DSC	Differing site condition	Geo.Wash.L.Rev.	
D.Me.	District Court of Maine (federal trial court)	Gonz.L.Rev.	
D.N.H.	District Court of New Hampshire (federal trial court)	H.R.Doc.	

D0085356

DSC	Differing site condition
D.Me.	District Court of Maine (federal trial court)
D.N.H.	District Court of New Hampshire (federal trial court)
DOTBCA	Department of Transportation Board of Contract Appeals[10]
DOTCAB	Department of Transportation Civil Aeronautics Appeal Board[10]
D.R.I.	District Court of Rhode Island (federal trial court)
Del. Super.	Delaware Superior Court
DePaul L.Rev.	DePaul Law Review
Det.C.L.Rev.	Detroit College Law Review[6]
Duke L.J.	Duke Law Journal
Duq.L.Rev.	Duquesne Law Review[6]
E.D.N.C.	Eastern District of North Carolina (federal trial court)
E.D.VA	Eastern District of Virginia (federal trial court)
EEO	Equal Employment Opportunity (federal)
EJCDC	Engineers Joint Contract Documents Committee
EPA	Environmental Protection Agency (federal)
Emory L.J.	Emory Law Journal[6]
ENG BCA	Corps of Engineers, Board of Contract Appeals[10]
Eng.Rep.	English Reports
Exec.Order	Executive Order (federal)
F.	Federal Reporter
F.2d	Federal Reporter, Second Series
F.3d	Federal Reporter, Third Series
FAACAB	Federal Aviation Agency, Contract Appeals Board[10]
FAR	Federal Acquisition Regulations
FHA	Federal Housing Authority
FIDIC	Federal Internationale Des Ingenieurs–Conseils
Fed.Cir.	United States Circuit Court, Federal Circuit
Fed.Reg.	Federal Register
Fed.Rules	Federal Rules
Fla.	Florida Supreme Court[1]
Fla.Dist.Ct.App.	Florida District Court of Appeals
Forum	Forum[6]
F.Supp.	Federal Supplement
GAAP	generally accepted accounting principles
GAO	General Accounting Office (federal)
GMC	guaranteed maximum cost
GMP	guaranteed maximum price
GSA	General Services Administration (federal)
GSBCA	General Services Board of Contract Appeals[10]
Ga.	Georgia Reports[1]
Ga.App.	Georgia Appeals Reports
Ga.Code Ann.	Georgia Code Annotated[8]
Ga.L.Rev.	Georgia Law Review[6]

Geo.Wash.L.Rev.	
Gonz.L.Rev.	
H.R.Doc.	
HUD	Department of Housing and Urban Development (federal)
Harv.Bus.Rev.	Harvard Business Review[6]
Harv.L.Rev.	Harvard Law Review[6]
Hastings L.J.	Hastings Law Journal[6]
Haw.Adm.Rules	Hawaii Administrative Rules
Hawaii	Hawaii Reports[1]
Hous.L.Rev.	Houston Law Review[6]
IBCA	Interior Board of Contract Appeals[10]
Idaho	Idaho Reports[1]
Ill.	Illinois Reports[1]
Ill.2d	Illinois Reports, Second Series[1]
Ill.Ann.Stat.	Illinois Annotated Statutes[8]
Ill.App.2d	Illinois Appellate Court Reports, Second Series
Ill.App.3d	Illinois Appellate Court Reports, Third Series
Ill.Comp.Stat.Ann.	Illinois Compiled Statutes Annotated[8]
Ill.Rev.Stat.	Illinois Revised Statutes[8]
Ind.	Indiana Reports[1]
Ind.Ct.App.	Indiana Appellate Court Reports
Ins.Couns.J.	Insurance Counsel Journal[6]
Int'lConstr.L.Rev. (U.K.)	International Construction Law Review[6]
Iowa	Iowa Reports[1]
Iowa Code Ann.	Iowa Code Annotated[8]
Iowa L.Rev.	Iowa Law Review[8]
J ACCL	Journal of the American College of Construction Lawyers[6]
J.Disp.Resol.	Journal of Dispute Resolution[6]
J.of Constr.Eng'g & Mgmt.	Journal of Construction Engineering and Management[6]
J.Mar.L.Rev.	John Marshall Law Review[6]
J.Pat.Off.Soc.	Journal of Patent Officials Society[6]
K.B.	King's Bench (England)
KC News	Kellogg Corp. News
Kan.	Kansas Reports[1]
Kan.App.2d	Kansas Appellate Reports, Second Series
Kan.L.Rev.	Kansas Law Review[6]
Kan.Stat.Ann	Kansas Statutes Annotated[8]
Ky.	Kentucky Reports[1]
Ky.L.J.	Kentucky Law Journal[6]
Ky.Rev.Stat.	Kentucky Revised Statutes[8]
LDA	local development authority
L.R–H.L.	Law Reports–House of Lords (England)
La.	Louisiana Reports[1]
La.App.	Louisiana Court of Appeals
Law & Contemp.Probs.	Law and Contemporary Problems[6]
Law Div.	Law Division (New Jersey)
Law Week	United States Law Week
Loy.L.A.L.Rev.	Loyola (Los Angeles) Law Review[6]
Loy.L.Rev.	Loyola Law Review[6]

(continued on back endsheets)

Legal Aspects of Architecture, Engineering and the Construction Process

EIGHTH EDITION

Justin Sweet

John H. Boalt Professor of Law Emeritus
University of California (Berkeley)

Marc M. Schneier

Editor, Construction Litigation Reporter

CENGAGE
Learning™

Australia • Canada • Mexico • Singapore • Spain • United Kingdom • United States

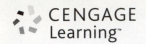
CENGAGE
Learning™

Legal Aspects of Architecture, Engineering and the Construction Process, Eighth Edition
Justin Sweet and Marc M. Schneier

Director, Global Engineering Program:
Chris Carson

Senior Developmental Editor: Hilda Gowans

Editorial Assistant: Jennifer Dinsmore

Marketing Specialist: Lauren Betsos

Director, Content and Media Production:
Barbara Fuller-Jacobsen

Content Project Manager: Emily Nesheim

Production Service: RPK Editorial Services, Inc.

Copyeditor: Fred Dahl

Proofreader: Martha McMaster

Indexer: Shelly Gerger-Knechtl

Compositor: Integra

Senior Art Director: Michelle Kunkler

Cover Designer: Andrew Adams

Text Permissions Researcher: Natalie Barrington

Photo Permissions Researcher: Natalie Barrington

Senior First Print Buyer: Doug Wilke

For product information and technology assistance, contact us at
Cengage Learning Customer & Sales Support, 1-800-354-9706.

For permission to use material from this text or product,
submit all requests online at **www.cengage.com/permissions**
Further permissions questions can be emailed to
permissionrequest@cengage.com.

Library of Congress Control Number:
U.S. Student Edition: 2008933517

ISBN-13: 978-0-495-41121-5

ISBN-10: 0-495-411213

Cengage Learning
200 First Stamford Place, Suite 400
Stamford, CT 06902
USA

Cengage Learning is a leading provider of customized learning solutions with office locations around the globe, including Singapore, the United Kingdom, Australia, Mexico, Brazil, and Japan. Locate your local office at: **international.cengage.com/region.**

Cengage Learning products are represented in Canada by
Nelson Education, Ltd.

For your course and learning solutions, visit
academic.cengage.com/engineering.

Purchase any of our products at your local college store or at our preferred online store **www.ichapters.com.**

Printed in the United States of America
1 2 3 4 5 6 7 12 11 10 09 08

To my wife, Sheba
  —*Justin Sweet*

To my family
  —*Marc M. Schneier*

Contents

CHAPTER FOUR

The Agency Relationship: A Legal Concept Essential to Contract Making   27

CHAPTER FIVE

Contracts and Their Formation: Connectors for Construction Participants   32

CHAPTER SIX

Remedies for Contract Breach: Emphasis on Flexibility   44

CHAPTER SEVEN

Losses, Conduct, and the Tort System: Principles and Trends   52

CHAPTER EIGHT

Introduction to the Construction Process: Ingredients for Disputes   85

CHAPTER NINE

Limits on Ownership: Land Use Controls   93

CHAPTER THIRTEEN

Compensation and Other Owner Obligations   224

CHAPTER FOURTEEN

Professional Liability: Process or Product?   237

CHAPTER FIFTEEN

Risk Management: A Variety of Techniques 302

CHAPTER SIXTEEN

Intellectual Property: Ideas, Copyrights, Patents, and Trade Secrets 317

CHAPTER SEVENTEEN

Planning the Project: Compensation and Organization Variations 339

CHAPTER EIGHTEEN

Competitive Bidding: Theory, Realities, and Legal Pitfalls 381

CHAPTER NINETEEN

Sources of Construction Contract Rights and Duties: Contract Documents and Legal Rules 412

CHAPTER TWENTY

Contract Interpretation: Chronic Confusion 429

CHAPTER TWENTY-ONE

Changes: Complex Construction Centerpiece 449

CHAPTER TWENTY-SEVEN

Claims: By-Products of Construction Process   592

CHAPTER TWENTY-EIGHT

The Subcontracting Process: An "Achilles Heel"   633

CHAPTER TWENTY-NINE

The Design Professional as Judge: A Tradition Under Attack   662

CHAPTER THIRTY

Construction Disputes: Arbitration and Other Methods to Reduce Costs and Save Time   674

CHAPTER THIRTY-THREE

Terminating a Construction Contract: Sometimes Necessary but Always Costly   752

Preface

The primary focus of this edition, as in editions that preceded it, is to provide a bridge for students, mainly architectural and engineering students, but increasingly, those in business schools and law schools, between the academic world and the real world. We hope to provide a cushion for the inevitable shock such a transition generates. The world of the classroom, with its teachers and its books, is not the same as the world of construction with its developers, owners, design professionals, contractors, building inspectors, loan officers, and public officials that regulate the construction process.

This cushion requires that readers understand what is law, how it is created, how it affects almost every activity of human conduct, and how legal institutions operate. This cannot be accomplished through simply stating "the law." It requires clear, concise, jargon-free text that probes beneath the surface of legal rules to uncover why these rules developed as they did, outline arguments for and against these rules, and examine how they work in practice.

This is a national textbook. We do not concentrate on the law of any particular state. It is difficult to state an American rule with fifty states and a federal jurisdiction. A few areas of law that are relevant to the Construction Process are exclusively regulated by federal law.[1] Yet most disputes are governed by the law of each state. This does not mean that there are fifty rules. Those who survey the cases will find that there is a majority rule and one or more minority rules. We will try to emphasize the majority rule without ignoring that some states have provided a different legal solution.

We provide many illustrations of how the legal rules operate through summaries of actual cases and reproduction of some cases. We do this to demonstrate how legal rules work in practice, how most disputes are resolved by the facts in the case, and, in the case of actual judicial opinions, how judges decide cases.

Describing legal rules and how they operate is not an in-depth or exhaustive treatment of something as complex as Construction Law. Footnotes are one way to at least hint at greater complexity. With this in mind, let us express our footnote philosophy.

The primary role of footnotes is to provide authority for textual statements that we believe require documentation. Another function is to indicate a deeper level of complexity of a legal problem without detracting from the primary explanation provided in the text. This leads to a third function of footnotes: to provide references that will enable the student, the researcher, or the practicing attorney to dig more deeply into a particular legal problem.

Footnote references can be statutes, regulations, or cases. In addition to these primary sources of law, references can include the burgeoning secondary literature, texts and legal journals, that examine Construction Law in depth and from a practical perspective. Finally, references can include techniques used by Standard Construction Documents that have suggested solutions for many legal and practical problems.

Each of us has taught courses in Construction Law. We have found that students often wish to do research into legal problems. They are helped by footnote references we have provided. Legal research is not limited to law students. We have found that students in architecture and engineering schools also engage in legal research and

---

[1]As illustrations see Sections 14.08G (Occupational Safety and Health Act) and 16.03–16.06 (Copyright and Patents).

that footnote references help them. In addition, lawyers or design professionals handling legal materials in practice have informed us that these references have helped them.

The main impetus for the Eighth Edition is the publication by the American Institute of Architects (AIA) of new editions of its principal documents for design services (B101–2007), prime construction services (A101–2007 and A201–2007), and subcontracting services (A401–2007). The dual linchpins in the AIA panoply of standard form documents—for design and prime construction services—were greatly changed from their 1997 predecessors. Both documents were affected by the AIA's introduction of a new project participant: the Initial Decision Maker (IDM). Unusually, the numbering of the design services document was changed from B141 to B101. Stylistically, the AIA in 2007 began referring to the parts of its documents as sections, whereas previously they were referred to as paragraphs. This style change is reflected in the text of this book.

AIA documents are the backbone of the contractual aspects of the American construction industry. Any book dealing with construction law should refer to the most current AIA documents. These documents reflect the AIA thinking on risk distribution and contract administration. The newest versions of these documents—the agreement between owner and architect (B101), the agreement between owner and prime contractor (A101), the "general conditions" (A201), the performance and payments bonds (A312), and the subcontract (A401)—are reproduced respectively in Appendices A, B, C, D and E.

While a major revision to the AIA standard form documents by itself justifies a new edition, new sets of standard form documents were published by two other organizations.

The Engineers Joint Contracts Documents Committee (EJCDC) publishes documents dealing with engineering services. As this book is going to press, the EJCDC has published some new documents, whereas others are still under revision. The appendix of this book contains a new "general conditions" document, C-700 (2007), and a new owner/contractor agreement, C-520 (2007). However, the existing version of the agreement between owner and engineer, E-500 (2002), is included in the appendix because its successor document is still under revision. These standard form documents are reproduced in Appendices G (E-500), H (C-520), and I (C-700).

Finally, in 2007 the Associated General Contractors of America (AGC) for the first time has refused to endorse the new AIA documents. Instead, the AGC, in combination with several other trade associations, has produced a new series of standard form documents, called ConsensusDOCS. At this point, both the market impact and judicial treatment of the ConsensusDOCS are unknown. While these new standard form documents are mentioned,[2] they are not reproduced in the appendix and are not discussed in depth.

Alternative dispute resolution (ADR) is of increasing importance to the construction industry. The most significant document governing ADR is the Construction Industry Arbitration Rules and Mediation Procedures of the American Arbitration Association, revised and issued on September 1, 2007. This document is reproduced in Appendix F.

Reasons other than the issuance of new standard construction documents also prompt this new edition. Law, though it seeks stability, is not static. New cases come pouring out of the courts daily. Legislatures enact new statutes in every new session.

The construction industry is also constantly changing in response to market and technological forces. This Eighth Edition amplifies on alternative project delivery methods previously discussed, such as construction management[3] and design-build.[4] More importantly, this edition introduces new project delivery methods: lean project delivery,[5] project alliance,[6] program management,[7] and building information modeling (BIM).[8]

A plethora of changes have had an impact in disparate ways on the roles and rights of design professionals and on other important actors in the construction industry, particularly contractors. Changes in the roles and rights of design professionals impact on contractors and others.

As noted, one of the most significant changes made by the AIA to its standard form documents was the introduction of a new project participant, the Initial Decision Maker. The nature of this new role and its impact upon

---

[2]Sections 20.02F and 28.01.
[3]Section 17.04D.
[4]Section 17.04F.
[5]Section 17.04I.
[6]Section 17.04J.
[7]Section 17.04K.
[8]Section 17.04M.

the architect are reviewed.[9] We explore how a new United States Supreme Court decision on patent infringement impacts an architect's intellectual property rights.[10] California has passed a series of statutes directly addressing an architect's indemnity rights.[11] Limitation of liability clauses, most commonly used by engineers, receives expanded attention.[12] The question of whether the relationship between an architect and her client is a fiduciary one is newly reexamined.[13]

Substantive construction law is a complex amalgam of judge-made and statutory law. Looking at the latter factor first, statutory regulation of the construction industry continues unabated, as does the judicial interpretation of those laws. An unlicensed contractor's right of recovery against the client, as well as the client's right of reimbursement of payments already made, are given an updated treatment.[14] A growing trend is for state legislatures to remove certain classes of disputes—in particular, those involving homeowners—from the litigation process, either by giving owners specified warranty rights or contractors a right to remedy the defects before litigation may be started.[15] Risk transfer through the use of indemnity agreements is essential to the construction process. An updated review of statutory regulation of these agreements is provided.[16]

Changes in the general, judge-made law impacts on the law applicable to the construction industry. The material dealing with products liability,[17] the economic loss rule,[18] and contribution and noncontractual indemnity[19] have been updated to take into consideration these broader legal developments. In addition, construction law and general contract law are intimately related. Discussion of the "plain meaning" rule of contract interpretation is greatly amplified, and a new case has been added.[20]

The ever essential *Spearin* doctrine, imposing upon the owner an implied warranty for certain design defects, receives an updated and expanded analysis, including a discussion of a United States Supreme Court decision.[21]

The concept of the "stigma" effect of defective construction and the effect that it may have on the owner's damages are introduced.[22]

As the holder of the contract funds, the construction lender's liability to unpaid subcontractors for allegedly negligent disbursement of the funds receives updated treatment.[23]

Much of the dispute resolution process in the construction industry is through alternative dispute resolution (ADR) and, in particular, arbitration. The concern that such agreements deprive consumers and homeowners of their rights remains a hotly disputed topic.[24] A corollary issue is whether a judge or the arbitrator decides whether the parties must arbitrate their claim.[25] The newest AIA documents give a greater role to mediation.[26]

The regulatory scheme surrounding the construction industry is also updated. A leading United States Supreme Court case upheld from constitutional challenge a municipality's taking of private property for the purpose of economic development.[27] The environmental movement, broadly defined, affects the construction industry in a variety of ways. Statutory developments have been updated and also expanded to include so-called brownfields legislation.[28] A new discussion addresses the industry's own attempts to create environmentally friendly design and construction, including the creation of LEED standards.[29]

Although federal procurement law as a general matter is not covered, federal legislation, regulations, and decisions in federal procurement construction disputes often spearhead changes among the states. For this reason, a new Section discussing the Federal False Claims Act is added.[30] In addition, creation of the new Civilian Board of Contract Appeals is noted.[31]

---

[9] Sections 29.01A, 29.05, 29.08A, and 29.11.
[10] Section 16.03B.
[11] Section 31.05D.
[12] Section 15.03D.
[13] Section 11.04B.
[14] Section 10.07.
[15] Section 23.02K.
[16] Section 31.05D.
[17] Section 7.09.
[18] Section 14.08E.
[19] Sections 31.02, 31.03A and 31.03E.
[20] Section 20.02A.

[21] Section 23.05E.
[22] Section 27.03D.
[23] Section 22.02I.
[24] Section 30.02.
[25] Section 30.04.
[26] Section 30.18B.
[27] Section 9.07.
[28] Section 9.13.
[29] Section 9.13E.
[30] Section 27.16.
[31] Section 30.20A.

We seek not only to accurately describe the construction industry and the applicable law, but also to provide suggestions for reform. A proposal is made for the major project participants, before commencement of the work, to enter into a Defect Response Agreement, under which they agree in advance to divide the cost of inevitable defects, thereby precluding this issue from impeding performance.[32] Also, mechanics' liens are a time-honored mainstay of the construction process, meant to guarantee payment of subcontractors, yet the question is asked whether subcontractors would be better off if such liens were simply abolished.[33]

Now, for our nonlawyer readers, let us outline the mechanics of citing case decisions. Although legal citations seem complicated, they are, in reality, quite simple. There are four elements to a citation. A simple citation would be *Sniadach v. Family Finance Corp.*, 395 U.S. 337 (1969). First, the name of the case is given, usually with the plaintiff (the person starting the lawsuit) listed first, followed by the defendant. "U.S." is an abbreviation of the reporter system from which the case is taken, in this case the United States Supreme Court Reports. The number preceding the abbreviation of the reporter system (395) is the volume in which the case is located. The number following the abbreviation of the reporter system (337) states the page on which the case begins. The citation concludes with the year that the court announced the decision.

Many significant reported appellate opinions come from the Federal Circuit Courts of Appeal. *Moorehead Construction Co. v. City of Grand Forks*, 508 F.2d 1008 (8th Cir.1975) is illustrative. Following the name of the parties, an abbreviation of the series of reports in which the opinion is contained (F.2d) (Federal Reporter, Second Series) is given. Again, it is preceded by the volume (508) of the report. The reporter abbreviation is followed by the page on which the case begins (1008). The particular federal circuit court (8th) deciding the case follows. The citation is completed by the year (1975) of the decision.

A third type of federal court is the district court, the trial court of general jurisdiction. Although state trial court opinions (except those from New York) are not collected and published, some trial court opinions of the federal district courts are published. An example is *Gevyn Construction Corp. v. United States*, 357 F.Supp. 18 (S.D.N.Y.1972). The opinion is collected in the Federal Supplement (F.Supp.). The name of the reporter series is preceded by the volume number of the report and followed by the page on which the opinion begins. This citation shows that the district court was the Southern District of New York (S.D.N.Y.). Larger states have different courts located in different cities. Again, the citation concludes with the date (1972) the opinion was issued.

To understand state court citations, it is essential to understand the role of the West Regional Reporter System. At one time, most state citations had two citations—one to the official reporter system selected by the court and the other to the unofficial reporter system, usually the West Regional Reporter System. West divided the states regionally and published sets of reporters based on these regional classifications. If a state still has an official report system, a typical citation would be *Anco Construction Co. v. City of Wichita*, 233 Kan. 132, 660 P.2d 560 (1983). This citation indicates that the case came from the Kansas Supreme Court (Kan.) and is also collected in the Pacific Reporter (P.2d).

States are increasingly abandoning the official reporter system and cite only to the West Regional Reporter System. An example is *Smith v. Gilmer*, 488 So.2d 1143 (La.App.1986). This citation, from the Southern Reporter, indicates that this was a decision of the Louisiana Court of Appeals (its intermediate appellate court) and that there is no official reporter system in Louisiana.

Finally, two of the more populous states—California and New York—have an unofficial reporter published by West that collects all the reported opinions from those states. In California, it is the California Reporter; in New York, it is the New York Supplement. Decisions by the highest appellate court—in California, the Supreme Court, and in New York, the Court of Appeals—are found in the California Reporter and the New York Supplement, respectively, as well in the official and regional reports. For example, the California Supreme Court case of *Pollard v. Saxe and Yolles Dev. Co.* is cited 12 Cal.3d 374, 525 P.2d 88, 115 Cal.Rptr. 648 (1974). "Cal" is the official reporter, Cal.2d being the second series of that reporter. "P.2d" is the second series of the Pacific Reporter. "Cal.Rptr." collects all California appellate cases, Supreme Court decisions as well as those of the intermediate California Courts of Appeal. Similarly, a decision by the New York Court of Appeals would be found in its official reporter (N.Y. or N.Y.2d) and the West

---

[32]Section 24.11.
[33]Section 28.07.

Regional Reports, the North Eastern Reporter (N.E. or N.E.2d) and the New York Supplement (N.Y.S., N.Y.S.2d, or N.Y.S.3d). The New York Supplement also contains decisions of the New York intermediate courts of appeal (the Appellate Division of the Supreme Court) and the Supreme Court (the New York trial court). Intermediate appellate court decisions from these states would have only two citations—one to the official reports and the other to the California Reporter and the New York Supplement. Intermediate appellate court decisions from these states are not included in the regional reports. (The Table of Abbreviations lists citation references—see the endpapers of this book.)

The number of reported appellate decisions has increased greatly. Most reporter systems have gone into a second, third, or even fourth series of reports. This is indicated by the citation in the *Pollard* case, which shows that the official reporter is in the third series, the unofficial regional reporter is in the second series, and the California Reporter was at that time still in its first series. (It now is in its third series.)

The hierarchical nature of the court system means that parties receiving a decision by a lower-level court may try to appeal the decision. The higher-level court has several options. If it accepts the appeal, it usually either affirms the lower court opinion or reverses it. This is indicated by the terms "aff'd" (affirmed) or "rev'd" (reversed) appearing after the citation of the lower court decision. If the higher-level court refuses to hear the appeal, that is indicated in various ways: "review denied," "appeal denied," or "cert. denied," where "cert." is short for the term "certiorari." An example is *McDowell-Purcell, Inc. v. Manhattan Constr. Co.*, 383 F.Supp. 802 (N.D.Ala.1974), aff'd, 515 F.2d 1181 (5th Cir.1975), cert. denied, 424 U.S. 915 (1976). If the higher court opinion appears in the same year as the lower court opinion, the date appears only once and at the end. An example from the New York court system is: *J. P. Stevens & Co. v. Rytex Corp.*, 41 A.D.2d 15, 340 N.Y.S.2d 933, aff'd, 34 N.Y.2d 123, 312 N.E.2d 466, 356 N.Y.S.2d 278 (1974). This later case history is provided primarily for purposes of accuracy and should not concern the student reader.

On the very different stylistic matter of pronous, we avoid the awkward "he or she" format. In many instances, the actor is a corporation or government entity and the neutral pronoun ("it") is used. Otherwise, we alternate chapters: the male pronoun is used in odd chapters and the female pronoun in even chapters.

Finally, special thanks are given to several professors who reviewed the Seventh Edition and made suggestions for the Eighth: Lansford C. Bell, Clemson University; Steven M. Goldblatt, University of Washington; Dana Sherman, University of Southern California; Kelly Strong, Iowa State University; and Blake E. Wentz, Milwaukee School of Engineering. Preparation of this Eighth Edition was greatly enhanced by their input. These reviewers have contributed ideas, suggestions, and new perspectives. The result is an improved, newer edition.

Justin Sweet
Marc M. Schneier

Credits

This page constitutes an extension of the copyright page. We have made every effort to trace the ownership of all copyrighted material and to secure permission from copyright holders. In the event of any question arising as to the use of any material, we will be pleased to make the necessary corrections in future printings. Thanks are due to the following authors, publishers, and agents for permission to use the material indicated.

A-1: (c) American Institute of Architects B-1: (c) American Institute of Architects C-1: (c) American Institute of Architects D-1: (c) American Institute of Architects E-1: American Institute of Architects F-1: Used with permission of American Arbitration Association. G-1: Reprinted with permission of the National Society of Professional Engineers (www.nspe.org) H-1: Reprinted with permission of the National Society of Professional Engineers (www.nspe.org) I-1: Reprinted with permission of the National Society of Professional Engineers (www.nspe.org)

Chapter 12. 171: From "Organic Transactions: Contract, Frank Lloyd Wright, and the Johnson Building," by Stewart Macaulay [1996], Wisconsin Law Review 75. Copyright (c) 1996 by the Board of Regents of the University of Wisconsin System. Reprinted by permission of the Wisconsin Law Review. Most notes have been omitted. Those not omitted have been renumbered. 172: Courtesy of SC Johnson 172: Courtesy of SC Johnson 173: Courtesy of SC Johnson

CHAPTER ONE

Sources of Law: Varied and Dynamic

SECTION 1.01 Relevance

Law consists of coercive rules created and enforced by the state to regulate the citizens of the state and provide for the general welfare of the state and its citizens. Law is an integral part of modern society and plays a major role in the construction process. Because this text examines the intersection between law and the construction process, it is important to be aware of the various sources of law and the characteristics and functions of the law.

Many illustrations can be provided; suppose a man who owns property wishes to build a house to provide shelter for his family. Without assurance that stronger people will not use force to seize materials with which he is building or throw him out of the house after it is built, it would take an adventurous or powerful person to invest time and materials to build the house. Similarly, workers would be reluctant to pound nails or pour concrete if they were fearful of being attacked by armed gangs. Here, criminal law protects both the property owner from those who might take away his property and the workers from those who might harm them.

Similarly, contractors would hesitate to invest their time or money to build houses if they did not believe they could use the civil courts to enforce their contracts and help them collect for their work if owners did not pay them. Workers would be less inclined to work on a house if they were not confident they could use the civil courts to collect for work they had done, or to compensate them if they were injured on the job.

Finally, some would be unwilling to engage in construction activity if they were not confident that an impartial forum would be available if disputes arose over performance.

Were the state not to provide such a forum, participants might settle their disputes by force.

Today various sources of law seek to fill these needs. These sources of law are spotlighted in this chapter. Chapter 2 focuses on the U.S. judicial system.

SECTION 1.02 The Federal System

Very large countries, such as the United States, Canada, Australia, Nigeria, and India, employ a federal system of government. Even smaller countries, particularly those with distinct religious, linguistic, ethnic, or national communities (for example, Switzerland), may choose to live under a federal system. A federal system of government gives local entities limited autonomy to deal with cultural and other activities. The alternative would be domination by the majority or by small, often economically inefficient or weak nation-states.

In a federal system, power is shared; the exact division of power between the central government and constituent members varies. For example, Canada has a looser federal system than the United States, with Canadian provinces having more autonomy than U.S. states.

A federal system may be created in a large country where people living in one part of the country hold political, social, or economic views distinct from the rest of the country. For example, New York might choose to provide greater social benefits for its citizens than Texas. Similarly, some states may wish to execute murderers, whereas others may not. In a federal system, diverse views may be accommodated within one country.

Also, depending on its characteristics, a federal system can bring citizens closer to those who govern them. A rancher in Wyoming may resent being controlled by a legislature in Washington but may be more willing to submit to laws that come from Cheyenne. When the entities of a federal system are themselves large, such as Canadian provinces or U.S. states, citizens may see even the state capital as being too remote and may prefer to be governed by those they elect in their cities or towns.

The federal system recognizes the need for a central government to deal with certain issues on behalf of all citizens. For example, the U.S. federal government controls currency, foreign relations, and defense, to mention a few of the areas it controls exclusively. The constitutional framers did not want each state to have its own currency, foreign policy, or military forces.

Other functions, such as the enactment of tax laws, can be shared by the federal government and state governments. Similarly, federal and state laws deal with crime, labor relations, and—to use an illustration more germane to construction—worksite safety. To avoid duplication of enforcement efforts and to relieve construction contractors from inconsistent regulations, the federal government, though dominant, can delegate workplace safety standards and their enforcement to the states so long as the states meet federal standards. Similar delegations are found in environmental protection laws.

Despite general federal supremacy over the states, the U.S. Constitution reserved some authority to the states. For example, except for contracts made by the federal government or those affected with a strong federal interest, state law determines which contracts will be enforced, the remedies granted for breach of contract, the conduct that gives rise to civil liability, and the laws that relate to the ownership of property—all core legal concepts in the construction process. As a result, most law that regulates construction is determined by the state in which the project is located or in which the activities in question are performed.

This can and sometimes does lead to variations in legal rules that relate to construction. However, the dominance of standard contract forms, created by national associations for nationwide use; the willingness of courts in one state to look at and often follow the decisions from another state; and the unification of areas of private law, such as the sale of goods, have all minimized the actual variation in state laws that relate to construction. Some laws vary greatly, such as mechanics' lien laws and licensing, both of which are regulated by state statutes.

SECTION 1.03 Constitutions

When organizations are created, whether political, economic, or social, they usually attract members because of their goals, the methods they choose to achieve the goals, and the procedures by which they operate. Often these factors are set forth in a basic set of principles or rules providing a constitutional framework that induces people to join and regulates the operations of the organization. Although this basic set of rules may be referred to by different labels, it provides a constitutional framework intended to be durable and, although not immutable, amendable only with the consent of most members of the organization.

Most organizations need rules to govern day-to-day operations. These rules, although created with less formality than constitutional rules and more responsive to changing conditions, must stay within the basic framework of the constitutional rules. For example, a business corporation will usually have articles of incorporation that operate as a constitution of the corporation and are agreed on by those who create the corporation and available to those who wish to purchase shares. The corporation is also likely to have bylaws that regulate corporate activities and that are often more detailed and more easily changed. In addition, a large corporate organization may need standard operating procedures and hierarchical authority mechanisms to control the corporate operation.

For two centuries the U.S. Constitution has regulated the power among the branches of the federal government, the federal government and the states, and all governments and their citizens. Despite universal admiration for this durable document, the wealth of federal laws that have been enacted attests to the importance of legislation as a method of dealing with changing circumstances and the varying allocation of political power.

On the whole, state constitutions are longer than the federal Constitution and are changed more frequently. Although not identical, they share common characteristics with the U.S. Constitution, mainly the separation of powers among the legislative, executive, and judicial branches of government and the importance of protecting citizens from abuse of power by the states.

With a few exceptions, constitutional law does not play a significant role in the construction process.[1]

SECTION 1.04 Legislation

Legislation, mainly at the state level, plays an increasingly large role in the construction process. Those who wish to understand how law affects design and construction must pay increasingly careful attention to legislation. Legislation is the most democratic lawmaking process. It expresses the will of the majority of the citizens of the state. (Perhaps the only source of law more democratic is the law that contracting parties create to regulate their private rights and duties.)

Legislatures are political instrumentalities. Persons and organizations seek to influence lawmakers. Although often denigrated as an area where power and money control, legislation often reflects popular attitudes. Lawmakers who vote in ways not popular with their constituents are likely to be removed from office. Also, legislation can be enacted quickly to respond to what the legislature believes to be important social and economic needs of its citizens.

The legislative process functions at different governmental levels, such as federal, state, county, and city legislative bodies.

Legislators do not have absolute freedom to enact legislation. The checks and balances so central to the American political process bar legislators from absolute power even though the legislatures do represent the political will of the citizens. Laws must be constitutionally enacted, and the constitutionality of a law is determined by the courts. In their role of interpreting legislation, the courts can indirectly control the legislature and the political desires of the majority. Similarly, local units such as counties, cities, or special districts are limited by the legislation that has created them.

Legislation—again, mainly at the state level—affects numerous aspects of the construction industry, from the planning stage through the resolution of postperformance disputes. Local legislative bodies, exercising their power to protect the public and regulate land use, determine whether and which types of projects may be built.[2] Housing and building codes control the quality of construction.[3] Licensing and registration laws determine who may design and who may build.[4]

Even though American law values freedom of contract, legislation increasingly regulates the content of construction contracts. Statutes preclude certain types of indemnification provisions in construction contracts.[5] The payment process is also regulated, with statutes addressing payment conditions,[6] payment promptness[7] and retainage.[8]

Mechanics' liens[9] and trust fund[10] laws bestow statutory rights upon unpaid subcontractors and suppliers. Statutes affect the litigation process by encouraging arbitration.[11] Residential building contracts are especially regulated, with statutes specifying contract language,[12] dispute resolution procedures[13] and warranties.[14] In short, while construction law originated in the common (or judge-made) law, today legislation is an integral part of the legal framework governing the building industry.[15]

The Uniform Commercial Code (UCC), particularly Article 2, deals with transactions in goods. The UCC was developed by the American Law Institute and the Commissioners on Uniform State Laws, both

---

[2]Chapter 9.

[3]Section 19.02C.

[4]Chapter 10.

[5]Section 31.05D.

[6]Section 28.06.

[7]Section 22.02F.

[8]Section 22.03.

[9]Section 28.07D.

[10]Section 28.07F.

[11]Section 30.02.

[12]West Ann.Cal.Civ.Code § 7159; *Ragucci v. Professional Constr. Serv.*, 25 A.D.3d 43, 803 N.Y.S.2d 139 (2005) (New York statute, prohibiting mandatory arbitration clauses in consumer contracts, applies to an architect's contract with a homeowner).

[13]Section 23.02K.

[14]Ibid.

[15]For a survey discussion, see George, Gerber & Montez, *Legislative Update: The Kansas Fairness in Private Construction Contract Act and Other Legislation Regulating "Unfair" Provisions in Construction Contracts*, 26 Constr. Lawyer, No. 3, Summer 2006, p. 33.

---

[1]Sections 18.03P (affirmative action programs), 23.03G (statutes of repose), 22.02E (withholds from payments), 28.07D (mechanics' liens) and 33.03J (termination of public contracts).

private organizations devoted to unification of private law. At present, all states but Louisiana have adopted the UCC. (Louisiana, influenced by its civil law tradition, has adopted only some parts of the UCC.) Article 2 regulates the sale of materials and supplies used in a project. Although it does not regulate construction itself (inasmuch as it does not govern services), it can be influential in a construction dispute. Reference is made to it in this book.

SECTION 1.05   The Executive Branch

The separation of powers central to the U.S. political system is designed to prohibit any of the three branches—executive, legislative, or judicial—from wielding dominant power. Although laws come out of the legislature, the executive and judicial branches participate in lawmaking. The executive branches—the president at the federal level, the governor at the state level, and the mayor in some municipal systems—are elected by all the citizens. Federal and state chief executives and sometimes local chief executives can veto legislation. To override a veto requires more than a simple majority of the legislature, usually two-thirds.

The chief executive can issue executive orders to those under his control. But more important, the chief executive, such as the president or the governor, has the power to appoint the heads of administrative agencies or administrative boards that play a significant role in construction. Examples of such regulatory instrumentalities are those that deal with workplace safety, registration of design professionals, and licensing of contractors, to name only some of the most important.

SECTION 1.06   Administrative Agencies

One great change in the U.S. governmental structure has been the emergence of administrative regulatory agencies at every level of government, but particularly at federal and state levels. Such agencies developed for a variety of reasons. First, special activities and industries were thought best regulated by experts in those activities and industries. Second, legislatures were often unable or unwilling to involve themselves in the details of regulation. Third, regulation through

agencies was thought to be better than no regulation (this changed in the 1980s, with some industries being deregulated) or government operation of these activities and industries.

The constitution for a regulatory agency is the legislation creating it. In that sense, the agency is a creation of the legislature. However, the chief executive usually appoints key agency officials, often with the advice and consent of the legislature. Legislatures, particularly Congress, monitor agency performance through exercising an oversight function. This is accomplished through committee hearings during which complaints are heard and agency officials are asked to give explanations.

Agencies operate through issuance of regulations and through disciplinary actions. Such activities are subject to judicial review. During the period when many of these agencies were created, courts extended almost total deference to such agencies because of presumed agency expertise. As the agencies became more active and as complaints became more vocal about their aggressiveness and preoccupation with problems that some thought trivial, the courts and legislatures looked at agencies' powers and activities more carefully.

Regulation through administrative agencies has generated intense controversy. Some agencies are attacked as being under the control of those they are supposed to regulate. This is asserted by pointing to agency employees being selected from the regulated industries or to agency employees looking forward to being hired by members of the industry after they leave agency employment. However, agencies have also sometimes been attacked for overzealous regulation, and such attacks had some success in the 1980s.

Administrative agencies play an important role in some aspects of the construction process. They are often given responsibility for implementing laws enacted by the legislatures, such as laws on licensing and registration of professionals, workplace safety (especially the federal Occupational Safety and Health Administration (OSHA)), environmental protection (especially the federal Environmental Protection Agency (EPA), discussed in Section 9.13B), and social insurance for workplace injuries (worker's compensation, discussed in Section 7.04C). At local levels, administrators play important roles in land use control and construction quality. In areas affecting public safety, the most important source of law is regulations issued by administrative agencies.

SECTION 1.07 Courts: The Common Law

Courts are an important part of the lawmaking process. Exercising their power of judicial review, they determine whether legislation is constitutional and interpret the legislative enactments. They have the principal responsibility for granting remedies under legislative systems. Courts also pass on and interpret administrative regulations and grant remedies for violations.

The sense in which courts are a source of law central to the construction process is principally based on the dual function of U.S. appellate courts. First, the appellate courts review trial court decisions by providing a forum for litigants dissatisfied with judgments of the trial courts. Second, and more important, the process by which appellate courts judge the correctness of decisions made by trial courts has had an immense impact on design and construction. To resolve disputes, American courts follow precedent. An understanding of how courts make law and resolve disputes requires an understanding of the precedent system discussed in Section 2.14.

SECTION 1.08 Contracting Parties

Sources of law usually are public bodies. Yet contract law, with the broad autonomy granted the contracting parties to determine the terms of their exchange, grants lawmaking power to those who make contracts.

Even though autonomy is seen as an inherent liberty in a free society, the state still plays a role—enough of a role to create a "partnership" between contracting parties and the state. Despite broad autonomy given the parties, the state still determines who can contract, creates the formal requirements, and provides remedies when contracts are not performed.

If the contracting parties can make their contract a source of law, this is the most democratic form of lawmaking. The contracting parties freely determine the terms of their exchange and voluntarily limit their freedom of action by agreeing to perform in the future, with the coercive arm of the state operating if they do not.

Chapter 5 is devoted to some aspects of contract law, and Chapter 6 to contract remedies. At this point, it is sufficient to mention the emergence of the bipolar analysis of contract law, which sharply distinguishes between negotiated contracts and contracts of adhesion. The former emphasize autonomy or freedom of contract based on consent being freely given by the contracting parties. The latter involve contracts made with the terms, or at least most of them, largely dictated by one of the parties and merely adhered to by the other. This distinction is discussed in Section 5.04C.

SECTION 1.09 Publishers of Standardized Documents

Adhesion contracts, represented by printed contracts, are prepared by one party to the contract largely, if not exclusively, with its own interests in mind. Yet many printed forms used in contracts for design and construction should not be considered adhesion contracts in the sense that the term is used in Sections 1.08 and 5.04C. Contracts for design and construction services often are based largely on documents published by professional associations such as the American Institute of Architects (AIA), the Associated General Contractors of America (AGC), and the Engineers Joint Contract Documents Committee (EJCDC), a consortium of professional engineering associations, to mention some of the most important.[16]

These associations have no official status, and some would contend they cannot be considered sources of law. Certainly contracting parties need not use the standardized documents or, if they do use them, are free to make drastic changes. But the frequency with which these documents are used largely unchanged justifies classifying these associations as sources of law. Increasing legislative regulation of the construction contracting process may at some point necessitate the promulgation of state-specific standard documents; however, that day has not as yet arrived.[17] Even if in a technical sense they should not be ranked alongside courts, the realities of construction contracting make them even more important.

---

[16]Sweet, *The American Institute of Architects: Dominant Actor in the Construction Documents Market*, 1991 Wis.L.Rev., 317 (1991). Some of the more important AIA and EJCDC standard documents are included as appendices to this book.

[17]Section 1.04.

SECTION 1.10 Restatements of the Law

Many of the common law rules, together with comments and examples, have been codified in a series of volumes called Restatements of the Law. They are published by the American Law Institute, a private organization made up of lawyers, judges, and legal scholars. The institute's function is to collect case law from all the states and distill it into rules. Because the common law is a dynamic and robust disputes resolution system, the legal rules generated change over time. As a result, the Restatements of the Law have been updated; for example, the Restatement (Second) of Contracts was published in 1981, while a Restatement (Third) of Torts is currently being issued.

Unless adopted by a legislature or followed by a court, Restatement rules are not law. The principal restatements for the purposes of this book are those of torts, contracts, and agency.

SECTION 1.11 Summary

Understanding how law bears on design and construction requires a recognition of the wide variety of law sources. People involved in design and construction should be aware of the laws and legal institutions that have placed their mark on these fields. And people in these fields must appreciate which sources of law predominate and the forms that legal intervention takes. A half-century ago, it could be confidently stated that the principal sources of law in design and construction were the contracting parties and the courts. Increasingly, however, other instrumentalities such as legislatures and regulatory agencies are leaving their imprint on design and construction.

CHAPTER TWO

The American Judicial System: A Forum for Dispute Resolution

SECTION 2.01 State Court Systems: Trial and Appellate Courts

Each U.S. state has its own judicial system. Courts are divided into the basic categories of trial and appellate courts. Within each category may be a subclassification based on amount or type of relief sought or on the nature of the matter being litigated.

The basic trial court, frequently called the court of general jurisdiction, hears all types of cases. Depending on the state, it may be called a superior, district, or circuit court. The bulk of the work before such courts consists of criminal cases, personal injury cases, commercial disputes, domestic matters (divorce, custody, and adoption), and probate (transfer of property at death).

A court of general jurisdiction may review determinations of administrative agencies (zoning, employment injuries, licensing, etc.). The presiding official will be a judge, and in certain matters there may also be a jury. The division of functions between judge and jury is discussed in Sections 2.10 and 2.11. Courts of general jurisdiction are usually located at county seats.

Many states have established subordinate courts of limited jurisdiction. Municipal or city courts often have jurisdiction to hear disputes that involve less money than a minimum figure set for courts of general jurisdiction. There can also be a jury in such cases. As a rule, the dockets are less congested in these courts of limited jurisdiction than are the dockets in the courts of general jurisdiction. The procedures in municipal or city courts are essentially the same as those in the courts of general jurisdiction.

Some state legislatures have established small claims courts that provide expeditious and inexpensive procedures for disputes involving small sums. Usually the party seeking a remedy from such a court can start the legal proceeding by paying a small filing fee and filling out a form provided by the clerk of the small claims court. The procedures are usually informal. There are no juries. Although the judges are lawyers, attorneys are less important in such disputes, and in some states attorneys are not permitted to represent the parties.

Some states still use justices of the peace, a vestige of a rural, dispersed society. These judges often have had little or no legal training. Although they are being phased out in favor of small claims, municipal, or city courts, they still exist in sparsely populated areas of some states. The future may witness their elimination.

At least one appeal is usually possible from a decision of a trial court. Generally the appeal is to the next highest court. For example, a party losing a decision in a municipal court may have a right to appeal to the court of general jurisdiction or to an appellate division consisting of judges of the court of general jurisdiction. Appeals from courts of general jurisdiction are made to an appellate court. Beginning in the more populous states and increasingly in other states, appeals first must go to an intermediate appellate court. In states with such courts, the state supreme court generally is given the discretion to decide whether it will hear an appeal from an intermediate appellate court. In states without intermediate appellate courts, an appeal is made from the court of general jurisdiction to the supreme court of the state.

State court judges are either appointed by the governor or elected. Even states that elect judges have many judges who were initially appointed. Generally, judges are reelected and leave office only when they retire or die. Such vacancies are filled by interim appointments, and appointees are usually elected when they go before the voters.

Some states use an elective process in which all judges—or at least appellate judges—periodically submit their records to the electorate but do not run against other candidates.

SECTION 2.02 The Federal Court System

The U.S. judicial system includes two systems—federal and state—which to a degree exist side by side. Although each state has its own judicial system, the federal courts also operate in each state. The federal courts have jurisdiction to decide federal questions; that is, disputes involving the federal Constitution or federal statutes. They also have jurisdiction to hear civil disputes between citizens of different states, often called *diversity of citizenship* jurisdiction. In theory, the amount in controversy in each type of case must exceed $50,000 exclusive of interest and costs, but the amount in controversy is important principally in diversity of citizenship cases.

Many claims brought before federal courts can also be brought in state courts, creating concurrent jurisdiction. Some matters, however, can be brought only in the federal courts. The principal areas of exclusive federal jurisdiction are as follows:

1. admiralty
2. bankruptcy
3. patent and copyright
4. actions involving the United States
5. violations of federal criminal statutes

Where exclusive federal jurisdiction exists, there is no requirement for amount in controversy.

The federal courts operate under more modern and less formal procedural rules than do most state courts. For substantive law (laws that establish legal rights and duties), the federal courts use the substantive law of the state in which they sit or some other applicable state law, unless the case involves federal law, such as a federal constitutional provision or a federal statute.

The basic trial court in the federal court system is the district court. Each state has at least one, and the populous states have a number of district courts located in the principal metropolitan centers.

The district court is presided over by a federal district court judge, and juries are used in certain cases. A party may appeal a decision of the district court to one of the eleven circuit courts that hear appeals from that district. Another circuit court hears appeals mainly from decisions of U.S. administrative agencies. Another hears appeals from specialized federal courts.

A party dissatisfied with the result of a decision by the circuit court of appeals may ask that the U.S. Supreme Court review the case. In general, the Supreme Court determines which cases it will review. If it decides not to hear the case, the matter is ended. If it decides to hear the case, briefs and oral arguments are presented to the Court. The Supreme Court usually rejects petitions for hearings unless the matter involves either a conflict between federal circuit courts or an important legal issue.

Federal court judges are appointed by the president of the United States and are confirmed by the Senate. In essence, the judges serve for life or until voluntary retirement.

Within the federal system, a few courts deal with specialized matters. For example, the U.S. Court of Federal Claims (formerly the Claims Court) hears claims against the U.S. government, appeals from which are taken to the U.S. Court of Appeals for the Federal Circuit. Usually these cases relate to tax disputes or to disputes between government contractors and government agencies that award government contracts. Another specialized court is the Tax Court, which hears disputes between taxpayers and the government.

SECTION 2.03 Statute of Limitation: Time to Bring the Lawsuit

Statutory provisions require that legal action be commenced within a specified period of time. Such statutes in the context of the construction process are discussed in Sections 14.09C and 23.03G.

SECTION 2.04 Hiring an Attorney: Role and Compensation

Usually a person with problems that may involve the law or the legal system consults an attorney. After an inquiry into the facts and a study of the law, the attorney advises a client of her legal rights and responsibilities. Although the attorney may also give an opinion on the desirability of instituting legal action or defending against any action that has been asserted against a client, the litigation choice is usually made by the client.

Unless the law sets the fee, which is rare, attorney and client can determine the fee. Sometimes the fee is a specified amount for the entire service to be performed, such as uncontested divorces or simple incorporations. The fee may be a designated percentage of what is at stake, such as in the probate of an estate. Often the fee is based on time spent by the attorney, typically computed on an hourly basis. If no specific agreement on the fee is reached, the client must pay a reasonable amount.

Hourly rates charged depend on the attorney's skill, the demands on the attorney's time, the amount at stake, the complexity of the case, what the client can afford, the locality in which the attorney practices, and the outcome in the event of litigation. Although hourly rates vary greatly, at present an experienced attorney located in a large city is likely to charge from $250 to $500 an hour. (The second edition of this book, published in 1977, stated the hourly rate as $50 to $75.)

Outstanding lawyers with a great deal of experience are encouraged to use "value" billing instead of an hourly rate. This is based on the possibility that in a very short period of time they can give advice that is worth much more than would be compensated under even a high hourly rate.

A potential client should ask a prospective attorney the likely charges for legal services. Such advance inquiry can avert possible later misunderstandings or disputes over the hourly rate. The client may wish to set a maximum figure for a particular legal service. Attorneys are reluctant to accept maximum figures, because it is often difficult to predict the amount of time needed to provide proper legal service.

Commonly in personal injury or death cases and occasionally in commercial disputes, attorneys use contingent fee contracts. The lawyers are not paid for legal services if they do not obtain a recovery for the client. Usually a client agrees to reimburse the attorney for out-of-pocket costs, such as deposition expenses, filing fees, and witness fees. For taking the risk of collecting nothing for time spent, the attorney will receive a specified percentage of any recovery. In personal injury cases, the percentage can range from 25 percent to 50 percent, depending on the locality, the difficulty of the case, and the reputation of the attorney. In some metropolitan areas, attorneys will charge 25 percent to 33 percent if they obtain a settlement without trial and 33 percent to 40 percent if they go to trial.

The contingent fee system is much criticized, especially where the attorney obtains an astronomical sum in a tragic injury or death case. Such a system gives the attorney an entrepreneurial stake in the claim. Some feel this is unprofessional and may influence the attorney's advice as well as raise questions when attorney and client disagree over settlement. Yet some defend the contingent fee as the only method by which poor clients can have their claims properly presented. States are beginning to regulate contingent fees.

While contingent fee contracts tend to be used in claims for personal injury, a variant on this technique has been used in commercial disputes. This variant is sometimes called a "success" or "blended or hybrid" fee. In such a contract, the attorney agrees to reduce her normal fee, usually an hourly fee, in exchange for a percentage of anything recovered. In one case, the attorney reduced her fees by 20 percent in exchange for 10 percent of any final recovery, less fees paid by the client. In such a method, the other party, rather than the client, must pay a contingent fee to the client's attorney.

A retainer is an amount paid by a client to an attorney either at the beginning of or periodically during the attorney–client relationship. Although the retainer can be an agreed-on value for the legal services performed by the attorney or an amount to pay for having a call on the attorney's services unrelated to fees for services, the retainer payment is more commonly an advance payment on any fees that the client is obliged to pay the attorney. Sometimes it covers routine services but does not include extraordinary services, such as litigation. The attorney and the client should agree on the function and operation of any retainer.

Some attorneys, particularly in criminal or matrimonial matters, label the retainer as nonrefundable. If the attorney's services are not requested under the terms of such a retainer, no refund is available. Attorneys who use nonrefundable retainers assert that once having been retained, they are barred from being retained by the other party, because retention can generate conflict-of-interest problems later. Such attorneys assert that the client has an option to use the attorney whose services she purchases by paying the retainer.

Opponents of the nonrefundable retainer assert that such a payment limits the right of the client to choose her own attorney. They also contend that it is unconscionable and unethical for attorneys to receive compensation when they do not perform services.

In the past, bar associations published schedules of recommended or minimum fees. Such schedules violate the antitrust laws and are no longer used. For a discussion of architectural and engineering fee schedules, see Section 11.03.

The high cost of legal services and an increasing use of prepaid group health plans has led some to suggest prepaid group legal service plans for those in construction, such as those available to members of many trade unions.

The law recognizes the need for a client to be as candid in communicating with an attorney as with a member of the clergy, a doctor, or a spouse. For this reason, the attorney must keep confidential any communication made by the client, unless the communication was to plan or commit a crime or if the client challenges the competence of the attorney's performance.

SECTION 2.05 Jurisdiction of Courts

Jurisdiction—the power to grant the remedy sought—means power over the person being sued and over the subject matter in question. This section discusses state court jurisdiction.

The U.S. Constitution requires that state and federal governments ensure that a defendant receives due process, the opportunity of knowing what the suit is about, who is suing, and where and when the trial will take place. Usually due process is accomplished by having a process server hand to the defendant, in the state where the trial is to take place, a summons ordering the defendant to appear and a complaint stating the reasons for the lawsuit. The power of the state to compel the defendant to appear does not extend beyond the border of the state. If the defendant does not come within the state, the state court in which the lawsuit has been brought would traditionally not have jurisdiction over the person or entity against whom the claim has been made. The plaintiff would have to sue in the state where the defendant actually could be handed the legal papers.

The requirement of physical presence in the state where the court sits can place great hardship on a plaintiff who may be forced to begin a lawsuit in a state that is far from the evidence and witnesses. For this reason, most states have enacted long-arm statutes that permit plaintiffs in certain cases to sue in their home state even though they cannot hand the legal papers to the defendant within the plaintiff's state.

Long-arm statutes are often used to sue defendant motorists who reside in a state other than the one where the accident occurred or where the injured party resides. Suppose an Iowa driver were involved in an accident in Illinois that injured an Illinois resident. The Iowa driver returns to Iowa after the accident. The injured Illinois resident can always sue in Iowa. Usually the injured party would also be given the opportunity to sue in Illinois by mailing a notice of the lawsuit to the other driver in Iowa or by filing legal papers with a designated state official in Iowa or Illinois.

Sometimes long-arm statutes are used to sue businesses that have their legal residence in a state other than that of the plaintiff. To illustrate, suppose a New York company sells, directly or indirectly, electric drills to Wyoming purchasers. A Wyoming buyer is likely to be able to sue the New York seller in a Wyoming court for injuries resulting from the purchase and use of a defective drill.

Jurisdiction may also involve the subject matter of the lawsuit. For example, if land is located in State A and there is a dispute as to the ownership of the land, only State A will have jurisdiction, even though persons who may claim an interest in the land may not be residents of State A and may not be served with the legal papers in State A.

States usually limit their jurisdiction to matters of concern to that state. A state may have jurisdiction because property is located in the state, the injury occurred in the state, the plaintiff or defendant is a resident of the state, or a contract was made or performed within the state.

Another aspect of jurisdiction relates to the type of remedy or decree sought. Early in English legal history, two sets of courts existed—law courts and equity courts. Juries were used in the former but not in the latter. In addition, equity court judges could award remedies not available in the law courts. Parties seeking equitable relief had to show that their remedy in the law courts was inadequate. As a result of this dual set of courts, two sets of procedural and substantive rules emerged. Generally, the procedures and remedies tended to be more flexible in the equity courts.

The division between law and equity was incorporated into the U.S. judicial system. However, a gradual merger of the two systems has occurred in most states. For all practical purposes, in most states only one set of courts exists, although some differences remain. The most important is that certain remedies may be given only by an equity court. The most important of these remedies are specific performance (under which one party is ordered by a court to perform as promised in a contract), injunctions (orders by the court that persons do or do not do certain things), and reformation (rewriting of a written document to make it accord with the parties' actual intention).

As Section 2.10 explains, there are no juries when an equitable remedy is requested.

Contractual provisions sometimes specify the judicial forum for resolving disputes under the contract. Such clauses are generally upheld if they result from negotiation between contracting parties who are aware of the nature of the provision and are able to protect themselves. In the absence of any forum selection clause, the courts would apply their jurisdictional rules.

A distinction must be drawn between clauses that seek to bar a forum from one that would be available under the jurisdictional rules of the state, and clauses that seek to create jurisdiction where the state laws would not do so. The law will more likely allow a party to waive its right to use its courts if this was done deliberately and knowingly than it would allow contracting parties to create jurisdiction that would not already be present. The latter can place an unwelcome burden on the courts. Yet forum selection clauses can be the result of an abuse of power.

Increasingly, legislatures have addressed the problem of abusive forum selection clauses. For example, a Wisconsin law enacted in 1991 invalidates clauses in contracts for improving land in Wisconsin that require "that any litigation, arbitration or other dispute–resolution process on the contract occur in another state."[1] Under such a statute, a contract for building a structure in Wisconsin could not contain a provision requiring that disputes be resolved in New York.

SECTION 2.06 Parties to the Litigation

Generally, the party commencing the action is called the plaintiff, and the party against whom the action is commenced is called the defendant. In construction disputes, it is common for the defendant to assert a counterclaim

---

[1]West Wis. Stat. Ann. § 779.135(2). Virginia also limits the place of arbitration to Virginia. Va. Code Ann. § 8.01-262.1(B). For judicial intervention into the contractual arbitration forum selection process, see *Player v. George M. Brewster & Son, Inc.*, 18 Cal.App.3d 526, 96 Cal.Rptr. 149 (1971) discussed in Section 30.02. See Lyon & Ackerman, *Controlling Disputes by Controlling the Forum: Forum Selection Clauses in Construction Contracts,* 22 Constr. Lawyer, No. 4, Fall 2002, p. 15.

against the plaintiff or to make claims against third parties arising from the same transaction. For example, suppose an employee of a subcontractor sues the prime contractor based on a claim that the prime contractor has not lived up to the legal standard of conduct and this caused a loss to the claimant. The prime contractor may, in addition to defending the claim made by the employee of the subcontractor, sue the architect and the owner, claiming that the former was negligent and the latter was responsible for failure of the architect to live up to the legal standard, or sue the subcontractor employer of the claimant based on indemnification.

The prime contractor in asserting these claims is called a cross-complainant in most state courts and a third-party plaintiff in the federal courts. Those against whom these claims are made would be called cross-defendants in most state courts and third-party defendants in the federal courts.

Cross-claims by defendants were difficult to maintain in the law courts in England and under early American procedural rules. However, equity courts freely permitted a number of different parties to be involved in one lawsuit as long as the issues did not become too confusing. They sought to resolve in one lawsuit disputes relating to the same transaction. Ultimately, the rules developed in the equity courts prevailed, and it is generally possible to have multiparty lawsuits if jurisdiction can be obtained over all the parties and if litigation can proceed without undue confusion or difficulty.

SECTION 2.07 Prejudgment Remedies

Not uncommonly, the defendant against whom a judgment has been obtained has insufficient assets to pay the judgment. A judgment may be uncollectible if assets are hidden or heavily encumbered by prior rights of other creditors. Most states have statutes creating prejudgment or, as they are sometimes called, *provisional* remedies. Plaintiffs (or defendants who are asserting counterclaims) may be able, in advance of litigation, to seize or tie up specific assets of the opposing party to ensure that if they prevail, they can collect the judgment. Such assets may be attached if in the possession of the defendant or garnished if in the hands of third parties. It is not uncommon for a plaintiff to attempt to tie up assets such as bank accounts or other liquid assets in advance of the litigation.

The party whose assets are tied up can dissolve (have removed) the attachment or garnishment by posting a sufficient bond. Parties whose assets have been tied up may ultimately prevail in the litigation. If so, they may have unfairly suffered damage from the seizure or tying up of their assets. To protect against the risk of not finding assets out of which the prevailing party can be indemnified for such losses, the law frequently requires that the attachment or garnishment be accompanied by a bond.

Even if, by posting a bond, a party can avoid having assets tied up in advance of litigation, such a protective device is often not feasible where the party whose assets are being tied up or seized is poor. As a result, great hardship can fall on poor people and their families if the wage earner's wages are seized or if before the court hearing the sheriff repossesses a car, a television set, or furniture being purchased under a conditional sales contract. Under such arrangements, the buyer takes possession, but the seller retains ownership until all payments have been made. As a result of this, the U.S. Supreme Court has looked closely at prejudgment remedies to ensure that they give the defendant wage earner or debtor due process.[2]

Another prejudgment remedy is a preliminary injunction—an equitable decree ordering a defendant to cease doing something that the plaintiff claims is causing or will wrongfully cause injury to the plaintiff. Such an injunction can be issued before a full trial if the plaintiff can show that it is needed to preserve the status quo until the trial, that the plaintiff would suffer irreparable loss if it were not issued, and that there is a strong likelihood that the plaintiff will prevail at the trial. The party seeking the preliminary injunction must post a bond. This provides security to the other party for damages that may result from a subsequent determination that the injunction should not have been issued.

SECTION 2.08 Pleadings

The lawsuit is usually begun by handing to the defendant a summons and complaint, called *service of process*. The defendant has a specified time to answer. The defendant's answer may assert that the plaintiff has not stated the facts correctly or that defenses exist even if the allegations of the plaintiff's complaint are true. In some states, the

legal sufficiency of the complaint may be tested by filing a demurrer—a statement that even if the facts as alleged by the plaintiff are true, the plaintiff has no valid claim.

In addition to answering the plaintiff's complaint, the defendant in some cases may assert a claim against a third party by filing a cross-complaint or may file a counterclaim asserting a claim against the plaintiff.

In some states, the plaintiff is required to submit a reply to the answer. Most states require an answer to a counterclaim. In early American legal practice, a substantial number of other pleadings might be filed. The tendency in American procedural law today is to reduce the number of pleadings.

The pleadings should inform each party of the other party's contentions and eliminate from the trial matters on which there is no disagreement. By examining the pleadings, an attorney or a judge should be able to determine the salient issues to explore in the litigation. Pleadings should streamline the litigation and avoid the proof of unnecessary matters. Trial preparation should be more efficient and settlement expedited if each side knows the issues on which the other intends to present evidence at the trial.

The complaint should be a concise statement of facts on which a plaintiff is basing a claim and the specific remedy sought. Unfortunately, complaints are often prolix and contain factual statements and serious charges that the plaintiff's attorney may not be able to prove or that have not been checked out carefully. The inexperienced litigant should understand some reasons for this.

Frequently the lawsuit is begun before the plaintiff's attorney has had enough time to investigate thoroughly and use the discovery procedures described in Section 2.09. Unsure of the facts, an attorney may plead a number of different legal theories as a protective measure. The attorney wants to be certain that the complaint will be legally sufficient; that is, that the judge will not uphold a claim made by the opposing attorney that no legal claim has been stated.

The plaintiff's attorney may decide to use the identical language of pleadings that have been held sufficient in the past. Such language may not be tailor-made to the particular case in which it is used. If the language is taken verbatim from an old pleading form book, it may contain archaic and exaggerated language.

The plaintiff's attorney may believe that the only way to get the defendant or the defendant's attorney to consider seriously a client's claim is to start a lawsuit immediately,

---

[2]*Fuentes v. Shevin*, 407 U.S. 67 (1972).

make strong accusations, and ask for a large amount of money. Unfortunately, exaggeration and overstatement are frequently used legal weapons. But when litigants are not made aware of the reasons for such language, its use may simply increase the already existing hostility between the litigants and make settlement more difficult.

If the defendant has been served with a summons and complaint, failure to answer or to receive an extension of time to answer within the time specified by law allows the plaintiff to take what is called a *default judgment:* The case is lost by default.

The plaintiff who obtains a default judgment can collect on this judgment in the manner described in Section 2.13. Most states permit the judge to set aside the default judgment. To do so, the defendant must present cogent reasons why it would be unfair to enforce the judgment despite failure to respond to the pleadings in the designated time. Modern courts are more willing than courts a generation ago to set aside default judgments if the defendant can show that there is a valid defense, that failure to answer was the result of an excusable mistake, and that the amount of the judgment is substantial. Still, default judgments are rarely set aside. For this reason, a defendant who wishes to contest a claim made in a complaint should have an attorney answer the complaint within the required time.

SECTION 2.09 Pretrial Activities: Discovery

The parties and their attorneys wish to discover all facts material to the lawsuit. Discovery is the legal process to uncover information in the hands of the other party. One method of discovery is written interrogatories. One party's attorney sends a series of written questions to the other party, who has a specified period of time in which to respond.

The other method is to take a deposition; that is, to compel the other party or its agent to appear at a certain time and place and answer questions asked by the attorney seeking discovery. The attorney is also likely to demand that the person being questioned bring all relevant documents relating to the matter in dispute. Unlimited discovery demands are being curbed in the federal courts and in some states because of time and expense burdens. Limits can include restricting the scope of the inquiry, the number of persons that can be deposed, and the durations of the deposition. Discovery, though designed to promote

settlement and avoid surprise, has become enormously expensive.

The person being questioned usually brings an attorney. Questions are asked under oath. Questions and answers are transcribed by a reporter, and a transcript is made available to both parties.

Generally the transcript does not substitute for testimony at the trial. However, the transcript can be used to impeach (that is, to contradict) the witness if the testimony at the trial is inconsistent with the statements made under oath in the discovery process.

A proper use of discovery should avoid surprise at the trial, reduce trial time, and encourage settlement. However, the sweeping range of inquiry and the wide latitude given to demand that documents be produced has made the process time consuming and very costly.

Witnesses who will be out of jurisdiction at the time of trial or who are very sick and may not be alive when the case is tried can also be examined under oath and their testimony recorded and transcribed. This deposition process can substitute for testimony that would otherwise be unavailable at the time of trial.

In many states, the pretrial conference is another step in the litigation process. Usually such a conference is conducted in the judge's chambers in the presence of the judge, the attorneys, and sometimes the parties. The main purposes of the pretrial conference are, like those of the pleadings, to narrow the issues, avoid surprise, and encourage settlement.

SECTION 2.10 The Jury

A dispute not settled or abandoned before trial will be submitted to a judge and sometimes to a jury. Generally, the right to a jury trial is constitutionally guaranteed, except in claims where decrees by an equity court are sought. The parties can agree to try the case without a jury. The principal present use of juries is in criminal and personal injury trials.

The use of juries in commercial disputes, although available, is less common. In commercial disputes, the use of juries is criticized when the case is extraordinarily complex. This problem comes up frequently in construction disputes, which often can involve jury trials of up to six months. Yet U.S. courts as a rule have been unwilling to deprive a claimant of the right to a jury trial as guaranteed by the Constitution, despite the complexity of the case and the difficulty the jurors might have in deciding the case properly.

Historically, juries have consisted of twelve laypersons selected from the community to pass on guilt and sometimes to sentence in criminal matters and to decide disputed factual questions in civil matters. Juries have been attacked as inefficient. Jury trials take longer, cost more, and have a higher degree of reversals by appellate courts than those cases resolved by a judge. Some states reduce jury size from twelve to six members. In criminal matters, the decision must be unanimous. Some states require only a five-sixths decision in civil disputes.

In many state courts, attorneys are allowed to question prospective jurors to determine their impartiality. An attorney dissatisfied with a particular juror can ask that the juror be excused for cause. The judge rules on whether there is proper cause to excuse the juror. Usually the attorneys can strike (that is, excuse) a designated number of jurors peremptorily (that is, for no reason). But a peremptory challenge cannot be based on race. In the federal courts, jury examination is typically conducted by the judge.

Jurors need give no reason for their decision in a case. Although the judge instructs the jury as to the law, jurors are, for all practical purposes, free to do as they choose. Nullification is the name given to this informal power. Some criticize nullification; others feel it is a useful safety valve, especially in criminal matters, to allow members of the community to excuse law violators.

Until recently, jurors did not make the details of their deliberations public. But of late jurors, especially in controversial cases, openly describe the jury process to representatives of the media and anyone else who may be interested or willing to pay.

Some feel that juries are easily manipulated by clever and persuasive attorneys. Others feel that jurors are sensible and generally come to the right result. Those who prefer juries often point to the unfortunate phenomenon of judges who are incompetent but who cannot be removed because of the great difficulty of impeaching a judge. (Some states allow one peremptory challenge of a judge to whom a case has been assigned.)

SECTION 2.11 Trials: The Adversary System

In some jurisdictions, the parties are required before the actual trial to submit their dispute to nonbinding arbitration. Usually the arbitrator is a volunteer local attorney, and usually the requirement exists only for claims within a certain dollar amount.

After the results of the nonbinding arbitration, parties generally do not seek an actual trial. They are either satisfied with the outcome, are principally concerned in having had a neutral look at their dispute, or are deterred by the cost of going to trial.

The trial is usually conducted in public and is begun with an opening statement by the plaintiff's attorney. Sometimes the defendant's attorney also makes an opening statement. In the opening statement, the attorney usually states what she intends to prove and also seeks to convince the jury that the client's case is meritorious. Opening statements are less common in trials without a jury.

After opening statements, the plaintiff's attorney may call witnesses who give testimony. Also, during this phase physical exhibits and documents can be offered into evidence. In civil actions, the other party can be called as a witness.

An attorney cannot ask leading questions of the witnesses except to establish less important preliminary matters or where the witness is of low mentality or is very young. A leading question is one that tends to suggest the answer to the witness. Such questions can be and usually are asked of witnesses called by the other party.

Witnesses are supposed to testify only to matters they have perceived through their own senses. Witnesses usually cannot express opinions on technical questions unless they are qualified as experts. Expert testimony is discussed in greater detail in Section 14.06.

The hearsay rule precludes witnesses from testifying as to what they have been told if the purpose of the testimony is to prove the truth of the statement. The danger in hearsay testimony is that the party who made the statement is not in court and cannot be cross-examined as to the basis on which the statement was made. However, there are many exceptions to the hearsay rule.

After the attorney has finished questioning the witness, the other party's attorney will cross-examine and try to bring out additional facts favorable to her client or to discredit the testimony given on direct examination. Cross-examination can be an effective tool to catch a perjurer or show that a witness is mistaken. However, used improperly it can create sympathy for the witness or reinforce testimony of the witness.

Documents play a large role in litigation. Under modern rules of evidence, it is relatively easy to introduce into evidence documents for the court to consider. Documents are hearsay testimony. They are usually writings made by persons who are not in court. However, exceptions to the hearsay rule permit the admission of documents such as official records and business entries. Often the attorneys will stipulate to the admissibility of certain documents.

After the plaintiff's attorney has presented her client's case, the defendant's attorney will present the defendant's case. Then the plaintiff is given the opportunity to present rebuttal evidence. After all the evidence has been presented, the judge submits most disputed matters to the jury (when one is used). The judge instructs the jury on the law, using instructions sometimes difficult for a juror to understand. The jury meets in private, discusses the case, takes ballots, and decides who prevails and how much should be awarded to the winning litigant. (In controversial cases, this is often followed by a press conference. See Section 2.10.)

The adversary system, though much criticized because of its expense, its often needless consumption of time, and the hostility it can engender, is central to the American judicial process. Each party, through its attorney, determines how its case is to be presented and can present its case vigorously and persuasively. Also, each party, mainly through cross-examination and oral argument, has considerable freedom to attack the other's case. The judge generally acts as an umpire to see that procedural rules are followed.

One justification offered for the adversary system is that when the smoke has cleared, the truth will emerge. This assumes well-trained advocates of relatively equal skill and a judge who ensures that procedural rules are followed.

Another justification is that the adversary system gives the litigants the feeling they are being honestly represented by someone of their choosing and are not simply being judged by an official of the state. The system can give the litigants a sense of personal involvement in the process rather than simply being passive recipients of decisions handed them by the state. However, excessive partisanship can generate bitter wrangles and delays. Also, it can make a trial resemble military combat rather than a reasoned pursuit of the truth.

Even under the adversary system there are limits on advocacy. The law prescribes rules of trial decorum administered by the judge, who can punish those who violate these rules by citing violators for contempt. The person cited can be fined and, in some cases, imprisoned. In some court systems, an attorney can be sanctioned for certain types of conduct that do not rise to the level required to hold the attorney for contempt of court. Usually the sanction carries a fine.

The legal profession can discipline its members who violate professional rules of conduct. As obvious illustrations, an attorney must not bribe witnesses, encourage or permit perjured testimony, or mislead the judge on legal issues. An attorney is the champion for her client yet part of the system of the administration of justice, and must not do anything that would subvert or dishonor that system.

In hotly contested litigation, attorneys frequently make objections. This is always irksome to participants, but it is often, though not always, necessary. If objections are not made at the proper time, the right to complain of errors may be lost. The judge should be given the chance to correct judicial mistakes on the spot.

The possibility of drama and tension in the litigation process can increase when a jury is used. Some attorneys employ dramatic tactics to influence the jury. Although some enjoy the combat of litigation, litigation is at best unpleasant and at worst traumatic for the litigant or witness who is not accustomed to the legal setting, the legal jargon, or the adversary process. This can be intensified if the trial attracts members of the public or is reported in the media.

A trial is an expensive way to settle disputes. In addition to attorneys' fees, witness fees, court costs, and stenographic expenses, there are less obvious expenses to the litigant. Much time must be spent preparing for and attending the trial. The litigant may have to disrupt business operations by searching through records. For these and other reasons, most lawsuits are settled out of court.

Yet extensive preparation costs are often incurred even if a case is settled out of court. The litigation process can generate huge costs. This has led to the search for and the creation of the Alternative Dispute Resolution (ADR) discussed in Chapter 30.

SECTION 2.12 Judgments

A judgment is an order by the court stating that one of the parties is entitled to a specified amount of money or to another type of remedy. Some judges rule immediately after the trial. Usually the judge takes the matter under consideration. Often judges write opinions giving reasons for their decisions.

SECTION 2.13 Enforcement of Judgments

If the plaintiff obtains a money award, the defendant should pay the amount of money specified in the judgment. However, if the defendant does not pay voluntarily, the plaintiff's attorney will deliver the judgment to a sheriff and ask that property of the defendant in the hands of the defendant or any third party be seized and sold to pay the judgment. It is often difficult to find a defendant's property. In some states, defendants can be compelled to answer questions about their assets.

Even if assets can be found, exemption laws may mean that certain assets may not be taken by the plaintiff to satisfy the judgment. Legislatures declare certain property exempt from seizure. Statutes usually contain a long list of items of property that cannot be seized by the sheriff to satisfy a judgment. Such items are considered necessary to basic existence and include automobiles, personal clothing, television sets, tools of trade, the family Bible, and other items that vary depending on the state and the time in which the exemption laws were passed.

Homestead laws exempt the house in which the defendant lives from execution to satisfy judgments. Financial misfortunes notwithstanding, a debtor and those dependent on the debtor should have shelter. Sometimes homestead rights are limited to a certain amount. Under certain circumstances, a homestead can be ordered sold.

In many cases, judgments are not satisfied, because the defendant has no property or the property in the defendant's possession is exempt. Collecting on a judgment when the defendant does not pay voluntarily can be difficult, uncertain, and costly—another reason for settlement.

SECTION 2.14 Appeals: The Use of Precedent

The party who appeals the trial court's decision is called the appellant. The party seeking to uphold the trial court decision is called either the appellee or the respondent.

The attorney for each party will submit a printed brief to the appellate court. The court can permit a typewritten brief. Courts increasingly limit the number of pages in a brief. The brief of the appellant seeks to persuade the appellate court that the trial judge has made errors, and the brief for the respondent seeks to persuade the appellate court to the contrary or that any errors committed were not serious enough to warrant reversal.

Usually the attorneys make brief oral arguments before the appellate court. To save time, some judges do not allow oral arguments in simple cases. Sometimes oral arguments consist of a summary of the briefs. Sometimes the attorneys are questioned by the judges on specific points that appear in the briefs or trouble the judge. Introducing evidence is not permitted in appeals.

In the American system, most appellate judges write opinions stating their reasons for the way they have decided the case. Decisions vary in length. Usually the legal basis for a decision in a trial court or appellate court is a statute, a regulation, or a prior case precedent. The latter requires amplification.

All decision makers seek guidance from the past. Nonjudicial decision makers, such as committees, boards of directors, and organizations of all types, often seek to determine what they have done in the past as a guide for what they should do in the present. Similarly, a refusal to make a particular decision may be based on the fear that it will be looked on as a precedent and the basis of a claim by others for similar treatment. However, in such decision making it is not likely that the precedents of the past must be followed. English and American courts must follow precedent, subject to an exception described later in this section. Within a particular judicial system, judges must decide matters in accordance with earlier decisions of higher courts within the system.

For example, all judges in the state of California must follow the precedents set by the highest member of the system—the state supreme court. Trial judges and those who serve on the intermediate appellate courts must follow decisions of the intermediate appellate courts.

As stated, most appellate judges give reasons for their decisions in written opinions. All members of the system are expected to follow the reasoning as well as the results in earlier cases decided by a higher court. However, California judges need not follow decisions from other state courts, although they may look to decisions of other state courts for guidance. Nor are they compelled to follow the decisions of federal courts except the decisions of the U.S. Supreme Court, which involve federal law.

Despite reverence for precedent, some European countries forbid following precedent, and some decision makers in this country, such as labor and commercial arbitrators, avoid it. This suggests that arguments can be made for and against following precedent.

A legal system should be reasonably predictable. To plan their activities, persons and organizations wish to know the law in advance. If they can feel confident that future judges will follow earlier decisions in similar cases, they can predict and plan more efficiently. Knowing in advance what a judge or court will do encourages settlement of disputed matters. A fair judicial system treats like cases alike.

Another argument for following precedent relates to conservation of judicial energy. If a matter has been thoroughly reviewed, analyzed, and reasoned in an earlier decision, there is no reason for a later court to replow the same ground. This reason for following precedent assumes great confidence in the decision makers of the past and assumes a relatively static political, economic, and social order in which the judicial system operates.

Those who attack the precedent system maintain that the sought-after certainty is illusory, because courts can avoid precedent. They also argue that a precedent system tends to become rigid and unresponsive to changing needs. Finally, they contend that the precedent system can be an excuse for judges avoiding hard questions that should be examined.

Precedents must be followed only in similar cases. If the facts are different, precedents can be distinguished by judges and held not to control the current case. Even under a system of following precedents, at times earlier precedents become so outmoded that they are overruled. When American judges cannot accept the result that applying precedent compels, they create exceptions to the precedent. The greater degree of dissatisfaction, the greater the number of exceptions likely to be created. When the exceptions seem to swallow up the precedent, an activist court may recognize the realities and overrule the precedent.

Good courts seek to avoid unthinking rigidity at one extreme and whimsical decision making at the other. The important feature of a precedent system is reasonable predictability that can accommodate change. The skillful attorney reads the precedents but also looks at the facts and at any relevant political, social, and economic factors that may bear on the likelihood that the precedent will be distinguished or overruled.

Usually all the members of the appellate court agree. Sometimes a dissenting judge gives reasons for not agreeing with the majority. Sometimes the majority of the court that agrees on the disposition of the case cannot agree on the reasons for the decision, and one or more majority judges may write a concurring opinion.

SECTION 2.15 International Contracts

Increased worldwide competition exists for building and engineering contracts. As a result, those who design and construct may be engaged by a foreign national, often the sovereign itself, in a foreign country where the work is to be performed. These international contracts raise special problems discussed in Sections 8.09 and 30.21. One topic deserves mention here, however.

Legal systems vary not only as to substantive law but also as to the independence of their judiciary and dispute resolution processes. In some countries, the judiciary is simply an arm of the state, and follows to a large degree the will of the head of state or the agency with whom the contract has been made.

A foreigner may not have confidence in the impartiality of such a dispute resolution process. This lack of confidence can even be a problem in the United States or within a particular state. The U.S. Constitution recognized this by allowing the removal of a case from a state court to the federal court if there is diversity of citizenship. This problem can be particularly difficult when the designer or contractor is a company from an industrialized country doing business in a less developed country.

Lack of confidence in the judiciary often necessitates a contract clause under which disputes will be resolved by some international arbitration process. Even where there is more confidence in the independence of the judiciary, such as in a transaction where disputes would normally be held before a court in the United States or a western European country, unfamiliarity with the processes and the possible application of unfamiliar law may lead to contractual provisions dealing with disputes. For example, European legal systems do not use the adversary system. They use a more inquisitorial system under which the judge is likely to be a professional civil servant who plays a more active role in resolving disputes. Similarly, laws in western European countries are often based on brief yet comprehensive civil codes quite different from those an American lawyer may encounter in domestic practice. Under such conditions, it is common for the parties to provide by contract that disputes will be resolved by international arbitration with the applicable law of a neutral and respected legal system.

CHAPTER THREE

Forms of Association: Organizing to Accomplish Objectives

SECTION 3.01 Relevance

People engaged in design or construction should have a basic understanding of the ways in which individuals associate to accomplish particular objectives. The professional in private practice should know the basic elements of the forms of associations that are professionally available. Anyone who becomes an employee or executive of a large organization should understand the basic legal structure of that organization. The design professional who deals with a corporate contractor on a project should understand corporate organization. Those who are shareholders in or transact business with a corporation should understand the concept that insulates shareholders from almost all liabilities of the corporation.

The following are the most important organizational forms:

1. sole proprietorship
2. partnership
3. corporation
4. limited liability entity
5. joint venture
6. unincorporated association
7. loose association

SECTION 3.02 Sole Proprietorships

The sole proprietorship, although not a form of association as such, is the logical business organization with which to begin the discussion. It is the simplest form and is the form used by many private practicing design professionals.

The creation and operation of the sole proprietorship is informal. By the nature of the business, the sole proprietor need not arrange with anyone else for operating the business. Generally, no state regulations apply except for those requiring registration of fictitious names or for having a license in certain businesses or professions. Sole proprietors need not maintain records on the business operation except those necessary for tax purposes. Sole proprietors have complete control over the business operation, taking the profits and absorbing the losses. They rent or buy space for operating the business, hire employees, and may buy or rent personal property used in the business.

The proprietorship continues until abandonment or death of the sole proprietor. Continuity of operation in the event of death sometimes can be achieved through a direction in the proprietor's will that an executor continue the business until it can be taken over by the person to whom it is sold or given by will. This ensures that the business is continued during the handling of the estate and that there is no costly hiatus in operation.

The proprietor may hold title to property in his own name or in any fictitious name. Interest in the business can be transferred, the only exception being that in some states the spouse of the proprietor may have an interest in certain types of property used in the enterprise. Capital must be raised by the proprietor or by obtaining someone to guarantee the indebtedness. This differs from a corporation, which can issue shares as a method of raising capital.

SECTION 3.03 Partnerships

A. Generally: Uniform Partnership Act

A partnership is an association of two or more persons to carry on a business for profit as co-owners. Unlike a corporation, it is not a legal entity. All states except

Louisiana, many making minor variations, adopted the original Uniform Partnership Act (UPA), issued in 1914. The Act was revised in 1997. As of 2005, 34 states have adopted the Revised Uniform Partnership Act (RUPA). The Acts set forth the rights and duties between the partners themselves if the partners have not specified to the contrary in the partnership agreement. They also deal with claims of third parties.

A new form of partnership, limited liability partnership (LLP), is discussed in Section 3.06.

B. Creation

Although the partnership agreement need not be evidenced by a sufficient written memorandum, such a memo is advisable. In addition to reasons for having any agreement in writing, other reasons exist for expressing partnership agreements in writing. The transfer of an interest in land must be evidenced by a written memorandum, as must agreements that by their terms cannot be performed within a year. If the creation of the partnership is accompanied by a transfer of land, that portion of the partnership agreement dealing with the land should be evidenced by a written memorandum.

If the duration of the partnership expressly exceeds a year, the partnership agreement should be evidenced by a written memorandum. Agreements to answer for the debts of another must, under certain circumstances, be evidenced by a written memorandum. Partnership agreements with provisions of this type should be evidenced by a written memorandum.

In stating that certain agreements must be evidenced by a written memorandum, a distinction is frequently misunderstood. Suppose a partnership is created orally for a period of more than a year. The rights and duties of the partners will still be governed by the oral agreement and by the applicable Uniform Partnership Act. However, partners have no legally enforceable obligation to perform in the future, although they will have to perform those obligations that accrued before the decision to terminate the oral partnership.

Third parties will still have whatever rights the law would give them against the partners despite the requirement that the partnership agreement be evidenced by a written memorandum. The fact that the law required the partnership agreement to be so evidenced because it was to last over a year will have a limited impact on the partners and, in most cases, no impact on third parties.

C. Operation

Generally the partners decide who is to exercise control. Most matters can be decided by majority vote. However, certain important matters, such as amendment of the partnership agreement, must be by unanimous vote. Sometimes partners want something other than majority rule. In a large partnership, a small group of partners may be given authority to decide certain matters without requiring that a majority of the partners approve such decisions.

D. Fiduciary Duties

Partnership is an important illustration of the fiduciary relationship. As fiduciaries, each partner must be able to trust and have confidence in the other. Neither must take advantage of this trust for selfish reasons or in any way betray the partnership. For example, a partner must account to the partnership for any secret profits made. A partner must not harm the partnership because of any undisclosed conflict of interest. For example, where a partnership composed of design professionals represents the owner in dealing with a construction company, it would be improper for one of the partners to have a financial interest in the construction company unless it was disclosed in advance to the partners and to the owner-principal.

Because the fiduciary obligation is very important, it may be helpful in the partnership agreement to spell out the permissible scope of outside activities of individual partners. In the absence of an understanding among the partners as to permissible outside activities, any activity that raises a conflict of interest or competes in any way with the partnership is not proper. For further discussion, see Section 11.04B.

E. Performance Obligations, Profits, Losses, Withdrawal of Capital and Interest

In the absence of any agreement to the contrary, the partners are to devote full time to the operation of the business and to share profits and losses equally even if one works more than the others. If a specified proportion of the profits is allocated by partnership contract to specified partners, that same percentage will apply to losses. If some other arrangement is desired, it must be expressed in the partnership agreement.

Normally partners cannot withdraw capital during the life of the partnership unless permitted by the partnership agreement. A partner may collect interest for money or property lent to the partnership as well as for capital contributions advanced on request of the partnership.

F. Authority of Partner

Section 301 of the 1997 RUPA makes each partner the agent of the partnership. Acts that apparently carry on ordinary partnership business bind the partnership. If the partner had no actual authority and the person dealing with the partner knew or was notified of this, the acts do not bind the partnership. Acts that apparently do not carry on the partnership business bind the partnership only if the act was authorized by the other partners.

Apparent authority can charge the partnership with unauthorized acts by a partner. The partnership can create apparent authority by making it appear to third parties that a partner has authority he does not actually possess (see Section 4.06). Certain extraordinary acts, such as criminal or other illegal acts, are not charged to the partnership. However, because of the vast range of authority—both actual and apparent—given to the partners and the fiduciary obligation owed by each partner to the other, the character and integrity of prospective partners are of great importance.

G. Liability of General Partnership and Individual Partners

The liability of the partnership and the partners for debts of the partnership and individual debts of the partners is complicated and confusing. Clearly, creditors of the partnership can look to specific partnership property and, if this is insufficient, to the property of the individual partners. Creditors of individual partners who obtain a court judgment may satisfy the judgment out of the partner's interest in the partnership property, but the partnership can avoid losing the property by paying the judgment.

Even more complicated is the liability of an incoming partner for obligations created before becoming a partner, and of an outgoing partner for obligations incurred after leaving the partnership. Generally, the partner is not liable for obligations incurred before becoming a partner. Partners can eliminate liability for obligations that occur after leaving the partnership by informing those who have dealt with the partnership that they have left. But because of the uncertainty of the law in this area, it is best for those who enter existing partnerships to determine what obligations exist at the time they enter the partnership and for those who leave partnerships to so notify those who have dealt with the partnership.

Because of the ease with which partnerships can be dissolved and reconstituted, considerable confusion exists as to the right of creditors of predecessor partnerships to hold successor partnerships liable for the obligations of the predecessor.

A partner who incurs liability or pays more than a proper share of a debt should receive contribution from the other partners. Similarly, a partner who incurs liability or pays a partnership debt should be indemnified by the partnership. Usually these instances are expressly dealt with in the partnership agreement.

H. Transferability of Partnership Agreement

A partner has some transferable interests. Profits can be assigned, but not specific partnership property or management and control.

I. Dissolution and Winding Up

Partnerships do not have the legal stability of corporations. Sections 601 and 602 of the RUPA list events that can dissolve the partnership or give a court the power to dissolve a partnership on application by a partner or by a purchaser of a partner's interest. Sections 801–805 of the RUPA provide rules for winding up a partnership.

J. Limited Partnership

Most states permit the creation of a limited partnership made up of one or more general partners and one or more limited partners. Such an organization permits investors to receive profits but limits their liability to their investment without having to use the corporate form. Generally, the general partner manages or controls the corporation while the limited partners are investors. The general partner owes fiduciary obligations to the limited partners.

Most states adopted the Uniform Limited Partnership Act of 1916. Most of those states adopted a new version, the Uniform Limited Partnership Act of 1976, with modifications made in 1985. Currently all states but

Louisiana have adopted one of the two versions of the Uniform Limited Partnership Act, and many states have modified the act in adopting. The details of such statutes are beyond the scope of this book.

SECTION 3.04 Profit Corporations

A. Use

The corporate form is used by most large and medium-sized businesses in the United States and has become the vehicle by which many small businesses are conducted. In addition, practicing design professionals are increasingly choosing the corporate form, where possible, for their business organization. The corporate form takes on even greater significance in light of the increasing number of design professionals employed by corporations. Because many complexities in corporation law cannot be discussed in this book, the discussion here must be brief and simple.

B. General Attributes

Although the partnership is merely an aggregate of individuals who join together for a specific purpose, the corporation is itself a legal entity. It exists as a legal person. It can take, hold, and convey property and sue or be sued in its corporate name. As shall be seen, the other important corporation attributes are management centralized in the board of directors, free transferability of interests, and perpetual duration. In addition, the corporation offers the advantage of limiting shareholder liability for debts of the corporation to the obligation to pay for corporate shares purchased.[1] Although not all these attributes are available to all types of corporations, as a general introductory statement these attributes set the corporation apart from the sole proprietorship and partnership.

C. Preincorporation Problems: Promoters

Often persons called *promoters* set into motion the creation of a corporation. Although they may continue in control, sometimes they merely organize the corporation and turn control over to others. Usually they are compensated by the corporation after the corporation has been organized.

This compensation may be cash, shares of stock, stock options, or positions within the corporation.

During the promotion phase, promoters often make contracts with third parties, who should consider the possibility that the corporation will not be formed or will not adopt the promoter's contract. Considering such possibilities may induce the third party to insist that the promoter be held individually liable if the corporation is not formed or does not adopt the promoter's contract. A third party who has doubts about the finances of the prospective corporation may wish to hold the promoter liable even if the contract is adopted by the corporation.

D. Share Ownership

The corporation is a separate legal entity owned by the shareholders, who do not own any part of specific corporation property. Shareholders have a right against the corporation that is governed by statutes, the articles of incorporation, and the wording of the shares. One great strength of modern corporation law is the variety of types of shares that can be employed by the corporation—preferred and common, par and no-par, and voting and nonvoting.

Ordinarily, shares of stock in a corporation are freely transferable (a major advantage of incorporation). Usually restrictions exist on transferability of shares in closely held corporations whose shares are not sold to the general public. This lets shareholders keep control of the company by preventing outsiders from becoming shareholders. Restrictions on transferability sometimes accompany shares purchased by executives or employees as part of a stock option plan.

E. Piercing the Corporate Veil

A principal function of the corporation is to shield shareholders from corporate obligations. Ordinarily, the liability of shareholders is limited to paying for shares they have purchased. This limitation of liability can operate unfairly where a third party relies on what appears to be a solvent corporation. The corporation thought to be solvent may be merely a shell. The assets of the corporation may not be sufficient to pay the corporate obligations. Someone injured by the acts or failure to act of a corporation may find the corporation cannot pay for damages because the amount of capital paid into the corporation was very small.

---

[1]*Saltiel v. GSI Consultants, Inc.*, 170 N.J. 297, 788 A.2d 268 (2002).

The corporate form can be disregarded and, to use a picturesque phrase, the corporate veil pierced and shareholders held liable. A court may do so if unjust or undesirable consequences would result by interposing the corporation as an entity between the injured party (or a creditor) and the shareholders. This is more likely to be done where a person has suffered physical harm through corporate activities, where a creditor could not reasonably have been expected to check on the credit of the corporation, and where the one-person, family, or closely held corporation is used. The latter types of corporations have been popular because they combine control with limitation of liability. Ordinarily, such corporations are valid and protect the shareholders from liability. However, this is so only if the corporation is used for legitimate purposes and the business is conducted on a corporate basis. Also, the enterprise must be established on an adequate financial basis so that the corporation can respond to a substantial degree to its obligations. If not, circumstances such as those mentioned can result in piercing the corporate veil. Similar problems can arise when a corporation organizes a subsidiary corporation and holds all the shares of the subsidiary.

Sometimes owners on small projects (usually homeowners), who deal with a corporation consisting of a very limited number of individuals, seek to impose liability not just on the corporation, but also on the individual corporate officers who either did the actual work or supervised it. These owners sue the corporation for breach of contract and the officers for negligence. In *Greg Allen Construction Company, Inc. v. Estelle*, the Indiana Supreme Court refused to find the corporation's president and sole owner to be individually liable for his own negligent work. The court pointed out that, because the president was working as the corporation's agent, he could be personally liable to the homeowners only if he had committed an independent tort. (Agency is discussed in Chapter 4.) Instead, the president was simply fulfilling the corporation's contract obligations, albeit in a negligent manner.[2]

---

[2]798 N.E.2d 171 (Ind.2003). However, corporate officers were found liable in *Michaelis v. Benavides*, 61 Cal.App.4th 681, 71 Cal.Rptr.2d 776 (1998) (defective work created hazardous conditions) and *Fontana v. TLD Builders, Inc.*, 362 Ill.App.3d 491, 840 N.E.2d 767 (2005) (court pierced the corporate veil).

F. Activities, Management, and Control: Insider Misconduct

The state law under which the corporation was created determines the outer limits of corporate activities and organization. Because state laws generally extend considerable latitude to the corporation on such matters, as a rule, activities and organization are governed by the articles of incorporation.

The articles of incorporation are the constitution of the corporation. They generally set forth the permissible activities of the corporation, the organization and management of the corporation, and the rights of the shareholders. Sometimes these matters are phrased in general terms in the articles of incorporation and articulated more specifically in the corporate bylaws. In large corporations, a set of corporate documents is likely to delineate the chain of command and state the authority of corporate officers and employees to handle particular corporate matters. The ultimate power within the corporation lies with the shareholders, who delegate this power to an elected board of directors. In theory, the board of directors controls long-term corporate policies, the day-to-day operations being handled by the corporate officers.

This model of corporate control may vary depending on the type of corporation. In a smaller corporation, the board of directors, or even large shareholders, may influence day-to-day operations. If the corporation is a large, publicly held corporation with thousands of shareholders, the actual power rests largely with the management. Even though theoretically the shareholders can displace the board of directors, the diffusion of share ownership often makes such displacement difficult, and gives the board of directors and the officers of the corporation effective power, with the shareholders having little control over policy or corporate acts. However, this does not make the corporation immune to takeover bids by other corporations. Increasingly, investors with large blocks of shares, such as pension funds, are playing a more active role in monitoring management decisions.

Some corporations have cumulative voting of shareholders for directors. A shareholder asked to choose a slate of seven members of the board may cast all seven votes for one candidate. Cumulative voting is required in some states and has the effect of protecting minority interests in the corporation.

The directors choose the officers of the corporation. Usually they will select a president, vice president, secretary, and treasurer. Larger corporations might also, through the board of directors, designate persons to serve as general counsel or controller or as other corporate officers.

The officers are in charge of the day-to-day running of the corporation. The larger the corporation, the more likely that many details will be delegated to other employees of the corporation.

There has been a marked development in U.S. law toward extending fiduciary obligations of fair dealing to directors and to officers. Directors and officers owe each other fiduciary duties of fair dealing and owe a fiduciary duty to the shareholders. It is not proper, for example, for a member of the board or for an officer to use inside information to purchase shares of corporate stock from shareholders who are not aware of the same inside information. Nor would it be proper (and very likely be illegal) for corporate insiders to disclose information, such as impending mergers or takeovers or other news that would inflate or depress the value of the stock not available to the general public.

Directors and officers are not permitted to take advantage of economic opportunities that should be made available to the corporation.

The articles of incorporation usually provide for annual shareholder meetings as well as for periodic meetings of the board of directors. Minutes must be kept of all board proceedings. The administrative burden of operating a corporation can be formidable.

G. Profits and Losses

Usually the board of directors determines the disposition or distribution of profits made by the corporation. Sometimes control is limited by the articles of incorporation, especially the rights of preferred shareholders. In addition, the corporation may have obligations to the creditors of the corporation based on contracts or on agreements made between the corporation and lenders such as shareholders or banks. Profits may be reinvested in the corporation and used for corporate purposes.

Subject to statutory regulations, the board determines when (and how large) a share of the profits is to be paid as dividends to the shareholders. The board may pay dividends only out of certain specified funds. Sometimes the board issues a stock dividend instead of a cash dividend. In such cases, the shareholders receive additional shares in the corporation instead of money. Statutory or

other limitations may exist relating to the redemption of shares by the corporation and to the repurchase by the corporation of its own shares.

If the board unlawfully issues dividends, the board members are liable to the corporation and to those creditors of the corporation harmed by the unlawful declaration of dividends. Normally, individual shareholders are not responsible for losses of the corporation because of the insulation from personal liability given corporate shareholders. A shareholder who paid for shares in accordance with the purchase agreement with the corporation is not liable for any obligations of the corporation.

H. Life of the Corporation

One advantage of the corporation is its perpetual life. Sole proprietorships end with the death of the sole proprietor. Often partnerships end with the death of any of the partners. The corporation will continue despite the death of shareholders. A court can dissolve a corporation in the event of a deadlock in the board of directors or under other circumstances.

The dissolution of a corporation is complicated and is governed largely by statute. The bankruptcy laws play a large role in disposing of corporation assets on dissolution. This book does not discuss dissolution.

SECTION 3.05 Nonprofit Corporations

Nonprofit corporations are similar in organization and operation to profit corporations. The major difference is that a nonprofit corporation cannot distribute profits to shareholders. Examples of nonprofit corporations are hospitals, most educational institutions, and charities.

Capital for a nonprofit corporation is raised by donations, grants, and, occasionally, the sale of shares. Usually there are members instead of shareholders. The members elect the board of directors or trustees; the board selects officers to run the corporation. Nonprofits have articles of incorporation, bylaws, meetings, and other similarities to profit corporations. Generally, the shareholders are insulated from personal liability for the debts of the corporation. Nonprofit corporations are exempt from taxes because they have no profit. Nonprofit corporations that engage in certain types of political or profit-making activities can lose their tax exemptions.

SECTION 3.06 Professional Corporations, Limited Liability Companies (LLC), and Limited Liability Partnerships (LLP)

Although design professionals traditionally have practiced as sole proprietors or partnerships, states have increasingly permitted them to practice through a professional corporation. The corporation form has been used mainly to take advantage of tax laws that allow employees of corporations to receive fringe benefits without having them included within taxable income.

Some states that have allowed professionals to incorporate have made clear that the professionals cannot use the corporate form to shield their individual assets—a principal reason for using business corporations.

Design professionals have begun to take advantage of two forms of business association that became prominent in the 1990s: the limited liability company (LLC) and the limited liability partnership (LLP). The LLC is a hybrid of the corporate and partnership forms, containing the pass-through income tax benefits of a partnership (in contrast to the two-tier tax treatment of corporations) with the limited liability protections of a corporation. The limited liability company is a legal entity, separate from its owners, who have no liability for the entity's debts. An LLP alters the rule of joint and several liability for general partners, replacing it with protection from liability of the LLP itself.

State law varies in permitting design professionals to take advantage of these new business entities. Pennsylvania allows architects to form either LLCs or LLPs, if certain requirements are met.[3] By contrast, California permits design professionals to form LLPs, but not LLCs.[4]

SECTION 3.07 Joint Ventures

Joint ventures can be created by two or more separate entities who associate, usually to engage in one specific project or transaction. Such arrangements are contractual and can be expressed by written contracts or implied by acts. In large construction projects, two contractors may find they cannot handle a particular project individually but can if they associate in a joint venture. Another type of joint venture can be created by an organization that performs design and an organization that constructs. Sometimes joint ventures are needed to bid on "design/build" projects.[5]

Usually the agreement under which such a joint venture is created is complex and sets forth in detail the rights and duties of the joint venturers. Joint ventures created informally by acts have gaps as to specific rights and duties that are filled in by the law. When such gaps exist, it is likely that principles of partnership will be applied (see Section 3.03).

Suppose a dispute arises as to the existence of a joint venture. Because joint ventures are considered very much like partnerships for one transaction, the relationship is often examined to determine whether partnership-like attributes are present. For example, sometimes it is stated that there must be a community of interest, a common proprietary interest in the subject matter, the right of each to govern policy, and a sharing of profits and losses.

SECTION 3.08 Unincorporated Associations

Individuals sometimes band together to accomplish a collective objective without using any of the forms of association described thus far, such as partnerships or corporations. Instead they may organize an unincorporated association, such as a fraternal lodge, a social club, a labor union, or a church. Design professionals may perform professional services for these associations or may become involved in the associations themselves.

Generally, unincorporated associations are not legal entities. For this reason, early American law did not allow them to hold property in the association name, make contracts, sue in the name of the association, or be sued as a group. They were merely a group of individuals who banded together to accomplish a particular purpose.

In most states, statutes have removed many of the former procedural disabilities. Although they are still not entities, unincorporated associations are often permitted

---

[3]63 Pa.Stat.Ann. § 34.13(a)(6) & (7). The requirements pertain to the minimum number of members or managers of the LLC, or partners of the LLP, who must be licensed to practice architecture. Id., § 34.13(f) & (g).

[4]West Ann.Cal.Bus. & Prof. Code § 5535 (architects may form LLPs) and West Ann.Cal.Corp. Code § 17375 (LLCs may not render professional services). For a state-by-state survey, see Lurie & Anderson, *The Practice of Architecture and Engineering by Limited Liability Entities*, 24 Constr. Lawyer, No.1, Winter 2004, p. 24.

[5]See Section 17.04F.

to contract, to hold property, and to sue or be sued in the name of the association.

Usually such groups have constitutions, bylaws, and other group-related rules that govern the rights and duties of the members. They elect officers who have specified authority, such as to hire employees and run the activities of the association. All the members generally are responsible for those contracts entered into by the officers who were authorized by the members. Liability rests on agency principles.[6]

Establishing the authority of the officers is often difficult because formalities are often disregarded in these organizations. It may be necessary to show evidence of which members voted for resolutions authorizing the officers to make a particular contract. Although the officers may have certain inherent authority by virtue of their positions, important projects, such as engaging a design professional or a contractor, usually are beyond the scope of their inherent authority.

Officers who make the contract may be individually liable if the contract was clearly made by them as individuals. It may be possible to hold the officers for having misrepresented their authority if the contract was not authorized. Because of the potential risk to officers and to members, many contracts contain provisions that limit liability to certain designated property held in trust for the association in those states where they are not permitted to own property in the association name. Where the association is permitted to hold property in its own name, liability is often limited to that property.

SECTION 3.09 Loose Associations: Share-Office Arrangement

Sometimes design professionals use the term *association* to describe an arrangement under which they are independent sole proprietors but join together to share offices, equipment, and clerical help. They may also do work for each other. Sometimes such loose associations are known as *share-office arrangements*.

During the sharing arrangement, several legal problems can develop. First, suppose one associate performs services for another and the latter is not paid by his client. In the absence of any specific agreement or well-documented

understanding dealing with this risk, the associate who performs services at the request of another should be paid.

Second, suppose the associate performing services at the other associate's request does not perform properly and causes a loss to the client. The client would have a claim against the associate who did the work and the associate with whom it dealt. If the latter settled the claim or paid a court judgment, he would have a valid claim for indemnification against the associate who did not perform properly.

Third, if those in the share-office arrangement create an impression to the outside world that they are a partnership, they will be treated as such. For example, if the letterhead, building directory, and telephone directory list them as Smith, Brown, and Jones, such listings may create an apparent partnership. If so, each member of the association can create partnership-like contracts and tort obligations.

When a loose association terminates or one participant withdraws, all associates in the first case and the withdrawing associate in the case of withdrawal can freely compete with former associates or those remaining. However, the associates in the first case and the withdrawing associate in the second cannot do any of the following:

1. take association records that belong to another associate
2. represent that they still are associated with former associates if this is not the case
3. wrongfully interfere with contractual or stable economic relationships existing at the time the association ended or one associate withdrew

SECTION 3.10 Professional Associations

Professional associations of participants in the construction process, such as the American Institute of Architects (AIA), the National Society of Professional Engineers (NSPE), and the Associated General Contractors (AGC), have had a substantial impact on design and construction. Although they are not organizations like the others mentioned in this chapter created for the purposes of engaging in design and construction, some mention should be made of their activity and some of the legal restraints on them.

These professional associations engage in many activities. They speak for their members and the professions or industries associated with them. In addition, they seek to educate their members in matters that relate to their

---

[6]See Sections 4.05 and 4.06.

activities. One of their most important activities is publishing standard documents for design and construction that are used not only as the basis for contracts for design and construction but also to implement the construction administrative process. Keep in mind that membership in these professional associations is not required for someone to design or build.

By and large, professional associations can choose their members. However, as these associations become more important, the law is beginning to limit their power to determine who will be admitted to membership and what the legitimate grounds are for discipline. Restraints placed on professional associations have related to attempts by public authorities to encourage competition in the professional marketplace. This has resulted in attacks by public officials on what were called the rules of ethical conduct, which determine whether an applicant would be admitted to the association and can be disciplined for violating ethical rules of the association.

Often such ethical rules sought to limit the power of one professional to supplant another and bar or discipline members, based on "ungentlemanly conduct such as competing with a fellow member on the basis of price." Disciplinary action based on violating these rules was found to violate agreements between an association and federal antitrust officials that the association would not limit competition.[7]

---

[7]*United States v. American Society of Civil Engineers*, 446 F.Supp. 803 (S.D.N.Y.1977).

CHAPTER FOUR

The Agency Relationship: A Legal Concept Essential to Contract Making

SECTION 4.01 Relevance

Agency rules and their application determine when the acts of one person bind another. In the typical agency problem, there is a principal, an agent, and a third party. The agent is the person whose acts are asserted by the third party to bind the principal. See Figure 4.1. There can be legal problems relating to the rights and duties of the principal and agent as between themselves as well as disputes between the agent and the third party. But because in the agency triangle the third party versus principal part is most important—and most troublesome—these problems will be used to demonstrate the relevance of agency law.

The design professional, whether in private practice or working as an employee, may be in any of the three positions of the agency triangle.

For example, suppose the design professional is a principal (partner) of a large office. The office manager orders an expensive computer. In this illustration, the design professional, as a partner, is a principal who may be responsible for the acts of the agent office manager.

Suppose the design professional is retained by an owner to design a large structure. The design professional is also engaged to perform certain functions on behalf of the owner in the construction process itself. Suppose the design professional orders certain changes in the work that will increase the cost of the project. A dispute may arise between the owner and the contractor relating to the power of the design professional to bind the owner. Here the design professional falls into the agent category, and the issue is the extent of her authority.

Suppose X approaches a design professional in private practice regarding a commission to design a structure.

X states that she is the vice president of T Corporation. The design professional and X come to an agreement. Does the design professional have a contract with T Corporation? Here the design professional is in the position of the third party in the agency triangle.

The agency concept is basic to understanding the different forms by which persons conduct their business affairs, such as partnerships and corporations.

SECTION 4.02 Policies Behind Agency Concept

A. Commercial Efficiency and Protection of Reasonable Expectations

As the commercial economy expanded beyond simple person-to-person dealings, commercial necessity required that people be able to act through others. Principals needed to employ agents with whom third parties would deal. Third parties will deal with an agent if they feel assured that they can look to the principal. The agent may be a person of doubtful financial responsibility. The concept of agency filled the need for giving third persons some assurance that they can hold the principal responsible.

Agency exposes the principal to risks. The principal may be liable for an unauthorized commitment made by the agent. Suppose the principal authorizes its agent to make purchases of up to $1,000, but the agent orders $5,000 worth of goods. From the third party's standpoint, this $5,000 purchase may be reasonable in light of the agent's position or what the agent had been ordering in the past. The law protects the principal from unauthorized commitments but not at the expense of reasonable expectations of

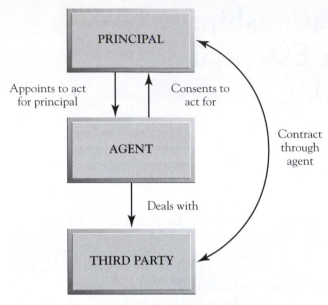

FIGURE 4.1   Agency concept illustrated.

the third person. That such problems can develop does not destroy the unquestioned usefulness of the agency concept. Such problems require the law to create rules and solutions to handle such questions that accord with commercial necessity and common sense.

Two recent cases illustrate the agency concept from the perspective of the owner. In *Shoals v. Home Depot, Inc.*, a homeowner was swindled by a dishonest employee of Home Depot, a large home improvements chain. The court found that the employee's representations made to the homeowner were binding on Home Depot, the principal. In *Shoals*, the owner was the third party who personally benefitted from application of the agency concept.[1] By contrast, in *Ciraulo v. City of Newport Beach*, the contractor assured the homeowners that it would obtain a building code variance from the city, then went ahead with the construction even though it never talked to the city. After completion, the city ordered the owners to remove the unauthorized building. The court upheld the order, because the owners, as principals, were bound by the fraudulent acts of their agent, the contractor.[2]

B. Relationships Between Principal and Agent

As a rule, the relationship of principal and agent is created by a contract expressed in words or manifested by acts. The agent may be a regular employee of the principal or an independent person hired for a specific purpose and not controlled as to the details of requested activities. Several aspects of the agency relationship distinguish it from the ordinary commercial, arm's-length relationship.

An arm's-length transaction is one wherein the parties are expected to protect themselves. In such a transaction, no general duty is imposed on one party to protect the other party, nor is any duty imposed to disclose essential facts to the other party. Although there are some exceptions, generally commercial dealings are at arm's length. On the other hand, principal and agent have a fiduciary relationship—one of trust and loyalty.[3] In such a relationship, one person relies on the integrity and fidelity of the other. The latter must not take unfair advantage of the trust in her by benefiting at the expense of the former.

Often the arrangement under which an agent performs services is sketchy and does not specifically delineate all the rights and duties of principal and agent. In such a case, terms often must be implied by law.

SECTION 4.03   Other Related Legal Concepts

Confusion often occurs because of the overlapping nature of terms such as principal–agent, master–servant, and employer–employee. This chapter deals only with the principal–agent relationship and the power of agency under which one person may bind another to a contractual obligation.

SECTION 4.04   Creation of Agency Relationship

Generally, agency relationships are created by a manifestation by principal to agent that the agent can act on the principal's behalf and by some manifestation by agent to principal that the agent will so act. However, these elements of consent are often informal. They need not

---

[1]422 F.Supp.2d 1183 (E.D.Cal.2006).
[2]147 Cal.App.4th 838, 54 Cal.Rptr.3d 515 (2007).

[3]See Section 11.04B.

be written, except in cases where statutes require that the agent's authority be expressed in writing if the transaction to be consummated by the agent would require a writing.

SECTION 4.05 Actual Authority

The agent is ordinarily authorized to do only what it is reasonable to believe the principal wants done. In determining this, the agent must look at the surrounding facts and circumstances. If, for example, the principal has authorized the agent to purchase raw materials to be used in a particular manufacturing process and the agent learns that the principal has decided not to proceed with the project, it is unreasonable for the agent to believe that the authority to buy the materials still exists.

The agent must consider the situation of the principal, the general usages of the business and trade, the object the principal wishes to accomplish, and any other surrounding facts and circumstances that would reasonably lead to a belief that the agent can or cannot do something. This is largely a matter of common sense. Sometimes the agent is given specific authority to do those incidental acts that are reasonably necessary to accomplish the primary act. For example, when an agent is given authority to purchase a car for the principal, it is likely that the agent also has authority to buy liability insurance for the principal. The agent, when possible, should seek authorization from the principal to perform acts not expressly authorized.

Sometimes emergencies arise that make it difficult or impracticable for the agent to communicate with the principal. In such cases, the agent has authority for necessary acts to prevent loss to the principal with respect to the interests committed to the agent's charge.

SECTION 4.06 Apparent Authority

Most difficult principal–agency cases arise under the doctrine of apparent authority, which exists when the principal's conduct reasonably leads a third party to believe that the principal consents to acts done on its behalf by the person purporting to act for it.

It may be useful to draw some initial distinctions essential to an understanding of apparent authority problems. First, suppose the asserted agent neither is an employee of the principal nor has been hired by the principal to perform any specific task. This can be illustrated by *Amritt v. Paragon Homes, Inc.*[4] The plaintiff homeowner sued for damages caused by faulty construction of a home built for her by Romero, a contractor. Clearly, Romero was liable but evidently could not pay a court judgment. The homeowner sued Paragon, the manufacturer of the prefabricated home erected by Romero, and Sewer, the manufacturer's local distributor.

The homeowner had arranged through Sewer to purchase the materials, plans, and drawings for such a home. Evidently Sewer represented himself to the homeowner as Paragon's agent and implied that Paragon would not be responsible unless the homeowner used Romero to construct the home. The agreement was on a form supplied by Paragon and filled in by Sewer. The homeowner borrowed money from and gave a mortgage to Paragon for the construction costs. The homeowner made payments by depositing funds in escrow with Paragon, and Paragon demanded a completion certificate before disbursing the funds. Sewer inspected the construction work for Paragon.

Paragon, Sewer, and Romero were all separate legal entities. Romero worked for himself, as did Sewer. The court held, however, that Sewer and Romero were agents for Paragon because both had "the outward trappings of apparent authority to be Paragon's agents."[5] The court concluded that the homeowner was "reasonably led to believe they were Paragon agents."[6]

The court, quoting the Restatement of the Law of Agency,[7] pointed to Paragon's conduct—supplying all materials and forms through Sewer, requiring that the house be inspected by Sewer and built to its plans, and disbursing all construction payments—and stated that combining of these activities could reasonably have led the homeowner to believe Paragon had authorized Sewer and Romero to be its agents. Although the court spoke in terms of apparent authority, it actually found an apparent agency.

Because of the many legal entities involved in a construction project, one party not infrequently seeks to bind an entity other than the one with whom it has dealt, on the same theory as that used in *Amritt v. Paragon Homes, Inc.*

---

[4] 474 F.2d 1251 (3d Cir. 1973).

[5] Id. at 1252.

[6] Ibid.

[7] For an explanation of the Restatements of the Law, see Section 1.10. The *Amritt* court cited to the Restatement (Second) of Agency, published in 1958. There is now a Restatement (Third) of Agency, published in 2006.

This is especially likely when consumers find that the people with whom they have dealt are unable to respond when those people have not performed in accordance with their legal obligation. The law is more likely to permit a claimant, as in the *Amritt* decision, to recover from the only solvent defendant who can be connected to the transaction if that solvent party has organized the transaction, controlled it, and sought to eliminate responsibility for a risk legitimately a part of its enterprise by using a separate legal entity that lacks the financial capability to respond for product defects. The conclusion reached in the *Amritt* decision is also more justified if the party dealing with the contractor is a consumer of limited commercial experience.

A second problem can stem from a client's engaging a professional person. For example, the owner may engage a design professional to perform services relating to design and contract administration for a particular project. The latter is rarely an employee of the owner but has certain designated authority. The contractor may claim that communications it had with the architect should be treated as if coming directly from the owner, while the owner may contest that conclusion. This matter is treated more fully in Section 17.05B.

The third and perhaps most difficult problem relates to whether an admitted employee of an employer has the authority to engage in certain acts and thereby to bind the employer. Here there is an agency relationship in that the agent has some authority to bind the principal. The problems that develop relate to the extent of that authority.

Clearly, acts of the agent that are authorized expressly or implicitly (authority implicit in the actual authority) will bind the principal. In addition, acts of the principal can cloak the agent with apparent authority beyond that actually possessed. A third party who relies reasonably on the appearance of authority can hold the principal responsible as if the acts were authorized.

SECTION 4.07   Termination of Agency

If authority is conferred for a specified period, it will terminate at the end of that period. If no time is specified, authority continues for a reasonable time. If authority is limited to performing a specified act or to accomplishing a certain result, it terminates when the act or result is completed.

Sometimes the terms of the authorization specify that the authorization is to continue until a certain event occurs. If so, the occurrence of the event will terminate the agency unless the event would not come to the attention of the agent. Sometimes the loss or destruction of certain subject matter terminates the agency.

Sometimes the agency terminates when principal and agent consent to terminate the relationship. This can occur if either principal or agent manifests to the other that the relationship will no longer continue. In some cases, the agency can also be terminated by death of the principal, and almost always by death of the agent. However, agency relationships sometimes are created by contract for fixed terms. The principal may, despite a fixed-term contract, terminate the agency and terminate the power of the agent to bind it, but such a termination will be a breach of contract with the agent.

These events effect actual authority of the agent. If termination of authority has taken place but manifestations to third parties that the agent has authority still exist, the agent may be able to bind the principal by the doctrine of apparent authority, which will continue only so long as the third party should not realize that the agent no longer has authority to bind the principal.

SECTION 4.08   Disputes Between Principal and Third Party

Sometimes the principal wishes to be bound by a transaction between the principal's agent and the third party. The latter may refuse to deal with the principal, claiming it did not know it was dealing with an agent. The agent may not have informed the third party that she was an agent, and the third party may not have had reason to know that the agent was acting on behalf of someone else. The undisclosed principal may disclose its status and hold the third party, unless the facts would make it unjust to permit the principal to assert its status. (Once the principal is disclosed, the third party can hold the principal.)

Suppose the agent did not have actual authority to enter into a transaction but the principal discovers the transaction. In such a case, the principal may ratify the transaction within a reasonable time and bind the third party (and itself) by notifying the third party and affirming the agent's unauthorized acts.

SECTION 4.09 Disputes Between Agent and Third Party

Disputes between an agent and a third party are relatively rare. If the agent is acting on behalf of an undisclosed principal, the third party has the right to sue the agent individually.[8] The third party may lose this right if it pursues its remedy against the principal, once the principal becomes disclosed. If the agent has misrepresented her authority and the third party is unable to hold the principal, the third party may have an action (be able to press a lawsuit) against the agent for misrepresenting authority.[9]

Within the construction industry, third party claims against the agent of a disclosed principal most commonly occur when a contractor sues the project architect or engineer. The contractor usually claims that the design professional committed a tort for which she is personally liable. For example, the contractor may contend that the design professional negligently administered the project or intentionally interfered with the contractor's contractual relationship with the owner. A design professional who acts within the scope of her agency and is not motivated by improper motive will not be liable to the contractor.[10]

SECTION 4.10 Nontraditional Project Delivery Systems

The discussion of agency principles to this point focused on a traditional construction project arrangement, in which the owner hires a design professional and prime contractor under different contracts, and the design and construction services are separately and sequentially performed.[11] An increasing number of large commercial and public projects use newer, nontraditional project delivery systems. These modern variations might include the introduction of new entities (such as construction managers or project managers) or entities that integrate the delivery of design and construction services (such as design/build).[12] The agency principles discussed in this chapter also apply to these new project delivery systems, although sorting out the principal/agency relationships may be difficult.

---

[8]*Mitchell v. Linville*, 148 N.C.App.71, 557 S.E.2d 620, 626 (2001).
[9]See Section 17.05B.

[10]*Wiekhorst Bros. Excavating & Equipment Co. v. Ludewig*, 247 Neb. 547, 529 N.W.2d 33 (1995). See also Section 14.08F.
[11]This "eternal triangle" is described in Section 8.02.
[12]Section 17.04.

CHAPTER FIVE

Contracts and Their Formation: Connectors for Construction Participants

SECTION 5.01 Relevance

The private design professional makes contracts with, among others, clients, consultants, employees, landlords, and sellers of goods. In addition, contracts are made between owners and prime contractors, prime contractors and subcontractors, contractors and suppliers, employers and employees, buyers and sellers of land, and brokers and property owners.

Contracts for design and for construction are analyzed in other parts of the text. This chapter provides a framework for such analysis.

SECTION 5.02 The Function of Enforcing Contracts: Freedom of Contract

The principal function of enforcing contracts in the commercial world is to encourage economic exchanges that lead to economic efficiency and greater productivity. This is accomplished by protecting the reasonable expectations of contracting parties that each will perform as promised. Although many and perhaps most contracts are and would be performed without resort to court enforcement, the availability of legal sanctions plays an important role in obtaining performance.

Generally, American law gives autonomy to contracting parties to choose the substantive content of their contracts. Because most contracts are economic exchanges, giving parties autonomy allows each to value the other's performance.

To a large degree, autonomy assumes and supports a marketplace where market participants are free to pick the parties with whom they will deal and the terms on which they will deal. Not only are the parties in the best position to determine terms of exchange, but the alternative—state-prescribed rules for economic exchanges—would lead to rigidity and heavily burden the state. Also, parties are more likely to perform in accordance with their promises if they have participated freely in making the exchange and determined its terms. Finally, such autonomy—often called *freedom of contract*—fits well in a free society that encourages individual enterprise. However, broad grants of autonomy assume contracting parties of relatively equal bargaining power, equal accessibility to information, and a relatively free marketplace. Such conditions place a check on overreaching. A party who believes the other party's terms to be unreasonable can deal with others. If the parties do arrive at an agreement under such conditions, the give-and-take of bargaining should ensure a contract that falls within the boundaries of reasonableness.

The development of mass-produced contracts and the emergence of large blocs of economic power often dealing with parties with limited or no bargaining power have made this earlier model of the negotiated contract the exception. If the state, through its courts, enforces adhesion contracts (contracts presented on a take-it-or-leave-it basis), the state is according almost sovereign power to those who have the economic power to dictate contract terms. For this reason, many inroads have been made on contractual freedom by federal and state legislation, regulations of administrative agencies, and courts through their power to interpret contracts and determine their validity.

SECTION 5.03 Preliminary Definitions

In the simplest terms, a contract is an enforceable bargain. A valid contract requires agreement by the parties, sometimes called manifestations of mutual assent (discussed in Section 5.06). In addition to agreement, the parties usually exchange promises. This bargained-for exchange is called consideration, another requirement for a valid contract (discussed in Section 5.08). Often there are formal requirements for a valid contract (discussed in Section 5.10). Breach of contract generates a remedy, usually the award of damages, but in rare cases an award of specific performance (discussed in Chapter 6).

The promisor is a person who makes a promise, and the promisee is the person to whom the promise is made. In most two-party contracts, each party is a promisee and promisor, both making and receiving promises.

The offeror is a person who makes an offer, and the offeree is a person to whom the offer is made. All other definitions will be given as the particular term is discussed.

SECTION 5.04 Contract Classifications

A. Express and Implied

Sometimes contracts are classified according to the method by which they are created. Using this classification, there are express contracts and implied-in-fact contracts. In express contracts, the parties manifest their assent or agreement by oral or written words. In implied-in-fact contracts, assent is manifested by acts rather than by words.

U.S. v. Young Lumber illustrates this in a construction project context.[1] The employee of a subcontractor brought equipment he had rented to the job site. The employer admitted that the employee's equipment was on the job site and was used but contended there was no express agreement under which it would pay the employee rent for using the equipment.

In ruling for the employee, the court stated,

> even if there was no express oral agreement for equipment rental, an implied contract may be inferred from the conduct of the parties. Implied contracts arise ... where there is circumstantial evidence showing that the parties intended to make a contract. Where one performs for another a useful service of a character that is usually charged for, and such service is rendered with the knowledge and approval of the recipient who either expresses no dissent or avails himself of the service rendered, the law raises an implied

promise on the part of the recipient to pay the reasonable value of such service.[2]

Unless the agreement is required to be in writing, the implied-in-fact agreement is as valid as an express contract.

B. Subject Matter

Sometimes contracts are classified by the transaction involved, such as sales of land, sales of goods, loans of money, leases, service contracts, professional service contracts, insurance contracts, and family contracts. Although American contract law traditionally has been thought of as a unitary system, different transactions are treated differently. For example, the seller of goods is in many instances held to a successful outcome standard, usually referred to as "implied warranty of fitness," whereas those who sell professional services are usually held to a less strict standard[3] comparing what was done with what others would have done. Subject matter variations are manifested by regulatory legislation over certain types of contracts. Judicial opinions often treat one transaction differently from another.

C. Bargain and Adhesion

Contracts are sometimes classified as negotiated or adhered to. The latter are referred to as *contracts of adhesion*.

A negotiated contract arises when two parties with reasonably equivalent bargaining power enter into negotiations, give and take, and jointly work out a mutually satisfactory agreement.

The adhesion contract has no or minimal bargaining. The dominant party hands the contract to the weaker party on a take-it-or-leave-it basis. At its extreme, the weaker party who wishes to enter into the transaction must accept all the terms of the stronger party.

Sometimes important terms can be negotiated, although the balance will be dictated by one party, with little opportunity for bargaining. For example, the purchaser of a new automobile may be able to bargain on price but will have to accept the dealer's standardized terms on all or almost all other aspects of the purchase. The rare buyer who reads the standardized terms and

---

[1]376 F.Supp. 1290 (D.S.C.1974).

[2]Id. at 1298. Where such an arrangement is considered nonconsensual, recovery can be based on unjust enrichment.
[3]See Section 14.05.

objects to them will have great difficulty in persuading the dealer to change the terms.

Sometimes the adhesion contract is accompanied by monopoly power. In a competitive economy, a weaker party who does not want to accept harsh terms may deal with others. However, in many transactions the weaker party will find the same terms used by the competitors of the party whose terms were unpalatable, or will find no competitor. Frequently, such adhesion contracts are printed, and the person with whom the weaker power is dealing (such as a salesperson or clerk) lacks authority to vary the printed terms of the contract. Modern courts recognize the difference between the negotiated contract and the contract of adhesion.

The weaker party has the burden of proving the contract it entered into was adhesive. The difficulty of that task is illustrated by a recent case. Buyers of a home in a large subdivision near St. Louis asserted that their purchase agreement was adhesive. The Missouri Supreme Court was not convinced. It pointed to testimony by the developer's attorney, who stated that all the contract terms were negotiable. In the court's view, the developer's attorney's admission—that the objectionable contract clause (the arbitration provision) had never been negotiated by the developer—did not prove the negative (that negotiation would have been refused if the buyers had tried). The court also observed that there was no evidence that *all* St. Louis area builders used the same arbitration term. The court concluded the buyers did not prove their purchase agreement was a contract of adhesion.[4]

Interpretation of contracts between clients and their design professionals[5] and between contractors and owners[6] is discussed later in this book. It is useful, however, to note how courts have treated adhesion contracts. At the very least, ambiguous terms will be interpreted against the party who supplied the adhesion contract.[7] Even more important are legal rules that seek to determine the reasonable expectation of the party who was presented a contract on a take-it-or-leave-it basis and that interpret the contract in accordance with that expectation. This may even be done if the party to whom such a contract has been presented was aware of the terms and of the likely intention of the party who supplied them.[8] Finally, it is more likely that terms in an adhesion contract than in a negotiated contract will be considered unconscionable and will not be enforced by a court before whom such a contract has been presented.[9]

SECTION 5.05 Capacity to Contract

Capacity usually relates to age and mental awareness. Because it rarely affects contracts for design or construction, it is not discussed in this book.

SECTION 5.06 Mutual Assent

A. Objective Theory of Contracts: Manifestations of Mutual Assent

Early English contract law stated there had to be a meeting of the minds—actual agreement as to the existence of a contract and its terms—before there could be a valid contract.

As a rule, each party believes it has made a contract, though perhaps the parties differ as to the exact nature of performance. Suppose one person harbors a secret intention not to be bound and yet manifests to the other party an intention to be bound. The party who relied on the objective manifestation should be protected.

The law protects the reasonable expectations of the innocent party through the objective theory. A party is bound by what it manifests to the other party. Secret intentions are not relevant. If one party, innocently or otherwise, misleads the other into thinking it has serious contractual intention, a contract exists although the parties do not actually agree. The same principle holds true if one party is only joking when entering into negotiations and making an agreement with the other party. Unless the other party should reasonably have realized that the negotiations were not serious, the party who is not serious will be held to the agreement.

---

[4]*State ex rel. Vincent v. Schneider*, 194 S.W.3d 853 (Mo.2006). Sections 30.02 and 30.03D discuss whether arbitration clauses are unconscionable, a concept distinct from that of adhesion contracts.

[5]Section 11.04D.

[6]Section 20.02.

[7]Sections 11.04D, 20.02.

[8]*Graham v. Scissor Tail, Inc.*, 28 Cal.3d 807, 623 P.2d 165, 171 Cal. Rptr. 604 (1981).

[9]Section 5.07D.

B. Offer and Acceptance in the Assent Process

Typically the process by which contracting parties make an agreement involves communications, oral and written, that culminate with one party making an offer and the other party accepting it.

Where parties exchange communications at the same time and place and an agreement is reached, rarely is it necessary to determine whether there has been an offer and acceptance. But where parties communicate with each other at a distance, it is often necessary to determine whether the parties have arrived at an agreement, and this requires an examination of whether there has been an offer and acceptance.

Offers are different from preliminary negotiations or proposals for offers. Although several abstract definitions of an offer exist, it is more useful to consider the *effect* of an offer. An offer creates a "power of acceptance in the offeree." The offeree can create a legally enforceable obligation without any further act of the offeror. How is it determined whether the offeree has a reasonable belief that he has a power of acceptance?

Some cases scrutinize the language of the offer, especially a written offer, looking for definite words of commitment on the part of the offeror, such as "I offer" or "I promise." However, parties only rarely express themselves in legal terms.

The entire written proposal is examined to see whether a reasonable person receiving it would think he could "close the deal." Factors include the certainty of the terms, any indication that the proposer will not have to take further action, the past dealings between the parties, and the person to whom the offer is made. For example, if nothing is stated on essential terms such as price, quantity, or quality, the bargaining is probably in a preliminary stage. The first proposal is intended merely to start the negotiating mechanism, and more negotiating will take place before agreement is reached.

If the proposer is negotiating with a number of other people at the same time and the person to whom a particular proposal is directed knows of this, it is likely that the latter realizes, or should realize, that the proposer wants the last word rather than risk being obligated to a number of people. In such a case, the communication is not an offer.[10]

---

[10]*Fletcher-Harlee Corp. v. Pote Concrete Contractors, Inc.*, 482 F.3d 247 (3d Cir.2007) (prime contractor's solicitation of bids from several subcontractors; the solicitation is not an offer and the bid is not an acceptance).

The common law concluded that offers are revocable even if stated to be irrevocable. At any time before acceptance, an offeror can withdraw the offer, provided the offeror communicates the revocation directly or indirectly to the offeree. This rule of revocability has generated many exceptions, the most important being reliance on the offer and the firm offer under Section 2-205 of the Uniform Commercial Code dealing with the sale of goods. The latter makes a firm written offer in a signed writing by a merchant irrevocable for the time stated or, if not stated, for a reasonable time not to exceed three months.

Suppose a written offer states that it will be open for a specified period; say, ten days. At the end of this period, the power of acceptance ends. If the offer is made by letter, it may not be clear when the time period begins. Does the period begin on the date of the letter offer, at the time the letter is mailed, at the time it would normally be received, or at the time it is actually received? Usually the time period begins when the letter is actually received. It is best to use a specific terminal date, such as "You have 10 days from May 1, 2008." See subsection C later in this section.

If the duration of the offer is not stated specifically, the offer remains open for a reasonable time. Facts considered are the state of the market for the goods or services in question, the offeror's need to be able to deal with others if the offeree does not accept, and custom and usage in the particular transaction. For example, the reasonable time period in which to accept an offer for the sale of shares of stock in a fluctuating market is usually shorter than the reasonable time to accept an offer for the sale of land. The time allowed to accept an offer to sell perishable goods is shorter than the time to accept an offer to sell durable goods. The amount of time is judged from the viewpoint of the offeree, whose reasonable belief as to the duration of the offer governs.

An offer can be revoked by the offeror. Generally, revocation must be communicated to the offeree. A few states make a revocation effective when placed in the means of communication, such as mail, telegram, fax machine, or when "send" is clicked dispatching an e-mail on its electronic way, similar to an acceptance (discussed in Section 5.06C). After a valid revocation, an offeree who is still interested must make another proposal to the original offeror.

Generally an offer terminates if the offeree rejects it. Sometimes the rejection is explicit. For example, the offeree may communicate a lack of interest. In such a case, the offeror is free to deal with others unless the power of acceptance has been created by an option that has been purchased.[11] Usually rejection is implied if the offeree makes a counteroffer. The normal expectation of the offeror in the event of a counteroffer is that the offeree is no longer interested in doing business on the basis of the original offer. This implication may be negated if the counteroffer makes clear that the offeree is still considering the original offer. A counteroffer both creates a power of acceptance in the original offeror and usually terminates the power of acceptance of the original offeree.

The acceptance must be sent or communicated while the power of acceptance still exists. Any acceptance after that time does not operate to create the contract without a further act by the offeror.

Under the common law's *mirror image* rule, the acceptance must be unequivocal and must not propose new or different terms. When the offeree (the person to whom the offer is made) proposes different terms, it usually means no agreement has been reached as yet.[12] This rule often frustrated the common intentions of the parties. For example, an acceptance letter on a company standard form may contain preprinted language. This new language may defeat the formation of a contract, even though the variance in contract terms was minor and not of importance to the parties themselves.

Another complication is devising a rule of acceptance between a buyer and seller with their own standard form contracts, which rarely agree. To remedy this problem and to deal with transactions where buyer and seller in the sale of goods will sign only their own forms that rarely agree, the Uniform Commercial Code adopted Section 2-207. Including additional or different terms does not necessarily preclude the communication from being an acceptance if it appears the offeree is accepting the terms of the offer and does not condition his acceptance expressly on the offeror agreeing to additional or different terms. Those additional terms are considered proposals for additions to the contract that are sometimes binding without fur-

ther communication of the offeror.[13] Section 2-207 also provides that if the conduct of the parties indicates the parties have made a contract although the writings of the parties do not match, a contract has been concluded on the terms that do match, with the supplemental terms being supplied by law.

As offerors usually wish to know whether their offers have been accepted, acceptance, as a rule, must be communicated to the offeror. The offeror can deprive himself of the right to receive actual notice of acceptance by stating, "If you accept, sign the letter and you need not communicate any further with me."

C. Contracts by Correspondence

Generally, to be valid, offers, revocations, and rejections must be communicated to the other party. Because contract law protects reasonable expectations, contracting parties must know where they stand. However, a special rule has been developed for acceptances.

When parties began to make contracts by correspondence, English courts adopted the mailbox, or dispatch, rule. If the offeree used the same means of communication as the offeror had employed, the acceptance was effective when placed in the means of communication. It did not have to be actually received. For example, if the offeree mails an acceptance to the offeror in response to an offer received by mail, the address on the letter of acceptance is correct, and the proper postage is placed on the letter, the contract is formed at the time the letter is mailed. This places the risk of a delayed or lost letter on the offeror and protects the offeree's expectation that when the letter has been mailed, a contract has been formed.

Although early cases scrutinized the means of communication used by the offeree, generally the mailbox rule applies if a reasonable means of communication is employed. As noted in Section 5.06F, the means of communication can include mail, fax, or electronic mail.

One aspect of this rule is often ignored. The rule applies only if the offeror has not specifically required that the communication actually be received. For example, if the offeror makes an offer by mail but states that

---

[11]Restatement (Second) of Contracts § 37 (1981).

[12]*Poel v. Brunswick-Balke-Collender Co. of N.Y.*, 216 N.Y. 310, 110 N.E. 619 (1915) (letter of acceptance, containing printed provisions along the margins that differed from the offer, was not an acceptance).

[13]*Wachter Management Co. v. Dexter & Chaney, Inc.*, 282 Kan. 365, 144 P.3d 747 (2006) (construction manager's negotiated purchase of construction software became a contract when the software was ordered; a venue provision in the shrinkwrap's license agreement was disregarded as an attempt to modify the contract terms).

word must be received by a specified date as to whether the offeree accepts, the contract is not formed unless the offeror receives the offeree's acceptance by the stated date.

D. Reasonable Certainty of Terms

As mentioned in Section 5.06B, one test for determining whether an offer has been made is the clarity and completeness of the terms of the proposal. Even when an agreement on terms has been reached and no issue exists as to the existence of an offer, agreed terms that lack reasonable certainty of meaning preclude a valid contract.

Why require that even agreed terms be reasonably clear in meaning before a valid contract can exist? First, vagueness or incompleteness of terms may indicate that the parties are still in the bargaining stage. Second, a third party, such as a judge or an arbitrator, should be able to determine without undue difficulty whether the contract has been performed. But reasonable certainty does not require terms so clear that there can be no doubt as to their meaning.

Janzen v. Phillips[14] involved an action by a landscape architect to recover his fee. The architect and his client were somewhat hazy in their initial dealings. There had been no definite discussion of costs; the defendant stated that he wanted a "first-class job," and work went ahead on this basis.

Difficulties developed when the interim billings exceeded what the client thought he would have to pay. After discussions regarding costs, the architect sent a letter to the client in which he stated that he would "substantially landscape most of your property" for $6,500, or "nearly so."[15] The court was concerned with the terms "substantially" and "nearly so." Without these terms, clearly, assent to this letter would have created a valid contract. The court stated that where it appears the parties have intended to make a contract, courts will tread lightly in concluding that their efforts were not successful because of the absence of reasonable certainty. The terms "substantially" and "nearly so," according to the court, have a definite meaning and are frequently used to indicate that the cost would be about $6,500 and the work would be essentially completed for that price. Some latitude was expected, and the architect would have some leeway from the $6,500 figure.

_____
[14]73 Wash.2d 174, 437 P.2d 189 (1968).
[15]437 P.2d at 191.

Cases often appear to be inconsistent. Many cases that involve contract formation problems, such as the certainty requirement and the effect of preliminary agreements (see Section 5.06E and I), depend on the particular facts in the case before the court. The court will examine the language of any agreement, the intention of the parties, and the surrounding facts and circumstances, such as any performance by one or both parties. In addition, the availability of any restitution remedy based on unjust enrichment (one party's performance having benefited the other party) may be a basis for concluding that a valid contract was not made.

E. A More Formal Document: Preliminaries

Suppose an informal writing, sometimes called or even labeled as a preliminary or tentative agreement, a letter of intent (see Section 5.06I), or a purchase order, is concluded, but the parties expect that a more formal document will be drawn up. Have they intended to bind themselves, with the formal written document only a memorial of their actual agreement, or did they not intend to bind themselves until assent to the more formal writing? This depends on the intention of the parties.

An expression of their intention will control the decision. But suppose specific statements of intentions do not exist. If the transaction is complicated and nonroutine and the basic agreement is sketchy, it is likely that the parties do not intend to bind themselves until the more formal writing is prepared and signed. Commencement of performance by both parties very likely means that a contract has been formed without a formal agreement. Performance by one party is more ambiguous.

Modern courts tend to find a contract at the earliest possible stage if the court is convinced that the parties intended to be bound and only minor gaps need to be filled.

F. Agreements to Agree

Parties to a contract are often unable to specifically define every aspect of their relationship in the contract, yet they may wish to create a legally enforceable agreement. To accomplish these two objectives, they may specify that certain terms will be agreed on by the parties at a later date. For example, a long-term supply-of-goods contract may specify that the shipping arrangements or delivery dates will be agreed on later. The parties have agreed to agree.

In older cases, courts refused to enforce contracts that contained such provisions. Judges saw themselves as simply enforcing agreements that the parties made, rather than as making agreements for the parties. Modern courts are more willing to enforce contracts that contain agreements to agree as long as the parties are not still in the negotiation stage and the matters to be agreed on are not essential. For example, Section 2-204(3) of the Uniform Commercial Code states that a contract for sale of goods is valid even though certain terms are left open "if the parties have intended to make a contract and there is a reasonably certain basis for giving an appropriate remedy."

Where courts have faced contracts that contain agreements to agree, as a rule the parties are not disputing the to-be-agreed-to provisions. One party contends that it is not bound, because a fatal defect rendered the contract invalid. If performance has begun and the matter to be agreed to is not so essential that it indicated the parties were still in the bargaining stage, courts are likely to enforce the agreement. The agreed-to provision will be supplied by a standard of reasonableness that will often take trade usage into account.

G. Agreements to Negotiate in Good Faith

Sometimes an agreement to agree is interpreted to be an agreement to negotiate in good faith. The parties do not state they will agree but state they will negotiate in good faith on certain matters. Although some courts will not enforce such provisions, there is an increasing tendency to enforce them by requiring the parties to negotiate in good faith. If the parties in good faith seek to agree but cannot, further performance should not continue, and any performance that has been rendered should be dealt with by principles of restitution based on unjust enrichment. However, if one of the parties does not negotiate in good faith, the court may determine what good-faith negotiation would have been produced by way of agreement.[16]

H. Memorandum of Understanding

In the course of a complicated and often lengthy negotiation, the parties may draw up and agree to what they call a *memorandum of understanding*, which indicates that they have reached agreement on certain terms but are still negotiating over others. Generally, such memoranda are not binding. If the negotiations have not been concluded, all matters are open to negotiation.

The purpose of a memorandum of understanding is to give the parties some indication of those things that have come to the stage of tentative agreement, to aid the parties in seeing what still remains to be negotiated. Although one party may assert that the other party is not negotiating in good faith if it seeks to reopen terms on which there has been agreement, the agreement process should not be placed in this straitjacket.

I. Letter of Intent

During the course of a complex negotiation, the parties may wish to begin performance before there has been a final agreement. The party asked to commence performance may wish some assurance that if the negotiations are not successfully concluded, it will still be paid for what it has done. A letter of intent issued by the other party will be a directive to proceed and will provide a compensation formula for work performed that will be binding even if the parties do not reach a final agreement. If one party commences performance after the other party issues a letter of intent, the letter of intent will govern their relationship, even if they do not successfully conclude the negotiations.

Quake Construction, Inc. v. American Airlines, Inc.[17] involved an attempt, ultimately unsuccessful, to employ the classic use of a letter of intent to get performance started while negotiations continued. The letter, while noting many of the details of the proposed subcontract, included a provision stating that the prime reserved the right to cancel the subcontract if the prime and sub could not agree on all of the terms.

The Supreme Court of Illinois concluded that the case should be sent back to the trial court. It found that other factors, such as the prime informing the sub that it had been awarded the subcontract, announcing this at a pre-construction conference and promising to send a final draft of mutually agreed terms to the sub for its signature (this was never done) made the language of the letter of intent ambiguous and allowed the use of extrinsic evidence to show the intent of the parties.

---

[16]*Purvis v. United States*, 344 F.2d 867 (9th Cir.1965).

[17]141 Ill.2d 281, 565 N.E.2d 990 (1990).

The *Quake* case shows that the language of the letter of intent coupled with the surrounding facts and circumstances can create a binding contract, not merely an authorization to proceed.

Yet without a letter of intent, the party performing services before the contract has been finalized, such as the design professional or the contractor, may be doing work on a risk or contingency basis or "on the come." Was he taking his chances that the contract will never be finalized?[18]

SECTION 5.07 Defects in the Mutual Assent Process

A. Fraud and Misrepresentation: Duty to Disclose

The negotiation process frequently consists of promises and factual representations made by each party. If important factual representations or promises are made to deceive the other party, the deceived party may cancel the transaction or receive damages if it has reasonably relied on the representations or promises.

Sometimes such representations are made not with the intention of deceiving but are negligent or even innocent. As the conduct becomes less morally reprehensible, the remedies to the deceived party will be fewer. For example, in fraudulent representation the deceived party may receive punitive damages designed to punish the conduct and not to compensate the victim. Although the victim of negligent misrepresentation will be able to recover damages to compensate for the loss or cancel the transaction, the only relief likely to be accorded someone who relies on an innocent misrepresentation is cancellation of the contract.

Suppose one party to a contract wishes to cancel the transaction because of a claim that the other party should have disclosed important facts that, had they been known, would have persuaded the former not to enter into the transaction. Although early cases rarely placed a duty to disclose on contracting parties, there is an increasing tendency to do so where the matter that is not disclosed is important and where the party knowing of the facts should have realized that the other party was not likely to ascertain them.

B. Economic Duress

A valid contract requires consent freely given. Consent cannot be obtained by duress or compulsion. Early common law duress was physical, such as obtaining a deed or a contract by threatening to kill or injure the other party.

Modern duress in the commercial world principally involves economic duress or (as it is sometimes called) business compulsion, which can exist if one party exerts excessive pressure beyond permissible bargaining and the other party consents because it has no real choice. Courts have found economic duress to exist in very few cases. Their hesitance undoubtedly relates to the inherent pressures involved in the bargaining process and the fear that many contracts could be upset if the economic duress concept were used too frequently. Yet, as indicated by *Rich & Whillock v. Ashton Development* (reproduced in Section 27.13), today some courts will set aside contractual adjustments or releases even in the hard-bargaining commercial world.[19]

C. Mistake

The doctrine used most often in attacking the formation process is mistake. Here again, there are no fixed or absolute rules. There may be mistakes as to the terms of the contract. One party may not read the agreement, or because of mistake or fraud the agreement may not reflect the earlier understanding of the parties. Also, parties are often mistaken as to the basic assumptions on which the contract was made. Everyone who makes a contract holds certain underlying assumptions that, if untrue, render the contract undesirable.

A few generalizations can be made. Although there is always some carelessness in the making of a mistake, the greater the carelessness the less likely that the party making the mistake will be given relief. A comparison of the values exchanged is relevant. If one party is getting something for

---

[18]*GSGSB v. New York Yankees*, 862 F.Supp 1160 (S.D.N.Y. 1994). The architect may be saved if he can persuade the jury that, despite there not having been a contract, the architect could recover based on *quantum meruit*. To do this, he must show that he rendered services, the owner accepted those services, the architect expected compensation, and submit the reasonable value of the services. 862 F.Supp. at 1170. In light of the conclusion that no contract had been made and that the architect did the work "on the come," it may be hard to persuade the jury that compensation was expected. The opinion is very instructive on the issues in Section 5.06.

[19]*Centric Corp. v. Morrison-Knudsen Co.*, 731 P.2d 411 (Okla.1986) (collecting many cases).

almost nothing, the argument of mistake is more likely to be employed to relieve the other party from the contract. The question of when the mistake is uncovered is also important. If the mistake is discovered before the other party has relied on or performed under the contract, there is a greater likelihood that the contract will not be enforced.

In extreme cases, courts will relieve a party from certain risks. However, the law also seeks to prevent parties from avoiding performance of their promises by dishonest claims of mistake that are economically advantageous. Steering a course between these two policies as well as supporting the policy that people should be able to rely on agreements has meant that few fixed legal rules can be used to predict probable results in such cases. A determination of whether relief will be granted requires a careful evaluation of the facts, with principal scrutiny paid to the relative degree of negligence, any assumption of risk, the disparity in the exchange values in the transaction, and the likelihood that the status quo can be restored.

D. Unconscionability

Equity courts refused to enforce contracts that were unconscionable—contracts that shocked the conscience of the judge. Borrowing this concept, the Uniform Commercial Code in Section 2-302 gave the trial judge the power to strike out all or part of an unconscionable contract for the sale of goods. Gradually, the concept is being accepted in all contracts. But contracts or clauses that have been struck down (and few have) have tended to

1. be consumer rather than commercial transactions
2. involve sharp dealing in the bargaining process rather than harsh terms
3. involve clauses that exculpate a party from its obligation or make vindication of the other party's legal rights very difficult rather than price terms

As yet, the doctrine has not made much of a direct impact on contracts for design and construction except for agreements to arbitrate as discussed in Sections 30.02 and 30.03A. Yet it has been held that placing a duty on the owner to seek verification of a bid if the bid appears to have been a mistake, is done to avoid an unconscionable contract.[20] As noted in Section 19.02E, the doctrine's

development has, however, encouraged courts to examine all the circumstances that surround the making of a contract and to intervene more frequently.

E. Formation Defects and Restitution

An important use of restitution relates to the conferring of benefits when for a variety of reasons a claim cannot be based on a valid contract. There may be no one with whom a valid contract can be made. For example, suppose a doctor passing a construction site sees a passerby struck by a falling piece of lumber and rendered unconscious. The doctor administers medical aid in an attempt to save the victim. Because the doctor is a person trained in such emergencies, he should be encouraged to act. His training, the emergency, and the fact that he is not likely to be rendering services gratuitously takes him out of the volunteer category and allows him to recover.[21]

Such cases are rare. More commonly, restitution claims are made where parties intended to make a contract but did not do so. In the context of design or construction services, two principal formation defects require a person who has performed services to look to restitution. First, the contract fails because the terms lack sufficient certainty (usually in contracts for design services), the parties could not agree on terms while work progressed, or the parties did not get around to formalizing the contract by executing a formal writing where it was their intention to formalize it that way or there was a misunderstanding over the terms of the contract. In all these instances, the parties believed they had or would have a valid contract. Where benefit is conferred under these circumstances, the party who has conferred the benefit will very likely be able to recover in restitution.

The measure of recovery may vary, from the value of the expenditures to the benefit actually conferred, depending on the equities. It may even, in rare cases, be based on expenditures that did not benefit the defendant.[22] In a close case as to whether a valid contract has been made, performance by both parties or even one party

---

[20]*Sulzer Bingham Pumps, Inc. v. Lockheed Missiles & Space Co., Inc.,* 947 F.2d 1362 (9th Cir.1991), reproduced in Section 18.04(E).

[21]*Cotnam v. Wisdom,* 83 Ark. 601, 104 S.W. 164 (1907). Admiralty law allowed a recovery to the owner of a ship that altered its course to help a stricken ship that lacked a medical person. *Peninsular & Oriental Steam Navigation Co. v. Overseas Oil Carriers, Inc.,* 553 F.2d 830 (2d Cir.1977), cert. denied, 434 U.S. 859 (1977).

[22]*Minsky's Follies of Florida, Inc. v. Seenes,* 206 F.2d 1 (5th Cir.1953) (failure to have a written memorandum).

may tip the scales in favor of a conclusion that a valid contract did exist. In these illustrations, neither party had breached. As a result, recovery here is more likely to be limited to actual benefit conferred.

The second defect requiring restitution that is relevant to design and construction services (more common in construction) relates to contracts made illegally, principally where public contracts require that awards be made competitively. Because of the varying judicial attitude toward the importance of rigorously enforcing those rules and the various levels of illegality (compare a bribed official awarding a contract[23] with a technical irregularity by well-meaning officials[24]), court decisions may not appear consistent. If there is any trend, it is toward granting recovery in favor of those who have performed in good faith whose contracts were invalid because of technical irregularities.[25]

SECTION 5.08 Consideration as a Contract Requirement

A. Definition

While an offer and acceptance result in a bargain, the added element of *consideration* is required to transform that bargain into an enforceable contract. Consideration is loosely defined as bargained-for exchange: one person giving up something of value in return for receiving a benefit from the other person to the bargain. Using more technical language, consideration is found to exist if the promisee (the recipient of the promise) suffers a legal detriment (for example, pays money) and this detriment induces the promisor to act (give the promisee an apple).

Early cases required benefit to the promisor. Later decisions, however, enforced a promise where the promisor did not receive pecuniary benefit but the promisee suffered a detriment. Nor need the detriment be economic in nature; it may include the promisee refraining from doing what he has the legal right to do. In general, courts do not look into the sufficiency of the consideration, instead leaving that valuation up to the judgment of the parties. However, the detriment must consist of more than a moral obligation to act.

The *preexisting duty* rule is a corollary to the element of detriment.[26] One who already has a legal duty to perform an act cannot claim that performing that act was a legal detriment. A firefighter is paid to save people from burning buildings and so cannot charge someone he saved from a burning building on a theory of that a contract existed between the firefighter and the saved person.

Yet another way of understanding the consideration requirement is the principle that promises motivated by benefits received *in the past* are not enforceable. If an owner and contractor have entered into a contract, the owner cannot use the already existing consideration to compel the contractor to perform new and additional work at no additional pay. In the commercial world, exceptions to the *past consideration* rule include promises to pay a debt discharged by bankruptcy[27] or barred by statutes of limitation.[28]

B. Reliance

Suppose the promisor's promise foreseeably induces reliance by the promisee, but consideration is found lacking. While no contract has been formed, many courts allow the promisee to recover under the doctrine of promissory estoppel. Section 90(1) of the Restatement (Second) of Contracts states, "A promise which the promisor should reasonably expect to induce action or forbearance on the part of the promisee or a third person and which does induce such action or forbearance is binding if injustice can be avoided only by enforcement of the promise." In the construction process, promissory estoppel has primarily been used as a means of enforcing bids made by subcontractors to prime contractors, as further described in Section 28.02.

SECTION 5.09 Promises Under Seal

When most people were unable to write, a method was needed to accomplish conveyances and other legal acts. The method adopted in early English legal history was the seal, an impression made by a signet ring or other

---

[23]*Manning Eng'g, Inc. v. Hudson County Park Comm'n*, 74 N.J. 113, 376 A.2d 1194 (1977) (no recovery).

[24]*Layne Minnesota Co. v. Town of Stuntz*, 257 N.W.2d 295 (Minn. 1977) (no recovery but based on absence of benefit).

[25]*Blum v. City of Hillsboro*, 49 Wis.2d 667, 183 N.W.2d 47 (1971).

[26]Restatement (Second) of Contracts § 73 (1981).

[27]Restatement (Second) of Contracts § 83 (1981). Formal requirements to pay debts barred by bankruptcy are contained in 11 U.S.C.A. § 524.

[28]Restatement (Second) of Contracts § 82 (1981). Many states require a written promise.

similar instrument in soft wax that had been placed on the document. The seal emphasized the seriousness of the act being performed. Consideration was not necessary. A sealed instrument was a powerful document.

In the United States, the seal began to lose its power when, instead of being a formal impression in wax, it was reduced to either putting the word *seal* on the document or using some abbreviation such as *LS*. This did not have the trappings of formality possessed by the old seal. As a result, in many states the seal was denied any function in preparing documents. In other states, a document under seal creates enforceability or creates a presumption of consideration.

SECTION 5.10 Writing Requirement: Statute of Frauds

A. History

In 1677, the English Parliament passed the Statute of Frauds, which requires that a sufficient written memorandum be signed by the defendant before there can be judicial enforcement of certain specified transactions. One reason for the writing requirement is to protect litigants from dishonest claims and protect the courts from the burden of hearing claims of questionable merit. Another is that the requirement of a writing acts as a cautionary device. It warns people that they are undertaking serious legal obligations when they assent to a written agreement. The Statute of Frauds has been adopted in all American jurisdictions in one form or another. The tendency has been to expand the classifications that require a writing. See also Section 11.04F.

B. Transactions Required to Be Evidenced by a Sufficient Memorandum

Transactions singled out in the original statute were promises by an executor or administrator to pay damages out of his own estate; promises to answer for the debt, default, or miscarriages of another; agreements made on consideration of marriage; contracts for the sale of land or an interest in land; agreements not to be performed within a year from their making;[29] and contracts for the sale of goods over a specified value. Some states require

contracts for performing real estate brokerage services and contracts to leave property by will to be evidenced by a written memorandum. Statutes regulating home improvements often require that there be a written contract.[30]

Classification of the transaction has been one device by which courts have cut down the effectiveness of this statute. For example, a contract to construct a twenty-story building will not require a writing if the court concludes that the contract could be performed within a year. Many other limitations and exceptions on the various transactions have been made by the courts.

C. Sufficiency of Memorandum

The memorandum must be signed by the party to be charged. The proliferation of electronic communications led Congress to enact the Electronic Records and Signatures in Global and National Commerce Act that provides for a method of validating signatures on electronic communications.[31] Obtaining digital signatures is simple and inexpensive. It must contain the basic terms but need not express the entire agreement. It need not have been signed at the time the agreement was made but can be signed later. For example, a letter written and signed by a party after a dispute has arisen will provide the needed memorandum if the letter contains sufficient terms. In goods transactions, the Uniform Commercial Code Section 2-201(3b) allows the writing requirement to be satisfied by an admission in the course of litigation. As technology changes, the law will have to decide whether audio or videotapes or e-mail will satisfy the requirement.[32]

---

[29]*J.R. Loftus, Inc. v. White*, 85 N.Y.2d 874, 649 N.E.2d 1196, 626 N.Y.S.2d 52 (1995) (oral agreement to build house *and* to provide a one-year warranty is subject to the statute).

[30]West Ann.Cal.Bus. & Prof.Code § 7159 (the contract and any changes to it must be evidenced by a writing and signed by all parties). Courts have allowed both contractors and homeowners to get around § 7159's writing requirement. *Arya Group, Inc. v. Cher*, 77 Cal.App.4th 610, 91 Cal.Rptr.2d 815 (2000) involved a $4 million oral contract to build a house for Cher, the movie star; the contractor was allowed to sue under a theory of *quantum meruit*. In *Shoals v. Home Depot, Inc.*, 422 F.Supp.2d 1183 (E.D.Cal.2006), a blind homeowner who was swindled by a Home Depot employee was able to enforce the oral agreement against the company.

[31]15 U.S.C.A. § 7001.

[32]*Ellis Canning Co. v. Bernstein*, 348 F.Supp. 1212 (D.Colo.1972) held that an audio tape satisfied the Statute of Frauds. While *Swink & Co., Inc. v. Carroll McEntee & McGinley, Inc.*, 266 Ark. 279, 584 S.W.2d 393 (1979) held that even assuming a tape recording is a writing, the tape recording of a conversation did not satisfy the Statute as it had not been signed. See Misner, *Tape Recordings, Business Transactions via Telephone and the Statute of Frauds*, 61 Iowa L. Rev. 941 (1976).

D. Avoiding the Writing Requirement

Sometimes oral agreements that generally require a memorandum are enforced despite the absence of a memorandum. Section 2-201 of the Uniform Commercial Code states that acceptance of part payment for the goods is sufficient to make enforceable an oral contract for those goods.

In cases involving the sale of land, the part performance doctrine developed as a substitute for a writing. For example, if the buyer took possession and made improvements, the court could order the seller to convey despite the absence of a writing. However, part payment by the purchaser was not sufficient part performance.

In service contracts, full performance by the parties would bar the use of the Statute of Frauds as a defense. However, this exception was not applied in a home improvement contract required to be in writing.[33]

Some jurisdictions enforce oral agreements if one party has reasonably relied on and changed its position based on the promised written agreement or if the other party has been unjustly enriched by the performance.[34]

The New York Appellate Court saved an oral contract for pre-construction construction management services, and possibly the construction itself, by reversing the trial court decision that had struck down the contract claim on the basis of the one-year Statute of Frauds. It held that the case should be sent back to the trial court to determine whether the two parts of the services could be separated. If so, the part relating to pre-construction construction management services could be performed in nine months and was enforceable.[35]

The Statute of Frauds has become eroded in many states. If a claim seems genuine, and if the claimant has relied reasonably on the oral agreement, a strong likelihood exists that the claimant will be permitted to prove the agreement.

---

[33]*Caulkins v. Petrillo*, 200 Conn. 713, 513 A.2d 43 (1986).

[34]*Monarco v. Lo Greco*, 35 Cal.2d 621, 220 P.2d 737 (1950).

[35]*Lehrer McGovern Bovis, Inc. v. New York Yankees*, 207 A.D.2d 256, 615 N.Y.S.2d 31 (1994). The contractor sent a letter of intent and a written contract to the owner. It was never signed and returned by the owner.

CHAPTER SIX

Remedies for Contract Breach: Emphasis on Flexibility

SECTION 6.01 An Overview

Determining the remedies available for breach of contract is much more difficult than determining whether a valid contract has been made. It is exasperatingly difficult to predict before trial what the law will require that a breaching party must pay. Even after all the evidence has been produced at the trial, it can be very difficult to determine the precise remedy that the party entitled to relief should receive. This sometimes leads to a trial procedure under which remedies are considered before going into the issues of whether a valid contract has been formed and whether it has been breached. If very little can be accomplished remedially, it makes little sense to have a long, protracted lawsuit to establish the claim.

Although the function of awarding a remedy for contract breach is to compensate the injured party (the exceptions to this are discussed in Section 6.04), measuring the losses incurred and gains prevented can be very difficult. There may be agreement on general objectives, yet implementing these objectives can be difficult because of the great variety of fact situations, the different times at which a breach can occur in the history of a contract, the variety of causes that may generate the breach, and the different judicial attitudes toward breach of contract itself.

The common law has developed conventional formulas that are applied in particular cases designed to implement the basic compensation objective. These formulas may not appear to achieve a just result when they are applied. Borderline cases make it difficult to determine which formula should be applied. The range of remedies, both as to type and amount, can give discretion to juries, trial judges, and appellate courts. For example, if one party seems to have taken unfair advantage of the other, doubts may be resolved in favor of the latter. The reason for the breach, though perhaps not exculpating the breaching party, can be influential in measuring the award. The relative abilities of plaintiff and defendant to bear the loss can be an influential factor, although often an unstated one.

Generally, when juries are used, they are given considerable latitude to determine the award. The jury award is likely to be upheld if it is based on substantial evidence or reasonable inferences from the evidence, unless the award seems to result from passion or prejudice. Although compromise can play a strong part in jury determinations, the failure to require a jury to be specific as to the reasons for its award can give the jury a mechanism for achieving what it believes to be a just result even if it would differ from what the law appears to require.

SECTION 6.02 Relationship to Other Chapters

This chapter explores generally the judicial remedies awarded for breach of contract. Succeeding chapters apply these basic principles to construction process disputes.

Usually claims by a design professional are for services rendered for which payment has not been made. Claims by the owner against the design professional usually relate to a delay caused by the design professional, defective design, or a failure to monitor the contractor's performance. These topics are considered in greater detail in Sections 12.14B and 14.10A. Claims by owners against contractors usually involve losses asserted because of unexcused contractor delay (covered in Chapter 26) or failure by the contractor to build the project as required (covered in Chapter 24).

Claims by the contractor against the owner are usually based on payment for work performed that has not been received or on increased cost of performance attributable to the owner (covered in Chapters 22, 23, and 25). Measuring the value of the claim is discussed in Chapter 27.

SECTION 6.03 Judicial Remedies: Money Awards, Specific Decrees and Declaratory Judgments

Judicial remedies can be divided into judgments that simply state that the defendant owes the plaintiff a designated amount of money (sometimes called the *money award*); judgments that specifically order the defendant to do something (specific performance) or to stop doing something (injunction); and a judgment declaring the contract rights of the parties (declaratory judgment). It is important to recognize the essential differences between money awards, specific decrees and declaratory judgments.

The ordinary court judgment (the money award) is not a specific order to the defendant to pay this amount to the plaintiff. If the defendant does not voluntarily comply with the court decree, the plaintiff must ask law enforcement officials to seize property of the defendant (now a judgment debtor) that the law does not exempt. If this is done—often a costly and frustrating process—the property is sold to satisfy the judgment. Any amount remaining after paying the judgment and costs of sale is paid to the defendant. Enforcing court judgments in this fashion is often costly where successful and often is unsuccessful.

The specific decree, a personal order or command by the judge, is a much more effective remedy. Failure by the defendant to comply with the court order can be the basis for citing the defendant for contempt of court. The defendant is brought before the judge to explain why she has not complied with the decree. If the explanation is not satisfactory, the judge can punish the defendant by fine or imprisonment or coerce the defendant into performing by stating that the defendant must pay a designated amount or stay in jail until she performs.

A declaratory judgment delineates the contract rights of the parties under stipulated facts. A party seeking a declaratory judgment is asking the court to assess the party's rights under the contract, usually as a means to preclude further litigation. Most commonly, declaratory judgment actions are sought in insurance disputes to determine whether insurance coverage exists under the terms of the insurance policy. In one construction case, an owner's request for a declaratory judgment—that its contractor's invocation of the contract's *force majeure* clause was untimely—was disallowed.[1]

With some exceptions, Sections 11.02B and 28.07D (mechanics' liens), and Section 30.12 (specific performance of arbitration award), the claims central to this book will, if successful, result in money awards—the principal remedy sought when claims are made that someone has not performed a contract for design or construction services. Courts have been very reluctant to issue orders (specific performance) requiring design professionals or contractors to perform in accordance with their contractual obligations.

The principal reason for refusing to grant specific performance is that the remedy "at law" (the ordinary money judgment) is adequate. Adequacy assumes that the party with such an award can use the money to obtain the promised performance from a third party. The party seeking specific performance must show that the remedy at law is inadequate.

This is demonstrated by *Sokoloff v. Harriman Estates Development Corp.*,[2] a decision of the highest court of New York. In this case, the owner sought specific performance from the contractor who had agreed to perform pre-construction services that included supplying a set of plans and specifications for a house. (The contractor had retained an architect.)

When negotiations broke down over the price to build the house, the owner sought a court decree ordering the contractor to hand over the plans and specifications to the owner. He had paid most of the promised compensation and offered to pay the balance. (The owner had participated in creating the design.) While the law usually will not order specific performance of a personal service contract—too much like court-compelled slavery and a recipe for more problems that will place a substantial burden on the court—the New York Court of Appeals sent the case back to the trial court to see whether the money award would be adequate. If the owner cannot procure a similar design from a third party, the court will issue a decree of specific performance, ordering the contractor to turn over the plans to the owner.

---

[1] *Milford Power Co., LLC v. Alstom Power, Inc.*, 263 Conn. 616, 822 A.2d 196 (2003).

[2] 96 N.Y.2d 409, 754 N.E.2d 184, 729 N.Y.S.2d 425 (2001).

SECTION 6.04 Compensation and Punishment: Emergence of Punitive Damages

As stated in Section 6.01, the basic purpose in awarding damages for breach of contract is to compensate a party who has suffered losses or who has been prevented from making gains. The common law did not view contract breach as an immoral act that might justify punishment. This view may have been based on the importance, in a market-oriented society, of people engaging in economic exchanges. Excessive sanctions may discourage people from making contracts. In the past few years, courts have been willing to award punitive damages for certain types of egregious contract breach, justifying this by concluding that a particular breach was tortious. Wrongful dismissal claims, those made by employees against their employers, have begun to generate punitive damages, which are awarded to the claimant to prevent the defendant from repeating the conduct or to make an example of the defendant to deter others. As most punitive damages are awarded in tort claims, they will be discussed again in Section 7.10C.

Punitive damage awards have been quite rare in construction disputes. However, awarding punitive damages in the construction contract context, particularly against contractors, has been increasing. Specific illustrations are given in Section 27.10.

SECTION 6.05 Protected Interests

Increasingly, law looks on the varying ways in which the party who makes a contract can be protected from breach by the other party. This protection is frequently described as encompassing restitution, reliance, or expectation.

The most protected of the three interests, restitution, seeks to restore the status quo that existed before the contract was made by awarding the plaintiff any benefit the plaintiff has conferred on the defendant. It is most protected because it is relatively easy to establish and involves both a loss to the plaintiff and a gain to the defendant. Its importance in measuring claims by design professionals and contractors for services performed justifies its inclusion in Sections 12.14C and 27.02E.

A plaintiff protects its reliance interest by obtaining reimbursement of expenses from the defendant that have been incurred either in reliance on the contract or incurred before the contract was made and that have become valueless because of the breach. Such a remedy looks backward to a point in time even before the contract was made (somewhat similar to restitution), but also looks forward. If the defendant can establish that the expenditure sought would never have been reimbursed in the venture engaged in by the plaintiff, the plaintiff cannot recover from the defendant.

An architect brought a successful claim for reliance damages in *Designer Direct, Inc. v. DeForest Redevelopment Authority*.[3] A town's redevelopment agency teamed with an architect on a three-phase project to revise the downtown area. The redevelopment agency dragged its feet through the first two phases, causing the architect to terminate the agreement before work could begin on the third phase. Finding the termination well justified, the court permitted the architect to recover the cost of his design services performed in preparation for the third-phase work as a form of reliance damages.

Suppose a contractor failed to build a building in which the owner intended to manufacture trendy, stylish pants suits created by a well-known clothing designer for the very wealthy. Suppose further that the owner invested a large amount of money to promote these suits. If the contractor can establish that market saturation or changing fashions meant not one suit would have been sold, the owner could not recover the marketing costs, as they would not have been reimbursed from sales proceeds.

The third protected interest, expectation, looks forward and seeks to place the plaintiff in the position it would have found itself had the defendant performed. One way of accomplishing this objective is to order the defendant to perform as promised, a remedy generally unavailable in contracts to perform design or construction services. The same objectives can be achieved by a money award, however.

Suppose a design professional or contractor refuses to perform a contract. Awarding an amount that would enable the owner to hire someone to perform identical design or construction services would protect the owner's expectation interest. As owners use this formula very frequently in claims, it will be discussed in greater detail in Chapter 27.

---

[3] 313 F.3d 1036 (7th Cir.2002), rehearing denied, Jan. 29, 2003 (applying Wisconsin law).

SECTION 6.06 Limits on Recovery

Using rules described in this section, the law limits recovery by a claimant who has suffered losses that can be connected to the other contracting party's failure to perform in accordance with the contract. These rules often provide insurmountable obstacles to the claimant, prompting some to contend that the law does not adequately protect those who suffer losses from contract breach. Undoubtedly these obstacles can make recovery difficult. Yet they can also be looked on as rules that encourage parties to make contracts without inordinate fear that their nonperformance will expose them to unpredictable or devastating damage claims.

These doctrines can limit recovery but do not invariably do so. To the extent that any generalization can be made, modern law seems more willing to grant greater compensatory damages than in earlier periods in American legal history.

A. Causation

The claimant must show that the defendant's breach has caused the loss. Losses may be caused by more than one actor, and on occasion, causal factors can include other events and conditions. For example, the contractor's performance may be delayed by the owner, by strikes, by material shortages, and by the contractor's poor planning. It may be difficult if not impossible to establish the amount of the loss caused by contributing causes or conditions.

The defendant will be responsible for the loss if its breach was a substantial factor in bringing about the loss. It need not be the sole cause of the loss. Generally this question is determined by the finder of fact, sometimes the jury, and sometimes the trial judge. Any judicial finding regarding causation not based on guesswork will be upheld.

Causation problems can be particularly difficult when claims are made for breach of a contract to design or to construct, mainly because of the number of different entities that participate in the process and also because of the variety of conditions that may affect causation.

One common fact pattern involves tracing responsibility for a defect both to defective design and improper workmanship. Similarly, delayed performance by the contractor may be traceable to poor management by the contractor and excessive design changes made by

the design professional. These difficult problems are treated in Sections 24.06 and 26.06, respectively.

B. Certainty

A claimant must prove the extent of losses with reasonable certainty. One court described the certainty rule in these terms:

> Courts have modified the "certainty" rule into a more flexible one of "reasonable certainty." In such instances, recovery may often be based on opinion evidence, in the legal sense of that term, from which liberal inferences may be drawn. Generally, proof of actual or even estimated costs is all that is required with certainty.
>
> Some of the modifications which have been aimed at avoiding the harsh requirements of the "certainty" rule include: (a) if the fact of damage is proven with certainty, the extent or the amount thereof may be left to reasonable inference; (b) where a defendant's wrong has caused the difficulty of proving damage, he cannot complain of the resulting uncertainty; (c) mere difficulty in ascertaining the amount of damage is not fatal; (d) mathematical precision in fixing the exact amount is not required; (e) it is sufficient if the best evidence of the damage which is available is produced; and (f) the plaintiff is entitled to recover the value of his contract as measured by the value of his profits.[4]

Certainty in the context of construction claims is examined in Section 27.04.

C. Foreseeability: Freak Events and Disproportionate Losses

A series of improbable events can combine to lead to large losses. In the contract context, suppose a contractor is hired to build a high-rise office building. Shortly after the work is completed and accepted, the electrical system fails because of improper workmanship, causing the elevators to be out of service for two hours. During those two hours, the building owner has scheduled an interview with a prospective tenant who is thinking of leasing three floors of office space at a rental very attractive to the owner. Because the elevators are out of service, the prospective tenant decides to rent elsewhere. Should the contractor

---

[4]*M & R Contractors & Builders, Inc. v. Michael,* 215 Md. 340, 138 A.2d 350, 355 (1958).

be liable to the building owner for the extraordinary lease profits that the owner lost?

Clearly, the contractor's breach, perhaps a minor one, set off a chain of events that culminated in the loss of extraordinary rental profits. The improper installation caused an electrical failure, causing the elevators to malfunction. The prospective tenant was in the building at that very time. The prospective tenant's decision not to rent may have been caused by the tenant's irascibility or ignorance. This series of possible but improbable events has caused the building owner to suffer a loss. Is it fair to transfer this loss to the contractor?

The loss was an extraordinary loss and one that could not reasonably have been foreseen at the time the contract was made. Contrast this with an electrical malfunction that causes a fire. Such a loss falls more easily into the type of loss that can be reasonably foreseen at the time a contract is made. Also, such a risk is commonly covered by property insurance taken out by the owner and public liability insurance maintained by the contractor.

The leading common-law case dealing with what are sometimes called consequential damages is the English case, *Hadley v. Baxendale*,[5] which gave rise to the requirement that losses be reasonably foreseeable as a probable result of the breach. A shipper claimed against a carrier when the latter's delay in returning a shaft that had been sent to the factory for repair caused the shipper's plant to be shut down. The amount paid for shipping the shaft was disproportionately small compared with the cost of an entire plant shutdown. The court concluded that the carrier is liable for losses naturally resulting from the breach and other losses, the possibility of which is brought to the carrier's attention at the time a contract is made.

Advance awareness of the risk gives the carrier the chance to adjust its rates for performance or decide to forgo the transaction. Similarly, is it fair to place the responsibility for a building shutdown on an electrician who is called to make a minor repair in a high-rise building but who fails to perform in accordance with the contract obligations? This may be too onerous in light of the small profit earned. Care must be taken, however, to distinguish risks that are insurable from the premiums paid to transfer the risk to the insurer as part of a performing party's overhead.

---

[5]156 Eng.Rep. 145 (1854). See Restatement (Second) of Contracts § 351 (1981).

D. Avoidable Consequences (The Concept of Mitigation)

The rule of avoidable consequences is another limitation. A claimant cannot recover those damages that the claimant could have reasonably avoided. Sometimes this rule is expressed as one that requires the victim of a contract breach to do what is reasonable to mitigate or reduce the damages. This limiting rule relates to the requirement that the breaching party is responsible only for those losses that its breach has caused. If the loss could have been reasonably avoided or reduced by the claimant, the claimant cannot transfer that loss to the breaching party.

This limitation has not been a favored one. The law not only has placed the burden of establishing that the loss could have been avoided by the claimant on the breaching party but also has been hesitant to give the rule much scope. For illustrations in the construction context, see Section 27.07.

E. Lost Profits

Claims for lost profits, extraordinary or ordinary, have caused particular difficulty. They involve the preceding limitations on contract recovery, causation, certainty, foreseeability, and avoidable consequences.

In the construction contract, owner claims are usually based upon delays. Delay can create owner liability on other related contracts, such as leases with prospective tenants and lost opportunities for resale profits (discussed in Section 27.06B). Contractor claims for profits lost on the contract for construction breached by the owner are routine parts of the basic damage formula treated in Sections 27.02B and C. Profits on future construction contracts that may be lost because of the diminished reputation or reduction in bonding capacity caused by the breach are covered in Section 27.06C.

Some states deny recovery of lost profits by new businesses, proof being too uncertain that it would have earned profits. The modern tendency has been to treat this issue like any other factual issue. Even a new business can try to prove it would have earned additional profits from other contracts.

F. Collateral Source Rule

Suppose the defendant has breached but the plaintiff has been compensated by a third party. Can the defendant show that the plaintiff has not suffered a loss?

The issue in these cases is whether the collateral source rule applies. This rule denies the right of a defendant to deduct the amounts recovered from third parties if the third party who has compensated the plaintiff is considered a collateral source. If so, the defendant cannot deduct from its obligation the amount paid by the third party. The contribution simply is not taken into account.

While some courts refuse to invoke the collateral source rule in a contract as opposed to a tort claim, there is a trend toward treating contract and tort claims similarly and basing the application of the rule on other criteria. Still, most of the cases where the collateral source rule is invoked are tort claims. The rule will be discussed in Section 7.10B dealing with tort remedies. As shall be noted there, the common-law collateral source rule increasingly is modified by statute.[6] In the construction context, the rule is treated in Section 27.08.

G. Contractual Control: A Look at the UCC

The preceding discussion has assumed an absence of any controlling contract clauses regulating the remedy. Yet it is becoming more common for contracts to regulate the remedy. Some contracts specify or liquidate the damages when establishing actual damages would be difficult, if not impossible, to prove. Some contracts contain language insisted on by one of the contracting parties to either exculpate itself from responsibility or limit its exposure.

When remedies are specified, it is common to state in the contract that they are not exclusive. For example, the American Institute of Architects (AIA) provides remedies for termination. Yet the 2007 A201, Section 13.4.1, states that remedies specified are not exclusive. Similarly, evidence must be clear that a specified remedy is to be exclusive before it is given that effect. Contractual control of remedies is treated in other parts of this book.[7] The Uniform Commercial Code, which governs transactions in goods, has been having increasing impact on construc-

tion contract disputes. Section 2-719 allows the parties to provide for remedies by contract but states, "Where circumstances cause an exclusive or limited remedy to fail of its essential purpose, remedy may be had as provided in this Act."

Sellers of goods frequently seek to limit their obligation for breach to the repair and replacement of defective goods. In *Coastal Modular Corp. v. Laminators, Inc.*,[8] a contractor was allowed to recover damages from a supplier of defective panels for a construction project despite a contract providing that the remedy was limited to repair and replacement. The court held that the remedy had failed of its "essential purpose" under Section 2-719. The contractual remedy would apply only if the defect had been discovered while the work was in progress. Here the defect had been discovered after completion. Repair and replacement would not have been a viable remedy.

H. Noneconomic Losses

To this point, with the exception of the discussion relating to punitive damages, this chapter has focused on compensatory economic losses—losses that can be established precisely or roughly in the marketplace. Increasingly, claims for breach of contract, sometimes tied with tort claims, are made for noneconomic losses, such as emotional distress, caused by a breach of contract.

Generally contract law does not grant recovery for such losses. Denial has usually been based on the lack of foreseeability. However, a more acceptable rationale is that contracting parties should not be liable for potentially open-ended and freak losses that are extremely difficult to measure in economic terms. A homeowner who cannot take possession of a new residence when promised can certainly suffer emotional distress. Similarly, a homeowner may suffer emotional distress when she occupies a very badly built house. (*Erlich v. Menezes* reproduced in Section 27.09 involves such a claim.) But the likelihood and gravity of such distress generally depends on the emotional and psychological makeup of the homeowner.

Reasons that have been given for denying recovery for such losses, even if they are established as genuine and are reasonably foreseeable, are that recovery is likely to result

---

[6]A description, history, and application of the New York statute that in part eliminated this common-law rule can be found in *Fisher v. Qualico Contracting Corp.*, 98 N.Y.2d 534, 779 N.E.2d 178, 749 N.Y.S.2d 467 (2002) (insurance proceeds offset amounts received from negligent contractors).

[7]Sections 15.03, 26.09B, and 26.10A.

[8]635 F.2d 1102 (4th Cir.1980).

in disproportionate compensation.[9] Another commentator sought to explain the variant decisions by drawing a line between personal and commercial contracts, though he was not satisfied with this dichotomy.[10]

SECTION 6.07   Cost of Dispute Resolution: Attorneys' Fees

This text has suggested that contract remedies should compensate the injured party. Yet full compensation is rarely achieved. The rules that determine whether and how much damages are awarded sometimes make it difficult to achieve full compensation. Even more important, the winner cannot recover its cost of litigation, an item that has risen to staggering proportions. American common law (judge-made) does not transfer the prevailing party's attorneys' fees to the other party unless the claim is based on a contract that contains a provision providing for a recovery[11] or recovery is based on a statute granting attorneys' fees. Under extraordinary circumstances, such as commission of an intentional tort or the losing party's claim having been vexatious or frivolous, such costs can be recovered.

Denial of attorneys' fees is based on the reluctance of American law to discourage citizens from using the legal system. Although an unsuccessful claimant may have to bear the cost of its own fees, the law absolves the claimant of the responsibility of bearing the other party's costs.

This rule has led to increased legislation authorizing attorneys' fees as part of costs in consumer and civil rights cases. As shall be shown in Section 23.02J, claims by owners for defective work are often based on deceptive trade practices statutes, to expand remedies to treble damages and award attorneys' fees. Some states have enacted reciprocal attorneys' fees legislation. These statutes provide that if one party can recover attorneys' fees under a contract, the other party can do so even though the latter is not specifically granted this in the contract.

Some statutes grant attorneys' fees to the prevailing party where a mechanics' lien has been sought. Some give attorneys' fees when claims are made on surety bonds. A few statutes allow attorneys' fees *generally* for contract actions.[12]

Documents published by the AIA do not, except for indemnification, provide for attorneys' fees, each party bearing its own litigation costs.

AIA documents allow the parties to elect arbitration under the arbitration rules of the Construction Industry Dispute Resolution Procedures (CIDRP) (see Appendix F) administered by the American Arbitration Association. These rules do not specifically provide for attorneys' fees. A decision by the Supreme Court of Tennessee reviewed an arbitration award made under the CIDRP that included attorneys' fees.[13] The court pointed to two provisions in the contract that dealt specifically with attorneys' fees while the CIDRP Rules do not, as noted earlier in this Section, provide for attorneys' fees. While as a rule (see Section 30.14), the courts give wide scope to the arbitration award, the court refused to confirm that portion of the award that awarded attorneys' fees. This issue is discussed in Section 30.12.

Local law, currently in a state of ferment, must be consulted by those who draft contracts or those who must counsel as to the availability of attorneys' fees to the prevailing party. Although discussion has centered on attorneys' fees, the other formidable costs of construction litigation—particularly complex litigation that involves defects and impact claims—cannot be ignored. These claims generate immense costs to reproduce, classify, analyze, and store documents as well as the staggering costs involved in conducting pretrial discovery. To these costs must be added the costs of preparing exhibits and retaining expert witnesses and the nonproductive costs incurred by personnel in preparing for the lawsuit. Those who draft contracts may wish to deal with these costs. Such costs should remind the parties they must make every effort to avoid litigation.

---

[9]E. A. FARNSWORTH, CONTRACTS, § 12.17 (3d ed.1999).

[10]Sebert, *Punitive and Non-pecuniary Damages in Actions Based on Contract: Toward Achieving the Objective of Full Compensation*, 33 U.C.L.A. L. Rev. 1565, 1594 (1986).

[11]*Caldwell Tanks, Inc. v. Haley & Ward, Inc.*, 471 F.3d 210 (1st Cir.2006) (attorney fee clause in indemnity agreement enforced).

[12]Toomey & Brown, *The Incredible Shrinking "American Rule": Navigating the Changing Rules Governing Attorneys' Fee Awards in Today's Construction Litigation*, 27 Constr. Lawyer, No. 2, Spring 2007, p. 34.

[13]*D. & E. Constr. Co. v. Robert J. Denley Co., Inc.*, 38 S.W.3d 513 (Tenn.2001).

SECTION 6.08 Interest

Many cases have involved recoverability of prejudgment interest. Claimants often seek interest from a time earlier than entry of court judgment. Perhaps this is due to the ease with which a claim for interest can be tacked onto a claim for work performed or defective work. In any event, states vary considerably in their treatment of claims for prejudgment interest.

The varying legal rules dealing with prejudgment interest are caused by the differing judicial attitudes toward the desirability of complete compensation on one hand and on the other hand punishing a defendant when the latter exercised a good-faith judgment not to pay that turned out to be incorrect.

Although case holdings and statutes have many slight variations, they follow three principal rules. One limits prejudgment interest to liquidated (specific) amounts or unliquidated amounts that are easily determinable by computation with reference to a fixed standard contained in the contract without reliance on opinion or discretion. A party against whom a claim is made should be able to avoid prejudgment interest by tendering the amount due. Unless the amount is known or easily determined, tender cannot be made. This ignores the likely possibility that payment is not made because of a dispute over the validity of the claim, not the amount due were the claim held to be valid.

Some courts and statutes give the judge or jury discretion to determine whether interest should be awarded.[14] Often discretion takes into account whether the refusal by the defendant to pay was vexatious (without sufficient grounds, solely to cause annoyance), what the inflation rate is, and how important money use is to the party entitled to the payment.

Some jurisdictions, either by court decision or increasingly by statutes, mandate prejudgment interest under certain circumstances.[15] Undoubtedly this reflects increased legislative recognition that delay in payment causes a serious loss that should be compensated through interest. Again, local law must be consulted.

Special rules frequently deal with claims against public authorities. Until recently, interest could not be recovered against the federal government. This changed in 1978.[16]

With costs of financing now more recognizable as an important element in construction costs, contracts should specify that payments will bear interest from the time they are due. As to claims that are difficult to evaluate, the issue is not so clear. Suppose each party has a good-faith belief in the merit of its position and for that reason refuses to settle. Ultimate determination that refusal to pay was unjustified, permitting recovery of prejudgment interest, can place a heavy burden on the party who has asserted an honest reason for not paying. On the other hand, undoubtedly the loss should have been paid, and not awarding prejudgment interest denies full compensation to the party whose position was ultimately vindicated. Balancing compensation against punishment makes resolution of this issue difficult.

Unless the contract or an applicable statute states otherwise, in most states the interest rate will be the amount specified by law to be paid on legal judgments. See Section 22.02L.

---

[14]West Ann.Cal.Civ.Code § 3287(b) (grants judge discretion to award interest from date no earlier than date legal action filed).

---

[15]Penn.Consol.Stat.Ann., tit. 73, § 512(b) (prompt payment act), interpreted in *John B. Conomos, Inc. v. Sun Co., Inc. (R&M)*, 2003 PA Super. 310, 831 A.2d 696 (2003).

[16]41 U.S.C.A. § 611 allows interest from the date the claim is received until payment. The rate is set by the secretary of the treasury. From January 1, 2008 to June 30, 2008, the rate was 4.75 percent. See 72 Fed.Reg. 74408 (2007). (In 1982, it was 15.5 percent.)

Losses, Conduct, and the Tort System: Principles and Trends

SECTION 7.01 Relevance to the Construction Process

During the construction process, events can occur that might harm persons, property, or economic interests. Workers or others who enter a construction site may be injured or killed. The owner or adjacent landowners may suffer damage to land or improvements. Participants in the process may incur damage to or destruction of their equipment or machinery. The owner or other participants in the project may incur expenses greater than anticipated. Investors in the project or those who execute bonds on participants may also suffer financial losses.

After completion of a project, people who enter or live in the project might be injured or killed because of defective design, poor workmanship, or improper materials. Those who invest in the project may find investment value reduced for similar reasons.

The construction project is a complex undertaking involving many participants. It presents high risk of physical harm to those actively engaged in it. Sometimes such harm is caused by participants failing to live up to the standards of conduct required by law. Losses sometimes occur by human error that does not constitute wrongful conduct. Losses sometimes occur because of unpredictable and unavoidable events for which no one can be held accountable. Because of the varying causes of losses, the many participants in the project, and the complex network of laws, regulations, and contracts, placing responsibility is difficult.

SECTION 7.02 Tort Law: Background

A. Definition

A *tort* has been defined as a civil wrong, other than a breach of contract, for which the law will grant a remedy, typically a money award. This definition, though not very helpful, mirrors the difficulty of making broad generalizations about tort law in the United States. One reason is the incremental or piecemeal development of tort law necessitated by new activities causing harm. For this reason, much American tort law consists of a collection of wrongs called by particular terms, which were given legal recognition in order to deal with particular problems.

Some basic distinctions are essential. Although tort law and criminal law have features in common—each regulating human conduct—they operate independently. Crimes are offenses against the public for which the state brings legal action in the form of criminal prosecution. Prosecution is designed to protect the public by punishing wrongdoers through fines or imprisonment and to deter criminal conduct.

The tort system is essentially private. Only individual victims can use the system. However, under certain circumstances a class can maintain a class action. A class allowed to bring a class action consists of those who have suffered similar harm by the same cause (silicone breast transplants, asbestosis caused by working with asbestos). Any sanctions imposed against those who do not live up to the standard of tort conduct are for the benefit of the victim. For example, the automobile driver who violates the criminal law may be fined or imprisoned. Those who are injured because of such criminal conduct are likely to institute a civil action to transfer their losses. This civil action is part of the tort system.

B. Function

Tort law has different functions. The particular function most emphasized at any given time depends on social and economic conditions in which the system operates.

The principal functions are to compensate accident victims, to deter unsafe or uneconomic behavior (to encourage investment of up to but not beyond the point at which incremental safety costs equal incremental injury costs), to punish wrongful conduct, or to protect social norms from a sense of outrage generated by perceived injustice, by providing a dispute resolution mechanism, or a combination of any of these goals.[1]

In seeking to implement these functions, the law must often choose between goals and interests of individuals and groups. One person's desires may be filled at the expense of another. One person may wish to drive a car at a high speed, exposing others to danger. Property owners may wish the freedom to maintain their property as they wish. But this freedom may come at the expense of those who enter the land and are injured.[2] A manufacturer may wish complete freedom to design a product that will earn the highest profit. But this freedom may come at the expense of buyers who suffer harm from using the product.[3] Adjusting these conflicts can reflect conscious decisions to select or favor one competing interest or goal over another—a form of social engineering.

Tort rules may severely limit a property owner's freedom or the freedom of a manufacturer to give greater protection to those who are injured on unsafe property or by defective goods. The historical description in Section 7.02D suggests trends in social engineering.

C. Threefold Classifications

Two important threshold concepts are threefold. The first concept describes the interests considered sufficiently important to merit tort protection and often described as follows:

1. *personal*, sometimes defined to include psychic or emotional interests
2. *property*, tangible and intangible
3. *economic*, unconnected to harm to a person or damage to property

The second concept classifies the conduct of the person causing the loss:

1. *intentional*, including not only the desire to cause the harm but also the realization that the conduct will almost certainly cause the harm
2. *negligent*, usually defined as failure to live up to the standard prescribed by law
3. *nonculpable*, though in a sense wrongful, in which the actor neither intends harm nor is negligent

These threefold classifications play important roles in determining which victims will receive reparation from those causing the loss. On the whole, harm to a person is most deserving of protection, with harm to property being considered second in importance. At the conduct end, intentional conduct that causes harm is least worthy of protection, followed by negligent conduct. These classifications are gross, and many subtle distinctions must be made.

D. Historical Patterns

Earliest English private law developed during the feudal period, in which land dominated society. As a result, property law developed before any significant developments in tort or contract law. The feudal period was dominated by agriculture and a largely illiterate population, with little need for a developed contract or tort law system.

Although early English legal history saw the development of laws that dealt with finance, banking, and maritime matters, what is known today as *tort law*—as well as much of what is known today as *contract law*—did not develop until the Industrial Revolution in the late eighteenth and early nineteenth centuries. The Industrial Revolution moved manufacturing out of cottages and into factories, necessitating a transportation system and migration of workers from the farms and villages to the towns and cities.

The changes brought significant developments in tort law. The preindustrial agrarian society, with its emphasis on property, was most concerned with property rights and keeping the peace. The important torts were those that were intentional, as they could invite retribution, and those that invaded property interests. As a result, although persons may not have always acted at their peril in the sense that they would have to account for any damages their activities caused, much liability was "strict."

---

[1]Smith, *The Critics and the "Crisis": A Reassessment of Current Conceptions of Tort Law*, 72 Cornell L.Rev. 765 (1987). A general discussion of the aims, policies, and methods of tort law is found in D. DOBBS, TORTS, 12–25 (2000).

[2]See Section 7.08.

[3]See Section 7.09.

Trespass, an invasion of a property owner's right to exclusive possession of property, did not require any showing of fault. Any trespass, whether innocent or deliberate, was wrongful. Both the importance of property rights and the inability to deal with subtle concepts such as negligence and fault contributed to the rather simple and often harsh strictures of early tort law.

Keeping the peace required protection against serious intentional torts, such as trespass, assault (apprehension of harm), battery (harmful or offensive touching), or false imprisonment (deprivation of freedom of movement). These serious matters could lead to breaches of the peace, and such conduct had to be eliminated or at least minimized through making the actor pay for the harm caused.

Unintentional or negligent conduct did not rate very high on the interest protection scale, both for reasons mentioned and because more important matters required attention. Matters such as harsh words, offensive conduct, or careless jostling were not sufficiently important in such a society to receive protection.

The Industrial Revolution generated factory and transportation accidents as well as migration to population centers. Now the law had to deal with conduct that became serious in crowded towns and cities. Many matters of an earlier day that were too trivial to be dealt with now required attention.

Even more important, difficult choices had to be made when commercial activity caused harm. The law chose to protect new and useful commercial and industrial activities from potentially crushing liability by not making liability as "strict" as it had been in preindustrial times. With some important exceptions,[4] a person who suffered a loss could not transfer the loss to the person causing it unless the injured person was free of negligence and could establish that the person causing the harm did not live up to the negligence standard of conduct. Transferring the loss only on a showing of negligence and the development of other legal doctrines were designed to free useful activities from responsibility. These rules were less concerned with compensating victims of these activities.

Liberalization, as the term was then used, was designed to free economic activity from the shackles of heavy state mercantilistic controls. Rules that protected industrial and commercial activity may also have been generated by the belief that such activities brought long-run social and economic advantages. Society in the nineteenth century emphasized the moral aspects of individual responsibility, and shifting a loss required wrongdoing. But as the toll in human misery and economic deprivation rose because of industrial, commercial, and transportation activities and as the automobile replaced the horse, changes were inevitable.

At the beginning of the twentieth century, many industrial countries sought to remove industrial accidents from the tort system. Workers' compensation is an example of a change to tort law brought about by legislative action.[5]

Common law judges also began to fundamentally expand the nature of tort liability. One of the most important defenses to liability created during the "liberalization" period was the *privity doctrine*, under which a party who negligently performs a contract obligation is liable only to the other party to that contract, not to third parties injured by the defendant's conduct. In the 1920s, judges began to reject the privity doctrine, making liability dependent upon the foreseeability of injury to the plaintiff arising from the defendant's conduct.

The new foreseeability rules immediately expanded liability to persons injured by negligently manufactured products. Manufacturer liability for negligence was later replaced with strict products liability. Compensating victims rather than unshackling enterprises became the predominant goal. The shift is sometimes described as *enterprise liability*. Under it an enterprise can and should bear the normal risks of its activity. These risks can be predicted, computed, and insured. The social costs of the activity can, through pricing, be spread to all who benefit from the enterprise by using its products.[6] (Workers' compensation is an example of enterprise liability, but it is limited to injury to workers and coupled with tort immunity to employers.)

Beginning in the mid-twentieth century, the common law expanded liability to property owners for injuries to those who entered upon their property.[7] Liability also expanded to those who furnished services, such as architects and engineers.[8]

---

[4]See Section 7.04.

[5]See Section 7.04C.

[6]Keating, *The Theory of Enterprise Liability and Common Law Strict Liability*, 54 Vand. L. Rev. 1285 (2001); Section 7.09 (strict products liability).

[7]See Section 7.08.

[8]See Chapter 14.

What of the postindustrial tort law? Some have advocated its abolition. They would replace it with social insurance, under which all victims of accidents would be compensated by the state or by private insurers.

Some have advocated, and many states have adopted, no-fault handling of road accidents. The victim recovers up to a threshold amount of medical expenses (which is quite low) without showing any fault. Above this threshold, the victim can use tort law. (Low thresholds have meant inflated medical expenses and very little reduction in lawsuits.)

Some have felt that enterprise liability has placed too heavy a burden on manufacturers and professionals. Much has been written about the "malpractice crisis," and legislative activity has moderated the harsh treatment given professionals and manufacturers by the courts. In addition, some have questioned the expansion of liability without fault and have suggested a return to a more negligence-oriented tort standard. Such calls for a brake on expanded liability are based on assertions either that liability cannot be insured against or that insurance premiums and claims expenses make useful services or products unprofitable. This has been seen in decisions by drug companies to pull particular drugs off the market and increased reluctance by medical professionals to pursue certain specialties, such as obstetrics and gynecology.

As noted in Section 7.02B, deterring unsafe and inefficient conduct is one goal of tort law. With the advent of liability insurance, some believe that the deterrent function is no longer accomplished by tort law. They would prefer direct control by legislation, similar to the Occupational Safety and Health Act enacted by Congress. Although much criticized mainly for obsessive interest in detailed rules, this act is a direct attempt to make industrial activities safer than would the indirect method of the tort system.

Although predictions are dangerous, the tort system in some form will likely continue to serve, if not the primary, at least an ancillary role in compensating victims and regulating activity.

E. General Factors in Determining Tort Liability

Some have examined the unruly and disparate thousands of tort cases and have attempted to articulate factors that affect tort liability. One scholar listed the following items as important:[9]

1. the moral aspect of the defendant's conduct
2. the burden of recognizing a legal right on the judicial system
3. the capacity of each party to bear or spread the loss
4. the extent to which liability will prevent future harm

F. Coverage of Chapter

Tort law is simply too diverse and immense to cover in this text. By and large, there will be no discussion of intentional torts[10] such as trespass,[11] assault, battery, false imprisonment, intentional infliction of emotional distress,[12] defamation,[13] invasion of privacy, and interference with contract or prospective advantage.[14] Emphasis is placed on negligence and strict liability, as those concepts relate to the construction process, starting with design and culminating with the finished project.[15]

---

[9]See W. PROSSER, TORTS, 16–23 (4th ed. 1971). A subsequent edition added "a recognized need for compensation" and "historical development." W. P. KEETON et al, TORTS, 20–26 (5th ed. 1984).

[10]Comprehensive treatment of intentional torts can be found in D. DOBBS, supra note 1 at 47–154.

[11]In re Catalano, 29 Cal.3d 1, 623 P.2d 228, 171 Cal.Rptr. 667 (1981), held that a union official did not violate the criminal trespass statute when he refused to leave the site when ordered to do so by the owner. The official was conducting a safety inspection, a power given the union under its contract.

[12]See Section 6.06H.

[13]The tort of defamation, usually subdivided into libel and slander, sometimes arises in the construction context. See, for example, Diplomat Elec., Inc. v. Westinghouse Elec. Supply Co., 378 F.2d 377 (5th Cir.1967) (subcontractor may sue its supplier for defamation); Tutor-Saliba Corp. v. Herrera, 136 Cal.App.4th 604, 39 Cal.Rptr.3d 21 (2006) (city attorney's allegedly defamatory remarks, made about a public contractor with whom the city was in litigation, are subject to the "official duty" privilege); and Mishler v. MAC Systems, Inc., 771 N.E.2d 92 (Ind.App.2002) (sign posted by owners on their property, complaining of their contractor's poor workmanship, is protected speech).

[14]This intentional tort is noted briefly in Section 14.08F. See Custom Roofing Co. v. Alling, 146 Ariz. 388, 706 P.2d 400 (App.1985), which upheld a punitive damage award against a supplier in a claim by a contractor. The award was based on the supplier's wanton conduct and indifference to the rights of others. The supplier had failed to supply material, knowing it would cause the contractor to lose its contract.

[15]Most law relating to torts is called common law in that the principal sources of law are reported appellate decisions. Legislative bodies increasingly bar certain conduct, such as improper reasons for refusing to sell, rent, or hire.

SECTION 7.03 Negligence: The "Fault" Concept

A. Emergence of the Negligence Concept

The Industrial Revolution was the precipitating factor for tort law moving from principal emphasis on intentional torts and the need to keep the peace, to a system that would deal with increased accidents brought about by industrialization. This movement culminated with the recognition of negligence as the principal basis for liability. To legitimately transfer the plaintiff's loss to the defendant, the plaintiff was required to establish that the defendant had not performed in accordance with the legal standard of conduct that, although somewhat inaccurately, was called the "fault" system.

Today the negligence concept largely governs road accident losses, losses caused by the possessor of land failing to keep the land reasonably safe, losses that occur in the home, losses caused by the activities of professionals, and, for the most part, losses caused by participants in the construction process.

B. Elements of Negligence

To justify a conclusion that the defendant was negligent, the plaintiff must establish the following:

1. The defendant owed a duty to the plaintiff to conform to a certain standard of conduct in order to protect the plaintiff against unreasonable risk of harm.[16]
2. The defendant did not conform to the standard required.[17]
3. A reasonably close causal connection existed between the conduct of the defendant and the injury to the plaintiff.[18]
4. The defendant invaded a legally protected interest of the plaintiff.[19]

C. Standard of Conduct: The Reasonable Person

The Objective Standard and Some Exceptions. Nineteenth-century English and American courts rejected a standard based on the subjective ability of the defendant. It was

---

[16]See Section 7.03E.
[17]See Section 7.03C.
[18]See Section 7.03D.
[19]See Section 7.03F.

not sufficient for the defendant to show that he did the best he could. To protect the community, its members are held to a standard that can exceed what they are able to do. In this sense, negligence is not synonymous with fault. The community standard requires that the defendant do what the reasonable person of ordinary prudence would have done. Such a standard can hold the defendant liable despite his having done the best he could. Negligence, then, or much of it, is not congruent with morality. The standard holds people who live in the community to an average community standard. For example, an inexperienced driver is expected to drive as well as the average driver.

Exceptions do exist, and a person can be held to a lower or higher standard than that of the community. Usually such exceptions are created by designating special subcommunities smaller than the general community and then applying an objective subcommunity standard. For example, children generally are not held to the adult standard but are held only to the standard of children of similar age and experience. Persons who, because of special training or innate skill, are expected to do better than the average person. Physicians are held to a higher standard in dealing with medical matters than are ordinary members of the community. Architects and engineers are held, as a rule, to the standards of their subcommunity (discussed in greater detail in Chapter 14). Professional truck drivers are expected to drive better than ordinary drivers. The combination of objective and subjective standards in such cases is reflected by statements that defendants are judged by what they knew or should have known or by what they did or should have done.

Because these standards are generally applied by a jury, some tolerance of human weakness and some exceptions to the objective standard other than those mentioned may find their way into jury decisions.

Unreasonable Risk of Harm: Some Formulas. Courts and commentators seek to refine vague community or reasonable person standards by articulating factors that should sharpen the inquiry into whether the standard of conduct has been met. One approach is to evaluate the magnitude of the risk, the utility of the conduct, and the burden of eliminating risk.

To determine the magnitude of risk, the relevant factors are the gravity, the frequency, and the imminence of the risk. Clearly the likelihood and the severity of harm that can result are important factors in determining the type of conduct that, to avoid these risks, should be expected. Driving at an excessive speed on the highway clearly creates a high risk because accidents often result from excessive speed and their consequences are usually serious. Railroad crossings are dangerous because, though the likelihood of a train's striking an automobile may be small, the consequences of such an occurrence are serious. Conversely, although throwing a soft rubber ball into a crowd may not cause serious harm to anyone, it is likely to cause some harm to someone.

The other factors recognize that imposing liability on actors restricts human freedom. The social utility of the conduct being regulated is an important criterion in determining whether the legal standard has been met. Suppose a bank robber carelessly jostles a bank patron in the course of robbing the bank. If the law holds the bank robber responsible for the harm caused the patron, one factor is likely to be that bank robbing is not considered a useful activity and can be regulated by tort law as well as by criminal law. Conversely, the same carelessness by a bank security guard in performing a socially useful activity, such as organizing the patrons of the bank so that they can make an orderly retreat from a fire, would not expose the bank or guard to the same liability as the bank robber.

The burden of eliminating the risk recognizes that almost all risks can be eliminated or minimized if sufficient resources are mobilized. Undoubtedly the impact of road accidents would be minimized if guardrails were installed on all public highways. Yet doing so would involve an immense expenditure that might not be commensurate with the gain thus realized. Likewise, the burden of eliminating the risk in this manner would take into account not only the cost but also the aesthetic deprivation caused by universal installation of guardrails.

Where serious harm can be avoided by minimal effort or expenditure, failure to do so will very likely be negligent. For example, if serious burns can be avoided by installing a $5 mixer valve in bathroom fixtures, failure to do so would very likely be negligent.[20]

Common Practice: Custom. Suppose the defendant conformed to or deviated from common practice, or what is sometimes called "the custom in the community."

_____
[20]*Schipper v. Levitt & Sons, Inc.*, 44 N.J. 70, 207 A.2d 314 (1965) (builder-vendor strictly liable without need to show negligence).

Defendants frequently attempt to exculpate themselves by showing that they performed as others do. This defense arises most frequently in claims against a manufacturer. Often the manufacturer establishes that it performs in accordance with industry practices.

Although compliance with customary practices is evidence of compliance with the legal standard of care, it is not conclusive. The customary standard itself may be careless and create unreasonable risk of harm. For example, suppose that most pedestrians jaywalk or that all workers refuse to wear hard hats in a hard-hat area. Similarly, failure to conform to customary practices is not conclusive, and a defendant may be exonerated despite deviation from customary practices if good reasons existed for deviation and the defendant conducted activities with reasonable care.

One important exception relates to the standard of conduct expected of professionals. As this is more appropriately dealt with in examining liability of design professionals, major discussion is postponed until Chapter 14. It is enough to state here that when defendants are judged by the subcommunity of their profession, customary practices of the profession become the standard.

Violations of or Compliance with Statutes. In exercising its responsibility to protect all citizens, government frequently prohibits certain conduct and attaches civil or criminal sanctions for violations. The construction process is governed by a multitude of laws dealing with land use, design, construction methods, and worker safety. What effect do violations of those statutes have on civil liability?

It is possible, though uncommon, for the statute to expressly declare that violations of the statute determine civil liability. More commonly, the statute expressly imposes criminal sanctions only. In such cases, courts can and do look at the statute as a legislative declaration of proper community conduct. To have any relevance, however, some preliminary questions must be addressed.

First, the person suffering the harm must be in the class of people that the legislature intended the statute to protect. Many statutes are intended to protect members of the community at large by achieving public peace and order rather than to protect any particular group or individual. For example, statutes sometimes prohibit certain businesses from operating on Sunday. Such statutes are not designed to protect those who suffer physical harm while a business operates in violation of the Sunday closing laws. Similarly, although statutes require that automobiles be registered, the purpose of such laws is to raise revenue and not to impose liability on the driver of an unregistered car even though driving properly.

Sometimes the legislation is designed to protect an extremely limited class of people that may not include the injured party. For example, legislation requiring that dangerous machinery be shielded may be designed for the benefit of employees and not of those who enter the plant for other purposes. Similarly, some states hold that safety regulations imposed on a contractor-employer are not designed to protect employees of other employers on the site.

Second, did the statute deal with the particular risk that caused the injury? For example, suppose a statute limits the time a train may obstruct a street crossing. This is designed to deal with traffic delays, not with the risk of personal harm caused by the delaying train. On the whole, the tendency has been to broadly define the particular risk.

The violation of law can be excused in most instances by showing extraordinary circumstances that made compliance more dangerous than violation. For example, a statute may require that drivers always drive on the right unless they are passing another vehicle or making a left turn. Suppose the driver veers to the left lane to avoid hitting a child. This technical violation will be excused and have no bearing on negligence.

Where the policy expressed by the statute is particularly strong, the statute may expressly eliminate any possibility of a violation being excused. Violations of such statutes are conclusive on the question of negligence. Such statutes often impose liability despite assumption of risk or contributory negligence by the injured party, which, as shall be seen in Section 7.03G often is a defense. This form of strict liability may result if the statute was intended to protect someone from his own immaturity or carelessness. Illustrations are those prohibiting child labor or requiring safety measures in construction work.

Much can depend on the particular law violated. Some laws seem anachronistic and continue to exist only because the legislature lacks the energy to modernize rules. Violation of such laws may have very little impact.

Suppose it is concluded that the preliminary requirements have been met. The plaintiff is in the class of people to be protected, the statute was intended to cover the risk in question, and the violation was not excused. Most courts hold the violation to be negligence *per se* and conclusive on the question of negligence. The trier of fact,

whether judge or jury, need not decide whether there had been negligent conduct.

With the exception of special protective statutes of the type described earlier, a *per se* violation may or may not preclude a defense such as assumption of risk or contributory negligence. It does not preclude the defendant from showing that violating the statute did not cause the harm. For example, *Hazelwood v. Gordon*[21] concerns a case in which an employee fell down a flight of stairs, which were too narrow at the bottom. The narrowness of the stairs, together with an inadequate handrail, violated a city ordinance. However, the court held that the injured party could not recover from the property owner, because her injury was not caused by a violation of the ordinance but was caused by her negligently placing her foot on the top step of the staircase, knowing the stairs were dangerous.

Some jurisdictions, however, find that a statutory violation is simply evidence of negligence to be given to the jury and weighed along with other evidence to determine whether the defendant lived up to the legal standard of care. Some states that employ negligence *per se* hold violations of local ordinances, traffic regulations, or administrative regulations to be only evidence of negligence. For example, *Bostic v. East Construction Co.*[22] dealt with administrative regulations for fire safety. The court held that administrative regulations can be the basis for negligence *per se* but are less likely to be. The court also held that the regulations must be understandable, and the regulations involved in this case were not clear enough. (In any event, the court seemed to believe that the failure to comply with the fire regulations did not cause the injury.)

Compliance with the statutory standard does not necessarily preclude a finding of negligence. The statutory standard is a minimum. Additional precautions can be required. For example, it may not be a statutory violation to park a car on the shoulder of a highway as long as a taillight functions. But under certain circumstances, such as an extremely foggy night, this conduct may be below the legal standard and be negligent. Such instances are rare.

Res Ipsa Loquitur. Proof of negligence can be by direct testimony of witnesses who testify based on their own observations of the defendant's conduct. However, sometimes this evidence is not available. Absence of

---

[21]253 Cal.App.2d 179, 61 Cal.Rptr. 115 (1967).
[22]497 F.2d 712 (6th Cir.1974).

direct evidence does not preclude the plaintiff from establishing indirectly (that is, by circumstantial evidence) that the defendant was negligent. This can be accomplished by showing facts relating to the accident that tend to show, in the absence of an explanation by the defendant, that the defendant's negligence probably caused the accident. For example, suppose a tool falls from a scaffold and injures a passerby. Once the facts are established and it is known that the contractor's workers were working on the scaffold, it is more likely than not that the accident was caused by the contractor's negligence. The unfortunate Latin term *res ipsa loquitur*, used to describe this process of indirect proof of negligence, was employed in an English case in which a passerby was struck by a flour barrel falling from a warehouse window.[23]

Much controversy has developed regarding the *res ipsa* concept, mainly centered around judicial and scholarly statements of *res ipsa* requirements. In reality, the supposed requirements are not truly requirements, because the doctrine is sometimes applied despite absence of some of them. However, the stated requirements may provide some assistance in understanding when the concept will be used. It is usually stated that *res ipsa* requires the following:

1. The event is one that ordinarily does not occur in the absence of someone's negligence.
2. The event must be caused by an agency or instrumentality within the exclusive control of the defendant.
3. The accident must not be due to any voluntary action or contribution by the plaintiff.

Some courts have suggested that the evidence must be more readily accessible to the defendant than to the plaintiff. But as stated, these requirements cannot be taken as conclusive, because they are not always found in cases where the doctrine is applied.

The use of circumstantial evidence does not, as a rule, shift the burden of proof from plaintiff to defendant. But if the facts indicate that the defendant's negligence probably caused the accident, the matter will be submitted to the jury. Applying the doctrine does not preclude defendants from introducing evidence that they

were not negligent. For example, in the illustration given earlier, the contractor can show that the tool fell because of a strong gust of wind for which it was not responsible. But the doctrine helps plaintiffs get their cases before the jury.

Sometimes the inference of negligence by the defendant is so strong it may persuade the jury that the plaintiff has met his burden of proof. Similarly, a very strong inference of negligence may justify the judge directing a verdict for the plaintiff.[24]

D. Legal Cause: Cause in Fact and Proximate Cause

A reasonably close connection must exist between conduct of the defendant and the harm to the plaintiff. Legal cause is divided into two separate though related questions:

1. Has the defendant's conduct caused the harm to the plaintiff? (This is usually referred to as "cause in fact.")
2. Has the defendant's conduct been the "proximate cause" of the harm to the plaintiff?

Cause in Fact. The first, considered less complicated, is a factual question decided by the finder of fact, usually the jury. Even this supposedly simple question of causation can raise difficult issues. First, the defendant's conduct need not be the sole cause of the loss. Many acts and conditions join together to produce a particular event. Suppose an employee of a subcontractor suffers a fatal fall while working on a scaffold high above the ground. Any one of the following events could be considered a cause of the death in the sense that without any of these events the fatal fall would not have occurred:

1. the worker's need to pay medical bills, causing the worker to take this risky job
2. defective scaffolding supplied by a scaffolding supplier
3. failure by the subcontractor or prime contractor to remove the scaffolding when complaints were made about its unsafe condition
4. weather conditions that made the scaffold particularly slippery on the day of the accident

---

[23]*Byrne v. Boadle*, 159 Eng.Rep. 299 (Exch.1863); Webb, *The Law of Falling Objects:* Byrne v. Boadle *and the Birth of Res Ipsa Loquitur*, 59 Stanf. L. Rev 1065 (2007).

[24]*Morejon v. Rais Constr. Co.*, 7 N.Y.3d 203, 851 N.E.2d 1143, 818 N.Y.S.2d 792 (2006).

5. a low-flying plane that momentarily distracted the worker
6. the worker's refusal to wear a safety belt
7. the subcontractor's or prime contractor's failure to enforce safety belt rules

Although the list can be amplified, remove any link in the causation chain and the worker would not have been killed. Yet it would be unfair to relieve any actor whose failure to live up to the legal standard played a significant role in the injury simply because other actors or conditions also played a part in causing the fall.

Liability in such a case would depend on a conclusion that any of the defendants substantially caused the injury. Suppose a claim had been made against the scaffold supplier and the prime contractor. Each could have been a substantial cause of the injury. In an indivisible injury, each would be liable for the entire loss, and whether the party paying the claim would be entitled to recovery from the other party would depend on whether a right to contribution or indemnity existed.[25]

Cause in fact requires that the harm would not have occurred without the defendant's failure to live up to the legal standard. Put another way, the defendant will usually be exonerated if the injury would have happened even if the defendant had lived up to the legal standard of conduct. For example, suppose the prime contractor did not supply a safety belt to the worker as required by law but the worker would have been killed because he would have refused to wear it. Under these conditions, it is likely that the prime contractor would not be held liable because the contractor's negligence did not cause the harm.

The "but for" defense has one important exception. Suppose two builders are constructing houses on adjacent lots. Each is simultaneously negligent, causing fires to begin on each building site. The fires join together, roar down the street, and burn a number of homes. Either fire would have been sufficient to burn the houses. Yet neither builder will be able to point to the "but for" rule as a defense. Each will be liable for the entire harm.[26]

Proximate Cause. Proximate cause, though related to cause in fact, serves a different function. Cause-in-fact judgments are factual and best made by commonsense decisions of juries. Proximate cause, in contrast, involves a legal policy that draws liability lines to relieve those whose failure to live up to the legal standard of conduct causes harm. Proximate cause serves a function similar to the requirement that there be a duty on the defendant to act to protect the plaintiff, as discussed in Section 7.03E. Both duty and proximate cause can minimize crushing liability burdens on those engaged in useful activities and can prevent liability from going "too far."

Proximate cause can involve the following:

1. harm of a different type than reasonably anticipated
2. harm caused to an unforeseeable person
3. harm caused by the operation of intervening forces

As an example of the first, suppose a contractor installs a sheltered walkway around a project. It can foresee that defective planking could cause a sprained ankle, but will it be liable if a pedestrian pushing a baby stroller falls on a defective plank, causing the baby to tumble from the stroller and fracture its skull? Suppose defective wiring in a high-rise building causes a power failure and shuts down all the elevators. A person who intends to submit a bid on a public project on the top floor cannot reach the awarding authority's office in time to submit the bid. Can the person recover the lost profits on the contract from the supplier of the electric wire or the owner of the building?

These freak accidents, though they occur infrequently, are dramatic enough to excite scholarly interest when they reach appellate courts. The most famous illustration of the second type of case was *Palsgraf v. Long Island Railroad Co.*[27] It involved a passenger running to catch one of the defendant's trains. Some employees of the defendant sought to help the passenger board the train but did so carelessly, dislodging a package from the passenger's arms, which fell on the rail. The package contained fireworks that exploded with some violence. The concussion overturned some scales many feet down the platform. This was unfortunate for the plaintiff, who was struck by a scale, but fortunate for legal scholars and generations of law students who dissected the subsequent appellate court decision. In a four-to-three opinion, the court held that the plaintiff was outside the zone of risk.

---

[25]A discussion of concurrent liability for indivisible loss is found in Sections 24.06 and 27.14. A claim by one concurrent wrongdoer against the other is discussed in Sections 31.03 and 31.05.

[26]See Section 24.06.

[27]248 N.Y. 339, 162 N.E. 99 (1928).

As an unforeseeable plaintiff, the railroad company did not owe any duty toward her despite its employees being careless toward someone else.

A case illustrating the third type was *Petition of Kinsman Transit Co.*, which involved two claims and two appeals. A negligently moored ship in the Buffalo River was set adrift by floating ice, dragging along another ship on the way. Bridge attendants were warned but inexplicably failed to lift a bridge. Both ships collided with the bridge, and the bridge collapsed. The ships and ice blocked the river channel by creating a dam. Water and ice backed up, damaging factories on the bank as far up the river as the original mooring. One claim involved flood damage to the factories, and the other involved pecuniary loss caused by the necessity of transporting ship cargoes around the blockage. The court granted a recovery to the factory owners[28] but denied recovery for those ships incurring additional expenses to go around the blockage.[29]

It would serve no useful function to explore the many formulas used in these cases, such as direct cause, foreseeability, hindsight, and superseding causes. It is sufficient to indicate that freak accidents cause unusual harm, and lines must be drawn. One treatise, after cataloging the various formulas, suggested that those who propose formulas are groping for something that it is difficult if not impossible to put into words. It suggested the need for a method "of limiting liability to those consequences which have some reasonably close connection with the defendant's conduct and the harm which it originally threatened, and are in themselves not so remarkable and unusual as to lead one to stop short of them."[30] Even though proximate cause is analytically considered an issue of law because it deals with a policy of limiting liability, juries are usually given a vague instruction relating to proximate cause similar to the instruction they are given relating to the standard of conduct. It is hoped that the jury will use common sense in deciding such freak cases.[31]

E. Duty

Under negligence law, every person has a duty to use the level of care of a reasonable and prudent person. That standard of conduct is explained in Section 7.03C. While this standard is a universal one that applies without regard to the relationship between the defendant and plaintiff, this was not always so. In particular, the common law has struggled with when a duty should be applied to a defendant whose injurious conduct occurred during the performance of a contract obligation.

The first duty cases were decided in the mid-nineteenth century in a legal climate favorable to new industries developing after the beginning of the Industrial Revolution. Like proximate cause, it was an attempt to draw a line beyond which recovery would not be granted. Unlike proximate cause, it tended to emphasize the relationship between individuals that imposes on one a legal obligation to watch out for the other. As examples, innkeepers owed legal obligations to their guests, as did landowners and business owners to their visitors.

However, as more fully explained in Section 7.09B, the privity doctrine was the primary device used by nineteenth-century courts to limit a defendant's duty of care. Under this doctrine, first devised in England, a defendant who performed a contract obligation negligently owed a duty of care only to the other parties to the contract: those with whom it was in privity. Absent privity, a defendant was not liable for injury it had caused to the plaintiff by its negligence or breach of contract with another. Recovery was denied regardless of whether the plaintiff's injury consisted of physical harm or financial losses.[32] Generally speaking, the privity doctrine protected manufacturing and commercial activity.

The privity doctrine was rejected early in the twentieth century, replaced by a general rule that a defendant may be liable in negligence if its lack of due care proximately caused foreseeable harm to a foreseeable victim.[33] Otherwise stated, a victim cannot recover against the defendant under a theory of negligence unless the defendant could have reasonably anticipated that someone would be injured in *this* way by *this* careless conduct of the defendant.[34]

---

[28]*Petition of Kinsman Transit Co.*, 338 F.2d 708 (2d Cir.1964), cert. denied, 380 U.S. 944 (1965).

[29]*Petitions of Kinsman Transit Co. v. City of Buffalo*, 388 F.2d 821 (2d Cir.1968).

[30]W. P. KEETON, supra note 9 at 300.

[31]Perhaps a layperson's view based on a gut reaction of what is going too far is the best solution to these vexatious problems. In commenting favorably on a hindsight test, one commentator suggested a line be drawn: "short of the remarkable, the preposterous, the highly unlikely, in the language of the street, the cock-eyed and far-fetched, even when we look at the event, as we must, after it has occurred." (Id. at p. 299.)

[32]*Winterbottom v. Wright*, 152 Eng. Rep. 402 (1842) (physical harm).

[33]*MacPherson v. Buick Motor Co.*, 217 N.Y. 382, 111 N.E. 1050 (1916); D. DOBBS, supra note 1 at 334–36.

[34]If the defendant's careless behavior results in an entirely unanticipated injury, the defendant does not owe a duty of care and is not liable. *Pulawa v. GTE Hawaiian Tel*, 112 Haw. 3, 143 P.3d 1205 (2006).

Replacement of the privity doctrine with the foreseeability doctrine led to a dramatic expansion in potential tort liability. Contracting parties have a duty not only to perform in accordance with the contract terms (both express and implied), but also to perform their obligations in a non-negligent manner. The tort duty of workmanlike conduct requires a contractor to build not only in conformity with the design, but also (where the design is silent) as would a prudent and careful contractor.[35] If the design specifies a manner of building the contractor believes to be dangerous or inadequate, it must bring this to the attention of the owner.[36]

The open-ended nature of potential tort liability resulting from the adoption of the foreseeability doctrine is not always healthy for society as a whole. While expansive tort liability should ensure greater compensation to victims and encourage careful conduct, absolute safety is neither possible nor affordable and excessive risk aversion may stifle product innovation. For these reasons, courts have struggled to devise objective rules by which to limit this premise of open-ended liability.

One major limitation is to impose a narrowed duty of care when the plaintiff suffered pecuniary or financial loss, in contrast to physical harm such as death, personal injury, or damage to property.[37] Absent privity between the parties, liability for economic losses negligently caused was limited to those with whom the defendant had a special relationship or to whom the defendant had made representations upon which the plaintiff had relied.[38]

Courts may also invoke amorphous policy considerations as a limitation on tort liability. Even if the plaintiff's injury was foreseeably caused by the defendant's negligent conduct, courts may disallow recovery if this would place an unreasonable burden on the defendant or create tort exposure lacking a sensible or objective cutoff point.[39]

Courts are also sensitive to the financial burdens that expanded tort liability may impose upon state and local governments and, by extension, taxpayers. Under the public duty doctrine, a city, whose building inspector is charged with negligence, owes no duty to individual owners above that duty owed to the general public.[40] Local law must be consulted.

F. Protected Interests and Emotional Distress

As Section 7.02C indicates, the particular loss suffered often controls liability. Even if all the requirements for negligence are met, the particular harm that has resulted may not receive judicial protection. Courts speak of whether particular interests are protectable in the process of deciding whether the defendant must respond for certain losses caused the plaintiff.

Harm to the person is most worthy of protection. Death not only ends one's life but also can have a severe financial and emotional impact on the deceased's survivors. Physical injury often means medical expenses and diminished earnings as well as pain and suffering. Often those who suffer physical harm are low-income people who may not procure insurance to protect themselves and their dependents. As a result, for these as well as for humanitarian reasons, it may be important to extend protection to those who suffer physical harm.

Harm to property such as damage or destruction is also considered worthy of protection because of the importance placed on property in modern society. But as the harm moves away from personal and property losses, the interest receives less protection. Economic harm such as diminished commercial contractual expectations, lost profits, or additional expenses to perform contractual obligations, although sometimes protected, is considered less

---

[35]*Stonegate Homeowners Ass'n v. T.A. Staben*, 144 Cal.App.4th 740, 50 Cal.Rptr.3d 709, 716–17 (2006) (subcontractor has a tort duty to perform with due care).

[36]*George B. Gilmore Co. v. Garrett*, 582 So.2d 387, 394–95 (Miss.1991) (duty to warn of soil conditions); Section 25.02 discusses a contractor's liability for defective soil conditions.

[37]The closely related economic loss rule is discussed in Section 14.08E.

[38]*Plourde Sand & Gravel Co. v. JGI Eastern, Inc.*, 154 N.H. 791, 917 A.2d 1250 (2007) (no special relationship between a testing company hired by the engineer and a gravel supplier); Section 14.08D discusses negligent misrepresentation.

[39]For examples of courts using policy considerations to limit tort liability in construction cases, see 532 *Madison Ave. Gourmet Foods, Inc. v. Finlandia Center, Inc.*, 96 N.Y.2d 280, 750 N.E.2d 1097, 1103, 727 N.Y.S.2d 49 (2001); *Butler v. Advanced Drainage Systems, Inc.*, 294 Wis.2d 397, 717 N.W.2d 760, 767–69 (2006); and *Hoida, Inc. v. M&I Midstate Bank*, 291 Wis.2d 283, 717 N.W.2d 17, 31–34 (2006).

[40]*Sinning v. Clark*, 119 N.C.App. 515, 459 S.E.2d 71, review denied, 342 N.C. 194, 463 S.E.2d 242 (1995); D. DOBBS, supra note 1 at 723–27.

worthy of protection than harm to person or to property. Even less protection is accorded emotional distress and psychic harm, sometimes called *noneconomic losses*.

A number of reasons for caution exist when claims are made for economic loss, and even more when claims are made for emotional distress. Economic losses result from many causes other than the conduct of the defendant, and are difficult to prove. Liability can place crushing burdens on the defendant. Courts have only cautiously extended protection for economic harm, and they have been even less willing to extend protection for emotional distress. This caution is partly grounded in the undoubted difficulty of establishing the genuineness of the claim. Another is the difficulty of placing an economic value on the loss. Because of these difficulties, the law has been unsettled in this area, and decisions vary from jurisdiction to jurisdiction.

With some exceptions,[41] the plaintiff cannot recover for mental disturbance caused by the defendant's negligence in the absence of accompanying physical injury or consequences. Clearly, however, there can be recovery for mental disturbance if the negligence of the defendant has inflicted an immediate physical injury and the mental disturbance such as pain and suffering was caused by the physical injury. Suppose physical harm follows fright or shock of the plaintiff, such as a miscarriage suffered after negligent conduct by the defendant caused a mental disturbance. Many American cases have required the plaintiff to show that some physical impact on the plaintiff was caused by the defendant's negligent acts. Others have not. The recent cases have eliminated the requirement of impact, and the impact rule is likely to disappear.

Suppose physical injury resulting from fright occurs when the injured person feared for his own safety. Here most states allow recovery. But fear for someone else's safety has divided the courts. The classic example has involved a mother, though not in danger, seeing her child seriously harmed. Some courts allow recovery. Others do not.[42]

Emotional distress and mental disturbance can arise in the context of construction work. Suppose a worker is on a roof. The roof partially collapses, but the worker is not injured. The worker may have suffered emotional distress from the fear that he was about to fall or that he might have fallen. Also, suppose a worker experiences traumatic shock when he sees someone else fall from the roof of a building.

G. Defenses

Acceptance Doctrine. What is the liability of a contractor if a construction defect caused injury to a third party (whether an occupant, visitor, or passerby), but only after performance was completed and the building had been accepted by the owner? In the nineteenth century, the privity doctrine (discussed in Section 7.03E) meant that the injured person could not sue the contractor, just as an injured consumer could not sue the manufacturer of a defective product.

In the first half of the twentieth century, as privity gave way to an expanded view of tort liability, different rationales were used to continue the acceptance doctrine. One rationale was that the owner's acceptance of the completed building was an intervening cause that severed the causal link between the contractor's negligence and the injury, especially in light of the owner's affirmative duty to maintain the structure in a reasonably safe condition. Courts also feared that abandonment of the acceptance defense would subject contractors to unlimited, potential liability, since the statute of limitations for tort actions begins to run on the date of the injury.

The acceptance doctrine was criticized for shifting liability from a negligent contractor to an innocent owner. Critics pointed out that the rationale in favor of the doctrine was especially weak where the injury was caused by a latent construction defect, which the owner could not have been expected to discover and remedy. Passage of statutes of repose—which cut off liability for contractors a specified number of years after completion of their work—meant that the acceptance doctrine was no longer needed to prevent unlimited contractor liability.[43] Eventually, many courts began to entirely reject the doctrine as incompatible with modern tort jurisprudence. Those who continued adherence to the doctrine usually ameliorated its harsher results by creating a variety of exceptions to it; for example, that the defense would not apply to defects that were imminently dangerous to others, that were latent, or where the design was so obviously defective that no reasonable contractor would follow it. The competing reasons in favor and against continued

---

[41]See Sections 6.06H and 27.09 for claims for emotional distress based on contract breach. See also W. P. KEETON, *supra* note 9 at 361–62.

[42]W. P. KEETON, *supra* note 9 at 365–66.

[43]Statutes of repose are discussed in Section 23.03G.

adherence to the rule mean the common law remains in a state of flux.[44]

Assumption of Risk. Assumption of risk completely bars recovery even if the defendant has been negligent. Advance consent by the plaintiff relieves the defendant of any obligation toward the consenting party. The plaintiff has chosen to take a chance. Suppose the plaintiff voluntarily entered into a relationship with the defendant with knowledge that the defendant would not protect the plaintiff from the risk. In such cases, the plaintiff implicitly assumed the risk. Sometimes the plaintiff is aware of a risk created by the negligence of the defendant but proceeds voluntarily to encounter it.

Express agreements to assume the risk are often given effect if knowingly and freely made by parties of relatively equal bargaining power.[45] But sometimes the relationship is one regulated by law and this freedom is denied. For example, workers are frequently prohibited from signing agreements to assume the risk of physical harm.

Most cases involve implied assumption of risk. Did the plaintiff know and understand the risk? Was the choice free and voluntary? Voluntariness has generated considerable controversy where workers take risks under the threat that they will be discharged if they do not continue working.[46] Under such circumstances, some American courts conclude that the worker even under such pressure has assumed the risk of performing dangerous work.[47] Although statutes often preclude assumption of risk in employment relationships, the concept can be applied in third-party actions brought by workers against persons *other* than their employers.

The assumption-of-risk defense is not favored. As a result, the defendant must plead and prove that the plaintiff assumed the risk.

Contributory Negligence. Until quite recently, most states would bar the plaintiff from recovering if the defendant established that the plaintiff's negligence played a significant role in causing the injury. Juries were thought to use techniques to mitigate the harshness of this defense. Where the plaintiff's negligence was slight and much less than that of the defendant, juries sometimes decided that the plaintiff was not contributorily negligent or that the contributory negligence did not cause the harm. Where both parties were negligent but the defendant much more so, the jury may have compromised by finding that there had not been contributory negligence but by diminishing the amount of the plaintiff's award to take the plaintiff's negligence roughly into account.

The contributory negligence rule has been criticized because slight negligence bars what would otherwise appear to be a just claim. A number of states, notably Wisconsin and Louisiana, enacted legislation early in the century that modified this rule by requiring that the negligence of the plaintiff and the defendant be compared rather than automatically barring the plaintiff from recovery if the plaintiff has been negligent. Most states have enacted statutes creating comparative negligence, and a few courts have created comparative negligence without the benefit of legislation.

The details of comparative negligence laws vary considerably. The essential feature is that the plaintiff's negligence does not automatically bar recovery but only diminishes recovery. A plaintiff whose negligence reaches a specific percentage, such as 50 percent, is barred in some states. In other states, the comparison is "pure." An 80 percent negligent plaintiff would be entitled to recover 20 percent of the loss from a 20 percent negligent defendant. Comparative negligence has been used in admiralty law and in claims by employees of certain common carriers subject to the Federal Employers Liability Act.

Independent Contractor Rule. Unlike the two preceding defenses, the defense of the independent contractor rule does not involve conduct or risk taken by the party who suffered the loss. But because of its importance in construction legal problems, it merits brief mention at this point.

Suppose a homeowner hires a contractor to do remodeling work and the contractor negligently drops tools from the roof, injuring a passerby. As a general rule, the homeowner will be not liable to the passerby because the injury was

---

[44]Compare *Moglia v. McNeil Co., Inc.*, 270 Neb. 241, 700 N.W.2d 608 (2005) (recognizing the imminently dangerous and latent defect exceptions, but finding neither applicable) with *Davis v. Baugh Industrial Contractors, Inc.*, 159 Wash.2d 413, 150 P.3d 545 (2007) (rejecting the doctrine). For a compilation of the cases, see Annot., 75 A.L.R.5th 413 (2000).

[45]*Delta Air Lines, Inc. v. Douglas Aircraft Co.*, 238 Cal.App.2d 95, 47 Cal.Rptr. 518 (1966). However, a release signed by a person seeking admission to a university hospital was not upheld in *Tunkl v. Regents of the Univ. of California*, 60 Cal.2d 92, 383 P.2d 441, 32 Cal.Rptr. 33 (1963).

[46]*Fonseca v. Orange County*, 28 Cal.App.3d 361, 104 Cal.Rptr. 566 (1972).

[47]*Mitchell v. Young Refining Corp.*, 517 F.2d 1036 (5th Cir.1975).

caused by an independent contractor.[48] An independent contractor is asked to achieve a result and is not controlled as to means by which this is accomplished. The homeowner's lack of control over the contractor's means and methods of performance defeats imposition of vicarious liability upon the owner (as would happen if the injury was caused by the owner's employee). The hirer of an independent contractor can be either the owner (who hires the prime contractor) or the prime contractor (who hires subcontractors).

Many exceptions to the independent contractor rule exist. Where the work to be performed involves an "inherently dangerous" activity, an injury arising out of that work will subject the hirer to vicarious liability.[49] Vicarious liability will also be imposed if the injury involved a "nondelegable duty," such as violation of a workplace safety statute.[50] (The term "nondelegable duty" is something of a misnomer; the duty to provide a safe workplace may be delegated; however, liability arising out of violation of that duty remains on the hirer.) Many other exceptions impose liability based on the hirer's own acts of negligence, including retention of control over the contractor's work.[51] In the past quarter century, an increasing number of courts have refused to impose vicarious liability—and have narrowed the exceptions based upon the hirer's own negligence—where the victim was a construction worker entitled to workers' compensation benefits.[52]

SECTION 7.04 Nonintentional Nonnegligent Wrongs: Strict Liability

A. Abnormally Dangerous Things and Activities

As indicated earlier,[53] preindustrial law was often strict in the sense that liability did not require negligence. Despite the emergence of negligence as the dominant basis for liability that developed in the nineteenth century, some

pockets of law found liability "strictly." One—keeping of animals—is of little importance to construction. (But watch strict liability for keeping a vicious dog on the site to deter vandals.) The other—strict liability for abnormally dangerous things and activities—does find application to modern construction problems and merits some comment.[54]

Rylands v. Fletcher,[55] an 1868 English decision, held that a mill owner not shown to have been negligent was liable when he built a reservoir whose waters broke through to an abandoned coal mine and flooded parts of the plaintiff's adjoining mine.

Ultimately, most American decisions in some form or another found strict liability where the defendant owned dangerous things or engaged in dangerous activities. The prototype abnormally dangerous activity is often stated to be blasting. The activity may not be negligent, because the social utility and the high cost of avoiding the harm makes the conduct reasonable despite the high risk of serious harm. The modern rationale for strict liability for such activities is that such enterprises should pay their way and are better risk bearers than the persons who have been harmed by the activity.

As modern tort law emphasizes victim compensation and loss spreading through insurance and pricing goods and services, strict liability (liability without a need to establish negligence) has become more important in construction-related activities. Strict liability for defective products plays an increasingly important role in construction-related litigation (discussed in Section 7.09). (Also, the professional standard to which design professionals are held is increasingly under attack, the assertion being that they should be held to a "stricter" standard.)[56] This trend is reflected in federal laws that make those who generate, transport, or dispose of hazardous waste strictly liable for violating statutes and regulations that regulate these activities.[57]

B. Vicarious Liability

Sometimes one person is responsible for the negligence of another. The most common illustration is the employment relationship. Despite the absence of negligence by the employer, the employer is liable for harm caused by the negligent conduct of the employee as long as that conduct

---

[48]Restatement (Second) of Torts § 409 (1965).

[49]Id., §§ 416 and 427.

[50]E.g., N.Y. Labor Law § 240(1).

[51]Restatement (Second) of Torts § 414 (1965).

[52]*Privette v. Superior Court*, 5 Cal.4th 689, 854 P.2d 721, 21 Cal. Rptr.2d 72 (1993) (Restatement § 416); *Toland v. Sunland Housing Group, Inc.*, 18 Cal.4th 253, 955 P.2d 504, 74 Cal.Rptr.2d 878 (1998) (Restatement § 414); M. SCHNEIER, CONSTRUCTION ACCIDENT LAW: A COMPREHENSIVE GUIDE TO LEGAL LIABILITY AND INSURANCE CLAIMS, 131–147 (1999).

[53]See Section 7.02.

---

[54]See supra note 49.

[55]L.R.-3 H.L. 330 (1868).

[56]See Section 14.07.

[57]See also Section 9.13A.

is within the scope of the employment. Sometimes the employer can be held liable for intentional torts committed by the employee. For example, a subcontractor was held liable for injuries sustained by two employees of the prime contractor as a result of a physical attack committed on them by employees of the subcontractor.[58] The ruling was made despite the physical attack having occurred after the attacking employees had completed their work shift.

One reason sometimes given for vicarious liability, or as it is sometimes called, *respondeat superior,* is the incentive it gives to an employer to choose employees carefully. The real basis for vicarious liability is that an enterprise should pay the full cost of its activities or the products it produces, such as the cost of those injured by its activities or products. Also, vicarious liability is premised on the ability to pay, or the "deep pocket." The pocket of the employer is likely to be deeper than that of the employee.

C. Employment Accidents and Workers' Compensation

Although the principal thrust of this chapter is the role of tort law in distributing losses, one exception is the workers' compensation law, which was to a large degree to supplant tort law in workplace accidents. The employer's liability under workers' compensation is strict. Recovery against the employer does not require the latter's negligence. For these reasons, it may be useful to briefly describe certain doctrines that deal with employment injuries.

The injured worker did not fare well under English or American common law. Expansion of the Industrial Revolution generated employment injuries but only limited legal protection to those injured. The employer's duties were limited: to furnish the worker a safe place to work and safe appliances, tools, and equipment and to warn the worker of any known danger connected with the work. The injured worker faced, in addition to limited employer duties, obstacles to compensation. First, the worker had to prove that the employer was negligent. Many accidents occurred because of the nature of the work and not the negligence of the employer.

Even if negligence were established, the worker faced the "unholy trinity" defenses of contributory negligence,[59]

assumption of risk,[60] and the fellow servant rule. The first barred a worker whose injury was substantially caused by the worker's negligence unless the employer's conduct had been willful or wanton. The second barred the worker from recovery if taking the job was an assumption of the risk of injuries normally incident to the employment. A worker who stayed on a job under protest after the worker knew or appreciated the danger assumed the risk. The third defense meant that with some exceptions, a worker could not recover for injuries caused by the negligence of a fellow servant.

These formidable barriers often meant hardship to injured workers and their families. Following German social insurance law, American states began in the early twentieth century to enact what were then called workmen's compensation laws designed to replace tort law with social insurance for industrial accidents. Although some early statutes were held unconstitutional, currently all states have workers' compensation laws.

Although there are considerable variations among the states, certain common questions have arisen. For example, are the employer and the worker covered under the workers' compensation law? Many statutes exclude agricultural workers and domestics as well as employers who have only a few employees. The injury must arise out of the employment. Do injuries that occur on company picnics or while parking the worker's car in the company lot arise out of the employment? Doubts generally are resolved in favor of the employee.

Some states cover occupational diseases. Others do not. The worker need not show negligence by the employer. Nor are workers precluded by their own negligence from recovering, by their having assumed the risk, or by the injury having been caused by their fellow workers.

Some states deny recovery if the worker was guilty of willful misconduct or intoxication and such misconduct caused the injury.

Most states require that the employer obtain compensation insurance. A few states have set up state funds to pay compensation awards. In some states, an employer can be a self-insurer if it makes adequate proof of financial responsibility. All these requirements are designed to ensure that there will be a financially responsible entity.

The particular problems of the construction industry have been recognized. Many states enacted "subcontractor

---

[58]*Rodgers v. Kemper Constr. Co.,* 50 Cal.App.3d 608, 124 Cal.Rptr. 143 (1975).

[59]See Section 7.03G.

[60]Ibid.

under" or "statutory employer" statutes under which a prime contractor is the employer of subcontractor employees under certain circumstances, such as a showing that the subcontractor did not procure the required insurance. This allows recovery against the prime contractor's compensation insurer. Special rules for the construction industry recognize that many subcontractors are not financially sound and may not obtain the requisite insurance. Compensation coverage may be diluted if a prime contractor subcontracted out work that would normally be performed by the prime contractor. This may be done to reduce the number of employees for whom the prime contractor would have to obtain insurance coverage or be exempt as having too few employees.

Typically, the award consists of a proportionate amount of the employee's wages as well as reimbursement for medical expenses incurred. Sometimes the employee is also able to receive a specific monetary award for designated injuries. More intangible, noneconomic losses—such as amounts to compensate for the emotional distress caused by disfigurement or for pain and suffering—are generally not recoverable. Workers' compensation awards provide part (estimated at from one-third to three-quarters of economic loss) and not full compensation. The recovery is intended to ensure that the injured worker does not become a burden on others. Compensation recoveries are less, and in many states much less, than can be recovered in a tort action. This has led to demands for federal minimum standards for compensation awards.

Workers' compensation remedies generally supplant whatever tort remedy the worker may have had against the employer. This is accomplished by statutory immunization of the employer from tort claims by the workers. Immunity gave the employer something in exchange for giving up existing legal protection in employment accidents.

Eliminating all tort suits would create a total compensation system and keep disputes within the administrative agencies charged with the responsibility for handling such claims. This has not been accomplished. Most states permit injured workers, or compensation insurers who have paid them, to institute tort claims against most third parties whose negligence caused the injury. A factory worker may be able to recover in tort against the manufacturer of a defective product. A worker on a construction project may recover in tort against one of the many entities involved in the project other than his own employer. Because of the limited recovery available under work-

ers' compensation, it is becoming increasingly common for injured construction workers to institute third-party actions against anyone they can connect with their injuries other than their employers.

Workers' compensation claims are handled by an administrative agency rather than a court. Hearings are informal and usually conducted by a hearing officer or examiner. The employee can represent himself or be represented by a layperson or lawyer. Fees for representation are usually regulated by law. Although awards by the agency can be appealed to a court, judicial review is extremely limited, and very few awards are overturned. Third-party actions, on the other hand, because they involve tort claims, are brought to court, and incredibly complicated lawsuits often result.

Workers' compensation took on political and economic dimensions in the 1990s. In many states, insurance rates jumped drastically because of the expansion of claims falling within the workers' compensation system to claims involving, for example, job-connected stress. Employers in some states where the insurance rates were high complained that they could not compete with employers in states with lower rates and often threatened to move their operations to states that did not generate such expensive claims. Others complained that many claims were fraudulent, that too much of the ultimate payout went into the pockets of attorneys and health care providers, and that insurers charged exorbitant administration overhead. These factors have led to attempts to legislate changes designed to cure some of the problems, for better or worse. However, the efforts of those who would be adversely affected by change often thwart such attempts.

D. Product Liability

Another and perhaps more spectacular illustration of strict liability is imposed on manufacturers for defective products (discussed in greater detail in Section 7.09).

SECTION 7.05 Claims by Third Parties

A. Lost Consortium

Suppose a spouse is seriously injured. In addition to causing economic loss, the injury can cause losses of an intangible nature to the other spouse or a child of the injured person. Such losses are sometimes called *consortium*, a term that encompasses the services and society lost and, in the case of the spouse, sexual relations.

Although the husband could recover for loss of consortium when his wife was injured, the law generally did not give corresponding rights to his wife. Slowly the courts have equalized consortium rights of husband and wife. However, a child cannot recover for lost consortium of either parent.

B. Survival and Wrongful Death Statutes

Before the mid–nineteenth century, the death of the wrongdoer precluded any claim being made against the wrongdoer's estate. The wrong died with the wrongdoer. Similarly, death of the party harmed also barred any claim he would have had. However, statutes generally changed this common law rule. The easiest were those that allowed claims to be made against the estate of the wrongdoer.

More problems developed because of the patchwork of statutes affecting the right of the claimant. Those statutes are divided into survival statutes and wrongful death statutes. The former gave rights—often limited ones—to the estate of the deceased for harm suffered between the wrongful act and the death. The second provided compensation to the deceased person's dependents, for their economic losses. Local statutes must be consulted.

The claim is usually measured by any loss the deceased has suffered prior to death resulting from the defendant's negligence and, more important, any loss to the estate or survivors. Most statutes limit recovery to pecuniary losses, although some permit a limited amount of noneconomic losses to be recovered. Recovery for economic losses to the survivors is based on potential earnings of the deceased during his working life that would have been available to the survivors. The amount recovered for the death of a person with high earnings or high earning capacity is often large. A few states limit the amount of recovery for wrongful death.

SECTION 7.06 Immunity

A. Charitable Organizations

Initially, American law granted immunity from tort liability to charitable organizations. Immunity was originally based on the charitable and nonprofit characteristics of the organization. The availability of public liability insurance, the recognition that even charitable institutions take on some of the characteristics of commercial enterprises, and the injured party's needs are factors that led to virtual abolition of this immunity.

B. Employers and Workers' Compensation

The worker's sole remedy against the employer is under workers' compensation. Employers have immunity, subject to a few exceptions, from tort claims made by their workers.

C. Public Officials

Judges and high public officials are usually granted absolute or qualified immunity from claims against them.

D. Sovereign Immunity

For various reasons—some metaphysical and some practical—English law immunized the sovereign from being sued in royal courts. This doctrine was adopted early in the nineteenth century by the U.S. federal courts, and it soon became established that the federal government could not be sued without its consent.

In 1946, the Federal Tort Claims Act (FTCA)[61] was adopted. This legislation gave individuals the right to sue the United States for certain wrongs it committed. The act had two important exceptions: (1) the federal government could not be sued for certain intentional torts, and (2) certain discretionary functions or duties performed by government officials could not give rise to tort liability. *Dalehite v. United States* clarified this situation by denying liability when the conduct was a policy or planning decision, also holding that the government could not be held liable unless it was shown to have been negligent.[62] *Feres v. United States* barred a claim by a member of the armed services against the United States.[63] This has been extended to a manufacturer who followed government specifications.[64] Strict liability cannot be the basis for any claim against the federal government. Negligence

---

[61]28 U.S.C.A §§ 2671–80.
[62]346 U.S. 15 (1953).
[63]340 U.S. 135 (1950).
[64]See Section 7.09J.

under applicable state law must be established. However, government conduct that is negligent under state law may also fall within the "discretionary function" exception to the FTCA. In that situation, the federal government would be immune even though a private party might be liable under state law.[65]

States generally adopted the English rule of sovereign immunity, although many states have given consent to be sued for specific claims.[66] Municipal corporations, such as counties and cities, receive immunity for governmental acts but not for those they have performed in a private or proprietary capacity. Sometimes immunity is waived for specified activities or operation or control of real property.[67]

Modern justifications are usually based either on relieving already burdened public entities of serious financial responsibilities or on precluding judicial intrusion into the running of government. Those who oppose immunity contend it is better to spread the loss among all the taxpayers in the public entity than to concentrate loss on the person who has suffered it. As to the fear of judicial intrusion, opponents of immunity contend that the current trend is toward increased accountability rather than relieving persons from their negligent acts. Yet expanded liability of public entities has generated a drastic increase in the cost of the insurance they procure, and in some instances has made insurance unavailable. Although it is unlikely that the law will reverse its field and restore sovereign immunity wholesale, these concerns are likely to limit expansion of governmental liability or, in special cases, restore immunity.[68]

Even when immunity has been eliminated, claims against governmental units require particular attention. Often such claims must be made within a shorter period of time than claims against private persons. In addition, such claims must be presented, as a rule, to legislative bodies of the governmental unit for their review before court action can be begun.

For sovereign immunity from claims for breach of contract, see Section 23.03I.

SECTION 7.07 Misrepresentation

A. Scope of Discussion

Although misrepresentation problems can occur in the contract formation process, this section treats the liability of people whose business it is to make representations. A surveyor makes representations as to boundaries, a geotechnical engineer as to soil conditions, a design professional as to costs and the amount of payment due a contractor.

B. Representation or Opinion

Representations should be distinguished from opinions. For example, an architect may give his best considered judgment on what a particular project will cost. The prediction, however, may not be intended by him or understood by the client to be a factual representation that will give the client a legal claim in the event the prediction turns out to be inaccurate. If the statement is merely an opinion and not a representation of fact, it is reasonably clear that the person making the representation will not be liable simply because he is wrong.

C. Conduct Classified

The person making the misrepresentation may have had a fraudulent intent. He may have made the representation knowing that it was false, with the intention of deceiving the person to whom the representation was made. The representation may not have been made with the intention to deceive but may have been made negligently. Finally, the representation may have been made with due care but turned out to be wrong. This is sometimes referred to as an *innocent misrepresentation*.

D. Person Suffering the Loss

Another classification relates to the person who was harmed, the person to whom the representation was made, or a third party. For example, the geotechnical engineer may make a representation of soil conditions to a client.

---

[65]*Wood v. United States*, 290 F.3d 29 (1st Cir.2002) (worker injury claim brought on federal construction project); Schminky, *The Liability of the Government Under the Federal Tort Claims Act for the Breach of a Nondelegable Duty Arising from the Performance of a Government Procurement Contract*, 36 A.F.L.Rev. 1 (1992).

[66]D. DOBBS, supra note 1 at 715–18.

[67]Id. at 718–23.

[68]Krent, *Reconceptualizing Sovereign Immunity*, 45 Vand.L.Rev. 1529 (1992).

If the representation is incorrect, the harm may be suffered by the client or, in some cases, by third parties, such as a contractor or a subsequent purchaser or occupant.

E. Type of Loss

Cases can also be classified by the harm that resulted from the misrepresentation. A misrepresentation of soil conditions may result in a cave-in that kills or injures workers. It may also cause damage to property or economic loss unrelated to personal harm or damage to the client's property. The client may have to pay for damage caused to an adjacent landowner's property. Because of the misrepresentation, a subcontractor may incur additional costs during the excavation.

F. Reliance

In addition to the representation having to be material or serious, it must have been relied on reasonably by the person suffering the loss. If there is no reliance or if the reliance is not reasonable, there is no liability for misrepresentation. A principal application of this doctrine is discussed in Chapter 25.

Often the owner makes representations as to soil conditions to the contractor and then attempts to disclaim responsibility for the accuracy of the representation. The disclaimer is an attempt to transfer the risk of loss for any inaccurate representations to the contractor. It is intended to negate the element of reliance, a basic requirement of misrepresentation. Generally, but not invariably, as seen in Section 25.05, such disclaimers successfully place the risk of loss on the contractor. However, they cannot relieve the person making the representation from liability for fraud, and they may not be effective if the representations were negligently made.

G. Generalizations

Generally, the more wrongful the conduct by the person making the representation, the greater the likelihood of recovery against the party. For example, a fraudulent misrepresentation will always create liability to the party to whom it was made or to third parties. A negligent one, in addition to providing the basis for a claim by the party who has paid for the representation, may be the basis for a claim by third parties. An innocent misrepresentation,

being least culpable, is the most difficult on which to base a claim. Third parties are rarely able to recover, and even the other party to the recovery will be able to recover damages only if there is a warranty of accuracy.

Liability to third parties often depends on the type of harm suffered. If personal harm such as death or injury results, the absence of a contractual relationship between the person suffering the harm and the person making the misrepresentation is not likely to constitute a defense.

Where harm is economic, lack of privity (contractual relationship) between the claimant and the person who made the misrepresentation causes the greatest difficulty. One reason is the wide range of individuals who are affected by the representations of people who are in the business of making them. Such persons can be exposed to enormous liability, often disproportionately high to the remuneration paid for the services. For example, a certified public accountant may make an audit report that causes thousands of investors to buy shares in a particular company. Were the accountant accountable to all these investors, he would face enormous risk exposure.

In the construction context, the misrepresentation cases typically involve claims against surveyors and those who provide geotechnical information—but see Section 14.08D for expanded use of misrepresentation against architects and engineers. Those who provide services often are liable to third parties for economic losses when the representations made were found to be negligent, provided the professional making the representation can reasonably foresee the type of harm that is likely to occur and the people who may suffer losses. This trend is reflected in *Rozny v. Marnul*,[69] in which a surveyor was held liable to a homeowner who had built a house and garage relying on a survey the surveyor had prepared for a developer.

SECTION 7.08 Premises Liability: Duty of the Possessor of Land

A. Relevance

Tort law determines the duty owed by the possessor of land—that is, the one with operative control—to those people who pass by the land or enter on it. Before, during, or after completion of a construction project, members

---

[69]43 Ill.2d 54, 250 N.E.2d 656 (1969).

of the public will pass by the land or, with or without permission, enter on the land with the potential of being injured or killed by a condition on the land or by an activity engaged in by the person "in control" of the land. Workers also may suffer injury or death because of the condition of the land or activities on it. People who live in or enter a completed project may suffer injury or death because of something related to the land or to the construction process. Liability for such harm depends on the particular nature of the obligation owed to the plaintiff by the possessor of land near which or on which the physical harm was suffered. It can depend on the injured party's permission to be near or on the land and the purpose for being there.

The term *possessor of land* is used in this section without exploring the troublesome question of whether the owner, the prime contractor, or the subcontractors fall into this category during the construction process.

As the following discussion makes clear, the courts are nearly evenly split on the duty a possessor owes to those who enter on to the land. Under the traditional common law rules, followed by a slight majority of states, the possessor's duty varies depending on the status of the entrant: trespasser, child, licensee, or invitee. However, under the minority modern rule, a duty of reasonable care extends to all entrants, regardless of their common law status.[70]

B. To Passersby

Passersby can expect that the conditions of the land and activities on it will not expose them to unreasonable risk of harm. Whether the possessor has measured up to the legal standard depends on factors discussed in Section 7.03C.

C. To Trespassing Adults

The trespasser enters the land of another without permission. In so doing, the trespasser is invading the owner's exclusive right to possess the land. Veneration for landowner rights led to a very limited protection for trespassers by English and American law. The possessor was not liable if trespassers were injured by the possessor's

failure to keep the land reasonably safe or by the possessor's activities on the land.

Exceptions developed as human rights took precedence over property rights. For example, possessors who know that trespassers use limited areas of their land must conduct their activities in such a way as to discover and protect trespassers from unreasonable risk of harm. Railroads were required to be aware of people who crossed the tracks at particular places. Another exception was applied frequently to railroads for dangerous activities conducted on the land.

Discovered trespassers are entitled to protection. The landowner must avoid exposing such trespassers to unreasonable risk of harm.

Despite their unfavored position, trespassers are not outlaws. Possessors cannot shoot them or inflict physical harm on them to protect their property. What steps can possessors take to protect their land from trespassers? Suppose a contractor puts up a barbed-wire fence. Suppose a contractor keeps savage dogs on a fenced-in site at night to protect the site and construction work from vandals.

A case that excited controversy was *Katko v. Briney*.[71] The defendant owned an uninhabited farmhouse containing some antiques and old jars. The house had been broken into and some of the contents removed several times. After requests to law enforcement authorities were unproductive, the defendant installed a spring gun aimed at the legs of anyone who entered the house and sought to enter a particular room. The plaintiff, thinking the house uninhabited and looking for old canning jars, entered the house and was severely injured by the spring gun's discharge. Charged with a felony, the plaintiff pleaded guilty to a misdemeanor and received a sixty-day suspended jail term.

The plaintiff then sued the defendant for having caused the injury. A jury award against the landowner for compensatory and punitive damages was upheld because the privilege to protect property did not extend to the infliction of serious bodily harm.

California has sought to immunize the possessor of land, even from intentional use of a deadly weapon when such force is justifiable, by enacting Section 847 of the California Civil Code.[72] This approach would produce an outcome different from *Katko v. Briney* described in this Section.

---

[70]D. DOBBS, supra note 1 at 591–620. Professor Dobbs's treatise, published in 2000, lists 21 states following the modern rule, although some of those make an exception for trespassers. Id. at 666, notes 4 & 5. See also Section 7.08G.

---

[71]183 N.W.2d 657 (Iowa 1971).

[72]The California attempt to immunize the possessor who uses force is described in *Calvillo-Silva v. Home Grocery*, 19 Cal.4th 714, 968 P.2d 65, 80 Cal.Rptr. 2d 506 (1998).

D. To Trespassing Children

Special rules have developed for trespassing children. Children frequently do not realize that they are entering the land of another. Sometimes they are not aware of the dangerous characteristics of natural and artificial conditions on the land that they enter. Possessors must conduct their activities in such a way as to avoid unreasonable risk of harm to trespassing children. However, controversy frequently develops regarding the extent to which limited protection given trespassers as to artificial conditions on the land should be applied to trespassing children.

The law's strong protection for landowners and their rights has, for the most part, been qualified by humanitarian concerns for children of tender years injured or killed when they confront or deal with dangerous conditions on the land of another. The Restatement (Second) of Torts reflects this diminution of landowner protection and states that the possessor of land is liable for injuries to trespassing children caused by artificial conditions on the land if

(a) the place where the condition exists is one upon which the possessor knows or has reason to know that children are likely to trespass, and

(b) the condition is one which the possessor knows or has reason to know and which he realizes or should realize will involve an unreasonable risk of death or serious bodily harm to such children, and

(c) the children because of their youth do not discover the condition or realize the risk involved in intermeddling with it or in coming within the area made dangerous by it, and

(d) the utility to the possessor of maintaining the condition and the burden of eliminating the danger are slight as compared with the risk of children involved, and

(e) the possessor fails to exercise reasonable care to eliminate the danger or otherwise protect the children.[73]

Children often trespass on construction sites. In doing so, they may engage in an activity that can result in injury or death.

Judges less sympathetic to trespassing children also emphasize the responsibility of even young children to know what is dangerous and of parents to keep their children away from construction sites. In many cases, part of

the responsibility for the child's injury must fall on parents who do not supervise the child properly.

E. To Licensees

A licensee has a privilege of entering or remaining on the land of another because of the latter's consent. Licensees come for their own purposes rather than for the interest or purposes of the possessor of the land. Examples of licensees are people who take shortcuts over property with permission, people who come into a building to avoid inclement weather or to look for their children, door-to-door salespeople, and social guests.

Some anomalous exceptions exist, such as the firefighter who enters a building at night to put out a fire or the police officer who enters to apprehend a burglar. Logically, such people should be considered as benefiting the possessor of land, but many cases hold that they are simply licensees.

Early cases held that the only limitation on the possessor's activity was to refrain from intentionally or recklessly injuring a licensee. However, most courts today require that the possessor of land conduct activities in such a way as to avoid unreasonable risk of harm.

The possessor has a duty to repair known defects or dangerous conditions or to warn licensees of nonobvious dangerous conditions. Licensees cannot demand that the land be made reasonably safe for them. The possessor need not inspect the premises, discover dangers unknown to the possessor, or warn the licensee about conditions that are known or should have been known to the licensee.

F. To Invitees

An invitee receives the greatest protection. The possessor must protect the invitee not only against dangers of which the possessor is aware but also against those that could have been discovered with reasonable care. Although not an insurer of the safety of invitees, the possessor is under an affirmative duty to inspect and take reasonable care to see that the premises are safe. Sometimes the possessor can satisfy the obligation by warning the invitees of nonobvious dangers.

Who qualifies for such protection? Some cases have limited invitees to those people who furnish an economic benefit to the possessor. More jurisdictions, however, seek to determine whether the facts imply an invitation to the

---

[73]Section 339 (1965). For a recent case collecting many cases, see *Kessler v. Mortenson*, 16 P.3d 1225 (Utah 2000). The case involved a six-year-old child playing in a partially completed house. The court emphasized these child trespasser cases are decided on a case-by-case basis.

entrant. However, the invitation concept does not include those invited as social guests.

The line between licensee and invitee is difficult to draw and sometimes seems arbitrary. However, among the states following the common law "status" rule of possessor liability, contractors and their employees are universally viewed as invitees. This means that the owner, as possessor, owes the highest duty of care to the contractor and its employees. If the owner vacates the land and possession is assumed by the prime contractor, then the contractor owes the highest duty of care to its invitees: the subcontractors and their employees.

G. Movement Toward General Standard of Care

Undoubtedly, the various categories that determine the standard of care are difficult to administer. Exceptions develop within the categories, and the application of the categories is often uneven. For this reason, there is clear movement toward a rule that would require the possessor of land to avoid unreasonable risk of harm to all who enter on the land.[74] Those courts following the modern rule often adopt the test for possessor liability found in the Restatement (Second) of Torts: that a possessor has a general duty to take reasonable care to protect entrants from hazardous conditions of which the entrant either is not aware or, even if aware, would likely encounter.[75]

H. Defenses to Premises Liability

Underlying premises liability law is a balancing of responsibility between the land possessor and the entrant. Just as the possessor must take reasonable steps to protect entrants from hazardous conditions of the premises, so too must the entrant exercise prudence while on the property. It is sometimes said that the land possessor's liability is premised upon his *superior knowledge* of the hazard. If the entrant's knowledge is equal to that of the possessor—or if a reasonable person in the entrant's position would have discovered the hazard and appreciated the risk—then the possessor's liability is either eliminated altogether or at least reduced by the degree of the entrant's negligence.[76]

This onus upon the entrant to exercise care and prudence while on the premises has congealed into a number of defenses a land possessor may assert. One defense is that the entrant was injured by a patent defect—one the entrant knew of or should have discovered. Another defense is that the invitee exceeded the scope of the invitation—that he wandered off into an area into which he was not authorized to enter.

Several states have passed "recreational use" statutes, whose purpose is to encourage property owners, both private and public, to provide the public with free access to their property for recreational use. With narrow exceptions, these statutes immunize owners from liability to entrants injured while engaging in recreational activities.[77]

Of importance to the construction industry are various defenses that apply to injuries to independent contractors. An owner who turns the premises entirely over to the control of a prime contractor may argue that the contractor is now the possessor and that the landowner is therefore no longer subject to premises liability. (This defense may not be successful if the hazardous condition preexisted the prime contractor's assumption of temporary possession.) Another defense is that the hazardous condition was itself the object of the contractor's work or arose from its activities.[78] The reason for this defense arises directly from the "superior knowledge" rationale of premises liability law: a contractor hired to repair a hazardous condition, or who created that condition while doing its work, was obviously aware of the dangerous condition; so its knowledge was at least equal to that of the landowner.

SECTION 7.09 Product Liability

A. Relevance

The historical development of legal rules relating to the liability of a manufacturer for harm caused by its products manifests a shift from protection of commercial ventures toward compensating victims and making enterprises bear the normal enterprise risks. Manufacturer's liability has become important in the construction process, as harm can be caused by defective equipment or materials.

---

[74]*Rowland v. Christian*, 69 Cal.2d 108, 443 P.2d 561, 567–68, 70 Cal. Rptr. 97, 103–4 (1968); Annot., 22 A.L.R.4th 294 (1983).

[75]*Jordan v. National Steel Corp.*, 183 Ill.2d 448, 701 N.E.2d 1092 (1998), adopting the Restatement (Second) of Torts §§ 343 and 343A (1965).

[76]The California Supreme Court recently reformulated this rationale under the doctrine of "delegation"; see *Kinsman v. Unocal Corp.*, 37 Cal.4th 659, 123 P.3d 931, 937–38, 36 Cal.Rptr.3d 495 (2005).

[77]West Ann.Cal.Civ.Code § 846; Wash.Rev.Code Ann. § 4.24.210.

[78]*Blair v. Campbell*, 924 S.W.2d 75 (Tenn.1996) (contractor, hired to perform repairs on a decrepit building, knew the work site was unstable); *Baber v. Dill*, 531 N.W.2d 493 (Minn.1995) (worker fell on protruding reinforcing steel rods—a hazardous condition the contractor had created).

B. History: From Near Immunity to Strict Liability

In 1842, the English case of *Winterbottom v. Wright*[79] held that an injured party could not recover from the maker of a defective product in the absence of privity between the injured party and maker. As noted in Section 7.03E, privity was usually a contractual relationship, but it could also be based on a status that placed a duty on one party to watch out for another.

The privity requirement protected an infant manufacturing industry developing during the Industrial Revolution. It allowed contracting parties to know their liability exposure and deal with inordinate risks by contract. Privity permitted the manufacturer to be secure in the belief that it would not be held liable to people other than those with whom it dealt. It could relieve itself, by contract disclaimers, from inordinate risks to those with whom it dealt. It could not do so with the many third parties who might be injured by its products or activities.

Protection was no more tolerable than the sad plight of the uncompensated industrial accident victim who ultimately received protection through workers' compensation. By the early twentieth century, the privity rule was no longer acceptable. Injured parties needed a solvent defendant from whom they could recover. Ultimate responsibility should be on the manufacturer. The latter can insure against predictable losses and pass on the cost to those who benefited from the enterprise, such as owners or users of the enterprise's activities.

The major turning point occurred in New York in 1916. Before 1916, New York had held that a negligent manufacturer could be liable despite the absence of privity if there was a latent defect in the goods sold or if the goods sold were inherently dangerous. In 1916, the New York Court of Appeals held in *MacPherson v. Buick Motor Co.*[80] that this exception included goods that were dangerous if made defectively. The *MacPherson* case, involving an automobile, led to abolition of the privity rule in the United States.

The plaintiff still had the difficult task of establishing that goods were negligently made by the manufacturer. But the *res ipsa loquitur* doctrine proved of great assistance. It permitted the plaintiff to present its case to the jury even though the plaintiff introduced no direct evidence of negligent conduct by the manufacturer.

Res ipsa, as noted in Section 7.03C, did not as a rule deprive the defendant of the opportunity of introducing evidence that it had not been negligent. This was typically done by seeking to establish that the defendant had followed common industry practices and had used a system designed to ensure that its products were safe. Yet juries typically ruled for victims of adulterated food or beverages or defective products.

A more efficient way of placing this risk on the manufacturer was needed. The concept that first accomplished this purpose was implied warranty. This doctrine, borrowed largely from commercial law, held sellers liable when their goods were not merchantable or, under certain circumstances, not fit for the purposes for which buyers bought them. It eliminated the need to establish negligence. Implied warranty would under certain circumstances extend protection to third parties.

Warranty first was used in food and drug cases. It then began to be employed when harm was caused by manufactured goods. This concept, at least for a time, put normal enterprise risks on the enterprise.

Early in the history of implied warranty, some courts recognized that though useful, this essentially commercial doctrine was inappropriate for determining who should bear the risk of physical harm caused by defective products. In addition, warranty carried with it technical rules more appropriate to commercial transactions. As a result, a few courts began to treat claims by injured parties against manufacturers as involving strict liability in tort.[81] Soon other courts fell into line, and the strict liability concept gradually supplanted implied warranty as a risk distribution device.

C. Restatement (Second) of Torts Section 402A and Restatement (Third) of Torts: Products Liability Section 2

One factor that led to the replacement of implied warranty by strict liability in tort was the decision by the American Law Institute—a private group of judges, scholars, and lawyers—to restate the law of torts by publishing Section 402A in 1965. This section, found in the Restatement (Second) of Torts, states,

---

[79] Supra note 32.
[80] 217 N.Y. 382, 111 N.E. 1050 (1916).

[81] The leading case is *Greenman v. Yuba Power Products, Inc.*, 59 Cal.2d 57, 377 P.2d 897, 27 Cal.Rptr. 697 (1963).

1. One who sells any product in a defective condition unreasonably dangerous to the user or consumer or to his property is subject to liability for physical harm thereby caused to the ultimate user or consumer, or to his property, if
   a. the seller is engaged in the business of selling such a product, and
   b. it is expected to and does reach the user or consumer without substantial change in the condition in which it is sold.
2. The rule stated in Subsection (1) applies although
   a. the seller has exercised all possible care in the preparation and sale of his product, and
   b. the user or consumer has not bought the product from or entered into any contractual relation with the seller.

Section 402A has had a significant impact. Courts following that section held manufacturers of defective products liable without any privity between the injured party and the manufacturer and without the injured party having to establish that the manufacturer had been negligent.

Courts have broken down the linchpin of Section 402A—"a defective condition unreasonably dangerous"—into three types of product defect: in the manufacturing, design, or marketing of the product. A manufacturing defect is a production flaw; an example is the sale of contaminated food. A design defect occurs when the design of the product is needlessly dangerous, such as a car that is prone to roll over. A marketing defect means a lack of instructions or warnings explaining to the end user how to operate the product safely. All of these defects are measured against the reasonable expectations of the ultimate consumer of the product. The precise nature of each of these defects, as well as of the consumer expectation test against which they are measured, is further discussed in Section 7.09I; however, a detailed analysis of these issues is beyond the scope of this book.[82]

Under the common law system, plaintiffs are free to advance new theories of product defect not found in the Restatement of Torts. In *Slisze v. Stanley-Bostitch*,[83] a construction worker was injured by a nondefective pneumatic nail gun. At the time of the accident, the manufacturer had another, safer model, which made it harder for nails

to be discharged unintentionally. The court found the manufacturer not liable in negligence or strict products liability, based either on a duty to refrain from marketing a nondefective product when a safer model was available or on a duty to inform the consumer of the availability of the safer model.

Over the years, use of strict liability for design and marketing defects came under increasing criticism, with the result that these two defects had come to be evaluated under a negligence standard. In 1998, the American Law Institute published the Restatement (Third) of Torts: Products Liability.[84] This Restatement changed the terminology of a defective product from one that is "unreasonably dangerous" to one that is "not reasonably safe." Section 2, titled "Categories of Product Defect," states,

> A product is defective when, at the time of sale or distribution, it contains a manufacturing defect, is defective in design, or is defective because of inadequate instructions or warnings. A product:
> (a) contains a manufacturing defect when the product departs from its intended design even though all possible care was exercised in the preparation and marketing of the product;
> (b) is defective in design when the foreseeable risks of harm posed by the product could have been reduced or avoided by the adoption of a reasonable alternative design by the seller or other distributor, or a predecessor in the commercial chain of distribution, and the omission of the alternative design renders the product not reasonably safe;
> (c) is defective because of inadequate instructions or warnings when the foreseeable risks of harm posed by the product could have been reduced or avoided by the provision of reasonable instructions or warnings by the seller or other distributor, or a predecessor in the commercial chain of distribution, and the omission of the instructions or warnings renders the product not reasonably safe.

Because the impact of the Restatement of Products Liability lies largely in the future, the following discussion is directed to Section 402A, unless otherwise indicated.[85]

---

[82]For a thorough introduction, see D. DOBBS, supra note 1 at 977–1020.

[83]979 P.2d 317 (Utah 1999).

[84]The first two Restatements of Torts were published as a single, multivolume work. The third Restatement of Torts is being published in separate volumes, each addressing individual topics. Hence, the Restatement (Third) of Torts: Products Liability is a stand-alone volume, with its own, separate title.

[85]For an overview of the changes made by the Third Restatement, see Schwartz, *The Restatement (Third) of Torts: Products Liability: A Guide to Its Highlights*, 34 Tort & Ins. L.J. 85 (Fall 1998).

D. Product Misuse

A product is defective in design or marketing only if the harm was reasonably foreseeable. The manufacturer will not be liable if the consumer's unforeseeable misuse of the product was the sole cause of the harm. Unforeseeable misuse may show that the product was not defective, that the defect did not cause the injury, or that the plaintiff's recovery should be reduced by his comparative negligence. Some courts impose upon manufacturers a duty to design or warn against foreseeable product misuse, which one court has defined as "any danger inherent in normal use which is not known or obvious to the ordinary user. Normal use includes misuse which might reasonably have been anticipated by the manufacturer."[86]

In *Romito v. Red Plastic Company*,[87] the California Court of Appeal addressed the policy reasons for not subjecting a manufacturer to liability for unforeseeable product misuse. In that case, a construction worker, while working on a roof, stumbled backward and fell through a plastic skylight to his death. Plaintiff provided evidence that the skylight, at the same cost to the manufacturer, could have been made from stronger plastics. The court found the manufacturer not liable, noting that it had no control over the work-place and that the stronger plastic might not have worked against a heavier worker. For policy reasons, liability should not be imposed where so many variables uncontrollable by the manufacturer contribute to a worker's death.[88]

E. Parties

Many entities play significant roles in the manufacturing and distribution of products. Manufacturers buy component parts and materials from other suppliers. They may obtain independent design and testing services. The product itself may be sold or installed by independent retailers or installers. Sometimes products are distributed through wholesalers who sell to retailers. See Figure 7.1.

It is generally assumed that the manufacturer is best able to spread the loss and avoid harm. But some retailers also have this capacity. For example, many products are distributed through large national retail chains that design or set performance standards for products in contracts with smaller manufacturers.

The purchaser of the product is not the only one who may be injured by a defective product. Members of the purchaser's family may use the product. Social guests may be injured if a television set explodes in the living room. A defectively designed car can injure drivers, passengers, other vehicles, or pedestrians.

Section 402A of the Restatement of Torts took no position as to whether people other than users or consumers can use strict liability to recover against the manufacturer. Nor did it take a position as to whether the seller of a component part would be liable. The trend seems toward holding responsible all who play significant roles in product manufacture, sale or leasing. The Restatement (Third) of Torts: Products Liability makes clear that liability extends to all those who are in the business of selling or distributing products, including retailers and lessors.[89]

Although there has been great expansion of manufacturer's liability, at the same time there has been some hesitation to place liability on *everyone* who can be connected in some way to the injury-creating product. This is demonstrated in the construction process by *Bay Summit Community Association v. Shell Oil Co.*[90] The developer of a condominium installed polybutylene plumbing in its individual condominium units and in the common areas. When numerous leaks appeared, the owners of condominiums and the condominium association sued U.S. Brass, a polybutylene manufacturer, and Shell, a supplier of resin for the pipes.

Shell's defense was that it merely had supplied nondefective material. But the court pointed to Shell's marketing activities, which consisted of

1. direct marketing to manufacturers
2. committing resources to obtain code approval
3. employing marketing "blitzes"

The court held Shell's liability was not based on its role as a supplier of resin. Rather, liability depended on

---

[86]*Reilly v. Dynamic Exploration, Inc.*, 571 So.2d 140, 144 (La.1990). Under the Third Restatement, there is no duty to design or warn against unforeseeable risks and uses; see the Restatement (Third) of Torts: Products Liability § 2, comment m.

[87]38 Cal.App.4th 59, 44 Cal.Rptr.2d 834 (1995).

[88]Id., 44 Cal.Rptr.2d at 835. For other applications of the unforeseeable misconduct defense, see *Jackson v. Tennessee Valley Auth.*, 413 F.Supp. 1050, 1058–59 (M.D.Tenn.1976), aff'd, 595 F.2d 1120 (6th Cir.1979); *Colvin v. Red Steel Co.*, 682 S.W.2d 243, 246 (Tex.1984); Annot., 65 A.L.R.4th 263 (1988).

[89]Restatement (Third) of Torts: Products Liability § 1, comment c.

[90]51 Cal.App.4th 762, 59 Cal.Rptr.2d 322 (1996).

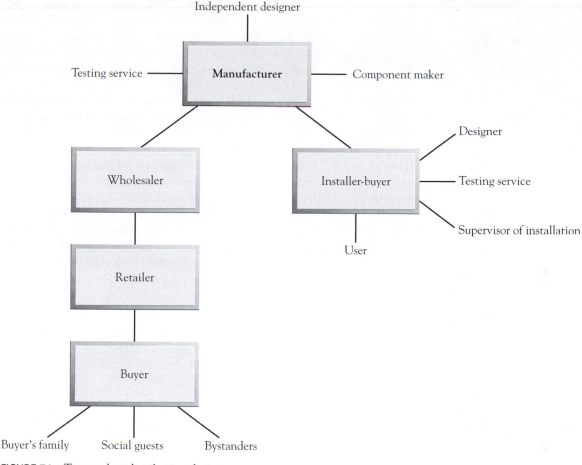

FIGURE 7.1   Two product distribution chains.

Shell's participation in marketing and distributing the plumbing system.

The *Bay Summit* case may not be followed by every court dealing with a similar issue. Different judges may draw the outer limits of responsibility for defective products differently. Therefore outcomes cannot be easily predicted and apparently inconsistent outcomes may result. Section 7.09K discusses whether strict liability extends to design professionals and contractors.

F. Defenses

The plaintiff's contributory negligence, where it still exists, is generally not a defense available to a manufacturer. If the manufacturer establishes that the injured party voluntarily assumed the risk, however, the plaintiff cannot recover. Where comparative negligence applies, it has been applied in strict liability claims.[91]

Sometimes product manufacturers expect or require that those to whom they sell will take steps to ensure that the product will be used safely. For example, suppose user instructions make clear that guards are to be used around dangerous machinery. Similarly, suppose a manufacturer of automobiles requires the car dealer to prepare the car in a designated way for the customer. Some courts have held that the manufacturer's duty to make a product

---

[91]*Daly v. General Motors Corp.*, 20 Cal.3d 725, 575 P.2d 1162, 144 Cal.Rptr. 380 (1978); *Barry v. Quality Steel Products, Inc.*, 263 Conn. 424, 820 A.2d 258 (2003) (rejecting the doctrine of superseding cause in favor of comparative fault).

that is reasonably safe cannot be delegated to others.[92] However, negligent conduct by the purchaser under some circumstances can be a superseding cause that may relieve the manufacturer. This superseding cause may consist of the purchaser making changes that would affect the way the machine was used.[93]

Another defense manufacturers may assert is that federal government regulations preempt (that is, displace) state product liability law. For example, the U.S. Supreme Court held that a cigarette manufacturer who complied with the health warnings required by a federal statute cannot be liable under state tort law for failure to give a better warning, although liability based on express warranty or conspiracy was not preempted.[94] The New Jersey Supreme Court similarly ruled that a construction worker, injured by a forklift manufactured in conformity with federal standards, could not sue the manufacturer for failure to include in the forklift additional warning devices.[95] However, preemption will not be found if the federal regulations do not apply to the particular safety device the plaintiff claims was defective and caused the accident.[96]

G. Economic Losses

Is the manufacturer liable for economic losses such as delay damages, lost profits, or injury to the product itself? Clearly, a manufacturer is responsible if there is an express warranty. After some early uncertainty, the trend is to bar recovery for economic losses in claims based on strict liability.[97]

As shown in Section 7.09E, which deals with which actors are responsible for defective products, it will be equally difficult to implement any rule defining the types of losses for which defendants are responsible. This difficulty is demonstrated by *Fieldstone Co. v. Briggs Plumbing Products, Inc.*[98]

The *Fieldstone* case involved a claim by a developer against manufacturers of sinks it installed in its developments. The defendants manufactured inexpensive bathroom sinks. The developer sought to recover the cost of replacing sinks that very early rusted and chipped because of spot welding and inadequate coating around the steel overflow outlets. The manufacturers claimed strict liability should not apply, because the sinks only damaged themselves and caused purely noncompensable economic losses.

The economic loss rule will be discussed in Section 14.08E. However, note that at the outset the court stated,

> [T]he line between physical injury to property and economic loss reflects the line of demarcation between tort theory and contract theory. [Citation.] "Economic" loss or harm has been defined as "damages for inadequate value, costs of repair and replacement of the defective product or consequent loss of profits—without any claim of personal injury or damages to other property."[99]

Next the court noted the variety of outcomes when this rule is applied by stating,

> Jurisdictions differ as to whether tort recovery is available where the sole physical injury is to the product itself. . . . A number of courts have allowed such recovery, finding the rationales behind the adoption of strict liability apply whether damages are to the same or other property; a large number of other courts have ruled otherwise, reasoning that warranty theories provide the exclusive remedy; and, yet "[o]ther courts have recognized that there may be particular exceptions to the rule of nonrecovery for mere economic damage to the product itself, based on an analysis of the nature of the defect and the risks involved. Accordingly, these courts have ruled that strict liability in tort could serve as a basis of recovery where the damage occurred in a sudden or calamitous manner, since this was akin to ordinary tort claims which ordinarily involve sudden injuries or damage, as opposed to mere deterioration over a length of time."[100]

---

[92]*Vandermark v. Ford Motor Co.*, 61 Cal.2d 256, 391 P.2d 168, 37 Cal. Rptr. 896 (1964); *Bexiga v. Havir Mfg. Corp.*, 60 N.J. 402, 290 A.2d 281 (1972).

[93]*Schreffler v. Birdsboro Corp.*, 490 F.2d 1148 (3d Cir.1974).

[94]*Cipollone v. Liggett Group, Inc.*, 505 U.S. 504 (1992).

[95]*Gonzalez v. Ideal Tile Importing Co., Inc.*, 184 N.J. 415, 877 A.2d 1247 (2005), cert. denied sub nom., *Gonzalez v. Komatsu Forklift, U.S.A., Inc.*, 546 U.S. 1092 (2006). The Restatement (Third) of Torts: Products Liability § 4(b) states that a product's compliance with a product safety statute does not preclude a finding of product defect, but presumably the Restatement here is referring to a state (not federal) statute.

[96]*Lindsey v. Caterpillar, Inc.*, 480 F.3d 202 (3d Cir.2007).

[97]Compare *Santor v. A & M Karagheusian, Inc.*, 44 N.J. 52, 207 A.2d 305 (1965) (permitting recovery of economic losses in strict products liability) with *Moorman Mfg. Co. v. National Tank Co.*, 91 Ill.2d 69, 435 N.E.2d 443 (1982) (disallowing a strict liability claim for purely economic harm). However, *Santor* remained a minority position and was overruled in *Alloway v. General Marine Industries, L.P.*, 149 N.J. 620, 695 A.2d 264, 275 (1997).

[98]54 Cal.App.4th 357, 62 Cal.Rptr.2d 701 (1997).

[99]62 Cal.Rptr.2d at 704.

[100]Id. at 705.

After canvassing California cases, the court stated,

> We conclude that here there was no injury to "other property" for purposes of imposing tort liability. The spot welding and inadequate coating were latent defects which made the sinks prone to rusting, chipping and premature deterioration. In other words, this case presents a routine situation in which a purchaser seeks replacement costs because a poorly designed and built product failed to meet its expectations. The doctrine of strict liability, however, is not a substitute for contract and warranty law where the purchaser's loss is the benefit of the bargain, and unless the parties specifically agree the product will perform in a certain way, the manufacturer is not responsible for its failure. . . . Certainly, Fieldstone is a sophisticated consumer and could have specified a higher quality product; but . . . there is no justification here for imposing tort liability on manufacturers who guaranteed their products for only one year. We reject Fieldstone's analysis, under which virtually every defective product evidencing deterioration of any nature would constitute "other property" for purposes of tort recovery. Such is not the law.[101]

The Restatement (Third) of Torts: Products Liability § 21 adopts the economic loss rule, but without using that term. Instead, it states that strict liability does not apply where the damage is only to the product itself. A comment to § 21 explains that recovery under tort law is unnecessary if there has been no physical harm to person or other property, because "the law governing commercial transactions sets forth a comprehensive scheme governing the rights of the buyer and seller."[102]

H. Disclaimers

In the commercial world, sellers frequently seek to limit their risk by disclaiming responsibility for certain losses or by limiting the remedy. One reason for shifting from implied warranty to strict liability was the desire to avoid disclaimers frequently part of the commercial transaction where the party injured was not a real participant in the transaction.

Disclaimers are likely to be given effect between the parties if the parties are business entities of relatively equal bargaining strength. However, disclaimers are not likely to be given effect in consumer purchases.

I. Design Defects and Duty to Warn: Two Restatements

Early product liability cases involved manufacturing defects. To determine whether a product is defective, the product causing the injury is compared to either the design plans or other products made from the same design. If there is a deviation, the product is defective.

Even in the absence of negligence, some products will have unintended manufacturing defects. It is not economically feasible to eliminate all risks of randomly defective products. It is better to predict the likelihood of defects, calculate the liability exposure, insure or self-insure against these risks, and include the cost in the product price. In addition to the exasperating question of whether there was a manufacturing or design defect (was it a bad weld or cheap material that made welding difficult?), judging the design has no neat test such as does manufacturing.

The law has struggled painfully with design defect definitions. California had in 1972 rejected the Restatement definition of "unreasonably dangerous" as reinstituting the discarded negligence test.[103] Then in 1978 it decided *Barker v. Lull Engineering Co.*,[104] a design defect case. The court held that two tests must be applied. First, did the product meet the expectations of an ordinary consumer as to product safety? Second, even if it did, the manufacturer is liable if it cannot establish (using negligence-like criteria) that on balance, benefits of the design outweighed the risks. The court allowed a hindsight look at the design and reserved judgment on whether the state of the art being followed would be a defense. The court did not decide whether products are defective because they lack adequate warnings or directions.

New Jersey has dealt with some of the issues unresolved in the *Barker* case. In 1982 the New Jersey Supreme Court decided *Beshada v. Johns-Manville Product Corp.*[105] The case involved the duty of a manufacturer to provide adequate warning of the dangers of its product. Duty to warn, a third type of liability in addition to an assembly-line defect and

---

[101]Id. at 707.

[102]Restatement (Third) of Torts: Products Liability § 21, comment d, p. 294.

[103]*Cronin v. J. B. E. Olson Corp.*, 8 Cal.3d 121, 501 P.2d 1153, 104 Cal.Rptr. 433 (1972).

[104]20 Cal.3d 413, 573 P.2d 443, 143 Cal.Rptr. 225 (1978).

[105]90 N.J. 191, 447 A.2d 539 (1982).

defective design, resembles negligence. The manufacturer is judged by the standard of reasonableness. However, in negligence claims the claimant must prove negligence. But in claims against a manufacturer based on inadequate warning, the burden is on the manufacturer to prove its conduct was reasonable.

In the *Beshada* case the court, as in the *Barker* case, employed a "hindsight" test to determine whether there had been adequate warning. It also rejected the "state of the art" as a defense. It held the manufacturer must reduce the risk of harm to the greatest extent possible consistent with the product's utility.

Yet two years later the court decided *Feldman v. Lederle Laboratories*.[106] This was a duty-to-warn case against a drug manufacturer. Although not expressly overruling the *Beshada* case, the court refused to use the hindsight test. It limited the applicability of the *Beshada* case to its facts. These cases reflect the confusion and uncertainty in manufacturers' liability, particularly those who make drugs, a product with great utility and high risk.

Courts are favoring a risk-utility analysis in design-defect cases. The court, usually done through a jury, balances the risks inherent in product design against the utility of product so designed. The inquiry requires determining whether the manufacturer acted reasonably in choosing a particular product design, given the probability and seriousness of the risk imposed by the design, usefulness of the product in that condition, and the burden on the manufacturer to take necessary steps to eliminate the risk.[107] In effect, this is a negligence test: the issue thought to have been eliminated in claims based upon defective products. But design-defect claims are different from those that involve manufacturing.

The Third Restatement changed the tests for both design and marketing defects. For design defects, the Restatement introduces a requirement that the plaintiff prove that a "reasonable alternative design" was available to the product manufacturer for the design to be defective.[108] By replacing the "reasonable consumer expectation" test with a "reasonable alternative design" requirement, the Third Restatement arguably replaces strict liability with a standard of reasonableness—a negligence concept. The existence of a reasonable alternative design is determined using a risk/utility analysis.[109] According to one scholar, the view underlying the Third Restatement is that, in the law of torts, negligence law is the norm, and strict (or enterprise) liability is the rare exception.[110]

With regard to marketing defects, the Third Restatement appears to expand liability. Under the Second Restatement, a manufacturer's warning or directions rendered an otherwise defective product nondefective.[111] The Third Restatement no longer states that instructions or warnings will make a product nondefective. Finally, the new Restatement mandates a safer design (where possible) over the use of warnings.[112]

Both Restatements allow an "assumption of the risk" defense to claim of inadequate warning. While the Second Restatement provides that a user who "discovers the defect and is aware of the danger . . . is barred from recovery,"[113] the courts have split on whether the plaintiff's recovery actually is barred or only diminished under principles of comparative negligence.[114] The Third Restatement says that the plaintiff's fault results only in a reduction in his recovery.[115] In either event, this "sophisticated user" defense is applicable particularly to the construction industry, which is characterized by highly trained workers using complicated, and potentially dangerous, machinery.[116]

J. Government-Furnished Design

Suppose someone suffers physical harm because of a defectively designed product procured by the federal government, the design compelled by the United States in its contract with the manufacturer. The federal government is immune from tort liability except to the extent that it has deprived itself of that immunity in the

---

[106]97 N.J. 429, 479 A.2d 394 (1984).

[107]*Banks v. ICI Americas, Inc.* 264 Ga. 732, 450 S.E.2d 671 (1994).

[108]Restatement (Third) of Torts: Products Liability § 2(b) and comments d & g.

[109]Id., comment f.

[110]Keating, supra note 6 at 1333–35.

[111]Restatement (Second) of Torts § 402A, comment j.

[112]Restatement (Third) of Torts: Products Liability § 2, comment l, followed in *Rogers v. Ingersoll-Rand Co.*, 144 F.3d 841 (D.C.Cir.1998) and *Uniroyal Goodrich Tire Co. v. Martinez*, 977 S.W.2d 328 (Tex.1998), cert. denied, 526 U.S. 1040 (1999).

[113]Restatement (Second) of Torts § 402, comment n.

[114]Annot., 46 A.L.R.3d 240 (1972).

[115]Restatement (Third) of Torts: Products Liability § 17, comment d.

[116]For applications of the defense, see *Johnson v. American Standard, Inc.*, 43 Cal.4th 56, 179 P.3d 905, 74 Cal.Rptr.3d 108 (2008); *Antcliff v. States Employees Credit Union*, 414 Mich. 624, 327 N.W.2d 814, 820–21 (1982); *Blackwell Burner Co., Inc. v. Cerda*, 644 S.W.2d 512 (Tex.App. Ct.1982).

Federal Tort Claims Act. The act requires that the government be negligent, and negligence is not required to establish the liability of the manufacturer for a defective product. Inasmuch as the federal government is likely to be immune, those who have suffered personal harm, or their survivors, frequently sue the manufacturer of the product.

The manufacturer is likely to assert two defenses: (1) the government specification defense if it followed nonobviously defective government specifications[117] and (2) the government contract defense, which gives the manufacturer the same immunity that would be given to the government.[118]

These defenses were relatively uncontroversial until the explosion of the environmental movement in the 1970s. The realization that acts committed many years ago can create serious risks led to many claims based on exposure to unsafe chemicals and hazardous wastes. The most controversial have been those that have related to Agent Orange, a defoliant used by the U.S. Armed Forces in Vietnam and manufactured in accordance with government specifications. A federal court held that the manufacturer will be given a defense if it can show that it followed government specifications in manufacturing the defoliant and that the United States knew as much as or more than the manufacturer of the hazards.[119]

K. Beyond Products: Sellers of Services

The discussion to this point addresses mass-produced products made by large-scale, commercial manufacturers. Does this paradigm extend to the construction industry, and does strict products liability also apply to individual products made by service providers?

A distinction should be drawn between design professionals and contractors. Attempts by injured persons to hold architects and engineers strictly liable or liable based on implied warranties have been largely unsuccessful.[120] Suppose a claimant sues an independent designer of a defectively designed product, or suppose the manufacturer sues the designer. In either case, the designer would not be held to a standard of strict liability or implied warranty. Some have advocated holding designers to enlarged, more strict liability.[121]

Developers who mass-produce homes[122] or lots[123] have been held strictly liable in some states. Contractors (or subcontractors) may be strictly liable for the delivery of defective appliances or products.[124] In some states, hybrid transactions, consisting of both services and the supply of products, will subject a contractor to strict liability if the delivery of the product is the predominate purpose of the transaction.[125]

The Restatement (Third) of Torts: Products Liability § 19(b) states that services, even when provided commercially, are not products.

L. Future Developments

Considerable dissatisfaction has been expressed with the evolution of product liability law. Manufacturers have complained that the options available to them are all unsatisfactory. One alternative is simply not to insure when the cost of premiums makes the price uncompetitive or is beyond the financial capacity of the manufacturer. Another is to overdesign a product that will pass even hindsight judicial or jury review evaluation. The product

---

[117]The U.S. Supreme Court held (5–4) that the contractor who drew attention to the problem was not liable. *Boyle v. United Technologies Corp.*, 487 U.S. 500 (1988). See Gleason, *In the Name of Boyle: Congress's Overexpansion of the Government Contract Defense*, 36 Pub. Cont. L.J. 249 (2007).

[118]See *Smith v. Rogers Group, Inc.*, 348 Ark. 241, 72 S.W.3d 450 (2000) (defense applied in state contract).

[119]*In re "Agent Orange" Product Liability Litigation*, 818 F.2d 179 (2d Cir.1987), cert. denied sub nom. *Krupkin v. Dow Chem. Co.*, 487 U.S. 1234 (1988).

[120]*LaRossa v. Scientific Design Co.*, 402 F.2d 937 (3d Cir.1968). See also Section 14.07.

[121]See Comment, 55 Calif.L.Rev. 1361 (1967); Comment, 15 Cal. W.L.Rev. 305 (1979).

[122]*Schipper v. Levitt & Sons, Inc.*, 44 N.J. 70, 207 A.2d 314 (1965); *Kriegler v. Eichler Homes, Inc.*, 269 Cal.App.2d 224, 74 Cal.Rptr. 749 (1969). But see *Wright v. Creative Corp.*, 30 Colo.App. 575, 498 P.2d 1179 (1972). For a thorough discussion, see Comment, 33 Emory L.J. 175 (1984). See Section 24.10.

[123]*Avner v. Longridges Estates*, 272 Cal.App.2d 607, 77 Cal.Rptr. 633 (1969).

[124]*State Stove Mfg. Co. v. Hodges*, 189 So.2d 113 (Miss.1966), cert. denied sub nom. *Yates v. Hodges*, 386 U.S. 912 (1967). In *Jimenez v. Superior Court*, 29 Cal.4th 473, 58 P.3d 450, 127 Cal.Rptr.2d 614 (2002), the California Supreme Court, by overruling two courts of appeal decisions, made clear that strict products liability applies to subcontractors who provide defective products.

[125]*Hill v. Rieth-Riley Constr. Co., Inc.*, 670 N.E.2d 940, 943 (Ind. App.1996), overruled on other grounds, *Peters v. Forster*, 804 N.E.2d 736 (Ind.2004).

line can be dropped, which can mean diminished competition, fewer consumer choices, and higher prices.

In addition, the cost of defense is staggering, including attorneys' fees, costs of testing, and experts. As if this were not enough, courts are beginning to award punitive damages when a jury decides that design choices did not take safety or public needs into account.[126]

In the 1990s, attacks were made on expanded product liability law by those who contended that such restraints on American industry made American products uncompetitive internationally. Attackers pointed to the costs incurred by manufacturers—such as increased cost of testing, ballooning insurance premiums, and the need to set aside reserves for uninsured losses and claims overhead—and compared American manufacturers to international competitors who did not face these same costs. They also pointed to manufacturers' decisions to withdraw products from the market because the costs associated with liability made the products unprofitable.

Defenders of product liability law asserted that competing manufacturers in other countries were highly regulated by government agencies. Costs to U.S. manufacturers are not much higher than those of foreign competitors, the defenders claimed. If production costs in the United States are truly higher than elsewhere, defenders said, U.S. prices should be higher and U.S. manufacturers less profitable. Yet, they claimed, this is not the case.

Defenders also pointed to the policy rationale for product liability law: that the need for safe products spurs manufacturers to develop safe products that would be attractive to consumers and acts as a *dis*incentive to produce unsafe products. Defenders of U.S. product liability law further claimed that no evidence supports the argument that such laws stifle innovation.

Clearly, emphasis on competitiveness points to product liability law as a form of government regulation. When that formulization is revealed, we can see more clearly which groups are struggling over this issue. Plaintiffs' lawyers and consumer organizations argue for retaining the expanded liability of manufacturers, whereas manufacturers and many economists stress the need for a free market.

This debate has reached the attention of the courts. Although the law may not be rolled back by the courts, it is not likely to be expanded.

Many legislatures have enacted statutes that have limited common law liability. Recommendations have been made for a uniform federal statute dealing with manufacturer's liability for defective products.

SECTION 7.10 Remedies

A. Compensation

The principal function of awarding tort damages is to compensate the plaintiff for the loss. In the ordinary injury case, the plaintiff is entitled to recover economic losses and certain noneconomic losses.[127] Economic losses are, for example, lost earnings and medical expenses. The principal noneconomic loss is pain and suffering.

Recovery for emotional distress has always been given hesitantly. Where there is physical injury, however, there has been no difficulty in allowing recovery for pain and suffering. Often the plaintiff's attorney will seek to obtain a large award for pain and suffering by asking the jury to use a per diem or even per hour method to compute the pain and suffering award. Breaking down the period of pain and suffering into small units can generate a large award.

Many have suggested that pain and suffering not be recoverable or that limits be placed on recovery for pain and suffering. This has been especially attractive to those seeking to minimize malpractice liability of doctors. In the 1970s, many states reduced the liability of health care providers. One method was to cap noneconomic losses.

Recovery of pain and suffering damages, which are often open-ended, has been justified by the fact that a large amount of the damage award usually goes to pay the victim's attorney. Plaintiff advocates emphasize that pain and suffering are real and that placing an economic value on them, though difficult, can give victims a sense that the legal system has taken adequate account of the harm they have suffered.

B. Collateral Source Rule

Often an accident victim's hospitalization costs are paid by an employer, a health insurer, or the government. Many victims recover lost wages through disability insurance.

---

[126]*Grimshaw v. Ford Motor Co.*, 119 Cal.App.3d 757, 174 Cal.Rptr. 348 (1981).

[127]See Section 27.09 for this in the context of a contract breach.

Although the many types of benefits create varied results, on the whole most sources of compensation are considered collateral and are not taken into account when determining the victim's loss.[128]

What is known as the collateral source rule has been considerably criticized because it can overcompensate. However, in tort cases, three principal justifications have been made for the rule. First, the defendant is a wrongdoer who should not receive any credit for benefits provided by third parties. Second, accident victims frequently must use a large share of their award to pay their attorneys. Third, accident victims often receive benefits because they have planned for them or because they are part of payment for their services.

Increasingly, state statutes reduce the scope or eliminate the common-law collateral source rule.[129] These statutes usually are enacted as part of an overall tort reform effort. This demonstrates the importance of benefits provided by third parties.

C. Punitive Damages

Tortious conduct that is intentional and deliberate—bordering on the criminal—can be punished by awarding punitive damages. Such damages are designed not to compensate the victim but to punish and make an example of the wrongdoer, to deter others from committing similar wrongs. In some areas of tort law, such as defamation, punitive damages play an important role because compensatory damages are often difficult to measure. There has been a tendency to award punitive damages for wrongful refusal to settle claims by insurance companies with their own insureds to ensure fair dealing in claim settlement. A few courts have awarded punitive damages in claims against manufacturers of defective products where the manufacturer seemed unwilling to place a high value on human life.[130]

Punitive damages have become extremely controversial. Supporters of such awards say large organizations trample on weak consumers and that the threat of actual imposition of punitive damages is the only way such organizations will take their legal obligations seriously. That is why the awards must hurt, or they will not affect conduct. As a result, the wealth of the defendant or its revenues have to be taken into account. To those that say that award by juries are often irrational and out-of-line, defenders say such awards can be upset or modified by appellate courts on review.

Attackers say that even if out-sized awards can be challenged on appeal, the threat of a horrendous punitive award will be a powerful tool in settlements. Attackers claim that there are other mechanisms to police bad conduct. They also complain that the jury can take into account the wealth of the defendants. Finally, they assert that the use of class actions (uniting claims of many people in a similar position) can generate huge awards.

Challenges by appeal also require an appeal bond that is often beyond the resources of the defendant. As a result, many large companies have filed for Chapter 11 bankruptcy. This allows the companies to reorganize and continue to operate while providing a shield against claims.

Most of the constitutional challenges to punitive damages awards have not been successful. But in 2003, the United States Supreme Court held in *State Farm Mut. Auto Insurance Co. v. Campbell* that the due process clause of the United States Constitution was violated when a state awarded 145 million dollars in punitive damages where the compensatory damages were one million dollars.[131]

The Court found that the award was grossly excessive or arbitrary. This violated the constitution. Such an award serves no legitimate purpose and constitutes an arbitrary deprivation of property. Also, an excessive award violates due process as the defendant is deprived of fair notice that his conduct will subject him to a defined punishment.

The amount must bear some relationship to compensatory damages. In practice, said the Court, few awards exceeding a single digit multiplier (the amount of punitive damages times compensatory damages) will satisfy due process. More than four times is close to the line, even to the point of presumptively being invalid.

But the Court created no fixed rule. If the harm is personal and not merely economic, a higher multiplier may be appropriate. If the amount of compensatory damages is substantial, a lesser ratio, even equal to the amount of compensatory damages, may be a limit on punitive

---

[128] See this rule in the contract claim context in Section 6.06F.

[129] As an illustration, see *Fisher v. Qualico Contracting Corp.*, 98 N.Y.2d 534, 779 N.E.2d 178, 749 N.Y.S.2d 467 (2002) (New York statute applied to the issue of whether insurance proceeds should offset a negligence claim against contractor).

[130] See supra note 126.

[131] 538 U.S. 408 (2003) (6 to 3). See note 132 infra for a recent important U.S. Supreme Court decision.

damages.[132] Particularly reprehensible conduct that results in a small amount of economic losses may justify a higher ratio.

While all will depend on the circumstances, the language of the Court should place some practical limits on punitive damages by the multiplier comparison to compensatory damages.

The trial court looked at the defendant insurer's settlement practices on a national basis. This condemned the insurer for its nationwide practices. The Court found this to be an error. The award of punitive damages must be for conduct directed toward the plaintiffs.

The Court also held that the amount of punitive damages must take into account the civil fines and penalties that the conduct would have justified.

Finally, the Court held that the wealth of the defendant cannot cure any failure to satisfy other requirements for the award of punitive damages.

While this case may slow the growth of punitive damages, this decision will have to be applied in the state courts. Some will see this case as a justification to slow the advance or even roll back punitive damages. Others, though, sympathetic to the award of substantial punitive damages awards, may find ways to continue liberal awards. It remains to be seen how the controls will work out in practice.

One criticism of punitive damages is that these damages are awarded to a claimant over and above his actual losses. Because the purpose of punitive damages is essentially a public one—that is, to deter wrongful conduct by making an example of the defendant—some have suggested that the damages should go into the public coffers and not into the pocket of the claimant. Seven states have enacted legislation requiring that a designated percentage—usually between 30 percent and 75 percent—be paid to the state. An unprecedented recent decision by the Supreme Court of Ohio held that part of a punitive damages award against a health insurer must go to a cancer research fund named after the deceased insured. The fund is to be administered by a hospital and research institute at the Ohio State University.[133]

Punitive damages in the construction context will be discussed in Section 27.10.

D. Attorneys' Fees: Cost of Litigation

Generally, accident victims are not able to recover their attorneys' fees or other costs of litigation from the wrongdoer. This has led to expanded damages through pain and suffering and the collateral source rule as well as the contingency fee contract, under which attorneys risk their time if they do not obtain a recovery.[134]

E. Interest

Because tort damages are rarely liquidatable, it is difficult to receive interest from any period of time before award of judgment. However, as indicated earlier,[135] statutes vary considerably, and in some states the trial judge has discretion to award interest from the date legal action commenced.

---

[132]As this book went to press, the United States Supreme Court made an important decision in reviewing an award of punitive damages that resulted from a disasterous Alaska oil spill in 1989. The jury awarded five billion dollars for punitive damages (later reduced to two and one-half billion dollars) against Exxon. The Court reduced the award to one-half billion dollars as the award was excessive under common law Maritime Law. Its reduction was based on the award of one-half billion dollars as compensatory damages. *Exxon Shipping Co. v. Baker*, 128 S.Ct. 2605 (2008).

---

[133]*Dardinger v. Anthem Blue Cross & Blue Shield*, 98 Ohio St.3d 77, 781 N.E.2d 121 (2002).

[134]See Section 2.04.

[135]See Section 6.08.

CHAPTER EIGHT

Introduction to the Construction Process: Ingredients for Disputes

SECTION 8.01 Relationship to Rest of Book

The rest of this book deals with the impact of law on the construction process. To provide a deeper understanding of such interaction, it is important at the outset to have an overall picture of certain salient characteristics of the construction industry, with special reference to the main participants in the construction process. This chapter provides such an introduction. Figure 8.1 provides a simplified chart of the participants in a building project organized in the traditional (design–bid–build) delivery system. Figure 8.2 is a more detailed breakdown of direct construction participants. For variants, see Section 17.04.

SECTION 8.02 The Eternal Triangle: Main Actors

A. The Owner

As a rule, the owner is the entity that provides the site, the design, the organizational process, and the money for the project. With some exceptions, such as real estate developers, owners tend to be "one-shot" players, not the repeat players found in the contractor and design professional segments of the industry. There is a wide range of those who commission construction; some differentiation between them is thus essential.

Most important is the differentiation between public and private entities. A private owner can select its design professional (by competition, competitive bid, or negotiation), its contractor (by competitive bid or by negotiation), and its contracting system (single contract or multiple prime, separate contracts) in any manner it chooses. Public agencies, in contrast, are limited by statute or regulation. As a rule, they must hire their designers principally on the basis of design skill and design reputation rather than fee. Construction services generally must be awarded to the lowest responsible bidder through competitive bidding. Often a public entity must use separate or multiple prime contracts because of successful efforts in the legislatures by specialty trade contractors.

Other important differentiations exist between public and private owners. Public contracts have traditionally been used to accomplish goals that go beyond simply getting the best project built at the best price in the optimal period of time. Contracts to build public projects have often been influenced by the desire to improve the status of disadvantaged citizens, to remedy past discrimination, to give preferences to small businesses, to place a floor on labor wage rates, and to improve economic conditions in depressed geographical areas. Considerations exist that are not likely to play a significant role in awarding private contracts.

Public entities are more likely than are their private counterparts to be required to deal fairly with those from whom they procure design and construction services. Yet public entities, having responsibility for public monies, usually impose tight controls on how that money is to be spent. As a result, such transactions have often generated intense monitoring by public officials and by the press to avoid the possibility that public contracts will be awarded for corrupt motives or favoritism. In addition, public projects are more controversial. To whom the project is awarded, the nature of the project, and the project's location often excite fierce public debate and occasional treks to the courthouse. Public owners are expected not to allow those from whom they procure goods and services to make profits on unperformed work or to make excessive profits.

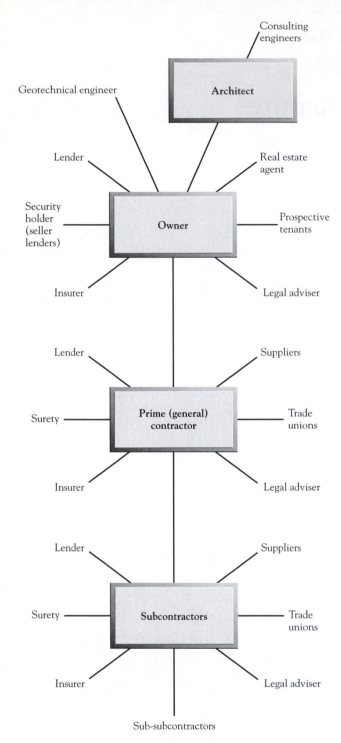

FIGURE 8.1   Organization of building project (traditional D–B–B system).

Public owners often seek to control dispute resolution, by contract or by law. Some public owners now require arbitration. Many experienced public owners, such as federal contracting agencies and similar agencies in large states, have developed a specialized dispute resolution mechanism, often using specialized regulatory, arbitral, or judicial forums.

Another differentiation is between experienced and inexperienced owners. An experienced owner engages in construction, if not routinely, at least repeatedly. This owner is familiar with common legal problems that arise, construction legal and technical terminology, and standard construction documents. It may also have a skilled internal infrastructure of attorneys, engineers, risk managers, and accountants.

An inexperienced owner, though often experienced in its business, is likely to find the construction world strange and often bewildering. Such an owner lacks the internal infrastructure and may, even if it has resources to hire such skill, not even know whether it should do so and, if so, how it can be done. The prototype, of course, is an owner building a residence for her own use.

Many other owners are "inexperienced." Although as a general rule, public owners are more experienced than private ones, care must be taken to differentiate between, for instance, the U.S. Corps of Engineers and a small local school district. The former has experienced contracting officers, contract administrators, and legal counsel and operates through comprehensive agency regulations, standard contracts, and an internal dispute resolution mechanism. The small local school district, in contrast, may be governed by a school board composed of volunteer citizens, be run by a modest administrative staff, and have a part-time legal counsel who may be unfamiliar with construction or the complicated legislation that regulates the school district.

Similar comparisons can be made between private owners. Compare General Motors building a new plant with a group of doctors building a medical clinic or a limited partnership composed of professionals seeking tax shelters through building or renting out commercial space. The clinic or limited partnership at least can buy the skill needed to pilot through the shoals of the construction process. But a greater dichotomy can be seen if a comparison is made between General Motors and a private person building a residence she intends to occupy. The latter as a rule does not obtain technical assistance because of costs.

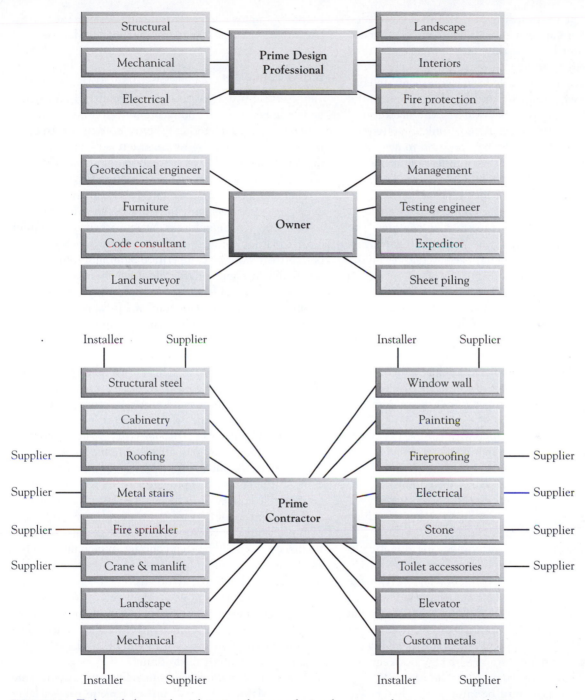

FIGURE 8.2 Technical players along the prime design professional, owner, and prime contractor chains in a building project.

Looking next to awarding construction contracts, inexperienced private owners will likely prefer competitive bidding, not because they must but because they will not know the construction market well enough to sit across the negotiating table from a contractor. In contrast, an experienced owner may be able to review the contractor's proposal and be aware of the market and other factors necessary to negotiate. (Ironically, often the public agency, which is in the best position to negotiate, must use the competitive process, a reflection of the distrust in public officials.)

Much more than an experienced owner, an inexperienced owner will need to engage an architect or engineer to design and administer the construction contract. The experienced owner may have sufficient skill within its own internal organization and not need an outside adviser experienced in construction.

An inexperienced owner will more likely use standard construction contracts such as published by the American Institute of Architects (see Appendices A through E) or the Engineers Joint Contracts Documents Committee (see Appendices G through I). It will do so because it does not wish to spend the money for an individualized contract, does not have an attorney who can draft such a contract, or wishes to acquiesce to the suggestions of its architect or engineer.

An owner's lack of experience with complex standard contract terms can generate many legal problems. The "form" may not "fit," leading to interpretation problems. In addition, if people operating under the contract are not familiar with or do not understand the contract, they are likely to disregard provisions, leading to claims that those provisions have been waived.

This differentiation may also affect performance in other ways. Inexperienced owners may make many design changes that raise the cost of construction and increase the likelihood of disputes over additional charges. They may also refuse legitimate contractor requests for additional compensation because they are unaware of those provisions of a contract that may provide the basis for additional compensation. Inexperienced owners may not keep the careful records that are so crucial when disputes arise.

Where the language must be interpreted, doubts will very likely be resolved in favor of the inexperienced owner. An experienced owner could have made the contract language clear. A claim made by a contractor that it be excused from default or be given additional compensation because of unforeseen events during performance is less likely to be successful against an inexperienced owner.

B. The Contractor

The contracting industry is very decentralized, with a large number of small and medium-sized firms. Although concentration characterizes other industries, construction is largely local, with most contractors serving a single metropolitan area. Few construction companies are even regional, let alone national or international.

Half a million companies engage in construction. The average company is family owned, with an average of five to ten permanent employees. Most workers are hired for a particular job through unions or otherwise. Contractors obtain their work by competitive bidding. Profit margins are usually low and bankruptcies high. Because construction requires outside sources of funds, it is often at the mercy of the changing monetary and fiscal policies.

Two out of three contractors are specialty contractors, and one out of two workers plies a specialized trade, such as plumbing, electrical work, masonry, carpentry, plastering, and excavation. This means that in many construction projects, the contractor acts principally as a coordinator rather than as a builder. Its principal function is to select a group of specialty contractors who will do the job, schedule the work, police specialty trades for compliance with schedule and quality requirements, and act as a conduit for the money flow. Moreover, in a fixed-price contract the contractors provide security to the owner by giving a fixed price.

The volatility of the construction industry adds to the high probability of construction project disputes. Because the fixed-price or lump-sum contract is so common, a few bad bids can mean financial disaster. Contractors are often underfinanced. They may not have adequate financial capability or equipment when they enter into a project. They spread their money over a number of projects. They expect to construct a project with finances furnished by the owner through progress payments and with loans obtained from lending institutions.

Labor problems, especially jurisdictional disputes, are common. Many of the trade unions have restrictive labor practices that can control construction methods. Some contractors are union; others are nonunion; still others are double-breasted, having different entities for union and nonunion jobs.

Some contractors do not have the technological skill necessary for a successful construction project. Often the technological skill, if there is any, rests with a few key employees or officers. The skill is often spread thinly over a number of projects and can be effectively diminished by the departure of key employees or officers for better-paying jobs.

The construction industry has attracted a few contractors of questionable integrity and honesty. These contractors will try to avoid their contractual obligations and conceal inefficient or defective performance. Such contractors are skillful at diverting funds intended for one project to a different project.

C. The Design Professional

Design professionals, the third element of the construction triangle, find themselves in the often uncomfortable position of working for the owner yet being expected to make impartial decisions during construction. They, like the contractors, have financial problems, because there is usually not enough work to go around.

D. The Industry

All the main actors—owners, contractors, and design professionals—suffer from a chronically sick building industry particularly affected by rapid movements of the economy, changes in public spending policies, and swings of monetary policy, all of which can affect the interest rates, a formidable factor in most construction.

SECTION 8.03 The Supporting Cast on a Crowded Stage

A. The Owner Chain: Spotlight on Lender

Because owners rarely have funds to provide for major (or even minor) construction work, an important actor along the owner chain is the entity providing lending for the project. Availability of funds is essential to the participants who provide services and materials. Often the owner's inability to obtain funds is the reason a project does not proceed, despite the design professional's having spent time on the design. Similarly, available funding may not be adequate to deal with such common events in con-

struction as design adjustments, escalation of prices, changed conditions, and claims. Inadequate funding can create a breakdown in the relationship among the project's participants, which in turn leads to claims.

With regard to funding, it is important again to differentiate between public and private projects. Public projects are usually funded by appropriations made by a legislature or administrative agency. In controversial projects, challenges may be made to the project itself, to the method by which the contract was awarded, or to compliance with other legal requirements. If challenges are made to the project, funds may never become available or may generate delay in payment.

Another problem generated in public projects relates to the project's being funded by a mixture of funds from various public agencies or even a combination of public agencies and private entities. For example, the interstate road system was funded 90 percent by the federal government and 10 percent by the states, while wastewater projects were funded 75 percent by the federal government, 12.5 percent by the state, and 12.5 percent by local entities. Mixed funding can generate problems because one funding entity requires certain contract clauses that may differ from those required by another funding entity.

In private projects, funds are usually provided by construction loans made by lenders for short periods, usually only for the time necessary to complete the project. The interest rates are short term, usually floating, and generally higher than those charged by permanent lenders (most commonly insurance companies and occasionally ordinary lending sources).

The private construction lender wants the loan to be repaid and usually takes a security interest in the project that can be used to obtain repayment if the loan is not repaid. The lender makes an economic evaluation to determine whether a commercial borrower will derive enough revenue to repay the loan. Also, such decisions are made by permanent lenders.

In addition, the lender will wish to be certain that the borrower has put up enough of its own money so that if the project runs into problems, the borrower will not walk away from the project and avoid the loan obligation by going bankrupt—leaving the lender with only the security interest and the unpleasant prospect of taking over a defaulted project.

Similarly, the lender wants to be certain that advanced funds go into the project and enhance the value of the project by the amount of money advanced. It does not wish to see the borrower diverting funds for other purposes.

The lender also wishes to be assured that there are no delays, which can often be caused by faulty design or poor workmanship. Costly design changes, claims for extras by the contractor, and additional expenses due to unforeseen conditions are events the lender wishes to avoid. Similarly, the lender wishes to be assured that those who provide services, such as contractors and subcontractors and those who provide materials and equipment, such as suppliers, are paid and do not assert liens against the project. If liens are asserted, it may be difficult for the construction lender to obtain a permanent "takeout" lender.

Finally, the lender usually wants to know how the project is proceeding and may, in addition to approving the construction contracts made at the outset, insist on receiving copies of contract addenda, modifications, requests for changes, change orders, claims, and even routine correspondence.

The seller often retains a security interest in the land. The owner may be constructing a commercial structure in which space has been leased in advance to tenants. The owner's creditors may have an interest in the construction project. They may hope to collect their debts from profits made by the project or by having the land seized or sold to pay the owner's debts.

B. The Contractor Chain

The contractor chain in the single-contract system involves, in addition to the prime contractor, a large number of subcontractors and possibly sub-subcontractors. Each one of these contractors, as well as the prime contractor itself, purchases supplies and rents equipment. The material and equipment suppliers also have a substantial stake in the construction project.

Using surety bonds for prime contractors, and often for subcontractors, brings a number of surety bond companies into the picture. The creditors of the contractors, other than suppliers, are often involved. Contractors may owe taxes to the authorities, and people may have lent money to the contractors. The contract chain would not be complete without reference to the trade unions, which have a substantial stake in the construction project.

C. The Design Professional Chain

In addition to the owner and contractor chains, there is a somewhat shorter chain beginning with the design professional, who may hire consultants. Some consultants, such as geotechnical engineers, may have been hired by the owner and be in the owner's chain. In larger projects, an employee of the design professional might be on the site daily.

Consultants also have a stake in the construction project. They wish to be paid, and their negligence may cause damage to any number of people who are also affected by or involved in the construction process.

D. Insurers

Insurers are also important actors on this crowded stage. All the major participants, such as design professionals, owners, and contractors, are likely to have various types of liability insurance. In addition, either the owner or the prime contractor will carry some form of property insurance on the work as it proceeds and on materials and equipment not yet incorporated into the project.

Figures 8.1 and 8.2 illustrate the participants in a complex building project.

SECTION 8.04 The Construction Contract

Building a construction project is a complicated undertaking. The construction documents, particularly plans and specifications, should be clear and complete. At their best, they should give a good indication of the contractor's duties. Unfortunately, even the best design professionals cannot do a perfect job of drafting the construction documents that encompass the complete construction obligation.

In American construction the contract documents must often be interpreted by the person who designed the project and who was selected and is paid by one party. The inevitable interpretation issues can induce corner-cutting contractors to bid low and submit a large bill for extras. The variety of contract documents creates additional dispute possibilities, inconsistencies, and ambiguities.

The competitive bidding process so often used to select a contractor and the frequent use of the fixed-price contract play a significant role in dispute generation. The former emphasizes price rather than quality, and the latter places tensions on the relationship by placing the risk of many unknown and abruptly shifting factors on the contractor.

SECTION 8.05 The Delivery Systems

Although there have always been variations in construction delivery systems, construction in America in the twentieth century historically has been dominated by what is sometimes called the *traditional construction method,* also known as "design–award–build" or "design–bid–build." Under this system, the owner acquires the right to improve land, commissions an architect or engineer to prepare the design, and engages the contractor to agree to execute the completed design, either through competitive bidding or through negotiation, under a fixed-price or lump-sum contract. The American construction delivery system commonly employs the design professional to monitor construction as it proceeds and to play a central role in interpreting the contract, resolving disputes, and issuing certificates of payment and completion. Depending on the type of project, some parts of the design are prepared by consultant design professionals, and much of the actual construction is performed by subcontractors.

Although variations on the system have always existed, increasing emphasis on the variations began after World War II, particularly in the 1970s. Principal variations include awarding the contracts for construction to separate contractors or multiple prime contractors, commencing construction before design is completed (known as "phased construction" or "fast-tracking"), inserting a new professional called a construction manager into both design and construction, and combining design and construction through the use of what has been called a turn-key or design–build. Probably the majority of large-scale public and private commercial projects currently use nontraditional project delivery systems, especially design-build and construction management. In the past twenty years, newer project delivery methods include a more collaborative approach by the main project participants ("partnering" and "project alliance"), importing modernized manufacturing delivery systems to the construction industry ("lean project delivery") and a fundamental revamping of the design and implementation methods based upon advanced computer modeling technology ("building information modeling"). Section 17.04 deals in greater detail with these variations.

SECTION 8.06 The Applicable Law

The law that regulates the rights of parties along any of the chains should be the contracts which connect the various persons on the chain. For example, the basic law between contractor and owner should be the prime contract. Similarly, the law that regulates the relationship between design professional and owner and prime contractor and subcontractor should be found, explicitly or implicitly, in their contracts.

There is an increasing use of standard contracts such as those published by the American Institute of Architects (AIA) or the Engineers Joint Contract Documents Committee (EJCDC). It is impossible to anticipate all the problems and deal with them properly in each individual construction contract. Standard contracts rely heavily on the experiences of the past and on the expertise of people who have wide experience in construction projects.

Good standard contracts are planned carefully. However, their existence does not solve contract problems for lawyers or for design professionals. First, some standard contracts acquire the reputation for being heavily slanted in favor of one of the parties. When asked to pass judgment on these contracts, lawyers may reject them completely or substantially modify them. Second, a standard contract is often unread or misunderstood by the other party if the party to whom the contract is presented is not represented by a lawyer. A possibility exists that the contract will not represent all the law regulating the relationship between the parties.

A number of other laws regulate the construction project. Building codes and industry standards are often incorporated, expressly or implicitly, in the construction documents. Building codes lack uniformity and consist of complicated and often cumbersome rules that regulate the construction process. There are zoning laws and subdivision laws. Tort doctrines, such as those relating to nuisance and soil support, affect the use of land. Title and security problems are often difficult. Because of history and the archaic language of surety bonds, interpreting a surety's obligation is difficult. The rights of injured persons or injured property owners are governed by tort law and the bewildering process of indemnification. If more were needed, ultimate responsibility for some losses seems to require a "slug fest" between insurance carriers, all armed with unreadable policies with hordes of special endorsements. Is it any wonder that disputes are common and litigation time consuming and costly?

SECTION 8.07 The Construction Site

The owner's choice of location for the project is important. Designing for an area where craft unions are strong is different from localities where nonunion labor is used. If the site is in a city with environmentally sensitive citizens, the project will have characteristics different from those designed for localities where this scrutiny is less intense. The site may affect the likelihood that bribes must be paid to receive contracts or permits. The site may also affect weather conditions, which can influence design and performance. The site may determine the availability of an efficient and honest judicial system to resolve disputes.

The physical site itself contributes to the likelihood of disputes and construction process difficulties. No two pieces of land are exactly alike, generating a high probability of subsurface surprises. Testing methods for soil conditions are expensive and often do not give an accurate picture of the entire site. In addition, the physical limitations of the site, together with the large number of persons and contracting parties who must perform within this limited physical area, increase the probability of difficulties.

SECTION 8.08 Contract Administration

Even if the general terms and conditions of the contract documents are well expressed, difficulties often develop because contracting parties often are sloppy in contract administration. Decisions are made on the site, modifications are agreed to, changes are ordered—all without the formal requirements frequently expressed in the general terms and conditions of the written contract documents. Telephone conversations are often used to resolve difficulties and continue the work, but a dispute may arise later as to what was said in such conversations.

In the process of the dispute, one party often points to the contract clauses requiring that certain directions be given in writing or that certain modifications be expressed in writing. The other party then states that throughout the entire course of administration these formal requirements were disregarded. From these seeds develop disputes and lawsuits.

SECTION 8.09 International Transactions

Crucial differences exist between foreign and domestic owners. Contracts made with foreign owners are most likely to involve the sovereign or one of its agencies. Under such circumstances, the people contracting to provide design or construction services must be aware of the sovereign's power to regulate foreign exchange rates, import or export of goods or money, local labor conditions, the necessity for bribes, the lack of an independent judiciary, and the risk of expropriation.

Even contracting with a private owner in a foreign country involves risks that are not significant when contracting with domestic owners. Problems may arise from unfamiliar laws, different subcontracting practices, different laws and customs regulating the labor market, and different legal solutions.

SECTION 8.10 Unresolved Disputes and Litigation

Disputes between parties may be resolved without litigation when they want to maintain goodwill in their future dealings. However, this element—the necessity of future relations—may be missing in many construction projects. A dispute may involve a number of parties, such as insurance companies and sureties, that must consent to any settlement. The uncertainty of both the law and the facts and variety of legal issues discourages settlement.

Unless the parties value the goodwill of the owner, as is not common in much construction, they need not compromise—another factor discouraging settlement and leading to the courtroom. If there is an arbitration provision, one or both of the parties may not trust the arbitration process. Even if the arbitrator makes an award, the party against whom the award is made may not perform. Such awards must be confirmed by a court.

In summary, parties in a dispute-prone process such as construction have the propensity to call on the legal system to enforce contracts or obtain compensation for losses. Participants in the process must be aware of this. They must do all they can to avoid disputes, seek to settle those that do develop, and be aware of the role law plays in the process.

CHAPTER NINE

Limits on Ownership: Land Use Controls

This chapter examines a variety of controls or limitations on the nature of the project. Sections 9.01 through 9.05 look at private land use controls created by contract or tort law and enforced by courts. Sections 9.06 through 9.15 deal with public land use controls enacted mainly by local entities through powers given them by the state. Administration is handled largely through local regulatory commissions or legislative bodies. Courts, though less important, play mainly a passive, oversight role, determining whether local laws or administrations meet the state legislative and constitutional requirements. Controls on the design itself, imposed by building codes and the permitting process, are discussed in Section 19.02C.

SECTION 9.01 Nuisance: Unreasonable Land Use

The law protects the rights of landowners (those protected can also include family members or tenants or employees) to enjoy their land free of unreasonable interference by landowner activities in the same vicinity. This law regulates land use by prohibiting uses that unreasonably interfere with others' land use. The unfortunate term *nuisance* has developed as the means of describing such land uses.

A landowner (as used here, *landowner* can include a tenant) must make reasonable use of the land and not deprive neighboring landowners of the reasonable use and enjoyment of their land. To a large degree, granting of legal protection through nuisance controls depends on the activity or use complained of and on its effect on other landowners.

Looked at first from the perspective of the complaining landowner, the interference can take many forms. At one extreme, the land can actually have physical damage. For

example, a neighbor's discharging solid, liquid, or gaseous matter on the land may change the physical characteristics and shape of the complaining landowner's property and seriously affect the land's use.

The offensive activity may disturb the comfort and convenience of the complaining landowner. For example, activities on the offending land can create loud noise or the emission of noxious odors that may disturb the occupants of adjacent or nearby land. Moving to even more intangible interference, the neighbor's complaint may be based on simply knowing the nature of the activities on the adjacent or nearby land. For example, a neighbor's mental tranquility may be disturbed by the knowledge that an adjacent landowner maintains a house of prostitution or a meeting place for people whose activities are bizarre or unconventional.

Moving to the outer extreme of intangible interference, one neighbor may be extremely disturbed by the appearance of a neighbor's house.

An objective standard is used. Would the acts or activities in question have disturbed a reasonable person under such circumstances? Could the interference have been avoided had the complaining landowner taken reasonable measures? For example, would closing the windows have reduced or eliminated excess noise caused by the neighbor? If so, would such protective measures unduly interfere with the use and enjoyment of the land?

Generally, the more the activity in question causes an actual physical result, such as changing the characteristics or contours of the land or making a physical impact on the land, the more likely the law will give relief. As the interference moves toward more intangible matters, such as peace of mind and aesthetic judgments, the less likely the law will accord protection. Although such interests may be protected, the law would require that there be a significant interference with the use and enjoyment of the complaining landowner's property.

The likelihood of legal protection may depend on the extent and duration of the interference. Temporary interference or interference that occurs only at long intervals is less likely to justify relief.

Legal protection does not require the complaining party to show that the acts or activities in question were intended to interfere with the use and enjoyment of adjacent or nearby landowners. The conduct need not be negligent. However, the more intentional and more negligent the conduct, the more likely the activity in question is a nuisance.

The social utility of the activity is important. Certain commercial and industrial activities are necessary and are encouraged by society. The necessity for these activities may overcome minor inconveniences to those in the vicinity. If the activity has little social value, slight interference with the use and enjoyment of the adjacent landowner's land can constitute a nuisance.

Often the crucial issue is the remedy to be given the complaining owner when the defendant has made an unreasonable use of its land. The choice is between giving a money award for damages to the complaining party and ordering the offending party to cease the activities—an equitable injunction.

The money damage award must be inadequate before an injunction can be awarded. However, it is often difficult to establish the precise value of the damages caused in such cases. This is especially true if the offensive activity consists of noise or emission of noxious fumes or if the activity has caused losses other than physical harm. As a result, it has become relatively easy to obtain injunctive relief in nuisance cases, especially if the nuisance is a continuing one.

Suppose an industrial activity is being conducted on a large scale and has adversely affected the occupancy and use of nearby land. Some early court decisions affirmed orders that required cessation of the offensive activities. Such an order can close industrial plants on which may depend the economy of the town where the plant is located. As a result, some courts award what are called *permanent damages,* based on the diminution in the value of the land affected by the industrial activity, rather than to order that the offending activity cease.[1]

Class actions are increasingly being used, and some landowners maintain the action for the benefit of all landowners similarly situated. By aggregating small claims against the offending landowner, class actions can justify the often large expenses needed to mount a lawsuit against the industrial activity adversely affecting a large number of landowners in the vicinity.

SECTION 9.02   Soil Support

The owner of unimproved land has the right to lateral support by adjacent land. Because this is an absolute property right, the owner of unimproved land can receive legal relief if lateral soil support was withdrawn by excavation, and the owner need not show that the excavating landowner did not use proper methods.

This rule of property—an absolute rule as contrasted to the flexible one used in nuisance—does not apply if the affected land has been improved, adding weight to the soil. In such a case, the excavator is liable only if the adjacent landowner can show that the excavator did not comply with the legal standard of care. If the adjacent landowner can show that the soil support would have been withdrawn by the neighboring excavation regardless of the added weight caused by the improvement, the adjacent landowner can recover without showing that the excavator did not comply with the legal standard.

State legislation frequently controls the result in excavation cases. For example, in Illinois, the excavator must notify adjacent landowners of its intent to excavate, the proposed depth of excavation, and when excavation will begin. Legal rights depend on the depth of the excavation.[2]

The excavator should comply with any public controls such as the Illinois statute mentioned, give notice of excavation to the other party, seek to work out an advance agreement on protection and responsibility, and use reasonable care in excavating.

SECTION 9.03   Drainage and Surface Waters

Construction frequently affects and is affected by drainage at the site and adjacent land. Drainage changes can cause troublesome collection of surface waters. Three surface

[1]*Boomer v. Atlantic Cement Co.,* 26 N.Y.2d 219, 257 N.E.2d 870, 309 N.Y.S.2d 312 (1970).

[2]765 I.L.C.S. 140/0.01.

water rules have emerged in the United States.[3] The first is the "common enemy" rule: Each landowner can treat surface water as a common enemy, with the right to deal with it as the landowner pleases, without regard to neighbors. The second is the "civil law" rule: A landowner who interferes with the natural flow of water is strictly liable for any damage to neighbors.

The preceding rules are absolute. One gave great freedom to deal with surface water. Another created absolute liability. A third rule, that of "reasonable use," is gradually supplanting the absolute rules. Each case in jurisdictions with such a rule is decided on its own facts. The landowner affecting the water flow must act reasonably. In making this determination, the law considers the following:

1. Was there a reasonable necessity for such drainage?
2. Was reasonable care taken to avoid unnecessary damage to the land receiving the water?
3. Did the benefit accruing to the land drained reasonably outweigh the resulting harm?
4. When practicable, was the diversion accomplished by reasonably improving the normal and natural system of drainage, or if such a procedure was not practicable, was a reasonable and feasible artificial drainage system installed?

SECTION 9.04 Easements for Light, Air, and View

Light, air, and view are important considerations in constructing buildings and residences. *Fontainebleau Hotel Corp. v. Forty-Five Twenty-Five, Inc.*[4] involved an addition built by one luxury hotel in such a way as to cut off sun, light, and view to an adjacent hotel. The hotel on which an addition was being built—the Fontainebleau—was constructed in 1955; its adjacent competitor, the Eden Roc, was constructed in 1956. The addition was to be a fourteen-story tower. During the winter months from two o'clock in the afternoon and for the remainder of the day, the tower would cast a shadow over the cabana, swimming pool, and sunbathing areas of the Eden Roc.

When the addition was roughly eight stories high, the Eden Roc sought to obtain an order restraining the Fontainebleau from proceeding with the construction of the addition.

The trial court granted the requested order, but the appellate court reversed. The court stated that no American decision, in the absence of some contractual or statutory obligation, had held that a landowner "has a legal right to the free flow of light and air across the adjoining land of his neighbor."[5] The court concluded that construction of a structure that serves a useful and beneficial purpose could not be restrained by the court even though it cut off light and air and interfered with the view of a neighbor. Despite the partially spiteful motivation for the structure, relief was denied the Eden Roc.

Wisconsin refused to follow the absolute property rule. It applied the flexible nuisance doctrine to preclude an owner from building in such a way as to reduce the effectiveness of a neighbor's solar heating panels.[6] An owner who constructs to take advantage of view should consider either obtaining easements from adjacent landowners or buying adjacent land.

Some California communities have adopted tree ordinances. An owner whose view is impeded by a tree on a neighbor's land can request the tree be removed or trimmed. If this is done, the costs are divided. If the tree owner is unwilling, a local "tree commission" decides the dispute.

Suppose a neighbor puts up a "spite fence"—a high, unsightly fence that interferes with a neighbor's view and the entrance of light to the neighbor's land—and creates an eyesore. States have not been uniform in their treatment of spite fences. Some emphasize the right of a land-owner to use its land as it wishes, and permit spite fences. Other states hold that if the dominant motive in erecting the fence is malicious and if the fence serves no useful purpose, the neighbor can recover damages and be granted a court decree ordering the fence removed. In some states, statutes limit the height of any fence erected maliciously or for the purpose of annoying a neighbor. The increasing tendency is to restrict the use of spite fences.

---

[3]*Butler v. Bruno*, 115 R.I. 264, 341 A.2d 735 (1975) presents a complete discussion of the history of the three American water law rules.

[4]114 So.2d 357 (Fla.Dist.Ct.App.1959), cert. denied, 117 So.2d 842 (Fla.1960).

[5]Id. at 359.

[6]*Prah v. Maretti*, 108 Wis.2d 223, 321 N.W.2d 182 (1982). This case is noted in 21 Duq.L.Rev. 1159 (1983) and 48 Mo.L.Rev. 769 (1983).

SECTION 9.05 Restrictive Covenants: Common Interest Communities (CIC)

An individual grantor (one who conveys) of land generally has no interest in how the land is used after the deed has transferred ownership. But suppose the grantor lives or will live in the vicinity of the land transferred. Or, more important, suppose the grantor is a residential developer who is selling lots or houses to a large number of buyers. In either case, the grantor may wish to control land use after the title is transferred by deed. The grantor who is not a developer may wish to control the land use to enhance or maintain the use or value of land it owns in the vicinity of the land transferred. The developer-grantor may wish to assure buyers in the development that the land will retain its residential character to protect the enjoyment of and investment in the land. Although restrictive covenants can be created in agricultural or commercial property, this section emphasizes their use in developing residential property.

To accomplish the land-planning objectives, the developer-grantor can obtain an express promise relating to land use. There are no particular legal problems between the developer and the original buyer in accomplishing this result by contract, with the exception of racial or religious restrictive covenants (mentioned later in the section). The contract in such cases can and usually does contain promises that limit the buyer's right to use the land. In land law, such promises are called *restrictive covenants*.

Land ownership in a housing development is likely to be transferred. Because use affects value, the developer wishes to assure buyers that restrictions on use will bind future buyers and to assure buyers or their successors that they will have enforcement rights if any buyers violate the restrictive covenants. As a marketing device, the protection of tightly drawn enforceable restrictive covenants can persuade buyers to buy. To accomplish this, a developer will include in the deeds restrictions that give buyers of lots or residences in the development the right to enforce these restrictions when the covenants are violated.

To be effective over a long period of time, such restrictions must be "tied to" or "run with" the land. All buyers, present and future, must be bound by the restrictive covenants. All owners, present and future, must be able to obtain judicial enforcement of these restrictive covenants. Early English land law that sought to preserve property values by limiting land use through such restrictive covenants became unsatisfactory. A number of technical requirements often frustrated attempts to enforce these covenants.

The ineffectiveness of the system led to the nineteenth-century development of equitable servitudes in England and subsequently in the United States. In the famous case of *Tulk v. Moxhay*,[7] the English Court of Equity held that a purchaser of land who knew of the restrictions on the use of the land would be ordered by the Court of Equity not to violate these restrictions even though nothing in the deed restricted the land use. Equitable servitudes avoided technical requirements of covenants "running with the land." They created a viable private system of land use controls that had great impact on the development of cities and that still exist despite the proliferation of public land use controls.

In the United States, an equitable servitude requires an intent to benefit the adjoining land and notice of the servitude or restriction to buyers of the burdened land; that is, the land that is limited in use. Intent to benefit is usually shown by a uniform, common scheme of restrictions. Generally this scheme is shown by a legal document recorded in the land records, although in some cases it can be shown by a uniform pattern of use throughout the development.

Courts have generally given considerable freedom to developers and buyers in creating restrictive covenants. However, contractual freedom is not absolute. In *Shelley v. Kraemer,* the U.S. Supreme Court held that state enforcement of racial restrictive covenants violated the Equal Protection Clause of the Fourteenth Amendment to the federal Constitution.[8]

Should the constitutional tests that determine the validity of *public* land use controls (as seen in Section 9.15) be applied to private restrictive covenants? Some contend that buyers do not have adequate notice of the restrictions, the agreements are contracts of adhesion and are not consensual, and *in effect,* broad-scale enforcement of private restrictive covenants is tantamount to private zoning. Private restrictions, especially those that limit construction to single-family residences, can frustrate plans to have integrated housing in urban and suburban areas.

One writer urged that constitutional requirements for zoning ordinances be applied to decisions of those who administer such systems, such as homeowner associations

---

[7]41 Eng.Rep. 1143 (Ch.1848).
[8]334 U.S. 1 (1948).

and architectural committees, as they often wield power analogous to local authorities.[9] However, courts generally regard restrictive covenants as a useful private planning device and have not as yet been willing to apply public land use standards to reviewing restrictive covenants and actions of homeowner associations.

Private restrictive covenants can be rigid. Restrictions must be effective for a sufficiently long time to protect buyer investments. Yet the future is difficult to predict. To deal with the need for flexibility, many planned developments create homeowner associations composed of owners in the development. These associations can modify existing restrictions under certain circumstances and pass judgment on requests to deviate from the restrictive covenants. They are often given the power to seek injunctions for violations even though they do not own land in the affected area.

While the original reason for restrictive covenants was to protect property values, increasingly, they are created for other reasons. This is demonstrated by the names given to them. One way of referring to them is as gated or walled communities. These names show the emphasis on a sense of security to those who live in them. Those who live in them want a feeling that they will be safe in their homes and in their communities.[10]

At the beginning, these communities created by private restrictive communities were the domain of the more affluent. But now, they are no longer reserved for the rich but are created for different class and economic groups as long as all members share cultural bonds. As a result, they are currently called *common interest communities* (CIC). They include not only gated or walled communities but also condominiums and cooperatives. In a condominium, owners own their apartments, with common areas owned by a homeowners or community association. In a cooperative, the property is owned by a cooperative association with shares in the cooperative owned by those who live there.

CICs seek to create an authentic sense of community much like the villages in preindustrial times. While they are still interested in property values, they also want to create and maintain neighborhoods of people with like interests. They want to create a community. Those who create CICs and those who live in them seek to develop a connectedness that many feel is absent from modern life.

The administration of these communities is done by an ownership or community association created at the outset by those who organize CICs. Ultimately, they are elected by those who live in them.

Community objectives are sought through regulations. Professor Paula A. Franzese has chronicled the life, legal and social, of these CICs. She states:

> In the CIC setting, covenants have been devised to regulate everything from whether pets are limited or prohibited, to the permissibility and style of one's screen and storm doors, to the ration of grass, trees and shrubs allowed on one's property. Restrictions are imposed to regulate the mounting of basketball hoops, the retrieval of dog droppings, the posting of for-sale signs, the trimming of bushes and the color of window curtains. [Regulations] exist mandating that any doghouse must be made of the same material as the master house (either wood or brick) and hidden from view by a six-foot fence or prescribed greenery. Rules exist to prohibit wok cooking, compel 'poorly dressed guests' to ride in service elevators and ban those wearing 'flip-flops' from sitting in common-area chairs.[11]

She chronicles the conflict and litigation that often resulted from disagreements between neighbors and between residents and the association.

This was illustrated by conflicts that developed in the exclusive enclave of Edgewater Point on a spit of land called Satanstoe on Long Island.[12] One owner replaced a wooden pier with a $2,500 cement one. This angered one neighbor. This dispute "expanded to include everything from the placement of trash cans in the neighborhood to whether one's in-laws can live in the community."

Hundreds of thousands of dollars and a dozen or so lawsuits later the dispute continued and expanded to include allegations that garbage sheds and outbuildings violated local building rules.

The creating of a CIC through detailed regulations may not achieve its objectives.

---

[9]Comment, 21 UCLA L.Rev. 1655 (1974).

[10]Much of this draws upon the work of Professor Paula A. Franzese in *Does It Take a Village? Privatization, Patterns of Restrictiveness and the Demise of Community*, 47 Vill.L.Rev. 553 (2002).

[11]Supra note 10 at 555–556 (footnotes omitted).

[12]Mr. Millstein's Backyard Brawl—Weil Gotshal Partner Tried to Broker Peace Among His Neighbors but Got Burned," *Wall Street Journal*, June 20, 2000, p. C-1.

Returning to restrictive covenants, they have begun to deal with aesthetic and architectural aspects of a development. In *Hanson v. Salishan Properties, Inc.*,[13] leases for beachfront lots contained provisions dealing with design aspects that affected the view of others in the development. In addition to being required to comply with specific covenants set forth in the lease relating to these matters, tenants were referred to an architectural checklist drafted by the architectural committee that set forth general concepts relating to aesthetics and view for the development. The lease required that plans be approved by the architectural committee.

A tenant submitted preliminary plans for a home that were approved by the architectural committee. Some upland tenants sought an injunction against construction of the home that they claimed violated the restrictive covenants. The Oregon court concluded that the committee acted properly and that its decision would not be disturbed. The court did not have to articulate standards for judicial review of decisions of architectural committees because the court concluded that the architectural committee's decision was correct.

The association or committee generally must act reasonably and in good faith. It must act in accordance with predetermined standards. If there are no predetermined standards, its decisions must be consistent with the standards set forth in the original declaration of restrictions and with existing neighborhood conditions.

Sometimes private restrictive covenants prohibit owners from modifying or changing the exterior in such a way as to detract from architectural harmony or historic tradition. For example, *Gaskin v. Harris*[14] involved a restrictive covenant that required all exteriors to be Old Santa Fe or Pueblo-Spanish architecture. One homeowner wished to build a modern Oriental swimming pool, and other homeowners objected. New Mexico upheld the restrictive covenant despite the pool builder's claim that the pool had already been constructed and that the restrictive covenant was too vague to be enforceable.

Other techniques, though used less often, can soften some of the rigidity of restrictive covenants. A landowner burdened by a restrictive covenant can seek relief or may resist an injunction sought by other landowners by showing that conditions have so changed that the servitude should be removed or modified. The court will not order an injunction if the conditions in the neighborhood have so changed that enforcement of the restrictive covenant would be unreasonable. Denying enforcement on the basis of changed conditions requires a showing that enforcing the covenant would harm the burdened land much more than nonenforcement would harm the other land in the development. Reasons for denial can include a significant increase in noise and traffic, the unsuitability of the burdened land for the restricted purpose, and rezoning of the land for a more permissive use.

Servitudes can be eliminated by agreement of all the affected landowners or occasionally by benefited landowners failing to seek judicial relief for past violations. The servitude can also be eliminated if a public agency takes the land for a use inconsistent with the restriction. This taking is accomplished by condemnation, and property owners that would have benefited by continued enforcement generally receive compensation from the agency that exercises the power of eminent domain.

Despite the proliferation of public land use controls, private restrictive covenants are still important. They are often the principal land use controls in many existing developments and are used in new developments to some degree as protection from the risks of zoning law changes.

SECTION 9.06 Development of Land: Expanded Public Role

In the individualistic system preceding restrictive covenants and after the development and enforcement of restrictive covenants, the law had a minimal and mainly passive role. Individual judgment and the free market predominated. The present dominant system of land use control has expanded the law's role. Today the state itself is the biggest player in deciding the permissible use of land.

It is essential to compare the passive approval of private restrictive covenants and the more active public developmental control to understand why the latter has dominated modern land use controls. The essentially political nature of public control is discussed in Sections 9.07 and 9.08.

---

[13]267 Or. 199, 515 P.2d 1325 (1973).
[14]82 N.M. 336, 481 P.2d 698 (1971).

Other differentiations exist between private restrictive covenants and public land use controls. The principal objective of restrictive covenants is to preserve property values and maintain the characteristics of the neighborhood. Although public land use controls take these factors into account, local authorities must also concern themselves with fiscal matters. Because property taxes have been the principal revenue-raising measure available to local authorities—at least until the taxpayer revolt against property taxes in the late 1970s—permitting industrial, commercial, or expensive residential use can increase the tax base and raise revenue.

Local authorities must furnish fire and police protection, educational facilities, and parks as well as water and sewage facilities. Local authorities are also responsible for the overall quality of life within and around the community. They must consider community-wide problems such as housing and crime. Public land use planning is more complicated and difficult than private controls.

Flexibility is another differentiating aspect of public land use controls. Although some flexibility is needed in private restrictive covenants, the private system emphasizes stability. In contrast, public controls must take into account and devise methods for dealing with the changes that occur at an astounding pace in urban life.

The developer thinks mainly of profits and creates restrictive covenants to sell lots or homes. The public planning process should marshal the best thinking of many disciplines to accomplish legitimate planning of objectives. One judge described the public planning process as bringing to bear on planning decisions "the insights and the learning of the philosopher, the city planner, the economist, the sociologist, the public health expert and all the other professions concerned with urban problems."[15] For these and other reasons, the passive role did not seem adequate. As a result, the government took primary responsibility for urban land development, which then became a political process.

Legislatures, both state and local, consist of people elected by and responsible to voters. The public land use control system that evolved in the twentieth century not only was representatively democratic but also often became directly democratic through decisions made by the voters themselves.

SECTION 9.07   Limitations on Land Use Controls: Takings

The power to control public land use essentially belongs to each state. But exercise of state powers is limited because property rights have received special protection in the United States. Owners cannot be deprived of their property without due process of law. If public regulation has placed too great a limit on an owner's right to use his land, such a limitation can constitute a "taking" of the property.

Takings can take two forms. There can be a physical taking. The state may use its powers of eminent domain and condemn, that is, take over the property. If it does, the Fifth Amendment of the U.S. Constitution requires that it pay compensation. The only issues in such physical takings are whether the "takeover" is for a public purpose and the value of the land taken.

The first, the requirement of a public purpose, presents no difficulty if the state appropriated the land for schools, roads, or parks. But increasingly, the state, particularly cities, seeks to take away land from one private person and hands it over to another who it believes will make better economic use of the land. When a city seeks to take away land from one private owner and give it to another, the city will contend this appropriation is made for a "public purpose." In *Kelo v. City of New London, Connecticut,* the U.S. Supreme Court interpreted the term "public purpose" broadly to encompass the destruction of nonblighted private homes as part of a city's economic development plan.[16] A backlash to the decision by property rights activists has led to local proposals to restrict municipalities from engaging in similar economic development plans.[17] Some state courts have rejected the premise, accepted in *Kelo,* that economic benefit alone satisfies the "public use" requirement.[18]

Most cases in the past involved physical takings. The more recent cases have involved regulatory takings. If the

---

[15]*Udall v. Haas,* 21 N.Y.2d 463, 235 N.E.2d 897, 900, 288 N.Y.S.2d 888, 893 (1968).

[16]545 U.S. 469, reh'g denied, 545 U.S. 1158 (2005). The Court relied in part upon *Berman v. Parker,* discussed in Section 9.11.

[17]A criticism of proposed legislative restrictions on *Kelo* is provided in Burke, *Much Ado About Nothing*: Kelo v. City of New London, Babbitt v. Sweet Home, *and Other Tales from the Supreme Court,* 75 U. Cincinnati L. Rev. 663 (2006).

[18]*City of Norwood v. Horney,* 110 Ohio St.3d 353, 853 N.E.2d 1115 (2006); see also *County of Wayne v. Hathcock,* 471 Mich. 445, 684 N.W.2d 765 (2004) (preceding *Kelo*). See generally Annot., 21 A.L.R.6th 261 (2007).

state regulates the use of land in such a way as to deprive the owner of any significant use of his land, he may assert that the state has taken his land and must pay compensation. Essentially, these regulatory taking cases have been decided on a case-by-case basis, not by any fixed rules, but by an evaluation of the facts in the case. Because of the complexity of this subject, this section will look at illustrative cases of what constitutes a taking and some recent cases on temporary takings and whether that will give the owner a right to any compensation.

In 1992 the U.S. Supreme Court examined the issue of whether land use control or environmental rules have so reduced the value of the land to the owner that they constitute a taking under the Fifth and Fourteenth Amendments to the U.S. Constitution, entitling the landowner to just compensation. *Lucas v. South Carolina Coastal Council*[19] involved lots on the Isle of Palms, east of Charleston, South Carolina. Although the beachfront had been a critical area since 1977 and subject to regulation, the lots were landward of the restricted area when Lucas bought them, and were zoned for development as residential home sites.

In 1988, pursuant to a new beachfront statute, construction of improvements, with some exceptions, was prohibited in such a way as to bar Lucas from developing his property. The statute was designed to protect life and property by creating a storm barrier that dissipates wave energy and contributes to shoreline stability, both to protect the tourist industry and to provide a habitat for plants and animals. Lucas conceded that the ban was necessary to protect life and property against serious harm but asserted that the new law had completely extinguished his property's value, entitling him to compensation. The trial court found that the ban had made Lucas's lots valueless, but the South Carolina Supreme Court reversed the determination.

By a five-member majority, the U.S. Supreme Court held that compensation would be required unless "A law or decree … do no more than duplicate the result that would have been achieved in the courts … under the State's law of private nuisance or by the State's power to abate a public nuisance."[20] The U.S. Court remanded the case to the South Carolina Supreme Court to determine what existing law permitted.

The U.S. Supreme Court in the *Lucas* case was criticized for its failure to articulate rules that would determine whether there had been a taking. The implementation of its holding was somewhat clarified in *Reahard v. Lee County*,[21] which involved the creation of open space (also discussed in Section 9.10H). As that section will show, public entities sometimes seek to preserve open space by barring any development of particular space. Reahard's parents had purchased 540 acres in 1944. During the mid-1970s, the family had subdivided, developed, and sold tracts of the parcel, retaining approximately forty acres. Reahard inherited the site in 1984 and sought to continue development for single-family homes. However, in December 1984 Lee County (Florida) classified the forty acres as part of a resource protection area, which limited the forty acres to a single residence or for uses of a "recreational, open space, or conservation nature."[22]

Reahard filed a complaint, contending that this had been a taking, entitling him to compensation. Among the trial court's findings was one that simply stated that as a result of the adoption of the plan, there was "a substantial deprivation of the value" of Reahard's property, resulting in a taking.[23] The federal appellate court for the Eleventh Circuit initially noted the difficulty of determining whether a regulation "goes too far."[24] The court must determine whether the regulation has denied an owner economically viable use of his own property. To do so, the court must determine the economic impact of the regulation on the property owner and the extent to which the regulation has interfered with investment-backed expectations. The appellate court held that the trial court neither analyzed the factors nor set forth findings necessary for such analysis. A proper taking analysis, according to the appellate court, would address certain questions:

In this case, those questions are: (1) the history of the property—when was it purchased? How much land was purchased? Where was the land located? What was the nature of title? What was the composition of the land and how was it initially used? (2) the history of development—what was built on the property and by whom? How was it subdivided and to whom was it sold? What plats were filed? What roads were dedicated? (3) the history of zoning and regulation—how and when was the land classified? How was use proscribed? What changes in classifications occurred? (4) how did development

---

[19]505 U.S. 1003 (1992).
[20]505 U.S. at 1029.

[21]968 F.2d at 1131, supplemented, 978 F.2d 1212 (11th Cir.1992).
[22]968 F.2d at 1133.
[23]Id. at 1134.
[24]Id. at 1135.

change when title passed? (5) what is the present nature and extent of the property? (6) what were the reasonable expectations of the landowner under state common law? (7) what were the reasonable expectations of the neighboring landowners under state common law? and (8) perhaps most importantly, what was the diminution in the investment-backed expectations of the landowner, if any, after passage of the regulation?[25]

The court then remanded the case to the trial court to address these issues and determine whether there had been a taking. Further analysis of these questions ended when the federal court of appeals, in a subsequent appeal following the remand, decided that the owner's remedy lay with the Florida courts and that the federal courts therefore lacked jurisdiction (or authority) to rule on the matter.[26]

Regulators must consider whether the regulations proposed will constitute a taking. If so, enacting the regulations will be costly. Yet these standards still leave a considerable amount of room for regulators to enact land use and environmental controls.

The risks of the controls so diminishing the value of the property so as to constitute a "taking" have led to new techniques, such as transferable developmental rights, known as TDRs, a technique to be noted in Section 9.12. The operation of TDRs was before the U.S. Supreme Court in *Suitum v. Tahoe Regional Planning Authority*.[27]

In the *Suitum* case the Tahoe Regional Planning Agency barred the development of Suitum's undeveloped land near Lake Tahoe to protect runoff into the watershed. But owners whose land was barred from development for environmental reasons could receive a TDR. Such rights make it easier to develop other undeveloped land. TDRs can be transferred. In effect, they have economic value. Granting TDRs to those who are not allowed to build on their land is designed to avoid the restriction being a "taking." (The majority held the case was "ripe" for adjudication but did not rule on the validity of the TDR method to avoid the control amounting to a "taking" entitling the owner to compensation.)

Lake Tahoe generated another important U.S. Supreme Court decision. This issue before the court in *Tahoe-Sierra*

Preservation Council, Inc. v. Tahoe Regional Planning Agency[28] was whether two moratoria totaling thirty-two months on any development in the Lake Tahoe basin (the dissent concluded that the land was tied up for almost six years) was a taking requiring that the landowners be compensated. The moratoria were to give the regional planning authority time to formulate a land use plan for the area.

The challengers claimed the taking question was governed by *Lucas v. South Carolina Coastal Council*[29] discussed earlier in this section. In the *Lucas* case the Court found there had been a total taking of Lucas's land. The challengers in the *Tahoe-Sierra* case asserted that the *Lucas* case established a categorical *per se* rule. If a regulation permanently denies the owner all productive use of an entire parcel, this is a taking. Even if this denial is temporary, as in the *Tahoe-Sierra* case, the owner must be compensated. Under a categorical *per se* rule, there is no inquiry into the owners' investment-backed expectations, the actual impact on the individual landowner, the importance of the public interest served by the regulation, or the reasons for the temporary restriction.[30]

The Court, in a six-to-three opinion, rejected this contention. Instead, it chose to follow an *ad hoc* balancing approach used by the court in the *Penn Central* case.[31] While the Court did not overrule the *Lucas* case, it limited the categorical *per se* rule announced there to extraordinary circumstances when no productive or economically beneficial use is permitted. This *ad hoc* balancing approach is illustrated by the *Reahard* case[32] discussed earlier in this section.

The Court rejected the landowners' contention that the 32-month moratoria removed a segment of the landowner's fee simple (total ownership). If any use during such a segment is barred by the moratoria, there is a taking, claimed the landowners. The Court stated that the temporary nature of the restriction does not control the taking question one way or the other.[33] The Court thought thirty-two months a long time, but the trial court found that period reasonable, especially as the regional authority was considering a comprehensive plan for an environmentally sensitive area.[34]

[25]Id. at 1136.
[26]*Reahard v. Lee County*, 30 F.3d 1412 (11th Cir.1994), cert. denied, 514 U.S. 1064 (1995).
[27]520 U.S. 725 (1997).
[28]535 U.S. 302 (2002).
[29]505 U.S. 1003 (1992).
[30]535 U.S. at 320–21.
[31]*Penn Central Transp. Co. v. New York City*, 438 U.S. 104 (1978).
[32]Supra note 21.
[33]535 U.S. at 331–32.
[34]Id. at 341–42.

The case-by-case approach, the immense variety that regulation can take, and the different impact such regulation can have on land ownership means that "taking" questions will continue to be a complex, controversial area of constitutional law.[35]

To avoid these complicated taking issues, most public control over land use is predicated on the police powers of the state to protect its citizens and the state's concern for the general health, safety, and welfare of the community. To bring legislated land use controls within the police powers of the state, the state legislation stresses objectives such as reduced street congestion, better fire and police protection, and promotion of health and general welfare as the reasons for restricting land use.

SECTION 9.08 Local Land Use Control: The Process

Land use control was authorized by state enabling acts. They expressed goals in general terms and allowed local authorities to make specific rules and administer them. Generally, local authorities exercised this power. They enacted and administered ordinances to regulate land development. The traditional method for accomplishing this was to create districts in which only certain activities are permitted.

The enabling acts passed in the 1920s recognized that much of the work needed to create and administer public land use controls could not be done by the local legislative governing bodies. These bodies generally are composed of unpaid or modestly paid citizens elected to local legislative bodies. The elected officials would not have the time or expertise to actually draft and administer a development system.

The enabling acts allowed two agencies to be created to deal with these matters. A planning commission was authorized to draft a master plan and detailed ordinances. A board of adjustment was authorized to deal with appeals from decisions by the local administrator or public officials. The structures, personnel, and names of these agencies varied considerably from locality to locality.

Members were generally given tenure for a specific number of years to insulate them from improper pressures. Beyond tenure requirements, localities varied in the attributes of such agencies. Compensation, qualifications, staff, and support largely depended on the size of the locality. Even in larger cities, however, members of planning commissions frequently were interested citizens who served without pay and often with minimal staff. The final local authority in enacting ordinances and appeals from decisions was usually vested in the local governing body such as the city council.

At this point and in other parts in this chapter, reference is made to the Model Land Development Code (MLDC),[36] a model enabling act approved in 1975 by the American Law Institute, a group of scholars, lawyers, and judges. The MLDC gives greater responsibility to the local development authority (LDA), the agency implementing and operating the development program. The code removed the local governing body, such as the city council, from decisions relating to specific development plan proposals and day-to-day operation of the system given to the LDA.

Development control is given to local authorities because land use is largely a local matter. Local decision makers are expected to be knowledgeable about the community and its needs and represent the people who will be most affected by the development rules. Should an increasingly interdependent urban society give such controls to often small political units? Insularity and unwillingness to take metropolitan social needs into account have led to increased demands for regional or statewide controls. The MLDC has recognized this, and certain provisions are designed to encourage increased activity by regional and statewide authorities.

In contrast, criticism has been made of control by governing bodies of large cities that can be insensitive to the needs of local neighborhoods. Although local neighborhoods were not given decision-making powers in the MLDC, neighborhood organizations were recognized and given greater stature.

The Copley Place project in Boston, a private project costing half a billion dollars, demonstrated the increased

---

[35]*Lingle v. Chevron U.S.A. Inc.*, 544 U.S. 548 (2005) is the most recent Supreme Court takings case. It dealt with an attempt by Hawaii to limit the rent dealers paid Chevron to 15% of the dealer's gross profits on gasoline sales and a similar cap on other sales. The Court held that a test used in many cases, "substantially advances legitimate state interests," would no longer be used in takings cases. The Court stated that earlier takings cases would not be affected by its decision.

[36]This will also be referred to as "the code." This code is not law. It can be adopted in whole or in part by state legislatures. Sometimes such model codes have great influence; sometimes they are ignored.

activity of neighborhood groups in the 1970s. Before the state authorities would grant needed air rights over a freeway, they required the developer to negotiate agreements with surrounding neighborhoods. From 1977 to 1980, fifty public hearings were held with neighborhood groups representing a wide range of interests. Issues thrashed out related to the scale of the project and its effect on neighborhoods, traffic, parking, shadows, wind generation, and jobs. Such processes are repeated in many cities where large-scale development is proposed.

Land use legislation and administration engender controversy. Much of land value depends on use, and land can involve large economic stakes. Such decisions often polarize the community. Land use debates frequently feature clashes between those who favor growth and those who oppose it and between those who advocate the rights of landowners and those who champion social control through the political process. Land use controls can exclude certain people and activities (discussed in greater detail in Section 9.15). From a business vantage point, controls can restrict competition.

Udall v. Haas[37] illustrates some of these features. The municipality had enacted a building zoning ordinance that reclassified a landowner's property from business to residential. The new classification would have permitted only public and religious buildings and residences with a designated minimum size. The event that led to the rezoning appears to have been the submission to local authorities of a preliminary sketch for a bowling alley combined with either a supermarket or a discount house to be built on a vacant lot zoned for business use. The powerful economic effect of zoning on property value can be demonstrated by the observation made by the New York Court of Appeals that more than 60 percent of the value of the land, or over $260,000, was wiped out by the rezoning. Describing the process by which the rezoning was accomplished as a "race to the statute books," the court invalidated the rezoning as not being in compliance with the master plan.

Although the basic power to regulate land use resides in local legislative bodies, citizens are given greater direct power over such decisions, a power used increasingly to limit growth. Sometimes local voters can use the initiative process. This is direct enactment of law by the voters. The process requires the filing of a designated

number of signatures—usually 5 percent to 25 percent of the voters—on petitions to place a proposed local ordinance on the ballot. Voters then determine whether the proposed ordinance will be adopted.

Sometimes after a local ordinance is passed by the local legislative body, direct voter power can be exercised by a referendum that submits the ordinance to the voters for review. The process can begin by presenting a petition with a designated number of voter signatures within a certain number of days after the passage of the challenged ordinance or permit decision. Typically, the petition must be filed before the effective date of the ordinance or permit. The petition compels the local governing body to repeal the ordinance, change its decision on the permit, or place the issue on the ballot for voter decision.

Both initiative and referendum can be costly and not likely to be undertaken unless stakes are high or citizens are greatly distressed with decisions made by their city council or board of supervisors.

Increasingly, attacks are made on land use controls as unconstitutional. See "takings" discussed in Section 9.07.

SECTION 9.09 Original Enabling Acts and Euclidean Zones: The MLDC

Most zoning enabling statutes passed in the 1920s dealt mainly with the physical characteristics of development. They tended to divide the territory of the municipality into districts of different contemplated uses. Each district was to have uniform regulations. This system was sometimes called Euclidean zoning after the U.S. Supreme Court case that first validated a zoning plan.[38] Euclidean zones required homogeneous use. Only particular uses were permitted in each district. For example, residential districts permitted only residences, commercial districts only commercial activities, and industrial districts only industrial activity. Major categories were further divided. For example, industry would be divided into heavy industry and light industry. Commercial use was divided into different categories. Residential use was divided into single-family homes, two-family homes, and multiple-family uses. Homogeneous districts were based on the assumption that differing uses within a district would harm property values.

---

[37]Supra note 15.

[38]*Village of Euclid v. Ambler Realty Co.*, 272 U.S. 365 (1926).

Some cities allow multiuse districts, or what is sometimes called *cumulative zoning*. For example, residences only might be permitted in the residential zones, but commercial zones would also permit residences and industrial zones would permit residential and commercial uses as well.

In addition to creating districts of permitted use, the enabling acts of the 1920s regulated matters such as height, bulk, and setback lines. Rigid patterns of land and development were assumed. Single homes were to be placed on gridlike lots.

The assumption of most early enabling acts was that landowners could develop their land as they wished provided they did not contravene specific restrictions expressed in the ordinances. The statutes authorized but did not compel the local authority to control development decisions. There was no exhortation or incentive to owners to undertake desirable development. The statutes expressed prohibitions designed to avoid undesirable development.

The original enabling acts were relatively indifferent to the effect of no or poor local planning on areas outside the municipality. Local public interest dominated.

Much has changed since the 1920s, and the development of the MLDC was a response to such changes. The code seeks to go beyond simply prohibiting poor planning and establish guidelines for desirable planning. The code recognizes the need for regional and statewide planning. It seeks to develop flexibility and new planning methods (discussed in Section 9.10) to replace the rigid district techniques of the original enabling statutes. The code more openly recognized aesthetics, environmental problems, and the preservation of historical sites as proper planning and developmental factors (discussed in Sections 9.11, 9.12, and 9.13).

SECTION 9.10 Flexibility: Old Tools and New Ones

As stated in the preceding section, the controls of early zoning rarely went beyond height, bulk, and setback regulations. But the ugly, inefficient, and expensive explosion of urban sprawl just before and after World War II compelled planners to find methods to improve the quality of urban life. Planners were asked to enact land use controls that would create and preserve a healthy, aesthetic community with proper regard for the finite quality of resources and the historical patrimony of the community. This and the following two sections deal with traditional

and new methods to accomplish these more ambitious objectives.

A. Variances and Special Use Permits

A variance permits the parcel of land to be used otherwise than prescribed in the zoning ordinance. Issuance requires the landowner to meet requirements that will appear in the cases discussed later in this subsection. Usually the landowner asserts that the use permitted by the ordinance was not economically feasible and this would cause great hardship.

Zoning ordinances often allow permits to be issued for a special use. These permits can be issued by the local land development authority only if the special uses are set forth in the ordinance. Sometimes the zoning ordinances spell out specific uses that can justify the issuance of a special use permit. Sometimes the standards for issuing these permits are described in general terms.

Ordinances often distinguish use and area variances. To avoid an unconstitutional confiscatory ordinance, use variances are given if the land as zoned cannot yield a reasonable return. But because a use variance is a more drastic disruption of planning, a stronger burden is placed on the landowner who seeks a use variance than where an area variance is sought.

B. Nonconforming Uses

Although not considered a technique for avoiding the overrigidity of the Euclidean grid, the nonconforming use has had that effect. Most zoning enactments created carefully segregated districts, but the cities on which these grids were to be imposed did not follow such a neat arrangement. Junkyards existed in residential or shopping areas. Stores were found in residential areas and factories next to retail sales stores. To ensure similarity of use within districts, early planners sought to eliminate nonconforming uses. Although planners felt that police powers of the state could justify elimination of existing uses as well as prohibition of future uses, the political process dictated that the two be treated differently. To avoid entire zoning plans being turned down by local governing bodies, a system for protecting existing lawful uses was developed.

At first rules were adopted based on the hope that time would eliminate nonconforming uses. These rules precluded nonconforming uses from being changed. Nonconforming structures could not be altered, repaired, or restored, and a use could not be reinstituted after it had been abandoned. Implementing these rules was difficult. Defining a nonconforming use was not easy. Things became more complex when courts sought to differentiate a nonconforming use in a conforming building, from a nonconforming use in a nonconforming building.

In any event, nonconforming uses did not wither away. Permission for a nonconforming use created a monopoly, such as a small grocery store in a residential neighborhood. The next attack was to enact amortization ordinances ordering nonconforming uses to cease after a period of time during which the nonconforming user could recover its investment. Amortization periods ranged from one year for billboards to twenty-five years for gas stations to from fifty to sixty years for substantial buildings.

Although courts generally enforced amortization ordinances, judicial opinions expressed skepticism about the fairness of such ordinances. Research revealed a patchwork quilt created by variances and special use permits rather than homogeneity of uses. Some expressed doubt regarding the desirability of homogeneous zones. As a result of these factors, elimination of nonconforming uses by amortization looked less attractive. Put another way, can there really be nonconforming uses in an urban environment that essentially was nonconformist?

C. Rezoning

The only alternatives for a landowner who cannot meet the often stringent requirements for a variance and who does not wish or is not able to obtain a conditional use permit is to seek rezoning. Rezoning can cause a hardship to the neighboring property owners, and under rigid Euclidean theory, an all-or-nothing choice must made between the landowner and the neighboring property owners. As to conditions for rezoning, see Section 9.10D.

D. Contract and Conditional Zoning

Some have proposed increased use of contract or conditional zoning. Suppose the developer wishes to build a shopping center. The governing body or administrative agency can rezone the land or issue a permit. Contract zoning allows the governing body to obtain in exchange for its action the applicant's promise to do certain things. In conditional zoning or issuance of a special permit, the effectiveness of rezoning or the permit is conditional on the applicant's doing certain things required by local authorities. The promised acts or conditions can deal with noise abatement, traffic control, setback lines, erection of fences, or any other devices to enable the center to blend into the surrounding neighborhood. As an alternative to making promises to do these things, the developer can promise to pay to have these designated things done by the local authority.

Currently, conditions for approval of a large-scale project such as a mixed-use development or a high-rise office building may include protection of historic views, employment preferences to local citizens and minorities, inclusion of low-income housing units or subsidy for them elsewhere, or a promise of child care facilities at the proposed project or elsewhere.

Although contract or conditional zoning gives flexibility, courts have divided as to its validity. Contract zoning is especially vulnerable to attack, because it appears the lawmaking power is being traded away. As a result, conditional zoning is more likely to be the method selected.

As to development conditions, the U.S. Supreme Court requires that the condition be directly related to a detrimental aspect of the project and that it must substantially advance a legitimate state interest.[39] This ruling has raised concern among city planners that cities could be liable for monetary damages to developers if they impose conditions that the court subsequently finds to have no substantial connection to the burdens caused by the development project. Other recent cases have passed judgment on this "connection" issue. *Dolan v. City of Tigard* involved approval of an owner's plan to expand her store and pave her parking lot conditioned on her dedicating some of her land for a public greenway and for a pedestrian/bicycle pathway.[40] *Ehrlich v. City of Culver City* involved imposing a mitigation fee as a condition to rezone a private tennis club and recreational facility to permit construction of a condominium project. The fee would be to provide recreation facilities and to provide for art in public places.[41] *Town of Flower Mound v. Stafford Estates Ltd. Partnership*

---

[39]*Nollan v. California Coastal Commission*, 483 U.S. 825 (1987) (California Coastal Commission granted a permit to an owner of a beachfront lot with a small house on it to build a larger house on condition the owner allowed the public an easement to pass across its beach, which was located between two beaches: action of commission constituted a "taking," permit condition struck).

[40]512 U.S. 374 (1994) (required dedication an unconstitutional taking.)

[41]12 Cal.4th 854, 911 P.2d 429, 50 Cal.Rptr.2d 242, cert. denied, 519 U.S. 929 (1996) (art exaction upheld but not recreation fund fee).

involved requiring a builder of a housing development to destroy a functioning, asphalt abutting road and rebuild it as a concrete road.[42]

Section 2-103 of the MLDC allows the imposition of conditions for issuing a special development permit. These conditions can include the requirement that the developer set up mechanisms for future maintenance of the property, such as a homeowner association.

E. Floating Zones

Floating zones provide flexibility within fixed ones. The boundaries of such districts are not determined by ordinance but are fixed by approval of a petition by a property owner to develop a specific tract for a specifically designated use.

F. Bonus Zoning

The developer may claim its apartment has special features that justify an area variance. In some cities, a developer can receive dispensation from normal development requirements by providing bonus features. For example, in 1971, the Bankers Trust Building in New York City was given certain planning dispensations in tower height and floor area in exchange for including a large, elevated open plaza with desirable architectural features and a two-level covered arcade of shops. The bonus features were attractive to the developer not only because it could get dispensation mainly in floor area but also because those features provided amenities that could make the project more attractive to tenants and obtain a higher rental.

G. Planned Unit Development

The gridlike Euclidean zoning, with its emphasis on individual lots, is inappropriate for planning large developments on unimproved land. Section 2-211 of the MLDC authorizes the designation of specialized planned areas in which development will be permitted only in accordance with a plan of development for the entire area. The drafters of the code contemplated relatively undeveloped land where local planners anticipate some development in the future but wish to discourage small scattered uncontrolled developments. The designation can also be applied to urban areas anticipating major development.

If a specially planned area has been designated, no development can take place until the local development agency adopts a precise plan for the area, which may include street locations, utilities, dimensions and grading of parcels, and siting of structures as well as the location and characteristics of permissible types of development. Developers owning land in the area can obtain a development permit if their development is consistent with the plan specified in the planned area ordinance.

H. Open Space

Various techniques have evolved to preserve open space. Greenbelts, developed in England, are buffer zones of open space between developed areas. Cluster zoning allows more than concentrated density, usually in the form of high-rise multiple dwellings, in exchange for commonly used open space. The overall density does not exceed density limits.

A Seattle greenbelt ordinance constituted a taking (later overruled) of private property without just compensation, violating the Washington constitution and the Fifth Amendment of the U.S. Constitution. The ordinance deprived landowners of all profitable use of a substantial portion of their land. Under the ordinance, 50 percent to 70 percent of greenbelt zone lots had to be preserved in or returned to a natural state. The owner of a lot located within the greenbelt zone could not make any profitable use of that portion of land required to be preserved under the ordinance.[43]

For a decision involving an attempt to create open space, refer to *Reahard v. Lee County*,[44] discussed in Section 9.07. The *Reahard* case noted the desire on the part of local authorities to limit land development and to preserve open space, often at the expense of property owners. As in many of these taking cases, the issue is whether the loss in value should be borne by the property owner or whether the city should exercise its power of eminent domain and pay just compensation. What cannot be

---

[42]135 S.W.3d 620 (Tex.2004) (rebuilding road using concrete was not proportionate to the town's legitimate interests and so was an unconstitutional taking).

[43]*Allingham v. City of Seattle*, 109 Wash.2d 947, 749 P.2d 160 (1988). Overruled by *Presbytery of Seattle v. King County*, 114 Wash.2d 320, 787 P.2d 907 (1980) (holding that barring use of part of a parcel was not a "taking"), cert. denied, 498 U.S. 911 (1990).

[44]Supra note 21.

ignored, however, is the potential impact that a requirement of just compensation would place on environmental protection.

SECTION 9.11 Aesthetics and Control

Early in the history of public land use control, doubts existed as to the constitutionality of limits on property rights designed to create a more aesthetic and pleasing environment. Police powers could be used to restrict development property rights to protect public health, safety, and welfare. But could property rights be limited to accomplish aesthetic objectives?

Berman v. Parker, a 1954 decision of the U.S. Supreme Court that upheld an urban redevelopment program, ended any doubt over the validity of regulation designed to create a more aesthetically pleasing environment. Although aesthetics were not directly involved, the opinion included language that was later used to justify aesthetics as a land use control objective: "It is within the power of the legislature to determine that the community should be beautiful as well as healthy, spacious as well as clean, well-balanced as well as carefully patrolled."[45] In residential developments, regulations have been enacted to promote uniformity by specifying certain design requirements, such as attached garages, two-level houses, or styles of exteriors. Sometimes ordinances sought to avoid subdivision monotony by prohibiting a house from looking too much like the surrounding houses.

An ordinance passed by the Village of Olympia Fields, Illinois, combined both uniformity and nonuniformity by requiring that a permit would not be issued in the case of a design that was excessively similar, dissimilar, or inappropriate in relation to nearby property.[46]

An increasing number of cities have sought to protect views from developers' projects. Illustrations of views protected have been Seattle's nearby bay and mountains, the state capitol in Austin, Texas, the rolling hills outside Cincinnati, the view of the Capitol in Washington, D.C., and the view of the Rocky Mountains from downtown Denver.[47] But the Scottsdale, Arizona, attempt to ban building from a nearby mountain partly within the city was successfully challenged.[48]

SECTION 9.12 Historic and Landmark Preservation

National concern has been increasing over the destruction of historically significant buildings. A survey of historic buildings made in 1933 included 12,000 buildings. By 1970, over half had been razed. The pressure to demolish or substantially alter historic structures is especially strong in fast-changing urban areas. The response has been activity at every governmental level, but especially at the local level of government.

The principal steps taken have been to survey historic landmarks, create historic preservation districts, acquire ownership of historically significant structures or easements over their facades, and grant tax concessions to owners who will preserve the historic character of their structures. Illustrations of historic district designations are Vieux Carré in the city of New Orleans;[49] a four-block area surrounding the Lincoln house in Springfield, Illinois;[50] and the Old Town district of San Diego, California.[51]

In addition to creating historic districts, some local entities have designated specific buildings as historic landmarks despite the possibility that the designation could be considered spot zoning. Illustrations of buildings that could justify such a designation are New York City's Grand Central Terminal;[52] Lincoln's home in Springfield, Illinois; and Monticello, Jefferson's home outside Charlottesville, Virginia.

---

[45]348 U.S. 26, 33 (1954).

[46]The ordinance was found to be invalid as an improper delegation of decision making to the architectural advisory committee without providing adequate standards. See *Pacesetter Homes, Inc. v. Village of Olympia Fields,* 104 Ill.App.2d 218, 244 N.E.2d 369 (1968).

[47]Upheld in *Landmark Land Co., Inc. v. City and County of Denver,* 728 P.2d 1281 (Colo.1986), appeal dismissed sub nom. *Harsh Inv. Corp. v. City and County of Denver,* 483 U.S. 1001 (1987).

[48]*Corrigan v. City of Scottsdale,* 149 Ariz. 538, 720 P.2d 513 (1986) (owner entitled to damages as well as invalidation of ordinance), cert. denied, 479 U.S. 986 (1986).

[49]Upheld in *Maher v. City of New Orleans,* 516 F.2d 1051 (5th Cir.1975), cert. denied, 426 U.S. 905 (1976).

[50]Upheld in *Rebman v. City of Springfield,* 111 Ill.App.2d 430, 250 N.E.2d 282 (1969).

[51]Upheld in *Bohannan v. City of San Diego,* 30 Cal.App.3d 416, 106 Cal.Rptr. 333 (1973).

[52]Upheld in *Penn Central Transp. Co. v. New York City,* 438 U.S. 104 (1978).

Designation usually means that the landowner may not demolish or alter the existing structure. To alleviate possible hardship from such limitations, the ordinances can provide that the owner is expected to realize at least a designated return on the property. If the designation proves an economic hardship because of a lower return, a commission is given discretion to ease the hardship by rebating the real estate tax or the commission is afforded the additional right of producing a buyer or lessee who can profitably use the premises without the sought-for alteration or demolition. If these remedies prove unrealistic or unobtainable, the city is given the power to condemn the property.

The most serious constitutional attack on ordinances creating historical districts or designating landmark buildings has been the claim by the landowner that the use limitation created by the ordinance or designation violates the constitutional prohibition against taking private property without paying just compensation. If the action by local authorities is considered a taking, the owner must be compensated. Because compensation can be costly, especially at a time when local government faces continuing fiscal problems, ordinances and designations are usually based on the police powers. Courts have been hesitant to consider anything short of actual appropriation to be a taking as long as the law can be justified as an exercise of police powers.

Yet the Pennsylvania Supreme Court, in *United Artists Theatre Circuit, Inc. v. City of Philadelphia*,[53] showed some dissatisfaction with the flexibility given public entities to designate a particular building as historic. The Boyd Theater had been designated historic because it was an example of art deco architecture, was done by an important architectural firm, and "represent[ed] a significant phase in American cultural history and in the history of Philadelphia."[54] The four-judge majority of the court asked whether the cost of designation—that is, the reduction in the property owner's value—should be borne by all taxpayers or by the owner. This particular ordinance, according to the court, would prohibit any change except painting or papering—including even the moving of a mirror from one wall to another—without a permit. The majority held that this constituted an unconstitutional taking under the Pennsylvania Constitution. Three concurring judges would

have focused attention on the radical nature of the ordinance and would have limited the ordinance to requiring a permit only for changes to the exterior and to the interior only if they affected the exterior.

After the case was reargued, the court held that designation of a historic building without the owner's consent was not a taking; the court followed the decision of the U.S. Supreme Court in the *Penn Central* case.[55] But the court held that the Philadelphia Historic Commission acted outside its statutory authority when it designated the interior of the Boyd theatre as "historic". The case does reflect concern that the power to declare property as historic can be abused.

A method of relieving against the hardship of placing the entire burden on the owner of the landmark building has been the development in New York City of transferable development rights (TDRs) as a means of preserving historic buildings without incurring the high cost of condemnation. Under this plan, a landmark owner who is prevented from using its property to its full permitted use, such as the maximum floor area ratio permitted by law, is compensated by being given a transferable developmental right for a specific transfer district. The developmental right equals the excess potential the landowner was precluded from using because its building had been designated a historical landmark. The landowner can use this excess potential in the transfer district building even if it exceeds normal density limits for the building. An owner who does not wish to use the development right can transfer it. See Section 9.07 for discussion of this method to protect Lake Tahoe.

SECTION 9.13 The Environmental Movement and Owner Liability

During the late 1960s, the American public began to demand that government make strong efforts to protect the natural environment. The goals of the environmental movement are evident in the preamble of the National Environmental Policy Act (NEPA),[56] enacted by Congress in 1969:

The Congress, recognizing the profound impact of man's activity on the interrelations of all components of the natural environment, particularly the profound influences of popula-

---

[53]528 Pa. 12, 595 A.2d 6 (1991).
[54]595 A.2d at 8.

[55]535 Pa. 370, 635 A.2d 612 (1993).
[56]42 U.S.C.A. §§ 4321–4375.

tion growth, high-density urbanization, industrial expansion, resource exploitation, and new and expanding technological advances and recognizing further the critical importance of restoring and maintaining environmental quality to the overall welfare and development of man, declares that it is the continuing policy of the Federal Government, in cooperation with State and local governments, and other concerned public and private organizations, to use all practicable means and measures, including financial and technical assistance, in a manner calculated to foster and promote the general welfare, to create and maintain conditions under which man and nature can exist in productive harmony, and fulfill the social, economic, and other requirements of present and future generations of Americans.[57]

The strength of this movement has led Congress to enact several key pieces of environmental legislation, including the Clean Air Act,[58] the Clean Water Act,[59] the Resource Conservation and Recovery Act (RCRA),[60] and the Comprehensive Environmental Response, Compensation, and Liability Act (CERCLA).[61] Although these laws provide much-needed protection of our natural resources, they also create significant obligations and liabilities for landowners and developers.

Given their complexity, it would be impractical—if not impossible—to provide a comprehensive discussion of environmental laws and regulations. The following statutes provide insight into the kinds of obligations and liabilities that environmental laws impose on owners and developers.[62]

A. National Environmental Policy Act (NEPA)

NEPA is a broad-based environmental statute that imposes continuing responsibility on the federal government "to improve and coordinate Federal plans, functions, programs, and resources" to protect the natural and human environment.[63] As a means to that end, Section 102 of NEPA requires all federal agencies to

include in every recommendation or report on proposals for legislation and other major Federal actions significantly affecting the quality of the human environment, a detailed statement by the responsible official on—

(i)   the environmental impact of the proposed action;

(ii) any adverse environmental effects which cannot be avoided should the proposal be implemented;

(iii) alternatives to the proposed action;

(iv) the relationship between local short-term uses of man's environment and the maintenance and enhancement of long-term productivity; and

(v)   any irreversible and irretrievable commitments of resources which would be involved in the proposed action should it be implemented.[64]

This detailed statement—commonly known as an environmental impact statement (EIS)—must be made available to the public, the president, and the Council on Environmental Quality and must accompany the proposal through the agency review process.[65]

An owner or developer making a proposal that requires approval by a federal agency faces a number of potential hurdles under NEPA. First, to determine whether the action will "significantly" affect the environment, the appropriate federal agency typically prepares a shortened EIS—commonly known as an environmental assessment (EA). If the agency determines that the proposed action will not significantly affect the environment, the project may proceed without a full-scale EIS. Citizens' groups or other parties may, however, challenge the agency's determination, thereby delaying the project.

Second, if the agency determines that the proposed action *will* significantly affect the environment, the owner or developer must await completion of a comprehensive EIS. The EIS process can be extremely costly and time consuming. Moreover, once the EIS is completed, the federal agency may decide to cancel or significantly alter the proposed action. In addition, if the agency decides to proceed with the action as proposed, citizens' groups or other parties may once again challenge the decision in court.

[57]Id. § 4331(a).

[58]42 U.S.C.A. §§ 7401–7671q (also known as the Federal Air Pollution and Control Act).

[59]33 U.S.C.A. §§ 1251–1387 (also known as the Federal Water Pollution and Control Act).

[60]42 U.S.C.A. §§ 6901–6992k, amending the Solid Waste Disposal Act.

[61]42 U.S.C.A. §§ 9601–9675.

[62]For a comprehensive survey of environmental risks facing contractors, see P. BRUNER & P. O'CONNOR, JR., BRUNER & O'CONNOR ON CONSTRUCTION LAW, §§ 7:89–:105 (1992).

[63]42 U.S.C.A. § 4331(b).

[64]Id. § 4332(C).

[65]Id.

Since its enactment in 1969, NEPA has created considerable litigation; still, very few projects have been enjoined under the statute. When agency determinations are challenged, courts usually extend great deference to agencies, based on the latter's expertise and experience. Therefore, the most significant burden of NEPA appears to be the time and expense of preparing the EA and EIS.

B. Comprehensive Environmental Response, Compensation, and Liability Act (CERCLA)

Overview of Statute. CERCLA, commonly known as Superfund, governs the cleanup of hazardous waste sites. Sections 104 and 106 of CERCLA give the federal government broad power to clean up contaminated sites. Under Section 104, the president is authorized to take removal and remediation action whenever

(A) any hazardous substance is released or there is a substantial threat of such a release into the environment, or
(B) there is a release or substantial threat of release into the environment of any pollutant or contaminant which may present an imminent and substantial danger to the public health or welfare.[66]

Similarly, under Section 106, the president is authorized to "require the Attorney General of the United States to secure such relief as may be necessary to abate such danger or threat" and to "take other action . . . as may be necessary to protect public health and welfare and the environment."[67]

Section 107 of CERCLA imposes liability on four categories of "potentially responsible parties" (also known as PRPs):

(1) the owner and operator of a vessel or a facility,
(2) any person who at the time of disposal of any hazardous substance owned or operated any facility at which such hazardous substances were disposed of,
(3) any person who by contract, agreement, or otherwise arranged for disposal or treatment . . . of hazardous substances . . . and
(4) any person who accepts or accepted any hazardous substances for transport to disposal or treatment facilities, incineration vessels or sites selected by such person, from

which there is a release, or a threatened release which causes the incurrence of response costs, of a hazardous substance. . . .[68]

Courts have interpreted these categories to make up a broad range of parties. For example, the first category of PRPs—owner and operator—has been held to include "current owners and operators of a facility . . . bankruptcy estates, absent landowners/lessors, lessees, foreclosing banks, corporate officers, parents, and successors."[69] As potentially liable parties, owners and developers should be aware of any hazardous substances that may exist on their property.

The costs of liability under CERCLA can be extremely high. Potentially responsible parties are liable for the removal or remedial costs incurred by the federal government, other necessary response costs, damages for injury to or destruction or loss of natural resources, and the costs of certain health assessments.[70] In addition, parties who fail to provide for proper removal or remedial action may be liable for punitive damages of up to three times the amount of costs incurred by the federal government as a result of such failure.[71] Moreover, parties may be subject to costly civil and criminal penalties for violating CERCLA's notification, recordkeeping, or financial responsibility requirements and for violating or refusing to obey settlement agreements, administrative orders, consent decrees, or interagency agreements.[72]

CERCLA provides few defenses to liability. The statutory defenses—an act of God, an act of war, and certain acts or omissions of a third party other than an employee or agent[73]—are very narrowly interpreted. Liability is both joint and several, and traditional notions of causation carry little weight in CERCLA case law. Consequently, even *de minimis* (very little) contributors may be held liable for the entire cost of cleaning up a hazardous waste site. Moreover, because liability is strict and retroactive, "liability attaches regardless of the time when the material was deposited."[74]

---

[66]42 U.S.C.A. § 9604(a).
[67]Id. § 9606(a).

[68]42 U.S.C.A. § 9607(a).
[69]McSlarrow et al., *A Decade of Superfund Litigation: CERCLA Case Law from 1981–1991*, Environmental Law Reporter 10367, 10390 (1991).
[70]42 U.S.C.A. § 9607(a).
[71]Id. § 9607(c)(3).
[72]Id. § 9609.
[73]Id. § 9607(b).
[74]Nash, *An Economic Approach to the Availability of Hazardous Waste Insurance*, Annual Survey of American Law 455, 463 (1991).

May a PRP who remediates a site seek to recoup those costs from other PRPs? Under the Superfund Amendments and Reauthorization Act of 1986, which amended CERCLA, a PRP who is subject to a civil action under Section 106 has a right of contribution against other PRPs.[75] Under CERCLA itself, a PRP who voluntarily remediates the site may seek to recoup those expenses from other PRPs.[76]

Construction Industry Participants. The wide-ranging scope of CERCLA's strict liability has ensnared prime contractors who unknowingly worked on contaminated land. In a 1992 decision, a federal appeals court ruled that an excavating and grading contractor, who quite unknowingly moved some contaminated soils in the course of his work and who stopped work as soon as he noticed something amiss, was nevertheless a responsible party under CERCLA. Liability was imposed because the contractor was found to be both a "transporter" of hazardous substances and as an "operator" of a facility (the construction site), and his dispersal of the contaminated soil was a "disposal" within the meaning of the statute.[77]

In contrast, design professionals have not been found to be PRPs because they lack control over actual disposal or movement of the hazardous materials. For example, an architect who designed a wood treatment plant that released hazardous wastes when operated, was found not to be a responsible party.[78] Another court found that a consulting engineer, who provided not only the design, but also substantial technical assistance during the construction phase, still lacked the requisite control to merit PRP status.[79]

Insurance Coverage. Are owners and developers, compelled to perform cleanup under CERCLA entitled to coverage (reimbursement) of those costs under their commercial general liability (CGL) policies? Under these policies, the insurer generally provides coverage for "all sums which [the insured] shall become legally obligated to pay as *damages* because of … property damage to which this policy applies." [Ed.: Italics added.] Owners and developers, who were ordered by the EPA or equivalent state agencies to clean up hazardous wastes, sought coverage under their CGL policies. Insurers denied coverage, claiming that their insured's cleanup costs were not "damages" within the meaning of the policy. In a leading 1990 decision, the California Supreme Court, in *AUI Insurance Company v. Superior Court*, ruled that the costs of complying with a government remediation order are "damages" within the meaning of CGL policies.[80] This ruling has been universally adopted.

Must policyholders wait to be sued by the EPA to obtain coverage? The question here is whether an owner's or developer's voluntary remediation cost are "sums … legally obligated to pay as damages" within the meaning of the policy. Limiting the term "damages" to money that a court orders an insured to pay, the California Supreme Court found no coverage in this situation.[81] However, most states have held that an owner or developer, who receives a PRP letter from the EPA[82] and in response undertakes cleanup efforts, is entitled to coverage under their CGL policies. In *Johnson Controls, Inc. v. Employers Insurance of Wausau*, the Wisconsin Supreme Court explained why receipt of an EPA letter marks the beginning of an adversarial relationship between the insured and the EPA, in effect exposing the insured to liability for "damages" within the meaning of the policy:

> PRP notice letters expose an insured to agency action that is not inconsequential to its liability interests. In this case, if Johnson Controls had refused to respond to these letters and refused to become involved in remediation efforts with the EPA or comparable state agencies, then, inevitably, the cleanup and remediation work would have been done by the EPA, the state agencies, or by settling responsible parties, any of whom could have sued Johnson Controls for

---

[75]*Cooper Industries, Inc. v. Aviall Services, Inc.*, 543 U.S. 157 (2004).

[76]*United States v. Atlantic Research Corp.*, 127 S.Ct. 2331 (2007).

[77] *Kaiser Aluminum & Chemical Corp. v. Catellus Dev. Corp.*, 976 F.2d 1338 (9th Cir.1992). This decision has been followed in other federal circuits; see *Tanglewood E. Homeowners v. Charles-Thomas, Inc.*, 849 F.2d 1568 (5th Cir.1988) and *Redwing Carriers, Inc. v. Saraland Apartments*, 94 F.3d 1489 (11th Cir.1996).

[78]*Edward Hines Lumber Co. v. Vulcan Materials Co.*, 861 F.2d 155 (7th Cir.1988).

[79]*City of North Miami, Fla. v. Berger*, 828 F.Supp. 401 (E.D.Va.1993). See also *Blasland, Bouck & Lee, Inc. v. City of North Miami*, 96 F.Supp.2d 1375 (S.D.Fla.2000) (environmental engineering firm, which negligently implemented a remediation plan, is not a PRP under CERCLA).

[80]51 Cal.3d 807, 799 P.2d 1253, 274 Cal.Rptr. 820 (1990).

[81]*Certain Underwriters at Lloyd's of London v. Superior Court (Powerine Oil Co.)*, 24 Cal.4th 945, 16 P.3d 94, 103 Cal.Rptr.2d 672 (2001).

[82]A PRP letter is a letter issued by the EPA notifying the recipient that the EPA considers it to be a potentially responsible party for contamination at a given site. 42 U.S.C.A. § 9622(e).

its share of the costs. PRP letters, which are more analogous to a civil complaint than a traditional demand letter, alerted Johnson Controls that the EPA had begun a legal process to conclusively and legally determine the appropriate "response activities" that liable parties must perform or pay for to abate the pollution at the sites in question. [Citation omitted.]

This is why many courts have concluded that a PRP letter is so adversarial that it constitutes the functional equivalent of a suit and triggers the insurer's duty to defend. In the absence of such a conclusion, the insured has a perverse incentive not to cooperate with government remedial actions until the EPA or a state agency files a civil action in court "to force the insured's compliance with CERCLA." [Citation omitted.] Deliberate non-compliance for the purpose of obtaining a defense from the insurer is completely contrary to public policy.[83]

Brownfields. CERCLA's onerous liability provisions had the unintended effect of causing developers to shy away from potentially contaminated urban sites—popularly called "brownfields"—opting instead to pursue unpolluted green space at the outskirts of towns and cities. In response, CERCLA was amended through passage of the Small Business Liability Relief and Brownfields Revitalization Act of 2001, also known as the Brownfields Act.[84] The Brownfields Act seeks to promote development of these urban sites by creating defenses ("safe harbors") to CERCLA's strict liability.

Three significant safe harbors exist: for innocent owners, for contiguous (neighboring) owners, and for prospective purchasers. Under the first safe harbor provision, an owner will be exempted from CERCLA's joint and several liability scheme if it acquired the land not knowing it was polluted, even after having conducted an appropriate inquiry into its history.[85] The contiguous owner defense applies to owners and developers of property that is adjacent to contaminated land. It protects these persons from responsibility for contamination that was not originally on the property, but rather migrated there from another contaminated site.[86] Both innocent owners and contiguous owners, in order to partake of the safe harbor provisions, must cooperate with EPA cleanup programs, including granting access to the land.

The final safe harbor provision seeks to exempt future purchasers of contaminated property so long as they do not hinder existing or future cleanup operations. This bona fide prospective purchaser defense does not require the purchaser to be unaware of the contamination in the same way that the innocent owner and contiguous owner defenses do.[87] Moreover, a prospective purchaser who discovers contamination during its due diligence inquiry remains protected so long as it does not hinder any EPA cleanup response.[88] By allowing a prospective purchaser to develop a contaminated site without threat of CERCLA liability, this final safe harbor provision may do the most to stimulate brownfield development.[89]

C. Resource Conservation and Recovery Act (RCRA)

Enacted as amendments to the Solid Waste Disposal Act of 1965, RCRA regulates solid waste—primarily hazardous solid waste—"from cradle to grave" (from generation to disposal). RCRA provides specific, albeit confusing, definitions of solid waste[90] and hazardous waste.[91] RCRA directs the EPA to promulgate regulations for these hazardous wastes but provides exemptions for domestic sewage, legal source point discharges, irrigation return flows, and other potentially hazardous materials.

RCRA regulates three categories of hazardous waste handlers: (1) generators, (2) transporters, and (3) owners and operators of treatment, storage, and disposal (TSD) facilities. Hazardous waste generators are required to keep accurate records, store the waste in particular containers, label the waste in a specific manner, provide information about the waste and its characteristics, and use a manifest system to track the hazardous waste until it is delivered to a TSD facility. Similarly, transporters are required to ensure that the manifest system is accurate, that spills are minimized and properly handled, and that hazardous wastes are not mixed or stored in violation of RCRA.

---

[83]264 Wis.2d 60, 665 N.W.2d 257, 284 (2003).

[84] Public Law No. 107-118, H.R. 2869 (2002), codified at 42 U.S.C.A. § 9601 et seq.

[85]42 U.S.C.A. § 9601(35).

[86]42 U.S.C.A. § 9607(q).

[87]42 U.S.C.A. § 9601(40).

[88]42 U.S.C.A. § 9607(r)(1).

[89]Vanderberg, *The Brownfields Revitalization Act of 2001: New Hope for Urban Development*, 23 Constr. Lawyer, No. 3, Summer 2003, p. 39.

[90]42 U.S.C.A. § 6903(27).

[91]Id. § 6903(5).

Owners and operators of TSD facilities are the most closely regulated parties under RCRA. In addition to recordkeeping, storage, labeling, information, and manifest requirements, TSD owners and operators must meet financial responsibility standards and provide records of past regulatory compliance. These additional criteria are intended to ensure that the owners will not become insolvent, leaving behind "orphan" sites for the government to clean up.

Like CERCLA, RCRA presents owners and developers with potentially significant costs and penalties. RCRA authorizes the EPA to issue administrative orders to comply with the law and to assess civil penalties of up to $25,000 per day for violations of the statute. The exact penalty is determined based on a variety of factors, including the potential for harm and the extent of deviation. The scope and severity of liability often make it difficult for businesses to obtain liability insurance. Moreover, federal prosecutors have become increasingly willing to pursue criminal convictions for violations of RCRA and other environmental laws. Although RCRA is primarily targeted at owners and operators of TSD facilities, contractors are not immune from statutory liability. In one case, a demolition contractor soon after beginning work found two canisters in the building. The canisters were marked with a skull and crossbones and had the word "poison" written on them. The contractor did nothing about the canisters for three weeks, at which point they were stolen. The contractor was found criminally liable under RCRA as a storer of hazardous waste without a permit.[92]

D. State Law

The environmental movement has also led state legislatures to enact environmental laws often based on, if not more protective than, federal environmental laws. For example, roughly three-fourths of the states have enacted NEPA–type laws. In California, the law includes detailed requirements for agency procedures, substantive requirements for environmental impact reports (EIRs) and agency review, and judicial review standards.[93] Similarly, many states have enacted stringent hazardous waste laws, modeled after CERCLA and RCRA, that may create additional regulatory burdens for owners and developers. For example, under Massachusetts state law parties liable for hazardous waste cleanup are required to pay the Commonwealth of Massachusetts for costs of assessment, containment, and removal of hazardous wastes and for damages for injury to and destruction of natural resources. Unlike CERCLA, however, Massachusetts law also holds such parties liable to any third party for damages to its real or personal property.[94]

Some states have moved beyond the federal legislation in their regulation of environmental risks. California has responded to "sick building" complaints through enactment of the Toxic Mold Protection Act of 2001.[95] The Act requires the state Department of Health Services to convene a task force to develop permissible exposure levels to mold and to devise remediation standards. A second, new law requires a different task force to study and publish its findings on fungal contamination of indoor environments.[96]

E. Environmentally Friendly Design and Construction

The environmental movement is not limited to imposition of statutory liability, but encompasses as well promotion of environmentally friendly design and construction. This movement first came to national attention with passage—in response to an oil embargo by the Organization of Petroleum Exporting Countries (OPEC) in the early 1970s—of the federal Energy Policy and Conservation Act.[97] California followed suit through amendment of its building codes.[98]

While these statutes focused on individual consumer products (including many found in buildings), a newer approach is to promote efficiency through manipulation of the design itself. The United States Green Building

---

[92]*United States v. Sims Brothers Constr., Inc.*, 277 F.3d 734 (5th Cir. 2001). For an empirical analysis of environmental crime prosecutions (including RCRA), see Brickey, *Charging Practices in Hazardous Waste Crime Prosecutions*, 62 Ohio St. L.J. 1077 (2001).

[93]California Environmental Quality Act, West Ann.Cal.Pub.Res. Code §21000 et seq.

[94]Mass.Gen.Laws Ann. ch. 21E, § 5.

[95]West Ann. Cal. Health & Safety Code §§ 26100 et seq.

[96]Id. §§ 26200 et seq.

[97]42 U.S.C.A. §§ 6201 et seq., passed in 1975. Sections 6291–6295 addresses the energy efficiency of consumer products other than cars; for example, § 6294(a)(E) advocates energy-efficient products, and § 6295(k) addresses low-flush toilets.

[98]West Ann. Cal. Health & Safety Code § 17921.3 mandates the use of low-flush toilets.

Council (USGBC), an organization of industry participants, has established a Leadership in Energy and Environmental Design (LEED) rating system establishing design components for environmentally friendly buildings.[99] A Connecticut statute requires most new state facilities to be designed to comply with or exceed compliance with the LEED rating system, or an equivalent system.[100] Thus, the promoters of this new approach are industry participants, with government mandates coming later.

SECTION 9.14 Judicial Review

Although action or inaction of a local governing body can be challenged politically, such as by initiative, referendum, or the election process, the opponents of an ordinance or those who wish to contest a permit decision often go to court. A preliminary and often serious obstacle is whether the challenger has "standing" to challenge the ordinance judicially.

If there is standing, the principal grounds for legal challenge have been the following:

1. The enabling act did not authorize the act in question.
2. The ordinance or permit decision was unconstitutional.
3. The making of decisions by nonelected officials, such as an architectural review commission, was invalid because the local governing body delegated power without proper standards. Successful judicial challenges to decisions of local authorities have been rare.[101]

In *Village of Belle Terre v. Boraas*,[102] Belle Terre, New York, a village of less than one square mile consisting of 700 people and 220 houses, restricted the entire village to single-family dwellings. The village prohibited three or more unrelated people from living together. The ordinance was challenged by six unrelated students from a nearby university who wished to live together. In upholding the challenged ordinance, the Supreme Court stated,

> A quiet place where yards are wide, people few, and motor vehicles restricted are legitimate guidelines in a land-use project addressed to family needs. . . . The police power is not confined to elimination of filth, stench, and unhealthy places. It is ample to lay out zones where family values, youth values, and the blessings of quiet seclusion and clean air make the area a sanctuary for people.[103]

The court concluded that the ordinance was a rational means of achieving these objectives. This decision will make constitutional attacks on local development control even more difficult than in the past.

SECTION 9.15 Housing and Land Use Controls

A. Residential Zones

When zoning took center stage in the 1920s, one justification for local control was the need to develop and maintain residential neighborhoods that would keep their value and enable people to live among their own kind. In the early days of zoning, this was accomplished principally by limiting residences in certain districts to single-family detached homes and severely curtailing the multifamily living unit.

Social aspects of residential zoning were revealed in the first opinion of the U.S. Supreme Court that passed on the validity of zoning laws. In *Village of Euclid v. Ambler Realty Co.*,[104] the U.S. Supreme Court, after justifying the limitation of property rights by the police power to provide a safe and more pleasant community, expressed the American deification of the single-family home and disdain for apartment living.

Home ownership has always been a prized goal for upward mobile classes and a status symbol of achievement. Before the advent of modern public land use controls, private mechanisms existed to develop and maintain quality residential neighborhoods of individually owned

---

[99]According to the USGBC's website, "LEED promotes a whole-building approach to sustainability by recognizing performance in five key areas of human and environmental health: sustainable site development, water savings, energy efficiency, materials selection, and indoor environmental quality." See http://www.usgbc.org/DisplayPage. aspx?CMSPageID=222, last visited April 2, 2008. An alternative to the LEED system is the Green Globes assessment and rating system, promulgated by the Green Building Initiative. See 31 Constr. Contracts L.Rep., No. 2, Jan. 19, 2007, ¶ 21.

[100]Conn.Gen.Stat.Ann. § 16a-38k(a). See generally, Montez & Olsen, *The LEED™ Green Building Rating System and Related Legislation and Governmental Standards Concerning Sustainable Construction*, 25 Constr. Lawyer, No. 3, Summer 2005, p. 38.

[101]The most successful attacks have been in New Jersey and Pennsylvania. See Section 9.15.

[102]416 U.S. 1 (1974).

[103]416 U.S. at 9.

[104]Supra note 38.

homes. Affluent Americans wanted and were able to buy homes in districts exclusively devoted to attractive, well-built homes on spacious lots. Such home buyers generally sought to live with people of their own socioeconomic background. They built or bought in neighborhoods where their children could play and go to school with children of their own cultural background and social class. The free market encouraged the development and maintenance of such neighborhoods. Only the affluent could afford fine residential homes because, at least in normal times, the demand for such homes was large and the supply limited. As a result, high prices kept out people of low or moderate income.

Buyers wanted assurance that there would be no change in the residential environment. To induce them to buy in such neighborhoods, developers used restrictive covenants. Buyers would be willing to pay higher prices to live in districts that were carefully restricted to quality single-family homes. Buyers wanted assurance that property values would be maintained, that schools would be kept high quality, and that neighbors shared their interests and values.

Political power of local communities to control land use enabled communities to resist market forces that might otherwise have impinged on the model of the single-family detached home. Unlike restrictive covenants, public land use controls clash with free-market concepts. This is demonstrated by the frequent alliance of developers and low-income groups who unite to attack zoning laws.

From its inception, residential zoning excluded certain types of people from designated districts. This was justified as a proper planning control designed to encourage investment and the development of good residential neighborhoods. Not until the 1960s were strong attacks made on zoning laws as excluding low and moderate income families from districts reserved to the more affluent, as discussed in greater detail in Section 9.15C.

B. Subdivision Controls

Many communities grew when developers subdivided raw land for homes. Subdividing increased population and placed added burdens on the community to provide streets, fire and police protection, schools, parks, and other municipal services. Although many of these costs could have been provided by special property assessments, many local communities often exacted conditions for approving subdivisions. These conditions could include dedication of part of the raw land for public use for streets, schools, and recreation. Sometimes in-lieu payments were conditions for subdivision approval. These payments—sometimes called "exactions" or "impact fees"—were used to reimburse the community for providing facilities and services in the subdivision or services that the community had to furnish to residents of the subdivision.

Some states have gone further and have permitted local communities to require land or a money exaction for parks and recreation facilities "to bear a reasonable relationship to the use of the park and recreational facilities by the future inhabitants of the subdivision." Such an exaction was upheld with the suggestion that a city could exact a fee to purchase park land some distance from the subdivision but that could be used by subdivision residents.[105]

Subdivision exactions place on new subdivision residents the cost for additional municipal services they will require. But if exactions substantially increase construction costs and prices of subdivision houses, they can, in addition, be exclusionary, (as discussed in greater detail in Section 9.15C).

C. Exclusionary and Inclusionary Zoning

Attacks on zoning laws began with changing demographic patterns during the mid-1950s. In *Oakwood at Madison, Inc. v. Madison Township*, the court recognized this and stated,

Madison Township, among other municipalities, is encouraging new industry. Industry is moving into the county and region from the central cities. Population continues to expand rapidly. New housing is in short supply. Congestion is worsening under deplorable living conditions in the central cities, both of the county and nearby. The ghetto population to an increasing extent is trapped, unable to find or afford adequate housing in the suburbs because of restrictive zoning.[106]

Focusing on zoning laws as a cause for social injustice took the form of describing white suburbs surrounding central cities as "tight little islands" that refused to do their fair share of housing the poor.[107]

[105]*Associated Home Builders etc., Inc. v. City of Walnut Creek*, 4 Cal.3d 633, 484 P.2d 606, 94 Cal.Rptr. 630, appeal dismissed, 404 U.S. 878 (1971). See also Annot., 97 A.L.R.5th 123 (2002).

[106]117 N.J. Super. 11, 17, 283 A.2d 353, 356 (Law. Div. 1971).

[107]Sager, *Tight Little Islands: Exclusionary Zoning, Equal Protection, and the Indigent*, 21 Stan.L.Rev. 767, 791–92 (1969).

The decline of central cities and the flight to the suburbs caused fiscal problems for central cities. (Beginning in the 1980s, and continuing into the 21st century, some cities saw a reversal of this trend as the middle classes moved back and gentrified decaying inner-city neighborhoods.) Cities were left with the increased costs of social services for the poor and a declining tax base because the middle classes, along with many businesses, moved to the suburbs. Central cities looked on the suburbs as parasite communities composed of people who earned their livings in central cities and used the central cities' cultural and recreational facilities. These parasite communities, however, did not bear their proper share of the urban costs that were increasing because of the need to care for poor people.

Suburbs sought to attract light industry that paid high property taxes but did not require many employees of low or moderate income. The latter were to be avoided. They lived in modest homes or apartments that brought in less revenue than the expenditures—mainly in social services and education—required to take care of such people. People of middle income and above wanted to live with others who shared their social and cultural values. Some techniques used to accomplish these purposes were as follows:

1. large lot requirements
2. minimum house size requirements
3. exclusion of multiple dwellings
4. exclusion of mobile homes
5. unnecessarily high subdivision requirements or in-lieu exactions[108]

Judicial attacks have been made on many ordinances by those who wish to build or those who wish to be able to live in suburbs. Usually these attacks are based on claims that such ordinances violated federal and state constitutions.

In 1965, the Pennsylvania Supreme Court, in *National Land and Investment Co. v. Kohn*,[109] invalidated a four-acre minimum lot requirement. Noting that evaluating the constitutionality of a local zoning ordinance was not easy and declining to be "a super board of adjustment" or "a planning commission of the last resort," the court nevertheless saw itself as a judicial overseer "drawing the limits beyond which local regulation may not go."[110]

Pennsylvania and New Jersey have gone far toward what some have called "socioeconomic public land use controls." Courts in these states have taken an active role in attempting to compel local communities to confront and deal with housing problems. This activistic approach has been criticized as an undue interference in local affairs and an unwarranted attempt by courts to venture into complicated social and fiscal problems more appropriate for legislative treatment. There has been some question regarding the right of courts and legislative bodies to frustrate the desires of people to choose the types of people near whom they wish to live.

Other courts have not yet been as willing as Pennsylvania and New Jersey courts to step into local land use decisions. This is demonstrated by cases discussed in Section 9.15D that permit local communities to limit growth through the land use control process.[111]

The introduction of the term *exclusionary* has been accompanied by increasing use of what is called *inclusionary zoning*. New Jersey compelled each community to do its fair share to house persons of low or moderate income.[112] Failure by local communities in New Jersey to do so can result in the communities' zoning laws being invalidated. This, in a sense, is similar to affirmative action employment programs. The court is asking local communities not simply to refrain from excluding certain people but to affirmatively use efforts to include them. In 1969, Massachusetts attempted to require municipalities to provide for such housing by an "anti-snob" zoning law.[113]

Efforts have been made at local levels of government to include those often excluded. Typically, such ordinances require that developers of more than a certain number of units include a designated percentage for low- and moderate-income tenants. Developers challenge these ordinances as a taking or an unconstitutional deprivation of their due process rights. Courts that uphold inclusionary zoning view these ordinances as a valid exercise of the local government's police powers.[114]

---

[108]*Building the American City*, Report of National Commission on Urban Problems, H.R.Doc. No. 91-34, 91st Cong., 1st Sess. 211–16.

[109]419 Pa. 504, 215 A.2d 597 (1965). For a more recent case, see *C & M Developers, Inc. v. Bedminster Township Zoning Hearing Bd.*, 573 Pa. 2, 820 A.2d 143 (2002).

---

[110]215 A.2d at 607.

[111]Also see *Village of Belle Terre v. Boraas*, discussed in Section 9.14.

[112]*Southern Burlington County N.A.A.C.P. v. Township of Mt. Laurel*, 67 N.J. 151, 336 A.2d 713 (1975).

[113]Mass.Gen.Laws Ann. ch. 40B, §§ 20 *et seq.*

[114] *Home Builders Ass'n of Northern Cal. v. City of Napa*, 90 Cal. App.4th 188, 108 Cal.Rptr.2d 60 (2001) (upholding the ordinance); Annot., 22 A.L.R.6th 295 (2007).

D. Phased Growth

In the 1960s, the once accepted goal of growth as the proven road to prosperity and social mobility began to be questioned seriously. Some people advocated goals that would look less at material measurements such as the gross national product and more at measurements of the quality of life. Different measurement processes and the realization of the finite nature of resources were factors that led to the development of the environmental movement discussed in Section 9.13.

From the standpoint of urban planning, growth became synonymous with urban sprawl—the often uncontrolled development of land at the outskirts of major American cities. What often resulted was an unplanned, market-oriented development of isolated and scattered parcels of land on the fringes of suburbia followed by the gradual urbanization of the intervening undeveloped areas. Urban sprawl has been criticized as unaesthetic, wasteful of valuable land resources, and unduly increasing the cost of providing municipal services.

Rethinking of goals and priorities reflected themselves in public land use planning. Communities sought to avoid the disruptive effect of an influx of people into their communities.

Plans attacked as exclusionary and defended as phased growth often have common characteristics. However, a crude comparison can be made between these two classifications. Suburbs whose plans are exclusionary seek to be tight little islands. They wish to keep out low- and moderate-income families for social and fiscal reasons. However, phased growth plans such as those discussed in this subsection are designed to keep out all people, not simply undesirable ones.

Another preliminary caution is that all land use controls tend to exclude, because density control is a legitimate and frequent objective. For this reason, terms such as *exclusionary* and *phased growth* must be looked at in light of their value-laden characteristics. *Exclusionary zoning* has been the label that critics of suburbs developed when they attacked homogeneous suburbs as tight little islands. Yet *phased growth*, though frequently a label used to justify an exclusionary plan, carries with it socially desirable goals of protecting nature and wildlife and preserving the values and attributes of small-town life.

The first significant case to pass on a phased growth plan involved a system devised by Ramapo, New York, a suburb of New York City some twenty-five miles from downtown Manhattan. The master plan adopted in 1966 contemplated an eighteen-year period during which public facilities would be built and after which the town would be fully developed. In 1969, an ordinance was passed to control residential development on any vacant lots or parcels. Applications for permits were to be made to the town board, which was to take into account the availability of major public improvements and services such as sewers or approved substitutes, drainage facilities, parks, or recreational facilities including public school sites, improved roads, and firehouses. The degree of availability of each facility to the site was to be measured and scored on a scale of 0 to 5, and no permit could be issued unless a minimum of fifteen points were obtained. A developer could advance the date for development by providing facilities itself or by obtaining a variance. Defenders of the Ramapo approach pointed to the advantage of each parcel owner's knowing the eventual use and density classification of every parcel. The plan only postponed development. Backers of the Ramapo approach pointed to specific criteria governing planning board decision making rather than the fuzziness of most permit-issuing criteria.

The Ramapo approach was upheld by the New York Court of Appeals.[115] The court held that the plan was within the powers given the town by the state enabling laws, with the effective end to be served by the time controls within the state's police powers, and was not so unreasonable as to constitute a taking without compensation. (By the 1980s, the economic slowdown forced Ramapo to substantially modify the plan to encourage development.)

A second case involved an attempt by a suburban community to limit growth by controlling the number of units that could be built. The plan was upheld,[116] but again by the 1980s the plan was too effective and changes had to be made to encourage growth.

A third case citing the fragile ecology of a small rural town upheld an ordinance drastically increasing the minimum lot size to block a large development of second homes for urban residents.[117]

---

[115]*Golden v. Planning Bd. of Town of Ramapo*, 30 N.Y.2d 359, 285 N.E.2d 291, 334 N.Y.S.2d 138, appeal dismissed, 409 U.S. 1003 (1972). Both majority and dissent called for regional land use planning.

[116]*Construction Indus. Ass'n v. City of Petaluma*, 522 F.2d 897 (9th Cir.1975), cert. denied, 424 U.S. 934 (1976).

[117]*Steel Hill Dev., Inc. v. Town of Sanbornton*, 469 F.2d 956 (1st Cir.1972).

While these three phased growth plans were upheld by the courts, the Massachusetts high court struck down a town zoning bylaw of unlimited duration that regulated the number of building permits issued annually for the construction of single family homes. The court found that restrictions of unlimited duration on a municipality's rate of development are in derogation of the general welfare and therefor unconstitutional.[118]

---

[118]*Zuckerman v. Town of Hadley,* 442 Mass. 511, 813 N.E.2d 843 (2004).

CHAPTER TEN

Professional Registration and Contractor Licensing: Evidence of Competence or Needless Entry Barrier?

SECTION 10.01 Overview

Legal requirements for professional practice is the logical place to begin a study of how law intersects with design and the design professions. This chapter deals with professional registration laws. It also examines state occupational licensing that can and often does include contractors, a latecomer into the field of occupational regulation.

Registration and licensing laws are enacted by state legislatures. They are implemented and enforced by administrative agencies. Because of the political nature of these laws, they often are changed. A Tennessee decision handed down in 1995 stated that the Tennessee contractor licensing law had been changed seventeen times since its original enactment in 1931.[1] This is not uncommon.

Similarly, as shall be seen in Section 10.02, the controversial nature of these laws and regulations means that court decisions can seem contradictory and occasionally are reversed by later decisions.

For these reasons, textual statements must be regarded as general and even tentative. Local laws must be consulted when planning to perform design and construction services or in actual cases that involve the legality of performing these services.

SECTION 10.02 Public Regulation: A Controversial Policy

Public regulation of professional activity can take many forms. One form of indirect public regulation—professional liability—is discussed in Chapter 14. This chapter discusses the direct legal requirements that determine whether a particular person or business entity may use a particular title or perform design or construction services.

Statutes enacted by legislative bodies and regulations promulgated by state administrative agencies set criteria for professional practice and administer systems that determine who may legally perform these services.

In the United States, the states control registration. Generally, the factors considered by a state before granting the contractor a license are similar to those considered by the federal government when it selects contractors to perform services. But state licensing laws are not applied to those contractors who work exclusively for the federal government.[2] To give effect to the supremacy clause of the U.S. Constitution, the federal government must operate free of interference by the states.

There is no federal regulation system for registering and licensing design professionals. Local regulation, where it exists, is designed principally to raise revenue and not to regulate the professions. A few states have given local authorities the power to impose competence requirements on contractors.

A. Justification for Regulation

The police and public welfare powers granted to states by their constitutions permit states to regulate who may practice professions and occupations. States exercise this power by setting requirements for those who wish to

---

[1]*Winter v. Smith*, 914 S.W.2d 527, 536 (Tenn.App.1995), appeal denied Dec. 18, 1995

[2]*Gartrell Constr. Inc. v. Aubry*, 940 F.2d 437 (9th Cir.1991), following *Leslie Miller, Inc. v. Arkansas*, 352 U.S. 187 (1956).

practice professions or occupations. Such requirements are usually expressed in licensing or registration laws.[3]

Some members of the public who seek to have design or construction work performed are unable to judge whether those who offer to perform design or construction services have the necessary competence and integrity. One purpose of licensing laws, whether they apply to architects, engineers, surveyors, or contractors, is to give members of the public some assurance that those with whom they deal will have at least minimal competence and integrity. The public generally may suffer property damage or personal harm because of poor design or construction. In denying recovery to an architect licensed where he practiced but not where the project was located, one court refused to allow an exception for an isolated transaction and stated, "One instance of untrained, unqualified, or unauthorized practice of architecture or professional engineering—be it an isolated transaction or one act in a continuing series of transactions—may be devastating to life, health or property."[4] In passing judgment on a contractor licensing law, one court stated that the law was designed to

> prevent unscrupulous or financially irresponsible contractors from deceiving and taking advantage of those who engage them to build. . . . It often happens that fly-by-night organizations begin a job and, standing in danger of losing money, leave it unfinished to the owner's detriment. Or they may do unsatisfactory work, failing to comply with the terms of their agreement. The licensing requirement is designed to curb these evils; the license itself is some evidence to the owner that he is dealing with an honest and qualified builder.[5]

Society seeks to accomplish these undeniably desirable objectives by requiring that those who wish to perform certain services have had specified education and experience and are able to demonstrate competence by passing examinations. People in business occupations may also have to establish financial responsibility. Evidence of immoral or illegal conduct goes to the issue of integrity

and provides an independent reason to deny licensing.[6] In addition, after entry into the profession, professional misconduct or gross incompetence may justify suspending or revoking a license.

Another objective was revealed when states began to require comprehensive occupational licensing. When this occurred, licenses were needed not only by doctors, lawyers, architects, and engineers but also by barbers, beauticians, auto mechanics, and shoe repairers, to name but a few. Such laws were designed to bring status and dignity to those in useful occupations.

B. Criticism of Licensing Laws

Occupational licensing laws can deny some people their only means of livelihood. As a result, imposition or tightening of licensing laws is often accompanied by "grandfathering in" those who already are in those occupations. This can relieve the hardship caused by denying people the right to perform their best and often only skill. It can also dilute standards.

More important, even in the early history of occupational licensing, critics attacked the state's role of determining who and how many people would be allowed to enter professions or occupations. Occupational licensing laws can be viewed as "fence me in" legislation designed to limit competition and protect those already in the profession or occupation.[7] This criticism became even more strident when the states went, as noted in Section 10.02A, into "wholesale" licensing. Licensing as a method of "turf protection" surfaced again in the 1980s as the architectural profession sought to stop interior designers from being licensed. Any group with sufficient organization and political strength to demand that standards be set for its group can generate a new licensing law.

The competition-inhibiting aspect of licensing laws was demonstrated in 1973 when 2,149 general contractors took and failed the Florida Construction Industry Licensing Board examination. Either all Florida applicants were incompetent, or the board was seeking to limit competition

---

[3]The term *registration* is commonly used for the design professions and is used in this chapter interchangeably with the term *licensing,* except in the sections dealing with contractors.

[4]*Food Management, Inc. v. Blue Ribbon Beef Pack, Inc.*, 413 F.2d 716, 723–24 (8th Cir.1969).

[5]*Sobel v. Jones*, 96 Ariz. 297, 394 P.2d 415, 417 (1964).

[6]*Hughes v. Board of Architectural Examiners*, 17 Cal.4th 763, 952 P.2d 641, 72 Cal.Rptr.2d 624 (1998); *Nye v. Ohio Bd. of Examiners of Architects*, 165 Ohio App.3d 502, 847 N.E.2d 46 (2006).

[7]*People v. Johnson*, 68 Ill.2d 441, 369 N.E.2d 898 (1977) (registration rules, which have the practical effect of perpetuating a monopoly of those already licensed, are unconstitutional).

by barring new entrants to the field. After indignant protest from builders who had failed, the board abruptly reversed itself and curved the grades so that 88 percent were given passing marks and contractors' licenses.[8]

In 1987 Florida again moved into the spotlight. After many complaints about shoddy workmanship performed by contractors, the state overhauled its licensing examinations, upgraded tests, and added new sections in accounting, business administration, and insurance. Some 93 percent of contractors failed the new examination.[9]

Critics also note that the agencies that administer licensing laws are often dominated by members of the profession being regulated. This can generate practices that keep down the number of practitioners, to improve the economic status of those already in their profession.

The Florida experiences also demonstrate the difficulty of determining the proper level of competence necessary to receive state permission to engage in a particular occupation. This issue has triggered protest and lawsuits from minority groups, who contend that their low representation in the professions results, among other things, from examination questions and grading that either are designed to limit their entry or have that effect. The long experience requirements for designers have been criticized as either inadequate for proper training or as solely designed to provide registered designers with cheap labor.

New criticism of licensing and registration laws surfaced during the galloping inflation of the mid-1970s. Many pointed to professional licensing laws, among other protections given to the professions, as causing the high cost of professional services. Educational and experience preexamination requirements mean many unproductive years at great expense. This can lead to fees designed to recoup these losses after entry. Similarly, education, practice, and testing requirements limit the number of people who can perform these professional services. This reduces supply and increases fees. Another undesirable by-product is that high professional fees tend to limit design services to those clients who can afford to pay these fees and to high-cost projects.

Yet the regulatory process, despite such criticism, continues to proliferate.

In Arizona the winds of deregulation blew fiercely, if fitfully. In 1981, Arizona concluded that licensing laws were not needed in commercial and industrial construction, where the market could provide a cheaper and more effective regulatory mechanism. The legislature stated,

> The legislature finds that regulation of the commercial and industrial construction business, including public works, . . . is not necessary for the protection of the public health, safety and welfare, and that it is in the public interest to deregulate such business. It is the purpose and intent of the legislature to continue the registrar of contractors agency in order to protect the public health, safety and welfare by providing for the continuing licensing, bonding and regulation of contractors engaged in residential construction.[10]

Yet in 1987 Arizona restored much of the regulation eliminated in 1981.[11] The Arizona experience illustrates not only the controversial nature of occupational licensing, but also the apparently inexorable movement toward greater state control.

Finally, wholesale occupational licensing laws can be a method by which the state can exercise pervasive control over its economy and its citizens. Enterprises or individuals thought to be operating against the public interest but not reachable by the penal laws can be put out of business by license suspension or revocation.

C. Importance of Attitude Toward the Regulatory Process

The attitude of lawmakers toward the regulation process clearly is influential. Legislators who take a beneficent view are likely to enact more licensing laws. Similarly, such legislators may seek to bar judicial doctrines—such as substantial compliance, discussed in Section 10.10C—that reduce the effectiveness of such laws. Of course, the attitude of courts that must often pass on these laws will influence decisions. For example, courts are often called on to determine the constitutionality and meaning of the legislation. They also decide whether particular conduct has violated the statute, whether substantial compliance with licensing laws is adequate to excuse a violation, and whether a party who has performed work, though unlicensed, will be able to recover compensation.

---

[8]See *Wall Street Journal*, Jan. 8, 1975, at 1.
[9]11 Constr. Contractor 333 (1987).

[10]1981 Ariz.Sess.Laws, ch. 221, § 1 (repealed by Laws 1986, ch. 318, § 21).
[11]Ariz.Rev.Stat.Ann. § 32–1101 *et seq.*

D. Judicial Attitudes Toward Registration Laws

Courts in different states express variant attitudes. For example, one court passing judgment on a particular licensing practicing system noted that it expressed "grave policy."[12] Yet in that same year, another granted recovery to a contractor who through technical default had allowed his license to lapse, stating, "It performed in all other respects competently and without injury to any person. . . . We are not involved in aiding an incompetent or dishonest artisan. . . . The defendant received full value under the terms of the contract. The licensing law should not be used as a shield for the avoidance of a just obligation."[13]

Even a particular state over time can change its attitude toward licensing laws. For example, in 1957, California precluded an unlicensed subcontractor from collecting from the prime contractor who knew it was unlicensed despite the latter's having been paid by the owner.[14] Yet cases decided by the California courts in 1966, 1973, and 1985 were more tolerant toward technical contractor noncompliance.[15] As noted in Section 10.10C, in 1989 the legislature responded negatively toward these pro-contractor decisions. Finally, in 2001, it added Section 7031(b) to its Business and Professions Code. That section permits a person who has dealt with an unlicensed contractor to recover any payments that she made.

On the whole, licensing laws still are considered to express important and desirable policy. Criticism that has been made, however, has begun to be reflected in judicial decisions.

SECTION 10.03 Administration of Licensing Laws

Modern legislatures articulate rules of conduct by statute and create administrative agencies to administer and implement the laws. Agencies created to regulate the professions can make rules and regulations to fill deliberate gaps left by the statutes and particularize the general concepts articulated by the legislature. For example, the licensing laws state that there must be examinations to determine competence. The details of the examinations, such as the type, duration, and frequency, are determined by the agency.

In addition to having quasi-legislative powers, the agencies have quasi-judicial power. They may, subject to judicial review, decide disputed questions, such as whether a particular school's degree will qualify an applicant to take the examination or whether certain conduct merits disciplinary sanction. They also have power to seek court orders ordering that people cease violating the licensing laws.

These agencies generally possess expertise in the fields of activity being regulated. Members of these agencies who must perform quasi-judicial functions, such as deciding whether an architect or engineer should have her license suspended or revoked, traditionally are members of the professions being regulated, such as architects or engineers. But there is an increasing tendency to appoint some lay persons to these boards.

Martin v Sizemore[16] is instructive on the issue of whether certain charges made by the agency and contested by the person whose registration or license is being challenged must be supported by expert testimony. As shall be seen in Sections 14.05B and 14.06, a claim made in a judicial proceeding against a professional person must usually be supported by the testimony of experts unless common knowledge is sufficient to determine whether the professional has acted properly.

The case catalogued the different bases for disciplinary action. It noted that breach of contract under Tennessee law may be the basis for a license suspension or revocation. If breach of contract is the basis for the disciplinary action, expert testimony to sustain the charge is not required.

But if the basis for the disciplinary action is gross negligence, incompetence, malpractice, commission of excessive errors or misconduct, expert testimony must be submitted by the agency to justify the disciplinary action. Though the court extends considerable deference to agencies on technical matters within their expertise, the court held that due process requires a level of fairness to the person whose livelihood is at stake.

The court recognized that members of the agency performing quasi-judicial functions are professionals, though

---

[12]*Hedla v. McCool*, 476 F.2d 1223, 1228 (9th Cir.1973).

[13]*Vitek, Inc. v. Alvarado Ice Palace, Inc.*, 34 Cal.App.3d 586, 110 Cal. Rptr. 86, 92 (1973).

[14]*Lewis & Queen v. N. M. Ball Sons*, 48 Cal.2d 141, 308 P.2d 713 (1957).

[15]*Latipac, Inc. v. Superior Court*, 64 Cal.2d 278, 411 P.2d 564, 49 Cal.Rptr. 676 (1966); *Vitek, Inc. v. Alvarado Ice Palace, Inc.*, supra note 13; *Asdourian v. Araj*, 38 Cal.3d 276, 696 P.2d 95, 211 Cal.Rptr. 703 (1985).

[16]78 S.W.3d 249 (Tenn.Ct.App.2001).

in Tennessee, as well as in many states, lay members serve as well. This was not enough to satisfy due process requirements where the basis of the challenge relates to professional competence. But, as in litigation, expert testimony is not needed if the issue can be resolved on the basis of common knowledge.

Grounds for disciplinary action will be discussed in Section 10.04B.

SECTION 10.04   The Licensing Process

A. Admission to Practice

Requirements imposed by states or territories vary considerably. Some states and territories require citizenship and residency. Most have minimum age requirements, usually ranging from 21 to 25. All require a designated number of years of practical experience that can be substantially reduced if the applicant has received professional training in recognized professional schools. All states require at least one examination, and some require two. Most inquire into character and honesty. Some require

interviews. As to interstate practice, see Section 10.06G. Section 10.08A covers contractors.

B. Postadmission Discipline:
Duncan v. Missouri Board for Architects, Professional Engineers and Land Surveyors

Although the regulatory emphasis has been on carefully screening those who seek to enter the professions, all states can discipline people who have been admitted. Discipline can be a reprimand or suspension or revocation of the license.

Grounds for disciplinary action vary considerably from state to state, but they are generally based on wrongful conduct in the admissions process, such as submitting inaccurate or misleading information, or conduct after admission that can be classified as unprofessional or grossly incompetent.

Yet the disciplinary action in the aftermath of the Hyatt Regency tragedy of 1981 in which many people were killed and injured shows that in spectacular accidents, courts will back up strong sanctions imposed by a licensing authority. The decision of the intermediate Missouri court follows.

DUNCAN V. MISSOURI BOARD FOR ARCHITECTS, PROFESSIONAL ENGINEERS AND LAND SURVEYORS

Missouri Court, Eastern District, Division Three, 1988. 744 S.W.2d 524.
[Ed. note: Footnotes have been renumbered and some omitted.]

SMITH, Judge.
On July 17, 1981, the second and fourth floor walkways of the Hyatt Regency Hotel in Kansas City collapsed and fell to the floor of the main lobby. Approximately 1500 to 2000 people were in the lobby. The walkways together weighed 142,000 pounds. One hundred and fourteen people died and at least 186 were injured. In terms of loss of life and injuries, the National Bureau of Standards concluded this was the most devastating structural collapse ever to take place in this country. That Bureau conducted an investigation of the tragedy and made its report in May 1982. In February 1984, the Missouri Board for Architects, Professional Engineers and Land Surveyors filed its complaint seeking a determination that the engineering certificates of registration of Daniel Duncan and Jack Gillum and the engineering certificate of authority of G.C.E. International were subject to dis-

cipline pursuant to Sec. 327.441 RSMo 1978. The Commission, after hearing, found that such certificates were subject to suspension or revocation. Upon remand for assessment of appropriate disciplinary action, the Board ordered all three certificates revoked. Upon appeal the trial court affirmed. We do likewise.

G.C.E. is a Missouri corporation holding a certificate of authority to perform professional engineering services in Missouri. Gillum is a practicing structural engineer holding a license to practice professional engineering in Missouri. He is president of G.C.E. Duncan is a practicing structural engineer holding a license to practice professional engineering in Missouri and is an employee of G.C.E.

Gillum-Colaco, Inc., a Texas corporation, contracted with the architects of the Hyatt construction to perform structural engineering services in connection with the erection of that

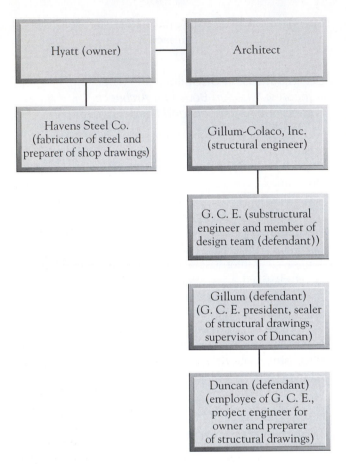

FIGURE 10.1 Participants in structural design, Hyatt Regency project.

building. By subcontract the responsibility for performing all of such engineering services was assumed by G.C.E. The structural engineer, G.C.E., was part of the "Design Team" which also included the architect, and mechanical and electrical engineers. Gillum was identified pursuant to Sec. 327.401.2(2) RSMo 1978, as the individual personally in charge of and supervisory (sic) of professional engineering activities of G.C.E. in Missouri. His professional seal was utilized on structural engineering plans for the Hyatt. Duncan was the project engineer for the Hyatt construction in direct charge of the actual structural engineering work on the project. He was under the direct supervision of Gillum. [Ed. note: See Figure 10.1.]

* * *

We will not attempt to set forth in detail the extensive evidence before the Commission. Some review of that evidence is,

however, required. The atrium of the Hyatt was located between the 40 story tower section of the hotel and the function block. Connecting the tower and function block were three walkways, suspended from the atrium ceiling above the atrium lobby. The fourth floor walkway was positioned directly above the second floor walkway. The third floor walkway was to the east of the other two walkways. As originally designed the fourth and second floor walkways were to be supported by what is referred to as a "one rod" design. This consisted of six one and one quarter inch steel rods, three on each side, connected to the atrium roof and running down through the two walkways. Under this design each walkway would receive its support from the steel rods and the second floor walkway would not be supported by the fourth floor walkway. At each junction of the rods and the walkways was a box beam–hanger rod connection. These were steel to steel connections and the design of such connections is an engineering function, because the design includes the performance of engineering calculations to determine the adequacy of the connection to carry the loads for which it is designed. [Ed. note: See Figure 10.2.]

Connections are basically of three kinds, simple, complex, and special. All connections are the responsibility of the structural engineer. Simple connections are those which have no unusual loads or forces. They may be designed by looking up the design in the American Institute of Steel Construction (AISC) Manual of Steel Construction and following directions found therein. This can be done by a steel fabricator utilizing non-engineering personnel. Complex connections are those where extreme or unusual loads are exerted upon the connection or where the loads are transferred to the connection from several directions. These connections cannot be designed from the AISC manual and require engineering expertise to design.

Special connections are a hybrid having characteristics of each of the other two. A simple connection becomes special where concentrated loads are placed thereon and the AISC manual no longer provides all the information necessary to properly design the connection. Such connections may also become special where the connections are "non-redundant." A "redundant" connection is one where failure of the connection will not cause failure of the entire system because the loads will be carried by other connections. A "non-redundant" connection which fails will cause collapse of the structure. The box beam–hanger rod connections were "non-redundant." The Commission found the box beam–hanger rod connections to be special connections.

The steel fabricator on the Hyatt project, Havens Steel Company, had engineers capable of designing simple, complex or special connections. The structural engineer on a project

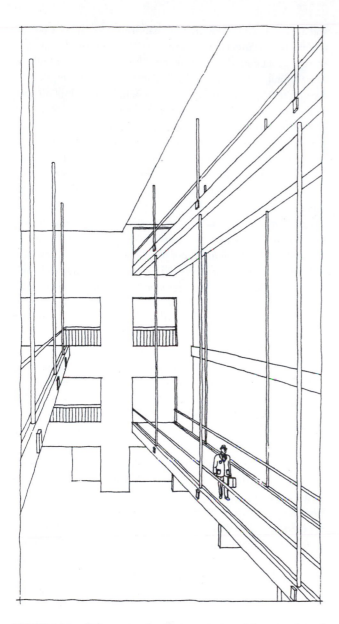

FIGURE 10.2   Schematic of walkways as viewed from north wall of atrium.

may, as a matter of custom, elect to have connections designed by the fabricator. To do this, he communicates this information to the fabricator by the manner in which he portrays the connection on his structural drawings. The adequacy of the connection design remains the responsibility of the structural engineer. The Commission found that the structural drawings (S405.1 Secs. 10 and 11) did not communicate to the fabricator that

it was to design the box beam–hanger rod connection, and did communicate to the fabricator that those connections had been designed by the engineer. Duncan testified that he intended for the fabricator to design the connections. Havens prepared its shop drawings on the basis that the connections shown on the design drawings had been designed by the structural engineer. Certain information concerning loads and other aspects of the box beam–hanger rod connections which appeared on Duncan's preliminary sketches was not included on the final structural drawings sent to the fabricator. The Commission also found that Duncan's structural drawings did not reflect the need for a special weld, did not reflect the need for stiffeners and bearing plates, and reflected that the hanger rods should be of regular strength steel rather than high strength. These factual findings are not contested. The hanger rods and the box beam–hanger rod connections shown on the structural drawings did not meet the design specifications of the Kansas City Building Code. That finding of fact by the Commission is also not contested.

Because of certain fabricating problems Havens proposed to Duncan the use of a "double rod" system to suspend the second and fourth floor walkways. [Ed. note: See Figure 10.3.] Under this system the original six rods would be connected only to the fourth floor walkway. A second set of rods would then connect the second floor walkway to the fourth floor walkway. The effect of this change was to double the load on the fourth floor walkway and the box beam–hanger rod connections on that walkway. There was evidence that one of the architects contacted Duncan to verify that the double rod arrangement was structurally sound and was advised by Duncan that it was. Appellants dispute that the architect's testimony clearly establishes such an inquiry and contend that the conversation dealt rather with the aesthetic nature of the change. Our review of the record causes us to conclude that the architect did testify to receiving assurances that the new design was structurally safe. It is difficult to understand why the architect would consult the structural engineer if his only concern was the aesthetics of the new design. The Commission further found that the records of G.C.E. failed to contain a record of a web shear calculation which Duncan testified he made and which would normally be a part of the G.C.E. records. Duncan's testimony reflected the need for such a calculation before approval of the double rod arrangement. The Commission also found certain additional necessary tests or calculations were not made. Appellants do not challenge that finding. It is a reasonable inference from the evidence that Duncan did not make the engineering calculations and tests necessary to determine the structural soundness of the double rod design.

Havens prepared the shop drawings of the structural steel fabrication. These drawings were returned to Duncan for review

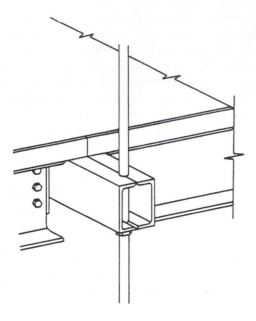

Original detail

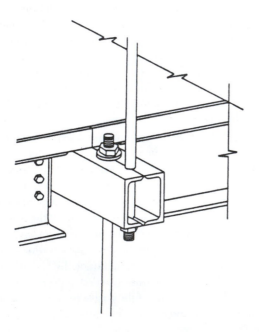

As built

FIGURE 10.3 Comparison of interrupted and continuous hanger rod details.

and approval. They contained the fabrication of the box beam–hanger rod connections based upon the structural drawings previously submitted by Duncan and bearing the seal of Gillum. The Commission found, and appellants do not dispute, that its own internal procedures called for a detailed check of all special connections. The primary reason for such a procedure is to provide assurance for the owner that the fabricator is conforming to the contract and that any engineering work conforms to acceptable standards. A technician employed by G.C.E. checked the sizes and materials of the structural members for compliance with design drawings. He called to Duncan's attention questions concerning the strength of the rods and the change from one rod to two. Duncan stated to the technician that the change to two rods was "basically the same as the one rod concept." Duncan did not "review" the fourth floor box beam connection shown on the Havens shop drawings nor did he, in accord with usual engineering practice, assemble its components to determine what the connection looked like in detail. The Commission found, again not disputed, that appellants did not review the shop drawings for compliance of the box beam–hanger rod connection with design specifications of the Kansas City Building Code; did not review the shop drawings for conformance with the design concept as required by the contract of G.C.E. and the specifications on the Hyatt project nor for compliance with the information given in the contract documents. Duncan and Gillum approved the shop drawings.

While construction of the Hyatt was in progress the atrium roof collapsed. Investigation into that collapse established that the cause was poor construction workmanship. During the course of their investigation of the atrium roof collapse, appellants discovered that they had made certain errors in their design drawings and that they had failed to find discrepancies in their review of shop drawings involving the atrium. These errors and discrepancies were in areas other than the walkway design. The owner and architect directed G.C.E., for an additional fee, to check the design of the entire atrium. G.C.E. undertook that review. Gillum assured the owner's representative that "he would personally look at every connection in the hotel." Appellants were also specifically requested by the construction manager to inspect the steel in the bridges including the connections. Duncan subsequently advised him that had been done. In their report to the architects, appellants advised "we then checked the suspended bridges and found them to be satisfactory." This report was a culmination of a design check of the "structural steel framing in the atrium as per the request of Crown Center [the owner]." Appellants did not do a complete check of the design of all steel in the atrium nor a complete check of the suspended bridges. Gillum reviewed the report prepared by Duncan and took no exception. At a meeting with the owner and architect, Gillum stated that his company had "run a detailed, thorough re-analysis of all of the structure. And to determine if there was any other areas that were critical or had any kind of a design deficiency or detail deficiency." Duncan reported at that meeting: "We went back, myself and another engineer, and checked all the atrium steel. . . . Everything in the atrium checked out very well [with one non-relevant exception]." Appellants checked only the atrium roof steel.

Approximately a year after completion of the Hyatt Regency the second and fourth floor walkways collapsed. The cause of the walkway collapse was the failure of the fourth floor box beam–hanger rod connections.

[Ed. note: The engineers challenged the constitutionality of the licensing laws. They contended that the "gross negligence" standard was so vague as to deny them due process under the U.S. Constitution. The court rejected this contention, concluding that the "phrase provides a guideline sufficient to preclude arbitrary and discriminatory application." See Section 10.06A.]

The Commission rejected the definition utilized in the first category of cases (difference in degree) and utilized a definition recognizing that gross negligence is different in kind from ordinary negligence. Appellants do not disagree with this selection of category. The Commission defined the phrase in the licensing context as "an act or course of conduct which demonstrates a conscious indifference to a professional duty." This definition, the Commission found, requires at least some inferred mental state, which inference may arise from the conduct of the licensee in light of all surrounding circumstances. Appellants have posited a definition purportedly different that would define the phrase as "reckless conduct done with knowledge that there is a strong probability of harm, and indifference as to that likely harm." We are not persuaded that the two definitions are in fact different. An act which demonstrates a conscious indifference to a professional duty would appear to be a reckless act or more seriously a willful and wanton abrogation of professional responsibility. The very nature of the obligations and responsibility of a professional engineer should appear to make evident to him the probability of harm from his conscious indifference to professional duty and conscious indifference includes indifference to the harm as well as to the duty. The structural engineer's duty is to determine that the structural plans which he designs or approves will provide structural safety because if they do not a strong probability of harm exists. Indifference to the duty is indifference to the harm. We find no error in the definition utilized by the Commission. It imposes discipline for more than mere inadvertence and requires a finding that the conduct is so egregious as to warrant an inference of a mental state unacceptable in a professional engineer.

The appellants also challenge the Commission's findings that each of several different acts or omissions constituted gross negligence justifying discipline. These findings were apparently made by the Commission in an abundance of caution for in a footnote it recognized that "it is only after a complete analysis of their overall performance within the system that any judgment of their conduct can be made under the terms of the licensing statute." This is clearly true. It is the combination of a series of acts and omissions which created the structurally unsound walkways. Any one of those acts or omissions alone might well not have compromised the structural integrity of the walkways if the series of acts and omissions had not existed in combination. For the Commission to require a finding that each of these acts or omissions had to be grossly negligent to support discipline (if in fact the Commission so required) would place upon the Board a greater burden than was required. It is apparent, however, that the Commission found the overall conduct of appellants grossly negligent and if that finding is supported by competent and substantial evidence we are bound by it.

* * *

Appellants in this connection challenge one of the Commission's findings on the basis that the collapse of the walkways was not caused by a certain specific failure. This is raised in connection with the finding that the failure to delineate special strength steel for the rods was grossly negligent. The rods themselves did not fail. In making this assertion appellants rely on the elements of a common law cause of action for negligence, i.e., duty, breach, proximate causation, and injury or damage. They assert that proximate causation is not present. In the first place we are not dealing with a civil cause of action for negligence. We are dealing instead with a determination of whether appellants negligently breached their duty in the design of the walkways. That breach occurred at the latest when their design was incorporated into the building with their approval and they were subject to discipline whether or not any collapse subsequently occurred. It is the appellants' conscious disregard of their duty for which discipline is being imposed not the result of that breach. . . . Damage or injury is not an element of this disciplinary proceeding and proximate cause is the legal concept that authorizes civil recovery for damage resulting from negligence. It is not in and of itself an aspect of "negligence," only an aspect of a civil cause of action for negligence. Related to that concept is the fact that indeed there was damage caused by the breach. By statute and under the contract the owner of the building was entitled to a building structurally safe and sound. . . . The owner did not receive such a building because of appellants' breach of their professional responsibility. The owner received a defective building. Whether the walkways collapsed or not, the owner was damaged because it received less than it was entitled to and that damage was proximately caused by appellants' acts and omissions. Further we have previously stated that gross negligence is not required in each act of appellants; it is their overall conduct in regard to the Hyatt construction which justifies discipline.

* * *

The statutory provisions make clear that Missouri has established a stringent set of requirements for professional engineers practicing in the state. The thrust of those requirements is professional accountability by a specific individual certified engineer. These requirements establish the public policy of the state for the protection of the public. They require that plans for construction of structures in this state which require engineering expertise be prepared by or under the direct supervision of a specified certified engineer and that that engineer bear personal and professional responsibility for those plans. The affixing of his seal on the plans makes him responsible for the entire engineering project and all documents connected therewith unless he specifically disclaims responsibility for some document relating to or intended to be used for any part of the engineering project. It would be difficult to imagine statutory language more clearly evidencing the total responsibility imposed upon the engineer, and accepted by him when he contracts to provide his services. The statutory statement that the right to engage in the profession is a personal right based upon the individual's qualifications in no way impacts upon the responsibilities imposed upon an engineer. Rather the assessment of the individual "qualifications" of the engineer include his willingness and ability to accept the responsibilities imposed on him by the statutes.

[Ed. note: The court rejected the engineers' contention that the custom of relying on the fabricators to design the connections precluded any finding they were grossly negligent. The court noted that the employees of the fabricator are exempt from licensing requirements.]

* * *

The public policy of this state as it pertains to the responsibility of engineers has been established by the General Assembly in Chapter 327. That Chapter imposes upon the engineer a nondelegable duty of responsibility for projects to which he affixes his seal. . . . The purpose of disciplinary action against licensed professionals is not the infliction of punishment but rather the protection of the public. . . . Chapter 327 has established the responsibility a certified engineer bears when he undertakes a contract in his professional capacity. Sec. 327.191 authorizes noncertificated engineers to perform engineering work "under the direction and continuing supervision of and is checked by" a certified engineer. It is a misdemeanor for a certified engineer to affix his seal to plans which have not been prepared "by him or under his immediate personal supervision." Sec. 327.201. A corporation may engage in engineering activities if it has assigned responsibility for proper conduct of its professional engineering to a registered professional engineer. Sec. 327.401. Gillum was the engineer designated by G.C.E. as having that responsibility. An engineer affixing his seal to plans is personally and professionally responsible therefor. Sec. 327.401. Affixing his seal to plans imposes upon the engineer responsibility for the whole engineering project unless he, under seal, disclaims such responsibility. Gillum made no such disclaimer here. The entire thrust of Chapter 327 is to place individual personal and professional responsibility upon a known and identified certificated engineer. This is the responsibility the engineer assumes in exchange for the right to practice his profession. It is the assumption of this responsibility for which he is compensated. The statutory framework is established to protect the public and to hold responsible licensed engineers who fail to afford that protection. It is clear that the statute expresses the intent to impose disciplinary sanctions on the engineer responsible for the project whether the improper conduct is that of himself or attributable to the employees or others upon whom he relies. This case differs, therefore, from the cases relied upon by Gillum and G.C.E. where the statute did not impose such non-delegable responsibility. The Commission did not err in finding that Gillum and G.C.E. were subject to discipline for the acts or omissions of Duncan.

* * *

We now turn to the sufficiency of the evidence to support the Commission's findings that discipline was warranted. In so doing we note again the concession of appellants that except for five findings the findings of fact of the Commission are supported by evidence. We have previously stated our analysis of the five disputed findings. We review the sufficiency within the legal principles and framework heretofore explicated.

We look first to Duncan. He was the project engineer for the Hyatt and as such had primary responsibility within his company for designing and approving those aspects of the Hyatt which required structural engineering expertise. The design of the connections in the walkways and the design of the walkways themselves were included in that responsibility. The walkways were intended to carry pedestrian traffic. They were suspended above the main lobby of the hotel, recognized to be the main point of congregation within the hotel. The walkways each weighed approximately 35 tons and were comprised of heavy and largely non-malleable materials such as steel, concrete, glass and wood. The connections in the walkway were nonredundant so that if any one within a single walkway failed they all would fail and the walkway would collapse. Duncan had never designed a system similar to the Hyatt walkways. It is self-evident that the walkways offered a potential of great danger to human life if defectively designed. The Commission could properly consider the potential of danger in determining the question of gross negligence. That which might constitute inadvertence where no danger exists may well rise to conscious indifference where the potential danger to human life is great. This is simply to say that the level of care required of a professional engineer is directly proportional to the potential for harm arising from his design and as we have previously stated indifference to harm and indifference to duty are closely related if not identical.

The structural drawings of Duncan furnished to the fabricator contained several serious errors. Under standard engineering practice Duncan could either design the box beam–hanger rod connections or cause the drawings to reflect his intention that they be designed by the fabricator. These drawings did neither. They appeared to be connections fully designed by the engineer and were reasonably so interpreted by the fabricator. Duncan testified that he intended the fabricator to design the connections. The drawings did not contain information indicating that the connections were to be designed by the fabricator and omitted important engineering load calculations necessary to enable the fabricator to design the connections. The drawings failed to properly identify the type of weld required, the need for bearing plates and/or stiffeners, and erroneously identified the hanger rods as standard rather than high-strength steel. The box beam–hanger rod connections and the hanger rods themselves on all three walkways, as shown by the structural drawings, did not meet the design specifications of the Kansas City Building Code. That Code is intended to provide a required level of safety for buildings within the City. It is difficult to conclude that gross failure to comply with that Code can constitute other than conscious indifference to duty by a structural engineer.

Because of certain difficulties in fabrication Havens requested a change to the double rod configuration. This request was transmitted to Duncan who approved it and verified its structural soundness and safety to the architect. He did so without having conducted all necessary engineering tests and calculations to determine the soundness and safety of the double rod arrangement. His concern was with its architectural acceptability not its structural acceptability. The result of this change was to double the load on the fourth floor walkway and impose a similar increase on the connections which were already substantially below Code requirements.

Havens supplied Duncan with its shop drawings. Under the contract, and under the statute, review and approval of the shop drawings is an engineering function. Appellants' normal in-house procedures called for detailed check of all special connections during shop drawing review. Duncan was aware of the change to the two-rod system but did not review the box beam–hanger rod connection on the fourth floor walkway. Duncan did not, as is standard practice, look for an assembled detail of the connection and did not assemble the components, either in his mind or on a sketch, to determine what the connection looked like in detail. The shop drawings did not reflect the use of stiffeners or bearing plates necessary to bring the connections within Code requirements. No review was made nor calculations performed to determine whether the box beam–hanger rod connection shown on the shop drawings met Code requirements. Shop drawing review by the engineer is contractually required, universally accepted and always done as part of the design engineer's responsibility. The box beam–hanger rod connections and the hanger rod shown on the shop drawings did not meet design specifications of the Code.

Following the atrium roof collapse appellants were requested by the architect and owner to recheck all the steel in the atrium. They reported that they had done so and included in that report was the statement "we then checked the suspended bridges and found them to be satisfactory." In fact appellants did not do a complete check of the design of all steel in the atrium and did not do a complete check of the suspended "bridges," i.e., walkways. As finally built, the hanger rods and the box beam–hanger rod connections did not meet the requirements of the Code. The walkway collapse was the result of the failure of the fourth floor box rod connections. The third floor walkway, which did not collapse, had a "high probability" of failure during the life of the building.

The determination of conscious indifference to a professional duty, i.e., gross negligence, is a determination of fact. The conduct of Duncan from initial design through shop drawing review and through the subsequent requested connection review following the atrium roof collapse fully supports the Commission's finding of conscious indifference to professional duty. The responsibility for the structural integrity and safety of the walkway connections was Duncan's and that responsibility was non-delegable. . . .

He breached that duty in continuing fashion. His reliance upon others to perform that duty serves as no justification for his indifference to his obligations and responsibility. The findings of the Commission as to Duncan's gross negligence are fully supported by the record.

The Commission also found Duncan subject to discipline for misconduct in misrepresenting to the architects the engineering acceptability of the double rod configuration when he performed no engineering calculations or other engineering activities to support his representation. The Commission found such representation to have been made either knowing of its falsity or without knowledge of the truth or falsity. The Commission found Duncan's misrepresentation to be the willful doing of an act with wrongful intention which it had defined as misconduct. We find no error in either the factual findings or legal conclusion of the Commission. Duncan's representation to the architect concerning a material fact, without a basis for knowledge of its truth or falsity, could properly be viewed as either misconduct as an engineer or gross negligence. In either event it subjected Duncan to disciplinary action.

The Commission found Gillum subject to discipline for gross negligence under the vicarious liability theory and also personally grossly negligent in failing to assure that the Hyatt engineering designs and drawings were structurally sound from an engineering standpoint prior to impressing thereon his seal and in failing to assure adequate shop drawing review. It further found Gillum to be subject to discipline for unprofessional conduct and gross negligence in his refusal to accept his responsibility as mandated by Chapter 327 and his denial that such responsibility existed. All of these findings arise from the same basic attitude of Gillum that the responsibility imposed by Chapter 327 is not in keeping with usual and customary engineering practices and that that responsibility did not mandate his personal involvement in the design of the Hyatt. In essence he placed the responsibility for the improper design of the connections on Havens and took the position that the structural engineer was entitled to rely on Havens' expertise. What we have heretofore said in regard to the requirements of Chapter 327 and the responsibility imposed upon an engineer thereby sufficiently deals with Gillum's contentions.

His argument here that utilization of his seal without disclaimer could not impose responsibility upon him for the shop drawings of another entity prepared after impression of the seal is clearly rejected by the language of the statute. By Section 327.411.2 the owner of the seal is responsible for the "whole . . . engineering project" when he places his seal on "any plans" unless he expressly disclaims responsibility and specifies the documents which he disclaims. The shop drawings were part of the documents comprising the engineering project and were "intended to be used for any part or parts of the . . . engineering project. . . ." Gillum was by statute responsible for those drawings and he accepted such responsibility when he entered into the contract and utilized his seal. His refusal to accept a responsibility so clearly imposed by the statute manifests both the gross negligence and unprofessional conduct found by the Commission. These findings are further bolstered by the evidence of Gillum's participation in the misrepresentations concerning, and nonperformance of, a review of the atrium design upon direct request of the architect and owner. Although we have found that a specific finding of misconduct and discipline therefor cannot be based upon the atrium design review because not charged in the complaint, the evidence is relevant and persuasive on Gillum's overall mental approach to his responsibilities as an engineer and the cavalier attitude he adopted concerning the Hyatt project.

Appellant G.C.E. is, for reasons heretofore stated, subject to discipline for the conduct of its employees and particularly for the conduct of the engineer assigned the responsibility for the "proper conduct of all its . . . professional engineering . . . in this state. . . ."

The finding of misconduct against Gillum arising from the "atrium design review" is reversed.[17] In all other respects the order of the Commission and the discipline imposed by the Board is affirmed.

KAROHL, P. J., and KELLY, J., concur.

---

[17][Ed. note: This ground for discipline was not alleged in the complaint.]

Other cases have also involved sufficiently flagrant misconduct or incompetence to justify a drastic agency decision. For example, a court affirmed an agency decision revoking the license of a professional engineer who had performed welding without being certified as required by the state administrative code.[18] In the same case, however, the court would not affirm the agency's revocation where the professional engineer designed and supervised the construction of a garage that collapsed. The court held that the incompetence did not have to consist of continued and repeated acts. However, incompetence must refer to some demonstrated lack of ability to perform professional functions. Although recognizing that there was an admitted error in the design of the roof supports for the garage, the court noted that the error was not obvious and

---

[18]*Vivian v. Examining Bd. of Architects, etc.*, 61 Wis.2d 627, 213 N.W.2d 359 (1974).

that this was the first failure the engineer had experienced in eleven years of practice.

A revocation of an architect's license was affirmed based on serious design errors leading to the failure of the basement wall.[19] The architect also caused construction delay, failed to obtain a building permit, misplaced the building in reference to the lot line, and secured the owner's endorsement of payment without informing him of the facts. Another court affirmed a six-month suspension based on a deficient ventilation plan, a superficial inspection of the premises, a superficial scanning of the architectural plan, and reliance on the judgment of two relatively inexperienced employees.[20] In addition, the engineer, after finding that the original certifications were in error and serious defects existed, did not disclose this to appropriate city officials.

Many members of the design professions operate under economically unstable conditions. Swings in the economic cycle can prove devastating. As a result, individual bankruptcies of architects and engineers are not uncommon when the economy turns for the worse. Those in the contracting business face similar risks.

The California experience illustrates how difficult it can be to reconcile the state's right to protect the public through its licensing laws with the interest of the federal government in effectuating the bankruptcy laws designed in part to give the bankrupt a fresh start. Section 7113.5 of the California Business and Professions Code had stated that grounds for disciplinary action existed if the contractor had settled a lawful obligation for less than the full amount of the obligation by having all or part of the debt discharged in bankruptcy. Yet *Grimes v. Hoschler*[21] held that this would frustrate the fresh-start purpose of the federal bankruptcy law. Risking one's license could discourage a contractor from discharging its business debts in bankruptcy. The contractor would very likely pay its discharged business debts to avoid losing its license. The licensing law would not only discourage contractors from availing themselves of the provisions of the Bankruptcy Act but also deny a fresh start to those licensees who are adjudicated bankrupts.

Shortly after this decision, the California legislature bowed to the federal law and the court decision by modifying Section 7113.5, barring disciplinary action being taken against a contractor who settled his debts for less than the full amount by bankruptcy.

The issue surfaced again in 1986. In *Parker v. Contractors State License Board*,[22] the licensing board sought to suspend a bankrupt contractor who had deducted benefit contributions from his employees' wages but did not forward them to the union. However, the union trust fund claim against him had been discharged by bankruptcy. The court held that the board could not discipline the contractor for going bankrupt, as that would prevent him from obtaining a fresh start. But disciplinary action could be based on fraud or a violation of the California Labor Code. Factors other than the bankruptcy could indicate the contractor was unsuitable to hold a license. The court sent the case back to the trial court to determine whether the activity of the contractor was fraudulent or simply failure to pay for lack of money.

Usually, attempts to discipline an architect or engineer relate to conduct after the license was issued. But California held that conduct before admission to practice could be misconduct that justified revocation of the license.[23] The architect in this case had difficulties with Virginia architectural board authorities that led to criminal charges. The architect did not report this when he applied for admission in California. The Supreme Court rejected the contention by the architect that the registration statutes were penal and must be strictly construed. The court held that the registration statutes were not penal but were designed to protect the public.

The effectiveness of postadmission disciplinary powers has been limited. Attempts to suspend or revoke are almost always challenged by the design professional or contractor whose means of livelihood are being taken away. Challenges often mean costly appeals. Often administrative agencies charged with responsibility for regulating the profession or occupation are underfunded and understaffed. Suspension and revocation are unpleasant tasks. Even when action is taken, courts closely scrutinize

---

[19]*Kuehnel v. Wisconsin Registration Bd. of Architects & Professional Engr's*, 243 Wis. 188, 9 N.W.2d 630 (1943).

[20]*Shapiro v. Bd. of Regents*, 29 A.D.2d 801, 286 N.Y.S.2d 1001 (1968). See also *Martin v. Sizemore*, supra note 16. It sustained a holding that omission of fire alarms, exit lights, and emergency lighting were grounds for disciplinary action. This case was discussed in Section 10.03. Cases involving architect license revocations and suspensions are collected in Annot. 58 A.L.R.3d 543 (1974) and those involving engineer licenses in Annot. 64 A.L.R.3d 509 (1975).

[21]12 Cal.3d 305, 525 P.2d 65, 115 Cal.Rptr. 625 (1974), cert. denied, 420 U.S. 973 (1975).

[22]187 Cal.App.3d 205, 231 Cal.Rptr. 577 (1986).

[23]*Hughes v. Board of Architectural Examiners*, supra note 6.

decisions of administrative agencies. As an example, Washington held that the board findings must be based on "clear, cogent, and convincing evidence", a more demanding standard than the usual civil law standard of preponderance of the evidence.[24]

Courts *should* extend considerable deference to the agency decision. In matters as important as these, however, courts seem to redetermine what is proper. The combination of agency lethargy and overextensive judicial scrutiny when agencies do act may be one reason, among many, for increased professional liability. The latter can supplement or even replace the regulatory licensing process as a means by which incompetent practitioners are eliminated.

SECTION 10.05   Types of Licensing Laws

It must be emphasized again that each state determines licensing rules. As a result, considerable variety exists in the regulatory controls in both prohibited conduct and sanctions for violations. Also, legislation frequently changes. Case decisions cannot always be relied on because subsequent legislation may change the result in a particular case. However, some broad patterns have emerged as to the types of controls enacted.

A. Licensing of Architects and Engineers: Holding Out and Practice Statutes

Licensing laws fall into two main categories. Some statutes regulate the professional title and are called "holding out" statutes. For example, in 1965 Ohio enacted such a statute stating that "[n]o person shall use the title 'landscape architect' . . . unless he is registered . . . or holds a permit." A court passing on the validity of this legislation stated,

> A practitioner, upon qualifying, becomes entitled to use the label, "landscape architect." Only the title, not the practice or profession, is restricted to licensees. Unregistered members may continue to practice, but without employing the title. Thus, appellant's allegation that the law prohibits him from practicing his profession is not well taken; only his use of the title "landscape architect" is proscribed.[25]

Projects that affect the public generally are more likely to require the services of a licensed design professional. A licensed design professional is more likely to be required where the project is one for human habitation or one in which large numbers of persons will gather. Some states exclude less complex structures, such as single-family residences, agricultural buildings, or warehouses and commercial buildings below specified sizes and heights.[26] See Section 10.06E. Local laws must be checked.

B. Contractor Licensing

The proliferation of occupational licensing has led to an increased number of state licensing statutes for contractors. Until recently, only half of the states had such statutes, although the number seems to be increasing steadily. The statutes raise different problems and are discussed in Sections 10.08 through 10.11.

C. Variations on the Traditional Contracting System

The traditional system for delivering construction services, discussed in Section 8.05, will be discussed again in Section 17.03A. Modern variations on this system are covered in Section 17.04. Some of these variations, such as the use of construction management (Section 17.04D) and combining designing and building (Section 17.04F), have raised difficult licensing and registration problems, which are discussed in those sections.

SECTION 10.06   Statutory Violations

A. Preliminary Issue: Constitutionality

Because of the controversial nature of professional regulation, constitutional attacks are common. By and large, such legislation is upheld based on the police powers granted by state constitutions and constitutional provisions permitting the state to legislate in the interest of public welfare.[27] Some successful attacks have been made that have usually involved language or procedural problems and

---

[24]*Nims v. Washington Bd. of Registration*, 113 Wash.App. 499, 53 P.3d 52 (2002).

[25]*Garono v. State Bd. of Landscape Architect Examiners*, 35 Ohio St.2d 44, 298 N.E.2d 565, 567 (1973).

[26]Iowa Code Ann. § 544A.18.

[27]*Richmond v. Florida State Bd. of Architecture*, 163 So.2d 262 (Fla.1964); *State v. Beck*, 156 Me. 403, 165 A.2d 433 (1960); *State v. Knutson*, 178 Neb. 375, 133 N.W.2d 577 (1965); *Pine v. Leavitt*, 84 Nev. 507, 445 P.2d 942 (1968); *Chapdelaine v. Tennessee State Bd.*, 541 S.W.2d 786 (Tenn.1976) (state required surveyors but not engineers to register).

not the power of the state to regulate the professions and occupations.[28]

B. Holding Out

Rodgers v. Kelley involved an action by the plaintiff for design services for a project that was abandoned because of excessive costs. The clients claimed the plaintiff violated the Vermont holding-out statute. The plaintiff contended that he had never signed his name or in any way represented that he was an architect. He testified that he had twelve years of experience in the architectural field but contended that so long as he did not label himself, his plans, or his business with the title "architect" he had not violated the statute. However, the court stated,

> But "holding oneself out as" an architect does not limit itself to avoiding the use of the label. The evidence is clear that, in the community of Stowe, this plaintiff was known as a proficient practitioner of all of the architectural arts with respect to homebuilding, at least. It was a business operation from which he received fees. He presented himself to the public as one who does the work of an architect. This constitutes holding oneself out as an architect, and is part of the very activity sought to be regulated through registration.[29]

C. Practicing

Which activities fall within architectural practice? Design services can range from simply sketching a floor plan or planning to place a residence on a designated site all the way to construction documents with sufficient detail to obtain a bid from the contractor. In addition to the varying design activities, a differentiation can be made between those services that are part of the design process and those services performed by a design professional during construction.

The differentiation between services performed during the designing phase and those performed during construction can be difficult. It has been stated that services performed during the design phase are more likely to be considered within the licensing laws than those performed during the construction phase.[30] This tendency is undoubtedly based on the specialized training and professional skill possessed by a design professional being manifested mainly in the design phase.

The design professional does bring skill to the construction process by interpreting documents, advising on changes, and judging performance. Some of these skills are intimately connected to other design process skills. The generalization may be true if emphasis is placed on construction methods and organization, more properly the province of the contractor.

This differentiation may be of little practical importance. Design professionals rarely perform services connected only with construction. They perform services connected either with design and construction or with design alone. However, as shall be seen in Section 17.04D, the problem may come to a head when the law is asked to determine whether construction managers must have a design professional license, a contractor's license, or neither.

Sometimes the person performs design services for registered design professionals and does not deal directly with the public. For example, a Mississippi case involved an unsuccessful attempt by the state registration board to restrain Rogers from performing engineering services or holding himself out as a mechanical designer.[31] Rogers was not a registered engineer. He operated a small office and did all his work for other architects and engineers and never in any way dealt with members of the public. Mississippi exempted people who performed design services, such as drafters, under the direct control of a registered design professional.

Sometimes the design services are incidental to the selling of equipment or mechanical systems. For example, in *Dick Weatherston's Associated Mechanical Services, Inc. v. Minnesota Mutual Life Insurance Co.*,[32] an insurance company was dissatisfied with advice received from its architects regarding the air-conditioning system. It approached Weatherston, a mechanical contractor with a degree in engineering, for advice. Weatherston made some design

---

[28]*H&V Eng'g, Inc. v. Idaho State Bd. of Professional Eng'r and Land Surveyors*, 747 P.2d 55 (Idaho 1987) ("gross negligence" standard for disciplinary action too vague). But see *Duncan v. Missouri Bd. for Architects, Professional Eng'r and Land Surveyors* reproduced in Section 10.04B (upheld such a standard against constitutional challenge).

[29]128 Vt. 146, 259 A.2d 784, 785 (1969). See also *Carlson v. SALA Architects, Inc.*, 732 N.W.2d 324 (Minn.App.2007), review denied Aug 31, 2007; Annot., 13 A.L.R.4th 676 (1982).

[30]Annot., 82 A.L.R.2d 1013 (1962).

[31]*State Bd. of Registration v. Rogers*, 239 Miss. 35, 120 So.2d 772 (1960).

[32]257 Minn. 184, 100 N.W.2d 819 (1960).

changes but made clear that he was not a registered engineer. These changes were ultimately accepted by architects and engineers retained by the insurance company.

Weatherston claimed and proved he had a valid contract with the insurance company to supply air-conditioning systems. He brought legal action, claiming the insurance company had repudiated the contract. The insurance company claimed Weatherston had been practicing engineering without a license. However, the court concluded after noting that the company knew he was not licensed that his action was brought on the contract to supply the air-conditioning equipment and not a contract for engineering services. The court seemed to assume that those who sell or install mechanical equipment often provide some engineering advice to buyers. When it appears that the principal basis for the claim is the sale of the equipment, the absence of a license under these circumstances should have no effect.

Does testifying as an expert constitute professional practice? For example, in one case a party challenged a lower court verdict because an unlicensed architect had been permitted to testify.[33] The claim was made that testifying was practicing, and violated the licensing laws.

Possession of a license generally does not, however, determine whether a witness will be permitted to testify as an expert. The court permitted the testimony by concluding that testifying was not practicing. However, registration or absence of it goes to the weight of the testimony (how persuasive it is to judge or jury) and not to its admissibility. Expert testimony is discussed in Section 14.06.

D. Architecture and Engineering Compared

States generally regulate architecture and engineering separately. Each profession and the agencies regulating them sometimes differ over where one profession begins and the other ends. These conflicts demonstrate the economic importance of registration as well as the secondary role sometimes played by the public interest.

The two professions can be differentiated by project types and their use. One court stated,

> One prominent architect, in explaining the difference between architecture and engineering, said in effect that

the entire structure and all of its component parts is architecture, if such structure is to be utilized by human beings as a place of work or assembly. He pointed out that, if the authorities were going to erect a courthouse as the building in which the [case] was being tried, they would obtain the service of an architect; but, if it was proposed to construct a power plant . . . they should employ an engineering firm. . . .

> All of the architects and those who were registered as both architect and engineer agreed that the overall plan of a building and its contents and accessories is that of the architect and that he has full responsibility therefor. As one witness answered it, he is the commander in chief.[34]

In *State v. Beck*,[35] the Maine Supreme Court stated aesthetics to be the principal difference between engineering and architecture. To that court an architect was "basically an engineer with training in art." The court also stated,

> While categorically an engineer, the architect—without disparagement toward the professional engineer—is required to demonstrate that he possesses and utilizes a particular talent in his engineering, to wit, art or aesthetics, not only theoretically but practically, also, in coordination with basic engineering.[36]

The court then cited a Louisiana case that stated that an engineer "designs and supervises the construction of bridges and great buildings, tunnels, dams, reservoirs and aqueducts."[37] The Maine court then described architectural projects:

> Architects are commonly engaged to project and supervise the erection of costly residences, schools, hospitals, factories, office and industrial buildings and to plan and contain urban and suburban development. Health, safety, utility, efficiency, stabilization of property values, sociology and psychology are only some of the integrants involved intimately. Banking quarters, commercial office suites, building lobbies, store merchandising salons and display atmospheres, motels, restaurants and hotels eloquently and universally attest to the decisive importance in competitive business of architectural science, skill and taste. A synthesis of the utilitarian, the efficient, the economical, the healthful, the alluring and the blandished is often the difference between

---

[33]*W. W. White Co. v. LeClaire*, 25 Mich.App. 562, 181 N.W.2d 790 (1970).

[34]*State Bd. of Registration v. Rogers*, supra note 31, 120 So.2d at 774.
[35]Supra note 27.
[36]165 A.2d at 435.
[37]*State v. Beck* quoted from *Rabinowitz v. Hurwitz-Mintz Furniture Co.*, 19 La.App.811, 133 So. 498, 499 (1931).

employment and unemployment, thriving commerce and a low standard of existence. Basic engineering no longer suffices to satisfy many demands of American health, wealth or prosperity.[38]

Statutes sometimes use definitions that recognize some differences between architecture and engineering. For example, in New York the practice of architecture is defined as "rendering or offering to render services which require the application of the art, science, aesthetics of design and construction of buildings, groups of buildings, including their components and appurtenances and the spaces around them wherein the safeguarding of life, health, property, and public welfare is concerned."[39] The practice of engineering is defined as "performing professional service such as consultation, investigation, evaluation, planning, design or supervision of construction or operation in connection with any utilities, structures, buildings, machines, equipment, processes, works, or projects wherein the safeguarding of life, health and property is concerned, when such service or work requires the application of engineering principles and data."[40]

Some statutes permit engineers to perform architectural services incident to engineering work. For example, one case allowed an engineer to seal drawings that, in the words of the Kansas statute, he is qualified by education, training and experience to prepare, in effect recognizing the overlap between architecture and engineering.[41]

Yet another case involved an attempt by the state licensing authorities to restrain two officers of a construction company, who were licensed engineers, from designing a seventy-eight-bed nursing home.[42] The engineers contended that the design work they were performing was incident to their engineering services to build the structure. The court recognized that some design services are required for any structure and that this is the proper function of the exemption for architectural services incident to engineering work. However, the building of a nursing home, with both the aesthetic considerations of the human beings who would live there and the aesthetic

aspects of positioning the project, was architectural, and the defendants could not perform these services.

The lack of a clear demarcation between the practice of architecture and the practice of engineering means that friction is endemic to the design professions. By far the most common scenario is for an architect to complain to the architectural licensing board that a nonarchitect (usually an engineer) is engaged in the unauthorized practice of architecture on a particular project. If the board finds against the engineer and assesses a penalty, the engineer challenges the administrative action in court.

The vast majority of courts view these complaints as a turf war between the professions and rule in favor of the accused. Little deference is given to the licensing board, which is not a disinterested examiner. Given the overlap between the two professions, expert testimony is rarely helpful. As observed by one court:

> We are concerned that, on the testimony entered in this record, had Mr. Lomax [the architect who brought the initial complaint] been awarded the project, the Engineers' Board could have assessed civil penalties against him for the unauthorized practice of engineering. It is noted that the engineering expert witnesses testified that the project comprised 80% engineering and 20% architecture, even though the architectural expert witnesses testified that the project was 80% architectural and 20% engineering.[43]

However, the Arkansas Supreme Court came to the opposite conclusion in *Holloway v. Arkansas State Board of Architects*.[44] It rejected the argument of Mr. Holloway, the accused engineer, that the statutes distinguishing architecture from engineering were unconstitutionally vague. It then upheld the board of architecture's assessment of the maximum civil penalty, finding that Holloway's conduct threatened public health and safety.

Sometimes these disputes are resolved by the legislatures, with some intervention by the courts and regulatory agencies. For example, the Michigan constitution required that a majority of the members of any regulatory board be members of the profession being regulated. Subsequently, Michigan passed a statute that required only three members of the seven-member board regulating architects

---

[38]165 A.2d at 437.

[39]N.Y.Educ.Law § 7301.

[40]Id. § 7201.

[41]*Schmidt v. Kansas State Bd. of Technical Professions*, 271 Kan. 206, 21 P.3d 542 (2001).

[42]*Dahlem Constr. Co. v. State Bd. of Examiners*, 459 S.W.2d 169 (Ky.1970). The result was undoubtedly assisted by a special Kentucky statute requiring that architects design nursing homes.

[43]*Rosen v. Bureau of Professional and Occupational Affairs*, 763 A.2d 962, 969 (Pa.Commw.2000), appeal denied, 566 Pa. 654, 781 A.2d 150 (2001). Other courts reaching the same conclusion include *Schmidt v. Kansas Bd. of Examiners*, supra note 41.

[44]352 Ark. 427, 101 S.W.3d 805 (2003).

to be architects. To deal with the prior constitutional requirement, the legislature simply declared architects and engineers to be members of the same profession.

Such a legislative declaration was rejected by a Michigan court that stated the legislature could not declare something that is not so.[45] The court noted that architects and engineers have different educational requirements and serve different functions.

That these turf fights can be resolved harmoniously between the two professions was demonstrated in New Mexico in 1992 by an agreement made among the agencies regulating engineers, surveyors, and architects. The agreement recognized that architecture and engineering overlap and that in some projects either an architect or an engineer can meet the owner's needs and provide a safe, serviceable building. The agreement defined the incidental practice of architecture and engineering "as those services performed on buildings which have a construction valuation of no more than $250,000 or an occupant load of no more than 50." According to the agreement, a single professional seal would meet the requirement on building plans submitted for permits within these limits. The agreement resulted from an order by a court in 1985 that an architect–engineer joint practice committee be established to resolve these turf disputes.

For several years, the committee worked on an agreement, which resulted in the agreement made in 1992 and adopted by the regulatory agencies on March 2, 1992. The agreement was also designed to permit the regulatory boards to focus their efforts on unlicensed practice rather than interdisciplinary disputes.[46]

The issue of the relationship of architecture and engineering, along with other allied professions, surfaces in the issue of who can testify in malpractice claims against certain design professionals. This will be discussed in Section 14.06B.

E. Statutory Exemptions

Small projects do not necessarily merit an architect. In the 1960s, several states by statute exempted projects that did not cost more than a designated amount of money, from a require-

ment to be designed by a licensed architect.[47] Such statutes generated interpretation problems because the cost of a project evolves through various stages. The statutes must also be periodically updated to take inflation into consideration.[48]

Rather than exempt projects based upon their cost, contemporary statutes more commonly exempt projects based upon their size (square footage), height, and type (single-family dwellings are more likely to be exempted). For example, California Business and Professions Code § 5537 exempts "single-family dwellings of woodframe construction not more than two stories and basements in height," multiple-family dwellings with no more than four units of a similar size and construction unless the units form apartment and condominium complexes of over four units, garages and structures appurtenant to exempt dwellings, and agricultural or ranch buildings—unless public officials deem "that an undue risk to the public health, safety or welfare is involved."[49]

F. Possessor of License

Design professionals traditionally performed as sole proprietors or partners. Increasingly, for tax and other reasons, design professionals are being permitted to perform design services through the corporate structure, a concept discussed in Section 3.06.

Most regulation of the design professions assumes individual practitioners. Yet increasingly, design professionals practice in what is known as "firm practice," such as partnerships, corporations, and limited liability companies. Compliance within a state can be difficult, as the rules and regulations often do not keep up with new forms of practice. But since many firms practice in more than one state, the problems become extraordinarily complex. While there is general similarity of regulations of individual practitioners, there are confusingly complex and often contradictory rules for firm practice.

Although a number of issues must be addressed, one commentator states that many states "require licensees

---

[45]*Nemer v. Michigan State Bd.*, 20 Mich.App. 429, 174 N.W.2d 293 (1969).

[46]Press release, Office of the Governor, Santa Fe, New Mexico (May 13, 1992).

[47]A collection of these statutes, now virtually all repealed, may be found in *State v. Spann*, 270 Ala. 396, 118 So.2d 740 (1960).

[48]For example, as of October 1, 2007, North Carolina increased from $700,000 to $1,000,000 the value of a project that may be built by a general contractor holding an "intermediate license." N.C.Gen.Stat.Ann. § 87–10, amended, 2007 N.C. Laws, S.L. 2007–247.

[49]Similar statutes are Iowa Code Ann. § 544A.18 and Tenn. Code Ann. § 62-6-103(a)(2)-(5).

to hold certain defined positions with the firm." This, he states, "is the area of greatest variation." He states that the issue that requires closest attention is whether the right people in the firm have the right licenses, with particular reference to who is in charge of services and who can seal and sign drawings.[50]

G. Out-of-State Practice

The practice of architecture and engineering as well as the performance of construction services increasingly crosses over state lines. As a result, a person licensed in one state often performs services in another.

Sometimes design professionals who perform services on a multistate basis obtain licenses in each state in which they perform services or in which projects for which they perform services are located. This has become easier because of standardized examinations and increased reciprocity.

Some registration laws do not require professionals licensed in a foreign state to become registered if they perform work only for an isolated transaction or perform work not to exceed a designated number of days. Often such exemption requires an easily obtained temporary license.

As interstate practice has become more common, an increasing number of cases have dealt with attempts by design professionals licensed in one state to recover for services related to a project in another. For example, in *Johnson v. Delane*[51] an engineer licensed in the state of Washington obtained a commission to prepare plans and specifications for a project to be built in Idaho. The engineer was not to perform any supervisory function. He obtained the commission while visiting in Idaho. He performed the requisite design services in Washington and delivered the plans and specifications to the client in Idaho. The court held that he was not practicing architecture in Idaho, and he recovered for his services.

Some state laws do not require a license for performance of services by out-of-state design professionals if they are licensed in the state where they practice principally and as long as there is a licensed design professional in overall charge of the project. However, the licensed local design professional must not be a figurehead but must actually perform the usual design professional functions. Sometimes, as in *Food Management, Inc. v. Blue Ribbon Beef Packing, Inc.*, out-of-state architects or engineers "associate" a local architect as a means of ensuring that they will be able to collect their fee.[52]

In *Hedla v. McCool*,[53] a Washington architect performed design services for a project to be built in Alaska and had the plans approved by an Alaskan engineer. The court emphasized the importance of the policy expressed in Alaska's licensing law. The court distinguished this case from *Johnson v. Delane*[54] (discussed earlier) by noting that the Alaska statute was broader, that this was not an isolated transaction, and that conditions in Alaska were different from those in the state of Washington. The court was more persuaded by the holding in the *Food Management* case, which had not looked kindly on out-of-state architects associating a local architect when the out-of-state architects were principally responsible for design decisions.[55]

One important consideration permeating these cases is whether the unlicensed person seems to have performed properly. If so, denial of recovery, despite the importance of license compliance, may seem unjust. For example, in *Delane* the work appears to have been performed correctly. But in *Hedla v. McCool*, where recovery was denied, the costs overran considerably and the design was never used—factors that make recovery less attractive.

The expansion of reciprocity and the ease with which out-of-state design professionals who are registered in their home states can be allowed to practice in other states may make it more difficult for design professionals to recover if they do not use these techniques for legitimating their projects in states where they are not licensed. In any event, any design professionals considering performing design services either in another state or for a project that will be built in another state should receive legal advice on the proper process for ensuring that they are not violating the laws of the state where the services are being performed or the project is located.

---

[50]Noble, "DON'T LEAVE HOME WITHOUT IT—A Strategy for Complying with Interstate Architectural Practice Rules," *UNDER CONSTRUCTION*, Newsletter of the ABA Forum on the Construction Industry, p. 1 (Dec. 2001).

[51]77 Idaho 172, 290 P.2d 213 (1955).

[52]413 F.2d 716 (8th Cir.1969).

[53]476 F.2d 1223 (9th Cir.1973). Similarly, a local association failed in *O'Kon and Co., Inc. v. Riedel*, 588 So.2d 1025 (Fla.Dist.Ct.App.1991).

[54]Supra note 51.

[55]Supra note 52.

H. Substantial Compliance

Increasingly, attempts by unlicensed contractors to recover have been based on asserted "technical" noncompliance with the law. This is not, strictly speaking, a contention that the law has not been violated. As a result, this doctrine is discussed in Section 10.07B and, inasmuch as contractors have been the principal users of this approach, in Section 10.10C.

SECTION 10.07 Sanctions for Licensing Law Violations

A. Criminal and Quasi-Criminal Sanctions

Licensing laws usually carry criminal sanctions. Violations can be punished by fine or imprisonment. However, use of the criminal sanction is relatively rare. Perhaps occasional instances have occurred where fines have been imposed, although it is quite unlikely, though possible, that imprisonment will be ordered. Where a penal violation is found, it is likely that sentence will be suspended or probation granted.[56]

Increasingly, states are permitting victims of crimes to receive restitution for losses caused by those who commit crimes. For example, Washington has allowed restitution as a condition of probation.[57] Where restitution is denied, denial is based upon the defendants not having protections available in civil actions, such as pleading requirements, discovery, and a jury trial. The Washington court only allowed restitution to correct dangerous defects.

Arizona provides a constitutional right to restitution to the victim of crime and has implemented statutes to provide the details. The statute allows lost earnings and interest but not recovery for pain and suffering, punitive damages, or consequential damages.[58]

In *State v. Wilkinson*,[59] the Arizona Supreme Court applied this statute in a criminal action against an unlicensed contractor (criminal fine was $2,500 while the claims for restitution were for $45,000). The court held the statute was designed to force the criminal to yield up to the victim the fruits of his crime. To do this, the victim can recover payments made but not the cost of remedying the defective work or completing the work.

A quasi-criminal sanction can be invoked. If the licensing authorities obtain a judgment from the court ordering that the unlicensed design professional cease practice and the order is disobeyed, failure to comply will be contempt of court and punishable by a fine or imprisonment. Just as law enforcement officials rarely have the staff or resolve to seek criminal sanctions, the regulatory agencies rarely seek a judicial order that the violator is in contempt.

In 1985, the California Code of Civil Procedure Section 1029.8 added another quasi-criminal sanction. It *trebles* any actual damages (additional damages cannot exceed $10,000) and allows the court to award attorneys' fees if an unlicensed person negligently "causes injury or damage to another person." This sanction does not apply to people who believe in good faith that they are licensed and to people who have failed to renew their license when renewal would have been automatic.

Although the statute appears directed to personal injury claims, it can be read broadly to include damage to property and even economic losses. (Section 253 of the California Insurance Code was added at the same time, stating that coverage of such a treble damage award cannot be insured against.) Such a statute will encourage smaller claims to be taken to court.

B. Recovery for Work Performed and Payment Refunds

The infrequent use of criminal or quasi-criminal sanctions means that the principal sanction for unlicensed practice of architecture, engineering, or construction work has been to deny recovery for work performed. Attempts by unlicensed persons to recover for their work has been the main legal battleground.

Generally, neither party can enforce an illegal contract. The law provides no help if the parties are equally guilty. If a citizen bribes a public official, the citizen cannot sue to have the promised performance made nor can the public official sue to recover the promised bribe.

Courts do not view parties who hire unlicensed contractors as analogous to those who bribe public officials. In the former situation, not only are the parties not equally guilty, they might both be ignorant of the violation (for example, if the contractor lets its license lapse through inadvertence). Also, the illegality of the two situations is

---

[56]However, in *Huffman v. State*, 356 Md. 622, 741 A.2d 1088 (1999) the unlicensed contractor was sentenced to jail, given probation, and also fined.

[57]*State v. Bedker*, 35 Wash.App. 490, 667 P.2d 1113 (1983) (crime was failure to procure a bond).

[58]Ariz.Rev.Stat.Ann. § 13–105.14.

[59]202 Ariz. 27, 39 P.3d 1131 (2002).

not comparable in terms of intention of the parties, morality, or public policy. In legal parlance, bribing officials is a *malum in se*—intrinsically illegal—while violating a licensing law is a *malum prohibitum*: admittedly a prohibited act, but disallowing enforcement of the contract would be entirely disproportionate to the illegality. For these reasons, the client can enforce its contract against the unlicensed architect or contractor and maintain an action for faulty performance.[60] The more difficult questions are whether the contractor may recover for work properly performed and whether the client may recoup payments already made under the contract.

An unlicensed contractor's right of recovery *on the contract* for work properly performed depends on the wording of the statute the contractor violated. Three types of statutes exist: those that specify the contractor's right to compensation; those that are silent on the contractor's right to compensation; and those that expressly prohibit the contractor's right to compensation. Tennessee Code Section 62-6-103(b) is an example of the first type of statute, one that specifies a remedy for the contractor. It allows an unlicensed contractor to recover in a court of equity if it can show clear and convincing proof of actual documented expenses. Profits cannot be recovered. Very likely overhead would not be recoverable.[61]

Statute Silent on Contractor's Right to Compensation.
Where the statute does not specify that its violation will deprive the contractor of its right to compensation, courts generally view the illegality as a *malum prohibitum* and will not impose upon the contractor the penalty of forfeiture. One case from California and another from Massachusetts illustrate the courts' reasoning.

In *MW Erectors, Inc. v. Niederhauser Ornamental and Metal Works Co., Inc.*,[62] a sub-subcontractor on a hotel construction project entered into an *ornamental steel* contract with the defendant, a metal works company. (The sub-subcontractor's second, *structural steel* contract is discussed later in this Section.) The sub-subcontractor was unlicensed when it *signed* the ornamental steel contract but licensed when it *performed* the work.

The defendant argued that it was illegal for the sub-subcontractor to sign the contract while unlicensed. The defendant relied upon California Business and Professions Code Section 7028(a), which makes it a misdemeanor for a person to "engage in the business or act in the capacity of a contractor within this state without having a license therefor." The defendant took the position that the contract was illegal and therefore void; that is, entirely unenforceable by the sub-subcontractor.

The California Supreme Court allowed the sub-subcontractor recovery under the contract. It agreed that, as a general rule, a contract made in violation of a regulatory statute is void. The contractors' licensing law is regulatory because its purpose is to protect the public. For this reason, courts will not enforce an agreement for contractor services executed by a person who was not licensed to perform them. But that rule, the court continued, did not compel the conclusion that a contractor should be denied any recovery where it was licensed at all times during performance, even though not licensed when it signed the construction contract. In addition, a statute which makes conduct illegal and also provides for a fine or administrative discipline (as do the licensing laws) by implication precludes the court from imposing additional penalties. Also, where a contract's illegality was not a *malum in se*, it is voidable but not necessarily void; courts will declare the contract void only if public policy compels such a harsh measure. The court concluded:

> We see no such necessity [for forfeiture] here. As indicated, the CSLL [Editor: Contractors' State License Law] expressly provides multiple means of enforcing the general ban on acting as an unlicensed contractor, insofar as that prohibition includes the mere execution of contracting agreements while unlicensed. Though the Legislature barred recovery of compensation by unlicensed contractors under certain circumstances, it did not impose this bar against contractors who, though licensed at all times during performance of contracting work, had executed agreements for the work while unlicensed. No compelling reason exists to conclude that the public protective purposes of the CSLL can only be served by deeming such contracts illegal, void, and unenforceable on that basis alone.[63]

Town Planning & Engineering Associates, Inc. v. Amesbury Specialty Co., Inc. involved a contract for design services

---

[60]*Hedla v. McCool*, supra note 53; *Domach v. Spencer*, 101 Cal. App.3d 308, 161 Cal.Rptr. 459 (1980) (unlicensed contractor); *Cohen v. Mayflower Corp.*, 196 Va. 1153, 86 S.E.2d 860 (1955).

[61]The statute was applied in *Roberts v. Houston*, 970 S.W.2d 488 (Tenn.App.1997).

[62]36 Cal.4th 412, 115 P.3d 41, 30 Cal.Rptr.3d 755 (2005).

[63]Id., 115 P.3d at 59.

made by an unlicensed designer who had been approached by a client. The unlicensed designer retained two registered professional engineers as well as a registered land surveyor to perform services related to the project. However, much of the work was done by the designer and his six drafters. The project was ultimately abandoned when the bids were too high. At first the client told the unlicensed designer to renegotiate the bids or get lower ones, but shortly thereafter the client terminated the contract, claiming it was illegal. Subsequently, the client negotiated directly with a contractor, obtained a reduced bid for a somewhat scaled-down project, and to some degree used the unlicensed designer's plans.

The court noted that the plaintiff's work had been properly performed. In holding that the plaintiff could recover, the court stated,

> If there was a violation here, it was punishable as a misdemeanor under the statute. Violation of the statute, aimed in part at least at enhancing public safety, should not be condoned. But we have to ask whether a consequence, beyond the one prescribed by statute, should attach, inhibiting recovery of compensation, and we agree with the judge in his negative answer to the question in the present case. To find a proper answer, all the circumstances are to be considered and evaluated: what was the nature of the subject matter of the contract; what was the extent of the illegal behavior; was that behavior a material or only an incidental part of the performance of the contract (were "the characteristics which gave the plaintiff's act its value to the defendant . . . the same as those which made it a violation . . . of law"); what was the strength of the public policy underlying the prohibition; how far would effectuation of the policy be defeated by denial of an added sanction; how serious or deserved would be the forfeiture suffered by the plaintiff, how gross or undeserved the defendant's windfall. The vector of considerations here points in the plaintiff's favor.
>
> Our cases warn against the sentimental fallacy of piling on sanctions unthinkingly once an illegality is found. As was said in *Nussenbaum v. Chambers & Chambers, Inc.,* 322 Mass. 419, 422, 77 N.E.2d 780, 782 (1948): "Courts do not go out of their way to discover some illegal element in a contract or to impose hardship upon the parties beyond that which is necessary to uphold the policy of the law." Again the court said in *Buccella v. Schuster,* 340 Mass. 323, 326, 164 N.E.2d 141, 143 (1960), where the plaintiff was allowed to recover his compensation for blasting ledge on the defendant's property although he had not given bond or secured a blasting permit

as required by law: "We do not reach the conclusion that blasting without complying with the requirements . . . is so repugnant to public policy that the defendant should receive a gift of the plaintiff's services." Professor Corbin adds: "The statute may be clearly for protection against fraud and incompetence; but in very many cases the statute breaker is neither fraudulent nor incompetent. He may have rendered excellent service or delivered goods of the highest quality, his non-compliance with the statute seems nearly harmless, and the real defrauder seems to be the defendant who is enriching himself at the plaintiff's expense. Although many courts yearn for a mechanically applicable rule, they have not made one in the present instance. Justice requires that the penalty should fit the crime; and justice and sound policy do not always require the enforcement of licensing statutes by large forfeitures going not to the state but to repudiating defendants." 6A A. Corbin, Contracts § 1512, at 713 (1962) [footnote omitted].[64]

Statute Bars Contractor's Right to Compensation. California Business and Professions Code § 7031(a) is an example of a statute that prohibits an unlicensed contractor from suing the client for compensation for work performed. Section 7031(a) states that no contractor

> may bring or maintain any action, or recover in law or equity in any action, in any court of this state for the collection of compensation for the performance of any act or contract where a license is required by this chapter without alleging that he or she was a duly licensed contractor *at all times during the performance* of that act or contract, regardless of the merits of the cause of action brought by the person . . . [italics added].

The California Supreme Court in the *MW Erectors* case also addressed Section 7031(a), this time when discussing the *structural steel* contract signed by the sub-subcontractor. Here, the sub-subcontractor was *unlicensed* when it signed the structural steel contract *and also during the first month of work.* So part of the actual work was done while the sub-subcontractor was unlicensed. Applying the plain wording of § 7031(a), the court ruled that, because the sub-subcontractor was not licensed "at all times" during performance of the structural steel contract, it could not recover *any* compensation under that contract.

---

[64]369 Mass. 737, 342 N.E.2d 706, 711–712 (1976).

The effect of the court's decision is that a contractor who becomes unlicensed any time during performance faces the threat of forfeiture. The *MW Erectors* court had before it briefs from amici curiae ("friends of the court" who file briefs in addition to those filed by the parties). These amici curiae cautioned the court that its ruling would encourage contractors to abandon a project once they discovered even a momentary gap in licensure, because they will have forfeited the right to any compensation under the contract. In response, the court noted that Section 7031 does not prevent the owner from paying for work voluntarily and that "[b]usiness considerations may persuade the beneficiary to ignore license lapses it deems insignificant, and to continue compensating the contractor, in order to avoid disruption of progress on the project."[65]

The court's reasoning is brought into question by the legislature's passage in 2001 of Section 7031(b), which allows the owner to obtain reimbursement of payments it had made to the unlicensed contractor. Under § 7031(b), a devious owner may coax the contractor to continue performance by paying it. Once the work is nearly complete, the owner could then recoup these payments based upon the contractor's lack of a license for even a brief time (assuming the substantial compliance doctrine does not apply).

The complications from the court's ruling do not end there. One argument not mentioned by the court is that a contractor who abandons a project fearing it has lost its right to compensation would subject itself to loss of its license.[66] The threat of losing its license should prevent most contractors from choosing abandonment.

When faced with a statute that expressly bars an unlicensed contractor's right to compensation (such as California Business and Professions Code § 7031(a)), the contractor may attempt recovery not under the contract, but based on restitution. The contractor claims it has conferred a benefit on the other party and that leaving the parties in status quo could cause unjust enrichment.[67] The unlicensed contractor might also claim that the hiring party knew of its unlicensed status but represented that

this would not be a problem. Under these facts, denial of restitution could be viewed as the courts assisting in the furtherance of a fraud.

In *Hydrotech Systems, Ltd. v. Oasis Waterpark*,[68] the California Supreme Court rejected both the restitution and fraud arguments. A California corporation wished to build a waterpark in California using a New York vendor's wave-generating equipment. The vendor wanted to sell but not install the equipment, because it was not licensed in California. The prime contractor insisted the vendor install the equipment, and it did so. When the vendor later sued for payment, the owner and prime contractor raised its unlicensed status as a defense. Even under these compelling facts, the California high court refused to recognize either a fraud or an "isolated transaction" exception to Section 7031. Nor was an exception made because the vendor was a subcontractor who dealt with a prime contractor, not with a member of the public.

In *Stokes v. Millen Roofing Company*,[69] homeowners knowingly hired an unlicensed contractor to install a roof. When the unpaid contractor filed a mechanics' lien, the owners asked the court to have it discharged because the contractor was unlicensed. The licensing statute provided that a "person . . . shall not bring or maintain an action . . . for the collection of compensation for the performance of an act or contract for which a license is required by this article. . . ."[70]

The court denied even reimbursement of the costs of the contractor's supplies, finding that such a recovery would constitute "compensation" within the meaning of the statute. But the case produced four opinions: the majority, two concurrences, and a dissent. All four opinions addressed the same issue: the equity of preventing payment based solely on the contractor's unlicensed status, where the contractor produced quality workmanship and the owners hired him knowing he was unlicensed. Most troubling was evidence that in a *separate* case involving the same improvement, the owners had hired another unlicensed contractor and then refused to pay him because he was unlicensed. It appeared that the owners were abusing the statutory scheme: hiring unlicensed contractors and then hiding behind their lack of a license in order to deny them payment.

---

[65]Supra note 62, 115 P.3d at 51 n. 10.

[66]West Ann.Cal.Bus. & Prof. Code § 7107, discussed later in this Section.

[67]*C. B. Jackson & Sons Constr. Co. v. Davis,* 365 So.2d 207 (Fla.Dist. Ct.App.1978) (recovery based on unjust enrichment); *West Baton Rouge Parish School Bd. v. T. R. Ray, Inc.,* 367 So.2d 332 (La.1979) (corporate architect's licensing employee quit after contract made; corporation cannot arbitrate but can recover based on unjust enrichment).

[68]52 Cal.3d 988, 803 P.2d 370, 277 Cal.Rptr. 517 (1991).

[69]466 Mich. 660, 649 N.W.2d 371, reh'g denied, 467 Mich. 1202, 651 N.W.2d 920 (2002).

[70]Mich.Comp.Laws § 339.2412(1).

Sophisticated Parties. While claims by and against unlicensed contractors will be discussed in Section 10.10, one difficulty faced by the law is revealed when the claim is made by an unlicensed subcontractor against a prime contractor who knew the subcontractor was unlicensed. If the purpose of the law is to protect homeowners who may not be able to determine the competence of contractors, the unlicensed subcontractor should recover. But if the state enacts licensing laws to limit those who perform construction services to those who have met the competence required by the state and to encourage contractors to become licensed, the unlicensed subcontractor should not recover for her work even if the prime knew she was not licensed.

The difficulty of reconciling these objectives is shown by the different judicial attitudes when the contract is made between contractors rather than between a contractor and an owner.[71] Although this factor may not be determinative, recovery by the unlicensed contractor is more likely if the other party was also a contractor.

Despite some recent criticism of licensing laws noted in Section 10.02B, the laws generally are still received favorably by the courts. Even in the absence of an express statutory provision dealing with the right to collect, unlicensed persons generally are not able to recover either on the contract or on unjust enrichment.[72]

Payment Reimbursement. May a client recoup payments already made during performance to an unlicensed contractor? As always, analysis of this question must begin with the governing statute. In 2001, the California legislature added subsection (b) to California Business and Professions Code § 7031, which allows "a person who utilizes the services of an unlicensed contractor [to] bring an action . . . to recover all compensation paid to the unlicensed contractor for performance of any act or contract." Although no appellate court has interpreted this statute, its intention to grant the client a right of reimbursement seems clear.

Some states allow the owner to recover payments it already made based on common law principles.[73] Also, as seen in Section 10.07A, another method to recover payments made has been in those states that allow a victim of crime to recover restitution from the criminal.[74]

Taken together, these statutes and cases may signal the increased importance of registration and licensing laws and a tougher attitude toward those who perform under such illegal contracts. However, the Washington Supreme Court has taken a contrary view, holding that an unlicensed contractor "may assert completion of the work under the contract as a defense to a claim by a customer for reimbursement."[75]

Some courts give partial relief to the unlicensed person, permitting her to set off compensation for work that has been performed against any claim made against the unlicensed person. For example, suppose an unlicensed architect had furnished services for which she was still owed $10,000. Suppose that the client on an unrelated matter had $15,000 coming from the unlicensed architect. In such a case, if the client brought a claim for the $15,000, some courts would permit the setoff of $10,000 for work performed under the illegal contract to reduce the claim against the unlicensed architect to $5,000.[76]

A claim usually seeks a money award. But other uses can be made of the claim. For example, suppose the unlicensed person files a mechanics' lien based on a statute giving security to the person who has improved another's property. Although it can be asserted that this remedy is given to avoid unjust enrichment, some courts, pointing to the need for a valid contract, have denied the right to impose a lien.[77] Yet one court allowed a lien despite the illegality of the contract when the obligation was restitutionary based on unjust enrichment.[78] If the Mechanics'

---

[71]Compare *Enlow & Son, Inc. v. Higgerson,* 201 Va. 780, 113 S.E.2d 855 (1960); (granted recovery) with *Opp v. St. Paul Fire & Marine Ins. Co.,* 154 Cal. App.4th 71, 64 Cal.Rptr.3d 260 (2007); *Triple B Corp. v. Brown & Root, Inc.,* 106 N.M. 99, 739 P.2d 968 (1987) and *Frank v. Fischer,* 108 Wash.2d 468, 739 P.2d 1145 (1987) (denied recovery). See Section 10.10A.

[72]*Food Management, Inc. v. Blue Ribbon Beef Pack, Inc.,* supra note 52. The decision cites many cases from a number of jurisdictions denying recovery. See also *Southern Metal Treating Co. v. Goodner,* 271 Ala. 510, 125 So.2d 268 (1960). An earlier edition of this book is cited in *Gerry Potter's Store Fixtures, Inc. v. Cohen,* 46 Md.App. 131, 416 A.2d 283, 285 (1980).

[73]*Kansas City Community Center v. Heritage Industries, Inc.,* 972 F.2d 185 (8th Cir.1992); *Cevern, Inc. v. Ferbish,* 666 A.2d 17 (D.C.1995); *Ransburg v. Haase,* 224 Ill.App.3d 681, 586 N.E.2d 1295, appeal denied, 145 Ill.2d 644, 596 N.E.2d 637 (1992) (to protect the public); *Mascarenas v. Jaramillo,* 111 N.M. 410, 806 P.2d 59 (1991) (consumer knew contractor was not licensed).

[74]*State of Arizona v. Wilkinson,* supra note 59.

[75]*Davidson v. Hensen,* 135 Wash.2d 112, 954 P.2d 1327, 1336 (1998).

[76]*Sumner Dev. Corp. v. Shivers,* 517 P.2d 757 (Alaska 1974); *S & Q Constr. Co. v. Palma Ceia Dev. Org'n,* 179 Cal.App.2d 364, 3 Cal.Rptr. 690 (1960).

[77]*Sumner Dev. Corp. v. Shivers,* supra note 76; *Chickering v. George R. Ogonowski Constr. Co., Inc.,* 18 Ariz.App. 324, 501 P.2d 952 (1972).

[78]*Expert Drywall, Inc. v. Brain,* 17 Wash.App. 529, 564 P.2d 803 (1977). Cf. *Bastian v. Gafford,* 98 Idaho 324, 563 P.2d 48 (1977).

Lien statute itself requires that the lien claimant be licensed, the court will deny a lien even if there are "equities" in favor of the lien claimant.[79]

Suppose the parties to the contract arbitrate the dispute and the arbitrator grants an award to the unlicensed person. Will a court confirm it? One court, stressing the importance of the licensing statute (the case was decided in 1949) refused to confirm the award.[80] Another case, decided in 1982, stressing that the arbitrator could resolve the legal issues, confirmed the arbitration award.[81]

In a 1988 case, the owners discovered soon after the arbitration hearing had ended that their contractor was unlicensed. The arbitrator denied the owners' request to reopen the hearing and entered an award in favor of the contractor. The court affirmed the award, stressing the need for finality in the arbitration process and observing that the contract, while illegal, was not so repugnant to public policy as to be void *ab initio* ("from the start").[82]

Suppose the claimant presents its claim to the bankruptcy court when the other party to the contract has been adjudged a bankrupt. An opinion allowed the claim, for whatever it will be worth, based on the broad discretion given to the bankruptcy court.[83] The court was influenced by the bankrupt having been enriched and having acquiesced in continued performance by the claimant after the bankrupt knew the performing party was not licensed. (This claim will come at the expense of the other unsecured creditors, not of the bankrupt.)

C. Summary

Although it must still be generally stated that unlicensed persons are not able to recover based on either the contract they have made or unjust enrichment, some factors may motivate a court to determine that the statute was not violated, substantial compliance is sufficient, or recovery will be had despite a licensing violation. Such factors are as follows:

1. The work for which recovery is sought conforms to the contract requirements.
2. The party seeking a defense based on a violation of the licensing law is in the same business or profession as the unlicensed person.
3. The unlicensed person apparently had the qualifications to receive the license.

SECTION 10.08 Should Contractors Be Licensed?

A. Purpose of Licensing Laws

To protect the public, the state regulates professions, occupations, and businesses. Those who wish to build must retain a contractor and often a designer. Licensing laws should provide a representation to the public that the holder meets a minimal level of competence, honesty, and financial capacity. The state wishes to protect the general public from being harmed by poor construction work and seeks to accomplish this by allowing only those who meet state requirements to design and to build. Yet as seen in this chapter, these efforts are not costless. Educational and experience requirements and the administration of these regulatory systems increase the cost of providing these services. They also reduce the pool of contractors.

Professional designers usually must meet certain educational and training requirements before they are allowed to take the examination that will determine whether they will be registered. Although these requirements do raise the cost of entering the profession, it can be contended that they improve the quality of designers.

As a rule, contractors need meet only experience requirements and not educational ones. (The California application fee is $300.)[84] In California, for example, they must have four years of experience in the last ten years in the trade they wish to enter. Under certain circumstances, particular education courses taken can fulfill up to three years of this four-year experience requirement.[85] Applicants in California take a one-day written examination, half of which is devoted to law and the other half to trade practices. As seen in the Florida experiences described in Section 10.02B, the "pass" line can be drawn

---

[79]*Stokes v. Millen Roofing Co.*, supra note 69.

[80]*Loving & Evans v. Blick*, 33 Cal.2d 603, 204 P.2d 23 (1949).

[81]*Parking Unlimited, Inc. v. Monsour Medical Found.*, 299 Pa.Super. 289, 445 A.2d 758 (1982).

[82]*Davidson v. Hensen*, supra note 75; see also *Migdal Plumbing & Heating Corp. v. Dakar Developers, Inc.*, 232 A.D.2d 62, 662 N.Y.S.2d 106 (1997), lv. denied, 91 N.Y.2d 808, 692 N.E.2d 130, 669 N.Y.S.2d 261 (1998).

[83]*In re Spanish Trails Lanes, Inc.*, 16 B.R. 304 (Bkrtcy.Ariz.1981).

---

[84]West Ann.Cal.Bus. & Prof.Code § 7137. Legislation enacted in 1996 allows fees to be reduced if revenues are adequate for certain purposes. Id. § 7138.1.

[85]16 Cal.Code Regs. § 825.

arbitrarily. Can examiners determine what is a reasonable level of classroom performance? Can the classroom simulate actual construction? Even these minimal examination requirements can be waived under certain circumstances under the California Business and Professions Code Section 7065.1.

How will the examiners determine the integrity of the person who applies for a license? This is undoubtedly accomplished by letters of reference (of dubious value) and perhaps other investigations. Yet will investigations uncover much that will reflect the integrity of the contractor? How will the regulatory agency determine financial responsibility? In California financial solvency currently requires a working capital that exceeds $2,500.[86] An applicant must post a bond for $12,500 in California.[87] Is this bond, which covers more than complaints by homeowners, adequate to establish financial capacity?

California has augmented these relatively modest bond requirements. The California Business and Professions Code requires a contractor to post a bond that equals any unsatisfied final judgment against the contractor that is substantially related to the qualifications, functions, or duties of the license applied for.[88] Again, as seen in Section 10.04B, respecting the supremacy of the federal bankruptcy law, if there has been a discharge of the unsatisfied judgment in a bankruptcy proceeding, the bond need not be furnished.[89] Finally, a contractor who has been disciplined by having had its license suspended or revoked can be forced to file an additional bond of not less than $15,000 or ten times the ordinary bond if it seeks to have its license restored.[90]

What must a contractor do to keep a license in effect? In California, the license must be renewed every two years.[91] The cost of renewal for an active contractor is $360.[92] An inactive contractor can renew for $180.[93] There is no provision for periodic examination or continuing education.

---

[86]West Ann.Cal.Bus. & Prof.Code § 7067.5.

[87]Id. § 7071.6.

[88]Id. § 7071.17.

[89]Ibid.

[90]Id. § 7071.8.

[91]Id. § 7140.

[92]Id. § 7137.

[93]Ibid.

Are the undeniable costs that such a system imposes more than balanced by helping *those who need help* to determine who is competent? This raises additional questions. Who needs the protection, and who gets it? Clearly, such legislation was intended to protect those who are inexperienced in the world of construction or who cannot without great cost obtain sufficient information to make judgments. Owners that fall into either category are more likely to make judgments based on whether a contractor is licensed and bonded.

The inexperienced owner building a residence, a duplex, or a four-unit apartment or the naive owner about to engage in home improvement may be helped by the existence of licensing laws. But does that justify an across-the-board licensing requirement and an expensive licensing apparatus? Particular abuses can be dealt with by already abundant consumer legislation.

In addition to these questions, does this system deter unlicensed persons from building? As seen in Section 10.10, often an unlicensed contractor can use the courts to recover for services performed.

B. Harmful Effects

Suppose contractor licensing laws neither screen out the incompetent nor only protect those who need protection. Do they do any harm?

Contractor licensing laws can artificially reduce the pool of contractors, as demonstrated by the Florida experience described in Section 10.02B.

Contractor licensing statutes usually draw lines one cannot exceed between prime contractors and the specialized trades. In addition to the difficulty of drawing these lines, the segregation of specialty trades by licensing tends to reduce the likelihood of better organized, more efficient organizational structures—something so needed in construction.

Other dangers exist. It is not uncommon for licenses to be "bought" and "sold" like any other valuable commodity. An owner who wishes to build a residence or small commercial structure and avoid the cost of a full-time prime contractor sometimes engages an individual with a license, gives that person very little overall control, and limits her activity to periodic visits. This is done to avoid the owner's being liable for violating the licensing laws and gives a licensed contractor a saleable commodity for a relatively small investment.

The issuance of a license can be a false representation by the state that may deceive the unwary consumer, that is, state-sponsored "consumer fraud." Consumers may believe they are dealing with a competent, honest, and financially reliable builder because the builder is licensed. To obtain business from a gullible public, many builders advertise that they are licensed and bonded. Eliminating contractor licensing laws might induce consumers to protect themselves.

SECTION 10.09 Contractor Licensing Laws

Along with the general proliferation in occupational licensing, there has been an increase in contractor licensing laws. At present about half of the states have such laws. Perhaps the deregulation spirit seen briefly in Arizona[94] may signal a slowdown or even a rollback. But ideological movements such as deregulation often lack the staying power of organized groups, such as consumer groups and contractors, both prime movers in occupational licensing.

Yet inevitably, licensing laws demonstrate the tradeoffs inherent in the political process. Even though in 1987 Arizona went back to a traditional contractor licensing law, it specified sixteen exemptions. For example, owners who improve their own property by themselves or jointly with a licensed contractor are exempt if the structure is not intended for sale or rental. Similarly, employees of owners of condominiums, townhouses, cooperatives, or apartments of four units or fewer need not have a contractor's license. Also, a surety company that undertakes to complete a contract under the terms of a bond needs no license, provided the work is performed by licensed contractors.[95]

Many states exempt certain projects from the ambit of the contractor licensing laws. For example, North Carolina exempts "furnishing or erecting industrial equipment, power-plant equipment, radial-brick chimneys, and monuments"[96] Alabama exempts emergency work for public utility or power activity as long as the work is performed under the supervision of a licensed architect or engineer.[97]

Evidently, these states believe that owners in these projects can protect themselves from incompetent contractors

or have persons with construction skills who can supervise the work.

The sanctions imposed on the unlicensed contractor are discussed in detail in Section 10.10. However, many states have statutes that specifically bar an unlicensed contractor from recovering for work performed, something usually not found in architectural and engineering registration laws.

Yet some legislatures have sought to induce contractors to become licensed by granting benefits only to licensed contractors. For example, California permits only a licensed contractor to hold a supplier of goods to any orally communicated price proposal made under certain circumstances.[98]

Contractor licensing laws have also regulated the formation and performance of construction contracts. For home improvement contracts, California requires registration of door-to-door salespersons[99] and prescribes the contracts' essential terms. For example, these contracts must identify the contractor, list a start and estimated ending date, notify the owner of lien provisions and methods to avoid liens, and provide notice of a right to cancel within three days after contract signing.[100] Violation of these provisions is grounds for disciplinary action against the contractor by the Contractors Licensing Board.[101]

More important, California licensing laws include grounds for disciplinary action against the contractor that appear to be no more than simple breach of contract. For example, the following can be grounds for disciplinary action:

1. abandonment of project[102]
2. willful departure from accepted trade standards or "from or disregard of plans and specifications in any material respect."[103]
3. material failure to complete project for price stated[104]
4. failure to prosecute work diligently[105]

These provisions can cause public intervention into relations between owner and contractor (discussed in greater detail in Section 10.11).

---

[94]Refer to Section 10.02B.
[95]Ariz.Rev.Stat.Ann. § 32-1121.
[96]N.C. Gen. Stats § 87–1.
[97]Ala. Code § 34–8–7.

[98]West Ann.Cal.Comm.Code § 2205.
[99]West Ann.Cal.Bus. & Prof.Code §§ 7152–58.
[100]Id. § 7159.
[101]*Handyman Connection of Sacramento, Inc. v. Sands*, 123 Cal. App.4th 867, 20 Cal.Rptr.3d 727 (2004).
[102]West Ann.Cal.Bus.&Prof.Code § 7107.
[103]Id. § 7109.
[104]Id. § 7113.
[105]Id. § 7119.

SECTION 10.10 The Unlicensed Contractor: Civil Sanctions

A. Recovery for Work Performed

The chief sanction for violating licensing laws has been to bar the use of the courts to collect for services.[106] Yet for reasons noted in Section 10.10D, denials of recovery, although still common, are less likely when a contractor seeks judicial relief than when such relief is sought by a design professional. This occurs despite the specific legislative direction in many contractor licensing laws to bar the state courts to unlicensed contractors.

These specific legislative directives, such as found in California or Washington, do not mean contractors never recover. Such a harsh sanction has led these states to create a substantial compliance doctrine to avoid unjust enrichment (discussed in Section 10.10C). Courts have divided on the use of such laws as a defense by a knowledgeable participant in the construction industry.[107] The cases show the usual variant results when unlicensed contractors seek recovery.[108]

Unwillingness by the courts in some states to impose the harsh sanction of barring recovery for work performed may have led Tennessee to enact a statute that permits an unlicensed contractor to recover if it can show clear and convincing proof of actual documented expenses.[109]

The construction industry, particularly the portion that consists of contractors and subcontractors, witnesses chronic financial difficulties and frequent bankruptcies. The tension between protecting consumers from financially insecure or irresponsible contractors and the policy in the federal bankruptcy law to give debtors a fresh start has led to disputes over whether bankruptcy is grounds for taking away the contractor's license. As noted in Section 10.04B, this conflict has been resolved in favor of the bankruptcy policy's permitting a fresh start.

B. Exceptions

Barring the unlicensed contractor from using the courts to recover for services it has performed does not mean that the unlicensed contractor cannot use other avenues.

In states that bar the unlicensed contractor from using the courts to recover under the contract, unlicensed subcontractors have been allowed to sue a prime contractor (or vice versa) for delay damages, defective work or negligent administration of the contract, which caused the unlicensed contractor to suffer increased costs of performance.[110] The reasoning of these cases—that recovery is not sought for performance—should allow an unlicensed prime contractor to recover from an owner for delay damages.

Similarly, an unlicensed minority subcontractor could not recover for work it performed but was not barred from bringing a civil rights claim based upon what it contended were illegal racial conditions in the contract and racially motivated retaliation.[111]

Other cases demonstrate the ambivalence of courts when asked to deny recovery to unlicensed contractors who have done their work properly. For example, courts have divided as to the effect of noncompliance when the dispute does not involve an ordinary member of the public but involves those knowledgeable about construction.[112]

Another evidence of the greater solicitude shown unlicensed contractors is reflected by *Moore v. Breeden.*[113] The applicable law required that the owner give notice to the contractor that a license is required before the owner can defend any claim made against it based on the contractor's having been unlicensed. It was held that this notice requirement applies even if the contractor knew it was required to be licensed. This interpretation allowed the court to grant a recovery of almost $300,000 to an unlicensed contractor.

A similar judgment in favor of the claimant was based on the claimant's not being a contractor at all but simply

---

[106]Another California sanction is trebling the award against an unlicensed contractor and giving the court discretion to award attorneys' fees. See Section 10.07A.

[107]See Section 10.07B; note 71.

[108]Compare *C. B. Jackson & Sons Constr. Co. v. Davis,* supra note 67 (recovery), and *Lignell v. Berg,* infra note 115 (sophisticated owner), with *Cochran v. Ozark Country Club, Inc.,* 339 So.2d 1023 (Ala.1976) (had been paid $35,000, sought $28,000 more: denied); *United Stage Equipment, Inc. v. Charles Carter & Co.,* 342 So.2d 1153 (La.App.1977) (paid all but $1,300: denied); *Revis Sand & Stone, Inc. v. King,* 49 N.C.App. 168, 270 S.E.2d 580 (1980) (need to protect public).

[109]Tenn.Code Ann. § 62-6-103.

[110]*E. C. Ernst, Inc. v. Contra Costa County,* 555 F.Supp. 122 (N.D.Cal.1982) (subcontractor suffered delay damages); *American Sheet Metal, Inc. v. Em-Kay Eng'g Co., Inc.,* 478 F.Supp. 809 (E.D.Cal.1979) (recovery by an unlicensed contractor for the defective work of a subcontractor); and *Gaines v. Eastern Pac.,* 136 Cal.App.3d 679, 186 Cal.Rptr. 421 (1982) (same). The continued viability of these cases is in doubt; see *Pacific Custom Pools, Inc. v. Turner Constr. Co.,* 79 Cal.App.4th 1254, 94 Cal.Rptr.2d 756, 764–65 (2000).

[111]*Holland v. Diesel-Morse Intern., Inc.,* 86 Cal.App.4th 1443, 104 Cal. Rptr. 2d 239 (2001).

[112]See Section 10.07B; note 71.

[113]209 Va. 111, 161 S.E.2d 729 (1968).

a foreman who had been engaged by the owner to perform construction work.[114] These cases should not convey the impression that courts simply disregard contractor licensing laws. They do show the difficulty some courts have had in reconciling the legislative requirement for a license with the merits of a particular claim, the denial of which would appear to create unjust enrichment.

C. Substantial Compliance

A number of states recognize the substantial compliance doctrine. In these states, excusable errors by the contractor relating to obtaining or renewing a license will not preclude a recovery.[115] The California experience is instructive.

In *Latipac, Inc. v. Superior Court*,[116] failure to renew a license was due to the inadvertence of an office manager who had had a mental breakdown. California has struggled with this doctrine but on the whole has used it to allow recovery where it appears that the work was properly performed and failure to obtain a license did not relate to competence but to bureaucratic error.[117]

Finally, the California Supreme Court decided *Asdourian v. Araj*[118] in 1985. The contractor had made a contract in his own name instead of the business under which he had obtained his license. Nor had he made a written contract for home improvements as required by California law. Yet the court allowed him to recover by a liberal use of the substantial compliance doctrine and a narrow reading of the statutory requirements for a written contract. The court, over one dissenting opinion, showed little respect for the public policy supposedly behind contractor licensing laws. It also scolded intermediate appellate courts, which had not made broad use of the substantial compliance doctrine and which were more inclined to follow the literal meaning of the statutes.

The court noted that the substantial compliance doctrine had been followed for almost five decades without any response by the legislature. The legislature took up the challenge in 1989, amending Section 7031 of the California Business and Professions Code to make clear that the doctrine of substantial compliance would not permit an unlicensed contractor to recover for its services.

But an amendment to Section 7031 enacted in 1994 shows the ambivalence of the California legislature toward contractor licensing laws: This attitude is shared by many legislatures and courts. After stating in subsection (e) that the substantial compliance doctrine cannot be applied by the courts if the contractor has never been licensed, the statute allows a court to apply the doctrine under specified conditions. The contractor must show that it had been licensed prior to doing the work, that it acted "reasonably and in good faith" to maintain its license, and that it did not or should not have known that it was unlicensed.[119]

California's inability to decide what to do when an unlicensed contractor performs work is again demonstrated by the amendment to Section 7031 enacted in 2001 and noted in Section 10.07B. That amendment allows the person who has paid for work done by an unlicensed contractor to recover payments it has made.[120] This goes beyond barring the unlicensed contractor from using the courts to collect for its work.

To sum up, in 1989 the Legislature stated it did not want substantial compliance to be used. In 1994, it decided to allow it under limited circumstances. In 2001, it went farther in punishing an unlicensed contractor by not only barring it from recovering for work done but allowing the person who has paid for work to recover payments made. Such are the twists and turns of contractor licensing laws.

The California Supreme Court's most recent analysis of the substantial compliance doctrine is *MW Erectors, Inc. v. Niederhauser Ornamental and Metal Works Co., Inc.*[121] Recall that a sub-subcontractor entered into two contracts: a structural steel contract followed, one month

---

[114]*Sobel v. Jones*, 96 Ariz. 297, 394 P.2d 415 (1964).

[115]*McCormick v. Reliance Ins. Co.*, 46 P.3d 1009 (Alaska 2002); *Lignell v. Berg*, 593 P.2d 800 (Utah 1979); *Murphy v. Campbell Investment Co.*, 79 Wash.2d 417, 486 P.2d 1080 (1971); *Expert Drywall, Inc. v. Brain*, supra note 78. For Washington law, see Comment, 14 Gonz. L.Rev. 647, 659–61 (1979). For a case vehemently rejecting the substantial compliance doctrine, see *Brady v. Fulghum*, 309 N.C. 580, 308 S.E.2d 327 (1983).

[116]64 Cal.2d 278, 411 P.2d 564, 49 Cal.Rptr. 676 (1966).

[117]See *Vitek, Inc. v. Alvarado Ice Palace, Inc.*, 34 Cal.App.3d 586, 110 Cal.Rptr. 86 (1973).

[118]38 Cal.3d 276, 696 P.2d 95, 211 Cal.Rptr. 703 (1985).

[119]Two cases came to opposite conclusions on the question of whether the contractor met the stringent statutory definition of substantial compliance. See *ICF Kaiser Engineers, Inc. v. Superior Court*, 75 Cal.App.4th 226, 89 Cal. Rptr.2d 88 (1999) (substantial compliance was found where a computer snafu kept even the state licensing board from knowing the contractor's license suspended) and *Pacific Custom Pools, Inc. v. Turner Construction Co.*, supra note 110 (substantial compliance not found where contractor knew for five months its license was suspended).

[120]West Ann.Cal.Bus.&Prof. Code § 7031(b).

[121]Supra note 62.

later, by an ornamental steel contract. It was unlicensed when it signed both contracts and obtained its license only after it had begun work on the structural steel contract. The court ruled that the sub-subcontractor may not apply the substantial compliance doctrine to the structural steel contract because it did not hold a California license prior to performance. Under the statutory definition, the substantial compliance doctrine is limited to contractors who, prior to commencement of the job, had previously held a valid license.[122]

That the substantial compliance doctrine still is troublesome is illustrated by *Crow v. Hickman's Egg Ranch, Inc.*[123] The Arizona Statute was much like California's 7031 discussed above. It required that the contractor have a license when the contract is made. The contractor (an experienced contractor) was not licensed when the contract was made. The owner knew this. The contractor performed a small amount of work. Then it became licensed.

A dispute over quality then developed. The contractor sued only for the work done after it was licensed. In denying recovery, the majority stated that substantial compliance requires that the contractor attempt to become licensed before entering into the contract. In this case it did not, although it was aware of the license requirement.

Again, showing the controversial nature of the substantial compliance excuse for not being licensed, the dissenting judge stated that substantial compliance should be applied where it satisfies the general purpose of the licensing laws. The judge noted that the contractor sought recovery only for work done after it became licensed.

Those jurisdictions that are quick to employ the substantial compliance doctrine are inclined to see contractor licensing laws as traps for unwary contractors who perform proper work and then face what appears to be a technicality that may bar them from recovering for their work. Yet those same jurisdictions are often emphatic in their desire to protect consumers from those who might prey on them. Again, as noted in Section 10.07B, the absence of a license may not bar a contractor from recovering where it appears that it did not mislead the owner and did its work properly.

---

[122]This same reasoning was applied to find the substantial compliance doctrine not applicable in *Opp v. St. Paul Fire & Marine Ins. Co.*, supra note 71.

[123]202 Ariz.113, 41 P.3d 651 (App.2002), review denied Jun. 25, 2002.

D. Observations

The contractor licensing laws and cases that interpret them demonstrate different attitudes from those that involve architects, engineers, and surveyors. More appears to be expected of design professionals than of contractors. Contractors are often small businesses that may not be aware of the existence of licensing laws and may not have the staff to ensure that they are always in compliance. The recent enactment of many contractor licensing laws and the difficulty of formulating standards for determining competence may play a part in some judicial reluctance to bar a contractor. Even in cases where contractors are precluded from recovery, generally all that has been lost has been part payment, with contractors often having collected most of their compensation.

SECTION 10.11 Indirect Effect: Forum for Consumer Complaints

Some make a case for contractor licensing laws by pointing to the power they give the consumer-homeowner when a contractor is not responsive to complaints that work has been done poorly. A criminal sanction, though rarely imposed, can be the basis for restitution, at least in aggravated cases. More important, a threat to go to the licensing authorities may provide an incentive for the contractor to take the complaint seriously. If the dispute cannot be resolved amicably, there is likely to be an informal hearing before a representative of the state licensing agency. In effect, these agencies have become small claims courts.

The consumer protection aspect of contractor licensing would be more effective if licensing agencies made available a record of complaints against particular contractors. Although perhaps not typical, California's experience is instructive. In 1986, California permitted the registrar of contractors to make public information related to complaints on file against a licensed contractor that have been referred for legal action (very few are referred) but not those complaints that were resolved in favor of the contractor.[124] The registrar may set reasonable limits on the number of requests for information per month from any one requestor.[125]

---

[124]West Ann.Cal.Bus. & Prof.Code § 7124.6.

[125]16 Cal.Code Regs. § 863

But a true consumer protection system would allow a consumer to learn of the number of complaints against a contractor whom the consumer plans to use. The California public information system does not approach this.

To be sure, this indirect function has dangers. It can provide too powerful a tool to an owner who abuses it to obtain more than promised under the contract or required by law. The system can be costly to operate, particularly where a high level of consumer dissatisfaction exists. This induced California to require arbitration of certain disputes.[126]

SECTION 10.12 The Trained but Unregistered Design Professional: Moonlighting

A. Unlicensed Persons: A Differentiation

This chapter has spoken of persons who violate the registration for licensing laws as if they all can be placed in the same category. Yet differentiation can be made between those who violate these laws who have been educated and trained as design professionals and others, such as contractors, developers, designer-builders, or self-styled handypersons.

A long period ordinarily elapses between the beginning of architecture or engineering training (architecture will be used as an illustration) and registration. As a result, there are many people with substantial education and training as design professionals but who for various reasons do not work under the supervision of a registered architect yet engage in the full spectrum of design services. To be sure, the law does not differentiate these people from those without design education and training. Yet that these people often do perform these services in violation of the law demonstrates not only the financial burden involved in education and training before registration but also that a market exists for such services.

This section is directed toward some of the legal problems faced by what is referred to as the *moonlighter*, who is likely to be a student, a recent graduate, or a teacher. Some of the discussion anticipates material to be discussed in greater detail in the balance of the book, such as professional liability, exculpation, and indemnification.

---

[126]West Ann. Cal.Bus. & Prof.Code § 7085.

However, a focus on the range of legal problems that such individuals face is also useful.

B. Ethical and Legal Questions

A threshold issue before the discussion of collection for services performed by moonlighters (Section 10.12C) and liability of moonlighters (Section 10.12D) is whether such work should be performed. A differentiation will be drawn between work by a moonlighter that does not violate the law because the project is exempt and the requirement that a registered designer perform services that do violate the registration laws. The former course raises no ethical or legal questions; the latter does.

The discussion in this section should not be taken as a suggestion that moonlighters should violate the law by engaging in design services prohibited by law. However, two factors necessitate discussion of moonlighting. First, arrangements between clients and moonlighters demonstrate that each group feels it will benefit from such an arrangement or such arrangements would not be made. That there is a good deal of moonlighting demonstrates either that the registration laws only make illegal conduct that should not be prohibited or that there is great disrespect for the registration laws.

Second, the many cases that involve attempts by unlicensed people to recover for services that they have performed, and the periodic success of their claims, indicate some dissatisfaction with the registration laws even by those charged with the responsibility of enforcing them indirectly by denying recovery for services performed in violation of them.

Anyone tempted to moonlight and violate registration laws must take into account—in addition to the risk of criminal and quasi-criminal sanctions—the risks discussed in the balance of this section: going unpaid for services rendered or being wiped out financially because of claims by the client or third parties.

C. Recovery for Services Performed

Clearly, the safest path is to design only those projects exempt from the registration laws. Undertaking nonexempt projects creates substantial risks. This subsection looks at these risks and ways of minimizing the likelihood that the moonlighter will not be able to recover for services performed.

Because payments made generally cannot be recovered by the client,[127] it is best to specifically provide that interim fee payments will be made and to make certain that they are paid. If payment has not been made in full, it is possible at least in some jurisdictions to recover for services in a restitutionary claim based on unjust enrichment. Such a claim is more likely to be successful if the client is aware of the moonlighter's nonregistered status, if the moonlighter has had substantial education and training in design, if the work appears to have been done properly, and if it appears that the client has benefited by a lower fee. The stated purpose of registration laws is to deter unqualified people from performing design services and to protect the public from hiring unqualified professionals. A client who is aware of the status of the designer engaged has given up protection accorded by law, and to allow the client to avoid payment where the services have been properly rendered can create unjust enrichment.

If the project will expose the general public to the risk of physical harm, it is more difficult to justify either entering into such an arrangement or allowing recovery for services performed under one. But as a practical matter, this is not the type of project that moonlighters will design. The uncertainty of recovery and the high cost of litigation even if there is a recovery should deter moonlighters as a whole from doing nonexempt work. If the deterrent function will be accomplished, assuming that is one of the desirable results of these laws, there appears to be no reason to deny recovery where services have been performed by someone with the requisite education and training.

D. Liability Problems

This discussion assumes that the moonlighter will not be carrying professional liability insurance. As a result, any liability that moonlighters incur will be taken out of their pockets if there is anything in their pockets to take.

As shall be seen in Section 14.05, the professional is generally expected to perform as others in the profession would have performed. Should this standard be applied when a moonlighter is knowingly engaged by a client?

Differentiation must be made between an express agreement dealing with a standard of performance and an agreement implied by law. Suppose the parties agree that the standard of performance will be greater or lesser than

that usually required of a registered design professional. Would the court enforce any agreement under which the standard would be less than that required by a registered professional? This depends on a balancing of laws favoring autonomy, which would enforce such an agreement, and the strength of the policy expressed in the registration laws, which would deny enforcement of such an agreement. Although resolving such a question would not be easy, the law would probably not give effect to an express agreement under which someone performing services in violation of law would be judged by some standard less than would be used in the event that the person doing the design were registered.

If no express agreement is made regarding the standard of performing, a case can be made for judging the performance by what the parties are likely to have intended. If the client knowingly engaged a moonlighter who received a lower fee, should the client get what she pays for—a lower level of performance for a lower price? However, again it is unlikely that the court in determining the implied terms of such an illegal agreement would prefer autonomy to a result that might appear to frustrate the registration laws.

Another approach, related to but somewhat different from the one discussed in the preceding paragraphs, would be for the moonlighter to seek to persuade the client to exculpate the moonlighter from the responsibility for any performance that would violate the contract or to limit the moonlighter's exposure to a designated portion of the entire fee. Because by definition the moonlighter is not insured, the moonlighter may be able to persuade the client that she would not be in a position to pay for any losses she caused, and this is taken into account in setting the fee.

Suppose the client will agree to exculpation or liability limitation. The moonlighter must take particular care to express these provisions very clearly, as courts will at the very least construe them against the moonlighter. Even more, there is a risk that a court that feels strongly about the registration laws will not enforce such a clause if it would frustrate those laws. Keep in mind, however, that because the moonlighter is not likely to have either insurance or assets, a claim by the client is likely to be rare.

Third-party liability raises different problems, inasmuch as any arrangement for exculpation or a liability limitation between the moonlighter and the client will not affect the rights of third parties. But again, the moonlighter without insurance or assets is unlikely to be sued by

---

[127]Section 10.07B.

a third party. A moonlighter concerned about the liability to third parties either because of moral considerations or because the moonlighter does have assets should attempt to seek indemnification from the owner. But as shall be seen, such clauses are also narrowly interpreted and in some states are unenforceable.[128]

---

[128]See Sections 31.05D and E.

From the standpoint of liability either to the client or to third parties, it does not appear that there is a great risk if moonlighters do not have assets of their own. However, if moonlighters do have assets or if they are concerned about causing loss to someone for which they ought to pay, the liability problems of moonlighters are likely to mean that the moonlighters will be judged by the standard of registered professionals.

CHAPTER ELEVEN

Contracting for Design Services: Pitfalls and Advice

SECTION 11.01 Authority Problems

A. Private Owners

Negotiations to provide design services are always made with individuals. It is important to determine whether the individual with whom a design professional negotiates has authority to make binding representations and to enter into a contract. The general agency doctrines that regulate these issues were discussed in Chapter 4. Reference should be made to that chapter, particularly the material in Section 4.06 that deals with apparent authority.

Design professionals will be motivated by many considerations when they decide whether to enter into a contract to perform design services. One factor will be representations made by the prospective client. For example, a design professional may be influenced by representations that relate to the likelihood that the project will be approved by appropriate authorities, that adequate financing can be secured, and that the design professional will work out the design with a particular representative of the client or to other facts that will influence the design professional to contract to perform design services for the client.

In the best of all worlds, the people at the top of the client's organization would be contacted to determine the authority of the person with whom the design professional is directly dealing. However, it may not be easy to determine who is at the top of the client's organization, and it may not always be politic to take this approach. The realities of negotiation mean that the design professional will have to rely largely on appearances and common sense to determine whether the person with whom he is dealing can make such representations. This may depend on the position of the person in the corporate structure, the importance of the representation, and the size of the

corporation. The more important the representation and the more important the contract, the more likely it is that only a person high up in the organizational structure will have authority to make representations that will bind the corporation. If the corporation is particularly large, a person in a relatively lower corporate position may have authority to make representations. For example, in a large national corporation, the head of the purchasing department may be authorized to make representations that would require a vice president's approval in a smaller organization.

One useful technique is to request that important representations be incorporated in the final agreement. If the person with whom the design professional is dealing is unwilling to do so, that person may not have authority to make the representations that have been made.

This approach has an additional advantage. In the event of a dispute that goes to court, the parol evidence rule (discussed in Section 11.04E) may make it difficult to introduce evidence of representations that are not included in the final agreement. Even if such evidence is admitted, failure to include such representations in the written agreement may make it difficult to persuade judge or jury that the representations were made if the client denies making them.

Next it is important to examine the authority to contract for design services. This problem will be approached by looking at the principal forms of private organizations that are likely to commission design services.

Sole Proprietors. The sole proprietor clearly has authority to enter into the contract. Dealings with an agent of the sole proprietor are controlled by concepts of agency and scope of actual or apparent authority. Generally, sole proprietorships are small business operations. For a sole proprietorship, a contract for design services or construction is probably a serious and important transaction. For that reason, agents of a sole proprietor are not likely to have such authority.

Partnerships. Partnerships are usually not large businesses. Generally only the partners have authority to enter into contracts for design services or construction. The partners may have designated certain partners to enter into contracts. Unless this comes to the attention of the design professional, it is unlikely that such a division of authority between the partners will affect him. Although partners have authority to enter into most contracts, it is advisable to get all the partners to sign the contract.

If the partnership is a large organization, agents may have authority to enter into a contract for performance of design services and construction. It is probably best, however, to have some written authorization from a partner that the agent has this authority.

Corporations. The articles of incorporation or the bylaws of a corporation generally specify who has authority to make designated contracts and how such authorization is to be manifested. The more unusual the contract or the more money involved, the higher the authority needed. Contracts that deal with land, loans, or sales or purchases not in the normal course of business frequently must be authorized by the board of directors. Lesser contracts may require authorization by higher officials, whereas for contracts involving smaller amounts of money or those in the usual course of business, subordinate officials may be authorized to contract.

Contracts needing board approval are passed by resolution and entered into the minutes of the board meeting. In addition, bylaws or any chain-of-authority directive frequently states which corporate officials must actually sign the contract. Again, the importance of the contract usually determines at what echelon the contract must be signed and how many officials must sign it. The corporate bylaws or state statutes may require that the corporate seal be affixed to certain contracts.

For maximum protection, the design professional should check the articles of incorporation, bylaws, and any chain-of-authority directive of the corporation. The design professional should see who has authority to authorize the contract and whether the proper mechanism,

such as the appropriate resolution and its entry into the minutes, was used. Then the design professional should determine whether the person who wants to sign or has already signed for the corporation has authority to do so.

In very important contracts, it may be wise to take all these steps. It is not unreasonable for the design professional to request that the corporation attach to the contract a copy of its articles and bylaws and a copy of the resolution authorizing the particular project in question. The design professional can request that the contract itself be signed by the appropriate officers of the corporation, such as the president and secretary of a smaller corporation or the vice president and secretary of a larger corporation.

Sometimes such precautions need not be taken. The project may not seem important enough to warrant this extra caution. The design professional may have dealt with this corporation before and is reasonably assured there will be no difficulty over authority to contract. However, laziness or fear of antagonizing the client is not a justifiable excuse.

For the risks involved in dealing with the promoter of a corporation not yet formed, see Section 3.04C.

Unincorporated Associations. Dealing with an unincorporated association seems simple but may involve many legal traps. To hold the members of the association liable, it is necessary to show that the people with whom the contract was made were authorized to make the contract. In such cases, it is vital to examine the constitution or bylaws of the unincorporated association and to attach a copy of the resolution of the governing board authorizing the contract. The persons signing the contract should be the authorized officers of the association. It may be wise to obtain legal advice when dealing with an unincorporated association, unless the unincorporated association is a client for whom the design professional has worked in the past and in whom the design professional has confidence.

Spouses or Unmarried Cohabitants. A design professional may deal with a married or unmarried couple living together. Even with spouses, no presumption exists that one is agent of the other.[1] But one can bind the other if

---

[1]*Oldham & Worth, Inc. v. Bratton,* 263 N.C. 307, 139 S.E.2d 653 (1965).

the former acts to further a common purpose or if unjust enrichment would result without a finding of agency.[2] It is advisable to have both spouses or members of an unmarried couple sign as parties to the contract.

A recent case from Utah demonstrates the potential risks of dealing with spouses.[3] The wife owned a house that had been damaged. She contracted with a contractor to repair the damage and to make some improvements. Both husband and wife negotiated with the contractor. A written contract for the work was made that contained a general arbitration clause. The house was listed in the title page as the house of husband and wife, but the signature page contained one line and that was signed only by the wife.

Problems developed and the contractor demanded arbitration with the husband and the wife. The husband refused, claiming he was not a party to the contract with its arbitration clause.

The Utah Supreme Court agreed with the husband. It held that the contract was ambiguous and would be resolved against the contractor that had prepared it. The court recognized that there are exceptions to the usual requirement that before a party could be compelled to arbitrate, it must be clear that he had agreed to arbitrate.[4] Illustrations of these exceptions under which a nonsigner can be required to arbitrate are seeking to enforce the contract or receiving a direct benefit from the contract, but the court concluded that none applied here.

Most important, the court held that the husband could not be held to the contract based upon the wife being the agent of the husband. It concluded that the mere status of husband and wife does not entitle one to bind the other on principles of agency.[5] More facts need to be shown, and none was shown here.

The court did recognize the special requirements for an agreement to arbitrate, stating the husband could be held to be a party to the contract but not to the agreement to arbitrate. For that reason the decision could be limited to agreements to arbitrate. Yet the court's holding that there is no presumption that one spouse can bind the other is in accord with general agency law.

B. Public Owners

See Section 8.02A, which deals with public owners and describes some of the special rules for dealing with public entities. Authority issues arise with some frequency because of the need to protect public funds and preclude improper activity by public officials.

The law has generally provided little protection to those who deal with public entities, rarely invoking the apparent authority doctrine. Similarly, it also rarely applies the concepts of estoppel (a party whose representations are reasonably relied upon by the other party is barred from asserting a defense to the other party's claim), which would bar the public entity from establishing that representations were unauthorized or that contracts were not made by proper persons or in accordance with law.[6] Yet some relaxation of the protections has been accorded to public entities. Restitution claims based on unjust enrichment are beginning to have some success,[7] as are assertions that a public entity cannot rely on provisions of a contract where it has misled the contractor.[8]

This relaxation should not encourage carelessness in dealing with public entities. Substantial risks still attend reliance on representations by government officials and entering into contracts with public entities. Immunity of public entities as a defense to a claim for breach of contract will be discussed in Section 23.03I.

---

[2]*Capital Plumbing & Heating Supply Co. v. Snyder*, 2 Ill.App.3d 660, 275 N.E.2d 663 (1971) (wife agent for lien purposes).

[3]*Ellsworth v. American Arbitration Ass'n*, 148 P.3d 983 (Utah 2006).

[4]See Section 33.03A.

[5]See Section 4.06.

---

[6]*School Dist. of Phila. v. Framlau Corp.*, 15 Pa.Comwlth. 621, 328 A.2d 866 (1974) (school board president); *Stephenson County v. Bradley & Bradley, Inc.*, 2 Ill.App.3d 421, 275 N.E.2d 675 (1971) (chairman of board of supervisors) distinguished in *Commonwealth Dep't of Transp. v. Limestone Products & Supply Co.*, 72 Pa. Commw. 360, 456 A.2d 706 (1981). See also *M.A.T.H., Inc. v. Housing Auth. of East St. Louis*, 34 Ill.App.3d 884, 341 N.E.2d 51 (1976), which held a promise to renegotiate after completion made by the president of the housing authority to be beyond the latter's authority and unenforceable.

[7]*Coffin v. Dist. of Columbia*, 320 A.2d 301 (D.C.App.1974) (dictum: recovery limited to the amount of money that the contracting officer had authority to commit); *Saul Bass & Assoc. v. United States*, 205 Ct.Cl. 214, 505 F.2d 1386 (1974) (recovery based on an informal agreement despite failure to execute a formal contract).

[8]*Emeco Indus., Inc. v. United States*, 202 Ct.Cl. 1006, 485 F.2d 652 (1973) distinguished in *Kaeper Machine, Inc. v. United States*, 74 Fed. Cl. 1 (2006) (questioning *Emeco* case as precedent in light of "more recent decisions"; while no *per se* rule treats government differently, claimant bears "heavy burden." Id. at 9; *Manloading & Management Assoc., Inc. v. United States*, 198 Ct.Cl. 628, 461 F.2d 1299 (1972) distinguished in *RGW Communications, Inc. dba Watson Cable Co.*, ASBCA Nos.54495, 54497, 05–1 BCA ¶32,972.

SECTION 11.02 Financial Capacity

A. Importance

A crucial factor in determining whether to undertake design work is the client's financial capacity. The relationship between the design professional and client can deteriorate when the latter cannot or will not pay interim fee payments as required, is unable or unwilling to obtain funds for the project, or lacks the financial capacity to absorb strains on the budget caused by design changes or market conditions that increase the cost of labor and materials. If the client will not have the financial resources to pay any court judgment that may be awarded, the contract is risky and should be avoided.

Despite constant warnings, many design professionals enter into contracts, perform work, and then find they cannot collect for services. Professionals do not like to confront financial responsibility openly. They prefer to assume that clients are honorable people who will meet their obligations, and usually this is the case. However, there are too many instances of uncollected fees to justify cavalier disregard of this problem. Design professionals must seriously confront the problem of financial capacity.

B. Private Clients

Retainers and Interim Fees. Most standard contracts published by the professional associations provide for initial retainers to be paid by the client at the time the contract is signed and for interim fee payments to be made during performance. A major purpose of such provisions is to give the design professional working capital.

Another purpose is to limit the scope of the financial risk taken by the design professional. If services and efforts do not run very far beyond the money paid, risk of nonpayment is substantially reduced. The difficulty is that design professionals frequently do not insist that the client comply with these contract terms. This is absolutely essential. If the matter is explained properly to the client, there should be no difficulty. It should alert the design professional to possible danger when the client seems to be insulted when businesslike requests are made for advance retainers and interim fee payments when due. Clients who react adversely to such requests are often clients who either do not have the money or will not pay even if they do have the money.

Client Resources. Until 1997 (more later on the change), AIA Document B141, Paragraph 4.3, permitted the architect to ask the client to furnish evidence that financial arrangements had been made to fulfill the owner's obligations. The power to request this evidence could be exercised at any time during the contract period. Such a provision created both a promise and a condition to the architect's continued performance. Failure by the client to comply permitted the architect to suspend performance. If suspension continued or if the client indicated an inability or unwillingness to furnish this information, the architect could treat it as a material breach terminating the obligation to perform and giving remedies for breach of contract. (See Sections 27.02 and 33.04.) The design professional can and should ask for this information before deciding whether to perform services. If the client is new or if the information furnished does not appear trustworthy, it may be useful to run a credit check.

In 1997, the AIA dropped this provision from B141-1997. Although the Institute realized that such a power was exercised on occasion, it believed the standard contract should reflect "positive" aspects of the architect–client relationship. Of course, architects are free to add such a provision to B101-2007. Very likely many will, though exercise of the power will still be rather infrequent.

One-Person and Closely Held Corporations: Individual Liability of Officers. Many small businesses are incorporated, and the shares of the business are held entirely by the proprietor of the business. The proprietor is permitted to do this by law. One purpose is to insulate personal assets from the liabilities of the corporation. Some corporations are small, closely held corporations, with the shares owned by a family or by the persons actually running the business. As noted in Section 3.04E, the law can pierce the corporate veil and treat the inadequately capitalized one-person or closely held corporation as a sole proprietorship or partnership. If this is done, the design professional can recover from the shareholder or shareholders. Piercing a corporate veil is rarely done, however. If a credit check reveals that the corporation is merely a shell with very few assets, it may be necessary to demand that the shareholders assume personal liability. To do so, the sole shareholder or shareholders should sign both as representatives of the corporation and as individuals. If those signing as individuals are solvent, this is a reasonably secure method of assuring payment if the corporation is unable to pay the contractual obligations. If the individuals signing are not solvent, individual liability will be of little value. A refusal to sign as an individual may be a warning that the corporation is in serious financial trouble.

A Surety or Guarantor. Chapter 32 discusses the role of the surety in the construction phase of the project. If the design professional believes the person with whom he is dealing is not financially sound, the design professional may wish to obtain a financially sound person to guarantee the obligation of the corporation or the individual before entering into the contract. Subject to some exceptions, promises to pay the debt, default, or miscarriages of another must be in writing. The design professional who requests and obtains a third person to act as surety for the client should make certain that the surety signs the contract or a separate surety contract. When a surety or guarantor is involved, legal advice should be obtained.

Real or Personal Property Security. Another method of securing the design professional against the risk that the client will not pay is to obtain a security interest in real or personal property. This can be done by obtaining a mortgage or a deed of trust on the land on which the project is to be constructed or on other assets owned by the client. This measure may seem drastic. If that much insecurity is involved, it may be advisable not to deal with the client at all. The law regarding creating and perfecting security interests is beyond the scope of this book. If such protection is needed, legal advice should be obtained.

Client Identity. Some clients move quickly in and out of various different but related legal forms. What appears to be a partnership turns out to be a corporation of limited resources. The design professional may find that the corporation with whom he dealt is insolvent, whereas other, solvent, related corporations are controlled by the client. It can be difficult for the design professional to know for whom he is working or who has legal responsibility. This can be complicated by the different entities described in the preliminary correspondence, the contracts, and the communications during performance. Design professionals should know the exact identity and legal status of the client.

Spouses or Unmarried Cohabitants. As mentioned in Section 11.01A, dealing with spouses or cohabiting persons can raise agency questions. Design professionals may discover that the person with whom they dealt has no assets, whereas that person's spouse or other cohabitant owns all the assets. It is important to have both spouses or cohabitants sign the contract.

Mechanics' Liens. Mechanics' liens—more important to contractors and suppliers—are treated in Section 28.07D. Their occasional utility to professional designers can tempt those who are about to perform design services to avoid some of the techniques previously mentioned for fear of losing a client. For that reason, they are discussed here briefly.

State statutes often give persons a security interest in property that they have improved to the extent of any debt owed to the improver by the owner of the property or someone who has authority to bind the owner. The lien holder can demand a judicial foreclosure of the property and satisfy the obligation out of the proceeds.

Design professionals face two problems when they assert a mechanics' lien. Both problems relate to improvement of the owner's property as the basis for awarding a specific remedy of foreclosure against the land itself rather than simply giving the design professional a money award as described in Sections 2.12 and 2.13.

The first problem—a problem not faced by contractors—involves whether or not design services have improved the land. This problem generally has been surmounted by mechanics' lien legislation that specifically names design professionals as lien recipients. However, not all states have done this.

The second issue recognizes the unstable world of design. It is not uncommon for the design professional to perform design services for a project that is abandoned before site work is initiated. (The beginning of the site work warns those who might advance credit to the owner that there may be a lien that will take precedence against any subsequent security interest.) As noted, improvement of the land is the basis for awarding mechanics' liens.

Courts have not been uniform when faced with the issue of whether a lien can be asserted despite the project's having been abandoned. Most courts have ruled that no lien has been created. But some cases go in the other direction.

The cases from Kansas are instructive, showing the complexities of lien laws and the uncertain efficacy of liens as a reliable security interest for those who perform services, such as architects or engineers, before construction begins. Decided in 1990 in *Mark Twain Kansas City Bank v. Kroh Bros. Development Co. et.al.*,[9] a Kansas intermediate court of appeals held that a claimant who had performed off-site architectural and engineering services was not entitled to a lien where no construction ever commenced. The court required that there be visible improvements so that the world would know that a lien was a possibility.

But in 1996 the Kansas Supreme Court in *Haz-Mat Response Inc. v. Certified Waste Services, Inc.* concluded that the opinion of the Kansas intermediate court in the *Mark Twain* case had been incorrect.[10] The Supreme Court held that there need not be a physical addition or actual construction. The court concluded that the lien claimant did not succeed because the work was part of routine maintenance.

The Kansas intermediate appeals court in 2003 decided *Mutual Savings Association v. Res/Com Properties LLC*.[11] The court stated that the *Mark Twain* case had been in effect overruled by subsequent cases, that there need not be physical addition, and that preliminary stakes and surveying could be the basis for a lien.

To demonstrate the complexity of lien laws, the court in the *Mutual Savings* case stated that mechanics' liens were a creature of statute and that the claimant must clearly bring itself within the statute. In other words, the statute must be strictly followed. But if this standard were met, the remedial nature of the statute meant that the law is construed liberally in favor of the claimant.

In a number of states, the mechanics' lien laws have been amended to grant liens even though the project is never begun.[12] Such laws are usually supported by professional associations of architects and engineers who assert that design professionals perform services and should be paid. Clearly, design professionals would have a right to recover an ordinary court judgment unless they have assumed the risk that they would not be paid if the project did not go forward.[13] However, the professional associations seek to go further and create a specific remedy, that is, a lien against the property for which the design was commissioned. Often ignored is the fact that this specific remedy places design professionals ahead of those ordinary unsecured creditors who may have also performed services or supplied goods and have not been paid. Participants in the construction industry have sought to enlarge their lien rights, but it should not be ignored that this enlargement often comes at the expense of unsecured creditors.

Lien rights cannot be asserted against public improvements. Many technical requirements exist to create a valid lien. Notices have to be given, filings have to be made, and foreclosure actions must be taken within specified times. Without strict compliance, there is no lien. Other persons may have equal or prior security rights in the land. The land value may not be enough to pay the lien claims in their entirety.

Mechanics' lien statutes vary considerably from state to state and are frequently changed by legislatures. Design professionals should seek legal advice to see whether they are within the class of persons accorded liens and to ascertain the steps needed to perfect a lien. It is unwise to undertake work for a client who may not be able to pay, hoping that in the event of nonpayment, there will be a right to a mechanics' lien. The possibility of being able to assert a mechanics' lien is never a substitute for a careful consideration of the financial responsibility of the client and collection of interim fee payments.

C. Public Owners

Those who deal with public entities run the risk that money will not be appropriated or bonds not issued to pay for the project or design services. An Illinois decision held that an architect could not recover for his services where the evidence showed he was risking compensation on the passage of a bond issue.[14] The court pointed to a statement made by the architect during negotiations that he was gambling with the county, which convinced the court that he took this risk.

The best protection against risks is to check carefully on the availability of funds and make certain that work does not begin until it is relatively certain that funds will be available to compensate the design professional.

---

[9]14 Kan.App.2d 714, 798 P.2d 511 (1990). For similar cases see *Brownstein v. Rhomberg-Haglin & Assocs.*, 824 S.W.2d (Mo.1992), *Anthony & Assocs.etc. v. Muller*, 598 A.2d 1378 (R.I.1991), and *Goebel v. National Exchangers, Inc.* 88 Wis.2d 596, 277 N.W.2d 755 (1979).

[10]259 Kan. 166, 910 P.2d 839 (1996). See also *Design Assoc. Inc. v. Powers*, 86 N.C.App. 216, 356 S.E.2d 819 (1987).

[11]32 Kan.App.2d 48, 79 P.3d 184 (2003), review denied Feb. 10, 2004.

[12]West Ann.Fla.Stat.Ann, § 713.03 (a design professional who has a direct contract to perform services relating to a specific parcel of real property has a lien for the amount owing to him regardless of whether such real property is actually improved). Similarly see West Ann.Cal. Civ. Code § 3081.2 enacted in 1990. A permit must have been issued. Under § 3081.10, there are no liens by design professionals against single-family residences that cost less than $100,000.

[13]See Section 12.13C.

[14]*Stephenson County v. Bradley & Bradley, Inc.*, supra note 6.

SECTION 11.03 Competing for the Commission: Ethical and Legal Considerations

A. Competition Between Design Professionals: Ethics

Associations of design professionals historically have sought to discipline their members for ungentlemanly competition, both in obtaining a commission for design services and in replacing a fellow member who has been performing design services for a client. As an illustration, AIA Document J330 in effect in 1958 listed, among its mandatory standards of ethics,

> 9. An Architect shall not attempt to supplant another Architect after definite steps have been taken by a client toward the latter's employment.

> 10. An Architect shall not undertake a commission for which he knows another Architect has been employed until he has notified such other Architect of the fact in writing and has conclusively determined that the original employment has been terminated.

In addition, J330 stated that the architect would not compete on the basis of professional charges.

These restraints were considered part of ethics. Undoubtedly they are traceable to the desire to preserve these professions as professions and to avoid certain practices that would be accepted in the commercial marketplace. Although the associations must have recognized that there will be competition for work, it was hoped that competition would be conducted in a gentlemanly fashion and would emphasize professional skill rather than price.

Beginning in the 1970s, attacks were made on these ethical standards by public officials charged with the responsibility of preserving competition. One by one, association activities that were thought to impede competition came under attack, particularly those that dealt with fees. Fee schedules, whether required or suggested, were found to be illegal.[15] Disciplinary actions for competitive bidding were also forbidden.[16]

Standard 9 in J330 attempted to inhibit competition before any valid contract had been formed.[17] The extent to which the law will protect a valid contract or a prospective contractual relationship by imposing liability on those who seek to or do interfere with such legitimate expectations will be discussed in Section 11.03D. Standard 10, though less susceptible to the charge that it is anticompetitive, could have had the effect of limiting the client's power to replace one architect with another.

B. Brooks Act

Design professionals do not like to compete on the basis of price. Although their associations cannot use their disciplinary power to attack this practice, current federal legislation precludes head-to-head fee competition.

In 1972, Congress enacted the Brooks Act.[18] The Act determines how contracts for design services can be awarded by federal agencies. It declares that the policy of the federal government is to negotiate on the basis of "demonstrated competence and qualification for the type of professional services required and at fair and reasonable prices."

Those who perform design services submit annual statements of qualifications and performance data. The agency evaluates the statements, together with those submitted by other firms requiring a proposed product, and discusses with no fewer than three firms "anticipated concepts and the relative utility of alternative methods." The agency then ranks the three most qualified to perform the services. It attempts to negotiate a contract with the highest qualified firm and takes into account "the estimated value of the services to be rendered, the scope, complexity, and professional nature thereof."

If a satisfactory contract cannot be negotiated with the most qualified, the agency will undertake negotiations with the second most qualified, and if that fails, with the third. The U.S. Justice Department has proposed repeal, arguing that the statute restricts competition. Many states have enacted comparable legislation regulating the award of contracts for design services by state agencies.

Despite the Brooks Act and various state statutes adopting the same approach, some procurement agencies have sought to incorporate price at the first level—that of determining competence and qualification. A well-reasoned and researched opinion of the Colorado attorney general concluded that a public entity cannot require cost proposals in connection with selecting the most qualified architects,

---

[15]*Goldfarb v. Virginia State Bar*, 421 U.S. 773 (1975) distinguished in *Arizona. v. Maricopa County Medical Society*, 457 U.S. 332 (1982).

[16]*Nat'l Soc'y of Professional Eng'rs v. United States*, 435 U.S. 679 (1978).

[17]*United States v. American Soc'y of Civil Eng'rs*, 446 F.Supp. 803 (S.D.N.Y.1977) (barred society from disciplining members who supplanted another member).

[18]40 U.S.C.A. §§ 1101-04.

engineers, or land surveyors for a project.[19] However, an Ohio intermediate appellate court upheld a decision by the governor of Ohio that there can be two most qualified firms and that a tie in such a case may be broken at the first stage by using competitive price proposals. The low bidder then becomes the most qualified firm.[20]

One commentator, although obviously despairing at the prospect, made the following suggestions to design professionals forced to compete on price:

1. Make bids only on projects where there is a clearly defined scope of work.
2. Select your clients carefully.
3. Bid only on projects where your experience would favor selection in the absence of price competition.
4. Evaluate your competition.
5. Plan to use the most competent and qualified people you have.
6. Define your work with great precision.
7. Make detailed cost estimates.

The commentator also suggested that it is crucial to know when to walk away from a prospective commission and to be prepared to do so.[21]

One troubling question that arose as a result of the increased use of construction management in the 1970s related to whether contracts for the performance of these services were more like those for design, which did not require competitive bidding, or more like contracting, which did. This aspect of construction management is discussed in Section 17.04D.

C. Federal Art-in-Architecture Program

The General Services Administration (GSA) administers the design and construction of buildings occupied by many federal employees. GSA regulations require that a designated percentage of the budget for federal buildings (originally one and one-half percent) be set aside for artistic work for new buildings. This became known as the Art-in-Architecture Program (AIA). Many public entities have similar programs.

The GSA has created a complex competitive process to select the artist and approve the work. This process involves a delicate balancing of the views of the community and the artistic freedom of the artist selected.[22]

D. Interference With Contract or Prospective Advantage

In the competitive market system, persons are allowed and even encouraged to seek commercial advantages by claiming they can do better than their competitors. The classic example of competition is the auction. Each bidder seeks to win the item being auctioned by offering more than others or to perform the work for less than others, such as in competitive bidding. The latter method is often used in construction and will be discussed in detail in Chapter 18.

But at a certain point, attempting to persuade someone to use one's services or buy one's goods will run afoul of the protection tort law accords valid contracts: a commercial relationship, such as a stable supplier-customer relationship, or a negotiation that has proceeded so far that it is very likely it will be concluded successfully. In such cases, the person who makes the claim has the reasonable expectation of commercial gain that has been frustrated wrongfully.

Whether a third party has interfered wrongfully requires balancing protection given a commercial relationship with the freedom given persons to seek to persuade others that they can perform services better than the one who has been retained to perform them.

The two issues that surface when claims are made against third parties are whether the interest affected merits protection of the law and whether the conduct in question has risen to the necessary level of wrongdoing.

As to the first, a federal trial court opinion demonstrates this in the context of a construction contract.[23] A catastrophic discharge of water occurred during performance. The contractor sought to find out what had caused the problem and to fix it. The defendant, an architect who was not the original architect in the project, urged

---

[19]Op.Colo.Att'y Gen., Sept. 2, 1992.

[20]*Ohio Ass'n of Consulting Eng'rs v. Voinovich.*, 83 Ohio App.3d 601, 615 N.E.2d 635 (1992). The state and national professional societies filed an appeal to the Supreme Court of Ohio based on the Ohio mini-Brooks Act, which they asserted bars a procedure under which price is a factor in the selection of the most qualified firm. This appeal was denied. 66 Ohio St.3d 1459, 610 N.E.2d 423 (1993) (over three dissents).

[21]Lekamp, *Professional Liability Perspective*, vol. 10, no. 11 (November 1991).

[22]Gandhi, *The Pendulum of the Art-in-Architecture Program's Struggle to Balance Artistic Freedom and Public Acceptance*, 31 Pub.Cont.L.J. 535 (2002) (describing procedures and case studies).

[23]*RCDI Constr., Inc., v. Spaceplan/Architecture, Planning & Interiors, P.A.*, 148 F.Supp.2d 607 (W.D.N.C. 2001) (affirmed by the Court of Appeals for the 4th Circuit in an unpublished opinion. 2002 WL 53927).

the owner to terminate the plaintiff contractor so that the defendant architect could provide a method of repairing the damage. The owner terminated the contractor.

The plaintiff contractor settled with the owner and brought a claim against the defendant for interfering with its contract.[24]

The court held that the contract made by the contractor and the owner was illegal as the contractor was not licensed. Since the contractor could not bring legal action based upon this contract, it had no reasonable expectation for the law to protect. Also, the court held that the contractor could not create a legal right, that is, to have its contract protected from third parties, based upon its having committed an unlawful act.

As to the second type of conduct, early cases required malicious intent or bad faith.[25] But the trend has been toward requiring merely an intent to do the act that will have the effect of interfering with a legally protected interest of the plaintiff.[26]

SECTION 11.04   Professional Service Contracts: Some Remarks

A. Profits and Risk

Ordinarily, professional advisers do not have potentially high profit returns. They serve their client and expect to be paid for their services. This does not mean that advisers are *always* paid for their work. Nonpayment because the client does not have the financial resources is a risk taken by all who perform services.

This section is directed toward defenses by the client to any claim for compensation for services rendered based on an assertion that no money is due. Such defenses are usually predicated on an asserted understanding that if the project did not go forward—usually because public approval or funds could not be obtained—the professional designer would not be paid for the work.

Although autonomy gives contracting parties the power to make an arrangement under which the professional

designer may risk the fee, any conclusion that this risk has been taken should be arrived at only if weighty evidence supports this conclusion. This is based not only on customary practices but also on the conclusion that professional designers do not as a rule recover profits but recover only payment for services they render.[27]

B. Good Faith, Fair Dealing, and Fiduciary Relationships: Confidentiality

The law increasingly finds that all contracting parties owe each other the responsibility of good faith and fair dealing.[28] This is particularly so when objectives sought by the design professional and the client require close cooperation. Each party should help the other achieve its goals under the contract.

While the obligations of good faith and fair dealing have become more or less universal in contract law and are applied in all contracts, the fiduciary concept is much more particular, one that is created in specific and limited contexts.

The fiduciary relationship is a complex doctrine created originally by English Courts of Equity. There are many facets to this doctrine, and it can occur in many contexts. Generally speaking it involves the duty of the finest loyalty.[29] Put another way, the fiduciary owes certain persons strict fidelity. One court stated that a relationship is fiduciary when one party has superior knowledge and authority and that party is in a position of trust and confidence over the weaker party.[30] Another way of putting it is that the relationship places one party in the hands of another. The fiduciary is expected to act in the best interests of the other party. Obviously, this concept is fluid and expresses bare bones concepts. The normal rules of conduct in the commercial world allow parties to act at arm's length. Each actor in that world can think mainly of its own interest.

---

[24]Had the defendant been the architect of the project under which he could advise the owner on termination, the architect would have had the advisor's privilege as this was his function. See Section 14.08F.

[25]*Dehnert v. Arrow Sprinklers*, 705 P.2d. 846 (Wyo.1985).

[26]*Blivas Page, Inc. v. Klein*, 5 Ill.App.3d 280, 282 N.E.2d 210 (1972); *Williams v. Chittenden Trust Co.*, 145 Vt. 76, 484 A.2d 911 (1984).

[27]*Designer Direct, Inc. v. DeForest Redevelopment Auth.*, 313 F.3d 1036 (7th Cir.2002), rehearing denied Jan. 29, 2003, appeal after remand, 368 F.3d 751 (7th Cir.2004).

[28]Restatement (Second) of Contracts § 205 (1981): Uniform Commercial Code § 1–304 (formerly 1–203). See Section 19.02D.

[29]*Meinhart v. Salmon*, 164 N.E. 545 (N.Y.1928). Though this case was decided in 1928, many cases have continued to cite it. See Rosen, *Meador Lecture Series: Fiduciaries*, 58 Ala. L.Rev. 1041 (2007).

[30]*Carlson v. SALA Architects, Inc.* 732 N.W.2d 324 (Minn.App. 2007), review denied (Minn. Aug. 21, 2007). The court held that the relationship between architect and his client does not create *per se* a fiduciary relationship. More is needed to show such a relationship.

But this changes drastically when the fiduciary relationship is created. These general concepts must be fleshed out with examples that show the concept in operation.

The best and clearest example of a fiduciary relationship is between the trustee of a trust and its beneficiaries. The trustee is to act with the interests of the beneficiaries in mind, not the interests of the trustee. Similarly, the attorney-client relationship automatically (*per se*) creates a fiduciary relationship. For example, the attorney's advice must be based solely on the best interests of the client, not his own. There can be dealings between the attorney and client, but the law will look very carefully at any transaction to make sure it is fair to the client in light of the greater knowledge of the attorney and the trust the client places in his attorney. Similarly, the trustee of an express trust can contract with the trust, but again the law examines any such transaction carefully to make sure it is a fair one.

Of course, parties to a contract can create a fiduciary relationship or one that resembles one. The AIA cost-type construction contract documents[31] include Article 3. While it does not state that such a relationship has been created, it states that "The Contractor accepts the relationship of trust and confidence established by this Agreement." The contractor also promises to cooperate with the architect and to "exercise the Contractor's skill and judgment in furthering the interests of the Owner. . . ." The contractor also promises "to perform the Work in an expeditious and economical manner consistent with the Owner's interests." In light of the contractor's control of the performance, this language is crucial in a contract in which the owner promises to pay the cost of the work.[32]

Yet in the design and construction context, the issues usually relate to the relationship between the design professional (more commonly the architect) and the client. As noted, one court held that the relationship between architect and client does not in itself (*per se*) create a fiduciary relationship.[33] Nor is it an arm's length relationship.

The difficulty of fitting the architect or engineer into the fiduciary relationship is shown by the unwillingness of the law to scrutinize the contract between a design professional and the client at the same intensity as it would a contract between the attorney and client. Even without a fiduciary relationship between the parties, the law would most likely construe any ambiguities against the architect both because the architect in most cases drafted the contract and because the balance of knowledge tilts toward the architect despite the owner often having superior bargaining power.[34]

This question is complicated by the various roles played by the architect in both the design and the construction phases. In the design phase the architect can act as the agent of the client. The architect who acts as agent for the client owes the client a fiduciary obligation when playing that role. He must act in the best interests of the client, not his own. The architect, as a rule, is not the general agent of the client.[35] In AIA document A201–2007 the architect is not the general agent of the client, but an agent whose authority is limited to the extent provided in the contract documents.[36]

The architect can also be a professional advisor, particularly in design matters that concern cost. For example, it has been held that the architect must warn his client if costs are likely to overrun.[37] Another case, while concluding that the trial court properly refused to give an instruction that in essence was one based on a fiduciary relationship, held that the architect must disclose information that the client would want to know.[38] Some contracts require the architect to warn the owner of possible cost overruns, but the outcome would be similar without such a provision based on the concept of implied terms.

During the construction phase, the architect or engineer has other roles. He may be given the power by the parties to the construction contract to certify payment certificates[39] or certificates of completion.[40] They may also give him the power to initially resolve disputes.[41] In these

---

[31]A102-2007 (with Guaranteed Maximum Price (GMP)) and A103-2007 (without GMP).

[32]That this does not always create such a relationship is shown in Section 17.02B.

[33]See supra note 30.

[34]See Section 20.02.

[35]*Incorporated Town of Bono v. Universal Tank & Iron Works, Inc.*, 239 Ark. 924, 395 S.W.2d 30 (1965).

[36]A201-2007, §4.2.1. It is interesting to note that B101-2007 deleted from the new § 3.6.2.1, dealing with site visits, the phrase "as a representative of the Owner."

[37]*Zannoth v. Booth Radio Stations, Inc.*, 333 Mich. 233, 52 N.W.2d 678 (1952).

[38]*Getzschman v. Miller Chemical Co.*, 232 Neb. 885, 443 N.W.2d 260, 268, 273 (1989) distinguished on a different ground, prejudgment interest, in *Folgers Arch. Ltd. v. Kerns*, 262 Neb. 530, 633 N.W.2d 114 (2001).

[39]AIA Doc. A201-2007, §§ 9.4, 9.5.

[40]Id. at § 9.8.4.

[41]AIA Doc. A201-1997, ¶ 4.4. In 2007 the AIA changed this. It now gives this power to an Initial Decision Maker, which can be the architect or someone else. AIA Doc. A201-2007, § 15.2. This will be discussed in Section 29.05.

capacities he acts in a quasi judicial role, not as agent for the client. Certainly there is no place for any fiduciary duties even if there were such a relationship.

Two other issues that can be part of or at least related to the fiduciary concept concern confidentiality and conflict of interest. These ancillary issues often arise in construction. The parties may learn information that one party wishes to stay confidential. Also, any decision making can be affected by a conflict of interest.

The client should be able to trust its professional adviser and receive honest professional advice. Both parties—design professional and client—should be candid and open in their discussions, and each should feel confident that the other will not divulge confidential information to third parties.

In 2007 the AIA changed B141-1997 to B101-2007, renumbered old Paragraph 1.2.3.4 and made it Section 10.8 (Article 10 captioned "Miscellaneous"). The new B101-2007 made the promise of confidentiality reciprocal. The new Section 10.8 expanded the core obligation. It added "business proprietary" to "confidential" and required that such information be kept strictly confidential compared to the earlier requirement of maintaining the confidentiality of information.

The confidentiality obligation in B101-2007 was also given greater scope by reducing the exceptions to the obligations found in B141-1997. In 1997 there were exceptions that permitted disclosure of confidential information that would violate the law or confidential information that "would create the risk of significant harm to the public or prevent the Architect from establishing a claim or defense in an adjudicatory proceeding." These were deleted in 2007. In their place the receiver of the information can disclose it to its employees, to those who need it to perform services or construction on the project, and to its consultants whose contracts "include similar restrictions on the use of confidential information."

B101-2007, Section 10.7 continued the permission to the architect to use certain representations of the design in promotional and professional material. The architect can include the owner's confidential or proprietary information unless the owner "had previously advised the Architect in writing of the specific material considered by the Owner to be confidential or proprietary." This had been Paragraph 1.3.7.7 in B141-1997. Putting this burden on the owner limits the confidentiality protection of the owner, but requiring advance notice does protect the

architect, who may not realize what information is proprietary or confidential.

These changes show the importance of careful review of new AIA documents. The change to reciprocal obligations gives protection to the architect not found in B141-1997. Yet the tightening up of the core obligation of confidentiality shows the increasing stress on protection of intellectual property, often at the heart of nondisclosure commitments.

Do the deletions in B101-2007 of exceptions that had been included in B141-1997, those that had allowed disclosures that would violate the law or create a risk of significant harm to the public, adversely affect any architect that uses B101-2007? This is at best uncertain. In any event, even without the 1997 exceptions, any architect who plans to withhold confidential information or proprietary data or withhold information that could violate the law or harm the public should seek competent legal advice before taking this step.

Some aspects of the fiduciary relationship are obvious. Design professionals should not take kickbacks or bribes. They should not profit from professional services other than by receiving compensation from the client.

Funds held by the design professional that belong to the client should be kept separate. Commingling is a breach of the fiduciary obligation. In such a case, any doubts about to whom the money belongs or for whom profitable investments were made is resolved in favor of the client.

Financial opportunities that come to the attention of the design professional as a result of the services he is performing for the client should be disclosed to the client if the services would be an opportunity falling within the client's business.

One of the most troubling concepts in law relates to conflict of interest. A person cannot serve two masters. The client should be able to trust its design professional to make judgments based solely on the best interests of the client. Advice or decisions by the design professional should be untainted by any real or apparent conflict of interest. The client must believe the design professional serves it, and it alone.

Design professionals should not have a financial interest in anyone bidding on a project for which they are furnishing professional advice. Likewise, they should not have a financial interest in any contractor or subcontractor who is engaged in a project for which the professionals have been engaged. Design professionals should not have any significant financial interest in manufacturers, suppliers, or distributors whose products might be specified by them. Products should not be endorsed that could affect specification writing, nor should designated products be specified because manufacturers or distributors of those products have furnished free engineering. The purpose of these restrictions is to avoid conflict of interest. Design professionals cannot serve their clients loyally if they might personally profit by their advice.

Generally, a client who is fully aware of a potential conflict of interest can nevertheless choose to continue to use the design professional. Consent to a conflict of interest should be binding only if it is clear that the client knows all the facts and has sufficient understanding to make a choice. A design professional who intends to rely on consent by the client must be certain that such requirements are met.

The client clearly can avoid responsibility for any act by the design professional that is tainted by a breach of the fiduciary obligation. For example, if a contract is awarded to a bidder with whom the architect or engineer colluded, the contract can be set aside by the client. If the architect or engineer issued a certificate for payment dishonestly or in violation of his fiduciary obligation, the certificate can be set aside if the client so desires.

While the mere retention of a design professional may not establish a fiduciary relationship, the design professional–client relationship is not one at arm's length. Though neither extremity fits this relationship, the nature of the relationship has many attributes of a fiduciary relationship because of the client's lack of the specialized skills possessed by the design professional and the power the design professional has to affect the interests of his client. The client must have confidence in the loyalty of his design professional and the two must trust each other.

The breach of a fiduciary obligation can give the client grounds to dismiss the design professional. Any bribes or gifts taken by the design professional can be recovered by the client. Any profit the design professional has made as a result of breaching the fiduciary obligation must be given to the client even if the profit was generated principally by the skill of the design professional. Obviously, the design professional must take his fiduciary obligation seriously and make every effort to avoid its breach or the appearance of such a breach.

The AIA recognized conflict of interest in its B101-2007, Section 2.4. It provides that, unless the owner knows and consents, "the Architect shall not engage in any activity, or accept any employment, interest or contribution that would reasonably appear to compromise the Architect's professional judgment with respect to this Project."

This demonstrates the fluidity of these concepts. Clearly these restraints would bind the architect were there a fiduciary relationship. Yet even without one, such conflict of interest can be barred by the contract or would be implied by the law.

C. Variety of Contracts: Purchase Orders

Some professional relationships are not accompanied by careful planning in a contractual sense. Such relationships sometimes are created by a handshake without any exploration of important attributes of the relationship. If such attributes are discussed and resolved, the resolution is often not expressed in tangible form.

At the other extreme, the relationship often is cemented by assent to a preprepared standardized form frequently supplied by architect or engineer and occasionally by the client. Often the client does not understand these standardized agreements. On occasion the actual agreement of client and design professional may differ from provisions of the standardized form. Under any of the circumstances mentioned, difficulties can exist if disagreements arise between client and design professional over the services the latter is to perform.

Another contracting system that can create problems, particularly for the design professional, involves the use of purchase orders as a method for commissioning professional services by architects and engineers. A client will send a purchase order to the design professional and ask that a copy be signed and returned to the client. Purchase orders ordinarily contain a set of standard terms and conditions that relate more to delivery of "off-the-shelf" goods than to professional services. Their use may cause the design professional and his client to fail to achieve a clear mutual understanding as to the services to be performed, the compensation to be received, and responsibility if things go wrong.

From a substantive standpoint, the use of a purchase order more geared to the delivery of goods may impose on the design professional stricter warranties as to the outcome of the services than are required normally of design professionals. As shall be seen in Sections 14.05 and 14.07, such warranties and guarantees may be excluded from professional liability insurance coverage.

The best approach for a design professional who receives a standard purchase order is to send the form back unsigned and begin discussions with the client over scope and responsibilities before commencing performance.

D. Interpretation

Contracts between owners and contractors and between contractors and subcontractors generate more interpretation disputes than those between design professionals and clients. For this reason, the bulk of the discussion in this book relating to contract interpretation is found in Chapters 19 and 20. However, a few interpretation guides relating to contracts between design professionals and clients may be useful here.

One attribute of the design professional–client relationship that sets it apart from ordinary commercial contracts is the relative inexperience of many clients in design and design services. Another is the way such agreements are made. Not uncommonly, such relationships are created by a vague, informal agreement sometimes followed by assent to a preprepared contract form supplied by the design professional. These two attributes frequently generate honest misunderstandings between design professionals and clients.

The most important lodestar—the common intention of the parties at the time the contract was made—is determined by examining any discussions the parties may have had before entering into the agreement, the language in any written contract to which both parties have assented, the facts and circumstances surrounding the making of the agreement, the conduct of the parties after the relationship has begun, and any custom or usage of the trade of which both parties were aware or should have been aware.

Although discussion between the parties—the admissibility of which is determined by the parol evidence rule described in Section 11.04E—and surrounding facts and circumstances are important, emphasis here is on the language of the contract. Although courts still invoke the "plain meaning" rule discussed in Section 20.02—the search for evidence outside the writing requiring a preliminary determination that the meaning is not "plain" on its face—by and large courts today will look at any relevant evidence to determine the meaning of language selected by the parties.

As a rule, the most important evidence of the intention of the parties is any written agreement to which the parties have assented. This is true whether the agreement is a carefully negotiated contract between client and design professional of relatively equal bargaining power and experience, or an agreement between an unsophisticated client and an experienced design professional where the latter has supplied a preprepared form contract.

In the former, there is likely to be a neutral reading of the language. In the latter, the interpretation guide that contract terms are generally interpreted against the party who has prepared the agreement will be used (*contra proferentum*).[42] This interpretation guide can be justified either because the drafter carelessly or deliberately caused the language ambiguity or because the law assumes that the party who prepares the agreement is in a position to force unfair or onerous terms on the other party. This doctrine may be an implicit recognition that the party who does not prepare the language may not take the time or have the ability to examine carefully the language selected by the other party.

This discussion assumes that either both or at least one party had an intention as to specific language. When the parties use lengthy standard contracts published by a third party such as the AIA, it is possible that neither party had any idea of the purpose of a particular clause, often an important datum invoked to interpret the clause. As this "absence of intention" can arise in construction contracts, discussion of this topic is postponed until Section 20.02.

The interpretation guide that resolves the dispute against the party who supplied the ambiguous terms will also be applied to contracts published by the professional association of the design professional.[43] Ambiguous language is likely to be interpreted to be consistent with the reasonable expectations of the client rather than with the literal interpretation that the language might otherwise bear.

Suppose a design professional is retained by a large institutional client such as a public agency or large private corporation. In such a case, the client often prepares the contract for professional services and presents

---

[42]*Williams Eng'g, Inc. v. Goodyear*, 496 So.2d. 1012 (La.1986).

[43]*Malo v. Gilman*, 177 Ind.App. 365, 379 N.E.2d 554 (1978); *Durand Assoc., Inc. v. Guardian Inv. Co.*, 186 Neb. 349, 183 N.W.2d 246 (1971).

it to the design professional on a take-it-or-leave-it basis. Ambiguous language in such a case should be interpreted in favor of the design professional.

As to other interpretation guides, handwritten portions of a contract are preferred to typewritten portions or printed portions, and typewritten portions are given more weight than printed provisions in a form contract. Where parties choose language in connection with a particular transaction, the language is more likely to reflect their common intention.

Specific provisions are given more weight than general provisions. For example, suppose a contract between architect and client states that the architect will perform services in accordance with normal architectural professional standards. Suppose further that the same agreement provides that the architect will visit the site at least once a day. If normal professional standards would not require that the architect visit the site this often, a conflict exists between the two provisions. In such a case, the provision specifically relating to the number of site visits will control the general provision requiring that the architect perform in accordance with accepted professional standards.

Words are generally used in their normal meanings unless both parties know or should know of trade usages that have grown up around the use of certain terms. This rule is important in the design professional's relationship to the client. Often a word used in an agreement will have a definite meaning to the design professional but have a different meaning, or at least an indefinable meaning, to the client. For example, suppose that the agreement between design professional and client states that the design professional will make periodic visits to the site during construction. Suppose further that the design professional shows evidence of a custom that visits customarily occur at weekly intervals during a certain stage of the project. If the client does not know or have reason to know of this custom, the custom cannot be used to interpret the phrase "periodic visits." Although parties can be said to contract with reference to established usages in a trade, if one party is not or should not have been aware of the usage, it would be unfair to permit evidence of usage to be used to interpret the language in question.

Another interpretation guide—that of practical interpretation—looks at the practices of the parties. Such practices may give good evidence of what the parties intended. For example, suppose the language to be interpreted is "periodic visits." Suppose the architect visited the site weekly and this was known by and acquiesced to by the client. A court would consider this persuasive evidence that the parties agreed that the visits would be weekly.

E. The Parol Evidence Rule and Contract Completeness

One aspect of the parol evidence rule relates to the provability of asserted prior oral agreements when the parties have assented to a written agreement. A client may contend that the design professional agreed to perform certain services that were not specified in the written agreement. The latter's attorney may contend that the writing expressed the entire agreement and that parol evidence or oral evidence is not admissible to "add to, vary, or contradict a written document." Sometimes such oral agreements are provable. Sometimes they are not.

Courts generally do not consider written agreements between clients and design professionals to be complete expressions of the entire agreement.[44] Undoubtedly, one reason why clients are usually allowed to testify as to promises not found in the written contract is the close relationship between the design professional and the client. Under such circumstances, the client may have been reasonable in not insisting that oral promises be included in the writing.[45]

This emphasis on the close relationship between the contracting parties can be demonstrated by a Mississippi case in which the principal issue was whether a contractor should have read an addendum that contained a provision that the architect told the contractor would be deleted.[46] In excusing the contractor's failure to read the addendum and his relying solely on the statement of the architect, the court noted that the contractor had been assured that the provision had been removed by the senior architect in the firm, "a man in whom he reposed considerable

---

[44]*Spitz v. Brickhouse*, 3 Ill.App.2d 536, 123 N.E.2d 117 (1954); *Malo v. Gilman*, supra note 43.

[45]*Lee v. Joseph E. Seagram & Sons, Inc.*, 552 F.2d 447 (2d Cir.1977).

[46]*Godfrey, Bassett & Kuykendall, Architects, Ltd. v. Huntington Lumber & Supply Co.*, 584 So.2d 1254 (Miss.1991). The *Godfrey* case has been distinguished in a number of insurance cases on a variety of grounds. See *Ross v. Citifinancial, Inc.*, 344 F.3d 458 (5th Cir.2003); *Reed v. American Soc. Group, Inc.*, 324 F. Supp.2d 798 (S.D.Miss.2004). This indicates that the *Godfrey* case was one of a kind and has little value as a precedent.

confidence following over a decade of business dealings."[47] The contractor, according to the court, expected to have been told the truth and not to have been misled. The court concluded stating that this was "a case of dealings between gentlemen, two honorable men."[48]

Sometimes a court will refuse to listen to any evidence outside the writing if it is convinced the writing is so clear in meaning it needs no outside assistance. Most courts will permit either party, especially the client, to show prior oral agreements not included in the writing.

Although attorneys for design professionals frequently rely on the parol evidence rule, such reliance is often misplaced. This is especially so if the agreement is sketchy and does not spell out the details adequately and if the client was not represented by an attorney during the negotiations.

Reducing an arrangement to writing does not necessarily protect the design professional from assertions of additional oral agreements. However, the more detail included in the agreement and the greater the likelihood that the client understood the terms or had legal counsel, the greater the probability that the client will not be allowed to prove the claimed oral agreement. The parol evidence rule does not apply to agreements made after the written agreement has been signed by the parties.

The parol evidence rule relates only to the provability of antecedent or contemporaneous agreements. If such agreements are admitted into evidence, the trial court or the jury—depending on who makes the determination of fact—must decide whether the evidence shows that such an agreement was made. Very often the attorney for the design professional places heavy reliance on the parol evidence rule and does not adequately prepare for the more important question of whether the asserted agreement took place.

Most parol evidence cases require a determination of whether the writing was intended to be the complete and final repository of the entire agreement of the parties. To preclude the court from determining that a particular writing was not intended to be complete, contracts prepared by attorneys or by professional associations often contain integration or merger clauses.

Before 1966, the AIA Standard Documents for professional services did not contain an integration clause,

reflecting the then emphasis on the close professional relationship between architect and client. Beginning in 1966, the AIA documents tended to become more concerned with liability and litigation, and this undoubtedly was the reason for including B141, Paragraph 9.6, which had stated, "This Agreement represents the entire and integrated agreement between the Owner and the Architect and supersedes all prior negotiations, representations or agreements, either written or oral." In B141-1997 this became Paragraph 1.4.1 and in B101-2007 it became Section 13.1. In commercial contracts, such clauses are generally successful in accomplishing the drafter's objective unless the party attacking the clause asserts that the other party fraudulently induced the agreement.

F. Formal Requirements: Recent California Legislation

Formal requirements for contracts in general were discussed in Section 5.10. Statutes of Frauds, though with some local variations, have been enacted in all states. They require a written memorandum for certain transactions. As a practical matter, none of the transactions required by the Statute of Frauds include transactions between a design professional and his client.

Yet increasingly, legislatures single out for special treatment other transactions that have generated disputes often ending in court, usually requiring that they be in writing. In addition, statutes often require conspicuousness, such as requiring a particular size type or all capital letters.

It was inevitable that legislatures would turn their attention to contracts between architects and their clients. The culmination of fragmentary discussions between a professional and his client, the likelihood that the client has never made a transaction like this one, and the complexity of the transaction and the lack of clarity of scope of services—these factors and undoubtedly others make it rare for clear and complete contracts to be made by architect and client.

The most complete contracts available are standard contracts published by professional associations of professional designers, such as the American Institute of Architects (AIA). Though the AIA claims its contracts are fair to both parties and reflect existing practices, such pre-prepared contracts are designed with the best interests of architects in mind. Such contracts rarely are gone over in detail by the architect and his client. Even language

---

[47]*Godfrey, Bassett & Kuykendall, Architects, Ltd. v. Huntington Lumber & Supply Co.*, supra note 46 at 1257.

[48]Ibid.

that deals with issues that come up frequently, such as cost estimates, scope of services, additional services, ownership of documents, and compensation, may not be reviewed carefully. Clients, as a rule, do not understand the terms used, are not given much time to review the contract, and rarely engage a lawyer to review the contract.

The realities of the contract making process led the California legislature in 1995 to add Section 5536.22 to its Business and Professions Code. It requires that an architect use a written contract executed by both architect and client before the architect begins work, unless the client (in writing) directs that the work can begin before the contract is made. At the very least, the contract must contain language dealing with the scope of services, basis for compensation, method of payment, procedures for additional services, and procedures for termination. There are some exceptions.

Another legislative interference with contracts is the Connecticut Home Improvement Act.[49] A valid contract for a home improvement requires a writing signed by both parties. This amplifies the Statute of Frauds.[50] The Act also requires written change orders. But it made an exception for a claim for the reasonable value of services if the court decides it would be inequitable to deny recovery to the claimant.

Enactment of this statute raises many questions. Are transactions that do not comply with this statute valid? Is failure to comply only a breach of the architect's professional responsibility and the basis for professional discipline? Will the AIA or local architects develop a standard contract that will meet the statutory requirements? What effect will this statute have on existing AIA documents, such as AIA Document B101-2007 reproduced in Appendix A? Will other states enact similar legislation?

Yet enactment of these statutes does reflect public concern over the way architects and their clients create the architect–client relationship.

---

[49] Conn.Gen.Stat. § 20–429(a).

[50] See Section 5.10.

CHAPTER TWELVE

Professional Design Services: The Sensitive Issues

SECTION 12.01 Range of Possible Professional Services: Fees and Insurance

The range of potential professional services, already substantial because of the complexity of design and construction and the centrality of the design professional's position, has expanded. The range of potential services performed by the design professional has expanded because of:

1. Increased public control over design and construction. See Section 12.06.
2. Greater participation by the design professional in financial and economic aspects of the project. See Sections 12.04 and 12.05.
3. Greater demand by the owner that the design professional play a role in selecting or drafting the construction documents. See Section 12.07.
4. Greater participation by the design professional in disputes that culminate in arbitration or litigation. See Section 29.04.
5. Greater desire by the client to involve the design professional in activities prior to actual design or those which are performed after completing construction. See Section 12.02.

The reaction of the AIA to this expanded range of professional services will be discussed in Section 12.02.

Compensation is discussed in Chapter 13, but one aspect of compensation relates to the potential list of professional services. Although the percentage-of-construction-cost method of compensating the design professional is beginning to be replaced by time-based compensation, it is still common to compute the fee by the percentage formula. However, and not without occasionally surprising the

client, this amount is called the *basic fee*. Other services are frequently designated as "additional services," which entitle the design professional to compensation in addition to the basic fee. In many cases, the design professional is willing to perform any services the law allows and within her professional skill, provided she is compensated for performing these services. But disputes may arise over whether the services requested fall under the basic fee or entitle the design professional to additional compensation. As shall be seen in Section 13.01G, the American Institute of Architects (AIA) in its B141-1997 eliminated the concept of additional services and substituted Paragraph 1.3.3, "Change in Services."

Ten years later, the AIA brought back the term "additional services." Document B101-2007 Article 3 lists the architect's basic services, and Article 4 lists the architect's additional services. This book uses the terms *basic services* and *additional services*.

Another factor—professional liability insurance (discussed in Section 15.05)—must be taken into account in planning services. Design professionals should be aware of which services can be covered by professional liability insurance and which are generally excluded. As a general rule, insurers wish to insure only those services routinely performed and connected to the professional training and experience of the insured design professional. Such services include analyzing the client's needs, preparing a design solution, and performing site services to see that the design is executed properly. Services that go beyond these may very well be excluded unless there is a special endorsement.

Similarly, the client should be aware of which services are included. Although it is possible to agree that services be rendered that are excluded from insurance coverage, this should be a deliberate choice made by the contracting parties, not inadvertent.

At the outset of the professional relationship or as soon as possible thereafter, the client and design professional should determine in advance which of the possible professional services the design professional is to perform and whether those services fall within the basic fee. After this is done, the written agreement should reflect the actual agreement, and performance should accord with the contract requirements or any subsequent modifications.

SECTION 12.02 Traditional Roles of Design Professional: B101-2007

Before proceeding to the 2007 AIA changes to the "Basic Agreement for Design Services," it is advisable to look at the traditional model of design professional services.[1]

Usually the client comes to the designer with a problem it hopes the designer can solve. The client describes its needs and, as a rule, what it wishes to spend. After consulting the client, the designer develops a schematic design and may revise the client's budget.

Next the designer studies the design and prepares drawings and possible models that illustrate the plan, site development, features of construction equipment, and appearance. The designer is also likely to prepare outline specifications and possibly again revise the predicted costs. In small projects, schematic design and design development are still designated as "preliminary studies."

After the client approves the design development, the designer prepares working drawings and specifications that cover in detail the general construction, structure, mechanical systems, materials, workmanship, site development, and responsibility of the parties. Often the designer supplies or drafts general conditions and bidding information (discussed in Section 12.07, dealing with services of a legal nature).

The final preconstruction phase is generally called the *bidding* or *negotiation phase*. The designer helps the client obtain a construction contractor through bidding or negotiation. These phases are discussed again in Section 13.02, as phases often determine timing of interim fee payments.

During construction, the designer interprets the contract documents, checks on the progress of the work to issue payment certificates, participates in the change order process, and initially resolves disputes. As shall be noted in Section 29.01, A201-2007 Article 15 introduces a new actor, the Initial Decision Maker (IDM). She makes initial decisions. If an IDM is not specifically designated, then initial decisions are made by the architect.

In creating B141-1997, the AIA sought to emphasize the expanding range of services that the architect might perform. It divided B141-1997 into two components. Both were labeled as B141-1997 but were physically divided: the "Standard Form of Agreement Between Owner and Architect" (Articles 1.1-1.5) and the "Standard Form of Architect's Services: Design and Construction Administration" (Articles 2.1-2.9). The latter document listed seven categories of architect's services.

With B101-2007, the AIA returns to the traditional one-document "Standard Form of Agreement Between Owner and Architect." Article 3, dealing with basic services, divides these services into five components:

1. schematic design phase (§ 3.2)
2. design development phase (§ 3.3)
3. construction documents phase (§ 3.4)
4. bidding or negotiation phase (§ 3.4)
5. construction phase (§ 3.6)

SECTION 12.03 Cost Predictions

A. Inaccurate Cost Prediction: A Source of Misunderstanding

The relative accuracy of cost predictions is crucial to both public and private clients. Clients may be limited to bond issues, appropriations, grants, loans, or other available capital. Unfortunately, many clients think that cost estimating is a scientific process by which accurate estimates can be ground out mechanically by the design professional. For this reason, design professionals should start out with the assumption that cost predictions are vital to the client and that the client does not realize the difficulty in accurately predicting costs.

The close relationship between design professional and client often deteriorates when the low bid substantially exceeds any cost figures discussed at the beginning of the relationship or even the last cost prediction made by the

---

[1]The traditional model described here contrasts with the impact on design services brought about by digital information technology. See Section 19.01.

design professional. Not uncommonly the project is abandoned, and the client may claim that it should not have to pay the design professional. The design professional contends that cost predictions are educated guesses and that their accuracy depends in large part on events beyond the control of the design professional. Clients frequently request cost predictions before many details of the project have been worked out. Clients frequently change the design specifications without realizing the impact such changes can have on earlier cost predictions. Design professionals point to unstable labor and material costs. The amount a contractor is willing to bid often depends on supply and demand factors that cannot be predicted far in advance. Design professionals contend they are willing to redesign to try to bring costs down but unless it can be shown they have not exercised the professional skill that can be expected of people situated as they were, they should be paid for their work.

The terminology relating to *cost predictions* indicates this is a sensitive area. The most commonly used expression, though used less frequently in contracts made by professional associations, is *cost estimates*. The term *estimate* is itself troublesome. In some contexts it means a firm proposal intended to be binding, and in others it is only an educated guess. Probably many clients believe that cost estimates will be in the ballpark, whereas design professionals may look on such estimates as educated guesses.

Professional associations seek to minimize the likelihood that their members will go unpaid if a project is abandoned because of excessive costs (explored in greater detail in Section 12.03C). Part of this effort is reflected in the terminology chosen. For example, in its 1977 B141 the AIA sought to differentiate between "statements of probable construction costs, detailed estimates of construction costs, and fixed limit of construction costs." Only the latter, according to the AIA, created a risk assumption by the architects that they will go unpaid if the project is abandoned because of excessive costs. In 1987, the AIA changed to *preliminary cost estimates*. In B141-1997, Paragraph 2.1.7.1 stated that the architect will prepare "a preliminary estimate of the Cost of the Work."

In B101-2007, Section 3.2.6 states that the architect will submit to the owner "an estimate of the Cost of the Work prepared in accordance with Section 6.3." Section 6.3, in turn, states that the estimate may be based on "current area, volume, or similar conceptual estimating techniques." If the owner asks for "detailed cost estimating services," this will be charged as an additional service.

The Engineers Joint Contracts Document Committee (EJCDC) in EJCDC E-500 (2002), Paragraph 5.01, refers to "opinions of probable construction costs." The potential for misunderstanding is demonstrated in the reported appellate decision reproduced in Section 12.03C.

B. Two Models of Cost Predictions

The many reported appellate cases demonstrate not only the frequency of misunderstanding but also the difficulties many design professionals have predicting costs. Recognition of this, along with other factors explored in Chapter 17, has led the sophisticated client to seek more refined ways to control and predict costs. It is important at the outset to distinguish between what can be called the *traditional method* and these more refined methods.

The traditional method usually involves the design professional's using rough rules of thumb based on projected square or cubic footage, modulated to some degree by a skillful design professional's sense of the types of design choices a particular client will make. Cost predictions are likely to be given throughout the development of the design, but they are usually based on these rough formulas, refined somewhat as the design proceeds toward completion.

As time for obtaining bids or negotiating with a contractor draws near, the cost predictions should become more accurate. In this model, there is a great deal of suspense when bids are opened or when negotiations become serious. Under this system, a greater likelihood exists of substantial if not catastrophic differences between the costs expected by the client and the likely costs of construction as reflected through bids or negotiations. This model gives rise to the bulk of litigation.

The other model—that of more efficient techniques—is likely to be used by sophisticated clients who are aware of the difficulties design professionals have using the model just described. Clients are likely to engage someone who can more accurately predict costs as the design evolves. Sometimes they hire a skilled cost estimator—someone close to the quantity surveyor used in England—as a separate consultant. Sometimes they hire a construction manager (CM), who is supposed to have a better understanding of the labor market, the materials and equipment market, and the construction industry, as well as of the construction process itself. Using a CM is intended not only to free the designer from major responsibility for cost predictions but also to keep an accurate, ongoing cost prediction. Sometimes the CM agrees to give a guaranteed maximum price (GMP) that may vary as the design evolves. To do so, the CM may obtain firm price commitments from the specialty contractors.

This second model—a fine-tuned model—is designed to avoid the devastating surprises common under the traditional model. To determine whether the design professional bears the risk of losing the fee, differentiating between the two models is vital, the risk being greater in the traditional model.

C. Creation of a Cost Condition: Frank Lloyd Wright and the Johnson Building and *Griswold & Rauma v. Aesculapius*

When clients assert they have no obligation to pay the design professional because the low bid exceeded predicted costs, they are asserting a cost condition. They are claiming their obligation to pay was conditioned on the accuracy of a design professional's cost prediction. A cost condition is a gamble by the design professional that the cost prediction will be reasonably accurate. If it is not, the fee is lost unless the client prevented the condition from occurring, such as making excessive design changes, or is willing to dispense with the cost condition. In such a case, the client need not show that the design professional has not lived up to the professional standard of care.

Clients sometimes contend the design professional not only gambled the fee but also promised that the project will be brought in at or below a designated price. Failure to perform as promised makes the design professional responsible for any damage to the client that was reasonably foreseeable as a probable result of the breach at the time the contract was made. Section 12.03C emphasizes the question of whether a cost condition has been created, while Section 12.03F discusses the question of damages for breach.

FRANK LLOYD WRIGHT AND THE JOHNSON BUILDING: A CASE STUDY

Much can be learned from case studies. They provide a picture of how the design professional and the client interact over costs. The case studies in a book dealing with law are usually actual cases that have reached an appellate court. One such case, *Griswold*, is reproduced in Section 12.03C. It very likely involved the use of AIA documents.

The Frank Lloyd Wright case study did *not* culminate in a written contract, a lawsuit, and a reported appellate decision. Yet much can be learned from it. The human side of the architect–client relationship is not always depicted in actual cases. This case study involved Frank Lloyd Wright and one of his most famous projects, the Johnson Building in Racine, Wisconsin (see photos A–C). In 1986 it ranked ninth on the list of best architectural designs by the College of Fellows of the American Institute of Architects. For the portion of this case study that deals with costs, we are indebted to Professor Stewart Macaulay,[2] who gained access to records and obtained testimony from witnesses.

Macaulay writes that at the time Wright received his commission, his practice was fallow due to scandals involving Wright and the onset of the Great Depression. Wright managed to survive this period by lecturing, writing, and forming the Taliesin Fellowship. There young architects and artists paid to work with Wright. Macaulay continues,

The Johnson Building was Wright's next major commission. He was eager to get it, and it marked a major turning point for him. Wright was nearing seventy, and after more than five years when he had no significant commissions, the publicity provoked by "Fallingwater" and the Johnson Building reminded people that Frank Lloyd Wright was not merely an important figure in the history of architecture. He began about twenty years more of highly productive work. This last phase of Mr. Wright's career ended in 1959 when he died at age ninety-one with New York's Guggenheim Museum under construction.

. . .

The company needed a new administration building. It hired a local architect named Matson who offered a traditional design. Jack Ramsey [general manager of Johnson] was dissatisfied. Several people suggested Frank Lloyd Wright. Ramsey and Bill Connolly, the Advertising Manager, went to Taliesin to meet Wright on July 17, 1936. Ramsey knew Wright's reputation as an

[2]From "Organic Transactions: Contract, Frank Lloyd Wright, and the Johnson Building," by Stewart Macaulay [1996], *Wisconsin Law Review* 75. Copyright © 1996 by the Board of Regents of the University of Wisconsin System. Reprinted by permission of the Wisconsin Law Review. Most notes have been omitted. Those not omittted have been renumbered.

(A)

(B)

PHOTOS A AND B   Frank Lloyd Wright's Johnson Building—Research Tower (A) and Administration Building (B) exteriors. Photos courtesy of SC Johnson, A Family Company.

(C)

PHOTO C   Frank Lloyd Wright's Johnson Building—The Great Workroom, located inside the Administration Building. Photo courtesy of SC Johnson, A Family Company.

architect from Ramsey's experience in Europe, but he also knew Wright's negative reputation in Wisconsin. People there thought about the scandals related to Wright's domestic situation, how seldom Wright paid bills on time, his unconventional houses, and the cost of his work. The commission could be a great opportunity for Wright, after having so little work for a number of years. Wright was at his most persuasive, and Ramsey was impressed. He wrote a memorandum to Hibbard Johnson who was at his cottage in Northern Wisconsin. Ramsey's memo strongly recommended that Johnson meet Wright. A Frank Lloyd Wright building became Ramsey's cause within the company. In effect, he committed his reputation to a project by the controversial architect. Ramsey said:

Regarding the new building, I had a day Friday that so confirmed and crystallized my feeling about Matson's present offering and that at the same time so inspired me as to what can be done that I was on the point of sending you wild telegrams Friday night when I got home, or getting you out of bed on the telephone.

. . . Honest, Hib, I haven't had such an inspiration from a person in years. And I won't feel satisfied about your getting what you want until you talk to him—to say nothing of not feeling justified in letting $300,000 be clothed in Matson's designs.

. . .

He's an artist and a little bit "different," of course, but aside form his wearing a Windsor tie, he was perfectly human and very easy to talk to and most interested in our problem and understood that we were not committing ourselves, but, gosh, he could tell us what we were after when we couldn't explain it ourselves.

. . .

And he asked about what we thought this building would cost us. I said, when we got through with the building, landscaping, furnishings, etc., we'd be investing around $300,000. He asked how many people it would house. I said about 200. He snorted and said it was too damn much money for the job and he could do a better functional job in a more appropriate manner for a lot less. . . .

He is very easy to talk to, much interested in our job whether he has anything to do with it or not, because it hits his ideas of modern building, because it is a Wisconsin native proposition, and because it seems to hurt his artistic conscience to see so much money spent on anything ordinary. . . . Will you see him?

On July 21, 1936, Johnson drove to Taliesin to meet Wright. At first, the two men argued. Johnson later said, "I showed him pictures of the old office, and he said it was awful. . . . He had a Lincoln-Zephyr, and I had one—that was the only thing we agreed on. On all other matters we were at each other's throat." Johnson described his goals for the new building. He wanted it to symbolize the progressive company that his grandfather and father had built. Wright then "described the kind of building he would design, unconventional, imaginative, trend-setting, a visual symbol of a great company."

On July 23rd, Johnson wrote Wright:

I am now asking you to proceed with plans and sketches of a $200,000 office building for us in Racine on the basis of 2½% or $5,000 to be paid you when sketches and plans are submitted. . . .

It is my understanding that the remaining commission of 7½% or $15,000 will not be paid to you unless your plans are used wholly and under your supervision. Also, that we are free to use any or all the ideas you offer—either ourselves, or other architect . . . .

I want to take this opportunity of expressing my appreciation, as well as Mr. Ramsey's, for your gracious hospitality, and for the inspiration and education we received.

. . .

Wright and his associates then worked around the clock to produce his proposal. He presented it to Johnson, Ramsey and several other executives on September 15, 1936. On that same day, Johnson and Wright presented the plan to the firm's Board of Directors. The Board approved the project.[3]

---

[3]On August 18, 1936, Johnson wrote Wright,

Some time ago the Directors approved a sum of $200,000 for a new office building. No mention was made of furnishings, fees, etc. At the next meeting I will advise them of your goal—the building complete at $250,000—which I feel will be acceptable to them, considering the plus value we will receive by having you do it for us.

[The author of a book on Wright and the Johnson Building] also quotes John Howe, Mr. Wright's chief draftsman, as saying, "From the start, the money they were talking about wouldn't have done the most ordinary kind of building. Mr. Wright always started doing what he thought was right for the building. He didn't burden himself with undue considerations of cost."

The company hired a contractor on the basis of cost plus an agreed upon percentage for overhead and profit. There is no record of any detailed written contract signed by Wright and the Johnson firm.[4]

Macaulay then describes the building of the structure. Problems developed, such as

1. Would the structure serve its business needs?
2. Discovery of two major structural problems
3. Greater than expected delay and
4. "The building cost far more than Mr. Wright's various estimates."

As to the latter, Macaulay continued, summarizing the facts already given:

Responsibility for delay strained the relationship, but the real arguments focused on the cost of the building. When Jack Ramsey had first visited Wright, Ramsey said that the Johnson company was planning to spend about $300,000 on the local architect's plan. Ramsey reported to Hib Johnson that Wright "snorted and said it was too damn much money for the job and he could do a better functional job in more appropriate manner for a lot less. . . ." Johnson's letter offering the commission to Wright suggested a $200,000 building. Later, Johnson wrote to Wright and quoted a total cost of "$250,000 including furnishings, fees, etc." Wright's fee was to be ten percent of the total cost of the building. As the cost of the building increased so did the architect's fee. Although the final cost was not announced by S. C. Johnson & Son, it was clearly many times Mr. Wright's original target of $250,000. One speculation was $750,000, Mr. Wright's figure was $850,000, and another estimate was $900,000.

---

[4]Wright wrote Ramsey:

I am sorry you had to break off the thread of continuity so abruptly next day. I tried to get . . . you to stay until we could get formalities over with so we might proceed with Mr. Johnson's decision to build immediately. But we are so proceeding anyway without formalities so that no time will be lost. When you return we can get things straight. The first part of the service according to our agreement is practically rendered and a letter of acceptance from the company closing the preliminary episode and opening the second phase is in order when you get around to it.

. . . Later, Jack Ramsey wrote Frank Lloyd Wright about the Johnson firm's arrangement with Wright. He mentioned Hibbard Johnson confirming "our verbal agreement" in a letter of July 23, and Mr. Wright's "long-hand note" of August 15 " 'assuring and driving at the Building complete at a cost of $250,000 including an appropriation of $20,000 for furnishings. Architect's fee is included and also the Clerk of the Works fee.' Letter from Ramsey to Wright, Dec. 11, 1936."

Wright and Ramsey first debated fees in December of 1936. Wright wrote to Ramsey, addressing him as "My dear Jack," with a "proposition" about designing the furniture for the building. He said that he had charged others twenty percent for designs of furniture and equipment. However, he offered do it for ten percent if Ramsey would send him $3,000 "on furniture designs you have not yet seen . . ." Wright noted that "Christmas is coming and the best way for me to get a good one is to pay up the thousand and one petty accounts nagging my footsteps."

Ramsey responded the next day, noting that he was "the Scotchman in this picture." This letter was the only time that either Johnson or Ramsey turned to the language of their contract with Wright. Ramsey referred to the parties' agreement for designing the building:

[A]fter various conferences, you finally wrote him [Hib Johnson] on August 15 (long-hand note so maybe not in your files), "I am assuring and driving at building complete at a cost of $250,000 including an appropriation of $20,000 for furnishings. Architect's fee is included and also the Clerk of the Works fee."

Ramsey said that they had not added to the original plans except a squash court over the garage. "But now you tell me it will run about $300,000 and that apparently exclusive of furnishing and fees!" Ramsey thought that Johnson's Wax might not be able to "splurge" on new furnishings, and they could not commit to Wright's designs without seeing them. "Money is an irritating part of this world, but we've got to take it into account—not for piling up gold for its own sake, but just so that this business continues to run properly and serve the very human destiny that it has for fifty years." He concluded:

In any case, it seems to me that there are a lot of things about the building itself that have to be completed first [before the Johnson firm agrees to Wright-designed furniture]. Do you realize that Hib has advanced, to be exact $20,964 on an extreme expectation of something under $25,000 [in total architect's fees] and we have not yet the completed construction plans, to say nothing of final interior layout and approved plans on heating, ventilating, lighting—even the glass to be used in wall construction? That is confidence beyond anything I can say in words, so I know you will not take my plain words wrong.

Wright responded over a month later, beginning, sarcastically: "Dear Mr. Ramsey. Thanks a Lot Anyway." He noted that he had had pneumonia, and Ramsey's letter had been kept from him because it was "considered disturbing." Wright justified the increased costs by pointing to what had been added to the structure. He wrote of what he had saved the project by battling state regulators. He said that commercial buildings usually carried separate fees for a "structural engineer, sanitary engineer, heating and ventilating engineer as well as architect." He argued that the creative work of planning had been completed. "I would be then entitled to $7\frac{1}{2}\%$ of the revised est. of $300,000 or $22,000." He noted that he felt free "to throw away details . . . and make others when I find I can improve the structure or save something. . . . This will never . . . [stop] until it is finished if I can keep my form." He ended, "So you see, Jack, Scotch though you may be, your architect is no longer trespassing on his client."

Shortly before Christmas in 1937, Hibbard Johnson indicated his concern about the mounting costs of the building. Wright responded in a long letter. He pointed out that he was not profiteering from the project. "No architect creating anything worth naming as creative work ever made or can make any money on what he does." He then asserted that the company could afford what Wright was providing. Johnson had "the privilege of paying for something way beyond money value." Wright had saved money on the project in various ways. The building would benefit the company as a symbol that could be advertised.

He then turned to the costs. He said, "if the office building runs to $450,000 (as it will) including furnishing:—it will have cost the Company about 33 cents per cubic foot, which is the price of any ordinary well-built, fire proof, air-conditioned factory building." The structure was bigger and built under more expensive conditions than Wright had imagined. The total costs were not extravagant "considering the resources of the owners and what they are getting for their outlay. . . . The labor scale and shorter hours and prices for materials, all these are higher than any previous work of mine." Wright said that the demands on him in supervising construction had been excessive. Wright ends by asking Johnson to send him a check "to help get started in the [Arizona] desert."

Hibbard Johnson was surprised by the new cost estimate of $450,000, and he responded with some heat:

I know it does no good to complain as you are an artist so in love with your work that nothing will make you change your ideas of what the . . . [building] ought to be, even though it works a hardship on your client. You would rather tell the client whatever comes into your head as to the cost and the time to construct, at the start, just to sell the job and give satisfaction to your art to create something worthwhile, rather than be accurate in cost estimates. Why didn't you put me wise long ago as to the true costs and time to construct? Would that be unreasonable to ask? That is water over the dam now and I am going to have to take it, but I will never like it. That is, the way you have handled me; the buildings . . . I am going to love. . . .

Now, Frankie, this reply to your letter is no complaint as it would do no good to complain. You have us hooked and we can't get away. Rather, it is written to show you how I feel and, if possible, spur you on to economize on matters still undecided in the building. . . .

Most people called Frank Lloyd Wright, "Mr. Wright." Hib Johnson was a special client and called him, "Frank." "Frankie" should be taken as an expression of annoyance.

A writer on the staff of Architectural Forum was visiting Taliesin in 1938. He overheard a meeting between Mr. Wright and Johnson officials.

[We could hear the loud voices raised on the client's side, and afterward Mr. Wright came out . . . and said that the client was unhappy . . . [because] the building was going over budget. . . . [H]e said . . . "You know, they really don't understand this building at all." He said, "They're acting as if this were a normal office building and you calculate this the way you would a normal office building. But they have forgotten what they told me initially, which was that this was a memorial to Grandpa, the founder of this great industrial enterprise, and you don't build memorials with the same materials, or the same spirit, or the same budget, you know as you do speculative office buildings." He said, "One of these days . . . you're going to see . . . tourists from all over the country . . . come and see this building."

Almost a year later, Jack Ramsey and Frank Lloyd Wright again debated the delays and cost of the building. Mr. Wright wrote Ramsey while traveling on the train to Arizona. He remarked that the original plans were only "a crude unfinished sketch" of what was in process in Racine. He said, "I realize fully the strain the growth of this great landmark in new-world architecture has thrown on you—and do not resent the breakdown of good feeling and consequently of good sense."

Ramsey replied that he appreciated the letter. However, "I can't subscribe to the statement that we ever lost 'good sense'; but I freely admit that 'good nature' took an awful long vacation." Then Ramsey sought to justify the Johnson company's concern with the costs of the building:

Cost, as measured in money, is a most difficult thing to argue with you. Idealistically you despise the idea of money as a measure of anything. It probably has not occurred to you that Hib and I and probably 90% of the rest of the world also realize the imperfection of such a measuring stick. Nevertheless, it remains a fact that it is a universal yardstick used even to measure happiness. Wastefulness of dollars in the construction of our building did not grieve us because each dollar came off of a cold figure concerning a bank balance, but offended our sense of justice in that such wasted dollars were a measure of some other constructive accomplishment that thereby must be omitted from the scheme of things.

It is, I believe, only a matter of proportion on which we have differed with you. If a farmer has a hundred dollars and has a certain aim in view concerning the raising of poultry, he might be justified in spending twenty dollars on a chicken coop, or even twenty-five dollars with the extra five dollars as a measure of additional content and happiness afforded the chickens and the eyes of all beholders, but he would be morally unjustified in spending ninety-nine dollars on his chicken coop and thereby starving his horses and cows. And if he hired a chicken-coop specialist to build the finest coop in the world at an estimated cost of thirty dollars and the cost ran up to ninety-nine dollars, moral responsibility would be upon aforesaid specialist.

That is not a pretty example and it is probably crude and exaggerated but I am impelled to try to illustrate our side somehow.

Mr. Wright did not accept Ramsey's position. He said that if the chicken house simile were apt, "[w]e would all be written off for damn'd fools and sent over the hills to the poor house. I've felt (as I know you and Hib have felt) that there were human values involved, in this building way beyond any that could be measured by money."

. . . .

One sanction usually available to a dissatisfied party to a contract is refusing to interact again with the other party. Wright and Johnson had become friends, but Hib Johnson felt that Mr. Wright had manipulated him with an unreasonably low estimate to get the job. This strained the relationship. Wright wrote Ramsey after Wright had prepared a revised edition of his autobiography: "I've heard nothing from Hib since I sent on the piece on the building now appearing (soon) in An Autobiography . . . I thought he would like the piece very much. But I guess he didn't. . . . We expected Hib to invite us to dinner sometime this winter—but no."

After the Johnson Administration Building was completed in 1939, the firm decided to build facilities for research and development of new products. World War II delayed the project. In October of 1943, Hibbard Johnson wrote Frank Lloyd Wright, asking for comments. Johnson said:

To be frank, Frank, we simply will not consider a financial and construction nightmare like the office building. It is a plain factory kind of job that should be built by an engineer or contractor like our other factory buildings. Yet because of its proximity to your masterpiece, it should have a relationship thereto and we feel it would be unfair to you and a mistake on our part if we didn't ask how you think you would want to fit into such a picture.

After several letters and a proposed design, Johnson changed his mind and hired Wright to build the Research Tower. Mr. Wright's plan was not for "a plain factory kind of job." The Company attempted to be more cautious and formal in its dealings with Frank Lloyd Wright. Hib Johnson wrote Wright: "We want this building built on a contract basis, if at all possible; if not that, then on a basis where the cost would not vary 10% over estimates." Nonetheless, they went ahead on a cost-plus basis without the ceiling that Mr. Johnson wanted. Mr. Wright was to be paid his usual commission of 10% of the costs of the project.

The tower's estimated cost rose from $750,000 to over $2 million. In May of 1948, Johnson accepted this estimate but again bargained to cap Mr. Wright's commission at $200,000 to be paid over two years.

Contract Completeness and the Parol Evidence Rule. The issue that has arisen most frequently in cost cases relates to the parol evidence rule (discussed in Section 11.04E). This rule determines whether a writing assented to by contracting parties is the sole and final repository of the parties' agreement. If so, testimony relating to agreements made before or at the time of the written contract is not admitted into evidence by the judge. In the context of a cost condition, the issue is whether the client will be permitted to testify that it had made an earlier oral agreement or had an understanding that it could abandon the project and not pay for design services if the low construction bid substantially exceeded the cost prediction of the design professional.

The client generally will be permitted to testify that such an agreement had been made if the agreement between the design professional and the client is oral, if the agreement is written but nothing is stated as to the effect of accurate cost predictions, or in some cases even if this problem is dealt with in the agreement. Permitting such testimony is based on the conclusion that such agreements are not, as a rule, the final and complete repository of the entire agreement between design professional and client.[5]

Standardized contracts prepared by the AIA or EJCDC include language that seeks to protect the design professional from assertions of the existence of a cost condition. Although such language has been useful to design professionals,[6] its presence is not ironclad protection against clients being permitted to testify to their understanding that if the costs were excessive they could abandon the project and not pay the design professional for services.[7] The case reproduced later in this section involved a clause that sought to make the writing complete. Yet, as shall be seen, testimony seems to have been freely admitted. (The current AIA language is discussed later in this section.)

Keep in mind that permitting the testimony does not end the matter. Issues are still likely to exist as to whether the agreement took place, the nature of the agreement, and whether the condition has occurred or been excused.

Preliminary Issues. Before discussing cost conditions and their legal effect, other issues should be addressed. Are cost and cost control essential elements of professional service? What is the legal effect of giving

---

[5]*Stevens v. Fanning*, 59 Ill.App.2d 285, 207 N.E.2d 136 (1965). Many cases are collected in Annot., 20 A.L.R.3d 778 (1968).

[6]*Anderzhon Architects v. 57 Oxbow II*, 250 Neb. 768, 553 N.W.2d 157 (1996) (AIA language requiring a writing to create a cost condition justified summary judgment in favor of the architect).

[7]*Kahn v. Terry*, 628 So.2d 390 (Ala.1993) (despite the use of an AIA contract both parties permitted to testify and issue sent to jury: architect prevailed).

a cost estimate? Do other obligations exist relating to cost and cost control?

Two cases, each involving identical AIA documents that required the architect to give preliminary cost estimates when requested, came to different conclusions. Texas held that the fiduciary relationship did not require the architect to take cost into account and advise the owner as to estimated cost.[8] Michigan held giving cost estimates and, presumably, monitoring costs are inherent in the client–design professional relationship.[9] The unusual facts in the Texas case, the increased responsibility placed on fiduciaries, and the usual expectations of the client make it likely that the Michigan result will be followed.[10]

Although a cost estimate usually is simply an opinion and not a *guarantee*,[11] it can be a factual representation by the design professional. (This may be the reason the AIA in 1987 changed from "a statement of probable construction cost" to "a preliminary cost estimate," in 1997 to "a preliminary estimate of the Cost of the Work," and in 2007 to "an estimate of the Cost of the Work.") If it is sufficiently certain to be relied on reasonably by the client, an inaccurate estimate is a misrepresentation by the design professional. New York held that a cost prediction could constitute an intentional misrepresentation if it was guaranteed or if the client relied on the architect's opinion as an expert.[12] A Florida court ruled that a contract disclaimer, stating that the architect's estimate of probable cost was not a guarantee, meant the trial court erred in instructing the jury that "it is the general duty of an engineer to be reasonably accurate in providing estimated costs of a project."[13]

An innocent misrepresentation that induces the making of the contract permits rescission of the contract. If the contract is made, the design professional cannot recover for services and must repay any amounts received. If her design

was used by the client, the design professional can recover in restitution based on unjust enrichment. If the representation was negligently made, the client would also have a claim for damages (discussed in Section 12.03F).

Even if an estimate is not a guarantee or even if a cost condition is not created, there can be other cost-related obligations. For example, the Louisiana Supreme Court held that the engineer who designed a water slide was liable for damages despite a contractual provision stating that the engineer was simply giving opinions to be used only as a guide.[14] The court concluded that the engineer had breached the contract "not by giving an inaccurate initial estimate, but by failing to employ a professional estimator, failing to look at other water slides, failing to advise the owners about other contractual possibilities and failing to provide revised cost estimates."[15]

Existence of Cost Condition. Although cost condition cases generally involve the client claiming an express agreement based on a cost agreed to or set forth in the contract under which it could abandon the project and not pay a fee, a cost condition can be created by implication without any specific agreement as to the effect of inaccurate costs.[16]

Certainly design professionals do not operate in the dark. If they know what funds are available and are aware of the remoteness of obtaining additional funds, any cost specified may be "hard." In such a case, a cost condition may be created despite the absence of an express agreement under which this risk is taken.

Evidence that bears on the softness or hardness of any projected costs discussed by the design professional and the client or expressed in their agreement is crucial. This evidence, such as labeling the amounts as merely an estimate or using a cost range, may indicate that the amount or range specified is what is hoped for rather than a fixed-cost limitation. Where the amount is

---

[8]*Baylor Univ. v. Carlander*, 316 S.W.2d 277 (Tex.Ct.App.1958).

[9]*Zannoth v. Booth Radio Stations*, 333 Mich. 233, 52 N.W.2d 678 (1952).

[10]See *Williams Eng'g, Inc. v. Goodyear, Inc.*, 496 So.2d 1012, 1017 (La.1986), which followed the Michigan case. Similarly, see *Getzschman v. Miller Chemical Co., Inc.*, 232 Neb. 885, 443 N.W.2d 260 (1989) (duty to warn of potential overruns). See Section 11.04B.

[11]AIA Doc. B101-2007, § 3.2.6.

[12]*Pickard & Anderson v. Young Men's Christian Ass'n*, 119 A.D.2d 976, 500 N.Y.S.2d 874 (1986).

[13]*Post, Buckley, Schuh & Jernigan, Inc. v. Monroe County*, 851 So.2d 908 (Fla.App.Dist.Ct.2003).

---

[14]*Williams Eng'g v. Goodyear, Inc.*, supra note 10. This case is discussed in Section 12.03F.

[15]Id. at 1017.

[16]*Stanley Consultants, Inc. v. H. Kalicak Constr. Co.*, 383 F.Supp. 315. (E.D.Mo.1974) (dictum). In *George Wagschal Assoc., Inc. v. West*, 362 Mich. 676, 107 N.W.2d 874 (1961), a cost condition was found in a consultant contract because the consultant knew of the client's budget limit. But if the contract expressly negates a cost condition, it will not be implied. *Kurtz v. Quincy Post Number 37, Am. Legion*, 5 Ill.App.3d 412, 283 N.E.2d 8 (1972).

"soft," design professionals are being exhorted to use their professional skill to bring the project in for the amount specified. Where it is "hard," the client may be informing the design professional that the latter is risking the fee on ability to accomplish this objective. This distillation of appellate cases appeared in a legal journal dealing with architectural cost predictions:[17]

> Courts have admitted evidence of custom in the profession. Architects have been permitted to introduce evidence that customarily architects do not assume the risk of the accuracy of their cost predictions. Also, courts have been more favorably disposed toward holding for the architect if the project in question has involved remodeling rather than new construction, because estimating costs in remodeling is extremely difficult. The same result should follow if the type of construction involves experimental techniques or materials.
>
> Courts sometimes distinguish between cases and justify varying results on the basis of the amount of detail given to the architect by the client in advance. Generally, the greater the detail, the easier it should be for the architect to predict accurately. However, it is much more difficult for the architect to fulfill the desires of the client within a specified cost figure if the client retains a great deal of control over details, especially if these controls are exercised throughout the architect's performance. For this reason, some courts have held that a cost condition is not created where the architect is not given much flexibility in designs or materials.
>
> Some courts have looked at the stage of the architect's performance in which the cost condition was created. If it is created at an early stage, it is more difficult for the architect to be accurate in his cost predictions. Generally, the later the cost limit is imposed in good faith, the more likely it is to be a cost condition. But courts should recognize that if it is imposed later, creation—or, more realistically, imposition—may be an unfair attempt by the client to deprive the architect of his fee.
>
> Occasionally the courts have applied the rule that an ambiguous contract should be interpreted against the person who drew it up and thus created the ambiguity. If the client is a private party, the contract is usually drafted or supplied by the architect. Courts have looked at the building and business experience of the client. If the client is experienced, he should be more aware of the difficulty of making accurate cost estimates. If he has building experience, the client is more likely to be aware of the custom that architects usually do not risk their fee on the accuracy of their cost estimates.
>
> Courts have sometimes cited provisions for interim payments as an indication that the architect is not assuming the risk of losing his fees on the accuracy of his cost estimates. However, standard printed clauses buried in a contract are not always an accurate reflection of the understanding of the party not familiar with the customs or the forms. If payments have actually been made during the architect's performance, this is a clearer indication that the client is not laboring under the belief that he will not have to pay any fee if the low bid substantially exceeds the final cost estimate. A few cases have looked for good faith on the part of the client. For example, if the client has offered some payment to the architect for his services, this may impress a court as a show of fairness and good faith. [Ed. note: Footnotes omitted.]

Standard Contracts and Disclaimers: A Look at AIA Standard Contracts. Professional associations have dealt with cost problems by including language in their standard contracts to protect design professionals from losing fees when cost problems develop. The protective language has ranged from the brief statement that cost estimates cannot be guaranteed to the elaborate contract language beginning with the 1987 AIA Document B141 and carried forward to the present.

To understand the changes the AIA made in 1997, it is important to look at the treatment this sensitive topic received in 1987.

Paragraph 5.2 required the client to include in its budget amounts that take into account the possibility that the bids may substantially exceed the budget. Paragraph 5.2.1 required the architect to use her best judgment to evaluate any client-supplied project budget or architect-created cost estimates but stated that the architect does not warrant accuracy. In addition to Paragraph 9.6 stating that the writing is complete, Paragraph 5.2.2 stated that a project budget is not a fixed-cost limit and *requires that any such limit be in writing and signed by the parties.* If such a *fixed-cost limit* (not an ordinary budget) is established, the architect has the right to determine "materials, equipment, component systems, and types of construction" to be included and to make reasonable adjustments in

---

[17]Sweet & Sweet, *Architectural Cost Predictions: A Legal and Institutional Analysis,* 56 Calif.L.Rev. 996, 1006–1007 (1968).

project scope. The architect can include pricing contingencies and alternate bids.

Paragraph 5.2.3 required an adjustment if bid or negotiation is delayed more than ninety days after submission of construction documents. Finally, Paragraph 5.2.4 is a contractual expression of the obligation of good faith and fair dealing. It states that if the lowest bona fide bid exceeds any fixed-cost limit, the owner must do one of the following:

1. approve an increase in the fixed limit
2. authorize rebidding or renegotiating
3. terminate and pay for work performed and termination expenses if the project is abandoned
4. cooperate in project revision to reduce cost

If option 4 is chosen, under Paragraph 5.2.5 the architect must bear the cost of redesign, specified as the limit of the architect's responsibility, and she is to be paid for any other service rendered.

In 1997 the AIA revised its treatment of cost predictions in B141-1997. It linked the owner's budget with preliminary cost estimates furnished by the architect. Now the owner must specify its budget for the project and the cost of the work at the time the architect is engaged[18] and must keep the budget current as the architect works.[19] It may not change the budget or budget contingencies "without the agreement of the Architect to a corresponding change in the Project scope or quality."[20] If the budget is exceeded by the lowest bona fide bid or negotiated proposal, the owner must take one of the steps set forth in the 1987 Paragraph 5.2.4, the most important being to cooperate in contract revision to reduce the cost.[21]

Also, in 1997 the AIA deleted the phrase "fixed limit of construction cost," which had required a separate writing signed by the parties. The trigger for the four methods of resolving the problem was no longer the hard-to-establish fixed-cost limitation (cost condition) but the budget being exceeded by the bids or proposal. This emphasis on the budget accented the need for cooperation of the owner and the architect. It also eliminated reference to the possibility of a cost condition being created under which the architect

would risk her fee on the cost coming close to the cost estimates.

Presumably the AIA felt that the contract integration clause[22] would protect against assertions by the owner that an oral cost condition had been created, often the chief battleground of these cost prediction cases. That such clauses can help the architect is shown by *Torres v. Jarmon*,[23] a Texas case decided in 1973. Earlier cases, such as *Stevens v. Fanning*,[24] and some others reflect a different attitude toward such protective clauses. Much will depend on the facts surrounding the transaction, especially the degree of variance, the extent of client changes in the design, and the way each party behaved before and after the problem surfaced.

In 2007, the AIA sought to shift responsibility for the budget more squarely onto the owner. The owner's Initial Information must include its "budget for the Cost of the Work."[25] The Cost of the Work includes not only the cost to build, but also the contractor's overhead and profit. The Cost of the Work does not encompass the architect's compensation or costs that are the owner's responsibility, including a contingency.[26]

Section 6.2 enigmatically says that this budget provided as part of the Initial Information "*may* be adjusted throughout the Project as *required* under Sections 5.2, 6.4 and 6.5." Notwithstanding Section 6.2's use of the permissive "may," Sections 5.2 and 6.4 make clear that periodic updating of the budget is a mandatory duty of the owner.

The AIA envisions the owner updates occurring with input from the architect.[27] If the architect's estimate exceeds the owner's budget, the architect will make recommendations to adjust the project's size or quality, and the owner will cooperate with the architect.[28] At the same time, the AIA emphasizes that the architect's inputs are estimates, not guarantees. Section 6.3 allows the architect to base her estimates on "conceptual estimating techniques" without providing a detailed cost estimate. More

---

[18] AIA Doc. B141-1997, ¶1.1.2.5.

[19] Id. at ¶1.2.2.2.

[20] Ibid.

[21] AIA Doc. B141-1997, ¶2.1.7.5.

[22] Id. at ¶1.4.1. For a detailed review of the B141-1997 scheme, see Melton & Autry, *Beyond His Power to Build It: Who Is to Blame for the Overbudget Project?* 25 Constr. Lawyer, No. 1, Winter 2005, p. 20.

[23] 501 S.W.2d 369 (Tex.Ct.App. 1973).

[24] Supra note 6.

[25] AIA Doc. B101-2007, § 1.1, cross-referencing to B101-2007, Exhibit A, § A.1.3.

[26] AIA Doc. B101-2007, § 6.1.

[27] Id. at § 6.2.

[28] Id. at § 6.5.

directly, Section 6.2 warns that "the Architect cannot and does not warrant or represent" that her cost evaluations will not vary from bids received.[29]

As with the 1987 and 1997 versions of B141, if the lowest bona fide bid or negotiated price exceeds the budget, the owner must either take one of the four steps previously allowed or, as of 2007, may also "implement any other mutually acceptable alternative."[30] The integration clause seeks to prevent the owner from using parol evidence to establish oral agreements of a cost condition.[31]

The current EJCDC treatment of a "construction cost limit" is found in its 2002 E-500, optional Exhibit F, paragraph F5.02. The owner and engineer agree to a "construction cost limit." If the bids exceed the limit, the owner will either increase the limit, rebid, or renegotiate the project, or work with the engineer to revise the project's scope, extent, or character, consistent with sound engineering practices.

An Illustrative Case: **Griswold and Rauma, Architects, Inc. v. Aesculapius Corp.** A case very likely involving an earlier AIA standard document is reproduced here.

GRISWOLD AND RAUMA, ARCHITECTS, INC. v. AESCULAPIUS CORP.

Supreme Court of Minnesota, 1974. 301 Minn. 121, 211 N.W.2d 556.
[Ed. note: Footnotes omitted.]

PETERSON, Justice.

Plaintiff brought this action to enforce and foreclose a lien for $19,438.65 for architectural services provided defendant. Defendant answered by denying that it owed plaintiff anything and filed a counterclaim to recover $17,436.04 already paid plaintiff, defendant's theory being that plaintiff was entitled to no fee because it breached its contract by grossly underestimating the probable cost of construction of the as-yet-unbuilt building. The trial court found for defendant. We reverse, for reasons requiring an extended recital of the factual setting out of which the litigation arose.

Plaintiff is a corporation engaged in providing architectural services. Defendant is also a corporation, the principal stockholders of which are Drs. James Ponterio, P. J. Adams, and A. A. Spagnolo. Defendant owns a medical building in Shakopee which it rents to the Shakopee Clinic, which in turn is operated by Drs. Ponterio, Adams, and Spagnolo.

In early 1970 defendant decided to expand the Shakopee Clinic to allow for a larger staff of doctors. As a result the members of the corporation and their business manager, Frank Schneider, contacted various architectural firms and selected plaintiff.

At his first meeting with the doctors in February or March 1970 David Griswold, one of plaintiff's senior architects, was shown a rough draft of the proposed addition and given a very general idea of what the doctors wanted. Although the evidence is conflicting, it appears that at this meeting the doctors talked in general terms of a budget of about $300,000 to $325,000.

After a number of subsequent conferences with the doctors and Mr. Schneider, plaintiff prepared and delivered to the doctors on May 8, 1970, a document entitled "Program of Requirements." This document, which outlined and discussed the requirements of the project as then contemplated, contained the following final section:

BUDGET:

The design to evolve from this program will indicate a certain construction volume that can be projected to a project cost by the application of unit (per square foot and per cubic foot) costs; and eventually, as the design is developed in detail, by an actual materials take-off. Inevitably the projected cost must be compatible with a budget determined by available funds. It is obvious that adjustment of either the program or the budget may be necessary and that possibility must be recognized.

The project budget established, as currently understood, is $300,000. It has not been stated if this is intended to include non-building costs such as furnishings, equipment and fees—which may be approximately 25% of the total expenditure—as well as construction costs. Advice in this respect will eventually be necessary.

The essential principle to be considered in the design development is as previously stated in the paragraphs of the section titled PROJECT OBJECTIVES.

---

[29]This disclaimer language has been repeated virtually verbatim since B141-1977, ¶ 3.2.1.

[30]AIA Doc. B101-2007, § 6.6.5.

[31]Id. at § 13.1.

The construction shall be as economical as possible within the limitations imposed by the desire to build well and provide all of the facility required for a medical service.

In spite of the suggestion at the end of the second paragraph quoted above, neither party at any time thereafter sought to define more particularly what the budget was intended to include. Mr. Schneider testified, however, that he believed the original budget figure included the cost of construction, architects' fees, and the remodeling of the old building. In contrast, Dr. Ponterio testified that it was his belief that the original budget figure did not include a communications system valued at $12,755, architects' fees, or the $20,000 remodeling of the existing building.

On or about May 22, 1970, plaintiff submitted to defendant two alternate preliminary plans, designated SK-1 and SK-2. Plan SK-1 projected the programmed services to be housed partly in the existing building and partly in the new building. Plan SK-2 projected the programmed services as being housed entirely in the new building. Plan SK-2, as specifically shown on the plans, involved a larger plan in terms of area than SK-1. Defendant indicated its preference for plan SK-2, the larger and more elaborate of the two.

On June 1, 1970, plaintiff provided defendant with a "Cost Analysis" of the plan chosen by defendant. This cost analysis showed the dimensions of the project in square feet as then contemplated, and computed the cost of the project at two different rates per square foot. At the higher rate per square foot, the cost came to $322,140, plus an estimated $20,000 for remodeling the old building, totaling $342,140. At the lesser rate per square foot, the cost came to $284,575, plus $15,000 for the remodeling of the old building, totaling $299,575. The cost analysis memorandum also noted that "the best procedure for projecting costs is by a materials take-off" which was to be done "when sufficient information is available."

It is undisputed that subsequent to the June 1, 1970, cost estimate, no further cost estimates were ever conveyed to defendant. What is disputed is whether in the ensuing months there was any discussion as to whether the project was coming within the budget. According to the testimony of Mr. Schneider, defendant was assured at all times during the preparation of the building plans and in all discussions with Griswold that the construction would come within the budget. Dr. Ponterio also emphasized that Griswold constantly mentioned the budget figure of $300,000 at their meetings. Griswold, however, denied that he had ever assured defendant that the project was coming within the budget.

Although the architectural services began in March and the first billing was May 6, 1970, no written contract was forwarded until June 23, 1970. At that time a standard American Institute of Architects (AIA) contract was forwarded, calling for payment at plaintiff's standard hourly rate and for reimbursement of expenses and recognizing that a lump sum fee for the construction phase would be negotiated prior to its commencement. The following provisions of the contract have relevance to this case:

SCHEMATIC DESIGN PHASE

1.1.3 The Architect shall submit to the Owner a Statement of Probable Construction Cost based on current area, volume or other unit costs.

DESIGN DEVELOPMENT PHASE

1.1.5 The Architect shall submit to the Owner a further Statement of Probable Construction Cost.

CONSTRUCTION DOCUMENTS PHASE

1.1.7 The Architect shall advise the Owner of any adjustments to previous Statements of Probable Construction Cost indicated by changes in requirements or general market conditions.

THE OWNER'S RESPONSIBILITIES

2.8 If the Owner observes or otherwise becomes aware of any fault or defect in the Project, or nonconformance with the Contract Documents, he shall give prompt written notice thereof to the Architect.

CONSTRUCTION COST

3.4 . . . Accordingly, the Architect cannot and does not guarantee that bids will not vary from any Statement of Probable Construction Cost or other cost estimate prepared by him.

3.5 When a fixed limit of Construction Cost is established as a condition of this Agreement, it shall include a bidding contingency of ten per cent unless another amount is agreed on in writing.

3.5.1 If the lowest bona fide bid . . . exceeds such fixed limit of Construction Cost (including the bidding contingency) established as a condition of this Agreement, the Owner shall

(1) give written approval of an increase in such fixed limit,
(2) authorize rebidding the Project within a reasonable time, or
(3) cooperate in revising the Project scope and quality as required to reduce the Probable Construction Cost.

In the case of (3) the Architect, without additional charge, shall modify the Drawings and Specifications as necessary to bring the Construction Cost within the fixed limit. The providing of this service shall be the limit of the Architect's responsibility in this regard, and having done so, the Architect shall be entitled to his fees in accordance with this Agreement.

PAYMENTS TO THE ARCHITECT

6.3 If the Project is suspended for more than three months or abandoned in whole or in part, the Architect shall be paid his compensation for services performed prior to receipt of written notice from the Owner of such suspension or abandonment, together with Reimbursable Expenses then due and all terminal expenses resulting from such suspension or abandonment.

Between June 1, 1970, and November 25, 1970, when the bids were opened, plaintiff worked actively with defendant both in the design development and construction documents phases of the project, plaintiff continuing to bill defendant on an hourly basis without objection by defendant. Although the project was substantially increased in size and scope during this period, plaintiff did not furnish and defendant did not request any up-to-date cost projections. Significantly important changes in the project made during this period include:

1. Replacement of offices with examining rooms necessitating additional plumbing;
2. More extensive X-ray space;
3. A doubling of the size of the laboratory;
4. Addition of a sophisticated communications system;
5. Addition of 2,100 feet of finished space in the basement (to provide facilities originally projected for the remodeled old building);
6. Enlargement of the structure as follows:
   waiting area 1710' to 1724'
   first floor 5880' to 6650'
   basement 6260' to 6814'

All of the changes were discussed, approved, and understood by defendant. Defendant alleges, however, that it was under the impression that all such changes would be included in the original budget figure.

Bids were opened on November 25, 1970. The low construction bid was Kratochvil Construction Company at $423,380. Deductive alternates agreed to by defendant would bring the total low bid cost, including carpeting, down to $413,037.

Subsequent to the opening of the bids, the doctors called a meeting with Griswold at which the doctors informed Griswold that they could not complete the building according to the cost evidenced by the bids. Thereafter, Griswold met with the doctors and offered suggestions as to how the low bid figure could be reduced. Approximately $42,000 of reductions were projected, so that the final bid as reduced totaled $370,897. This figure included construction, carpeting, and remodeling of the old building and would meet the program of requirements without reducing size in any way. Griswold also pointed out to the doctors that the project cost could be further reduced by eliminating "bays" (series of examination rooms) from the building at a saving of approximately $35,000 per bay. Defendant was willing to accept the $42,000 reduction but was not in favor of eliminating any bays from the proposed project.

From November 25, 1970, when the bids were opened, until October 12, 1971, plaintiff and defendant met and corresponded many times in connection with various possible revisions of the project. During this time defendant never indicated that the project was abandoned and in fact, in January 1971 made a payment of $12,000 on its bill. During this time defendant never asked to be excused from the balance of the bill, and defendant offered to help plaintiff by paying interest on the open account if plaintiff required bank financing by reason of nonpayment. During this time defendant advised plaintiff that ground breaking would be deferred for 6 months, and that request by plaintiff for a further payment was "well taken" but that no further payment would be recommended until construction began.

The building, in fact, was never constructed. Plaintiff filed its lien on April 30, 1971, and commenced this action to enforce it in October 1971.

In analyzing the facts of this case we have considered Minnesota decisions as well as decisions from other jurisdictions. From these decisions we have extracted a number of factors which we believe are relevant to a determination of what the effect on compensation of an architect or building contractor should be when the actual or, as here, probable cost of construction exceeds an agreed maximum cost figure.

One very significant factor is whether the agreed maximum cost figure was expressed in terms of an approximation or estimate rather than a guarantee. Where the figure was merely an approximation or estimate and not a guarantee, courts generally permit the architect to recover compensation provided the actual or probable cost of construction does not substantially exceed the agreed figure.

Another significant factor is whether the excess of the actual or probable cost resulted from orders by the client to change the plans. Where the client ordered changes which increased the actual or probable construction costs, courts are more likely to permit the architect to recover compensation notwithstanding a cost overrun.

A third factor is whether the client has waived his right to object either by accepting the architect's performance without objecting or by failing to make a timely objection to that performance.

A fourth factor, applicable in a case such as this where the planned building was never constructed, is whether the architect, after receiving excessive bids, suggested reasonable revisions in plans which would reduce the probable cost. Courts have held that if the architect made such suggestions and the proposed revisions would not materially alter the agreed general design, then the architect is entitled to his fee, again provided that the then probable cost does not substantially exceed the agreed maximum cost figure.

Considering this case in light of these factors, we conclude that the trial court erred in denying plaintiff's motion for amended findings of fact, conclusions of law, and order for judgment.

First, it does not appear that plaintiff guaranteed the maximum cost figure. The trial court did not expressly state whether the agreed cost figure of $300,000 to $325,000 was an established estimate or a guarantee. However, the intention of the parties, as evidenced by the record, especially by contract provision 3.4, quoted earlier, more reasonably supports an established cost estimate than a guaranteed cost figure. Secondly, the probable cost of the project in our view did not substantially exceed the cost figure. At trial architect Griswold testified and Mr. Schneider agreed, that the project could be completed for approximately $360,000 to $370,000 without reducing the square footage of the project. A probable construction cost of $370,000 exceeds the agreed cost estimate maximum found by the trial court, $325,000, by only 13 percent. It seems difficult to classify such a degree of cost excess as substantial. A review of cases cited in Annotation, 20 A.L.R.3d 778, 804 to 805, suggests that most courts would not consider such a degree of cost excess substantial.

Thirdly, we think it relevant that defendant approved substantial changes beyond the original plans. Defendant not only adopted and acknowledged these changes but was often active in advocating them (especially in expanding the X-ray facilities and changing the doctors' offices into extra exam rooms). It is true that defendant contends that it was under the impression at all times that such changes were within the original cost estimate. However, this impression seems unreasonable and unjustified in view of the scope of the changes.

Finally, we think it relevant that plaintiff showed defendant how they could reduce the cost of the lowest bid below $370,000. For example, defendant would have reduced the cost drastically by simply agreeing to eliminate one of the "bays" or a fraction of a bay from the project. Provision 3.5.1 of the contract required the parties to revise the project scope and quality if necessary to reduce the probable construction cost. While it might be against public policy to allow an architect, under such a provision, to reduce substantially the area of a proposed project, it seems that, barring a specifically guaranteed area, a reasonable reduction in the size of the project should be allowed when necessary to meet the construction cost. Reversed and remanded.

SHERAN, C. J., took no part in the consideration or decision of this case.

Negligent Cost Predictions. If a cost condition is created, it is not necessary to determine whether the design professional met the appropriate standards of performance. However, in many cases where the cost estimate is wide of the mark, it is likely that a design professional did not live up to the legal standard of performance.

If the client does show that the design professional has been negligent in preparing the cost estimates, it is not necessary to establish that a cost condition has been created. It is implied that the design professional will perform in accordance with the standards of her profession. If she does not do so and if the project is abandoned, the client can recover any fees it has paid and need not pay any additional fees. However, if the client uses the design prepared by the design professional, the client must pay the net benefit of the architect's work. This is computed by subtracting any damages caused by the breach from the amount the client has been benefited by use of the design.

Even if the state law will permit the owner to maintain a tort claim based on professional negligence, one court said that "the contract may control the scope of the duty undertaken" by the architect.[32] The contract, according to the court, will determine whether there has been a breach. In that case the court concluded that the architect was to be relieved from responsibility for costs (the owner intended to use a contractor to check costs) in exchange for the architect making more frequent inspections than usual.

---

[32]*Getzschman v. Miller Chemical Co.,* supra note 10, 443 N.W.2d at 270.

Stanley Consultants, Inc. v. H. Kalicak Construction Co.,[33] illustrates a clear case of a failure to estimate properly. The project was a sixty-one-unit housing project to be built in Zaire, Africa. The cost estimate was $8 million, and the only bidder submitted a bid of $16 million. The design professional had prepared cost estimates without data from Zaire when such data were available. The driveways contained impassable grades. The sewer lines were twenty feet above the surface with eight-foot supports, and the sewer lines were to be fifty feet above a river without any supports being designated. Structures were located outside the property lines, and the design professional failed to take into account an easement to a religious shrine and created an encroachment. The court found ample evidence that the design professional had not lived up to the reasonable standard of professional skill. Such a finding could be the basis not only for denying the design professional compensation for services rendered but also for holding the design professional responsible for damages. See Section 12.03F.

D. Interpretation of Cost Condition

Is the design professional allowed some tolerance in determining whether the cost condition has been fulfilled? One court required a reasonable approximation.[34] Roughly 10 percent seems to be accepted, although this tolerance figure may be reduced if the project is large and the fee justified continual detailed pricing takeoffs. B141-1997, Paragraph 1.2.2.2 required contingencies in the budget.[35] The degree of tolerance permitted may also depend on the language used to create the cost limitation. The more specific the amount, the more likely a small tolerances figure will be applied.

One court spoke of a "relaxed version," presumably there also being a "strict version."[36] The former apparently requires that the amount be *substantially* exceeded before the architect loses her fee. Presumably the "strict" version requires the amount not be exceeded *at all*. The

relaxed version, much like the doctrine of substantial performance,[37] resembles the common use of some amount of tolerance.[38]

The AIA requires the owner to include a contingency in its budget. While one purpose of a contingency is to accommodate changes in the work made during performance, a contingency may also cushion a discrepancy between the owner's and architect's anticipated cost of construction and the bids or negotiated offers the owner receives.[39] The architect too may include contingencies in her cost estimates.[40] The size of the contingency is not specified.

E. Dispensing with the Cost Condition

A cost condition generally is created for the benefit of the client. If it so chooses, the client can dispense with this protection. Courts that conclude the client has dispensed with the cost condition usually state that the condition has been waived. Where this occurs, the condition is excused and the design professional is entitled to be paid even if the cost condition has not been fulfilled.

Excusing a condition can occur in a number of ways. A condition is excused if the client had prevented or unreasonably hindered its occurrence. For example, if the client does not permit bidding by contractors or limits bidding to an unrepresentative group of bidders, the condition is excused. The most common basis for excusing the condition has been the client's making excessive changes during the design phase.[41]

The client proceeding with the project despite the awareness of a marked disparity between cost estimates and the construction contract price can excuse the condition. By proceeding, the client may be indicating it is willing to dispense with the originally created cost conditions. However, proceeding with the project should not automatically excuse the condition. It may be economically disadvantageous to abandon the project. Proceeding in such cases may not indicate a willingness to dispense with

---

[33]383 F.Supp. 315 (E.D.Mo.1974).

[34]*Durand Assoc., Inc. v. Guardian Inv. Co.*, 186 Neb. 349, 183 N.W.2d 246 (1971).

[35]Despite a 10 percent tolerance figure, a court allowed the architect to recover where there was a 13 percent difference in *Griswold and Rauma, Architects. Inc. v. Aesculapius Corp.* (reproduced in Section 12.03C). But this factor was taken with others in ruling for the architect.

[36]*Peteet v. Fogarty*, 297 S.C. 226, 375 S.E.2d 527 (App.1988).

[37]See Section 22.06B.

[38]In *Peteet v. Fogarty*, supra note 36, the court found that the relaxed version had been applied. (The maximum cost was $150,000 and the low bid $307,000.)

[39]AIA B141-1997, ¶ 1.2.2.2; AIA B101-2007, § 6.1.

[40]AIA B141-1997, ¶ 2.1.7.3; AIA B101-2007, § 6.3.

[41]*Koerber v. Middlesex College*, 128 Vt. 11, 258 A.2d 572 (1969). See also *Griswold and Rauma, Architects, Inc. v. Aesculapius Corp.* (reproduced in Section 12.03C).

the condition. In such cases, any recovery of the design professional should be based on restitution measured by any benefit conferred.

F. Nonperformance as a Breach: Recovery of Damages

The client occasionally seeks damages based on the breach of a promise of accuracy or negligence in making the cost prediction. An understanding of the basis of such claims requires that promises be differentiated from conditions. See Section 12.03C.

Creating a cost condition does not necessarily mean the design professional promises to fulfill it. Design professionals can risk their fees on the accuracy of the cost prediction, although they may not wish to be responsible for losses caused by nonperformance.

Courts seem to assume, however, that a fixed-cost limit constitutes a promise by the design professional that the project would cost no more than the designated amount. Under this assumption, if costs substantially exceed predicted costs, the design professional has breached even though she has lived up to the professional standard in making cost predictions. The breach entitles the non-breaching party to recover foreseeable losses caused by the breach that could not have been reasonably avoided by the nonbreaching party.

Suppose the project is abandoned. The client can recover any interim fee payments based on restitution.[42] Despite this, restitution of fees paid does not occur often. Denial of a design professional's claim does not necessarily mean services have been performed without any remuneration.

Abandoning the project may cause other client losses, such as wasted expenditures in reliance on the design professional's promise to bring the project in within a designated cost.[43]

Redesign followed by construction very likely causes delay. Delay damages are recoverable if they can be proved with reasonable certainty and were reasonably foreseeable at the time the agreement was made.[44]

Suppose the project is constructed and the client seeks to recover the difference between the cost prediction and the actual cost. The property may be worth its cost. If so, the client has suffered no loss. Proceeding with the project knowing the costs would substantially exceed the predictions may show that the loss could have been reasonably avoided by abandoning the project.

Kellogg v. Pizza Oven, Inc.,[45] involved a client who rented space for a restaurant. The lease provided that the landlord would pay up to $60,000 for the cost of an improvement to the landlord's building. The architect had negligently estimated costs at $62,000, but the project cost $92,000. The tenant-client had to pay the balance of approximately $30,000. The tenant recovered this amount less a 10 percent tolerance for errors from the design professional because of the excess cost.[46]

Where the project is built, the architect should not be held for the excess of actual costs over predicted costs. This is not based on the client's having proceeded despite its knowledge that the costs will be more than anticipated. The client should not be required to give up the project to reduce damages for the design professional. However, the client benefits by ownership of property presumably of a value equal to what the client has paid for the improvement. Damages should not be awarded unless the client can prove the economic utility of the project was reduced in some ascertainable manner because of the excessive costs.

Williams Engineering, Inc. v. Goodyear, Inc.,[47] also involved a claim for damages in a commercial context similar to *Kellogg*, noted earlier in this section. Williams, the engineer, had been retained to design a recreational water slide. Time was important, as the client hoped to

---

[42]*Peteet v. Fogarty*, supra note 36 (plans worthless as project abandoned).

[43]The client's claim in *Durand Assoc., Inc. v. Guardian Inv. Co.*, supra note 34, included excavation costs and losses suffered on a steel prepurchase made to avoid an anticipated price rise. These losses should have been recovered if they were reasonably foreseeable and not avoidable. Because they were incurred by an affiliated company of the client and not by the client, they were denied.

[44]*Impastato v. Senner*, 190 So.2d 111 (La.App.), appeal denied, 249 La. 833, 191 So.2d 639 (1966), denied recovery for delay, but *Hedla v. McCool*, 476 F.2d 1223 (9th Cir.1973), allowed recovery. AIA Doc. B101-2007, § 6.7, states that redesign is the limit of the architect's responsibility. This should not relieve the architect from responsibility for delay damages caused by negligence.

[45]57 Colo. 295, 402 P.2d 633 (1965). Similarly, see *Kaufman v. Leard*, 356 Mass. 163, 248 N.E.2d 480 (1969).

[46]The landlord ended up with improvements probably worth considerably more than the $60,000 he spent. Perhaps the design professional should have claimed against the landlord based on unjust enrichment. The unjust enrichment argument was rejected in *Kaufman v. Leard*, supra note 45.

[47]Supra note 10.

open for business in the summer of 1979. Williams gave a preliminary estimate of some $409,000 but suggested that the client proceed on a "fast track" basis to expedite completion.[48] Also, he suggested the contractor be hired on a cost-plus basis.

The design phase was completed on April 18, 1979. During design and construction, the engineer submitted written invoices for his fee based on a percentage of the estimated cost of $409,000. On August 1, 1979, three days before the slide opened, the engineer's bill still showed the cost as $409,000. But twenty days after the water slide was opened for business, the client was billed for construction costs of almost $888,000. The project was still only 82 percent complete. Even worse, the engineer submitted a new bill based on the projected cost of almost $1,000,000. The client paid $824,000 for the water slide. The water slide turned out to be a financial failure; the client lost its lease and had to pay to have the slide removed. Although the basic reason for the failure was lack of customers, the delayed opening and some design features contributed to the financial disaster.

In addition to resisting the engineer's claim for additional fees, the client claimed and testimony supported that the engineer breached by failing to employ a professional estimator, failing to look at other water slides, failing to advise the client about the other contract possibilities, and failing to provide revised cost estimates. The client contended that it would have modified the design or given up the venture had it been apprised of ultimate cost.

The client initially sought damages of $634,000. As the trial proceeded, it reduced its claim to $409,000, by chance the amount it thought it was going to have to pay for the completed water slide. After an eight-day trial, the jury awarded the engineer additional professional fees of $25,000, attorneys' fees of $23,000, $2,800 in litigation costs, and expenses of expert witnesses of $6,000. It also decided the engineer had breached the contract and the client should have damages of $125,000 plus expert witness fees of $3,000. This created a net award of $71,200 in favor of the client.

The intermediate court of appeals reversed the jury's award granting the engineer additional fees and increased the damages to the client to $205,000.[49] The Louisiana Supreme Court, though noting the engineer did not operate in bad faith, held he had committed a misrepresenta-

tion by his failure to reveal "even an approximation of the true cost of the project until after the construction was completed."[50] The court could not use the frequently used damage measure of the difference between the market value of the project and the excess cost, as the water slide had become worthless and had to be removed. The court affirmed the decision of the intermediate court, denying any additional engineering fees to the engineer and awarding the client over $205,000 in damages.

The intermediate appellate court's decision induced representatives of the design professions to submit briefs as "friends of the court." The briefs contended the engineer did not guarantee the cost and could not be responsible for damages for the overrun. In response, the Louisiana Supreme Court stated damages were awarded not because of any guarantee of estimates but because the engineer was negligent, as shown by the testimony at the trial.

The case illustrates not only the difference between a cost condition and a promise but also the risk involved in designing for commercial ventures. Also, the protective language in the contract, much the same as the court commonly found in contracts prepared by design profession associations or by design professionals with good bargaining power, did not protect the design professional when he did not watch costs.

What damages might an owner suffer if the design professional overestimates the cost of the project? In *Post, Buckley, Schuh & Jernigan, Inc. v. Monroe County*,[51] a county, relying on its engineer's estimate of the cost of a public improvement, floated a bond issue to pay for the work. The bids received were so low that the county could have financed the project with its existing funds. The county then released its bond obligations and sought reimbursement from the engineer for the unnecessary financing expenses.

G. Relationship Between Principal Design Professional and Consultant

Section 12.10B discusses the use of consultants by the prime design professional. One problem that can arise is whether the consultant bears the risk that the prime design

---

[48]See Section 17.04B.

[49]480 So.2d 772 (La.App.1985).

[50]496 So.2d at 1018.

[51]Supra note 13. The judgment in favor of the county was reversed because of an erroneous jury instruction addressing the engineer's duty of care.

professional will not be paid by the client. Although non-payment can result from many causes, often this occurs when a project is abandoned because of excessive costs. *George Wagschal Associates, Inc. v. West*[52] involved a legal action by the consulting engineer against the principal design professional (the architect) for services he had performed in the design of a school that was never built to these plans because of excessive cost. The cost overrun apparently was due to engineering overdesign.

The architect evidently had not been paid by the school district and contended that the engineer should share this loss with him. The court upheld this contention and seemed to hold that the principal design professional and consulting engineer were jointly engaged in the project. The architect risked his fee by his contract with the school district, and the engineer risked his fee because he knew of the budgetary constraints.[53]

The AIA, representing mainly prime design professionals, and the Engineers Joint Contract Documents Committee (EJCDC), representing professional engineers, have differed as to the right of a consulting engineer to be paid if the architect has not been paid. The EJCDC argues that consultants generally do not take this risk. It notes that the prime design professional selects the client and has the best opportunity to evaluate its capacity to pay,[54] also points to AIA Document B141, Paragraph 4.3, which allows the architect to request that the client provide a statement of funds available for the project and of the source of funds. (This was deleted in 1997.)

The AIA contends that a long-term relationship frequently exists between prime design professionals and consultants that resembles a partnership even if not cast in legal terms. It believes that under such conditions, the risk of nonpayment should be shared.

The chronic instability of the construction industry means this problem will occur repeatedly. Most prime design professionals are aware of the difficulties that may arise if the project is abandoned for excessive costs. They should discuss this potential problem with their consultants and seek a commonly accepted allocation

of risk. When this is successfully accomplished, the contract should make the result *quite* clear. As shall be seen in Section 28.06, dealing with subcontracts, the risk of nonpayment taken by the consultant must be very clearly expressed or the court will not enforce it.

H. Advice to Design Professionals

Many design professional–client relationships deteriorate because of excessive costs. Design professionals should try to make the cost prediction process more accurate. The chief method chosen by the professional associations has been to use the contract to protect their members from losing their fees where their cost predictions are inaccurate. The professional associations have included provisions in their contracts that are supposed to ensure that fees will not be lost when cost predictions are inaccurate and that fees will be lost only where the cost predictions are made negligently.

Protective language should be explained to the client. Design professionals who give a reasonable explanation to a client are not likely to incur difficulty over this problem. They should inform the client how cost predictions are made and how difficult it is to achieve accuracy when balancing uncontrollable factors. They should state that best efforts will be made but that for various specific reasons, the low bids from the contractors may be substantially in excess of the statement of probable construction costs. The suggestion should be made that under such circumstances, the design professional and the client should join to work toward a design solution that will satisfy the client's needs. In helping the client be realistic about desires and funds, the design professional should ask the client to be as specific as possible as to expectations about the project.

The design professional who takes these steps may lose some clients. It may be better to lose them at the outset than to work for many hours and then either not be paid or be forced to go to court to try to collect. Without honest discussion with the client at the outset, the design professional takes these risks.

Design professionals should also consider greater flexibility in fee arrangements. If the stated percentage of construction costs is used, it may be advisable to reduce or eliminate any fee based on construction costs that exceed cost predictions.

During performance, the design professional should state what effect any changes made by the client will have

[52]Supra note 16.

[53]If the lost fee resulted from engineering overdesign, should the architect be able to transfer his lost fee to the engineer?

[54]Until 1997, the EJCDC could point to AIA Doc. B141-1987, ¶ 4.3, which allowed the architect to request that the client provide a statement of funds available for the project and of the source of funds. This was deleted in 1997 and not renewed in 2007.

on any existing cost predictions. It is hoped that not every change will require an increase in the cost predictions. If the client approves any design work, the request for approval should state whether any change has occurred in cost predictions.

SECTION 12.04 Assistance in Obtaining Financing

A building project frequently requires lender financing. To persuade a lender that a loan should be granted, the client generally submits schematic designs or even design development, economic feasibility studies, cost estimates, and sometimes the contract documents and proposed contractor. This submission includes materials prepared by the design professional.

Must the design professional do more than permit the use of design work for such a purpose? Does the basic fee cover such services as appearing before prospective lenders, advising the client as to who might be willing to lend the client money, or helping the client prepare any information that the lender may require? Must such work be done only if requested as additional service and paid for accordingly?

The professional education and training of a design professional do not include techniques for obtaining financing for a project. Nor is it likely that the design professional will be examined on this activity when seeking to become registered. It should not be considered part of basic design services.

Clearly this is true if the design professional has been retained by a large institutional client with personnel experienced in financial matters. The result does not change, however, even if the client does not have the skill within its own organization. Rarely does the client engage a professional designer because the client expects the designer to have and use skills relating to obtaining funds for the project.

Most firms give financing advice to some clients, many furnish financial contacts, and almost all occasionally arrange financing. Yet extra charges for such services are rare. Perhaps the activity is not burdensome, or merely reflects economic realities of practice.

Clearly the AIA does not consider services related to project financing to be part of basic services. They should also not be viewed as additional services. As noted in Section 12.01 and more fully discussed in Section 13.01G, B101-2007 distinguishes between "basic" and "additional"

services. Additional services are addressed in Article 4, with specific examples listed in Sections 4.1, 4.3.1, and 4.3.2. Financing advice is not in any of these lists. The closest comparable service is "site evaluation and planning" in 4.1.5. This language should not be viewed as including a duty to help procure financing.

An architect who agrees to provide financing advice creates a risk of liability exposure if she does the job badly.[55] For this reason it is better to view such a service as provided pursuant to a separate agreement, not as an additional service. If financing advice is an additional service under Section 4.1, this would mean the architect would have to perform the service if the owner requests, albeit for additional compensation.

SECTION 12.05 Economic Feasibility of Project

AIA B141-1997 Paragraph 2.8.3.6 included "economic feasibility studies" as a service that the architect must provide if requested by the owner. This was dropped from B101-2007 which, as noted,[56] does not include any form of financing advice in the list of additional services.

The AIA's decision to drop economic feasibility studies from B101-2007 is wise. The determination by a lender will be affected by the general economic feasibility of the venture. But if the architect should hesitate before making representations as to the availability of funds, she should certainly avoid venturing into economic feasibility studies. These are not part of basic services. If such studies are requested and made, and the venture is unsuccessful, the design professional will be exposed to claims by the client. Such predictions are treacherous, ones for which design professionals are rarely trained by education and experience.

In *Martin Bloom Associates, Inc. v. Manzie*,[57] the plaintiff architect sued the defendant clients for design services he rendered to the defendants before the project was abandoned by the clients because they could not obtain financing. The clients owned land in Las Vegas, Nevada, land that they wished to develop for investment purposes.

---

[55]See *Herkert v. Stauber*, 106 Wis.2d 545, 317 N.W.2d 834 (1982) (citing and following earlier edition of this book).
[56]See Section 12.04.
[57]389 F.Supp. 848, 852 (D.Nev.1975).

The architect produced a written contract under which he was to provide design services and contract administration for a designated compensation fee. Here the issue was not the architect's right to additional compensation but his responsibility for the accuracy of his representations as to financing and profitability.

The clients claimed that the architect had represented that the clients would have no difficulty in obtaining financing and that the project would be profitable. The architect's version differed. He contended that the entire agreement was in the letter and that no representation had been made as to profitability.

First, the trial judge concluded that the written agreement was not the complete contract between the parties. The judge then stated,

> Throughout the two days of trial, the Court had the opportunity to observe the manner of the witnesses while they testified, the consistency of their versions, both internally and compared to those of other witnesses, the probability or improbability of their versions and their interest in the outcome of the lawsuit. Based on these factors, the Court credits the testimony of Manzie [the client] that representations of Bloom [the architect], both express and implied, induced Manzie to enter into the agreement and also constituted part of the agreement itself.

The case illustrates the danger of exceeding one's professional capabilities. If the architect had, as determined by the trial judge, made representations relating to financial feasibility and profitability, he may very well have ventured into an unpredictable area beyond his professional skill, services excluded from coverage under his professional liability policy. As noted in Section 12.03C, architects and engineers often lose fees when their cost predictions are inaccurate. But at least cost predictions, though certainly difficult, should be within the professional competence of a design professional. Economic feasibility goes beyond this boundary and should be avoided.

SECTION 12.06 Approval of Public Authorities: Dispute Resolution

Greater governmental control and participation in all forms of economic activity have meant that the design professional increasingly deals with federal, state, and local agencies. Also, it is more likely that the design professional

will help the owner prepare for third-party dispute resolution such as arbitration or litigation. Must the design professional help the client or its attorney prepare a presentation for the planning commission, zoning board, or city council? Must she appear at such a hearing and act as a witness if requested? Is the design professional who does these things entitled to compensation in addition to the basic fee? Must she help in arbitration or litigation?

Reasonable cooperation by the design professional in matters relating to public land use control can be expected by the client. Design professionals are expected to have expertise in matters that are often at issue in these public hearings. It may be within the client's reasonable expectations that the architect or engineer will render reasonable assistance and advise the client on these matters. Each contracting party should do all that is reasonably necessary to help achieve the objectives of the other. The same can be said for dispute resolution. But are these part of basic services?

The increased likelihood that the owner will ask for help on permit matters or on dispute resolution has led the associations of design professionals to seek to make clear that these services are not part of basic services and merit additional compensation. Each new issue of standard documents makes more clear that such services require added compensation. For example, B101-2007, Section 4.3.1.7 states that "preparation for, or attendance at, a public presentation, meeting or hearing" is an additional service for which the architect is entitled to additional compensation beyond the basic fee.

Reflecting the different nature of engineering projects and their funding sources, EJCDC E-500 (2002) states in Exhibit A (Engineer's services) Paragraph A2.01A(1), the following are additional services requiring advance authorization:

> Preparation of applications and supporting documents (in addition to those furnished under Basic Services) for private or governmental grants, loans or advances in connection with the Project; preparation or review of environmental assessments and impact statements; review and evaluation of the effects on the design requirements for the Project of any such statements and documents prepared by others; and assistance in obtaining approvals of authorities having jurisdiction over the anticipated environmental impact of the Project.

* * *

20 Preparing to serve or serving as a consultant or witness for Owner in any litigation, arbitration or other dispute resolution process related to the Project.

Again, it may be useful or desirable to perform such services without requesting additional compensation. This section deals solely with the questions of whether design professionals are obligated to perform the services and whether they are legally entitled to additional pay for doing so.

SECTION 12.07 Services of a Legal Nature

Some design professionals volunteer or are asked to perform services for which legal education, training, and licensure may be required. This may be traceable to the high cost of legal services, and the uncertainty as to which services can be performed only by a lawyer.

Undoubtedly, one of the most troublesome activities is drafting or providing the construction contract. It may not be wise for laypeople to draw contracts for themselves, although the law allows professional designers to supply or draft contracts for their own services. Suppose the designer drafts or selects the construction contract for a client. Lawyers often contend that when design professionals perform such activities, they are practicing law. To meet such complaints, the AIA includes language on its standard forms of agreement stating that the document has important legal consequences and that "consultation with an attorney is encouraged with respect to its completion or modification."

B101-2007 states,

§ 3.4.3. During the development of the Construction Documents, the Architect shall assist the Owner in the development and preparation of: (1) bidding and procurement information that describes the time, place and conditions of bidding; including bidding or proposal forms; (2) the form of agreement between the Owner and Contractor; and (3) the Conditions of the Contract for Construction (General, Supplementary and other Conditions). The Architect shall also compile a project manual that includes the Conditions of the Contract for Construction and Specifications and may include bidding requirements and sample forms.

In addition, Section 5.8 requires the owner to furnish legal services as necessary. The AIA clearly wants architects not to perform legal, insurance, and accounting services. Some provisions in construction contracts can be considered "architectural" or "engineering" in that architectural or engineering training and experience are essential to reviewing or even drafting such provisions. This may tempt professional designers to play an aggressive role rather than realize that their *advice* will certainly be useful to the client or its lawyer.

A blurred border exists between legal and nonlegal services in the land use area. Although land use matters can and do involve legal skills, the architect may be knowledgeable about the politics and procedures in land use matters, particularly if she has had experience in dealing with land use agencies.

Another illustration of the overlap in matters relating to land use can be demonstrated by the hearings required before public agencies charged with the responsibility of issuing permits. On the whole, such hearings tend to be political rather than legal. Such a hearing may be best orchestrated by someone with political skill and experience. This is the responsibility of the owner, although the architect or engineer can be of great value in advising the owner or even in orchestrating the hearing and the strategy for it.

The line between legal and nonlegal services can also be demonstrated by the not uncommon phenomenon of the design professional being asked to advise the client as to whether a surety bond should be required. Because of their experience, design professionals may be put in the position of being able to determine whether a particular contractor should be bonded. Their experience may also lead them to have strong opinions on the general desirability of surety bonds. Repeated involvement with such matters may give design professionals confidence that they can handle them.

Here the balance tips strongly in favor of considering these legal services. Sometimes bonds are required for public projects. Using surety bonds may largely depend on other legal remedies given subcontractors and suppliers, such as the right to assert a mechanics' lien or to stop payments. The design professional should simply answer specific questions of a nonlegal nature, rather than advise generally on such matters. If asked to advise on nonlegal issues, the services a design professional provides in response to such a request are additional (such services are excluded from professional liability insurance coverage).

SECTION 12.08 Site Services

A. Relation to Chapter 14

Section 12.08 concentrates on the reasonable expectations of the client as to the design professional's role on the site while construction proceeds. The section focuses on the belief that the client may have that the design professional has been paid to ensure that the client receives all it is entitled to receive under the construction contract. The client may expect the design professional to accomplish this by watching over the job, seeing that the contractor performs properly, and being responsible if the contractor does not so perform. As noted in Section 12.08B, a design professional may take a different view of her function. The purpose of Section 12.08 is to explore this problem by examining what the law has done when faced with resolving the often different expectations of client and design professional.

Site services are but a part of the total professional services performed by the design professional. Disputes over design performance may involve determining whether the design professional has performed in accordance with her professional or contractual obligation.

The architect's site services and the frequently used disclaimers of responsibility for the work of the contractor or for safety often play a significant part in the architect's liability to clients or third parties who have suffered losses that they can trace to the failure of the design professional to perform site services properly. Liability will be discussed in Chapter 14.

B. Supervision to Observation:
Watson, Watson, Rutland/Architects v.
Montgomery County Board of Education

Contracts between design professionals and their clients include, as noted in Section 12.01, a lengthy list of services. This section treats the general role of the design professional during construction with special reference to site visits. The details of site visits are often handled in the contract between the design professional and her client. Provisions dealing with this topic can include requiring the design professional to take an active role in the process, performing such tasks as solving execution problems, directing how work is to be done, searching carefully to determine failure to follow the design, and seeing that safety regulations are followed. They can also limit the design professional to a passive role under which she "walks the site" periodically to check to see how things are going, perhaps measures progress for payment purposes, and reports to the client about work in progress. If the design professional takes an active role, she will be diluting the authority and responsibility of the contractor and exposing herself and her client to liability. Yet some clients may prefer this route because they see it as a means of ensuring a more successful project.

The professional associations representing design professionals have favored placing the authority and responsibility for executing the design solely on the contractor and limiting the architect or engineer to a more passive role. To implement this separation, the standard documents published by professional associations include language that clearly gives sole responsibility for executing the design to the contractor. Such language disclaims any responsibility on the part of the design professionals for how the work is to be done or for the contractor's failure to comply with the contract documents. The effect of such disclaimers, discussed later, can generate concern both by clients ("What am I getting for my money?") and by courts ("Is the design professional taking any responsibility?").

As noted, the AIA has, like the EJCDC, selected a passive mode for architects and engineers. This can be demonstrated by a brief look at the history of AIA documents.

Before 1961, the architect had general supervision of the work. In response to the specter of the expanded liability to third parties, the AIA dropped the phrase "general supervision" in favor of language under which the architect observed rather than supervised, made visits "at intervals appropriate to the stage of construction or as otherwise agreed"[58] (in B141-1997 this became "the Contractor's operations") rather than conducted exhaustive on-site inspections, and did her best. She was not responsible, however, for the contractor's failure to perform in accordance with the contract documents or for methods of executing the design.

The AIA justified these language changes by suggesting that the architect was no longer the master builder exercising almost total domination of the construction

---

[58]AIA Doc. B141-1987, ¶2.6.5; B141-1997, ¶2.6.2.1; B101-2007, § 3.6.2 (changed to "as otherwise required in Section 4.2.3").

process from beginning to end.[59] Instead, said the AIA, the architect has turned over the responsibility for executing the design to the contractor, who presumably has the necessary skill to accomplish this properly and safely. Even more, in sophisticated construction the architect may simply be part of the management team that may consist of a project manager, a construction manager, and a field representative of owner or lender. This shift in role and function, though principally a response to increased liability to third parties, was also designed to recognize the shift in organization for most construction, the architect no longer being master builder but being called in to interpret or decide disputes and making periodic observations to check on the progress of the work, mainly to issue certificates for progress payments.

The design professional associations contend that such terminological changes simply reflect actual practices and were not intended to change design professional site responsibilities. Yet it seems clear that such changes were directed toward the courts with the hope they would provide defenses against what the professional associations thought were unmeritorious claims against design professionals. The AIA's approach is explored in the *Watson* case (reproduced in this section).

The following issues can arise as to the specific obligation to visit the site:

1. When must the design professional be on the site (continuous presence versus periodic visits)?
2. What is the intensity level of checking for compliance (intense inspection to ferret out deficiencies versus casual observation)?
3. If a deficiency or noncompliance is discovered, what must the design professional do (direct work be corrected versus report to owner)?

The following issues can arise as to the contractual disclaimers:

1. Do they totally exculpate the design professional from any responsibility for failure by the contractor to comply with the design documents? (Are they a basis for summary judgment resolving the issue of responsibility without a full trial?)

2. Do they exculpate the design professional even if she discovers the noncompliance by the contractor?
3. Do they exculpate the design professional if she has failed to comply with her contractual obligation? (Is she responsible for what she would have detected had she complied with her contractual commitment?)

The exact nature of the design professional's site visits depends, of course, on the contract, both its express and implied terms. The contract may specify continuous versus noncontinuous presence, the frequency and timing of visits, the intensity of inspection, and action to be taken on discovery of deficiencies.

Factors likely to be examined to determine whether the duration, frequency, and timing of site visits are adequate are as follows:

1. owner's absence during construction and contractor's reputation for corner-cutting[60]
2. size of the project[61]
3. distance between the site and the design professional's home office[62]
4. when crucial steps are undertaken, such as pouring concrete or covering works[63]
5. type of construction contract (cost contracts require more monitoring)
6. experimental design or unusual materials specified
7. extent to which owner has a technical staff that will take over some of these responsibilities
8. observation of contractor's performance during visits[64]
9. contractor's record of performance on the project

The *Watson* case (reproduced here) notes the important cases dealing with the responsibility of the design professional for failure of the contractor to comply with the contract documents, with special reference to the effect of disclaimers. *Watson* also demonstrates the complexity of a defect claim with multiple parties and multiple claims.

---

[59]See Sweet, *The Architectural Profession Responds to Construction Management and Design-Build: The Spotlight on AIA Documents*, 46 Law & Contemp.Probs. 69 (1983).

[60]*Kleb v. Wendling*, 67 Ill.App.3d 1016, 385 N.E.2d 346 (1979).

[61]*First Nat'l Bank of Akron v. Cann*, 503 F.Supp. 419, 437 (N.D.Ohio 1980) (citing earlier edition of book), aff'd, 669 F.2d 415 (6th Cir.1982).

[62]*Warde v. Davis*, 494 F.2d 655 (10th Cir.1974).

[63]*Tectonics*, PSBCA No. 2417, 89-3 BCA ¶ 22,119.

[64]*Chiaverini v. Vail*, 61 R.I. 117, 200 A. 462 (1938) (visits when contractor not on site inadequate).

WATSON, WATSON, RUTLAND/ARCHITECTS, INC. v. MONTGOMERY COUNTY BOARD OF EDUCATION, et al.

Supreme Court of Alabama, 1990. 559 So.2d 168.

MADDOX, Justice

This case arises out of property damage incurred when the roof of Brewbaker Junior High School in Montgomery leaked. The Montgomery County Board of Education (hereinafter "the School Board") filed this action against Bear Brothers, Inc., the general contractor; United States Mineral Products Company (hereinafter "U.S. Mineral"), the manufacturer of the roofing membrane; and W. Murray Watson, W. Michael Watson, and J. Michael Rutland, the architects for the school project. Bear Brothers joined Dixie Roof Decks, Inc. (hereinafter "Dixie"), the roofing subcontractor, as a third-party defendant. The trial court substituted the corporate entity Watson, Watson, Rutland/Architects, Inc. (hereinafter "the Architect"), for the individually named architects.

The School Board alleged negligence and breach of contract against the Architect and breach of contract and breach of guaranty against Bear Brothers and U.S. Mineral. The Architect filed a cross-claim against U.S. Mineral and Dixie for indemnity in case the Architect were held liable to the Board. The trial court entered summary judgment for U.S. Mineral and Dixie on that cross-claim, holding that a one-year statute of limitations applied to that cross-claim and that the statute barred the claim. [Ed. note: The legal complications as to parties in this simple claim are shown in Figure 12.1.]

The School Board settled with Bear Brothers and U.S. Mineral for a total of $100,000. At trial against the Architect, the court granted the Architect's motion for a directed verdict as to the School Board's negligence claim, on the ground that the statute of limitations had expired. The breach of contract claim was submitted to the jury, and the jury returned a verdict of $24,813.08 against the Architect. The court entered a judgment based on that verdict.

* * *

At the center of this dispute is the architectural agreement between the Architect and the School Board; it included the following language:

> ARTICLE 8. Administration of the Construction Contract. The Architect will endeavor to require the Contractor to strictly adhere to the plans and specifications, to guard the Owner against defects and deficiencies in the work of Contractors, and shall promptly notify the Owner in writing of any significant departure in the quality of material or workmanship from the requirements of the plans and specifications, but he does not guarantee the performance of the contracts.

* * *

> The Architect shall make periodic visits to the site and as hereinafter defined to familiarize himself generally with the progress and quality of the Work and to determine in general if the Work is proceeding in accordance with the Contract Documents. On the basis of his on-site observations as an Architect, he shall endeavor to guard the Owner against defect and deficiencies in the work of the Contractor. The Architect shall not be required to make continuous on-site inspections to check the quality of the Work. Architect shall not be responsible for construction means, methods, techniques, sequences or procedures, or for safety precautions and programs in connection with the Work, unless spelled out in the Contract Documents, and he shall not be liable for results of Contractor's failure to carry out the work in accordance with the Contract Documents.

* * *

> The Architect shall not be responsible for the acts or omissions of the Contractor, or any Subcontractors, or any of the Contractor's or Subcontractor's agents or employees, or any other persons performing any of the Work.

The following issues respectively have been argued orally and in the parties' briefs:

On the Architect's appeal:

1. Does the exculpatory language in the architectural agreement absolve the Architect from liability for damages arising from the failure of the contractor to follow the plans and specifications?
2. If the answer to Issue 1 is no, then is the Architect entitled to cross-claim for indemnity against the roofing subcontractor and the manufacturer of the roofing membrane?

On the School Board's cross-appeal:

3. When did the statutory period of limitations applicable to the school board's negligence claim begin to run?

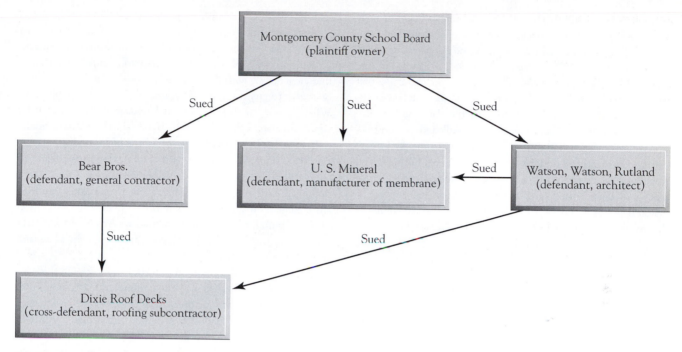

FIGURE 12.1   Original parties in *Watson, Watson, Rutland* case.

I

The Architect argues that the exculpatory language in Article 8 of the architectural agreement absolves the Architect from liability to the School Board because all roof leaks involved were attributable to the faulty workmanship of the contractor. The Architect points out that Article 8 provides for two types of inspection services by the Architect; that is, the owner could elect to receive only general site inspection by the Architect, or the owner could elect to pay an additional fee for continuous on-site inspections (known as the "clerk of the works" alternative); the School Board elected not to pay for the second options.[65]

This Court has construed language virtually identical to that at issue here, in *Sheetz, Aiken & Aiken, Inc. v. Spann, Hall, Ritchie, Inc.*, 512 So.2d 99 (Ala.1987). In that case, the contract in question contained the following language:

The ARCHITECT shall not be responsible for the acts or omission of the contractor, or any subcontractors, or any other contractor or subcontractors, agents or employees, or any person performing any work.

The Court, in *Sheetz* stated: "The contract expressly states that Spann is not responsible in any fashion for the acts or omission of the contractor, subcontractors, agents, or employees performing the work." 512 So.2d at 102.

A case analogous to this one is *Moundsview Indep. School Dist. No. 621 v. Buetow & Associates, Inc.*, 253 N.W.2d 836 (Minn.1977), where the contract language regarding inspection duties was virtually identical to that used here; the trial court entered a summary judgment for the architect, and the Minnesota Supreme Court affirmed. The supreme court found it significant that the school district had elected not to obtain continuous supervisory services from the architect through the "clerk of the works" clause. The court held that the contractual provisions "absolved [the architect] from any liability, as a matter of law, for a contractor's failure to fasten the roof to the building with washers and nuts." 253 N.W.2d at 839.

Other courts have also held that similar language absolved the architect from liability as a matter of law. [Citing cases.]

The Architect contends that imposition of liability on it here would be nothing short of a holding that the Architect was the guarantor of the contractor's work.

The School Board responds to the Architect's argument by saying that it has never contended that the Architect should be a

---

[65]Article 8 is modeled after language found in American Institute of Architects (AIA) contracts used throughout the country.

guarantor of the contractor's work, but the School Board strongly contends that where a contract is for work and services, there is an implied duty to perform with an ordinary and reasonable degree of skill and care, citing, *C. P. Robbins & Associates v. Stevens*, 53 Ala. App. 432, 301 So.2d 196 (1974). The School Board also argues that its claim was covered under the terms of the agreement because its claim was limited to those deviations from the plans and specifications that should have been obvious to one skilled in the construction industry. According to the School Board, if the Architect's construction of the contract language is accepted, the Architect would not be accountable to anyone for its failure to make a reasonably adequate inspection so long as it made an "inspection," no matter how cursory, on a weekly basis. The School Board points out that the contract also included the following language:

> On the basis of his on-site observations as an Architect, he shall endeavor to guard the Owner against defects and deficiencies in the work of the Contractor.
>
> The Architect shall have authority to reject Work which does not conform to the Contract Documents.
>
> The administration of the contract by the Architect is not normally to be construed as meaning the furnishing of continuous inspection which may be obtained by the employment of a Clerk of the Works. However, the administration shall be consistent with the size and nature of the work and must include, at least, one inspection each week, a final inspection, and an inspection at the end of the one year guarantee period shall be required on all projects.

It is apparent that our focus must be drawn to the exculpatory language contained in Article 8 of the agreement. Most courts in other jurisdictions that have considered similar exculpatory clauses have recognized that while such clauses do not absolve an architect from all liability, an architect is under no duty to perform continuous inspections that could be obtained by the employment of a "clerk of the works."

The critical question in most of the cases we have reviewed from other jurisdictions seems to focus on the extent of the obligation owed when an architect agrees to perform the type of inspection that the architect agreed to perform in this case. Some of the courts hold, as a matter of law, that the agreement does not cover particular factual situations, and at least one court makes a distinction based on whether a failure of a contractor to follow plans and specifications is known *to the architect during the course of the construction*.

A fair reading of Article 8 obviously operates to impose certain inspection responsibilities on the architect to view the ongoing construction progress. The frequency and number of such inspections is made somewhat specific by the contractual requirement that these visits to the site be conducted at least once each week. The difficulty arises in construing the particular terminology used, such as "*endeavor* to require," "familiarize himself generally with the progress and quality of the work," and "*endeavor* to guard the owner against defect and deficiencies." When coupled with the contract language stating that "[t]he administration of the contract by the Architect is not normally to be construed as meaning the furnishing of continuous inspection which may be obtained by the employment of a Clerk of the Works," these phrases make it obvious that the Architect's duty to inspect is somewhat limited, but we cannot agree with the argument on appeal that there could never be an imposition of liability under an agreement similar to Article 8 no matter how serious the deviation of the contractor from the plans and specifications.

We begin our interpretation of Article 8 by stating the general rule that one must look at the contract as a whole in order to determine the intent of the parties. In this case, the School Board presented some evidence by persons knowledgeable of construction practices, and they testified concerning the cause of the leaks and gave general testimony concerning the conditions at the construction site, which, they allege, should have put the Architect on notice that the plans and specifications were not being followed. However, the only testimony in the record we find concerning the obligation of the Architect to conduct an inspection pursuant to the agreement was to the effect that, according to one architect witness who had read the agreement, it did not call for an inspection to discover the defect alleged in this case.

Some courts have determined, as a matter of law, that certain factual situations are not covered under an agreement to inspect like that agreement involved here, and some courts have emphasized that the contract is one with a professional and have appeared to apply a rule requiring the presentation of expert testimony in order to prove the contract meaning. For example, the *Moundsview* court said, "An architect, as a professional, is required to perform his services with reasonable care and competence and will be liable in damages for any failure to do so." 253 N.W.2d at 839. In *Mayor v. City Council of Columbus v. Clark-Dietz & Associates-Engineers, Inc.*, 550 F.Supp. 610 (N.D.Miss. 1982) the court held that the architect's liability was limited by the contract language, but also held that the architect still had a duty to supervise construction by observing the general progress of the work. The court did not impose liability on the architect, because it explicitly found that the architect had not been negligent in its limited duties.

The same AIA contract language was involved in *First Nat'l Bank of Akron v. Cann*, 503 F.Supp. 419 (N.D.Ohio 1980), aff'd, 669 F.2d 415 (6th Cir.1982), where the court wrote:

> That exhaustive, continuous on-site inspections were not required, however, does not allow the architect to close his eyes on the construction site, refrain from engaging in any inspection procedure whatsoever, and then disclaim liability for construction defects that even the most perfunctory monitoring would have prevented. 503 F.Supp. at 436.

Although the contract here clearly made the Architect's inspection duty a limited one, we cannot hold that it absolved the Architect from all possible liability or relieved it of the duty to perform reasonably the limited contractual duties that it agreed to undertake. Otherwise, as the School Board argues, the owner would have bought nothing from the Architect. While the agreement may have absolved the Architect of liability for any negligent acts or omissions of the contractor and subcontractors, it did not absolve the Architect of liability arising out of its own failure to inspect reasonably. Nor could the Architect close its eyes on the construction site and not engage in any inspection procedure, and then disclaim liability for construction defects that even the most perfunctory monitoring would have prevented, or fail to advise the owner of a known failure of the contractor to follow the plans and specifications.

The issue here is whether the Architect can be held liable for its failure to inspect and to discover the acts or omissions of the contractor or subcontractors in failing to follow the plans and specifications. Under the terms of the contract, the Architect had at least a duty to perform reasonable inspections, and the School Board had a right to a remedy for any failure to perform that duty. There is no question that the Architect performed inspections; the thrust of the School Board's argument is that these inspections were not as thorough as the Architect agreed they would be.

As we have already stated, the only evidence concerning the Architect's duty under the agreement was from an architect who stated that under his interpretation of the contract the Architect was not under a duty to inspect for the specific defect that was alleged in this case.

As was pointed out in *Moundsview*, an architect is a "professional," and we are of the opinion that expert testimony was needed in order to show whether the defects here should have been obvious to the Architect during the weekly inspections. Just as in cases dealing with an alleged breach of a duty by an attorney, a doctor, or any other professional, unless the breach is so obvious that any reasonable person would see it, then expert testimony is necessary in order to establish the alleged breach. The nature and extent of the duty of an architect who agrees to conduct the inspection called for by the subject agreement are not matters of common knowledge. The rule of law in Alabama concerning the use of expert testimony is as follows: "[E]xpert opinion testimony should not be admitted unless it is clear that the jurors themselves are not capable, from want of experience or knowledge of the subject, to draw correct conclusions from the facts. The opinion of the expert is inadmissible on matters of common knowledge." *Wal-Mart Stores, Inc. v. White*, 476 So.2d 614, 617 (Ala.1985) (quoting—C. Gamble, *McElroy's Alabama Evidence* § 127.01(5) (3d ed.1977)).

The breach alleged in this case involved architectural matters that would not be within the common knowledge of the jurors, yet the School Board presented no expert testimony regarding the Architect's inspections and any deficiencies in those inspections. The School Board presented no expert testimony regarding the standard of care imposed within the architectural profession by the weekly inspection provision contained in this contract, and there was no expert evidence that that standard was breached by the Architect. In fact, the only expert that testified concerning the "weep holes" (which were the source of the leaks) was an architect who stated that it was not within the standard of care under the weekly inspection provision to keep track of each and every weep hole.

In a case startlingly similar to this one, a New York court, while holding that an architect could be found liable under the AIA contract for a failure to notify the board of education about leaks in a school's roof, did so only after finding that there was evidence that the architect had knowledge during the course of the construction that the contractor was not following the plans and specifications. In *Board of Educ. of Hudson City School Dist. v. Sargent, Webster, Crenshaw & Folley*, 146 A.D.2d 190, 539 N.Y.S.2d 814, 817–18 (1989), the court stated:

> While this very clause in the standard AIA architect/owner contract [the clause absolving the architect from responsibility for the contractor's failures] has been given exculpatory effect in this State and other jurisdictions [citations omitted] *none of the cases involved defects known by the architect during the course of construction,* which he failed to apprise the owner of under the contractual duty to keep the [School District] informed of the progress of the work. We decline to extend the application of the clause in question to an instance such as this, *where the trier of facts could find that the architect was aware of the defect and failed to notify the owner of it.* [Emphasis added.]

The key distinction between that case and this one is that in *Sargent, Webster* there was evidence that the architect knew of the defects during construction and failed to notify the owner. Clearly, the architect there would have breached his contractual duty if he failed to notify the owner of a known defect. It is undisputed in this case that during construction the Architect did not know of the defects with the weep holes. In fact, the School Board does not suggest that the Architect did know, only that under the provisions of the contract the trier of fact could have found that the Architect should have conducted an inspection that would have discovered the defect.

We conclude that, although the Architect had a duty under the contract to inspect, exhaustive, continuous on-site inspections were not required. We also hold, however, that an architect has a legal duty, under such an agreement, to notify the owner of a known defect. Furthermore, an architect cannot close his eyes on the construction site and refuse to engage in any inspection procedure whatsoever and then disclaim liability for construction defects that even the most perfunctory monitoring would have been prevented. In this case, we hold, as a matter of law, that the School Board failed to prove that the Architect breached the agreement.

In making this judgment, we would point out the value, and in some cases, the necessity, for expert testimony to aid the court and the trier of fact in resolving conflicts that might arise.

II

We now address the School Board's argument on its cross-appeal that the negligence claim should have been submitted to the jury. The School Board stated at oral argument that its negligence claim was asserted as an alternate theory of recovery in case its contract claim was not allowed. The trial court ruled that this negligence claim was barred by the statute of limitations.

* * *

The trial court was correct, therefore, in directing a verdict for the Architect on the School Board's negligence claim.

III

Because we hold for the Architect on the School Board's contract and negligence claims, we determine that the issue of whether the Architect should be allowed to assert its cross-claim against the roofing subcontractor and the manufacturer of the roofing membrane is moot.

IV

For the above-stated reasons, that portion of the judgment based on the jury's verdict against the Architect on the School Board's contract claim is reversed, and a judgment is rendered for the Architect on that claim. That portion of the judgment based on the directed verdict in the Architect's favor on the School Board's negligence claim is affirmed.

88-220 REVERSED AND JUDGMENT RENDERED.

88-273 AFFIRMED.

HORNSBY, C. J., and JONES, SHORES, HOUSTON and STEAGALL, J. J., concur.

Some observations on the *Watson* decision and some discussion of other important decisions (some of which were noted in *Watson*) may be useful in understanding the legal aspects of the general obligation of the design professional to perform site services.

After concluding that the exculpatory language did not completely shield the architect, the court faced the question of whether the inspections were as thorough as the architect agreed they would be. First, note that the AIA documents do not specifically require the architect to make inspections until the contractor claims that the work has been substantially completed. Prior to that point, the architect's site visits are observations, not inspections.[66]

The *Watson* decision spoke of the duty to *inspect* and discover the acts or omissions of the contractor in failing to follow the plans and specifications. Yet the court was not able to determine whether the architect had conducted a proper inspection in the absence of expert testimony showing whether the defects should have been obvious to the architect during his weekly inspections. Because the school board had the burden of providing this expert testimony, but did not, it failed to prove that the architect had breached the contract.

The *Watson* case illustrates the interrelationship that often exists between a breach of contract claim involving a design professional and the professional standard applicable to a claim of professional negligence (malpractice).[67]

---

[66]AIA Doc. B101-2007, §§ 3.6.2.1, 3.6.6.1.

[67]The professional standard is discussed in Section 14.05.

An architect or professional engineer owes her client a duty, not only to act in conformity with the contract requirements, but also in a non-negligent manner. For this reason, an owner who has hired a design professional to inspect the project during the construction phase, but who ends up with a defective building, can sue the design professional for breach of contract and for tort.

Whether the claim is approached as one based on breach of contract or professional malpractice, evidence will have to be introduced that bears on whether the design professional did what she promised or what the law requires. Depending on the nature of the contractual obligation, the evidence necessary to establish breach of contract or malpractice may be identical,[68] or it may be very different.

In comparing the proof necessary to establish a breach of contract and that necessary to establish a tort, an initial distinction must be drawn between two types of contract terms. Where the standard of performance dictated by the contract is unambiguous and readily discernible, expert testimony is not necessary to establish its breach. The reason is that the contract creates its own standard of expected performance, so that failure to meet that standard (absent excuse) results in a breach of contract. A party who breaches an unambiguous contract term cannot raise as a defense that it acted in a non-negligent manner.[69] The modified AIA contract addressed in *Watson*, for example, required the architect to provide "at least one inspection each week, a final inspection, and an inspection at the end of the one year guarantee period . . ." An architect who fails to inspect at least weekly would be in breach of contract, even if the professional standard did not require that many inspections. Here, a corollary exists with tort law: Expert testimony is not necessary when a design professional's negligence is so obvious that it is a matter of common knowledge. For example, expert testimony would not be necessary to show that a supervising

professional engineer who never showed up at the project while it was being built was negligent.[70]

Sometimes a contract establishes a standard of performance which is not easily discernable. For example, the contract addressed in *Watson* requires not only weekly inspections, but also that "the administration [of the contract by the Architect] shall be consistent with the size and nature of the work" and that the architect through her inspections shall "endeavor to guard the Owner against defects and deficiencies in the work of the Contractor." These contract requirements create no objective, readily discernible, and unambiguous standards of conduct. Must the court turn to the professional standard and ask what inspections would the reasonable architect make in an effort to protect the owner from defective work by the contractor? A corollary to that inquiry is: Must that contractual standard of conduct be proved by expert testimony?

Here, the *Watson* court created a distinction by way of its citation to *Board of Education of the Hudson School District v. Sargent, Webster, Crenshaw & Folley*.[71] The *Sargent* case also involved an architect's duty of inspection under an AIA contract. A newly installed roof failed, and the school district sued the inspecting architect. Evidence revealed that the architect's representative saw the roof being installed improperly; nonetheless, at the close of the evidence, the trial court had entered a judgment for the architect. The appellate court reversed and entered judgment for the school board, ruling that the architect breached the contract when she did not inform the owner of a defect known to the architect from her inspections.

Citing *Sargent*, the *Watson* court adopted the rule that an architect with a contractual duty of inspection breaches that contract when it fails to notify the owner of a known defect. The fact that the contractual duty itself does not contain a readily discernible, objective standard of conduct does not mean that in *all* cases the owner must present

---

[68]In *Adobe Masters, Inc. v. Downey*, 118 N.M. 547, 883 P.2d 133 (1994), the court stated that the evidence the owner needed to prove either breach of contract or negligence against its architect was identical.

[69]*Jowett, Inc. v. United States*, 234 F.3d 1365 (Fed.Cir.2000) (contractor must insulate ceiling ducts as required by the contract, notwithstanding industry custom of not doing so) and *R.B. Wright Constr. Co. v. United States*, 919 F.2d 1569 (Fed.Cir.1990) (contract calling for three coats of paint is enforced, notwithstanding a trade custom of applying two coats of paint).

[70]*Seven Tree Manor, Inc. v. Kallberg*, 688 A.2d 916 (Me.1997). Accord, *Town of Breckenridge v. Golforce, Inc.*, 851 P.2d 214, 216 (Colo. App. 1992), cert. denied, (Colo. May 17, 1993) (court refuses to allow expert testimony on whether golf course designer breached the contract, as the contract itself establishes the standard of performance) and *Turney Media Fuel, Inc. v. Toll Bros., Inc.*, 725 Pa.Super. 836, 725 A.2d 836 (1999) (in breach of contract action, expert's testimony as to why the work did not meet industry standards was properly excluded).

[71]146 A.D.2d 190, 539 N.Y.S.2d 814, appeal denied, 75 N.Y.2d 702, 551 N.E.2d 107, 551 N.Y.S.2d 906 (1989).

expert testimony to prove a breach of contract. However, excluding this narrow exception of a known defect in the contractor's work, the owner cannot prove the architect's breach of her duty of inspection absent expert testimony.

One more point needs to be made concerning breach of a contract obligation which is readily discernible. As noted above, an architect's failure to make *any* inspections would constitute a breach of contract without any need to resort to expert testimony as to the professional standard of care. However, even when the design professional performs an inspection in conformity with the contract requirement—or with the professional standard—expert testimony still may be necessary to establish whether the inspection itself was carried out properly. Such testimony must prove that a reasonable architect or engineer, performing that inspection, would have perceived that the contractor was performing in an improper manner and that such improper conduct caused the defect.[72]

A second issue relates to the troublesome question of the interrelation of claims based on breach of contract and those based on the tort of negligence. In the *Watson* case, the school board alleged negligence and breach of contract against the architect. It lost its claim based on negligence because the claim had not been filed within the time required by the statute of limitations. Typically, the claimant has a longer period to file a claim based on breach of contract than one based on negligence. But a claim based on tort may provide a larger judgment than one based on breach of contract. Although this distinction is relatively unimportant in defect cases, it would not be if the claimant sought recovery for emotional distress or if the claim were argued under a legal theory that might provide for punitive damages.[73]

Third, *Watson* seeks to impose an objective standard in determining whether there has been a breach of the obligation to inspect and discover. The court's call for expert testimony is designed to determine what other architects would have done under similar circumstances. The court noted that the language could not exculpate the architect "no matter how serious the deviation of the contractor from the plans and specifications."

The AIA sought to employ a subjective standard to measure performance even if the disclaimers do not exculpate the architect. Yet the court was frustrated by the AIA's attempt to provide a loose standard as exemplified by phrases such as "endeavor to require," "familiarize himself generally with the progress and quality of the work," and "endeavor to guard the owner." Frustration with the imprecision of these terms undoubtedly motivated the court to employ an objective standard. (In B141-1997 the phrases are slightly modified in Paragraph 2.6.2.1 but are essentially the same. In B101-2007, Section 3.6.2.1, the "endeavor to guard the owner" language has been dropped.)

Fourth, the contract in *Watson*, a modified AIA Document B141, specified that the "inspection" would be made at least once a week. Although the B141 in effect at the time this contract was made used somewhat different language, it is useful to note that in 1987, B141, Paragraph 1.5.4, required the architect to visit the site "at intervals appropriate to the stage of construction or as otherwise agreed." (In 1997 this became in Paragraph 2.6.2.1 "the stage of the Contractor's operations" and in 2007 this became in Section 3.6.2.1 "the stage of construction.") Parties commonly reject this formulation and include a requirement that visits be made at specific times. However, requiring weekly visits can generate excessive costs because of visits being made when they are not necessary. Document B101-2007, Section 4.3.3.2, seeks to solve this number-of-visits schedule problem by providing a blank that sets a limit on the number of visits as part of basic services.

At the time of the *Watson* decision, other courts were dealing with these issues. *Hunt v. Ellisor & Tanner, Inc.*,[74] cited in many subsequent decisions, held that the disclaimer is simply designed to ensure that the architect is not held to be a guarantor for the mistakes of the contractor. It rejected the *Moundsview* decision and its conclusion that the disclaimers, as a matter of law, relieved the architect. *Moundsview* held that it was proper to grant the architect a summary judgment (a resolution of the case without a full trial). In doing this, *Hunt* pointed to the responsibility of the architect "to visit, to familiarize, to determine, to inform and to endeavor to guard" and found these obligations to be nonconstruction responsibilities of the architect to provide information and not to perform work. It then concluded that the exculpatory paragraph emphasized the architect's nonconstruction responsibility and that the architect as a provider of information does not ensure or guarantee the contractor's work.

---

[72]*City of York v. Turner-Murphy Co., Inc.*, 317 S.C. 194, 452 S.E.2d 615 (App.1994).

[73]See Sections 27.09 and 27.10.

---

[74]739 S.W.2d 933 (Tex.Ct.App.1987), writ denied Mar. 23, 1988.

Although it is likely that the AIA intended to provide a contractual method by which the architect would obtain summary judgment solely by virtue of the exculpatory language, the method of arriving at the result in *Hunt* is an imaginative but strained interpretation of the contract provisions. Undoubtedly, reactions such as those shown in *Watson* and *Hunt,* as well as others to be noted shortly, reflect a negative reaction against this attempt by the AIA to shield the architect from responsibility for the work of the contractor.

A New York case demonstrates an attitude toward disclaimers similar to that in *Hunt.* The *Sargent* case[75] (noted earlier in this section) concluded that the exculpatory language was designed to avoid the architect's being held as a guarantor for the work of the contractor. It conceded that cases had concluded that the exculpatory language could be the basis for granting a summary judgment to the architect, but distinguished them by noting they did not involve cases where the architect had known of the defects during construction and where the architect failed to keep the owner informed of the progress of the work.

Note that *Sargent* did not go as far as *Watson.* It did not hold the architect responsible where she has not been paid for the type of close supervision that would reveal most if not all defects in the contractor's performance. It noted that the owner bargained and paid for the architect to make periodic inspections and to convey information relating to defects the architect discovered as a result of her visits that would enable the owner to ameliorate the problems.

These decisions indicate a trend toward making the issue of the architect or engineer's performance a factual one that must be resolved by evidence. This trend appears to reverse the tendency of some earlier decisions, which had used exculpatory language to justify awarding the architect a summary judgment, making a trial unnecessary.

The cases reproduced and summarized provide some generalizations regarding the site role of design professionals who perform services under contracts such as published by the AIA. First, the design professional breaches her contract if she does not do as she has promised regarding site visits, their frequency, and their intensity. Second, if her breach is a substantial cause of harm to the client, she must compensate the client for its losses. (Third-party claims are discussed in Section 14.08.) Third, if she *exceeds* her contractual commitment, such as visiting more frequently,

more intensively examining the contractor's work, directing the contractor as to how work should be performed, or acting positively as to safety, she has led the client to believe she would continue to perform those services even if not contractually obligated to do so. She must compensate the client if she is negligent in performing these services or if the client relies to its detriment on her performance by not having others perform those services. Fourth, the exculpatory provisions will not protect the design professional if she became aware of the contractor's failure to perform properly—performance including both compliance with the contract and laws such as building or safety laws—and did not take reasonable action to eliminate or minimize the likelihood that the finished project would not comply with the contractual requirements, that third persons on or about the site might suffer physical harm, or that others nearby would suffer damage to their property.

Fifth, language protecting the design professional in the contract between the design professional and her client will not be effective in a claim by a contractor against an architect if the specifications in the construction contract provided that the architect would supervise the work. This has been interpreted to include giving instructions when problems develop and inspecting the work as it is being performed.[76]

The cases discussed here demonstrate the inherent difficulties of standard form contracts. It is difficult to quarrel with the basic allocation expressed in AIA documents of responsibility for executing the design. That the architect checks the work periodically and issues certificates of payment should not disturb the division of authority and responsibility under which the architect designs and the contractor builds. However, general language often cannot take into account unforeseen circumstances. If the architect has not breached her obligation to observe (or inspect) and has not learned of defects, the exculpatory language should be sufficient to grant her a summary judgment. But where she has learned of defects and not reported them or where she has not complied with her contractual obligations, the exculpatory language should not provide a shield. Although it must be recognized that contracts drafted by the professional associations will inevitably be designed to favor members of those associations, such standard contracts cannot anticipate every problem.

---

[75]Supra note 71.

[76]*Colbert v. B. F. Carvin Constr. Co.*, 600 So.2d 719 (La.App.), writ denied, 604 So.2d 1309, 1311 (La.1992).

Also, the AIA's attempt to employ a subjective standard and to avoid the requirement of inspections until substantial completion has clearly not been successful in court. This may be due to the recognition that such contracts as those published by the AIA seek to give marginal advantages to the architect or due to the feeling that the owner pays for a service and should receive it. But also note that the courts recognize the owner should not receive more than the services for which it has paid. If the owner wishes a continuous presence on site, it will have to pay for it.

C. Submittals: Design Delegation

The contractor is usually required to submit information generally known as *submittals* to the design professional. The best-known submittals are shop drawings. (The term comes from the fabrication process: Shop drawings instruct the fabrication "shop" how to fabricate component parts.) AIA Document A201-2007, Section 3.12.1, defines shop drawings as data especially prepared for the work that is "to illustrate some portion of the Work." Submittals also include product data, defined in Section 3.12.2 as including "illustrations, standard schedules, performance charts, instructions, brochures, diagrams and other information" that illustrate the materials or equipment the contractor proposes to use. Finally, Section 3.12.3 defines samples as physical examples that illustrate "materials, equipment or workmanship and establish standards by which the Work will be judged."

An owner who is aware of the submittal system sees the process as a double- and even triple-checking system that should reduce the likelihood of defects or accidents. Often submittals are prepared by the subcontractor and reviewed by the contractor and the design professional or her consultant. See Figure 12.2. Although submittals should reduce potential loss, they also generate additional expense.

The prime design professional is even more concerned about submittals. Submittal review exposes the design professional to claims and even liability when people are injured, often the result of construction methods that are not within her expertise and not her responsibility in many if not most construction contracts. In addition, project failure often can be traced to submittal review.[77]

The prime design professional must rely heavily on the expertise of her consultants and will be held accountable to her client or to third parties if the consultant does not perform properly.[78]

Finally, the submittal process can expose the prime design professional to claims by the client and the contractor that improper delay in reviewing submittals delayed the project or disrupted the contractor's planned sequence of performance.[79]

The consultant who reviews submittals, such as a structural engineer, sees the submittal process as one that exposes her to the risk of license revocation or other disciplinary action. The *Duncan* case reproduced in Section 10.04B dealt with disciplinary action against a licensed engineer based on shop drawing review, among other things.

In addition, the consultant may, like the prime design professional, be exposed to liability to the owner or to third parties. She may also have to indemnify the prime design professional against whom a claim has been made.

The contractor often relies on the subcontractor to prepare submittals. Just as the prime design professional will be held responsible for the acts of her consultant, the contractor will be held responsible for the acts of the subcontractors.[80] More important, the contractor may believe that approval of its submittals transfers certain risks and responsibilities to the owner, a position often contrary to contract language.

Subcontractors and prime contractors see submittals as forcing them to fill in gaps in the design created by the design professional as well as exposing them to liability.

It has always been recognized that specialty subcontractors often design parts of the project within their expertise, in addition to supplying and installing parts of the structure. This is most commonly done in steel fabrication, window walls, and vertical transportation. In effect, part of the design has been delegated to contractors. The question of allocating risk when this is done has troubled the construction industry.

The controversy surfaced when the AIA issued its 1987 A201. Paragraph 3.12.11 briefly recognized the possibility of the contractor, through its specialty subcontractors or separate contractors (multiple primes) being given a set of performance specifications and asked to design in a way

---

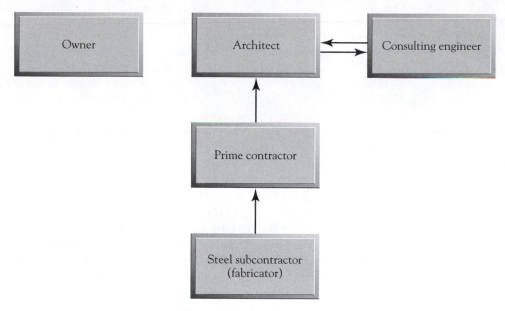

FIGURE 12.2   Path of a steel fabrication submittal.

that would accomplish these goals. That paragraph stated that when this is required by the contract, the architect can rely on the accuracy of the certification; in essence, on the professional quality of the design furnished by others.

This brief provision was the center of controversy.[81] Many attacked it as allowing the architect to relieve herself from design responsibility, often contrary to the understanding of the owner; that design is often excluded by the contractor's liability insurance; and that dual responsibility will create risk allocation problems. The AIA defended the provision, stating that it simply reflected industry practices that some skills, such as providing prefabricated trusses, rest with specialty subcontractors, and that specialty contractors compete by promising to deliver complete service.

This controversy was addressed in A201-1997, Paragraph 3.12.10, and the 1997 language remained virtually unchanged in A201-2007, Section 3.12.10. It states that the contractor is not required to provide professional architectural or engineering services, "unless such services are specifically required . . . for a portion of the Work." The contractor need not provide professional services that violate law. When such services are required, the contractor must be given performance and design criteria, and the contractor must see to it that "a properly licensed design professional signs and seals the drawings." The owner and architect can rely on the services and the certifications and the architect's review is for "the limited purpose of checking for conformance with the information given and the design concept expressed in the Contract Documents."

In essence this provision allows the owner (and the architect) to delegate the responsibility for the design quality of specified parts of the project to the contractor. The latter must use a licensed design professional to do the design and be responsible for it.

It remains to be seen how the division of responsibility for design will work, particularly with respect to the understanding of the client, licensing, insurance, risk allocation, and indemnification.

Before noting the varied purposes of the submittal process, imagine a building project without a submittal process. The contractor would build in accordance with its beliefs as to its contractual obligations. It would not have to state in advance what it proposes to do. It would simply be judged by what it does. If it fulfilled its contract commitment, it would earn the contract price. If it did not,

[81]J. SWEET & J. SWEET, SWEET ON CONSTRUCTION INDUSTRY CONTRACTS, § 10.08 (4th ed.1999).

the owner could exercise legal remedies. But submittals, like the site monitoring performed for the owner by the design professional, should reduce the likelihood that the project will not be built as required.

Although submittals have ancillary purposes, such as to persuade a vendor to process a purchase order[82] or allow different manufacturers to show their products and how such products join and relate to other products,[83] the principal purpose of a submittal is to obtain a representation from the contractor as to how it plans to execute the aspects of the design for which submittals are required.

In traditional construction projects,[84] such as those for which AIA documents are used, the contractor determines how the design is to be executed. The architect does not control the contractor's methods of executing the design. The architect monitors performance by site visits mainly to determine progress in order to issue certificates for payments. Until substantial completion, however, the architect makes visits to determine whether she sees anything to indicate that the completed project will not comply with contract requirements. Similarly, as to submittals, AIA Document A201-2007, Section 4.2.7, describes the architect's review to be only for "the limited purpose of checking for conformance with information given and the design concept." The architect does not check the methods by which the contractor intends to execute the design in accordance with Section 3.3.1. It states that the contractor is solely responsible for and controls the "construction means, methods, techniques, sequences and procedures." (In 1997 and continuing in 2007, A201 adds language to Section 3.3.1 that recognizes the possibility that the contract may give the contractor specific instructions as to execution.)

The submittals do not, however, lend themselves to a neat breakdown between design and its execution. As a result, information may be submitted (or omitted) that relates to execution of the completed project or the temporary work needed to accomplish the completed project. *Waggoner v. W. & W. Steel Co.*[85] demonstrates this. An accident resulted from inadequate temporary connections while the structural steel was being assembled and erected,

a choice made by the contractor. The court commenced its opinion by stating the issue to be whether the architect was responsible for ensuring "that the contractor employ safe methods and procedures in performing his work."[86] Under the AIA contract used in that project, this was the responsibility of the contractor. The site visits by the architect did not change this.

The plaintiffs contended the shop drawings that were submitted and approved by the architect "should have included specifications for temporary bracing and connections."[87] They contended the architect was negligent when he approved shop drawings that did not provide for proper temporary connections on the expansion joints. They asserted the architect or his structural engineer reviewing the submittal should have refused to approve it until the reviewer determined that the contractor would use proper temporary connections.

The court pointed to contract language stating that the *contractor* was solely responsible for construction methods and that it was not the architect's responsibility to require this information be furnished.

The court stated it was the duty of the contractor, not of the architect, to see that the shop drawings included how the temporary connections were to be made. It would be more correct to state there was no requirement that the contractor communicate its plan as to temporary connections to the architect through submittals. The architect could assume that a failure to include provisions in the shop drawings for temporary connections did not indicate that the contractor would use unsafe connections but indicated only that this was not information required by the submittal process. (Had the architect known unsafe methods were to be employed, she should have taken some affirmative action to avoid structural collapse and harm to workers.)

Another less well-known function of the submittal process is to make more definite the contract document requirements for which submittals are required. Language and even graphic depiction have communication limitations. (Try to describe or draw even a simple house!) As the contractor plans its actual performance rather than review the design prior to making its bid, its submittal indicates its interpretation of the contract, particularly in sensitive areas such as steel detailing. Submittal review

---

[82]*Day v. National U.S. Radiator Corp.*, 241 La. 288, 128 So.2d 660 (1961).

[83]*Alabama Society for Crippled Children & Adults, Inc. v. Still Constr. Co., Inc.*, 54 Ala.App. 390, 309 So.2d 102, 104 (1975).

[84]See Section 17.03.

[85]657 P.2d 147 (Okla.1982).

[86]Id. at 148.

[87]Id. at 151.

can be the process by which some potential interpretation disputes are exposed.

In addition, submittals as to product data and samples may be the method by which the contractor proposes substitutions if they are permitted by the contract. For example, AIA Document A201-1987, Paragraph 3.12.8, stated that the contractor will be relieved of responsibility for deviations if it specifically informs the architect of the deviation at the time of the submittal and the architect approves the specific deviation. In 1997 language was added making clear that the deviation must be a minor change under Paragraph 7.4.1, or a change order or a construction change directive, the latter two requiring consent of the owner. This language was continued unchanged in 2007.

Finally, although the contractor is not entitled to technical advice from the design professional on how to execute the design, the submittal process can be the method by which the contractor seeks information from the prime design professional or her consultants, particularly as to engineering techniques. (More on this later in this section.)

Much of what has been suggested regarding the submittal process and its use in construction is reflected in AIA documents. For example, B101-2007, Section 3.6.4.2 contained in the AIA document for design services, states that the architect's review and approval are only for the "limited purpose of checking for conformance with information given and the design concept." It also states that the review "shall not constitute approval of safety precautions or, unless otherwise specifically stated by the Architect, of construction means, methods, techniques, sequences or procedures." Similar language appears in A201-2007, Section 4.2.7.

B101 and A201 reflect the AIA's desire to avoid exposing the architect to liability. One liability exposure generated by the submittal process is the architect being responsible for defects or accidents that can be traced to her approval of submittals, as in *Waggoner*.

Another method the AIA uses to avoid liability exposure is A201-2007, Section 3.12.4, which states that submittals are not contract documents but merely demonstrate how the contractor proposes to conform to the design concept expressed in the documents. Yet Section 3.12.8 states that the contractor is relieved from responsibility for deviations if it informs the architect in writing of these deviations and the architect approves them. If after receiving the architect's approval the contractor

performs as it has proposed, the owner should not be able to claim a breach by pointing to Section 3.12.4 and its declaration that submittals are not contract documents. (The troublesome question of the legal effect of compliance with approved submittals is discussed later in this section.)

Finally, A201-2007, Section 3.12.4, recognizes that some submittals are not intended to be reviewed and approved but are simply for information.

B101 and A201 also focus on other liability exposure. The AIA seeks to minimize the likelihood of successful contractor claims that it has suffered losses because the architect unreasonably delayed passing on submittals. B101-2007, Section 3.6.4.1, states that the architect shall act with reasonable promptness and should be given "sufficient time" in her professional judgment "to permit adequate review." Similar language is found in A201-2007, Sections 4.2.7 and 3.12.5. More important, A201-2007, Section 3.10.2, requires that the contractor and architect agree to a schedule of submittals that allows the architect reasonable time for review.

The effect of review and approval of submittals depends on the purpose of the submittal. In making a submittal, the contractor can be asking one or a number of different questions. It can be asking whether what it proposes to do "is okay." If the contractor performs as it says it will, has it performed its contractual obligation?

As mentioned earlier, this can create an interpretation issue. The submittal process can flush out problems and make the contractual obligations of the contractor more concrete or specific. Relieving the contractor if it complies can be based on A201-2007, Section 4.2.11, which gives the architect the power to interpret the requirements of the contract documents on request of either party. To be sure, Section 4.2.11 looks to interpretation issues that arise on the site and not in the submittal process. Also, interpretations during the submittal process may bypass the process in Article 15 for resolving disputes. Yet the submittal process can also serve a concretization function. The architect has no power to change the contract, except under those contracts such as A201 that give her the power to make minor changes that do not affect time or price.[88] If concretization is accomplished and the contractor performs in accordance with the proposal it makes in the submittal, it has complied with the contract documents.

---

[88]AIA Doc. A201-2007, § 7.4.

Yet the preceding conclusion—that the contractor who complies with an approved submittal has performed in accordance with its contract commitments in A201-2007—is troublesome. Section 3.12.8 states that the contractor is not relieved from responsibility for deviations in the contract documents by the architect's approval of submittals unless the architect has been given specific notice of the proposed deviation and gives written approval to it. It also states that the contractor shall not be relieved of responsibility for errors or omissions in the submittals even if approved by the architect.

Other provisions of A201-2007 are relevant, however. Section 3.1.2 states that the contractor will perform the work in accordance with the contract documents. Yet Section 3.1.3 states that the contractor will not be relieved from its obligation to perform in accordance with the contract documents by approvals required, presumably approved submittals. Section 3.12.7 states that the contractor will not perform any work that requires submittal and review until the architect has given approval.

These provisions place the contractor on the horns of a dilemma. The contractor cannot perform certain work until submittals have been approved, because it must perform in accordance with approved submittals. Yet apparently the contractor's noncompliance with the contract documents will be a breach even if its submittals have been approved unless it can point to the deviation under Section 3.12.8 and the architect has given written approval to the specific deviation. Other than that, the contractor's work can be challenged even if its submittals indicate how it proposes to do the work and those submittals have been approved.

The specific exception of Section 3.12.8 undoubtedly recognizes that review of submittals can often be perfunctory. Yet it does not seem fair to deny any legal effect to approval and allow the owner to challenge the work as not complying with the contract documents only if the limited exception of Section 3.12.8 applies.

This analysis assumes that the information submitted is data that the contractor must furnish and the architect must review. (The architect should, however, inform the owner of any approval that is in effect an interpretation ruling of which the owner would want to be aware.)

A submittal may be a request by the contractor for technical advice from the design professional. The contractor may be stating what it proposes to do and asking the design professional, "What do you think?" or "Can you help me out?" If this request relates to construction means, methods, or techniques, the design professional need not respond. But if the design professional knows that the contractor proposes to proceed in a way that will lead to project failure or accidents, the design professional must draw attention to this danger. This duty to act is similar to her obligation when she visits the site. During site visits, the design professional need not search for potential trouble in operations under the contractor's control, but she cannot close her eyes to obviously dangerous conditions or to information that clearly indicates project failure.[89] Similarly, she should not be able to close her eyes to such information revealed during the submittal process even though she has no obligation to review that information. For example, had the submittal in the *Waggoner* case discussed earlier in this section been a clear signal to the architect that unsafe temporary connections were to be used or if industry custom required this information be shown in the submittals and it was not, the architect should have directed attention to this problem.

Whether such a duty to speak out exists depends on the balance of expertise. For example, if the subject of the submittals relates to steel detailing largely within the expertise of the structural engineer, there is a clear duty to take some action.[90] However, if the submittal relates to an activity in which the balance of expertise lies on the contractor's side, such as vertical transportation, approval by the design professional should not relieve the contractor from responsibility. Similarly, the design professional need not answer any questions on such matters that the submittals may ask.

Finally, some information contained in submittals is simply to inform the design professional of what the contractor intends to do but is not a request for approval or even a request for information or advice. This is what is contemplated by A201-2007, Section 3.12.4, which speaks of information submittals. In such a case, only a clear indication of the almost certain likelihood of project failure or harm to third parties would place on the design professional a duty to take action.

---

[89] See Sections 14.08G and 14.11B.

[90] *Duncan v. Missouri Bd. for Architects, Professional Eng'rs and Land Surveyors*, reproduced in Section 10.04B.

D. Use of Project Representative

The standard contracts published by the professional associations do not require the design professional to have a continuous presence on the site. If such presence is required, parties can agree to have a permanent representative such as a project representative or a resident engineer on the site continually. Although the principal problems involved in using a project representative relate to the representative's authority,[91] such use can bear on the responsibility of the design professional.[92]

At the very least, the presence of a full-time project representative means the design professional is not expected to be on the site continuously. It should also mean that frequency of visits may be diminished. However, this should not mean the design professional need not visit the site at times when it would otherwise be appropriate.[93] Also, the design professional must select competent representatives and must monitor their work.[94]

E. Use of Construction Manager

Although newer construction organization methods are discussed in Chapter 17, mainly in Section 17.04, it may be useful at this point to note the development of construction management. The role of the design professional has been substantially changed by standard contracts, particularly in the area of predicting costs, scheduling, safety, visiting the site, and passing on submittals. This has left a vacuum that in some projects has been filled by a construction manager (CM).

A CM should provide construction experience and skill, particularly at the design stage but also during design execution. There are two models of cost estimating: One uses rough formulas, and the other does a continuous evaluation of those elements that construction costs comprise. Similarly, during construction the CM should bring greater skill than the design professional in monitoring contract compliance and schedules. Is work being done in accordance with safety regulations? Will or does the end product meet the contract document requirements? It is likely that the future will see increased use of specialized professionals, unconnected with design creation, to perform many services mentioned in this chapter.

F. Statutory and Administrative Regulations

Increasingly, state statutes and regulations have dealt with design professional services, mainly those related to site services. Florida requires that a structural inspection plan be submitted to public authorities prior to issuing a building permit for any building greater than three stories or 50 feet in height or that has an assembly occupancy classification that exceeds 5,000 square feet in area and an occupant content of greater than 500 people.[95] The plan must be prepared by the engineer or architect of record. Its purpose is to provide specific inspection procedures and schedules so the building can be adequately inspected for compliance with the permitted documents.[96] The public agency must require that a state-certified inspector perform structural inspections of the shoring and reshoring for conformance with plans submitted to public authorities. The architect or engineer of record can perform these special inspections if she is on the list of those qualified. The costs of the special inspector are borne by the owner.

California enacted statutes in 1971 for engineers,[97] and in 1985 for architects[98] that deal with site services. The former currently requires that where supervision by a licensed engineer is required, the engineer must perform periodic observations "to determine general compliance." Supervision does not include responsibility "for the superintendence of construction processes, site conditions, operations, equipment, personnel, or the maintenance of a safe place to work or any safety in, on, or about the site." Periodic observation by an engineer is defined by the statute as a visit by the engineer to the site of the work. The statute regulating architects states that those who sign plans are not obligated to observe the construction of the work but the architect and client are not precluded from making a contract dealing with these services. The statute defines construction observation services as "periodic observation of completed work

---

[91]See Section 17.05B.

[92]*Deyo v. County of Broome*, 225 A.D.2d 865, 638 N.Y.S.2d 802 (1996) (use of project representative with authority to report on safety can make architect supervisor and liable to injured worker).

[93]*Central School Dist. No. 2 v. Flintkote Co.*, 56 A.D.2d 642, 391 N.Y.S.2d 887 (1977).

[94]*Town of Winnsboro v. Barnard & Burk, Inc.*, 294 So.2d 867 (La. App.), cert. denied, 295 So.2d 445 (La.1974).

[95]West Ann.Fla.Stat. § 553.71(7).

[96]Id. § 553.79(5)(a).

[97]West Ann.Cal.Bus.&Prof.Code § 6703.1.

[98]Id. § 5536.25(b).

to determine general compliance with the plans, specifications, reports, or other contract documents."[99] The statute states that these services do not encompass construction processes, site conditions, and activities similar to those set forth in the statute dealing with engineers.

Unlike California, which uses a statutory approach, Hawaii requires, by administrative regulation, an application made for a building or construction permit "involving the public safety or health" to state that all plans and specifications shall include a warranty by the certificate holder stating that construction would be under her supervision. If she cannot make such a warranty, she must notify the registration board within fifteen days and include the name of another licensed professional who will continue with construction supervision.[100]

State control over every aspect of the construction process is increasing.[101] People who engage in that process should obtain legal advice on the requirements for performing particular activities. In addition, the legislative definitions such as those found in the California statutes show an attempt by the design professions to obtain statutory language that seeks to limit the obligations of design professionals. However, those statutory definitions relate only to the public law aspects of construction. Those statutes do harmonize with language in standard contracts published by associations of design professionals. (Very likely the statutes were "pushed" by those associations.) But those statutes would not bar design professionals and their clients from making contracts that place a greater responsibility on the design professional than found in the statutes.

SECTION 12.09 Hazardous Materials

The environmental movement noted in Section 9.13 has had a significant effect on the construction process. A proposed site might be found to have been contaminated by hazardous or even toxic materials stored there years ago or leaking from an adjacent site. Similarly, during a renovation project, asbestos or other hazardous materials might be encountered that could prove dangerous to workers or others in the building. Even in new projects, greater emphasis on sensitivity to environmental factors could generate concern over materials and equipment to be incorporated in the project.

As the public has become more concerned over environmental matters, regulatory efforts have been directed toward this problem. Although this is not the place to discuss these regulations in detail, we can point to some of the factors involved in such regulation.

Regulation has often included requirements that reports be submitted before certain work involving hazardous materials can be performed and that only people certified by state or local authorities can perform dangerous work, such as asbestos removal. Also, the uncertain liability connected with environmental problems has led insurers to exclude many hazardous material activities from liability coverage. Much of the difficulty encountered by insurers relates to the uncertainty of what could be considered safe exposure levels and the fact that the claims may come many years after work has been performed.

These factors have led those actively involved in the construction process, such as design professionals and contractors, to seek methods to exculpate themselves from responsibility for these environmental concerns, in the hope that liability would fall on the owner. Illustrations can be found in the AIA Document B101-2007. For example, Section 5.5 requires the owner on request of the architect to furnish geotechnical information that includes "evaluations of hazardous materials." In addition, Section 5.7 requires the owner to furnish tests for air and water pollution and tests for hazardous materials. Finally, Section 10.6 states that "unless otherwise required in this Agreement, the Architect shall have no responsibility for the discovery, presence, handling, removal or disposal of or exposure of persons to, hazardous materials or toxic substances in any form at the Project site."

The uncertainties created by the uncovering of environmental risks has led participants in the construction process to use contract language to avoid responsibility. The hope is that these clauses will be effective, at least to contracted connected parties, even though liability exposure of other sorts, such as claims by third parties or public officials, may not be affected. Because the owner has both the upside and downside risks of ownership, it is not unreasonable to place ultimate responsibility on the owner and let the owner deal with transferring or distributing this risk through insurance.[102]

---

[99]Id. at § 5536.25(c).

[100]Hawaii Adm.Rules § 16-115-9(b), (c).

[101]For discussion of building codes, see Section 14.05C.

[102] AIA Doc. A201-2007, § 10.3.3. See Section 23.05I for more on hazardous material.

SECTION 12.10 Who Actually Performs Services: Use of and Responsibility for Consultants

A. Within Design Professional's Organization

The design professional may operate through a corporation or a partnership or be a sole proprietor.[103] The actual performance of the work may be done by the sole proprietor, a principal[104] with whom the client discussed the project, another principal, employees of the contracting party (whether the contracting party is a sole proprietor, a partnership, or a corporation), or (as discussed in Section 12.10B), a consultant hired by the design professional. Are the client's obligations conditioned on the performance being rendered by any particular person? Can certain portions of the performance be rendered by people other than the design professional without affecting the obligation of the client to pay the fee?

Services of design professionals, especially those relating to design, are generally considered personal. A client who retains a design professional usually does so because it is impressed with the professional skill of the person with whom it is dealing or the firm that person represents. The client is likely to realize that licensing or registration laws may require that certain work be done by or approved by people who have designated licenses.[105] Yet the client is also likely to realize that some parts of the performance will be delegated to other principals in the firm with whom the client is dealing, employees of that firm, or consultants retained by the design professional organization.

As to design, unless indicated otherwise in the negotiations or in the contract, the client probably expects the design professional with whom it has dealt to assemble and maintain a design "team" and to control and be responsible for the design. The fleshing out of basic design concepts, such as the construction drawings and specifications, is likely to be actually executed by other employees of the design professional's organization. Although contract administration is probably less personal than design, the client will still probably expect that the principal contract administration decisions will be made, if not by the person with whom it has dealt, at least by another principal in the design professional organization. However, just as in design, the client is likely to realize that the person who has overall responsibility will not actually perform every aspect of contract administration. (As to interpretations and disputes, see Section 29.06.)

These conclusions can depend on the client's knowledge of the size and nature of the design professional's organization. They may also depend on whether the client has ever dealt with design professionals before and whether it knew of the division of labor among principals, employees, and consultants.

B. Outside Design Professional's Organization: Consultants

Because of the complexity of modern construction, the high degree of specialization, and the proliferation of licensing laws,[106] design professionals frequently retain consultants to perform certain portions of the services they have agreed to provide. Although the principal legal issues have been whether consultant fees are additional services not covered by the basic fee[107] and whether the design professionals are responsible for the consultants they hire—a point to be mentioned later in this subsection—inquiry should be directed initially to whether the design professional can use a consultant to fulfill the obligations owed the client.

Clients generally prefer that highly specialized work be performed by highly qualified specialists, often outside the design professional's organization. Suppose, though, the client insists that all design services be performed within the design professional's organization. The client may believe that it would not be able to hold the design professional accountable if the consultant did not perform properly. In such a case, the client might feel at a serious disadvantage if forced to deal with or institute legal action against a consultant it has not selected and with whom it has no direct contractual relationship. Clearly, the client can insist on this in the negotiations, and the contract could so provide. But suppose nothing is discussed or stated specifically in the agreement.

It is unlikely that any client who is or should be aware of the customary use of consultants for certain types of

---

[103]New variations on these business organizations are limited liability partnerships (LLPs) and limited liability companies (LLCs). See Section 3.06.

[104]The term *principal* is defined as a partner in a partnership or a person with equivalent training, experience, and managerial control in a corporation.

[105]For discussion of the effect of licensing laws on who is permitted to do the work, refer to Section 10.05A.

[106]See Section 10.02.

[107]See Sections 12.01 and 13.01G.

work can insist on consultants not being used. Standard contracts published by professional associations contain language relating to professional consultants. Such language usually does not directly relate to the question of whether consultants may be used but deals indirectly by providing that certain consulting services are covered under the basic fee and others are additional services. Inclusion of such language should indicate to the client the likelihood that some consultants will be used and that possibly others will be used if the client consents.

One person can be held liable for the wrongdoing of another. The most common example is the liability of an employer for the negligent acts of its employee committed in the scope of employment. Although many reasons exist for vicarious liability, one reason sometimes given is that the employer controls or has the right to control the details of the employee's activities.

If the negligent actor is not controlled or subject to the control of the person who has hired the actor, the actor is an independent contractor and not an employee. Subject to many exceptions, the employer of an independent contractor is generally not liable for the negligence of the latter. For example, the businessperson who hires an independent garage to service its fleet of trucks generally will not be held liable for the negligence of the garage. An opposite result would follow if the businessperson had a repair service as part of its organization.

Generally the consultant is an independent contractor. As a rule, the consultant is asked to accomplish a certain result but can control the details of how it is to be accomplished. If an architect retains a structural engineer as a consultant, the latter's negligent conduct that causes injuries to third persons will generally not be chargeable to the architect.

The client who retained the architect is not a third party in the same sense. Permitting the architect to use a consulting engineer as a substitute to perform certain portions of the work should not relieve the architect of responsibility to the client for proper performance of that work. The architect's obligation to the owner is based on their contract.[108]

A trilogy of consultant cases decided by the Supreme Court of Oregon are instructive. The first, *Scott & Payne v. Potomac Insurance Co.*,[109] involved a claim by an owner against an architect based on failure in the heating system that the owner claimed was caused by defective design. The architect's insurer refused to defend, claiming that the negligent act occurred in a period not covered by the policy. The architect settled the claim and brought an action against his insurance company when it refused to reimburse him. To recover, the architect found himself in the strange position of claiming he was liable. One argument made by the insurer was that the architect was not negligent because he relied on the advice of a heating engineer. In rejecting this argument, the court stated that the architect should possess skill in all aspects of the building process and cannot shift responsibility to a consultant.

The second case in the trilogy was *Johnson v. Salem Title Co.*[110] in which an injured third party sued an architect when a wall defectively designed by the consulting engineer collapsed. The court concluded that the engineer was an independent contractor, which would normally give the architect a defense. However, the court held the architect liable because of an exception to the independent contractor rule for nondelegable duties created by safety statutes. In this case, the architect was required to comply with building codes, and this duty could not be delegated to the engineer. By nondelegation, the court did not mean that the architect could not use the engineer to fulfill code requirements but meant that the architect could not divest himself of ultimate responsibility if the code were not followed.

The action was brought not by the client against the architect but by an injured party. Had the action been brought by the client for its losses, the independent contractor rule would not have been relevant.

The third case in the Oregon trilogy, *Owings v. Rosé*,[111] involved the owner suffering a loss when the floor cracked because of defective design by the consulting engineer. In passing on the indemnity claim, the court stated that the architect was liable to the owner on the basis of *Scott & Payne v. Potomac Insurance Co.*[112]

---

[108]*Harold A. Newman Co. v. Nero,* 31 Cal.App.3d 490, 107 Cal. Rptr. 464 (1973). Similarly, see *South Dakota Bldg. Auth. v. Geiger-Berger Assoc.,* 414 N.W.2d 15 (S.D.1987) (architect denied indemnification against consulting engineer because it fixed its seal to the engineer's drawings and it failed to comply with its contractual obligations to prepare architecturally sound plans), and *Brooks v. Hayes,* 133 Wis.2d 228, 395 N.W.2d 167 (1986) (prime liable to owner for negligence of subcontractor even though latter was independent contractor).

[109]17 Or. 323, 341 P.2d 1083 (1959).
[110]246 Or. 409, 425 P.2d 519 (1967).
[111]262 Or. 247, 497 P.2d 1183 (1972).
[112]Supra note 109.

Prime design professionals can exculpate themselves from liability to clients for errors of their consultants. One method—a novation—is a tripartite contract under which the consultant is substituted for the prime design professional for that part of the work. Another is to seek and obtain from the client exculpation from any responsibility for the conduct of the consultants. This is more likely to be obtainable if the client designates that particular consultants be used. It does not seem unreasonable for the principal design professional in such a case to seek exculpation from the client.

Following this approach to its logical extreme, the principal design professional can suggest or insist that the client contract directly with consultants. Clearly, this would relieve the principal design professional unless she had information regarding the consultant that should have been communicated to the owner. Although this approach is often used in retaining geotechnical engineers, principal design professionals frequently prefer to keep overall professional control and are not anxious for clients to contract separately with consultants.[113]

It is likely that the principal design professional will not be relieved from liability to the client for the acts of consultants. Probably the best the principal design professional can do is to ensure that the consultant is obligated to perform in an identical manner to the principal design professional's obligation to the client. The principal design professional must consider the financial responsibility of consultants. Having a good claim against a consultant may be meaningless if the consultant is not able to pay the claim. The principal design professional should undertake an investigation of the financial capacity of consultants employed. If the consultant is a small corporation, the principal design professional should bind the individual shareholders to the contract so that they are personally liable. The principal design professional probably should require that the consultant carry and maintain adequate professional liability insurance.

Does the consulting engineer assume the risk of not being paid if the principal design professional—the architect—is not paid by the client? See Section 12.03G.

SECTION 12.11 Ownership of Drawings and Specifications

Who owns—or, more properly, who has use rights of—the tangible manifestations, usually written but increasingly computer generated, created by the design professional and necessary to build the project?

Clients sometimes contend, and many agencies of state and local government insist, that the party who pays for the production of drawings and specifications should have exclusive right to their use.[114]

However, design professionals contend they are selling their ideas and not the tangible manifestations of these ideas as reflected in drawings and specifications. This, along with the desire to avoid implied warranties to which sellers of goods are held, is the basis for calling the tangible manifestations "instruments of service." Design professionals contend that the subsequent use of their drawings and specifications may expose them to liability claims. If another design professional completes the project, the original designer may be denied the opportunity to correct design errors as they surface during construction. Design professionals contend that most projects are one of a kind and that in reality design is a trial-and-error process. Similarly, liability exposure can result if the design is used for an addition to the project, or a new project, for which it may not be suitable. Even if the design professional is absolved, absolution may not come until after a lengthy and costly trial.[115] (As discussed later in this section, this risk can be dealt with by indemnification.) In addition, reuse without adaption may compromise the aesthetics or structural integrity of the original design.

Another reason a design professional may wish to retain ownership of the design is to receive credit as an author and to protect her professional reputation whenever her

---

[113]Even if the owner retained the consultant, the architect must exercise due care in relying on that consultant's report. *Kerry, Inc. v. Angus-Young Assocs.*, 280 Wis.2d 418, 694 N.W.2d 407 (App.), review denied, 286 Wis.2d 98, 705 N.W.2d 659 (2005).

[114]A Louisiana statute specifies that all plans, designs, and specifications "resulting from professional services paid for by any public entity shall remain the property of the public entity." La.Rev.Stat § 38:2317. The State of Georgia has considered creating "stock" plans to be used on all school construction projects at an anticipated savings of nine million dollars annually in architectural fees. See Roberts, *After the Ball: Subsequent Use of Construction Documents After the Project for Which They Were Originally Prepared: A Sketchy Area of the Law*, 17 Constr. Lawyer No. 2, April 1997, p. 35.

[115]See *Karna v. Byron Reed Syndicate*, 374 F.Supp. 687 (D.Neb.1974) (designer absolved when project's use unforeseeably changed). See also West Ann.Cal.Bus.&Prof.Code § 5536.25(a), which gives the architect a defense if her design is changed without her approval.

creative works are displayed to the public.[116] In addition, employees of the architect want recognition for their work. Draftspeople, project designers, and others often wish to make copies of their work to show prospective employers and clients.[117] Perhaps most important to design professionals is that the actual use by the owner is an important factor in determining the value of design professionals' services and, correspondingly, their fee—a factor recognized by the frequent use of the percentage of construction cost to determine compensation. (See Section 13.01B.) Had the design professional known that drawings and specifications would be used again, a larger fee would have been justified.[118]

In the absence of any specific provision in the contract dealing with reuse, the client who has paid for the services has the exclusive right to use the tangible manifestations of the design services performed by design professional. Although cases are rare, this result stems from the analogy to a sale of goods. In ownership terms, the client "owns" the drawings and specifications.[119]

Suppose a design professional establishes a custom that drawings and specifications belong to the person creating them and that the client is allowed their use only for the particular project. The client would not be bound by it unless it knew or should have known of such a custom.[120] It has been held that an AIA design services document supports the establishment of a custom that in the architectural profession the architect retains ownership of the plans unless there is an express agreement to the contrary.[121]

In both of the cases referred to in the preceding paragraph, the architect retained the original blueprints, and the owner (in *Meltzer*) or contractor (in *Aitken*) were provided copies. Retention by the architect of the physical plans may indicate that they are her personal property, and a person is not entitled to take or use the plans without the architect's consent. However, in *Sokoloff v. Harriman Estates Development Corporation*,[122] the New York Court of Appeals (that state's highest court) granted the homeowners "specific performance"—a court order—requiring their design/build contractor to turn over to them the design the contractor had obtained from an architect. (The owners had decided not to proceed with construction with the contractor when its low bid greatly exceeded their budget.) The owners had argued successfully that the architectural plans were unique, and that specific performance was warranted because the plans were based on a design conceived by them.

Consecutive editions of the AIA documents have shown increasing attention to the issue of ownership of the design documents. In B141-1987, Paragraph 6.1, the design and other documents prepared by the architect are instruments of service and the architect is described as the author. The owner could not use the design on other projects without agreement with and compensation of the architect.

This simple scheme was greatly expanded in B141-1997, Paragraph 1.3.2. The documents covered include those prepared by consultants and created in electronic format. The architect and consultants are owners of their respective instruments of service. The owner is given a nonexclusive license to reproduce the documents for the purpose of building, using, and maintaining the project. The owner cannot use the documents on another project without the architect's permission. Unauthorized use of the documents is at the owner's risk and without liability to the architect. A201-1997, Paragraph 1.6.1, provides similar protection to the architect and her consultants and provides methods for handling the instruments of service.

---

[116]By way of contrast, Peter Eisenman, architect of the Wexner Center for the Arts in Columbus, Ohio, has reportedly expressed another view. "My feelings about buildings is this: You make it and you give it to the client and it is theirs. They can tear it down for all I care." Wall Street Journal, p. W14 (July 13, 2001).

[117]Ellickson, *Ownership of Documents: Does It Matter Who Owns the Drawing or the Design?* AIA Documents Supplement Service, July 1991, at 1–4.

[118]*Garcia v. Cosicher*, 504 So.2d 462 (Fla.Dist.Ct.App.), review denied, 513 So.2d 1060 (Fla.1987) (contract provided for repeat fees in event of owner reuse).

[119]Leading commentators early in the twentieth century were in agreement that architectural plans, paid for by the owner, are the property of the owner, notwithstanding assertions of a contrary custom in the architectural community. See J. WAIT, ENGINEERING AND ARCHITECTURAL JURISPRUDENCE § 815 (1907) and C. BLAKE, THE LAW OF ARCHITECTURE AND BUILDING 90 (1916) ("[U]nless there be a specific provision in the contract whereby it is agreed that the plans are to be and remain the property of the architect, they must be legally considered, it seems, as the property of the employer who has ordered, accepted, and paid for them."). However, both commentators cite rather obscure case law in support of this proposition, such as *Windrim v. City of Philadelphia*, 9 Phila. 550, 552, 1872 Westlaw 15115 (Pa.Com.Pl.1872).

[120]*Meltzer v. Zoller*, 520 F.Supp. 847, 856 (D.N.J.1981).

[121]*Aitken, Hazen, Hoffman, Miller, P.C. v. Empire Constr. Co.*, 542 F.Supp. 252, 261 (D.Neb.1982).

[122]96 N.Y.2d 409, 754 N.E.2d 184,729 N.Y.S.2d 425 (2001)

B101-2007 addresses instruments of service in Article 7. Under Section 7.2, the architect and her consultants are deemed the authors and owners of their respective instruments of service. Documents in electronic format, mentioned in 1997, are not referred to in 2007. Under Section 7.3, the architect grants the owner a nonexclusive license to build, alter, or maintain the project. If the owner uses the documents without employing the architect, the owner will hold the architect harmless against any liability. The owner cannot assign or grant the nonexclusive license to another party without the architect's prior written consent, and unauthorized use is at the owner's sole risk. A201-2007, Section 1.5.2, addresses the use of the instruments of services by the contractor, subcontractors, and material or equipment suppliers.

The relationship between contractually created use rights and copyright law will be discussed in Section 16.04D.

Legitimate reasons exist for giving the design professional exclusive right to reuse the drawings and specifications, mainly on the basis of use determining value of the professional services. Yet the prohibition against the client's using the materials for additions to or for completing the project can be looked on as a device to discourage the client from retaining a new architect or at least to make the client pay compensation if it replaces the original architect. It is as if an implied term of the original retention agreement gave the design professional an option to perform any additional design services required by an addition to the original project. Hiding such "options" in the paragraph dealing with ownership of drawings and specifications can make courts suspicious of the fairness of such standardized contracts.[123]

The Engineers Joint Contracts Documents Committee (EJCDC), a consortium made up of the Professional Engineers in Private Practice, a division of the National Society of Professional Engineers, the American Council of Engineering Companies, and the American Society of Civil Engineers, published an edition of its standard form of agreement between owner and engineer for professional services in 2002. In this document, E-500, Paragraph 6.03 is headed "Use of Documents." Paragraph 6.03E states that the documents are not intended to be suitable for reuse on extensions of the project or on any other project. The owner who uses the documents without written verification or adaptation by the engineer does so at its own risk and will indemnify and hold harmless the engineer or her consultants from claims and other expenses, including attorneys' fees relating to this use. Any verification or adoption entitles the engineer under Paragraph 6.03F to further compensation at "rates or in an amount to be agreed upon by Owner and Engineer."

The EJCDC also deals more extensively with electronic data generated. For example, Paragraphs 6.03B and C deal with electronic data furnished by the owner to the engineer and the engineer to the owner. Such data are furnished for the convenience of the recipient but only the hard copies may be relied upon. Any conclusion or information from such data is "at the user's sole risk." Under Paragraph 6.03B if there is a discrepancy between "electronic files and hard copies, the hard copy governs." Finally, Paragraph 6.03C deals with the possible deterioration of data stored in electronic media format, and Paragraph 6.03D covers transferring of documents in electronic media format.

The EJCDC is slated to publish a new version of E-500 in 2008. That new edition must he examined to determine the EJCDC's most current approach to this issue.

The EJCDC's concern with the interface between electronic and written data is well merited. Digital technologies that increasingly are used to create the design may well have the effect of redefining the basic relationships between designer, contractor, and owner. For example, when subcontractors and suppliers electronically participate in the creation of the original design components, who at that point is the overall "designer" of the project? In addition, the ease with which electronic data may be altered (without leaving a "paper trail") raises concerns as to what constitutes the actual design. (The EJCDC's response is that the hard copy governs if it conflicts with the electronic version.) These issues (and others) undermine the basic concept of ownership so vital to the AIA's idea of the design as the architect's "instruments of service." A recent article compared the adequacy of the AIA A201–1997 Paragraph 1.6.1 in light of these technological changes:

> The "Ownership of Documents" clause will have to be rethought in the digital age. For one thing, it completely ignores intellectual property rights in those portions of project documentation that are furnished by or through the Contractor or subs. In a collaborative design process, is the entire project database a joint work, in which the copyright is

---

[123]*Foad Consulting Group, Inc. v. Azzalino*, 270 F.3d 821, 829, n. 12 (9thCir.2001) and *Eiben v. A. Epstein & Sons International, Inc.*, 57 F.Supp.2d 607, 612, n. 5 (N.D.Ill.1999) (interpreting earlier AIA language).

owned by many participants or are different parts of the database owned by different parties? If portions of the Architect's contribution consist of digital "objects" imported from third party sources, to what extent is the Architect the compiler of a collective work rather than the author of an original work? With respect to control of multiple copies, there will be myriad digital files maintained in multiple storage media, not simply "one record set" retained by the Contractor. The Contractor will probably not be able to "suitably account for" all of these files to the Architect at the end of the Project. What will the "statutory copyright notice" look like in a collaboratively-generated 3D or 4D electronic file?[124]

Returning to more conventional technologies, even if the design professional obtains ownership of the design documents, the owner may have a statutory right to use them. For example, in 1992 the California legislature enacted legislation providing that in the event of damage covered by insurance to a single-family home as a result of a declared natural disaster, the architect must release a copy of the plans to the homeowner's insurer or to the homeowner on request and verification that the plans will be used solely for verifying the fact and amount of damage for insurance purposes. Under this legislation, the homeowner cannot use the plans to rebuild without the prior written consent of the architect. If the architect does not consent, the architect who has drafted the original plans and released them is not liable for any property damage or personal harm if the plans are subsequently used by the owner to rebuild all or a part of the residence.[125]

Are there any limitations on the design professional's right to reuse the drawings and specifications? Suppose the design professional plans to use the documents to build an identical residence near the completed residence that would diminish the exclusivity of the original residence. Although the owner could obtain exclusivity by contractual protection, as done in the Trump Plaza contract noted later in this section, this event is rarely anticipated in the negotiations. It is likely that the law would imply a promise by the design professional not to reuse the drawings and specifications in any way that would significantly diminish the value of the original residence.

The AIA reports that Donald Trump obtained an out-of-court settlement against his architect barring the architect and a competing developer from using the design of Trump Plaza in developing a site one block south of the plaza. According to the AIA, the written agreement stated that the drawings were not to be used for any other buildings, although it allowed the architect to retain ownership of the drawings. The settlement required changes in the façade of the second building to distinguish it from Trump Plaza.[126]

This section has dealt with contractual provisions designed to protect the design professional's exclusive right to use the documents she has prepared. Claims by design professionals may have other substantive bases. As noted in Sections 16.03 and 16.04, a design professional, as author, can be given federal copyright protection. In some jurisdictions, unauthorized use by a third party may be a tortious conversion, which can be the basis for a claim.[127] Remedially, generous protection is provided by the federal copyright law, as seen in Section 16.04D. If the claim is based on a breach of contract, in addition to a possible injunction (also permitted under the federal copyright law), the design professional would be able to recover the market value of her services that have been used improperly by the defendant—very likely the cost of producing the documents. A claim based on the tort of conversion would be more expansive than one based on breach of contract. The remedy could include not only the cost to prepare the documents but also, in the case of an intentional conversion, any enrichment or benefit the defendant has received as a result of the conversion and, under extreme circumstances, punitive damages.

The discussion to this point has assumed that the architect or engineer was an independent contractor hired by the owner. But on the increasingly popular design-build projects, discussed in Section 17.04F, the contractor, not the owner, retains the architect. Who owns the design documents on a design-build project, the designer or the contractor? This question is resolved under the work for hire doctrine. If the designer was an employee of the contractor, and the design was created within the scope of the employment, then the design would be a work for hire and the employer/contractor would be the owner of the documents.[128]

---

[124]Stein, Alexander & Noble, *AIA General Conditions in the Digital Age: Does the Square "New Technology" Peg Fit Into the Round A201 Hole?* Construction Contracts Law Rep., vol. 25, no. 25 ¶367 at p. 7 (Dec. 14, 2001). A thorough discussion of the new technologies affecting design professionals is found in Ashcraft, *New Paradigms for Design Professionals: New Issues for Construction Lawyers*, an unpublished paper presented at an American Bar Association meeting. The author can be contacted at hashcraft@hansonbridgett.com.

[125]West Ann.Cal.Bus.&Prof.Code § 5536.3.

[126]Supra note 117.

[127]*Williams v. Chittenden Trust Co.*, 145 Vt. 76, 484 A.2d 911 (1984).

[128]*Trek Leasing, Inc. v. United States*, 62 Fed.Cl. 673 (2004).

SECTION 12.12   Time

If no specific provision deals with time for performance, the parties must perform within a reasonable time. However, the cooperative nature of design, the client providing a program, and the design professional creating a design subject to client approval make it difficult to determine who is responsible for delay. Delay may be caused by public officials failing to move the administrative process along or by lenders deciding whether to make a loan. Even if a schedule is created, it is likely to require frequent adjustment.

For these reasons, claims for a delay during design are difficult to sustain. For delays in contract administration, claims are more likely to be made by contractors. This is dealt with in Section 26.10.

Starting in 1977, the AIA included language in B141 dealing with scheduling of design services. Document B101-2007, Section 2.2, requires the architect to perform services as "expeditiously as is consistent with the professional skill and care and the orderly progress of the project." Section 3.1.3 requires the architect to submit a schedule for the owner's approval. The schedule may be adjusted as necessary and must allow time for owner review, performance by the owner's consultants, and approvals by public authorities. Time limits may not be exceeded by either party "except for reasonable cause."

There is much fluidity in performing design work. It is unlikely that any schedule will be the basis for a successful claim. Yet the mere existence of a schedule approved by both parties should help move the process along.

SECTION 12.13   Cessation of Services: Special Problems of the Client–Design Professional Relationship

A. Coverage

Contracts to perform design or construction services are service contracts, and many of the same legal rules apply. Principles discussed in Chapter 33 dealing with construction contracts also apply to contracts for design services. However, some aspects of the client–design professional relationship have generated special legal rules discussed in this section.

B. Specific Contract Provision as to Term

Often contracts specifically define the duration of the contract relationships. However, this rarely exists in contracts to perform design services because of the high likelihood of delay traceable to the large number of participants in the construction process, including not only owners, designers, and contractors but also lenders and public authorities. (See Section 12.12, dealing with time and schedules.)

At the outset, a number of issues must be differentiated. First, when do the contractual obligations between client and design professional end? Second, when does delay in the performance of design services entitle the design professional to additional compensation? Third, when has the design professional's dispute resolution power ended? The first is discussed in this section, the second in Section 13.01G, and the third in Section 29.05.

Usually the design professional's services are divided into phases. Because the construction phase is last, it is the focal point for determining when the client–design professional relationship has terminated. Also, the construction phase is itself divided into segments, principally to enable the contractor to be paid as it works. Phases are important for other purposes as well.

Generally, the owner would like the design professional to perform until the project is "completed." At this point, the owner takes possession of the work. After that point, except for postcompletion services such as furnishing or reviewing as-built drawings, the owner no longer needs the design professional's services.

In most construction contracts executed on standard documents published by the professional associations, completion has two stages: substantial and final. AIA Document B101-2007, Section 3.6.1.3, states that the obligation of the architect to provide administration "terminates on the date the Architect issues the final Certificate for Payment." This end date corresponds to A201-2007, Section 4.2.1, which provides that the architect will perform until she issues the final certificate for payment. Compare these two provisions with B101-2007, Section 4.3.2.6, which entitles the architect to additional compensation if services extend sixty days after substantial completion.

Clearly some time will pass between the dates of substantial and final completion. It is during this time that the contractor performs a so-called punch list of relatively minor items to wrap up the project. But the AIA recognizes that issuance of a final certificate may be delayed because of the contractor's unwillingness to correct the punch list items. That services performed after the sixty-day period are considered additional may surprise the client.

C. Conditions

A condition is an event that must occur or be excused before a party is obligated to begin or continue performance. Often contracting parties do not wish to begin or continue performance unless certain events occur or do not occur. For example, an owner may not wish to start construction until it has obtained a loan or permit. Design professionals may wish to condition their obligations on their ability to rent additional space or hire an adequate staff. Yet each party may wish to make a binding contract in the sense that neither can withdraw at its own discretion.

A condition that is within the *sole* power of one of the parties may prevent a valid contract from being formed. Such a conclusion is avoided by implying an obligation to use good faith to seek occurrence of the condition. However, the creation of a condition does not affect the validity of the contract as long as the condition is described with reasonable certainty.

In contracts for design services, the client frequently asserts conditions the nonoccurrence of which gives the client power to end the relationship and, depending on the language of the condition, precludes the design professional from being compensated for services rendered before termination.

Sometimes the client asserts that a condition was created though not expressed in the written contract (discussed in Section 12.03C, dealing with costs). Generally, the client can testify as to an oral condition so long as the condition does not directly contradict the written contract or unless the written contract is clearly the final and complete repository of the entire agreement. Even if the condition is expressed in the written agreement, problems of reconciling it with other contractual provisions may exist. *Parsons v. Bristol Development Co.*[129] is instructive in this regard. The contract provided that a condition precedent to payment by the owner was the owner's obtaining a loan to finance the project. The architect commenced work and was paid part of his fee. He then began to draft

final plans and specifications for the building. However, the owner was unable to obtain a construction loan and abandoned the project.

The owner pointed to the condition in the contract as a basis for its refusal to pay for any unpaid services that the architect had rendered. The architect pointed to language in the contract stating that if any work designed by the architect is abandoned, the architect "is to be paid forthwith to the extent that his services have been rendered." Such language frequently appears in documents published by the AIA and is known as a "savings clause." (In B101-2007, see Sections 9.3 and 9.6.) However, the court interpreted the two apparently inconsistent provisions in favor of the more specific one dealing with the condition of obtaining a construction loan.

The case demonstrates the tendency of courts to construe language against the parties that supplied the language and to protect clients from having to pay for design services when the project is abandoned. Such client protection is based on the understandable reluctance of the client to pay for services that it ultimately does not use. But professionals generally expect to be paid for their work. Fee risks can be taken, but any conclusion that the risk was taken should be supported by strong evidence that the design professional assumed the risk.

Courts facing claims of finance conditions often come to different results depending on the language, the surrounding circumstances, and judicial attitude toward outcomes that either deny any payment for work performed or force a party to pay for services it cannot use.[130]

D. Suspension

Suspension of performance in contracts for design and construction is not uncommon. The owner may wish to suspend performance of either design professional or contractor if it is having financial problems or it wishes to rethink the wisdom of the project. Those who perform services in exchange for money, such as the design professional and the contractor, may wish to have the power to suspend performance if they are not paid. Suspending performance can provide a powerful weapon to obtain

---

[129]62 Cal.2d 861, 402 P.2d 839, 44 Cal.Rptr. 767 (1965).

[130]Compare *Campisano v. Phillips*, 26 Ariz.App. 174, 547 P.2d 26 (1976) (architect assumed risk), with *Vrla v. Western Mortgage* Co., 263 Or. 421, 502 P.2d 593 (1972) (architect recovered).

payment for work that has been performed. Also, suspending performance reduces the risk that further performance will go uncompensated.

The common law did not develop clear rules that allowed suspension. The performing party either must continue performing or under proper circumstances could discharge its obligation to perform.

An interim remedy—suspension—was not well recognized until states adopted the Uniform Commercial Code, beginning in the early 1950s. Section 2-609 of the code allowed a party to a contract involving goods to demand assurance under certain circumstances and withhold its performance until reasonable assurance was provided. This doctrine has carried over to contracts that do not fall within the jurisdiction of the Uniform Commercial Code, such as those that involve design or construction.[131]

Contracts for design services frequently include provisions granting the client the power to suspend the performance of the design professional. Usually such clauses give the design professional an immediate right to compensation for past services and a provision stating that suspension will become a termination if it continues beyond a certain period. If the design professional has enough bargaining power, she will seek to include a provision stating that she can suspend if she is not paid and that if suspension continues for a designated period, she can terminate her performance under the contract. Because of the uncertainty of the common law as to suspension, it is desirable to include specific provisions granting this power to either party.

Document B101-2007, Section 9.2, gives the owner the power to suspend the architect's performance. If suspension continues for more than thirty consecutive days, the architect is paid for past work. If work is resumed, she is paid for "expenses incurred during the interruption and resumption." Also, fees and schedules are "equitably adjusted." Under Section 9.3, suspension for over ninety consecutive days gives the architect the power to terminate.

B101-2007, Section 9.1, gives the architect the power to suspend or terminate if the owner fails to make payment. The election to suspend is effective after seven days' written notice to the owner. The seven-day period allows for cure by the owner and continuation of performance. Before resuming services, she must be paid "all sums due." Fees and time are "equitably adjusted."

E. Abandonment

Because of the uncertainties inherent in the construction process, project abandonment is not rare. When it occurs, some clients refuse to pay for services that have been performed by their design professional. Sometimes, as seen in the *Bristol* case described in Section 12.13C, the client points to language in the contract or an oral agreement under which the design professional forfeits her compensation in case of project abandonment.

In the absence of any contract provision giving the owner the power to abandon or any common law power to abandon based on changed circumstances or frustration of purpose, abandonment by the client is a breach of contract. The design professional is entitled to recover for the work she has performed and to any profit that would have been earned had the work been performed, if that can be established. See Section 27.02E. For that reason, contracts usually deal with abandonment, the client usually wishing to reduce its exposure if it does abandon and the design professional wishing to protect her right to be paid for work performed if there is an abandonment.

Document B101-2007 eliminated abandonment in favor of more detailed coverage of suspension and insertion of Section 9.5, which gives the owner the power to terminate for its own convenience.

F. Termination Clauses

Contracts frequently contain provisions under which one or both parties can terminate their contractual obligations to perform. Termination does not necessarily—nor usually—extinguish any claim either party may have for the other's failure to perform. Termination clauses vary. Some provide that one party can terminate if the other commits a serious breach of the contract. Some allow termination powers for any breach, a method intended to foreclose any inquiry into the seriousness of the breach. However, such a power can be abused. For that reason, careful judicial inquiry is likely to be made into the way in which the contract was made. If such a clause would operate unfairly, at the very least it will be interpreted against the stronger party to the contract and may be found unenforceable.

Some contracts provide that either party can terminate by giving a specified notice without any need to show that the other party has breached the contract. This

---

[131]Restatement (Second) of Contracts, § 237, illustration 1, §§ 251, 252 (1981).

codifies what the common law called "contracts at will," under which the contract regulates rights and duties of the parties only as long as both wish to continue. For a close, confidential relationship, such as one created when a design professional is retained by a client, such a provision may be reasonable.

However, AIA Document B101-2007, Section 9.4, is a "default" termination clause requiring a substantial failure to perform through no fault of the terminating party. As shown in Section 12.13H, the owner's failure to make payments is considered substantial nonperformance and cause for termination. Section 9.5 allows the owner to terminate for its convenience.[132]

Remedies for breach of contract will be discussed in Section 12.14C. Generally the law allows the parties to provide contractual remedies for breach of contract. AIA Document B101-2007 provides for remedies in the event of a breach by the owner. Section 9.6 gives the architect payment for services performed and not yet paid for and reimbursement for certain expenses incurred prior to termination. It also allows termination expenses set forth in Section 9.7. These include expenses for which the architect has not been compensated and profit on services not performed.

Termination of a contract can result from a material breach without any specific power to terminate. Although the same result may not be reached in every jurisdiction, it is likely that a contractual termination clause will not be exclusive and will not limit any common law right to terminate.[133]

In that regard, note that AIA Document A201-2007, Section 13.4.1 states that remedies provided in the contract do not supplant or preempt common law remedies. There is no comparable provision in B101-2007. This could mean that termination remedies provided by contract in B101-2007 *do* preempt common law remedies. Usually, though, preemption requires clear language to that effect. But the expansion of remedies seen in B101-2007, Section 9.7, seems to finesse the preemption possibility.

Default termination clauses often specify that the breaching party be allowed time to cure any default. As this opportunity to cure arises more frequently in construction contracts, it is discussed in Section 33.03F.

Termination clauses often require written notice of termination. For example, AIA Document B101-2007, Section 9.4, allows termination "upon not less than seven days' written notice." Although it is clear that termination does not actually become effective until expiration of the notice period, it is not always clear what the rights and duties of the contracting parties are during the period between receipt of notice and the effective date of termination. This may depend on the purpose of the notice.

The notice period can serve as a cooling-off device. Termination is a serious step for both parties. If it is ultimately determined that there were insufficient grounds for termination, the terminating party has committed a serious and costly breach. A short notice period can enable the party who has terminated to obtain legal advice and to rethink its position.

If the notice period provides a cooling-off period, performance should continue during the notice period but cease when the notice period expires. Only if the termination is retracted during the notice period should the parties continue performance after the effective date of termination.

If the right to terminate requires a contract breach, notice can have an additional function. It may be designed to give the breaching party time to cure past defaults and provide assurances that there will be no future defaults. If cure is the function of the notice period, actual termination should occur only if the defaults are not cured by the expiration of the notice period. During the notice period, the parties should continue performance. If it appears there is no reasonable likelihood that past defaults can be cured and reasonable assurances given, performance by the defaulting party should continue only at the option of the party terminating the contract. The latter should not be forced to receive and perhaps pay for substandard performance.

Probably the principal purpose of a notice is to wind down the work to allow the parties to plan new arrangements made necessary by the termination. A short continuation period can avoid a costly shutdown of the project or the unavoidable expenses that can result if the design professional must stop performance immediately. The notice period can enable the client to obtain a successor design professional while retaining the original

---

[132]This is dealt with in a construction contract context in Section 33.03B.

[133]*North Harris County Junior College Dist. v. Fleetwood Constr. Co.*, 604 S.W.2d 247 (Tex.Ct.App.1980) (dictum). Cf. *Glantz Contracting Co. v. General Elec.*, 379 So.2d 912 (Miss.1980).

professional for a short period. It can also enable the design professional to make workforce adjustments, get employees back to home base, cancel arrangements made with third parties, and allow time to line up new work for employees.

If making adjustments is the principal reason for the notice period, each party should be able to continue performing during the notice period. However, if relations have so deteriorated that continued performance would likely mean deliberately poor performance by either or both parties during the notice period, neither should be compelled to perform during the notice period. Certainly work that cannot be finished before the effective date of termination should not be begun.

Contracting parties should decide in advance what function the notice period is to serve and what will be the rights and duties of the parties during this period. Once this determination is made, the contract should reflect the common understanding of the parties, and any standardized contracts should be modified accordingly.

G. Material Breach

Commission of a material (that is, serious) breach empowers the other party to terminate the contract. Whether a breach is material will depend on a number of factors discussed in Section 33.04A. Generally, a material breach by the client is an unexcused and persistent failure to pay compensation or cooperate in creating the design. A material breach by the design professional is likely to be negligent performance or excessive delays. As in construction contracts, termination of contracts to perform design services—the principal problem involving client abandonment of the project—is relatively rare.

H. Subsequent Events

Contract law normally places the risk of performance being more difficult or expensive than planned, on the party promising performance. But sometimes after the contract is made events go far beyond the assumptions of the parties at the time of contract making. If so, the law will relieve a party who is affected by these events unless the contract clearly allocates this risk to the performing party. This is more commonly a problem in construction contracts and is dealt with in Section 23.05A. However, the highly personal nature of the performance of design

services makes it important to discuss one problem that infrequently surfaces in the performance of construction contracts, as follows.

Suppose a key person is no longer available to perform professional services. Is that person's continued availability so important that her inability to perform—because of either disability, death, or an employment change—will terminate the contract? The issue usually arises if the client wishes to terminate its obligation because a key design person is no longer available. That key person can be the sole proprietor, a partner, or an important employee of a partnership or professional corporation.

Contract obligations generally continue despite the death, disability, or unavailability of people who are expected to perform. Only in clear cases of highly personal services will performance be excused.

Unavailability of key design people can frustrate contract expectations. For example, in the absence of a contrary contractual provision, the death of a design professional who is a party to the contract, such as a sole proprietor or partner, will terminate the obligation of each party. The personal performance of that particular design professional was very likely a fundamental assumption on which the contract was made. A successor to the design professional can, of course, offer to continue performance, and this may be acceptable to the client. However, continuation depends on the consent of both successor and owner. Without agreement, each party is relieved from further performance obligations.

Suppose the person expected to actually perform design services is an employee of a large partnership or professional corporation. That person's unavailability may still release each party, but it would take stronger showing that the unavailable design professional was crucial to the project and that her continued performance was a fundamental assumption on which the contract was made.

The parties should consider the effect of the unavailability of key design personnel and include a provision that states clearly whether the contract continues if that person dies, becomes disabled, or for any other reason becomes unavailable.

AIA Document B101-2007, Section 10.3, binds owner and architect and their successors to the contract. Under this obscure language it appears that the parties contemplate successors stepping in if for some reason a contracting party, such as the architect, can no longer perform. This appears to require that the client continue dealing with the partnership if the partner with whom the client had originally dealt dies, becomes disabled, or leaves the partnership.

Continuity may be desirable. However, the close relationship required between design professional and client may mean that the client does not wish to continue using the partnership if the person in whom it had confidence and with whom it dealt is no longer available. Similarly, a successor may not want to work for the client. Specific language should be included dealing with this issue.

I. The Lender's Perspective

In the course of constructing the project, the owner may have difficulty paying its construction loan. If so, two events can occur. First, the owner defaults and the lender may decide to take over the project to salvage something from the financial disaster. Second, cutting off the money flow may give the design professional, usually the architect, the power to terminate her obligation to perform further under the contract.

It may be important to the lender to keep the architect on the project. The architect may no longer wish to continue performing or may believe she is entitled to additional compensation. Even if the lender can persuade the architect to continue and even if the renegotiation is successfully completed, costly delay is likely to result.

The lender would prefer as part of the takeover to receive an assignment of the owner's rights to the architect's services without the need for renegotiation. However, either the law or a contractual provision barring assignment without the architect's consent may make this course of action unavailable.

In 1997 the AIA sought to meet these lender concerns. It added to its general nonassignment clauses, B141-1997, Paragraph 1.3.7.9, an exception for an "institutional lender providing financing for the Project." This exception was continued in B101-2007, Section 10.3, "if the lender agrees to assume the Owner's rights and obligations under this Agreement." Such a lender can take an assignment by the owner of the architect's contract without the consent of the architect. In such a case the lender agrees to assume all the owner's obligations under the contract. A similar provision in the construction contract is found in A201-2007, Section 13.2.2. Neither provision makes clear that the assignee (the lender) must pay any amounts owed the architect or the contractor. Although the lender must

agree to assume the obligations of the owner, it is not clear whether this covers only future obligations but not those that arose in the past. It is likely that neither the architect nor the contractor will work for the lender without being paid unpaid bills.

SECTION 12.14 Judicial Remedy for Breach: Special Problems of the Client–Design Professional Relationship

A. Coverage

Basic judicial remedies for contract breach were discussed in Chapter 6. Chapter 27 discusses claims in the context of the construction contract. This section applies basic legal doctrines to special problems found in the relationship between client and design professional.

B. Client Claims

The principal claims that clients make against professionals relate to defective design. A breach of contract by the design professional entitles the client to protect its restitution, reliance, and expectation interests. (Refer to Section 6.05.)

Although clients occasionally seek to protect their restitutionary interest by demanding return of any payments made, the principal problem relates to the client's expectation interest. If the project is designed defectively, the client is entitled to be put in the same position it would have been had the design professional prepared a proper design. The first issue that can arise is whether the client can measure its expectation loss by proving the cost of correcting the defective work or whether it is limited to the difference between the project as it should have been designed and the project as it was designed. Claims against contractors that involve this issue are discussed in Sections 22.06B and 27.03D.

This issue can be demonstrated by looking at *Bayuk v. Edson*.[134] In this case, the owner complained about faulty design consisting of, among other things, an improperly designed floor, closets too small, outside doors constructed for a milder climate than where the house was built and of an unusual type that could not be constructed by artisans

---

[134]236 Cal.App.2d 309, 46 Cal.Rptr. 49 (1965).

in the area, unaesthetic kitchen tile, sliding doors that did not fit properly in their tracks, and a fireplace that became permanently cracked.

A number of witnesses testified that it would not have made economic sense to repair the defects. One witness testified that tearing out and repairing would cost more than the cost of rebuilding the house in its entirety. The plaintiff produced an expert real estate appraiser who fixed the value of the house without the defects at $50,000 to $60,000 and with the defects at $27,500 to $31,500. The trial court awarded a judgment of $18,500, the least of the possible remainders. This was affirmed by the appellate court.

Suppose, however, that it would not have been economically wasteful to correct the defective work. This would entitle the owner to the cost of correction. In claims against a design professional for improper design, application of this standard involves the betterment rule, based on the cost of correction sometimes unjustly enriching the owner. For example, in *St. Joseph Hospital v. Corbetta Construction Co., Inc.*,[135] the hospital sued its architect, the contractor, and the supplier of wall paneling that had been installed when it was disclosed that the wall paneling had a flame spread rating some seventeen times the maximum permitted under the Chicago Building Code.

After the hospital had been substantially completed, it was advised that it could not receive a license because of the improper wall paneling. The city threatened criminal action against the hospital for operating without a license. The hospital removed the paneling and installed paneling that met code standards. The jury awarded $300,000 for removal of the original paneling and its replacement by code-complying paneling and an additional $20,000 for architectural services performed in connection with removal and replacement.

In reviewing the jury award of $320,000, the appellate court noted that had the architect complied with his obligation to specify wall paneling that would have met code standards, the construction contract price for both paneling and cost of installation would have been substantially higher. The court stated that the hospital should not receive a windfall of the more expensive paneling for a contract price that assumed less expensive paneling.

The paneling that should have been specified together with installation would have cost $186,000, whereas the paneling specified with installation cost $91,000. This, according to the court, should have reduced the judgment by $95,000. The court reduced the award an additional $21,000 for items installed when the panels were replaced that were not called for under the original contract. As a result, the judgment was reduced some $116,000.

That awarding the full cost of correction would put the client in a better position than it would have been had there been full performance is an affirmative defense that must be established by the design professional. The design professional must show that the owner would have proceeded with the project at the same site and under the same conditions regardless of the increased cost.

A variation of the betterment rule is one sometimes called the *extended life rule* or *added value*. For example, if a roof must be replaced after the repairs turn out to be ineffective, the client has received a new roof instead of a roof that would have used up some of its useful life value. If the design professional can prove that the client now has an asset with an extended life value, this can be deducted from the damages.

Under either the betterment or the extended life rule, the client is entitled to any additional cost incurred because the corrected work has taken place after the initial work has been performed if the price for the cost of correction has increased.

Although the betterment and extended life rules have a logical attractiveness, they are not always applied, particularly if the party who must pay for the cost of correction is a consumer or a client of limited financial capacity. In addition, any close questions as to whether the work would have been done and cost incurred originally or as to the extended life value may be resolved in favor of such a client. Similarly, the law sometimes offsets the cost to the client of having to correct a defective condition many years before it would have had to correct this condition had there not originally been a design defect.

Another claim sometimes made by clients is unexcused delay in preparing the design or performing administrative work during construction. Delay can harm the contractor, and most delay disputes are between an owner and a contractor. Claims for delay during the design phase are difficult to establish because of the likelihood of multiple causes. But suppose the project is completed late because of negligence by the architect in passing on submittals of

---

[135]21 Ill.App.3d 925, 316 N.E.2d 51 (1974). See also *Lochrane Eng'g, Inc. v. Willingham Realgrowth Investment Fund, Ltd.*, 552 So.2d 228 (Fla.Dist.Ct.App.1989), rev. denied, 563 So.2d 631 (Fla.1990); Bales, O'Meara & Azman, *The "Betterment" or Added Benefit Defense*, 26 Constr. Lawyer, No. 2, Spring 2006, p. 14.

the contractor. There are two main damage items. First, the owner will lose the use of its project for the period of unexcused delay. Second, the contractor may make a claim against the owner or design professional for any loss suffered because of delay wrongfully caused by the design professional. (The second claim is discussed in Section 26.10.)

The first damage item, that of loss of use caused by the delay, was before the court in *Miami Heart Institute v. Heery Architects & Engineers, Inc.*[136] This case demonstrates the liability exposure of design professionals when they have delayed completion of a project without justification. In the case, the architect had agreed to design plans for the building of a new hospital structure that would house patients. During construction, the patients were housed in older buildings on the premises. The architect's failure to comply with code requirements caused a ten-month delay in the issuance of a certificate of occupancy.[137]

The court rejected the contention by the architect that damages for loss of use can be recovered only when there is a delay by the contractor. It also rejected the contention that lost use value can be recovered only if the project itself was to be rented to others. The court concluded that the proper measure of recovery should be measured by the reasonable rental value of the structure during the period of delay.

The court was faced with the fact that the patients who would have been in the new structure were housed in the old structure for the period of the delay. With this in mind, the court concluded that the proper measure of recovery was the difference between the reasonable rental value of the new structure and that of the old structure for the period of delay. The patients were undoubtedly inconvenienced in still being housed in the old structure, and the quality of the medical services may have been poorer than if they had been in the new structure. Yet to establish the actual economic value of such losses would be almost impossible. Of course, some of this may be factored into the reasonable rental value amounts. Clearly, this is one case where it would have been helpful to have

had liquidated damages—(discussed in Section 26.09B, dealing with unexcused delay by the contractor).[138]

There were other claimed damages. The court held that the betterment rule would prevent the owner from recovering for additional expenses incurred to make the building meet code requirements. It noted that the plans called for 2 feet of electrical wire, but proper plans would have required 12 feet of wire. But the owner's expense in purchasing the additional 10 feet cannot be chargeable to the architect. The owner would have incurred this expense regardless of the architect's defective plans.

Finally, the court concluded that the owner could also recover any delay damage claims it paid to the contractors and subcontractors as a result of the architect's breach.

Delay can also cause less direct, or what are sometimes called *consequential*, damages. For example, delay in completing an industrial plant may generate losses caused by the inability of the plant owner to fill the orders of its customers, exposing the plant owner to claims by the customers and causing it to lose profits on the transactions. (Such losses were discussed in Section 6.06.)

AIA Document B101-2007, Section 8.1.3, provides that each party "waives consequential damages . . . due to either party's termination." This is also included in A201-2007, Section 15.1.6. Since Section 15.1.6 is more complex and involved greater controversy, detailed discussion will be postponed until Section 27.06. But it should be noted that in the *Miami Heart Institute* case damages for lost use would not be recoverable nor would the losses caused by delay in completing the industrial plant were such a waiver clause in the contract.

As noted in Section 15.03C, design professionals faced with such losses and having sufficient bargaining power may seek to deal with this problem by limiting the liability of the design professional in such a way as to preclude consequential damages from being recoverable. Clients with strong bargaining power sometimes insist that contracts contain provisions stating that the design professional will be liable for consequential damages. In the absence

---

[136]765 F.Supp. 1083 (S.D.Fla.1991), aff'd, 44 F.3d 1007 (11th Cir.1994).

[137]The court's opinion in this case was in response to a preliminary dispute relating to the measure of recovery. The opinion made the assumption that the owner can establish that the architect was liable. This had not yet been established.

[138]See *E.C. Ernst, Inc. v. Manhattan Constr. Co. of Texas*, 387 F. Supp. 1001 (S.D.Ala.1974) aff'd in part, vacated in part, 551 F.2d 1026, reh'g granted in part and opinion modified, 559 F.2d 268 (5th Cir.1977), cert. denied sub nom. *Providence Hosp. v. Manhattan Constr. Co. of Texas*, 434 U.S. 1067 (1978) (court measured damages for delay by the liquidated damages clause in the contract between the owner and the contractor, the court noting that the architect had participated in selecting the liquidated damages amount).

of such a provision, the law would determine whether consequential damages can be recovered. The inclusion of such a provision will create greater liability exposure for the design professional.

Suppose the client claims its design professional exceeded her authority in ordering changes in the work or accepting defective work without authority of the client. As this usually involves claims by the contractor as well, this topic and the measure of recovery for a valid claim are discussed in Section 21.04D.

Whether a client can invoke state consumer protection legislation with its expanded remedies is discussed in Section 23.02J.

C. Design Professional Claims

The principal claims made by a design professional against the client relate to the latter's failure to pay for services performed. These claims have not raised difficult valuation questions. Design professionals commonly seek to protect their restitution interests and recover the reasonable value of their services.[139] Occasionally, clients have resisted this claim by contending that they did not use the plans and specifications drafted and thereby have not been enriched. Such defenses have been generally unsuccessful.[140] (The more difficult problem—the problem of a contractor who seeks to protect its restitution interest when the owner has breached—is discussed in Section 27.02E.)

Suppose the design professional seeks to protect her expectation interest. To do so, she must be put in the position she would have been had the client performed as promised. This could be computed by what she could have earned had she fully performed less the expense she has saved or her expenditures in part performance plus her profit, less what she has already been paid.

---

[139]*Getzschman v. Miller Chemical Co.*, supra note 10.

[140]*Barnes v. Lozoff*, 20 Wis. 2d 644, 123 N.W.2d 543 (1963) (measured by rate of pay in community).

CHAPTER THIRTEEN

Compensation and Other Owner Obligations

SECTION 13.01 Contractual Fee Arrangements

A. Limited Role of Law: Methods of Compensation

A number of different fee arrangements are available to design professionals and owners. The choice among these arrangements is principally guided by criteria that are professional, not legal. Put another way, design professionals are generally in a better position to determine the type of fee structure than are their attorneys. Nevertheless, the law does play a limited role.

Principally, the law interprets any contractual terms that bear on fee computation when the contracting parties disagree. If no fee arrangement is specified in the contract and the parties cannot determine an agreed-on fee subsequent to performance, the law may be called on to make this decision.

Despite this limited role, certain legal principles must be taken into account in choosing a fee arrangement or in predicting the legal result if a dispute arises. For example, faced with the question of whether certain services come within the basic design fee, the law may choose to protect the reasonable expectation of the client if the design professional selected the language.

Likewise, any fee arrangement that measures compensation by a stated percentage of construction cost may be interpreted to favor the client. This can result from the belief that such a fee formula can be unfair to the client, given the design professional's incentive to run up costs. In rare cases, the contractual method selected will be disregarded because supervening events occur that neither party contemplated.

In its instructions to B101-2007, the AIA states there are at least ten ways to compute compensation for architectural services. Four are based on cost and time:

1. multiple of direct salary expense
2. multiple of direct personnel expense
3. professional fee plus expenses
4. hourly billing rates

In the first the multiple represents benefits, overhead, and profit. In the second the multiple represents overhead and profit. In the third salaries, benefits, and overhead are expenses and the fee represents profit. In the fourth, salaries, benefits, overhead, and profit are included in the hourly rates.

The other methods, though indirectly related to time and expenses spent, are related more directly to the project costs or attributes of the project. They are

1. stipulated sum
2. percentage of the cost of the work
3. multiple of consultants' billing
4. square footage
5. unit cost (apartment, rooms, acres, etc., times a price factor)
6. royalty (share of owner income or profit)

These methods may be used in combination. This chapter comments on those used most often.

B. Stated Percentage of Construction Costs: Square Footage

Although no longer universal (fixed-fee and cost methods of compensation being used increasingly), the stated percentage of construction costs is still a common method of fee computation. In such a method, the fee is determined by multiplying the construction costs by a designated percentage set forth in the contract. (A close variant of this method used in real estate residential developments is a stated dollar amount per square foot of the residence.) Because the stated percentage of construction cost typically covers only basic design services, this amount is often augmented by payments for additional services and reimbursable expenses. (However, see Section 13.01I on fee limits.)

This method has been criticized. It can be a disincentive to cut costs and may reward the design professional who is less cost conscious. It can be too rigid, because projects and time spent can vary considerably. It may not reflect time spent. It also tends to subsidize the inefficient client at the expense of the efficient one.

Despite constant criticism, the fee method is still used. Clients seem accustomed to it. The method can avoid bargaining over fee. In normal projects it may accurately reflect the work performed. Although it may undercompensate on some projects and overcompensate on others, some design professionals feel the fees average out. The fee method can avoid extensive recordkeeping. It also is much less likely to generate a client demand to examine the design professional's records, a common feature of a cost type of fee formula.

Percentages vary, with the figure selected in major part reflecting the amount of work the design professional must perform and, increasingly these days, the risk of failure. Such risk is addressed in greater detail in Chapter 14. The amount of work depends on a number of factors.

If the design professional has worked for the client before, past experience may be relevant in determining the stated percentage. An inexperienced, inefficient client may require more work than an efficient client who has dealt with construction before. Sometimes the percentage is based on whether the project is residential or commercial, whether the construction contracts are single or separate, and whether the construction contract price is fixed or a cost type. Smaller projects may have a minimum fee.

To avoid criticism that the fee method encourages high costs and discourages cost reduction, some design professionals use a flexible percentage. One method is a percentage that declines as the costs increase. The actual percentage for the entire project is determined from a schedule that has variable percentages depending on the ultimate construction cost. For example, a fee schedule may provide for a 5 percent fee if the costs do not exceed $1 million, a fee of 4 percent if the ultimate cost is between $1 million and $1.5 million, and 3 percent if the cost is over $1.5 million.

Another method to encourage cost consciousness is to employ a sliding scale under which the highest percentage is applied to a cost up to a specified amount and then the percentage reduces on succeeding amounts. For example, the fee can be 8 percent on the first $1 million of cost, 7 percent on the next $4 million, and 6 percent on all amounts over $5 million.

As to the construction cost that is multiplied by the percentage, see AIA Document B101-2007, Section 6.1, in Appendix A.

Suppose responsibility for an increase in construction cost is shared by the design professional and the contractor, based on the contractor's not having directed attention to obvious design errors. Apportionment of responsibility may be appropriate, with the design professional's fee increased by that portion of the corrected work cost chargeable to the contractor. However, the difficulty of making such allocation, and primary responsibility for design being that of the design professional, will likely preclude such an apportionment. The major issue is likely to be who bears responsibility for the cost of correction, the design professional or the contractor. If an apportionment is made for this purpose, that apportionment can be used to determine the design professional's fee.

Courts have interpreted fee provisions, but because of the different provisions that can be or are employed, generalizations are perilous. In close cases, courts are likely to favor the position of the client if the design professional selected the contract language.

Techniques for imparting flexibility to what otherwise can be a rigid fee method are mentioned in Section 13.01J.

C. Multiple of Direct Personnel Expense: Daily or Hourly Rates

Personnel multipliers determine the fee for basic and additional services by multiplying direct personnel expense by a designated multiple ranging from 2 to 4, the average being 2.5 to 2.7. AIA Document B141-1997, Paragraph 1.3.9.4, defined direct personnel expense as "the direct salaries of the Architect's personnel engaged on the Project and the portion of the cost of their mandatory and customary contributions and benefits related thereto, such as employment taxes and other statutory employee benefits, insurance, sick leave, holidays, vacations, employee retirement plans, and similar contributions." Once fringe benefits were truly on the "fringe." Today, they can constitute as much as 25 to 40 percent of the total employee cost. The contract must clearly specify personnel compensation cost beyond the actual salaries or wages.

In AIA Document B101-2007 this paragraph has been deleted. The AIA believes that this method is less significant. In its Instructions to B101-2007, it merely defines Multiple of Direct Personnel Expense as "salaries plus benefits . . . multiplied by a factor representing overhead and profits." Multiples are still common in contracts for engineering services.

There are obvious disadvantages to daily or hourly rates. Because a day is a more imprecise measurement than an hour, if either method is used it is likely to be the hourly rate. Such a method requires detailed cost records that set forth the following:

1. the exact amount of time spent
2. the precise project on which the work was performed
3. the exact nature of the work
4. who did the work

Different hourly rates may be used for work by personnel of different skills.

When compensation is based on cost incurred by the design professional, the client often prescribes the records that must be kept, how long they must be kept, and that they be made available to the client. The B141 published in 1977 stated that the architect would keep records based on GAAP (generally accepted accounting principles).[1]

B101-2007 does not prescribe accounting standards for records that must be kept. Many owners will insist on contract language that requires the architect to keep any relevant cost records based on GAAP standards, as in earlier editions of AIA Documents. They are also likely to require that language deal with where records will be maintained, the power to inspect and make copies, and the period of record retention.

B101-2007, Section 11.10.4, simply provides that "Records of Reimbursable Expenses, expenses pertaining to Additional Services and services performed on the basis of hourly rates shall be available to the Owner at mutually convenient times."

D. Professional Fee Plus Expenses

The form of fee arrangement known as "professional fee plus expenses" is analogous to the cost type of contracts discussed in greater detail in Section 17.02B. One advantage of a cost type of contract for design services is that the compensation is not tied to actual construction costs and there should be incentive to reduce construction costs. It can be a disincentive, however, to reduce the cost of design services.

The cost type of contract, or what the AIA calls "professional fee plus expenses," necessitates careful definition of recoverable costs. Costs, direct and indirect, can be an accounting nightmare. Disputes can arise over whether certain costs were excessive or necessary. In cost contracts, advance client approval can be required on the size of the design professional staff, salaries, and other important cost factors. Cost contracts require detailed recordkeeping. (As noted in Section 13.01I, a ceiling can be placed on costs.)

Suppose the design professional estimates what the costs are likely to be in such a contract. Although the client may wish to know approximately how much the design services are likely to cost, an estimate can easily become a cost ceiling. If an estimate is given that is not intended as a ceiling, it should be accompanied by language that indicates the assumptions on which the estimate is based and that it is not a fixed ceiling or a promise that design costs will not exceed a designated amount.

E. Fixed Fee

Design professional and client can agree that compensation will be a fixed fee determined in advance and incorporated in the contract. Before such a method is employed, a design professional should have a clear idea of direct cost, overhead, and profit as well as appreciate the possibility that contingencies may arise that will affect performance costs. A fixed fee should be used only where the scope of design services is clearly defined and the construction project well planned. It works best in repetitive work for the same client.

---

[1]For a criticism of this gap and a listing of the ways this topic is dealt with in other contracts, see J. SWEET & J. SWEET, 1 SWEET ON CONSTRUCTION INDUSTRY CONTRACTS § 6.06 (4th ed. 1999), Aspen Law & Business.

Does the fixed fee cover only basic design services? Does it include additional services and reimbursables? Standard contracts published by professional associations usually limit the fixed fee to basic design services. A design professional who intends to limit fixed fees to basic services should make this clear to the client. See Section 13.01I for discussion of fee provisions that place an absolute ceiling on compensation.

A recent study indicates that there has been a significant rise in the number of design professionals using what it calls "lump-sum/fixed-fee contracts" between 1998 and 2002.[2] In all markets, except industrial facilities and private environmental, the percentages of those markets using these contracts rose from 48 percent in 1998 to 68 percent in 2002.

This increase reflects the greater desire on the part of owners and those providing financial resources for price certainty "up-front" and a great fear of uncertainty caused by cost overruns and claims. This also is reflected in the use of no-damage clauses (see Section 26.10A), saddling design/build contractors with responsibility for cost overruns due to owner mistakes (see Sections 17.04F and I) and the greater use of caps on fees of design professionals (see Section 13.01I). While cogent arguments can be made for not placing these risks on contractors and architects, the seemingly insatiable need by some for "front-end" certainty can trump them.

A recent unreported federal district court case demonstrates the risk of such a contract made by a project manager (PM). It gave a fixed price without any provisions for additional services. Despite an eighteen month delay in completion and more delay after completion when a flood resulted from a burst sprinkler head, the PM received no additional compensation.[3]

F. Reasonable Value of Services or a Fee to Be Agreed On

The fee will be the reasonable value of the services where the parties do not agree on a compensation method. If there is no agreed valuation method for additional services, compensation is the reasonable value of the services. The reasonable value of a design professional's services will take into account the nature of the work, the degree of risk to

---

[2]Zweig White, *2002 Fee & Billing Survey of A/E/P & Environmental Consulting Firms.*

[3]*Plante & Moran Cresa, L.L.C. v. Kappa Enterprises, L.L.C.,* 2006 Westlaw 1676411 (E.D.Mich.2006).

---

the design professional, the novelty of the work, the hours performed, the experience and training of the design professional, and any other factors that bear on the value of these services, including overhead and a reasonable profit. Proving the reasonable value of services requires detailed cost records. Leaving the fee open is generally inadvisable.

Where this issue does arise, each party usually introduces evidence of customary charges made by other design professionals in the locality as well as evidence that bears on factors outlined in the preceding paragraph. In many cases, great variation exists among the testimony given by the expert witnesses for each party. It is not unusual for the court or jury to make a determination that falls somewhere in between.

The parties can agree to jointly determine the fee at the completion of performance. Where the project has gone well, where the parties wish to work with each other again, and where adequate records have been kept, agreement on fees may be reached easily. However, when such fortunate events have not occurred, an agreement on fees may prove difficult.

At one time, as noted in Section 5.06F, such agreements were considered unenforceable as simply "agreements to agree." However, if the work has been performed, it is likely that the parties must negotiate in good faith to determine a fee or, more likely, to determine what the parties would have agreed to had they bargained in good faith and made an agreement on compensation. The same result can follow if the parties cannot agree and the matter is submitted to arbitration.

G. Additional Services

The scope of services is a central component of the design professional–client relationship. What does the contract require as to scope of services, sometimes called basic services, performed by the design professional and encompassed by the basic fee? Do these match the expectations of design professional and client? These issues are at the heart of the concept of additional services.

Sections 12.01 and 12.02 referred to the distinctions between basic and additional services. The importance of this distinction depends on the compensation system used. If the system is tied to costs incurred, such as multiples of direct personnel expenses, a daily rate, or an hourly rate, the distinction is less important. When those systems are used, the only issues are whether the work was required under the contract and the expenses incurred.

But if the system used is a percentage of construction cost or per-unit formula, the question of basic versus additional becomes a central issue and often a contentious one. The client may believe that the services in question are part of basic services. If there are too many additional services that require compensation in addition to the basic fee, the client may believe that it is not getting much for the basic fee and that the basic fee is only a "sticker" price unrelated to the ultimate payout. Finally, as shall be seen, procedures required by additional compensation clauses, such as written orders and notices, are often ignored by the parties. This generates claims by the design professional that these formal requirements were met or waived by the client.[4]

If the services are part of basic services under compensation systems, such as a percentage of construction cost or a fixed fee, the design professional receives no additional compensation. These services are compensated under the basic fee. But if the services are not part of basic services, if authorized, they become additional services and earn the design professional additional compensation.

The principal difficulty is generated by the increasing number of services that must be performed by someone, whether the design professional, the owner, or a consultant retained by either. One reason for the expansion of services is the increased public controls over construction, the emphasis on the environment, and the increasing likelihood of disputes that often end up in arbitration or litigation. These new services can be needed in the design or construction phase. Additionally, each project can be handled differently, depending upon the client's capacity to do them itself and the specific nature of the project. In this regard in AIA Document B101-2007, Section 4.1 lists twenty-seven specific services that can be done by the architect, done by the owner, or not be provided.

Contracts for design services generally set forth the basic services in the scope of services section. As a rule they specify services that are additional in a separate section labeled "Additional Services." Such services are listed to recognize that these listed services are often but not always performed. If ordered by the owner, they must be performed, and the design professional earns additional compensation. In that sense the list of additional services is similar to a "changes clause" in a contract for construction services. Both grant unilateral power to the owner to demand that certain services be performed. In the case of a contract for design services, they are designated as additional services. In the case of construction services, they must be within the general scope of the work.

The list of additional services in a contract for design services serves another function. Listing a service as "additional" makes clear that it is not part of basic services. Even if the contract seeks to describe basic services in the "scope" section, it is not always clear what falls within the ambit of such services.

Two cases demonstrate the effect of not having a clause detailing additional services. In a case noted in Section 13.01E, the absence of an additional services clause was pointed to by the court when it denied any additional compensation to a project manager (PM) when inordinate delay and a flood drastically increased the PM's cost of performance.[5]

Another involved a claim by the client that its architect should have alerted the client that a report by an engineer retained by the client was inadequate. It did not show underwater deficiencies in the substructure.[6]

The contract with the architect required the architect to provide normal structural services. The architect pointed out that one of the optional services that could have been performed by the architect but was not chosen by the client was to investigate existing conditions or facilities for which an extra fee would be paid.

Yet this argument was rejected in favor of the client's contention that, even though the architect had not breached his contract, he had been negligent. He should have recognized the inadequacy of the engineer's report. This conclusion was also supported based upon expert testimony presented by the client. That expert testified that the requirement to provide normal structural services included the obligation to check the adequacy of the engineer's report.

Usually the basic-versus-additional issue controls the compensation to the architect but not whether liability is transferred from the architect to the owner. Also, the architect knows what information is needed to design properly. The report did not relieve the architect from his responsibility to determine whether the information needed to

[4]*Belot v. Unified School Dist. No. 497*, 27 Kan.App.2d 367, 4 P.3d 626 (2000) (additional services claim was denied because architect failed to segregate costs of additional services from costs incurred for basic services); *Newman Marchive Partnership, Inc. v. City of Shreveport*, 944 So.2d 703 (La. App.2006), writ denied, 940 So.2d 448, 452 (La.2007) (failure to get advance authorization barred claim because this showed that the architect viewed the services as basic).

[5]*Plante & Moran Cresa, L.L.C. v. Kappa Enterprises, L.L.C.* supra note 3.

[6]*Kerry, Inc. v. Angus-Young Assocs.*, 280 Wis.2d 418, 694 N.W.2d 407, review denied, 286 Wis.2d 98, 705 N.W.2d 659 (2005).

execute the design was provided properly. Had the issue been whether the architect was entitled to additional compensation for doing work that the owner chose not to have him perform, the outcome might have been different.

Some of these difficulties in distinguishing basic from additional services are reflected in the history of AIA design service documents. Until 1958 there was no list of additional services. From 1958 to 1963 there were nine specified additional services. Steady increases culminated in twenty-two additional services in 1977. The AIA is not alone in listing many additional services. In E-500 Exhibit A (2002), the EJCDC lists twenty-two *optional* additional services that require owner's written authorization (Paragraph A2.01) and seven *required* additional services not requiring owner's written authorization (Paragraph A 2.02).

In 1987 the AIA divided additional services into nine *contingent* and twenty *optional* additional services. The latter were classic additional services, requiring a written order in advance of performance. Contingent additional services had a different mechanism, as follows. The architect would inform the owner he was going to perform these services. The need for such services had to be caused by events beyond control of the architect. Unless the owner ordered the architect *in writing* not to perform the services, he performed them. These services were contingent in the sense that they were to be performed *contingent* on the owner writing that it *did not* want them performed. This division was made to help the architect who was ordered to perform such services in writing but was unable to supply the paperwork when time came for billing. The AIA also noted that it would be rare for an owner not to want these services performed.

In B141-1997 additional services, at least in name, were eliminated. They were replaced by provisions scattered around the document that could be the basis for a bill for services in addition to the basic fee.[7]

In 2007 additional services were reorganized. Provisions that had been scattered around B141-1997 were consolidated into Article 4 entitled "Additional Services." There were some substantive changes, mostly classifying more services as additional.

Services were divided into those that were agreed upon at the time the contract for design services is made (Section 4.1) and those made during project performance (Section 4.3). Section 4.2 provides a blank space to describe additional services itemized in Section 4.1 unless they are described in an exhibit to the contract. Section 4.1 lists twenty-seven specific services (up from twenty-two) that can be performed by the architect, by the owner, or not be provided. If the architect performs these services, he is paid additional compensation.

Some services are not included in B101-2007 that had been included in B141-1997, such as land survey services (the owner provides an expanded survey, including legal limitations under Section 5.4), geotechnical services (the owner provides geotechnical engineering services under Section 5.5), economic feasibility studies, special bidding and negotiations, and construction management. If something appears to be deleted, it can, though not here, show up elsewhere.

As noted earlier in this section, listing additional services distinguishes those services from basic services. Another less apparent purpose is demonstrated when services are removed from the list of additional services, such as removal of economic feasibility services from the list of additional services in 2007. Removal makes clear that the architect need not perform these services even if requested by the owner.

The list of additional services is similar to changes in construction contracts. Just as the contractor must perform changes that are properly ordered under the changes clause, the architect must perform additional services. To be sure, he is compensated beyond the basic fee when he performs the additional services. But he is not compelled to perform services not listed as additional.

The purpose of removing specified services from the list of additional services reflects the AIA's view that the architect should not perform these services. As noted in Section 12.05, it is best for the architect not to make economic feasibility studies because they are beyond his training and experience.

Some services were added, such as environmental-related services,[8] BIM,[9] as-designed and as-constructed

---

[7] AIA Doc. B141-1997, ¶ 1.1.6 (information changes require adjustments in compensation); ¶¶ 1.3.3.1, 1.3.3.2 (circumstances beyond architect's control, such as code changes, owner changes of mind, poor administration by owner, significant project changes, testifying at hearings or dispute resolution); ¶ 1.5.9 (delay in completion of design services contract through no fault of architect); ¶ 2.8.1 (number of specified services, such as submittal reviews, site visits, inspections, exceeding the designated number specified in the contract); ¶ 2.8.2 (designated services beyond normal services, such as submittal reviews out of sequence, evaluating an extensive number of claims, evaluation of substitutions, and administrative services provided sixty days after substantial completion); ¶ 2.8.3 (twenty-two specified services [increased in 2007 to twenty-seven] ordered, much like earlier optional services).

---

[8] AIA Doc. B101-2007, §§ 4.1.2.3, 4.1.2.4.
[9] Id. at § 4.1.6.

record drawings,[10] coordination of owner consultants,[11] security design,[12] and fast-track design services.[13]

Section 4.3.1 lists eleven specific services that are provided after the agreement has been made. When the architect recognizes the need to perform these services, he must "notify the Owner with reasonable promptness and explain the facts and circumstances giving rise to the need." The architect does not perform these services until he receives "the Owner's written authorization."

Some new items found in Section 4.3.1 deal with environment-related services,[14] BIM,[15] dealing with bidding alternates,[16] preparing for and attending a variety of public hearing or meetings,[17] preparing and attending dispute hearings,[18] evaluating bidders,[19] and assisting the Initial Decision Maker (IDM) if he is not the architect.[20]

Another six services are listed under Section 4.3.2. The architect performs these services "To avoid delay in the Construction Phase." Concurrent with performance of these services, he must "notify the Owner with reasonable promptness, and explain the facts and circumstances giving rise to the need." If the owner subsequently determines that these services "are not required the Owner shall give prompt written notice to the Architect, and the Owner shall have no further obligation to compensate the Architect for those services." Because Section 4.3.2 includes "no *further* obligation" (emphasis added), services performed before notification that they not be performed are compensated as additional services.

The complete list of services in Sections 4.1 and 4.3 can be found in Appendix A. In Sections 4.3, both Sections 4.3.1, which requires advance authorization before performance, and 4.3.2, relating to services that are performed subject to not receiving notice not to perform them, contain changes from earlier documents.

Some changes expand or clarify the list of additional services. For example, Section 4.3.1.1 states that services caused by material changes in the project are not limited to "size, quality or complexity, the Owner's schedule or budget." Section 4.3.1.6 makes services that relate to alternate bids or proposal requests additional services. Similarly, Section 4.3.1.9 makes services that evaluate bidders additional services. "Assistance to the Initial Decision Maker, if other than the Architect," under Section 4.3.1.11 is additional.

Similarly, changes of this nature can be found in Section 4.3.2, the section that allows the architect to perform and then notify the owner. For example, preparing change orders or construction change directives that require evaluation of the proposal and supporting data are additional services under Section 4.3.2.3. Services that affect basic construction phase services by the architect performed sixty days after substantial completion or the "anticipated date of Substantial Completion" are additional under Section 4.3.2.6.

Section 4.3.3 sets up a system to control the frequency of performance of certain services performed by the architect. A blank is provided for limits as to the number of shop drawing reviews,[21] site visits[22] and inspections.[23] When the limit is reached, the architect must notify the owner. Services that exceed the limit are additional.

Finally, Section 4.3.4 makes services performed after a designated date (a blank is provided) not the fault of the architect additional services.

To sum up this intricate method of providing for additional services, Section 4.1 requires an agreement in advance of who is responsible for specified services. Section 4.3.1 creates a list of services that require advance authorization by the owner. Section 4.3.2 lists services that the architect can perform until ordered not to do so. Section 4.3.3 makes additional those services that exceed a designated number.

The AIA system dealing with additional services demonstrates that there can be a substantial difference between the basic fee and the ultimate fee payout where the compensation is not one based on cost. Inefficient or incompetent contractors, unforeseeable or even foreseeable events that affect performance adversely, and dithering or mind-changing owners can, under B101-2007, turn a percentage of construction cost or a fixed price contract into a cost contract.

---

[10]Id. at §§ 4.1.14, 4.1.15

[11]Id. at § 4.1.19.

[12]Id. at § 4.1.21.

[13]Id. at § 4.1.25.

[14]Id. at § 4.3.1.2.

[15]Id. at § 4.3.1.5.

[16]Id. at § 4.3.1.6.

[17]Id. at § 4.3.1.7.

[18]Id. at § 4.3.1.8.

[19]Id. at § 4.3.1.9.

[20]Id. at § 4.3.1.11.

[21]Id. at § 4.3.3.1.

[22]Id. at § 4.3.3.2.

[23]Id. at §§ 4.3.3.3 (substantial completion) and 4.3.3.4 (final completion).

H. Reimbursables

B101-2007 Section 11.8.1 illustrates reimbursable expenses. As for changes, B101-2007 makes reimbursable taxes levied on professional services and on reimbursables,[24] as well as site office expenses.[25] The catch-all "Other similar Project-related expenses"[26] makes clear that the list is not exclusive. The list always seems to grow.

Incurring obligations for the client and paying them can impose an administrative burden on the design professional. Sometimes design professionals charge the client a markup for handling reimbursables. For example, suppose the design professional incurred expenses of $1,000 for traveling in connection with the project and long-distance calls. Under a markup system, the design professional might bill the client $1,000 plus an additional 10 percent or $100, making a total of $1,100. The markup percentage can depend on the number of reimbursables and the administrative overhead incurred in handling them. A design professional who wishes to add an overhead markup should explain this to the client in advance and obtain client approval.

B101-2007, Section 11.8.2, added to the amounts incurred by the architect a blank percentage for expenses incurred, in effect a markup.

I. Fee Ceilings

Owners are often concerned about the total fee, particularly if the fee is cost based. But even in a compensation plan under which the design professional is paid a fixed fee or a percentage of compensation, the client may be concerned about additional services, reimbursables, or increases based on an unusual jump in the construction cost. Public owners with a specified appropriation for design services may seek to limit the fee to a specified amount. It may be useful to look at two cases that have dealt with fee limits.

In *Hueber Hares Glavin Partnership v. State*,[27] the contract limited the fee. It also excluded recovery for work to correct design errors. The Appellate Division held the language unambiguous and the fee limit not an estimate. The fee limit could not be exceeded by costs attributed to design errors. By dictum, the court stated that the city would have

been precluded from asserting the fee limit had it ordered extra services knowing the fee limit had been reached.

Harris County v. Howard[28] involved an AIA document. The upset price was also held to unambiguously include additional services and reimbursables. The public owner inserted a detailed recital on the fee limit and what it included. The court rejected the architect's contention that his having been paid $20,000 over the limit showed the limit did not include additional services or reimbursables.

Substantial changes in project scope should eliminate any fee ceiling that may have been established.[29]

J. Adjustment of Fee

Generally, the law places the risk that performance will cost more than planned on the party who has promised to perform. Careful planners with strong bargaining power build a contingency into their contract price that takes this risk into account. Suppose a client directs significant and frequent changes in the design. Suppose for any reason the design services must be performed over a substantially longer period than planned. It is unlikely that the law will give the design professional a price adjustment under a fixed-price contract. However, the AIA has sought to protect architects from these risks and others in B101-2007, Article 4, dealing with additional services. This has been discussed in Section 13.01G.

K. Deductions from the Fee: Deductive Changes

Acts of the design professional may cause the client to incur expense or liability, and the client may wish to deduct expenses incurred or likely to be incurred from the fee to be paid to the design professional. Suppose a design professional commits design errors that cause a claim to be made by an adjacent landowner or by the contractor against the client. Suppose the client settles any claims or wishes to deduct an amount to reimburse itself in the event it must pay the claims.

The right to take deductions or offsets in such cases can be created either by the contract or by law. An illustration of the first is the frequent inclusion of provisions in construction contracts that give the owner the right to

---

[24]§ 11.8.1.9.
[25]§ 11.8.1.10.
[26]§ 11.8.1.11.
[27]75 A.D.2d 464, 429 N.Y.S.2d 956 (1980).

[28]494 S.W.2d 250 (Tex.Ct.App.1973).
[29]*Herbert Shaffer Assoc., Inc. v. First Bank of Oak Park*, 30 Ill.App.3d 647, 332 N.E.2d 703 (1975).

make deductions and offsets against the contractor. This is usually not found in contracts between design professionals and clients. However, even in the absence of such provisions, the client may be able to take deductions for expenses incurred or likely to be incurred by the client as a result of any contractual breach by the design professional. The amount deducted must not be disproportionate to the actual or potential liability of the client.

It is likely that the construction contract price will be increased or decreased during actual performance through issuance of change orders. AIA Document B141-1997, Paragraph 1.3.9.1, had stated that no payments will be withheld from the architect "on account of the cost of changes" unless the architect is "adjudged to be liable." Presumably, this would encompass only judicially determined negligent design. Deductive changes, though they may reduce the construction contract price, do not affect the architect's fee even if the fee is based on a stated percentage of construction costs. Deductive change orders not only are not likely to reduce the extent of the architect's services but also may increase it.

In addition, B141-1997 Paragraph 1.3.9.1 had stated that there will be no reduction from the architect's compensation if the total contractor payout is reduced because of deductions for liquidated damages or other sums withheld from the contractor, such as deductions for damages suffered because of the contractor's breach. Such deductions do not reduce the contract price; they only reimburse the client for its losses when it does not receive performance it has been promised by the contractor.

B101-2007 Section 11.10.3 continued and clarified this policy. There is to be no deduction in the architect's compensation because of a reduction in the payout to the contractor because of deductive changes. In a change from 1997, a reduction of the architect's fee results if "the Architect agrees or has been found liable for the amounts in a binding dispute resolution procedure." This makes clear that a binding arbitration award that the architect has breached, rather than as in 1997 the need for a court judgment, can reduce his compensation.

The effect on the architect's fee of a reduction in the payout to the contractor because of the contractor's breach of contract is puzzling. In B101-2007, instead of dealing with breach by the *contractor* that reduced the amount paid to the contractor when this is used to compute the

architect's fee, Section 11.10.3 speaks of a "penalty of liquidated damages on the *architect*" (emphasis added). What is the purpose of this change?

Suppose the architect is guilty of a breach of contract that delayed completion. This would be rare because usually delayed completion is caused by the contractor. Delay is often dealt with by liquidating damages in the construction contract because actual damages are very difficult to prove.[30] Also, unlike the frequent use of damage liquidation in the construction contract, it is almost unheard of for a contract between the architect and his client to include a liquidation of damages clause.

The only possible justification in B101-2007 for this language is to prevent the owner from reducing the architect's fee by arbitrarily liquidating any claim it has against the architect. This would be rare. It is important to include language making clear that the architect's fee will not be reduced if amounts are deducted from the payout to the contractor to compensate the owner for the contractor's breach. This makes a return to the language of B141-1997 useful.

L. Project Risks

In determining appropriate compensation, the design professional should consider the projected risk the commission creates. Although risk management is discussed in greater detail in Chapter 15, it is important to note in this chapter that the scope of risk should play an influential if not dominant role in determining whether a project should be undertaken and the appropriate compensation for it. Risk means the likelihood that claims will result and the cost of dealing with such claims.

One factor that should be taken into account, particularly in a fee structure not based on cost, is the likelihood that the client will be cooperative and efficient in performing its part of the contract for design services. Another relates to the technical problems that may be encountered relating to both design and administration. Design professionals should be aware of the Architecture and Engineering Performance Information Center headquartered at the University of Maryland. The center collects, analyzes, and disseminates information about the

---

[30]See Section 26.10C.

performance of buildings, civil structures, and other constructed facilities. It uses its data to analyze trends over time relating to insurance claims. It also notes the likelihood particular clients will press claims and the organizational structures (the traditional method, cost projects, design–build) that seem prone to increased claims. It can also give information on the likelihood of particular types of problems in particular projects.

M. The Fee as a Limitation of Liability

Looking ahead to professional liability, the fee can serve another function. Some design professionals seek to limit or actually limit their liability exposure to their client, to the amount of their fee.[31]

SECTION 13.02 Time for Payment

A. Service Contracts and the Right to Be Paid as One Performs

In service contracts, the promises exchanged are payment of money for performance of services. Unless such contracts specifically deal with time for payment to occur, the performance of all services must precede the payment of any money. Put another way, the promise to pay compensation is conditioned on the services being performed.

Such a rule operates harshly to the person performing services. First, if the performance of services spans a lengthy time period, the party performing these services may need a source of financing to perform. Second, the greater the performance without being paid, the greater the risk of being unpaid. For these reasons, the law protects manufacturers and sellers of goods by giving them the right to payment as installments are delivered. However, this protection was not accorded to people performing services. If the hardships and risks described are to be avoided, contracts for professional services must contain provisions giving design professionals the right to be paid as they perform.

State statutes generally provide that employees are to be paid at designated periodic intervals. However, such statutes do not protect those who perform design services who are not employees of the owner.

B. Interim Fee Payments

Design professionals commonly include contract clauses giving them the right to interim fee payments. This avoids the problems described in the preceding section. From the client's standpoint, interim fee payments can create an incentive for the design professional to begin and continue working on the project.

Usually, interim payments in design professional contracts become due as certain defined portions of the work are completed. Although AIA Document B101-2007, Section 11.5, leaves blanks for interim fee payments, the 1977 B141a ("Instruction Sheet") suggested interim fee payments as follows:

- schematic design phase 15 percent
- design development phase 35 percent
- construction documents phase 75 percent
- bidding or negotiation phase 80 percent
- construction phase 100 percent

Any schedule used should depend on the breakdown of professional services and the predicted work involved in each phase.

Dividing the design services and allocating a designated percentage of the fee to each service can make the contract appear divisible. A divisible contract matches specified phases of the work to specified compensation or a specified portion of the total compensation. In a truly divisible contract, the amount designated is earned at the completion of each phase and the value of the work for each completed phase cannot be revalued.

As an illustration, suppose the contract were considered divisible and the design professional unjustifiably discharged after completing the construction documents phase. In such a case, the design professional will recover 75 percent of the fee if this were the amount specified even if the reasonable value of services exceeded this amount. Conversely, the client would not be permitted to show that the reasonable value of the services was less than 75 percent of the fee. Interim fee payments provisions should not make these contracts divisible. The amounts chosen are usually rough approximations and not

---

[31]See Section 15.03D.

agreed final valuations for each phase. This is especially true if the standard form contract specifies the phases and allocates a percentage of the fee for each phase. Such payment should be considered only provisional.[32]

C. Monthly Billings

In many projects, months may elapse before a particular phase is completed. To avoid overly long periods between payments, contracts for such projects should provide for monthly billings within the designated phases.[33]

D. Late Payments

Financing costs are an increasingly important part of performing design professional services. Late payments and reduced cash flow can compel design professionals to borrow to meet payrolls and pay expenses. The contract should provide that a specified rate tied to the actual cost of money be paid on delayed payments.[34]

In the absence of a contractually specified rate or formula, the interest is the "legal rate." In most states, this is the amount of interest payable on court judgments, see Section 6.08. During the inflationary period of the 1970s, the actual cost of borrowing was substantially in excess of the legal rate. As a result, many states increased the legal rate to better reflect the actual costs of borrowing money. But in the 1990s inflation dropped drastically, and as a result the legal rate currently may be substantially greater than the actual cost of borrowing money. (This is not likely to be the case for short-term commercial loans.)

Among the welter of consumer protection legislation that has spewed forth from Congress is the federal Truth in Lending Act. This act requires those who lend money to disclose the details of the cost of the loan. The AIA has expressed concern over this law. It has pointed to the law's possible application in a note to AIA Document B141-1997, Paragraph 1.5.8, which provides for late payments. The note states that certain consumer protection laws may affect the validity of the late payment provision and

suggests that specific legal advice be obtained if changes are made in the clause or if disclosure must be made.

An early draft of B101-2007, Section 11.10.2, continued the cautionary note. This was dropped in the final version. But design professionals who face tardy payments and the need to "carry" the client should be aware of this risk.

It is possible to conceive of the architect or the engineer—although generally not thought of as being a lender—as extending credit by allowing payments to be made after services are performed. If the amounts specified are considered a finance charge rather than a charge for late payment, disclosure requirements must be made in accordance with the Truth in Lending Act.

This situation is illustrated by *Porter v. Hill*,[35] a case decided by the Oregon Supreme Court in 1992. An attorney had included at the end of a billing statement a provision stating that a late payment charge of 1.5 percent per month would be added to the balance due if the amounts were more than thirty days overdue. The client challenged this provision, claiming it was a financing charge and required disclosures in accordance with the Truth in Lending Act.

The court held that two factors are examined to determine whether the charge is a late payment charge to which the disclosure requirements do not apply: (1) whether the terms of account require the consumer to pay the balance in full each month and (2) whether the creditor acquiesces in the extension of credit by allowing the consumer to pay the account over time without demanding payment in full.

The court concluded here that under these standards, this was a late payment charge and not a financing charge. It pointed to the fact that the amount was described as a late payment charge and not a finance charge. It also noted that the written agreement required the client to pay the balance in full on billing and that the attorney had not given his client the option of paying over time, subject to the 1.5 percent payment charge. Instead, the court noted, the attorney demanded full payment each month and, when he was not paid, brought legal action to collect the full amount due.

Design professionals should make sure their bills state that the client is required to pay the balance in full each month. Also, if the design professional allows clients to pay the accounts over time without demanding payment

---

[32]*Herbert Shaffer Assoc., Inc. v. First Bank of Oak Park*, supra note 29. But see *May v. Morganelli-Heumann & Assoc.*, 618 F.2d 1363 (9th Cir. 1980) (architect contract divisible).

[33]B101-2007, §11.10.2.

[34]B101-2007, §11.10.2, provides blank spaces to be filled in for the rate of interest and when payable. If the blanks are not filled, late payments invoke the legal rate at the architect's principal place of business.

---

[35] 314 Or. 86, 838 P.2d 45 (1992).

in full and no effort is made to collect the full amount due, any late charge may be considered a finance charge, invoking the onerous disclosure requirements of the Truth in Lending Act. Legal advice should be sought on this issue.

E. Suggestions Regarding Interim Fee Payments

When clients delay payments, the design professional should make a polite and sometimes strong suggestion that payments should be made when due. If the design professional, as noted in Section 13.02D, acquiesces to a pattern of delayed payments, he may find he has extended credit and is subject to the Truth in Lending Act. Also, a pattern of delayed payments should make the design professional seriously consider exercising any power to suspend further performance until payments are made.[36] If the suspension continues for a substantial time period, the design professional should consider terminating the contract.

In cases of suspension or termination of performance, it is desirable to notify the client of an intention to either suspend or terminate unless payment is received within a specified period of time. This gives the client an opportunity to make the payment. It also shows the client that failure to make interim fee payments as promised will not be tolerated.

SECTION 13.03 Payment Despite Nonperformance

Denying a contracting party recovery for services performed unless it has performed all the obligations under the contract can create forfeiture (loss of contractual payment rights substantially exceeding harm caused by breach) or unjust enrichment (the other party retaining and using the performance without paying for it). Legal doctrines have developed that minimize the likelihood of forfeiture and unjust enrichment.

Where it is not clear from the contract whether exact performance is a promise or a condition, the law is likely to classify the nonperformance as a breach creating a right to recover damages, rather than as a failure of a condition that bars recovery for work performed under the contract. The party who has breached can recover if it has substantially performed. Failure to fully perform does not bar recovery if caused by the other party's prevention or hindrance of performance or if the latter has failed to extend reasonable cooperation necessary to performance. Sometimes the party to whom performance is due may have waived its right to performance by indicating that it was satisfied with less than exact performance.

Repudiating the contract, denying its validity, or communicating an unwillingness or inability to perform excuses full performance. It would make little sense for the party to complete performance when the other party has indicated it does not wish performance and will not accept it.

Increasingly, the law is recognizing the right of a party in default to recover despite failure to perform under the contract. These problems arise more commonly in construction contracts, and legal doctrines have been developed with those contracts in mind. For this reason, a detailed discussion of these doctrines is postponed until Section 22.06D.

SECTION 13.04 Other Client Obligations

Contracts for design usually place a number of obligations on the owner. These usually relate to furnishing information that is within its knowledge as owner and obligations that it is in the best position to perform.

Sometimes the owner's role in the construction process is forgotten. It is clear that in a design–bid–build plan the design is the responsibility of the design professional, and construction is the responsibility of the contractor. When design-build is selected, the contractor both designs and builds.

Yet the owner plays a large role in the design and in construction. This is recognized in B101-2007, Article 5, captioned "Owner's Responsibilities." The owner is required to provide information through a written program.[37] It establishes and updates its budget.[38] It designates a representative to act on its behalf.[39] It provides surveys that "describe physical characteristics, legal limitations . . . and a written legal description of the site."[40]

---

[36]Restatement (Second) of Contracts § 237, illustration 1; §§ 251, 252 (1981) (power to suspend until adequate assurance given). See Section 12.13D.

[37]B101-2007 § 5.1.
[38]Id. at § 5.2.
[39]Id. at § 5.3.
[40]Id. at § 5.4.

It furnishes the services of geotechnical engineers,[41] furnishes tests and inspections required by law,[42] as well as legal, insurance, and accounting services.[43] Article 5 includes other operational responsibilities, but the items listed show the significant role played by the owner.

In addition to express provisions, obligations can be implied into contracts. The client impliedly promises not to interfere with the design professional's performance and to cooperate. For example, the client should not refuse the design professional access to information necessary for performing the work. Refusal to permit the design professional to inspect the site would be prevention and a breach of the implied obligation owed by the client to the design professional.

Positive duties are owed by the client. The client should exercise good faith and expedition in passing judgment on the work of the design professional and in approving work at the various stages of the latter's performance. It should request bids from a reasonable number of contractors and should use best efforts to obtain a competent bidder who will agree to do the work at the best possible price. If conditions exist that will require acts of the client, such as obtaining a variance or obtaining financing, the client impliedly promises to use best efforts to cause the condition to occur.

---

[41]Id. at § 5.5.

[42]Id. at § 5.7.

[43]Id. at § 5.8.

CHAPTER FOURTEEN

Professional Liability: Process or Product?

SECTION 14.01 Claims Against Design Professionals: On the Increase

The possibility that a design professional will find herself in court has increased dramatically. One out of every three practicing architects is likely to find herself in litigation. Sometimes litigation is necessitated by the client's failure to pay fees for professional services. This chapter concentrates on claims made by clients and others that they have suffered losses that should be transferred to the design professional. This section explores the many reasons for increased claims against design professionals.

A. Changes in Substantive Law

Some defenses that had proved useful when claims were made by parties other than the client (collectively referred to as *third parties*) have proved of diminished value or have largely disappeared. For example, the requirement of privity between claimant and design professional and acceptance of the project terminating liability of the design professional—both valuable in avoiding third-party claims—have proved much less effective. Courts have loosened the requirement that expert testimony be introduced to establish that a professional has not lived up to the standards of her profession. These substantive changes have resulted from increasing emphasis on ensuring that victims receive compensation and on deterring wrongful conduct, rather than on protecting socially useful activities.

B. Procedural Changes

With little difficulty and minimal costs, a claimant can bring legal action against a number of defendants in the same lawsuit. Similarly, those against whom claims have been brought can bring other defendants into that lawsuit with relative ease and without much expense. The result has been complicated lawsuits with a host of parties defending and asserting claims. Also, statutes of limitations designed to protect defendants from stale claims, based on activities that took place many years before the claims, have provided increasingly less protection.

C. Ability of Design Professionals to Pay Court Judgments

Claimants usually do not assert legal action against people who they believe will be unable to pay for a court judgment or are not insured. Although expanded liability and increased cost of insurance premiums have begun to reduce the percentage of design professionals who carry insurance, many will have professional liability insurance or sufficient resources to respond to court judgments. As a result, claims increase.

D. Access to Legal System

A person who seeks relief through the legal system usually engages a lawyer. Easier access to legal services will mean more claims and litigation. Increased accessibility to lawyers began by giving those charged with major crimes a lawyer even if they cannot afford to hire one. Programs were later developed to give legal representation to the poor when they sought to use the legal system. This development was accompanied by increased emphasis on informing people of their legal rights. For example, construction trade unions routinely inform their members of their legal rights and encourage them to use the legal process. Increasingly, prepaid legal insurance plans provide legal services to union members.

Another reason for easier access to the legal system is the much-maligned contingent fee contract (discussed in Section 2.04). Under such a contract, the client pays the lawyer for her time only out of any recovery obtained, and the client's investment is generally limited to expenses (in injury cases often paid by the attorney).

Another reason for more claims is that the prevailing party does not recover its costs of defense, including attorneys' fees, unless it can point to a contract providing for such recovery or to a statute granting attorneys' fees to the prevailing party. This can encourage legal claims by assuring the claimant that it does not run the risk of having to pay the other party's legal expenses if the claim fails.

E. Societal Changes

Americans today are less willing to accept their grievances silently. They are more inclined to use the legal system if they feel they have a grievance. The high ratio of lawyers to population in the United States, perhaps the highest in the world, demonstrates this.

American society has become increasingly urban and impersonal. There is much less likelihood that disputants will be part of a cohesive social unit. Such units usually provide informal mechanisms for adjustment of rights and discourage resort to outside processes.

A sense of alienation in a large impersonal society makes people feel powerless, with no one to protect them or help them. This was undoubtedly one element prompting the rise of consumerism in the 1960s. Aggressive use of the legal system responded to this phenomenon.

F. Enterprise Liability: Consumerism

Compensating victims rather than protecting enterprises has been the dominant modern tort motif. This has led to liability rules based on the belief that it is best that victims recover from the enterprise that is in the best position to avoid or spread the losses to those who benefit from the enterprise. Much of this tendency drew unconsciously from the emphasis on security that became dominant after World War II. Much was accomplished through social welfare legislation, such as unemployment insurance, public housing, public welfare, and job security, and compensating victims by expanding tort rights was a useful adjunct.

Liability expansion is also traceable to the consumer movement of the 1960s. Those who felt consumers were being supplied shoddy goods and services advocated increased liability as a means of bringing home to the business sector the importance of dealing fairly with consumers. Consumer protection legislation with its expanded remedies is discussed in Section 23.02J.

In the 1990s, a backlash resulted from the belief by some that expanded liability and increased litigation placed too heavy a burden on commercial enterprises and insurers. As noted in Section 7.08L, much of this backlash was cloaked in the garb of competitiveness as American enterprises increasingly competed with products and services performed by enterprises in foreign countries. This backlash in the product liability field was demonstrated by the increased tendency of state legislatures to enact legislation that would limit the exposure of those who manufacture products. In the field of professional services, some states enacted legislation designed to limit the liability of health care providers. Still the strong tendency toward compensating victims continues to play a significant (perhaps even dominant) role in American tort law.

G. Design and Social Policy

Societal concerns with the quality and nature of the built environment naturally focuses on the design as a promoter of public safety and an allocator of resources. Emergency service agencies view design as the first step in promoting safety. Public officials have stated that much of the responsibility for avoiding fires or minimizing fire losses falls on those who design structures. One fire chief, for example, commented that "good fire protection in high-rise buildings begins on the architect's drawing board." He also said it was the builder's responsibility to design "a safe building, not a firetrap."[1]

Law enforcement officials have asserted that assaults can be reduced if those who design business and residential areas plan properly. A director of the National Institute of Law Enforcement and Criminal Justice said,

> Better environmental design can do much more [to reduce crime]. New housing projects, schools, shopping centers and other areas can be designed, for example, with more

---

[1]*New York Times*, August 4, 1974, p. 11.

windows looking out on streets, fewer hidden corridors, and other crime discouraging features.[2]

Law enforcement concerns recently have shifted from everyday crimes to designing buildings to withstand terrorist attacks.[3] Claims alleging the design of unsafe buildings also include prisons, where both prison guards and families of suicidal inmates have sought to hold the design professional responsible.[4]

As discussed in Section 9.13E, environmental concerns have spawned industry and statutory efforts to implement "green" design and construction techniques.

H. Codes

Liability has also expanded because detailed building and housing codes have proliferated. Violating these codes, although not conclusive on the question of negligence, makes it relatively easy to establish that the design professional is liable not only to the client but also to third parties who suffer foreseeable harm because of the violation.

Although most building codes are enacted by local authorities, the enactment of the federal Americans with Disabilities Act will also be used as the basis for extending liability to design professionals. See Section 14.05C.

I. Expansion of Professional Services

Design professionals are expected to either provide or volunteer to provide services that go beyond core design services. These services can relate to availability of funds for the project, likely profitability of the project, and likely approval by public officials and agencies who control building and construction. This can increase claims when clients suffer disappointments or losses.

J. Site Services

A design professional's varied site services discussed in Section 12.08 and in the balance of the book make the design professional more vulnerable to a claim traceable not only to design but also to the way in which the design was executed, both as to compliance with the design and the methods of accomplishing it.

SECTION 14.02 Overview of Chapters 14 and 15

The preceding section outlined the reasons why professional liability has expanded and the likelihood that more claims will therefore be made against design professionals. Chapter 14 deals with professional liability claims mainly by clients but also by third parties. Chapter 15 discusses risk management techniques to avoid or reduce liability risks.

A. Applicable Law

Professional liability is usually determined by state law. Depending on various factors, professional liability rules will be those of the state in which the design professional has its principal place of business, where the actual design is created, or where the project is located. Because both legal rules and actual outcomes may vary depending on the applicable state law, discussion of liability must be general and not directed to a particular state except to the extent that illustrative cases may do so.

B. Types of Harm

It is useful to divide claims into those that involve personal harm, those that involve harm to property, and those that involve economic loss not connected to personal harm or property damage. As a general rule, the law is more likely to sustain claims based on personal harm than claims based on harm to property or economic loss, and more likely to sustain claims related to property harm than those involving economic loss. Traditionally, though less so today, tort law dealt with harm to person or property.

---

[2]Ibid. See also O. NEWMAN, DEFENSIBLE SPACE: CRIME PREVENTION THROUGH URBAN DESIGN (1973).

[3]*Clark Concrete Contractors, Inc. v. General Services Admin.*, GSBCA No. 14340, 99-1 BCA ¶30,280, recon. denied, GSBCA No. 14340-R, 99-2 BCA ¶30,393; appeal dismissed, 230 F.3d 1372 (Fed.Cir.1999) (redesign of new Federal Bureau of Investigation building in Washington D.C. in response to the Oklahoma City bombing of the Alfred P. Murrah Federal Office Building); The Times of London, September 6, 2002, p. 1 (Scotland Yard is working with Israeli architects to design shopping centers to be less vulnerable to suicide bombings after the September 11, 2001 attacks on the World Trade Center and the Pentagon).

[4]*State v. Gathman-Matotan*, 98 N.M. 740, 653 P.2d 166 (1982) (a claim by security guards confronted by a prison riot). The following cases involved inmates who committed suicide while in their cells: *Easterday v. Masiello*, 518 So.2d 260 (Fla.1988); *La Bombarbe v. Phillips Swager Assoc.*, 130 Ill.App.3d 896, 474 N.E.2d 942 (1985); *Tittle v. Giattina, Fisher & Co., Architects, Inc.*, 597 So.2d 679 (Ala.1992).

This hierarchy of protection can help predict the outcome of any lawsuit, particularly where a court is asked to veer from established rules and recognize new legal rights. This recognition is more likely when a claim is based on personal harm and the need for compensation is more urgent. Once this inroad has been made, some courts stop at that point. But more often, courts decide that there is no particular reason to limit the extension to cases involving personal harm. Over time, what was once an established rule may disappear through this process. Disappearance is usually gradual, however. This preserves the appearance of stability while accommodating the need for change.

SECTION 14.03 Claims Against Design Professionals: Some Illustrations

Another way to demonstrate the liability explosion described in Section 14.01 is to note the types of claims that have been made against design professionals. Although most claims noted were successful, some cautionary remarks are essential.

Any conclusion that particular conduct gave rise to a valid claim was made in the context of a particular contract and particular facts. Conduct held to fall below the legal standard required in a particular case does not mean such conduct will always fall below.

Many appellate opinions simply conclude that the determination made by the finder of the facts—either the trial judge or jury—was within its discretion. An appellate court reviewing a lower court decision may not agree with the decision but will respect the differentiation between the role of trial and appellate courts, the former resolving factual disputes and the latter, as a rule, deciding questions of law.

As to negligence in the design phase, one writer stated,

Architects might fail to use due care in various ways. The architect may inadequately consider the nature of the soil under the building; he may design an inadequate foundation; he may design a roof too weak to support the weight it will foreseeably have to bear; he may insulate or sound-proof the building inadequately. The architect may negligently design a sewer so that waste is carried toward rather than away from the house, he may design windows too small or too large, he may fail to put a handrail on a stairway, or he may specify that nails rather than bolts be used

to secure a sundeck. . . . In addition the architect may negligently fail to notice a defect in the work of a consultant he has hired to help prepare the plans and specifications. The architect would also probably be liable for damage caused by his failure to hire a consultant where a reasonable architect would have done so.

Negligence in design can be based on negligently incomplete specifications as well as on complete but erroneous ones. The plans and specifications must be complete and unambiguous. For example, specifications are negligently prepared if they are so indefinite that a contractor can bid as if he were going to use first-class materials and then build using inferior materials. If measurement of a material is involved, the specification must distinguish between dry and liquid states, or loose or tight packing, where there is any chance of ambiguity.[5]

Since that analysis, judicial opinions have provided other illustrations. For example, cases have involved the following claims, many of which were successful:

1. misrepresenting existing topography[6]
2. relying on an out-of-date map and building on land not owned by the owner[7]
3. specifying material that did not comply with building codes[8]
4. positioning the building in violation of setback requirements[9]
5. failing to inform client of potential risks of using certain materials[10]
6. failing to advise client about potential problems with new product[11]

---

[5]Comment, 55 Calif.L.Rev. 1361, 1370–71 (1967). The author's footnotes to cases are omitted. Cases involving surveyors' mistakes are collected in Annot., 117 A.L.R. 5ᵗʰ 23 (2004).

[6]*Mississippi Meadows, Inc. v. Hodson*, 13 Ill.App.3d 24, 299 N.E.2d 359 (1973) (dictum).

[7]*Jacka v. Ouachita Parish School Bd.*, 249 La. 223, 186 So.2d 571 (1966). Here the architect was relieved of responsibility because the client was obligated to and did furnish the out-of-date map.

[8]*St. Joseph Hosp. v. Corbetta Constr. Co.*, 21 Ill.App.3d 925, 316 N.E.2d 51 (1974).

[9]*Armstrong Constr. Co. v. Thomson*, 64 Wash.2d 191, 390 P.2d 976 (1964).

[10]*Banner v. Town of Dayton*, 474 P.2d 300 (Wyo.1970).

[11]*White Budd Van Ness Partnership v. Major-Gladys Drive Joint Venture*, 798 S.W.2d 805 (Tex.Ct.App.1990), cert. denied, 502 U.S. 861 (1991).

7. failing to inform the client that it could not make a reliable judgment as to materials, and failing to collect information about such materials[12]

8. failing to inform the client that the design professional had acquired information that water temperature was too low for operating a pump[13]

9. drafting ambiguous sketches, which caused extra work[14]

10. designing a house that could not be built by tradespeople in the community where the project was to be built[15]

11. designing closets too small for the clothing to be contained in them[16]

12. designing a project that greatly exceeded the client's budget[17]

13. specifying untested material solely because of seller's representations[18]

14. designing inadequate solar heating system[19]

15. failing to consider energy costs[20]

16. failing to disclose an underground high-voltage live wire[21]

17. failing to include owner as named insured and omitting indemnity clause[22]

18. failing to advise a need for use permit[23]

19. failing to design a prison that would make it "takeover proof"[24]

20. failing to design a prison that would avoid prisoner suicide[25]

21. failing to know of local building codes and safety laws[26]

22. specifying a competition diving board for a grade school and failing to warn users of dangers[27]

23. failing to design a road sign which would cause lesser injuries if crashed into by a motorist[28]

24. failing to identify a deficiency in an engineer's report obtained by the owner[29]

25. failing to design an apartment building so that it is accessible by the disabled[30]

26. failing to design a preengineered metal building so that it can be built using an autowelder, the standard industry equipment[31]

Claims, relating to the construction process phase, again mostly successful, were as follows:

1. allowing material not approved by code to be installed[32]

2. ordering excess fill to be placed without consulting a soil tester[33]

3. failing to make changes needed to comply with codes[34]

4. failing to condemn defective work[35]

5. scheduling and coordinating incompetently[36]

---

[12]*Richard Roberts Holdings, Ltd. v. Douglas Smith Stimson Partnership and Others*—decision by official referee in the United Kingdom and summarized in 5 Constr.L.J. 223 (1989).

[13]*Green Island Assoc. v. Lawler, Matusky & Skelly Eng'rs*, 170 A.D.2d 854, 566 N.Y.S.2d 715 (1991).

[14]*General Trading Corp. v. Burnup & Sims*, 523 F.2d 98 (3d Cir.1975).

[15]*Bayuk v. Edson*, 236 Cal.App.2d 309, 46 Cal.Rptr. 49 (1965).

[16]Ibid.

[17]*Stanley Consultants, Inc. v. H. Kalicak Constr. Co.*, 383 F.Supp. 315 (E.D.Mo.1974).

[18]*New Orleans Unity Soc'y v. Standard Roofing Co.*, 224 So.2d 60 (La. App.1969) (dictum).

[19]*Keel v. Titan Constr. Corp.*, 639 P.2d 1228 (Okla.1981).

[20]*Board of Educ. v. Hueber*, 90 A.D.2d 685, 456 N.Y.S.2d 283 (1982) (unsuccessful).

[21]*Mallow v. Tucker, Sadler & Bennett, Architects & Eng'rs., Inc.*, 245 Cal.App.2d 700, 54 Cal.Rptr. 174 (1966).

[22]*Transit Cas. Co. v. Spink*, 94 Cal.App.3d 124, 156 Cal.Rptr. 360 (1979) (trial court finding for client not challenged on appeal), disapproved on other issues in *Commercial Union Assurance Co. v. Safeway Stores*, 26 Cal.3d 912, 610 P.2d 1038, 164 Cal.Rptr. 709 (1980).

[23]*Chaplis v. County of Monterey*, 97 Cal.App.3d 249, 158 Cal.Rptr. 395 (1979).

[24]*State v. Gathman-Matotan*, supra note 4.

[25]*Easterday v. Masiello*, supra note 4; *La Bombarbe v. Phillips Swager Assoc.*, supra note 4; *Tittle v. Giattina, Fisher & Co. Architects, Inc.*, supra note 4.

[26]*Insurance Co. of North Am. v. G.M.R., Ltd.*, 499 A.2d 878 (D.C.App. 1985).

[27]*Francisco v. Manson, Jackson & Kane, Inc.*, 145 Mich.App. 255, 377 N.W.2d 313 (1985).

[28]*Polak v. Person*, 232 Ill.App.3d 505, 597 N.E.2d 810 (1992).

[29]*Kerry, Inc. v. Angus-Young Assocs.*, 280 Wis.2d 418, 694 N.W.2d 407 (App.), review denied, 286 Wis.2d 98, 705 N.W.2d 659 (2005).

[30]*Options Center for Independent Living v. G & V Dev. Co.*, 229 F.R.D. 149 (C.D.Ill.2005). See Section 14.05C.

[31]*C. H. Guernsey & Co. v. United States*, 65 Fed.Cl. 582 (2005) (not architectural negligence).

[32]*St. Joseph Hosp. v. Corbetta Constr. Co.*, supra note 8.

[33]*First Ins. Co. of Hawaii v. Continental Casualty Co.*, 466 F.2d 807 (9th Cir.1972).

[34]*Mississippi Meadows, Inc. v. Hodson*, supra note 6.

[35]*Skidmore, Owings & Merrill v. Connecticut General Life Ins. Co.*, 25 Conn.Sup. 76, 197 A.2d 83 (1963).

[36]*Peter Kiewit Sons' Co. v. Iowa Southern Utilities Co.*, 355 F.Supp. 376 (S.D.Iowa 1973).

6. failing to properly exercise supervisory powers[37]
7. failing to warn an experienced contractor of general precautions not known in the industry[38]
8. failing to engage and check with a consultant[39]
9. failing to stop work after discovering contractor using unsafe methods[40]
10. issuing payments or certificates negligently[41]
11. failing to warn of bankruptcy when paying for materials in contractor's possession[42]
12. failing to observe design deviation when checking shop drawings[43]
13. failing to give instructions, issue change orders, and conduct inspections[44]
14. allegations of gross negligence: allowing trusses to be erected without first inspecting the welds, not fixing discrepancies between the shop drawings and the trusses as fabricated, not ensuring that the trusses were actually tested, and not discovering that the fabricator had never fabricated steel trusses before[45]

The remainder of this chapter discusses the different types of claims brought against design professionals. Before reading about these different theories of recovery, it is helpful to know the nature of claims brought against design professionals. Victor O. Schinnerer & Company, the largest insurer of design professionals, has reviewed its internal files dating back to the late 1950s to analyze the type, frequency, and severity of claims filed. It determined that slightly over 75 percent of claims are for property damage or economic loss, with nearly 55 percent of those brought by the owner/ client. Of the remaining quarter consisting of personal injury claims, one-third are brought by construction workers and the other two-thirds involve nonworker injury.[46]

The frequency of claims peaked between 1978 and 1985 and has leveled off between 1998 and 2004 (the last year available). The frequency and severity of claims are greatest on projects involving houses or townhouses, schools or colleges, and condominiums. About half of all claims are brought before project completion, and 95 percent are brought within five years of completion.[47]

SECTION 14.04   Specific Contract Standard

A. Likelihood of Specific Standard

Contracting parties can by agreement determine the standard of performance. Because the client–design professional relationship is created by agreement, a primary source of any agreed-on standard is the contract itself. Most disputes between the client and the design professional do not involve a specific contractually designated standard. Rather, they involve a general standard not set forth in the contract, such as the professional standard to be described in Section 14.05, or an outcome-oriented standard, such as implied warranty described in Section 14.07.

Why do most contracts fail to specifically state how the design professional is to perform? First, many relationships are created without any written agreement—by handshake arrangements. Second, many are made by casual letter agreements drafted by the design professional that are not likely to describe specific standards. Third, those relationships created by assent to standard contracts published by professional associations such as the American Institute of Architects (AIA) do not, despite their completeness as to services and what the design professional is not responsible for, specifically describe how the work will be done. However, in 2007, the AIA for the first time by contract specified that the architect is subject to the professional standard. Document B101, Section 2.2, reproduced in Appendix A, states that the "Architect shall perform its services consistent with the professional

[37]*Aetna Ins. Co. v. Hellmuth, Obata & Kassabaum, Inc.*, 392 F.2d 472 (8th Cir.1968).

[38]*Vonasek v. Hirsch & Stevens, Inc.*, 65 Wis.2d 1, 221 N.W.2d 815 (1974) (dictum).

[39]*Cutlip v. Lucky Stores, Inc.*, 22 Md.App. 673, 325 A.2d 432 (1974).

[40]*Associated Eng'rs, Inc. v. Job*, 370 F.2d 633 (8th Cir.1966). See Annot., 59 A.L.R.3d 869 (1974).

[41]*Aetna Ins. Co. v. Hellmuth, Obata & Kassabaum, Inc.*, supra note 37.

[42]*Travelers Indem. Co. v. Ewing, Cole, Erdman & Eubank*, 711 F.2d 14 (3d Cir.1983) (unsuccessful).

[43]*Jaeger v. Henningson, Durham & Richardson, Inc.*, 714 F.2d 773 (8th Cir.1983).

[44]*Colbert v. B. F. Carvin Constr. Co.*, 600 So.2d 719 (La.App. 1992).

[45]*Travelers Indemnity Co. of Conn. v. Losco Group, Inc.*, 136 F.Supp.2d 253 (S.D.N.Y.2001), overruled sub silentio, *St. Paul Fire & Marine Ins. Co. v. Universal Builders Supply*, 409 F.3d 73 (2d Cir.2005).

[46]M. SCHNEIER, CONSTRUCTION ACCIDENT LAW: A COMPREHENSIVE GUIDE TO LEGAL LIABILITY AND INSURANCE CLAIMS, 262 (1999).

[47]Jones, Jr., *Looking Back to Look Ahead: Benchmarking Design Professional Liability Claims*, 28 Constr.Litg.Rep., No. 3, March 2007, p. 103.

skill and care ordinarily provided by architects practicing in the same or similar locality under the same or similar circumstances." Similarly, the Engineers Joint Contracts Documents Committee (EJCDC) in Paragraph 6.01A of its 2002 E-500 (owner–engineer contract), reproduced in Appendix G, states that the standard of care "will be the care and skill ordinarily used by members of the subject profession practicing [engineering] under similar circumstances at the same time and in the same locality," and that the engineer "makes no warranties, express or implied, under this Agreement or otherwise. . . ."

Even if the standard of performance is discussed in advance—something that is rare—it may not be included in the written contract: The design professional may not want it included, or the client may think it unimportant. Both may believe that any assurances as to outcome are simply nonbinding opinions or expectations.

Large architectural and engineering firms commonly include in their standard agreements language specifying the professional standard described in Section 14.05. They do so for two reasons. First, the design professional may wish to refine elements of the professional standard that may operate to her advantage. For example, she may include a provision stating that she is to be compared to other design professionals *in her community,* to avoid any contention that she is to be compared to design professionals *with whom she competes.* Second, and more commonly, the specification of the professional standard is accompanied by language stating that the architect will not be held to any express or implied warranty standard, a standard often more rigorous than the professional standard.[48] Clients who are in a powerful bargaining position and experienced in design work, such as public entities, often resist such clauses, seeking to preserve the possibility of later claiming the design professional should be held to a stricter standard.

If challenged, the design professional will justify such a clause by stating that its professional liability insurance coverage will not include contractual risks that deviate from the professional standard. If the matter cannot be resolved one way or the other, the language may be omitted, leaving the standard to that applied by law. Yet sometimes contractually specified standards of performance are created. Looking at some of them can provide a useful backdrop to the nonspecific standards described in Sections 14.05 through 14.07.

B. Client Satisfaction

Sometimes the client's obligation to pay arises only if the client is satisfied with the work of the design professional. Such a contract may be interpreted to be a promise by the design professional to satisfy the client. Although design professionals generally seek to avoid such relatively one-sided agreements, if clear evidence exists of such an agreement it will be enforced.

If satisfaction is a condition of the client's obligation to proceed and to pay, the client need not pay unless it is satisfied or waives this performance measurement. Any legal obligation that may arise must then be based on unjust enrichment created by the client using the work of the design professional.

If satisfaction has been promised, failure to perform requires the design professional to compensate the client for any losses the latter may have suffered because of the breach. For example, if a breach caused the project to be delayed or abandoned, the design professional is accountable for any foreseeable losses that can be established with reasonable certainty and that could not have been reasonably avoided.

Although the design professional may, however unwisely, risk the fee, strong evidence of such a risk assumption should be produced before she must respond for losses caused by her failure to satisfy the client.

There are two standards of satisfaction.[49] If performance can be measured objectively, the standard is reasonable satisfaction. Would a reasonable person have been satisfied? Objective standards are more likely to be applied where performance can be measured mechanically. For example, if an engineer agreed with a manufacturer that the manufacturer would pay if satisfied with the performance of a particular machine designed by the engineer, the obligation to pay would require that a reasonable manufacturer be satisfied.

More personal performance invokes a subjective standard. Suppose an artist agrees to paint a portrait that will satisfy the person commissioning the portrait. The latter must exercise a good-faith judgment and must be genuinely dissatisfied before she is relieved of the obligation to pay. If, for example, the person refused to view the portrait

---

[48]See Section 14.07.

[49]*First Nat'l Realty v. Warren-Ehret Co.,* 247 Md. 652, 233 A.2d 811 (1967) (collecting authorities). For a careful analysis, see *Morin Building Products Co. v. Baystone Constr. Co.,* 717 F.2d 413 (7th Cir.1983).

or give it sufficient light to judge its quality, any judgment was not exercised in good faith.

In practice, the two standards may not operate differently. In the preceding example, if judge or jury thought the performance satisfactory it would take a strong showing by the person commissioning the portrait that she was genuinely dissatisfied. As a rule, the subjective satisfaction standard arises in the design phase, particularly in aesthetic matters. However, in standard commercial projects an objective standard may be invoked.

What about a design professional's performance during construction? If the performance related to delicate matters, such as how the design professional handled the contractor or public officials or how she dealt with site conflicts, a subjective standard may be applied. Roughly speaking, though, design is more likely to be measured subjectively, whereas contract administration is more likely to be measured objectively—another illustration of the important difference between design and nondesign site services.[50]

Suppose the client is justified in refusing to pay. This may create a forfeiture (performance by one party for which she is not compensated). In such a case, the loss to the design professional may substantially exceed the loss to the client that would occur if the client, though dissatisfied, were required to accept the design professional's work. Various legal doctrines, among them waiver (party entitled to performance accepted less than full or proper performance) and estoppel (party misled into not performing in accordance with contract by other party's representation that it would accept performance other than that promised), can be employed to avoid forfeiture. But if the language makes clear that the risk of forfeiture was clearly assumed by one party to the contract, the clause will be enforced even if this would create a forfeiture. (In some jurisdictions exceptions to this rule exist in transactions involving the purchase of land.)

C. Fitness Standard

A more specific performance standard than satisfaction can be contractually created: The parties may agree that the completed project will be suitable or fit for those purposes for which the client entered into the project.

For example, the client who plans a luxury residence usually wants a house suitable for a person of her means and taste. In addition to wanting the normal requirements for any residence, such as structural stability, shelter from the elements, and compliance with safety and sanitation standards, the client may want a house that is admired by those who enter it or a residence that can facilitate closing business deals or making business contracts. The client may hope that the opulence of the residence will make social events successful.

The client who plans a commercial office building wishes to profit from renting space. To do so, suitable tenants at an economically adequate rent must be found. Such a client assumes that the planned use of the structure will be permitted under zoning laws and that the structure will comply with the applicable building codes and zoning regulations relating to materials, safety, density, setback regulations, and other land use controls. In addition, the client who builds an unusually designed office building may hope the structure will attract national architectural interest.

The client who wishes to build an industrial plant generally assumes that the plant when completed will be adequate to perform anticipated plant activities. The building is expected to comply with applicable laws relating to public health and safety.

Proper design requires that the design professional consider client objectives such as those discussed in the preceding paragraphs. Some items mentioned in those paragraphs will be discussed and included in the client's program. Some of the matters discussed in the preceding paragraphs would be assumed and probably not discussed. One would not expect client and design professional to discuss the necessity of complying with building codes or regulations dealing with health and safety. Yet beyond these basic objectives, discussions may have taken place of economic and social goals less directly connected with basic design objectives.

One possible standard to measure the performance of the design professional is whether the project accomplishes the client's objectives. Put another way, is the structure suitable for the client's anticipated needs, or is it fit for the purpose for which it was built? Is the building an architectural success? Has the client attracted good tenants? Has plant production increased? Are the social events successful?

To determine whether suitability or fitness performance standards will be used to measure the design professional's obligation, it is important to look at any antecedent

---

[50]*Jaeger v. Henningson, Durham & Richardson, Inc.*, supra note 43. (expert testimony not needed for claim based on site services).

negotiations, discussions, or understandings that may have preceded the client–design professional contract or may have occurred during the course of the design professional's activities.

Suppose client objectives were discussed during precontract negotiations or during the design professional's design performance. Did the design professional give any assurances relating to the fitness or suitability of her design to accomplish particular client objectives, promises that the design would accomplish the objectives or just statements of opinion that these objectives would be achieved?

Suppose the design professional made statements relating to such matters. The design professional may have stated that a particular luxury residence would create an artistic stir within the client's social circle. The design professional may have expressed a belief that suitable tenants could be found or that someday a particularly unusual office building would be considered an architectural landmark. She may have assured the client that the client could conduct certain activities on the premises.

To determine whether a statement makes a promise or expresses an opinion, the law looks at the definiteness with which the statement was made ("I am certain your cost per unit will decrease" versus "It's my considered opinion you will improve productivity"), the degree to which the design professional's performance can bring about that objective ("People will like the exterior design" versus "Your parties will be great successes"), and the degree and reasonableness of any reliance by the client on the statement (using certain types of machinery in a plant versus redecorating the interior of a house at great expense for the new social season). The more definite the statement, the more the outcome is within the control or professional expertise of the design professional, and the more likely the client has justifiably relied on the statement, the more likely it is that the law will find there has been a promise.

If the statement was made before a contract was formed, a design professional may contend that because of the parol evidence rule, such statements cannot be proved. Although results are not always consistent, by and large the client will be permitted to testify as to these statements.

The design professional should avoid assuring the client that particular objectives will be achieved unless she is willing to risk the possibility of being held accountable if they are not. Assurances of certain matters should be given; for example, that the design will meet public land use controls, such as zoning laws and building codes. The design professional should avoid venturing into areas that

are beyond her expertise and require difficult predictions of the future.

D. Quantitative or Qualitative Performance Standards

Sometimes the contract between design professional and client contains a specific performance standard. For example, an engineer may make a contract with a manufacturer under which it was specifically agreed that the machine designed by the engineer would produce a designated number of units of a particular quality within a designated period of time.

Suppose the performing party finds that meeting the performance standard is extremely difficult or that the performance standard will require more time and money than it anticipated. In some extreme cases, the performance standard may be impossible to meet.

In such cases, two legal issues may arise. First, is the design professional entitled to be paid for the effort made in trying to perform to specifications? The contract may be onerous, but if this risk is assumed, the professional will not recover. As in satisfaction contracts, to avoid forfeiture interpretation doubts are resolved in favor of the performing party. If this cannot be done, the performing party will go uncompensated unless she can show that efforts she expended, although not fulfilling the performance standards, have benefited the other party. In most cases there will be no unjust enrichment and no recovery for the performing party—although it may benefit the other party to be shown that the performance standards were not possible.

Second, has the performing party breached by not accomplishing the objective? Again, this is a question of whether accomplishing the objective is not merely a condition on the client's obligation to pay but also a promise by the performing party.

Obviously, performance standards place a heavy risk on the performing party. Yet they are attractive to clients because they objectively measure whether the client is getting what was promised. Such a standard can be an effective sales device for the designer.

E. Indemnification

The frequency of indemnification (process by which indemnitor promises indemnitee that indemnitor will see to it that the indemnitee will not suffer a loss when a claim is made against indemnitee which should be responsibility of indemnitor) in construction contracts necessitates more complete treatment, which Chapter 31 provides. For the purposes of this section, note that clients increasingly demand that the design professional indemnify them against claims the client believes to be the responsibility of the design professional. Any design professional indemnification creates another specific contractual obligation, one potentially broader than the professional standard.

F. Cost Overruns Caused by Design

An Ohio public entity made a proposal requiring design professionals to pay the cost of construction change orders caused by the design professional. The professional societies and professional liability insurers objected, saying that the policy erroneously assumed that change orders are necessarily caused by negligence of the design professional and that the professional liability insurance policies would not cover such a form of strict liability. After these objections were made, the Ohio agency withdrew the proposal. However, this is probably not the last such attempt to charge cost overruns to the design professional.

Attempts have been made—mainly by clients but also by design professionals—to recognize the likelihood of cost overruns due to design and to deal with them in various ways in the contract. One public contract in Virginia specified that the design professional would be responsible for cost overruns related to design if the overruns exceeded three percent of the contract price, without any need to show that the design professional had not performed in accordance with the professional standard. Even more, as noted in the preceding paragraph, some clients insist that all overruns due to design be the responsibility of the design professional whether or not the designer has failed to perform in accordance with the professional standard. This form of strict liability would almost certainly not be insurable under normal professional liability policies.

It would be unfair to the contractor if the contractor submitted its bid without knowing that the architect was strictly responsible for all design overruns. If the contractor knew of such strict liability, it could infer that all questions of interpretation that arise during performance would be resolved against the contractor. This would affect its bid.

Looking at this problem from the standpoint of a design professional, people interested in limiting design professional liability have suggested including provisions stating that the design will inevitably contain errors, omissions, conflicts, and ambiguity, all of which will require clarification and correction during construction. They suggest the client be advised that producing perfect documents is impossible and that some design decisions are more efficiently deferred for the benefit of the client until construction is underway and actual field conditions exist. Finally, such people advise that a contract contain a provision relieving the design professional from liability for all cost overruns up to a stated percentage of the construction cost.

G. Contractual Diminution of Legal Standard

Specific contractual standards are usually higher than the professional standard. Suppose, however, that the agreement between the client and the design professional specifies a lower standard. Because this situation more directly involves the extent to which the legal standard can be varied, it is discussed later, in Section 14.05D.

SECTION 14.05   The Professional Standard: What Would Others Have Done?

A. Defined and Justified: *City of Mounds View v. Walijarvi*

In *City of Eveleth v. Ruble*[51] the Minnesota Supreme Court stated,

> (1) In an action against a design engineer for negligence, the applicable legal principles are held to be:
>
> (a) One who undertakes to render professional services is under a duty to the person for whom the service is to be performed to exercise such care, skill, and diligence as men in that profession ordinarily exercise under like circumstances.
>
> (b) The circumstances to be considered in determining the standard of care, skill, and diligence to be required include the terms of the employment agreement, the nature of the problem which the supplier of the service represented himself as being competent to solve, and the effect reasonably to be anticipated from the proposed remedies on the balance of the [water] system.

---

[51]302 Minn. 249, 225 N.W.2d 521 (1974).

(c) Ordinarily, a determination that the care, skill, and diligence exercised by a professional engaged in furnishing skilled services for compensation was less than that normally possessed and exercised by members of that profession in good standing and that the damage sustained resulted from the variance requires expert testimony to establish the prevailing standard and the consequences of departure from it in the case under consideration.[52]

Four years later, the same court was invited to jettison this standard and replace it with the implied warranty standard, an outcome-oriented standard (discussed in Section 14.07). A portion of that second opinion is reproduced here.

_____

[52]225 N.W.2d at 522.

CITY OF MOUNDS VIEW v. WALIJARVI

Supreme Court of Minnesota, 1978. 263 N.W.2d 420.
[Ed. note: Footnotes omitted.]

TODD, Justice.

[Ed. note: The city became apprehensive because of dampness in the basement of an addition that was being added to a city building. The architect wrote to the city that its design would, if executed properly, generate a "water-tight and damp-free" basement. But problems grew worse, and corrective work was needed.

The city sued the architect, based on claims of negligence, express warranty, and implied warranty.

The trial court held that the language in the letter asserted to constitute a warranty to be merely an expression of opinion and that Minnesota did not recognize implied warranty of a perfect plan or an entirely satisfactory result in an architectural service contract. The trial court granted the architect's motion for summary judgment (no trial needed) on the warranty claims. The Minnesota Supreme Court held that the express warranty claim failed because the contract required that modifications be written and signed by both parties and no evidence had been introduced of any written agreement by the city. The opinion then dealt with implied warranty.]

. . . As an alternative basis for recovering damages from the architects, the city urges that we adopt a rule of implied warranty of fitness when architectural services are provided. Under this rule, as articulated in the city's brief, an architect who contracts to design a building of any sort is deemed to impliedly warrant that the structure which is completed in accordance with his plans will be fit for its intended purpose.

As the city candidly observes, the theory of liability which it proposes is clearly contrary to the prevailing rule in a solid majority of jurisdictions. The majority position limits the liability of architects and others rendering "professional" services to those situations in which the professional is negligent in the provision of his or her services. With respect to architects, the rule was stated as early as 1896 by the Supreme Court of Maine (*Coombs v. Beede*, 89 Me. 187, 188, 36 A.104 [1896]):

In an examination of the merits of the controversy between these parties, we must bear in mind that the [architect] was not a contractor who had entered into an agreement to construct a house for the [owner], but was merely an agent of the [owner] to assist him in building one. The responsibility resting on an architect is essentially the same as that which rests on the lawyer to his client, or on the physician to his patient, or which rests on anyone to another where such person pretends to possess some skill and ability in some special employment, and offers his services to the public on account of his fitness to act in the line of business for which he may be employed. The undertaking of an architect implies that he possesses skill and ability, including taste, sufficient to enable him to perform the required services at least ordinarily and reasonably well; and that he will exercise and apply in the given case his skill and ability, his judgment and taste, reasonably and without neglect. But the undertaking does not imply or warrant a satisfactory result.

The reasoning underlying the general rule as it applies both to architects and other vendors of professional services is relatively straightforward. Architects, doctors, engineers, attorneys, and others deal in somewhat inexact sciences and are continually called on to exercise their skilled judgment in order to anticipate and provide for random factors which are incapable of precise measurement. The indeterminate nature of these factors makes it impossible for professional service people to gauge them with complete accuracy in every instance. Thus, doctors cannot promise that every operation will be successful; a lawyer can never be certain that a contract he drafts is without latent ambiguity; and an architect cannot be certain that a structural design will interact with natural forces as anticipated. Because of the inescapable possibility of error which inheres in these services, the law has traditionally required, not perfect results, but rather the exercise of that skill and judgment which can be reasonably expected from similarly situated professionals. As we stated in *City of Eveleth v. Ruble*, 302 Minn. 249, 253, 225 N.W.2d 521, 524 (1974): "One who undertakes to render professional services is under a duty to the person for whom the service is to be performed to exercise such care, skill, and diligence as men in that profession ordinarily exercise under like circumstances." See, also, *Kostohryz v. McGuire*, 298 Minn. 513, 212 N.W.2d 850 (1973).

We have reexamined our case law on the subject of professional services and are not persuaded that the time has yet arrived for the abrogation of the traditional rule. Adoption of the city's implied warranty theory would in effect impose strict liability on architects for latent defects in the structures they design. That is, once a court or jury has made the threshold finding that a structure was somehow unfit for its intended purpose, liability would be imposed on the responsible architect in spite of his diligent application of state-of-the-art design techniques. If every facet of structural design consisted of little more than the mechanical application of immutable physical principles, we could accept the rule of strict liability which the city proposes. But even in the present state of relative technological enlightenment, the keenest engineering minds can err in their most searching assessment of the natural factors which determine whether structural components will adequately serve their intended purpose. Until the random element is eliminated in the application of architectural sciences, we think it fairer than [sic] the purchaser of the architect's services bear the risk of such unforeseeable difficulties.[53]

The city suggests that many of the design-related tasks performed by modern architects are routine and carry no risk of error

if they are performed with professional due care. It is argued that with respect to such tasks, the premise on which the traditional rule rests is inoperative, making the adoption of the implied warranty theory fully proper. We note, however, that architectural errors in relatively simple matters are quite easily handled under the existing cause of action for professional negligence.

Moreover, if implied warranties are held to accompany only uncomplicated architectural endeavors, the finder of fact will be forced in every case to determine, as a preliminary matter, whether the alleged architectural error was made in the performance of a sufficiently simplistic task. Defects which are found to be more esoteric would presumably continue to be tried under the traditional rule. It seems apparent, however, that the making of any such threshold determination would require the taking of expert testimony and necessitate an inquiry strikingly similar to that which is presently made under the prevailing negligence standard. We think the net effect would be the interjection of substantive ambiguity into the law of professional malpractice without a favorable tradeoff in procedural expedience.

In addition, we observe that the ills which spurred the creation and expansion of the implied warranty/strict liability doctrine are not really present in this case or in the architect–client relationship generally. The implied warranty of fitness originated primarily as a means of facilitating the legitimate interests of the consuming public and bringing common-law remedies into step with the practicalities of modern industrialism. The outmoded requirement of contractual privity, coupled with manufacturers' sweeping disclaimers of liability, frequently operated to deny effective remedies to those who purchased commercial products at the t of a multi-tiered production and distribution network. See, generally, Prosser, Torts (4 ed.) §§ 97, 98. The introduction of the implied warranty doctrine created an effective remedy by allowing plaintiffs to proceed directly against the offending party without reliance on express contractual warranties.

The relationship between architect and client is markedly different. For a client, architectural services are hardly produced by a faceless business entity, insulated by a network of distributors, wholesalers, and retailers. Architects and clients normally enjoy a one-to-one relationship and communicate fairly extensively during the course of the relationship. When a legal dispute arises, the client has no trouble locating the source of his problem, and a remedial device like the implied warranty is largely unnecessary.

Finally, while it is undoubtedly fair to impose strict liability on manufacturers who have ample opportunity to test their products for defects before marketing them, the same cannot be said of architects. Normally, an architect has but a single chance to create a design for a client which will produce a defect-free

---

[53]Our decision in *Robertson Lumber Co. v. Stephen Farmers Co-op Elev. Co.*, 274 Minn. 17, 143 N.W.2d 622 (1966), does not compel a different result. In that case, we affirmed an award of damages for a defectively constructed grain storage building on a theory of implied warranty. The contract in question, however, was treated as a construction contract and not an architectural contract. This distinction for warranty purposes between "professional" and general contracting services is well established in other jurisdictions. See, e.g., *Kriegler v. Eichler Homes, Inc.*, 269 Cal.App.2d 224, 74 Cal.Rptr. 749 (1969); *Stuart v. Crestview Mutual Water Co.*, 34 Cal.App.3d 802, 110 Cal.Rptr. 543 (1973); *Schipper v. Leavitt & Sons, Inc.*, 44 N.J. 70, 207 A.2d 314 (1965); *Weeks v. Slavick Builders, Inc.*, 24 Mich.App. 621, 180 N.W.2d 503, aff'd, 384 Mich. 257, 181 N.W.2d 271 (1970).

structure. Accordingly, we do not think it just that architects should be forced to bear the same burden of liability for their products as that which has been imposed on manufacturers generally.

For these reasons, we decline to extend the implied warranty/strict liability doctrine to cover vendors of professional services. Our conclusion does not, of course, preclude the city from pursuing its standard malpractice action against the architects and proving that the basement area of the new addition was negligently designed. That issue remains for the trier of fact in the district court.

Affirmed.

OTIS, J., took no part in the consideration or decision of this case.

That the court felt compelled to justify at some length its decision made four years earlier in *City of Eveleth v. Ruble* demonstrates some dissatisfaction with the professional standard. This is reflected in Section 14.06A, dealing with exceptions to the requirement of expert testimony; in Section 14.07, comparing the professional standard with that of implied warranty (including claims under consumer protection laws); and in Section 14.11A, analyzing some of the current controversies relating to professional liability.

Some refinements should be made in the professional standard rule. First, against whom is the conduct of the professional measured? Usually it is assumed that such conduct is measured against the conduct of others in the professional's locality. For example, if the professional practices in a small town, it is usually assumed that she should not be compared with professionals who practice in a large city. To some degree, this is based on the reasonable expectations of the client, and to some degree, on the likelihood that advances in the profession come first to large urban areas. The AIA Document B101-2007, Section 2.2, uses the locality standard.

It is more appropriate, however, to measure the conduct against those professionals with whom the defendant competes. Some design professionals in smaller cities compete against design professional firms in large metropolitan areas. Similarly, the choice made by the client may be one based on a comparison of those design professionals who are established in a particular specialty, and those specialty firms may be scattered in small towns, cities, and large urban areas. Some courts have held that an architect would not be measured solely against those in the locality in which she practices.[54] This may set a trend for a more refined application of the professional standard.

Similarly, differentiations are not always made regarding the precise nature of the standard to which the design professional will be held. Two authors developed a classification system for determining whether engineers are liable for design errors. They divided the state of the art into the cutting edge, the open literature, acceptance by the profession, and what they refer to as "the undergraduate horizon." The authors concluded that in the ordinary case, the challenged engineer should be measured against the "informed" engineer, who, they assert, practices in accordance with the state of the art accepted by the profession.[55]

The burden of establishing a failure to comply with the professional standard is generally on the claimant who seeks to transfer her loss to the professional.[56]

B. Expert Testimony and the Professional Standard

As noted in the preceding section, expert testimony is usually required to support a finding that the professional standard has not been met.[57] Because this has been a central issue in professional liability claims, it is discussed in Section 14.06.

C. Building and Housing Codes: The Americans with Disabilities Act

Design professionals are generally expected to consider and comply with building and housing codes.[58] Failure

---

[54]*Hill Constr. Co. v. Bragg*, 291 Ark. 382, 725 S.W.2d 538 (1987), opinion after remand on different issue, 297 Ark. 537, 764 S.W.2d 44 (1989); *Underwood v. WaterSlides of Mid-Am., Inc.*, 823 S.W.2d 171 (Tenn.App.1991); *Parsons Main, Inc.*, ASBCA Nos. 51355, 51717, 02-2 BCA ¶31,886.

[55]Peck & Hoch, *Liability of Engineers for Structural Design Errors: State of the Art Considerations in Defining the Standard of Care*, 30 Vill.L.Rev. 403 (1985).

[56]*Coulson & C.A.E., Inc. v. Lake L.B.J. Mun. Util. Dist.*, 734 S.W.2d 649 (Tex.1987). For subsequent opinions on different issues, see 771 S.W.2d 145 (Tex.Ct.App.1988) and 781 S.W.2d 594 (Tex.1990).

[57]See Farrug, *The Necessity of Expert Testimony in Establishing the Standard of Care for Design Professionals*, 38 DePaul L.Rev. 873 (1989) (attacking Illinois law, which does not require expert testimony in cases of professional malpractice).

[58]See Sections 14.03, 14.05C.

to design in accordance with such codes is almost always considered to violate the obligation the design professional owes her client. Code compliance is considered one aspect of meeting the professional standard imposed by law on design professionals. In addition, the code violation may grant private rights to those who are damaged by the violation. This occurs if the statute specifically designates that a violation gives private rights to individuals damaged by the violation or if the statute is so interpreted.[59]

Building codes can determine design, materials, and construction methods. Design specifications, performance standards, or a combination of the two is used. Codes provide minimum standards to protect against structural failures, fire, and unsanitary conditions.

Housing codes are of relatively recent vintage. They have gone beyond structure, fire, and basic sanitation to include light, air, modern sanitation facilities, maintenance standards, and occupancy density rules. Sometimes they are used to measure the implied warranty of habitability owed tenants by landlords.

Although there is some movement toward statewide and even federal building codes, local codes predominate. In addition to reflecting extreme local variations in subsurface conditions and climate, local control reflects strong democratic roots in the local community. (Yet, as noted in this subsection, the federal government recently enacted the Americans with Disabilities Act, part of which deals with building requirements.)

Codes generally fall into one of three major types developed by various private associations, such as fire underwriters and local building officials. Choice within these types is often regional, with one part of the country choosing a particular type. Yet frequent local modification and variant local interpretation have made for variations. These, along with unrealistic and unneeded standards, often dictated by local special interest groups, have undoubtedly raised construction costs.

As shall be seen in some of the cases noted in this section, codes are often difficult to understand and are subject to uneven if not whimsical local interpretation. Unrealistically high code standards often exclude low- and moderate-income people from certain communities.

Moreover, the codes are not always enforced and can lead to bribery.

Some of the strengths and weaknesses of local building codes can be demonstrated by two cases. *Greenhaven Corp. v. Hutchcraft & Associates*, an Indiana case,[60] involved a claim by an architect against a client for fees for providing plans for remodeling a building. The preliminary plans called for two exits from the top floor, but the owner's representative, a contractor, requested that the plans be altered to provide for only one exit. The architect complied with this request. This would have violated the building code, but the project was abandoned.

The court noted that the architect impliedly promises to draw plans and specifications that comply with local codes. But it held that parties can contract for nonconforming plans—that contracting parties can make any agreement, as long as it is not illegal or contrary to public policy.

The court stated that neither party contended that an agreement permitting nonconforming plans was contrary to public policy. Also, the court stated that the public was not harmed by such an agreement. Before the building could have been occupied, the fire marshall would have had to approve by issuing an occupancy permit. If the fire marshal had not issued such a permit, the parties might have been able to obtain a variance or change the plans to conform to the requirements. In either case, the court found the public was protected regardless of the agreement between the parties. It also noted that the availability of variance procedures encourages nonconforming plans.

The court showed little respect for the building codes, but its language also recognized the flexibility achieved through the variance process. Although variance provides for flexibility—very useful if codes become too rigid and out of touch with reality—such a system can also lead to corruption.

The second case, *Edward J. Seibert, A.I.A., Architect and Planner, P.A. v. Bayport Beach and Tennis Club Association, Inc.*,[61] arose in Florida. The issue was whether the architect had designed in accordance with the local building code. The dispute also involved a fire exit. The architect had designed each second-floor dwelling unit with one independent unenclosed set of stairs leading directly from the front door of each unit to the ground.

[59]*Huang v. Garner*, 157 Cal.App.3d 404, 203 Cal. Rptr. 800, 804–06 (1984), rev'd on other grounds, *Aas v. Superior Court*, 24 Cal.4th 627, 12 P.3d 1125, 101 Cal.Rptr.2d 718 (2000).

[60]463 N.E.2d 283 (Ind.Ct.App.1984).

[61]573 So.2d 889 (Fla.Dist.Ct.App.1990), review denied, 583 So.2d 1034 (Fla.1991).

The completed plans were submitted to the city's chief building inspector, who was also the chief code enforcement officer. The inspector's interpretation of the building code led him to conclude that because of the size of the units and because the exit was unenclosed, the single unenclosed exit design complied with the code. After the fire department approved the plans, the building inspector issued a building permit, and the units were ultimately built according to these plans.

Two years later, the individual condominium owners assumed control of the association and filed a claim against a number of parties, including the architect, for various defects. Although the jury exculpated the architect for many defects, it found him liable for the improperly designed fire exit.

To determine whether there had been a violation, the court extensively examined the testimony of experts called by the parties. The association presented an expert witness—a structural engineer who testified that in his interpretation the exit did not comply with the code.

The architect presented two experts who testified to the contrary. One was a professional engineer who testified he had been employed by the Southern Building Code Congress, which had promulgated the code, and that during his employment he was responsible for building code changes, hearings, plans, reviews, and code interpretation. The engineer testified that the architect's design complied with the code and was consistent with safe design as intended by the code. Yet the jury, as noted, found the architect liable for the cost of adding a second exit.

The appellate court noted that the jury needed to know what the code required and that such information could be presented to the jury by means of expert testimony. The court stated that expert testimony could be presented if the jury needed to understand the evidence better or to determine a key issue in the case, but experts could not testify as to how the code should be interpreted. This was a question of law. The court held that the trial court should have interpreted the meaning of the code and instructed the jury concerning that meaning.

Finally, the appellate court held the trial judge should have concluded that the architect had complied with the code, because the chief building inspector and chief code enforcement officer stated the design did comply. This was relied on by the architect, and the building was constructed with the approved design. The court noted that even though a code could be interpreted in more than one way, the law will follow the interpretation of the agency that has authority to implement the code. The appellate court concluded by stating a judgment should have been entered in favor of the architect.

It is difficult to quarrel with the conclusion that the architect can rely on the interpretation of the building inspector. However, the fact that codes can be interpreted in various ways can make compliance difficult and encourage corruption.

The Americans with Disabilities Act (ADA)[62] enacted in 1990 generated building standards as part of an overall policy to protect disabled persons, a group that has long been subjected to discrimination in employment, housing, education, transportation, public accommodations, recreation, and health services. The act comprehensively defines the treatment of disabled persons, imposes significant new obligations on employers, and (most importantly for design professionals) mandates that commercial and public accommodations be accessible to people with disabilities.

Title 3 of the ADA states that existing buildings, new constructions, and alterations are governed by the requirement that individuals with disabilities have access to almost all businesses and public places. The rules exempt private clubs, religious institutions, residential facilities covered by fair housing laws, and certain owner-occupied inns.

Many of the details will be expressed in regulations (Americans with Disabilities Accessibility Guidelines) provided by the U.S. Architectural and Transportation Barriers Compliance Board. The guidelines are based largely on existing American National Standards Institute (ANSI) standards. Construction completed after January 26, 1993, must be readily accessible to and usable by people with disabilities. Even more important, the ADA requires removal of architectural barriers in existing buildings if the structure is considered a place of public accommodation. There are some exceptions, such as removal not being readily achievable. Achievability is defined by the guidelines as "easily accomplishable and able to be carried out without much difficulty or expense." The regulations take into account the nature and cost of removal, the financial resources of the party involved, and the impact of removal on the operation of the site and on profitability. These factors permit a case-by-case approach.

---

[62]Pub.L. No. 101-336, 104 Stat.327 (1990) codified in 42 U.S.C.A. § 12101 et seq.

Alterations to a place of public accommodation built after January 26, 1992, must ensure to the maximum extent feasible that the altered portions are readily accessible to and usable by disabled people, including those who use wheelchairs. The ADA defines certain areas of a building as having a "primary function." In those areas there are additional requirements. Not only must the design conform to Title 3 requirements, but the path of travel to the altered area to the maximum extent feasible must be accessible to and usable by the disabled. But if alterations to the path of travel cost more than 20 percent of the cost of the alterations, this is considered disproportionate and not required. The barriers are not limited to mobility impairments but can include those that impair vision, hearing, and reading.

If there is a violation of Title 3, the court can issue an order requiring the owner to modify physical facilities. The court can award monetary but not punitive damages, civil penalties, and attorneys' fees. In considering civil penalties up to $100,000 under Title 3, the court can take into account any good-faith efforts to comply.

Many state and local laws deal with the same topic. A state or locality can ask the federal attorney general to certify that its code complies with the ADA. If the certification is given, compliance with local or state building codes will also comply with the federal act. If local and state requirements are less stringent, the federal act will supersede them.

In effect, the ADA has created a form of national building code, but no entity is obligated under the ADA to review and approve drawings and specifications for compliance. Thus parties seeking to comply may not know whether they have discharged their obligation until a complaint is filed.

This legislation should provide a significant market for the services of design professionals. Some clients may seek to shift the ultimate risk of compliance to the design professional by providing that the design professional must certify that the design complies with the law. This certification may be uninsurable. Although liability resulting from negligent failure to produce a design that complies with codes is clearly insurable, contract provisions that turn the duty of compliance into a guarantee of compliance can create the possibility of exclusion from coverage.

May disabled persons who believe a building is not in compliance with the ADA sue the architect for violation of the act? That straightforward question does not yield a simple answer because the statute is unclear as to: the category of persons potentially liable under the act; the types of buildings subject to the act; and the types of activities for which one may be liable.

Two "liability" provisions of the ADA are inconsistent as to both the category of persons potentially liable under the act and the types of buildings subject to it. Section 302 broadly prohibits discrimination on the basis of disability "by any person who owns, leases (or leases to), or operates a place of public accomodation."[63] A simple reading of this provision would exclude liability for architects (or contractors) by limiting liability to those who own, lease, or operate places of public accommodation.

However, § 303 (titled "Application") appears to create a wider net of liability. In describing the conduct which constitutes "discrimination" under § 302, § 303 states:

> As applied to *public accommodations and commercial facilities,* discrimination for purposes of Section 12182(a) of this title includes—
>
> (1) a failure to *design and construct* facilities for first occupancy . . . that are readily accessible to and usable by individuals with disabilities. . . .[64] [Emphasis added.]

Section 303 appears to expand the universe of potentially liable persons and buildings from that described in § 302 in two respects: (1) § 303 adds a reference to "commercial facilities" and (2) § 303 imposes liability upon those who fail to "design and construct" conforming buildings. The "design and construct" language itself is ambiguous: will liability attach only to those who both design *and* construct the buildings (in which case the act would not apply to architects), or does it apply to those who either design *or* construct the buildings (in which case architects who design buildings not in conformity with the act and its regulations could be liable)? Not surprisingly, the courts have reached differing conclusions on the question of design professional liability under the ADA.[65]

---

[63]42 U.S.C.A. § 12182(a).

[64]Id. § 12183(a), italics added.

[65]Compare *Paralyzed Veterans of Am. v. Ellerbe Becket Architects & Engineers, P.C.*, 945 F.Supp.1 (D.D.C. 1996), aff'd on other grounds, 117 F.3d 579 (D.C.Cir.1997), cert. denied sub nom. *Pollin v. Paralyzed Veterans of Am.*, 523 U.S. 1003 (1998) (architect not liable); *United States v. Days Inns of Am., Inc.*, 151 F.3d 822 (8th Cir.1998), cert. denied, 526 U.S. 1016 (1999) (architect cannot be liable but design/build contractor or owner may be); and *Lonberg v. Sanborn Theaters Inc.*, 259 F.3d 1029 (9th Cir.2001) as amended on denial of rehearing *en banc* 271 F.3d 953 (2001) (same, noting *in dicta* that statutory liability would also not extend to general contractors and subcontractors), with *Johanson v. Huizenga Holdings, Inc.*, 963 F.Supp. 1175 (S.D.Fla.1997) (architect may be liable under ADA) and *United States v. Ellerbe Becket, Inc.*, 976 F.Supp. 1262 (D.Minn.1997) (same). For a comprehensive discussion, see Powers & Berg, *Accessiblity: Not Just the Owner's Responsibility,* 21 Constr. Lawyer, No. 2, Spring 2001, p. 13.

D. Contractual Diminution of Standard: Informed Consent

As noted in Sections 14.04 and 14.05, the professional standard is residual, applying only where the parties have not contractually agreed to a different standard. Usually any specific standard is more strict; the fact that the design professional did what others would have done will not in itself relieve the design professional. Suppose the design professional can point to a lower specific standard in the contract. Although rare, exploration of this possibility uncovers difficult problems.

Suppose the design professional would have selected X, a material that would have been designated by other professional designers because of its combination of durability, low maintenance, cost, and appearance. But the client, being aware of the tradeoffs orders that Y be used simply because it is the least expensive material. Suppose the material selected needed replacement earlier than the client expected. The client should not be able to recover any correction costs against the designer. The client and the designer agreed to a standard different from the professional standard—in this case, a lower standard. Assuming that this choice does not expose others to unreasonable risk of harm, the law will let the contracting parties decide whether the professional standard or something less will satisfy the designer's contractual obligations.

But suppose an applicable building code requires X, and Y is thought to be unsafe. The design professional makes the client aware of the code but the client insists on Y. What are the consequences? Several issues are presented:

1. Will the design professional be denied recovery for services that relate to selection of this material?
2. Will the design professional be given a defense if the client sues her for the cost of corrective work?
3. Will this selection expose the design professional to negligence claims by any third parties who suffer losses because of this choice?
4. Will the client and the design professional who violated the building laws (and any contractor who knowingly violates the code) be exposed to criminal prosecution?
5. Will this selection be grounds for disciplining the design professional under the registration laws?

The answers to these questions should be yes. Because of the importance of design, the transaction cannot, like the preceding one, be simply regarded as a private one

between client and design professional. In matters of safety, public protection takes precedence. *Bowman v. Coursey*[66] illustrates the difficulty in determining whether a particular agreement required performance less than that called for by the professional standard and whether such an agreement is simply a private arrangement or one that involved public safety.

Coursey hired Bowman to design a warehouse. Bowman's original plans included pilings beneath the walls in the floor slab. Because Coursey wanted to reduce the cost, the plans were revised. Coursey wanted to eliminate the pilings beneath the floor area. Bowman told Coursey the revisions would very likely mean the floor would be subject to settlement. Coursey asked how long it would take for the expected settlement to occur and, when it occurred, how the settlement could best be handled. After consulting with the soils engineer and an engineer for the lender, Bowman revised the design so that the floor slab would be reinforced to minimize the effect of any settlement. The revised plans were submitted to and approved by the local regulatory agency.

After construction began, problems developed with the quality of the construction work. Experts criticized the wall construction and the foundation design. A dispute arose over whether the architect Bowman had been negligent in determining the capacity of the piles beneath the wall area. Work on the project ceased. Coursey brought legal action against Bowman. A number of experts testified at the trial, most of whom testified that they would not have designed as did Bowman. They agreed that some settlement was inevitable, but none would say that the design was necessarily wrong or that the building was inadequate. One testified that the owner should be made aware of the possibility of settlement. An expert testifying for Bowman stated that although this was not the best design possible, it was adequate. He indicated that as long as Coursey knew what he was getting, allowing the floor to float was not a bad condition. Settlement, according to that witness, would occur in a controlled fashion because of the stiffness of the walls. This might be acceptable to the owner.

The court concluded that if the warehouse were built according to the revised plans, it would be less than perfect but the imperfections were not caused by a breach by the architect Bowman but by the client Coursey's preferring the imperfections to further financial outlays. When warned that the structure would settle, Coursey's response

---

[66]433 So.2d 251 (La.App.), cert. denied, 440 So.2d 151 (La.1983).

had been that he could live with it. The court noted that additional pilings would have been necessary to prevent settlement, but none of the witnesses concluded the design was clearly wrong or engineeringly unsound. The design did not violate the building code, nor would it present a danger or hazard to people working within the building. The court concluded that the problems Coursey would face were a tradeoff for reduced building expenses.

Although the court concluded Bowman had not been negligent, its conclusion was based principally on Bowman's having pointed out all the problems to Coursey and Coursey's deciding to accept certain risks in exchange for a lower construction cost. The court emphasized that the controlled settlement that was likely would not present a danger or hazard to persons within the building and that the design did not violate the building codes. Yet the expert testimony indicated that the experts, though testifying in careful and guarded terms, indicated that they would not have designed the warehouse in the way chosen by Bowman with Coursey's concurrence. In essence, the parties agreed to a design below the standard other professionals would have used.

This problem also exposes another issue—one increasingly debated. Must any client's consent be informed? Taking a leaf from the law regulating the relationship between physician and patient, one writer advocated that architects be required to inform their clients of the costs and benefits of design choices.[67] This can, under certain circumstances, be related to the requirement of good faith and fair dealing, which augments the express obligations required of contracting parties. The uncertain dimensions of such a duty and the increasing concern for expanded liability even under the generally more protective professional standard may be why informed consent has not as yet received overt approval.

E. Tort and Contract

Often the claimant prefers tort law, with its more limited foreseeability defense and its more potent remedies. For example, if the claim is based on breach of contract it is very difficult to recover for emotional distress and almost impossible to recover punitive damages. However, a tort claim may justify recovery of damages for emotional distress and, if a breach of contract is also considered a tort, may justify recovery of punitive damages.

Yet there are advantages to bringing the claim for breach of contract if that alternative exists. The time limit for commencing a lawsuit is longer—often very much longer—when the claim is based on a breach of contract rather than on a tort theory. (However, if the period to start a tort claim does not begin until the claimant discovers it has a claim against the defendant, the time period may, in states with a short contract period, be longer than for a breach of contract.)

Unlike claims by patients against their doctors, claims by clients against their design professionals are more likely to involve specific provisions of a contract. If the client is permitted to bring its claim in tort, it may be able to bypass certain contract defenses, such as the parol evidence rule, the statute of frauds, or exculpatory provisions found in the contract between the client and its design professional.

In addition, claims based on commission of a tort sometimes run afoul of the economic loss rule. Although this is more commonly a problem when claims are made by third parties, as discussed in Section 14.08E, it can also be a problem when the claim is by one party to a contract against the other. Usually there is no difficulty when patients sue their doctors, because there will be a claim for physical harm to which claims for economic losses can be attached. Claims by owners or contractors against design professionals usually do not involve physical harm but are more likely to involve purely economic losses, those based on disappointed expectations. In such cases, a tort claim for economic losses may not be sustained.

Also, claims against design professionals are generally based on the failure of the design professional to perform in accordance with the professional standard discussed in Section 14.05A. Although there is some tendency toward incorporating into the contract language that deals with how the services are to be performed, claims are most commonly based on a term implied by law—the professional standard. In such cases, it has been relatively easy for courts to focus on the professional relationship, to assert that the relationship creates the duty, and to conclude that the contract is simply a way to implement the relationship. Although a tort claim requires that negligence be established, whereas a breach of contract claim is, at least in theory, "strict," claims by clients against their design professionals, as mentioned earlier, usually employ the same standard—the professional standard—whether based on tort or contract. Only if the contract contains a refinement of the professional standard or imposes some

---

[67]Note, 30 Hastings L.J. 729 (1979).

form of strict liability will liability be based on a standard other than the professional standard.

However, courts are careful not to allow clients to convert descriptions of the architect's services into either express warranties or independent representations. Otherwise, every breach-of-contract claim also would be a breach of warranty or a misrepresentaion.[68]

Although some courts seem to look for the gravamen, or essence, of the claim,[69] the client will likely be able to assert a claim based on the commission of a tort or the breach of a contract, whichever is more advantageous to the client. Very likely the client will be able to assert alternative claims. Only at the stage when the judge determines that the client must choose a remedy must the client choose the theory on which it is basing its claim.

Yet, as seen in Section 12.08B, many contracts between clients and design professionals, especially those published by the AIA, clearly express the rights and obligations of both parties and also provide contractual exculpation for the design professional. Where such exculpations appear unfair, the client will likely be given the option of maintaining the claim under tort law.

As noted earlier in this Section, the law gives the claimant the widest possible berth when bringing a claim against a professional whom it has retained, thus allowing the claimant to maintain an action either in contract or in tort, depending on which theory is most advantageous. Although this is accomplished by stating that the duty exists independent of the contract and the contract is only a means of executing the duty—a doubtful fiction at best—it seems clear that their contract is the underlying basis for claims by a client against the design professional. As indicated in Section 14.04A, the contract may be explicit as to specified duties and how they are to be performed or it may be explicit as to duties but silent as to the standard. Many retention arrangements are silent as to specific duties and how they are to be performed.

A contract that details the duties of each party and allocates the risks of likely losses is a plan under which the parties agree to exchange their performance and apportion risks in a specific manner. To disregard that plan by allowing tort claims exposes design professionals to risks they did not plan to undertake and for which they were not paid. Clients should not have a choice of bringing their actions in contract or tort. In planned transactions, they should be required to base their claims on breach of the contract.

This suggestion is compatible with a desire to encourage people to plan their transactions by freeing them from the risk of open-ended tort exposure. Contracting parties who have suffered emotional distress can be protected by classifying their claims (though based on contract) as torts. Similarly, if there is a need to punish or deter, the claim can be based on tort law. Other claims based on failure to perform under the design professional–client relationship should be based solely on the contract.

SECTION 14.06 Expert Testimony

One important component of the professional standard is the need, at least as a general rule, for expert testimony. Section 14.06A looks at the exceptions to that requirement, 14.06B examines expert testimony generally and particularly in the context of claims against design professionals, 14.06C looks at criticism of the current system, and 14.06D approaches the problem from the vantage point of the expert witness.

A. Purpose and Exceptions to General Rule

In the judicial system, judges and juries make decisions. To do so, they may have to hear, evaluate, and judge testimony and exhibits that relate to technical matters unfamiliar to them. To help them, the law permits evidence of opinion testimony as an exception to the rules of evidence that generally bar opinion testimony. But before such opinion testimony can be admitted, the issue must be too difficult for a judge or jury to decide without technical assistance. Also, the person permitted to give opinions must possess the necessary education and experience to

---

[68]See *Dickerson Internationale, Inc. v. Klockner*, 139 Ohio App.3d 371, 743 N.E.2d 984, 987-91 (2000) (architect's breach of contract is not also a misrepresentation, breach of warranty, or fraud); *SME Industries, Inc. v. Thompson, Ventulett, Stainback & Assocs., Inc.*, 28 P.3d 669, 676–77 (Utah 2001) (contract's description of architect's services is not an express warranty); *Howard v. Usiak*, 172 Vt. 227, 775 A.2d 909, 913–14 (2001) (no negligent misrepresentation); and *Rochester Fund Municipals v. Amsterdam Municipal Leasing Corp.*, 296 A.D.2d 785, 746 N.Y.S.2d 512, 515–16 (2002) (no express warranty or intentional misrepresentation).

[69]*Lesmeister v. Dilly*, 330 N.W.2d 95 (Minn.1983); *Mac-Fab Products, Inc. v. Bi-State Dev. Agency*, 726 S.W.2d 815 (Mo.Ct.App.1987). See generally, W. P. KEETON et al., TORTS, 664–667 (5th ed. 1984).

be an "expert" on the issue for which the judge or jury needs help. Yet the expert is generally not allowed to give opinions that could determine the outcome of the case. This power is still retained by the judge and (when used) the jury. The importance of limiting expert testimony can be seen by the *Seibert* case, discussed in Section 14.05C, where the court held that the expert would not be allowed to testify on the proper interpretation of a building code.

The professional standard—generally—requires expert testimony to support any conclusion that the design professional has not performed in accordance with ordinary professional standards. There are important exceptions to this general rule. For example, in *City of Eveleth v. Ruble*, discussed in Section 14.05A, the plaintiff city had retained defendant engineer to design a new water treatment plant. After completion, two difficulties developed. First, the intake system that took water from a lake and processed it for distribution into the city's transmission lines and storage reservoir proved inadequate. Second, pressure in the cast-iron distribution lines leading from the water treatment plant to users and storage facilities caused some leaded joints in the line to give way. Without any expert testimony, the trial court awarded the city damages against the engineer.

As to the diminished intake capacity, the court noted that the engineer knew the city had decided to build a new water plant to increase the intake capacity to a designated number of gallons per minute. The intake capacity was inadequate because there was a failure to anticipate changes in the lake level. The existing intake line that was laid on the bottom of the lake in approximately 40 feet of water was not the 18-inch line expected by the engineer but in some places was 16 inches and in other places only 12 inches in diameter.

The Minnesota Supreme Court concluded that the trial judge was able to assess the validity of these excuses "without the aid of expert testimony." In drawing this conclusion, the court stated,

> In our judgment, no expert opinion is needed to demonstrate that a design engineer charged with the responsibility of analyzing the piping and other structural characteristics of an existing plant should be as certain of the dimensions of the intake line as circumstances would possibly permit before recommending a plan the function of which depended on this critical measurement. It would seem clear that the examination of a photograph of the line would be of little value. Incomplete drawings made available to the Engineer at its request indicate that the intake line was 18 inches in diameter at the point of terminus with the old plant, but we believe that common knowledge would reject this as adequate basis for careful analysis. We find the explanation given for the failure of the Engineer's employees who entered the lake for the purpose of measuring the intake line unsatisfactory when the record shows that others employed for this same purpose by the City were able to obtain the true dimensions of the intake line without difficulties disproportionate to the importance of the task.[70]

As to the excess pressures in the distribution lines causing the joints to give way, the trial court found that the high-service pumps that distributed the clear water to the storage tanks caused sudden surges "with consequent water hammer," blowing out the lead-sealed joints. Evidently, the cast-iron distribution lines were buried in the ground and had been in use for approximately sixty years. The court concluded it would not be fair to hold the engineer to any "express or implied commitment that the City's transmission lines would be trouble-free following the installation of the new facility."[71] The court stated that "it would not be reasonable to expect the Engineer to guarantee the performance of these lines under these circumstances." The court relieved the engineer by concluding that expert testimony was required, because determining what type of pressure system should be used with pipes of this age was a technical question. In employing the professional standard, the court concluded,

> Would a design engineer, in the exercise of that degree of care, skill, and diligence to be expected from this profession, knowing the line pressures which would be created by the operation of the high service pumps and knowing the age of the line and the character of its construction, have reasonably anticipated that it would fail in use? If uncertain, would he have employed tests or techniques of inspection which could have been employed and which would have been employed by a design engineer applying the requisite standards of skill and care which should have revealed the deficiencies in the line? If not, would such an engineer, being unable to ascertain the facts, have recommended the installation of devices such as those now recommended as a precaution against possible but unpredictable ruptures? We do not think that common knowledge affords answers to these questions.[72]

---

[70]302 Minn. 249, 225 N.W.2d 521, 527–28 (1974).
[71]225 N.W.2d at 529.
[72]Id. at 530.

The wide range of services often performed by the design professional in both design and construction phases will inevitably raise questions as to which services are sufficiently technical to require expert testimony and which are sufficiently nontechnical that the lay judge or jury needs no assistance. In many cases, the plaintiff cannot or chooses not to offer expert testimony and claims it is not needed, whereas the defendant contends that the lack of expert testimony bars any judgment of professional malpractice.

Claims based on design services as a rule require expert testimony. Choices that involve excavation,[73] design,[74] foundation sufficiency, structural stability, equipment and components,[75] protection against the elements,[76] energy efficiency,[77] surface water disposal,[78] and designing a sidewalk for a slope[79] are all matters that require expert testimony. Often they are technical areas for which the prime design professional will retain consultants. Whether such services are performed in-house or by outside consultants, they require decisions that should be made by professionals with specialized education and experience. Often these services must be performed only by people registered by the state, although this is not determinative.

Yet even where the services seem to be technical in nature, such as a defective soils report, one case excused the requirement for expert testimony, stating it would not be needed

> when the conduct of [the professional] is so unprofessional, so clearly improper, and so manifestly below reasonable

standards dictated by ordinary intelligence, as to constitute a prima facie case of either a lack of the degree of skill and care exercised by others in the same general vicinity or failure to reasonably exercise such skill and care.[80]

Laypeople can infer negligence by applying common sense.

As to site services unconnected to design, the issue becomes more cloudy. Is expert testimony needed to help a judge or jury decide whether the design professional should have certified a particular amount to be paid, detected a design deviation in a shop drawing or submittal, noted a deviation from an approved schedule, verified a changed condition, or coordinated the work of separate contractors, to name some of the tasks often given to the design professional?

In two cases, courts held that no expert testimony was needed to support claims based on construction-phase services, which, according to the courts, did not require specialized technical skill possessed only by people who have particular education and training.[81] These courts lumped all construction-phase services into the uninformative "supervision" category, failing to differentiate activities that require professional skill, such as preparing change orders, interpreting contract documents, or resolving disputes.

A federal district court interpreting Louisiana law did not need expert testimony to find an engineering firm negligent in failing to discover overbilling and other fraudulent conduct by the contractor it was supervising.[82]

At the other extreme, Virginia held that expert testimony would be required to support a claim by the owner against its architect based on the architect's asserted failure to detect defects in a complex project and the latter's alleged delay in processing proposed change orders.[83]

The increasing number of claims against design professionals based on their site services should develop rules that separate services requiring professional education and training from services that can be and often are performed by people without such training, such as

---

[73]Harbor Ins. Co. v. Schnabel Foundation Co., 992 F.Supp. 419 (D.D.C.1997) (design of sheeting and shoring system intended to stabilize the excavation walls and protect adjacent properties), vacated after settlement by the parties, 992 F.Supp. 437 (D.D.C.1997).

[74]Nauman v. Harold K. Beecher & Assoc., 24 Utah 2d 172, 467 P.2d 610 (1970). But a claim based on a failure to discover a structural barrier in a renovation project did not require expert testimony. See Milton J. Womack, Inc. v. State House of Representatives, 509 So.2d 62 (La.App. 1987), cert. denied, 513 So.2d 1208 (La.1987).

[75]Dresco Mechanical Constructors, Inc. v. Todd-CEA, Inc., 531 F.2d 1292 (5th Cir.1976); John Grace & Co. v. State Univ. Constr. Fund, 64 N.Y.2d 709, 475 N.E.2d 105, 485 N.Y.S.2d 734 (1984).

[76]South Burlington School Dist. v. Calcagni-Frazier Zajchowski Architects, Inc., 138 Vt. 33, 410 A.2d 1359 (1980).

[77]Board of Educ. v. Hueber, 90 A.D.2d 685, 456 N.Y.S.2d 283 (1982).

[78]National Cash Register Co. v. Haak, 233 Pa.Super. 562, 335 A.2d 407 (1975), reproduced in part in Section 14.06B. For a collection of cases, see Annot. 3 A.L.R.4th 1023 (1981).

[79]Seaman Unified Sch. Dist. v. Casson Constr. Co., 3 Kan.App.2d 289, 594 P.2d 241 (1979).

[80]Nicholson & Loup., Inc. v. Carl E. Woodward, Inc., et al., 596 So.2d 374, 381 (La.App.1991), writ denied, 605 So.2d 1098 (La.1992). Accord, Zontelli & Sons, Inc. v. City of Nashwauk, 373 N.W.2d 744 (Minn. 1985) (massive errors meant that expert testimony not necessary to establish that a bid for municipal street work was negligently prepared).

[81]Jaeger v. Henningson, Durham & Richardson, Inc., supra note 43 (approval of shop drawings); Bartak v. Bell-Galyardt & Wells, Inc., 629 F.2d 523 (8th Cir.1980) (checking on contractor compliance).

[82]City of Houma, La. v. Municipal & Industrial Pipe Service, 884 F.2d 886, 890 (5th Cir.1989).

[83]Nelson v. Commonwealth, 235 Va. 228, 368 S.E.2d 239 (1988).

inspectors and even construction managers. The former services should require expert testimony; the latter should not. Care must be taken to distinguish activities that are part of the design professional's "judging" role. In attacking a decision made by a design professional, expert testimony may not be needed, because the issue is not whether the design professional did as others would have, but the finality of her decision. This is discussed later, in Section 29.09.

Claims by those who engage one entity to both design and build, or who buy from a builder-vendor, raise special problems. At the outset, a claim based on professional malpractice must be differentiated from one based on implied warranty, a standard that usually measures the obligation of the builder-vendor or developer. In the latter, negligence is not the standard.[84] As a result, no expert testimony is required.

B. Admissibility of Testimony: *National Cash Register Co. v. Haak*

Before proceeding to the principal focus of this section—who can testify and what type of testimony is needed—a few preliminary remarks relating to admissibility generally must be made.

First, a party intending to call an expert witness must notify the other party before trial of the identity and qualifications of the expert and the issues on which expert testimony will be elicited. This lets its opponent investigate or use the deposition process to check on the qualifications of the experts the other side plans to use.

Second, the expert must give her opinion based on firsthand knowledge (such as an expert medical witness who has conducted her own examination), on facts admitted into evidence, or on a combination of firsthand knowledge and evidence. If the expert has no firsthand knowledge, in many states the opinion of the expert is based on a set of hypothetical facts included in the question and based on evidence that has been admitted or evidence the party calling the expert plans to introduce. Sometimes the expert can observe the testimony of witnesses and be asked to assume the truth of previous testimony as a basis for her opinion. This can simplify the often bewilderingly complex hypothetical question.

Third, a witness who is not registered in accordance with the registration laws of the state very likely will be permitted to testify.[85] Local law must be consulted.

The qualifications of an expert to give an opinion—an increasingly difficult issue because of overlapping specialization in professions—are discussed in the following opinion written by an outstanding state court judge.

NATIONAL CASH REGISTER CO. v. HAAK

Superior Court of Pennsylvania, 1975. 233 Pa.Super. 562, 335 A.2d 407.
[Ed. note: Footnotes omitted.]

SPAETH, Judge.
[Ed. note: After construction of a manufacturing plant was completed, sinkholes adjacent to dry wells developed, threatening the integrity of the building. The problem was diagnosed by Gannett-Fleming, an engineering firm. Corrective work was done. The owner brought a claim against the architect for negligent design. The controversy on this appeal centers around the adequacy of the testimony of the four expert witnesses called by appellant owner].

Charles W. Pickering was admitted as an expert in "civil engineering in hydraulics" (the court's characterization) or as a "civil engineer familiar with hydraulics" (defense counsel's characterization). He was employed by Gannett-Fleming

was the one who had examined the site for that firm and had recommended the removal of the dry wells and the installation of the new system. His opinion on the cause of the sinkhole activity was unequivocal:

> It is my opinion that the Surface Water System as was installed on the NCR site has accelerated the formation of sinkhole activity on the site.

---

[84]See Section 24.10.

[85]*Corcoran v. Sanner*, 854 P.2d 1376 (Colo.App.1993); *Thompson v. Gordon*, 221 Ill.2d 414, 851 N.E.2d 1231 (2006); *Owens v. Payless Cashways, Inc.*, 670 A.2d 1240 (R.I.1996); *Martin v. Barge, Waggoner, Sumner & Cannon*, 894 S.W.2d 750 (Tenn.App.1994), appeal denied, Mar. 9, 1992. For a review of the rules of evidence as applied to construction litigation, see Cohen, *Obstacles to Admitting Evidence in Construction Cases*, 20 Constr. Lawyer, No. 1, Jan. 2000, p. 36.

[I]f the present system were to be continued in use, . . . the formation of sinkholes would continue and with the ultimate possibility or ultimate meaning at sometime, some point in time, because it is a natural phenomenal [sic] it cannot be predicted, that at sometime there may be serious damage caused to the major facilities on the site.
Q. Would that include the building?
A. Yes.

Timothy Saylor was admitted as an expert in geology. He was also employed by Gannett-Fleming, and his testimony, where relevant to the issues to be considered here, corroborated Mr. Pickering's.

Professor Jacob Freedman of Franklin & Marshall College also testified as an expert in geology. His qualifications indicated extensive experience in his field covering over 25 years, including being frequently called in as a consultant on the geology of Lancaster County, especially with regard to "water problems, foundations problems, [and] studies of quarries." His testimony corroborated Mr. Pickering's and Mr. Saylor's with regard to the causal relationship between the dry wells and the sinkholes. He then added the following at the end of re-direct examination:

I probably ought to say and I probably haven't said this in my discussion so far along these lines, that I have and I know no geologist who has ever recommended dry wells for construction in an area like this; that, in fact, we urge people not to use this kind of system; and if they have French drains—in fact, as I say, we frequently get calls at school; and anytime anybody mentions a French drain we tell them they are in for trouble, and French drains and dry wells are very similar in their properties. So these are systems that should not have been installed because they lead to trouble.

This brief statement (which was not given in direct response to any question, but which also was not objected to) was immediately explored on recross-examination:

Q. This is your opinion, Professor, is that correct?
A. Let's say it is not only opinion. It is an observation.
Q. And I understand you to say that no geologist that you know of recommends this system for this type of area?
A. I know all my colleagues invade [sic; "inveigh"?] against them.
Q. These are all of your colleagues at Franklin & Marshall?
A. Everybody I have talked to.
Q. At Franklin & Marshall?
A. No, at other places too.
Q. Pardon me?

A. Other places, other colleagues. We discuss these at meetings and I have been a consultant on a situation where I have seen the result of one of these things.

That was the complete recross-examination of this witness.

Appellant's final witness was F. James Knight, a registered engineer, who was qualified as an "engineering geologist." He was employed by Gannett-Fleming and had supervised the repairs of the sinkholes on appellant's site. He testified as to the extent of the damage and the extent of the repairs necessary, and expressed his professional opinion that the dry wells had "contributed significantly to the accelerated formation of sinkholes . . . on that tract."

At the close of appellant's case an oral motion for compulsory nonsuit was made on the ground that appellant had failed to present sufficient expert testimony to establish the standard of care required of an architect in that locality with respect to the design of surface water disposal systems. The court granted the motion, ruling that "while [appellant] has presented testimony of experts in the field of geology and engineering, they have not presented any testimony in the field of architecture tending to prove that [appellees'] professional services departed from accepted practice in this profession or that [appellees] failed to meet the standards of their professional duties." A motion to take off the nonsuit was denied by the court en banc with an opinion by Judge BUCHER, who was also the trial judge. In that opinion the court held that "testimony in the field of architecture" was necessary. It also suggested a second reason for the nonsuit, by asking, "[D]id [appellant] prove any negligence on the part of [appellees] that justified submitting the case to the jury?" Both of these issues are before us on this appeal.

I.

The failure to present an architect as an expert witness was not fatal to appellant's case.

The court below states the general rule as being that "expert testimony is necessary to establish negligent practice in any profession." *Wohlert v. Seibert,* 23 Pa.Super. 213 (1903), is cited for this proposition (as it has been in other opinions), but in fact its test is more analytical:

The crucial test of the competency of a witness offered as an expert to give testimony as such is the resolution of the question as to whether or not the jury or persons in general who are inexperienced in or unacquainted with the particular subject of inquiry would without the assistance of one who possesses a knowledge be capable of forming a correct judgment on it.
Id. at 216.

The opinion then goes on to restate the rules for the standard of care to which a physician is held; it does not discuss any other profession or professions in general. The same observation may be made of the section of Wigmore most cited by professional experts:

> On any and every topic, only a qualified witness can be received; and where the topic requires special experience, only a person of that special experience will be received [cross-reference omitted]. If therefore a topic requiring such special experience happens to form a main issue in the case, the evidence on that issue must contain expert testimony or it will not suffice. Wigmore on Evidence (3d ed. 1940) ¶2090A at 453.

We have no doubt that expert testimony was required in this case. The "subject of inquiry" was the standard to be applied to one who holds himself out as competent to design and supervise the construction of a surface water disposal system in Lancaster County. This is certainly a subject that "requires special experience." However, there is nothing inherent in the nature of that experience that makes it unique to architects. What the jury needed was not "the assistance of one who possesses a knowledge" of architecture but of surface water disposal systems. Whether the person offering that assistance happened to be an architect, or engineer, or geologist, or something else, was unimportant; what was important was what he knew.

The error committed by the court below was that it literally analogized the instant case to one of medical malpractice. The court reasoned that in medical malpractice cases there must be expert testimony from physicians as to the appropriate standards of medical practice. From this the court reasoned that in a suit against architects, only architects are competent to testify as to the appropriate architectural standards. However, in medical malpractice cases the expert generally must be a physician because only a physician is trained to perform the medical functions that are the subject matter in controversy. In the instant case the subject matter in controversy (the design and installation of a surface water disposal system on the site in question) is not within the exclusive realm of one profession. To the contrary, it is within the realm of at least three professions: architects, engineers, and geologists. Therefore, a member of any of these professions (if otherwise qualified) was competent to state what the appropriate design and installation standards were.

A case similar to this one is *Bloomsburg Mills, Inc. v. Sordoni Construction Co.*, 401 Pa. 358, 164 A.2d 201 (1960). There the plaintiff hired some architects to design and supervise the construction of a weaving mill for nylon and rayon. This type of manufacturing requires a constant temperature and humidity.

It was thus necessary to build the roof with a "vapor seal" to prevent condensation and leakage of moisture. The plaintiff alleged that the roof was defective in that the vapor seal did not function properly, and that the roof was otherwise inadequately sealed. The defendant appealed a verdict in favor of the plaintiff, claiming insufficient evidence of negligence. As part of this claim the defendant attacked the competency of the plaintiff's expert witness. The expert had had extensive experience with a large roofing manufacturing company and was at the time of the trial a professional roofing consultant. (He had, in fact, been requested by the defendants to submit a bid on the project in question but had declined.) He was not, however, an architect. Nevertheless the Supreme Court held him qualified to testify against the defendant architects. For other decisions in accord, see *Abbott v. Steel City Piping Co.*, 437 Pa. 412, 263 A.2d 881 (1970) (witness with extensive experience in masonry competent to testify as to how a certain wall should be built despite lack of formal engineering degree); *Willner v. Woodward*, 201 Va. 104, 109 S.E.2d 132 (1959) (heating engineer competent to testify against architect regarding a heating and air conditioning duct); *Cuttino v. Mimms*, 98 Ga.App. 198, 105 S.E.2d 343 (1958) (engineer and contractor both competent to testify against architect in case based on faulty construction); *Covil v. Robert & Co. Associates*, 112 Ga.App. 163, 144 S.E.2d 450 (1955) (engineer competent to testify against architect who had drawn plans for water works where issue was whether a certain pipe joint was properly secured).

The court below dismissed *Bloomsburg Mills*, saying that the improper design of a roof is a "common problem" that "would hardly require expert testimony." As suggested by our preceding statement, however, in fact the problem was quite complex and involved a special type of structure. This is further apparent from the Supreme Court's quite lengthy and detailed discussion of the construction problems presented, and of the expert witness's qualifications, none of which would have been appropriate had the case presented only a "common problem," requiring no expert testimony at all. We thus find *Bloomsburg Mills* persuasive authority, and conclude that the failure of appellant to present an architect as an expert witness was not fatal to its case.

II.

As noted above, the opinion of the court below asks the question, "[D]id [appellant] prove any negligence . . . that justified submitting the case to the jury?" In fact that question is not there addressed. Instead, the entire opinion deals only with whether the trial judge was correct in holding that in an action against an architect acting in his professional capacity, the plaintiff must produce testimony by another architect, which is the issue that we have just disposed of. In these circumstances we have made our own examination of the record, and have concluded that appellant did prove sufficient evidence of negligence to send the case to the jury.

Messrs. Pickering, Saylor, and Knight all testified unequivocally that in their respective professional opinions the sinkholes were aggravated by the dry well system, and that considerable damage to appellant's property resulted. However, none of them testified as to the standards of skill required of one who undertakes to design and supervise the installation of a surface water disposal system in the particular area in question, and, as observed in the preceding section of this opinion, it was essential to appellant's case that there be some testimony by a qualified expert on that point.

Appellees have contended that there was no such testimony. (They do not challenge the testimony of Messrs. Pickering, Saylor, and Knight.) This contention, however, overlooks the testimony of Professor Freedman, which we have already quoted in relevant part, ante at 409. It is indeed true that Professor Freedman's testimony was not developed in an orderly manner, and in fact it appears to have emerged almost by chance. It is, nevertheless, in the record, and the professor was cross-examined with respect to it. Summarized, the testimony was that neither the professor, nor any other geologist, nor anyone consulted about surface water disposal, "has ever recommended dry wells for construction in an area like this . . . [W]e frequently get calls . . . and . . . we tell them they are in for trouble . . . [T]hese are systems that should not have been installed because they lead to trouble." When we bear in mind that we must give appellant every reasonable benefit from the evidence, *Shirley v. Clark*, supra, this testimony may fairly be read as a statement of opinion, by a qualified expert, that appellants violated the professional standards of care to which they were obliged to conform. Order reversed.

Although other recent cases have also been liberal in admitting expert testimony,[86] three recent cases refused to admit expert testimony where the expert was not of the same discipline as the person charged with being negligent.[87]

As noted in Section 14.09D, this problem can also arise when a claimant must support his claim for professional negligence with a certificate from a fellow professional that the claim is meritorious.

C. Critique of System

The expert testimony system has been severely criticized. The complex "hypothetical" question has led to overtechnical appellate review and confusion of jurors. In addition, complexity has made errors more likely because the professions are increasingly specialized, a point demonstrated by *Haak*. When the system is administered too strictly, experts may not be found—a particular difficulty when expert testimony is required. The pool of experts being limited to local experts not only has made it more difficult to obtain experts but also has not taken into account the increasingly statewide or national standards of practice.

Even more criticisms have been made of the entire system itself, based largely on the adversary system discussed in Section 2.11. Each party looks not for the best-qualified experts but for the experts who will best support its case. This, coupled with the high compensation paid to experts, has led to skepticism about the professional honesty of many experts.

Not many years ago intense criticism was leveled against the "conspiracy of silence," the unwillingness of

---

[86] *Tomberlin Assoc., Architects, Inc. v. Free*, 174 Ga.App. 167, 329 S.E.2d 296 (1985), cert. denied May 1, 1985 (civil engineer allowed to testify in claim against architect based on soil erosion); *Keel v. Titan Constr. Corp.*, supra note 19 (physics professor could testify as to adequacy of design of solar system); *Edgewater Apartments, Inc. v. Flynn*, 216 A.D.2d 53, 627 N.Y.S.2d 385 (1995) (architect may testify in claim against engineer for water intrusion), subsequent appeal, 268 A.D.2d 227, 701 N.Y.S.2d 357 (2000); *Wessel v. Erickson Landscaping Co.*, 711 P.2d 250 (Utah 1985) (structural engineer allowed to testify in claim against landscape architect as to design of retaining wall); *Perlmutter v. Flickinger*, 520 P.2d 596 (Colo.App.1974) (engineer and contractor permitted to testify about skylight design); *White Budd Van Ness Partnership v. Major-Gladys Drive Joint Venture*, supra note 13 (noting overlap between engineering and architecture permitted engineer to testify as expert in a claim against architect).

[87] *IMR Corp. v. Hemphill*, 926 S.W.2d 542 (Mo.Ct.App.1996) (civil engineer not allowed to testify as to the standard of care required for a general contractor); *Brennan v. St. Louis Zoological Park*, 882 S.W.2d 271 (Mo.Ct.App.1994), transfer denied Sept. 20, 1994 (engineer not allowed to testify as to the standard for architects despite 35 years "of working closely with and overseeing architects"; member of one profession cannot testify against an entirely different profession based on special experience); and *Walker v. The Bluffs Apartment*, 324 S.C. 350, 477 S.E.2d 472 (1996). The latter case was a malpractice claim against an architect. A licensed residential builder and building inspector who taught building codes was not allowed to testify as to architectural design. He had never built a structure and had had no architectural experience or training.

professionals to testify against one another. This problem was particularly difficult where local standards were employed. This led to the development of professional expert witnesses (called forensic design professionals because of their ability to persuade judges and juries of the soundness of their professional conclusions). Although this development has certainly helped overcome the "conspiracy of silence," it has had some unfortunate results.

Such experts have been looked on as "hired guns"—too quick to find fault. Of course, the design professional charged with malpractice can produce other experts. But in the end, both judge and jury are often confused. How can experts with such outstanding credentials differ so sharply as to the cause of the harm and the standards of professional practice?

The junk science debate that surfaced in 1991 relates to the power of trial court judges to admit scientific testimony by witnesses whose work has not been subject to peer review or published in professional journals but is often generated solely for use in litigation, mainly in claims against pharmaceutical houses. In *Daubert v. Merrill Dow Pharmaceuticals, Inc.*, the trial judge refused to hear expert testimony, presented by the plaintiffs, that was based on research that had not been published or subjected to peer review. Refusal was based on the *Frye*[88] test for the admissibility of expert testimony: that to be admissible, the proffered evidence must be "generally accepted" by scientific experts in the field. This was affirmed by the Ninth Circuit Court of Appeals.

The U.S. Supreme Court reversed, ruling that the *Frye* test was too limiting and in any event had been supplanted by the subsequently enacted Federal Rules of Evidence. These rules permitted scientific evidence if the testimony will help the trier of fact, judge or jury.[89]

However, the trial judge must act as a gatekeeper to ensure that the scientific evidence admitted is not only relevant but reliable. She must determine whether the testimony rests on scientific principles. The testimony and methodology must be based on generating hypotheses and testing to see if they can be falsified. The theory or technique must be subject to peer review, which as a rule is required for publication. Actual publication is not a prerequisite to admissibility. The Court sought to leave some room for innovative theories yet at the same time placed some obstacles in the path of such testimony.

In *Kumho Tire Co. Ltd. v. Carmichael*, the Supreme Court held that a district court's duty to screen the reliability and relevance of expert testimony is not limited to "scientific" testimony but applies to all expert testimony, including that provided by engineers.[90] The Court gave great latitude to the district court's role as gatekeeper, stating that the factors for determining the reliability of scientific evidence listed in *Daubert* were neither a definitive checklist nor a test.

The Court buttressed the district court's gatekeeper role by limiting review of its decision to admit or exclude expert testimony by the appellate courts. In *General Electric Co. v. Joiner*, the Court ruled that the district court's decision to admit or exclude expert testimony is reviewable under the "abuse of discretion" standard, even when the decision to exclude such evidence is outcome-determinative.[91]

Although the junk science debate has centered around pharmaceutical drugs, it can also arise in construction litigation. The reliability of expert testimony is most likely to be challenged when the claim is *not* that defective construction caused immediate damage to the building, but rather that it sickened the building's occupants years later. These "sick building" or toxic-mold claims especially are prone to "junk science" charges in the absence of federal or state standards on permissible levels of mold within habitable spaces, and where the causal link between exposure to mold and personal injury has yet to be proven conclusively.[92]

This dispute and the high cost of producing expert witnesses have again revived calls for greater use of expert panels from which trial court judges can select expert

---

[90]526 U.S. 137 (1999).

[91]522 U.S. 136 (1997).

[92]In *Mondelli v. Kendel Homes Corp.*, 262 Neb. 263, 631 N.W.2d 846, opinion modified on denial of rehearing, 262 Neb. 663, 641 N.W.2d 624 (2001), the court ruled that the homeowners' expert could testify that their injuries were caused by exposure to mold in their home. California has taken the lead in establishing safe mold levels by enacting legislation to monitor mold-related health complaints and to devise exposure standards. The Toxic Mold Protection Act of 2001 requires the state Department of Health Services to convene a task force to develop permissible exposure levels to mold and to devise remediation standards. See West Ann. Cal. Health and Safety Code §§ 26100 *et seq.* A second, new law requires a different task force to study and publish its findings on fungal contamination in indoor environments, see Id., §§ 26200 *et seq.* For a general discussion of sick building litigation, see O'Neal, *Sick Building Claims*, 20 Const. Lawyer, No. 1, Jan. 2000, p.16.

---

[88]*Frye v. United States*, 293 F. 1013 (D.C.Cir.1923).

[89]509 U.S. 579 (1993).

witnesses rather than rely on the testimony of experts called by the parties. Yet a report indicates that only 20 percent of the federal court judges have appointed their own experts. A system of court-appointed experts also raises problems. Not only does it interfere with the adversary system, but questions arise as to how judges will select the experts, whether the parties will be able to cross-examine the experts or offer their own expert testimony, and who will pay the fees of the experts.

D. Advice to Expert Witnesses

Space does not permit a detailed discussion of all the problems seen from the perspective of a design professional asked to be an expert witness. Some brief comments can be made, however. A clear, written understanding should precede any services being performed. Such a writing should include the following:

1. Specific language making clear that the expert will give her best professional opinion.
2. Language that covers all aspects of compensation for time to prepare to testify, travel time, and actual time testifying before a court, board, or commission. (Many experts use an hourly rate for preparation time and a daily rate for travel and testimony time.)
3. Specification of expenses to be reimbursed, using a clear and administratively convenient formula for reimbursing costs of accommodations, meals, and transportation.
4. A minimum fee if the expert is not asked to testify. Some attorneys retain the best experts, use the experts whose opinions best suit their case, and, by having retained the others, preclude them from testifying for the other parties.

The attorney calling the expert usually provides details as to appearance, description of qualifications, methods of answering questions, explanations for opinions, and defending opinions on cross-examination. It is important to recognize who is being addressed and the reason for seeking expert opinions. The expert should help the judge or jury—people often inexpert in evaluating technical material. So opinions and explanations must be understandable by the people who must evaluate them. The expert should never speak down to judge and jurors.

A client who loses its case, or who recovers less damages than anticipated, may place the blame on the experts—whether its own or those of the opposing party. An expert sued by a client for malpractice may invoke "witness immunity"—a defense designed to ensure that witnesses testify freely and without fear of civil liability. Witness immunity has been extended both to adverse witnesses sued for statements made either in preparation of trial or during trial, as well as to a client's claim against its own expert for in-court testimony.[93] However, one court permitted a client to sue its expert, an engineer, for negligence in pretrial litigation support services.[94]

SECTION 14.07 Implied Warranty: An Outcome Standard

As noted in Section 14.05A, the law generally looks at the process by which the services were to be performed by a design professional—the professional standard. Where a professional is retained to provide design services, American law has concluded (not without criticism, as explored in Section 14.11A) that the likely understanding between the client and the professional designer is not that a successful outcome will be achieved when professional services are purchased, but that the professional will perform as would other professionals.[95]

A minority view has developed under which the design professional must design to achieve the communicated or understood expectations of the client. The issue in such cases is whether the design professional has met the client's purposes of which she was aware or should have been aware. Under this doctrine, liability—like the liability for any breach of contract or any manufactured product—is strict. There is no need to compare what the challenged design professional did to what other design professionals similarly situated would have done, and no need to introduce expert testimony.

---

[93]*Western Technologies, Inc. v. Sverdrup & Parcel, Inc.*, 154 Ariz. 1, 739 P.2d 1318 (App.1986) (claim against adverse witness); *Middlesex Concrete Products & Excavating Corp. v. Carteret Indus. Ass'n*, 68 N.J.Super. 85, 172 A.2d 22 (App.Div.1961) (same); and *Bruce v. Byrne-Stevens & Assocs. Engineers, Inc.*, 113 Wash.2d 123, 776 P.2d 666 (1989) (claim against own witness, an engineer, for negligence in testifying).

[94]*Murphy v. A.A. Mathews, a Div. of CRS Group*, 841 S.W.2d 671 (Mo.1992).

[95]Quoted and followed in *Konkel & Kemper Architects v. McFall*, 843 P.2d 1178, 1186 (Wyo.1992).

Courts invoke a variety of rationales in support of a decision to apply an implied warranty or result-oriented standard. One rationale is that the design professional is being sued by her client for breach of contract, not negligence. Courts invoke this rationale usually in the context of either a design-build contract or a contract in which the architect or engineer agreed to both design the building and supervise its construction.[96]

Another rationale is that, contrary to other professions such as medicine or the law, design professionals deal in exact sciences and so can be held to a strict standard of compliance. Courts employing this rationale either extend it to both architects and engineers[97] or distinguish between types of design professionals [98] or types of design services.[99]

A third rationale, usually limited to cases involving design defects, involves the consideration of fairness. These courts observe that the architect holds herself out as the expert, while the client (and sometimes even the contractor) is both ignorant and wholly dependent upon the design for the successful execution of the project.[100]

Finally, clients may invoke consumer protection and deceptive trade practices laws when suing their design professionals, thereby benefiting from the relaxed standard for recovery (and increased penalties) specified in those laws. A Texas court of appeals applied the Texas Deceptive Trade Practices Act (DTPA) with its strict liability to an architect.[101] Although the continued viability of that decision is in doubt in light of subsequent legislative and judicial developments,[102] the Kansas Supreme Court has recently ruled that its state's consumer protection law applies to a client's claim against his engineer,[103] and a New York court applied that state's consumer protection law to a homeowner's contract with an architect.[104]

Once a jurisdiction has adopted an implied warranty standard in the context of a claim between the client and its design professional, it may extend that warranty to third parties. South Carolina, for example, has extended the implied warranty of design to subsequent owners[105] and contractors.[106]

---

[96]*J. Ray McDermott & Co. v. Vessel Morning Star*, 431 F.2d 714, 721–22 (5th Cir. 1970), reversed on rehearing *en banc* on other grounds, 457 F.2d 815 (5th Cir.), cert. denied, 409 U.S. 948 (1972) (admiralty; breach of contract to design and build fishing boats); *Hill v. Polar Pantries*, 219 S.C. 263, 64 S.E.2d 885 (1951) (contract to design and supervise); and *Board of Educ. v. Del Biano & Assocs.*, 57 Ill.App.3d 302, 372 N.E.2d 953, 957–58 (1978) (same, court finds implied contractual terms). One court noted in *dicta* that the design professional could contract to perform a higher duty than that imposed by law. *Tamarac Dev. Co., Inc. v. Delamater Freund & Assocs., P.A.*, 234 Kan. 618, 675 P.2d 361, 364 (1984). But see *Adobe Masters, Inc. v. Downey*, 118 N.M. 547, 883 P.2d 133 (1994) and *SME Industries, Inc. v. Thompson, Ventulett, Stainback & Assocs., Inc.*, supra note 68, 28 P.3d at 678–80, as both cases distinguish breach of contract from implied warranty.

[97]*Tamarac Dev. Co., Inc. v. Delamater Freund & Assocs., P.A.*, supra note 96, 675 P.2d at 365.

[98]*Broyles v. Brown Eng'g Co.*, 275 Ala. 35, 151 So.2d 767 (1963) (court extends implied warranty to civil engineer who designed drainage system, while stating that architects would not be subject to an implied warranty). Alabama may be shifting away from an implied-warranty standard; in *K.B. Weygand & Assocs., P.C. v. Deerwood Lake Land Co.*, 812 So.2d 1165 (Ala.2001), rehearing denied, Jul. 6, 2001, a divided court found a civil engineer not strictly liable where a road failed because of unique soil conditions.

[99]The Eighth Circuit Court of Appeals has ruled that no expert testimony is necessary to support claims based on construction-phase services, thereby in effect creating a strict liability standard. *Jaeger v. Henningson, Durham & Richardson, Inc.*, supra note 43 and *Bartak v. Bell-Galyardt & Wells, Inc.*, supra note 81.

[100]*Hill v. Polar Pantries*, supra note 96 (claim by client); *Bloomsburg Mills, Inc. v. Sordoni Constr. Co.*, 401 Pa. 358, 164 A.2d 201 (1960) (claim by client); and *Eastern Steel Constructors, Inc. v. City of Salem*, 209 W.Va. 392, 549 S.E.2d 266, 276–77 (2001) (claim by contractor).

[101]*White Budd Van Ness Partnership v. Major-Gladys Drive Jt. Venture*, supra note 11.

[102]In *Chapman v. Paul R. Wilson, Jr., D.D.S.*, 826 S.W.2d 214, 217 (Tex.Ct.App.1992, writ denied), the court expressly rejected *White Budd's* holding that the DTPA apples to "purely professional services." Then, in 1995, the DTPA was amended to exclude damages based on the rendering of a professional service, defined as the providing of advice, opinion, or judgment, see Tex. Bus. & Com. Code § 17.49(c).

[103]In *Moore v. Bird Eng'g Co. P.A.*, 273 Kan. 2, 41 P.3d 755 (2002), the court stated that to recover under the Act, the homeowner/consumer was required to show only that the engineer had reason to know that its design did not comport with the consumer's expectations, not that the engineer had an intent to deceive. The court rejected the engineer's contention that consumer protection laws are limited to mass-produced products and do not apply to face-to-face transactions.

[104]*Ragucci v. Professional Constr. Services*, 25 A.D.3d 43, 803 N.Y.S.2d 139 (2005).

[105]*Beachwalk Villas Condominium Ass'n, Inc. v. Martin*, 305 S.C. 144, 406 S.E.2d 372 (1991) (claim by condominium association).

[106]*Tommy L. Griffin Plumbing & Heating Co. v. Jordan, Jones & Goulding, Inc.*, 320 S.C. 49, 463 S.E.2d 85 (1995), subsequent appeal, 351 S.C. 459, 570 S.E.2d 197 (App.2002) (issue of expert testimony in negligence claim), rehearing denied Aug. 22, 2002, and cert. denied Mar. 19, 2003. Accord, *Eastern Steel Constructors, Inc. v. City of Salem*, supra note 100, 549 S.E.2d at 276–77.

Until recently the standardized contracts prepared by the design professional associations did not specify how the design professional's performance would be measured. As noted in Section 14.04A, however, both the AIA and the Engineers Joint Contracts Documents Committee (EJCDC) have included language stating that the professional standard applied. This was done to exclude any outcome-oriented express or implied warranties that would measure the engineer's performance.[107]

Some clients, both public and private, have sought to include language under which the design professional will perform to some successful outcome standard, such as provisions stating that the design professional will perform in accordance with the "highest standards of professional service." This could be taken to create an express warranty of a successful outcome or at least to require that it be determined whether the challenged design professional had done better than others would have. Usually in such cases the design professional who is aware of the risk will seek to persuade the client not to include such language, by stating that the liability is contractually assumed and would not be covered under the professional liability insurance policy. However, the realities of the market often mean the design professional must accept this language if the client insists.

Some case decisions are confusing, mixing professional standard language and implied warranty language.[108] One case stated that the implied warranty standard did not require a favorable result but simply required a reasonable result.[109] Another case required expert testimony to establish breach of the engineer's implied warranty of design.[110] Neither of these cases differs much from the professional standard. Finally, some claims for implied warranty are rejected when the implied warranty sought was one over which the design professional had no control.[111]

But is there much difference in outcome between the two standards? Although this issue is discussed in greater detail in Section 14.11A, it should be noted here that the design professionals, their professional associations, and their insurers think there is a difference. Much of their polemical work is devoted to protecting the professional standard. In many cases, failure by the design professional to design in accordance with the known or foreseeable purposes of the client will be a breach of the professional standard. Yet the struggle between the process-oriented professional standard and the outcome-oriented implied warranty is likely to continue.[112]

SECTION 14.08 Third-Party Claims: Special Problems

Third-party claims raise problems that do not arise when the claimant is connected to the design professional by contract. The proliferation of third-party claims has generated more litigation, varying state rules, and judicial opinions of divided courts than have claims by clients against design professionals. This reflects rules in transition, with inevitable strains and contradictions.

A. Potential Third Parties

The centrality of the design professional's position in construction—both designing and monitoring performance—generates a wide range of potential third-party claimants. At the inner core are the other major direct participants in the process who work on or enter the site itself, such as contractors, construction workers and delivery people. Around the core cluster those who supply money, materials, or equipment, such as lenders and suppliers. Next are those who "backstop" direct participants, such as sureties (persons who promise to perform another's obligations) and insurers. Still farther from the core are those claimants who will ultimately take possession of the project, such as subsequent owners, tenants, and their employees. Situated farthest from the core are people who may enter

---

[107]AIA Doc. B101-2007, § 2.2 and EJCDC E-500, ¶ 6.01A (2002).

[108]*Bloomsburg Mills, Inc. v. Sordoni Constr. Co.*, supra note 100; *Eastern Steel Constructors, Inc. v. City of Salem*, supra note 100.

[109]*E. C. Ernst., Inc. v. Manhattan Const. Co. of Texas*, 551 F.2d 1026, 1035, rehearing denied in part, granted in part, 559 F.2d 268 (5th Cir.1977), cert. denied sub nom. *Providence Hosp. v. Manhattan Constr. Co. of Texas*, 434 U.S. 1067 (1978) (applying Alabama law).

[110]*Tommy L. Griffin Plumbing & Heating Co. v. Jordan, Jones & Goulding, Inc.*, supra note 106.

[111]*Allied Properties v. John A. Blume & Assocs.*, 25 Cal.App.3d 848, 102 Cal.Rptr. 259 (1972).

---

[112]See Jones, *Economic Loss Caused by Deficiencies: The Competing Regimes of Tort and Contract*, 59 U. of Cinn.L.Rev. 1051, 1070, 1073 (1991) (argues for result-oriented standard if probability of success is high); Note, 20 Mem.St.U.L.Rev. 611 (1990) (argues against implied warranty and strict liability).

or pass by the project during construction or after completion, such as members of the public or patrons. See Figure 14.1 and please review Figures 8.1 and 8.2.

The wide variety of potential claimants, the varying distance from core participants, the type of harm, the difference between those who have other sources of compensation, such as workers—all combine to ensure complexity.

B. Contracts for Benefit of Third Parties

Section 14.08 spotlights tort claims by third parties who suffer losses related to the construction process. Yet some claims by third parties are based on the assertion that the claimants are intended beneficiaries of contracts to which they are not parties. Until the mid–nineteenth century, American law generally did not permit persons who were not party to a contract to maintain legal action for the contract's breach, even though they may have suffered losses from the breach. But U.S. law has steadily expanded the rights of third parties to recover for contract breach.[113]

The use of this doctrine has found its way into construction claims. In the context of claims against the design professional, the doctrine is sometimes employed by contractors and subcontractors who assert that they are intended beneficiaries of the contract made between the owner and the design professional.

Such claims have had spotty success,[114] with most cases concluding that contracting parties usually intend

to benefit only themselves. Yet the possibility of such a theory being invoked successfully has led contracting parties to seek to use the contract itself to bar claims by third parties as intended beneficiaries.[115]

Even if a claimant can establish that it is an intended beneficiary, its claim may fail because of provisions in the contract.[116] To complicate matters further, an architect sued by a contractor was allowed to use a provision in the contract between the contractor and the owner as a defense.[117] (Many of these problems could be avoided if owner, contractor, and design professional were all parties to one contract.)

C. Tort Law: Privity and Duty

Sections 14.08C, 14.08D, and 14.08E all relate to attempts by third parties, usually contractors or subcontractors, to assert tort claims against design professionals. Section 14.08C approaches the problem from the perspective of privity, the absence of which has been the principal justification for denying such claims in the past and, to some degree, today. Section 14.08D examines negligent misrepresentation, one avenue by which the requirement of privity has been avoided. Section 14.08E discusses the economic loss rule, a traditional principle of tort law that

---

[113]Restatement (Second) of Contracts § 302 (1979).

[114]*John E. Green Plumbing & Heating Co. v. Turner Constr. Co.*, 742 F.2d 965 (6th Cir.1984), cert. denied, 471 U.S. 1102 (1985) (contractor v. construction manager). But see *A. R. Moyer v. Graham*, 285 So.2d 397 (Fla.1973) (contractor could sue architect in tort but not as intended contract beneficiary). The cases, including those involving construction contracts, are canvassed in Prince, *Perfecting the Third Party Beneficiary Standing Rule Under Sec. 302 of the Restatement (Second) of Contracts*, 25 B.C.L.Rev. 919 (1984). See also Burch, *Third-Party Beneficiaries to the Construction Contract Documents*, 8 Constr. Lawyer, No. 2, April 1988, p. 1. Accord, *A.H.A General Construction, Inc. v. Edelman Partnership*, 291 A.D.2d 239, 737 N.Y.S.2d 85 (2002) (contractor cannot sue architect as third-party beneficiary) and *Eastern Steel Constructors, Inc. v. City of Salem*, supra note 100, 549 S.E.2d at 277–78 (same). Third-party beneficiary claims are not limited to contractors and subcontractors. In *The Ratcliff Architects v. Vanir Construction Management, Inc.*, 88 Cal. App.4th 595, 106 Cal.Rptr.2d 1 (2001), a construction management contract expressly disavowed the creation of rights "in persons not party

to this agreement, whether third-party beneficiaries or otherwise." The court relied upon this clause to reject the architect's indemnity claim against the CM. In *Jenne v. Church & Tower, Inc.*, 814 So.2d 522 (Fla. Dist.Ct.App. 2002), the court ruled that the county sheriff was not a third-party beneficiary to a contract between the county and a design-build contractor to build a new detention facility. See also *F. H. Paschen/ S. N. Nielsen, Inc. v. Burnham Station, L.L.C.*, 372 Ill.App.3d 89, 865 N.E.2d 228, appeal denied sub nom. *Paschen v. Burnham Station, LLC*, 225 Ill.2d 631, 875 N.E.2d 1111 (2007) (unsuccessful claim by investor, who lost its investment in a real estate development designed by the architect).

[115]See AIA Docs. B101-2007, § 10.5 and A201-2007, § 1.1.2. The latter gives third-party rights to the architect. The court in *SME Industries, Inc. v. Thompson, Ventulett, Stainback & Assocs., Inc.*, supra note 68, 28 P.3d at 684–85, relied upon a clause similar to B141-1997, ¶1.3.7.5 (the predecessor of B101-2007, § 1.1.2) in rejecting a subcontractor's third-party beneficiary claim against the architect and engineer.

[116]*Prichard Bros., Inc. v. Grady Co.*, 407 N.W.2d 423 (Minn.App. 1987) reversed 428 N.W.2d 391 (Minn.1988) (contractor can bring tort claim against architect and is not limited to third-party beneficiary claim).

[117]*Bates & Rogers Constr. Corp. v. Greeley & Hanson*, 109 Ill.2d 225, 486 N.E.2d 902 (1985) (architect sued by contractors could invoke "no damage" clause in prime contract as defense).

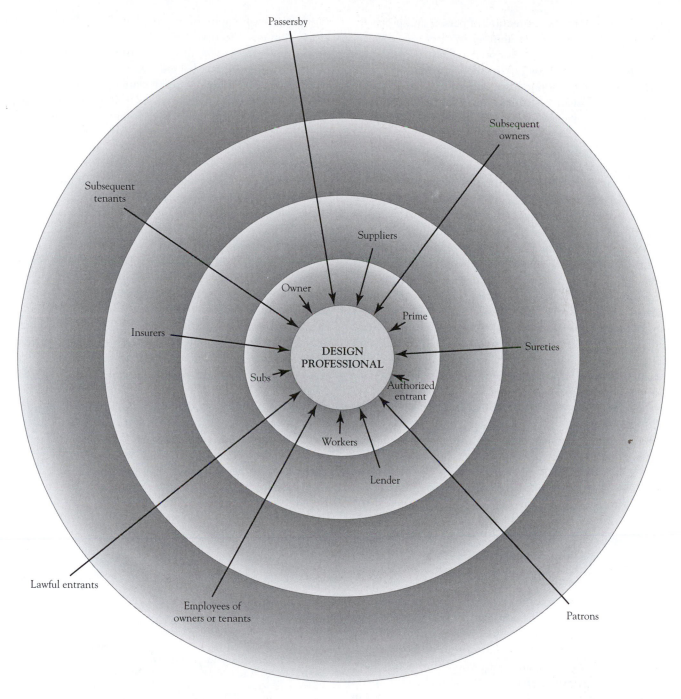

FIGURE 14.1   Potential third-party claimants against design professional.

limited tort law to the protection of person and property. Although this can also be applied in disputes between contract-connected parties, as noted in Section 14.05D, in most disputes today those against whom claims have been made assert the economic loss rule as a reason to bar the claimant from basing its claim on tort law. Because these doctrines interrelate, some overlap will occur in the three sections.

These sections all document the conflicting worlds of contract law, with its emphasis on planning and certainty, and tort law, with its emphasis on preventing harm and compensating victims.

A claimant employing tort law as a basis of transferring its loss to another must show that the latter owed a duty to protect the claimant from the unreasonable risk of harm. Often, particularly where the wrongful conduct was breach of contract, the duty concept was phrased as one requiring privity between claimant and the person against whom the claim was made. Usually, though not exclusively, privity was based on a contract between the claimant and the person against whom the claim was made. The defense of absence of duty or lack of privity seeks to avoid unlimited liability and to free people from the fear of being held accountable for any harm they might cause. As noted in Section 7.09B, in the nineteenth century and early twentieth century the requirement of privity was an insurmountable obstacle even to a claimant who suffered personal harm because of another's negligence. However, that section noted that the privity defense was eliminated in personal injury claims early in the twentieth century. The early cases doing so were claims against manufacturers by those injured from negligently manufactured products. By the mid–twentieth century, participants in the construction process, such as design professionals, could no longer invoke the privity doctrine as a defense to claims based on personal harm.[118]

Tort law is not limited to claims for personal harm or damage to property. In the construction process, claimants often suffer economic losses unconnected with personal harm or damage to property. Often they seek to recover such losses from participants with whom they do not have a contract. For example, contractors and sureties assert claims against design professionals, sometimes (as noted in Section 14.08B) by claiming to be intended beneficiaries of a contract that the design professional has breached,

and increasingly by invoking tort law. When tort law is invoked, in many states the economic loss rule is a barrier. While that rule will be discussed in Section 14.08E, it should be addressed briefly here.

States that permit third-party claims saw no reason to distinguish purely economic losses from harm to person or property. They were willing to compensate those who suffered economic losses because of the negligence of those against whom claims were made.[119] The leading case pointed to the life-or-death power the architect has over the contractor.[120] Requiring that the harm and the person suffering it be reasonably foreseeable was the method such courts felt would protect the person against whom a claim would be made from unlimited liability.

Not all states permit contractors or subcontractors to sue design professionals for economic losses. Some states find the design professional owes a contract duty solely to the owner and does not owe a tort duty of care to the contractor or subcontractor.[121] Other states still require privity in claims for purely economic losses.[122] They point to tort law's historic function of protecting personal and property rights and the need to draw some lines beyond which the law should not compel a person to pay for another's loss. Foreseeability was considered insufficient protection from third party claims. Also, decisions barring such claims frequently pointed to the claimant's often having a contractual right against the party with whom it contracted. One method of avoiding the privity rule has been to invoke the doctrine of negligent misrepresentation, which can apply to those who furnish information on which third parties have relied (discussed in Section 14.08D).

---

[118]*Miller v. DeWitt*, 37 Ill.2d 273, 226 N.E.2d 630 (1967).

[119]*E. C. Ernst. Inc. v. Manhattan Constr. Co. of Texas*, supra note 109 (subcontractor v. architect); *Donnelly Constr. Co. v. Oberg/Hunt/Gilleland*, 139 Ariz. 184, 677 P.2d 1292 (1984) (contractor v. architect); *Forte Bros., Inc. v. National Amusements, Inc.*, 525 A.2d 1301 (R.I.1987) (contractor v. architect). Many cases followed the influential opinion of the federal trial court judge in *United States v. Rogers & Rogers*, 161 F.Supp. 132 (S.D.Cal.1958). For surety claims, see Section 22.07. See also Annot., 65 A.L.R.3d 249 (1975).

[120]*United States v. Rogers & Rogers*, supra note 119.

[121]*Bernard Johnson, Inc. v. Continental Constructors, Inc.*, 630 S.W.2d 365, 371–76 (Tex.Ct.App.1982).

[122]*Bagwell Coating, Inc. v. Middle South Energy, Inc.*, 797 F.2d 1298 (5th Cir.1986) (Mississippi law) (contractor v. construction manager); *Peyronnin Constr. Co. v. Weiss*, 137 Ind.Ct.App. 417, 208 N.E.2d 489 (1965) (contractor v. engineer).

D. Negligent Misrepresentation: *Ossining Union Free School District v. Anderson LaRocca Anderson*

The design professional's performance consists in part of supplying information. Sometimes representations are made to the client, such as those involving cost estimates,[123] the availability of funds for the project,[124] and the likelihood that the project will meet the client's needs.[125] This Section emphasizes claims by third parties, such as contractors and subcontractors, that they relied on negligent misrepresentations made by the design professional.[126]

In 1991 the Supreme Court of Tennessee decided *John Martin Co., Inc. v. Morse/Diesel, Inc.*[127] The case was considered sufficiently important by participants in the construction process that an *amicus curiae* (friend of the court) brief was submitted on behalf of the defendant construction manager (CM) by the American Institute of Architects, the National Society of Professional Engineers, the American Consulting Engineers Council, and state chapters of those organizations. A similar brief was submitted on behalf of the contractor by two Tennessee contractor associations.

John Martin was engaged by the CM, Morse/Diesel, on behalf of the owner as a specialty concrete supplier. A dispute arose during performance over the concrete work. The issue was who had caused the need for additional concrete. John Martin and the CM blamed each other. After settling with the owner, John Martin sued the CM, charging negligent misrepresentation.

The CM, supported by briefs of architectural and engineering associations, asserted the economic loss rule as its principal defense. This rule, to be discussed later in Section 14.08E, bars tort claims unless the loss included harm to person or property. The CM admitted there was an exception for claims of negligent misrepresentation. But the CM and the associations supporting him claimed the exception applied only to a professional supplying information that induces a person to enter into a business transaction.

The court stated there was no clear majority of states on the scope of the exception. It then held the specialty trade concrete supplier could sue the CM for negligent misrepresentation, despite the absence of privity, provided

1. The defendant is acting in the course of his business, profession or employment . . .; *and*
2. The defendant supplies faulty information meant to guide others in their business transaction; *and*
3. The defendant fails to exercise reasonable care in obtaining or communicating the information; *and*
4. The plaintiff justifiably relies on the information.[128]

The court emphasized foreseeability of the plaintiff's use of the information. Further comment on this case will be made later in this section.

It is important to note that misrepresentations in the *John Martin* case are not limited to professional opinions such as surveyor reports or those made by geotechnical engineers. This decision would appear to allow the negligent misrepresentation exception to the privity requirement to be used in cases involving almost anything that a design professional does in the course of her professional services.

A person supplying information may be inhibited from doing so if she would be exposed to indeterminate liability to a large number of third-party claimants. Historically, the principal professionals who have sought legal protection have been accountants and others who supply financial information on which many may rely. As a result, liability for negligent misrepresentation has been more limited than claims based on ordinary negligence. Most important, the defense of foreseeability (i.e., the party supplying the information could not reasonably foresee that it would be relied on by particular persons) is more likely to be available if a claim is based on negligent misrepresentation than if it is based on ordinary negligence.

Some of the uncertainties created by the negligent misrepresentation exception to the privity rule are demonstrated in New York. *The Ossining Union* case addressed the policy issues in third party claims and clarified New York law. It is reproduced at this point.

---

[123]See Section 12.03.

[124]See Section 12.04.

[125]See Section 12.05.

[126]As to surety claims, see Section 22.07. For contractor claims against the owner for implied warranties created by the design, see Section 23.05E.

[127]819 S.W.2d 428 (Tenn.1991).

---

[128]819 S.W.2d at 431, following Section 552 of the Second Restatement of Torts.

OSSINING UNION FREE SCHOOL DISTRICT v. ANDERSON LaROCCA ANDERSON

Court of Appeals of New York, 1989. 73 N.Y.2d 417, 539 N.E.2d 91, 541 N.Y.S.2d 335.

OPINION

KAYE, Judge.

[Ed. note: Footnote omitted.]

At issue is a question that has long been a subject of litigation: in negligent misrepresentation cases, which produce only economic injury, is privity of contract required in order for plaintiff to state a cause of action? Whether defendants are accountants (as in several recent cases) or not (as here), our answer continues to be that such a cause of action requires that the underlying relationship between the parties be one of contract or the bond between them so close as to be the functional equivalent of contractual privity. Such a bond having been alleged in the present action against engineers, we reverse the Appellate Division order and deny defendants' motion to dismiss the complaint.

Viewing the facts presented in a light most favorable to plaintiff, as we must at this stage of the proceeding, plaintiff school district alleges that in 1984, it began a general study and structural evaluation of its buildings. To that end, it entered into a written agreement with an architectural firm, co-defendant Anderson LaRocca Anderson, whereby Anderson was hired to provide an evaluation and feasibility study of plaintiff's buildings; the contract authorized Anderson's retention of consultants. Anderson retained the defendants, Thune Associates Consulting Engineers and Geiger Associates, P.C., as engineering consultants to assist in various aspects of the work it had undertaken for the school district. Although the school board authorized the retention of Thune and Geiger, neither defendant had a contract with the school district.

[Ed. note: See Figure 14.2.]

This litigation arises from certain reports made by defendants following tests done on school district premises in order to determine the structural soundness of the high school annex. Specifically, defendant Thune and thereafter, at the school district's request to Anderson, second engineering firm—defendant Geiger—tested the concrete at various locations throughout the building. Both reported that there were serious weaknesses in the building, particularly the concrete slabs that formed the building's superstructure, and Anderson informed the school district of those findings.

It is alleged that defendants were aware that plaintiff would rely on their findings and that the intended purpose of defendant's reports was in fact to enable the school district to determine what measures should be taken to deal with structural problems in its buildings. For safety reasons, the school district closed the annex and, purportedly at substantial expense, obtained other facilities for the dislodged activities. The school district, however, later retained a third independent expert to check the results, and that expert advised plaintiff that the annex had been constructed with a lightweight concrete known as "Gritcrete" rather than the 2,500 pound per square inch cement defendants had assumed and reported. According to plaintiff, this information was available to defendants in the original building design drawings and specifications which had been furnished to them. Had defendants read these materials rather than acting on their mistaken assumption as to the type of concrete used, they would not have made the reports and rendered the advice that eventuated in the unnecessary and expensive closing of the annex.

The school district then began this lawsuit against Anderson and both engineering consultants. Claims of negligence and malpractice were asserted against all three. A claim for breach of contract was also asserted against Anderson, the only party with which plaintiff school district had a contract. [The court gave the procedural history and summarized the decision of the intermediate appellate court. The decision dealt with the tort claim by the school district against the consulting engineers.]

Courts have long struggled to define the ambit of duty or limits of liability for negligence, which in theory could be endless. While much of this struggle has been couched in the rhetoric of foreseeability of harm, under some circumstances foreseeability has appeared particularly inadequate for defining the scope of potential liability. In negligent misrepresentation cases especially, what is objectively foreseeable injury may be vast and unbounded, wholly disproportionate to a defendant's undertaking or wrongdoing. . . . In reaching the policy judgment called "duty," courts have therefore invoked a concept of privity of contract as a means of fixing fair, manageable bounds of liability in such cases.

[The court gave a history of the privity rule in cases involving personal harm, starting with *Winterbottom v. Wright* and highlighting *MacPherson v. Buick Motor Co.*, 217 N.Y. 382, 111 N.E. 1050 (1916). See Section 7.09B.]

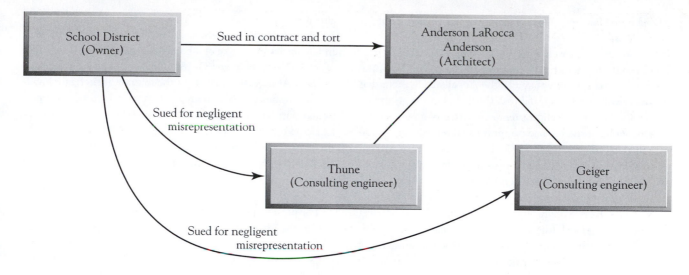

FIGURE 14.2 *Ossining Union Free School District v. Anderson LaRocca Anderson.*

In theory, there appeared to be no reason why the privity bar should be dispensed with in cases such as *MacPherson* but retained in certain other types of negligence cases, and in *Glanzer v. Shepard*, 233 N.Y. 236, 135 N.E. 275 we said as much. The defendants in *Glanzer* were public weighers hired by the sellers of beans to provide plaintiff with a certificate stating the weight of the beans. When defendants negligently misstated the weight, the buyers sued for the overpayment they had made in reliance on defendants' inaccurate statement. Rejecting defendants' claim that *MacPherson* applied only to products posing a risk of physical danger, Judge Cardozo wrote: "We do not need to state the duty in terms of contract or of privity. Growing out of a contract, it has nonetheless an origin not exclusively contractual. Given the contract and the relation, the duty is imposed by law." (*Id.*, at 239, 135 N.E. 275.)

In *Glanzer*, the particular relationship that was held to warrant imposition of a legal duty of care was found in the fact that "[t]he plaintiffs' use of the certificates was not an indirect or collateral consequence of the action of the weighers. It was a consequence which, to the weighers' knowledge, was the end and aim of the transaction" and that a copy of the certificate was sent to plaintiffs "for the very purpose of inducing action." (*Id.*, at 238–239, 135 N.E. 275.) While noting that the plaintiffs might be analogized to third-party beneficiaries of the contract between the bean sellers and the defendants, the opinion stressed that the result was reached more simply by analyzing the duty owed by the defendants under the circumstances: "The defendants, acting, not casually nor as mere servants, but in the pursuit of an independent calling, weighed and certified at the order of one with the very end and aim of shaping the conduct of another. Diligence was owing, not only to him who ordered, but to him also who relied." (*Id.*, at 242, 135 N.E. 275.)

The ambit of duty was not, however, determined simply by the class of persons who relied on the negligent misrepresentations. As Chief Judge Cardozo wrote in *Ultramares Corp. v. Touche*, 255 N.Y. 170, 174 N.E. 441, what was determinative in *Glanzer* was that there was a bond between plaintiff and defendant that was "so close as to approach that of privity, if not completely one with it." (*Id.*, at 182–183, 174 N.E. 441.) In *Ultramares*, by contrast, the range of potential plaintiffs was "as indefinite and wide as the possibilities of the business that was mirrored in the summary." (*Id.*, at 174, 174 N.E. 441.) That distinction led Chief Judge Cardozo to observe: "If liability for negligence exists, a thoughtless slip or blunder, the failure to detect a theft or forgery beneath the cover of deceptive entries, may expose accountants to a liability in an indeterminate amount for an indeterminate time to an indeterminate class. The hazards of a business conducted on these terms are so extreme as to enkindle doubt whether a flaw may not exist in the implication of a duty that exposes to these consequences." (*Id.*, at 179–180, 174 N.E. 441.)

While plaintiffs' reliance might have been objectively foreseeable both in *Glanzer* and in *Ultramares*, the court chose to circumscribe defendants' liability for negligent misstatements by privity of contract or its equivalent, because of concern for the indeterminate nature of the risk. That very concern, which has echoed in the law since *Winterbottom*, was also at the root of our recent decisions. . . .

In none of these cases, however, has the court erected a citadel of privity for negligent misrepresentation suits. Thus, the rule is not, as erroneously stated by the Appellate Division, that "recovery will not be granted to a third person for pecuniary loss arising from the negligent representations of a professional with whom he or she has had no contractual relationship." (135 A.D.2d at 520, 521 N.Y.S.2d 747.) The long-standing rule is that recovery may be had for pecuniary loss arising from negligent representations where there is actual privity of contract between the parties or a relationship so close as to approach that of privity.

Nor does the rule apply only to accountants. We have never drawn that categorical distinction, and see no basis for establishing such an arbitrary limitation now. It is true that in many of the cases involving claims for negligent misrepresentation, the defendants are accountants. Indeed, in attempting to fashion a rule that does not expose accountants to crippling liability, we have noted the central role played by that profession in the world of commercial credit. But while the rule has been developed in the context of cases involving accountants, it reflects our concern for fixing an appropriate ambit of duty, and there is no reason for excepting from it defendants other than accountants who fall within the narrow circumstances we have delineated. Notably, *Glanzer* itself did not involve a suit against accountants.

The remaining question, then, is whether under *Glanzer* and our subsequent cases, defendants owed a duty of care to plaintiff that was breached by their alleged negligent performance. We have defined this duty narrowly, more narrowly than other jurisdictions (see, e.g., *Rosenblum, Inc. v. Adler*, 93 N.J. 324, 461 A.2d 138). We have declined to adopt a rule permitting recovery by any "foreseeable" plaintiff who relied on the negligently prepared report, and have rejected even a somewhat narrower rule that would permit recovery where the reliant party or class of parties was actually known or foreseen by the defendants (*Credit Alliance Corp. v. Andersen & Co.*, 65 N.Y.2d at 553, n.11, 493 N.Y.S.2d 435, 483 N.E.2d 110). It is our belief that imposition of such broad liability is unwise as a matter of policy (see Siliciano, *Negligent Accounting and the Limits of Instrumental Tort Reform*, 86 Mich.L.Rev.1929) or, at the least, a matter for legislative rather than judicial reform.

Instead, we have required something more, and we have articulated the requirement in various ways. In *Glanzer*, it was described as reliance by the plaintiff that was "the end and aim of the transaction." (233 N.Y.2d, at 238–239, 135 N.E. 275; see

also *White v. Guarente*, 43 N.Y.2d 356, 362, 401 N.Y.S.2d 474, 372 N.E.2d 315 ["one of the ends and aims of the transaction"].) In *Ultramares*, we spoke of a bond "so close as to approach that of privity." (255 N.Y. at 182–182, 174 N.E. 441.) Most recently, in *Credit Alliance*, we spelled out the following criteria for liability: (1) awareness that the reports were to be used for a particular purpose or purposes; (2) reliance by a known party or parties in furtherance of that purpose; and (3) some conduct by the defendants linking them to the party or parties and evincing defendant's understanding of their reliance (*Credit Alliance Corp. v. Andersen & Co.*, 65 N.Y.2d at 551, 493 N.Y.S.2d 435, 483 N.E.2d 110).

For present purposes, the facts asserted in plaintiffs' submissions satisfy these prerequisites. Plaintiff alleges that through direct contact with defendants, information transmitted by Anderson, and the nature of the work, defendants were aware—indeed, could not possibly have failed to be aware—that the substance of the reports they furnished would be transmitted to and relied on by the school district. Plaintiff asserts that that was the very purpose of defendants' engagement.

Though under contract to Anderson, defendants allegedly undertook their work in the knowledge that it was for the school district alone, and that their findings would be reported to and relied on by the school district in the ongoing project—the evaluation of the structural soundness of the school buildings. Defendants were retained to visit plaintiffs premises, examine its buildings, and prepare reports of their findings on which action would be taken. The engagement of consultants was provided for in the contract between the school district and Anderson; the retention of defendants specifically was authorized by the school board, and they were so informed; in seeking compensation, Geiger itself wrote "we were hired by [the school district]," and it sent a bill directly to the school district. Given plaintiff's factual allegations of known reliance on the findings of both engineering consultants, Geiger's argument that it was not engaged until after the school district had decided to close the annex, and thus could not be liable for plaintiff's losses, does not at this juncture entitle it to dismissal of the complaint. Plaintiff further alleges that defendants had various types of contact directly with the school district. That, as well as the contents of some of Anderson's communications with defendants, constitutes conduct linking defendants to plaintiff and evidencing their understanding of plaintiff's reliance.

Not unlike the bean weighers in *Glanzer*, defendants allegedly rendered their reports with the objective of thereby shaping this plaintiff's conduct, and thus they owed a duty of diligence established in our law at least since *Glanzer* not only to Anderson who ordered but also to the school district who relied.

Accordingly, the order of the Appellate Division should be reversed, with costs, the motions to dismiss the complaint as against defendants Thune Associates Consulting Engineers and Geiger Associates, P.C. denied, and the certified question answered in the negative.

SIMONS, ALEXANDER, TITONE, HANCOCK and BELLACOSA, J. J., concur.

WACHTLER, C. J., taking no part.

Order reversed, etc.

Although both the *John Martin* and the *Ossining* decisions claim to be following the leading New York decisions, there is a clear distinction between them. The *John Martin* decision followed the Second Restatement of Torts, Section 552, requirements: first, that liability be restricted to the limited number of persons for whose benefit and guidance the defendant "intends to supply the information," and second, that supplier "intends the information to influence or knows that the recipient so intends." Yet the opinion itself appears to base protection principally on foreseeability of use, a relatively broad standard.

In the *Ossining* decision, the court took pains to reject both a rule permitting recovery by any foreseeable plaintiff and "a somewhat narrower rule that would permit recovery where the reliant party or class of parties was actually known or foreseen by the defendants." It thought these two standards would impose overly broad liability. Instead, the court required that the information be used for a particular purpose, that the reliance be by a known party in furtherance of that purpose, and that there be some conduct by the defendants linking them to the party or parties.

The court then found this necessary bonding in the direct contact between the engineers and the school district, in the undoubted awareness by the engineers that their reports would be submitted and relied on, that they were retained to visit the premises, examine the buildings, and repair reports, and finally that the retention by the architect of the consulting engineers was authorized by the owner. One of the consulting engineers himself wrote that he was hired by the school district, had sent a bill directly to the district, and had had various other types of direct contact with the district.

The differentiation noted between the *John Martin* and *Ossining* decisions is crucial in determining whether contractors and subcontractors will be allowed to sue the design professional for economic losses caused by a defective design. Those courts find Section 552 of the Second Restatement of Torts applicable to contractor and subcontractor claims for economic losses caused by defective design, which permits these parties to sue the architect or engineer for negligent misrepresentation,[129] while such a lawsuit is not allowed under the New York rule.[130]

Another position adopted by some courts is to refuse to apply Section 552 to third-party negligent design claims, even though these same jurisdictions apply the Restatement rule to claims against other professionals, such as accountants and attorneys. These courts refuse to apply Section 552 to the construction industry, which is characterized by contractual relationships in which the parties are free to negotiate the risk of economic losses caused by parties associated with the project, but with whom there is no privity.[131]

---

[129]*Donnelly Constr. Co. v. Oberg/Hunt/Gilleland*, 139 Ariz. 184, 677 P.2d 1292 (1984); *Moransais v. Heathman*, 744 So.2d 973, 983 (Fla.1999), rehearing denied Nov. 4, 1999; *Gulf Contracting v. Bibb County*, 795 F.2d 391 (11th Cir.1986) (Georgia law); *Prichard Bros., Inc. v. Grady Co.*, 428 N.W. 2d 391 (Minn.1988); *Jim's Excavating Service, Inc. v. HKM Assoc.*, 265 Mont 494, 878 P.2d 248 (1994); *Bilt-Rite Contractors, Inc. v. The Architectural Studio*, 581 Pa, 454, 866 A.2d 270 (2005); *Forte Bros. v. National Amusements, Inc.*, 525 A.2d 1301 (R.I.1987); and *Tommy L.Griffin Plumbing & Heating Co. v. Jordan, Jones & Goulding, Inc.*, supra note 110 (finding that a "special relationship" exists between the contractor and engineer such that the latter may be liable for economic damages).

[130]*Williams & Sons Erectors, Inc. v. South Carolina Steel Corp.*, 983 F.2d 1176 (2d Cir.1993). Ohio, which also follows the privity or near-privity rule, ruled that a contractor could not sue the architect; see *Floor Craft Floor Covering, Inc. v. Parma Community General Hosp.*, 54 Ohio St.3d 1, 560 N.E.2d 206 (1990).

[131]*Congregation of the Passion v. Touche Ross & Co.*, 159 Ill.2d 137, 636 N.E.2d 503, cert. denied, 513 U.S. 947 (1994); *SME Industries, Inc. v. Thompson, Ventulett, Stainback & Assocs., Inc.*, supra note 68, 28 P.3d at 682–84; *Sesenbrenner v. Rust, Orling & Neales, Architects, Inc.*, 236 Va. 419, 374 S.E.2d 55 (1988); and *Berschauer/Phillips Constr. Co. v. Seattle School Dist.*, 124 Wash.2d 816, 881 P.2d 986 (1994).

E. Economic Loss Rule: Losses Unconnected to Personal Harm or Damage to Property

As seen in Section 14.08C, in some states the requirement of privity can bar a third party's claim for economic loss. However, the economic loss rule not only bars third-party claims but also bars the use of tort law where claimant and defendant are contractually connected, such as in a claim by a client against its design professional. For convenience this doctrine is treated in this section dealing with third-party claims. Examples of economic losses relating to defective products (or buildings) are

1. damage to the defective product itself
2. diminution in the value of the product
3. natural deterioration of the product
4. cost of repair or replacement of the product
5. loss of profits caused by use of a defective product

Although the economic loss rule has long been a part of English and American tort law,[132] its modern development stems from the 1960s explosion of product liability. During that period, the courts had to face claims by purchasers of defective products against manufacturers when the product itself did not function properly and caused economic losses to the purchaser. In the leading case of *Seely v. White Motor Co.*, the court barred recovery of lost profits suffered by a purchaser of a defective truck.[133] In rejecting this claim, Chief Justice Traynor stated that the manufacturer can protect itself from liability for physical harm by meeting the tort standard. But it should not be charged for the performance level in the purchaser's business without showing a breach of warranty. The user cannot charge the success of her venture to the manufacturer.

In *East River Steamship Corp. v. Transmerical Delaval, Inc.*, an admiralty case, the U.S. Supreme Court held that a charterer of a supertanker could not recover in tort from the turbine manufacturer who had designed and manufactured the turbines installed in the charterer's vessels, for damage to the turbines themselves.[134] Speaking for the Court, Justice Blackmun stated that three views had developed in the state courts. The first would not allow tort to be used in such a claim, preserving a proper role for the

law of warranty if a defective product causes purely monetary harm. At the other end of the spectrum, a minority of courts would allow recovery where the product itself was injured; those courts saw no distinction between harm to person and property and purely economic loss. Those jurisdictions are not concerned about unlimited liability, because a manufacturer can predict and ensure against product failure. Some courts sought a compromise solution, allowing recovery if the defective product creates a situation potentially dangerous to people or property.

The Supreme Court followed the majority rule, concluding that the intermediate position was too indeterminate to allow manufacturers to structure their business behavior. When harm to the product itself occurs, the resulting loss due to repair costs, decreased value, and lost profit is the failure of the purchaser to receive the benefit of its bargain, a reason to reject the minority rule allowing such a claim to be brought in tort. Also, the minority view did not adequately keep tort and contract law in their proper compartments and failed to maintain a realistic limitation on damages.

Justice Blackmun stated that tort law was concerned with safety but not with injury to the product itself. In such a case, the commercial user

> stands to lose the value of the product, risks the displeasure of its customers who find that the product does not meet their needs, or, as in this case, experiences increased costs in performing a service. Losses like these can be insured. Society need not presume a customer needs special protection.[135]

In these transactions, as had been suggested by Chief Justice Traynor, contract law and the law of warranty function well.

The Supreme Court was not persuaded that the limitations set forth in the minority of cases granting recovery for economic losses—that of foreseeability—would be an adequate brake on unlimited liability. The Supreme Court stated, "Permitting recovery for all foreseeable claims for purely economic loss could make a manufacturer liable for vast sums. It would be difficult for a manufacturer to take into account the expectations of persons downstream who may encounter its product."[136] The *East River* Court's definition of economic loss, as involving damage to the

---

[132]*Robins Dry Dock & Repair Co. v. Flint*, 275 U.S. 303, 308–09 (1927).

[133]63 Cal.2d 9, 403 P.2d 145, 45 Cal.Rptr. 17 (1965).

[134]476 U.S. 858 (1986).

[135]Id. at 871–72.

[136]Id. at 874.

product itself, was adopted in the Restatement (Third) of Torts: Products Liability, Section 21(c).

While having its modern origin in products liability law, the economic loss rule has subsequently spread to virtually all aspects of construction defects litigation. The rule has been applied to claims (1) brought by third parties and between contracting parties and (2) by and against both commercial parties and unsophisticated homeowners. Thus, while this Chapter 14 addresses design professionals and this Section 14.08 discusses third-party claims, the following discussion is not limited to either category.

In truth, there is no one economic loss rule. While there is consensus as to what economic losses are (see the list at the beginning of this Section 14.08E), courts apply the rule (or find it inapplicable) in different situations and for different reasons. A comprehensive discussion of the economic loss rule is beyond the scope of this text, and what follows is intended simply as illustrations of this complex topic.[137]

The most expansive use of the economic loss rule is by states that apply the rule in a literal and mechanistic manner, flatly refusing to allow the recovery of economic losses under theories of nonintentional torts. In these states, such as Illinois, the rule applies whether the parties are in privity, regardless of whether the plaintiff has a contract remedy, and to claims against design professionals.[138] Other states permit third-party claims if the relationship between the parties approximates that of privity or if the parties have a "special relationship."[139] Colorado permits recovery of economic losses between parties in privity if the defendant breached a tort duty that was independent of its contract obligations.[140] Wisconsin goes so far as to bar a homeowner's claim against a subcontractor—that is, a claim by a noncommercial party not in a contractual relationship with the defendant.[141] At the same time, Wisconsin does not extend the economic loss rule to services contracts, including between commercial parties in privity.[142] One approach gaining increasing acceptance is to disallow third-party claims if there exists a scheme of interrelated contracts that affords the claimant a contract remedy, even if not against the actual tortfeasor. Under this rationale, for example, a subcontractor will be denied a tort claim against the project's architect, on the grounds that the subcontractor has a contract remedy against the prime contractor for the economic losses caused by the architect's negligence.[143]

Other jurisdictions are more willing to allow the recovery of economic losses in tort. These states apply the economic loss rule only under narrow circumstances. As a general rule, these courts refuse to allow the economic loss doctrine to swallow up the preexisting law of torts by eliminating a defendant's duty to act with due care.[144]

The California Supreme Court has repeatedly addressed the recovery of economic loss damages claims in construction defect cases. It has approached this issue *not* through a mechanistic application of the economic loss rule, but instead has viewed it through the lens of the 1958 *Biakanja v. Irving* decision (discussed shortly), which devised a test

---

[137]For a listing of the different rationales underlying the economic loss rule and its exceptions as applied to construction industry disputes, see Analysis, *Economic Loss Rule—Concentric Circles of Rationales*, 26 Constr.Litg.Rep., No. 5, May 2005, p. 216. For a generalized review of the doctrine from many perspectives, see *Symposium: Dan B. Dobbs Conference on Economic Tort Law*, 48 Ariz.L.Rev., No. 4, Winter 2006, pp. 689 et seq.

[138]*2314 Lincoln Park West Condominium Ass'n v. Mann, Gin, Ebel & Frazier, Ltd.*, 136 Ill.2d 302, 555 N.E.2d 346 (1990) (client must sue its architect in contract); *Anderson Elec., Inc. v. Ledbetter Erection Corp.*, 115 Ill.2d 146, 503 N.E.2d 246 (1986) (claim barred despite lack of contract remedy).

[139]*Ossining Union Free School Dist. v. Anderson LaRocca Anderson*, reproduced in Section 14.08D (relationship approximating privity); *Eastern Steel Constructors, Inc. v. City of Salem*, supra note 100 ("special relationship" between architect and contractor).

[140]*Town of Alma v. AZCO Construction, Inc.*, 10 P.3d 1256 (Colo.2000).

[141]*Linden v. Cascade Stone Co., Inc.*, 283 Wis.2d 606, 699 N.W.2d 189 (2005). The question of whether the economic loss rule applies to an owner's tort claim against a subcontractor is particularly divisive; compare *A.C. Excavating v. Yacht Club II Homeowners Ass'n, Inc.*, 114 P.3d 862 (Colo.2005) (claim permitted) with *Corporex Development & Construction Mgt., Inc. v. Shook, Inc.*, 106 Ohio St.3d 412, 835 N.E.2d 701 (2005) (claim denied).

[142]*Insurance Co. of North Am. v. Cease Elec. Inc.*, 276 Wis.2d 361, 688 N.W.2d 462 (2004).

[143]*BRW, Inc. v. Dufficy & Sons, Inc.*, 99 P.3d 66 (Colo.2004) (subcontractor has remedy against the prime contractor, who has a remedy against the owner under the *Spearin* doctrine discussed in Section 23.05); *SME Industries, Inc. v. Thompson, Ventulett, Stainback & Assocs., Inc.*, supra note 68. By the same logic, an owner's remedy lies in a contract claim against the prime contractor, not a tort claim against the negligent subcontractor; see *Corporex Development & Construction Mgt., Inc. v. Shook, Inc.*, supra note 141.

[144]*Indemnity Ins. Co. of North America v. American Aviation, Inc.*, 891 So.2d 532 (Fla.2004); *Kennedy v. Columbia Lumber & Mfg. Co., Inc.*, 299 S.C. 335, 384 S.E.2d 730 (1989).

for determining when a contracting party is liable to third parties for economic losses caused by his negligent performance of his contract obligations. In *Aas v. Superior Court*,[145] the court also confronted the vexing question of whether the economic loss rule should be suspended when a hazardous construction defect (which has not yet caused any personal injury) prompted the litigation.

In *Aas*, homeowners sued a developer and its subcontractors for multiple construction defects. Some of those defects—such as improper shear walls meant to stabilize the building in an earthquake and failure to build fire protection in party walls—implicated safety concerns. However, because the owners alleged no damage to other property or personal injury, their claim presented the issue of whether the economic loss rule should extend to hazardous defects. The trial court applied the economic loss rule, and a divided California Supreme Court affirmed.

The majority placed the issue before it—a builder's liability for construction defects which have not caused physical injury—in the wider context of circumstances under which a defendant's negligent performance of a contract obligation could subject it to liability in tort for economic losses foreseeably suffered by the plaintiff. The supreme court first addressed that issue in *Biakanja v. Irving*,[146] a case in which the plaintiff was denied a testamentary gift because the defendant, a notary public practicing law without a license, had negligently prepared a will by not having it properly solemnized. Applying a six-factor test, the *Biakanja* court concluded that the relationship between the defendant and plaintiff was such that the defendant owed plaintiff a duty to perform his contract obligation nonnegligently. These factors included the foreseeability of harm to the plaintiff, "the degree of certainty that the plaintiff suffered injury," and the public policy of preventing future harm.

In the early 1960s, the court extended the *Biakanja* test to the construction industry, ruling that both prime contractors and subcontractors owed a duty of care to homeowners for defective construction work.[147] However, in 1965, the court in the *Seely* decision, discussed earlier in this section, invoked the economic loss rule to bar a strict product liability claim. Whether the economic loss rule extended as well to negligence claims was brought into doubt nineteen years later. In *J'Aire Corp. v. Gregory*,[148] a restaurant owner who was a tenant in a building suffered lost business allegedly caused by construction delays by a contractor hired by the landlord. Applying *Biakanja*—especially the foreseeability of harm factor—the supreme court found that a "special relationship" existed between the tenant and contractor, allowing the tenant to sue the contractor for negligent interference with contract in order to recover lost profits.

Going back to the *Aas* case, the majority concluded that the homeowners did not satisfy the *Biajanka* test for tort liability. Plaintiffs did not meet the factor that looks to "the degree of certainty that the plaintiff suffered injury," because the builders' negligence caused only the threat of future harm. The homeowners also did not meet the foreseeability of harm factor, because they have suffered no actual harm. Most troublesome, though, was the *dissent's* view that under the final *Biakanja* factor—the public policy of preventing future harm—it is (in the words of the dissent) "economically efficient" to allow the homeowners to make the repairs, then charge that cost to the builders. The *majority* disagreed, stating:

> In some sense, that policy [of preventing future harm] might be served by a rule of tort liability making builders, in effect, the insurers of building code compliance, even as to defects that have not caused property damage or personal injury. Moreover, as plaintiffs argue, to require builders to pay to correct defects as soon as they are detected rather than after property damage or personal injury has occurred might be less expensive. On the other hand, such a rule would likely increase the cost of housing by an unforeseeable amount as builders raise prices to cover the increased risk of liability. Such a rule should also be unnecessary to the extent buyers timely enforce their contract, warranty and inspection rights, and to the extent building authorities vigorously enforce the applicable codes for new construction.[149]

An increasing number of courts have held, as did the *Aas* court, that the economic loss rule should bar both homeowners and commercial owners from a tort remedy when they perceive their building contains hazardous defects which have not (yet) caused either property

---

[145]24 Cal.4th 627, 12 P.3d 1125, 101 Cal.Rptr.2d 718 (2000).

[146]49 Cal.2d 647, 320 P.2d 16 (1958).

[147]*Stewart v. Cox*, 55 Cal.2d 857, 13 Cal.Rptr. 521 (1961) (subcontractor liable in negligence) and *Sabella v. Wisler*, 59 Cal.2d 21, 377 P.2d 889, 27 Cal.Rptr. 689 (1963) (prime contractor liable in negligence).

[148]24 Cal.3d 799, 598 P.2d 60, 157 Cal.Rptr. 407 (1979).

[149]12 P.3d at 1140.

damage or personal injury. As indicated in the discussion of *East River Steamship*, the Supreme Court rejected the so-called "intermediate position" (allowing recovery for hazardous defects) as too indeterminate a standard of manufacturer liability.[150] The experience of the Maryland Court of Appeals (that state's highest court) is instructive. In *Council of Co-Owners Atlantis Condominium, Inc. v. Whiting-Turner Construction Co.*,[151] the court held that the economic loss rule does not apply to hazardous defects: in that case, building code violations. However, less than a decade later, in *Morris v. Osmose Wood Preserving*,[152] it applied the economic loss doctrine and barred a builder's claim that deteriorating roof trusses, allegedly caused by the defendant manufacturer's wood preservative, led to the possibility of roof collapse.[153]

F. Interference with Contract or Prospective Advantage—The Adviser's Privilege

Intentional interference with a contract or prospective advantage is increasingly invoked in claims against design professionals. This intentional tort was discussed in Section 11.03 in the context of retaining a design professional. This section will refer to its use when the construction contract is awarded or during its performance.

In the award stage, the design professional can play a pivotal role. She may advise the owner as to which bid should be accepted when a contract is to be awarded by competitive bidding. She may also give advice when a contract is awarded by negotiation. During performance, the design professional plays a significant role both in advising the owner and in making decisions. The immunity sometimes granted to the design professional when she makes decisions is discussed in Section 14.09D. The design professional's role generally in the decision-making process is discussed in Chapter 29.

During performance, the design professional may be asked to advise the owner on subcontractor selection or removal. She may be asked to advise the owner during a negotiation with the contractor over compensation for changes. She may also be required to certify, as in AIA Document A201-2007, Section 14.2.2, that there is sufficient cause to terminate the contractor. Other illustrations can be given, but these are enough to show the design professional in her advising role and the capacity her advice may have to harm others. However, much of the complexity in claims against the design professional stems from the varied roles she plays—agent, quasi-arbitrator, and individual architect.[154]

Because of the open-ended exposure the tort can create, the law has tended to protect only against intentional interference. However, the important case of *J'Aire Corp. v. Gregory* discussed in Section 14.08E permitted the tenant in an airport to maintain an action for negligent interference with its restaurant business against a contractor whose negligence delayed the installation of an air-conditioning system.[155]

The substantive basis for an intentional interference claim can range from actual malice or bad faith to simply intending to perform an act that interferes with the contract or prospective advantage. As illustrations, a subcontractor sued the architect who advised the owner not to approve that particular subcontractor.[156] Similarly, a rejected bidder

---

[150]476 U.S. at 870.

[151]308 Md. 18, 517 A.2d 336 (1986).

[152]340 Md. 519, 667 A.2d 624 (1995).

[153]For other courts coming to the same conclusion, see *Moorman Mfg. Co. v. National Tank Co.*, 91 Ill.2d 69, 435 N.E.2d 443, 449–50 (1982); *Determan v. Johnson*, 613 N.W.2d 259 (Iowa 2000); and *Bellevue South Assocs. v. HRH Constr. Corp.*, 78 N.Y.2d 282, 579 N.E.2d 195, 574 N.Y.S.2d 165 (1991). Where the hazardous defect is the installation of asbestos-containing materials, application of the economic loss rule turns on whether the asbestos fibers have spread. If they have, courts find that the building has suffered property damage and so do not apply the economic loss rule. *City of Greenville v. W.R. Grace & Co.*, 827 F.2d 975 (1987), rehearing denied, 840 F.2d 219 (4th Cir.1988) (South Carolina law); *80 South Eighth Street Ltd. Partnership v. Carey-Canada, Inc.*, 486 N.W.2d 393, 397, rehearing denied Sep. 11, 1992, amended by 492 N.W.2d 256 (Minn. 1992); and *Northridge Co. v. W.R. Grace & Co.*, 162 Wis.2d 918, 471 N.W.2d 179, 186 (1991). Where there is no evidence that the asbestos fibers have escaped, the courts are less likely to find the defendant liable in tort since the harm remains speculative; see *Adams-Arapahoe School Dist. v. GAF Corp.*, 959 F.2d 868 (10th Cir. 1992) and *San Francisco Unified School Dist. v. W.R. Grace & Co.*, 37 Cal.App.4th 1318, 44 Cal.Rptr.2d 305, 309–12 (1995). A California court, analogizing to the law applicable to asbestos contamination, ruled that contamination of a utility's pipeline by polychlorinated biphenyls (PCBs) constitutes "property damage" for purposes of the economic loss rule. *Transwestern Pipeline Co. v. Monsanto Co.*, 46 Cal.App.4th 502, 53 Cal.Rptr.2d 887 (1996).

[154]*Lundgren v. Freeman*, 307 F.2d 104 (9th Cir.1962) (a leading case).

[155]Supra note 147.

[156]*Kecko Piping Co., Inc. v. Town of Monroe*, 172 Conn. 197, 374 A.2d 179 (1977) (architect given defense of adviser's privilege).

sued the architect.[157] Another case involved a claim by a contractor against an architect who did not confirm a preliminary approved product substitution by issuing a change order. When the architect ordered the contractor to remove the substituted product, the contractor refused, and on the architect's advice the owner terminated the prime contractor.[158] Another case involved a claim by a mechanical subcontractor and its supplier against an engineer for arbitrarily requiring equipment specifications the supplier could not have met.[159] Another involved a claim by the contractor that the engineer communicated directly with the contractor's suppliers, encouraged the suppliers to threaten to terminate or actually terminate their delivery of supplies, and encouraged the owner to treat nondelivery by the suppliers as grounds to terminate the contract.[160] Finally, an investigating engineer was sued by a soils engineer when he claimed that the report the investigating engineer gave the owner was false and interfered with anticipated future relations with the owner.[161]

By and large these claims have not been successful. Some defendants have been able to invoke the Restatement (Second) of Torts, Section 772,[162] which states that interference has not been improper if the person against whom the claim has been made has in good faith given honest advice when requested to do so. Another defense granted has been the privilege given to a person who is asked to prepare for litigation.[163]

---

[157]*Commercial Indus. Constr., Inc. v. Anderson*, 683 P.2d 378 (Colo. Ct.App.1984) (owner privileged to reject bid). See also *Riblet Tramway Co., Inc. v. Ericksen Assoc.*, 665 F.Supp.81 (D.N.H.1987) (engineer advised public entity that bidder was not qualified: defense of adviser's privilege granted).

[158]*Dehnert v. Arrow Sprinklers, Inc.*, 705 P.2d 846 (Wyo.1985) (claim unsuccessful, as bad faith not shown). See also *Victor M. Solis Underground Utility & Paving Co., Inc. v. City of Loredo*, 751 S.W.2d 532 (Tex.Ct.App.1988) (defense of engineer's superior contract right and duty to protect city granted).

[159]*Waldinger Corp. v. CRS Group Eng'rs, Inc.*, 775 F.2d 781 (7th Cir.1985) (engineer entitled to privilege unless it could be shown that it intended to harm the plaintiffs or further engineer's personal goals).

[160]*Santucci Constr. Co. v. Baxter & Woodman, Inc.*, 151 Ill.App. 3d 547, 502 N.E.2d 1134 (1986), appeal denied, 115 Ill.2d 550, 511 N.E.2d 437 (1987).

[161]*Western Technologies, Inc. v. Sverdrup & Parcel, Inc.*, supra note 93 (report absolutely privileged as made by potential witness in anticipation of litigation).

[162]*Kecko Piping Co., Inc. v. Town of Monroe*, supra note 156; *Williams v. Chittenden Trust Co.*, 145 Vt. 76, 484 A.2d 911 (1984) (adviser's privilege not a defense, as conduct went beyond advice).

[163]*Western Technologies v. Sverdrup & Parcel, Inc.*, supra note 93.

Despite the relative lack of success of such claims, the claims still expose design professionals to liability for an intentional tort with the possibility of punitive damages.[164]

G. Safety and the Design Professional: *Pfenninger v. Hunterdon Central Regional School District* and *CH2M Hill, Inc. v. Herman*

The responsibility of design professionals for safety has become a contentious issue. Does the design professional engaged to perform the normal site services have any responsibility for harm to people or to property when the principal cause of the harm has been negligence by the contractor or a subcontractor? This problem has surfaced principally in two forums. The first is the judicial system, where the problem is triggered by a claim by or on behalf of an injured worker, a member of the public, or an adjacent landowner that seeks to transfer to the design professional losses it has suffered, claiming negligence. This negligence usually consists of not taking reasonable steps to prevent contractors from performing work in a way that unreasonably exposes the claimant to personal harm or property damage.

The other forum in which this issue arises is workplace safety laws, such as the Occupational Safety and Health Act (OSHA), or equivalent state regulatory agencies. Agencies charged with administering such laws increasingly charge design professionals with responsibility for unsafe workplaces.

Claims for compensation by those who have suffered losses often involve the effect of contract language that seeks to exculpate the design professional from any responsibility for safety. This is similar to the language discussed in Section 12.08B that seeks to exculpate the design professional from responsibility for the contractor not performing properly under its contract. Tort claims brought through the judicial process on behalf of people who seek to transfer their losses to the design professional are examined first.

A tort claim requires that the defendant owe a duty to the claimant to act in a way that avoids exposing the claimant to unreasonable risk of harm. When the privity requirement was dropped in the late 1950s, design

---

[164]*Custom Roofing Co., Inc. v. Alling*, 146 Ariz. 388, 706 P.2d 400 (App.1985) (punitive damages justified by wanton conduct and indifference to the rights of others). The cases are cataloged and discussed by Schneier, *Tortious Interference with Contract Claims Against Architects and Engineers*, 10 Constr. Lawyer, No. 2, May 1990, p. 3.

professionals switched to another defense. They asserted that they owed no duty to third-party claimants, because their project monitoring was directed toward a project's fulfilling the contract obligations promised to the owner. (Of course, it was more complicated than that, as demonstrated in the cases reproduced next and in the analysis in Section 14.11B.)

Design professionals complained that courts were unjustifiably placing responsibility on them if anything went wrong in the construction process. Courts did this by focusing on the construction contract, with its many powers given the design professional, such as to reject work, stop the work, and provide general supervision. These powers, they contended, existed only to implement their "monitoring" function. The design professional, they asserted, had no right or duty to tell the contractor how the work was to be done. Nor was she paid, in the ordinary project, to be a "safety" engineer.

As a result of adverse court decisions, standard agreements made by the professional associations were changed in the 1960s to seek to make clear that the contractor, not the design professional, decided how the work was to be done and that the design professional did not provide continuous on-site observation and certainly did not supervise the work. These changes, designers claimed, simply reflected the true allocation of responsibility for work on the site.[165]

Most courts find that contracts which (1) reserve to the architect or engineer responsibility only to "observe" (not supervise) the work, (2) impose upon the contractor responsibility for the manner or method of performance and for job-site safety, and (3) grant the owner (not the architect) the authority to stop the work, do not impose a duty of care upon the design professional for the safety of the construction workers. For example, in *Krieger v. J. E. Greiner Co.*,[166] a construction worker was injured when a steel beam collapsed. He claimed that the erection subcontractor did not support the 780-pound reinforcing bars and the steel column. He asserted that the work was performed under the supervision of defendant Greiner, the prime engineer, and defendant Zollman, a consulting engineer. He sued Greiner and Zollman, claiming that they should have known that the subcontractor was

performing in a defective and dangerous manner and that the defendants previously had stopped the work when they perceived it being performed in a dangerous manner. The Maryland Court of Appeals (the state's highest court) rejected the worker's claim:

> We have carefully examined each of the contracts in question. We find no provisions in these contracts imposing any duty on the engineers to supervise the methods of construction. Some mathematics instructors have been heard to observe that there is more than one solution to a given problem and thus they are unable to say that any given method of solving a problem is the only correct solution, being able only to determine that the correct answer is produced. The same reasoning would apply to methods of construction. One skilled contractor may prefer one method for performing a given task while another such contractor may choose what seems to him a simpler, less expensive way of reaching the same end result, either of which procedures would be a proper method. It could well be, however, that one method might not have occurred to an engineer or another contractor.
>
> We likewise find nothing in the contracts imposing any duty on the engineers to supervise safety in connection with construction.
>
> The duty of the engineers under their contracts is to assure a certain end result, a completed bridge which complies with the plans and specifications previously prepared by Greiner. It will be observed that many of the cases which have held architects and engineers responsible for safety have done so on the basis of the construction by the courts of the contracts existing between the engineer or architect and the owner. We hold that a fair interpretation of the contracts between the Commission and Greiner and Zollman is that the duties of those engineers do not include supervision of construction methods or supervision of work for compliance with safety laws and regulations. Hence, the Kriegers may not recover from the engineers under the contracts between the owner and its engineers.[167]

In *Krieger*, the construction worker argued that the engineers "should have known" of the erection subcontractor's unsafe manner of performance. He essentially was saying that, while the engineer did not know of

---

[165]M. SCHNEIER, CONSTRUCTION ACCIDENT LAW, supra note 46 at 282–85 (detailing the changes made to AIA documents).

[166]282 Md. 50, 382 A.2d 1069 (1978).

[167]382 A.2d at 1079. Accord, *Baumeister v. Automated Products, Inc.*, 277 Wis.2d 21, 690 N.W.2d 1 (2004) (under AIA Doc. B141-1987, architect had no duty to supervise installation of roof trusses and was not liable to workers injured during that process).

the unsafe practice of the subcontractor, he was negligent in not knowing and should be liable for that negligence. Suppose, however, that the inspecting architect or engineer *knew* of the hazardous condition beforehand and did not act to have that condition ameliorated. Historically, courts were more inclined to find the inspecting design professional liable in such a situation.[168]

In the 1990s, courts began extending the rule of nonliability under the modern standard form contracts to "actual knowledge" cases. These courts reasoned that a design professional does not owe a duty to care to construction workers absent a duty assumed by contract or conduct. The engineer's mere knowledge of the hazardous condition, without more, does not create a duty of care where previously none had existed.[169]

New Jersey's experience with the question of an inspecting architect's or engineer's liability for workplace accidents is instructive. In *Carvalho v. Toll Brothers & Developers*,[170] an engineering firm designed a sewer line for a township, and provided a site representative to observe and inspect the construction on a daily basis. The various contracts made clear that the general contractor alone was responsible for worksite safety, although the engineer had the authority to stop the work. A subcontractor's employee was killed when a 13-foot deep trench in which he was standing collapsed. The engineering firm's representative knew before the accident that the trench lacked shoring and was unstable. The estate sued the engineering firm for wrongful death. The supreme court ruled that the engineer owed a duty of care to the construction workers, even though it had no contractual responsibility for site safety. Five years later, the New Jersey Supreme Court, in yet another case involving a trench cave-in, sought to restrict *Carvalho* to its facts. That decision is reproduced here.

PFENNINGER v. HUNTERDON CENTRAL REGIONAL HIGH SCHOOL

Supreme Court of New Jersey, 2001. 167 N.J. 230, 770 A.2d 1126.
[Ed. note: Footnotes omitted.]

PER CURIAM

[Ed. note: Pfenninger was killed in a trench collapse while installing drainage pipes on a school athletic field. The trench was eight feet deep and was not shored or braced, in violation of the Occupational Safety and Health Administration (OSHA) safety regulations, which Pfenninger was obligated by his contract to follow. The school board had hired defendant O'Sullivan to design and inspect the work. See Figure 14.3

Plaintiff, Pfenninger's wife, sued O'Sullivan for negligence. The trial court entered a summary judgment (a pre-trial ruling) in favor of the architect, but the appellate division reversed. The supreme court unanimously affirmed the trial court ruling in favor of O'Sullivan. Rather confusingly, it did so in a *per curiam* opinion (i.e., an opinion of the court as a whole, not authored by a particular judge) in which it adopted the dissenting opinion of Justice Coleman (a justice on the New Jersey Supreme Court) as the opinion of the entire court. (Justice Coleman dissented on a separate issue: whether the school board owed a duty of care to Pfenninger.) So the following is a reproduction of Justice Coleman's dissent, which—on the question of O'Sullivan's liability—is the opinion of the court as a whole.]

* * *

In 1994, Hunterdon Central Regional High School District Board of Education ("Board") commenced several construction projects at Hunterdon High School, including a drainage system, an underground electrical system, a new scoreboard, and renovation of the school's existing fountain. Although the

---

[168]*Erhart v. Hummonds*, 232 Ark. 133, 334 S.W.2d 869, 872 (1960) and *Balagna v. Shawnee County*, 233 Kan. 1068, 688 P.2d 157, 164 (1983), in effect overruled by Kan.Stat.Ann. § 44–501(f); see *infra* note 215. Scholars are in accord, see Sweet, *Site Architects and Construction Workers: Brothers and Keepers or Strangers?* 28 Emory L.J. 291 (1979).

[169]*Yow v. Hussey, Gay, Bell & DeYoung Internat'l, Inc.*, 201 Ga.App. 857, 412 S.E.2d 565, 568 (1991), cert denied Jan. 30, 1992; *Herczeg v. Hampton Municipal Auth.*, 2001 PA Super 10, 766 A.2d 866, 873–74, appeal denied, 567 Pa. 742, 788 A.2d 376 (2001); and *Peck v. Horrocks Engineers, Inc.*, 106 F.3d 949 (10th Cir.1997) (predicting the Utah Supreme Court would adopt the rule of non-liability).

[170]143 N.J. 565, 675 A.2d 209 (1996).

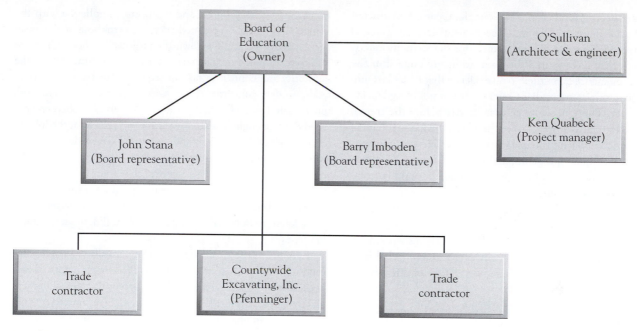

FIGURE 14.3   *Pfenninger v. Hunterdon Central Regional High School.*

Board hired multiple contractors to perform the various projects, it did not retain a general contractor. Responsibility for coordinating and scheduling the projects was shared by the Board and A.J. O'Sullivan Architects, P.A. ("O'Sullivan"), an architectural firm hired by the Board to provide architectural and engineering services.

The project involved in the present litigation involved the installation of a drainage system for the varsity baseball field. That required excavating an eight- to nine-foot deep trench to install drainage piping underneath the field. O'Sullivan contacted several excavating companies on the Board's behalf to solicit bids for the drainage project. The contract was ultimately awarded to Countywide Excavating, Inc. ("Countywide"), the owner and principal officer of which was the decedent, Matthew Pfenninger.

It was agreed that Countywide would supply labor and equipment and that the Board would supply the necessary materials, including the drainage pipe. According to the design specifications prepared by O'Sullivan, Countywide was "responsible for all safety precautions on the job." The design specifications also stated:

The contractor shall furnish, place and maintain all sheeting, bracing, lagging, shoring and miscellaneous supports, as required to support and prevent movement of earth which could injure persons in or around the work areas, and/or endanger any adjacent structures, tanks or utilities. . . .

. . . All preventive safety measures shall be in compliance with OSHA and the requirements of the local municipality and the owner.

The design plans further specified that O'Sullivan did "not have field inspection responsibilities for the job" and that Countywide was "responsible for the methods and means of construction."

The Board's representatives were John Stana, director of plants and facilities, and Barry Imboden, supervisor of grounds and maintenance. Although Stana was the self-acknowledged liaison between the Board, O'Sullivan, and Countywide, he did not visit the athletic field after Countywide commenced the excavation project. Barry Imboden, on the other hand, was on campus daily and recalls having at least two conversations with Matthew Pfenninger after the project began. Imboden peered into the trench on occasion, but never saw it in a fully exposed state because Countywide "backfilled" completed sections of the trench with gravel before moving on to the next section.

O'Sullivan appointed Ken Quabeck as its project manager. Quabeck communicated with Countywide on several occasions after the excavation began. For example, Quabeck instructed Countywide that it would have to erect a temporary fence around the exposed trench during the project for the students' safety. Quabeck may have visited the excavation site once after the project commenced, but there is no evidence that he looked into the trench or was aware that Countywide was not using bracing or shoring. There is evidence that shortly before the trench collapsed O'Sullivan was aware that Countywide was having difficulty completing the excavation work due to inclement weather, machinery problems, and soil conditions. O'Sullivan frequently requested that the project commence and be completed as soon as possible.

* * *

On August 24, 1994, Matthew Pfenninger was working in the trench when the walls suddenly caved in and killed him. The walls were not braced or shored at the time of the accident as required by the contract.

* * *

In an attempt to impose liability on the architect, plaintiff relies on the Court's discussion of an engineering firm's duty of care to an independent contractor in *Carvalho, supra,* 143 N.J. 565, 675 A.2d 209. In *Carvalho,* the township of West Windsor hired an engineering firm to prepare plans for a sewer construction project. Although the general contractor was responsible for safety measures and the methods of construction, the engineer was contractually required to maintain an on-site representative, who was to monitor work progress and ensure that the construction complied with the design specifications. To this end, the engineer's on-site representative had the contractual authority to stop construction. An excavation subcontractor's employee was killed when the trench in which he was working collapsed. Notwithstanding the fact that the contract required excavation protection systems, the trench was not shored or braced at the time of the accident. Notably, the engineering firm's on-site representative was standing near the trench on the day of the accident, observing the decedent perform his work.

The decedent's widow sued the engineering firm, arguing that it owed the decedent a duty to provide reasonably safe premises. The Court first determined that the accident was foreseeable because the contract provided for trench safety precautions and other trenches had previously collapsed at the construction site. The Court then observed that it was fair to impose a duty on the engineer because it had assumed a contractual responsibility

to monitor work progress and to ensure compliance with the design plans, which implicated safety concerns because the use of trench boxes would slow down the project. [Ed. note: Citations omitted.] The Court also stressed that the engineer "had the authority and control to take or require corrective measures to address safety concerns," when safety conditions affected work progress. [Ed. note: Citations omitted.] Finally, the Court emphasized that the engineer's representative was present at the job site on the date of the accident and had actual knowledge that work was being conducted in the trench without the protection of a trench box. For all of those reasons, the Court concluded that imposing a duty to exercise reasonable care to protect against the risk of injury on the construction site was consistent with considerations of fairness and public policy. [Ed. note: Citations omitted.]

Relying on the Court's decision in *Carvalho,* plaintiff contends that O'Sullivan owed Matthew Pfenninger a duty of care because it was foreseeable that the trench walls could collapse. I agree that there is a risk in nearly every excavation project that the walls of an unshored trench may cave in. Those who doubt this statement need only canvass New Jersey's case law, which is replete with lawsuits arising from trench collapses.
[Ed. note: Citations omitted.]

The risk of a cave-in can be eliminated, however, through the use of adequate protection systems, such as bracing, shoring, or trench boxes. Therefore, plaintiff is not correct that it was foreseeable that the walls would collapse on Matthew Pfenninger. O'Sullivan, like the engineering firm in *Carvalho,* inserted a clause into the design specifications requiring Countywide [the construction company employing Pfenninger] to adequately brace or shore the trench. In contrast to an unbraced trench, it is not foreseeable that a braced trench will collapse. Nor did O'Sullivan have reason to believe that Countywide would deviate from the design specifications and fail to secure the trench walls with a protection system.

The same can be said, of course, about the engineer in *Carvalho.* The critical difference, however, is that the engineer in *Carvalho* was contractually required to maintain an on-site representative who had the authority to ensure compliance with safety requirements and to stop work, if necessary, to ensure compliance. Furthermore, the engineer's representative in *Carvalho* was observing the decedent work in the trench at the time of the accident, and was, therefore, aware that neither a trench box nor any other safety system was being used.

O'Sullivan, in contrast to the engineer in *Carvalho,* was neither required to maintain an on-site representative nor did it have the authority to halt the project to remedy safety violations. The design specifications explained that O'Sullivan did not have "field inspection responsibilities" and that Countywide was "responsible for safety precautions on the job." An architect's duty to foresee and prevent harm is generally "commensurate with the degree of responsibility which the engineer [or architect] has agreed to undertake." *Sykes v. Propane Power Corp.*, 224 N.J.Super. 686, 694, 541 A.2d 271 (App.Div.1988). Furthermore, the summary judgment record lacks evidence that O'Sullivan, or the Board for that matter, had actual knowledge that Countywide

failed to brace the walls of the trench. Although O'Sullivan's project manager, Ken Quabeck, visited the excavation site at least once, there is no competent evidence in the record establishing that he looked into the trench or was told that the walls were not braced. Absent contractual responsibility for on-site safety and actual knowledge that the trench was not braced, O'Sullivan could not foresee that the trench would cave in on Matthew Pfenninger.

* * *

Future New Jersey cases involving claims by injured construction workers or their estates will have to contend with the state legislature's response to the *Carvalho* decision. N.J.S.A. § 2A:29B-1 provides that a professional engineer or engineering firm is not liable for a worksite accident unless (1) the engineer expressly assumed responsibility for worksite safety, or (2) on a multi-prime project, the engineer is the owner's representative and no contractor has been designated responsible for site safety, or (3) the engineer "had actual knowledge" of site conditions involving "an imminent danger" and failed, within a reasonable time, to notify the contractor or construction workers. Section 2A:29B-2 creates an exception for acts of willful misconduct or gross negligence. While the statute refers only to engineers, it presumably applies to architects as well. It modifies the common law "actual knowledge" rule by imposing upon the design professional a duty to warn of hazardous conditions involving an imminent danger.[171]

The *Krieger*, *Carvalho* and *Pfenninger* decisions all involved a design professional's liability to construction workers arising out of the architect's or engineer's presence on the job site. An architect who only provided the design and had no involvement in the actual construction would not be liable under such a theory. However, workers may also allege liability arising out of the architect's design duties. In *Waggoner v. W&W Steel Co.*,[172] the death of two workers and the injury of another resulted from the

contractor's failure to provide interior columns and cross beams that would have provided lateral bracing for the outside columns. While the steel was being erected, a gust of wind hit the unsecured and unbraced steel, causing it to collapse. The claimants asserted that the architect owed a duty to workers. They claimed the architect was negligent in approving shop drawings that did not provide for temporary connections on the expansion joints.

The court held that although privity was not required before a tort claim could be brought in claims for physical injuries to third parties, the actual liability depends on whether a duty has been created, and this depends on the nature of the architect's undertaking and his conduct.

The court then quoted extensively from general conditions of the contract for construction, very likely those of the American Institute of Architects (AIA). The contract provided that the responsibility for executing the contract and reasonable precautions for safety belonged to the contractor, not to the architect. The fact that the architect periodically visited the site did not remove the responsibility from the shoulders of the contractor. As to the plaintiffs' contention that the architect's approval of shop drawings that did not provide for temporary connections on the expansion joints constituted negligence, the court pointed to language in the contract stating that the contractor represented that he had checked all shop drawings for the requirements of the contract and also noted that the contract makes the contractor solely responsible for "construction means, methods, techniques, sequences and procedures."[173] But the contract specified that the

---

[171]Other statutory protections of design professionals against injured worker claims are discussed in Section 14.09E.

[172]657 P.2d 147 (Okla.1982).

[173]Id. at 151.

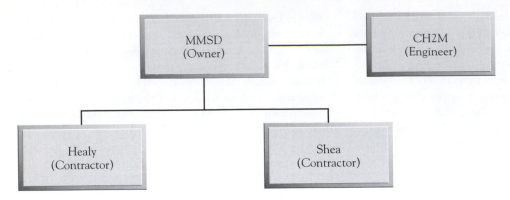

FIGURE 14.4   *CH2M Hill, Inc. v. Herman.*

architect's approval was only for conformance with the design concept and the information given in the contract documents.

Most important from the perspective of design professionals, relying mainly on the exculpatory provisions in the contract the court found there was no question of fact for the jury to decide. The design professional was able to receive a summary judgment and avoid the cost of a full trial and the uncertain outcome of a jury verdict.

As mentioned earlier, another forum where this question has become controversial is the Occupational Safety and Health Review Commission (OSHRC), which passes on violations of OSHA. In 1977, the commission held that an architect who does not perform construction work is exempt from OSHA job site regulations and extended this to a consulting structural engineer in 1992.[174]

However, in 1997, the OSHRC for the first time extended the OSHA regulations (called "construction standards") to an engineering firm. *Secretary of Labor v. CH2M Hill Central, Inc.*[175] involved a massive project by the Milwaukee Metropolitan Sewerage District (MMSD) to upgrade its sewer and treatment facilities. MMSD hired two general contractors—Healy and J. F. Shea—to build

one tunnel each. MMSD also hired *CH2M Hill Central, Inc.*, an engineering firm, to perform administrative and supervisory duties. See Figure 14.4. CH2M's contract with MMSD specified that the engineering firm's supervisory duties did not relieve a construction contractor of its sole responsibility for the method or means of construction. If an engineer discovered defects or nonconformance, MMSD alone had the right to stop the work.

When a contractor discovered methane gas in one of its tunnels, it filed a differing site conditions ("DSC") claim (seeking guidance for an unanticipated site condition). CH2M worked with the contractor to devise a safety program. However, when methane was again discovered, three employees went back into the tunnel after only a quarter of an hour (in violation of the safety program, which required a one-hour wait) and were killed when the gas exploded, presumably because the workers attempted to operate their equipment.

CH2M was cited for violation of the OSHA construction standards. The administrative law judge, applying the *Skidmore* and *Simpson* decisions cited in footnote 159, found the engineering firm not subject to the safety regulations because it did not engage in actual construction work. The Review Commission found CH2M liable only by devising a new test, holding that an architect or engineer may be subject to statutory liability if it (1) possessed broad responsibilities, both contractual and *de facto*, over the construction activities and (2) was directly and substantially engaged in activities that were integrally connected with safety issues. CH2M appealed to the federal Court of Appeals, and that decision is reproduced here.

[174]*Secretary of Labor v. Skidmore, Owings & Merrill*, 5 O.S.H.Cas. (BNA) 1762 (O.S.H.Rev.Comm'n.1977); *Secretary of Labor v. Simpson, Gumpertz & Heger, Inc.*, 15 O.S.H.Cas. (BNA) 1851 (O.S.H.Rev.Comm'n.1992), aff'd on other grounds, *Reich v. Simpson, Gumpertz & Heger, Inc.*, 3 F.3d 1 (1st Cir.1993).

[175]17 O.S.H. Cas. (BNA) 1961 (O.S.H. Rev.Comm'n.1997).

CH2M HILL, INC. v. HERMAN

United States Court of Appeals, Seventh
Circuit, 1999. 192 F.3d 711.

Before BAUER, MANION and KANNE, Circuit Judges
KANNE, Circuit Judge
[Ed. note: Some footnotes omitted and footnotes renumbered.]
Whenever accidental death occurs, it is human nature to place
blame. During a massive construction project on the Milwaukee
sewer system, three men died when methane gas located in
the tunnel in which they had been working exploded. During
the aftermath, the Secretary of Labor ("Secretary") issued a citation to CH2M Hill Central, Inc. ("CH2M Hill"), the firm that
Milwaukee had hired as consulting engineer. On administrative
review, CH2M Hill argued vigorously that the construction
standards under which it had been cited did not apply to it as a
professional firm not "engaged in construction" and the initial
Occupational Safety and Health administrative law judge agreed
with CH2M Hill. However, on review, the Occupational Safety
and Health Review Commission ("Commission") concluded
the standards did apply to CH2M Hill. On remand, a second
Occupational Safety and Health administrative law judge found
CH2M Hill had violated the regulations and imposed a series of
fines. We agree with the initial administrative law judge's conclusions and reverse.

* * *

II. Analysis

The primary issue presented by this case is whether OSHA's
construction standards apply to professional firms with responsibilities similar to those exercised by CH2M Hill in regard to
tunnel CT-7. While we offer no opinion as to the "new" test proposed by the Commission and that amici [Ed. note: "friends of the
court"] urge us to find invalid, we do conclude that all professionals operating in the field of construction are not per se exempt
from complying with these construction standards. Under the facts
of this case, we find that because CH2M Hill's responsibilities did
not rise to a level that constituted being engaged in construction
work, the regulations do not apply to it. Accordingly, we need not
determine whether its actions conformed to these standards.

* * *

A. Scope of the Construction Standards
CH2M Hill first urges us to reject the Secretary's interpretation of the construction standards that permits their
application to professionals, such as engineers and architects.
It contends that the plain meaning of the term "construction"
creates a per se exclusion of professionals working on or as part
of a construction project. In its view, "construction" should be
strictly construed according to its dictionary meaning of building, erecting, or putting together. [Ed. note: The court rejects
that argument, concluding its discussion as follows:]

While a blanket per se exclusion of professionals entering into
construction contracts might promote efficiency by avoiding this
often expensive and always time consuming process, the broader
approach is also more consistent with congressional intent. The
Act's purpose is "to assure so far as possible every working man
and woman in the Nation safe and healthful working conditions and to preserve our human resources. . . ." 29 U.S.C.
§ 651 (b). Congress designed the Act to enhance compliance
with standards and to reduce safety hazards in the workplace,
not to punish employers. See *Anning-Johnson Co. v. OSHRC*,
516 F.2d 1081, 1088 (7th Cir.1975). With these points in mind,
this Court has concluded that "[t]he underlying rationale in
effectuating these purposes by placing primary responsibility
on employers is that employers have primary control of the
work environment and should therefore insure that it is safe
and healthful." *Id.* (citing S.Rep. No. 91-1282, at 9 (1970),
U.S. Code Cong. & Admin. News at 5177, 5186; H.R.Rep.
No. 91-1291, at 21 (1970)). Thus, Congress did not intend the
Act, or the regulations flowing from it, to apply only to some
employers and not others, but rather to those employers who
were best suited to alleviate hazards at the construction site.
A per se exception excluding professionals, regardless of their
duties, from liability under the Act and its regulations would
diminish the aims of Congress in enacting this legislation.
Therefore, the Secretary's goal of adopting a balance as to whom
the regulations apply is a reasonable approach. We do not find
the Commission's decisions regarding the general applicability
to professionals in some cases arbitrary or capricious.

B. Applicability of the Construction
Standards to CH2M Hill
Prior to its decision in regard to CH2M Hill, the Commission had explained that it would apply the construction
standards "to employers who perform no physical trade labor,

such as [the respondent], only to the extent that such employers have actual and direct responsibility for the specific working conditions at the jobsite and for any hazards resulting from the actions of any trade contractor." *SGH*, 15 O.S.H. Cas. (BNA) at 1867. [Ed. note: "SGH" refers to the *Simpson* decision in footnote 174.] Or, stated more succinctly, professionals only have to comply with the standards if they exercise "substantial supervision over actual construction." *SOM*, 5 O.S.H. Cas. (BNA) at 1764. [Ed. note: "SOM" refers to the *Skidmore* decision in footnote 174.] This inquiry has become known as the "substantial supervision" test.

As part of its evaluation of CH2M Hill's responsibilities with regard to the Program, however, the Commission adopted a new test. It stated that an engineering or architectural firm would "engage in construction work" if it:

(1) possesses broad responsibilities in relation to construction activities, including both contractual and de facto authority directly to the work of the trade contractors, and (2) is directly and substantially engaged in activities that are integrally connected with safety issues . . . notwithstanding contract language expressly disclaiming safety responsibility.

As we noted in an earlier consideration of this case, this test means:

architects, engineers, and similar professionals should be treated as joint employers with the firms actually carrying out the construction, even if the contracts assign to the project owner full responsibility for directing the work, and to the general contractor sole responsibility for implementing the owner's decisions.

CH2M Hill Central, Inc. v. Herman, 131 F.3d 1244, 1245 (7th Cir.1997). However, in *Secretary of Labor v. Foit-Albert Assoc.*, 17 O.S.H. Cas. (BNA) 1975, 1978 (1997), the Commission admitted that it could not always hold professionals responsible for compliance with construction standards.

In an amicus brief, several professional engineering and architecture associations and societies urge us to accept CH2M Hill's arguments and reject the Commission's expanded interpretation of "engaged in construction work." They claim the new test is arbitrary, vague and too broad, making it inconsistent with the Act. They argue for a return to the "old" test under which the regulations applied only to engineers and architects who exercised "substantial supervision over actual construction." See *SOM*, 5 O.S.H. Cas. (BNA) at 1764. Their views certainly are well supported; however, we need not reach this issue. Even if this "new" test were appropriate, OSHA still fails to establish that CH2M Hill contractually or on a de facto basis exercised direct

authority and control over, or substantially engaged in activities integrally connected with, the safety measures for CT-7.

Even though we refrain from basing our decision on the validity of the "new" test, we, however, note our concern as to at least one aspect of the new test—specifically, the Commission's decision to ignore contract language in evaluating to whom the regulations apply. While perfunctory language that does not represent the true responsibilities of a particular employer should not absolve it from complying with the regulations, language exempting an employer from particular responsibilities that the facts confirm that the employer does not actually retain control cannot be casually thrown aside. Contracts represent an agreed upon bargain in which the parties allocate responsibilities based on a variety of factors. To ignore the manner in which the parties distributed the burdens and benefits is contrary to our notion of contract law. When an owner, such as MMSD, contracts for outside services relating to a construction contract, it's usually the owner who bargains from the position of strength, making it easier for it to shift liability away from itself. *Cf. Anning-Johnson Co. v. OSHRC*, 516 F.2d 1081, 1089 (7th Cir. 1975) (stating that general contractors often bargain from the position of strength when contracting with subcontractors). Thus, the owner's retention of such liability may, in fact, be a significant indicator of the relationship between the parties. The Secretary and Commission cannot ignore the fact that parties to a contract bargained for one to maintain safety responsibilities and for the other to refrain from such responsibility. As one of our colleagues aptly noted:

Congress, whose underlying rationale in adopting the statute . . . was that the employer has control of the work environment and should therefore be responsible for making it safe, could not have intended that the employer be sanctioned for failure to correct conditions he could not correct.

Anning-Johnson, 516 F.2d at 1092 (Tone, J. concurring). Ignoring the language of the contract or an attempt to create blanket liability for professionals regardless of their ability to control the safety aspects of a construction site would be, as Judge Tone explained, contrary to the very intent of Congress when it drafted OSHA. The question as to what responsibilities a particular defendant maintained should turn on a factual inquiry based on a review of the record, including the language of the contract.

In this case, no significant evidence in the record supports the factual findings of the Commission that CH2M Hill exercised substantial supervision or control such that it was "directly and substantially engaged in activities that [were] integrally connected with safety issues." . . .

A review of the Commission's prior considerations of this matter supports this conclusion as well. The cases in which the Commission has concluded that a "professional" employer is engaged in construction work it has found that the employer, either contractually or in actuality, had substantial control over the safety program at the construction site. In *Secretary of Labor v. Bechtel Power Corp.*, 4 O.S.H. Cas. (BNA) 1005, 1007 (1976), the Commission found that the construction standards applied to a construction manager who, among other things, coordinated the safety program, was an integral part of the total construction system and functioned in a manner "inextricably intertwined" with the actual physical labor of the construction. Similarly, the Commission concluded that the construction standards applied to an architectural firm that also managed the construction project because it had the authority to stop work until problems were resolved. See *Secretary of Labor v. Bertrand Goldberg Assoc.*, 4 O.S.H. Cas. (BNA) 1587, 1588 (1976). In *Secretary of Labor v. Cauldwell-Wingate Corp.*, 6 O.S.H. Cas. (BNA) 1619, 1621 (1978), the Commission stated that the construction standards applied to a firm employed as a construction manager had significant responsibilities that gave it "substantial supervision over actual construction," including the authority to act on behalf of the owner of the project. Finally, the Commission, in *Secretary of Labor v. Kulka Constr. Management Corp.*, 15 O.S.H. Cas. (BNA) 1870, 1871-73 (1992), decided that the construction standards applied to a management corporation that exercised substantial supervisory authority over the construction work because it reviewed the safety programs of each contractor, was responsible for giving general instructions to contractors as to how the work would proceed and its own employee assured compliance officers that he would make sure the safety measures would be completed. In all of these cases, the employer in question had authority to direct and control what was occurring at the construction site.

On the other hand, the Commission has also concluded that the construction standards do not apply to professional firms who do not maintain such authority or control. In *SOM*, the Commission vacated a citation issued to a firm finding that the construction standards did not apply to it due to its lack of "substantial supervision over actual construction" because of its limited functions and authority over the work, 5 O.S.H. Cas. (BNA) at 1764. While the firm inspected the work of various contractors to ensure that they met design specifications, employed field representatives to observe the work and could require that work be redone or repaired if it did not conform, the contract expressly denied the firm "the responsibility for or the authority to direct or supervise construction methods, techniques, procedures or

safety methods." *Id.* at 1762-63. Similarly, the Commission, again applying the "substantial supervision test," concluded that an architectural/engineering firm was not subject to the construction standards because it did not engage in construction work. See *SGH*, 15 O.S.H. Cas. (BNA) at 1869. The firm's duties included: preparing contract drawings and specifications to conform to local regulations and codes subject to approval by the owner, assisting the owner with the bidding process, undertaking measures to ensure that contractors conformed with the necessary codes and regulations, making on-site inspections and assisting the owner with resolving disputes between it and the contractors. Its responsibilities, like those of SOM, were limited by a contractual disclaimer, which stated it would not "be responsible for, construction means, methods, techniques, sequences or procedures, for safety precautions and programs in connection with the Work . . ." *Id.* at 1854. The Commission concluded that, while the firm was informed of the safety hazard, it "did not assume substantial supervisory authority over" the project, and, thus, it did not come within the scope of the regulation. *Id.* at 1870.

The responsibilities of CH2M Hill are more like those in these latter cases—SOM and SGH—than they are like those in which the Commission found the employers to be engaged in construction work. Like those of SOM and SGH, CH2M Hill's contract severely limited its authority and responsibility over safety programs. CH2M Hill, while it exercised some authority, had only limited authority that was always subject to final approval by MMSD. While CH2M Hill drafted the DSC changes, MMSD (whether it exercised its responsibility attentively or not) had to approve any changes made. CH2M Hill was required to consult with MMSD before such decisions were final. Unlike the employers found to come within the domain of the regulations, CH2M Hill lacked the necessary authority or supervisory responsibilities. CH2M Hill did not function as a coordinator of the safety program; the contract specifically removed this responsibility from CH2M Hill. Nor did it make representations that it would ensure the safety regulations were met. The firm through its drafting of contract modifications did not function in a manner that was "inextricably intertwined" with the actual construction. It could not instruct Healy how to perform the construction work, nor halt the work if the required regulatory safety measures were not met. In fact, in its explanation to Healy, it told the company to seek the advice of the manufacturer of the equipment for compliance. It did not accept that role itself. Thus, based on the Commission's own line of cases, CH2M Hill should not be subject to the construction standards for its work on the Program.

The Commission, in its evaluation of CH2M Hill, appears to have not only departed for no apparent reason from the "substantial supervision" test but also from its own precedent, which clearly supports the original ALJ's findings that CH2M Hill was not engaged in construction work. We find with all due deference to the Commission that its findings cannot be supported by the record, especially in light of its previous decisions on the subject.

III. Conclusion

We conclude that while the construction standards may apply to some professionals working on construction projects, they do not apply to CH2M Hill in this case because the firm did not engage in construction work based upon its contractual and actual responsibilities. Accordingly, the findings of violations and the imposition of fines are VACATED.

Efforts by public regulatory agencies to place more responsibility for workplace safety on design professionals have not been successful where the design professional performs site services in the traditional manner, such as that outlined in AIA or EJCDC documents. But it is likely that we will see an increasing number of attempts to place responsibility on design professionals.

That the design professional's duty to workers continues to be hotly debated is demonstrated not only by scholarly comment,[176] but also in the varied conclusions in the case law. As emphasized elsewhere in this book, however, *factual* differences, such as different contract terms, different actual practices, and whether *obviously* unsafe practices were occurring, may account for different conclusions. Yet even when these *are* taken into account, judicial opinions can still reflect different emphases, whether on the contract language and disclaimers of responsibility or on tort concepts of compensation, foreseeability, and avoiding harm to persons.

Cases have tended to exonerate the design professional when the contractor, not the design professional, is responsible for means, method, and sequences by which the design is accomplished or for safety.[177] In addition,

design professionals whose site activities are of a more passive nature, such as those outlined in AIA documents, will not be held responsible for personal harm occurring to workers. If nothing further is alleged, the design professional is entitled to a summary judgment without a trial on the issue of negligence. But if the claimant alleges that (1) the contract obligated the design professional to take an active role in overseeing construction safety, (2) she undertook these functions even though her contract did not require her to do so, or (3) she was aware of obviously unsafe practices, the dispute must proceed to trial. If those allegations are established, the trier of fact—jury or judge—must determine whether the design professional has been negligent.[178]

H. Action Taken on Site

Pastorelli v. Associated Engineers, Inc.[179] involved an injury to an employee of a racetrack caused by a falling heating duct. The accident occurred after the work had been completed and accepted by the injured employee's employer.[180] The injured employee sued the prime contractor, the sheet metal subcontractor who installed the duct, and engineers who prepared the plans and who agreed to "supervise the contractor's work throughout the job."

---

[176] See Sweet, supra note 168; Goldberg, *Liability of Architects and Engineers for Construction Site Accidents in Maryland—Krieger v. J. E. Greiner Co.; Background and Unanswered Questions*, 39 Md.L.Rev. 475 (1980); Comment, 30 U.Kan.L.Rev. 429 (1982); and Precella, *Architect Liability: Should an Architect's Status Create a Duty to Protect Construction Workers from Job-Site Hazards?* 11 Constr. Lawyer, No. 3, August 1991, at p. 11. The last author suggests the best theories for claimants are assumption of duty and knowledge of the dangerous condition. She does not approve of finding a duty simply on the basis of status.

[177] As examples, see *Ramey Constr. Co., Inc. v. Apache Tribe of Mescalero Reservation*, 673 F.2d 315 (10th Cir.1982) (no duty to contractor); *Yow v. Hussey, Gay, Bell & DeYoung Int'l, Inc.*, supra note 168; *Becker v. Tallamy, Van Kuren, Gertis & Assoc.*, 221 A.D.2d 1014, 634 N.Y.S.2d 282 (1995).

[178] *Phillips v. United Eng'rs and Constructors, Inc.*, 500 N.E.2d 1265 (Ind.App.1986) (CM assumed responsibility by conducting safety meetings, touring the site, and noting safety violations and unsafe practices.); Annot., 59 A.L.R.3d 869 (1974).

[179] 176 F.Supp. 159 (D.R.I.1959). This federal trial court case, interpreting Rhode Island law, was cited with approval by the state's high court in *Maggi v. De Fusco*, 107 R.I. 278, 267 A.2d 424, 427 (1970).

[180] Earlier American law immunized architect and contractor from most accidents that occurred after the project was completed and accepted by the owner. In *Pastorelli v. Associated Eng'rs, Inc.*, supra note 178, the court rejected this rule and held that completion and acceptance did not furnish a defense to the architect. This is discussed in Section 14.09B.

The duct was 20 feet long and weighed 500 pounds. It had not been attached directly to the roof itself or to the joists of the clubhouse but had been suspended from the ceiling of the clubhouse by the attachment of semirigid strips of metal called *hangers*, which were then attached to the ceiling. The ceiling was of sheathing and was nailed to the joists, leaving a considerable air space between sheathing and roof. The specification required that sheet metal work be erected "in a first class and workman-like manner" and that "the ducts be securely supported from the building construction in an approved manner."[181]

The trial judge concluded that the duct had not been properly installed. The engineer prepared and submitted periodic inspection reports while the work was in progress. The trial judge stated that the engineer's employee who prepared the reports

> testified that his employer assigned to him the task of supervising the installation of said systems, and that in pursuance of his duties he visited the job site on one, two or three occasions each week to inspect the work of the contractor as it was being done. He also testified, however, that he never observed any of the ducts being hung from the ceiling in said clubhouse, stating that whenever he visited the clubhouse the ducts were either on the floor or already installed. He also admitted that he never climbed a ladder to determine whether the hangers by which they were suspended were attached by nails or lag screws and never tested any of the hangers to see how securely they were attached.[182]

After holding prime and subcontractor negligent, the judge noted that the engineer's employee knew the safety of people in the clubhouse required that the ducts be attached to the joists, and that he made no attempt to ascertain whether they were so installed. The judge also noted that the employee made no visits at a time when he could determine how they were being installed. Holding the engineer negligent, the judge stated, "In other words, he failed to see that they were properly installed and took no steps after their installation to ascertain how and by what means they were secured. In my opinion he failed to use due care in carrying out his undertaking of general supervision."[183]

---

[181] 176 F.Supp. at 162.
[182] Id. at 162–63.
[183] Id. at 167.

Suppose the engineer's employee had determined that the ducts were not properly secured. What should she have done? Should she have directed the employees of the sheet metal subcontractor to correct the work? Should she have gone to the superintendent of the prime contractor with a similar request? Should she have ordered the work terminated until proper corrective measures were taken? Should she have suggested to a building inspector that the latter order the work corrected? These questions show the difficult position in which the design professional can find herself when she does determine that work is not being properly performed, especially when people could be injured or killed as a result.

If the design professional simply observes, one would think that in a case such as *Pastorelli* her principal responsibility would be to call the defective work to the attention of the contractor, and have the contractor transmit this to the subcontractor. Yet if it is likely that correction will not come quickly enough and if it is likely that persons will be injured, registering a complaint indirectly would not be sufficient. The owner should be given the information, along with advice as to what should be done. The owner can then determine the course to be followed. If the danger was imminent and the situation urgent, perhaps the building inspector should be called, or if the design professional has the power, she should stop the work. (Until 1970 the AIA empowered the architect to stop the work. Since then this power has been given only to the owner. See AIA Document A201-2007, Section 2.3.1.)

I. Safety Legislation

Often legislation is enacted after a tragic construction site accident. The L'Ambiance Plaza in Hartford, Connecticut, collapsed during construction in 1987. Twenty-eight workers died and sixteen were injured. As a result, Connecticut enacted Section 29-276b of its general statutes. After setting forth criteria for significant structures, it requires that for those structures local building officials hire an independent engineering consultant to review the plans and specifications for code compliance. The prime contractor and major subcontractors are required to keep a daily log and keep it available to both local building officials and engineers or architects involved with the building. All design professionals and general contractors must give signed approval that completed construction substantially complies with the plans. If a building has more than three stories, Section 29-276c requires the design professional to review the implementation of her design and observe the construction.

J. Summary

Increasingly, third parties, including public safety officials, have been asserting claims against design professionals. Claims based on the assertion that the claimant is an intended beneficiary of the contract between the design professional and the owner, though occasionally successful, have not proved as successful as tort claims. But although third-party claims in tort have met obstacles (as noted in this section), the trend except in site accident claims is clearly toward increased liability exposure for design professionals. Finally, safety legislation is becoming more common, and this poses new legal problems for design professionals.

SECTION 14.09 Special Legal Defenses

The principal issue in claims against design professionals relates to the standard of performance and whether there has been compliance. This section briefly outlines a number of defenses that have been used by the design professional when claims are made by the client or third parties. In addition to the special defenses discussed here, a design professional, of course, has available standard defenses to any claim, such as lack of causation.[184]

A. Approval by Client

Suppose the design professional asserts that the design has been approved by the client and this relieves her of any liability, even for negligent design. Ordinarily, such a defense is not successful.[185]

The client retains a design professional because of the latter's skill in design, a skill not usually possessed by the client. The very purpose of engaging an expert would be defeated if the approval by the nonexpert client relieved the expert design professional from her negligence. Approval does authorize the design professional to proceed with the next phase of services, and client-directed changes after approval may justify the design professional who is compensated by a method unrelated to costs receiving additional compensation.

There may be an unusual circumstance that could justify approval relieving the design professional. Suppose the design professional points out a design dilemma to the client. The designer may inform the client that particular material may prove unsatisfactory for specified reasons but asks the client's approval of that material because it is less costly. If the client is apprised of all the risks and authorizes that particular material to be used, the client assumes the risk. Even if the design professional has not done what others would have done, approval by the client with full knowledge of the risks and with the ability to evaluate them should relieve the design professional.[186]

Claims by third parties raise other problems. Even if approval by the client would bar its claim, this would not bar a third-party claim against the design professional. If approval by the client were itself negligent, any third party who suffered harm could recover against the client as well and may give the design professional an indemnification claim against the client.

B. Acceptance of the Project

Acceptance of the project usually involves the owner's taking possession of the completed project. It can—although under standard contracts it usually does not[187]—imply that the owner is satisfied with the work and bar any claim for existing defects or defects that may be discovered in the future.

For purposes of this chapter, acceptance can affect any claim third parties have against those who have participated in design. Although some early cases[188]—and

---

[184]*Outlaw v. Airtech Air Conditioning and Heating, Inc.*, 366 U.S.App. D.C. 374, 412 F.3d 156, 164 (D.C.Cir.2005) (Roberts, J., later Chief Justice of the United States Supreme Court) ("When there is a construction defect, the architect is one of the usual suspects, but [the architect's] proximity to the problem and [the owner's] accusation alone are not enough to survive summary judgment.")

[185]*C.H. Guernsey & Co. v. United States*, 65 Fed.Cl. 582 (2005); *Eichler Homes, Inc. v. County of Marin*, 208 Cal.App.2d 653, 25 Cal. Rptr. 394 (1962); *Bloomsburg Mills, Inc. v. Sordoni Constr. Co.*, supra note 100.

[186]*Bowman v. Coursey*, supra note 66, discussed in Section 14.05D.

[187]AIA Doc. A201-2007, § 9.10.4 (final payment does not bar most owner claims).

[188]*Sherman v. Miller Constr. Co.*, 90 Ind.App. 462, 158 N.E. 255 (1927).

occasionally recent ones[189]—have barred some such claims after acceptance, the modern tendency is to hold that acceptance does not bar third-party claims against design professionals.[190] If acceptance bars claims, it is because the intervening act of the owner—that of acceptance—is a superseding cause relieving even negligent participants of liability. This rationale is particularly weak when the claim is asserted against the design professional—the person who often decides whether the project has been completed and should be accepted. Even if the owner decides to accept, acceptance may not be negligent. Although acceptance by the owner can be looked on as an intervening cause, it is in most cases a nonnegligent one that should not immunize a design professional from her negligence.

Suppose, though, that acceptance by the owner precluded the design professional from correcting design errors. Should the design professional be immunized from third-party claims? If the owner knew that there were defects and barred their correction, the owner's intentional acts operate as an intervening cause to immunize the design professional from a third-party claim. The owner's acceptance should give the design professional a claim for contribution or indemnity against the owner if the latter's acceptance precluded correction of defects.

Immunization because of the acts of others—even negligent acts—is not favored today. Acceptance rarely bars third-party claims against a design professional.

C. Passage of Time: Statutes of Limitations

Sometimes the design professional can defend by establishing that legal action was not started within the time required by law. This is usually accomplished by invoking the statute of limitations as a bar to the claim. Because this bar can apply to all claims—not only to those against design professionals—and because most construction-related claims are brought against other participants such as the owner or contractors, full discussion of this defense is postponed until Section 23.03G. A few observations, dealing with claims against design professionals, should be made in this section.

Statutes of limitations usually prescribe a designated period of time within which certain claims must be brought. One of the troublesome areas in construction claims relates to the point at which that period begins. This can be a formidable problem in all construction-related claims, inasmuch as the defect may be discovered long after the design is created and executed. As a result, even in claims that do not involve the design professional, the law has had difficulty selecting from a number of base points, such as project completion, the wrongful act, the occurrence of damage, discovery of a substantial defect, or discovery of any defect.

This troublesome question is exacerbated when claims are made against the design professional. The latter usually develops the design, obtains approval by the client, gives it to the contractor to execute, and often issues a certificate that there has been substantial or final completion. In addition, the changes process often means that the design is changed during construction. All this means there are additional base points that can be used to determine when the period commences.

Essentially, design is a trial-and-error process. Although some states hold that the period commences when the negligence occurs, the trial-and-error aspects of design could, along with the continuation of the professional relationship between client and design professional, lead to a determination that the period commences on completion of the project.[191]

New York refused to apply this rule in a claim by a third party who brought a tort action based on professional negligence against a designer, with the court holding that the period began when the wrongful act was discovered (usually the time of injury).[192] The court held that the completion base point is proper in dealing with claims between client and design professional but not those brought by third parties.

In addition to the complexity resulting from a differentiation between claimants drawn in New York, additional confusion can result because the period for beginning action is usually longer for claims based on breach of contract than for those based on other wrongful conduct,

[189]*Phifer v. T. L. James & Co., Inc.*, 513 F.2d 323 (5th Cir. 1975); *Easterday v. Masiello*, 518 So.2d 260 (Fla.1988) (defense as to patent defects); *McBride v. Cole Assocs., Inc.*, 753 N.E.2d 730 (Ind.App. 2001).

[190]*Pastorelli v. Associated Engr's., Inc.*, supra note 179; *Theis v. Heuer*, 264 Ind. 1, 280 N.E.2d 300 (1972); *Totten v. Gruzen*, 52 N.J. 202, 245 A.2d 1 (1968).

[191]*Sears, Roebuck & Co. v. Enco Assocs.*, 43 N.Y.2d 389, 372 N.E.2d 555, 401 N.Y.S.2d 767 (1977). The court's choice of the six-year statute of limitations for contract claims was almost immediately replaced by the legislature with a three-year limitation period; see N.Y. Civ. Prac. Law & Rules § 214(6).

[192]*Cubito v. Kreisberg*, 69 A.D.2d 738, 419 N.Y.S.2d 578 (1979), aff'd, 51 N.Y.2d 900, 415 N.E.2d 979, 434 N.Y.S.2d 991 (1980).

such as negligence. This becomes a serious issue because claimants against design professionals are often given the option of bringing actions in tort or in contract.[193]

Another complication is when the statute of limitations should begin for an owner who continues to use the project architect in dealing with defects that are discovered after completion of the project. Owners often turn to their architects and contractors for initial help in determining the source of the problem. Yet an owner faced with a defective building must know that the architect's design may be the cause of the defect, even if the architect places the blame elsewhere. A court which finds that the owner reasonably relied upon the architect's advice so as to delay filing suit against her may invoke the "continuous treatment" doctrine to "toll" (temporarily suspend) running of the statue of limitations.[194]

The discussion to this point has been focused primarily on owner claims against its design professional. When does the limitations period begin when a contractor sues the design professional for economic losses caused by a defective design? The design professional would argue that the limitations period commenced when the contractor signed the contract and began to use the design. However, courts instead find that the limitations period begins when the contractor incurs economic loss with certainty.[195]

Generally, a defense based on the passage of time—though still of value to the design professional—has provided limited protection. This has led to the enactment of statutes of repose that seek to cut off liability after a designated period of time following substantial completion of the project. (The statutes are discussed in Section 23.03G.)

The dubious utility and confusion engendered by application of the statutes determining when an action must be commenced have led to contractual attempts to create a private statute of limitations. If reasonable, these provisions can regulate the relationship between client and design professional. Contractual provisions dealing with this problem can, as do AIA Documents B141-1997, Paragraph 1.3.7.3, and A201-1997, Paragraph 13.7, seek to control the commencement of the statutory period.[196] These 1997 AIA provisions, which started the limitations period from the date of either substantial or final completion, were dropped in the 2007 AIA documents. Under the new AIA Documents B101-2007, Section 8.1.1 and A201-2007, Section 13.7, suit must begin within the period specified by applicable law (usually the law of the state where the project is located), but in any case not more than 10 years after the date of substantial completion. In addition, some contract clauses stipulate the period during which the claim must be asserted as well as the commencement of the period. If the clause is reasonable, it will be enforced.[197]

D. Decisions and Immunity

The design professional is frequently given the power to interpret the contract documents, resolve disputes, and monitor performance. When claims are brought against a design professional by the client or others for activity that can be said to resemble judicial dispute resolution (such as deciding disputes or issuing certificates),[198] design professionals sometimes assert they should receive quasi-judicial immunity.

This is based on the analogy sometimes drawn between judges and the design professionals performing judge-like functions. For example, the judge is given absolute immunity from civil action, even for fraudulent or corrupt

---

[193]*In re R.M. Kliment & Frances Halsband, Architects (McKinsey & Co., Inc.)*, 3 N.Y.3d 538, 821 N.E.2d 952, 788 N.Y.S.2d 648 (2004), the court applied N.Y. Civ. Prac. Law & Rules § 214(6), the statute of limitations for malpractice claims, even though the owner claimed its architect breached a contract duty to comply with the building code. See Section 14.05E.

[194]Compare *Greater Johnston City School Dist. v. Cataldo & Waters, Architects, P.C.* 159 A.D.2d 784, 551 N.Y.S.2d 1003 (1990) (tolling doctrine applied where architect worked with the owner for several years to identify cause of insulation problem and devise a solution) with *Saint Alexander's Church v. McKenna*, 294 A.D.2d 695, 742 N.Y.S.2d 165 (2002) (doctrine not applied absent evidence the architect remained actively involved with the owner to rectify problems). See also *Lake Superior Center Auth. v. Hammel, Green & Abrahamson, Inc.*, 715 N.W.2d 458 (Minn.App.2006), review denied Aug 23, 2006 (architect who participated in repair efforts is precluded from arguing the owner's claim against it is time-barred).

[195]*Hardaway Co. v. Parsons, Brinckerhoff, Quade & Douglas, Inc.*, 267 Ga. 424, 479 S.E.2d 727 (1997); *MBA Commercial Constr., Inc. v. Roy J. Hannaford Co.*, 818 P.2d 469 (Okla.1991).

[196]*Harbor Court Assocs. v. Leo A. Daly Co.*, 179 F.3d 147 (4th Cir.1999) (beginning commencement from the date of substantial completion, even though the state statute specifies the limitations period begins when the harm is discovered or should have been discovered).

[197]*Therma-Coustics Mfg., Inc. v. Borden, Inc.*, 167 Cal.App.3d 282, 213 Cal.Rptr. 611 (1985). See also U.C.C. § 2-725(1), which allows the parties to reduce the four-year period to "not less than one year." The parties cannot extend the period.

[198]See Section 22.07.

decisions. The corrupt judge may be removed from office or subject to criminal sanctions. But a disappointed party cannot institute civil action against a judge. Immunity protects judges from being harassed by vexatious litigants and encourages them to decide cases without fear of civil action being brought against them.

Quasi-judicial immunity for design professionals has had a troubled history both in England and in the United States. The English House of Lords reversed an earlier decision and held that the architect can be sued by the owner for a negligently issued certificate.[199]

American decisions have not been consistent. Some have granted quasi-judicial immunity, but others have not.[200] Where immunity is given, the design professional cannot be sued for decisions made unless they were made corruptly, dishonestly, or fraudulently. However, immunity does not protect against negligent delay in making a decision.[201]

Again, the peculiar position of the design professional—independent contractor as designer, agent of the owner, decider of disputes, and an individual participating in construction projects—has caused difficulty. If immunity is granted, it should be based on both parties to the contract—owner and contractor—agreeing to give certain judging functions to the design professional.[202] Yet the cases in which courts have been quickest to deny immunity have been actions instituted against a design professional by the client—the party who has selected and paid the design professional. Courts that have denied immunity in such cases seem more impressed with the client's selection of and payment to the design professional as indicating that the design professional's principal responsibility is to protect the owner. This, in addition to the increasing tendency to hold professional people accountable, may not mean that immunity will never be available when the design professional claims it but reduces the likelihood that it will succeed.

There has been criticism of immunity even where limited to good-faith decisions. Can the design professional truly be neutral in rendering a decision when she has been selected and paid by the client? Also, as indicated earlier, the contractor rarely has much choice in these matters. The English House of Lords in *Sutcliffe v. Thackrah*[203] dealt with a claim by the client that the architect had negligently overcertified. Lord Reid drew a distinction between a dispute resolver who is a judge or arbitrator and an architect. According to Lord Reid, a true dispute resolver is a passive recipient of information and arguments submitted by the parties. An architect, in contrast, is a professional engaged to act at her client's instructions and to give her own opinions. The judge or true arbitrator does not investigate but simply decides matters submitted to her. Lord Reid concluded that deciding whether work is defective is not judicial. He noted there had been no dispute, the architect was not jointly engaged by the parties, the parties did not submit evidence to the architect, and the architect made his own investigation and came to his own decisions.

Lord Reid's reasoning may not be applicable in the United States because of different, though perhaps marginal, American practices. First, in the United States, the architect is "approved" by both parties, but engaged by the owner. Although the owner retains the design professional, when the contractor enters its bid or negotiates to perform the work, the contractor knows who the design professional will be. Under AIA Document A201-2007, the contractor has a limited power of veto over any successor architect.[204] Also, A201-2007 seems to contemplate the parties submitting evidence (supporting data) to the Initial Decision Maker (who by default is the architect).[205]

The desirability of immunity was also passed on by an American court in *E. C. Ernst, Inc. v. Manhattan Construction Co. of Texas*.[206] The case involved a claim against the architect based on his rejection of certain equipment proposed by the contractor. The court stated,

> The arbitrator's "quasijudicial" immunity arises from his resemblance to a judge. [The court here is speaking of the architect as arbitrator.] The scope of his immunity should be no broader than this resemblance. The arbitrator serves as a private vehicle for the ordering of economic relationships.

---

[199]*Sutcliffe v. Thackrah*, 1974 A.C. 727.

[200]Authorities are collected in *City of Durham v. Reidsville Eng'g Co.*, 255 N.C. 98, 120 S.E.2d 564 (1961). See also *Blecick v. School Dist. No. 18*, 2 Ariz.App. 115, 406 P.2d 750, 755–56 (1965), overruled on other grounds, *Donnelly Constr. Co. v. Oberg/Hunt/Gilleland*, 139 Ariz, 184, 677 P.2d 1292 (1984); Annot., 43 A.L.R.2d 1227 (1955).

[201]*E. C. Ernst, Inc. v. Manhattan Constr. Co. of Texas*, 551 F.2d 1026, rehearing denied in part and granted in part, 559 F.2d 268 (5th Cir.1977), cert. denied sub nom. *Providence Hosp. v. Manhattan Constr. Co. of Texas*, 434 U.S. 1067 (1978).

[202]*Newton Inv. Co. v. Barnard & Burk, Inc.*, 220 So.2d 822 (Miss.1969).

---

[203]Supra note 198.

[204]AIA Doc. A201-2007, § 4.1.3.

[205]Id. §15.2.4.

[206]Supra note 201.

He is a creature of contract, paid by the parties to perform a duty, and his decision binds the parties because they make a specific, private decision to be bound. His decision is not socially momentous except to those who pay him to decide. The judge, however, is an official governmental instrumentality for resolving societal disputes. The parties submit their disputes to him through the structure of the judicial system, at mostly public expense. His decisions may be glossed with public policy considerations and fraught with the consequences of stare decisis [a Latin phrase for precedent]. When in discharging his function the arbitrator resembles a judge, we protect the integrity of his decision-making by guarding his fear of being mulcted in damages. . . . But he should be immune from liability only to the extent that his action is functionally judge-like. Otherwise we become mesmerized by words.[207]

The court then concluded that such immunity as possessed by the architect as arbitrator did not extend to unexcused delay or failure to decide, with immunity limited to "judging."

Determining whether to grant immunity must also take into account the finality of the design professional's decision (taken up in greater detail in Section 29.09). Under many contracts, the initial decision by the design professional can be taken to arbitration. If so, any wrong or even negligent decision can be corrected by the arbitrators. However, if arbitration is not used and the dispute goes to court, the decision by the design professional is likely to have a certain degree of finality. Giving the decision substantial finality and the decision maker immunity may be granting too much power to the design professional—power that can be abused. In Section 29.10, the suggestion is made that the process under which the design professional interprets the contract and decides disputes can be justified principally by expediency–the need to move construction along. This justification will not be adversely affected even if the decision were given very little finality and even if the design professional is stripped of any immunity.

It is often difficult to determine whether the design professional is acting as agent or judge. Perhaps it is simply better to jettison immunity, as the English have done. Such immunity as exists protects only against the negligent decision and not against one made in bad faith. Also, as noted in Section 15.03E, one way of dealing with this problem is to provide immunity by contract.

---

[207]Id. at 1033.

E. Recent Legislative Activities

Legislatures have been more receptive to pleas by associations representing design professionals that expanding liability has had an undesirable effect not only on the professions but also on the public. Much of the legislative effort came in the backwash of similar legislative protection given health care providers in the 1970s and 1980s. Some, such as the requirement specified in Section 1029.5 of the California Code of Civil Procedure, that the claimant must post bond if a malpractice claim is made against a design professional under certain circumstances, may not extend much protection. A $500 bond must be posted for each defendant, with a total bond limit of $3,000.

Legislative protections for design professionals fall into two broad categories. Some statutes do not limit liability but impose special procedural or evidentiary requirements on the claimant with the intended goal of preventing frivolous claims. The second category of statutes limits the design professional's liability.

Certificate of merit statutes are the primary device for weeding out frivolous lawsuits through the imposition of special pleading burdens on claimants. These laws generally require the plaintiff to attach to the complaint a certificate from her attorney declaring that the attorney has consulted with an expert and that the attorney or expert has concluded that the suit is meritorious.[208] For example, California Code of Civil Procedure Section 411.35, enacted in 1979, requires the attorney for a claimant who files a claim for damages or indemnity arising out of the professional negligence of a licensed architect, a licensed engineer, or a licensed land surveyor, to file a certificate stating that the attorney has reviewed the facts of the case; that she has consulted and received an opinion from a licensed design professional in the same discipline as the defendant,[209] who she reasonably believes is knowledgeable in the relevant issues; and that she has concluded "that there is reasonable and meritorious cause for the filing of this action and that the person consulted gave an opinion that the person against whom the claim

---

[208]Hawaii and Kansas also require the plaintiff to present her "case" to an independent panel to obtain an advisory opinion. For a discussion of both types of statutes, see Davis, *Certificates of Merit and Review Panels: Conditions Precedent to Civil Actions Against Design Professionals,* 15 Constr. Lawyer, No. 2, April 1995, p. 88.

[209]*Ponderosa Center Partners v. McLellan/Cruz/Gaylord & Assoc.,* 45 Cal.App.4th 913, 53 Cal.Rptr.2d 64 (1996), review denied Aug. 14, 1996, held that a certificate by a structural engineer in a claim made against an architect in a roof collapse was sufficient.

was made was professionally negligent." Provisions seek to protect the identity of the design professional who has been consulted.

Georgia has gone farther, enacting legislation requiring that a complaint charging professional malpractice be accompanied by an affidavit of an expert setting forth at least one negligent act or omission.[210]

Pennsylvania's certificate of merit statute has been interpreted to apply only to negligence claims, not to intentional torts of fraud or misrepresentation.[211] A unique Pennsylvania law permits a design professional sued for negligence to file an "affidavit of noninvolvement," in which she asserts that she is "misidentified or otherwise was not involved with regard to the cause of the injury or damage" alleged by the plaintiff.[212]

Other statutes seek, at least to a limited degree, to limit the liability of design professionals. A significant number of states have enacted "Good Samaritan" laws to protect design professionals from potential liability when they provide expert services in emergencies. The statutes vary as to the type of emergency that will invoke the protective legislation and who can benefit from such legislation. As an illustration, under California law an architect or engineer who, at the request of a public official but without compensation or expectation of compensation, voluntarily provides structural inspection services at the scene of a declared emergency "caused by a major earthquake, flood, riot, or fire" is not liable for any harm to person or property caused by her good faith but negligent inspection of the structure. Protection does not extend to gross negligence or willful misconduct. The inspection must occur within thirty days of the "declared emergency."[213] Some states have extended this protection to those who provide inspection or other services following any natural disaster.[214]

Most significantly, beginning in the mid-1980s associations of design professionals were successful in persuading a number of states to grant immunity to design professionals from third-party claims by injured workers who are covered by workers' compensation for site services unless the design professional has contracted to oversee safety or undertook to do so.[215] The injured worker has rights under workers' compensation laws. Limitation of third-party actions—that is, tort claims that can be made by the injured worker against those who are not immunized by workers' compensation laws—may be of considerable importance in reducing liability.

SECTION 14.10 Remedies

A. Against Design Professionals

The remedies available when the design professional does not perform in accordance with the contract or tort obligations were set forth in Sections 7.10 and 12.14B.

B. Against Co-Wrongdoers

In construction losses are commonly caused by a number of participants and conditions. As to defective work, see Section 24.06. However, a few generalizations can be made in this section.

First, suppose the client asserts a claim against the design professional, but the loss has been caused in part by the client's negligence. If the client is allowed to pursue a tort claim, generally the negligence of the client and that of the design professional will be compared, and the amount recoverable by the client will be reduced by the amount attributable to its own negligence, as noted in Section 7.03G. For example, if the loss was $100,000 and a determination was made that 30 percent of the loss should be chargeable to the client, the client would

---

[210]The Georgia statute was interpreted in *Kneip v. Southern Eng'g Co.*, 260 Ga. 409, 395 S.E.2d 809 (1990).

[211]*McElwee Group, LLC v. Municipal Auth. of Borough of Elverson, Pa.*, 476 F.Supp.2d 472 (E.D.Pa.2007).

[212]Pa. Consolidated Stat. Ann. Tit. 42 § 7502(e), which the superior court likened to a motion for summary judgment but permitted earlier in the litigation. *Herczeg v. Hampton Municipal Auth.*, supra note 169, 766 A.2d at 869–70.

[213]West Cal.Ann.Bus.&Prof.Code §§ 5536.27, 6706.

[214]Tenn.Code Ann. § 62-2-109; Mo.Rev.Stat. § 44.023. Protection under the Missouri statute does not extend to those guilty of willful misconduct or gross negligence. The latter also provides that volunteers can receive their incidental expenses for up to three days' performance of voluntary services.

[215]Conn.Gen.Stat.Ann. § 31-293(c); West Fla.Stat.Ann. § 440.09(6); Kan.Stat.Ann. § 44-501(f); West Okla.Stat.Ann.Tit. 85, § 12; West Wash.Ann.Rev.Code Ann. § 51.24.035 (immunity not granted if responsibility is specifically assumed by contract terms that were mutually negotiated or the design professional actually exercised control over the portion of the premises where the worker was injured). None of these statutes grant immunity for negligent design. In *Edwards v. Anderson Eng'g, Inc.*, 284 Kan.892, 166 P.3d 1047 (2007), the court interpreted Kan.Stat.Ann. § 44–501(f) as overruling *Balagna v. Shawnee County*, supra note 168, an actual knowledge case, and as not applying to a negligent design claim.

recover $70,000. A few states apply the contributory negligence rule to bar the entire claim.

Suppose the claim by the client is based on breach of contract. Some states would still make an apportionment, whereas other states would not.[216] Those that do not apportion place the loss on the party whose breach substantially caused the loss.

Apportionment becomes more complicated when responsibility falls on a number of defendants and, in some instances, on the claimant itself. If a particular loss can be connected principally to a particular defendant, that defendant will pay the loss that it has caused. However, if the loss is indivisible and cannot be apportioned among the wrongdoers, it is likely that a tort claimant will be able to recover the entire loss from any of the defendants, with the ultimate responsibility determined by contribution laws and indemnification.[217]

SECTION 14.11 Current Controversies: Some Observations

It would be impossible to comment at length on the issues related to professional liability addressed in this chapter. But this section highlights issues that have generated heated controversies.

A. The Professional Standard: Should Professionals Be Treated Differently?

The professional standard in essence permits local professional practice, at the time the defendant's services were rendered, to be the legal standard. It has been subjected to intense criticism. Manufacturers of products and those who build and sell homes[218] are held to strict liability. Owners warrant to the contractor that the design they have supplied will be sufficient.[219] Should professionals be given special dispensation when they have caused harm?

This specialized treatment may have had some justification, according to its attackers, when professionals were practicing in one-person offices or in small firms and in times when the professionals largely learned from those practicing in their localities. Today, professionals such as doctors, lawyers, and many of the design professionals practice in large organizations, make large profits, compete nationally, and receive education and training that use state or national, not local, practices. Those who attack the professional standard point to this as a reason why professionals today do not deserve special treatment. They argue that design professionals must accomplish the objective for which they are retained. Any increased liability that may result can be handled by insurance and the cost spread to those who use the professional services. Even if insurers generally exclude contractual risks that go beyond negligence, those who insure are part of a competitive industry that will respond to market pressures. In any event, critics argue that insurance practices should no more determine the appropriate legal standard than professional practices. Attackers also stress the following:

1. Antitrust laws are increasingly applied to the professions. The law views their members as businesspeople rather than professionals.
2. The relationship between design professional and client has taken on a more commercial character. This should permit design professionals to reduce what they perceive to be open-ended liability by allowing them to contract for a standard lower than the professional standard, by limiting their liability, and by exculpating themselves from certain liability. See Section 15.03D.
3. The current standard greatly increases the cost of litigation by requiring a parade of well-paid experts.
4. The line between design, to which the professional standard applies, and site monitoring, where it may not, is blurred and creates an added dimension of uncertainty.
5. To obtain a commission, the design professional often stresses that it is better than others with whom it competes yet asserts it should be measured by the professional standard that looks to average local practices.
6. The professional standard is often unrefined, with insufficient attention devoted to clear analysis that takes into account the differences among parties on the cutting edge of the profession, the material

---

[216]For cases allowing an apportionment, see *S. J. Groves & Sons Co. v. Warner Co.*, 576 F.2d 524 (3d Cir.1978); *Grow Constr. Co. v. State*, 56 A.D.2d 95, 391 N.Y.S.2d 726 (1977). For cases refusing to employ apportionment, see *Hunt v. Ellisor & Tanner, Inc.*, 739 S.W.2d 933 (Tex. Ct.App.1987), writ denied Mar. 23, 1988.

[217]*Northern Petrochemical Co. v. Thorsen & Thorshov, Inc.*, 297 Minn. 118, 211 N.W.2d 159 (1973).

[218]See Sections 7.09 and 24.10.

[219]See Section 23.05E

available in the literature, what becomes accepted in the profession as part of the knowledge possessed by an informed professional, and what is taught at the undergraduate level.[220]

Indictment of the professional standard has had an effect on the law, such as the loosened requirements for expert witnesses[221] and the beginning of a tendency to hold that some design professionals are engaged in ultrahazardous activities for which they are strictly liable.[222] Counterarguments must exist that have persuaded courts to hold to the professional standard in the face of more strict standards being applied to others. Even modern cases in an atmosphere in which compensation is emphasized, have still applied the professional standard. Why? Perhaps most important is the belief that professionals operate amid great uncertainty. Will the professional be judged harshly in the event of an unsuccessful outcome when even the best professional services would not have been able to create a satisfactory outcome? Does any outcome standard run the risk of holding the professional to professional performance that was more appropriate at the time of trial than at the time the professional services were performed?[223] Defenders contend that the client expects good professional service rather than insurance. Those who defend the professional standard fear that any warranty standard will expose design professionals to unreasonable expectations of the client that will be resolved in favor of the client by a jury, particularly if the client is unsophisticated in the world of design and construction. Also, what will be the exact nature of the warranty if an outcome standard is used? Might it not extend far beyond the function of design services?

Defenders of the professional standard argue that an outcome standard will generate overdesign at unneeded costs to reduce or avoid the risk of liability. Increased liability also means higher cost of service, resulting in fewer practitioners, higher prices, and more uninsured professionals.

Those who defend the design professions state that the high compensation some professionals receive is rare in the design professions. Only "strong" design professionals will be able to "contract out" of any outcome standard. The ordinary professional designer lacks the bargaining power to obtain a more limited standard, or believes it is inappropriate to begin a professional relationship by demanding or requesting one.

So it stands. The reason for announcements of the professional standard in clear terms by courts means it is unlikely that the standard will be abolished in the near future. Dissatisfaction at the privileged position accorded professionals is likely to result in chipping away at its protections as well as in sophisticated clients demanding contractual protection beyond the standard.

B. The Design Professional's Duty to Workers

The cases cited in Section 14.08G demonstrate the contrary perspectives from which this hotly contested issue can be viewed. It is important to recognize that this is a multilevel problem. The *Pfenninger* decision reflects the majority rule, that the issue in the ordinary retention should be viewed almost exclusively as contractual. Design professionals are retained and paid to use their best efforts to see that the project is built in accordance with the contract documents. The contract makes clear that the contractor, not the design professional, is responsible for how the design is executed. This approach requires an examination of the contracts, mainly the ones for design services but also for construction. If they do not reveal that the design professional was accepting responsibility for construction methods, the design professional has no duty to workers, and there need be no inquiry into her conduct. As noted, some courts adhere to this contract-based perspective even in the face of allegations that the design professional knew of the hazardous condition which caused the worker's injury.[224]

---

[220]Peck & Hoch, *Liability of Engineers for Structural Design Errors: State of the Art Considerations in Defining the Standard of Care*, 30 Vill. L.Rev. 403 (1985). Prof. Jones, whose paper is cited at supra note 112, argues for a result-oriented liability if the probability of success is high. He contends that most services performed by a design professional fall into this category. Most design professionals would dispute this.

[221] See Section 14.06A.

[222]See Section 7.04A

[223]One study concluded that, even when applying the professional standard, defendants are at risk of being judged by current standards, rather than the standard in existence when they provided the services. The reason lies in the bias of witnesses to recall accurately the prior standard of care in light of current practices. See Miller, *What is the Standard of Care?*, 41 J. of Management in Engineering, No. 6, Nov./ Dec. 1996, p. 40.

---

[224]See supra note 169.

The other perspective is to start with tort law, with its function of compensating victims, deterring wrongful conduct, and avoiding harm. That body of law jettisoned the privity rule to ensure that claimants can find a person from whom compensation can be recovered and to avoid technical defenses that bar scrutiny into the conduct of those people who very likely caused the harm. Why reinstitute this barrier by looking mainly at the purpose for engaging a design professional and focusing almost exclusively on the contract?

To be sure, the contract for design services is important. Without it, vague or detailed, the design professional would have no business on the site and would not be in a position to look out for danger to workers or anyone else. The contract plays a significant role in determining what the design professional should have seen. The design professional cannot be expected to look for problems—that is not her function. If she sees unsafe practices or would have seen them had she done what the contract obligated her to do, her conduct will be judged. Did she act reasonably? She may satisfy this obligation by complaining to the contractor's superintendent, by bringing this matter to the client's attention, or even by inviting public officials who deal with safety matters to deal with the question. If the issue is viewed as a tort problem, however, she has a responsibility to act reasonably.[225]

Although most material dealing with this troublesome problem is found in case decisions and legal journals, there is other relevant material. For example, a California attorney general opinion addresses the duty of a registered engineer, retained to investigate the integrity of a building, who determines there are structural deficiencies but who is advised by the owner that no disclosure or corrective action is intended and that the professional information is to remain confidential. Does the professional engineer have a duty to warn occupants or notify local building officials that she has uncovered evidence of structural deficiencies that create imminent risk of serious injury to the occupants? The attorney general could not find any duty to warn based on the registration laws but concluded that there was such a duty, because the common law has imposed a duty on the part of professionals to protect third parties whose lives may be endangered when the professional becomes aware of this risk.[226]

A divergence—at first glance, counter-intuitive in nature—appears to be emerging in the law governing design professional liability to third parties. Tort law traditionally affords greater protection where personal harm rather than commercial interests are involved. However, a review of the cases cited in Sections 14.08C, 14.08D, and 14.08E dealing with claims for economic losses suffered by contractors shows an increased willingness to find the designer liable notwithstanding the economic-loss rule, while at the same time courts are less willing to find architects or engineers liable to injured construction workers.

This divergence may be more apparent than real. A common thread in both types of cases is the willingness of courts to place liability on the party in the best position to prevent the harm. The architect controls the quality of the design, and if that design is gravely defective, the resulting economic losses to the contractor and subcontractors are immediate and real. By contrast, the contractor—not the designer—is in the best position to ensure a safe project; accordingly, imposition of liability upon the design professional for a workplace accident is not merited. That basic allocation of responsibility is not altered by the designer's awareness of a dangerous site condition.

A second level looks at judicial procedures. Those who support the "no duty to workers" rule seek a defense that will bar a full trial and have the issue decided by looking solely at the pleading or the pleadings and affidavits submitted to support a motion for summary judgment. They wish to escape the high cost of trial even if they believe that the design professional will ultimately prevail if her conduct is judged.

Those who support the tort orientation see no reason why the design professional should escape being judged. They contend that if the facts clearly show that she could not have been expected to know of the unsafe practices or that she did what was clearly adequate, she will be able to avoid a full-scale trial. But they object to a rule of law

---

[225]A fuller expression of one author's views can be found in Sweet, supra note 168.

[226]68 Op.Cal.Att'y Gen. 250 (1985). But see *Burg v. Shannon & Wilson, Inc.*, 110 Wash.App. 798, 43 P.3d 526 (2002) (engineering firm, hired by the city to analyze the stability of a cliff, had no duty to warn homeowners living under the cliff of its results) and *Butler v. Advanced Drainage Systems, Inc.*, 294 Wis.2d 397, 717 N.W.2d 760 (2006) (no duty to warn of flooding hazard to owners of lakefront properties threatened by rising lake levels).

that relieves her. They prefer this question be treated as a factual question like any other.

A third level relates to workers' compensation law. As noted, a main function of modern tort law is to see that victims are compensated. Most workers will be compensated under workers' compensation law—a social insurance system that pays a certain portion of the economic losses through an administrative process that does not focus on wrongful conduct. Workers cannot sue their employers in tort; workers' compensation is their exclusive remedy. Those who feel the compensation system is inadequate seek to encourage the development of third-party claims that the injured worker can make to fully recover for the harm suffered. This is usually more than the often inadequate amounts recovered under workers' compensation law.

However, workers' compensation is sometimes used as an argument by those who oppose the design professional's having a duty to workers.[227] Those opposing such a duty argue that the worker will not be uncompensated, and they see no reason why a worker in a construction-related accident should be placed in a better position than a worker in an industrial accident who may not have as many third parties from whom she can seek tort recovery. They contend that the broadening of third-party claims has hopelessly overcomplicated often simple accident cases and has generated lawsuits with horrendous costs to all participants.

At a fourth level, operations on a construction site are stressed. Those who support the duty state that safety is everyone's business, and the more people concerned, the less likely injuries will occur.

Those who oppose a "duty to workers" rule claim it will induce design professionals to venture into areas where they do not have expertise, which can only cause blurred lines of responsibility as well as expose the design professional and the client to claims that intervention into these matters breaches the contract between owner and contractor.

Probably the best solution in an ideal world would be to have a total, enclosed social insurance system under which all workers receive adequate compensation without the necessity of going to court but without having rights against third parties. (Currently workers' compensation

laws are extraordinarily controversial, in light of the high workers' compensation premiums, the large percentage of awards and costs that go to attorneys and doctors, the large overhead and profit of some insurers, and the relatively low amounts that end up in the pockets of injured workers or dependents of those killed.) Until this utopia is achieved (it may be a long way off), it is hard to support a result that would revive the dead privity doctrine and put professional designers in a favored position under which their principal function precludes their conduct being judged.

C. Injection of Tort Law into the Commercial World: A Wild Card

The commercial world—the world of business dealings between merchants where much of the world's work is accomplished—certainly does not escape tort law intervention. For example, under certain limited circumstances the tort law will enter when a person has wrongfully induced another to breach a contract or impede a prospective economic advantage. On the whole, tort law interferes here only when there has been intentional wrongdoing or where there has been harm to people or property.[228] Tort law, at least in some states (see Sections 14.08C and 14.08E), has been reluctant to shift economic losses caused by negligent conduct and has left that largely to contract and commercial law.

In the construction world, as in others, tort law has begun to play an increasingly important role in allocating purely economic losses. For example, as noted in Section 14.08D, negligent representation has been the basis of claims made by people who have relied on the representations of those in the business of furnishing information, such as surveyors or geotechnical engineers and even architects, engineers, and construction managers.[229] Even more, there has been the tendency to expose the design professional, among other participants, to claims by other participants, particularly contractors, sureties, and even prospective occupiers of projects, such as buyers or tenants.

Perhaps these tendencies cannot be rolled back in a legal world where tort law and accountability have become

---

[227]*Balagna v. Shawnee County*, supra note 168, 668 P.2d at 170–72 (dissent).

[228]A subsequent tenant was allowed to bring a tort claim against the architect when the floor settled, walls became damaged, and premises became untenantable. See *A. E. Inv. Corp. v. Link Builders, Inc.*, 62 Wis.2d 479, 214 N.W.2d 764 (1974).

[229]*Rozny v. Marnul*, 43 Ill.2d 54, 250 N.E.2d 656 (1969).

dominant factors in private law even in commercial disputes. Those courts faced with these decisions or asked to expand existing rules should consider the effect of introducing this wild card.

For example, some courts allow a contractor to maintain a tort action against a geotechnical engineer for misrepresentation.[230] The context was a transaction in which the construction contract clearly placed the entire risk of unforeseen subsurface conditions on the contractor.[231] Allowing the tort action induces the geotechnical engineer to request indemnity from the client, to increase her contract price to take this risk into account, or to price her work to encompass performance designed to ensure that her representations are accurate even if "excessive" caution would not be justified. Either way, the system of allocating risks is frustrated, with the client perhaps paying twice for the same risk by increased contractor bids (tort recovery is too uncertain to permit the prudent contractor to reduce the bid because of potential tort recovery) and the higher compensation to the geotechnical engineer.

Perhaps even worse, one court allowed a tenant who lost full use of its premises because of delays caused by the contractor, to establish negligence and to use that as the basis for a claim against the contractor.[232] What effect does this holding have on the potential risk of construction contract participants?

Delay is dealt with at length in the construction contract with its time extensions, damage liquidations, or no-damage clauses. Will these clauses apply to tort claims? How can contractors or design professionals faced with this risk deal with it? They can hope the law will protect them. They may decide to build a contingency into their contract price to deal with this risk or demand indemnification if they have the bargaining power. They may, if they can identify potential claimants, seek exculpation from them or demand that the owner do so. All these approaches add transaction costs and must increase contract prices. It must also be kept in mind that the party who has suffered economic losses usually can transfer such losses to the party with whom it has contracted—the tenant to the landlord or the contractor to the owner.

The tort wild card has caused chaos in the construction legal world. It is bad enough to allow major participants, such as the design professional, owner, and contractors, to sue each other in tort. It is much worse to allow more remote participants, such as suppliers, potential buyers, tenants, lenders, or even sureties, to use tort law as a means of shifting their economic losses to the major participants in the construction process.

The law should encourage people to enter into commercial transactions. One way is to limit the exposure for consequential damages suffered by the other party. However, this protection is diminished in claims by third parties in tort. The fact that tort in this area usually requires negligence does not compensate for this added exposure, particularly when claims against design professionals usually use a tortlike standard anyway.[233]

D. The Effect of Expanded Professional Liability

This section has outlined some of the arguments for and against expanded professional liability. The focus of this subsection is on the effect of this expansion.

Expanded liability can be looked on as a method of eliminating incompetent practitioners from the professions, to supplement the unarguably ineffective registration laws or, at least in this area, the inefficient marketplace. But does it have this effect?

Does expanded liability drive out the practitioner who should be removed from the profession? Are incompetents sued more often than those who are competent? Are they likely to be forced out by increased insurance rates, decisions by insurers not to insure them, or unwillingness of prospective clients to engage them if they cannot be insured? At best, these are unprovable. Very likely the answer to all three is no.

Has expanded liability improved professional practice? Undoubtedly, design professionals are now more careful—perhaps too careful. The result can be overdesign, an unwillingness to take design risks, and mediocre design. Again, it will be difficult to assemble anything beyond anecdotal evidence and polemics to uncover the truth.

---

[230]M. Miller Co. v. Central Contra Costa Sanitary Dist., 198 Cal. App.2d 305, 18 Cal.Rptr. 13 (1961). But see Texas Tunneling Co. v. City of Chattanooga, 329 F.2d 402 (6th Cir.1964).

[231]See Section 25.05.

[232]J'Aire Corp. v. Gregory, supra note 148. The Aas decision, supra note 145, appears to limit the reach of J'Aire.

[233]See Sections 14.05–14.07. Jones, supra note 112 at 1070–73, 1092–93, also suggests tort may not be available to these claimants.

Of course, expanded liability can be and has been justified as a process for allocating responsibility to the people who are responsible and who can best spread the loss. The construction process, with its wealth of participants, its overlapping functions, and its unclear lines of responsibility, makes unlikely that expanding professional liability will place responsibility on the party who should take it. All that can be certain is that expanded liability has led to hopelessly complicated and unpredictable lawsuits, with an inevitable rise in the overhead of performing professional services.

What is also certain is that expanded liability has generated overprotective contract language that may simply drive prospective clients to others, such as those who design and build or manage construction. Along with other causes of high operational overhead, such as taxes and other regulations, expanded liability has slowly reduced the ranks of sole practitioners and small partnerships and has led to increased specialization. Expanded liability has led to an emphasis on risk management (explored in Chapter 15).

CHAPTER FIFTEEN

Risk Management: A Variety of Techniques

Chapter 14 chronicled the expansion of professional liability. Although often ignored, one technique for avoiding claims is to cultivate a good client relationship. Honesty in approach, respect for the client's intelligence, appreciation of the proper role of a professional adviser, and common courtesy (answering phone calls, letters, and e-mails) are perhaps the best techniques to avoid claims. These are nonlegal considerations. This chapter suggests legal and planning approaches to deal with risk management.

SECTION 15.01 Sound Economic Basis: Bargaining Power

This book is not an appropriate place to discuss in detail the economics of the design professions. To implement some if not most of the approaches suggested in this chapter, the design professional must be able to choose which commissions to accept and to request or even demand that particular contract language be excluded or included. Doing so requires sufficient economic strength to pick and choose among projects and contracts. Although this chapter cannot deal with methods to achieve this power, the approaches suggested will be of no value unless economic strength can be attained and mobilized.

SECTION 15.02 Evaluating the Commission: Participants and Project

As a rule, there are more design professionals than commissions. Therefore, most design professionals usually take whatever work they can get. Some design profes-

sionals can pick among projects, and this section is directed to them.

It is very important to evaluate the client and its financial resources. A client with limited resources, particularly one with extravagant expectations, may not be able to withstand the shocks of added costs generated by design changes, delays, claims, or other circumstances that will increase the ultimate contract payout. Such a client will be more inclined to abandon a project before construction. If construction does begin, the client may be quicker to point at the design professional and other participants if the project does not proceed as planned. From 1966 to 1997 (not in 2007), the AIA gave architects the power to request information about the financial resources of the owner at any time during their performance. Yet probably most architects did not exercise this power, particularly at the start of the professional relationship. Although an architect may understandably wish to make these inquiries at the beginning of performance, such inquiries may be crucial during performance if financial troubles appear imminent.

Another important client criterion is experience in design and construction. An inexperienced client may be more likely to make claims, because it does not realize the uncertainties inherent in construction and the likelihood that adjustments will have to be made. Such a client may also be mesmerized by a fixed-price contract and be unduly rigid as to price adjustments. Although construction contract pricing disputes principally affect the relationship between owner and contractor, when that relationship sours, more administrative burdens will be placed on the design professional and claims will be more likely.

The project, too, must be evaluated. Projects that involve new materials, untested equipment, and novel construction techniques must be viewed as creating special risks. If disappointments develop, claims (including those against the design professional) are more likely.[1]

It is important to evaluate the other key participants, such as the prime contractor, the principal subcontractors, and consultants, for technical skills, financial capacity, and integrity. The construction contract is another factor that must be taken into account. A tight fixed-price contract, a rigid time schedule with a stiff liquidation of damage clause for delay, a multiple prime contractor arrangement, and a fast-track sequence all contain the seeds for controversy and, possibly, liability exposure.

It may be helpful to develop a point system for evaluating these factors. If the points reach a certain level, the project should not be undertaken without careful executive review. Beyond the next numerical benchmark, the commission should be refused unless changes are made. At a point beyond even that, the commission should be refused.

SECTION 15.03 Contractual Risk Control

Section 15.02 noted the importance of a general appraisal of the contract. This section looks at specific contract clauses that can be useful in risk management.

A. Scope of Services

The contract should make clear exactly what the design professional is expected to do. Sections 12.01 and 13.01G referred to the difference between basic and additional services. Here the emphasis is on services the client may expect the design professional to perform that the design professional does not feel are part of his undertaking. Perhaps most important is the design professional's role in determining how the work is being performed and the responsibility of the design professional for the contractor not complying with the contract documents.

B. Standard of Performance

Usually the design professional wishes to be held to the professional standard discussed in Section 14.05. This is

demonstrated by the AIA's inserting the professional standard for the first time in AIA Document B101-2007, Section 2.2. When this is the case, the contract language as well as any other communications should not use words such as *assure, ensure, guarantee, achieve, accomplish, fitness,* or *suitability* or any language that appears to promise a specific result or achievement of the client's objectives. It is even better to include language that specifically incorporates the professional standard and, where possible, language that justifies it.

Section 15.03D deals with liability limitations, but one suggestion that relates to the standard of performance should be mentioned here. Many cost overruns result from design changes. Some have suggested a toleration figure—an amount usually based on some percentage of the construction contract for which the design professional would not be responsible. This suggestion is based on the assumption that design is essentially a trial-and-error process and some things truly cannot be discovered until the design is actually being executed. The design professional would be responsible over any such toleration figure if performance does not measure up to the obligation imposed on him by the contract or by law.[2] (Usually the amount specified is approximately 3 percent.)

C. Exclusion of Consequential Damages

The law does not charge a breaching party with all the losses caused by its breach. Usually the breaching party is not chargeable with what are sometimes called *consequential* or *less direct* damages. This limits a contracting party's responsibility.

Yet the foreseeability requirement—the standard used most frequently in determining whether consequential damages can be recovered—has been applied by modern courts in such a way as to diminish protection given by earlier courts. For that reason, the design professional should seek to exclude his liability for consequential damages in the contract.[3] This can be done expressly or by limiting

---

[1]The resources of the Architectural and Engineering Performance Information Center at the University of Maryland might be useful in evaluating problems that may arise on a project.

[2]EJCDC E-500, Exhibit I, ¶ I.6.11(B)(3)(2002), which can be added to ¶6.11, is entitled "Agreement Not to Claim for Cost of Certain Change Orders."

[3]Upheld in *Dept. of Water and Power of City of Los Angeles v. ABB Power, T. & D. Co.* 902 F.Supp. 1178 (C.D.Cal.1995) and *Wood River Pipeline Co. v. Willbros Energy Services Co.,* 241 Kan. 580, 738 P.2d 866 (1987).

the responsibility of the design professional to correcting work caused by defective design or, when that is not economically feasible, to the diminished value of the project.

In AIA Documents A201-2007, Section 15.1.6, and B101-2007, Section 8.1.3, its major standard documents for design and for construction, provisions under which each contracting party waives consequential damages against the other. The EJCDC in its E-500 dealing with engineering services provides a similar provision in its Paragraph 6.10(E) (2002). It can be supplemented by Exhibit I, Paragraph I.6.11(B)(2).

Undoubtedly general judicial expansion of the recovery of consequential damages and the judicial tendency to consider all AIA documents as adhesion contracts[4] will mean likely judicial challenge to the validity of such waivers.

D. Limiting Liability to Client

Beginning in the 1970s, professional associations of engineers, particularly the American Society of Foundation Engineers and the National Society of Professional Engineers, advocated that their members seek to get their clients to agree to provisions in their contracts that would limit the liability of the engineer. Efforts to obtain such provisions were also spurred on by professional liability insurers that offered a reduced premium for those who obtained these limitations in their contracts.

Before discussing techniques to obtain approval by the client and the types of clauses, traditionally client claims were the only claims that the design professional had to anticipate. But the growing risk of third-party claims has generated a new issue. From whom does the design professional seek a limitation of liability?

There are two ways of dealing with this problem. The first is to anticipate the likelihood and identity of third-party claimants. This can be a problem in the world of complex real estate development transactions. A good example is an Illinois case decided in 2007. Although the architect ultimately was absolved from a third-party claim, the facts show the difficulty of anticipating third-party claims.[5]

The developer, JDL, and its partner sought to develop condominiums and townhouses for speculative sales. JDL orally retained TMA, the architect, to design the project. To develop, market, and sell the units, JDL and its partner formed Burnham Station, a limited liability corporation. JDL and its partner were officers in Burnham. The project failed, and a major investor in Burnham sued TMA for breach of contract and in tort to recoup its lost investment.

As JDL and TMA had made an oral retention agreement, it is likely that TMA did not consider a liability limitation. But had it sought to limit its exposure, from whom should it have sought such protection in this tangled web of entities?

A second approach for a design professional to manage third-party claims is to seek and obtain an agreement with its client under which the latter will indemnify the design professional from third-party claims, to obtain a total cap on the design professional's liability exposure. Some refer to such an approach as a hybrid clause as it covers both direct liability to the client and protection from third-party claims. This total cap approach was used in a recent California case.[6]

From an operational standpoint, the next obstacle to this method of managing the risk is to persuade clients to agree to such a provision. Private clients may be reluctant to accept this risk. They may be persuaded by insistence upon such a clause by the design professional only if they think the design professional can do the job properly. They may also be persuaded if inclusion merits a fee reduction

---

[4]See Section 5.04 C.

[5]*F. H. Paschen/S. N. Nielson, Inc. v. Burnham Station L.L.C.*, 372 Ill. App.3d 89, 865 N.E.2d 228 (2007), rehearing denied, Apr. 18, 2007. The court held the third-party claimants, investors in a failed venture, were not third-party beneficiaries of the architect's contract, and the tort claim was barred by the economic loss rule. See Sections 14.08B, D, and E.

[6]*TSI Seismic Tenant Space, Inc. v Superior Court*, 149 Cal.App.4th 159, 56 Cal.Rptr.3d 751 (2007). In this case the owner of an apartment complex sued the prime, the structural engineer, and the geotechnical engineer. The prime and the structural engineer cross-complained against the geotechnical engineer. The owner settled with the geotechnical engineer for the amount of the liability limitation of $50,000. The geotechnical engineer waived its claim for $259,000 of attorneys' fees. Under California law if the settlement is in good faith, the settling defendant, here the geotechnical engineer, is shielded from claims by the cross-complainants, here the prime and structural engineer. The trial court held that the liability limitation was valid. It also held that the settlement was in good faith. But the appellate court held that the settlement was not in good faith because it did not bear a reasonable relationship to the geotechnical engineer's proportionate share of liability. The loss claimed by the owner was $6.4 million, and experts testified that the geotechnical engineer's share was over $3 million. The statute is designed to advance the goal of equitable allocation among wrongdoers. See Section 27.14 for claims against multiple parties.

based upon any saving in insurance premiums paid by the design professional. Finally, they may see the allocation of risk as a fair one if the design professional is exposed to very large risks for a modest fee. Acceptance may get the professional relationship off to a positive start, but public clients, with their emphasis on accountability, are unlikely to accept this risk.

The enforceability of such clauses has stirred controversy. Many recent cases have passed upon the validity of such clauses. Some will be noted in this section. But before that discussion, it is crucial to recognize the many variables that present themselves in any individual case that makes the law murky and enhances the difficulty of predicting outcomes.

The limitation can take many forms.[7] Some limit exposure to the design professional's fee, to his professional liability coverage or to a designated amount. The amounts may be nominal or substantial. If two methods are selected, the clause may state that the amount is the greater or the lesser, usually the greater.

Some clauses allow the client to buy out of the liability limitation by giving the client the choice of paying a premium to obtain full liability or to accept the liability limitation. It will be rare for the client to buy out, and the principal reason to give the client an alternative is to enforce the liability limitation.[8]

As noted earlier in this section, the cap may be a total cap, under which the design professional is indemnified by the owner from any third-party claims.

Some courts examine the relationship between the amount of the fee and the predicted or actual damages.[9] The greater the discrepancy between the liability limit and actual or potential damages, the less likely the clause will be enforced.

Clauses can vary as to the type of conduct controlled by the liability limitation clauses. They apply to a claim for breach of contract, negligence, gross negligence, or willful misconduct. The more culpable the conduct, the less likely the court will uphold the clause. Here the tort principle of deterring wrongful conduct that is more culpable than simple negligence takes precedence over freedom of contract to apportion risks by contracting parties.

Another variable is the nature of the transaction, commercial or consumer. Even within this variable there can be differences in the ability of the client to know what risk it is taking. For example, a small business owner may not be able to evaluate the risk of limiting the amount it can recover, in effect taking this risk upon itself. This variable will be analyzed later in this section.

Another factor that must be taken into account is the scope of services provided by the design professional. It may be a discrete one-off transaction (though, as will be seen, with the potential for damages greatly exceeding the fee) or one where the scope of services provided is broad and diverse, such as an architect who designs, predicts costs, and plays a significant role during construction. Many types of losses may occur in such a transaction.

Many arguments are made for and against the enforceability of liability limitation clauses. Focus upon them as a means of controlling risk to design professionals has stirred great debate. The powerful principle of freedom of contract, a pillar of any free market economy, collides with the tort principles of deterring wrongful conduct and protecting victims. This collision has made the law uncertain and controversial.

Proponents also give other reasons for enforcing such clauses. The centrality of design, as noted in Section 14.01G, exposes the design professional to great risks for which he does not receive compensation comparable to the risk.

Proponents also assert that judicial unwillingness to enforce such clauses can affect the willingness of engineers to perform services in quick, high-risk–low-cost transactions. For example, in *Estey v. Mackenzie Engineering, Inc.* a structural engineer was retained to make a limited, visual review of a house his client was considering purchasing. His fee was $200, and the contract limited his liability for negligence to that amount. The client claimed the engineer had been negligent. At the time of trial, the client proved losses of $190,000 due to needed repairs with an estimated cost of $150,000 yet to be incurred. The engineer pointed to the limit of liability.

---

[7]EJCDC E-500 (2002) includes optional Exhibit I, ¶1.6.11(B)(1). It lists alternative formulas. They include the engineer's compensation, insurance proceeds or a specified amount. The limitation can be added to ¶6.10.

[8]See *Cregg v. Ministor Ventures*, 148 Cal. App.3d 1107, 196 Cal. Rptr. 724 (1983) (upheld in a consumer rental space contract).

[9]*Fort Knox Self Storage, Inc. v. Western Technologies, Inc.*, 140 N.M. 233, 142 P.3d 1 (2006) (cap 28 times fee enforced). The court compared this ratio to another case which enforced a cap where the loss was seven times the fee. *Valhal Corp. v. Sullivan Assocs., Inc.*, 44 F.3d 195 (3d Cir. 1995).

Although the intermediate Oregon court enforced the clause,[10] the Oregon Supreme Court would not.[11] It stated that the client probably did not intend to take this risk.

Although it would seem unfair for the innocent client to suffer this loss, refusing to enforce such a clause in this context may mean that fewer engineers will perform this service even if they are insured, or will drastically increase the fee. Without enforcement, a quick, cheap inspection is very likely not possible.

Ricciardi v. Frank reflects the uncertainty of enforcing liability limitations in consumer transactions that involve discrete but crucial services. The facts are similar to the *Estey* case already noted. A consumer sued an inspecting engineer who performed a house inspection for $375. The New York trial court in a lengthy opinion refused enforcement of the limitation of liability to the fee when the actual damages were close to $3,000.[12] The ratio of $375 to $3,000 in the *Ricciardi* case was not as disproportionate as in the *Estey* case, a $200 fee compared to $190,000 proven losses and another $150,000 yet to be incurred. The appellate division within the New York trial system enforced the liability limit.[13] It should be clear from these cases that prediction in consumer cases with low fees and large exposure will be hazardous.

If the clause specifies a small sum, such as the $200 in the *Estey* case, it may appear that the clause exculpates the user from its negligence, not simply places a cap on liability exposure where it would seem reasonable to do so.

An indemnification clause must be compared to a liability limitation clause. Indemnification is a process by which one party, the indemnitee, receives a promise from the other party, the indemnitor, that the latter will take care of any claim against the indemnitee that results from conduct, usually negligent but not necessarily, of the indemnitor.[14] Often the conduct of the indemnitee plays some role in causing the loss. When this happens some such clauses may exculpate the indemnitee from its own negligence. To avoid this most states have statutes that place limits on such clauses, particularly if they indemnify the indemnitee when its negligence has been the sole cause of the loss.[15]

The similarities between an exculpatory clause, an indemnification clause, or one that limits liability to a specified amount have caused difficulties.[16] To be sure, all have an element of exculpation; the exculpation clause exonerates entirely, the indemnity clause may in part or completely exonerate, and the liability limitation can exonerate in part, for any amount of damage in excess of the liability amount.

With this similarity in mind, those who seek to bar enforcement of a limitation of liability often claim that a statute that bars indemnification for certain negligence precludes enforcement of a limitation of liability clause. While such a claim has had some success,[17] most courts see the difference and hold that the statute does not affect the limitation of liability.[18]

Cases have enforced[19] and rejected[20] such clauses. This is understandable in light of the newness of such clauses,

---

[10]137 Or. App. 1, 902 P.2d 1220 (1995).

[11]324 Or. 373, 927 P.2d 86 (1997).

[12]163 Misc.2d 337, 620 N.Y.S.2d 918 (City Court of N.Y., Yonkers, Westchester Cty.1994).

[13]170 Misc.2d 777, 655 N.Y.S.2d 242 (1996). See also *Fort Knox Self Storage System, Inc. v. Western Technologies, Inc.*, supra note 9 (enforced clause in commercial context but court issued a dictum that stated there might be a different result in a consumer transaction). See 142 P.3d at 6.

[14]See Section 31.05C.

[15]See Section 31.05D.

[16]*Mistry Prabhudas Manji Eng. Pvt. Ltd. v. Raytheon Engineers & Constructors, Inc.*, 213 F.Supp.2d 20 (D.Mass. 2002)

[17]*Dillingham v. CH2M Hill Northwest*, 873 P. 2d 1271 (Alaska 1994).

[18]*Mistry Prabhudas Manji Eng. Pvt. Ltd. v. Raytheon Engineers & Constructors, Inc.*, supra note 16 (clause enforced if made by business-people in a commercial setting unless amount is so low that it removes any incentive to perform with due care).

[19]*Valhal Corp. v. Sullivan Assocs. Inc.* supra note 9; *Mistry Prabhudas Manji Eng. Pvt. Ltd. v. Raytheon Engineers & Constructors, Inc.*, supra note 16; *Long Island Lighting Co. v. IMO Delaval Inc.*, 668 F.Supp. 237 (S.D.N.Y. (1987), aff'd, *Long Island Lighting Co. v. IMO Industries, Inc.*, 6 F.3d 826 (2d Cir.1993) (complex litigation culminated with this case); *Burns & Roe v. Central Maine Power Co.*, 659 F.Supp. 141 (D.Me.1987): *Markborough Cal. Inc. v. Superior Court*, 227 Cal.App.3d 705, 277 Cal.Rptr. 919 (1991) (aided by California statute): *Cregg Ministors Ventures*, supra note 8 (aided by availability of buying out of clause): *Fort Knox Self Storage, Inc. v. Western Technologies, Inc.*, supra note 9: *Ricciardi v. Frank*, supra note 13.

[20]*Dillingham v. CH2M Hill Northwest*, supra note 17. *Estey v. Mackenzie Engineering, Inc.*, supra note 11; *Garbish v. Malvern Fed. Sav, & Loan Ass'n.*, 358 Pa.Super. 282, 517 A.2d 547 (1986). A California case refused to enforce a liability limitation clause despite the client having the help of legal representation and having negotiated a change in the amount of the cap. *Viner v. Brockway*, 36 Cal. Rptr.2d 718 (Cal. App.1994). The intermediate appeals court held that the trial court was correct in not enforcing the clause despite a statute allowing such clauses if there was an opportunity to negotiate. The California Supreme Court would not grant a hearing to review this case but ordered the case not to be published. This case cannot be cited in California litigation. The court must have been dissatisfied but still unwilling to take the case.

the wide variety of contexts in which they are used, and the collision of powerful legal principles.

Opponents of enforcement claim these risks can be insured against. The added cost of insurance, if obtainable at a price justified by the revenues generated, may not be easy to pass on. Also, the insurance often includes high deductibles and limited protection. Finally, even if there is insurance coverage, the design professional will have to spend uncompensated time to defend the claim.

Those who oppose enforcing such clauses assert that enforcement will create a moral hazard by encouraging carelessness, particularly if the limitation is merely nominal. If so, the clause is actually an exculpation that relieves the design professional from tort liability. Opponents point out that design professionals are licensed by the state and perform a public service. Such an argument assumes that people who deal with these professionals usually lack the ability to assess the risks and make any meaningful decision, often the victims of adhesion contracts.[21] They see refusal to enforce as a needed form of consumer protection.

Yet the strong policies of protecting consumers and discouraging negligent performance are very powerful. They may trump freedom of contract. For example, in 1996 California enacted Business and Professions Code Section 7198 that made liability limitations invalid in contracts to inspect a home that limits liability to "the cost of the home inspection report."

Some struggles over this method of risk management are reflected in the enactment of statutes dealing with liability limitations, mainly in the context of design professional services. These statutes are in addition to the many anti-indemnification statutes found in most states. The statutes take a wide range of approaches. Wisconsin forbids eliminating tort liability in a construction contract.[22] California allows them if there is an opportunity to bargain.[23] Texas enforces them.[24] The statutes should settle the question in states that have them, but courts may inject their own ideas when interpreting the statutes, again leading to uncertainty.

Liability limitation clauses will continue to play a significant role in risk management for design professionals.

If the specified amount does not amount to complete exculpation, if the parties appear to know what they are doing and if the context is truly commercial, such clauses will very likely be enforced.

But other factors can play a role in deciding whether such a clause will be enforced. If there is a chance to buy out of the clause, there is a greater likelihood of enforcement. But suppose the ratio between the liability limitation created by the clause and the likely loss is so extraordinarily great as to constitute an almost total exculpation that can create an incentive not to perform properly. This can lead to rejection of the clause. Finally, if the clause is used in a consumer context, this seriously diminishes the likelihood of enforcement. But if the proponent of the clause makes a strong showing that important consumer needs will not be met unless such a clause is enforced, the clause has a good chance of enforcement.

E. Immunity: Decision Making

As noted in Section 14.09D, some American courts grant design professionals quasi-judicial immunity when they decide disputes under the terms of the construction contract. Many standard agreements published by the professional associations incorporate language that relieves the design professional from any responsibility if decisions are made in good faith.[25] They attempt to incorporate into the contract limited quasi-judicial immunity. To be effective, such clauses must be incorporated in the contracts both for design services and for construction services.

F. Contractual Statute of Limitations

As noted earlier, judicial claims can be lost simply by the passage of time, via statutes of limitations. Parties to a contract can include a private statute of limitations as long as it is reasonable and does not deprive one party of any viable judicial remedy. Some design professionals incorporate in their contracts language stating when the period begins and specifying the period itself.[26] Such

---

[21]See Section 5.04C.

[22]West Ann.Wis.Stat. § 895.49.

[23]West Ann.Cal.Civ.Code § 2782.5.

[24]Tex.Civ.Prac. & Rem.Code Ann. § 130.004.

---

[25]AIA Doc. B101-2007, § 3.6.2.4 grants the architect immunity for his interpretations and decisions made in good faith.

[26]AIA Docs. B141-1997, ¶1.3.7.3, and A201-1997, ¶13.7.1, sought to control commencement of the statutory period of limitations. In 2007 this was changed. Now all claims must be brought within the time specified by law but in no case more than 10 years after substantial completion. AIA Docs. B101-2007, § 8.1.1, and A201-2007, § 13.7. See Section 23.03G.

statutes of limitations can be useful as risk management tools.

G. Third-Party Claims

Third-party claims are sometimes based on the assertion that the claimant is an intended beneficiary of the contract. This contention can be negated by appropriate contract language.[27] Such language is unlikely to provide a defense to a tort claim.

H. Dispute Resolution

Some believe that the most important risk management tool is to control the process by which disputes will be resolved. Many American standardized construction contracts give first-instance dispute resolution to the design professional and frequently provide for an appeal to arbitration. Because arbitration has become such an important feature of construction contract dispute resolution, it is covered in detail in Chapter 30.

I. The Residue

Although a wish list could be created,[28] the realities of contract bargaining necessitate that emphasis be placed on those tools most useful in risk management. It makes little sense to spend precious negotiation time and bargaining power on seeking to obtain concessions that in actuality mean very little.

J. Some Suggestions

The clauses noted in this section are more likely to be enforced if they are drafted clearly and express specific reasons for their inclusion, if the client's attention is directed to them, and if suggestions are made to the client to seek legal advice if it has doubts or questions about them.

SECTION 15.04 Indemnity: Risk Shifting or Sharing

The important risk management device of risk shifting usually involves all key participants in the construction process. From the vantage point of the design professional, this tool seeks to shift any or a part of any loss he suffers related to claims by third parties—such as workers, members of the public, or owners of adjacent land—to another participant, usually the contractor, but possibly the client. See Figure 15.1. This process is discussed in Chapter 31.

The indemnification in AIA Document A201-2007 that will be discussed in greater detail in Section 31.05G stemmed from attempts by owners and architects to receive indemnification from contractors in the construction contract. AIA Document B101-2007—the standard form of agreement for design services—does not contain any indemnification provisions. Yet owners increasingly demand indemnification from their design professionals, and design professionals, where they have the bargaining power to do so, increasingly seek indemnification from the owner in the contract for design services.

In 1992, the EJCDC inserted indemnification clauses in its 1910-1, Standard Form of Agreement Between Owner and Engineer, at Paragraphs 8.7.1 and 8.7.2. Under the EJCDC approach, the owner indemnifies the engineer and the engineer indemnifies the owner. Indemnification applies only if the claim was caused solely by the negligent act or omissions of the indemnitor.

In its 1992 Guide Sheet to 1910-1, the EJCDC did not recommend that the engineer indemnify the owner as it would extend the engineer's liability "beyond that it would be at common law." But indemnity was included as strong owners demanded it. The EJCDC urged engineers not to go beyond the limited indemnity found in 1910-1. It can be found in EJCDC E-500 issued in 2002 at Paragraph 6.10. See Appendix G.

Many owner-drafted indemnity clauses expand indemnification far beyond that found in EJCDC agreements. Indemnification in construction contracts will be treated in Chapter 31.

---

[27]AIA Docs. B101-2007, § 10.5, and A201-2007, § 1.1.2, (with a limited exception for the architect), seek to bar those not parties to the contract from asserting rights as intended beneficiaries. Usually these clauses are successful. See Section 14.08B. But in *Gilbane Bldg. Co. v. Nemours Foundation*, 606 F.Supp.995 (D.Del.1985), a tort claim was allowed despite contractual negation of third-party rights. See Section 14.08E.

[28]Examples are an exculpation for consultants, a favorable choice of applicable law, waiver of a jury trial, a power to suspend work for nonpayment, insurance premiums as a reimbursable, and a stiff late payment formula, to mention a few.

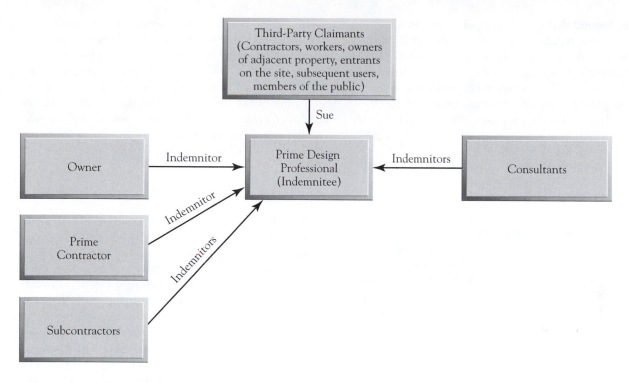

FIGURE 15.1   Indemnification to prime design professional.

SECTION 15.05 Professional Liability Insurance: Risk Spreading

A. Requirement of Professional Liability Insurance

The law does not require design professionals to carry professional liability insurance.[29] However, clients increasingly require that design professionals have and maintain professional liability insurance. Until 2007, the AIA did not specifically require insurance but stated that "expense of professional liability insurance dedicated exclusively to this Project or the expense of additional insurance coverage or limits requested by the Owner in excess of that normally carried by the Architect and the Architect's consultants" is a reimbursable.[30]

AIA Doc. B101-2007, Section 2.5, currently requires that the architect maintain general liability, automobile, workers' compensation, and professional liability insurance. This change may be in response to client demands for requiring such coverage and the hope that this would reduce the large number of architects "going bare." Architects that go bare may do so not only to avoid the cost of professional liability insurance but in the hope that lack of such insurance will mean they will not be sued. Even without insurance their nonexempt property, if substantial, will make them a target for claims. If they do not carry insurance, they run the risk of seeing those who have suffered losses caused by their negligence uncompensated. This is not a pleasant prospect for anyone, let alone professional architects.

Section 2.5 also states that "if any of the requirements set forth below exceed the types and limits the Architect normally maintains" any additional cost is reimbursable by the owner. Section 2.5 also requires that the types of limits and coverage carried by the architect be identified. This should avoid surprises when the architect requests that any coverage not normally carried be reimbursed by his client.

As noted earlier, before 2007 the cost of any insurance coverage dedicated exclusively to a specific

---

[29]This term will be used instead of the more commonly used "errors and omissions insurance."

[30]AIA Doc. B141-1997, ¶1.3.9.2.6.

project or beyond that usually carried by the architect or his consultants is reimbursable by the owner. In 2007 this issue relating to the normal insurance coverage of the *architect* is dealt with in Section 2.5. Any coverage dedicated exclusively to this project or beyond that usually carried by any *consultants* is reimbursable under Section 11.8.1.8.

To avoid any misunderstandings, the architect should provide information as the insurance coverage of his consultants as well as his own insurance coverage.

Clients may have a claim against the design professional for losses relating to the project or because they have satisfied claims of third parties that are directly traceable to the design professional's failure to perform in accordance with the legal standards. To make any claim collectible, they may require the design professional to carry professional liability insurance. If the design professional has adequate professional liability insurance, third parties injured as a result of his conduct may choose to bring legal action against the design professional directly rather than against the owner.

Even if not required to, many design professionals carry such insurance. One reason is to protect their nonexempt assets from being seized if a judgment is obtained against them. Another is that many design professionals do not wish to see people go uncompensated who suffer losses because of the design professional's failure to live up to the legal standard.

B. Volatility of Insurance Market

This section describes certain characteristics of professional liability insurance policies. Yet these characteristics, particularly coverage, exclusions, and premiums, are not static. Sometimes many entrepreneurs want to enter the insurance market. At these times, entrants are tempted by the investment of front-end premiums that can earn large returns. Payout by such entrants can be postponed to far into the future. In such a market, premiums are cut to obtain business.

However, when the investment portfolio does not generate large returns and claims begin to come in, the insurance business looks much less attractive. When this occurs, some insurers drop out of the market, and those that remain limit coverage, broaden exclusions, and raise premium rates. In addition, these insurers carefully select those for whom they will issue insurance and they defend claims more vigorously.

The cyclical nature of the insurance market means that after such a period of retrenchment and caution, entrants will again be tempted into the insurance market. Thus it is difficult to describe with any certainty the types of policies available over an extended period. Those who want to insure must obtain expert advice on the particular state of the market, coverage, and premiums. Similarly, those who impose contractual requirements for insurance must seek to determine whether any such insurance is available.

C. Regulation

All states regulate insurance, with varying details. State laws usually provide requirements of capital and financial capacity to ensure the solvency of insurers, and they increasingly regulate claims settlement practices. State insurance regulatory agencies determine which insurers will be permitted to do business in the state. In some states, regulators can determine coverage, exclusions, and premiums.

Courts have also regulated insurance, mainly through interpreting insurance policies when disputes arise. The law interprets ambiguities against the insurer unless the insurance policy is negotiated between a strong insured, such as a group of hospitals, and an insurer. Some courts seek to determine the reasonable expectations of the insured or some average insured. They will protect that expectation, some courts even disregarding insurance policy language in order to protect that expectation. Finally, there are many technical contractual requirements before the insurer's obligation to defend and indemnify matures, such as complying with warranties or representations as to activities, giving notice of claims, furnishing proof of loss, or cooperating with the insured in defending claims, to name some. Technically, a failure to comply with any of these requirements can result in loss of coverage. But the law often, though not always, protects the insured by requiring the insurer to show that the failure to comply prejudiced the insurer.

D. Premiums

Premiums have become an increasingly important overhead cost for design professionals. For example, a recent study stated that architects who carry insurance pay approximately 3 percent of their gross billings for insurance premiums and structural engineers currently pay some 6 percent of their gross billings for insurance premiums. The amount of premiums for any individual insured is determined by a number of factors, such as type of services performed, experience of the insured, locality in which work or projects are located, gross billings of the insured, contracts under which services are performed (one insurer states it reduces the premium if the contractual limitations of liability discussed in Section 15.03D are incorporated in a certain percentage of the insured's contracts), and claims record of the insured.

Premiums can rise because of underwriting predictions (selection of risks) by insurers, a decline in the value of the insurer's portfolio, a downturn in the economy that can induce insureds to cancel coverage, high claims payouts, and increased cost of defending claims.

E. Policy Types: Occurrence or Claims Made

In the construction process, a long time lag can exist between the act or omission claimed to be the basis for liability and the making of the claim. Coverage in cases depends on whether the policy is a "claims-made" or an "occurrence" policy. Claims-made policies cover only claims made during the policy period, regardless of when the act giving rise to the claim occurred. An occurrence policy, in contrast, gives coverage if the act or omission occurs during the policy period.

Professional liability insurance is currently written on a claims-made basis. This avoids the "tail" at the end of any occurrence policy. Proper insurance underwriting and rate making require the insurer to predict payouts in a designated period. Particularly in states that do not begin the statutory period for bringing claims until discovery that there is a claim and against whom it can be made, occurrence policies can result in coverage many years after the premium has been fixed. This led the professional liability insurers to base policies on claims made. It is easier to predict the value of claims that will be made during the period of the policy than to look far into the future, a prediction needed for setting premiums under an occurrence policy.

At any given time, a variety of claims-made policies can be available. Some are hybrid. Such policies specifically protect against claims made during the policy period and require that the act giving rise to the claim occur during the policy period. California rejected a challenge to such a policy, noting that the policy terms were

conspicuous and that the insured could expect only the coverage provided.[31]

Insurers usually offer retroactive, or prior acts, coverage, giving coverage regardless of when the act occurs. Retroactive coverage usually requires the insured to represent that he is unaware of any facts that could give rise to a claim or that he was insured when the act occurred and that he has carried insurance throughout his career.

Although most courts have sustained claims-made policies,[32] one case struck down a claims-made clause because its retroactive coverage applied only if the insured had carried a policy for an earlier period with the company that was insuring the insured at the time the claim was made.[33]

F. Coverage and Exclusions: Professional Services

In the absence of a special endorsement, professional liability insurance policies generally cover liability for performing normal professional services.[34] Conversely, coverage under commercial general liability (CGL) policies carried by contractors usually excludes professional services. Often, claims against the professional liability carrier are denied because the insurer asserts that the services in question were not professional.

Similarly, claims against a CGL carrier are denied because the carrier asserts that the services were professional and not covered. If one insurer pays the claim, it often brings a claim against the other carrier based upon its having paid a claim that should have been paid by the other. Examples will be seen in the cases described ahead.

---

[31]*Merrill & Seeley, Inc. v. Admiral Ins. Co.*, 225 Cal.App.3d 624, 275 Cal.Rptr. 280 (1990). In that regard, the California Insurance Code Section 11580.01 requires that an application or proposal for a claims-made policy must "recite prominently and conspicuously at the heading thereof that it is an application or proposal for a claims-made policy." Also, each policy must contain a prominent and conspicuous statement to that effect, along with language suggesting that the insured carefully study the policy and discuss coverage with his insurance agent or broker.

[32]*Zuckerman v. National Union Fire Ins.*, 100 N.J. 304, 495 A.2d 395 (1985) (describing history and analysis of claims-made policies). Similarly, see *Guidrey v. Lee Consulting Eng'g Inc.*, 945 So.2d 785 (La.App.2006) (enforcing hybrid claims made and occurrence policy). Both cases enforced clauses that required both the claimed act and claim to be within policy period.

[33]*Jones v. Continental Cas. Co.*, 123 N.J.Super. 353, 303 A.2d 91 (Ch.1973).

[34]See Annot., 83 A.L.R.3d 539 (1978).

At one time "professional" referred to the learned professions, such as medicine, law, and the clergy. Because this proved too limiting, the law began to look at persons who are licensed by the state. But since the state has gone wholesale into licensing, these days the law requires almost everyone to be licensed; so that test is useless. This classification requires that actual cases be examined.

The professional activities covered are core design services, such as preparing drawings and specifications.[35] Coverage can also encompass site services, such as monitoring the work as it proceeds for issuing payments and completion certificates.

American Motorists Insurance Co. v. Republic Insurance Co.[36] involved the question of whether the architect's preparation and submission of a competitive design–build bid is a professional service and covered under the architect's professional liability policy. The case involved a competitive design system, with the insured being the successful bidder. After the award of the bid, an unsuccessful bidder sued the insured for misrepresentations and other tortious conduct. The insured's professional liability insurer refused to defend, but the action was defended successfully by the insured's CGL insurer. The latter then sought to recover a pro rata share of its defense costs from the professional liability insurer.

The professional liability insurer argued that bid preparation and submission are merely preparation to render professional services and not the actual rendering of such services. However, the Alaska Supreme Court did not accept this contention, noting that only an architect using his specialized knowledge, labor, or skills could have prepared the bid. The bid consisted of two booklets approximating 160 pages in length and considerable detail as to the design. The court noted that the policy did not define professional services and that in such cases the term was considered ambiguous, requiring a construction that favored coverage.

Although coming under an exclusion for professional services in a CGL policy, the same issue came before the court in *Camp Dresser & McKee, Inc. v. Home Insurance Co.*[37] The original claim was made by an injured worker at a plant who attempted to throw ash onto a head pulley

of a conveyor that was missing a safety guard, thus exposing the worker's hand and arm to injury. Camp Dresser, the insured, acted as a project manager to coordinate the work and perform services similar to that of a construction manager. Camp Dresser's professional liability insurer defended the case and paid for the settlement in excess of the $150,000 deductible under the professional liability policy. But Camp Dresser also notified its CGL insurer because that policy had no deductible.

The court held in favor of Camp Dresser, noting that the fact that the services performed by Camp Dresser are usually performed by engineers or other professionals does not compel the conclusion that the contracts are those for professional services. The court held that professional services require specialized knowledge and mental rather than physical skills. Claims of ordinary negligence or negligent management and control were not expressly precluded, and the word "supervisory" in the exclusion is reasonably susceptible to ambiguous interpretation. The word can be construed narrowly, as describing supervision of purely professional activities, or broadly, as describing management or control of the aspects of a project involving professional and nonprofessional activities. Because the policy did not make clear in what sense the term was used, Camp Dresser received coverage.

These cases demonstrate that the borderline between professional and nonprofessional services can mean that there is coverage in both types of policies. Yet it can also mean that under certain circumstances, neither policy will cover. An insured who can anticipate this event and can be concerned that it will have to pay the formidable costs of defense might consider procuring business legal expense insurance.

As new actors enter the construction process, issues may arise as to whether their services are professional. For example, in a recent Louisiana case[38] a project manager (PM) was hired to act as a liaison between the owner and the prime contractor. He hired an engineer to perform engineering services because the PM was not a licensed engineer. The PM was to provide professional services required for design, engineering, construction management, and operations management for construction and operation of the plant being built. A subcontractor was negligent and damages occurred.

---

[35]*Stone v. Hartford Cas.Co.*, 470 F.Supp.2d 1088 (C.D.Cal.2006) (California law).

[36]830 P.2d 785 (Alaska 1992).

[37]30 Mass.App.Ct. 318, 568 N.E.2d 631 (1991).

---

[38]*North American Treatment Systems, Inc. v. Scottsdale Ins. Co.*, 943 So.2d 429 (La.App.2006), writ denied 949 So.2d 423, 424 (La.2007).

The owner's property damage insurer compensated the owner, and that insurer brought a subrogation claim against the PM. The PM sought a defense from its CGL insurer. The issue was whether the PM's services were professional and not covered because it was excluded by the CGL policy. The court on appeal held that the PM was performing professional services because its work took particular skill and training. Clearly, the PM should have gotten professional liability insurance.

Another court stated that professional services are those that are predominantly mental, not physical or manual. It held that drafting involved specialized intellectual labor and that the exclusion for professional services applied.[39]

More important are the long lists of exclusions incorporated in policies. Coverage excludes contractually assumed risks. One policy excluded work not customarily performed by an architect as well as activities relating to boundary surveys, subsurface conditions, ground testing, tunnels, bridges, and dams. Also excluded were failure to advise on or require insurance or surety bonds and failure to complete construction documents or to act on submittals in the time promised unless those losses were due to improper design. In addition, the policy excluded liability for guarantees, estimates of probable construction costs (currently, one leading professional liability insurer does not exclude this, thereby demonstrating the changing nature of the insurance market and liability coverage), and copyright, trademark, and patent infringements.

Coverage for activity related to the handling of hazardous materials is fluid. Initially, it was almost impossible to obtain such coverage. Currently, insurers are increasingly willing to cover pollution claims relating to alleged design negligence of a design professional. Again, consultation with a competent insurance counselor is vital in such a market.

The insured should review his contractual commitments to determine whether they are covered by his professional liability insurance policy. Some insureds submit unusual contracts to the insurer for its approval or to discuss coverage.

The usual professional liability insurance is not likely to cover many services performed by a construction manager or by those who both design and build. Special endorsements or specially tailored policies for these activities will be needed.

A complicated coverage problem can develop when several causes contribute to the loss. In some states, if one cause is covered and the other excluded, the insured receives coverage. This is demonstrated by *Comstock Insurance Co. v. Thomas A. Hanson & Associates, Inc.*,[40] in which the architect was held liable to the owner because of negligent design, which was covered, and negligent cost estimates, which was excluded under that policy. (As noted earlier, some professional liability insurers currently do not exclude this.) Under the Illinois law, which governed this transaction, if one cause is covered and the other excluded, there is coverage.

Increasingly, the contractor and specialty subcontractors participate in collaborative-designed projects. If so, these contractors should insure against claims based upon design. Usually, this risk is excluded from liability insurance coverage carried by contractors. To insure against design-based claims, the contractor should procure separate insurance or obtain a special endorsement on its liability insurance to cover design-based claims.[41]

G. Deductible Policies

Insurance companies increasingly seek to reduce their risk by excluding from coverage claims, settlements, or court awards below a specified amount. Policies that exclude smaller claims are called *deductible policies*. Generally, the higher the deductible, the lower the premium cost. A high deductible means that the insured bears substantial risks. The recent tendency to raise deductible amounts makes the insured increasingly a self-insurer for small claims. Some policies include in the deductible the cost of defense. For example, if the deductible is $5,000 and a claim of $3,000 is paid to a claimant, the insured bears any cost of defense up to the deductible amount.

Under some policies, the insured must pay all claim expenses until the deductible is reached. Other policies do not require payment until the claim is resolved.

Deductible policies can create a conflict of interest between insurer and insured. When small claims are made, the insurer may prefer to settle the claim rather than incur

---

[39]*Stone v. Hartford Cas. Co.*, supra note 35.

[40]77 Md.App. 431, 550 A.2d 731 (1988). Some states use a proximate or dominant cause analysis.

[41]*Harbor Ins. Co. v. OMNI Constr. Inc.*, 912 F.2d 1250 (D.C.Cir.1990) (exemption for professional services precluded coverage under a CGL policy despite design and construction responsibilities given to a single subcontractor).

the cost of litigation. The insured may oppose such a settlement, believing that it is an admission of negligence and that payment to the claimant will come out of the insured's pocket. One insurer gives the insured the right to veto a settlement recommended by the insurer but provides that if the insured's ultimate liability exceeds that settlement proposal, the insurer's liability will not exceed the amount of the proposed settlement and any expense costs incurred prior to the settlement. In effect, such a provision gives the insurer the right to determine settlement.

Usually the deductible amount applies to each occurrence. Where the policy was not clear in this regard, however, one court applied the deductible of $10,000 to each claim, and in an accident with eight claimants, the total deductible amounted to $80,000.[42]

H. Policy Limits

American policies generally limit insurance liability.[43] Suppose a claim for $50,000 is made and the policy limits are $100,000. The claimant offers to settle for $40,000. The insurance company exercises its right to veto settlement and refuses to settle. The claim is litigated, and the claimant recovers $125,000. The insured may contend that had the insurer settled, it would not have had to pay amounts in excess of the $100,000 policy limit. Insurance companies are generally liable for amounts over the policy limit if their refusal to settle was unreasonable in light of all the circumstances.[44]

Policies also contain an aggregate limit that applies to all claims made during the policy period. The maximum amount available to settle all claims arising out of one negligent act is usually the limit of liability per claim, but the maximum amount for all claims is the policy's aggregate limit of liability.

Some policies apply cost of defense to the aggregate limit. With the cost of defense so high, an insured who wishes to continue reasonable coverage might find it useful to increase its aggregate limit if defense costs are charged off against the aggregate limit.

---

[42]*Lamberton v. Travelers Indem. Co.*, 325 A.2d 104 (Del.Super.Ct. 1974), aff'd, 346 A.2d 167 (Del.1975).

[43]For discussion of whether the surety can be liable for more than the bond limits, see Section 32.10E.

[44]*Comunale v. Traders & General Inv. Co.*, 50 Cal.2d. 654, 328 P.2d 198 (1958); *Crisci v. Security Ins. Co.*, 66 Cal.2d 425, 426 P.2d 173, 58 Cal.Rptr. 13 (1967) (insured also recovered for her emotional distress).

I. Notice of Claim: Cooperation

Insurance policies usually state that the insured must notify the insurance company when an accident has occurred or when a claim has been made. The notice is to enable the insurer to evaluate the claim and gather evidence for a possible lawsuit.

In addition to requiring that the insured notify the insurer, policies usually require the insured to cooperate with the insurer in handling the claim. The insured must give honest statements (a false swearing clause may make any dishonest statement the basis for denying coverage). He must also make reasonable efforts to identify and locate witnesses and supply the insurer with material that may be important in defending the claim. The insured must also comply with any notice to attend hearings that have to do with the claim, such as depositions and trials.

Although insurance policies often state that coverage will be denied if there is a failure to cooperate (as stated in Section 15.05C and 15.05I), courts are not always willing to make failure of cooperation a sufficient basis to deny coverage unless the insurer can prove it was prejudiced by the lack of cooperation. Note, however, that the insured design professional has a disincentive for failure to cooperate. The design professional may be liable to the extent of the deductible and in certain unusual cases may be liable for any amount that exceeds the policy limits.

J. Duty to Defend

In most liability insurance policies, the insurer promises to defend and indemnify. One of the most difficult areas of law surrounds the question of when the insured has a duty to defend. The duty to defend is broader than the duty to indemnify, as policies usually state that the insurer will defend claims that are groundless, false, or fraudulent. Whether or not there is a duty to defend first depends on the complaint made against the insured. If coverage appears likely from the complaint, the insurer must defend. However, the complaint does not limit the duty to defend. The insured will have to defend if facts brought to its attention indicate a possibility of coverage.

Suppose a claim is made or liability determined that is less than the deductible specified in the policy. Some policies provide for the cost of defense in such a case, but if the amount paid on the claim is less than the deductible amount, defense costs are considered part of the deductible up to the deductible amount.

Suppose the policy limit is $100,000 and a claim is made for $200,000. Any recovery over $100,000 must be paid by the insured. If the insurer is willing to pay the policy limits, will the insurer be obligated to defend the claim?

Professional liability policies generally require the insurer to defend even though the latter is willing to pay the policy limit. The professional liability policy is designed to furnish the dual protection of paying the claim and defending the claim, subject to policy limits and deductibles.

Policies may differ as to claim expenses. Some provide "first-dollar defense" coverage. Under such provisions, the company will pay all claim expenses. In some of these policies, the amount of claim expenses reduces the aggregate limit of liability coverage. (As noted in Section 15.05G, some policies require the insured to pay for the claims defense costs up to the amount of any deductible.) Some policies provide that the insurer will pay 80 percent of claim expenses and the design professional 20 percent. It has been held that the cost of defense can be prorated between acts that are covered under the policy and acts that are not covered.[45] But a court held that where proration of defense costs between those claims covered and those not covered cannot be made, the insurer must pay the entire cost of defense.[46] The insured must evaluate any options available as to the cost of defense, which is often high.

In addition to the often formidable attorneys' fees involved in defending claims, there are other expenses. Exhibits must be prepared, and expert witness fees must be paid. Sometimes transcripts must be made of testimony taken before or at the trial. Bonds sometimes may have to be provided at stages of the legal action. The insured should determine whether the insurer is obligated to pay for these expenses.

Claim defense can raise questions about conflict of interest between insurer and insured. As noted in Section 15.05H, claims that can be settled within the policy limits can create a conflict of interest. The insured may wish to settle to avoid the possibility of a judgment that exceeds his insurance coverage. Similarly, as noted in Section 15.05G, a claim within any deductible can also raise conflict of interest as to settlement.

In addition, the insurer need not indemnify the insured if the trial reveals liability based on acts excluded from coverage. For example, suppose the claimant's loss could be attributable either to defective plans and specifications, which would be covered, or to an express warranty of a successful outcome, which would not. As the defense is within the control of the insurer, the insurer's attorney could use that control to obtain a judicial conclusion that placed liability on conduct not covered. (Also, the attorney is ordinarily more concerned with preserving his relationship with the insurer than with the interests of the insured.) This led a California court to conclude that if conflict of interest arises, the insurer must pay for an independent attorney for the insured.[47] Even if the insurer need not pay for the cost of an independent attorney for the insured, it may be advisable for the insured to retain independent counsel at his own expense.

K. Settlement

As mentioned in Section 15.05I, the typical professional liability policy, although granting the power to settle to the insured, places sharp restraints on that power by making the insured take certain risks if he refuses to settle when the insurer suggests he do so.

Settlement provisions typically state that the insurer will not settle without consent of the insured. If the insured fails to consent to a recommended settlement and elects to contest the claim and continue legal proceedings, the insurer's liability for the claim does not exceed the amount for which the claim would have been settled, plus claims expense incurred up to the date of such refusal. In other words, the insured will take the risk of the settlement having been a good one.

L. Multiparty Policies

In a transaction as complex as construction, with its host of participants, inefficiencies can develop if each party carries its own liability insurance. This situation has led some owners to require "wrap-up" policies for those engaged in construction and projectwide professional liability insurance for all the professionals involved in the project. Such insurance is relatively rare.

---

[45]*Insurance Co. of North America v. Forty-Eight Insulations, Inc.*, 633 F.2d 1212 (6th Cir. 1980).

[46]*National Steel Constr. Co. v. National Union Fire Ins. Co. of Pittsburgh*, 14 Wash.App. 573, 543 P.2d 642 (1975).

[47]*San Diego Federal Credit Union v. Cumis Ins. Soc'y, Inc.*, 162 Cal. App.3d 358, 208 Cal.Rptr. 494 (1984). Shortly after this decision, the California legislature enacted Civil Code Section 2860, which codifies the right to independent counsel where there is certain conflict of interest but also regulates who can be appointed as independent counsel and his fees.

M. Termination

Most insurance policies permit the insurer to terminate by giving a designated notice, often as short as thirty or forty-five days. Increasingly, legislatures and courts deny the insurer an absolute right to terminate.

SECTION 15.06 Preparing to Face Claims

Design professionals should anticipate the likelihood that claims will be made against them. With this in mind, the design professional must be able to document in the clearest and most objective way that a proper job was done. For example, expanded liability should not deter design professionals from using new designs, materials, or products. Design professionals must, however, prepare for the possibility that if things go wrong, they will be asked to explain their choice.

Using new materials as an example, the design professional should accumulate information directed toward predicting the performance of any contemplated new materials. Information should be obtained from unbiased people who have used the materials on comparable projects. A list of such people can be requested from manufacturers, whose representatives should be questioned about any bad results. The manufacturer can be notified as to intended use of the project, and asked for technical data that include limitations of the materials. Sometimes it is possible to have a manufacturer's representative present when new material is being installed to verify installation procedures. Any representations or warranties obtained should be kept readily accessible.

Design professionals should be able to reconstruct the past quickly and efficiently. A system for efficient making, storing, and retrieving of memoranda, letters, e-mails, and contracts is essential. Legal advice should determine the proper time to preserve records. If major design decisions have to be made, the design professional should indicate the advantages and disadvantages and obtain a final written approval from the client. Records should show when all communications are received and responses made. If work is to be rejected, the design professional should support his decision in communications to client and contractor. Similarly, if any previous approvals are to be withdrawn, written notice should be given to all interested parties. Records should be kept of all conferences, telephone calls, and discussions that may later need to be reconstructed in the event of a dispute.

In regard to recordkeeping, the instability of many design professional relationships can be troublesome. Design professionals dissolve partnerships frequently. Where dissolution occurs, records that should be kept are often lost or destroyed. When rearrangements occur, those involved should separate records and see that those who may need them, have them.

At all stages of their practice, design professionals need competent legal services at prices they can afford. Certainly, legal advice only after disputes have arisen is insufficient. Younger groups of design professionals should consider negotiating with those who provide legal services, for prepaid legal services plans.

Finally, the operations of the design professional, including contracts used, records kept, and compliance with laws regulating employers, should be evaluated periodically.

CHAPTER SIXTEEN

Intellectual Property: Ideas, Copyrights, Patents, and Trade Secrets

SECTION 16.01 Relevance to Design Professional

Design professionals use their training, intellect, and experience to solve design, construction, and manufacturing problems of their clients and employers. Usually design professionals reduce the proposed design solution to tangible form. These forms, whether sketches, renderings, diagrams, drawings, specifications, computer software, or models, communicate the design solution to the client or employer and others concerned. Often the design solution is followed by completion of the product being designed, such as the construction project, the industrial process, or the machine. In addition, some aspects of the design solution, such as the floor plans, sketches, diagrams, or pictures, may be used to advertise the project or the product. To sum up, the three steps are as follows:

1. the intellectual effort by which the solution is conceived
2. communication of the solution
3. development of the end product[1]

If the client owns the tangible manifestation of the design solution or the end product itself, the client may wish that the manifestation not be copied or used without permission. To preserve uniqueness of the project, whether a residence or a commercial building, the client

may not want another project constructed to an identical design. The client may feel wronged if the construction documents are used by others without payment if the client paid for and received exclusive ownership rights.

Similarly, the manufacturer who invests funds to develop a product or process may not want others to copy it without permission. The manufacturer may wish to recoup the money invested to develop the process or product, either directly, through royalties, or by retaining a competitive advantage the research investment has given.

The design professional may want to obtain similar protection. Dealings with the client were discussed in Section 12.11. Likewise, the design professional may want protection against third parties copying the construction documents, diagrams, or drawings to be used in developing a product or copying the end products themselves.

The subject of this chapter is the protection accorded those who create or hire others to create tangible manifestations of intellectual effort.

SECTION 16.02 An Overview

A. Specificity of Discussion

In this section, certain legal concepts relating to intellectual ideas will be explored in greater detail than others. For example, patents, though of great importance to engineers, will be discussed only briefly. Patent law is a highly technical area. Inventors who wish to obtain or enforce a patent require a patent lawyer. For this reason, only the basic principles and certain salient features of patent law will be mentioned.

Obtaining copyright protection, in contrast, is relatively simple. People who want copyright protection, in contrast

---

[1]For a thorough and provocative examination of the application of copyright law to architecture, see Newsam, *Architecture and Copyright— Separating the Poetic from the Prosaic,* 71 Tul.L.Rev. 1073 (1997). The author examines the design process and separates the unprotectible design ideas from the protectible expression of them. He suggests that the core of protectible design expression is what he calls the "poetic language." His approach would narrow protection but would, he says, assist the creative design process and advance design.

to legal enforcement of copyright remedies, can generally get it without the help of an attorney. For this reason, more detail will be given to copyrights than to patents.

B. Purpose of Protection

Copyrights and patents are given to authors and inventors for their writings and discoveries. The primary purpose of granting them is to foster social and industrial development for the public good. This development is accomplished by granting people monopoly rights to reward them for their contributions, rights that would otherwise be antithetical in a competitive system. One judge stated,

> The economic philosophy behind the clause empowering Congress to grant patents and copyrights is the conviction that encouragement of individual effort by personal gain is the best way to advance public welfare through the talents of authors and inventors. . . . Sacrificial days devoted to such creative activities deserve rewards commensurate with the services rendered.[2]

Yet society can suffer from excessive protection. Much intellectual and industrial progress depends on free interchange of ideas and free use of others' work. Commercial and industrial ventures can be frustrated or impeded if entrepreneurs are compelled to pay tribute to others who claim that their ideas, designs, or inventions have been used in some way by those entrepreneurs. The law seeks to reward truly creative and inventive work without unduly limiting the free flow of ideas and use of industrial and scientific technology. Patent law, for example, gives a twenty-year monopoly to the inventor of a novel, original, and not obvious invention, in exchange for disclosure to the public. This period was chosen as a compromise that adequately rewards an inventor but does not unduly perpetuate the stagnation that can accompany monopoly.

C. Exclusions from Discussion: Trademarks and Shop Rights

Creating an effective and universally recognized trademark or trade name is an intellectual act. However, design professionals are less concerned with trademarks and trade names than they are with copyrights and patents. For this

reason there will be no discussion of common law or statutory trademarks or trade names.[3]

The doctrine under which an employer under certain circumstances has shop rights—limited rights in the inventions of employees—will also not be discussed. For all practical purposes it has been preempted by near-universal use of standard form employment contracts.

SECTION 16.03 The Copyright Law of 1976[4]

A. Common Law Copyright Abolished

Section 301 of the federal Copyright Act of 1976 has preempted state laws "that are the equivalent to any of the exclusive rights within the general scope of copyright."[5] Preemption was intended to promote uniformity both by replacing state law with federal law and by preempting common law copyright, eliminating the frequently difficult question of when the work had become dedicated to the public, an act that deprived the author of a common law copyright.

Common law copyright remains only for works of authorship not fixed in a tangible medium of expression that can nevertheless be copyrighted. Such works include choreography that has never been filmed or put into notation, extemporaneous speech, original works of authorship communicated solely through conversations or live broadcasts, and dramatic sketches or musical compositions improvised or developed from memory and without being recorded or written down.

B. Statutory Copyright

Classification of Copyrightable Works. The classification of works that can be copyrighted was changed by the Copyright Law of 1976 from the close-ended thirteen to

---

[3]The Lanham Act, 15 U.S.C.A. § 1051 *et seq.*, a federal statute, permits registration of trade names, trademarks, and service marks as well provides for remedies. State law also deals with trademarks.

[4]In 1976 Congress enacted Pub.L. No. 94-553, 90 Stat. 2541, which replaced the 1909 Copyright Act and went into effect in 1978. References are to sections in the current act. The statute is also found in 17 U.S.C.A. § 101 *et seq.*

[5]17 U.S.C.A. § 301(a). Preemption has become more difficult than anticipated. For example, *Balsamol Olson Group v. Bradley Place Ltd. Partnersh.*, 950 F.Supp. 896 (C.D.Ill.1997), though citing authorities to the contrary, held that the Copyright Act preempted the Illinois Uniform Deceptive Trade Practices Act and Illinois tort law, which protects prospective economic advantage.

---

[2]*Mazer v. Stein*, 347 U.S. 201, 219 (1954).

an open-ended seven. Section 102 permits copyright of the following categories:

1. literary works
2. musical works, including any accompanying words
3. dramatic works, including any accompanying music
4. pantomimes and choreographic works
5. pictorial, graphic, and sculptural works
6. motion pictures and other audiovisual works
7. sound recordings; and
8. architectural works

Drawings or plans fall in category 5, whereas specifications fall under category 1. In that regard, literary works need not be "literary" as long as they express concepts in words, numbers, or other symbols of expression. "Architectural works" refers to copyright protection of the built structure. This category was added with passage of the Architectural Works Copyright Protection Act of 1990, discussed in Section 16.04D.

Copyright Duration: More Protection. The Copyright Act of 1909 protected the copyright holder from others reproducing, creating derivative works, and distributing copyrightable works. Protection lasted for twenty-eight years, with the right to renew for an additional twenty-eight years. Much criticism had been made of copyright duration, and longer life expectancy made it inadequate. To bring American law in line with that of most foreign countries, Section 302 of the 1976 act gives copyright protection for the life of the author plus fifty years thereafter. However, if a work was made for hire (discussed in greater detail in Section 16.04D), the duration of the copyright is seventy-five years after the year of its first publication or one hundred years from the year of its creation, whichever expires first.

In 1998, the Congress again extended copyright duration by enacting the Copyright Term Extension Act.[6] It extended copyright duration twenty years for all copyrights, existing and future. The period commences from copyright creation. As extended, the duration for individual copyrights is the life of the author plus seventy years. Works for hire are now ninety-five years from time of creation or one hundred and twenty years from time of

publication, whichever occurs first. This harmonizes with directives of the European Union issued in 1993 and puts the United States in conformity with international norms.

Codification of Fair Use Doctrine. Section 107 of the Copyright Act of 1976 expressly recognizes fair use. It permits reproduction "for purposes such as criticism, comment, news reporting, teaching (including multiple copies for classroom use), scholarship or research." Factors described as bearing on whether the use is a fair one include the following:

1. the purpose and character of the use, including whether such use is of a commercial nature or is for nonprofit educational purposes
2. the nature of the copyrighted work
3. the amount and substantiality of the portion used in relation to the copyrighted work as a whole
4. the effect of the use on the potential market for or value of the copyrighted work

Obtaining a Copyright. Although the form of copyright notice was not changed, Section 405(a) of the 1976 act gives some relief where the notice has been omitted in certain circumstances,[7] and Section 406 gives relief if the notice contains an error in name or date.

Section 411(a) requires registration with the Copyright Office before commencement of an infringement action. In addition, Section 412 precludes recovery of statutory damages or attorneys' fees if the copyrighted work is not registered within three months after first publication of the work.[8] Registration need not be difficult or expensive. Even the pre-1976 copyright regulations permitted substitutions for the original materials themselves if they were bulky or if it would be expensive to require deposit.

Section 407 stiffens the 1909 act requirement that copyrighted works be deposited with the Library of Congress. The act requires that two complete copies of the best edition be deposited, although failure to deposit will not affect the validity of the copyright. A person who fails to deposit within three months after a demand is subject to a fine of

---

[6]17 U.S.C.A § 302(a), upheld by seven to two decision in *Eldred v. Ashcroft*, 537 U.S. 186, rehearing denied, 538 U.S. 916 (2003) (within legislative power granted by U.S. Constitution).

[7]Relief from the requirement of attaching a copyright notice was denied because § 405(a) grants relief only if the notice is omitted from a relatively small number of copies. *Donald Frederick Evans & Assoc., Inc. v. Continental Homes, Inc.*, 785 F.2d 897 (11th Cir.1986).

[8]Section 412 provides authors with a practical incentive to register and also encourages potential infringers to check the federal register. *Johnson v. Jones*, 149 F.3d 494, 505 (6th Cir.1998).

not more than $250 for each work and a fine of $2,500 if refusal is wilful or persistent.

However, the 1976 act also gives authority to the Register of Copyrights to specify by regulation the nature of copies required to be deposited. These regulations can allow deposit of identifying material rather than copies. One rather than two copies can be permitted.

Remedies for Infringement. The 1976 act provides copyright authors with two broad remedies: injunctive relief (ordering the performance of a specific act) and damages.

Under the common law, an injunction is an extraordinary remedy given only in exceptional circumstances. However, Congress authorized the federal courts to grant both copyright and patent owners injunctive relief. Section 502 of the 1976 act allows a court to grant an injunction "to prevent or restrain infringement of a copyright." Under Section 503, a court may order impounding or destruction of infringing copies and articles by which infringement has been accomplished.

Copyright and patent owners traditionally have argued that, if they establish infringement, they should automatically be entitled to injunctive relief. In a patent infringement case, the United States Supreme Court rejected that view. In *eBay, Inc. v. MercExchange, L.L.C.*,[9] the Court ruled that the patent holder must prove the same four-part test of entitlement to injunctive relief as is required under the common law: (1) that the patent holder suffered an irreparable injury; (2) that a monetary award will not adequately compensate it; (3) that the balance of hardships between the plaintiff and defendant militates in favor of injunctive relief; and (4) that the public interest would not be disserved by a permanent injunction.

The *eBay* decision has been extended to copyright cases. In *Christopher Phelps & Associates, LLC v. Galloway*,[10] an owner built his house using infringed plans. The architect established the infringement and was awarded damages (the cost he would have charged to create the plans). The architect then sought a permanent injunction to prevent the owner from profiting from his infringement by either renting or selling the house within the period of plaintiff's copyright—95 years. The court refused to issue an injunction on the ground that the architect failed to prove the last two elements of the *eBay* test. With regard to the

third element (balance of hardships), an injunction would be unduly harsh because it would deprive the owner of his rights of ownership. With regard to the fourth element (the public interest), an injunction would undermine the public policy of allowing land to be productively used (including selling it). As the *Christopher Phelps* decision makes clear, an architect seeking injunctive relief faces an uphill battle after *eBay*.

Section 504 allows recovery of actual damages and profits made by the infringer attributable to the infringement. In establishing the infringer's profits, the copyright owner is required to present proof only of the gross revenue, and the infringer must prove deductible expenses and elements of profit attributable to factors other than the copyrighted work.[11] Statutory damages can now be awarded up to $30,000, with $150,000 for wilful infringement. The statutory damages for an innocent infringer can be reduced to $200.

A federal trial court opinion illustrates the difficulty of proving actual damges.[12] A developer paid an architect $13,000 (based on an hourly rate) to prepare plans and specifications for an apartment complex, which the developer built.

Without permission of the architect, the developer copied and revised the plans to produce another set of plans, from which he built another apartment complex. The architect sued the developer for copyright infringement, claiming $36,000 as his actual damages. The architect's damages claim was based on the estimated construction cost of the second apartment complex and a percentage fee of 7.5 percent.

The trial court rejected the architect's damages claim of $36,000, as the 7.5 percent was based not only on the preparation of the drawings and specifications but also on the "supervision" of the contract award and construction, which the architect did not do. The judge stated that the architect instead was entitled to the fair market value of the infringed architectural plans, which was defined as what the developer would have paid the architect and what the architect would have expected to receive as the fair market value for those plans. Plaintiffs' expert witnesses testified that the fair market value of the

---

[11]*Johnson v. Jones*, supra note 8, 149 F.3d at 506 (architect is entitled to infringing architect's entire profit of $16,500 where the defendant proved no expenses to deduct from that amount).

[12]*Aitken, Hazen, Hoffman, Miller, P.C. v. Empire Constr. Co.*, 542 F.Supp. 252 (D.Neb.1982).

---

[9]547 U.S. 388 (2006).

[10]492 F.3d 532 (4th Cir.2007).

architectural services ranged from $24,000 to $37,000. Each expert testified that he did not know that the plaintiff had received only $13,000 for the original drawings and specifications. The judge found that it would be inconceivable for the developer to pay more than the $13,000 he paid for the first set, and so concluded that the reasonable value was $13,000. The judge then reduced that amount by the cost of revising the original plans, as the architect had not incurred that expense. This reduced the actual damages to $10,000.

The architect also sought the developer's profits. The developer made a gross profit of $60,000 on the project. After deducting a portion of the project's administrative and general overhead expenses, his net profit was some $17,000.

The final judgment awarded the architect $10,000 actual damages and the profits of $17,000 earned by the developer. It is likely that this judgment came after a long and costly trial. The costs may have far exceeded the amount of the judgment. This should not be taken to mean that a copyright claim by a design professional will never produce a significant damage award. In another case, the architect recovered the fair market value of the plans ($11,968) plus the profit ($42,250) made by the party who infringed the architect's copyright in building a residence.[13] Nevertheless, a careful calculation must be made of the likely recovery, because those against whom infringement claims are brought often attack the validity of the claim. In addition, any settlement or recovery in a breach of contract action alleging unauthorized use of the architect's work product would be deducted from a copyright claim involving the same injury.[14]

Works Commissioned by U.S. Government. Some had advocated that there be no copyright in works commissioned by the U.S. government. However, Congress rejected this position and gave procuring agencies discretion to determine whether to give the design professional copyright ownership. Copyright protection will be denied only if the copyrighted work is authored by an employee of the government.

---

[13]*Eales v. Environmental Lifestyles, Inc.*, 958 F.2d 876 (9th Cir.), cert. denied sub nom. *Shotey v. Eales*, 506 U.S. 1001 (1992).

[14]*Sparaco v. Lawler, Matusky & Skelly, Engineers LLP*, 303 F.3d 460, 469 n.2 (2d Cir.2002), cert. denied, 538 U.S. 945 (2003). Upon remand, the district court ruled that the designer's settlement of his contract claim precluded the recovery of statutory damages. *Sparaco v. Lawler, Matusky & Skelly, Engineers LLP*, 313 F.Supp.2d 247 (S.D.N.Y.2004).

SECTION 16.04 Special Copyright Problems of Design Professionals

A. Attitude of Design Professionals Toward Copyright Protection

Design professionals vary in their attitude toward the importance of legal protection for their work. Some design professionals want their work to be imitated. Imitation may show professional respect and approval of work. When credit is given to the originator, imitation may also enhance the professional reputation of the person whose work is copied. Some design professionals are messianic about their design ideas and would be distressed if their work were not copied. Many design professionals believe that free exchange and use of architectural and engineering technology are essential to advancing the art of design.

Even design professionals who want imitation or who do not object to it draw some lines. Some design success is predicated on exclusivity. Copying the interior features and layout of a luxury residence or putting up an identical structure in the same neighborhood is not likely to please the architect or client. The same design professional who would want her ideas to become known and used might resent someone going to a public agency and without authorization copying construction documents required to be filed there. This same design professional is likely to be equally distressed if a contractor were to copy plans made available for the limited purpose of making a bid. Much depends on what is copied, who does the copying, and whether appropriate credit is given to the originator.

B. What Might Be Copied?

Design professionals may wish protection for ideas, sketches, schematic and design drawings, computer software, two-dimensional renderings, three-dimensional models, construction documents sufficiently detailed to enable contractors to bid and build, and the completed project itself. Ideas themselves cannot receive legal protection, and legal protection for the executed project, as noted in Section 16.04D, only recently has received limited protection. The principal problems relate to tangible manifestations of design solutions that are a step toward the project. These tangible manifestations vary considerably in the amount of time taken to create them and in the amount of time and money saved by the infringer who copies them. The tangible manifestations of the ideas in the architectural drawings must be sufficiently detailed to merit copyright protection.[15]

C. Common Law Copyright and Publication

Traditionally, design professionals sought protection through common law copyright. This may have been traceable to a lack of understanding of statutory copyright. Some design professionals believed, incorrectly, that statutory copyright protection against infringement requires the expensive registration of often formidable construction documents after they are created. Whatever the reason, the bulk of the cases involving design works involve claims of common law copyright. Common law copyright was abolished by the 1976 federal Copyright Act. For historical reasons, a brief comment on common law copyright is adequate.

The principal difficulty in perfecting a common law copyright had been the frequent claim made by the alleged infringer that the work copied had already been published. If dissemination of and the facts and circumstances surrounding the work indicated to a reasonable person that the creator had dedicated the work to the public, common law copyright was lost.

Common law copyright did not provide much protection. Design professionals should welcome its abolition. Now they are limited to statutory copyright, which, although it too has its weaknesses, is substantially better.

D. Statutory Copyright

If we look first at the beginning (creation of ideas) and end (execution of completed projects) of a design professional's services, it is clear that the former does not receive copyright protection. Before the 1976 act, the copyright holder of technical drawings had no exclusive right to complete the project. However, drawings or models for an unusual structure that could be classified as a work of art, such as the Washington Monument or the Eiffel Tower, gave the copyright holder the exclusive right "to complete, execute and finish it." On the one hand, the Copyright Office took the position that works of design professionals were copyrightable only as technical drawings and not as works of art. On the other hand, if a structure incorporated features such as artistic sculpture, carving, or pictorial representation that could be identified separately and were capable of existing independently as works of art, such features were eligible for registration. Thus, limited aspects of a building received copyright protection as a work of art. Under some circumstances, certain features of design could be sufficiently novel to justify a design patent.

Architectural Works Copyright Protection Act of 1990. Until the enactment of the Architectural Works Copyright Protection Act of 1990, the AWCPA,[16] architects were not given copyright protection of the actual three-dimensional design, that is, the completed project itself.[17] Change resulted from the desire of the Congress to bring U.S. law into line with the requirements of the Berne Convention, which provides the most comprehensive protection for copyright. Signatories to the convention agree to protect the copyright of citizens of other signatories to the convention. To bring America into line with the Berne Convention, it was necessary that U.S. law protect works of architecture and recognize the moral rights of artists, as noted in Section 16.05.

Under the AWCPA, an architectural work is the abstract three-dimensional design for a building. The law provides copyright protection for original design elements of such buildings. In addition, the new act makes clear that injunctive relief can be granted to enforce a copyright even if it means destruction of buildings.

Protection does not extend to all work designed by design professionals. It covers designs for buildings that are habitable, such as houses and office buildings.[18] It also

---

[15]Compare *Attia v. Society of the New York Hospital*, 201 F.3d 50 (2d Cir.1999), cert. denied, 531 U.S. 843 (2000) (highly preliminary design concepts do not receive copyright protection) with *Sparaco v. Lawler, Matusky & Skelly, Engineers LLP*, supra note 14 (site plan containing detailed proposed physical improvements receives copyright protection).

[16]Title VII of Pub.L. No. 101-650, 104 Stat.5089, codified at 17 U.S.C. §§ 101, 102, 106, 120, and 301. See J. DRATLER, JR., INTELLECTUAL PROPERTY LAW: COMMERCIAL, CREATIVE AND INDUSTRIAL PROPERTY, § 5.02(4) (1991). For more detailed discussion of the AWCPA, see MacMurray, *Trademarks or Copyrights: Which Intellectual Property Right Affords Its Owner the Greatest Protection of Architectural Ingenuity?* 3 Nw.J.Tech. & Intell.Prop. 111 (2005); Roberts, *There Goes My Baby: Buildings as Intellectual Property Under the Architectural Works Copyright Protection Act*, 21 Constr. Lawyer, No. 2, Spring 2001, p. 22; and Note, 2004 Utah L.Rev. 853 (2004).

[17]*Wright v. Eisle*, 86 A.D. 356, 83 N.Y.S. 887 (1903) and *Demetriades v. Kaufmann*, 680 F.Supp. 658 (S.D.N.Y. 1988).

[18]A store in a mall is not a "building" subject to copyright protection. *Yankee Candle Co., Inc. v. New England Candle Co., Inc.*, 14 F.Supp.2d 154, vacated pursuant to settlement, 29 F.Supp.2d 44 (D.Mass.1998).

includes structures not inhabited by human beings but used by them, such as churches, pergolas, gazebos, and garden pavilions. It does not cover structures designed by engineers such as interstate highway bridges, cloverleafs, canals, dams, and pedestrian walkways. The Berne Convention does not require such extended protection, and Congress determined that protection is not necessary to stimulate creativity in those fields. Extensive monuments or commemorative structures, such as the Washington Monument and Statue of Liberty, are protected if they permit entry and temporary use by people.

The protections accorded by the 1990 act have limitations. The act does not cover features of architectural works required by utilitarian function. An architectural work does not include "individual standard features." This would preclude protection for features such as common windows, doors, and other staple building components. Protection does not extend to forbidding use of pictures of buildings if the building is located in or ordinarily visible from a public place. As a result, people can use and reproduce photographs, posters, and other pictorial representations of architectural works.[19] The copyright owner cannot prevent the owner of a building embodying the work from altering or destroying the building or authorizing others to do so. Similarly, the new act ensures that local entities can exercise their police powers to enforce laws regarding landmarks, historical preservation, and zoning or building codes.

The new act applies to all architectural works created after December 1, 1990.[20] It also applies to architectural works created before that date that were unconstructed and embodied in unpublished plans or drawings on that date. However, such protection expires unless the building is constructed before December 31, 2002.

Copyright Owner: Work for Hire Doctrine. Before the 1976 act, it was assumed, in the absence of an agreement to the contrary, that the person commissioning copyrightable works was entitled to the copyright. This was one of the reasons for the frequent inclusion of clauses, in contracts between design professionals and their clients, giving ownership rights to the former.

Section 201 of the 1976 act gives copyright protection to the author of the work. However, if the work is made for hire, the employer or other person for whom the work was prepared is considered the author unless the parties have agreed otherwise in a signed written agreement. Section 101 defines a "work for hire" as prepared by an employee or a work specially ordered or commissioned. The legislative history did not include as work for hire a client commissioning a design professional.

Yet the law that has emerged since the 1976 act has made clear that the design professional who operates independently and is retained by a client to prepare the design, owns the copyright in the absence of an agreement to the contrary. The standard that applies is the common law of agency. Was the work prepared by an employee or by an independent contractor?[21] If the design professional is an independent contractor, she owns the copyright.

A number of factors are used in making this determination. The ones that seem to have emerged as most significant in the cases involving independent design professionals are whether the skills involved are beyond the capacity of a layperson; whether the client paid employee benefits such as health, unemployment, or life insurance benefits; and whether the client paid employment taxes and withheld federal and state income taxes.[22]

Ownership of Documents and Copyright: License to Use. Section 12.11 of this book discussed the ownership of drawings and specifications. There it was noted that associations of design professionals include language in their contracts reserving the ownership or re-use rights in the design professional. As seen in that section, the 2007 AIA Document B101, Section 7.3, states the architect grants the owner a nonexclusive license to use the plans and specifications to construct, use, and maintain the project. If the owner exceeds the limited use of the license by using the architect's design on other projects without

---

[19]*Leicester v. Warner Brothers*, 232 F.3d 1212 (9th Cir.2000).

[20]*Richard J. Zitz, Inc. v. Dos Santos Pereira*, 232 F.3d 290 (2d Cir.2000) (AWCPA not applicable to building substantially complete as of December 1, 1990).

[21]*Community for Creative Non-Violence v. Reid*, 490 U.S. 730 (1989).

[22]*Kunycia v. Melville Realty Co.*, 755 F.Supp. 566 (S.D.N.Y. 1990) (architect retained by client entitled to copyright, as copyrightable drawings were not made pursuant to a work for hire); M.G.B. *Homes, Inc. v. Ameron Homes, Inc.*, 903 F.2d 1486 (11th Cir.1990) (homebuilder is not the owner of a floor plan created by a draftsman, an independent contractor), appeal after remand, 30 F.3d 113 (11th Cir.1994); *Trek Leasing, Inc. v. United States*, 62 Fed.Cl. 673 (2004) (architect of design-build contractor was an employee and the architectural work was a work for hire).

permission and payment of a reuse fee, the architect may ask the court to order the owner to stop this conduct.[23]

But strong owners, particularly public owners, insist on owning the plans and specifications. In Section 12.11 it was also noted that if the contract did *not* deal with ownership or re-use rights, common law gives ownership to the person who commissioned and paid for them, unless a valid custom to the contrary can be established by the design professional.

What is the effect of copyright law on ownership rights? The copyright law gives an independent author the copyright. When federal copyright law and common law contract law clash, what is the outcome?

Clearly, contract law—at least that which deals with use of the design documents—has been preempted by federal copyright law. Yet as noted,[24] preemption issues are complicated and preemption should be invoked sparingly. Certainly Congress sought to avoid the confusion and uncertainty of state common law copyright. But it does not follow that Congress intended to preclude contractual control of re-use rights. Certainly the AIA and EJCDC do not believe so. If they did, they would not treat this topic so extensively in their standard contracts.[25]

Can we reconcile the two sources of law by dividing their functions? Copyright law bars reproducing, preparing derivative works, or distributing copies of copyrighted work.[26] That could be the ambit of copyright law. Infringement would invoke statutory remedies. Contract law could deal with *use* by others, invoking contract law remedies, including a restitution claim against a wrongful user. But this interpretation seems too artificial and confusing to implement. Use inevitably involves reproduction.

Another more promising method seeks to accommodate both sources of law. When two legitimate sources of law appear to clash, the law seeks to reconcile them. It is never assumed that one seeks to destroy the other. Clearly, an author can assign her copyright to another. In their contracts with their authors, publishers customarily require authors to assign the copyright. If the client knew that the Copyright Act gives the copyright to the design professional, it might also demand a written assignment of the

copyright as permitted by the Copyright Act in Section 204(a). The most effective way to preserve contract autonomy in the face of copyright law is to conclude that a contract that does not deal with ownership of documents gives a nonexclusive license to the owner to copy for a limited purpose, namely, execution of the design. This would also provide a legitimate reason for design professionals to include a provision preserving their ownership rights.

Since the mid-1990s, courts increasingly have grappled with the question of whether a "nonexclusive license" may be found in the absence of a writing and, if so, the scope of that license. A "nonexclusive license" is a grant by the copyright holder allowing a third person to use the copyrighted materials for a limited purpose. The nonexclusive license may be granted expressly—that is, in writing—or the grant may be implied from the circumstances of the relationship between the copyright holder and the third person.

The issue of whether a nonexclusive license was granted typically arises in the context of an architect who performed preliminary design work for an owner, only to be replaced by another architect who completes the design or administers the project's construction. May an owner who purchases the preliminary design work use those documents for further development and actual construction (in which case the owner has a nonexclusive right to the design) or must the owner compensate the original architect for the later use of the preliminary design (in which case a nonexclusive license has not been granted)?

The federal Fourth Circuit Court of Appeals in *Nelson-Salabes, Inc. v. Morningside Development, LLC*,[27] recently addressed that question as applied to architectural drawings. It stated that a key factor in determining the existence and scope of a nonexclusive license is whether the architect intended the owner to copy, distribute, or modify its work without further involvement of the architect. The court used a "totality of the circumstances" test to discern the architect's intent. It then reviewed application of that test by three other circuit courts—the Ninth Circuit in *Foad Consulting Group, Inc. v. Azzalino*,[28] the Sixth Circuit in *Johnson v. Jones*,[29] and the Seventh Circuit in *I.A.E., Inc. v. Shaver*,[30]—and summed up:

---

[23]*LGS Architects, Inc. v. Concordia Homes of Nev.*, 434 F.3d 1150 (9th Cir.2006).

[24]See supra note 5.

[25]Levin & Marshall, *Copyright Protection by Design: The B141 Document*, 21 Constr. Lawyer, No. 2, Spring 2001, p. 30.

[26]17 U.S.C.A. § 106.

---

[27]284 F.3d 505 (4th Cir.2002).

[28]270 F.3d 821 (9th Cir.2001).

[29]See supra note 8.

[30]74 F.3d 768 (7th Cir.1996).

Our analysis of these decisions thus suggests that the existence of an implied nonexclusive license in a particular situation turns on at least three factors: (1) whether the parties were engaged in a short-term discrete transaction as opposed to an ongoing relationship; (2) whether the creator utilized written contracts, such as the standard AIA contract, providing that copyrighted materials could only be used with the creator's future involvement or express permission; and (3) whether the creator's conduct during the creation or delivery of the copyrighted material indicated that use of the material without the creator's involvement or consent was permissible. In *Johnson,* the architect was retained with the understanding that he would develop the project to completion, and his contracts and conduct suggested that any use of his architectural work without his involvement or consent was impermissible. As a result, no implied license was found to exist. On the other hand, as we have pointed out, implied licenses were found to exist in *Foad Consulting* and *Shaver,* where the architects were hired for discrete tasks, with no indication of their further involvement in the project, and where they did not suggest in their proposed contracts or by their conduct that use of the copyrighted material without their involvement was impermissible.

The situation in this case, based on the record presented, falls somewhere between the situation in *Johnson,* on the one hand, and those in *Shaver* and *Foad Consulting,* on the other. Unlike the architect in *Johnson,* NSI [Ed. note: The architect] was not retained to develop plans for the entire Project and, like the architects in *Shaver* and *Foad Consulting,* NSI created the NSI Drawings pursuant to task-specific contracts. Moreover, as in *Shaver* and *Foad Consulting,* neither of those task-specific contracts contained language prohibiting future use of the architectural drawings without NSI's involvement or consent. In this instance, however, like the architect in *Johnson,* NSI and Strutt [Ed. note: The developer] plainly contemplated NSI's long-term involvement in Satyr Hill [Ed. note: The project], and they engaged in ongoing discussions for more than nine months concerning NSI's development of the Project. And during those discussions, NSI submitted contracts to Strutt that, similar to those in *Johnson,* contained the standard AIA prohibition against use of the NSI Drawings without NSI's future involvement or its express consent. Indeed, NSI never expressed to Strutt by its representations or conduct that Strutt could utilize NSI's plans without NSI's future involvement or express consent; in fact, NSI specifically advised Strutt to the contrary on at least two occasions.

Reviewed and evaluated in its totality, therefore, Strutt and NSI were engaged in an ongoing relationship that contemplated NSI's future involvement in Satyr Hill. Although the Project was performed in component parts, the facts found by the district court in its Opinion demonstrate that NSI created the NSI Drawings with the understanding that it would participate in the further development of Satyr Hill. Opinion at 17–18. Unlike the architects in *Foad Consulting* and *Shaver,* NSI eventually submitted an AIA agreement to Strutt, and it negotiated for more than nine months with Strutt concerning the work covered by the AIA agreement. In these circumstances, we agree that NSI did not intend for Strutt to utilize the NSI Drawings in the construction of Satyr Hill without NSI's future involvement in the Project or its express consent. Therefore, like the district court, we are "satisfied that these facts do not support a finding that [NSI] granted Strutt [an implied] nonexclusive license."[31]

In the *Nelson* case, no design contract was signed by the owner (Strutt); however, the court discerned the intent of the architect (NSI) by examining the AIA contract (apparently B141-1987) sent by NSI to Strutt for signature. Paragraph 6.1 of B141-1987 stated that the design documents created by the architect "are instruments of the Architect's service for use solely with respect to this Project" and that the architect is the owner of the documents and retains all rights, including copyright. Although Paragraph 6.1 did not expressly address the issue of license, it did prohibit the owner's use of the design to complete the project, if done without the architect's consent.

By contrast, B101-2007, Section 7.3, expressly addresses the issue of nonexclusive license. It states that the architect grants the owner "a nonexclusive license to use the Architect's Instruments of Service solely and exclusively for purposes of constructing, using, maintaining, altering and adding to the Project." The license remains in effect only if the owner substantially performs its contract obligations (including prompt payment of the architect) and is terminated if the architect "rightfully terminates this Agreement for cause as provided in Section 9.4."

Missing from Section 7.3, is an explanation of what would excuse the owner from continuing to be subject to the license's restrictions. That absence is conspicuous because AIA B141-1997, Paragraph 1.3.2.2, addressed

---

[31]*Nelson–Salabes, Inc. v. Morningside Development, LLC,* supra note 27, 284 F.3d at 516–17.

this question directly. Under the 1997 document, if the architect was "adjudged in default," the license would be terminated and replaced with a second license allowing the owner to use the design to complete, use, and maintain the project.

In the absence of contract language addressing this issue, an owner's rights would be determined under contract law. Suppose an owner believes the architect has breached the contract. The owner terminates the B101-2007 agreement and uses the design to complete the project. If a court later finds that the termination was justified, then the owner's use of the design to complete the project will also have been justified. But if the court finds in favor of the architect, then the owner will have violated Section 7.3 by continuing to use the design after the termination date.

What rights does the owner have under AIA B101-2007 to use the plans after project completion? A distinction must be drawn between the owner's use of the design on the same building for which the plans were made and its use of the plans to construct new buildings.

Section 7.3 allows the owner to use the design not only for the purpose of constructing the building, but also for purposes of "using, maintaining, altering and adding to the Project." This broad language should permit the owner to use the plans to modify or alter the completed project. In *Eiben v. A. Epstein & Sons, International, Inc.*,[32] the contract permitted the owner to retain the specifications "in connection with [the owner's] use and occupancy of the Project." The court ruled that this language authorized the owner, ten years after the original construction, to use the original architect's drawings on a new project that included renovating a small part of the project designed by that architect.

An owner could also argue that changes made in the 1997 and 2007 documents show that the AIA *intended* the owner to be able to use the architect's plans on future alterations to the completed project. AIA B141-1997 Paragraph 1.3.2.3 prohibited the owner from using the Instruments of Service "for future additions or alterations *to this Project*" without the architect's prior written agreement. (Emphasis added.) By deleting this language from AIA B101-2007, the AIA may have indicated it no longer intended to limit the owner's use in this manner.

As for the owner's use of the architect's design on other projects, AIA B101-2007, Section 7.3, is a nonexclusive license "solely and exclusively" for use on the project designed by the architect. Section 7.4 says that the owner may not transfer the license to any other person "without the prior written agreement of the Architect." This protects the architect from the owner of a store chain buying one design and paying for it once, but then using it multiple times. The owner would be in breach of the license if it did so without prior written agreement of the architect (presumably at least to determine a fee for reuse of her design). Finally, under Section 7.3.1, an owner who uses the design without retaining the architect who prepared it must indemnify and hold harmless the architect from any consequences.

Infringement: Sturdza v. United Arab Emirates. The owner's claim of a nonexclusive license is a defense to a copyright infringement claim. Even if the architect defeats that defense, she still must establish that the defendants—usually both the owner and the replacement architect—infringed her copyright. The following case examines proof of a copyright infringement claim involving design documents.

STURDZA v. UNITED ARAB EMIRATES

United States Court of Appeals, District of Columbia Circuit, 2002. 350 U.S.App.D.C. 154, 281 F.3d 1287
[Ed. note: Some footnotes omitted and footnotes renumbered.]

TATEL, Circuit Judge
[Ed. note: In 1993, the United Arab Emirates (UAE) held a competition for the architectural design of a new embassy in Washington D.C. The UAE informed the competitors that the building should express the "richness and variety of traditional

Arab motifs." Plaintiff Sturdza and defendant Demetriou both submitted designs. The UAE informed plaintiff that she had won.

Over the next two years, plaintiff and the UAE exchanged eight contract proposals. Although plaintiff agreed to the final proposal, the UAE ceased all communications with her. In 1997, plaintiff learned that the UAE had hired Demetriou to design the embassy. Plaintiff sued the UAE and Demetriou for

---

[32]57 F.Supp.2d 607 (N.D.Ill.1999).

copyright infringement. The district court granted summary judgment in favor of the defendants and the plaintiff appealed.]

* * *

III

To prevail on a copyright claim, a plaintiff must prove both ownership of a valid copyright and that the defendant copied original or "protectible" aspects of the copyrighted work. *Feist Publ'ns, Inc. v. Rural Tel. Serv. Co.*, 499 U.S. 340, 348, 361, 111 S.Ct. 1282, 1289, 1296, 113 L.Ed.2d 358 (1991). "Not all copying, however, is copyright infringement." *Id.* at 361, 111 S.Ct. at 1296. The plaintiff must show not only that the defendant actually copied the plaintiff's work, but also that the defendant's work is "substantially similar" to protectible elements of the plaintiff's work. *See generally* 4 MELVILLE B. NIMMER & DAVID NIMMER, NIMMER ON COPYRIGHT § 13.01 [B], at 13-8 to 13-10 (2001) (explaining that while few courts clearly differentiate between actual copying and substantial similarity, both are clearly required, and the latter inquiry concerns whether actual copying is illegally actionable). In their motions for summary judgment, the UAE and Demetriou disputed neither Sturdza's ownership of a valid copyright nor that Demetriou actually copied Sturdza's design. Instead they argued, as they do here, that Sturdza cannot prove substantial similarity.

The substantial similarity inquiry consists of two steps. The first requires identifying which aspects of the artist's work, if any, are protectible by copyright. "[N]o author may copyright facts or ideas. The copyright is limited to those aspects of the work—termed 'expression'—that display the stamp of the author's originality." *Feist Publ'ns*, 499 U.S. at 350, 111 S.Ct. at 1289 (internal quotation marks and citation omitted) (alteration in original); see also 17 U.S.C. § 102(b) ("In no case does copyright protection for an original work of authorship extend to any idea, procedure, process, system, method of operation, concept, principle, or discovery. . . . ."). Using Shakespeare as an example, Judge Learned Hand explained the distinction between protectible expression and unprotectible ideas:

> If Twelfth Night were copyrighted, it is quite possible that a second comer might so closely imitate Sir Toby Belch or Malvolio as to infringe, but it would not be enough that for one of his characters he cast a riotous knight who kept wassail to the discomfort of the household, or a vain and foppish steward who became amorous of his mistress. These would be no more than Shakespeare's "ideas" in the play, as little capable of monopoly

as Einstein's Doctrine of Relativity, or Darwin's theory of the Origin of Species.

Nichols v. Universal Pictures Corp., 45 F.2d 119, 121 (2d Cir.1930); *see also, e.g., Country Kids 'N City Slicks, Inc. v. Sheen*, 77 F.3d 1280, 1286 (10th Cir.1996) (holding that "wooden form of the traditional paper doll" is idea not expression). "[N]o principle," Judge Hand said, "can be stated as to when an imitator has gone beyond copying the 'idea,' and has borrowed its 'expression.' Decisions must therefore inevitably be ad hoc." *Peter Pan Fabrics, Inc. v. Martin Weiner Corp.*, 274 F.2d 487, 489 (2d Cir.1960).

Also relevant to this case, copyright protection does not extend to what are known as *scènes à faire*, i.e., "incidents, characters or settings which are as a practical matter indispensable, or at least standard, in the treatment of a given topic," *Atari, Inc. v. North Am. Philips Consumer Elecs. Corp.*, 672 F.2d 607, 616 (7th Cir.1982) (internal quotation marks and citation omitted), or elements that are "dictated by external factors such as particular business practices," *Computer Mgmt. Assistance Co. v. Robert F. DeCastro, Inc.*, 220 F.3d 396, 401 (5th Cir.2000). For example, because "[f]oot chases and the morale problems of policemen, not to mention the familiar figure of the Irish cop, are venerable and often-recurring themes of police fiction[,] . . . they are not copyrightable." *Walker v. Time Life Films, Inc.*, 784 F.2d 44, 50 (2d Cir.1986).

* * *

Once unprotectible elements such as ideas and *scènes à faire* are excluded, the next step of the inquiry involves determining whether the allegedly infringing work is "substantially similar" to protectible elements of the artist's work. "Substantial similarity" exists where "the accused work is so similar to the plaintiff's work that an ordinary reasonable person would conclude that the defendant unlawfully appropriated the plaintiff's protectible expression by taking material of substance and value." *Country Kids*, 77 F.3d at 1288 (internal quotation marks and citation omitted). Substantial similarity turns on the perception of the "ordinary reasonable person" or "ordinary observer," *id.* As the Second Circuit explained:

> [t]he plaintiff's legally protected interest is . . . his interest in the potential financial returns from his [work] which derive from the lay public's approbation of his efforts. The question, therefore, is whether defendant took from plaintiff's works so much of what is pleasing to the . . . lay . . . audience . . . that defendant wrongfully appropriated something which belongs to the plaintiff.

Arnstein v. Porter, 154 F.2d 464, 473 (2d Cir.1946).

The substantial similarity determination requires comparison not only of the two works' individual elements in isolation, but also of their "overall look and feel." *Biosson,* 273 F.3d at 272 (internal quotation marks and citation omitted).

* * *

Finally, and of particular importance to this case, "[b]ecause substantial similarity is customarily an extremely close question of fact, summary judgment has traditionally been frowned upon in copyright litigation." A.A. *Hoehling v. Universal City Studios, Inc.,* 618 F.2d 972, 977 (2d Cir.1980). . . .

In assessing whether Sturdza's claim of substantial similarity presents a genuine issue of material fact, the district court first eliminated from consideration those elements of Demetriou's 1997 design that were present in his 1993 competition submission. These include the building's overall volume, backyard garden, and atrium. *See Sturdza,* No. 98-2051, slip op. at 11 (Oct. 30, 2000) (listing elements). The district court excluded these features because Sturdza's amended complaint alleges that Demetriou copied her design after preparing his original, 1993 competition entry. Because Sturdza does not challenge this aspect of the summary judgment decision, we too will exclude these elements of Demetriou's 1997 design from our consideration.

The district court next "filter[ed] out" those elements of Sturdza's design it viewed as unprotectible ideas: "domes, wind-towers, parapets, arches, and Islamic patterns." *Id.* at 6, 12. According to the district court, Sturdza's expression of these elements, but not her use of them, is protectible. We agree with this aspect of the district court's decision. In and of themselves, domes, wind-towers, parapets, and arches represent ideas, not expression. . . . Indeed, to hold otherwise would render basic architectural elements unavailable to architects generally, thus running afoul of the very purpose of the idea/expression distinction: promoting incentives for authors to produce original work while protecting society's interest in the free flow of ideas. [Ed. note: Citation omitted.] We also agree that "Islamic" patterns are not protectible, though we would characterize them as *scènes à faire* dictated by the UAE's desire that the building "express[] the richness and variety of traditional Arab motifs." Particular shapes such as diamonds or circles that comprise a given pattern, however, do constitute ideas.

Proceeding item by item, the district court then meticulously compared how the concepts of domes, wind-towers, parapets, arches, and decorative patterns (referred to by the district court as "Islamic" patterns) are expressed in the two designs. "[A]t the level of protectable expression," the district court concluded,

"the designs are decidedly different." *Sturdza,* No. 98-2051, slip op. at 13 (Oct. 30, 2000).

Here we part company with the district court. Although we agree that Demetriou's design differs from Sturdza's, we think the district court overlooked several important respects in which Demetriou's design expresses particular architectural concepts quite similarly to Sturdza's. We also see significant similarities in the "overall look and feel" of the two designs. To help explain these two points, we attach as appendices to this opinion selected "elevations," i.e., views, of Sturdza's and Demetriou's designs. Appendices A and B show front and side elevations of Sturdza's design. Appendices C and D show front and side elevations of Demetriou's 1997 design.

We begin with the ways in which Demetriou's expression of architectural concepts mirrors Sturdza's. Consider the domes. Although we agree that Demetriou's dome differs from Sturdza's in some respects—Demetriou's is opaque and positioned toward the front of the building, while Sturdza's rises directly over the building's central section and is made of "glass[,] . . . allowing light in through the pattern," Pl.'s Suppl. Answer to Def. Demetriou's Interrogs. at 4—in other respects Demetriou's dome appears quite similar. Viewed from the front, both domes appear to rise from the center and toward the front of the buildings. Both domes rise to essentially the same height, correspond in width to the buildings' midsections, and taper gently upward to a point. Although the domes have different decorative patterns, the patterns create a similar effect. Sturdza encircles her dome with three bands of pointed arches: largest at the dome's base and becoming progressively smaller toward its top. Her arches' decreasing size and pointed shape create a feeling of upward movement from the dome's base toward its top. Demetriou creates a similar effect by covering his dome with diamonds whose upper points correspond to Sturdza's pointed arches and that (like Sturdza's arches) become progressively smaller toward the top of the dome. Finally, Sturdza gives her dome a ribbed effect by raising the edges of the arches above the dome's surface; Demetriou creates a similar effect by accenting his diamonds' edges.

Like the domes, Demetriou's wind-towers differ in some respect from Sturdza's: Sturdza's are three-dimensional, emerge from the building's roof, and are decorated with diamond patterning; Demetriou's are essentially two-dimensional extensions of the building's front facade and are decorated with three vertical bands. Viewed from the front, however, the wind-towers appear quite similar in terms of size and placement. Indeed, because the wind-towers are essentially the same height and width and rise on either side of the domes, they create extremely similar building contours. Moreover, by placing diamonds atop the three vertical bands, Demetriou creates a decorative effect similar to Sturdza's.

[Ed. note: The court then applies its analysis to the two architects' use of parapets and arches.]

The final concept—decorative patterning that covers the facades of the two buildings—is the idea Demetriou expresses most differently. As the district court pointed out, Demetriou's patterning has sixteen-sided stars on the upper levels and shapes somewhere between a circle and a diamond on the ground level; Sturdza's has diamond shapes throughout. Demetriou also uses significantly less patterning overall than Sturdza, who covers the entire facade of her building with decoration. Even with these differences, however, we see a significant similarity: like Sturdza, Demetriou covers his building's facade with a grid of diamonds that creates a diamond motif and emphasizes the facade's division into horizontal and vertical planes.

Moving on to our second basis for questioning the district court's conclusion that the designs are "decidedly different," *id.* at 13, we see no indication that, in addition to comparing the ways in which the two architects express individual concepts, the district court considered the two buildings' "overall look and feel." *Biosson*, 273 F.3d at 272. Examining the two designs ourselves, we are struck by the significant extent to which Demetriou's design resembles Sturdza's. The size, shape, and placement of Demetriou's wind-towers, parapets, and pointed domes, when viewed from the front, give his building a contour virtually identical to Sturdza's. Contributing to the similarity in overall look and feel, both buildings have a pyramid-like clustering of pointed arches around the front entrances, prominent horizontal bands and vertical

columns demarcating the windows, slightly protruding midsections, diamond grids, and similar latticework patterning inside the arches. Finally, Demetriou achieves the "Islamic" effect sought by the UAE by expressing and combining his wind-towers, arches, dome, parapet, and decorative patterning in ways quite similar to Sturdza's expression and combination of these elements.[33]

To sum up, we think Demetriou's design, though different in some ways from Sturdza's (as the district court thought), is sufficiently similar with respect to both individual elements and overall look and feel for a reasonable jury to conclude that the two are substantially similar. Unless the jury "set out to detect the disparities" between the two works, it might well "be disposed to overlook them, and regard their aesthetic appeal as the same." *Biosson*, 273 F.3d at 272 (internal quotation marks and citation omitted). Because Sturdza's copyright claim presents an extremely close question, and because "summary judgment has traditionally been frowned upon in copyright litigation, *A.A. Hoehling*, 618 F.2d at 977, we will reverse the grant of summary judgment.

At the beginning of its analysis, the *Sturdza* court noted that Demetriou and the UAE did not dispute either Sturdza's ownership of a valid copyright or that Demetriou actually copied Sturdza's design. Those elements of a copyright claim can, of course, provide the defendant with significant defenses to the plaintiff's claim. For example, in *LZT/Filliung Partnership, LLP v. Cody/Braun & Assocs., Inc.*,[34] the district court rejected the original architects' copyright claim against the successor architect on the same project. The court concluded that the similarity in the two architects' designs was explained by the owners' extensive input into both designs.

E. Advice to Design Professionals

If copyright protection is to be sought, design professionals should comply with the statutory copyright requirements. First, they should be certain they have not assigned to others their right to copyright ownership. There is still utility, as suggested in Section 16.04D, in including a contract clause giving ownership rights to the plans and specifications to the design professional.

Second, design professionals should comply with the copyright notice requirements. The word *Copyright* can be written out, or the notice can be communicated by abbreviation or symbol. The authorized abbreviation is "Copr.," and the authorized symbol is the letter "C" enclosed within a circle (©). The year of first publication

should be given, and the name of the copyright owner or an abbreviation by which the name can be recognized or generally known can be used. The notice must be affixed to the copies in such a manner and location as to give reasonable notice of the copyright claim. The Register of

---

[33][Ed. note: Other federal courts also state that proof of improper copying extends not only to particular details in the two designs, but also to the "overall form" or "look and feel" of the two designs. See *T-Peg, Inc. v. Vermont Timber Works, Inc.*, 459 F.3d 97 (1st Cir.2006) ("overall form" standard) and *Shine v. Childs*, 382 F.Supp.2d 602 (S.D.N.Y.2005) ("total concept and feel" standard).]

[34]117 F.Supp.2d 745 (N.D.Ill.2000.

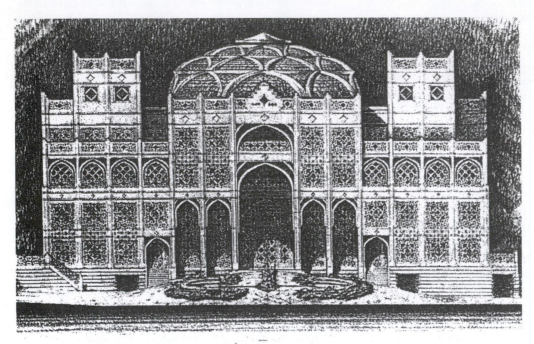

APPENDIX A Elena Sturdza

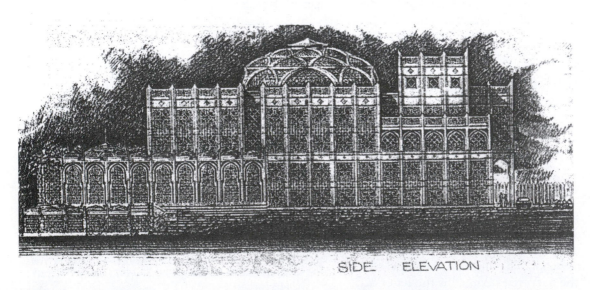

APPENDIX B Elena Sturdza

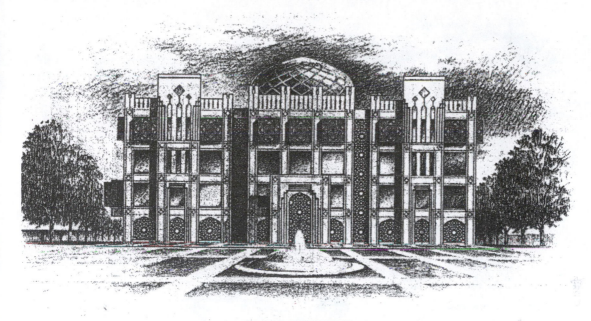

FRONT ELEVATION

APPENDIX C   Angelos Demetriou (1997)

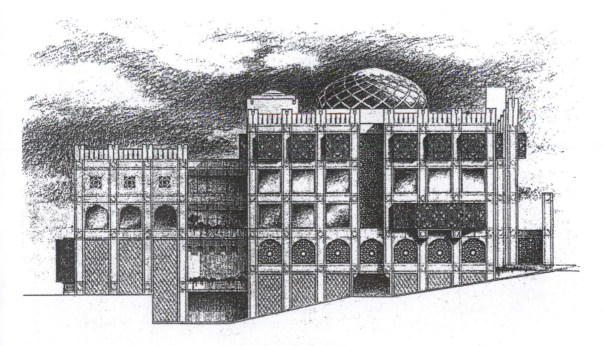

SIDE ELEVATION

APPENDIX D   Angelos Demetriou (1997)

APPENDIX E   Angelos Demetriou (1993)

Copyrights is given authority to specify methods by which copyright notice can be given.[35]

If the design professional wishes to take advantage of the statutory damage award and to recover attorneys' fees, she should register the copyrighted work within three months of publication. Methods are available to minimize this burden. The Copyright Office should be consulted.

SECTION 16.05 Moral Rights of Artists

The moral rights of artists were recognized by the enactment of the Visual Artist Rights Act of 1990, which was also designed to bring America into line with the Berne Convention.[36] Two kinds of rights are protected: the right of attribution and the right of integrity. Works protected are limited to graphic, sculptural, and photographic works.

Although generally this topic is beyond the scope of this book, one aspect should be mentioned.

The right of integrity consists of the artist's right to prevent certain distortions, mutilations, or other modifications of her work as well as the right to prevent destruction of work of recognized stature. This protection relates only to acts that will be prejudicial to the author's honor or reputation.

The act has a number of limitations, one of which relates to visual art incorporated in buildings. If an artist has consented to the incorporation of her visual art in a building, the owner of the building can remove the art even if it cannot be removed without destroying it or otherwise violating the artist's right of integrity. Consent given after June 1, 1991, must be evidenced by a written instrument signed by the artist and the owner and must specifically give the owner the right to remove even if removal destroys the work of art.

If the incorporated art can be removed without destruction, the building owner may still avoid infringing on the artist's moral rights by notifying the artist and giving her a chance to remove the art and pay for its removal within ninety days of receiving notice. The owner who is unable

---

[35]As indicated, 17 U.S.C.A. §§ 405 and 406 provide relief if the copyright notice is omitted or erroneously made. See Section 16.03B.

[36]17 U.S.C.A. §§ 101, 106A, 113(d). See DRATLER, supra note 16, at Section 6.01(6).

to notify the artist despite a good-faith attempt to do so can proceed to remove the incorporated visual art. The Register of Copyrights provides a method of recording names and addresses of artists and permits building owners to record their efforts to locate an artist. For the interface between statutes granting artists moral rights and modification of the design, see Section 21.04C.

SECTION 16.06 Patents: Some Observations and Comparisons

A. Scope of Coverage

A design professional who wishes to institute legal action for infringement of a patent or a copyright should retain an attorney. Although the steps for perfecting a copyright are simple, an inventor who wishes to obtain a patent must secure the services of a patent attorney. A patent attorney is needed to guide the inventor through the maze of patent law and the complexities of a patent search. As a rule, perfection of a copyright will not require the services of an attorney.

B. Patent and Copyright Compared

The subjects of patents generally are products, machines, processes, and designs.[37] Copyright generally protects writings.

The principal protection accorded by copyright law is the exclusive right to reproduce, prepare derivative works, or distribute the copyrighted material. Copyright law does not protect against someone who, without knowledge of the copyrighted work or access to it, creates a similar work. A patent gives the patent holder a monopoly. The patent holder can exclude anyone from the field covered by the patent even if the same invention has been developed independently and without any knowledge of the patented device.

The most important difference between patent and copyright law is the higher degree of creativity required for issuance of a patent. Copyright law requires only that the work have some originality and be the independent labor of the author. Patent law requires that the work

be original, inventive, useful, novel, and not obviously derived from the prior art in the particular field.

Protecting an invention begins with the issuance of a patent, which is supposed to presumptively establish that the patent is valid. In the bulk of patent infringement cases, the defendant attacks the validity of the patent. The defendant has a good chance of establishing that the patent is not valid.

As the United States economy has shifted over the past century from heavy manufacturing to information processing, a debate has emerged over the need to reform the Patent Act of 1952.[38] Critics are concerned with the ease with which patent claims may be filed in an information-based technology, in which a single "product" may include numerous patentable processes or inventions. Another criticism is that patent owners (in particular of software processes) do not use their patents to create a product, but bring an infringement action only when someone uses that process to make and market a product. In addition, statistics appear to support the view that increased infringement claims and escalating damages awards threaten to stifle industry innovation.[39]

Patent infringement suits are lengthy, complicated, and expensive. The plaintiff can recover damages and, in some cases, a compulsory royalty for patent infringement. In some flagrant infringement cases, the plaintiff can recover treble damages and attorneys' fees.

A patent's duration is twenty years from the date the patent application is filed. This is substantially less than copyright protection. The limited protection accorded copyright and the monopoly protection accorded a patent are probably why the patent protection is more limited in duration.

Generally, patent protection is harder to acquire than copyright but once acquired is worth much more.

SECTION 16.07 Trade Secrets

A. Definition

The First Restatement of Torts has defined a trade secret as

> any formula, pattern, device, or compilation of information which is used in one's business, and which gives him an opportunity to obtain an advantage over competitors who

---

[37]For an instructional design patent case in an architectural context, see *Blumcraft of Pittsburgh v. Citizens & Southern Nat'l Bank of South Carolina*, 407 F.2d 557 (4th Cir.1969).

[38]35 U.S.C.A. §§ 1–376 (2000).
[39]"Industries Brace for Tough Battle Over Patent Law," *Wall Street Journal*, June 6, 2007, at A1.

do not know or use it. It may be a formula for a chemical compound, a process of manufacturing, treating or preserving materials, a pattern for a machine or other device. A trade secret is a process or device for continuous use in the operation of the business. Generally it relates to the production of goods, as, for example, a machine or formula for the production of an article.

* * *

The subject matter of a trade secret must be secret. Matters of public knowledge or of general knowledge in an industry cannot be appropriated by one as his secret. Matters which are completely disclosed by the goods which one markets cannot be his secret. Substantially, a trade secret is known only in the particular business in which it is used. It is not requisite that only the proprietor of the business know it. He may, without losing his protection, communicate it to employees involved in its use. He may likewise communicate it to others pledged to secrecy.

The Restatement sets forth the following factors that are considered in determining whether particular information is a trade secret:

1. the extent to which the information is known outside of the proprietor's business
2. the extent to which it is known by employees and others involved in the proprietor's business
3. the extent of measures taken by the proprietor to guard the secrecy of the information
4. the value of the information to the proprietor and her competitors
5. the amount of effort or money expended by the proprietor in developing the information
6. the ease or difficulty with which the information could be properly acquired or duplicated by others[40]

B. Context of Trade Secret Litigation

Trade secret litigation can arise when an employee leaves an employer either to go into business or to work for a new, and frequently competing, employer. If the former employee has commercial or technical information, the prior employer may seek a court decree ordering the former employee not to

---

[40]Restatement (First) of Torts § 757 comment b. The Second Restatement of Torts omitted this topic. Although less authoritative, the First Restatement still provides a useful summary.

disclose any trade secrets "belonging" to the prior employer and a decree ordering the new employer not to use the secret information. Such a court order can be justified by a confidential relationship between the prior employer and the former employee or the breach of an employment contract between the prior employer and the former employee.

Trade secret litigation can result when the proprietor of a trade secret learns that someone to whom a trade secret has been disclosed on a basis of confidentiality intends to make or has made unauthorized use of the information. For example, the developer of a new product may give technical information relating to the product, to the contractor building the plant in which the product is to be manufactured, or to the manufacturer who is to build the machinery needed to make the product.

Similarly, a confidential disclosure of the information may be made to a manufacturer by an inventor who seeks to interest the manufacturer in a process or product developed by the inventor. The unauthorized use of such a precontract disclosure can lead to trade secret litigation.

Developers of technology sometimes try to recover research costs by licensing others to use the data. To protect the secrecy of the technology and to enable them to sell the data to others, developers usually obtain a promise from the licensee not to disclose the data to anyone else. Breach or a threatened breach of such a nondisclosure promise may cause the proprietor of the trade secret to seek a court decree forbidding any unauthorized use or disclosure.

C. Contrast to Patents: Disclosure vs. Secrecy

Patent law requires public disclosure of the process, design, or product that is the subject of a patent. In exchange for this public disclosure, the patent holder obtains a twenty-year monopoly. Trade secret protection, in contrast, requires that the data asserted to be a trade secret be kept relatively private and nonpublic.

A patent requires an invention to be original, inventive, useful, and not obvious from the prior art. The trade secret need not meet these formidable requirements. Although the courts are not unanimous on the point, it seems clear that the person who asserts ownership of a trade secret must show she has made some advance on what is generally known. If the information is generally known or generally available, the information is not a trade secret.

Roughly, trade secret protection has the same relationship to patent protection that common law copyright had to statutory copyright. Both the doctrine of common law copyright and the doctrine of trade secrets are predicated on extending legal protection to creative people by giving them the right to determine when, how, and if the fruits of their intellectual labor should be made generally available. Patent and statutory copyright are predicated on disclosure. Trade secret protection is accorded by state law and suffers from the same lack of uniformity that common law copyright had. In contrast, patent law is governed by federal law, resulting in general uniformity throughout the United States.

D. Adjusting Competing Social Values

The doctrine of trade secrets, like many other legal doctrines, must consider and adjust various desirable, yet often antithetical, objectives. This can be shown by examining these objectives from the viewpoints of the various persons affected.

Those who seek trade secret protection—primarily inventors and research-oriented organizations—want to be rewarded economically for their creativity. Restricting others from using the information and technology that they have developed can make their information more valuable. Without adequate economic incentives, scientific and industrial progress is likely to be impeded. Protection of trade secrets can discourage industrial espionage and corruption.

Trade secret protection can restrain the freedom of choice and action for research employees. Creative employees can be prevented from making the best economic use of their talents. An employer's failure to consider or develop an employee's research ideas can destroy the employee's creativity. Many high-tech industries developed when creative people banded together to start new companies. Had they been tied to older, established companies unwilling to engage in experimental research, many of these industries might not have developed or might have taken considerably longer to do so.

Overzealous protection of trade secrets can hamper commercial, scientific, and industrial progress. To a great extent, such progress is made possible by free dissemination of technical and scientific information. Disseminating such information can avoid costly duplication of research efforts.

Expansive protection of trade secrets can also have an anticompetitive effect. Protection of trade secrets can give the developer a virtual monopoly that can hinder the development of competitive products and can result in higher prices to consumers.

Trade secret law has had to consider and adjust all these competing objectives—not an easy task.

E. Availability of Legal Protection

Duty Not to Disclose or Use: Confidential Relationship and Contract. The circumstances surrounding the disclosure and the nature of the information disclosed are relevant in determining whether a duty exists not to use or disclose the information. If the disclosure is accompanied by an express promise not to use or disclose, a general duty exists not to disclose. It may still be necessary to interpret the agreement to determine what cannot be disclosed, to whom disclosure is prohibited, and the duration of the restraint. Section 10.8 of AIA Document B103-2007 states that if the architect or owner receives information designated as "confidential" or "business proprietary," the information will be kept strictly confidential. B103 is the owner/architect agreement for large or complex projects. Confidentiality is also discussed in Section 11.04B.

Suppose a licensing agreement exists by which the licensor permits the licensee to use technological data disclosed by the licensor to the licensee. Does the restraint on disclosure include information that the licensee knew before the disclosure? Does it include information developed by the licensee from the disclosed information? Does the restraint include parts of the technological data disclosed that are known at the time of disclosure or become generally known? Can disclosure be made to an affiliated or successor company? Is there a continuing obligation for either or both parties to communicate new technology? These questions should be and usually are covered in the licensing agreement. If not, courts must interpret the agreement and, if necessary, imply terms.

In some circumstances, no express provision prohibits unauthorized use or disclosure. The method by which the information is acquired often determines whether the disclosure is made in confidence and whether the person to whom it is made obligates herself not to disclose the information to others. This is similar to the process by which the law implies certain promises between contracting parties not expressed in the written contract. The communication may be part of a contractual arrangement. For example, the possessor of the information may communicate it to a consulting engineer who has been retained to advise the possessor on the type of machinery to be used in the process. If a written contract exists, the possessor will usually require a promise by the consulting engineer not to divulge certain specified information. Even without such an express promise not to disclose, the law would probably imply such a promise, based on surrounding facts and circumstances. The same is true if the disclosure is made to the manufacturer of the machine or to a building contractor.

In some circumstances, no contractual relationship exists between the possessor of the information and the person to whom it is disclosed. For example, an inventor may disclose information to a manufacturer in order to interest the manufacturer in buying the information. It is possible for the inventor to obtain a promise from the manufacturer not to disclose the information. Even without such a promise, if it is apparent from the surrounding facts and circumstances that the disclosure is made in confidence, any disclosure of the information by the manufacturer would be a breach of confidence and remedies would be available to the inventor.

Nature of Information.   Courts look at the type or nature of the information itself to help decide whether the information is a trade secret. A plaintiff who invested considerable time and money to create the information and then keep it secret from competitors is more likely to receive legal protection.

While trade secret litigation in the construction industry is rare, the recent case of *San Jose Construction, Inc. v. S.B.C.C., Inc.*,[41] presented an issue unique to the industry. The plaintiff and defendant, both prime contractors, were local competitors in the business of commercial construction. Mr. Foust, a project manager for the plaintiff, left that company to work for the defendant. Foust secretly copied and took with him the bidding documents on five prospective projects on which the plaintiff was ready to begin work if awarded the contracts.

Plaintiff sued the defendant and Foust, claiming its bidding information was a trade secret. One defense raised by the defendants was that the information had already been revealed to third parties, such as the project owners and subcontractors. The court rejected that defense, stating:

> We can readily infer . . . that the information contained in [plaintiff's] project binders, viewed as a whole, derived economic value from being kept secret from competitors such as [defendant]. As [plaintiff] describes it, "only [plaintiff] had the completed puzzle for each project, contained in the Project Binders. . . . No third party had it. The subcontractors each had a piece, and the owners had a piece, but no one except [plaintiff] had it all."[42]

With regard to the effort expended by plaintiff to create the bidding documents, the court accepted plaintiff's testimony that it can take from six months to over a year to create such information, at a cost of tens of thousands of dollars.

If the person to whom the information is disclosed—whether a contractor hired to build a plant, a manufacturer hired to build a machine, or a consulting engineer hired to furnish technical services—knows that the information is not generally known in the industry, this is likely to persuade a court that a confidential relationship was created or that a nondisclosure promise should be implied.

The nature of the information will also determine the legal remedy for a breach of confidence or a breach of contract. Under U.S. law, the normal remedy for a breach of contract is a judgment for money damages. Only if that remedy is inadequate will the law specifically order that a defendant do or not do something. This is crucial in trade secret cases. Typically, if the information is truly valuable and not generally known, the most important remedy is the court decree ordering the person who has the information to not disclose it to anyone else. Violating such an order is punishable by a fine, or even imprisonment, under the contempt powers of the court. Such a decree puts the plaintiff in a good position to demand a substantial royalty or settlement price if the defendant needs to use the trade secret information.

To obtain such an extraordinary remedy, the plaintiff must show that irreparable injury would occur without such a court order and, as mentioned, that a judgment for money damages would be inadequate. In a trade secret case, the plaintiff seeks to show that irreparable economic harm would be suffered if the information claimed to be a trade secret is broadly disseminated. The plaintiff usually asserts that such broad disclosure will enable competitors to "catch up" despite the plaintiff's research expenditure to turn out a better product or develop a better process. The plaintiff will also claim that it is difficult, if not impossible, to establish

---

[41]155 Cal.App.4th 1528, 67 Cal.Rptr.3d 54 (2007).

[42]Id., 67 Cal.Rptr.3d at 63.

the actual damages suffered by general dissemination of the secret information. A court concluding that the information is a trade secret usually gives injunctive relief.

The principal defense in trade secret cases is that the information was not secret. Often defendants point to the existing literature in a given scientific or technical area, with a view toward showing that a person diligently searching for this information could put it together and arrive at the process independently. Courts have not been particularly receptive to this defense. Usually the defendant has not gone through the literature to ferret out the secret. The information is often obtained from an employee of the trade secret possessor, paid for its disclosure by virtue of a licensing agreement, or received through a confidential, limited disclosure. On the whole, the defense has not succeeded. Part of the difficulty in arguing for this defense is that sometimes information and data are available, but not in a collected, organized, convenient, and usable form. These factors are the principal advantages of the trade secret. Sometimes the data are collected and organized in readily accessible form, but most people in the industry are unaware of this fact or unable to locate the material easily.

Employee Cases. There are special aspects to cases in which the information has been learned or developed by an employee and that employee goes into business herself, joins in a venture with others, or is hired by an existing or potential competitor of the prior employer. In addition to using the confidentiality theory, the former employer often points to an employment contract under which the employee agreed not to disclose the information after leaving the employment. Sometimes the limitation on disclosure far exceeds what is reasonable. Often the employee has little bargaining power in deciding whether to sign such an agreement. Some courts have recognized the adhesive (nonbargaining) nature of such agreements and have refused to give these agreements literal effect. However, the employer who has an agreement by the employee not to divulge information is in a better position to obtain a court decree ordering the employee not to disclose particular information.[43]

On the one hand, in addition to recognizing the take-it-or-leave-it nature of most employment contracts, some courts feel that agreements under which employees cannot practice their trade or profession or use the information that is their principal means of advancement are unduly oppressive to employees. Such courts are not likely to sympathize with claims for trade secret protection. On the other hand, other courts manifest great concern with immorality and disloyalty on the part of employees and look on employee attempts to cash in on information of this type as morally indefensible. These courts are likely to deal harshly with employees in trade secret cases.

To sum up, a former employee will be restrained from using confidential information if that restraint is reasonable, taking into account the legitimate needs of the former employer, the former employee, and the public.

F. Scope of Remedy

A trade secret claimant can recover damages suffered, profits made by the infringer resulting from the infringement, and a court decree prohibiting her from using or divulging the information. The injunctive relief usually does not exceed the protection needed by the plaintiff. An injunction may be only for a period of time commensurate with the advantage gained through the technological information improperly acquired or used. If the defendant could have ascertained the information within a designated period, the court decree may require that she not use the information for that period of time. Unless the defendant has made the information public, the court order for nondisclosure will apply only until the information is generally known. Some courts take a more punitive attitude and order that the trade secret not be used even if it becomes generally known. Generally, the more reprehensible the conduct by the defendant, the broader the injunction.

G. Duration of Protection

The trade secret is protectible as long as it is kept relatively secret. This unlimited time protection has caused some to advocate protecting trade secrets for a limited period of time by according a patentlike monopoly to the developer of a trade secret.

Some trade secrets are patentable. Unlimited duration of protection for a trade secret can frustrate the twenty-year patent monopoly policy. To the extent that states frustrate patent law by protecting trade secrets, such trade secret protection may be unconstitutional.

---

[43]In *San Jose Constr., Inc. v. S.B.C.C., Inc.*, supra note 41, the court pointed to the plaintiff's use of confidentiality agreements (which Mr. Foust had signed) to show that the employer made reasonable efforts to keep its information secret.

H. Advice to Design Professionals

Design professionals who invent processes, designs, or products should, wherever possible, use contracts to give them protection against the possibility that people to whom they divulge the information may disclose the information to others or use it themselves.

Design professionals who occupy managerial positions in companies where trade secrets are important should use all methods possible to keep the information secret. Only those who have an absolute need to use the information should be given access to it, and these people should expressly agree in writing not to disclose the information. Management should also realize that employee loyalty is probably the best protection against the loss of trade secrets. Reasonable treatment of employees is likely to be a better method of preserving trade secrets than litigation.

Design professionals who are technical employees and who wish to take their technological information to start their own businesses, join in a business venture, or work for a competitor of their present employers should realize that their departure under these circumstances may result in litigation, or at least the threat of litigation. Legal advice should be sought to examine the legality of any asserted restraints and to determine the scope of risk involved to the employee who chooses to leave her present employment.

CHAPTER SEVENTEEN

Planning the Project: Compensation and Organization Variations

SECTION 17.01 Overview

A. Some Attributes of the Construction Industry

Industry characteristics are central to this chapter, so please review Chapter 8, with special attention to Sections 8.01 and 8.02.

B. Owner's Objectives

In the design phase, quality, price, and completion date are interrelated. An owner who wishes the highest quality may have to make tradeoffs among quantity, price, and completion date. If early completion is crucial, the owner may need to sacrifice price and probably quality and quantity.

After these choices have been made, the owner may have to make additional choices when it decides how to select a contractor, which contractor or contractors to select, and the type of construction contract or contracts. These choices should be made in a way that maximizes the likelihood that the owner will receive quality that complies with the contract documents and on-time completion at the lowest ultimate cost. Compromises may be needed. The contractor who will do the highest-quality work is not likely to be cheapest and quickest. The quickest contractor may not be the one who will provide the best quality. The importance the owner attaches to these objectives will affect the process the owner uses to select the contractors and how construction contracts are organized. This chapter looks at pricing and organizational variations. Chapter 18 focuses on competitive bidding.

C. Blending Business and Legal Judgments

This book examines law in the context of the construction industry. A sharp line cannot always be drawn between business and legal considerations.

Clearly, choices as to the pricing of a construction contract and how the project is to be organized must seek to achieve the owner's objectives. Choices made in these crucial matters should seek to obtain the best on-time work at the best price. To a significant degree, this will depend on each participant's knowing what it is supposed to do and being able to do so in the most efficient way. Modern methods designed to bring efficiency to what can be a chaotic process are described in Section 17.04. Although Chapter 8 describes some characteristics of the construction industry, one characteristic—operational inefficiency—is relevant to this chapter. Section 17.05 looks at internal efficiency—mainly authority and communication.

An empirical study dealt with "the pervasive and distressing inefficiency" in construction work. One writer stated that the study indicated that only 32 percent of the total time spent on a construction site involved actual work on the project.[1] He quoted the study as showing that the remainder of the work was divided in the following manner:

1. 7 percent for equipment transportation delays
2. 13 percent for traveling on the job site
3. 29 percent consumed by waiting delays

---

[1]Foster, *Construction Management and Design-Build/Fast Track Construction: A Solution Which Uncovers a Problem for the Surety,* 46 Law and Contemp.Probs. 95, 116 n. 119 (1983).

4. 8 percent for late starts and early quits
5. 6 percent for receiving instructions
6. 5 percent for personnel breaks

Quoting these statistics, the writer concluded that much inefficiency is caused by the many contractual relationships and parties all working on the same structure, each "under a different management and each marching to the beat of a different drummer."[2]

Choices of the type described in this chapter that deal with these matters are to a great extent best made by people who know design and construction. What role does the law play? The answer can be divided into two categories: direct and indirect legal controls.

Direct legal controls involve such matters as registration and licensing laws, legal controls dealing with how a construction contract is awarded, the standard to which a contracting party is held, and the way in which the law will treat claims by a party who has suffered losses that it seeks to transfer to someone else.

Indirect legal considerations are often as important. For example, where lines of authority are blurred and risk allocations unclear, the law will frequently be called on to pass on claims. Similarly, inefficiency and other matters that cause losses are likely to lead to claims that, although usually settled, are done so against the backdrop of what the law would provide if the dispute ended up in court.

It is increasingly being recognized that one overhead cost in construction work that is incurred by all participants is the cost of making, avoiding, preparing for, and resolving claims. Choices of the type described in this chapter must be made carefully and intelligently. Failure to do so can only increase the cost of construction.

D. Public vs. Private Projects

An important criterion is the status of the owner—whether it is a private party or a public entity. A private party who wishes to build can choose any type of compensation plan it can persuade a contractor to accept. It can award a contract in any way it chooses. It can make one contract with a prime contractor or a number of separate contracts with individual contractors.

Public entities may be limited by laws and regulations in making these choices. Public entities commonly must award their construction contracts by competitive bidding

under which contractors are all given a chance to submit a bid for particular work. That design professionals for public projects do not—at least at the first round—compete on the basis of price[3] causes problems when a public entity wishes to use the design–build system, as noted in Section 17.04F.

Some states and cities require separate contracts for certain types of public work. Limitations are often placed on cost contracts used in public works. A public entity may be limited in the way it resolves disputes with its contractors. Early in this century, public entities frequently took the position that they could not arbitrate disputes, as this would delegate power to private arbitrators. Increasingly, however, public entities are being required, by law or regulation, to use arbitration or some other method of resolving disputes.

People working on public contracts must first examine the applicable statutes and regulations to determine what restraints are placed on them in awarding or organizing construction contracts. This chapter largely assumes that there are no restraints and that the owner and contractor can make any type of contract they wish and the owner is not limited how it chooses to organize participants contractually and administratively.[4]

SECTION 17.02   Pricing Variations

Selecting a compensation system must take into account the responsibility for certain risks. Although many variations are possible, each major category deals with risk allocation.

A. Fixed-Price or Lump-Sum Contracts: Some Variations

In American usage, fixed-price and lump-sum contracts are used interchangeably. Under such contracts, the contractor agrees to do the work for a fixed price. Almost all performance risks—events that make performance more costly than planned—fall on the contractor. For example, when the construction industry experienced a sudden and sharp increase in the price of steel, after many years of price stability, contractors who had entered into fixed-price

---

[2]Ibid.

[3]See Section 11.03.

[4]For an instructional case study on planning for construction of a hospital, see Macomber, *You Can Manage Construction Risks*, Harv.Bus. Rev., March–April 1989, p. 155.

contracts were unable to pass these unanticipated costs on to the owner.[5]

Only if the contract itself provides a mechanism for increasing the contract price can the contractor receive more than the contract price. Although not common in ordinary American contracts (English and international contracts often use fluctuations clauses), some contracts have price escalation clauses under which the contract price is adjusted upward (and sometimes downward), depending on market or actual costs of labor, equipment, or materials. Such a provision protects the contractor from the risks of any unusual costs that play a major part in its performance. The provision usually requires the contractor to use its best efforts to obtain the best prices.

Many construction contracts contain changed-conditions clauses that allow a price increase if conditions under the ground or in existing structures are discovered that are substantially different from those anticipated by the parties.

Sophisticated procurement systems sometimes provide variant fixed-price formulas even in fixed-price contracts. For example, one is the federal fixed-price incentive firm (or FPIF) contract used in negotiated contracts. Before contract finalization, owner and contractor negotiate the following items:

1. target cost estimate—against which to measure final costs
2. target profit—a reasonable profit for the work at target cost
3. ceiling price—the total dollar amount for which the owner will be liable
4. sharing formula—the arrangement for establishing final profit and price

After the work is completed, the contractor and the owner negotiate the final costs of the contract, sharing the overruns or underruns according to the agreed formula.

This form of contract has the advantage of establishing a price ceiling similar to the guaranteed cost to the owner under the fixed-price contract. By penalizing the contractor for cost overruns above the target estimate (which is always something less than the price ceiling) and by

rewarding it for cost savings, this type of contract provides financial motivation to the contractor to perform at the most economical cost.

A contract of this type is appropriate where the owner's plans are not sufficiently detailed to allow fixed-price bidding without excessive provision for contingencies yet are sufficiently advanced that a reasonably accurate target estimate can be made. Such a contract can also, if desired, include monetary incentive provisions to the contractor for early completion.

A fixed-price contract has the obvious advantage of letting the owner and those providing funds for the project know in advance what the project will cost. It works best when clear and complete plans and specifications are drawn. Incomplete contract documents are likely to cause interpretation questions that can lead to cost increases that under a "cost plus overhead and profit changes" clause can convert what appears to be a fixed-price contract into a cost contract. The fixed-price contract is used most efficiently when a reasonable number of experienced contractors are willing to bid for the work. This is less likely where the design is experimental or where work is abundant.

Another advantage to the fixed-price contract is that the owner need not be particularly concerned with the contractor's recordkeeping. If changed work or extra work is priced on a cost basis, there may have to be some inquiry into the contractor's cost. On the whole, the fixed-price contract avoids excessive owner concern with the contractor's cost records. Conversely, such a contract is attractive to the contractor, which need not expose its cost records, something that occurs in a cost contract, as noted in the next subsection.

The fixed-price contract has come under severe attack for its inherent adversarial nature. A contractor that reduces costs increases profits. As long as the cost is not reduced at the expense of the owner's right to receive performance specified in the construction contract, the owner cannot object.

In construction, however, the performance required and whether such performance has been rendered are often difficult to establish. In this gray area of compliance, the interests of owner and contractor can clash. Unless the contractor seeks a reputation for quality work or values goodwill, it is likely to perform no more than the contract demands. This has led to some of the variations described in Section 17.04, particularly the cost contract with a guaranteed maximum price (GMP), sometimes called a guaranteed maximum cost (GMC).

[5]See *Holder Constr. Group v. Georgia Tech Facilities, Inc.*, 282 Ga.App. 796, 640 S.E.2d 296 (2006), reconsideration denied, Dec. 8, 2006 (*force majeure* clause inapplicable); *Spindler Constr. Corp.*, ASBCA No. 55007, 06-2 BCA ¶33,376 (subcontract not commercially impracticable to perform); Guidry, *The Steel Price Explosion: What Is an Owner or a Contractor to Do?* 24 Constr. Lawyer, No. 5, Summer 2004, p. 5.

Another disadvantage to the fixed-price contract is that the risk of almost all performance cost increases falls on the contractor. A prudent contractor will price these risks and include them in its bid. However, in a highly competitive industry the prudent contractor may not receive the award because others may be more willing to gamble with a low price and either hope that problems will not develop or recoup any losses by asserting claims for extras and delays. Even if a prudent contractor does take these risks into account and does receive the award, if the risks do not materialize the owner may be paying more than if the risk had been taken out of the contractor's bid.

B. Cost Contracts

When prospective contractors cannot be relatively certain of what they will be expected to perform, or where they are uncertain as to the techniques needed to accomplish contractual requirements, they are likely to prefer to contract on a cost basis. For either reason, projects that involve experimental design, new materials, work at an unusual site, or in which the design has not been thoroughly worked out, are likely to be made on a cost basis.[6]

Usually a cost contract allows the contractor to be paid costs plus an additional amount for overhead and profit. This arrangement should be distinguished from what is sometimes called a "time and materials" contract, which at least in one case was held to preclude recovery by the contractor of overhead on direct labor costs.[7]

The cost contract has two principal disadvantages. First, at the time it engages the contractor the owner does not know what the work will cost. Second, as a general rule a cost contract does not give enough incentive for the contractor to reduce costs.[8] These factors have led, in

both public and private contracting systems, to variants on a pure cost contract.

Cost contracts often contain provisions that require the contractor to use its best efforts to perform the work at the lowest reasonable cost. Provisions are often included requiring the contractor who has reason to believe that the cost will overrun any projected costs to notify the owner or its representative and give a revised estimate of the total cost. Sometimes these provisions state that failure to give notice of prospective cost overruns will bar recovery of any amounts higher than any cost estimates given.

Another method of keeping costs down is to include provisions in the contract stating that a fiduciary relationship has been created between owner and contractor that requires that each use its best efforts to accomplish the objectives of the other and to disclose any relevant information to the other.

AIA Document A102-2007 (formerly A111-1997) is intended for a transaction where compensation is to be cost plus a fixed fee with a Guaranteed Maximum Price (GMP). Since the cost is, to a large degree, within the control of the contractor, the AIA sought to create a fiduciary relationship between the owner and the contractor; the contractor obligating itself to protect the owner's interest. Article 3 states in part that "the Contractor accepts the relationship of trust and confidence" and will seek to do the work "in an expeditious and economical manner consistent with the interests of the Owner."[9]

Yet this attempt by the AIA in A111 (now A102) to create a fiduciary obligation was not successful in *Eastover Ridge, LLC v. Metric Constructors, Inc.*[10] While the case did not deal with cost control, the usual issue in cost-type contracts, its analysis shows that the fiduciary concept can become blurred when the owner uses an independent architect.

The owner sought recovery against the contractor under the North Carolina Unfair and Deceptive Trade

---

[6]See Rosenfeld & Geltner, *Cost-Plus and Incentive Contracting: Some False Benefits and Inherent Drawbacks*, 9 Constr. Mgmt. & Econ. 481 (1991) (argues that at a macro level, widespread use of cost-plus contracts on the average and over time contributes to adverse selection, blunts incentives for production efficiency, leads to higher costs and prices in the industry, and helps mediocre contractors win jobs against efficient ones).

[7]*Colvin v. United States*, 549 F.2d 1338 (9th Cir.1977).

[8]In *Smith v. Preston Gates Ellis, LLP*, 135 Wash.App. 859, 147 P.3d 600 (2006), review denied, 161 Wash.2d 1011, 166 P.3d 1217 (2007), an owner who entered into a cost plus contract to build his dream home had such disastrous results that he (unsuccessfully) sued his lawyer for malpractice for allowing him to enter into the contract.

[9]In 2007, the AIA also published A103, a cost plus fixed fee without a GMP (formerly A114-2001). The language in A103 Article 3 is identical to that in A102 Article 3.

[10]139 N.C.App. 360, 533 S.E.2d 827, review denied, 353 N.C. 262, 546 S.E.2d 93 (2000). But see *Jones v. J.H. Hiser Constr. Co., Inc.*, 60 Md.App. 671, 484 A.2d 302 (1984), cert. denied, 303 Md. 114, 492 A.2d 616 (1985) (contractor in cost contract promises to inform owner of prospective overruns). See also Holloway, *Pitfalls in Cost-plus Contracts*, ABA Forum on the Construction Industry Newsletter, March 2003.

Practices Act. That statute allows treble damages. To use the statute as the basis for recovery, the owner must show constructive fraud based on trust and confidence, the usual standards in a fiduciary relationship. Constructive fraud, unlike actual fraud, does not require specific fraudulent representations. It only requires a transaction where the contractor took advantage of its position of trust in a confidential relationship.

The parties used A111 with the trust and confidence language noted previously. An employee of the contractor testified that it would look after the owner's interest.

Despite this, the court refused to find a relationship of trust and confidence. It pointed to the fact that Article 3.1 (now A102-2007, Article 3) was never discussed. More important, it directed attention to the parties having used A201: the AIA General Conditions that is to be used with A111. A201, as shown in Section 12.08, gives the architect administrative powers and responsibilities during construction that in part are to protect the owner. The architect's close involvement, according to the court, meant that the contractor did not owe the owner the obligation of trust and confidence. Therefore, there was no constructive fraud and no violation of the Deceptive Trade Practices Act.

To be sure, the breach of a cost-type contract should not automatically constitute a violation of the Trade Practices Act with its treble damages provision. Yet the AIA sought to create a fiduciary relationship in a cost-type contract mainly as a cost-control device. Its use of A201 as the general conditions with its architect involvement did not change the clear language of A111, Article 3.1.

The role of the architect as described in A201 should not eliminate the trust and confidence spelled out in A102-2007, Article 3. Were this a case involving an abuse of the contractor's power to determine costs, it is likely the court would have concluded that there has been an abuse of the relationship of trust and confidence created by A102.

Even without specific provisions designed to protect the owner from excessive costs, the design professional and the contractor should keep the owner informed of costs based either on the covenant of good faith and fair dealing noted in Section 19.02D, or on an obligation inherent in the fiduciary relationship created by such a contract.[11]

One type of cost contract gives the contractor cost plus a percentage of cost for overhead and profit. Obviously, such a contract not only creates little incentive to cut costs but also grants a reward for increasing costs. For this reason, it is not used in federal procurement. However, sometimes it is used to price changed work in private contracts. It provides a readily accepted guideline for determining the percentage in contrast to the less readily definable negotiated fixed fee—its alternative for compensating overhead and profit. With a contractor of the highest integrity, this type of cost contract is useful.

Another type is the "cost plus a fixed fee" contract. The parties agree that the contractor will be reimbursed for allowable costs and paid a fee that is fixed at the time the contract is made. The fee is normally not affected when actual cost exceeds or is less than the estimated cost. However, if the scope of the work is substantially changed, sometimes the fee is renegotiated. Because the contractor's fee is not affected by cost savings, the contractor has no compensation incentive to reduce costs. For this reason, in federal procurement this type of contract has largely been superseded by cost contracts that create incentives to reduce costs.

An incentive contract took a disastrous twist for the contractor in *Koppers Co. v. Inland Steel Co.*[12] The contractor agreed to design, procure materials for, and construct an industrial plant for the owner. At the time of the award, the contractor estimated that the project would cost $267 million. Because of the likely number and scope of changes, a fixed price was not feasible. As a result, the contract contained targeted costs and an adjustment provision for agreed changes. The contract also provided a bonus under which the contractor would receive 50 percent of any cost underrun up to $6.3 million. In the event of an overrun, however, the contractor would refund up to $4.8 million.

The project cost nearly $444 million. The owner brought suit, charging that the contractor should be liable for much of the excess cost. Although there had been numerous design changes, the owner alleged that the actual cost far exceeded what should have resulted from these changes, because the contractor failed to use reasonable care in design and construction.

The contractor contended that the incentive bonus/penalty provision was the owner's sole remedy for cost overruns, regardless of why or how they occurred. The issue was

---

[11]*Williams Eng'g, Inc. v. Goodyear, Inc.*, 496 So.2d 1012 (La.1986) (engineer responsible for cost overrun because he did not update estimates, did not advise of other types of contracts, and did not hire a cost estimator).

[12]498 N.E.2d 1247 (Ind.App.1986), transfer denied Nov. 25, 1987.

whether a distinction should be drawn between increased costs attributable to errors and omissions by the contractor and those that occurred for other reasons. The court held that only clear language can deprive the contracting party of normal rights accorded by law. The incentive bonus/penalty provision was not intended, according to the court, to supplant the owner's remedies for any failure by the contractor to use due diligence in performance. As a result, the owner was awarded some $64 million in damages from the contractor.

Another method of providing cost reduction incentives is value engineering, discussed in the next subsection.

These methods of seeking to keep costs down, although sometimes successful, still do not accomplish the objective of letting the owner or anyone supplying funds for the project know that the costs will not exceed the particular designated amount. To deal with this problem, owners sometimes insist that the contractor give a guaranteed maximum price (GMP) or that an "upset" price be included in the contract. These techniques are designed to give some assurance that the project will not cost more than a designated amount. These should be differentiated from any cost estimates given by the contractor, although there is always a risk that any cost figures discussed will end up being a GMP.

Construction managers are frequently asked to give a GMP if they engage the specialty trade contractors or perform some of the work with their own forces. For that reason, discussion of a GMP is postponed until Section 17.04D. However, a GMP may not be worth much if the design is quite incomplete at the time the GMP is given. If costs exceed the GMP, the contractor is likely to claim that the scope of the work has so changed that the GMP no longer applies.[13]

The owner has additional administrative costs in a cost contract. Usually the design professional will seek a higher fee than for a fixed-price contract because many more changes are made in a cost contract as the work progresses. The design professional may have additional responsibilities for checking on the amount of costs incurred by the contractor and for ensuring that the costs claimed actually went into the project and were required under the contract. Determining costs involves not only the often exasperating problems of cost accounting but also the creation of record management and management techniques for determining just what costs have been incurred.

Innumerable variations of allowable costs exist. Usually no question arises on certain items, such as material, labor, rental of equipment, transportation, and items of the contractor's overhead directly related to the project. However, sometimes disputes arise over such matters as whether the cost of visits to the project by the contractor's administrative officials, the cost of supervisory personnel employed by the contractor, and the preparatory expenses or delay claims by subcontractors, are allowable costs.

The drafter must try to anticipate all types of costs that can relate directly or indirectly to the project. A determination should be made as to which will be allowable costs for the purposes of the contract. Some troublesome areas can be highlighted by comparing Articles 7 and 8 of AIA Document A102, which deal with cost-plus arrangements. Article 7 lists reimbursable costs, and Article 8 specifies certain costs that are not to be reimbursed. EJCDC C-525, an agreement form for a cost-type contract, also illustrates cost issues. The years of experience of federal procurement have generated complicated allowable cost rules, yet problems still arise in this troublesome area.

C. Value Engineering

Owners are always seeking methods of reducing costs in both fixed-price and cost contracts. Value engineering, a method developed by the federal procurement system, attempts to provide an incentive to the contractor to analyze each contract item or task to ensure that its essential function is provided at the lowest overall lifetime cost. The federal method states that an owner who accepts a value engineering change proposal initiated and developed by the contractor grants the contractor a share in any decrease in the cost of performing the contract and in any reduced costs of ownership.[14]

Although it is difficult to quarrel with the concept and objectives of value engineering, the actual operation of the system has generated a significant amount of litigation. Two federal procurement cases are instructive. The first, *John J. Kirlin, Inc. v. United States*,[15] involved a contract for renovating the Pentagon's heating, ventilating, and air-conditioning system. The contractor was required,

---

[13]*C. Norman Peterson Co. v. Container Corp. of America*, 172 Cal. App.3d 628, 218 Cal.Rptr. 592 (1985).

[14]48 CFR § 52.248-1 (2007)
[15]827 F.2d 1538 (Fed.Cir.1987).

among other things, to replace certain minimum dampers but was not required to replace the "maximum" outside air dampers.

One month after the contract had been awarded, the contractor submitted a proposal suggesting that the maximum outside air dampers be replaced. The contractor asserted this would reduce the government's annual energy costs by $1.6 million. This proposal was rejected, the contracting officer believing that the estimate of energy savings was not valid.

Nevertheless, about three months after completion of the contract, the agency awarded another contract to a different contractor under which the maximum outside dampers were to be replaced. The original contractor then asserted it was entitled to share in the energy and maintenance cost savings the agency would realize. The U.S. Claims Court denied recovery, concluding that the value engineering proposal must be for the purpose of changing any requirement of the contract and does not deal with work beyond the scope of the contract that could not be accomplished by change order. Because the contractor could not be ordered to perform such a drastic change, the court rejected the claim.

The U.S. Circuit Court of Appeals agreed that value engineering does not cover cardinal changes, ones beyond the scope of the project. Yet it did not find necessary to determine whether this would have been a cardinal change. It concluded that the original contractor's claim required that its proposal be accepted. The court rejected the contractor's contention that the proposal had been constructively accepted when the government agency decided to do the work as the original contractor had suggested after the original contract had been completed. The original contractor had no right to savings based on a subsequent contract to which this contractor was not a party. The court would not follow a decision by the Armed Services Board of Contract Appeals (ASBCA), which had granted recovery on the theory of an implied contract, in effect a conclusion that denying the claim would create unjust enrichment.[16]

The second case, *ICSD Corp. v. United States*,[17] involved a contract to supply night vision field sights. The value engineering clause provided that the contractor would receive 50 percent of the government "contract"

savings, that is, those realized on future purchases of essentially the same item as that to be acquired under the contract. The contractor would receive only 20 percent of the savings, however, for cost reductions considered "collateral," that is, reductions of operation costs and government-furnished property costs. Under ICSD's contract for night vision gunsights, batteries, although a major component of the sights, were not to be manufactured or delivered to the army.

ICSD submitted a value engineering change proposal (VECP) to substitute alkaline batteries for the mercury batteries specified. The contracting officer determined that the savings were over $1 million per year and awarded ICSD a share of the savings. However, the contracting officer concluded that these savings were collateral to the contract and that the amount owed ICSD was 20 percent, not the 50 percent it had sought. Also, the contracting officer made no award for savings based on the increased safety of alkaline batteries, reduced disposal costs, reduced logistics costs, and the elimination of the need for a cold-weather adapter.

While ICSD's proposal was under consideration, another government contractor submitted a similar VECP. The government evaluated both proposals and decided that ICSD's was superior in three areas, while the other contractor's was superior in one. This led the contracting officer to split the savings share award, giving 75 percent to ICSD and 25 percent to the other contractor.

The court held that to be considered contract savings, there must be a reduction in the cost of the item or essentially the same item as that to be acquired under the contract. But because batteries were not acquired under the ICSD contract, the proper award was 20 percent of the government's savings. As to the other collateral savings not recognized by the government, the court found that the savings were not measurable, documentable, or ascertainable and that ICSD's estimates were neither reasonable nor credible. Finally, the court held that splitting savings awards between contractors is permissible.

The court rejected the argument that granting only 20 percent thwarts the policy of the incentive clause, noting that although the 50 percent award would provide greater incentives, less incentive to make proposals is needed where the reduced cost does not relate to items furnished under the contract.

In 1990, the Federal Acquisition Regulations (FAR), which regulate federal procurement, adopted value

---

[16]*Alan Scott Indus.*, ASBCA No. 24729, 82-1 BCA ¶15,494.
[17]934 F.2d 313 (Fed.Cir.1991).

engineering principles and procedures for architect/engineer contracts. When authorized by the contracting officer, such studies will be made after completion of 35 percent of the design stage or such other stages as the contracting officer determines. However, unlike value engineering for contractors, the government and the contractor do not share savings of costs. Instead, the design professional provides a fee breakdown schedule for the services relating to value engineering activities, and when approved, these services are compensated.[18]

D. Unit Pricing

One risk of a fixed-price contract relates to the number of work units to be performed. This risk can be removed by unit pricing. The contractor is paid a designated amount for each work unit performed.

A number of factors must be taken into account in planning unit pricing. First, the unit should be clearly described. The cost of a unit should be capable of accurate estimation. Best unit pricing involves repetitive work in which the contractor has achieved skill in cost predicting. Second, it must be clearly specified whether the unit prices include preparatory work such as cost of mobilizing and demobilizing apparatus needed to perform the particular work unit.

Third, it is important to decide whether the invitation to bidders will include an "upset" or maximum price or whether the invitation will include a minimum price. The reason for an upset price is obvious. However, the reason for a minimum price involves understanding the use of an unbalanced bid. Under such a bid, the contractor does not actually base its unit bid price on its prediction of the cost to be incurred for performing work on that unit. Commonly the contractor bids high for unit work that will be performed early, and low for later work. In a competitive bid evaluation, such a contractor suffers no disadvantage by this distortion. The award is based on the total unit prices multiplied by the total estimated quantities. As a result, such a bidder takes no apparent risk when it unbalances the bid.

The reasons for an unbalanced bid, sometimes called "pennying," were revealed in a New Jersey decision,

Boenning v. Brick Township Municipal Utilities Authority.[19] The case involved a bidding invitation that set minimum unit prices for certain portions of work connected with installing a municipal sewer system. The issue was whether the municipality could set minimum unit prices.

The court first distinguished *fixed* from *discretionary* items. It described a fixed unit as one that can be measured with certainty by reference to the plans and specifications, such as sewer pipe whose size and length are shown on the plans. (This type forms the basis of what the English call "measured contracts," with the price based on so many board feet of lumber, doors, windows, etc.)

Discretionary items (an unfortunate term)[20] are a type of work whose quantities cannot be accurately measured or ascertained in advance of the work itself being performed. The contract before the court involved two discretionary items: underground and restoration work. The underground work required laying sewer pipes in trenches ranging from 6 to 20 feet in depth. Although test borings had been taken before design, the soil conditions that would be encountered could not be determined with certainty. Some soil conditions would require specified material or a concrete cradle to support the pipe. Timber or steel shoring left in place might be needed to shore up the trench. Similarly, it could not be determined in advance how much restoration work, such as asphalt paving or landscaping, would be required after the sewers had been installed.

In what the court called "discretionary work," engineers for the awarding authority usually estimate units that will be required based on the borings, their knowledge of local conditions, and their experience. (The problem of mistaken estimates is taken up later in this subsection.) An unbalanced bid—or what the court referred to as *penny bidding*—can encourage the contractor who has pennied certain discretionary items to challenge an engineer's orders because the contractor may not want to incur the expenses of doing work for which it made a nominal bid. This can increase the number of disputes and interfere with timely completion of the project. Yet according to the court, it had become common

---

[18]48 CFR § 52.248-2 (2007).

[19]150 N.J.Super. 32, 374 A.2d 1214 (App.Div.), cert. denied, 75 N.J. 537, 384 A.2d 516 (1977).

[20]Most disputes involve excavation. The units are not, as a rule, discretionary or fixed. The issues usually are overruns or underruns (discussed later in this subsection).

practice in New Jersey bidding for utility work to penny units of discretionary work.

The court then dealt with how a pennying contractor can compete with one who does not penny units. The former may simply gamble that such work will not be needed or may believe that the engineer has overestimated the need for such items. Such a contractor will gain a competitive advantage by pennying and can afford to increase its bid on items that are certain, which can give it a windfall.

Another, more unsavory, reason for pennying may be the prospect of collusion between the contractor and the engineer, the latter finding the pennying units low and the excessively priced units high. Even without a specific "deal," the contractor may believe that the prospective gains might make it worthwhile to attempt to influence or even bribe the engineer.

The contractor may submit an unbalanced bid to obtain progress payments more quickly, called "front-end loading." This can mean a greater chance of late-finishing subcontractors and suppliers not being paid.

The court noted that the minimum prices were fixed below the estimated cost so that a contractor cannot reap a windfall if there is an overrun on any high-priced item. Establishing a minimum at a figure less than the likely cost will encourage competition among bidders. The minimum ensures that a contractor who has bid that price will be paid at least a part of its cost and can minimize gambling by contractors and potentially unnecessary disputes. It can also inhibit front-end loading by using the unbalanced bid.

The court rejected an argument that if the contractors gamble and lose, their sureties will complete the job and the local authority will receive a cheaper price. The court held that the local authority could use the minimum bid for units inasmuch as it could, if it chose, reserve the right to reject unbalanced bids.[21]

As to the effect of an unbalanced bid on pricing deleted work, the New Jersey Supreme Court refused to penalize a contractor who had submitted an unbalanced bid in an attempt to incorporate a last-minute subcontract offer into its overall bid proposal. Finding none of the evils associated with the unbalanced bid present in this case,

the court refused to allow the government to rely upon the bid values to calculate a work deletion. Instead, the contractor was entitled to an equitable adjustment, that is, the Department of Transportation must preserve the contractor's profit margin while making the deletion.[22]

The second legal issue, and perhaps one that arises most frequently, relates to inaccurate estimates of units to be performed. These usually involve excavation cases in which the actual units substantially overrun or underrun the estimates. Pricing the unit work usually assumes that there will not be a substantial deviation from the estimates. If the actual units substantially underrun, the cost cannot be spread over the number of units planned and will cost more per unit than the contractor expected. If the unit is overrun, the contractor may be expected to perform more unit work in the same period of time, another factor that can increase planned costs.

Often contracts specifically grant price changes if costs overrun or underrun more than a designated amount. For example, the federal procurement system grants an equitable adjustment if costs overrun or underrun more than 15 percent above or below the estimated quantity.[23]

AIA Document A201-2007 deals with this problem in a more limited way. Section 7.3.4 (formerly Paragraph 4.3.9) grants an equitable adjustment to either party if the quantities "are materially changed in a proposed Change Order or Constructive Change Directive" and applying the unit prices will cause "substantial inequity." This does not grant an automatic adjustment for overruns or underruns.

Many contracts provide equitable adjustments if the subsurface conditions actually encountered vary from the conditions stated to exist or usually encountered (see Section 25.06).

Some of the more difficult cases have involved claims where the owner simply gives an estimate that turns out to be inaccurate (most commonly grossly inaccurate) without any showing of negligence chargeable to the owner or any contractual adjustment method.

As a rule, the mere existence of a variation does not entitle the contractor to additional compensation or a change in the unit price. Such a result would be based on the contractor's having assumed the risk or, occasionally,

[21]See also *Department of Labor and Industries v. Boston Water and Sewer Comm'n*, 18 Mass.App.Ct. 621, 469 N.E.2d 64 (1984) (agency could not refuse a bid that had pennied some items, especially where bid was not unbalanced, front-end loaded, or otherwise inflated).

[22]*M. J. Paquet, Inc. v. New Jersey Dep't of Transp.*, 171 N.J. 378, 794 A.2d 141 (2002).
[23]48 CFR § 52.211-18 (2007).

on an express provision in the contract stating that estimates cannot be relied on. For example, in *Costanza Construction Corp. v. City of Rochester*,[24] the drawings showed 20 cubic yards of rock, and the specifications estimated 100 cubic yards. The actual amount of rock to be removed was 600 cubic yards. The contractor claimed it bid below cost because it expected to encounter only a small amount of rock. It tried to avoid the unit price by claiming that the actual amount of rock found constituted a cardinal change that altered the essence of the contract. However, it was not granted relief, because the contract included the city's disclaimer of responsibility as to the accuracy of the estimate and required the contractor to make its own inspection. The court brushed off a claim that there was no time for an independent inspection and that the cost would have been excessive.

The dissent stated that normal variations in quantities can be dealt with by a unit price. Here, however, the difference was so great (the contract price was $936,000, the actual cost for excavation was alleged to be $800,000, and the contractor's estimate was $2,500) that the estimates became meaningless. Applying the unit price, the dissent argued, could economically ruin a good-faith bidder through no fault of its own.

Reasonable reliance on the estimate, however, may justify reforming the contract to grant a price readjustment. For example, in *Peter Kiewit Sons' Co. v. United States*,[25] the contractor estimated its unit prices for three types of work of varying difficulty but then learned that the specifications required it to submit a combined unit price bid. It computed and submitted a unit price bid by averaging the three types of excavation and based its bid on the units estimated by the government.

When the work was in progress, the government ordered a reduction in one excavation type—a change that, because of the averaging used, the contractor claimed would increase the composite unit cost of the work. The claim was denied, on the contention that the estimates did not bind the government and could not be relied on, pointing to language to that effect in the specifications.

One basis for the contractor's claim was that each party had made a mutual mistake of fact as to the quantity of work required. The Court of Claims upheld the claim, stating that the contract was intended not to be speculative but to be capable of proper computation and conservative bid. The composite bid being required meant that a great variation in the actual quantities performed either could be ruinous to the contractor or could cause the government to pay far more than the work was worth. The court concluded that the government did not intend to contract on such an irrational basis. If the parties each believe the estimates to be accurate, a mutual mistake to that effect would be a basis for reforming the contract.

As to language stating that estimates are not guaranteed, the court first pointed to language granting an equitable adjustment in the event of changed conditions, noting that such a provision conflicted with the one stating that estimates are not guaranteed. Perhaps more important, the court stated that the disclaimer did not mean "that all considerations of equity and justice are to be disregarded, and that a contract to do a useful job for the Government is to be turned into a gambling transaction."[26]

In addition, this case demonstrates the not uncommon conflict between clauses that grant relief under certain circumstances and clauses that seek to place on the contractor risks closely related to those circumstances (explored in greater detail in Section 25.04). The case also reflects a recognition of the contractual doctrine of mutual mistake sometimes applied in extreme cases despite language that appears to place the risk of unexpected quantities on the contractor.

Davidson & Jones, Inc. v. North Carolina Department of Administration[27] employed the mutual mistake doctrine in a case involving a claim for delay damages. The contract stated that the unit price for rock excavation would apply to "all rock removed above or below these quantities." The contractor was required to include 800 cubic yards of rock excavation in its basic bid. However, early in the excavation it became clear this estimate was grossly inaccurate. The contractor actually excavated 3,714 cubic yards. It was also delayed six months while doing this excavation. It did not seek an adjustment in its unit price but sought delay damages.

[24]147 A.D.2d 929, 537 N.Y.S.2d 394, appeal dismissed, 74 N.Y.2d 714, 541 N.E.2d 429, 543 N.Y.S.2d 400 (1989).

[25]109 Ct.Cl. 517, 74 F.Supp. 165 (1947). See also *Timber Investors, Inc. v. United States*, 218 Ct.Cl. 408, 587 F.2d 472 (1978) (dictum: reformation allowed if estimated quantities "grossly and unreasonably inaccurate" and reliance reasonable, such as inability to verify).

[26]74 F.Supp. at 168. For a similar result based on misrepresentation that induced a particular composite bid, see *Acchione & Canuso, Inc. v. Pennsylvania Dept. of Transp.*, 501 Pa. 337, 461 A.2d 765 (1983).

[27]315 N.C. 144, 337 S.E.2d 463 (1985).

The court rejected the owner's argument that the contractor assumed the risk of delay and stated,

> The trial court found that the plaintiff had inspected the site as required in the bidding documents and had seen nothing to indicate the presence of such an excess. The court further found that it was neither customary nor reasonable for a contractor to order his own subsurface investigation. Contractors customarily relied on the State's figures; plaintiff here had actually relied on them. Some variation was to be expected. Based on the evidence presented at the trial, the trial court found that ten to fifteen percent was a reasonable variation. The court's findings of fact were based on competent evidence and may not be disturbed on appeal.[28]

E. Cash Allowance

Sometimes after the prime contract has been awarded the owner wants to select certain items. For example, it may wish the right to select particular hardware or fixtures after a contractor is selected. A cash allowance can be specified to cover the cost of the items selected. If the cost of the items ultimately selected varies from the cash allowance, an appropriate adjustment in the contract sum is made.

Disputes sometimes arise relating to what is encompassed within the allowance. The contract should specify whether the cash allowance encompasses only the net cost of the materials and equipment delivered and unloaded at the site, including taxes, or whether it also includes handling costs on the site, labor, installation costs, overhead and profits, and other expenses. Commonly, the allowance includes only the cost of the items selected.[29]

F. Contingencies

Contingencies are often included in contracts to deal with uncertainties. For example, in a bid contract the contractor might be told to include in its bid a specific sum to cover the cost of contingencies (defined as items necessary or desired to complete the project in accordance with the owner's wishes). These contingencies might be items related to latent conditions discovered after work begins, items omitted from the documents, or items the owner feels are necessary for a complete project. Under a contingency clause, if the contractor runs into no unexpected problems

and completes the job in accordance with the plans and specifications, it receives the contract price but not the contingency amount. If it encounters unexpected expenses or conditions, the contingency amount is expected to be spent on them.

As shall be seen in Section 17.04D, contingencies (as discussed in Section 17.02F) are often included in cost contracts with a guaranteed maximum price (GMP).

SECTION 17.03   Traditional Organization: Owner's Perspective

This chapter looks at compensation and organizational aspects of the construction process mainly from the owner's perspective. It is generally assumed that contractors who engage in the process are individual legal entities, usually corporations. However, some projects may be beyond the capacity of an individual contractor, who may seek to associate with other contractors in a joint venture. This organizational method—roughly, a partnership for one project—was discussed in Section 3.07.

A. Traditional System Reviewed

The traditional system separates design and construction, the former usually performed by an independent design professional and the latter by a contractor or contractors. The sequencing used in this system is creation of the design followed by contract award and execution. Two processes are available to the owner when choosing a contractor. Under design-bid-build (DBB), the owner provides the design to several contractors and they competitively bid to perform the work. The owner usually hires the contractor who offers to do the job at the lowest price. (Competitive bidding is discussed in detail in Chapter 18.) If the owner wants a particular contractor to perform the work, the owner will use the design-award-build (DAB) approach and give the design to a contractor it has already chosen.

The contractor to whom the contract is awarded is usually referred to as the *prime* or *general contractor*. It will perform some of the work with its forces but is likely to use specialized trades to perform other portions of the work. The prime contractor is both a manager of those whom it engages to perform work and a producer in the sense that it is likely to perform much or some of the work

---

[28]337 S.E.2d at 468.
[29]AIA Doc. A201-2007, § 3.8.2.1 (less applicable trade discounts).

with its own forces. The traditional system used in the United States usually gives certain site responsibilities to the design professional who has created the design (discussed in detail in Section 12.08B).

The principal advantages of this system are as follows:

1. The owner can select from a wide range of design professionals.
2. For inexperienced owners, an independent professional monitoring the work with the owner's interest in mind can protect the owner's contractual rights.
3. Not awarding the contract until the design is complete should enable the contractor to bid more accurately, making a fixed-price contract less likely to be adjusted upward except for design changes.
4. Subcontracting should produce highly skilled workers with its specialization of labor and should create a more competitive market because it takes less capital to enter.

B. Weaknesses

The modern variations to the traditional system described in Section 17.04 developed because, despite its strengths, the traditional system developed weaknesses. It is important to see these weaknesses as a backdrop to Section 17.04.

Separating design and construction deprives the owner of contractor skill during the design process, such as sensitivity to the labor and material markets, knowledge of construction techniques, and their advantages, disadvantages, and costs. A contractor would also have the ability to evaluate the coherence and completeness of the design and, most important, the likely costs of any design proposed.

Sequencing the work in the traditional system not only precludes work from being performed while the design is being worked out but also deprives the contractor of the opportunity of making forward purchases in a favorable market. Although the leisurely pace that often accompanies the traditional construction process at its best can produce high-quality work, it is almost inevitable that such a process will take more time than a process that allows construction to begin while the design is still being completed.

The frequent use of subcontractors selected and managed by the prime contractor causes difficult problems. Subcontractors complain that their profit margins are unjustifiably squeezed by prime contractors demanding they reduce their bids and, more important, reducing the price for which they will do the work after the prime contractor has been awarded the contract. Subcontractors also complain that contractually they are not connected to the owner—the source of authority and money. They also complain that prime contractors do not make enough effort to move along the money flow from owner to those who have performed work and that they withhold excessive amounts of money through retainage.

The traditional system tends to keep down the number of prospective prime contractors who could bid for work and thereby reduces the pool of competitors. The traditional system, with its emphasis on a fixed-price contract and competitive bidding, also can create an adversarial relationship between owner and contractor (described in Section 17.04A).

Also, the traditional system with its linked set of contracts—owner–design professional, owner–contractor, contractor–subcontractors—did not generate a collegial team joining together with a view toward accomplishing the objectives of all the parties. Some note that the designer and the contractor often generate a semiadversarial mood, which can generate accusatory positions when trouble develops. The designer acting as the owner's representative can generate administrative costs, papers having to pass through many hands. All this can increase the likelihood of disputes. Some of the variations noted in Section 17.04 emphasize better organization, the creation of a construction team, and the need for all involved to pull together to accomplish the objectives.

Another weakness of the traditional system relates to the role of the design professional during construction. For the reasons outlined in Section 12.08B, modern design professionals seek to exculpate themselves from responsibility for the contractor's work and to limit their liability exposure. Perhaps even more important, many design professionals lack the skill necessary to perform these services properly.

Under the traditional method, the managerial functions of the prime contractor may not be performed properly. The advent of increasing, pervasive, and complex governmental controls over safety often found many contractors unable to perform in accordance with legal requirements. In addition, the managerial function of scheduling, coordinating, and policing took on greater significance as pressure mounted to complete construction as early as possible and as claims for delay by those who participate in the project proliferated. A good prime contractor should be able to manage these functions, as its fee is paid to a large degree for performance of these services. Not all prime contractors were able to do this managerial work efficiently. Some owners believed the managerial fee included in the cost of the prime contract could be reduced.

The division between design and construction, although at least in theory creating better design and more efficient construction, had the unfortunate result of dividing responsibility. When defects develop, as noted in Chapter 24, the design professional frequently contends that such defects were caused by the contractor's failure to execute the design properly, whereas the contractor asserts that the design was defective. This led to bewildered owners not being certain who was responsible for the defect as well as to complex litigation.

SECTION 17.04 Modern Variations

A. Introductory Remarks

Modern project delivery systems are becoming the norm, especially in large commercial projects, but increasingly in public contracting as well. The most popular modern variations on the traditional system are construction management (CM) (Section 17.04D) and design-build (DB) (Section 17.04F), with these methods sometimes incorporating phased construction (fast-track) (Section 17.04B). Other new delivery systems, such as program management (PM) (Section 17.04K), teaming arrangements (Section 17.04H), and project alliance (Section 17.04J), have also made significant inroads, although to a lesser extent than CM and DB. Yet other new delivery systems, in particular building information modeling (BIM) (Section 17.04M), have yet to be fully utilized.

The traditional system, which separates design from construction, has relatively clear lines of responsibility. Risk allocation methods—insurance, indemnity, and contract disclaimers—are easy to devise. By contrast, all of the modern variations have as a common denominator: a blurring of the lines of responsibility. This blurring has led (and will continue to lead) to the creation of new risk allocation methods, including new insurance products and new forms of indemnity.

The law, too, must adapt to these changing commercial realities. Some changes may be accomplished through legislation, such as allowing the use of CM or DB on public contracts.[30] Common law changes occur more gradually. The common law functions to a large degree through crude categories aided by analogies, largely for administrative convenience. Law often lags behind organizational and functional shifts in the real world. New project delivery systems described in this section inevitably create temporary disharmony in the law. Over time, predictable legal rules emerge.

Much standardization in the traditional system is attributable to general acceptance of construction contract documents, in particular those published by the American Institute of Architects (AIA). As alternative project delivery systems have become more accepted, industry groups have begun issuing standard form documents for these newer systems as well. The AIA has published standard form documents for CM and DB, sometimes with endorsement from the Associated General Contractors of America (AGC), but not for the 2007 AIA documents. The AGC, in turn, has very recently created a standard form contract for PM. Even with these developments, descriptions of newer project delivery methods must be viewed as nothing more than generalizations. The reason is that these projects are almost always undertaken by sophisticated owners who are likely to tailor the agreement to meet their individual needs.

Many legal rules are premised on the traditional system, as described in Section 17.03. Modern variations change this system. Using separate contracts (multiple primes) (to be discussed in Section 17.04C) shifts responsibility for coordination from a prime contractor to someone else, often an independent professional adviser retained by the owner or a principal (not a prime) contractor who has no direct contracts with those contractors whose work it must, as a manager, coordinate.

Phased construction (fast-tracking, discussed in Section 17.04B) allows construction to proceed during design. When the spotlight is placed on construction management (in Section 17.04D), a new adviser or coordinating contractor is engaged (not, as a rule, after the design is completed but during design) by the owner, who also plays a significant monitoring role during construction. Here the owner, through at least some permutations of construction management, has a more active role in managing the project and is not simply turning over the design to the contractor, a businessperson engaged in the venture to earn a profit.

---

[30]N.C.Gen.Stat. § 143-128.1 (CM); Wash.Rev.Code § 47.20.780 (DB).

In turnkey and design–build (discussed in Sections 17.04E and 17.04F), the owner has not engaged an independent adviser to prepare the design but has turned over everything to a businessperson who represents that he has paid or can pay for the skill to both design and build, and in the case of a turnkey contract even more, such as providing financing, furnishing interior furnishings, and providing a computer system. Even more difficult, the DB owner can, as noted in Section 17.04F, span a wide range of owners.

Partnering (Section 17.04G) seeks to close the gaps among owner, designer, and contractor by having all three of these project participants work as a team to accomplish a project or series of projects. Teaming agreements (Section 17.04H) were promoted primarily by the federal government to allow would-be competitors with complementary skills to bid together on a project. While one team member is sometimes contractually described as the prime contractor and the other as the subcontractor, teaming agreements envisage more equal relationships between the two team members.

Lean project delivery (Section 17.04I) looks to the production industry—specifically the principles of the "Toyota Way" used by the Toyota Motor Company—to bring greater efficiency to the construction process. The project participants seek to deliver a high-value project through collaborative efforts. This is also true for project alliance (Section 17.04J), where a high degree of collaboration and commitment is required by the parties' agreement. As a general rule, partnering, lean project delivery, and project alliance all seek to convert the construction process into a collaborative venture, in which the major project participants subsume their own interests to the goals of the project itself.

Program management (Section 17.04K) is perhaps the next evolutionary stage following the CM as advisor. The PM provides the large commercial or public owner with the type of support the owner would have received from an in-house construction or engineering department.

The build-operate-transfer (BOT) system of project delivery (Section 17.04L) began with large infrastructure projects in developing countries. Here, a consortium with the financial means builds the project and operates it, with the ultimate goal of transferring it back to the country in which the project is located. Most BOT projects are turnkey projects.

Finally, building information modeling (BIM) (Section 17.04M) is unique among the many project delivery systems

in that it is a technology-driven organizational model. BIM is premised upon the creation of a computer model of the project that is both information-rich and information-integrative. Virtually all the project participants play an interactive role in the design and implementation of the building, thereby erasing the traditional system's sharp boundary between design and construction.

B. Phased Construction (Fast-Tracking)

Phased construction, or what has come to be known as "fast-tracking," is not an organizational variation. It differs from the traditional method in that construction can begin before design is completed. There is no reason it cannot be used in a traditional single contracting system, although it is likely to mean that the contract price will be tied to costs.

Construction can begin while design is still being worked out. Ideally this means that the project should be completed sooner.[31] If a contractor is engaged during the design phase and knows which material and equipment to use, it can make purchases or obtain future commitments earlier and often cut costs.

Another advantage to the owner is the early activation of the construction loan. In traditional construction, an owner, particularly a developer, does not receive loan disbursements until the construction begins. Yet it must pay for investigation and acquisition of the project site. It may have had to pay either the purchase price for the land or installments of ground rent, as well as insurance premiums, real estate taxes, design service fees, or legal expenses. If it must make these expenditures before it starts to receive construction loan funds, it must draw on its personal unsecured credit line, necessitating monthly interest payments and reducing its credit line, which could be used for other purposes. If construction can begin through fast-tracking, the owner need not use its personal funds but can employ loan disbursements.

---

[31]Some recent studies have indicated that these ideals do not always mature. One study indicates that very low percentages of design completion before commencement of construction may result in considerable construction delays. See Laufer & Cohenca, *Factors Affecting Construction-Planning Outcomes*, 116 J.of Constr.Eng'g.&Mgmt. 135 (1990). Similarly, another study states that fast-tracking is of much less value than it appears to be. The benefit of early completion, it stated, is more than half offset by the consequent shift in the timing of construction expenditures. It not only may wipe out any benefits gained by an early start but also may lead to the accumulation of losses. Rosenfeld & Geltner, supra note 6.

The principal disadvantage of fast-tracking is the incomplete design. The contractor will be asked to give some price—usually after the design has reached a certain stage of completion. Very likely the contract price will be cost plus overhead and profit, with the owner usually obtaining a guaranteed maximum price (GMP).[32] However, the evolution of the design through a fast-tracking system is likely to generate claims. Completing the design and redesigning can generate a claim that the contract has become one simply for cost, overhead, and profit and that any GMP has been eliminated.[33] The owner may be constrained in making design changes by the possibility that the contractor will assert that the completed drawings must be consistent with the incomplete drawings.[34]

Two other potential disadvantages need to be mentioned here. First, there is greater likelihood that there will be design omissions—items "falling between the cracks." Needed work is not incorporated in the design given to any of the specialty trade contractors or subcontractors. Although such omissions can occur in any design, they are more likely to occur when the design is being created piecemeal rather than prepared in its entirety for submission to a prime contractor.

Second, in fast-tracking there is a greater likelihood that one participant may not do what it has promised, and thus adversely affect the work of many other participants. For example, an opinion of a federal contract appeals board, discussing fast-tracking in the context of construction management and multiple primes, stated,

> "Phased" construction has been analogized to a procession of vehicles moving along a highway. Each vehicle represents a prime contractor whose place in the procession has been pre-determined. The progress of each vehicle, except that of the lead vehicle, is dependent on the progress of the vehicle ahead. The milestone dates have been likened to mileage markers posted along the highway. Each vehicle is required to pass the mileage markers at designated times in order to insure steady progress.[35]

When any of the vehicles do not pass the mileage markers assigned to them, claims for delays and complex causation problems are likely to result.[36]

C. Separate Contracts (Multiple Primes): *Broadway Maintenance Corp. v. Rutgers, State University*

Separate contracts (using multiple prime contractors) consist of the owner contracting directly with the principal subcontractor trades. It developed because of owner concern over the subcontracting process briefly described in Section 17.03B. See Figure 17.1. Separate contractors give the owner greater control in specialty contracting and avoid the subcontractor complaints that they are a contract away from the source of power and funds. Separate contracts were used if owners were not confident of prime contractor management skill. Their use was also spurred by successful legislative efforts by subcontractor trade associations to require that state and sometimes local construction procurement use separate contracts.

Another factor, though of less significance, was the hope that separate contracts could develop lower contract prices. Breaking up the project into smaller bidding units allows more contractors to bid. Owners hoped the managerial fees could be reduced by taking this function from the prime contractor. These hoped-for pricing gains can compensate for the additional expense of conducting a number of competitive bids.

In the traditional contracting system, the linked set of contracts determines communication and responsibility. If subcontractors have complaints, they can look to the prime contractor. If the prime has problems, it can look to the subcontractors or the owner.[37]

Separate contractors are required to work in sequence or side by side on the site, but they do not have contracts with one another. Disputes among them must be worked

---

[32]See Sections 17.02B, 17.04D, and 17.04F.

[33]*Armour & Co. v. Scott*, 360 F.Supp. 319 (W.D.Pa.1972), aff'd, 480 F.2d 611 (3d Cir.1973); see *C. Norman Peterson Co. v. Container Corp. of America*, supra note 13.

[34]*City Stores Co. v. Gervais F. Favrot Co.*, 359 So.2d 1031 (La.App. 1978), overruled on other grounds, *St. Tammany Manor, Inc. v. Spartan Bldg. Corp.*, 509 So.2d 424 (La.1987).

[35]*Pierce Assoc., Inc.*, GSBCA No. 4163, 77-2 BCA ¶12,746, citing *Paccon, Inc. v. United States*, 185 Ct.Cl. 24, 399 F.2d 162 (1968).

[36]The use of construction management often accompanied by fast-tracking in federal projects led to an inordinate number of delay claims. See U.S. General Accounting Office, *Federal Construction: Use of Construction Management Services*, U.S. General Accounting Office GAO/GGD-90-12 (January 1990).

[37]Even in the traditional system, lines were becoming blurred as owners and subcontractors sued each other. See Sections 28.05B, 28.07H, and 28.08B. AIA prime contracts give the owner power to intervene to a degree in the prime–subcontractor relationship. See AIA Doc. A201-2007, §§ 5.2.2, 5.3, 9.3.1.2, 9.6.2, and 9.6.3.

Traditional (design–bid–build or design–award–build)

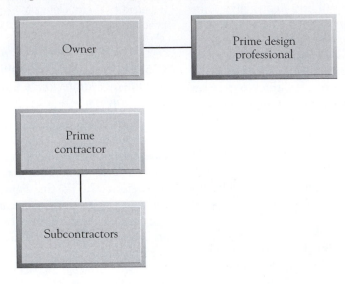

Separate contracts (multiple primes)

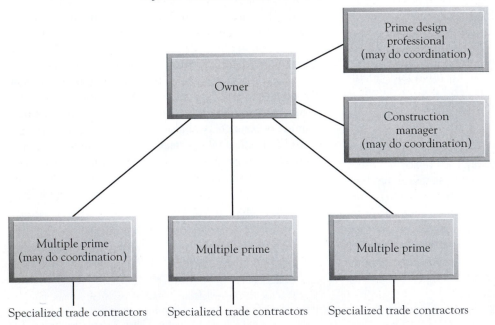

FIGURE 17.1   Traditional and separate contract systems compared.

out by the participant performing the coordination function, something that is not as neat in the separate contract system as in the single-contract system. To be sure, even in the traditional system this problem can arise if disputes develop between subcontractors who have no contract with each other. The traditional system handled this with relative clarity by requiring that the prime contractors deal with these matters as part of its managerial function and that the owner remove itself from these problems.

Who performs the managerial function? Who has legal responsibility? Under the construction management system, these often fall to the construction manager (CM). But the managerial function can be performed by the design professional if he has the skill and willingness to do so, by a staff representative of the owner, by a program manager, or by a managing or principal separate contractor. If it is not clear who will coordinate, police, and be responsible in any system, the project will be delayed and claims made. A case that generated all these problems is reproduced here.

BROADWAY MAINTENANCE CORP. v. RUTGERS, STATE UNIVERSITY

Supreme Court of New Jersey, 1982. 90 N.J. 253, 447 A.2d 906.
[Ed. note: Footnotes renumbered and some omitted.]

SCHREIBER, J.

Two contractors engaged in the construction of the Rutgers Medical School in Piscataway Township sought to recover damages from Rutgers, The State University (Rutgers), for its failure to coordinate the project and to compel timely performance by a third contractor. Rutgers asserts that it allocated the sole duty to coordinate the work of the prime contractors on the project and ensure their timely performance to Frank Briscoe Co., Inc. (Briscoe), its general contractor. Plaintiffs deny their right to sue Briscoe as third party beneficiaries of Briscoe's contract with Rutgers. We must determine first, whether plaintiffs were intended to be beneficiaries of that contract. Even if they were, we must determine whether Rutgers, as owner, retained any duty to coordinate or supervise which could give rise to a cause of action in plaintiffs. Finally, we must determine whether Rutgers is excused from any such liability by an exculpatory clause in the contract.

On October 31, 1966, Rutgers signed contracts for general construction work with Briscoe for $7,392,000, electrical work with plaintiff, Broadway Maintenance Corp. (Broadway), for $2,508,650, and plumbing and fire protection with plaintiff, Edwin J. Dobson, Jr., Inc. (Dobson), for $998,413. Six other contracts, also entered into, covered precast concrete; structural steel; elevators; heating, ventilating and air conditioning; laboratory furniture and independent inspection and testing. In contrast with construction projects in which the owner contracts with a general contractor who undertakes the entire project and coordinates operations of subcontractors, Rutgers entered into contracts with each of several prime contractors. In Rutgers'

agreement with Briscoe, Briscoe agreed to act as the supervisor on the job and coordinator of all the contractors. [Ed. note: See Figure 17.2.]

After the work was finished, at a date well beyond the scheduled time for completion, Dobson and Broadway filed separate complaints against Rutgers in the Superior Court, Law Division, asserting a variety of claims, including damages due to delays and disruptions caused by Rutgers' failure to coordinate the activities of the various contractors on the site. Rutgers filed third party complaints seeking indemnification from Briscoe and its surety. The two actions were consolidated for trial with a pending third suit brought by Briscoe against Rutgers for money due under the Rutgers-Briscoe contract. Dobson and Broadway never added Briscoe as a party defendant in their actions. The suit between Rutgers and Briscoe was settled before trial, except for two claims for indemnification. Briscoe is not a party to this appeal.

The non-jury trial proceeded for 43 days. Rutgers produced no evidence and rested at the end of plaintiffs' case. The trial court in an extensive written opinion granted plaintiffs judgments for some of their claims, but denied recovery against Rutgers for failure to coordinate the activities of the prime contractors, including Briscoe. 157 *N.J.Super.* 357, 384 A.2d 1121 (Law Div.1978). Dobson and Broadway appealed to the Appellate Division, which affirmed. 180 *N.J.Super.* 350, 434 A.2d 1125 (App.Div.1981).

Each plaintiff petitioned for certification. We granted both petitions to consider three questions: (1) in a multi-prime contract, is each prime contractor liable to the other, (2) in such a contract, does the owner have a duty to coordinate the work of

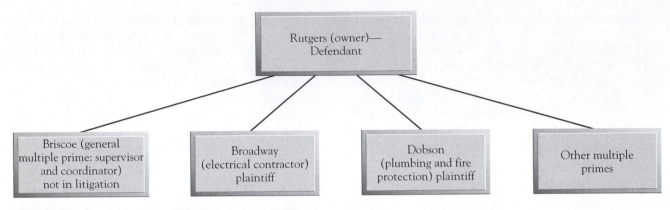

FIGURE 17.2 The relationships among parties in *Broadway Maintenance v. Rutgers*.

the contractors, and (3) does the exculpatory clause in the prime contracts at issue here shield Rutgers from liability for damages due to delay. The Mechanical Contractors Association of New Jersey, Inc. and the Building Contractors Association of New Jersey were granted leave to file briefs as amici curiae ["friends of the court," allowed to present their views by court permission].

[Ed. note: The court noted that jurisdictions had reached different results when faced with the question of whether one separate contractor could sue another. It concluded that the record supported the findings of the trial court holding that all parties had agreed the separate contractors could maintain legal action against each other for damage due to unjustifiable delay. In determining this question, the court held that the terms and conditions of the contract should be examined to determine the intent of the parties.]

The existence of a third party claim does not necessarily extinguish all claims between the parties to the contract. Here plaintiffs argue that Rutgers breached its agreement with them and is liable irrespective of any third party claims they may have against the general contractor, Briscoe. The plaintiff's contention is sound at least with respect to those matters that Rutgers had contractually obligated itself to do and for which it would be responsible. The trial court did award damages to the plaintiffs against Rutgers for certain contractual breaches. Rutgers has not appealed from those determinations.

The narrow questions before us on this appeal are whether Rutgers had agreed to synthesize the operations of the prime contractors, including Briscoe, and, if so, whether Rutgers breached that duty and would be liable for the delay flowing from that breach. The order granting the petitions for certification was limited to the subject matter of coordination of the operations of all

the prime contractors including Briscoe. Damages flowing therefrom involve only delay on the job and its consequential costs.

Plaintiffs urge various claims that they relate to delay such as additional expenses incurred because of lack of elevators and stairs. These particular claims were disallowed by the trial court, which pointed out, among other things, that Briscoe's delay did not cause the asserted damages. Though these items are not before us on this appeal, we have reviewed the record and are satisfied that the trial court's findings have adequate factual support.

Plaintiffs also continue to press before us matters such as Rutgers' alleged default in not withholding funds due Briscoe in order to satisfy plaintiff's claims for delay against Briscoe, and Rutgers' non-fulfillment of its supposed duty to place the site in condition so that plaintiffs could proceed. On this appeal we are concerned solely with the general supervision over the contractors. What duties Rutgers had in this respect depends on what obligations were imposed on it by the contract.

In the absence of any compelling public policy, an owner has the privilege to eliminate a general contractor and enter into several prime contracts governing the construction project.[38] In that event the owner could engage some third party or one of the contractors to perform all the coordinating functions. Where all the parties enter into such an arrangement the owner would have no supervisory function. The situation would be analogous to one where a general contractor had been engaged to construct

---

[38]Use of multiple prime contracts assumes the owner will benefit from savings that will accrue from eliminating the overhead and profits of the general contractor. Provisions for construction of any public buildings by the state provide for separate bids for different major aspects of the job and bids for all work in one contract, the award to be made to whichever method results in a lower cost. N.J.S.A. 52:32-2.

the project on a turnkey basis. Surely the subcontractors would have no claim against the owner for failure to coordinate.

If no one were designated to carry on the overall supervision, the reasonable implication would be that the owner would perform those duties. In so doing, the owner impliedly assumes the duty to coordinate the various contractors to prevent unreasonable delays on the project. See, e.g., *Born v. Malloy*, 64 Ill.App.3d 181, 184, 21 Ill.Dec. 117, 120, 381 N.E.2d 52, 55 (1978); *Carlstrom v. Independent School District No. 77*, Minn., 256 N.W.2d 479 (1977). That is a reasonable assumption because the contracting authority has the power to use its superior position and to invoke its contractual rights to compel cooperation among contractors. *Shea-S & M Ball v. Massman-Kiewit-Early*, 606 F.2d 1245, 1251 (D.C.Cir.1979). The owner is impliedly obligated to act in good faith and to do that which it reasonably can to ensure that the other contractors adhere to the time schedules established for the project. *Paccon, Inc. v. United States*, 399 F.2d 162, 169-70 (Ct.Cl.1968). An owner's failure to take action in the face of unnecessary and unreasonable delays by one of the contracting parties would ordinarily evidence bad faith and constitute a breach of its implied duty to coordinate.

It is, of course, also possible that the owner might fractionalize those supervisory functions. Where the owner has chosen to engage several contractors directly, the bottom line is to ascertain who the parties agreed would orchestrate and harmonize the work. The answers may be found in the contract language as illuminated by surrounding circumstances.

The complications that arise in this case are due in part to the unclear nature of who was to perform what supervisory function. Briscoe had agreed to supervise the job generally. It was entrusted with the "oversight, management, supervision, control and general direction" of the project. It was to control "the production and assembly management of the building construction process." Briscoe was obligated to have sufficient executive and supervisory staff in the field so as to handle these matters efficiently and expeditiously. The General Conditions also recited that Rutgers relied on Briscoe's management and skill to supervise, direct and manage Briscoe's own work and the efforts of the other contractors; indeed, the agreements stated that the contractors also relied on Briscoe's supervisory powers to deliver the building within the scheduled time.

However, Rutgers also designated its Department of New Facilities to represent it in technical and administrative negotiations with the contractors and the Department could stop the work if necessary. Rutgers also selected a critical path method consultant whose progress chart was to be followed, but could be amended with the consultant's approval. Further, Rutgers

engaged an architect. The architect agreed to "[s]upervise the construction of work by periodic inspections sufficient to verify the quality of the construction and the conformity of the construction to the plans and specifications . . . and by which supervision the architect shall coordinate the work of the various contractors and expedite the construction of the Project." If a contractor were delayed in completing the work whether due to the owner, any other contractor, the architect, or other causes beyond the contractor's control, the architect alone was authorized to determine extensions of time to which a prime contractor would be entitled. In this respect the trial court found that the architect acted as an impartial and independent umpire, and as such was not the agent of either Rutgers or the contractor.

Plaintiffs contend that Rutgers retained supervisory control because Briscoe had not been given the power to enforce its coordinating authority. They argue that Rutgers had the economic weapons of terminating the contracts and withholding payments, so that only Rutgers could effectively cause the prime contractors, particularly Briscoe, to keep up to schedule. Though this contention has some surface appeal, it fails to account for the entire contractual structure governing the project. Rutgers' power to terminate a contract or withhold funds did not alter its expressed intent to have someone else supervise the work. Rutgers had delegated that overall responsibility to Briscoe. Rutgers never intervened to coordinate the operations. It never assumed control.

The plaintiffs also claim that Rutgers retained coordination and supervision over the job because of the roles of the Department of New Facilities and critical path method consultant. These contentions are misconceived. The Department's functions were primarily quality control and only incidentally affected time of performance. The consultant plotted actual progress as against the scheduled performance, but did not supervise and coordinate the work. Nor does the architect's supervisory role support plaintiffs' position. First, the architect's agreement was not incorporated in the contractors' contracts. Second, Rutgers' delegation of coordinating functions to the architect confirms, if anything, that *Rutgers* itself was not engaged in any such undertaking. Lastly, plaintiffs rely on an indemnity clause in each contract that provided that if a contractor or subcontractor sued Rutgers, the contractor would defend, indemnify and save Rutgers harmless. However, that provision simply confirms the intent that Rutgers was not to be responsible for defaults of other prime contractors, including Briscoe.

When viewed in its entirety, the contractual scheme contemplated that if a contractor were adversely affected by delays, it could maintain an action for costs and expenses against the fellow contractor who was a wrongdoer. Furthermore, a contractor had a right to obtain extensions of time to complete the work when delayed by "any act or neglect" of any other contractor. Other than those remedies, the contractor could not look to Rutgers for recourse because of its failure to coordinate the work.

[Ed. note: The court reviewed the enforceability of a "no damage" (no pay for delay) clause and concluded that it would be given effect in this case. The court rejected an argument that the clause be given effect only when the delays are reasonable and concluded that Rutgers did not actively interfere with the progress of the work nor did it act in bad faith.]

In summary then, we hold that a third party beneficiary may sue on a contract when it is an intended and not an incidental beneficiary. Resolution of that issue depends on examination of the contractual provisions and the attendant circumstances. In a construction project where the owner directly hires the various contractors, the owner may engage a separate contractor to coordinate and supervise the project and agree with the several contractors that the general supervision will be carried out in that manner. Lastly, the owner may exculpate itself from liability for damages to the extent delineated in the contract, in the absence of any public policy reasons to the contrary.

The judgment is affirmed.

For affirmance—Justices PASHMAN, CLIFFORD, SCHREIBER, HANDLER and O'HERN—5.

For reversal—None.

The *Rutgers* decision exposes some of the administrative and legal difficulties inherent in the separate contracting system. Injecting a new "entity" into the process—in that case, a CPM consultant (perhaps a limited CM)—without a well-defined role or clearly delineated responsibility can create problems. The case demonstrates the administrative difficulties of giving one entity managerial responsibilities without the tools to effectively manage—one reason why the *Rutgers* decision has been criticized.[39]

This problem is recognized in North Carolina public contracts. The statute dealing with separate contracts gives the coordination to a project expediter selected by the state.[40] The expediter can be a contractor. But the expediter can make recommendations to the state for payments to the contractors and can prepare the schedule (with input from the contractors). This gives the expediter some power over payments and scheduling (both significant factors in policing coordination).

If managerial functions are given the managing contractor, what effect does this have on any owner responsibility? For example, had that power been given to the architect or even a CM, it is likely that the owner would be responsible if either had failed to coordinate properly.

AIA Document A201-2007, Section 6.1.1, allows the owner to award separate contracts but only on conditions "identical or substantially similar" to A201. That paragraph requires any separate contractor claim for delay or added cost be handled by Article 15, a complicated disputes process in A201-2007. The process will deal with two disputes: the claimant separate contractor against the owner and the owner against the separate contractor whose acts were alleged to have generated the claim.

A201-2007, Section 6.1.3, requires the owner to provide coordination.[41] Each separate contractor agrees to cooperate. A separate contractor can claim another has damaged or destroyed its work or property. Most important, a separate contractor can claim that another has failed to perform in accordance with the latter's contractual obligation and adversely affected the former's scheduling and performance.

A201-2007, Section 6.2.3, makes the owner responsible for a contractor's costs incurred because of delayed or defective performance by another contractor. The at-fault contractor must then reimburse the owner. Under this system, an injured contractor looks to the owner for relief instead of trying to sue the coprime contractor. This system contrasts with that of the EJCDC, which gives the separate contractors rights against each other.[42]

Many cases have involved who can sue whom when separate contracts or multiple prime contracts are used. Sometimes the owner, as in the *Rutgers* case, can persuade the court that the separate contract system insulates the owner from any claims by a separate contractor. As demonstrated in that case, much depends on the language of the

---

[39]Bynum, *Construction Management and Design-Build/Fast Track Construction from the Perspective of a General Contractor,* 46 Law & Contemp.Probs. 25, 33 (1983).

[40]N.C. Gen. Stat. § 143-128 (e).

[41]This coordination is not substantially different in A201 CMa-1992 ¶6.1.3 and B801 CMa-1992 ¶2.3.6, its CM agency documents.

[42]EJCDC Doc. C-700, ¶7.03 (2007).

contract and any right that one separate contractor has to maintain a direct action against the other. Although one case held that the owner is insulated from responsibility,[43] another did not.[44] Even if the owner is not absolved from responsibility, it is not clear whether the owner's obligation is "strict" or whether its obligation is simply to use its best efforts, in effect a negligence standard.[45] Finally, as to the right of one separate contractor to maintain a direct action against the other, although Illinois held that one cannot sue the other,[46] Tennessee, in *Moore Construction Co., Inc. v. Clarksville Dept. of Electricity*,[47] came to a contrary conclusion in a case involving AIA Document A201. The court reviewed the cases and pointed to what it saw as an emerging trend:

> Unless the construction contracts involved clearly provide otherwise, prime contractors on construction projects involving multiple prime contractors will be considered to be intended or third party beneficiaries of the contracts between the project's owner and other prime contractors. They have been permitted to recover when the courts have found that their fellow prime contractor assumed an obligation to the owner to them (the "duty owed" test) or

that their fellow contractor assumed an independent duty to them in their own contract with the owner (the "intent to benefit" test). The courts have generally relied on the following factors to support a prime contractor's third party claim: (1) the construction contracts contain substantially the same language; (2) all contracts provide that time is of the essence; (3) all contracts provide for prompt performance and completion; (4) each contract recognizes the other contractor's rights to performance; (5) each contract contains a non-interference provision; and (6) each contract obligates the prime contractor to pay for the damage it may cause to the work, materials or equipment of other contractors working on the project.[48]

The court found these requirements present in the A201 contract to which both separate contractors had assented.

Suppose one separate contractor can sue the other for the latter's breach of contract. Should this have any bearing on the separate contractor's right to sue the owner? Under A201-2007, Section 6.1.3, the owner provides coordination. If it does not do so, the separate contractor has a claim against the owner. Suppose, however, that an agreed schedule provides for coordination but one separate contractor does not perform as agreed. Under the *Moore* case, the separate contractor whose work has been disrupted can sue the separate contractor whose failure to perform properly has caused the disruption. But can the separate contractor also sue the owner? Can it assert that the owner has impliedly warranted that each separate contractor's work would be properly coordinated with the work of others? If so, the separate contractor has claims against the owner and the other separate contractor. Under A201-2007, these claims would then have to be processed through Article 15 dealing with disputes.

Wisconsin blends these problems by a statute regulating state public contracts. The state is not liable to a separate contractor for delay caused by another if the state took reasonable steps "to require the delaying prime contractor to comply with its contract." If the state is exonerated, the prime delayed can sue the delaying prime.[49]

The coordination complexities are revealed in *Beltrone Construction Co., Inc. v. State of New York*.[50] A subcontractor to a separate contractor asserted a claim against

---

[43]*Hanberry Corp. v. State Bldg. Comm'n,* 390 So.2d 277 (Miss.1980).

[44]*Shea-S & M Ball v. Massman-Kiewit-Early,* 606 F.2d 1245 (D.C.Cir. 1979).

[45]The Court of Claims was not able to resolve this issue squarely. Compare *Fruehauf Corp. v. United States,* 218 Ct.Cl. 456, 587 F.2d 486 (1978) (warranty based on suspension of work clause), with *Paccon, Inc. v. United States,* 185 Ct.Cl. 24, 399 F.2d 162 (1968) (not liable if owner takes reasonable steps). In *Blinderman Constr. Co. v. United States,* 695 F.2d 552 (Fed.Cir.1982), only best efforts were required. In *Amp-Rite Elec. Co. v. Wheaton Sanitary Dist.,* 220 Ill.App.3d 130, 580 N.E.2d 622 (1991), appeal denied, 143 Ill.2d 635, 587 N.E.2d 1011 (1992), the owner was held liable to two multiple primes, despite one of the multiple primes being required to coordinate the work. Although the court did not find that the owner had warranted (strict liability) that delays would not occur, the owner owed an implied duty to coordinate. This was breached when the owner actively created or passively permitted to continue a situation over which it had control. It alone had the power to control cooperation by withholding payments or terminating or taking over the work. This demonstrates that whatever pains the owner takes to relieve itself from responsibility for coordination, the owner is likely to be held responsible because of its other powers. See Goldberg, *The Owner's Duty to Coordinate Multi-Prime Construction Contractors, A Condition of Cooperation,* 28 Emory L.J. 377 (1979), for an exhaustive treatment.

[46]*J. F., Inc. v. S. M. Wilson & Co.,* 152 Ill.App.3d 873, 504 N.E.2d 1266, appeal denied, 115 Ill.2d 542, 511 N.E.2d 429 (1987).

[47]707 S.W.2d 1 (Tenn.Ct.App.1985), aff'd March 24, 1986 (thorough discussion).

[48]Id. at 10.

[49]West Ann.Wis.Stat. § 16.855(14)(b). Presumably, if the state is not exonerated, the state can still sue the delaying prime contractor.

[50]256 A.D.2d 992, 682 N.Y.S.2d 299 (1998).

a public agency (through its separate prime) that it had been delayed by two other separate contractors. In rejecting the claim, the court noted that the contract had stated that the state owner could not guaranty the unimpeded progress of contractors in this multiple prime contract project. The court held that the trial court ruled correctly in concluding that the state fulfilled its obligation, as it acted promptly to address disruptions. The state made significant and meaningful attempts to move the project along. The court stated that the state could not have been expected to terminate the contracts of those contractors causing delay to the claimant.

Suppose one separate contractor damages the work or property of the other separate contractor. Under the *Moore* case, a properly drafted construction contract would give the separate contractor harmed a legal claim against the separate contractor causing the harm. But does the former also have a claim against the owner? Such a claim could be based on the owner's breach of an express or implied provision to indemnify a separate contractor if its property is damaged by another separate contractor.

Finally, can a separate contractor maintain a direct action against the surety of the other separate contractor? New Jersey faced this in a state public contract.[51] State contracts usually require a bond be furnished to protect unpaid subcontractors, suppliers, or laborers. In this case, each separate contractor furnished a performance bond (one not required by statute) and a payment bond (one required by statute). The statute specifically gave rights to people who perform work under the contract of the contractor who furnished the bond, such as suppliers and subcontractors, but not to those who perform under their own contracts, such as other separate contractors. Consequently, New Jersey did not allow the separate contractor to sue on the bond. Also, it held that only the owner could maintain an action on the performance bond. As a result, the separate contractor was left with a claim against a bankrupt separate contractor. Although New Jersey stated it was following the majority rule, much depends on the language of any statute compelling that a bond be furnished, on the bond language itself, and on any language in the contract requiring the bond, as well as on the common law.

Owners contemplating a separate contract system should consider these problems. Clearly, a separate contract system

ought not to be used unless the managerial function can be better performed and at a lower cost than if performed by a prime contractor under a single contract system. If the separate contract system is selected, all contracts should make clear who has the administrative responsibility and who has the legal responsibility if a contractor has been unjustifiably delayed. More particularly, if this responsibility is to rest solely in the hands of the people who are given the responsibility for managing the contract, all contracts should make clear whether the owner is exculpated.

D. Construction Management

Before seeking to define construction management (CM), the reasons for its development must be outlined. Principally, CM developed because of perceived weaknesses and inefficiencies in the traditional construction process with special reference to the inability of design professionals and contractors to use efficient management skills. Design professionals were faulted for their casual attitude toward costs, their inability to predict costs, and their ignorance of the labor and materials market, as well as of the costs of employing construction techniques. Owners were also concerned about the tendency of design professionals to take less responsibility for quality control, policing schedules, and monitoring payments. Contractors also came in for their share of blame. Some lacked skills in construction techniques and the ability to work with new materials. Others did not have the infrastructure to comply with the increasingly onerous and detailed workplace safety regulations.

Construction management instead would be an *efficient* tool for obtaining higher-quality construction at the lowest possible price and in the quickest possible time. A limited study by the U.S. General Accounting Office of certain federal agencies' use of construction management records some problems in the past, mainly caused by delay claims in fast-track projects.[52] It notes the difficulty of using CM concepts in federal construction and reports that projects using the CM concept experienced time delays exceeding six months more often than projects that did not use the concept. From a cost standpoint, the CM projects less often had cost increases that exceeded 10 percent. Obviously, there are many ways of accounting for these differences other than simply the CM process.

---

[51]*MGM Constr. Corp. v. N.J. Educ. Facilities Auth.*, 220 N.J.Super. 483, 532 A.2d 764 (Law Div.1987). For a case that came to a different conclusion, see *Hanberry Corp. v. State Bldg. Comm'n*, supra note 43.

[52]U.S. General Accounting Office, supra note 36.

In any event, spotlighting the CM process should lead to more studies of its effectiveness.[53]

A great variety of approaches may be used to achieve the objectives promised by construction management. For this reason, this book can provide only generalizations. That having been said, a broad division of two types of CMs has been developed in response to the widespread use of standard form contracts: the CM as agent (CMa) and the CM as constructor (CMc).

Until 1992, the AIA and AGC published competing CM documents, the former emphasizing the CM as agent and the latter emphasizing the CM as constructor. This divide between the two organizations was bridged in 1992. The AIA published a set of CMa documents (A101/CMa, A201/CMa, B141/CMa, and B801/CMa) that are recommendations for how the CM fits into the contract system when acting as an adviser to the owner. At the same time, the AIA published A121/CMc, and the AGC published AGC 565, a joint product to be used when the CM acts as constructor.

In 1993, the AIA published B144/ARCH-CM, an amendment to B141 to be used when the architect also performs CM services. The cooperation between the AIA and AGC continued with the joint publication of AIA A121/CMc-2003 and AGC 565 (CMc with a GMP) and also A131/CMc-2003 and AGC 566 (CMc with no GMP). This cooperative period ended in 2007 when the AGC joined several other trade associations in the publication of ConsensusDOCS 500 (owner/CM agreement with a GMP) and 510 (owner/CM agreement without a GMP).

The divide between CMa and CMc is illustrated in Figure 17.3. The CMa contracts only with the owner and has no contract relationship with the subcontractors. (There is rarely any contract between design professional and CM.) By contrast, the CMc contracts with both the owner and the specialty trades. The CM acting as a professional adviser can be compensated by any of the various methods used to compensate a design professional, such as percentage of construction costs, personnel multiplier, cost plus fee, or fixed price.[54] As noted, the CM acting as

---

[53]For a series of articles on construction management, see 46 Law & Contemp.Probs., Winter 1983. See also Spangler & Hill, *The Evolving Liabilities of Construction Managers*, 19 Constr. Lawyer, No. 1, Jan. 1999, p. 30.

[54]For a case describing functions and compensation of a CM, see *Gibson v. Heiman*, 261 Ark. 236, 547 S.W.2d 111 (1977).

CM as Adviser (CMa)

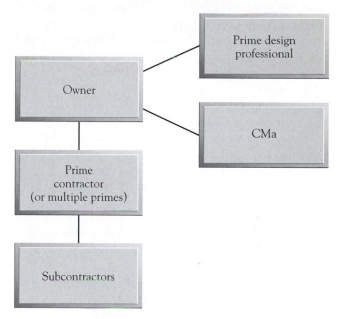

CM as Constructor (CMc)

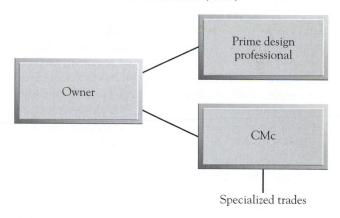

FIGURE 17.3   Construction management systems. (See Figure 14.2 for another CMa illustration.)

a constructor can be paid either by guaranteed maximum price or on a cost plus fee basis without a GMP. The CMc looks very much like a prime contractor.

While the risk of unforeseen subsurface conditions will be discussed in Chapter 25, it is helpful to note the effect of having a CMc rather than a prime contractor and its effect on the federal government's obligation when a differing site condition (DSC) is encountered. As will be noted in Section 25.06A, the contractor is entitled to an equitable adjustment if the actual conditions are materially different from those represented.

In *Whiting-Turner/A. L. Johnson Jt. Venture v. General Services Administration*,[55] the claimant acted as a CM during the design phase and later made a contract to build the project for a cost-plus a fixed fee with a GMP. One of the CMc's subcontractors encountered water all around the project rather than the limited places indicated in the subsurface information given the CMc. Even though the CMc was involved in assembling the information that turned out to be inaccurate, the board rejected the government's contention that the CMc made the representations. It held that they were made by the government and relied upon by the CMc. The use of the CMc did not relieve the government of its DSC obligations. The decision shows that the injection of the CM system could have had an effect upon the owner's obligations to furnish accurate subsurface information. But the board, though admitting that the CMc was involved in planning in a way that a normal contractor would not have been, concluded that the CMc was basically a contractor.

Another illustration of the ambiguous status of the CM can be shown by a recent case in which the CM sought to assert a mechanics' lien on property whose construction he had supervised. Were he a laborer, a subcontractor, or a design professional, he would have been entitled to a lien. But the court held that the CM had not designed, or built, but had merely supervised and was not entitled to a lien under the laws of Indiana.[56] In other states, the CM might have succeeded, as mechanics' lien laws vary greatly.

CMs who contract on their own usually guarantee a maximum price (GMP)—a cost contract with a "cap." The courts have taken a strict attitude toward the GMP. Rejecting a contention by the CM that the GMP was a "target figure," an Indiana court held the CM to the GMP, despite the fact that one of the trade contractors gave a price quotation substantially higher than planned and the CM decided to do the work itself.[57] In such arrangements,

CMs may also perform some construction with their own employees. As noted, in 2003 the AIA and the AGC jointly drafted CM documents in which the CM acts as constructor.[58]

When CMs give a GMP, they usually agree to give it after the design has been worked out to a designated portion of finality, say 75 percent. They protect themselves by getting as many fixed-price contracts from specialty trades as they can and usually build enough into their GMP to take into account not only their fees but also contingencies that may arise from both design changes and unforeseen circumstances.

Despite this caution, one risk inherent in a GMP is that the specialty trades or the CM who gives a GMP may claim drastic scope changes have eliminated any contractual price commitments made. Usually, though not invariably, a construction management system uses a fast-track approach, a principal justification for construction management, to complete the project in a shorter time. Yet it is possible to have a CM without having a fast-track system. The dangers of fast-tracking—sloppier work and uncontrolled costs—may persuade an owner who wishes to use the construction management system not to fast-track the job.

The legal issues have at their core whether the CM is more like a design professional or an entrepreneurial contractor. For example, must the CM be registered or licensed by the state, and if so, which type of registration or license is needed?[59] The answer may depend on which form of CM is used, whether the CM is engaged solely as a professional adviser, or whether the CM undertakes to perform some construction himself.

Must awards for CM services in public contracts be made in the same manner as other professional services? Most courts view CMs as providing primarily professional services and permit public agencies to hire them through the negotiation process.[60] Others require use of the competitive bidding statutes when hiring CMs.[61]

The role of the CM in awarding construction contracts also raises the issue of analogies. For example, one case held

[55]GSBCA No. 15401, 02-1 BCA ¶31,708.

[56]*Murdock Constr. Management, Inc. v. Eastern Star Missionary Baptist Church, Inc.*, 766 N.E.2d 759 (Ind.App.), transfer denied, 783 N.E.2d 694 (Ind.2002).

[57]*TRW, Inc. v. Fox Dev. Corp.*, 604 N.E.2d 626 (Ind.App.1992), transfer denied May 21, 1993.

[58]The AIA version is A121/CMc, and the AGC version is AGC 565.

[59]See Bynum, supra note 39 at 27–28; Lunch, infra note 74 at 86–87.

[60]*Shivley v. Belleville Township High School Dist. No. 201*, 329 Ill. App.3d 1156, 769 N.E.2d 1062 (2002) and *Malloy v. Boyertown Area School Bd.*, 540 Pa. 915, 657 A.2d 915 (1995).

[61]*City of Inglewood-L.A. County Civic Center Auth., v. Superior Court*, 7 Cal.3d 861, 500 P.2d 601, 103 Cal.Rptr. 689 (1972).

that the CM represented the owner when he awarded a contract, thereby allowing the trade contractor to demand arbitration with the owner.[62] If he is simply a professional adviser who, unlike most design professionals, has been given the authority to award contracts, the contractor has made a contract with the owner *through* the CM. Here he looks more like a design professional than a contractor.[63] (Usually the design professional does *not* have authority to award contracts.) However, if he contracts on his own with specialty trade contractors, he looks like a prime contractor.[64]

Liability problems also involve analogies. For example, it has been held that a contractor can bring a negligence claim against the CM just as a negligence claim can be instituted against a design professional.[65] If the CM is analogized to a design professional or performs services usually performed by a design professional,[66] will the CM be given the benefit of the professional standard described in Section 14.05 or quasi-judicial immunity noted in Section 14.09D?

As shown in Section 14.05, the professional standard has its greatest application when the design professional is performing design rather than administrative services. Inasmuch as the services performed by a CM are connected more to administration than to design, it is possible that the *ordinary* but not *professional* negligence standard will be applied, meaning that no expert testimony will be required. But if the CM is simply an adviser performing administrative tasks for the owner, he looks less like a design professional. As a result, the CM may be able to defend himself more successfully if sued by other participants by asserting that he was simply advising the owner and not acting as an independent professional.

Yet an architect was not allowed to assert a negligence claim for economic losses against the CM. The court held that the CM had no duty to the architect, that his sole duty was to the owner who had engaged him, and that the contract between owner and CM was not made for the benefit of the architect.[67] The court held that the architect could not satisfy the California requirements for a negligence claim for economic losses. Such claims are dealt with in detail in Section 14.08E.

A CM is also exposed to claims by construction workers injured on the job. Liability under the common law may be based on analogies to injured worker claims against design professionals.[68] If CMs instead look like prime contractors, either because they are managing the work of the specialty trades, have overall safety responsibility, or are performing some of the construction work with their own forces, they may be found to be contractors with contractor-like liability.[69] Liability may also arise under safety statutes. New York imposes a duty on all contractors and owners "and their agents" to make sure workers are furnished with safety equipment. A CM acting as the owner's advisor who has authority over the work may be liable under this statute as an agent of the owner.[70] Under federal law, a CM with a pervasive job site presence and authority to stop unsafe work may be liable for a worker's injury under the Occupational Safety and Health Act, even if the CM did no actual construction work.[71]

In some states, being a "contractor" CM can be advantageous in the event of worker claims. For example, such a CM may be considered a "statutory employer" entitled to immunity under the workers' compensation laws as described in Section 7.04C.[72] This was unsuccessful where

---

[62]*Seither & Cherry Co. v. Illinois Bank Bldg. Corp.*, 95 Ill.App.3d 191, 419 N.E.2d 940 (1981).

[63]See *Bagwell Coatings, Inc. v. Middle South Energy, Inc.*, 797 F.2d 1298 (5th Cir.1986) (contractor cannot sue CM who was owner's agent).

[64]But see *Turner Constr. Co.*, ASBCA No. 25171, 81-1 BCA ¶15,070 (trade contractor engaged by CM could sue public agency who maintained significant direct involvement in award and performance despite disclaimer of any contractual relationship between public agency and trade contractor).

[65]*Gateway Erectors Div. v. Lutheran Gen. Hosp.*, 102 Ill.App.3d 300, 430 N.E.2d 20 (1981).

[66]*Pierce Assoc., Inc.*, supra note 35 (CM interprets contract and judges performance).

[67]*The Ratcliff Architects v. Vanir Constr. Management, Inc.*, 88 Cal. App.4th 595, 106 Cal.Rptr.2d 1 (2001). The architect sought unsuccessfully to claim indemnity under the owner–CM contract. The court seemed to think the negligence claim was an "end run" around the attempt to claim indemnification.

[68]*Caldwell v. Bechtel, Inc.*, 631 F.2d 989 (D.C.Cir.1981). But see *Everette v. Alyeska Pipeline Service Co.*, 614 P.2d 1341 (Alaska 1980) (CM relinquished safety responsibilities and was not liable).

[69]*Lemmer v. IDS Properties, Inc.*, 304 N.W.2d 864 (Minn.1980) (CM assumed responsibility for job safety); *Farabaugh v. Pennsylvania Turnpike Comm'n*, 590 Pa. 46, 911 A.2d 1264 (2006) (same).

[70]*Walls v. Turner Constr. Co.*, 4 N.Y.3d 861, 831 N.E.2d 408, 798 N.Y.S.2d 351 (2005).

[71]*Bechtel Power Corp.*, 4 BNA OSHC 1005, 1975–1976 CCH OSHD ¶20,503 (OSHRC 1976), aff'd, 548 F.2d 248 (8th Cir.1977). As to OSHA, see Section 14.08G.

[72]*O'Brien v. J.C.A. Corp.*, 372 Pa.Super. 1, 538 A.2d 915 (1988).

the CM did not construct anything but was simply hired to manage and inspect, more like a design professional than a contractor.[73]

Those planning to engage a CM or those about to engage in CM work must also determine whether the insurance coverage of the CM will be adequate. The CM who is principally a contractor will find his commercial (formerly called "comprehensive") general liability (GGL) coverage does not include design, whereas a CM who is essentially a design professional may find his professional liability insurance excludes any coverage relating to the construction process or the work of the contractor.[74] For this reason, CMs should carry professional liability insurance (as should design professionals).[75]

Owners contemplating hiring CMs must consider the advantages and disadvantages of inserting this new actor into an already crowded assemblage of project participants. CMs claim to make the construction process more efficient (and hence faster and cheaper) by bringing sophistication and modern management techniques to the construction process. CMs hired as advisors, however, may be viewed by the courts as acting as the owner's agent. As explained in Section 4.02A, this could expose the owner (as principal) to liability to trade contractors who claim they were injured by the CM's actions.[76] In addition, employing a CMa, together with an architect and prime contractor, invariably will add cost to the owner and may also cause confusion as to the chain of command. On the other hand, an owner who hires a CMc may, if problems arise, find itself in the same adversarial relationship as it would with a prime contractor, especially if a GMP is involved.

E. Turnkey Contracts

The discussion in this subsection must be taken together with the material in subsection F dealing with design–build. Both subsections involve the owner contracting with an entity who agrees to both design and build. However, because turnkey contracts have other functions, they should be taken separately, both on their own and as an introduction to design–build.

There are a great variety of turnkey contracts. At its simplest, the contract is one in which the owner gives the turnkey builder some general directions as to what is wanted and the turnkey builder is expected to provide the design and construction that will fill the client's communicated or understood needs. In theory, once having given these general instructions, the owner can return when the project is completed, turn the key, and take over.

Many turnkey projects are not that simple. The instructions often go beyond simply giving a general indication of what is wanted. They can constitute detailed performance specifications.[77] Also, the obligation to design and build may depend on the owner furnishing essential information or completing work on which the turnkey contractor relies to create the design and to build. Finally, the owner who has commissioned a turnkey project is not likely to remain away until the time has come to turn the key. As in design–build, the owner may decide to check on the project as it is being built and is almost certain to be making progress payments while the project is being built. The most important attribute of the turnkey project is that one entity both designs and builds, a system discussed in greater detail in Section 17.04F.

Some turnkey contracts require the contractor not only to design and build the building but also to provide the land, the financing, and interior equipment and furnishings.

A turnkey contract looks more like a sale than a contract for services. As a result, one court held that a turnkey contract created warranties that made the seller-contractor responsible for any defects.[78] This is something that may also be found in any design–build contract.

---

[73]*Brady v. Ralph Parsons Co.*, 308 Md. 486, 520 A.2d 717 (1987), subsequent appeal, 327 Md. 275, 609 A.2d 297 (1992).

[74]See Lunch, *New Construction Methods and New Roles for Engineers,* 46 Law & Contemp.Probs. 83, 93–94 (1983).

[75]*1325 North Van Buren, LLC v. T-3 Group, Ltd.*, 293 Wis.2d 410, 716 N.W.2d 822 (2006).

[76]*Aladdin Constr. Co., Inc. v. John Hancock Life Ins. Co.*, 914 So.2d 169 (Miss.2005) (CM was agent of the owner for purposes of paying the trade contractors).

[77]Another definition is given in *Smithco Eng'g, Inc. v. Int'l Fabricators, Inc.*, 775 P.2d 1011, 1015 (Wyo.1989), citing Vol. 1 B. J. MCBRIDE, & I. WACHTEL, GOVERNMENT CONTRACTS, § 9.120[12] (1980) as follows: "A turnkey project is one in which the developer builds in accordance with plans and specifications of his own architect subject to performance specifications for quality and workmanship, and with limited guidance for features as style of house, number of bathrooms, etc."

[78]*Mobile Housing Environments v. Barton & Barton,* 432 F.Supp. 1343 (D.Colo.1977).

F. Combining Designing and Building (DB)

In addition to the weakness of the traditional design–bid–build method outlined in Section 17.03B, there are other weaknesses that led the explosion of the design-build (DB) system in the 1990s. One commentator states that it can no longer be assumed that "the most advanced construction technology and knowledge of the most construction methods lie with architects and engineers."[79] He goes on to state that this knowledge lies increasingly with "specialty contractors and building-product manufacturers." He also notes that it increasingly has become difficult to prepare complete and accurate drawings and specifications, hereby exposing owners to claims based on defective specifications. He also points to increased cost and delay in determining who is responsible for defects, the designer or the contractor, and the need for "single point" responsibility. Finally, he states that the traditional method simply takes too long and costs too much in preconstruction services.

As a result, it is increasingly popular for one entity to both design and build, an important variation from the traditional method of organizing for construction. See Figure 17.4. DB can encompass at one extreme the homeowner building a single-family home patterned on a house that the builder has already built and at the other a large engineering company agreeing to both design and build highly technical projects, such as petrochemical plants. The former is likely to be a builder with an in-house architect on its staff or one that engages an independent architect where it is required by law that design be accomplished by a registered architect or engineer. At the more complex extreme, the D/builder may employ a large number of construction personnel and licensed architects and engineers in-house to offer a total package for projects such as power plants, dams, chemical processing facilities, and oil refineries. Between these extremes, DB is often used for less technical repetitive work such as warehouses or small standard commercial buildings.

Organizationally, DB has variations. One method is an architect promising to design and build and employing a contractor to execute the design. Because of the capital needed, a more common technique is for the D/builder to be a contractor who engages a design professional to create the design. Finally, a DB project can be

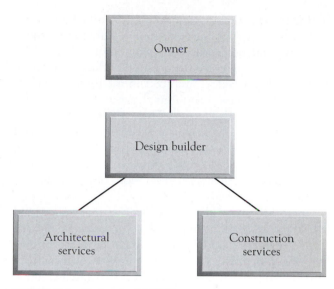

FIGURE 17.4 Design–build (DB) system.

a joint venture between a design professional and a contractor. As shall be noted, each form raises licensing and insurance issues.

The second method, that of a contractor engaging an architect to do the design work, is illustrated by the case of *C. L. Maddox v. Benham Group, Inc.*[80] That case involved a $10 million contract made by Maddox, the plaintiff D/builder. Maddox engaged the defendant architectural firm under a "subcontract" to complete the drawings and specifications (both used the same marketing agent). The architect must have had some site role. Language in the "subcontract," Paragraph 2.1.6, stated that the architect would keep the contractor "informed of the progress and quality of the Work, and shall endeavor to guard [Maddox] against defects and deficiencies in the Work of [Maddox]."[81]

The D/builder admitted that it and the owner had caused a substantial amount of the damages, but claimed that Paragraph 2.1.6 required the architect to guard the D/builder against its own deficiencies.

The court, however, pointed to Paragraph 2.1.7, which stated that the architect was not responsible for those performing construction, including the D/builder. This exonerated the architect.

[79]Hinchey, *Karl Marx and Design-Build*, 21 Constr. Lawyer, No. 1, Winter 2001, p. 46.

[80]88 F.3d 592 (8th Cir.1996).
[81]Id. at 602.

Both Paragraphs 2.1.6 and 2.1.7 are identical to language in AIA standard documents used for the performance of design services in the traditional design–bid–build system. This raises the question of the responsibility of the "subcontractor" architect in a DB contract. Is it the same as when he performs these site services under a contract with the owner? This, of course, should be controlled by the contract between the D/builder and its architectural "subcontractor" when the latter is not his employee. But the public interest, as noted later in this section, as well as the expectations of the owner may expand the architect's responsibility.

Increased use of the DB system led the professional and trade associations to publish standard documents for design as part of their DB document packages. Documents developed by these associations, such as the American Institute of Architects (AIA), the Associated General Contractors of America (AGC), the Design–Build Institute of America (DBIA), and the Engineers Joint Contract Documents Committee (EJCDC), reflect a wide variation in approaches often dictated by the policies of the associations and the persons they represent. A recent thorough and insightful study has been made of these documents.[82]

These standard documents (or any individualized contract between the D/builder and the designer) regulate the relationship between the D/builder and the person it engages to create the design.

But what is the relationship between the owner and the designer, whether an independent design professional retained by the D/builder or one on the D/builder's staff? Even though the designer works for the D/builder and owes his primary allegiance to the one who has selected him and pays him, the owner may believe the design professional has some obligation to the owner. The owner indirectly pays him, and as a licensed design professional, he has obligations to the public and to the owner even though the owner has no contract with him. In a vague and undefined way, the owner may believe that a registered architect or engineer will make independent decisions that take his interests and those of the public into account even though the D/builder has engaged him and pays him. This is demonstrated by some successful owner claims against design professionals with whom they had no contract.[83]

To give the owner more direct control over the design, some have advocated the "bridging" method. This technique, as described by one commentator,[84] blends the traditional design–bid–build (DBB) or the design–award–build (DAB) system with DB. Under it, the owner's consultant, generally an architect or an engineer, develops the schematic design and budget. He then develops the design and "prepares an extensive legal and technical request for proposal (RFP) for a design-build contract." Once the D/builder has been selected under whatever pricing is mutually acceptable, the D/builder provides the design through its own designer either as a subcontractor, as in the *Maddox* case, or as a member of its staff.

The D/builder's designer prepares the construction documents, "with the owner's consultant monitoring the work." The D/builder's A/E is the architect or engineer of record and is responsible for the design. But the owner's consultant administers the project. The D/builder's A/E checks submittals but submits them to the owner's consultant. The D/builder's A/E issues payments and completion certificates.

Such a "bridging" method may solve some problems but may create others. If the owner's consultant in effect controls the designing after the DB contract is made, the single-point responsibility, a main reasons for DB, may be lost. In addition, the owner may be found to have provided design specifications to the D/builder, not performance specifications as in a pure DB project. As explained in Section 23.05E, an owner impliedly warrants to the contractor the constructability of design specifications, but a warranty does not attach to performance specifications. So by using the "bridging" method, the owner shifts liability for

---

[82]Noble, *The Design-Builder-A/E Contract: A Comparison of Standard Forms*, 21 Constr. Lawyer, No. 1, Winter, 2001, p. 1. The most recent AIA documents are A141-2004 (owner and D/builder), A142-2004 (D/builder and contractor), B142-2004 (owner and consultant), and B143-2004 (D/builder and architect). For a discussion, see Quatman, *The AIA's New (and Improved) Design-Build Contracts*, 25 Constr. Lawyer, No. 2, Spring 2005, p. 37.

[83]*Keel v. Titan Constr. Corp*, 639 P.2d 1228 (Okla.1981); *Nicholson & Loup v. Carl E. Woodward, Inc.*, 596 So.2d 374 (La.App.), writ denied, 605 So.2d 1098 (La.1992) (though in-house architect did not select negligent soils engineer nor had input into the structural work, he did seal the documents representing that either he had done the design or it was done under his responsible supervision); and *Cunningham Hamilton Quiter v. M. L. of Miami*, 776 So.2d 940 (Fla.Dist.Ct.App.2000), rehearing denied Jan. 21, 2001 (owner's claim against D/builder's architect met with successful defense by latter that the claim must be arbitrated based on arbitration clause in DB contract).

[84]See Hinchey, supra note 79.

defects in the design from the D/builder to itself.[85] From the D/builder's perspective, this method may create uncertainty as to where the owner's design parameters end and where the D/builder's design responsibility begins.[86]

The relative newness of the DB concept can cause confusion as to whether the contract is a classic DB contract, with the D/builder given general parameters and providing the design. Such a classic DB contract makes the D/builder responsible for a successful implementation of the owner's expectations. But, as noted earlier, this risk allocation assumes that the owner has not taken control of the design.

FSEC, Inc.[87] demonstrates how difficult it may be to determine whether a classic DB contract has been created. The contract included drawings that required four exhaust fans and dust collectors. An amendment to the contract required the contractor to supply design drawings and project specifications. This left the contractor unsure of its obligation.

An amendment, in addition to describing the project as a DB contract, stated that the contractor was to complete design details for approval by the government. This labeling of the contract as a DB contract persuaded the contractor that it could design a system as long as it met performance criteria.

A specialty subcontractor determined that the performance specifications could be met with two exhaust fans. The contracting officer ordered the contractor to install four exhaust fans. The contractor's claim was rejected. The Armed Services Board of Contract Appeals held that all elements of the contract must be considered. It pointed to the bid documents requiring four fans and that design was subject to the government's approval. In essence, though described as a DB contract, it was not. It was a hybrid design–award–build contract with design specifications but labelled DB contract.

Even more than in construction management, the DB system has generated a large number of legal problems that must be confronted before the choice can be made to use such a system. In public projects, the principal problem is the usual requirement that construction work be competitively bid by the contractor. As seen in Section 11.03, the design professional is chosen either entirely on the basis of demonstrated competence and experience or in a two-stage process where the design professional competes with others as to competence and, once that has been determined, enters into a negotiation with the public entity.

The difficulty of classifying the project for determining whether a DB contract is a construction contract that must be competitively bid or one for professional services that can be negotiated is shown by *C&C/Manhattan v. Government of the Virgin Islands*.[88] The project was a prison to be done on a DB basis. The statutes allowed an exception for professional services, such as architecture design work. The government negotiated a DB contract relying on the design exception.

The contractor who had received the DB award pointed to cases that had held that DB contracts were an exception to the requirement for competitive bidding. The court stated that the exemption applied to highly technical contracts where the crux of the work was exempt services, the nonexempt services merely being incidental. The court would not make a blanket exception for DB contracts It noted that 70 percent of the price was for construction services. The court was fearful that exempting all DB projects could swallow up the requirement that, generally, public works contracts should be bid competitively.

In the traditional competitive bidding process by which contractors are selected (discussed in greater detail in Chapter 18), the design is completed and given to potential contractors to bid for the work. The lowest responsible bidder is usually selected. Bids of this sort cannot be made unless the design is relatively worked out.

Often a fast-track system is used for a design–build project. Thus prospective bidders cannot be given a detailed set of plans and specifications on which to make their bids. Although the DB system can be used competitively,[89] it does not fit comfortably with the requirement that the competitive bidding process be used to award construction work.[90] As a result, many barriers hinder the use of DB projects for public work. Some states have not allowed this method, whereas other states allow it, usually because it has been authorized for specific projects or because the state has enacted legislation that allows—or

---

[85]*Dillingham Constr., N.A., Inc. v. United States*, 33 Fed.Cl. 495 (1995), aff'd, 91 F.3d 167 (Fed.Cir.1996).

[86]*M. A. Mortenson Co.*, ASBCA No. 39,978, 93-3 BCA ¶26,189; *United Excel Corp.*, VABCA No. 6937, 04-1 BCA ¶32,485.

[87]ASBCA No. 49, 509, 99–2 BCA ¶30,512.

[88]40 V.I. 51, 1999 V.I. LEXIS 4, 1999, WL 117765 (Terr. V.I.1999).

[89]*Ogden Dev. Corp. v. Federal Ins. Co.*, 508 F.2d 583 (2d Cir.1974).

[90]*Sloan v. Department of Transp.*, 365 S.C. 299, 618 S.E.2d 876 (2005).

even requires—the design–build system to be used for certain types of construction.[91]

Because of the increased use of DB in federal procurement and the uncertainties of its legality in that system, in 1996 the Congress enacted the Federal Acquisitions Reform Act (FARA) of 1996.[92] It authorizes a two-step system under which the first step looks primarily at competence and qualifications. After the most qualified offerors are selected, they are asked to submit proposals for the second phase, which includes cost or price.

This two-step process is intended to avoid focus on low cost to the detriment of technical qualifications. This process resembles the Brooks Act, noted in Section 11.03, under which professional designers are selected by the federal government, and "little Brooks Acts," which are used by some states.

Another problem in public contracts is that the design for a design–build project may be made by a designer who has not been selected in accordance with the state laws for selecting design professionals for public works. Florida met this problem in 1989 by enacting legislation that sets forth specific standards when a public agency seeks to use the DB system.[93] These standards include a requirement that the agency engage a licensed professional to prepare a design criteria package. Also, the agency must select no fewer than three DB firms and specify criteria procedures and standards for evaluating the proposals. The agency must consult with the retained design criteria professional concerning the evaluation of bids.

A public entity wishing to use design–build must pilot through the shoals of the competitive bidding laws and the laws under which design professionals are selected before such a method can be used.[94]

In private contracting, the principal problem in design–build again relates to licensing laws. Under Missouri law, a contractor who did not have an architect's license could not recover for work performed under a DB contract, even though the architect it used was licensed.[95] As noted, the D/builder may be a business organization that will either furnish the design services needed in-house or retain an independent design professional to perform these design functions.[96] In addition, many states bar business corporations from performing architectural services. As a result, design–build contracts have been challenged, even if the design portion was to be fulfilled by the engaging of a licensed design professional.[97] Although the trend has been toward permitting such contracts despite the absence of a registered independent design professional, legal opinion should be sought if the design–build system is to be employed.

Finally, the liability insurance of a design professional usually excludes construction and work performed under a joint venture, whereas the liability insurance of a contractor usually excludes design. An entity planning to engage in a DB venture must check insurance carefully.

Despite the legal obstacles to the DB system, DB has clearly been found to be useful. The principal advantage of DB (called in England a "package" job), usually associated with fast-tracking, is speed. One entity replaces different entities who design and build with the inevitable delay caused by using two entities whose work intersects. Those who design and build frequently do repetitive work and acquire specialized expertise.[98] Perhaps most important from the standpoint of an unsophisticated owner, the system concentrates responsibility on the D/builder. Owners are often frustrated when they look to the designer who claims that the contractor did not follow the design, with the latter claiming that the problem was poor design.

---

[91]*State ex rel. Citizens Against Tolls (CAT) v. Murphy*, 151 Wash.2d 226, 88 P.3d 375 (2004).

[92]Enacted as part of the Defense Authorization Act of 1996, P.L. 104-106. See FARA § 5105. For a description, see Baltz & Morrissey, *Procuring Design-Build Construction Services: Federal Government's New Approach*, 31 Procur. Lawyer, No. 4, Summer 1996, p. 12.

[93]West Ann.Fla.Stat. § 287.055.

[94]According to the Design-Build Institute of America, as of April 2006, 45 states and the District of Columbia permit DB in some or all public contracts. For a survey, see THE DESIGN/BUILD DESKBOOK (Ed. Heisse, 2d ed., 2000). See also Busing, *The Law Struggles to Keep Pace with the Trend of State and Local Government Experience with Design/Build*, 11 Constr. Lawyer, No. 4, Oct. 1991, p. 22; Halsey & Quatman, *Design/Build Contracts: Valid or Invalid?* 9 Constr. Lawyer, No. 3, Aug. 1989, p. 1 (focusing on private contracts); Comment, 17 U.Dayton L.Rev. 109 (1991).

[95]*Kansas City Community Center v. Heritage Industries, Inc.*, 972 F.2d 185 (8th Cir.1992).

[96]Such a contract was upheld in New York in a 4–3 decision in *Charlebois v. J. M. Weller Assocs.*, 72 N.Y.2d 587, 531 N.E.2d 1288, 535 N.Y.S.2d 356 (1988). The dissent worried about the effect on the public and the registration laws when the design professional is engaged by the D/builder.

[97]See supra note 95.

[98]The Federal Highway Administration has reported a 14 percent reduction in project duration using DB, as well as slightly lower costs, with no reduction in quality. See *Design-Build Effectiveness Study* (Jan. 2006), found at http://www.fhwa.dot.gov/reports/designbuild/design build.htm.

Another impetus for the use of design–build occurred when the AIA, at first on a provisional basis, decided that its members could engage in business without running afoul of ethical constraints. The AIA did this because it felt economic realities in the construction industry dictated that architects be given the opportunity of engaging in DB ventures. After a study of how this was working, the AIA approved its members engaging in DB work and published a series of design–build documents.[99]

DB has weaknesses. The absence of an independent design professional selected by the owner can deprive the owner of the widest opportunities for good design. (One English architectural critic ascribed the recent low level of English design to increasing use of the package or DB system.) An unsophisticated owner often lacks the skill to determine whether the contractor is doing the job well or as promised. This can reflect itself not only in substandard work but also in excessive payments being made early in the project or in slow payment or nonpayment of subcontractors.

Owners, such as developers, do not like to make cost-plus contracts with D/builders. Clear and complete drawings and specifications protect a developer who uses a cost-plus contract. Although the price is not defined, unless there is an effective GMP the project is defined. In design–build, however, the owner may not even know what is to be built when it enters into the contract.

The contractor is reluctant to give a fixed price on a design–build project. It cannot know with any certainty what it will be expected to build. If the design has not been worked out at the time the contract is made, the design must be completed later. Very likely there will be redesign. As a result, the owner will prefer a fixed price, whereas the contractor would like an open-ended cost contract.

To ameliorate this possible impasse, the owner may find it helpful to prepare a set of performance specifications. Even if these specifications cannot define the elements of the building in detail, they may be able to prescribe intelligent criteria for performance in advance.

Another useful technique is to prepare a budget for each phase of the work and designate the budget estimate as a target price. If the actual cost is greater or less than the estimate, the contract price can be adjusted.

Progress payments also provide problems in design–build contracts. Under traditional methods, the architect certifies progress payments. Although some complain that the architect cannot be impartial when he does this, the architect's professional stature gives hope that he will be impartial.

In design–build, payment certificates issued by an architect who has been retained by the contractor may not be reliable. Owners who use design–build often take control of the progress payment process, a method that can operate to the disadvantage of the contractor. As a result, it may be useful for each party to agree on an independent certifier for progress payments.

One advantage of combining designing and building is the ability to fast-track the project. However, the two need not go together. One entity who both designs and builds may create the complete design before starting construction.[100]

An owner building a single-residence home, a warehouse, or simple commercial building relies entirely on the contractor for design and construction. Here warranties of fitness are appropriate. The owner is buying not services but a finished product. If the D/builder is held to have sold a "product" and given seller-like warranties, it may hold its design professional only to the professional standard as noted in Section 14.05A, unless the contract between the D/builder and the design professional it engages provides for a different standard.

The sophisticated owner who employs DB and has within its own organization the services of skilled professionals to control design or monitor performance has bought, not a product, but rather expert services to prepare and execute the design it chooses. Such an owner should not receive a sale-of-goods–like warranty.

The differentiation between the two types also reflects itself in pricing. It is likely that the homeowner commissioning a builder to design and build a house will contract on a fixed price. The sophisticated owner building a petrochemical plant will contract on a cost basis with a GMP, at least until the design is put in final form.[101] Such projects create immense exposure for consequential damages. It is likely that the D/builder will limit its obligation to correction of defects.

---

[99]See supra note 82.

[100]See Dibner, *Construction Management and Design-Build: An Owner's Experience in the Public Sector*, 46 Law & Contemp.Probs. 137, 143–144 (1983).

[101]See Foster, supra note 1 at 118–119.

G. Partnering

"Partnering" has been receiving increasing attention in the construction process world. Although it is not strictly speaking a form of organization, partnering is an approach that can remedy some of the stiffness and adversary tendencies that are so commonly part of the construction process.

Partnering is a team approach in which owners, contractors, architects, and engineers form harmonious relationships among themselves for the express purpose of completing a single project or group of projects. The partnering process usually is initiated when the principal project participants are assembled. It can also be used to steer a troubled project back on course. These relationships are to be structured on trust and dedication to the smooth operation of the project. Those who advocate this concept hope that it will accomplish the project successfully and minimize the likelihood of claims. Partnering, then, is a method of building a closeness that transcends organizational boundaries. This closeness should be premised on a shared culture and the desire to maximize the likelihood of attaining the common goals.

Although the objectives of partnering can be sought and achieved on an informal basis, focus on this concept has led to the designation of a facilitator and development of an agreement designed to highlight the goals of partnering and the methods by which they are to be achieved. These can be done after contracts have been awarded to the major participants: the owner, design professional, prime contractor, and key subcontractors. The agreement can make concrete the often nebulous partnering commitments.

What is important is that the parties express as clearly as possible what each party expects and what each party is willing to undertake. Clearly, expressions of this sort are difficult to make sufficiently specific and concrete. But some commitments, such as confidentiality and the agreement not to profit at the other "partner's" expense, may be worth expressing in the contract.

Partnering can suffer unless the entities involved in the effort are willing to make sacrifices to keep the team intact. This is particularly a problem in dealing with design professionals, because clients often express concern when design professionals with whom they have been working are assigned to other work. But it can also be a problem in large engineering firms and the organizations that employ them. Contacts among people working at different operational levels of the various entities must be developed, and a sense of confidence and trust among these workers must be built up and maintained.

It is likely that resort to litigation will be rare in partnering arrangements. Judicial intervention will almost never take the form of ordering people to continue such arrangements or compensating one party—particularly as to acts in the future—because of the other's refusal to continue. There may, however, be resort to legal action if one party has breached confidentiality or has stolen business opportunities that should go to the other. For this reason, it is useful to deal with the possibility of litigation in the contract. Finally, legal action may be taken if one party has incurred expenses that it feels the other should bear or at least share. Again, these matters may be appropriate for contractual resolution.[102]

One often unrecognized danger in partnering is the possibility the partnership can become "too successful." Understanding the other party's problems and desiring to implement common goals can make an employee forget for whom he works. At its worst, it can be an avenue for corruption. For example, an employee may allow an unmeritorious claim by the other party to gain favor with an employee of the latter or to take a bribe. These possibilities, however, do not destroy the value of the concept.

H. Teaming Agreements

Teaming agreements have been used when the federal government has embarked on programs to fulfill procurement needs, mainly in high-tech defense. A teaming agreement is a contract between a potential prime contractor and another firm that agrees to act as a potential subcontractor. The prime, sometimes with the assistance of the potential subcontractor, draws up a proposal for a federal agency contract in competition with other prospective contractors. Typically, the prime promises that if awarded the prime contract, it will reward its "teammate" with an implementing subcontract in return for the often unpaid support given in making the bid.

Most litigation has involved claims by aggrieved teammates against the prime when they have not been awarded the implementing subcontract. Such agreements generally

---

[102]See Roberts & Parisi, *Partnering: Prescriptions for Success*, 13 Constr. Litig.Rep., No. 11, Nov. 1992, p. 298 (contains suggestions and cites relevant literature). See also Stipanowich, *The Multi-Door Contract and Other Possibilities*, 13 Ohio St. J. on Disp. Resol. 303, 378-384 (1998). See also Section 30.18H.

use such phrases as "best efforts" and "reasonable efforts" as well as other nonspecific terminology. Often the question of whether the prime is actually committed to award the implementing subcontract to the potential subcontractor is left vague and uncertain. These arrangements resemble in some ways the normal prime contractor–subcontractor relationship discussed in Chapter 28. As noted in Section 28.02, the primes seek to retain as much flexibility as possible while still keeping their teammate interested in supporting the arrangement. However, the prime would like to ensure that any commitments made by the teammate can be legally enforced if necessary.

Moreover, similar to the prime–subcontract relationship discussed in Chapter 28, the terms that would govern the relationship if the prime wins the award and awards the subcontract to its teammate frequently have not been fully worked out. Sometimes such agreements include express covenants of good faith and fair dealing, but such terms do not necessarily mean the two parties will always agree on more substantive questions. Also, the twilight area between unenforceable agreements to agree and enforceable agreements to negotiate in good faith can create difficult legal questions.[103]

I. Lean Project Delivery

As described by one commentator,[104] lean project delivery involves extending the business principles of Toyota Motor Company to the construction process. The three principles underlying the "Toyota Way" are: (1) allowing anyone, including factory workers, to stop production in the face of a defect; (2) just-in-time delivery of materials; and (3) subsuming the individual production units within output of the entire project.

The organizational foundation for lean delivery is the Integrated Agreement, which is signed by the owner, architect, and construction manager (CM) or prime contractor *before* creation of the final design. These three participants are the project team. Major project-related decisions are made by consensus of these three members with the owner having the final word in the event of an impasse. Selection of additional project members is made by requests for proposal, rather than based strictly on price. New members must agree to participate in the project based on a level of responsibility and collaboration described in the Integrated Agreement:

> By forming an Integrated Team, the parties intend to gain the benefit of an open and creative learning environment, where team members are encouraged to share ideas freely in an atmosphere of mutual respect and tolerance. Team members shall work together and individually to achieve transparent and cooperative exchange of information in all matters relating to the Project, and to share ideas for improving project delivery as contemplated in the Project Evaluation Criteria. Team members shall actively promote harmony, collaboration and cooperation among all entities performing on the Project.[105]

Rather than the architect's presenting a completed design to the construction participants, the Integrated Agreement calls for the core group to create a "target value design." The target value design makes explicit the value, cost, schedule, and constructability of basic components of the design criteria. This process allows the major participants to have input into the design during its creation, where problems (whether with goals, constructability, or cost) can be identified and dealt with as early into the process as possible. Also, the major trade contractors (such as mechanical, electrical, and plumbing) participate in the design of their segments of the work.

The project team creates a "phase plan" for each segment of the work. A six-week look-ahead plan is prepared in which the team identifies the prerequisites each phase must meet for the work not to be defective. Each week, the team screens upcoming work assignments for defects (broadly including unanswered requests for information, incomplete prerequisite work, and missing materials or labor resources) and releases work to the field only if that work has no constraints. Work commitments are then obtained from the relevant subcontractors who are to perform the next segment of work. Because the Toyota Way values reliability over speed, the subcontractors are expected to decline a work assignment if the work is defective or if they lack confidence they can timely perform the work. By saying "no," the

---

[103]For further discussion, see M. MUTEK, CONTRACTOR TEAM ARRANGEMENTS—COMPETITIVE SOLUTION OR LEGAL LIABILITY (2006), Killian & Fazio, *Creating and Enforcing Teaming Agreements*, 25 Constr. Lawyer, No. 2, Spring 2005, p. 5, and Note, 91 Colum.L.Rev. 1990 (1991).

[104]Lichtig, *The Integrated Agreement for Lean Project Delivery*, 26 Constr. Lawyer, No. 3, Summer 2006, p. 25.

[105]Id. at 29.

subcontractors avoid the creation of defective work and allow the project team to replan the work and avoid the creation of waste. The subcontractors must stay in constant communication with each other and the project team to know whether their commitment to perform new work can be kept.

Compensation of the prime contractor (or CM) is on a cost plus fee basis with a guaranteed maximum price (GMP). The prime contractor and major subcontractors agree to limit the basis for requested change orders to a few, predesignated conditions: material changes in the scope of the work, changed site conditions and unforeseen regulatory or code interpretations. Disputes are decided not by the architect, but through participation of the project team. If resolution is still not possible, an outside expert is brought in.

The Integrated Agreement seeks to eliminate negligence as the measure of the architect's financial responsibility. Instead, the owner and core group members negotiate a deductible as a percentage of the construction costs and allocate it to design defects, without investigation of underlying fault. Above that deductible, the parties negotiate a percentage for which the designer is responsible without proof of negligence. Above that amount, the owner must prove negligence to recover.

In short, the Integrated Agreement seeks to create a system of shared risk, rather than shifting it. Value is increased by emphasizing collaboration, early detection and correction of defects, and early involvement by the project team to understand the project's goals and produce a target value design.[106]

J. Project Alliance

Project alliancing shares with lean project delivery the establishment of collaborative relationships among the owner and the major project participants. Unlike lean project delivery, project alliance does not seek to mimic the Toyota Way and does not draw its theoretical framework from the manufacturing sector.

In project alliance, the owner, design team, prime contractor and major trade contractors and suppliers all sign the same contract, the Project Alliance Agreement (PAA). The PAA has as its goal the benefit of the project as a whole. The financial success of the project participants is contingent upon the financial success of the project. Financial goals are set forth in the PAA; each party is financially rewarded or penalized based upon whether the project as a whole meets or fails to meet these predetermined goals. In addition, the parties agree in advance to release one another from all liability arising out of the project, except for willful misconduct. In short, the alliance participants:

- assume collective responsibility for delivering the project
- take collective ownership of all risks associated with the delivery of the project
- agree to performance targets
- share in the financial pain or gain, depending on how actual project outcomes compare with the preagreed targets that they have jointly committed to achieve, and
- release one another from negligence liability[107]

The owner chooses the major project participants through a rigorous selection process, in which their technical skill, performance capacity, and ability to get along are all taken into consideration. Because of the full commitment demanded of the alliance participants, senior management must be brought into the project early in the design process. They must be able to sign off on the performance and scheduling targets, then maintain a hands-on approach during the performance period so that disputes are quickly identified and resolved.

The project alliance delivery system is reserved for particularly difficult, large-scale projects. It was developed by the petrochemical industry, but it then spread to large public works and commercial projects. At present, it seems to be used primarily in Australia and New Zealand.[108] Initial results show that these projects have been brought in ahead of schedule and under budget.[109]

---

[106]For further discussion, see the website for the Lean Construction Institute, www.leanconstruction.org.

[107]Wilke, *Alliancing for Infrastructure Projects—Sharing Risks and Rewards with a "No Blame" Agreement,* J. A.C.C.L., May 2007, pp. 211, 212 (listing just the first four components).

[108]Construction of the National Museum of Australia in Canberra is discussed in Hauck et al., *Project Alliancing at National Museum of Australia — Collaborative Process,* 130 J. of Constr. Eng'g & Mgmt., No. 1, Jan–Feb. 2004, p. 143. The Department of Treasury & Finance, State of Victoria, Australia, has published a *Project Alliancing Practitioners' Guide* (Apr. 2006). Go to www.dtf.vic.gov.au and look under Publications.

[109]Wilke, supra note 107 at 216; Hauck, supra note 108 at 150.

K. Program Management

The demand for program management arises not from dissatisfaction with the traditional model for project delivery, but from company downsizing and budget constraints, which have led large commercial owners to eliminate their in-house engineering and construction departments.[110] A program manager (PM) promises to bring to commercial owners the expertise that these owners need to manage and expand already existing large facilities. The PM's role has been described as "doing for the owner what it would do for itself if it had the in-house capacity or if it chose to divert its resources to the project(s)."[111]

Akin to a CM as advisor,[112] the PM does not undertake responsibility for actual construction work. One commentator lists likely services a PM would offer as including:

- analysis of owner's current facilities and needs
- assisting in project site evaluation and acquisition
- identifying regulatory and permitting requirements
- documenting the existing site's conditions
- managing a public/community relations program
- developing risk management and insurance program
- prequalifying the design and construction team members
- developing design criteria
- developing and updating project budgets
- managing project communications, documenting flow, and record keeping during the construction phase
- monitoring quality assurance and safety programs[113]

Either large design firms or construction companies may market themselves to owners as qualified to provide program management services. Both AIA and EJCDC standard form documents allow either architects or engineers to offer many of these services, with the exception of assuming responsibility for site safety.[114] Contractors, by contrast, may have more experience with constructability, pricing, and scheduling services.

Is a PM's liability assessed under a standard negligence standard or under the professional standard? Where the PM is a licensed design firm, courts will probably conclude that the professional standard applies. Would the result change if the PM is not originally a design professional company?

Much like construction management thirty years ago, program management is still an evolving concept. One impediment to a clearer definition of PM has been the lack, until quite recently, of a standard form contract available for those providing such services. Only in 2007, with the Associated General Contractor's publication of ConsensusDOCS Document 800 (owner/PM agreement and general conditions), has a standard form contract been made available to the marketplace. The impact of this recent development remains to be seen.

L. Build-Operate-Transfer (BOT)

In the international field, most developing countries are not able to finance large infrastructure projects. As a result, after competitive bidding they may make an agreement with a consortium under which the latter operates the project and then ultimately transfers it back to the state. See Figure 17.5. The consortium that seeks to enter such a contract usually must provide the financial backing to build the project, get it operating, and keep it running so that it can be transferred. The consortium expects to be repaid mainly from revenues generated. BOT projects are used in large infrastructure projects such as energy generation, bridges, and transportation (roads, rails and waterways) systems. Most BOT projects are turnkey contracts, which makes financing more secure and easier to obtain. Increasingly, these privately financed projects are being considered and used by developed countries and some American states, mainly for toll roads and high-speed rail systems. When this is done, the government may put up a part of the funds needed.

The principal actors involved in a simple BOT project are

1. the granting authority (usually a sovereign state)
2. the project sponsors (the entity that has put together the principal actors)
3. the financiers
4. the users of the project

The granting authority wants to be certain the project is built properly, because its citizens are likely to use the

[110]Scotti, *Program Management: The Owner's Perspective*, 16 Constr. Lawyer, No. 4, Oct. 1996, p. 15.

[111]Terio, *Program Management: A New Role in Construction*, 16 Constr. Lawyer, No. 4, Oct. 1996, p. 4.

[112]See Figure 17.3.

[113]Noble, *Program Management: The Design Professional's Perspective*, 16 Constr. Lawyer, No. 4, Oct. 1996, p. 5.

[114]Id. at 6–7 (comparing PM services with AIA Doc. B141 and EJCDC Doc. 1910-1). B141 is now B101, and EJCDC 1910-1 is now E-500.

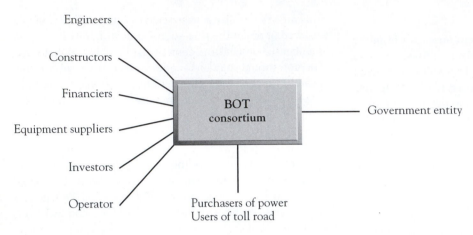

Engineers

Constructors

Financiers

Equipment suppliers

Investors

Operator

BOT consortium

Government entity

Purchasers of power
Users of toll road

FIGURE 17.5 Build–operate–transfer (BOT) organization.

project directly or indirectly and because it will be the recipient of the project at the end of the designated period. For their part, the project sponsors and financiers—who are operating in a highly regulated environment—want to ensure that the project finances are sufficient, through assurances from the granting authority that there will be no rate decreases for electricity or tolls to please its citizen users.

Increasingly, the BOT classification is being replaced by the term "privately financed infrastructure projects." This shift is due to the proliferation of variations on the same basic concept: The sponsors find the money, rather than the government. In addition to BOT, there are, to name some,

1. BOO (build-own-operate)
2. BOOT (build-own-operate-transfer)
3. BOLT (build-own-lease-transfer)
4. BRT (build-rent-transfer)

Also, the propensity for disputes in construction contracts noted in Chapter 8, the much increased dispute potential in international contracts, as seen in Section 8.09, the BOT contract can generate new problems.

There are business risks. Will users choose free roads rather than pay tolls? If there is no choice, such as a bridge connecting two parts of the city as in Istanbul, Turkey, will there be citizen user complaints, that the tolls are too high? Everyone uses energy. Will citizens say the electricity rates are exorbitant when furnished by a BOT power-generation plant? Some contracts contain guarantees by

the state that revenues will reach a certain amount, and this type of guaranty may also be unpopular.

There are political risks other than disgruntled users. The opposition party may charge that the agreement was a "sweetheart" deal and seek to have it annulled or cite it as the basis to oust the governing party.

Since those who build and operate are relying on revenue generated by users, they are subject to the risk of currency fluctuations.

Finally, those who finance the BOT project want the costs to be certain and "up front." They may insist upon the inclusion of language in the construction contract that places all risks on contractors, even those that normally would not be contractor risks, to protect the contract price by avoiding claims and cost overruns.

Yet despite these problems, contractor-financed projects, of which the BOT is the best known, are used increasingly. The United Nations Commission on International Trade Law (UNICITRAL) has been preparing voluminous and useful materials on these contracts.[115]

---

[115]For helpful articles see Nicklisch, *Realization of Privately Financed Infrastructure Projects—Economic Viability, Contract Structure, Risk Management,* [2003] Int'l Constr. L. Rev., pt 1, p. 80; Huse, *Use of the FIDIC Silver Book in the Context of a BOT Contract,* [2000] Int'l Constr. L. Rev., 384; and Merna *et al., Benefits of a Structured Concession Agreement for Build-Own-Operate-Transfer (BOT) Projects,* [1993] Int'l Constr. L. Rev., 32.

M. Building Information Modeling (BIM)

The modern variations discussed up to now involved organizational changes to the design-bid-build (DBB) and design-award-build (DAB) systems. The mechanics of the construction process remained the same. Efficiencies were promoted through the introduction of new entities (such as construction managers or program managers) or, more commonly, through changes in the contractual relations among the project participants (such as multiple prime, design-build, teaming, project alliances, etc.).

By contrast, building information modeling (BIM) is a technology-driven organizational model. The key to BIM is the creation of a computer model of the project that is both information-rich and information-integrative. It is information-rich in that, by clicking on any object in the model (such as a steel beam), the viewer is provided information about the object, such as the beam's size and structural characteristics (the structural forces acting on the beam and its capacity). It is information-integrative, because the object will automatically adjust to changes in other parts of the model. The design power of BIM, as compared with the traditional computer-aided design(CAD) software, was explained in a recent article:

> The difference between BIM and traditional design approaches is most striking when changes are made to the design. For example, if a steel structure is designed with traditional CAD tools, the drafted design might contain columns and beams with specific connections. If a column is removed to create a larger bay, the designer must recalculate the size of adjacent columns, resize beams, reanalyze load paths, and re-detail the connections. In object-oriented [i.e., BIM] design packages, . . . if a column is removed, the model will communicate with the remaining columns, adjust their size as necessary, change beam dimensions, and change the beam/column connections. . . . Thus, a change in the architectural requirements can ripple through the structural design without direct engineering involvement. The model can "design" itself based on rules embedded in the objects themselves. . . . Not only is this process efficient, it sharply reduces inconsistencies unforeseen when the design was modified.[116]

The goal of BIM is a paperless project in which changes to the specifications are seamlessly integrated with the rest of the design and simultaneously communicated to all the project participants. The BIM could also incorporate the project schedule (associating each phase of construction with when it is to be accomplished) and thus adjust the schedule if necessary (because of a delay in delivery of supplies or equipment, for example).

Widespread adoption of BIM by the construction industry faces serious legal impediments. BIM project delivery in many ways erases the distinctions among designer, builder, and component supplier. While the model is created in the architect's or engineer's office, it integrates contractor, vendor, and fabricator information into a seamless whole. BIM is a collaborative effort; yet the law allocates responsibility upon individuals. Who is responsible for design defects on a collaborative design? How is insurance coverage allocated?[117] Who owns the copyright in a BIM model? All of these fundamental questions are exacerbated by the absence of standard form contracts governing BIM projects. As noted in Section 24.01, existing standard form contracts make written documents paramount over electronic information—an approach entirely inimical to the BIM process.

Practical considerations must also be addressed. The level of detail in a BIM model means that the design process must start much earlier than on a design-bid-build (or even fast-track) projects, with the key trades joining in the design collaboration process. Each of the many participants must have the software and know-how for dealing with the BIM model. Financial considerations also must be addressed. A BIM model creates an informational database of such detail as to eliminate the need for shop drawings, thereby eliminating the primary compensation methods of most architectural offices.[118] For now, AIA B101-2007, Section 4.1.6, specifies that

---

[116]Ashcraft, Jr., *Building Information Modeling: Electronic Collaboration in Conflict with Traditional Project Delivery*, 27 Constr.Litig.Rep., Nos. 7–8, July–Aug. 2006, pp. 335, 336–37. For further discussion, see Wheatley & Brown, *An Introduction to Building Information Modeling*, 27 Constr. Lawyer, No. 4, Fall 2007, p. 33.

---

[117]For example, a contractor's commercial general liability policy typically excludes coverage for professional services performed by the contractor. Would the contractor's input into the BIM model constitute professional services?

[118]"The economic *lingua franca* of most professional firms is the billable hour. The economic foundation of many design firms is built around the hordes of souls hunched over drafting tables creating drawing after drawing. BIM will render this practice obsolete." P. O'Connor, Jr., *Productivity and Innovation in the Construction Industry: The Case for Building Information Modeling*, 1 J. A.C.C.L., No. 1, Winter 2007, pp. 135, 165.

architectural services performed on a BIM project are additional services deserving compensation separate from the basic fee.

To date, BIM projects have been limited to large owners, whether private or governmental.[119] As experience grows and larger numbers of design firms adopt the technology, BIM's spread to smaller projects appears likely.

N. Summary

The modern variations described in this section developed because of weaknesses in the traditional system, particularly leisurely performance, divided responsibility, and complaints of subcontractors. However, time and the courts will test these new systems and reveal their deficiencies. This does not in any way diminish the utility of these variations, although it does impose a serious burden on those who wish to engage in them, to anticipate the problems in deciding whether to use a variant from the traditional process and to plan in such a way so as to minimize the difficulties, both administrative and legal.

SECTION 17.05 Administrative Problems

A. Overview

This chapter has examined how a construction project can be organized, mainly the problems of selecting the type of contract (fixed-price or cost), the type of contract system to use (single, separate, DB), and methods to bring efficiency to the management of the process (prime contractor, design professional, or construction manager). Throughout this chapter, reference has been made to the need for efficient organization of any project, principally emphasizing coordination of the activities of the various participants.

Yet another factor essential to construction efficiency must be examined. A successful construction project recognizes the importance of clear and efficient lines of communication among the main participants. It is particularly important to know who has authority to bind the owner (with the hope that such a person can be identified and found), how communications (and there are many) are to be made, to whom they should be directed, and what the rules are for determining their effectiveness (effective when posted or when received?).

Communications are essential elements in the responsibility for defects (Chapter 24) and delays (Chapter 26) as well as equitable adjustments for unanticipated subsurface conditions (Chapter 25). (When those topics are discussed, communications problems will be addressed.) After a contract has been awarded and before performance begins, the participants must address these administrative issues openly and seek to make clear at the outset who speaks for whom, to whom and how communications are to be directed, and what the rules are for their implementation. Usually these issues are addressed at a preconstruction meeting attended by the representatives of the major participants in the process. This section deals with some of these problems.

B. Authority: Special Problems of Construction Contracts

Were the construction contract an ordinary one, it would be enough simply to read Chapter 4, dealing with agency and authority. However, special characteristics of the construction project make additional discussion essential. One factor relates to the multifaceted roles of an independent design professional engaged to design and monitor performance, to act as agent of the owner, to interpret and judge performance, and in general to be the central hub around which the process revolves. Using a construction manager does not clear up the problems, because the process revolves around two participants of often unclear responsibilities.

In construction projects that justify doing so, the owner may have a full-time site representative who observes, records, and reports the progress of the work. The representative (called a *clerk of the works*, *project representative*, *resident engineer*, or otherwise) can be a regular employee of the owner, an employee of the design professional, or CM and paid for either by the employer or by the owner, or an independent entity retained to perform this service. The representative's permanent presence invites difficulties, with the contractor often contending that he directed the work, knew of deviations, accepted defective work, or knew

---

[119]BIM has so far been used on large, complex projects such as stadiums, entertainment venues, and large commercial facilities. O'Connor, supra note 118 at 160. It was also used on the Denver Art Museum project. Cunz & Larson, *Building Information Modeling*, Under Construction, Dec. 2006, p. 1. The federal government's use of BIM is discussed in Silberman & McKee, *GSA's Building Information Modeling Program Raises Cutting-Edge Construction, Contracting, and Procurement Issues*, 41 Procur. Lawyer, No. 4, Summer 2006, p. 4.

of events that would be the basis for a contractor claim for a time extension or additional compensation.[120]

Communication is vital to construction. Problems often develop that should be brought to the attention of the appropriate party. For example, any design difficulties should be reported to the design professional. If there are interventions by public authorities, clashes among participants, material shortages, or strikes—to name only a few problems—the owner and design professional or construction manager should be made aware of them. Early discussion of problems or potential difficulties should bring prompt corrective action, efficient readjustments, and the gathering of necessary information for the ever-present "claim." A successful construction project requires clear lines of communication and their use.

Looking first at the traditional organization (CMs are discussed later), the design professional, though an agent of the owner, has limited authority;[121] the authority of a project representative is even more limited.[122] Which acts of the design professional or project representative will be chargeable to the owner? Before looking at that question, it is important to differentiate between acts of the design professional that bind the owner by principles of agency from those decisions made by the design professional as interpreter of the contract and judge of its performance. Although courts sometimes blur this distinction, this section concentrates on the first. Chapter 29 examines the second.

Generally the owner retains a design professional to provide professional advice and services, not to make or modify contracts or issue change orders. Although design professionals may enter into discussions with prospective contractors or even assist in administering a competitive bid, participants in the construction industry recognize that design professionals do not have authority to make contracts, modify them, or issue change orders on behalf of owners.

This does not mean that a design professional will never bind the owner. First, a design professional may be given express authority to perform any of these functions. For example, the design professional is frequently given authority to make minor changes in the work.[123] In addition, the design professional may be vested with certain authority that by implication includes other authority. For example, an architect who had the power to require that a contractor post a payment bond was authorized to represent to a subcontractor that such a bond would be required.[124]

Similarly, the owner may, by its acts, cloak the architect with apparent authority. For example, suppose the architect directs major changes for which he does not have express authority. If the owner stands by and does not intervene to deny such authority or pays for the changes, either may represent to the contractor that the architect has apparent authority to order that work be changed.[125]

Care must be taken to determine whether the principal—that is, the owner—has authority that can be exercised through its agent, the design professional. For example, suppose the engineer directs the contractor as to how the work is to be performed where this choice by contract is given the contractor. The owner would not have the authority to direct the methods. Any directions given by the engineer would not have to be followed any more than any direction by the owner. This is not an agency issue but one of contract interpretation.

Accepting defective work raises problems that are discussed in Section 24.05. A few points need to be made in this chapter, however. First, the design professional, when he is interpreter and judge of performance, decides whether work complies. If he decides the work is proper, he has not accepted defective work. Of course, his determination may be reversed by arbitrators or a court. However, if he determines that the work has not conformed, any acceptance by him constitutes changing the contract or waiving the owner's right to proper performance, neither of which he has authority to do, at least in the ordinary case.

Because of the failure to make this distinction, it is more likely that a design professional's acceptance of the

---

[120]See Section 21.04H.

[121]*Crown Constr. Co. v. Opelika Mfg. Corp.*, 480 F.2d 149 (5th Cir. 1973). But an actual employee may have more authority. *Grand Trunk Western R. Co. v. H. W. Nelson Co.*, 116 F.2d 823 (6th Cir.1941) (employee design professional authorized to make important contract).

[122]*Lemley v. United States*, 317 F.Supp. 350 (N.D.W.Va.1970), aff'd, 455 F.2d 522 (4th Cir.1971); *Samuel J. Creswell Iron Works, Inc. v. Housing Auth. of the City of Camden*, 449 F.2d 557 (3d Cir.1971); *Acoustics, Inc. v. Trepte Constr. Co.*, 14 Cal.App.3d 887, 92 Cal.Rptr. 723 (1971).

[123]AIA Doc. A201-2007 § 7.4.

[124]*Bethlehem Fabricators, Inc. v. British Overseas Airways Corp.*, 434 F.2d 840 (2d Cir.1970).

[125]See Section 21.04C.

work, even defective work, will be considered to bind the owner. To preclude this result, many contracts provide that acts of the design professional, such as issuing certificates, will not be construed as accepting defective work.[126]

As indicated earlier in this section, the construction process requires many communications among the participants. Clearly, a contract can provide that a notice can or must be given to the design professional. If this is the case, the notice requirement has been met if notice has been given to the design professional. The notice need not come to the owner's attention. In the absence of specific contract language of this type, will a notice that should go to the owner be effective if it is delivered to the design professional?

Because owner and design professional are closely related, and the law tends to be impatient with notice requirements if they appear to bar a meritorious claim, a notice given to the design professional will very likely be treated as if given to the owner. For example, in *Lindbrook Construction, Inc. v. Mukilteo School District Number 6*,[127] the contractor discovered unanticipated subsurface conditions substantially at variance with the contract documents. The contractor notified the architect and claimed an equitable adjustment in the price and additional time. The architect denied that the contractor was entitled additional compensation and directed it to proceed. The contractor completed the work and sued the owner.

The issue before the appellate court was whether the contract requirement of notice to the owner had been satisfied by giving notice to the architect. The court held that the architect was the only representative of the owner with whom the contractor had contact. It noted that the architect had complete knowledge of the changed condition and of the fact that the contractor was going to claim additional compensation. There was no evidence of a breakdown in communication between the architect and the owner, and the court noted that it would be unbelievable to assume that the architect did not notify the owner. The court held that the notice to the architect was imputed to the school district.

The dissenting judges argued that the architect's knowledge of the extra work could not be imputed to the owner because the architect had no authority to modify the con-

tract that required that notice be given to the owner. The architect was therefore, according to the dissenters, not operating within the scope of his authority, and the school district should not be bound. The dissenters also pointed to specific language stating that the architect could not commit the district to cost allowances and noted the importance of protecting public funds from unlawful expenditure.

Sometimes the notice problem is dealt with specifically by statute, such as provisions in mechanics' lien laws stating that notices can be given to the architect.

Are facts known to the design professional as if they were known by the owner? One case held that the notice of a limited warranty by a seller given to a mechanical engineer consultant bound the owner-buyer.[128] Similarly, a design professional's knowledge of the construction industry custom was held to be chargeable to the owner.[129]

As noted earlier in this section, if the design professional has limited authority, the project representative has even less. Usually the project representative is simply authorized to observe, keep records, and report.[130] Use of a project representative is often accompanied by a document that describes these functions and limits the authority of any project representative.[131] There is always a risk that a project representative will become overactive and seek to perform unauthorized activities, such as directing or accepting work. Although these are clearly unauthorized and the contractor should realize it, the contractor may submit and then later claim that these acts bind the owner. Usually, such assertions are not successful.[132]

This section has assumed the traditional construction delivery system. However, the problem of authority may have to take into account the way in which an architect or engineer, in a traditional system, may differ from a CM. Again, great caution must be exercised in making generalizations because construction management is fluid and imprecise.

---

[126]AIA Doc. A201-2007, §§ 9.6.6, 9.10.4.

[127]76 Wash.2d 539, 458 P.2d 1 (1969).

[128]*Trane Co. v. Gilbert*, 267 Cal.App.2d 720, 73 Cal.Rptr. 279 (1968).

[129]*Fifteenth Ave. Christian Church v. Moline Heating & Construction Co.*, 131 Ill.App.2d 766, 265 N.E.2d 405 (1970).

[130]But in *Town of Winnsboro v. Barnard & Burk, Inc.*, 294 So.2d 867 (La.App.), application denied, 295 So.2d 445 (1974) (knowledge of a project representative was chargeable to the architect).

[131]AIA Doc. B352.

[132]See supra note 122.

Nevertheless, the difference in professional status between a design professional and a CM will probably reflect itself in the extent to which acts of the CM may be held to bind the owner. In many instances, it may be difficult to distinguish the activities of regular employees of a sophisticated owner, such as members of an engineering department, from the activities of a program manager (PM) or CM. Even if the PM or CM is an independent entity, much of what either does, though resembling what design professionals were expected to do in the traditional system, is more geared toward the efficiency expected from the owner's own employees. As a result, it may be easier to establish apparent authority of a PM or CM than of a design professional.[133]

To sum up, independent advisers such as design professionals, PMs, CMs, and project representatives, though agents of the owner, do not have actual authority to make contracts, modify existing contracts, accept defective work, or waive contract requirements on behalf of the owner. Yet doctrines such as apparent authority or ratification may, in a proper case, justify a conclusion that acts of these professional advisers will be chargeable to the owner.[134]

C. Communications

A communication can be made personally, by telephone, or by a written communication. It is advisable to specify in the contract that a facsimile (fax) or electronic mail (e-mail) is a written communication. Generally, written communications should be required where possible. If it is not possible for a written communication to be made, at the very least a person making the oral communication should give a written confirmation as soon as possible. If the contract provision that deals with communications of notices does not state how communications are to be made, the communication can be made in any reasonable manner.

Usually, communications are signed. There may be challenges to fax communications, as the signature is not original. But the almost universal use of fax communications should not cause rejection of any fax as a valid communication on that basis.

Similar problems can accompany electronic mail. Electronic signatures can be made valid by compliance with the Electronic Records and Signatures in Global and National Commerce Act, effective October 1, 2000.[135] Digital signatures are simple to obtain and inexpensive.

Section 17.05B discussed the problems of authority and noted that it is essential to have designated persons with authority for each party to make decisions and to take responsibility. With regard to communications, each party should know to whom it should direct any particular communication. The authorized person should be designated by name in the contract and his address given. It is important to notify the other party if there has been a change in personnel and notices are to be sent to someone else.

If the contract does not designate time requirements for notices, any notice or communication required must be given within a reasonable time. Commonly the construction contract specifies time requirements. For example, AIA Document A201-2007, Section 15.1.2, requires a notice "within 21 days after the occurrence of the event giving rise to such Claim or within 21 days after the claimant first recognizes the condition giving rise to the Claim, whichever is later."

It is also important to designate whether days are working days or calendar days. Using working days raises problems because of weekends and various holidays and possible delays due to inclement weather. It is usually best, as AIA Document A201-2007, Section 8.1.4, states, to specify calendar days.

The absence of a base point, such as simply stating that a twenty-day notice must be given, can create difficulties. Does the notice period begin to run when the event occurred, when the contractor found out about its occurrence, when it could have found out about its occurrence, or when the event had sufficient impact on the work to cause the delay? Such a dispute would be best resolved by concluding that the period of time for giving notice began when it was reasonably clear that a delay would occur.

Suppose the owner justifiably notifies the contractor that the contract will be terminated unless the contractor pays certain subcontractors within ten days. When does the ten-day period begin? Suppose the letter was dated September 1, mailed September 2, and received September 4. Under such circumstances, when does the time period begin?

---

[133]*Turner Constr. Co.*, supra note 64 (owner bound by notice to CM).
[134]See Section 21.04C.

[135]15 U.S.C.A. § 7001 et seq.

It is likely that the time would begin when the letter is received. Doubts will be resolved against the letter sender because that party could have clarified any possible uncertainty. The receiver would probably expect the time period to begin to run when the letter was received, unless the receiver knew there had been an inordinate delay in the mail.

At the other end of the notice, suppose a contractor is asked to respond to a communication within ten days. The contractor posts the letter on the tenth day but the letter is not received until the twelfth. Has the contractor complied?

American law generally follows what is called the *mailbox,* or *dispatch,* rule. Posting a reply within the time period is effective unless it is made clear that actual receipt must be had within the time period. The mailbox rule will apply if the communication is sent by a reasonable means of communication. The mailbox concept, though developed in cases involving formation of contract, would very likely be applied to construction project communications.

Many difficulties discussed can be avoided by careful contract drafting. It is best to make absolutely clear when periods begin, when they conclude, and whether receipt will be effective when placed in the means of communication or whether a communication must be actually received.

Systems should be developed that make it likely that communications will be received and that proof exists that communications were sent and received. Although false claims of dispatch and receipt are rare, a carefully conceived and administered construction project should consider the importance not only of proving that notices were sent or received but also of knowing when these events occurred.

Communications received and copies of communications dispatched should be logged and kept in readily accessible files. Such records should be kept for a substantial time after project completion, to deal with the possibility of long-delayed claims being asserted.

Contracts frequently set forth formalities to avoid difficult proof problems. During the course of contract administration, parties often dispense with formal requirements. If they do, there is a serious risk that the formal requirements have been eliminated by the conduct of the parties. For this reason, formal requirements should be complied with throughout contract administration. If the formal requirements must be dispensed with, it is important for the party who wishes to rely on the formal requirements at a later date to notify the other party that the dispensation of requirements on this occasion will not eliminate the formal requirements for the rest of the contract. Sometimes general conditions contain provisions stating that waiver of formal requirements in one or more instances will not operate to eliminate formal requirements in the future. These clauses may not mean much if the parties, by their conduct, show an intention to generally dispense with formal requirements.

Sometimes legal rights, such as the right to a mechanics' lien, may require compliance with statutory provisions for notices. Many problems that have been discussed in this section can arise when statutory notices are required. Particularly in the area of mechanics' liens, it is necessary to obtain and follow legal advice regarding notice requirements.

CHAPTER EIGHTEEN

Competitive Bidding: Theory, Realities, and Legal Pitfalls

SECTION 18.01 Basic Objectives Reconsidered

The owner wishes to obtain a completed project that complies with the contract requirements as to quality, quantity, and timeliness at the lowest possible cost. Just as the organizational method and pricing discussed in Chapter 17 must take this into account, so must the method selected to designate the contractor.

In looking at cost—the factor most commonly emphasized in competitive bidding—attention must be directed not only to the bid price but also to other costs. For example, it is likely to cost more to conduct a competitive bid than to negotiate a contract. Similarly, the ultimate cost must take into account any administrative costs (called increasingly these days "claims overhead") incurred to obtain the promised performance and resolve disputes. These costs are likely to be greater under competitive bidding because the low bidder's performance and administration can increase the cost of monitoring performance and resolving disputes as well as create a greater likelihood of claims, thus increasing the ultimate contract price because of extra work.

The ultimate cost of a project should take into account the cost of maintenance and the durability of the project. Methods exist that take these factors into account.[1]

Although many of these costs are difficult to quantify, any owner who can *choose* to have competitive bidding should take them into account. Techniques within the competitive bidding system can reduce some of these risks.[2]

SECTION 18.02 Competitive Bidding: Theories and Some Pitfalls: Corruption

A succinct formulation of the principles that underlie competitive bidding are "competition in bidding sufficient to insure that the owner gets a fair price for the work; uniformity in the treatment of contractors to avoid favoritism; and the use of objective criteria and methods of evaluating contractor credentials."[3]

Another more fleshed out description of these principles employed by an Ohio court states that competitive bidding "gives everyone an equal chance to bid, eliminates collusion, and saves taxpayers money. . . . It fosters honest competition in order to obtain the best work and supplies at the lowest possible price because taxpayers' money is being used. It is also necessary to guard against favoritism, imprudence, extravagance, fraud and corruption."[4] This is a tall order.

Before examining competitive bidding as a mechanism to get the best work at the best price, it is helpful to look at the final objective noted by the Ohio court: that of guarding against corruption. As corruption can play a significant role in frustrating the goal of getting the best value for public funds, it strikes at the heart of public governance.

Competitive bidding is often used in private construction projects. This is done not only to get the best price but also because a private owner, often inexperienced in construction, may not be able to negotiate efficiently with a contractor whose business is to build.

---

[1]Vickrey & Nicol, *Total-Cost Bidding—A Revolution in Public Contracts?* 58 Iowa L.Rev. 1 (1972) (bidders bid initial cost, guarantee maintenance costs, and give a price to repurchase goods: variables totaled and bid made to lowest total cost bidder).

[2]Comment, 130 U.Pa.L.Rev. 179 (1981) (competitive negotiation).

[3]McMillan & Luschei, *Prequalification of Contractors by State and Local Agencies: Legal Standards and Procedural Traps*, 27 Constr. Lawyer, No.2, Spring, 2007, pp. 21, 27.

[4]*United States Constructors & Consultants, Inc. v. Cuyahoga Metro. Housing Auth.*, 35 Ohio App.2d 159, 300 N.E.2d 452, 454 (1973).

But emphasis will be placed in this discussion of corruption in public contracts. Here, the fear of corruption is greater and the need to account for public funds more crucial. Also, as a rule, construction work in the public sphere requires competitive bidding. Private projects can be made in any manner that the owner chooses.

Of late, much attention has been paid to corruption, particularly in international transactions. No longer is it winked at or easily dismissed as endemic to certain societies, as a method to bypass useless and burdensome rules, or as an efficient method to get public officials to do what they are supposed to do. Corruption is seen increasingly as a generator of useless infrastructure projects or senseless industrial projects. Corruption is expensive, adding greatly to the cost of all projects. It not only holds back economic progress in lesser developed countries, but can bring down governments and create regional instability.

Unfortunately, corruption is seen as endemic to construction, particularly public construction. It is, as seen above, one reason for competitive bidding. Why is corruption so readily linked to construction?

Construction is a risky business. Contractors operate on small profit margins. Yet there is, as a rule, no shortage of those who seek construction contracts. The press will announce that a particular contractor has "won" or been awarded a contract. Certainly, the bidders think that profit can be made despite a fixed-price contract placing the risk on the contractor of almost all events that increase construction cost. How does corruption enter the picture?

In the course of construction, many events occur that can be the basis of a contractor asking for more money. Design is an ongoing process. Often, there is the need for redesign coupled with a contractor request for more money through the changes process. Interpretation of construction documents raises issues that must be resolved by the parties. It is almost impossible to express clearly and completely what the designer seeks. Events occur that are unforeseen and raise costs. Compliance control requires inspections. These and other events often generate claims in fixed-price contracts. Even in cost-type contracts there can be disputes over allowable costs.

How claims are handled depends on the honesty and intelligence of those officials administering the contract. Corruption can generate interpretation decisions that favor the contractor at the expense of the public agency that result in over-generous settlements.

Of course, there can be corruption even when there is competitive bidding. There can be a corruption-generated decision to exempt a project from competitive bidding. The corrupt official may reveal submitted bids to a bidder before the deadline for bid submission. She may be induced by a bribe to decide whether a bid conforms or whether to waive an irregularity in the bidding. A particular alternative can be selected because an official has been bribed.

Despite the corruption in the competitive bidding process itself, competitive bidding creates transparency and openness that, along with a free and vigorous press, can create an atmosphere that discourages or exposes corruption.

Moving back to assumptions that underlie competitive bidding, it assumes that goods or services requested can be objectively evaluated or compared, preferably before award, or at least after. An award for pencils of a standardized type is an illustration. All pencils would be comparable, and all that would be needed would be to compare prices. If a number of competing sellers will bid, the price should be as low as can be obtained.

Even here, performance uncertainties exist. Will the party awarded the pencil contract deliver as promised? But if it does not, replacement pencils could be obtained and the excess procurement cost charged to the contractor or its surety.

As an example of different approaches, California held that a public award for a construction manager must be made competitively,[5] whereas Massachusetts held that competitive bidding was not required to purchase an existing vessel for use as a ferry boat.[6] The Massachusetts court noted that competitive bidding does not work well where the various properties offered for purchase are not identical, as in the sale of real property. This non-comparable factor is a reason given for holding that turnkey contracts for public housing are not subject to the requirement for competitive bidding.[7] However, the same

---

[5]*City of Inglewood—L.A. County Civic Center Auth. v. Superior Court,* 7 Cal.3d 861, 500 P.2d 601, 103 Cal.Rptr. 689 (1972). But see Section 17.04D.

[6]*Douglas v. Woods Hole, Martha's Vineyard, & Nantucket S.S. Auth.,* 366 Mass. 459, 319 N.E.2d 892 (1974).

[7]*United States Constructors & Consultants, Inc. v. Cuyahoga Metro. Housing Auth.,* supra note 4. See Section 17.04E.

argument can be made in favor of permitting a negotiated contract for a construction manager. Yet the California court was more impressed with the need for competition as a method of reducing costs.

A procurement for the award of police cars that would meet designated standardized performance specifications would fit the competitive bidding model. Either before or after award, testing can be done to determine whether the bidders have prototype vehicles that will comply or have complied with the specifications. Again, there can still be the problem of predicting whether all the cars will conform and be delivered on time.

The principal objective of the awarding agency will be frustrated if the performance standards are not met. Time has been lost and the objective is not yet accomplished. Even if the successful bidder is financially solvent or has furnished a surety bond, the objective has still not been achieved. Collecting for nonperformance certainly was not the objective of the procurement.

The difficulty of predicting whether a prospective bidder will be able to meet the performance standards is an important reason why competitive bidding is rarely used in research and development contracts. In these procurements, the agency must find a contractor who has the technological skill to perform properly. Unless accompanied by some preselection process, competitive bidding may not be the best method to select a process in such a procurement.

Although most construction work is not as sophisticated or experimental as building nuclear submarines or space capsules, some elements in construction work do not make it perfectly suitable for competitive bidding. First, how likely are prospective bidders to be willing and able to do the work they promise? Second, how difficult is it to determine whether they have done the work they promised? Third, what is the likelihood that the actual price will substantially exceed the contract price for reasons other than owner-directed changes or circumstances over which neither party had control?

The competitive bidding process can be structured to minimize some of these risks. The invitation may state that the contract will be awarded to the lowest *responsible* or lowest and *best* bidder. Preselection can screen out bidders who do not have the requisite competence or capacity. Nevertheless, despite these provisions, the likelihood of success depends on the integrity and ability

of the contractor—often difficult to measure in competitive bidding, where the tendency is to look solely at price.

Competitive bidding also assumes free bids and true competition. If bidders collude to "take turns" or submit fictitious bids, antitrust laws are violated and competitive bidding cannot accomplish its objective of obtaining the lowest price. Although construction is, on the whole, fiercely competitive, collusion between competitors in the competitive bidding process is not unknown.

James Cape & Sons Co. v. PCC Construction Co. was recently decided by a federal court of appeals.[8] It illustrates corruption in competitive bidding as well as the remedial difficulty for the contractor who has been victimized by the bid riggers.

A group of road contractors, not including the plaintiff Cape, created a group that rigged the state transportation department's bidding process from 1997 to 2004. The group discussed projects soon to be bid, shared information, discussed potential competitors, and "set bids amongst themselves in an attempt to allocate projects between them."[9] (All later pleaded guilty to criminal charges.)

One member of the group contacted Beaudoin, a key Cape official, the plaintiff in the case. Beaudoin gave confidential information to the group often just before bid submission. The group members used this information to make their bids and further the bid rigging. Cape sued the group members and Beaudoin.

The court rejected Cape's claim under the antitrust laws and the Racketeering Influenced and Corrupt Practices Act (RICO). The latter statute was aimed at organized crime but has been used in many civil claims. Other than a claim, probably worthless, against its employee Beaudoin, what avenues are open to Cape who should have received the award? Can it challenge the award as one obtained by corruption, though not the corruption of the awarding authority? It appears that Cape waited too long to do this. Can it make a claim for lost profits based upon unjust enrichment against any member of the bid rigging group that won contracts it should not have gotten?[10]

---

[8]453 F.3d 396 (7th Cir.2006).
[9]Id. at 398.
[10]This has had mixed success. See infra notes 159, 160.

Anticompetitive devices can be found in competitive bidding. If product specifications do not provide for alternative products and a viable method for substitutes, competitive pricing may be unduly restricted. Also, specifying a particular product may create a warranty of commercial availability by the awarding authority to the bidder. New Jersey prohibits specifications in local public contracts that limit free and open bidding, such as specifying brand names without allowing equivalent products to be substituted.[11]

Another weakness of competitive bidding is the difficulty of involving the contractor in the design process. This can be alleviated by using a construction manager[12] or, in the extreme case, the design–build process.[13] But the traditional contract system, where the owner supplies a design prepared by the design professional and submits it to the contractors for competitive bidding, generally fails to employ any design skills of the contractors.

Some competitive bidding systems permit *alternative bidding*.[14] The design professional prepares a design for one method of conventional construction and, in addition, prescribes parameters and other specific configurations or finish requirements for the alternate. Performance specifications for allowed alternates are usually included in the contract documents. The contractor can select the alternative design or bid on the design supplied. However, this method is limited to those projects where the design professional is aware of another viable design solution. Another limitation to alternative bidding is the greater likelihood of disputes as to responsibility and liability.

Prebid design, or what federal procurement calls "two-step" formal advertising (discussed in Section 18.03D), is another method of involving the contractor in the design. This employs a competition first on design, then on price, similar to the Brooks Act, under which design services are procured for federal projects.[15]

To sum up, the competitive bidding system is presently, and is likely to continue to be, the major method of obtaining construction contractors. However, competitive bidding has pitfalls, and in the proper circumstances,

serious thought should be given to another method of obtaining a contractor.

SECTION 18.03 The Competitive Bidding Process

A. Objectives

Competitive bidding should result in contract awards made impartially at the lowest price. Nonconforming bids are usually disregarded because conformity is needed for a proper comparison of bids and to give each bidder an equal opportunity. The competitive bidding system cannot function properly unless honest and capable bidders have enough confidence in the fairness of the system to submit bids. Submitting a bid proposal is expensive and time consuming. The bidders are entitled to reasonable assurance that they will be treated fairly and that the owner will follow its own rules.

No effort will be made here to cover every aspect of the competitive bidding process. Most public agencies have standard forms for the competitive bidding process, and they are often regulated by statutes, regulations, and ordinances. Private owners often have engaged in substantial construction work and have also developed their own forms and methods. Design professional associations have developed standard or recommended forms for bidding documents. They are usually available to the design professional. The principal objective of the following subsections is to present an overview of the process.

B. Invitation to Bidders

The initial step in conducting a competitive bid is the *invitation to bidders*. In federal procurement, this is known as *the request for bids (RFB)*. Public agencies are frequently required to invite bids by public advertising. By using the broadest dissemination, such as trade newspapers, professional journals, and government publications, the largest number of bidders can participate. This gives all bidders a chance and should obtain the lowest price.

Although inviting the maximum number of competitors should result in a lower price, dangers exist in having too many bidders. Good bidders may be discouraged if the large number of bidders makes their chances of winning quite remote.

---

[11]N.J.Stat.Ann. § 40A:11-13.
[12]See Section 17.04D.
[13]See Section 17.04F.
[14]See Section 18.03G.
[15]See Section 11.03.

Ordinarily, the invitation to bidders is not an offer and does not create a power of acceptance in the bidders. It is a request that bidders make offers to the owner that can be accepted or rejected.

But the invitation is an important document and must be drafted carefully. It sets up the ground rules for the competitive bid. If a contract is formed with one of the bidders or if the successful bidder refuses to enter the contract, the invitation may have legal significance. Of course, the invitation can be modified by formal agreement between the owner and the successful bidder, but as noted in Section 18.04G, the formal contract may simply create a memorandum of what has been agreed to in the invitation and the bid proposal.

C. Prequalification

Devices are available to avoid the competence risks inherent in the price-oriented competitive bid. In addition to a statement that the award will be made, if at all, to the lowest responsible or the lowest and best bidder, a prequalification system can be used. Increasingly prequalification is used in public projects, extremely large projects, high-profile projects, and projects that use new construction techniques.

The owner selects a group of bidders, all of whom are likely to have the capability of doing a competent job. The design professional or construction manager may request information from specific contractors on prior jobs completed, capital structure, machinery and equipment, and personnel (including supervisory personnel). After evaluating this information, the owner's professional adviser usually decides which contractors should be permitted to receive an invitation to bid.

Three choices need to be made to implement the prequalification system. One is to delineate exclusive from nonexclusive bidding for a particular project. On an exclusive project the owner receives bids only from prequalified bidders. Nonexclusive projects are those on which the owner will receive bids from both prequalified bidders and those not prequalified. The cost to bidders of the prequalification process may deter bidders that will have to face those who are not prequalified. But the openness of the nonexclusive system can avoid the charge of limiting competition and increasing the risk of corruption.

Another choice is between specific and programwide qualification. Specific means preselected for specific projects. Programwide means qualification for certain types, such as expressways or subways. Programwide can sharpen the criteria and also act as an incentive to incur the expense of participation.

The program may be full or limited. Full means unconditional prequalification. Limited prequalification may apply conditions to qualification, such as financial capacity, bonding capacity, or limits on the dollar amount.[16]

The advantage to prequalification is that there is a better chance of finding a good contractor. However, administrative time and expense will be expended in making this preselection or prequalification, and the competitive aspect is likely to be diminished. Similarly, experience requirements that are too narrowly drawn or drawn to favor a particular bidder can also be an indication of corruption and may infect the entire prequalification process.

D. Two-Step Process

The federal procurement regulations sometimes permit a two-step formal advertising procedure. This is used where definite specifications cannot be prepared and offered for fixed-price competitive bids. As a first step, unpriced technical proposals are solicited to meet certain requirements specified in the solicitation for proposals. Typically, these requirements are for an end product. All proposals received are evaluated, and those considered to be within the range of acceptability are discussed with the proposers to obtain clarification and more detailed definition.

The second step is a request for price bids from all those whose first-step proposals met the criteria specified in the original solicitation for proposals. Award is then made in accordance with the procedures for awarding a fixed-price contract, with each bidder pricing its own technical proposal previously approved as having met the specified criteria.

E. First Article

In procuring products or equipment, the federal government can protect itself from having to wait for a product that may not be adequate by inserting a "first article"

---

[16]For a thorough canvassing of prequalification, see McMillaan & Luschei, supra note 3.

clause. This clause requires the contractors to deliver for government testing a preproduction model of what the contractor has agreed to furnish.

F. Deposits

In public procurement and occasionally in private procurement, the invitation requires the bidder to deposit a bid bond, a cashier's check, or a certified check to provide security in the event that the bidder to whom the award is made does not enter into the contract. The amount is either a stated percentage of the bid or a *fixed* amount determined by a designated percentage of the estimated costs made by the design professional.

The invitation should also state how long the securities will be held, which is important to bidders. After award, the owner should release the securities of all but the lowest three or four bidders. The bidder to whom the award is made may not enter into the contract. It still may be possible under some circumstances to hold the next lowest bidder. However, there is no justification for holding the securities of bidders who are not likely to be awarded the contract. After the successful bidder signs the contract, all securities should be released.

Usually the bidders must make a small monetary deposit when they request information to study a possible proposal. This may discourage people who are not serious about making a bid proposal from obtaining the plans and specifications merely out of curiosity. The deposit is usually refunded when the bidding documents are returned.

G. Alternates

Sometimes larger projects can be divided into designated stages. The owner may decide not to build the entire project if it does not have the money or if bids are too high for certain portions of the work. For this reason, the invitation to bidders can be divided into the stages or project alternates. But alternates can mean that favoritism can be accomplished by the award. For example, there may be an invitation to bidders involving a hospital, housing for staff, and a parking structure. One alternate could be the hospital alone. A second could be the hospital with the parking structure, and a third the hospital with the staff housing. One bidder may be low on the total bid. Another

bidder may be low on the first alternate, another on the second, and another on the third. The determination of which alternate is to be awarded may be based on favoritism to one of the bidders. To avoid this difficulty, the invitation to bidders should state the preferred alternate choices.

The same type of award manipulation is possible if the alternates consist of different methods of construction or materials. A list of preferences within price limits should avoid suspicion of possible favoritism.

H. Information to Bidders

Information to bidders—called in federal procurement *the information for bids (IFB)*—accompanying the invitation usually consists of drawings, specifications, basic contract terms, general and supplementary conditions, and any other documents that will be part of the contract. Sometimes soils test reports are included. Alternatively, the information to bidders may state that designated reports are available in the office of a particular geotechnical engineer or the design professional, for examination by bidders. Subsurface information is discussed in Section 25.03.

Bidders should be given adequate opportunity to study the bidding information, to make tests, to inspect the site, and to obtain bids from subcontractors and suppliers. For example, under AIA Document A201-2007, Section 3.2.1, the contractor represents that it "has visited the site, become generally familiar with local conditions . . . and correlated personal observations with the requirements of the Contract Documents." Even when there is adequate time, bidders often wait to complete their bids until the bid closing deadline is imminent. Generally, this is due to reluctance on the part of subcontractors to give sub-bids to bidders until shortly before the deadline for bid submissions. (The reasons for this reluctance are explored in Section 28.02C.) Even if bidders do not make proper use of the time available to them, having allowed reasonable time for bid preparation can be helpful to the owner if a dispute arises over claimed computation errors or unforeseen subsurface conditions.

The drawings and specifications should be detailed and complete so that bidders can make an intelligent bid proposal. Imprecise contract documents and much discretion to the design professional may discourage honest bidders from submitting bids and may encourage bidders of doubtful integrity who make low proposals in the hope that they will later be able to point to ambiguities and make large claims for extras.

The information to bidders can specify that any uncertainties observed by the bidder must be resolved by a written request for a clarification to the design professional before bid opening. Any clarification issued should be in writing and sent to all people who have been invited to bid. The law increasingly requires that under certain circumstances the party conducting the competitive bid must disclose information in the bidding documents (discussed in Section 18.04B).

I. Bid Proposals: Changes

Bid proposals should be submitted on forms provided by the owner. To properly compare bids, all bidders must be proposing to do the same work under the same terms and conditions. The bid proposal form should include or refer to any important disclaimers contained in the bidding information. Such disclaimers include denying responsibility for subsurface information and specifying any rules that seek to govern the rights of the bidder to withdraw the bid, such as limiting withdrawal to clerical errors or setting forth other requirements, such as a deadline for claiming mistake.

The bid proposal should be signed by an authorized person. If the bidder is a partnership, the entire name and address of the partnership should be given. If the bidder is a corporation, the bid should be signed by appropriate officers and the corporate seal should be attached. It may be desirable to attach to the bid a resolution of the board of directors approving the bid.

Invitations can preclude bids from being changed, corrected, or withdrawn after submission. However, there is a tendency toward more flexibility. More commonly, invitations permit changes, corrections, and withdrawals of submitted bids before bid opening.

Permitting changes can help the owner by permitting and encouraging reduction of bids in a changing market. However, overliberality in permitting changes, corrections, and withdrawals may cause bidders to lose confidence in the honesty of the bidding competition. Where the right to change is given, it may be limited to reducing the bid. If changes, corrections, or withdrawals are to be permitted, the invitation can limit such a right to a designated time period and require that any change, correction, or withdrawal be expressed in writing and received by the owner by a designated time.

The owner should keep all the records connected with the bidding process. Time logs should be kept that show exactly when the bidder obtained the bidding information. Changes in bidding information should be sent to all people who have picked up bidding information, with copies of the changes kept for future reference. After the bidding information is disseminated, changes should be kept to a minimum.

J. Bid Opening

Bid proposals are sent or delivered in a sealed envelope to the owner or the person designated to administer the competitive bidding, such as a design professional or construction manager. In public contracts, bids are opened publicly at the time and place specified in the invitation to bidders. Usually the person administering the process announces the amount of the bids and the bidders. The invitation specifies a designated period of time in which the owner can evaluate the bids. At bid opening, care must be taken to avoid any impression that the low bidder has been awarded the contract. Often the person administering the process states that a particular bidder is the "apparent low bidder."[17]

In private competitive bids, the invitation can state that bids need not be opened in public. When the bid opening is private, the bidders, at least at that time, are not able to compare their bids with the others. This minimizes the likelihood that the low bidder will begin to suspect a bidding error if its bid is much lower than the others. One case involved the successful bidder not finding out that its bid had been much lower until it was well into the project. At that time, it threatened to walk off the project unless it was given a price adjustment, claiming that the owner must have known its bid was erroneous. Although the court did not grant relief to the contractor, its determination to grant the contractor relief on another theory was probably affected by the belief that the owner should have notified the contractor of the wide discrepancy between its bid and the others.[18] This may reflect the law beginning to impose on contracting parties, or even those in the process

---

[17]In *McCarty Corp. v. United States*, 204 Ct.Cl. 768, 499 F.2d 633 (1974), the contracting officer stated that the plaintiff was the "apparent low bidder." This, among other facts, precluded a finding that the plaintiff's bid had been accepted.

[18]*Paul Hardeman, Inc. v. Arkansas Power & Light Co.*, 380 F.Supp. 298 (E.D.Ark.1974). This case is discussed again in Section 33.01.

of making a contract, an obligation of good faith and fair dealing, such as drawing attention to a possible mistake. A private bid opening may tempt the owner not to notify a bidder when it is apparent that a mistake has been made.

K. Evaluations of Bids

Some legal problems relating to bid evaluations and awards are discussed in Sections 18.04E and F. This section outlines some steps that should be taken when evaluating bids.

First, the owner or its representative should follow the procedures set forth in the invitation to bidders. The bids should be checked to see whether they conform. A nonconforming bid is a proposal based on performance not called for in the invitation or information or offering performance different from that specified. A deviation is material if "it gives the bidder a substantial competitive advantage and/or prevents other bidders from competing on an equal footing."[19] Nonconformity may relate to the work, the time of completion, the bonds to be submitted, or any requirements of the contract documents. Bidding documents sometimes allow the owner to waive bidding technicalities. This allows the owner to accept the lowest bid despite minor irregularities. However, advance notice that technicalities can be waived can encourage careless proposals and may also create the impression that favoritism may be shown to certain bidders. If a bid is not in conformity and the defect cannot be waived, the bid must be rejected. This will discussed in Section 18.04D.

Second, the owner should examine the invitation to bidders to determine whether any bid must be accepted. Typically, the invitation reserves the right to reject all bids. The invitation must be reviewed to determine the basis for making an award if one is made. Commonly, the award is to be made to the lowest responsible or lowest and best bidder. Responsible means the bidder must have "the necessary technical, managerial and financial capability and integrity to perform the work."[20]

To determine the lowest responsible bidder, the owner can take into account the following factors as well as any others that bear on which bidder would be most likely to do the job properly:

1. expertise in type of work proposed
2. financial capability

3. organization, including key supervisory personnel
4. reputation for integrity
5. past performance

These examples relate to the basic objectives of obtaining a contractor who is likely to do the job properly at the lowest price and with the least administrative cost to the owner.

A recent federal procurement case demonstrates the inclusion of integrity and business ethics as an evaluation criterion applied to bidders. The regulations in effect at the time of the invitation required the contracting officer (CO) to make an affirmative determination that the bidder has "a satisfactory record of integrity and business ethics."[21] A federal appeals court ruled that this determination by the CO can be reviewed by the court under the difficult to overturn "rational basis" standard.[22]

Experience in the type of construction work being procured is an important selection criterion. However, overemphasis on experience or drawing the experience factor too narrowly carries great risk. It can result in a loss of qualified bidders that will increase the price. It can also generate a corrupt award and violate the competitive bidding laws.[23]

In contracts for the purchase of machinery, bid evaluation should take into account the following factors:

1. length and extent of warranty
2. availability of spare parts
3. service and maintenance
4. cost of replacement parts
5. cost of installation
6. durability

Although the owner should evaluate criteria other than price in selecting a contractor, awarding the bid to someone other than the low bidder raises a substantial risk of a lawsuit challenging the award. Sometimes bid awards in public contracts can be judicially challenged. This should not deter even a public owner from awarding the

---

[19]Construction Briefings, Jan. 2005, p. 7.
[20]Construction Briefings, Jan. 2006, p. 10.

[21]48 CFR § 9.104-1(d) (2007).
[22]*Impresa Construzioni Geom. Domenico Garufi v. United States*, 238 F.3d 1324 (Fed.Cir.2001).
[23] See *Gerzof v. Sweeney*, 16 N.Y.2d 206, 211 N.E.2d 826, 264 N.Y.2d 376 (1965), discussed in Section 18.04J. In that case, experience requirements were so tightly drawn that only one bidder qualified. This made the award illegal.

contract to someone other than the low bidder. However, compelling reasons for not awarding it to the low bidder must exist and be documented.

Third, if bidders were asked to bid separately on project alternates, the owner should determine whether any bidders had made "all-or-nothing" bids. Such a bid indicates an unwillingness on the part of the bidder to perform any part of the project except the entire project. An all-or-nothing bid is permitted unless the invitation specifically precludes it.

L. Notification to Bidders

When the successful bidder is selected, the successful and unsuccessful bidders should be officially notified. However, sometimes with or without legal justification, the successful bidder will not enter into the contract. For this reason, the notification to unsuccessful bidders should state that their bids still remain available for acceptance by the owner for the period of time specified in the invitation to bid. This is done to preclude any later contention that awarding the contract released the unsuccessful bidders.

M. Postaward Changes

If the owner changes the contract terms after the bid has been awarded and the change makes the obligation less burdensome to the successful bidder, this can be unfair to the other bidders. For this reason, such changes should be done with caution. (*See Conduit and Foundation Corp. v. Metropolitan Transportation Authority*, reproduced in Section 18.04D.)

N. Signing the Formal Contract

The successful bidder is sent to the formal contract for signature or requested to meet at a specified time and place to sign the formal documents. Records of all correspondence should be kept in the event that subsequent disputes arise between the owner and the contractor relating to the bidding process and awarding of the contract.

The successful bidder must enter into the formal contract awarded unless legal grounds exist for refusal (discussed in Section 18.04E). The effect of signing the formal contract, as well as such signing's effect on bidding documents and the function of the formal contract, is discussed

in Section 18.04G. Because the date the contract has been made can measure the time commitment, this is discussed in Sections 26.02 and 26.04.

O. Readvertising

Sometimes the bids are all too high and the owner decides to readvertise the project. In public contracts, there are procedures for readvertising, which require time and administrative expense. Some feel that readvertising without reducing the scope or quality of the project is unfair because it is an attempt to beat down the bids of the contractors, often resulting in deficient workmanship and substandard materials. Readvertising frequently means higher bids from all bidders because the less skillful bidders have seen what the more skilled bidders have bid.

P. Special Rules for Public Contracts

Owners generally seek a construction contract that will give them the balance they choose among quality, timely completion, and cost. On the basis of these criteria, the owner looks for a contractor who will do the best work at the best price.

Looking only at the project and these goals, the owner will not be concerned with the racial or gender characteristics of the contractor's labor force, the wages the contractor pays its employees, and the source of the supplies. Obviously, this lack of concern is not absolute. The owner could be concerned with the wages if it felt that workers who were not paid the prevailing rate would not perform properly. The source of the supplies could be important if the owner felt that supplies from certain sources were higher quality. But on the whole, the owner who wishes to obtain the best price through competitive bidding or through negotiation must give broad latitude to the contractor in these matters.

Public contracts, however, involve billions of taxpayer dollars. Elected officials frequently look to the procurement process and public spending as ways of accomplishing goals that go beyond the best quality at the best price. In doing so, they often respond to interest groups who also see procurement as a means of obtaining their objectives. The federal government has been the pioneer in using the procurement process to achieve social and economic objectives. State and local public agencies and even some private owners have also begun to see procurement in this light.

This book cannot deal in detail with the many rules that seek to achieve a variety of objectives through the public contract process. However, some goals sought to be achieved by the public contract process are as follows:

1. Providing employment opportunities to disadvantaged minorities, women, handicapped persons, or disabled war veterans.
2. Setting aside certain procurement awards for small businesses or disadvantaged minorities.[24]
3. Favoring contractors, suppliers, or workers who reside in a particular state or city.[25]
4. Awarding contracts to bidders located in economically depressed areas.
5. Ensuring that workers are paid at the local prevailing wage rate (Davis-Bacon Act).[26]
6. Avoiding corruption in procurement.
7. Protecting American manufacturers.[27]

8. Encouraging trade unions.[28]
9. Ensuring compliance with workplace, environmental, and civil rights laws.[29]

Another component of the public bidding process is to obtain the best price for the materials used on the project. Specifications which are overly restrictive as to the type of materials or products to be used can result in an individual manufacturer becoming the sole source for the project. To avoid placing manufacturers in a monopoly position, the federal government and many states require that the specifications on public works projects permit the supply of brands "equal" to those specified.[30]

Public entities, particularly cities, have sought and increasingly seek to influence the conduct of domestic and even foreign business through their procurement policies.

At the domestic level some public entities give preference to those with child care facilities on the premises, or refuse to do business with companies that do not give rights to domestic partners. As an unusual example, an Illinois county awarded a slightly more expensive food-service contract for inmates to a nonprofit agency that also provided food-service training for the mentally handicapped.[31] At the foreign level the invitation may bar those who deal with certain foreign countries.

Often these policies are challenged, sometimes successfully. In any event, these attempts to use public procurement to influence foreign and domestic policies make doing business with public entities more tenuous and fraught with uncertainty.

Clearly, such rules have their costs. In addition to higher bid prices, administrative costs are likely to be incurred by all participants to see that these rules are followed. These costs may be worth incurring if they accomplish the objectives sought. However, attempts to use the

---

[24]The leading case is *Adarand Constructors, Inc. v. Pena*, 515 U.S. 200 (1995) (racial classifications require strict judicial scrutiny). The *Adarand* case generated considerable litigation. The last opinion was *Adarand Constructors, Inc. v. Mineta*, 534 U.S. 103 (2001). The litigation is catalogued in 23 Constr.Litig.Rep. 26 (2002). State affirmative action programs that comply with federal regulation, such as the Federal Highway Act, are likely to be enforced. There is a compelling governmental interest in dealing with past discrimination in the national construction market. These programs will be allowed if they are narrowly tailored. *Northern Contracting, Inc. v. Illinois*, 473 F.3d 715 (7th Cir.2007). See generally, Day, *Retelling the Story of Affirmative Action: Reflections of a Decade of Federal Jurisprudence in the Public Workplace*, 89 Cal.L.Rev. 59 (2001).

[25]As to local workers on a state funded project, see *United Building & Constr. Trades Council v. City of Camden, et al.*, 465 U.S 208 (1984). The Court held that the city cannot discriminate against out-of-state residents on matters of fundamental concern. That the city is putting up some of the money is an important factor, that permits them to prefer locals. If it can demonstrate a substantial reason for different treatment (to reverse white flight, help depressed city, etc.), this might be lawful. The court sent the case back to the trial court for findings on this issue.

[26]Many states also have such laws. As an illustration, the Massachusetts law was interpreted in *Teamsters Joint Contract No.10 v. Director of Dep't of Labor and Workforce Development, etc.*, 447 Mass. 100, 849 N.E.2d 810 (2006).

[27]For a discussion of the Buy America Act and the difficulties of determining what is foreign content, see West & Handwerker, *How the BAA Affects Construction Contractors*, 32 Proc.Lawyer, No. 3, Spring 1997, p. 3. Many states require that preference be given to materials made in that state in state contracts.

[28]See Section 23.05B dealing with Project Labor Agreements (PLAs).

[29]Toward the end of the Clinton administration, the president issued regulations requiring federal contracting agencies to consider the bidder's history of compliance with federal laws before awarding the bidder any government contracts. The bidder would have to certify that for three years it had been in compliance with federal laws and had a satisfactory record of integrity and business ethics. President Bush rescinded these rules shortly after he entered office. See Sherrill & McQueen, *The High Price of Campaign Promises: Ill-Conceived Labor Responsibility Policy*, 30 Pub.Cont.L.J. 267 (2001).

[30]Federal Acquisition Regulation (FAR) §§ 52.211–216 (2007); West Ann. Cal. Pub. Cont. Code § 3400.

[31]*Court Street Steak House, Inc. v. County of Tazewell*, 163 Ill.2d 159, 643 N.E.2d 781 (1994) (within discretion of county).

public contract process for these objectives have generated intense controversy, both as to the fairness of the programs and as to whether the objectives can be accomplished in a different way at a lower cost.

Every aspect of awarding public contracts, especially competitively bid awards, must avoid even the appearance of impropriety. For example, public officials who make procurement decisions are expected to have the interests of their agencies in mind and avoid conflict of interest.[32]

As a result of public concern over contract awards, public officials often are rigid and unbending in such matters as bidding irregularities and withdrawal of bids.

SECTION 18.04 Some Legal Aspects of Competitive Bidding

Some sections in this chapter describe issues that are encountered before the actual construction process begins, such as the requirement that the award be made by competitive bidding[33] and the duty of the awarding authority to disclose information to prospective bidders.[34] Some deal with the bidding and award processes.[35] Others examine mistakes made by bidders (the most litigated issue) and whether they should be given some relief.[36] One section relates to the need for a formal contract.[37] Another discusses the legal nature of the bidding documents.[38] Others treat security given the awarding authority,[39] the process of judicial review,[40] and the effect of an illegal award.[41]

Where the owner has the choice of whether to select a contractor by competitive bidding or by negotiation, it must take into account the likelihood of disputes that may develop in competitive bidding. To those who must be a part of the competitive bidding process, the legal issues should be understood and methods developed to avoid litigation.

A. Obstacles to Competitive Bidding

Generally, public contracts must be awarded through competitive bidding. An improper use of the exceptions that allow negotiation can frustrate this. This can be an issue if change orders are given to the contractor performing the work. To get the best price, statutes often specify that changes over a specified amount must be awarded through competitive bidding.[42] (This can create site access, delay, and operation problems, just some of the reasons why awarding authorities often seek a way to avoid the need for competitive bidding on existing projects.)

As noted in Sections 18.02 and 18.03A, competitive bidding should enable the awarding authority to get the best price and the greatest assurance of a successful completion. To do this the process assumes that all qualified bidders will be given a chance to bid and that the award will be made on objective criteria and not favoritism.[43]

Opening the process to all qualified bidders can be frustrated if some potential bidders are excluded. A preselection process can unduly restrict the number of bids received.[44]

A relatively new challenge to this objective can be posed by the use of a Project Labor Agreement (PLA).[45] PLAs are designed to keep labor peace. All contractors and subcontractors, union or nonunion, are required to enter into the PLA. Uniform terms and conditions of employment are achieved by requiring that all contractors designate a particular union to be the collective bargaining representative of all workers on the project.

Nonunion contractors often claim, usually without success, that a PLA restricts competition to union contractors, reducing the pool of bidders and depriving the awarding entity of getting the best price.[46]

B. Duty to Disclose

The early common law rarely obligated a contracting party to disclose information that the other party would want to know. Contracting parties were generally expected to

---

[32]See *Conduit Foundation Corp. v. Metro. Transp. Auth.*, reproduced in Section 18.04D.

[33]Sections 18.02, 18.03A, 18.04A, and 18.04J.

[34]Section 18.04B.

[35]Section 18.04C and D.

[36]Section 18.04E.

[37]Section 18.04G.

[38]Section 18.04H.

[39]Section 18.04F.

[40]Section 18.04I.

[41]Section 18.04J.

---

[42]Section 21.04B.

[43]Section 18.03G (selection between alternates can be based on favoritism).

[44]Section 18.03B.

[45]For more discussion see Section 23.05B.

[46]Ibid.

look out for themselves. They could not make deliberate misrepresentations but could conceal matters within their knowledge, even if they knew that the other party would want to know these matters.[47] But currently the common law is likely to require disclosure of vital information that the other party is not likely to discover.[48]

This has been reflected in an increasing tendency to require that the party conducting a competitive bid disclose certain information to bidders. The Court of Claims has held that a federal procurement agency must disclose procurement plans of other federal agencies of which it knows and which may affect the pricing assumption of bidders.[49] Similarly, it has held that the procuring agency must disclose technical information that it had relating to the manufacturing process that it knew the contractor intended to use.[50] The court noted that on many occasions each party has an equal opportunity to uncover the facts. But where one party knew much more than the other and knew the other was proceeding in the wrong direction, it could not betray the contractor into a "ruinous course of action by silence." State courts have also recognized a duty to disclose.[51]

Finally, a Federal Court of Appeals decision held that a federal government agency breached its requirements contract when it failed to notify the contractor before the bid that it knew that the orders would be significantly below (about 10 percent) of the estimate.[52]

Yet contractors cannot rely too heavily on the possibility of legal protection. Generally parties who *can* protect themselves must do so.[53] Contracting parties, although expected to cooperate, do not owe each other fiduciary obligations to look out for each other.[54]

C. Bid Proposal

The bid proposal is an offer and creates a power of acceptance in the owner. The owner has a right to close a deal and bind a bidder without a further act of the bidder.[55] Subject to many exceptions,[56] offers are generally revocable even if stated to be irrevocable for a specified period of time. This raises an issue that does not usually surface in construction litigation. Suppose either *before* bid opening or *after*, but before the formal contract is signed, the bidder revokes its bid. It is generally assumed, at least in public contracts, that despite the common law rule of revocability, the bid is irrevocable for the period stated in the invitation unless the invitation permits withdrawal before bid opening. What is the justification for such an assumption?

Sometimes public contract competitive bidding is authorized by statutes or regulations that state bids to be irrevocable. This may supersede the common law rule of revocability.[57] Sometimes the bid is revocable, but any deposit made with it is forfeited.[58] One argument for irrevocability is that the bidder receives the benefit of having the bid considered by the owner. Another is the bid may become irrevocable if the owner has justifiably relied on the bid. The owner may have given up the opportunity of negotiating a contract, having relied on the bidders' stating that their bids would be irrevocable or, at the very least, expended considerable time and money in conducting the competitive bid process.[59] If

---

[47]*Swinton v. Whitinsville Sav. Bank*, 311 Mass. 677, 42 N.E.2d 808 (1942).

[48]*Obde v. Schlemeyer*, 56 Wash.2d 449, 353 P.2d 672 (1960).

[49]*J.A. Jones Constr. Co. v. United States*, 182 Ct.Cl. 615, 390 F.2d 886 (1968) (Corps of Engineers knew of Air Force plans); *Bateson-Stolte, Inc. v. United States*, 145 Ct.Cl. 387, 172 F.Supp. 454 (1959) (must prove Corps of Engineers knew of AEC plans). Cases cited in this note were distinguished in many subsequent cases that point to factual differences. Reliance on cases cited must be tempered with a comparison of the facts in later cases. The cases cited did establish a precedent, but one that is largely based on the particular facts of the case. See also *Hardeman-Monier-Hutcherson v. United States*, 198 Ct.Cl. 472, 458 F.2d 1364 (1972) (duty to disclose weather and sea conditions after contractor request).

[50]*Helene Curtis Indus., Inc. v. United States*, 160 Ct.Cl. 437, 312 F.2d 774, 778 (1963).

[51]*Welch v. California*, 139 Cal.App.3d 546, 188 Cal.Rptr. 726 (1983) (records of earlier attempt to repair that documented tidal difficulties); *Commonwealth Dept. of Highways, v. S. J. Groves & Sons Co.*, 20 Pa.Commw. 526, 343 A.2d 72 (1975) (another contractor would occupy site after access given). See Annot. 86 A.L.R.3d 182 (1978).

[52] *Rumsfeld v. Applied Cos.,Inc.*, 325 F.3d 1328 (Fed.Cir.), cert. denied, 540 U.S. 951 (2003).

[53]*H. N. Bailey & Assoc. v. United States*, 196 Ct.Cl. 166, 449 F.2d 376 (1971) (information obtainable elsewhere); *T.L. James & Co., Inc. v. Traylor Bros., Inc.*, 294 F.3d 743 (5th Cir.2002) (applying Louisiana law).

[54]See Section 19.02D.

[55]*Hadaller v. Port of Chehalis*, 97 Wash.App. 750, 986 P.2d 836 (1999).

[56]See Section 5.06B.

[57]*Powder Horn Constructors, Inc. v. City of Florence*, 754 P.2d 356 (Colo.1988).

[58]1 A.CORBIN, CONTRACTS § 2.18 (rev. ed.1993).

[59]*Drennan v. Star Paving Co.*, 51 Cal.2d 409, 333 P.2d 757 (1958) (subcontractor bid irrevocable because prime relied).

the transaction is governed by the Uniform Commercial Code (transactions in goods), the bid will be irrevocable as a "firm offer."[60] If the procurement involves the sale of goods with installation incidental, the Code would govern.[61] Even if it does not, it may be possible to use the code by analogy.[62]

Once the bid has been opened, it has become irrevocable and can be withdrawn only if legal grounds exist for doing so—usually, a mistake in preparing the bid.[63]

Can the owner hold bidders to whom the bid has not been awarded for the period specified in the invitation to bidders? The not uncommon refusal of the bidder to whom the contract has been awarded to enter into the contract makes it imperative that no expectation be created that the unsuccessful bids are no longer binding when an award has been made. Of course, the period of time cannot be extended, but it should not be shortened simply because the award appears to have been made to someone else.

D. Award and Waiving of Irregularities—Remedies— *Conduit and Foundation Corp. v. Metropolitan Transportation Authority*

Usually the owner promises to award the contract to the lowest responsible or lowest and best bidder, reserving the right to reject all bids. Although these two standards for determining to whom a contract should be awarded are often used as if they dictated the same outcome, a careful study of the language shows that this is not always the case. Being the *lowest responsible* bidder would not be the same as being the *lowest and best bidder*. The California Supreme Court emphasized that "responsible" simply means that the low bidder has the quality, fitness, and capacity to satisfactorily do the required work. In other words, according to the court, a contract must be awarded to the low bidder unless it has been found to be not

responsible, that is, "not qualified to do the particular work under consideration."[64]

This distinction is demonstrated by *KAT Excavating, Inc. v. City of Belton*.[65] The standard for making the award was the "lowest and best bidder." The city accepted the second lowest bid. It was $4,000 higher than the low bid in a contract of about $500,000. The court held the "best bidder" takes into account honesty and integrity, skill and business judgment, experience and facilities, and previous work. The city had discretion to select the best bidder, but discretion had to be exercised "in good faith, in the interests of the public, without collusion or fraud, nor corruptly, nor from motives of personal favor or ill will. . . ."[66]

Here, the bidder to whom the contract was awarded had a good track record, especially as to timely performance. That the bids were so close, along with this factor, meant that the city exercised its discretion properly. There might have been a different outcome had the standard been "lowest responsible bidder."

Before making the award, the owner must evaluate the bids. This is done between bid opening and award, if any. Yet this evaluation period is not open ended. Bids are irrevocable for the period specified in the invitation to bidders. This restriction is illustrated in *Hennepin Public Water District v. Petersen Construction Co.*,[67] where the invitation had stated that the awarding authority would have to obtain financing within sixty days from bid opening. There was a tentative acceptance before the awarding authority obtained financing. Then, sixty-seven days after bid opening, the awarding authority forwarded the formal contract to the successful bidder. The court held the acceptance to have been too late, and the bidder was released.

Before examining legal constraints on the awarding authority when it considers bids, the following case reviews some of the purposes of competitive bidding in the public sector discussed in Section 18.02 and passes on the conditions under which the awarding authority can reject all bids and rebid the project.

---

[60] U.C.C. § 2-205.

[61] *Bonebrake v. Cox*, 499 F.2d 951 (8th Cir.1974). See also *Wachter Management Co. v. Dexter & Cheney, Inc.* 282 Kan. 365, 144 P.3d 747 (2006) (construction software sale involved goods governed by U.C.C., even though support services were included).

[62] *Transatlantic Financing Corp. v. United States*, 363 F.2d 312 (D.C. Cir.1966).

[63] *Elsinore Union Elementary School Dist. v. Kastorff*, 54 Cal.2d 380, 353 P.2d 713, 6 Cal.Rptr. 1 (1960).

[64] *City of Inglewood—L.A. County Civic Center Auth. v. Superior Court*, supra note 5, 500 P.2d at 604, 103 Cal.Rptr. at 692.

[65] 996 S.W.2d 649 (Mo. App.1999), rehearing and/or transfer denied June 29, 1999, transfer denied Aug. 24, 1999.

[66] Id. at 651.

[67] 54 Ill.2d 327, 297 N.E.2d 131 (1973).

CONDUIT AND FOUNDATION CORP. v. METROPOLITAN TRANSPORTATION AUTHORITY

Court of Appeals of New York, 1985. 66 N.Y.2d 144, 485 N.E.2d 1005, 495 N.Y.S.2d 340.
[Ed. note: Footnotes omitted.]

JASEN, Judge.

The narrow issue presented on this appeal is whether the evidence on the record supports a holding that the Metropolitan Transportation Authority and the New York City Transit Authority acted unlawfully in rejecting all bids submitted in an initial round of bidding on a public works project.

The relevant facts are not in dispute. Respondents, the Metropolitan Transportation Authority and the New York City Transit Authority (Transit Authority), solicited bids for a public works contract for the massive rehabilitation of part of New York City's subway system. Three bids were received, including that of petitioner, a joint venture, whose bid was the lowest and fell within the advertised estimated cost between $120 [million] and $140 million. Shortly after opening the bids, the Transit Authority met with each of the three bidders for the stated purpose of determining why the prices submitted were as high as they were and whether the project costs might be reduced. Although the bidders were met separately, each was informed that the meetings with the others were taking place.

Subsequent to the meetings, the second lowest bidder advised the Transit Authority by letter that significant cost reductions were possible and suggested another meeting to discuss the possibility of preparing revised contract documents. Several days thereafter, it was petitioner that met again with the Transit Authority. Petitioner was informed that its bid, though the lowest, was far in excess of the revised estimated cost and that, therefore, all the bids might be rejected. Nevertheless, petitioner was told that it was deemed technically and financially qualified to undertake the project and, indeed, was afforded the opportunity to reduce its bid for consideration. Within the next two weeks, petitioner once again met with the Transit Authority and offered to reduce its bid price by two million dollars, a proportionately small amount. The reduction was deemed insufficient by the Transit Authority's Chief Engineer who, on the following day, recommended that all the bids be rejected and that the contract, with modifications, be readvertised for a second round of bidding. The recommendation was adopted by the President of the Transit Authority and the three bidders were notified. Thereafter, the Transit Authority circulated a notice soliciting new bids for the project and indicating an estimated cost range

between $100 [million] and $120 million, $20 million lower than that originally advertised.

Petitioner commenced this article 78 proceeding seeking an injunction against a second round of bidding and a judgment directing respondents to award the contract to petitioner as the lowest responsible bidder. Special Term granted the petition concluding that the Transit Authority's postbid communications with the three bidders, its revision of the advertised cost estimate and its purpose of obtaining bid prices lower than that originally received rendered the decision to reject all first round bids arbitrary and capricious. The Appellate Division agreed with Special Term, two justices dissenting, that the contract should be awarded to petitioner, but modified the judgment to reduce the contract price by two million dollars on the basis of petitioner's prior offer to the Transit Authority to so lower its bid. The court held that the record demonstrates an "appearance of impropriety" on the part of the respondents which "might have" a detrimental effect "on the broad interest of the public in the entire public bidding process." (111 A.D.2d 230, at p. 234, 489 N.Y.S.2d 265.) The two dissenters argued that the mere "appearance" of impropriety is not sufficient ground to disturb the decision of the Transit Authority absent a showing of actual favoritism, fraud or similar evil which competitive bidding is intended to prevent. We agree and now reverse the order below.

As this court has stated on prior occasion, the purpose of the laws in this State requiring competitive bidding in the letting of public contracts is "to guard against favoritism, improvidence, extravagance, fraud and corruption." (*Jered Contr. Corp. v. New York City Tr. Auth.*, 22 N.Y.2d 187, 193, 292 N.Y.S.2d 98, 239 N.E.2d 197; . . .

These laws were not enacted to help enrich the corporate bidders but, rather, were intended for the benefit of the taxpayers. They should, therefore, be construed and administered "with sole reference to the public interest." (10 McQuillin, Municipal Corporations § 29.29, at 302 [3d rev ed]; . . . This public interest, we have noted, is sought to be promoted by fostering honest competition in the belief that the best work and supplies might thereby be obtained at the lowest possible prices. . . .

Indeed, this policy is reflected explicitly in the statutory provisions which mandate that public work contracts be awarded "to the lowest responsible bidder." (Public Authorities Law § 1209[1]; General Municipal Law § 103[1]; *see also*, General Municipal Law § 100-a.) Dishonesty, favoritism and material or substantial irregularity in the bidding process, which undermines the fairness of the competition, impermissibly contravene this public interest in the prudent and economical use of public moneys. . . .

Nevertheless, where good reason exists, the low bid may be disapproved or, indeed, all the bids rejected. Neither the low bidder nor any other bidder has a vested property interest in a public works contract . . . and statutory law specifically authorizes the rejection of all bids and the readvertisement for new ones if deemed to be "for the public interest so to do" (Public Authorities Law § 1209[1]; . . .

Although the power to reject any or all bids may not be exercised arbitrarily or for the purpose of thwarting the public benefit intended to be served by the competitive process . . . the discretionary decision ought not to be disturbed by the courts unless irrational, dishonest or otherwise unlawful. . . .

Moreover, it cannot be gainsaid that a realistic expectation of obtaining lower contract prices on a second round of bidding constitutes a reasonable, bona fide ground for rejecting all first round bids and serves the public's interest in the economical use of public moneys. . . .

Nor is the decision to reject all bids rendered arbitrary and capricious solely on the basis of nondiscriminatory postbid changes in the contract specifications . . . or postbid communications with individual bidders for the bona fide purpose of ascertaining how contract costs might be reduced. . . .

Only on a showing of actual impropriety or unfair dealing—i.e., "favoritism, improvidence, extravagance, fraud and corruption"

(*Jered Contr. Corp. v. New York City Tr. Auth.*, *supra*, 22 N.Y.2d at p. 193, 292 N.Y.S.2d 98, 239 N.E.2d 197)—or other violation of the statutory requirements, can the decision to reject all bids and readvertise for a second round of bidding be deemed unlawful. . . .

Consequently, where the party challenging the decision does not satisfy the burden of making such a demonstration, that decision should remain undisturbed. . . .

Here, where the court below found only that the postbid activity of the Transit Authority created an "appearance of impropriety" and "cast a doubt" on its fair dealing with the bidders, the respondents' decision to seek a rebid ought not to have been disturbed. It may be true, as the majority at the Appellate Division found, that it would have been wiser for the Transit Authority to meet with all the bidders at the same time, instead of separately, in order to avoid the possible appearance of unfair dealing. Likewise, other bid interactions between the Transit Authority and certain bidders may have been more discreet. Nevertheless, as noted in the dissenting opinion of Justice Niehoff absent some finding that the respondents had, in fact, engaged in some unfair or unlawful practice tainting the impartiality of the competitive bidding process, it was error to grant the petition.

The Transit Authority clearly was empowered to reject all the first round bids, and its expectation of obtaining lower bid prices on readvertisement was a rational basis for deciding to do so. . . .

Accordingly, the order of the Appellate Division should be reversed, with costs, and the petition dismissed.
WACHTLER, C. J., and MEYER, SIMONS, KAYE, ALEXANDER and TITONE, J. J., concur.

The U.S. Court of Claims has held that a bidder is entitled to honest consideration of its bid. To show that the bid was not properly considered, the bidder must establish that the government acted arbitrarily and capriciously and that there was no reasonable basis for the government's decision.[68]

On the other hand, California rejected a bidder's contention that the awarding authority must consider bids in good faith.[69] The court stated,

> Such a promise would be contrary to a well established rule which allows a public body where it has expressly reserved the right to reject all bids, to do so for any reason and at any time before it accepts a bid. If the entity so decides, it may return all bids unopened. The courts have consistently refused to interfere with the exercise of a public body's right to reject bids, however arbitrary or capricious.[70]

---

[68]*Keco Indus. Inc. v. United States*, 192 Ct.Cl. 773, 428 F.2d 1233 (1970). Absence of fraud or palpable abuse of discretion is required in Mississippi. *Warren G. Kleban Eng'g Corp. v. Caldwell*, 361 F.Supp. 805 (N.D.Miss. 1973), vacated on other grounds, 490 F.2d 800 (5th Cir.), on remand dismissed petition, 492 F.2d 1200 (5th Cir. 773, 1974). New Jersey requires a bona fide judgment. *Mendez v. City of Newark*, 132 N.J.Super. 261, 333 A.2d 307 (Law Div.1975).

[69]*Universal By-Products, Inc. v. City of Modesto*, 43 Cal.App.3d 145, 117 Cal.Rptr. 525 (1974).
[70]117 Cal.Rptr. at 529.

The court also rejected a claim based on misrepresentation because of immunity granted the public entity. But the awarding authority cannot solicit a bid without any intention of considering it.[71]

Although the California rule gives maximum discretion to the awarding authority, there is an undeniable trend toward holding government agencies accountable for their acts.

Where the low bidder is rejected, procedures vary. For example, New Jersey precludes the local awarding authorities from rejecting the high bid for property or low bid for work without giving a hearing to the disappointed bidders.[72] California, however, does not require a full-fledged courtlike hearing if the low bid is rejected. Rather, the awarding authority must give the low bidder access to any evidence that has reflected on its responsibility received from others or produced as a result of an independent investigation. The bidder must be afforded an opportunity to rebut such adverse evidence and present evidence that it is qualified to perform the contract.[73]

California imposes stricter due process requirements on the awarding authority when the bidder is claimed by the awarding authority not to be responsible. It must receive some level of due process, such as notice and an opportunity to be heard. But if a dispute is whether the bid is responsive, the facts are generally undisputed and the decision does not involve agency discretion. In such a case, the bidder is not entitled to the same due process level.[74]

States vary as to the requirement for a hearing and its nature. But again, with the emphasis on greater accountability, it is likely that low bidders who are not awarded the contract can at least demand that reasons be given and that they be given an opportunity to rebut adverse evidence.

As for specific reasons for rejecting a bidder, New Jersey held that disputes over a previous job were insufficient to deny award to the low bidder.[75] But Mississippi upheld a provision in the invitation that allowed the awarding authority to reject a bidder for being in arrears on an existing contract or in litigation with the awarding authority for having defaulted on a previous contract.[76] The bidder's wholly owned subsidiary was suing the awarding authority on another matter, and on this basis the bidder's bid was rejected. The court recognized the provision as possibly coercing contractors from asserting their legal rights but held that the standard of rejecting bidders was one within the discretion of the authority. This holding allows an awarding authority to use the procurement process for an improper purpose.

Massachusetts upheld rejection of a subcontractor who had had a dispute with the prime contractor on an earlier job and had made misstatements of its previous work experience.[77] A federal court applying Mississippi law upheld rejection of a bidder because of the bidder's poor reputation for quality work.[78]

An awarding authority or owner conducting a competitive bid should be able to reject a bidder if a good-faith determination has been made that the bidder is not likely to be able to complete the required performance.

The evaluation may proceed even in the face of the awarding authority having informed the low bidder that it had been awarded the contract. This is demonstrated by *Hadaller v. Port of Chehalis*.[79] The plaintiff was one of three apparent low bidders. Next, the awarding authority, the Port, had to determine who was the lowest qualified bidder. The plaintiff was selected and informed of this. The Port then informed the plaintiff that it was awarded the contract.

But a formal contract was not drawn up. The Port requested more information about the metal building system. It asked the plaintiff to provide a list of satisfactory installations, a certificate from the manufacturer, and the manufacturer's 20-year warranty. The instructions to bidders specified that the successful bidder would furnish this information. When the required information was not submitted, the Port awarded the contract to another bidder.

---

[71]Supra notes 69, 70.

[72]*Mendez v. City of Newark*, supra note 68; *D. Stamato & Co. v. Township of Vernon*, 131 N.J.Super. 151, 329 A.2d 65 (App.Div.1974).

[73]*City of Inglewood—L.A. County Civic Center Auth. v. Superior Court*, supra note 5.

[74]*D. H. Williams Constr., Inc. v. Clovis Unified School Dist.*, 146 Cal. App.4th 757, 53 Cal.Rptr.3d 345 (2007), rehearing denied Jan. 31, 2007, review denied May 9, 2007 (bid responsive despite listing an unlicensed subcontractor, as the proposed subcontractor must be licensed only when it enters into the subcontract).

[75]*D. Stamato & Co. v. Township of Vernon*, supra note 72.

[76]*M.T. Reed Constr. Co. v. Jackson Mun. Airport*, 227 So.2d 466 (Miss.1969).

[77]*Kopelman v. Univ. of Massachusetts Building Auth.*, 363 Mass. 463, 295 N.E.2d 161 (1973).

[78]*Warren G. Kleban Eng'g Corp. v. Caldwell*, supra note 68.

[79]See supra note 55.

The plaintiff contended that it had made an oral contract with the Port. As no formal contract had been made and as the facts showed that the award process was not yet completed, this contention was rejected by the court.

Nevertheless, officials of the awarding authority should be careful to avoid any indication that the process had been completed and a contract awarded. It would have been better had the awarding authority made more clear that the award was conditioned on the requested information being furnished.

To the extent that the reason for rejection departs from evaluation of the bidder's experience, financial ability, integrity, and availability of facilities necessary to perform the contract, it is more likely that the rejection has not been made in good faith.[80] Bidder evaluation was discussed earlier.[81]

The preferred remedy by a disappointed bidder is an injunction barring the public entity from awarding the contract to another.[82] Sometimes, it is too late for an injunction, or it is not available for another reason. If this is the case, the only remedy left is damages.

The *Kajima* case dealt with the money award remedy accorded a bidder who should have received the award but did not because of errors of the awarding authority.[83] This case merits attention both because of its holding on the remedy issue and its thorough canvassing of the cases dealing with the money award remedy throughout the United States.

Kajima sought recovery for its first and second round bid and protest expenses totaling $143,000, its unabsorbed overhead expenses of $1,300,000 and $1,500,000 in lost profit on the contract, plus prejudgment interest. (All numbers approximate.) The trial court considerably reduced the amounts but awarded a judgment for all items claimed.

The intermediate appeals court, noting the equitable nature of the basis for the claim, concluded that each case should be judged on its merits. It held that the trial court acted within its discretion.

On appeal to the California Supreme Court, the issue was whether Kajima would be limited to its bid preparation costs. In essence, the issue was lost profits. (Bid protest costs and overhead were not contested by the Transit Authority.)

Kajima's claim was based upon promissory estoppel. The bidder relied on the Transit Authority's promise that when it solicited bids it would award the contract to the lowest responsible bidder. The court held there could be no claim based on tort nor one for breach of contract.

The court would not award Kajima lost profits, limiting Kajima's remedy to the costs of preparing its bid. It pointed to the uncertainty connected to the bidding process and the speculative nature of profits on such a contract. Finally, it was concerned that too expansive a remedy could "encourage frivolous litigation and further expend public resources."[84]

It rejected Kajima's contention that limiting the disappointed bidder to its bid preparation expenses would provide little incentive to challenge an improper award and would ultimately lead to a decline in bidding for public work. The court pointed to the many cases where awards were challenged in court.

The issue of the remedy for a reliance-based claim has always been a troublesome one. But the real basis for the holding, protection of public entities, is revealed by the court's refusal to follow a case that had awarded lost profits by stating it involved private parties.[85]

Its canvas of cases from other jurisdictions reveals that a few states bar any cause of action against a public entity. Most limit recovery to bid preparation expenses.[86] A few allow recovery of lost profits.[87]

The limited remedy and the broad scope accorded public agencies to award contracts reflects judicial hesitance to interfere with the operations of a public entity.[88]

---

[80]*D. Stamato & Co. v. Township of Vernon*, supra note 72.

[81]See Sections 18.03K and 18.04D.

[82]Injunctive relief was granted in *Keefe-Shea Joint Venture, Inc. v. City of Evanston*, 332 Ill.App.3d 163, 773 N.E.2d 1155, appeal denied, 201 Ill.2d 570, 786 N.E.2d 184 (2002), appeal after remand, 364 Ill.App.3d 48, 845 N.E.2d 689 (2005), appeal denied, 218 Ill.2d 541, 850 N.E.2d 808 (2006); *Modern Continental Constr. Co., Inc. v. City of Lowell*, 391 Mass. 829, 465 N.E.2d 1173 (1984).

[83]*Kajima/Ray Wilson v. Los Angeles County Metro.Trans.Auth.*, 23 Cal. 4th 305, 1 P.3d 63, 96 Cal.Rptr.2d 747 (2000), rehearing denied, Aug. 23, 2000.

[84]1 P.3d at 70.

[85]Id. at 73.

[86]*Cementech, Inc. v. City of Fairlawn*, 109 Ohio St.3d 475, 849 N.E.2d 24 (2006) (reversing an intermediate appellate court that had allowed lost profits, noting that awarding lost profits would punish the taxpayers and that injunctive relief is preferable to damages).

[87]1 P.3d at 70–72.

[88]See Section 18.04I on standing to judicially contest an award.

Suppose there are irregularities in the proposal or process. The awarding authority may believe them minor and wish to waive them and accept the bid. The legal issues have been whether the irregularity in question is minor and whether waiver would encourage carelessness, create opportunity for favoritism, and operate unfairly to other bidders.

The awarding authority's discretion to waive deviations in a nonresponsive bid depends on the importance of the deviation,[89] possible prejudice to the other bidders, and prejudice to the public authority. For example, in deciding whether an irregularity in the low bid was minor and could be waived, the court allowed the awarding authority to take into consideration that the award to the next lowest bidder would mean the same work was being done for nearly $8 million dollars more.[90]

Yet a different, less flexible method of dealing with nonresponsive bids can be seen in Louisiana. Earlier Louisiana decisions had sought to limit the power to waive irregularities to formal ones, not those of substance. Yet this did not satisfy the legislature. It constantly revised its statutes to make it clear that no waiver would ever be permitted.[91] Constant revision indicated a battle between some lower courts and the public agencies on one side and the legislature on the other.[92]

Many cases involve bid submission deadlines. *H. R. Johnson Construction Co. v. Board of Education of Painsville Twp., etc.*[93] held that an awarding authority could not accept a bid made one minute late. The court was not persuaded that one minute could not give a bidder a competitive advantage over other bidders. The court felt that if the awarding authority is given the discretion to allow a one-minute deviation, it could stretch this power to even fifteen minutes or an hour.

In *William F. Wilke, Inc. v. Department of Army*,[94] the bids were to be opened at 3 P.M. At that time, a representative of the low bidder was in the room, but he neglected to put the sealed bid in the receptacle designated for that purpose. At 3:04, the contracting officer gathered the receptacle and started to sort out the bids. During the sorting, the low bidder's representative added his sealed bid to the box of yet unopened bids. The awarding authority awarded the bid to the low bidder, and the next low bidder complained.

The court concluded that the bid should not have been accepted, as it had been deposited four minutes late. But because it did not appear that the low bidder had obtained any competitive advantage, the court would not order the bid to be awarded to the next low bidder but limited the disappointed bidder to recovery of its bidding expenses. Other cases have allowed the awarding agency discretion to waive such irregularity, especially if the bidding information permitted this.[95]

These time deadline cases, simple though they seem, are a microcosm of the tensions in the competitive bidding process set forth in Section 18.02. It is easy to look at one minute late as a technical deficiency that should not preclude an awarding authority from accepting the lowest bid, thus saving the public money.

But laxness in enforcement of this plain rule can lead to favoritism and even corruption. Waiver of the time deadline can be easy to justify, such as the discretion given the awarding authority to waive minor deficiencies, or by concluding that the late bidder was not at fault.

---

[89]*Albano Cleaners, Inc. v. United States*, 197 Ct.Cl. 450, 455 F.2d 556 (1972). The court held that a substantial deviation affects price, quality, or quantity. In *Rosetti Contracting Co. v. Brennan*, 508 F.2d 1039 (7th Cir.1975), the court held that nonconformity as to affirmative action hiring was not correctable. Here the court was facing a substantial deviation but one that the bidder offered to correct. The court was fearful that allowing a correction would give the bidder an option exercisable after seeing the other bids.

[90]*Hill Bros. Constr. & Eng'g Co. Inc. v. Mississippi Transp. Comm'n*, 909 So.2d 58, 69–70 (Miss.2005).

[91]La. R.S. § 33-2212A(1)(b).

[92]The Louisiana history is outlined in *Hamp's Constr. L.L.C. v. City of New Orleans*, 924 So.2d 104 (La.2006) (legislature has made its will clear: no waivers).

[93]16 Ohio Misc. 99, 241 N.E.2d 403 (1968). But see *Mickey O'Connor General Contractor, Inc. v. City of Westwego*, 804 So.2d 128 (La. App.2001), writ denied, 811 So.2d 908 (La. 2002) (bid deadline could be waived). See also *PHC, Inc. v. Kelleys Island*, 71 Ohio App.3d 277, 593 N.E.2d 386 (1991) (would not go by clock in public office that was 2–3 minutes fast).

[94]485 F.2d 180 (4th Cir.1973). See also *Mickey O'Connor General Contractor, Inc. v. City of Westwego*, supra note 93 (city could accept bid stamped one minute after deadline as bidder in line with three other bidders before deadline).

[95]*William M. Young & Co. v. West Orange Redevelopment Agency*, 125 N.J.Super. 440, 311 A.2d 390 (App.Div.1973) (two minutes late preceded by a telephone call from the bidder stating he would be a few minutes late because of inclement weather); *Gostovich v. City of West Richland*, 75 Wash.2d 583, 452 P.2d 737 (1969) (three days late due to mail mixup). *Quinn Constr. Co., L.L.C. v. King County Fire Protection Dist.*, 111 Wash.App. 19, 44 P.3d 865 (2002) (award to lowest bidder, who missed filing deadline by 5 to 10 seconds, upheld).

Also, accepting a late bid can mark the beginning of the tolerance of sloppy procedures throughout the performance of the contract. It can give the impression that the contractor need not perform "to the letter." It too can lead to favoritism and even corruption.

Finally, this can be unfair to other bidders and make them hesitant to submit bids in the future. It can, in the end, cost the awarding authority money.[96]

E. Withdrawal or Correction of Mistaken Bids: *Sulzer Bingham Pumps, Inc. v. Lockheed Missiles & Space Company, Inc.*

Sometimes a bidder will seek to withdraw or correct a bid. Such a request usually occurs just after bid opening or, more rarely, after the formal award. Usually the basis for the request is computation errors, such as omitting a large item, making a mathematical miscalculation in determining an item price, or making an error in adding bid price components.[97]

Early cases would not relieve a bidder for these mistakes.[98] A reason given for denying relief was that the mistake was "unilateral," one made only by the bidder and not shared in by the owner. There was, as a rule, negligence in computing the bid. The fear of false claims and the integrity of the bidding system were other reasons for denying relief.

Some courts began to moderate the strictness of this doctrine. They pointed to the rule that a person to whom an offer has been made cannot accept the offer if she knows or should know that the offer was made by mistake. This is known as the "snap-up" doctrine. For example, if the owner or design professional knew or should have known that an entire item had been left out of the bid or that there had been a mistake in adding the total, it would be unfair to accept the proposal and seek to bind the bidder. The courts usually focus on the knowledge or constructive (what it should have known) knowledge of the owner before the formal award and execution of the formal contract, most commonly at the time the bids are open. Some courts have required that there be actual knowledge on the part of the owner,[99] whereas others appear to require constructive knowledge that the mistake was known or ought to have been known.[100] In practice, the difference between actual and constructive knowledge will rarely control the outcome of a case.

The typical case involves the low bidder claiming that it had made a mistake at the time of bid opening or shortly thereafter. The "snap-up" doctrine noted earlier is demonstrated in *Santucci Construction Co. v. Cook County*[101] in which the awarding authority had estimated that the cost of drain work would be $1.9 million.[102] The cost of the drainpipe, including labor, was expected to be $1.4 million. Santucci submitted a total bid of $1.1 million, with $775,000 for the drainpipe, including labor. Three other contractors submitted total bids between $1.7 million and $1.8 million, with their bids for the drainpipe, including labor, being between $1.2 million and $1.4 million. The engineer for the awarding authority had thought that Santucci's bid was "cheap, low." A day after bid opening, Santucci claimed a clerical error and sought to withdraw his bid. The request was refused, and Santucci would not enter the contract. The awarding authority retained Santucci's bid deposit. Santucci brought legal action to recover it.

Noting that ultimately the work was let for $1.6 million, the appellate court affirmed the finding of the trial court that Santucci had made a mistake and that the awarding authority should have known of this mistake. It rescinded Santucci's bid and ordered that the deposit be returned to him. The court emphasized that Santucci's bid was $600,000 less than the next lowest bidder and over

---

[96]For further discussion of this topic, see Waagner & Evans, *Agency Discretion in Bid Timeliness Protests: The Case for Consistency*, 29 Pub. Cont.L.J. 713 (2000).

[97]Occasionally, claims for relief are based on a sub-bidder's refusal to contract at the price proposed or for an error of judgment relating to performance cost.

[98]*Steinmeyer v. Schroeppel*, 226 Ill. 9, 80 N.E. 564 (1907). This case was distinguished in *Rushlight Automatic Sprinkler Co. v. City of Portland*, 189 Or. 194, 219 P.2d 732 (1950) (reasonable excuse for error and awarding authority accepted the bid even if it thought the bid was too good to be true). At most, the *Steinmeyer* case is important today for historical reasons. Yet the concept of no relief has some attraction in light of the complexities and uncertainties generated by opening the gates to mistake claims.

---

[99]*Westinghouse Elec. Corp. v. New York City Transit Auth.*, 735 F.Supp. 1205 (S.D.N.Y.1990) (interpreting *Iversen Constr. Corp. v. Palmyra-Macedon Central Sch. Dist.*, 143 Misc.2d 36, 539 N.Y.S.2d 858 (1989) as requiring actual knowledge).

[100]*Bromley Contracting Co. v. United States*, 219 Ct.Cl. 517, 596 F.2d 448 (1979).

[101]21 Ill.App.3d 527, 315 N.E.2d 565 (1974).

[102]Amounts are approximations.

$800,000 less than the awarding authority's estimate for the project.

The "snap-up" concept used by the court in the *Santucci* case has been the vehicle for bidder relief where courts have thought it appropriate. Yet an appraisal of the cases and an awareness of the immense variation of the bids for many types of construction work leads inescapably to the conclusion that the courts have emphasized, not the fact that the awarding authorities should have known of the mistake, but the unfairness of holding a bidder who has made an honest mistake when the next bid can still be accepted.[103]

Increasingly fewer jurisdictions deny relief.[104] These jurisdictions emphasize the need to protect the process from possible favoritism that may accompany the power to allow withdrawal. Such decisions reflect skepticism that the fact-finding process can determine whether honest mistakes have been made. Even in these jurisdictions, relief can be granted if the facts appear to make it inequitable to hold the bidder to its bid.[105]

Most jurisdictions will relieve the bidder if the mistake is clerical rather than an error of judgment and involves a substantial portion of the total bid or a large amount of money and if the owner has not relied to its detriment on the mistaken bid.

The Federal Acquisitions Regulations (FAR) permit the awarding agency to correct clerical mistakes before an award is made.[106] The regulations provide a systematic method of dealing with clerical and other bidding mistakes "apparent on its face in the bid." A correction may be made if the bidder requests permission to correct a mistake and if there is "clear and convincing evidence" of a mistake and of the "bid actually intended." However, if the correction would displace one or more bids, correction "shall not be made unless the existence of the mistake and the bid actually intended are ascertainable substantially from the invitation and the bid itself."[107] If the evidence of the mistake is clear and convincing but the bidder does not provide clear and convincing evidence of the bid actually intended, the public agency may decide that the bidder may withdraw the bid.

Suppose the bidder wishes to withdraw the bid rather than correct it. If the evidence is clear and convincing, both as to the existence of a mistake and as to the bid actually intended, and if the bid as corrected would be the lowest received, the agency may correct the bid and not permit its withdrawal.

The regulations govern the relationship between bidders and the federal agency conducting the bidding competition. They may also control bidding mistakes in disputes between a subcontractor and a prime contractor. If the subcontract selects federal procurement law as governing, the FAR will be applied.[108]

State legislation increasingly regulates attempts by those who bid on public contracts to withdraw bids because of mistake after bid opening. In 1945, California adopted legislation that allowed a bidder to be relieved from a bid if it established to the satisfaction of the court that

1. "A mistake was made."
2. It gave the department "written notice within five working days . . . after the opening of the bids of the mistake, specifying in the notice in detail how the mistake occurred."

---

[103]Similarly, cases holding that a right to rescind exists and that the claimant met the standards required for rescission are *Marana Unified School Dist No.6 v. Aetna Cas. & Sur. Co.*, 144 Ariz. 159, 696 P.2d 711 (1984) (bidder did not need to forfeit bond despite statute); *Powder Horn Constructors, Inc. v. City of Florence*, supra note 57 (reversed intermediate court's requirement that bidder show it was not negligent). See Jones, *The Law of Mistaken Bids*, 48 U.Cin.L.Rev. 43 (1979); Cavico, *Relief for Unilateral Mistake in Construction Bids*, 10 Thurgood Marshall L.Rev. 1 (1985). The many cases are collected in Annot., 2 A.L.R.4th 991 (1980).

[104]*Alaska Int'l Constr., Inc. v. Earth Movers of Fairbanks, Inc.*, 697 P.2d 626 (Alaska 1985) (no loss on contract; strong dissent), called into doubt a year later by *Jensen & Reynolds Constr. Co. v. State Dep't of Transp. & Public Facilities*, 717 P.2d 844 (Alaska 1986); *Anco Constr. Co. Ltd. v. City of Wichita*, 233 Kan. 132, 660 P. 2d 560 (1983), disagreed with by *Marana Unified School Dist. No. 6 v. Aetna Cas. & Sur.*, supra note 103; *Nelson Inc. of Wis. v. Sewerage Comm'rs of Milwaukee*, 72 Wis.2d 400, 241 N.W.2d 390 (1976) (despite statute), disagreed with and for all practical purposes overruled by *James Cape & Sons v. Mulcahy*, 285 Wis.2d 200, 700 N.W.2d 243 (2005). Care must be taken to distinguish cases that refuse relief because the elements are not established from those that hold that no relief can be granted in any case.

[105]See *Travelers Indem. Co. v. Susquehanna County Commr's.*, 17 Pa. Commw. 209, 331 A.2d 918 (1975) (the court refused to forfeit the bid bond because the bidder had not actually withdrawn his bid but had simply requested to do so and the awarding authority had failed to present the bidder with contract papers for execution).

[106]48 CFR § 14.407-2 (2007).

[107]The Federal Procurement Regulations are discussed in Rudland, *Rationalizing the Bid Mistake Rules*, 16 Pub.Contract L.J. 446 (1987), and Hagberg, *Mistake in Bid, Including New Procedures Under Contract Disputes Act of 1978*, 13 Pub.Cont.L.J. 257 (1983).

[108]*Sulzer Bingham Pumps, Inc. v. Lockheed Missile & Space Co., Inc.*, 947 F.2d 1362 (9th Cir.1991) (reproduced in part later in this subsection).

3. "The mistake made the bid materially different" from what was intended.

4. "The mistake was made in filling out the bid and not due to error in judgment or to carelessness in inspecting the site of the work or in reading the plans or specifications."[109]

The statute reflects a more mechanical method of dealing with the problem, particularly in barring relief unless notice is given within five days after bid opening specifying in detail how the mistake occurred. There is no requirement under the statute that holding the bidder would be unconscionable[110] or that the city must have had knowledge before the bid was accepted that there had been a clerical mistake making it unjust and unfair for the city to take advantage of the bidder's error. Nor does the statute require any evaluation of the degree of negligence.

The evolution of legal rules dealing with bidding mistakes reflects a slow process, starting first in those transactions where it is determined that the awarding authority knew or should have known of the mistake but limiting mistake to clerical errors.[111] Yet suppose a mistake has been made that is not a simple clerical error. Cases have differed as to whether relief is confined to clerical mistakes. Some cases refuse to draw this distinction.[112]

Again the California experience is instructive. Early California cases relieved bidders from good faith, unilateral, clerical mistakes.[113] Implicit in those cases was that demand by the awarding authority that the bidder perform at the bid price was unconscionable.[114] Much depended on when the mistake claim was made, whether the next lowest bid could still be accepted , and the size and cause of the error.

In 2001, California faced the issue of unilateral mistake in a case involving a mistake in a newspaper advertisement.[115] The car agency making the mistake offered to pay the claimant the cost of his fuel, time, and effort expended in traveling to the agency to examine the car. The claimant refused and brought a lawsuit.

The court agreed with the trial court that had refused to allow the claim. The court specifically adopted Section 153 (a) of the Second Restatement of Contracts. It authorizes relief from the mistaken bid where the mistake pertains to an important element of the bargain, the mistake materially alters the bargain, the risk is not one borne by the party seeking relief, and enforcement would be unconscionable.

A commentator, while recognizing that unconscionability is an important element in these cases, worried about the subjective and difficult to define nature of this standard. She preferred relief for good faith unilateral clerical errors if "the mistaken party compensates the non-mistaken party for detrimental reliance."[116]

Were this applied in the usual bid mistake cases, the bidder who made a good faith unilateral clerical error would be relieved if the bidder compensated the awarding authority any costs it had incurred. If it were too late to accept the next low bidder, the bidder seeking relief would receive relief only if it paid for the difference in bid prices. If the mistake were caught early, more likely the case, and communicated in time to accept the next lowest bid, the awarding entity would be entitled to whatever administrative costs it incurred in processing the mistaken bid. This would raise troublesome proof questions, unlike the car ad case before the court.

The best solution lies in a standard like the California statute noted earlier in this section.

Yet another method of relieving from large forfeitures exists, though it is often ignored. It looks at the *security* that bidders are usually requested to deposit, such as a certified check, a cashier's check, or a bid bond, based on a designated percentage of the bid. Suppose a bidder who has posted security seeks to withdraw its bid based on an asserted clerical or mathematical error of the type discussed in this section.

---

[109]West Ann.Cal.Pub.Cont.Code § 5103.

[110]See Sections 5.07D and 19.02E for discussion of unconscionability.

[111]See also *Osberg Constr. Co. v. City of The Dalles*, 300 F.Supp. 442 (D.Or.1969); *Elsinore Union Elementary School Dist. v. Kastorff*, supra note 63.

[112]*Balaban-Gordon v. Brighton Sewerage Dist. No. 2*, 41 A.D.2d 246, 342 N.Y.S.2d 435 (1973); *White v. Berrenda Mesa Water Dist.*, 7 Cal. App.3d 894, 87 Cal.Rptr. 338 (1970).

[113]*M.F. Kemper Constr. Co. v. City of Los Angeles*, infra note 125: *Elsinore Union Elementary School Dist. v. Kastorff*, supra note 63.

[114]Please note the discussion of unconscionability in the *Sulzer* case reproduced in this section.

[115]*Donovan v. RRL Corp.*, 26 Cal.4th 261, 27 P.3d 702, 109 Cal. Rptr.2d 807 (2001), rehearing denied and modified, Sep. 12, 2001.

[116]*Recent Cases*, 115 Harv.L.Rev. 724 (2001).

In some of the early cases allowing relief, the courts pointed to the large discrepancy between the low bidder who claims a mistake and the next low bidder. When this amount was large, courts allowed relief for what was essentially unilateral mistake, noting that the difference, if awarded, would constitute a large forfeiture and was unconscionable. In the landmark California *Kemper* case, the successful bidder omitted a $301,000 item. Its bid was $780,000 and the next low bid was $1,049,000 (amounts approximations).[117]

The bidder sought to have its bid bond cancelled. The city sought to recover on the bond and recover damages as well, a remedy provided by the Los Angeles City Charter. The damages would be the difference between the two bids, here $257,000. The court reasoned that to allow damages here would be unconscionable and it relieved the bidder from its bid.

Suppose the city asked only for enforcement of the 10 percent bid bond or $78,000. This would have been one-third the damages sought by the city. Whether this would have led the court to refuse cancellation of the bond is, of course, uncertain. But the amount of the forfeiture certainly played a large role is the court's decision.

This is often the outcome under statutes regulating competitive bidding. If the successful bidder refuses to enter into the contract, its deposit, usually a bid bond, is forfeited. In effect the statute providing the forfeiture has created a liquidated damages clause, or more realistically, a limitation of liability.[118] By limiting the remedy to forfeiture of the bid bond, there might not be the need to relieve the bidder from its bid.

This would seem a fair solution to the vexatious bidding mistake cases. The contractor's exposure would be the deposit. Although one court drew this distinction,[119]

most generally have held that the contractor is relieved from its performance entirely if the requirements for relief are established. In such a case, the contractor is entitled to recover the deposit. Although the deposit may be only 5 to 10 percent of the bid, in large jobs this amount is substantial. Where the facts are sufficient to allow the mistake doctrine to be applied, most courts would prefer to relieve the bidder entirely.[120]

Courts generally have been reluctant to allow a bidder to correct a mistake, particularly if the claim to reform the contract is made after performance has begun or been completed.[121] Yet some cases, particularly in the federal procurement system, do permit the bidder to correct its mistake.[122] Some courts outside the federal procurement system have also permitted correction through the equitable doctrine of reformation.[123] Yet clearly rescission (cancellation) is easier to obtain than reformation.[124]

A case that involves the application of the Federal Acquisition Regulations (FAR) deals with what appears to be a mistaken bid created by errors of judgment and employs an equitable solution when the contract has been partly performed. This instructive case is reproduced here.

---

[117]*M. F. Kemper Constr. Co. v. City of Los Angeles,* infra note 125.

[118]This will be discussed in Section 18.04F. Also, please review Section 15.03D.

[119]*Triple A Contractors, Inc. v. Rural Water Dist. No. 4,* 226 Kan. 626, 603 P.2d 184 (1979). The precedent value of this case has been diminished. See *Marana Unified School Dist. No. 6 v. Aetna Cas & Sur.,* supra note 103 (disagreeing with the *Triple A* case): *Florence v. Powder Horn Constructors, Inc.* supra note 57 (claiming *Triple A* allowed this relief despite *Triple A* allowing the bond to be forfeited). But see *A. & A. Elec. Inc. v. King City,* infra note 129. The Canadian Supreme Court has used a similar analysis. See *The Queen in Right of Ontario v. Ron Eng'g and Constr. (Eastern), Ltd.,* [1981] S.C.R. 111.

[120]*Marana Unified School Dist. v. Aetna Cas. & Sur. Co.,* supra note 103; *Powder Horn Constructors, Inc. v. City of Florence,* supra note 57 (court might have done so had clause liquidated damages).

[121]*Lemoge Elec. v. County of San Mateo,* 46 Cal.2d 659, 297 P.2d 638 (1956). But *Martin Eng'g, Inc. v. Lexington School Dist. No. One,* 365 S.C.1, 615 S.E.2d 110 (2005) held that the bidder, though usually not allowed to correct, can correct its bid if the correction would still leave him the low bidder. It can be awarded the contract at the corrected price. Of course, the bidder must show grounds to allow correction, usually a mistake of some kind, usually clerical.

[122]*United States v. Hamilton Enters., Inc.,* 711 F.2d 1038 (Fed. Cir.1983) (contractor must establish claim by clear and convincing proof of clerical or arithmetic error or misreading of specifications).

[123]*Nat Harrison Assocs., Inc. v. Louisville Gas & Elec. Co.,* 512 F.2d 511 (6th Cir.1975) (dictum), cert. denied, 421 U.S. 988 (1975).

[124]*Liebherr Crane Corp. v. United States,* 810 F.2d 1153 (Fed. Cir.1987).

SULZER BINGHAM PUMPS, INC. v. LOCKHEED MISSILES & SPACE COMPANY, INC.

United States Court of Appeals, Ninth Circuit, 1991. 947 F.2d 1362.
Before KILKENNY, GOODWIN and SCHROEDER, Circuit Judges.

SCHROEDER, Circuit Judge.

This appeal arises out of an unusual dispute between a major government contractor and a subcontractor providing components for the United States Navy's Trident II nuclear submarines. The contractor, Lockheed Missiles & Space Company, awarded a subcontract to the low bidder for the subcontract, the appellee Sulzer Bingham Pumps, Inc. Sulzer Bingham's bid, however, was in fact millions of dollars lower than it would have been if Sulzer Bingham had not committed a series of errors in preparing the bid. According to the findings of the respected district judge who heard the evidence, Lockheed doubted that Sulzer Bingham could perform the contract at that price, but nevertheless awarded the contract for the bid price, resulting in an unconscionably low price. The district court concluded that Lockheed's conduct amounted to overreaching in violation of basic contractual principles, and that its failure to ask the subcontractor to verify its bid violated the terms of the subcontract. The district court declined to rescind the contract. The district court did, however, award Sulzer Bingham some equitable relief in the form of the actual costs it incurred above the contract price, in a total amount which was not to exceed the next lowest bid. Lockheed appeals.

* * *

FACTUAL BACKGROUND

The facts as determined by the district court are not seriously disputed on appeal. They can be summarized as follows.

Lockheed is the prime contractor for the Navy's Trident II nuclear submarines. The subcontract at issue involved the production of ballast cans. When the submarines are not carrying nuclear missiles, they need ballast cans for stability. Each ballast can weighs 64,000 pounds and stands 15 feet high.

In 1988, Lockheed sent out a request for quotation to potential subcontractors, seeking bids for the manufacture of 124 ballast cans. In February 1989, Lockheed received eight bids, including one from Sulzer Bingham. Sulzer Bingham was the lowest bidder at $6,544,055. The next lowest bid was $10,176,670, and the bids ranged up to $12,940,540, with one high bid at $17,766,327. Lockheed estimated that the job would cost about $8.5 million.

Lockheed's employees were shocked by Sulzer Bingham's bid and thought it was surprisingly low. Price extensions, submitted to Lockheed by Sulzer Bingham, revealed no arithmetic errors. Lockheed then asked Sulzer Bingham to verify that its bid included shipping charges and First Article Compatibility Testing, but did not ask for verification of the entire bid. Sulzer Bingham informed Lockheed that its bid was complete. Lockheed then inspected Sulzer Bingham's Portland facility to evaluate Sulzer Bingham's technical capabilities. The inspection revealed that Sulzer Bingham would have to make many modifications to its existing facility in order to complete the contract. The turntable Sulzer Bingham anticipated using for machining and assembling the ballast cans was inadequate, and an entirely new lead pouring facility needed to be constructed. None of these shortcomings were revealed to Sulzer Bingham by Lockheed.

At no time did Lockheed notify Sulzer Bingham that it suspected a mistake in Sulzer Bingham's bid. Lockheed did not inform Sulzer Bingham that its bid was significantly lower than the next lowest bid, and lower than Lockheed's own estimate of the cost of the job as well. Lockheed never informed Sulzer Bingham that it suspected that Sulzer Bingham would not be able to complete the contract at the bid price.

Sulzer Bingham made a variety of errors in its bid. It had underestimated the number of hours the job would require, and had used hourly labor rates that were below cost. Sulzer Bingham had not realized that its existing facilities were inadequate for the job, and had overlooked certain costs of the job. Sulzer Bingham's bid broke down to $30,707 per ballast can. The next lowest bid broke down to $58,137 per can, and Lockheed estimated at least $40,000 per can.

In late February 1989, Lockheed accepted Sulzer Bingham's bid and Sulzer Bingham started work. In November 1989, Sulzer Bingham revised its estimate of the cost of the job and asked Lockheed for additional $2,111,000 in compensation. Lockheed rejected Sulzer Bingham's request for additional compensation.

* * *

The district court . . . concluded that under section 14.406-3 [Ed. note: Now 14.407-3] of the Federal Acquisition Regulation,

Lockheed had a duty to notify Sulzer Bingham when it suspected a mistake in Sulzer Bingham's bid. The district court further concluded that Lockheed breached this duty, and Sulzer Bingham was therefore entitled to equitable relief.

The district court denied rescission because Sulzer Bingham had delayed its request for rescission and because Sulzer Bingham had already completed about half of the contract. The district court then concluded that it could award Sulzer Bingham other equitable relief because Lockheed had breached its duty to verify Sulzer Bingham's bid, and the resulting contract was unconscionable. The district court required Sulzer Bingham to complete the contract, and ordered that Sulzer Bingham could "recover its actual costs only, including a reasonable amount for depreciation and overhead. Under no circumstances may [Sulzer Bingham's] recovery exceed the amount of the next lowest bid."

THE APPLICABILITY OF FEDERAL ACQUISITION REGULATIONS REGARDING BID VERIFICATION

Lockheed argues that the district court erred in holding that the Federal Acquisition Regulations ("FAR") imposed specific verification duties on Lockheed. [Ed. note: The court held that the subcontract imposed these duties on Lockheed.]

Section 14.406-3 [Ed. note: now 14.407-3] of the Federal Acquisition Regulation requires a contracting officer to take the following steps if a bidding mistake is suspected:

1. The contracting officer shall immediately request the bidder to verify the bid. Action taken to verify bids must be sufficient to reasonably assure the contracting officer that the bid as confirmed is without error, or to elicit the allegation of a mistake by the bidder. To assure that the bidder will be put on notice of a mistake suspected by the contracting officer, the bidder should be advised as appropriate—

i. That its bid is so much lower than the other bids or the Government's estimate as to indicate a possibility of error;

* * *

iv. Of any other information, proper for disclosure, that leads the contracting officer to believe that there is a mistake in bid.

48 C.F.R. § 14.406-3G (1) [Ed. note: now § 14.407-3G(1)] The district court held that Lockheed breached its duty to Sulzer Bingham by not notifying Sulzer Bingham that it suspected a mistake, and not informing Sulzer Bingham that the bid was much lower than all other bids.

* * *

The district court correctly concluded that FAR governed the parties' conduct during the bid acceptance and award period, and that Lockheed breached its duty under FAR by failing to notify Sulzer Bingham that it suspected a mistake in the bid.

THE APPROPRIATE REMEDY FOR LOCKHEED'S BREACH

Lockheed argues that even if FAR applies and Lockheed breached its duty to properly verify Sulzer Bingham's bid, nevertheless Sulzer Bingham is not entitled to relief. In support of this argument Lockheed relies on cases holding that the terms of a substantially performed contract may not be changed through reformation if the bid mistake is not attributable to an arithmetic or clerical error.

It is apparently well settled government contract law that reformation, based on a mistake in bidding, is available to correct only "clear cut clerical or arithmetical error, or misreading of specifications." *Aydin Corp. v. United States*, 229 Ct.Cl. 309, 314, 669 F.2d 681, 685 (1982) (quoting *Ruggiero v. United States*, 190 Ct.Cl. 327, 335, 420 F.2d 709, 713 (1970)). According to the district court's findings, errors in judgment predominated in this case. It is apparently well settled government contract law that such errors in judgment do not, by themselves, justify the award of equitable relief in the form of a change in the contract price. . . . Such a conclusion is consistent with the understanding that contracting entities should live up to their contractual obligations when bidding errors are based on economic misjudgment, and not attributable to the other contracting party's conduct.

In this case, however, the district court did not fashion equitable relief solely on the basis of the bidder's economic misjudgments. Rather, the district court awarded relief because Lockheed, in failing to follow contractual provisions requiring bid verification, accepted an unconscionably low bid. The verification procedures Lockheed ignored were designed to ensure that such unconscionably priced contracts would not be awarded.

None of the authorities relied on by Lockheed involve such a situation. In *Aydin Corp.*, 229 Ct.Cl. at 318, 669 F.2d at 687, for example, the disparity in bids was not sufficient to put the contracting officer on constructive notice of any mistake in the bid, whereas in this case, Lockheed had actual notice. In *Hamilton Enterprises*, 711 F.2d at 1045, the bidder underestimated the number of hours needed to perform the contract, and the government failed to adequately verify the bid. The bidder defaulted on the contract. The court denied the claims of both parties, describing the case as one of "mutual fault to the extent that neither party is entitled to recover on the claims asserted against the other." There was no finding of unconscionability. In this case, unlike *Hamilton Enterprises*, Lockheed is reaping the rewards of Sulzer Bingham's performance at an unconscionably low price. The Defense Department's own adjudicatory

arm has itself recognized in contract disputes that equitable principles do apply to prevent the enforcement of an unconscionable contract.

* * *

AFFIRMED.

Suppose the invitation to bidders states that bidders will not be released for errors. One court interpreted this language to cover errors of judgment and not clerical errors.[125] Another refused to employ this interpretation technique to give relief for a clerical error.[126] The former approach is preferable. If the invitation *clearly* covers clerical errors, the risk of even clerical errors should be placed on the contractor. However, it is more likely that such language will not tie the hands of courts to grant relief for such mistakes if the enforcement of the mistaken bid would be unconscionable.

Mirroring increased judicial activism in other fields, constitutional principles have been invoked in bid mistake cases. In *Midway Excavators, Inc. v. Chandler*,[127] the court rejected the claim by the bidder that the failure by the public entity to develop guidelines determining the type of technical mistakes that would justify relief violated the bidder's constitutional right of not being deprived of property without due process of law.

On rare occasions unjustified refusal by the bidder to whom the contract has been awarded results in a benefit to the awarding authority. This occurred in *Macon-Bibb County Water & Sewer Authority v. Tuttle/White Constructors, Inc.*[128] The court invoked the concept of *offsetting benefits*. In this case the defendant submitted the low bid of $8.7 million for the incineration of waste sludge. The next low bid was $9.9 million or a difference of about $1.2 million. The defendant, without legal justification, refused to sign the contract. After it refused, a third party advised the awarding authority of a new type of boiler that would save the awarding authority $1.2 million

in capital costs and $155,000 in annual costs. The defendant was able to offset the benefits its breach had generated. Without its breach, the new information could not have been obtained and used.

F. Bid Deposit

The bidders are usually requested to submit deposits with their proposals. Is the deposit a security deposit out of which the owner can take whatever damages it has incurred? Does the payment limit the damages to a specified figure that still obligates the owner to prove damages up to that figure? Is the deposit submitted in an attempt to set damages in advance by agreement of the parties?

Suppose Bidder A submits a proposal for $1 million and the invitation to bidders requires it to submit a bid bond for 5 percent of its bid. A bid bond for $50,000 is deposited. The bids are opened, and A is lowest. The next lowest bidder has submitted a bid of $1.1 million. A is offered the contract but without any legal justification declines to enter into it. The contract is offered to the next lowest bidder, who bid $1.1 million. Is the owner entitled to $100,000 in damages, with $50,000 of that amount as a security deposit out of which it can assure itself that it will be able to collect at least a part of its damages? Or is the owner limited to $50,000, because this is what the parties have agreed will be the actual damage amount whether the actual damages are higher or lower?

Suppose the next lowest bidder had been $1,025,000 instead of $1.1 million. In such a case, can the owner keep the entire $50,000 or must it be limited to $25,000?

To a certain extent, the parties are free by their contract to determine whether the amount submitted or the bid bond deposited liquidates (agrees in advance on the amount of) damages, is a security deposit, or is a limitation of liability. A security deposit is an amount of money deposited with one party, out of which the latter can satisfy whatever damages to which it is entitled. If the 5 percent deposit is merely a security deposit, the owner

[125]*M.F. Kemper Constr. Co. v. City of Los Angeles*, 37 Cal.2d 696, 235 P.2d 7 (1951). See also *Jobco, Inc. v. Nassau County*, 129 A.D.2d 614, 514 N.Y.S.2d 108 (1987) (disclaimer in bid bond did not preclude rescission for mistake).

[126]*City of Newport News v. Doyle & Russell, Inc.*, 211 Va. 603, 179 S.E.2d 493 (1971).

[127]128 N.H. 654, 522 A.2d 982 (1986).

[128]530 F.Supp. 1048 (M.D.Ga.1981).

can retain this amount and sue for any balance to which it is entitled, or it must return the excess of the deposit over damages.

If the deposit is a valid liquidated damages clause, the parties have agreed in advance that whatever the amount of the actual damages, the breaching party will pay the amount stipulated in the clause. The owner could retain the $50,000 whether the damages were $200,000 or $1.

A valid liquidated damages clause requires that it be difficult to ascertain the damages at the time the contract is made and that the amount agreed be a genuine pre-estimate of the potential damages.

An amount disproportionate to the actual or antici-pated damage chosen merely to coerce performance is a penalty, and unenforceable. For example, if the deposit were 50 percent of the bid and if it were most unlikely that there would be damages approaching this amount, the clause would be a penalty and unenforceable. The owner would be entitled only to actual damages without regard for the amount of the deposit, and it would have to refund the excess of the deposit over its actual damages.

Legislation, state or local, may give the public agency damages based on the difference between the low bid and the next lowest bid if the low bidder does not enter into the contract awarded to it. In such a case, the deposit is for security and is not an attempt to liquidate damages. A liquidated damages clause *establishes* the damages.

Legislation sometimes provides that the public agency may retain the amount deposited but only to the extent of the difference between the defaulting bidder's bid and the amount for which the contract is ultimately awarded. For example, suppose the amount deposited was $50,000 or 5 percent of the $1 million bid and the next bidder was awarded the contract at $1,025,000. In such a case, the owner would be entitled to retain $25,000. Such statutes set up a liquidation of damages that will apply only if actual damages are greater than the amount deposited. If actual damages are less, the deposit is simply security.

Is it desirable to liquidate damages? From the owner's standpoint, the chances of collecting an amount in excess of the deposit from the contractor are remote. In addi-tion, the owner would like to retain the amount deposited without having to show actual damages. For these reasons, it is preferable to liquidate damages rather than use the deposit solely as security. The pure security deposit does allow the owner to seek to recover an amount beyond the deposit. This would occur if the discrepancy between the

defaulting bidder's bid and the next bidder is more than the deposit. In such a case, there is a strong likelihood of a mistake that would permit the bidder to withdraw its bid. Many contractors would not be able to satisfy a large court judgment. This is a reason for providing the security deposit. Liquidated damages protect the contractor from the risk of excessive damages and guarantee the owner a reasonable amount of collectible damages.

A properly drafted liquidated damages clause is likely to be enforced if created by legislation that specifically permits the awarding authority to forfeit the deposit.[129] In the absence of such legislation, courts divide. Some enforce such a clause; others do not.[130, 131]

The uncertainty as to the amount of damages must exist at the time the contract is made and at the time of the deposit. Some courts ignore this requirement and seem to look at whether the amount of damages can be easily ascertained at the time of breach. These courts seem unwilling to forfeit an amount in excess of actual damages. This approach does not take note of the long-haul aspects of denominating the forfeiture clause as stip-ulated damages. Over the long haul of many competitive bids, the losses to the public agency probably average per competition the amount stipulated in each competi-tive bid. Although the long haul may seem unfair to the particular bidder who must lose more than what it *appears* the agency has been damaged in this competitive bid, the *particular* bidder is relieved from any risk *beyond* the deposit amount.

There are administrative costs when the next low-est bidder is selected. Admittedly, that loss often seems much less than the amount forfeited. But again, the par-ties expect the amount to be deposited to be forfeited, and the bidder is relieved from the risk of loss beyond the deposit. As long as the amount selected is reasonable, the

---

[129]*A & A Elec., Inc. v. King City*, 54 Cal.App.3d 457, 126 Cal.Rptr. 585 (1976) (awarding authority limited to forfeiture of bid bond, no damages). See also *Emma Corp. v. Ingleside Unified School Dist.*, 114 Cal. App.4th 1018, 8 Cal.Rptr.3d 219 (2004), review denied, Apr. 21, 2004 (no damages, only forfeiture of bid bond). Cf. *Powder Horn Constructors, Inc. v. City of Florence*, supra note 57.

[130]*Jobco, Inc. v. Nassau County*, supra note 125 (dictum); *Bellefonte Borough Auth. v. Gateway Equip. & Supply Co.*, 442 Pa. 492, 277 A.2d 347 (1971); *City of Fargo v. Case Dev. Co.*, 401 N.W.2d 529 (N.D.1987) (failure to develop property).

[131]*Ogden Dev. Corp. v. Federal Ins. Co.*, 508 F.2d 583 (2d Cir.1974), held that the clause forfeiting the deposit was a penalty; *Petrovich v. City of Arcadia*, 36 Cal.2d 78, 222 P.2d 231 (1950).

forfeiture clause should be considered a valid liquidated damages clause.

The clause should be *clearly enforceable* if it is necessary to rebid the entire project. Rebidding entails substantial additional administrative expense.

Suppose both low bidder and the next lowest bidder unjustifiably refuse to enter into the contract. Can the owner retain the deposit by both bidders? Although it may seem unfair to retain both bidders' deposits, it is not logically indefensible. Each bidder has breached, and each has been to some degree relieved from the risk of excessive damages by using an agreed damages provision. However, goodwill and the avoidance of litigation may necessitate some solution, such as retaining one-half of each deposit rather than trying to retain both.[132]

G. The Formal Contract

The culmination of a successful competitive bidding process is the award by the awarding authority to the successful bidder. Usually, a formal contract is forwarded or given to the successful bidder for its execution.

Suppose the award is made *before* the expiration of the period during which the bid is irrevocable but the formal contract is not executed within that period. Although one case discussed earlier held that a tentative acceptance within the period was not a sufficient acceptance,[133] two Wisconsin cases held that the validity of the contract did not require execution of the contract. In one,[134] the awarding authority had voted, in the presence of the bidder, to accept the bid. In the other,[135] approval by a federal regulatory agency that conditioned the award was not received until the morning of the final day of the period during which the bid was irrevocable. On that morning (a Friday), the engineer notified the contractor that the formal contracts would be in the mail that day. The following day (Saturday), the contractor wrote that it was withdrawing its bid, because the forty-five-day period during which its bid was irrevocable had expired. The contracts were received on the following Monday. The court could have held that the acceptance took place on the forty-fifth day—when the contracts were put in the mail. But the court held for the awarding authority by concluding that contracts had been formed, and the formal contracts merely memorialized the agreement that had already been made.

An Iowa case, noting the contrary holdings as to the requirement that a formal contract was needed to form the contract, held that written, formal acceptance created a valid contract despite the failure to execute the formal written contract.[136]

The difficulties generated by this issue are shown in a recent Arizona case. The intermediate Arizona appellate court held the contract was valid even though the parties did not execute a formal contract.[137] Yet this opinion was vacated by the Arizona Supreme Court. It held there was no valid contract until the parties had executed a formal agreement.[138]

Cases may come out differently where the test is the intention of the parties and factual patterns vary. If the transaction is a "one-off" and not a routine transaction, the amounts at stake are substantial, the transaction is a complicated one, and attorneys are involved at some stage of the transaction, it is likely that the parties have not bound themselves until they have signed a formal, written agreement.

Suppose the formal contract is not consistent with earlier communications exchanged between the successful bidder and the awarding authority. The Court of Claims held that the formal contract, while typically superseding all previous negotiations, documents, etc., is merely a reduction to form of the actual agreement made by the advertisement, bid, and its acceptance.[139] This is another recognition of the formal contract often being simply a memorial of the agreement that has been made. However, the date of the formal contract *may* set into motion any time commitment of the contractor.

---

[132]West Ann.Cal.Pub.Cont.Code §§ 10181–10182 permit forfeiture of the security of lowest, second lowest, and third lowest if none will enter into the contract.

[133]*Hennepin Public Water Dist. v. Petersen Constr. Co.*, supra note 67, discussed in Section 18.04D .

[134]*Nelson, Inc. of Wisconsin v. Sewerage Comm'n*, supra note 104.

[135]*City of Merrill v. Wenzel Brothers, Inc.*, 88 Wis.2d 676, 277 N.W.2d 799 (1979). See also *Citizens Bank of Perry v. Harlie Lynch Constr. Co.*, 426 So.2d 52 (Fla.Dist.Ct.App.1983) (oral acceptance by owner's board valid acceptance).

[136]*Horsfield Constr., Inc. v. Dubuque County, Iowa*, 653 N.W.2d 563 (Iowa 2002). For an analysis see 24 Constr. Litig. Rep. 19 (2003).

[137]*Ry-Tan Constr. Inc. v. Washington Elementary School Dist. No. 6*, 208 Ariz. 379, 93 P.3d 1095 (App.2004) (collecting many authorities).

[138]210 Ariz. 419, 111 P.3d 1019 (2005).

[139]*Dana Corp. v. United States*, 200 Ct.Cl. 200, 470 F.2d 1032 (1972).

H. Bidding Documents

The culmination of the complex competitive process usually is assent by both parties to the construction contract. This process has generated a series of many long and complex writings, such as the invitation to bidders, information to bidders, and the bid proposal. Included among these are other materials, such as plans, specifications, general conditions, supplemental conditions, addenda, and the agreement forms. Chapters 19 and 20 deal with the problems generated by this wealth of written material. But what about the materials generated by the bid process itself?

AIA Document A201-2007, Section 1.1.1, seeks to deny the bidding materials any legal effect by excluding bidding requirements as contract documents.[140] Bidding materials have been superseded by execution of the construction contract. It is important to avoid contradiction in the voluminous contract documents. But a provision can state that in the event of conflict, the construction contract takes precedence over the bidding documents. Although implementing provisions of this type is not as simple as it appears, the solution AIA has selected—excluding bidding documents—is undesirable.

Most participants in the process believe that the bidding documents do have legal efficacy.[141] Information is given in the bidding documents that is relied on by the bidders. For example, subsurface information is frequently included in the information to bidders. Under AIA Document A201-2007, Section 3.7.4, unless this information is found in the specifications, it cannot be the basis of any claim by the contractor for an equitable adjustment for subsurface conditions different from those usually encountered or disclosed by the contract documents. Similarly, AIA Document A701-1997 (instructions to bidders) imposes many contractual terms intended to survive, such as liquidating damages for failure to enter into the construction contract and requiring that a particular type of surety bond be used. If the award is made and the construction contract documents are signed without these provisions, if taken literally, A201-2007 would discharge any obligations the contractor may have that are expressed in the instruction to bidders. There may be factual material in the bidding instructions or information to which the owner may wish to point at some later date if a claim has been made.[142]

The bidding material was useful to the contractor in *Village of Turtle Lake v. Orvedahl Construction, Inc.*[143] Here the bidding material was a contract document. Because the arbitration clause included disputes relating to the contract documents, the bidder was given a chance to submit its claim of mistake to arbitration even if it could not meet the requirements of the statute regulating bidding mistakes in public contracts.

Attempts to deny any legal effectiveness to the bidding documents may not be in the owner's best interest and may frustrate the reasonable expectations of both parties.

I. Judicial Review of Agency Action

Public procurement encompasses a vast number of social and economic goals. The result has been a complicated set of statutes and regulations with the increasing likelihood of irregularities. At the state and local level standing to contest the agency award in court does not seem to have been the problem it has been at the federal level.

Before 1970, disappointed bidders were not granted standing to challenge federal agency decisions based on *Perkins v. Lukens Steel Co.*,[144] which held that the government has the sole right to choose with whom and under what terms it will contract. A disappointed bidder has only a privilege and not a right to do business with the government. In addition to this wooden logic, the court was concerned that judicial interference with government procurement would cause delay and involve the courts in decision making beyond their competence.

But in *Scanwell Laboratories, Inc. v. Shaffer*,[145] a federal appeals court cited the Administrative Procedure Act,[146] a statute passed after *Perkins v. Lukens Steel Co.*, and held

---

[140]See also AIA Doc. A101-2007, § 9.1.7.

[141]*Jack B. Parson Constr. Co. v. State Dep't of Transp.*, 725 P.2d 614 (Utah 1986) (effect given to bidding information).

[142]*D.A. Collins Constr. Co., Inc. v. State*, 88 A.D.2d 698, 451 N.Y.S.2d 314 (1982) (information regarding possible delay barred contractor delay claim).

[143]135 Wis.2d 385, 400 N.W.2d 475 (App.1986).

[144]310 U.S. 113 (1940).

[145]424 F.2d 859 (D.C.Cir.1970).

[146]5 U.S.C.A. §§ 551–559, 701–706 (2000). The *Scanwell* court relied on § 702, which states that "[a] person suffering legal wrong because of agency action . . . is entitled to judicial relief thereof."

that a disappointed bidder had standing to seek a judicial order stopping an allegedly invalid procurement order.

The *Scanwell* case seemed to open some federal court-house doors to disappointed bidders seeking federal contracts. This would involve courts in difficult procurement problems. A year later, the same court had second thoughts. It held that judicial interference with a procurement award required the challenger to demonstrate that there was no "rational basis" for the award. In addition, trial courts were given broad discretion to refuse to interfere with the procurement.[147] It is easier for a disappointed bidder to recover damages that are usually limited to bidding expenses if it can establish a defect in procurement procedures.[148]

The authority of the federal district courts to hear protests by disappointed bidders became the subject of congressional action. In 1996, Congress gave both the U.S. Court of Federal Claims (formerly the United States Claims Court, located in Washington, D.C.) and the federal district courts concurrent jurisdiction over bid protests.[149] At the same time, however, it enacted a "sunset" provision which stated that the district courts' jurisdiction granted by this statute would terminate on January 1, 2001, unless extended by Congress.[150] Congress did not extend the district courts' jurisdiction by the announced deadline. One court has interpreted Congress's failure to act to mean that the federal district courts are entirely divested of jurisdiction over bid protests involving federal contracts.[151] A contrary view is that the district courts are divested of their jurisdiction under the 1996 law, but not under the Administrative Procedure Act—the statute relied upon by the *Scanwell* court.[152]

J. Illegal Contracts

The imposing number of legal controls on public contracts, the interest generated by public projects, and the staff inadequacies in small public entities all create a substantial risk that an award may be made illegally. Suppose the party to whom the award had been made partly or fully performed.

Clearly, recovery cannot be made under an illegal contract. But two other issues can arise. First, and most frequently, can the contractor who has performed under an illegal contract recover for the work it has performed based on restitution? Second, can any payments made to a contractor be recovered by the awarding authority?

These issues of unjust enrichment, like attempts by bidders to withdraw their bids, generate sharp differences of opinion. Those who would deny recovery or even require repayment stress the importance of an honest competitive bidding system and the need to protect public funds. Although a recognition exists of the occasional unfairness of denying, because of technical irregularities, a contractor recovery for work it has performed, those who take a hard line cite the difficulty of making these judgments and the importance of not allowing any loopholes in the laws regulating public contracts.

Those who take a softer approach are willing to concede that there should not be recovery or that there should even be repayment where there is venality or corruption, but they draw a distinction between those cases and ones that do not involve serious criminal misconduct. Where corruption is pervasive, they would concede that only harsh and unremitting punishment has a chance of deterring such corruption. In other jurisdictions, inefficiency is common and corruption rare. In such jurisdictions, there should be greater willingness to allow payment to a contractor where the award was not tainted with bad faith, fraud, or corruption.

Generally, illegally awarded contracts cannot be the basis for restitution. But in 1971, restitution was allowed in *Blum v. City of Hillsboro*.[153] Immediately after acceptance of the $47,000 contract, council members of the awarding authority asked the contractor if it would be willing to do additional work for a designated price. The contractor stated it would, and the awarding authority, through its mayor and city council, specified the work

---

[147]*M. Steinthal & Co. v. Seamans*, 455 F.2d 1289 (D.C.Cir.1971).

[148]*Keco Indus., Inc. v. United States*, supra note 68. Lost profits were not allowed in *Armstrong & Armstrong, Inc. v. United States*, 514 F.2d 402 (9th Cir.1975); *Swinerton & Walberg Co. v. City of Inglewood*, 40 Cal.App.3d 98, 114 Cal.Rptr. 834 (1974). See Rosengren & Librizzi, *Bid Protests: Substance and Procedure on Publicly Funded Construction Projects*, 7 Constr. Lawyer, No. 1, Jan. 1987, p. 1.

[149]28 U.S.C.A. § 1491(b)(1).

[150]Administrative Dispute Resolution Act of 1996, Pub.L. 104-320, § 12(d) 110 Stat. 3875.

[151]*Emery Worldwide Airlines, Inc. v. United States*, 264 F.3d 1071, 1078–80 (Fed.Cir.2001), rehearing and rehearing *en banc* denied Nov. 28, 2001.

[152]Note, 32 Pub. Cont. L.J. 393 (2003).

[153]49 Wis.2d 667, 183 N.W.2d 47 (1971).

to be done and drew up an amendment to the original contract.

The contractor performed the additional work, which increased the amount of the contract to $154,000. The awarding authority paid $82,000 but refused to pay the balance of $72,000. The contractor sought to recover the balance, and the awarding authority counterclaimed for the amount it paid in excess of the original contract price. This claim was based on the failure to follow state law, which required that the additional work be competitively bid.

The court held that recovery of restitution based on benefit conferred (though a minority view) was justified in this case. According to the court, failure to allow profits would be sufficient deterrence. If recovery is not granted, a claim will very likely be made on the municipality to use its discretionary power to pay moral claims. The court limited recovery to actual costs, including overhead not exceeding actual benefit, but denied profit. Nor could recovery exceed the unit cost of the original contract that had been properly awarded.

This issue continues to plague the courts. *Bozied v. City of Brookings*, decided by the Supreme Court of South Dakota in 2001, involved facts similar to those in the *Blum* case just described.[154] The court held that the additional work performed by the contractor should have been awarded by competitive bidding. This made the change order illegal. But the contractor had received most of what it would have been entitled to for performing the additional work. The court held the contractor cannot get any more but it need not return what it had received. Were that to be done, the city would have been unjustly enriched. The court held the parties should be left where they were.

But the court was quick to point out that this solution assumes that there was no evidence of fraud or collusion, that the public entity was authorized to make the contract but did so in an unlawful manner, that the payments were received, and the amounts paid were reasonable.

The remedy is crucial in these cases. The flexibility provided by remedial choices can provide a way of reaching a fair result. This will be seen again in the *Gerzof* case discussed shortly in this sub-section.

In 1977, the Minnesota Supreme Court followed the *Blum* case in a well-drilling contract that was illegally awarded because it violated the competitive bidding statute. Whether there had been a violation was a close question, inasmuch as the awarding authority thought there had been a sufficient emergency that granted it an exemption from the competitive bidding requirements. Yet because the contractor had not actually found water, the court denied recovery because there had been no benefit to the awarding authority.[155]

Other jurisdictions have taken a much harder line. *Manning Engineering, Inc. v. Hudson County Park Commission*[156] involved pervasive corruption in the awarding of contracts in Jersey City, New Jersey. The court not only denied the engineering company recovery for work that it had performed but also indicated that the awarding authority would have had a good claim had it sought repayment of funds that had been paid. It cited a New York case[157] that had involved a contractor convicted of conspiring to violate state bribery laws through a kickback system. When the contractor sued for the unpaid balance, the city successfully defended the claim and recovered payments that it had made.

Gerzof v. Sweeney[158] demonstrated not only the difficulty of fashioning an appropriate remedy but also how the "tough" New York court can be persuaded to relax its harsh rules. In this case, the Village of Freeport (New York) had advertised for bids for a 3,500-kilowatt generator. Enterprise bid $615,000, and Nordberg bid $674,000. After an advisory committee had recommended acceptance of Enterprise's bid, a new village election was held at which a new mayor and two new trustees were elected. Shortly thereafter, Nordberg's higher bid was accepted.

Enterprise obtained a court order setting the award aside. The board of trustees then drew up new specifications for a 5,000-kilowatt generator with the active participation of Nordberg. The specifications were so rigged that only Nordberg could comply. As expected, Nordberg was the only bidder, and its bid of $757,000 was accepted. Nordberg installed the generator and was paid.

---

[154]638 N.W.2d 264 (S.D.2001)

[155]*Layne Minnesota Co. v. Town of Stuntz*, 257 N.W.2d 295 (Minn. 1977).

[156]74 N.J. 113, 376 A.2d 1194 (1977).

[157]*S.T. Grand, Inc. v. City of New York*, 32 N.Y.2d 300, 298 N.E.2d 105, 344 N.Y.S.2d 938 (1973).

[158]22 N.Y.2d 297, 239 N.E.2d 521, 292 N.Y.S.2d 640 (1968).

After a court declared the second award invalid, the trial court held that the village should retain the generator and recover the $757,000 from Nordberg. The intermediate appellate court modified that judgment by providing that Nordberg could retake the machine on posting a bond for $357,000 to secure the village against damages from removal and replacement of equipment. New York's highest court, the Court of Appeals, first emphasized the importance of protecting the public against corruption and collusion between public officials and bidders.

In the *normal* case it would make no difference whether the village was defending a claim brought by Nordberg or was seeking to recover the money paid Nordberg. But this was not a normal case. Granting recovery of payments made would cost Nordberg three-quarters of a million dollars, and the village would have its generator. Motivated by the enormity of the forfeiture, the court awarded a remedy different from that awarded by the trial court or the intermediate appellate court. The Court of Appeals stated that the award should have been made to Enterprise. Had this been done, the village would have had a 3,500-kilowatt generator for $615,000. The court awarded judgment against Nordberg based on the difference between the $757,000 paid Nordberg and the $615,000 that the village would have paid Enterprise. To this was added $37,000, the difference between what it cost the village to install the Nordberg generator and what it would have cost to install the one offered by Enterprise. In addition, the village was awarded interest.

Suppose the contract should have been awarded to X but was awarded illegally to Y. Suppose X seeks the profits from Y that Y made on the contract. Although some cases have denied recovery,[159] a federal court decision applying Iowa law employed unjust enrichment to award a bidder who should have been awarded the contract the profit of the contractor who had been awarded the contract improperly.[160]

SECTION 18.05 Subcontractor Bids

The relationship between prime and subcontractor is discussed in greater detail in Section 28.02B. However, one aspect of that relationship should be mentioned briefly here. A legal problem that has surfaced frequently relates to the right of a prime contractor to hold a subcontractor to its bid after the former has used that bid in computing its own bid and submitting it to the owner. Although the cases are by no means unanimous,[161] the clear trend is toward holding the subcontractor's bid irrevocable after it has been used by the prime contractor.[162]

---

[159]*Savini Constr. Co. v. Crooks Bros. Constr. Co.*, 540 F.2d 1355 (9th Cir.1974); *Royal Services, Inc. v. Maintenance, Inc.*, 361 F.2d 86 (5th Cir.1966).

[160]*Iconco v. Jensen Constr. Co.*, 622 F.2d 1291 (8th Cir.1980).

[161]*Home Elec. Co. of Lenoir, Inc. v. Hall & Underdown Heating & Air Conditioning Co.*, 86 N.C.App. 540, 358 S.E.2d 539 (1987), aff'd, 322 N.C. 107, 366 S.E.2d 441 (1988). See Note, 10 Campbell L.Rev. 293 (1988). See also Kovars & Schollaert, *Truth and Consequences: Withdrawn Bids and Legal Remedies*, 26 Constr. Lawyer, No. 3, Summer 2006, p. 5.

[162]The leading case is *Drennan v. Star Paving Co.*, 51 Cal.2d 409, 333 P.2d 757 (1958). See Section 28.02B.

CHAPTER NINETEEN

Sources of Construction Contract Rights and Duties: Contract Documents and Legal Rules

SECTION 19.01 Contract Documents: An Electronic Age

As noted in Sections 17.03 and 17.04, there have been changes in the way construction delivery services are organized. Under traditional systems, design–bid–build (DBB) or design–award–build (DAB), the design was done by design professionals and constructed by constructors, usually called contractors. The law has been slow to respond to these new organizational arrangements.

Similarly, the design traditionally is expressed through plans and specifications on paper and is prepared or reviewed by the design professional. The legal documents that organize the project and allocate responsibilities are set forth on paper. Collectively, all of these papers are called the Contract Documents.

Figure 19.1 gives an idea of the complexity of construction contract documents.

But technological changes have created an electronic age.[1] Design is increasingly created and expressed electronically. Communications, including elements of and comments on the design, are electronic. This made it necessary for the Congress to enact legislation making electronic documents and signatures legally binding.[2]

But the stunningly rapid advances in technology far outstrip the willingness and ability of the legal world, including those who publish standard documents and the construction bar, to adapt to them.[3]

Major issues relevant to this chapter (others will surface in other chapters) are what is and who creates the design? The design will no longer be done solely by the design professional and expressed on paper. Specialty subcontractors, vendors, and software manufacturers will be involved in design. Design is fluid and collaborative. Much of the coordination and communication will be electronic.

Yet this will not eliminate paper. While some may disagree,[4] one commentator thinks that the electronic documents will be treated as drafts or copies. Paper has many advantages. It is better for archival purposes, more secure, and more easily used by those who must review it.[5]

Yet the electronic world is quicker and more efficient. There will have to be methods of blending the two worlds. These things must be taken into account in dealing with the sources of construction contract rights and duties, the subject of this section.

A. Bidding Documents

Section 18.04 discussed bidding documents.

---

[1] This is discussed in detail in Ashcraft, *New Paradigms for Design Professionals: New Issues for Construction Lawyers*. This unpublished paper was presented at a meeting of the American Bar Association Forum on the Construction Industry, Scottsdale, Arizona, October 12, 2000. It was not bound into the book of program materials. The author can be reached at hashcraft@hansonbridgett.com

[2] Electronic Records and Signatures in Global and National Commerce Act, effective October 1, 2000, 15 U.S.C.A. § 7001 et seq.

[3] See Stein, Alexander & Noble, *The AIA General Conditions in the Digital Age: Does the Square "New Technology Peg" Fit into the Round A201 Hole?* 25 Constr. Contracts Law Reports, No. 25, Dec. 14, 2001 at 3-20 (West Group).

[4] Stein, Alexander & Noble supra note 3, state "contracts for construction of Frank Gehry's buildings provide that the information in the 3D electronic files takes precedence over information contained in hard copies." Id. at 4. See Section 20.03B.

[5] Ashcraft, supra note 1 at 18–19.

B. Basic Agreement

The basic agreement is the culmination of competitive bidding or negotiation. As an illustration, AIA Document A101, set forth in Appendix B, identifies the parties and the architect and contains provisions dealing with the work to be performed, time of commencement and completion, the contract sum, and provisions for progress payments and final payment. In essence, the basic agreement sets forth the principal incentives of each contracting party. The owner seeks a project completed on schedule, and the contractor seeks agreed compensation for its performance.

The construction project is a complex undertaking. The basic agreement forms are but one part of the total package of construction documents. See Figure 19.1. Frequently the other documents are incorporated by reference in the basic agreement. For example, A101-2007, Article 9, includes as contract documents the general and supplementary conditions, drawings, specifications, addenda issued before execution of the basic agreement, and all modifications issued later. Section 9.1.7 provides space to list the documents incorporated by reference.

Contracts frequently incorporate industry standards. The parties can even incorporate a document not yet in existence. For example, in *Randolph Construction Co. v. Kings East Corp.*,[6] the basic agreement incorporated plans that had not yet been completed. But according to the court, if the completed plans differ substantially from those anticipated, no contract exists unless the parties agree to the completed plans.

Incorporation by reference—a technique for giving legal effectiveness to writings not physically attached to the contract—should be differentiated from reference to other writings. Often reference is made to other writings for informational purposes, without any intention of making them part of the contract obligations. Classifying a writing referred to in the contract as simply providing information can be a technique to avoid binding a party to a writing of which it was unaware. Although such binding can occur when an owner signs an A201, it is more of a problem for subcontractors. Subcontracts frequently refer to the prime contract, and serious questions can arise as to whether the prime contract provisions are incorporated into the subcontract as well as how to reconcile contrary provisions in prime and subcontracts. For that reason, incorporation into subcontracts is discussed in Section 28.04.

C. Drawings (Plans)

The drawings graphically depict the contractor's obligations. Together with the other contract documents, particularly the specifications, they define and measure the contract obligation.

Drawings are of great importance to the construction process. Complying with them, as shall be seen in Section 24.02, usually relieves the contractor of liability if the project is unsuccessful or is not in accord with the owner's expectations. Drawings that are incomplete or inconsistent with the specifications almost always generate increased construction costs. As discussed in Section 14.03, defective drawings or drawings that do not fit with the specifications usually result in liability to the design professional.

A common construction process problem is inconsistency between the drawings and specifications (dealt with in Section 20.03).

D. Specifications—*Fruin-Colnon v. Niagara Frontier* and *Blake v. United States*— Work Preservation Clauses

Specifications use words to describe the required quantity and quality of the project.[7] They also provide information that helps the contractor plan price and performance, such as subsurface conditions, site access for heavy equipment, and availability of temporary power. They should be clear and complete and should "fit together." Otherwise there are likely to be defects, disputes, increased costs, and litigation.

Specifications are classified by type. The most important types are design, performance, and purchase description.

Design specifications (sometimes called *materials and methods* or *detail specifications*) state precise measurements, tolerances, materials, construction methods, sequences, quality control, inspection requirements, and other information. They tell the contractor in detail the material it must furnish and how to perform the work.

Performance specifications state the performance characteristics required; for example, the pump will deliver fifty units per minute, a heating system will heat to 70°F within a designated time, or a wall will resist flames for a designated period. As long as performance requirements are met, design and measurements are not stated or considered important.

---

[6] 165 Conn. 269, 334 A.2d 464 (1973).

[7] See Sections 23.05E and F for more on specifications.

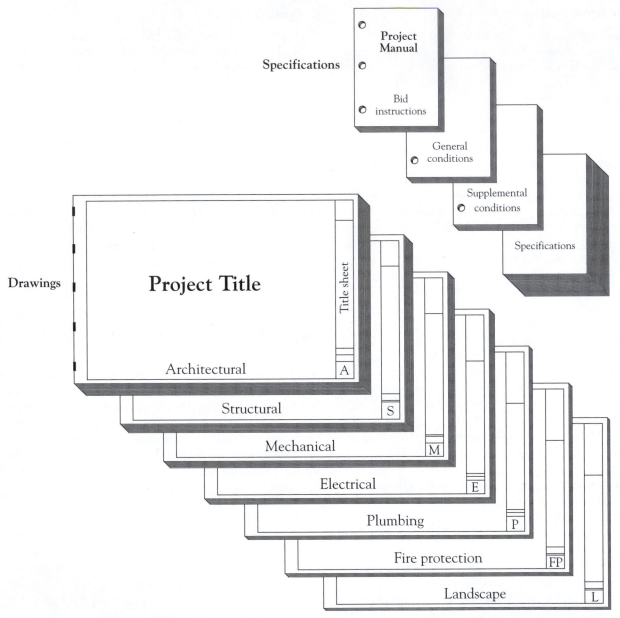

FIGURE 19.1   Construction documents.

Under a pure performance specification, the contractor accepts responsibility for design, engineering, and performance requirements, with general discretion as to how to accomplish the goal. Sometimes the contract documents give the contractor suggestions for foundation work. They may be clearly labeled as indicative of "general requirements" or accompanied by a statement that the foundation shall be redesigned to suit subsurface conditions.

Performance specifications are more common in large-scale industrial work where the contractor agrees to design and build a plant that will turn out a designated number of units of a particular quality in a specified period of time. However, performance specifications may also be used in residential or commercial work.

The following case excerpts differentiate performance from design specifications and indicates how a contract should be interpreted to determine whether performance or design specifications have been created. The first case involved tunnel construction and a claim by the contractor that it was entitled to additional compensation. The owner asserted that the contractual requirement of watertightness was a performance specification.

FRUIN-COLNON CORP. ET AL v. NIAGARA FRONTIER TRANSPORTATION AUTHORITY

Supreme Court of New York, Appellate Division, 4th Department, 1992.
180 A.D.2d 222, 585 N.Y.S.2d 248.

[Ed. Note: The contractor agreed to construct twin subway tunnels, each approximately two miles long, as a part of the Buffalo Light Rail Rapid Transit System.]

* * *

A performance specification requires a contractor to produce a specific result without specifying the particular method or means of achieving that result. . . . Under a performance specification, only an objective or standard of performance is set forth, and the contractor is free to choose the materials, methods and design necessary to meet the objective or standard of performance. . . . Concomitant with control over the choice of design, materials and methods is the corresponding responsibility to ensure that the end product performs as desired. . . . In other words, the contractual risk of nonperformance is on the contractor. . . . That is in contrast to a design specification, where the owner specifies the design, materials and methods and impliedly warrants their feasibility and sufficiency. . . . A contractor must follow a design specification without deviation and bears no responsibility if the design proves inadequate to achieve the intended result. . . . In that instance, the contractor's guarantee, even if framed in "absolute" terms, is limited to the quality of the materials and workmanship employed in following the owner's design. . . . Whether a provision is a performance specification or a design specification depends on the language of the contract as a whole. . . . Other factors to consider include the nature and degree of the contractor's involvement in the specification process, and the degree to which the contractor is allowed to exercise discretion in carrying out its performance under the contract.

* * *

In arguing that the watertightness requirement was a performance specification that plaintiff assumed the responsibility of

meeting, defendant relies primarily on the language of article 3.12 (§ 03300) of the contract, the watertightness clause. Read in isolation, article 3.12 appears to be a performance specification; it specifies the end objective (watertightness) and the standards for measuring that objective, but does not specify the methods of achieving watertightness. Nevertheless, the language and structure of the contract as a whole, as well as the parties' usage and course of performance under the contract, support the conclusion that a design specification was created.

Although the watertightness clause itself does not set forth a particular method for achieving watertightness, the contract as a whole establishes complex and exacting standards for design and construction of the tunnel. Plaintiff was to construct an unreinforced, cast-in-place concrete liner of precise dimension. The type and mix of the concrete was precisely specified, as were detailed requirements for placing, curing, protecting, and finishing the concrete. Plaintiff was given no discretion to deviate from those specifications, whether for the purpose of waterproofing or otherwise. For example, plaintiff had no discretion to install an impermeable outer limit to resist the hydrostatic pressure that both parties knew would exist following completion of construction.

Other provisions of the contract contemplate that waterproofing would be accomplished by means of fissure grouting, which also was to be carried out pursuant to detailed specifications. Additionally, the payment and warranty provisions of the contract support the conclusion that, as a whole, it created a design specification. The contract explicitly provides that "all measures necessary for achieving the degree of watertightness specified in . . . article 3.12, including remedial treatments to stem leaks," would be paid for at the contract unit prices. It is unlikely that defendant would have agreed to pay plaintiff on a per unit basis if, as defendant contends, plaintiff had assumed the responsibility of achieving watertightness. Further, unlike the general warranty set forth in the contract, the extended watertightness warranty did not provide that plaintiff would remedy water leaks at its own expense, as it would have if plaintiff had assumed the responsibility of achieving watertightness.

The parties' course of dealing prior to construction also supports the inference that a performance specification was not intended. Bidders had no input into the design of the tunnel, nor did plaintiff exercise any independent design judgment after it was awarded the contract. Defendant relies heavily on plaintiff's March 5, 1980 letter submitted in support of its Value Engineering Change Proposal (VECP), a type of bilateral change order. In that letter, plaintiff asserted, as a reason why it should be permitted to change the design of the tunnel liners to a "full circle pour," that plaintiff bore the risk of meeting the watertightness requirement. In relying on the language of that letter, defendant overlooks the context in which it was sent. If the contract had created a performance specification with respect to watertightness, it would have been unnecessary for plaintiff to obtain defendant's approval for a design change, and it would have been improper for defendant to withhold such approval.

Similarly, the parties' course of dealing during and following construction illustrates that the contract did not establish a performance specification. As evidence that plaintiff undertook the responsibility of waterproofing, defendant cites the fact that plaintiff, on its own initiative, developed and carried out a course of chemical grouting, a method not mentioned in the contract and for which plaintiff did not bill defendant. It is far more significant, however, that defendant routinely denied plaintiff's numerous requests to implement various other waterproofing methods. Even before the concrete liners were put in place, and while the water diversion system was operating, plaintiff repeatedly requested permission to grout the numerous water-bearing fissures that were present in the exposed rock. Those requests consistently were denied based on defendant's erroneous view that fissure grouting was not intended to achieve watertightness. After the tunnel was constructed and the water diversion system turned off, plaintiff unsuccessfully requested permission to fissure grout, plug the deep wells and piezometer testing holes, and construct a permanent dewatering system. Those requests were denied by defendant, which maintained an uncooperative and obstructive attitude. Plaintiff nonetheless proceeded to fissure grout under protest, based on its proper interpretation of the contract. It would have been unnecessary for plaintiff to seek defendant's consent for such measures, and contractually impermissible for defendant to withhold such approval, if plaintiff were in fact responsible for achieving watertightness and had discretion to choose the means to achieve that objective.

* * *

For the foregoing reasons, the court properly awarded judgment to plaintiff on its watertightness claim. Because the contract did not create a performance specification, plaintiff was not contractually responsible for making the tunnels watertight at its own expense. Consequently, plaintiff is entitled to reimbursement for the work performed in an attempt to achieve watertightness.

* * *

Accordingly, the judgment and subsequent order of the court should be affirmed.

Judgment unanimously affirmed without costs.

Another instructive look at the difference between design and performance specifications can be found in *Blake Construction Co., Inc. v. United States.* Part of the decision is reproduced at this point.

BLAKE CONSTRUCTION COMPANY, INC. v. UNITED STATES

United States Court of Appeals, Federal Circuit, 987 F.2d. 743, rehearing denied: suggestion for rehearing in banc declined, April 13, 1993, cert. denied, 510 U.S. 963 (1993).

[Ed. Note: The drawings depicted the electrical conduits as installed overhead. Notes on the drawings stated that the drawings were "diagrammatic." They also stated that the contractor would "relocate . . . conduits to coordinate with all others trades." Blake's subcontractor began to install the electrical feeder system in an underground concrete duct bank.

The Navy ordered that the conduits be installed overhead. Blake complied but claimed this order was a constructive change that entitled it to an equitable adjustment. The U.S. Court of Federal Claims agreed with Blake, and the Navy appealed.

Like the decision in the *Fruin-Colnon* case reproduced above, the court first distinguished design and performance specifications, noting that the design specifications do not allow deviation but must be followed as if they were a road map. Blake contended that it faced a performance specification. This would give discretion to Blake or its subcontractor to choose an underground location for installing the electrical feeder system. The conduit was to be installed to avoid interference with other trades. But it did not detail the manner that this was to be done. It also contended that the system could not be installed exactly as depicted on the drawings as alterations were needed to avoid conflict with other trades. Also, the specifications did not provide a road map "characteristically associated with a design specification."]

* * *

This reasoning obscures the real question in the case. Taking the second argument first, the mere fact that a specification cannot be followed precisely does not, in and of itself, indicate that it is "performance" and not "design." Were this true, any specification intended to be a design specification would be transformed into a performance specification if it were faulty. This is nonsensical; common sense dictates that the contractor does not acquire unfettered discretion to complete the contract in any manner it sees fit, just because one aspect of the specification might be defective. *See J.L. Simmons*, 412 F.2d 1360 (contractor brought flaws in design specifications to government's attention, proceeded under government's direction, and was entitled to equitable adjustment); *S.W. Elecs. & Mfg. Corp. v. United States*, 228 Ct.Cl. 333, 655 F.2d 1078, 1081 (Ct.Cl. 1981) (same). The fact that the electrical conduits could not be installed overhead in the precise manner depicted by the drawings, and at some points had to be installed outside the corridor itself, did not automatically relieve Blake of the obligation to install them overhead.

More generally, the problem with both of Blake's arguments is that the distinction between design and performance specifications is not absolute, and does not dictate the resolution of this case. Contracts may have both design and performance characteristic. *See, e.g., Utility Contractors, Inc. v. United States*, 8 Cl.Ct. 42, 50 n. 7 (1985) ("Certainly one can find numerous government contracts exhibiting both performance and design specifications."), *aff'd mem.*, 790 F.2d 90 (Fed.Cir.1986); *Aleutian Constructors v. United States*, 24 Cl.Ct. 372, 379 (1991)

("Government contracts not uncommonly contain both design and performance specifications."). It is not only possible, but likely that a contractor will be granted at least limited discretion to find the best way to achieve goals within the design parameters set by a contract. *See, e.g., Penguin Indus., Inc. v. United States*, 209 Ct.Cl. 121, 530 F.2d 934, 937 (1976). "On occasion the labels 'design specification' and 'performance specification' have been used to connote the degree to which the government has prescribed certain details of performance on which the contractor could rely. However, those labels do not independently create, limit, or remove a contractor's obligations." *Zinger Constr. Co. v. United States*, 807 F.2d 979, 981 (Fed.Cir.1986) (citations omitted). These labels merely help the court discuss the discretionary elements of a contract. It is the obligations imposed by the specification which determine the extent to which it is "performance" or "design," not the other way around.

The real issue is not whether the drawings and diagrammatic notes in their entirety should be labeled design specifications or performance specifications, but how much discretion the specifications gave Blake in the placement of the electrical feeder system. This is a question of contract interpretation which is a matter of law for this court to decide. *R.B. Wright Constr. Co. v. United States*, 919 F.2d 1569, 1571 (Fed.Cir.1990). Although the trial court's opinion may be helpful, we are not bound by it. *J.B. Williams, Co. v. United States*, 196 Ct.Cl. 491, 450 F.2d 1379, 1388 (1971). There is no question that the diagrammatic notes gave the electrical contractor some discretion to work around the other trades, but we think the Claims Court defined too broadly the amount of discretion permitted under the contract.

"Contracts are viewed in their entirety and given the meaning imputed to a 'reasonably intelligent contractor' acquainted with the involved circumstances, regardless of whether labeled 'design,' 'performance,' or both." *Zinger Constr.*, 807 F.2d at 981 (citing *J.B. Williams*, 450 F.2d at 1388). We believe that a reasonable contractor would understand that the contract required more than mere avoidance of conflict with the other trades. The specifications, viewed as a whole, additionally required installation of the conduits overhead within the confines of the corridor. This is the only conclusion that gives meaning to the drawings. An interpretation which gives reasonable meaning to all parts of a contract is preferred to one which renders part of it insignificant or useless. *Hill Materials Co. v. Rice*, 982 F.2d 514, 517 (Fed.Cir.1992). All the drawings depicted overhead installation of the electrical conduits and, more specifically, showed either an exposed or concealed installation depending on their position along the length of the corridor. An interpretation permitting underground installation renders these drawings meaningless. Accordingly, the government's interpretation of the specifications is the only reasonable one.

Contrary to Blake, we do not revisit the Claims Court's fact findings to reach our conclusion. The Claims Court thought the specifications were "performance" for several nondispositive reasons—first, the diagrammatic notes indicated the drawings need not be followed exactly; second, the evidence showed an underground installation was better than an overhead installation; and third, industry practice favored underground installation. As we have observed, the meaning of the diagrammatic notes is a matter of law. Because we conclude

the contract did not permit Blake to install the electrical conduits underground, whether or not this method would have been "better" is irrelevant. Finally, whether local trade custom was to install electrical conduits underground is also irrelevant here. Contracting parties may freely choose to have work performed in a specified manner; here, the government opted to install the electrical conduits overhead, even if this was not the usual mode.

CONCLUSION

Accordingly, the judgment of the Court of Federal Claims is reversed.

Purchase description specifications designate the required product or equipment by manufacturer, trade name, and type. Sometimes the contractor can select from an approved list. These specifications usually increase contract cost, because they limit the contractor's ability to use materials or equipment that may be just as good as specified and that may cost less. For this reason, public contracts and many private contracts frequently create a method by which the contractor can seek approval to use alternative products or materials.

When such methods are used, a number of legal problems can arise. First, as discussed in Section 24.03, does use of substitute materials or products transfer the risk of design failure? Second, does unreasonable delay in passing on such a request expose the design professional to liability to the owner or contractor?[8] Third, will the contractor have a claim for intentional interference with its contract if the design professional rigs the specifications and methods for approving alternates in favor of a particular manufacturer?[9]

The contractor's proposal to use an alternate is most commonly made before bid opening but occasionally is made during performance. When such proposals are made, the design professional should be prompt and fair in passing judgment on whether the alternate product or material is the equal or equivalent of the brand

specified. If the design professional has a reputation for intransigence or unreasonably delayed decisions, a contractor is likely to assume that alternates will not be available, and to bid accordingly. Unreasonable delay in passing on requests for alternates can result in liability to the design professional.[10]

Specifications should state that the determination of whether a proposed alternate is the equivalent of that specified can take into account not only function and performance but also aesthetics, manufacturer's warranty, and the reputation of the manufacturer for servicing the product and supplying spare parts.

Another question that can arise with purchase description specifications, particularly those that do not provide for authorized substitutions, is whether the owner who uses such specifications makes any warranties regarding the commercial availability of the designated materials or equipment. Although such a warranty has been found to have been created, as a rule, it is very limited. The owner warrants that the supplier is capable but does not warrant its willingness to meet time requirements, nor does the owner warrant that the supplier designated will accede to the terms and conditions the contractor insists on. Finally, the warranty does not include an assurance that the supplier will perform. In the case where this limited warranty was recognized, the court simply concluded that it would not change the normal rights and responsibilities that attend the use of subcontractors and suppliers.[11]

---

[8]*E. C. Ernst, Inc. v. Manhattan Constr. Co. of Texas,* 551 F.2d 1026, rehearing denied in part and granted in part, 559 F.2d 268, (5th Cir.1977), cert. denied sub nom. *Providence Hospital v. Manhattan Constr. Co. of Texas,* 434 U.S. 1067 (1978) (claim by owner). As to contractors, see Section 14.08.

[9]*Waldinger Corp. v. CRS Group Eng'rs, Inc.,* 775 F.2d 781 (7th Cir.1985).

[10]*E.C. Ernst, Inc. v. Manhattan Constr. Co. of Texas,* supra note 8.

[11]*Edward M. Crough, Inc. v. Dept. of Gen. Serv. of Dist. of Columbia,* 572 A.2d 457 (D.C.App.1990).

Specification writing can play an important part in relations between contractors and any labor union representing employees. Traditionally, most of the labor in a construction project is performed at the site. The relatively high wages paid to unionized craft workers, as noted in Section 23.05B, the frequent disruption caused by labor disputes at the site, and delays caused by weather conditions have encouraged systems under which more work is done in factories.

Construction specifications increasingly require integrated units or prefabricated materials, designed to reduce the amount of labor performed at the site.[12] But reducing work at the site affects job security of people in the construction craft trades. As a result, many craft unions have won what are called "work preservation" clauses in their collective bargaining agreements with employers, under which the employers, mainly subcontractors, agree not to handle products on which labor is performed at the factory when that labor traditionally was performed at the site. Under these work preservation clauses, unions are given the right to strike if there is a violation.

Enterprise Association of Steam, etc., Local Union Number 638 v. NLRB[13] involved the construction of the Norwegian Home for the Aged in New York City. The prime contractor prepared the specifications, which required that climate control units manufactured by a designated supplier be installed as integrated units. These units contained factory-installed internal piping. If the units were installed complete with factory prepiping, the supplier guaranteed all units for a year.

The subcontractor who had agreed to install the heating, ventilating, and air-conditioning had had a collective bargaining agreement for many years with the plumbers' local that contained a provision requiring the subcontractor employer to preserve certain cutting and threading work for performance at the job site by its own employees. That agreement would have required the internal piping already installed in the climate control units to be cut and threaded at the job site.

When the units arrived at the job site, the union's business agent inspected them and informed both prime contractor and subcontractor that the union employees would not install them. The dispute delayed completion, and the prime contractor filed an unfair labor practice charge with the National Labor Relations Board (NLRB), alleging that the union could not instruct members to refuse to handle the units, because their object would have been to force a "neutral"—the prime contractor—to cease using the supplier's products.

In dealing with work preservation cases, the NLRB and some of the federal circuit courts of appeals had adopted a "right to control" test. If the struck employer (here the subcontractor) had the right to control work assignments, the strike is legal. The pressure must be on the employer in such a case, as it can control work assignment. But if it does not have the right to control, the pressure must be on a "neutral"—the prime contractor. Then the strike is illegal.

Applying the right-to-control test to the Local 638 case, the NLRB concluded that the subcontractor never had the power to assign the disputed piping work to its union employees. This, according to the NLRB, made the subcontractor a neutral. The union's principal target was the prime contractor, inasmuch as the subcontractor had no right to control the work assignment. Pressure was being put on the subcontractor to force the prime contractor to stop buying the units and restore the work. Such pressure was a secondary boycott and an unfair labor practice.

However, the Federal Circuit Court of Appeals for the District of Columbia, following opinions in other federal circuit courts, refused to follow the right-to-control test and concluded that the subcontractor was not a neutral, inasmuch as the union was simply attempting to enforce its lawful work preservation clause with the employer subcontractor. The court also suggested that the pressure could have been placed on the subcontractor to negotiate a compromise or to terminate the contract with the prime contractor. Because the core of the union's grievances was with the subcontractor, the court held the strike lawful. However, the U.S. Supreme Court reversed the circuit court and concluded that this was an illegal secondary boycott.[14]

---

[12]The subject matter of work preservation clauses illustrates some of the technologically motivated decisions being made by those drafting specifications. For example, cases have involved precast walls, *NLRB v. Carpenters Dist. Council of Kansas City and Vicinity*, 439 F.2d 225 (8th Cir.1971); a prepiped sink unit, *Associated General Contractors of California, Inc. v. NLRB*, 514 F.2d 433 (9th Cir.1975); prefabricated fireplaces, *Western Monolithics Concrete Products, Inc. v. NLRB*, 446 F.2d 522 (9th Cir.1971); and prepiped heating and ventilating controls discussed in the text of this subsection.

[13]521 F.2d 885 (D.C.Cir.1975).

---

[14]*NLRB v. Enterprise Ass'n of Steam, etc., Local Union* No. 638, 429 U.S. 507 (1977).

Design professionals drafting specifications must take work preservation clauses into account. Craft unions will oppose specifications that reduce their work. Various steps, some legal and some probably illegal, are likely to be taken by the craft union if substantial work traditionally performed by their members will be performed by others. This does not mean the best and cheapest technology should never be specified. However, in taking into account the advantages of specifications that involve prefabricated products or sealed-at-the-factory components, the disadvantages of potential work stoppages and legal battles must be considered.

Specifications can also play a role in avoiding sales taxes on material and equipment needed for the project. A nonprofit owner may be exempt from sales taxes. If this is so, the specifications should state that the prime or subcontractors purchase as agents for the owner.[15]

E. Conditions: General and Supplementary

Construction is a complex and dispute-prone activity. Guidelines are needed to spell out clearly and completely the rights and duties of the parties. For these reasons, most construction documents include general conditions (often supplemented by supplementary conditions) of the contract. These ground rules, under which the project will be constructed, are often lengthy and deal with the following subjects:

1. scope of contract documents and resolution of conflicts between them
2. roles and responsibilities of the principal participants in the project
3. subcontractors and separate contractors
4. time
5. payments and completion
6. protection from and risk of loss to persons and property
7. changes
8. corrections
9. termination
10. disputes
11. insurance

---

[15]*F. Miller & Sons, Inc. v. Calcasieu Parish School Bd.*, 838 So.2d 1269 (La.2003) (successful in avoiding sales tax).

Differentiations should be made among a *legal condition*, *general conditions*, and *supplementary conditions*. The first is a legal classification. It is an event that must occur or be excused before an obligation to perform arises. Whether a particular event—sometimes all or part of a promised performance and sometimes an event not within the control of a contracting party—is a condition depends on any contractual language manifesting this conclusion, the probable intentions of the parties, and elements of fairness. However, this is not the sense in which the term *conditions* is used in this subsection.

General conditions are usually expressed in standardized prepared printed contract documents often published by professional associations such as the American Institute of Architects (AIA) or the Engineers Joint Contract Documents Committee (EJCDC). Usually the court seeks to determine how terms should be interpreted by ascertaining the intention of the parties. Because documents prepared by the AIA or EJCDC are in essence prepared by third parties—not the contracting parties—as noted in Section 20.02, they raise difficult interpretation questions.

Sometimes general conditions are prepared by owners who are "repeat players" in the world of construction. These owners may enter into many similar transactions. As shall be seen in Section 20.02, unclear general conditions are likely to be interpreted against the owner that has prepared them.

Although frequently called *general conditions*, its provisions are almost never automatically considered legal conditions, although nothing prevents language within a general condition from creating a legal condition. General conditions are prepared in a way that lets them be used in many types of transactions, either by different contract makers or by a single contract maker.

Any individual construction contract, however, may have attributes that make supplementary conditions necessary. For example, the indemnity provisions in general conditions prepared for national use may not be enforceable or desirable in states that have specific statutes regulating indemnification. Similarly, the frequent existence of specialized statutes dealing with arbitration may make it essential to add supplementary conditions if those requirements are to be met. Finally, because of the individualized aspects of insurance, specific insurance requirements are always found in supplementary conditions. Not all contract makers wish to use general conditions drafted by others, such as the AIA. As a result, these contract makers may use AIA

Document A201 but either modify it or attach their own supplementary conditions and make those conditions take precedence over the general conditions.

Two further observations must be made. First, some owners or design professionals may lull contractors into a false sense of security by prescribing general conditions with which the contractors are familiar and comfortable, but may make many changes in the supplementary conditions that destroy some of the protection accorded contractors by the general conditions.

Second, as noted in Section 20.03B, the law is frequently called on to sort out inconsistencies between general conditions and supplementary conditions. Although it is best to delete from general conditions any provisions supplemented by the supplementary conditions, often this deleting requires more work than the attorney wants to do. As a result, the attorney may simply state that the supplementary conditions take precedence over the general conditions. This ploy often requires a judge or arbitrator to try to reconcile apparently conflicting language. This problem is endemic to construction, with its wealth of contract documents.

Section 12.07 discussed the role of the design professional in drafting or suggesting that particular general conditions be used.

F. Site Subsurface Test Reports

Invitations to bidders frequently contain subsurface test reports or site data or a reference that such data are available in the office of a geotechnical consultant engaged by the owner. This information and data are of great importance in the construction project. Chapter 25 discusses the legal effect of furnishing this information.

G. Prior Negotiations and the Parol Evidence Rule

The parol evidence rule relates to the provability of oral agreements made before or at the same time as a written contract. This rule is based on the concept of "completeness of writings." If the written agreement and other written documents incorporated by reference or attached to it are the complete agreement of the parties, prior or contemporary oral agreements are not binding on the parties. Even if they were made, they were integrated, or merged, into the written, complete document.[16]

Most construction contracts are complete. Nevertheless, a possibility exists that oral agreements have been made before or at the same time as the final written agreement. If the written agreement contains an integration or merger clause—one that specifies that the written agreement is the complete and final agreement—it will be difficult for either party to prove a prior oral agreement.[17] It will not be impossible, however, because doctrines that attack the validity of the contract, such as fraud, permit proof of an oral agreement even if provisions specify that the written agreement is the complete document. Also, in one construction case the court permitted evidence of an antecedent oral agreement despite the presence of an integration clause by concluding the oral agreement was separate and not integrated into the written contract.[18] If no contract provision deals with the question of completeness, testimony claiming an oral agreement generally will be admitted.

It is important to incorporate the entire agreement into the writing. Even if the writing is prepared and ready for execution, the design professional, and certainly the owner's attorney, should insist on incorporating any changes or additions into the writing itself. The entire document need not be retyped if time does not permit (word processing has made this easy). If the oral agreement is not included in the writing, a substantial risk exists that it cannot be proved.

Suppose a dispute arises over meaning of terms used in the writing. If the words chosen by the parties are ambiguous, the court can look at the surrounding facts and circumstances to determine how the parties used those particular words. These circumstances include the setting of the transaction, the contracting parties' objectives in making the contract, and any conversations they may have had. Even if a writing is considered complete, antecedent or subsequent conversations may be admissible to interpret the writing.

Two cases are instructive, both involving the parol evidence rule in the design and construction context. The first, *Godfrey, Bassett & Kuykendall, Architects, Ltd. v. Huntington Lumber & Supply Co., Inc.*,[19] was discussed in Section 11.04E. That case noted the closeness

---

[16]See Sections 11.04E and 12.03C for further discussion.

[17]*Lower Kuskokwin School Dist. v. Alaska Diversified Contractors, Inc.*, 734 P.2d 62 (Alaska 1987), denied motion for reconsideration, 778 P.2d 581 (Alaska 1989), cert. denied, 493 U.S. 1022 (1990).

[18]*C. L. Maddox, Inc. v. Benham Group*, 88 F.3d 592 (8th Cir.1996).

[19]584 So.2d 1254 (Miss.1991).

of the relationship between the contracting parties as one basis for determining whether one party could claim there had been an antecedent agreement not expressed in the writing, despite a failure by that party to have read the agreement. In the case, the court looked carefully at the background and relationship between the parties to determine whether failure to include an asserted promise precluded the party from testifying as to its existence.

Before discussing the second case, some background is in order. There are many expectations to the parol evidence rule. Many scholars and important judges have expressed hostility toward the rule's continued existence. This situation has led some to believe that in no case will a party be prevented from testifying of an asserted antecedent oral agreement despite its not having been included in the writing.

These beliefs would certainly have been supported by cases decided in California in the 1960s.[20] However, *Brinderson-Newberg Joint Venture v. Pacific Erectors, Inc.,*[21]—a construction contract dispute decided in 1992 by the Ninth Federal Circuit Court applying California law—shows there is still vitality in the parol evidence rule.

H. Modifications

Modifications are different than changes. Changes occur frequently during a construction project. They are generally governed by carefully drawn and complete provisions under which the owner has the right to order changes and the contractor must perform them, with appropriate contract language dealing with compensation for deletions or additions caused by the changes. A modification is a change agreed on by both parties in the basic obligation not based on any contractual provision giving the owner the right to order changes.

A modification is a contract. It must meet the requirements of a valid contract, such as mutual assent (see Section 5.06B), consideration (see Section 5.08), and the Statute of Frauds (see Section 5.10). Service contracts, such as construction, raise special consideration problems. To demonstrate, suppose there is a construction project for $100,000. The obligations of the contractor are expressed in the contract documents. Suppose that during the term of the agree-

ment the parties mutually agree to increase the contract price to $110,000. The owner must receive something in exchange for the additional compensation. If the contractor agrees to improve the quality or increase the quantity of the building or to shorten the time for completion, there is an appropriate exchange and the modification agreement is binding. Difficulties arise when the modification agreement encompasses an increase in price without any change in the contractor's obligation. The *preexisting duty rule* can invalidate such an agreement. Unless one of many exceptions applies, the agreement is not binding if the party—in this case the contractor—is obligating itself to do no more than it was previously obligated to perform under the original contract.[22]

The preexisting duty rule has been criticized. It limits the autonomy of the parties by denying enforceability of agreements voluntarily made. Implicit in the rule is an assumption that an increased price for the same amount of work is likely to be the result of expressed or implied coercion by the contractor, as if the contractor were saying, "Pay me more money or I will quit and you will have to go to court to get damages." However, suppose the parties have voluntarily arrived at a modification of this type. There is no reason for not giving effect to their agreement. A number of exceptions can relieve the sometimes harsh effect of the preexisting duty rule.

In the construction contract, minor changes in the contractor's obligation have been held sufficient to avoid the rule even where the increase in price was not commensurate with the change in the contractor's obligation. This approach may permit a contract modification to be enforced if the parties have had enough foresight or legal knowledge to provide for some minor and relatively insignificant change in the contractor's obligation as a means of enforcing the increased price.

An exception that developed in construction contracts enforces the modification if it is fair and equitable, in view of circumstances not anticipated by the parties when the contract was made.[23] As this exception (frequently referred to as the "unforeseen circumstances exception") has occurred mainly in the area of subsurface conditions, it is discussed in greater detail in Section 25.01D.

---

[20]*Masterson v. Sine,* 68 Cal.2d 222, 436 P.2d 561, 65 Cal.Rptr. 545 (1968); *Pacific Gas & Elec. Co. v. G. W. Thomas Drayage & Rigging Co.,* 69 Cal.2d 33, 442 P.2d 641, 69 Cal.Rptr. 561 (1968).

[21]971 F.2d 272 (9th Cir.1992), cert. denied, 507 U.S. 914 (1993).

[22]*Crookham & Vessels, Inc. v. Larry Moyer Trucking, Inc.,* 16 Ark.App. 214, 699 S.W.2d 414 (1985); *Hiers-Wright Assoc., Inc. v. Manufacturer's Hanover Mortgage Corp.,* 182 Ga.App. 732, 356 S.E.2d 903 (1987).

[23]*Linz v. Schuck,* 106 Md. 220, 67 A. 286 (1907) followed in the Restatement (Second) of Contracts § 89 (1981).

Dissatisfaction with requiring consideration for a change is also shown by Section 2-209 of the Uniform Commercial Code, dealing with goods transactions. The section states that modifications are valid without consideration. However, a modification obtained in bad faith is not valid.

Curiously, despite the movement toward abolishing the consideration requirement for a modification, the rule is applied vigorously in federal public contracts. Unless the government receives something for an increase in price or a reduced price for a deletion, the modification is not valid. This rigid adherence to the consideration requirement may be justified as a means of preventing gifts of public funds, and corruption and collusion between the contractor and a public official. Giving some advantage to a contractor without the government's getting anything in return may be unfair to the other bidders.

Contracts frequently state that modifications must be in writing. Such provisions generally were not enforced by the common law.[24] In enforcing oral modifications despite such clauses, the courts have stated that by making a subsequent oral modification, the parties have changed the agreement requiring that the modification be in writing.

A New York court, applying a statute enforcing such clauses, held that a contract clause requiring a modification be in writing bars an oral agreement unless the oral modification is fully executed or partly performed and unequivocally referable to the oral modification.[25]

Section 2-209 of the Uniform Commercial Code controls transactions in goods. It provides that oral agreements modifying a written agreement are not effective if the written agreement contains a provision requiring that modifications be in writing. There are some exceptions, but the code expresses a policy that such contractual provisions requiring a writing as a condition to enforcement of an asserted modification agreement should be given more effect than courts have given in the past. It remains to be seen whether this change in the law relating to the sale of goods will have any impact on judicial thinking in other types of contracts.

SECTION 19.02 Judicially Determined Terms

A. Necessity to Imply Terms

When making a contract, parties frequently do not consider all problems that may arise. Some matters may have been considered, but the parties believed the resolution of the matters to be so obvious that contract coverage would be unnecessary. Matters have been discussed by the parties during negotiations, but the parties could not agree on a contract solution to the problem. Yet these parties may intend to have a binding contract despite their inability to resolve all problems during negotiation. In such cases, the parties may state in the contract that they will agree in the future on certain less important contract matters or omit the matter from the contract entirely. Courts may be asked to fill in the gaps not covered by the contract or decide matters left for future agreement where the parties cannot agree. Courts will be more likely to perform these functions if convinced that the parties intended to make an enforceable agreement, especially where performance has begun.

Until recently, common law judges hesitated to imply terms. These judges saw themselves as enforcing contracts made by the parties rather than making contracts for the parties. One court stated,

1. The implication must arise from the language used or it must be indispensable to effectuate the intention of the parties.
2. It must appear from the language used that it was so clearly within the contemplation of the parties that they deemed it unnecessary to express it.
3. Implied covenants can be justified only on the grounds of legal necessity.[26]
4. A promise can be implied only where it can be rightfully assumed that it would have been made if attention had been called to it.
5. There can be no implied covenant where the subject is completely covered by the contract.[27]

These requirements reflect judicial reluctance to imply terms. In addition, courts during the first half of the twentieth century hesitated to provide a term where the parties had agreed to agree but did not.[28]

---

[24]*Prince v. R. C. Tolman Constr. Co., Inc.*, 610 P.2d 1267 (Utah 1980). Four states, most notably California, require that modifications of written contracts be in writing. There are many exceptions.

[25]*F. Garofalo Elec. Co., Inc. v. New York University*, 270 A.D.2d 76, 705 N.Y.S.2d 327, leave to appeal dismissed, 95 N.Y.2d 825, 734 N.E.2d 762, 712 N.Y.S.2d 450 (2000).

[26]This means that courts will imply a promise if it is necessary in order to have a valid contract and the parties have so intended.

[27]*Stockton Dry Goods Co. v. Girsh*, 36 Cal.2d 677, 681, 227 P.2d 1, 3–4 (1951).

[28]See Section 5.06F.

Modern judges seem less insecure and more realistic about their role in contract disputes. They are beginning to recognize that they "make" contracts for the parties. Because they recognize the proliferation of standard form contracts often made in an adhesion context, they are more willing to imply terms to redress unequal bargaining than were courts a half century ago.

The law should exercise restraint in implying terms where the contract has been negotiated. In such cases, most of the major problems were probably considered and the absence of a promise may be deliberate. Implying a term in such cases could frustrate the bargain.

Even in negotiated contracts it may be necessary to imply terms that were so obvious that the parties did not think it necessary to express them. A court should "complete" the deal by filling in gaps where parties intended to make a contract and have left minor terms for future agreement but have not been able to agree.

Again, a word of caution. The existence of express contract terms generally precludes terms on that subject being implied.[29] Also, custom takes precedence over implied terms. First, the contract must be examined. If the contract deals with subject matter that relates to the proposed implication, a court is less likely to imply terms. However, sometimes the effect of implication is achieved by interpreting particular contract terms in a way consistent with the implication.

One court, discussing the owner's duty to furnish a site, stated,

> Each party to a contract is under an implied obligation to restrain from doing any act that would delay or prevent the other party's performance of the contract. . . . A party who is engaged to do work has a right to proceed free of let or hindrance of the other party, and if such other party interferes, hinders or prevents the doing of the work to such an extent as to render the performance difficult and largely diminish the profits, the first may treat the contract as broken and is not bound to proceed under the added burdens and increased expense.[30]

The owner should not prevent the contractor, nor should the prime contractor prevent the subcontractor from performing in a logical, orderly, and efficient manner.[31] Breach of this obligation can generate a claim for delay or disruption, or better, an inefficiency claim.[32]

Implied obligations can go further and, under certain circumstances, require that the owner perform positive acts to assist the contractor in performance. However, the law will be more reluctant to impliedly require affirmative acts of cooperation than to preclude negative acts of hindrance or prevention. Positive acts are more likely to have been thought about if they were important, and failure to express them in the contract may indicate that neither expected them to be done.

Any duty to cooperate should not require undertaking heavy burdens of cooperation that would very likely frustrate the contractual allocations of responsibility made by the parties. However, if one party can assist the other party's performance at a minimal cost, such cooperation should be required.[33]

Some specific illustrations of implied terms in the construction contract have been mentioned earlier, such as the standard for the design professional's performance[34] and, as noted in Section 17.04C, the obligation of the owner to coordinate work of separate contractors or stand behind the obligation of another entity that has that responsibility.

There are other illustrations. The owner impliedly promises that the site will be ready for the contractor to commence performance.[35] (Often, as Section 26.02 notes, this is handled by not starting the obligation to commence performance until a notice to proceed has been given.) The owner impliedly promises to obtain the necessary easements or rights to enter the project site or land of another necessary for the contractor's performance.

---

[29]*Weber v. Milpitas County Water Dist.*, 201 Cal.App.2d 666, 20 Cal. Rptr. 45 (1962).

[30]*United States v. Guy H. James Constr. Co.*, 390 F.Supp. 1193, 1206 (M.D.Tenn.1972), aff'd without opinion, 489 F.2d 756 (6th Cir. 1974). See also, *C. A. Davis, Inc. v. City of Miami*, 400 So.2d 536 (Fla. Dist.Ct.App.), review denied, 411 So.2d 380 (Fla.1981) (no duty to do contractor's work—only not to hinder or interfere with that work).

[31]*Howard P. Foley Co. v. J. L. Williams & Co., Inc.*, 622 F.2d 402 (8th Cir.1980) (Arkansas law) (work obstructed, interfered with, and delayed) and *Housing Auth. of City of Little Rock v. Forcum-Lannom, Inc.*, 248 Ark. 750, 454 S.W.2d 101 (1970) (claimant could not perform in an orderly manner and could not follow any normal sequence). See also Sweet, *Contract Regulation of Delay and Disruption Claims in America*, [2002] Int'l Constr. L. Rev. 284, 285-289.

[32]See also Sections 26.10A and 27.02F.

[33]See *Allied Fire & Safety Equip. Co. v. Dick Enterprises, Inc.*, 886 F.Supp. 491 (E.D.Pa.1995) (prime had duty to coordinate the work).

[34]See Section 14.05A.

[35]*North Harris County Junior College Dist. v. Fleetwood Constr. Co.*, 604 S.W.2d 247 (Tex.Civ.App.1980).

The prime contractor impliedly promises the owner to use proper workmanship and materials and to complete the project free and clear of liens.[36] A contractor in a cost contract promises to inform the owner of prospective overruns of cost estimates.[37] One court held that the prime contractor promised the subcontractors that the work would be coordinated and that the prime contractor would process subcontractor claims to the owner.[38] But another court refused to imply a promise that the prime contractor would create and maintain an efficient schedule when this was neither customary nor bargained for.[39]

The owner impliedly promises that it and its designated representative, usually the design professional, will perform in such a way as to reasonably expedite the contractor's performance. For example, the owner impliedly promises that the design professional will give contract interpretations and pass on sufficiency of shop drawings within a reasonable time.[40]

The law usually implies a promise by the owner to supply adequate drawings and specifications (discussed in greater detail in Section 23.05E). However, that such terms are implied does not eliminate the necessity of focusing more closely on the nature of the implication. For example, suppose the owner designates specified material. That will very likely mean that if the contractor uses that material, the contractor will not be responsible if the material proves unsuitable.[41] However, does specifying a material imply that the material is available? Does it imply that the material is stocked by local suppliers? Such questions expose the basic function of implying terms—that of allocating risk and responsibility.

A federal appeals board held that the owner does not warrant that a product specified will be in stock, inasmuch as parties do not usually guarantee the performance of third parties.[42] Such a result may also be based on the likelihood that owners know no more about the stock of suppliers than does the contractor.

Usually the owner reserves certain powers during the contractor's performance to protect its interests. This should not automatically convert into an affirmative duty for the benefit of the contractor or subcontractors. For example, it has been held that the power to monitor contractor performance and stop the work if necessary did not imply that the owner had promised to direct the contractor's workers.[43]

Courts are sometimes asked to imply completion times to construction contracts. The contract usually gives a specified time for completion. If no express provision deals with this question, courts hold that the contractor must complete performance within a reasonable time, given all the surrounding facts and circumstances.

Courts should not imply a reasonable time for performance if performance has not yet commenced and if the failure to agree on a time for performance indicates that the parties have not yet intended to conclude a contract. The absence of agreement on such an important question may mean the parties are still in a bargaining stage.

B. Custom

Customary practices are important in determining rights and duties of contracting parties, particularly in complex transactions such as construction, where not everything can be stipulated in the contract. Section 12.11 mentioned the frequent claim by design professionals that they customarily retain ownership of drawings and specifications. Custom plays other significant roles in construction. In placing such a heavy emphasis on customary practices when interpreting contract terms, courts often state that parties contract with reference to existing customs. In that sense, courts can be said to be simply giving effect to the actual intention of the parties. However, a contracting party may also be held to those customs that it knew or should have known. The law places a burden on contracting parties to learn customs that apply to the type of

---

[36]But see AIA Doc. A201-2007, § 9.3.3 (work previously paid for free of liens as far as contractor knows).

[37]*Jones v. J. H. Hiser Constr. Co., Inc.*, 60 Md.App. 671, 484 A.2d 302 (1984), cert. denied, 303 Md. 114, 492 A.2d 616 (1985).

[38]*Citizens Nat'l Bank of Orlando v. Vitt*, 367 F.2d 541 (5th Cir.1966), appeal after remand, 414 F.2d 696 (5th Cir.1969). As to duty to coordinate the work, see also *Allied Fire & Safety Equip. Co. v. Dick Enterprises, Inc.*, supra note 33.

[39]*Drew Brown, Ltd. v. Joseph Rugo, Inc.*, 436 F.2d 632 (1st Cir.1971).

[40]See AIA Doc. A201-2007, §§ 4.2.7, 4.2.11 (reasonable promptness).

[41]See Section 24.02.

[42]*James Walford Constr. Co.*, GSBCA No. 6498, 83-1, BCA ¶16,277, aff'd upon reconsideration, 83-1 BCA ¶16.342. See *Edward M. Crough, Inc. v. Dept. of Gen. Serv. of Dist. of Columbia*, supra note 11 (discussed in Section 19.01D).

[43]*Nat Harrison Assoc., Inc. v. Louisville Gas & Elec. Co.*, 512 F.2d 511 (6th Cir.), cert. denied, 421 U.S. 988 (1975).

transaction they are about to enter and in the place where they contract.

In any event, an established custom can be a more convenient and proper method of filling gaps than a court determination of what is reasonable. For example, if customarily the contractor obtains a building permit, the court is likely to place this responsibility on the contractor, but only where there is a contract gap on this point.[44] If the contract does not deal with the matter, custom is preferable to a judge's determination of who should obtain the permit. Custom can help the court interpret terms or conflicts in the contract documents.[45]

Custom is an important part of construction contracts. However, parties should never rely on custom when there are express provisions to the contrary. Custom can trap a contractor who performs a project in a locality where customs may differ from those where it customarily works. Custom can often be difficult to establish. In the absence of an express provision controlling the dispute, the party seeking to establish custom will be given an opportunity to do so.

C. Building Codes and Permits

Building codes and land use controls play a pervasive role in construction. The contract frequently implies or expresses that the contractor will comply with applicable laws. Further analysis is needed to separate several related but distinct problems.

Where design and construction are separate, the former is the owner's responsibility and is usually done by the design professional. Suppose the design violates building code requirements. The contractor may contend that its job is to build the design, but the contractor should not ignore violations of law. It should direct the design professional's attention to any obvious code violations in the design.

The contractor's responsibility in design code violations is dealt with by AIA Document A201-2007 Section 3.2.3, which states that a contractor, upon receipt of the design, is not required to determine whether the Contract Documents are in accord with legal requirements; however, it must promptly report to the architect any nonconformity discovered or made known to the contractor (presumably by a subcontractor). Under Section 3.7.3, if the contractor performs work knowing it to be contrary to such laws without notifying the architect, the contractor "shall assume appropriate responsibility for such Work and shall bear the costs attributable to correction."[46]

This solution effectuates a sound middle ground between putting an unreasonable responsibility on the contractor and allowing the contractor to close its eyes in the face of danger. A contractor who does not comply with the requirements of Section 3.7.3 is exposed to liability, not only for the cost of correction but also for losses suffered by third parties. (As noted in Section 27.06, it would not be responsible for consequential damages to the owner because Section 15.1.6 included in A201-2007 requires each party to waive consequential damages against the other.)

In many instances no *direct* clash exists between building codes and the design. In such instances, the owner and the design professional expect the contractor to follow building codes. In this sense, the contractor must build in conformity with legal requirements. Essentially, this is a gap-filling instrumentality. In addition, such code compliance provisions are intended to obtain compliance with worker safety rules.

Other legal requirements can consist of permits issued by public officials at various stages of the performance. These permits are not limited to building and occupancy permits. Permits must frequently be obtained from public utilities to connect the utilities of the project to public utility lines. Drainage rights of way may be needed, as well as permits from state highway officials for certain types of construction.

Who must obtain permits required by law? Construction documents should cover these matters.[47] If the contract does not, custom may allocate responsibility for obtaining permits. In the absence of an express contract provision or accepted custom, the law will be likely to imply that the owner will obtain the more important permanent permits, such as land use control permits,[48] but the contractor should

---

[44]*In Weber v. Milpitas County Water Dist.*, supra note 29, custom that the owner procured these permits was not relevant because an express provision in the contract required the contractor to obtain such permits.

[45]*Fifteenth Ave. Christian Church v. Moline Heating & Constr. Co.*, 131 Ill.App.2d 766, 265 N.E.2d 405 (1970).

[46]See Section 14.05D, dealing with design that violates local building codes.

[47]AIA Doc. A201-2007, § 3.7.1 (contractor obtains building permit).

[48]*COAC, Inc. v. Kennedy Eng'rs*, 67 Cal.App.3d 916, 136 Cal.Rptr. 890 (1977) (owner must obtain environmental impact report).

get operational permits, such as building permits[49] and occupancy permits. The contractor is likely to be required, by implication of law, to obtain from public officials permits that are associated with facilities and equipment. The contractor should obtain utility hookup permits and permits for temporary construction. Determining who should obtain particular construction permits will depend on the extent of experience the particular owner has had in construction projects.

Public authorities, inundated with permit requests, are not always able to process these requests expeditiously, thereby causing significant project delays. To ameliorate this bottleneck, large urban centers (beginning with New York City in 1976) created a process of "self-certification," under which architects, engineers, or certain trade contractors (such as plumbers and electricians) certify their own work as being in compliance with the building code. Because of the obvious conflict of interest, public authorities continue to spot-check self-certifications, and violators are subject to severe sanctions, including loss of the future ability to self-certify projects and loss of licensure.[50]

Generally, the contractor's failure to comply with any requirement to obtain a work permit or submit the plan does not affect its right to recover compensation.[51] However, it is dangerous to rely on a court's subsequent determination that the violation was technical.

D. Good Faith and Fair Dealing

The common law did not usually hold contracting parties to obligations of good faith and fair dealing. Refusal to do so reflected the common law's belief that the written contract was sacred, contracting parties should take care of themselves, and good faith is imprecise.

Beginning in the mid–twentieth century, however, and heavily influenced by the Uniform Commercial Code,[52] American contract law began to hold contracting parties to the covenant of good faith and fair dealing. One party cannot be expected to guarantee that the other party will receive the benefit it expects from a contract, nor must a contracting party make unreasonable sacrifices for the other party. However, each party should not only avoid deliberate and willful frustration of the other party's expectations but should also extend a helping hand where to do so would not be unreasonably burdensome. Contracting parties, although not partners in a legal sense, must recognize the interdependence of contractual relationships.

Some applications of the doctrine are simply recognition of implied terms, as discussed in 19.02A. But more expansive use of the doctrine emphasizes the unspoken objectives of the contracting parties, the spirit of the contract itself, and the need for elementary fairness. This more expansive use can even dictate an outcome contrary to the literal interpretation of the written contract. Finally, some breaches of this covenant, particularly in contracts of insurance and employment, violate public policy. When they do, they can become the basis for tort remedies, including punitive damages.

Because the doctrine is pervasive and the various parts of the construction process are particularly interdependent, illustrations of the covenant of good faith and fair dealing are sprinkled throughout this book.[53] It may be useful to note one case here.

Maier's Trucking Co. v. United Construction Co.[54] illustrates good faith and fair dealing in the context of a deductive change. The prime contractor suggested a modification that would essentially eliminate a subcontractor's work. Although the majority held that this did not breach the covenant of good faith and fair dealing, two judges strongly dissented. They would have ordered a trial to determine whether this suggestion had been made in bad faith.

Yet in *Scherer Construction LLC v. Hedquist Construction, Inc.*,[55] the Supreme Court of Wyoming held that the trial court was incorrect in awarding a summary judgment to the prime contractor in a claim by the subcontractor similar to the claim in the *Maier* case discussed in the preceding paragraph. In the *Scherer* case, the subcontractor claimed that the obligation of good faith and fair dealing was breached when the prime suggested a change to the owner that would have wiped out 75 percent of the value of the subcontract. The subcontractor had spent a substantial amount of money to prepare to perform as required by the original contract. The court held that the obligation of good faith and fair dealing applied to every contract, citing the Restatement of Contracts (Second) Section 205. The court sent the case

---

[49]*Drost v. Professional Bldg. Service Corp.*, 153 Ind.App. 273, 286 N.E.2d 846 (1972).

[50]Kubes, *The Design Professional's Project Self-Certification: A Key to Efficiency or Liability?* 26 Constr. Lawyer, No. 4, Fall 2006, p. 5. Wrongful certification may also expose the architect to liability to third parties. See *27 Jefferson Avenue, Inc. v. Emergi*, 18 Misc.3d 336, 846 N.Y.S.2d 868 (2007).

[51]See Annot., 26 A.L.R.3d 1395 (1969).

[52]U.C.C. § 1-304 (formerly 1-203).

---

[53]See Sections 11.04B, 14.04B, 21.04C, 24.04, 26.03C, and *Broadway Maintenance Corp. v. Rutgers*, reproduced at Section 17.04C.

[54]237 Kan. 692, 704 P.2d 2 (1985).

[55]18 P.3d 645 (Wyo.2001).

back to the trial court to see if the subcontractor could prove its claim, the very action suggested by the dissent in the *Maier* case.

The trend is clearly in the direction of the *Scherer* case. The trial court must decide whether there was a breach of the obligation. Much will depend on the prime's motive for suggesting the change. If it were done to "teach the sub a lesson" or to "get even with the sub," the obligation would have been breached. But if the suggestion were made in commercial good faith, made to help the owner cut costs, then the subcontractor will not have a valid claim of breach. These will be hard claims to sustain, but the door is open.

Another element of good faith and fair dealing can be seen in the need for communication. For example, AIA Document A201-2007, Sections 9.4.1 and 9.5.1, require an architect who cannot certify the full amount requested to notify the owner and contractor and give his reasons. Similarly, AIA Document A312, Section 3.1, states that the surety's obligation arises after the owner has notified the contractor and surety that it considers defaulting the contractor and "has requested . . . a conference with the Contractor and Surety . . . to discuss methods of performing the Construction Contract." Both elements illustrate specific requirements of obligations that would very likely be encompassed by the covenant of good faith and fair dealing.

The settlement process and the pressures that attend it are fertile ground for claims of bad faith. One case decision employed the concept of duress to void a settlement.[56] Another held that a prime contractor who was empowered by contract to negotiate the settlement of a delay claim on behalf of its subcontractors was required to do so in good faith.[57] Still another case held that a termination by the owner in accordance with the powers given it under the contract was improper, because it had not been made in good faith.[58]

Even this sampling shows the range of breaches. Some are considered tortious, some illustrate that contract language does not necessarily insulate a party from its obligation to act in good faith and to deal fairly, and some are based on implied terms. The outpouring of cases dealing with this doctrine demonstrates that the doctrine will be an important component of the construction contract obligation.[59]

E. Unconscionability

The common law generally allowed contracting parties to decide the terms of their contracts with relatively minimal judicial intervention. But again, as noted in Section 19.02D, the Uniform Commercial Code regulating transactions in goods led to the broadening of the unconscionability doctrine that had been used in the courts of equity.[60] The obligation of good faith and fair dealing placed limits on the conduct of contracting parties. The unconscionability doctrine authorizes an inquiry into the circumstances under which the contract was made and permits a judge to refuse to enforce unconscionable clauses or contracts.

The doctrine has rarely invalidated clauses in construction cases.[61] Clauses where it might have some utility, such as those of exculpation or indemnification, have been regulated by requiring a high degree of specificity or by statute.[62] Unconscionabiltiy was the basis for invalidating a provision in a subdivision purchase agreement that required, at the option of the developer, that disputes be referred to a court-appointed referee.[63] As seen in Section 30.03D, it can be the basis for challenging an agreement to arbitrate. The unconscionability doctrine is likely to be increasingly invoked in the future.

---

[59]See Section 18.04E for another illustration—requiring an owner to notify a bidder if it appears a bidding mistake has been made.

[60]U.C.C. § 2-302.

[61]*Arcwel Marine, Inc. v. Southwest Marine, Inc.*, 816 F.2d 468 (9th Cir.1987), cert. denied, 484 U.S. 1008 (1988) (exculpatory clause); State *Highway Admin. v. Greiner Eng'g Sciences, Inc.*, 83 Md.App. 621, 577 A.2d 363, cert. denied, 321 Md. 163, 582 A.2d 499 (1990) (no damage clause); *Curtis Elevator Co. v. Hampshire House, Inc.*, 142 N.J.Super. 537, 362 A.2d 73 (Law Div.1976) (strike clause); *S. Brooke Purll, Inc. v. Vailes*, 850 A.2d 1135 (D.C.2004) (liquidated damages clause); and *Mistry Prabhudas Manji Eng. Pvt. Ltd. v. Raytheon Engineers & Constructors, Inc.*, 213 F. Supp.2d 20 (D.Mass.2002) (Pennsylvania law) (combination of limitation of liability and waiver of consequential damages). The *Mistry* case is described in Section 15.03D.

[62]See Sections 31.05D and 31.05E.

[63]*Pardee Constr. Co. v. Superior Ct.*, 100 Cal.App.4th 1081, 123 Cal. Rptr.2d 288 (2002). But see *Woodside Homes of Calif. v. Superior Ct.*, 107 Cal.App.4th 723, 132 Cal.Rptr.2d 35 (2003) (came to the opposite conclusion in a similar case). See also *Lucier v. Williams*, 366 N.J.Super. 485, 841 A.2d 907 (App.Div.2004) (limitation of liability provision in home inspection contract unconscionable); *State ex rel. Vincent v. Schneider*, 194 S.W.3d 853 (Mo.2006) (arbitration clause unconscionable in part).

---

[56]*Rich & Whillock, Inc. v. Ashton Devt., Inc.*, 157 Cal.App.3d 1154, 204 Cal.Rptr. 86 (1984), reproduced in Section 27.13.

[57]*T.G.I. East Coast Constr. Co. v. Fireman's Fund Inc. Co.*, 534 F.Supp. 780 (S.D.N.Y.1982).

[58]*Paul Hardeman, Inc. v. Arkansas Power & Light Co.*, 380 F.Supp. 298 (E.D.Ark.1974). Cf. *Darwin Constr. Co., Inc. v. United States*, 811 F.2d 593 (Fed.Cir.1987), cert. denied, 484 U.S. 1008 (1988).

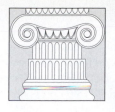

CHAPTER TWENTY

Contract Interpretation: Chronic Confusion

SECTION 20.01 Basic Objectives

The basic objective in contract interpretation is to determine the intention of the parties. However, within this relatively simple standard lurk many problems, some of which follow:

1. What can be examined to ascertain the intention of the parties?
2. Once the relevant sources are examined, how is the intention of the parties determined?
3. What if each party has different intentions?
4. What if one party knows of the other party's intention?
5. What if the parties have no particular intention about the matter in question?
6. Can the court go beyond these presumed intentions of the parties and interpret in accordance with what the court thinks the parties would have intended had they thought about it?
7. Can a court disregard the intention of the parties and base determination of the rights and duties of the parties on judicial notions of proper allocation of risk?

The problems can be even more complicated when the language has been selected not by any or all of the parties but by a third party such as the American Institute of Architects (AIA) or the Engineers Joint Contract Documents Committee (EJCDC). Should the intention of AIA personnel who selected the language be examined, if it can be ascertained? If it cannot, what should the law do when neither party had any intention whatsoever when it agreed to use AIA documents (not an uncommon phenomenon as to certain types of contract clauses)?

The preceding list is given merely to indicate that phrasing the test as the process of ascertaining the intention of the parties is deceptively simple, often hiding difficult interpretation problems.

SECTION 20.02 Language Interpretation

Words have no inherent meaning—they develop meanings because people who use them as tools of communication attach meanings to them. A judge asked to interpret contract terms could decide all interpretation questions simply by using a dictionary and choosing the definition that seemed most appropriate. However, even dictionary meanings are not exclusive. Choosing a definition can be a formidable task. Interpretation of contract terms should take into account the setting and function of the transaction and other matters not found in the contract or the dictionary. Courts seek to put themselves in the position of the contracting parties and determine what the contracting parties must have meant or intended when they used the language in question.

Yet judges and juries do not have absolute freedom to determine the meaning of words. This constraint may be traceable to the fear that juries, and sometimes trial judges, will be unduly sympathetic to a hard-luck story and too inclined to protect the party in the weaker bargaining position. Underlying this reluctance is the skepticism the law has toward the ability of the trial process to separate truth from falsehood, especially when parties differ in their testimony as to what transpired during the negotiations.

Reluctance to give absolute discretion to the fact finders may also be traceable in part to the fear that juries, and sometimes trial judges, may not realize the importance to the commercial world of attaching consistent, commercially accepted meanings to terms.

A. Plain Meaning Rule

The *plain meaning rule* has been the method employed to limit interpretation powers of trial judges and juries. A judge must first determine whether the words used by the parties have a plain meaning. If so, the judge cannot look beyond the document itself to determine what the parties meant when they used those particular words.

Of course, the language selected by the parties for inclusion in the contract is relevant, often crucial. It is the starting point for interpreting the contract. But unless the language has a plain meaning, the search is not limited to the contract language.

Most, though not all, courts invoke the plain meaning rule to bar evidence outside the writing (extrinsic evidence). Even if the plain meaning rule blocks the use of extrinsic evidence, the judge may look at the dictionary or dictionaries if she is not confident she can decide the issue based solely upon her linguistic experience.

Whether the plain meaning rule applies is a question of law. This issue is decided by the trial judge and not the fact finder, such as a jury impaneled to hear the case. Being a question of law allows any appellate court to decide this issue without extending any deference to the trial judge's decision.

If not barred by the plain meaning rule, extrinsic evidence can be used to determine the intention of the parties. This evidence highlights the contextual setting of the transaction. Such evidence can include the surrounding facts and circumstances, preliminary negotiations that do not run afoul of the parol evidence rule,[1] customs and practices of the industry, and the practices of the parties before the dispute arose. These will be discussed later in this Section.

There has been scholarly criticism of the plain meaning rule. Unbending use of the rule can generate a contract that neither party intended. Also, critics point to the deficiencies of human language. Even a search for dictionary meanings can involve a choice between different dictionaries and many dictionary meanings.

Yet the great majority of jurisdictions follow the plain meaning rule.[2] It is also followed more strictly than in private litigation by the United States Court of Appeals for the Federal Circuit.[3] This court hears appeals from the various federal boards of appeals in federal procurement cases and also from the United States Court of Federal Claims. In these disputes the courts are conscious of the need to protect public funds. Too much extrinsic evidence can upset the highly bureaucratic mechanisms that should protect the taxpayers.

The decision as to whether the plain meaning rule applies relates to the objective theory of contracts discussed in Section 5.06A. The plain meaning rule creates an objective standard. If it blocks examination of often subjective extrinsic evidence, the result should be a detached, objective interpretation. If it does not bar inquiry into the contextual meaning, the search is for the actual intention of the parties. This can be said to effectuate the subjective meaning of the parties. But that does not mean that undisclosed intentions prevail. The search is for the *common* intention of the parties, not some objective intention based upon the judge's interpretation, aided by the dictionary.

The plain meaning rule applies only to litigation. An arbitrator can examine extrinsic evidence, if she chooses, without first finding the language was not plain on its face.

B. Illustrative Federal Cases: *TEG-Paradigm Environmental, Inc. v. United States*

It is essential to see these rules of contract interpretation in actual cases. With that in mind, some cases from the federal procurement system will be described. They involve construction disputes that can also arise in state public contracts and in private construction projects. Finally, a case will be reproduced in part decided recently by the United States Court of Appeals for the Federal Circuit, the court that handles many federal procurement disputes.

The conflict between contractual provisions and trade practices can be shown in a construction contract context. In *Jowitt, Inc. v. United States*,[4] the contract provided that the contractor was to insulate all cold-air supply ducts. In a separate provision an exemption from the insulation requirement was provided for return ducts in ceiling spaces and ceilings which form plenums. The government contended that this exemption applied only to ceilings which form plenums.

---

[1] See Sections 11.04E and 19.01G.

[2] J. CALAMARI & J. PERILLO, CONTRACTS, § 3.10 at 151 (5th ed. 2003).

[3] *Travelers Cas. & Sur. of Am. v. United States*, 75 Fed.Cl. 696 (2007). For a critical discussion of the federal courts' adoption of the plain meaning rule, see Johnson, *Interpreting Government Contracts: Plain Meaning Precludes Extrinsic Evidence and Controls at the Federal Circuit*, 34 Pub. Cont.L.J. 635 (2005).

[4] 234 F.3d 1365 (Fed.Cir.2000).

To rebut the plain language of the specifications, the contractor introduced affidavits from other contractors that standard practice was to not insulate supply ducts in the above-ceiling space. The Court of Federal Claims (the trial court in claims against the federal government) refused to give any weight to the affidavits. Even if they were true as to industry practice, they could not be used to vary or contradict the plain language of the contract.

The Court of Appeals for the Federal Circuit affirmed. But it did state that the trade usage could be relevant to interpret a *term* used in the contract that differs from its ordinary meaning and that reasonable reliance on trade practice showed that the contract was susceptible to two interpretations.

This case shows the difference between interpreting what appears to be a clear, contractual *term* with the help of trade usage and simply saying that "this is not the way we do things in the trade." The court stated that the affidavits did not identify a specific term that needed trade usage to show how the parties were using it in their contract. There was no showing of an accepted industry meaning different from its ordinary meaning.

For example, in *Metric Constructors, Inc. v. National Aeronautics & Space Administration*,[5] trade usage testimony was permitted to interpret the contract requirement that "new lamps (i.e., light bulbs) shall be installed immediately prior to completion of the project." The dispute was whether all light bulbs (as contended by the government) or only broken or defective light bulbs (as argued by the contractor) needed to be replaced.

At this point, a case will be reproduced in part decided recently by the United States Court of Appeals for the Federal Circuit.

TEG-PARADIGM ENVIRONMENTAL, INC. v. UNITED STATES

United States Court of Appeals, Federal Circuit, 2006, 465 F.3d 1329

SCHALL, Circuit Judge.

TEG-Paradigm Environmental, Inc. ("TEG") entered into a contract with the United States Department of Housing and Urban Development ("HUD"). Pursuant to the contract, TEG agreed to perform asbestos abatement work at the Geneva Towers, an apartment complex, in San Francisco. After the contract work was completed, TEG submitted a claim to the contracting officer in which it sought an equitable adjustment in the contract price. In support of its claim, TEG asserted that it had been required to perform excessive cleaning and that it had been required to remove excessive quantities of asbestos. After the contracting officer denied the claim, TEG filed suit in the United States Court of Federal Claims under the Contract Disputes Act of 1978, 41 U.S.C. §§ 601–613 (2000).

[Ed. note: The court gave the procedural history of the case. The issue before the court was TEG's appeal from the Court of Federal Claims (the trial court) decision granting the government's motion for summary judgment on its two claims for breach of contract. It affirmed the trial court granting summary judgment in favor of the government.

The court then amplified the facts in the case as follows:

The Geneva Towers were two-high-rise apartment buildings in San Francisco. HUD acquired the buildings in 1991 and decided to implode them to make way for new development. . . . However the buildings contained asbestos, which had to be removed before implosion. . . . HUD solicited bids on a contract for asbestos abatement and TEG was awarded the contract on May 8, 1997, for a fixed price of $5,153,625.00.

Other salient facts will be given in the court's decision. The court then stated the rules for granting summary judgment. Contract interpretation in this case is a question of law and the appellate court need give no deference to the trial court's decision. It can decide the case *de novo* (from the beginning).

The court next gave the contentions of the parties. These will be referred to in the opinion that deals with the main issue, interpretation of the contract.]

---

[5]169 F.3d 747 (Fed.Cir.1999).

III.

When interpreting a contract, "'the language of [the] contract must be given that meaning that would be derived from the contract by a reasonably intelligent person acquainted with the contemporaneous circumstances.'" *Metric Constructors, Inc. v. Nat'l Aeronautics & Space Admin.*, 169 F.3d 747, 752 (Fed.Cir.1999) (quoting *Hol–Gar Mfg. Corp. v. United States*, 169 Ct.Cl. 384, 351 F.2d 972, 975 (Ct.Cl.1965)). When deriving this meaning, we begin with the contract's language. *Coast Fed. Bank, FSB v. United States*, 323 F.3d 1035, 1038 (Fed.Cir.2003) (en banc). When the contract's language is unambiguous it must be given its "plain and ordinary" meaning and the court may not look to extrinsic evidence to interpret its provisions. *Id.* at 1040; *McAbee Constr.*, 97 F.3d at 1435. Although extrinsic evidence may not be used to interpret an unambiguous contract provision, we have looked to it to confirm that the parties intended for the term to have its plain and ordinary meaning. *See Coast Fed. Bank.* 323 F.3d at 1040 (looking to contemporaneous evidence of the parties' understanding and "not[ing] that much of it is consistent with the [contract's] plain meaning"). When a provision in a contract is susceptible to more than one reasonable interpretation, it is ambiguous, *Edward R. Marden Corp. v. United States*, 803 F.2d 701, 705 (Fed.Cir.1986), and we may then resort to extrinsic evidence to resolve the ambiguity, *see McAbee*, 97 F.3d at 1435. We utilize extrinsic evidence to derive a construction that effectuates the parties' intent at the time they executed the contract. *See Dureiko v. United States*, 209 F.3d 1345, 1356 (Fed.Cir.2000).

Even when a contract is unambiguous, it may be appropriate to turn to one common form of extrinsic evidence—evidence of trade practice and custom. *Hunt Constr. Group, Inc. v. United States*, 281 F.3d 1369, 1373 (Fed.Cir.2002). We have stated that "evidence of trade practice may be useful in interpreting a contract term having an accepted industry meaning different from its ordinary meaning—even where the contract otherwise appears unambiguous—because the 'parties to a contract . . . can be their own lexicographers and . . . trade practice may serve that lexicographic function in some cases.'" *Id.* (quoting *Jowett, Inc. v. United States*, 234 F.3d 1365, 1368 (Fed. Cir.2000)). Trade practice and custom may not be used, however, "to create an ambiguity where a contract was not reasonably susceptible of differing interpretations at the time of contracting." *Metric Constructors*, 169 F.3d at 752.

The parol evidence rule provides a further limitation on the use of extrinsic evidence in interpreting contracts. Under the parol evidence rule, extrinsic evidence pre-dating a written agreement may not be used "to add to or otherwise modify the terms of a written agreement in instances where the written agreement has been adopted by the parties as an expression of their final understanding." *Barron Bancshares*, 366 F.3d at 1375 (citation and quotation marks omitted). However, extrinsic evidence such as prior agreements and documents will be considered part of a contract when they are incorporated into the contract. *See S. Cal. Fed. Sav. & Loan Ass'n v. United States*, 422 F.3d 1319, 1330 (Fed.Cir.2005). One common way to incorporate extrinsic evidence is through an integration clause that expressly incorporate the extrinsic evidence. *Id.; McAbee Constr. v. United States*, 97 F.3d 1431, 1434 (Fed.Cir.1996). Although the parol evidence rule bars the use of extrinsic evidence to supplement or modify a written agreement, the rule does not bar the use of extrinsic evidence to interpret the terms of a contract when the plain and ordinary meaning is not clear from the contract itself. *See* Restatement (Second) Contracts § 215 cmt. b (1981); 6–26 *Corbin on Contracts* § 579 (2006); Cibinic, Nash & Nagle, *supra*, at 199. Armed with these rules, we turn to the issues TEG raises on appeal.

IV.

We consider first TEG's claim that the Court of Federal Claims erred in holding that it was required to clean debris and residue from pores and cracks of the Geneva Towers under the contract's abatement standard. As seen, the provision of the contract containing the abatement standard for the Geneva Towers project, Section 2080, 4.3C, provided:

> Asbestos-containing materials applied to concrete, masonry, wood and nonporous surfaces, including, but not limited to, steel structural members (decks, beams and columns), pipes and tanks, shall be cleaned to a degree that no traces of debris or residue are visible by the Observation Services Contractor.

The Court of Federal Claims correctly identified two issues raised by the abatement standard. First, we must determine whether this standard requires the removal of asbestos within pores and cracks. Second, we must determine what asbestos-containing "debris or residue" means.

Based upon the plain language of the abatement standard, we conclude that the Court of Federal Claims did not err in ruling that TEG was required to remove asbestos within pores and cracks. The plain language of the contract indicates that it requires abatement to the point that there is no "debris or residue . . . visible." Thus, to the extent that "debris or residue" was "visible" within the pores and cracks of concrete or other porous surfaces, it had to be abated under the contract.

However, if the "debris or residue" was not "visible" within the pores and cracks, it was not required to be abated under the contract. Accordingly, we find that the plain and ordinary meaning of the abatement standard required TEG to remove visible asbestos within the pores and cracks of the towers.

As we did in *Coast Federal Bank,* we turn to extrinsic evidence, specifically, the course of dealing of the parties, to confirm that our interpretation of the plain and ordinary meaning was, in fact, the parties' understanding. *See Coast Fed. Bank,* 323 F.3d at 1040. The original specifications provided for two different abatement standards for friable and non-friable materials. As far as friable materials were concerned, the specifications expressly stated that materials must be cleaned "to a degree that no traces of debris or residue are visible." In contrast, the specifications provided that non-friable materials "shall be cleaned until no residue is visible other than that which is embedded in the pores, cracks, or other small voids below the surface of the non-friable material." Thus, the original specifications expressly allowed for the contractor to leave non-friable asbestos in pores and cracks. In a pre-bid conference call, TEG's representative stated that it was not clear which standard, friable or non-friable, would apply and that this was an important difference. TEG's representative noted, "It's a difference, because on one it has to be clean to a degree there's no trace; on the other, it's clean to a degree that material can still be embedded in pores, cracks and voids." In our view, the conference call demonstrates that TEG understood the visibility standard, which was eventually adopted for all asbestos abatement under the contract, to require that no asbestos remain in the pores and cracks.

We find unpersuasive TEG's argument that the Court of Federal Claims erred by failing to consider other pieces of evidence, including ATC's letter and Mr. Oberta's expert opinion. These documents could be considered evidence of trade practice and custom, which we have found appropriate to consider in some cases even when a contract is unambiguous. *See Hunt Constr.,* 281 F.3d at 1373. However, neither of these documents aids in the interpretation of a term of art in the asbestos abatement field. Rather each document offers an alternate explanation of the contract's abatement standard generally. Under *Hunt Construction,* it is not permissible to use these extrinsic sources to impart ambiguity into an otherwise unambiguous contract—they may only be used to interpret a term of art. *Id.* at 1369; *see also Metric Constructors,* 169 F.3d at 752. Given the clarity of the meaning from the language and the parties' pre-contractual negotiations, none of the extrinsic evidence cited by TEG carries weight.

We are also not swayed by TEG's argument that Section 2080, 4.3C requires only that "surfaces" be abated and that "surfaces" do not include pores and cracks. TEG turns to a common usage dictionary that defines a "surface" as "[t]he outer-face, outside, or exterior boundary of a thing; outermost or uppermost layer or area." Appellant's Br. at 21 (quoting *Ramdom House Webster's College Dictionary* 1314 (2nd ed.1999)). We reject TEG's argument based on this definition that a "surface," as used in Section 2080, 4.3C, is only the "outer-face" of the concrete and not the pores and cracks. The definition upon which TEG relies does not expressly state that "surfaces" do not include pores and cracks. Further, if we were to accept TEG's argument we would have to decide how small a crack or pore had to be in order to be excluded from "surfaces." The Court of Federal Claims correctly noted that we "would be left to quarreling over the depth of recess needed to differentiate a crack or pore from a smooth surface" if we adopted TEG's interpretation. *TEG–Paradigm* slip op. at 13. We also find that TEG's "surfaces" argument is weak in comparison to the evidence of the parties' understanding of the language in the contract as demonstrated by the conference call. Thus, we believe that the strict visibility abatement standard is more in line with the parties' contemporaneous understanding of Section 2080, 4.3C.

Turning to the second issue relevant to TEG's abatement standard claim, the meaning of "debris or residue," we see no error in the Court of Federal Claims's holding that, under the contract, any dust or powder found on inspection was assumed to be asbestos-containing "debris or residue" that had to be abated. "Debris" "residue" are not defined in the contract. As previously noted, evidence of trade custom may be used to interpret terms of art such as "debris" and "residue." *See Hunt Constr.,* 281 F.3d at 1373. The ASTM standard for asbestos abatement provides that debris and residue is "assumed" to contain asbestos. *TEG–Paradigm,* slip op. at 13. Therefore, we agree with the Court of Federal Claims that trade practice and custom demonstrates that in the asbestos abatement field any "debris and residue" found is assumed to contain asbestos. *Id.* Thus, we affirm the Court of Federal Claims's holding that the contract required TEG to clean all visible powder and dust found on inspection, including powder and dust in cracks and pores.

V

[Ed. note: The court rejected TEG's contention that its work plan was part of the contract and took precedence over the specifications. Bidders were required to submit a work plan "stating the details of the bidder's engineering controls and work procedures." The court noted that the contract did not state the work plan was integrated into the contract and superceded the specifications. Other documents were specifically incorporated into the contract. The mere fact that the work plan was physically attached to the contract did not in itself make it part of the contract. The court held the requirement of the work plan was a preaward submission "to aid the government in assessing TEG's ability to perform the contract."]

CONCLUSION

For the foregoing reasons, we therefore affirm the decision of the Court of Federal Claims granting summary judgment in favor of the government on TEG's claims for breach of contract.

COSTS

Each party shall bear its own costs.
AFFIRMED.

C. Extrinsic Evidence

Surrounding facts and circumstances often determine how language is interpreted. The relevant surrounding facts and circumstances typically are those that existed at the time the contract was made.[6] However, to interpret contract language sometimes courts look at circumstances that existed at some time during performance.[7]

Evidence of the surrounding facts and circumstances that courts will not examine is any undisclosed intentions of the parties they claim existed at the time they made the contract. If these intentions are made known to the other party, they may be relevant. Clearly disclosed intention is relevant.[8]

The Uniform Commercial Code (U.C.C.) encourages the use of trade usages.[9] But the question of whether such evidence can be used in the face of a writing that seems to point in another direction is not clear.[10] Cases governed by the U.C.C. have not been consistent when facing a conflict between trade usage and language in the contract with a clear meaning.[11]

As an illustration, *Fletcher-Harlee Corp. v. Pote Construction Contractors, Inc.* noted that increasingly common commercial practices aid interpretation of contracts. Yet, despite a finding that commonly subcontractors make firm offers that can be relied upon, a statement on the price quotation that the price is not firm has greater weight than commercial practices.[12] Here the plain language trumped commercial practices.

Such a clash is inevitable, because both contract language and trade usage are powerful sources of contractual obligations.

D. Practical Interpretation

Statements and, more important, acts of the parties before the dispute arose may indicate how the parties interpreted the language. Courts often invoke and give considerable weight to those acts under what is called *practical interpretation*. The practices of the parties often indicate their intention at the time the contract was made. For example, making two progress payments, without a showing that the work complied with the contract, was considered to manifest an intention that the contractor was entitled to progress payments despite noncompliance.[13] Likewise, the prime contractor periodically billing the owner in accordance with certain unit prices and the owner paying these billings indicated that the unit prices were correct.[14] The contractor doing what the owner had directed without complaint indicated the contractor's acquiescence in the owner's interpretation.[15] Note that the doctrine of practical interpretation applies only to language susceptible to more than one interpretation.[16] However, this rarely is a difficult obstacle.

---

[6]*Metric Constructors v. United States*, 314 F.3d 578 (Fed.Cir.2002) (interpreting a settlement release).

[7]*Contracting & Material Co. v. City of Chicago*, 20 Ill.App.3d 684, 314 N.E.2d 598 (1974). The holding was reversed, not on the admissibility of the evidence but on finding the evidence irrelevant because of a strict interpretation given to the contract clause in question. See 64 Ill.2d 21, 349 N.E.2d 389 (1976).

[8]*United States v. F. D. Rich Co.*, 434 F.2d 855 (9th Cir.1970).

[9]U.C.C. § 1-303 formerly § 1-205.

[10]Id. at §§ 1-303, 2-202.

[11]E. A. FARNSWORTH, CONTRACTS § 7.13 (4th ed. 2004).

[12]482 F.3d 247 (3d Cir.2007). See Section 28.02.

[13]*Giem v. Searles*, 470 S.W.2d 327 (Ky.1971).

[14]*Berry v. Blackard Constr. Co.*, 13 Ill.App.3d 768, 300 N.E.2d 627 (1973).

[15]*Bulley & Andrews, Inc. v. Symons Corp.*, 25 Ill.App.3d 696, 323 N.E.2d 806 (1975).

[16]*Dana Corp. v. United States*, 200 Ct.Cl. 200, 470 F.2d 1032 (1972).

E. Canons of Interpretation

Not infrequently the surrounding circumstances and the predispute conduct of the parties provide little assistance. When this occurs, courts sometimes resort to secondary assistance, called "canons of interpretation." These canons are interpretation guides. Some are used frequently.

***Expressio* Rule.** *Expressio unius est exclusio alterius,* "the expression of one thing excludes the others," excludes an item from relevance when there is a list of items and the item in question is not expressed. For example, suppose a party were excused from performance in the event of strikes, fire, explosion, storms, or war. Under the *expressio guide,* the occurrence of an event not mentioned, such as a drought, would not excuse performance. Where the parties have expressed five justifiable excuses, they must have intended to exclude all other excuses for failure to perform. This guide is sometimes harshly applied and does not give realistic recognition of the difficulty of drafting a complete list of events.

Ejusdem Generis. Another guide, *ejusdem generis,* states that the meaning of a general term in a contract is limited by the specific illustrations that accompany it. Anything not specifically mentioned must be "of that sort," that is, similar in meaning to those things that are. For example, damages payable for harm to crops, trees, fences, and premises would not include depreciation in market value of the land. The particular item in question is too unrelated to those items listed to be covered by the contract terms.

Reasonableness. As to other guides, one court stated, "where one interpretation makes a contract unreasonable or such that a prudent person would not normally contract under such circumstances, but another interpretation equally consistent with the language would make it reasonable, fair and just, the latter interpretation would apply."[17] This does not give the court the power to rewrite the language.[18] Nor does it preclude the contracting parties from agreeing to language that a court might consider unreasonable. However, for an unreasonable

interpretation to be selected, the language must make clear that this is what the parties intended.

Another illustration can be seen in *Medlin Construction Group, Ltd. v. United States.*[19] In this case the specifications gave two alternative performance options to the contractor but the drawings illustrated only one of the options. The government contended that the drawings narrowed the scope of the options given in the specifications. The contractor countered that the drawings merely illustrated one of the options. That did not eliminate its right to use the other option. The court concluded that the contractor's interpretation was the only reasonable interpretation because it preserved both options. The government's interpretation would have read one of the options out of the specifications.

Fairness. Another court required that the language be interpreted "so as not to put one side at the mere will or mercy of the other."[20] Another stated that a particular interpretation should be rejected because it "would certainly be both unconscionable and inequitable, and the recognized rule is that the interpretation that makes a contract fair and reasonable will be preferred to one leading to a harsh and unreasonable result."[21]

Contra Proferentem. Another important interpretation guide, *contra proferentem,* "against the one who puts it forward" (the drafter of the language) interprets ambiguous language against the party who selected the language or supplied the contract. Usually this guide is applied not to negotiated contracts but to those mainly prepared in advance by one party and presented to the other on a take-it-or-leave-it basis (adhesion contracts).

One basis for this guide is to penalize the party who created the ambiguity. Another and perhaps more important rationale is the necessity of protecting the reasonable expectations of the party who had no choice in preparing the contract or choosing the language. This rationale had its genesis in the interpretation of insurance contracts. Insureds frequently received protection despite language that appeared to preclude insurance coverage. This

---

[17]*Elte, Inc. v. S. S. Mullen, Inc.,* 469 F.2d 1127, 1131 (9th Cir.1972).

[18]As seen in Section 20.04, the court can reform the written contract if it did not express the intention of the parties.

[19]449 F.3d 1195 (Fed.Cir.2006).

[20]*Contra Costa County Flood Control & Water Conservation Dist. v. United States,* 206 Ct.Cl. 413, 512 F.2d 1094, 1098 (1975).

[21]*Glassman Constr. Co., Inc. v. Maryland City Plaza, Inc.,* 371 F.Supp. 1154, 1159 n.3 (D.Md.1974).

preference recognizes that insurance policies are difficult to read and understand and the insured's expectations as to protection are derived principally from advertising, sales literature, and sales people's representations. Giving preference to the insured's expectations may also rest on the judicial conclusion that insurance companies frequently exclude risks that should be covered.

The *contra proferentem* guide can break a tie when all other evidence is either inconclusive or unpersuasive. The court may mention it to bolster an interpretation that has already been determined by other evidence.

F. Industry Contracts

Design and construction work frequently are performed after parties have assented to a standard preprepared contract documents created by associations such as the American Institute of Architects (AIA) or the group of engineering associations who created the Engineers Joint Contract Documents Committee (EJCDC). Sometimes parties agree to such printed contracts without carefully considering the language. On other occasions, the parties carefully considered all or some of the language before entering into the contract. Sometimes the parties have dealt with and understood the language, and on other occasions one party may be unfamiliar with the terminology and concepts employed in the standard contract.

A look at a few cases in the construction context may be instructive. *Durand Associates, Inc. v. Guardian Investment Co.*[22] construed an AIA standard contract against the engineer who had supplied it despite the fact that the owner was an investment company about to build a medical clinic that later was changed to an apartment complex. Although the owners in the case seemed to be experienced business people, evidently the court felt that their knowledge and experience did not equip them to carefully appraise the language of the AIA document. Yet the owners read the document and made some revisions that were accepted by the engineer.

Other cases have involved disputes between owners and contractors, with varied outcomes. One case noted that the facts that the contractors were sophisticated business people and the owner was legally represented indicated that the document should not be construed against

either party.[23] Another case pointed to the contract's having been the result of arm's-length bargaining.[24] A third interpreted A201 against the owner.[25] Another interpreted a subcontract, probably an AIA Doc. A 401, against the prime contractor.[26]

Although the contract involved was not one published by a professional society, *W. C. James, Inc. v. Phillips Petroleum Co.*,[27] is instructive. The plaintiff, a large pipeline contractor, contracted with Phillips, a large gasoline supplier. The contractor contended that the contract had been preprepared, or boilerplated, and presented to bidders on a take-it-or-leave-it basis. The trial court noted that the clause in question—waiving damages for delay— was common in construction work. The trial court stated that the contractor entered into the contract voluntarily, intelligently, and knowingly. The trial court upheld the clause, stating that it was not unfair and that the contract was not one of adhesion. The Federal Circuit Court of Appeals affirmed the trial court, noting that the pipeline contractor was of sufficient size, even in relationship to Phillips, so that it could not seek relief on the grounds of an adhesion contract.

Probably the pipeline contractor in the *James* case had no choice. Yet if the contractor was aware that it was risking delay damages, it could adjust its price and take this risk into account. It is not in the same position as a consumer, who is usually not in the position to adjust to the risk that the contract language seeks to place on her.

The preceding discussion has demonstrated the need for an analysis that will recognize some of the particular problems of dealing with standard contracts published by the AIA or EJCDC. First, it is important to look at the surrounding facts and circumstances that led to the use of the standard agreement. One should not always assume that the architect dictates that an AIA document be used for either design or construction services. Nor should one assume that an owner dictates the use of an AIA document for construction services. An AIA document may be selected by each party to the contract because of the

---

[22]186 Neb. 349, 183 N.W.2d 246 (1971).

[23]*Robinhorne Constr. Corp. v. Snyder*, 113 Ill.App.2d 288, 251 N.E.2d 641 (1969), aff'd, 47 Ill.2d 349, 265 N.E.2d 670 (1970).

[24]*Cree Coaches, Inc. v. Panel Suppliers, Inc.*, 384 Mich. 646, 186 N.W.2d 335 (1971).

[25]*Osolo School Buildings, Inc. v. Thorleif Larsen and Son of Indiana, Inc.*, 473 N.E.2d 643 (Ind.App.1985).

[26]*Katzner v. Kelleher Constr.*, 545 N.W.2d 378 (Minn.1996).

[27]485 F.2d 22 (10th Cir.1973).

document's reputation for fairness or because of familiarity. This is more likely when AIA construction documents are used, unless the owner has dealt with design services before.

Second, changes are frequently made in AIA documents, principally in the areas of payments, changes, indemnification, arbitration, and the responsibilities of the design professional. It would be administratively inconvenient to use different standards of interpretation when substantial changes have taken place. Where changes have been substantial, the parties have likely considered the entire document, in some cases even jointly negotiating the agreement. If so, the agreement should be interpreted neutrally. Where one party has used some clauses from a standard agreement that favor it and then incorporated other clauses more favorable than those contained in the standard agreement, clearly the language should be interpreted against that party.

Third, it is possible to interpret any AIA B-series agreements—those that deal with design services—in favor of the client, because the AIA receives no input from other organizations for these documents and is likely to be drafting with the best interests of architects in mind.[28]

Interpretation of AIA A-series documents requires an understanding of the endorsement system used by the AIA. Originally, the AIA sought and obtained the endorsement of A101/201 from the many contractor associations that represented prime contractors and subcontractors. The involvement of the many participants in the construction process gave the documents the appearance of a fair and balanced contract.

This endorsement process was costly and time-consuming, and it deprived the AIA of total control. Beginning in the 1950s, the AIA sought endorsement of A101/201 *only* from the Associated General Contractors of America (AGCA, better known as the AGC). The AGC also published its own documents, but their use never rivaled those of the AIA.

A401, the A-series document for subcontracting, was endorsed by the American Subcontractor Association (ASA) and the Associated Specialty Contractors, Inc. (ASC), the principal umbrella organizations for the many subcontractor associations.

The A101/201 endorsement process involved exchange of drafts and discussions between AIA and AGC. While the AIA made some concessions to the AGC to obtain endorsement of A101/201, the principal bargaining power belonged to the AIA. A similar process led to endorsement of A401 by the ASA and the ASC. The AGC did not participate in the A401 process because it saw no reason for the AIA to be involved in the relationship between primes and subcontracts.

For reasons that cannot be explored here, in 2007 the AGC,[29] ASA, and ASC decided not to endorse AIA documents. (It continued to endorse construction documents published by the EJCDC.) Led by the AGC, a group of owner, contractor, subcontractor, and surety associations published ConsensusDOCS. This group states it has published more than 70 documents. It hopes to compete with the AIA. The future will tell whether it will dethrone the AIA from its dominant position.

Contracts drafted by trade associations can create surrogate bargaining. For example, contracts prepared by associations that represent buyers and sellers of certain products can be the result that would be reached if individual buyers and sellers met across the bargaining table. Such standard contracts would be interpreted neutrally or at least in a manner similar to individually made contracts.

During the endorsement period, that is, until 2007, AIA A-series documents could be looked upon as imperfect surrogate bargaining. The parties who use A-series documents are owners and prime contractors. The AGC could be said to represent the prime contractors, and the AIA could be said to represent and look after the interests of the owners. But the interests of architects and owners do not always coincide. On many issues the AIA could be said to be looking out for the owner. But on others, such as role and responsibility of the architect, the interests of the owner and the architect diverge. Yet, despite the imperfect model even in the pre-2007 period, the courts could and often did take a neutral position in resolving interpretation issues.

Since 2007 it is no longer possible to use even the imperfect surrogate bargaining model. Yet the prestige of the AIA and of the architectural profession and the appearance of neutrality will very likely lead to a neutral interpretation, though preferring an interpretation that

---

[28]Two cases in which this has been done are *Malo v. Gilman*, 177 Ind. App. 365, 379 N.E.2d 554 (1978) and *Kostohryz v. McGuire*, 298 Minn. 513, 212 N.W.2d 850 (1973).

[29]See Construction News for a press release of October 12, 2007 announcing refusal to endorse. See the AGC website at www.agc.org.

marginally would favor the contractor. For historical reasons the law might still see the AIA as favoring the owner.

In the pre-2007 period only the AIA, the ASC, and the ASA were involved in A401. The AGC was not part of the process. A401 could not be considered the result of surrogate bargaining. The architects certainly could not be said to represent the prime contractors. In A401 the AIA did act neutrally. Because primes usually possess stronger bargaining power than subcontractors (subcontractors often make up for this by their muscle in state legislatures), it is likely that the law would interpret A401 at least neutrally, if not in favour of the subcontractor.[30]

The subcontractors no longer participate or endorse. Now it can be said that the AIA is neutral in publishing A401. It seeks to find a fair solution. It will try to recognize the interests and needs of those who will use this document. If so, the law should be neutral in its interpretation of A401.

Whether the emergence of the ConcensusDOCS will result in fewer users of A401 remains to be seen. Yet the involvement of the AGC in the ConcensusDOCS but not in A401 may mean subcontractors will prefer A401 even if they are not part of the A401 process.

Fourth, courts usually seek to find the intention of the contracting parties as the lodestar of contract interpretation. However, it is not uncommon for neither party to an AIA document to have any intention whatsoever as to certain clauses and their function when they have used an AIA document as the basis for their agreement. Does that mean that parties should be allowed to introduce any evidence of the intention of AIA representatives, if any can be obtained? This is analogous to the use of legislative history by courts in interpreting statutes. In 1999, the AIA published its Commentary on AIA Document A201-1997.[31] As yet the AIA has not published a commentary on documents issued in 2007. Letters or affidavits by AIA officials can be used.

Such evidence is untrustworthy at best, and the parties would have no opportunity to question those who have sought to describe the AIA's intention. It is probably best for courts to ignore this evidence and decide intention questions based on common law principles and the hypothetical intention of the parties—what they would have intended.

A Florida court faced this problem when it was asked to interpret the meaning of an applicable building code. Expert testimony was submitted by both parties. The architect's expert had served as manager of engineering of the Southern Building Code Congress and had been responsible for building code changes, hearings, plan reviews, and code interpretations. In effect, that expert testified as to what the drafters had in mind when they created the code. The design, he stated, was consistent with the *intent* of the code. However, the court concluded that the most important evidence to interpret the code was the interpretation given by the code official who had authority to implement the code.[32]

G. Particular Clauses

Courts will apply particular scrutiny to contract clauses that appear to be one-sided or that could encourage careless conduct. Exculpatory clauses relieve one party from liability for its own negligence.[33] They are not enforced unless evidence of the other party's consent to accept the risk is sufficiently clear. Even if the language is clear, these clauses are not enforced if the conduct in question has been grossly negligent. Indemnification clauses that seek to indemnify the indemnitee against the consequences of its negligence must also convey that intent in the language selected.[34]

H. Bidding Process—Patent Ambiguity: *Newsom v. United States*

The competitive bidding process has developed special rules. Prospective bidders are given a complex set of construction documents that have been prepared over a long period of time. Bidders are asked to review the materials within the short period of time available to them before submitting a bid. They are also asked to report any errors or inconsistencies they have observed. This laudable

---

[30]But see *Katzner v. Kelleher Constr.*, supra note 26 (appeared to construe the pre-2007 A401 in favor of the subcontractor).

[31]It can be obtained on its Web site, www.aia.org. See also J. SWEET & J. SWEET, SWEET ON CONSTRUCTION INDUSTRY CONTRACTS: MAJOR AIA DOCUMENTS (4th ed. 1999). The 5th edition is scheduled to be published in the Fall of 2008.

[32]*Edward J. Seibert, A.I.A., Architect and Planner, P.A. v. Bayport Beach & Tennis Club Ass'n., Inc.*, 573 So.2d 889 (Fla.Dist.Ct.App.1990), review denied, 583 So.2d 1034 (Fla.1991).

[33]See Sections 15.03D, 31.05A.

[34]See Sections 31.05D and E.

attempt to catch errors and invoke the knowledge and skill of the contractor often generates problems. Many errors are detected through submittal review while the work is in progress.

One way of dealing with interpreting construction contracts in this context is to accept the interpretation that a reasonably prudent bidder would have given.[35] However, courts have had to struggle, sometimes painfully, with the question of whether, because the ambiguity was patent and glaring, the contractor should have sought clarification before bidding.[36] The following case explores this problem.

NEWSOM v. UNITED STATES

United States Court of Claims, 1982. 676 F.2d 647.
[Ed. note: Footnotes renumbered and some omitted.]

SMITH, Judge.

This case is an appeal by petitioner, George E. Newsom, of a decision of the Veterans Administration Board of Contract Appeals (board). The board found that certain parts of the contract for hospital improvements were patently ambiguous and that, having failed to consult with the contracting officer about the ambiguities, petitioner was barred from recovering for work done beyond that required under petitioner's interpretation of the contract. We affirm the decision of the board.

On August 28, 1978, the Veterans Administration (VA) issued an invitation for bids for building mediprep and janitor rooms in the VA hospital at Knoxville, Iowa. Drawings and specifications for the work to be done were supplied to the prospective bidders.

Paragraphs 4, 5, and 6 of the specifications described, respectively, buildings 81, 82, and 85. Each paragraph had two parts: the first described the first floor of the building and referenced page 7 of the drawings; the second described the second floor of the building and referenced page 8 of the drawings. Conversely, the caption block on page 7 of the drawings indicated that it described work for all three buildings, 81, 82, and 85. However, page 8 of the drawings indicated only building 85. Petitioner at no time inquired about this discrepancy.

As a consequence, petitioner included in his bid the costs of the second floor of building 85 only. He was the low bidder and the contract was awarded to him on October 13, 1978. It was not until March 29, 1979, that the parties realized that there was a discrepancy between what the VA had intended and what petitioner had understood. Petitioner then did the work as intended by the VA at an additional cost of $14,600, and he appealed the decision of the contracting officer denying relief to the Veterans Administration Board of Contract Appeals. The board held against petitioner on the ground that the error on page 8 of the drawings was a patent ambiguity which imposed on the contractor a duty to inquire about it. Petitioner now appeals that finding to this court under the Contract Disputes Act.

The doctrine of patent ambiguity is an exception to the general rule of *contra proferentem* which requires that a contract be construed against the party who wrote it. If a patent ambiguity is found in a contract, the contractor has a duty to inquire of the contracting officer the true meaning of the contract before submitting a bid.[37] This prevents contractors from taking advantage of the Government; it protects other bidders by ensuring that all bidders bid on the same specifications; and it materially aids the administration of Government contracts by requiring that ambiguities be raised before the contract is bid on, thus avoiding costly litigation after the fact. It is therefore important that we give effect to the patent ambiguity doctrine in appropriate situations.

The existence of a patent ambiguity is a question of contractual interpretation which must be decided de novo by the court. This determination cannot be made on the basis of a single general rule, however. Rather, it is a case-by-case judgment based

---

[35]*Corbetta Constr. Co. v. United States*, 198 Ct.Cl. 712, 461 F.2d 1330 (1972).

[36]*Zinger Constr. Co., Inc. v. United States*, 807 F.2d 979 (Fed.Cir. 1986) (design would not produce an operational system; duty to inquire). New Jersey adopted the patent ambiguity rule in *D'Annunzio Bros., Inc. v. New Jersey Transit Corp.*, 245 N.J.Super. 527, 586 A.2d 301 (App. Div.1991). A contractor's failure to seek clarification of a patent ambiguity will not bar its recovery if the government also knew of the ambiguity but did not notify the bidders of it. *Metcalf Constr. Co., Inc. v. United States*, 53 Fed. Cl. 617 (2002).

[37]*Beacon Constr. Co. v. United States*, 161 Ct.Cl. 1, 6, 314 F.2d 501, 504 (1963); *Blount Bros. Constr. Co. v. United States*, 171 Ct.Cl. 478, 495–96, 346 F.2d 962, 971–72 (1965).

on an objective standard. In coming to our decision, we are bound neither by the legal conclusions of the board, nor by the subjective beliefs of the contractor, subcontractors, or resident engineer as to the obviousness of the ambiguity.

The analytical framework for cases like the instant one was set out authoritatively in *Mountain Home Contractors v. United States*.[38] It mandated a two-step analysis. First, the court must ask whether the ambiguity was patent. This is not a simple yes–no proposition but involves placing the contractual language at a point along a spectrum: Is it so glaring as to raise a duty to inquire? Only if the court decides that the ambiguity was not patent does it reach the question whether a plaintiff's interpretation was reasonable. The existence of a patent ambiguity in itself raises the duty of inquiry, regardless of the reasonableness . . . of the contractor's interpretation. It is crucial to bear in mind this analytical framework. The court may not consider the reasonableness of the contractor's interpretation, if at all, until it has determined that a patent ambiguity did not exist.[39]

Examining the contract itself, we find that a patent ambiguity existed. Two parts of the contract said very different things: the specification required construction on the second floors of buildings 81, 82, and 85, whereas the drawings required construction on the second floor of only building 85. It is impossible from the words of the contract to determine what was really meant. The contractor speculated that it meant that part of the project had been dropped along the way. Looking at the same language, the Government can insist that it was clearly a drafting error. We do not consider which interpretation is correct; at this stage we determine only whether there was an ambiguity. What is significant about the differing interpretations is that neither does away with the contractor's ambiguity or internal contradiction. There is simply no way to decide what to do on the second floors of buildings 81 and 82 without recognizing that the contract also indicates otherwise.

Mountain Home, discussed above, involved a very similar ambiguity. The specifications ordered inclusion of kitchen fans in certain housing units, but the drawings appeared to indicate that kitchen fans were not to be installed. There is a crucial difference between that case and this, however. In *Mountain Home* the indication on the drawing was that the fans were to be under an alternate bid. Thus, the drawings indicated that the Government was reserving the option of either including the fans in the main contract or ordering them separately. The drawings did not state that the fans were simply not to be included. The *Mountain Home* contract, therefore, was susceptible of an interpretation which did not leave significant ambiguities or internal contradictions. Here, even petitioner's interpretation acknowledges that the contract is not internally consistent. Petitioner's interpretation explains the reason for the inconsistency but does not eliminate it.

We recognize that the instant case does not represent a difference in kind from the *Mountain Home* facts, but this area of the law involves a case-by-case determination of placement along a spectrum. In our opinion, this case is closer to *Beacon Construction*[40] than to *Mountain Home*. In *Beacon Construction*, the specifications stated only that "weatherstrip shall be provided for all doors," while the drawings describing the weatherstripping clearly indicated weatherstripping around the windows as well. The conflict between the specifications and the drawings was direct, as in the instant case. And the court was not swayed by the mere fact that the contractor was able to come up with a highly plausible interpretation of the ambiguity. No interpretation could in *Beacon Construction*, or can in the instant case, eliminate the substantial, obvious conflict between the drawings and the specifications.

Finally, we emphasize the negligible time and the ease of effort required to make inquiry of the contracting officer compared with the costs of erroneous interpretation, including protracted litigation. While the court by no means wishes to condone sloppy drafting by the Government, it must recognize the value and importance of a duty of inquiry in achieving fair and expeditious administration of Government contracts.

Accordingly, on consideration of the submissions, and after hearing oral argument, the decision of the Veterans Administration Board of Contract Appeals is
AFFIRMED.

---

[38]*Mountain Home Contractors v. United States*, 192 Ct.Cl. 16, 20–21, 425 F.2d 1260, 1263 (1970).

[39]If the court finds that a patent ambiguity did not exist, the reasonableness of the contractor's interpretation becomes crucial in deciding whether the normal contra proferentem rule applies.

[40]*Beacon Constr. Co. v. United States*, supra note 37, 161 Ct.Cl. at 4–5, 314 F.2d at 502–03.

Note the difference between the drawings and the specifications in the *Newsom* case. Often contracts seek to deal with this problem by incorporating a precedence-of-documents clause, which gives more weight or precedent to one document than to others. This method of dealing with ambiguity in the construction context is discussed in Section 20.03.

Inasmuch as many of the disputes that have dealt with the duty of the contractor to draw attention to defective design have, such as in the *Newsom* case, come before a federal contracting agency appeal board, it may be useful to look at criteria that have been developed by those boards to determine whether a defect should have been brought to the attention of the owner. Some of these criteria are as follows:

1. Did other bidders discover the error and seek clarification?
2. Did skilled professionals for the agency detect the error?
3. Were the construction documents so complicated that a small detail could have been easily missed?
4. Was the cost of correcting the defect a relatively small part of the contract price?
5. Did the defect occur in one item out of many items involved in the bid?
6. Was there a single prime contractor or many (a "multiprime" job)?
7. Will the contractor make a profit from a failure to inquire?

Interpretation guides are only guides. Courts look principally at the language, the surrounding facts and circumstances, and the acts of the parties. Nevertheless, interpretation guides are useful in close cases and may provide the court with a rationalization for a result already achieved.

SECTION 20.03 Resolving Conflicts and Inconsistencies

A. Within the Written Agreement

The difficult process of contract interpretation requires a court to put itself in the position of the parties and determine the parties' intentions at the time of the agreement. This process is complicated by the modern tendency for longer agreements. Often, long agreements are not carefully examined to avoid conflicts and inconsistencies. This is especially true where a set of standardized general conditions or contract terms is appended to a letter agreement that expresses the basic elements of the contract.

Courts generally interpret the contract as a whole and, wherever possible, seek to reconcile what may appear to be conflicting provisions. It is assumed that every provision was intended to have some effect. Yet this process of reconciliation may not accomplish its objective, and it may be necessary to prefer one provision over another.

Where one clause deals generally with a problem and another deals more specifically with the same problem, the specific takes precedence over the general. The specific clause is likely to better indicate the intention of the parties. For example, suppose a building contract specified that all disputes were subject to arbitration but also stated in a different paragraph that disputes as to aesthetic effect were to be resolved by the design professional whose decision was to be final. It is likely that the parties intended the specific clause dealing with nonreviewability of specific decisions to control the general clause.

If parties have expressed themselves specifically, their failure to change the general clause is likely due to a desire to avoid cluttering up the contract with exceptions and provisos. Also, the parties may not have noticed the discrepancy.

Later clauses take precedence over earlier ones. Although this is rarely used, it is premised on the assumption that parties sharpen their intentions as they proceed, much as an agreement made Tuesday displaces one made the preceding Monday.

Operative clauses take precedence over "whereas" clauses that seek to give the background of the transaction. As to inconsistency between printed, typed, and handwritten provisions, handwritten will be preferred over typed, and typed preferred to printed. These preferences are premised on the assumption that the parties express themselves more accurately when they take the trouble to express themselves by hand. Likewise, specially typed provisions on a printed form are more likely than the printed provisions to indicate the intention of the parties.

Some of these rules may have to be reevaluated in the light of the frequent use of contracts printed by word processors. Such agreements may appear to have been typed particularly for this transaction but may have actually been prepared for a variety of transactions, just like printed contract forms. The test in seeking to reconcile conflicting language within a written document should be whether some language was specially prepared for this transaction and indicated with greater accuracy the intention of the parties.

These priorities are simply guides to help the court resolve these questions. Guides to the meaning of language *within* a document are similar to the canons of interpretation that were explored in Section 20.02E. Each party can usually point to guides that support its position. What must be done is to focus on Section 20.01 and its emphasis on the intention of the parties. Which meaning is most likely to accomplish those objectives without grossly distorting the language?

B. Between Documents: *Unicon Management Corp. v. United States; Hensel Phelps Construction Co. v. United States*

Construction contracts present particularly difficult interpretation questions because of the number and complexity of contract documents and the frequent incorporation of bulky specifications by reference. Suppose one document calls for work that another does not specifically require. Suppose one document sets up procedures for changes different from those another document sets forth. At the outset, it may be instructive to reproduce a case dealing with this problem.

UNICON MANAGEMENT CORP. v. UNITED STATES

United States Court of Claims, 1967. 375 F.2d 804.
[Ed. note: Footnotes renumbered and some omitted.]

DAVIS, Judge.

In March 1959 the contractor agreed with the Corps of Engineers to construct, for a fixed price, two phases of the Missile Master Facilities near Pittsburgh. The current claim is that the Government required plaintiff to install in one room a steel-plate flooring which was not called for by the plans and specifications. The contracting officer and the Armed Services Board of Contract Appeals (65-1 BCA para. 4775) refused the demand for an equitable adjustment and this suit was brought. The problem arises because the most pertinent specification, if read alone, could be said to contemplate a wholly concrete rather than a partial steel-plate floor, while the most pertinent drawings, if read alone, direct the steel-plate covering. The contractor resolves the difficulty by relying, mainly, on the contractual clause that "in case of difference between drawings and specifications, the specifications shall govern." The Board and the defendant invoke another provision that "anything mentioned in the specifications and not shown on the drawings, or shown on the drawings and not mentioned in the specifications, shall be of like effect as if shown or mentioned in both."[41]

Since the question is one of contract interpretation, we are free to decide the matter for ourselves.

The room was Equipment Room No. 1 (also designated as Room 117) in one of the buildings being erected by plaintiff. To secure heavy equipment, this space was to have pallets along the floor with parallel cable trenches for bringing electric current to the machines. Steel beams were to be used in and along the floor in connection with these pallets. It is agreed that the top cover of the floor, apart from the portions devoted to the pallets and the trenches, was to be a resilient floor tile. There was also to be a concrete base. The dispute is whether the tile was to be laid directly on this concrete or over a quarter-inch steel plate above the concrete. The specifications, in a section of the Technical Provisions on miscellaneous metalwork, contain a paragraph dealing with the floor of this room and with the

---

[41]These sentences are from Article 2 of the General Provisions, which reads as follows:

The Contractor shall keep on the work a copy of the drawings and specifications and shall at all times give the Contracting Officer access thereto. Anything mentioned in the specifications and not shown on the drawings,

or shown on the drawings and not mentioned in the specifications, shall be of like effect as if shown or mentioned in both. In case of difference between drawings and specifications, the specifications shall govern. In any case of discrepancy either in the figures, in the drawings, or in the specifications, the matter shall be promptly submitted to the Contracting Officer, who shall promptly make a determination in writing. Any adjustment by the Contractor without this determination shall be at his own risk and expense. The Contracting Officer shall furnish from time to time such detail drawings and other information as he may consider necessary, unless otherwise provided.

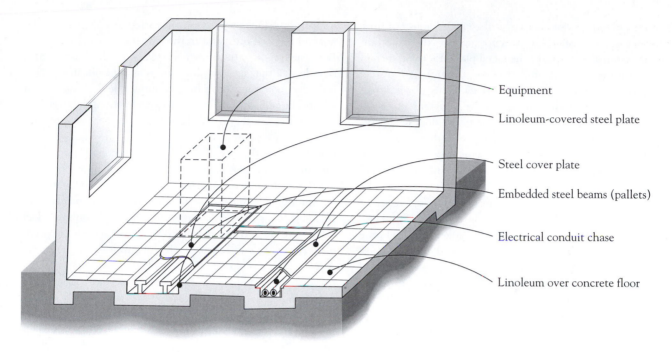

FIGURE 20.1 Disputed drawings in Unicon Management Corp. v. United States.

Equipment

Linoleum-covered steel plate

Steel cover plate

Embedded steel beams (pallets)

Electrical conduit chase

Linoleum over concrete floor

pallets (TP 17-23).[42] By itself, this provision can be interpreted as implicitly envisioning an all-concrete floor; there is no reference to steel plate but there are references to "the concrete floor," "a concrete floor," "pouring of the floor," "pouring of the floors," and steel beams being "embedded" in the concrete floor.

The relevant drawings, by themselves, give a very different impression. The most significant sketch—a cross-section of the floor of Equipment Room No. 1—shows the trench covered by a removable steel cover plate, with a depression into which the plate's handle is fitted flush with the floor level; the open-floor portion of the drawing shows the top of two steel beams covered by ¼" steel plate and the resilient floor tile on top of this plate. This is a specific directive to use steel plate immediately beneath the tile. To the same effect, another detail drawing of the trenches and beams in the room shows steel plate on top of an "I" beam and the resilient tile above the steel plate. The other drawings are uninformative or unclear on the point, but they do not contradict the sketches explicitly showing the steel plate. The result is that the drawings as a whole affirmatively support the Government's understanding as to the nature of the floor of Equipment Room No. 1.

[Ed. note: See Figure 20.1.]

We agree with the Board that these plans need not, and should not, be construed as in conflict with specification TP 17-23 footnote [42], supra but, instead, as supplementing the latter. The parties directed in the contract that "anything . . . shown on the drawings and not mentioned in the specifications, shall be of like effect as if shown or mentioned in both" (footnote [41] supra). This rule is peculiarly applicable here because

---

[42] 17-23 Equipment Room No. 1: AAOC Main Building. Floor shall have steel beams embedded in the concrete floor so that they run transversely to the equipment pallet lengths. The flange surfaces shall be flush with the finished cement. Two beam runs shall be used, one at each end of the pallets, as shown on the drawings. These beams will be used to anchor the equipment pallets, to level the equipment pallets and can be used as references for cement finishing tools during pouring of the floor. These beams shall be connected to the building ground system so that they will serve as a grounding means for the equipment pallets. The steel beams shall be on one continuous length, with portions removed to provide clearance for the cable troughs. The cross section area of the beams or jumpers, at the points where metal is removed shall provide at least as much conductivity as the ground cables used to connect the beams to the ground system. The grounding system must be installed prior to the pouring of the floors, as specified in Section 47, 'Electrical Work.'"

TP 17-23 obviously does not cover the entire subject of floor construction. For instance, it does not mention the important item of resilient floor tile, or the metal plates for the trenches, nor does it even describe fully the construction of the steel beams running transversely to the pallets. When the contract is viewed as a whole, the function of the paragraph is seen as calling attention to the steel beams set in the floor to support the pallets—the references to concrete are in that connection and in that special light—not as drawing together all of the requirements on the makeup of the floor. Certainly, as the Board pointed out, the specification does not provide in terms or by necessary implication that the resilient tile is to rest directly on concrete; nothing is said about the tile or its placing. It is therefore proper to read TP 17-23 as open to complementation by the drawings (or by other specifications) insofar as these cover items or aspects other than the features of the steel beams discussed in the former provision. Plaintiff seems to insist that, if the specification can be read as conflicting with the drawings, that reading must be adopted even though a more harmonious interpretation is also reasonably available. The rule, however, which the courts have always preferred is, where possible, to interpret the provisions of a contract as coordinate not contradictory. . . . Contractors, too, have long been on notice that in reading contract documents they should seek to find concord, rather than discord, if they properly can.

This brings us to another reason why plaintiff's position is weak. There is no evidence as to the view actually taken by the company's estimators before it submitted its bid (plaintiff did not call them to the stand in the administrative proceeding). But assuming . . . that they read TP 17-23 the way plaintiff now does, if they examined the plans and specifications carefully they could not have helped notice the drawings which specifically embodied the contrary requirement. If they were not aware of this fact they should have been. The contract provided that "in any case of discrepancy either in the figures, in the drawings, or in the specifications, the matter shall be promptly submitted to the Contracting Officer, who shall promptly make a determination in writing. Any adjustment by the Contractor without this determination shall be at his own risk and expense" (footnote [41] supra). A warning of this kind calls on the bidder to bring to the Government's attention any serious or patent discrepancy of significance, of which he is or should be cognizant. . . . The discrepancy here—if TP 17-23 was then thought to mean what plaintiff now contends—surely met that standard of importance. Yet there is no suggestion that plaintiff brought it to the contracting officer or sought the required guidance before bidding. If, on the other hand, plaintiff did not study the plans and specifications before bidding, it cannot complain that the Board and this court strive, in accordance with the established canon, to read the relevant contract provisions together rather than at odds.

The plaintiff is not entitled to recover. Its motion for summary judgment is denied and the defendant's is granted. The petition is dismissed.

Other decisions have dealt with conflicts in documents. First, as in the *Unicon* decision, the court attempts to reconcile the apparently conflicting language. When this cannot be done, the court often bases its decision on contract language that seeks to resolve potential conflicts.[43]

The last holding in the *Unicon* case reproduced above has caused difficulty in federal procurement cases. *Franchi Construction Co., Inc. v. United States* took a different attitude toward the duty to report "any serious or patent discrepancy of significance, of which he is or should be cognizant" and any precedence of documents clause. It held that the contractor could rely on the precedence of documents clause under certain circumstances.[44] The *Franchi* case is discussed and applied in the following case.

[43]*U.S. Fidelity & Guar. Co. v. West Rock Dev. Co.*, 50 F. Supp.2d 127 (D.Conn.1999) (HUD provisions over A201 incorporated into the contract); *Graham v. Com*, 206 Va. 431, 143 S.E.2d 831 (1965) (specifications over general conditions); *Dunlap v. Warmack-Fitts Steel Co.*, 370 F.2d 876 (8th Cir.1967) (specifications over bid proposal). See Stein, Alexander, & Noble, *The AIA General Conditions in the Digital Age: Does the Square "New Technology Peg" Fit Into the Round A201 Hole?*, 25 Constr. Contracts Law Reports, Dec. 14, 2001 (West Group) at p. 4 (contracts for construction of Frank Gehry's buildings provide that information in the 3D electronic files take precedence over information in hard copies), but see EJCDC C-700 ¶ 3.06. (2007). It prefers hard copy over electronic data. See Section 20.03.

[44]221 Ct.Cl. 796, 609 F.2d 984 (1979). Similarly, see *Shah Constr. Inc.*, ASBCA No. 50411, 01-1 BCA ¶31,330.

HENSEL PHELPS CONSTRUCTION CO. v. UNITED STATES

United States Court of Appeals, Federal Circuit, 1989. 886 F. 2d 1296

ARCHER, Circuit Judge.

[Ed. note: Footnotes renumbered and some omitted.]

Hensel Phelps Construction Co. (Hensel Phelps) appeals the decision of the Armed Services Board of Contract Appeals, ASBCA No. 35757, 88-2 BCA (CCH) ¶20,701, denying its claim for an equitable adjustment in the amount of $100,983.00 on Contract No. DACA 56-87-C-0002. We reverse and remand.

BACKGROUND

Hensel Phelps' contract with the United States, entered into on October 29, 1986, was for the construction of a jet engine blade repair facility at Tinker Air Force Base, Oklahoma. The initial contract price was $33,617,000.00.

On December 4, 1986, Hensel Phelps requested a contract interpretation, pointing out that the specifications called for a minimum of 18 inches of non-expansive fill under the concrete floor slabs, whereas a note on the drawings called for 36 inches of non-expansive fill.[45] The contracting officer was advised that Hensel Phelps and its subcontractor, C. Watts Construction Company (Watts), had used the requirement of the specifications in calculating the bid instead of the conflicting drawing note requirement because the contract provided that the specifications control over the drawings. The pertinent clause, which is commonly referred to as the "order of precedence" clause, provides:

> Anything mentioned in the specifications and not shown on the drawings, or shown on the drawings and not mentioned in the specifications, shall be of like effect as if shown or mentioned in both. *In case of difference between drawings and specifications, the specifications shall govern.* In case of discrepancy in the figures, in the drawings, or in the specifications, the matter shall be promptly submitted to the Contracting Officer, who shall promptly make a determination in writing. Any adjustment by the Contractor without such a determination shall be at its own risk and expense.

Contract Clause No. 47, entitled "52.236-21 SPECIFICATIONS AND DRAWINGS FOR CONSTRUCTION (Apr. 1984)" (emphasis added).

The contracting officer directed that 36 inches of non-expansive fill should be placed under the concrete floor slabs as required by the drawings and Hensel Phelps and Watts proceeded as instructed. Thereafter Hensel Phelps timely submitted a $100,983.00 claim, properly certified, for equitable adjustment in the contract price based on an asserted modification of contract terms. The claim was denied by the contracting officer.

The Board found that Watts, in preparing its subcontract bid to Hensel Phelps,

> recognized the clear conflict between the drawings and the specification. There was no way that the conflict could be resolved or harmonized. [Watts] relied on the "Order of Precedence" clause and prepared [its] bid based upon the 18 inches set out in the specification, rather than the 36 inches called for by the contract drawings.

88-2 BCA (CCH) ¶20,701, at 104,600. The board also found that Hensel Phelps "relied upon Watts' bid in preparing its own bid which it submitted to the government on this contract."[46]

The Board concluded however, that the decision of the United States Court of Claims in *Franchi Constr. Co. v. United States*, 609 F.2d 984 (Ct.Cl.1979), and the decision in *Shirley Contracting Corp.*, ASBCA No. 29028, 87-1 BCA (CCH) ¶19,389, applying the *Franchi* case, were controlling and precluded recovery by Hensel Phelps.

OPINION

The Court of Claims has held that when the requirements of the specifications of a government contract conflict with the drawings and the contract contains an order of precedence clause, the specifications shall control as the order of precedence clause provides. *Franchi Constr. Co. v. United States*, 609 F.2d at 989-90; *William F. Klingensmith, Inc. v. United States*, 505 F.2d 1257.

---

[45]The difference in the amount of the work is significant. According to the Board, "[t]he 18" requirement would involve 9750 cu. yds of fill and 530 cu. yds of excavation. The 36" requirement, on the other hand, would involve about 5000 cubic yards *more* of both fill and excavation." 88-2 BCA (CCH) ¶20,701, at 104,600-01.

[46]Hensel Phelps apparently learned of the discrepancy when Watts was advised by a government inspector during construction that 36 inches of fill would be required.

The government argues, however, that *Franchi* limits the applicability of the order of precedence clause to situations in which the contractor was not aware of the discrepancy prior to bidding and notes that Hensel Phelps' subcontractor knew of the discrepancy between the specification and the drawings before it submitted its bid. The Board accepted the government's position, stating that

> [i]t was clearly unreasonable for the appellant to presume that the Government intended to set out such blatantly conflicting requirements and leave it to the operation of the "boiler plate" clause to resolve and establish the Government's specific needs. In these circumstances, it was unreasonable for appellant to bid as it did, knowing as it must have, that a serious mistake had been made.

88-2 BCA (CCH) ¶20,701 at 104,602. Thus, the Board appeared to interpret *Franchi* as foreclosing reliance on an order of precedence clause whenever the contractor knows, or should know, of a discrepancy prior to bidding. *See also Shirley Contracting Corp.*, ASBCA No. 29028, 87-1 BCA (CCH) ¶19,389 (holding contractor should have sought clarification despite order of precedence clause when three of appellant's representatives knew of the difference between the specification and the drawing before bid). We disagree with the Board's holding and do not find that its interpretation is consistent with *Franchi*.

In *Franchi*, the trial judge found that there was a patent discrepancy between the specifications and drawings but held that the order of precedence clause could be relied on to resolve the discrepancy. In his opinion the trial judge stated:

> The Government authored the order of precedence clause as a mechanism to automatically remove conflict between specifications and drawings by assigning preeminence to the former. . . .
>
> The plaintiff is entitled to take the Government sponsored order of precedence clause at face value. Once its right to do so in the present situation is recognized, no conflict sufficient to occasion inquiry remains. . . .

Franchi Constr. Co. v. United States, 609 F.2d at 989-90 (citation omitted). In appealing the trial judge's ruling, the defendant argued that the clause should not apply when a discrepancy is patent. The Court of Claims said:

> We cannot in the circumstances say, in face of the precedence clause, our characterization of a discrepancy as patent automatically triggers an obligation to report. The clause itself seems designed to excuse such reporting, instances where equity would intervene aside.

Id. at 986. As the court in *Franchi* also noted, a critical distinction is made in the order of precedence clause itself. Precedence applies only in the case of "discrepancies between specifications and drawings, while one in figures, drawings, or specifications, each by themselves, must be promptly reported. . . ." *Id.*

The *Franchi* case, therefore, stands for the proposition that clarification must be sought where an internal discrepancy in the "figures, drawings, or specifications" is found, or should have been found, prior to bid, but that a discrepancy *between* the specifications and drawings, a matter covered by the order of precedence clause, will generally be resolved in the manner prescribed by that clause. We believe that this is the proper application of the order of precedence clause. Contractors should, as a general rule, be entitled to rely on the order of precedence clause and not be required to seek clarification of a putative inconsistency between the specifications and drawings. The order of precedence clause itself resolves that inconsistency. It is, after all, the government that is the author of this contract clause, as well as the specifications and drawings.

Despite this general rule, however, we recognize, as the *Franchi* court foresaw, that a strict application of the order of precedence clause in all situations where it literally applies could, under some circumstance, give rise to overreaching. In *Franchi*, the court in *dictum* noted:

> We would assume arguendo that a bidder, who noticed or should have noticed a serious mistake in the invitation or other of the contract documents, must divulge what he has or should have noticed to the government, and *will not in equity be allowed to profit by not doing so, as it would be an instance of overreaching.* That is not this case, whether the discrepancy be patent or latent. *Defendant does not accuse plaintiff of overreaching nor could it do so.*

Id. at 985-86 (emphasis added).

In this case, there is no evidence of overreaching by Hensel Phelps or Watts. The findings of the Board demonstrate that Watts' price included only an amount sufficient to perform the specification requirement and that this price was incorporated into Hensel Phelps' bid price. Reliance was properly placed on the order of precedence clause to resolve a discrepancy between the specifications and the drawings and this resolution was reflected in the bid. When the government insisted on 36 inches of fill, rather than the 18 inches called for in the specifications, the contractor was required to perform more work than the contract required and more than its bid price contemplated. Consequently, on the record here, neither Hensel Phelps nor Watts can be said to have profited or otherwise benefited by reliance on the order of precedence clause.[47]

In sum, we hold that an order of precedence clause may be relied on to resolve a discrepancy between the specifications and drawings even though the discrepancy is known to the contractor prior to bid or is patent. If the contractor is required to perform work in addition to that called for by application of the order of precedence clause, he may seek an *equitable* adjustment in the price of the contract for such work and, in that event, *equitable principles* would apply to prevent overreaching or profiteering.

The decision of the Board is reversed and the case remanded for a determination of an equitable adjustment.

COSTS

Costs to Hensel Phelps.
REVERSED AND REMANDED.

Another technique for dealing with potential conflict appears in AIA Document A201-2007, Section 1.2.1, which states,

The intent of the Contract Documents is to include all items necessary for the proper execution and completion of the Work by the Contractor. The Contract Documents are complementary, and what is required by one shall be as binding as if required by all; performance by the Contractor shall be required only to the extent consistent with the Contract Documents and reasonably inferable from them as being necessary to produce the indicated results.

This paragraph warns the contractor to assume that work called for under any document will be required even if apparently omitted from other documents. The paragraph requires work not specifically covered if that work is necessary to produce the intended results. Undoubtedly, such a provision gives considerable power to the architect, inasmuch as she may be called on to determine whether particular work is reasonably inferred though not specifically required. A cautious contractor must take this power to determine whether work is inferred into account when submitting a bid. Even though there are limits to the architect's power, the contractor may find itself at the mercy of an architect who wishes to cover up for poor document drafting. Understandably, a provision is needed to take into account that it is not possible to specifically include every aspect of work that should be done. However, if the architect does not exercise this power fairly, such a provision can generate performance difficulties.

The AIA has chosen not to include a precedence-of-documents clause such as discussed in the *Unicon* case. It believes, as has been seen in the *Hensel Phelps* case, that such a clause can be a disincentive for contractors to report errors, omissions, or inconsistencies in the design documents. Certainly, if such a clause is read to eliminate the need to make inquiry at the bidding or negotiations stage, the AIA's position can be justified. The AIA also believes that including such a clause can lead to "wooden" interpretation decisions that ignore the documents as a whole. As seen in the *Unicon* case, the reconciliation process is not always easy. Further, the AIA believes that no standardized precedence system can be the basis for a provision in a nationally used contract form. The AIA suggests that any parties that wish to include such a provision can simply add it.

Other arguments against a precedence-of-documents clause can be made. The common law has developed a method of reconciling apparently inconsistent clauses within a writing and among writings, which may do as good a job as a precedence-of-documents clause, particularly where the language can be easily supplemented by evidence of industry custom and a course of dealing between the parties. As can be seen in the Court of Claims decisions reproduced in this section, not only is it often difficult to determine whether a precedence clause diminishes or eliminates the obligation to inquire, but it also requires mental gymnastics to distinguish differences within a document from those differences that appear between documents. A precedence-of-documents clause does not always deal well with problems where the

---

[47]Obviously equitable principles would not permit a contractor to submit a bid priced to include the additional work required by the drawings and then seek additional compensation based on an order of precedence clause argument when the government required the work called for in the drawings, rather than in the specifications, to be performed. This would be overreaching in that the contractor would, in effect, be seeking payment twice for the same work, i.e., the amount included in the bid price for such work and the extra compensation claimed for said work under the order of precedence clause argument.

evidence of a lower-ranked document is strong and the evidence of a higher-ranked document is weak.

Yet the frequent inclusion of precedence-of-documents clauses must indicate that such clauses have value. In insoluble interpretations, they may provide some method of arriving at a solution. This is likely the principal reason they are used despite their weaknesses.

The Engineers Joint Contract Document Committee (EJCDC) C-700, Paragraph 3.06 (2007) deals with electronic data. Expressing concern over electronic data, Paragraph 3.06A states that only hard copy can be relied on. If there is a discrepancy between electronic files and hard copy, the hard copies govern. Paragraph 3.06B expresses concern over deterioration or modification of electronic files, whether inadvertent or otherwise.

SECTION 20.04 Reformation of Contracts

Sometimes parties to a contract reduce their agreement to a writing, and for various reasons, the writing does not correctly express the intention of the parties. The remedy of reformation granted by equity judges rewrites the contract to make the writing conform to the actual agreement of the parties. Because there were no juries in equity cases, courts felt freer to go into the difficult questions of intention and mistake.

Most cases have involved an improper description of land in a deed. These descriptions were complicated and often taken from old deeds and tax bills. It was not unusual for the person copying the old description to make a mistake. If it could be shown by reliable evidence that a mistake had been made, the court would reform the contract. This would be a judicial declaration that the contract covered the particular land the parties actually intended to buy and sell rather than the land covered in the description.

In addition to the cases of mutual mistake in description, other types of mistake would justify reformation. Usually these mistakes would involve the process by which the actual agreement was reduced to writing.

The reformation doctrine has been given an interesting application by the U.S. Court of Claims in federal procurement contracting.[48] In *National Presto Industries,*

Inc. v. United States,[49] the contractor under a fixed-price contract sought more money when its costs were substantially increased because a method that it had intended to use was not feasible for mass production. The court found mutual mistake in that the contractor and the government each assumed that the particular process in which the United States had interest could be used. Usually the remedy for mutually mistaken fundamental assumptions is to relieve the performing party—in this case the contractor—from the obligation to perform. But the Court of Claims extended reformation beyond its normal correction of mistakes function to change the fixed-price contract to one of a joint enterprise in which each joint enterpriser would share the unforeseen expenses.

Such a doctrine can impair the certainty and risk assumption features of procurement. As a result, two years later, the Court of Claims seemed to have had second thoughts about its decision in *National Presto* and sought to make clear that the reformation remedy awarded in *National Presto* could be justified only in a joint enterprise experimental situation in which neither party has assumed the particular risk, a great concern on the part of the government in the process and not merely the end product, and a distinct benefit to the government from the contractor's period of trial and error.[50] Nevertheless, in exceptional circumstances reformation may be the vehicle for redistributing risks, at least in federal procurement.[51]

Reformation can correct writings that do not reflect the actual agreement of the parties,[52] but the requirements for invoking this equitable remedy are formidable. Parties obviously cannot rely on the possibility that the law will correct their mistakes. They must make every effort to make the writing conform to their actual agreement.

---

[49]167 Ct.Cl. 749, 338 F.2d 99 (1964), cert. denied, 380 U.S. 962 (1965). The value of the *National Presto* case as a precedent is considerably weakened because it has been distinguished 19 times between 1964 and 2001. As an example, see *Northrup Grumman Corp. v. United States,* 47 Fed.Cl. 20 (2000).

[50]*Natus Corp. v. United States,* 178 Ct.Cl. 1, 371 F.2d 450 (1967).

[51]*Dynalectron Corp. v. United States,* 207 Ct.Cl. 349, 518 F.2d 594 (1975).

[52]*Timber Investors, Inc. v. United States,* 218 Ct.Cl. 408, 587 F.2d 472 (1978) (mistaken estimates in quantities). See Section 17.02D.

---

[48]This court was replaced by the Claims Court, a trial court, and the U.S. Circuit Court for the Federal Circuit, an appellate court. The former is currently called the U.S. Court of Federal Claims.

CHAPTER TWENTY-ONE

Changes: Complex Construction Centerpiece

SECTION 21.01 Definitions and Functions of a Changes Clause: *Watson Lumber Co. v. Guennewig*

After award of a construction project, the owner may find it necessary to order changes in the work. The contract documents are at best an imperfect expression of what the design professional and owner intend the contractor to execute. Circumstances develop during the construction process that may make it necessary or advisable to revise the drawings and specifications.

Design may prove inadequate. Methods specified become undesirable. Materials designated become scarce or excessively costly. From the owner's planning standpoint, program or budget may change. Natural events may necessitate changes. For any of many reasons, it often becomes necessary to direct changes after the contract has been awarded to the contractor.

"Changes" problems can arise in a number of different contexts. Most commonly, they involve claims by contractors that they have gone beyond the contract requirements and are entitled to additional compensation, sometimes called *disputed changes*. These are primarily matters of contract interpretation and of the implementation of any system under which the design professional or any Initial Decision Maker (IDM) resolves disputed questions. Although to some degree this chapter examines certain aspects of the changes process that bear on such disputes, the principal focus of this chapter is on the power of the owner to order a change, the process by which the change is ordered, and the effect on the contract price and time of the change.

Less commonly, disputes may arise when a contractor defends a claim brought against it by the owner for noncompliance with the contract requirements by asserting that the original contract requirements were changed and that the contractor has complied with the contractual obligations as changed.

Preliminarily, differentiations should be made between *changes, extras,* and *deletions,* although the AIA encompasses all of them in AIA Document A201-2007, Article 7, dealing with changes. However, this chapter refers to "extras," which, for example, is the way the court classified the problem involved in the *Watson* case reproduced in this section. This *type* of change highlights the difficulties that a long postcompletion list of extras can present to the owner or those furnishing funds for the project. Extras usually involve additional or more expensive items than those in the original contract documents. Conversely, a deletion, depending on how it is priced, can adversely affect planning. Although technically both are changes, additions or deletions that affect contract price or time are sensitive areas.

It can be helpful to compare *changes* with *additional* work. This can be demonstrated in *Eldeco Inc. v. Charleston County School District.*[1] The contract was for construction of a high school. The original design included an unfinished area to be used for vocational training. In the course of performance the school district decided to add some classrooms and a culinary arts uplift addition to the unfinished vocational training space. The school district requested a price quote from its prime contractor. The prime's bid included electrical work by the electrical subcontractor. The school district thought the price given by the subcontractor was too high, and another subcontractor was used for the new work.

One basis for the subcontractor's claim against its prime was that it should have been awarded the new work. As

---

[1] 372 S.C. 470, 642 S.E.2d 726 (2007).

449

shall be seen in Section 21.04E, one issue that can arise is the right of the existing contractor (or subcontractor) to be given changed work. The court concluded that the proposed work was additional work, not part of the changes process, as it was ordered and agreed upon in a supplemental agreement. Unlike change order work, the subcontractor had no contractual right to perform additional work.

The *Eldeco* case will be discussed in Sections 21.04B and 21.04E. But the case does illustrate that even if the existing contractor has some claim to changed work, if the work is additional, the changes process does not apply. It is simply a new contract and is made the way any other contract is made.

Changes must also be contrasted with *modifications* and *waivers*. Modifications are two-party agreements in which owner and contractor mutually agree to change portions of the work. They are discussed in various contexts in Sections 5.11C, 19.01H, and 25.01D. (One way to avoid formal requirements in a changes clause is to conclude that the work in question was a modification agreed on by the parties and not a change. Another is to classify the work as "repairs" rather than additional work requiring a change order. See Sections 21.03A and 21.04H.)

A *change* is the term used in construction contracts that allows the owner to unilaterally direct that changes be made *without* obtaining the contractor's consent to perform the work.

A *waiver* is generally based on the owner's acts that either manifest an intention to dispense with some of the contractual requirements or lead the contractor to reasonably believe that the owner is giving up its right to have required work performed. In such a case, the contractor can recover the full contract price despite not complying with the contract documents (discussed in Section 22.06E).

Reference has been made to the changes process and to the changes clause. In 1987, the AIA published a new edition of A201 that not only moved changes from Article 12 to Article 7 but also introduced new terminology. What had been called a *change order*—that is, the exercise of a contractual power to unilaterally direct changes in the work—was now designated under Paragraph 7.3 as a construction change directive (CCD). In 1987, under Paragraph 7.2, a *change order* (CO) became the wrap-up paperwork after there has been agreement as to price and time adjustment. This terminology change carried over to A201-2007, Sections 7.2 and 7.3. What is now a *change order* looks very much like a contract modification.

This change in terminology was designed to expedite the changes process, but it remains to be seen whether other standard contracts will adopt the change. This chapter uses traditional terminology, such as the changes process and the changes clause. However, when the AIA documents are discussed, AIA terminology is used.

As shall be seen in Section 21.04, the changes clause controls the scope of the unilateral power to order changes, the mechanism for doing so, and the pricing of changed work. Section 21.04I will show that the changes clause usually provides a complex process to price the changes. Unit prices, if any, are preferred. If not, the parties, the owner and the contractor, seek to reach an agreement as to any time or price adjustments that should be made because of the change. If that is not achieved, price and time adjustments are made by a third party. As shall be noted in Section 21.04I, if AIA documents are used, under A201-2007, Section 7.3.7, any price and time adjustments are made by the architect.[2]

The absence of a changes clause with its pricing formulas can make pricing drastic changes difficult. This is demonstrated in *Chong v. Reebaa Construction Company*.[3] The owner-lawyer and the contractor were both of Chinese ancestry. The original work was remodeling, and the contract price was $96,208.00. The contractor sought a written contract, but the owner-lawyer said they did not need one and they should base their agreement on honor and trust. There was no written contract.

The owner made many changes and upgrades. Though warned that these changes would increase the contract price, the owner said money was no object and he would pay what it cost. The owner made payments of $108,000. The contractor billed the owner for an additional $128,982.

The jury found for the contractor based upon a breach of the contract. But the Georgia intermediate appellate court held that the contractor could not recover for breach of contract as there had been no agreement on price. But it did allow the contractor to recover in *quantum meruit*, based on the reasonable value of the work.

However the Georgia Supreme Court disagreed. It concluded that breach of contract has been established as the contract was not so uncertain as to be unenforceable.[4] Uncertainty could be cured by subsequent words or

---

[2]A dissatisfied party can assert a claim under A201-2007, Section 15.2. Disputed claims are resolved initially by the Interim Decision Maker (IDM) under Section 15.2.1. For more discussion see Section 29.01.

[3]284 Ga.App. 830, 645 S.E2d. 47 (2007), reconsideration denied Apr. 10, 2007.

[4]283 Ga.222, 657 S.E.2d 826 (2008).

actions of the parties. When Chong requested upgrades, the contractor warned him that they would increase project cost. Chong stated that money was no object and he should be billed for the additional work.

The case teaches that the absence of a changes clause can create great difficulty when, as is common, the work is changed. (Though not mentioned by the supreme court, honor and trust can plunge into bitter acrimony, not an uncommon event in contracts for home construction.)

The beginning of this Section noted why changes are common in construction projects. Understanding the changes process requires that the process be looked at from the perspective of the principal parties to the construction contract.

Although the contractor recognizes that some changes are inevitable in construction work, it fears it will not receive adequate compensation for changed work or for unchanged work that is affected by the change. It also worries that it will not receive a proper time extension nor adequate compensation for delay or disruption caused by having to do work out of order. It is also concerned that poor administrative practices relating to changes will impede its cash flow and unduly burden its financial planning.

Contractors also express concern that unexpected changes may place a drain on their resources, divert capital they would like to use on other projects, and require more technical skills than they possess.

Contractors also complain about the unilateral nature of the changes clause. They must perform before they know how much they are going to receive in price and time adjustments. Although changes clauses usually provide a pricing mechanism through initial decisions by the

design professional, the contractor may fear that the design professional will not grant fair compensation and time adjustment, especially when the change will reflect on the professional competence of the design professional.

To the owner, the changes process looks quite different. As mentioned earlier, the changes process is needed to make those design changes required by subsequent experience and events. But the owner also fears that the changes process can expose it to large cost overruns that may seriously disrupt its financial planning and capacity. In that sense, the changes process is an important element of cost control. Owners and, perhaps more important, lenders fear a "loose" changes mechanism. They fear that bidders of questionable honesty and competence will bid low on a project with the hope that clever and skillful postaward scrutiny of the drawings and specifications will be rewarded by assertions that requested work is not required under the contract. A loose changes mechanism will generate claims for additional compensation. A changes clause, which gives either too much negotiating power to the contractor or too much discretion to the design professional, any IDM, or arbitrator, will convert a fixed-price contract into an open-ended cost type tied to a generous allowance for overhead and profit. This is why owners and lenders want tight, complete specifications, a mechanical pricing provision, such as unit pricing, and limits on overhead and profit. Their horror is the prospect that the end of the project will witness a long list of claimed extras that, if paid, will substantially increase the ultimate construction contract payout. The following case explores the changes mechanism mainly from the latter perspective.

WATSON LUMBER COMPANY v. GUENNEWIG

Appellate Court of Illinois, 1967. 79 Ill.App.2d 377, 226 N.E.2d 270.

EBERSPACHER, Justice.
[Ed. note: Footnotes renumbered and some omitted.]

The corporate plaintiff, Watson Lumber Company, the building contractor, obtained a judgment for $22,500.00 in a suit to recover the unpaid balance due under the terms of a written building contract, and additional compensation for extras, against the defendants William and Mary Guennewig. Plaintiff is engaged in the retail lumber business, and is managed by its president and

principal stockholder, Leeds Watson. It has been building several houses each year in the course of its lumber business.

* * *

[Ed. note: The project was a four-bedroom, two-bath house with air-conditioning for a contract price of $28,206. The total amount claimed as extras and awarded by the trial court was $3,840.09.]

The contractor claimed a right to extra compensation with respect to no less than 48 different and varied items of labor and/or materials. These items range all the way from $1.06 for extra plumbing pieces to $429.00 for an air-conditioner larger than plaintiff's evidence showed to be necessary, and $630.00 for extra brick work. The evidence, in support of each of these items and circumstances surrounding each being added, is pertinent to the items individually, and the evidence supporting recovery for one, does not necessarily support recovery for another.

* * *

Most of the extras claimed by the contractor were not stipulated in writing as required by the contract. The contractor claims that the requirement was waived. Prior to considering whether the parties, by agreement or conduct dispensed with the requirement that extras must be agreed to in writing, it should first be determined whether the extras claimed are genuine "extras." We believe this is an important area of dispute between these parties. Once it is determined that the work is an "extra" and its performance is justified, the cases frequently state that a presumption arises that it is to be paid for. . . .

No such presumption arises, however, where the contractor proceeds voluntarily; nor does such a presumption arise in cases like this one, where the contract makes requirements which any claim for extras must meet.

* * *

The law assigns to the contractor, seeking to recover for "extras," the burden of proving the essential elements. . . . That is, he must establish by the evidence that (a) the work was outside the scope of his contract promises; (b) the extra items were ordered by the owner, . . . (c) the owner agreed to pay extra, either by his words or conduct, . . . (d) the extras were not furnished by the contractor as his voluntary act, and (e) the extra items were not rendered necessary by any fault of the contractor. . . .

The proof that the items are extra, that the defendant ordered it as such, agreed to pay for it, and waived the necessity of a written stipulation, must be by clear and convincing evidence. The burden of establishing these matters is properly the plaintiff's. Evidence of general discussion cannot be said to supply all of these elements.

The evidence is clear that many of the items claimed as extras were not claimed as extras in advance of their being supplied. Indeed, there is little to refute the evidence that many of the extras were not the subject of any claim until after the contractor requested the balance of the contract price, and claimed the house was complete. This makes the evidence even less susceptible to the view that the owner knew ahead of time that he had ordered these

as extra items and less likely that any general conversation resulted in the contractor rightly believing extras had been ordered.

In a building and construction situation, both the owner and the contractor have interests that must be kept in mind and protected. The contractor should not be required to furnish items that were clearly beyond and outside of what the parties originally agreed that he would furnish. The owner has a right to full and good faith performance of the contractor's promise, but has no right to expand the nature and extent of the contractor's obligation. On the other hand, the owner has a right to know the nature and extent of his promise, and a right to know the extent of his liabilities before they are incurred. Thus, he has a right to be protected against the contractor voluntarily going ahead with extra work at his expense. He also has a right to control his own liabilities. Therefore, the law required his consent be evidenced before he can be charged for an extra . . . and here the contract provided his consent be evidenced in writing.

The amount of the judgment forces us to conclude that the plaintiff contractor was awarded most of the extra compensation he claims. We have examined the record concerning the evidence in support of each of these many items and are unable to find support for any "extras" approaching the $3,840.09 which plaintiff claims to have been awarded. In many instances the character of the item as an "extra" is assumed rather than established.[5] In order to recover for items as "extras," they must be shown to be items not required to be furnished under plaintiff's original promise as stated in the contract, including the items that the plans and specifications reasonably implied even though not mentioned. A promise to do or furnish that which the promisor is already bound to do or furnish, is not consideration for even an implied promise to pay additional for such performance or the furnishing of materials. The character of the item is one of the basic circumstances under which the owner's conduct and the contractor's conduct must be judged in determining whether or not that conduct amounts to an order for the extra.

The award obviously includes items which Watson plainly admits "there was no specific conversation." In other instances,

---

[5]We cite as some examples: An extra charge was made for kitchen and bathroom ceilings, concerning which Watson testified that he was going to give these as gifts "if she had paid her bill." According to the testimony, the ceilings were lowered to cover the duct work. We consider it unlikely that the parties intended to build a house without duct work or with duct work exposed. Likewise an "extra" charge was made for grading, although the contract clearly specifies that grading is the contractor's duty. An "extra" charge is sought for enclosing the basement stairs, although the plans show the basement stairs enclosed. An extra charge is sought for painting, apparently on the basis that more coats were necessary than were provided in the contract.

the only evidence to supply, even by inference, the essential element that the item was furnished pursuant to the owner's request and agreement to pay is Mr. Watson's statement that Mrs. Guennewig "wanted that." No specific conversation is testified to, or fixed in time or place. Thus it cannot be said from such testimony whether she expressed this desire before or after the particular item was furnished. If she said so afterward, the item wasn't furnished on her orders. Nor can such an expression of desire imply an agreement to pay extra. The fact that Mrs. Guennewig may have "wanted" an item and said so to the contractor falls far short of proving that the contractor has a right to extra compensation.

* * *

Many items seem to be included as "extras" merely because plaintiff had not figured them in the original cost figures.

It is clear that the contractor does not have the right to extra compensation for every deviation from the original specification on items that may cost more than originally estimated. The written contract fixes the scope of his undertaking. It fixed the price he is to be paid for carrying it out. The hazards of the undertaking are ordinarily his. . . .

* * *

If the construction of an entire work is called for at a fixed compensation, the hazards of the undertaking are assumed by the builder, and he cannot recover for increased cost, as extra work, on discovering that he has made a mistake on his estimate of the cost, or that the work is more difficult and expensive than he anticipated. 17A C.J.S. Contracts ¶371(6), p. 413.

Some so called "extras" were furnished, and thereafter the owner's agreement was sought.[6] Such an agreement has been held to be too late. . . .

The judge, by his remarks at the time of awarding judgment, shows that the definition of extras applied in this case was, indeed, broad. He said,

substantial deviation from the drawings or specifications were made—some deviations in writing signed by the parties, some in writing delivered but not signed, but nevertheless utilized and accepted, some delivered and not signed, utilized and not accepted, some made orally and accepted, some

made orally but not accepted, and some in the trade practice accepted or not accepted.

While the court does not state that he grants recovery for all extras claimed, he does not tell us which ones were and were not allowed. The amount of the judgment requires us to assume that most were part of the recovery. It can be said with certainty that the extras allowed exceeded those for which there is evidence in the record to establish the requirements pointed out.

Mere acceptance of the work by the owner as referred to by the court does not create liability for an extra. . . . In 13 Am.Jur.2d 60, "Building & Cont." § 56, it is stated that "The position taken by most courts considering the question is that the mere occupancy and use do not constitute an acceptance of the work as complying with the contract or amount to a waiver of defects therein." Conversation and conduct showing agreement for extra work or acquiescence in its performance after it has been furnished will not create liability. . . . More than mere acceptance is required even in cases where there is no doubt that the item is an "extra". . . .

The contractor must make his position clear at the time the owner has to decide whether or not he shall incur extra liability. Fairness requires that the owner should have the chance to make such a decision. He was not given that chance in this case in connection with all of these extras. Liability for extras, like all contract liability, is essentially a matter of consent; of promise based on consideration. . . .

The Illinois cases allow recovery for extra compensation only when the contractor has made his claim for an extra, clear and certain, before furnishing the item, not after. They are in accord with the comments to be found in 31 Ill.Law Rev. 791 (1937). There the author, after reviewing the cases, makes the following analysis:

The real issue in these cases is whether or not the contractor has, at the time the question of extra work arises, made his position clear to the owner or his agents and that would seem to be the true test in situations where a written order clause is sought to be disregarded. If he does expressly contend that work demanded is extra, the owner certainly cannot be said to be taken unawares, and if orders are given to go ahead it is with full knowledge of the possible consequences.

The contractor claims that the requirement of written stipulation covering extras was waived by the owner's conduct. The defendants quite agree that such a waiver is possible and common but claim this evidence fails to support a waiver of the requirement. There are many cases in which the owner's conduct has waived such a requirement. . . . In all the cases finding that

---

[6]The drain tile around the foundation of the house, according to the evidence, was already in place when the owner learned that it was more expensive material. Only then did the contractor secure the owner's consent to pay for one-half the cost of the more expensive material.

such a provision had been waived thus allowing a contractor to collect for extras, the nature and character of the item clearly showed it to be extra. Also, in most cases the owner's verbal consent of request for the item was clear beyond question and was proven to have been made at the time the question first arose while the work was still to be finished. The defendants' refusal to give a written order has in itself been held to negative the idea of a waiver of the contract requirements for a written order. . . .

We think the waiver of such a provision must be proved by clear and convincing evidence and the task of so proving rests on the party relying on the waiver. . . .

[Ed. note: The court ordered a new trial, stating that the contractor could recover only for those extras he could prove were ordered as such by the owner in the proper form, unless he could show that the owner waived the requirements of a writing by clear and convincing evidence.]

Undoubtedly, the court is correct in emphasizing the owner's right to know whether particular work will be asserted as extra. However, in a small construction project the Illinois court's requirements would place an inordinate administrative burden on the contractor. It would not only have to make clear its position that particular work was extra but also obtain the written change order executed by the owner. If the contractor did not obtain a written change order, it might be able to assert the doctrine of waiver if it could persuade the court that the work was extra and that the owner was made aware of this and allowed the work to proceed.

Rather than comply with the excessively formal requirements set forth by the court in the *Watson* case, a contractor in a small project, such as that involved in *Watson,* might better add a contingency in the contract price to cover small extras that are very likely to be requested.

The *Watson* case shows that courts often ignore the cost of complying with rules of law. Undoubtedly, larger projects will bear the administrative costs of doing things correctly and "according to the book." But smaller jobs may not permit "by the book" contract administration. The changes process involves the following:

1. the exercise of a power to order a change or a direction the contractor contends is a change in the work
2. methods for the contract or the parties to price the change and its effect on time requirements
3. a residuary provision that controls the price in the event the parties do not agree

SECTION 21.02 Shifts in Bargaining Power

To appreciate the centrality of the changes process to construction, the shifts in bargaining power because of changes must be appreciated. Some of them have been noted in the preceding section. But it is important that these factors be underlined. When a dispute develops, the resolution of the dispute, whether by design professional, interim decision maker (IDM), arbitrator, or court, may be influenced in some way by the bargaining power of the parties in structuring the changes process or at the time the dispute develops. In addition, the bargaining power is important at the time the owner must decide whether to issue a change order.

When preparing to engage a contractor, the owner, as a rule, has superior bargaining power. The hotly competitive construction industry and the frequent use of competitive bidding usually allow the owner to control many aspects of the construction contract terms.

A few contractors will deliberately bid low and drive out more prudent and experienced contractors with the hope that they will be able to demand and receive additional compensation by pointing to design ambiguities and amassing large claims toward the end of the project.

The contractor who is performing moves into a much stronger bargaining position. This is clearly so if the owner *must,* for either practical or legal reasons, order any additional work from the contractor. It would be in an even stronger bargaining position if it could refuse to execute the change unless there were a mutually satisfactory agreement on the effect of the change on price or time. Yet any bargaining advantages to the contractor by being in the position to refuse to do the work until there is an agreement are usually tempered by contract provisions that require the contractor to do the work even if there is no agreement on the price or time. (To counter this, a contractor can assert that the direction is a cardinal change, discussed in Section 21.03A.) This can be made even worse by the dominance some owners have over the changes process through their control over the purse strings. This power, exercised either directly or through the design professional, to withhold payment until the contractor agrees on price and time can exert intense pressure on the contractor to accept whatever the owner or design professional is willing to pay.

The changes mechanism can operate adversely to the contractor if the owner makes many small changes but is niggardly in its proposals for adjusting the price. (This can backfire, however, generating claims by the contractor, particularly in a losing contract, that the cumulative effect of any changes has created a cardinal change.)

To sum up, the changes mechanism on the whole favors the owner except if it is dealing with a clever, claims-conscious contractor who uses the changes mechanism to extract large amounts of money at the end of the job. Judicial resolution of disputes that involve changes may take into account the owner's strong position, particularly if the owner seems to have abused its power.

SECTION 21.03 Types of Changes

A number of classifications, in addition to those noted in Section 21.01, can be made that can help one understand the changes process.

A. Cardinal Change

Section 21.04B discusses the limits on the owner's power to order changes. Here it is important to look at the cardinal change, a concept developed in federal procurement law that relates to the power to order a change. The cardinal change concept originated not because of potential abuses created by a changes clause but because of jurisdictional aspects of federal procurement law.[7] Before 1978, a dispute between a contractor and a federal procurement agency would have to first go before an agency board of contract appeals if it arose under the contract. To bypass the board and bring the dispute before the U.S. Court of Claims (now the U.S. Court of Federal Claims), the contractor had to show a breach of the contract. To accomplish this, the contractor would seek to show a course of conduct—usually a large number of changes, drastic changes, or other wrongful agency conduct—that it could establish as a breach of contract.

Despite legislation enacted in 1978 that allows the contractor to choose which route to take,[8] the concept still has utility in both federal procurement and private disputes.

Clearly, any changes clause must be interpreted. Not any change can be ordered. A direction that goes beyond the scope of the work, if that is the standard, need not be obeyed. Such an order shall be called a "one-shot" scope change, to distinguish it from the more common "nibbling" or "aggregate of changes," currently the most common claim.

Suppose the contractor complies. Now there has been an agreement. But does it fall within the jurisdiction of the changes clause with its procedural and pricing mechanisms? The facts may indicate that the contractor, knowing it could not be compelled to perform because the order went beyond the scope of the work, agreed to waive any "beyond the scope" defense and allow the work to be governed by the changes clause. In the absence of such evidence, the parties have simply made a new agreement. If the parties cannot agree on price, the contractor is entitled to reasonable compensation.[9]

A cardinal change is considered a breach largely because of its origin in federal procurement jurisdictional issues. Can the contractor choose to treat it as a serious material breach and terminate any further contractual obligations it owes the owner? Section 33.04 discusses material breach.

The real nature of the direction or order can now emerge. It is simply a proposal, an order, that the contractor can choose to accept. It is not a breach of contract unless the owner asserts that it will not proceed further unless it is accepted. If so, this is a contract repudiation, and the contractor can terminate and recover any damages it may have suffered.

The other type of cardinal change does not occur all at once, as in the one-shot direction, but occurs throughout the performance of the contract. It consists of many changes,[10]

---

[7]The authoritative text on changes in federal procurement is R. NASH & S. FELDMAN, GOVERNMENT CONTRACT CHANGES (3d ed. 2007). It deals with the topics discussed in Sections 21.03A and B. See also Sweet, *The Amelco Case: California Bars Abandonment Claims in Public Contracts*, 32 Pub. Cont. L.J. 285, 288–89 (2003) (discussion of federal procurement).

[8]41 U.S.C.A. §§ 601 *et seq.*

[9]*Nat Harrison Assoc., Inc. v. Gulf States Util. Co.*, 491 F.2d 578, rehearing denied, 493 F.2d 1405 (5th Cir. 1974).

[10]*Wunderlich Contracting Co. v. United States*, 240 F.2d 201 (10th Cir.1957) (6,000 changes). Rejecting a contention by the defendant prime contractor that the cardinal change should be limited to federal procurement law, the court applying Kentucky law recognized the cardinal change. *L. K. Comstock & Co. v. Becon Constr. Co., Inc.*, 932 F.Supp. 906 (E.D.Ky.1993) (citing this book), aff'd, 73 F.3d 362 (6th Cir.1995).

drastic changes,[11] or other conduct that has gone beyond the reasonable expectations of the contractor and made the transaction different from what the parties had in mind when they made the contract.[12] This is a breach. But what are the remedies?

Here two scenarios emerge. One involves the contractor's deciding in the middle of the project that it has had enough, that it feels it can walk off the job because the total effect of the owner's conduct constitutes a material breach.[13] If upheld, the contractor can recover the reasonable value of its work, based on restitution or damages.[14] (The former is selected in losing contracts.) The second, and more common, scenario involves completion of the work by the contractor and a claim for the reasonable value of the services and materials, along with an allowance for overhead and profit, bypassing any contract price or guaranteed maximum price (GMP).[15]

Some states recognize the cardinal change doctrine; others do not.[16] It is important to recognize the utility of the cardinal change and its close cousin, abandonment, in jurisdictions other than the federal procurement system. If the owner has committed serious breaches, sometime called material breaches, this discharges the obligation of the contractor to proceed further and gives it, as a rule, a claim that either protects the contractor's expectation interest or prevents unjust enrichment.[17] This is also true if the contractor has fully performed and seeks postcompletion damages. Often the value of such a postcompletion claim can be based on crude global formulas, such as total cost or jury verdict.[18] These formulas avoid the need to connect each breach with specific damages.

There are some substantive differences between ordinary breaches and cardinal changes. But both are premised on errors (excessive changes) or poor administrative practices (unreasonable delays) by the owner or persons for whom it must take responsibility, such as the design professional, that increase the contractor's cost.[19] The same method of proving damages, ordinary contract damages with its strict certainty rules and the crude global formulas, are used in both types of claims. Whether the claim is based on ordinary contract damages or a global formula, what types of proof will a court demand?

The real use of cardinal change is to avoid any contract clauses that can affect the claim, such as exculpation clauses or those that limit damages. That is the reason contractors try to convince a court that the cardinal change should be recognized.

Federal procurement has witnessed the beginnings of a cardinal delay.[20] The term "cardinal" attached to a change or changes gets rid of the contract price and allows measurement of the claim by damages or the reasonable value of the work. A cardinal delay removes any remedial

---

[11]*Saddler v. United States*, 152 Ct.Cl. 557, 287 F.2d 411 (1961) (doubling excavation in small contract).

[12]*Allied Materials & Equip. Co. v. United States*, 215 Ct.Cl. 406, 569 F.2d 562 (1978). *Wunderlich Contracting Co. v. United States*, 173 Ct.Cl. 180, 351 F.2d 956 (1965) (cumulative effect of magnitude and quality of changes). In *Pellerin Constr. Inc. v. Witko Corp.*, 169 F.Supp.2d 568 (E.D. La.2001), the court found no Louisiana case recognizing the cardinal change doctrine. Even if it were allowed, the changes were not so drastic and profound that they altered the thing to be built. But see *Hensel Phelps Constr. Co. v. King County*, 57 Wash.App. 170, 787 P.2d 58 (1990). In this case, the painting subcontractor's claim for a cardinal change entitling it to quantum meruit (as much as it merits, as it is worth) was based on a drastic acceleration, a redoing of its work, and the stacking of trades (more than one trade working at the same time). The court pointed to the contractor having been compensated under the changes clause for its additional expenses and to other clauses indicating the contractor had taken certain risks. The court also pointed to the fact that there had been no fundamental alteration of the project itself, such as the shape or the square footage of the surfaces painted. While recognizing the cardinal change doctrine, the court stated that "this case does not involve the magnitude of changes that existed in the federal cases." 787 P.2d at 65.

[13]See Section 33.04A.

[14]See Sections 27.02D, E.

[15]*C. Norman Peterson Co. v. Container Corp. of Am.*, 172 Cal.App. 3d 628, 218 Cal.Rptr. 592 (1985) (cardinal change eliminated GMP). The California Supreme Court, while affirming *C. Norman Peterson* (involving a private sector project), ruled that the closely related doctrine of abandonment does not apply to public works contracts. *Amelco Electric v. City of Thousand Oaks*, 27 Cal.4th 228, 38 P.3d 1120, 115 Cal.Rptr.2d 900 (2002), rehearing denied Mar. 13, 2002. See *Rudd v. Anderson*, 153 Ind.App. 11, 285 N.E.2d 836 (1972). See also Section 27.02F.

[16]Silverman, *Abandonment and Cardinal Change Claims on "Projects from Hell,"* 25 Constr. Lawyer, No. 4, Fall 2005, p. 18. Nevada has recently recognized cardinal change and abandonment. *J. A. Jones Constr. Co. v. Lehrer McGovern Bovis, Inc.* 89 P.3d 1009 (Nev.2004). See also Sweet, supra note 7 (abandonment not allowed in California public contracts).

[17]See Sections 27.02D, F.

[18]See Section 27.02F.

[19]Sweet, supra note 7.

[20]*Godwin Equip., Inc.*, ASBCA No. 51939, 01-1 BCA ¶31,221. See Section 26.10A, note 87.

provisions in the contract that deal with government-caused delay.

B. Constructive Change

Again, reference must be made to federal procurement law. Claims for breach of contract could be taken only to the Court of Claims before 1978. A claim had to go before any procuring agency appeals board if it arose "under the contract." Relief had to be provided by statute or by a clause in the contract. To keep claims within the appeals board, the constructive change developed. Had a change order been issued, any claim clearly came under the contract.

In many cases, the contractor claimed that a direction or order was a change. The contracting officer acting for the procuring agency refused to issue a change order because he asserted that the work ordered was within the contract. The fiction of the constructive change allowed the boards of appeal to take jurisdiction by concluding that a change should have been issued if it were to later agree with the contractor. It is then as if a change order had been issued. The doctrine was also used if specifications were defective, requiring the contractor to do additional work. Even though the doctrine has no jurisdictional significance because of 1978 federal legislation,[21] it still merits comment.

Federal procurement disputes clauses require the contractor to keep working pending resolution of a dispute. If a contractor wishes to stop work, it may assert that its claim is based on breach and does not arise under the contract. Here the agency may assert there has been a constructive change and the contractor must continue under the contract. The changes clause precludes claims after final payment. Postfinal payment claims may be asserted to be based on breach, not "under the contract." In private contracts, the problem can still arise, although not in a jurisdictional sense.

Suppose that design professional and contractor dispute over work to be done. The design professional contends the work is called for under the contract, and the contractor claims it is not. Suppose the contractor performs the work but makes clear that it considers the position of the design professional unjustified and that it intends to claim additional compensation.[22]

Here, as in so many other aspects of the construction process, it is important to recognize the design professional's power to interpret the document and his quasi-judicial role if he is the interim decision maker (IDM). If the design professional determines that the work falls within the contract, many construction contracts make his decision final unless it is overturned by arbitration or litigation. If the contractor later requests additional compensation for the work and is met with the contention that no change order has been issued, the absence of a written change order should not bar the contractor's claim as long as it made clear that it intended to claim additional compensation. However, the claim should be denied if the design professional or the IDM's decision is binding and has not been overturned. A contractor dissatisfied with the design professional's decision should invoke any process under which that decision can be appealed. If it later is determined that the design professional or IDM had been incorrect and the decision is overturned, the absence of a written change order should not bar the claim.

Although no change order had been issued, there was a *constructive* change order—that is, one should have been issued.[23] (Some problems of this type are handled under "waiver," largely because the party directing the change had no authority to do so.)[24]

Another type of constructive change developed in federal procurement law—the constructive acceleration (discussed in Section 26.03B, dealing with time).

C. Deductive Change (Deletion)

Changes clauses usually permit the owner to delete a portion of the work, sometimes known as a *deductive change*. Although as a rule a deductive change is lumped together with changes that add to the contractor's contractual commitment, it raises special problems. First (as noted in Section 21.04B), the power to change the work typically is limited to those changes that fall within the scope of the work. Clearly a deductive change cannot be measured

---

[21]See supra note 8. Before 1968, constructive changes were also used to get around the *Rice* doctrine. In *United States v. Rice*, 317 U.S. 61 (1942), the court barred recovery for additional expense of *unchanged work*.

[22]See 48 CFR § 52.243-1(c) (2007); AIA Doc. A201-2007, § 15.1.4.
[23]*Chris Berg, Inc. v. United States*, 197 Ct.Cl. 503, 455 F.2d 1037 (1972).
[24]*Weeshoff Constr. Co. v. Los Angeles County Flood Control Dist.*, 88 Cal.App.3d 579, 152 Cal.Rptr. 19 (1979).

by that standard. Whether the prime suggesting to the owner that certain work be deleted that would wipe out much of the value of the subcontract was a violation of the obligation of good faith and fair dealing toward the subcontractor whose work was deleted was discussed in Section 19.02D.

Similarly, problems can develop over whether deleted work reduces the overhead and profit that are usually added in the event of additional work (discussed in Section 21.04I).

Finally, a problem similar to that discussed in Section 21.04E—the duty to order additional work from the contractor—can arise when there is a deductive change. Suppose the owner wishes to delete part of the work and offer it to another contractor. Unless the owner had good reason to do so, or at least acted in good faith, this would be a breach of the contract.

D. Minor Change

It sometimes becomes necessary to make minor changes in the drawings and specifications that are not intended to affect the contract price or completion date. AIA Document A201-2007, Section 7.4.1, gives the architect the authority to order minor changes as long as the changes are consistent with the intent of the contract documents. (Even if the architect is given authority to order minor changes, it is best that he not make changes when the owner is available.) In 2007 this Section added a requirement that the order be "signed by the architect."

Suppose the contractor believes the change is not minor and demands additional time and compensation. It should perform the work and make a claim under Sections 15.1.4 and 15.1.5 for more money and more time.

E. Tentative Change

As mentioned in Section 21.01, the classic changes clause requires the contractor to comply with the change order before any agreement is made on price. Contractors generally are not happy with this arrangement. But owners may not wish to be obligated to pay for the change without knowing what it will cost. For that reason, either by contract provisions or by practice, some owners will issue a tentative change, obtain a price from the contractor, and then decide whether to implement it. If time does not per-

mit the processing of such a tentative change order, the owner can direct the change unilaterally, and the contractor must comply.

One problem that sometimes arises when a tentative changes process is invoked is the contractor's claim that it should be compensated for its expenses in preparing a price quotation if the change is not ordered. Owners usually resist such claims, contending that the cost of preparing a price quotation is similar to the overhead costs of preparing a bid at the outset.

SECTION 21.04   Change Order Mechanisms

A. Judicial Attitude Toward Changes Mechanisms

From a planning standpoint, a changes mechanism is essential. Design flexibility and cost control cannot be accomplished without a system for changing work.

Yet judicial attitude toward the changes mechanism often determines how courts will interpret contract language and how quickly courts will find that the changes mechanism has been waived.

Judicial attitude is reflected by language in opinions that must pass judgment on these questions. For example, Massachusetts held that the contractor could not recover for extra work and concluded, "Although it seems a hardship for the [contractor] not to be able to recover for the extra work which apparently it performed in good faith, yet such failure results from its not obtaining from the architect or his agents written authority to perform the work."[25]

A federal court granted recovery to a subcontractor despite the absence of formal requirements by pointing to the fact that "[F]rom the beginning of the contract work the parties to the subcontract ignored the provisions as to written orders and proceeded with the work with little or no regard for them."[26] The court obviously thought the formal requirements simply technicalities that should not preclude a contractor from recovering for work performed beyond the contract requirements.

In a decision denying recovery for extras ordered orally by the city engineer, the Pennsylvania Supreme Court seemed unwilling to force the city of Philadelphia to pay for work not properly ordered and that could not be returned

---

[25]*Crane Constr. Co. v. Commonwealth*, 290 Mass. 249, 195 N.E. 110, 112 (1935).

[26]*Ross Eng'g Co. v. Pace*, 153 F.2d 35, 49 (4th Cir.1946).

by the city.[27] Here, control over public funds predominated. Likewise, the West Virginia Supreme Court denied recovery for work performed at the direction and with the knowledge of the county court president and other board members because of statutory requirements that all proceedings of the county court be entered into the record books of such court. In justifying its decision, the court stated,

> So the requirements of the statute and the decision that county courts must enter of record its orders for the expenditure of the funds must not be construed as meaningless, but must be enforced for the benefit of the whole public and not for the benefit of any particular individual who may suffer on account of a mistaken reliance on invalid acts of individual officers, however unfortunate, harmful or deceptive any such acts may have been, unless there is a predominant reason not to do so. To extend the doctrine of liability in every instance because of unjust enrichment is to open the door to all such claims as have not been properly authorized. . . . Too often is it apparent that the expenditures first authorized under a contract are enlarged by so-called "extras" without proper authorization, either intentionally or otherwise, and any such practice should certainly not be given judicial sanction. Nor should sloven management of county affairs be approved. In the absence of special reason, based on competent evidence, why the paramount consideration inherent in the legislative act requiring formal orders by county courts should not be made effective, there should be no digression therefrom as to its enforcement.[28]

It is more difficult to grant recovery to a contractor where there has not been compliance with formal requirements in contracts for public works. However, as seen in the *Watson Lumber* case (reproduced in Section 21.01), courts are also protective of inexperienced owners who are building their first home. Such owners must be made aware of price increases for proposed changed work.

The judicial attitude toward formal requirements must take into account the contract language, experience of the parties in construction, need to protect public funds, reasonable expectations of the owner, and potential unjust enrichment of the owner that can result if the contractor is unable to recover for admittedly extra work.

B. Limitation on Power to Order Changes

Although a changes clause is essential to construction projects, it is, as suggested earlier in this chapter, amenable to abuse. A contractor can expect some changes, but the changes clause should not be a blank check for the owner to order the contractor to do anything it wishes, to compel the contractor to perform before there has been an agreement on price, and to take care of compensation later. This behavior can be even more abusive if the owner seeks to avoid any increased cost of performing unchanged work or any cost incident to disruption of the contractor's method of performance.

At the outset, there can be a question in a public contract as to whether the additional work must be awarded by competitive bidding rather than as a change. This issue comes up frequently when there is a judicial challenge to the ordering of changed work to the contractor performing the contract. Many of the cases discussed in Section 18.04J deal with illegality caused by failure to award additional work through competitive bidding.[29]

A recent Illinois statute requires rebidding only if the amount in question is more than 50% of the original bid price. Originally, the limit was 25% but public entities objected. They contended that rebidding increased administrative costs, caused delays, and created a stacking of trades in that original and new contractors needed to work side by side.[30]

To determine whether a proposed change can be ordered, the overall character of the work must be considered. More particularly, the change must permit the work to retain its specific character, must be similar in nature to the work already contracted for, must not change the extent of the total performance in an intolerable manner, and must be within the capability of the contractor, taking into account the contractor's technical skill and financial capability.

Almost all construction contracts place some limit on the owner's power to direct changes unilaterally. Sometimes

---

[27]*Montgomery v. City of Philadelphia*, 391 Pa. 607, 139 A.2d 347 (1958).

[28]*Earl T. Browder, Inc. v. County Court of Webster County*, 143 W.Va. 406, 102 S.E.2d 425, 432–33 (1958). The California Supreme Court expressed similar sentiments in *Zottman v. San Francisco*, 20 Cal. 96, 104–05 (1862).

[29]*Bozied v. City of Brookings*, 638 N.W.2d 264 (S.D.2001); *Blum v. City of Hillsboro*, 49 Wis.2d 667, 183 N.W.2d 47 (1991).

[30]Montez, *Legislative Update: Illinois' Public Works Contract Change Order Act and Similar Statutes from Other Jurisdictions*, 25 Constr. Lawyer, No. 1, Winter 2005, p. 40 (listing 14 other states).

a limit is based on a certain percentage of the contract price. The difficulty is that the limitations may not be known until the contract is close to full performance. More commonly in the United States, the power to direct changes is limited to the scope of the work. For example, an owner may be able to order a 10 percent increase in the floor space of a residential home or that a carport be built or even a swimming pool, but not that a beach house be built twenty miles away.

The difficulty in applying the AIA "scope of the Work" limitation can be shown by *Eldeco Inc. v. Charleston County School District*,[31] discussed in Section 21.01. In that case the original contract was for a new high school. The court does not tell us the original price. The contract appeared to be an AIA contract with the power to order changes limited to the scope of the work.

The school district decided to finish an unfinished vocational training wing and to add some new classrooms. The court concluded that these were not simply changes but in effect new contracts made not on a change order but as supplementary agreements. The subcontractor plaintiff argued, unsuccessfully, that it had been given 109 change orders, indicating a pattern of how changes of all types were handled. The court spoke of the scope of the contract and referred to Paragraph 1.1.3 that defined the Work.

Clearly, the additional work was not part of the original contract. Changed work never is. The issue is whether the work order is so closely related to the original contract to be considered part of its scope. The court could have concluded that vocational training facilities and added classrooms are functionally part of the high school and could be demanded under the changes clause.

It is not clear what the price of the additional work bore to the original price. The court did not tell us the price quoted by the prime to the school district. But that quote included price quotes of the subcontractor for the additional work of $968,311. The school district thought the price too high. It requested the prime to get another bid. That bid was $578,152. It ordered the prime to accept that bid. (The successor subcontractor went bankrupt.)

It is important to separate two issues. The prime had no objection to doing the work. The issue was whether the electrical subcontractor should have been awarded the subcontract. That depended upon whether the school

district was required to award the additional work to the prime.[32] This will be discussed in Section 21.04E.

What is important for analyzing the limits on the change order power is whether the additional work was sufficiently connected to the project, in this case the high school, to be within the scope of the work. The amount for which the contractor would be paid was very likely within the 10 percent that often limits the power to order changes.

As mentioned in Section 21.03C, the scope limitation is difficult to apply when the owner wishes to delete part of the work. Here the only suitable guideline is the reasonable expectation of the contractor.

Suppose the owner points to the changes clause and directs that the contractor accelerate its performance. Must the contractor comply? For example, if the contract completion time is 500 days after the notice to proceed, can the owner direct that the work be done in 400 days?

AIA Document A201-2007 does not address this question directly. Section 7.3.1 simply provides that the owner may "order changes in the Work within the general scope of the Contract consisting of additions, deletions or other revisions." Work is defined in Section 1.1.3 as "the construction and services required by the Contract Documents," language contemplating the physical structure and not the time during which it must be completed. In the light of the drastic nature of acceleration, A201 should not empower the owner to direct an acceleration. A201 should be interpreted to allow only quantity and quality changes.[33]

Can the owner order a delay or a "stretch-out" by a change? For example, suppose the owner directs the 500-day time be increased to 600 days. It is unlikely that Section 7.3.1 would permit such a stretch-out. Yet Section 7.3.1 does allow the owner to suspend, delay, or interrupt performance without cause. If so, under Section 14.3.2 the contractor receives a price adjustment if the suspension increases its cost, plus profit and time.

Some may contend that if the owner can suspend, the owner can stretch out. And if the owner can stretch out, it should be able to speed up. But there is a difference between a suspension, with its generous remedy, and a stretch-out and certainly between a suspension and a speed-up, with

---

[31]Supra note 1.

[32]The large difference in price is a good reason for not giving a monopoly to the prime (and the sub) on changes. See Section 21.04E.

[33]*Mobil Chemical Co. v. Blount Bros. Corp.*, 809 F.2d 1175 (5th Cir.1987) (acceleration order assumed to be a breach of contract). Similarly, see *Hensel Phelps Constr. Co. v. King County*, supra note 12.

the indeterminate remedy (which some consider generous) of Section 7.3.7.

The federal procurement system speaks directly to this point, giving the federal agency the power to order accelerations.[34] Even if a pure acceleration were permitted, there is still a scope limit. The amount of the acceleration cannot exceed what could be reasonably expected and for which a reasonable price adjustment can be made.

Can the owner through the changes process take over control of means, methods, and techniques by which the contract is to be performed? Under AIA Document A201-2007, Section 3.3.1, this power belongs to the contractor. Such a change would not be within the owner's power. The changes process is limited to quantity and quality changes.

A limitation sometimes ignored is the duration of the power to order changed work. As noted in Section 26.04, some contracts use "substantial compliance" as the benchmark for determining compliance with the contractor's time commitment. If there has been substantial completion the owner can make effective use of the project even though some things still need to be completed. At that point the owner's power to direct changes unilaterally should terminate. Otherwise, the contractor might have to perform for a period of time not contemplated at the time the contract was made.

Another limitation that a contractor with strong bargaining power would like is the power to demand that the owner show evidence that it can pay for additional work it may want done. The AIA gives the contractor, before it begins to perform and under certain circumstances during performance, the power to determine whether a strong likelihood exists that the contractor will be paid for the work it performs.[35] A contractor without this contractual protection either at the beginning of the project or when substantial changes are ordered will have to take the owner's ability to pay into account at the time it decides the price for which it is willing to do the work.

Even if such protection is not found in the contract, if the owner directs a substantial change, the obligation of good faith and fair dealing should require that the owner satisfy the contractor that the owner can pay for the work that it has ordered.

For the effect of the law known as the Visual Artists Rights Act on the right of the owner to make changes, see Section 16.05.

C. Authority to Order Change

In the absence of any contract clause dealing with the question of authority to make changes, doctrines of agency control the question of who can order a change. The owner can order changes. Members of the owner's organization may also have the authority to order changes expressly, impliedly, or by the doctrine of apparent authority. Carefully thought-out construction contracts usually specify which members of the owner's organization have the power to order changes.

Neither the design professional,[36] the construction manager, nor, surely, the project representative[37] has inherent authority by virtue of his position to direct changes in the work. Some public contract cases are instructive. Sometimes the ultimate outcome is based on a contractor's claim that a *direction* had been given by someone clearly *without* the authority to direct the change when this unauthorized act was known by authorized officials. For example, *Chris Berg, Inc. v. United States*[38] involved a contract awarded under federal procurement rules. The contractor painted a stairway clearly not required under the contract to be painted. A field memo had been executed by a project engineer and a resident engineer, neither of whom had authority to issue a change order, which impliedly directed the contractor to paint the stairway in question.

Responsible officials in the agency knew the plaintiff was doing work not called for under the contract. The plaintiff had requested a ruling on whether the stairway should be painted and that it be advised of the government's interpretation of the field memo. The court stated that the agency had ample opportunity to warn the plaintiff that the plaintiff was painting a stairway not called for under the contract. The court concluded that this was a constructive change and the contractor was entitled to an equitable adjustment. In addition, the constructive change was based on the contractor's

---

[34] 48 CFR § 52.243–4(a)(4) (2007).
[35] AIA Doc. A201-2007, § 2.2.1.

[36] *F. Garofalo Elec. Co., Inc. v. New York Univ.*, 270 A.D.2d 76, 705 N.Y.S.2d 327, leave to appeal dismisssed, 95 N.Y.2d 825, 734 N.E.2d 762, 712 N.Y.S.2d 450 (2000). But see also Section 17.05B.
[37] See Section 17.05B.
[38] Supra note 23.

having painted certain areas after having received the tacit and undoubted oral approval of the project engineer.

Similarly, in *Weeshoff Construction Co. v. Los Angeles County Flood Control District*,[39] the contractor performed additional work in a road construction contract largely under the direction of the agency's site inspector. The court noted the problem of the site inspector's lack of authority to waive the requirement for a written change order. However, the court concluded the change order had already been issued, inasmuch as the procuring agency had threatened to do the work with its own workforce if the contractor did not do it. The court concluded the contractor was justified in relying on the site inspector's statements.

These two decisions demonstrate that directions by project representatives, coupled with other facts that would make denial of recovery unjust, may be the basis for the contractor's receiving additional compensation. (Excusing formal requirements is discussed in Section 21.04H.)

Yet a third case, one arising in federal procurement as did the *Chris Berg* case described earlier in this section, points in a different direction. *Winter v. Cath-dr/Balti Joint Venture*[40] took a strict position, limiting the authority to order changes to the Contracting Officer (CO) despite facts that might have led to another outcome.

At the preconstruction conference Navy officials instructed the contractor to get written authorization for contract changes from the resident officer in charge of contracts (ROICC). The CO, designated by the contract as the only person to authorize changes, was not present at the conference.

During performance the contractor followed the instructions given at the preconstruction conference, obtaining written authorization from the ROICC. But when the contractor sought an equitable adjustment to pay for the changed work, the Navy took the position that only the CO could authorize the changes.

The United States Court of Appeals for the Federal Circuit agreed with the government. It held that only the CO has authority to enter into or modify contracts.[41] In light of the clear contract language and the regulation, and notwithstanding the instructions given at the preconstruction conference, the ROICC had neither actual nor apparent authority to change the contract requirements.

These cases demonstrate the uncertainty outcome when this issue arises. The *Chris Berg* and *Weeshoff* cases looked at all the facts and were willing to hold for the contractor based upon its reasonable reliance. The *Winter* case looked only at the formal rules and not at what appears to be the reasonable reliance of the contractor. It shows the strong policy in federal procurement to protect the bureaucratic method devised by the government to control expenditures of public funds.

Private contracts frequently specify who has authority to issue change orders. As an illustration, AIA Document A201-2007, Section 7.2.1, states that a change order is prepared by the architect and signed by the architect, the contractor, and the owner. Section 7.3.1 states that the construction change directive is again prepared by the architect but signed only by the architect and the owner. Finally, for a minor change Section 7.4 requires a written order signed only by the architect.

This set of requirements raises an issue that is rarely addressed. All documents connected to the changes process are prepared by the architect. But Section 7.1.2 requires the architect's agreement, as well as those of the owner and the contractor, for a change order and under Section 7.3.1 for a construction change directive. Can the architect prevent a change either by refusing to prepare the change order or refusing to sign either the change order or the construction change directive?

Suppose AIA Document B101-2007 is used. Section 3.6.5.1 requires the architect to prepare change orders and construction change directives "for the Owner's approval." Is agreement by the architect a purely ministerial (nondiscretionary) act, one that does not give the architect the power to refuse to sign a change order or constructive change directive?

As seen in the *Duncan* case reproduced in Section 10.04B, the law places professional responsibility on the architect because he has been registered and qualified by the state, a responsibility that can take precedence over contract obligations. Surely the architect could refuse to sign if the change clearly would violate law. In addition, suppose his refusal to sign is based on his belief that the finished project would ruin his professional reputation. Would this belief justify his refusal?

To be sure, the architect who refuses to sign a change order or (using AIA current terminology) a construction

---

[39] Supra note 24.
[40] 497 F.3d 1339 (Fed.Cir.2007).
[41] 48 CFR § 43.102.

change directive authorized by the owner and agreed to by the contractor, may lose a client. Yet the owner should take seriously any objection the architect may have to signing these documents. This is part of the covenant of good faith and fair dealing the owner owes the architect. As noted, the architect would not be required to sign a change order or construction change directive that clearly violates building laws. Yet even if the matter involved only aesthetics, the owner should take into account the architect's professional standing. Admittedly, the money to be spent is the owner's, and the performance is the contractor's. Yet the owner should meet and confer with the architect and seriously consider the architect's professional judgment. In the end, the owner may decide to go ahead with the change, but its decision to do so must be made in good faith. If the architect cannot continue without either violating the law or compromising his professional integrity, he should terminate the contractual relationship. Neither party has breached, but the architect should be paid for work he has performed for which he has not yet been paid, based on restitution.

Just as the owner can give the design professional express authority, as seen in Section 4.06, so also the owner may cloak the design professional with apparent authority. The owner's acts can reasonably lead the contractor to believe that the design professional has the authority to change the work. Suppose the owner knows the design professional is executing change orders and makes no objection or, even further, pays the contractor, based on change orders issued by the design professional. This activity could reasonably lead the contractor to believe that the design professional has authority to change the work. The contractor is less likely to successfully invoke the apparent authority doctrine if the project involves public work.[42]

Fletcher v. Laguna Vista Corp.[43] illustrates enlarged authority. Changes were required to be ordered in writing by owner and architect. However, in concluding that the architect had authority to execute a written change order, the court stated,

> the manner in which the parties themselves have interpreted the contract through their course of dealings is of utmost importance. The record in this case is filled with

testimony to the accord that both [owners] and [contractor] had relied on architect Frye to make adjustments in the contract sum and had abided by his decision. [Owners] knew that there would be at least a slight overage in the sums spent by [contractor] for overhead but had never objected. [Owners] accepted decreases in the cost of millwork which were incorporated into a change order signed only by architect Frye and the contractor. The parties themselves have interpreted the contract to allow an increase and a decrease in the contract sum with only the written signature of architect Frye. Even if the contract does not grant this authority to architect Frye, the parties through their course of dealings have interpreted and modified the document so as to place in the hands of architect Frye the final authority to authorize increases and decreases in the contract sum.[44]

It is possible for the owner to give the architect this authority. However, if the basis for enlarging the architect's normal authority is the architect's power to interpret the contract documents, the case is incorrect. The court failed to distinguish the normal power given to the design professional to interpret the contract documents from the power to order changes.

Sometimes there is insufficient time to obtain authorization from the owner for work needed immediately, because there is impending danger to person or property. The contractor should have authority to do such emergency work. Contracts frequently provide that extra work in emergencies can be performed without authorization.[45] Even without express or implied authority, the contractor should be able to recover for emergency work, based on the principle of unjust enrichment.

D. Misrepresentation of Authority

If the contractor performed the additional work at the order of the design professional and this order was beyond the latter's actual authority, what recourse does the contractor have?

First, the contractor would seek to establish that the design professional had apparent authority to order the work. But in the absence of the owner's having led the contractor to believe that the design professional had this authority, the contractor will not be successful. Likewise, any claim that

---

[42]This is discussed in greater detail in Section 21.04H.
[43]275 So.2d 579 (Fla.Dist.Ct.App.), cert. denied, 281 So.2d 213 (Fla.1973).

[44]Id. at 580–81 (the court's footnotes omitted).
[45]AIA Doc. A201-2007, § 10.4 (contractor can act in emergency affecting safety of persons or property).

the work has unjustly enriched the owner would not be successful. The contractor would be considered a volunteer, and denied recovery.

Next, the contractor would look to the design professional. The threshold question would be whether the contractor reasonably relied on misrepresented authority. In many cases, reliance would not be reasonable, because design professionals typically are not given this authority and because the contractor should have checked with the owner. But the reliance element is often minimized or even ignored. In any event, if compliance with the order had been reasonable under the circumstances, the contractor can recover, from the design professional,[46] the cost of performing the unauthorized work and any cost of correction made necessary to make the work conform to the contract documents. A design professional held to have misrepresented his authority should not be able to recover from the owner on the theory of unjust enrichment.

Suppose the contractor had recovered from the owner because the latter's acts cloaked the design professional with apparent authority. Remember, the design professional acted without authority even though the acts of the owner may have created apparent authority. Suppose the owner seeks to transfer this loss to the design professional because of the latter's misrepresentation of authority. Passing by the question of whether the owner had suffered any real loss other than having work it did not choose, the loss was caused both by the design professional's misrepresentation of authority and by the owner's acts making it appear that the design professional had the authority. Because the more culpable act appears to be the misrepresentation of authority, this loss should be borne by the design professional.

Yet if the misrepresentation resulted in the value of the property having been increased, the owner has received a benefit. Although the law does not explicitly divide losses in such cases, it would be fair to do so here. Another method of sharing the loss would be to give the owner any difference between what the owner had to pay the contractor and the enhanced value of the project because of the unauthorized work.

At this point, a distinction should be made between the contractor's recovery having been based on implied authority or on apparent authority. In the former case, there has been no wrongdoing by the agent, who was authorized to act even though authority was not expressly given. For example, the architect's representation to the subcontractor in the *Bethlehem Fabricators* case mentioned in the notes to Sections 17.05B and 28.07J was authorized implicitly, though not explicitly. Had the architect's statement been *beyond* his authority—express or implied—recovery against the owner would have to have been based on apparent authority. If so, the owner would have had a claim against the architect. In the *Bethlehem Fabricators* case, there was no unjust enrichment of the type present in the preceding discussion. As a result, if the subcontractor's recovery against the owner had been based on an unauthorized representation by the architect, the owner should be reimbursed for what it paid the subcontractor.

An owner who knows of the unauthorized conduct of its agent but who then expressly approves the agent's decision ratifies the decision. The owner can no longer claim that the agent's decision is not binding on it.[47]

E. Duty to Order Change from Contractor

The changes clause clearly gives the owner the power to order that work within the general scope of the contract be performed by the contractor. Suppose the owner asserts the right to award work within the general scope of the contract to a third party or perform that work with its own forces. The power to do this would give the owner some bargaining advantage in negotiating the price for the changed work with the original contractor.

Most changes clauses do not deal with this question specifically. The contractor could contend that giving the owner this power would be unfair, putting the contractor at the mercy of a changes clause without the compensating advantage of being able to perform all work within the general scope of the contract. Giving the contractor the right to do the changed work, it could assert, would not put the owner at a disadvantage. Usually pricing formulas apply to the changed work if the parties do not agree on a price. Although as a practical matter the owner may find it inexpedient to have the original contractor working side by side with a substitute or its own employees, the express language of most changes clauses does not give the contractor the power to perform changed work.

---

[46]*Brown v. Maryland Cas. Co.*, 246 Ark. 1074, 442 S.W.2d 187 (1969) (dictum).

[47]See Section 4.08.

In *Hunkin Conkey Construction Co. v. United States*,[48] the contractor had entered into a contract with the Army Corps of Engineers for the construction of a dam. During performance, subsurface problems developed that necessitated a significant change in design. The contractor and the Corps discussed alternative designs but were unable to agree on a price for an alternative design desired by the Corps. As a result, the Corps negotiated a contract with a third party to perform the alternative design work at a cost of $200,000 less than proposed by the original contractor.

After completion of the project, the contractor contended that the work should have been awarded to it. The Court of Claims was not persuaded that the power to direct changes meant that the government had a duty to order those changes from the original contractor. But the court focused on another provision of the contract. This provision allowed the government to undertake or award other contracts for "additional work" and required the original contractor to cooperate with other contractors or government employees. This, according to the Court of Claims, had to be read together with the changes clause. When it was so read, the government was clearly not obligated to award the additional work to the contractor.

In *Eldeco, Inc. v. Charleston County School District*,[49] discussed in Sections 21.01 and 21.04B, there was a substantial difference in price between the original electrical subcontractor and a competing subcontractor, and the owner ordered the prime to engage the lower bidding subcontractor. This meant employees of two different subs might be working side by side, perhaps one union and one nonunion. But the original subcontractor's high bid may have been based on its belief that it was entitled to the work.

It should be noted that, just as in the *Hunkin Conkey* case, the contract in the *Eldeco* case was an AIA contract that stated that there might be separate contractors and the original contractor promised to cooperate. The owner may not want to be "locked in" to the contractor for "extra" work. Conversely, though, the contractor may assert an "expectation" that its price was predicated on a "monopoly" on "extras."

F. Formal Requirements

Before proceeding to the formal requirements for a change order, the work in question may fall outside the changes clause. It may be beyond the power granted the owner by the changes clause. As noted in Section 21.04B it may be work needed to correct errors of the design professional or to repair leaks to bring the water and sewer line project into conformity with the contract.[50] These are not within the jurisdiction of the changes clause and formal requirements specified in that clause do not apply.

As noted in Section 21.04C, construction contracts generally require that change orders be written and signed by the person or persons authorized to execute change orders. Obviously, it is best to have the change order issued and price agreed on before the work is started. If price cannot be agreed on before the work is begun, issuing a change order does assure the owner that if no agreement is reached, the changed work will be compensated in accordance with a changes clause formula. Issuance also assures the contractor that the owner will not contend the work in question falls within the contract requirements.

Should the contract require that the change order be issued before the work is begun? If the change order is issued after the work is begun or completed, issuance generally forecloses any question of whether the work was within the contract requirements. However, not requiring that the change order be issued before the work is begun invites issuance of oral change orders, with the assurance by the owner that a written change order will follow. Although oral orders sometimes may be necessary, suppose the owner denies issuing an oral change order. (In many cases, the owner admits giving a particular direction but contends it was not a change.) In such cases, the owner will insist that no additional compensation should be paid, and the contractor will contend that the issuance of an oral change order has dispensed with a requirement for a writing.

The contract should require that the change order be issued before the start of work even if this means some delay while the written change order is issued.[51]

Under AIA Document A201-2007, Section 7.3, the sequence is the issuance of a written construction change

[48]198 Ct.Cl. 638, 461 F.2d 1270 (1972).
[49]Supra note 1.

[50]*Town of Palm Beach v. Ryan Inc. Eastern*, 786 So.2d 665 (Fla.Dist. Ct.App.), review dismissed, 794 So.2d 608 (Fla.2001).
[51]In *Uhlhorn v. Reid*, 398 S.W.2d 169, 175 (Tex.Ct.App.1965), the clause stated that "No extra work or charges under this contract will be recognized or paid for unless agreed to in writing before the work is done or the charges made."

directive, commencement of the work with a simultaneous attempt to agree on price and time, and the residual power in the architect to decide price or time changes if the parties cannot agree. Some contracts do not require a written change order in advance of the work being done. The contractor can recover if a written change order is subsequently issued. Here the obvious advantage is speed; the disadvantage is the possibility of subsequent disputes over whether the work was extra. At the very least, having a written change order eliminates this difficulty. Selecting a less rigorous changes mechanism may reflect lack of confidence that a more structured clause will be effective. This doubt may be attributed to the belief that courts will not deny compensation to the contractor despite the absence of a written change order if the owner has apparently directed work to be performed that the court decides was beyond the contract documents.[52]

Another problem that regularly surfaces relates to the conflict between a changes clause, with its formal requirements, and other provisions of the contract that may have different formal requirements. This problem arises most frequently when the contractor encounters subsurface conditions different from those specified or normally expected and seeks to receive an equitable adjustment. (This special problem is discussed in Section 25.06.)

Any suggestions for formal requirements must face two paradoxical observations. In the helter-skelter world of construction, habits of sloppiness and the felt need for rapid action mean there will always be transactions in which the formal requirements fall to the wayside. This invokes the frequent claims of waiver to be noted in Section 21.04H. At the same time, courts often deny claims despite claims of waiver with the hope that demands for strict compliance with the formal requirements of the contract will spare the courts, the arbitrators, and the parties the frustrations of seeking to resolve these waiver claims.[53]

Finally, increased attention must be paid to the role of the legislatures in regulating consumer construction transactions. The Connecticut Home Improvement Act

requires that a change order be in writing and signed by the owner and the contractor. Yet the statute provides a restitution-like recovery for the reasonable value of the work done if the judge thinks it would be inequitable to deny recovery.[54] This multipolar approach seeks to get those who deal with home improvements to provide a written memorial of their agreement while providing a "way out" when it seems fair to do so.

G. Intention to Claim a Change

A claim entailing a lengthy list of extra work submitted after completion is a sad, though not infrequent, part of construction work. Section 21.03B, which treated constructive changes, noted that a vehicle for such claims is the contention that particular directions or instructions given by owner or design professional were changes entitling the contractor to extra compensation. One method of minimizing this problem is to require the contractor to give a written notice within a designated number of days after any event that the contractor will claim as the basis for an increase in the contract price. AIA Document A201-2007 requires such a notice under Sections 15.1.4 (more money) and 15.1.5.1 (more time). Except in an emergency, the notice must be given before beginning the work.[55]

A similar provision is used by the California Department of Public Works. A general contractor doing work for the state must submit a written protest to the state architect within thirty days after receiving a *written* order from the state architect to perform any disputed work. Failure to do so precludes compensation for the work.[56] The California provision also requires that the protest notice specify in detail how requirements were exceeded and the resulting appropriate change in cost. This content requirement goes considerably beyond A201-2007, Section 15.1.4. The latter states that a contractor who "wishes to make a Claim" for more money must give "written notice . . . before proceeding to execute the Work." There is an exception for emergencies "endangering life or property."

---

[52]As examples, see *Universal Builders, Inc. v. Moon Motor Lodge, Inc.*, 430 Pa. 550, 244 A.2d 10 (1968), and *Ecko Enter., Inc. v. Remi Fortin Constr., Inc.*, 118 N.H. 37, 382 A.2d 368 (1978). See also Section 21.04H.

[53]*Cameo Homes v. Kraus-Anderson Constr. Co.*, 394 F.3d 1084 (8th Cir.2005) (claim barred for failure to submit claim to architect as required by the contract); *Newmam Marchive Partnership, Inc. v. City of Shreveport*, 944 So.2d 703 (La.App.2006), writ denied, 949 So.2d 448 (La.2007) (failure to get advance authorization for additional work barred claim).

---

[54]Conn.Gen.Stat. § 20-429(a).

[55]*Pioneer Roofing Co. v. Mardian Constr. Co.*, 152 Ariz. 455, 733 P.2d 652 (1986) (oral order by prime to subcontractor in emergency)

[56]This provision was interpreted in *Acoustics, Inc. v. Trepte Constr. Co.* and *Weeshoff Constr. Co. v. Los Angeles County Flood Control Dist.*, discussed in Section 21.04H.

H. Excusing Formal Requirements

Not uncommonly, the change order mechanism is disregarded by the parties. This may be because of unrealistically high expectations by the drafters, time pressures, or unwillingness by the parties to make and keep records. When the owner denies contractor claims for extras because of the absence of a written change order or intention to claim an extra, contractors frequently assert the requirements have been waived. Waiver questions can be divided into three issues:

1. Is the requirement waivable?
2. Who has the authority to waive the requirement?
3. Did the facts claimed to create waiver lead the contractor to reasonably believe that the requirements have been eliminated or indicate that the owner intended to eliminate the requirements?

Except where the waiver concept cannot be applied to public contracts,[57] formal requirements can be waived. They are not considered an important element of the exchange and are often viewed as simply technical requirements.[58]

As a rule, only parties who have authority to order changes have authority to waive the formal requirements.[59] Usually only the owner or its authorized agent can waive the formal requirements, although a design professional with authority to order changes should have authority to waive the writing requirement.

Waiver can generally be based on acts such as conduct by the owner that indicated no written change order would be required[60] or oral orders by the owner or by its authorized representative.[61] Generally, if the owner apparently gave oral orders for the changed work, the court will assume the writing requirement has been waived.

Unfortunately, courts do not differentiate between cases where the owner admits ordering the work but claims it was within the contract requirements, from cases where the owner admits the work was extra and was ordered. One purpose for requiring a written change order is to eliminate the issue of whether the work was extra. The absence of a writing indicates at the very least that the owner did not consider the work extra. However, courts seem to disregard this factor. If they determine that the work was extra and that the contractor was ordered to perform it, lack of a written change order is not likely to prevent the contractor from recovering. But in a public contract, recovery could be denied even in such a case unless a constructive change is found.

Another factor that must be taken into account is whether the owner or design professional knew of the facts that would be the basis for a the contractor's claim of a change. For example, in *Moore Construction Co. v. Clarksville Department of Electricity*,[62] one issue was whether one separate contractor could claim delay damages caused by another separate contractor. However, the claimant did not give notice of intention to claim additional compensation as required by the predecessor of AIA Document A201-2007, Section 15.1.4. The court held this notice could be waived. It noted that all the participants knew that the claimant was being delayed and that the delays were not the claimant's fault. The claimant had been given a time extension through a job site memo, but no format change order had been issued.

The court noted that the claimant could have reasonably believed that the owner would not demand that the claimant submit a written notice of a claim for additional compensation because it had been granted a time extension. In addition, throughout this project the requirement that the contractor submit written notices had not been followed. (In fact, most of the formal requirements appear to have been abandoned.) In addition, the court concluded the owner had not shown itself prejudiced by the contractor's failure to give the required written notice.

---

[57]In *Delta Constr. Co. of Jackson v. City of Jackson*, 198 So.2d 592 (Miss.1967), appeal after remand 228 So.2d 606 (Miss.1969), the court held that waiver could operate against a municipality but not where the formality in question was a supplemental agreement on price. In *Metropolitan Sanitary Dist. of Greater Chicago v. Anthony Pontarelli & Sons, Inc.*, 7 Ill.App.3d 829, 288 N.E.2d 905 (1972), the court held that a provision giving the public agency the right to recover illegal and excessive payments meant that there was no way the required approval by the board of trustees could be waived.

[58]See Section 21.04A.

[59]See Section 21.04C.

[60]*Huang Int'l Co. v. Foose Constr. Co.*, 734 P.2d 975 (Wyo.1987). But see *Hall Contracting Corp. v. Entergy Services, Inc.*, 309 F.3d 468 (8th Cir.2002) (under Arkansas law, no waiver of written change order requirement where the owner had paid for several written change orders, but did not pay a purported oral request for additional work). Two recent cases found waiver by conduct, the unpublished case of *Spraungel Constr. Co. v. West Bloomington Motel, Inc.*, 2005 WL 832063 (Minn. Ct.App.2005) and *H. E. Contracting v. Franklin Pierce College*, 360 F.Supp.2d 289 (D.N.H.2005). See Annot., 2 A.L.R.3d 620 (1965).

[61]*City of Mound Bayou v. Roy Collins Constr. Co.*, 499 So.2d 1354 (Miss.1986); *T. Lippia & Son, Inc., v. Jorson*, 32 Conn.Supp. 529, 342 A.2d 910 (1975).

[62]707 S.W.2d 1 (Tenn.App.1985).

Owner payment based on oral change orders by the design professional or conceivably by a resident engineer or project representative can be a waiver. Payment can lead the contractor to reasonably believe the design professional has the authority to order changes orally.[63] If payment is accompanied by a notice making clear to the contractor that the formal requirements are not being waived, no waiver should be found.

It may be useful to examine language in some waiver cases. In *Rivercliff Co. v. Linebarger,*[64] the court found the writing requirement had been waived, stating,

> For a second ground, appellant contends that . . . the trial court should not have made any allowance to the contractor because the extra work was not authorized in accordance with the terms of the contract. This contention appears to be supported by the terms of the contract, which provides that extras must be approved in writing prior to execution. This provision was not complied with but it does not constitute a defense available to appellant, because, as we hold, a strict compliance with this provision of the contract was waived by appellant in this instance. It is not disputed that the extra excavation was done with the knowledge and at the direction of Smith who was not only the architect supervising the work for Rivercliff but was also a part owner of the appellant corporation. From his testimony we gather that he refused to approve an allowance for extras mainly because he did not think the contractor was entitled to anything as a result of the changed method of constructing the foundation. It appears that other changes in construction had been made and paid for where no written change order had been previously issued. Although it was shown that several such changes had been made and paid for during the construction of the four buildings, yet Mr. Smith testified that only one written change order had been made.[65]

Another court stated,

> Several situations may form the basis for waiver: (1) when the extra work was necessary and had not been foreseen; (2) when the changes were of such magnitude that they could not be supposed to have been made without the knowledge of the owner; (3) when the owner was aware of the additional work and made no objection to it; and (4) when there was a subsequent verbal agreement authorizing the work.[66]

Finally, in a subcontractor context, a court stated,

> the court finds that plaintiff is entitled to recover for several of the jobs which it performed that were not incidental to the building of a cofferdam and that were required to be done due to the insistence or inaction of James. Where a party is aware that extra work is being done without proper authorization but stands by without protest while extra work is being incorporated into the project, there is an implied promise that he will pay for the extra work. See *United States v. Klefstad Engineering Co.,* 324 F.Supp. 972 (W.D. Pa.1971). In *Klefstad,* the prime contractor was given the right to recover for work it performed which was the responsibility of the subcontractor; whereas in the present case the subcontractor, E & R, is entitled to recover for work it performed which was the responsibility of the prime contractor, James. Whether the theory of recovery is considered as quasi-contract, implied-in-fact, or promissory estoppel, a subcontractor is entitled to be compensated for extra work it performed as the result of inducing statements and conduct by the prime contractor.[67]

Two California cases demonstrate the difficulty in predicting when a court will find that formal requirements have been excused. In *Acoustics, Inc. v. Trepte Construction Co.,*[68] the court held that compliance with contractual provisions for written orders is indispensable, and denied recovery to a contractor who had been verbally ordered to perform changed work. The court noted that the state inspector who had ordered the changes had no authority to waive the formal requirement and that the contractor erred in expecting payment without a written change order from the state architect as required in the contract.

Eight years later, in *Weeshoff Construction Co. v. Los Angeles County Flood Control District,*[69] which involved a similar claim, the only differentiation being the apparent knowledge by the awarding agency that a site inspector

---

[63]*Oxford Dev. Corp. v. Rausauer Builders, Inc.,* 158 Ind.App. 622, 304 N.E.2d 211 (1973).

[64]223 Ark. 105, 264 S.W.2d 842, cert. denied, 348 U.S. 834 (1954).

[65]264 S.W.2d at 846.

[66]*Nat Harrison Assoc., Inc. v. Gulf States Util., Co.* supra note 9 at 583.

[67]*United States v. Guy H. James Constr. Co.,* 390 F.Supp. 1193, 1223 (M.D.Tenn.1972), aff'd, 489 F.2d 756 (6th Cir.1974).

[68]14 Cal.App.3d 887, 92 Cal.Rptr. 723 (1971).

[69]See supra note 24.

was directing changes to be made, the court came to the opposite result based on, though not explicitly, the constructive change rationale.

Admittedly, case decisions can appear to reach inconsistent results on similar facts. Often this is due to a failure by the court to apply a sensible and accepted reliance standard. When the issues are who can waive the condition and whether it has been waived, the court should consider whether the contractor (or subcontractor) has been reasonable in its belief that acts of the owner indicate that a written change order would not be a condition to payment.

Two New York lawyers suggest that New York courts draw a distinction between public and private contracts.[70] They suggest that contractor claims are routinely barred unless the notice is given as required in public contracts, but waiver may be found in private contracts. (In that regard, the *Acoustic* and *Weeshoff* cases discussed earlier in this section were both California public contract cases. Clearly, the courts examined the facts and routinely did not bar the claim because of a failure to give the contractually required notice.)

After canvassing the recent New York cases, these lawyers state that the reason for the distinction between public and private contracts is based on the need to protect "the public fisc," to uphold the integrity of the competitive bidding process, and the effect of the requirement that public contracts must be awarded to the low responsible bidder. The latter requirement may generate bids by "unsavory" contractors.

Private owners do not need the same protection, they say. They can pick their bidders without the restraints faced by public entities. They may pay more for protection against claims-prone contractors by choosing only reputable ones and paying for this by a higher contract price. But many private contracts contain such notice provisions that are often relied upon by private owners.

It is clear that the public–private distinction is real and plays a role in judicial decisions. But the reliance factor and a scrutiny of the relevant facts are the key factors.

Another way of avoiding the formalities of the changes clause is to conclude the work was not a "change." Section 21.03A noted that work may be considered outside the changes clause. Likewise, work made necessary because of errors of the design professional must be paid for despite the absence of the formal requirements set forth in the changes clause. Such work can be regarded as a remedy for defective work given to the contractor.

Similarly, as noted Section 21.04F, it has been held that work needed to conform to contract requirements is not a change requiring a change order.[71]

I. Pricing Changed Work

Pricing changed work is an important part of the changes mechanism. A tightly drawn pricing formula, along with clear and complete contract documents, can discourage a deliberately low bid made with the intention of asserting a long list of claims for extra work.

Where possible, work should be compensated by any unit prices specified in the agreement. If no specifically applicable unit price is specified in the contract, compensation should be based on an analogous unit item, taking into account the difference between it and the required work. It is also important, however, to take into account the possibility of great variations in units of work requested and the effect on contractor costs. (See Section 17.02D.)

When the contractor receives the directive under A201-2007, Section 7.3.5, it must proceed with the change in the work and indicate whether it agrees or disagrees with the method set forth in the construction change directive of adjusting price or time. If it agrees, under Section 7.3.6 it signs the construction change directive. If it does not respond promptly or disagrees with the proposed method of adjustment, under Section 7.3.7 the price adjustment will be determined by the architect "on the basis of reasonable expenditures and savings . . . including, in the case of an increase in the Contract Sum an amount for overhead and profit." It appears that Section 7.3.10 gives the architect the power to make time as well as cost adjustments.

Some contracts, including A201-2007, Section 7.3.7 provide that if the parties cannot agree, the contractor will be paid cost plus a designated percentage of cost in lieu of overhead and profit. Pricing extra work in this fashion can discourage the contractor from reducing costs. However, it may be difficult to arrive at a fixed fee for overhead and profit when the nature of the extra work cannot be

---

[70]Postner & Cruz, *The Public/Private Distinction in Enforcement of Contractual Notice and Claim Provisions*, Construction and the Law, Vol. 17, No. 1, Spring 2003 (published by Postner and Rubin).

[71]*Town of Palm Beach v. Ryan Inc. Eastern,* supra note 50.

determined until the work is ordered. For this reason, cost plus a percentage of cost will probably continue to be used to price changed work.[72]

When a cost formula is used, the design professional must examine the cost items to see whether they are reasonable and required under the change order. The changes clause can specify that only material, equipment, and labor costs or direct overhead costs are to be included as cost items.[73]

Public contracts—federal and state—frequently provide that if the parties cannot agree, there will be an "equitable" adjustment of the price.[74]

Suppose the change reduces the work. Does deductive change require that overhead and profit also be deducted for deleted work? Section 7.3.7 of AIA Document A201-2007 permits the architect to add a reasonable allowance for overhead and profit for an increase in the work but Section 7.3.8 appears to preclude reduction of overhead and profit if work is deleted. In contrast, a federal procurement decision held that overhead and profit would be deducted when work is deleted.[75]

Suppose a changes clause specifies that the parties will agree on compensation for changed work and the parties cannot agree. Early cases concluded that such an agreement was not enforceable as simply "an agreement to agree."[76] Today, courts would very likely determine a reasonable price for the work where the parties do not agree.[77] One case involving changed work on a subcontract held that where the parties could not agree on a price, the subcontractor could receive cost plus overhead and profit.[78]

A changes clause that specifies that the parties will agree should also provide an alternative if the parties do not agree, such as a pricing formula or a broadly drawn arbitration clause. Earlier federal procurement cases seemed to prohibit contractors from recovering for added costs of doing unchanged work caused by the change order.[79] However, federal procurement regulations have been changed to permit the contractor to recover additional costs of performing unchanged work.[80]

In addition to claims involving the cost of performing unchanged work, an excessive number of change orders of the type that can be considered cardinal changes are, along with other acts of the owner, often the basis for the now frequent delay damage claim (discussed in Section 26.10).

SECTION 21.05 Effect of Changes on Performance Bonds

When sureties were usually uncompensated individuals, courts held that any change in the contract between the principal and the obligee would discharge the surety. This would be unjust where a professional surety bond company is used, especially as changes and modifications are common in construction contracts. Most surety bonds provide that modifications made in the basic construction contract will not discharge the surety. Some changes clauses permit changes up to a designated percentage of the contract price without notifying the surety.

---

[72]Federal procurement regulations have developed weighted guidelines for determining profit. The factors taken into account are degree of risk, relative difficulty of the work, size of job, period of performance, contractor's investment, assistance by government, and subcontracting. Each of these factors is weighted depending on the particular procurement. An illustration of how these guidelines are used can be found in *Norair Eng'g Corp.*, ASBCA No. 10856, 67-2 BCA ¶6619.

[73]AIA Doc. A201-2007, § 7.3.7.

[74]48 CFR § 52.243-1(b) (2007). For a municipal contract containing a similar clause, see *Fattore Co. v. Metropolitan Sewerage Comm.*, 454 F.2d 537 (7th Cir.1971), cert. denied, 406 U.S. 921 (1972).

[75]*Algernon Blair, Inc.*, ASBCA No. 10738, 65-2 BCA ¶5127.

[76]See Section 5.06F.

[77]*Purvis v. United States*, 344 F.2d 867 (9th Cir.1965).

[78]*Hensel Phelps Constr. v. United States*, 413 F.2d 701 (10th Cir.1969).

[79]*United States v. Rice*, supra note 21.

[80]48 CFR § 52.243-1(b) (2007).

CHAPTER TWENTY-TWO

Payment: Money Flow as Lifeline

SECTION 22.01 The Doctrine of Conditions

This chapter deals principally with the process by which the contractor is paid for performing work required by the construction documents. The rules that control this process are derived principally from the contract documents, most notably the basic agreement and the general conditions. An important backdrop to these contract provisions is the part of the legal doctrine of conditions that deals with the order of performance.

This important legal doctrine seeks to protect the actual exchange of performance specified in the contract. A party should not have to perform its promise without obtaining the other party's promised performance. For example, suppose a contract is made under which a supplier agrees to deliver supplies to a small manufacturer. The supplies arrive by truck at the warehouse. However, the seller's truck driver refuses to unload the supplies until payment is made. The buyer's employee refuses to pay until the supplies are unloaded and placed on the buyer's receiving dock. Obviously, such a dispute could have been dealt with in the first instance by appropriate contract language dealing with the question of whether delivery precedes payment. When the contract does not deal with this question, the law must determine the sequence of performance.

Before proceeding to the legal resolution of this question, it should be noted that had the seller unloaded the supplies and not been paid, the seller would have had a valid claim for payment. Conversely, had the buyer paid but the supplies not been unloaded, the buyer would have had a valid claim for the value of the supplies or return of the money. However, neither party wishes to exchange its actual performance for a legal claim. Each would prefer to receive the other's performance before rendering its own.

Likewise, buyer and seller of real estate wish to avoid performance without obtaining the other's performance. The seller does not wish to transfer ownership by deed without obtaining the money. Conversely, the buyer does not wish to pay without receiving the deed. In a real estate purchase, protecting the desire of each is usually accomplished by using a third party or escrow holder. The seller will transfer the deed, and the buyer will deliver the money to the third party. Each believes the third party will effectuate the exchange.

In the absence of a third-party system or specific contract clause dealing with sequence of performance, the law must determine whether performance or payment must come first. The common law required that the performance of services precede the payment for those services.[1] This protected the party receiving the services by permitting that party to withhold payment until all the services were performed. Not only did this rule allow the paying party to avoid the risk of paying and then not receiving performance, but it also allowed the paying party to dangle payment before the performing party, a powerful incentive for rendering performance.

However, the "Work first and then be paid" rule is disadvantageous to the performing party. The latter must finance the entire cost of performance. The party performing services must take the risk that any deviation, however trivial, would enable the paying party to withhold money greatly in excess of the damages caused by the deviation. The performing party must assume the risk that the paying party might not pay after complete performance, leaving the performing party with a legal claim.

---

[1] E. A. FARNSWORTH, CONTRACTS § 8.11 at 614 (4th ed. 2004).

The doctrine of conditions is central to the discussion of progress payments (discussed in Section 22.02) and the right of the contractor to recover despite noncompliance (discussed in Section 22.06). Although less important, it is also relevant to retainage and final payment, discussed in Sections 22.03 and 22.05, respectively.

SECTION 22.02 Progress Payments

A. Function

The doctrine of conditions requires that work be performed before any obligation to pay arises. Such a rule places severe financial obligations on the contractor and creates a substantial risk of nonpayment. To avoid these problems, construction contracts generally provide for periodic progress payments made monthly or at designated phases of the work. This section examines the process by which progress payments are made and some common problems involved in this process.[2]

B. Schedule of Values

To facilitate computation of the amount to be paid under progress payments in a fixed-price contract, the contractor is generally required to submit to the design professional a schedule of values before the first application for payment. This schedule, when approved, constitutes an agreed valuation of designated portions of the work. The aggregate of the schedule should be the contract price. Contracts sometimes permit adjustments as work proceeds.

C. Application for Payment Certificate

The contractor submits an application for a progress payment a designated number of days before payment is due. This application is usually accompanied by documentation that supports the contractor's right to be paid the amount requested. In addition to the work itself, AIA Document A201-2007, Section 9.3.2 allows payment for goods stored at the site or, with advance approval by the owner, off the site for subsequent incorporation in the work. If stored off

site, the location must be agreed upon in writing. Payment for goods stored on or off the site is conditioned upon the contractor complying with procedures satisfactory to the owner to establish the owner's title to the goods.

Allowing the contractor to retain possession of material for which the owner has paid can create legal problems, mainly claims by creditors of the contractor or the trustee in bankruptcy if the contractor is declared bankrupt.[3] To avoid this, some owners require that material be stored only in bonded warehouses.

Some contracts require the contractor to give assurance or proof that it has paid its subcontractors and suppliers when progress payments applications are made. Alternatively, such contracts allow the prime contractor to submit documents executed by subcontractors or suppliers that give up their rights to any mechanics' liens.

Neither the standard contract published by the American Institute of Architects (AIA) nor the one published by the Engineers Joint Contracts Documents Committee (EJCDC) requires such assurance or proof of payment or submission of lien waivers at the time *progress* payments are made. This is justified in part by the fact that administrative requirements can be formidable in large projects if partial lien waivers or evidence of partial payment is required. These steps may not be necessary if other techniques exist that protect against such risks. One function of a retainage (discussed in Section 22.03) is to protect against liens. Also, Section 9.3.3 in AIA Document A201-2007 requires the contractor to warrant to the best of its "knowledge, information, and belief" that all work and materials will be "free and clear of liens." This will not preclude liens but may create a right against the contractor or surety under a payment bond.

Allowing design professionals to withhold progress payments when they learn that subcontractors are not being paid or that liens have been or are likely to be filed may be sufficient owner protection. Usually such information is communicated quickly to the design professional. The AIA acceded to requests by subcontractor associations to include more recognition of and protection for subcontractors in A201-1997. For example, Paragraph 9.3.1.2 barred the contractor from including in the payment application "requests for payment for portions of the Work for which

---

[2]This section examines the AIA payment process. For a comparison of the AIA approach with that of the Associated General Contractors of America (AGC) and the Engineers Joint Contract Documents Committee (EJCDC), see Nielsen, *Payment Provisions: Form Contract Approaches and Alternative Perspectives*, 24 Constr. Lawyer, No. 4, Fall 2004, p. 33.

[3]See *Traveler's Indem. Co. v. Ewing, Cole, Erdman & Eubank*, 711 F.2d 14 (3d Cir.1983), cert. denied, 464 U.S. 1041 (1984) (architect not liable for not warning owner of bankruptcy risk in paying for materials stored off site).

the Contractor does not intend to pay a Subcontractor or material supplier unless such Work has been performed by others whom the Contractor intends to pay." Similarly, Paragraph 9.6.3 required the architect, on request, "if practicable" to furnish information to a subcontractor regarding the percentages of completion or amounts applied for by the contractor and action taken thereon on account of work performed by the subcontractor. This language has been incorporated in A201-2007, Sections 9.3.1.2 and 9.6.3. The 2007 A201 adds new language to Section 9.6.4, giving the owner "the right to request written evidence from the Contractor that the Contractor has properly paid" subcontractors and suppliers out of amounts paid by the owner to the contractor. If the contractor does not furnish the written evidence in seven days, the owner has the right to contact the subcontractors and suppliers directly to ascertain whether they have been properly paid.

Also, the AIA recognizes the increasingly difficult problems of cash flow for the contractor. Section 9.3.1.1 allows the contractor to include in its application amounts that had been authorized by construction change directives or by interim determinations of the architect though not formally included in change orders.

To speed the cash flow, Section 7.3.9 allows the contractor to request payment "for Work completed" under a Construction Change Directive (CCD), even if the total cost of the CCD has yet to be determined. The architect then makes "an interim determination for purposes of monthly certification for payment" of the cost of the completed work and certifies it for payment. Either the owner or contractor may challenge the interim determination in accordance with the "claims and disputes" process set forth in Article 15. However, if they agree to the interim determination, then Section 7.3.10 requires the architect to prepare a change order to that effect. Section 7.3.10 makes explicit that "Change Orders may be issued for all or any part of a Construction Change Directive."

D. Observations and Inspections: Representations from Certificate Issuance

Before issuing a payment certificate, the design professional visits the site to determine how far the work has progressed. This is the basis on which progress payments are made.

The inspection should uncover whatever an inspection principally designed to determine the progress of the work would have uncovered. Each new edition of A201 has

sought to limit the architect's responsibility during the progress payment certification process. Under A201-2007, Section 9.4.2, issuance of a certificate is "based on the Architect's evaluation of the Work and the data comprising the Application for Payment." Section 9.4.2 also states that the certificate warrants only that the work complies with the contract requirements "to the best of the Architect's knowledge, information and belief." The architect's representations are "subject to an evaluation of the Work for conformance with the Contract Documents on Substantial Completion, to results of subsequent tests and inspections, to correction of minor deviations from the Contract Documents prior to completion and to specific qualifications expressed by the Architect." Finally, Section 9.4.2 states that the issuance of a certificate does not represent that the architect has made exhaustive or continuous on-site inspections, reviewed the construction methods, reviewed copies of requisitions from subcontractors and suppliers, or made any examination to determine how previous payments have been used.

Some owners, particularly developers influenced by their lenders, require more intensive inspections and place more responsibility on the architect. To accomplish these aims, such owners require the architect to give warranties that go beyond the experience of the design professional and the services she has contractually committed to perform. To protect architects, the AIA in its B101-2007, Section 10.4 requires that the architect receive certificate language, for her "review," "at least 14 days prior to the requested dates of execution" and prevents the owner from requiring certifications "that would require knowledge services, or responsibilities beyond" the scope of B101-2007.

E. Amount Certified for Payment: Incorporation into the Project

The amount certified for payment depends on the pricing provisions in the construction contract. In fixed-price contracts, the pricing benchmark is the contract price. In cost contracts, the principal reference point for determining payments is the allowable costs incurred by the contractor. In unit-priced contracts, the progress payments are based on the number of units of designated work performed.

Each pricing provision commonly uses a retainage system under which a designated amount is withheld from progress payments to provide security to the owner. (Retainage is discussed in Section 22.03.)

The computation of the amount to be paid is facilitated by using a schedule of values, discussed in Section 22.02B. Such an agreed schedule determines the extent to which the work has progressed. Smaller contracts without a schedule of values sometimes provide that specified payments are to be made at designated phases of the work with an appropriate allowance for retainage where used.

Despite a properly prepared schedule of values, measurement problems can develop. Much time elapses between the contractor's ordering material and equipment and incorporating that material and equipment into the work. The contractor prefers payment as early as possible. The owner prefers not to have to pay until the material and equipment have been incorporated into the project. However, as noted in Section 22.02C, AIA Document A201-2007, Section 9.3.2, allows for payment for materials and equipment not incorporated but stored on and off the site. Also, as noted in 22.02C, payment prior to incorporation into the project raises legal and insurance questions.

Construction contracts allow the design professional to make *partial* certification for payment. Partial certificates authorize payments of an amount less than requested by the contractor. Usually such certificates are based on the design professional's determination that the work has not progressed to the extent claimed by the contractor or on a determination that the work does not meet the requirements in the contract documents. If an architect intends to withhold the certificate either in whole or in part, Section 9.4.1 of AIA Document A201-2007 requires the architect to notify the contractor and the owner in writing of such a decision and to give reasons for this action. Under Section 9.5.1, failure by the contractor and architect to agree on payment amount allows the architect to issue a certificate for payment for the amount "for which the Architect is able to make such representations to the Owner."

Construction contracts frequently give the design professional the power to revoke a previously issued certificate. Revocation can be effectuated by a partial certificate or by withholding a certificate for work that has been performed. One reason for revocation is the discovery of defective work that had been the basis of a previously issued certificate.

Section 9.5.1 of AIA Document A201-2007 gives the architect power to make "withholds" from payments due.[4]

Withholds are designed to protect the owner from losses that have occurred or may occur in the future. Losses can relate to nonconforming work, nonpayment of subcontractors and suppliers, or claims made by other contractors or other third parties against the owner for which the contractor may be responsible.

Although the law would give the owner certain offset rights under these circumstances without a contractual right of offset, it is advisable to expressly recognize this right. Doing so avoids the necessity of establishing a right to offset and can exceed the protection given by law. However, the design professional should not be unreasonable or arbitrary in determining when and how much to withhold from payment amounts earned by the contractor.[5] Issuing partial certificates or, more important, withholding certificates can result in contractor default. For this reason, language that requires design professional and contractor to discuss and negotiate on these matters is useful.[6]

Offsets or withholdings from public works contractors raise constitutional issues. For example, state prevailing wage acts require that all workers employed on public construction projects of a certain minimum size be paid the "prevailing wage" (which is essentially the wage and benefits paid to unionized workers). The state department of labor is charged with investigating workers' complaints that they are not receiving the prevailing wage. If the department determines that underpayments have been made, it may order the public agency in charge of the project to withhold from the contractor or subcontractor the amount of the alleged wage underpayments, plus statutory penalties, without first giving the contractor or subcontractor a chance to contest the workers' allegations. In *Lujan v. G & G Fire Sprinklers, Inc.*,[7] the Supreme Court ruled that California's Prevailing Wage Act, which mandated the immediate withholding of wage underpayments and statutory penalties without a prior hearing from payments otherwise due a subcontractor, does not violate due process.

---

[4]See *Howard S. Lease Constr. Co. v. Holly*, 725 P.2d 712 (Alaska 1986) (need not notify contractor before withholding). But see AIA Doc. A201-2007, §§ 9.4.1, 9.5.1.

---

[5]*City of Mound Bayou v. Roy Collin Constr. Co.*, 499 So.2d 1354 (Miss.1986) (altering requests without proof of poor or incomplete work); *S.I.E.M.E., S. r. 1*, ASBCA No. 25642, 81-2 BCA ¶15,377 (improper to withhold $10,000 when $1,000 would have been sufficient to complete the few remaining items).

[6]Nevada by statute has severely limited the owner's right to withhold. See Nev.Rev.Stat.§ 624.609, discussed in Sebastian, *Legislative Update*, 27 Constr. Lawyer, No. 2, Spring 2007, p. 38.

[7]532 U.S. 189 (2001). California has since amended its Prevailing Wage Act to grant contractors and subcontractors a post-deprivation hearing as to the propriety of the withholding; see West Ann.Cal.Labor Code § 1742.

A federal court of appeals similarly ruled that due process does not require that a public works contractor receive a formal hearing before a design professional withholds progress payments.[8] Informal hearings are sufficient. The court was concerned that such a requirement could cause delay and unreasonably burden the public entity. Increasingly, contractors invoke constitutional rights in their disputes with state entities.[9]

Federal procurement policy allows reduction or suspension of progress payments despite the absence of any present contract breach by the contractor where the latter's financial position makes future nonperformance likely.[10]

F. Time of Payment

The federal government[11] and the state legislatures have been increasingly responsive to complaints by contractors and subcontractors that payments are not made promptly to them. For example, California legislation establishes deadlines for making progress payments, places a cap on withholding where amounts are disputed, and imposes penalties for noncompliance.[12] In addition, a prime contractor who violates the law is subject to licensing disciplinary action and must pay the subcontractor a penalty of 2 percent per month in addition to normal interest.

Care must be taken to check for legislation dealing with the payment process. Some such legislation covers public contracts, some private contracts, and some both. Also, such legislation should be checked to see whether it

provides that parties can, by contract, change the legislative rules.

Another payment-timing issue is whether a prime contractor may, by contract, delay payment of subcontractors contingent upon the contractor's receipt of payment by the owner. These "pay when paid" and "pay if paid" clauses are discussed in Section 28.06.

G. Passage of Title

As work proceeds, the materials and equipment are procured and installed in the building or attached to the land. Problems can develop that relate to when title passes from the contractor to the owner. There may be a preliminary question as to when title passes between the vendor of the product and the purchaser, in the construction context, the contractor or subcontractors. But for purposes of the construction contract, the principal question relates to when the materials and equipment belong to the owner.

Passage of title is important. If the materials and equipment still belong to the contractor, the contractor has the risk of loss, and the creditors of the contractor can seize the property or equipment in payment for the contractor's debts. Conversely, once the title has passed to the owner the owner has the risk of loss, and its creditors may have rights in the property.

The risk of loss is usually dealt with by insurance. Either the contractor or, more commonly, the owner takes out insurance on all materials and equipment that reach the site or, in the case of materials and equipment stored, at an earlier point. But typically, the owner would like to have title pass to it as soon as possible to avoid the possibility that creditors of the contractor will claim that they can seize its property in payment for the contractor's debts or because of unsatisfied court judgments.

By law, title passes when the materials and equipment are incorporated into the project. Other solutions can be expressed in the contract, such as when materials are delivered to the site[13] or when payment is made.

---

[8]*Signet Constr. Corp. v. Borg*, 775 F.2d 486 (2d Cir.1985).

[9]See Sections 18.04E and 33.03J.

[10]This power was exercised in *National Eastern Corp. v. United States*, 201 Ct.Cl. 776, 477 F.2d 1347 (1973).

[11]Federal Prompt Payment Statutes, 31 U.S.C.A. § 3901 et seq.

[12]West Ann.Cal.Bus. & Prof.Code § 7108.5 (applies to payments by contractors on private and local public contracts); West Ann.Cal. Civ. Code § 3260.1 (applies to private contracts); West Ann.Cal.Pub. Cont. Code § 10261.5 (applies to payment of contractors by state agencies); id. at 10262.5 (applies to payment by contractors on projects for state public works agencies). These statutes are interpreted in *Morton Eng'g & Constr., Inc. v. Patscheck*, 87 Cal.App.4th 712, 104 Cal.Rptr.2d 815 (2001) (subcontractor's claim successful) and *Denver D. Darling, Inc. v. Controlled Environments Constr., Inc.*, 89 Cal.App.4th 1221, 108 Cal. Rptr.2d 213 (2001) (subcontractor's claim denied because a bona fide dispute existed with the contractor). For a state-by-state compilation, see Hays, *Prompt Payment Acts: Recent Developments and Trends*, 22 Constr. Lawyer, No. 3, Summer 2002, p. 29.

---

[13]*In Franklin Pavkov Constr. Co. v. Roche*, 279 F.3d 989 (Fed.Cir. 2002), rehearing denied Mar. 22, 2002, the court ruled that the contractor, not the government, bore the risk of loss of government-furnished property stored in a fenced-in, but unlocked, enclosure at the project site. The court looked to the U.C.C.'s definition of "delivery" to determine the contractor's responsibility for discovering that the promised items were missing.

H. Assignment of Payments

Contractors or subcontractors often must borrow funds to operate their businesses. Lenders commonly require collateral to secure them against the possibility that the borrower will not repay the loan. Sometimes collateral consists of funds to be earned under specific construction contracts. Lenders often seek information from owners regarding the construction contracts whose payments are to be used as security and assurances by the owner that payments will be made to the lender or that checks will be issued jointly to lender and contractor.

Rather than rely on a promise, lenders more commonly demand that assignments be made to them of the payments. An assignment transfers the right to receive payment and effectuates a change of ownership in the rights transferred. It is a more substantial security than a promise. The party making the transfer is the *assignor*. The party to whom ownership is transferred is the *assignee*. The party owing the obligation being transferred is the *obligor*. Using the fact pattern of the prime contractor seeking a loan, the prime is the assignor, the bank the assignee, and the owner the obligor.

Such assignments were difficult if not impossible to accomplish early in English legal history. Modern law not only makes such assignments possible but also encourages them. However, these assignments should not put the obligor in a substantially worse position. Encouragement of assignments has gone to the extreme of invalidating contract clauses that prohibit assignment of such rights where the assignments are given as collateral to obtain loans.[14] Terms in contracts precluding assignment of payments for such purposes are invalid.

Although the validity of such assignments no longer raises serious questions, other legal problems exist. They center principally on the extent to which the original contract obligation can be changed because a third party (the assignee) now owns contract rights.

Normally, the obligor must pay the assignee after it has received notice of the assignment. But sometimes the obligor, such as the owner, can pay the assignor contractor to enable it to complete the contract.[15]

Modification of existing contracts whose rights have been assigned can be made "in good faith and in accordance with reasonable commercial standards" as long as payments assigned have not as yet been earned.[16] The obligor (owner) can offset against the assignee (lender) amounts owed it by the assignor (prime contractor).[17]

I. Lender Involvement

The assumption in this section has been that the owner is making payments to its prime contractor. Sometimes lenders involve themselves directly in the payment process. However, this can create the type of problem that was presented in *Davis v. Nevada National Bank*.[18] The lender was held liable for paying directly to a prime contractor despite warning by the owner that there were structural defects. The owner successfully pursued a claim against the contractor, but the contractor's bankruptcy wiped out that claim. As a result, the owner brought a claim against the lender. Recognizing that this was a "one of a kind" case and seeking to reassure lenders that it would not expose them to unreasonable risk, the court noted,

> Nothing we have said should be interpreted beyond the comparatively narrow confines of the instant case. Specifically, under usual construction loan terms and conditions, no lender should consider itself at risk if it elects not to generally inspect the progress of the construction of a project financed by the lender. Nor is a lender to consider itself at risk if it volitionally elects to inspect and does so negligently or ineffectively. A lender also has no duty, under our instant holding, either to withhold payment at borrowers' requests or to inspect on such requests, for construction deficiencies or omissions of a type that inevitably will occur in all projects and that commonly are remedied by a contractor as part of a "punch list" prior to project completion and the release of retained funds.[19]

The *Davis* case involved an owner claim against the lender. Subcontractors too may look to the lender for relief when the prime contractor does not pay them.

---

[14]*Mississippi Bank v. Nickles & Wells Constr. Co.*, 421 So.2d 1056 (Miss.1982); *Aetna Cas. & Sur. Co. v. Bedford-Stuyvesant Restoration Constr. Corp.*, 90 A.D.2d 474, 455 N.Y.S.2d 265 (1982). See U.C.C. § 9-406(d).

[15]*Fricker v. Uddo & Taormina Co.*, 48 Cal.2d 696, 312 P.2d 1085 (1957).

[16]U.C.C. § 9-405.

[17]Id. at § 9-404.

[18]737 P.2d 503 (Nev.1987).

[19]Id. at 506. A construction lender was found to have a fiduciary relationship with the owner in *First Nat'l Bank & Trust Co. of the Treasurer Coast v. Pack*, 789 So.2d 411 (Fla.Dist.Ct.App.2001), review denied, 817 So.2d 846 (Fla.2002).

Suppose the prime contractor obtains a loan from a lender. As explained in Section 22.02H, the contractor would provide security for the loan by assigning its future payments on a project to the lender. If the lender applies these payments from the owner to pay off the contractor's outstanding debt, there may be insufficient funds left to pay the subcontractors. For this reason, AIA A201–2007, Section 9.6.7, specifies that payments received by the contractor for work properly performed by subcontractors and suppliers "shall be held by the Contractor for those Subcontractors or suppliers." This language has been interpreted to create an express trust, which means the owner's payments (the trust funds) can only be used to pay the beneficiaries of the trust (the subcontractors and suppliers). The lender is liable for using the owner's payments to offset the contractor's debt if the bank should have known the funds were for a construction project and may be trust funds.[20]

Subcontractors may also claim that the lender had a duty to shield them from the contractor's misconduct. In *Hoida, Inc. v. M&I Midstate Bank*,[21] an unpaid subcontractor, unable to recover from the prime contractor, sued the project's construction lender. The subcontractor argued that the lender's disbursement of progress payments to the prime contractor were negligent because the lender did not first verify that the prime contractor had obtained lien waivers from the subcontractors. (The lien waivers would have meant the subcontractors had been paid.) A divided Wisconsin Supreme Court instead ruled that construction lenders do not breach a common law duty of care by making disbursements, authorized by the owner, without first verifying that the subcontractors had been paid. The court feared that imposing such a duty would place too great an administrative responsibility upon lenders.[22] It also pointed out that making the lender liable would violate the public policy of the mechanics' lien act, which gives liens by construction lenders priority over liens filed by subcontractors.[23]

J. Joint Checks

One of the principal aims of the payment process is to make certain that payments go to those who have provided labor or materials. Those who have provided labor or materials for private projects, such as contractors, subcontractors, and suppliers, usually have the right to assert a mechanics' lien on the property they have improved when they are not paid. This gives them a security interest in the property and a right to have the property sold to pay their claims.

To avoid this and also to ensure that those who are doing the work will continue to have the incentive to do the work, those who make payments, such as owners and prime contractors, seek to avoid diversion of the funds from those to whom the funds should go. One method sometimes used to avoid diversion is to have a payment issued by a joint check. For example, an owner may issue a joint check to a prime contractor for work that includes work by a subcontractor or materials furnished by a supplier. The names of both payees appear on the check. Each will have to endorse the check for it to be converted into cash. This should avoid the possibility that the prime contractor will take the funds and not pay those whose services or materials have provided the basis for the payment. Similarly, prime contractors sometimes issue payments to subcontractors by using a joint check, to ensure that the subcontractor pays sub-subcontractors or suppliers.[24]

This method, though apparently simple, creates legal problems. The first is whether the contract allows the payor to make payments by joint check. Using the owner as payor as an illustration, the power to pay by joint check initially depends on whether the prime contract allows this method of payment. Imposing the joint check method may generate opposition by the prime contractor who does not wish to have to obtain cooperation from the joint payee to cash the check. (Of course, this is the purpose for the joint check method.) Without a contractual power allowing the owner to use the joint check method, its unilateral use would be a breach of contract.[25] The contractor can claim with justification that the contract requires payment to the contractor and that any attempt to interject

---

[20]*Chang v. Redding Bank of Commerce*, 29 Cal.App.4th 673, 35 Cal. Rptr.2d 64 (1994); *Westview Investments, Ltd. v. U.S. Bank Nat. Ass'n*, 133 Wash.App. 835, 138 P.3d 638 (2006). As discussed in Section 28.07F, a trust in favor of the subcontractors and suppliers may also be created by statute.

[21]291 Wis.2d 283, 717 N.W.2d 17 (2006).

[22]Id., 717 N.W.2d at 32.

[23]Id. at 33–34.

---

[24]Barrett, *Joint Check Arrangements: A Release for the General Contractor and Its Surety*, 8 Constr. Lawyer, No. 2, April 1988, p. 7.

[25]*Piedmont Eng'g & Constr. Corp. v. Amps Elec. Co.*, 162 Ga.App. 564, 292 S.E.2d 411 (1982).

a third party interferes with its management prerogatives. Where this is discussed in advance and jointly resolved, it is likely that the clause will allow the owner to use the joint check process if the contractor is in default in paying subcontractors or suppliers.

Another contractual solution is to bar the issuance of joint checks if the contractor requests the owner not to do so because of a good-faith dispute it has with a subcontractor or supplier. The more difficult problem is whether the owner would impliedly have the power to use the joint check process if it had reasonable grounds to believe that the payee would not pay those to whom payments should be made. The obligation of good faith and fair dealing would likely give the owner this power under such circumstances, but the outcome of such a dispute would be difficult to predict.

Even if the payor has the power to pay by joint checks, problems can develop if the process misfires, principally because the payee obtains an endorsement from the copayee but the latter does not actually receive payment or the amount owed the copayee and the amount of the payment are not the same. In a California case,[26] the owner drew a check payable jointly to the contractor and a supplier, after which the supplier waived its mechanics' lien rights. Both parties endorsed the check. Proceeds were paid to the prime contractor, who gave its personal check to the supplier. Unfortunately for the supplier, the check was dishonored because the contractor did not have sufficient funds in its account.

The supplier then sued the contractor and received a judgment for the amount of the debt but could satisfy only a portion of it from the assets of the prime contractor. Then the supplier brought a lawsuit against the surety of the prime contractor for the balance. The court held that the unpaid supplier did not lose its right to claim on the bond despite having waived its lien and endorsing the check. Yet the joint check process clearly failed because the supplier did not insist on being paid at the time it endorsed the check but relied on a personal check issued by the prime.

K. Surety Requests That Payment Be Withheld

Payment bonds require the surety to pay unpaid subcontractors and suppliers. To recover any losses caused by having to make such payments, sureties often demand

that the owner stop paying the prime contractor when the latter has defaulted in its payments to subcontractors or suppliers.

In public work, unpaid subcontractors and suppliers do not have lien rights. As a result, the owner may decide to pay the prime contractor to enable the latter to continue performance and complete the work. In doing so, the surety's right to reimbursement can be affected adversely. Cases decided in the U.S. Court of Claims, now the U.S. Court of Appeals for the Federal Circuit, have given the contracting officer broad discretion to pay earned progress payments to the contractor despite some minor defaults.[27] The court has recognized the government's interest in obtaining a completed project by giving discretion to make payments even though payment can harm the surety.

Although the contracting officer is given considerable discretion, a subsequent opinion by the Court of Claims requires that that discretion be exercised "responsibly" and that the surety's interest be considered.[28] In one case, a summary judgment had been granted to the government by the trial court based solely on the contractor's having been on schedule and having completed 91 percent of the project. The appellate court held that the summary judgment in favor of the government was improper where the surety contended that progress payments and retainage should not have been released when it notified the government of the contractor's impending default.[29] Other factors must be examined to determine whether the government exercised reasonable discretion in disbursing the funds. Even if the government acted unreasonably in paying the contractor, the surety must still establish that it was damaged by the improper payments.[30]

Suppose the owner accedes to the surety's requests or demands. In such a case, the unpaid prime contractor may, in addition to claiming a right to the payment withheld, assert a claim against the surety for wrongful interference with the contract that the prime contractor has made with the owner.

---

[26]*Ferry v. Ohio Farmers Ins. Co.*, 211 Cal.App.2d 651, 27 Cal.Rptr. 471 (1963).

[27]*Argonaut Ins. Co. v. United States*, 193 Ct.Cl. 483, 434 F.2d 1362 (1970).

[28]*United States Fid. & Guar. Co. v. United States*, 201 Ct.Cl. 1, 475 F.2d 1377 (1973).

[29]*Balboa Ins. Co. v. United States*, 775 F.2d 1158 (Fed.Cir.1985).

[30]*National Surety Corp. v. United States*, 118 F.3d 1542 (Fed.Cir.1997). See also Section 32.10H.

L. Remedies for Nonpayment

For convenience, the discussion will assume an unpaid prime contractor. However, any conclusions expressed in that context are likely to apply to unpaid first- or second-tier subcontractors. The law, as noted in Section 22.02F, recognizes the importance of prompt payment. Yet under certain circumstances, drastic remedies for nonpayment may not be appropriate. Sometimes delay in payment is unavoidable. Delay may not harm the contractor. Care must be taken to avoid an unpaid party's using minor delay as an excuse to terminate an unprofitable contract and causing economic dislocation problems. Despite these possibilities, on the whole nonpayment is—and should be—considered a serious matter.

At the outset, there must be a determination that failure to make payment is a breach of contract. Because construction contracts are detailed and contract procedures often ignored, those accused of not paying in accordance with the contract often assert that prompt payment has been waived.

If payment is not made as promised, the contractor is entitled to interest on the payment. Because as a rule the payments are liquidated or relatively certain in amount, the interest runs from the date payment was due.

AIA documents state that in the absence of a specified rate in the contract, late payments are paid at the "legal rate" of interest.[31] This term generates confusion. Generally, the legal rate is the amount specified by law that is paid on unpaid court judgments. Yet some states hold that the legal rate is the highest rate that can be lawfully exacted for the particular transaction. Until the explosive inflation of the 1970s, the interest rate on unpaid judgments ran roughly between 5 and 8 percent. During that inflationary period, the actual market rate for a commercial loan was between 15 and 20 percent. In states that used the unpaid judgment rate, owners were often tempted to delay payment and pocket the difference between the two rates.

In the 1980s, inflation subsided and some states increased the rate of interest on unpaid judgments. For that reason, generalizations at any given time as to the differential between these two rates can be perilous. To add complexity, in many states commercial loans have no limits on interest. Contracting parties should agree on an interest rate for late payments related to the market rate, rather than leave it to the legal rate.

As noted in Section 22.02F, state prompt payment statutes increasingly provide stiff penalties for violations. For example, in California, failure to pay a subcontractor triggers a penalty of 2 percent of the amount due per month for each month the payment is not made in addition to whatever interest is provided by law or by contract.[32] Also, if the subcontractor must resort to legal action and it prevails, it is entitled to its attorneys' fees and costs.[33] An unpaid prime contractor can recover the 2 percent penalty, but this is in lieu of any other interest it would be entitled to recover. If it prevails, it can also recover attorneys' fees.[34] Local law must be consulted.

Contractors *unable* to borrow money may suffer other losses, such as the ability to bid on other projects, loss of key personnel who leave when they are not paid, or eviction caused by nonpayment of rent. If such losses are reasonably foreseeable at the time the contract is made and could not have been reasonably avoided by the contractor, they can be recovered by the contractor if they can be proven with reasonable certainty. The obstacles to recovering for losses of this sort are formidable, and as a rule, interest is all that can be recovered for delayed payment or nonpayment.

A formidable weapon in the event of nonpayment is the power to suspend work. In addition to placing heavy pressure on the owner, suspension avoids the risk of further uncompensated work. The availability of a mechanics' lien is a pale substitute when there is a substantial risk of nonpayment. Until recently, the law was not willing to recognize a remedy short of termination for such breaches. Some cases now hold that the contractor can suspend work if it is not paid.[35]

Nonpayment certainly should not automatically give the right to suspend performance. Shutting down and starting up a construction project is costly. It would be unfair to allow the contractor to shut down the job simply because payment is not made absolutely on schedule.

---

[31]AIA Docs. A101-2007, § 8.2; A201-2007, § 13.6; A401-2007, § 15.2; and B101-2007, § 11.10.2. See Section 6.08 for further discussion of interest.

[32]West Ann.Cal.Bus. & Prof.Code § 7108.5.

[33]Ibid.

[34]West Ann.Cal.Civ.Code § 3260.1(b), § 3260(g).

[35]*Hart & Son Hauling, Inc. v. MacHaffie*, 706 S.W.2d 586 (Mo.Ct. App.1986); *Zulla Steel, Inc. v. A. & M. Gregos*, 174 N.J.Super. 124, 415 A.2d 1183 (App.Div.1980). See also Restatement (Second) of Contracts § 237 (1981).

Under AIA Document A201-2007, Section 9.7, seven days after a progress payment should have been made the contractor can give a seven-day notice of an intention to stop work unless payment is made. Failure to pay after expiration of the notice period permits suspension. Section 9.7 also states that the contractor shall receive a price increase for its reasonable costs of shutdown, delay, and startup. Interest is added to these amounts. Undoubtedly, when it is not paid after a reasonable period of time the contractor should be able to suspend work. However, the two seven-day periods may be too short.

Can an unpaid contractor terminate its obligation to perform? Suspension is temporary. Termination relieves the contractor from the legal obligation of having to perform in the future. In the absence of any contract provision dealing with this question, termination is proper if the breach is classified as "material."

Although materiality of breach is discussed in greater detail in Section 33.04, it may be useful to look at some factors that relate to nonpayment as a material breach. The most important are effect on the contractor's ability to perform and the likelihood of future nonpayment. Persistent nonpayment may indicate that the problem is a serious one. A clear statement that performance will not be made is a repudiation and clearly gives the right—and probably the obligation—to stop performance. Termination is an important decision. The law looks at many factors before deciding whether there has been a material breach. Many cases, however, have concluded that a failure to pay (often together with other breaches) gives the contractor the right to terminate its obligation to perform under the contract.[36]

Emphasis in the discussion and in the cases has been on the effect of nonpayment on the contractor's ability to perform. This can operate to the disadvantage of a financially sound contractor. Another approach, and perhaps a better one, is to permit termination if a changed cash flow will cause the contractor to finance the project to a larger degree than anticipated. If, however, the facts clearly demonstrate that a particular contractor was selected for, among other factors, its capacity to absorb payment delay, this may show an intention by the contractor to accept the risk of substantial alteration of financing the work.

Suppose the contractor continues performance. Continued performance may indicate that nonpayment was not sufficiently serious to constitute termination. Continued performance may manifest an intention to continue performance that is relied on by the owner. However, continued performance should not invariably preclude nonpayment from justifying termination. An unpaid contractor may choose to continue work for a short period while awaiting performance.[37] This would be especially true if the contractor clearly indicated that if payment were not forthcoming, work would cease.

AIA Document A201-2007, Section 14.1.1.3, permits the contractor to terminate if the work has been suspended for thirty days for nonpayment. The contractor must first give a seven-day written notice that it intends to terminate. Does this express termination provision affect any common law power to terminate for nonpayment? That depends on whether the express power to terminate is exclusive. Generally, the law hesitates to make specified remedies in a contract exclusive unless there is a clear indication that this is the intention of the parties.[38] This has been codified in A201-2007, Section 13.4.1, which states that remedies specified are in addition to any remedies otherwise imposed or available by law.

Can the owner preclude termination by paying all unpaid progress payments with interest during the seven-day period? As noted in Section 33.03F, the notice period may allow the owner to cure past defaults. Termination is a drastic remedy, and the obligation of good faith and fair dealing support the conclusion that the owner can cure past defaults during the notice period. Section 27.02 discusses remedies to a contractor who has justifiably ceased performance or who has been wrongfully removed from the site. Some courts allow the contractor to recover only for work performed and not for lost profits.[39] However, the general tendency is to treat this breach as no different from any other.

---

[36]*Guerini Stone Co. v. P. J. Carlin Constr. Co.*, 248 U.S. 334 (1919); *Integrated, Inc. v. Alec Fergusson Elec. Contractors*, 250 Cal.App.2d 287, 58 Cal.Rptr. 503 (1967) (many factors in addition to nonpayment); *Leto v. Cypress Builders, Inc.*, 428 So.2d 819 (La.App.1983) (nonpayment made work precarious); *Aiello Constr., Inc. v. Nationwide Tractor Trailer, etc.*, 413 A.2d 85 (R.I.1980) (substantial underpayment for prolonged period).

[37]*Darrell J. Didericksen & Sons, Inc. v. Magna Water & Sewer Improvement Dist.*, 613 P.2d 1116 (Utah 1980). But see *Drew Brown Ltd. v. Joseph Rugo, Inc.*, 436 F.2d 632 (1st Cir.1971).

[38]*Glantz Contracting Co. v. General Elec.*, 379 So.2d 912 (Miss. 1980); *Bender-Miller Co. v. Thomwood Farms, Inc.*, 211 Va. 585, 179 S.E.2d 636 (1971).

[39]*Palmer v. Watson Constr. Co.*, 265 Minn. 195, 121 N.W.2d 62 (1963).

The effect of owner nonpayment on the right of a subcontractor to recover for its performance is treated in Section 28.06.

M. Payment as Waiver of Defects

Sometimes contractors contend that making progress payments waives any deviations from contract document requirements. Generally, payment in and of itself does not waive defects.[40] Suppose the owner knows or should know of the defect and makes payment. This may lead the contractor reasonably to believe that the owner intends to pay despite the defect. If so, minor defects are waived.

To preclude waiver in such cases, construction contracts frequently contain provisions stating that progress payments do not waive claims for defects. For example, Section 9.6.6 of AIA Document A201-2007 states that progress payments are not an acceptance of work that is not in compliance with the contract documents. The payment certificate should make clear that payment does not constitute waiver.

Acts that can constitute waiver of defects more commonly arise when the project is accepted or completed and final payment is made. Detailed discussion of this problem is found in Section 24.05.

N. Progress Payments and the Concept of Divisibility

Divisibility, sometimes called *severability*, is a multipurpose legal concept sometimes applied to a contract where performance is made in installments or is divided into designated items. As an illustration of the first, the contract for the sale of goods could allow twelve monthly deliveries with payment at the time of each delivery. An illustration of the second would be a contract for the sale of goods that consisted of five different items to be delivered with different prices for each item.

If a contract is considered divisible, it would be as if there were twelve separate contracts in the first illustration and five in the second. Each partial performance,

whether a monthly delivery or delivery of less than all the five items, would be considered the equivalent of the money promised for each installment or item. Without describing the many legal issues that can sometimes depend on the divisibility classification, the progress payment mechanism should not result in the construction contract being considered divisible. The amounts certified are approximations and not agreed valuations for the work as it proceeds. The amounts can be—and frequently are—adjusted at the time of final payment.[41]

O. Payment as Preference

Bankruptcy law prohibits payments to creditors a certain number of days before filing of bankruptcy. Allowing such a payment would prefer some creditors over others. See Section 33.04C.[42]

SECTION 22.03 Retainage

Retainage (retention) is a contractually created security system under which the owner retains a specified portion of earned progress payments to secure itself against certain risks. For example, suppose the schedule of values establishes that the contractor is entitled to a progress payment of $50,000. Construction contracts commonly provide that a portion of this $50,000 will be retained by the owner and paid on completion. Consistent with the AIA position that to a large degree equates substantial with final completion as shown in Sections 22.04 and 26.04, Section 9.8.5 of AIA Document A201-2007 requires that any retainage, with the consent of any surety, be released to the contractor when the owner and contractor have accepted the responsibilities given them in the Certificate of Substantial Completion. The owner may retain an amount "for Work that is incomplete or not in accordance" with the contract requirements.

The purpose of retainage is to provide money out of which claims that the owner has against the contractor can be collected without the necessity of a lawsuit.

[40]*Annen v. Trump*, 913 S.W.2d 16 (Mo.Ct.App.1995) transfer denied, May 2, 1995 (acceptance and payment did not bar claim). See Annot., 66 A.L.R.2d 570 (1959). Waiver generally is discussed in Section 22.06E.

[41]*Dravo Corp. v. Litton Systems, Inc.*, 379 F.Supp. 37 (S.D.Miss. 1974); *Kirkland v. Archbold*, 113 N.E.2d 496 (Ohio App.1953).

[42]See Hughes & Dunning, *Holding on to What You Have Been Paid: Defending a Preference Action Against a Contractor, Subcontractor or Supplier*, 16 Constr. Lawyer, No. 1, Jan. 1996, p. 15.

Retainage is not an agreed damage or damage limit.[43] Some states require retainage in public contracts.

Retainage has become controversial. Contractors and subcontractors increasingly request legislators to help them. They contend that retention of money they have earned is unfair to them and costly and unnecessary to the owner. They assert that the reasons for creating the security, such as defective work or the prime contractor not paying subcontractors or suppliers, are dealt with by the contractor's furnishing performance and payment bonds. Contractors and subcontractors also contend that financing costs that are inevitably incurred because of retention are ultimately transferred to the owner through the contract price. They then argue that the owner is paying twice for the same risk—once through financing costs included in the contract price and once by the cost of surety bond premiums.

This argument equates the efficacy of retainage with bonds. Obviously, it is better for the owner to actually have funds within its control that it can use to secure the owner against these risks, rather than have to deal with or send unpaid subcontractors or suppliers to the surety. Also, owners can earn interest on the funds retained. Some owners keep the retainage after completion to address postcompletion defects. As shall be seen, some legislatures have required that retention be released on substantial completion, retaining only an amount related to the estimated cost of punch list items.

Early finishing subcontractors, such as excavating subcontractors, make another attack on retainage. They would like line item retention, in essence decoupling their work from that of the other contractors on the project and granting them the right to the retainage for their work when their work is finished.

As one purpose of retainage is to pay lien claimants, subcontractors and suppliers can benefit from retainage, something shown by state statutes to be discussed later in this section.

Contractors also complain that they should be able to substitute other securities for the retainage and obtain the use of the retainage as quickly as possible. They also assert that owners unfairly earn interest on the retainage. California, by statute, provides that a contractor on a state project is entitled to interest on any securities deposited in lieu of retainage, and that same right is conferred on subcontractors performing more than five percent of the total bid.[44] In a Missouri case, a city allowed bidders to establish an escrow account consisting of securities acceptable to the city in lieu of the city withholding retainage. The city's specification did not state who would receive interest generated by the account during the construction period. The court ruled that the contractor, who had established the account, was entitled to the interest on it.[45]

Contractors and subcontractors may worry about the safety of the retainage. Will the owner have the money to pay the retainage at the end of the project? Finally, in the progress payment system, retainage acts as an incentive to move the work along. It is a spur to the contractor to finish the project and obtain the retained amounts.

There has been some tendency to reduce retainage, and some contracts eliminate it entirely. Sometimes retainage is limited to the first 50 percent of the work. This should give the owner an indication of whether the contractor will do a good job and finish on time. After 50 percent has been completed, the amount retained is held until there has been substantial completion. At that point, as has been noted, the owner retains an amount related to the estimated amount needed to complete the work (punch list items). Some contracts allow the contractor to receive the retention at the 50 percent mark.

Complaints of contractors led to a decision by the Office of Federal Procurement Policy[46] to effectuate a uniform government-wide retainage policy. Retainage must not be used as a substitute for good contract management. The agency cannot withhold funds without good cause. Determinations concerning retainage should be based on an assessment of the contractor's past performance and the likelihood that such performance will continue. The office suggests that retainage not exceed 10 percent and that it be adjusted downward as the contract approaches completion, particularly if there is better than expected

---

[43]An earlier edition of this book was cited and followed in *Van Knight Steel Erection, Inc. v. Housing & Redevelopment Auth. of the City of St. Paul*, 430 N.W.2d 1, 3 (Minn.App.1988).

[44]West Ann.Cal.Pub.Cont.Code §§ 10263(d) (contractors) and 10263(e) (subcontractors). Accord, Fla.Stat.Ann. § 255.052 (permits substitution of securities in lieu of retainage on public contracts).

[45]*McCarthy Building Cos. v. City of St. Louis*, 81 S.W.3d 139 (Mo. App.2002), transfer denied Aug. 27, 2002.

[46]OFPP Policy Letter 83-1, 48 Fed.Reg. 22,832 (1983). This has been implemented in Federal Acquisition Regulation § 32.103; 48 CFR § 32.103 (2007).

performance or alternate safeguards.[47] Once all contract requirements have been completed, all retained amounts should be paid promptly to the contractor.

Contractor complaints, as noted, have increasingly led to legislation.[48] Statutes in some states recognize the problems described earlier in this section and attempt to deal with them. Texas, for example, has long allowed a mechanics' lien on retainage. In contracts for which liens by subcontractors and suppliers may be filed, the owner must retain a 10 percent retainage as security for lien claimants. In addition, lien claimants are given a preference on the retainage that may entitle them to payment ahead of the prime contractor. The owner's failure to comply with its requirement for retainage gives lien claimants preferential mechanics' liens on the property itself.[49]

Michigan limits the retainage to 10 percent until 50 percent of the work is in place. At that point retention stops unless the contractor is not making substantial progress. In that case more than 10 percent can be retained. Michigan prohibits state owners from commingling retainage with other funds. With some exceptions for federally funded projects, retainage must be placed in an interest-bearing account. At final payment, the contractor receives the retainage and the interest it earned. After 94 percent of the work has been completed, the contractor can post an irrevocable letter of credit and receive the retainage.[50]

In contracts over $500,000, Tennessee requires the owner, public or private (or contractor who holds retainage for subcontractors), to deposit retainage in escrow. The escrow holder must give security for the retainage. The interest earned by the retainage goes to the contractor or subcontractor. Rights granted by the statute cannot be waived.[51] Tennessee allows the prime contractor, by posting security, to obtain the retainage held by local entities.[52]

California has enacted legislation that requires local entities to include provisions in any invitation to bid, with certain exceptions, that allow the contractor to substitute certain securities for retainage held by the agency or an escrow agent. When this system is invoked, the owner makes full progress payments to the contractor from which retainage normally would have been withheld. In the event of a default by the contractor, the owner can draw on the securities by notifying the escrow agent, who converts the securities to cash and distributes the cash as instructed by the owner. Alternatively, at the written request of the contractor the owner makes payments of the retainage directly to the escrow agent, who invests it on behalf of the contractor. On satisfactory completion of the contract, the escrow agent distributes to the contractor the securities and any payments made to the escrow agent by the owner. The contractor must pay each subcontractor, no later than twenty days after receipt of this payment, the respective amount of interest earned attributable to funds withheld from each subcontractor. All expenses of the escrow arrangement are paid by the contractor.[53]

Another indication of legislative activity can be seen by legislation enacted in Idaho in 1990.[54] The legislation covers private works of improvement. Under the statute, the retainage must not exceed 5 percent of any payment, and the total withheld can never exceed 5 percent of the contract price, but only if the contractor or subcontractor posts a performance bond. The 5 percent maximum does not apply to contracts for the improvement of residential property consisting of from one to four units (one of which is occupied by the owner).

Thirty-five days after substantial completion, the amount retained must be reduced to whichever is less: 150 percent of the estimated value of the work to be completed or the amount retained by the owner—again not to exceed 5 percent. Within thirty-five days from final completion, the retainage must be released except in the event of a dispute. If there is a dispute, the amount that can be withheld from the final payment cannot exceed 150 percent of the estimated value in dispute.

Within ten days from the time of receipt, the prime contractor must pay its subcontractors any retainage plus interest that is received. The prime contractor can deduct from the interest paid a 1 percent fee for administration. However, the retainage need not be paid to the subcontractor if a bona fide dispute exists between the prime

---

[47]A contracting officer may also release the retainage early in an effort to alleviate the contractor's cash flow problems. *Fireman's Fund Ins. Co. v. United States*, 909 F.2d 495 (Fed.Cir.1990).

[48]A recent canvassing of legislation is found in Stockenberg & Limbaugh, *Fifty-State Review of Retainage Laws*, 22 Constr. Lawyer, No. 1, Spring 2002, p. 24.

[49]Tex.Prop.Code Ann. §§ 53.102–105.

[50]Mich.Comp.Laws Ann. § 125.1563.

[51]Tenn.Code Ann. § 66-11-144.

[52]Id. at § 12-4-108.

[53]West Ann.Cal.Pub.Cont.Code § 10263.

[54]Idaho Code § 29-115.

and the subcontractor. The amount withheld in such case must not be more than 150 percent of the estimated value of the work to be completed or amount in dispute. If the retainage is not paid within the time limits required, the payments will be subject to an additional charge over and above interest of 1½ percent per month. For segregated accounts created by the owner or lender, provisions also exist under which funds are not mixed with other funds of the owner. Retained funds are deposited into these segregated accounts when withheld by the owner, and the accounts bear interest. The interest earned must be paid to the prime contractor, but the owner can deduct from the interest an amount to cover administration. Most important, the Idaho legislation states that the parties cannot waive any provisions of this statute.

It can be seen that increased legislative activity in the field of payments with regard to their promptness and with regard to retainage has meant that parties who are about to engage in a construction contract must check local laws to see whether these laws can be varied by contract and to ensure that these laws are followed. Particular reference must be paid to whether the laws are limited to public contracts, are limited to private contracts, or cover both types. Typically, a regulation of this sort begins in public contracts and often is extended to private contracts. However, as seen in the Idaho statute, smaller projects are sometimes exempt.

One problem incident to retainage can develop at the end of the project. Suppose the retained amount is $50,000 and the owner is entitled to take $20,000 from it to remedy defects in the work. A solvent contractor will be paid the balance. However, in the volatile construction industry, it is not uncommon for a contractor to run into financial problems. In such a case, a horde of claimants descends on the owner and demands the money. The claimants can be lenders who have received assignments of contract payments as security for a loan, sureties who have had to discharge the obligations of the contractor, taxing authorities who claim the funds when the contractor has not paid its taxes, and, in the event the contractor has gone bankrupt, the trustee in bankruptcy.

The scramble for funds usually is dealt with by the owner's initiating what is called an "interpleader" action. The owner pays the disputed funds into court, files a lawsuit in which it names as parties all claimants to the funds, and withdraws from the fray. The court must unscramble the claims.

Although the possible multiple claimants' confusion at the end of the job may not in itself be a reason to limit or eliminate retention, the troublesome disputes that relate to retainage ownership should be taken into account in determining whether and to what extent retainage should be used.

SECTION 22.04 Substantial Completion

Sometimes construction contracts provide that a payment will be made to the contractor on substantial completion. A201-2007 defines substantial completion in Section 9.8.1 as "sufficiently complete . . . so that the Owner can occupy or utilize the Work for its intended use." Perhaps more important, the "end of the job" process, which begins with the contractor's representation that it has substantially completed the project and concludes with a certificate of final completion, is one about which a number of legal issues cluster (described in this section and in Section 22.05).

Although time problems are discussed in Chapter 26, it should be noted here that under AIA Document A101-2007, Section 3.3, and A201-2007, Section 8.2.3, compliance with the contractor's time obligation is measured by substantial completion.

The process for determining substantial completion under A201-2007, Section 9.8.2, begins with a submission to the architect by the contractor of a list of items to be completed or corrected. Under Section 9.8.3 this is followed by an inspection by the architect to determine whether she should issue a certificate of substantial completion. Such inspections are more carefully made than are inspections for ordinary progress payments. More important issues are involved. For example, this may be the point at which the owner retakes possession of the site and begins to use the project.[55] This can have important insurance consequences. Also, statutes of limitation that cut off liability a designated number of years after completion frequently use substantial completion as the event that triggers the commencement of the period.[56]

A201-2007, Section 9.9.1, permits the owner to occupy or use a portion or all of the project prior to substantial completion.

---

[55]AIA Doc. A201-2007, § 9.8.1.
[56]See Section 23.03G.

SECTION 22.05 Completion and Final Payment

The AIA process begins with an act by the contractor, a notice that the work is ready for final inspection and acceptance, and an application for final payment. The architect inspects the work to determine whether the certificate should be issued—an inspection that should be undertaken with great care.

This stage in the construction process is a benchmark that has serious implications for major participants in the process. If the owner has not taken possession of the project at the time of substantial completion, it will certainly do so at this point.

Possession by the owner involves legal responsibilities that rest on the possessor of land.[57] Perhaps more important, the end of the job often has an effect on allocation of risks, such as those that relate to defects, and the continued existence of any claim. Because completion is sometimes equated with acceptance—an important legal doctrine that relates to defects—it is discussed in Sections 14.09B and 24.05.

Contractors also have claims that may be affected by completion and final payment. As noted in Section 27.13, the law looks for benchmarks that can put an end to disputes. In the case discussed in that section the benchmark under discussion—payment for changed work—reflected some of the pressure contractors face when seeking final payment.

The effect of final payment accompanied by a release and its effect on contractor claims surfaced in *Mingus Constructors, Inc. v. United States*.[58] Mingus substantially completed his work on July 30, 1982. In July and August, he sent letters to the contracting officer stating that he intended to file a claim and giving some indication as to the basis of the claim.

On October 28, 1982, the contracting officer inquired about the status of the claims release form the government had sent to Mingus. This was part of the final payment process. Mingus was advised that in the space provided he could "except" his claims against the government and that the exception would not hold up processing of his final payment. On the following day, he executed the claims release form, which stated that he released all claims. In the space provided for exceptions,

Mingus stated that pursuant to correspondence, he would file a claim, the amount of which was undetermined. He was paid on November 2, 1982. On November 5, 1982, he wrote to the government seeking confirmation of his understanding that by inserting the statement in the form he had not waived his claims. The contracting officer replied on November 10, 1982, stating that he could "except" any claims in the space provided without waiving his rights.

Mingus finally submitted his claim on January 5, 1984. When the claim was returned to Mingus, Mingus brought legal action in the Claims Court. The government took the position that Mingus had not protected his claim when he executed the release. The Claims Court held in favor of the government. This decision was affirmed by the federal appellate court based on a contractual provision barring claims after final payment. Preserving the claim, according to the court, can be done only by specifically excepting claims at the time of the general release.

Mingus was barred because of his release. Releases are strictly construed against the contractor. The exception Mingus included in the release, even if supplemented by his prior letters, was a "blunderbuss exception," which "does nothing to inform the government as to the source, substance, or scope of the contractor's specific contentions."[59] The court felt it was important that releases should put an end to controversies. Although the contractor need not include a final certified claim when it executes the release, merely stating the claim in general terms does nothing to preserve the claim.

Failure to provide a contractually required notice at the time of final payment was held to have waived not only the public contractor's claims against the public entity but also its negligence claim against the public entity's engineer. This was based on the acceptance of the final payment form stating that acceptance released the owner and others for any claims or any liability to the contractor arising out of the work.[60]

A201-2007 deals with release of contractor claims in Section 9.10.5. To protect its claim when it has accepted final payment, the contractor must have made the claim previously in writing and must have identified that

[57]See Section 7.08.
[58]812 F.2d 1387 (Fed.Cir.1987).

[59]Id. at 1394.
[60]*McKeny Constr. Co., Inc. v. Town of Rowlesburg*, 187 W.Va. 521, 420 S.E.2d 281 (1992).

claim as unsettled at the time of the application for final payment. Again, the needed specificity of the previous written claim can generate disputes. For the owner to be aware of the nature of the contractor's claim—both to investigate it and to have some idea of the contingencies that must be set aside to deal with it[61]—the contractor should give as much information as is reasonably available. As noted in the *Mingus* case, blunderbuss reservations of rights may not be sufficient.

On the other hand, courts will not construe even a broadly worded release to apply to a contractor's pre-existing claim unless the release explicitly refers to that claim. To borrow the *Mingus* court's terminology, even a blunderbuss release will be limited as to its scope in accordance with the parties' intent. Determination of that intent may require the admission of extrinsic evidence.[62]

Mechanics' liens can surface at the end of the job. Although they are discussed in greater detail in Section 28.07D, dealing with subcontracts, this section looks briefly at liens from the vantage point of the owner, with particular reference to the end of the job.

Owners employ a variety of techniques to avoid liens. Some relate to the payment process. In some contracts, though not those of the AIA, owners seek to avoid liens by requiring evidence that subcontractors and suppliers who have lien rights either have been paid or have executed lien waivers before the contractor receives progress payments. Even if this is not required for a progress payment, it is almost certainly required for final payment. For example, AIA Document A201-2007, Section 9.10.2, conditions final payment on the contractor's submitting an affidavit that it has paid all its bills and gives the owner the right to require other data "such as receipts, releases and waivers of liens . . . arising out of the Contract, to the extent and in such form as may be designated by the Owner." Because of concern that lien waivers or—even worse—"no lien" contracts (under which the prime contractor waives all liens in advance for itself, its subcontractors, and suppliers) may be unfair to subcontractors and suppliers,[63] legal controls have

been enacted by the states that regulate these methods.[64] Where this is the case, as noted in Section 22.02J, owners may choose to issue final payments as well as progress payments through joint checks.

Although these legislative activities have been designed to protect subcontractors and suppliers, other legislative activity protects owners from liens. For example, an Iowa statute provides that the owner need not make final payment until ninety days after completion of the project if the project would be subject to a lien.[65] This does not apply if the prime contractor has filed a payment bond or supplied signed receipts or lien waivers. Generally, liens must be filed a designated number of days after substantial or final completion.[66] One way of avoiding liens is to delay final payment until the time for filing liens has expired.

Owner protection does not eliminate the requirement that contracting parties behave fairly toward one another. For example, one case involved a contract with lien avoidance conditions before final payment. Yet the contractor was awarded final payment without complying because the owner had insisted that the contractor waive its delay damage claim before final payment would be made. The court found this tactic unconscionable.[67]

SECTION 22.06 Payment for Work Despite Noncompliance

A. The Doctrine of Conditions

The promise by the owner to make any payments is conditioned on the contractor's compliance with the contract documents. During the course of the work, the contractor is not entitled to be paid unless, as discussed in Section 22.02A, provisions are made for progress payments. At the end of the project, the doctrine of conditions in its strictest sense requires absolute compliance with the contract documents before the contractor is entitled to any unpaid part of the contract price.

---

[61]*Bornstein v. City of New York*, 94 A.D.2d 683, 463 N.Y.S.2d 198 (1983) (contractor had filed claim before accepting final payment).

[62]*Metric Constructors, Inc. v. United States*, 314 F.3d 578 (Fed.Cir. 2002); *Sedona Contracting, Inc.*, ASBCA No. 52093, 99-2 BCA ¶30,466.

[63]West Ann.Wis.Stat. § 779.03(1) prohibits general contractors from waiving the lien rights of subcontractors. However, § 779.03(2) allows the owner to eliminate those rights by requiring the contractor to obtain a payment bond.

[64]O.C.G.A. § 44-14-366, enacted in Georgia, makes lien waivers made in advance of furnishing labor and materials null and void and sets forth statutory interim and final lien waiver forms. Similarly, see West Ann.Cal.Civ.Code § 3262. See also Section 28.07D.

[65]Iowa Code Ann. § 572.13.

[66]West Ann.Cal.Civ.Code § 3116 (ninety days after completion or cessation of work or thirty days after notice of cessation or completion filed by owner).

[67]*North Harris County Junior College Dist. v. Fleetwood Constr. Co.*, 604 S.W.2d 247 (Tex.Ct.App.1980).

Although the owner would be permitted to hold back all the money until it receives what has been promised in construction contracts, strict application of the doctrine can cause hardship to the contractor. During performance, progress payments are essential to finance the job and avoid the risk of going unpaid for work. At the end of the job, a strict application of the doctrine of conditions can cause a loss to the contractor disproportionate to the loss caused by nonperformance. Such loss can create unjust enrichment where the owner is occupying the project. This section outlines some of the legal doctrines that have developed to relieve the contractor when performance has not been in exact compliance with contract documents.

A subcontractor's promise to complete by a designated time was held to be a promise and not a condition to the prime contractor's obligation to pay the contract price.[68] The court noted that conditions are not favored, because they can create forfeiture and unjust enrichment. The contractor can recover the unpaid balance, but it must pay any damages caused by delay.

In the absence of a valid liquidated damages clause, delay damages are very difficult to prove. The ultimate outcome is likely to be that the contractor will collect for the work despite its late performance. However, were timely performance a condition for final payment—unlike defects of quality that can be cured—the prime contractor would have the benefit of the subcontractor's work without paying fully for it.

B. Substantial Performance: *Plante v. Jacobs*

The substantial performance doctrine developed early in English law and made its way into American law. The dimensions of the doctrine will be sketched later in this subsection; in its simplest form, it requires the owner to pay the balance of the construction contract price if the contractor has substantially performed, leaving the owner a claim for damages based on failure of the contractor to perform strictly in accordance with the contract requirements.

Most substantial performance cases involve construction contracts. Construction is a complex undertaking, with detailed contract requirements. A strong likelihood exists that minor deviations will surface at the end of the job. Work performed by prime contractor employees may escape the scrutiny of the prime's superintendent. Much of the work is performed by subcontractors, and the exact quantity and quality may be difficult to determine while the work is being performed.

The complexity of contract documents frequently generates interpretation questions, and it is often difficult to get an authorized interpretation by the design professional while work is being performed. Likewise, it may be difficult to find someone with authority to direct a change or approve substituted materials or products claimed by the contractor to be equal or equivalent to those specified.

Unlike in other types of contracts, the breaching party—that is, the contractor—cannot take back its performance. The contractor's performance has become incorporated into the owner's land, and very frequently the owner has taken possession of the project.

With some limited exceptions,[69] substantial performance issues usually arise at the end of the job. The prime contractor contends that the owner has essentially what it bargained for and often points to the owner's occupying and using the project. One court defined substantial performance as "when construction has progressed to the point that the building can be put to the use for which it was intended, even though comparatively minor items remain to be furnished or performed in order to conform to the plans and specifications of the completed building."[70]

Another case added that the defects must not "so pervade the whole work that a deduction in damage will not be fair compensation."[71] Yet another case concluded there had not been substantial performance when there was an accumulation of small defects, even though the owner was occupying the premises.[72]

The ratio of defect correction costs to contract price and the determination of whether the building has met its essential purpose are relevant.[73] But one court affirmed a

[68]*Landscape Design & Constr., Inc. v. Harold Thomas Excavating, Inc.*, 604 S.W.2d 374 (Tex.Ct.App.1980).

[69]*Nordin Constr. Co. v. City of Nome*, 489 P.2d 455 (Alaska 1971) (owner sued to recover payments).

[70]*Southwest Eng'g Co. v. Reorganized School Dist. R-9*, 434 S.W.2d 743, 751 (Mo.Ct.App.1968). Accord, *All Seasons Constr., Inc. v. Mansfield Housing Auth.*, 920 So.2d 413 (La.App.2006) (contractor substantially performed where the building was occupied and the punch list items were mostly cosmetic).

[71]*Jardine Estates, Inc. v. Donna Brook Corp.*, 42 N.J.Super. 332, 126 A.2d 372, 375 (App.Div.1956).

[72]*Tolstoy Constr. Co. v. Minter*, 78 Cal.App.3d 665, 143 Cal.Rptr. 570 (1978).

[73]*Stevens Constr. Corp. v. Carolina Corp.*, 63 Wis.2d 342, 217 N.W.2d 291 (1974).

judgment of substantial performance despite cost of completion and correction of defective work being 31 percent of the contract price.[74]

Jacob & Youngs, Inc. v. Kent,[75] a leading case, found that there had been substantial performance when it was discovered after completion that one brand of pipe had been substituted for the one specified. The court noted that the pipe substituted was of equal quality to that designated.

Kirk Reid Co. v. Fine[76] involved the installation of an air-conditioning system. The owner proved that the following deviations existed:

1. The system was 46 to 50 tons short of capacity.
2. The primary air unit was of a lower rating by 7,850 cubic feet per minute.
3. The condenser water pipe was 120 gallons per minute short of capacity.

Yet the court held that the contractor had substantially performed, stating, "It appears from the . . . finding that there was a workable air conditioning plant installed and operating by May 1, 1959, the time called for by the contract. It further appears from the record that the system so installed and operating has been used, since that time, by the defendant without complaint by him that it is insufficient to perform its task."[77]

A case dealing with construction of a church held that the contractor had substantially performed when the deviation consisted of a two-foot lower ceiling, windows shorter and narrower than called for, and seats narrower than designated by the specifications.[78] Another court held that the contractor had substantially performed when it was discovered at the completion of the project that the roof shingles were discolored.[79] The court noted that the shingles could not be seen from the street and did not constitute a functional defect. However, another roof case came to a contrary result.[80]

Finally, another court concluded that the contractor had substantially performed when it buried pipe six and one-half feet into the ground instead of the required seven.[81]

Early cases required the contractor to show that it was free from fault and that the breach was not willful, such as by establishing that the deviation was caused by a subcontractor or was not done knowingly.[82] However, recent cases have reflected ambivalence on this issue. One case held that substantial performance could be used despite a finding that the contractor knowingly deviated from the contract.[83] Other cases have held that the reason for the breach must be taken into account with other factors to determine whether there has been substantial performance.[84]

Although the substantial performance doctrine is designed to avoid forfeiture and unjust enrichment, it would be difficult to conclude there was unjust enrichment if a contractor testified that it had substituted one brand of pipe for another because "it was just as good" or because the contractor did not feel like going to the warehouse to get the right kind.[85]

It has been stated that explicit contract language precludes substantial performance from determining the contractor's rights to the contract price.[86] Where such statements are made, they are likely to be in cases where the contractor has been found to have substantially performed.[87] The dilemma posed by this question requires that the essential nature of the substantial performance doctrine be determined. If the doctrine merely reflects a common intention that minor deviations will not preclude recovery, a contract clause negating this intention should be controlling. However, if the doctrine is based on the avoidance of forfeiture by the contractor and corresponding unjust enrichment of the owner, an express contract clause should not bar use of this doctrine. Resolving this difficult question will be aided by an evaluation of *Plante v. Jacobs* (reproduced next).

---

[74]*Jardine Estates, Inc. v. Donna Brook Corp.*, supra note 71.

[75]230 N.Y. 239, 129 N.E. 889 (1921).

[76]205 Va. 778, 139 S.E.2d 829 (1965).

[77]139 S.E.2d at 837.

[78]*Pinches v. Swedish Evangelical Lutheran Church*, 55 Conn. 183, 10 A. 264 (1887).

[79]*Salem Towne Apartments, Inc. v. McDaniel & Sons Roofing Co.*, 330 F.Supp. 906 (E.D.N.C.1970).

[80]*O. W. Grun Roofing & Constr. Co. v. Cope*, 529 S.W.2d 258 (Tex. Ct.App.1975) (streaks in a structurally sound roof precluded substantial performance and required the roof to be replaced).

[81]*All Seasons Water User Ass'n. v. Northern Improvement Co.*, 399 N.W.2d 278 (N.D.1987).

[82]*Jacob & Youngs, Inc. v. Kent*, supra note 75.

[83]*Kirk Reid Co. v. Fine*, supra note 76.

[84]*Hadden v. Consol. Edison Co. of N.Y., Inc.*, 34 N.Y.2d 88, 312 N.E.2d 445, 356 N.Y.S.2d 249 (1974); *Nordin Constr. Co. v. City of Nome*, supra note 69; *Watson Lumber Co. v. Mouser*, 30 Ill.App.3d 100, 133 N.E.2d 19 (1975).

[85]*O. W. Grun Roofing & Constr. Co. v. Cope*, supra note 80.

[86]*Jacob & Youngs, Inc. v. Kent*, supra note 75.

[87]Ibid. But see *Winn v. Aleda Constr. Co., Inc.*, 227 Va. 304, 315 S.E.2d 193 (1984) (language barred substantial performance).

Following this case will be a further critique of the substantial performance doctrine and its relationship to contract language dealing with strict compliance.

Usually the contractor must prove it has substantially performed.[88] However, once substantial performance has been established, as a rule the owner must prove its damages.[89] Substantial performance, though entitling the contractor to recover the balance of the contract price, is still a breach. Because damages reduce the amount to which the contractor is entitled, once substantial performance has been established the law places the burden of proving

damages on the owner. The extent of damages will likely be an important factor in the owner's attempt to establish that there has not been substantial performance.

A few states place the burden of establishing damages on the contractor, an anomalous result, because the contractor's first line of attack usually is that it has fully performed.[90]

Damages are sometimes measured by the cost of correcting the defect, sometimes by the diminished value of the project, and sometimes by a combination of the two.[91] The interrelation of various damage measurements can be seen in *Plante v. Jacobs*, reproduced here.

PLANTE v. JACOBS

Supreme Court of Wisconsin, 1960. 10 Wis.2d 567, 103 N.W.2d 296.

Suit to establish a lien to recover the unpaid balance of the contract price plus extras of building a house for the defendants, Frank M. and Carol H. Jacobs, who in their answer allege no substantial performance and breach of the contract by the plaintiff and counterclaim for damages due to faulty workmanship and incomplete construction. . . . After a trial [before the judge without a jury] judgment was entered for the plaintiff in the amount of $4,152.90[92] plus interest and costs, from which the defendants, Jacobs, appealed and the plaintiff petitioned for a review. . . .

The Jacobses, on or about January 6, 1956, entered into a written contract with the plaintiff to furnish the materials and construct a house on their lot in Brookfield, Waukesha County, in accordance with plans and specifications, for the sum of $26,765. During the course of construction the plaintiff was paid $20,000. Disputes arose between the parties, the defendants refused to continue payment, and the plaintiff did not complete the house. On January 12, 1957, the plaintiff duly filed his lien.

The trial court found the contract was substantially performed and was modified in respect to lengthening the house two feet and the reasonable value of this extra was $960. The court disallowed extras amounting to $1,748.92 claimed by the plaintiff because they were not agreed on in writing in accordance with the terms of the agreement. In respect to defective workmanship the

court allowed the cost of repairing the following items: $1,550 for the patio wall; $100 for the patio floor; $300 for cracks in the ceiling of the living room and kitchen; and $20.15 credit balance for hardware. The court also found the defendants were not damaged

---

[90] *Vance v. My Apartment Steak House of San Antonio, Inc.*, 677 S.W.2d 480 (Tex.1985). See Note, 22 Hous.L.Rev.1093 (1985), attacking the Texas rule.

[91] These damage measurements are discussed in Section 27.03D. In *Jacob & Youngs, Inc. v. Kent*, supra note 75, the measure of recovery for substitution of pipe was the diminished value of the project that the court concluded was nominal. But in *Salem Towne Apartments, Inc. v. McDaniel & Sons Roofing Co.*, supra note 79, the diminished value of the project was found to have been about one half the cost of correction. But see *Kaiser v. Fishman*, 187 A.D.2d 623, 590 N.Y.S.2d 230 (1992), appeal denied, 81 N.Y.2d 711, 619 N.E.2d 658, 601 N.Y.S.2d 580 (1993) (approved award of almost $215,000 as the cost of correction, despite what court called quick-fix correction work of $60,000).

[92] [Ed. note: The computation of the judgment must have been as follows:

Original contract price	$26,765.00
Extras	$ 960.00
Contract price as adjusted	$27,725.00
Payments received by plaintiff	$20,000.00
Balance unpaid	$ 7,725.00
Minus: (1) Amount conceded by plaintiff for omissions (kitchen cabinets, gutters and downspout, sidewalk, closet clothes poles, and entrance seat)	$ 1,601.95
(2) Cost of correction for patio wall ($1,550), Patio floor ($100), cracks in living room and kitchen ceiling ($300), and credit balance for hardware ($20.15).	$ 3,572.10
Amount of Court Judgment	$ 4,152.90]

---

[88] *A. W. Therrien Co. v. H. K. Ferguson Co.*, 470 F.2d 912 (1st Cir.1972); *Teramo & Co., Inc. v. O'Brien-Sheipe Funeral Home, Inc.* 283 A.D.2d 635, 725 N.Y.S.2d 87 (2001).

[89] *Maloney v. Oak Builders, Inc.*, 224 So.2d 161 (La.App.1969), rev'd on other grounds, 256 La. 85, 235 So.2d 386 (1970); *Hopkins Constr. Co. v. Reliance Ins. Co.*, 475 P.2d 223 (Alaska 1970).

by the misplacement of a wall between the kitchen and the living room, and the other items of defective workmanship and incompleteness were not proven. The amount of these credits allowed the defendants was deducted from the gross amount found owing the plaintiff, and the judgment was entered for the difference and made a lien on the premises. . . .

HALLOWS, Justice. The defendants argue the plaintiff cannot recover any amount because he has failed to substantially perform the contract. The plaintiff conceded he failed to furnish the kitchen cabinets, gutters and downspouts, sidewalk, closet clothes poles, and entrance seat amounting to $1,601.95. This amount was allowed to the defendants. The defendants claim some 20 other items of incomplete or faulty performance by the plaintiff and no substantial performance because the cost of completing the house in strict compliance with the plans and specifications would amount to 25 or 30 percent of the contract price. The defendants especially stress the misplacing of the wall between the living room and the kitchen, which narrowed the living room in excess of one foot. The cost of tearing down this wall and rebuilding it would be approximately $4,000. The record is not clear why and when this wall was misplaced, but the wall is completely built and the house decorated and the defendants are living therein. Real estate experts testified that the smaller width of the living room would not affect the market price of the house.

The defendants rely on *Manitowoc Steam Boiler Works v. Manitowoc Glue Co.*, . . . for the proposition there can be no recovery on the contract . . . unless there is substantial performance. This is undoubtedly the correct rule at common law. . . . The question here is whether there has been substantial performance. The test of what amounts to substantial performance seems to be whether the performance meets the essential purpose of the contract. In the *Manitowoc* case the contract called for a boiler having a capacity of 150 percent of the existing boiler. The court held there was no substantial performance because the boiler furnished had a capacity of only 82 percent of the old boiler and only approximately one-half of the boiler capacity contemplated by the contract. In *Houlahan v. Clark*, . . . the contract provided the plaintiff was to drive pilings in the lake and place a boat house thereon parallel and in line with a neighbor's dock. This was not done and the contractor so positioned the boat house that it was practically useless to the owner. *Manthey v. Stock*, . . . involved a contract to paint a house and to do a good job, including the removal of the old paint where necessary. The plaintiff did not remove the old paint, and blistering and roughness of the new paint resulted. The court held that the plaintiff failed to show substantial performance. The defendants also cite *Manning v. School District No. 6*, . . . However, this case involved a contract to install a heating and ventilating plant in the school building which would meet certain tests which the heating apparatus failed to do. The heating plant was practically a total failure to accomplish the purposes of the contract. See also *Nees v. Weaver*, . . . (roof on a garage).

Substantial performance as applied to construction of a house does not mean that every detail must be in strict compliance with the specifications and the plans. Something less than perfection is the test of specific performance unless all details are made the essence of the contract. This was not done here. There may be situations in which features or details of construction of special or of great personal importance, which if not performed, would prevent a finding of substantial performance of the contract. In this case the plan was a stock floor plan. No detailed construction of the house was shown on the plan. There were no blueprints. The specifications were standard printed forms with some modifications and additions written in by the parties. Many of the problems that arose during the construction had to be solved on the basis of practical experience. No mathematical rule relating to the percentage of the price, of cost of completion or of completeness can be laid down to determine substantial performance of a building contract. Although the defendants received a house with which they are dissatisfied in many respects, the trial court was not in error in finding the contract was substantially performed.

The next question is what is the amount of recovery when the plaintiff has substantially, but incompletely, performed. For substantial performance the plaintiff should recover the contract price less the damages caused the defendant by the incomplete performance. Both parties agree. *Venzke v. Magdanz*, . . . states the correct rule for damages due to faulty construction amounting to such incomplete performance, which is the difference between the value of the house as it stands with faulty and incomplete construction and the value of the house if it had been constructed in strict accordance with the plans and specifications. This is the diminished-value rule. The cost of replacement or repair is not the measure of such damage, but is an element to take into consideration in arriving at value under some circumstances. The cost of replacement or the cost to make whole the omissions may equal or be less than the difference in value in some cases and, likewise, the cost to rectify a defect may greatly exceed the added value to the structure as corrected. The defendants argue that under the Venzke rule their damages are $10,000. The plaintiff on review argues the defendants' damages are only $650. Both parties agree the trial court applied the wrong rule to the facts.

The trial court applied the cost-of-repair or replacement rule as to several items, relying on *Stern v. Schlafer*, . . . wherein it was stated that when there are a number of small items of defect or omission which can be remedied without the reconstruction of a substantial part of the building or a great sacrifice of work or material already wrought in the building, the reasonable cost of correcting the defect should be allowed. However, in *Mohs v. Quarton*, . . . the court held when the separation of defects would lead to confusion, the rule of diminished value could apply to all defects.

In this case no such confusion arises in separating the defects. The trial court disallowed certain claimed defects because they were not proven. This finding was not against the great weight and clear preponderance of the evidence and will not be disturbed on appeal. Of the remaining defects claimed by the defendants, the court allowed the cost of replacement or repair except as to the misplacement of the living-room wall. Whether a defect should fall under the cost-of-replacement rule or be considered under the diminished-value rule depends on the nature and magnitude of the defect. This court has not allowed items of such magnitude under the cost-of-repair rule as the trial court did. Viewing the construction of the house as a whole and its cost we cannot say, however, that the trial court was in error in allowing the cost of repairing the plaster cracks in the ceilings, the cost of mud jacking and repairing the patio floor, and the cost of reconstructing the nonweight-bearing and nonstructural patio wall. Such reconstruction did not involve an unreasonable economic waste.

The item of misplacing the living room wall under the facts of this case was clearly under the diminished-value rule. There is no evidence that defendants requested or demanded the replacement of the wall in the place called for by the specifications during the course of construction. To tear down the wall now and rebuild it in its proper place would involve a substantial destruction of the work, if not all of it, which was put into the wall and would cause additional damage to other parts of the house and require replastering and redecorating the walls and ceilings of at least two rooms. Such economic waste is unreasonable and unjustified. The rule of diminished value contemplates the wall is not going to be moved. Expert witnesses for both parties, testifying as to the value of the house, agreed that the misplacement of the wall had no effect on the market price. The trial court properly found that the defendants suffered no legal damage, although the defendants' particular desire for specific room size was not satisfied. . . .

On review the plaintiff raises two questions: Whether he should have been allowed compensation for the disallowed extras, and whether the cost of reconstructing the patio wall was proper. The trial court was not in error in disallowing the claimed extras. None of them was agreed to in writing as provided by the contract, and the evidence is conflicting whether some were in fact extras or that the defendants waived the applicable requirements of the contract. The plaintiff had the burden of proof on these items. The second question raised by the plaintiff has already been disposed of in considering the cost-of-replacement rule.

It would unduly prolong this opinion to detail and discuss all the disputed items of defects of workmanship or omissions. We have reviewed the entire record and considered the points of law raised and believe the findings are supported by the great weight and clear preponderance of the evidence and the law properly applied to the facts.

Judgment affirmed.

Returning to the question posed before *Plante v. Jacobs*, to what extent can contract language preclude use of the substantial performance doctrine? In *Plante v. Jacobs* the court stated:

> Substantial performance as applied to construction of a house does not mean that every detail must be in strict compliance with the specifications and the plans. Something less than perfection is the test of specific performance *unless all details are made the essence of the contract.* [Editor's emphasis.] This was not done here. There may be situations in which features or details of construction of special or of great personal importance, which if not performed, would prevent a finding of substantial performance of the contract.

A distinction must be drawn between different methods by which the contract itself might preclude use of the doctrine. First, a clause could simply state that strict compliance with all requirements of the contract documents is required. Such a clause is likely to be found among the many clauses included in the general conditions of the contract.

Second, the contract could create, as does AIA in A201-2007, Section 9.8.5, a system dealing with minor defects discovered when an inspection is made to determine whether there has been substantial completion and a mechanism for dealing with these defects between substantial and final completion.

Third, interpretation of specific language in the contract documents and the surrounding facts and circumstances may indicate that exact compliance with certain contract requirements was expected. For example, suppose in *Plante v. Jacobs* that the contractor knew or should have known that the Jacobses had purchased expensive furniture that could not be used because the living room wall had been misplaced. Alternatively, suppose the Jacobses had informed Plante that for aesthetic reasons the living room must comply in all respects to the floor plan dimensions and no deviation would be allowed. In either case, it could be argued that the substantial performance doctrine should not be used, as the owners expected strict performance and this was communicated to the contractor at the time the contract was made.

If the doctrine of substantial performance is based on the implication that "close to perfect" compliance is all the owner could reasonably expect, the doctrine should not be applied to contracts using any of the methods previously outlined. Any method could indicate that the owner expected full compliance and that the contractor was aware of this responsibility. Perhaps a clause of the first type would be less persuasive, as it could be boilerplate (preprinted clauses) and would not sufficiently bring to the contractor's attention the need for strict compliance. Even here the contractor should be aware of the importance of strict compliance.

A misplaced wall could be troublesome even if the contract appeared to negate use of the substantial performance doctrine. This is the very type of defect for which the doctrine is designed. If so, another objective of the doctrine is to avoid economic waste, a goal that originally helped generate the doctrine.

In *Plante v. Jacobs*, correcting the misplaced wall would have cost $4,000, but the misplaced wall was found not to have diminished the value of the house. Although the court did not face a contract method for avoiding substantial performance, it is doubtful that the court would have either ordered the contractor to correct the mistake or permitted the owner to deduct $4,000 from the final payment for damages. Actual replacement of the wall either by court order or by allowing the Jacobses to reimburse themselves out of the final payment after they corrected the work themselves would cause or endorse an uneconomic expenditure of $4,000 that would not increase the value of the house.

Suppose the Jacobses were allowed to deduct the $4,000 without the need to show they were sufficiently concerned to correct the mistake. Would the Jacobses simply take the money and not make the correction? In the likely event they would not make the correction, they would be given a windfall. But this windfall possibility assumes that the Jacobses built their house principally as an investment and not as a place to live. Even assuming that the Jacobses would not invest the $4,000 to correct the mistake, some recognition should be given to the possibility that the Jacobses might have been damaged in some difficult-to-measure way by having to live in a house with a living room one foot shorter than they had planned even though the misplaced wall gave them an extra foot in the kitchen. Although some might doubt that a mistake of this kind could cause emotional distress, as noted in Section 6.06H the law is slowly moving toward the recognition that certain contracts are made not simply for commercial reasons but for reasons of personal solicitude. Were this the case here, such a loss could have been taken into account in determining the damages caused by the misplaced wall.

The substantial performance doctrine straddles concepts of contract and unjust enrichment. There can be risk assumption by contract, although economic waste and windfall cannot be excluded. As evidence becomes clear that the risk was specifically assumed by the contractor, there may be less room for the doctrine. But as it becomes apparent that there will be economic waste or windfall, the doctrine has more scope despite contract language that might seem to preclude its use.

Plante v. Jacobs illustrates yet another aspect of the interrelationship between the contract and substantial performance. The court stated,

> In this case the plan was a stock floor plan, no detailed construction of the house was shown on the plan. There were no blueprints. The specifications were standard printed forms with some modifications and additions written in by the parties. Many of the problems that arose during construction had to be solved on the basis of practical experience.

In this type of transaction, the contract documents, such as they were, may have only been starting points from which the parties work together toward a particular solution. In such a case, it would be unfair to hold the contractor to the original requirements of the contracts, because contractor and owners were essentially designing the house as it was being built.

Similarly, this type of transaction is not likely to involve a detailed set of general terms and conditions that provide a way to determine when there has been substantial completion and ultimately final completion. As a result, it is much easier to insert the common law substantial performance doctrine, because there is no contractual method to deal with disputes where the owner has received largely what it had bargained for but had not received everything.

Plante v. Jacobs also illustrates possible abuse of the doctrine. The defects at the end of the job consisted of the contractor's failure to furnish the kitchen cabinets, gutters and downspouts, sidewalk, closet clothes poles, and entrance seat. The estimated cost of furnishing these items was $1,610.95. The amount retained was $7,725. Superficially, it would seem as if Plante had substantially performed and should be entitled to the outstanding balance less the cost of correction. In theory, the deduction from the outstanding balance would compensate the owner for the omissions. In actuality, the legal measure of recovery does not accomplish this. There are additional costs, such as obtaining a substitute contractor, the delay in effectuating the necessary corrections, and the risk of a substitute contractor not performing, which would be difficult to deduct from the outstanding balance.

Even if the contractor *had* removed its workers from the project, it would not impose any serious burden on it to return and complete the job as promised. Would it be unreasonable for the owner to withhold the entire $7,725 as an unashamedly coercive device to get the contractor to finish the work if the owner chose this route rather than engage a substitute contractor? If the amount withheld appears excessive, it could be reduced to two or three times the projected cost of correction (often statutes use 150 percent). But the amount retained must be enough to make it worthwhile for the contractor to finish the job properly. If not, some contractors will refuse to perform and will invoke the substantial performance doctrine to recover the outstanding balance less the amount deducted as damages. In such a case, a court would likely conclude there had been substantial performance. The court noted that the house had been decorated and occupied by the owners even though the owners were dissatisfied with the contractor's work. Certainly the house, with the omissions, was fit for its primary purpose, and close to strict completion had been accomplished.

The preceding criteria were generated from cases that clearly involved economic waste or windfall because the defects were discovered at the end of the project and would have involved substantial redoing of the project at expenditures greatly in excess of the diminished value caused by the deviation. However, where these elements are not present, the substantial performance doctrine should not be a device by which contractors can walk away without having completed the project and expect to be paid the balance less the damages allowed by law. Yet one court expressed this view:

> The doctrine of substantial performance is a necessary inroad on the pure concept of freedom of contracts. The doctrine recognizes countervailing interests of private individuals and society; and, to some extent, it sacrifices the preciseness of the individual's contractual expectations to society's need for facilitating economic exchange. This is not to say that the rule of substantial performance constitutes a moral or ethical compromise; rather, the wisdom of its application adds legal efficacy to promises by enforcing the essential purposes of contracts and by eliminating trivial excuses for nonperformance.[93]

This court views the doctrine as one to protect contractors from owners who seek an excuse to avoid paying. This attitude can lead a contractor to walk away from a commitment and be able to force the court to perform an "accounting job," compute the balance owed by deducting the cost of correction or completion, or use the diminished value, a result not bargained for by the owner. It can promote a "just as good" philosophy and encourage sloppy, incomplete work. Whether it promotes promise making is, at best, debatable.

C. Divisible Contract

For various reasons, the law sometimes artificially treats one contract as if it were composed of a number of separate contracts. Section 22.02N discussed whether progress payments in construction contracts make the contract divisible.

One function that can be served by the divisibility fiction is to enable a contractor to recover despite not having fully performed the contract. For example, suppose a construction project has five well-defined phases and designated payments for each phase. Suppose the contractor completes three phases but omits the remaining two.

---

[93]*Bruner v. Hines*, 295 Ala. 111, 324 So.2d 265, 269–70 (1975).

In such a case, classifying this essentially single contract into a series of five contracts would enable the contractor to recover for the three phases completed, the owner being left with a deduction for the damages caused by failure to complete the final two.

The presence of progress payments does not mean that the parties have agreed that each segment of work is an agreed equivalent for the amount of the progress payment being made. The progress payments are approximations and not intended as agreed figures. Although it is not inconceivable that a court would classify a construction contract as divisible in order to permit the contractor to recover despite default, the doctrine of substantial performance (discussed in Section 22.06B), and the doctrine of restitution (to be discussed in Section 22.06D), are more appropriate vehicles for giving relief to a contracting party who has not fully performed yet has benefited the other party.

D. Restitution: Unjust Enrichment

The defaulting building contractor may not have performed sufficiently to take advantage of the substantial performance doctrine. Another concept sometimes available that can enable the contractor to recover for any net benefit that its performance has conferred on the owner is restitution, a concept sometimes called *quasi-contract*, or *quantum meruit*.

Restitution, unlike its substantial performance counterpart, has not been uniformly accepted. Early cases were not sympathetic to defaulting parties. Some courts continue to hold that a party who has not substantially performed may not recover for work performed.[94] However, there has been some relaxation of this attitude, and in many jurisdictions, contractors are able to recover restitution despite their own breach, in order to avoid unjust enrichment of the owner.[95]

Suppose there is a construction contract in which a promise to pay $100,000 is exchanged for the promise to construct a building. Suppose the contractor unjustifiably leaves the project during the middle of performance. The defaulting contractor may have conferred a net benefit despite its having breached by abandoning the project. For example, suppose the defaulting contractor has received $20,000 in progress payments but the work has been sufficiently advanced so that it can be completed by a successor contractor for $75,000. Absent any delay damages, the defaulting contractor has conferred a net benefit of $5,000 on the owner and should be entitled to this amount.

Some of the same problems as those discussed in relation to substantial performance apply when the contractor seeks to use restitution to recover for the benefit that has been conferred. Such recovery may be precluded if the contractor's breach is considered "willful," and similar problems of measuring the damages caused by the breach arise when restitution rather than substantial performance is the basis of recovery.

Restitution cases do not arise as frequently as do "end of the job" substantial performance disputes. Part of this infrequency may be due to the unlikeliness of a contractor's walking off a project that would be profitable. Only in these cases is there likely to be a net benefit to the owner that exceeds progress payments that have been made.

E. Waiver

During performance or at the end of the project, the owner may manifest a willingness to pay the full contract price despite the contractor's not having fully complied with the contract documents. Such manifestation is often described as a *waiver*. If the deviation is an important aspect of the contract, any promise by the owner to pay despite noncompliance would have to be supported by consideration. Consideration, as discussed in Section 5.08, usually consists of something given in exchange for accepting the deviation, such as a price reduction. Alternatively, the waiver of even important matters would be enforced if the contractor had reasonably relied on the statement. Examples are stating, "Do the extra work, and I won't insist on a written change order," or misleading the contractor by a course of conduct such as paying despite noncompliance.[96]

---

[94] Under Massachusetts common law, a contractor must show complete and strict performance to recover under the contract, see *Peabody, N.E., Inc. v. Town of Marshfield*, 426 Mass. 436, 689 N.E.2d 774 (1998), and must show substantial performance to recover in quantum meruit, see *J.A. Sullivan Corp. v. Commonwealth*, 397 Mass. 789, 494 N.E.2d 374 (1986).

[95] *American Sur. Co. of N.Y. v. United States*, 368 F.2d 475 (9th Cir.1966); *United Coastal Industries, Inc. v. Clearheart Constr. Co., Inc.*, 71 Conn.App. 506, 802 A.2d 901 (2002); *Starling v. Housing Auth. of City of Atlanta*, 162 Ga.App. 852, 293 S.E.2d 392 (1982); *R. J. Berke & Co. v. J. P. Griffin, Inc.*, 116 N.H. 760, 367 A.2d 583 (1976); *Kreyer v. Driscoll*, 39 Wis.2d 540, 159 N.W.2d 680 (1968).

[96] See Section 21.04H.

Waivers of less important matters are effective despite the absence of any consideration or reliance. All that is needed in such cases is evidence that an intention to waive or give up the requirement was communicated to the other party.

A waiver can be directly expressed ("You told me you would pay even though I omitted the doorstops"). It can be expressed indirectly by implication ("When you made the final payment knowing I omitted the doorstops, I assumed you were waiving the defect"). Waiver is more likely to be found when the defect is small (doorstops) or technical (failure to obtain a written change order). It is more likely to be found where it is directly expressed ("I'll pay anyway" or "Change orders are a bother, and I know you did the extra work"). It is less likely to be found when expressed by implication ("When you paid, I thought you were waiving those defects of which you were aware" or "When you didn't complain, I assumed you waived the defects"). These factors recognize that small things can be given up if the evidence is clear that the party entitled to them was willing to do so and that acts are often ambiguous.

The act frequently contended to indicate waiver of defects is the owner's accepting by using or occupying the project; particularly those defects apparent at the time the owner used or occupied the project will be asserted to have been waived. Although this topic is treated in greater detail in Section 24.05, it should be noted that court decisions are not consistent, though the trend is against using acceptance as waiver of a claim.[97]

General conditions of construction contracts sometimes seek to eliminate the possibility of waiver. For example, Section 9.10.4 of AIA Document A201-2007 states that payment is not an acceptance of nonconforming work. Sometimes clauses more generally state that when one party does not insist on its full contract rights, it shall not be precluded from insisting on these rights in the future. Clauses of this type, though supportive of a conclusion that there has been no waiver,[98] are not absolutely waiver proof. Where it appears there has been a course of conduct that deviates from contract requirements or clear evidence that a defect has been given up, waiver can be found. The clause is removed by concluding that the nonwaiver clause itself has been waived.

Waiver, unlike substantial performance or restitution, allows recovery of the entire contract balance without any deduction for damages. Waiver usually deals with technical requirements such as written change orders[99] for minor defects in the work.

The person asserted to have waived contract rights must have been authorized to do so. Authority is determined by agency concepts and is often dealt with in the contract.[100] However, it is likely that it would take less authority to waive a minor contract defect than to make a contract in the first instance or to modify an existing contract. Waiver is less likely to be applied in public contracts.[101]

SECTION 22.07   The Certification and Payment Process: Some Liability Problems

The design professional's role in the certification and payment process can generate a variety of claims, such as failure to discover defects, issuing certificates without determining whether the contractor has paid subcontractors and suppliers or whether liens have been or will be filed, and failure to issue certificates within a reasonable time. But this section looks at the design professional's liability for certifying an incorrect amount.

As discussed earlier,[102] one of the multifaceted aspects of the design professional's performance is to interpret the contract, judge performance, and decide disputes. The design professional's role can be analogized to that of a judge when performing some of these functions. When the design professional looks like a judge, in some states she will be accorded quasi-judicial immunity—being liable for only corrupt or dishonest decisions or those not made in good faith. Quasi-judicial immunity where granted by common law is not likely to be asserted as a defense when

---

[97]Although moving into the house and signing a form of acceptance for a federal agency was held to waive defects in *Cantrell v. Woodhill Enter., Inc.*, 273 N.C. 490, 160 S.E.2d 476 (1968), *Honolulu Roofing Co. v. Felix*, 49 Haw. 578, 426 P.2d 298 (1967), held that mere occupancy is not a waiver, particularly when the owners had to move in because they had vacated their own home.

[98]*Armour & Co. v. Nard*, 463 F.2d 8 (8th Cir. 1972).

[99]See Section 21.04H, dealing with changes and Section 26.08, dealing with waiving formalities to obtain time extensions.

[100]On the question of authority, see Section 17.05B.

[101]See Section 11.01B. But in *Kenny Constr. Co. v. Metro. San. Dist. of Greater Chicago*, 52 Ill.2d 187, 288 N.E.2d 1 (1972), the court treated waiver in a public contract much as if it had been a private contract.

[102]See Section 14.09D.

the design professional has been sued for improper certification. In fact, the landmark English decision reversing earlier decisions granting immunity was based on a claim of improper certification.[103]

Quasi-judicial immunity can also be created by contract. The AIA does not view the architect's role in the certification process as creating quasi-judicial immunity. Under AIA A201-2007, Article 15, if the amount certified is disputed by either owner or contractor, the dispute must be submitted to the architect (assuming the architect is the Initial Decision Maker under Section 15.2) before it can proceed to mediation under Section 15.3, to arbitration under Section 15.4, or to litigation. A201-2007 does not create quasi-judicial immunity for the architect's participation in the disputes process.[104] It does grant immunity under Section 4.2.12 for "interpretations or decisions . . . rendered in good faith." This is part of the administration of the contract under Section 4.2, not the architect's part in the disputes process.

The principal legal issue has been the extent to which persons other than the owner can pursue a claim for improper certification. Potential claimants are the contractor, subcontractors, or sureties. Overcertification shrinks the retainage and can cause harm to the owner. It can also cause harm to the surety. The surety looks to the retainage for security if the contractor defaults and the surety must take over. Undercertification can harm the contractor and the surety as well. It can adversely affect the contractor's capacity to continue or complete performance.

Case decisions have tended to permit third parties such as contractors and sureties to maintain legal action against the design professional for improper certification.[105] Those cases that have protected the architect have pointed to a lack of privity.[106] The principal protection against such claims in transactions that involve AIA documents is the multitiered disputes process under which the architect's certificate, if challenged, must (ironically) go back to the architect (or IDM) and ultimately to arbitration, if selected. Any mistakes can be corrected by the arbitrators. As to the finality of the decision if challenged in court, see Section 29.09.

---

[103]*Sutcliffe v. Thackrah* [1974] A.C. 727.

[104]See Chapter 29 for discussion of the design professional as judge.

---

[105]*Aetna v. Hellmuth, Obata & Kassabaum, Inc.*, 392 F.2d 472 (8th Cir.1968); *State v. Malvaney*, 221 Miss. 190, 72 So.2d 822 (1954); *Westerhold v. Carroll*, 419 S.W.2d 73 (Mo.1967); *Peerless Ins. Co. v. Cerny & Assoc., Inc.*, 199 F.Supp. 951 (D.Minn. 1961); *Designed Ventures, Inc. v. Auth. of the City of Newport*, 132 B.R. 677 (D.R.I.1991).

[106]*Engle Acoustic & Tile, Inc. v. Grenfell*, 223 So.2d 613 (Miss.1969) (sharply limiting *State v. Malvaney*, supra note 105). See Sections 14.08C, D, and E.

CHAPTER TWENTY-THREE

Expectations and Disappointments: Some Performance Problems

SECTION 23.01 Introduction to Chapters 23 through 28

When the contract is created, each party to a construction contract (the contract between owner and contractor will be illustrative) has expectations. The owner expects to receive the project specified in the construction contract at the time promised. It also expects to pay the contract price as adjusted for change orders.

The contractor's expectations are more complex. Its performance entails a variety of tasks that depend on cooperation by many entities as well as on weather conditions. The contractor expects that it will be able to purchase labor and materials at or within certain prices and that work will proceed without excessive slowdowns or stoppages due to weather, labor difficulties, acts of public authorities, or failure by the owner to cooperate. The contractor does not anticipate that construction will be impeded or destroyed by fire, earthquake, or vandalism. It expects to be paid as it performs.

Yet either or both may be disappointed. The owner may not receive the project it expected for a variety of reasons, explored in Chapter 24, dealing with defects, and in Chapter 25, dealing with subsurface problems. The owner may not receive the completed project at the time promised, a disappointment discussed in Chapter 26. Chapter 27 explores the methods of measuring any claims the owner may have against the contractor for not performing.

The contractor's disappointments that relate to failure to receive progress payments treated in Chapter 22 are explored in greater detail in Chapter 27. This chapter deals with the disappointments that relate to obstacles to the contractor's performance or to insecurity about the owner's performance. The special problems of increases in the cost of performance due to unexpected subsurface conditions are treated in Chapter 25. The contractor's disappointed expectations as to the pace with which it will work are treated in Chapter 26.

Disappointments can encompass losses caused by subcontractors, whose special problems are discussed in Chapter 28.

This chapter introduces performance problems discussed in Chapters 24 through 28. Initially, it describes and briefly illustrates the essential attributes of legal doctrines central to performance problems. In addition, the chapter notes some performance problems that do not justify full chapter treatment.

SECTION 23.02 Affirmative Legal Doctrines: The Bases for Claims

A. Introduction: The Shopping List

This section summarizes in mostly nontechnical terms the legal theories that can be the basis for construction-related claims between owner and contractor. Unfortunately, it has become common for claimants to assert a variety of claims based on assorted legal theories, with the law generously permitting, at least up to a point in a lawsuit, the use of such a shotgun approach. This chapter deals only with claims by contract-connected parties. As will be seen, the variety of doctrines is astounding.

B. Fraud

Fraud is the intentional deception of one contracting party by the other. The deception induces the former to make a contract that it would not have made had it not been deceived. Usually the deception consists of giving false information or half-truths, or occasionally, of concealing information. The deception must relate to important matters. The party defrauded relies by making the contract. For example, an owner may defraud a contractor who is concerned about the owner's resources by falsely telling the contractor it has received a loan commitment. The contractor may defraud the owner by falsely representing it has access to equipment needed to perform particular work. In most states, fraud can consist of making a promise to induce the making of a contract without intending to perform it.

The defrauded party can rescind the contract and recover any benefit it has conferred on the other party based on restitution. It can affirm the contract, cease or continue performance, and recover damages. Punitive damages are generally available.

Fraud should not play a significant role in construction performance disputes. Although on rare occasions fraud can be committed during performance, such as deceiving by telling an owner that certain work has been performed or has been approved by a city inspector, fraud usually consists of preperformance conduct. Performance disputes over unanticipated subsurface conditions sometimes involve allegations that an owner deliberately gave false information on which the contractor relied.

Fraud is easy to allege, particularly in a free-swinging construction dispute. Although fraud claims have a nuisance and settlement value, they are rarely successful.

C. Concealment of Information

Concealing information, though not always fraudulent, can be the basis for a claim. The early common law placed no duty on a contracting party to disclose information to the other party during negotiation. But as noted in Section 18.04B, there is an increasing tendency to require a party to disclose information to the other party that the latter would want to have known and that it is not likely to discover on its own.

Concealing information, like fraud and misrepresentation (to be discussed in Sections 23.02B and D), relates mainly to subsurface problems, as discussed in Section 25.03. Failure to disclose when disclosure should have been made permits the innocent party to rescind the contract and recover for any benefit it may have conferred on the other party based on restitution. Alternatively, it can

be the basis for a claim for damages, based on either the additional expenses incurred or the difference between the expenses that should have been incurred and those that were incurred.

D. Misrepresentation

Misrepresentation (discussed in Section 7.07) is intentionally, negligently, or innocently furnishing inaccurate information relied on by the other party in deciding to make the contract and its terms—principally price and time. Although, like fraud and concealment, it is principally a preperformance claim, it can be based on conduct during the performance of the contract, such as misrepresenting the value or the quality of work done. But like fraud and concealment, misrepresentations occurring before making the contract are usually discovered during performance. Again, this arises mainly, though not exclusively, in subsurface disputes.

E. Negligence

Negligence claims generally involve parties not connected by contract, such as a claim by a person delivering materials to a site against the contractor based on the latter's negligent conduct. But as seen in Section 14.05E, negligence can be the basis of a claim by one party to a contract against the other party. A negligence claim requires a duty owed to the claimant, a violation of that duty, and a suffering of harm caused factually and proximately by the negligent party.[1]

Negligence claims have become more common in construction contract disputes that relate to defects or delays, usually with the hope of involving the contractor's public liability insurer, avoiding exculpatory clauses in the contract, or obtaining a more expansive remedy than one that could be obtained for breach of contract.[2]

Negligence claims require that the claimant establish that the defendant's conduct did not live up to the standard required by law, something not required in a simple breach-of-contract claim. Negligence claims are barred after a shorter period than breach-of-contract claims. The remedy for negligent conduct related to construction is usually the additional expense incurred. Although this

---

[1]See Section 7.03.
[2]See Sections 14.05E and 24.08.

remedy is roughly similar to damages for breach of contract, certain defenses used in breach-of-contract claims, such as foreseeability,[3] offer less protection to a defendant when the claim is based on negligence. Negligence as a tort can more easily serve as the basis for a claim for emotional distress.[4] If wrongdoing goes beyond simple negligence and involves intentional misconduct or recklessness, the wrongful party cannot defend on the basis that the loss that was incurred was not reasonably foreseeable, and punitive damages may be recovered.

F. Strict Liability

As noted in Section 7.09, strict liability has its greatest modern use in claims against manufacturers who market defective products. It does not require proof of negligence. In that sense, it is like a breach of contract or its near relative, breach of an implied warranty. Although strict liability is often the basis of claims by those who buy homes or lots built by developers,[5] it is rarely used as the basis for a claim between owner and contractor in the ordinary construction contract.

G. Breach of Contract

To determine whether a contract breach has occurred, the following must be interpreted:

1. the written contract, if any
2. antecedent negotiations not superseded by a written contract
3. terms implied by law

The contract may prescribe conditions that relate to claims, set forth exculpatory provisions, or control remedies. Breach does not require fault, although doctrines described in Section 23.03 may provide a defense. Remedies were discussed generally in Chapter 6 and are discussed more particularly in Chapter 27. Basically, the claimant can recover gains prevented and losses incurred that are established with reasonable certainty, were caused substantially by the breach, could not have been reasonably avoided, and were reasonably foreseeable at the time the contract was made.

H. Express Warranty

A warranty is an assurance by one party relied on by the other party to the contract that a particular outcome will be achieved by the warrantor. For example, a manufacturer may expressly warrant that a tank it promises to build will travel 100 miles on 100 gallons of fuel, will not be inoperable more than ten days a month, and will hit a moving target ten miles away while the tank is traveling at 25 miles per hour.

Failure to accomplish the promised objectives is a breach. Fault is not required. (Defenses, though rare, are discussed in Section 23.03.) The method in which the promised outcome is to be achieved is frequently determined by the warrantor. Courts have been reluctant to hold the warrantor to its warranty if the warrantor has not been given the freedom to determine the method by which the warranted outcome is to be achieved.

Express warranty for various purposes has been considered a special form of contract breach. It can be considered simply an outcome promise by the warrantor. Perhaps express warranty continues to be used to emphasize its nonfault nature. It is less susceptible to contract doctrines that can excuse nonperformance by a contracting party. In some jurisdictions, breach of a contract to perform services must be based on negligence.[6] Warranty, express or implied, in such jurisdictions may simply bring the law back to the normal nonfault contract standard. (Express warranty can create rights for a third party. It was used before the modern law of product liability allowed a user to sue a manufacturer directly without privity and based on nonfault conduct in manufacturing the product.)[7] Its use in construction relates mainly to warranty or guarantee clauses discussed in Section 24.09. Its remedy is the same as for breach of contract.

I. Implied Warranty

Implied warranty is another "subspecie" of contract breach that developed historically as a separate basis for a claim. Like express warranty, implied warranty must be relied on and an outcome is warranted. Liability is strict.

---

[3]See Section 6.06C.
[4]See Sections 7.10C and 27.10.
[5]See Sections 7.09K and 24.09.

[6]*Samuelson v. Chutich*, 187 Colo. 155, 529 P.2d 631 (1974).
[7]See *Baxter v. Ford Motor Co.*, 168 Wash. 456, 12 P.2d 409 (1932). It is still used when the product was not defective. *Collins v. Uniroyal, Inc.*, 64 N.J. 260, 315 A.2d 16 (1974).

No negligence need be shown. Remedy is the same as for breach of an express warranty.

In sale of goods, implied warranty usually relates to merchantability or fitness for the buyer's known purposes. Implied warranty developed as an offshoot of modern tort law, mainly to allow a claimant who had suffered harm—usually personal harm—to maintain an action against a manufacturer with whom it was not in privity, without the necessity of establishing negligence. It was also used to avoid real property doctrines that often barred an occupier of a home from maintaining a legal claim against a builder who had sold the home to it or to a predecessor owner—again, with no need to show negligence. Implied warranties in the sale of homes is discussed in Section 24.10.

In construction contract disputes, as shall be seen,[8] a contractor often claims that the owner impliedly warrants the sufficiency of the design. This could have been described as an implied term, that the owner "promises" that executing the design will achieve the expected outcome of both owner (a successful project) and contractor (performance in a reasonably efficient manner).

Where the construction contract does not specify the quality of the contractor's services,[9] the law usually implies a warranty that the contractor will perform its services in a workmanlike manner, free of defects of workmanship.[10] In operation, this implied warranty will closely resemble the negligence standard. However, because warranty is more result oriented, application of a warranty is likely to place the burden on the contractor where the work has not measured up to the ordinary expectation of the owner. The closely related implied warranty of habitability applicable to residences is discussed in Section 24.10.

Although sometimes implied terms are justified as simply reflecting the intention of the parties,[11] warranties are often implied because of "fairness," a conscious judicial allocation of risk. In sales transactions, the express warranty may be one that is intended to limit or negate the warranties that would be provided by the law in the absence of contrary provisions in the contract. Such a disclaimer warranty (it takes away when it appears to give) must be conspicuous to be effective[12] and even if conspicuous may not be enforced if such a warranty is unconscionable.[13] An implied warranty is preempted (displaced) by an express warranty, just as any express contract term takes precedence over or may bar an implied term.

An implied warranty is not only preempted by an express warranty but may not have been relied on by the other party, an essential element in a claim for breach of warranty. It is an issue in Section 25.05 (which deals with subsurface claims), with the owner often giving information but disclaiming responsibility for the information's accuracy.

J. Consumer Protection Legislation

Most states have enacted consumer protection legislation.[14] Such laws bar misleading or fraudulent business practices in consumer transactions. They provide a list of forbidden activities and enlarge the remedies available to consumers, such as permitting treble damages and awarding attorneys' fees.

In some states, these statutes have been used against contractors who build for consumers. As illustrations, the following acts have been found to constitute unfair trade practices:

1. using substandard material[15]
2. failing to complete work within a specified period of time[16]
3. failing to adequately supervise the construction project[17]
4. intentionally supplying low estimates[18]

---

[8]See Sections 23.05E and 24.02.

[9]See AIA Doc. A201-2007, § 3.5, for a warranty as to labor, material and equipment.

[10]*New Zion Baptist Church v. Mecco*, 478 So.2d 1364 (La.App.1985) and *Mascarenas v. Jaramillo*, 111 N.M 410, 806 P.2d 59 (1991). But see *Milau Assoc. v. North Avenue Dev. Corp.*, 42 N.Y.2d 482, 368 N.E.2d 1247, 398 N.Y.S.2d 882 (1977), which held that there was no implied warranty of fitness in a service contract.

[11]See Section 19.02A.

---

[12]U.C.C. § 2-316(2).

[13]U.C.C. § 2-302.

[14]The closely related federal Magnuson-Moss Warranty Act, governing warranties of consumer products, is described in Section 24.09B.

[15]*New Mea Constr. Corp. v. Harper*, 203 N.J.Super. 486, 497 A.2d 534 (App.Div.1985); *Tang v. Bou-Fakhreddine*, 75 Conn.App. 334, 815 A.2d 1276 (2003).

[16]*Watson v. Bettinger*, 658 S.W.2d 756 (Tex.Ct.App.1983).

[17]*Building Concepts, Inc. v. Duncan*, 667 S.W.2d 897 (Tex.Ct.App. 1984).

[18]*Hannon v. Original Gunite Aquatech Pools, Inc.*, 385 Mass. 813, 434 N.E.2d 611 (1982) (dictum); *Eastlake Constr. Co. v. Hess*, 102 Wash.2d 30, 686 P.2d 465 (1984).

5. providing defective workmanship[19]
6. misrepresentation as to experience and qualifications of builder[20]

Consumer protection laws generally govern the relationships between clients (usually homeowners) and the contractor or design professional they directly hire. For example, clients may invoke state consumer protection or deceptive trade practices acts in their claims against design professionals so as to extend these statutes' strict liability provisions to professional services.[21]

Generally deceptive trade practices acts are not applicable to disputes between prime contractors and subcontractors[22] or against a supplier who sold to a builder.[23] But the Connecticut Unfair Trade Practices Act was violated when a prime contractor made an oral contract with a subcontractor, listed the subcontractor as a disadvantaged business enterprise in its bid with a public entity, and then sought to renegotiate the contract. In addition to being awarded lost profits, the subcontractor

was awarded its attorneys' fees under the act.[24] However, these laws do not apply to more remote actors, because there is no "consumer transaction" between subcontractors and the homeowner.[25]

As noted in Section 10.11, consumer dissatisfaction sometimes results in the charge that the licensed contractor should be disciplined because of its failure to perform properly. Such complaints have generated another method of consumer protection, structuring an informal complaint process, with the contractor's license at stake.

K. Residential Construction Defects Legislation

Several states have consumer protection legislation aimed specifically at buyers of new homes. These statutes are of two types: new home warranty acts and right to cure laws.

New home warranty acts (NHWA) create statutory warranties that apply to the sale of new homes. They require contractors to guarantee that the home is free of major defects. For example, Connecticut's NHWA implies in every sale of a new residence one-year warranties that the residence is free of faulty materials, constructed according to sound engineering standards, constructed in a workmanlike manner, and fit for habitation.[26] These statutory warranties are in addition to any warranties implied by law.[27]

By contrast, Louisiana's NHWA displaces any rights the builder of a new residence and its owner would have under general law. Instead, the act provides these parties with the exclusive remedies, warranties, and limitations periods.[28] For example, an owner could sue the contractor for breach of the statutory warranties, but not for breach of contract and negligence.[29]

Right to cure laws do not give owners new, substantive rights. Instead they seek to prevent unnecessary litigation by giving contractors the opportunity to cure defects before the homeowner is allowed to sue. Homeowners who discover what they believe to be construction defects must

---

[19]*Jim Walter Homes, Inc. v. White,* 617 S.W.2d 767 (Tex.Ct.App. 1981).

[20]*Gennari v. Weichert Co. Realtors,* 148 N.J. 582, 691 A.2d 350 (1997) (claim against real estate agent-developer); *Plath v. Schonrock,* 314 Mont. 101, 64 P.3d 984 (2003), rehearing denied Mar. 6, 2003 (containing excellent discussion of the history of consumer protection laws and types of legislative relief granted).

[21]*Ragucci v Professional Constr. Services,* 25 A.D.3d 43, 803 N.Y.S.2d 139 (2005). In *White Budd Van Ness Partnership v. Major-Gladys Drive Jt. Venture,* 798 S.W.2d 805 (Tex.Ct.App.1990), cert. denied, 502 U.S. 861 (1991), the Texas Court of Appeals applied the Texas Deceptive Trade Practices Act's (DTPA) strict liability provisions to a client's claim against an architect. The continued viability of that decision is in doubt in light of subsequent legislative and judicial developments. In *Chapman v. Paul R. Wilson, Jr., D.D.S.,* 826 S.W.2d 214, 217 (Tex.Ct.App.1992), the court expressly rejected *White Budd's* holding that the DTPA applies to "purely professional services." Then, in 1995, the DTPA was amended to exclude damages based on the rendering of a professional service, defined as the providing of advice, opinion, or judgment (see Tex.Bus. & Prof.Code Ann. § 17.49(c)). The Kansas Supreme Court extended its state's consumer protection law to a client's claim against his engineer. *Moore v. Bird Eng'g Co.,* 273 Kan.2, 41 P.3d 755 (2002). By dictum, Louisiana seems to allow a client to sue an architect for "unreasonably high fees for services rendered." *Abbyad v. Mathes Group,* 671 So.2d 958, 962 (La.App.1996) (not awarded, because no damages were shown).

[22]*Lake County Grading Co. of Libertyville, Inc. v. Advance Mechanical Contractors, Inc.,* 275 Ill.App.3d 452, 654 N.E.2d 1109 (1995) (did not involve consumer protection).

[23]*Morris v. Osmose E. Wood Preserving,* 340 Md. 519, 667 A.2d 624 (1995). In this case the claim was brought by homeowners for defective plywood. But the deception occurred in the sale to the developer-builder.

[24]*Bridgeport Restoration Co., Inc. v. A. Petrucci Constr.,* 211 Conn. 230, 557 A.2d 1263 (1989).

[25]*Messeka Sheet Metal Co., Inc. v. Hodder,* 368 N.J.Super. 116, 845 A.2d 646 (App.Div.2004) (subcontractors); *Amstadt v. United States Brass Corp.,* 919 S.W.2d 644 (Tex.1996) (upstream suppliers of raw materials and component parts of defective plumbing system).

[26]Conn.Gen.Stat.Ann. § 47-118.

[27]Id. § 47-120.

[28]La.Rev.Stat. § 9:3150.

[29]*Carter v. Duhe,* 921 So.2d 963 (La.2006). For further discussion of new home warranty acts, see Annot., 101 A.L.R.5th 447 (2002).

comply with a statutory prelitigation procedure. They must provide the contractor with written notice of the alleged defects and then give the builder the opportunity to inspect the building and attempt repairs. If the contractor believes the home is not defective or if the owner is dissatisfied with the repairs, California and Hawaii require the parties to submit to mediation.[30] In other states, the owner may proceed to litigation without going first to mediation.

California's right to cure law, known generally as Senate Bill (SB) 800, is the most ambitious of these types of laws.[31] SB 800 applies to purchases of new homes after January 1, 2003. As with most such statutes, it requires the homeowner to provide the builder with written notice of the claimed defects. The contractor has a relatively short time to inspect and then effectuate repairs. Should the contractor fail to cure in a timely manner and the dispute is not resolved by mediation, the homeowner is released from the statute's requirements and may proceed to court.[32]

What makes SB 800 unique among right to cure laws is its detailed enumeration of "actionable defects." As described by one commentator:

> [V]ery little has been overlooked. In Section 896, the Act defines "water issues" with respect to door and window systems, roofing systems, decks, foundations and slabs, stucco and exterior wall systems, plumbing systems, shower and bath enclosures, countertops, and exterior landscaping, irrigation, and drainage. Section 896 covers structural issues, soil issues, fire protection, plumbing, sewer and electrical issues, and numerous miscellaneous matters right down to the dryer ducts.[33]

Although right to cure laws do not grant homeowners new, substantive rights, they do deprive contractors of one common law defense. By defining certain construction defects as "actionable," these statutes preclude the builder from arguing that the economic loss rule (discussed in Section 14.08E) immunizes it from liability.[34]

---

[30]West Ann.Cal.Civ.Code § 919; Haw.Rev.Stat. § 672E-7.

[31]West Ann.Cal.Civ.Code § 895 et seq.

[32]Id. § 920.

[33]Coven, *California Attempts to Resolve Residential Construction Defect Claims Without Litigation*, 23 Constr. Lawyer, No. 2, Spring 2003, p. 35. The author nonetheless predicts that "[t]he comprehensiveness of the Act . . . will not put construction lawyers and experts out of business." Ibid.

[34]SB 800 statutorily overrules *Aas v. Superior Court*, 24 Cal.4th 627, 12 P.3d 1125, 101 Cal.Rptr.2d 718 (2000); see also Nev.Rev.Stat.

SECTION 23.03 Defenses to Claims

A. Introduction: Causation and Fault

On the whole, the doctrines described in this section relate to the occurrence or discovery of a supervening event that one contracting party asserts should relieve it from its obligation to perform. A party asserting such a defense must show that the event has seriously disrupted performance planning or performance itself. If the occurrence of the event could have been prevented or if the effect of the event on performance could have been minimized or avoided, the occurrence of the event does not provide a defense. For example, suppose laborers struck because of illegal conduct by the contractor who employed them. The contractor will not receive relief, because its own conduct caused the disruptive event. Likewise, if work were shut down because of a court order issued because the owner did not comply with land use controls, the owner will not be given relief if a contractor makes a claim.

Similarly, a party who does not appear "innocent" is less likely to be given relief, either because its conduct has caused the event or because its fault denies it the right to obtain relief from its contractual obligations.

Generally, the doctrines described in this section deal with events that affect performance and that cannot be said to have been the fault of either party.

B. Contractual Risk Assumption

Some of the doctrines discussed in this section were developed by the common law. They involve claims for relief based on doctrines that courts develop to deal with drastic changes of circumstances. The willingness or even the power of courts to use these common law doctrines is often affected negatively by specific contract clauses. For example, an owner may defend against a contractor's claim for additional compensation for delays caused by wrongful acts of the owner by pointing to a clause under which the contractor is limited to a time extension.

Similarly, the common law doctrine of impossibility is not likely to be employed if there is a *force majeure*

---

§ 40.600 et seq., interpreted in *Olson v. Richard*, 120 Nev. 240, 89 P.3d 31 (2004). For further discussion of right to cure laws, see Allen, *Construction Defects Litigation and the "Right to Cure" Revolution*, Constr. Briefings (Thomson/West March 2006), and Quatman & Gonzalez, *Right-to-Cure Laws Try to Cool Off Condo's Hottest Claims*, 27 Constr. Lawyer, No. 3, Summer 2007, p. 13 (listing thirty states that have adopted such laws).

provision, which grants relief under specified circumstances to the performing party. For example, contract clauses frequently allow for a time extension if certain events occur. The contractor may defend a claim by the owner that the contractor has not performed on time by pointing to contract language that justified a time extension.

A fixed-price contract itself is a contractual assumption of risk that in many instances bars the contractor from receiving additional compensation in the event its performance costs have increased because of events over which it may have had no control.

Contractual risk assumption is not limited to express contract terms. Courts often hold that the owner warrants the "sufficiency" of its design. The law will protect a contractor who follows the design if the owner asserts a claim for a defect by concluding that the owner impliedly warranted the design.

As contracts become more detailed, the potential for judicial intervention may narrow. Yet the more open recognition of the adhesive nature of many construction contracts may tempt a court to judge the fairness of contract clauses dealing with these claims for relief.[35]

C. Mutual Mistake

Each contracting party, particularly the party agreeing to perform services, has fundamental assumptions often not expressed in the contract. For example, often the contracting parties assume that the subsurface conditions that will be encountered will not vary greatly from those expected by the design professional or the contractor. Although some deviation may be expected, drastic deviation that has a tremendous effect on the contractor's performance may be beyond any mutual assumption of the parties. In such a case, the contractor may be relieved from further performance (discussed in Section 25.01E).

Gevyn Construction Corp. v. United States[36] involved a contract with the U.S. Postal Service. The specifications referred to a letter that Michigan state officials sent to the Post Office, stating that the contractor could tap a drain into a designated storm sewer of the state of Michigan. Relying on this letter, the contractor made its bid. But the state of Michigan refused permission, and the contractor was forced to connect the drain to a more distant outlet.

Although the court to some degree emphasized that the U.S. Postal Service invited the contractor to rely on the letter, the main basis for relief was the parties' having based their agreement on the false assumption that the drain could be tapped into the state's storm sewer.

Another illustration of the occasional use of mutual mistake as a basis for relief is seen in Section 23.05B, dealing with disruptive labor activities.

D. Impossibility: Commercial Impracticability

Despite the generally harsh attitude taken by nineteenth-century common law courts toward claims of impossibility, in that century contracting parties could be relieved if the court concluded that their performance had become impossible.[37] However, it soon became recognized that true impossibility is rare, and what is usually asserted as a defense by a performing party is that performance cannot be accomplished without excessive and unreasonable cost. This recognition emphasized "commercial impracticability" rather than actual impossibility.[38]

Despite the apparently increased willingness to grant relief to a contracting party, the need for commercial certainty meant that relief would be granted sparingly. First, something unexpected must occur. Second, the risk of this unexpected occurrence must not have been allocated to the performing party by agreement or by custom. The occurrence of the event must have rendered performance commercially impracticable.[39] These are formidable requirements, and relief requires extraordinary circumstances.

In addition to the differentiation between impossibility and impracticability, other differentiations are useful. First, *objective* impossibility is sometimes differentiated from *subjective* impossibility. To grant relief, it must usually be shown not simply that the contracting party could not perform (subjective impossibility) but also that other contractors similarly situated (objective impossibility) would not have been able to do so.

---

[35]See Sections 5.04C and 19.02E.

[36]357 F.Supp. 18 (S.D.N.Y.1972).

[37]*Taylor v. Caldwell*, 122 Eng.Rep. 309 (Q.B. 1863) (theater owner exonerated when theater burned).

[38]U.C.C. § 2-615(a).

[39]*Transatlantic Financing Corp. v. United States*, 363 F.2d 312 (D.C. Cir.1966) (charterer not relieved when Suez Canal closed in 1956); *M. J. Paquet, Inc. v. New Jersey Dept. of Transp.*, 171 N.J. 378, 794 A.2d 141, 148–49 (2002) (doctrine of impracticability used to allow a state transportation department to delete work made more expensive to perform by new OSHA regulations, where the parties could not agree upon a price adjustment for that work).

Second, *temporary* impossibility should be differentiated from *permanent* impossibility. For example, suppose a severe material shortage makes it impossible for the contractor to continue performance. This shortage may be relieved by eliminating the event that caused the shortage, such as a transportation strike. In such a case, impossibility is only temporary. Unless the contractor is held to assume these risks, the only relief it should obtain is a time extension, not termination. However, if the event will apparently continue for an indefinite time or for so long that it will drastically affect the cost of performance, termination may be appropriate.

Third, *partial* impossibility must be differentiated from *total* impossibility. Suppose a material shortage is caused by a strike. As a result, a supplier who has contract and customer commitments of 1,000 units has only 500 in stock and cannot obtain others. It will be allowed relief if it makes a good-faith allocation of its available supply to all its customers.[40] Suppose it has a contractual commitment to deliver ten units to a contractor. If it would be an exercise of commercial good faith to supply all customers—those with contracts and those with commercial commitments—alike, the contractual obligation would be discharged if the supplier delivered five units to the contractor.

E. Frustration

Frustration of purpose developed early in the twentieth century. It looked at the effect of subsequent events not on performance but on desirability of performance. For example, the leading English case[41] giving rise to this doctrine involved a contract under which the plaintiff rented rooms to the defendant from which the defendant would have a good view of the coronation of King Edward VII. The coronation was postponed because the king became ill, and the defendant was relieved from his obligation to pay for the rooms. Clearly, this case did not involve impossibility. The coronation parade postponement made the contract much less attractive than originally. This doctrine has great similarity to mutual mistake. Each party probably assumed that the coronation would take place as originally scheduled, and each was equally mistaken.

One can see that relief under this doctrine must be awarded sparingly or contracts would lose much of their effectiveness. The leading American case, *Lloyd v. Murphy*,[42] involved a claim by a commercial tenant at the onset of World War II that it should be relieved of its obligation to pay rent on premises from which it intended to sell new cars that became unavailable because of the war. The court held that relief under this doctrine required that the value of the contract be almost totally destroyed by an event that was not reasonably foreseeable by the party seeking relief. The court held that the tenant did not meet these requirements.

Suppose during construction of a racetrack a law were passed that made horse racing illegal in the state. Suppose the owner sought relief from contractual obligation by pointing to this event. If the project can be used only for an activity that is now illegal, perhaps the owner should be freed from its construction contract. However, much would depend on whether the legislative action were reasonably foreseeable and whether the project could be used for other purposes, such as a go-cart track, a tennis court, commercial exhibitions, a nine-hole golf course, or auto racing.

Although frustration claims are rare in construction projects, the Massachusetts Supreme Judicial Court employed this doctrine in a case that involved a contract to resurface and improve a public road.[43] The contract required the contractor to replace a grass median strip with precast concrete barriers. The prime contractor entered into subcontracts under which a designated subcontractor would provide the concrete barriers for the median strips. After work began, a citizens' group filed a lawsuit protesting removal of the grass median strip. The public entity settled the case by agreeing not to install any additional barriers for the median strips and deleting the concrete barriers from the project. As a result, the prime contractor canceled its subcontract with the subcontractor to whom the work had been awarded, generating a claim for lost profits by the canceled subcontractor. (The subcontractor had been paid at the contract price for the barriers it had delivered.)

The court held that the prime contractor was not responsible for the elimination of the barriers and that the risk of elimination of the barriers was not allocated by the subcontract. The prime contractor was excused because of the frustration-of-purpose rule, because neither party to the subcontract anticipated that the public entity would

---

[40]U.C.C. § 2-615(b).
[41]*Krell v. Henry*, 2 K.B. 740 (C.A.1903).

[42]25 Cal.2d 48, 153 P.2d 47 (1944).
[43]*Chase Precast Corp. v. John J. Paonessa Co., Inc.*, 409 Mass. 371, 566 N.E.2d 603 (1991).

cancel a major portion of the job. Unlike earlier cases and commentaries, the court's analysis did not draw any significant distinction between frustration and commercial impracticability.

F. Acceptance

Sometimes the contractor claims a defense when a claim is asserted against it for defective work, based on the owner's having accepted the project. Inasmuch as this defense is principally related to defects, it is discussed in Section 24.05.

G. Passage of Time: Statutes of Limitation and Repose

Sometimes a claimant is met with a defense that it did not begin legal action on the claim within the time required. Time limits may be created by contract, by judge-made law, or (most commonly) by statutes. Time limits created by contract or judge-made law are discussed at the end of this Section.

Statutory limits on the time available for bringing a lawsuit has generated immense complexity and must be looked at in any treatment of claims generally. Statutes barring claims based on the passage of time are of two types: statutes of limitations (discussed also elsewhere)[44] and statutes of repose. Both types of statutes deal with the effect of the passage of time on the maintainability of claims. They are designed to protect defendants from false or fraudulent claims that may be difficult to disprove if not brought until relevant evidence has been lost or destroyed and witnesses become unavailable. In addition, entirely apart from the fault aspect of delay in bringing legal action, the law can also seek to promote certainty and finality in transactions, especially commercial transactions, by terminating contingent liabilities at specific points in time. This second function is accomplished by statutes of repose. The statute can bar a claim of which the claimant was never aware.

Although statutes of limitation are necessary, their implementation has generated great difficulty in construction performance claims, claims often discovered long after the breach by the contractor or completion of the work. Such statutes frequently are unclear as to the time the statutory period begins. In construction, the period can begin when the wrongful act occurred, when the contract performance had been completed, when the defective

work had been discovered, when the owner knew of the defect and its cause, or when the owner knew or should have known all the elements of its claim, including against whom the claim can be made. For example, suppose a contract is made to build a commercial building in 1990. Work proceeds during that year. The contractor does not use proper workmanship in applying adhesive materials when constructing the roof. The building is accepted and occupied in 1990. In 2000, severe roof leaks occur that are traceable to improper workmanship by the contractor.

Suppose the applicable statute of limitations states that a claim for breach of contract must be begun within four years from the time the "cause of action accrued." If the cause of action accrued at the time of the improper workmanship or the acceptance of the building, the claim has been barred by the passage of time. If the claim accrued at the time of damage, should the period begin at the time of completion (at that time the improper workmanship could not have been corrected) or at the time the poor workmanship was discovered? If the cause of action accrued in 2000, legal action can be brought until 2004. If so, the contractor will be forced to defend a claim based on events that occurred fourteen years earlier. Its defense may be hampered by its inability to produce witnesses and documentary evidence. Yet if the earlier time is selected, the owner has lost a claim before it becomes aware it had one, unless an objective standard is used and a reasonable inspection would have discovered the poor workmanship during the time of performance.[45]

Most states would find that the owner's breach of contract claim against the contractor was barred by the four-year statute of limitations because the right to sue the contractor accrued in 1990, when the completed construction was accepted. (Remember that the statute begins to run at the time the "cause of action accrued.") A cause of action accrues at the earliest time a plaintiff may bring a lawsuit under a particular legal theory. A breach of contract action may be brought immediately upon project completion because the owner is deemed

---

[44]See Sections 2.03 and 14.09C.

[45]See *D. J. Witherspoon v. Sides Constr. Co.,* 219 Neb. 117, 362 N.W.2d 35 (1985) (where architect merely designs, period begins when design documents delivered; where he supervises, it begins at completion of his services); *Holy Family Catholic Congregation v. Stubenrauch Assoc.,* 136 Wis.2d 515, 402 N.W.2d 382 (App.1987) (period commences when contractor loses significant control and owner can occupy or use the building for its intended purpose; not determined by architect certificate or when contractor stopped work).

to have suffered injury, even if it did not yet know the construction was defective. If four years go by before the owner, despite reasonable investigations, discovers the defective workmanship, its contract claim would be barred.

Some states have ameliorated the harsh effect of this rule. They employ the "discovery rule" to delay accrual for latent (hidden) defects until the time when a reasonable and diligent owner would have discovered the defect.[46] Other states refuse to apply the discovery rule to breach of contract claims.[47]

Unlike a breach of contract claim, which accrues immediately upon the defendant's wrongful act, a negligence cause of action accrues only when a plaintiff suffers appreciable injury. Deciding when the plaintiff suffered appreciable injury can also be difficult to determine. Suppose a contractor claims that defects in the architect's design caused the contractor to incur higher-than-anticipated performance costs. The contractor's negligence or misrepresentation action would not accrue until it suffered economic losses with certainty, caused by the defective design.[48]

Participants in the construction industry persuaded all states but New York and Vermont to enact *statutes of repose* that cut off liability a designated number of years after substantial completion.[49] Some statutes dealt only with claims based on breach of contract, whereas others included tort claims.[50] Some differentiated between latent and patent defects, the former being defects not discoverable by reasonable inspection and the latter being defects that were reasonably discoverable. The statutes varied as to the

people who could take advantage of them. Typically, those protected were design professionals and contractors, with protection not accorded owners or suppliers. The time period selected by the legislatures also varied.

The statutes were attacked in many states, principally based on the unfairness of depriving a party of the claim that it did not realize it had. More technically, statutes were attacked based on constitutional grounds, mainly as violating the guarantee of equal protection of the law. Some people were given legislative protection, whereas others were not.

Most courts upheld such statutes. Some were held unconstitutional. Those held unconstitutional were usually redrafted to meet constitutional objections. Even some redrafted were found to be unconstitutional.[51] In addition, the statutes generated a great amount of litigation, not only passing on their constitutionality but also seeking to integrate them with the normal statutes of limitations of the state.[52] As this book cannot provide a detailed description and analysis of these statutes, local law must be consulted.

One observation that relates to risk management (discussed in greater detail in Section 15.03F) merits mention at this point. Racing to the legislatures may have distracted from another approach to the long-delayed claim. It is possible for the contracting parties to regulate this problem. Although contractual regulation will not affect claims of third parties, such as those who may be injured if a building collapses years after it is completed, the bulk of the exposure in this area usually involves claims by the owner against the contractor or design professional.

Parties to a construction contract can regulate the period of time in which the claim may be made. These agreed-upon limits will be enforced by the courts if the time periods are not unreasonably short. AIA Document

---

[46]*Erenhaft v. Malcolm Price, Inc.*, 483 A.2d 1192 (D.C.1984) is a leading case. See also West Ann.Cal.Code Civ.Proc. §§ 337.1 and 337.15 (specifying different limitations periods based on whether the defect was patent or latent); Md.Real Prop.Code § 10-203, interpreted in *Lumsden v. Design Tech Builders, Inc.*, 358 Md. 435, 749 A.2d 796 (2000); Annot., 33 A.L.R.5th 1 (1995).

[47]*Stephens v. Creel*, 429 So.2d 278 (Ala.1983); West Ann.Wash.Rev. Code § 4.16.326(1)(g).

[48]*Hardaway Co. v. Parsons, Brinckerhoff, Quade & Douglas, Inc.*, 267 Ga. 424, 479 S.E.2d 727 (1997) (contractor sued engineer for misrepresentation based on defective design); *MBA Commercial Constr., Inc. v. Roy J. Hannaford Co.*, 818 P.2d 469 (Okla.1991) (subcontractor's action did not accrue until it knew it would not be paid its delay costs).

[49]Gwyn & Davis, *Statutes of Repose*, 21 Constr. Lawyer, No. 3, Summer 2001, p. 33.

[50]See *Martinez v. Traubner*, 32 Cal.3d 755, 653 P.2d 1046, 187 Cal. Rptr. 251 (1982) (did not apply to tort claim for personal injury) and *Stoneson Dev. Corp. v. Superior Court*, 197 Cal.App.3d 178, 242 Cal. Rptr. 721 (1987) (applied to claim based on strict liability). See also *Hagerstown Elderly Assocs. Ltd. Partnership v. Hagerstown Elderly Bldg. Associates Ltd. Partnership*, 368 Md. 351, 793 A.2d 579 (2002) (applies to contract and tort actions).

[51]The leading case refusing to uphold such a statute is *Skinner v. Anderson*, 38 Ill.2d 455, 231 N.E.2d 588 (1967) (statute subsequently revised). The leading case upholding such a statute is *Freezer Storage, Inc. v. Armstrong Cork Co.*, 234 Pa.Super. 441, 341 A.2d 184 (1975). Currently, most courts find statutes of repose constitutional, see *Weston v. McWilliams & Assocs., Inc.*, 716 N.W.2d 634 (Minn.2006) and *Winnisquam Regional School Dist. v. Levine*, 152 N.H. 537, 880 A.2d 369 (2005). Care must be taken not to rely on a court decision holding the statute unconstitutional. The legislature may have subsequently amended the statute to meet the objections of the court decision.

[52]This topic was dealt with extensively in *Northern Indiana Pub. Service Co. v. Fattore Constr. Co.*, 486 N.E.2d 633 (Ind.App.1985). The court held that a special statute takes preference over a general statute, especially if the former has been more recently enacted. It also held that if either of two statutes can apply, the court is to apply the longer statute.

B101-2007, Section 8.1.1, and A201-2007, Section 13.7, require legal action to be brought "within the time period specified by applicable law, but in any case not more than 10 years after the date of Substantial Completion of the Work." The 10-year outer limit functions as a contractual statute of repose. Another example of regulation of the time in which the claim can be brought is Section 2-725(1) of the Uniform Commercial Code, which governs transactions in goods and can, by analogy, be applied to service transactions. Section 2-725(1) creates a four-year period of limitation but allows a contractually created period of not less than one year but not longer than the four-year statutory period.

Parties to a construction contract can also regulate when the limitation period begins. AIA Document B141-1997, Paragraph 1.3.7.3, and A201-1997, Paragraph 13.7.1, specified that the limitation period began to run from the time of substantial completion. Although enforced by the courts,[53] these provisions were dropped from the 2007 AIA documents in response to owner dissatisfaction.

The doctrine of laches is a judge-made limitation on the time available for bringing a lawsuit. Laches was developed in England by the equity courts and applies to plaintiffs seeking equitable remedies, such as injunctions and specific performance. It generally prevents a person who is aware of his injury from "sitting on his rights" and delaying to bring a claim, to the disadvantage of the defendant. For example, in *Chirco v. Crosswinds Communities, Inc.*,[54] a copyright holder of architectural plans waited to bring suit for two and one-half years while the defendant built a condominium using the infringed plans. The court ruled that laches barred the copyright holder's request that the infringing building be destroyed.

Sometimes contractors contend that any warranty or guarantee clause is a private period of limitation (discussed in Section 24.09B, dealing with these clauses).

H. Release

In some disputes, the party against whom a claim is made asserts the claim has been settled. Section 27.13 discusses this defense, and Section 22.05 discusses it in the context of final payment.

---

[53]*Gustine Uniontown Assocs., Ltd. v. Anthony Crane Rental, Inc.*, 2006 PA Super 12, 892 A.2d 830 (2006) (interpreting identical language in B141-1987, Paragraph 9.3).

[54]474 F.3d 227 (6th Cir.2007). For further discussion of laches, see D. DOBBS, REMEDIES § 2.4(4) (2d ed. 1993).

I. Sovereign Immunity: Federal and State

When the United States enters the marketplace, as a general rule its contracts are treated as would be those made by private parties. To be sure, such contracts are heavily regulated, to some degree by statutes but mainly by administrative regulations. Specified clauses must often be included in contracts. It also means that claims must be brought before the U.S. Court of Federal Claims, the United States having given up its sovereign immunity by the Tucker Act.[55]

The special status of the federal government became an issue that culminated in *United States v. Winstar Corp.*[56] This case developed from the attempt by the U.S. agencies to sort out the mess that resulted when many savings and loan associations (S&Ls) failed in the 1980s. These agencies urged solvent S&Ls to acquire the defunct S&Ls. To encourage the solvent ones, government officials promised a favorable accounting ruling on the acquisitions. Yet after the acquisitions took place, the United States enacted laws that required less favorable accounting rules. Parties that acquired the defunct S&Ls brought breach-of-contract claims against the United States.

The government asserted a number of defenses, but the main defenses centered on the sovereign status of the United States. These defenses are the doctrines of sovereign acts and of unmistakability. The doctrine of sovereign acts was developed in a 1925 case that involved enactment of a law making illegal the performance of a contract the United States had made before enactment.[57] The unmistakability doctrine requires unmistakable evidence, at the time of the transaction, showing that the government was surrendering its immunity. Surrender must be shown in unmistakable terms.[58]

In the *Winstar* case, the U.S. Supreme Court—sharply divided both as to outcome and rationale—narrowly applied these defenses and ruled against the government. The opinions tended to prefer *congruence* (that the U.S. government contracts should be treated as would any other) over *exceptionalism* (that government contracts are treated differently).[59] The sharp debate within the Court

---

[55]The core of the Tucker Act is found in 28 U.S.C.A. § 1491(a)(1).
[56]518 U.S. 839 (1996).
[57]*Horowitz v. United States*, 267 U.S. 458 (1925).
[58]*Bowen v. Public Agencies Opposed to Social Security Entrapment*, 477 U.S. 41 (1986).
[59]For an analysis of this doctrine and for a survey of all the law on this topic, see Schwartz, *Assembling* Winstar: *Triumph of the Ideal of Congruence in Government Contracts Law?* 26 Pub.Cont.L.J. 481 (1997). See also Citron, *Lessons From the Damages Decisions Following* United States v. Winstar Corp., 32 Pub.Cont.L.J. 1 (2002).

shows that making an agreement with the federal government carries some risks not present in normal private contracts. It should be noted, however, that this sharp debate did not involve *making* contracts or asserting claims but whether the government could defend its failure to perform by pointing to a law it enacted after the contract was made and that made performance illegal.[60] See Section 23.05D.

Although sovereign immunity has been abolished in many states, some states use immunity to provide a defense to a public owner. Despite the many cases abolishing immunity, a 1985 case stated that a statute abolishing immunity must be strictly construed so as not to undermine sovereignty.[61] A number of cases with varied outcomes have recently passed on whether a public entity waives its immunity by entering into a construction contract.[62]

SECTION 23.04 Restitution

Some of the doctrines described in Section 23.03 must be looked at in connection with restitution. If a contractor is relieved because of common law doctrines described in Section 23.03, it may seek affirmatively to recover for the benefit it has conferred before the time its obligation to perform was terminated. One aspect of this, destruction of the work in progress, is discussed in Section 23.05C. For two reasons, this problem has not generated much difficulty in construction disputes. First, the contractor is usually being paid as its performance progresses, and any restitutionary claim it might have would be reduced by these progress payments. This means the amount at stake is likely to be small. Second,

courts have used these relief doctrines sparingly in construction cases. However, in a series of Massachusetts cases, subcontractors were granted broad restitution remedies when the prime contract was found to have been illegally awarded.[63]

Restitution, however, has another use. Section 23.02 discussed claims and their bases. Sometimes a party wishes to measure its recovery for its claim not by the other party's promised performance but by the benefit it has given the other party (discussed in greater detail in contractor claims analyzed in Section 27.02).

SECTION 23.05 Specific Applications of General Principles

A. Increased Cost of Performance

In the course of performance, the contractor may find that its cost of performance has risen dramatically. Illustrations can be drastic increases in the cost of materials, equipment, supplies, utilities, labor, or transportation. These risks are generally assumed by the contractor under a fixed-price contract.

The common law provided little protection to a contractor who performs under a fixed price. To be sure, extraordinary and unanticipated events may occur that drastically affect the cost of the contractor's performance and may be the basis for a claim for relief. Here again the remedy sought may be influential if not determinative. Additional time is easiest to obtain, although this depends on the language of the clause that grants time extensions. Additional compensation and termination, if sought, are more difficult to obtain.

There are exceptions, most relating to specific contract provisions or terms implied into the contract. Examples of the former are equitable adjustment provisions that relate to subsurface work, as discussed in Section 25.06; escalation clauses that can grant price increases under certain market conditions; and time extension provisions, as discussed in Section 26.08. Examples of the latter are implied warranties by the owner that the design can be constructed economically and efficiently as discussed in Section 23.05E.

---

[60]Predictions of huge damages judgments being awarded against the federal government in the wake of *Winstar* have, according to two recent articles, failed to materialize. See Graves, "Winstar *Wars*"—*Revenge of the Thrift: A Viable Model to Right a Decade of Wrongs*, 36 Pub. Cont. L. J. 361 (2007); Rizzo & Gomez, *Erosion of the Sovereign Acts Doctrine? How Recent* Winstar *and Spent Nuclear Fuel Litigation Impacts Government Contractors*, 42 Procur. Lawyer, No. 3, Spring 2007, p. 3.

[61]*De Fonce Constr. v. State*, 198 Conn. 185, 501 A.2d 745 (1985).

[62]Sovereign immunity was applied in *County of Brevard v. Miorelli Eng'g, Inc.*, 703 So.2d 1049 (Fla.1997), rehearing denied Dec. 12, 1997, Jan. 7, 1998; *C & M Constr. Co., Inc. v. Com.*, 396 Mass. 390, 486 N.E.2d 54 (1985); *Travis County, Texas v. Pelzel & Assocs., Inc.*, 77 S.W.3d 246 (Tex.2002); *G. M. McCrossin, Inc. v. W. Va. Bd. of Regents*, 177 W.Va. 539, 355 S.E.2d 32 (1987). A waiver of sovereign immunity was found in *Architectural Woods, Inc. v. State*, 92 Wash.2d 521, 598 P.2d 1372 (1979). A waiver of tribal immunity was found in *C & L Enterprises, Inc. v. Citizen Band Potawatomi Indian Tribe of Okla.*, 532 U.S. 411 (2001).

[63]*Albre Marble & Tile Co. v. John Bowen Co.*, 338 Mass. 394, 155 N.E.2d 437 (1959); *Boston Plate & Window Glass Co. v. John Bowen Co.*, 335 Mass. 697, 141 N.E.2d 715 (1957); *M. Ahern Co. v. John Bowen Co.*, 334 Mass. 36, 133 N.E.2d 484 (1956).

B. Labor Disruptions: The Picket Line and Project Labor Agreements

The labor dispute is one of the most disruptive events in a construction project. Such disputes can cause frequent and lengthy work stoppages. Craft unions sometimes dispute who has the right to perform certain work. Not uncommonly, employees of unionized contractors refuse to work on a site with nonunion employees of another contractor. Workers, as noted in Section 19.01D, sometimes refuse to work on a project because prefabricated units are introduced on the site in violation of a "work preservation" clause in a subcontract or collective bargaining agreement. Any of these situations can result in a strike, a picket line, or both. Workers frequently refuse to cross a picket line. Such refusals can shut down a job directly (because no workers are willing to work on the site) or indirectly (through the refusal of workers to deliver materials to a picketed site).

Pressure and even coercion are common tactics in the struggles between employers and groups that are attempting to organize the workers as well as represent them in negotiating wages, hours, job security, and working conditions. Each of the disputants—employer and union—not only seek to persuade but also use economic weapons to obtain a favorable outcome.

Disputants direct pressure at each other by use of economic weapons such as strikes or lockouts. Even these direct primary pressures affect neutrals not involved in the labor dispute. A strike affects nonstriking workers, the families of the strikers, and those who deal with the struck employer, such as those supplying it goods or purchasing its products or services. Likewise, a lockout by the employer affects neutrals.

Economic warfare often expands to include pressure and coercion on third parties important to the employer, such as suppliers or customers. The pressure can be direct, as with communicated coercive threats, or indirect, as with a picket line whose signs describe the union's reason for its grievance against the employer and ask that certain activity or nonactivity be taken.

As these tactics broaden the field of economic warfare, they may begin to seriously harm neutrals and become what are called *secondary boycotts*. Such impermissible activities are unfair labor practices, giving the party injured by such activity the right to damages and, more important, the right to obtain an injunction ordering that such activities cease or be modified. The U.S. Supreme Court has stated that the law reflects "the dual congressional objectives of preserving the right of labor organizations to bring

pressure to bear on offending employers in primary labor disputes and of shielding unoffending employers and others from pressures and controversies not their own."[64]

In addition, the Court recognized that the law was concerned not only with pressure brought to bear on the other disputants but also with pressure brought to force third parties "to bring pressure on the employer to agree to the union's demands."[65]

The line between permitted primary and prohibited secondary boycotts is difficult to draw in industrial collective bargaining warfare. It is even more difficult in construction work because of the transient nature of the workers, the seasonal nature of the work, the proliferation of craft unions, the frequent occupation of the construction site by employees of many bargaining entities such as contractors and subcontractors, and the special rules that govern collective bargaining in the construction industry.

One of these factors merits additional comment. As indicated, the construction project usually involves work by a number of contractors, sometimes at different stages of the work, but often at the same time. Some of the employers may have collective bargaining agreements with a union, whereas others may not. Some employers may have collective bargaining agreements with one union, and others may have collective bargaining agreements with a different one. For example, a nonunion prime contractor may be working alongside union subcontractors. Sometimes union subcontractors are working on a site with employees of a subcontractor who does not have a collective bargaining agreement with any union. A union engaged in an economic struggle with an employer can maximize its bargaining power if it can shut down the entire project by putting up a picket line that no union workers will cross. The very effectiveness of common situs picketing (picketing an entrance used by all workers) is the reason it can also enmesh many neutrals and cause frequent and costly work stoppages.

The law has limited common situs picketing by allowing the prime contractor to set up separate gates, one for the employees involved in the labor dispute and the other for those not involved in the dispute. This two-gate system prevents the project from being totally shut down. Picketing can be done at the first gate but not at

---

[64]*NLRB v. Denver Building & Constr. Trades Council*, 341 U.S. 675 (1951).

[65]*NLRB v. Local 825, International Union of Operating Eng'rs*, 400 U.S. 297 (1971).

the second. In *NLRB v. International Union of Elevator Constructors*,[66] an employee of a neutral subcontractor refused to use the neutral gate when another gate at the site was being picketed as a result of a labor dispute involving another subcontractor. The union attempted to enforce a contractual picket line clause that protected its employees from discipline when they refused to use a neutral gate. The Court of Appeals upheld a decision of the National Labor Relations Board (NLRB) that the union was guilty of violating the National Labor Relations Act when it sought to enforce such a clause, because the employee's refusal to work during picketing by another subcontractor's employees was not a protected activity.

Owners on large projects may circumvent problems of both union and nonunion trades working together *and* jurisdictional disputes among different unions by having the prime contractor and all subcontractors enter into a Project Labor Agreement (PLA). A PLA is a project-specific labor agreement. It ensures uniform terms and conditions of employment for all workers, whether union or nonunion, covering such matters as working hours, shift times, scheduling, holidays, overtime, and premium pay. All contractors, unions and employees are subject to the same collective bargaining agreement. Nonunion contractors must hire workers through the union hiring hall (although an exception may exist for so-called core employees), but the employees need not actually join the union. Of even greater importance for owners, the PLA establishes a dispute resolution mechanism and prohibits strikes. In short, PLAs are intended to guarantee labor peace through the life of the contract.

Legal challenges to PLAs have arisen on public works projects. In the *Boston Harbor* case, the Supreme Court ruled that PLAs are not preempted by the National Labor Relations Act.[67] Legal attacks then shifted to state laws. The most frequent complaint by nonunion contractors is that PLAs violate the state competitive bidding laws by restricting competition for public works jobs to union contractors. This narrowing of the number of contractors willing to bid on advertised projects, the opponents to PLAs contend, deprives the state from obtaining the lowest bid for the work. (Competitive bidding statutes are discussed in

Chapter 18.) Most courts apply a balancing test. They will uphold use of PLAs on public projects if the size and complexity of the project are such that the danger of labor disruptions without a PLA would be very high.[68] Increasingly courts reject the premise of opponents that PLAs discourage bidding by nonunion contractors.[69] The legality of PLAs notwithstanding, the propriety of public agencies entering into such agreements has generated heated debate.[70]

But suppose a labor dispute does shut down or curtail a project. What effect will this have on the construction contract obligations of participants in the project?

Labor disputes, like damage to a partially completed project, can involve three separate but related questions:

1. Is the contractor entitled to a time extension for the period that work is shut down or curtailed?
2. Can the contractor collect additional compensation for expenses caused by the shutdowns or curtailment of work?
3. Can the contractor whose work is severely disrupted by a labor dispute to which it is not a party terminate its obligation to perform under the contract?

As for the first, construction projects typically provide time extensions for work disrupted by labor difficulties. For example, Section 8.3.1 of AIA Document A201-2007 awards the contractor a time extension for delays caused "by labor disputes."[71] Without such a provision granting relief, a contractor will have a difficult time obtaining a time extension in the event of labor difficulties. The contractor generally assumes the risk of events that would increase the cost and time of performance.

---

[66]902 F.2d 1297 (8th Cir.1990).

[67]*Building & Constr. Trades Council of the Metro. Dist. v. Associated Builders & Contractors of Mass./R.I., Inc.*, 507 U.S. 218 (1993) (involving a project to clean up the polluted Boston Harbor).

[68]A leading case upholding use of a PLA is *New York State Chapter, Inc. v. New York State Thruway Auth.*, 88 N.Y.2d 56, 666 N.E.2d 185, 643 N.Y.S.2d 480 (1996) (renovation of the Tappan Zee bridge, spanning the Hudson River in New York). Use of a PLA on a routine library construction project was rejected in *Tormee Constr., Inc. v. Mercer County Improvement Auth.*, 143 N.J. 143, 669 A.2d 1369 (1995).

[69]Statistical evidence that PLAs do not discriminate against non-union contractors is cited in *John T. Callahan & Sons, Inc. v. City of Malden*, 430 Mass. 124, 713 N.E.2d 955, 964 (1999) and *Associated Builders & Contractors, Inc. v. Southern Nev. Water Auth.*, 979 P.2d 224, 229 n. 1 (Nev.1999).

[70]A point-counterpoint pair of articles is Baskin, *The Case Against Union-Only Labor Project Agreements*, 19 Constr. Lawyer, No. 1, Jan. 1999, p. 14 and Kopp & Gaal, *The Case for Project Labor Agreements*, 19 Constr. Lawyer, No. 1, Jan. 1999, p. 5.

[71]New Jersey held a strike clause used in all elevator installer contracts not unconscionable. *Curtis Elevator Co. v. Hampshire House, Inc.*, 142 N.J.Super. 537, 362 A.2d 73 (Law.Div.1976).

Moving to the second question, it is usually difficult for the contractor to recover additional compensation for costs increased markedly by unusual labor disruptions, because these are, on the whole, assumed risks.[72]

The third question, that of termination, will depend principally on the construction contract. Interestingly, the AIA standard documents, although providing for time extensions, do not provide for termination.[73] The absence of a clause specifically allowing termination should not necessarily preclude termination from being a proper remedy under certain circumstances.[74] Some courts hold that a contract that specifically mentioned labor problems and did not make them the basis for termination, indicated that the parties contemplated continued performance, with time extensions being the only remedy.

It may be useful to examine a case that involved a defense of impossibility asserted to be grounds for terminating performance obligations where the contract did not treat labor difficulties. In *Mishara Construction Co. v. Transit-Mixed Concrete*,[75] a prime contractor brought legal action against a supplier who claimed it could not deliver because of a picket line at the site. The court rejected the prime contractor's contention that a labor dispute making performance more difficult never constitutes an excuse for nonperformance. The court instead held that whether a labor dispute excuses the supplier's obligation to perform would be a fact question that must be determined by a jury. The court noted that a picket line might constitute a mere inconvenience and not make performance impracticable. Also, according to the court, a contract made in the context of an industry with a long record of labor difficulties shows that the parties assumed the risk of labor disputes. For example, if the supplier knew it was agreeing to deliver concrete to an employer who had had chronic and bitter labor difficulties, having made a contract without providing for contractual protection could indicate that it was assuming this risk. The court stated, "Where the probability of a labor dispute appears to be practically nil, and where the occurrence of such a dispute provides unusual difficulty, the excuse of impracticability might well be applicable."[76]

The court concluded by noting that the tendency has been to recognize strikes as an excuse for nonperformance. Although the court indicated that under certain circumstances, absence of protective language might be considered an assumption of risk, the Uniform Commercial Code, with its emphasis on commercial impracticability, might not find the absence of a clause *conclusive* on the question of whether a particular risk was assumed.[77]

C. Partial or Total Destruction of Project: Insurance and Subrogation Waivers

During project construction, the project may be partially or totally destroyed by circumstances for which neither party is chargeable. The work may be destroyed by fire, unstable subsurface conditions, or violent natural acts such as earthquakes or hurricanes. Although destruction is often discussed in terms of impossibility or assumption of risk, it is essential to recognize that at least three issues can arise when such events occur:

1. Are the parties relieved from any further obligation to perform?
2. Must the contractor repair and restore damage to the work?
3. Is the contractor entitled to be paid for work incorporated into the structure before destruction?

As for the first question, the performance obligation is not terminated in the absence of a clause relieving the contractor from its performance obligation. Similarly, with regard to the second question the contractor is responsible for repairing and restoring damaged work. Courts have stated that a party who in unqualified terms promises to perform is itself at fault when it does not expressly protect itself from these contingencies.[78] If the contract is for repairs or additions to an existing structure or for part of

---

[72]*McNamara Constr. of Manitoba, Ltd. v. United States*, 206 Ct.Cl. 1, 509 F.2d 1166 (1975).

[73]See AIA Doc. A201-2007, Art. 14.

[74]See AIA Doc. A201-2007, § 13.4.1, which states contractual remedies are not exclusive. See also Section 33.03A, which discusses this in the context of termination.

[75]365 Mass. 122, 310 N.E.2d 363 (1974).

[76]310 N.E.2d at 368.

[77]New Jersey stated that even without a clause excusing performance hindered by labor disputes, the doctrine of commercial impracticability might give the performing party relief. *Curtis Elevator Co. v. Hampshire House, Inc.*, supra note 71.

[78]*Stees v. Leonard*, 20 Minn. 494 (1874). Although the principal cases articulating this rule were decided in the nineteenth century, the rule was applied in *Dravo Corp. v. Litton Systems, Inc.*, 379 F.Supp. 37 (S.D.Miss.1974).

a new structure and the work is destroyed in whole or in part without the fault of either party, each party is relieved from further contract obligations.[79]

As to the third question—that of recovery for work performed before destruction—the result depends, as a rule, on the termination issue. If the contractor is not discharged from its obligation to perform, it cannot recover for work performed. If it is discharged, as where it had agreed to build only part of the structure or to repair an existing structure, generally the contractor can recover for work performed prior to destruction.[80]

Usually each participant in the construction process insures its property from loss of the type discussed in this subsection. Typically, the owner insures the work in progress, whereas the contractor insures its equipment and other property that will not go into the project. When the project is destroyed during construction, the owner receives proceeds from the insurance company. It holds these proceeds as fiduciary for those who have suffered property damage covered by the proceeds. Because much of the work has been paid for by progress payments, the proceeds held for the contractor are typically work for which progress payments have not yet been received and work for which progress payments were received but from which the retainage has been deducted. A well-planned property insurance system should reimburse the parties for the losses suffered. If this occurs, an argument can be made for the rule that does not discharge the parties from their obligation to perform further. This result follows whether the construction was for an entire structure or part of one or for the repair of an existing structure.

This appears to be the solution under AIA Document A201-2007, Article 14, dealing with the termination of the contract. Article 14 does not include destruction of the project as a ground for termination. The 1987 AIA Document A201 appeared to assume that work destroyed or damaged would be replaced.

Changes were made in 1997. Prior to 1997 after the loss was adjusted, unless there were a "special agreement," presumably by the owner and the contractor, the damaged property would be replaced. This provision assumed that there would be no discharge of the contract unless the parties agreed to do so. But in 1997 Paragraph 11.4.9 (replacing old Paragraph 11.3.9) gave the owner a right to terminate "for convenience." See Section 33.03B. This provision, renewed in A201-2007 Section 11.3.9, gives the owner the freedom to terminate if the project looks much different after destruction. But no corresponding option is given to the contractor if it believes it made a bad contract or things have occurred that make the contract much less desirable.

If a substantial period of time has elapsed from the making of the contract until the destruction of the project, it would be unfair to require each party to begin over. Even if property losses have been reimbursed through insurance, the original bid was based on prices and conditions existing at the time bids were made. The prime contractor might not be able to hold subcontractors to the original subcontracts in states where those who build only part of a whole structure are relieved from further performance obligations. In these states, new subcontracts would have to be negotiated. Where it would be an essentially different contract than originally made, holding that the contractor must still perform would be unfair. Likewise, the owner's perspective may be different after the project has been destroyed while being constructed.

New information may bear on the economic viability of the project, and it might be unreasonable to compel the owner to continue as if nothing had happened.

It is advisable to incorporate a provision in the construction contract that would give either party the right to terminate if the project is totally or nearly destroyed after a designated period of time has expired from start of performance.

As indicated throughout this book, legislatures are increasingly enacting laws that regulate the effect of accidental destruction of the work in progress. For example, in 1990 California enacted a law that requires that public contracts "not require the contractor to be responsible for the cost of repairing and restoring damage to the work, which damage is determined to have been proximately caused by an Act of God, in excess of 5% of the contracted amount."[81]

The statute provides this protection only if the contractor has built in accordance with accepted building standards and the project's plans and specifications. However, the contracts may contain provisions for terminating the contract. The statute defines acts of God to include only tidal waves and earthquakes that exceed 3.5 on the Richter scale.

Although discussion has emphasized destruction for which neither party can be held accountable, another issue

---

[79]*Fowler v. Ins. Co. of North Am.*, 155 Ga.App. 439, 270 S.E.2d 845 (1980).

[80]Annot., 28 A.L.R.3d 788 (1969).

[81]West Ann.Cal.Pub.Cont.Code § 7105.

often arises over responsibility. The most frequent cause of destruction—fire—is usually an insured risk. Fire losses are often traceable to human failings. When the insurer pays, it often looks for reimbursement from a party, other than the insured, who it can claim negligently caused the fire.

Assignment and the equitable doctrine of subrogation are methods insurers use to obtain reimbursement. Insurance policies often provide that on being paid for the loss, the insured assigns to its insurer its claims against third parties who caused the loss. Subrogation allows the insurer to "step into the shoes" of the insured. The insurer can assert any claims the insured may have against third parties (not its insured) who caused the loss. But construction projects, with their wealth of entities, usually provide ample third parties for the insurer to pursue.

Construction contracts frequently use techniques to bar subrogation. One is to require that the property insurance designate all the active participants as additional insureds. The EJCDC Standard General Conditions of the Construction Contract, C-700, Paragraph 5.04(B)1 (2007), uses this technique.

Another technique, often used together with the former method, is to require all active participants to waive any rights they have against other participants. The AIA uses such waiver of subrogation clauses in all AIA documents.[82] Similarly, the EJCDC, in its C-700, includes a waiver of subrogation in Paragraph 5.07. To illustrate, suppose a fire has been caused by the prime contractor and the insurer pays the owner for the damage. The insurer will not be able to pursue the prime for negligently causing the fire if the owner has waived its rights of subrogation against the prime. This method seeks to place the entire loss on the insurer and frees the participants in the construction project or, more realistically, their public liability insurers from responsibility.

An issue can arise as to whether the AIA waiver of subrogation clause, A201-2007, Section 11.3.7, protects against a subrogation claim for a loss that occurs *after* the project has been completed. That section states that the waiver applies to property damage "to the extent covered by property insurance obtained pursuant to this Section 11.3." The insurance required by Section 11.3.1

must cover losses until final payment has been made or until no entity other than the owner has an insurable interest, whichever is later. This would preclude the waiver being effective after coverage expires. However, Section 11.3.5 also extends the waiver agreement to property insurance obtained by the owner and "provided on the completed Project through a policy or policies other than those insuring the Project during the construction period." This Section has been interpreted to extend waiver to post-completion claims.[83]

Some insurers do not look favorably on such waivers. To them, waivers give away their claims, much like a motorist admitting he was wrong at the scene of an auto accident. An insured lost its coverage by waiving subrogation, held to be a violation of the failure-to-cooperate provision in the insurance policy.[84] For that reason A201-2007, Section 11.3.7, gives the insurer notice by requiring that the waivers of subrogation be included as an endorsement on the policy.

Insurers defend subrogation claims by contending that as between the blameless insurance company and a negligent third party, the latter or its liability insurer should pay the loss. In addition, they suggest that relieving all the parties from negligent conduct can encourage carelessness. Waivers also create an externality, a cost borne by others that should be allocated to the party whose activity has caused the loss.

Those who oppose subrogation claims point to their transaction costs, a second lawsuit often needed to complete the process. They also state that subrogation recoveries are insurer windfalls, as they are too uncertain to be included in the data used to determine premiums. Finally, those who structure a waiver of subrogation system feel that such a system best accords with the intention of all the parties to the construction project. For this reason, some courts deny the owner's property insurer's subrogation claim even in the absence of a waiver provision. These courts reason that the owner's decision to insure against a risk implied an intent to limit its remedy to insurance coverage.[85]

---

[82]See AIA Docs. A201-2007, §§ 5.3, 11.3.7; A401-2007, § 13.9; and B101-2007, § 8.1.2. See *Davlar Corp. v. Superior Court*, 53 Cal. App.4th 1121, 62 Cal.Rptr.2d 199 (1997) (insurer not permitted subrogation against subcontractor based on prime contract incorporated into subcontract).

[83]*Colonial Properties Realty Ltd. Partnership v. Lowder Constr. Co., Inc.*, 256 Ga.App. 106, 567 S.E.2d 389, 391–92 (2002). See also, *Touchet Valley Grain Growers, Inc. v. Opp & Seibold General Constr., Inc.*, 119 Wash.2d 334, 831 P.2d 724, 728 (1992) (not involving an AIA contract).

[84]*Liberty Mutual Ins. Co. v. Altfillisch Constr. Co.*, 70 Cal.App.3d 789, 139 Cal.Rptr. 91 (1977).

[85]*Morsches Lumber, Inc. v. Probst*, 180 Ind.App. 202, 388 N.E.2d 284, 287 (1979); *Acadia Ins. Co. v. Buck Constr. Co.*, 756 A.2d 515 (Me.2000).

A clear majority of courts have adopted the latter viewpoint and have enforced waiver agreements and disallowed subrogation claims under a variety of circumstances. These courts interpret the waiver clause broadly to encompass damage to all property covered by the owner's property insurance, even if the contractor damaged property not included within its scope of work.[86] Waiver clauses have also been enforced notwithstanding an insurer's claim that it had no notice of its insured's agreement.[87] Perhaps most controversial is the willingness of some courts to deny subrogation claims even when the owner alleges gross negligence by the defendant.[88]

Other courts interpret waiver of subrogation provisions more narrowly. These courts are most likely to find the waiver does not extend to damage to property not included in the scope of the contract, an issue that most commonly arises on renovation projects.[89] Courts may also refuse, on public policy grounds, to apply the waiver provision to claims of gross negligence.[90]

D. Governmental Acts

Government interference can take different forms. After the contract has been made, a law may be passed that makes performance of the contract illegal. For example, suppose a law is passed prohibiting the construction of any nuclear facility. Certainly enactment of such legislation would terminate any contract to build a nuclear plant. Performance would require an illegal act. The contractor's right would be limited to recovery for any work performed before the enactment of the legislation. Such recovery would be restitutionary and based on unjust enrichment. If the party whose performance became illegal was a government entity, special problems develop if that government enacted the law that made the performance by the government entity illegal. In effect it is the act of the contracting party, here the government, that made performance illegal. This situation was discussed in Section 23.03I, dealing with sovereign immunity.

A change in the law, made after the contract was signed and performance begun, may also upset the parties' pricing calculations. For example, in 1972, President Richard M. Nixon discontinued existing wage and price controls, causing the price of materials to escalate. Contractors who at the time were performing fixed price contracts, calculated on the assumption that the price controls would remain in place, had to cope with this unexpected price hike.[91] In a New Jersey public works case, new OSHA regulations made certain work under the contract more expensive to perform. When the parties were unable to agree upon a price increase, the state deleted the affected work.[92]

Another form of governmental interference would be the issuance of a judicial or administrative order shutting the project down. For example, a project might be shut down because of improper construction methods, an invalid building permit, or a design or use that violated land use controls. If the project is shut down for an appreciable period of time, the party not responsible for the shutdown should be relieved from further performance obligations and may have a cause of action against the other party. For example, if a project is shut down because of improper design or failure to comply with land use controls, the contractor may have a valid claim for

---

[86]*Tokio Marine & Fire Ins. Co. v. Employers Ins. of Wausau*, 786 F.2d 101, 104–05 (2d Cir.1986) (Florida law); *Lloyd's Underwriters v. Craig & Rush, Inc.*, 26 Cal.App.4th 1194, 32 Cal.Rptr.2d 144, 148 (1994); *Village of Rosemont v. Lentin Lumber Co.*, 144 Ill.App.3d 651, 494 N.E.2d 592, 598 (1986); *South Tippecanoe School Building Corp. v. Shambaugh & Son, Inc.*, 182 Ind.App. 350, 395 N.E.2d 320, 332–33 (1979); *Employers Mutual Cas. Co. v. A.C.C.T., Inc.*, 580 N.W.2d 490, 493 (Minn.1998) (broad waiver if the owner relies on its property insurance, but waiver is limited to damage to property included in the contractor's scope of work if the owner obtained a separate all-risk policy just to insure against damage to the work); *Chadwick v. CSI, Ltd.*, 137 N.H. 515, 629 A.2d 820, 826 (1993); *Trinity Universal Ins. Co. v. Bill Cox Constr., Inc.*, 75 S.W.3d 6, 11–14 (Tex.App.Ct.2001). See Singer, *AIA Waivers of Subrogation Continue to Benefit Architects Even While the Facts Are Stretching the Analysis*, 19 Constr.Litig.Rptr., No. 5, May 1998, p. 147.

[87]*Universal Underwriters Ins. Co. v. A. Richard Kacin, Inc.*, 2007 PA Super 13, 916 A.2d 686 (2007); *Bakowski v. Mountain States Steel, Inc.*, 52 P. 3d 1179 (Utah 2002).

[88]*St. Paul Fire & Marine Ins. Co. v. Universal Builders Supply*, 409 F.3d 73 (2d Cir.2005) (distinguishing between waiver of subrogation clauses and exculpatory clauses); *Behr v. Hook*, 173 Vt. 122, 787 A.2d 499 (2001) (AIA waiver clause applies to owner's gross negligence claim against subcontractor).

[89]*Independent School Dist. 833 v. Bor-Son Constr., Inc.*, 631 N.W.2d 437 (Minn.App.2001), review denied Oct. 16, 2001; *S.S.D.W. Co. v. Brisk Waterproofing Co., Inc.*, 76 N.Y.2d 228, 556 N.E.2d 1097, 1099–1100, 557 N.Y.S.2d 290 (1990); *Travelers Ins. Cos. v. Dickey*, 799 P.2d 625, 630–31 (Okla.1990) (not involving an AIA contract).

[90]*Colonial Properties Realty v. Lowder Constr. Co.*, supra note 83, 567 S.E.2d at 394 (owner's claim of gross negligence against contractor).

[91]*Blake Constr. Co.*, GSBCA No. 4118, 75-1 BCA ¶11,278 (invoking the sovereign immunity doctrine and rejecting the contractor's attempt to hold the government responsible for the price increases).

[92]*M.J. Pacquet, Inc. v. New Jersey Dept. of Transp.*, 171 N.J. 378, 794 A.2d 141 (2002).

damages as well as be relieved from further performance obligations.[93] If the shutdown is due to poor construction methods or failure to comply with the contract documents, the owner should have a valid legal claim against the contractor for breach as well as should be able to terminate any further obligations owed the contractor. The mere fact that performance was stopped by a government official does not necessarily absolve a party from contract breach even if the work is shut down. If the shutdown was due to the unexcused nonperformance by one of the parties, there can be breach as well as a possible termination.

Despite the generalization made in the preceding paragraph, that government intervention may be traceable to the fault of one of the parties or someone for whose acts the parties are responsible, the difficulty of placing clear responsibility on one of the parties for acts of government intervention makes it unlikely that government intervention in the ordinary case will be chargeable to one of the contracting parties.

Suppose the governmental acts that stop performance are not the responsibility of *either* contracting party. The site may have been condemned by the state's exercise of its power of eminent domain. In wartime, the state may have drafted workers, requisitioned essential materials, or commandeered all transportation facilities. Such acts would relieve the contractor from further contractual obligations. (They might also discharge the owner's obligations through frustration of purpose. See Section 23.03E.) Any claims for work performed by the contractor prior to shutdown would be restitutionary.

Suppose a work stoppage results because important equipment being used by the contractor has been repossessed by the owner of the equipment or someone with a security interest in it. Repossession is often accomplished by court order. However, such a court order is a risk clearly assumed by the contractor. That the actual act that interferes with performance is an order by a public official should not relieve the contractor from its obligation to perform.

Suppose the job was shut down because a vital piece of *subcontractor* equipment was seized by court order. The result would be the same as if the equipment were owned by the prime contractor. The prime contractor would not be entitled to a time extension, either because this was an assumed risk, as in the preceding paragraph, or because the job was not really shut down by the state.

Suppose legislation is enacted that would bar a method of performance contemplated at the time the contract was made. It is likely that the performing party will not receive relief if the legislation was reasonably foreseeable and did not increase costs astronomically.[94]

E. Misrepresentaton Through Defective Specifications: *United States v. Spearin*

Contractor allegations that the specifications were defective are common in construction performance disputes. Unfortunately, such allegations often produce more heat than light and obscure the real issues in the dispute. Before looking at these issues, it is useful to reproduce the fountainhead case that is often cited to support a "*Spearin*" claim.

UNITED STATES v. SPEARIN

Supreme Court of the United States, 1918. 248 U.S. 132.
[Ed. note: Footnotes renumbered.]

BRANDEIS, Justice.

Spearin brought this suit in the Court of Claims, demanding a balance alleged to be due for work done under a contract to construct a dry-dock and also damages for its annulment. Judgment was entered for him. . . .

First. The decision to be made on the Government's appeal depends on whether or not it was entitled to annul the contract.

The facts essential to a determination of the question are these: Spearin contracted to build for $757,800 a dry-dock at the Brooklyn Navy Yard in accordance with plans and specifications which had been prepared by the Government. The site selected by it was intersected by a 6-foot brick sewer; and it was

---

[93]*Gordon v. Indusco Management Corp.*, 164 Conn. 262, 320 A.2d 811 (1973).

[94]See Section 23.03E. For a review of the different legal doctrines that may apply in the event a change in the law impacts the project, see Hinchey & Queen, *Anticipating and Managing Projects: Changes in Law*, 26 Constr. Lawyer, No. 4, Fall 2006, p. 26.

necessary to divert and relocate a section thereof before the work of constructing the dry-dock could begin. The plans and specifications provided that the contractor should do the work and prescribed the dimensions, material, and location of the section to be substituted. All the prescribed requirements were fully complied with by Spearin; and the substituted section was accepted by the Government as satisfactory. It was located about 37 to 50 feet from the proposed excavation for the dry-dock; but a large part of the new section was within the area set aside as space within which the contractor's operations were to be carried on. Both before and after the diversion of the 6-foot sewer, it connected, within the Navy Yard but outside the space reserved for work on the dry-dock, with a 7-foot sewer which emptied into Wallabout Basin.

About a year after this relocation of the 6-foot sewer there occurred a sudden and heavy downpour of rain coincident with a high tide. This forced the water up the sewer for a considerable distance to a depth of 2 feet or more. Internal pressure broke the 6-foot sewer as so relocated, at several places; and the excavation of the dry-dock was flooded. On investigation, it was discovered that there was a dam from 5 to 5½ feet high in the 7-foot sewer; and that dam, by diverting to the 6-foot sewer the greater part of the water, had caused the internal pressure which broke it. Both sewers were a part of the city sewerage system; but the dam was not shown either on the city's plan, nor on the Government's plans and blue-prints, which were submitted to Spearin. On them the 7-foot sewer appeared as unobstructed. The Government officials concerned with the letting of the contract and construction of the dry-dock did not know of the existence of the dam. The site selected for the dry-dock was low ground; and during some years prior to making the contract sued on, the sewers had, from time to time, overflowed to the knowledge of these Government officials and others. But the fact had not been communicated to Spearin by anyone. [Spearin] had, before entering into the contract, made a superficial examination of the premises and sought from the civil engineer's office at the Navy Yard information concerning the conditions and probable cost of the work; but he had made no special examination of the sewers nor special enquiry into the possibility of the work being flooded thereby; and had no information on the subject.

Promptly after the breaking of the sewer Spearin notified the Government that he considered the sewers under existing plans a menace to the work and that he would not resume operations unless the Government either made good or assumed responsibility for the damage that had already occurred and either made such changes in the sewer system as would remove the danger or assumed responsibility for the damage which might thereafter be occasioned by the insufficient capacity and the location and design of the existing sewers. The estimated cost of restoring the sewer was $3,875. But it was unsafe to both Spearin and the Government's property to proceed with the work with the 6-foot sewer in its then condition. The Government insisted that the responsibility for remedying existing conditions rested with the contractor. After fifteen months spent in investigation and fruitless correspondence, the Secretary of the Navy annulled the contract and took possession of the plant and materials on the site. Later the dry-dock, under radically changed and enlarged plans, was completed by other contractors, the Government having first discontinued the use of the 6-foot intersecting sewer and then reconstructed it by modifying size, shape and material so as to remove all danger of its breaking from internal pressure. . . .

The general rules of law applicable to these facts are well settled. Where one agrees to do, for a fixed sum, a thing possible to be performed, he will not be excused, or become entitled to additional compensation, because unforeseen difficulties are encountered. *Day v. United States*, 245 U.S. 159; *Phoenix Bridge Co. v. United States*, 211 U.S. 188. Thus one who undertakes to erect a structure on a particular site, assumes ordinarily the risk of subsidence of the soil. *Simpson v. United States*, 172 U.S. 372; *Dermott v. Jones*, 2 Wall. 1. But if the contractor is bound to build according to plans and specifications prepared by the owner, the contractor will not be responsible for the consequences of defects in the plans and specifications. *MacKnight Flintic Stone Co. v. The Mayor*, 160 N.Y. 72; *Filbert v. Philadelphia*, 181 Pa.St. 530; *Bentley v. State*, 73 Wisconsin, 416. See *Sundstrom v. New York*, 213 N.Y. 68. This responsibility of the owner is not overcome by the usual clauses requiring builders to visit the site, to check the plans, and to inform themselves of the requirements of the work, as is shown by *Christie v. United States*, 237 U.S. 234; *Hollerbach v. United States*, 233 U.S. 165, and *United States v. Utah & C. Stage Co.*, 199 U.S. 414, 424, where it was held that the contractor should be relieved, if he was misled by erroneous statements in the specifications.

In the case at bar, the sewer, as well as the other structures, was to be built in accordance with the plans and specifications furnished by the Government. The construction of the sewer constituted as much an integral part of the contract as did the construction of any part of the dry-dock proper. It was as necessary as any other work in the preparation for the foundation. It involved no separate contract and no separate consideration. The contention of the Government that the present case is to be distinguished from the *Bentley Case, supra*, other similar cases, on the ground that the contract with reference to the sewer is purely collateral, is clearly without merit. The risk of the existing system proving adequate might have rested on Spearin, if the contract for the dry-dock had not contained the provision for relocation of the 6-foot sewer. But the insertion of the articles prescribing the character, dimensions and location of the sewer imported a warranty that, if the specifications were complied with, the sewer would be adequate. This implied warranty is not overcome by the general clauses requiring the contractor, to examine the site,[95] to check up the plans,[96] and to assume responsibility for the work until completion and acceptance.[97] The obligation to examine the site did not impose on [the

contractor] the duty of making a diligent enquiry into the history of the locality with a view to determining, at his peril, whether the sewer specifically prescribed by the Government would prove adequate. The duty to check plans did not impose the obligation to pass on their adequacy to accomplish the purpose in view. And the provision concerning contractor's responsibility cannot be construed as abridging rights arising under specific provisions of the contract.

* * *

The judgment of the Court of Claims is, therefore, Affirmed.

The *Spearin* doctrine consists of three components: the owner's issuance of design specifications, misrepresentation to the contractor caused by a defect in those specifications, and injury in the form of the contractor's inability to build the project successfully. All three components must be satisfied for the contractor to recover.

Design Specifications. As explained in Section 19.01D, the distinction between design and performance specifications lies in the amount of discretion given to the contractor in choosing how to build the project. An owner who issues design specifications, which dictate how the contractor is to go about its work, impliedly warrants that following the design will lead to a result acceptable to the owner.[98]

Misrepresentation and Defect. A contractor claiming breach of an owner's implied warranty of design is assert-

ing a *misrepresentation* claim. As the *Spearin* Court explained, "the contractor should be relieved, if he was misled by erroneous statements in the specifications."[99] Specifications can convey a wide variety of information, all of which provide the potential basis for a *Spearin* claim. According to one treatise:

> The owner's breach of its implied warranty of design has been found in factual situations such as: (1) noncompatible soils; (2) structural defects; (3) fire damage; (4) dredging difficulties; (5) highway concrete; (6) sewer design/water infiltration; (7) roof leaks; (8) survey errors; (9) concrete design mix; (10) sealant; and (11) excavation quantity error.[100]

A contractor who entered into a contract knowing the specifications were defective was not misled and so cannot assert a claim for misrepresentation.[101] More commonly, a court denying a claim will find that the contractor had a duty to investigate the condition and failed to do so. In denying the claim, the court is in effect finding that the contractor's reliance on the misrepresentation was not

---

[95]"271. Examination of site.—Intending bidders are expected to examine the site of the proposed dry-dock and inform themselves thoroughly of the actual conditions and requirements before submitting proposals."

[96]"25. Checking plans and dimensions; lines and levels.—The contractor shall check all plans furnished him immediately on their receipt and promptly notify the civil engineer in charge of any discrepancies discovered therein. . . . The contractor will be held responsible for the lines and levels of his work, and he must combine all materials properly, so that the completed structure shall conform to the true intent and meaning of the plans and specifications."

[97]"21. Contractor's responsibility.—The contractor shall be responsible for the entire work and every part thereof, until completion and final acceptance by the Chief of Bureau of Yards and Docks, and for all tools, appliances, and property of every description used in connection therewith."

[98]*Daewoo Eng'g and Constr. Co., Ltd. v. United States*, 73 Fed.Cl. 547, 566-68 (2006).

[99]248 U.S. at 136. See also *Souza & McCue Constr. Co. v. Superior Court*, 57 Cal.2d 508, 370 P.2d 338, 339–40, 20 Cal.Rptr. 634 (1962). Nearly all states have adopted the *Spearin* doctrine; see Annot., 6 A.L.R.3d 1394 (1966).

[100]3 BRUNER & O'CONNOR ON CONSTRUCTION LAW § 9:81 at 669–70 (2002) (footnotes omitted).

[101]*Robins Maintenance, Inc. v. United States*, 265 F.3d 1254 (Fed.Cir. 2001) (grounds maintenance contractor entered into a contract renewal knowing the specification erred in the description of the size of the grounds to be maintained); *Helene Curtis Industries, Inc. v. United States*, 160 Ct.Cl. 437, 312 F.2d 774, 779 (1963) (contractor producing a disinfectant, who renewed the contract knowing the specifications describing the manufacturing process were defective, was not misled).

justified, because it should have learned of the true conditions itself.[102] On the other hand, a contractor's failure to investigate will not bar a *Spearin* claim if a reasonable investigation would not have uncovered the erroneous information.[103]

May an owner disclaim the warranty of design adequacy? *Spearin* makes clear that general, nonspecific disclaimers are ineffective.[104] Some courts are sensitive to the need to protect the integrity of the implied warranty and so place high standards on the specificity required for a disclaimer to be enforced. For example, the Federal Circuit Court of Appeals upheld a *Spearin* claim, finding the attempted disclaimer "does not clearly alert the contractor that the design may contain substantive flaws requiring correction and approval before bidding."[105] However, other courts are willing to enforce disclaimers of specific representations, such as soils reports[106] or estimated quantities.[107]

The *Spearin* doctrine fits within the broader duty of an owner—and in particular a public owner—to act in good faith with a contractor it hires. It complements a public owner's duty to disclose to a contractor relevant information in its possession which the contractor is not likely to discover.[108]

Before attempting to break down the imprecise term *defective specifications* into more workable categories, it is important to note the different legal issues affected by the quality of specifications. The risk of building defects may depend on whether the defect was caused by the design. If so, the entity who supplied design specifications (materials and methods)—usually the owner—will not be able to transfer the cost of correction to the contractor who has executed the design. (Whether it can transfer this loss to the design professional was discussed in Chapter 14.) Put another way, the contractor who executes the required design is not liable for the cost of correction, as seen in *United States v. Spearin*.

Suppose the contractor claims that it has expended more than it anticipated because of defective specifications. Sections 25.03 and 25.05 deal with this in the context of subsurface conditions. This subsection looks briefly at this issue in other contexts.

If the design authorizes a *means or method* of performance, the owner impliedly warrants that the contractor's use of that means will achieve an acceptable result. In *Ace Constructors, Inc. v. United States*,[109] the specifications permitted the contractor to pave an airplane runway using either slip-form or fixed-form types of pavers. However, the runway could be built only using slip-form pavers. The contractor's wasted cost of trying to build the runway using fixed-form pavers was compensable under the implied warranty.[110]

A contractor who cannot comply with the specifications sometimes asserts that the specifications were "impossible." Sometimes specifications are impossible in the sense that they are not coherent, that they do not "fit." Compliance with one part makes compliance with another part an "impossibility." In Section 23.05F,

---

[102]*Green Constr. Co. v. Kansas Power & Light Co.*, 1 F.3d 1005, 1008-10 (10th Cir.1993) (stating that "where a contractor has a duty to make an independent investigation, reliance on an owner's specifications may very well be unreasonable"); *Allied Contractors, Inc. v. United States*, 180 Ct.Cl. 1057, 381 F.2d 995, 999 (1967) ("[I]t was not true that plaintiff [the contractor] was justified in blithely proceeding with its work in the face of obvious and recognized errors. The obligation was cast upon plaintiff to do something about it.")

[103]*Sherman R. Smoot Co. of Ohio v. Ohio Dept. of Adm. Serv.*, 136 Ohio App.3d 166, 736 N.E.2d 69 (2000) (a contractor's failure to perform a pre-bid inspection did not preclude it from being able to assert an implied warranty of design, where an inspection would not have revealed the problem with the site) and *Zontelli & Sons, Inc. v. City of Nashwauk*, 373 N.W.2d 744 (Minn.1985).

[104]Contract provisions requiring bidders to examine the site, verify the plans and dimensions, and be responsible for the entire work does not negate the owner's implied warranty. 248 U.S. at 137.

[105]*White v. Edsall Constr. Co., Inc.* 296 F.3d 1081, 1086 (Fed.Cir. 2002). Accord, *W. H. Lyman Constr. Co. v. Village of Gurnee*, 84 Ill.App. 3d 32, 403 N.E.2d 1325, 1332 (1980). The *W. H. Lyman* case is discussed more fully in Section 24.03.

[106]*Green Constr. Co. v. Kansas Power & Light Co.*, supra note 102; *McDevitt & Street Co. v. Marriott Corp.*, 713 F.Supp. 906 (E.D.Va.1989), aff'd in relevant part, 911 F.2d 723 (4th Cir.1990); *Anderson v. Golden*, 569 F.Supp. 122, 142–43 (S.D.Ga.1982); and *Mooney's, Inc. v. South Dakota Dept. of Transp.*, 482 N.W.2d 43 (S.D.1992) (disclaimer of quality of aggregate in gravel pits).

[107]*J.F. White Contracting Co. v. Massachusetts Bay Transp. Auth.*, 40 Mass.App.Ct. 937, 666 N.E.2d 518 (1996).

[108]See Section 18.04B.

[109]70 Fed.Cl. 253 (2006), aff'd, 499 F.3d 1357 (Fed.Cir.2007).

[110]Id. at 284–285. See also *Miller v. City of Broken Arrow, Okla.*, 660 F.2d 450, 456–58 (10th Cir.1981), cert. denied sub nom. *Benham-Blair & Affiliates, Inc. v. City of Broken Arrow, Okla.*, 455 U.S. 1020 (1982) (engineer's insistence that contractor use crushed rock to handle unstable soil rendered performance impossible) and *McCree & Co. v. State*, 253 Minn. 295, 91 N.W.2d 713 (1958) (on highway project, it was impossible for the contractor to achieve soil compaction to the specified density using the prescribed method). In these cases the government was found liable under the *Spearin* doctrine.

impossibility as applied to performance specifications means beyond the state of the art, while impossibility as applied to design specifications is more akin to physical impossibility.

It may be helpful to start with the attributes of good or high-quality specifications. This requires an understanding of what specifications include and what they are designed to accomplish.

In the sense that specifications measure the contractor's contract obligations, the specification should make clear to the contractor what it will be expected to do. A complete failure of the specifications to convey to the contractor the result intended by the owner may, in an extreme case, result in the parties not having had a "meeting of the minds," which is a prerequisite to the existence of a contract.[111]

The nature of the information conveyed to the contractor varies by the type of specification. A design specification tells the contractor what it is to do and how it is to do it. A purchase description specification is even more limiting, telling the contractor that the materials and equipment it must furnish will be made by a particular manufacturer and be of a designated model or type.[112] A performance specification describes a specific outcome. When the performance specification is combined with a design specification, the owner has sought to tell the contractor what it must do, how it must do it, and the result it must achieve. As noted in Section 24.03, this can cause problems.

In addition, as seen in *United States v. Spearin* specifications can provide relevant information that the contractor needs and uses to determine whether it will enter into the contract, what will be its price, and how it expects to perform. In that case, the United States did not tell the contractor there was a dam in the 7-foot sewer. Put another way, it failed to provide complete information. It described the sewer but did not inform the contractor that there was a dam in it or that there had been prior flooding problems. Information can relate to the site, its subsurface characteristics, its accesses, and any conditions that would be helpful in planning performance.

High-quality specifications clearly inform the contractor what it will be expected to do and the conditions under which it will perform. They enable the contractor to plan its performance and price with the expectation of predictable and efficient work sequences. Similarly, a subcontractor relies on information supplied by the owner or the prime.[113] From the owner's standpoint, high-quality specifications will advance the owner's anticipated objectives, particularly those objectives expressed in or implied by the contract documents.

Defective specifications do not accomplish these objectives. More detailed classification helps one understand some of the legal issues that surround this term, one that hides a multitude of sins.

Erroneous specifications contain factual errors of the type described in *United States v. Spearin* and subsurface data errors, to be discussed in Section 25.03. They also include *legal* errors, such as noncompliance with applicable laws such as building and housing codes and environmental regulations.[114] Usually legal errors are the responsibility of the owner, with ultimate responsibility belonging to the design professional.[115] Only if the contractor knew of the errors may this risk have been shifted or shared.[116]

Specifications may contain errors of a mechanical nature, such as water intrusion, foundation settling, insufficient heating or cooling, or a partial or total collapse of the structure. Although such problems are often classified as defects (the subject of Chapter 24), the party responsible for defects of this type that are caused by the design is the party who created the design or had the design created for it, usually the party with a preponderance of expertise in the element of design that failed. Typically,

---

[111]*Denver D. Darling, Inc. v. Controlled Environments Constr., Inc.,* 89 Cal.App.4th 1221, 108 Cal.Rptr.2d 213 (2001) (no meeting of the minds as to whether the floor flatness standard applied only to the freezer floor or also to the loading dock).

[112]See Section 24.02.

[113]See *W. F. Magann Corp. v. Diamond Mfg. Co.,* 775 F.2d 1202 (4th Cir.1985) (prime did not disclose information it received from owner to subcontractor). But in *Gillingham Constr., Inc. v. Newby-Wiggins Constr., Inc.,* 136 Idaho 887, 42 P.3d 680 (2002), the court ruled that the implied warranty of design did not extend from a prime contractor to the subcontractor.

[114]*St. Joseph Hosp. v. Corbetta Constr. Co.,* 21 Ill.App.3d 925, 316 N.E.2d 51 (1974) (paneling violated code); *Atlantic Nat. Bank of Jacksonville v. Modular Age, Inc.,* 363 So.2d 1152 (Fla.Dist.Ct.App.1978) (wall violated code).

[115]But see *Green v. City of New York,* 283 A.D. 485, 128 N.Y.S.2d 715 (1954) and *Quedding v. Arisumi Bros., Inc.,* 66 Haw. 335, 661 P.2d 706 (1983), where contractors who designed and built were liable where specified materials violated code because of express or implied promise to comply with the law.

[116]AIA Doc. A201-1997, ¶3.7.4. But this language was cut from A201-2007.

this is the owner, unless the owner has transferred this risk to the contractor.

Errors can be functional, such as a project failing to accomplish the owner's desired objectives. This can result from bad business judgment or changed circumstances—both owner risks. The owner may be able to transfer this risk to the design professional if the latter warranted a successful outcome or if the design professional's failure to perform as other design professionals would have performed caused the project to fail.

Defective specifications can "fail" as a method of communication. For example, specifications that do not describe what will be demanded or that fail to give sufficient design detail to enable the contractor to accomplish the desired objective are "incomplete."

Where the design is incomplete and the contract does not clearly state who will provide the missing part of the design, it is likely to be the responsibility of the owner.[117] It drafted the contract and hired a design professional to provide the design.

Sometimes the specifications are confusing or contradictory, again a problem of failing to communicate the performance that will be demanded. In *Jasper Construction, Inc. v. Foothill Junior College District*,[118] the specification stated,

11. CONSTRUCTION JOINTS

A) Locations and details of construction joints shall be as indicated on the structural drawings, or as approved by the Architect. Relate required vertical joints in walls to joints in finish. In general, approved joints shall be located to least impair the strength of the structure.[119]

The locations of the construction joints were not shown on the drawings. However, Jasper contended that certain structural drawings indicated to him that the steel was from "floor to floor" and therefore the concrete would be poured in the same manner. He began to pour the basement "floor to floor," but the architect informed him that the joints would have to be "wall to wall." The contractor complained but did the work as directed and made a claim.

The jury was instructed that a public entity that issues plans and specifications *impliedly warrants* that they are

free of defects, and complete, and will, if followed, result in the project intended.

The appellate court found this jury instruction erroneous. The court recognized that warranties do attach to owner-supplied specifications. However, the court limited the implied warranty doctrine to *affirmative misrepresentations* or concealment of material facts that misled the contractor. The court concluded that the contractor had not been misled. It could have cleared up any ambiguities or incompleteness in advance by seeking a clarification from the architect.

The result in the *Jasper* case in no way undermines the importance or existence of implied warranties. Warranties are implied either where the issue is who bears the risk of inaccurate information, such as in the subsurface cases, or in disputes that involve the outcome of compliance with specifications. In either case, an implied warranty may be found it if is more equitable to make the owner responsible for inaccurate information relied on reasonably by the contractor or to make the owner pay for defects caused by the design or unanticipated expenses of complying with that design. Implied warranties do not depend on fault. They are found if this is what the parties are very likely to have intended or, more commonly, if this is the fairest way of allocating the risk for particular losses.

Most implied warranty cases involve owners who were public entities. There is no reason to differentiate public from private owners. Private owners who have the same resources and expertise of public entities should be held to similar implied warranties of quality specifications. If a private entity does not have these resources and expertise (and there may be cases where public entities do not either) and relies on the contractor, the owner may be the beneficiary of an implied warranty by the contractor.

Injury. The *Spearin* Court invoked the implied warranty to relieve a contractor from liability for a failed project. In addition to this defensive use of the doctrine, contractors may also invoke the warranty in an attempt to recover higher than anticipated construction costs. However, in the latter situation the doctrine applies only if the higher costs were incurred by the contractor in trying to implement unbuildable design specifications. This is the injury addressed by *Spearin*; the warranty does not guarantee a perfect design and trouble-free performance.

A buildable design does not violate the implied warranty, even if the construction process was more expensive

[117]*Tibshraeny Bros. Constr. v. United States*, 6 Cl.Ct. 463 (1984).
[118]91 Cal.App.3d 1, 153 Cal.Rptr. 767 (1979).
[119]153 Cal.Rptr. at 769.

and took longer than the contractor anticipated. A large number of requests for information, while perhaps evidence that the design team did a sloppy job, does not establish an injury within the meaning of the *Spearin* warranty, as long as the design is buildable.[120]

In *Hercules Inc. v. United States*,[121] the Supreme Court addressed the question of whether the *Spearin* doctrine applies to third-party personal injury claims against the contractor. Vietnam War veterans, claiming injury from exposure to the defoliant Agent Orange, sued manufacturers of the chemical. (Agent Orange litigation is discussed in Section 7.09J.) The manufacturers settled, then sued the government to recoup the cost of settlement. The manufacturers contended that the Agent Orange design, which had been created by the government, was defective and resulted in their settlement and defense costs. The Supreme Court refused to extend *Spearin* to the manufacturers' claims:

> When the Government provides specifications directing how a contract is to be performed, the Government warrants that the contractor will be able to perform the contract satisfactorily if it follows the specifications. The specifications will not frustrate performance or make it impossible. It is quite logical to infer from the circumstance of one party providing specifications for performance that that party warrants the capability of performance. But this circumstance alone does not support a further inference that would extend the warranty beyond performance to third-party claims against the contractor. In this case, for example, it would be strange to conclude that the United States, understanding the herbicide's military use, actually contemplated a warranty that would extend to sums a manufacturer paid to a third party to settle claims such as are involved in the present action. It seems more likely that the Government would avoid such an obligation, because reimbursement through contract would provide a contractor with what is denied to it through tort law.[122]

F. "Impossible" Specifications

Performance specifications require the performing party to accomplish a designated objective.[123] For example, an aircraft company might agree to manufacture an airplane that will fly twice the speed of sound or a machinery manufacturer might agree to build a system for a plant that would turn out 1,000 units of a particular quality per hour. Suppose the airplane manufacturer or machinery maker fails. Ordinarily, the failure to comply with performance specifications is a breach of contract.[124] But suppose the party promising to meet these specifications asserts that it was "impossible" to do so. Does proof of impossibility provide a defense?[125] Does such proof entitle the performing party to reimbursement for the expenses incurred while seeking to meet the performance specification?

The answers to these questions in the first instance depend on the contract. The contract can place such risks on one or both parties. Often contracts are not clear on this point, and the law must determine the answers to these difficult questions.

Unfortunately, such problems are classified as "impossible" specification cases because the performing party, who will be referred to as the contractor, claims that it was "impossible" to meet these performance specifications. But the word *impossible* has a number of subtle shadings that often complicate cases and make prediction uncertain.

Another difficulty with the "impossible" label is that it can obscure the crucial issue of risk assumption. Clearly, a party can promise to do the impossible, although evidence that such a foolish risk was taken should be clear.[126] The fact that it is not physically possible—that is, beyond the state of the art—should not invariably resolve the matter in favor of the contractor.

---

[120]*Caddell Constr. Co. Inc. v. United States*, 78 Fed.Cl. 406 (2007); *Dugan & Meyers Constr. Co., Inc. v. Ohio Dept. of Adm. Servs.*, 113 Ohio St.3d 226, 864 N.E.2d 68 (2007) (delays due to excessive changes in the plans are not compensable under *Spearin*).

[121]516 U.S. 417 (1996).

[122]*Id.* at 425. The *Hercules* decision is discussed in Leaderman, *The Spearin Doctrine: It Isn't What It Used to Be*, 16 Const. Lawyer, No. 4, Oct. 1996, p. 46.

---

[123]For a cavalier judicial treatment of performance specifications, see *Kurland v. United Pacific Ins. Co.*, discussed in Section 24.03.

[124]*Gurney Indus., Inc. v. St. Paul Fire & Marine Ins. Co.*, 467 F.2d 588 (4th Cir.1972).

[125]The discussion in this subsection emphasizes attempts by the performing party to be reimbursed for its efforts, as this has been the principal issue in cases raising this problem. If there were a sufficient degree of impossibility to justify reimbursement, it seems obvious that the performing party would have a defense if sued by the other party for not complying with the performance specifications. The contractor was given a defense when the plans were considered "impossible" in *City of Littleton v. Employers Fire Ins. Co.*, 169 Colo. 104, 453 P.2d 810 (1969).

[126]*J. C. Penney Co. v. Davis & Davis, Inc.*, 158 Ga.App. 169, 279 S.E.2d 461 (1981).

A key factor in determining if the contractor assumed the risk of performing "impossible" specifications is whether it created the design. *Foster Wheeler Corp. v. United States*,[127] a Court of Claims opinion, dealt at length with an impossible specifications problem. Foster Wheeler Corporation (FWC) entered into a fixed-price supply contract under which it agreed to design, fabricate, and deliver within thirteen months two boilers and perform a "dynamic shock analysis" study called a DDAM (dynamic design analysis method) that would demonstrate that the boilers could withstand shock up to certain designated intensities set forth in the contract specifications. The total contract price was $280,000. The boilers were ultimately to be installed in naval ships. Because the boilers could not be subjected to actual shock testing, the DDAM—a mathematical model to represent a piece of equipment and the use of dynamic inputs to substitute for physical stresses and failure criteria—was to substitute.

After many months of design work and creation of mathematical models, the contractor ceased performance and sought an equitable adjustment of $192,000, claiming the "impossibility" of meeting the specifications. This claim was based on the impossibility of meeting the performance specifications of "shock hardness." Given other design requirements, this degree of hardness could not be demonstrated by the DDAM.

The court recognized that the term *impossibility* does not require absolute impossibility but encompasses impracticability, a type of commercial impossibility caused by extremely unreasonable expense to perform. *Absolute* impossibility, in the sense of requiring performance beyond the state of the art, would entitle the contractor to recover its cost in attempting to perform unless it assumed this risk. The court concluded that demonstrating the boiler to be shock hard by the DDAM method was both "commercially and absolutely impossible."

After giving facts that supported a conclusion of absolute impossibility, the court went on to the more controversial "commercial impossibility," stating,

Under this theory, it is contended that the construction of a shock-hard boiler, even if ultimately possible, could not be accomplished without commercially unacceptable costs and time input far beyond that contemplated in the contract. To design a shock-hard boiler by means of a mathematical model and dynamic analysis could . . . take an infinite amount of time. . . . The evidence shows . . . that the . . . contract contained specifications which were impossible to meet, either commercially or within the state of the art.[128]

This did not end the matter. The government argued that FWC assumed the contractual responsibility for performing the impossible. To determine who should assume this risk, the court examined which party had the greater expertise in the subject matter of the contract and which party took the initiative in drawing up specifications and promoting a particular method or design. The court ruled for FWC.[129]

The preceding discussion addressees the doctrine of impossibility in the context of performance specifications. The doctrine may also extend to design specifications. Here, the contractor usually argues that the design is physically impossible to implement.[130] Sometimes, the specifications are impossible to implement because of time constraints.[131]

_____

[127]206 Ct.Cl. 533, 513 F.2d 588 (1975).

[128]513 F.2d at 598.

[129]For additional cases finding that the contractor assumed the risk of impossible specifications, see *Noslo Eng'g Corp.*, ASBCA No. 27120 86-3 BCA ¶19,168 (contractor created the design); *Austin Co. v. United States*, 161 Ct.Cl. 76, 314 F.2d 518, cert. denied, 375 U.S. 830 (1963) (the contractor convinced the government to revise the specifications before the contract was signed); and *Bethlehem Corp. v. United States*, 199 Ct.Cl. 247, 462 F.2d 1400 (1972) (the contractor did not prepare the specifications but assured the government that the specifications were reasonable and results obtainable). In *National Presto Indus. v. United States*, 167 Ct.Cl. 749, 338 F.2d 99 (1964), cert. denied, 380 U.S. 962 (1965), the court applied the doctrine of mutual mistake where the parties, with equal expertise, agreed on an impossible method of performance.

[130]*R.P Wallace, Inc. v. United States*, 63 Fed.Cl. 402 (2004) (window model that was impossible to manufacture); *O'Neal Eng'g, Inc.*, ASBCA No. 31804, 86-2 BCA ¶18,906 (design called for installation of a four-inch cable in an existing conduit less than four inches in diameter) and *James W. Sprayberry Constr.*, IBCA No. 2130, 87-1 BCA ¶19,645 (on a re-roofing contract, the specifications were written on the assumption the roof was level; however, the roof was uneven, so the specifications were impossible to implement).

[131]*Triax Pacific, Inc. v. West*, 130 F.3d 1496 (Fed.Cir.1997) (the specifications' painting schedule and curing (or drying) requirement, when read together, made timely completion of the project impossible). See also *Southland Enterprises, Inc. v. Newton County*, 838 So.2d 286 (Miss.2003) (county mandated that road resurfacing job be performed

G. Weather

An empirical study reports that weather is "one of the most important causes of delay in construction, its influence being felt through lost or non-productive working days, idle equipment, spoiled materials, contingency overheads and consequently higher prices."[132] Generally, the weather is considered a risk borne by the contractor.[133] Time extensions are granted only if the weather is severe and abnormal for the time and place.[134]

H. Financial Problems

After the construction contract is made and before completion of each party's performance, either party can suffer severe financial reverses. Such reverses may manifest themselves in difficulties that range from short delay in paying bills to insolvency and even bankruptcy. Financial reverses of this sort can raise two related but separate legal questions. First, the party in financial difficulty may contend that it should be relieved from further performance because it does not have the financial capacity to continue performance. Second, one party may be unwilling to continue performance if it appears the other party's financial difficulties will make it unlikely that the latter will perform as promised. Under certain circumstances, the law allows a party who has legitimate concern over the other party's ability to perform to refuse to continue performance until it is reassured that the other party will perform its promise.[135] AIA, in its A201-2007 relating to construction services, gives the performing party—the contractor in A201-2007—the right to demand certain information regarding the owner's financial resources.[136]

Suppose the contractor asserts that the owner has made a *promise* to supply adequate information as to its financial resources and that its failure to do so constitutes a breach. Such assertion would be based on the furnishing of information not only being a condition to the contractor's obligation to execute the contract but also being a promise by

the owner that it will furnish the information if requested. If so, the contractor is entitled to damages. At the very least, damages should encompass its expenses incurred in negotiating or submitting a competitive bid. However, the contractor should not recover lost profits. To avoid such recovery, such a provision would likely be held to create a condition rather than a promise.

Suppose the required information is not provided. As already noted, the common law would give the contractor the right to demand adequate assurance and to suspend performance until it is furnished. In transactions involving goods, a similar right is given the insecure party under the Uniform Commercial Code, Section 2-609. If the information is crucial or if the requested information is not furnished within a reasonable time not exceeding thirty days, the party who is suffering insecurity can terminate. Under AIA Document A201-2007, Section 2.2.1, failure to furnish the requested information on financial resources gives the contractor the right to refuse to begin work or to continue working. A201-2007, Section 14.1.1.4, gives the contractor the right to terminate its obligation if it has ceased performance for thirty days because the owner has not furnished the requested financial information.

Some owners may object to such a provision, particularly the part added in 1997 (and continued in 2007) that bars the owner from materially varying its financial commitments "without prior notice to the Contractor." (Such a clause can give the contractor an opportunity to get out of the contract if it finds that it has bid too low.) Such owners may justify their refusal by noting that contractors in private construction have a right to mechanics' liens if they are not paid. However, the contractor would prefer a battery of weapons to deal with financial insecurity and nonpayment.

Suppose an A201-2007 is used but the portion giving the contractor this power is deleted. Would a court use common law doctrines that might otherwise be available to the contractor who has reasonable insecurity as to the owner's ability to pay?[137] Deletion appears to indicate the parties did not intend that this "gap filler" to be part of this transaction.

---

in December; however, winter temperatures prevent the asphalt from adhering to the rock bed).

[132]Laufer & Cohenca, *Factors Affecting Construction-Planning Outcomes*, 116 J.of Constr.Eng'g & Mgmt 135, 147 (1990).

[133]See Sections 26.07 and 26.08.

[134]*Hardeman-Monier-Hutcherson v. United States*, 198 Ct.Cl. 472, 458 F.2d 1364 (1972) (no additional compensation in absence of a clause granting it).

[135]Restatement (Second) of Contracts §§ 251, 252 (1981).

[136]AIA Doc. A201-2007, § 2.2.1.

---

[137]See Section 33.04B. West Ann.Cal.Civ. Code § 3110.5, effective January 1, 2002, may partially displace the common law in California. It requires certain private owners on large projects to provide the prime contractor with either a payment bond, an irrevocable letter of credit, or an escrow account, so as to increase the prime's financial security.

Suppose the contractor's power is limited to that period between award and execution of the formal contract. Would this limitation preclude a court from using any common law financial insecurity doctrines that might otherwise be available during performance?[138] Does a contract that deals with a particular problem signal a court that the law should not imply terms? The detailed treatment of this problem would make legal intervention unlikely.

Financial problems encountered by the contractor are not likely to give it any justification for refusal to continue performance. However, construction contracts usually grant the owner certain remedies if the contractor runs into financial difficulties. This grant should make it unnecessary for the law to employ any common law doctrines dealing with financial insecurity. To use provisions giving the owner remedies, such as suspension or termination, if the contractor runs into serious financial problems is unlikely, given the frequent use of surety bonds to provide financial security for the owner if the contractor does not perform as promised. As to the effect of bankruptcy law on contract provisions, see Section 33.04C.

I. Asbestos and Other Hazardous Materials

Section 9.13 described some of the contours of the environmental movement and the increasing liability exposure for those who are involved in any way with hazardous materials. Also, Section 12.09 treated this issue from the perspective of design services. This subsection focuses on the discovery of a condition involving hazardous materials during construction and the effect of this discovery on the contractual obligations of the parties to a construction contract.

How does this discovery affect the obligation of the contractor to continue performance or to resume performance? AIA Document A201-2007, Section 10.3.1, mandates cessation of work if asbestos or polychloride biphenyl (PCB)—given as nonexclusive examples—are encountered and if "reasonable precautions will be inadequate to prevent foreseeable bodily injury or death."

Under Section 10.3.2, first added in 1997, the owner must use a licensed laboratory to determine whether such danger exists and, if so, the owner shall "cause it to be rendered harmless," presumably through the use of a licensed

remediation contractor. The architect and contractor may raise reasonable objections over who will do the testing and remediation. If the materials have been rendered harmless, work resumes if the owner and contractor so agree. Also added in 1997, the contractor is given a time extension and a price adjustment amounting to its cost of shutdown, delay, and start-up.

Indemnity will be treated in Chapter 31, but here it should be noted that indemnity for hazardous materials has been created in Sections 10.3.3, 10.3.5 and 10.3.6. Section 10.3.3 requires the owner to indemnify the design and construction participants (architect and contractors) for all risks relating to hazardous materials except to the extent the damage is due to the negligence of an indemnitee. Under Section 10.3.5 the contractor indemnifies the owner if the owner is held liable for costs of remediation caused by the contractor's negligence or breach. Under Section 10.3.6, the owner indemnifies the contractor if the contractor "is held liable by a government agency for the cost of remediation of a hazardous material or substance solely by reason of performing Work …."

Cessation and resumption of work raise other legal issues. Suppose the work is stopped, an asbestos abatement contractor performs asbestos removal, and the work is resumed. Who must pay for the cost of asbestos abatement and removal? Can the contractor receive an equitable adjustment if discovery of asbestos or other hazardous material has adversely affected its performance cost? If its work has been drastically affected, does the contractor have the power to terminate the contract? Similarly, if the project looks very different to the owner after the high cost of abatement, can the owner use the impossibility or frustration doctrines to terminate its obligation to proceed further under the contract? Finally, if the owner chooses not to resume the work, has it breached its contract with the contractor?

Some, though not all of these problems, have been addressed in AIA Document A201-2007, Section 10.3. These questions should be addressed in the contract. If the contractor encounters such materials, it should receive an equitable adjustment if it can continue to perform any work. The owner should be allowed to terminate the contract if in good faith it believes it would not be economical to continue working under such conditions.

Hazardous materials at the site may injure the contractor's workers or, on a renovation project, the building's occupants. In the latter situation, the hazardous materials

---

[138]Ibid.

usually were introduced as part of the construction process. For example, a contractor's application of sealants, curative agents, paints, and glues often releases fumes which may be toxic to some people.

The causal connection between the hazardous materials and the claimed personal injuries is not always clear-cut. If the injury is immediate upon exposure, the connection appears more evident, and the contractor is informed of the possibility of a liability claim. However, as in the case of asbestos, workers may allege that their illness manifested itself many years after the project was over.

Worker injury claims should be covered by workers' compensation insurance. Contractors look to their liability insurance for coverage of claims by third parties, such as building occupants. Unfortunately for contractors, commercial general liability policies increasingly include a so-called "absolute" pollution exclusion. The exclusion purports to negate coverage for any claims caused by the exposure to "pollutants," a term broadly defined to include "any solid, liquid, gaseous, or thermal irritant or contaminant, including smoke, vapor, soot, fumes, acids, alkalis, chemicals and waste."

The courts are deeply split on the scope of the pollution exclusion. Some interpret it literally to apply to all exposures to hazardous materials.[139] Other courts find that the intent of the exclusion is to apply to environmental pollution claims. Those courts adopting this narrower view of the exclusion's scope continue to find coverage where the plaintiff was injured by exposure to hazardous substances that are part-and-parcel of the construction process.[140]

Another problem relates to the discovery of radon, either during construction or after construction has been completed. Increasingly, radon will create legal problems, both of a public and of a private nature.[141]

---

[139]*Certain Underwriters at Lloyd's, London v. C. A. Turner Constr. Co., Inc.,* 112 F.3d 184 (5th Cir.1997) (Texas law) and *Madison Constr. Co. v. Harleysville Mutual Ins. Co.,* 557 Pa. 595, 735 A.2d 100 (1999), reargument denied Oct. 8, 1999.

[140]*Meridian Mutual Ins. Co. v. Kellman,* 197 F.3d 1178 (6th Cir.1999) (Michigan law; fumes from floor sealant); *MacKinnon v. Truck Ins. Exchange,* 31 Cal.4th 635, 73 P.3d 1205, 3 Cal.Rptr.3d 228 (2003) (pesticide spray); *Freidline v. Shelby Ins. Co.,* 774 N.E.2d 37 (Ind.2002) (fumes from carpet glue); *Clendenin Bros., Inc. v. United States Fire Ins. Co.,* 390 Md. 449, 889 A.2d 387 (2006) (welding fumes). See also M. SCHNEIER, CONSTRUCTION ACCIDENT LAW: A COMPREHENSIVE GUIDE TO LEGAL LIABILITY AND INSURANCE CLAIMS § 9:B:VII (1999) and Annot., 39 A.L.R.4th 1047 (1985).

[141]Note, 15 Seton Hall Legisl.J. 171 (1991); Note, 37 Wash.J.Urb. & Contemp.L. 135 (1990).

CHAPTER TWENTY-FOUR

Defects: Design, Execution, and Blurred Roles

SECTION 24.01 Introduction: The Partnership

The sad but not uncommon discovery that the project has defects often generates a claim by the owner against parties it holds responsible for having caused the defect. Claims against the design professional were discussed in Chapter 14. This chapter concentrates on owner claims against the contractor.[1]

Defects in a house can include a leaky roof, a sagging floor, structural instability, and an inadequate heating or plumbing system. A commercial structure can include these defects as well as an escalator that is unsafe or inefficient or that requires excessive repairs. An industrial plant can include the preceding defects as well as inadequate space to install machinery or the inability of the computer system to operate the assembly line.

A differentiation must be made between temporary and permanent work. Usually defects deal with permanent work—the finished product that the owner intends to use. Under AIA Document A201-2007, Section 3.3.1, how the permanent work is to be accomplished—such as means, methods, and sequences, including methods for accomplishing the temporary work, such as temporary shoring, bracing, formwork, or coffer dams—is usually the responsibility of the contractor.[2] Yet the AIA recognizes that this model of responsibility is not universal. In 1997 it added language to Paragraph 3.3.1, which spoke specifically to the possibility that some owners may wish to "give

specific instructions concerning construction methods." If so, the contractor must evaluate "job site safety" and be responsible for "job site safety of such means." If the contractor believes the methods specified are not safe, it must notify the owner and architect in writing and must not proceed until instructed to do so. This language is preserved in A201-2007, Section 3.3.1. Of course, AIA assumptions, that of giving execution responsibility to the contractor, could always be changed by the parties. They can add language recognizing that projects can and do use different delivery and risk allocation methods.

It is important to differentiate the different types of owners discussed in Section 8.02A. The most important is a differentiation between an owner who supplies a design—usually by an independent design professional—and one who hires a contractor to both design and build. (Design–build (DB) was discussed in Section 17.04F.)

It is also important to review certain types of specifications noted in Sections 19.01D and 23.05E. They are *design* specifications, *performance* specifications, and *purchase description* specifications. The first requires the contractor to use designated methods and materials. The second gives the contractor particular goals, with the contractor often but not always able to determine how the goals are to be achieved. The third designates materials and equipment by name of manufacturer, trade name, and type. Sometimes purchase description specifications let the contractor request authorization to substitute something equal to or the equivalent of the designated product. Specifications can combine types, such as design and performance. Also, they vary as to completeness of materials and methods requirements.

Often it is difficult to determine what causes a defect. A defect can be caused by the design, by workmanship, by

---

[1]For claims against public liability insurers, see Section 24.08.

[2]AIA Doc. A201-2007, §1.1.3, defines "Work" as any activity performed by the contractor. This can include temporary work, such as shoring, bracing, or formwork.

extraneous factors such as weather, or by a combination of factors. Perhaps the greatest difficulty involves the defects traceable to materials (discussed in Section 24.02 and 24.09C).

One rough classification is to charge the owner with defects caused by design and hold the contractor accountable for defects caused by failure to follow the design or by poor workmanship. This classification can be deceptive unless account is taken of the increasingly blurred roles related both to design and to its execution. For example, Section 12.08C, dealing with submittals, noted that A201-2007 recognized, by inserting Section 3.12.10, the increasing use of what has been called "design delegation." Section 12.08C deals with this topic in great detail.

Although design is principally the responsibility of the owner when it has not been delegated to the contractor (the owner usually acting through the design professional), the contractor plays a role in design, particularly a contractor retained because it has specialized skill in certain work or because it retains subcontractors with those skills. Illustrations of this shared responsibility can be contracts that require the contractor to submit drawings indicating how it proposes to do the work (often prepared by specialized subcontractors) and that require the design professional to approve the submittals. (The submittal process was discussed in Section 12.08C.)

Shared design responsibility also should increase as digital technologies play a larger and larger role in design creation. As discussed in Section 12.11, ownership of the design becomes blurred when subcontractors and suppliers contribute electronically to the formation of the original design. As discussed in Section 17.04M, Building Information Modeling contemplates a collaboratively created design, in which the design author cannot easily be traced or determined. Moreover, making electronic media part of the design complicates the contractor's design review role. AIA A201-2007, Section 3.2.2, requires the contractor, before starting the work, to "carefully study and compare the various Contract Documents" and to report to the architect "any errors, omissions, or inconsistencies" that it discovers. A recent article (commenting on the same language in A201-1997, Paragraph 3.2.1) questions how that review would work in a digitally-based design:

> The process of "carefully studying and comparing" interoperable object-oriented CAD files may be quite different from the process of "carefully studying and comparing" paper

Drawings and Specifications. For one thing, there may be "plan-checking" software introduced to the market that the "careful study" standard would obligate the Contractor to use. And the sheer amount of information in an object-oriented interoperable file (including links to other documents and web sites) may be impossible for the Contractor to absorb in detail "before starting each portion of the Work."[3]

Another complication is the standard by which the contractor's review of the design is measured. Section 3.2.2 states that the contractor's review is made in its capacity as a contractor, not as a design professional. However, as the contractor becomes a more active participant in the collaborative design process, a professional license may become necessary.[4] At this point, must the contractor meet the professional standard? What types of design defects would the contractor then be expected to discover? Would it, like an architect, have an obligation to determine whether the design complies with applicable laws? In sum, as the sharp line between architect and contractor blurs in the creation and review of the design, so does the demarcation blur as to responsibility for design defects.

Sometimes bidders are asked to provide design alternates, or do so voluntarily. Before or after award or during performance, the contractor may request to substitute different equipment or material from that specified. Although approval by the design professional is usually required, approval often is based on representations or even warranties by the contractor that the proposed substitution will be at least as good as that specified or will accomplish the desired result. The contractor, though clearly subordinate to the design professional, plays an important role in design.

Similarly, the design professional frequently monitors the contractor's performance during the work and at the end of the job. Some design professionals may even direct how the work is to be done, although this role is typically not within their power or responsibility.

As shall be seen in Section 24.02, control is a key factor determining responsibility for defects. Yet the rough "partnership" between owner and contractor makes it difficult to neatly divide responsibility by design and execution.

[3]Stein, Alexander & Noble, AIA *General Conditions in the Digital Age: Does the Square "New Technology" Peg Fit Into the Round A201 Hole?*, 25 Construction Contracts Law Rep., No. 25, Dec. 14, 2001, ¶367, pp. 3, 8. "CAD" refers to computer-assisted design.

[4]Id. at p. 9.

SECTION 24.02 Basic Principle: Responsibility Follows Control

As a basic principle, responsibility for a defect rests on the party to the construction contract who essentially controls and represents that it possesses skill in that phase of the overall construction process that substantially caused the defect. Usually, defects caused by design are the responsibility of the owner in a traditional construction project and the contractor who both designs and builds. Control does not mean simply the power to make design choices. Every owner usually has the power to determine design choices. For example, an owner may require a particular type of tile to be used, a power within its contract rights. But the control needed to invoke the basic principle means a skilled choice, either one made by an owner who has professional skill in tile selection or an adviser such as an architect with those skills.

This principle recognizes that the owner who supplies the design is responsible for design that does not accomplish the owner's objective and yet may not have a claim against the design professional. Usually the standard to which the design professional is held is whether she would have performed as would have other design professionals similarly situated.

Under this principle, the owner, though not at fault, bears the cost of correcting defects. The owner has the principal economic stake in the project and will benefit from a successful project. There is no reason why it cannot be responsible for project failures even though it is blameless and cannot transfer the loss.

Many cases have held that the contractor who follows the design is not responsible for a defect unless it warrants the design or was negligent.[5] This establishes the principle that the owner is responsible for any design it has furnished. Similarly, the owner impliedly warrants the accuracy of specific information it furnishes that is reasonably relied on by the contractor.[6] It also warrants that any required materials, design features, or construction methods will create a satisfactory end product within the completed time and without extraordinary unanticipated expense.[7]

Cases supporting this principle usually involve traditional construction in which the owner furnishes and monitors the design but is not primarily responsible for its execution. Risk allocation is based on the probable intention of the parties, the greater skill possessed or supplied by the owner, the contractor's lack of discretion, and the owner's being in the best position to avoid the harm, as well as the owner's ability to spread, absorb, or shift the risk to the design professional. Similarly, contractors are generally held to have impliedly warranted the quality of their workmanship.[8]

Contractors who both design and build usually warrant that the finished product will be fit for the owner's purposes of which the contractor knew or should have known, for the same reasons placing risk on the owner in a traditional method. Also, these warranties are similar to those placed on sellers of goods. For example, one court found an implied warranty[9] where

1. The contractor holds itself out, expressly or by implication, as competent to undertake the contract; and the owner

2. has no particular expertise in the kind of work contemplated;

3. furnishes no plans, designs, specifications, details, or blueprints; and

4. tacitly or specifically indicates its reliance on the experience and skill of the contractor, after making known to him the specific purposes for which the building is intended.

Defective material generates the most difficult problems. The huge variety of available materials (often

---

[5]Annot., 6 A.L.R.3d 1394 (1966). See also *John Grace & Co., Inc. v. State Univ. Constr. Fund*, 99 A.D.2d 860, 472 N.Y.S.2d 757, aff'd as modified, 64 N.Y.2d 709, 475 N.E.2d 105, 485 N.Y.S.2d 734 (1984) and *Bunkers v. Jacobson*, 653 N.W.2d 732, 741–42 (S.D.2002). By statute, contractors in Louisiana are not liable to third parties injured by design defects if the contractor followed the design and did not have reason to know that deficiencies in the design would create the hazardous condition. See La.Rev.Stat. § 9:2771, applied in *Dumas v. Angus Chemical Co.*, 729 So.2d 624 (La.App.1999). But see *Calcasieu Parish School Bd. v. Lewing Constr. Co., Inc.*, 931 So.2d 492 (La.App.2006) (architect's specification that floor be installed in accordance with the manufacturer's instructions shifted to the flooring subcontractor, as the party with the greatest expertise, the duty to perform moisture tests).

[6]See Section 25.03.

[7]See Section 23.05E.

[8]*Northern Pac. Ry. Co. v. Goss*, 203 F. 904 (8th Cir.1913); *Trahan v. Broussard*, 399 So.2d 782 (La.App.1981); *Smith v. Erftmier*, 210 Neb. 486, 315 N.W.2d 445 (1982). But see *Samuelson v. Chutich*, 187 Colo. 155, 529 P.2d 631 (1974) (fault required). Sklar, Filer & Bird, *Implied Duties of Contractors: Wait a Minute, Where Is That in My Contract?*, 21 Constr. Lawyer, No. 3, Summer 2001, p. 11.

[9]*Dobler v. Malloy*, 214 N.W.2d 510, 516 (N.D.1973).

untested), the pressure to cut costs or weight, and the inability to test or rely on manufacturers all combine to make defective materials a prime cause of defects in the project.

Materials specified may be unsuitable and may never accomplish the purpose for which they have been specified. As specification is part of design, responsibility for such materials falls on the person who controls the design or to whom the risk is transferred.[10]

Materials may be suitable yet fail either because they were installed improperly—in which case the contractor clearly would be responsible for the failure—or because they were manufactured improperly. Should the owner's control over the design mean that it is also responsible for faulty materials, or is the manufacturer more akin to a subcontractor, for whose defective performance the contractor is responsible? This problem is addressed in greater detail in Section 24.09C, dealing with the AIA warranty clauses. A few observations can be made here.

Arguments can support placing the risk of faulty materials on either owner or contractor. As to putting the risk on the owner, it is often difficult to determine whether the materials were unsuitable or whether they were faulty. It is more efficient to place responsibility for any specified materials on the party that controls the design—the owner, in the case of traditional construction, and the contractor who both designs and builds. It is unfair to use purchase description specifications and then seek to hold the contractor accountable. When a purchase description specification is used, the contractor is not a seller but is simply a procurer of goods ordered by someone else. The contractor should not be held to the warranties of sellers or manufacturers.

Cogent arguments, however, can be made for holding the contractor responsible for faulty materials. It is difficult to determine whether a defect is caused by unsuitable or faulty materials or by the contractor's failure to install them properly. Very likely a contractor is responsible for faulty materials if it knew or should have known they

were faulty. If the contractor is responsible for faulty material that it knew was faulty, it is administratively more efficient to make the contractor responsible for faulty materials, particularly as the contractor is likely to have a better claim against the supplier or manufacturer than the owner.

The federal boards of contracts appeals adopt the latter approach. They reason that, in the event a delivered product is manufactured improperly, the contractor, rather than the government, is in a better position to seek relief from the manufacturer. This is because the contractor is in privity with the manufacturer and so can protect itself contractually, while the government is not in privity with the manufacturer. The boards' reasoning follows the basic principle of responsibility following control, but in this case "control" means the party in the best position to obtain a remedy.[11]

This deceptively simple problem should be handled specifically in any contract that uses design or purchase description specifications. An owner successfully dealt with this issue through contract language in *Rhone Poulenc Rorer Pharmaceuticals, Inc. v. Newman Glass Works*.[12] In that case, a federal court of appeals ruled that a subcontractor's express warranty that the work will be "free from faults and defects" shifted to it responsibility for latent, manufacturing defects in products designated in the specifications.

SECTION 24.03 Displacing the Basic Principle: Unconscionability

Autonomy (freedom of contract) generally allows contracting parties to determine how particular risks will be borne. But as noted in Sections 5.07D and 19.02E, a clause or contract that is unconscionable will not be enforced. As a rule, the owner is the party who usually seeks to take advantage of autonomy.

---

[10]*Trustees of Indiana University v. Aetna Cas. & Sur. Co.*, 920 F.2d 429 (7th Cir.1990); *Teufel v. Wienir*, 68 Wash.2d 31, 411 P.2d 151 (1966). The *Aetna* case is discussed in great detail by Reynolds, *What Is a Contractor's Warranty Responsibility for Owner-Specified Materials*—Trustees of Indiana University v. Aetna Casualty & Surety, 11 Constr. Lawyer, No. 4, Oct. 1991, p. 1. Although the author represented Aetna in the case and admits that he might be biased, the article is a careful and perceptive treatment of the problem.

[11]*DeLaval Turbine, Inc.*, ASBCA No. 21797, 78-2 BCA ¶13,521; *Cascade Electric Co.*, ASBCA No. 28674, 84-1 BCA ¶17,210; Baltz, *The Spearin Doctrine: How Far Does It Go?*, 33 Procur. Lawyer, No. 1, Fall 1997, p. 11.

[12]112 F.3d 695 (3rd Cir.1997), rehearing and rehearing *in banc* denied Jun 4, 1997 (Pennsylvania law). See also *Graham Constr. Co., Inc. v. Earl*, 362 Ark. 220, 208 S.W.3d 106 (2005) (contractor's express warranty, coupled with its poor workmanship, made it responsible for leaking skylights specified by the owner).

Displacement should be differentiated from clauses that clarify or augment the basic principle. Section 24.02 defines the basic principle for allocating the risk for defects. It places responsibility on the party that controlled or had the right to control the activity causing the defect. A warranty clause (discussed in Section 24.09) can transfer design risks to the contractor. Such a clause would displace what would otherwise be the common law rule. However, such a clause may make clear that the contractor is responsible for poor workmanship and specify a remedy. The latter warranty clause seeks not to displace but to augment the basic principle.

Any clauses that seek to displace the basic principle will be scrutinized carefully to determine both whether the contracting party on whom the risk is placed was aware of the risk allocation and the fairness of displacing the basic principle. Although cases do not always openly recognize the need to scrutinize the clause, the courts are more willing to determine the fairness of such clauses than to simply enforce the clause as written.

W. H. Lyman Construction Co. v. Village of Gurnee[13] involved a specification requiring that the contractor use a particular manhole base and seal. The specifications also stated that the contractor assumed the risk of complying with infiltration limits. If the contractor thought the design would be inadequate, it was to direct attention to this inadequacy in writing at the time it submitted its bid.

The court held that this was an "impermissible attempt on the part of the Village to shift the responsibility for the sufficiency and adequacy of the plans to the contractor, without providing the contractor the corresponding benefit of something to say about the plans that he is strictly bound to follow."[14] After noting the "possible" unconscionability of such a clause, the court concluded that the clause would not shield the village from its negligence because of public policy. Shielding the village would discourage bidders, and the public interest would suffer in the long run.

Kurland v. United Pacific Insurance Co.[15] demonstrates the unwillingness of courts to simply apply language that seeks to displace the basic principle that risk follows control, as set forth in Section 24.02. It involved a claim by an owner against a subcontractor's surety. The subcontractor had undertaken to install an air-conditioning system in an apartment building. The plans and specifications designated equipment to be used and required the contractor to meet performance standards for cooling and heating. The subcontractor did as required, but the air-conditioning system did not function as required.

In affirming a judgment for the surety, the court noted that the plans and specifications had been prepared by the architect. The air-conditioning unit was an owner design choice, perhaps a negligent one. The court held that it could not be reasonably concluded that the subcontractor would assume the responsibility for the adequacy of the plans and specifications. The court concluded that warranty-like language was simply an undertaking by the subcontractor that it would work as effectively as possible to achieve the desired result. The language was not a warranty of a successful outcome.

Nevertheless, courts do not always relieve the contractor if it follows the plans and specifications dictated by the owner. In *United States Fidelity & Guaranty Co. v. Jacksonville State University*,[16] the contractor did as the specifications required. However, the court held the contractor responsible for wall leaks caused by unsuitable sealing material, because it found language in the contract to be a guarantee by the contractor.

The *Kurland* and *Jacksonville* cases, along with *Teufel v. Wienir*,[17] raise a problem that relates to subcontracting. Often the prime contract contains language requiring the prime contractor to obtain a warranty from the subcontractor of the subcontractor's work. In the case of purchase description specifications, the enforceability of the subcontractor's warranty may turn on the nature of the defect. In *Kurland*, the failure of the air-conditioning system to work may well have been a design defect, which is the responsibility of the owner. However, in the *Rhone* case,[18] discussed in Section 24.02, the specified product contained a manufacturing defect, and the court interpreted the warranty as shifting to the subcontractor the risk of such defects. As noted earlier,[19] good reasons exist to shift to the party in privity with the manufacturer (the contractor or subcontractor) responsibility for manufacturing defects.

---

[13]84 Ill.App.3d 28, 403 N.E.2d 1325 (1980).
[14]403 N.E.2d at 1332.
[15]251 Cal.App.2d 112, 59 Cal.Rptr. 258 (1967).

[16]357 So.2d 952 (Ala.1978).
[17]Supra note 10.
[18]Supra note 12.
[19]See Section 24.02.

Subcontractor warranties raise another issue reflected in the *Jacksonville* and *Teufel* cases. Do these warranties affect any prime contractor obligation? Contract language commonly makes the prime contractor responsible for the subcontractor's work. However, the contract may indicate that the owner was exclusively relying on the *subcontractor's* warranty. This exonerates the prime contractor.

That a subcontractor can be or has been sued should not change the legal obligation of owner and prime contractor unless the owner, by either demanding or accepting the subcontractor's warranty, manifests an intent to release the prime. For example, the trial court judge gave the prime contractor a defense in *Teufel v. Wienir* because the specifications indicated that the owner agreed to look only to the subcontractor.[20]

Suppose the design is created by the subcontractor. As between owner and prime, responsibility falls on the contractor.[21]

Another problem of risk shifting results from proposals by the prime contractor for substitutions of materials or equipment for that specified. Although substitutions must be approved by the design professional, to protect herself and to recognize that she may be relying on representations of the contractor, the design professional often obtains a guarantee by the contractor.

The effectiveness of the contractor's warranty will depend on the language of the warranty and surrounding facts and circumstances. Although a few court decisions have upheld warranties in the context of a substitution request,[22] the outcome of such a dispute cannot be easily predicted.[23] Review Section 12.08C for more on submittals.

Another method of displacing the basic principle is the frequent inclusion in construction contracts of provisions requiring that the contractor study and compare the design documents and report any observed errors as well as any violation of building laws that it observes.[24] These provisions should be strictly interpreted. If given effect, they displace the basic principle set forth in Section 24.02 by relieving the owner and ultimately the design professional from responsibility for design. The contractor should be required to use whatever design skills it possesses as long as doing so does not place unreasonable burdens on the contractor. Yet design in the traditional method of construction is still the responsibility of the owner. These objectives can be accommodated by requiring the breaching contractor to share the cost of correction and any consequential damages with the owner. This approach is discussed in Section 24.06.

Varying results show that language will not *automatically* displace the basic principle of control outlined in Section 24.02. Yet the outcome can often be predicted. Which party really made the design choice? Which party had the greater experience and skill? Which party relied on the other? The answers do not always implicate the owner, even in a traditional system. Even with these questions answered, account must be taken of the different attitudes toward the use of contract language to achieve what appears to be an unconscionable result.

SECTION 24.04 Good Faith and Fair Dealing: A Supplemental Principle

Parties who plan to enter into a contract are not, as a general rule, expected to look out for each other. Although exceptions arise, this principle is still strong in American contract law.

Once a contract has been made, the law—led by the Uniform Commercial Code (U.C.C.), Section 1-304, dealing with certain commercial transactions—increasingly expects parties to act in good faith and to deal fairly with one another.[25] In construction contracts, either party—though most commonly the contractor—must warn the other when the other is proceeding in a way that will cause failure. Sometimes this attitude is reflected in contract clauses requiring that the contractor bring to the attention of the owner or design professional design or other problems that can adversely affect the project.[26]

---

[20]The appellate court affirmed a judgment for the contractor by finding that the prime contractor simply followed the design. *Teufel v. Wienir,* supra note 10.

[21]*Stevens Constr. Corp. v. Carolina Corp.,* 63 Wis.2d 342, 217 N.W.2d 291 (1974).

[22]*Urania v. M. P. Constr. Co.,* 492 So.2d 888 (La.App.1986); *New Orleans Unity Soc. v. Standard Roofing,* 224 So.2d 60 (La.App.), cert. denied, 254 La. 811, 227 So.2d 146 (1969).

[23]A warranty whose meaning was not clear did not shift the risk to the contractor in *Habenicht & Howlett v. Jones-Allen-Dillingham,* Cal. Ct.App., 1 Civ. 46449 (1981) (an unpublished opinion that cannot be cited in California).

---

[24]AIA Doc. A201-2007, §§3.2.2 and 3.7.3.

[25]See Section 19.02D.

[26]Supra note 24.

Suppose there is no specific contract obligation. In this case, what does the law demand of the contractor? Must it examine the contract documents

1. but only to prepare its bid, design errors being none of its business?
2. principally to prepare its bid but also to note and report any errors observed?
3. both to prepare its bid and to check for errors, being held for errors it should have observed?

Even making allowance for factual variations, the cases display diverse results. This is common when a new doctrine of a vague nature limits the powerful principle of autonomy. A few seem to permit the contractor to ignore any design errors it may even observe, with its responsibility simply to build and not design.[27] The better reasoned cases and, as indicated, the AIA require the contractor to warn the owner if the contractor believes a suitable result cannot be obtained from the design.[28]

Is the contractor's conduct measured objectively? Is it responsible for an examination based on what the contractor should have discovered and what it should have reported?

Cases that have given a defense to the contractor who has followed the design have stated that this defense will be lost if the contractor has been negligent—an apparently objective standard.[29]

However, the AIA has used a subjective standard in A201-2007, Section 3.2.2. To further clarify the standard of review, Section 3.2.2 states that "the Contractor's review is made in the Contractor's capacity as a contractor and not as a licensed design professional." The owner is paying for whatever design skill the contractor possesses and uses. The principal purpose for the contractor's reviewing the contract documents is to prepare its bid. The owner should expect attention to be drawn only to those errors that the contractor does discover.

The owner often has years to prepare the design, contrasted to the thirty days or so given the contractor to review the design and prepare its bid.[30] An objective approach may operate in a way as to place design risks unfairly on the contractor.

The contractor who *does* notify the owner of errors is not charged for defects that result despite its warnings, and is entitled to be paid for what it has done.[31] The contractor who does not report obvious errors should not be given advantage of the "following the plans" defense.[32] If it performs work *knowing* that the work violates building codes or technical competence, the contractor will not be able to recover for the work performed.

If the contractor's failure to report errors causes a loss, as such a failure will undoubtedly do, should the contractor be responsible for the entire loss? Sharing responsibility rather than placing it all on the contractor is fairer, particularly if the contractor did not have actual knowledge of the error. This is discussed in Section 24.06.

SECTION 24.05 Acceptance of Project

If the owner communicates a clear intention to relinquish a claim for obvious defects, the claim is barred.[33] Acts frequently asserted to have indicated such an intention are payment, particularly the final payment, acceptance, and taking possession of the project.

---

[27]*Lewis v. Anchorage Asphalt Paving Co.*, 535 P.2d 1188 (Alaska 1975); *Luxurious Swimming Pools, Inc. v. Tepe*, 379 N.E.2d 992 (Ind. Ct.App.1978), overruled on other grounds, *Berns Constr. Co., Inc. v. Miller*, 516 N.E.2d 1053 (Ind.1987); *Associated Builders, Inc. v. Oczkowski*, 801 A.2d 1008 (Me.2002); *Lebreton v. Brown*, 260 So.2d 767 (La.App.1972); *Hutchinson v. Bohnsack School Dist.*, 51 N.D. 165, 199 N.W. 484 (1924); *Home Furniture, Inc. v. Brunzell Constr. Co.*, 84 Nev. 309, 440 P.2d 398 (1968).

[28]*American & Foreign Ins. Co. v. Bolt*, 106 F.3d 155 (6th Cir.1997) (Michigan law; contractor was liable in negligence for installing additional purlins to strengthen the roof in a manner it knew to be improper, albeit pursuant to the specifications); *Eichberger v. Folliard*, 169 Ill. App.3d 145, 523 N.E.2d 389, appeal denied, 122 Ill.2d 573, 530 N.E.2d 243 (1988) (contractor did not perform in a workmanlike manner when it built the house in conformity with the plans, although knowing that the foundation was inadequate because of soil conditions); *Rubin v. Coles*, 142 Misc. 139, 253 N.Y.S. 808 (Cty.Ct.1931).

[29]See supra note 5.

[30]For cases directing attention to this, see *Southern New England Contracting Co. v. State*, 165 Conn. 644, 345 A.2d 550 (1974); *Pittman Constr. Co. v. Housing Auth. of New Orleans*, 169 So.2d 122 (La.App. 1964). See also *T. H. Taylor, Inc.*, ASBCA No. 27699, 86-2 BCA ¶18,743, which required the contractor to report only obvious errors.

[31]Architects who pointed out design difficulties can recover for their work if the owner still proceeds. *Greenhaven Corp. v. Hutchcraft & Assoc.*, 463 N.E.2d 283 (Ind.App.Ct.1984) (design violated fire code); *Bowman v. Coursey*, 433 So.2d 251 (La.App.), cert. denied, 440 So.2d 151 (La.1983) (design below acceptable professional practice).

[32]*Allied Contractors, Inc.v.United States*, 180 Ct.Cl. 1057, 381 F.2d 995 (1967); *Davis v. Henderlong Lumber Co.*, 221 F.Supp. 129 (N.D.Ind.1963).

[33]*Stevens Constr. Corp. v. Carolina Corp.*, supra note 21.

Standard documents increasingly make clear that these acts do not constitute a waiver of any claim for defective work.[34] The universal presence of warranty (guarantee) clauses makes it quite unlikely that such a claim will be lost because of the occurrence of these acts.[35]

Generally, the owner by final payment,[36] acceptance or taking possession of the project[37] does not waive its claim for defective work. Either act alone does not unambiguously indicate the owner's intention to give up a claim for work to which it was entitled and for which it paid. Possession may be taken for reasons other than satisfaction with the work. Such holdings are frequently supported by contract clauses denying that such acts have waived claims for defects. Cases that have *found* waiver have involved disputes over particular work followed by some act of the owner, such as payment or taking possession, which communicated satisfaction with the work and an intention not to assert any claim.[38]

SECTION 24.06 Owner Claims and Divided Responsibility

If proof establishes the cause of the defect and that cause is clearly the responsibility of one of the parties to the construction contract, the loss is chargeable to that party. But in a venture as complicated as construction, it is not uncommon for a construction defect to be traceable to multiple causes. For example, suppose the owner asserts a claim against the contractor based on the contractor's not having followed the plans and specifications. Suppose the contractor admits having breached the contract but points to other possible causes, such as abnormal weather conditions or third parties for whom the contractor is not responsible. Legal responsibility for breach of contract does not require the claimant to eliminate all causes except acts or omissions by the party against whom the claim has been made. The defendant's breach must only be a substantial factor in causing the harm.[39] If other conditions or actors played a minor or trivial part in causing the loss, the contractor is responsible for the entire loss.

Multiple causation becomes more complicated when the defect is traceable both to the party against whom the claim has been made and to the claimant itself. For example, Section 24.02 stated the basic principle that in the traditional construction project, the owner is responsible for the design and the contractor for its execution. Yet as noted in Section 24.01, these activities are not watertight compartments. The owner plays a role in execution because of the design professional's site responsibilities. The contractor plays a role in the design because through a specialty subcontractor, it may supply the design or it may be obliged to study and compare the contract documents and report any errors observed.

Another complicating factor is that the cause of the defect may be traceable to wrongful acts of the design professional, some of which may be chargeable to the owner. This section deals with defect claims by the owner against the contractor, with the defect having been caused in whole or in part by acts of the owner or someone for whom the owner is responsible. More commonly in construction, the owner asserts a claim for defective work against both the contractor and a third party, usually the design professional. This claim is discussed in Section 27.14.

Before analyzing shared responsibility, a threshold question must be addressed. Can a contractor who has not followed the plans and specifications point to defective design? A few cases have precluded such a contractor from pointing to a design defect.[40] This conclusion is too punitive and does not take into account the role the design

---

[34] AIA Doc. A201-2007, §§ 9.6.6 and 9.10.4. See Section 22.02M. A federal regulation that stated that taking possession or use is not acceptance was influential in concluding that the project had not been accepted in *M. C. & D. Capital Corp. v. United States*, 948 F.2d 1251 (Fed.Cir.1991).

[35] AIA Doc. A201-2007, §§ 3.5, 9.10.4.3, and 12.2.2.

[36] *Metro. Sanitary Dist. of Greater Chicago v. Anthony Pontarelli & Sons, Inc.*, 7 Ill.App.3d 829, 288 N.E.2d 905 (1972); *Parsons v. Beaulieu*, 429 A.2d 214 (Me.1981); *Quin Blair Enterprises, Inc. v. Julien Constr. Co.*, 597 P.2d 945, 955 (Wyo.1979) (citing earlier edition of book).

[37] *M. C. & D. Capital Corp. v. United States*, supra note 34 (U.S. not precluded from terminating for default despite the fact that it had taken possession); *Kangas v. Trust*, 110 Ill.App.3d 876, 441 N.E.2d 1271 (1982); *Annen v. Trump*, 913 S.W.2d 16 (Mo.Ct.App.1995), transfer denied May 2, 1995 (acceptance does not bar claim for defective work).

[38] *Sentell Bros., Inc.*, DOTCAB No. 1824, 89-3 BCA ¶21, 904, reconsideration denied, 89-3 BCA ¶22,219; *Grass Range High School Dist. v. Wallace Diteman, Inc.*, 155 Mont. 10, 465 P.2d 814 (1970). See generally, Sweet, *Completion, Acceptance and Waiver of Claims: Back to Basics*, 17 Forum 1312 (1982).

[39] *Krauss v. Greenbarg*, 137 F.2d 569 (3d Cir.), cert. denied, 320 U.S. 791 (1943).

[40] *Valley Constr. Co. v. Lake Hills Sewer Dist.*, 67 Wash.2d 910, 410 P.2d 796 (1965); *Robert G. Regan Co. v. Fiocchi*, Ill.App.2d 336, 194 N.E.2d 665 (1963), cert. denied, 379 U.S. 828 (1964).

professional played in causing the loss. A federal court, discussing an Arkansas project but applying California law by agreement of the parties, refused to bar the contractor's claim against the owner for a defective design simply because the contractor failed to execute the design, where the contractor's defective performance was not the cause of the project's failure.[41]

Where defects can be traced to both the owner and the contractor, the law has taken two approaches. Some cases have held the loss will be shared, with the owner's claim being reduced by a rough formula comparison such as used in those states that use comparative negligence as part of their tort law.[42] This formula takes into account the complexity of causation and the desire to avoid all-or-nothing outcomes.

Yet courts in other cases have not been willing to apply comparative fault in a contract claim for economic losses.[43] They seek to preserve a sharp line between contract and tort claims, a difficult objective given that the owner in some jurisdictions may be able to choose whether to maintain its claim in contract or tort.[44]

Contract language can control the apportionment problem. For example, in 1987, AIA Document A201 provided that if the contractor did not fulfill its obligation under Paragraph 3.2.1 to report errors it discovers, the contractor "shall assume appropriate responsibility . . . and shall bear an appropriate amount of the attributable costs for correction." This would appear to invite shared responsibility.

But in 1997, the AIA made a change. A201-1997, Paragraph 3.2.3 stated that the contractor "shall pay such costs and damages to the Owner as would have been avoided if the Contractor had performed such obligations." This new language is retained in A201-2007, Section 3.2.4. This would appear to relieve the design professional even though she committed the design error.

Similarly, under A201-2007, Section 3.7.3, if the contractor performs work knowing it to be in violation of the building laws, it must "assume full responsibility for such Work and shall bear the costs attributable to correction." This would appear to place the *entire* responsibility on the contractor. Such an outcome would relieve the owner and ultimately the architect from any responsibility for supplying a design that did not comply with building laws. Admittedly the intervening cause of the contractor's failure to report errors of which it is aware is an intentional act, which can cut off responsibility for a negligent act of the architect, which is charged to the owner. Yet this outcome exculpates the owner (and the architect) from responsibility for the design.

SECTION 24.07   Third-Party Claims

This chapter has emphasized the discovery of defects that harm the owner. However, defects caused either by design, for which the owner is responsible, or by execution, for which the contractor is responsible, or by a combination of the two can also harm third parties. As noted in Section 14.08, third-party claims create complex legal issues.

Damage to the property of a third party—a tort claim—is dealt with as would be any tort claim.[45]

Suppose a third party suffers property damage that can be traced to defective design. The complexity of such third-party tort claims is demonstrated by *Cincinnati Riverfront Coliseum, Inc. v. McNulty Co.*[46] Five years after completion of construction, an elevated walkway for pedestrian traffic suffered extensive damage and deterioration. The owner repaired the structure and brought legal action against eleven defendants. All settled except the consulting engineer who designed the walkway and the city, which had obligated itself to maintain it.

The jury found that 40 percent of the losses were attributable to the negligence of the consulting engineer and 5 percent to the city. The consulting engineer claimed that he was not responsible for construction defects, because the structure had not been constructed in accordance with the design, and that his tort liability was cut off by the intervening acts of the contractor.

---

[41]*Arkansas Rice Growers Coop. Ass'n v. Alchemy Indus., Inc.*, 797 F.2d 565 (8th Cir.1986).

[42]*Calcasieu Parish School Bd. v. Lewing Constr. Co., Inc.*, supra note 5 (architect responsible for 20 percent of owner's damages); *Grow Constr. Co. v. State*, 56 A.D.2d 95, 391 N.Y.S.2d 726 (1977); *Bunkers v. Jacobson*, supra note 5, 653 N.W.2d at 743; *Circle Elec. Contractors, Inc.*, DOTCAB 76-27, 77-BCA ¶12, 339.

[43]*Broce-O'Dell Concrete Products, Inc. v. Mel Jarvis Constr. Co.*, 6 Kan.App.2d 757, 634 P.2d 1142 (1981) (claim by subcontractor against concrete supplier).

[44]See Section 14.05E.

---

[45]See Section 7.03.

[46]28 Ohio St.3d 333, 504 N.E.2d 415 (1986).

The court stated that the engineer is responsible for the foreseeable consequences of his failure to exercise reasonable care. The jury could have found it was reasonably foreseeable that the contractor would alter the design without approval of the engineer and fail to execute the design properly. The engineer can escape responsibility if the deviations are material and were the "proximate cause" of the loss. The breach by the contractor must independently break the causal connection between the negligent design and the damage. If it does, the breach supersedes the negligent design as the cause.

As noted in Sections 14.08C, D, and E, if the losses suffered by third parties are purely economic losses unconnected to personal harm or damage to property, some jurisdictions do not allow tort claims.

The *Cincinnati Riverfront* case dealt with one party's liability for negligence being cut off by the acts of another. Other courts have looked on design negligence and failure to follow the design as concurrent and not independent causes. If so, the claimant (in defect cases, the owner) can contend that both caused the loss and can recover an entire, indivisible loss from either party, the ultimate responsibility dealt with by indemnity or contribution.[47]

SECTION 24.08 Claims Against Liability Insurer

Commonly, the contractor is required to carry commercial (formerly called "comprehensive") general liability (CGL) insurance. A primary purpose of CGL insurance is to indemnify the insured contractor against claims by third parties who assert they have suffered losses because the insured contractor has not acted in accordance with tort law. Usually claimants are workers on the job who are not employees of the contractor (the latter are usually covered by workers' compensation) or members of the public. Claims covered are those that involve personal harm or property damage.

In the 1970s, contractors began to seek insurance coverage for defective workmanship claims brought against them by the owner. (The owner often alleged that the contractor was negligent in an attempt to bring its claim within the scope of the insurance policy.) Insurance

coverage of owner claims is advantageous to the contractor. The insurer, unlike a surety, cannot claim against its insured.[48]

Generally, CGL policies provide the insured with a legal defense and indemnification (up to the policy limits) against any claims of liability brought against the insured that assert "bodily injury" or "property damage" caused by an "occurrence." An "occurrence" is defined as an "accident." Hence, the claim must have been in some way unexpected and unintended by the insured. Coverage for claims alleging "bodily injury" is relatively straightforward.[49] A more difficult question is whether an owner's claim of defective workmanship constitutes "property damage" caused by an "occurrence."

There is no doubt that liability insurance is intended primarily (both by the insurer and the insured) to cover unusual, unexpected losses, such as a wall collapsing and destroying a Porsche automobile parked nearby. Conversely, CGL policies are not intended to cover ordinary business losses; defective work being an example. However, sometimes courts confuse the *purpose* of CGL insurance with the *scope of coverage*. Rather than examine whether the claim against the insured involves "property damage" caused by an "occurrence," they broadly exclude from coverage all breach of contract claims alleging economic losses, including owner claims for defective workmanship.[50] However, a more sophisticated analysis is to determine coverage based on the policy language, not on the theory of liability or the broad purpose of commercial liability insurance.[51]

---

[47]*Northern Petrochemical Co. v. Thorsen & Thorshov, Inc.,* 297 Minn. 118, 211 N.W.2d 159 (1973). See Section 27.14.

[48]See Section 32.02.

[49]This is not to say that coverage is guaranteed, or even likely, since bodily injury claims, especially by construction workers, are subject to numerous exceptions. See M. SCHNEIER, CONSTRUCTION ACCIDENT LAW: A COMPREHENSIVE GUIDE TO LEGAL LIABILITY AND INSURANCE CLAIMS §§ 9:B:II–IX (1999).

[50]*Heile v. Herrmann,* 136 Ohio App.3d 351, 736 N.E.2d 566 (1999); *Century Indemnity Co. v. Golden Hills Builders, Inc.,* 348 S.C. 559, 561 S.E.2d 355 (2002); *Erie Ins. Property & Cas. Co. v. Pioneer Home Improvement, Inc.,* 206 W.Va. 506, 526 S.E.2d 28 (1999).

[51]*Fejes v. Alaska Ins. Co., Inc.* 984 P.2d 519, 523–24 (Alaska 1999), rehearing denied Sept. 29, 1999 (contract liability claims are not *per se* excluded from CGL coverage); *Vandenberg v. Superior Court,* 21 Cal.4th 815, 982 P.2d 229, 243-46, 88 Cal.Rptr.2d 366 (1999), rehearing denied Oct. 20, 1999 (same); *Travelers Indemnity Co. of Am. v. Moore & Associates, Inc.,* 216 S.W.3d 302 (Tenn.2007); *Lamar Homes, Inc. v. Mid-Continent Cas. Co.,* 242 S.W.3d 1 (Tex.2007), rehearing denied Dec. 14, 2007.

Even courts willing to apply the policy language often conclude that there is no coverage for defective workmanship claims. Many find that such claims either were not caused by an "occurrence" or did not result in "property damage."[52] Courts which do decide that these threshold requirements for coverage exist may then find coverage defeated through application of the policy's "business risk" exclusions. Under these standard exclusions, CGL policies usually exclude work products, property in control of the insured (this should be handled by property insurance), or work performed by or on behalf of the contractor, such as work performed by a subcontractor. (But this last exclusion is subject to an exception, discussed *infra*.) In sum, unless the defective workmanship caused collateral damage (such as water damage to the owner's personal property),[53] defective workmanship claims generally fall within the scope of the exclusions.[54]

Where the insured is a prime contractor and the defective workmanship is caused by a subcontractor, it is less clear whether the standard exclusions apply. Does the insured's "work" include that of its subcontractors, or do the subcontractors produce separate "work"? Analysis of this question reveals a fundamental fact of insurance law: Policy language is constantly being changed by the insurance industry, and apparently, slight changes in wording can impact coverage significantly. Simplifying this issue greatly, earlier policies excluded coverage for work performed "by or on behalf of" the insured, thus including within its scope the work of subcontractors. However, prime contractors could purchase a separate endorsement—the broad form property damage endorsement—which removed the "on behalf of" language and thereby restored coverage for defective workmanship caused by subcontractors.[55] Nor did the matter end there: Recent policies either have dropped the "on behalf of" language (thereby obviating the need for a prime contractor to purchase a special endorsement) or have included exceptions to certain "work product" exclusions for claims alleging defective work by subcontractors.[56]

Another problem relates to the duration of coverage. The *standard* CGL policy terminates coverage after final payment. This would bar coverage for defects discovered after that point. It is possible to obtain "completed operations" coverage to cover such claims.[57]

A more detailed analysis of insurance coverage for defective workmanship claims is beyond the scope of this book.[58]

SECTION 24.09 Warranty (Guarantee) Clauses

A. Relationship to Acceptance

Acceptance of the project is an important benchmark in the relationship between owner and contractor. Contractors would like to know when the law can no longer call on them to perform further work or to respond

---

[52]*Oak Crest Constr. Co. v. Austin Mutual Ins. Co.*, 329 Or. 620, 998 P.2d 1254 (2000) (breach of contractual obligation cannot be an "occurrence"); *Firemen's Ins. Co. of Newark v. National Union Fire Ins. Co.*, 387 N.J.Super. 434, 904 A.2d 754 (App.Div.2006) (contractor's installation of defective firewalls does not constitute "property damage"); *L-J, Inc. v. Bituminous Fire & Marine Ins. Co.*, 366 S.C. 117, 621 S.E.2d 33 (2005) (deterioration of road is not an "occurrence"); *Vogel v. Russo*, 236 Wis.2d 504, 613 N.W.2d 177 (2000) (diminution in value of building, caused by subcontractor's defective masonry work, is not "property damage").

[53]*Canal Indemnity Co. v. Blackshear Farmers Tobacco Warehouse, Inc.*, 227 Ga.App. 637, 490 S.E.2d 129 (1997), reconsideration denied Jul. 28, 1997; *Pekin Ins. Co. v. Richard Marker Assocs., Inc.*, 289 Ill.App.3d 819, 682 N.E.2d 362, appeal denied, 175 Ill.2d 531, 689 N.E.2d 1140 (1997); *Iberia Parish School Bd. v. Sandifer & Son Constr. Co, Inc.*, 721 So.2d 1021 (La.App.1998); *ACUITY v. Bird & Smith Constr., Inc.*, 2006 ND 187, 721 N.W.2d 33 (2006). For a review of what constitutes an "accident" under a CGL policy, see Annot., 7 A.L.R.3d 1262 (1966).

[54]The following courts applied various business risk exclusions: *American Equity Ins. Co. v. Van Ginhoven*, 788 So.2d 388 (Fla.Dist. Ct.App. 2001); *Sapp v. State Farm Fire & Cas. Co.* 226 Ga.App. 200, 486 S.E.2d 71 (1997); *Supreme Services and Specialty Co., Inc. v. Sonny Greer, Inc.*, 958 So.2d 634 (La.2007); *Alverson v. Northwestern Nat. Cas. Co.*, 559 N.W.2d 234 (S.D.1997), rehearing denied Mar. 20, 1997.

[55]*Fireguard Sprinkler Systems, Inc. v. Scottsdale Ins. Co.*, 864 F.2d 648, 652 (9th Cir.1988) (citing insurance industry bulletins as to the purpose of the broad form endorsement); *Fejes v. Alaska Ins. Co., Inc.*, supra note 51, 984 P.2d at 525; *Maryland Cas. Co. v. Reeder*, 221 Cal.App.3d 961, 270 Cal.Rptr. 719 (1990), review denied Sept. 19, 1990; but see *Knutson v. St. Paul Fire & Marine Ins. Co.*, 396 N.W.2d 229 (Minn.1986) (broad form endorsement does not change loss from business cost to insurance risk). See also Casamassima & Jerles, *Defining Insurable Risk in the Commercial General Liability Insurance Policy: Guidelines for Interpreting the Work Product Exclusion*, 12 Constr. Lawyer, No. 1, Jan. 1992, p. 3 (providing a useful, historical approach).

[56]*Kalchthaler v. Keller Constr. Co.*, 224 Wis.2d 387, 591 N.W.2d 169, 174 (App.1999).

[57]*Pardee Constr. Co. v. Insurance Co. of the West*, 77 Cal.App.4th 1340, 92 Cal.Rptr.2d 443(2000), review denied April 26, 2000 (extending completed operations coverage in subcontractors' policies to benefit a prime contractor, who was an "additional insured" on those policies, even though the additional insured endorsement was obtained only after the subcontractors had completed their work).

[58]See INSURANCE COVERAGE FOR DEFECTIVE CONSTRUCTION (American Bar Association 1997) and S. TURNER, INSURANCE COVERAGE OF CONSTRUCTION DISPUTES (2d ed. 2002).

to a legal claim. For that reason, they frequently contend that acceptance of the project manifests owner satisfaction and the owner cannot make any further complaints about the work.

The law has been reluctant to find that "acceptance" has been sufficient to bar any future claims for latent or non-discoverable defects against the contractor.[59] Undoubtedly this reluctance stems from the difficulty of discovering defects at the time the project is turned over. Many defects will not be apparent until the owner has taken over the project and used it. One method of dealing with the risk

that acceptance will be found and claims barred is to insert a provision making clear that liability does not end on the project's being turned over to the owner.[60] The exact nature of that liability is discussed in Section 24.09B.

B. Purposes: *St. Andrew's Episcopal Day School v. Walsh Plumbing Co.*

The deceptively simple warranty clause obscures a variety of possible purposes. Before discussing the many possible purposes, a case involving such a clause is reproduced.

ST. ANDREW'S EPISCOPAL DAY SCHOOL v. WALSH PLUMBING CO.

Supreme Court of Mississippi, 1970. 234 So.2d 922.

ROBERTSON, Justice.

The appellant, St. Andrew's Episcopal Day School, a charitable corporation, brought suit in the Chancery Court of the First Judicial District of Hinds County, Mississippi, against Appellee Walsh Plumbing Company, the contractor of the mechanical work, and Appellee The Trane Company, the manufacturer of the major portion of the air conditioning system installed in the new Day School building. The suit was one for breach of warranty or guaranty. The chancellor, after a full trial, dismissed the bill of complaint, and the Day School appeals from this judgment.

On April 8, 1965, appellant entered into a contract with Walsh Plumbing Company whereby, in consideration of $149,420, Walsh agreed to:

> furnish all of the materials and perform all of the work shown on the Drawings and described in the Specifications entitled: Item II, Mechanical Construction, St. Andrews Episcopal Day School, Old Canton Road, Jackson, Mississippi.

The General Conditions of the Contract provided in Article 20:

> Correction of the Work After Substantial Completion
> The Contractor shall remedy any defects due to faulty materials or workmanship and pay for any damage to other work resulting therefrom, which shall appear within a period of one year from the date of Substantial Completion as defined in these General Conditions, and in accordance with the terms of any special guarantees provided in the Contract. The Owner shall

give notice of observed defects with reasonable promptness. All questions arising under this Article shall be decided by the Architect subject to arbitration, notwithstanding final payment.

The Construction Specifications, in Paragraph 22 of Mechanical Construction, required:

> This contractor shall guarantee each and every part of all apparatus entering into this work to be *the best of its respective kind and he shall replace within one year from date of completion all parts which during that time prove to be defective and he must replace these parts at his own expense.*
> He shall guarantee to install each and every portion of the work in strict accordance with the plans and specifications and to the satisfaction of the owner.
> Guarantee to include replacement of refrigerant loss from air conditioning and refrigerant system. [Emphasis added.]

On August 19, 1966, Lomax, North and Beasley, consulting engineers, by letter, advised Biggs, Weir, Neal & Chastain, architects, that the mechanical contractor should furnish "as built drawings which locate all underground piping and clean-outs, framed operating instructions in the boiler rooms, CFM figures for all air units, and certification that all safety valves and devices have been tested." The engineers ended their letter with this comment:

> Other than the above, we feel that the mechanical work is ready for final certification, subject to contract guarantee provisions.

---

[59]See Section 24.05.

[60]AIA Doc. A201-2007, §§ 3.5, 9.10.4.3 and 12.2.2.

On August 30, 1966, the engineer wrote the architects:

Final inspections of the subject project have been completed and we recommend final certification of the mechanical contract, subject to contract guarantee provisions.

In accordance with the requirements of Paragraph 22 of the Construction Specifications, on August 16, 1966, Walsh Plumbing Company wrote St. Andrew's Episcopal Day School:

We hereby guarantee all work performed by us on the above captioned project to be free from defective materials and workmanship for a period of one (1) year, unless called for in the specifications to be a longer period of time.

Between August 16, 1966, and July 17, 1967, Appellee Walsh was called on several times to remedy defects in the air conditioning system, and Walsh always responded promptly. On July 17, 1967, the air conditioning system broke down completely and ceased to function. The headmaster, the Reverend James, immediately tried to contact Ray Walsh, only to find that he was out of town. Mrs. Walsh suggested that the School call somebody else.

During the first two weeks of July, 1967, James E. Davis, Jr., operator of Davis-Trane Service Agency and also a salesman and representative of The Trane Company, was contacted by John B. Walsh and together with Mr. Walsh attempted to determine why the air conditioning system was not cooling. Mr. Davis described the meeting that took place on July 18, 1967, in these words:

[T]he meeting that we had on a particular day at the school, in which Mr. Nicholson was there, and Mr. Ray Walsh was there, and Mr. Nicholson said, "We want to get this thing fixed," and Mr. Walsh told me to fix it. He said, "You have the people, the personnel and the know-how, you go ahead and do it and I will just stay out of it," so Mr. Nicholson said, "Okay, Davis-Trane Service Agency, go ahead and fix this machine." [Emphasis added.]

Forrest G. North, the mechanical engineer, testified that the failure of two safety devices, the flow control switch and the freeze protection thermostat caused the copper tubes inside the chiller shell and the shell itself to rupture. The purpose of the flow control switch was to prevent the operation of the machine when there was no circulation of water in the chiller shell. The flow control switch was installed by Walsh outside the chiller and was not Trane equipment. The freeze protection thermostat was a Trane part, and was installed inside the chiller unit by Trane at its factory.

The repairs made by Davis-Trane Service Agency pursuant to Ray Walsh's instructions to Davis to go ahead and fix it amounted to $6,813.05. One of the major items of expense was a new chiller unit purchased from the Trane Company. The Trane Company and Davis-Trane billed the appellant, and the appellant paid Trane separately for the new chiller unit, and Davis-Trane for all the repairs.

The appellant is in the business of running a Christian day school; it is not in the air conditioning business. Appellant does not profess to have any knowledge or expertise about air conditioning systems or equipment. That was the main reason for the provision in the Construction Specifications that the mechanical contractor "shall guarantee each and every part of all apparatus entering into this work to be *the best of its respective kind*, and he shall replace within one year from date of completion all parts which during that time prove to be defective and he must replace these parts at his own expense."

Walsh was a reputable, responsible and knowledgeable contractor of mechanical work; and when Lomax, North and Beasley, consulting engineers for the Day School, recommended to the appellant that the bid of Walsh Plumbing Company be accepted, the duty and responsibility was placed squarely on Walsh's shoulders to purchase and properly install the best air conditioning system on the market. Not only was Walsh to purchase and install, he was to guarantee the system and its installation for one year. This was what Walsh contracted to do, and this was what the appellant paid Walsh to do. Not knowing anything about air conditioning systems, the appellant employed experts in this field and reposed full confidence in these experts to look after its interests.

The chancellor was correct in finding that the breakdown of the air conditioning system occurred "within the time of the warranty by Walsh," and that the repairs were made necessary to properly repair the system.

The chancellor was in error in holding that:

St. Andrews never gave any written notice or made any demand on Walsh to comply with his warranty to fix the machine, which would be as I hold a condition precedent to hiring someone else to do the work.

The sole purpose of notice is to give the contractor who selected, purchased and installed the system the first opportunity to remedy the defects at the least possible expense to him. Appellee Walsh was given this opportunity.

The evidence is undisputed that Walsh was at the Day School building on July 18, 1967, the day after the breakdown, with James E. Davis, Jr., of Davis-Trane Service Agency and John W. Nicholson of the Day School. Davis, called as an adverse witness by the appellant, testified that Ray Walsh said at that time:

"You have the people, the personnel and the know-how, you go ahead and do it and I will just stay out of it."

Walsh was afforded the opportunity of doing the work himself or employing somebody else to do it. With full knowledge and full notice he chose to employ Davis to go ahead and remedy the defects and make the necessary repairs.

Appellee Trane was the major supplier of items and equipment going into the air conditioning system and Trane was well paid for these. It is unfortunate that Trane's one-year guaranty to Walsh had run out at the time of the complete breakdown of the air conditioning system. Trane guaranteed the items and equipment furnished by it for one year from the date of shipment; these parts and equipment were shipped in March, 1966. The complete breakdown did not occur until July, 1967. The chancellor was correct in holding that Trane's warranty had expired.

The chancellor found that $6,813.05 was "a reasonable amount to make the repairs" and put the air conditioning system back in operation.

The judgment of the chancery court is affirmed as the Appellee The Trane Company, but the judgment is reversed as to Appellee Walsh Plumbing Company, and judgment is rendered against Walsh Plumbing Company on its warranty for $6,813.05, together with 6% interest from July 17, 1967.

Judgment affirmed as to The Trane Company, but reversed and rendered as to Walsh Plumbing Company.

GILLESPIE, P. J., and JONES, BRADY, and INZER, J. J., concur.

From the owner's vantage point, one disadvantage of the traditional contracting system is the possibility of the owner being "whipsawed" between the design professional and the contractor when a defect is discovered. The design professional contends the defect resulted from improper execution of the design. The contractor contends the defect existed because of design inadequacy. To make matters worse, each or both may contend the defect was caused by conditions over which neither had control, such as unusual weather, misuse, or poor maintenance by the owner. One way of dealing with divided responsibility is to hire one entity to both design and build.

Another method of dealing with this problem is to incorporate a provision under which one entity is responsible for defects however and by whomever caused. The owner is likely to place this risk on the contractor. For example, the owner could require the contractor to give a full warranty on a roof that would cover defects however caused.

Despite the undoubted advantage in placing the risks of defects *however caused* on the contractor, this is not often done. Contractors may respond to such a guarantee on a roof by drastically increasing the contract price to take into account the possibility of design errors, the additional cost of checking on the quality of the design, the cost of any special endorsement to be obtained to cover design risks, or the cost of a roofing bond (a bond by a surety that guarantees the roof for a certain number of years). Nevertheless, the owner may decide these additional costs are worth the advantage of centralizing responsibility. Although enforcement requires very clear

language (even that may not be sufficient), such clauses have been enforced.[61]

Many issues lurk in a simple warranty clause. Earlier discussion emphasized the owner's purpose in having such a clause. The contractor is also concerned with how such a clause will function. Although it would prefer to be relieved from responsibility on acceptance, it may be willing to accept a warranty clause if it believes that expiration of the period will terminate its obligation. Similarly, the contractor may be concerned with exposure created by such a clause even within the warranty period. Failure to take immediate steps to deal with a defect may increase the loss. The cost of correcting a defect is likely to be greater if the correction is done by someone other than the contractor. For that reason, contractors sometimes contend that the clause is for their benefit, that its function is to give the contractor notice as quickly as possible, to enable the contractor to cut the losses and repair the defect as inexpensively as possible.[62] Or the owner may contend that the

---

[61]*Potler v. MCP Facilities Corp.*, 471 F.Supp. 1344 (E.D.N.Y.1979); *Bryson v. McCone*, 121 Cal. 153, 53 P. 637 (1898) (construction of industrial plant).

[62]*See St. Andrew's Episcopal Day School v. Walsh Plumbing Co.*, reproduced in this section; *Barrack v. Kolea*, 438 Pa.Super. 11, 651 A.2d 149 (1994), appeal denied, 542 Pa. 671, 668 A.2d 1134 (1995); *Simek v. Rocky Mountain, Inc.*, 977 P.2d 687 (Wyo.1999). If it is clear that the contractor will not correct the defect, giving the notice is excused. *Orto v. Jackson*, 413 N.E.2d 273 (Ind.Ct.App.1980). See, generally, Slutzky, *Fully Understanding and Utilizing the Call Back Warranty*, 23 Constr. Lawyer, No. 1, Winter, 2003, p. 13; and Senter, *Construction Warranties and Guarantees: A Primer*, 23 Constr. Lawyer, No. 1, Winter 2003, p. 17.

clause is for its benefit. If the owner has lost confidence in the original contractor or wants the right to bring in a successor for other reasons, it need not call back the original contractor (discussed in Section 24.09C).

Another possible purpose is to bar the contractor from contesting how the owner has chosen to correct the defect if the contractor refuses to attempt to correct the work when requested to do so. The clause may place the burden on the warrantor to establish that other causes, such as abnormal weather or owner misuse, caused the defect.

In the consumer context, warranties can mislead the consumer. The consumer may believe that what looks like a very broad warranty has been so hedged about with restrictions that it may not amount to very much. To deal with this problem, Congress enacted the Magnuson-Moss Warranty Act, which carefully regulates consumer warranties.[63] The act also specifies remedies for breach of such a warranty and allows the consumer who prevails in a claim based on the act to recover his or her attorneys' fees.

Although the act was thought principally to deal with consumer products, a case held that it also applies to a contract to roof a house.[64] In this case, the court looked at the legislative history and the desire to protect consumers and concluded that the act was not limited solely to products (the award of attorneys' fees exceeded the damages).

Moving back to the warranty clause as putting a time limit on liability, although a few decisions have held the clause to have created a private (created by the contract) period of limitation,[65] most have held that such a provision is not intended to cut off liability at the end of the warranty period.[66] Clearly, the AIA has not made its warranty clause a period of limitation.[67] The law will *not* bar a claim

by a contractual provision of this type unless the clause clearly shows this intention.[68]

Does the clause affect the remedy? Usually the clause states that the contractor will correct any defect within the warranty period. As discussed in Section 24.09C, will this provision justify the issuance of a court decree *specifically ordering* the contractor to come back and correct the work? If the contractor fails to correct the work, the owner can do so, and charge the contractor. This recognizes the adequacy of the money award, usually a bar to judicially ordered specific performance. A warranty clause should not affect the availability of specific performance.

Remedy can be approached another way. Suppose there were no warranty clause. The owner can recover the cost of correction unless it would be disproportionately high in relation to the diminished value caused by the defect. If the latter, the measure is the difference in value between the project as built and as it should have been built. In addition, the owner may be able to recover consequential damages, such as lost use or profits if reasonably foreseeable and proved with reasonable certainty.

A warranty clause is a promise by the contractor to correct defects within the warranty period. Does this make correction cost the only remedy? Is diminished value barred? Does the clause limit liability to cost of correction? Very likely the clause was not intended to deal with any of these subtle issues. The law should not give the clause any remedial effect.[69]

Although historically such clauses may have been justified as a method to avoid acceptance barring claims, is there a current justification for including a warranty clause? An unfortunate problem that often plagues the owner is the unwillingness of the contractor to return and correct defects for which it is responsible. Including a specific provision under which the contractor promises to return and correct defects if notified to do so within a

---

[63]15 U.S.C.A. § 2301 et seq.

[64]*Muchisky v. Frederic Roofing Co., Inc.*, 838 S.W.2d 74 (Mo.Ct.App. 1992). See also *Atkinson v. Elk Corp. of Texas*, 142 Cal.App.4th 212, 48 Cal.Rptr.3d 247 (2006) and Schneier, *The Magnuson-Moss Warranty Act: Federalizing Homeowner Construction Defect Cases*, 13 Constr. Lawyer, No. 4, Oct. 1993, p. 1.

[65]*Cree Coaches, Inc. v. Panel Suppliers, Inc.*, 384 Mich. 646, 186 N.W.2d 335 (1971); *Independent Consol. School Dist. No. 24 v. Carlstrom*, 277 Minn. 117, 151 N.W.2d 784 (1967); *Mountain View/Evergreen Imp. & Service Dist. v. Casper Concrete Co.*, 912 P.2d 529 (Wyo.1996).

[66]*First Nat'l Bank of Akron v. Cann*, 503 F.Supp. 419 (N.D.Ohio 1980), aff'd, 669 F.2d 415 (6th Cir.1982) (citing earlier edition of book); *Norair Eng'g Corp. v. St. Joseph's Hosp., Inc.*, 147 Ga.App. 595, 249 S.E.2d 642 (1978); *Board of Regents v. Wilson*, 27 Ill.App.3d 26, 326 N.E.2d 216 (1975) (citing and following earlier edition of book).

[67]AIA Doc. A201-2007, § 12.2.5.

[68]*Glantz Contracting Co. v. General Elec.*, 379 So.2d 912 (Miss. 1980); *Bender-Miller Co. v. Thomwood Farms, Inc.*, 211 Va. 585, 179 S.E.2d 636 (1971) (remedy not exclusive).

[69]*United States v. Franklin Steel Products, Inc.*, 482 F.2d 400 (9th Cir.1973), cert. denied, 415 U.S. 918 (1974) (consequential damages recoverable); *Oliver B. Cannon & Son, Inc. v. Dorr-Oliver, Inc.*, 336 A.2d 211 (Del.1975); *New Zion Baptist Church v. Mecco, Inc.*, 478 So.2d 1364 (La.App.1985). But see Heckman, *Drafting the "Perfect" One-Year Warranty*, 27 Constr. Lawyer, No. 3, Summer 2007, p. 5 (recommending contract language to make the one-year warranty the owner's exclusive remedy).

designated period may persuade a contractor that it must do as it has promised.

Exhortation, of course, may not be the sole function of such a clause or even a function of a particular clause. A warranty clause may be designed to shift design risks to a contractor or to affect the legal standard of conduct. Nevertheless, on the average, modern warranty clauses are principally exhortations to get defects corrected quickly before they generate greater losses. This does not mean that the drafter may not want a different purpose. If so, the objective should be expressed clearly in the contract. Even so, clauses will always need interpretation as to coverage.

C. Warranty and Correction of Work Under A201

A201-2007, Section 3.5, creates an express warranty as to the quality of the contractor's work. It must be read together with Section 12.2, which deals with correction of work that does not meet the quality warranty.

Section 3.5 creates three warranties from the contractor to the owner and to the architect:

1. "[M]aterials and equipment . . . will be of good quality and new unless the Contract Documents require or permit otherwise."
2. "The Work will conform to the requirements of the Contract Documents."
3. "The Work . . . will be free from defects, except for those inherent in the quality of the Work."

Section 24.02 discussed the difference between unsuitable and faulty (defective) materials. Suppose purchase description specifications that require the contractor to use particular materials or equipment by manufacturer and model number are not used. Section 3.5 makes the contractor responsible for faulty materials or equipment. Suppose, though, purchase description specifications are used. If the contractor did as ordered, would the contractor be liable under Section 3.5 for defects caused by faulty materials? It failed to supply good quality. But it can be contended that the contractor is relieved because it installed what was "required or permitted." This contention would be supported by the strong body of case law that relieves the contractor when it does as it has been ordered.[70] That legal doctrine does not apply if the contractor has given a warranty or violated its obligation of good faith and fair dealing, such as by failing to communicate information that would have warned the architect not to proceed. Yet Section 3.5 is not sufficiently *specific* to create a warranty that can overcome the basic rule that the contractor is relieved if it does what it has been required to do.

Section 3.5 exculpates the contractor from its warranty if the defect is caused by "abuse, alterations to the Work not executed by the Contractor, improper or insufficient maintenance, improper operation, or normal wear and tear and normal usage." The loss of control and the possibility of such exculpatory events are reasons why contractors seek to be relieved from any responsibility after expiration of the warranty period.

If there is a breach of the quality requirement under Section 3.5, Section 12.2 deals with correction of work needed before substantial completion (Section 12.2.1) or after substantial completion (Section 12.2.2). The latter is generally known as the "one year correction-of-work" clause.

Although the Associated General Contractors (AGC) has sought to cut off the contractor's liability for patent (discoverable) defects after the expiration of the one-year period, the AIA has made clear in Section 12.2.5 that expiration of the period does not terminate the contractor's liability, which is terminated only by the expiration of the appropriate statute of limitation. Other issues are not so clearly resolved.

Does the contractor have "first crack" at repairing defective work that is discovered within the first year?[71] Section 12.2.2.1 requires that the contractor correct the defect promptly after receipt of written notice. The owner must give notice promptly after discovery of the condition. In 1997 language was added (continued in 2007) under which the contractor is relieved from responsibility if the owner "fails to notify the Contractor" and give it "an opportunity to make the correction."

If the correction-of-work clause is intended for the owner's benefit, the owner need not give the contractor the first chance to make the corrections.[72] Applying the obligation of good faith and fair dealing, however, it would not be fair for the owner to have the absolute right to determine whether to give the contractor the first chance to correct the work. The owner can refuse to do so

---

[70]See Section 24.02. See also *Teufel v. Wienir*, supra note 10.

[71]*St. Andrew's Episcopal Day School v. Walsh Plumbing Co.*, reproduced in Section 24.09B.

[72]*Baker Pool Co. v. Bennett*, 411 S.W.2d 335 (Ky.1967).

only if the owner has had good reason to lose faith in the contractor, such as prior assurances by the contractor having proved unreliable.

What if the defect is discovered after the one-year correction-of-work period expires? Must the owner notify the contractor? Notification provisions should apply only to defects discovered within one year. One court troubled by this problem pointed to Paragraph 7.1.4 (now Section 10.2.8), which requires either party who suffers injury or damage to make a claim in writing to the other party within a reasonable time.[73] This was very likely intended to apply only to harm suffered during performance and not to place a notice requirement as a condition precedent for any claim.

The one-year correction-of-work requirement expressed in Section 12.2.2.1 can be expanded if "an applicable special warranty [is] required by the Contract Documents." This provision recognizes that some specific warranties may exceed the one-year period. For example, roof warranties are often given for lengthy periods. If such a warranty is considered a specific warranty required by the contract, the one-year correction-of-work period should be extended to the period covered in the specific warranty clause.[74]

In 1987, the AIA took the position that the purpose of the warranty clause is remedial, to give the owner a right to specific performance (judicial order to the contractor to correct the defect).[75] It supported this position by pointing to its language in Paragraph 12.2.1.1 stating that the contractor has a specific obligation to correct the work.

The AIA Commentary on AIA Document A201-1997 issued in 1999 appeared to have retreated from this stance. The remedial purpose was not mentioned. Instead, it stated that Paragraph 12.2.2 was a separate remedy for nonconforming work from the general warranty clause, Paragraph 3.5, and the clause allowed the owner to correct nonconforming work during performance, Paragraph 2.4.1. According to the Commentary, the purpose of the one-year clause was to give the contractor "first crack" at correcting the work.

SECTION 24.10 Implied Warranties in the Sale of Homes: Strict Liability

A. Home Buyers and Their Legal Problems

In the preceding section, project success assumes construction documents that set forth the contractor's obligations and are the basis for judging the design professional's performance. Yet people often buy partially or totally completed homes without the protection of plans that must be met before final payment is made. They purchase homes from developers or builder-vendors, each of whom builds homes for immediate sale to home buyers. Suppose after the purchase the basement leaks, the walls crack, or the heating system malfunctions. What recourse does the buyer have?

Before the mid-1960s, home buyers who faced such problems had little recourse against the sellers, or for that matter against anyone. Part of this situation was traceable to legal rules designed to avoid uncertainties in real property transfers and to transfer certain risks by ownership change. In addition, the common law expected buyers to protect themselves, a concept expressed in the maxim *Caveat emptor*, "Let the buyer beware." Even if these barriers were surmounted, the homeowner discovering these defects could only, as a rule, pursue the party who had sold her the house. Recovery could not be obtained from those more responsible, such as the builder, the designer, or the developer who had not sold the house to the person discovering the defect. In a fast turnover market, such requirements of privity often left the discoverer of the defect with a claim only against the person who had sold the discoverer the house.

B. The Implied Warranty Explosion of the 1960s

Dissatisfaction with the traditional denial of protection meshed with the general consumer movement that began in the 1960s. Consumers of goods began to demand products that worked, and home buyers demanded similar protection. In response to this demand, courts recognized an implied warranty of habitability in the sale of new homes.[76] Some courts have gone beyond implied warranty and have held

---

[73]*First Nat'l Bank of Akron v. Cann*, 669 F.2d 415 (6th Cir.1982).

[74]*Hillcrest Country Club v. N. D. Judd Co.*, 236 Neb. 233, 461 N.W.2d 55 (1990).

[75]3 Architect's Handbook of Professional Practice, A201 Commentary p. 69 (1987). But *Gerety v. Poitras*, 126 Vt. 153, 224 A.2d 919 (1966) refused to order specific performance of a guarantee clause.

[76]One of the most interesting of the many opinions is *Humber v. Morton*, 426 S.W.2d 554 (Tex.1968). In *Centex Homes v. Buecher*, 95 S.W.3d 266 (Tex.2002), rehearing denied Feb. 20, 2003, a divided Texas Supreme Court reaffirmed *Humber* and also carefully distinguished between the warranty of habitability and the implied warranty of workmanlike conduct.

that a builder of mass-produced homes can be held strictly liable for injuries caused by the conditions in the home.[77]

C. Current Problems

The breakthrough consisted of overthrowing the old rules that had denied any recovery. With the exception of a few jurisdictions,[78] most modern courts have recognized the implied warranty of habitability. But the courts and legislatures must still work out details. Courts have articulated similar but somewhat variant formulas for describing the nature of the implied warranty.[79]

The exact nature of the warranty will depend on the price of the house, the customary standards of the community, and the reasonable expectations of the buyer.

For whose protection does the implied warranty exist, and who is subject to its requirements? Clearly, a buyer can sue the seller. Beyond that, the law varies in the different jurisdictions. Some courts allow a subsequent purchaser to maintain an action,[80] whereas some do not.[81] One case held that the infant son of a tenant could bring an action against the developer.[82]

As to other potential defendants, successful actions have been brought against a lender who took more than a normal lender's role in creating the development[83] and an engineer who conducted soil studies.[84]

Courts seem hesitant to imply a warranty in the sale of an existing home not built for immediate sale.[85] Here it is more difficult to justify warranty implication as part of the normal enterprise risk of developers or builder-vendors. The *caveat emptor* doctrine, with the modern qualification that sellers disclose serious defects of which they know and that the buyer would not likely be able to discover, seems proper.

Most plaintiffs seek damages based either on the cost of correcting the defect or on the diminished value of the property.[86] A host of structural defects and incurable water problems should enable the buyer to call the deal off and receive any money paid less any benefit received by occupying the premises. Like cars, houses can be lemons.

Can the seller disclaim the implied warranty by a contract clause? Express contract language usually takes precedence over implied warranties. It is likely that exculpatory clauses will not be effective where the plaintiff has suffered personal harm. Exculpation may be effective where there is damage to property or other economic loss if the exculpatory language was brought to the buyer's attention in such a way as to clearly indicate that the buyer assumed this risk.[87] Marketing methods of most developers and vendor-builders are not likely to employ this open approach.

---

[77]*Schipper v. Leavitt & Sons, Inc.*, 44 N.J. 70, 207 A.2d 314 (1965). See Comment, 10 Ohio N.U.L.Rev. 103 (1983). But in *Calloway v. City of Reno*, 116 Nev. 250, 993 P.2d 1259 (2000), the court rejected application of strict products liability law to development housing.

[78]*Bruce Farms, Inc. v. Coupe*, 219 Va. 287, 247 S.E.2d 400 (1978). This decision effectively has been overruled by Virginia's adoption of statutory warranties; see infra note 89.

[79]*Hartley v. Ballou*, 20 N.C.App. 493, 201 S.E.2d 712 (1974) (suitable for habitation); *Padula v. Deb-Cin Homes, Inc.*, 111 R.I. 29, 298 A.2d 529 (1973) (fit for human habitation); *Gable v. Silver*, 258 So.2d 11 (Fla.Dist.Ct.App.1972) (fit and merchantable); *Crawley v. Terhune*, 437 S.W.2d 743 (Ky.1969) (major structural features constructed in a workmanlike manner with suitable materials); *Albrecht v. Clifford*, 436 Mass. 706, 767 N.E.2d 42 (2002) (implied warranty of habitability and safety); but see *Stuart v. Coldwell Banker Commercial Group, Inc.*, 109 Wash.2d 406, 745 P.2d 1284 (1987) (warranty of habitability does not extend to exterior, nonstructural defect). See generally, Annot., 25 A.L.R.3d 383 (1969). See Conn.Gen.Stat.Ann. § 47-121 (built in accordance with building codes). Implied warranties are discussed in detail in Note, 42 S.C.L.Rev. 503, 514–524 (1991).

[80]*Wells v. Clowers Constr. Co.*, 476 So.2d 105 (Ala.1985); *Kriegler v. Eichler Homes, Inc.*, 269 Cal.App.2d 224, 74 Cal.Rptr. 749 (1969); *Speight v. Walters Development Co., Ltd.*, 744 N.W.2d 108 (Iowa 2008).

[81]*Redarowicz v. Ohlendorf*, 92 Ill.2d 171, 441 N.E.2d 324 (1982); *Barnes v. MacBrown & Co.*, 264 Ind. 227, 342 N.E.2d 619 (1976); *Evans v. Mitchell*, 77 N.C.App. 598, 335 S.E.2d 758 (1985), review denied, 316 N.C. 376, 342 S.E.2d 89 (1986). See Note, 35 Baylor L.Rev. 670 (1983).

[82]*Schipper v. Leavitt & Sons, Inc.*, supra note 77.

[83]*Connor v. Great Western Sav. & Loan Assn.*, 69 Cal.2d 850, 447 P.2d 609, 73 Cal.Rptr. 369 (1968). But see West Ann.Cal.Civ.Code § 3434, which states that construction lender, acting only in the capacity of a lender, is not liable to third persons for construction defects. See Ferguson, *Lender's Liability for Construction Defects*, 11 Real Est.L.J. 310 (1983). Lender liability is also discussed in Note, S.C.L.Rev. 503, 510–512, 519–520 (1991).

[84]*Avner v. Longridge Estates*, 272 Cal.App.2d 607, 77 Cal.Rptr. 633 (1969).

[85]*Pollard v. Saxe & Yolles Dev. Co.*, 12 Cal.3d 374, 525 P.2d 88, 91, 115 Cal.Rptr. 648, 651 (1974).

[86]These measures of recovery are discussed in more detail in Section 27.03.

[87]Compare *Albrecht v. Clifford*, supra note 79 (warranty may not be waived); *Centex Homes v. Buecher*, supra note 76 (warranty may be waived only as to known defects); *Board of Managers of the Village Centre Condominium Ass'n, Inc. v. Wilmette Partners*, 198 Ill.2d 132, 760 N.E.2d 976 (2001) (waiver only if warranty of habitability referred to by name in the disclaimer).

While the implied warranty of habitability has its origin in the common law, state legislatures increasingly are addressing the issue of defective residential construction.[88] The relationship between the judicial doctrine and the statutory schemes will have to be worked out.

D. Insurance Protection

The preceding discussion may have overemphasized the value of the implied warranty to buyers. Although the 1960s did see an explosion of protection for home buyers, the law in any given jurisdiction may still be unclear, especially as to the particular nature of the warranty. Often the defects complained of are not large, and instituting individual legal action to enforce the warranty may cost more than what can be recovered in court. This is a reason for increased use of the class action in which a number of buyers "similarly situated" join together in one lawsuit. In any event, there are often long delays in litigating these cases.

Developers and vendor-builders are engaged in a hazardous business and often are either out of business or unable to pay for damages by the time a claim is brought. It is also more difficult for builders to obtain liability insurance coverage because of the increasing prevalence of homeowners suits.[89] As a result, private industry has been offering homeowner's warranties to cover the types of defects discussed in this section.

E. Deceptive Practices Statutes

For a review of statutes governing deceptive practices, see Section 23.02J.

SECTION 24.11 A Suggestion: Defect Response Agreements

This section is directed to owners who are about to engage in projects of a substantial nature, particularly those with the likelihood of defects developing. It is an attempt to outline a skeletal proposal for dealing with defects.

---

[88]See Section 23.02K.
[89]*Wall Street J.*, Feb. 27, 2002, p. B-7 (builders from Colorado to California having difficulty obtaining liability insurance).

Dealing with defects involves three stages. Stage I, the *diagnostic* stage, involves correcting the defect. Stage II is the *preparation* to resolve the allocation of legal responsibility for the losses caused by the defect. Stage III, the actual trial, involves *judicial* resolution of legal responsibility.

Although Stage III is the most dramatic (and costly), Stages I and II are the most crucial. The defect must be fixed for the owner to receive the project for which it has paid and to avoid large losses. Stage II, the gathering of information that can enable the parties to try to voluntarily settle a dispute, involves a large expenditure of time and money before the parties are in a position to settle the dispute. Stage III is rarely reached. Regardless of how expensive Stages I and II are, the expenses of Stage III with a trial of from one to six months add immense direct and indirect costs and often generate unsatisfactory outcomes. As a result, most disputes on large projects never go beyond Stage II, with the dispute being settled out of court.

Before the parties are seriously in the position to discuss settlement, they must have some idea of what caused the defect, who is legally responsible, and how the losses should be apportioned. To prepare for serious negotiations, much expense must be incurred, such as the expenses of lawyers, experts, and testers, as well as the often ignored indirect expense of officers and key employees of the major participants having to spend time that does not earn their employers any revenue. In disputes of this nature, the stakes are sufficiently high so that no participant feels that it can "cut costs."

When all the major participants in the dispute—and there are many—have sufficient pressure to seriously discuss negotiations, it is likely that there is enough blame to spread around, enough technical uncertainty and disagreements among experts, and uncertainty as to the legal outcome. As a result, the major participants—the owner, contractors, design professionals, major suppliers, manufacturers, sureties, insurers, and funding agencies, to name the most visible—each decides it will pitch some money into a settlement "pot."

How do participants determine how much they are willing to pitch into the pot? It will very likely result from hard bargaining with all the variable factors that make negotiation an art, such as willingness or reluctance to continue the fight, financial strength and weakness, appraisal of the strength and weakness of each participant's position, and a prediction of how the matter will be resolved, whether decided by arbitration or litigation.

All of this takes place after a large expenditure of time and money needed to amass the information that each party feels it needs before it can enter into meaningful negotiations. Is there any way that this can be avoided?

One method is to simulate in a rough way the type of negotiations that result at Stage II but to do it *before* the project begins. This can be accomplished by a Defect Response Agreement (DRA) made among *all* the major participants. To accomplish this, the major participants enter into the negotiation as they "sign on" the project, all under the direction and supervision of the owner. The DRA deals solely with the response to defects, how to correct the defects, and who is responsible for the "immediate" response as well as with the formula for determining the ultimate loss distribution and possibly the security to back up any such formula.

Because all the major parties are signatories, there would be none of the increasingly complex third-party problems. The formula itself would be determined by negotiation, influenced very likely by the same matters that affect any negotiation—a calculation of risk and profits. Fear of the open-ended consequential damages would lead to each participant's giving up claims it would have for such damages.

The DRA would have to be imposed by a strong private owner[90] that recognizes that *it* bears the large burden of the wasteful costs incurred in the present system. It can "persuade" other participants that they will benefit in the long run by a system that will avoid the worst aspect of the current dispute resolution process.[91]

---

[90]Public owners would be too fearful of accountability requirements to try such a no-fault approach.

[91]Another suggestion is to use a property damage policy that would cover defects, however caused, with *all* participants waiving subrogation claims and consequential damages. Defects would then be an insurance problem.

CHAPTER TWENTY-FIVE

Subsurface Problems: Predictable Uncertainty

SECTION 25.01 Discovery of Unforeseen Conditions

A. Effect on Performance

A contractor during performance may discover unexpected subsurface conditions. On renovation projects, these subsurface conditions exist within the building. For example, removal of drywall may reveal the presence of asbestos. On new projects, the contractor may encounter subsurface *soil* conditions that differ from what was anticipated. While analysis of these two types of subsurface conditions sometimes involve the same legal analysis (for example, the federal differing site conditions clause discussed in Section 25.06 applies equally to both types of conditions), this chapter focuses on subsurface soil conditions.

For reasons explored in Section 25.01B, the discovery of unforeseen subsurface conditions is not unusual in the construction process. When such conditions are discovered, they usually have an adverse effect on the contractor's planned performance and prediction of performance costs.

Sometimes subsurface materials encountered are more difficult to excavate or extract, with many cases involving the discovery of hardpan or more rock than expected. When this occurs, performance is likely to take more time and money. Sometimes the subsurface conditions encountered generate a great increase or decrease in the quantities to be excavated. This also can affect the time and costs, particularly if the contractor has bid a composite unit price that may be adversely affected by finding that certain work runs over and other work runs under the bid. (This was discussed in Section 17.02D, dealing with unit pricing.) Materials (borrow pits) that the contractor expects to use for fill or compaction may turn out to

be unsuitable. If so, the contractor must obtain materials from a site more distant than planned or more costly to extract. When these conditions are discovered, the contractor may request more time and more money.

B. Causes

Why are unexpected subsurface conditions frequently encountered? Soil testing is expensive, and usually the number of borings is limited.[1] As a result, the data reported may not reflect subsurface conditions throughout the entire area. Subsurface conditions even within small areas may vary greatly.

As part of the design preparation process, the owner hires a geotechnical engineer to perform a geotechnical soil report. The engineer determines the load-bearing capacity of soil and rock, and the plasticity (or expansion potential) of different soils or clays. The geotechnical engineer may use this information to recommend the design of the building's foundation.

Geotechnical engineers may be exposed to great liability if the information they report is incorrect or their suggestions do not accomplish the owner's expectations. Although this chapter concentrates mainly on subsurface

---

[1] A soil boring is a three-inch diameter metal tube drilled into the soil to a specified depth. The soil content of the core is examined to determine the nature of the subsurface soil. As one court observed, "[t]he narrowness of a soil boring limits the information that can be inferred from it because it is incapable of giving information of the existence of anything larger than three inches in diameter. However, soil borings can indicate cobbles, and the presence of hard materials can be detected by the resistance encountered when the drilling is pushed through the ground." *Travelers Casualty and Surety Co. of Am. v. United States*, 75 Fed.Cl. 696, 700 n. 6 (2007).

conditions encountered during construction, problems such as structural instability, settling, or cracking can develop later. They are often traceable to the subsurface conditions, often leading to claims against geotechnical engineers. As a result, geotechnical engineers may not perform services where liability exposure greatly exceeds anticipated profit, or may sharply increase their fees. They may also use "limitation of liability" clauses to limit their liability to the owner and require indemnification of any third-party liability.[2] Refusal to perform geotechnical services, etc., can eliminate testing in smaller projects or make less extensive testing more likely.

Information, however reliable, must be gathered and used by designers and contractors. When real conditions are encountered that vary from what was anticipated, the risk must be borne by someone. This chapter examines contractor claims for additional compensation when contractors' costs are more than expected because of unforeseen subsurface conditions.

C. Two Models

It is important to recognize the different methods by which people organize construction. First, consider construction done by a contractor who is given a site and asked both to design and to build a particular project. The case of *Stees v. Leonard,*[3] cited many times for the proposition that the contractor bears all risks of subsurface conditions, involved a contract under which the contractor both designed and built a house for the owner. As discussed in Section 24.02, in such cases, it is easy to place the risk of success on the contractor.

The traditional American construction process—design, bid, build (DBB)—divides design and construction, with the owner engaging a design professional to design, a geotechnical engineer to gather data, and a contractor to build. This organizational structure makes it more difficult to determine who will bear the risk of unforeseen subsurface conditions.

Although there are abundant factual variations, typically the owner—private or public—calls for bids on a construction project that will involve subsurface excavation. Building specifications may be furnished, but methods of excavation and construction will probably be left to the contractor.

Depending on owner identity and project size, an independent geotechnical engineer makes soil tests for cost estimation, design, and scheduling. The reports are made available to the design professional and the owner. Although the information is likely to be available to the contractor, owners take different approaches as to whether they will take responsibility for the accuracy of the information given to the contractor.

Whichever approach is taken, the bidder will usually be warned that it must inspect the site and, under some contracts, conduct its own soil testing. However, often the bidder will not make independent soil tests because the profit potential may be too small to justify such expenditure or because the bids are due in a relatively short period of time. As a result, contractors frequently bid without knowledge of actual conditions that will be encountered. Bids are calculated on the basis of expected conditions, and if the unexpected is encountered, the actual cost will vary widely from that anticipated. This chapter focuses on who will bear this risk.[4]

D. Enforceability of a Promise to Pay More Money

When unforeseen subsurface conditions are discovered, the owner may promise to pay additional compensation. The enforceability of such a promise depends on the application of the preexisting duty rule. The traditional application of this rule would deny enforceability of such a promise because the owner is getting for its promise nothing more than it was entitled to get under the contract.[5] However, special rules have developed in subsurface cases that can in some jurisdictions permit enforcement of the promise. Sometimes the promise is enforced because the contractor gave up its right to rescind the contract for a mutual

---

[2]*Lanier at McEver, L.P. v. Planners and Engineers Collaborative, Inc.,* 285 Ga.App. 411, 646 S.E.2d 505 (2007) and *Fort Knox Self Storage, Inc. v. Western Technologies, Inc.,* 140 N.M. 233, 142 P.3d 1 (App.2006). Sample limitation of liability clauses are listed in EJCDC Doc. E-500, Exhibit I (2002). See Section 15.03D.

[3]20 Minn. 494 (1874), cited in *Dravo Corp. v. Litton Systems, Inc.,* 379 F.Supp. 37 (S.D.Miss.1974).

[4]For a comparative study, see Wiegand, *Allocation of the Soil Risk in Construction Contracts: A Legal Comparison,* [1989] Int'l. Constr. L. Rev. 282.

[5]For a case indicating that the traditional rule still retains some of its vigor, see *Crookham & Vessels, Inc. v. Larry Moyer Trucking, Inc.,* 16 Ark. App. 214, 699 S.W.2d 414 (1985). For a collection of cases, see Annot., 85 A.L.R.3d 259, 315–36 (1978).

mistake as consideration for the promise.[6] More commonly, the law enforces such a promise so long as "the modification is fair and equitable in view of circumstances not anticipated by the parties when the contract was made."[7]

E. Supervening Geotechnical Conditions and Mistake Claims

Most problems involve encountering unexpected *preexisting* subsurface conditions. Additional costs can result from a change in geological conditions because of unexpected weather (unseasonable rains or frost) or third-party interference (flooding from adjacent lands). A distinction can be drawn between the types of risks involved. If the parties want to spend enough time and money, the risk of a preexisting geological condition can be eliminated. However, the likelihood of a future occurrence, whether from third-party conduct or atmospheric conditions, can rarely be predicted with any accuracy.

Supervening changes of geological conditions are not discussed in this chapter. Usually when such events occur, the contractor seeks a time extension but not additional compensation. The differing site conditions clause used by the federal government and discussed in Section 25.06A generally does not apply to atmospheric difficulties or third-party interference.[8]

If contracting parties share a fundamental mistake as to material assumptions at the time the contract is made, the parties may be discharged from any further obligation to perform because of mutual mistake unless the risk has been allocated by contract or by custom to the party whose performance has been adversely affected. For example, if both parties hold fundamental assumptions as to particular subsurface characteristics, the discovery of substantially different subsurface characteristics could entitle the party adversely affected, usually the contractor, to terminate its obligation under the contract and be paid for what it has performed, based on restitution.

Most subsurface disputes involve claims by the contractor for additional compensation, not for the right to terminate its obligation. As a rule, these claims rely on misrepresentation, as discussed in Section 25.03, or on contractual provisions, as discussed in Section 25.06. As a result, mutual mistake does not play a significant role in this chapter.

Mistake can be relevant in two instances. As noted in Section 25.01D, mistaken assumptions may be part of the basis for enforcing a subsequent promise to pay additional compensation. Also, the differing site conditions method under which the contractor receives additional compensation as discussed in Section 25.06 can be substituted for any common law right to terminate because of mutual mistake (discussed in that section).

SECTION 25.02 Common Law Rule

The traditional common law rule, universally followed up to the beginning of the twentieth century, placed the risk of project failure caused by defective soil squarely upon the contractor. The owner had a duty not to intentionally deceive the contractor but had no duty to disclose to the contractor soil conditions that the owner knew were relevant to the contractor. Rather, the risks of contract performance—including soil conditions and design adequacy—belonged to the contractor.[9]

In 1918, the Supreme Court announced the modern rule in *United States v. Spearin*,[10] reproduced in Section 23.05E. Under the *Spearin* doctrine, the owner, not the contractor, assumes the risk of defects in design specifications that prevent the contractor from building an acceptable project. Because a building must necessarily be built on a particular plot of land, the owner's implied warranty of design includes the risk that the design failed to accommodate the soil conditions. As explained by one court:

> Plaintiff [owner] contends here that the court below failed to distinguish between defects inherent in the plans and specifications and defects extrinsic to such specifications, *such as a latent defect in the soil.* This argument is untenable, since plans and specifications do not exist in a vacuum; they are made for a particular building at a particular place.

---

[6]*Healy v. Brewster,* 251 Cal.App.2d 541, 59 Cal.Rptr. 752 (1967).

[7]*Angel v. Murray,* 113 R.I. 482, 322 A.2d 630, 636 (1974), quoting what is now Restatement (Second) of Contracts § 89A. See also *Linz v. Schuck,* 106 Md. 220, 67 A. 286 (1907).

[8]*Hardeman-Monier-Hutcherson v. United States,* 198 Ct.Cl. 472, 458 F.2d 1364 (1972). It can apply if representations are made to the contractor regarding sea or climate conditions.

[9]*Dermott v. Jones,* 69 U.S. 1 (1864); *Thorn v. Mayor etc. of London,* 1 Appeal Cases 120 (H.L.1876); *Ford & Denning v. Shepard Co.,* 36 R.I. 497, 90 A. 805 (1914) (owner who did not disclose the presence of quicksand was not liable to the contractor).

[10]248 U.S. 132 (1918).

The defect in the plans and specifications for the building in question was the failure to make provision for adequate pilings and other support for the floor; the fact that these plans and specifications might provide for an adequate building in some other place does not render the plans and specifications less defective for the location in question. (Emphasis added.)[11]

The owner's implied warranty as to the adequacy of the design specifications is counterbalanced by its right to rely upon the contractor's expertise in building. As discussed in Section 23.02I, a contractor has an implied duty to perform in a workmanlike manner. Courts include within the contractor's implied warranty of workmanlike conduct a duty to warn the owner of adverse subsurface conditions known or reasonably known to the contractor.

As an illustration, the Mississippi Supreme Court held a builder responsible for severe cracking because the builder was found to have been negligent in not warning the buyer of potential soil problems and not recommending that soil tests be made. The court held in favor of a subsequent purchaser, based on the builder's duty to use its technical knowledge to build a house on yazoo clay in a way that would protect users from the house settling. The dissenting judge would have exonerated the builder because he concluded that no such duty was imposed on the builder and because the builder had built according to plans and specifications required by the Veterans Administration.[12]

As a general proposition, then, the contractor will bear the risk of unforeseen subsurface conditions unless (1) it can establish that it has relied on information furnished by the owner (discussed in Sections 25.03 and 25.05), (2) the contract itself provided protection (discussed in Section 25.06), (3) the owner did not disclose information it should have disclosed,[13] or (4) the cost of performance was extraordinarily higher than could have been anticipated, a fact that would justify applying the doctrine of mutual mistake or impossibility, as discussed in Sections 23.03C and D.

The number of exceptions to the basic rule should not convey the impression that the policy of placing these risks on the performing party no longer exists or does not reflect, at least to some courts, an important legal principle in fixed-price contracts. For example, in *W. H. Lyman Construction Co. v. Village of Gurnee*,[14] a contractor had been engaged to perform a sanitary sewer project. One basis for a claim against the public entity was that the sewer had to be constructed through subsurface soil that was for the most part water-bearing sand and silt rather than clay, as indicated by the soil-boring log shown on the plans. A high groundwater table was also discovered. This required the contractor to install numerous dewatering wells.

It is important to note the judicial attitude in the *Lyman* case, which dealt with the claim by the contractor that the public entity impliedly warranted that the plans and specifications would enable it to accomplish its promised performance in the manner anticipated. In rejecting the contractor's claim, the court stated "It is well settled that a contractor cannot claim it is entitled to additional compensation simply because the task it has undertaken turns out to be more difficult due to weather conditions, the subsidence of the soil, etc. To find otherwise would be contrary to public policy and detrimental to the public interest."[15]

The court looked on the common law rule as expressing a principle of great importance, one needed to protect public entities and public funds. It is not clear that the court would have felt as strongly as it did were the contract a private one. Yet private owners and those who supply funds for the project are also greatly concerned with the ultimate cost of the project and rely heavily on the contract price in their planning.

Different rules apply to developers and subdividers. Developers both own the land and build the structures. As discussed in Section 24.10, the developer of a residential development may be liable to the homeowners under the implied warranty of habitability or fitness or under a theory of strict products liability, if the homes are damaged by defective soils.

A subdivider prepares empty land for construction, but does not do the construction work itself. The subdivider grades the land and subdivides it into lots in preparation for a home development. The subdivider then sells these

---

[11]*Ridley Investment Co. v. Croll*, 56 Del. 209, 192 A.2d 925, 926–27 (1963). See also *Gaybis v. Palm*, 201 Md. 78, 93 A.2d 269, 272 (1952).

[12]*George B. Gilmore Co. v. Garrett*, 582 So.2d 387 (Miss.1991). Accord, *Lewis v. Anchorage Asphalt Paving Co.*, 535 P.2d 1188, 1199 (Alaska 1975); *Farmer v. Rickard*, 150 P.3d 1185 (Wyo.2007) (contractor's duty to warn extends only to latent soil conditions); Annot., 73 A.L.R.3d 1213 (1976).

[13]See *P. T. & L. Constr. Co., Inc. v. State of New Jersey Dep't of Transp.*, 108 N.J. 559, 531 A.2d 330 (1987).

[14]84 Ill.App.3d 28, 403 N.E.2d 1325 (1980).

[15]403 N.E.2d at 1328.

lots to contractors who build the houses. If the houses later fail because of defective soils, may either the homeowners or contractors shift responsibility to the subdivider? A subdivider has a duty to disclose to the contractors defects in the soil it knew or should have known through the exercise of reasonable care. Disclosure of these defects to the contractors will shield the subdivider from liability to either the contractors or home purchasers.[16]

SECTION 25.03   Information Furnished by Owner

Owners who intend to build substantial projects often make subsurface and soil reports available to prospective contractors. One exception to the general allocation of unexpected costs to the contractor relates to the owner furnishing information relied on by the contractor. The owner may make this information available but may disclaim responsibility for its accuracy. This section emphasizes the effect of providing this information with the full realization that the contractor is likely to rely on it and without any technique to shield the owner from responsibility.

At the outset, differentiation must be made among types of information that may be given to the contractor. Reports of any tests that have been taken are usually included in such information. The reports may also contain opinions or inferences that the geotechnical engineer may have drawn from observation and tests taken. The information may include estimates as to the type and amount of material to be excavated or needed for fill or compaction. (This information may also be included in the specifications.) Although less common, the reports may recommend particular subsurface operational techniques.

Misrepresentation is the basic theory on which the contractor bases its claim. A threshold question involves what constitutes a misrepresentation. Facts and opinions must be differentiated. Reporting the result of tests is clearly a factual representation, whereas professional judgments that seek to draw inferences from this information may be simply opinions.

A misrepresentation claim may be based on improperly selected test sites, the inference being that a contractor may believe that the test sites selected will generally represent the site. Misrepresentation can consist of a combination of providing some information but not disclosing all the information that qualifies the information given. Half-truths can be just as misleading as complete falsehoods. Misrepresentation is distinguished from simply failing to disclose any information or information that the owner knows would be valuable to the contractor and that the contractor is not likely to be able to discover for itself.

Representations may be fraudulent, that is, made with the intention of deceiving the contractor. Fraud claims, though having a heavy burden of proof, are most valuable to a contractor. The contractor's negligence in failing to check the data generally does not bar recovery.[17] Recovery would be barred only if the contractor relied on information it gathered. Fraud may extend the time a contractor has for bringing suit. Under the "discovery" rule for determining when a cause of action accrues, a distinction exists between when the contractor knew that the subsurface conditions differed from what it expected (a claim grounded in misrepresentation) and the contractor's knowledge that the owner or architect fraudulently misrepresented to it what the soil conditions were.[18] Fraud also offers the contractor a variety of remedies. The contractor can rescind the contract and refuse to perform further, raise fraud as a defense if sued for nonperformance, or—most important—complete the contract and recover additional compensation in a claim for damages.

More commonly, though, misrepresentations are not made with the intention of deceiving the contractor. If the misrepresentations were made negligently, some added complexities develop. Very likely the negligence is that of the geotechnical engineer, and owner recovery against the geotechnical engineer may not always be available.[19]

---

[16]Stepanov v. Gavrilovich, 594 P.2d 30, 35 (Alaska 1979) (subdivider is not liable for failure to disclose the existence of permafrost, where the geological testing did not reveal that danger); Anderson v. Bauer, 681 P.2d 1316, 1322–23 (Wyo.1984) (subdivider warned contractors of the presence of a water table and so was not liable to homeowners); and Smith v. Frandsen, 94 P.3d 919 (Utah 2004) (subdivider not liable to homeowners, because the contractors had an independent duty to apprise themselves of the soil conditions).

[17]Seeger v. Odell, 18 Cal.2d 409, 115 P.2d 977 (1941).

[18]Trafalgar House Constr., Inc. v. ZMM, Inc., 211 W.Va. 578, 567 S.E.2d 294 (2002). The discovery rule is discussed in Section 23.03G.

[19]See Section 14.05. Compare Texas Tunneling Co. v. City of Chattanooga, 329 F.2d 402 (6th Cir.1964) (claim denied), with M. Miller Co. v. Central Contra Costa Sanitary Dist., 198 Cal.App.2d 305, 18 Cal. Rptr. 13 (1961) (claim allowed). The latter has been followed in the Restatement (Second) of Torts, § 525 (1977).

Clearly, if the negligence had been that of the owner, the contractor can recover any losses it suffers caused by the negligence that it could not have reasonably avoided.

What is the responsibility of the owner for any negligent representations in the soil information generated by the geotechnical engineer? If the geotechnical engineer is an independent contractor—which is likely to be the case—the owner would not be chargeable with the latter's negligence. However, very likely the owner will be chargeable either for breach of contract (that the breach is caused by a person whom it engages to perform services does not relieve it of responsibility),[20] or implied warranty of the accuracy of the information supplied by the owner that is relied on by the contractor.[21]

The least culpable conduct is that of innocent misrepresentation. In such cases there is neither intention to deceive nor negligence. Innocent misrepresentation generally allows the contracting party who was misled to rescind the contract. Some states allow a restitutionary damage remedy—the difference between what was received and what was paid out[22]—an ineffective remedy in subsurface cases.

Actual rescission in these subsurface cases is rare. Many legal and factual issues may make it difficult to predict whether the right to rescind is available. Walking off the job under these conditions exposes the contractor to liability. Even if its rescission were determined to be justified, the contractor can recover only the reasonable value of its services. This recovery formula should, at least in theory, put the contractor in the position it was in when it made the contract.

It may be difficult to recover lost overhead and profit using this restitutionary remedy.[23] More likely the contractor will continue performance and later claim additional compensation. Because innocent misrepresentation is not a contract breach, however, contractors will shift to implied warranty to justify their claims for additional compensation. The principal problem with the implied warranty theory is the owner's attempt to exonerate itself by using contract language negating the necessary reliance.[24]

---

[20]*Harold A. Newman Co. v. Nero*, 31 Cal.App.3d 490, 107 Cal.Rptr. 464 (1973); *Brooks v. Hayes*, 133 Wis.2d 228, 395 N.W.2d 167 (1986).
[21]See Section 23.05E.
[22]See Section 7.07.
[23]See Section 27.02E.
[24]See Section 25.05.

SECTION 25.04 Risk Allocation Plans

One of the principal planning decisions that those who prepare construction contracts must make relates to subsurface problems, broadly defined. As noted in Section 25.02, in the absence of any contractual risk assumption system incorporated into the contract, the contractor will bear the risk of additional expenses attributable to having to perform work under subsurface conditions that differ from that anticipated. Thus the contractor must bear the cost of any additional expenses and cannot be relieved from any further obligation to perform.

The exceptions to this basic risk allocation were noted in Section 25.02. But those who plan risk allocation for such contracts must keep in mind that generally the law will permit the contract to distribute the risks in any way chosen by the parties. That it is usually the owner who makes this determination, and that the only choice the contractor has is to enter into the contract or not, rarely affects the law's respect for the contract provisions.

Assuming that the owner chooses the risk distribution, the owner has essentially three options:

1. Gather no information, and let the contractor proceed based on its own evaluation.
2. Gather the necessary information, make it available to the contractor, but disclaim any responsibility for its accuracy (disclaimer system).
3. Gather the information, make it available to the contractor, and promise the contractor to equitably adjust the contract price if the actual conditions turn out to differ from those represented or anticipated (the differing site conditions (DSC) system).

The preceding techniques are discussed in Sections 25.05 and 25.06. In those sections, however, the assumptions are that a particular risk distribution choice has been made and how the law deals with that choice. In this section, the emphasis is on the advantages and disadvantages of the three systems.

Emphasis in this section will be not on the first system but on the second and the third. In construction projects of any size, gathering subsurface information is crucial for design choices, and the owner rarely just allows the contractor to deal with the problem. Contractors are not always geared to gather the information and make these choices, and any attempt to use the first system runs a great risk of construction failure.

With the important exceptions noted in Section 25.06, most private and many state and local public owners make information available to the bidders but use contractual disclaimers in an attempt to relieve themselves from any responsibility for the information's accuracy.

Those who adopt the disclaimer system recognize the possibility or even the likelihood that the contractor will encounter physical conditions different from those represented or anticipated. They expect the contractor to calculate the risks and include contingencies in the bid price, which takes this risk into account. Those who prefer this system also think that if the contractor will have to pay for the added expenses for corrective work, this system will encourage contractors to more carefully evaluate the information and inspect sites. Advocates of this system want the pricing of such uncertainty to go into the contract at the front end—rather than at the back end, through claims. To sum up, the principal justification for the disclaimer system is the need many public entities and many private owners have to know at the outset what the project will cost.

But the disclaimer system has drawbacks. First, as shall be seen in Section 25.05, the disclaimer does not always work. A study by the American Society of Civil Engineers concluded that despite owners seeking to place these risks on the contractor through disclaimer language, contractors will make claims that in the end the owner will compensate. The claims will be based on other contract clauses or different legal theories.[25]

Second, if the disclaimer system does not protect the geotechnical engineer who furnishes the information, the contractor may be able to make a successful claim against the geotechnical engineer based on negligence.[26] This will frustrate the risk allocation system by giving the contractor a windfall if it has priced the contingency in its contract price and can still recover against the engineer. This uncertainty may also force the engineer to protect himself by demanding higher compensation or indemnification.

Third, the ruthlessly competitive construction market may mean that contractors do not include contingencies for subsurface conditions in their bid prices. Although this may appear beneficial to the owner, the contractor who loses money is likely to make a claim and may win it; in any event, all the parties will suffer extensive claims overhead.

Despite these undoubted disadvantages, the need for certainty, particularly when budgets are fixed and tight, motivates many owners to seek to use the disclaimer system.

The third system—the differing site conditions (DSC) method—assures the contractor that it can bid on what it believes it is likely to encounter and that it will receive an equitable adjustment if actual conditions do not turn out that way. This system is intended to generate lower bid prices because the contractor need not attempt the often difficult task of providing a contingency in its pricing for the subsurface uncertainties and need not incur extensive costs connected with its own testing if it is not certain that the information furnished by the owner under the disclaimer system is accurate.[27]

Obviously, the DSC method has disadvantages. The most important is the uncertainty of the ultimate contract price. Yet over the long run—and this point is made by many owners who are repeat players—the prices will be lower. Also, the cost of administering the DSC system can be formidable, and a contractor who is not convinced it will be treated fairly may include a contingency price anyway. Determining the amount of the equitable adjustment can generate difficult problems and expensive claims overhead.

Another argument sometimes made against the DSC system is the uncertainty as to whether it actually induces lower bids. Some contend that bids are based on workload, the desirability of keeping the workforce together, and the prospects of a well-administered construction project. The presence of the DSC system may be a relatively insignificant item in determining the actual bid price. Also, the bid price may not be reduced unless the contractor is confident that obtaining an equitable adjustment will not be extraordinarily difficult or administratively expensive.

Finally, another factor in favor of using the DSC system is that it should provide an incentive for the owner to furnish the best subsurface information by hiring competent soil testers and giving them enough time and compensation to develop accurate information.

---

[25]*Managing Unforeseen Site Conditions*, 113 J.of Constr.Eng'g & Mgmt, No. 2, June 1987, quoted by Jones, *The U.S. Perspective on Procedures for Subsurface Ground Conditions Claims*, [1990] Int'l Constr. L. Rev. 169.

[26]See Stein & Popovsky, *Design Professional Liability for Differing Site Conditions and the Risk-Sharing Philosophy*, 20 Constr. Lawyer, No. 2, April 2000, p. 13.

[27]See *Foster Constr. C. A. & Williams Bros. Co. v. United States*, 193 Ct.Cl. 587, 435 F.2d 873, 887 (1970), for an articulation for the reasons for a differing site conditions clause.

Clearly, no system is perfect, and whoever must plan the risk allocation faces difficult choices. There is an increased tendency, as demonstrated in Section 25.06, to use the DSC system. But in the greater number of construction projects, the disclaimer system will likely still be employed.

SECTION 25.05 Disclaimers— Putting Risk on Contractor

Owners use a variety of techniques to relieve themselves of responsibility for the accuracy of subsurface information they have obtained and made available. Although emphasis is on making the information available and then seeking to disclaim responsibility for it, there is the possibility that the owner will choose simply not to make this information available. Owners may seek to do so if they believe that no sure technique can relieve them of the responsibility for the information's accuracy. Owners are increasingly expected to disclose information that would be valuable to the contractor if the contractor would probably not be able to discover this information for itself. Yet an opinion of the Pennsylvania Commonwealth Court[28] held that the owner's failure to include information on subsurface conditions was not constructive fraud. Even more, an opinion of the Post Office Board of Contract Appeals held that the public entity was not required to make a subsurface investigation if it did not want to do so.[29]

In these cases, other factors made it difficult for the contractor to recover. For example, in the first case, the court noted that the contractor was experienced and had access to public documents describing the mine water levels in the project area. In the second case, the court noted that the contractor was aware of the existence of subsurface rock, that before bidding the contractor anticipated that some rock would be encountered in excavation, and that the amount of rock actually encountered was not unusual for the area. These factors might have been sufficient to enforce any disclaimer if the information given and responsibility for it were disclaimed.

More commonly, the owner needs the information, commissions it, and makes it available to the contractor.

Some owners use a number of techniques to avoid responsibility for the accuracy of the information. Sometimes owners place the responsibility on the contractor to check the site and make its own tests. As seen in *United States v. Spearin* (reproduced in Section 23.05E), these generalized disclaimers are not always successful.

Some owners use a different approach: They do not include subsurface data in the information given bidders but state where the contractor can inspect such information available elsewhere.

Another approach is to give the information but state that it is for information only and is not intended to be part of the contract. This is another way of stating the contractor cannot rely on the accuracy of the information, in the hope of shielding the owner from any claim based on misrepresentation or warranty. These techniques generate varied outcomes, reflecting the difficulties courts face in deciding whether a party can use the contract to shift a risk to the other party.[30]

Judicial ambivalence when facing disclaimers can be demonstrated by comparing *Wiechmann Engineers v. State Department of Public Works*,[31] which *enforced* the disclaimer and *barred* the claim, with *Stenerson v. City of Kalispell*,[32] which *did not enforce* the disclaimer and *allowed* the claim. To be sure, the *Wiechmann* and *Stenerson* decisions can be reconciled. The contractor in the *Wiechmann* case had the opportunity to examine the information but did not. It also observed the site and knew that boulders (the basis for its claim) were present. It would be difficult to justify a claim for additional expenses under these conditions. The public entity made available all it knew, and the information was readily accessible to the contractor. The contractor lost its claim because it did not rely on the information. This case invoked the common law rule that places on the contractor the risk of performance that proves to be more expensive than anticipated.

---

[28]*Tri-County Excavating, Inc. v. Borough of Kingston*, 46 Pa.Commw. 315, 407 A.2d 462 (1979).

[29]*Wyman Constr. Inc.*, PSBCA No. 611, 80-1 BCA ¶14,215.

[30]Compare *City of Columbia, Mo. v. Paul N. Howard Co.*, 707 F.2d 338 (8th Cir.), cert. denied, 464 U.S. 893 (1983) (disclaimer not enforced) with *R. Zoppo Co. v. City of Dover*, 124 N.H. 666, 475 A.2d 12 (1984) (disclaimer enforced).

[31]31 Cal.App.3d 741, 107 Cal.Rptr. 529 (1973). See also *Cook v. Oklahoma Bd. of Public Affairs*, 736 P.2d 140 (Okla.1987). The contractor was denied recovery even where the owner knew of the condition, which it did not mention in bid information. A reasonably prudent bidder would have known of the condition.

[32]629 P.2d 773 (Mont.1981). See also *Jack B. Parson Constr. Co. v. State*, 725 P.2d 614 (Utah 1986) (bidder could reasonably rely on specific misrepresentation despite disclaimer).

In the *Stenerson* case, the court concluded that the public entity either knew or should have known that the information would be relied on by the bidders. It also held that no on-site inspection would have revealed any information that was not on the plans. The *Wiechmann* case condemned the contractor because it did not perform as the court felt contractors should perform. The court in *Stenerson* concluded that the contractor did as other contractors would have done—relied on the expertise of the geotechnical engineers.

The fact remains that the *Wiechmann* decision reflects the court's unwillingness to deprive the public entity of the opportunity of disclaiming responsibility and placing the risk on the contractor. The *Stenerson* decision, in contrast, showed an unwillingness to allow the public entity to exonerate itself when it knew that the contractor did rely on the information, and that reliance in the long run was beneficial to the public entity because it generated a lower price.[33] Although all judges are likely to give great autonomy to contracting parties to apportion risks as they choose, they often differ when one party has *dictated* a risk allocation plan, even though the contractor can adjust its bid price to take into account the risks it is being asked to bear.

Judges view the disclaimer process differently. Judges who are unwilling to allow owners to furnish information and disclaim responsibility for its accuracy (the issue here is not fraud or negligent misrepresentation) seem more influenced by the apparent unfairness of allowing the owner to place the risk on a party who is often in a poorer position to distribute or shift that risk. This is even more persuasive if the owner knows that the information will be relied on and derives a benefit through lower bid prices when the contractor does not make its own tests. They are also influenced by the belief that the contracting parties owe each other the duties of good faith and fair dealing. They believe autonomy should not be used to place subsurface risks on the contractor when the contractor did as others and relied.[34] Judges favoring broad autonomy are less influenced by these considerations.

They seek only to determine how the risk was apportioned by the contract.[35]

SECTION 25.06 Contractual Protection to Contractor

A. Public Contracts: The Federal Approach

In the absence of contract protection, misrepresentation, or breach of warranty, the contractor must bear the risk of unforeseen site and subsurface conditions. If disclaimer language is chosen carefully and the bidder given a reasonable opportunity to observe and test, the disclaimer will likely be effective. Even where the disclaimer language is clear and the contractor could observe and test, the owner will be held responsible for factual representations, not opinions or inferences.

The owner may find that shifting risk to the contractor is not in the owner's best interest. In such a case, the owner may choose to accept the risk of unforeseen site and subsurface conditions. This acceptance is accomplished in federal construction through what was formerly called the *changed-conditions clause* and is now known as the *differing site conditions (DSC) clause*.[36]

The differing site conditions system plays a significant part in federal procurement. In 1982, an EPA study identified DSC claims as representing 50 percent of the dollar amount in contract modifications.[37] A 1984 study showed about 34 percent of all federal contract modifications were based on the DSC.[38]

---

[33]See Parvin & Araps, *Highway Construction Claims*, 12 Pub.Cont. L.J. 255 (1982).

[34]In *Sherman R. Smoot Co. v. Ohio Dep't of Adm. Serv.*, 136 Ohio App.3d 166, 736 N.E.2d 69 (2000), the court invoked the *Spearin* doctrine to allow the contractor to rely upon a soils report provided "for information only."

[35]See *Berkel & Co. Contractors, Inc. v. Providence Hosp.*, 454 So.2d 496 (Ala.1984); *Empire Paving, Inc. v. City of Milford*, 57 Conn.App. 261, 747 A.2d 1063 (2000); *Green Constr. Co. v. Kansas Power & Light Co.*, 1 F.3d 1005 (10th Cir.1993) (Kansas law); *Millgard Corp. v. McKee/Mays*, 49 F.3d 1070 (5th Cir.1995) (Texas law); *McDevitt & Street Co. v. Marriott Corp.*, 713 F.Supp. 906 (E.D.Va.1989), aff'd in relevant part, 911 F.2d 723 (4th Cir.1990) (Virginia law).

[36]For an in-depth discussion of the federal DSC clause, see J. CIBINIC, JR., R. NASH, JR. & J. NAGLE, ADMINISTRATION OF GOVERNMENT CONTRACTS, Ch. 5 (4th ed. 2006) and Chu, *Differing Site Conditions: Whose Risk Are They?*, 20 Constr. Lawyer, No. 2, April 2000, p. 5.

[37]*Report of Internal and Management Audit, Environmental Protection Agency, Office of the Inspector General*.

[38]*Contract and Change Order Summary Reports, Naval Facilities Engineering Command*.

The federal DSC clause states,

(a) The Contractor shall promptly, and before the conditions are disturbed, give a written notice to the Contracting Officer of (1) subsurface or latent physical conditions at the site which differ materially from those indicated in this contract, or (2) unknown physical conditions at the site, of an unusual nature, which differ materially from those ordinarily encountered and generally recognized as inhering in work of the character provided for in the contract.

(b) The Contracting Officer shall investigate the site conditions promptly after receiving the notice. If the conditions do materially so differ and cause an increase or decrease in the Contractor's cost of, or the time required for, performing any part of the work under this contract, whether or not changed as a result of the conditions, an equitable adjustment shall be made under this clause and the contract modified in writing accordingly.

(c) No request by the Contractor for an equitable adjustment to the contract under this clause shall be allowed, unless the Contractor has given the written notice required; *provided,* that the time prescribed in (a) above for giving written notice may be extended by the Contracting Officer.

(d) No request by the Contractor for an equitable adjustment to the contract for differing site conditions shall be allowed if made after final payment under this contract.[39]

A DSC (differing site condition) creates two methods of obtaining an equitable adjustment, conditions different from those represented (Type I) and unanticipated conditions (Type II).

A Type I DSC requires that there be an actual physical (subsurface or latent) condition encountered at the site that differs materially from the conditions "indicated" in the contract documents. First the contract documents must be examined. In federal procurement, such documents are broadly defined and include the invitation for bids, drawings, specifications, soil-boring data, representations of the type of work to be done, and the geographical area of construction. Included among the contract documents are not only specific information as to the subsurface characteristics but also the description of the nature of the project and the physical conditions that

relate to the work. Physical conditions include the details of excavation or construction work, as they may include representations of the physical conditions.[40] The contractor, when preparing its bid, must consider not only the representations contained in the physically attached documents, but also information located elsewhere but made available for inspection by the bidders, and even the contractor's own experience on similar projects in the same location.[41] The contractor can draw reasonable inferences as to the physical conditions, and there need not be express representations of them. The important thing is whether the representations are sufficient to provide a reasonably prudent contractor with a sufficient basis on which to rely when the contractor prepared its bid.[42] Where the contract documents are silent as to subsurface conditions, no Type I DSC claim can be brought.[43]

Actions by third parties, including economic, governmental, political, or labor conditions, do not constitute physical conditions.[44] Moreover, natural conditions, such as abnormally heavy rainfall, rough seas, strong winds, hurricanes, severe temperatures, frozen ground, or flooding by themselves do not constitute differing site conditions. However, an abnormal weather condition may interact with a latent physical condition in such a way as

---

[39]48 CFR § 52.236-2 (2007).

[40]*Ace Constructors, Inc. v. United States,* 70 Fed.Cl. 253, 269–72 (2006), aff'd, 499 F.3d 1357 (Fed.Cir.2007) (government's erroneous topological survey, indicating a "balanced" project, when instead the project lacked adequate fill from the cut areas, is a Type I DSC).

[41]See *Randa/Madison Jt. Venture III v. Dahlberg,* 239 F.3d 1264 (Fed. Cir.2001) (geotechnical reports made available for inspection) and *T. L. James & Co., Inc. v. Traylor Bros., Inc.,* 294 F.3d 743 (5th Cir.2002) (Louisiana law; dredging contractor is bound by information found in maps in possession of the public owner indicating subsurface obstructions in a river). In *Massman Constr. Co. v. Missouri Highways & Transp. Com'n,* 31 S.W.3d 109 (Mo.App.2000), transfer denied Dec. 5, 2000, the court allowed the contractor to recover under a *Spearin* claim; even though the contractor had placed the rock revetment, it did not know the revetment would interfere with the current project.

[42]In *H. B. Mac, Inc. v. United States,* 153 F.3d 1338 (Fed.Cir.1998), the court rejected the argument that the "reasonably prudent contractor" standard should be relaxed where the contractor was a "small, disadvantaged business."

[43]*Conner Brothers Constr. Co., Inc. v. United States,* 65 Fed.Cl. 657, 679–82 (2005).

[44]*Olympus Corp. v. United States,* 98 F.3d 1314 (Fed.Cir.1996) (strike by co-prime's employees, which prevented the plaintiff from accessing the site, not a DSC).

to create a DSC.[45] Physical conditions can also include artificial or manufactured conditions, such as underground electric power lines, sewer lines, or gas pipelines that are unexpectedly encountered at the site.

The condition must be subsurface or latent. Most claims involve encountering something different than anticipated in the subsurface materials or structure—usually physical or mechanical properties, behavioral characteristics,[46] quantities, etc. But a DSC need not always be a subsurface condition; it can also be a latent physical condition at the site, such as undisclosed concrete piles, the thickness of a concrete wall, or the height of ceilings above a suspended ceiling.

The DSC must be encountered at the site, which usually is the place where construction will be undertaken but can include off-site pits from which soil is to be borrowed for fill or disposal sites if the contractor is directed to obtain or dispose of materials off-site.

The physical conditions encountered must be materially different from those indicated—an essentially factual inquiry. Was there a substantial variance from what a reasonably prudent contractor would have expected to encounter, based on its review of the contract documents and what it would have encountered had it complied with the site investigation clause in the contract?[47] The "reasonably prudent contractor" test imposes upon the contractor not only a duty to examine all the information stated in the contract documents or separately made available, but also to evaluate this information in a balanced manner. As explained in one case:

> For example, the contractor must consider whether the borings are numerous and well-spaced, or whether they are few and far between. In addition, the contractor must consider the general description of the site and any warnings of

conditions which might be encountered. . . . [A] reasonable bidder will consider the information provided by the boring logs and then consider how other available information sheds light upon the results of the test borings and upon the extent to which the test borings are representative of conditions throughout the site.[48]

The DSC clause makes the contractor's actual reliance upon the contract documents when preparing its bid a precondition for bringing a claim. A contractor who did not rely upon certain contract documents when preparing its bid may not then use those documents as the basis for a claim.[49] Moreover, even without such a clause, a contractor would be expected to make at least a minimal inspection to familiarize itself with the site. Typically, site inspection clauses obligate the contractor to inspect and familiarize itself with the conditions at the site, but the contractor is obligated only to discover conditions apparent through a reasonable investigation. It is not expected to perform burdensome, extensive, or detailed tests or analysis. The contractor can rely on information received from the owner, particularly when it does not have adequate time or opportunity to conduct a thorough investigation. What other bidders did or what they thought they would encounter will bear heavily on the question of whether the site investigation was properly performed.

A Type I DSC clause is in essence a contractual mechanism for dealing with misrepresentation claims concerning subsurface conditions. Within federal procurement jurisprudence, the DSC clause is a subset of the *Spearin* doctrine, discussed in Section 23.05E. In the case of an overlap of these two theories—such as when the specifications are allegedly defective in their misrepresentation of subsurface conditions—the contractor must pursue its claim under the more specific, contractual mechanism.[50]

---

[45]*Baldi Bros. Constructors v. United States*, 50 Fed.Cl. 74 (2001).

[46]Behavioral DSC claims were raised in two tunneling cases: *Mergentime Corp. H/T Constr., Inc.* (JV), ENG BCA No. 5756, 94-3 BCA ¶27,119 and *Municipality of Anchorage v. Frank Coluccio Constr. Co.*, 826 P.2d 316 (Alaska 1992). See Smyth, *Behavioral Differing Site Conditions Claims*, 11 Constr.Litig.Rep., No. 6, June 1990, p. 158.

[47]*H. B. Mac, Inc. v. United States*, supra note 42 (in geologically diverse soil, borings logs located 300 yards from the construction site would not be viewed as reliable by a reasonable contractor); *Neal & Co., Inc. v. United States*, 36 Fed.Cl. 600 (1996), aff'd, 121 F.3d 683 (Fed.Cir. 1997) (groundwater was not a Type I DSC where the contract did not indicate dry soil conditions).

[48]*PCL Constr. Services, Inc. v. GSA*, GSBCA No. 16588, 06-2 BCA ¶33,403, p. 165,616.

[49]*Comtrol, Inc. v. United States*, 294 F.3d 1357, 1363–64 (Fed.Cir. 2002); *Fru-Con Constr. Corp. v. United States*, 43 Fed.Cl. 306, 319–21, motion for recon. denied, 44 Fed.Cl. 298 (1999).

[50]*Comtrol, Inc. v. United States*, supra note 49, 294 F.3d at 1362. However, an Ohio court ruled that a contractor who was not entitled to relief under the DSC clause could then bring the same claim under the *Spearin* doctrine; *Sherman R. Smoot Co. v. Ohio Dep't. of Adm. Serv.*, supra note 34.

The DSC clause is also akin to the *Spearin* doctrine in that courts are loath to allow disclaimers to undermine either theory of recovery.[51]

A Type II DSC requires a variance between the site condition actually encountered and that which would be reasonably expected at the time the contract was made. Expectations look at the information furnished to the contractor or information acquired from other sources. Were the conditions encountered common, usual, and customary for that geographical area? The contractor is expected to be aware of conditions under which the work will be performed. For example, a contractor who is working in winter in a mountainous area where snow is common should expect to encounter wet conditions.[52]

To obtain an equitable adjustment, the contractor must notify the contracting officer promptly in writing. This generates the inevitable difficulty over substantial compliance and waiver.[53] Courts seem more willing to waive strict notice requirements if it appears that the owner knew the contractor had encountered unforeseen subsurface conditions and that a claim would likely be made.[54] In *Kenny Construction Co. v. Metropolitan Sanitary District of Greater Chicago*,[55] the contractor notified the engineer, who told the contractor to go ahead and they would

figure costs later. Despite the absence of a written order by the engineer, the court granted recovery, because the contractor relied on the engineer's statement and the public entity was estopped (condition cannot be asserted as a defense, as representation that it would not be asserted as a defense had been relied on) to assert the condition.

In *Centex Construction Co.*,[56] the written notice requirement was waived because the project representative was aware of the oil dump initially at the prebid site investigation and again when the contractor orally advised him of the condition when work commenced. The agency Board of Appeals also noted that failure to provide written notice did not prejudice the government's position.

The argument in favor of enforcement of the notice requirement is made in *Fru-Con Construction Corp. v. United States*.[57] On a dam renovation project, the blasting method employed by the contractor caused "overbreak" of the existing concrete walls. This delayed the project because the contractor was required to rebuild the walls. The contractor provided no DSC notice, oral or written, even though the evidence of overbreak was immediate and obvious. Indeed, the contractor did not even make a DSC Type II claim until three years after project completion and on the eve of trial. The court rejected the contractor's arguments that late notice was excused because it was on a tight construction schedule and because it did not know the cause of the overbreak (that it was caused by a DSC, rather than its blasting subcontractor's improper work) until it consulted an expert. The court observed that the government was prejudiced by the late notice because it was deprived of the opportunity to consider alternative construction methods.[58]

The use of a DSC is becoming more common in public contracts made by entities other than the federal government. This usage can create difficulties. For example, one case involved a clause that incorporated a Type I DSC but not a Type II DSC.[59] Similarly, confusion can develop because a local public entity may be compelled to use a DSC because the project is being partially funded with federal money granted by the Environmental Protection Agency (EPA). But the local entity making the contract

---

[51]In *Whiting-Turner/A. L. Johnson Jt. Venture v. General Services Admin.*, GSBCA No. 15401, 02-1 BCA ¶31,708 the Board of Contract Appeals refused to enforce disclaimers providing that boring logs are for information only and not part of the contract documents, as doing so would render the Type I DSC clause meaningless. The Board also ruled that the government could not negate its DSC liability to a subcontractor by hiring a construction manager. See also *Condon-Johnson & Assocs., Inc. v. Sacramento Mun. Utility Dist.*, 149 Cal.App.4th 1384, 57 Cal.Rptr.3d 849 (2007), review denied July 25, 2007.

[52]*Housatonic Valley Constr. Co, Inc.*, AGBCA No. 1999-181-1, 00-1 BCA ¶30,869 (wet soil in the Oregon forest not a Type II condition).

[53]*Moorhead Constr. Co. v. City of Grand Forks*, 508 F.2d 1008 (8th Cir.1975); *Metropolitan Paving Co. v. City of Aurora, Colo.*, 449 F.2d 177 (10th Cir.1971) (notice waived).

[54]*Brinderson Corp. v. Hampton Roads Sanitation Dist.*, 825 F.2d 41 (4th Cir.1987) (actual or constructive notice of condition by public owner and its opportunity to investigate waives requirement that contractor give formal notice); *Roger J. Au & Son, Inc. v. Northeast Ohio Regional Sewer Dist.*, 29 Ohio App.3d 284, 504 N.E.2d 1209 (1986) (waiver of written notice depends on whether owner already knew of the condition and was not prejudiced by contractor's failure to give it notice of the condition); *Parker Excavating, Inc.*, ASBCA No. 54637, 06-1 BCA ¶33,217.

[55]52 Ill.2d 187, 288 N.E.2d 1 (1972). A subsequent opinion is found in 56 Ill.2d 516, 309 N.E.2d 221 (1974).

[56]ASBCA No. 26830–26849, 83-1 BCA ¶16,525.

[57]Supra note 49.

[58]43 Fed.Cl. at 327.

[59]*Metropolitan Sewerage Comm'n v. R. W. Constr., Inc.*, 72 Wis.2d 365, 241 N.W.2d 37 (1976).

may have had a practice of using disclaimers, and these disclaimers are sometimes still included in the contract documents. One court faced with this question concluded that the DSC provision took precedence.[60]

Similarly, in 1989 DSC clauses were required to be incorporated into road-building contracts for federally subsidized highway programs under which roads are constructed by private contractors under contract with state highway agencies. However, under the applicable regulations the DSC provisions need not be included if the state law requires a different method of risk distribution.[61] Also, statutes can mandate slight deviations from classic federal DSCs. For example, in 1989 California enacted a statute dealing with public works contracts of local entities that involve digging trenches or other excavations that extend deeper than four feet below the surface. Such contracts are required to include a truncated Type I DSC and a Type II DSC clause.[62]

If the necessary substantive requirements are met under the DSC, the contractor or the owner receives an equitable adjustment. The legal systems that handle such claims, particularly in federal procurement, have had to struggle with a variety of measurement techniques—all of which tend to employ approximations rather than precise measurements. Using the contractor's claim as an illustration, the best proof is item-by-item proof of the difference (1) between the cost it would have incurred had it not encountered the differing site condition and (2) what it actually incurred. However, often it is impossible to establish those additional costs with anything resembling precision, and the contractor is likely to seek to employ a crude formula such as total cost, modified total cost, or jury verdict.[63] These are discussed in greater detail in Section 27.02F.

Sometimes contractors claim that encountering the subsurface conditions required them to perform the work out of sequence or delayed completion of the project. In such cases, contractors sometimes seek additional compensation in addition to time extensions. As shall be seen in Section 26.10A, claims for additional compensation may be barred by a "no damage" (no pay for delay) clause. Although the law has had considerable difficulty when such clauses are attacked, the current trend seems to be toward increased scope for such exculpatory clauses.

Finally, the law has struggled with contractor claims that it is also entitled to additional overhead and profit.[64]

Suggestions have been made for contract provisions that will deal in some way with the amount of the equitable adjustment. One commentator suggests the possibility of a contractual provision that requires the contractor to include a per-diem calculation of delay damages.[65] The contractor would include within the stipulated amount field overhead, home office overhead, idle labor and equipment, and any loss of efficiency or impact on later performance. These figures would be included in the bid. Determining the low bidder would also involve an assumed number of delay days appropriate to the project multiplied by the bidder's per-diem daily delay price, which would be added to the bid.

Another method is to include a provision under which the contractor agrees that certain items would not be included in any DSC claim and that any dispute resolution system would not include in its award payment for profit, labor and efficiencies, cost of vital equipment, or project and home office overhead.

It is likely that these attempts to structure remedies for a DSC could be used only in less regulated public contracts or in private contracts.

Does the right to receive an equitable adjustment preclude the contractor from terminating its obligation if the condition discovered would have granted it that power? In addition to taking this potentially disastrous risk away from the contractor, the clause is designed to keep the job going. It should substitute for any common law rights that the contractor might be accorded.

---

[60]*Fattore Co. v. Metropolitan Sewerage Comm'n*, 505 F.2d 1 (7th Cir.1974). See also *Baltimore Contractors, Inc. v. United States*, 12 Cl.Ct. 328 (1987).

[61]23 CFR § 635.109(b) (2007).

[62]West Ann.Cal.Pub.Cont.Code § 7104.

[63]See *Baldi Bros. Constructors v. United States*, supra note 45, employing a modified total-cost measure to compensate a contractor faced with a massive differing site condition. See Section 27.02F.

[64]*Kenny Constr. Co. v. Metro. Sanitary Dist. of Greater Chicago*, 52 Ill.2d 187, 288 N.E.2d 1 (1972) (allowed as part of changes process); *Fattore Co. v. Metropolitan Sewerage Comm'n*, supra note 60 (profit on unperformed work denied). See Beh, *Allocating the Risk of the Unforeseen Subsurface and Latent Conditions in Construction Contracts: Is There Room for the Common Law?* 46 Kan.L.Rev. 115, 144–47 (1997) suggesting profit not be awarded.

[65]Ashcraft, *Avoiding and Managing Risk of Differing Site Conditions*, a chapter in DIFFERING SITE CONDITIONS CLAIMS (1992).

B. AIA Approach: Concealed Conditions

AIA Document A201-2007, Section 3.7.4, uses language similar to the federal DSC clause, with its Type I and II substantive bases for additional compensation. Whether the occasional owner who builds once in a lifetime is better off with such a provision is debatable. One commentator questions the use of a DSC when the owner is a small, "one time" player, private or public.[66] Its use *may* generate a lower bid price. If so, where it is not so necessary to protect the contract price it might be useful. Yet most small owners cannot manage the disputes process and deal with possible contractor claim manipulation. Also, unlike large "repeat players" in construction, such an owner may not have nor intend to create a long-term relationship with the contractor. Similarly, the contractor may have no long-term relationship with the owner to nurture. Thus, the commentator says, "the incentive to promote future relationships is nonexistent."[67] Nor is the small owner concerned, as are "repeat players" such as federal agencies, with preserving a pool of contractors able to build. Finally, she notes that a small owner often lacks the knowledge to assess the risks, and the contractor often has superior knowledge of potentially adverse subsurface conditions. In such a case it may be more important to protect the contract price by not having a DSC and by placing this risk on the contractor through the fixed price. She suggests that at the very least the AIA should give such an owner a choice by alternative provisions. This choice would direct attention to and force negotiation on this important and frequently litigated provision.[68]

Just as it is questionable where a small one-shot owner should use a DSC, it is questionable whether lenders will want a provision that can substantially expand the ultimate cost of the project.

In AIA Document A201-2007, Section 3.2.1, the contractor represents that it has visited the site and familiarized itself with local conditions under which the work is to be performed and "correlated personal observations with the requirements of the Contract Documents." Conditions that could have been observed at a normal site visit cannot justify additional compensation.

In 1987, the AIA substantially expanded the administrative features of the then Paragraph 4.3.6 (now Section 3.7.4). The observing party must give a notice "promptly before conditions are disturbed and in no event later than 21 days after first observance of the conditions." Although many notices required by AIA documents can be given to the architect, this notice must go to both the owner and the architect if the contractor is the observing party. The architect can be considered the owner's agent for this purpose, and a notice to the architect may be sufficient.[69] The law has been hesitant to give literal effect to notice provisions.[70]

A201-2007 is honeycombed with notice provisions. For example, if the contractor wishes to receive additional compensation, it must give a written notice under Section 15.1.4. If the contractor wishes additional time, it must give a written notice under Section 15.1.5.1. If it has suffered injury to property, the contractor must give a written notice under Section 10.2.8. As can be seen, each of the three previously mentioned sections requires a *written* notice. Yet Section 3.7.4 apparently requires only that notice be given. Certainly it is safer for a contractor discovering such conditions to give notices under Sections 3.7.4, 15.1.4, 15.1.5.1, and 10.2.8. However, if an oral notice were given, apparently sufficient under Section 3.7.4, failure to give the other notices might be waived if the owner knew from the notice given that claims for additional compensation, additional time, and even damage to property are likely to be made.

Returning to the substantive basis for a claim, the Type I claim under Section 3.7.4 requires that conditions be encountered that "differ materially from those indicated in the Contract Documents." AIA Document A201-2007, Section 1.1.1, excludes bidding requirements from the contract documents. (A101-2007, Section 9.1.7, recognizes that some may wish to include them.)

Because specifications are contract documents, the contractor will not face any obstacles if the information is included in the specifications. But suppose the information is included in the instructions to bidders or is simply made available at the geotechnical engineer's office. A literal interpretation of Sections 1.1.1 and 3.7.4 could bar the contractor's claim.

---

[66]Beh, supra note 64.
[67]Id. at 153.
[68]Id. at 153–54.

[69]See Section 17.05B.
[70]See Section 21.04H.

However, it can be contended that including a DSC, yet giving the contractor relevant information in something *other* than a contract document, would be at least a violation of the covenant of good faith and fair dealing, and possibly would be fraud. Put another way, assuring the contractor it will receive an equitable adjustment if it encounters something unexpected and then refusing to grant an equitable adjustment in a Type I claim—the most common basis for equitable adjustments—would violate the obligation of good faith and fair dealing discussed in Section 19.02D.

A201-2007, Section 3.7.4, creates a complicated system to determine the existence of a DSC and the extent of any equitable adjustment. Either party observing an apparent DSC must provide prompt notice to the architect. The architect investigates.

If the architect determines that there is a DSC, he recommends an equitable adjustment. If the owner and contractor cannot agree on the adjustment, this question is referred to the Initial Decision Maker (who could be the architect) under Article 15. If the architect determines that there is no DSC and that no adjustment is warranted, he notifies the owner and contractor in writing, giving his reasons. If either party objects, then the claims resolution procedure under Article 15 is again invoked.

Section 3.7.4 provides no review of the architect's decision that a DSC exists. This decision, favorable to the contractor, is apparently binding on the owner. In addition, A201 gives no guidance as to the amount of the equitable adjustment, but presumably the vast federal jurisprudence can be looked to, especially in light of the identical language used to define a DSC.

As noted in Section 25.06A, the delay that may be caused by the discovery and treatment of the DSC may be compensated by using a liquidated-damages clause. Some have suggested language limiting the remedy to which the contractor is entitled.

C. EJCDC Approach

The Engineers Joint Contracts Documents Committee (EJCDC) employs a more complex methodology in its 2007 "Standard General Conditions of the Construction Contract," C-700. It divides coverage into subsurface (¶4.03) and underground facilities physical conditions (¶4.04). Paragraph 4.02(A) states that information will be given in the Supplementary Conditions. Under Paragraph 4.02(B), the contractor may rely on "the accuracy of the 'technical data'" even though the data are not contract documents. But Paragraph 4.02(B)(1) states that the contractor cannot rely on the completeness of the information for purposes of execution or safety nor under Paragraph 4.02(B)(2) on "other data, interpretations, opinions and information."

Paragraph 4.02(B)(3) precludes a contractor's claim against the owner or engineer with respect to the contractor's interpretation or conclusions drawn from the technical data. This is a sensible line of demarcation and fits the DSC system.

To claim an equitable adjustment, under Paragraph 4.03(A) a written notice must be given describing the technical data the contractor claims it intends to rely on and why the data are materially inaccurate or stating that the condition differs materially from that shown or indicated in the contract documents. The Type II language of Paragraph 4.03(A)(4) is similar to that in the federal procurement system and the AIA.

After a claim is made, under Paragraph 4.03(B) the engineer can review the condition and determine whether the owner should obtain "additional explorations and tests." Under that paragraph, he advises the owner of his findings and conclusions. If he concludes a change is required, he issues a work change directive. But this does not guarantee price or time changes in the contract. Under Paragraph 4.03(C), an equitable adjustment is granted only to the extent that the condition causes an increase or decrease in the cost or time. The inability of the parties to agree on the amount of adjustment in price or time under Paragraph 4.03(C)(3) is a dispute that falls into the disputes clause, Paragraph 10.05, under which the engineer plays roughly the same role as does the architect under A201-1997.

Underground facilities are treated differently. Information is given by the owners as to these facilities. According to Paragraph 4.04(A)(1), neither the owner nor the engineer is responsible for its "accuracy or completeness." The contractor must review and check such information. But if uncovering reveals an underground facility "not shown or indicated or indicated with reasonable accuracy," Paragraph 4.04(B)(2) allows an equitable adjustment "to the extent that they are attributable to the existence or location of any Underground Facility that was not shown or indicated or not shown or indicated with reasonable accuracy . . . and the Contractor did not know of and could not have reasonably expected to be aware of or to have anticipated." Again claims are handled under Paragraph 10.05.

EJCDC documents are more detailed than those of the AIA, largely because of the type of engineering projects for which they are designed. But their classifications and detail would be useful in many building contracts.

D. The FIDIC Approach

Although this book emphasizes domestic construction contracts, not those made in an international context, occasionally it is useful to look at a particular problem from the vantage point of a standardized contract used in international transactions. The most commonly used standard contract is the one generally referred to as the FIDIC contract, published by the Federation Internationale des Ingenieurs-Conseil (International Federation of Consulting Engineers).

The FIDIC issued its Conditions of Contract for Construction for Building and Engineering Works Designed by the Employer in 1999, generally known as the Redbook. It demonstrates that there are many ways to deal with unforeseen subsurface conditions encountered during the work.

Section 25.04 sets forth different risk allocation plans to deal with unforeseen subsurface conditions. One is to furnish information to the contractor but disclaim responsibility for its accuracy. Another is to use differing site conditions (DSC) clauses. These are used in federal procurement[71] and by AIA[72] and EJCDC.[73] There are two types of DSCs that justify the contractor receiving an equitable adjustment. Type I requires a comparison of what was represented and what was encountered. Type II requires an inquiry into whether the contractor encountered unusual and unforeseeable subsurface conditions. The FIDIC blends these approaches. It uses a modified-disclaimer method and a Type II DSC.

Subclause 4.10 is a modified-disclaimer provision. It requires the employer (owner) to make available to the contractor all relevant data in its possession "on subsurface and hydrological conditions at the Site, including environmental aspects" 28 days prior to tender (bid) submissions as well as any data that comes into the employer's possession after that date. The contractor must interpret this data.[74]

This places all of the risks on the contractor. It includes an assumption that it has inspected and examined the site. The availability of this data means that the contractor is satisfied as to "the form and nature of the Site, including subsurface conditions," as well as hydrological and climatic conditions.

But Subclause 4.10 limits the disclaimer to the "extent which was practicable (taking into account the cost and time)." This exception to the disclaimer recalls the attacks on such disclaimers by contractors based on the time and cost of verifying the information made available to it and making its own tests.[75] Many said it was too expensive. The limitation in Subclause 4.10 takes into account the practicalities of prebid activities.

FIDIC contracts are likely to be used in large infrastructure and engineering projects. The contractors are likely to be well financed. But these are often megaprojects in strange, inhospitable, and unexplored places where it is difficult to gather reliable information.

Subclause 4.11 hammers away at the disclaimer, stating the contractor is satisfied with the contract price and that it has based its price on the site data available.

Subclause 4.12 enacts a Type II DSC. If "the contractor encounters adverse physical conditions which he considers to have been unforeseeable," he must notify the engineer. Physical conditions include subsurface and hydrological conditions but exclude climatic conditions. The engineer inspects, investigates, and determines whether the conditions were unforeseeable and the adjustments to be made. If unforeseen conditions cause the contractor to suffer delay or increased cost, the contract price and time are adjusted.

There are two other interesting aspects to the FIDIC approach. First, under Subclause 4.12, if the engineer finds that there were more favorable conditions than could have been expected, this can offset any cost increase. The net effect of any adjustment may not reduce the contract price. Second, the contract defines cost to include overhead but not profit.[76]

---

[71]See Section 25.06A.
[72]See Section 25.06B.
[73]See Section 25.06C.

---

[74]FIDIC Conditions of Contract for Construction for Building and Engineering Works Designed by the Employer, Subclause 1.1.3.1 (1999).
[75]See Section 25.05.
[76]See supra note 74 at 1.1.4.3.

SECTION 25.07   Some Advice to Courts

Undoubtedly, arguments can be made for a disclaimer system or a system that uses a DSC clause. Courts should effectuate whichever choice has been made. If it appears clear the contractor has, happily or not, accepted the risk of unforeseen subsurface conditions, the court should support that risk allocation and not seek to destroy it by tortured interpretation generated by a belief that it is unconscionable for the owner to place these risks on the contractor. Only if the contractor clearly was not made aware of the risk or if imposing the risk violates the obligation of good faith and fair dealing should the court entertain disturbing the loss distribution selected and expressed in the contract. An illustration would be the owner deliberately misleading the contractor or not giving the contractor sufficient time to examine the data, visit the site, or take its own tests. What is unclear is whether use of a disclaimer system knowing the contractor is aware of the data is a violation of the obligation of good faith and fair dealing.

Similarly, courts should not let a contractor maintain a tort action against a geotechnical engineer if the contractor clearly was expected to assume this risk. Allowing the contractor to transfer this risk to the geotechnical engineer through a tort action frustrates the efficiency of the system selected.

CHAPTER TWENTY-SIX

Time: A Different but Important Dimension

SECTION 26.01 An Overview

The law has not looked at time as part of the basic construction contract exchange—that is, money in exchange for the project. This may be because of the frequency of delayed construction projects. Timely completion depends on proper performance by the many participants as well as on optimal conditions for performance, such as weather and anticipated subsurface conditions.

Delayed performance is less likely to be a valid ground for termination. Delayed performance is also less likely to create legal justification for the owner's refusal to pay the promised compensation, with the owner left to the often inadequate damage remedy. Similarly, delayed payment by the owner is less likely to automatically give the contractor a right to stop the work or terminate its performance. Finally, a strong possibility exists that a performance bond that does not expressly speak of delay will not cover delay damages.

Computation of damages is also different. If the owner does not pay or the contractor does not build properly, measuring the value of the claim is, relatively speaking, simple. If the owner does not pay, at the very least contractors are entitled to interest. If the contractor does not build properly, the owner is entitled to cost of correction or diminished value of the project.[1]

Delay creates serious measurement problems. The owner's basic measure of recovery for unexcused contractor delay is lost use of the project. The contractor's basic measure of recovery for owner-caused delay is added expense. Lost use is difficult to establish in noncommercial projects. Added expense is even more difficult to measure.

---

[1]See Section 27.03D.

Because of measurement problems, each contracting party—whether it pictures itself the potential claimant or the party against whom a claim will be made—would like a contractual method to deal with delay claims, either to limit them or to agree in advance on amount. This does not mean time is not important in construction. The desire to speed up completion to minimize financing costs and accelerate revenue-producing activities was a large factor in leading to techniques intended to generate efficient methods of organizing construction (discussed in detail in Section 17.04). Time is another dimension, however.

SECTION 26.02 Commencement

The very nature of construction sometimes makes complicated what in other contracts is simple. In an ordinary contract, such as an employment contract, or in a long-term contract for the sale of goods, as a rule, the commencement dates are simple to establish. In construction, the date when the performance can begin in earnest (procurement can precede site access) is usually the date when site access is given. This cannot always be precisely forecast. The owner may need to obtain permits, easements, and financing before the contractor can be given site access. To avoid responsibility for site access delay and to measure the contractor's time obligation fairly, the commencement of the time commitment period, when measured in days, is often triggered by the contractor's being given access to the site, usually by a notice to proceed (NTP). When an NTP is used, the contractor assumes the risk of ordinary delays in site access but not those that go beyond the reasonable expectations of the contracting parties.

Commencement raises other problems. For example, must the contractor actually begin work at the site when an NTP is given? Is the actual commencement date of the NTP a date specified in the NTP, or the date when the NTP is received? The contract should deal with these issues but frequently does not. What are reasonable expectations of the parties? The custom in the industry? What is fair?

If the NTP does not expressly specify the commencement date, the date should begin when the NTP is received. Any date specified in the NTP should be effective only if it meets the standard of good faith and fair dealing. For example, the notice should not specify a date that the owner knows the contractor cannot meet or that fails to take into account realistic commencement requirements.

If an NTP system is employed, the owner will know when the contractor has commenced performance. However, if it is not employed, the contractor could conceivably commence performance before the commencement date. This could disadvantage the lender who wishes to perfect its security interest before work begins, in order to have priority over any mechanics' liens. Similarly, early commencement before insurance is in place can create a coverage gap.

With perfection of security interests in mind, in the absence of an NTP, the 1987 AIA Document A101, Paragraph 3.1, stated the contractor must notify the owner in writing not less than five days before beginning work to allow time for filing security interests. In A101-1997 the five-day period was deleted and a blank inserted for a date that correlates with any owner need to file security interests. Yet the AIA Document A201-1997, Paragraph 8.2.2 barred the contractor from premature commencement prior to the effective date of insurance and for some reason retained the five-day notice to file security interests.

This system was simplified in the 2007 documents. A101-2007, Section 3.1, remains unchanged from 1997. However, A201-2007, Section 8.2.2, simply states that the contractor should not begin work before the effective date of insurance, unless with the owner's written consent. The language from 1997 requiring the five-day notice has been deleted.[2]

Suppose there are delays between bid opening and bid award, and between award and execution of a formal contract. These issues were involved in two instructive cases.

The case of *Quin Blair Enterprises, Inc. v. Julien Construction Co.*[3] involved a competitively bid contract to build a motel. Julien's bid stated that it would complete 240 days from the date the contract was signed. Julien was awarded the contract and was asked to prepare the agreement. He used AIA Document A101 and filled in the blank with 240 days, without specifying as to when the period began. But A101 stated the contract was made on October 8, 1971.

AIA Document A201, Paragraph 8.1.2, in effect at that time, stated that the date of commencement, if there is no notice to proceed, begins on the date of the agreement "or such other date as established therein."

Blair signed on October 22 and Julien on October 25. Julien could not start until Blair cleared the site. That was not completed until November 18. When did the 240-day period commence? On October 8? On October 25? On November 18?[4] The trial court chose October 8, but the Wyoming Supreme Court disagreed. Recognizing the ambiguities and seeking to harmonize all the writings, the court concluded the time began on October 25.

Because no date was specified in A101, the court referred to A201. Because there was no NTP, the commencement date should have been the date of the agreement. But was that date October 8, the date on the agreement, or October 25, when the agreement was signed by the contractor? Because A201 states the date of the agreement, one would think it was the date on the agreement, not the date when it was made. According to the court, each party agreed the bid was part of the contract. The bid stated that time would begin on signing the contract on October 25, and although parties can designate a retroactive date, nothing indicated that "either party ever intended . . . . [a] retroactive date."[5]

In *Bloomfield Reorganized School District No. R-14, Stoddard County v. Stites*,[6] the court did find a retroactive date. The contract was dated August 8, 1955, and provided that the contract was to be substantially completed in 395 calendar days. The architect mailed the contract to the contractor on August 17, 1955. The contractor signed the contract and returned it to the architect, who delivered it to the school superintendent for execution by the school board. School board officials signed the contract,

---

[2]For a neater and sharper method of determining commencement of contract time, see EJCDC C-700, ¶2.03 (2007).

[3]597 P.2d 945 (Wyo.1979).

[4]Julien's failure to submit the required notice of an intention to ask for an extension barred a time extension.

[5]597 P.2d at 951.

[6]336 S.W.2d 95 (Mo.Ct.App.1960).

and the superintendent mailed it to the architect on September 14. The latter forwarded the signed copy to the contractor on September 22, six weeks after the contract date. Yet the court looked solely at the language of the contract, which stated that the agreement had been made on August 8, 1955.

SECTION 26.03 Acceleration

A. Specific: The Changes Clause

One way to accelerate the completion date is a specific directive by the owner that the contractor must complete in a time shorter than originally agreed. Power to accelerate is usually determined by the changes clause. See Section 21.04B.

B. Constructive Acceleration

Constructive acceleration originated in federal procurement law. Although the original jurisdictional basis for its development is no longer applicable,[7] it can be applied to all construction contracts.

Constructive acceleration is based on the owner's unjustified refusal to grant a time extension. To establish constructive acceleration, a contractor must prove five elements:

1. the contractor experienced an excusable delay and is entitled to an extension;
2. the contractor properly requests a time extension;
3. the owner denies the time extension;
4. the owner demands completion by the original completion date; and
5. the contractor incurs reasonable increased costs caused by its actual acceleration.[8]

The current justification for constructive acceleration is that denying a deserved time extension can force additional expenses when work is not performed in the order planned. Suppose, though, that the contractor continues to perform as it would have performed had an extension been granted. This will very likely lead to untimely completion. If the time extension should have been granted and it is granted later (by agreement, by an arbitrator, or by a court), any attempt by the owner to recover actual or liquidated damages would not succeed. The constructive acceleration doctrine allows the contractor to speed up its performance and recover any additional expenses it can establish, or to use the wrongful denial of the time extension as a defense against any claim that the owner might bring against the contractor for late completion.

C. Voluntary: Early Completion

Delays are so common in construction that attention is rarely paid to the legal effect of the contractor's completing early or claiming it would have completed early had it not been delayed by the owner.

Some owners may find early completion desirable. This can be evidenced by a penalty/bonus clause, as discussed in Section 26.09. Yet early completion may, if unexpected, also frustrate owner plans. For example, suppose a contractor building a factory finishes substantially earlier than planned. The owner may have to take possession before it can install its machinery. Early completion can require payments in advance of resource capabilities. It can be as disruptive as late completion.

AIA Document A101-2007, Section 3.3, requires the contractor to substantially complete the project "not later than" a specified date. This appears to give the contractor the freedom to complete early even if it disrupts owner plans.

Construction contracts of any magnitude usually have schedules. It is unlikely the owner will be greatly surprised by early completion. Yet even awareness during construction that performance will be completed earlier than required may not enable the owner to make the adjustments needed to avoid economic losses.

The obligations of good faith and fair dealing, discussed in Section 19.02D, require that a contractor notify an owner if it intends to finish much earlier than expected or when it appears this is likely. If this notification is made or the owner is aware of that prospect, the contractor should

---

[7]Like constructive changes, constructive acceleration was a claim based on the contract and not its breach. This gave jurisdiction to the agency appeals board. Since 1978, a claimant can choose to bring a claim before either an appeals board or the U.S. Court of Federal Claims. See Sections 21.03A and B.

[8]*Fraser Constr. Co. v. United States*, 384 F.3d 1354, 1361 (Fed. Cir.2004) (claim denied); *Ace Constructors, Inc. v. United States*, 70 Fed.Cl. 253, 280–81 (2006), aff'd, 499 F.3d 1357 (Fed.Cir.2007) (claim granted); *Clark Constr. Group, Inc.*, JCL BCA No. 2003-1, 05-1 BCA ¶32,843 at pp. 162,559–562 (government's cure notice issued to the contractor constituted an order to accelerate, where the contractor was at the same time stymied by a design defect).

receive additional compensation if the owner interferes with any realistic schedule under which the contractor would have completed earlier than required by the contract.[9] To establish an early completion claim, a contractor must prove three elements:

1. from the outset of the contract, the contractor intended to complete early;

2. the contractor had the capacity to complete early; and

3. the contractor would have completed early but for the delay caused by the owner.[10]

SECTION 26.04   Completion

Construction contracts generally provide a way to measure compliance with the contractor's time commitment. For example, as discussed in Section 22.04, the AIA has selected substantial completion as the benchmark for determining compliance with the time commitment. Some construction contracts use actual or final completion.

Completion is an important benchmark in the history of a construction project, similar to agreement as to changed work or final payment. The effect of completion, often linked to final payment and acceptance, was discussed in Section 24.05.

SECTION 26.05   Schedules: Simple and Critical Path Method (CPM)

A project schedule is a formal summary of the planned activities, their sequence, and the time required and the conditions necessary for their performance. A schedule alerts the major participants of the tasks they must accomplish to keep the project on schedule. It can reduce project cost by increasing productivity and efficiency, facilitates monitoring of the project, and can support or disprove delay claims.

The schedule for a very simple project, such as the construction of a garage, may simply be starting and completion dates. A somewhat more complex project, such as

a residence, may add designated stages of completion, mainly as benchmarks for progress payments. When construction moves upscale, for example, from a simple commercial structure to a nuclear energy plant, the schedule will take on more complex characteristics. Until the past twenty-five years, schedules in anything but the simplest project would be a bar chart, sometimes referred to as a Gantt chart, after its inventor. One such bar chart[11] is shown in Figure 26.1. Bar charts continue to be used where feasible because they are easy and inexpensive to prepare and simple to understand.

Bar charts have deficiencies. They provide no logical relationship between work packages. There are limits to the number of work packages that can be represented in a bar chart—perhaps thirty to fifty—until the level of detail becomes unwieldy. Rates of progress within a package may not be uniform. The different activities are represented equally. In case of a delay, management cannot determine which activities have priority over others and the courts cannot assess the significance of the delay.

On complex projects, a critical path method (CPM) schedule is used.[12] A CPM schedule lacks the intuitive simplicity of a bar chart but is capable of showing many more activities and, even more importantly, the logical relationship between the different activities: how a delay on one activity affects other activities.

A variety of legal issues have surfaced with respect to delay scheduling. Section 26.05A compares the AIA's handling of scheduling with other methods. Section 26.05B discusses CPM scheduling. Section 26.05C concludes with some description of the legal issues involved.

A. Approaches to Scheduling

The approach taken by the AIA is reflected in AIA Document A201-2007, Section 3.10.1. It requires the contractor to submit its construction schedule for the information of the owner and the architect. The schedule must provide "for expeditious and practicable execution of the Work." Although the contractor's failure to conform to the most recent schedule constituted a breach under Paragraph 3.10.3, issued in 1987, both A201-1997 and A201-2007 relaxed this commitment by inserting

---

[9]BECO Corp., ASBCA No. 27090, 82-2 BCA ¶16,124.

[10]Interstate General Government Contractors, Inc. v. West, 12 F.3d 1053, 1059 (Fed.Cir.1993) (claim denied); Fru-Con Constr. Corp., ASBCA No. 53544, 05-1 BCA ¶32,936 at pp. 163,160–64 (claim denied).

[11]B. BRAMBLE & M. CALLAHAN, CONSTRUCTION DELAY CLAIMS 11-5 (3d ed. 2000).

[12]Critical path method (CPM) is also discussed in Section 26.05B.

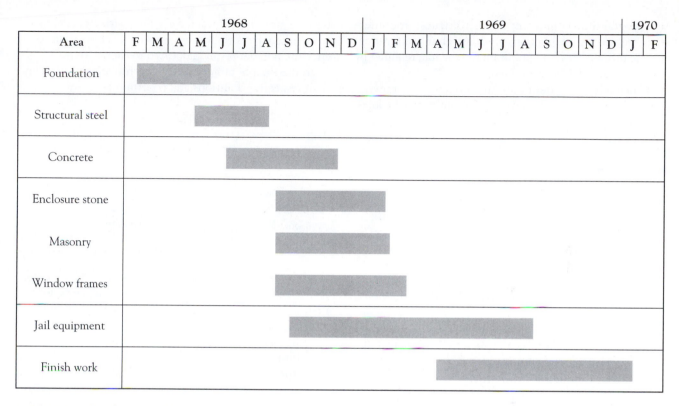

FIGURE 26.1   Bar chart schedule.

"in general accordance" with the "most recent schedules submitted to the Owner and Architect." This issue will be discussed in Section 26.05C.

Neither details nor schedule type are required. All is left to the contractor. Of course, a contractor is also interested in completing the project as promised. It should develop and meet a schedule that will accomplish that objective. But should responsibility be put solely in the hands of the contractor? Failure to meet the completion date, though giving a claim against the contractor, is not getting an on-time project. Of course, as seen in Section 22.02, payments are keyed to work progress and can be an incentive to move the work along. Failure to supply enough skilled workers may be grounds for termination,[13] though rarely will this be sought. Taken as a whole, A201 seems to ignore delay claims and contains few levers to obtain timely completion.

Contracts prepared by experienced public or private owners, particularly private owners under the influence of their lenders, usually prescribe much greater detail and take greater control over the contractor's schedule. This can manifest itself in language requiring that the schedule be on a form approved by the owner or the owner's lender; that each monthly schedule specify whether the project is on schedule (and if not, the reasons therefore); that monthly schedule reports include a complete list of suppliers and fabricators, the items that they will furnish, the time required for fabrication, and scheduled delivery dates for all suppliers; and that the contractor hold weekly progress meetings and report in detail as to schedule compliance.

Similarly, the Engineers Joint Contract Documents Committee (EJCDC) takes progress much more seriously than does the AIA. For example, the EJCDC's C-700 (2007), Paragraph 2.05(A), requires the contractor to submit within ten days after the effective date of the contract a *preliminary* progress schedule, a *preliminary* schedule of values, and a *preliminary* schedule of submittals.

---

[13]AIA Doc. A201-2007, § 14.2.1.1.

Paragraph 2.06 requires a preconstruction conference before any work is started. At that conference, among other topics, the preliminary schedule required by Paragraph 2.05A is discussed.

Paragraph 2.07 requires another conference to review "for acceptability to Engineer" the preliminary schedules. The contractor has ten days to make corrections and adjustments and "to complete and resubmit the schedules." No progress payments are made "until acceptable schedules are submitted to Engineer."

B. The Critical Path Method (CPM)

This system has generated burgeoning literature.[14] The description here must be simple, its goal mainly to point out the essential characteristics of the process and note the effect of float or slack time. The CPM process also relates to measuring claims (discussed in Section 27.02F and 27.03E).

To show how a CPM schedule operates, a very simple construction project will be used as illustration, without all the complexities of arrow diagrams, precedence diagrams, and nodes.

First, the contractor divides the total project into different activities or work packages. A major project may have thousands of activities, with each subcontractor generally performing a different activity.

Next, the contractor determines the activities that must be completed before other activities can be started. These constraints are the key to the CPM schedule. For example, usually excavation must be completed before foundation work can be begun. Conversely, plumbing and electrical work can usually be performed at the same time, as neither depends on the other. Subcontractors performing this work can work side by side.

Finally, the contractor estimates how long it will take subcontractors to complete their activities. This estimate

is made after the contractor consults with its subcontractors and analyzes the design drawings. These data influence the number of days allocated to each activity.

In the sample project, the following are activities and their respective durations and constraints:

Activity	Duration (days)	Constraint
1. Excavation	7	None
2. Formwork	5	Excavation
3. Plumbing	4	Excavation
4. Electrical	2	Excavation
5. Concrete pour	5	Formwork and plumbing
6. Roof	4	All

Constraints dictate the form of the CPM schedule. Because excavation has no constraints, it can be performed first. Once it is completed, the formwork, plumbing, and electrical activities can be performed. The concrete pour activity cannot be performed until the formwork and plumbing are completed. The roof cannot be installed until the concrete pour and electrical have been completed. Figure 26.2 illustrates the CPM schedule for this project. The total project under this schedule should be completed in twenty-one days.

The critical path, the longest path on this simple schedule, consists of activities that will delay the total project if they are held up. In the preceding example, excavation, formwork, concrete pour, and roof work are on the critical path. A delay to any of these activities will hold up the entire project.

In contrast, plumbing and electrical activities are not on the critical path. Their delay, up to a point, will not hold up the total project. If electrical work is delayed seven days, the total project will not be held up. The number of days each noncritical path activity can be delayed before the total project is affected is called *float* or *slack time*. In the illustration, plumbing and electrical work have one day and eight days of float, respectively.

If a noncritical path activity is delayed beyond its float period, it becomes part of the critical path. Moreover, some activities that were previously on the critical path will no longer be there. Suppose there is a three-day delay to plumbing. Originally, plumbing had one day of float. Now the CPM must be adjusted as shown in Figure 26.3.

---

[14]B. BRAMBLE & M. CALLAHAN, supra note 11, at 11-6 to 11-13; Wickwire & Smith, *The Use of Critical Path Method Techniques in Contract Claims*, 7 Pub.Cont.L.J. 1 (1974). The latter seminal paper was updated in Wickwire, Hurlbut & Lerman, *Critical Path Method Techniques in Contract Claims: Issues and Developments, 1974–1988*, 18 Pub.Cont.L.J. 338 (1989). Wickwire's latest discussion on this topic includes a critical analysis of modern scheduling software. Wickwire & Ockman, *Use of Critical Path Method on Contract Claims—2000*, 19 Constr. Lawyer, No. 4, Oct. 1999, p. 12. For a more skeptical look at CPM, see Laufer & Tucker, *Is Construction Project Planning Really Doing Its Job? A Critical Examination of Focus, Role and Process*, 5 Constr.Mgmt. & Econ. 243 (1987).

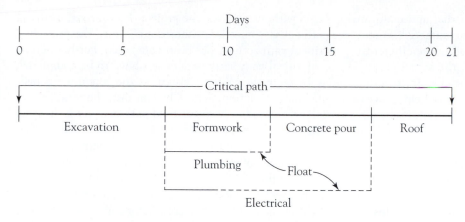

FIGURE 26.2   CPM chart schedule.

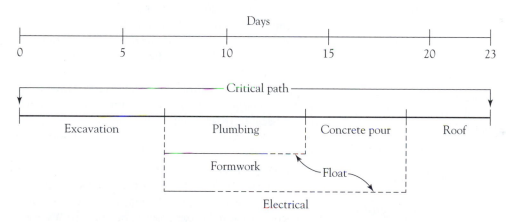

FIGURE 26.3   CPM chart schedule as adjusted.

The total project has now been delayed two days. Plumbing has become part of the critical path, and formwork has moved off the path.

This method can be illustrated by *Morris Mechanical Enterprises, Inc. v. United States.*[15] The contractor was to deliver and install a chiller within 120 days. The contractor delivered the chiller 231 days late. As a result, the government withheld $23,100 as liquidated damages from the final payment.

In ruling for the contractor, the court pointed to the CPM. The schedule showed that delivery and installa-

tion of the chiller were originally on the critical path. However, they were later taken off because of delays to other activities for which the contractor was not responsible. The chiller was to be installed in an equipment room. Another contractor who was responsible for completing the equipment room had difficulty procuring materials. When the chiller was actually delivered, the room had not yet been completed. Even though the chiller was delivered 231 days late, the contractor's breach did not delay the total project, inasmuch as its performance was no longer on the critical path. The court relieved the contractor by concluding it should have been given a time extension, precluding the agency from deducting from the unpaid contract balance.

---

[15]1 Cl.Ct. 50, 554 F.Supp. 433 (1982), aff'd, 728 F.2d 497 (Fed.Cir.), cert. denied, 469 U.S. 1033 (1984)

Requiring the contractor to construct and maintain a CPM schedule has at least three advantages.[16] First, it should require the contractor to work more efficiently. Second, it gives the owner notice of the actual progress of the work. Third, from a litigation standpoint requiring the contractor to maintain a CPM schedule helps prove or disprove the impact of an owner-caused delay. (Schedules are more persuasive evidence if they are actually used during the construction of the project. By contrast, a schedule compiled after completion and with an eye toward litigation will be given little probative value by the courts.)

A CPM schedule has disadvantages. First, it will increase the total contract price. Such schedules are expensive to create and maintain. Second, the contractor may believe such a requirement an unnecessary intrusion into its work. In such a case, the contractor's creation and maintenance of the schedule during construction may be haphazard.

Authoritative commentators state that the following questions must be asked in evaluating a delay claim in which CPM is going to be employed:

1. How was the project actually constructed?
2. What are the differences between the project as planned and as constructed with respect to activities, sequences, durations, manpower, and other resources?
3. What are the causes of the differences or variations between the project as planned and the actual performance?
4. What are the effects of the variances in activities, sequence, duration, manpower, and other resources as they relate to the costs experienced, both by the contractor and by the owner?[17]

As discussed earlier, float is the number of days a noncritical path activity can be delayed before it becomes part of the critical path. Both owner and contractor prefer that the project have float periods. The float periods reflect that every project has some flexibility. A noncritical path activity can be started a few days after it theoretically can

begin without delaying the project. By contrast, a critical path activity must start immediately once the preceding critical path activity has been completed. Furthermore, a noncritical path activity does not have to be completed, as shown in the *Morris* decision, by the date it was scheduled for completion. Again, by contrast, the total project will be delayed only if a critical path activity is delayed. The only constraint on a noncritical path activity is that it cannot be delayed longer than its float period.

A number of cases have explored the problem of "who owns the float."[18] Suppose activity A has a float period of thirty days and is the only activity delayed on the project. If either the owner or the contractor causes activity A to be delayed for fewer than thirty days, neither is responsible for delay damages. The delay did not result in delaying the total project. The project can still be completed by the contract completion date. Conversely, if one of the parties causes activity A to be delayed beyond thirty days, that party is liable for delay damages because activity A has become part of the critical path and the project has been delayed.

Suppose owner and contractor each cause a twenty-day delay to activity A. The total project time will be delayed ten days because this activity originally had a thirty-day float. But who is liable for the ten-day delay? The project would not have been delayed at all if either owner or contractor had not delayed the activity.

If the contractor owned the float, the owner could not charge the contractor for project delay and would be liable for any contractor delay expenses. The contractor's argument that it owns the float can stress that it is responsible for scheduling the activities of the project. In a sense, it created the float because it created the CPM schedule.

But what if the owner requires a CPM schedule or demands ownership of the float? The contractor could change the float period by changing the critical path. Also, owner and contractor rarely bargain for ownership of the float. Usually the contract provides that the contractor must complete by a certain date and the owner must pay a designated sum of money for the completed project. Implicit is that the owner will not interfere with

---

[16]It is misleading to speak of a CPM schedule. In any project where CPMs are used to prove delay claims, it is likely that the claimant will have to show a reasonable "as planned" CPM, an "as-built" CPM reflecting all delays—government, contractor, and excusable—and an adjusted CPM to establish completion of the project absent government delays. See Wickwire, Hurlbut & Lerman, supra note 14.

[17]Wickwire, Hurlbut & Lerman, supra note 14, at 341, analyze these issues in detail in their paper.

---

[18]Compare *Brooks Towers Corp. v. Hunkin-Conkey Constr. Co.*, 454 F.2d 1203 (10th Cir.1972) with *Natkin & Co. v. George A. Fuller Co.*, 347 F.Supp. 17 (W.D.Mo.1972). See generally, Wickwire, Hurlbut, & Lerman, supra note 14; Finke, *The Burden of Proof in Government Contract Delay Claims*, 22 Pub.Cont.L.J. 125 (1992); Lifschitz & Scott, *Who Owns the Float?* Constr. Briefings (Feb. 2005).

the contractor's performance. If there were interference, the contractor would not be able to complete the project by the completion date (this seems to be the AIA hands-off approach). If the owner takes part of the float period, it interferes with the contractor's performance. If interference causes the contractor to incur delay damages, the owner should be liable for the time-related added costs. These are contractor arguments to claim the float.

Some owners take float ownership by contract.[19] Where this is done, the contractor does not receive a time extension even if delayed by the owner. In a noncritical path activity, if the delay can be absorbed, completion is not delayed.[20]

A responsible contractor will inflate its bid price to cover its potential liability for a delayed project. Also, it must absorb its delay costs. At the bidding stage, the contractor cannot estimate the number of days of float the owner will use. Depending on the complexity of the project, the total increase in the bid can be substantial. In addition, initial enforcement of float ownership clauses may be uncertain, principally because of the owner's superior bargaining position.

The *project* can also own the float, using one of two methods. First, the party actually causing the delay to the project is liable. Suppose, in the preceding example, the owner caused the first twenty-day delay. Then the contractor caused another twenty-day delay. The contractor is liable for the ten-day delay of the total project. The owner's delay had used up twenty days of the total thirty-day float, leaving ten days of float remaining. The contractor's delay used up the balance of that period but then caused the activity to be delayed beyond its float period.

This method works fairest if both parties innocently or negligently caused delays to the same activity. However, if the owner's delay had been caused intentionally or willfully, the owner rather than the contractor should be liable.

The second method is borrowed from tort law. Liability for the delayed project depends on the comparative fault of the parties. In the preceding example, because both parties had caused an equal delay to activity A, each is liable for one-half of the other's time-related losses. This method can become complicated if liquidated damages are used (see Section 26.09B).

Each of these solutions has advantages and disadvantages. Giving either the contractor or the owner the float provides some certainty. If it exclusively owns the float, a party is exonerated for delays that use up the float of a noncritical path activity.

In complex projects, it may be best to agree that neither owns the float.[21] In such projects, it is equally likely that both parties will delay the project. The owner will anticipate that it is likely to delay the project but plan on using some of the float. Nevertheless, the contract price may increase if the project owns the float. A responsible contractor will increase its bid when some of its previously recognized rights are taken away but not as much as if the owner exclusively owned the float.

Where the project is simple, it may be best that the contractor own the float. This should develop the lowest contract price. Fewer delays by owner and contractor are likely to occur. If the owner does cause a delay and uses part of the float, the contractor's delay damages may be less than the addition to the contract price that would result if the owner took the entire float for itself.

Generally, the CPM system is practical for only major construction projects. Contractors that bid for these projects already appreciate the usefulness of the schedule. The amount of money spent to construct and maintain the schedule is nominal, compared to the total cost of the project. Liability exposure in such projects can be vast.

C. Other Legal Issues

Sometimes contractors, mainly subcontractors, contend they are entitled to have their work scheduled properly.

---

[19]For example, "The inclusion of float time in the activity listing of the Contractor's Construction Schedule shall be owned entirely by the Owner. The Contractor shall not be entitled to any adjustment in the Contract Time, the Contractor's Construction Schedule, or the Contract Sum, or to any additional payment of any sort by reason of the loss or use of any float time. . . ." This language is found in an unpublished decision, *Construction Enterprises & Constructors, Inc. v. Orting School Dist. No. 344*, 121 Wash.App. 1012, 2004 WL 837912, review denied, 152 Wash. 2d 1034, 103 P.3d 201 (2004). Float is discussed in detail by Wickwire, Hurlbut & Lerman, supra note 14, at 361–62; Lifschitz & Scott, supra note 18, at 7–8.

[20]*E. C. Ernst, Inc. v. Manhattan Constr. Co. of Tex.*, 387 F.Supp. 1001 (S.D.Ala.1974), aff'd 551 F.2d 1026, rehearing denied in part, granted in part, 559 F.2d 268 (5th Cir.1977), cert. denied sub. nom. *Providence Hosp. v. Manhattan Constr. Co. of Tex.*, 434 U.S. 1067 (1978).

[21]See Wickwire, Hurlbut, & Lerman, supra note 14, at 361–62. The authors state that in significant public procurements, contract clauses are included that provide that the float is not for the exclusive benefit of either party to the project. Such clauses permit the individual that gets to the float first to gain the benefit of the float.

One court would not imply an obligation by the prime contractor that it would supply a schedule for subcontractor work because this was not customarily done and the subcontractor did not ask for it in negotiations.[22] The court also observed that the subcontractor had not made a showing that scheduling was important in that particular project. However, the increasing use of obligations of good faith and fair dealing may be the basis for implying that there would be a schedule, particularly if this were customarily done.[23]

An often-ignored issue is whether failure to comply with a progress schedule is a breach of contract. AIA Document A201-2007, Section 3.10.3, as stated earlier, requires the contractor to perform the work in general accordance with the most recent schedules. EJCDC C-700, Paragraph 6.04 (2007), also requires the contractor to "adhere to the Progress Schedule . . . as it may be adjusted from time to time." To underline the greater concern with schedules in documents published by the EJCDC, its C-700, Paragraph 4.02(A)(1), states that the owner can terminate if the contractor does not adhere to the current schedule.

Although schedules are important and the contractor's performance can be terminated for consistent failure to meet the progress schedule, schedule dates must be more flexible than completion dates. The contractor must be given some latitude both in creating and in maintaining the schedule. Yet changing the schedule, not complying with it, and even abandoning it can be serious enough to be a breach, perhaps even a material one.

SECTION 26.06 Causation: Concurrent Causes

A variety of participants and events can cause delays. Delay can be caused by acts of the owner or someone for whose acts the owner is responsible, such as the design professional. Sometimes delays are caused by the contractor or someone for whose acts the contractor is responsible, such as a subcontractor. Sometimes delays are caused by events not chargeable to either owner or contractor, such as nonnegligent fires, unpreventable labor difficulties, or unforeseeably extreme weather. Delays can be caused by third parties, such as a union shutting down the project over a labor dispute.

The multiple causes for delay create a number of legal problems. For example, delays caused by events not chargeable to either party will not justify additional compensation unless there is a warranty by one party that those events will not occur. If a delay is caused by both owner and contractor, it may be extremely difficult to determine which portion of the delay is caused by either party. This can mean that neither party will be able to recover losses it suffers because of the delay,[24] or that any clause liquidating damages will not be upheld, or that the contractor will be limited to a time extension.[25]

Where a clear apportionment can be made between owner-caused delay and contractor-caused delay, as where those delays are sequential rather than concurrent, the delay damages caused by each party may be divided accordingly. The apportionment also can be used to determine the amount of any time extension as noted in Section 26.08.[26]

One case illustrates an attempt to use a "total time" theory much like the total cost method of measuring a contractor's post-completion claim discussed in Section 27.02F.[27] The expert witness for the contractor in a concurrent delay case compared the original and extended completion dates, pointed to a number of individual delay incidents for which the government was responsible that contributed to the overall extended time, and sought to conclude that the entire overrun period was attributable to the government. The court concluded that

---

[22]*Drew Brown Ltd. v. Joseph Rugo, Inc.*, 436 F.2d 632 (1st Cir.1971).

[23]See Section 19.02D.

[24]*Singleton Contracting Corp. v. Harvey*, 395 F.3d 1353 (Fed.Cir.2005) (government's failure to correct the design and contractor's failure to provide a certificate of insurance were concurrent delays); *Westinghouse Elec. Corp. v. Garrett Corp.*, 601 F.2d 155 (4th Cir. 1979) (discretion in trial judge to deny both damages); *Hartford Elec. Applicators of Thermalux, Inc. v. Alden*, 169 Conn. 177, 363 A.2d 135 (1975) (breach by both converts completion date to a reasonable time).

[25]*Hartford Elec. Applicators of Thermalux, Inc. v. Alden*, supra note 24. But see Section 26.09B.

[26]*Robinson v. United States*, 261 U.S. 486, 488–89 (1923); *Essex Electro Engineers, Inc. v. Danzig*, 224 F.3d 1283 (Fed.Cir.2000) (delays by government and contractor in submissions-approval process is subject to apportionment); *R.P. Wallace, Inc. v. United States*, 63 Fed.Cl. 402 (2004) (sequential delays by government and contractor are apportioned to reduce liquidated damages). For a general discussion, see Kutil & Ness, *Concurrent Delay: The Challenge to Unravel Competing Causes of Delay*, 17 Constr. Lawyer, No. 4, Oct. 1997, p. 18. For the view in England and other common law jurisdictions, see Marrin, *Concurrent Delay*, 18 Constr.L.J. 436 (2002).

[27]*Morganti Nat'l, Inc. v. United States*, 49 Fed.Cl. 110 (2001).

this "total time" theory was virtually of no value. It did not prove that the government-caused delay actually delayed the overall completion of the project.[28]

SECTION 26.07 Allocation of Delay Risks

A. Compensable Vs. Noncompensable

At the outset, it is important to divide risk allocation relating to delay into two categories. First, the party whose performance has been delayed may seek relief from its obligation to perform by a particular time. For example, a contractor who has agreed to complete the project by a given date may justify its failure to do so by pointing to the causes that it claims excuse its obligation to complete by the time specified. Second, this contractor may claim additional compensation based on an increase in its anticipated spending because events occurred that delayed its performance. As a general rule, the law has been more willing to excuse performance than to grant additional compensation. Because of this differentiation, speaking simply of risk allocation of events that impede performance can be misleading. This is demonstrated in this section.

B. Common Law

A party who has agreed to perform by a specific time generally assumes the risks of most events that may delay its performance. In the absence of any common law defenses, such as impossibility or mutual mistake, a contractor will not be relieved of its obligation to perform as promised, let alone receive any additional compensation. Relief, if any, must come from the contract.

The contractor normally does not assume the risk that it will be unreasonably delayed by the owner or someone for whom the owner is responsible. However, under certain special circumstances, a contractor may have assumed even this impediment to its performance. For example, suppose an owner constructs an addition to a functioning plant. The contract provides that the owner can order the contractor off the site if the owner's manufacturing operation requires. The contractor has assumed the risk of these delays, though not delays that are beyond those normally contemplated, such as constant or excessive delays or those caused by the owner's bad faith.

C. Fault

The law can take into account the blameworthiness of the party causing the delay. For example, in *Broome Construction, Inc. v. United States*,[29] the Court of Claims held the government not liable for delay in making a work site available where it sought to do so in good faith. The contractor assumed the risk of this delay but would not have assumed the risk of negligently caused delay. The contractor's request for additional compensation requires a clause expressly warranting that the government was making itself strictly liable for its failure to furnish the site.

Fault can also play a role when a party who has been delayed seeks to be relieved from its responsibility. For example, in *J. D. Hedin Construction Co. v. United States*,[30] the Court of Claims held that a contractor was entitled to a time extension because of a cement shortage. The court criticized findings by the Agency Appeals Board that the delays were foreseeable and were the fault of the contractor. The court stated that the contractor could not be expected to show prophetic insight but need only resort "to the usual and long-established methods employed by the commercial world in general."[31]

D. *Force Majeure* Clauses

Many events can occur that cause delay in construction. The common law placing almost all these risks on the contractor has led to the frequent use of *force majeure* clauses, which single out specific events and general causes as justifying relief to the contractor.[32]

The list of specified events justifying a time extension can invoke the canon of interpretation, or better, interpretation guide, of *expressio unius est exclusio alterius* noted in Section 20.02E. Stripped of its Latin phraseology, it means that if one event is specified, those not specified are excluded.

Legal systems can differ in their handling of a catalog of events. Some look upon them as illustrations, and like events can be encompassed within such a clause as justifying relief. Others, such as the common law, look upon the list as exclusive. The common law assumes that if the parties made the effort of specifying some events, they

---

[28]Id. at 134.

[29]203 Ct.Cl. 521, 492 F.2d 829 (1974).

[30]187 Ct.Cl. 45, 408 F.2d 424 (1969)

[31]408 F.2d at 429.

[32]Wright, *Force Majeure Delays*, 26 Constr. Lawyer, No. 4, Fall 2006, p. 33.

must have intended to exclude those not specified.[33] For example, suppose there is an abrupt, sharp rise in the cost of certain material, an event not included in the catalog of events. The *expressio* guide would exclude this event. But such an event might be included were a general "catch-all" phrase included in the clause.

Often, a *force majeure* clause will include a "catch-all" phrase, such as "any other event beyond the control of the contractor." Such a phrase would grant relief if events, though unspecified, occur, have the designated effect on the contractor's performance, and are beyond the control of the contractor. Use of such a "catch-all" invokes another interpretation guide: *ejusdem generis*.[34]

Events claimed to fall within the general or "catch-all" phrase must be *similar* to those events specified. For example, the clause may not encompass an extreme rise in labor costs caused by the outbreak of war, no specific events relating to war being included. But it may encompass delay in transportation of workers to the site because of a strike of public transportation workers when strikes of site workers are specified.

E. Weather

In 1987, the AIA moved weather from its *force majeure* clause (A201, Paragraph 8.3.1) to Paragraph 4.3.8.2, dealing with claims. The latter—currently A201–2007, Section 15.1.5.2—requires the contractor to document "by data substantiating that weather conditions were abnormal for the period of time, could not be reasonably anticipated, and had an adverse effect on the scheduled construction," before it can receive a time extension for weather conditions.

These requirements can make it difficult for a contractor who does not keep detailed weather records to claim a time extension for adverse weather conditions. They reflect a belief by the AIA that weather generally is a risk assumed by the contractor and that only in extraordinary circumstances should weather be the basis for a time extension.[35]

Because a contractor will frequently contend that severe weather should be the basis for a time extension, it is useful to look at *Fortec Constructors v. United States*.[36] The contractor contended that it should have been given a fifty-seven-day time extension because of unusually severe weather. The contract had provided that the contractor would be relieved if it were delayed for causes "other than normal weather," including but not restricted to severe weather. The court then stated,

> Unusually severe weather is "adverse weather which at the time of year in which it occurred is unusual for the place in which it occurred." [Citing case.] Proof of unusually severe weather is generally accomplished by comparing previous years' weather with the weather experienced by the contractor. [Citing cases.] In the present case, contract provision 1A-06(b), a meteorological chart of past weather averages, established the usual weather conditions to be expected during contract performance. Notwithstanding the occurrences of unusually severe weather, however, a plaintiff is only entitled to an extension of contract time if such unusually severe weather has an adverse impact on the construction being performed. [Citing cases.] On a daily basis, the contractor completed a Daily Inspection Report (DIR) and the Government completed a Quality Assurance Report (QAR). These reports record daily rainfall, temperature extremes, and a rating of how the weather affected work that day. While certain testimony indicated that the parties did not always complete the reports on a daily basis, the Court believes that the DIRs and QARs represent the most reliable documents presented regarding both the actual weather at the job site and its effect on job performance. Accordingly, the Court has utilized these documents in assessing the merits of the plaintiff's weather claims.[37]

The court compared the DIRs and the QARs for each month against the contract's chart of bad weather conditions to see whether the contractor had experienced more days of severe weather than expected. The court concluded that for some months, the contractor had experienced less severe weather than expected and awarded it a one-day time extension.

---

[33]*Holder Constr. Group v. Georgia Tech Facilities, Inc.*, 282 Ga.App. 796, 640 S.E.2d 296 (2006), reconsideration denied Dec. 8, 2006 (*force majeure* clause—applicable to delayed delivery of materials caused by government order, riot or other civil disorder, and extreme weather conditions—did not apply to a spike in the price of steel). See also Section 20.02E.

[34]Section 20.02E.

[35]See Annot., 85 A.L.R.3d 1085 (1978).

[36]8 Cl.Ct. 490 (1985), aff'd, 804 F.2d 141 (Fed.Cir.1986). This case was also considered a seminal one for determining the effect of a CPM schedule.

[37]8 Cl.Ct. at 492–93.

F. Subcontractor-Caused Delay

Suppose the prime contractor seeks to excuse its responsibility for delay by pointing to the delay having been caused by a subcontractor. Unlike some contracts, A201-2007, Section 8.3.1, does *not* expressly include subcontractor-caused delay among those that justify a time extension. It does give the architect the right to grant a time extension for any cause that justifies the delay. However, failure to specifically include a common cause of delay, such as subcontractor-caused delay, should mean that the subcontractor-caused delay should not be encompassed within the "catchall" or within the broad grant of power given the architect.

Another reason for not granting a time extension is the single contract system's objective of centralizing administration and responsibility in the prime contractor. Only if subcontractor-caused delay is *specifically* included should it excuse the prime contractor.

The independent contractor rule, although subject to many exceptions, relieves the employer of an independent contractor for the losses wrongfully caused by the latter.[38] Contractors sometimes assert that the subcontractor is an independent contractor inasmuch as the subcontractor is usually an independent business entity and can, to a large extent, control the details of how the work is performed. Even so, the independent contractor rule does not relieve the employer of an independent contractor when the independent contractor has been hired to perform a contract obligation and the party who suffers the loss caused by the independent contractor is the party to whom the contract obligation was owed.[39] In the construction contract context, the owner usually permits the prime contractor to perform obligations through subcontractors. This does not usually mean that the prime contractor is relieved of its obligation to the owner for subcontractor-caused delay unless the owner specifically agrees to exonerate the prime contractor for that delay.

G. Role of Architect

The residual power granted to the architect to grant time extensions in AIA documents has been criticized as leading to a deterioration of any fixed completion date. It does have the virtue of not forcing the drafter to think of every possible event and to include it in the "catalog of events" justifying a time extension.

Delays caused by a separate contractor to another separate contractor were discussed in Section 17.04C.

SECTION 26.08   Time Extensions

Construction contracts usually provide a mechanism under which the contractor will receive a time extension if it is delayed by the owner or by other designated events such as those described in Section 26.07. This section examines the time extension process.

A. Role of Design Professional

Ideally, owner and contractor should agree on the issuance and extent of a time extension. In many construction contracts, the resolution of these issues, at least in the first instance, is given to the design professional or construction manager. The finality of her decision in the event of subsequent arbitration or litigation is discussed later.[40] Whether a time extension should be granted usually requires that the *force majeure* clause be applied to the facts that are asserted to justify a time extension.

B. Duration of Extension

Suppose a time extension is justified. How is the amount of time extension determined? To some degree, this was discussed in Section 26.05, dealing with scheduling. Depending on the existence of float and who can take the benefit of it, delay in performance of a particular activity may not justify any time extension.

Suppose abnormal weather conditions not normally anticipatable precluded work from October 1 to October 14. Suppose that period contained ten working days. Unless the contractor can show that it would have worked on other than normal working days, the time extension should be ten days.[41]

The extent of time extension need not necessarily, however, be the same as the number of days of delay. Suppose

---

[38]See Section 7.03G.

[39]*Harold A. Newman Co. v. Nero,* 31 Cal.App.3d 490, 107 Cal.Rptr. 464 (1973); *Brooks v. Hayes,* 133 Wis.2d 228, 395 N.W.2d 167 (1986).

[40]See Section 29.09.

[41]*Missouri Roofing Co. v. United States,* 357 F.Supp. 918 (E.D.Mo. 1973) appears to support this conclusion.

a fourteen-day delay caused the contractor to work during a period when the weather was more rigorous than during the period of delay. Under such circumstances, a time extension of fifteen days is proper when the contractor was actually precluded from working for only ten days.

One court faced the imprecision of measuring the exact impact of delay. That court arbitrarily granted a time extension of 65 days for a 131-day delay because this amount was "as accurate an estimate as can be made from the actual resulting delay."[42] Whoever determines the amount of delay will be given considerable latitude. However, it is vital for both owner and contractor to keep careful and detailed records. (See Sections 26.10E and 27.05.)

C. Notices

Usually time extension mechanisms provide that the contractor must give notice of the occurrence of an event that is to be the basis for a time extension claim and its probable effect.[43] Some courts seem to consider such notices as technicalities. For this reason, these courts seem quick to find that the notice condition has been waived if it appears that the owner—or someone with actual or apparent authority—knew of the delay-causing event and that a claim would be made and was not harmed by failure to give the notice. Likewise, the requirement will be waived if in any way the owner has misled the contractor into believing that the notice will not be required.[44]

Notice conditions, however, serve a useful function. In dealing with the requirement that a home buyer give notice of a breach of warranty within a reasonable time after discovery of the breach, a court stated, "The requirement of notice of breach is based on a sound commercial rule designed to allow the defendant opportunity for repairing the defective item, reducing damages, avoiding defective products in the future, and negotiating settlements. The notice requirement also protects against stale claims."[45]

Likewise, in the context of a time extension mechanism the notice of an intention to claim a time extension serves a number of useful functions. The notice informs the design professional or owner that people for whom it is responsible, such as other separate contractors, consultants, or the design professional, are delaying the contractor. This can enable the owner or design professional to eliminate the cause of the delay and minimize future delays or damage claims. A timely notice should permit the design professional to determine what has occurred while the evidence is still fresh and witnesses remember what actually transpired.

The notice shows that the contractor has been adversely affected and can eliminate long-delayed, sometimes spurious contractor claims made after completion of the work. If the owner or design professional knew of the event causing the delay, the impact of the event on the contractor's performance, and the fact that the contractor would quite likely ask for a time extension, waiver may be proper.[46] However, the value of the notice requirement declines substantially if courts too frequently ignore the requirement.

SECTION 26.09 Unexcused Contractor Delay

Delay as justification for termination is discussed in Section 33.04A. This section deals with the recovery of damages for the contractor's delayed performance. Although Chapter 27 discusses in greater detail measurement of claims that owners and contractors have against each other, this section deals with damage liquidation—the contractual method used most frequently to deal with contractor delay.

A. Actual Damages

The damage formula applied most frequently by the common law to delayed contractor performance is the value of the lost use of the project caused by the delay. For example, delayed completion of a residence to be occupied

---

[42]*E. C. Ernst, Inc. v. Manhattan Constr. Co. of Tex.*, supra note 20, 387 F.Supp. at 1012–1013.

[43]AIA Doc. A201-2007, §§ 15.1.2, 15.1.5.1.

[44]*Travelers Indem. Co. v. West Georgia Nat'l Bank*, 387 F.Supp. 1090 (N.D.Ga.1974) (provision that HUD approve time extension waived). For additional discussion on waiver of technical requirements, see Section 21.04H.

[45]*Pollard v. Saxe & Yolles Dev. Co.*, 12 Cal.3d 374, 525 P.2d 88, 92, 115 Cal.Rptr. 648, 652 (1974). See also Section 27.05C.

[46]*Southwest Eng'g Co. v. Reorganized School Dist. R-9*, 434 S.W.2d 743 (Mo.Ct.App.1968).

by the owner will be measured by the lost rental value.[47] Although the owner may have suffered other losses, such as the inconvenience of living in a motel or with relatives or having to transport a child to a more distant school, such consequential damages are generally difficult to recover.

In commercial construction, the owner will very likely be able to recover the lost use value in the event of unexcused delay by the contractor.[48] Even in projects that have readily ascertainable commercial value, losses are often suffered that may be difficult to recover. This problem is even greater when the project is built for a public entity that intends to use it as a school, an office building, or a freeway. Although some public projects have a readily ascertainable use value, most do not.

B. Liquidated Damages: *Bethlehem Steel Corp. v. Chicago* and *Rohlin Constr. Co. v. City of Hinton*

Because proving delay damages is very difficult, particularly in public projects, construction contracts commonly include provisions under which the parties agree that certain types of unexcused delay will result in damages of a specific amount. They are usually known as *liquidated damages clauses*. One court stated,

> There was a time when the courts were quite strong in their view that almost every contract clause containing a liquidated damage provision was, in fact, a forfeiture provision which equity abhorred, and therefore, nothing but actual damages sustained by the aggrieved party could be recovered in case of contract breach caused by delay past the proposed completion date. But, in modern times, the courts have become more tolerant of such provisions, probably because of the Anglo-Saxon reliance on the importance of keeping one's word, and have become more strongly inclined to allow parties to make their own contracts and to carry out their own intentions, free of judicial interference, even when such non-intervention would result in the recovery of a prestated amount as liquidated damages, on

proof of a violation of the contract, and without proof of actual damages.[49]

The early common law also felt that contractually stipulated damages could be used unconscionably by parties possessing strong bargaining power. The common law saw its role in awarding contract damages as compensating losses and not punishing or effectuating abuse of power.

The earliest use of such clauses was penal bonds, a precursor of surety bonds explored in Chapter 32. Penal bonds were absolute promises by the maker to pay a designated sum (the penal sum) followed by a condition that this promise would be null and void *if* certain things occurred, such as the maker, usually a debtor or a surety, paying the full amount of the debt. Frequently the penal sum greatly exceeded the harm caused. Equity courts would not enforce penalties. (This is the historical basis for differentiating liquidated damages clauses from penalties.) In addition, English courts felt that damages were the exclusive province of the courts. They would not enforce clauses that would "oust the court of jurisdiction."

Despite the nineteenth-century common law courts' belief in liberty and autonomy, the common law courts followed equity courts and placed strict limits on enforcement of these clauses. Courts were willing under limited circumstances to enforce liquidated damages clauses. They would do so if the party seeking to enforce the clause showed that the damages were extremely difficult to ascertain at the time the contract was made and that the amount selected was a genuine pre-estimate of the damages likely to occur as a probable result of the breach. In addition, some courts would not enforce these clauses unless it were shown that the parties intended the clause to compensate and not to punish. The amount selected must be the sole money award. The proponent should not have the option of choosing liquidated or actual damages.

Modern courts, particularly in public contracts, recognize the difficulty of proving damages generally, particularly those relating to delay, and the certainty that these clauses can provide both parties. As indicated, the law

---

[47]*Muller v. Light*, 538 S.W.2d 487 (Tex.Ct.App.1976) ($100-a-day clause held a penalty). See also *Miami Heart Institute v. Heery Archs. & Eng'rs*, 765 F.Supp. 1083 (S.D.Fla.1991), aff'd, 44 F.3d 1007 (11th Cir.1994) (delay claim by owner against architect) (discussed in detail in Section 12.14B).

[48]*Ryan v. Thurmond*, 481 S.W.2d 199 (Tex.Ct.App.1972).

[49]*Sides Constr. Co. v. City of Scott City*, 581 S.W.2d 443, 446 (Mo. App.1979).

is more willing to be relieved of the burden of measuring damages.[50] Although the results are by no means unanimous, clauses liquidating damages for construction delay are generally enforced if they are reasonable as judged by the circumstances existing at the time the contract was made.[51] Often a comparison is made between the amount stipulated and the contract price.[52] Some states even require such clauses in public construction contracts.[53] A case that typifies the modern judicial attitude toward such clauses is reproduced next.

BETHLEHEM STEEL CORP. v. CHICAGO

United States Court of Appeals, Seventh Circuit, 1965. 350 F.2d 649.
[Ed. note: Footnotes omitted.]

GRANT, District Judge.

Plaintiff-Appellant (Bethlehem) brought this action to recover an item of $52,000.00 together with certain items of interest, etc., withheld by the Defendant (City), as liquidated damages for delay in furnishing, erecting, and painting of the structural steel for a portion of the South Route Superhighway, now the "Dan Ryan Expressway," in the City of Chicago. . . . [T]he District Court concluded that Plaintiff's claims on the items in controversy should be denied and entered judgment accordingly. We agree and we affirm.

The trial court's findings included the following uncontroverted facts:

* * *

The work which Bethlehem undertook was the erection in Chicago of structural steel for a 22-span steel stringer elevated highway structure, approximately 1,815 feet long, to carry the South Route Superhighway from South Canal Street to the South Branch of the Chicago River. Bethlehem's work was preceded and followed by the work of other contractors on the same section.

The "Proposal and Acceptance" in the instructions to bidders required the bidders to ". . . complete . . . within the specified time the work required. . . ." Time was expressly stated to be the essence of the contract and specified provisions were made for delivery of the steel within 105 days thereafter, which was to be not later than 15 days from notification. The successful bidder was to submit to the Commissioner of Public Works a "Time Schedule" for his work and if "less than the amount . . . specified to be completed" were accomplished "the City may declare this contract forfeited. . . ." The work had to be completed irrespective of weather conditions.

The all important provision specifying $1,000 a day "liquidated damages" for delay is as follows:

The work under this contract covers a very important section of the South Route Superhighway, and any delay in the completion of this work will materially delay the completion of and opening of the South Route Superhighway thereby causing great inconvenience to the public, added cost of engineering and supervision, maintenance of detours, and other tangible and intangible losses. Therefore, if any work shall remain uncompleted after the time specified in the Contract Documents for the completion of the work or after any authorized extension of such stipulated time, the Contractor shall pay to the City the sum listed in the following schedule for each and every day that such work remains uncompleted, and such moneys shall be paid as liquidated damages, not a penalty, to partially cover losses and expenses to the City.

Amount of Liquidated Damages per Day . . . $1,000.00.

The City shall recover said liquidated damages by deducting the amount thereof out of any moneys due or that may become due the Contractor. . . .

[50]*Sides Constr. Co. v. City of Scott City,* supra note 49 (avoids laborious item-by-item damage recitations); *Osceola County v. Bumble Bee Constr., Inc.,* 479 So.2d 310 (Fla.Dist.Ct.App.1985) (city tourist information center).

[51]*Dahlstrom Corp. v. State Highway Comm'n of State of Miss.,* 590 F.2d 614 (5th Cir.1979); *Pembroke v. Gulf Oil Corp.,* 454 F.2d 606 (5th Cir.1971); *Dave Gustafson & Co. v. State,* 83 S.D. 160, 156 N.W.2d 185 (1968). Cases are collected in Annot., 12 A.L.R.4th 891 (1982). The Uniform Commercial Code (U.C.C.) enforces them if reasonable in the light of actual or anticipated damages. See U.C.C. § 2-718.

[52] *Dahlstrom Corp. v. State Highway Comm'n of State of Miss.,* supra note 51.

[53]West Ann.Cal.Pub.Cont.Code § 10226. The predecessor of this statute, West Ann.Cal.Govt.Code § 14376, is cited in *Westinghouse Elec. Corp. v. County of Los Angeles,* 129 Cal.App.3d 771, 181 Cal.Rptr. 332, 339 (1982).

Provision was made to cover delay in a contractor's starting due to preceding contractor's delay. Unavoidable delays by the contractor were also covered, and extensions therefor accordingly granted.

Bethlehem's work on this project followed the construction of the foundation and piers of the superhighway by another contractor. Bethlehem, in turn, was followed by still another contractor who constructed the deck and the roadway.

Following successive requests for extensions of its own agreed completion date, Bethlehem was granted a total of 63 days' additional time within which to perform its contract. Actual completion by Bethlehem, however, was 52 days after the extended date, which delay the City assessed at $1,000.00 per day, or a total of $52,000.00 as liquidated damages.

Bethlehem contends it is entitled to the $52,000.00 on the ground that the City actually sustained no damages. Bethlehem contends that the above-quoted provision for liquidated damages is, in fact, an invalid penalty provision. It points out that notwithstanding the fact that it admittedly was responsible for 52 days of unexcused delay in the completion of its contract, the superhighway was actually opened to the public on the date scheduled.

In other words, Bethlehem now seeks to re-write the contract and to relieve itself from the stipulated delivery dates for the purposes of liquidated damages, and to substitute therefor the City's target date for the scheduled opening of the superhighway. This the Plaintiff cannot do.

In *Wise v. United States*, . . . the Supreme Court said:

> . . . [T]he result of the modern decisions was determined to be that . . . courts will endeavor, by a construction of the agreement which the parties have made, to ascertain what their intention was when they inserted such a stipulation for payment, of a designated sum or on a designated basis, for a breach of a covenant of their contract. . . . When that intention is clearly ascertainable from the writing, effect will be given to the provision, as freely as to any other, where the damages are uncertain in nature or amount or are difficult of ascertainment or where the amount stipulated for is not so extravagant, or disproportionate to the amount of property loss, as to show that compensation was not the object aimed at or as to imply fraud, mistake, circumvention or oppression. *There is no sound reason why persons competent and free to contract may not agree on this subject as fully as on any other, or why their agreement, when fairly and understandingly entered into with a view to just compensation for the anticipated loss, should not be enforced.*
>
> . . . *The later rule, however, is to look with candor, if not with favor, on such provisions in contracts when deliberately entered into between parties who have equality of opportunity for understanding and insisting on their rights, as promoting prompt performance of contracts and because adjusting in advance, and amicably, matters the settlement of which through courts would often involve difficulty, uncertainty, delay and expense. . . .*
>
> . . . It is obvious that the extent of the loss which would result to the Government from delay in performance must be uncertain and difficult to determine and it is clear that the amount stipulated for is not excessive. . . .
>
> The parties . . . were much more competent to justly determine what the amount of damage would be, an amount necessarily largely conjectural and resting in estimate, than a court or jury would be, directed to a conclusion, as either must be, after the event, by views and testimony derived from witnesses who would be unusual to a degree if their conclusions were not, in a measure, colored and partisan. [Italics supplied.]

* * *

Affirmed.

The *Bethlehem* decision illustrates increased judicial cordiality toward liquidated damages clauses in the construction context, particularly public contracts. The law is more willing to enforce such contracts if they have been bargained for and if actual damages would be difficult to establish in court. Although the amount stipulated is rarely the result of bargaining in a competitive bid contract, the contractor can adjust its bid to take this risk into account.[54]

In another road-building contract, however, the Supreme Court of Iowa expressed concern with the arbitrary way in which the liquidated damages amount was selected. That case is reproduced next.

---

[54]See Sweet, *Liquidated Damages in California*, 60 Calif.L.Rev. 84, 118–23 (1972). But see *Space Master Internat'l, Inc. v. City of Worcester*, 940 F.2d 16 (1st Cir.1991), to which reference was made in the *Rohlin* case reproduced in this section. The contract involved modular classrooms, and delay meant that children had to have classes in hallways, gymnasiums, auditoriums, and libraries. Educational programs were forced to be compromised, and morale suffered. These facts would certainly have been good enough to allow damages that could be liquidated. The court pointed to the fact that the amount selected was not the result of arm's-length bargaining but was simply put out to competitive bidding. It used this fact to support its holding that a summary judgment should not be granted. It did not recognize the difficulties of negotiating a liquidated damages clause in competitive bidding for public work.

ROHLIN CONSTRUCTION CO., INC. v. CITY OF HINTON

Supreme Court of Iowa, 1991. 476 N.W.2d 78.

Considered by HARRIS, P.J., and SCHULTZ, CARTER, LAVORATO and SNELL, J.J.

SCHULTZ, Justice.

The issues in this appeal center on the question of whether the liquidated damage clauses in three road construction contracts were actually penalty clauses and therefore void. The trial court refused to impose liquidated damages on the contractor. Our court of appeals affirmed. We affirm both courts.

In the spring of 1988, defendants Plymouth County (county) and the City of Hinton (city) entered a joint project for resurfacing certain county and city roads. Defendants jointly planned and advertised for bids on one city and two county resurfacing projects. Plaintiff Rohlin Construction Co., Inc. (Rohlin) was the successful bidder and entered into two contracts with the county for total prices of $221,588.39 and $251,696.99, and one contract with the city for a price of $37,957. The two county contracts were dated May 31, 1988, and the city contract was dated June 16, 1988.

All three contracts contained a provision requiring that work be completed within forty "working days." The contracts specified a completion date of September 2, 1988, but did not specify a starting date. The contract language was prepared by the Iowa Department of Transportation on "proposal forms" which contained the following language:

If this bid is accepted, Bidder agrees . . . to either complete the work within the contract period or pay liquidated damages, which shall accrue at the daily rate specified below, for each additional working day the work remains uncompleted.

Each contract established $400.00 per day as the amount of liquidated damages.

Rohlin commenced work on September 17 and completed the project in less than thirty days. Rohlin completed the project 25.5 days past the September 2 deadline on the city contract and 27.5 days and 28 days late on the county contracts. The city withheld $10,200 and the county withheld $22,200 as liquidated damages for late completion of the project.

Rohlin then commenced separate law suits against the city and county seeking judgments against both defendants for the sums withheld, plus interest and attorney fees. Following trial, the court ruled in favor of Rohlin, allowing recovery for the amount of the withholdings plus interest at eight percent from November 15, but denying attorney fees.

* * *

In the past, we disfavored the use of liquidated damage clauses and favored interpretation of contracts that make stipulated sums penalties. . . . Later, we relaxed this penalty rule and recognized that parties may fix damages by contract when the amount of damages is uncertain and the amount fixed is fair. . . . This change in contractual interpretations is consistent with the trend of favoring liquidated damage clauses.

* * *

We often turn to Restatements of the Law and believe it is appropriate to do so in this case. The American Law Institute adopts a more conservative approach as follows:

Damages for breach by either party may be liquidated in the agreement but only at an amount that is reasonable in the light of the anticipated or actual loss caused by the breach and the difficulties of proof of loss. A term fixing unreasonably large liquidated damages is unenforceable on grounds of public policy as a penalty.

Restatement (Second) of Contracts § 356(1) (1981). However, the American Law Institute shows no hostility toward liquidated damages by stating:

The parties to a contract may effectively provide in advance the damages that are to be payable in the event of breach as long as the provision does not disregard the principle of compensation. The enforcement of such provisions for liquidated damages saves the time of courts, juries, parties and witnesses and reduces the expense of litigation. This is especially important if the amount in controversy is small. However, the parties to a contract are not free to provide a penalty for its breach. The central objective behind the system of contract remedies is compensatory, not punitive. Punishment of a promisor for having broken his promise has no justification on either economic or other grounds and a term providing such a penalty is unenforceable on grounds of public policy.

Id. comment a. This Restatement section also sets out the test for a penalty:

Under the test stated in Subsection (1), two factors combine in determining whether an amount of money fixed as damages is so unreasonably large as to be a penalty. The first factor is the anticipated or actual loss caused by the breach. The amount fixed is reasonable to the extent that it approximates the actual loss that has resulted from the particular breach, even though it may not approximate the loss that might have been anticipated under other possible breaches. Furthermore, the amount fixed is reasonable to the extent that it approximates the loss anticipated at the time of the making of the contract, even though it may not approximate the actual loss. The second factor is the difficulty of proof of loss. The greater the difficulty either of proving that loss has occurred or of establishing its amount with the requisite certainty, the easier it is to show that the amount fixed is reasonable. To the extent that there is uncertainty as to the harm, the estimate of the court or jury may not accord with the principle of compensation any more than does the advance estimate of the parties. A determination whether the amount fixed is a penalty turns on a combination of these two factors. If the difficulty of proof of loss is great, considerable latitude is allowed in the approximation of anticipated or actual harm. If, on the other hand, the difficulty of proof of loss is slight, less latitude is allowed in that approximation. If, to take an extreme case, it is clear that no loss at all has occurred, a provision fixing a substantial sum as damages is unenforceable.

Id. comment b (citations omitted). We believe that application of these principles is appropriate to our determination in this case.

A review of the record reveals uncertainty of how the sum of $400 per day for liquidated damages was derived. This liquidated damage amount was placed in the specifications as a result of the county engineer's consultation with someone in the office of the Department of Transportation (DOT). A DOT construction manual contained a schedule of suggested rates for liquidated damages based strictly on the engineer's estimate of the contract price. According to the schedule in the manual, the city's contract called for $100 per day in liquidated damages and the two county contracts called for liquidated damages of $200 and $300 per day respectively, for a total of $600. As noted previously, there were three separate contracts but the resurfacing project was a joint city-county project involving connected highways. Thus, if the total amount rather than the individual amounts of the three contracts is used, then the manual's suggested rate is $400 per day in liquidated damages.

The county engineer indicated that the reason for deviating from the manual's suggested rates was due to the city and county's desire for a completion date prior to the increased traffic that would accompany the start of school and the grain-hauling season in Hinton. In addition, the engineer stated that "we wanted the liquidated damage amount to be sufficient to make the contractor aware that we need that project completed."

There is no valid justification for the individual liquidated damage amounts contained in each of the three contracts. Under the record of this case, the person who set the $400-per-day amount in each contract is unknown and was not called as a witness. Additionally, no witness was called to justify the suggested liquidated damage amounts contained in the DOT manual schedule. The county engineer did not conduct studies or present any other data suggesting that defendants anticipated that the government entities and the public could sustain damages equivalent to the $400-per-day liquidated damage amount contained in each of the three contracts. Furthermore, plaintiff called the school superintendent as a witness to give evidence that the school experienced no problems due to the road work. The county engineer also indicated that a Hinton grain elevator company had not complained that delayed completion of the road work caused the company or its patrons any damages or losses.

Plaintiffs seem to contend that Rohlin's delayed completion of the project caused no damages. The county did sustain damages, however, due to erosion because it could not seed the highway shoulders because of the delay. Undoubtedly, there was some inconvenience to school bus drivers and to grain haulers because of the late completion.

We agree with the trial court that the provisions for liquidated damages in the three contracts were penalties rather than reasonable amounts for liquidated damages. "Liquidated damages must compensate for loss rather than punish for breach. . . ." *Space Master Int'l, Inc. v. City of Worcester*, 940 F.2d 16, 18 (1st Cir.1991). We recognize that proving the amount of loss with any degree of certainty is difficult; nevertheless, the amount of liquidated damages set in each contract appears to be unreasonably large and goes far beyond the anticipated loss caused by delay in performance of the contract. The road project would inevitably cause some inconvenience to the parties and the public for a certain number of days regardless of when performed. Therefore, we conclude that the $400-per-day liquidated damage clause contained in each of the three contracts is an unrealistic amount and is therefore a penalty that should not be enforced.

The court of appeals affirmed the trial court's denial of an award of attorney fees to plaintiff. We agree that this determination is a matter within the discretion of the trial court and it did not abuse its discretion in denying plaintiff attorney fees.

Costs on appeal are taxed two-thirds to Plymouth County and one-third to the City of Hinton.

DECISION OF COURT OF APPEALS AND JUDGMENT OF DISTRICT COURT AFFIRMED.

The court's concern over the absence of evidence indicating why the particular amount was selected, with particular reference to the DOT manual schedule, will place unreasonable obstacles to the enforcement of such clauses. The difficulty of amassing proof of this sort is one of the reasons why parties liquidate and why modern courts as exemplified in the *Bethlehem* decision are more inclined to enforce such clauses. Had the public entity introduced testimony indicating why it felt it necessary to deviate from the manual schedule, the outcome might have been different. But if all the bidders had taken into account the likely enforcement of the liquidated damages clauses and had seen the importance the public entity attached to timely completion, they should have included in their bids an amount to take care of the possibility that additional resources would have to be used to accomplish timely completion. If Rohlin did not, and thereby obtained the contract, this would not have been fair to the other bidders who bid rationally. If Rohlin had included these pricing contingencies, it unjustly enriched itself in this contract. To be sure, there would not have been unjust enrichment if Rohlin had signed the contracts on the assumption that the liquidated damages clause would never be enforced.

The court's decision also reflects its belief that if a party's delayed performance cannot be said to have caused monetary losses, the liquidated damages clause should not be enforced. It is clear that actual losses in such cases cannot be established at the time the contract is made or even after it is performed. This is the reason for using liquidated damages clauses.

While both *Bethlehem Steel* and *Rohlin* involved public contracts, liquidated damages are widely enforced in private works agreements as well. Delayed completion causes a private owner to suffer a variety of losses, both direct and indirect, including lost rental income, higher financing costs, higher administrative expenses, the cost of an idled workforce, and other damages. An owner faces a daunting task of proving all of these damages with reasonable certainty. In addition, the cascading consequences of delayed completion means that the owner would have to establish that its claimed damages were direct, rather than consequential.[55] Supplanting proof of actual damages with a stipulated damages sum provides many advantages to a private owner, whether commercial or residential.

Although there is a greater willingness of modern courts to enforce liquidated damages clauses, additional legal issues remain. For example, if both the owner and contractor delay the project, liquidated damages will be assessed only to that period of delay apportioned solely to the contractor.[56] Judicial acceptance of such clauses also means that the party challenging the clause, not the party seeking to enforce the clause, bears the burden of establishing that it is a penalty.[57]

One reason courts enforce liquidated damages clauses is that the application of such clauses precludes the need for a lengthy and difficult trial on the issue of the owner's damages. Owners must understand, however, that these clauses also preclude them from recovering actual damages, even if those damages are apparently well in excess of the liquidated sum.[58] On the other hand, contractors are precluded from arguing that the owner's actual damages were less or even nonexistent.[59] As explained by two commentators, "liquidated damages become a ceiling as well as a floor for establishing an owner's recovery for contractor-caused delay."[60]

Should a liquidation clause be applied if the contractor *abandons* the project? It can be contended that parties drafting such clauses are thinking principally of *delayed* completion by the contractor and not *abandonment,* particularly if followed by abandonment of the project by the owner.[61] Another reason for not applying the clause is the fear that the owner may be tempted to delay completion and thereby increase the liquidated damages amount.

Although application can appear to create open-ended liability, the clause can be applied for a reasonable period

---

[55]Heckman & Edwards, *Time Is Money: Recovery of Liquidated Damages by the Owner,* 24 Constr. Lawyer, No. 4, Fall 2004, p. 28.

[56]*Robinson v. United States,* supra note 24; *William F. Klingensmith, Inc. v. United States,* 731 F.2d 805 (Fed.Cir.1984); *Jasper Constr. Inc. v. Foothill Junior College Dist.,* 91 Cal.App.3d 1, 153 Cal.Rtpr. 767 (1979)

[57]*S. Brooke Purll, Inc. v. Vailes,* 850 A.2d 1135 (D.C.2004) (unusual case involving a contractor's use of a liquidated damages clause); *Commercial Union Ins. v. La Villa Indep. School Dist.,* 779 S.W.2d 102 (Tex.Ct.App.1989) (party opposing liquidated of damages must prove actual damages to show liquidation amount was not a reasonable approximation); West Ann.Cal.Civ.Code § 1671(b) (burden on opponent, but not in certain consumer transactions).

[58]*Worthington Corp. v. Consolidated Aluminum Corp.,* 544 F.2d 227 (5th Cir.1976).

[59]*Southwest Eng'g Co. v. United States,* 341 F.2d 998 (8th Cir.), cert. denied, 382 U.S. 819 (1965).

[60]Heckman & Edwards, supra note 55, at 31.

[61]*City of Elmira v. Larry Walter, Inc.,* 76 N.Y.2d 912, 564 N.E.2d 655, 563 N.Y.S.2d 45 (1990).

of time, not for infinity. Another way of dealing with this problem is to contractually "cap" the liquidation amount and thereby avoid the open-ended liability that can generate a forfeiture vastly disproportionate to the actual damages.[62]

Unjustified abandonment by the contractor or proper termination by the owner can cause the owner *two* harms: additional cost to complete the project and additional cost caused by delayed completion. Damage liquidation can apply to the second, delay in completion that is determined by the actual completion by a successor contractor or when the project could have been reasonably completed by a successor.[63]

Another problem relates to the difference between substantial and final completion of the project.[64] This issue was treated in *Hungerford Construction Co. v. Florida Citrus Exposition, Inc.*[65] The contract involved an exhibition center for the citrus industry. One important feature of the project was a concrete dome 170 feet in diameter that was to operate as the center's roof. The roof was to be waterproof without independent waterproof covering. Completion time was 180 calendar days, and the liquidated damages were specified to be $200 per calendar day of delay.

The project was completed within 180 days, and the owner moved into the project. However, from the beginning the roof leaked, and corrective work was necessary. The corrective work did not require the owner to leave the premises, but it did preclude the owner from making the premises available to those who might want to rent it for exhibitions. The court referred to this use as a secondary use. Even after the corrective work, the secondary use was diminished because the leaking and correction had caused unsightly discolored plaster across the roof.

The court held that the liquidated damages clause would not be applied in this case. The court stated that the building was available to the owner for its primary use and that the loss of secondary use was entirely speculative.

To deal with this problem, the contract can create a two-tier liquidation system, with one tier dealing with substantial completion and the other with final completion. This would avoid the *Hungerford* case result, in which it is likely that the owner will be unable to establish losses with sufficient certainty for the reduced use value of the structure while corrective work was being performed.

Applicability of the clause also was an issue in *Northern Petrochemical Co. v. Thorsen & Thorshov, Inc.*[66] After completion, serious structural defects were discovered. This necessitated large-scale redesign and reconstruction. Correcting the defects took eight months. The court affirmed an award based on lost profits because of delayed occupancy and excess operating costs while awaiting occupancy. The court rejected application of the contractual liquidated damages clause, noting that it was intended to apply to normal delay in the "rate of construction," not to an extraordinary eight months' delay due to redesign and reconstruction.

The use of fast-tracking[67] can generate particular problems, especially if the project is being built by a number of separate contractors or multiple prime contractors. In *Casson Construction Co.*,[68] the contractor, though finishing on time, failed to complete a phase of its work that delayed a follow-on contractor. The follow-on contractor was ordered to accelerate. It did so by the use of overtime and double shifts for several weeks. This acceleration cost $174,000.

The breaching contractor claimed its liability was limited to the liquidation of a damages amount of $240 a day, as contained in its contract. The cost of acceleration was $644 a day. The board pointed to another provision in the contract stating that the contractor would indemnify the public agency for acceleration payments made to other contractors. The court held that the latter clause controlled and that the same breach can invoke different clauses. The board would not apply the liquidated damages clause to milestone date delays, only to delay in completing the entire contract performance. The board noted that the standardized language used in the contract was adopted long before the advent of fast-track construction.

[62]*Reed & Martin, Inc. v. Westinghouse Elec. Corp.*, 439 F.2d 1268 (2d Cir.1971).

[63]*Construction Contracting & Mgmt., Inc. v. McConnell*, 112 N.M. 371, 815 P.2d 1161 (1991). Compare *City of Boston v. New England Sales & Mfg. Corp.*, 386 Mass. 820, 438 N.E.2d 68 (1982) and *Austin-Griffith, Inc. v. Goldberg*, 224 S.C. 372, 79 S.E.2d 447 (1953) (limited to abandonment after completion date) with *Continental Realty Corp. v. Andrew J. Crevolin Co.*, 380 F.Supp. 246 (S.D.W.Va.1974) (refusal to apply clause). For a collection of cases, see Annot., 15 A.L.R.5th 376 (1993).

[64]AIA Doc. A101-2007, § 3.3, measures the time commitment by substantial completion.

[65]410 F.2d 1229 (5th Cir.), cert. denied, 396 U.S. 928 (1969).

[66]297 Minn. 118, 211 N.W.2d 159 (1973).

[67]See Section 17.04B.

[68]GSBCA No. 4884, 78-1 BCA ¶13,032.

Subcontracting raises special problems. Sometimes prime contractors seek to charge all subcontractors a ratable amount of any liquidated damages that the prime contractor must pay the owner. This avoids the often difficult question of which subcontractors have caused the delay, yet it requires a subcontractor who has not caused the delay to share the liquidation loss.[69] As a result, AIA Document A401-2007, Section 3.3.1, expressly provides that liquidated damages against the subcontractor are assessed "only to the extent caused by the Subcontractor" or those for whom it is responsible. Unfortunately, this language could justify a claim by the prime contractor against a subcontractor who has caused the delay even if the subcontractor has not breached its contract.

Actual damages suffered by the prime contractor itself can be separated from amounts that the prime contractor must pay the owner.[70] There is no reason why the liquidated damages clause cannot encompass both. Although sometimes the amount that the prime contractor may have to pay the owner can be ascertained at the time the contract is made (the stipulated amount in the owner–prime contract), the *other* damages that the prime may suffer in *addition* may be difficult to establish. Two provisions can be included: one making clear that the subcontractor will reimburse the prime for any amounts it must pay the owner attributable to the subcontractor's breach, and another to liquidate damages for other harm suffered by the prime. Alternatively, the contract can specify that the liquidation amount applies only to the liquidation amounts that must be paid by the prime contractor to the owner, not precluding the recovery of actual damages suffered by the contractor as well. But the parties should be able to include both in one clause.

Project completion dates can be "hard" or "soft," depending on the importance attached to timely completion. When a high liquidation amount is selected, the contractor should increase its contract price either to ensure timely completion (double time, expedited deliveries, more workers) or to pay damage liquidation. Coupling a soft completion date with a stiff damage liquidation amount will generate timely completion at a high cost when timely completion is not crucial.

When the completion date is hard, the amount selected must be sufficiently high (but not too high so as to risk finding that it is a penalty) to make it more profitable for the contractor to finish on time than to delay and pay liquidated damages. Selecting an amount to accomplish this objective requires a sophisticated understanding of contractor costs. It involves awareness of the increased administrative costs of such a clause. Invariably, there will be more requests for time extensions. Any stiff clause should be accompanied with a contractual justification for the amount, as in the *Bethlehem* case.

This advice is inconsistent with the stated requirement that liquidated damages be based on a genuine pre-estimate of actual damages that will result in the event of delay. In many construction contracts—mainly public but also private—the amount selected does not in reality represent an attempt to estimate damages. In such contracts, it is almost impossible to estimate the economic loss caused by delay. For example, the $1,000 a day selected in the *Bethlehem* case probably did not reflect the City of Chicago's judgment as to the economic loss that it or its citizens would suffer if the project was delayed. It is more likely that the amount was selected to make it more economical for the contractor to perform on time than to delay and pay damages. Although courts do not overtly concede that they are doing so, their enforcement of liquidated damages clauses for a delay in many construction contracts amounts to enforcing reasonable penalties.

Sometimes liquidation clauses are joined with bonus provisions. Under such clauses, the contractor forfeits a designated amount for each day of unexcused delay but gains a designated daily bonus if it completes the project before the completion date. Although enforceability does not require that a bonus be attached to a damage liquidation, it may be tactically desirable to do so. The bonus clause may help to enforce the liquidation clause, because "mutuality" is attractive. A bonus clause may also make it appear that the amount had actually been bargained by the contractor and owner. However, a bonus clause should not be used unless it is very important to obtain early completion.

Section 27.06D will discuss contractual methods of excluding consequential damages. The relationship of such contractual provisions and liquidated damages will be discussed in that section. However, it should be noted here that the AIA's effort to exclude consequential damages could, according to the AIA, be frustrated by the owner's use of harsh liquidated damages clauses. As a

---

[69]*United States v. Arundel Corp.*, 814 F.2d 193, decision clarified on denial of rehearing, 826 F.2d 298 (5th Cir.1987), appeal after remand, 896 F.2d 143 (5th Cir.1990), rehearing denied Apr. 11, 1990.

[70]*United States v. Foster Constr. (Panama) S.A.*, 456 F.2d 250 (5th Cir.1972).

result, the two subjects were treated together in AIA Document A201-2007, Section 15.1.6.

Damage liquidation for unexcused contractor delay is usually desirable even though claims based on liquidated damages clauses are often traded away in a final settlement. Liquidated damages clauses must be drafted carefully. First, the applicable law must be determined. Second, the clause must be tailored to the particular type of delay to which it is expected to apply. Third, the amount selected should take into account the importance of timely completion, the likely lost use value, and the likelihood that the amount selected will actually achieve the objective.

SECTION 26.10 Owner-Caused Delay

A. Assumed Contractual Risk: No-Damage Clauses: *Triple R Paving, Inc. v. Broward County*

Increasingly, contractors make large claims for delay damages. The substantive basis for what are sometimes called "delay and disruption" or inefficiency claims is the implied obligation not to prevent the contractor from performing its obligations in a logical, orderly, and efficient manner.[71] Delay can mean not being able to work, causing the project not to be completed on time. Disruption can mean being prevented from working efficiently. This section deals with the response of the owner to the possibility of delay and disruption claims being made.

Before looking at these responses, it is important to examine the nature of these claims. Some are based on the owner, or those for whom the owner is responsible,[72] not doing a proper job of *communicating information (Spearin* claims),[73] *designing* (excessive changes or changes beyond the power granted the owner by the changes clause),[74] or *administering* the project.[75]

As illustrations, cases have pointed to failure to schedule the work so that it could be done smoothly and quickly;[76] creating site congestion, failing to coordinate the project (where that was the owner's responsibility);[77]

failure to deliver owner-supplied material on time;[78] acceleration, which generates excessive working hours and overcrowded conditions;[79] and constant revision of drawings causing confusion and interruption of the orderly progress of the work.[80] These adversely affect productivity and impede "job rhythm," a necessary condition for good productivity.[81]

Returning to owner responses to the threat of such contractor claims, public entities are often limited by appropriations and bond issues. As a result, they must know in advance the ultimate cost of a construction project. To do so, many public entities use the disclaimer system for unforeseen subsurface conditions described in Section 25.05. Similarly, they wish to avoid end-of-project claims based on allegations that they have delayed completion or required the contractor to perform work out of sequence. These entities recognize that barring claims may cause higher bids, but they would prefer to see bidders increase their bids to take this risk into account rather than face claims at the end of the job.

Before looking at the troublesome question of the validity of no-damage clauses, it is important to note other methods that can be used to eliminate or reduce the likelihood of such claims. The contract may specify that the owner has the right to delay the contractor and that the interference is not a contract breach. The surrounding facts and circumstances may indicate that the claimant, prime, or subcontractor could expect to have its performance interrupted. For example, a subcontractor knew that it was working on an existing, functioning hospital and that it could not expect to hold the prime contractor responsible for the inevitable delays.[82] Even in such a case, there would be limits to the extent of interference.[83]

---

[71]See Section 19.02A.

[72]See Section 17.05B.

[73]See Section 23.05E.

[74]See Section 21.04B.

[75]See Section 17.05C.

[76]*U.S. Industries, Inc. v. Blake Constr. Co.*, 671 F.2d 539 (D.C. Cir.1982), rehearing denied March 23, 1982.

[77]*Howard P. Foley Co. v. J. L. Williams & Co. Inc.*, 622 F.2d 402 (8th Cir.1980) (Arkansas law).

[78]Ibid.

[79]*S. Leo Harmonay, Inc. v. Binks Mfg. Co.*, 597 F.Supp. 1014 (S.D.N.Y.1984), aff'd without opinion, 762 F.2d 990 (2d Cir.1985) (New York law).

[80]Ibid; Jones, *Lost Productivity: Claims for the Cumulative Impact of Multiple Change Orders,* 31 Pub.Cont.L.J. 1 (Fall 2001).

[81]Burke, *Productivity Loss Claims,* a paper presented to the Surety and Fidelity Claims Conference Ass'n, April 18, 1991.

[82]*Port Chester Elec. Constr. Corp. v. HBE Corp.*, 978 F.2d 820, 821 (2d Cir.1992) (New York law).

[83]But see *Keeney Constr. v. James Talcott Constr. Co., Inc.*, 45 P.3d 19 (Mont.2002), rehearing denied Apr. 30, 2002 (court did not seek to place reasonable limits on the contractor's power to determine completion date where work was to be done as directed by the contractor).

Another technique is to contend that any time extension mechanism is the exclusive remedy. Generally, the availability and use of the time extension mechanism does not impliedly preclude delay damages.[84] The AIA's time extension mechanism specifically states that it does not bar the contractor from recovering delay damages.[85]

Many public and some private construction contracts meet the delay damage problem head on. These contracts contain no-damage or no-pay-for-delay clauses. Such clauses attempt to place the entire risk for delay damages on the contractor and to limit the contractor to time extensions.

Such clauses have become very controversial. Owners—usually public entities, but increasingly private owners—justify these clauses as a means of fiscal control. Public owners are often limited as to what they can spend by appropriation bills or bond issues. Also, cost overruns can be politically devastating. Public owners want *all* costs, including delay and disruption, to be put "up front" in the contract price. They do not want potentially open-ended delay or disruption claims at the end of the project.

Contractors oppose such clauses. First, they claim that in public contracts they have no choice but to take these clauses. They also say that such costs cannot be rationally priced and included in the contract price. Most such claims result from poor administrative practices, such as excessive change orders, delay in furnishing necessary information, and dilatory submittal approvals. (Other illustrations will be given later in this section.) How can a contractor know, at the time it bids, the likelihood of such delay and disruption and what they will cost? Also, contractors say if they do price these risks and put them in their bid prices, in the cutthroat competition of the construction world, they will lose the awards to contractors who will not. The latter will take their chances that they will not suffer such losses or that they can avoid these clauses in court.

Finally, contractors point to the moral hazard such clauses create. An owner and its design professional insulated by such clauses will not do their best to administer the project efficiently.

These objections have not gone unnoticed by courts and legislatures. As shall be seen, although a few courts

will interpret the clauses "as written,"[86] most will find exceptions that will justify not applying such a clause literally.[87] *U.S. for Use and Benefit of Williams Electric Co., Inc. v. Metric Constructors, Inc.,*[88] after stating that generally such clauses are enforced, stated,

> A majority of jurisdictions, however, recognize certain exceptions to such clauses. . . . Among the recognized exceptions are (a) delay caused by fraud, misrepresentation, or other bad faith; (b) delay caused by active interference; (c) delay which has extended such an unreasonable length of time that the party delayed would have been justified in abandoning the contract; (d) delay that was not contemplated by the parties and (e) delay caused by gross negligence.[89]

It also noted,

> The most contested of the exceptions is that for "delay not contemplated by the parties."[90] Under this exception, a number of courts find that a "no damage" provision will not bar claims resulting from delays caused by the contractee if the delays "were not within the contemplation of the parties at the time they entered into the contract." *Corinno Civetta Constr. v. City of New York,* 67 N.Y.2d 297, 502 N.Y.S.2d 681, 686, 493 N.E.2d 905, 910 (1986). The rationale for this exception, as stated by the *Corinno Civetta* court, is that "[i]t can hardly be presumed . . . that the contractor bargained away his right to bring a claim for damages resulting from delays which the parties did not contemplate at the time."

---

[84]*Selden Breck Constr. Co. v. Regents of Univ. of Michigan,* 274 F. 982 (E.D.Mich.1921).

[85]AIA Doc. A201-2007, § 8.3.3.

[86]*I. L. & B. Constr. Co. v. Ragan Enterprises,* 367 Ga. 809, 482 S.E.2d 279 (1997).

[87]*J.A. Jones Constr. Co. v. Lehrer McGovern Bovis, Inc.,* 120 Nev. 277, 89 P.3d 1009 (2004). See Parvin & Araps, *Highway Construction Claims,* 12 Pub.Cont.L.J. 255, 277–81 (1982); Vance, *Fully Compensating the Contractor for Delay Damages in Washington Public Works Contracts,* 13 Gonz.L.Rev. 410 (1978); Comment, 28 Loy.L.Rev. 129 (1982). For a review of the American experience directed at an international audience, see Sweet, *Contract Regulation of Delay and Disruption Claims in America* [2002] Int'l Constr. L. Rev. 284.

[88]325 S.C. 129, 480 S.E.2d 447 (1997).

[89]480 S.E.2d at 448.

[90]This exception resembles the cardinal delay that has been recognized by the Armed Services Board of Contract Appeals. *Godwin Equip. Inc.,* ASBCA No. 51939, 01-1 BCA ¶31,221. Cardinal delay, in turn, resembles the cardinal change; a change or changes so different than could have been expected that the contract price is removed and the contractor entitled to recover its damages or the reasonable value of its services. See Section 21.03A. If the delays fall into the cardinal category, increased cost and lost profit can be recovered despite contractual provisions that limit the contractor's remedy for government-caused delay.

However, this view is not universally accepted and has recently been questioned by a number of courts. *See e.g. State Highway Administrator v. Greiner*, 83 Md.App. 621, 577 A.2d 363 (1990); *Gregory and Son, Inc. v. Guenther and Sons*, 147 Wis.2d 298, 432 N.W.2d 584 (1988). These courts hold that a clear, unambiguous clause which precludes a contractor's recovery of damages for "any delays" is binding, notwithstanding uncontemplated delays, absent some allegation of intentional wrongdoing, gross negligence, fraud, or misrepresentation. *Greiner, supra.* In *Gregory,* the Wisconsin Supreme Court stated:

> Indeed, the adoption of a no-damage-for-delay clause shows that the parties realize that some delays cannot be contemplated at the time of the drafting of the contract.

. . . The parties can deal with delays they contemplate by adjusting the start and completion dates or by including particular provisions in the contract. "[I]t is the unforeseen events which occasion the broad language of the clause since foreseeable ones could be readily provided for by specific language." *City of Houston v. R.F. Ball Constr. Co., Inc.,* 570 S.W.2d 75, 78 (Tex.Civ.App.1978).[91]

Judicial application of the exceptions to enforcement of no damages clauses outlined in the *Williams* case described earlier in this section is demonstrated by *Triple R Paving, Inc. v. Broward County.* This case dealt with exceptions (a) and (b) noted in that case: fraud, misrepresentation, bad faith, and active interference. It is reproduced in part at this point.

TRIPLE R PAVING, INC. v. BROWARD COUNTY

District Court of Appeal of Florida, Fourth District, 2000. 774 So.2d 50, Rehearing denied Jan. 12, 2001.

STONE, J.

Triple R Paving, Inc. ("Triple R"), plaintiff below, and Frederic R. Harris, Inc. ("Harris"), third-party defendant, appeal from a judgment awarding damages to Triple R resulting from construction delays in the performance of its road construction contract with Broward County, Harris' indemnitee. The appeals have been consolidated for review.

Triple R successfully bid on a road construction contract, the design for which was prepared by Harris. The contract called for widening a portion of Rock Island Road, including widening a bridge which spanned a canal. During construction, delays resulted from a horizontal sight distance design flaw, a Florida Power & Light (FP & L) utility relocation, and detention pond elevation problems. Triple R filed suit against Broward County for delay damages, which included lost home office overhead and lost efficiency. The county, in turn, filed a third-party complaint against Harris for indemnification. [Ed. note: Lost home office overhead claim omitted. See Section 27.02B.]

The standard form contract, used by the county, contained the following pertinent [clause]:

43: No Damages for Delay:
NO CLAIM FOR DAMAGES OR ANY CLAIM OTHER THAN FOR AN EXTENSION OF TIME SHALL BE MADE OR ASSERTED AGAINST THE COUNTY BY REASON OF ANY DELAYS. The

CONTRACTOR shall not be entitled to an increase in the Contract Sum or payment or compensation of any kind from the COUNTY for direct, indirect, consequential, impact or other costs, expenses or damages, including but not limited to costs of acceleration or inefficiency, arising because of delay, disruption, interference or hindrance from any cause whatsoever, . . .; provided, however, that this provision shall not preclude recovery or damages by the CONTRACTOR for hindrances or delays due solely to fraud, bad faith or active interference on the part of the COUNTY or its agents. Otherwise, the CONTRACTOR shall be entitled only to extensions of the Contract Time as the sole an sic [and] exclusive remedy for such resulting delay, in accordance with and to the extent specifically provided above.

* * *

During the construction, Triple R determined that it would be more cost efficient to build the bridge with a single span rather than two spans and submitted what is known as a value engineering contract proposal (VECP) to make that change. The county indicated it was interested in pursuing the VECP and asked Triple R to have its engineer design a single span bridge.

---

[91]*U.S. for Use and Benefit of Williams Electric Co., Inc. v. Metric Constructors, Inc.*, supra note 88, 480 S.E.2d at 450.

Triple R retained engineer Joe Roles to design the single span bridge. He testified that his design of the bridge for the VECP did not change any element of the horizontal geometry, and he did not check horizontal sight distances because his work did not affect them. Both Roles and Hawks, Triple R's president, stated that Harris' representative, John Wise, knew that the plans did not meet horizontal sight distance standards. According to Hawks, John Wise was aware of the problem as early as 1992, but never mentioned throughout the VECP process, or in any meetings regarding the VECP, that the horizontal sight distance should be checked, despite the fact that he agreed to personally check the sight distance against standards at the time the VECP was under discussion for approval.

In September 1994, during construction, it became apparent that the bridge was too close to the Inverrary driveway and Triple R sent a letter to the county advising them of the problem. Broward County then refused to allow the bridge to open in the condition it was in on September 2, 1994. In September 1994, Triple R advised Harris that it would move its manpower and equipment off the job until a solution to the horizontal sight distance problem was discovered; however, it was directed not to do so.

While a solution was being worked out, Triple R worked on a portion of the project north of the bridge, but the inability to open the bridge and switch traffic to the other lane impeded its ability to proceed efficiently with its work.

A second delay occurred once the driveway was completed and construction was to be resumed. Terry Opdyke, Harris' chief inspector for the job, stated from the beginning of the job that he would coordinate all utilities because he wanted to control them. FP & L was scheduled to remove the power lines that went over the existing bridge. Opdyke told Triple R that the utility relocation would take only a matter of hours and that work could be resumed the same day or the next day. Triple R's subcontractor went out to the job, but FP & L did not show up to remove the power lines because of maintenance problems with its equipment. This delay lasted for several days.

The third delay resulted from problems involving the detention pond, as designed by Harris, which was to drain into a canal controlled by the City of Tamarac. The canal's elevation was higher than depicted in the design and higher than the pond; however, no proof was elicited to show that either the county or Harris was aware of this design flaw prior to actual construction. Had the construction proceeded as designed, a backward flow would have resulted. When the problem was recognized, Opdyke was told to ensure that the canal was at a specified elevation; however, when he approached the city of Tamarac to request

that it drop the elevation for the canal, the city refused. The detention pond design delay also led to extended performance costs for Triple R.

The entire project was finished in late August of 1995, within the extension period granted for the delays, but not within the original contract period. The work was never completely suspended on the project. According to Triple R, it was never able to become more efficient in performing the remainder of the work.

* * *

Harris moved for directed verdict at the close of Triple R's case claiming that Section 43 of the contract precluded delay damages inasmuch as Triple R had failed to prove that the construction delays were the result of fraud, bad faith, or active interference. The trial judge reserved ruling.

* * *

The jury returned a verdict which awarded damages to Triple R for loss of efficiency in the amount of $112,929.31, ...

Harris raises several points, only one of which merits discussion: whether the trial court erred in denying its motion for directed verdict as to Triple R's delay damages claim in the face of Section 43 of the contract, which prohibits damages for delay absent fraud, bad faith, or active interference.

Clauses providing for "no damages for delay," except in the case of fraud, bad faith, or active interference by the owner, are legal and enforceable. *See Newberry Square Dev. Corp. v. Southern Landmark, Inc.*, 578 So.2d 750 (Fla. 1st DCA 1991); *Southern Gulf Util., Inc. v. Boca Ciega Sanitary Dist.*, 238 So.2d 458 (Fla. 2d DCA 1970); *see also McIntire v. Green-Tree Communities, Inc.*, 318 So.2d 197 (Fla. 2d DCA 1975); *Peter Kiewit Sons' Co. v. Iowa S. Util. Co.*, 355 F.Supp. 376 (S.D.Iowa 1973); *Williams Elec. Co. v. Metric Constructors, Inc.*, 325 S.C. 129, 480 S.E.2d 447 (1997). *See generally* Susan Sisskind Dunne, *"No Damage for Delay"* Clauses, CONSTR. LAW, Apr. 1999 at 38.

It is Harris' assertion that, as a matter of law, Triple R failed to establish sufficient proof of either fraud, bad faith, or active interference to overcome the contract prohibition against delay damages. We agree with Harris with respect to the delays occasioned by the FP & L utility relocation and the detention pond elevation. *See Southern Gulf Util.*, 238 So.2d at 458 ("mere lethargy or bureaucratic bungling" would not overcome no damage for delay clause). However, we find that the facts surrounding the delay which resulted from the horizontal sight distance design flaw were sufficient to allow a jury to decide the question of fraud, bad faith, or active interference.

* * *

In *Newberry Square*, the court recognized that a "no damage for delay" clause would be vitiated by "delays resulting from a party's fraud, concealment, or active interference with performance under the contract." *Id.* Furthermore, "willful concealment of foreseeable circumstances which impact timely performance" will also limit applicability of a "no damages for delay clause." *Id.* Active interference was found in *Newberry Square* where the property owner delayed approving plans and change orders and ordered that construction not proceed absent such orders. Furthermore, the president of the corporation failed to make timely payments required by the contract and had threatened to "break" the contractor before he would pay him. *See id.*

We recognize that the facts of *Newberry Square* are more egregious than the circumstances surrounding the horizontal sight distance design flaw which resulted in the first construction delay at issue here; however, evidence of Harris' knowledge of the design flaw and the subsequent failure to apprise Triple R of the problem is sufficient to constitute "willful concealment of foreseeable circumstances which impact timely performance," such that the "no damages for delay" clause may be overcome. *See also McIntire*, 318 So.2d at 200 ("[C]ircumstances which caused the delay were brought about by appellee and were even foreseen but concealed by appellee when the contract was made.").

It is undisputed that, where a public authority does not willfully or knowingly delay job progress, it is protected by a "no damage for delay clause." *See C.A. Davis, Inc. v. City of Miami*, 400 So.2d 536, 539 (Fla. 3d DCA1981). In this case, however, the evidence showed a "knowing delay" and silence when assigned the responsibility of verifying compliance with standards on this point.

We note that Black's Law Dictionary 134 (7th ed. 1999) defines "bad faith" as "Dishonesty of belief or purpose <the lawyer filed the pleading in bad faith>" [sic] and includes the following quotation from the Restatement (Second) of Contracts § 205 cmt. d (1981):

A complete catalogue of types of bad faith is impossible, but the following types are among those which have been recognized in judicial decisions: . . . interference with or failure to cooperate in the other party's performance.

Black's at 134. The evidence of Harris' knowledge of the plan defect and failure to apprise Triple R of the problem while simultaneously agreeing to check the horizontal sight distance against standards during the VECP approval process is evidence of interference with, or at the very least, "failure to cooperate in the other party's performance," and, therefore, was sufficient proof

of bad faith to allow the jury to decide the question of Triple R's entitlement to delay damages.

Decisions of other jurisdictions lend further support to our conclusion. *See, e.g., Port Chester Elec. Constr. Corp. v. HBE Corp.* 894 F.2d 47 (2d Cir.1990); *P.T. & L. Constr. Co. v. New Jersey, Dep't of Transp.*, 108 N.J. 539, 531 A.2d 1330 (1987); *Buckley & Co. v. State of New Jersey*, 140 N.J.Super. 289, 356 A.2d 56 (1975). In *Port Chester Electrical*, the court of appeals premised its holding on the "universally accepted proposition that contract provisions aimed at relieving a party from the consequences of his own fault are not viewed with favor by the courts." *Id.* at 48 (citations omitted).

In *P.T. & L.*, the court recognized:

When the government makes a positive statement of fact about the character of work to be performed, upon which the contractor may reasonably rely, it is binding on the government notwithstanding the inclusion of exculpatory clauses in the contract.

P.T. & L., 531 A.2d at 1335. Accordingly, the court refused to apply a no damages for delay clause where the watery condition of the construction site caused construction delay after the state had misled the contractor into believing that the conditions of the site would be dry or normal. *See id.* Similarly, in this case, the evidence adduced by Triple R in its case-in-chief established that Harris was well aware of the design flaw in the bridge construction, but withheld that information from Triple R. Thus, there is sufficient evidence to allow a jury to determine whether fraud, bad faith, or active interference vitiated the no damages for delay clause. *See also Buckley*; 391 PLI/REAL ESTATE LAW AND PRACTICE COURSE HANDBOOK SERIES, "Defending Against the Contractor's Delay Damages Claim" at 395-97.

Accordingly, we affirm the trial court's denial of Harris' motion for directed verdict on the "no damages for delay" clause with respect to the horizontal sight distance delay. However, in view of the county's and Harris' lack of control over FP & L and the lack of proof of knowledge as to the detention pond flaw, we find the evidence insufficient, as a matter of law, to show fraud, bad faith, or active interference with respect to the FP & L relocation delay and the detention pond delay and, accordingly, reverse and remand for judgment in favor of the county and Harris on those damage claims. As to all other issues raised by Harris, we affirm.

* * *

REVERSE AND REMAND for proceedings consistent with this opinion.

KLEIN, J. and OWEN, WILLIAM C., Jr., Senior Judge, concur.

The ceaseless struggle between private autonomy (freedom of contract) and judicial control of what can be overreaching and abuse of contract power are demonstrated in the varying attitude of the courts toward "no-damage" clauses.[92] It is also seen in increasing legislative efforts to limit the use of such clauses.[93] To develop an understanding of this struggle, it is useful to look at other cases to see how these exceptions are applied.

The court in *Broadway Maintenance Corp. v. Rutgers*[94] (reproduced in part in Section 17.04C) enforced such a clause. The court stated, "No-damage provisions are a part of the economic package upon which the parties agree. The contractor who chooses to accept these risks will reflect the accompanying responsibility in his price."[95]

The contractor had complained of backfill problems, slow-pouring concrete, and lack of temporary heat. The court affirmed a finding by the intermediate court that the situations were not sufficiently exceptional to fall outside the no-damage clause. Such situations are ordinary and usual types of damage that most contractors frequently encounter.

Another court[96] held that such a clause would bar *delay* damages, which it defined as time lost when work cannot be performed because materials have not been delivered or preliminary work done. But it does not bar a claim based on hindering the work, such as failure to coordinate the work or supply temporary heat. The court's holding that hindering is not barred by the clause may only illustrate active interference. Another illustration of the exception for active interference was based on the owner's testimony that he would "break" the contractor before he would pay.[97]

Pennsylvania held that such clauses will not be enforced if there is affirmative or positive interference or a failure to act in some essential matter necessary to prosecution of the work.[98] The court held that failure to keep a lake drained actively interfered with a contractor's lake excavation contract. The court looked at the facts liberally from the perspective of the contractor and concluded that this was not the type of interference contemplated by the parties.

Similarly, the bad-faith or active interference exception was applied in *United States Steel Corp. v. Missouri Pacific Railroad Co.*[99] The claim of active interference was based on the owner's directing the contractor to proceed before requisite work of an earlier contractor had been completed. Access to the work was denied during completion of the earlier contractor's work, causing a delay of 170 days. The owner granted a time extension for this delay but, based on a no-damage clause, refused to pay any delay damages.

The court noted that the owner had an implied obligation to refrain from anything that would reasonably interfere with the contractor's opportunity to proceed with its work and to allow the contractor to carry on that work "with reasonable economy and dispatch." The court concluded that issuing the notice to proceed was an affirmative, willful act and that the owner's bad faith was demonstrated by its knowledge of circumstances that would prevent the contractor from proceeding timely with its work.

The preceding discussion has assumed a clause that *bars any damages* for owner-caused delay. It is possible to employ a targeted clause that does not have the sweep of a classic no-damage clause. Such a clause can target specific causes and specific losses. It can bar recovery for specific acts of the public entity, such as issuing a permit, scheduling adjustments, or a delay in the delivery of owner-supplied material.

---

[92]*U.S. for Use and Benefit of Williams Electric Co., Inc. v. Metric Constructors, Inc.*, supra note 88, 480 S.E.2d at 450. Many cases are collected in *State Hwy. Admin. v. Greiner Eng'g Sciences, Inc.*, 83 Md.App. 621, 577 A.2d 363, cert. denied, 321 Md. 63, 582 A.2d 499 (1990). See also Lesser & Wallach, *Risky Business: The Active Interference Exception to No-Damage-for-Delay Clauses*, 23 Constr. Lawyer, No. 1, Winter 2003, p. 26.

[93] West Ann.Cal.Pub.Cont. Code § 7102 limits no-damage clauses to unreasonable delays or those not contemplated by the parties. The statute provides that no public contract can waive this legislative protection. Oregon and Virginia provide that a contractor cannot waive claims for unreasonable delay in public contracts. See Or.Rev.Stat. § 279C.315 and Va. Code Ann. § 2.2-4335. Colorado bars no-damage clauses in public contracts. See Colo.Rev.Stat. § 24-91-103.5. Washington bans all no-damages clauses. See West Ann.Rev.Wash.Code § 4.24.360, interpreted broadly in *Scoccolo Constr., Inc. ex rel. Curb One, Inc. v. City of Renton*, 158 Wash.2d 506, 145 P.3d 371 (2006). The statutes are canvassed in Dunne, *Legislative Update: No Damage for Delay Clauses*, 19 Constr. Lawyer, No. 2, Apr. 1999, p. 38.

[94]490 N.J. 253, 447 A.2d 906 (1982).

[95]447 A.2d at 914.

[96]*John E. Green Plumbing & Heating Co., Inc. v. Turner Constr. Co.*, 742 F.2d 965 (6th Cir.1984), cert. denied, 471 U.S. 1102 (1985). See also Lesser & Wallach, supra note 92.

[97]*Newberry Square Dev. Corp. v. Southern Landmark, Inc.*, 578 So.2d 750 (Fla.Dist.Ct.App.), case dismissed, 584 So.2d 999 (Fla.1991).

[98]*Coatesville Contractors & Eng'rs, Inc. v. Borough of Ridley Park*, 509 Pa. 553, 506 A.2d 862 (1986).

[99]668 F.2d 435 (8th Cir.), cert. denied, 459 U.S. 836 (1982).

As to specific losses, contracts can allow recovery only for direct additional expenses, excluding additional overhead and profit. Such a clause can limit recovery to designated costs, such as for premiums paid on its bonds, for wages and salaries needed to maintain the work, or for plant and equipment when work is stopped.[100]

Another targeted approach is to provide a corridor period of losses being chargeable against the owner after a designated number of days of delay caused by the owner have elapsed.

Another method, though more uncertain of enforcement, to control owner exposure is to bar use of the broad global formulas, such as total cost, modified total cost, or jury verdict.[101] These formulas can create a large damage award and frighten owners.

Another approach is to limit the use of exceptions to enforcement of classic "no-damage" clauses that have been noted in this section.

It is unclear whether attempts to bar a court from using a global formula or to limit the exceptions to enforcement will be successful. As noted, courts do extend considerable freedom (private autonomy) to contracting parties. Yet attempts to tie the hands of courts when they administer remedies for breach of contract or determine the enforceability of clauses that exculpate a contracting party will be examined carefully by the courts, much as they look carefully at clauses that liquidate damages for delay.[102]

The advantage of such targeted clauses is that they are more likely to be interpreted neutrally, unlike the broad, "no-damage" or "no-pay-for delay" clause.[103] But the popularity of delay or disruption claims by contractors will force courts to pass on the validity of such exculpatory clauses. Undoubtedly, case decisions in different jurisdictions will apply different tests and will come to different outcomes.

B. Subcontractor Claims

Delays caused by the owner not only harm the prime contractor but also may harm subcontractors. The absence of a contract between subcontractors and the owner generally precludes direct legal action, and as a rule precludes direct negotiations between subcontractors and owners over delay claims. Often the prime contractor processes the subcontractor's claim. This processing is dealt with in Section 28.08B.[104]

C. Liquidated Damages

Until recently, it was rare for owners to liquidate damages for delays they caused. Public owners usually protected themselves, or at least hoped to, by using no-damage clauses. However, it is becoming more common to liquidate damages as an alternative to exposure to the often open-ended and difficult to establish or to disprove delay damage claims. Although no judicial analysis of these clauses has been found, very likely courts will enforce them. The process of establishing the additional costs incurred by the contractor for delay is even *more* complex and *more* difficult than establishing lost use by owners, even public owners. This should encourage enforcement, provided the amount bears some reasonable relationship to the anticipated or actual damages.

D. Measurement

Section 27.02F deals with the measurement of the value of the contractor's claim when without excuse the owner delays the contractor's performance.

E. Records

In delay disputes, the party with the best records has a great advantage. Each party should keep job records such as the site representative's daily field reports, correspondence, memoranda, photographs or video recordings, and change orders. These records should include data on labor, equipment, and materials used for each activity and should document the cause and impact of every delay. See Section 27.05.

---

[100]*Calumet Constr. Corp. v. Metropolitan San. Dist.*, 222 Ill. App.3d 374, 581 N.E.2d 206 (1991), appeal denied, 143 Ill.2d 636, 587 N.E.2d 1012 (1992).

[101]See Section 27.02F.

[102]See Section 26.09B.

[103]See Section 20.02.

[104]Sometimes the prime and the subcontractor use liquidating agreements under which the prime confesses liability to the subcontractor for owner-caused delay and the subcontractor releases the prime for all other liability. Under such agreements, the subcontractor is relegated to whatever delay damages the prime can recover from the owner (discussed in Section 28.08B).

CHAPTER TWENTY-SEVEN

Claims: By-Products of Construction Process

SECTION 27.01 Introduction

Review Chapter 6, the basic building block dealing with remedies for breach of contract.

The law seeks to compensate the contracting party who has suffered losses because of the other party's breach of contract. Generally it does not punish the party that has breached the contract.[1] The court judgment seeks to put the injured party in the position it would have reached had there been no breach (expectation), or seeks to restore the party to the position it occupied before performance began (restitution).

The law developed conventional formulas to implement the compensation objective. At times, a choice can be made between different formulas. Judicial attempts to determine what would have happened can lead to unsupported speculation and guesswork. The complexity of construction performance, especially in large projects, makes it difficult to determine what has happened.

As a result, claims measurement uses rough approximations that may be all that is available. This chapter reveals a constant tension between demanding specific and solid proof and the willingness to use formulas or approximations.

Measurement of claims and remedies for breach of the construction contract have generated a sea of reported case decisions from the fifty U.S. jurisdictions and the federal courts. The differing judicial attitudes toward claims, the multitude of American jurisdictions, and the inevitable factual variations make generalizations treacherous, except at the most abstract levels. Although the basic principle of compensation is the foundation of claims

measurement (except for the increasing use of punitive damages), implementing the compensation objective can invoke different formulas. This chapter describes the varying rules, discusses the reasons behind them, provides illustrations, and notes trends.

SECTION 27.02 Measurement: Contractor Versus Owner

A. Illustrations

Claims by contractors against owners can arise in many contexts. The principal claims are based on the following:

1. refusal to award the contract to the successful bidder
2. refusal of the owner to permit the contractor to commence performance after the contract has been awarded
3. wrongfully terminating the contractor's right to perform during performance
4. committing acts that justify the contractor in ceasing performance
5. failing to pay the contractor for work performed under the contract
6. committing acts that increase the contractor's cost of performance

B. Cost Contracts

This section does not deal with claims by the contractor for additional compensation under cost contracts. Usually such claims are based on assertions that costs were incurred and that the owner's payment did not cover those costs. Those questions generally involve factual

---

[1]See Section 6.04.

determinations of whether particular work was done or whether it was called for under the contract and the value of the work performed. Although the contractor must establish the particular work done and its cost, the owner has the burden of showing that the cost of particular work was unreasonable.[2] The purpose of the cost contract is to place the risk of costs incurred on the owner.

C. Project Never Commenced

Section 27.02 describes the measure of recovery granted a claimant contractor that begins performance but does not complete performance (D and E) or contractor claims after it has completed (F). This Section examines contractor claims when it has not begun to perform. Failure to commence performance may result from not being awarded the contract to which it was entitled in a competitively bid process. Failure to begin performance may also result from the owner repudiating the contract without legal justification. The owner may simply state it does not wish to receive performance.

The contractor is entitled to protection of its expectation interest.[3] This would put it in the position it would have been had it been allowed to complete performance. Had it performed, it would have received the contract price. But from that would be subtracted the expense saved, the cost of performance. The difference would have been the contractor's profit on the project.

Horsfield Construction, Inc. v. Dubuque County, Iowa[4] involved a competitively bid repaving contract. The contract price was $1,260,000 (all amounts approximations). The contractor claimed its costs would have been $750,000 resulting in a gross profit of $510,000. It sought to protect its expectation interest, its profit on the contract.

The trial judge (this is a trial court opinion) rejected the defendant's claim that this should be reduced by any amount

that the contractor earned on other work it performed during the time it would have spent on this contract. Profit on the other work performed was not caused by the breach as the contractor could have performed both the original and the additional work.

The gross payments would be reduced by the contractor's direct cost of performance, such as "material, subcontract, labor and equipment costs." There was no deduction for overhead as this "would be constant whether the contract was performed or not."[5] There was no showing that overhead costs would be increased had the contractor performed this contract.

The judge also rejected the defendant's claim that it failed to mitigate its damages by not submitting the same bid on a re-bid.[6] The contractor had no obligation to lower its bid on the re-bid. Whether the contractor would have won the re-bid is too speculative.

But not all arrows shot by the defendant missed the mark. It claimed that the contractor's price prediction did not take into account "a single adverse circumstance." The contractor's vice president testified that it would add a 15 to 20% contingency for unforeseen circumstances.[7]

The judge also had difficulty with the 40.1% gross profit submitted by the contractor.[8] The evidence showed that in the past three years its profit margin was 27%. But the contractor responded that in paving contracts it made an average of 35% profit. The judge accepted the contractor's explanation for the greater profitability in paving contracts.

The defendant's attempt to reduce the award by taxes that would been paid was rejected by the judge.

The judge used expert testimony as a basis for concluding that the gross profit would have been 35%. The judge reduced the damage award to $442,000, plus interest from the time it should have been awarded the contract.

D. Project Partially Completed: Damages

The damages award should place the contractor in the position it *would* have been in had the owner performed in accordance with the contract. Three possible formulas,

---

[2]*Sloane v. Malcolm Price, Inc.*, 339 A.2d 43 (D.C.App.1975).

[3]See Section 6.05.

[4]2003 Westlaw 24141325 (Iowa Dist.2003). An earlier stage of this dispute involved the issue of whether the contractor had a valid claim despite no formal contract having been made. The Iowa Supreme Court held that the contractor had a valid claim and sent the case back to the trial court to determine damages. 653 N.W.2d 563 (Iowa 2002). This case is noted in Section 18.04G. The opinion of the trial court judge on remand is described in the text. A trial court decision does not create a precedent. But it provides a valuable illustration of how damages are computed in a real case. It is not reported in the *Northwest Regional Reporter*.

[5]See Section 27.03F for an analysis of extended home office overhead in a claim by a contractor who has completed performance.

[6]See Section 27.07.

[7]See Section 25.04.

[8]Profits on contracts other than the one in dispute will be discussed in Section 27.06C.

which for convenience are referred to as Formulas 1, 2, and 3, can determine the amount of recovery.

Formula 1 determines what *would* have happened by awarding the contractor the contract price. This is the amount the contractor *would* have had on completion. But from this amount the cost of completion must be deducted.[9] This expense is saved by the breach. Progress payments received also must be deducted. Completion cost is what a reasonably prudent contractor in the contractor's position would have had to spend to complete the work.[10]

A contractor who finds it difficult to prove what its costs *would have* been can use the costs of a successor, provided those costs were reasonable.[11] If the owner can establish that the contractor would have sustained a loss had it completed the project, this loss will be deducted from the recovery.[12] (As seen in Section 27.02E, a deduction need not be made if the claim brought by the contractor is based on restitution.)

Formula 2 is the contractor's expenditures in part performance, including preparation, and, if the contractor can establish them, profits on the entire project.[13] To illustrate Formulas 1 and 2, assume that a project has the following figures:

1. contract price—$100,000
2. expenditures in part performance—$60,000
3. cost of completion—$30,000
4. progress payments received—$50,000

Under Formula 1, the contractor would receive the following:

$ 100,000	(contract price)
–30,000	(cost of completion)
$ 70,000	
–50,000	(progress payments received)
$ 20,000	(net recovery)

Under Formula 2, the recovery would be as follows:

$ 60,000	(expenditures in part performance)
+10,000	(profits—$100,000 contract price less $90,000, total cost of project)
$ 70,000	
–50,000	(progress payments received)
$ 20,000	(net recovery)

The result using Formulas 1 and 2 is identical. But suppose expenditures in part performance were $90,000. Had the contract been completely performed, the contractor would have lost $20,000. Using Formula 2, the contractor would still receive $20,000, inasmuch as the two principal items of the formula are not changed. However, the contractor has already expended $90,000 and has received $70,000 through progress payments ($50,000) and the money award ($20,000). This would create the same loss of $20,000 that the contractor would have suffered had it completed the contract.

Formula 2—that is, expenditures in part performance plus profits or minus losses—would compute recovery as follows:

$ 90,000	(expenditures in part performance)
–20,000	(loss that would have been suffered)
70,000	
–50,000	(progress payments received)
$ 20,000	(net recovery)

Under Formula 2, the contractor would be in the same position as under Formula 1. The contractor expended $90,000 and received $70,000, leaving the same $20,000 loss.

Formula 3 entitles the contractor to such proportion of the contract price as the cost of the work done bears to the entire cost of completing performance and, for the remaining portion of the work, the profit that would have been made on that work.[14] Formula 3 should produce the same recovery on profitable contracts but will produce a different result in losing contracts. Formula 3 is rarely used, however, so detailed illustrations of its application will not be given.

Compensation can involve the difficult question of establishing lost profits. Some contracting systems make a sharp differentiation between profits on performed and unperformed work. For example, federal procurement policies rarely give recognition to the profit on unperformed work.[15] However, the ordinary contract

---

[9]*Tull v. Gundersons, Inc.*, 709 P.2d 940 (Colo.1985). See Kirksey & Smedley, *Protecting the Contractor's Expectation Interest After the Owner's Substantial Breach*, 6 Constr. Lawyer, No. 1, Oct. 1985, p. 1.

[10]*Watson v. Auburn Iron Works, Inc.*, 23 Ill.App.3d 265, 318 N.E.2d 508 (1974).

[11]*Carchia v. United States*, 202 Ct.Cl. 723, 485 F.2d 622 (1973). But in *Edward Elec. Co. v. Metropolitan Sanitary Dist.*, 16 Ill.App.3d 521, 306 N.E.2d 733 (1973), the contractor's attempt to base the cost of completion on the completed work was unsuccessful.

[12]*Watson v. Auburn Iron Works, Inc.*, supra note 10.

[13]*Bensch v. Davidson*, 354 S.C. 173, 580 S.E.2d 128 (2003); Restatement (Second) of Contracts § 347, Comment d (1981).

[14]*Kehoe v. Borough of Rutherford*, 56 N.J.L. 23, 27 A. 912 (1893).

[15]*General Builders Supply Co. v. United States*, 187 Ct.Cl. 447, 409 F.2d 246 (1969). A weighting formula developed by the chief of engineers for determining profit on performed work was cited and applied in *Norair Eng'g Corp.*, ASBCA No. 10856, 67-2 BCA ¶6619.

measurement formulas that have been set forth do not make this differentiation. In such cases, the principal problem has been the standard of certainty required to recover lost profits.

Early case decisions took a negative view toward accepting lost profits as part of contract damages. Many cases held that profits could not be established for a new business.[16] Even lost profits by someone in an existing business were closely scrutinized. Was the profit reasonably foreseeable by the contract parties at the time the contract was made? In contracts for the sale of goods, this often excluded recovery of unusual resale profits.

In construction contracts, however, the principal problem has been certainty. This is discussed in Section 27.04. How does the contractor establish not only the profits on work performed but also the profits on unperformed work? If it uses Formula 1—contract price less cost of completion—lost profits need not be established directly. However, what it would have cost to complete the project can often be difficult to show. As a result, more commonly, contractors use Formula 2—expenditures in part performance, coupled with profits. Although the former can be difficult to establish if the contractor has not kept good records, it is generally easier to prove than cost of completion. However, the contractor will have to establish profits.

E. Project Partially Completed: Restitution[17]

The law generally permits a contracting party who has not fully performed to use an alternative measure of recovery where the other party has committed a serious breach.[18] Breaches sufficiently serious to justify restitution have been failure to make progress payments,[19] excessive changes,[20] and failure to perform those acts during performance that would allow the contractor to perform in the most expeditious way.[21]

Sometimes restitution through *quantum meruit* provides the basis for recovery to a contractor whose claim based upon its construction contract failed due to a lack of required certainty.[22]

Expectation (discussed in Section 27.02D) looks forward and seeks to determine what would have happened had the parties completed performance. Restitution looks backward to the position the parties were in at the time they entered into the contract. In some contracts, this is accomplished by a party's returning any performance the other party may have conferred on it. In construction, actual restitution is ordinarily impracticable, as the work has been attached to the owner's land. Instead, the contractor is entitled to receive the value that its performance has benefited the owner, generally the reasonable value of the materials and labor it contributed to improving the owner's land.

Restitution should be measured objectively, with the contractor recovering an amount that equals what the owner would have had to pay to purchase the materials and services from one in the contractor's position at the time and place the services were rendered.[23] Ordinarily, the contractor seeks to introduce evidence of its actual costs incurred in performing its contractual obligations.

---

[16]J. FISCHER, UNDERSTANDING REMEDIES 62 (1999).

[17]One confusing aspect of restitution is the frequent mention of the term *quantum meruit*. Often courts state that a particular action is brought on a *quantum meruit*. Early in English legal history, a party seeking relief had to bring itself within the language of writs issued by public officials commanding persons to appear before one of the King's courts. One frequently used writ was *Assumpsit*. This writ was broken down into a number of subwrits, one of which was called *General Assumpsit*. Because of common usage of certain subwrits such as General Assumpsit, many of them became known as *Common Counts*. One of the Common Counts was given the name of *quantum meruit*, which usually dealt with a method of recovering for the reasonable value of services furnished the defendant at his request. It was used as a method of recovering for breach of contract and could be used when the claim was based on unjust enrichment. This book refers to restitution rather than *quantum meruit*, even though some courts still use that term in describing the process by which the plaintiff recovers based on benefit conferred rather than damages.

[18]*Oliver v. Campbell*, 43 Cal.2d 298, 273 P.2d 15 (1954); *John T. Brady & Co. v. City of Stamford*, 220 Conn. 432, 599 A.2d 370 (1991); *Kass v. Todd*, 362 Mass. 169, 284 N.E.2d 590 (1972) (dictum). See Restatement (Second) of Contracts § 373. The seriousness is similar to that needed for a power to terminate. See Section 33.04A.

[19]*United States v. Algernon Blair, Inc.*, 479 F.2d 638 (4th Cir.1973); *United States v. Safeco Ins. Co.*, 555 F.2d 535 (5th Cir.1977) (nonpayment factor in determining material breach); *John T. Brady & Co. v. City of Stamford*, supra note 18; *Salo Landscaping & Constr. Co. v. Liberty Elec. Co.*, 119 R.I. 269, 376 A.2d 1379 (1977) (nonpayment goes to essence).

[20]*Glassman Constr. Co. v. Maryland City Piazza, Inc.*, 371 F.Supp. 1154 (D.Md.1974); *Rudd v. Anderson*, 153 Ind.App. 11, 285 N.E.2d 836 (1972).

[21]*Leo Spear Constr. Co. v. Fidelity & Cas. Co. of New York*, 446 F.2d 439 (2d Cir.1971).

[22]*Troutwine Estates Dev. Co. LLC v. Comsub Design & Eng'g, Inc.* 854 N.E.2d 890 (Ind.App.2006), rehearing denied Nov. 21, 2006, transfer denied April 17, 2007. The court held that failure to agree upon one major term deprived the contract of certainty. Some courts would not reach this conclusion. See Section 5.06D.

[23]*United States v. Algernon Blair, Inc.*, supra note 19.

Although this evidence is not conclusive, the court is likely to accept it unless the owner can establish that they were not market rate costs or that they were costs incurred to correct deficiencies chargeable to the contractor. One case affirmed a judgment for a subcontractor in its claim against the prime contractor by pointing to the opinion testimony of the subcontractor as to the direct costs and the court's examination of the site.[24] Another court looked at the bids submitted by unsuccessful bidders in determining the reasonable value of the services.[25]

The contractor should be able to recover its overhead costs, as they are incurred by the contractor in improving the owner's land and should be considered as benefiting the owner.[26]

Clearly, the contractor who seeks restitution cannot recover profit on unperformed work. The contractor has been allowed to recover profit on work that has been performed.[27] This profit would be included in what other contractors would have charged.[28]

The most difficult problem relates to the effect of the contract price on the contractor's recovery. Should the court ignore the contract price, use it as evidence of the value of the benefit conferred, use the pro rata contract price to actually measure the recovery, or use the contract price to put a limit on the contractor's restitutionary claim? Although the general tendency has been to use the contract price only as evidence of value,[29] some cases have limited the contractor to the contract price.[30] Florida strikes a middle ground. It limits the claimant to the contract price unless the breach by the defendant is willful. If so, recovery can exceed the contract price.[31]

The effect of contract price on recovery has stirred great debate. The majority rule—that of giving the contract price only evidentiary effect—can allow a contractor who has entered into a losing contract to at least break even. One argument given to justify this result has been that the restitutionary principle (that is, looking backward and avoiding unjust enrichment) has as much validity as expectation (that is, looking forward to what would have happened had the contract been performed properly). Another justification for the majority rule is that an owner who has breached a contract should not be able to take advantage of the contract price. A reason frequently given is that the expenses most likely exceed the contract price because of breaches by the owner. Perhaps—most persuasive—restitution not limited to the contract price is relatively easy to administer. If expectation (discussed in Section 27.02D) were used, losses would be deducted. This computation would require a determination of cost of completion, often at best a guess. If the contract price is not controlling, all that is needed is to determine the reasonable value of the work performed.

Those who support using the contract price either to measure the value or to limit recovery point to the anomaly that can result if the contractor does complete performance. Restitution cannot be used if the contractor has completed performance and all that is left is for the owner to pay the balance of the contract price. If the owner's breach has caused the contractor to lose money, the contractor must establish such losses. (As shall be seen, various crude formulas are sometimes allowed, making proof easier. See Section 27.02F.) Barring the owner from using the contract price because it has breached does not take into account the possibility or even the likelihood that the question of breach has been a close one, with much fault attributable to both parties. Why take away the benefit of the bargain made by the owner because it is ultimately determined that the owner has breached?

The difficulty with the majority rule is that it can encourage a contractor in a losing contract to claim that it has adequate grounds for termination and to stop work. If it continues to perform, it will almost certainly lose money and face difficult problems if it seeks to recover compensation based on breaches by the owner. If it stops and claims the right to terminate, it runs the risk that it will be found to be in default and have to pay heavy damages. These damages may be no greater, however, than the contractor will suffer if it continues to perform a losing contract when there is very little hope of turning things around.

---

[24]*Leo Spear Constr. Co. v. Fidelity & Cas. Co of New York*, supra note 21.

[25]*Paul Hardeman, Inc. v. Arkansas Power & Light Co.*, 380 F.Supp. 298 (E.D.Ark.1974).

[26]*Leo Spear Constr. Co. v. Fidelity & Cas. Co. of New York*, supra note 21.

[27]Ibid. See also *C. Norman Peterson Co. v. Container Corp. of Am.*, 172 Cal.App.3d 628, 218 Cal.Rptr. 592 (1985) (overhead and profit).

[28]*W. F. Magann Corp. v. Diamond Mfg. Co.*, 775 F.2d 1202 (4th Cir.1985) (profit not recoverable *per se* but can be recovered if relevant to reasonable value).

[29]*United States v. Algernon Blair, Inc.*, supra note 19; *Paul Hardeman, Inc. v. Arkansas Power & Light Co.*, supra note 25; *Kass v. Todd*, supra note 18.

[30]*United States v. Mountain States Constr. Co.*, 588 F.2d 259 (9th Cir.1978); *Johnson v. Bovee*, 40 Colo.App. 317, 574 P.2d 513 (1978).

[31]*Ballard v. Krause*, 248 So.2d 233 (Fla.Dist.Ct.App.1971).

In the confusion of a construction dispute in court, a good chance exists that the contractor may be exonerated.

One method for handling this use of restitution is to examine the nature of the conduct of each party. If the contractor's claim of material breach by the owner is a tactic to allow the contractor to bail out of a losing contract that resulted from causes that are the responsibility of the contractor, restitution does not seem attractive. Since restitution requires that the contractor show that the owner has committed a serious, and in some states even a willful breach, restitution gives the contractor the reasonable value of its services. This allows the contractor to avoid having to prove its damages caused by such breach.[32]

It is important to consider the reasons for the contract being a losing one. A losing contract can result from incompetent estimating by the contractor, its taking on a losing contract deliberately, its inability to perform efficiently, or the occurrence of events that fall within the contractor's zone of risk. If so, it is difficult to justify restitution and its disregard of the contract price.

Restitution requires that the contractor show the reasonable value of its services that benefited the owner. A loss that resulted from the contractor's poor performance should be reflected in the proof as to the reasonable value of the services. A loss that resulted from poor estimating or deliberately making a losing contract (to keep the workforce together or learn new skills) also will be reflected in the evidence that establishes the reasonable value of the contractor's services. This operates as a check against the unjust use of restitution.

Most restitution claims are based upon the owner's failure to pay the contractor as required by the contract.[33] It is the most important owner obligation. The importance of payment is reflected in the remedies granted the contractor when payment is not made, such as the invocation of prompt payment statutes[34] or the power to suspend[35] or terminate.[36] Another powerful incentive for the owner to pay is to avoid granting the contractor the remedy of restitution, including the ability to by-pass the contract price.

If full performance precludes use of restitution and the loss resulted from breaches by the owner, the contractor must, as indicated, show the specific damages resulting from each breach. As shall be seen in Section 27.02F, in some cases, the contractor may be able to surmount this obstacle by use of one of the global formulas: such as total cost, modified total cost, or the jury verdict.

This section has discussed restitution as a method for measuring the recovery for breach of contract. In *C. Norman Peterson Co. v. Container Corp. of America*,[37] restitution, including an allowance for overhead and profit, was granted a contractor based on an "implied" agreement to abandon the original cost contract with its guaranteed maximum price coupled with the subsequent "implicit" understanding by the parties that the parties would proceed on a *quantum meruit* basis. These implications resulted from an incomplete design and vast number of changes.[38]

F. Project Completed: *Complete General Construction Co. v. Ohio Department of Transportation* and *New Pueblo Constructors v. State of Arizona*

Denial of Restitution. A contractor who has fully performed but has not been paid the balance of the contract price cannot measure its recovery by the reasonable value of its services.[39] This contractor recovers only the unpaid balance of the contract price and any other losses it can prove.

Limitation on the restitutionary remedy is sometimes justified on historic grounds. However, the modern justification is the ease of using the contract price as a measure compared with making the more complicated factual determination of the reasonable value of the services. Here, unlike the past performance discussed in Section 27.02E, there is no need to determine value.

Site Chaos and Productivity. Most complex construction contract disputes involve claims by the contractor who has completed the project are not, as a rule, for the unpaid balance of the contract price (it has probably been paid), but instead for the additional costs it incurred. The contractor asserts that defective specifications required it to perform

---

[32]*Ballard v. Krause*, supra note 31.

[33]See supra note 19.

[34]See Section 22.02L.

[35]AIA Doc. 201-2007, § 9.7.1.

[36]Id. at § 14.1.1.3.

[37]Supra note 27. But see *Hensel Phelps v. King County*, 57 Wash.App. 170, 787 P.2d 58 (1990) (*quantum meruit* not allowed where contract exculpated owner or provided relief such as equitable adjustments and time extensions).

[38]This case and the California cases decided subsequent to it are discussed in Sweet, *The Amelco Case: California Bars Abandonment Claims in Public Contracts*, 32 Pub.Cont.L.J. 285 (2003).

[39]See supra note 18.

in an inefficient and costly manner, and, most important, that unexcused delays caused by the owner increased its cost of performance.

The problems of defective specifications and unforeseen subsurface conditions, to the extent that they do not involve delay, cause less difficulty. It is not difficult, comparatively speaking, to establish the value of additional services and materials that were necessary because of defective specifications or unforeseen subsurface conditions. Even more, in the case of claims based on unforeseen subsurface conditions, the justification for additional compensation is usually based on a differing site conditions clause, which either may contain a formula for measuring the additional compensation or has developed legal rules that provide a solution for measuring compensation. (Where delay does result, the crude formulas for measuring damages such as total cost, modified total cost, jury verdict, and extended home office overhead may have to be applied.)

However, the greatest difficulty arises in cases where the contractor claims it had to perform work in an inefficient manner mainly because of excessive changes by the owner or administrative incompetence of the owner or someone for whose conduct it is responsible. To set the scene, a court facing such a claim stated,

> We note parenthetically and at the outset that, except in the middle of a battlefield, nowhere must men coordinate the movement of other men and all materials in the midst of such chaos and with such limited certainty of present facts and future occurrences as in a huge construction project such as the building of this 100 million dollar hospital. Even the most painstaking planning frequently turns out to be mere conjecture and accommodation to changes must necessarily be of the rough, quick and *ad hoc* sort, analogous to ever-changing commands on the battlefield. Further, it is a difficult task for a court to be able to examine testimony and evidence in the quiet of a courtroom several years later concerning such confusion and then extract from them a determination of precisely when the disorder and constant readjustment, which is to be expected by any subcontractor on a job site, become so extreme, so debilitating and so unreasonable as to constitute a breach of contract between a contractor and a subcontractor. This was the formidable undertaking faced by the trial judge in the instant case and which we now review on the record made by the parties before him.[40]

Walter Kidde Constructors, Inc. v. State[41] involved a contractor (Kidde-Briscoe) claim in a dispute over a project that finished almost 900 days later than planned. Much of the responsibility fell on the public owner. The owner failed to make certain parts of the site available as promised. It issued many hold orders that substantially interfered with and disrupted construction while plans were being redesigned. It was late in processing change orders and approving shop drawings while still insisting that the contractors accelerate their work under threat of imposing $250-per-day penalty for not completing within the time requirements of the contract. In describing the effect of these delays, the court stated,

> The long delay in completing construction had a devastating impact on Kidde-Briscoe. They were not only required to provide labor and materials far in excess of and different from that called for in the original contract but to perform work out of sequence at an accelerated rate and in a manner not planned and not utilized under normal conditions. The unanticipated almost two and a half years of delayed construction extended through a period of increasing inflation and escalation in labor rates and material costs, through two winters with adverse building conditions and into a time when Kidde-Briscoe was faced with an 89 day regional strike by the sheet metal workers and 137 day regional strike by the plumbers union.
>
> A computation of the consequent damages to Kidde-Briscoe involves damages of two sorts: (a) delay damages due to the extended periods of field and home office overhead and (b) damages due to disruption, loss of productivity, inefficiency, acceleration and escalation.

In addition to the items noted in the *Walter Kidde* decision, contractors often claim and are sometimes awarded the following:[42]

1. idleness and underemployment of facilities, equipment, and labor
2. increased cost and scarcity of labor and materials
3. use of more expensive modes of operation
4. stopgap work needed to prevent deterioration

---

[40]*Blake Constr. Co. v. C. J. Coakley Co.*, 431 A.2d 569, 575 (D.C. App.1981).

[41]37 Conn.Supp. 50, 434 A.2d 962, 977 (1981).

[42]For an opinion that illustrates a claim involving a large number of these items, see *Contracting & Material Co. v. City of Chicago*, 20 Ill. App.3d 684, 314 N.E.2d 598 (1974). The opinion was reversed by the Illinois Supreme Court for failure by the contractor to comply with a contractual condition precedent of working double shifts, 64 Ill.2d 21, 349 N.E.2d 389 (1976), but the facts provide a good illustration.

5. shutdown and restarting costs
6. maintenance
7. supervision
8. equipment and machinery rentals and cost of handling and moving
9. travel
10. bond and insurance premiums
11. interest

In the *Walter Kidde* decision just discussed, the court stated that damages could include loss of productivity and inefficiency. *Luria Bros. & Co., Inc. v. United States*,[43] discussed a claim by a contractor that the defendant's delay had caused diminished productivity of its workforce. The workers were required to work during severe winter conditions as well as under adverse water conditions as a result of constant revisions in the contract drawings that resulted in confusion and interruption of the orderly progress of the work.

The court noted that loss of productivity can rarely be proven by books and records and that it almost always must be proven by opinions of expert witnesses. But any expression of opinion will be looked at skeptically if no further evidence supports that opinion or provides a sufficient basis for making a reasonable approximation of damages.

The plaintiff presented testimony of an expert witness. Although the court found the witness was competent and well qualified to express an opinion, it did note that the witness had been a former employee of the contractor and that he might have "a certain predilection for his old employer"[44] and might have "wanted to 'help them out' all he could."[45] The witness testified that productivity was reduced, because men had to work outside on trench excavations and foundation construction in winter weather. This required the men to wear gloves and warmer clothing and work on frozen or extremely wet ground. The witness estimated that the loss of productivity during this period was 33⅓ percent, and as to other periods, he estimated the loss of productivity between 25 and 20 percent.

The court, although noting that the testimony had not been rebutted, observed that these percentages were merely estimates based on observation and experience. The court reduced the estimates from 33⅓ percent to 20 percent, from 25 percent to 10 percent, and from 20 percent to 10 percent, respectively.

Extended Home Office Overhead: Eichleay Formula. The *Luria* case also involved an award for extended home office overhead. This has been one of the more complicated questions in measuring damages for owner breach, particularly in the field of federal procurement.

Contractors seek to use the *Eichleay* formula to measure extended home office overhead. The claimant must submit proof that (1) a government-caused delay extended the contract completion date; (2) the contractor was required to be on standby for an uncertain period of time; and (3) the contractor was unable to take on replacement or substitute work during the delay.[46] Proof of the elements is, to a large degree, in the hands of accountants employed as expert witnesses and requires careful understanding of job site overhead, home office overhead, fixed overhead costs, variable overhead costs, and unabsorbed home office overhead.

When a contractor bids for construction work, it usually takes into account not only job site overhead but also home office overhead. Many decisions of the federal boards of appeal and the old Court of Claims dealt with the question of home office overhead costs incurred after the original contract completion date and caused by compensable delays during the project. Home office overhead includes costs that are incurred to the mutual benefit of all contracts and cannot be tied to a specific project. One commentator stated,

> Costs which are normally included are: executive and clerical salaries, outside legal and accounting expense, mortgage expense, rent, depreciation, property taxes, insurance, utilities/telephone, auto/travel, professional and trade licenses and fees, employee recruitment, relocation, training and education, photocopying, data processing, office supplies, postage, books and periodicals, miscellaneous general and administrative expenses, advertising, interest on borrowing and other financial costs, entertainment, contributions and donations, bad debts, losses on other contracts, and bid and proposal costs.[47]

Under various federal regulations, some of these costs cannot be included in home office overhead. Illustrations are advertising, interest on borrowing, entertainment costs, and contributions and donations.

---

[43]177 Ct.Cl. 676, 369 F.2d 701 (1966).
[44]369 F.2d at 713.
[45]Ibid.

[46]*P. J. Dick, Inc. v. Principi*, 324 F.3d 1364, 1370 (Fed. Cir.2003). For the earlier Department of Veterans Affairs Board of Contract Appeals decision, see infra note 52.
[47]Long, *Extended Home Office Overhead Damages*, VI KC News, No. 2, June 1989, p. 1.

The *Complete General* case reproduced in this section explains and analyzes the *Eichleay* formula for allocating extended home office overhead when the government agency causes an indefinite suspension in the contractor's performance. But it important to recognize at the outset that an *Eichleay* claim is based on unabsorbed home office overhead, an indirect cost. When work is suspended indefinitely, the contractor has been placed on "standby." Though its indirect costs continue to run, it is not receiving payments against which these indirect costs can be charged. But if the contractor can use its resources to take on other work, that work can absorb the indirect costs. In that case there are no unabsorbed overhead costs and no *Eichleay* claim. But if the replacement work absorbs part of the indirect overhead, the contractor can recover the balance of the unabsorbed overhead incurred during the standby period under the *Eichleay* formula.[48]

A recent Ohio case, distinguishing overhead costs and providing an approved formula that can be used as an alternative to the *Eichleay* formula, is reproduced in part at this point.

COMPLETE GENERAL CONSTRUCTION CO.
v. OHIO DEPARTMENT OF TRANSPORTATION

Supreme Court of Ohio, 2002. 94 Ohio St.3d 54, 760 N.E.2d 364.

PFEIFER, J.

This case evolved out of a contractor's claim for additional costs on a highway construction project as a result of delays caused by the Ohio Department of Transportation ("ODOT"). The bulk of this case concerns the method of calculating unabsorbed home office overhead—the cost of running a contractor's home office during the delay period—and whether Ohio should adopt the "*Eichleay* formula," an equation employed by federal courts in determining such costs. The formula acquired its name from the Armed Services Board of Contract Appeals decision in *Eichleay Corp.* (1960), ASBCA No. 5183, 60-2 BCA ¶2688, 1960 WL 538, and "is the most well-known formula for calculating unabsorbed overhead" costs arising out of government-caused delay. Shapiro & Worthington, Use of the *Eichleay* Formula to Calculate Unabsorbed Overhead for Government Caused Delay Under Manufacturing Contracts (1996), 25 Pub.Contr.L.J. 513, 514. We hold that Ohio courts may use the *Eichleay* formula, with certain important modifications, in calculating such costs. We do not find that the *Eichleay* formula is the exclusive manner of determining unabsorbed home office overhead.

Factual Background

This action arises out of the construction of a stretch of I-670 from just north of downtown Columbus to Port Columbus International Airport. Construction of that portion of I-670 was broken into five sections, with each section being bid as a separate project. Appellee and cross-appellant, Complete General Construction Company ("Complete General"), successfully bid on four of those five projects. One of those four, Project 56-91 ("the Project"), is the focus of this action.

The Project provided for the construction of I-670 from the Greater Columbus Convention Center to a point just west of I-71. The contract called for the construction of that stretch of highway, including the erection of three new bridges and the widening of another. Work on the Project began on March 15, 1991, with a slated completion date of August 31, 1992.

However, early on in the Project, design errors relating to the bridges and attributable to ODOT caused a seven-month delay. Due to the delay, on May 13, 1992, ODOT granted Complete General a twelve-month work extension, moving the completion date to August 31, 1993. While the actual delay was seven months, an extension to March 31, 1993, automatically triggered an additional five-month extension under the contract because the March date fell within the winter shutdown period.

Later, ODOT granted Complete General an additional sixty-day extension for other work not contemplated by the original contract, moving the completion date to October 31, 1993.

Following the completion of the Project, the parties entered into negotiations to compensate Complete General for costs it incurred as a result of the extension of the completion date. On October 31, 1996, they agreed that ODOT would pay Complete General $177,662.47 as final compensation for all costs related to the contract extensions, except "home office overhead,

---

[48]*Melka Marine, Inc. v. United States*, 187 F.3d 1370 (Fed.Cir.1999), cert. denied, 529 U.S. 1053 (2000).

interest, major equipment costs and bond costs." The settlement was a part of Change Order 39. The parties continued negotiating on the unresolved issues.

On January 7, 1997, Complete General sued ODOT to recover unabsorbed home office overhead, idle equipment costs, extended equipment costs, and additional bond costs, as well as interest on all these costs from the time of completion of the Project. ODOT offered to pay Complete General $196,410.34: $182,500 for unabsorbed overhead, $888.31 for bond costs, and $13,022.03 for interest. Compete General accepted the amount as partial payment for the disputed claims, and continued on with its lawsuit. This partial settlement was memorialized in Change Order 40.

The parties tried the case before the Court of Claims beginning on April 13, 1998. On November 18, 1998, the court awarded Complete General $374,231.08. The award broke down as follows: $184,947 for unabsorbed home office overhead, $62,622.50 for idle equipment costs, $115,171.49 in interest on the overhead and idle equipment awards, and $11,490.09 in additional bond costs. The court found for ODOT on Complete General's claim for extended equipment costs, i.e., costs for additional equipment time required beyond that originally allocated in Complete General's bid.

[Ed. note: Discussion of idle equipment costs, interest, additional bond costs, and extended equipment costs omitted.]

Both parties appealed the decision of the Court of Claims. In its decision, the Franklin County Court of Appeals affirmed the judgment of the Court of Claims in part, reversed it in part, and remanded the cause for further proceedings.

The cause is before this court upon the allowance of a discretionary appeal and cross-appeal.

Law and Analysis

Both parties appeal aspects of the court of appeals' decision. We resolve those issues below.

Unabsorbed Home Office Overhead

ODOT appeals this part of Complete General's award based upon the lower courts' reliance on the *Eichleay* formula for the calculation of home office overhead during the delay period.

Bids on construction projects incorporate two different kinds of costs. The first type, direct costs, include construction wages and equipment expenses and are attributed to specific projects. The second type, indirect costs, are the expenses involved in generally running a business, not attributable to any one project. The most significant indirect cost is home office overhead. Such costs typically include salaries of executive or administrative personnel, general insurance, rent, utilities, telephone, depreciation, professional fees, legal and accounting expenses, advertising, and

interest on loans. See *Interstate Gen. Govt. Contrs., Inc. v. West* (Fed.Cir.1993), 12 F.3d 1053, 1058.

Each project a contractor undertakes derives benefits from the home office, and each contributes to paying for home office overhead. Contractors typically do not apportion overhead costs among individual projects Each project in some degree is responsible for the contractor's costs of simply doing business, and each project plays its proportionate part in paying those costs. When a delay occurs on a particular construction project, that particular project ceases to carry its weight in regard to running the business, which can result in damages to the contractor. See Kauffman & Holman, *The Eichleay Formula: A Resilient Means for Recovering Unabsorbed Overhead* (1995), 24 Pub.Contr.L.J. 319, 320–321.

Assigning a value to a delayed project's effect on home office overhead can be difficult. Calculating overhead costs allocable to a delay on a given project is generally achieved through the employment of a mathematical formula. The most prominent of those formulas, especially in the federal government context, is the *Eichleay* formula.

* * *

The *Eichleay* formula " 'seeks to equitably determine allocation of unabsorbed [home office] overhead to allow fair compensation of a contractor for government delay.' " *Satellite Elec. Co. v. Dalton* (Fed.Cir.1997), 105 F.3d 1418, 1421, quoting *Wickham Contracting Co., Inc. v. Fischer* (Fed.Cir.1994), 12 F.3d 1574, 1578. The formula was developed in the federal court system, beginning in 1960 with *Eichleay Corp., supra,* ASBCA No. 5183, 60-2 CBA ¶2688, and has been adopted by the Federal Circuit Court of Appeals as the prevailing method for calculating home office overhead expenses attributable to owner-caused delay on federal contracts. *Wickham,* 12 F.3d at 1579–1581.

The *Eichleay* formula creates a *per diem* rate for overhead costs attributable to a single project, multiplying that rate by the number of days of delay to arrive at a total home office overhead award. *Wickham,* 12 F.3d at 1577 fn. 3. The formula is calculated as follows:

1. (Total billings for the contract at issue / Total billings from all contracts during the original contract period) × (Total overhead during the original contract period) = Overhead Allocable to the Contract.

2. (Overhead Allocable to the Contract) / (Original planned length of the contract in days) = Daily Contract Overhead Rate.

3. (Daily Contract Overhead Rate) × (Compensable period in days) = Unabsorbed Overhead Damages.

See *West v. All State Boiler, Inc.* (Fed.Cir.1998), 146 F.3d 1368, 1379, fn. 4.

[Ed. note: In 1 and 2 the slash mark (/) means "divided by."]

The above is simply the mathematics for the *Eichleay* formula— an owner-caused delay in construction does not necessarily lead to an award of damages for home office overhead. Indeed, "recovery under the *Eichleay* formula is an extraordinary remedy designed to compensate a contractor for unabsorbed overhead costs that accrue when contract completion requires more time than originally anticipated because of government-caused delay." *All State Boiler*, 146 F.3d at 1377.

Before the *Eichleay* formula may be applied, the contractor must demonstrate two important elements in order to establish a prima facie case for the award of damages. First, the contractor must demonstrate that it was on "standby." *Interstate Gen. Govt. Contractors*, 12 F.3d at 1056. A contractor is on standby "when work on a project is suspended for a period of uncertain duration and the contractor can at any time be required to return to work immediately." *All State Boiler*, 146 F.3d at 1373. In effect, the contractor is not working on the project, yet remains bound to the project. The contractor must be ready to immediately resume performance at any time.

The second element in a prima facie case is that the contractor must prove that it was unable to take on other work while on standby. *Id.* That is, the contractor must show that the uncertainty of the duration of the delay made it unable to commit to replacement work on another project. Impracticability, rather than impossibility, of other work is the standard, and the contractor is entitled to damages "*only* if its inability to take on additional work results from its standby status, *i.e.*, is attributable to the government."' (Emphasis *sic*) *Id.*, 146 F.3d at 1375, quoting *Satellite Elec. Co.*, 105 F.3d at 1421.

In establishing a prima facie case, then, a contractor demonstrates that it has committed a portion of its overhead costs to a particular project and that not only has the project's suspension left those costs unabsorbed, but that the character of the government-caused delay is such that it is impractical for the contractor to obtain other work to fill the gap. Once the contractor commits resources to a project, the resources remain committed whether the project moves forward or not. The contractor is all geared up with nowhere to go.

That problem results in damages once the original contract period runs out and the extension period begins. At that point the contractor begins expending home office overhead on the project beyond what the contract had contemplated. It is important to note that a contractor may recover under *Eichleay* only if the suspension of the project results in the extension of the completion date. If the suspension does not affect the completion date, the contractor cannot claim damages because he has not suffered any injury, *i.e.*, he spent the time he had originally allocated on the project. *All State Boiler*, 146 F.3d at 1379. Thus, as the court holds in *All State Boiler*, damages are measured based on the number of days the contractor continues to expend home office overhead on the project beyond what was allocated:

"Once the contract performance period extends beyond the initial deadline, indirect costs continue to accrue but the contractor has neither allocated them to the newly-extended contract nor is able to begin a new contract to absorb the next portion of these continuing costs.

* * *

The ordinary course of the contractor's business is thus interrupted by the suspension; where normally the contractor would begin the next contract, to which a new portion of its indirect costs would be attributable, it is forced to extend performance on the old, suspended contract, while additional indirect costs accrue with no additional revenue to support them." *All State Boiler*, 146 F.3d at 1379.

The government can rebut the contractor's prima facie case for unabsorbed overhead damages by demonstrating either "(1) that it was not impractical for the contractor to obtain 'replacement work' during the delay, or (2) that the contractor's inability to obtain such work, or to perform it, was not caused by the government's suspension." *Melka Marine, Inc. v. United States* (Fed.Cir.1999), 187 F.3d 1370, 1375.

ODOT argues that the *Eichleay* formula allows contractors to recover for breach of contract without establishing causation. To the contrary, we find that before the formula can be applied, a contractor must prove a rather extraordinary set of circumstances that by their very nature demonstrate causation and damages.

The *Eichleay* formula goes nowhere without causation. A contractor may recover only if there is an *owner-caused* construction delay. Moreover, the "standby" character of the delay must also be caused by the owner, and must prevent the contractor from finding replacement projects to cover the overhead.

The fact that a delay that creates an uncertain extension period causes damages for a contractor is axiomatic. The outlay of overhead on a delayed project increases as the time allotted for the project is extended. *Eichleay* starts with the proposition that all of a contractor's projects share in a contractor's home office overhead. It only follows that the suspension of a particular project creates a gap in the coverage of overhead costs. The fact that damages are caused by an owner's breach is self-evident—the very nature of the formula requires that overhead costs are not replaced by another job.

Finally, as with any other contract claim, the contractor also has the duty to mitigate damages. Central to *Eichleay* is the requirement that, if able, the contractor must take on other work to absorb the overhead allotted to the delayed project.

Thus, *Eichleay* does no violence to contract law. However, we agree with ODOT that differences between federal and Ohio public contracting law allow contractors to recover inappropriate costs when the *Eichleay* formula is applied in Ohio. Comparing federal and Ohio highway contracting systems is like comparing apples and orange barrels. The federal government has adopted the Federal Acquisition Regulations ("FARs"), which set forth "uniform policies and procedures for acquisition by all executive agencies." Section 1.101, Title 48, C.F.R. Section 31.205-1 *et seq.* governs which costs are allowable and which are not. FAR prohibitions include interest on borrowings, *id.* at Section 31.205-20, entertainment expenses, Section 31.205-14, contributions and/or donations, Section 31.205-8, bid and proposal costs, 31.205-18, and bad debts, 31.205-3. See, generally, Section 31.205-1 *et seq.*, Title 48, C.F.R. Therefore, those types of costs go uncompensated in federal cases applying the *Eichleay* formula.

In general, these unrecognized costs are of the variety that do not bestow any benefit on the government owner in regard to the project at issue. The idea that the government should fund a contractor's parties, sports tickets, political contributions, or other expenses that bring nothing tangible to the government's project is unreasonable. Under the federal system, such costs are not included in recoverable overhead costs and should not be recoverable in an Ohio case applying the *Eichleay* formula.

Thus, we modify the use of the *Eichleay* formula in Ohio. Courts applying the formula must allow owners the opportunity to dispute particular items a contractor submits in an overall overhead cost presentation. Government agencies would do well to consider the FAR's dissection of allowable and unallowable indirect costs, Sections 31.205-1 through 31.205-52, Title 48, C.F.R., for guidance.

We find today that the *Eichleay* formula is *one* way of determining unabsorbed home office overhead damages in public construction delay cases. Once it is determined that an owner-caused

delay has caused a contractor to suffer unabsorbed overhead costs, then the *Eichleay* formula can be employed, but not necessarily exclusively. For instance, a court could utilize the direct cost formula. The direct cost method compares the direct costs actually attributed to a project as a portion of all of the direct costs incurred by the business over a particular period. The result is a ratio by which the percentage of indirect costs can be calculated, including home office overhead applicable to a particular project. *Royal Elec. Constr. Co. v. Ohio State Univ.* (Dec. 21, 1993), Franklin App. Nos. 93AP-399 and 93AP-424, unreported, 1993 WL 532013, reversed on other grounds (1995), 73 Ohio St.3d 110, 652 N.E.2d 687.

We find that the trial court acted within its discretion in applying the *Eichleay* formula in this case and we defer to the trial court's finding that Complete General did suffer unabsorbed home office overhead as a result of ODOT's delay on the project at issue. We find that the court correctly measured Complete General's damages stemming from the extension period. However, the court erred in applying the *Eichleay* formula without allowing ODOT to dispute items of overhead that did not bestow any benefit to the project at issue. Accordingly, we affirm the judgment of the court of appeals in part, reverse it in part, and remand the matter to the trial court for a determination of damages that makes specific findings regarding specific items of overhead disputed by ODOT. . . .

Conclusion

Accordingly, we reverse the judgment of the appellate court on the issue of home office overhead, idle equipment costs, and interest, and affirm the appellate court's judgment on the issue of extended equipment costs. We remand the cause to the trial court for a determination consistent with this opinion.

Judgment affirmed in part, reversed in part and cause remanded.

MOYER, C. J., DOUGLAS, RESNICK, FRANCIS E. SWEENEY, SR. and LUNDBERG STRATTON, J. J., concur.

COOK, J., concurs in part and dissents in part.

[Ed. note: Concurring and dissenting opinion omitted.]

Courts have not been uniform in their treatment of methods by which a contractor can establish its damages for extended home office overhead. Some are strict in demanding accurate proof of the loss caused by the delay and are unwilling to allow formulas to be employed in place of strict proof requirements.[49] Others are more

willing to employ formulas such as the *Eichleay* formula, recognizing the difficulties of proof and being willing to give the contractor the benefit of the doubt.[50] (As noted in Section 27.06D, the documents the AIA issued in 1997 and continued in A201-2007, Section 15.1.6.2, would bar

---

[49] *W. G. Cornell Co. etc. v. Ceramic Coating Co., Inc.*, 626 F.2d 990 (D.C.Cir. 1980); *Berley Industries, Inc. v. City of New York*, 45 N.Y.2d 683, 385 N.E.2d 281, 412 N.Y.S.2d 589 (1978).

[50] *Southern New England Contracting Co. v. State*, 165 Conn. 644, 345 A.2d 550 (1974), followed in *Walter Kidde Constructors, Inc. v. State*, supra note 41. *Broward County v. Russell, Inc.*, 589 So.2d 983 (Fla.Dist. Ct.App.1991).

the contractor from recovering for extended home office overhead.) Strict application of certainty rules can make it difficult if not impossible for the contractor to establish the exact amount of its loss.[51] To counter this, some courts have been willing to relax these rules by using formulas, such as the *Eichleay* formula, that, though not perfect, provide a rough way of measuring an undoubted but difficult-to-establish loss caused by the delay.

Productivity Loss Preferred Formulas: Measured Mile and Industry Productivity Studies. Before proceeding to the more controversial global formulas, two other techniques have been approved to measure productivity losses. The measured mile compares contractor productivity "in an undisputed area of work with the contractor's productivity on a similar task during a disrupted work period."[52] One commentator suggests that while comparisons usually are made of activities on the same project, comparisons also can be made of work on a different project "involving the same or similar type of work."[53] This method had been used on different, though similar, activities of the same project.[54]

Another preferred method of proving diminished productivity is industry or trade productivity studies.[55]

Total Cost. Some contractors seek to use an even rougher formula than those that have been noted in this section. A contractor faced with the almost insurmountable obstacle of establishing actual losses knows it has lost money and feels its losses were attributable, not to its own poor estimating or inefficient performance, but to acts of the owner or to those for whom the owner is responsible. As a result, the contractor will sometimes seek to use what is called the *total cost method*, a comparison of the actual costs of performance with what the contractor contends should have been the cost of the project.

Some courts have not been willing to accept the total cost formula because it assumes that the defendant's breaches caused all the expenditures in excess of the original estimate, that the original estimate accurately represented the cost of performance, and that the contractor efficiently performed its work.[56] Although the Court of Claims allowed this method in *WRB Corporation v. United States*, it placed limits on possible abuse by requiring proof that

(1) The nature of the particular losses make it impossible or highly impracticable to determine them with a reasonable degree of accuracy;
(2) The plaintiff's bid or estimate was realistic;
(3) Its actual costs were reasonable; and
(4) It was not responsible for the added expenses.[57]

Pennsylvania will apply the total cost *only* if there are no other means of determining damages and the claimant has presented reasonably accurate evidence of the various costs incurred.[58]

The *modified* total cost method seeks to avoid the crudeness of the total cost method. It focuses on the impacted work activities[59] and adjusts the original estimate to remove mistakes, inaccuracies, and work items not affected.

As to the reasonableness of the contractor's original estimate, the second item noted in the *WRB* case noted above, one commentator stated that the original bid could be considered realistic by factors such as:

The experience of the person responsible for estimating the bid, the failure of a competitor to challenge the bid (on a public project), the relative similarity of competing bids, and the track record of the bidder on similar jobs. The very acceptance of the contractor's bid should be seen as some proof of its reasonableness.[60]

---

[51]See Sections 6.06B and 27.04.

[52]Jones, *Lost Productivity: Claims for the Cumulative Impact of Multiple Change Orders*, 31 Pub. Cont.L.J.1, 34 (2001). For cases allowing this method, see *U.S. Industries, Inc. v. Blake Constr. Co, Inc.*, 671 F.2d 539 (D.C.Cir.1982); *P. J. Dick, Inc.*, VABCA No. 5597, *et al.* 01-2 BCA ¶31,647 ("most reliable though not exact, methodology to quantify labor inefficiency"), aff'd in part, vacated in part, reversed in part by *P. J. Dick, Inc. v. Principi*, 324 F.3d 1364 (Fed.Cir.2003) (issue mainly extended home office overhead); *W. G. Yates & Sons Constr. Co.*, ASBCA Nos. 49398, 49399, 01-2 BCA ¶31,348. See also Patton & Gatlin, *Claims for Lost Labor Productivity*, 20 Constr. Lawyer, No 2,April 2000, pp. 21, 24–25.

[53]Jones, supra note 52 at 34.

[54]*Clark Constr. Group, Inc.*, VABCA No. 5674, 00-1 BCA ¶30,870, reconsideration denied, 00-2 BCA ¶30,997.

[55]Ibid; *P. J. Dick, Inc.*, supra note 46 (approving manual on efficiency losses published by the Mechanical Contractors Association of America). See also Patton & Gatlin, supra note 52 at 26–27.

[56]*Namekagon Dev. Co. v. Bois Forte Reservation Housing Auth.*, 395 F.Supp. 23 (D.Minn.1974), aff'd, 517 F.2d 508 (8th Cir.1975).

[57]183 Ct.Cl. 409, 426 (1968). *Propellex Corp. v. Brownlee*, 342 F.3d 1335 (Fed.Cir.2003) failed to prove first element of *WRB* test.

[58]*John F. Harkins Co., Inc. v. School Dist. of Philadelphia*, 313 Pa.Super. 425, 460 A.2d 260 (1983) (contractor did not meet burden).

[59]*Servidone Constr. Corp. v. United States*, 931 F.2d 860 (Fed.Cir. 1991) (reduced award because bid too low and substituted a reasonable bid); *E. C. Ernst, Inc. v. Koppers Co.*, 626 F.2d 324 (3d Cir.1980); *Seattle Western Indus., Inc. v. David A. Mowat Co.*, 110 Wash.2d 1, 750 P.2d 245 (1988) (deducted cost for which subcontractor claimant responsible).

[60]Sobel, *The Modified Total Cost Method of Determining Damages*, 21 Constr. Lawyer, No 4, Fall 2001, p.p.3, 7

Yet the often insurmountable proof problems have led an increasing number of courts, often in road construction disputes, to recognize total cost where appropriate limitations are observed.[61]

Jury Verdict. Another rough measurement sometimes employed either in conjunction with a total cost theory or by itself is the jury verdict. The jury verdict seeks to employ the same educated guesswork used by a jury. This formula recognizes the inherently imprecise nature of the proof that can be produced in such cases. However, a jury verdict can be used only where there is a clear proof of injury, there is no more reliable method for computing damages, and the evidence is sufficient to make a fair and reasonable approximation of damages.[62] Inasmuch as these crude formulas depend heavily on the facts in a particular case, understanding their use requires that case decisions be examined. The following case, reproduced in part, involves these crude formulas and comes to the conclusion that using these formulas was appropriate.

NEW PUEBLO CONSTRUCTORS, INC. v. STATE OF ARIZONA

Supreme Court of Arizona, En Banc, 1985. 144 Ariz. 95, 696 P.2d 185.

[Ed. note: A catastrophic storm and unusually heavy rainfall caused extensive damage to road work. The contractor (NPC) performed corrective work and sued the state (ADOT) to recover its added expenses under a clause that relieved the contractor for "Acts of God."]

* * *

III. THE MEASURE OF DAMAGES

The jury awarded damages of approximately $201,000 in actual damages (without overhead) including the following amounts:

1. Cleaning and refurbishing medians and ditches $ 3,500
2. Replacing eroded material in the Anamax pit lost in the flood, including costs of hauling substitute material from the Duval pit to the Anamax pit $ 28,500
3. Restoring eroded and lost special backfill material $ 8,000
4. Restoring mineral aggregate from new area after flooding of the Agua Linda pit $ 102,000
5. Additional stripping of the Duval mineral aggregate pit $ 19,000
6. Dewatering of the Duval and Agua Linda mineral aggregate pit $ 40,000
$ 201,000

The contractor kept records of the actual costs of the entire project but did not keep separate records of the actual costs of the rebuilding and rework caused by the weather, except for item #6.

* * *

When additional work is performed on construction projects, there are traditionally at least three ways of proving costs. Specifically:

1. *Actual cost.* Keeping separate records of the actual costs of additional work is the most reliable method of qualifying costs.

* * *

[61]*Moorhead Constr. Co. v. City of Grand Forks*, 508 F.2d 1008 (8th Cir.1975); *C. Norman Peterson Co. v. Container Corp. of Am.*, supra note 27, but see *Amelco Elec. v. City of Thousand Oaks*, 27 Cal. 4th 228, 38 P.3d 1120, 115 Cal.Rptr.2d 900 (2000) (rejected total cost in claim against public entity); *Amp-Rite Elec. Co., Inc. v. Wheaton San. Dist.*, 220 Ill.App.3d 130, 580 N.E.2d 622, appeal denied, 143 Ill.2d 635, 587 N.E.2d 1011 (1992); *McKie v. Huntley*, 620 N.W.2d 599 (S.D.2000). For a comprehensive analysis, see Aaen, *The Total Cost Method of Calculating Damages in Construction Cases*, 22 Pac.L.J. 1185 (1991) (noting that majority of appellate cases accept total cost and illustrating various formulas).

[62]*Fattore Co. v. Metropolitan Sewerage Comm'n*, 505 F.2d 1 (7th Cir.1974) (approved jury verdict but warned that such formulas must require clear proof that the contractor has suffered a loss and that there was no more reliable method for computing damages); *Meva Corp. v. United States*, 206 Ct.Cl. 203, 511 F.2d 548 (1975); *State Department of Transp. v. Guy F. Atkinson Co.*, 187 Cal.App.3d 25, 231 Cal.Rptr. 382 (1986) (approved decision by arbitrator in a required judicial-like arbitration in claim against state entity who arbitrarily reduced award 35 percent from the highest claim estimates submitted by contractor); *State Highway Comm'n of Wyo. v. Brasel & Sims Constr. Co.*, 688 P.2d 871 (Wyo.1984); *Bechtel National Inc.*, NASA BCA No. 1186-7, 90-1 BCA ¶22,549.

2. *Jury verdict.* Where the contractor cannot prove actual costs, the contractor may present evidence of the cost of additional work to the finder of fact including any actual cost data, accounting records, estimates by law and expert witnesses, and calculations from similar projects.

* * *

3. *Total cost.* Under certain circumstances, the contractor may subtract the estimated cost or bid of the entire project from the final cost of the entire project. The resulting figure is the amount claimed as damages.

* * *

Courts have resorted to the total cost method only under exceptional circumstances and then only as a last resort method. . . . This method of measuring costs suffers from many defects. By simply subtracting the bid estimate from the cost of the overall project, the total cost method:

(a) presumes that the bid estimate was realistic;
(b) can pass along costs to the state which might have been incurred despite the act of God or the other party's breach;
(c) can reward the contractor's inefficiency, managerial ineptitude, financial difficulties and other failings by passing these costs along to the state.

* * *

To avoid these defects of the total cost method, NPC proposes a fourth method of measuring costs in construction contract cases.

4. *Modified total cost (or cost variance).* The original bid for a particular item of work is subtracted from the actual costs for this item of work, though both the bid and actual costs are limited and adjusted in the following manner. Damages are limited to only certain time intervals wherein the work was adversely affected by the weather. The actual costs are totaled only for these work activities affected by the weather and to only those cost accounts recording such work. The employee and machinery time used for a particular item of work is determined by reference to the superintendent's reports describing the weekly work performed, engineer's diaries, labor distribution reports, and equipment distribution reports. Actual costs which were otherwise compensated or unrelated to such work activity are eliminated. The original bid estimate is redetermined in the light of the actual unit costs for the month before and after the months adversely affected. The unit cost overrun for a particular work item during the rainy months would be multiplied by the total actual work performed to arrive at the cost of rework. The adjusted estimate of the particular items of work in dispute is subtracted from the actual adjusted cost of this work only if: (1) the nature of the particular losses makes it impossible or highly impracticable to determine them with a reasonable degree of accuracy; (2) the contractor's bid was realistic; (3) his actual costs were reasonable; (4) the added expense was not caused by the contractor or by some cause for which he assumed the risk, but was proximately caused by the unforeseen circumstance or the other party's breach.

It is misleading to refer to this as a separate "method" for determining damages. Most courts refuse to apply the total cost method in the absence of the aforementioned circumstances. . . . Additionally, the acceptability of such a determination of damages involves a factual question as to how carefully the total cost method was modified to restrict its deficiencies. The trial judge must play an active role in this fact-bound inquiry in determining that the measure of damages is appropriate to the nature of the harm involved and that the specific estimates have been appropriately adjusted to avoid recovery of unrelated costs by the contractor.

* * *

First, a sufficient foundation for the use of the cost variance method has been established in this case. Most of the additional expenses were caused by an invisible, rising subterranean water table, so that segregating and precisely recording rework costs and original work costs was impracticable. Other cases have applied a similarly modified total method as a last resort method on similar facts. See *J. D. Hedin Construction Co. v. United States*, 171 Ct.Cl. 70, 85–88, 347 F.2d 235, 246–47 (1965) (contractor suffered weather-related damage during extended period of work caused by faulty and changed government specifications); *State Highway Comm'n v. Brasel & Sims Construction*, supra, 688 P.2d at 878–79 (contractor suffered increased expenses and delay due to failure of a state to supply adequate water and inferior quality of state supplied gravel); *Moorhead Construction Co. v. City of Grand Forks* [508 F.2d 1008 (8th Cir.1975)] (excess moisture and lack of soil compaction caused additional expense and are different than represented); *Thorn Construction Co. v. Utah Dept. of Transp.*, 598 P.2d 365 (Utah 1979) (additional expense is caused contractor by inferior quality of road material and it is necessary to excavate other materials at another pit).

Because the additional work was best performed concurrently with the principal contract work, it was not feasible in this situation to quantify the actual costs of rework.

* * *

Nor has ADOT shown that NPC's adjusted estimates of the costs of rework were unrealistic. NPC submitted the low bids on both the Tubac and Carmen projects and increased these estimates to reflect the costs actually experienced on the project. The jury obviously concluded that the weather was the cause of the contractor's cost overruns during the months in question. NPC did keep some actual cost records where it was feasible to do so, namely, for item #6 of the damages, the dewatering of the mineral aggregate pits.

* * *

NPC used the jury verdict method to determine damages for . . . additional stripping of the Duval aggregate pit. The use of the jury verdict method is appropriate when (1) the state is liable for a changed condition that increases the contractor's expenses, (2) due to circumstances beyond his control, the claimant cannot feasibly prove specific damages by a more reliable method, and (3) when there is sufficient evidence in the record to provide a reasonable basis for approximating the damages.

* * *

We also believe that there is sufficient foundation for the use of the jury verdict method. Courts have approved this method of damages on similar facts. *Metro Sewerage Comm'n v. R. W. Constr.*, supra, 78 Wis. at 466, 255 N.W.2d at 302 (contractor suffered increased costs due to dewatering of tunnel caused by artesian water not shown on specifications); *Foster Constr. v. United States*, 193 Ct.Cl. 587, 435 F.2d 873 (1970) (contractor caused additional expense of uncertain amount by misrepresented subsurface conditions in constructing pier). We believe

that the use of the jury verdict was appropriate in this case for much the same reasons that the use of the modified total cost method was warranted. We are far from unqualifiedly endorsing either the jury verdict or the modified total cost measure of damages. The availability of both measures of damage must be proven by the contractor. Neither measure of damages can be used where there is no excuse for the failure to keep track of actual costs. *Pickard's Sons Co. v. United States*, 209 Ct.Cl. 643, 532 F.2d 739 (1976); *Appeal of Soledad Enterprises, Inc.*, ASBCA 20376, 77-2 BCA 12757 (1977).

We do not believe that this is a case wherein the contractor inexcusably failed to keep track of costs because this was not feasible due to circumstances beyond his control. The case at bar is quite different from the cases cited in this respect. Id.

Nor can the contractor use the modified total cost method when there is sufficient evidence to use the jury verdict method.

* * *

The contractor cannot use the jury verdict method if he can prove only that the state caused part of the damages and cannot make a reasonable approximation of those damages. See *Electronic & Missile Facilities, Inc. v. United States*, 189 Ct.Cl. 237, 256–57, 416 F.2d 1345, 1357–58 (1969) (contractor was caused additional expense to remove contaminated gravel, but cannot reasonably approximate government causation of damages). Even on facts where the use of both these measures of damages is appropriate, they are subject to close judicial scrutiny to insure that the contractor does not receive a windfall. We find no error, however, on the facts of this case.

The *New Pueblo* case's apparent preference for the jury verdict over the one that measured the loss to the contractor by the modified total cost formula would not be shared by other courts. While the modified total cost formula even in its refined form is "rough," the jury verdict is even "rougher." Only if the modified total cost formula cannot be employed is it likely that the jury verdict would be the formula used by the court.

SECTION 27.03 Measurement: Owner Vs. Contractor

A. Illustrations

The principal measurement problems relate to the contractor's unjustified failure to start or complete the

project, to complete the project as specified, or to complete the project on time. Because the issues are similar—except for the uncompleted or defective work being on the owner's land, which raises unjust enrichment problems—this section will encompass claims by primes against subcontractors.

B. Project Never Begun

Damages are determined by subtracting the contract price from the market price of the work. Market price is usually based on the best competitive price that can be obtained from a successor contractor for the same work. Suppose the contract price were $100,000 and the successor cost $120,000. The owner would be entitled to $20,000. This would protect its contract bargain.

Claims for lost bargains do not receive the protection that claims based on out-of-pocket losses do. It is conceivable that in a case of this type it would be concluded that there was no difference between contract and market price, no lost bargain, and no recovery for that element of the breach.

In addition to the expectation interest (loss of bargain), the owner would be entitled to any losses caused by the delay (discussed in Section 27.03E).

C. Project Partially Completed

Suppose the contractor ceases performance unjustifiably or is properly terminated by the owner. Restitution is one way of measuring the owner's claim. This measure seeks to restore the status quo as it existed at the time the contract was made. This can be accomplished by restoring to the owner any progress payments it has made to the contractor. From this is deducted the benefit to the owner of the partially completed project. However, the difficulty of measuring the benefit conferred may lead a court to conclude that restitution is not proper. For example, one case denied the owner restitution of its payments in a well-drilling contract in which the well produced 80 percent of the required amount of water.[63] In such a case, diminished expectation would be more appropriate. The owner would be entitled to the cost of performance that would generate the promised performance by the contractor, less the payments it had made.

When a fallout shelter leaked and finally caved in, the owner was allowed to recover progress payments.[64] To restore the status quo, the contractor would be entitled to deduct the extent to which its work added value to the land. The more likely event is that it added no value, with the owner being entitled to recover the cost of removing the shelter and restoring the land to its precontract condition.[65]

Restitution is rarely used to measure owner claims in construction contract disputes. More commonly, failure to complete generates two types of damages: excess cost of reprocurement and damages for omitted or defective work. First, the owner very likely will have to pay more than the balance of the contract price to have the work completed by a successor.

Usually the cost of engaging a successor contractor or subcontractor will exceed the balance of the contract price outstanding. The successor will not, as a rule, be selected by competitive bidding, because of time constraints. Also, a successor will need to incur expenses to bring its own equipment and personnel to the site. It may also determine that it will need a premium to take into account that it may in the end be held liable for defective work performed by the original contractor. Similarly, it may worry about working with an owner with whom the original contractor had problems, no matter who may have been at fault in causing the termination. Finally, the refusal by the original contractor to continue work may be based on the belief that market forces make it advisable to use its resources elsewhere. The market may have risen.

The law would award the owner any excess costs in using a successor for doing the work required to be performed under the contract. This would put the owner in the position it would have been had the work been completed by the original contractor. The AIA in Document A201-2007, Section 14.2.4, codified this common law outcome. But that section also takes into account the unlikely possibility that the cost of completion by a successor may be less than the amount outstanding. In such a case Section 14.2.4 states "such excess shall be paid to the Contractor." A similar result would occur under A401-2007, Section 7.2.1, governing subcontracts.

In *Saxon Construction and Management Corp. v. Masterclean of North Carolina, Inc.*,[66] the cost of engaging a successor subcontractor was *much less* than the amount outstanding. The court does not state why this occurred. It may have been due to a blown-up original price, a shift in the market, a desire by the successor to get into a new market or sharp bargaining by the prime. In any event the contract required the prime to give the excess to the defaulting subcontractor. But the court found this outcome promotes waste, provides an incentive to breach, creates unjust enrichment, is unconscionable, and violates public policy.

Although arguments can be made for either outcome, the court refused unfortunately to enforce the clause. The clause properly reflected the concept that the breaching contractor is not a criminal deserving of punishment and

[63]*Village of Wells v. Layne-Minnesota Co.*, 240 Minn. 132, 60 N.W.2d 621 (1953).

[64]*Economy Swimming Pool Co., Inc. v. Freeling*, 236 Ark. 888, 370 S.W.2d 438 (1963).

[65]*Bourgeois v. Arrow Fence Co.*, 592 So.2d 445 (La.App.1991) (addition so defective so as to be irreparable permitted owner to recover payments made and cost of demolition), writ denied, 596 So.2d 214 (La.1992).

[66]273 N.J.Super. 231, 641 A.2d 1056 (App.Div.), cert. denied, 137 N.J. 314, 645 A.2d 142 (1994). For the trial court opinion, see 273 N.J.Super. 374, 641 A.2d 1129 (Law.Div.1992).

the common law rule that the damages should simply put the claimant—here, the prime—in the position it would have been in had the contract been performed but not in a *better* position.

Second, in many cases, the work will have been performed defectively, and this too may justify additional compensation (discussed in Section 27.03D). Completion is likely to be delayed (delay damages are discussed in Section 27.03E).

D. Defective Performance: Correction Cost or Diminished Value?

Claims relating to defective performance may arise when a contractor has performed improperly, never having reached the level of substantial performance. This situation has been discussed in Section 22.06B. More commonly, defective performance cases involve attempts by the contractor to recover the balance of the contract price by alleging either substantial or full performance, the secondary issue being the deduction available to the owner if the work was only substantially performed.

The law seeks to place the owner in the position it would have occupied had the contractor performed properly. One way of doing so is by giving the owner the *cost of correction*—the amount necessary to correct the defective work or complete the work to bring it to the state required under the contract.

Another method of giving the owner what it was promised is to award the *diminished value*—the difference between the value of the project as defectively built and what it would have been worth had it been completed as promised. This measure gives the owner's balance sheet what it would have had by the time of full performance. This aim is achieved by combining the value of the project as it sits with the diminished-value measure of recovery because performance has been less than complete.

Each method presents problems. Suppose cost of correction is used, and the correction costs greatly exceed the value correction would add to the property. The likelihood that the owner will not use the money for this purpose presents the possibility that the money will be a windfall. The award would put the owner in a better position than if the contractor had performed properly. If the owner *did* correct defective work or complete the work when it would not be economically sound to do so, this would waste scarce societal resources.

But awarding diminished value deprives the owner of the performance bargained for and forces the owner to take an amount that is usually determined by the testimony of expert witnesses. Such testimony is expensive to procure and may not take into account the subjective expectations of the owner when it made the contract. In addition, this evidence is "softer" than the evidence of the cost of correcting the work. Finally, if the award is less than the cost of correction and the owner elects to correct or complete, the award will be inadequate.

Some courts have begun to speak of a "stigma" effect, a term taken from environmental remediation cases.[67] These cases involve mold or environmental hazards that may diminish the resale value of a sound and defect-free building.[68] The stigma effect is more likely to relate to cost of correction. In such cases, despite correcting the defect, the value of the property may be reduced, much like a car that has been in an accident or a house where a heinous murder has been committed. Buyers may be hesitant to buy because of the stigma attached to the house. The stigma factor means that the claim combines cost of correction with diminished value caused by the stigma.

Events that occur after the claim has arisen can create problems. Suppose the owner sells the house and cannot effectuate the repairs. One case held that the jury properly awarded the owner the costs of correction that exceeded diminished value despite the owner's having sold the house and being unable to correct the defects.[69] The court in effect held that the owner's right vested at the time of the wrongful act and was unaffected by subsequent events. Yet another case limited the owner to the diminished value as she could no longer correct the defects.[70]

As a rule, the owner is entitled to the quality spelled out in the contract and no more. It is not entitled to a better or different house than it bargained for. But one case held the corrected work need not be the same as that called for under the contract.[71] The case involved the failure of an earthen trench system. The owner was entitled

[67]See Section 9.13.

[68]*Vista Resorts, Inc. v. Goodyear Tire & Rubber Co.*, 117 P.3d 60 (Colo.App.2004), cert. denied, 2005 Westlaw 1864133 (Colo.2005) (defective radiant heating systems).

[69] *St. Louis L.L.C. v. Final Touch Glass & Mirror, Inc.*, 386 N.J.Super.177, 899 A.2d 1018 (App.Div.2006). Similarly see *Vaughn v. Dame Constr. Co.*, 223 Cal.App.3d 144, 272 Cal.Rptr. 261 (1990) (owner did not lose her right to recover by subsequent sale).

[70]*Wentworth v. Air Line Pilots Ass'n*, 336 A.2d 542 (D.C.App.1975).

[71]*Stovall v. Reliance Ins. Co.*, 278 Kan.777, 107 P.3d 1219 (2005).

to replacement damages to install the superior concrete shallow trench system where that cost would be less than installing a new earthen trench system.

Predictably, different jurisdictions will go off in different directions; there may even be contradictions within a particular jurisdiction. A few states employ only the cost of completion or correction. Some hold that diminished value is the only proper measure. Some states hold that the owner can elect to invoke either measure. A few hold that the proper measure of recovery is cost of correction or diminution in value, whichever is less.[72] As shown in *Plante v. Jacobs*,[73] (reproduced in Section 22.06B), one formula can be applied to some items and the other to other items. If cost of correction will diminish the value, both can be applied.[74] Most states prefer the cost of correction measure unless it would generate "economic waste."[75] One author stated,

"Economic waste" is primarily a result-oriented concept, not a fiscal one. Economic waste comes into play in those cases in which the defective building is still serviceable and useful to society. If repairs are possible but would completely destroy a substantial portion of the work, damage or injure good parts of the building, impair the building as a whole, or involve substantial tearing down and rebuilding, then that is "economic waste."[76]

The choice of an appropriate measure may also depend on whether either party, particularly the contractor, has acted in good faith. If there does not appear to have been a good reason for the deviation, there is a stronger likelihood that a measure preferable to the owner will be selected.[77]

This topic was dealt with in Section 22.06B, which examined substantial performance. If the contractor has substantially performed, the owner can offset against the balance of the contract price the damages caused by the contractor's failure to perform the contract as promised. The offset sometimes is the cost of correction and sometimes the diminution in value or a combination of the two. Please review Section 22.06B.

The most difficult cases involve a residence constructed for the use of the owner. In one case, the contractor had installed discolored roofing shingles. The cost to correct this properly functioning roof by replacing the roofing material was estimated at between $14,000 and $25,000. The contractor contended that the shingles would ultimately weather to a uniform color and that the roof was not visible to passersby. The court concluded the measure of recovery to be $7,500, based on diminution in value.[78]

Another case involving the installation of a shingled roof concluded that the contractor had not substantially performed, because the roof had yellow streaks, although it was structurally sound and would protect the owner from the elements. The court concluded that in matters relating to homes and their decoration, taste or preference "almost approaching whimsy" may be controlling with the homeowner so that variations that under other circumstances might be considered trifling would preclude a finding that there had been substantial performance.[79] This gave the owner a defense to any contractor claim for

---

[72]*Thompson v. King Feed & Nutrition Service, Inc.*, 153 Wash.2d 447, 105 P.3d 378 (2005) (unless the building was entirely destroyed, rather than merely damaged).

[73]10 Wis.2d 567, 103 N.W.2d 296 (1960).

[74]*Italian Economic Corp. v. Community Eng'rs, Inc.*, 135 Misc.2d 209, 514 N.Y.S.2d 630 (1987) (structural repairs needed to comply with code reduced floor space and windows).

[75]The many cases are collected in Annot., 41 A.L.R.4th 131 (1985). For some recent cases, see *Fisher v. Qualico Contracting Corp.*, 98 N.Y.2d 534, 779 N.E.2d 178, 749 N.Y.S.2d 467 (2002) (New York statute changing common law collateral source rule applied to property insurance payments and the New York rule using the measure that produces the lower amount); *Granite Constr. Co. v. United States*, 962 F.2d 998 (Fed. Cir.1992) (though material did not meet technical requirements of the specifications, it exceeded by 20 the safety factor for the overall project: diminution in value applied), cert. denied, 506 U.S. 1048 (1993); *City of Charlotte v. Skidmore, Owings, & Merrill*, 103 N.C.App. 667, 407 S.E.2d 571 (1991) (sidewalk deteriorated because of negligent design by architect; replacement value applied). See also Abney, *Determining Damages for Breach of Implied Warranties in Construction Defect Cases*, 16 Real Est. L.J. 210 (1988); Chomsky, *Of Spoil Pits and Swimming Pools: Reconsidering the Measure of Damages for Construction Contracts*, 75 Minn.L.Rev. 1445 (1991) (suggests that measure exceed diminution of market value if cost to complete too burdensome; wants all factors considered that relate to owner's loss and uses highest level consistent with fairness to the contractor).

[76]Abney, supra note 75 at 218 (footnotes omitted).

[77]In *Kaiser v. Fishman*, 187 A.D.2d 623, 590 N.Y.S.2d 230 (1992), leave to appeal denied, 81 N.Y.2d 711, 619 N.E.2d 658, 601 N.Y.S.2d 580 (1993), the court affirmed an award for cost of correction that involved physically lifting the residence off its pilings and performing substantial corrective work, mainly because the contractor had deliberately failed to comply with the contract terms.

[78]*Salem Towne Apartments, Inc. v. McDaniel & Sons Roofing Co.*, 330 F.Supp. 906 (E.D.N.C.1970). In *Orndorff v. Christiana Community Builders*, 217 Cal. App.3d 683, 266 Cal.Rptr. 193 (1990) homeowners were awarded repairs costs that exceeded the loss of market value by 2.5 percent.

[79]*O. W. Grun Roofing & Constr. Co. v. Cope*, 529 S.W.2d 258 (Tex. Ct.App.1975).

the balance and a basis for an owner claim measured by the cost of replacing the roof.

As noted earlier, a combination of the two measures is appropriate. For example, some deviations can be measured by diminution in value, and others by cost of correction.[80] The two formulas can be applied together if the cost of correction will not achieve the outcome promised under the contract. For example, in one case, settling and slanting had been caused by a failure to sink piles far enough. The court held that the owner could recover not only the cost of doing as well as could be done to correct the problem but also the diminished value, as the cost of correction could never accomplish the contractual requirements.[81]

When cost of correction is used, several other factors must be considered. First, the cost of correction cannot be recovered if the corrective work includes work not called for under the original contract. In one case, replacing leaky gutters and downspouts with galvanized gutters was not the proper measure of recovery when aluminum had originally been called for.[82] The measure should have been the cost of installing aluminum gutters that would have stopped the leaking, assuming this was the contractor's responsibility.

Suppose the original construction is in some way inherently defective or inadequate. Here, an award using the cost of repair standard may not require simply repairs, but the actual removal and replacement of the structure, with the new structure often being of better quality than the original. For example, in *Parsons v. Beaulieu* the defendant had contracted to build a septic tank for the plaintiff. The completed work did not comply with legal requirements. The plaintiff replaced the tank with a larger one that did comply. The court held that the defendant contractor had to pay for the cost of the larger septic tank because the original one would not function in a way that would meet the contractor's implied warranty. The court noted that the owner had installed the least expensive one that would meet the legal requirements. In the *Parsons* case,

the owner was not being put in a better position, as it had contracted for a tank that met the legal requirements.[83]

Suppose the correction takes place a year after completion. Installing new materials and equipment can be said to put the owner in a better position than that bargained for. The old materials and equipment have used up some of their useful life, as they have depreciated. Generally, the law takes into account any extended life that replacement may give the owner.[84] But in *Dickerson Construction Co. v. Process Engineering Co., Inc.*,[85] the court refused to deduct any amount for extending the life of the building, inasmuch as the defects had been discovered only six months after completion.

At what point is the cost of correction determined? Usually it is at the time of the breach, assuming the owner knows of the defect and is in a position to have the defective work corrected. However, courts in some cases have used the time of trial as the benchmark for determining when the cost of correction should be computed. These courts looked at the highly inflationary period of the 1970s and noted that an award based on time of breach rendered years after the work has been completed would not put the owner in the position it would have been had the work been done correctly in the first place.[86]

A similar problem can arise when the issue is diminution in value. A Florida case is instructive on this and other points: *Hourihan v. Grossman Holdings, Limited* involved a claim by the owners against a contractor for failure to build a house as promised. The owners contracted with the developer to purchase a house to be built in a planned development. Both the model and the office drawings showed the house with a southeast exposure, and the contract stated that the developer would construct the house "substantially the same" as in the plans and specifications or as in the model. A short time later, a new model and map went on display that showed the owners'

---

[80]*Plante v. Jacobs,* supra note 73, reproduced in Section 22.06B.

[81]*Kahn v. Prahl,* 414 S.W.2d 269 (Mo.1967). See also *Northern Petrochemical Co. v. Thorsen & Thorshov, Inc.,* 297 Minn. 118, 211 N.W.2d 159 (1973); *Italian Economic Corp. v. Community Engineers, Inc.,* supra note 74.

[82]*St. Joseph Hosp. v. Corbetta Constr. Co., Inc.,* 21 Ill.App.3d 925, 316 N.E.2d 51 (1974); *Steinbrecher v. Jones,* 151 W.Va. 462, 153 S.E.2d 295 (1967).

[83]429 A.2d 214 (Me.1981). See also *Kangas v. Trust,* 110 Ill.App.3d 876, 441 N.E.2d 1271 (1982) (better material needed because contractor's breach made it more costly to repair), and *Hendrie v. Board of County Comm'rs,* 153 Colo. 432, 387 P.2d 266 (1963) (pool removed and replaced with a different model compatible with soil conditions).

[84]*Freeport Sulphur Co. v. S. S. Hermosa,* 526 F.2d 300 (5th Cir.1976).

[85]341 So.2d 646 (Miss.1977). Similarly, see *Five M. Palmer Trust v. Clover Contractors, Inc.,* 513 So.2d 364 (La.App.1987).

[86]*Anchorage Asphalt Paving Co. v. Lewis,* 629 P.2d 65 (Alaska 1981); *Corbetta Constr. Co. of Illinois, Inc. v. Lake County Public Bldg. Comm'n.,* 64 Ill.App.3d 313, 381 N.E.2d 758 (1978) (contractor denied breach and refused to correct).

lot and to-be-built house facing the opposite direction from what they expected and wanted. The owners brought this discrepancy to the attention of the developer and demanded the construction of the house that they had contracted for. (They wanted the house facing a particular direction for aesthetic and energy-saving reasons.) The developer refused to change the plans and began constructing the house facing in the direction opposite of that expected by the owner.

The trial court refused to award any damages, finding that the cost of turning the house around would have been economically wasteful and out of proportion to the good to be attained. It also noted that the value of the house had risen above the contract price since the date of the contract.

The intermediate court of appeals applied a different measure of damages.[87] That court held that the economic waste doctrine does not apply to residential construction. It also found that the developer's willful and intentional failure to perform as promised barred it from using the substantial compliance doctrine. The court held that the proper damages would have been an amount necessary to reconstruct the dwelling to make it conform to the plans and specifications.

The Florida Supreme Court did not agree.[88] First, it noted that the proper remedy for breach of a construction contract is usually the reasonable cost of correction if this is possible and does not involve unreasonable economic waste. Where it does, the measure is the difference between the value of the product contracted for and the value of the performance received. The court saw no reason to separate residential buildings from other construction. The court concluded that repositioning the house would result in economic waste but held that the trial court had incorrectly applied the proper formula—diminution in value. The court stated that the measurement is to be determined at the date of breach. Fluctuations in value after breach do not affect the measure of recovery. (The house would have been worth even more had it been built properly.)

The *Grossman* case also reflects the difference between objective and subjective measurements of value. The impersonal marketplace may not have found any difference in the house facing in one direction or the other. But the Hourihans did.

There is something to be said for the intermediate appeals court decision that refused to use the economic waste argument in residential construction.[89] Balance sheet gains or losses are appropriate for certain types of commercial ventures but hardly make sense when a couple is buying a home in which they expect to live.

The case also rejects, or at least seems to reject, the statement made in other cases that allows the economic waste measurement to be used only if the breach is not willful. Here the owners complained as soon as they noticed that the developer was starting to turn the house around, and the developer simply refused to make any change. To be sure, the developer might have wanted all the houses facing in a particular way and might have thought it had sufficient discretion to make that sort of change inasmuch as the contract stated that the house that was built would be *substantially* the same as the model and house as shown on the maps. Yet this could be classified as a willful and deliberate breach.

The assumption generally is that there are two possibilities, *diminished value* with the possibility of the idiosyncratic owner not getting what it wanted and *cost of correction*, with the possibility of excessive damages and economic waste. There is a third, though not as yet a formula that has achieved judicial recognition. Where either of the two accepted formulas would not achieve a just outcome, the court could award the owner the difference between what the work *should have cost* had it been done properly and the *actual cost*. This would prevent the contractor from being unjustly enriched when the owner cannot establish damages under either of the two accepted formulas.[90]

E. Delay

Chapter 26 discussed the problems of delay, and Section 26.09B looked at contractual attempts to agree on damages in advance, liquidating damages. Delay was noted in Section 27.02F, dealing with contractor claims for

---

[87]396 So.2d 753 (Fla.Dist.Ct.App.1981).

[88]*Grossman Holdings, Ltd. v. Hourihan,* 414 So.2d 1037 (Fla.1982). The case before the intermediate appellate court was captioned (titled) *Hourihan v. Grossman Holdings, Ltd.* When the case was accepted for review by the Florida Supreme Court, the first party in the caption became the party seeking review. That was Grossman Holdings, Ltd.

[89]*Edenfield v. Woodlawn Manor, Inc.,* 62 Tenn.Ct.App. 280, 462 S.W.2d 237 (1970). See *Orndorff v. Christiana Community Builders,* supra note 78.

[90]For a discussion of this damages measure in England, see Wallace, *Cost of Repair or Diminution in Value: An Intermediate Measure? Or, the Too Shallow Deep End and How Both Sides Lost,* [1996] Int'l.Constr.L.Rev. 238.

unabsorbed home office costs. This section looks briefly at owner claims for unexcused contractor delay, delay claims that were not liquidated in the contract.

The basic measurement formula used for delay is lost use, measured as a rule by rental value.[91] Suppose the owners made leases with tenants who could not be put into possession because of delayed construction. One case allowed the owner to recover for such rentals that were not paid.[92] Other cases have concluded that no damages were suffered when the leases were extended to the full lease period, as long as the tenants made no claims against the owner for the delay.[93]

The uncertainties over the lost rental value and other related expenses provide an incentive to liquidate damages for contractor-caused delay.

As noted in Section 27.06D, AIA documents issued in 1997 and continued in A201-2007, Section 15.1.6.2, contain waivers of consequential damages and limitations on liquidation of damages. Under these documents some of the losses suffered by the owner for lost use would not be recoverable and liquidation might prove difficult.

SECTION 27.04 Certainty

Please review Section 6.06B for background on the certainty requirement. Construction litigation can require adequate proof of the following:

1. contractor cost of performing the work
2. owner or contractor projected cost of completing the work
3. additional cost incurred by contractor because of owner disruption or delay
4. contractor profit on performed work, unperformed work, or other consequential damages
5. owner lost profit for delayed performance or faulty work

As to item 1, a contractor who kept time records but who did not segregate them by jobs was able to prove how much should have been attributed to the defendant's project by testimony of his foreman.[94] Similarly, subcontractors are sometimes relieved from strict certainty proof.[95]

However, a prime contractor was limited to the amount that it could prove by time cards for labor work performed after the defendant subcontractor had defaulted.[96] Similarly, an estimate by the owner as to the cost of correction was held inadequate to prove damages.[97]

Item 2 is usually established by introducing evidence of estimates by other contractors or, even better, contracts for completion and correction.

As to item 3, see Section 27.02F.

As to item 4, although one court refused to award profits in a construction contract claim because of the lack of experience, personnel, equipment, and background of the contractor,[98] the principal questions have been the type of proof that will support a claim for lost profits. One court allowed a subcontractor to establish general profitability of the job.[99] Another submitted lost profits to the jury based on testimony of the president of the plaintiff contractor as to profits on other similar work, his method of estimating profits, and his expectation as to profits.[100] In the latter case, another witness had testified as to a range of profits. However, the appellate court reduced the judgment to the smallest profit that could be supported by the testimony.[101]

Court decisions are not always consistent as to the amount of certainty required to establish a loss. Some courts will not hold the claimant to a high standard of certainty if they are convinced that a loss has occurred,

[91]*Ryan v. Thurmond*, 481 S.W.2d 199 (Tex.Ct.App.1972). See *Miami Heart Institute v. Heery Architects & Engineers, Inc.*, 765 F.Supp. 1083 (S.D.Fla.), aff'd, 44 F.3d 1007 (11th Cir.1991), discussed in Section 12.14B, in which the court employed lost use value in a claim by an owner against its architect.

[92]*Herbert & Brooner Constr. Co. v. Golden*, 499 S.W.2d 541 (Mo. App.1973).

[93]*Ryan v. Thurmond*, supra note 91; *Ralph D. Nelson Co. v. Beil*, 671 P.2d 85 (Okla.App.1983) (owner may not keep property for full term of twenty-five-year lease).

[94]*Don Lloyd Builders, Inc. v. Paltrow*, 133 Vt. 79, 330 A.2d 82 (1974).

[95]*St. Paul-Mercury Indem. Co. v. United States*, 238 F.2d 917 (10th Cir.1956); *McDowell-Purcell, Inc. v. Manhattan Constr. Co.*, 383 F.Supp. 802 (N.D.Ala.1974), aff'd, 515 F.2d 1181 (5th Cir.1975), cert. denied, 424 U.S. 915 (1976); *Certain-Teed Products Corp. v. Goslee Roofing & Sheet Metal, Inc.*, 26 Md.App. 452, 339 A.2d 302, cert. denied, 276 Md. 739, 741 (1975).

[96]*Welch & Corr Constr. Corp. v. Wheeler*, 470 F.2d 140 (1st Cir. 1972).

[97]*Gross v. Breaux*, 144 So.2d 763 (La.App.1962).

[98]*Electric Service Co. of Duluth v. Lakehead Elec. Co.*, 291 Minn. 22, 189 N.W.2d 489 (1971).

[99]*Construction, Ltd. v. Brooks-Skinner Bldg. Co.*, 488 F.2d 427 (3d Cir.1973).

[100]*Natco, Inc. v. Williams Bros. Eng'g Co.*, 489 F.2d 639 (5th Cir. 1974).

[101]See Section 27.02D for formulas that include profit. For lost profit unconnected to this contract, see Section 27.06C.

that it has been caused by the defendant's breach, and that the claimant has marshaled the best available evidence. But other courts, particularly in cases in which the issue of breach is a close one and it appears that the defendant has performed in good faith, will use the certainty requirement to either limit or bar recovery.[102]

In arbitration or other nonjudicial systems, certainty will be less of an obstacle to obtaining the amount claimed. Nonjudicial dispute resolution involves practical people aware of the realities of recordkeeping and more willing to guess, compromise, or use rough formulas in these intractable proof problems.

SECTION 27.05 Records and Notices

A. During Performance

The certainty requirement for establishing damages discussed in Section 27.04 demonstrates the importance of recordkeeping during performance of the contract. The contractor who seeks to prove the reasonable value of the labor and materials it has furnished will be in a substantially better position if it can produce accurate, detailed records of what it spent in performing the work. If it seeks to prove additional expenditures that it claims resulted from wrongful acts of the owner, it will be in a substantially better position if it can establish those costs that resulted from the breach by the owner. If it cannot do so, the contractor may be able to take advantage of crude formulas, such as total cost, modified total cost, or the jury verdict discussed in Section 27.02F. But the party that goes into settlement negotiations, arbitration, or litigation with the most complete and detailed records is in the best position to obtain an optimal settlement, award, or court judgment.

In addition, the contractor should keep and maintain its records on *other* contracts, particularly records that establish its worker productivity, scheduling, and profit margins.

A contractor who wishes to obtain a time extension for bad weather should maintain careful records of the weather conditions at the place of performance. This is particularly important to those who use AIA documents.

A201-2007, Section 15.1.5.2, requires the contractor to produce records relating to weather and establish the effect weather had on its performance.

The importance of records is demonstrated in *Clark Construction Group, Inc.*[103] In dealing with an inefficiency claim by a subcontractor, the board pointed to the absence of daily logs, CPM fragments, correspondence, and other documentation that would support the disruption claim.

In another recent case failure to set up special accounts for investigative work barred a claim based upon the "total cost" method discussed in Section 27.02F.[104]

One commentator indicated that this demonstrates a new trend in the Federal Boards of Contract Appeals to require documentary evidence supporting the alleged inefficiencies. This would consist of contemporary documents such as "daily reports, change order requests, RFIs [Ed. note: Request for Information.], project logs or journals, correspondence, and other types of project materials." He states that this evidence "must identify the impacts on which the claim is based, as well as the cause and even the effect of the impact on the contractor's work." He concludes that without this "contemporaneous documentary evidence" the contractor will have difficulty recovering its costs.[105]

Recordkeeping is a prime motivation for having a full-time site representative.[106]

B. After Dispute Arises

After a dispute has arisen, each party seeks to augment its records with the records of the other party to the dispute. Usually this is handled through discovery prior to trial.[107]

Those who deal with public owners may be able to invoke the federal or any state freedom-of-information laws[108] or any federal or state constitutional due process

---

[102]*In Huber, Hunt & Nichols, Inc. v. Moore*, 67 Cal.App.3d 278, 136 Cal.Rptr. 603 (1977), the court refused to admit computer costs records. as they did not segregate costs incurred because of the breach.

[103]Supra note 54.

[104]*Propellex Corp. v. Brownlee*, supra note 57.

[105]Nelson, *Lost Productivity Claims and Documentation*, 15 Common Sense Contracting, No. 13, Smith, Currie & Hancock, ¶410.

[106]See Sections 12.08D and 17.05B.

[107]See Section 2.09.

[108]*De Maria Bldg. Co., Inc. v. Dept. of Management and Budget*, 159 Mich.App. 729, 407 N.W.2d 72 (1987). But see Sea Crest Constr. Corp. v. Stubbing, 82 A.D.2d 546, 442 N.Y.S.2d 130 (1981), which held that the correspondence between the public agency and its consultants was, with some exceptions, exempt and did not have to be disclosed. See also *Suffolk Constr. Co., Inc. v. Division of Capital Asset Management*, 449 Mass. 444, 870 N.E.2d 33 (2007) (Massachusetts public records law does not impliedly repeal the government's attorney-client privilege).

rights as a basis for obtaining relevant records of the public owner.[109]

C. Notices

Notices play a crucial role in construction contract administration. As examples, Section 21.04G treats the frequent contractual requirement that the contractor must give a notice that it intends to request additional compensation, usually because it was ordered to do work not part of its contractual obligation. As noted in Section 25.06, if the contractor discovers unexpected subsurface conditions that justify additional compensation under a Differing Sites Condition (DSC) clause, it must notify the owner before conditions are disturbed. If the contractor seeks a time extension as discussed in Section 26.08C, it must give a notice. Other illustrations can be given, but it should be clear that notice requirements are often central to claims. Not giving a required notice can bar the claim.[110]

While the function of a notice depends upon the particular issue, all notices have a basic function. The notice system allows the person in control, usually the owner or the design professional, to make needed adjustments in its program to eliminate or lessen further losses. Notices are also needed to prepare for future claims and disputes, such as disputed changes, time extensions, or DSC claims. Notices should trigger an investigation while facts are fresh and before the claimant can create conditions to justify a spurious claim. Also, the absence of a notice may reflect on the honesty of the claimant and the legitimacy of the claim.

In specific cases the defense by an owner that the required notice was not given may seem like an excuse to avoid paying a legitimate claim. When this occurs, the person resolving the dispute may search for evidence that the notice requirement has been waived. While waiver is a handy tool to avoid barring what appears to be a just claim, those resolving disputes should be aware of the crucial role notices play.

SECTION 27.06 Consequential Damages: Lost Profits

A. Defined

Direct losses can occur in any construction contract. They are usually measured by conventional objective formulas, such as lost use for unexcused contractor completion delay,[111] cost of correction or diminished value for other contractor defaults,[112] interest for owner-delayed payment,[113] the balance of the purchase price for nonpayment, and the additional costs caused by owner disruption or delay.[114]

Before examining specific consequential damages claims, lost profits on the contract that has been breached must be distinguished from profits on other transactions that the claimant asserts it would have made had there been no breach of contract. The first were covered in Section 27.02D. There lost profits are part of the conventional formula for measuring the contractor's claim when the owner has breached. In this section profits lost on *other transactions* are examined.

Among the difficulties encountered in this Section are the different labels given to these claims by courts in different states. Some courts distinguish general from specific damages. Others differentiate direct from indirect damages. Some contrast generally accepted conventional formulas, such as part performance plus profits, from consequential damages.

Whatever the formula selcted by the courts, the law must recognize the difference between normal, routine damages from unusual or unanticipated damages. If the damages are normal, routine damages, the claimant can plead them generally and proof of them is not subject to an elevated level of proof. If the damages claimed are special, indirect, or consequential, the law will establish hurdles that must be surmounted before there can be recovery. Consequential damages are more difficult to establish than direct losses.

One reason for carefully scrutinizing claims for consequential damages is the difficult issue of whether something that did not happen would have happened. Would the contractor have won the contracts it claims it lost because of the owner's breach? In such claims the law

[109]*Zurn Eng'rs v. State of California ex. rel. Dept. of Water Resources,* 69 Cal.App.3d 798, 138 Cal.Rptr. 478, cert. denied, 434 U.S. 985 (1977).

[110]Section 17.05 treats how notices are to be prepared, to whom they must be given, and when notice requirements are waived.

[111]See Section 26.09A.
[112]See Section 27.03D.
[113]See Sections 6.08 and 22.02L.
[114]See Section 27.02F.

must try to decide what *would* have happened. These are difficult issues to resolve.

In addition, the law seeks rules that enable contracting parties to know the extent of their exposure. They must know this if they are deciding whether to enter into the contract. Normal, general, routine damages can be expected and have to be taken into account when a party decides to make the contract and agree to its terms. Special, unusual, unforeseeable, consequential damages do not go into the planning calculus.

For these reasons, the legal doctrines relevant to consequential damages are foreseeability, discussed in Section 6.06C, and certainty, discussed in Sections 6.06B and 27.04.

B. Owner Claims

Generally, owners' consequential damage claims are for lost profits caused by delay in completing the project.[115] Traditionally, recovery of lost profits was denied if the business in which the owner intended to engage was a new one, an inflexible application of the certainty rule.[116] However, the modern tendency has been to treat this issue as any other issue and regard it as a question of fact that should be resolved based on relevant evidence.[117]

C. Contractor Claims

Generally, contractor claims are based on the loss of other business opportunities or business goodwill in general.[118] Often claims for lost profits are made because the owner's conduct has reduced or impaired the contractor's bonding capacity. Most cases have denied these claims as being too

speculative.[119] Some claims have been allowed in cases where the contractor could clearly establish a history of making consistent profits.[120] The contractor must also establish that these damages were reasonably foreseeable by the parties at the time the contract was made.[121]

As noted, many lost profit claims by contractors involve loss or reduction of their bonding capacity. Cases from California show the difficulty of predicting outcomes in these cases.

California faced this issue in *Lewis Jorge Construction Management, Inc. v. Pomona Unified School District*.[122] The contractor claimed that the breach by the owner adversely affected its bonding capacity. It claimed that this caused it to lose profits on unidentified future projects. The California Supreme Court held that these were not general damages, that is, not routine, expected losses. They were special or consequential damages. Here the contractor did not establish the requirements for special damages as its proof was too general. Had the contractor showed it lost specific projects because of its loss of its bonding capacity, it could have recovered lost profits.

The court had its attention directed to a Montana case that had awarded damages in such a case.[123] The court was not obligated to follow a case from Montana. But it did note that that decision had not been followed outside Montana for 33 years and was distinguished even in Montana.[124] It was "one of kind" and at best a questionable precedent.

Yet three years later a California intermediate appeals court decided *BEGL Construction Co., Inc. v. Los Angeles Unified School District*.[125] The court distinguished the *Jorge* case, one basis being that the contractor in the *Jorge* case only showed evidence of unidentified projects, not specific projects lost.

---

[115]*Ambrogio v. Beaver Road Assocs.*, 267 Conn. 148, 836 A.2d 1183 (2003).

[116]*Drs. Sellke & Conlon, Ltd. v. Twin Oaks Realty, Inc.*, 143 Ill.App.3d 168, 491 N.E.2d 912 (1986); *Exton Drive-In v. Home Indem. Co.*, 436 Pa. 480, 261 A.2d 319 (1969), cert. denied, 400 U.S. 819 (1970).

[117]*Mechanical Wholesale, Inc. v. Universal-Rundle Corp.*, 432 F.2d 228 (5th Cir.1970) (recovery not automatically excluded for new businesses); *Grossman v. Sea Air Towers, Ltd.*, 513 So.2d 686 (Fla.Dist.Ct. App.1987), review denied, 520 So.2d 584 (Fla.1988) (lost rents are reasonably certain in nature).

[118]*Larry Armbruster & Sons, Inc. v. State Public School Bldg. Auth.*, 505 A.2d 395 (Pa.Commw.1986), appeal denied, 115 Pa. 636, 520 A.2d 1386 (1987) (not established).

[119]*Indiana & Michigan Elec. Co. v. Terre Haute Indus., Inc.*, 507 N.E.2d 588 (Ind.Ct.App.1987).

[120]*Tempo, Inc. v. Rapid Elec. Sales and Service, Inc.*, 132 Mich.App. 93, 347 N.W.2d 728 (1984).

[121]*Texas Power & Light Co. v. Barnhill*, 639 S.W.2d 331 (Tex.Ct.App. 1982) (owner aware of existence of collateral contracts).

[122]34 Cal. 4th 960, 102 P.3d 257, 22 Cal.Rptr.3d 340 (2004), rehearing denied, Feb. 16, 2005.

[123]*Laas v. Montana State Highway Comm'n*, 157 Mont. 121, 483 P.2d 699 (1971) (contractor had earned profit for past twenty-two years).

[124]*Zook Bros. Constr. Co. v. State*, 171 Mont. 64, 556 P.2d 911(1976).

[125]154 Cal.App.4th 970, 66 Cal.Rptr.3d 110 (2007), rehearing denied Sep. 18, 2007, rev. denied and ordered not to be published Nov. 28, 2007. The history of this case demonstrates the uncertainty of the law in California as to lost profits.

BEGL introduced expert testimony that it is well-known that a surety will not bond a contractor that is having a dispute with its prior surety. While BEGL could not show specific projects lost because of its inability to become bonded, experts testified as to BEGL's profits.

Despite the *Jorge* case a contractor that has been wrongfully terminated and loses it bonding capacity has some chance of recovering lost profits on future projects if its case is properly presented.

The two cases do show that this issue is unsettled in California and probably elsewhere.

D. Waiver of Consequential Damages: AIA Approach

The uncertainty of liability exposure caused by the specter of large and often unquantifiable consequential damages often leads to contractual provisions that seek to bar such claims. In a commercial context, in dealings between commercial people who know what they are doing, such a clause will be very likely to be upheld.[126]

Beginning in 1997 the American Institute of Architects included such waiver provisions (continued in 2007) in its principal documents for design and construction.[127] This decision was generated in part by a case in which a construction manager who received a fee of some $600,000 was held liable in an arbitration award in favor of a casino for consequential damages of $14.5 million.[128] But the AIA went farther than simply requiring waiver of such claims. It also gives *examples*:

1. damages incurred by the Owner for rental expenses, for losses of use, income, profit, financing, business and reputation, and for loss of management or employee productivity or of the services of such persons; and
2. damages incurred by the Contractor for principal office expenses including the compensation of personnel stationed there, for losses of financing, business and reputation, and for loss of profit except anticipated profit arising directly from the Work.[129]

Most important, A201-2007 excludes the much-debated home office overhead that generated the *Eichleay* formula.

During discussions with the Associated General Contractors (AGC), the latter contended that strong owners would bypass the prohibition by including harsh liquidated damages clauses. This discussion led to language in A201-1997, Paragraph 4.3.10, which, although allowing damage liquidation of "direct" damages, presumably would not allow liquidation of damages for breaches that could be classified as indirect or consequential.

This approach does not take into account the difficulties of determining what are *direct* damages that can be liquidated, the fact that it is in the area of consequential damages that liquidation is most needed, and that strong owners who want to reserve their right to recover consequential damages will simply strike Paragraph 4.3.10 (and now A201-2007, Section 15.1.6), assuming they know it is there.

SECTION 27.07 Avoidable Consequences: The Concept of Mitigation

The general dimensions of the mitigation doctrine (the claimant cannot recover damages that it could have reasonably avoided, sometimes called the "duty to mitigate damages") were discussed in Section 6.06D. Some defendants have invoked this concept in construction disputes. In *C.A. Davis, Inc. v. City of Miami*,[130] a defendant contractor sought to reduce the recovery by the owner by asserting that the owner spent more than was necessary to correct the contractor's work. The court held that the contractor could challenge the completion cost only if it could show waste, extravagance, or lack of good faith. It will be unusual for a court to conclude that expenses incurred by the claimant to correct defective work by the contractor were out of line and not recoverable. Yet one court did reduce an award because it concluded that overtime was not necessary and the claimant's refusal to allow

---

[126]For a trial court decision upholding such a clause in a commercial context, see *Department of Water and Power of L.A. v. ABB Power*, 902 F.Supp. 1178 (C.D.Cal.1995).

[127]A201-2007, § 15.1.6., A401-2007 § 15.4, and B101-2007 § 8.1.3.

[128]*Perini Corp. v. Greate Bay Hotel & Casino, Inc.*, 129 N.J. 479, 610 A.2d 364 (1992). Two years later the Supreme Court of New Jersey had second thoughts about its decision that had allowed a bit more room for judicial review of an arbitration award, even though it stated the normal rule that mistakes of law are not enough to upset the award. In *Tretina Printing, Inc. v. Fitzpatrick & Assoc. Inc.*, 135 N.J. 349, 640 A.2d 788 (1994), the court adopted a concurring opinion in the *Perini* case that limited the scope of judicial review. This case is discussed in Section 30.14.

[129]A201-2007, §§ 15.1.6.1, 15.1.6.2.

[130]400 So.2d 536 (Fla.Dist.Ct.App.1981), petition for review denied, 411 So.2d 380 (Fla.1981).

the designer–builder to provide free engineering services was not justified.[131]

This doctrine was invoked, strangely, in *S. J. Groves & Sons Co. v. Warner Co.*, by a subcontractor in default who claimed the prime contractor should have fired the subcontractor.[132] The prime contractor sought damages against the subcontractor, who had failed to supply ready-mixed concrete.

Trouble developed almost at the outset. The owner was forced repeatedly to reject the concrete supplied by the subcontractor. In addition, the subcontractor frequently failed to make deliveries in accordance with the prime contractor's instructions. The prime contractor considered using other sources but felt it had no real alternatives. Building its own plant would cost too much. The only other concrete source had not been certified to do state work, and its price was higher than that of the subcontractor. In addition, the only alternative source had limited facilities and trucks and the subcontractor continued to assure the prime contractor that things would improve.

Despite these promises, the subcontractor's performance continued to be erratic, and the public entity ordered all construction halted until the subcontractor's performance could be discussed at a conference. Again after renewed assurances that things would improve, the public entity allowed work to resume. For succeeding months, the subcontractor's performance continued to be uneven and unpredictable.

During performance, the prime contractor approached the alternate source, which by then had been certified by the state. The alternate source agreed to reduce its price to the same price as the subcontractor, but the prime contractor continued to use its subcontractor as its sole supplier.

Had the prime contractor acted reasonably in continuing to use the subcontractor despite its poor performance? The trial court concluded that the prime contractor had not, but the appellate court did not agree.

After noting that the breaching party had the burden of proving that the losses could have been avoided by reasonable effort, the court looked at the alternatives available to the prime contractor. One alternative was to simply terminate the subcontractor, an alternative the court did not find realistic. Another alternative was for the prime contractor to set up its own cement-batching plant, an alternative the court found impractical because of time and expense. Another alternative was to accept the subcontractor's assurances that performance would be satisfactory in the future, the alternative selected. Another alternative was to use the alternate supplier as a supplemental source or as a substitute.

The appellate court concluded that all the alternatives had their drawbacks. Even if the alternate supplier had been engaged as a supplemental source, there was still no guarantee that the subcontractor would perform properly. The use of two suppliers might raise other problems, and there was a question as to whether the alternate supplier would have been able to perform.

The court concluded that confronted with these choices, the prime contractor's decision to stay with the subcontractor may not have been the best choice. The test is, however, whether the course chosen was *reasonable*, not whether it was necessarily the best. The court was not willing to engage in hypercritical examination of the choice made. It concluded that staying with the subcontractor may have been not only reasonable but also the best choice under the circumstances. The court noted that the breaching subcontractor, who sought to second-guess the choice made by the prime contractor, could also have engaged a supplemental supplier and that, where each party had the equal alternative to reduce the damages, the defendant was in no position to contend that the plaintiff failed to mitigate.

Another aspect of the rule of avoidable consequences is often ignored. Can a person who takes action to avoid a foreseeable loss *before* it occurs recover the costs incurred from the person who would have been responsible? Put another way, suppose the owner performs a prophylactic replacement of questionable material to mitigate likely future damages that would be much more costly. Can it recover its costs from the manufacturer to prevent (mitigate) greater losses in the future if nothing were done?

This issue is illustrated by *Toll Bros., Inc. v. Dryvit Systems, Inc.*[133] Toll Bros., a national developer, built a housing project using synthetic stucco exterior. Afterward, the media began to publicize defects with synthetic stucco—that it permits moisture to enter the building. Although the development's houses showed no evidence

---

[131]*First Nat'l Bank of Akron v. Cann*, 503 F.Supp. 419 (N. D. Ohio 1980), aff'd, 669 F.2d 415 (6th Cir.1982).

[132]576 F.2d 524 (3d Cir.1978).

[133]432 F.3d 564 (4th Cir.2005) (Connecticut law).

of water infiltration, Toll Bros. replaced the exterior stucco and sought to recoup that expense from the manufacturer. A divided court allowed the developer to recover from the manufacturer in tort the cost of prophylactically replacing the stucco in order to avoid future water damage to its houses.

SECTION 27.08 Collateral Source Rule: Off-Setting Benefits

The collateral source rule was described generally in Sections 6.06F and 7.10B. There it was noted that a good number of states have modified the common law rule by statute. These statutes require that reimbursement for losses from third parties be taken into account in measuring damages.

The collateral source rule must be seen against the backdrop of what can be called the "single recovery" rule. A claimant should not recover more than it has lost. The claimant would be unjustly enriched if this occurs. As shall be seen in some of the cases described in this Section, a claimant who is paid by one defendant must reduce its recovery against another defendant so it is not paid twice for the same loss.

The collateral source rule is an exception to the single recovery rule. If the person or entity that has paid the claimant is considered *collateral*, such as an insurer or public entity through some form of social insurance, that payment does not reduce the claim.

A review of some cases helps illustrate applications of the collateral source rule in construction disputes. In *New Foundation Baptist Church v. Davis*[134] a church sued the contractor for defects in the sanctuary floor that caused collapse during a funeral three years after the church was completed. A church member carpenter donated his labor and repaired the damage for a total cost to the church of $3,000. Yet the jury's award of $6,500 to the church was upheld. The court noted that the church member had no wish to benefit the contractor, and it would be unfair for the contractor to receive the advantage of the church member's generosity.

One commentator states that gifts and friendly help are considered collateral and do not reduce the recovery to the claimant.[135] This may explain the result in the *New Foundation* case discussed earlier in this Section.

The issue of whether "the single recovery" rule or the "collateral source" rule applied also arose in *Huber, Hunt & Nichols, Inc. v. Moore*.[136] A contractor claimed against the owner and the architect. The contractor settled with the owner before trial. One reason given for denying recovery against the architect was that the owner had paid the contractor substantial amounts. This was a tort claim, there being no contract between contractor and architect. Although this would make the payments come from a collateral source that cannot be used to reduce the claim, the court held that the owner and the architect were sufficiently associated with each other to preclude the owner from being considered a collateral source.

RPR & Associates v. University of N.C.—Chapel Hill also involved the triangle of owner, architect, and contractor.[137] In this case, the contractor brought a legal claim based upon breach of contract against the owner and the owner's architect. It settled with the architect and pursued its claim against the owner. The owner asserted that it was entitled to offset the amount received in the settlement with the architect in the claim against it by the contractor.

The court agreed, stating that refusal to do so would overcompensate the contractor. No mention was made of the collateral source rule.

Putting together the *Huber, Hunt* and *RPR* cases, it would appear that the owner and its architect are considered one entity for determining the scope of the contractor's recovery. Any amount received from the owner (the *Huber, Hunt* case) would offset any liability of the architect; while any amount received from the architect (*RPR* case) would offset any liability against the owner. There appears to be no role for the collateral source rule in disputes between owner, architect, and contractor, whether the claim is brought in tort or contract.

On rare occasion, the breach may *benefit* the non-breaching party. If so, a benefit that is direct and that clearly would not have occurred were it not for the breach, can be offset against any losses.

Benefit conferred by the breaching party can arise in a number of different contexts. Professional liability claims against a design professional based on negligent cost estimates are treated in Section 12.03F. Items that should have been included in the design in a professional

---

[134]257 S.C. 443, 186 S.E.2d 247 (1972).
[135]D. DOBBS REMEDIES, § 3.8(1), p. 267 (2d ed. 1993).

[136]Supra note 102.
[137]153 N.C.App. 342, 570 S.E.2d 510 (2002), cert. dismissed and review denied, 357 N.C. 166, 579 S.E.2d 882 (2003).

liability claim are discussed in Section 12.14B. Benefits to an awarding authority when the sucessful bidder refuses to enter into the contract are discussed in Section 18.04E. Cases for a reduction of damages because cost of correction extended the life of the original item are discussed in Section 27.03D. Correction of the defective work using a more expensive item is also covered in Section 27.03D.

SECTION 27.09 Noneconomic Losses: *Erlich v. Menezes*

The recovery of noneconomic losses for breach of contract was discussed in general in Section 6.06H. There it was noted that the general rule was that such losses could not be recovered for a breach of contract. But there are exceptions. One is a contract breach that can be classified as an independent tort. (*Erlich v. Menezes*,[138] a case that is reproduced in part in this section, dealt with this exception.) A second is a personal contract or one made to provide mental solicitude. It can be difficult to determine which contracts fall into the category of those that lie outside of the basic rule that such damages cannot be recovered for breach of contract.

In addition, presenting a frightening residential construction nightmare, a case is reproduced in part at this point that sought to canvas the many cases that have dealt with this issue. It also stated why the law should venture very cautiously, if at all, into granting recovery for emotional distress in the context of a residential construction contract.

ERLICH v. MENEZES

Supreme Court of California, 1999. 21 Cal. 4th 543, 981 P.2d 978, 87 Cal.Rptr.2d 886.

BROWN, J.

We granted review in this case to determine whether emotional distress damages are recoverable for the negligent breach of a contract to construct a house. A jury awarded the homeowners the full cost necessary to repair their home as well as damages for emotional distress caused by the contractor's negligent performance. Since the contractor's negligence directly caused only economic injury and property damage, and breached no duty independent of the contract, we conclude the homeowners may not recover damages for emotional distress based upon breach of a contract to build a house.

I. FACTUAL AND PROCEDURAL BACKGROUND

Both parties agree with the facts as ascertained by the Court of Appeal. Barry and Sandra Erlich contracted with John Menezes, a licensed general contractor, to build a "dreamhouse" on their ocean-view lot. The Erlichs moved into their house in December 1990. In February 1991, the rains came. "[T]he house leaked from every conceivable location. Walls were saturated in [an upstairs bedroom], two bedrooms downstairs, and the pool room. Nearly every window in the house leaked. The living room filled with three inches of standing water. In several locations water 'poured in [] streams' from the ceilings and walls. The ceiling in the garage became so saturated . . . the plaster liquefied and fell in chunks to the floor."

Menezes attempts to stop the leaks proved ineffectual. Caulking placed around the windows melted, " 'ran down [the] windows and stained them and ran across the driveway and ran down the house [until it] . . . looked like someone threw balloons with paint in them at the house.'" Despite several repair efforts, which included using sledgehammers and jackhammers to cut holes in the exterior walls and ceilings, application of new waterproofing materials on portions of the roof and exterior walls, and more caulk, the house continued to leak—from the windows, from the roofs, and water seeped between the floors. Fluorescent light fixtures in the garage filled with water and had to be removed.

"The Erlichs eventually had their home inspected by another general contractor and a structural engineer. In addition to confirming defects in the roof, exterior stucco, windows and waterproofing, the inspection revealed serious errors in the

---

[138]21 Cal. 4th 543, 981 P.2d 978, 87 Cal.Rptr.2d 886 (1999). This case has generated controversy, particularly in California. Nine California cases refused to extend it in unpublished opinions. Five California cases distinguished it, four in unpublished opinions and one in a published opinion. *Gu v. BMW of North Amercia, LLC*, 132 Cal.App.4th 195, 33 Cal.Rptr.3d 617 (2005). Other cases show that claims for emotional distress for what is essentially breach of contract may, in extreme cases, have a chance of succeeding. For example, Colorado refused to follow the *Erlich* case. *Giampapa v. American Family Mut. Ins. Co.*, 64 P.3d 230 (Colo.2003). See also *Kishmarton v. William Bailey Constr. Co.*, infra note 139, which allowed recovery for emotional distress.

construction of the home's structural components. None of the 20 shear, or load-bearing walls specified in the plans were properly installed. The three turrets on the roof were inadequately connected to the roof beams and, as a result, had begun to collapse. Other connections in the roof framing were also improperly constructed. Three decks were in danger of 'catastrophic collapse' because they had been finished with mortar and ceramic tile, rather than with the light-weight roofing material originally specified. Finally, the foundation of the main beam for the two-story living room was poured by digging a shallow hole, dumping in 'two sacks of dry concrete mix, putting some water in the hole and mixing it up with a shovel.'" This foundation, required to carry a load of 12,000 pounds, could only support about 2,000. The beam is settling and the surrounding concrete is cracking.

According to the Erlichs' expert, problems were major and pervasive, concerning everything "related to a window or water-proofing, everywhere that there was something related to framing," stucco, or the walking deck.

Both of the Erlichs testified that they suffered emotional distress as a result of the defective condition of the house and Menezes invasive and unsuccessful repair attempts. Barry Erlich testified he felt "absolutely sick" and had to be "carted away in an ambulance" when he learned the full extent of the structural problems. He has a permanent heart condition, known as super-ventricular tachyarrhythmia, attributable, in part, to excessive stress. Although the condition can be controlled with medication, it has forced him to resign his positions as athletic director, department head and track coach.

Sandra Erlich feared the house would collapse in an earth-quake and feared for her daughter's safety. Stickers were placed on her bedroom windows, and alarms and emergency lights installed so rescue crews would find her room first in an emergency.

Plaintiffs sought recovery on several theories, including breach of contract, fraud, negligent misrepresentation, and negligent construction. Both the breach of contract claim and the negligence claim alleged numerous construction defects.

Menezes prevailed on the fraud and negligent misrepresentation claims. The jury found he breached his contract with the Erlichs by negligently constructing their home and awarded $406,700 as the cost of repairs. Each spouse was awarded $50,000 for emotional distress, and Barry Erlich received an additional $50,000 for physical pain and suffering and $15,000 for lost earnings.

By a two-to-one majority, the Court of Appeal affirmed the judgment, including the emotional distress award. The majority noted the breach of a contractual duty may support an action in tort. The jury found Menezes was negligent. Since his negligence exposed the Erlichs to "intolerable living conditions and a constant, justifiable fear about the safety of their home," the majority decided the Erlichs were properly compensated for their emotional distress.

The dissent pointed out that no reported California case has upheld an award of emotional distress damages based upon simple breach of a contract to build a house. Since Menezes negligence directly caused only economic injury and property damage, the Erlichs were not entitled to recover damages for their emotional distress.

We granted review to resolve the question.

[Ed. note: The court noted that damages for breach of contract are limited to those losses within the contemplation of the parties when the contract was made. Then it discussed the difference between contract (enforce the intention of the parties and encourage contractual relations by enabling parties to plan the "financial risks" of their enterprise) and tort (vindicate social policy). The court pointed to the intermediate appellate court's conclusion that the same wrongful act can be both a breach of contract and a tort. But the court stated that breach of contract can be a tort only if it violates "a duty independent of the contract arising from principles of tort law." Cases where tort damages have been permitted in contract cases require an independent tort or conduct that is intended to harm. That mental distress is foreseeable does not alone create an independent duty.

The court held that "mere negligent breach of a contract is not a tort." While California has held that a breach of contract can be a tort in the context of an insurance policy, this is based on the special relationship between insured and insurer. Such a special relationship does not apply to single-family construction contracts. The law, as a rule, keeps tort and contract apart. Tort, with its expansive remedies, could adversely affect predictability in commercial relationships.

Even if the contractor's negligence were linked to a sufficient independent duty to the plaintiffs, these facts would not allow recovery for emotional distress. Property damage and economic injury cannot be the basis for damages for negligent infliction of emotional distress. This also holds true for construction of a home. There was no physical injury here.]

The Erlichs may have hoped to build their dream home and live happily ever after, but there is a reason that tag line belongs only in fairy tales. Building a house may turn out to be a stress-free project; it is much more likely to be the stuff of urban legends—the cause of bankruptcy, marital dissolution, hypertension and fleeting fantasies ranging from homicide to suicide. As Justice Yegan noted below, "No reasonable homeowner can embark on a building project with certainty that the project will be completed to perfection. Indeed, errors are so likely to occur that few if any homeowners would be justified in resting their peace of mind on [its] timely or correct completion. . . ." The connection between the service sought and the aggravation and distress resulting from incompetence may be somewhat less tenuous than in a malpractice case, but the emotional suffering still derives from an inherently economic concern.

* * *

. . . . [D]amages for mental suffering and emotional distress are generally not recoverable in an action for breach of an ordinary commercial contract in California. (*Kwan v. Mercedes-Benz of North America, Inc.*, 23 Cal.App.4th 174, 188, 28 Cal.Rptr.2d 371 (1994) (*Kwan*); *Sawyer v. Bank of America* (1978) 83 Cal. App.3d 135, 139, 145 Cal.Rptr. 623). "Recovery for emotional disturbance will be excluded unless the breach also caused bodily harm or the contract or the breach is of such a kind that serious emotional disturbance was a particularly likely result." (Rest.2d Contracts, § 353.) The Restatement specifically notes the breach of a contract to build a home is not "particularly likely" to result in "serious emotional disturbance." (*Ibid.*)

Cases permitting recovery for emotional distress typically involve mental anguish stemming from more personal undertakings the traumatic results of which were unavoidable. (See, e.g., *Burgess v. Superior Court*, *supra*, 2 Cal.4th 1064, 9 Cal.Rptr.2d 615, 831 P.2d 1197 [infant injured during childbirth]; *Molien v. Kaiser Foundation Hospitals* (1980) 27 Cal.3d 916, 167 Cal.Rptr. 831, 616 P.2d 813 [misdiagnosed venereal disease and subsequent failure of marriage]; *Kately v. Wilkinson* (1983) 148 Cal.App.3d 576, 195 Cal.Rptr. 902 [fatal waterskiing accident]; *Chelini v. Nieri* (1948) 32 Cal.2d 480, 196 P.2d 915 [failure to adequately preserve a corpse].) Thus, when the express object of the contract is the mental and emotional well-being of one of the contracting parties, the breach of the contract may give rise to damages for mental suffering or emotional distress. (See *Wynn v. Monterey Club* (1980) 111 Cal.App.3d 789, 799-801, 168 Cal.Rptr. 878 [agreement of two gambling clubs to exclude husband's gambling-addicted wife from clubs and not to cash her checks]; *Ross v. Forest Lawn Memorial Park* (1984) 153 Cal.App.3d 988, 992-996, 203 Cal.Rptr. 468 [cemetery's agreement to keep burial service private and to protect grave from vandalism]; *Windeler v. Scheers Jewelers* (1970) 8 Cal.App.3d 844, 851-852, 88 Cal.Rptr. 39 [bailment for heirloom jewelry where jewelry's great sentimental value was made known to bailee].)

Cases from other jurisdictions have formulated a similar rule, barring recovery of emotional distress damages for breach of contract except in cases involving contracts in which emotional concerns are the essence of the contract. (See, e.g., *Hancock v. Northcutt* (Alaska 1991) 808 P.2d 251, 258 ["contracts pertaining to one's dwelling are not among those contracts which, if breached, are particularly likely to result in serious emotional disturbance"; typical damages for breach of house construction contracts can appropriately be calculated in terms of monetary loss]; *McMeakin v. Roofing & Sheet Metal Supply* (Okla.Ct. App.1990) 807 P.2d 288 [affirming order granting summary judgment in favor of defendant roofing company after it negligently stacked too many brick tiles on roof, causing roof to collapse and completely destroy home, leading to plaintiff's heart attack one month later]; *Day v. Montana Power Company* (1990) 242 Mont. 195, 789 P.2d 1224 [owner of restaurant that was destroyed in gas explosion allegedly caused by negligence of utility company employee not entitled to recover damages for emotional distress]; *Creger v. Robertson* (La.Ct.App.1989) 542 So.2d 1090 [reversing award for emotional distress damages caused by foul odor emanating from a faulty foundation, preventing plaintiff from entertaining guests in her residence]; *Groh v. Broadland Builders, Inc.* (Mich. Ct.App.1982) 120 Mich. App. 214, 327 N.W.2d 443 [reversing order denying motion to strike allegations of mental anguish in case involving malfunctioning septic tank system, and noting adequacy of monetary damages to compensate for pecuniary loss of "having to do the job over," as distinguished from cases allowing recovery because situation could never be adequately correctd].)

Plaintiffs argue strenuously that a broader notion of damages is appropriate when the contract is for the construction of a home. *Amicus curiae* [Ed. note: friends of the court] urge us to permit emotional distress damages in cases of negligent construction of a personal residence when the negligent construction causes gross interference with the normal use and habitability of the residence.

Such a rule would make the financial risks of construction agreements difficult to predict. Contract damages must be clearly ascertainable in both nature and origin. (Civ.Code, § 3301.) A contracting party cannot be required to assume limitless responsibility for all consequences of a breach and must be advised of any special harm that might result in order to determine whether or not to accept the risk of contracting. (1 Witkin, Summary of Cal. Law, *supra*, Contracts, § 815, p. 733.)

Moreover, adding an emotional distress component to recovery for construction defects could increase the already prohibitively high cost of housing in California, affect the availability of insurance for builders, and greatly diminish the supply of affordable housing. The potential for such broad-ranging economic consequences—costs likely to be paid by the public generally—means the task of fashioning appropriate limits on the availability of emotional distress claims should be left to the Legislature. (See Tex.Prop.Code Ann., § 27.001 et seq. (1999); Hawaii Rev. Stat., § 663-8.9 (1998).)

Permitting damages for emotional distress on the theory that certain contracts carry a lot of emotional freight provides no useful guidance. Courts have carved out a narrow range of exceptions to the general rule of exclusion where emotional tranquility is the contract's essence. Refusal to broaden the bases for recovery reflects a fundamental policy choice. A rule which focuses not on the risks contracting parties voluntarily assume but on one party's reaction to inadequate performance, cannot provide any principled limit on liability.

The discussion in *Kwan*, a case dealing with the breach of a sales contract for the purchase of a car, is instructive. "[A] contract for [the] sale of an automobile is not essentially tied to the buyer's mental or emotional well-being. Personal as the choice of a car may be, the central reason for buying one is usually transportation. . . . [¶] In spite of America's much-discussed 'love affair with the automobile,' disruption of an owner's relationship with his or her car is not, in the normal case, comparable to the loss or mistreatment of a family member's remains [citation], an invasion of one's privacy [citation], or the loss of one's spouse to a gambling addiction [citation]. In the latter situations, the contract exists primarily to further or protect emotional interests; the direct and foreseeable injuries resulting from a breach are also primarily emotional. In contrast, the undeniable aggravation, irritation and anxiety that may result from [the] breach of an automobile warranty are secondary effects deriving from the decreased usefulness of the car and the frequently frustrating process of having an automobile repaired. While [the] purchase of an automobile may sometimes lead to severe emotional distress, such a result is not ordinarily foreseeable from the nature of the contract." (*Kwan*, supra, 23 Cal.App.4th at p. 190, 28 Cal. Rptr.2d 371.)

Most other jurisdictions have reached the same conclusion. (See *Sanders v. Zeagler* (La.1997) 686 So.2d 819, 822–823 [principal object of a contract for the construction of a house was to obtain a place to live and emotional distress damages were not recoverable]; *Hancock v. Northcutt, supra*, 808 P.2d at pp. 258–259 [no recovery for emotional distress as a result of defective construction; typical damages for breach of house construction contracts can appropriately be calculated in terms of monetary loss]; *City of Tyler v. Likes* (Tex.1997) 962 S.W.2d 489, 497 [mental anguish based solely on property damage is not compensable as a matter of law].)

We agree. The available damages for defective construction are limited to the cost of repairing the home, including lost use or relocation expenses, or the diminution in value. (*Orndorff v. Christiana Community Builders* (1990) 217 Cal.App.3d 683, 266 Cal.Rptr. 193.) The Erlichs received more than $400,000 in traditional contract damages to correct the defects in their home. While their distress was undoubtedly real and serious, we conclude the balance of policy considerations—the potential for significant increases in liability in amounts disproportionate to culpability, the court's inability to formulate appropriate limits on the availability of claims, and the magnitude of the impact on stability and predictability in commercial affairs—counsel against expanding contract damages to include mental claims in negligent construction cases.

DISPOSITION

The judgement of the Court of appeal is reversed and the matter is remanded for further proceedings consistent with this opinion.

GEORGE, C.J., KENNARD, J., BAXTER, J., and CHIN, J., concur.

Concurring and Dissenting Opinion by WERDEGAR, J. [omitted].

While *Erlich v. Menezes* represents the majority rule (that emotional distress cannot be recovered for negligent breach of a construction contract), a few cases have allowed such claims to succeed.[139] One was a tort claim based upon intentional infliction of emotional distress.[140] This would have been allowed under the decision in *Erlich v. Menezes*.

---

[139]A recent case came to the opposite conclusion in a claim by a homebuyer against a builder–vendor. It would allow recovery for emotional distress if there were physical harm or if serious emotional distress is likely to result from contract breach. *Kishmarton v. William Bailey Constr., Inc.* 93 Ohio St.3d 226, 754 N.E.2d 785 (2001).

[140]*Randa v. United States Homes, Inc.*, 325 N.W.2d 905 (Iowa App.1982) (wife of buyer spent time in hospital for nervous breakdown, petrified when she was told she couldn't put mirrors in bedroom, got sick, and started to cry when liens filed).

Another was *B & M Homes, Inc. v. Hogan*[141] involving a negligence claim by the Hogans against Morrow (the owner of B & M Homes) based on the contract under which the Hogans agreed to buy a lot and a house to be built by Morrow. During construction, Mrs. Hogan discovered a hairline crack in the concrete slab. Morrow told her such cracks were common and not to worry. After the Hogans moved in, the crack widened and caused extended damage. Morrow was notified and dealt with the damage but did not attempt to repair the slab himself. The claim was based in part on mental anguish that the Hogans suffered. The Hogans introduced evidence that they were concerned over their safety because they believed the house to be structurally defective, that the condition of the house might cause gas and water lines to burst, and that they were forced to live in a defective house because they could not afford to move.

The court noted that as a general rule, damages for mental anguish are not recoverable for breach of contract. However, an exception is made for contracts that involve "mental concern or solicitude." The court held that this contract fell into that category and that it was reasonably foreseeable that faulty construction of a house would cause the homeowners to suffer severe mental anguish. The court noted that a home is the largest single investment for most families and one that places the family in debt for many years. The court pointed to an earlier decision that had allowed recovery for mental anguish when a builder performed improperly under a contract to build a home, emphasizing the homeowner's view of her home as her castle and a place to protect her against the elements and to shelter her belongings. One commentator described *B & M Homes v. Hogan* as "an unusually liberal decision."[142]

SECTION 27.10 Punitive Damages

Punitive damages were discussed in the contract context in Section 6.04, in the tort context in Section 7.10C, and will be discussed in the arbitration context in Section 30.12.

---

[141]376 So.2d 667 (Ala.1979). See Annot., 7 A.L.R.4th 1178 (1981). That this was a controversial holding is shown by one court refusing to follow it, *Hancock v. Northcutt*, 808 P.2d 251 (Alaska 1991), rehearing denied April 8, 1991, and two cases distinguishing it. See also *Orto v. Jackson*, 413 N.E.2d 273, 278 (Ind.Ct.App.1980) (damages for aggravation and inconvenience).

[142]E. A. FARNSWORTH, CONTRACTS, § 12.17 at note 18 (1982). This language was deleted in subsequent editions.

Punitive damages, almost unheard of in construction contract disputes, cannot be ignored today. To justify punitive damages the conduct must go beyond a simple or even negligent breach of contract. It must be at least grossly negligent, outrageous, or malicious.

Despite the windfall aspects of punitive damages where the plaintiff recovers an amount in excess of its losses, both contract law and tort law have begun to look at punitive damages as a way of deterring outrageous conduct, particularly in that part of the construction world that involves ordinary consumers rather than knowledgeable businesspeople.

Again, particular cases are instructive. In *F. D. Borkholder Co. v. Sandock*,[143] the contractor deviated from the plans. The court justified the punitive damages award by concluding that the defendant was guilty of intentional and wrongful acts that constituted fraud, misrepresentation, deceit, and gross negligence in its dealings. The court stated,

> The Court of Appeals cited our decision in *Hibschman Pontiac, Inc. v. Batchelor*, 266 Ind. 310, 362 N.E.2d 845 (1977) for the proposition that punitive damages are recoverable in breach of contract actions only when a separate tort accompanies the breach or tort-like conduct mingles in the breach. Here, prior to the execution of the contract, Sandock representatives expressed their concern about moisture on the walls. Under the terms of the contract, they were to pay $200 for plans to be drawn up by Borkholder's architect. The contract provided that all labor and material would be furnished in accordance with specifications. Sandock was given a copy of the plans. However, contrary to these plans, the top and bottom courses of block forming the one wall were not filled with concrete, thus constituting latent variances. Furthermore, the roofline was shortened which represented an additional deviation from the plans.
>
> There was testimony that the cut-off roofline enabled water to leak down into the top of the block wall. Other evidence indicated that the wetness problem resulted from this water percolating down through the inside of the wall, collecting at the bottom, and then rising again by capillary action. Sandock made numerous complaints but was constantly reassured by several Borkholder representatives that the problem was caused by simple condensation, a

---

[143]274 Ind. 612, 413 N.E.2d 567, 570–71 (1980). Similarly, see *Jeffers v. Nysse*, 98 Wis.2d 543, 297 N.W.2d 495 (1980) (misrepresentation of insulation and heating costs).

theory ultimately disproved by an on-site test conducted by the Borkholder firm. Sam Sandock testified that Freeman Borkholder, president of the company, promised that the situation would be remedied whereupon Sandock tendered all but $1,000 of the contract price. The problem was never corrected. The Borkholder people knew, of course, that the blocks in the wall were not filled with concrete. Also, Borkholder himself conceded that the roofline adjustment increased the likelihood of water running down into the core of the wall.

We believe that there is cogent and convincing proof that the Borkholder firm engaged in intentional wrongful acts constituting fraud, misrepresentation, deceit, and gross negligence in its dealings with Sandock. *Hibschman Pontiac, Inc.,* supra. Accordingly, we agree with the Court of Appeals that the trial court could have concluded that separate torts accompanied the breach. Next, relying on *Hibschman,* the Court of Appeals attempted to identify the public interest to be served by imposing punitive damages. However, the majority could not perceive any such interest and refused to let the award stand. We disagree. As Judge Garrard stated in his dissent:

"I have no problem identifying the public interest to be served in requiring that the builders of public buildings be deterred from fraudulently disregarding building code requirements or those contained in the plans and specifications they have agreed to comply with." *Sandock v. Borkholder,* supra, at 959.

The purpose of punitive damages generally is to punish the wrongdoer and to deter him and others from engaging in similar conduct in the future. . . . An award of such damages is particularly appropriate in proper cases involving consumer fraud. . . .

The building contractor occupies a position of trust with members of the public for whom he agrees to do the desired construction. Few people are knowledgeable about this industry, and most are not aware of the techniques that must be employed to produce a sound structure. Necessarily, they rely on the expertise of the builder. Here, the builder has been found to have engaged in fraudulent or deceptive practices by constructing a building with latent deviations from the plans which resulted in damage to the owner. Further, the builder has attempted to disclaim responsibility for such damage when it may be inferred that it knew or should have known that its work was the cause. Under these circumstances, certainly the imposition of punitive damages furthers the public interest.

The cases described have involved performance problems. However, *Brant Construction Co., Inc. v. Lumen Construction, Inc.,*[144] involved a smoke screen to hide evasion of the Minority Business Enterprise (MBE) requirements of a public contract. The prime contractor deceived the subcontractor and used it as a "front" to obtain award of a federally funded contract that required good-faith effort toward employing a designated percentage of minority contractors.

The MBE program is designed to give experience to minority contractors. The prime contractor entered into a contract with a minority subcontractor but for much of the contract period used another subcontractor. When it finally gave some work to the designated subcontractor, the prime did not offer the required assistance and prevented it from performing properly. This attempted circumvention of the MBE program justified awarding punitive damages to the subcontractor to deter other primes from similar conduct by making an example of the prime.

But noting cases in which punitive damages were awarded should not be taken to mean that they are common in construction contract disputes. For example, one case refused to award punitive damages when there had been a wrongful termination, because there had been no showing of malicious or oppressive conduct. The court did not want to open the claims floodgates in the wake of bitterly disputed construction contracts.[145]

Finally, punitive damages were denied in a breach of contract case involving only delay or nonperformance.[146] In that case, the intentional breach of an underpriced contract was held not to subject the breaching party to punitive damages. The court was unwilling to force a contractor faced with a substantial financial loss to perform when it made a sound business judgment not to perform, knowing it would be subject only to compensatory damages.

Although punitive damages will still be rare in the ordinary construction contract dispute, the increased willingness of courts to award such open-ended damages determined by a jury means that punitive damages will play an increasingly important role in construction dis-

---

[144]515 N.E.2d 868 (Ind.Ct.App.1987).

[145]*Indiana & Michigan Elec. Co. v. Terre Haute Indus.,* supra note 119.

[146]*Construction Contracting & Mgmt, Inc. v. McConnell,* 112 N.M. 371, 815 P.2d 1161 (1991).

putes, particularly where contractors prey on unsuspecting owners and where ordinary compensatory damages will not be sufficient to deter such conduct.[147]

SECTION 27.11 Cost of Dispute Resolution: Attorneys' Fees

See Section 6.07.

SECTION 27.12 Interest

See Section 6.08.

SECTION 27.13 Disputes and Settlements: *Rich & Whillock v. Ashton Development, Inc.*

When important benchmarks in a transaction are reached, the law must deal with the effect of those benchmarks on claims. Illustration of such benchmarks in the construction process are the issuance of progress payment certificates, the issuance of certificates of substantial and final

completion, and the acceptance of the project by the owner.

At these important benchmarks, either party, the owner for claims for defect or delay and the contractor for delay and disruption or inefficiency claims, may lose its claim unless it has been reserved.[148] It is also at these crucial points that the parties will seek to settle their claims.

Settlement negotiations inevitably involve pressure, mostly on the contractor (or subcontractors). They often face excruciatingly difficult cash flow problems. Creditors demand payment or refuse to extend loans. Government officials insist on being paid back taxes. Workers expect and demand wages.

Under these conditions, some owners will seek to take advantage of the financial stress on contractors by offering amounts that contractors desperately need but are far below what is justified. Under this pressure, the contractor may accept payment, usually by check, with endorsements stating that cashing the check extinguishes the claim.[149] Extinguishing the claim was accomplished through the doctrine of accord and satisfaction.

To deal with the effect of payment on its claim, the following case employed economic duress to set aside a settlement that had been made between a prime and a subcontractor.

RICH & WHILLOCK, INC. v. ASHTON DEVELOPMENT, INC.

California Court of Appeal, 1984. 157 Cal.App.3d 1154, 204 Cal.Rptr. 86.
[Ed. note: Footnotes omitted.]

WIENER, Associate Justice.

Ashton Development, Inc. and Bob Britton, Inc. appeal from the judgment awarding Rich & Whillock, Inc. $22,286.45 for the balance due under a grading and excavating contract. Following a non-jury trial the court entered judgment after it found a settlement agreement and release signed by Rich & Whillock, Inc. were the products of economic duress and thus provided no defense to its contract claim. We conclude substantial evidence supports the court's finding and affirm the judgment.

FACTUAL AND PROCEDURAL BACKGROUND

On February 17, 1981, Bob Britton, president of Bob Britton, Inc., signed a contract for grading and excavating services to be provided by Rich & Whillock, Inc. at a price of $112,990. The work was to be done on a project by Ashton Development, Inc. Bob Britton, Inc. was general contractor on the project and the agent for Ashton Development, Inc. in all dealings with Rich & Whillock, Inc. Work began the day the contract was signed.

In late March 1981 Rich & Whillock, Inc. encountered rock on the project site. A meeting was held at the site to discuss the problem. In attendance were Greg Whillock and Jim Rich,

---

[147]See *Gennari v. Weichert Co. Realtors*, 148 N.J. 582, 691 A.2d 350 (1997) (punitive damages awarded against real estate agent in favor of purchasers based on fraudulent representation as to builder's experience and quality).

[148]See Sections 22.02M (payment), 24.05 (acceptance of project), and 22.05 (completion and final payment).

president and vice-president of Rich & Whillock, Inc., Bob Britton, Berj Aghadjian, president of Ashton Development, Inc., and a man from a blasting company. Everyone agreed the rock would have to be blasted. The $112,990 contract price expressly excluded blasting. The contract also stated "[a]ny rock encountered will be considered an extra at current rental rates." In response to Britton's inquiry, Whillock and Rich estimated the extra cost to remove the rock would be about $60,000, for a total contract price of approximately $172,000. They also emphasized, however, the estimate was not firm and the actual cost could go much higher due to the unpredictable nature of rock work.

Britton directed Whillock and Rich to go ahead with the rock work and bill him for the extra costs and said they would be paid. Rich & Whillock, Inc. proceeded accordingly, submitting invoices and receiving payments every other week. The invoices separately stated the charges for the regular contract work and the extra rock work and were supported by attached employee time sheets. Toward the end of April Whillock asked Britton if he had any questions and [he] told Whillock to continue with the rock work because it had to be done.

By June 17, 1981, after receiving payments totaling $190,363.50, Rich & Whillock, Inc. submitted a final billing for an additional $72,286.45. After consulting with Aghadjian, Britton refused to pay. When Whillock asked why, Britton explained he and Aghadjian were short on funds for the project and had no money left to pay the final billing. Up until he received that billing, Britton had no complaints about the work done or the invoices submitted by Rich & Whillock, Inc. and had never asked for any accounting of charges in addition to that already provided. Whillock told Britton he and Rich would "go broke" if not paid because they were a new company, the project was a big job for them, they had rented most of their equipment and they had numerous subcontractors waiting to be paid. Britton replied he and Aghadjian would pay them $50,000 or nothing, and they could sue for the full amount if unsatisfied with the compromise.

On July 10, 1981, Britton presented Rich with an agreement for a final compromise payment of $50,000. The agreement provided $25,000 would be paid "upon receipt of this signed agreement," to be followed by a second $25,000 payment on August 10, 1981 "upon receipt of full and unconditional releases for all labor, material, equipment, supplies, etc., purchased, acquired or furnished for this contract up to and including August 10, 1981." Rich repeated Whillock's earlier statements about the probable effects of nonpayment on their business. Britton replied: "I have a check for you, and just take it or leave it, this is all you get. If you don't want this, you have got to sue me." Rich then signed the agreement and received a $25,000 check after telling Britton the agreement was "blackmail" and he was signing it only because he had to in order to survive. Rich & Whillock, Inc. received the second $25,000 payment on August 20, 1981, at which time Whillock signed a standard release form.

In December 1981 Rich & Whillock, Inc. filed this action for damages for breach of contract. The court found Ashton Development, Inc. and Bob Britton, Inc. were liable for the $22,286.45 balance due under the contract, and that the July 10 agreement and August 20 release were unenforceable due to economic duress. On the latter point the court found Britton and Aghadjian "never really disputed the amount of plaintiff's charge in that they never asked for an accounting nor documentation concerning the extra work." The court also stated it disbelieved Britton when he testified Rich & Whillock, Inc. had agreed to do the extra work for a sum not to exceed $90,000. By disbelieving Britton and finding no dispute about the actual amount owed, the court impliedly found Britton and Aghadjian acted in bad faith when they refused to pay Rich & Whillock, Inc.'s final billing and offered instead to pay a compromise amount of $50,000. Based on its finding of bad faith, the court concluded the July 10 agreement and August 20 release were signed "under duress in that plaintiff felt they would face financial ruin if they did not accept the lesser sum and that defendants, knowing this, threatened no further payment unless plaintiff accepted the lesser sum."

DISCUSSION

"At the outset it is helpful to acknowledge the various policy considerations which are involved in cases involving economic duress. Typically, those claiming such coercion are attempting to avoid the consequences of a modification of an original contract or of a settlement and release agreement. On the one hand, courts are reluctant to set aside agreements because of the notion of freedom of contract and because of the desirability of having private dispute resolutions be final. On the other hand, there is an increasing recognition of the law's role in correcting inequitable or unequal exchanges between parties of disproportionate bargaining power and a greater willingness to not enforce agreements which were entered into under coercive circumstances." (*Totem Marine T. & B. v. Alyeska Pipeline, Etc.* (Alaska 1978) 584 P.2d 15, 21, fn. omitted.)

---

[149]Usually, it is the owner who seeks to use the check endorsement as a means of extinguishing the contractor's claim by accord and satisfaction. But in an Alabama case, a customer of a termite inspection company sent in its renewal check with a statement that it no longer agreed to arbitrate disputes as provided for in the contract. The customer succeeded. *Cook's Pest Control v. Rebar*, 852 So.2d 730 (Ala.2002).

California courts have recognized the economic duress doctrine in private sector cases for at least 50 years. (*Young v. Hoagland* (1931) 212 Cal. 426, 430–432, 298 P. 996.) The doctrine is equitably based (*Burke v. Gould*, supra, 105 Cal. at p. 281, 38 P. 733) and represents "but an expansion by courts of equity of the old common law doctrine of duress." (*Sistrom v. Anderson* (1942) 51 Cal.App.2d 213, 220, 124 P.2d 372.) As it has evolved to the present day, the economic duress doctrine is not limited by early statutory and judicial expressions requiring an unlawful act in the nature of a tort or a crime. (Civ.Code, § 1569, subd. 2; . . . .)

Instead, the doctrine now may come into play on the doing of a wrongful act which is sufficiently coercive to cause a reasonably prudent person faced with no reasonable alternative to succumb to the perpetrator's pressure. . . . The assertion of a claim known to be false or a bad faith threat to breach a contract or to withhold a payment may constitute a wrongful act for purposes of the economic duress doctrine. . . .

Further, a reasonably prudent person subject to such an act may have no reasonable alternative but to succumb when the only other alternative is bankruptcy or financial ruin. . . .

The underlying concern of the economic duress doctrine is the enforcement in the marketplace of certain minimal standards of business ethics. Hard bargaining, "efficient" breaches and reasonable settlements of good faith disputes are all acceptable, even desirable, in our economic system. That system can be viewed as a game in which everybody wins, to one degree or another, so long as everyone plays by the common rules. Those rules are not limited to precepts or rationality and self-interest. They include equitable notions of fairness and propriety which preclude the wrongful exploitation of business exigencies to obtain disproportionate exchanges of value. Such exchanges make a mockery of freedom of contract and undermine the proper functioning of our economic system. The economic duress doctrine serves as a last resort to correct these aberrations when conventional alternatives and remedies are unavailing. The necessity for the doctrine in cases such as this has been graphically described:

> Nowadays, a wait of even a few weeks in collecting on a contract claim is sometimes serious or fatal for an enterprise at a crisis in its history. The business of a creditor in financial straits is at the mercy of an unscrupulous debtor, who need only suggest that if the creditor does not care to settle on the debtor's own hard terms, he can sue. This situation, in which promptness in payment is vastly more important than even approximate justice in the settlement terms, is

too common in modern business relations to be ignored by society and the courts. (Dalzell, Duress by Economic Pressure II (1942) 20 N. Carolina L.Rev. 340, 370.)

Totem Marine T.&B. v. Alyeska Pipeline, Etc., supra, 584 P.2d 15, presents an example of economic duress remarkably parallel to the circumstances of this case. Totem, a new corporation, contracted with Alyeska to transport pipeline construction materials from Houston, Texas to a port in southern Alaska, with the possibility of one or two cargo stops along the way. Totem chartered the equipment necessary to perform the contract. Unfortunately, numerous unanticipated problems arose from the outset which impeded Totem's performance. When Totem's chartered tugs and barge arrived in the port of Long Beach, California, Alyeska caused the barge to be unloaded and unilaterally terminated the contract. Totem then submitted termination invoices totaling somewhere between $260,000 and $300,000. At the same time, Totem notified Alyeska it was in urgent need of cash to pay creditors and that without immediate payment it would go bankrupt. After some negotiations, Alyeska offered to settle Totem's account for $97,500. In order to avoid bankruptcy, Totem accepted Alyeska's compromise offer and signed an agreement releasing Alyeska from all claims under the contract. (Id, at pp. 17–19.)

About four months after signing the release agreement Totem sued Alyeska for the balance due under the contract. The trial court entered summary judgment for Alyeska based on the release agreement. (*Totem Marine T. & B. v. Alyeska Pipeline, Etc.*, supra, 584 P.2d at p. 19.) The Supreme Court of Alaska reversed, explaining:

> [W]e believe that Totem's allegations, if proved, would support a finding that it executed a release of its contract claims against Alyeska under economic duress. Totem has alleged that Alyeska deliberately withheld payment of an acknowledged debt, knowing that Totem had no choice but to accept an inadequate sum in settlement of that debt; that Totem was faced with impending bankruptcy; that Totem was unable to meet its pressing debts other than by accepting the immediate cash payment offered by Alyeska; and that through necessity, Totem thus involuntarily accepted an inadequate settlement offer from Alyeska and executed a release of all claims under the contract. If the release was in fact executed under these circumstances, we think that under the legal principles discussed above that this would constitute the type of wrongful conduct and lack of alternatives that would render the release voidable by Totem on the ground of economic duress. (Id., at pp. 23–24, fn. omitted.)

Here, Britton and Aghadjian acted in bad faith when they refused to pay Rich & Whillock, Inc.'s final billing and offered instead to pay a compromise amount of $50,000. At the time of their bad faith breach and settlement offer, Britton, and through him, Aghadjian, knew Rich & Whillock, Inc. was a new company overextended to creditors and subcontractors and faced with imminent bankruptcy if not paid its final billing. Whillock and Rich strenuously protested Britton's and Aghadjian's coercive tactics, and succumbed to them only to avoid economic disaster to themselves and the adverse ripple effects of their bankruptcy on those to whom they were indebted. Under these circumstances, the trial court found the July 10 agreement and August 20 release were the products of economic duress. That finding is consistent with the legal principles discussed above and is supported by substantial evidence. Accordingly, the court correctly concluded Ashton Development, Inc. and Bob Britton, Inc. were liable for the $22,286.45 balance due under the contract.

DISPOSITION

Judgment affirmed.

COLOGNE, Acting P.J., and STANIFORTH, J., concur.

While contractors occasionally have prevailed on their claims of economic duress,[150] economic duress claims are difficult to sustain.[151] The law will look at the extent of the pressure on the contractor (threat of bankruptcy), the lack of practical choice (the existence of realistic, practical alternatives), whether the pressure was compounded by actual or threatened breach of contract by the owner, and the state of mind of the owner (lack of good faith) as in the *Rich & Whillock* case.

Yet the reason for difficulty, the crucial role of private autonomy (freedom of contract) in American law, makes prediction of the outcome of these duress cases treacherous. One English judge stated that the aim is "to distinguish between contracts which are entered into as a result of illegitimate pressure from those entered into under the rough and tumble of the pressures of normal commercial bargaining. . . ."[152]

As noted, an accord and satisfaction can bar claims. Yet there have been inroads on this venerable doctrine, such as case decisions emphasizing economic duress as in the *Rich & Whillock* case, judicial outrage at unconscionable conduct by the stronger party,[153] and state legislation.[154] Although a party that accepts tender of an amount under these circumstances still runs a substantial risk that its claim will be barred, increasingly the law is willing to allow the claim to be pursued despite the claimant's having accepted the tendered payment with the restrictive conditions.[155]

SECTION 27.14 Claims Against Multiple Parties

Chapter 24—particularly Sections 24.06 and 24.07—noted the not uncommon phenomenon of defects that can be traceable to design and to failure by the contractor to execute the design properly. That chapter looked at disputes between owner and contractor. In this section, emphasis is on a claim made by the owner against both its independent design professional and the contractor. How is recovery measured when there are multiple causes?

---

[150]*Centric Corp. v. Morrison-Knudsen Co.*, 731 P.2d 411 (Okla.1986) (CM took advantage of the trade contractor's precarious financial condition).

[151]*Selmer Co. v. Blakeslee-Midwest Co.*, 704 F.2d 924 (7th Cir.1983) (financial difficulty not enough); *Pellerin Constr. Co. v. Witko Corp.*, 169 F. Supp.2d 568 (E.D.La.2001) (stress of business conditions does not create duress unless the defendant engaged in conduct designed to produce that distress); *Turner v. Low Rent Housing Agency of the City of Des Moines*, 387 N.W.2d 596 (Iowa1986).

[152]*DSND Subsea v. Petroleum Geo-Services*, [2000] Build.L. R. 530, 545. The recent English cases are analyzed in Tan, *Constructing a Doctrine of Economic Duress*, 18 Constr.L.J. 87 (2002). Tan provides a "guided" three-stage test at 95–96.

[153]*City of Mound Bayou v. Roy Collins Constr. Co.*, 499 So.2d 1354 (Miss.1986); *North Harris County Junior College Dist. v. Fleetwood Constr. Co.*, 604 S.W.2d 247 (Tex.Ct.App.1980).

[154]West Ann.Cal.Pub.Cont.Code § 7100 (barred in state contracts). In 1987, California drastically reduced the effectiveness of accord and satisfaction in the check-cashing situation. See West Ann.Cal.Civ. Code § 1526. A creditor can strike out a "paid in full" notation, cash the check, and still preserve a disputed claim.

[155]*John Grier Constr. Co. v. Jones Welding & Repair, Inc.*, 238 Va. 270, 383 S.E.2d 719 (1989) (subcontractor's claim not barred despite its endorsement of a check marked "paid in full" because its lack of knowledge meant no meeting of the minds).

The problem is best illustrated by *Northern Petrochemical Co. v. Thorsen & Thorshov, Inc.*[156] The owner contracted with an architect to design its new headquarters, which contained offices, a warehouse, and a manufacturing plant. It also contracted with a prime contractor under a fixed-price contract. The construction began in the fall of 1967 and was virtually completed in April 1968. At that time, it became clear that there were major structural flaws in the building. Large cracks were developing. Walls and columns were out of plumb. An investigation led to remedial action, but the deterioration continued. It became apparent that the building was moving in fits and starts. An agreement was made by the major participants to perform corrective work.

Until actual reconstruction began, it was believed that the sole reason for the building's failure was a structural defect in a support wall. When corrective work began, it was determined that the reinforcing steel for the concrete flooring had not been imbedded in the concrete and the underlying fill had not been compacted as required. These omissions left large voids under the floor that caused cracking and uneven settling. Correcting these faults required a massive reconstruction process that took approximately eight months. The trial court concluded that the owner's damages were approximately $750,000. It determined that some of the problem was traceable to design, some to faulty construction, and the balance to both.

Interestingly, the court treated this as a tort case, finding that both the architect and the contractor had been negligent. It then shifted to tort principles applicable when there are co-wrongdoers and there is no difference in the culpability of wrongdoing. It concluded that where it is not reasonably possible to make a division of the damage caused by the separate acts of negligence closely related in point of time, the negligent parties, even though they acted independently, are jointly and severally liable. Each was responsible for the entire indivisible loss, even though they did not act together. The court adopted a rule that puts the burden on either architect or contractor to limit its liability by providing a method of apportionment. In other words, either the architect or the contractor must prove that its negligence caused a particular harm for which it is responsible. In the absence of a rational method of apportionment, each is responsible for the entire loss. (Rights and duties between the co-wrongdoers are discussed in Section 31.02.)

The issue of joint and several liability is controversial. If two or more defendants are responsible for an indivisible loss, one may be much more at fault than the other. For example, if one is 90 percent at fault and the other 10 percent, and the party 90 percent at fault cannot pay a court judgment, the entire loss may fall on the party who was 10 percent at fault.

This possibility led the Iowa legislature to enact a law allowing joint and several liability only if a party is at least 50 percent negligent. Under such a statute, a case involving a claim against a contractor and a design professional concluded that the contractor was 86 percent responsible and the design professional 14 percent. However, the contractor was judgment proof; that is, it could not respond to a judgment. As a result, the owner desperately sought to show that the design professional was 50 percent responsible so that it could saddle the design professional with the entire loss. This was not successful, and the design professional was responsible for only 14 percent of the loss.[157]

The increased use of comparative negligence and shared responsibility can lead to another approach. The loss can be divided by comparing negligence or, in the absence of a tort standard being appropriate, the level or intensity of wrongdoing in a nontort sense.[158]

Suppose the architect were looked on as having caused 60 percent of the loss and the contractor 40 percent. This would place 60 percent of the loss on the architect and 40 percent on the contractor. Here cause would be replaced by "fault." This formula could also bypass complicated contribution and indemnity issues. But it would expose the owner to a greater likelihood of being uncompensated if either defendant could not pay the judgment.

This Section deals with co-wrongdoers and the amount each must pay the claimant and the effect of joint and several liability. Another issue can arise when claims are made against multiple parties. How will the ultimate

---

[156]297 Minn. 118, 211 N.W.2d 159 (1973).

[157]*Eventide Lutheran Home for the Aged v. Smithson Elec. & Gen. Constr., Inc.*, 445 N.W.2d 789 (Iowa 1989), applying Iowa Code Ann. § 668.4. In 1986 California passed Proposition 51, which changed the joint and several liability rule. In claims for noneconomic losses the defendants pay only for the percentage of their fault. This was enacted as West Ann.Cal. Civ. Code §§ 1431–1431.5. In *Thomas v. Duggins Constr. Co., Inc.*, 139 Cal. App.4th 1105, 44 Cal.Rptr.3d 66 (2006), the court held that Proposition 51 does not protect a defendant whose employee intentionally injured an innocent person.

[158]*Shepard v. City of Palatka*, 414 So.2d 1077 (Fla.Dist.Ct.App.1981) (dictum).

liability be determined? This is particularly difficult when one defendant settles and seeks to use that settlement as a defense when another wrongdoer seeks to pin the entire loss on it or at least seeks to apportion some of the loss to the party that has settled.

California law provides that, if a settlement is made by one defendant in "good faith," it cannot be sued for indemnity or contribution by defendants who have not settled.[159] This is done to encourage settlements.

The lawsuits with multiple claimants, claims, and defendants generate complexity that cannot be discussed in this book. Also, such lawsuits have generated many legislative solutions to the substantive and procedural problems created by these lawsuits.

SECTION 27.15 Security for Claims

A. Owner Claims

The owner can secure its claim against a contractor by withholding payment of funds for work performed[160] or refusing to release retainage.[161] Another method is to require that the contractor furnish either individual guarantees, unsecured or backed-up security interests in property, or a performance or warranty bond.

B. Prime Contractor Claims

The contractor indirectly can secure its claim against the owner by the pressure of a threatened or actual stoppage or even termination until payment is made. It can file a mechanics' lien against any private property that the prime contractor has improved.[162]

C. Subcontractor Claims

In addition to the methods described in Section 27.15B, in public projects, the subcontractor can secure its claim by looking to any payment bond that the prime contractor

has been required to furnish. In some states, stop notice rights (unpaid subcontractor can stop flow payments from owner or lender to prime) can be effective to obtain payment.[163]

D. Summary

Claims related to the construction project are often worth very little unless they can be collected. Obtaining a court judgment in many instances does not provide actual reparation for the loss. For that reason, participants in the construction project should plan their transactions to give them as much security as they can obtain by their contracts and should use every effort to perfect any security that is given to them under the contract, such as a surety bond, or perfect security provided by law, such as a mechanics' lien or stop notice.

SECTION 27.16 Claims Against Public Entities: Federal False Claims Act

The negotiation process that follows the making of a claim has been noted in Section 27.13. There focus was upon taking advantage of the claimant's financial difficulties. As noted in that Section the parties are given a good deal of room to negotiate the best settlement they can. But as seen in that Section even in the freewheeling negotiation atmosphere a few settlements will be set aside. Setting aside the settlement has been based upon economic duress. Usually this consists of taking undue advantage of the financial difficulties of the other party.

Rough tactics are often part of the negotiation process. The air can be filled with threats, bluster, and wild exaggerations. But the rules are different when the claim is made against a public entity, whether federal, state, or local. Such claims are governed by special statutes that seek to protect public funds.

This is demonstrated by *Daewoo Engineering & Construction Co. v. United States*.[164] Daewoo was a Korean company that contracted with the U.S. Corps of Engineers to build a 53-mile road on the Island of Palau. Daewoo presented a claim against the Corps for $64 million

---

[159] West Ann.Cal. Code of Civ.Proc. § 877.6. This statute was discussed in connection with a construction dispute in *TSI Seismic etc. v. Superior Court,* 149 Cal.App.4th 159, 56 Cal.Rptr.3d 751 (2007). This case is discussed in Section 15.03D at note 6.

[160] See Section 22.02E.

[161] See Section 22.03.

[162] See Section 28.07D.

[163] See Section 28.07E. See also Sections 28.07F, G, H, and J for additional methods available to subcontractors.

[164] 73 Fed.Cl.547 (2006), appeal filed Jun. 13, 2007.

(the original contract price was $73 million) but ended up being ordered to pay the Corps more than $50 million. The trial judge found Daewoo had violated the Contract Disputes Act's fraud provisions and the Federal False Claims Act (FCA).[165]

The government did not make its claim until Daewoo had rested its case. It claimed that there had been unexpected testimony. A number of factors persuaded the trial judge to make this order against Daewoo. He found that some of the legal arguments by Daewoo were not credible. He concluded that its witnesses testified in a vague and unreliable manner. Most important for purposes of this Section there had been testimony by one of Daewoo's executives that it made the large claim as a negotiating ploy to make the government pay attention. The judge pointed to Daewoo having misled the Corps as to its actual costs and its having lowered its claim from $50 million to $29 million. It used heightened theoretical projections instead of actual acquisition costs, claimed equipment that had been fully depreciated, and relied on an incorrect production rate to calculate its productivity loss. These can constitute fraud and an intention to deceive.[166]

What is interesting about the *Daewoo case* is that some of its tactics would have been normal in claims against private entities. But its claim against the Corps of Engineers invoked a number of federal statutes that deal with false claims and was the basis for a huge fine. Also, in another case the claimant forfeited its claim for $53,534,679 under the Forfeiture of Fraudulent Claims Act because its requests for reimbursement for bond premiums were false.[167]

The FCA originated in the Civil War period. What is unusual about this and related statutes is that the person (the whistleblower) who alerts the public agency to the wrongdoing under certain circumstances can pursue the claim itself (a *qui tam* claim) and recover a generous award for bringing this matter to the attention of public authorities. Because this disclosure to the federal entity, usually by an employee, will not be appreciated by the employer, the FCA protects the whistleblower from retaliation. A significant number of states have enacted similar statutes.[168]

Those who engage in public projects must be aware of these statutes and their requirements. They must also emphasize to its employees that these special rules must be followed.

---

[165]31 U.S.C.A.§ 3729. Each amendment strengthens the Act. For a detailed analysis of the FCA and related statutes, see FALSE CLAIMS IN CONSTRUCTION CONTRACTS: FEDERAL, STATE and LOCAL (ed. Sink & Pages, 2007).

[166]The discussion of the *Daewoo* case owes much to Littlejohn, *Turnabout at Trial: Contractor Ordered to Pay $50 Million for Fraudulent Claims*, Update Construction Law, Spring 2007, pp. 1, 3.

---

[167]*Morse-Diesel Intern. Inc. v. United States*, 74 Fed.Cl. 601 (2007). In a connected case the government was awarded $7,292,213 for violations of the Anti-Kickback Act and the FCA. 79 Fed.Cl. 116 (2007), reconsideration denied, 81 Fed.Cl. 311 (2008).

[168]At present 14 states have similar statutes and the list is growing.

CHAPTER TWENTY-EIGHT

The Subcontracting Process: An "Achilles Heel"

SECTION 28.01 An Overview of the Process

At the risk of oversimplifying, the subcontracting process, as used in this chapter, is defined as the method of construction organization under which the prime contractor is allowed to perform some or even much of its contract obligations through other contracting entities. The latter contracting entities are first-tier subcontractors. Likewise, the process in a large construction project can involve first-tier subcontractors performing their contract obligations through other contracting entities called *second-tier subcontractors*, or *sub-subcontractors*.

Other business entities frequently furnish equipment, machinery, products, supplies, or materials incorporated into the project or used to construct the project. These entities—collectively called *suppliers*—usually make contracts with prime contractors or subcontractors. Although the line between subcontractors and suppliers is sometimes difficult to draw, for purposes of this discussion subcontractors are defined as people who perform significant services at the site.[1]

The subcontracting process results in a chain of contracts that runs from owner to prime contractor or separate contractors, from prime or separate contractors (multiple primes) to subcontractors, and from subcontractors to sub-subcontractors. Likewise, there are direct contract lines between contractors that for purposes of discussion include prime contractors and subcontractors and their suppliers.

Some legal problems generated by the subcontracting process have already been discussed, and others are discussed later in this book. But subcontracting is the legal Achilles heel of the construction process. It generates many legal problems and therefore merits a separate chapter.

The principal advantage of a subcontracting system is improved efficiency, accomplished by breaking down work into categories that require a small number of related skills and the development of those skills by repetition. People who perform services, whether laborers working on the site, cost estimators making bid proposals, or managers making procurement decisions, should become more skilled as they repeatedly perform these specified services.

The subcontracting process, if working properly, can reduce costs not only by allowing more efficient work but also by creating many competitive prime contractors and subcontractors. Entry into the prime contract field is facilitated by allowing contractors to conserve capital, by relieving them of investment and financial burdens to the extent that subcontractors are used. The subcontracting system enables prime contractors to shift over much of the contract risks to subcontractors. If these risks involve the particular skill of the subcontractor and are ones over which the subcontractor has direct control, this can be an efficient allocation of risk.

As for subcontractors, the subcontracting system should encourage many smaller, highly competitive subcontractors to enter the construction field. The subcontracting system may be one reason why the construction industry, unlike the automobile industry, is made up of many contractors with specialized talents who operate mainly in limited localities. This has meant vigorous competition for work that, although it has the disadvantage of economic instability, frequent financial failures, and disputes, should, through competition, reduce construction cost.

---

[1] *U.S. Industries v. Blake Constr. Co.*, 671 F.2d 539, 543 (D.C.Cir. 1982) (citing earlier edition of this book).

An often ignored advantage of the subcontracting system is that many subcontracting entities can create social mobility and avoid the rigidity of class structures that are more closed. This is accomplished by permitting individuals or small business units to enter the field with a minimum amount of capital.

These undoubted potential and actual advantages have their cost. The subcontracting system generates a large share of construction legal problems.

Subcontractors are normally the financially weakest participants in the project and so are subject to the harshest contract obligations.[2] The principal subcontract problem deals with payments. As shall be seen in Section 28.06, the subcontracting system creates risk of delayed payment and nonpayment to subcontractors. Typically, the prime contractor on any substantial project is paid monthly as work progresses, with a customary retainage of from 5 to 10 percent. This delays cash flow from prime contractor to those to whom it owes payment. Even in no-retainage contracts, the prime contractor often faces cash flow problems caused by the lag between payments to it and its obligations. When there are retainages, the cash flow problem is more serious. To the extent of payment delay, the prime contractor is providing the owner with financing services that it seeks to transfer to subcontractors.

Prime contractors generally use subcontract payment provisions to minimize cash flow problems. These provisions frequently permit the prime contractor to delay paying subcontractors until the former have been paid by the owner. Such delayed payment provisions may squeeze subcontractors who must pay for their labor and supplies. Even those with credit face financial hardship when credit is withdrawn or limited.

Cash flow problems are increased as a greater percentage of work is performed by subcontractors. Increased subcontracting means a greater likelihood that nonperformance by one subcontractor will delay payment to others who have performed properly.

The construction industry is composed of many small businesses with limited financial capacity and credit. This is particularly true with regard to subcontractors, many of whom started as tradesmen and many of whose businesses are family owned. A short delay in the flow-through of

funds from the owner to those contractors performing services on the site can cause financial hardship that may deprive a contractor, especially at the lowest tier, of funds needed to continue performance. The subcontracting system heightens the financial stress inherent in the flow-through process because it increases the distance of the money flow. See Figure 28.1.

The extent of work subcontracted also plays a role in creating legal problems. Subcontractors frequently assert that prime contractors are merely assemblers or brokers for the services of others. Prime contractors deny this and claim that they supply much of the materials and services themselves. Prime contractors and subcontractors agree that the *amount* of work subcontracted depends on the *type* of construction. Whether or not prime contractors or subcontractors are correct on this controversial question, it is clear that the greater the percentage of subcontracted work, the greater the likelihood of severe cash flow problems. A prime contractor who does not have much money tied up in the project is less concerned about the swiftness of the cash flow. This is especially true if the prime contractor's obligation to pay subcontractors is conditioned on receiving payments from the owner.

Moving from cash flow delay to nonpayment problems, the volatility of the construction industry must again be emphasized. Business failures and bankruptcies frequently occur in the construction industry. The higher the proportion of work subcontracted, the greater the risk that unpaid subcontractors will seek some type of legal relief when their work has benefited the owner. One type of relief often sought by unpaid subcontractors and suppliers is a mechanics' lien. Valid liens usually result in owners paying lien claimants to remove the liens. To avoid liens, owners create payment structures to minimize payment diversion by the prime contractors. Owners—private and especially public—frequently require payment bonds to give an effective remedy to unpaid subcontractors and suppliers. Bond requirements inject the complexity of surety bonds and additional parties into the already complicated construction structure.

Problems of delayed payment and nonpayment highlight another significant feature of the subcontracting process. The subcontractor is "a contract away" from the major source of power and control over the project—the owner. Because the subcontractor has no direct contractual relationship with the owner, it must look to the prime contractor for payment, for dealing with disputes over

---

[2]Sklar, *A Subcontractor's View of Construction Contracts*, 8 Constr. Lawyer, No. 1, Jan. 1988, p. 1.

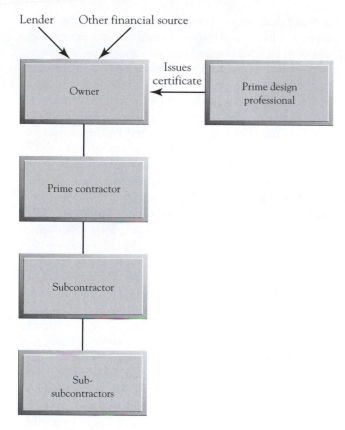

FIGURE 28.1 Cash flow in traditional system.

performance, and to process claims against the owner. This can create a sense of powerlessness in subcontractors, leading to friction and lawsuits at worst and poor communication at best.

Subcontracting compounds the already difficult problems caused by the multitude of construction documents regulating construction relationships. One principal cause of legal problems is the wealth of potentially conflicting documents that regulate the relationship between owner and contractor. Adding first- and second-tier subcontracts, which frequently refer *generally* to contract provisions in contracts above them on the contract chain, dumps in additional ingredients for disputes.

The subcontracting process generates a potentially large number of construction contractors—all working on the same site and often at the same time. For purposes of this discussion, contractors include prime contractors, separate contractors, and the various tiers of subcontractors. Contractors may disagree on who will

have access to a particular part of the site at a particular time. One contractor's performance often depends on another contractor's work being completed or at least in a certain state of readiness. One contractor's work may be disturbed or ruined by another contractor's workers, and disputes may develop over which contractor is responsible. One contractor may employ workers who belong to a construction trade union, whereas another uses non-union workers. One contractor's employee may be injured or killed, and the responsibility may be asserted against or shared by a number of other contractors. The sheer number of different contractors at work can create immense administrative and, ultimately, legal problems.

The illustrations given do not exhaust the legal problems incident to subcontracting. In addition, there is frequent bargaining disparity present in construction contracts and especially in subcontracts. Generally, the dominant bargaining strengths parallel the money flow. Lenders can exact terms from owners because the total number of borrowers seek more money than lenders have to lend. The owner who has funds or can borrow them seeks a contractor from among the many willing to perform the work. This position gives the owner substantial bargaining advantage over the prime contractor. Although the prime contractor awarded the contract does not as yet have funds, it has contract rights. These rights generally give the prime contractor bargaining advantage over subcontractors, many of whom are looking for work. As a result, at this stage, the prime contractor usually has the bargaining power to demand, and often obtain, favorable terms from subcontractors, sometimes terms more favorable than those in the prime contract. Likewise, subcontractors frequently exert parallel bargaining advantage over sub-subcontractors.

Lawyers do not play a significant role drafting and reviewing construction contracts generally, and particularly with subcontractor contracts. Lawyers are frequently not brought into the picture until the need for legal action becomes imminent. This can mean that extraordinarily one-sided contracts may be forced on subcontractors, who are generally unaware of legislation or case decisions that might protect them.

The bargaining position of suppliers and contractors cannot be so easily generalized. Suppliers range from large manufacturing companies with strong financial positions to small distributors with limited financial capacity. As a result, some suppliers are in the position to dictate terms to subcontractors and even to prime contractors. Yet even suppliers in this advantageous position must sell and must often extend credit to do so. Often their extension of credit is predicated on statutory lien rights or the existence of payment bonds.

In the typical construction project, on the lowest tier is the subcontractor with the weakest bargaining power. This bargaining pattern is illustrated by the frequent passing down of increasingly harsh and one-sided indemnity agreements. In addition to strong bargaining pressures from prime contractors or higher-tier subcontractors, lower-tier subcontractors also face the strong bargaining power of large suppliers and construction trade unions. Subcontractors, especially those at the lowest tier, often find themselves squeezed on all sides because of their poor bargaining position.

Institutional bargaining disparities generate legal problems. Harsh terms exacted at the bargaining table may be resisted by the weaker party when disputes develop. The law looks with disfavor on harsh terms, even though the legal system generally allows parties to write their own contracts. As a result, the uncertain enforcement of harsh terms exacted by the dominant party generates legal disputes that often require judicial resolution.

Those in vulnerable bargaining positions may seek other avenues of relief. For example, as seen in Section 31.05D, industry associations sometimes obtain legislative enactments such as anti-indemnity legislation to overturn contract clauses exacted from them by parties in a stronger bargaining position. Injecting legislative rules in a field regulated largely by private contracts adds a further complication to construction.

Associations composed of members who are often in a weak bargaining position often seek to remedy this weakness by participating in the creation of standard construction contracts such as those published by the American Institute of Architects (AIA) and the Engineers' Joint Contracts Documents Committee (EJCDC). In the past, prime contractors, through the Associated General Contractors of America (AGC), sought to persuade the AIA to incorporate or eliminate language in the AIA standard documents that they might not otherwise be able to do in dealing with an owner. Likewise, associations representing subcontractors frequently asked the AIA to incorporate provisions in the AIA prime contract requiring that certain rights be accorded subcontractors when these rights might not be obtainable in normal bargaining between prime contractors and subcontractors. When the prime contract deals with subcontracting and subcontractors, an increased likelihood exists of conflicting contract documents and a more generally complicated prime contract.

Subcontractor associations had sought protective language not only in prime contract documents published by the AIA but also in the AIA's standard subcontract, A401. Just as prime contractors prefer an AIA document over a construction contract drafted by the owner, subcontractors are likely to prefer a subcontract drafted by the AIA to one prepared by the prime contractor.

Past attempts by the AGC and subcontractor associations to influence the AIA gave way in 2007 to creation of their own standard documents, called ConsensusDOCS. ConsensusDOCS 750 is the standard form subcontract. Some provisions are written to the advantage of the subcontractor. Disavowing the use of pay-when-paid clauses (see Section 28.06), Section 8.2.5 states that if the owner fails to pay the contractor for work satisfactorily performed by the subcontractor, the contractor must pay the subcontractor "within a reasonable time." Under Section 8.2.6, if the subcontractor has not been paid within a reasonable time, then, upon giving seven days' written notice to the contractor "and without prejudice to and in addition to any other legal remedies," the subcontractor "may stop work until payment of the full amount owing to the Subcontractor has been received." The impact of the ConsensusDOCS at this point is unknown.

One final aspect of the subcontracting process adds complexity. Contracts for the performance of design and construction services are basically regulated by common law rules, that is, rules that have evolved through court decisions. However, contracts for the sale of goods are governed by the Uniform Commercial Code (U.C.C.), a comprehensive statutory regulation. When a subcontractor, as is usually the case, provides both goods and services, which rules apply—the common law or the U.C.C.? Generally the test is whether goods or services are the predominant aspect of the transaction.[3] This can mean that a particular subcontract or a severable part of a particular subcontract is governed by the U.C.C., whereas the rest may be governed by the common law.

To sum up, the subcontracting system has undoubted advantages, but generates a host of legal problems. Bear this in mind as some particular subcontract problems are addressed next.

---

[3]*Bonebrake v. Cox,* 499 F.2d 951 (8th Cir.1974).

SECTION 28.02 The Subcontractor Bidding Process

A. Using Sub-Bids

In technical projects requiring specialized skills, the prime contractor may do very little work itself. In such projects, the cost to the prime contractor of work to be done by subcontractors is not likely to be known until subcontractors submit sub-bids. The prime contractor cannot bid until hearing from all the prospective subcontractors, something that does not usually occur until close to time for submitting the bid to the owner. In submitting its own bid, the prime contractor relies on the subcontractors' bids.

The prime contractor's reliance on subcontractors is most acute in the mechanical specialty trades (e.g., electrical, plumbing, air conditioning). Prime contractors also use a large number of subcontractors from the non-mechanical specialty trades (e.g., masons, roofers). Reliance upon subcontractors' bids provides the prime contractor with a relatively accurate means of calculating its own bid to the owner, while the sub-bids provide it with downside protection. These subcontractors, in addition, provide the prime contractor with a skilled workforce, which the prime contractor could not possibly afford to employ between projects.

Prime contractors generally use the bids given by subcontractors in computing their bids. Suppose a subcontractor withdraws its bid, usually because it contends that its bid had been inaccurately computed or communicated. Can the prime contractor hold the subcontractor to its bid by contending that the prime had relied on the sub-bid in making its own bid?

B. Irrevocable Sub-Bids

The early cases dealing with this problem used traditional contract analysis to allow the subcontractor to revoke its bid. Generally the prime contractor does not contract with the subcontractor, conditioned on the prime contractor's being awarded the bid. The prime contractor wishes to preserve maximum freedom to renegotiate with the low bidder or negotiate with other bidders. Ignoring the prime contractor having relied on the sub-bid in making its own bid, *James Baird Co. v. Gimbel Brothers*[4] did not hold the subcontractor to its bid.

However, contract law, under the concept of "promissory estoppel," has expanded reliance as a basis for making an offer irrevocable. This carried over into subcontractor bid cases, the leading case being *Drennan v. Star Paving Co.*[5] The subcontractor had submitted the lowest sub-bid for the paving portion of the work, a sub-bid the prime contractor used in computing its overall bid. The prime contractor listed the defendant on the owner's bid form as required by statute.[6] The prime contractor was awarded the contract and stopped by the subcontractor's office the next day to firm up the "subcontract." On arrival, the subcontractor immediately informed the prime contractor that it had made a mistake in preparing its bid and would not honor it. The prime contractor sued and was awarded the difference in cost between the subcontractor's sub-bid and the cost of a replacement.

The court concluded that using the sub-bid did not create a bilateral—or two-sided—contract between the plaintiff and the defendant.[7] But the court held the subcontractor to its bid because the bid was a promise relied on reasonably by the prime contractor when it submitted its own bid. The possible uncertainty of subcontract terms (discussed later in this section) was brushed aside.

To satisfy the doctrine of promissory estoppel as articulated in the Restatement (Second) of Contracts Section 90, there must be a clear and definite offer; a reasonable expectation that the offer will induce reliance; actual and reasonable reliance by the offeree; and an "injustice" that can be avoided only by enforcement of the offer. To be clear and definite, the offer must be more than a mere estimate or price quote. Courts rely upon industry custom

---

[4]64 F.2d 344 (2d Cir.1933).

[5]51 Cal.2d 409, 333 P.2d 757 (1958), followed in the Restatement (Second) of Contracts, § 90 (1981). But *Drennan* was not followed in *Home Elec. Co. of Lenoir, Inc. v. Hall and Underdown Heating and Air Conditioning Co.*, 86 N.C.App. 540, 358 S.E.2d 539 (1987) (rejected as one-sided), aff'd, 322 N.C. 107, 366 S.E.2d 441 (1988), noted in 10 Campbell L.Rev. 293 (1988).

[6]As to listing laws, see Section 28.03. In *Southern California Acoustics Co. v. C. V. Holder, Inc.*, 71 Cal.2d 719, 456 P.2d 975, 79 Cal.Rptr. 319 (1969), the court held that an improper substitution created a claim based on the statutory violation against the prime contractor in favor of the improperly substituted subcontractor. This approximates binding both parties. But listing the subcontractor to comply with the statute generally does not create a bilateral (two-sided) contract. *Holman Erection Co. v. Orville E. Madsen & Sons, Inc.*, 330 N.W.2d 693 (Minn.1983).

[7]For a case holding that using the bid is not an acceptance, see *Mitchell v. Siqueiros*, 99 Idaho 396, 582 P.2d 1074 (1978). Similarly, see *Four Nines Gold, Inc. v. 71 Constr., Inc.*, 809 P.2d 236, 239 (Wyo.1991) (citing an earlier edition of this book).

to conclude that a prime contractor's reliance is reasonable. The final element of injustice is met where the prime contractor will suffer a detriment if the subcontractor's offer is not enforced. However, promissory estoppel is an equitable doctrine (meaning the court promotes a fair outcome), and misconduct by the contractor (such as bid shopping, snatching up an unreasonably low bid, or unreasonable terms of acceptance) will release the subcontractor from being held to its offer.[8] The prime contractor's damages is the difference in price between the withdrawn bid and the replacement subcontractor's cost. A recent article concludes that a majority of states follow promissory estoppel, a very few reject it outright, and in several the issue remains undecided.[9]

C. Bargaining Situation: Shopping and Peddling

The *Drennan* rule improves the prime contractor's already powerful bargaining position. Although the prime contractor under the *Drennan* rule is not free to delay its acceptance or to reopen bargaining with the subcontractor and still claim a right to accept the original bid, it can at least for a short period seek or receive lower bid proposals from other subcontractors.

Under the *Drennan* rule, until the prime contractor is ready to sign a contract with the subcontractor whose bid it has used, the subcontractor is not assured of getting the job. Before the award of the prime contract, the plurality of competing prime contractors' bidding on a project tends to diffuse their bargaining power over subcontractors. This competition before the award of the contract should result in lower sub-bids and consequently lower overall bids by the prime contractors, a definite benefit to the owner. Although subcontractors often wait until the last minute to submit their sub-bids in an effort to minimize the prime contractor's superior bargaining position, a substantial amount of competition still exists among the subcontractors themselves.

After award of the contract, the relative bargaining strengths of the successful prime contractor and the competing subcontractors change drastically. The prime contractor now has a "monopoly" and a substantially superior bargaining position over the subcontractors under the *Drennan* rule. The sub-bids used provide the prime

contractor with a protective ceiling on the cost of the work with no obligation to use the subcontractors. The prime contractor is therefore free to look elsewhere for yet a better price and is able to increase its profits by engaging in postaward negotiations.

Postaward negotiations have become controversial. Sometimes they are called "bid shopping"; the prime contractor uses the lowest sub-bid to "shop around" with the hope of getting still lower sub-bids. "Bid peddling" is the converse, with other subcontractors attempting to undercut the sub-bid to the prime, in essence engaging in a second round of bidding. Subcontractors often refer to these postaward negotiations as "bid chopping" and "bid chiseling." (These tactics can also be used before prime bids are submitted.)

Both subcontractors and owners have reasons to condemn postaward competition. The subcontractors assert that preparing a bid involves considerable expense. Subcontractors who "bid peddle" may not even prepare their own bids, thus saving overhead expenses. The subcontractor who went to the expense of calculating a bid subsidizes the bid-peddling subcontractor's overhead costs as well as the prime contractor's costs of bid preparation.[10]

Subcontractors who fear bid shopping often wait until the last minute to submit their sub-bids to the prime contractor to give the prime contractor as little time as possible to bid shop. This last-minute rush is the cause for many mistakes by both subcontractors and prime contractors. Some subcontractors simply refrain from bidding on jobs where bid shopping is anticipated, to save the expense of preparing a bid. To that extent, competition among subcontractors is diminished and higher prices can result.

Subcontractors feel they must pad their bids to make allowance for the eventual postaward negotiations. This "puffing" raises the cost to the owner, as the inflated bid is the bid the prime contractor uses to compute its overall bid. Any subsequent negotiations that result in reducing the price benefit only the prime contractor.

The superior bargaining position of a successful prime contractor spurs cutthroat competition among subcontractors, resulting in lost profits that can upset industry stability. Prime contractors respond by noting that

---

[8]Sections 28.02C & D.

[9]Kovars & Schollaert, *Truth and Consequences: Withdrawn Bids and Legal Remedies*, 26 Constr. Lawyer, No. 3, Summer 2006, p. 5.

[10]One court has held that bid shopping by a prime contractor constitutes an unfair trade practice; see *Johnson Elec. Co., Inc. v. Salce Contracting Assocs., Inc.*, 72 Conn.App. 342, 805 A.2d 735, certification denied, 262 Conn. 922, 812 A.2d 864 (2002).

sub-bids are often unresponsive to the specifications and require further clarification and negotiation. This may be especially true when prime contractors are dealing with subcontractors with whom they have never dealt. They must investigate the subcontractor's reputation and work experience before making a firm contract. A prime contractor who goes through the effort of deciding upon a specific subcontractor is, at the very least, inconvenienced by a withdrawal of a sub-bid on the eve of project commencement.[11]

Prime contractors state that bids are often requested for alternative proposals, and they lack the time to evaluate all the alternatives in the short time available between receipt of the sub-bids and bid closing. They assert that negotiations are sometimes required to decide on the specific alternative to be chosen. Prime contractors also justify postaward negotiation by stating that estimating a job is a normal cost of overhead in the construction industry.

D. Avoiding *Drennan*

Subcontractors can avoid their sub-bids' being firm offers that bind them and not the contractor. They can call their bids "requests for the prime to make offers" to them, or "quotations," given only for the prime contractor's convenience.[12] They can state in their bids that the bid is provided for information only and is not a firm offer.[13] They may also try to annex language stating that using the bid constitutes an acceptance that ties the prime contractor to them. They may refuse to submit bids unless they receive a promise by the prime contractor to accept the bid if it is low and the prime contractor is awarded the contract.[14] They may refuse to begin work without receiving a letter of intent from the prime contractor that provides that they will be paid if the parties

cannot agree on the terms of the subcontract. Finally, one subcontractor notified the public owner that it had made a mistake in its bid to the prime, and the owner canceled the procurement.[15]

The difficulty with most methods of avoiding the *Drennan* rule is that either the subcontractors do not have the bargaining power to implement them or the process does not make it convenient to use them.

Courts will not apply promissory estoppel if the conditions for finding a contract are not present. The *Drennan* rule can be avoided if many crucial areas have been left for further negotiation[16] or if there is no "meeting of the minds" as to the scope of work.[17] Promissory estoppel is inappropriate if the prime contractor does not accept the bid within a reasonable time[18] or proposes a subcontract that contains new unreasonable or onerous terms.[19] Although this defense was unsuccessful in the *Drennan* case, other courts have upheld the subcontractor's claim of mistake.[20]

Finally, a subcontractor who provides an oral bid may try to invoke the Statute of Frauds as a defense.[21] However, the Restatement (Second) of Contracts, Section 139, views promissory estoppel as an exception to the Statute of Frauds, and most courts find oral bids enforceable.[22]

---

[11]In *Arkansas Contractors Licensing Bd. v. Pegasus Renovation Co.*, 347 Ark. 320, 64 S.W.3d 241 (2001), the court affirmed a licensing board ruling that a subcontractor who repeatedly backed out of bids engaged in "misconduct in the conduct of the contractor's business," which actions justified revocation of the sub's license.

[12]*Leo F. Piazza Paving Co. v. Bebek & Brkich*, 141 Cal.App.2d 226, 296 P.2d 368 (1956); *Cannavino & Shea, Inc. v. Water Works Supply Corp.*, 361 Mass. 363, 280 N.E.2d 147 (1972).

[13]*Fletcher-Harlee Corp. v. Pote Concrete Contractors, Inc.*, 482 F.3d 247 (3d Cir.2007).

[14]*Electrical Constr. & Maintenance Co., Inc. v. Maeda Pacific Corp.*, 764 F.2d 619 (9th Cir.1985) (sub sued prime when latter used a different sub).

[15]*Four Nines Gold, Inc. v. 71 Constr., Inc.*, supra note 7.

[16]*Preload Technology, Inc. v. A. B. & J. Constr. Co., Inc.*, 696 F.2d 1080, rehearing denied, 703 F.2d 557 (5th Cir.1983) (Texas law) (dictum). But the court rejected this in *Arango Constr. Co. v. Success Roofing, Inc.*, 46 Wash.App. 314, 730 P.2d 720 (1986), pointing to the sub's reason for not making the contract relating to price and not other terms.

[17]*Camosy, Inc. v. River Steel, Inc.*, 253 Ill.App.3d 670, 624 N.E.2d 894 (1993) (no promissory estoppel because the steel erection bid was too ambiguous to have been relied upon).

[18]*Pickus Constr. & Equipment v. American Overhead Door*, 326 Ill. App.3d 518, 761 N.E.2d 356 (2001) (no acceptance within 30 days).

[19]*Hawkins Constr. Co. v. Reiman Corp.*, 245 Neb. 131, 511 N.W.2d 113 (1994) (subcontract contained new terms, including "no damages for delay" clause); *Lichtenberg Constr. & Dev., Inc. v. Paul W. Wilson, Inc.*, 2001 Ohio App. LEXIS 4372, 2001 Westlaw 1141236 (2001) (subcontract contained new "time of the essence" clause).

[20]*B. D. Holt Co. v. OCE, Inc.*, 971 S.W.2d 618 (Tex.Ct.App.1998), review denied Jun 25, 1998, rehearing for petition for review overruled, Aug. 25, 1998; *Tolboe Constr. Co. v. Staker Paving & Constr. Co.*, 682 P.2d 843 (Utah 1984). See Section 18.04E.

[21]Section 5.10.

[22]*SKB Indus., Inc. v. Insite*, 250 Ga.App. 574, 551 S.E.2d 380, 384 (2001), cert. denied Jan. 9, 2002 (the purpose of the statute is to prevent fraud, not "to prevent the use of the equitable principle of promissory estoppel to enforce a promise which was expected to and did induce detrimental reliance").

E. Uniform Commercial Code

Offers generally at common law are revocable even if stated to be irrevocable. Dissatisfaction with this rule has generated solutions taking many forms, including the *Drennan* rule making the bids irrevocable. Another method adopted is Section 2-205 of the Uniform Commercial Code (U.C.C.). Section 2-205 recognizes firm offers and enforces them if certain requirements are met.

The U.C.C. does not apply to service contracts or those that involve interests in land. The U.C.C. has also had limited use in making sub-bids by suppliers irrevocable. Section 2-205 requires a written offer stating that the bid will be held open. Bids are often communicated by telephone, facsimile, or electronic medium.

The U.C.C. is not a viable solution to the problem of sub-bids for another reason. Rarely will all the needed terms be expressed by the offer, however made. Even if the bid is communicated by a writing with all the terms, rarely do prime contractors intend to be bound to those terms, choosing to either use terms used before or dictate terms later. The firm offer of Section 2-205 does not fit the sub-bid, another reason why the *Drennan* rule, a substitute for the U.C.C. firm offer, was incorrectly decided.

California revised its version of the U.C.C. in 1980.[23] A written or oral sub-bid for goods made to a licensed contractor that the bidder knows or should know will be relied on is irrevocable for ten days after the prime contractor is awarded the contract but not later than ninety days after the bid. Oral bids of over $2,500 must be confirmed in writing within forty-eight hours. The bid can limit the duration of the offer.

Again the problem of terms can arise. The statute does not preclude the sub-bidder from asserting that the terms needed to cure any completeness requirement were not included in the offer, particularly if it were oral. Although the California statute as revised expands the enforcement of oral sub-bids for goods, it has the same defects as the *Drennan* rule (noted in Section 28.02B).

F. A Suggestion

The *Drennan* rule, now firmly in the saddle in most American jurisdictions, was a laudable attempt to avoid the common law rule of revocability and expand reliance as a method of enforcing *promises that should be enforced*. It is singularly inappropriate for construction sub-bids. First, it gives advantage to the prime contractors that already have a strong bargaining position. Second, it creates a binding offer before there has been agreement on important terms such as bonds, payment, indemnification, and dispute resolution.[24] The traditional common law rule expressed in the *Baird* case[25] is more appropriate. One-sided contracts should be found only when this is clearly intended or where absolutely necessary. Neither occurs here.

Using the bid should be an acceptance of the offer, and a contract concluded if there are sufficient terms supplied by well-established custom or a course of dealings between the parties which can supply the ancillary terms in addition to the work to be performed and payment. When such customs do not exist, neither party should be bound until both parties work out the terms needed to make a binding contract.[26]

G. Teaming Agreements

Teaming agreements are, as a rule, informal arrangements between prime contractors who wish to obtain federal procurement high-technology contracts and subcontractors who help devise proposals designed to help the prime contractor win the competition. The informality of these arrangements can often lead to litigation if the prime contractor wins the competition and does not award subcontract work to its "teaming partner." This situation was discussed in Section 17.04H.

SECTION 28.03 Subcontractor Selection and Approval: The Owner's Perspective

Owners wish to have competent contractors building their projects. They want subcontractors treated fairly and given an incentive to perform the work expedi-

---

[23]West Ann.Cal.Comm.Code § 2205.

[24]This problem was recognized in *Saliba-Kringlen Corp. v. Allen Eng'g Co.*, 15 Cal.App.3d 95, 92 Cal.Rptr. 799 (1971). But too much emphasis on this would frustrate the *Drennan* rule, an outcome the court was not willing to endorse.

[25]Supra note 4.

[26]This approach has been suggested in *Loranger Constr. Corp. v. E. F. Hauserman Co.*, 376 Mass. 757, 384 N.E.2d 176 (1978). See, generally, Closen & Weiland, *The Construction Industry Bidding Cases*, 13 J.Mar.L.Rev. 565 (1980).

tiously. Some owners contract directly with the specialty trades to avoid the prime contractor as an "intermediary" between owner and specialized trades. Some contract directly with the specialized trades and assign those contracts to a main or prime contractor. Some owners dictate to the prime contractor which subcontractors will be used, known in England as the *nominated subcontractor* system.

The American version used in some public contract systems is called "prefiled" bids. Prospective subcontractors file sub-bids with the public entity. The successful prime bidder must contract with the lowest subcontract bidders.[27]

These methods of direct intervention are not common in the traditional contracting system. Some owners leave subcontracting exclusively to the prime contractor. Others take a role that gives them *some* control but does not involve the owner *directly* with the subcontractor. Using the prime contractor as a buffer is done for administrative and legal reasons. The subcontracting system requires a well-defined organizational and communication structure under which each participant knows what it must do and with whom it must deal. From a legal standpoint, the owner does not want to be responsible for subcontract work, does want the prime contractor to be responsible for defective subcontract work, and does not want to be responsible if subcontractors are not paid.

One "part-way" control is to require that prime contractors list their subcontractors at the time they make their bids. Statutes in some states, called *listing laws*,[28] impose this on prime contractors in public projects. Although undoubtedly some impetus for such laws came from subcontractor trade associations,[29] one reason for the listing laws is to assure the awarding authority that only competent subcontractors will perform on the project.[30]

Legislatures frequently justify listing statutes by condemning bid shopping and bid peddling that, they assert, cause poor quality of materials and workmanship to the detriment of the public and also deny "the public of . . . the full benefits of fair competition among prime contractors and subcontractors, and lead to insolvencies, loss of wages to employees, and other evils."[31]

Listing laws usually regulate substitution of listed subcontractors by providing specific justifications for substitutions and a procedure for determining the grounds for replacing one subcontractor with another. Complications have developed not only as to grounds for substitution but also as to who has a claim for violating the statute.[32] From the subcontractor's viewpoint, listing laws dampen any attempt by the prime contractor to reopen negotiations.

Listing is used by the American Institute of Architects (AIA) in a more limited way. AIA Document A201-2007, Section 5.2.1, requires the contractor "as soon as practicable after award of the contract" to furnish the owner and architect the names of subcontractors and those who will furnish materials and equipment fabricated to a special design. Before 1976, the architect had to *approve* subcontractors. To make the architect's role more passive and to minimize the likelihood of liability, A201-2007, Section 5.2.1, states that the architect "may reply" to the contractor within 14 days stating (1) whether the owner or contractor has "reasonable objection to any such proposed person or entity" or (2) that the architect needs additional time for review. Failure to "reply within the 14-day period shall constitute notice of no reasonable objection." If the architect has reasonable objection, Section 5.2.3 requires an increase or decrease in the contract price caused by any substitution. Responding to criticism, the AIA in A201-1997, Paragraph 5.2.3, allowed the price change only if the rejected subcontractor "was reasonably capable of performing the Work," and this language is continued in A201-2007, Section 5.2.3.

Some control over subcontractors, though desirable, has negative features. Requiring that a subcontractor not

---

[27]This is described in *J. F. White Contracting Co. v. Dept. of Public Works*, 24 Mass.Ct.App. 932, 508 N.E.2d 637, review denied, 400 Mass. 1104, 511 N.E.2d 620 (1987).

[28]West Ann.Cal.Pub.Cont.Code §§ 4100 *et seq.*

[29]*Clark Pacific v. Krump Constr., Inc.*, 942 F.Supp. 1324 (D.Nev. 1996) (citing earlier edition of this book).

[30]Ibid. The court stated at 1338 that listing laws were also enacted "because of the indirect effects of ruthless bid-shopping, which include poor workmanship provided by subcontractors who, desperate to retain their subcontracts, shave their profits and expenses below a level which guarantees quality work, subcontractors' insolvencies, and construction workers' lost wages."

[31]West Ann.Cal.Pub. Cont. Code § 4101.

[32]See *Golden State Boring & Pipe Jacking, Inc. v. Orange County Water Dist.*, 143 Cal.App.4th 718, 49 Cal.Rptr.3d 447 (2006); *R. J. Land & Assocs. Constr. Co. v. Kiewit-Shea*, 69 Cal.App.4th 416, 81 Cal.Rptr.2d 615 (1999). See also *Southern California Acoustics Co. v. C. V. Holder, Inc.*, supra note 6; *Clark Pacific v. Krump Constr. Co.*, supra note 29. The latter opinion contains a good discussion of listing laws and their purposes.

be removed unless there is "due investigation" can invite a subcontractor who has been removed to assert a claim that the architect or owner has intentionally interfered with its actual or prospective contract. As noted, to get a price increase if a proposed subcontractor has been rejected under Section 5.2.3 the contractor must show that its proposed subcontractor was reasonably capable of performing the work. It is difficult to tell whether this 1997 change will have any effect on a claim by a rejected subcontractor against the owner or architect. It can be contended that proposing a subcontractor was a representation by the prime that the proposed subcontractor could do the work. At the very least this change may shift the burden of showing there were good reasons for the rejection to those who are rejecting the proposed subcontractor. The change made in 1997 will likely mean fewer rejections of proposed subcontractors. Although there may be defenses,[33] this does increase potential liability exposure to the owner or architect.

The Engineers Joint Contracts Documents Committee (EJCDC), in its C-700, Paragraph 6.06B (2007), does not require that the identity of subcontractors be furnished. Only if the supplementary conditions require the identity of subcontractors and suppliers are their names to be submitted to the owner for acceptance by the owner. The owner can revoke any acceptance "on the basis of reasonable objection after due investigation." A price adjustment is made if the contractor must use a replacement.

But Paragraph 6.06A states that the prime will not use a subcontractor "against whom Owner may have reasonable objection." That paragraph also provides, as does A201-2007, Section 5.2.2, that the prime cannot be forced to use a subcontractor against whom the prime has "reasonable objection."

Suppose the contractor wishes to remove an approved subcontractor? AIA Document A201-2007, Section 5.2.4, bars a contractor from doing so if the owner or architect makes a reasonable objection. But objections to proposed substitutions must be exercised carefully. In *Meva Corp. v. United States*,[34] the contracting officer denied the contractor permission to substitute, and the contractor was forced to keep the original subcontractor. The subcontractor did a poor job, however, and the contractor recovered from the public agency losses caused by the subcontractor.

To sum up, owner intervention into the relationship between prime contractor and subcontractor may be essential, but carries risks. If the risks are so great but the need for intervention so strong, the owner should consider a method other than the traditional contracting system, such as separate contractors (multiple primes). (The owner's attempt to keep subcontractors "on board" if the prime contract is terminated is discussed in Section 33.06.)

SECTION 28.04   Sources of Subcontract Rights and Duties: Flow-Through Clauses

The principal source of contract rights and duties between prime contractor and first-tier subcontractor is the subcontract itself. Similarly, the principal source of contract rights and duties between first- and second-tier subcontractors is the sub-subcontract, and so on down the subcontract chain. However, this source is not exclusive. As with any contract, the express terms will be supplemented by terms implied judicially into the subcontract relationship.

In addition, each contract on the subcontract chain can be and frequently is affected by contracts higher up the chain. The subcontract relationship is usually affected and may be controlled by terms in the prime contract. Correspondingly, second-tier subcontract relationships are affected and may be controlled by both first-tier subcontract and prime contract provisions. For convenience, discussion focuses on the relationship between prime contractor and first-tier subcontractor and the effect of the prime contract to which the subcontractor is not a party on that relationship.

The discussion in this section centers principally on the subcontractor's being bound to provisions of the prime contract because the latter is referred to or incorporated in the subcontract. However, it has been contended that a series of interrelated contracts can create a single contract binding all parties to the entire series even though each party may not have signed each contract in the series. Although such a contention was rejected in the context of an architect–owner contract followed by an owner–prime contractor contract,[35] a California decision intimated that

---

[33]*Commercial Indus. Constr. Co. v. Anderson*, 683 P.2d 378 (Colo. App.1984) (architect given adviser's privilege). See Section 14.08F.
[34]206 Ct.Cl. 203, 511 F.2d 548 (1975).

---

[35]*C. H. Leavell & Co. v. Glantz Contracting Corp.*, 322 F.Supp. 779 (E.D.La.1971).

such a conclusion might be sustained in a case involving a subcontract followed shortly by a prime contract.[36]

Even absent a one-contract theory, a court may look to the terms of the prime contract to discern the intent of the subcontract terms. In a Wyoming case, the city was required to give the prime contractor seven days to cure any defects; however, the subcontract permitted the contractor to repair the subcontractor's defective work only "after reasonable notice." The court, reading the two contracts together, ruled that the contractor breached the subcontract when it gave the subcontractor less than seven days to repair defective work.[37]

The owner and the prime contractor generally seek to bind the subcontractors at least to the performance provisions of the prime contract. The prime contractor's objective is to ensure that the entity it has selected to perform its obligations is obligated to do so. The owner's reason is less obvious. It could simply bind the contractor and let the prime take care of ensuring consistent performance obligation down the line. But the owner seeks proper performance and not claims against its prime contractor. It is most likely to obtain proper performance if the subcontractors down the line have committed themselves to do so. As an ancillary but much less important reason for obtaining such commitments from subcontractors, some owners might be seeking to perfect a claim against their subcontractors by asserting that they are intended beneficiaries of the subcontracts. This is discussed in Section 28.05B.

To tie subcontractors, the prime contract will usually contain a "flow-through," or conduit, clause. Such a clause requires the prime contractor to tie the subcontractors to provisions of the prime contract that affect their work. But flow-through clauses are not self-executing, the objective being accomplished only if the prime contractor does incorporate the prime contract into the subcontract. It is crucial to be aware of the maze of different contract documents and those administrative provisions, such as those that deal with disputes, that go beyond substantive performance obligations. To implement the objectives of owner and prime, the latter must obtain a promise by the subcontractor to perform the performance commitments of the prime and be bound to the administrative provisions, such as design professional decisions and arbitration. When a

total system is used, such as AIA Documents A101-201 and A401, this is done automatically. But if the prime uses its own subcontract or one negotiated with a subcontractor, the owner should check the ultimate subcontracts to be certain the prime contractor has done what it has obligated itself to do in the prime contract.

Implementation usually is accomplished by incorporation by reference. This must be distinguished from simply referring to the prime contract. If there is merely a reference to the prime contract, the subcontract will be interpreted in the light of the prime contract, or the prime will take precedence in the event of a conflict.[38] Although the prime contract—principally the general and supplementary conditions and technical writings—need not be physically attached to the subcontract, the subcontract should clearly state that the prime contract is incorporated by reference. If properly incorporated, it is part of the subcontract.[39]

Except for interpreting specifications or what work is included in the work of a particular specialty, debates over whether the subcontractor must comply with the performance aspects of the prime contract are rare. Most reported appellate decisions involve dispute resolution—principally the issue of whether particular disputes between prime and subcontractor must be resolved by arbitration. The cases have struggled with conflicting arbitration clauses in prime and subcontract, with transactions where arbitration is included in the prime but not in the subcontract, and even with contracts that do not contain an arbitration clause and do not even refer to another contract that does contain one.

Before examining specific cases, it is important to look at flow-through, or conduit, clauses. As noted, the principal reason for such a clause is to ensure that subcontractors commit themselves to the performance requirements of the prime contract. As an illustration, AIA Document A201-2007, Section 5.3, states that the contractor will "require each Subcontractor, to the extent of the Work to be performed by the Subcontractor, to be bound to the Contractor by the terms of the Contract Documents, and to assume toward the Contractor all the obligations and

---

[36]*Varco-Pruden, Inc. v. Hampshire Constr. Co.*, 50 Cal.App.3d 654, 123 Cal.Rptr. 606 (1975) (dictum).

[37]*Scherer Constr., LLC v. Hedquist Constr., Inc.*, 18 P.3d 645, 658 (Wyo.2001).

[38]*Oxford Dev. Corp. v. Rausauer Builders, Inc.*, 158 Ind.App. 622, 304 N.E.2d 211 (1973) (conflict over work to be performed); *United States v. Foster Constr. (Panama) S.A.*, 456 F.2d 250 (5th Cir.1972) (conflict over terms and conditions).

[39]*West Bank Steel Erectors Corp. v. Charles Carter & Co.*, 248 So.2d 52 (La.App.1971), held that a subcontractor was required to perform in accordance with specifications in the prime contract.

responsibilities which the Contractor . . . assumes toward the Owner and Architect."

But this clause goes beyond that objective. Under it, the *benefits* given the prime in A201 flow down to the subcontractor. For example Section 5.3 gives the subcontractor all the "rights, remedies and redress" against the contractor that the contractor has against the owner unless the subcontract specifically provides otherwise. For example, prime contracts usually contain *force majeure* provisions that excuse delayed performance if certain events occur. Such a provision for flow-through of benefits would give identical rights to the subcontractor if a claim is made against it by the prime contractor for delay.

A201's flow-through of benefits applies only if the subcontract does not specify otherwise. The flow-through provision may not be of much value to the subcontractor, because subcontracts frequently contain provisions less favorable to the subcontractor than the prime contractor has in its contract with the owner. Note that Section 5.3 requires the contractor to make available to proposed subcontractors, before execution of the contract, copies of the general conditions to which the subcontractor will be bound and "upon written request of the Subcontractor, identify to the Subcontractor terms and conditions of the proposed subcontract agreement which may be at variance with the Contract Documents." The AIA hopes to give the subcontractor some bargaining muscle when the prime contractor seeks to force provisions on the subcontractor that are harsher than those in the prime contract. Whether this will have any effect on the bargaining power in the actual subcontracts is debatable.

Provisions for flow-through of benefits raise problems. For example, AIA Document A201-2007, Section 2.2.1, gives the prime contractor a right to inquire into the owner's financial arrangements. Does the benefits flow-through provision give a similar right to the subcontractor to inquire into the financial sources of the prime contractor? The prime contractor may with justification contend that this power constitutes a trap for unwary prime contractors who may not realize how using a boilerplated flow-through provision can compel it to disclose sensitive information to subcontractors.

The attempt to give negotiation power to the subcontractor to augment its weak position at the bargaining table has other difficulties. For example, A201-2007, Article 15, sets up a dispute resolution system under which initially disputes are decided by the Initial Decision Maker,

subject to arbitration if chosen by the parties. If this is a *burden* under the prime contract, the subcontractor must accept this method of dispute resolution. If, however, it is a *benefit*, the subcontractor is entitled to it only if there is nothing specific in the subcontract to the contrary.

Similarly, the benefits and burdens dilemma can be illustrated by *Davlar Corp. v. Superior Court of Los Angeles County*.[40] In this case an insurer paid a loss to the prime and brought a subrogation claim against the subcontractor. The subcontractor defended by pointing to the waiver-of-subrogation clause in the prime contract and claiming that the prime contract was incorporated into the subcontract by reference. The insurer pointed to the absence of a waiver-of-subrogation clause in the subcontract and the inclusion of an indemnity clause in the subcontract. In essence it claimed the waiver was inconsistent with the indemnity clause and should not be incorporated into the subcontract.

The court did not find inconsistency. It held that the waiver covered insured losses, whereas the indemnity clause applied to noncovered claims, presumably property damage claims.

Structuring the provision for flow-through of benefits to apply only if nothing to the contrary is found in the subcontract raises the inevitable problem of seeking to determine when a benefit has been specifically excluded by the subcontract. Not only that, this attempt to give bargaining power to the subcontractor, as noted earlier in this section, seems to depend on a requirement in Section 5.3 that the contractor make available to proposed subcontractors copies of the prime contract and identify to the subcontractor "terms or conditions" of the subcontract that may be "at variance" with the prime contract.

Some prime contractors complain that requiring them to identify to the subcontractor provisions in the proposed subcontract, which may vary from the prime contract, requires that they act as the subcontractor's attorney.

This benefits flow-through provision demonstrates some of the difficulties in construction contract drafting. It seems attractive to include a clause that helps the subcontractor in its often difficult negotiations, if there are any at all, with the prime contractor. However, often drafters do not think about the possible applications of a general clause both as to substance and procedures. This

---

[40]53 Cal.App.4th 1121, 62 Cal.Rptr.2d 199 (1997).

well-meaning attempt to aid the subcontractors may only create more problems in construction contract administration and in dispute resolution.

As to specific cases, *Pioneer Industries v. Gevyn Construction Corp.*[41] involved legal action by a subcontractor against the prime contractor and its surety for work performed before the public awarding authority terminated the prime contract. The prime contractor disputed the termination and sought arbitration in accordance with the prime contract arbitration provision. However, the subcontractor sought to have the dispute decided by the court rather than by arbitrators. The prime contractor pointed to a provision in the subcontract that incorporated the arbitration clause of the prime contract into the subcontract. The court pointed to subcontract provisions, stating that any conflict between prime contract and subcontract would be controlled by the subcontract and that disputes arising out of termination would be arbitrated only if the dispute involved less than $3,500. Because the dispute in the lawsuit involved over $3,500, the court concluded that the dispute did not fall within the arbitration clause of the subcontract. This holding illustrates that the specific subcontract arbitration clause takes precedence over the incorporation by reference of the prime contract arbitration clause. The probably unintended result was an arbitration between prime contractor and owner and litigation between prime contractor and subcontractor.

The *Pioneer* case enforced the subcontract arbitration clause. However, the court in *John F. Harkins Co., Inc. v. The Waldinger Corp.*,[42] chose the arbitration clause in the prime contract. The subcontract contained a broad arbitration clause but also contained a clause stating that the subcontractor would be bound by all applicable provisions of the prime contract, the latter containing a much narrower arbitration clause.

After a dispute arose, the subcontractor sought to invoke the arbitration clause in the subcontract. The prime contractor contended it did not have to arbitrate, as the dispute was not arbitrable under the prime contract. The appellate court upheld a decision of the trial court that the dispute was not subject to arbitration.

The court pointed to preliminary negotiations that indicated that the subcontractor realized that the subcontract

was favorable to the prime contractor and sought unsuccessfully to obtain language that would give it the benefit of provisions in the prime contract. Evidence also showed that both parties intended that incorporating the prime contract would not give the subcontractor greater rights against the prime than the prime had under the prime contract. This conclusion, over a vigorous dissent, came in the face of federal law generally extending great scope to arbitration clauses.

Two cases held that arbitration would be required even though the subcontract did not contain an arbitration clause. The first, *Maxum Foundations, Inc. v. Salus Corporation*,[43] involved AIA Document A201, which incorporated the general conditions into the subcontract and also contained a flow-through provision. The court held that this was sufficient to incorporate the arbitration clause of A201 into the subcontract. Despite the prime contractor not having "passed through" prime contract obligations, the court concluded that all the evidence indicated that the clause was intended to be part of the subcontract, even though not expressly included in it. The court was influenced by the policy of federal courts favoring arbitration.

The second case, *John Ashe Associates v. Envirogenics Co.*,[44] involved a purchase order subcontract that did not even refer to the prime general conditions with its arbitration clause. Yet the subcontractor was bound to arbitrate because

1. The prime normally sent its prime general conditions with an arbitration clause to all prospective subcontractors.
2. The specifications and special conditions of the specifications referred to the prime general conditions.
3. The subcontractor referred to the prime general conditions to prepare its bid.

These reasons were held sufficient to put the subcontractor on notice that the arbitration clause of the general conditions was incorporated into the purchase order executed by the subcontractor.

Most of the cases noted have involved dispute resolution. As provisions seeking to limit claims—such as waivers

---

[41] 458 F.2d 582 (1st Cir. 1972).

[42] 796 F.2d 657 (3d Cir. 1986), cert. denied sub nom. *TWC Holdings, Inc. v. John F. Harkins Co.*, 479 U.S. 1059 (1987).

[43] 779 F.2d 974 (4th Cir. 1985), appeal after remand, 817 F.2d 1086 (4th Cir. 1987).

[44] 425 F.Supp. 238 (E.D.Pa. 1977).

of consequential damages as noted in Section 27.06D and "no-damage" clauses, discussed in Section 26.10A—become more common, there will be more attempts to pass them through to the subcontractor through a "flow-through" clause. The devastating effect such clauses can have on subcontractor claims, particularly when the subcontractor may not be aware of them or have participated in their creation, may mean courts will carefully scrutinize such "flow-through" provisions.[45]

These cases reveal some of the chaos that can occur in dispute resolution because of the subcontract system. Although, as indicated earlier in this section, there is no requirement that the prime and subcontracts be perfectly parallel, the risk of contradictory dispute resolution provisions or the absence of such provisions in one contract and their presence in another generates complicated issues. This makes disputes between prime and subcontractors extraordinarily difficult to settle and resolve.

SECTION 28.05   Subcontractor Defaults

A. Claims by Contract-Connected Parties

A subcontractor's unexcused refusal or failure to perform in accordance with its contract obligations can damage those with whom the subcontractor has contracted, such as the prime contractor or a lower-tier subcontractor. As shall be seen in Section 28.05C, a prime contractor may be accountable to the owner for defaults of the subcontractor. This can mean that the extent of the claim and the method of resolving any dispute over it may be affected by provisions in contracts other than the subcontract.

B. Claims by Third Parties

Some tort claims against a subcontractor are made by those who suffer personal harm. This section looks at claims against a subcontractor based on economic losses suffered by third parties, principally the owner.

Owners employ three theories to recover losses they have suffered that were caused by a subcontractor breach. First, an owner may contend that the subcontractor has breached its contract with the prime contractor—the contract intended for the benefit of the owner. Second, the owner may assert that the subcontractor's breach was tortious in that it failed to live up to the legal standard of care.[46] Third, the owner may contend that the prime contractor was merely a conduit between owner and subcontractor—essentially a contention based on the prime contractor's contracting as an agent of the owner.[47]

Generally, attempts have met with mixed success, with some courts allowing the owner to assert a claim as an intended beneficiary,[48] whereas others, taking a realistic look at the intention-to-benefit test, have not.[49] However, a prime contractor who began dealing with a sub-subcontractor because the subcontractor was having financial problems, was allowed to sue the sub-subcontractor for defective work.[50]

Because of the unsettled state of the law, the third party's principal claim is against the party with whom it has a contractual relationship and that party institutes action against the subcontractor. For example, if the owner has suffered a loss it is likely that the owner's action will be against the prime contractor and the prime contractor will bring an action against the subcontractor.

The third-party claim—that is, the direct claim against the subcontractor—may also be asserted. But the uncertainty of its success makes it ancillary to the principal claim unless the claim against the contract-connected party would be uncollectible.

---

[46]See cases cited in Sections 14.08C, D, and E.

[47]*National Cash Register Co. v. UNARCO Industries, Inc.*, 490 F.2d 285 (7th Cir.1974) (dictum that owner could sue as principal of an agency relationship but claim allowed under subrogation); *Seither & Cherry Co. v. Illinois Bank Bldg. Co.*, infra note 56.

[48]*Chestnut Hill Dev. Corp. v. Otis Elevator*, 653 F.Supp. 927 (D.Mass.1987) (ambiguity because clause that allowed the owner to assume the contractor's rights did not appear in the subcontract, but subcontract required sub to perform strictly in accordance with prime contract); *United States v. Ogden Technologies Laboratories, Inc.*, 406 F.Supp. 1090 (E.D.N.Y.1973); *Gilbert Financial Corp. v. Steelform Contracting Co.*, 82 Cal.App.3d 65, 145 Cal.Rptr. 448 (1978).

[49]*Vogel Bros. Building Co. v. Scarborough Constructors, Inc.*, 513 So.2d 260 (Fla.Dist.Ct.App.1987) (owner could not compel prime and sub to arbitrate); *Vogel v. Reed Supply Co.*, 277 N.C. 119, 177 S.E.2d 273 (1970); *Manor Junior College v. Kaller's, Inc.*, 352 Pa.Super. 310, 507 A.2d 1245 (1986).

[50]*White Constr. Co., Inc. v. Sauter Constr. Co.*, 731 P.2d 784 (Colo. App.1987).

---

[45]But see *L. & B. Constr. Co. v. Regan Enterp., Inc.*, 267 Ga. 809, 482 S.E.2d 279 (1997), which held a subcontractor bound to a "no-damage" clause that flowed through to its subcontract and barred its claim.

C. Responsibility of Prime Contractor

The subcontracting process permits the prime contractor to discharge its obligations by using a subcontractor. However, this power to perform through another does not relieve the prime contractor of responsibility for subcontractor nonperformance.[51] Most standard construction contracts make the prime contractor responsible for subcontractor defaults.[52]

In *Norair Engineering Corp. v. St. Joseph's Hospital, Inc.*,[53] one issue involved the responsibility for defective work performed by a subcontractor who was hired at the owner's demand by the contractor who had been brought in by the bonding company to replace the original prime. The court concluded that the contractor was well aware that it was taking the risk of the subcontractor's default when it entered into the prime contract, that it was not without remedies against the subcontractor, and that it would never have been awarded the contract to complete the work unless it had agreed to the provisions under which the owner could select and approve subcontractors.

Owners can exercise a variety of controls over the selection of subcontractors.[54] That the owner approves a proposed subcontractor should not relieve the prime contractor of responsibility. Clearly, this is a matter for negotiation, and the law should enforce the choice that the contracting parties have made. Where there is no clear evidence one way or the other, a strong argument can be made that the owner *dictating* the use of a particular subcontractor should relieve the prime contractor of responsibility.[55] In such a case, the prime contractor may be acting as an agent of the owner to engage a particular subcontractor and should not be responsible unless it knew or should have known that the subcontractor was incompetent.[56] If the owner steps in and takes away the prime contractor's power to manage the subcontractors, the prime contractor cannot be held accountable for the work of the subcontractors. Its obligation is conditioned on its right to manage the job without unreasonable interference.[57]

Sometimes the prime contractor contends that it should not be held responsible, pointing to the independent contractor rule. This rule, though subject to many exceptions, relieves the employer of an independent contractor from responsibility from the latter's improper performance.[58] However, this defense, where available, does not relieve the employer of the independent contractor when the independent contractor's failure to perform properly has damaged a party to whom the employer of the independent contractor owed a contract right of proper performance.[59]

This section has looked largely at judicial claims against the prime contractor based on conduct of the subcontractor. However, increasing attention is being directed toward claims by public safety officials against the prime contractor under powers granted by the Occupational Safety and Health Act (OSHA), a federal law that regulates the workplace. An OSHA citation was issued against a prime contractor for defective work performed by its electrical subcontractor.[60] The citation was based on the conclusion that the prime should have general knowledge of electrical requirements. Although the prime was entitled to rely on the knowledge and expertise of its subcontractors, the OSHA Review Commission concluded that the prime did not show evidence of its reasonable reliance on the subcontractor. This may mean that a prime contractor will be accountable under OSHA for defective work performed by specialty trades unless it obtains some written acknowledgment from the subcontractors that requirements are being complied with and that the prime is relying on the subcontractor's expertise.

---

[51]*Kahn v. Prahl*, 414 S.W.2d 269 (Mo.1967); *Waterway Terminals Co. v. P. S. Lord Mechanical Contractors*, 242 Or. 1, 406 P.2d 556 (1965) (prime responsible for subcontractors).

[52]AIA Document A201-2007, § 5.3. This is aided by requiring that the prime contractor obtain agreements from subcontractors that preserve and protect the rights of the owner "so that the subcontracting thereof will not prejudice such rights."

[53]147 Ga.App. 595, 249 S.E.2d 642 (1978).

[54]See Section 28.03.

[55]*Cf. National Cash Register Co. v. UNARCO Industries, Inc.*, supra note 47. Here the owner was allowed to sue the subcontractor that it ordered the prime contractor to use. Very likely the prime would be relieved of responsibility.

[56]*Seither & Cherry Co. v. Illinois Bldg. Corp.*, 95 Ill.App.3d 191, 419 N.E.2d 940 (1981) (CM agent of owner in hiring contractor).

[57]See Section 19.02A.

[58]See Section 7.03G.

[59]*Harold A. Newman Co. v. Nero*, 31 Cal.App.3d 490, 107 Cal.Rptr. 464 (1973); *Brooks v. Hayes*, 133 Wis.2d 228, 395 N.W.2d 167 (1986) (quoting and following earlier edition of this book).

[60]*Secretary of Labor v. Blount Int. Ltd.*, 15 O.S.H.Cas. (BNA) ¶ 1897, 1992 O.S.H.D. (CCH) ¶ 29,854 (O.S.H.Rev.Comm'n 1992).

SECTION 28.06 Payment Claims Against Prime Contractor

One particularly sensitive area in the subcontract relationship relates to money flow, with subcontractors frequently contending that they invest substantial funds in their performance and are entitled to be paid as they work and to be completely paid when their work is completed. This issue surfaces around two concepts: line item retention and payment conditions. The subcontractors wish to divorce their work and payment for it from the prime contract. They argue *for* line item retention and *against* payment conditions.

Line item retention gives the subcontractor the right to be paid after it has fully performed. Delay usually occurs because the owner holds back retainage—a designated amount of the contract price for the entire performance, that of prime contractor and all subcontractors. When all the work is completed and the project accepted, the owner will pay the retainage. However, early finishing subcontractors may have to wait a substantial period of time after they have completed their work for the retention allocated to their contract because the entire project is not yet completed. They would like to disassociate their contract from the rest of the subcontracts and the prime contract.

A payment condition ("pay when paid" clause or more realistically, from the position of the prime, "pay *only* if paid") makes payment to the prime contractor a condition of the prime contractor's obligation to pay the subcontractor. Prime contractors seek to create such a condition by including language in the subcontract stating that the prime contractor will pay "if paid by the owner," "when paid by the owner," or "as paid by the owner." An endless number of cases have interpreted this language. Does the language create a condition to payment or simply indicate that the payment flow contemplated some delay, with payment in any event being required after a reasonable time expires?

All courts agree, or at least so they state, that the parties can make a payment condition under which the subcontractor assumes the risk that it will not be paid for its work. The legal issue has centered around the requisite degree of specificity needed to create such a condition. On the surface, the courts simply examine the language and seek to determine the intention of the parties. Other factors are operating, however. This section looks at what

courts have done, discusses some reasons for such clauses, and concludes by looking briefly at the system used in the AIA standard documents.

In 1933, New York held that language of the type that appears to tie the subcontractor's right to payment to the contractor's receiving payment created a payment condition precluding the subcontractor from recovering when the prime contractor had not been paid.[61] In 1962 the court in the leading case of *Thomas J. Dyer Co. v. Bishop International Engineering Co.*,[62] came to a different conclusion. The court stated that performing parties usually expect to be paid for their work. Any result under which a party would suffer a forfeiture (that is, performing the work and not being paid) must be supported by clear and specific language that this risk has been taken. The court stated that the subcontractor, in addition to its mechanics' lien protection, contracts mainly on the basis of the solvency of the prime contractor. To change this normal credit risk, the contract should "contain an express condition clearly showing that to be the intention of the parties."[63] The court concluded that the language dealing with linkage of payment to payment to the prime contractor was designed "to postpone payment for a reasonable period of time after the work was completed, during which the general contractor would be afforded the opportunity of procuring from the owner the funds necessary to pay the subcontractor."[64]

Most recent cases have followed the reasoning in the *Dyer* case.[65] But some decisions have concluded that the language showed a payment condition had been created.[66] Courts here are not simply construing contract language. The language used usually indicates that the prime contractor need not pay *until* or even *if* it is paid. Courts have

---

[61]*Mascioni v. I. B. Miller, Inc.*, 261 N.Y. 1, 184 N.E. 473 (1933).

[62]303 F.2d 655 (6th Cir.1962) (Ohio law).

[63]Id. at 661.

[64]Ibid.

[65]Alsbrook, *Contracting Away an Honest Day's Pay: An Examination of Conditional Payment Clauses in Construction Contracts*, 58 Ark.L.Rev. 353 (2005) and Hendricks, Spangler & Wedge, *Battling for the Bucks: The Great Contingency Clause Debate*, 16 Constr. Lawyer, No. 3, July 1996, pp. 12, 16–20. See also *Federal Ins. Co. v. I. Kruger, Inc.*, 829 So.2d 732 (Ala.2002) (clause is only a timing mechanism).

[66]Hendricks, supra note 65, pp. 23–24; *Associated Mechanical Corp., Inc. v. Martin K. Eby Constr. Co., Inc.*, 67 F.Supp.2d 1375 (M.D.Ga. 1999), aff'd, 271 F.3d 1309 (11th Cir.2001) (pay-when-paid clause prevents subcontractor from receiving interest on retainage, without regard to prime's multi-year delay in suing the owner).

tilted the balance toward the subcontractor because of the feeling that the subcontractor has little choice as to the contract language, that the subcontractor has no opportunity to evaluate the solvency of the owner, and that it would be unfair for subcontractors to have to bear this risk when the prime contractor is in the best position to evaluate the credit of the owner.

The *reason* for nonpayment is important. *Some* cases should be set aside. First, if the nonpayment to the prime contractor results from the *subcontractor's* failure to perform properly, the subcontractor is not entitled to recovery. Second, if the reason for nonpayment is improper performance by the prime contractor or an unwillingness to make reasonable efforts to obtain payment, the subcontractor is entitled to be paid. Even if a condition to payment *has* been created, preventing its occurrence or not taking reasonable efforts to make it occur excuses the condition, allowing the subcontractor to recover.[67]

The more difficult questions involve the owner's nonpayment under circumstances for which neither the prime contractor nor the subcontractor can be held directly responsible. Suppose the nonpayment results from the design professional's refusal to issue a payment certificate or revocation of a previously issued certificate.[68] For example, suppose work by *another* subcontractor had been previously accepted but was discovered to be defective after payment to the subcontractor. In such a case, the latter's correction of the work will start the money flow again. Under these circumstances, the prime contractor should be required to pay the unpaid subcontractor who *had* performed properly. The prime contractor hired the subcontractor whose performance was improper and is in the best position to put pressure on *that* subcontractor to correct the defective work. In addition, the prime contractor could have required the subcontractor whose work had been found to be defective, to post a performance bond. This would have entitled the prime contractor to look to the surety for the nonperformance. Allowing the prime contractor to refuse to pay the unpaid subcontractor whose work was properly performed would also strengthen the already strong bargaining position the prime contractor is likely to

have. This can give the latter additional leverage to force settlement of disputes between prime and unpaid contractor even for projects unaffected by the payment dispute.

Another difficult problem involves nonpayment due to owner insolvency or bankruptcy. Generally, such risks should be borne by the prime contractor.[69] The prime has entered into the contract with the owner and is in the best position to make financial capacity judgments.[70] Although the subcontractor's willingness to perform work may have rested to some degree on its evaluation of the financial position of the owner, it would not be sufficient to conclude that the prime contractor has transferred the risk to its subcontractors in the absence of a clear showing that this was intended.

Other factors exist that may determine whether a payment condition has been created. If the prime contractor has invested substantial amounts of money into the project itself, it can more persuasively contend that it and the unpaid subcontractor should share the loss. If the prime contractor has invested very little of its own money, a stronger argument can be made for refusing to find a payment condition, the prime contractor in such a case acting as a broker who should take the risk of the owner's insolvency.

Suppose the unpaid subcontractor would have a right to a mechanics' lien if it has improved the owner's property. Some recent cases have held that even clear language indicating a condition precedent has been created will not provide the prime contractor a defense. To do so would frustrate mechanics' lien laws.[71]

Although it can be argued that a mechanics' lien is a security interest based on a valid contract claim, these courts have been emphasizing the lien as being based on unjust enrichment. It takes precedence over a payment

---

[67]*Urban Masonry v. N & N Contractors, Inc.*, 676 A.2d 26 (D.C.App.1996) (sub recovers as prime and owner made a "walkaway settlement" that frustrated the condition).

[68]*R. C. Small & Assoc., Inc. v. Southern Mechanical, Inc.*, 730 S.W.2d 100 (Tex.Ct.App.1986) (certificate not a condition precedent).

[69]*Pacific Lining Co. v. Algernon Blair Constr. Co.*, 819 F.2d 602 (5th Cir.1987) (payment to prime not a condition precedent).

[70]This conclusion would be strengthened under AIA Document A201-2007, § 2.2.1. Under it, a contractor can require the owner to furnish the contractor reasonable evidence that the owner has made financial arrangements to fulfill its obligation.

[71]Hendricks, supra note 65 pp. 20–22; *Wm. R. Clarke Corp. v. Safeco Ins. Co. of Am.*, 15 Cal.4th 882, 938 P.2d 372, 376–77, 64 Cal.Rptr.2d 578 (1997), adopting the reasoning of *West-Fair Elec. Contractors v. Aetna Cas. & Sur. Co.*, 87 N.Y.2d 148, 661 N.E.2d 967, 971, 638 N.Y.S.2d 394 (1995). But if the parties choose that a project located in New York will be governed by the law of another state that *does* enforce pay-when-paid clauses, New York courts will enforce the clause. *Welsbach Electric Corp. v. MasTec North Am., Inc.*, 7 N.Y.3d 624, 859 N.E.2d 498, 825 N.Y.S.2d 692 (2006).

condition clause. In effect, the mechanics' lien has a life of its own.

How does language that seeks to control the subcontractor's right to be paid despite nonpayment to the prime interact with any surety payment bonds that have been issued? Of course, if the prime has the legal obligation to pay, the surety is responsible. But the more difficult cases have been ones involving a claim against a surety where the prime has a defense. As usual, the cases are divided, although much will depend on the language of the bond and whether the bond has been incorporated into the subcontract.[72] On federal projects, courts view the Miller Act payment bond as providing subcontractors with a separate and independent guarantee of payment. They allow the subcontractor recovery against the bond even if the subcontract contains a pay when paid clause shielding the prime contractor from liability.[73]

The AIA is not entirely clear on whether it allows the prime contractor to delay payment to the subcontractor until the prime receives payment from the owner. AIA Document A201-2007, Section 9.6.2, requires the contractor to pay each subcontractor "no later than seven days after receipt of payment from the Owner." This language leaves unanswered the subcontractor's rights if the owner fails to pay the prime contractor through no fault of the subcontractor.

Yet A401-2007, the AIA's standard construction subcontract, appears to provide more succor to subcontractors. Sections 11.3 and 12.1 both state that if the contractor does not receive payment "for any cause which is not the fault of the Subcontractor," then the contractor shall pay the subcontractor "on demand." (Under Section 11.3, the subcontractor would receive a progress payment; under Section 12.1 it would receive final payment.) In addition,

Section 4.7 provides that if the contractor is seven days late in paying the subcontractor through no fault of the subcontractor, then the subcontractor, "upon seven additional days' written notice to the Contractor," may stop the work until it has received the payment due, plus "reasonable costs of demobilization, delay and remobilization."

In light of the A401 provisions, a prime contractor who has not received timely payment from the owner pertaining to the subcontractor's work has little option but to pay the subcontractor out of the contractor's own funds. Most prime contractors would rather advance payment to a fault-free subcontractor than risk that subcontractor's stopping work.

The struggle between primes and subcontractors in some states has moved to the legislative arena.

The statutes take many forms. North Carolina will not enforce either a "pay when paid" or "pay if paid" clause.[74] Wisconsin voids any payment condition (i.e., "pay if paid" clause) while also permitting clauses which delay payment of the subcontractor until the contractor receives payment.[75] Missouri does not directly invalidate payment conditions but limits their effectiveness by stating that they are not a defense to a mechanics' lien.[76] Maryland, in addition to protecting the lien rights in the face of the nonoccurrence of the payment condition, also bars the payment condition defense to the surety.[77]

SECTION 28.07 Payment Claims Against Property, Funds, or Entities Other Than Prime Contractor

A. Court Judgments and Specific Remedies

Legal relief generally comes in the form of court-issued judgments. Most judgments simply state that the judgment creditor (the person who seeks and obtains the judgment) is entitled to a specific amount of money from the judgment debtor (the person against whom the judgment is issued). The judgment creditor is given methods of collecting on these judgments, but as mentioned in Section 2.13, such remedies are often cumbersome and ineffective. If the

---

[72]*Hendricks*, supra note 65, pp. 25–27; see also *Moore Bros. Co. v. Brown & Root, Inc.*, 207 F.3d 717 (4th Cir.2000) (Virginia law), permitting a subcontractor claim against the payment bond even though the pay-when-paid clause would have shielded the prime contractor from liability. But see *Wellington Power Corp. v. CAN Surety Corp.*, 217 W.Va. 33, 614 S.E.2d 680 (2005), permitting a surety to invoke a pay-when-paid clause as a defense to a subcontractor's claim.

[73]See *United States v. Weststar Eng'g, Inc.*, 290 F.3d 1199 (9th Cir.2002) and Kanzler & McCarter, *The Effect of a Disputes Clause and Pay-If-Paid Provisions on Miller Act Claims and the Types of Damages Recoverable Under the Act*, 21 Constr. Lawyer, No. 3, Summer 2001, p. 28. The Miller Act requires prime contractors on federal projects to post a payment bond for the benefit of subcontractors, who cannot file mechanics' liens on federal buildings. See Section 28.07I.

[74]N.C.Gen.Stat. § 22C-2.

[75]West Ann.Wis.Stat. § 799.135(3).

[76]Mo. Rev.Stat. § 431.183.

[77]Md.Real Prop. Ann.Code § 9-113(b).

judgment debtor does not pay voluntarily, collection problems sometimes make the judgment worthless.

Specific remedies, in contrast, are remedies that either command the defendant to do or not do a particular act (specific performance decrees or injunctions) or operate against specific property (liens against particular funds, goods, or property). Specific remedies are equitable decrees backed up by the contempt power of the court. The judge can punish a person adjudged to be in contempt of court by a fine or imprisonment. See Section 6.03. Specific remedies are more effective, as a rule, than the ordinary money award court judgment.

One reason unpaid subcontractors seek mechanics' liens[78] or other specific remedies is that they operate against particular property and are more effective than ordinary judgments, whether obtained against the prime contractor or third parties such as the owner. Sometimes even specific remedies are ineffective, but on the whole they are preferable to ordinary court judgments.

B. Statutory and Nonstatutory Remedies

Most of the material discussed in this section, such as mechanics' liens, stop notices, and trust fund protection, have been created by the state legislatures. Many bonds are required on public work because of legislative enactments. The statutory nature of these remedies has added complications.

Reported appellate cases that seem contradictory are often traceable to the variant statutes before the courts. Statutes change frequently. Compliance is difficult, and the law becomes murky. Although state statutes follow broad patterns of similarity, variance in details places a heavy burden on those contractors who operate in different states and may in part account for the local character of the construction industry.

Much of the law in this area involves interpretation of these frequently complex state statutes. Such state statutes create many requirements for the creation of lien rights or stop notice protection. Who will be accorded statutory protection, how such protection is achieved, and the nature of the protection are often resolved by reference to the complicated, almost unreadable statutes. Frequently, such interpretation questions are resolved by holding that the statutes are designed to prevent unjust enrichment by protecting unpaid subcontractors, the intended beneficiaries of the legislative protection. Matters can become even more complicated in jurisdictions that state that the standards for perfecting a mechanics' lien will be *strictly* required, but once the lien has been perfected, the remedy will be administered *liberally*.[79] Strained interpretations and language distortion often result, yet construing these statutes to protect subcontractors does not invariably result in lien protection. The principal legacy of this approach is legal uncertainty and unpredictability.

The extensiveness of the statutory system can make it more difficult for subcontractors to assert nonstatutory claims. Where the subcontractor has been given statutory protection but did not take the necessary steps to obtain it, courts sometimes deny the nonstatutory claim.[80] Although the statutory systems are not exclusive, their existence can persuade courts that they offer sufficient protection to justify denial of a nonstatutory claim.

C. Public and Private Work

The nature of the remedy available may depend on whether the work is public or private. Mechanics' liens, to the extent that they permit foreclosure rights against public buildings, are not available. As a result, the federal government enacted the Miller Act, a compulsory prime bonding system. Many states have enacted comparable legislation for state construction projects.

The demarcation between public and private work is not always clear. Municipalities may confer public benefits to lure private businesses to build within the city limits. These benefits may include direct cash grants, reduced property taxes on the improved property, allowing the

---

[78]The many mechanics' lien laws make generalizations perilous. For example, a mechanics' lien generally creates only a security interest in the property improved by the lien claimant. However, Florida permits a lien claimant a personal judgment against the owner even if there is no contract relationship between them. West Ann.Fla.Stat. § 713.75. Similarly, what are called "liens" against public buildings may only be liens against funds. See Section 28.07E. Also, Texas gives a lien on the retainage. Tex.Prop.Code Ann. § 53.103.

[79]*Talco Capital Corp. v. State Underground Parking Comm'n*, 41 Ohio App.2d 171, 324 N.E.2d 762 (1974).

[80]*Engle Acoustic & Tile, Inc. v. Grenfell*, 223 So.2d 613 (Miss.1969); *Banks v. City of Cincinnati*, 31 Ohio App.3d 54, 508 N.E.2d 966 (1987). See, generally, Comment, *Mississippi Law Governing Private Construction Projects: Some Problems and Proposals*, 47 Miss.L.J. 437 (1976).

proprietor to divert sales taxes to pay off its construction costs, or publicly built infrastructure whose sole purpose is to benefit the new development. These arrangements raise the legal question of whether the project is a private or public work.[81] In *North Bay Construction, Inc. v. City of Petaluma*,[82] a city leased its land to a developer to build a sports complex. The court ruled that, because the underlying land was publicly owned, sovereign immunity barred the filing of mechanics' liens.

Sometimes systems developed in public work spill over into private construction. For example, stop notices (discussed in Section 28.07E) were originally developed to compensate contractors for public work because they were not accorded lien rights. However, in those states that have stop notices, the stop notice is often available for private work.

D. Mechanics' Liens

Mechanics' lien laws are complicated and vary considerably from state to state.[83] For that reason, it would be inadvisable to attempt a summary of all aspects of these statutory protections accorded certain participants in the construction process. Instead, this discussion focuses on rationales for such protection and salient features and current criticisms of lien laws.

Participants in the construction process who can in various ways trace their labor and materials into property improvements of another are given lien rights against the property in the event they are not paid by the party who has promised to pay them. The remedy accorded a lien holder is the right to demand a judicial foreclosure or sale of the property and be paid out of the proceeds, including, in some states, the legal costs of perfecting the lien, such as attorneys' fees. Section 28.06 treated the effect of liens on payment conditions in subcontracts.

Among the many lien recipients are prime contractors, subcontractors, suppliers, laborers, and design professionals.[84] Lien claimants are divided into two principal categories, those who have direct contract relations with the owner and those who do not. Typical illustrations of the first are design professionals and prime contractors. Illustrations of the second are subcontractors, laborers, and suppliers to prime contractors and subcontractors. Owners can avoid liens by paying their design professionals and prime contractors. But because the majority of the difficult problems are generated by lien claimants not connected by contract with the owner and because this chapter focuses on subcontractor problems, the discussion centers around the second class of lien claimants. The usual justification given for granting these liens is unjust enrichment. Those whose labor or materials have gone into the property of another should have lien rights in the property when they are not paid as promised.

The unjust enrichment rationale loses some of its attractiveness when "double payment" is considered. An owner who pays the prime contractor may have to pay again to an unpaid subcontractor if it wants to remove the lien. States increasingly protect residential owners from double payment. Unpaid Michigan subcontractors are given a claim against a state-administered fund created by payments of licensed contractors.[85]

Mechanics' lien laws provide a quick and effective remedy for unpaid workers who cannot wait until a full trial to collect their wages. Quick and certain remedies can induce workers to work on construction by assuring them they will be paid.[86]

Amplification of this inducement so vital to a developing country's lien laws were first enacted to spur development of Washington D.C. more than 200 years ago could and did lead to expansion of lien beneficiaries

---

[81]For example, in *City of Long Beach v. Department of Industrial Relations*, 34 Cal.4th 942, 102 P.3d 904, 22 Cal.Rptr.3d 518 (2004), a city gave the Society for the Prevention of Cruelty to Animals money to cover preconstruction costs for the building of a new animal control center. The court found the construction was not a public work and so not subject to the prevailing wage act.

[82]143 Cal.App.4th 552, 49 Cal.Rptr.3d 455 (2006).

[83]The National Conference of Commissioners of Uniform State Laws has proposed a new "Uniform Construction Lien Act." One of its features would be to protect against double payment by the owner, a failing of many mechanics' lien laws. Only one state has adopted it in full. See Neb.Rev.Stat. § 52-125 to 52-159. An earlier attempt at uniformity failed, no state adopting the proposed act. The commissioners withdrew the earlier act. See Benfield, *The Uniform Construction Lien Act: What, Whither, and Why*, 27 Wake Forest L. Rev. 527 (1991).

[84]Architects or engineers who are involved in the planning stage of a project may find their services are not lienable, see *New England Sav. Bank v. Meadow Lakes Realty Co.*, 243 Conn. 601, 706 A.2d 465 (1998) (services lienable after a statutory amendment). For a review of the lien rights of design professionals, see Annot., 31 A.L.R.5th 664 (1995). See Section 11.02B.

[85]See Section 28.07G.

[86]*Judd Fire Protection, Inc. v. Davidson*, 138 Md.App. 654, 773 A.2d 573 (2001).

to include not only laborers but also all those who participate directly in the construction process. The state gives credit to prime contractors by granting subcontractors lien rights, which encourages people to furnish labor and materials for construction. This state credit was especially needed to bolster an unstable construction industry composed of many contractors unwilling or unable to pay subcontractors and suppliers. This support is probably the principal reason for giving lien rights today.

Expansion of lien laws is undoubtedly traceable to the realities of the political process. Once some participants in the construction process had received lien rights on a frequently asserted unjust enrichment theory, it was relatively easy to expand the list of lien beneficiaries. Those who might oppose lien expansion, such as owners, are often unrepresented as an organized group in the legislatures. This too may have accounted for expansion of lien beneficiaries and lien rights.

The desire by participants to expand mechanics' lien rights is understandable, because the mechanics' lien is a more effective remedy than a money award. However, legislatures that respond to such pressures often ignore the fact that these liens come at the expense of others: In the case of subcontractor liens, they come at the expense of the unsecured creditors of prime contractors; in the case of prime contractor liens, they come at the expense of the unsecured creditors of the owner.

Some salient characteristics of lien laws have been mentioned. A lien is a security interest in the property and can be foreclosed on by the lien claimant. Some states known as having New York-type statutes limit the amount of the lien to the unpaid balance owed the prime contractor by the owner. Some states known as having Pennsylvania-type statutes do not place a limit of this type on the lien. States with an open-ended lien permit the owner to limit the lien to the contract price by filing a copy of the contract and posting a bond—something rarely done.

Although owners typically set the lien-creating events into motion by contracting with prime contractors, tenants can create liens by hiring contractors. Property owners can avoid such liens by posting a notice of nonresponsibility within a designated time after the owners learn the improvement is being made.[87]

Those entitled to liens are usually set forth in the statute, and the list typically is lengthy and expanding. Because of the lien statutes, whether particular work qualifies for a lien is often unclear. Some statutes use generic terms such as *improvement*, *building*, or *structure*. Others that attempt to be detailed do not always keep up with changes in the construction process. For example, liens have been denied where the lien claimant had placed engineering stakes and markers;[88] where a claimant had graded and installed storm and sanitary sewers, paving, curbing, and seating;[89] where a claimant had performed demolition work;[90] where a claimant had installed a swimming pool;[91] where a claimant had performed electrical work on a modular home erected at a factory;[92] and where an entity had helped subcontractors assemble a workforce, had dealt with payroll, and had advanced funds.[93] Most states deny liens to lessors of equipment used in construction unless the items are consumed in the process of use.[94] Not all work that would be considered part of the

---

[87]A notice of nonresponsibility was found ineffective under the "participating owner" *doctrine in Howard S. Wright Constr. Co. v. Superior Court,* 106 Cal.App.4th 314, 130 Cal.Rptr.2d 641 (2003), review denied

Jun. 11, 2003. In subsequent litigation, the court ruled that the tenant's anticipatory breach of the construction contract entitled the contractor to immediately file a mechanics' lien. *Howard S. Wright Constr. Co. v. BBIC Investors, LLC,* 136 Cal.App.4th 228, 38 Cal.Rptr.3d 769 (2006), review denied Apr. 26, 2006.

[88]*South Bay Eng'g Corp. v. Citizens Sav. & Loan Ass'n,* 51 Cal.App.3d 453, 124 Cal.Rptr. 221 (1975).

[89]*Sampson-Miller Assoc. Companies v. Landmark Realty Co.,* 224 Pa.Super. 25, 303 A.2d 43 (1973). The court reached its result reluctantly and urged that the legislature liberalize the statute to permit liens for work similar to that done by the claimant. The court noted a number of states where liens are available for preliminary work, such as California, Hawaii, Texas, and Illinois.

[90]*John F. Bushelman Co. v. Troxell,* 44 Ohio App.2d 365, 338 N.E.2d 780 (1975). The court noted that it could not extend the right to demolition work but left the problem for legislative action.

[91]*Freeform Pools, Inc. v. Strawbridge Home for Boys, Inc.,* 228 Md. 297, 179 A.2d 683 (1962). The statute gave liens for buildings, and the court concluded that a swimming pool was not a building.

[92]*C & W Elec., Inc. v. Casa Dorado Corp.,* 34 Colo.App. 117, 523 P.2d 137 (1974). The court noted that no owner of real property had requested that the work be performed, and this was a requirement of the statute.

[93]*Primo Team v. Blake Constr. Co.,* 3 Cal.App.4th 801, 4 Cal.Rptr.2d 701 (1992) and *Onsite Eng'g & Management, Inc. v. Illinois Tool Works, Inc.,* 319 Ill.App.3d 362, 744 N.E.2d 928 (2001).

[94]Annot., 3 A.L.R.3d 573, 578 (1965). In California, the seller of equipment who retains a security interest cannot obtain a mechanics' lien. *Davies Machinery Co. v. Pine Mountain Club, Inc.,* 39 Cal.App.3d 18, 113 Cal.Rptr. 784 (1974). But Louisiana granted a lien for nails, lumber, and plyform consumed in temporary work. *Slagle-Johnson Lumber Co. v. Landis Constr. Co.,* 379 So.2d 479 (La.1979).

construction process can qualify for a lien.[95] In short, both the identity of the claimant and the type of service or item that may be liened can be subject to dispute.[96]

Lien laws typically create an obstacle course of technical requirements for claimants. Generally, any failure to comply with these requirements will invalidate the lien. Substantial compliance is insufficient.[97]

If the lien amounts exceed the value of the property after parties with security interests that take priority are paid, all claimants are treated equally. In some states, liens of prime contractors are subordinated to other lien claimants, whereas in other states, laborers are sometimes given preference.

As between lien claimants and others with security interests in the property, such as the seller of the property who retains a security interest or a construction lender, the party who perfects its interest first takes priority. For this reason, lenders will not make construction loans if work has begun on the project, for fear that their security interest will not take priority over those who have already begun work. As a rule, lenders and other security holders in the land perfect their security interest before work begins. As a result, lien claims can become valueless if trouble develops and prior security holders foreclose on the property. This occurs frequently because of market imperfections. The lender is typically able to buy in at less than the amount owing on the construction loan because the liens of other claimants are not extensive enough to justify bidding in or they may not have sufficient funds to be able to compete with the lender.

To deal with this problem, New Hampshire enacted a statute that with some exceptions gives mechanics' lien claimants priority over construction lenders.[98]

Lien claimants must establish that they have performed work under the terms of a valid contract. As has been emphasized throughout this book, there has been a proliferation of legal controls on the construction process. (Licensing laws, land use controls, building and housing codes, and the controls imposed on projects built in part with public funds are illustrations.) As a result, a lien may be denied because of a technical violation of a law or regulation.[99]

Mechanics' lien laws have been attacked as violating the U.S. Constitution. Attacks were based on decisions of the U.S. Supreme Court that had invalidated certain remedies granted before a full trial as depriving the defendants against whom the remedy was allowed the use of their property without due process of law. Generally, such attacks have been unsuccessful in part because the Supreme Court in *Spielman-Fond, Inc. v. Hanson's, Inc.*[100] summarily upheld Arizona's mechanics' lien laws from constitutional atttack.[101] Courts have held that the owner's deprivation of the use of its property is not substantial. Although marketability may be somewhat impaired, there is no interference with the owner's possession. Courts that have not sustained these attacks have emphasized that the work of the claimants has improved the value of the property and that the owner can force an expeditious adjudication of the lien claim within a short period.

The question of the constitutionality of mechanics' lien laws was brought into renewed focus with the Supreme Court's issuance of *Connecticut v. Doehr*.[102] A plaintiff brought a civil action for assault and battery against Mr. Doehr and, as permitted by Connecticut law, also filed a prejudgment "attachment"—a form of lien—on Doehr's home, even though the plaintiff did not have a preexisting interest in Doehr's real estate. The

---

[95]Mechanics' liens were denied in the following cases: *Nickel Mine Brook Assocs. v. Joseph E. Sakal, P.C.*, 217 Conn. 361, 585 A.2d 1210 (1991) (legal services for developer); *Thompson & Peck, Inc. v. Division Drywall, Inc.*, 241 Conn. 370, 696 A.2d 326 (1997) (sub's unpaid insurance premiums); *CIT Group/Equipment Financing, Inc. v. Horizon Potash Corp.*, 118 N.M. 665, 884 P.2d 821 (App.1994) (workers' compensation premium); and *Integon Indemnity Corp. v. Bull*, 311 Ark. 61, 842 S.W.2d 1 (1992) (creditor who paid subs directly on behalf of prime contractor).

[96]Zimmerman & Orien, *Can I Lien That?* 27 Constr. Lawyer, No. 4, Fall 2007, p. 28.

[97]*IGA Aluminum Products, Inc. v. Manufacturers Bank*, 130 Cal. App.3d 699, 181 Cal.Rptr. 859 (1982); *Ng Bros. Constr., Inc. v. Cranney*, 436 Mass. 638, 766 N.E.2d 864 (2002); *Ryan Contracting, Inc. v. JAG Investments, Inc.*, 634 N.W.2d 176 (Minn.2001).

[98]N.H.Rev.Stat.Ann. 447:12-a.

[99]In *Stokes v. Millen Roofing Co.*, 466 Mich. 660, 649 N.W.2d 371 (2002), the court denied a mechanics' lien to an unlicensed contractor. The case produced four opinions—the majority, two concurrences and a dissent—all addressing the same issue: the equity of preventing payment based solely on the contractor's unlicensed status, where the contractor produced quality workmanship and the owners hired it knowing it was unlicensed.

[100]417 U.S. 901 (1974).

[101]*Columbia Group, Inc. v. Jackson*, 151 Ariz. 76, 725 P.2d 1110 (1986); *Connolly Dev., Inc. v. Superior Court*, 17 Cal.3d 803, 553 P.2d 637, 132 Cal.Rptr. 477 (1976); and *York Roofing, Inc. v. Adcock*, 333 Md. 158, 634 A.2d 39 (1993). All three cases upheld mechanics' lien laws, although sometimes only after those laws were first declared unconstitutional and then amended.

[102]501 U.S. 1 (1991).

Doehr court ruled that Connecticut's prejudgment attachment lien procedures violate due process. The court found that the lien deprived the owner of significant property interests because it: clouds title, impairs the owner's ability to sell, taints the owner's credit rating, reduces the owner's chances of getting a new mortgage or home equity loan, and can even constitute a technical default of an existing mortgage which has an "insecurity" clause.[103] (These same interests would be impaired by the filing of a mechanics' lien.) The court, in addition, announced that the *Spielman* decision, which had been decided summarily and not after argument on the merits, was no longer to be accorded full precedential value.[104] However, both the majority in its reference to *Spielman* and a concurrence by Chief Justice Rehnquist pointed out that mechanics' liens differ from attachment liens because the mechanics' lien claimants, unlike a general judgment creditor, have a pre-existing interest in the property they helped improve.[105] Over a dozen years later, state courts consistently have refused to view *Doehr* as a basis for finding mechanics' lien laws unconstituional.[106]

The brief examination of some of the salient characteristics of mechanics' lien protection has revealed the weaknesses of the remedy. Most important, the statutes are complex and change frequently. Carelessness in compliance can result in the lien's being lost. On the other hand, strict compliance is costly—perhaps, in smaller jobs, more than the value of the lien. The lien is most important in construction projects that fail. It is in such situations that it is most likely that lien claimants will find their claims wiped out because prior security holders have foreclosed and the funds left over for lien claimants are nonexistent. Because mechanics' liens differ from state to state, a large contractor with operations in several states must deal with different lien perfection and enforcement requirements.

These deficiencies have led some critics to contend that the mechanics' lien system gives subcontractors a false belief that they will receive compensation even in the case of default by the owner or contractor. In the view of these critics, abolition of mechanics' lien acts would not imperil subcontractors' payment options and might well compel them to demand contractual protections, such as bonding.[107] In any event, subcontractors have sought other forms of legislative and judicial relief—the focal point of the balance of Section 28.07.

Discussion has centered on the weaknesses of mechanics' lien laws as protection for unpaid subcontractors and suppliers. However, as noted earlier in this section, another frequently made criticism of such laws is that they can compel an inexperienced owner to pay twice for the same work. The owner may pay the prime contractor and then have to pay an unpaid subcontractor to remove a lien. The possibility of double payment has led to some legislative change to protect homeowners.[108]

Another obstacle to asserting a mechanics' lien is the presence in the prime contract of a provision under which the prime contractor gives up liens for itself and its subcontractors. In some states, such no-lien contracts are common. Whether a subcontractor will be precluded from asserting a mechanics' lien under such circumstances depends on whether the law requires that it consent to giving up its lien.[109]

Because of the circumstances under which such lien waivers are given, legislatures have started to regulate these transactions.[110]

---

[103]Id. at 11.

[104]Id. at 12, n. 4.

[105]Ibid (majority opinion) and id. at 28 (concurrence). See generally, Mills & Paskert, *What Lies Behind the Doehr? A Review of the Constitutionality of Mechanics' Lien Laws*, 14 Constr. Lawyer, No. 4, Oct. 1994, p. 3.

[106]See *Haimbaugh Landscaping, Inc. v. Jegen*, 653 N.E.2d 95 (Ind. App.1995), rehearing denied Aug. 18, 1995, transfer denied Feb. 29, 1996; *Vernon Hills III Ltd. Partnership v. St. Paul Fire and Marine Ins. Co.*, 287 Ill.App.3d 303, 678 N.E.2d 374 (1997); and *Gem Plumbing & Heating Co., Inc. v. Rossi*, 867 A.2d 796 (R.I.2005).

[107]See Sweet, *A View From the Tower*, 18 Constr. Lawyer, No. 1, Jan. 1998, p. 47.

[108]Mich.Comp.Laws Ann. § 570.1203. See also Md. Code, Real Prop. § 9-104(f)(3), expansively interpreted to apply to homeowner renovations in *Ridge Heating, Air Conditioning & Plumbing, Inc. v. Brennen*, 366 Md. 336, 783 A.2d 691 (2001). See also Benfield, supra note 83, 27 Wake Forest L. Rev. pp. 540–41.

[109]*Pero Bldg. Co., Inc. v. Donald H. Smith*, 6 Conn.App. 180, 504 A.2d 524 (1986) (contractor cannot bind subs and suppliers). If a subcontractor knows the prime contract is a no-lien contract, it may be barred. *Baker Sand & Gravel Co. v. Rogers Plumbing and Heating Co.*, 228 Ala. 533, 154 So. 591 (1934).

[110]West Ann.Cal.Civ.Code § 3262 (bars owner or prime from waiving liens of others, except with their written consent, and differentiates between conditional release, which is given in the advance of payment, and unconditional release, given when payment has been made) interpreted in *Tesco Controls, Inc. v. Monterey Mechanical Co.*, 124 Cal. App.4th 780, 21 Cal.Rptr.3d 751 (2004), rehearing denied Dec. 30, 2004; West Wis. Ann.Stat. § 779.135(1) (lien waivers are void). But lien waivers were upheld in *First American Bank of Va. v. J.S.C. Concrete Constr., Inc.*, 259 Va. 60, 523 S.E.2d 496 (2000).

E. Stop Notices

The ineffectiveness of mechanics' liens in private work and the unavailability in public work has led some states to supplement mechanics' lien protection by enacting stop notice laws. A stop notice imposes a lien not on the improved property but on the unpaid contract funds. Like mechanics' lien laws, these statutes vary from state to state.[111] Essentially, compliance stops the payment flow.

In California, if the funds withheld are insufficient to pay all stop notice claimants, each claimant shares pro rata in the funds without regard to when the stop notices were filed.[112] An owner, lender, or prime contractor who questions the amount of the notice can file a bond for 1.25 times the amount stated with the party served with a stop notice. When such a bond is filed, the funds withheld must be released.[113] California statutes authorizing stop notices on public works specify a detailed, speedy procedure to deal with disputed stop notices.[114]

The stop notice is more effective than a mechanics' lien in obtaining payment, as it can stop the flow of funds crucial to the project.

F. Trust Fund Legislation: Criminal and Civil Penalties

A number of states have enacted legislation designed to prevent prime contractors from diverting funds received from the owner and meant for payment of subcontractors and suppliers. These trust fund statutes make the prime contractor a trustee with a statutory duty to use the trust funds (the owner's payment) to pay the trust fund beneficiaries (the subcontractors and suppliers). Some trust fund statutes apply to public works, some to private works, and some to both.

If the funds paid can be located—ordinarily a difficult task—the statutes typically give the claimant priority over bankruptcy trustees and creditors.[115] To enforce the trust, the court can order an accounting, set aside as a diversion any unauthorized payments, award damages for breach of trust, or issue an order terminating or eliminating the authority of the contractor to apply trust assets.

Because such funds often disappear, the main effect of trust fund statutes is to provide harsh penalties for those who violate the trust. Sanctions for violation of the trust vary. Frequently, the breach of trust caused by diversion is a crime. For example, in New York, a diverting prime contractor is guilty of larceny.[116] In New Jersey the contractor can be convicted of theft.[117] In Oklahoma, the managing officers of a corporate contractor are guilty of embezzlement.[118] Although the penal sanction is rarely used,[119] its existence should deter diversion and result in payments to subcontractors and suppliers.

Some statutes give the trust beneficiaries (the subcontractors and suppliers) enhanced civil remedies. New York law gives beneficiaries[120] a right to punitive damages against the prime contractor.[121] Wisconsin also allows a civil remedy against an officer of the diverting corporation, but requires the subcontractor to prove by a preponderance of the evidence (more than 50 percent) a "specific intent" on the part of the diverter.[122] Some states allow a civil remedy without express language in the statute creating such a remedy.[123]

While trust fund statues normally impose trustee status on the contractor or subcontractor, some claimants have sought to extend the statute's reach to the project lender under a theory of "involuntary trustee." Courts have been

---

[111]For a listing of such state statutes, see Blanton & Diak, *Legislation Division Update*, 12 Constr. Lawyer, No. 1, Jan. 1992, p. 31.

[112]West Ann.Cal.Civ. Code § 3167.

[113]Id. at § 3171.

[114]Id. at §§ 3197–3205.

[115]*In re Munton*, 352 B.R. 707 (9th Cir.BAP 2006) (developer who violated Texas trust fund statute cannot use bankruptcy law to erase its debt to a subcontractor).

[116]N.Y.—McKinney's Lien Law § 79-a.

[117]N.J.Stat.Ann. 2C:20-9.

[118]Okla.Stat.Ann. tit. 42, § 153.

[119]But criminal convictions were successfully obtained in *State v. Cohn*, 783 So.2d 1269 (La. 2001); *State v. Spears*, 929 So.2d 1219 (La.2006); and *People v. Brooks*, 249 A.D.2d 572, 670 N.Y.S.2d 934 (1998), aff'd for reasons stated below, 93 N.Y.2d 862, 711 N.E.2d 198, 689 N.Y.S.2d 13 (1999).

[120]N.Y.—McKinney's Lien Law § 77(1).

[121]*Sabol & Rice, Inc. v. Poughkeepsie Galleria*, 175 A.D.2d 555, 572 N.Y.S. 811 (1991). In this case, the owner diverted $28 million to its partners and did not pay its contractors. Even though punitive damages are not specified in the statute, they were allowed in this case. The court noted that the conduct would have been sufficient to justify criminal charges of larceny.

[122]*Tri-Tech Corp. of Am. v. Americomp Services, Inc.*, 254 Wis.2d 418, 646 N.W.2d 822 (2002).

[123]*B. F. Farnell Co. v. Monahan*, 377 Mich. 552, 141 N.W.2d 58 (1966); *Hiller & Skoglund, Inc. v. Atlantic Creosoting Co.*, 40 N.J. 6, 190 A.2d 380 (1963).

amenable to this argument when the lender was actively involved in the payment process.[124]

California does not have a trust fund statute, but does make it a crime to divert funds intended for subcontractors and suppliers and to submit a false voucher to receive payment from a construction lender.[125]

G. Michigan Homeowners Construction Lien Recovery Fund

In 1982 Michigan enacted legislation that avoids a residential owner paying twice (owner pays prime and then pays the unpaid subcontractor to remove the latter's lien).[126] At the same time, Michigan allows a licensed subcontractor who could not receive a lien because the prime contractor had been paid, to recover from a state-administered fund created by fees paid by all licensed contractors. The subcontractor who seeks to recover from the fund must establish that it would have had a lien, that payments were made to its prime, that the prime retained the funds, and that the subcontractor demanded payment from the prime. A cap of $100,000 was placed on claims against the fund for any residence.[127]

H. Texas Trapping Statute

The Texas "trapping statute" is designed to aid subcontractors and suppliers. It allows an owner to withhold and pay a subcontractor directly if the latter notifies the owner that it has not been paid and the prime does not object.[128]

I. Compulsory Bonding Legislation

There are no lien remedies against public structures. To compensate for this and to encourage work on public projects, most public work—federal (Miller Act) and state (Little Miller Acts)—requires that prime contractors post payment bonds to protect subcontractors and suppliers. Statutes and case decisions vary as to how far down the line such protection exists.

Like mechanics' lien laws, the compulsory bonding legislation has its pitfalls. Again, rules determine who must file notices, to whom they must be sent, and what they must state, as well as requirements relating to when the lawsuit must be filed. On the whole, with their deficiencies, payment bonds on public work are a more effective remedy for subcontractors and suppliers than are mechanics' liens.[129]

Suppose a public agency fails to require the posting of a bond, or accepts a bond without investigation and the surety named on the bond is found to be a sham company. May the unpaid subcontractors sue the public agency for failure to enforce the bonding statute?

Most claims against the federal government, brought under the Federal Tort Claims Act, have proved unsuccessful.[130] The states are divided on whether a public entity's failure to require its prime contractor to obtain a proper bond subjects the government entity to liability to the unpaid subcontractors.[131]

J. Nonstatutory Claims

Subcontractors and suppliers who do not receive payment for work performed sometimes seek remedies against third parties not based on statutes of the type described earlier in the section. Instead or alternatively, they seek recovery based on a variety of theories against a number of parties other than the party with whom they have made the construction contract. To illustrate,

---

[124]See *Phillips Way, Inc. v. Presidential Financial Corp. of the Chesapeake*, 137 Md.App. 209, 768 A.2d 94 (2001) and *Sandpiper North Apartments, Ltd. v. American Nat'l Bank & Trust Co.*, 680 P.2d 983, 988 (Okla.1984).

[125]West Ann.Cal.Penal Code §§ 484b, 484c. Enforcement of this statute was made more difficult by *People v. Butcher*, 185 Cal.App.3d 929, 229 Cal.Rptr. 910 (1986), in which the court required the state to prove that the diversion caused damage to the project—a requirement not found in the statute itself.

[126]Mich.Comp.LawsAnn. § 570.1203.

[127]Id. at § 570.1204. See also *Mercantile Nat'l Bank of Indiana v. First Builders of Indiana, Inc.*, 774 N.E.2d 488 (Ind.2002), rehearing denied Nov. 18, 2002 (under Indiana's Personal Liability statute, an owner may be personally liable to an unpaid sub, limited, however, to the amount the owner owed the prime contractor).

[128]Tex.Prop.Code Ann. §§ 53.081–53.084, applied in *Don Hill Constr. Co. v. Dealers Elec. Supply Co.*, 790 S.W.2d 805 (Tex.Cr.App.1990).

[129]See Yarbrough, *Rights and Remedies Under Mississippi's New Public Construction Bond Statute*, 51 Miss. L.J. 351 (1980–81).

[130]Cann, *What to Do When There Is No Miller Act Payment Bond*, 14 Constr. Lawyer, No. 2, April 1994, p. 1.

[131]Compare *Imperial Mfg. Ice Cold Coolers, Inc. v. Shannon*, 101 P.3d 627 (Alaska 2004) (subcontractor has no private right of action under the Little Miller Act) with *N.V. Heathorn, Inc. v. County of San Mateo*, 126 Cal.App.4th 1526, 25 Cal.Rptr.3d 400 (2005) (city liability under the California Tort Claims Act). See also Mercier, *Little Miller Acts: Liability of Public Owners for Failure to Obtain Payment Bonds on Public Construction Projects*, 14 Constr. Lawyer, No. 4, Oct. 1994, p. 7; Annot., 54 A.L.R.5th 649 (1997).

an unpaid subcontractor may seek recovery for work performed from the owner, the lender, or a design professional. Similarly, a claim may be made by another party farther down on the subcontract chain, such as a second-tier subcontractor against all parties on the contract chain other than the first-tier subcontractor with whom the contract was made.

Nonstatutory claims can be further divided into specific remedies and ordinary court judgments. An illustration of a specific remedy is a claim against funds, in the hands of the owner, or lender earmarked for the project. An illustration of other remedies is a claim generally against the owner, lender, or design professional that would result in an ordinary judgment and not a claim against specific property.

Claims to specific funds such as funds in the hands of a lender or owner are usually based on assertion of an equitable lien—an equitable remedy based on unjust enrichment. Such claims generally point to the claimant's having improved the property of the defendant owner or having improved the land in which the lender has a security interest. Some courts have granted equitable liens on construction loan funds held by owners,[132] by lenders,[133] or on the retainage.[134]

Although the law is somewhat unsettled, it seems likely that claimants such as unpaid subcontractors or suppliers or those sureties that have had to pay such parties will be able to recover from the design professional if the latter did not perform in accordance with the professional standards established by the law when approving payments to the prime contractor.[135] Alternatively, if the claimant can show that the design professional breached the contract with the owner and that this contract was in part for the claimant's benefit, recovery is possible. These burdens have been difficult to sustain.[136] Claims against owners and lenders are based on similar theories but have been difficult to sustain.[137]

One differentiation between claims brought against design professionals and those brought against owners is that claims against the latter can be based on unjust enrichment. Claimants assert that their work or materials have benefited the owner and that it would be unjust for the owner to retain this benefit without paying the claimants.

As a rule, such claims have not been successful because the owner can show it has paid someone, usually the prime contractor, or that the retention of benefit was not unjust because the claimant could have protected itself by using the statutory remedies.[138] One case granted recovery where the owner had not paid anyone, with the prime contractor's having left town before payment.[139]

The subcontractor's reliance on a promise made by the owner's authorized agent that the prime contractor would be required to file a surety bond was the basis for a successful subcontractor claim against the owner.[140]

With the exception of an occasional recovery based on reliance on a direct promise or unjust enrichment, claimants such as subcontractors and suppliers have had limited success against owners. Subcontractors must use statutory remedies or bonds or they will be left to whatever claim they have against the party with whom they dealt directly.

Second-tier subcontractors sometimes, like first-tier subcontractors, look up the subcontract chain for a solvent party to pay for work when the party who has promised to do so does not. In *Friendly Ice Cream Corp. v. Andrew E. Mitchell & Sons, Inc.*,[141] an unpaid second-tier

---

[132]*Avco Delta Corp. v. United States*, 484 F.2d 692 (7th Cir.1973), cert. denied sub nom. *Canadian Parkhill Pipe Stringing Ltd. v. U.S.*, 415 U.S. 931 (1974).

[133]*Trans-Bay Eng'rs & Builders, Inc. v. Hills*, 551 F.2d 370 (D.C.Cir. 1976) (prime contractor recovered from lender and HUD). See generally Reitz, *Construction Lenders' Liability to Contractors, Subcontractors, and Materialmen*, 130 U.Pa.L.Rev. 416 (1981).

[134]*Williard, Inc. v. Powertherm Corp.*, 497 Pa. 628, 444 A.2d 93 (1982).

[135]See *Boren v. Thompson & Assocs.*, 999 P.2d 438 (Okla.2000) (architect on public works project liable to subs for failing to verify that prime obtained a payment bond) and *Cullum Mechanical Constr., Inc. v. South Carolina Baptist Hosp.*, 344 S.C. 426, 544 S.E.2d 838 (2001) (architect may have a duty of care to ensure a prime pays its subs). See also Section 22.07; but see Sections 14.08B through F.

[136]*Engle Acoustic & Tile, Inc. v. Grenfell*, supra note 80, denied recovery against owner and architect.

[137]Subcontractors were unsuccessful in *Helash v. Ballard*, 638 F.2d 74 (9th Cir.1980) (against owner); *Urban Systems Dev. Corp. v. NCNB Mortgage Corp.*, 513 F.2d 1304 (4th Cir.1975) (against lender). For further discussion of claims against lenders, see Section 22.02I.

[138]*Blum v. Dawkins*, 683 So.2d 163 (Fla.Dist.Ct.App.1996) (sub cannot recover from owner where lien failed and owner paid out the entire contract price: no unjust enrichment); *DJ Painting, Inc. v. Baraw Enterprises, Inc.*, 172 Vt. 239, 776 A.2d 413 (2001) (no unjust enrichment of owner by sub, where owner had paid prime).

[139]*Costanzo v. Stewart*, 9 Ariz.App. 430, 453 P.2d 526 (1969).

[140]*Bethlehem Fabricators, Inc. v. British Overseas Airways Corp.*, 434 F.2d 840 (2d Cir.1970) (promise made by architect).

[141]340 A.2d 168 (Del.Super.1975).

subcontractor asserted claims based on a number of theories against a prime contractor but was unsuccessful. But a third-tier subcontractor was allowed to recover against a first-tier sub based on the latter's promise to the prime contractor to pay all suppliers.[142]

The preceding discussion should indicate the importance of dealing with a party who has the desire and financial capability to pay for work that is ordered. Obviously, a subcontractor's principal concern would be the financial responsibility of the prime contractor with whom it has dealt. Yet the panoply of legislative protection and the occasional nonstatutory relief given indicates that such parties frequently are not paid and seek other methods to collect.

K. Joint Checks

One way owners and prime contractors seek to avoid liens and other claims is to issue joint checks. For example, the owner may issue a joint check to the prime and to the subcontractors, and the prime contractor may issue joint checks to subcontractors and their sub-subcontractors and suppliers. This was discussed in Section 22.02J.

SECTION 28.08 Other Subcontractor Claims

A. Against Prime Contractor

Suppose a subcontractor asserts that its cost of performance was wrongfully increased because of acts or omissions by the prime contractor or someone for whom the prime contractor is responsible. Generally, the law implies an obligation on the part of the prime contractor to take reasonable measures to ensure that the subcontractor can perform expeditiously and is not unreasonably delayed.[143] However, terms will not be implied if express provisions in the subcontract directly deal with this matter. Commonly subcontracts, amplified by relevant provisions in prime contracts incorporated into the subcontract, deal in some way with this problem. Express provisions can place such a responsibility on the prime contractor, such as the contract requiring that the site be ready by a particular date or that the work be in a sufficient state of readiness by a designated time to enable the subcontractor to perform specific work.

Conversely, contract provisions may indicate that the subcontractor has assumed certain risks regarding the sequence of performance. For example, the subcontract might require that the work be performed "as directed by the prime contractor," and such a provision would give the prime contractor wide latitude to determine when the subcontractor will be permitted to work.[144]

Additional clauses in the prime contract to which the subcontract refers or that are incorporated into the subcontract are also relevant. The contract can specify that certain delay-causing risks were contractually assumed risks or grant only a time extension.

In addition, contract clauses may control by denying recoverability of delay damages. For example, a no-damage clause that limited the subcontractor to a time extension in the subcontract, especially if tracked with a no-damage clause in the prime contract, would very likely preclude recovery of delay damages by the subcontractor against the prime contractor for delays caused by owner or prime contractor.[145]

The problems of tracking or parallelism in prime contracts or subcontracts complicate these claims. As a general rule, prime contracts and subcontracts are parallel in terms of rights and responsibilities. For example, should a prime contractor be held liable for delay to the subcontractor for acts caused by the owner when the prime contractor is precluded from recovering from the owner because of a no-damage clause?

The bargaining situation sometimes permits the prime contractor to better its position through the subcontract. For example, the prime contractor might be able to include a no-damage clause in the subcontract when the prime contract allows the prime contractor its delay damages against the owner. Although a careful subcontractor might be able to preclude this possibility, time or realities of the process often frustrate parallel rights. An abuse of prime contractor bargaining power can result in discontented subcontractors and poor performance. It is becoming increasingly common for owners to insist that

---

[142]*Aetna Cas. & Sur. Co. v. Kemp Smith Co.*, 208 A.2d 737 (D.C. App.1965).

[143]Annot., 16 A.L.R.3d 1252, 1254 (1967). See also Section 19.02A.

[144]*Keeney Constr. v. James Talcott Constr. Co., Inc.*, 309 Mont. 226, 45 P.3d 19 (2002), rehearing denied Apr. 30, 2002 ("as directed" clause precluded sub's delay claim).

[145]*McDaniel v. Ashton-Mardian Co.*, 357 F.2d 511 (9th Cir.1966) (subcontractor assumed the risk of government-caused delays). But in *J. J. Brown Co. v. J. L. Simmons Co.*, 2 Ill.App.2d 132, 118 N.E.2d 781 (1954), a subcontractor recovered delay damages from the prime contractor when the former was delayed by another subcontractor.

prime contracts contain provisions requiring the prime contractor to give the subcontractor benefits parallel to those given the prime contractor in its contract with the owner. (See Section 28.04.)

Other disputes between prime and subcontractor can arise. For example, a case that sharply divided the Kansas Supreme Court involved a claim by a subcontractor against the prime based on the latter having suggested a modification in the contract with the owner that would have essentially eliminated all of the subcontractor's performance. The court, over a dissent, held that this suggestion did not violate the obligation of good faith and fair dealing owed by the prime to the subcontractor.[146] However, the Wyoming Supreme Court, when faced with near-identical circumstances, remanded the case to the trial court to determine whether the prime contractor's conduct violated the implied covenant of good faith and fair dealing.[147]

B. Pass-Through Claims Against Owner: Liquidating Agreements and *Severin* Doctrine

When a project is disrupted or delayed by the owner or other responsible entity, all participants incur expenses. Frequently, they seek to transfer these expenses to the owner. Because of the high costs of pursuing claims, subcontractors often pool their claims with those of the prime, and the prime will present them against the owner. However, even where there is an agreement under which the subcontractors take whatever the prime gets for them, the prime must pursue the subcontractors' claims in good faith.[148]

One mechanism for accomplishing this is a liquidating agreement. Under such an agreement, the prime confesses liability to the subcontractor for owner-caused delay, the subcontractor releases the prime from all other liability, and the subcontractor is relegated to whatever delay damages the prime can recover from the owner.

In addition to saving costs in authorizing the prime to handle the subcontractor's claim, such pass-through or liquidating agreements avoid the problem of the subcontractor

not being in privity of contract with the owner. Usually, to make a claim against the owner the prime must show it has suffered a loss that it seeks to transfer to the owner. Usually the prime does this by showing either that it has reimbursed the subcontractor for the latter's damages or that it remains liable for such reimbursement. Sometimes, as noted, the liquidating agreement is drafted so as to make the contractor liable to the subcontractor but only as, when, and to the extent that the contractor recovers from the owner based on the subcontractor's claim.

Such pass-through claims have generated litigation. For example, in one case the liquidating agreement obligated the contractor to pay the subcontractor 9.5 percent of the first $20 million recovered and 5 percent of any excess. Combined with a partial payment, this formula left the contractor responsible for passing through less than the face amount of the subcontractor's claim. The court rejected the owner's attempt to limit its liability to the prime to the amount the prime would pay its subcontractor under the liquidating agreement. The court noted that the prime's liability to its subcontractor and therefore the prime contractor's right to recover on the subcontractor's claims are not connected to the agreed division of the recovery.[149]

Although these cases should encourage such pass-through or liquidating claims, a decision by the New York Court of Appeals created grave doubts about the utility of liquidating agreements. In that case, the subcontractor sued a separate prime contractor for delays caused primarily by the owner, its engineer, other prime contractors, and bad weather. At the end of the subcontractor's case, the court granted the separate prime contractor's motion to dismiss, concluding that the subcontractor would not be able to recover damages from the prime and that the prime did not cause the loss over which it had no control.[150]

This result is in no way unusual, but it must be looked at in the light of liquidating agreements under which the prime confesses liability for owner-caused delay and the subcontractor releases the prime from all other liability and relegates itself to whatever delay damages the prime

---

[146]*Meier's Trucking Co. v. United Constr. Co.*, 237 Kan. 692, 704 P.2d 2 (1985).

[147]*Scherer Constr., LLC v. Hedquist Constr., Inc.*, supra note 37.

[148]*T. G. I. East Coast Constr. Corp. v. Fireman's Fund Ins. Co.*, 534 F.Supp. 780 (S.D.N.Y.1982).

[149]*Frank Briscoe Co. v. County of Clark*, 772 F.Supp. 513 (D.Nev. 1991). See Calvert & Ingwalson, *Pass Through Claims and Liquidation Agreements*. 18 Constr. Lawyer, No. 4, Oct. 1998, p. 29.

[150]*Triangle Sheet Metal Works, Inc. v. James H. Merritt & Co.*, 79 N.Y.2d 801, 588 N.E.2d 69, 580 N.Y.S.2d 171 (1991).

can recover from the owner. If the subcontractor cannot recover from its prime, it is difficult to see how the prime can "confess" liability to its subcontractor and, if so, how a prime can recover from an owner for subcontractor claims for which the prime is not independently responsible to the subcontractor.

Special rules apply on public works projects. The federal government is immune from contract liability to parties with whom it does not have an express or implied contract.[151] For this reason, any subcontractor who wishes to bring a claim against the government must enter into a pass-through agreement with the prime contractor. However, under the *Severin* doctrine,[152] the prime cannot prosecute the claim unless it is at least potentially liable to the subcontractor. While the *Severin* court placed upon the contractor the burden of proving its potential liability to the subcontractor,[153] the Federal Circuit Court of Appeals has relaxed the standard for subcontractor claims by placing upon the government the burden of asserting and proving that the prime contractor is not responsible for the subcontractor's costs.[154] State courts have generally adopted the *Severin* doctrine.[155]

C. Against Third Parties

Suppose the costs of performance are increased because of acts of third parties such as owners or design professionals. Suppose a subcontractor, in addition to making any claim that might be available against the prime contractor, seeks legal relief directly against those third parties who it asserts caused the loss. For example, Section 28.07J stated that subcontractors sometimes seek payment against third parties such as owners or design professionals when the prime contractor does not pay them. Claims for additional cost of performance are not treated as favorably as claims to be paid for work performed. Even the latter claims are difficult to sustain against third parties. Very likely, claims against third parties for increased cost of performance will meet more severe resistance. However, a subcontractor who can persuade a court that the third party did not live up to the legal obligation imposed by law and thereby committed a tort or that the third party promised to perform in a certain way to the prime contractor and that this promise was for the benefit of the subcontractor can recover against third parties.[156]

---

[151]28 U.S.C.A. § 1491.

[152]See *Severin v. United States*, 99 Ct.Cl. 435 (1943), cert. denied, 322 U.S. 733 (1944).

[153]99 Ct.Cl. at 443.

[154]*E. R. Mitchell Constr. Co. v. Danzig*, 175 F.3d 1369, 1371 (Fed. Cir.1999), rehearing denied, en banc suggestion declined, Aug. 25, 1999. See generally, Thrasher, *Subcontractor Dispute Remedies: Asserting Subcontractor Claims against the Federal Government*, 23 Pub.Cont.L.J. 39 (1993).

[155]See *Howard Contracting, Inc. v. G. A. MacDonald Constr. Co., Inc.*, 71 Cal.App.4th 38, 83 Cal.Rptr.2d 590 (1998), review denied

March 31, 1999; *Department of Transp. v. Claussen Paving Co.*, 246 Ga. 807, 273 S.E.2d 161, 164 (1980) (burden remains on prime contractor); *Board of County Comm'rs of Frederick County v. Cam Constr. Co.*, 300 Md. 643, 480 A.2d 795 (1984); *Frank Coluccio Constr. Co. v. City of Springfield*, 779 S.W.2d 550 (Mo.1989).

[156]The owner who had negotiated a contract with subcontractors and then assigned the contract to a prime contractor was held liable to a subcontractor for the owner's change of the critical path method. *Natkin & Co. v. George A. Fuller Co.*, 347 F.Supp. 17 (W.D.Mo.1972).

CHAPTER TWENTY-NINE

The Design Professional as Judge: A Tradition Under Attack

SECTION 29.01 Overview

A. Introducing the Initial Decision Maker

In many construction projects, the design professional makes decisions affecting the rights and duties of owners, contractors, and subcontractors. Usually the power to make these decisions is created by the contract between owner and contractor.

Using A201-1997 as a starting point for discussion, the architect's role was divided into two broad categories. Under the rubric of contract administration,[1] the architect interpreted the contract documents and determined whether the contractor's performance complied with the design requirements. The architect was given the right to make inspections and uncover work. He had the right to grant time extensions, determine price adjustments to changed work if the parties did not agree, approve substitutions of subcontractors, and pass judgment on the sufficiency of the contractor's submittals. The architect reviewed the contractor's applications for payment and determined the dates of substantial and final completion. The architect was the owner's representative and acted as the conduit through which the owner and contractor communicated with each other. This administrative function continues unchanged under A201-2007.[2]

The second broad category of architect responsibilities under A201-1997 was as initial judge. (The term "judge" is not found in the AIA documents.) Contractor claims, including those alleging negligence by the architect, were first sent to the architect for an initial decision.[3] If either

the owner or contractor was dissatisfied with the architect's decision, that person could seek mediation and then either arbitration or litigation.[4] In addition, although the owner had the right to terminate the contractor, the architect played an important role by certifying that grounds existed under the contract to terminate.[5]

This second category of architect responsibility—the design professional as judge—has come under attack by contractors. Contractors expressed concern that architects could not be objective in evaluating their claims, especially since the architect is paid by the owner and the contractor's claim may require the architect to find his own design to be defective.

In response, the American Institute of Architects in A201-2007 created a new project *position:* the Initial Decision Maker (IDM).[6] The IDM is a new position, not necessarily a new participant. If the parties do not designate an IDM, the architect is the IDM. The parties need not agree that the IDM is to be the architect. He is the default choice. If the architect acts as IDM, then the design professional's judging function largely mirrors that found in A201-1997. Regardless of who occupies the position, the IDM's primary roles are to issue an initial decision when presented with a contractor or owner claim[7] and also to certify that grounds exist for the owner to terminate the contractor for cause.[8]

If the parties elect to have a different person (or different persons) fulfill the role of IDM, then they must resolve several issues—such as selection process and payment

---

[1]A201-1997, ¶ 4.2.
[2]A201-2007, § 4.2.
[3]A201-1997, ¶ 4.4.1.

[4]A201-1997, ¶¶ 4.5 and 4.6.
[5]A201-1997, ¶ 14.2.2. See Section 33.03D for analysis of this power.
[6]A201-2007, § 1.1.8.
[7]A201-2007, § 15.2.
[8]A201-2007, § 4.2.2.

mechanism—for which A201-2007 provides no guidance. Complications arising from the use of an IDM other than the architect are discussed in Section 29.11.

B. Relation to Chapter 30

This chapter must also be seen against the backdrop of the material in Chapter 30 that deals with dispute resolution outside of the courtroom. Sections 30.01 through 30.17 deal with arbitration. Often, but not always, any decision made by the design professional can be taken to arbitration.

More important, the construction industry has been seeking alternatives to the traditional method under which disputes are decided initially by the design professional with the right to challenge that decision by seeking arbitration or going to court. These alternatives are canvassed in Sections 30.18 through 30.20 in the context of both private and public projects. The methods discussed there reflect dissatisfaction with the traditional method under which the architect or engineer initially resolves disputes. Some of the difficulties noted in this chapter were instrumental in generating the search for alternative methods.

Also, methods discussed in Section 17.04 that deviated from the traditional design-bid-build (DBB) system, such as Design-Build, did not use the designer in the same way as in the DBB method.

Yet the role of the architect or engineer in the A201-1997 DBB system, as the person who both designs and administers the contract, including resolving disputes in the first instance, is still a prominent feature in the American construction industry. It is the initial subject of this chapter.

SECTION 29.02 The Doctrine of Conditions

Parties to a contract can condition specific obligations on the occurrence of designated events. Normally, unless these events occur or are excused, the duty to perform does not arise or ceases to exist.

Commonly, parties to a construction contract give the design professional power to make certain decisions. Such third-party decisions can be expressed as conditions. For example, if a certificate issued by the design professional conditions payment, issuance of that certificate—or the existence of facts that excuse issuance—is necessary before the owner is obligated to pay.

Although a decision of a third party can be a valid condition, there are two aspects of the circumstances under which third-party dispute resolution is created in construction contracts that bear on legal treatment of such conditions. First, the owner pays, and for all practical purposes selects, the design professional. Under these circumstances, one basic element of a decision-making process—that of an impartial judge—may not be present in the construction contract disputes process. Second, in many construction contracts the contractor, and even more so the subcontractor, has no real choice but to accept the decision-making process under which the design professional is given broad powers.

As shall be seen,[9] the law accords a degree of finality to decisions made by the design professional. But the degree to which these two factors cast doubt on the impartiality of the design professional may cause courts to scrutinize these decisions carefully.

SECTION 29.03 Excusing the Condition

Under certain circumstances, the condition can be excused. If so, any disputes between owner and contractor would have to be negotiated by the parties, submitted to arbitrators if the parties have agreed to use this process, or decided by a court.

The condition is excused if the design professional becomes unavailable or unable[10] to interpret the contract or judge its performance. The construction contract can provide for a successor design professional when the one originally designated cannot or will not perform this function. For example, AIA Document A201-2007, Section 4.1.3, allows the owner to appoint a successor "as to whom the Contractor has no reasonable objection." Even if no successor mechanism is specified in the contract, the parties can agree to use a replacement design professional. But if there were neither an agreement nor a mechanism for a successor design professional, any obligation conditioned on the issuance of a certificate or the resolution of a dispute by the design professional would be unconditional and the condition no longer be part of the contract.

---

[9]See Section 29.09.

[10]*United States v. Klefstad Engineering Co.*, 324 F.Supp. 972 (W.D.Pa. 1971) (surveyor lost records); *Grenier v. Compratt Constr. Co.*, 189 Conn. 144, 454 A.2d 1289 (1983); *Manalili v. Commercial Mowing & Grading*, 442 So.2d 411 (Fla.Dist.Ct.App.1983).

Other acts can excuse the condition. For example, one court excused the condition when the architect examined the work, found it satisfactory, but still refused to issue the certificate.[11] Similarly, another court held that the requirement of the issuance of a payment certificate had been waived because the architect failed to reinspect the work and specify which terms remained to be corrected.[12] The condition was excused in a private contract for developing a road when the city engineer, whose approval was a condition, simply refused to act.[13]

Clearly, collusion between the design professional and either owner or contractor excuses the condition.[14] A condition can be excused if the parties agree to eliminate it.[15]

The condition can be excused if the party for whom the condition is principally inserted manifests an intention to perform its obligations despite nonoccurrence of the condition. Although one case held that the condition of a certificate is for the owner's benefit and the owner can waive it,[16] there is no reason why such a condition cannot be considered for the benefit of *both* parties, particularly where the contractor establishes that it entered into the contract on the assumption that a *particular* design professional would make these decisions.

SECTION 29.04 The Design Professional as Judge: Reasons

To a continental European, the design professional as judge seems incongruous. With the exception of Great Britain, European construction administration might include the design professional giving his interpretation of design documents he drafted, but he would not be given a "judging role." The close association between owner and design professional based on the former's selection and payment of the latter precludes this role being given to the latter.

Despite the close association between owner and design professional, most American construction contracts, both public and private, give the design professional broad decision-making powers.[17] A number of reasons can be given for the development of this system.

First, the stature and integrity of the design professions may give both parties to the construction contract confidence that the decisions will reflect technical skill and basic elements of fairness.[18]

Second, the design professional's role in design before construction equips the design professional with the skill to make decisions that will successfully implement the project objectives of the owner. In a sense, the role as interpreter and judge is a continuation of design.

Third, owners are often unsophisticated in matters of construction and need the protection of a design professional to obtain what they have been promised in the construction documents. Implicit is the assumption that without a champion and protector, the owner might be taken advantage of by the contractor. Couple this with the owner's bargaining strength, and the present system results.

Fourth, even assuming that complete objectivity is lacking and that the contractor rarely has much choice, the alternative would be worse.

Suppose the design professional did not act as interpreter and judge. Such matters would have to be resolved by owner and contractor, which for many owners would require professional advice. If owner and contractor cannot agree, the complexity of construction documents and performance will necessitate many costly delays, because the alternative forums for owners and contractors who cannot agree—litigation and arbitration—still involve time and expense.

Fifth, despite the dangers of partiality and conflict of interest (the design professional may overlook defective workmanship to induce the contractor not to press a delay claim or a claim for extras based on the design professional's negligence, or the design professional may find a defect due to poor workmanship rather than expose himself to a claim for defective design), the system seems to have kept the project going notwithstanding inevitable disputes. Perhaps a quick decision that may at times be

---

[11]*Anderson-Ross Floors, Inc. v. Scherrer Constr. Co.*, 62 Ill.App.3d 713, 379 N.E.2d 786 (1978).

[12]*Hartford Elec. Applicators of Thermalux, Inc. v. Alden*, 169 Conn. 177, 363 A.2d 135 (1975).

[13]*Grenier v. Compratt Constr. Co.*, supra note 10.

[14]*Metropolitan Sanitary Dist. of Greater Chicago v. Anthony Pontarelli & Sons, Inc.*, 7 Ill.App.3d 829, 288 N.E.2d 905 (1972).

[15]*Steffek v. Wichers*, 211 Kan. 342, 507 P.2d 274, 281–82 (1973).

[16]*Halvorson v. Blue Mountain Prune Growers Co-op*, 188 Or. 661, 214 P.2d 986, rehearing denied, 217 P.2d 254 (1950).

[17]The history is described in Dreifus, *The "Engineer Decision" in California Public Contract Law*, 11 Pub.Cont.L.J. 1 (1979).

[18]See *Zurn Eng'rs v. State Dep't of Water Resources*, 69 Cal.App.3d 798, 138 Cal.Rptr. 478, cert. denied, 434 U.S. 985 (1977).

unfair was better than a more costly, cumbersome system that might give better and more impartial decisions.[19]

The potential for bias toward the owner[20] and decision making often involving the design professional's own prior design work are bound to reflect themselves in the judicial treatment of the design professional's decisions and whether the design professional will be held responsible for decisions (discussed in Section 14.09D).

SECTION 29.05 Jurisdiction of Decision-Making Powers

The design professional's interpreting and judging functions are created by the contract between the owner and the contractor. As a result, interpretation of the contract determines whether the design professional has jurisdiction to resolve the dispute. The contract will also govern the finality of the design professional's decision.[21]

Generally, jurisdictional grants to decide disputes are likely to be broader in matters that involve the experience and expertise of the design professional. Jurisdictional grants are likely to be broad where an on-the-spot decision during the operational phases, as opposed to post-completion disputes, must be made.[22] Despite these basic guidelines, decisions are likely to vary, depending on the particular language of the contract and judicial attitude toward this decision-making process.

In 1987, the AIA drastically revised the part of AIA Document A201 that dealt with the role of the architect in interpreting the contract documents and resolving disputes. First, a line was drawn between the architect's interpreting and deciding matters concerning performance and the architect's resolving claims and disputes. The architect's jurisdiction to interpret and decide performance matters under Paragraph 4.2.11 related to the requirements of the contract documents. This jurisdiction was very broad, because the contract documents covered many matters,

some technical, some administrative, and some legal. The architect's jurisdiction to pass on claims and resolve disputes under Paragraph 4.3.1 was even more broad, covering almost any dispute related to the project that could involve the owner and the contractor. Under Paragraph 4.3.2, disputes must first (subject to some exceptions) go to the architect.

A201-1997 retained the broad jurisdiction language. Jurisdiction was enlarged by deleting the exceptions to jurisdiction that had existed in 1987. (These exceptions related to there being no architect, delay in making the decision and the dispute relating to a mechanics' lien.) Most important, the AIA retained language in Paragraph 4.4.1 making clear that the architect's jurisdiction was not affected by the claim involving "an error or omission by the Architect." But Paragraph 4.4.2 allowed the architect in his own discretion to advise the parties that it would be "inappropriate" for him to resolve the claim. This was likely intended to allow the architect to remove himself if he believed there was too great a conflict of interest, as noted in Section 29.05.

Two limits to jurisdiction were added in 1997. Paragraph 4.4.1 excluded claims that relate to hazardous material and to disputes between the contractor and entities other than the owner.

As noted in Section 29.01A, A201-2007 retains the architect's jurisdiction over operational decisions during the construction phase. A new Section 4.2.14 specifies that the architect reviews and responds to requests for information forwarded by the contractor. This Section formalizes a process, no doubt already in existence on an informal basis, by which a contractor may seek clarification of questions it has with the design.

"Claims" must now be presented for review first to the Initial Decision Maker (IDM), who can be either the architect or a new project participant.[23] In Section 15.1.1, a claim is defined as a "demand or assertion by one of the parties seeking, as a matter of right, payment of money or other relief with respect to the terms of the Contract." As compared with the A201-1997 definition, A201-2007, Section 15.1.1, excludes from the definition of a claim "adjustment or interpretation of Contract terms" and "extension of time." However, jurisdiction over claims for

---

[19]*Cofell's Plumbing & Heating, Inc. v. Stumpf,* 290 N.W.2d 230 (N.D.1980).

[20]To avoid the false impression that the cases generally involve the design professional's refusing to award a certificate, note that many cases involve the owner's refusal to pay despite the design professional's having issued a certificate.

[21]See Sections 29.09, 29.10.

[22]*Cofell's Plumbing & Heating, Inc. v. Stumpf,* supra note 19. See also *Rockland County v. Primiano Constr. Co.,* 51 N.Y.2d 1, 409 N.E.2d 951, 431 N.Y.S.2d 478 (1980).

---

[23]A201-2007, § 15.1.2, states claims by either the owner or contractor must be initiated by written notice to the other party and to the IDM, "with a copy to the Architect, if the Architect is not serving as the Initial Decision Maker."

additional time is granted in Section 15.1.5. In addition, under Section 14.2.2, the IDM, not the architect, must certify to the owner that grounds exist for termination of the contractor for cause.

Some matters are outside the IDM's jurisdiction. Under Section 15.2.1, the IDM has no jurisdiction over claims concerning hazardous materials, actions by the contractor in response to an emergency situation, the owner's bonding against insured losses and the owner's settlement of property insurance claims. Also under Section 15.2.1, the IDM may decide disputes only between the contractor and owner, unless "all affected parties" agree otherwise. Under Section 4.2.13, the architect's decision on aesthetic matters is final "if consistent with the intent expressed in the Contract Documents." Should either the owner or contractor disagree with the architect as to aesthetic matters, the IDM's jurisdiction is limited to whether the decision was inconsistent with the design intent—a narrow basis for review.

SECTION 29.06 Who Can Make the Decision?

The interpreting and judging powers given the design professional by the contract are important, and the parties—particularly the owner—can expect those powers to be exercised by the particular person or persons in whom the parties have confidence. If a design professional partnership is named, any principal partner should be able to make such interpretations and judgments. However, the owner may expect that the particular partner with whom it dealt or the partner with principal responsibility for the design would be the appropriate and suitable person to interpret and judge. An individual design professional who is empowered to make such determinations cannot delegate this power without the permission of the contracting parties.[24]

AIA Document A201-1997, Paragraph 4.1.1, assumed the architect had an authorized representative. But this should not permit the designated architect to delegate

the responsibility of interpreting the contract or resolving disputes, even to a person in his firm. A representative may be appropriate for some administrative purposes, such as receiving notices or even executing change orders. But delegating the highly personal service of giving interpretations or resolving disputes requires consent of owner and contractor. A201-2007, Section 4.1.1 eliminates reference to an architect's authorized representative.

SECTION 29.07 The Contract as a Control on Decision-Making Powers

In addition to jurisdiction, there are other limits to the decision-making powers of design professionals or IDMs. The contract may limit the *grounds* upon which a claim may be made. For example, A201-1997, Paragraph 8.3.1, gave a laundry list of grounds upon which a claim for additional time may be made, including changes ordered in the work, labor disputes, fire and unusual delay in deliveries, as well as "other causes which the Architect determines may justify delay." (But adverse weather conditions is not one of the enumerated causes.) In A201-2007, Section 15.1.5.1, this enumeration of events is dropped, so that the IDM has jurisdiction over any claim for additional time, regardless of cause. However, Section 15.1.5.2 then states that if adverse weather is the basis for the claim, then the claim "shall be documented by data substantiating that weather conditions were abnormal for the period of time, could not have been reasonably anticipated and had an adverse effect on the scheduled construction."

Similarly, the power to decide disputes does not give the design professional authority to change the specifications.[25]

SECTION 29.08 Procedural Matters

A. Requirements of Elemental Fairness: A201

The IDM's role as interpreter or judge invites comparison with arbitration and litigation. Should the IDM conduct a hearing similar to that used in arbitration or litigation? Clearly, the formalities of the courtroom would be

---

[24]*Huggins v. Atlanta Tile & Marble Co.*, 98 Ga.App. 597, 106 S.E.2d 191 (1958). But see *Atlantic Nat'l Bank of Jacksonville v. Modular Age, Inc.*, 363 So.2d 1152 (Fla.Dist.Ct.App.1978), cert. denied, 372 So.2d 466 (Fla.1979), where a different architect judged the quality of modular units, the delegation issue not having been raised. An A201 was used.

[25]*Northwestern Marble & Tile Co. v. Megrath*, 72 Wash. 441, 130 P. 484 (1913).

inappropriate and unnecessary. Even the informal hearings conducted by an arbitrator would not be required. Unless a clause that confers jurisdiction requires a hearing, no hearing at all is necessary.

Cogent reasons exist for some semblance of elemental fairness to both parties. First, continued good relations on the project necessitate a feeling on the part of the participants that they have been treated fairly. Each party should feel it has been given a fair chance to state its case and be informed of the other party's position. A fair chance need not necessarily include even an informal hearing. The IDM should listen to the positions of each party where feasible before making a decision.

The exact nature of what is fair will depend on the facts and circumstances existing at the time the matter is submitted to the design professional. Small matters and those that require quick decisions may not justify the procedural caution that would be necessary where large amounts of money are at stake or where an urgent decision is less important.

The second reason for elemental fairness is the likelihood that a decision made without it will not be accorded much finality. For example, *John W. Johnson, Inc. v. Basic Construction Co.*,[26] involved a dispute in which the architect had ordered a prime contractor to terminate a particular subcontractor. The trial judge stated,

> This amazing directive was issued by the architect's office without notice to the plaintiff and without giving the plaintiff any opportunity to be heard, orally or in writing, formally or informally. This action on the part of the architect's office was contrary to the fundamental ideas of justice and fair play. The suggestion belatedly made at the trial that it was not appropriate for the architect to maintain any contacts with subcontractors is fallacious in this connection. Any such principle as that did not bar the architect's representative from according a hearing to the subcontractor before directing that his subcontract be cancelled.[27]

Termination is a serious matter, and more process fairness in such matters can be expected. But the case also reflects judicial unwillingness to uphold rash, impetuous decisions.

The third reason relates to the immunity sometimes given to design professionals when they act in a quasi-judicial role

(explored in Section 14.09D). The more design professionals act like judges or arbitrators, the more likely they will be given judicial protection from a lawsuit by someone dissatisfied with the decision.

As indicated in Section 29.04, in 1987, the AIA greatly amplified the jurisdictional rules in A201 for resolving disputes. It also provided detailed procedural directions and time deadlines. This was done to give greater guidance to architects and to resolve disputes as they develop, to avoid their being "blown up" at the end of the project. Yet it is likely these provisions will not be followed in many transactions—either because parties become impatient with the burdens and paperwork or because the transaction does not involve enough money to make it worthwhile incurring the overhead expenses such a system would inevitably entail.

Despite the criticism of the overdetailed procedures outlined in the 1987 A201, A201-1997 amplified the procedural instructions and A201-2007 preserved the changes made in 1997 but transferred them to the IDM. For example, Section 15.2 (corresponding to A201-1997, Paragraph 4.4.3) gives the IDM the power to consult outside experts if they will help him make the decision. Section 15.2.2 (Paragraph 4.4.2) states he can advise the parties that he cannot resolve the claim because he lacks sufficient information. Section 15.2.4 (Paragraph 4.4.4) details how he seeks additional supporting data. Section 15.2.5 (Paragraph 4.4.5) makes clear that his decision must be in writing and must state his reasons for the decision.

B. Standard of Interpretation

Section 29.09 deals with the finality to be accorded decisions by the design professional. This section emphasizes the process from the vantage point of the design professional. How should relevant contract language be interpreted?

This book has dealt extensively with methods by which contracts—especially construction contracts—are to be interpreted.[28] Were it not for the fact that design professionals are often called on to interpret construction documents that they either have prepared or have had a role in preparing, the earlier discussion would suffice. However,

---

[26]292 F.Supp. 300 (D.D.C.1968), aff'd, 429 F.2d 764 (D.C.Cir.1970).
[27]Id. at 304.

[28] See Section 20.02.

the complexity of interpreting and judging one's own work requires some additional comments in this section.

In addition to the limits imposed by the contract language being interpreted, contracts that give interpreting and judging powers to the design professional often specify general standards of interpretation for the design professional to use. For example, AIA Document A201-2007, Section 4.2.12, requires that all "interpretations and decisions of the Architect . . . be consistent with the intent of and reasonably inferable from the Contract Documents." Despite the limits and guides, the design professional often faces the formidable task of deciding disputes over the meaning of contract document language. The prior participation in developing contract documents makes this task even more difficult.

N. E. Redlon Co. v. Franklin Square Corp.,[29] held that the architect must not take into account the intention of the owner or contractor or his intention at the time he drafted the documents in question. According to the court, the architect should first look at the contract documents to see whether they have provided guidance in interpretation matters. In *Redlon*, the contract specifications stated that terms were to be used in their trade or technical sense. After applying any contractual guides given, the architect was to use an objective standard to determine the meaning of contract language.

If an objective standard is desired, design professionals should not consider their or the owner's intention.[30] However, the court may have gone too far when it directed the architect to ignore any intention the contractor may have had when it made the contract.

The design professional should judge the contract documents from the perspective of an honest contractor examining them before bid or negotiation. The owner's preparation of the contract documents, through the design professional, takes a long time, much longer than the contractor has to examine and bid. Any ambiguities the contractor should not have been expected to notice, and to which attention should not have been directed before bid submission or negotiation, should be resolved in favor of the contractor. Conversely, unclear language to which the contractor should have directed attention should be resolved against it.

This conclusion should not be eliminated by a printed clause in the general conditions or specifications that seeks to put the risk of all unclear language in the contract documents on the contractor unless the contractor draws this to the design professional's attention. For reasons described elsewhere,[31] this is an unfair burden to place on the contractor.

Admittedly, a standard that looks at the honest contractor can place design professionals in a difficult position where the language in question was derived from the drawings and specifications they prepared.[32] Openly acknowledging that the specifications were unclear can be a confession of professional failure. The standard of interpretation suggested—that of favoring the contractor under certain circumstances—can inhibit the design professional from ever finding language unclear for fear that it would reflect on his work. An honest design professional should be fair to both owner and contractor despite this possibility.

Perhaps it will be expecting too much for the design professional to step back from his own work and judge it objectively. This possibility may be a reason to accord less finality to the decision. It is certainly a reason the 2007 A201 uses an IDM to review contractor claims. In any event, if unwilling to construe unclear language in favor of the contractor, the design professional will likely use or claim to use the objective standard required in the *Redlon* decision.

C. Form of Decision

The contract clause giving the design professional the power to interpret documents can require that a particular form be followed when a decision is made. AIA Document A201-2007, Section 4.2.12, requires that interpretations be in writing or in the form of drawings. As noted in Section 29.08A, Section 15.2.5 requires that decisions resolving claims must be in writing with the reasons for the decision stated. For this and other reasons, it is generally advisable for decisions to be made in writing and communicated as soon as possible to each party. Where it is not feasible to make a decision in writing on the spot, any

---

[29]89 N.H. 137, 195 A. 348 (1937), aff'd, 197 A. 329 (1938).

[30]The architect's intention was overcome by other evidence in *Alabama Society for Crippled Children & Adults v. Still Constr. Co., Inc.,* 54 Ala.App. 390, 309 So.2d 102 (1975).

[31]See Section 24.03.

[32]The discussion assumes that the language in question was not part of the basic contract or general conditions, writings that should have been drafted or supplied by an attorney. See Sections 12.07, 19.01B, and 19.01D.

oral decision should be confirmed in writing and sent by a reliable means of communication to each party. A written communication giving the design professional's interpretation of his decision need not give reasons to support the decision. The essential requirement is that the parties know that a decision has been made and that they know the nature of the decision. However, the process will work more smoothly if the participants are given a reasoned explanation for the decision. The decision need not be elaborate or detailed but should specify the relevant contract language and facts and the process by which the decision has been made.

Under A201-2007, Section 15.2.6.1, either party may, within 30 days from the date of the initial decision, demand that the other party file for mediation within 60 days of the initial decision. If that demand is made, and the other party does not ask for mediation within the time allowed, then both parties waive their rights to mediate or to pursue "binding dispute resolution proceedings with respect to the initial decision."

D. Appeal

Many construction contracts specify a method of resolving disputes, such as mediation or arbitration. These will be discussed in Chapter 30. By and large the IDM's decision-making powers are designed to create an initial decision that can be appealed. The finality of his decision will be discussed in Section 29.09.

E. Costs

The architect's costs during the construction phase—including interpreting the contract, reviewing submittals, inspecting the contractor's work, and responding to the contractor's requests for information—are included within his basic fee. If the architect is not the Initial Decision Maker, then the architect's cost in assisting the IDM is an additional service.[33]

The AIA documents do not specify how the IDM is to be paid. One reason for creation of the IDM position was the concern of contractors that the architect, because he is paid by the owner, would be biased in favor of the owner. However, if the cost of the IDM is borne by the owner, then the possibility of bias in favor of the owner remains.

If, instead, the cost of the IDM is borne equally by the owner and contractor, then this expense must be budgeted in advance by the contractor when entering into the contract. In either case, the payment mechanism of the IDM should be addressed by separate agreement between the owner and contractor.

The Engineers Joint Contracts Documents Committee (EJCDC) provides in its E-500, A2.02(6) (2002), that evaluating "an unreasonable claim or an excessive number of claims" is a *required* additional service justifying additional compensation above the basic fee. It parallels A201-1997, Paragraph 2.8.2.5, which provided that the architect's "evaluation of an extensive number of claims" was an additional service.

The costs incurred by the parties, such as transporting witnesses to any informal hearing, obtaining any expert testimony, or attending any informal hearings, will be borne by the parties who incur them unless the contract provides otherwise. As most dispute resolution done by the design professional is informal, costs of this sort are not likely to be comparable to those incurred in an arbitration or litigation. Under A201-2007, Section 15.1.3, if the IDM elects to consult with experts or persons with specialized knowledge, then the owner must cover any cost involved.

Suppose the IDM orders that work be uncovered. Uncovering work is costly. Often contracts specify who will pay for the cost of uncovering and recovering. AIA Document A201-2007, Section 12.1.1, states that if work had been *improperly* covered by the contractor, such as covering work despite a request by the architect that it not be covered, the cost must be borne by the contractor even if the work had been properly performed. Section 12.1.2 deals with work properly covered. If the work is found "in accordance with the Contract Documents," costs are borne by the owner. If the work did not comply, the contractor must pay the cost of uncovering and recovering unless the deviation was caused by the owner or a separate contractor (on a project involving multiple primes). Similarly, Section 13.5.3 places the entire cost of "testing, inspection or approval" on the contractor if these processes reveal that the contractor has not complied with the contract documents.

Should the contractor be required to pay the entire cost if any deviation is discovered? One purpose of having the architect visit the site periodically is to observe work before it is covered. The assumption under Section 12.1.2

---

[33] B101-2007, § 4.3.1.11.

is that the architect did not see the defective work before it was covered or request the contractor not to cover it until he had a chance to inspect it. Under such circumstances, the architect may be fearful of a claim being made against him unless some defective work is found.

The all-or-nothing solution is justified only if there were major deviations or if the work was covered to hide defective work. If the defect is slight or inadvertent, it is better to share the cost of covering and uncovering.

SECTION 29.09   Finality of Initial Decision

Parties to a contract can create a mechanism by which disputed matters or other matters that require judgment can be submitted to a third party for a decision. The finality of that decision—that is, whether it can be challenged and the extent of the challenge—can range from 0 to 100 percent. It can be absolutely unchallengeable. At the other extreme, a decision by a third party can be simply advisory.

Issuance of an initial decision by the Initial Decision Maker is a condition precedent to mediation.[34] The AIA does not specify whether the mediator should give deference to the IDM's decision. However, the purpose of a mediator is not to decide issues but to help the parties settle. Mediation should be viewed as a fresh start to resolution of the dispute, and the mediator and parties are not in any way bound by the IDM's decision.

Outside the context of modern AIA contracts, most impartial third-party decisions have some finality but not the finality of an arbitral award, as noted in Section 30.14, or a court ruling. But such decisions are clearly more than advisory. Unless they are shown either to have been dishonestly made or to be clearly wrong, they are likely to be upheld if challenged in court.[35]

The preceding discussion has dealt with decisions by impartial third parties. A design professional cannot be said to be a disinterested third party. In discussing the reasons for limiting the effect of the refusal by an architect to issue a certificate for payment, a New York judge stated, "The rule is based upon the fact that the architect, in contracts of this sort, rarely a disinterested arbiter, is usually the representative of the party, often the owner, who must ultimately bear the cost of the work."[36] This factor—the dubious impartiality of the decision maker—and others noted in this section make the law uncertain and apparently contradictory.

Disputes submitted to the design professional can range from purely factual (Did the work meet certain specific standards?) to matters that though sometimes called legal are really factual (How should this clause be interpreted?) to legal questions (Is the substantial performance doctrine applicable where the design professional is to judge performance?).[37]

The wide range of issues, along with the ambivalent status of the design professional and possible conflict of interest, has led to uncertainty over finality of the design professional's decision. Finality in the first instance depends on the language of the contract giving the design professional the power to make decisions. If it says nothing about conclusiveness or finality, the decision would be purely advisory. But clauses giving this power generally state that the decision shall be "final and binding."

Suppose the architect's decision is final and binding unless arbitration is invoked. If the arbitration clause has been deleted or arbitration waived, the court will extend a considerable amount of deference to the architect's decision.[38] (If arbitration is sought, the arbitrators need pay no attention to the architect's decision. This can encourage arbitration by the party that is dissatisfied with the architect's decision.)

As noted earlier in this section, however, language of finality does not actually mean that the decision is final. One court held the decision to be binding unless there was fraud or bad faith.[39] The Restatement (Second) of Contracts makes the decision binding as long as it is made

---

[34]A201-2007, § 15.2.1.

[35]Earlier edition of this book cited and followed in *Bolton Corp. v. T. A. Loving Co.*, 94 N.C.App. 392, 380 S.E.2d 796, 801, review denied, 325 N.C. 545, 385 S.E.2d (1989).

[36]*Arc Elec. Constr. Co. v. George A. Fuller Co.*, 24 N.Y.2d 99, 247 N.E.2d 111, 113, n. 2, 299 N.Y.S.2d 129, 133 n. 2 (1969). See also *Martel v. Bulotti*, 65 P.3d 192 (Idaho 2003) (refused to apply arbitration standard of review to architect's decision).

[37]Earlier edition of this book cited in *In the Matter of Dutchess Community College*, 57 A.D.2d 555, 393 N.Y.S.2d 77, 78 (1977).

[38]*Martel v. Bulotti*, supra note 36.

[39]*Laurel Race Course, Inc. v. Regal Constr. Co.*, 274 Md. 142, 333 A.2d 319 (1975).

honestly and not on the basis of gross mistake as to the facts.[40]

New York gives greater finality to a certificate that has been issued than to the absence of a certificate. In the absence of a certificate, the contractor can recover if it can show substantial performance, with refusal to issue the certificate being unreasonable.[41]

These varying, though related, degrees of finality make the decision conclusive if honestly made unless it is clear that the design professional made a serious mistake.[42] But the standards do not tell the entire story. Other relevant factors determine the degree of finality.

The particular nature of the dispute is important. If the dispute is more technical and less legal, the decision will be given more finality.[43]

AIA Document A201-2007, Section 4.2.13, states that architect decisions as to aesthetic effect "will be final if consistent with the intent expressed in the Contract Documents." While A201-1997, Paragraph 4.6.1 made clear that such decisions could not be arbitrated, this language was deleted from A201-2007, Section 15.4.1. Nonetheless, the arbitrator's review is narrowly limited to whether the architect's decision was inconsistent with the design intent. The very subjectivity of aesthetic effect and the power for abuse that such a clause can create will inevitably lead to a difference of opinion among courts as to the enforceability of such a clause.[44]

If a certificate is issued and the *owner* refuses to pay, it is likely that the certificate will be considered final.[45] Undoubtedly, this recognizes that the owner for all practical purposes has selected the design professional and should be given less opportunity to challenge the decision. If a subcontractor's rights are at stake, less finality will be given.[46] The subcontractor had even less of a role in selecting the design professional. Giving power to the design professional is often accomplished by incorporating prime contract general terms by reference into the subcontract. The subcontractor may not have had much opportunity to present its case to the design professional.[47]

The availability of arbitration may also play a role. If the decision can be appealed to arbitration, arguably more finality can be given.[48] Another factor that may bear on the degree of finality to be given design professional decisions is the process by which the decision is made. If it appears to have been made precipitously and without elemental notions of fairness, less finality, if any, will be accorded.[49]

Another problem relates to the interaction between language giving finality to a design professional decision and language that bars acceptance of the project from waiving claims for defective work subsequently discovered.[50] One case held that the architect's power to judge performance did not give his decisions finality because of a clause stating that neither the issuance of a final certificate nor final payment relieved the contractor from responsibility for faulty materials or workmanship.[51] Yet another decision more sensibly reconciled these two clauses by concluding that the issuance of a certificate is conclusive where defects are patent or obvious but not where defects are latent, that is, not reasonably discoverable.[52]

---

[40]Restatement (Second) of Contracts § 227, comment c, illustrations 7 and 8 (1981). Other somewhat variant standards are expressed in *Perini Corp. v. Massachusetts Port Auth.*, 2 Mass.App.Ct. 34, 308 N.E.2d 562 (1974) (binding unless arbitrary or in bad faith); *City of Mound Bayou v. Roy Collins Constr. Co.*, 499 So.2d 1354 (Miss.1986) (binding if made in good faith, an honest judgment after a fair consideration of the facts); and *E. C. Ernst, Inc. v. Manhattan Constr. Co. of Texas*, 387 F.Supp. 1001 (S.D.Ala.1974), aff'd, 551 F.2d 1026, rehearing denied in part and granted in part, 559 F.2d 268 (5th Cir.1977) (engineer must use good faith), cert. denied, 434 U.S. 1067 (1978).

[41]*Arc Elec. Constr. Co. v. George A. Fuller Co.*, supra note 36.

[42]Text cited in *Ingrassia Constr. Inc. v. Vernon Tp. Bd. of Educ.*, 345 N.J. Super 130, 784 A.2d 73, 80 (App.Div. 2001).

[43]*Yonkers Contracting Co. v. New York State Thruway Auth.*, 25 N.Y.2d 1, 250 N.E.2d 27, 302 N.Y.S.2d 521 (1969), opinion amended by 26 N.Y.2d 969, 259 N.E.2d 483, 311 N.Y.S.2d 14 (1970); *John W. Johnson, Inc. v. J. A. Jones Constr. Co.*, 369 F.Supp. 484 (E.D.Va.1973).

[44]Compare *Baker v. Keller Constr. Corp.*, 219 So.2d 569 (La.App. 1969) (reviewed decision), with *Mississippi Coast Coliseum Comm'n v. Stuart Constr. Co.*, 417 So.2d 541 (Miss.1982) (refused to review).

---

[45]*Hines v. Farr*, 235 S.C. 436, 112 S.E.2d 33 (1960).

[46]*Walnut Creek Elec. v. Reynolds Constr. Co.*, 263 Cal.App.2d 511, 69 Cal.Rptr. 667 (1968).

[47]*John W. Johnson, Inc. v. Basic Constr. Co.*, supra note 26.

[48]*Roosevelt Univ. v. Mayfair Constr. Co.*, 28 Ill.App.3d 1045, 331 N.E.2d 835 (1975).

[49]*John W. Johnson, Inc. v. Basic Constr. Co.*, supra note 26.

[50]See Section 24.05.

[51]*Flour Mills of America, Inc. v. American Steel Bldg. Co.*, 449 P.2d 861 (Okla.1968).

[52]*City of Midland v. Waller*, 430 S.W.2d 473 (Tex.1968).

SECTION 29.10 Finality: A Comment

If the parties to a contract voluntarily designate a third party to make certain determinations or decide certain disputes and agree that these determinations or decisions shall be binding on both parties, the law should give effect to such an agreement. Parties who genuinely agree to abide by such third-party decisions and who are satisfied that the decision was honestly made are likely to perform in accordance with the decision. In the view of the contracting parties, the third party may be better equipped to give a fair and quick decision. Failure to make such agreements final may discourage parties from agreeing to submit determinations and disputes to third parties or may encourage them not to live up to such agreements. The result can be an increasing burden on an already burdened judicial system.

Contract language and courts often state that in making certain determinations, the design professional is acting in a quasi-judicial function and not as representative of one of the contracting parties. This does not change the realities. The design professional is not a neutral judge, and the contractor has little choice but to accept the current system.

Generally, determinations and decisions by the design professional are followed by the parties. This may be because the parties are satisfied, because it is too costly to arbitrate or litigate, or because of the need to retain the goodwill of the design professional. Giving the design professional initial but reviewable decision-making powers will provide a quick method of handling construction disputes. According such decisions some degree of finality simply gives effect to the superior bargaining position the owner frequently enjoys at the time a construction contract is made.

In *Cofell's Plumbing & Heating, Inc. v. Stumpf*,[53] one issue related to the finality of an engineer's decision under a contract that gave the engineer "binding authority to determine all questions concerning specification interpretations in the execution of the contract." After completion, a dispute arose that related to how particular work should be priced. The project engineer had concluded the work should be compensated under a particular provision in the contract. The owner refused to pay, but the trial court concluded the contractor was entitled to be paid in accordance with the engineer's decision.

The appellate court held that a line must be drawn between "disputes or specification interpretations which deal with work performance and require immediate resolution and those which deal with other contractual disputes, such as rates, method, or time of payment, and do not require prompt on-the-site determination."[54] The court held the contract should be interpreted to give finality to the former but not to the latter. The project will continue whatever the determination by the engineer, at least in the view of the majority.

The dissenting judge wished to expand the finality provision to *both* types of decisions, to prevent unnecessary delays in the construction process while disputes are settled. He contended that the majority's decision will invite delays by contractors who will be forced to stop work while resolving other types of disputes that the engineer, under the interpretation of the majority, cannot resolve with any degree of finality.

It is important to keep the project moving. Decisions needed to accomplish this goal may be given finality. Yet the project can likely be expedited by getting a decision but still allowing a dissatisfied party to challenge it later.

Some American jurisdictions give immunity to the design professional when that party is acting as a judge.[55] In those jurisdictions, to both give the design professional immunity and accord substantial finality to his decisions concentrates too much power in the hands of the design professional. Because this power will occasionally be abused, in those jurisdictions very little, if any, finality should be given to the design professional's decision. Even where the design professional has *no* immunity, it is better to accord no finality to his decisions. The system will work without it, and a needless issue can thus be removed from construction litigation.

SECTION 29.11 The Initial Decision Maker: Some Observations

Creation of the Initial Decision Maker (IDM) position is one of the most significant changes made in A201-2007. This change was made in response to several contractor concerns. Historically, contractors have viewed the architect as biased because the architect was both selected and

---

[53]See supra note 19.

[54]290 N.W.2d at 234.

[55]See Section 14.09D.

paid by the owner. Many also believed the architect could not be impartial in response to allegations of negligent design or failure to timely respond to requests made during the construction phase. The architect would likely be reluctant to render an initial decision blaming himself, both for psychological reasons and in the event of future legal disputes with the owner.

While good reasons may exist for the creation of an IDM position, A201-2007 suffers from a lack of detail as to how the new system would work, especially if the IDM is someone other than the architect. There is no explanation of how the IDM is selected. While the architect must be licensed,[56] there is no similar requirement for the IDM; indeed, the IDM might be a retired judge, lawyer, contractor, or scientist, rather than a design professional. Unless the IDM is put "on staff" from the beginning of the project waiting for disputes to arise (essentially a one-person

dispute resolution board as discussed in Section 30.18E), there may be a significant delay while the IDM is "brought up to speed" on the facts leading up to the present dispute. In any event, significant costs may be involved with employment of an outside IDM; yet A201-2007 gives no indication as to how the IDM is to be paid.[57] There is also no provision shielding the IDM from potential liability and no insurance requirement. If the IDM is a contractor and does not carry professional liability insurance, he may not be covered against a claim. In short, the general lack of guidance in A201 as to the nature of this new arrangement will likely foster disputes.[58]

---

[56] A201-2007, § 4.1.1.

[57] By comparison, the owner and contractor must share equally in the cost of the mediator; see A201-2007, § 15.3.3.

[58] Several of these points are made in Lesser & Bacon, *Meet the 2007 A201 "General Conditions of the Contract for Construction"—Part I: Letting Go of Paper and Making Way for the IDM*, 29 Constr.Litg.Rep., No. 4, Apr. 2008, p. 143.

CHAPTER THIRTY

Construction Disputes: Arbitration and Other Methods to Reduce Costs and Save Time

SECTION 30.01 Introduction

This chapter deals principally with a voluntary method of resolving disputes by submitting disputes to a third party and agreeing to be bound by that party's decision. Third-party resolution can be used to determine narrow issues such as the strength of concrete or the value of property. The former would be classified as third-party testing and the latter as third-party appraisal. The parties can go further and authorize a third party to decide any dispute that might arise between them in the performance of a contract. Such a general referral to the third party can encompass specific disputes narrow in character or broad disputes that can encompass any matter that might be resolved by a court. This latter system, frequently called *arbitration*, is the focal point of this chapter.

This chapter also looks at some methods that are implemented by the judicial system, such as the use of special referees or masters or a summary jury trial. In addition, the chapter looks at techniques by which a third party will facilitate an agreement by the disputing parties. What links these methods is that they all seek to avoid the traditional lawsuit as a means of resolving construction disputes.

Although this chapter deals with intervention by third parties in some form, it must be emphasized that about 95 percent of the disputes in the construction process are resolved through negotiation. It has become fashionable to speak of alternative dispute resolution (ADR) and its various methodologies, but it cannot be overemphasized that the most important initial step toward resolving disputes is good-faith negotiations by the parties to try to solve the problem. This means that the parties must gather information, look at the dispute from the other

party's position as well as their own, and recognize the horrendous costs that can be incurred in terms of time, money, and irrational outcomes because of the intervention of third parties, either through ADR or through the litigation system.

SECTION 30.02 The Law and Arbitration

Until the 1920s, the law was openly hostile to arbitration. Although courts would enforce an arbitral award after it was made, they frustrated agreements to arbitrate disputes that might arise in the future. Some courts found such agreements invalid, some allowed a party to revoke such an agreement prior to award, and some would give only nominal damages for breaching a contract to arbitrate future disputes. In such a legal system, arbitration could not thrive.

In the 1920s, commercial arbitration was greatly encouraged by statutory enactments in a substantial number of states, including commercially important states, that sought to remedy some of the deficiencies in the legal treatment of arbitration. Principally, these arbitration statutes accomplished the following:

1. made agreements to submit future disputes to arbitration irrevocable
2. gave the party seeking arbitration the power to obtain a court order compelling the other party to arbitrate
3. required courts to stop any litigation where there had been a valid agreement to arbitrate a pending arbitration
4. authorized courts to appoint arbitrators and fill vacancies when one party would not designate the arbitrator or arbitrators withdrew or were unable to serve

5. limited the court's power to review findings of fact by the arbitrator and her application of the law
6. set forth specific procedural defects that could invalidate arbitral awards and gave time limits for challenges

Almost all states and the federal government currently have modern arbitration statutes. Many of the modern state statutes were adopted from the Uniform Arbitration Act (UAA) revised in 2000.[1] Clearly, this has greatly encouraged the use of arbitration. Yet this has also created concurrent jurisdiction, where a party is often free to invoke either state or federal arbitration laws. If this does occur, one party may contend that the Federal Arbitration Act (FAA) preempts state arbitration law. As noted in Section 30.03C, this potential conflict has led to recent U.S. Supreme Court cases that have struck down state laws in conflict with the FAA. These disputes can slow down the arbitration process, as such an issue must be resolved in court. The more opportunities to seek judicial rulings, the less desirable it is to include an arbitration clause.

The more favorable attitude toward arbitration in the legislatures and the courts has been tempered by the modern judicial recognition that many agreements to arbitrate are forced on the weaker party through an adhesion contract. This can be demonstrated by two cases. The first case, *Spence v. Omnibus Industries*,[2] involved a dispute between a homeowner and a remodeling contractor over a remodeling contract. The contract—a standardized contract—included a clause requiring arbitration in accordance with the rules of the American Arbitration Association (AAA).

After the dispute arose, the homeowner brought a lawsuit against the remodeling contractor, seeking damages of $37,000. The remodeling contractor filed a petition for arbitration. The petition was granted, and the court ordered that the homeowner pay the arbitration filing fee of $720. The homeowner was willing to arbitrate but appealed the court's decision requiring her to pay the filing fee.

The court reversed the decision regarding the filing fee and in doing so compared the cost of beginning an action in court with the cost of submitting the dispute to arbitration. The court noted that the filing fee for commencing

an action in court would have been $50.50, but it would cost the homeowner $720 to submit the matter to arbitration. The court stated,

> The reason for this disparity is obvious. Courts are established and supported by the State in order to afford forums to which all, rich and poor alike, may present controversies at minimum cost to the parties. Arbitration is supported by the parties. If the parties are equal in bargaining power, arbitration is good. If the parties are not equal, arbitration may deny a forum to the weaker.[3]

After characterizing the contract as one of adhesion (see Section 5.04C), imposed by the party of greater bargaining power on the weaker, the court noted the contract was over 2,000 words jammed into a tightly printed jumble of "terms and conditions." The court noted it was quite unlikely that one homeowner in one hundred would ever read the massive information on the reverse page. Despite the judicial policy favoring arbitration, the court pointed to the strong judicial policy to protect the weaker party to the bargain. The court felt that a $720 fee could discourage a homeowner from presenting a claim against a builder.

Concluding that the homeowner waived her right to arbitrate by filing an action in court,[4] the court held that the contractor seeking arbitration became the initiating party and required the latter to pay the filing fee. Had the homeowner wanted arbitration, according to the court, she would have had to pay the filing fee.

The second case, *Player v. George M. Brewster & Son, Inc.*,[5] involved an arbitration clause in a subcontract. The work was to be performed in California, and the arbitration clause required that arbitration be governed by New Jersey law and be held in New Jersey. The arbitration clause specified that each party would select one arbitrator and the third would be picked by a New Jersey trial judge.[6] The prime contractor, though having its home office in New Jersey, performed much of its work in the western United States.

---

[1] The UAA is discussed in Section 30.14. For a discussion of the Revised UAA, see Ness, *Legislative Update: The Revised Uniform Arbitration Act of 2000*, 21 Constr. Lawyer, No. 4, Fall 2001, p. 35. See also Section 30.09.

[2] 44 Cal.App.3d 970, 119 Cal.Rptr. 171 (1975).

[3] 119 Cal.Rptr. at 172. But, applying Kentucky law, *Stutler v. T.K. Constructors, Inc.*, 448 F.3d 343 (6th Cir.2006) rejected the claim that up-front costs were so prohibitive that arbitration should not be compelled.

[4] See Section 30.05.

[5] 18 Cal.App.3d 526, 96 Cal.Rptr. 149 (1971).

[6] By the time the matter went to court, the prime contractor was willing to have the arbitration heard in California but insisted that the third arbitrator be picked by the New Jersey trial court judge.

The court interpreted the clause to determine whether it covered the particular dispute. Like the court in *Spence v. Omnibus Industries*, this court noted that this was a contract of adhesion and should be construed in favor of the subcontractor. The court stated, "As a whole it appears to be a 'house attorney' prepared form intended by Brewster to be submitted to all of its subcontractors on a take-it-or-leave-it basis."[7] The court then paid some tribute to arbitration, stating,

> The law favors contracts for arbitration of disputes between parties. They are binding when they are openly and fairly entered into and when they accomplish the purpose for which they are intended.
>
> * * *
>
> Our trial courts are clogged with cases, many of them involving disputes between contracting parties. One of the principal purposes which arbitration proceedings accomplish is to relieve that congestion and to obviate the delays of litigation.[8]

The court was concerned that a strong prime contractor that made all its subcontractors agree to such clauses could deprive subcontractors access to the courts in their states. The prime contractor would possess a powerful weapon enabling it to force the subcontractors to arbitrate thousands of miles away from the subcontractor's place of business and to have the third arbitrator, the neutral, be selected by a hometown judge where the prime contractor had its offices. The court concluded,

> A skepticism is born when we read paragraph 13 as to whether it was written as Brewster wrote it for the purpose of expeditious disposition of controversies with its subcontractors. Its plan, it is suggested, may have been designed to effectuate a more unilateral benefit to itself.
>
> * * *
>
> We think the courts would and should scan closely contracts which bear facial resemblance to contracts of adhesion and which contain cross-country arbitration clauses before giving them approval.[9]

The *Player* and *Spence* cases were decided in 1971 and 1975, respectively. They showed the tension between encouraging arbitration as a fast, inexpensive method that avoids the courthouse and relieves an overburdened judicial system and making certain that the right to judicial resolution, including the right to a jury, of the dispute in a convenient forum is given up willingly and knowingly. Two cases decided by two different divisions of the same California intermediate appeals court, that came to different outcomes on slightly different facts, show that this tension has not abated.[10]

Returning to the *Player* case discussed earlier, the court would not permit a stronger party to force the weaker to agree to an inconvenient forum. Yet a different attitude toward arbitration led to another court to uphold a clause that required arbitration in California despite the subcontractor being a Mississippi company and the project a federal one in Mississippi.[11]

Despite increased recognition of realities in the contract-making process, arbitration continues to be looked on favorably as a way to avoid the courthouse.[12]

Still, arbitration is exceedingly controversial in contracts that deal with employment, brokerage, franchises, in addition to consumer contracts.

SECTION 30.03 Agreements to Arbitrate and Their Validity

Arbitration is a voluntary system based on a valid contract to arbitrate. For that reason, analysis of the validity of a general arbitration clause involves the requisite elements for a valid contract as well as an awareness of the special treatment the courts accord agreements to arbitrate.

A. Legal Controls: Submissions and Agreements to Arbitrate Future Disputes

Arbitration requires a valid agreement to arbitrate. As noted, modern law looks favorably upon arbitration. Arbitration relieves courts from having to resolve many disputes. This enables courts to use their facilities and

---

[7]Supra note 5, 96 Cal.Rptr. at 154.
[8]Ibid.
[9]Id. at 156.

[10]*Pardee Constr. Co. v. Superior Court*, 100 Cal.App.4th 1081, 123 Cal.Rptr.2d 288 (2002) and *Woodside Homes of California, Inc. v. Superior Court*, 107 Cal.App.4th 723, 132 Cal.Rptr.2d 35 (2003).

[11]*Ellefson Plumbing Co. v. Holmes & Narver Constructors, Inc.*, 143 F.Supp.2d 652 (N.D.Miss.2000).

[12]See West Ann.Cal.Bus.&Prof.Code § 7085 (consumer complaints to California Contractors Licensing Board resolved by arbitration). See Section 10.11.

skills to resolve other disputes that cannot be handled by arbitration, such as criminal cases, family disputes, or accident claims. Classically, and in the ideal, a properly structured arbitration can provide skilled dispute resolution. When arbitration first developed arbitrators were selected from a pool of persons who were experts in the context of the transaction and possessed knowledge of the customs and risks of the relevant market.

Ideally, arbitration should be quicker and cheaper than judicial resolution of disputes. Also, arbitration is private. Most disputants dislike public airing of their disputes.

The parties can agree to arbitrate after a dispute has arisen, in what are called submission agreements. They do not raise the difficulties encountered when parties agree to arbitrate all future disputes. These latter agreements have generated complexity and controversy.

The legal requirements for a valid agreement to arbitrate clearly go beyond those needed to make an ordinary contract. Section 30.02 illustrated that arbitration can be abused, usually by the party with the economic power to require that disputes be resolved by arbitration. As a rule, this is the party that provides goods and services to another. Arbitration can be the method by which it can deny the party with whom it dealt the right to have its claims addressed in a court presided over by a neutral judge and decided by an impartial jury. Often the party that agrees to arbitrate does not realize what it is getting and what it is giving up.

Section 30.02 noted that the early common law was not friendly toward arbitration. Its rules made arbitration unworkable. This led to statutes, such as the Federal Arbitration Act (FAA) and similar state statutes that encouraged arbitration.

Since the enactment of this legislation, the law became concerned with the possibility of abuse of the dispute resolution process. The economically more powerful party could use arbitration to control the disputes process. The arbitration clause could control the jurisdiction of the arbitrator, the identity of the arbitrator, the place and structure of the hearing, the nature of the award, the remedies allowed, and the finality of the award.

Ironically, in their haste to encourage arbitration, the early arbitration statutes made it almost impossible to challenge the award. As shall be seen in Section 30.14, virtually unchallengeable awards can generate unjust outcomes and citizen discontent.

B. State Statutes

Concern with the one-sided nature of some arbitration clauses generated an outpouring of state statutes. They were designed to ensure that parties asked to assent to arbitration clauses knew what they were getting and to prevent the system from operating in such a way as to bar a dispute from being resolved in a convenient forum by a neutral person after a fair hearing. While the details varied from state to state, these statutes sought to make certain that all the parties, particularly the ones inexperienced and lacking bargaining power, were aware of what they were getting and what they were giving up.

Some statutes compelled arbitration clauses to be in capital letters, be of a different print and a different color than the rest of the contract, be signed separately, and be signed by both the parties and their attorneys, even that the signer knew it was giving up its "day in court."[13]

Statutes took other forms. One bars arbitration in adhesion transactions.[14] Another bars agreements to arbitrate future disputes in consumer transactions.[15] Texas bars agreements to arbitrate that are unconscionable.[16] Others barred clauses that required that the arbitration be held in an inconvenient place, as was seen in the *Player* case discussed in Section 30.02.[17]

These notice statutes sought to avoid what came to be known as procedural unconscionability, a contract that

---

[13]As examples, Montana had required that an arbitration clause in a home improvement contract be in special type and placed just before the signature line. This was held to have been preempted by the FAA in *Doctor's Assocs., Inc. v. Casarotto*, 517 U.S. 681 (1996) and later repealed by Mont.Code Ann. §29-5-114(4). California requires that arbitration clauses in residential projects of four or less units be printed in a certain type, in red, in capital letters and placed before the signature line. It also requires that the clause include language stating that the parties give up their right to go to court. West Ann. Cal.Bus.&Prof. Code §7191. Texas requires in certain transaction of $50,000 or less that there be a special agreement signed by the parties and their attorneys.Tex.Civ.Prac.&Rem. Code Ann. §171.002.

[14]Mo.Rev.Stat.§ 435.350. This statute was interpreted and applied in *State ex.rel. Vincent v. Schneider*, 194 S.W.3d 853 (Mo.2006). Iowa requires a separately signed writing in adhesion contracts. Iowa Ann. Code § 697A-1.

[15]N.Y.Gen.Bus.Code § 399-c. This was interpreted to control a contract between an architect and her client in *Ragucci v. Professional Constr. Services*, 25 A.D.3d 43, 803 N.Y.S.2d 139 (2005).

[16]Tex.Civ.Prac.&Rem. Code Ann. § 171.022. Unconscionability is discussed in Section 30.03D.

[17]West Ann. Cal. Code Civ.Proc. § 410.42a; La.Rev.Stat. § 9-2779; Va. Code Ann. § 8.01-262.1(b); West Wis.Stat.Ann. § 779.135.

did not bring home to the inexperienced party what it was doing.[18]

C. Federal Preemption

This outpouring of state statutes led to a new problem, that of federal preemption. This doctrine bars the application of state law that burdens interstate commerce or frustrates the policy of encouraging arbitration expressed in the FAA. Since most construction transactions affect interstate commerce,[19] a number of cases held that certain state statutes were preempted by federal law.[20] One writer stated that the states must enforce the substantive provisions of the FAA but that state laws that were procedural in nature could be enforced.[21] In any event, some states found that their attempts to regulate arbitration were swept aside by the preemption doctrine. But not every preemption attack succeeded.[22] The uncertainty of the application of preemption added another complication to arbitration.

D. Unconscionability

There was another obstacle to arbitration, the common law doctrine of unconscionability. This was discussed in Section 5.07D. Unconscionability is divided into procedural and substantive elements. As with the state statutes, one purpose of this doctrine is to notify contracting parties that they were agreeing to arbitrate, how it would work and that they were giving up their right to go to court. The doctrine also refuses to enforce unfair or one-sided clauses.

Procedural unconscionability examines how the contract was put together. It seeks to prevent oppression or surprise. Oppression could result from vast disparity of bargaining power, the lack of an opportunity to negotiate, and

the absence of meaningful choice.[23] Surprise directs the inquiry toward hidden terms, ones that were not pointed out to the weaker party and to language that could not be understood.

Substantive unconscionability directs attention to the effect of enforcing a challenged clause. The purpose of the doctrine is to bar enforcement of unfair, one-sided clauses and those that control the outcome by the structure of the arbitration process.

The unconscionability doctrine gives the judge the power to rewrite a contract. This means the judicial attitude will vary from state to state and even from judge to judge. California combines both types of unconscionability and requires that both be met. But a close case of procedural unconscionability can be made up for by a strong case of substantive unconscionability.[24]

Generally, the law has been reluctant to refuse to enforce a contract or a contract clause. The dimensions of the doctrine are imprecise. The doctrine limits freedom of contract. In rare cases courts have set aside contracts or contractual clauses, mainly in consumer or quasi consumer transactions.[25] Where it has been applied to arbitration clauses it has been used mainly in the choice of the arbitrators and the location of the arbitration.[26] This common law doctrine, while operating as a brake on some who would take advantage of arbitration, created additional uncertainty in the enforcement of agreements to arbitrate.

---

[18]See Sections 5.07D and 19.02E.

[19]*Citizens Bank v. Alafabco, Inc.*, 539 U.S. 52 (2003) (commerce protected to the fullest extent possible).

[20]*Doctor's Assocs., Inc. v. Casarotto*, supra note 13: *Shepard v. Edward Mackey Enterprises, Inc.*, 144 Cal.App.4th 1092, 56 Cal.Rptr 3d 326 (2007) (statute allowed defect claims to go directly to court despite arbitration clause in purchase agreement).

[21]Turner, *Under Construction*, ABA Forum Newsletter, March 2007, p. 6.

[22]*Woolls v. Superior Court*, 127 Cal.App.4th 197, 25 Cal.Rptr.3d 426 (2005), review denied June 5, 2005. (Single-family residence did not affect interstate commerce); *Joseph v. Advest, Inc.*, 2006 PA Super 213, 906 A.2d 1205 (2006) (statute providing shorter time to appeal award than the FAA not preempted by FAA).

[23]*Avid Eng'g, Inc. v. Orlando Market Place, Ltd.*, 809 So.2d 1 (Fla. Dist.Ct.App.2001), rehearing denied Feb. 7, 2002 (enforced as negotiated in good faith by sophisticated parties).

[24]*Nagrampa v. Mailcoups, Inc.*, 469 F.3d 1257 (9th Cir.2006). This case applied California law. The issue divided the court and resulted in an *en banc* court (nine judges rather than the usual three). It also noted the division in the federal circuit courts on whether the judge or the arbitrator would decide the issue of unconscionability. This issue will be noted in Section 30.04C.

[25]Ibid. The *Nagrampa* case was a dealer franchise dispute that the court treated as a consumer transaction.

[26]*Sehulster Tunnels/Pre-Con v. Traylor Bros., Inc./Obayashi Corp.*, 111 Cal.App.4th 1328, 4 Cal.Rptr.3d 655 (2003) (dispute review board to which the subcontractor was to submit claim consisted of members approved by the city and the prime). Dispute review boards will be discussed in Section 30.18E. *State ex.rel. Vincent v. Schroeder*, supra note 14 (dispute with developer; arbitrator was president of the Homebuilders Association of Greater St. Louis, who was the defendant seller at that time). In *D. R. Horton, Inc. Green*, 120 Nev. 549, 96 P.3d 1159 (2004) the court refused to enforce an arbitration clause. The clause entitled the developer to $10,000 if the buyer went to court and did not arbitrate as agreed. Also, the clause did not notify the buyer that it was giving up its legal rights.

Remza Drywall, Inc. v. W. G. Yates & Sons Construction[27] provides a good example of the attacks on arbitration based upon unconscionability. It is an opinion by a trial court judge in a federal court applying Mississippi law. A trial court opinion does not create a legal precedent.

The dispute was between a prime and subcontractor over amounts owed the latter on six different projects. The parties had agreed to a broad arbitration clause that required arbitration of all disputes under the rules of the American Arbitration Association. But if the dispute involved third parties, such as the owner or the architect, the prime had the sole discretion to remove the dispute to another forum, arbitral or judicial "to promote economy and avoid inconsistent results."[28] This power could be used to enable consolidation of claims and parties in one hearing. The clause also required that the arbitration be held in a specific city in Mississippi unless the prime designated another city to facilitate the policy of having all claims in one hearing. Also, the subcontractor agreed to waive special, consequential, or punitive damages.

The subcontractor sought to avoid arbitration by asserting both procedural and substantive unconscionability. The assertion of procedural unconscionability was based on the subcontractor's claim that the clauses were "hidden within the contract with no heading," that the clauses were so "complex and legalistic" that it did not have the opportunity "to present the subcontracts to legal counsel," and that the primary language of its principal who signed the agreements was Spanish.

The court rejected this contention. It stated the clause was in the same font and same size as the rest of the contract. (Some statutes require that the clause be larger, in different color and in all caps.[29]) The court found the language clear and and not "complex nor legalistic." The court stated that the subcontractor had time to consult with its attorney. It concluded by noting that inability to understand a contract is not a basis to conclude the contract is unconscionable. Most importantly, the court pointed to this not being a consumer transaction but one between two corporations contracting for over 6 million dollars worth of work.

The next attack was that the contract was substantively unconscionable. Such assertions are often based upon one-sidedness or rigged systems, such as in the appointment of arbitrators. The court found the power given the prime to remove the case from arbitration and take it to court in order to effectuate consolidation was not unconscionable. It concluded that mutuality is not needed, that the power served a useful business function, and that it had to be exercised in good faith. The power given the prime over venue was not, according to the court, too burdensome.

But barring the subcontractor but not the prime from recovering special, consequential, or punitive damages was indeed unconscionable. However, the unconscionability doctrine allows severance of the offending clause. The court severed the clause from the rest of the contract and enforced the arbitration clause.

This case is presented as an example. Other judges might have come to different conclusions. The *Remza* case demonstrates that, in addition to the statutory control over arbitration, the common law also plays a role. But the common law protections tend to be used in noncommercial transactions, such as those that involve consumers who buy goods and services for their own use. But even in the construction world unconscionability can play a role, though a minor one, in arbitration.

E. Mutuality

Another attack often made on arbitration is based upon the clause forcing one party to arbitrate but giving the option to arbitrate to the other. Sometimes attacks on these one-sided clauses are based upon unconscionability.[30] However, often such attacks are premised on the common law doctrine of lack of mutuality, a doctrine that is often linked to consideration.[31] They will be dealt with in Section 30.03H.

F. Underlying Contract

Discussion has centered upon the arbitration clause itself and the laws that seek to prevent abuse of arbitration. On occasion there is a validity attack on the underlying contract containing the arbitration clause. If sustained, the arbitration clause is not enforced. The clause can rise no higher than the underlying contract in which it

---

[27]2007 U. S. *Lexis* 50287, 2007 *Westlaw* 2033047 (S.D. Miss. 2007).
[28]See Section 30.09 for discussion of multiparty arbitrations.
[29]See supra note 13.

[30]*Avid Eng'g Inc. v. Orlando Marketplace Ltd.*, supra note 23 (exchanged one forum for another).
[31]See Section 5.08.

is contained. For example, the arbitration clause was not effective when the underlying contract violated public procurement laws.[32]

The discussion until this point has assumed that an agreement to arbitrate is easily found and the issue is the validity of this agreement. In construction disputes this is often not the case. The search for the relevant language is made difficult because of the set of complex contracts that connect the participants in the construction project. Some complexity results from the common practice of incorporating by reference parts or entire contracts into another contract or at the very least referring provisions to another contract. This is done to save space in an already lengthy contract, to provide a consistent set of rules, and to pass risks down from higher participants to those lower on the chain. These problems are most common in subcontracts.[33]

G. Contract Formalities and Nonsignatories

Usually, the contract that contains an arbitration clause is written and signed by the parties. But those who make contracts do not always follow the method intended to show they assented to the contract. For example, a court concluded that there was a valid contract to arbitrate even though an officer of the contractor did not sign in the blank provided in the contract for his signature.[34]

The court concluded that a signature was not the only way that assent could be manifested. Here, conduct of the parties can make up for the lack of a signature. In this case, the contractor's performance of the contract requirements clearly showed it was bound to the contract and expected the owner to perform as well.

Because of the inherently ambiguous nature of acts, cases often come to different conclusions when the process of making the contract is not followed yet work proceeds.[35]

It is much better to follow the formalities (sign and return the contract) by which the contract is supposed to be made. Sloppiness may be rescued by the court's willingness to consider the total picture. But contracting parties should not rely on being saved by the court.

Usually, the person demanding arbitration and the person against whom it is demanded are parties to the same contract. But on rare occasions, they are not. For example, in *Cuningham Hamilton Quiter P.A. v. M.L. of Miami*, the owner in a design–build (DB) contract sued the architect engaged by the contractor. The architect successfully invoked arbitration based on the arbitration clause in the DB contract.[36]

But the reverse may not be true. In one case the contractor sought to arbitrate with the husband who had participated in the negotiations but had not been a party to the construction contract made by his wife. The contractor claimed unsuccessfully that the wife was the agent of the husband.[37]

The usual method of showing an agreement to arbitrate is a signature on a written contract. But the increasing use of electronic means of communication has led assent to arbitration being done electronically. The FAA requires that agreements to arbitrate must be in writing. However, a federal circuit court held that electronic mail can satisfy the requirement.[38]

The employee claimed he had been terminated in violation of the Americans with Disabilities Act (ADA). The employer sent e-mails to its employees. The e-mail did mention that disputes resolution had a four-step approach, the final being arbitration before a qualified and independent arbitrator. Included was an embedded link to its brochure and handbook that stated all, including ADA disputes, would be arbitrated. The employee claimed he never saw the e-mail. The employer set up a tracking log to monitor whether the employees opened the e-mail, but it did not require a response.

---

[32]*C. R. Klewin Northeast LLC v. City of Bridgeport*, 282 Conn. 54, 919 A.2d 1002 (2007).

[33]See Section 28.04.

[34]*Stinson v. America's Home Place, Inc.*, 108 F.Supp.2d 1278 (M.D.Ala.2000). Among the many cases holding the signature is not always needed, see *Medical Dev. Corp. v. Industrial Molding Corp.*, 479 F.2d 345 (10th Cir.1973). Cases are collected in Windle, *Arbitration and the Unsigned Contract*, 15 Common Sense Contracting, No. 13, ¶408 (Smith Currie & Hancock).

[35]Compare *Landmark Properties, Inc. v. Architects International-Chicago*, 172 Ill.App.3d 379, 526 N.E.2d 603 (1988) (developer bound to arbitration clause despite not having signed or returned the contract:

having referred to the contract and not objecting showed an intent to be bound) with *Brooks & Co. General Contractors, Inc. v. Randy Robinson Contracting, Inc.*, 257 Va. 240, 513 S.E.2d 858 (1999) (contract not signed or returned and starting work without objection did not show agreement with terms).

[36]776 So.2d 940 (Fla.Dist.Ct.App.2000), rehearing denied Jan. 21, 2001.

[37]*Ellsworth v. American Arbitration Ass'n*, 148 P.3d 983 (Utah 2006).

[38]*Campbell v. General Dynamics Government Systems Corp.*, 407 F.3d 546 (1st Cir.2005) (Massachusetts law).

The court held under the right circumstances e-mail can satisfy the FAA writing requirement. But there must be some minimal level of notice to the employee. There was no actual notice. Then the issue was whether a reasonable person would have recognized this new method of resolving disputes. Here, according to the court, this communication involved a term of employment, and the e-mail itself did not make clear what the employee was being asked to do: to submit his dispute to an arbitrator and give up his right to go to court to assert his rights under the ADA. An e-mail can be effective if the proponent shows minimally sufficient notice.[39]

Another case with a different outcome involved a notice of arbitration on the seller's website that had become incorporated by reference into the purchase agreement. The website notice was in blue type.[40] This made it conspicuous. The purchaser purchased online and was not a novice in using computers.

Electronic communications can be effective but the facts in each case will determine whether the needed notice was present.

H. Fraud, Mutuality, Termination of Contract and Conditions Precedent

Section 30.03 deals with the validity of the agreement to arbitrate. Other attacks surface in construction contract disputes. The main ones are fraud, mutuality (one-sided clauses), the effect of termination of the underlying construction contract, or frustration of the disputes process due to nonoccurrence of a condition precedent to arbitration.

Fraud can be a challenge to any agreement to arbitrate. Where fraud is claimed, the principal issue is who decides whether the challenge is meritorious. For that reason fraud will be examined in Section 30.04C. It treats who decides certain issues. That section also introduces the seminal *Prima Paint* doctrine. It is central to the issue of who decides when these attacks are made. Other legal attacks on arbitration are discussed here.

One attack on an arbitration clause is that it lacks mutuality—when one party is required to arbitrate but the other is not.

A Florida case shed light on this method of avoiding arbitration.[41] The contract was between a shopping center and an engineer. The contract provided that all disputes would be arbitrated under the Construction Industry Rules of the American Arbitration Association "at the sole discretion of the Engineer."[42] The shopping center brought a lawsuit against the engineer. The engineer sought arbitration. The trial court refused to order arbitration as the arbitration clause lacked mutuality since only one party, the engineer, could demand arbitration.

The mutuality concept is an offshoot of the consideration requirement for enforcement of a contract discussed in Section 5.08. Here there was adequate consideration for the contract. The appellate court concluded the issue was one of unconscionability.[43] The court held this clause was not procedurally unconscionable. The contract had been negotiated at arms' length by sophisticated parties. Also, the clause was not substantively unconscionable. It merely exchanged one forum for another.

There may be contrary outcomes in cases that involve one-sided clauses, depending on the facts. But the fact that one party can choose to arbitrate will not as a rule automatically invalidate the arbitration even in consumer transactions.[44]

Discussion in this Section has centered on the validity of agreements to arbitrate. Suppose a valid contract with an arbitration clause is terminated? What effect does termination have on the agreement to arbitrate?

Termination can result from a serious breach by one of the parties,[45] unforeseen events that cause termination

---

[39]*Caley v. Gulfstream Aerospace Corp.*, 428 F.3d 1359 (11th Cir.2005) (Georgia law) declining to follow *Campbell*, supra note 38.

[40]*Hubbert v. Dell Corp.*, 359 Ill.App.3d 976, 835 N.E.2d 113 (2006).

[41]*Avid Eng'g, Inc. v. Orlando Marketplace, Ltd.*, supra note 23.

[42]Id. at 2.

[43]See Sections 5.07D, 19.02E, 30.02 and 30.03D.

[44]*Kalman Floor Co. v. Jos. L. Muscarelle, Inc.*, 196 N.J.Super 16, 481 A.2d 553 (App.Div.1984), aff'd, 98 N.J. 266, 486 A.2d 334 (1985) (after reviewing decisions with different conclusions upheld clause). See also *Albert M. Higley Co. v. N/S Corp.*, 445 F.3d 861 (6th Cir.2006) (enforced even if prime had discretion in prime–subcontractor dispute); *State ex. rel. Vincent v. Schroeder*, supra note 14 (enforced even if seller could go to court to obtain an injunction to protect its intellectual property and buyer had to submit all of its claims to arbitration). See also Nahmias, *The Enforceability of Contract Clauses Giving One Party the Unilateral Right to Choose Between Arbitration and Litigation*, 21 Constr. Lawyer, No. 3, Summer 2001, p. 36 (trend toward enforcement but no majority rule or consensus in courts).

[45]See Section 33.04.

(force majeure),[46] or an exercise of a power to terminate for convenience.[47] Termination will be discussed more fully in Chapter 33.

Termination not based on a formation defect should not and does not abrogate any broad arbitration clause.[48] This can become cloudy because some dispute resolution systems include steps that must be taken prior to arbitration; for example, the AIA system requires an initial decision by a design professional or any Initial Decision Maker (IDM). These steps are usually a condition precedent to arbitration. But the power of the architect or any IDM under A201-2007, Section 15.2.1 terminates 60 days after final payment is due.

Suppose the IDM or any design professional given the power to initially resolve disputes is no longer "on board" because her power expired. This should mean that the dispute can proceed directly to arbitration because the condition precedent had been excused.[49] One case held that the failure to have an interim decision eliminates the arbitration requirement.[50] This frustrates the intention of the parties and the agreed dispute resolution method.

This issue raises the question of whether the arbitration clause covers post-completion claims. This will be discussed in Section 30.04A.

SECTION 30.04 Specific Arbitration Clauses: Jurisdiction of Arbitrator and Timeliness of Arbitration Requests

A. Jurisdiction Conferred by Clause

A frequently disputed issue relates to whether the arbitration clause covers the particular matter in dispute. Obviously, much depends on the language of the clause,

and a broadly drafted arbitration clause can cover almost anything that relates to the contract between the parties. Alternatively, the clause can specify that it applies only to claims that do not exceed a certain amount or claims that involve *factual* as opposed to *legal* disputes.

Whether the clause has conferred jurisdiction on the arbitrator depends on the language of the arbitration clause as well as judicial attitude toward arbitration. A few examples illustrate this. Courts have differed as to whether delay damages fall within an arbitration clause.[51] A court seeking to encourage arbitration would not limit arbitration to damages solely to person or property.[52] Another court less favorable to arbitration held that a general arbitration clause did not cover the owner's possible liability for water damage.[53] That court required that the clause be "crystal clear" before it would confer jurisdiction on the arbitrator.

Another court held that a general arbitration clause in a subcontract did not confer jurisdiction on the arbitrator to determine which portion of the funds the prime contractor received to train minority workers should go to the subcontractor.[54] That same decision refused to permit the architect to arbitrate a dispute between prime contractor and subcontractor when the subcontractor's principal claim was that the architect had committed design errors.

Courts have differed in their willingness to encompass implied terms under the arbitration clause.[55] Most courts hold that the arbitration clause will be applied to disputes that arise after the work is completed despite a provision

---

[46]See Section 23.03D.

[47]See Section 33.03B.

[48]*Middlesex County v. Gevyn Constr. Co.*, 50 F.2d 53 (1st Cir.1971), cert. denied, 405 U.S. 955 (1972); *Reiss v. Murchison*, 384 F.2d 277 (9th Cir.1967); *Auchter Co. v. Zagloul*, 949 So.2d 1189 (Fla.Dist. Ct.App.2007). But see *Tri-Star Petroleum Co. v. Tipperary Corp.*, 107 S.W.3d 607 (Tex.Civ.App.2003), rehearing overruled July 2, 2003, review denied April 7, 2004, mandamus denied April 7, 2004, rehearing for petition for review denied June 25, 2004, rehearing for petition for mandamus overruled July 2, 2004 (refusing to follow *Middlesex* case).

[49]*Auchter Co. v. Zagloul*, supra note 48.

[50]*Lopez v. 14th Street Development LLC*, 40 A.D.3d 313, 835 N.Y.S.2d 186 (2007).

[51]*Harrison F. Blades, Inc. v. Jarman Memorial Hosp. Bldg. Fund, Inc.*, infra note 55, held a delay damage claim beyond the scope of arbitration, while *Aberthaw Constr. Co. v. Centre County Hosp.*, 366 F.Supp. 513 (M.D.Pa.1973), aff'd, 503 F.2d 1398 (3d Cir.1974), held that arbitration was required. See *Smay v. E. R. Stuebner, Inc.*, infra note 62.

[52]*Muhlenberg Township School Dist. Auth. v. Pennsylvania Fortunato Constr. Co.*, 460 Pa.260, 333 A.2d 184 (1975).

[53]*Silver Cross Hosp. v. S. N. Nielsen Co.*, 8 Ill.App.3d 1000, 291 N.E.2d 247 (1972).

[54]*Paschen Contractors, Inc. v. John J. Calnan, Co.*, 13 Ill.App.3d 485, 300 N.E.2d 795 (1973).

[55]*Roosevelt Univ. v. Mayfair Constr. Co.*, 28 Ill.App.3d 1045, 331 N.E.2d 835 (1975) and *Allentown Supply Corp. v. Hamburg Municipal Auth.*, 463 Pa. 167, 344 A.2d 477 (1975), held that arbitration encompassed implied terms, whereas *Harrison F. Blades, Inc. v. Jarman Memorial Hosp. Bldg. Fund, Inc.*, 109 Ill.App.2d 224, 248 N.E.2d 289 (1969), did not.

in the arbitration clause stating that work will continue while the dispute is being arbitrated.[56]

Prior to 2007, AIA Document A201-1997, Paragraph 4.6.1, dealing with arbitration was very broad. It covered any claim "arising out of or relating to the Contract." It excluded only claims relating to aesthetic effect, claims waived by final payment, and those dealing with consequential damages.

But in 2007 changes were made in dispute resolution. A101/201 set up a three-step process and gave more scope to mediation. The first was a decision by an Initial Decision Maker (IDM). She could be a named individual designated in A101-2007, Section 6.1. In the absence of a named IDM the architect acted in this role.

Mediation was the next step and the third, arbitration, if selected in A101-2007, Section 6.2. Since the steps are interlocked, attention to each step must be noted to determine the jurisdiction of the arbitrator. Special attention will be directed to disputes relating to aesthetic effect because this was clearly excluded in 1997 and in prior editions of A201.

Before looking at A201-2007 Article 15 dealing with disputes, it should be noted that A201-1997 Paragraph 4.2.13 stated that "decisions on matters relating to aesthetic effect will be final. . . ." This was unchanged in A201-2007. But what was clear in 1997 became less clear in 2007. Section 15.2.1 excludes claims that must be sent to the IDM, including those related to hazardous materials[57] and distribution of funds received from an insurer for property damage.[58] There was no exclusion for aesthetic effect disputes as in 1997.

The second step is mediation. Section 15.2.1 requires that all claims and disputes be mediated except those waived by final payment and those that affect consequential damages. Disputes that involve hazardous materials or distribution of insurance proceeds are not excluded.

Arbitration under Section 15.4 is the third step. All disputes not resolved by mediation must be arbitrated if that method is selected in A101, but mediation is a condition precedent to arbitration.[59] If a claim need not be mediated, apparently it need not be arbitrated. But not all claims that go to mediation need first go to the IDM.

In this Byzantine system what happens to matters relating to aesthetic effect clearly excluded in A201-1997? All that remains is Section 4.2.13, which states that the architect's decision is final "if consistent with the intent expressed in the Contract Documents." If there is a dispute over consistency, it must be mediated and possibly arbitrated. That the AIA removed the exclusion in 2007 from both mediation and arbitration that had been included in A201-1997 would support this conclusion.

Another jurisdictional issue relates to the arbitrability of tort claims.[60] Florida has allowed tort claims to be arbitrated, even ones for noneconomic losses.[61] Similarly, an indemnity claim triggered by a tort claim for personal harm fell within the arbitration clause.[62] As noted earlier, judicial resolution of the jurisdictional question is likely to be influenced by the court's attitude toward arbitration, the relative bargaining power of the parties, and the apparent appropriateness of arbitration for a particular dispute.

A recent Utah case showed the difficulty that jurisdiction of the arbitrator can generate. In this case, a developer and a subcontractor had a dispute that involved two parcels of land. They settled the dispute over the first and agreed to arbitrate their dispute over the second. At the arbitration, both introduced evidence that related to the first parcel that was not before the arbitrator. The arbitrator considered this a modification of the submission agreement and ruled on both parcels.

The Utah intermediate court of appeals agreed, concluding that the parties could expand the written submission agreement by their conduct.[63]

---

[56]*Warren Bros. Co. v. Cardi Corp.*, 471 F.2d 1304 (1st Cir.1973); *Auchter v. Zagloul*, supra note 48; *Hudik-Ross, Inc. v. 1530 Palisade Ave. Corp.*, 131 N.J.Super. 159, 329 A.2d 70 (App.Div.1974); *Warwick Tp. Water & Sewer Auth. v. Boucher & James*, 2004 PA Super 201, 851 A.2d 953 (2004). But see *Lopez v. 14th Street Development LLC*, supra note 50 (absence of a decision by the architect precluded arbitration). A case to the contrary was *Hussey Metal Div. v. Lectromelt Furnace Div.*, 471 F.2d 556 (3d Cir.1972) (clause stating that no demand for arbitration could be made after final payment). See Section 33.03I.

[57]A201-2007, §§ 10.3, 10.4.

[58]Id. at §§ 11.3.9, 11.3.10.

[59]Id. at § 15.3.2.

[60]*Harman Elec. Constr. Co. v. Consolidated Eng'g Co.*, 347 F.Supp. 392 (D.Del.1972); *Morton Z. Levine & Assoc., Chartered v. Van Deree*, 334 So.2d 287 (Fla.Dist.Ct.App.1976).

[61]*Seifert v. U.S. Home Corp.*, 750 So.2d 633 (Fla. 1999); *Kaplan v. Kimball Hills Homes Florida, Inc.*, 915 So.2d 755 (Fla.Dist.Ct.App.2006). (claim for emotional distress).

[62]*Smay v. E.R. Stuebner, Inc.*, 2004 PA Super 493, 864 A.2d 1266 (2004).

[63]*Pacific Dev. L.C. v. Orton*, 982 P.2d 94 (Utah App.1999) cert. granted, 20 P.3d 403 (table) 1999.

The Utah Supreme Court reversed, concluding that there could be no implied expansion of a written submission agreement.[64] It expressed concern that this sort of informal expansion would discourage persons from arbitrating, that the written agreement should set firm boundaries, and that Utah law required a written agreement.

This holding shows that the sometimes casual approach the law takes to informal modification of contracts will not be applied where the issue is the jurisdictional aspects of agreements to arbitrate.[65]

Although arbitration typically involves performance problems, it has been held that a claim by a contractor that it should be relieved from a bidding mistake should be decided by arbitration.[66]

Although it will be discussed again in Section 30.12, dealing with arbitral remedies, *Advanced Micro Devices v. Intel Corp.*[67] can be looked at in a jurisdictional sense as well. In that case the arbitrator ruled for the claimant, AMD, on its claim that Intel violated its obligation of good faith and fair dealing. The arbitrator ruled that AMD could use certain Intel intellectual property, in effect requiring that Intel forfeit a defense in a separate federal copyright suit *not* before the arbitrator. Although the intermediate California appellate court found this ruling to be beyond the arbitrator's jurisdiction, the California Supreme Court confirmed it in a 4-to-3 decision. The majority held that the remedy bore a rational relationship to the underlying contract as interpreted by the arbitrator or the breach of contract found by the arbitrator. While all the judges agreed that this award went *beyond* what a court could do, the dissenters did not agree with the broad interpretation of jurisdiction given by the majority.

The arbitrator must have found Intel guilty of extremely oppressive conduct and held that the only way to "make things right" was to bar it from using a defense in another dispute not before the arbitrator—in effect punishing Intel.

This decision may lead some parties to contractually specify very sharp limitations on the remedies the arbitrator can award, to state in the contract that the award must be in accord with state law, or to give up arbitration in favor of another method of resolving the dispute.

The increasing attempt of nonparties to enforce contracts by claiming that they are intended beneficiaries of contracts made by others has added to arbitration complexity. This increases the likehood that contracting parties in this context will exclude third-party claims by appropriate contract language. Courts should use caution in granting enforcement rights to a third party as an intended beneficiary.

Expiration of any time limit specified in the contract or statute for making the award generally ends jurisdiction of the arbitrator, and any awards made after expiration are invalid. This rule has been eroded where an award made after expiration of the time limit is enforced on the basis of waiver, such as proceeding with the arbitration without protest after expiration or failure to protest after a late award has been rendered. Waiver can also be predicated on a course of conduct that shows the parties did not consider time of the essence, especially where no prejudice is shown. If the expiration period for making the award has been waived, the award must be made within a reasonable time.[68] One of the more difficult jurisdictional questions relates to the application of an arbitration clause to a disputed termination (discussed in Section 33.03I).

Jurisdiction problems highlight drafting approaches between no arbitration and a general arbitration clause. They are discussed in Section 30.17.

B. Timeliness of Arbitration Demand

AIA Document A201-1997 had set two standards for timeliness. Decisions by the architect that state that they are final and subject to appeal must be appealed to arbitration within thirty days.[69] If the decision did not contain such a statement, arbitration must be requested within a reasonable time.[70] This method was criticized in the seventh edition of this book because it gave the architect the power to create a fixed period in her award. Failure to do so would inject a reasonable time into the formula for deciding the time limit on submission of claims. This can raise proof issues.

AIA Document A201-2007 made a change. Section 15.3.1 makes mediation a condition precedent to

---

[64]23 P.3d 1035 (Utah 2001).

[65]See Section 19.01H.

[66]*Village of Turtle Lake v. Orvedahl Constr. Co.*, 135 Wis.2d 385, 400 N.W.2d 475 (App.1986).

[67]9 Cal.4th 362, 885 P.2d 994, 36 Cal.Rptr.2d 581, cert. denied, 512 U.S. 1205 (1994).

---

[68]Annot., 56 A.L.R.3d 815 (1974).

[69]AIA Doc. A201-1997, ¶4.4.6.

[70]Id. at ¶4.6.3.

arbitration, and Section 15.4.1.1 requires that a demand for arbitration be "made no earlier" than the filing of a request to mediate "but in no event shall it be made after the date" when a legal action "would be barred by the applicable statute of limitations." This would make local law the standard, and, as shown in Section 23.03G, this can be a difficult standard to apply.

Another method of handling this seemingly simple issue is shown by the EJCDC. Dealing with the issue of when a claim must be made to the engineer, EJCDC C-700 (2007), Paragraph 10.05B, states the claim must be made "promptly (but in no event later than 30 days) after the start of the event giving rise" to the claim. Here the general standard is promptness with a thirty-day cap.

Paragraph 11.05E states the decision of the engineer becomes final unless one of the parties elects to mediate the dispute under Article 16. The engineer's decision is final and binding thirty days after termination of the mediation unless one party submits the claim to the disputes process selected in the Supplementary Conditions. Fixed time limits set forth in the contract are preferable.

C. Who Decides Jurisdiction and Timeliness: *Prima Paint* Doctrine

One question has generated a large number of cases: whether the court or the arbitrator decides preliminary issues.[71] These issues do not relate to the substantive disputes that determine who should prevail. Instead they deal with the validity of the arbitration clause and its underlying contract. These issues include compliance with any formal requirements and attacks on the formation of the contract, such as claims of fraud, unconscionability, or illegality of the underlying contract. Other issues that precede the actual hearing can include whether any conditions precedent have occurred, such as submitting the dispute to a third party or mediators. There are other illustrations, but even this list brief list shows that many disputes can arise before the actual arbitration hearing.

The question of who decides these preliminary issues uncovers some strategic characteristics of dispute resolution. Often arbitrating parties admit they agreed to

arbitrate but choose strategies to either avoid the process they have selected or hope to delay the actual arbitration for tactical reasons.

They may have second thoughts about arbitration and now prefer to go to court. Alternatively, they may believe that delay will help them make a better settlement. Some parties have financial problems and need the money they hope to receive. Delay then benefits the other party, who may able to use this to make a better settlement.

These tactics often lead parties to allege that the underlying agreement was procured by fraud. Fraud is easy to allege but difficult to prove. This led to *Prima Paint Corp. v. Flood & Conklin Manufacturing Co.*[72] and the *Prima Paint* doctrine. The United States Supreme Court used the divisibility fiction that artificially severs the arbitration clause from the underlying contract. If the fraud attack challenges the arbitration clause itself, the issue goes to the judge. If it asserts other challenges to the underlying contract, the issue is decided by the arbitrator. This doctrine made it more difficult to attack the arbitration clause and strengthened arbitration.

The *Prima Paint* doctrine was recently affirmed in *Buckeye Check Cashing, Inc. v. Cardegna.*[73] The *Buckeye* case held that a charge that the underlying contract was an illegal usurious contract was to be decided by the arbitrator. The entire contract, not the arbitration clause, was challenged as illegal.

But it is important to see what the Court stated it was not deciding. It was not deciding whether any valid agreement to arbitrate was made, such as whether the party signed the agreement to arbitrate, whether the signer had authority, and whether the signer had the mental capacity to make a contract.[74]

Implementation of the *Prima Paint* doctrine has not been easy. For example, in the *Buckeye* case the trial court held the issue was to be decided by the court. The Florida intermediate court of appeals disagreed and held it should be decided by the arbitrator.[75] The Florida Supreme Court held it should be decided by the court.[76] And as seen, the United States Supreme Court held that the arbitrator decided this issue.

---

[71]The singular will be used. Of course, there may be a panel of arbitrators. Three is common where the claims are for very large sums. See Section 30.18A.

[72]388 U.S. 395 (1967).
[73]546 U.S. 440 (2006)
[74]Id. at 444, n. 1.
[75]824 So.2d 228 (Fla.Dist.Ct.App.2005).
[76]894 So.2d 860 (Fla.2006).

Another illustration of the vexatious nature of the *Prima Paint* doctrine can be seen in a recent decision of a federal appeals court.[77] In that case the trial court held that the issue of unconscionability would be decided by the arbitrator; the federal appeals court, sitting in a three-judge panel, agreed, but the court meeting *en banc* (an expanded panel of nine judges) in a divided opinion disagreed and held that the issue would be decided by the court. The dissent noted the disagreements in the federal circuit courts over this issue.[78]

At the extremes there is no confusion. All would agree that the issue of whether there was valid agreement to arbitrate, not the validity of the underlying contract, should be decided by the court. At the other extreme, procedural issues, such as how the arbitrator conducts the hearing, should be decided by the arbitrator.[79] But on issues in between these extremes there can be wide variation in judicial outcomes. While some courts give these issues to the trial court to decide,[80] the tendency has been to give more preliminary issues to the arbitrator.[81]

The complexities of who decides can be shown by *Wagner Construction Co. v. Pacific Mechanical Corp.*, a recent California case.[82] The court held that the arbitrator decides whether the underlying claim is barred by the statute of limitations. The court decides whether the claim has been filed within the period specified in the arbitration clause or set by the association handling the arbitration. Waiver of arbitration is decided by the court.

Arbitration works best when all or almost all of the issues are decided by the arbitrator and not the judge. Repeated trips to the courthouse defeat the main purpose of arbitration, that of an alternate process that is cheaper, quicker, and more expert than litigation.

Yet it must be recognized that arbitrators are not perfect. For that reason, Section 30.14 describes the constant struggle over the scope of judicial review. But in this section, the issue is who will decide preliminary questions, not the substantive outcome of the arbitration.

To be sure, some of these preliminary questions are important, such as the validity of the agreement to arbitrate future disputes, any submission agreement to submit existing disputes to arbitration, and the jurisdiction of the arbitrator. Yet in addition to the *Prima Paint* doctrine of severability, recent cases cited in note 81 show that the law increasingly allows arbitrators to decide these preliminary questions. This smoothes the path to arbitration.

---

[77]*Nagrampa v. Mailcoups, Inc.*, supra note 24.

[78]469 F.3d at 1299.

[79]*Industra/Matrix Jt. Venture v. Pope & Talbot, Inc.*, 341 Or. 321, 142 P.3d 1044 (2006). As shall be seen in Section 30.14 serious misconduct can be grounds to overturn the award.

[80]*Nagrampa v. Mailcoups Inc.*, supra note 24 (court decides unconscionability); *Wagner Constr. Co. v. Pacific Mechanical Corp.*, infra note 82 (court decides waiver of arbitration); *Steven L. Messersmith v. Barclay Townhouse Assoc.*, 313 Md. 652, 547 A.2d 1048 (1988) (court decides jurisdiction and timeliness).

[81]*Howsam v. Dean Witter Reynolds, Inc.* 537 U.S. 79 (2002) (arbitrator decides statute of limitation issue); *John H. Goodman Ltd. Partnership v. THF Constr. Inc.*, 321 F.3d 1094 (11th Cir.2003) (arbitrator decides whether unlicensed contractor can recover for work performed); *C. R. Klewin Northeast LLC v. City of Bridgeport*, supra note 32 (illegality of underlying contract to arbitrator); *O'Keefe Architects, Inc. v. CED Constr. Partners Ltd.*, 944 So.2d 181 (Fla.2006) (arbitrator decides timeliness of arbitration demand); *City of Lenexa v. C. L. Fairly Constr. Co.*, 15 Kan.App.2d 207, 805 P.2d 507 (1991), review denied Apr. 23, 1991 (arbitrator decides timeliness of arbitration demand); *Rockland County v. Primiano Constr. Co.*, 51 N.Y.2d 1, 409 N.E.2d 951, 431 N.Y.S.2d 478 (1980) (arbitrator decides timeliness of demand and conditions precedent); *Industra/Matrix Jt. Venture v. Pope & Talbot, Inc.*, supra note 79 (arbitrator decides timeliness of demand, conditions precedents met, and effect of lack of license); *Ross Development Co. v. Advanced Bldg Development Inc.*, 2002 Pa. Super. 194, 803 A.2d 194 (2002) (arbitrator decides whether condition precedents met or excused).

SECTION 30.05   Waiver of Arbitration

The law generally favors arbitration. Yet a party may indicate by its acts that it chooses to litigate even though it had agreed to arbitrate. If it does so, the other party has a choice. It can compel arbitration by filing a motion in court to hold off litigation until the dispute is submitted to arbitration. Alternatively, it can decide to have the dispute handled in court.

The issue of wavier arises when the party who took steps inconsistent with the desire to arbitrate changes its mind and seeks to arbitrate. The court, or in some cases, the arbitrator must decide whether that party has waived or lost its right to arbitrate.

In deciding whether the right to arbitrate has been waived, two factors may be relevant. The first is whether the acts by one party in pursuing litigation were so inconsistent with the desire to arbitrate that it is clear that party no longer wishes to arbitrate. The second is whether

---

[82]41 Cal.4th 19, 157 P.3d 1029, 58 Cal.Rptr.3d 434 (2007).

"blowing hot and cold" on arbitration, first seeking another way of resolving the dispute, and then seeking to go back to arbitration, prejudiced the other party.

Some courts emphasize the first factor, the issue of intent to waive arbitration. Was the evidence clear that the party who now seeks to arbitrate manifested a desire to waive it? If so, arbitration has been waived.

Other courts, more supportive of arbitration, require that there be evidence not only of an intent to waive arbitration but that the other party has been prejudiced by this change of course. Has it been put in a worse position by the "blowing hot and cold" by the other party? Only then has there been waiver of arbitration.

Regarding inconsistent acts as a waiver, two writers state:

> Actions that have been held to be inconsistent with the continued right to arbitrate include the initiation of litigation, participation in discovery on claims subject to an arbitration agreement, or, more generally, taking actions "adverse to the arbitration process." The courts generally recognize exceptions for pretrial actions or motions designed to avoid litigation or to stay proceedings, where they are potentially consistent with enforcement of the arbitration provision. Similarly "a party can also 'conduct discovery with respect to non-arbitrable claims without waiving their right to arbitrate.'" The Fifth Circuit stated that inconsistent activity can be marked by acts that indicate intent to repudiate the right to arbitrate. This can occur either prior to or during litigation, but most frequently involves participation in a lawsuit without seeking to invoke the arbitration agreement.[83]

As to a prejudicial change of position these commentators state:

> For example, prejudice may arise from delay in asserting the right to arbitrate and the consequential risk of lost evidence, the duplication of effort in litigating first and then arbitrating, the use of discovery methods in litigation that would be unavailable in arbitration, or, more generally, "when a party instigates litigation of substantial issues going to the merits."[84]

Just as these waiver cases may be influenced by the favorable attitude toward arbitration, the AAA Construction Industry Arbitration (CIA) Rule R-49(a) states that "No judicial proceeding by a party relating to the subject matter of the arbitration shall be deemed a waiver of the party's right to arbitrate." The AIA more specifically seeks to avoid the filing of a lien as a waiver of contractual dispute resolution system.[85] These efforts to avoid waiver of arbitration will not always be effective.[86] One court held that the court decides waiver of arbitration.[87] As to who decides whether the right to arbitration has been waived, see Section 30.04C.

Although cases will continue to seem contradictory because of the many factual situations, the general trend may be indicated by a case that stated that arbitration was favored and that waiver would not be lightly inferred. The court noted the necessity for showing prejudice rather than simply showing inconsistent acts and indicated that recent waiver cases had involved demands for arbitration long after suits had been started and after discovery proceedings had taken place.[88]

A number of recent cases continue this trend.[89] But if a lien claimant wishes to protect its rights to arbitrate, legal advice should be obtained.

SECTION 30.06 Prehearing Activities: Discovery

The party initiating arbitration usually files a notice of an intention to arbitrate and, under the AAA Construction Industry Arbitration Rules (CIA Rules) (there are also Fast-Track and Large/Complex Rules) (see Appendix F),

---

[83]Ness & Peden, *Arbitration Developments: Defects and Solutions*, 22 Constr. Lawyer, No. 3, Summer 2002, pp. 10, 12 (footnotes omitted).

[84]Id. pp. 12–13 (footnotes omitted).

[85]A201-2007, § 15.2.8; B101-2007, § 8.2.1.

[86]See *Joba Constr. Co., Inc. v. Monroe County Drain Comm'r*, 150 Mich.App. 173, 388 N.W.2d 251 (1986), appeal denied June 20, 1986 (despite waiver language in predecessor of CIA Rule R-49(a)).

[87]*Zedot Constr., Inc. v. Red Sullivan's Conditioned Air Services*, infra note 89. See also Section 30.04C, notes 80, 81.

[88]*Gavlik Constr. Co. v. H. F. Campbell Co.*, 526 F.2d 777 (3d Cir. 1975).

[89]*Nagrampa v. Mailcoups, Inc.*, supra note 24 (limited involvement in pretrial matters not a waiver: strong dissent); *Zedot Constr. Inc., Red Sullivan's Conditioned Air Services*, 947 So.2d 396 (Ala.2006) (filing of motion for summary judgment not a waiver: no hearings, discovery or, date for trial set); *Brendsel v. Winchester Constr. Co. Inc.*, 392 Md. 601, 898 A.2d 472 (2006) (obtaining and filing lien not a waiver). The *Winchester* court emphasized intent to waive and noted the trend in favor of not waiving arbitration when the contractor seeks to protect its lien rights. But see *Kaneko Ford Design v. Citipark, Inc.*, 202 Cal.App.3d 1220, 249 Cal.Rptr. 554 (1988) (arbitration waived when lien filed and no demand for arbitration for five months while seeking to settle).

must pay a fee based on the amount of the claim. The notice usually contains a statement setting forth the nature of the dispute, the amount involved, and any remedies sought.[90]

A much-debated issue is whether the parties to an arbitration should have the right to examine people who might be witnesses for the other party and examine documents in the other's possession to evaluate the other party's case, to prepare for trial, and possibly to settle. In judicial proceedings, these activities are classified as "discovery."[91]

Discovery itself, one should add, has become controversial. Originally created to preclude the trial from taking unexpected twists and turns because of surprises, it has become largely uncontrolled, often with excessive demands to produce documents and lengthy and often irrelevant questioning of many potential witnesses. To a large degree it has been run by the lawyers, although increasingly courts are beginning to take control to prevent costly and time-consuming excesses.

Yet despite the increasing concern over full-blown, lawyer-controlled discovery, it seems clear that parties cannot begin to negotiate to settle disputes or even seek to mediate them without having some idea of the merits of the case. Obtaining basic information need not require full-blown discovery. The latter would change the informal nature of arbitration. (Yet lawyers used to this system often insert provisions in contract arbitration clauses that require judicial-like discovery.)

It can be useful to conduct a restricted, limited discovery to get to the point where the parties realistically can appraise the costs and benefits of settlement. This can be accomplished if the parties must preserve and exchange those routinely kept records that can present a true picture of the dispute, such as contract documents, bidding documents, accounting records, correspondence, site meeting minutes logs, diaries, weather reports, lab reports, and inspection records. If settlement efforts fail, then full-blown discovery may be useful, although many oppose it as destructive of the informality and speed needed for arbitration.

In complicated arbitrations that involve stakes high enough to justify pre-hearing activities, it may be useful to permit or even suggest that the arbitrating parties submit in advance a written statement giving in detail the facts and legal justification for the contentions of the arbitrating parties. Such an advance submission can eliminate much irrelevant testimony at the hearing. In addition, advance statements can give each party some idea of what the other party will seek to assert. Like limited discovery, advance submissions can help the parties prepare for the hearing, can make the hearing more expeditious, and may lead to settlement.[92]

When alternative dispute resolution (ADR) systems are discussed in Section 30.18, the right to discovery again will be addressed.

SECTION 30.07   Selecting Arbitrators

A leading treatise states, "The most important tactical step in an arbitration proceeding is the selection of the arbitrator."[93]

Usually the method for selecting arbitrators is specified in the arbitration clause. Sometimes a particular arbitrator or specific panel of arbitrators is designated by the parties in advance and is incorporated in the arbitration clause. Advance agreement on the arbitrators or a panel of arbitrators should build confidence in the arbitration process. However, in construction contracts such advance agreement is uncommon.

Generally a procedure to select arbitrators rather than a designation of particular individuals is used. For example, the procedure can require each party to name an arbitrator, with the two-party–appointed arbitrators designating a third or neutral arbitrator. Some arbitration clauses provide that each party will appoint an arbitrator and only if they cannot agree on the disposition of the dispute do they appoint a third arbitrator, who makes the decision. The methods used in private arbitration systems, such as the American Arbitration Association, are discussed in Section 30.18A.

---

[90]CIA Rule R–4 (Regular Track rules are preceded by R; Fast-Track by F; and Large/Complex by L.).

[91]For a debate (Hinchey arguing the affirmative and House & Corgan arguing the negative), see *Do We Need Special ADR Rules for Complex Construction Cases?* 11 Constr. Lawyer, No. 3, Aug. 1991, p. 1. A trial lawyer bemoans the lack of statutory or common law authority that grants full litigation-type discovery. He also complains that arbitration rules, such as the CIA Rules of the AAA, make any right to discovery within the discretion of the arbitrator. He offers suggestions how such rights can be written into the contract. See Moseley, *What Do You Mean I Can't Get That? Discovery in Arbitration Proceedings*, 26 Constr.Lawyer, No. 4, Fall 2006, p. 18.

[92]See Section 30.18A for discussion of the AAA Large/Complex arbitration.

[93]J. ACRET CONSTRUCTION ARBITRATION HANDBOOK, §6.1 at 6-1 (2d ed. 2006).

In *Commonwealth Coatings Corp. v. Continental Casualty Co.*,[94] the U.S. Supreme Court dealt with the extent to which the neutral arbitrator must disclose facts that may bear on the arbitrator's impartiality. This case has stirred controversy and influenced arbitration law. The arbitration involved a dispute between a subcontractor and the surety on a prime contractor's bond. The neutral member performed services as an engineering consultant for owners and building contractors. One regular customer was the prime contractor with whom the subcontractor had the dispute in question. That relationship was sporadic. The arbitrator's services were used only from time to time, and there had been no dealings between the arbitrator and the prime contractor for about a year before the arbitration. The prime contractor had paid fees of about $12,000 to the arbitrator over a four- or five-year period. These facts were not revealed by the arbitrator until after the award had been made. When this was disclosed, the subcontractor sought to invalidate the award.

The court noted the resemblance between a judge and an arbitrator exercising quasi-judicial powers. Clearly, the judge must avoid any appearance of partiality. But the court felt it even more important for an arbitrator to avoid any appearance of partiality because the arbitrator's decision is only subject to limited review.

The court refused to confirm the arbitration award because the arbitrator had not disclosed this relationship. Although the court recognized that arbitrators cannot sever all their ties from the business world, it held that failure to disclose prior activities of this type would create the impression of possible bias.

The concurring justices emphasized that arbitrators are people of affairs and part of the marketplace and cannot be compared strictly to judges. The concurring justices were concerned that too great a burden of disclosure would disqualify the best informed and most capable arbitrators. These justices felt that the arbitrator cannot be expected to provide the parties with a complete and unexpurgated business biography. Nevertheless, the concurring justices felt that in this case the relationship was more than trivial and for that reason agreed that the arbitration award should be set aside.

The dissenting justices felt that the losing party was simply grasping at straws and such a requirement of disclosure would discourage arbitrators from serving and would render arbitration less effective because it would be too easily challengeable.

The *Commonwealth Coatings* decision has come in for criticism for reasons given by the concurring and dissenting justices. This is also demonstrated by a recent decision of an *en banc* (16 judges rather than the usual three) court of the Fifth U.S. Circuit Court of Appeals.[95] It gave a narrow reading of *Commonwealth Coatings*. It refused to vacate an award because of what it saw as a trivial and insubstantial relationship. The contacts between the arbitrator challenged and an attorney in the arbitration were tangential, limited and stale. The majority stated that arbitration needed finality, that the standard sought was greater than for a judge (appearance of bias) and many skilled arbitrators would not serve rather than expose themselves to the risk of blemishing their reputations by a post-award charge of bias. Five judges dissented.

Other subsequent cases have reflected ambivalence toward the *Commonwealth Coatings* requirement of disclosure, especially in the context of a trade association or closely knit industry.[96] Yet the spirit of *Commonwealth Coatings* has been followed in other decisions.[97]

The issue of arbitrator disclosure and qualification to serve as arbitrator reveals an important potential conflict. Many states' arbitration statutes provide solutions to these issues. At the same time private arbitration organizations, such as the American Arbitration Association (AAA), create detailed rules that govern the arbitration.[98] The state statutes and the rules, such as the AAA's Construction Industry Arbitration Rules, can conflict.

---

[94]393 U.S. 145 (1968).

[95]*Positive Software Solutions, Inc. v. New Century Mortgage Corp.*, 476 F.3d 278 (5th Cir.2007).

[96]*Garfield & Co. v. Wiest*, 432 F.2d 849 (2d Cir.1970), cert. denied, 401 U.S. 940 (1971) (stock exchange arbitration: waiver); *Baar & Beards, Inc. v. Oleg Cassini, Inc.*, 30 N.Y.2d 649, 282 N.E.2d 624, 331 N.Y.S.2d 670 (1972) (need for specialized knowledge and skill); *Reed & Martin, Inc. v. Westinghouse Elec. Corp.*, 439 F.2d 1268 (2d Cir.1971) (arbitrator need not provide complete and unexpurgated business biography); *William B. Lucke, Inc. v. Spiegel*, 131 Ill.App.2d 532, 266 N.E.2d 504 (1970) (bias too remote).

[97]*Sanko S. S. Co. Ltd. v. Cook Indus., Inc.*, 495 F.2d 1260 (2d Cir.1973); *J. P. Stevens & Co. v. Rytex Corp.*, 41 A.D.2d 15, 340 N.Y.S.2d 933, aff'd 34 N.Y.2d 123, 312 N.E.2d 466, 356 N.Y.S.2d 278 (1974) (stating major burden of disclosure properly falls on arbitrator). The cases are discussed generally in Annot., 56 A.L.R.3d 697 (1974). For CIA Rules, see Appendix F, Rule R-20.

[98]The Rules are sponsored by over twenty organizations that make up the National Construction Dispute Resolution Committee. It is likely that AAA plays the major rule in drafting these rules but provides drafts to all committee members for comment and suggestions.

This was demonstrated in a California case, *Azteca Construction, Inc. v. ADR Consulting, Inc.*[99]

In this case the court was faced with a conflict between the California statute and the CIA Rules over disqualification of a potential arbitrator. The statute provides a more strict test for arbitrator disclosure than does the CIA Rules incorporated into the contract.[100] Also, the California statute states that a party shall be disqualified on the basis of the disclosure statement if either party serves a notice of disqualification within 15 days.[101] The CIA Rules give the power to disqualify to the AAA and its decision is conclusive.[102]

The trial court held that the agreement of the parties to use the CIA Rules took precedence over the statute. But the appellate court disagreed. It held that the partiality statute was intended to counterbalance the limited right to appeal an arbitration award. The court also stated that neutrality is a structural aspect of arbitration and is so crucial to the process that the statute cannot be displaced by a private agreement.

Despite the attitude toward arbitrator bias being so crucial in the *Azteca* case, a California court took a more relaxed attitude in a recent case.[103] This reveals the tension inherent in seeking to reconcile the different policies highlighted by the *Commonwealth Coatings* case. It also demonstrates increased legislative activity in arbitration, as shown in Section 30.03B.

The conflicting policies must be accommodated. A pool of skilled arbitrators is clearly needed. But the parties must believe that they were given a fair hearing if they are expected not to challenge the award. Finally, as shall be seen in Section 30.14, the virtual finality of the award makes the choice of a skilled and unbiased arbitrator crucial. It remains to be seen what the future cases will hold as to the duty of a prospective arbitrator to disclose and the standard that will be applied to determine her qualifications to serve as an arbitrator.

---

[99]121 Cal.App.4th 1156, 18 Cal.Rptr.3d 142 (2004).

[100]West Ann.Cal.Code Civ.Pro. § 1281.9(a). The AIA incorporates the CIA Rules into its documents. See A201-2007, § 15.4.1, B101-2007, § 8.3.1.

[101]West Ann.Cal.Code Civ.Pro. § 1281.9(b)(1).

[102]CIA Rules R-20(b).

[103]*Fininen v. Barlow*, 142 Cal.App.4th 185, 47 Cal.Rptr.3d 687 (2006) (award not vacated when arbitrator failed to disclose participation in prior mediation with one of the parties where he was recognized before the hearing and all participants consented to his participation).

SECTION 30.08 Place of Arbitration

The arbitration clause need not specify where the arbitration is to take place. Any contractual designation of locale will be given effect as long as the selection is reasonable.[104] If hearing site selection is based on factors that will expedite the process and make it less expensive, the selection will certainly be considered reasonable. However, if the place is designated by the stronger party in order to frustrate the weaker party's right to have disputes heard, such a clause has not been given effect.[105] See Section 30.18A for discussion of private arbitral systems.

Increasingly, as noted in Section 30.03B, state statutes place limitations on the place of arbitration.

SECTION 30.09 Multiple-Party Arbitrations: Joinder and Consolidation

Frequently, a linked set of construction contracts contains identical arbitration provisions. For example, there are identical arbitration clauses in AIA standard contracts between owner and contractor and between owner and architect. Likewise, identical arbitration clauses are contained in the AIA standard contracts between architect and consulting engineer and between prime contractor and subcontractor.

Suppose a building collapses, and the owner wants to assert a claim. It may not be certain whether the collapse resulted from poor design or poor workmanship. The first would be chargeable primarily to the design professional and the second to the contractor. The owner may wish to arbitrate. Suppose there are identical arbitration clauses in the owner's contracts with the design professional and contractor. The owner can arbitrate separately with each, but there are disadvantages. Two arbitrations will very likely take longer and cost more than one. The owner may lose both arbitrations. (Although this loss may be unpalatable to the owner, the result may be correct if the design professional performed in accordance with the standards required of design professionals and the contractor executed the design properly.)

---

[104]*Central Contracting Co. v. Maryland Cas. Co.*, 367 F.2d 341 (3d Cir.1966). Similarly, in an international context, see *Republic Int'l Corp. v. Amco Eng'rs, Inc.*, 516 F.2d 161 (9th Cir.1975).

[105]*Player v. George M. Brewster & Son, Inc.*, supra note 5.

Inconsistent findings may result. For example, one arbitrator may conclude that the design professional performed in accordance with the professional standard, whereas the other may not. One may conclude that the contractor followed the design, and the other may not. To avoid this inconsistency and save time and costs, the owner may wish to consolidate the two arbitrations.

Suppose the owner demands arbitration only with the contractor. The contractor may believe that the principal responsibility for the collapse was defective design. In such a case, it might wish to add the design professional as a party to the arbitration or, in legal terminology, "join" the design professional in the arbitration proceedings.[106]

A contractor can be caught in a similar dilemma if it felt the responsibility for the owner's claim against it was work by a subcontractor. In such a case, the prime contractor may wish to seek arbitration with the subcontractor and consolidate the two arbitrations. Similarly, the architect may find it advisable to consolidate any arbitration she might have with her consulting engineer and any arbitration proceedings with the owner.

There is general agreement that the courts will respect any contract language that deals with consolidating existing arbitrations or adding a new party (joinder) to an existing arbitration. But where the parties have not spoken on this issue, there is great variation in court decisions in state and federal courts where one party seeks to consolidate or join while others object.

Some states preclude consolidation unless there is express statutory authority allowing it. Other states hold that, even in the absence of statutory authority, a court has the judicial authority to do so. Some states have enacted statutes granting the court authority to consolidate, while others give the arbitrator this authority. Some states permit a court to order a stay in the arbitration until related litigation has been completed.[107]

While early cases in the federal courts seemed to favor consolidation, the most recent cases permit consolidation only if the parties expressly or by implication have consented.[108]

This picture appears to be one of confusion. But the variation in outcomes simply shows that there is a constant struggle between legislatures and courts as well as different opinions as to the desirability of consolidation and joinder by the various actors in the construction industry. In any given state, the answer can be found. This picture also shows how there are strong arguments for and against consolidation and the need to treat this issue in the contract.[109]

As noted earlier, arguments of economy and efficiency support the use of consolidation. Many of those involved in construction dispute resolution favor consolidation. A survey revealed that 83 percent of construction arbitrators favor consolidation and that 82 percent of lawyers surveyed were also in favor of resolving multiparty disputes through consolidation.[110]

CIA Rule R-7 states that if the contract provides for "consolidation or joinder of related arbitrations" the parties will seek to agree on a process to effectuate it. If the parties cannot agree, the AAA will appoint a single arbitrator to decide whether there should be consolidation or joinder and if so, a fair process to accomplish it. The AAA has always been in favor of consolidation or joinder, and this rule seeks to accomplish it.

Finally, the Revised Uniform Arbitration Act (RUAA) was published in 2000. It is the successor to the Uniform Arbitration Act that was adopted in whole or in part in many states. The RUAA has been or will be presented to state legislatures for adoption.

It has taken a positive stance toward consolidation and joinder. Section 10 specifically empowers courts to consolidate in appropriate cases. It has adopted consolidation

---

[106]This assumes that contractor and design professional have agreed to arbitrate under the same rules as owner and contractor. Ordinarily, contractor and design professional do not have a contract, but they can agree to arbitrate their disputes. "Joining" the design professional as a party to the original arbitration is similar to but procedurally slightly different from consolidation. The latter merges two existing arbitrations into one.

[107]West Ann.Cal.Code Civ.Pro. § 1281.2(c). The California Supreme Court held that this statute did not run afoul of the FAA. *Cronus Investments, Inc. v. Concierge Services*, 35 Cal. 4th 376, 107 P.3d 217, 25 Cal.Rptr.3d 540 (2005). See also *Elizabethtown Co. v. Watchung Square Assoc., LLC*, 376 N.J.Super. 571, 871 A.2d 140 (App. Div.2005) (litigation stayed pending resolution of arbitration involving the same

construction project). For a thorough review see McCurnin, *Two-Party Arbitrations in a Multiple Party World*, 26 Constr. Lawyer, No.1, Winter 2006, p. 5.

[108]See Ness & Peden, supra note 83, pp. 10–11 (collecting and summarizing cases).

[109]See Stipanowich, *Arbitration and the Multi-Party Dispute: The Search for Workable Solutions*, 72 Iowa L. Rev. 473 (1987).

[110]Thomson, *Arbitration Theory and Practice: A Survey of Construction Arbitrators*, 23 Hofstra L. Rev. 137, 165–167 (1994) and Thomson, *The Forum's Survey on the Current and Proposed AIA A201 Dispute Resolution Provisions*, 16 Constr. Lawyer, No. 3, July 1996, pp. 3, 5.

as its "default" position. It permits consolidation, with limitations, in the absence of a clause expressly barring consolidation.

There can be reasons for refusing to add parties and consolidating existing arbitrations. Arbitrator selection is a crucial element of arbitration. Consolidation should not require a party to go before an arbitrator it has not selected. It should not have to arbitrate under evidentiary rules, or no rules of evidence, if it has not agreed to do so. Section 10(a)(4) of the RUAA states that if prejudice can be shown, such as undue delay or other hardship, this can overcome the efficiency and economy of consolidation and preclude consolidation.

Recognition of potential prejudice should not overcome the generally favorable attitude toward consolidation. For many years the AIA did not permit the architect to be a party to a multiparty arbitration without her written consent. After years of criticism of this policy, in 2007 the AIA permitted joinder and consolidation.[111]

Why treat the two consolidation questions differently? One reason given is that an arbitration between architect and owner will involve determining whether the architect has lived up to the professional standard, whereas the issue between owner and contractor is whether the contractor has performed in accordance with the contract documents. Although the legal standards may slightly differ, this is not sufficient justification to bar consolidation.

For reasons mentioned, consolidation and joinder are generally desirable. The possibility of confusion because of the potentially large number of parties in a consolidated arbitration can be handled by according the arbitrator the power to decide the number of parties and issues that would make consolidation or joinder too confusing. If so, a request to do so could be denied.

If a party would be seriously disadvantaged by consolidation, adjustments can be made to overcome them. If not, consolidation can be refused. This is recognized in Section 10(a)4 of the RUUA, mentioned earlier. That it may be necessary to make adjustments should not deprive the disputing parties of an efficient way of resolving multiparty disputes.

Another method by which participants in the linked set of contracts can be involved in arbitration is what is called the "vouching in" system. This is a method by which a person involved in the arbitration—although not a party—is given the opportunity to participate and to seek to persuade the arbitrators to make an award in accordance with its contentions.[112]

SECTION 30.10 The Hearing

A. A Differentiation of Issues: Desirable Vs. Required

Section 30.14 discusses attempts by the party satisfied with the award to have it confirmed in court or by the party dissatisfied with the award to have it vacated—that is, upset. That section refers principally to defects in the process that are important enough to justify not confirming or vacating the award. Even though complaints are made regarding the arbitration hearing that will not justify upsetting the award, examination of complaints might provide a blueprint for conducting a fair hearing. The arbitrator should consider not only what is compelled—which, as shall be seen, is relatively minimal—but also the type of hearing that will persuade the parties they have been treated fairly.

The exact type of hearing will largely depend on the intensity of the parties' feelings, the amount at stake, and the need for an expeditious decision. The discussion in this section assumes that a serious matter is brought before the arbitrators, one with sufficient economic importance to justify a careful and fair hearing. Because attorneys are frequently present in these arbitrations, unless otherwise indicated it is assumed the parties will be represented by legal counsel. This assumption does not negate the possibility or even likelihood that many arbitrations do not justify some of the steps suggested because of matters mentioned earlier in this paragraph. Nor does this assumption ignore the important differentiation between arbitration and litigation as to speed. Overjudicializing the arbitration process, by giving it those attributes of the legal process that led to arbitration in the first place, is clearly undesirable.

---

[111]AIA Doc. A201-2007, § 15.4.4, B101-2007, § 8.3.4.

[112]Such a system is described in detail in *Perkins & Will Partnership v. Syska & Hennessy & Garfinkel, etc.*, 50 A.D.2d 226, 376 N.Y.S.2d 533 (1975), aff'd, 41 N.Y.2d 1045, 364 N.E.2d 832, 396 N.Y.S.2d 167 (1977).

B. Waiver

Parties can agree to submit the dispute to arbitration based solely on written statements, the contract documents, or any other written data the parties feel relevant. Such a paper submission can save time and can be valuable in minor disputes. However, such submissions are often deceptive and may provide insufficient information to make an award. Even if the parties have agreed no hearing is necessary, an arbitrator can request a hearing in the presence of both parties, if resolving the dispute requires.

C. Time

The arbitrators should schedule the hearing as early as possible, but must take into account the time needed to prepare for the hearing. A complicated dispute may require the opinions or services of engineers, architects, accountants, attorneys, and photographers, among others. If adequate time is not allowed for preparation, continual requests will be made for recesses after the hearing process has begun.

Arbitrators should grant reasonable party requests to postpone a scheduled hearing or recess a hearing early. However, arbitrators must take into account that requests for delays and recesses are sometimes bargaining tactics used by the party who feels the other party's financial position will not tolerate delay. Arbitrators should avoid disrupting the process for their own convenience.

D. Proceeding Without the Presence of One of the Parties

Suppose one party to the arbitration indicates that it will not participate in the hearings. Under such conditions, should the arbitrator proceed with the hearings, or can the party who does attend be awarded the amount of the claim? It is better practice for the arbitrators to hear the evidence submitted by the party attending before making the award. Obviously, doubts will be resolved against the party who has chosen not to attend the hearings.

Rule R-32 of the CIA Rules states that the arbitration may proceed if a party despite "due notice, fails to be present or fails to obtain a postponement." But an award under this paragraph cannot be made "solely on the default of a party." The party attending must present such evidence as the arbitrator "may require."

E. The Arbitrators

At the outset, the parties should be permitted to question the arbitrators relating to any matters that could affect their impartiality. The arbitrators must disclose matters that could affect their impartiality.

The arbitrators should comply with any state arbitration laws requiring arbitrators to take an oath at the beginning of a hearing. Even if an oath is not required, it is good practice for arbitrators to take an oath that they will conduct the hearing and render their award impartially and to the best of their ability.

F. Rules for Conducting the Hearing

Arbitration clauses generally do not prescribe detailed rules relating to the method of conducting the hearing. Certain arbitration statutes give general directives, such as requiring that the arbitrator permit each party to present its case and to cross-examine witnesses for the other party. Arbitration associations or trade groups that conduct arbitration often have simple rules relating to the conduct of the hearing. In the absence of rules, the arbitrator determines how the hearing is to be conducted.

G. Opening Statements

In a complicated arbitration, or even in matters that may not appear complicated, it is often helpful to permit the parties or their attorneys to make a brief opening statement. The statement can help the arbitrator determine which evidence is relevant and can reduce the number of issues by having the parties agree to certain facts and issues.

H. Production of Evidence: Subpoena Powers

Most states give arbitrators the power to issue subpoenas that compel witnesses to appear and testify and requires people to bring in relevant records. Without such power, the arbitrator cannot compel witnesses to appear or documents to be produced.

Usually the arbitrating parties produce witnesses and supply whatever records are advantageous to their positions. If issues can be resolved more easily if certain witnesses are produced or certain documents are presented, the arbitrator can resolve those questions against the party who refuses to produce the witnesses or records within its control. If the arbitrator indicates this likelihood to a reluctant party, the latter is likely to produce the witnesses or the records.

I. Legal Rules of Evidence

The arbitrator need not follow the rules of evidence applied in courts. Principally, the rules of evidence that can be dispensed with relate to the form evidence must take to be admissible. However, certain principles that are part of the legal rules of evidence should guide the arbitrator when determining whether material submitted by the parties should be considered. These principles relate to relevance and administrative expediency.

One experienced arbitrator states that even though the strict rules of evidence do not apply, there should be some standards. He suggests that the rules of evidence be a guideline and that the "ultimate target is to take only evidence upon which a reasonable business person would rely to establish facts in the everyday conduct of business." If the evidence is not probative of the issue or the time and expense would greatly outweigh its probative value or add nothing to evidence already introduced, the arbitrator should exclude the evidence.[113]

The arbitrator should not go into matters not germane to the dispute. Germaneness, however, is often difficult to determine. At the outset, the arbitrator should not cut off a line of testimony that may not appear relevant at the moment it is presented. The relevance of this testimony may become clear as the hearing proceeds. However, the arbitrator should ask the party presenting the evidence what it intends to thereby establish. If then the arbitrator determines that what the party intends to establish is not germane, the evidence should be disregarded and testimony cut off. Evidence presented should be relevant and should not be cumulative. Once a fact has been established, it is usually not necessary to establish that fact again.

J. Documentary Evidence

As stated, the arbitrator need not follow the many technical rules of evidence that relate to the admissibility of documentary evidence. Any documentary evidence submitted by the parties can be examined provided that it is relevant and not cumulative. The authenticity of the document can be taken into account. An excessive preoccupation with form, notarial seals, and witnesses to the document can slow down the hearing. Unless a party

questions the authenticity of a document, the arbitrator should consider it.

K. Questioning Witnesses

One important constituent of a hearing is the testimony of witnesses. Failure to hear testimony of witnesses is likely to be procedural misconduct that can vitiate any award made by the arbitrator. Arbitrators should follow any state laws requiring the witnesses to be placed under oath. Even where the law does not require this, a simple oath adds to the dignity of the hearing and may induce truthful testimony.

Legal rules of evidence generally prohibit an attorney from asking witnesses leading questions (questions that suggest the answer in the question). However, leading questions are permitted by the opposing attorney. This process, called *cross-examination*, is designed to test the credibility of the witness and expose dishonest or inaccurate statements.

Arbitrators need not follow these courtroom rules. However, cross-examination, whether by the opposing attorney or by the arbitrators, can be a useful device to test the veracity of the witness. Sometimes the narrative method permits the witness to testify in a logical and understandable fashion. However, the form of examining witnesses lies largely within the discretion of the arbitrators.

L. Visiting the Site

The hearing is often conducted in a hearing room or in an office. It is possible, and often helpful, to conduct a hearing at the site where the evidence is available to the arbitrator or arbitrators. Even when the hearing is not conducted at the site, the arbitrator—either on her own motion or when requested by a party or the parties—may view the premises. It is best that the viewing of the premises be done in the presence of the arbitrating parties or their attorneys.

M. *Ex Parte* Communications

Ex parte communications are information or arguments communicated by one party to a dispute or by a third party to the person deciding the dispute without the knowledge of the other party. It would be improper for a judge to receive privately communicated information relative to a pending case from one of the parties, one of the attorneys

---

[113]*Myers, Ten Techniques for Managing Arbitration Hearings*, 51 Dispute Res.J, No. 1, Jan–March 1996, p. 28 (Am.Arb.Ass'n) (many other valuable suggestions).

for the parties, or a third party not connected with the case. The attorneys who represent the litigating parties should know what communications are being made to the judge, in order to respond to them or to point out inaccuracies. This openness is one reason why arbitrators should notify the parties if they plan to view the premises and set a time that will enable the parties to be present.

CIA Rule R-19(a) prohibits a party from communicating *ex parte* (that is, without knowledge or consent of the other party) with an arbitrator or candidate for direct appointment concerning the arbitration. A party may do so with "a candidate for direct appointment . . . in order to advise the candidate of the general nature of the controversy . . ." and to discuss the candidate's qualifications, availability, or independence or the suitability of other candidates under certain circumstances. Under R-19(b) this limitation does not apply to arbitrators directly appointed by the parties.

N. Transcript

Normally, there is no requirement that testimony be transcribed or that a written transcript be made. Some arbitrators prefer to have the testimony transcribed and reproduced for their own use as well as for settling any disputes that may arise between the arbitrators and the parties over the exact testimony of witnesses.

O. Reopening Hearing

After the hearing has been closed, either party may request to reopen the hearing to introduce additional evidence. The arbitrator can determine whether to reopen the hearing. Newly discovered evidence usually will be sufficient basis to reopen the hearing as long as the arbitrators have not yet made their award. If the evidence proposed to be introduced at an additional hearing could have been available for the original hearing or is merely cumulative, the arbitrator should not reopen the hearing. However, such matters are largely within the discretion of the arbitrator.

SECTION 30.11 Substantive Standards

Sometimes the arbitration clause or the rules under which the arbitration is to be held give general guidelines to the arbitrator regarding standards by which to decide the dispute.

Some arbitration clauses or rules permit the arbitrator to do almost anything regardless of the language of the contract as long as what is done accords with justice or fairness. As a rule, however, arbitrators decide the dispute based on the evidence, and in a contract dispute the contract terms play a central role.

Arbitrators need not follow case precedents. One reason frequently given for arbitrating is that the arbitrator is free to make a proper decision without the constraint of earlier precedent or rules of law. However, in complicated arbitrations, attorneys may present legal precedents in an attempt to persuade the arbitrators. Although the arbitrators would be free to consider these precedents, clearly they would not be required to follow them.

Must the arbitrator apply legal rules that would provide a defense to the claim were it litigated? For example, suppose the claim would be barred because of the statute of limitations[114] or because the claimant is not licensed in accordance with state law.[115] Usually such defenses are asserted when the demand for arbitration is made. The party opposing arbitration goes to court contending that the contract including the arbitration clause no longer has any legal validity because the claim has been made beyond the period allowed by law. Similarly, the party may seek to oppose arbitration by contending that the party demanding arbitration cannot be allowed to recover because it does not have the requisite license.

Section 30.04C dealt with the issue of whether certain preliminary issues are decided by the arbitrator or the court. Footnotes to the discussion there cited recent cases dealing with the contention by the party opposing arbitration that the arbitration is barred by the statute of limitations or that the contractor was unlicensed. Those cases held that these issues were to be decided by the arbitrator.[116]

These decisions may have been based in part on the desirability of having these issues resolved by the arbitrator. The arbitral system gives the arbitrator the flexibility to

---

[114]As to statutes of limitation, see Annot., 94 A.L.R.3d 533 (1979).

[115]As to licensing, see *Merkle v. Rice Constr. Co.*, 271 So.2d 220 (Fla. Dist.Ct.App.), cert. denied, 274 So.2d 234 (Fla.1973) (issue for the arbitrator); *Parking Unlimited, Inc. v. Monsour Medical Foundation*, 299 Pa.Super. 289, 445 A.2d 758 (Fla. 1982) (arbitrator upheld). But see *Loving & Evans v. Blick*, 33 Cal.2d 603, 204 P.2d 23 (1949) (refused to confirm award, as party not licensed). (For a discussion of the rights of unlicensed contractors, see Section 10.07B.)

[116]See supra note 81.

arrive at a just decision based on the merits unhampered by what may appear to be technical impediments that would have to be applied by a judge.

This admonition to the arbitrator to make a just and fair award is one reason why some do not wish to have arbitration clauses in their contracts while others see this a great strength of arbitration. This difference of opinion repeats itself through all aspects of arbitration: whether to use it and how the arbitrator's award should be treated when challenged in court. The latter will be dealt with in Section 30.14.

SECTION 30.12 Remedies

The arbitration clause or rules under which the arbitration is being held commonly state remedies that can be awarded. For example, R-44(a) of the CIA Rules allows the arbitrator to "grant any remedy or relief that the arbitrator deems just and equitable and within the scope of the agreement of the parties, including, but not limited to, equitable relief and specific performance of a contract." Also, Rule R-35 allows the arbitrator to take interim measures "he or she deems necessary, including injunctive relief and measures for the protection or conservation of property and disposition of perishable goods." It also states that security can be required. It concludes that a request by a party to a court is not a waiver of arbitration.

Usually the arbitrator issues a money award, stating that one party owes the other party a designated amount of money.

Yet there have been increasing efforts by arbitrators, generally sustained by the courts, to do more than simply award the winner a specified amount of money. Often these expansionary remedies could not be awarded by a court. But such awards are generally sustained by the courts, either because the arbitration agreement gave the arbitrator this power or because a final, just solution required such a remedy.

For example, *Grayson-Robinson Stores, Inc. v. Iris Construction Corp.*,[117] sustained an award that ordered the developer to build a shopping center. This was an award that the court admitted could not be awarded by a court. It was an equitable award that many courts would not make

because it would involve the court in supervising a complex transaction.

As noted, CIA Rule R-44(a) gives the arbitrator the power to award specific performance of a contract. But the arbitrator cannot do what a court can do: order specific performance and have the coercive power to issue a contempt order if its order is not obeyed. A finding of contempt can be the basis for a fine or jail sentence. The *Grayson-Robinson* court did sustain the award, but there is no showing that it would back up the decree with its broad contempt powers.

Sometimes arbitrators order that the party perform and state in the award that failure to perform in accordance with the order will entitle the other party to a specified amount of damages. This technique, where enforced, can give a method of enforcement without seeking court confirmation. However, such a technique is not likely to be effective unless the alternative damage claim is high enough to coerce performance. Although this may obtain enforcement of the award without going to court, California has held that its arbitration law did not authorize imposition of such an economic sanction. But it saw no legal impediment to enforcing an agreement by the parties giving the arbitrator this power. However, the agreement did not give the arbitrator this power.[118]

The arbitrator's determination of whether damages or a specific order are appropriate should take into account whether the parties will continue to be able to work together after the arbitration. One reason why courts are hesitant to order parties to perform in accordance with promises is that by the time the matter has reached court, the parties will no longer cooperate with each other. The same can be true in arbitration. However, if there is a cooperative attitude between the parties and if the work continues to be performed during arbitration, it may be useful to specifically order that work be performed rather than to award damages.

In *David Co. v. Jim W. Miller Construction Co.*,[119] the arbitrator made an award that was not sought by either party, something a court would not do. He ordered the builder to buy back from the developer buildings the

---

[117]8 N.Y.2d 133, 168 N.E.2d 377, 202 N.Y.S.2d 303 (1960).

[118]*Luster v. Collins*, 15 Cal.App.4th 1338, 19 Cal.Rptr. 2d 215 (1993) (arbitrator ordered wrongdoer to pay $50 a day for each day wrongful act continued).

[119]444 N.W.2d 836 (Minn.1989).

builder had built. There were many serious defects and numerous code violations. These exposed the developer to liability.

In *Advanced Micro Devices, Inc. v. Intel Corp.*,[120] discussed in Section 30.04A, the court, in a 4-to-3 decision, concluded that the limited scope of review of an arbitrator's decision will also be applied to the *remedy* awarded by the arbitrator. The court pointed to the broad remedies permitted under the agreement to arbitrate (identical to current CIA Rule R-44a). The majority approved the award, which in essence barred Intel from asserting a defense in a *separate* federal case between AMD and Intel and granted a license to AMD that it would not have received under full performance of the contract. The majority held that the award must and did bear "a rational relationship to the underlying contract as interpreted, expressly or impliedly, by the arbitrator and to the breach of contract found, expressly or impliedly, by the arbitrator."[121] The only apparent limitations are an express provision in the contract prohibiting such a remedy or the use of a source outside the contract. The majority refused to limit the remedy to one that could be awarded by the court, the position taken by the three dissenting judges.

The majority emphasized the need to end the dispute by limiting the scope of judicial review. Judicial review is discussed in Section 30.14. But some contend that giving the arbitrator so much power can act as a disincentive to arbitrate.

Under CIA Rule R-50 the filing fees must be advanced by the party or parties or can be apportioned by the arbitrator in the award. The AAA can "in the event of extreme hardship . . . defer or reduce the administrative fees." Under R-51 each party must pay its own witnesses. Other expenses shall be borne equally by the parties, unless they agree otherwise, or unless the arbitrator assesses expenses against any specified party or parties. Attorneys' fees are not in the items listed of expenses.

Rule R-44(d) allows the arbitrator in the final award to include interest at "such rate and from such date as the arbitrator may deem appropriate." That rule also allows the award to include attorneys' fees "if all parties have requested such an award or it is authorized by law or their arbitration agreement." The AIA, in its standard documents, does not provide for attorneys' fees,

but parties often add such a provision. *Harris v. Dyer*,[122] which involved an earlier AIA Document A201, awarded the winning party who had filed a mechanics' lien its attorneys' fees incurred in arbitration.

A leading Illinois case categorized the judicial attitudes toward the power of the arbitrator to award punitive damages.[123] One group, led by the federal courts, permits the arbitrator to award punitive damages, unless the agreement states otherwise. A second group reserves punitive damages to the state and denies the arbitrator this power. A third group allows the arbitrator to award punitive damages only if an express contract provision grants the arbitrator this power.

The centrality accorded remedies in all legal disputes has led to increased focus on remedies in arbitration. Some believe that an arbitrator, often not being a judge, should require great judicial scrutiny over the remedies awarded. Others see the great need to accord the arbitrator remedial power to accomplish a just outcome.

This uncertainty of judicial outcome has led one writer to propose remedial language in the agreement to arbitrate.[124] He suggests language that deals with all significant aspects of remedies such as consequential damages,[125] home office overhead,[126] profits,[127] interest,[128] interim relief,[129] counterclaims, punitive damages,[130] arbitration

---

[120]See supra note 67.

[121]885 P.2d at 996, 36 Cal.Rptr.2d at 583.

[122]292 Or. 233, 637 P.2d 918 (1981).

[123]*Edward Elec. Co. v. Automation, Inc.*, 229 Ill.App.3d 89, 593 N.E.2d 833, 842 (1992). In *Stark v. Sandberg, Phoenix & von Gontard, P.C.*, 381 F.3d 793 (8th Cir.2004) the clause stated punitive damages were waived to the fullest extent permitted by law. The law does not permit waiver of punitive damages. An award of $6 million punitive damages by the arbitrator was affirmed despite the compensatory damages being $4,000. The arbitrator also awarded $22,780 for attorneys' fees and $9,300 for the cost of arbitration. For a thorough analysis of this issue see Stipanowich, *Punitive Damages in Arbitration: Lyle Stuart, Inc. Reconsidered*, 66 B.U.L.Rev. 953 (1986). The *Lyle Stuart* case, the leading case refusing to allow punitive damages in arbitration, was effectively overruled by the United States Supreme Court on the ground that New York law was preempted by the FAA. *Mastrobuono v. Shearson Lehman Hutton, Inc.*, 514 U.S. 52 (1995).

[124]Sink, *Negotiating Dispute Clauses That Affect Damage Recovery in Arbitration*, 22 Constr. Lawyer, No. 3, Summer 2002, p. 5.

[125]See Section 27.06.

[126]See Section 27.02F.

[127]See Section 6.06E.

[128]See Sections 6.08, 30.12.

[129]See Section 30.12.

[130]See Sections 6.04, 7.10C, and 27.10.

fees and costs,[131] attorneys' fees and costs,[132] limitation of damages,[133] and liquidated damages.[134]

SECTION 30.13 Award

Before making the award, the arbitrators review any documents submitted and listen to or read any transcription of the hearings. They consider any briefs that may have been submitted by the parties or their attorneys. Submission of briefs is uncommon except for disputes involving large amounts of money. The decision need not be unanimous unless the arbitration clause or the rules under which the arbitration is being held so require.

The form of the award can be simple. The arbitrator need not give reasons for the award. There are arguments for and against a reasoned explanation of the decision accompanying the award. An explanation may persuade the parties that the arbitrators have considered the case carefully. This may lead to voluntary compliance, which is obviously better than costly court confirmation.

A disadvantage of giving an explanation is the possibility that the dissatisfied party or parties may refuse to comply and seek to reopen the matter by objecting to the reasons given.

A more persuasive argument against reasons accompanying the award is the additional time and expense entailed. Making the arbitration too much like a court trial can lose some of the advantages of arbitration.

Taking all this into account, a short, reasoned explanation accompanying the award is advisable even if not required. This will be discussed in the context of specific arbitral systems in Section 30.18.

The rules under which an arbitration is conducted often specify when the award must be made. For example, R-42 of the CIA Rules requires an award be promptly made and, unless otherwise agreed or required by law, "no later than thirty (30) calendar days from the date of closing the hearing, or, if oral hearings have been waived, from the date of the AAA's transmittal of the final statements and proofs to the arbitrator." Failure to make the award by the designated time may terminate the jurisdiction of the arbitrators, although more commonly the parties agree to waive any time deadlines. More important, failure to make the award in the time required may deprive the arbitrators of any quasi-judicial immunity (discussed in Section 30.16). If there are no specific time requirements, the award must be made within a reasonable time.

SECTION 30.14 Enforcement and Limited Judicial Review

Failure to comply with an arbitration award may necessitate judicial involvement and review. A party wishing enforcement may have to go to court to obtain confirmation. The party seeking to challenge the award may go to court and ask that the award be vacated.

Most state arbitration statutes specify grounds for reviewing an arbitrator's award. As to grounds, Section 12 of the Uniform Arbitration Act, enacted in whole or with minor variations in a substantial number of states, permits an award to be vacated (set aside) if there has been corruption or fraud, partiality by the arbitrator, taking excess jurisdiction, procedural misconduct, or lack of a valid agreement to arbitrate.

Section 13 of the act allows modification or correction of an award within ninety days after delivery of a copy of the award where any of the following exists:

1. There was an evident miscalculation of figures or an evident mistake in the description of any person, thing, or property referred to in the award.
2. The arbitrators have awarded on a matter not submitted to them and the award may be corrected without affecting the merits of the decision on the issues submitted.
3. The award is imperfect in a matter of form, not affecting the merits of the controversy.

Similar language is contained in the Federal Arbitration Act.[135] Grounds for vacating are limited and principally look to serious procedural misconduct on the part of the arbitrators. Courts are reluctant to upset the award because the arbitrator refuses to admit evidence, particularly after the hearing has been closed.

---

[131]See Section 30.12.
[132]See Sections 6.07, 30.12.
[133]See Section 13.01M.
[134]See Section 26.09B.

[135]9 U.S.C.A. §§ 10, 11.

In one case misconduct was found where the arbitrator refused to reopen the hearing to receive evidence. That evidence indicated that the contract had been procured as part of a bribery scheme. The court found that the evidence was relevant, not cumulative, and central to the plaintiff city's case. While sympathetic to the need for finality and reluctance to interfere, this decision of the arbitrator was misconduct that required that the award be vacated.[136]

Similarly, case decisions have employed language indicating a very limited judicial review of arbitration awards, mainly the manner of holding the arbitration. One court stated that the court will not inquire whether the determination was right or wrong.[137] Another stated that errors of fact or law are not sufficient to set aside the award.[138] Courts have held that an error of law was not reviewable unless the arbitrator gave a completely irrational construction to the provision in dispute.[139] Another stated that honest errors were not reviewable.[140] Another stated that arbitrators may apply their own sense of justice and make an award reflecting the spirit rather than the letter of the agreement.[141] Lawyers who have lavished a great deal of attention to specifically allocating risks and remedies in the contract may not appreciate arbitrators ignoring the contract in their attempt to "do the right thing." Finally, California, pointing to the statutory grounds as exclusive and overruling earlier precedents, held that with certain exceptions, an arbitrator's decision is not reviewable for errors of fact or law whether or not such error appears on the face of the award and causes substantial injustice to the parties.[142] It stated that, if the scope of review can create too much judicial interference, it will ruin the arbitration process. Errors of law are a tolerable part of the arbitration process.

Despite a pattern emerging under which there would be relatively little review of an arbitration award, the issue continues to divide the appellate courts. For example, this issue so fragmented the New Jersey Supreme Court in *Perini Corp. v. Greate Bay Hotel & Casino, Inc.*,[143] that no majority opinion could be written.

The dispute involved the contract between a casino in Atlantic City and a construction manager under which the casino would be renovated. The arbitration consumed sixty-four days of hearings, involved twenty-one witnesses, and resulted in almost 11,000 pages of transcript. The arbitrators (two to one) awarded the casino lost profits of $14.5 million against the construction manager. The CM's fee was $600,000 plus reimbursables.

The arbitration was followed by three and a half years of litigation, first at the trial level, then before the intermediate appellate court, and then before the Supreme Court of New Jersey. The litigation produced five judicial opinions (excluding the concurring opinion of sixteen double-columned pages) of over 150 pages. Except for two dissenting judges, the rest of the court was willing to affirm the arbitrator's award, but the three-judge plurality and the two-judge concurring opinion (of a seven-person court) strongly differed on the proper scope of judicial review.

The plurality of three judges devoted an extensive part of its opinion to examining the legal issues that had been resolved in favor of the casino. Although recognizing that a simple mistake of law is not a sufficient reason to overturn the award—that ground for reversal not being in the New Jersey statute—the plurality was willing to extensively review the award to see whether it complied with New Jersey law. It noted that an award would be sustained unless the mistake or error of law or fact resulted from a failure of intent or error so gross as to suggest fraud or misconduct.

---

[136]*City of Bridgeport v. The Kemper Group, Inc.*, 278 Conn. 466, 899 A.2d 523 (2006).

[137]*Drake, O'Meara & Assoc. v. American Testing & Eng'g Corp.*, 459 S.W.2d 362 (Mo.1970). See also *Seither & Cherry Co. v. Illinois Bank Bldg. Corp.*, 95 Ill.App.3d 191, 419 N.E.2d 940 (1981), noted in Annot. 22 A.L.R.4th 366 (1983).

[138]*Mars Constructors, Inc. v. Tropical Enterprises, Ltd.*, 51 Haw. 332, 460 P.2d 317 (1969). But see *West v. Jamison*, 182 Ga.App. 565, 356 S.E.2d 659 (1987) (refused to confirm award because of obvious mistake of law).

[139]*Firmin v. Garber*, 353 So.2d 975 (La.1977) (award upheld, not grossly irrational but simply debatable); *Maross Constr., Inc. v. Central N.Y. Regional Transp. Auth.*, 66 N.Y.2d 341, 488 N.E.2d 67, 497 N.Y.S.2d 321 (1985) (upheld unless totally irrational).

[140]*Reith v. Wynhoff*, 28 Wis.2d 336, 137 N.W.2d 33 (1965).

[141]*Matter of J. M. Weller Assoc., Inc. v. Charlesbois*, 169 A.D.2d 958, 564 N.Y.S.2d 854, appeal denied, 78 N.Y. 851, 577 N.E.2d 60, 573 N.Y.S.2d 69 (1991).

[142]*Moncharsh v. Heily & Blase*, 3 Cal.4th 1, 832 P.2d 899, 10 Cal. Rptr.2d 183 (1992), rehearing denied Sept. 24, 1992. As to the scope of review of the remedy awarded by the arbitrator in California, see *Advanced Micro Devices, Inc. v. Intel Corp.*, supra note 67, discussed in Sections 30.04A and 30.12.

[143]129 N.J. 479, 610 A.2d 364 (1992).

The concurring judges felt this would open up too many arbitration awards to judicial scrutiny and concluded that the only basis for overturning an arbitration award should be those reasons set forth in the arbitration statute (principally those dealing with corruption, fraud, or undue means).

The instability of the scope of judicial review is demonstrated by the fact that the concurring opinion, one that limited the grounds for reversal to the statutory grounds, was adopted in New Jersey just two years later.[144]

The concept of limited judicial review was undoubtedly designed to encourage arbitration by limiting the likelihood that an award will be overturned in court. However, limited judicial review almost to the point of no review can make contracting parties reluctant to use the arbitration process. As shall be seen in Section 30.17, one reason sometimes given for reluctance to enter into arbitration agreements is the absence of any meaningful review of the arbitrator's decision.

Yet it is hard to escape the logic of the concurring opinion in the *Perini* decision. Even opening the door as little as is done by the plurality to include gross mistakes of law creates sufficient uncertainty and encourages appeals to the courts.

Can parties in their agreement to arbitrate provide a broad scope of judicial review? Such a provision came before the California intermediate appellate court in *Crowell v. Downey Community Hospital Foundation.*[145] The contract required that the arbitrator issue written findings of fact and conclusions of law that are supported by law and substantial evidence. The contract provided that "a court shall have the authority . . . to vacate the arbitrator's award . . . on the basis that the award is not supported by substantial evidence or is based on an error or law. . . ."[146]

The court cited and followed a case decided by the California Supreme Court in 1992.[147] That case held that the statutory grounds for review were exclusive. Even though arbitration is a system created by contract, there are limits to what the parties can do. In the *Crowell* case, they were not allowed to create a standard of judicial review that mirrored the review given by an appellate court to a trial court decision.

The outcome in the *Crowell* case suggests two questions. First, will this conclusion be reached in other states? The varying outcomes in these cases suggest some states will and some will not. This issue of whether the parties can by their contract change the review standards set forth in the FAA has divided the courts.[148] One held there was no right to change the standards. Another held this could be done by clear language.[149]

Second, what affect will this have on the use of arbitration in California? Will it encourage or discourage arbitration? It is likely to discourage arbitration.

The limited review does not mean arbitration awards are never upset. For example, a court that articulated the standard of complete irrationality nevertheless upset an arbitrator's decision by concluding that the words of the contract were so clear there was nothing left to interpret.[150] For all practical purposes, however, an arbitrator's decision is final.

---

[144]*Tretina Printing Inc. v. Fitzpatrick & Assoc. Inc,* 135 N.J. 349, 640 A.2d 788 (1994). The judge who had concurred in the *Perini* case, supra note 143, explains why he changed his mind in this richly revealing statement: "Much as I would prefer to announce that my change of position is attributable to some epiphany, to some deeply moving event that produced a sudden startling cerebral awakening, to some lightning bolt of cognitive awareness and intellectual enrichment, the plain truth of the matter is that I have thought more about it and have changed my mind. My awakening, however belated, puts me squarely in the Chief Justice's camp. For whatever ambivalence that progression demonstrates I refuse to commit myself to the psychiatrist's couch, content to resurrect—as apparently I must every couple of decades—that reassuring old turkey, 'The matter does not appear to me now as it appears to have appeared to me then.' (quoting an English judge of a statement made in 1872) 640 A.2d at 797."

[145]95 Cal.App. 4th 730, 115 Cal.Rptr.2d 810 (2002). Similarly, see *Azteca Constr. Inc. v. ADR Consulting, Inc.,* supra note 99. It held the California statute on qualification of arbitrators displaced a contract provision that dealt differently with this issue. For a thorough discussion and citation to many cases, see *John T. Jones Constr. Co. v. City of Grand Forks,* 665 N.W.2d 698 (N.D.2003). For an expansion of the issues raised in this case, see Anderson, *Stepping on the Judiciary's Toes: Can Arbitration Agreements Modify the Standard of Review That the Judiciary Must Apply to Arbitration Decisions?* 24 Constr. Lawyer, No. 3, Summer 2004, p. 13. On this issue see also Goldman, *Contractually Expanded Review of Arbitration Awards,* 8 Harv. Negot. L. Rev. 171 (2003).

[146]115 Cal.Rptr.2d. at 812.

[147]*Moncharsh v. Heily & Blase,* supra note 142.

[148]*Kyocera Corp. v. Prudential-Bache Trade Services, Inc.,* 341 F.3d 987 (9th Cir.2003), cert. denied, 540 U.S.1098 (2004).

[149]*Jacada(Europe) Ltd. v. International Marketing Strategies, Inc.,* 401 F.3d 701 (6th Cir.), cert.denied, 546 U.S.1031 (2005).

[150]*O-S Corp. v. Samuel A. Kroll, Inc.,* 29 Md.App. 406, 348 A.2d 870 (1975), cert. denied, 277 Md. 740 (1976). Similarly, an award was set aside because it would violate state public bidding law in *State v. R. A. Civitello Co., Inc.,* 6 Conn.App. 438, 505 A.2d 1277, certification denied, 199 Conn. 810, 508 A.2d 770 (1986).

SECTION 30.15 Insurers and Sureties

A commercial general liability (CGL) insurance policy carried by a contractor or a professional liability insurance policy carried by a design professional generally indemnifies the insured if it incurs liability to a third party. Insurers in fixing their rates must be able to predict their losses. The professional liability insurer assumes its liability will be based on the normal professional activity performed by its insured. Similarly, an insurer who issues public liability insurance to a contractor expects to indemnify the insured only if accidents arise out of its normal construction activities.

To standardize risks, insurers traditionally excluded liability assumed by contract. An insured who wished to do so, such as a contractor who has contractually agreed to indemnify the owner or architect, would often obtain a special endorsement. This gives the insurer the chance to examine the risk and decide whether to accept it, refuse it, or accept it with a premium adjustment. Under current underwriting practices, the contractor need not obtain a special endorsement in its CGL policy. The "contractual liability" exclusion contains an exception for liability assumed under an "insured contract," basically an indemnity agreement.

Arbitration, though not *imposing* liability by contract, substitutes one form of dispute resolution for another. This problem is more serious for professional liability insurers. Unlike claims against a contractor, claims against a design professional are likely to be made by parties with contracts (often containing arbitration clauses) against the insured design professional.

Insurers are usually at least *wary* of arbitration. Although most do not specifically *exclude* arbitration, they may counsel that it not be used, suggest that only a certain clause be used, or, in rare cases, deny coverage for a design professional who intends to use or uses a general arbitration clause. If insurers are aware of such a clause at the time they insure or make no objection if they are asked their opinion, they have consented. Design professionals should check their policies, bring them to the insurers if they are in doubt as to the advisability of arbitration or its effect on coverage, and notify their insurer if they plan to submit an existing dispute to arbitration.

Sureties are discussed in greater detail in Chapter 32. The surety issues a bond to the prime contractor to protect the obligee—usually the owner—from the risk that any claim the owner will have against the prime contractor cannot be collected. Usually the surety bond either incorporates the construction contract by reference or refers to it.

Some cases have involved attempts by the surety to compel arbitration when the principal parties to the dispute, such as owner and prime contractor or prime contractor and subcontractor, do not wish to arbitrate. Usually the surety cannot compel arbitration, as it is not a party to the construction contracts that contain arbitration clauses.[151]

The more difficult problems are attempts by sureties to distance themselves from the arbitration and then refuse to pay the arbitration award. This may be done simply to delay payment by invoking a technical defense. But the surety in good faith may also feel it cannot participate in the arbitration because it does not have sufficient information or records.

As a rule, the obligee (owner on a prime's bond or prime on a subcontractor's bond) prefers the surety to participate in the arbitration. Participation will ensure that the surety will pay or at least be liable for the award. This overcomes any disadvantage to the obligee of possibly having to face two opponents and two attorneys. Most often, though, the surety does not participate and then seeks to deny its responsibility to pay the award.

The surety must pay the award if it has involved itself in some way in the process, such as taking an assignment of the construction contract, being subrogated (placed in the contractor's position) to the contractor's rights, completing the project, or participating in the arbitration. It must pay the award if it has expressly promised to participate or be bound by the award.[152] A surety will generally be obligated to pay the arbitration award if it had notice of the arbitration and an opportunity to defend. Even if the surety is not obligated to participate, its decision to

---

[151]*Aetna Cas. & Sur. Co. v. Jelac Corp.*, 505 So.2d 37 (Fla.Dist.Ct. App.1987) (surety only incidental, not intended, beneficiary). But see *Henderson Investment Corp. v. International Fidelity Ins. Co.*, 575 So.2d 770 (Fla.Dist.Ct.App.1987) (incorporated arbitration clause into the surety bond) and see *J&S Constr. Co., Inc. v. Travelers Indem. Co.*, 520 F.2d 809 (1st Cir.1975) (surety could invoke arbitration, as construction contract incorporated in bond).

[152]*Town of Melville v. Safeco Ins. Co. of America*, 589 So.2d 625 (La. App.1991). The summary judgment in this case was reversed and the case remanded to the trial court because there were genuine issues of fact that may not have been precluded by the arbitration award. 593 So.2d 376 (La.1992).

forgo participation when it knows of the arbitration and has had the opportunity to defend should result in it being bound to pay any arbitration award.[153]

To bind the surety to any arbitration award, the owner should incorporate the construction contract into the bond, notify the surety of any default, notify the surety of any arbitration, and invite the surety to participate. If the surety chooses not to participate, it should still have to pay the award.

Increasing legislative regulation of the settlement practices of sureties may make it more difficult for sureties to refuse to participate in any arbitration and to use this as the basis for refusing to pay an arbitration award against its principal debtor.[154]

SECTION 30.16 Arbitrator Immunity

As noted in Section 14.09D, some American states grant the design professional quasi-judicial immunity when acting as judge. More clearly, arbitrators are granted quasi-judicial immunity.[155] Arbitrators are even more like judges than are design professionals. However, *Baar v. Tigerman*[156] stripped an arbitrator of his quasi-judicial immunity when he did not make an award in accordance with the time requirements of the arbitration rules. In addition, the court refused to grant immunity to the American Arbitration Association (AAA).

Although immunity was restored by subsequent legislation,[157] the case is a warning to arbitrators and those who run arbitral systems to set up reasonable deadlines for making the award and then to comply with them or obtain extensions.

SECTION 30.17 Arbitration and Litigation Compared

Arbitration is voluntary. Parties can choose to employ it. Choice involves a number of considerations, some apparent and some not.

A lawsuit begins with the filing of pleadings by a lawyer. An arbitration process can start by the filing of a claim, which does not require a lawyer. However, the claimant may wish legal advice to decide whether arbitration or litigation can better handle the dispute. It is usually quicker and less expensive to begin arbitration than litigation.

However, the filing fees for beginning litigation are modest compared with those required to initiate arbitration (see Section 30.02). The CIA Rules employ a sliding scale of filing fees. These fees apply to both claims and counterclaims, something common in construction disputes.

Arbitration and litigation differ in the ease with which basically one dispute involving a large number of claimants and claims resistors can be handled. Generally, joinder of parties and consolidation of disputes is more easily accomplished in a lawsuit than in arbitration (discussed in Section 30.09). This may change because of AIA changes made in 2007.

Until recently arbitral systems did not provide for compulsory discovery, the process by which attorneys in litigation can obtain evidence from the other party to prepare for the hearing. Although the arbitrator can frequently compel the production of evidence at the hearing, this is not the same as judicial discovery. Of course, an arbitration system can and, as noted in Section 30.18A, increasingly does contractually compel discovery.

Discovery availability demonstrates the balance between process fairness and a quick resolution of disputes. Its availability can make arbitration more desirable

---

[153]*Sheffield Assembly of God Church v. American Ins. Co.*, 870 S.W.2d 926 (Mo.Ct.App.1994) (need actual but not formal notice); *Raymond Internat'l Builders, Inc. v. First Indem. of America Ins. Co.*, 104 N.J. 182, 516 A.2d 620 (1986). But see West Ann.Cal.Civ.Code § 2855, which appears to require that the surety be a party to the arbitration before it can be compelled to pay the award. It has been held that the surety must participate in the arbitration if the prime contract with an arbitration clause was incorporated into the bond. *United States Fid. & Guar. Co. v. West Point Constr. Co., Inc.*, 837 F.2d 1507 (11th Cir.1988). See generally Ruck, *Can a Contract Bond Surety Be Compelled to Arbitrate Claims Against It?* 16 Forum 765 (1981).

[154]See Section 32.10I.

[155]*Baar v. Tigerman*, 140 Cal.App.3d 979, 189 Cal.Rptr. 834 (1983). The case is noted in 67 Marq.L.Rev. 147 (1983). *Pullara v. American Arbitration Ass'n, Inc.*, 191 S.W.3d 903 (Tex.App.Ct.2006), rehearing overruled May 31, 2006, review denied Oct. 29, 2006, rehearing for petition to review overruled Dec. 15, 2006 (arbitrator and association conducting arbitration granted absolute immunity similar to judge).

[156]Supra note 155.

[157]West Ann.Cal.Civ.Proc.Code § 1280.1. This was repealed by its own terms in 1997. Now see § 1297.119. It grants immunity to an arbitrator.

to attorneys. Yet its use can be costly, and its availability can slow down the process and allow one party to delay for bargaining purposes.

Two advantages frequently claimed for arbitration are informality and speed of the process. Although lawyers are generally required in litigation, arbitration does not require parties to have legal representation. However, parties arbitrating important matters usually retain lawyers. This can hamper the arbitration, as arbitrators need not be lawyers. Having nonlawyer arbitrators deciding legal questions argued by attorneys can be undesirable. In addition, difficult legal questions may have to be decided by nonlawyer arbitrators.

The skill with which a judge or arbitrator conducts a hearing not only determines how quickly the hearing can be completed but also affects a party's perception of the fairness of the process. Because judges generally possess greater hearing experience, the hearing process in court will be conducted by a person with greater hearing experience. Although some arbitrators who work permanently or frequently can develop the skill to conduct good hearings, in the construction industry hearing expertise is rarer than in other classes of arbitration. Arbitrators who decide construction disputes rarely make their living at arbitration. Not only does this result in less experienced arbitrators, but the part-time and, even more important, volunteer arbitrators must often take lengthy recesses in the hearings because of having to work at their profession or business.

One advantage of arbitration in terms of hearing speed is the freedom the arbitrator has to move the hearing along unhampered by technical rules of evidence that, when accompanied by contentious attorneys, can make court hearings excessively long. The arbitrator, unlike the judge, need not make a transcript of the hearing; although helpful, a transcript is costly and time consuming.

Looking at all aspects of this problem, arbitration is likely to be quicker than judicial dispute resolution. Speedy hearings should not only result in faster decisions but also avoid the indirect costs of lengthy hearings, such as the unproductivity of witnesses who must attend the hearing, and direct costs, such as travel expense and attorneys' fees.

Hearing location is likely to be different. A court hearing is likely to take place at the principal city in the county where the lawsuit was commenced and whose court has jurisdiction over the dispute. This does not necessarily mean the trial will be close to the project or convenient for witnesses. Often lawyers seek to gain tactical advantages by having the matter heard in a particular court. In contrast, the arbitrator has more flexibility to schedule the hearing at a place more convenient for the witnesses. However, sometimes the arbitration clause can be deliberately designed to make it inconvenient for one party to demand arbitration.

Another differentiation relates to the public nature of the hearing. Although arbitration hearings are private, the judicial process is public. Sometimes trials are reported to local newspapers. The privacy of the arbitration not only may avoid unwanted publicity but also can be a more sympathetic setting for witnesses.

One differentiation is often ignored: The courtroom and the services of judge and clerk are furnished free to the litigants, whereas the room for the hearing and the arbitrator's fee are expenses that must be paid by the parties.

Viewing the project, often a helpful activity, is more easily accomplished in arbitration.

Quality comparisons are difficult. Arbitration has frequently been supported because of the expertise (often absent in judges) that arbitrators bring to the disputes. Doubtless, construction experience is useful. However, the dispute resolver who is too expert may conceive of her role not as providing a hearing and then making a decision but as simply deciding the dispute based on her own knowledge. On the assumption that experience in construction is likely to produce a better decision, it is worth comparing the experience in construction between an arbitrator and a judge.

Judges have legal training and as a rule have spent ten to twenty years practicing law before they are appointed or elected to the judiciary. This experience may have involved construction matters, but on the average, most judges have probably had little construction experience before becoming a judge. They often learn on the job. A trial judge who has been on the bench for five or ten years in a jurisdiction that has a wide range of cases may have gathered enough experience to be knowledgeable about construction matters. On the whole, though, trial judges do not bring great expertise to resolving construction disputes. Yet in some states, special masters or referees can be appointed to handle certain parts of the lawsuit. Often they are experienced.

Arbitrators frequently have experience in the types of matters arbitrated before them, particularly those who arbitrate under collective bargaining agreements or who arbitrate highly specialized disputes, such as disputes between members of the stock exchange, the diamond industry, or textile trade associations. Most disputes in these industries are between members of these associations, are repetitive, deal with technical matters, and are handled well through arbitration.

However, the construction industry arbitration panels, though made up of people with experience in construction, such as attorneys, architects, engineers, and contractors, do not have enough cases to justify full-time arbitrators or even part-time arbitrators who work with sufficient regularity to develop specialized knowledge. As noted in Section 30.18A, private arbitral systems are seeking to improve the quality of the panels, the JAMS system by using only retired judges and experienced retired attorneys and the American Arbitration Association by seeking to reduce the size of the panels and training those on them. The segments of the construction industry, although having some elements in common, are highly specialized, and an arbitrator's experience in electrical contracting may not prepare the arbitrator to handle a dispute involving road building. On the whole, though, arbitrators are more likely to have had more experience in construction matters than have judges.

It is beyond the scope of this book to attempt a detailed comparison from a substantive standpoint between arbitration awards and court judgments. However, two frequently raised issues merit brief comment.

It has been asserted that a dispute that runs through the arbitral process is less likely to be decided by the plain meaning of contract language than one proceeding through the judicial process. To some degree, this trait is inherent in the concept of arbitration, which is supposed to be less formal and technical than the judicial process. Also, the arbitral process is generally not designed to create binding precedents, something that can be created if a dispute culminates with a written opinion by an appellate court. Anecdotal evidence suggests that a party who seeks to base its claim or defense on language of the construction contract is more likely to succeed in a judicial proceeding than in an arbitration.

The second criticism made of arbitration is that arbitrators often refuse to decide wholly on the merits but seek to award each party something, so that no one should walk away from the arbitration feeling demolished.

Although opinions may differ on the desirability of accomplishing the latter objective, again anecdotal evidence indicates that lawyers generally prefer arbitrators to decide strictly on the merits, and lawyers

are not pleased when the arbitrator "splits the difference." An AAA study sought to determine whether this is the case. The study reviewed all construction arbitrations administered by the AAA in 1990 and concluded that in the vast majority of cases arbitrators decide clearly in favor of one party or the other. According to this study, in only 12 percent of the cases was the award split between the parties within the 40 percent to 59 percent range.

Another comparison of great importance is the degree of finality in the decision. A decision by a trial judge can be appealed to an appellate court. Although appeals on the whole have a low probability of success, a litigant who appeals will succeed if it can show that the trial judge has made an error of law or that the evidence did not support factual findings. However, for all practical purposes an arbitrator's decision is final. To some, this is a great advantage of arbitration, as it ends the dispute quickly. To others, it is a disadvantage because the arbitrator possesses immense power—including the power to make completely wrong decisions.

Comparing arbitration and litigation should not assume that the only alternatives are a general arbitration clause or none at all. Different clauses can be employed that may be preferable to either a general arbitration clause or no arbitration clause at all. Although the variations suggested are by no means exclusive, they do demonstrate that an arbitration clause can be "tailored" to make it preferable to either a general arbitration clause or none at all. Variations can include the following:

1. limiting arbitration to factual disputes such as those involving technical performance standards or eliminating arbitration of other types of more "legal" disputes such as termination
2. specifying the place of arbitration
3. providing a designated person or persons as arbitrator or arbitrators
4. limiting arbitration to claims not exceeding a designated amount or percentage of the contract price[158]
5. limiting disputes to those that occur while the work is proceeding with an expedited one-person panel

---

[158]The Engineers Joint Contract Documents Committee (EJCDC) has limited arbitration to claims under an amount set forth in EJCDC E-500 (2002), Exhibit H at H.6.09(A2) with a blank provided for the amount. E-500 is its Standard Form of Agreement Between Owner and Engineer for Professional Services. Currently, EJCDC is working on a new set of E-series documents. Earlier documents had limited arbitration to claims not to exceed $200,000.

(see CIA Rules F-1 through F-13, the fast-track system to be discussed in Section 30.18A)

6. permitting consolidation of separate arbitrations
7. providing a right to discovery
8. limiting the award to the most fair of the last proposals or an amount between the two final proposals of the parties (baseball arbitration)
9. eliminating the use of attorneys
10. making the award "nonbinding"[159]

The more variables, obviously, the greater the cost of obtaining agreement and expressing it. This probably inhibits specialized arbitration clauses, at least where one party cannot dictate the clause to the other. A look at the variables and attendant complexity may persuade an exasperated drafter or negotiator that it may be simpler where possible to agree to waive a jury and have all disputes tried before the judge.

Choosing a general arbitration clause may be influenced by those who issue surety bonds and professional liability insurance. If the insurer or surety will not cover work done under a contract with a general arbitration clause, this factor can be significant in the choice.

Whether to agree to a general arbitration clause at a particular time and place requires that the choice go beyond simply comparing abstract models of dispute resolution. For example, if the choice is between taking a dispute to an efficient and competent legal system or an unknown panel of AAA arbitrators, the former is preferable. However, an inefficient court system with questionable judgment is much less preferable than a highly skilled arbitration system.

As shall be seen in Section 30.19, many state dispute resolution systems have developed techniques that have eliminated the worst aspects of incompetence and delay that can cause parties to choose arbitration.

The *Perini* case, noted in Section 30.14, demonstrated that arbitration is not always a simple, expeditious, or inexpensive method of adjudicating commercial controversies. Any system will have its "horror story" cases, and perhaps the quality of the arbitrator's award could conceivably have been better than what would have resulted in court. It is unlikely, however, that arbitration in this case saved time or money.

Another important aspect of arbitration choice that favors arbitration relates to the likelihood that the parties will accept the arbitrator's decision without repeated and costly trips to the courthouse. This is becoming even more complicated by the expansion of jurisdiction under the Federal Arbitration Act and the increasing likelihood that arbitration may be sought in either state or federal court and under either a state arbitration statute or the federal act. This will undoubtedly add to the complexity of legal issues and may in the long run discourage arbitration.

This section has pointed to advantages and disadvantages of arbitration. The contracting parties and their attorneys should carefully consider such factors when choosing contract language dealing with dispute resolution. Choosing among a general arbitration clause, a limited arbitration clause, or no arbitration clause is an important decision that should be made with care.

SECTION 30.18 Private Systems

The cost, the time consumed, and the quality of litigation outcome has discouraged taking disputes to court. As a result, new nonjudicial methods have mushroomed. These other methods are collectively referred to as alternative dispute resolution (ADR). But as arbitration has become more like litigation as to time, expense, and formalities, emphasis has shifted to methods that seek techniques designed to prevent disputes, such as partnering (Section 17.04G), Project Counsel (Section 30.18G), Conflict Manager (Section 30.18I), Project Neutral (Section 30.18F), and methods to help the parties resolve disputes such as mediation (Section 30.18B), minitrials (Section 30.18D), and dispute review boards (Section 30.18E).

As ADR consists of fluid methodologies, so discussion must concentrate on the basic elements of these systems. See the notes for sources that can be consulted.[160]

---

[159]For another attempt at tailoring a clause see Riggs, *Update on Arbitration* "Fixes," 16 Common Sense Contracting, No. 3, p. 4 (Smith, Currie & Hancock). Riggs worries about enforceability of clauses that seem to reject arbitration policy. See *Crowell v. Downey Community Foundation Hosp.*, supra note 145, discussed in Section 30.14.

[160]For a comprehensive survey of ADR, see *Stipanowich, Beyond Arbitration: Innovation and Evolution in the United States Construction Industry*, 31 Wake Forest L. Rev. 65 (1996). For a survey of ADR methods, including many described briefly in Sections 30.18–30.20, see Stipanowich, *The Multi-Door Contract and Other Possibilities*, 13 Ohio St.J. on Disp. Res. 303 (1998). It will be referred to as Stipanowich, *Multi-Door*. For an incisive analysis, see *Hinchey, Evolution of ADR Techniques for Major Construction Projects in the Nineties and Beyond: A United States Perspective*, 12 Constr. L.J. 14 (1996).

A. Arbitration

Arbitration has generally been thought of as the "alternate" part of ADR. Yet, as noted in the preceding paragraph, the controversy over arbitration engendered by the change from simple to an excruciatingly more complex process has led both to a search for other methods and attempts by those who operate arbitral systems such as the American Arbitration Association (AAA) and JAMS to search for better methods to attract users by responding to customer complaints.

Although AAA is a nonprofit organization whereas JAMS is "for profit," both seek to attract the same users and can be compared. As AAA is built into AIA documents and has a much greater caseload, the emphasis here will be on AAA.

The movement from simple to complex requires consideration of classic, simple arbitration. The scene is a district where cloth manufacturers are located. The buyer of a bolt of cloth complains to the seller about the quality. Each walks over to the shop of a trusted and experienced clothing manufacturer. Buyer and seller agree to accept her decision. (This might even be implied.) She examines the cloth and pronounces that it would pass in the trade as acceptable. Even if the buyer disagrees, the buyer accepts the decision. The buyer pays the seller. The dispute is concluded.

A normal construction dispute, such as one over the quality of workmanship, *can* be that simple. But the normal construction dispute will be much more complex. A construction project is much more complex than the sale of cloth. The contract is much longer, consisting of pages of technical and contractual provisions. Resolving the dispute may require the re-creation of events that transpired months or even years before. Chapter 8 explored many of the reasons both for the likelihood of disputes and difficulty resolving them.

The difference in the nature of the dispute between the classic paradigm and most construction disputes, plus the tendency of parties involved in disputes to use lawyers, has led to the judicializing of the process. This shift often creates the same problems that caused ADR to develop as a substitute for litigation: cost, time, and quality of the decision.

The recognition by AAA that its users were unhappy with certain aspects of its process led AAA to convene a group of experts to evaluate its system and make suggestions for its improvement. The task force assembled and produced a report.[161] Its main criticisms were

1. the need for different "tracks" of differing complexity depending on the dispute
2. the need for a more expert panel
3. lack of clear standards for discovery
4. the form of the decision accompanying the award

The AAA responded both internally and in the Construction Industry Arbitration (CIA) Rules published by the National Construction Dispute Resolution Committee. The latter has representatives from twenty-two national professional and trade associations. The rules are a component of the Construction Industry Dispute Resolution Procedures (CIDRP).

Before looking at efforts made by the AAA to improve the quality of its arbitrators, it is helpful to see the method by which arbitrators are selected in the AAA system.

Over the years, the AAA has changed its method of appointing arbitrators. Rule R-3 states that the AAA will establish and maintain a National Roster of Construction Arbitrators (National Roster). The CIA Rules are in Appendix F.

If the parties have not appointed an arbitrator or provided a method of appointment, Rule R-11 provides that the AAA shall send to each party a list of names chosen from the panel. If the parties cannot agree upon an arbitrator, the parties have fifteen days to strike names on the national roster and number the remaining names in order of preference. Taking this into account, the AAA appoints the arbitrators. If this process does not work, the AAA can select the arbitrator from the panel "without submission of additional lists."

Rule R-14 permits the parties to appoint party-appointed arbitrators. These arbitrators select a neutral third arbitrator, called a chairperson. If this method does not work, the AAA selects the chairperson.

Administratively, the AAA sought to improve the quality of its arbitrators by slimming down the panel, vetting the applicants more carefully, and requiring panelists to develop greater skill by using arbitrators more frequently. Also, training programs were created with criteria for evaluating trainee arbitrators.

---

[161]CONSTRUCTION ADR TASK FORCE REPORT (Am.Arb. Ass'n, Oct. 26, 1995).

In addition to developing a panel of skilled and experienced panelists, those admitted to the panel must agree to set aside consecutive days to hear important cases or at least to hear them in large blocks of consecutive days. This provision is to avoid the too frequent "stop and go" nature of private arbitration.

The JAMS panel originally consisted of retired judges. JAMS recently added retired experienced attorneys. Currently it describes its panels as Resolution Experts. Its panelists have had considerable experience and are likely to hear the matter from beginning to end without interruption.

To provide rules that are appropriate to the dispute, the AAA created three tracks. The Fast Track (Rules F-1 through F-13) applies to claims which do not exceed $75,000. Highlights of this track are a single arbitrator, informal notices, a sixty-day time limit from appointment of arbitrator to award, a single-day hearing, fourteen days from hearing to award, and allowing a "paper only" arbitration for claims, none of which exceeds $10,000, "unless any party requests an oral hearing or conference call, or the arbitrator determines that an oral hearing or conference call is necessary."[162]

The Regular Track Rules (R-1 through R-55) apply to claims of from $75,000 to $500,000.[163] Unless the parties have selected party appointed arbitrators, the dispute is heard by a single neutral arbitrator with an increased role in her selection given to the parties.[164] Greater authority is given the arbitrator to control the discovery process.[165] This will be discussed later in this section.

The Large/Complex Track (Rules L-1 through L-4) applies to claims of at least $500,000.[166] Claims can be heard by one to three arbitrators depending on the agreement of the parties. If they cannot agree, the claim is heard by three arbitrators if the claim is at least $1 million.[167] If the claim is for less than $1 million, a single arbitrator hears the case similar to the regular track.[168] There is a mandatory preliminary hearing to discuss the issues, exchange documents, identify and schedule witnesses, schedule hearings, discuss the extent of discovery, and take up other matters relevant to the arbitration.[169]

Because discovery (as noted in Section 30.06) has been one of those sensitive issues in arbitration, it may be useful to look at those rules that deal with discovery. It must be remembered that many lawyers want full judicial-like discovery, whereas many oppose it because of the cost and time. Most believe some form of information exchange is needed to prepare for any settlement discussions. Also, complaints have been made about the lawyers in judicial discovery being essentially "in control."

Fast-Track rules provide that the parties must "exchange copies of all exhibits they intend to submit at the hearing."[170] This exchange must be made at least two days before the hearing. There is no discovery, unless it is ordered by the arbitrator "in extraordinary cases when the demands of justice require it."[171]

The Regular Track requires exchange of information. The arbitrator is empowered to direct the production of documents and other information and "the identification of any witnesses to be called."[172] The arbitrator resolves disputes concerning "the exchange of information." There is no discovery under Rule R-22. But the arbitrator, as in Rule F-7, can order it "in extraordinary cases when the demands of justice require it."[173]

Rule L-4 for Large/Complex cases requires the parties to participate in a preliminary hearing at which items "to be considered" are "the extent to which discovery shall be conducted." Information as to witnesses is exchanged. These rules incorporate the Regular-Track rules. If there is a conflict, the Large/Complex Rules apply.

Because Rule L-4 lists the extent of discovery among matters to be discussed at the preliminary hearing, apparently discovery can be demanded. The arbitrator decides disputes over discovery.

As to the award, the AAA has made similar moves to a more judicialized arbitration. Nothing is stated as to the award in the Fast-Track rules. Under the Regular rules, the arbitrator must "provide a concise, written breakdown of the award."[174] If requested by all parties prior to the

---

[162]Id. at Rule F-8.
[163]Id. at Rule R-1(c).
[164]Id. at Rule R-18.
[165]Id. at Rule F-7, R-22(d), L-4.
[166]Id. at Rule R-1(c).
[167]Id. at Rule L-2.
[168]Ibid.

[169]Id. at Rule L-3.
[170]Id. at Rule F-6.
[171]Id. at Rule F-7.
[172]Id. at Rule R-22(d).
[173]Ibid.
[174]Id. at R-43.

appointment of the arbitrator "or if the arbitrator believes it is appropriate to do so" she must provide "a written explanation of the award."[175]

JAMS also has moved to a more judicialized arbitration. In addition to a good faith exchange of information under Rule 17(a), Rule 17(c) permits each party to take one deposition of an opposing party "or individual under the control of the opposing party." Rule 17(c) continues "The necessity of additional depositions shall be determined by the Arbitrator based on the reasonable need for the requested information, the availability of other discovery options and the burdensomeness of the request on the opposing Parties and the witness."[176]

JAMS has also provided for methods of making the award that go beyond simply letting the arbitrator decide. Rule 32 allows the parties to choose a Bracketed (or High-Low) Arbitration Option. The parties may agree on minimum and maximum damages. This agreement is given to the case manager, but not to the arbitrator without consent of the parties. If the award is between the agreed minimum and maximum, the award stands. If the award is below the agreed minimum, the award is corrected to the agreed minimum. If it exceeds the agreed maximum, the award is corrected to reflect the agreed maximum.

Rule 33 allows use of the Final Offer Award Option, known as the "baseball" award. Under this option, at least seven days before the hearing, the parties exchange proposals for the amount they would offer or demand. This is given to the case manager and to the arbitrator, unless the parties do not wish the arbitrator to receive it. These proposals can be revised until the close of the hearing.

If the arbitrator has received the proposals, she must choose the proposal she finds most fair and appropriate. If the arbitrator is not given the proposals, she renders the award. The award is corrected to conform to the closest of the last proposals and becomes the award.

Finally, Rule 34 allows the parties to agree to the JAMS Operational Arbitration Appeal Procedure before the award. This would resemble an appeal to an appellate court. Few have chosen this option.

These rules seem to respond to complaints of lawyers for discovery and the right to appeal and to publicity given baseball-arbitration disputes, as well as the need to curb outrageous demands and "stonewalling" positions so common in negotiations.

The relentless move toward making arbitration more like litigation may entice some lawyers to suggest arbitration to their clients (or relieve lawyers from having to insert the right to discovery in the contracts to arbitrate). But it may induce parties not to arbitrate and either litigate or look for another ADR method.

B. Mediation

Mediation has become a popular ADR technique. Although it takes a number of forms, essentially it involves the use of a neutral, third party, usually an individual but on occasion a team of two co-mediators, to seek to bring disputants together and settle the dispute. Hinchey described three types of mediation as follows:

1. *Rights-based* mediation seeks to achieve a settlement focusing on the legal rights of the parties.
2. *Interest-based* mediation is more "freewheeling," directing attention less to legal rights and more to "the parties' legal interests or compelling issues of the dispute."
3. *Therapeutic* mediation emphasizes "the emotional dimensions of the dispute" and seeks methods to handle future conflicts.[177]

The first is the type with which most lawyers are familiar. The second type resembles what the arbitrator did in *Advanced Micro Devices, Inc. v. Intel Corp.*, discussed in Sections 30.04A and 30.12. In that case the arbitrator sought to wrap up the entire disagreement between the parties even if this effort went beyond the narrow contract dispute submitted to arbitration. The third type resembles the function of a labor mediator engaged by the parties to a collective bargaining agreement or to a marital dispute involving the nature of the a marital relationship, child custody, or visitation rights.

Hinchey states the process can be stopped by any party at any time.[178] As mediation is a voluntary process, Hinchey states that it is important for the disputants to seek mediation as a solution and not have mediation thrust on them.

---

[175]Ibid. The changes in AAA procedures are discussed in Stipanowich, *MultiDoor*, supra note 160 at 343–347.

[176]All references are to the JAMS Comprehensive Arbitration Rules and Procedures, Jan. 2007. JAMS also publishes streamlined Rules.

[177]Hinchey, supra note 160 at 17–18.

[178]Id. at 18.

There is a difference of opinion as to the desirability of standard contracts containing an obligation to mediate. Objection is based upon the concept of voluntariness crucial to mediation. There were voluntary mediations long before it began to receive the attention currently given to it.

Yet others believe a structured system spelled out in advance and agreed upon by the parties can advance mediation. This is the basis for inclusion of mediation in AIA Documents. AIA gave mediations a power impetus when it included them in 1997.[179]

In 2007 the AIA went further, making mediation a central part of its disputes process. A201-2007, Section 15.2.5, states that a decision by the Initial Decision Maker (IDM) is "final and binding but subject to mediation." If mediation does not dispose of the dispute, it goes to binding dispute resolution (arbitration), if selected by the parties. Section 15.2.6 allows each party to demand mediation of an initial decision at any time. Section 15.2.6.1 states a party can demand in writing that the other party file for mediation within sixty days of the initial decision. If it does not, "both parties waive their rights to mediate or pursue dispute resolution." Failure to mediate precludes arbitration or litigation and makes the initial decision binding. Under Section 15.3.2 mediation is governed by the AAA's Construction Industry Mediation Procedures, a formidable set of seventeen rules. See Appendix F. B101-2007, Section 8.2, provides a similar mechanism for mediation, except there is no requirement for an initial decision by an IDM.

The assumption has been that the parties have made a valid agreement to mediate. Unlike arbitration that resolves the dispute, mediation does not attract the attention given to arbitration as a system that can be abused. In fact mediation currently occupies a benign position in dispute resolution.

As seen in Section 30.03B, one aspect of arbitration that has received the attention of legislatures is the location of the arbitration. Increasingly, states are placing limits on agreements to arbitrate that require the arbitration be held in a locality inconvenient to one of the parties.[180]

One such statute is California Code of Civil Procedure Section 410.42(a). It bars enforcement of a clause that requires arbitration or litigation outside California in certain contracts. It also adds the phrase "or otherwise." A California court held that this can be interpreted to include mediation.[181] It held that a clause requiring mediation in Nevada would not be enforced. The issue will arise more often as mediation becomes increasingly highlighted.

Hinchey suggests it is important to make certain mediators are not compelled to testify in any litigation and that any material the mediator develops is not used if mediation fails.[182] Mediators must be knowledgeable in the subject matter and appreciate the subtleties needed to persuade each party to see the strengths and weaknesses of its respective position. Also, mediators need the trust and confidence of the disputants. The mediator does not render a decision and has no power to compel a party to submit any proposal or counterproposal, but seeks to catalyze, to induce the parties to reach a voluntary settlement. When mediation is done successfully, the parties themselves negotiate a settlement that allows them to control the terms.

Mediator styles vary. Some operate as messengers. They simply communicate the positions of the parties but in a less hostile way. Others are more active. These mediators inject their own view and seek to persuade each party of the strengths of the other party's position, the weakness of their own, and the cost of unreasonable rigidity.[183]

Section 30.19 examines approaches used within the legal system to deal with construction disputes. Because there is greater focus on mediation of construction disputes, it is inevitable that courts will try to use mediation of disputes brought before them. As a result, cases are beginning to reach appellate courts that deal with the power of the courts to order mediation.

One court held that the court could not compel litigants to attend and pay for private mediation.[184] Another held that the twelve parties to the lawsuit could be ordered to mediate before a private mediator and share the costs. The court was not able to find specific authority to give

---

[179]A201-1997, ¶ 4.5, B141-1997, ¶ 1.3.4, A401-1997, ¶ 6.1.

[180]See statutes noted in Section 30.03A.

[181]*Templeton Development Corp. v. Superior Court,* 144 Cal.App.4th 1073, 52 Cal.Rptr.3d 19 (2006).

[182]Hinchey, supra note 160 at 18.

[183]For discussion of mediation, see Stipanowich, *Multi-Door,* supra note 160 at 316–318. For a survey of attitudes toward mediation, see Thomson, *A Disconnect of Supply and Demand: Survey of Forum Members' Mediation Preferences,* 21 Constr. Lawyer, No. 4, Fall 2001, p. 17.

[184]*Jeld-Wen, Inc. v. Superior Court,* 146 Cal.App.4th 536, 53 Cal. Rptr.3d 115 (2007).

this order but held the inherent power of the court to manage and control its calendar was sufficient.[185] Clearly, this issue will arise more frequently.

C. Mediation-Arbitration

Another method that has been proposed is mediation-arbitration, which has been developed in some collective bargaining contexts and places the third party in two roles. The third party seeks to mediate disputes between the parties by aiding in the negotiation process with a view toward settlement, but if parties do not settle a dispute the third party can use normal arbitration power to decide the dispute.

Because the third party must possess a variety of skills and would have considerable power, this mediator-arbitrator would likely be selected in advance by the parties. For this reason, the system contemplates the use of the mediator-arbitrator throughout the performance period. Although this arrangement is technically available in ordinary arbitration, experience has shown that most disputes during performance are decided by the design professional. A mediator-arbitrator would very likely be called on more frequently during performance and may in practice supplant the dispute resolution role of the design professional.

Mediation-arbitration is rarely used in construction disputes. One difficulty is finding a third party in whom the contracting parties have confidence and who has not only the skill but also the time to play an ongoing role during construction. Opposition can come from design professionals or construction managers who see the injection of yet another major participant as a threat to their status and as an additional complicating factor in an already complicated system. Yet it does have the advantage of moving *automatically* from one method, mediation, to another, arbitration.

D. Mini-Trials

The mini-trial has received attention in the popular press and is beginning to receive attention in the scholarly journals.[186] Although there can be many variations (the

process being essentially a private one made by a contract), attention has been directed toward a particular mini-trial and the process it used.

After a lengthy process of negotiation, the disputants worked out an agreement for a mini-trial. The parties agreed on a judge, or a neutral adviser. They were allowed an expedited discovery procedure and exchanged briefs. They were given a designated time to present their positions before executives of each disputant not *directly* connected with the dispute who had authority to settle. The hearing itself was limited to two days, moderated by the adviser. If the executives were unable to reach a settlement after the hearing, the adviser would issue a nonbinding opinion. The executives would meet again with the hope of settling. If they could not, the parties could go to court, with any admissions made or the tentative opinion of the adviser not admissible in any subsequent trial.

E. Dispute Review Board

Although dispute review boards (DRB) originated in tunneling contracts, their use has extended to complex contracts, often of an international nature.[187] A DRB is essentially a committee that familiarizes itself with the site and work in progress, meets periodically, gets to know the major actors, learns of problems as they arise, and suggests interim solutions to claims and disputes.

Members are construction industry experts selected by the parties. Two commentators state that as a rule, "the owner selects one board member (subject to approval by the contractor) and the contractor appoints another (subject to approval by the owner), and the first two appoint a third (subject to approval by the owner and the contractor). This last appointee customarily serves as the chairperson. Each of the three members is to be impartial, neutral, and not an advocate for either side.[188]

---

[185]*In re Atlantic Pipe Corp.*, 304 F.3d 135 (1st Cir.2002). The court held the order should have put limits on duration and fees.

[186]For discussions of mini-trials, see Green, Marks, & Olson, *Settling Large Case Litigation: An Alternative Approach*, 11 Loy.L.A.L.Rev. 493 (1978); Nilsson, *A Litigation Settling Experiment*, 65 ABAJ 1818 (1979);

Anderson & Snipes, *Stretching the Concept of Mini-Trials: The Case of Bechtel and the Corps of Engineers*, 9 Constr. Lawyer, No. 2, April 1989, p. 3; Klitgaard & Mussman, *High Technology Disputes: The Mini-Trial as the Emerging Solution*, 8 Santa Clara Computer & High Tech.L.J. 1 (1992).

[187]For a thorough analysis by a DRB member in international contexts, see Renton, *The Role of the DRB in Long Term Contracts*, 18 Constr. L.J. 8 (2002).

[188]McMillan & Rubin, *Dispute Resolution Boards: Key Issues, Recent Case Law and Standard Agreements*, 25 Constr. Lawyer, No. 2, Spring 2005, p. 14. This paper contains a thorough, thoughtful review of DRBs.

The construction contract establishes the DRB. Provisions deal with selection of members, the board's operation, meetings, hearings, recommendations, and admissibility of recommendations if litigation develops. Usually a party must submit a claim to the board before initiating litigation. But the board does not make a binding decision. Yet its recommendations can aid in getting the parties to settle.

In addition, a tripartite agreement is executed by the board members, the owner, and the contractor. It deals with board responsibilities, removal, compensation, and methods of resolving disputes among the members.

Some issues that have come before the courts are the removal of members[189] and the effect of such boards on other participants, such as subcontractors.

The DRB process can be expensive but if operated properly can save a great deal of claims overhead.[190]

The composition of the DRB has begun to raise legal issues. Most issues have been the contractual obligations of the members and those that appoint them. But in a recent case a contract required a subcontractor to submit its claim to a DRB as a condition precedent to litigation. The DRB consisted of members approved by the owner and the prime contractor. This method of selection was found unconscionable.[191] That method would be appropriate for relations between the owner and the prime but not one that should pass upon subcontractor claims even to the extent of submission to the DRB being a condition precedent to litigation. The condition precedent was excused.

F. Project Neutral

The project neutral resembles the dispute review board. Each seeks to establish an "on the ground" presence during construction. Its use creates a rapid, expert technique for sorting out problems and helping parties resolve disputes before they fester and grow.

The project neutral usually is a *group* of experts much like the early neutral evaluation groups used in federal courts. Although developed principally for ongoing disputes, it can be adapted for use during performance. It uses expert skills to find the facts, evaluate claims, and make recommendations to the parties.

G. Project Counsel

Some have advocated the use of a project lawyer who does not represent any of the parties but represents the project. Such a lawyer seeks to get the project built properly on time, without claims.

The Project Counsel would work with the owner and other participants to help select a delivery system for design and construction. She would advise the owner on a process to select other participants. She would help produce a draft set of construction documents. She would help the participants affirm their relationship and produce a set of enforceable agreements. She would work with the participants to develop a conflict resolution system that could be administered by the Project Counsel. The Project Counsel would harmonize the contractual arrangements with insurance, bonding, and risk management devices. She would play an important role in the close-out process and be available to address post-completion disputes.

Any proponents would have to see if there is a market for such a person, determine whether there would be funding, how compensation would be defined and structured, analyze the effect of being a project lawyer from the perspective of ethical and professional responsibility, evaluate liability risks, and determine the best association for performance of such services.[192]

H. Partnering

Partnering has been discussed in Section 17.04G. There it was treated as part of the system for organizing the project. But it can be considered a part of dispute resolution. Proper implementation of partnering can diminish the incidence of claims by reducing friction and adversarial postures.[193]

[189]*Los Angeles County Metro. Transp. Auth. v. Shea-Kiewit-Kenny*, 59 Cal. App.4th 676, 69 Cal.Rptr.2d 431 (1997).

[190]For general background on DRBs see R. MATYAS, A. MATTHEWS, R. SMITH, & P. SPERRY, CONSTRUCTION DISPUTE BOARD MANUAL (1996): Carr, Rubin & Smith, *Dispute Review Boards*, Chapter 5 WILEY CONSTRUCTION LAW UPDATE p. 111 (1992).

[191]*Sehulster Tunnels Pre-Con v. Traylor Bros., Inc./Obayashi Corp.*, supra note 26.

[192]Noble, *Friend of the Project—A New Paradigm for Construction Law Services in a "Partnered" Construction Industry*, [1998] Int'l Constr.L.Rev., 79, 81–84.

[193]See Stipanowich, *MultiDoor*, supra note 160 at 378–384.

I. Conflict Manager

In addition to describing the Dispute Resolution Advisor used in a Hong Kong hospital project,[194] Professor Stipanowich has proposed use of a Conflict Manager.[195] Her role would be much the same as the Project Counsel noted in Section 30.18G. She would facilitate initial partnering workshops. She would assist the key actors in conducting a "conflict diagnosis" for the project and in developing a conflict management system.

Her role would continue during the project. She would make monthly visits, confer with the key participants, and mediate any conflicts, much like a Project Neutral or Dispute Review Board. She might suggest sending the failed negotiation to any Project Neutral or Dispute Review Board. She could "help the parties identify the issues to be addressed, assist in structuring limited discovery to elicit relevant information and fine-tune ADR procedures."[196] While her role would be facilitative, she might be authorized to decide the dispute.

Stipanowich recognizes the cost concerns, the enforceability of the program, the qualification and standards for Conflict Managers, and ethical issues.[197] But he, like others who propose these interactive systems,[198] sees the need for an active hands-on role for the neutral.

J. Court-Appointed Arbitrators: Adjudicator (Great Britain)

The Adjudication system is required in most British construction contracts made after March 1, 1998.[199] It allows a party to send disputes to an adjudicator. Within a short and defined term, the adjudicator resolves the dispute. Her decision is final unless reversed by an arbitrator or a judge. If taken up, the arbitrator or judge resolves the dispute without regard to the adjudicator's decision.[200]

The system originated in payment disputes between prime and subcontractor where there was no architect or engineer to resolve such disputes. It received a big push when the first edition of the New Engineering Contract split up the functions of the engineer so that a new party unconnected with the project adjudicated disputes.

It is clear that this system developed because of dissatisfaction with the traditional method of having the architect or engineer resolve disputes, as described in Chapter 29. But it reflects the need for a quick, though not necessarily final, decision by a neutral.

While this need is best reflected in the need to resolve disputes while the project is proceeding, experience shows that most adjudication takes place after the project has been completed. If that is the case, the system mimics court-appointed arbitrators used in many American courts. In these courts disputes under a certain amount must go to an arbitrator. Her decision can be appealed but few are. This demonstrates that parties want a neutral person to look at their dispute and are satisfied with this method or the cost of taking the dispute to court discourages that step.

K. Multi-Tier Systems

Some contracts use a multi-tier system of resolving disputes. As an example, the dispute would be referred to a mediator. If there is no agreement, the dispute might be sent to a mini-trial. If that does not produce a settlement, the dispute is referred to arbitration. Another multi-tiered system is the Dispute Review Board–Mediation-Arbitration.[201] In some public contracts, there may be an agency internal review process before the multi-tier system is invoked. Multi-tier systems are designed to start with less expensive, informal procedures; move to more expensive, somewhat more formal procedures; and conclude with the most formal, and most expensive, arbitration or litigation.

L. Architect/Engineer Resolution

These systems must be seen against the backdrop of the traditional process for resolving, at least initially, disputes. This process requires that the dispute be submitted to the

---

[194]Id. at 387–389.

[195]Id. at 392–394.

[196]Id. at 393.

[197]Id. at 394–403.

[198]See Noble, supra note 192.

[199]Housing Grants, Construction and Regeneration Act of 1996.

[200]Adjudication has spawned a great deal of literature in Great Britain. See Aeberli, *Binning the Black Bag (What Material Can an Adjudicator Consider?)*, 23 Constr.L.J. 399 (2007); Sheridan & Helps, *Construction Act Review*, 19 Constr.L.J. 25 (2003); Tweeddale, *Challenging Jurisdiction in Adjudication Proceedings*, 17 Constr.L.J. 3 (2001); Kennedy & Milligan, *Research Analysis of the Progress of Adjudication Based on Questionnaires*

Returned from Adjudicator Nominating Bodies (ANBS) and Practicing Adjudicators, 17 Constr.L.J. 231 (2001) and Blunt, *Adjudicators' Time Defaults*, 17 Constr.L.J. 371 (2001).

[201]Renton, supra note 187, p. 13.

architect or engineer. In addition to the objections made to this method noted in Chapter 29, particularly Section 29.04, these alternatives show the need for ongoing neutral monitoring of the project from beginning to end.

The relationship between resolution by the design professional and arbitration is demonstrated in *Martel v. Bulotti*.[202] The dispute in this case was resolved by the architect. The issue was whether the standard of review was the limited one provided by the arbitration statutes[203] or the somewhat broader common law review of third-party decisions.[204] The court held determination by the architect was not arbitration, and so the arbitration standard of review did not apply. Instead, the standard was the common law review standard of third-party decisions.

SECTION 30.19   Adjuncts of Judicial Systems

Just as outsiders saw problems in the judicial system, people operating the system were well aware of the system's flaws. Although wholesale restructuring was not considered possible, some judicial systems at federal and state court levels developed techniques to reduce the time and expense of providing a method of resolving private disputes. Two of these systems are noted in this section.

A. Special Masters and Referees

As courts have struggled with better methods of dealing with construction disputes, some have appointed individuals—called *masters* or *referees*—who are given certain powers to expedite construction litigation. Sometimes these individuals are given authority to set rules for the deposition process, which has become a costly method of obtaining information. Sometimes masters or referees informally act as mediators with a view toward persuading the litigants to settle the dispute. Finally, such individuals are sometimes authorized by the parties to resolve the dispute and are sometimes authorized by the judge to make findings of fact and conclusions of law, which can then be passed on and adopted by the judge if the parties have given up their rights to a jury trial. Sometimes a presiding judge designates a referee or master as a temporary judge

or a judge *pro tem*. Where the determination by the master or referee has been approved by the court or where the referee or master is a temporary judge or judge *pro tem*, the judgment is final and subject to the same appeal as an ordinary trial court judgment.

B. Summary Jury Trials

Summary jury trials have been developed in the federal trial courts. Although the Seventh Federal Circuit Court of Appeals has limited the trial judge's authority to compel use of such a method,[205] this process will likely be used increasingly, by agreement or in federal circuits that disagree with the Seventh.[206]

A jury is selected as would be a real jury. The judge informs the jury that the parties have agreed in advance to an abbreviated procedure, to save time and money. Not until after the proceedings is the jury told that its determinations are not binding.

Each attorney makes what in a real trial would be a combined opening and closing statement. Attorneys may use charts, graphs, or other visual aids that would be used in a normal closing argument. However, no witnesses offer testimony. The other attorney responds, and the first attorney is given a limited amount of rebuttal time. The time allotted for the arguments runs between one-half and one full day. After the attorneys make the presentation, the judge instructs the jury as at the conclusion of a normal jury trial. After instructions, the jury retires and then presents its verdict. Because the verdict is not binding, either party can demand a regular trial. But if the parties believe that the verdict is very likely what a real jury would determine, the mock verdict should encourage settlement. The obvious advantages to such a method are avoiding the expensive marshaling of documents, preparing a lengthy pretrial order, participating in a pretrial

---

[202]65 P.3d 192 (Idaho 2003).

[203]See Section 30.14.

[204]See Section 29.09.

[205]*Strandell v. Jackson County, Ill.*, 838 F.2d 884 (7th Cir.1988), noted in 40 Case W.Res.L. Rev. 491 (1989–1990). See *Metzloff, Reconfiguring the Summary Jury Trial*, 41 Duke L.J. 806 (1992).

[206]*In re Atlantic Pipe Corp.*, supra note 185, disagreed with the *Strandell* case and allowed a trial judge to order private mediation. This was based on the inherent power of a trial court to manage and control its calendar. The scholars are divided on this issue. See Ponte, *Putting Mandatory Summary Trial Back on the Docket: Recommendations on the Exercise of Judicial Power*, 63 Fordham L.Rev. 1069, 1094 (1995) (favoring such power). But see O'Hearne, *Compelled Participation in Innovative Pretrial Proceedings*, 84 Nw.U.L.Rev. 290, 317 (1989) (opposing such power).

conference, and preparing and presenting the witnesses. If the parties take their obligation to present their cases in good faith and then negotiate in good faith after the jury verdict, this method should save time and money.

However, some believe that a real jury is so different that what an advisory jury concludes may not be what a real jury would determine. Also, those who are not in favor of the summary jury trial emphasize the great difference between hearing summaries of the type that would be presented in a summary jury trial and actually seeing witnesses and listening to their testimony.

C. Mediation

Court-ordered mediation is being used more frequently. This has been discussed in Section 30.18B.

SECTION 30.20 Public Contracts

Public construction contracts involve considerations not found in private contracts. Many statutes, rules, and regulations govern the award of contracts, and the resolution of disputes under those contracts. This book cannot examine the details of these legal restraints, because of their complexity and variations. However, the increased and often intense spotlighting of disputes in construction work necessitates some observations regarding dispute resolution in the context of public construction contracts.

A. Federal Procurement Contracts

Before 1978, the disputes process in federal procurement contracts was based on the disputes clause, which gave the contracting officer of the federal agency awarding the contract the power to decide disputed questions that arose during performance or thereafter. The contracting officer is usually a high administrative official of the agency awarding the contract whose decisions were conclusive unless appealed to the head of the agency within thirty days from the decision.

The agency appeals boards are appointed by the head of the agency. Their hearings and decisions are very much like those of a court. Before 1978, contractors had to go before the agency appeals boards if the dispute arose under the contract. If their claims were based on a breach of contract, they could appeal to the then Court of Claims.

Alternatively, in claims under $10,000 they could appeal to a federal district court. They could appeal from board decisions, but a board's decision was final on questions of fact if supported by substantial evidence. The Court of Claims could make its own determination on legal questions. Many federal procurement doctrines were developed that had as their objective either to keep a dispute before an agency appeals board or to allow the board to be bypassed in favor of the Court of Claims.

In 1978, Congress enacted the Federal Contract Disputes Act,[207] which gave *legislative* authorization for a disputes process that up to that time had been created solely by contract. In addition, changes were made. Appeals cannot be taken as in the past to the federal district courts for claims under $10,000. Also, Section 605 of the act requires contracting officers to issue decisions within sixty days on claims of $100,000 or less. Claims over that amount require a decision within sixty days or a notification of the time within which a decision will be issued. Failure to issue a decision within the time required permits the contractor to bypass the contracting officer and go directly to an agency board of appeals or to court. Under Section 606 the contractor is given ninety days from the contracting officer's decision to appeal to the agency appeals board. Section 609 allows a contractor to bring an action directly to the U.S. Court of Federal Claims.

The Federal Courts Improvement Act made drastic changes in 1982.[208] It created a new U.S. Court of Appeals for the Federal Circuit and a new U.S. Claims Court[209] (to be distinguished from the earlier Court of Claims). On January 6, 2007 the Civilian Board of Contract Appeals was established. It combines a number of nonmilitary agencies, such as Departments of Agriculture, Housing and Urban Development, Interior, Labor, Transportation, Veterans Affairs, and the General Services Administration. The new board will follow the decisions of the Court of Federal Claims and the Court of Appeals of the Federal Circuit, as well as the predecessor boards of appeals. Claims are initially presented either to the board of contract appeals or the U.S. Court of Federal

---

[207]41 U.S.C.A. §§ 601 et seq.

[208]96 Stat. 25, Pub.L. 97-164. See Anthony & Smith, *The Federal Courts Improvement Act of 1982*, 13 Pub.Cont. L.J. 201 (1983); Miller, *The New United States Claims Court*, 32 Clev.St.L.Rev. 7 (1983–84).

[209]This court is now called the U.S. Court of Federal Claims.

Claims. Appeals go to the new Court of Appeals for the Federal Circuit.

The federal procurement system has been a pioneer in the more inventive forms of ADR.[210]

B. State and Local Contracts

A detailed treatment of the many state and local laws and regulations cannot be given in this book. It is important, though, to recognize the importance of complying with specialized requirements for disputes under such public contracts of this type. Special rules exist for making claims. Also, there is a great variety of contractual methods for resolving claims. Some make a decision of an employee of the public entity a condition precedent to litigation.[211] Some give broad dispute resolution power to an official of the public entity, with her decision final unless the decision was arbitrary or capricious.[212] Some states have created special courts to deal with these claims. Early decisions in some states precluded arbitration of public contract disputes, as it would place in the hands of private parties the power to decide public matters and expend public funds.[213] However, hostility has been fading, and contracts to arbitrate future disputes have been upheld. Some states even require the arbitration of these disputes.[214]

California's experience demonstrates the movement from a system where an official of the public entity decided the dispute with limited judicial review, to a judicialized form of arbitration. Before 1978, California state agencies followed the traditional pattern of having disputes resolved initially by a high official of the agency—the state engineer. The state engineer could issue a decision that had a substantial amount of finality, being final and conclusive unless fraudulent, capricious, arbitrary, or so grossly erroneous as to necessarily imply bad faith, a standard much like that applied to decisions by independent design professionals.

In 1977, a California appeals court decided a case that led to drastic overhaul of the dispute resolution system. In that case, the contractor appealed to the trial court after an adverse decision by the state engineer. The trial court found that errors had been committed, disregarded the decision of the state engineer, and retried the case.

The Court of Appeals agreed that the state engineer committed procedural errors[215] but did not set aside the decision as had the trial court. It sent the claim back to the state engineer for reconsideration. In doing so, the court emphasized that the parties had "by voluntary contract" agreed the disputes would be resolved by a designated person. This conclusion was disputed by state contractors, who stated they had no choice but to agree to such a provision.

The furor caused by this case led to an executive order and later to a statute that required major state procuring agencies to use a judicialized arbitration.[216] Disputes would be decided by the persons certified as competent arbitrators by an arbitration committee composed of representatives of the state agencies and the construction industry. Those approved were placed on the State Construction Arbitration Panel. Those who wished to be placed on the panel had to submit information that indicated their education and experience and set a rate at which they were willing to serve. The disputants selected from this certified list.

Discovery is required. The arbitrator can use expert technical or legal advisers, depending on whether the arbitrator is an attorney or a technically trained person. Consolidation and joinder are permitted. Hearings are

---

[210]For an article dealing with how the federal procurement agencies can use new methods of ADR, see Crowell & Pou, *Appealing Government Contract Decisions: Reducing Cost and Delay of Procurement Litigation with Alternative Dispute Resolution Techniques*, 49 Md.L.Rev. 183 (1990). But see Mackey, *ADR Can Banish Due Process From Public Contract Disputes*, 30 Procur. Lawyer, No. 4, Summer 1995, p. 3.

[211]*Yonkers Contracting Co., Inc. v. Port Authority Trans-Hudson Corp.*, 87 N.Y.2d 927, 663 N.E.2d 907, 640 N.Y.S.2d 866 (1996).

[212]*Westinghouse Elec. Corp. v. New York City Transit Auth.*, 14 F.3d 818 (2d Cir.1994).

[213]*City of Madison v. Frank Lloyd Wright Found.*, 20 Wis.2d 361, 122 N.W.2d 409 (1963).

[214]North Dakota requires arbitration of highway construction contract disputes. *Gratech Co., Ltd. v. Wohl Eng'g., P.C.*, 672 N.W.2d 672 (N.D.2003) (statute expansively interpreted to require arbitration of a highway contractor's tort claims against the project engineer). Arbitration is also required for certain public contracts in Rhode Island. See *Sterling Eng'g & Constr. Co. v. Town of Burrillville Housing Auth.*, 108 R.I. 723, 279 A.2d 445 (1971). Likewise, Pennsylvania requires that public contracts contain arbitration clauses. See *U.S. Fid. & Guar. Co. v. Bangor Area Joint School Auth.*, 355 F.Supp. 913 (E.D.Pa.1973).

[215]*Zurn Eng'rs v. State Dep't of Water Resources*, 69 Cal.App.3d 798, 138 Cal.Rptr. 478, cert. denied, 434 U.S. 985 (1977).

[216]See West Ann.Cal.Pub.Cont.Code § 10240 *et seq.* See also West Ann.Cal.Civ.Code § 1670, which requires local entities to either arbitrate or litigate. Employees of the agency cannot, as before, decide the dispute.

open to the public. The award must contain findings of fact, a summary of the evidence, and reasons underlying the award as well as conclusions of law. If the award is not supported by substantial evidence or is based on an error of law, it can be vacated (set aside).

California continued its innovation in 1990, enacting legislation that regulates disputes between contractors and local public entities for claims of $375,000 or less. This process uses informal conferences, nonbinding judicially supervised mediation, and judicial arbitration. The agencies were required to incorporate this process in plans and specifications that could give rise to a claim.[217]

SECTION 30.21 International Arbitration: FIDIC

Sections 2.15 and 8.09 mentioned the difference between domestic construction contracts and those that involve nationals of different countries, particularly contracts made by American construction companies requiring them to build projects in a foreign country. (Increasingly, foreign contractors build in the U.S.) Transactions of this type generate some issues of minimal or no importance in domestic contracts.

The contracts themselves may be expressed in more than one language, often generating problems that result from imprecise translation. They may also involve payment in currency that varies greatly in value. Contractors in such transactions may often have to deal with tight and often changing laws relating to repatriation of profits and import of personnel and materials. Perhaps most important, neither party may trust the other's legal system, and the contractor may believe it will not obtain an impartial hearing if forced to bring disputes to courts in the foreign country, particularly if the owner is, as so common in lesser developed countries, an instrumentality of the government.

Contractors making these contracts commonly insist on international arbitration to resolve disputes. Such arbitrations are usually held in neutral countries or in centers of respected commercial arbitration. Such awards are enforceable worldwide because of the Convention of the Recognition and Enforcement of Foreign Arbitration Awards, known as the New York Convention. Because of the complexities generated by international arbitration, this book does not discuss the subject in detail. Yet the emphasis on dispute resolution worldwide is a justification for some information regarding systems used for international transactions.[218]

The most commonly used contract for international engineering is the one published by the Federation Internationale des Ingenieurs-Conseils (International Federation of Consulting Engineers). The federation's contract, generally known as the FIDIC contract, for civil engineering construction provides for international arbitration.

Traditionally, the FDIC had a two-step procedure. First, a dispute would be submitted to the consulting engineer who administered the project and may have played a significant role in design. Her decision could be taken to international arbitration if either or both parties were dissatisfied.

The same dissatisfaction that led the United Kingdom to inject an adjudicator in place of the architect or engineer,[219] and led some Americans to use the Disputes Review Board to monitor the project and give an opinion as to existing disputes.[220] This generated a change in the FIDIC issued in 1999. In its Red Book, subclause 20 creates an impure three-tiered system for handling disputes.[221] Subclause 20.2 requires that disputes be submitted first to a Dispute Adjudication Board (DAB). The DAB consists of one to three "suitably qualified" persons. Members are selected by the parties.

Subclause 20.4 requires the DAB to give a reasoned decision within eighty-four days after the dispute has been submitted to it. If either party is dissatisfied, it has twenty-eight days to express its dissatisfaction and ask for arbitration. If there is no notice of dissatisfaction, the decision is binding and the parties must give it effect,

---

[217]West Ann.Cal.Pub.Cont.Code § 20104 *et seq.*

[218]Black, Venoit, & Pierson, *Arbitration of Cross-Border Disputes,* 27 Constr. Lawyer, No. 2, Spring 2006, p. 5 (helpful introduction). See also Paterson, *Canadian Developments in International Arbitration Law: A Step Beyond Mauro Rubino-Sammartano's International Arbitration Law,* 27 Willamette L.Rev. 573 (1991); Tiefenbrun, *A Comparison of International Arbitral Rules,* 15 B.C. Int'l & Comp.L.Rev. 25 (1992).

[219]See Section 30.18J.

[220]See Section 30.18E. Of course, there were other reasons for the DRB.

[221]*Conditions of Contract for Construction for Building and Engineering Works Designed by the Employer* (1999) (known as the Red Book).

unless and until it is revised in an amicable settlement or by arbitration.

Subclause 20.6 requires the parties to "attempt to settle the dispute amicably." Unless settled amicably under subclause 20.6, the dispute is sent to international arbitration. But arbitration can begin fifty-six days from notice of dissatisfaction even if there has been an attempt to settle amicably.

This creates a three-step procedure, although not a pure one. There is first the DAB. Then there is an attempt to settle. Then there is submission to arbitration. It is not pure because the arbitration is not postponed while there is an attempt to settle.

Most important, the FIDIC eliminated the engineer in favor of the DAB, a person or panel unconnected with the project who gives a final and binding decision unless it is taken to arbitration. The attempt to settle amicably is traditional in international contracts. It probably is present in all ADR systems.

Another system has been proposed by the World Bank. In its instruction to bidders, contained in its sample bidding documents for smaller (under $10 million) projects, Paragraph 35.1 provides that the owner will propose that a particular person be appointed as an adjudicator at a designated hourly fee. If the bidder disagrees, it must so state in its bid and make a counterproposal. If there is disagreement, the adjudicator will be appointed by a designated appointing authority. The rules under which the adjudicator would operate are specified in Paragraph 25 of the general conditions of the contract. The adjudicator must give her decision within twenty-eight days. The costs are divided. Either party may refer a decision of the adjudicator to an arbitration within twenty-eight days of the adjudicator's written decision. In essence, the World Bank suggests that one role that has typically been performed by the consulting engineer—that of initially resolving disputes—be given to a different entity whose decision can be taken to arbitration.[222]

---

[222]This is derived from the New Engineering Contract issued by the Institution of Civil Engineers in England. There have been two new editions, NEC-2 and NEC-3.

CHAPTER THIRTY-ONE

Indemnification and Other Forms of Shifting and Sharing Risks: Who Ultimately Pays?

SECTION 31.01 First Instance and Ultimate Responsibility Compared

Today there is increasing likelihood that those who suffer losses incident to the construction process will be compensated, especially where those losses involve personal harm. Although claimants do not always receive judicial awards, the tendency has been to expand liability to ensure that those who suffer losses are compensated.

Lawsuits today generally begin with claims against a number of defendants, something that is legally permissible and relatively inexpensive. Typically, these defendants make claims against each other as well as claims against parties who have not been sued by the original claimant. The result is a multiparty lawsuit that generally involves as many as a half-dozen interested participants as well as an almost equal number of sureties and insurers. For example, the injured employee of a prime contractor is likely to sue all contractors who are not immune, such as other separate contractors (multiple primes) and subcontractors, the owner, the design professional, any construction manager, and, depending on the facts, those who have supplied equipment or materials. Each of the defendants will very likely bring claims against the others based on indemnification. If there is a building defect, such as the failure of an air-conditioning system, the owner is likely to assert claims against the design professional, the contractor, and the manufacturers and sellers of the system. Again, those against whom claims have been asserted are likely to assert claims against each other.

These lawsuits show two levels of responsibility. There is *first-instance* responsibility to the original claimant, such as the injured worker or the owner. Resolving this issue depends on whether any of the defendants or someone for whom they are responsible had a duty to the claimant, failed to live up to the standard required by the contract or by the law, and was the legal and proximate cause of the claimant's injury. After this determination is made, the next level, *ultimate* responsibility, must be addressed. Who among those responsible will ultimately bear the loss? This inquiry—the focal point of this chapter—has developed unbelievably complex and costly legal controversies.

SECTION 31.02 Contribution Among Wrongdoers

Suppose A and B, *acting together* in pursuing a common plan or design, injure C. C can recover its loss from either A or B if each has committed a wrong. Because each has committed the wrong and because they have acted together, neither A nor B can receive contribution from the other if either has paid more than half of the total judgment. In this particular instance, A and B are joint wrongdoers. For example, if the design professional and owner acted together to destroy the contractor's business or reputation, the contractor could sue either or both and recover its loss from either or both. In such a case, neither design professional nor owner would have a claim against the other if either paid more than half of the judgment.

But a number of defendants may be sued in the same legal action even though they have not acted together. They are co-defendants because each may have played a substantial role in causing the injury and for procedural convenience, all the claims are decided in one lawsuit. The defendants are concurrent wrongdoers, not in the sense that their wrongdoing occurred at the same time but in the sense that each played a substantial role in causing an indivisible loss to the claimant.

Suppose three defendants are held liable to the plaintiff and one defendant pays the award. Can the one defendant seek reimbursement or "contribution" from the other defendants? The basis for awarding contribution is the unfairness of one defendant paying the entire judgment when all are responsible. Most American courts do not require contribution among wrongdoers. But half of the states have created contribution by statute. The statutes vary considerably, but where they exist, two factors have reduced the effectiveness of contribution statutes. First, some states require that there be a joint judgment against the defendants before contribution can be compelled. This requirement has obvious disadvantages, because it inhibits settlement.[1] Second, and more important for purposes of construction accidents, courts frequently deny contribution against a party who was immune from the original claim. For example, it would not be available against the employer (actual or statutory, as described in Section 7.04C) of the injured party if the employer were immune from liability because of workers' compensation laws. To encourage settlement, some states have enacted legislation under which a defendant who settles with the claimant and seeks to be relieved from any claim that other claimants may make can ask the court to determine whether the settlement was made in good faith. If the court so determines, the settling defendant is released from any claims that may be made against it by other defendants.[2]

Where contribution does exist, either by judicial rule or legislation, how is liability among the wrongdoers apportioned? Two basic approaches exist. The minority view is that each wrongdoer contributes a pro rata share. Where there are three wrongdoers, each would contribute one-third. Pro rata apportionment disregards the comparative fault of the individual defendants.

The clear majority approach, consistent with the doctrine of comparative negligence discussed in Section 31.03E, is apportionment based on *fault*.[3] Suppose there are three wrongdoers: A, B, and C. A is 50 percent at fault, B is 30 percent at fault, and C is 20 percent at fault. If A pays the plaintiff the entire judgment, then A may recover 30 percent of that amount from B and 20 percent of that amount from C. (Under the pro rata system, B and C are each liable to A for one-third of the settlement amount.) The "comparative fault" approach has been adopted by the Restatement (Third) of Torts: Apportionment of Liability.[4]

Of particular significance to the construction industry, contribution applies where the indivisible injury caused by the multiple defendants involved physical harm, whether death, personal injury or property damage. In addition, the defendants, in inflicting the injury upon the plaintiff, must have each committed a tort. This means that no right of contribution exists if an owner sues the architect and prime contractor for the same financial injury caused by each defendant's breach of contract.[5]

SECTION 31.03 Noncontractual Indemnity[6]

A. Basic Principle: Unjust Enrichment

Like contribution, noncontractual indemnity transfers responsibility for the plaintiff's injury from the defendant who pays the claim to another person who in fairness should bear that cost. Contribution and noncontractual indemnity differ in three respects.

First, while contribution applies to an indivisible, physical injury caused by joint tortfeasors, indemnity may apply whether the harm is physical or financial and whether the injury was caused by a tort or by breach of contract. Second, contribution arises when the plaintiff sues multiple defendants; in indemnity, the plaintiff may

---

[1] The Restatement (Third) of Torts: Apportionment of Liability § 23 (2000), extends a right of contribution to a co-defendant who "discharges the liability of another by settlement or discharge of judgment." Section 23 represents the majority rule.

[2] West Ann.Cal.Civ.Proc. Code § 877.6.

[3] *Pacesetter Pools, Inc. v. Pierce Homes, Inc.*, 86 S.W.3d 827 (Tex.App. Ct.2002).

[4] Restatement (Third) of Torts: Apportionment of Liability § 23(b), which states: "A person entitled to recover contribution may recover no more than the amount paid to the plaintiff in excess of the person's comparative share of responsibility."

[5] *Board of Educ. of Hudson City School Dist. v. Sargent, Webster, Crenshaw & Folley*, 71 N.Y.2d 21, 517 N.E.2d 1360, 523 N.Y.S.2d 475 (1987).

[6] This categorization separates indemnification based on express or implied-from-the-contract agreements from other forms of indemnification. The latter have various designations, such as common law indemnity, quasi-contractual indemnity, and equitable indemnity. Rather than attempt to sort out these terms and their implications, indemnification is divided into noncontractual and contractual.

sue only one defendant, who then brings a third-party claim for indemnity against another party. Third, while contribution sought to distribute the liability among the many defendants, indemnity historically sought to transfer the *entire* liability, not just part of it. For such an all-or-nothing transfer to occur, the culpability of the two parties in an indemnity action must be qualitatively different.

Suppose a worker employed by a prime contractor is injured because of deliberate safety violations by his employer. Under certain circumstances, the worker can recover from the owner. Liability in such a case may be based on the owner's nondelegable duty to furnish a safe workplace. Alternatively, liability may be based on the failure to determine whether safe practices were being followed or on the failure to discharge the contractor after becoming aware of the violations.

In this example, the *degree* of wrongdoing between the owner and contractor is qualitatively different. The owner's liability to the employee is *passive* or secondary and that of the contractor is *active* or primary. The contractor would be unjustly enriched and the owner unjustly impoverished if financial responsibility for the employee's injury stayed with the owner. To rectify this situation, the owner can receive indemnification from the contractor.

Similarly, suppose the prime contractor violated safety orders and a claim was made against the design professional based on the latter's failure to detect the violation or to exercise corrective power given by the contract. The contractor's conduct can be considered active if it orders an employee to work under dangerous conditions. However, the conduct of the design professional can be considered passive, his liability based on failure to act. Although judicial determination of active and passive conduct is sometimes at variance with ordinary meaning (a point to be explored later in this section), courts sometimes make this differentiation the basis for giving the design professional indemnity from the contractor or from someone else more directly connected with the injury or loss.[7] Indemnity in such a case is based on the unjust enrichment of the more culpable wrongdoer that would result if it were not required to bear this loss.

B. Noncontract and Contract Indemnity Differentiated

Sections 31.04 and 31.05 treat indemnification implied from or expressed in a construction contract. As indicated, noncontractual indemnity is based principally on unjust enrichment and the concept that losses should be shifted from one wrongdoer to another based on qualitative comparative fault. For historical reasons, however, courts have tended to treat noncontractual indemnity as more analogous to a contract claim than to a tort claim. For this reason, courts tend to classify this form of indemnity as quasi-contractual (something like a contract) even though it is not based on consent. Also, in some states such indemnity is called *equitable* as it can be based on equitable notions of fairness.

C. Some Classifications

Courts have articulated various tests to determine whether noncontractual indemnification will be awarded. Although one court recognized an indemnification obligation in order to impose an ultimate burden on one who was the "active delinquent" in bringing about the injury rather than the "lesser delinquent,"[8] most courts have employed the primary–secondary or passive–active differentiation, singly or together. See Figure 31.1. For example, one court stated that indemnity would be granted "to a person who, without active fault on his own part, has been compelled, by reason of some legal obligation, to pay damages occasioned by the initial negligence of another, and for which he himself is only secondarily liable."[9] The court stated that both vicarious liability—that is, liability for the wrongs of another—that arises out of a positive rule of common or statutory law, and liability imposed for failure to discover or correct a defect or remedy a dangerous condition caused by the act of the one primarily responsible, are illustrations of *secondary* liability. The passive–active differentiation, perhaps used more frequently, looks to similar factors. Again, some positive acts create liability, whereas sometimes negative or passive inaction, though creating liability, is less morally objectionable though enough to justify indemnification.

---

[7]*In Owings v. Rosé*, 262 Or. 247, 497 P.2d 1183 (1972), an architect was given indemnity against a consulting engineer when the architect was held vicariously liable essentially for the negligence of the engineer.

[8]*Miller v. De Witt*, 37 Ill.2d 273, 226 N.E.2d 630, 642 (1967).
[9]*Builders Supply Co. v. McCabe*, 366 Pa. 322, 77 A.2d 368, 370 (1951).

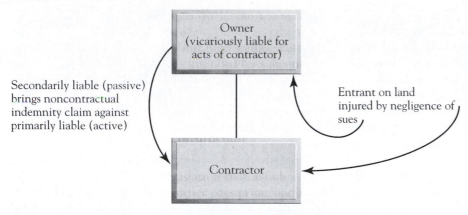

FIGURE 31.1   Noncontractual (equitable) indemnity illustrated.

It may be useful to look at a few cases that have dealt with noncontractual indemnity. In *Adams v. Combs*,[10] a road contractor sought indemnity from the city for whom the road work had been performed. An accident occurred two years after completion of the contract. The court denied the claim, however, holding that any negligence on the part of the city in not maintaining the road was secondary and passive, whereas the contractor's negligence was primary and active.

Architects sought indemnity in *St. Joseph Hospital v. Corbetta Constuction Co.*,[11] and *Owings v. Rosé*.[12] In the first case, the architect was *denied* indemnity from a supplier of defective tile because the architect knew the tile was defective. In the second case, the architect was *granted* indemnification against a negligent consulting engineer when the architect, who was *not* negligent, was liable because he had a nondelegable duty. Where each party has violated the same duty, indemnity is usually denied. For example, in *Harris v. Algonquin Ready Mix, Inc.*,[13] the electric company and the contractor it had hired each failed to warn employees of certain dangers. The absence of any qualitative difference in their liability precluded indemnity.

It is generally stated that failure to discover or remedy the defect caused by another is merely passive negligence.[14] However, in *Becker v. Black & Veatch Consulting*

Engineers[15] an engineer's failure to inspect was held to be *active* negligence when the engineer was employed for this express purpose. This active negligence precluded the engineer from receiving indemnity from owner or contractor, as all were considered actively negligent.

Contribution would have been permitted if such right existed and, generally, if all could have been held liable. If the contractor was immune because the injured person was its employee, however, the contractor could not be compelled to contribute to the judgment unless there was contractual indemnity. The judgment would be shared by owner and engineer unless there was contractual indemnity.

The perils of generalizations in this area are emphasized in an Illinois decision that, in discussing active and passive liability, stated,

> Determination of this question is not a matter of proceeding according to the usual dictionary definitions of the words "active" and "passive." These words are terms of art and they must be applied in accordance with concepts worked out by courts of review on a case-by-case basis. Under appropriate circumstances, inaction or passivity in the ordinary sense may well constitute the primary cause of a mishap or active negligence. . . . It has been appropriately stated that "mere motion does not define the distinction between active and passive negligence."[16]

---

[10]465 S.W.2d 288 (Ky.1971).

[11]21 Ill.App.3d 925, 316 N.E.2d 51 (1974).

[12]See supra note 7.

[13]59 Ill.2d 445, 322 N.E.2d 58 (1974).

[14]*See Builders Supply Co. v. McCabe*, supra note 9.

[15]509 F.2d 42 (8th Cir.1974). See *Associated Eng'rs, Inc. v. Job*, 370 F.2d 633 (8th Cir.1966), cert. denied sub nom. *Troy Cannon Constr. Co. v. Job*, 389 U.S. 823 (1967).

[16]*Moody v. Chicago Transit Auth.*, 17 Ill.App.3d 113, 117, 307 N.E.2d 789, 792–93 (1974).

Noting that certain terms have become words of art is a signal that a particular doctrine has developed difficulty, and it is therefore important to read between the lines of judicial opinions dealing with that doctrine. It is also an indication of considerable uncertainty in the law and the importance of using contract language to specifically deal with the problem rather than leave the matter to the vagaries of court decisions when either vague concepts must be applied[17] or legal terminology varies from ordinary meaning.

Noncontractual indemnity may be appropriate not under the active-passive test, but where one party has delegated to another exclusive responsibility to perform a specific duty. In *17 Vista Fee Associates v. Teachers Insurance & Annuity Association of America*,[18] an owner built a new building and hired an engineer to create the mechanical design. Building inspectors refused to approve the completed building unless the owner made upgrades to the mechanical system, which the owner did. The court permitted the owner to recover the cost of the upgrade from the engineer under a theory of noncontractual indemnity because the owner "was compelled to discharge a duty that it had delegated fully to, and that should have been discharged by, the engineer, whose negligence was the actual cause of the loss."[19]

D. Employer Indemnification

Sometimes the propriety of shifting liability depends on substantive law policies that tend to either protect a particular person from this responsibility or place it on him. Many noncontractual indemnity claims are brought against employers of injured persons. For example, suppose a subcontractor's employee is injured and the primary responsibility for the injury was the employer's failure to comply with safety rules. Because the injured employee cannot sue his own employer,[20] he institutes legal action against the prime contractor. Suppose the prime contractor can be sued as a third party; liability against the prime contractor is based on its responsibility as the employer to provide safe working conditions, and this duty cannot be delegated. This would be a classic case for granting the prime contractor noncontractual indemnity. The prime contractor's liability is based on a nondelegable duty, and clearly the subcontractor's negligence is active and primary.

Nevertheless, allowing indemnity would place *ultimate* liability on the subcontractor. Would this deny the subcontractor immunity from tort action granted by workers' compensation law? One of the tradeoffs in workers' compensation was granting immunity to the employer from tort liability in exchange for denying the employer rights possessed before workers' compensation statutes were enacted. Indemnification can, by indirection, frustrate this.

But there are cogent arguments for allowing indemnity in these cases. The third party, here the prime contractor, received nothing in the "trade" and arguably should not have existing rights taken away. This argument, however, ignores the likelihood that third parties are also employers who are part of some workers' compensation legislative trade. In addition to this argument, it would appear to be unjust to exonerate the more culpable employer when liability against the third party, such as a prime contractor or owner, may be vicarious, based on passive negligence or a statutory violation. It might also be argued that denial of indemnity may encourage carelessness by the employer.

The difficulty of this question is reflected in the case law. It has been considered the most evenly balanced issue in workers' compensation law.[21] But now it appears that a slight majority precludes noncontractual indemnity against an employer.[22] Those that grant indemnity usually find a separate duty owed by the employer to the third party seeking indemnity.[23]

---

[17]In *Santisteven v. Dow Chemical Co.*, 506 F.2d 1216 (9th Cir.1974), the court stated that quasi-contractual indemnity is based on whether it is "just" to shift liability.

[18]259 A.D.2d 75, 693 N.Y.S.2d 554 (1999).

[19]Id., 693 N.Y.S.2d at 556. Because the owner and engineer were in privity, this result could also have been reached under a theory of implied contractual indemnity, discussed in Section 31.04.

[20]The employee cannot sue his employer because the employee received workers' compensation benefits to cover his injury; in return, the employer is immune from tort liability. See Section 7.04C.

[21]See A. LARSON, 2A WORKMEN'S COMPENSATION § 76.11 (2002). Larson, a leading scholar in this field, advocates abolition of noncontractual indemnification against the employer. See Larson, *Third-Party Action Over Against Worker's Compensation Employer* [1982] Duke L.J. 483.

[22]See Annot., 100 A.L.R.3d 350 (1980). California has effectively barred noncontractual indemnity against the employer of the claimant by requiring an express indemnity agreement. West Ann.Cal.Lab.Code § 3864. Note that express indemnification is permitted.

[23]*Santisteven v. Dow Chemical Co.*, supra note 17.

E. Comparative Negligence

Currently, the great majority of states, whether through legislative or judicial initiative, have embraced comparative negligence. Comparative negligence, in which liability is proportionate to the degree of the defendant's fault, is fundamentally inconsistent with the traditional form of noncontractual indemnity, which sought to transfer the entirety of the loss from one tortfeasor to another. One commentator has concluded that the "active-passive kind of indemnity is now in disrepute and probably has little or no future in apportioning responsibility among tortfeasors."[24] While courts continue to use terms such as "equitable indemnity,"[25] the new regime of loss apportionment among multiple tortfeasors who have caused an indivisible injury is, in actuality, fault-based contribution.[26]

F. Preemption

Although express indemnification is discussed in detail in Section 31.05, one problem relating to indemnity clauses must be mentioned here. Does the presence of an express indemnification clause bar noncontractual indemnity?

If noncontractual indemnification is based on the likely intention of the parties, the matter should be resolved simply by interpreting the language of the indemnity clause. If no language deals specifically with preemption, it is likely that the parties, by focusing on contractual indemnification, have demonstrated an intention to exclude other forms of indemnification. If noncontractual indemnification is based on unjust enrichment, however, the problem becomes more difficult. On the one hand, it can be contended that there should be no inquiry into unjust enrichment where the parties have dealt specifically with the issue. On the other hand, unjust enrichment can have a life of its own, based on what the law considers fair and just.

The difficulty is reflected in case decisions that have passed on this problem. Most cases have concluded that the presence of an express indemnification provision bars indemnification based on a comparison of culpability.[27] But another case, *E. L. White, Inc. v. City of Huntington Beach*,[28] not only came to a different conclusion but also demonstrated some of the complexity raised by indemnification. For that reason, it may be useful to look at the facts in that case.

The city of Huntington Beach contracted with White to build a drain sewer. Under the contract, White agreed to indemnify the city. Problems developed after completion. White hired a subcontractor to make repairs. One employee of the subcontractor was killed and another injured in a cave-in caused by failure to shore or slope the trenches. The legal action by the workers resulted in judgments against the city and White. Before the legal claim was filed by the workers, the city filed a separate claim against White and its insurer asking, among other things, that a declaration be made that the city be entitled to indemnification from White under the indemnification clause contained in the prime contract. California law does not grant indemnity when the contractual indemnity clause is written in general terms and the party seeking indemnification is actively negligent. The city was found to have been actively negligent, which barred its indemnification claim.

After the judgment had been rendered, White and its insurer brought a legal action against the city seeking *equitable* indemnity against the city for amounts the insurer paid to satisfy the judgments.

The city contended that equitable indemnity was barred by the contractual indemnification clause, but the court did not agree. The court held that the two forms of indemnification—contractual and equitable—are separate bases for transferring a loss. If the express contractual provision does not apply to the factual setting before the court, equitable indemnification can come into play.

---

[24]D. DOBBS, TORTS, 1079 (2000) (footnote omitted).

[25]*American Motorcycle Ass'n v. Superior Court*, 20 Cal.3d 578, 578 P.2d 899, 146 Cal.Rptr. 182 (1978).

[26]*Frank v. Meadowlakes Development Corp.*, 6 N.Y.3d 687, 849 N.E.2d 938, 816 N.Y.S.2d 715 (2006) (under contribution statute, subcontractor who was less than 50 percent at fault was liable only for its proportionate share of fault when sued by the developer for noncontractual indemnity).

[27]*Felker v. Corning, Inc.*, 90 N.Y.2d 219, 682 N.E.2d 950, 660 N.Y.S.2d 349 (1997); *Southern Pac. Co. v. Morrison-Knudsen Co.*, 216 Or. 398, 338 P.2d 665 (1959); *Wyoming Johnson, Inc. v. Stag Indus., Inc.*, 662 P.2d 96 (Wyo.1983).

[28]21 Cal.3d 497, 579 P.2d 505, 146 Cal.Rptr. 614 (1978). See also *Ranchwood Communities, Ltd. Partnership v. Jim Beat Constr. Co.*, 49 Cal.App.4th 1397, 57 Cal.Rptr.2d 386 (1996) (indemnitee allowed to bring an equitable indemnity claim but not a contractual indemnity claim as it was not licensed).

The indemnity clause must be interpreted to see whether it expresses any intention by the parties that all equitable indemnity is to be eliminated. The indemnification clause protected only the city. The court permitted equitable indemnity when a claim was made by the *contractor*, because the clause did not preclude equitable indemnity.[29]

G. Settlements

When there are a number of joint wrongdoers the law seeks to provide incentives to settle. Some states have enacted statutes that allow a settling defendant to petition the court to conclude that the settlement was made in good faith, as a means of relieving the settling defendant from claims by other defendants. See Section 31.02.

SECTION 31.04 Implied Contractual Indemnity

As mentioned in Section 31.03F, one form of contractual indemnity rests not on express language but on the presumed intention of the contracting parties that one party will respond for a loss it causes to the other party. For example, suppose the prime contractor negligently leaves a hole uncovered. A mail deliverer who enters the site during construction with permission is injured. Suppose the latter successfully asserts a claim against the owner based on the owner's obligation, as a possessor of land, to keep the premises reasonably safe. Even in the absence of a specific indemnity clause in the prime contract, one can contend that the prime contractor has impliedly promised the owner that it will indemnify the owner if its negligent conduct harms a third party who asserts a claim against the owner and obtains a court judgment.

This form of implied indemnity has had a complicated history in admiralty law where longshoremen have been injured. The relatively uniform presence of indemnifica-

tion clauses and the expansive use of noncontractual indemnity, however, have made this form of indemnification relatively unimportant in modern construction disputes.[30]

SECTION 31.05 Contractual Indemnity

A. Indemnification Compared to Exculpation, Liability Limitation, and Liquidated Damages

This subsection compares indemnification clauses to other clauses that seek to distribute losses. Distinct differences exist between the various loss distributing clauses, but there is also much overlap. Indemnification—the principal topic in this chapter—has become controversial and highly regulated because it can be in effect exculpatory; that is, it can relieve one party of the cost of liability that it would otherwise have to bear. For that reason, it is instructive to look carefully at both exculpation and indemnification.

One of the most simple yet effective ways of relieving a party from responsibility is by a contractual exculpation under which a party that may suffer losses agrees it will not pursue the party that is legally responsible for such losses. For example, suppose a patient, on being admitted to a hospital, signs a form in which he agrees the hospital will not be responsible if he is harmed as a result of the hospital's negligence. The hospital would be seeking to relieve itself from liability. Although the law accords considerable freedom to contracting parties to make their own rules and distribute losses as they wish,[31] the hospital admissions room is hardly the place for the contract process. For that reason, such an agreement was held invalid.[32] Similarly, some courts and legislatures do not

---

[29]The Alaska Supreme Court has adopted the preemption approach expressed in *E. L. White, Inc. v. City of Huntington Beach*, supra note 24. See *Fairbanks North Star Borough v. Roen Design Associates, Inc.*, 727 P.2d 758 (Alaska 1986). For subsequent history not involving the preemption issue, see 795 P.2d 793 (1990) (measure of recovery) and 823 P.2d 632 (1991) (reversing jury's rejection of the borough's implied contractual indemnity claim against the project designer and remanding).

[30]See *Fall River Housing Auth. v. H.V. Collins Co.*, 414 Mass. 10, 604 N.E.2d 1310, 1313–14 (1992) (denying implied contractual indemnity where the parties could sue for breach of contract); but see *Kemper Architects, P.C. v. McFall, Konkel & Kimball Consulting Eng'rs*, 843 P.2d 1178 (Wyo.1992) (permitting the indemnity claim). A history of implied contractual indemnity in California is found in *Garlock Sealing Technologies, LLC v. NAK Sealing Technologies Corp.*, 148 Cal.App.4th 937, 56 Cal.Rptr.3d 177, 202–04 (2007).

[31]*Rutter v. Arlington Park Jockey Club*, 510 F.2d 1065 (7th Cir.1975) (exculpation in race track stabling contract enforced).

[32]*Tunkl v. Regents of the Univ. of California*, 60 Cal.2d 92, 383 P.2d 441, 32 Cal.Rptr. 33 (1963).

allow exculpation in a residential lease[33] or one in which the tenant is a small business.[34]

In addition to the potential for abuse in the examples just given, an exculpatory clause also eliminates the exculpated party's incentive to act with due care, because the clause deprives the plaintiff of his right to use the legal process when he would otherwise have had the opportunity to do so. For these reasons, clauses that exculpate one party from the consequences of its future negligence are often found to violate public policy.[35]

Indemnification, in contrast, does not preclude first-instance liability. It deals with ultimate responsibility by shifting the loss from one party to another. For example, if an injured worker recovers against the owner, permitting the owner to shift the loss to the contractor by indemnification in no way precludes the injured worker from recovering. A shift of ultimate responsibility from owner to contractor occurs. If, however, the owner is not able to recover against the contractor, for example because the contractor has filed for bankruptcy, the owner remains liable to the worker. The contingent nature of the owner's contractual right to be indemnified means that it continues to have an incentive to act with due care toward the worker. These twin facts—that the worker receives a remedy and the owner has an incentive to act with due care—means that indemnity agreements, unlike exculpatory clauses, do not violate public policy.[36]

If the owner was in any way at fault in causing the worker's injury, then risk shifting through indemnification resembles exculpation. Not only is the owner relieved of the consequences of its negligence, but indemnity agreements are often written in favor of the stronger party and against the weaker. Concern with the fairness of indemnity clauses, in particular when used in the construction industry, has led about half the states to regulate such clauses.[37] In addition, courts often interpret such clauses narrowly where the indemnified party was itself negligent.[38]

Another contract clause that can be compared with an indemnity clause is one that seeks to limit the legal remedy. For example, sellers of machinery sometimes seek to limit their liability to repair and replace defective parts. Similarly, some design professionals seek to limit their liability to their client to a designated amount of money or a certain percentage of the fee.[39] If the actual damages in the latter illustration are less than the specified amount, only actual damages can be recovered. But a liability limitation sets a ceiling on the damages.

Many of the same considerations that have been discussed with regard to exculpatory clauses apply to liability limitations. Where they are determined in a proper setting, by parties of relatively equal bargaining power, and where the language clearly expresses an intention to limit the liability, they are given effect.[40]

Contract clauses sometimes stipulate the amount of damages in advance. As seen in Section 26.09B, such clauses are frequently used for unexcused time delay. They are generally given effect as long as they are reasonable. However, some of the same considerations that relate to bargaining power and appropriateness of advance agreement will be taken into account. Liquidated damages clauses, unlike indemnity clauses, do not shift losses.

If the surrounding circumstances justify enforcement of an exculpatory clause or one that limits liability, those same circumstances may justify enforcement of an indemnification clause that has an exculpatory element.

B. Indemnity Clauses Classified

This section uses the prime contract as an illustration. The prime contractor is the *indemnitor,* that is, the party promising to indemnify. The owner is the *indemnitee,* that is, the party to whom indemnification has been promised. The analysis in this section also applies to a subcontract where the prime contractor is the indemnitee and the subcontractor the indemnitor.

Looking first at a claim as it relates to the indemnitor's conduct, the indemnity clause can be "work related." Such a clause covers a broad variety of claims that third parties such as injured workers or adjacent landowners may assert against the indemnitee owners. These claims may be

---

[33]*Crowell v. Housing Auth. of the City of Dallas,* 495 S.W.2d 887 (Tex.1973).

[34]*McLean v. L. P. W. Realty Corp.,* 507 F.2d 1032 (2d Cir.1974).

[35]*Hiett v. Lake Barcroft Community Ass'n,* 244 Va. 191, 418 S.E.2d 894 (1992).

[36]*Estes Express Lines, Inc. v. Chopper Express, Inc.,* 273 Va. 358, 641 S.E.2d 476 (2007).

[37]See Section 31.05D.

[38]See Section 31.05E

---

[39]See Section 15.03D.

[40]*Delta Air Lines, Inc. v. Douglas Aircraft Co.,* 238 Cal.App.2d 95, 47 Cal.Rptr. 518 (1966).

predicated on fault of the indemnitor prime. However, the only indemnification requirement is that the claim "arise out of," "be occasioned by," or "be due to," to use common indemnity phrases, the work or activity of the prime.[41]

Suppose the indemnitor was not at fault, and the indemnitee was instead the sole cause of the accident. Enforcing a "work related" indemnity agreement would then mean that the mere presence of the indemnitor at the work site is sufficient to impose upon it a contractual duty to indemnify. Courts are reluctant to reach this conclusion and instead require some casual connection between the indemnitor's conduct and the injury or damage. Unusually, these cases involve a solely negligent prime contractor who is the indemnitor suing the fault-free subcontractor indemnitee.[42]

The clause can be more limited. It may cover only claims based on wrongful conduct of the prime. This is employed by AIA Document A201-2007, Section 3.18 (discussed in Section 31.05G).

Focusing on the *indemnitee*, the claim usually asserts one or a number of bases for owner liability. The claimant may point to acts or failures to act by the owner itself or someone for whom it is responsible, such as design professional or prime contractor. This can raise questions about whose fault is passive versus active or primary versus secondary (as discussed in Section 31.03). The claim may be based on *status*, such as the owner being the possessor of land or the employer with common law or statutory responsibilities often strict in nature. Also, the claim against the indemnitee may be based *in whole or in part* on the indemnitee's own wrongful acts. Clauses that cover claims solely based on the negligence of the indemnitee are called "broad-form" indemnity clauses, and those that cover claims based in part on the negligence of the indemnitee are called "intermediate-form" indemnity clauses.

The exculpatory feature of many clauses, along with work-related clauses that do not require any legal wrong by the indemnitor prime, has caused courts to scrutinize such clauses. Also, the bargaining power that often enables apparently one-sided clauses to be passed "down the line" from owner to prime, from prime to subcontractor, and so

on, also has generated judicial concern, making it hazardous to try to predict the outcome of litigation.

C. Functions of Indemnity Clauses

The hostility the law has shown toward clauses under which one party has promised to indemnify another often ignores the important function served by the indemnification process. As an illustration, suppose an owner plans to make a construction contract with a prime contractor. The owner recognizes the increased likelihood that claims will be made against it that are based on the contractor's performance, sometimes but not always negligent. Recognizing that the law increasingly makes the owner responsible to third parties or at least that the law has made it more likely that third parties may make claims against the owner, the owner may say to the contractor,

> I have turned over the site to you. It is your responsibility to see to it that the building is constructed properly. You must not expose others to unreasonable risk of harm. The increasing likelihood that I will be sued for what you do makes it fair that you relieve me of ultimate responsibility for these claims by your agreeing to hold me harmless or to indemnify me.

Alternatively, a reassuring proposal may come from the contractor, who may say to the owner,

> I know you may be concerned about the possibility that a claim will be made against you by a third party during the course of my performance and that you will have to defend against that claim and either negotiate a settlement or even pay a court judgment. I always conduct my work in accordance with the best construction practices, and I have promised in my contract to do the work in a proper manner. I am so confident that I will do this that I am willing to relieve your anxiety by holding you harmless or by indemnifying you if any claim is made against you by third parties relating to my work. You will have nothing to worry about, as I will stand behind my work. If you are concerned about my ability to pay you, I will agree to back it up by public liability insurance coverage.

In this context, indemnification acts to seal a deal when one party is anxious. The same scenario can be played but in a slightly different way. Suppose the architect asks the owner to obtain indemnification for him from the prime through the prime contract in the manner accomplished

---

[41]*Nabholz Constr. Corp. v. Graham*, 319 Ark. 396, 892 S.W.2d 456 (1995).

[42]*Jones v. Strom Constr. Co., Inc.*, 84 Wash.2d 518, 527 P.2d 1115 (1974).

by AIA Document A201-2007, Section 3.18. The architect may be saying to the contractor through the owner,

> The law may hold me accountable for injury to your workers or to employees of your subcontractors because they may connect their injury with something they claim I did or should have done. You are being paid for your expertise in construction methods and your knowledge of safety rules. These are not activities in which I have been trained or in which I claim to have great skill or experience. For that reason, if a claim is made against me for conduct that is your responsibility, I want you to hold me harmless and indemnify me.

In addition, either owner or architect may back up its request for indemnification from the prime contractor by noting that it is exposed to potentially open-ended tort liability if claims are made by employees of the contractor or other subcontractors, whereas the actual employer, either the prime contractor or subcontractor, who is most directly responsible, need only pay the more limited liability that workers' compensation law imposes on it.

Insurance plays a significant part in the indemnification process, because the promise to indemnify may be worthless unless it is backed up by a solvent insurer. As explained in Section 31.05H, modern standard form liability insurance policies provide coverage for obligations assumed by a contractor under an indemnity agreement with the owner (although contractually assumed liability is generally excluded). Since the contractor is also the party in the best position to prevent a claim from being made under the indemnity clause, distribution of risks through indemnification can facilitate insurance at the cheapest possible cost.[43] (An architect who is asked to indemnify, as noted in Section 15.05D, may not be persuaded of this rationale if his professional liability insurance has a high deductible.)

Insurance facilitation assumes some accidents will occur and seeks an efficient method of distributing the ultimate loss to an insurer. Yet indemnification can be looked on from the perspective of worker safety and avoidance of accidents.[44] Here the emphasis is on the exculpatory aspects of indemnification. This escape from consequences is particularly a problem with an indemnification provision, such as a broad form or intermediate form, which can exculpate the indemnitee from its own wrongdoing. Even more, a prime contractor who extracts indemnification provisions from all its subcontractors may have little incentive to control project operations in a way that minimizes the likelihood of injuries to workers.

A federal judge, focusing on safety, stated that indemnification seeks to place the ultimate responsibility on the party who can most cheaply avoid the harm.[45] An application of this principle would extend broad scope to clauses under which subcontractors (closest to the actual operation) indemnify prime contractors and prime contractors (overall control) indemnify owners and design professionals. Despite this, some legislatures, as seen in Section 31.05D, and some courts, as seen in Section 31.05E, have not permitted full contractual freedom to the parties to decide how risks should be allocated by indemnification. Undoubtedly, much of this unwillingness is influenced by the belief that indemnification freely permitted will encourage carelessness. Yet it is unlikely that unregulated indemnification (enforcing clauses as written) would be a disincentive for creating and monitoring a safety program.

The factor that bears most heavily on how the contractor performs is its insurance rates, which may in part be predicated on the number of claims made against it. But the insurance premium is more likely to be predicated on the type of work, the locality, and the volume of the prime contractor's business. Other factors can deter careless performance, such as the possibility that public officials will take action against the prime contractor or the contractor's license will be placed in jeopardy.

In any event, the law allows parties to procure insurance even though the existence of insurance can encourage carelessness. Indemnification is a form of insurance, and there is no reason why the law should be hostile to this process. Undoubtedly, some of the hostility is also generated by the way in which indemnification is forced on the weaker party, which must usually accept such a clause on a take-it-or-leave-it basis. Again, if the risk

---

[43]*Guy F. Atkinson Co. v. Schatz*, 102 Cal.App.3d 351, 161 Cal. Rptr. 436 (1980) (citing earlier edition of this book); *Leitao v. Damon G. Douglas Co.*, 301 N.J.Super. 187, 693 A.2d 1209, 1211-12 (App. Div.), certification denied, 151 N.J. 466, 700 A.2d 879 (1997); *Di Lonardo v. Gilbane Bldg. Co.*, 114 R.I. 469, 334 A.2d 422 (1975).

---

[44]*Ft. Wayne Cablevision v. Indiana & Michigan Elec. Co.*, 443 N.E.2d 863 (Ind.App.1983); *Bosio v. Branigar Org., Inc.*, 154 Ill.App.3d 611, 506 N.E.2d 996 (1987).

[45]*McMunn v. Hertz*, 791 F.2d 88 (7th Cir.1986) (per Judge Posner).

is clearly brought to the attention of the weaker party and that party can insure against that risk and pass the cost of insurance on to the stronger party in its contract price, hostility is not justified.[46] Courts, adopting the view that indemnity clauses are simply insurance-procurement mechanisms and are enforceable just like any other contract provision, are discussed in Section 31.05E.

D. Statutory Regulation

The concern over indemnification that has been noted in Section 31.05C has led to frequent statutory regulation and, as shall be seen in Section 31.05E, to court decisions that in effect regulate indemnification clauses. Regulation does not view indemnification as a method of reassuring a nervous contract maker, obtaining insurance at the best possible cost, or placing the risk on the party that can avoid the harm most cheaply. Those who regulate indemnification look principally at the exculpatory aspects of all-or-nothing indemnification and the means by which the stronger party obtains indemnification from the weaker party.

Legislative intervention initially resulted from a struggle between the American Institute of Architects (AIA) and the Associated General Contractors of America (AGC)—really a struggle between professional liability insurers of design professionals and general liability insurers of primes and subcontractors. In 1966, the AIA for the first time included an indemnification provision in A201. The AGC, using the muscle of general liability insurers, fought this and ultimately obtained a modification in 1967. But the turf war extended to state legislatures. Contractors' associations were instrumental in obtaining legislation that limited enforceability of construction contract indemnification clauses. Even after the associations settled their dispute in 1967, parties continued to demand that legislatures limit indemnification. Finally, in the 1970s, long after the battle between the AIA and the AGC was over, some legislatures were persuaded to eliminate the exculpatory features of such clauses with their all-or-nothing solutions. A New York court explained the legislature's motivation in passing that state's anti-indemnity statute:

> The [anti-indemnity statute] was enacted in 1975 to prevent a practice prevalent in the construction industry of requiring

contractors and subcontractors to assume liability by contract for the negligence of others. The Legislature believed that such "coercive" bidding requirements restricted the number of contractors able to obtain or afford the necessary hold harmless insurance and that it unfairly imposed liability on a contractor or subcontractor for the fault of others over whom it had no control. Moreover, such insurance raised the costs of construction unnecessarily, since the cost of the insurance was added to the bid price, and it also resulted in double coverage in many cases by requiring both hold harmless insurance and protective liability insurance.[47]

Although a detailed examination of anti-indemnity statutes is beyond the scope of this book,[48] these statutes generally are of two types: "sole negligence" and "own negligence." A "sole negligence" statute prohibits indemnification of losses caused by the indemnitee's sole negligence. Under these statutes, an indemnitor who is fault-free is not required to indemnify, regardless of the wording of the indemnity clause. These statutes were motivated by the view that the stronger party who had also caused the loss should not be allowed to transfer its liability to a weaker party who was without fault. Still, an indemnitor who was only 1 percent at fault may (if the indemnity clause so provides) be liable for the entire loss. While most anti-indemnity legislation consists of "sole negligence" statutes, they are rarely applied to preclude enforcement of the indemnity clause. Most claims for which indemnification is sought do not result solely from the activity or inactivity of the indemnitee. To the contrary, the interdependent nature of elements in the construction project means that claims are usually attributable to both the indemnitee and the indemnitor.[49]

---

[46]*Di Lonardo v. Gilbane Bldg. Co.*, supra note 43.

[47]*Brown v. Two Exchange Plaza Partners*, 76 N.Y.2d 172, 556 N.E.2d 430, 434, 556 N.Y.S.2d 991 (1990).

[48]The statutes are categorized and discussed in M. SCHNEIER, CONSTRUCTION ACCIDENT LAW: A COMPREHENSIVE GUIDE TO LEGAL LIABILITY AND INSURANCE CLAIMS, 462-82 (1999) and Gwyn & Davis, *Fifty-State Survey of Anti-Indemnity Statutes and Related Case Law*, 23 Constr. Lawyer, No. 3, Summer 2003, p. 26.

[49]See, for example, N.J.Stat.Ann. § 2A:40A-1 and Mich.Comp.Laws Ann. § 691.991. In *Secallus v. Muscarelle*, 245 N.J.Super. 535, 586 A.2d 305 (App.Div.), aff'd, 126 N.J. 288, 597 A.2d 1083 (1991), the court refused to invalidate an overly broad indemnity clause where the indemnitor was, in fact, negligent. California bars indemnification for both sole negligence and willful misconduct of the indemnitee; see West Ann. Cal.Civ.Code § 2782(a).

"Own negligence" statutes bar clauses under which the indemnitor would have to pay for part of the loss caused by the indemnitee's negligence. Each party must bear the cost of the damage it had caused. The parties are not allowed by agreement to circumvent comparative negligence.[50]

Many in the construction industry deal with risk management. Professional and trade associations, lenders, insurers and sureties, to name some, play important roles in the legislative process. This being a political process insures the need for compromises and trade-offs to get the legislation enacted. Exceptions and complex statues are inherent in that process. Numerous statutes do not apply to agreements to procure general liability insurance, workers' compensation insurance or construction bonds.[51] Minnesota permits an owner to indemnify a contractor against strict liability under environmental laws.[52] California creates exceptions from its "sole negligence" statute for inspecting engineers who obtain indemnification from clients that have a designated financial capacity, indemnification for design defects, and indemnification for engineers dealing with hazardous materials, if liability did not arise out of the gross negligence or willful misconduct of the engineer.[53]

As already noted, one motivation for anti-indemnity statutes is to prevent prime contractors from imposing expensive insurance requirements on subcontractors—a cost that ultimately is transferred to the owner. Perhaps to bring down the cost of home construction, California bars residential builders from demanding indemnification from their subcontractors for losses that did not arise out of the subcontractor's work.[54]

Some statutes apply to public works. California prohibits public agencies from seeking indemnification against design professionals who were fault-free,[55] and Colorado prohibits public agencies from being indemnified for the losses caused by their own negligence.[56] Indiana does not apply its "sole negligence" statute to highway construction contracts.[57]

Florida bars indemnity agreements covering the indemnitee's negligence unless the contract contains a monetary cap on the amount of indemnification which bears a reasonable commercial relationship to the contract and is contained in the bid documents. Even if these conditions are fulfilled, the promise to indemnify cannot extend to damages resulting from the indemnitee's gross negligence; willful, wanton or intentional misconduct; or for statutory violations or punitive damages (unless these last two damages were caused by the indemnitor).[58] The effect of this language is to permit indemnity agreements covering the indemnitee's negligence, but only if the amount of indemnification is capped at a reasonable amount and the indemnitor had warning of the indemnity provision before entering into the contract. One effect of this type of legislation may be to lower the cost to the contractor of obtaining "contractual liability" insurance coverage, since the extent of potential liability is not open-ended.

The patchwork of state anti-indemnity statutes creates a number of problems. First, contractors whose work crosses state lines must be certain their liability insurance will cover the variety of indemnification clauses they may face. Second, additional legal difficulties can be generated if the contract was valid in state A where it was made but a dispute culminated in a lawsuit in state B. Usually the courts in state B will enforce the clause if it is valid where it was made. However, Maryland refused to enforce an indemnification clause that was not harmonious with its own anti-indemnification statute. It looked on enforcement of such a clause as violative of Maryland public policy.[59]

E. Common Law Regulation: Specificity Requirements

Long before the enactment of the anti-indemnification legislation just described, the common law looked with hostility at such clauses. Such clauses were enforced only

---

[50]Minn.Stat.Ann. § 337.02; N.Y.McKinney's Gen.Oblig.Law § 5-322.1; West Ann.Wash.Rev.Code § 4.24.115. The Minnesota statute is attacked in Kleinberger, *No Risk Allocation Need Apply: The Twisted Minnesota Law of Indemnification*, 17 Wm.Mitchell L.Rev. 775 (1987). The New York statute is interpreted in *Itri Brick & Concrete Corp. v. Aetna Cas. & Sur. Co.*, 89 N.Y.2d 786, 680 N.E.2d 1200, 658 N.Y.S.2d 903 (1997).

[51]740 Ill.Comp.Stat.Ann. § 35/3, interpreted in *Bosio v. Branigar Org., Inc.*, supra note 44.

[52]Minn.Stat.Ann. § 337.02(2)

[53]West Ann.Cal.Civ.Code §§ 2782.2, 2782.5, 2782.6.

[54]Id., § 2782(c).

[55]Id., § 2782.8. See also § 2782(b), which shields a contractor from liability for a public agency's active negligence.

[56]West Ann. Colo.Rev.Stat. § 13-50.5-102(8).

[57]West Ann.Ind.Code § 26-2-5-1.

[58]West Ann.Fla.Stat. § 725.06.

[59]*Bethlehem Steel Corp. v. G. C. Zarnas & Co.*, 304 Md. 183, 498 A.2d 605 (1985). Illinois follows the same rule; see *Lyons v. Turner Constr. Co.*, 195 Ill.App.3d 36, 551 N.E.2d 1062 (1990).

if the language clearly stated that indemnification would be made even if the loss had been caused in whole or in part by the indemnitee's negligence.[60] Some even required that negligence be mentioned specifically.[61] Regulating such clauses reflected the belief that indemnification often exculpated the indemnitee and could be a disincentive for good safety practices, as well as the belief that such clauses were forced on weaker parties. In operation, the effectiveness of such an approach depended on the skill of the person drafting the indemnification clause.

Although insurance is discussed in Section 31.05H, it is important to note here that many courts look at the contract in which the indemnification clause is contained to see whether the indemnitor is required to purchase and maintain liability insurance.[62] This focus may indicate an intention that indemnification is to cover claims caused at least in part by the indemnitee's negligence.

In recent years, there has been a modest tendency to look on indemnification clauses neutrally as representing a rational attempt to distribute losses efficiently by facilitating insurance coverage. Courts espousing a more favorable attitude toward such clauses seek to determine the intention of the parties as they would when interpreting any other clause.[63]

As noted in Section 20.02, finding the intention of the contracting parties is often difficult if the parties have contracted on the basis of a standard contract published by the American Institute of Architects (AIA) or the Engineers Joint Contracts Documents Committee (EJCDC).

Indemnification is often a key contract clause (especially for insurers), but very frequently, contracting parties do not understand such provisions. They may have selected an AIA or EJCDC document without carefully analyzing its terms. The drafts by these organizations (often influenced or dominated by liability insurers) makes the process of interpretation difficult. In these contracts, a more mechanical and predictable rule may be preferable than focus on the intention of the contracting parties.

The modest tendency toward interpreting indemnification clauses "as written" has by no means become majority rule. This is demonstrated by *Ethyl Corp. v. Daniel Construction Co.*[64] Earlier Texas decisions had required that the language be clear and unequivocal as to the obligation of the indemnitor to indemnify the indemnitee against the consequences of its own negligence. The court cited Texas decisions that had indicated a trend toward strict construction of such clauses, decisions that came close to adopting the requirement that negligence be expressly mentioned.

However, the court was concerned because

[t]he scriveners of indemnity agreements have devised novel ways of writing provisions which failed to expressly state the true intent of those provisions. The intent of the scriveners is to indemnify the indemnitee for its negligence, yet be just ambiguous enough to conceal that intent from the indemnitor. The result has been a plethora of lawsuits to construe those ambiguous contracts. We hold the better policy is to cut through the ambiguity of those provisions and adopt the express negligence doctrine.

. . .

Under the doctrine of express negligence, the intent of the parties must be specifically stated within the four corners of the contract.[65]

---

[60]For a few of the many cases, see *Becker v. Black & Veatch Consulting Eng'rs*, 509 F.2d 42 (8th Cir.1974); *Warburton v. Phoenix Steel Corp.*, 321 A.2d 345 (Del.Super.1974), aff'd, 334 A.2d 225 (Del.1975). Pennsylvania, in *Ruzzi v. Butler Petroleum Co.*, 527 Pa. 1, 588 A.2d 1 (1991), followed the majority rule of requiring that a clause be clear and unequivocal if it is to cover negligence of the indemnitee. Although *Ruzzi* was criticized as inflexible and mechanical, depriving the court of the opportunity of hearing testimony outside the writing itself that can indicate the intention of the parties; see *Pennsylvania Supreme Court Review*, 1991, 65 Temp.L.Rev. 679 (1992); the decision was reaffirmed in *Greer v. City of Philadelphia*, 568 Pa. 244, 795 A.2d 376 (2002).

[61]*Burns & Roe, Inc. v. Central Maine Power Co.*, 659 F.Supp. 141 (D.Me.1987); *Ethyl Corp. v. Daniel Constr. Co.*, 725 S.W.2d 705 (Tex. 1987).

[62]*Pennsylvania Supreme Court Review*, 1991, 65 Temp.L.Rev. 679 at 696, note 114 (1992) (citing cases).

[63]*New England Tel. & Tel. Co. v. Central Vermont Pub. Serv. Corp.*, 391 F.Supp. 420 (D.Vt.1975) (emphasizing two substantial corporations of equal bargaining power who knew what they were doing when they made the contract); *C. J. M. Constr., Inc. v. Chandler Plumbing & Heating*, 708 P.2d 60 (Alaska 1985) (two dissenting); *Washington Elementary School Dist. No. 6 v. Baglino Corp.*, 169 Ariz. 58, 817 P.2d 3, 6 (1991) (interpreting AIA contract); *Morton Thiokol, Inc., v. Metal Bldg. Alteration Co.*, 193 Cal.App.3d 1025, 238 Cal.Rptr. 722 (1987), review denied Oct. 17, 1987 (focuses on intention of the parties rather than applying prior California rule that general indemnification clauses

do not cover active negligence); *Columbus v. Alden E. Stilson & Assocs.*, 90 Ohio App.3d 608, 630 N.E.2d 59, 63 (1993) (enforcing work-related indemnity clause), motion to certify overruled, 68 Ohio St.3d 1461, 627 N.E.2d 1002 (1994).

[64]Supra note 61.

[65]Id. at 707–708.

The often inelegantly drafted anti-indemnification statutes and the ambivalent attitude of judges toward indemnification make this process one of the most difficult in construction law.

Complexity is confounded by some jurisdictions that attempt to distinguish between active and passive negligence, concluding that the requisite degree of specificity applies to active but not to passive negligence.[66] Other courts do not make the active–passive distinction for this purpose.[67]

The plethora of cases that have interpreted indemnity clauses and have come to variant results caused the Illinois Supreme Court to state in despair,

> We have examined the authorities cited by the parties and many of those collected at 27 A.L.R.3d 663, and conclude that the contractual provisions involved are so varied that each must stand on its own language and little is to be gained by an attempt to analyze, distinguish or reconcile the decisions. The only guidance afforded is found in the accepted rule of interpretation which requires that the agreement be given a fair and reasonable interpretation based on a consideration of all of its language and provisions.[68]

Despite this neutral approach, the court held that the language did not cover negligence of the indemnitee, as such intention was not expressed clearly and explicitly.

F. Interpretation Issues

Losses and Indemnity Coverage. Clauses can cover certain losses but not others. For example, indemnity clauses can be drawn broadly enough to cover any loss, even those relating to property damage that the indemnitee has suffered. Indemnity, however, is generally designed to transfer losses relating to claims third parties make against the indemnitee.[69]

Drafters of indemnification clauses should anticipate the types of losses that can occur. Typically, such clauses cover claims for harm to person or damage to property made by third parties. But clauses drafted with these claims in mind will not cover claims for purely economic losses unrelated to personal harm or damage to property.[70] Sometimes assertions of indemnity appear to be a grasping at straws. For example, an indemnity claim was made for damage resulting to window frames built faultily by the indemnitee prime contractor and then installed by the indemnitor subcontractor.[71] The court saw no specific language covering this loss and refused to include it within the general language.

Similarly, an indemnification clause between prime contractor and owner was held not to cover a loss incurred by the owner to a third party based on the prime contractor having trespassed on the third party's land while doing the work.[72] The owner did not comply with its contract requirement to obtain an easement. The court held that the loss did not arise out of the prime contractor's performance even though the trespass was caused by its performance. A claim made based on an injury that occurred after the work had been completed, however, falls within the ambit of the indemnity clause. Injuries, whether they occur during or after performance, are the type of loss typically covered by insurance and part of the indemnification process.[73]

Work-Relatedness of Injury. The indemnitee who uses a work-related indemnity clause usually seeks protection against claims made incident to or arising out of the indemnitor's performance. Interpretation problems develop when a claimant is injured while at work, yet the principal cause of the injury is not him doing the work but the activity of the indemnitee, who later seeks indemnification from, as a rule, the indemnitor employer

---

[66]*Morgan v. Stubblefield*, 6 Cal.3d 606, 493 P.2d 465, 100 Cal.Rptr. 1 (1972). But see *Morton Thiokol, Inc. v. Metal Bldg. Alteration Co.*, supra note 63; *Wrobel v. Trapani*, 129 Ill.App.3d 306, 264 N.E.2d 240 (1970).

[67]*Becker v. Black & Veatch Consulting Eng'rs*, supra note 60.

[68]*Tatar v. Maxon Constr. Co.*, 54 Ill.2d 64, 294 N.E.2d 272, 273–74 (1973). However, the court has recently interpreted similar indemnity language to cover an indemnitee's negligence; see *Buenz v. Frontline Transp. Co.*, 227 Ill.2d 302, 882 N.E.2d 525 (2008).

[69]In *Pacific Gas & Elec. Co. v. G. W. Thomas Drayage & Rigging Co.*, 69 Cal.2d 33, 442 P.2d 641, 69 Cal.Rptr. 561 (1968), the indemnitee sought recovery for damage to its own property. The court held that

evidence submitted by the indemnitor that tended to show that the parties intended to cover only claims made by third parties should have been admitted into evidence.

[70]*Mobil Chemical Co. v. Blount Bros. Corp.*, 809 F.2d 1175 (5th Cir.1987); *Fairbanks North Star Borough v. Roen Design Assoc., Inc.*, supra note 29.

[71]*Mesker Bros. Iron Co. v. Des Lauriers Column Mould Co.*, 8 Ill.App. 3d 113, 289 N.E.2d 223 (1972).

[72]*Serafine v. Metropolitan Sanitary Dist.*, 133 Ill.App.2d 93, 272 N.E.2d 716 (1971).

[73]*Becker v. Black & Veatch Consulting Eng'rs*, supra note 60.

of the claimant. The only connection that can be made between the accident and the activity of the indemnitor is that the accident would not have happened had the indemnitor not been on the job. These problems usually involve an accident to an employee of the subcontractor, a work-related clause, and a demand for indemnification from the indemnitee prime to the indemnitor subcontractor.[74]

Another interpretation issue arose in *General Accident Fire & Life Assurance Corp. v. Finegan & Burgess.*[75] A sign subcontract included an indemnity provision. After completion of the sign subcontractor's work, the owner's project engineer went to inspect the sign, accompanied by an employee of the sign subcontractor. After inspection, the engineer indicated he wished to check a particular switch that had been installed by another subcontractor. The employee of the sign subcontractor indicated the location of the switch and then left.

On his way to inspect the switch, the engineer fell from a walkway, which had no railing because of the prime contractor's negligence. The injured engineer sued both prime and sign subcontractor. The jury found that the prime contractor had been negligent but that the sign subcontractor had not. The prime contractor's insurer paid the claim and then brought an action against the sign subcontractor based on the indemnity provision. The purpose of inspection was to enable a tenant to move in earlier and was not directly related to the sign subcontractor's performance. The area where the injury had occurred was under the general contract of the prime contractor. For these reasons, the court held that the clause did not cover this accident. It seems that the engineer was no longer dealing with the sign subcontractor's work.

Amount Payable. Usually the indemnitee seeks to transfer its entire loss to the indemnitor. This, as a rule, includes any money paid to the claimant, any costs of investigating the claim, any legal, investigative, or expert witness costs, and interest from the time the payment was made to the third party.[76] Most clauses deal with these issues. The most troublesome questions relate to amounts paid under a settlement and costs to defend when it is determined that no liability existed.

As to the first, although one court seemed to require that the indemnitee establish that it would have been *liable*,[77] it is better to require indemnity if a settlement were made in *good faith*.[78] It should not be necessary for the indemnitee to have to either litigate or settle and then establish legal responsibility.

Invoking the insurance contract, which also promises to defend and indemnify, some courts have held only if the indemnitor was given notice and an opportunity to defend is the indemnitor bound to a settlement made by the indemnitee who has undertaken the defense. But even here the indemnitee must establish that the settlement was reasonable and prudent under all circumstances. It need not establish that it would have lost the case but need only establish that the settlement was reasonable.[79]

That indemnification may require the indemnitor to indemnify (pay off the claimant) and *defend* or pay the *cost of defense* raises another issue. The promise to defend may be broader than the promise to indemnify. The indemnitor may have the obligation to defend even if it is later determined that the indemnitor was not negligent. For example, suppose the indemnification is "work related" as defined in Section 31.05B. Work-relatedness is the threshold to indemnification. So long as the loss arose out of the indemnitor's work, it has a duty to defend the indemnitee, regardless of whether the indemnitee will ultimately be found entitled to indemnification.[80] Conversely, if it determined that the indemnitee was indeed negligent but

---

[74]*Twin City Fire Ins. Co., Inc. v. Ohio Casualty Ins. Co., Inc.,* 480 F.3d 1254, 1264 (11th Cir.2007) (Alabama law; contractor must indemnify solely negligent owner under a "work related" indemnity clause).

[75]351 F.2d 168 (6th Cir.1965).

[76]*Larive v. United States,* 449 F.2d 150 (8th Cir.1971).

[77]*Ford Motor Co. v. W. F. Holt & Sons, Inc.,* 453 F.2d 116 (6th Cir.1971), cert. denied, 405 U.S. 1067 (1972). As shall be seen in Section 31.05G, this is one of the deficiencies of the AIA clause.

[78]*Miller v. Shugart,* 316 N.W.2d 729 (Minn.1982) (indemnitor must pay settlement made by indemnitee if reasonable and prudent). But see *Peter Culley & Assocs. v. Superior Court,* 10 Cal.App.4th 1484, 13 Cal. Rptr.2d 624 (1992), which held that if the indemnitee settles without a trial, it must show that the liability is covered by the contract and that liability existed and the extent thereof. The court also held that the settlement is presumptive evidence of liability and the amount of liability but may be overcome by proof from the indemnitor that there was no liability or the settlement amount was unreasonable. This decision places too great a burden on the indemnitee.

[79]*United States Auto Ass'n v. Morris,* 154 Ariz. 113, 741 P.2d 246 (1987).

[80]*Hoffman Constr. Co. of Alaska v. U.S. Fabrication & Erection, Inc.,* 32 P.3d 464 (Alaska 2001); *Crawford v. Weather Shield Mfg. Inc.,* 9 Cal. Rptr.3d 721, 187 P.3d 424 (Cal. 2008).

the accident did not relate to the work, the indemnitor would like to recover its cost of defense on the basis of there never having been an obligation to indemnify.[81]

The question of who pays defense costs when there is no liability requires that liability insurance be addressed. Indemnification has attributes similar to liability insurance. The obligation to defend is determined by the claimant's allegations, augmented in some states by the discovery of facts as they evolve in the investigation. If there is *potential* liability, the insurer must defend. The insurer's refusal to defend may not stem from its belief that its insured was not liable. The refusal to defend may stem from the insurer's belief that the policy excluded coverage of the claim, that there had been misrepresentation by the insured, or that the insured did not cooperate.

When the insurer doubts its obligation to defend and indemnify, it may wish to assert that defense *later* but not lose control of the defense or be subject to huge liability for failure to defend. As a result the insurer will undertake the defense under a "reservation of rights." This means the insurer reserves the right to assert it had no obligation to the insured after the claim has been resolved. If later it is determined the insurer was not obligated to defend, it may seek to recover its cost of defense from its insured.

Returning to the illustration of the work-related indemnification clause, unless the indemnitor reserved its rights to assert that the claim was not work related, it would probably not be able to assert later that there was no obligation to indemnify and defend on that ground. Taking over the defense without such a reservation would preclude the indemnitor from recovering its defense costs. In such a case its obligation to defend was broader than its obligation to indemnify.

But suppose it was determined that the indemnitee was not negligent, and negligence triggers the indemnification obligation. Probably the same result would follow. The trigger to the obligation to defend is the *claim* of negligence. If there is a *bona fide* claim that the indemnitee was negligent, the indemnitor must defend. If it does so successfully, it need not pay the claimant, because the duty to indemnify does not arise. But the indemnitor cannot recover its cost of defense any more than an insurer can.

The AIA has chosen not to make the indemnitor defend, as in liability insurance, but to make the indemnitor *pay* the *cost of defense* provided by the indemnitee.

The issue of the successful defense is applied to an AIA indemnity clause, as discussed in Section 31.05G.

Who May Enforce Indemnity Clause. Usually, the question of who may enforce the indemnity agreement is straightforward, since the indemnitee is clearly identified. However, protection usually is extended also to persons acting under the direction of the indemnitee, such as its employees, agents, and contractors. At the same time, construction contracts typically state that they do not confer rights or benefits upon parties not signatory to the agreement.

The interplay between these various clauses was the subject of a California case, *The Ratcliff Architects v. Vanir Construction Management, Inc.*[82] A public owner entered into separate contracts with a construction manager (CM) and an architect. After project completion, the architect asserted that the CM, through its negligent performance, caused the architect to incur higher costs than anticipated and to suffer lost profits. The architect argued that it was an "agent" of the owner within the meaning of the indemnity clause and so had the right to enforce the clause against the CM.

The court rejected the architect's argument, finding that it would conflict with the contract clauses which denied the conferral of rights or benefits upon third parties. The court interpreted the contract as giving only the owner the right to enforce an indemnity clause against the CM on behalf of itself and any of its agents or employees.[83] For similar reasons, courts have refused to allow an independent contractor of the owner to enforce an indemnity agreement between the owner and prime contractor.[84]

G. The AIA Indemnity Clause

Before 1966, AIA Document A201 did not contain an indemnification clause. But the liability explosion of the 1960s persuaded the AIA and its insurance counsel that the sting from increased liability could be reduced if

---

[81]*Titan Steel Corp. v. Walton*, 365 F.2d 542 (10th Cir.1966).

[82]88 Cal.App.4th 595, 106 Cal.Rptr.2d 1 (2001).

[83]Id., 106 Cal.Rptr.2d at 7. The court also found the architect's negligence claim was barred by the economic loss rule; see id. at 8–9.

[84]*Pepe v. Township of Plainsboro,* 337 N.J.Super. 209, 766 A.2d 837 (App.Div.2001) (owner's contract administrator cannot sue the prime for contractual indemnity); *Tonking v. Port Auth. of N.Y. and N.J.,* 3 N.Y.3d 486, 821 N.E.2d 133, 787 N.Y.S.2d 708 (2004) (owner's construction manager cannot sue prime).

the contractor was required to indemnify the owner and the architect against certain losses. As noted in Section 31.05D, this attempt generated a struggle with the AGC and liability insurers of contractors. The issue was resolved in 1967.[85] Both groups and the insurers approved a clause, then designated Paragraph 4.18. The clause required the contractor to indemnify owner and architect against certain losses (harm to person or damage to property other than the work itself) that arose out of the work and was caused by the negligence of the contractor, despite the loss's having been caused in part by the indemnitee. However, indemnification was excluded for losses arising out of specified design activities of the architect or "the giving of or the failure to give directions or instructions by the Architect . . . provided such giving or failure to give is the primary cause of the injury or damage." The 1967 Paragraph 4.18.1 was an all-or-nothing intermediate indemnification clause. Noncontractual indemnification was not preempted by the indemnification clause. Unlike many indemnity clauses, this does not require the indemnitor to *defend* as well as indemnify. Instead, the indemnitor pays the cost of defense.

The drafts for the A201 published in 1987 had included language that would have created comparative indemnity, an approach the AIA had taken in its standard subcontract, A401. This would have complied with the increasing number of state statutes that require comparative indemnification.[86] Insurers resisted this approach; the troubled insurance industry was not in an adventurous mood. What emerged in 1987 was Paragraph 3.18.1. It stated that the indemnification applies "only to the extent caused in whole or in part" by the negligence of the contractor or those for whom it was responsible. This statement appears to change the clause from an all-or-nothing to a comparative clause. Yet the clause also states that indemnification applies whether the claim is caused in whole or in part by the indemnitee. The latter would not be necessary were the clause simply one under which the contractor indemnifies for that portion of the loss caused by its negligence. It is unlikely that the AIA intended Paragraph 3.18.1 to be a comparative indemnification clause. The phrase "only to the extent caused" may have been included inadvertently.

Yet most of the cases[87] and the commentators[88] looked at the language as creating a comparative indemnity clause. The offending language was deleted in 1997 and that deletion was continued in 2007. Now it is clear that Section 3.18 is a comparative indemnity clause. It also appears that the cost of defense, as well as indemnification, will be apportioned by the percentage of fault.[89] (See Figure 31.2.)

This apportionment raises the issue discussed in Section 31.05F: Who pays the defense costs if the claim is defended successfully? Before the clarification in 1997, it was held that if the architect (the indemnitee) defended the claim successfully, the contractor (the indemnitor) would still have to pay the cost of the defense.[90] However, these cases looked at Paragraph 3.18.3, which limited the broad indemnification required by Paragraph 3.18.1 by excluding design and site services of the architect under certain conditions. This exclusion generated claims, generally unsuccessful, by contractors that if the claim was based on activities excluded from indemnification, there was no obligation to indemnify and the costs of defense should not be borne by the indemnitor contractor.[91]

Because that exclusion was deleted in 1997, this threshold to indemnification issue should not arise. In A201-2007, indemnification covers only claims based on the negligence of the indemnitor contractor. Claims based on negligence of the architect or the owner will need to be defended and paid for by them.

A few other aspects of the indemnification provision in A201-2007 merit comment. First, Section 3.18.1 states that the indemnification is done to "the fullest extent permitted by the law." This can simply signal to users that local legislation must be checked. Or the statement can be

---

[85]Changes in 1970 and 1976 were marginal. In A201-1997, the clause is ¶3.18.

[86]See Section 31.05D.

---

[87]*MSI Constr. Managers, Inc. v. Corvo Iron Works, Inc.,* 208 Mich. App. 340, 527 N.W.2d 79 (1995); *Dillard v. Shaughnessy, Fickel & Scott Architects, Inc.,* 884 S.W.2d 722 (Mo.Ct.App.1994) (Kansas law); *Greer v. City of Philadelphia,* supra note 60; *Sullivan v. Scoular Grain Co.,* 853 P.2d 877 (Utah 1993).

[88]J. SWEET & J. SWEET, 2 SWEET ON CONSTRUCTION INDUSTRY CONTRACTS § 19.06[A] (4th ed.1999); Sklar, *An Overview of the New AIA Contact Documents as They Affect Subcontractors,* 3 Focus, No. 3 (Joseph E. Manzi & Assoc., Sept. 1988).

[89]*Dillard v. Shaughnessy, Fickel & Scott,* supra note 87.

[90]*Hillman v. Leland Burns,* 209 Cal.App.3d 860, 257 Cal.Rptr. 535 (1989); *Cuhaci & Peterson Architects, Inc. v. Huber Constr. Co.,* 516 So.2d 1096 (Fla.Dist.Ct.App.1987), review denied, 525 So.2d 878 (Fla. 1988).

[91]Ibid.

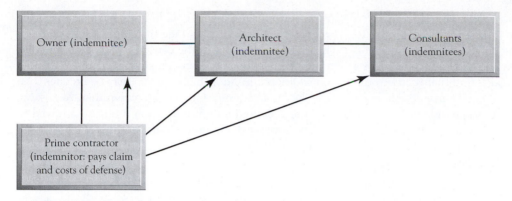

FIGURE 31.2   AIA Document A201-1997, Paragraph 3.18: Comparative indemnity clause.

interpreted to invite the court to scale down or "reform" any clauses that state legislation would invalidate.[92]

Indemnification applies only if the contractor is negligent. This requirement could preclude indemnity if there were a settlement in which no one admitted having been negligent. Indemnification, as noted in Section 31.05F, should apply to settlements made in good faith.

Finally, the architect given indemnification under Section 3.18 is not a party to the construction contract. His right to enforce indemnification depends on Section 1.1.2 or common law third-party beneficiary law.[93] Section 1.1.2 states that the architect shall be entitled to enforcement of those obligations "intended to facilitate performance of the Architect's duties."

H. Insurance

Owners frequently require that contractors procure insurance covering the risks specified in the indemnification clause. Even without this requirement, a prudent contractor will be certain that its insurance will cover this risk. Generally, liability policies cover only liability imposed by law and not that imposed or assumed by contract.[94]

At one time, it was necessary for the contractor to obtain a specific endorsement covering the liability assumed by these indemnification clauses. However, modern standard form commercial liability policies provide coverage for liability assumed under an "insured contract," which has been interpreted to mean an indemnity agreement.[95] Contractors should be certain, as should owners, that the liability policy covers this risk.

As noted, agreements to procure insurance to back up an indemnity provision are generally not invalidated by anti-indemnity legislation. Sometimes clarification that insurance does not run afoul of anti-indemnity legislation is done expressly by statute.[96] When courts have faced this issue without the benefit of a statute, they have recognized the difference between an indemnity clause with exculpatory features and a liability insurance policy.[97] The former is often thought to induce carelessness on the site, whereas the latter is looked on as a proper method of distributing risks.

---

[92]*Robertson v. Swindell-Dressler Co.*, 82 Mich.App. 382, 267 N.W.2d 131 (1978).

[93]Owners have been able to recover on indemnification clauses in subcontracts to which they are not a party. Schroeder v. C. F. Braun & Co., 502 F.2d 235 (7th Cir.1974); *Titan Steel Corp. v. Walton*, supra note 81.

[94]But in *Vandenberg v. Superior Court*, 21 Cal.4th 815, 982 P.2d 229, 88 Cal.Rptr.2d 366 (1999), the court ruled that a commercial liability policy may cover contract claims, so long as other requirements for coverage are met. The requirements for coverage, and not the plaintiff's theory of recovery, determine whether the policy applies. See also *Roger H. Proulx & Co. v. Crest-Liners, Inc.*, 98 Cal.App.4th 182, 119 Cal.Rptr.2d 442 (2002), as modified on denial of rehearing, May 28, 2002, review denied July 17, 2002.

[95]For coverage granted under the "insured contract" language, see *Gibson & Assocs., Inc. v. Home Ins. Co.*, 966 F.Supp. 468, 475–76 (N.D.Tex.1997).

[96]West Ann.Minn.Stat. § 337.05(1). The statutes are collected in 14 Constr.Litig.Rep., No. 1, Jan. 1993, p. 9.

[97]*Holmes v. Watson-Forsberg Co.*, 488 N.W.2d 473 (Minn.1992), M. SCHNEIER, CONSTRUCTION ACCIDENT LAW, supra note 48, at 473–75

Another difficulty that arises when an indemnity obligation is backed up by insurance is whether the insurance obligation remains in force if the indemnity clause is found invalid, either under an anti-indemnity statute or a common law analysis. Some courts find that the two obligations rise and fall together, so that a party not liable for indemnification also has no duty to provide insurance to cover the loss.[98] Other courts find the insurer's duty to provide coverage remains, even if its insured's liability under the indemnity agreement has been negated. These courts point out that the owner could hardly have intended the requested insurance coverage to fail precisely when its indemnity rights against the contractor are lost. They also note that the insurance coverage trigger (usually under an additional insured endorsement) may be broader than the duty to indemnify. Courts also reason that the insurer, having accepted a premium, should not be allowed to escape coverage for reasons having nothing to do with the policy itself.[99]

---

[98]*Hurlburt v. Northern States Power Co.*, 549 N.W.2d 919 (Minn.1996).

[99]*Shell Oil Co. v. National Union Fire Ins. Co.*, 44 Cal.App.4th 1633, 52 Cal.Rptr.2d 580, 585 (1996), review denied Aug. 14, 1996 (intent of owner); *Acceptance Ins. Co. v. Syufy Enterprises*, 69 Cal.App.4th 321, 81 Cal.Rptr.2d 557 (1999), review denied Apr. 14, 1999 (insurance coverage for owner's own negligence involving a premises defect not part of the contract work); *Heat & Power Corp. v. Air Products & Chemicals, Inc.*, 320 Md. 584, 578 A.2d 1202, 1208 (1990) (insurer collected premium).

CHAPTER THIRTY-TWO

Surety Bonds:
Backstopping Contractors

SECTION 32.01 Mechanics and Terminology

The surety bond transaction is a peculiar arrangement and differs procedurally from most contracts. The typical surety arrangement is essentially triangular. The "surety" obligates itself to perform or to pay a specified amount of money if the "principal debtor" (usually called the "principal") does not perform. The person to whom this performance is promised is usually called the "obligee" (sometimes called the "creditor"). In the building contract context, the surety is usually a professional bonding company. The principal is the prime contractor or, in the case of subcontractor bonds, a subcontractor. The obligee is the owner or, in the case of a subcontractor bond, the prime contractor.[1]

If the owner requires, the prime contractor (the principal) applies for a bond from a bonding company. Usually the owner indirectly pays the cost of the bond, because the bidder adds the bond premium to its costs when computing its bid. The bond is issued to the owner (the obligee). The bond "runs to" the owner in that performance by the surety has been promised to the owner even though the prime contractor applied for the bond.

Another problem in dealing with surety bonds results from the antiquated way in which bonds are written. The earliest bonds were called *penal bonds*. The surety made an absolute promise to render a certain performance or to pay a specified amount of money. This was followed by a paragraph stating that the bond would be void if the principal promptly and properly performed all the obligations under the contract between principal and obligee. As seen later in this section, use of this antiquated language has continued to the present.

This format can cause problems. The transaction for which the surety provides financial security can be simple or complex. As an illustration of the former, the bail bondsperson will pay if the accused fails to appear at the hearing, or a banker's blanket bond requires the surety to pay if the bank official embezzles funds. However, a prime contractor backed by a surety has a wide variety of obligations set forth in the construction contract. For example, the prime contractor promises to build the project properly, not to damage the land of adjacent landowners,[2] to perform the work in such a way as to avoid exposing workers and others to unreasonable risk of harm, and to indemnify the owner and design professional if certain claims are made.

Often bond language does not state which duties the bond covers. For example, AIA Document A311 (issued in 1970) refers to the construction contract and simply states that "if Contractor shall promptly and faithfully perform said Contract then this obligation shall be null and void; otherwise it shall remain in full force and effect." Which aspects of the contractor's performance does the surety back up? This will be discussed in Section 32.10D.

---

[1]The Restatement (Third) of Torts: Suretyship and Guaranty (1996) uses somewhat different terminology: the principal is the "principal obligor," the surety is the "secondary obligor," and the term "obligee" is unchanged.

[2]Generally the surety is not responsible for tort claims against its principal debtor, often those made by adjacent landowners. See Barker, *Third-Party Tort Claimants and the Contract Bond Surety,* 5 Constr. Lawyer, No. 1, Spring 1984, p. 7. But if the bond exceeds a statutory amount and has broad language, one commentator believes a tort claimant may be able to recover from the surety. Perry, *Third Party Tort Claimants Can Recover Against Sureties Under Construction Bonds: Is the Bond a Comprehensive General Liability Policy?* 5 Constr. Lawyer, No. 4, Apr. 1985, p. 5.

Also, bond language often does not specifically control problems of coverage, problems of notices, and other things that are normally part of any contract. Improvement in this regard has been made, and some bonds such as AIA Document 312 (issued in 1984) today do contain specific provisions that better inform parties of rights and duties under the bond. See Appendix D.

Another terminological distinction is between the conditional nature of bonds used in the American construction industry and "on demand" or unconditional bonds commonly used in British and international construction projects. These will be discussed in Section 32.10C.

SECTION 32.02 Function of Surety: Insurer Compared

A surety's function is to assure one party that the entity with whom it is dealing will be backed up by someone who is financially responsible. Sureties are used in transactions where people deal with individuals or organizations of doubtful financial capacity. They provide credit. Sureties must be distinguished from insurers, though each provides financial security.

Yet the similarities between insurers and sureties, as noted in Section 32.10I, mean that for regulatory purposes they are often lumped together. For example, the California Supreme Court found invalid a legislative attempt to exempt sureties from the rate control that Proposition 103 adopted in California to regulate insurance rates, as altering the proposition.[3]

An insured is concerned that unusual, unexpected events will cause it to suffer losses or expose it to liability. Although it can self-insure—that is, bear the risk itself—it usually chooses to indemnify itself against this risk by buying a promise from an insurance company, in exchange for paying a premium. The insurer distributes this risk among its policyholders.

In public liability insurance, the insured itself may be at fault and cause a loss to the insurer. But the insurer cannot recover its loss from its own insured. Although it may seek to recover its losses from third parties through subrogation (stepping into the position of the person it

has paid—its insured—and thereby acquiring any claims of its insured against those who caused the loss), it cannot recover from those named as insureds in the policy.[4]

Sureties, in contrast, in addition to dealing with ordinary construction contract performance problems, which can be considered business losses (not the "accidents" central to insurance), seek, through indemnity agreements, to recover any losses they have suffered from the principal party on whom it has written a bond as well as individual shareholders of a corporate principal. In the late 1970s, contractors who were liable because of defective workmanship sought to recover from their commercial (formerly called "comprehensive") general liability (CGL) insurers, preferring this arrangement to having their sureties pay the loss.[5] Their sureties would seek to recover from them; their CGL insurers could not.

SECTION 32.03 Judicial Treatment of Sureties

Judicial attitude toward sureties has changed. Before professional sureties developed, the surety would be a private person (perhaps a relative of the principal) who sought to aid the principal to obtain a contract or to stave off a creditor by obligating himself to perform if the principal did not. Such a surety was frequently not paid and received no direct benefit for taking this risk. For these reasons, the surety was considered a "favorite" of the law.

One illustration of this favored position was the Statute of Frauds, which requires that certain types of promises must be evidenced by a written memorandum.[6] The original Statute of Frauds, enacted in 1677, included promises to answer for the debts, defaults, or miscarriages of another. Without a writing, the surety was not held liable. This requirement was to protect her from the enforcement of an impulsive oral promise often made without due deliberation.[7] Also, any minor change in the contract between the principal and the obligee would discharge the surety (relieve her of liability).[8] The personal, unpaid

---

[3]*Amwest Surety v. Wilson*, 11 Cal.4th 1243, 906 P.2d 1112, 48 Cal. Rptr. 12 (1995).

[4]Subrogation is discussed in Section 23.05C.

[5]See Section 24.08.

[6]See Section 5.10.

[7]A written memorandum was not required where the main purpose or leading object of the surety was to benefit herself.

[8]See Section 32.10C for additional applications of this doctrine (called the *stricti juris* rule).

surety could not handle assurance needs in a commercial economy; the personal surety herself might not be financially responsible. For this reason, the professional paid surety has developed as an important institution in both economic life generally and in building contracts.

The development of professional sureties casts doubt on protective rules developed largely when sureties were uncompensated. Although some protective rules are still applied, the professional surety is not regarded with the tender solicitude accorded the personal, uncompensated surety. This changeover from a legally favored position to one of neutrality—and perhaps even to one of disfavor— creates uncertainty in the law. Older cases are sometimes cited as precedents to protect sureties, but these precedents may be of limited value, because being a surety is now a business.

The ambivalence toward sureties is illustrated by *Winston Corp. v. Continental Casualty Co.*,[9] which involved a claim by an owner against the surety on a performance bond. The bond incorporated all the provisions of the construction contract.

The contractor ran into financial difficulty causing delays, all known to the surety. Five months after the scheduled completion date, the owner met with the contractor and invited the surety to meet also. The surety, however, refused to attend. (The performance bond part of AIA Document 312, Paragraphs 2 and 3.1, requires the surety to participate in conferences when the owner considers declaring the contractor "in default." See Appendix D.) At this meeting, the owner and the contractor entered into an agreement designed to accelerate construction. Under the agreement, the contractor assigned the construction contract to the owner, permitted the owner to take possession of the premises, and assigned the contractor's subcontracts to the owner. However, the contractor's continued participation was expected. The owner immediately telephoned the surety, notifying it of the new arrangement, and mailed a copy of the letter agreement to the surety. A year after the scheduled completion date, the project was completed.

The owner sued on performance and payment bonds. The surety's defense was that the agreement between owner and contractor modified the original construction

contract and discharged the surety. Also, the surety contended that it was released by the owner's failure to give the surety seven days' written notice before terminating the contractor. The trial court, noting that sureties were favorites of the law, sustained the surety. The appellate court reversed, concluding that the owner had an absolute right to terminate the contract if the contractor was in default and that taking over the contract merely exercised this right.

As for the failure to give notice before taking over construction, the court agreed that historically, any slight deviation from the contract terms would discharge the surety. But, noted the court, applying this doctrine often caused harsh and unjust results, especially when compensated sureties were relieved of their obligations because technical breaches of the construction contract were incorporated as part of the bond. This harshness induced most courts to deviate from the doctrine under which the surety was the favorite of the law and to require the compensated surety to show that the change in the original agreement was material and prejudicial to the surety. The surety must show that the change increased the surety's risk or changed the risk to the surety's detriment.

In addition to no longer considering the surety as a favorite, courts generally construe ambiguities against the surety for the same reasons that most insurance policies are construed against the insurance company.[10]

SECTION 32.04 Surety Bonds in Construction Contracts

Surety bonds play a vital part in the construction process. The contracting industry is volatile: Bankruptcies are not uncommon, and a few unsuccessful projects can cause financial catastrophe. Estimating costs is difficult and requires much skill. Fixed-price contracts place many risks on the contractor, such as price increases, labor difficulties, subsurface conditions, and changing governmental policy.

Some construction companies are poorly managed and supervised. Often they are undercapitalized and rely heavily on the technological skill of a few individuals. If

---

[9]508 F.2d 1298 (6th Cir.), cert. denied, 423 U.S. 914 (1975). But see *In re Liquidation of Union*, 220 A.D.2d 339, 632 N.Y.S.2d 788 (1995) (surety discharged by change orders without consent of surety).

[10]*United States v. Algernon Blair, Inc.*, 329 F.Supp. 1360 (D.S.C.1971); *School Dist. No. 65R of Lincoln County v. Universal Sur. Co., Lincoln, Neb.*, 178 Neb. 746, 135 N.W.2d 232 (1965).

these people become unavailable, difficulties will likely arise. For many construction companies, credit may be difficult to obtain. Some contractors do not insure against the risks and calamities that can be covered by insurance. Finally, anti-inflationary government policies such as tight money policies almost always hit the building industry first.

In most construction projects, bonds are needed to protect the owner.[11] The owner in most projects would like to have a financially solvent surety if the successful bidder does not enter into the construction contract (bid bond), the prime contractor does not perform its work properly (performance bond), or the prime contractor does not pay its subcontractors or suppliers (payment bond). (Bond requirements can act as preliminary screen for contractor selection.)

If these events do not occur, the amount paid for a surety bond may seem wasted. Some institutional owners believe there is no need for a surety bond system if the prime contractor is chosen carefully and if a well-administered payment system eliminates the risk of unpaid subcontractors and suppliers. Such owners may choose to be self-insurers and not obtain bonds. They realize there may be losses, but they believe the losses over a long period will be less than the cost of bond premiums.

Even where a bond is not required at the outset, it is best to include a provision in the prime contract that will require the prime contractor to obtain a bond before or during performance if the owner so requests. Usually the owner pays the cost of a bond issued after the price of the project is agreed on.

Public construction frequently requires performance and payment bonds. The latter are required to protect subcontractors and suppliers who have no lien rights on public work. The Miller Act[12] requires federal prime contractors to obtain performance bonds and payment bonds based on the contract price, and similar requirements exist in state public contracting under "Little Miller Acts." In addition, often local housing development legislation requires that the developer furnish bonds to protect the local government if improvements the developer has promised are not made. Legal advice should be obtained to determine whether the project requires bonds.

The invitation to bid usually states whether the contractor is required to obtain a surety bond, the type or types of bonds, and the amount of the bonds. In some cases, the owner wishes to approve the form of bond and surety used. Also, the owner may want to have the right to refuse any substitution of the surety without its express written consent given before substitution.

Should the owner specify which surety bond company must be used? The choice depends on rates, bond provisions, and the reputation of particular bonding companies for efficient operation and for fairness in adjusting claims. If relative standardization exists on these matters, it is advisable to let the contractor choose the bonding company. The contractor may have an established relationship with a particular bonding company. In most cases, it is probably sufficient to permit the contractor to choose the bonding company, as long as the bonding company selected is licensed to operate in the state where the project will be built.

Occasionally a designated bonding company is unsatisfactory to the owner. This dissatisfaction may be based on suspicion of financial instability or a past record of arbitrariness in claims handling. One method of exercising some control over the selection of the bonding company is to provide in the contract that the contractor submit to the owner the bond of a proposed bonding company for the owner's review. If the owner, in exercising its best judgment, determines that it is inadvisable to use that bonding company, the owner can veto the proposed bonding company and designate another.[13]

SECTION 32.05 Bid Bond

The function of a bid bond is to provide the owner with a financially responsible party who will pay all or a portion of the damages caused if the bidder to whom a contract is awarded refuses to enter into it.[14]

---

[11]The relationship between the surety's obligation and any obligation to arbitrate was discussed in Section 30.15.

[12]40 U.S.C.A. § 270a–f, revised by Pub. L. No. 107-217, 116 Stat. 1062 (Aug. 21, 2002), codified at 40 U.S.C.A. §§ 3131–3133 (West Supp. 2003).

[13]In *Weisz Trucking Co. v. Emil R. Wohl Constr.*, 13 Cal.App.3d 256, 91 Cal.Rptr. 489 (1970), the standard for approval was held to be objective. However, a good-faith standard can be inserted in the contract.

[14]See Section 18.04E. Bid bonds are rare in Europe because bidders are frequently prequalified. This is also the reason why European bonds are for 5 to 10 percent of the contract price, compared with the 50 to 100 percent in the United States.

SECTION 32.06 Performance Bond

The performance bond provides a financially responsible party to stand behind some aspects of the contractor's performance. If a payment bond is furnished, the performance bond will not include payment of subcontractors, their suppliers, and suppliers of the prime contractor. See Figure 32.1. The AIA performance bond is reproduced in Appendix D.

Bonds usually place a designated dollar limit on the surety's liability. Typically, bond limits are 50 percent or 100 percent of the contract price. Some statutory bonds are required to be 50 percent of the contract price.

SECTION 32.07 Payment Bond

The payment bond is an undertaking by the surety to pay unpaid subcontractors and suppliers. See Figure 32.2. Appendix D reproduces the AIA payment bond. An understanding of the function of a payment bond requires a differentiation between private and public construction work. All states give unpaid subcontractors and suppliers liens if they improve private construction projects. This process was described in Section 28.07D.

Although there are various ways to avoid liens, one method has been to require prime contractors to obtain payment bonds. A payment bond obligates a surety to pay subcontractors and suppliers if the prime contractor does not pay them.[15] The owner seeks to avoid liens filed against its property. Although the owner generally has no contractual relationship with unpaid subcontractors or suppliers, the latter parties if unpaid can assert liens against the owner's property they have improved. The owner would prefer to direct them to the bonding company for payment.

Also, subcontractors are more likely to make bids when they can be assured of a surety if they are unpaid. The competent subcontractor who deals with a prime contractor of uncertain financial responsibility should add a contingency to its bid to cover possible collection costs and the risk of not collecting. Having a payment bond should eliminate the need for this cost factor. In addition, subcontractors and suppliers should be more willing to perform properly and deliver materials as quickly as

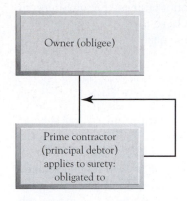

FIGURE 32.1 Performance bond.

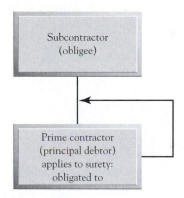

FIGURE 32.2 Payment bond.

possible when they are assured they will be paid. Although they have a right to a mechanics' lien, the procedures for perfecting the lien and satisfying the unpaid obligation out of foreclosure proceeds are cumbersome and often ineffective. Payment bonds are preferable to mechanics' liens.

Generally, subcontractors and suppliers cannot impose liens on public work. In some states, they can file a stop notice, which informs the owner that a subcontractor has not been paid and requires the owner to hold up payments to the prime contractor. However, the stop notice applies only to the unpaid balance still held by the owner when the notice was filed. A payment bond for the full contract price provides subcontractors with greater fiscal protection.

Competent subcontractors and willing suppliers are essential to the construction industry. These important components should have a mechanism that lets them collect for their work. Payment bonds do this. Without a reliable payment mechanism, a substantial number of

[15]Whether an unpaid subcontractor or supplier can recover on a payment bond if it has endorsed a joint check but did not receive payment is discussed in Section 22.02J.

subcontractors might go out of business. Eliminating competent subcontractors can have the unfortunate effect of reducing competition and the quality of construction work.

SECTION 32.08 Subcontractor Bonds

Sometimes the prime contractor requires that subcontractors also obtain payment and performance bonds. If a subcontractor does not perform as obligated, the prime contractor (or its surety looking toward reimbursement) wants to have a financially responsible person to stand behind the subcontractor. This is the justification for a subcontractor performance bond. See Figure 32.3. The justification for the subcontractor payment bond is similar to the justifications given for prime contractor payment bonds. If a subcontractor does not pay its sub-subcontractors or suppliers, the prime contractor (or its surety) is likely to be responsible, as it usually obligates itself to erect the project free and clear of liens. In many cases, unpaid sub-subcontractors and suppliers of subcontractors have lien rights against projects. In a large construction project, a substantial number of bonding companies may have written bonds on the various contractors in the project. This web makes the litigation in such cases complicated. Often the principal participants in a litigation are bonding companies, each seeking to shift the responsibility to the other.

SECTION 32.09 Other Bonds

Bonds also serve other purposes in the construction process. For example, in some states owners can post a bond that can preclude a lien from being filed or dissolve a lien that has been filed. The party filing the mechanics' lien must pursue its rights against the bond and the owner's title (or ownership rights) for the property to remain free and clear. Sometimes warranty bonds are used to back up the owner's claim under a warranty given by the contractor.

SECTION 32.10 Some Legal Problems

A. Who Can Sue on the Bond?

The most troublesome legal issue has been the seemingly simple question of whether unpaid subcontractors and suppliers can sue on a surety bond. Unpaid

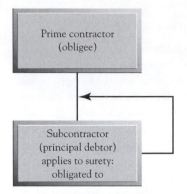

FIGURE 32.3 Subcontractor bond.

subcontractors and suppliers are not obligees under the bond. That is, the bond is not written to them, and the surety does not specifically oblige itself to them. Owners on prime contractor bonds or prime contractors on subcontractor bonds have no difficulty instituting legal action, because they are obligees and the bonds are written to them.

Early legal problems were complicated by using a single bond called the "faithful performance bond." Because the legal standard applied to determine the right of someone other than the obligee to sue on the bond was whether the owner as obligee intended to benefit unpaid subcontractors and suppliers, some courts denied unpaid subcontractors and suppliers the right to sue on the bond. These courts reasoned that the owner must have intended to benefit itself. The bonds covered aspects of the prime contractor's performance other than nonpayment of subcontractors and suppliers. Also, the interests of owner and unpaid subcontractors and suppliers could conflict if each had claims against the prime contractor and if the amount of the bond could not satisfy all claims. As a result, the practice changed to encompass two bonds. The payment bond was to cover default consisting of nonpayment of subcontractors and suppliers, whereas the performance bond covered all other aspects of nonperformance by the prime contractor. Yet even where two bonds are issued, some courts still deny unpaid subcontractors and suppliers the right to sue on the bond.[16]

---

[16]See cases at infra note 17.

This already complicated area was muddled further by courts that differentiated between bonds on public works and those for private projects. These courts permitted subcontractors and suppliers to sue on bonds executed for public projects. They noted that liens could not be asserted on public projects and the intention of the public agency requiring the bonds must have been to benefit the unpaid subcontractors and suppliers. But on private projects that were lienable, an unpaid subcontractor or supplier could not sue on the bond, because the intention must have been to benefit the owner. Courts deciding this question often focus solely on a private owner's desire to avoid liens being filed against its property. This ignores the other functions of surety bonds, such as those mentioned in Section 32.07.

Although the cases are not unanimous (bond language will vary), a strong modern tendency allows subcontractors and suppliers to sue directly on payment bonds if the bond states that the surety will pay unpaid subcontractors or suppliers.[17] This outcome is reflected in AIA bonds (in Appendix D), which clearly give unpaid subcontractors and suppliers a right to sue the surety. Courts interpreting unclear language in a bond should recognize that the owner wants to be able to tell an unpaid subcontractor or supplier that it will be paid by the bonding company and that if the bonding company wrongfully refuses to pay, the subcontractor or supplier will be able to institute legal action itself. A bond purchases this right. The party paying for the bond intends that unpaid subcontractors and suppliers have this right, and this intention should be controlling.

Where bonds are required for public work or where bonds are filed on private work under statutes that allow such filing as a substitute for lien rights, the language of the statute frequently determines who can sue on the bond.[18] Typically the claimant has furnished labor or materials that have gone into the project. Yet cases have permitted a contractor's lender,[19] a pension fund,[20] an unpaid architect,[21] a subcontractor's assignee,[22] and state and federal governments[23] to institute action on a payment bond.

Coverage under payment bonds sometimes turns on where the claimant is located in the chain of contracts ultimately leading to the prime contractor. The farther down a claimant is on the chain, the less likely it will be covered by the bond. A further complication is whether a remote claimant's contract is with a subcontractor or a supplier (also called a materialman). Because subcontractors provide on-site services, whereas suppliers do not, those who contract with a subcontractor are more likely to come within the scope of a payment bond than those who contract with a supplier.

The Miller Act—the bonding statute for federal construction projects over a given size—reflects this dichotomy between subcontractor and supplier. The furthest extent of payment bond coverage under the Miller Act is to one who has "a direct contractual relationship with a subcontractor."[24] Under this language, a subcontractor to a subcontractor may be a claimant, but a subcontractor to a supplier may not be a claimant. The Miller Act does not define the term "subcontractor," but it has been defined by the Supreme Court as one who assumes from the prime contractor a specific part of the contract requirements.[25] The courts split on whether a party that supplies a significant portion of the project's materials but that does not

---

[17]*Socony-Vacuum Oil Co. v. Continental Cas. Co.*, 219 F.2d 645 (2d Cir.1955) (a leading case); *Jacobs Assoc. v. Argonaut Ins. Co.*, 282 Or. 551, 580 P.2d 529 (1978) (reversing earlier opinion denying right of direct action), noted in 58 Or.L.Rev. 252 (1979). But in *Wyoming Machinery Co. v. United States Fid. & Guar. Co.*, 614 P.2d 716 (Wyo.1980), subcontractors were denied third-party beneficiary status on a prime contractor's payment bond. Material suppliers were denied bond coverage in *State of Florida v. Wesley Constr. Co.*, 316 F.Supp. 490 (S.D.Fla.1970), aff'd, 453 F.2d 1366 (5th Cir.1972) and *Day & Night Mfg. Co. v. Fidelity & Cas. Co. of New York*, 85 Nev. 227, 452 P.2d 906 (1969) (right to sue only if language clear).

[18]*Houdaille Indus., Inc. v. United Bonding Ins. Co.*, 453 F.2d 1048 (5th Cir.1972).

[19]*First Nat'l Bank of South Carolina v. United States Fid. & Guar. Co.*, 373 F.Supp. 235 (D.S.C.1974).

[20]*Trustees, Fla. West Coast Trowel Trades Pension Fund v. Quality Concrete Co.*, 385 So.2d 1163 (Fla.Dist.Ct.App.1980).

[21]*Herbert S. Newman and Partners, P.C. v. CFC Constr. Ltd. Partnership*, 236 Conn. 750, 674 A.2d 1313 (1996).

[22]*Quantum Corporate Funding, Ltd. v. Westway Industries, Inc.*, 4 N.Y.3d 211, 825 N.E.2d 117, 791 N.Y.S.2d 876 (2005).

[23]*Island Ins. Co., Ltd. v. Hawaiian Foliage & Landscape, Inc.*, 288 F.3d 1161 (9th Cir.2002). The bond was acquired by a subcontractor so the prime contractor, not the governments, was the obligee (the party for whose protection the bond was acquired). Nonetheless, the court ruled that the governments were third-party beneficiaries of the bond and could recover from the surety the employment taxes the subcontractor had failed to pay.

[24]40 U.S.C.A. § 3133(b)(2).

[25]*Clifford F. MacEvoy Co. v. United States ex rel. Calvin Tomkins Co.*, 322 U.S. 102, 109 (1944).

actually install these supplies qualifies as a subcontractor.[26] Claimants who contracted with a supplier are too remote from the prime contractor to come within the protection of the Miller Act.

As construction has become more complicated, with linked sets of contracts connecting a variety of parties, problems as to who can sue on surety bonds take new forms. For example, an Illinois case involved an attempt by the ultimate user of a project (a school district) to bring a claim on the performance bond given by the prime contractor. The construction contract clearly indicated that the ultimate user was to be an intended beneficiary of the construction contract, and that contract was incorporated into the performance bond. But the school district was denied any direct rights on the bond because the bond itself made clear that only the obligee—the public entity acting as owner—could maintain an action on the bond.[27] Similarly, a number of cases have involved unsuccessful attempts by separate contractors (multiple primes) to bring claims on bonds where the principal was a separate contractor and the bond was issued to the owner as obligee.[28]

Planning in the context of a traditional construction organization should include ensuring that unpaid subcontractors and suppliers can bring action on these bonds. This can be accomplished relatively easily by including language similar to that in AIA bonds. Where a nontraditional system is used, such as separate contracts (multiple primes) or the party most interested in performance is not a nominal owner but an ultimate user, planning must consider who can bring legal action on any bonds that are supplied by contractors and that choice expressed in the contract and the bond.

B. Validity of Bond

Suppose the contractor misrepresents its resources when it applies for the bond. This would be misrepresentation by the applicant, not by the obligee-owner. In such a case, the bond is generally valid, and the surety bond company must pursue any remedies it has against the contractor-applicant. Fraud on the part of the obligee-owner, however, gives the surety the power to avoid having to perform under the bond. If the obligee-owner participates in or knows about the fraudulent statements made by the applicant contractor or misleads the bonding company in some other way, the bond would not have been validly obtained and cannot be legally enforced.

For various reasons, the principal contract—that is, between owner and contractor—may not be enforceable. For example, suppose the building contract involved constructing a hideout for leaders of organized crime. To permit the owner to sue on the surety bond in such a case would further an illegal activity and would involve the court in that activity. For that reason, the bond would not be enforced.

Legal infirmities of contracts of a less serious nature may exist, however. Suppose the contractor is not licensed. In *Cohen v. Mayflower Corp.*,[29] the court held that the owner can recover on a bond written on an unlicensed contractor. The court held that the licensing law was designed to protect owners and the owner could have sued the contractor even though the unlicensed contractor could not have sued the owner. The court concluded that the surety would be held on the bond because the principal debtor, the contractor, could have been held liable.

The stronger the public policy making the contract illegal, the less likely the bond will be enforced. If the policy is designed to protect the owner, however, as in the *Cohen* case, it is more likely that the bond will be enforced.

C. Surety Defenses: On Demand Bonds

Suppose the owner does not make progress payments despite the issuance of progress payment certificates. Suppose the design professional unjustifiably interferes with the work of the contractor or does not approve shop drawings in sufficient time to permit proper performance. The contractor's performance may be rendered impossible because of a court order, the death of the contractor, or some natural catastrophe. The performance may be rendered impracticable due to the discovery of unforeseen subsurface conditions. Most defenses the contractor has against the owner would be available to the surety who has issued a performance bond.

---

[26]*United States ex rel. E & H Steel Corp. v. C. Pyramid Enterprises, Inc.*, 509 F.3d 184 (3d Cir.2007) (neither installation nor special fabrication of the product is a requirement for subcontractor status).

[27]*Board of Educ., School Dist. No. 15 DuPage County v. Fred L. Ockerlund, Jr. & Assoc., Inc.*, 165 Ill.App.3d 439, 519 N.E.2d 95 (1988).

[28]*M. G. M. Constr. Corp. v. New Jersey Educ. Facilities Auth.*, 220 N.J.Super. 483, 532 A.2d 764 (Law Div.1987); *Moore Constr. Co., Inc. v. Clarksville Dep't. of Elec.*, 707 S.W.2d 1 (Tenn.Ct.App.1985), aff'd Mar. 24, 1986.

[29]196 Va. 1153, 86 S.E.2d 860 (1955).

Likewise, defenses that the prime contractor could assert against a subcontractor or supplier claimant can generally be asserted by the surety who has issued a payment bond.[30] The surety's function is to provide financial responsibility for the acts of the prime contractor. Generally, the surety's obligation is coextensive with that of the principal debtor—that is, the prime contractor—but only to the extent of the bond limit.[31]

That the construction contract and the bond are linked together is illustrated by AIA Doc. 312, Paragraph 6. It states that the surety's obligation to the owner is not greater than those of the contractor under the construction contract and that the obligations of the owner to the surety are not greater than the owner's obligation under the construction contract.

The dependent nature of the bond in American construction practices may be contrasted to the security instruments that the contractor usually is required to furnish the employer (owner) in construction projects in other countries and in international engineering projects. In those transactions, the contractor usually is expected to furnish an unconditional bank guarantee, a standby letter of credit, or an "on demand" bond. These are unconditional promises by the issuing bank or surety that are not tied to default by the contractor. These are powerful securities that can be used in an abusive way. They can be unfair to the contractor who will be expected to indemnify the bank or surety. As a result, they have spawned considerable litigation in Great Britain and in international transactions.[32]

Suppose the claimant is an unpaid subcontractor or supplier under a payment bond. In *Houdaille Industries, Inc. v. United Bonding Insurance Co.*,[33] the surety was denied a defense based on an asserted claim that the owner had breached its contract with the contractor who was the principal on the bond. The court noted that although normally the surety can assert such defenses against the owner, this defense could not be asserted against a claimant "so long as it is not participated in or authorized by the materialman."[34]

Suppose the owner and the contractor modify the construction contract or the owner directs changes in the work. The surety's commitment can be limited by a fixed limit on the bond. Any change in the basic agreement, however, traditionally released the surety. The advent of the paid surety has made inroads on this rule. Bonds frequently provide that the surety "waives notice of any alteration or extension of time made by the owner." Without such a provision, the surety could be released because the principal obligation has been changed.[35]

A financially troubled contractor presents difficult problems for both the owner and surety. To keep the contractor afloat and the project moving forward, the owner may advance monies to the contractor for work the contractor has not yet performed. However, any diminishment in the contract balance reduces the funds from which the surety may obtain payment if, for example, it is later called upon to complete the contractor's work. The surety may argue that the owner's prepayment increased the surety's risk and so discharged the surety from its obligations under the bond. However, so long as the payments are used to build the project, thereby reducing the performance bond surety's ultimate liability, the fact that the payments were made ahead of schedule should not discharge the surety. The owner must have reasonable latitude to deal with a financially distressed contractor without fear of losing its protections under the bond.[36]

Suppose the construction process concludes with claims made by the participants. Two issues that may arise are the method of resolving disputes and whether claims have been barred by the passage of time. The first issue was discussed in Section 30.15, which deals with arbitration.

The second issue—that of the passage of time barring a claim—has caused difficulty. For example, in

---

[30]See Section 28.06 for a discussion of pay-when-paid clauses and whether they insure to the benefit of a payment bond surety.

[31]Restatement (Third) of Torts: Suretyship and Guaranty § 34. That the bond limits will not always limit the surety's obligation is discussed in Section 32.10E.

[32]See 2 I. N. DUNCAN WALLACE, HUDSON'S BUILDING & ENGINEERING CONTRACTS, §§ 17-054–17-079 (11th ed. 1995); *Bailey, Unconditional Bank Guarantees* [2003] Int'l Constr.L.Rev., Part 2, p. 240.

[33]Supra note 18.

[34]453 F.2d at 1053 n. 4. But see *Chicago Bridge & Iron Co. v. Reliance Ins. Co.*, 46 Ill.2d 522, 264 N.E.2d 134 (1970) (surety was given a defense when the claimant subcontractor submitted false lien waivers that allowed the prime to dissipate progress payments).

[35]*In re Liquidation of Union*, supra note 9. But see *Winston Corp. v. Continental Cas. Co.*, supra note 9, discussed in Section 32.03.

[36]*United States Fid. & Guar. Co. v. Braspetro Oil Services, Co.*, 369 F.3d 34, 61–66 (2d Cir.2004) (efforts by owner to keep project afloat did not discharge surety); *John T. Callahan & Sons, Inc. v. Dykeman Elec. Co., Inc.*, 266 F.Supp.2d 208 (D.Mass.2003) (surety is not discharged by prime's delay in declaring a struggling subcontractor in default).

State v. Bi-States Construction Co.,[37] the state of Iowa brought a claim against a contractor who had abandoned a project and its surety. The contractor, a Nebraska corporation, had been dissolved under the laws of Nebraska on March 26, 1969. Iowa law required that any claim against a dissolved corporation be made within two years. The owner's claim was barred because it did not institute the claim until almost five years after the contractor had been dissolved. The court held that the claim against the bonding company was also barred. The court recognized that divided authority on the question exists, but Iowa "adheres to the rule [that] a surety may assert as a defense the statute of limitation if available to the principal."[38]

Once disputes develop, both obligee and principal debtor should notify the surety and keep the surety informed as to the posture and process of the dispute. In this regard, Section 14.2.2 of AIA Document A201-2007 requires the surety to be notified in writing if the owner has terminated the contractor's performance.

D. Surety Responsibility

Some background must be kept in mind in understanding the responsibility of the surety. First, the surety's responsibility cannot exceed that owed by the principal debtor (the prime contractor on a prime contract performance bond and the subcontractor on a subcontractor performance bond) to the obligee (the owner on a prime bond and the prime on a subcontractor bond). Second, subject to some exceptions to be discussed in Section 32.10E, the surety's obligation is limited to the penal sum on the bond (the bond limit). Third, much depends on the language of the bond.

Fourth, with the prime contractor performance bond, the surety has no obligation to perform until the owner declares the principal prime contractor in default and communicates that declaration to the surety.[39] That communication is the dividing line from the surety's perspective between merely a troubled project and one under

which its bond obligations are triggered. The surety's options, at this point, are spelled out in the bond.

Suppose the prime contractor's default is that it abandoned the project without justification. Here, the principal item of damages will be the excess cost of reprocurement. AIA Document A312, Paragraph 4, allows the surety to arrange for the original contractor to return and complete the work, usually by pumping additional resources into the original contractor. Alternatively, the surety can complete the project itself (here the surety must be aware of contractor licensing laws) or get a substitute to do so. Finally, Paragraph 4.4 appears to let the surety pay whatever damages the owner would be entitled to recover from the defaulting contractor.[40] This will be the cost of correction or the diminished value of the project. As seen in Section 27.03D, the latter formulas will also be used if the prime contractor has completed the project but has not performed in accordance with its contractual commitment.

More complicated problems result when there has been delayed completion by the contractor or after its default by a successor contractor. To determine whether delay comes within the bond commitment, the language of the bond is crucial. One court, pointing to the bond language and emphasizing that the most important responsibility of the prime contractor is to build the project, did not allow the obligee to recover delay damages from the surety.[41] Its decision may have been traceable to a belief that delays are inherent in all construction projects. But other courts have allowed recovery against the surety for the contractor's delayed performance.[42]

---

[37]269 N.W.2d 455 (Iowa 1978).

[38]269 N.W.2d at 457. See also *Housing Auth. of City of Huntsville v. Hartford Accident & Indem. Co.*, 954 So.2d 577 (Ala.2006) (performance bond surety may assert statute of limitations defense available to its principal).

[39]See *Bank of Brewton, Inc. v. International Fidelity Ins. Co.*, 827 So.2d 747 (Ala.2002) (surety not liable where the owner did not unequivocally declare the principal in default).

[40]One option *not* available to the surety is to do nothing, force the owner to complete the performance itself, then offer to reimburse the owner for its reasonable costs. *National Fire Ins. Co. of Hartford v. Fortune Constr. Co.*, 320 F.3d 1260 (11th Cir.), cert. denied, 540 U.S. 873 (2003).

[41]*American Home Assur. v. Larkin General Hosp.*, 593 So.2d 195 (Fla.1992). Similarly, see *Downington Area School Dist. v. International Fidelity Ins. Co.*, 769 A.2d 560 (Pa.Cmwlth.), appeal denied, 567 Pa. 731, 786 A.2d 991 (2001).

[42]*MAI Steel Service, Inc. v. Blake Constr. Co.*, 981 F.2d 414 (9th Cir.1992); *National Fire Ins. Co. of Hartford v. Fortune Constr. Co.*, supra note 40 (applying Florida law and distinguishing *American Home Assur. v. Larkin General Hosp.*, supra note 41, on the ground that the subcontracts in *National Fire* contained liquidated damages clauses, which were incorporated into the bond); *Cates Constr., Inc. v. Talbot Partners*, 21 Cal.4th 28, 980 P.2d 407, 86 Cal.Rptr.2d 855 (1999); *New Amsterdam Cas. Co. v. Bettes*, 407 S.W.2d 307 (Tex.Ct.App.1966). See Douglas, *Delay Claims Against the Surety*, 17 Constr. Lawyer, No. 3, July 1997, p. 4.

The owner should be able to recover its delay damages from the surety, based on a liquidated damages clause or actual damages. Time, though not as important as the work itself, still merits bond protection. Such protection would be reasonably expected by the owner-obligee. This is reflected in AIA Doc. 312, Paragraph 6.3. It requires the surety to pay any liquidated or actual damages caused by the contractor's delay. The recovery against the surety should be the same as the amount that would be recovered from the prime contractor.

Does the scope of the surety's obligation extend to the contractor's liability for defective work discovered during performance where there is no termination or discovered after completion where the contractor has completed performance? In the latter situation, there is no possibility of termination and default.

Of course, much would depend on the language of the bond. The difficulty is the consistent use of the term "default" in bonds as well as case decisions that speak almost universally about default triggering the surety's obligation. Generally, default is thought to mean a contractor who quits without justification or an owner's justifiable decision to terminate the contractor and prevent it from completion. Yet it can go farther.

The surety's obligation can include failure to complete on time by a contractor or late ultimate completion caused by the contractor's default and the work being finished late by a successor. Default does not fit delay cases where the contractor simply has not finished on time. Yet there is authority for bringing this delay into the ambit of the surety's obligation.

What does default mean? Is it simply a synonym for any breach or one serious enough to remove the contractor from the project? If a contractor stops work without justification, this is a serious breach that discharges the owner's obligation to allow it to work any more. Similarly, if the owner orders the contractor to leave, this decision requires that the contractor have committed a serious uncurable breach as seen in Sections 33.04 and 33.05.

It can be argued that the surety's obligation does not include defective work that would not be serious enough to justify termination. This would be based on the failure to find a serious enough breach to be a default if that triggers the surety's obligation. Default works well if the issue is the failure of the contractor to complete the work.

But does that mean the surety's obligation does not extend to defective work where there has been no termination?

If the defective work is discovered during performance and the owner chose not to terminate, or it is discovered after completion and no termination is possible, if default is the trigger, then defective work does not fall within the surety's obligation under the bond.

There are arguments for extending defective work to the surety's obligation. If the purpose of the bond is to provide a financially sound party to stand behind the contractor, why should this be limited to failure to finish or even failure to finish on time?

Of course, the bond language would be examined and could be crucial. Still, if the language is unclear and focus is upon the reasonable expectation of the owner, the surety could lose.[43] Yet it must be admitted that the slavish use of "default" and the surety usually being called in when there is failure to finish would be substantial hurdles for the owner to overcome.

Generally, the surety will not be responsible for punitive damages that might be awarded against the prime contractor.[44] But the surety might, as noted in Section 32.10I, be liable for punitive damages for its bad-faith refusal to settle the claim made on the bond.

Special problems can occur when one separate prime contractor has been damaged by the acts of another separate prime contractor. As noted earlier, it is unlikely that the surety on a prime contract bond would be liable for a claim made by another separate contractor unless that contractor could sustain a claim that it was an intended beneficiary.[45] But if the claim by the separate contractor were against the owner, based on the owner's responsibility for coordinating the work properly, and the owner sought to transfer this loss to the separate prime contractor, who had caused the additional expense to the claimant, the surety for the latter should be liable.

As to payment bonds, much depends on any statute under which the payment bond was compelled or on the language of the bond itself. Clearly, an unpaid subcontractor or supplier can recover on the payment bond for the reasonable value of the work it has performed or materials it has supplied. Unless profit can be encompassed in the preceding formula, however, some courts have difficulty

---

[43]*DCC Constructors, Inc. v. Randall Mechanical, Inc.*, 791 So.2d 575 (Fla.Dist.Ct.App.2001) (obligee need not terminate to declare a default in a warranty breach case likely discovered after completion).

[44]Annot. 2 A.L.R.4th 1254 (1980).

[45]See supra note 28.

awarding profits against the surety—probably a reflection of the view that a payment bond is a substitute for lien rights.[46] Others have allowed a claim against the surety for increased costs of performance incurred by the subcontractor.[47]

A surety's obligations under its bond are triggered when the obligee declares the principal in default.[48] In the case of a performance bond, what is the effect on the surety's liability if it delays paying the owner? The surety may be liable for interest on the unpaid money, even if this results in an amount in excess of the bond's penal sum.[49] If the surety's failure to pay promptly delays completion of the project, resulting in higher performance costs, the surety will be liable for these increased costs, again without regard to the bond limit.[50]

A prime contractor who is having financial difficulties may ask its surety to advance funds to complete the project. (The effect of an owner advancing funds to a financially troubled prime is discussed in Section 32.10C.) Usually, at the time the bond is issued the surety obtains an indemnification agreement from the prime contractor in which the surety carefully disclaims any responsibility to advance funds. Claims of prime contractors have been based on an assertion that sureties have a good-faith obligation to investigate the request and provide the financing in order to minimize the prime's liability under the indemnification agreement, despite the latter's contract usually providing that advance of funds by the surety will be within the sole discretion of the surety. Although two commentators contend there should be no obligation to investigate in good faith, they also note that some court decisions have tended to go in that direction.[51]

E. Bond Limits and Surety Bad Faith

As noted, surety bonds generally contain what is called a *penal sum*, which limits the obligation of the surety. Generally, this bond limit is effective.[52] However, as noted in Section 32.10I, the law has begun to inquire into the settlement practices of sureties, just as it has examined the settlement practices of insurance companies. As a result, some exceptions have emerged to the general rule that the penal sum limits the surety's obligation. An important trial court opinion held that the surety's liability would not be limited by the bond limit, mainly because of improper settlement tactics of the surety.[53] Similarly, in another case where the surety was not cooperative in settlement discussions concerning an award of prejudgment interest, the bond limit did not place a ceiling on the liability of the surety.[54]

Bad faith settlement practices by the surety includes its duty to investigate once the obligee declares the principal debtor (the prime contractor) in default. In one case the owner properly declared the contractor in default. In bad faith the performance bond surety threatened the owner with a lengthy investigation before performing its bond obligation. This threat forced the owner to complete performance with its own funds. The court held the owner was entitled to prejudgment interest even if this exceeded the penal sum of the bond.[55]

F. Bankruptcy of Contractor

If, as discussed in Section 33.04C, during the course of the contractor's performance, the contractor is adjudicated a bankrupt, the trustee in bankruptcy (the person who takes

[46]*MAI Steel Service, Inc. v. Blake Constr. Co.*, supra note 42; *Lite-Air Products, Inc. v. Fidelity & Deposit of Maryland*, 437 F.Supp. 801 (E.D.Pa. 1977).

[47]*MAI Steel Service, Inc. v. Blake Constr. Co.*, supra note 42; *Tremack Co. v. Homestead Paving Co.*, 582 So.2d 26 (Fla.Dist.Ct.App.), review denied, 592 So.2d 680 (Fla.1991) (acceleration damages).

[48]*Insurance Co. of North America v. United States*, 951 F.2d 1244 (Fed. Cir. 1991) (surety's obligation to pay arises when principal defaults on contract).

[49]Ibid.

[50]*Republic Ins. Co. v. Prince George's County*, 92 Md.App. 528, 608 A.2d 1301(1992), cert. dismissed as improvidently granted, 329 Md. 349, 619 A.2d 553 (1993).

[51]Toomey & Fisher, *Is a Surety Obligated to Investigate Financing a Contractor Who Requests Financial Assistance?* 12 Constr. Lawyer, No. 4, Nov. 1992, p. 11.

[52]*Dawson Corp. v. National Union Fire Ins. Co. of Pittsburgh, Pa.*, 285 N.J.Super. 137, 666 A.2d 604, (App.Div.1995), cert. denied, 143 N.J. 517, 673 A.2d 276 (1996) (surety liability limited to original contract price and did not include unauthorized correction of work).

[53]*Continental Realty Corp. v. Andrew J. Crevolin Co.*, 380 F.Supp. 246 (S.D.W.Va.1974). As the case was settled, there was no appeal. The opinion is reviewed and criticized in Wisner, *Liability in Excess of the Contract Bond Penalty*, 43 Ins.Couns.J. 105 (1976).

[54]*Insurance Co. of North America v. United States*, supra note 48.

[55]*United States Fid. & Guar. Co. v. Braspetro Oil Services, Co.*, supra note 36, 369 F.3d at 78–81, enforcing McKinney's N.Y.Gen.Obl.Law § 7–301, which states that the obligee's recovery against the surety "shall not exceed the amount specified in the undertaking except that interest in addition to this amount shall be awarded from the time of default by the surety."

over the affairs of the bankrupt contractor) can determine whether to continue the contract. Usually it does not. If the contract is not continued, the bankrupt contractor has no further obligation to perform under the contract. The owner has a claim against the bankrupt contractor but is not likely to recover much. Ending the contractor's obligation, however, should not release the surety. This is the risk contemplated when the surety bond is purchased.[56]

G. Asserting Claims: Time Requirements

The surety's obligation usually requires that claimants give certain notices.[57] Likewise, statutes requiring public work to be bonded, such as the federal Miller Act and state "Little Miller Acts" (noted in Section 32.04), often specify that notices must be given within certain periods of time to designated persons.

Frequently, bonds require that legal action on the bond be brought within a designated time, usually shorter than the period specified by law. Such shortened periods to begin legal action are enforceable if reasonable.[58] In *Gateway Communications, Inc. v. John R. Hess, Inc.*, the court enforced the bond's two year limitations period and also refused to apply the "discovery rule" so as to toll the running of the period.[59] (The "discovery rule" prevents the limitations period from beginning to run until the owner knew or should have known that the building contained defects.)

Bonds often specify the court in which legal action must be brought, usually to courts—state or federal—in the state in which the project is located.[60]

Suppose a claimant does not comply with all the many requirements specified by bonds or statutes. A surety that denies responsibility because of some failure to comply appears to be using a technicality to avoid an obligation it was paid to perform. Courts generally interpret these requirements liberally in favor of claimants[61] and often require the surety to show prejudice caused by noncompliance before being given a defense.[62]

Courts have held that under certain circumstances the surety will be estopped to assert these provisions or be found to have waived them.[63]

The bond may also impose time limits on a surety's response to an obligee's claim. AIA Doc. 312, Paragraph 6.1, requires the surety to respond to a claim "within 45 days after receipt of the claim, stating the amounts that are undisputed and the basis for challenging any amounts that are disputed." Where a surety failed to make any response within the forty-five-day period, a court ruled the surety was precluded from contesting the claim.[64]

H. Reimbursement of Surety

It is sometimes said that sureties do not expect to take a loss. Sureties do not see themselves as insurers. As a result, they assert defenses that the principal debtor could have asserted. They seek bond language protection. Finally, they seek to recoup payments made under the bond, first from the principal and then, if unsuccessful, from third parties, including the obligee.

As a condition to receiving bonding, the prime contractor and often also the individuals who control it agree to indemnify the surety for any payments made in good faith under the bond. The agreement to indemnify extends also to the surety's settlement of any claims brought against the bond, whether or not the prime contractor turns out to have been liable to the bond

---

[56]Restatement (Third) of Torts: Suretyship and Guaranty § 34(1)(a).

[57]*Seaboard Sur. Co. v. Town of Greenfield*, 370 F.3d 215 (1st Cir.2004) (town's failure to give the performance bond surety a bond-mandated fifteen-day cure notice, before undertaking performance itself, discharged the surety).

[58]In *Rumsey Elec. Co. v. University of Delaware*, 358 A.2d 712 (Del. 1976), a one-year bond provision was held valid where the statutory period was three years. Sometimes statutes of limitation are excessively long. But *City of Weippe v. Yarno*, 94 Idaho 257, 486 P.2d 268 (1971), held that the period for bringing the action cannot be shortened by the parties, as it would be unconscionable. The court stated that stipulating a longer period would be enforced if reasonable.

[59]208 W.Va. 505, 541 S.E.2d 595 (2000).

[60]AIA Doc. A312, ¶¶ 9, 11, found in Appendix D.

---

[61]*United States v. Merle A. Patnode Co.*, 457 F.2d 116 (7th Cir.1972); *American Bridge Div. United States Steel Corp. v. Brinkley*, 255 N.C. 162, 120 S.E.2d 529 (1961).

[62]*Winston Corp. v. Continental Cas. Co.*, supra note 9.

[63]*Contee Sand & Gravel Co. v. Reliance Ins. Co.*, 209 Va. 672, 166 S.E.2d 290 (1969) (surety estopped to plead a one-year period of limitations in the bond because subcontractor was misled by surety statement that its prime contractor did not have a surety bond).

[64]*National Union Fire Ins. Co. of Pittsburgh v. David A. Bramble, Inc.*, 388 Md. 195, 879 A.2d 101 (2005). The intermediate court's decision in this case is discussed in Dranoff & Barthet, *Payment Bond Payout: The Forty-Five-Day Countdown*, 25 Constr. Lawyer, No. 1, Winter 2005, p. 17.

claimants paid by the surety.[65] While courts will enforce such clauses, a surety who acts in bad faith will either lose its right of indemnification or have that right reduced to the extent of losses caused by improper settlement practices.[66]

The surety's indemnity right will be of little use if, as often happens, a defaulting principal is not in a position to reimburse the surety. The surety may turn next to the obligee as a source of recovery. When the prime contractor defaults and the surety takes over, the surety usually notifies the owner that it should be paid all payments that would have gone to the prime contractor.[67] In addition, the surety usually demands at the end of the job any retainage that the owner has withheld to secure the owner against claims.[68]

In seeking the retainage, the surety usually competes with other creditors of the prime contractor, the taxing authorities, and the trustee in bankruptcy if the prime has been declared bankrupt. As a rule, many more claims than can be satisfied exist, and the result is a complicated lawsuit. Typically, the owner pays the retainage into court and notifies all claimants, and the court determines how the fund is to be distributed.

The surety's priority over competing claims of the government (for example, a claim that the principal owes taxes) and the owner depends upon whether the surety is seeking reimbursement of costs under a performance or payment bond. Under a performance bond, the surety may be required to complete the project, either performing the work itself, hiring a replacement contractor, or funding the original contractor. Such a surety acquires subrogation rights of—or, to use a colloquialism, "steps into the shoes of"—both the prime contractor (whose obligations the surety took over) and the owner (for whom the surety completed the project). This surety has priority to the retainage over all competing claims except those of the government.[69]

By contrast, a payment bond surety steps into the shoes of the subcontractors and suppliers to the extent it has paid them. That surety's claim to the retainage is superior to the claims of both the owner and prime contractor but is inferior to that of the government.[70]

In addition, the surety who takes over after default usually succeeds to any claims the prime contractor may have against the owner or third parties. It is common for the prime contractor to ascribe its difficulties to the owner, the design professional, subcontractors, or other third parties.

Sureties make strong efforts to be reimbursed and often succeed in salvaging a substantial amount of their loss when they are called on to respond for their principals' default.

I. Regulation: Bad-Faith Claims

Surety companies are regulated by the states in which they operate. In addition, sureties who wish to write bonds for federal projects must qualify under regulations of the U.S. Treasury Department. The financial capability of a surety limits the size of the projects a surety can bond. Bond dollar limits place a ceiling on exposure. In larger projects, there may be co-sureties or the surety may be required to reinsure a portion with another surety. Surety rates are usually regulated and are based on a specified percentage of the limit of the surety bond.

Section 32.10E noted that the surety's obligation is usually limited to the amount of the bond. It also cited cases in which the surety's obligation extended beyond the bond limit, largely because of the settlement tactics of the surety.

---

[65]*Gulf Ins. Co. v. AMSCO, Inc.*, 153 N.H. 28, 889 A.2d 1040 (2005).

[66]*PSE Consulting, Inc. v. Frank Mercede and Sons, Inc.*, 267 Conn. 279, 838 A.2d 135 (2004); *Republic Ins. Co. v. Prince George's County*, supra note 50.

[67]In *Gerstner Elec., Inc. v. American Ins. Co.*, 520 F.2d 790 (8th Cir. 1975), the court rejected a contention of the contractor that the request for payments by the surety to the owner was a wrongful interference in the contract between contractor and owner.

[68]Retainage is discussed in Section 22.03.

[69]While a completing surety's subrogation claim may be inferior to the federal government's tax lien, New York's high court found the surety's claim superior to the State's claim that the public works principal underpaid prevailing wages on a different project. See *RLI Ins. Co. v. New York State Dept. of Labor*, 97 N.Y.2d 256, 766 N.E.2d 934, 740 N.Y.S.2d 272 (2002).

[70]*National Fire Ins. Co. of Hartford v. Fortune Constr. Co*, supra note 40 (payment bond surety's claim is superior to owner's claim of set-off for delay damages). For a delineation of these subrogation rights, see the Restatement (Third) of Torts: Suretyship and Guaranty §§ 27–29.

The law has taken steps to deter insurers from conduct unfair to their policyholders. Most states allow claims based on tortious bad-faith conduct by the insurer to policyholders and in some states to those who have claims against policyholders. These techniques have enabled claimants to obtain awards that exceed the policy limits, usually through awards for emotional distress or punitive damages.

Building on increased judicial regulation of insurers, some states have allowed similar claims against sureties.[71] Where bad-faith claims are allowed and where the sureties' conduct falls below that required by law, the sureties' obligations can exceed the stated limit of the bond. This is accomplished by finding that the surety has committed a tort that in some instances may justify the award of punitive damages as well as normal tort damages.[72] However, some courts, fearful of blurring tort and contract distinctions, have not allowed tort claims of bad faith against sureties.[73]

In addition to statutes regulating claims settlement practices of insurers, courts may also extend a state's Deceptive Trade Practices Act to the surety. This too may result in the surety paying damages in excess of the bond's penal sum.[74]

---

[71]*United States v. Atul Constr. Co.*, 85 F.Supp.2d 414 (D.N.J.2000) (subcontractor may sue prime contractor's payment bond surety for bad faith delay in responding to claim); *Farmer's Union Central Exch., Inc. v. Reliance Ins. Co.*, 626 F.Supp. 583 (D.N.D.1985) (stalling, deceiving, and then refusing to pay based on failure to bring legal action); *Tonkin v. Bob Eldridge Constr. Co.*, 808 S.W.2d 849 (Mo.Ct.App.1991) (surety liable to owner under state statute for attorneys' fees and interest for a bad-faith failure to investigate the claim). For a case holding that the conduct did not constitute bad faith but only constituted litigating tactics, see *United States v. Seaboard Ins. Co.*, 817 F.2d 956 (2d Cir.1987) (surety denied liability, knowing the contractor was in default, failed to properly investigate, and told consultants not to put comments in writing).

[72]*Riva Ridge Apartments v. Roger J. Fisher Co.*, 745 P.2d 1034 (Colo. App.1987), cert. denied Nov. 9, 1987.

[73]*Shannon R. Ginn Constr. Co. v. Reliance Ins. Co.*, 51 F.Supp.2d 1347 (S.D.Fla.1999) (principal is not an insured and does not have a claim against its surety for bad-faith failure to settle); *Cates Constr., Inc. v. Talbot Partners*, supra note 42 (obligee may not sue performance bond surety for breach of the implied covenant of good faith and fair dealing); *Republic Ins. Co. v. Bd. of County Comm'rs of Saint Mary's County*, 68 Md.App. 248, 511 A.2d 1136 (1986) (court fearful bad-faith claims will be "boilerplate" and destroy differentiation between tort and contract); *Insurance Co. of the West v. Gibson Tile Co., Inc.*, 122 Nev. 455, 134 P.3d 698 (2006) (prime contractor may not sue performance bond surety for breach of the covenant of good faith and fair dealing to recover punitive damages); *Masterclean, Inc. v. Star Ins. Co.*, 347 S.C. 405, 556 S.E.2d 371 (2001) (prime contractor may not sue its performance bond surety for bad faith); *Great American Ins. Co. v. North Austin Util.*, 908 S.W.2d 415 (Tex.1995) (surety not an insurer and no duty of good faith and fair dealing to obligee).

[74]*R. W. Granger & Sons, Inc. v. J & S Insulation, Inc.*, 435 Mass. 66, 754 N.E.2d 668 (2001).

CHAPTER THIRTY-THREE

Terminating a Construction Contract: Sometimes Necessary but Always Costly

SECTION 33.01 Termination: A Drastic Step

Termination does not occur frequently in construction contracts. One reason for this is the difficulty of determining whether a legal right to terminate exists, a point discussed principally in Sections 33.03 and 33.04. Often each party can correctly claim the other has breached. It may be difficult to determine whether a party wishing to terminate is sufficiently free from fault and can find a serious deviation on the part of the other. Another reason is the often troublesome question of whether the right to terminate has been lost (to be discussed in Section 33.03E). A third reason, the serious consequences of terminating without proper cause, is treated in this section.

Two cases illustrate the danger of an improper termination. The first, *Paul Hardeman, Inc. v. Arkansas Power & Light Co.*,[1] involved a contract under which Hardeman was to construct electric transmission lines for the Power & Light Company for a contract price of $2.7 million. Hardeman's bid had been much lower than the other bids. This was known only to the owner because this private project did not use a public bid opening. Despite the likelihood that there had been a mistake, the Power & Light Company proceeded to award the contract without discussing the mistake possibilities.

Many difficulties developed, mainly because of Hardeman's inexperience. Several months before the completion date, at a time when somewhat less than one-half of the work had been completed, Hardeman discovered that its bid had been substantially lower than the other bids because of a calculation mistake. It immediately communicated this to the Power & Light Company and accused it of wrongfully awarding the contract while knowing of the likelihood of the mistake. It threatened to leave the project unless some equitable adjustment was made. A conference was held, but no resolution was accomplished. Finally the Power & Light Company and its engineer terminated Hardeman's contract, claiming that there had been unexcused delays as well as defective workmanship and materials. Evidently, the real reason for the termination was that the Power & Light Company sought the best tactical position for the inevitable lawsuit. Before the termination, Hardeman had claimed to have spent $3 million on the work. After termination, a successor contractor completed performance on a cost-plus basis and received $8 million.

Hardeman brought legal action claiming that the Power & Light Company wrongfully accepted its bid and that the termination had been improper. It sought restitution based on its asserted expenditures of $3 million minus payments made of $600,000. The Power & Light Company sought $5.8 million on its part as damages.

Although the court seemed sympathetic to Hardeman's first claim, it felt it could not grant relief, because Hardeman's estimator had been grossly negligent.[2] A convenient alternative solution, however, arose. The termination had not been made in good faith. According to the court, such a drastic step had been taken prematurely at best.

Termination proved useful to Hardeman. The court concluded that improper termination entitled Hardeman to recover the reasonable value of its services, which the court found to be $2 million, less the amount paid of $600,000. The defendant's counterclaim for the excess cost of correction was denied.

---

[1]380 F.Supp. 298 (E.D.Ark.1974).

[2]For a further discussion of relief from mistaken bids, see Section 18.04E.

Perhaps the case is not typical. The court seemed sympathetic to Hardeman's bidding mistake claim. Yet the case demonstrates that a precipitous termination by the owner can be very costly when the contractor has made a losing contract.

The second case, *Indiana & Michigan Electric Co. v. Terre Haute Industries*,[3] was tried before a jury. It lasted eighty-one trial days and generated a 36,000-page transcript, which was bound into 138 volumes. The reported appellate court decision filled thirty-three double-columned pages in the regional reports. The contract was for the installation of pollution control equipment for an energy plant using coal-fired generators. The state pollution control board had been exerting pressure on the utility to install an electrostatic precipitator and fix strict deadlines for its installation.

The contractor submitted the low bid of approximately $7 million. At a point at which approximately 80 percent of the by now contract price of $8 million had been paid, the utility and the contractor disagreed as to whether the contractor was on schedule. Claiming that the contractor was not and that it was the contractor's fault, the utility demanded an acceleration. Concluding that the contract would not be completed on time, the utility terminated the contractor because of its alleged failure to maintain the schedule. Evidence at the trial indicated that at the time of termination the project was 60 percent complete. The utility spent an additional $5 million to complete the work. At the time of termination, the utility took possession of construction equipment and used it for approximately four months, based on a power given to it in the termination clause of the construction contract.

The contractor's claim and the judgment it obtained in the trial court indicate the high degree of risk that termination can entail. The judgment was for $17 million (all figures approximations), broken down as follows:

1. retainage and interest, lost profits on the project, extras, and additional expenses—$2 million
2. loss of future business—$3 million
3. punitive damages—$12 million

In addition, the trial court did not grant the utility's counterclaim, because it concluded the termination by the utility had not been proper.

On appeal, the court reduced the award drastically, concluding that the contractor was not entitled to one small expense item, future profits, and punitive damages, leaving an award of $2 million.

The case demonstrates the difficult position in which the utility found itself when it decided to terminate the contractor's performance. It was difficult to determine who was responsible for the delay. Acrimonious disputes had arisen throughout the entire performance. Although the court ultimately concluded that punitive damages were not appropriate for this wrongful termination, certainly at the time of the termination an award of open-ended punitive damages could not be excluded as a possibility. To be sure, not terminating carried risks as well. But undoubtedly the economic dislocation and legal exposure make termination in construction disputes relatively rare.

The *Hardeman* and *Indiana Power* cases show the risks of termination. Were there adequate grounds to terminate? What is the cost of being wrong?

Other factors must be taken into account, even when there are grounds to terminate. What happens after termination? What will be the effect of termination on the contractor?

This is shown by the federal procurement guidelines that apply to termination even when there is a legal right to terminate.

Among the factors the contracting officer must take into account when he considers whether to declare a default termination are: "The availability of the supplies or services from other sources" and "The urgency of the need for the supplies or services and the period of time required to obtain them from other sources, as compared with the time delivery could be obtained from the delinquent contractor."

These factors stress what happens after termination. The contracting officer also should consider whether the contractor is important to the acquisition program and the effect termination would have on other contracts with the delinquent contractor.

Finally, the contracting officer should take into account the effect of termination on the contractor's ability to service loans and pay back progress and advance payments to which the government might be entitled.[4]

---

[3]507 N.E.2d 588 (Ind.Ct.App.1987).

[4]48 CFR § 49.402-3 (2007). This was applied in *Marshall Associated Contractors, Inc. and Columbia Excavating, Inc.* (JV), IBCA Nos. 1901 et.al., 01-1 BCA ¶31,248. Even though the contracting officer has considerable discretion, the board held he did not apply these criteria correctly.

It may sometimes be necessary to terminate a construction contract. Performance may be going so badly and relations may be so strained that continued performance would be a disaster. But because of the reasons mentioned, the drastic step of termination should not be taken precipitously.

SECTION 33.02 Termination by Agreement of the Parties

Just as parties have the power to make a contract, they can "unmake" it. In legal parlance, exercising this power may be described as *rescission, cancellation, mutual termination,* or some other synonym. However described, the parties have agreed that each is to be relieved from any further performance obligations. In lay terms, they have "called the deal off."

The legal requirements for such an arrangement are generally the same as those for making a contract: manifestations of mutual assent, consideration, a lawful purpose, and compliance with any formal requirements. By and large, the mutual assent requirement does not present unusual difficulty. Suppose, however, one party makes a proposal to cancel but the other party does not expressly accept. In most cases, silence is not acceptance, but silence can be sufficient coupled with other acts that lead the proposer to believe there has been an agreement.

If each party has obligations yet to perform, the consideration consists of each party relieving the other of performance obligations. If one party has fully performed, however, enforcement problems may arise. For example, suppose a contractor has been paid the full contract price but has not finished performance. If the parties agree to cancel remaining obligations of the contractor, the owner is not receiving anything for its promise. Courts generally relax consideration requirements somewhat when parties adjust or cancel existing contracts. Such an agreement can be enforced by calling it a waiver or completed gift. Although problems can arise, agreements under which each party agrees to relieve the other are generally enforced.

Because most construction contracts need not be expressed by a written memorandum, formal requirements rarely impede enforcement of contracts of mutual termination. Because proof is desirable, however, such agreements will usually be expressed in writing.

SECTION 33.03 Contractual Power to Terminate

Construction contracts frequently contain provisions giving one or both parties the power to terminate the contract. These provisions are a backdrop for material to be discussed in Section 33.04—the common law right to terminate a contract. Although contracts are not always clear on this point, often an interrelationship exists between specific termination provisions and the common law. Specific provisions can be considered illustrations or amplifications of common law doctrines, with the common law doctrines still applicable. Alternatively, contractual termination provisions can be said to have supplanted common law doctrines. The interaction between contractual and common law termination will be discussed in Section 33.03D. Generally, as noted in Section 22.02L, common law termination rights have not been eliminated by express contract termination provisions.

A. Default Termination

Construction contracts drafted by an owner generally give explicit termination rights only to the owner. In contrast, construction contracts published by professional associations provide that either owner or contractor can terminate for certain designated defaults by the other. Although great variations exist, it may be useful to begin discussion with AIA documents.

AIA Document A201-2007, Section 14.2.1.1, permits the owner to terminate "for cause" if the contractor "repeatedly refuses or fails to supply enough properly skilled workers or proper materials." Failure to pay subcontractors is grounds for termination under Section 14.2.1.2. Section 14.2.1.3 makes repeated disregard of laws grounds for termination. Finally, Section 14.2.1.4 is a "catch-all." It allows the owner to terminate the contract if the contractor is "guilty of a substantial breach of a provision of the Contract Documents."

There are two problems with Section 14.2.1.4. As shall be seen in Section 33.04A, the common law specifies that for a breach to permit termination, it must be material. Section 14.2.1.4 speaks of a substantial breach of presumably any contract provision. The common law looks more at the importance of the provision breached, not whether the breach is substantial. In most cases, though, the result will be similar under A201-2007 or the common law.

The second question relates to whether the provision referred to in Section 14.2.1.4 would encompass implied terms as well as those expressed in the contract. The duties of good faith and fair dealing and those of good workmanship are illustrations of implied terms. They are part of the contractual obligations, though rarely expressed in the written contract.

Is good workmanship expressly included in A201-2007? Section 14.2.1.1 mentions properly skilled workers. This relates to their training and experience, not to the quality of their work. Section 3.5 constitutes the warranty given to the owner by the contractor. It states that the work will conform to the contract and will be free of defects. But the well accepted quality standard of good workmanship is not expressly covered. Nevertheless, it is likely that this standard would be implied and should be included in Section 14.2.1.4 as a possible basis for termination.

This termination does not specifically allow partial termination, something that may be useful if the owner wishes to retain the contractor for part of the work and find a successor for the rest.

The Federal Acquisition Regulations (FAR) control federal procurement. They provide that the government may "terminate the contract completely or partially" if the contractor does not perform the required services within the specified time, does not perform any other contract provision, or does not make progress "and that failure endangers performance of the contract."[5] This resembles the common law requirement that discharge of the contract requires a material breach.[6]

Partial termination may be useful from the contractor's perspective. In a decision by the General Services Administration Board of Contract Appeals, the contractor had properly performed part (35 percent) of the work and improperly performed another part (35 percent). The board held that the agency could terminate only for the improperly performed part, based on the fiction of divisibility.[7]

Can the owner terminate if the contractor falls behind its schedule? As noted, federal construction contracts permit termination if the contractor fails to make progress and that failure endangers performance of the contract.[8]

Suppose the owner wishes to bring in a successor contractor when the original contractor is falling far behind schedule. AIA Document A201-2007, Section 14.2.1.1, as seen above, does permit the owner to terminate for repeated failure to supply proper workers or materials. But delays, even unexcused ones, may be caused by other factors. Does Section 8.3, with its provision for time extensions, preclude termination? Does presence of a liquidated damages clause indicate an intention that termination is not appropriate for failure to comply with schedule or completion requirements? Again, the nonexclusivity of the termination clause[9] can support a conclusion that protracted delay—particularly if it appears completion will not be "on time"—will create a power to terminate under common law principles.[10] Any exercise of such a power should be preceded by warnings and, where appropriate, an opportunity to cure.

A termination may be wrongful if the owner has not used good faith in exercising its contractual power to terminate. In *Paul Hardeman, Inc. v. Arkansas Power & Light Co.*, the court admitted that contractual grounds to terminate existed but held that the termination had not been made in good faith.[11]

Also, while not mentioning good faith specifically, a decision of a federal board of contract appeals pointed to elements that could indicate lack of good faith as a basis for holding that the contracting officer's default termination was an abuse of his discretion.[12] The board noted that the contracting officer:

1. Did not fully consider the contractor's claim
2. Did not review the contractor's expert analyses
3. Failed to share relevant information
4. Did not review analysis of his own personnel
5. Did not seek independent evaluations of borrow pit conditions
6. Authorized reprocurement that substantially relaxed the specifications
7. Authorized reprocurement that provided more time and information to bidders
8. Did not assess liquidated damages when the reprocurement contractor fell seriously behind schedule

---

[5]48 CFR § 49.402-1 (2007).

[6]See Section 33.04A.

[7]*Nestos Painting Co.*, GSBCA No. 6945, 86-2 BCA ¶18,993.

[8]48 CFR § 49.402-1 (2007). See also EJCDC, Standard General Conditions of the Construction Contract, C-700, ¶1.4.02(A)(1) (2007).

[9]A201-2007 §§ 13.4.1, 14.2.2.

[10]Whether the architect's certificate or that of any IDM needed under ¶14.2.2 would be required is discussed in Section 33.03D.

[11]See supra note 1.

[12]*Marshall Associated Contractors, Inc. and Columbia Excavating, Inc.* (JV), supra note 4.

The board converted the default termination to one for convenience, as discussed in Section 33.03B.

The law has been reluctant to inject the good-faith principle as a limitation on the exercise of a contractual power to terminate. This reluctance is due in part to the possibility that termination not made in good faith would be sufficiently wrongful to be considered a tort and sufficiently intentional to be considered the basis for punitive damages. This position is illustrated by the unwillingness of the court in *Indiana & Michigan Elec. Co. v. Terre Haute Industries*[13] to impose punitive damages after it concluded that the owner did not have grounds for terminating the contract. The award of punitive damages had been based on a finding that termination had been malicious and oppressive. The appellate court found that at worst the conduct was "substandard business practice, and arrogance."[14] It noted that the law did not impose punitive damages simply because "the contracting party or his agents are disagreeable people."[15] Awarding punitive damages, according to the court, would "let all disputes and quarrels over broken contracts and disappointed business ventures become the subject of acrimonious litigation over punitive damages."[16] Section 19.02D chronicles the increasing use of the good-faith concept in American contract law. Its use in most termination disputes should be limited to whether the termination was proper and should not be the basis for invoking tort law.[17]

Often contracts provide that the default termination does not take effect until the contractor has been given a notice that it must cure the default within a designated period of time. Curing provisions are useful. They can avoid a costly termination where the contractor has performance problems that can be cured.

AIA Document A201-2007 does not include a cure provision, though the seven-day notice period required by Section 14.2.2 can be considered a period to cure. The EJCDC C-700, Paragraph 14.02D precludes termination if the contractor begins to correct its failure to perform within the seven-day notice period required by Paragraph 14.02B and completes cure within 30 days from the receipt of the notice. Federal Procurement requires a ten-day curing

period before there can be a default termination.[18] Even without a specific curing provision, it has been held that before there can be a termination, good faith requires that the contractor be given an opportunity to cure.[19] Of course, if the defect is not curable within the cure period, the termination can take place in accordance with the termination clause without waiting for any cure period to elapse. As noted in Section 33.03B, a wrongful default termination may be converted into a termination for convenience.

It has been held that the owner cannot exercise its contractual power to terminate if the owner has caused or is responsible for the occurrence that is the basis for termination.[20] AIA A201-2007, Section 14.1, gives the contractor specific rights to suspend or terminate. But Section 14.1 must be looked at carefully. Four subsections—14.1.1.1 (court order stopping work), 14.1.1.2 (declaration of national emergency ordering work be stopped), 14.1.1.3 (nonpayment), and 14.1.1.4 (failure to receive evidence showing owner solvency under Section 2.2.1)—allow the contractor to terminate. But occurrence of these events does not automatically give the contractor a power to terminate. The contractor can terminate only "if the Work is stopped for a period of 30 consecutive days . . . for any of the following reasons: . . . ." This precludes hasty termination if the contractor is looking for an excuse to stop its performance.

Section 14.1.2 gives another ground to terminate to the contractor. If the work is delayed, interrupted, or suspended through no fault of the contractor for "more than 100 percent of the total number of days scheduled for completion, or 120 days in any 365-day period, whichever is less," the contractor can terminate.

Section 14.1.3 requires a seven-day written notice of termination. The contractor can recover for work done, reasonable overhead and profit, termination costs, and "damages." In 2007 the AIA deleted "proven loss with respect to materials, equipment, tools, and construction equipment and machinery" that had been in A201-1997. Presumably these losses will be compensated as termination costs.

---

[13]See supra note 3.

[14]507 N.E.2d at 617.

[15]Ibid.

[16]Ibid.

[17]*McClain v. Kimbrough Constr. Co.*, 806 S.W.2d 194 (Tenn.App. 1990), appeal denied Mar. 11, 1991.

[18]48 CFR § 52.249-8 (2007). Curing in federal procurement is canvassed in *Empire Energy Management Systems, Inc.v. Roche*, 362 F.3d 1343 (Fed. Cir.2004).

[19]*Bensch v. Davidson*, 354 S.C. 173, 580 S.E.2d 128 (2003).

[20]*Department of Transp. v. Arapaho Constr., Inc.*, 257 Ga. 269, 357 S.E.2d 593 (1987).

Section 14.1.4 adds another ground: work stoppage for sixty consecutive days because the owner repeatedly fails to fulfill its obligations "with respect to matters important to the progress of the Work." This recognizes the materiality requirement for a common law termination. Again the contractor must give a seven-day written notice. Because of A201-2007, Section 15.1.6, as discussed in Section 27.06D, damages do not include consequential damages. But Section 13.4.1 implicitly allows the contractor common law damages for wrongful termination by stating that specifying remedies in the contract does not exclude common law remedies.

The sparse grounds for termination and, as noted in Section 23.05C, the absence of a contractual right to terminate when the work in progress is destroyed show that AIA documents seek to continue performance and avoid the economic disruption caused by a contract termination.

Also, as noted in Section 33.01 under federal procurement regulations, authority to declare a default termination requires the evaluation of factors that go beyond grounds for termination.

B. Termination or Suspension for Convenience

Private construction contracts increasingly give private owners the right (pioneered by federal procurement regulations)[21] to terminate for convenience. This right is also found in subcontracts, particularly those tied to prime contracts where the owner has this power. An example is EJCDC C-700, Standard General Conditions of the Construction Contract, Paragraph 15.03, and as shall be seen later in this section, AIA Document A201-2007, Section 14.4. Invoking such a clause under federal procurement requires the contractor to stop work, place no further orders, cancel orders that have been placed, and perform other acts designed to terminate performance and protect the interests of the government. The contractor is reimbursed for work performed, unavoidable losses suffered, and expenditures incurred to preserve and protect government property. The contractor is also paid a designated profit for *work performed*.

In addition, federal procurement law has developed the concept of *constructive* convenience termination.[22] Under it,

a contract with a convenience termination clause converts a *wrongful* government default termination into a *convenience* termination.

On its face (convenience, without cause), the power given to the owner by such contracts appears unlimited. As illustrated, the owner may terminate if the project is no longer needed or has become outmoded or uneconomical. Yet, as demonstrated by federal procurement law, this power is not completely unrestricted. *Torncello v. United States*[23] outlined the history of such clauses and noted the importance of the government's ability to change its procurement objectives. Yet the court did not permit the agency to terminate when the agency used another contractor to do the work for less. (When it made the contract, the agency knew a cheaper source existed.) A plurality of the judges held that a termination for convenience can be used only when the circumstances of the bargain have changed. The changed-circumstances standard, according to the plurality opinion[24] in the *Torncello* case, was needed to save the validity of the contract. The prior limitation—bad faith and abuse of discretion—would be an insufficient limitation on the government and would make the government's contractual promise "illusory" and the contract invalid. One concurring judge disagreed, agreeing that in this case the termination was in bad faith, but seeing no reason for the more limiting "change-of-circumstances" standard.

The plurality standard significantly limited the use of the termination-for-convenience clause. As a result, later decisions tended to limit the *Torncello* plurality holding. Finally, in *Krygoski Construction Co. v. United States*,[25] a 1996 decision by a panel of the Court of Appeals for the Federal Circuit, although not overruling the *Torncello* plurality standard,

---

[21] 48 CFR § 52.249.1,.2, et. seq. (2007).

[22] *Torncello v. United States*, infra note 23. The constructive convenience termination was rejected in a case involving a private contract. *Rogerson Aircraft Corp. v. Fairchild Ind., Inc.*, 632 F.Supp. 1494 (C.D.Cal.

1986). It was adopted in New York by *Fruin-Colnon Corp. Traylor Bros., Inc. & Onyx Constr. & Equipment Inc. v. Niagara Frontier Transp. Auth.*, 180 A.D.2d 222, 585 N.Y.S.2d 248 (1992). The concept was attacked in Note, 52 Geo.Wash.L.Rev. 892 (1984).

[23] 231 Ct.Cl. 20, 681 F.2d 756 (1982), overruling *Colonial Metals Co. v. United States*, 204 Ct.Cl. 320, 494 F.2d 1355 (1974). The *Colonial Metals* case employed the bad faith/abuse of discretion standard and had allowed the United States to terminate when it found a cheaper source it should have known of when it made the contract.

[24] A plurality opinion is one not signed by the majority of the judges hearing the appeal. But the number signing the plurality opinion, amplified by judges concurring (agreeing with the outcome of the plurality though not their reasoning), decides the appeal.

[25] 94 F.3d 1537 (Fed.Cir.), cert. denied, 520 U.S. 1210 (1997). This subject is canvassed in Garson, *Krygoski and the Termination for Convenience: Have Circumstances Really Changed?* 27 Pub.Cont.L.J. 117 (1997).

applied the standard of prior bad faith/abuse of discretion. This is less limiting than the plurality in the *Torncello* case. Under these circumstances it cannot be stated with clarity which standard applies in federal procurement law.

Those owners who use a termination-for-convenience clause may have, at least judging by federal procurement, a false impression that the owner's rights are unlimited.

The early federal need for such clauses related to wartime or end-of-the war needs to cancel the procurement of unneeded items. Also, they were needed to adjust weapons needs as technology changed. This war procurement flexibility remained even while procurement circumstances changed.

State courts have different opinions regarding the requirements to terminate for convenience. States do not have weapons procurement policies that must be changed in the middle or at the end of a war. Nor do they deal with new weapons technology. Their current use is to protect public funds when circumstances change. Not only do states use such clauses, but, as shall be seen, they are found in private contracts. Some state courts are willing to allow the state to use them in much the same way as does the federal government. Others are more wary of them. Two cases demonstrate differing attitudes.

Capital Safety, Inc. v. State Division of Buildings & Construction involved a specially drafted clause.[26] It stated that the termination for convenience clause could be used if the contractor is unable to complete the project or if termination would be "in the public interest." The public owner could not relocate some of its employees. This prevented the contractor from working. The court held that the termination clause gave broad power to the state agency as long as it did not act in bad faith. The court cited federal procurement policies and followed the *Krygoski* case. It noted the heavy burden on the attacker of the clause and that rarely can bad faith be shown.

But another attitude was reflected in the same year by an Arizona intermediate appellate court in *Ry-Tan Construction, Inc. v. Washington Elementary School District No. 6*.[27] The case did not involve the usual attempt to terminate in the middle of performance. Instead, the principal issue was whether a contract had been formed when the school board accepted the low bid or whether no contract was formed

until the formal contract was signed.[28] The court (later reversed) concluded that there was a valid contract. The school district sought to avoid the claim by claiming (two years after the litigation began) that it was "constructively" terminating for its convenience. This would have severely limited the amount that could be recovered as damages.

The court, though not bound by federal law, looked to federal authorities. The court noted that the doctrine first was used in wartime and then later only when there was a change in circumstances. The court found no changed circumstances in this case. It concluded that the clause did not give unbounded discretion to the public entity and was not a license to dishonor contract obligations.

A series of cases from New York demonstrate the relationship between default terminations as well as another limitation on the power to terminate for convenience. *Paragon Restoration Group, Inc. v. Cambridge Square Condominiums*,[29] the most recent of the three cases, held that a party who terminates for convenience cannot make a claim for damages caused by defective construction. Usually, the contractor receives what it has spent, compensation for the losses it has suffered, and under some clauses profit on unperformed work. (In the federal system the contractor does not recover profit on unperformed work.) At first blush it seems strange that the owner should lose its claim when it uses its power to terminate for convenience. Generally the law is hesitant to bar a claim unless there is strong evidence that the claimant intended to give up its claim.

But this case shows the relationship between default termination and termination for convenience. The power to terminate for default requires a serious breach. In construction it is often difficult to predict whether there is the power to terminate for default because the facts may be hotly contested and application of the law may be hard to predict. To get the contractor off the project, the owner may decide to terminate for convenience and hope to deal with the performance dispute by offsetting its losses against the contractor's compensation under the clause. (This proved unsuccessful.)

---

[26]369 N.J.Super 295, 848 A.2d 863 (App.Div.2004).

[27]208 Ariz. 379, 93 P.3d 1095 (App. 2004), reversed on the contract issue, 210 Ariz. 419, 111 P.3d 1019 (2005).

[28]See Section 18.04G.

[29]42 A.D.3d 905, 839 N.Y.S.2d 658 (2007). The court cited *Tishman Constr. Corp. v. City of New York*, 228 A.D.2d 292, 643 N.Y.S.2d 589 (1996) and *Fruin-Colnon Corp. Traylor Bros., Inc. Onyx Constr. and Equipment, Inc. v. Niagara Frontier Transp. Auth.*, supra note 22. The latter case involved a wrongful default termination because it did not give the contractor a chance to cure. But this was constructively converted into a termination for convenience, with no right to recover for payments made to third parties to perform remedial work.

Also, a termination for default usually requires that the contractor be given a chance to cure, as noted in Section 33.03A. If the owner wants to be compensated for the contractor's breach, it should terminate for default, not convenience. The *Paragon* case is another limitation on termination for convenience. If the owner seeks to recover damages for breach in New York, it cannot use a convenience termination.

A201-2007 includes such a power in Section 14.4. Under Section 14.4.3 the contractor's remedy is payment for work performed, compensation for losses it cannot reasonably avoid because of the cancellation, reasonable overhead, and profits on *performed* and *unperformed* work. This remedy is more liberal than the federal formula, the federal government being unwilling to give the contractor profit on unperformed work. It remains to be seen whether the vast body of federal jurisprudence, including cases such *Torncello*, will be used when termination for convenience occurs under A201-2007. The federal system has its own unique characteristics, tightly regulated and with its own special disputes process. The availability of the well-developed federal jurisprudence, even if premised on federal administrative mechanisms, will mean continued use of federal sources by the states. Most states have not developed a detailed systematic jurisprudence to deal with these issues.[30]

A201-2007, Section 14.3.1, provides that the owner can *suspend, delay* or *interrupt the Work* without cause. Section 14.3.2 grants the contractor an adjustment "for increases in the cost and time caused by the suspension, delay or interruption" as well as profit on those adjustments.

As noted above, Section 14.1.2 permits the contractor to *terminate* if there are repeated suspensions, delays, or interruptions that aggregate more than 100 percent of the total number of days scheduled for completion, or 120 days in any 365-day period, whichever is less.

C. Events for Which Neither Party Is Responsible

Some contracts specifically allow termination when events have a devastating effect on performance, such as would justify common law relief for impracticability or impossibility. AIA documents do not specifically provide that these events permit termination. The party whose performance has been adversely affected receives time extensions and occasionally an equitable adjustment.

D. Role of Design Professional

Despite the AIA's movement toward reducing the activities and responsibilities of the architect, the owner's power to terminate under A201-2007, Section 14.2.2, requires the architect or any Initial Decision Maker (IDM) to certify "that sufficient cause exists to justify such action." No similar requirement exists for a termination by the contractor.

Termination is a drastic step. Why require the architect to certify that there are adequate grounds?[31] Such a decision, although involving some issues for which the architect may be trained, requires legal expertise rather than design skill. The owner may want the architect's advice. That need not require giving power to the architect to decide a sensitive and liability-exposing issue. The only possible justification is that such a preliminary step can act as a brake on any hasty, ill-conceived decision by the owner to terminate. It would be better to have the owner's attorney perform this function, especially if the architect's conduct is itself an issue in the termination.

Would an architect's certificate be needed for common law termination? Although it is unlikely such a procedure would apply to a common law termination, the common law can add grounds but should not destroy any agreed-on conditions precedent.

The requirement for such an architect certification and the interrelationship between contractual and common law termination came before the court in *Ingrassia Construction Co. v. Vernon Township Board of Education*.[32] The parties used A201, very likely the edition issued in 1987. However, the parties deleted the arbitration provision.

The board sought to terminate the contract based upon the contractor's failure to meet time milestones and to comply with quality requirements. It requested that the architect provide the certification under old Paragraph 14.2.2 that sufficient cause exists to justify termination. The architect was licensed in Canada but not in the United States. He refused to certify that there were sufficient

---

[30]*New Pueblo Constructors, Inc. v. State*, 144 Ariz. 95, 696 P.2d 185 (1985), reproduced in part in Section 27.02F; *Ry-Tan Constr. Inc. v. Washington Elementary School Dist. No. 6*, supra note 27, and *Capital Safety, Inc. v. State, Division of Buildings & Constr.*, supra note 26, both applied state law but looked at the federal procurement system for guidance.

[31]When "architect" is used, it includes the IDM. It remains to be seen whether this role will be played mainly by the architect or, as in England, a new specialist will be developed. In England he is called the adjudicator. See Section 30.18J.

[32]345 N.J.Super. 130, 784 A.2d 73 (App.Div.2001).

grounds to terminate. He stated that he had not been on the job daily and he "would not be able to certify from his own knowledge and observations that there was sufficient cause for termination."[33] He did certify that milestone dates had not been met, but not that this was a substantial breach or sufficient ground for termination.

The court upheld the determination by the trial judge that the certificate was defective, as it did not certify what was required under old Paragraph 14.2.2 (The trial judge pointed to the fact that the architect was unlicensed.) The appellate court did not comment on lack of a license. It simply concluded that the certificate did not meet the requirement of old Paragraph 14.2.2. That, stated the court, deprived the board of "any finality, presumption of correctness, or obligation of judicial deference that would otherwise attach to a proper architect's certification."[34]

That the certificate was defective, held the court, did not deprive the board of its common law right to claim that it had proper grounds to terminate the contractor. It pointed to old Paragraphs 13.4.1 and 14.2.2 that state the contractual remedies are not exclusive. The court concluded that a proper certificate is not a condition precedent to the board's exercise of its common law rights. The board's traditional burden to establish a material breach would have been largely exempted by the issuance of a proper certificate. The court did not have to decide how much finality to extend to a proper certificate, but it seems clear that issuance would have been very beneficial to the board.

The case raises many questions. Would a proper certificate by an unlicensed architect have been valid? The trial court seemed troubled by this, but the appellate court was not. It did not have to address this issue, as it was clear that the certificate was not adequate. A201-2007, Section 4.1.1, requires the owner to state that the architect is licensed. If both parties knew the architect was not licensed, that could have waived old Paragraph 4.1.1. A proper certificate would have been valid. (The architect would still be guilty of violating the licensing law.) But if the contractor did not know this, the certificate was not valid. The agreement to have the certification requirement in the contract was premised on the architect being licensed.

What of the architect's refusal to give the certificate because he had no personal knowledge? This would not accord with the intention of the parties. They knew the architect was not on the site every day. More likely, the architect should have conducted an investigation of the type specified in old Paragraph 4.3 dealing with claims. (Perhaps his unwillingness to do so was based upon his unlicensed status.)

This raises the issue of the interaction between old Paragraphs 14.2.2 dealing with termination and 4.3 dealing with all claims and disputes. If there had been no provision such as 14.2.2 and the board terminated, it would seem that the contractor could contest this by invoking Paragraph 4.3. The architect's decision, it might be noted, could not be taken to arbitration as the arbitration clause had been deleted.

Yet, as shall be suggested in Section 33.03I, termination is a special procedure and may not be treated the same as any other dispute. In that case, the issue would simply be one for the common law to resolve. The board would have to prove that the contractor committed a material breach to justify termination.

Finally, the court did not have to face the issue of the proper weight to be accorded an architect decision. But certainly, it would be given some finality, as the court noted.[35]

In sum, in addition to the cooling-off function noted earlier in this section, Section 14.2.2 gives a tactical advantage to the owner. It can relieve it of the burden of establishing a common law termination.

E. Waiver of Termination and Reinstatement of Completion Date

Suppose one party has the power to terminate but does not exercise it. Has the power to terminate been lost? Under what conditions can it be revived?

Cases that have dealt with this problem have usually involved a performing party—usually the contractor—who has not met the completion date, but for various reasons the owner decides not to terminate. The contractor continues to perform, believing the power to terminate will not be exercised. At some point during continued performance, the owner wishes either to terminate immediately or to set a firm date for completion, which, if not met, will then be grounds for termination.

Clearly, a termination after the contractor has been led to believe there will be no termination would be improper.[36]

---

[33]784 A.2d at 77.
[34]Ibid.

[35]See Section 29.09.
[36]*DeVito v. United States*, 188 Ct.Cl. 979, 413 F.2d 1147 (1969); *United States v. Zara Contracting Co.*, 146 F.2d 606 (2d Cir.1944).

The more difficult question relates to the requirements for reinstating a firm deadline that will allow termination if performance is not met by that time. The parties can agree to extend the time for completion, with the clear understanding that failure to comply would entitle the owner to terminate. The most difficult question involves unilateral attempts by the owner to reinstate a firm completion date and revive the right to terminate.

In *DeVito v. United States*,[37] the contractor did not meet a revised completion date of November 29, 1960, because of performance problems. Having anticipated default, on November 25, 1960, the contracting officer requested permission from higher headquarters to terminate the contract. The request moved slowly through various offices of the agency, and for unexplained reasons, authority to terminate was not granted until January 16, 1961. This was communicated to the contractor on January 17, 1961. During this period, the contractor continued to perform, made commitments, and delivered some units that were accepted by the government. At the time of termination, the contractor had many assemblies in various stages of completion and claimed to be on the verge of reaching full production.

In a legal action brought on behalf of the contractor, the court first noted that the government is habitually lenient in granting reasonable extensions, "for it is more interested in production than in litigation." Moreover, the court said, "default terminations—as a species of forfeiture—are strictly construed."[38] The court noted that permitting a delinquent contractor to continue performance can preclude the government from terminating if its actions or nonactions have led the contractor to believe no termination would occur; the contractor relied on this belief. The court held the government had waived its right to terminate by allowing the contractor to continue performance under these circumstances.

As to reinstatement, the court stated,

When a due date has passed and the contract has not been terminated for default within a reasonable time, the inference is created that time is no longer of the essence so long as the constructive election not to terminate continues and the contractor proceeds with performance. The proper way thereafter for time to again become of the essence is for the Government to issue a notice under the Default clause setting a reasonable but specific time for performance on pain

of default termination. The election to waive performance remains in force until the time specified in the notice, and thereupon time is reinstated as being of the essence. The notice must set a new time for performance that is both reasonable and specific from the standpoint of the performance capabilities of the contractor at the time the notice is given.[39]

The court held that such a process would not be required if the contractor had renounced its contract or was incapable of performance.

In the *DeVito* case, the court emphasized that there could be no fixed rules regarding the government's waiver of its right to terminate. The decision in *H. N. Bailey & Associates v. United States*,[40] decided by the Court of Claims two years after *DeVito*, illustrated this fluidity. The contractor had been awarded a contract set aside for small businesses and was relatively inexperienced in manufacturing the products sought by the procurement. The delivery date was July 27, 1966, and it seemed clear to the government that the contractor would not be able to perform. As a result, on August 10, 1966, the government notified the contractor that there had been a default and that any assistance given the contractor, or acceptance of delinquent goods, would be solely for the purpose of mitigating damages and not to be construed as an indication that the government was waiving its rights. The notice also gave the contractor ten days to advise the procuring agency of any reason why the contract should not be terminated.

On August 19, 1966, the contractor responded by stating it was confident it could produce the goods by the end of the following month. Yet production was still not made, and the government terminated for default on September 6, 1966.

The contractor claimed the conduct of the government was inconsistent with its right to terminate, and constituted a waiver. The court did not agree, and concluded that the government did not encourage or induce the contractor to continue performance; it characterized the government's conduct as "displaying a benevolent attitude towards the defaulting contractor."[41]

Sometimes two issues arise in the context of waiver. This is illustrated by *Consolidated Engineering Co., Inc. v. Southern Steel Co.*,[42] a dispute between a prime and subcontractor.

[37]See supra note 36.
[38]413 F.2d at 1153.

[39]Id. at 1154.
[40]196 Ct.Cl. 166, 449 F.2d 376 (1971).
[41]Id. at 385.
[42]699 S.W.2d 188 (Tex.1985).

The subcontractor was to install steel furnished by the prime contractor. It was unable to do so because of the prime's delay in furnishing the materials. Although the subcontractor contended it had the right to terminate, it continued to work for three more months. Then it abandoned the contract.

The prime claimed damages, contending that the subcontractor wrongfully abandoned the contract. The subcontractor claimed damages for not being able to perform its contract. The court concluded that the prime contractor had indeed breached and that the subcontractor was justified in stopping performance. This gave the subcontractor a defense to the claim by the prime contractor. It also gave the subcontractor a claim for damages against the prime contractor.

The prime did not contend that the subcontractor remaining on the job waived its rights to terminate; it contended that remaining on the job waived the subcontractor's claim for damages. The court disagreed, noting that the subcontractor reserved its rights at the time it continued performance. It in no way indicated it was giving up its right to recover damages for the prime contractor's breach.

F. Notice of Termination

Termination clauses often require that a notice of termination be sent by the terminating party to the party whose performance is being terminated and, in many contracts, to the lender or surety. The notice usually states that termination will become effective a designated number of days after dispatch or receipt of the notice.

In addition to problems common to all notices and communications,[43] such notice requirements create other legal problems. What are the rights and duties of the parties during the notice period? Can the defaulting party cure any defaults specified in the notice during the notice period and thereby keep the contract in effect? Can the party giving the notice to terminate demand continued performance or even accelerated performance during the notice period?[44] Answers to such questions often depend on the reason for requiring a notice to terminate.[45]

The notice period may be designed to allow the terminating party to "cool off." Construction performance problems often generate animosity, and before the important step of termination is effective, the terminating party may wish to rethink its position. The notice period can permit a defaulting party to cure defaults in order to keep the contract in effect for the benefit of both parties. Whether the notice period is designed to "cure" may depend on the facts that give rise to termination and on the notice period.

A recent case, noted in Section 33.03A, held that termination was wrongful unless the owner gave the contractor an opportunity to cure defects.[46] While this may have been an application of the obligation to act in good faith, it shows the importance of cure, especially when something as serious as termination is being considered.

Finally, the notice period can be used to wind down and protect the site. This allows each party to cut losses and make new arrangements. Such a position does not allow cure.[47] But if the owner terminates, it should have the option of ordering that work be done that can be completed by the effective date of termination.[48]

Such questions should be, but rarely are, answered by the termination clause.

The termination clause should clearly indicate whether it permits *cure*, provides a *cooling-off period*, or sets into motion a *winding down* of the project. To illustrate a cure provision, standard construction contracts used in Canada allow the owner to terminate if the contractor fails to perform after the owner notifies the contractor in writing that it is in default, based on a certificate from the architect. The notice must instruct the contractor to correct the default within five working days from receipt of the notice. If the default cannot be cured within the five working days, the owner cannot terminate if the contractor begins correction of the default within the specified time, provides the owner with an acceptable schedule for such correction, and completes the correction in accordance with that schedule. This is a sensible procedure. Termination is the last desperate step in a troubled contract.

---

[43]See Sections 12.13F and 17.05C.

[44]*New England Structures, Inc. v. Loranger*, 354 Mass 62, 234 N.E.2d 888 (1968) (can demand continued performance).

[45]*New England Structures, Inc. v. Loranger*, supra note 44 (five days too short to cure).

[46]*Bensch v. Davidson*, supra note 19.

[47]*New England Structures, Inc. v. Loranger*, supra note 44 (cut losses and make new arrangements).

[48]See supra note 40.

G. Taking Over Materials and Equipment

AIA Document A201-2007, Section 14.2.2, permits the owner to "exclude the Contractor from the site and take possession of all materials, equipment, tools, construction equipment and machinery thereon owned by the Contractor . . . [and] . . . finish the Work by whatever reasonable method the Owner may deem expedient." Construction contracts commonly include these provisions. The following reasons are given for such provisions:

1. to provide incentive to the contractor to take away its property from the site so that the owner can efficiently bring in a successor
2. to provide the owner with material and equipment by which it can expeditiously continue the work with a successor (note the conflict with item 1)
3. to give the owner property that can be sold and the proceeds used to pay for any claim it may have against the contractor

The owner, however, must first give a seven-day termination notice. Will this encourage the contractor to take away its property before the owner does? If so, the first justification will be accomplished, but the second will be frustrated. Also, suppose the successor contractor does not want to use the materials and equipment. Does the contractor have the right to take away the materials and equipment during the seven-day period? (It is unlikely that the contractor will stand by passively when the owner takes possession of its property.) Can the owner take materials or equipment owned by a subcontractor? If not, the clause may be of little value. This may depend on the implementation of any flow-through clause, as discussed in Section 28.04.

In addition, who actually owns the equipment? Much equipment is leased or purchased under conditional sales contracts. Even if the contractor owns the property, any attempt to use the equipment as security may force a struggle with others who contend that they have security interests that take precedence over any right created by such a clause or any other lien rights. If the contractor goes bankrupt, the trustee in bankruptcy surely will enter the fray. Also, if the contractor's property is used, a subsequent dispute may arise as to its use value.

As if these problems were not enough, suppose the termination is wrongful. Taking the contractor's property constitutes the tort of conversion. The owner must pay the reasonable value of property converted, restore any gains made through converting the property, and may have to pay punitive damages.[49]

The AIA did not expressly create a security interest. There is no express power to sell the materials and equipment and retain any amounts obtained to set off against claims it has against the contractor.

The 1987 AIA Document A201, Paragraph 12.2.4, dealing with failure to correct and remove defective work, did create such a security interest. This was deleted from A201-1997. But its earlier inclusion shows the AIA knew how to create a security interest when it wanted to do so.

Some justify the clause permitting takeover of materials and equipment as a basis to back up emergency measures under which it is essential that the materials and equipment be used to keep the project performance going. They would deal with legal subtleties later. The risks are greater if equipment, particularly construction machinery, rather than materials is taken. In any event, continued inclusion of such clauses in construction contracts may indicate that they do have utility.[50]

H. Effect on Existing Claims for Delay

Termination of the contract is usually accompanied by claims by each party against the other. If the owner terminated, the contractor, as a rule, claims not only that the termination was wrongful but also that the owner had committed earlier breaches that caused it losses. The terminating owner also is likely to have claims for damages based on delay and improper workmanship. Sometimes the terminated party will contend that invoking the termination remedy was a waiver of any claims that existed before termination. In *Armour & Co. v. Nard*,[51] this contention was rejected. After pointing to a provision stating that failure to exercise a right shall not be considered a waiver, the court noted that a contractual remedy for breach generally

---

[49] See *Indiana & Michigan Elec. Co. v. Terre Haute Indus., Inc.*, supra note 3 (punitive damages not awarded despite wrongful termination followed by takeover of contractor's equipment; independent tort required).

[50] In *Northway Decking & Sheet Metal Corp. v. Inland Ryerson Constr. Products Co.*, 426 F.Supp. 417 (D.R.I.1977), the court appears to have enforced such a clause. It denied a terminated subcontractor an injunction permitting him to remove unique hanging scaffolding.

[51] 463 F.2d 8 (8th Cir.1972).

does not exclude other remedies unless the clause shows a clear intention of the party to do so. Rarely will a terminating party implicitly give up damage claims. But as noted in Section 33.03B, New York holds that an owner who terminates for convenience cannot recover from the contractor for its costs to correct defective work.

I. Disputed Terminations

Suppose the owner terminates the contract in accordance with the termination clause and orders the contractor to leave the site. The contractor contends the termination is not justified and demands arbitration under a provision similar to AIA Document A201-2007, Section 15.1.4. This provides for arbitration for all claims, disputes, and other matters arising out of or relating to the contract documents or the breach thereof. Section 15.1.3 states that pending final resolution of a claim, both parties continue to perform. The following issues can arise:

1. Does a contractor's demand for arbitration give it the right to remain on the site pending the arbitrator's decision?
2. If the *contractor* wishes to terminate, may the owner demand arbitration and insist that the contractor continue work?
3. Is the disputed termination subject to arbitration?
4. If arbitration takes place, should the arbitrator order that work be resumed?

Requiring the contractor under Section 15.1.3 to continue working while the dispute is being resolved assumes performance disputes that do not involve termination.[52] Termination is a special dispute. The specific provisions of a termination clause should take precedence over the general arbitration clause. Section 15.1.3 does not require continued performance during arbitration *when agreed otherwise in writing*. This can refer to the termination clause with its specific mechanism. Also, that Section can be for the owner's benefit. The owner can waive its right to continued performance during arbitration by ordering the contractor to leave the site. Finally, the owner's ownership and control of the site should empower the owner to remove

the contractor. Whether the termination was proper can be resolved by any applicable arbitration clause.[53] The termination has not attacked the validity of the arbitration clause.[54]

Generally, arbitrators have remedial discretion. Any award to continue performance, at least in New York,[55] would be upheld. But termination is usually acrimonious and the last desperate step. Rarely will reinstatement be chosen as the remedy; this is another reason to force the contractor to leave the site even if termination is challenged. If termination were improper, a damage award would be adequate.

Suppose the contractor wishes to terminate and the owner seeks arbitration. The issue here is less clear. The provision for removing the contractor from the site can be for the benefit of the owner. This interpretation would allow the owner to order the contractor to continue working. It is more likely, however, that the conclusion will not depend on which party terminates.

J. Public Contracts and Constitutional Protection

The increasing use of constitutional doctrines has been noted earlier.[56] In the context of terminating a public construction contract, the contractor may contend that the contract cannot be terminated without extending it due process of law, usually taken to mean some basic procedural rights. In *Riblet Tramway Co., Inc. v. Stickney*,[57] the court held that the contract could be terminated despite the lack of a prior hearing. The court recognized that the contractor's interest in the contract could not be taken away without cause. But what process would be appropriate? Using the standard of fundamental fairness, the court held that the contractor could sue for breach. This was sufficient to satisfy the contractor's due process rights.

---

[52]But *Keyway Contractors, Inc. v. Leek Corp., Inc.*, 189 Ga.App. 467, 376 S.E.2d 212 (1988), cert. denied Jan. 11, 1989, held that a subcontractor who left the site when a dispute arose for which arbitration was required had breached the subcontract.

[53]Many cases hold that arbitration survives termination. See *Middlesex County v. Gevyn Constr. Corp.*, 450 F.2d 53 (1st Cir.1971), cert. denied, 405 U.S. 955 (1972); *State v. Lombard Co.*, 106 Ill.App.3d 307, 436 N.E.2d 566 (1982). But see *G&N Constr. Co. v. Kirpatovsky*, 181 So.2d 664 (Fla.Dist.Ct.App.1966). See Section 30.03B.

[54]*Riess v. Murchison*, 384 F.2d 727 (9th Cir.1976).

[55]*Grayson-Robinson Stores, Inc. v. Iris Constr. Corp.*, 8 N.Y.2d 133, 168 N.E.2d 377, 202 N.Y.S.2d 303 (1960).

[56]See Sections 18.04E and 22.02E.

[57]129 N.H. 140, 523 A.2d 107 (1987).

SECTION 33.04 Termination by Law

A. Material Breach

If the contract does not expressly create a power to terminate, termination is allowed in the event of a material breach. Even with a contract clause dealing with termination, many of the factors that determine materiality can be influential when such clauses are interpreted. Also, as indicated, the termination clause may not be the exclusive source for determining when a party has the power to terminate.

Rather than establishing fixed rules, such as the importance of the clause breached, the law examines all the facts and circumstances surrounding the breach to determine whether it would be fair to permit termination. For example, the Restatement (Second) of Contracts articulates factors in Section 241 that are significant in determining whether a particular breach is material.[58] They are

(a) the extent to which the injured party will be deprived of the benefit which he reasonably expected;

(b) the extent to which the injured party can be adequately compensated for the part of that benefit of which he will be deprived;

(c) the extent to which the party failing to perform or to offer to perform will suffer forfeiture;

(d) the likelihood that the party failing to perform or to offer to perform will cure his failure, taking account of all the circumstances including any reasonable assurances;

(e) the extent to which the behavior of the party failing to perform or to offer to perform comports with standards of good faith and fair dealing.

Looking first at breaches by the contractor, item (a) seeks to determine the importance of the deviation and the likelihood of future nonperformance. Item (b) examines whether the owner can be easily compensated for nonperformance. This factor would depend on whether the defect could be easily corrected or whether compensation for an uncorrected defect would be easy to measure. In construction contracts, the latter can be difficult. Item (c) looks for uncompensated losses that the contractor would suffer from termination. If the contractor had ordered materials

that had not yet been used and were not usable in other projects, this would militate against termination. Item (d) examines the likelihood that the contractor will be able to cure defects, and item (e) examines the reason for the breach. An example of the latter—a breach by a subcontractor that the prime contractor could not have reasonably prevented—might militate against materiality.

An owner breach would probably be nonpayment of money, failure to furnish the site, or noncooperation. Item (a) would look at the extent of the breach and the likelihood of future performance, while item (b) would look at whether the breach was easily compensable, such as interest for a breach consisting of not making a progress payment.

Item (c) examines the harm termination would cause the owner, such as lost loan commitments, liability to prospective tenants, or other lost business opportunities. Item (d) seeks to determine whether the breaches are likely to be cured. Finally, item (e) examines the reasons for nonperformance. Financial reverses or a steep rise in interest rates makes it less likely that the contractor can terminate for failure to pay. If, however, there was "bad blood" between the design professional and the contractor and the latter seized on the breach by the owner to injure the design professional, termination would be less likely. Likewise, if it appears the contractor sought an excuse to end the contract, the breach would less likely be material.

The case of *Oak Ridge Construction Co. v. Tolley*[59] involved a dispute between Tolley, the owner of land, and Oak Ridge, the contractor who had agreed to build a house and dig a well. Tolley disputed and refused to pay the invoice for extra drilling and casing. Oak Ridge stated that Tolley's failure to pay the invoice was a breach entitling Oak Ridge to terminate under the contract's termination provisions. It also notified Tolley that it was stopping work and would arbitrate the dispute. Tolley refused to arbitrate at that time, and the work proceeded no further.

After concluding Tolley had not committed a breach by anticipatory repudiation (see Section 33.04B), the court found that the work stoppage by Oak Ridge was a breach. To determine whether the breach was material, it applied the sections of the Restatement noted earlier. It stated,

We must therefore consider whether that breach was material in light of the factors enumerated above. The Tolleys, as the injured party, were deprived of the expected benefits of their

---

[58]Section 242 dealing with delay expands the list to include the extent to which the delay will hinder the injured party in making substitute arrangements and the extent to which the agreement provides for performance without delay.

[59]351 Pa.Super. 32, 504 A.2d 1343 (1985).

contract (i.e., receiving a completed home) by Oak Ridge's work stoppage, and could be adequately compensated for that deprivation by an award of damages. Furthermore, Oak Ridge's letter of September 7 gave no indication that the company would cure its failure to perform, and, in fact, the record indicates that Oak Ridge never did so act. Of course, a finding of material breach will result in forfeiture for Oak Ridge (i.e., the Tolleys will be discharged from all liability on the contract), however, Oak Ridge will be entitled to restitution for any benefit conferred upon the Tolleys by part performance or reliance (i.e., the cost of digging the well) in excess of the loss Oak Ridge caused by its own breach. See Restatement, supra, §§ 241 comment d, 374(1). We note that the record contains no evidence concerning whether Oak Ridge's conduct "comports with standards of good faith and fair dealing." Under these circumstances, we find that Oak Ridge's breach constituted a material failure of performance thereby discharging the Tolleys from all liability under the contract.[60]

Many factors are relevant. Perhaps the most relevant factors are the particular nature of the nonperformance, the likelihood of future breaches, and the possibility of forfeiture. Some courts may look at good faith not simply of the party failing to perform but of the party who seeks to exercise the power to terminate.[61] For example, Tennessee held that in the absence of a clause permitting the prime contractor to take over the subcontractor's work (the court noting that such clauses usually require a notice and an opportunity to cure), the prime contractor could not terminate the contract without notifying the subcontractor of its intention to do so and giving the subcontractor an opportunity to cure.[62]

Nonpayment—perhaps the most frequently asserted justification for termination—was discussed in Section 22.02L. The owner's failure to make an equitable adjustment when unforeseen underground conditions were discovered and the contractor was in financial trouble allowed the prime contractor to terminate.[63] A prime contractor's failure to have the site ready for the flooring subcontractor

was a material breach.[64] Failure to make satisfactory progress justified termination.[65] Hindering the subcontractor's operation was a material breach when coupled with the prime contractor's having stopped payment on a check that the subcontractor was about to negotiate.[66] Another court held that the collapse of a fallout shelter was sufficient grounds for the owner to terminate the contractor's performance and recover a down payment.[67] These obviously incomplete illustrations are not designed to indicate that breaches of the type described will always be considered material. As emphasized in this subsection, determining whether a breach is material requires a careful evaluation of the facts and circumstances surrounding the breach as well as the effect of termination.

B. Future Breach: Prospective Inability and Breach by Anticipatory Repudiation

Section 33.04A dealt with breaches that have occurred. This subsection deals with breaches that may occur in the future. It may appear that one party may not be able to perform when the time for performance arrives, or one party may state it will not perform when the time for performance arrives. The contractor may discharge some of its employees, or a number of employees may quit. The contractor may cancel orders for supplies, or its suppliers may indicate they will not perform at the time for performance. In such cases, the owner may realize the contractor will be unable to perform.

Such inability to perform—whether on the part of the owner, or the contractor, or subcontractors—is likely to be a breach. Each party owes the other party the duty to appear to be ready to perform when the time for performance arises. Even if no such promise is implied, under certain circumstances events may permit one party to suspend its performance or terminate the contract unless the other party who appears unable to perform can give assurance or security that when the time comes for performance, it

---

[60]504 A.2d at 1348–1349.

[61]See Section 33.01.

[62]*McClain v. Kimbrough Constr. Co.*, supra note 17. See also Restatement (Second) of Contracts § 242 (1981).

[63]*Metropolitan Sewerage Comm'n v. R. W. Constr., Inc.*, 72 Wis.2d 365, 241 N.W.2d 371 (1976), appeal after remand, 78 Wis.2d 451, 255 N.W.2d 293 (1977) (jury verdict approved).

[64]*Great Lakes Constr. Co. v. Republic Creosoting Co.*, 139 F.2d 456 (8th Cir.1943).

[65]*Aptus Co. v. United States*, 61 Fed.Cl. 638 (2004); *Mustang Pipeline Co., Inc. v. Driver Pipeline Co., Inc.*, 134 S.W.3d 195 (Tex.2004).

[66]*Citizens Nat'l Bank of Orlando v. Vitt*, 367 F.2d 541 (5th Cir.1966).

[67]*Economy Swimming Pool Co. v. Freeling*, 236 Ark. 888, 370 S.W.2d 438 (1963).

will perform.[68] Insolvency as a basis for insecurity may be affected in a construction dispute by AIA Document A201-2007, Section 2.2.1, which gives the contractor the power to ask the owner for evidence of its financial arrangements. Failure to exercise this power may preclude the contractor's suspending its obligation to perform if the owner later becomes insolvent. The owner's failure to make payments, as indicated in Section 22.02L, gives the contractor a right under A201-2007, Section 9.7.1, to suspend performance.

Prospective inability deals with probabilities. In the examples given, the question is whether the owner must wait to see whether actual performance or defective performance will occur or whether it can demand assurance and, in the absence of this assurance, legally terminate any obligation to use the contractor.

Contracts often deal with termination rights. In the absence of such provisions, it will take a strong showing on the part of the owner to terminate the contractor's performance on the grounds that the contractor may not be able to perform in the future. In many cases, a combination of present and prospective nonperformance exists. If present nonperformance exists, such as the installation of defective materials or poor workmanship, the likelihood of this continuing will strongly influence the court to let the owner terminate the contractor's performance. Pure prospective inability without present breach is likely to be considered insufficient grounds unless the probabilities are very strong or unless a contract provision exists giving the owner this right.

A breach by anticipatory repudiation occurs when one party indicates to the other that it cannot or will not perform.[69] Each party is entitled to reasonable assurance that the other will perform in accordance with the contract. If one party indicates that it will not or cannot perform, the other party loses this assurance. Should the latter be required to wait and see whether the threat or the indication of inability to perform will come to fruition?

The rights of a party to terminate a contract because of the other party's repudiation have expanded. A feeling of assurance is important, as this is one purpose for making a contract. Sometimes a party who indicates it will not perform is jockeying for position. The owner may be trying to pay less or obtain more than the called-for performance. If either party does repudiate, the other party will be given the right to terminate the obligation. The party to whom the repudiation is made need not terminate the obligation immediately. It may state that it intends to hold the other party to the contract. If the repudiator relies on the statement that the contract will be continued, the nonrepudiating party will lose its right to terminate based on the earlier repudiation.

The right to continue one's performance despite repudiation by the other party is qualified by the rule against enhancing damages. One party cannot recover damages caused by the other party's breach when those damages could have been avoided by taking reasonable steps to cut down or eliminate the loss. If the contractor repudiates the contract, for example, the owner cannot recover for damages caused by breach that could have been avoided by hiring a replacement contractor.

Suppose the contractor states unequivocally without justification that it will leave the job in three days. It might be reasonable for the owner to insist that it will hold the contractor to the contract for a short period, such as until the date of the walkout or even for a few days after the walkout. But when it is clear the contractor will not return to the project, the owner should take reasonable steps to replace the contractor if the owner wishes to continue the project. A replacement should be obtained when the repudiating contractor has committed its workers and machinery to another project and apparently has neither the willingness nor the capacity to return to the job. Any damages that could have been avoided by hiring a replacement will not be assessable against the repudiating contractor.

Repudiation may accompany present or prospective breach. The greater the scope of any present or prospective breach, the greater the likelihood that the court will release the innocent party by reason of both the present breach and the repudiation. Even a small breach, coupled with a repudiation, may be enough to terminate the innocent party's obligation to perform further.

C. Bankruptcy

Bankruptcy can affect contracts. Two types of bankruptcy are important for the purposes of this chapter. The first is liquidation under Chapter 7 and is known as *straight bankruptcy*.

---

[68] Restatement (Second) of Contracts §§ 251, 252 (1981) followed in *Danzig v. AEC Corp.*, 224 F.3d 1333 (Fed.Cir.2000), cert. denied sub nom. *ABC Corp. v. Pirie*, 532 U.S. 995 (2001).

[69] As to the existence of a repudiation, compare *Oak Ridge Constr. Co. v. Tolley*, supra note 59 (owner questioning invoice charges and saying they were in dispute, not a repudiation), with *Twenty-Four Collection, Inc. v. M. Weinbaum Constr., Inc.*, 427 So.2d 1110 (Fla.Dist.Ct.App. 1983) (contractor's mailgram threatening to shut down project unless a certain clause was eliminated constituted anticipatory repudiation).

It has two principal objectives. First, the bankrupt can wipe the slate clean, or mostly clean, by discharging most of the bankrupt's debts. Second, bankruptcy should provide a fair and efficient liquidation of the bankrupt's estate. The trustee for the bankrupt takes over the bankrupt's estate, collects any money owed the bankrupt, compels repayment of any preferential payments made to creditors,[70] turns over specific property in the hands of the bankrupt to those who have security interests in it, and, after paying expenses of administering the estate, distributes any amount remaining pro rata to unsecured creditors. For example, if the amount of unsecured debts is $1 million and the amount remaining is $100,000, each unsecured creditor receives ten cents for each dollar of unsecured debt. Payments to unsecured creditors, if made at all, are usually only a small fraction of the debts.

The second type is a petition for reorganization under Chapter 11. Such a petition protects the petitioner from creditor claims while it seeks to reorganize in order to put its business on a more sound financial basis. The petitioner submits for judicial approval a plan that involves reorganizing the interests of owners and creditors, rescheduling debts, and (one hopes), infusing cash and credit.

Bankruptcy law gives the trustee a choice to assume (continue performance) or reject (refuse further performance) existing contracts. The U.S. Congress precluded any contract clause from automatically barring this choice or election.[71] The trustee can assume any contract if it has the capacity to continue performance, or can pay compensation if it does not.

It was common for contracts to permit automatic termination if a party filed for bankruptcy. In construction contracts, the owner was usually given the power to terminate if the contractor filed for bankruptcy or had serious financial problems. Despite the change in the law, some contracts still contain such provisions, either because of ignorance as to their ineffectiveness or simply because of inertia. In any event, assumptions of construction contracts are rare. The trustee seldom has the resources to continue performance.

More important and more controversial were well-publicized Chapter 11 petitions by troubled airlines seeking to rid themselves of onerous labor contracts or asbestos manufacturers facing thousands of large tort claims by people who claimed that asbestos gave them cancer.

In *NLRB v. Bildisco and Bildisco*,[72] the contractor filed a Chapter 11 petition and immediately stopped paying pension, health plan, and wage increases agreed on in a recent collective bargaining agreement with the union representing its workers. The U.S. Supreme Court held that this labor practice would not be unfair if the contractor could show that the contract burdened its financial affairs and the equities balanced in favor of rejection. Congress responded to complaints by trade unions by creating difficult procedural requirements before a collective bargaining agreement could be set aside under Chapter 11.[73]

SECTION 33.05 Restitution When a Contract Is Terminated

When a contract is terminated, each party, as a rule, has conferred benefit on the other. The owner has made progress payments, and the contractor has performed work. What effect does termination have on the right of either to recover the value of what it has conferred on the other?

If termination has been made for the convenience of the owner, the clause permitting termination usually provides a formula for compensating the contractor for the work performed. Terminations for default made by the owner usually involve claims by the latter that exceed the value of any unpaid work performed by the contractor. If they do not, the contractor, though in default, in most jurisdictions would be entitled to recover the net benefit conferred on the owner.[74] If the contractor terminated based on an owner default, the contractor can use a restitutionary recovery that permits it in most cases to recover the reasonable value of the services performed.[75] If termination has been accomplished by mutual consent, the agreement will usually deal with compensation for benefits conferred. If it does not, either party is entitled to recover the net benefit it has conferred on the other. For example, if the work performed has a value of $100,000 and the contractor has been paid $90,000, the latter should recover $10,000. Conversely, if the figures were reversed, the owner would be entitled to that amount.

---

[70]Broadly speaking, a preference favors one creditor over the others without any legitimate business reason. For details, see 11 U.S.C.A. § 547.

[71]Id. at § 365(e)(1).

[72]465 U.S. 513 (1984).

[73]11 U.S.C.A. § 1113.

[74]See Section 22.06D.

[75]See Section 27.02E.

SECTION 33.06 Keeping Subcontractors After Termination

An owner (or its financial backers) who exercises its power to terminate the prime contract may want to employ a successor to continue the project. Alternatively, the surety on a performance bond may exercise its option to complete the project with a successor.

Default by the prime contractor is likely to result in failure to pay the subcontractors. This would give subcontractors the power to terminate their obligations under their subcontracts. The owner or surety may wish to take an assignment of some subcontracts (those of subcontractors who are performing well and are not owed an excessive amount of money) without having to renegotiate with them. Renegotiation always causes delay and may result in a higher contract price. Yet most subcontracts contain nonassignment clauses that preclude assignment of any rights under the contract without the consent of the party whose performance is being assigned—in this case, the subcontractors.

To avoid this tangle, it has become common for the prime contract to include an assignment conditioned on default by the prime to the owner of those subcontracts that the owner wishes to take over, with a promise by the prime contractor to obtain the subcontractors' consent to these assignments.[76]

---

[76]AIA Doc. A201-2007, § 5.4. See also § 14.2.2. For a comment on this and other approaches, see J. SWEET & J. SWEET, 2 SWEET ON CONSTRUCTION INDUSTRY CONTRACTS § 17.07 (4th ed. 1999). A fifth edition is planned for publication in the Fall of 2008.

Standard Form of Agreement Between Owner and Architect with Standard Form of Architect's Services

AIA DOCUMENT B101–2007

▓AIA® Document B101™ – 2007

Standard Form of Agreement Between Owner and Architect

AGREEMENT made as of the   day of
in the year of
(In words, indicate day, month and year)

BETWEEN the Architect's client identified as the Owner:
(Name, address and other information)

This document has important legal
consequences. Consultation with
an attorney is encouraged with
respect to its completion or
modification.

and the Architect:
(Name, address and other information)

for the following Project:
(Name, location and detailed description)

The Owner and Architect agree as follows.

TABLE OF ARTICLES

ARTICLE 1 INITIAL INFORMATION
§ 1.1 This Agreement is based on the Initial Information set forth in this Article 1 and in optional Exhibit A, Initial Information:
(*Complete Exhibit A, Initial Information, and incorporate it into the Agreement at Section 13.2, or state below Initial Information such as details of the Project's site and program, Owner's contractors and consultants, Architect's consultants, Owner's budget for the Cost of the Work, authorized representatives, anticipated procurement method, and other information relevant to the Project.*)

§ 1.2 The Owner's anticipated dates for commencement of construction and Substantial Completion of the Work are set forth below:
.1   Commencement of construction date:

.2   Substantial Completion date:

§ 1.3 The Owner and Architect may rely on the Initial Information. Both parties, however, recognize that such information may materially change and, in that event, the Owner and the Architect shall appropriately adjust the schedule, the Architect's services and the Architect's compensation.

ARTICLE 2 ARCHITECT'S RESPONSIBILITIES
§ 2.1 The Architect shall provide the professional services as set forth in this Agreement.

§ 2.2 The Architect shall perform its services consistent with the professional skill and care ordinarily provided by architects practicing in the same or similar locality under the same or similar circumstances. The Architect shall perform its services as expeditiously as is consistent with such professional skill and care and the orderly progress of the Project.

§ 2.3 The Architect shall identify a representative authorized to act on behalf of the Architect with respect to the Project.

§ 2.4 Except with the Owner's knowledge and consent, the Architect shall not engage in any activity, or accept any employment, interest or contribution that would reasonably appear to compromise the Architect's professional judgment with respect to this Project.

§ 2.5 The Architect shall maintain the following insurance for the duration of this Agreement. If any of the requirements set forth below exceed the types and limits the Architect normally maintains, the Owner shall reimburse the Architect for any additional cost:
(Identify types and limits of insurance coverage, and other insurance requirements applicable to the Agreement, if any.)

.1   General Liability

.2   Automobile Liability

.3   Workers' Compensation

.4   Professional Liability

ARTICLE 3   SCOPE OF ARCHITECT'S BASIC SERVICES
§ 3.1 The Architect's Basic Services consist of those described in Article 3 and include usual and customary structural, mechanical, and electrical engineering services. Services not set forth in Article 3 are Additional Services.

§ 3.1.1 The Architect shall manage the Architect's services, consult with the Owner, research applicable design criteria, attend Project meetings, communicate with members of the Project team and report progress to the Owner.

§ 3.1.2 The Architect shall coordinate its services with those services provided by the Owner and the Owner's consultants. The Architect shall be entitled to rely on the accuracy and completeness of services and information furnished by the Owner and the Owner's consultants. The Architect shall provide prompt written notice to the Owner if the Architect becomes aware of any error, omission or inconsistency in such services or information.

§ 3.1.3 As soon as practicable after the date of this Agreement, the Architect shall submit for the Owner's approval a schedule for the performance of the Architect's services. The schedule initially shall include anticipated dates for the commencement of construction and for Substantial Completion of the Work as set forth in the Initial Information. The schedule shall include allowances for periods of time required for the Owner's review, for the performance of the Owner's consultants, and for approval of submissions by authorities having jurisdiction over the Project. Once approved by the Owner, time limits established by the schedule shall not, except for reasonable cause, be exceeded by the Architect or Owner. With the Owner's approval, the Architect shall adjust the schedule, if necessary as the Project proceeds until the commencement of construction.

§ 3.1.4 The Architect shall not be responsible for an Owner's directive or substitution made without the Architect's approval.

§ 3.1.5 The Architect shall, at appropriate times, contact the governmental authorities required to approve the Construction Documents and the entities providing utility services to the Project. In designing the Project, the Architect shall respond to applicable design requirements imposed by such governmental authorities and by such entities providing utility services.

§ 3.1.6 The Architect shall assist the Owner in connection with the Owner's responsibility for filing documents required for the approval of governmental authorities having jurisdiction over the Project.

§ 3.2 SCHEMATIC DESIGN PHASE SERVICES

§ 3.2.1 The Architect shall review the program and other information furnished by the Owner, and shall review laws, codes, and regulations applicable to the Architect's services.

§ 3.2.2 The Architect shall prepare a preliminary evaluation of the Owner's program, schedule, budget for the Cost of the Work, Project site, and the proposed procurement or delivery method and other Initial Information, each in terms of the other, to ascertain the requirements of the Project. The Architect shall notify the Owner of (1) any inconsistencies discovered in the information, and (2) other information or consulting services that may be reasonably needed for the Project.

§ 3.2.3 The Architect shall present its preliminary evaluation to the Owner and shall discuss with the Owner alternative approaches to design and construction of the Project, including the feasibility of incorporating environmentally responsible design approaches. The Architect shall reach an understanding with the Owner regarding the requirements of the Project.

§ 3.2.4 Based on the Project's requirements agreed upon with the Owner, the Architect shall prepare and present for the Owner's approval a preliminary design illustrating the scale and relationship of the Project components.

§ 3.2.5 Based on the Owner's approval of the preliminary design, the Architect shall prepare Schematic Design Documents for the Owner's approval. The Schematic Design Documents shall consist of drawings and other documents including a site plan, if appropriate, and preliminary building plans, sections and elevations; and may include some combination of study models, perspective sketches, or digital modeling. Preliminary selections of major building systems and construction materials shall be noted on the drawings or described in writing.

§ 3.2.5.1 The Architect shall consider environmentally responsible design alternatives, such as material choices and building orientation, together with other considerations based on program and aesthetics, in developing a design that is consistent with the Owner's program, schedule and budget for the Cost of the Work. The Owner may obtain other environmentally responsible design services under Article 4.

§ 3.2.5.2 The Architect shall consider the value of alternative materials, building systems and equipment, together with other considerations based on program and aesthetics in developing a design for the Project that is consistent with the Owner's program, schedule and budget for the Cost of the Work.

§ 3.2.6 The Architect shall submit to the Owner an estimate of the Cost of the Work prepared in accordance with Section 6.3.

§ 3.2.7 The Architect shall submit the Schematic Design Documents to the Owner, and request the Owner's approval.

§ 3.3 DESIGN DEVELOPMENT PHASE SERVICES

§ 3.3.1 Based on the Owner's approval of the Schematic Design Documents, and on the Owner's authorization of any adjustments in the Project requirements and the budget for the Cost of the Work, the Architect shall prepare Design Development Documents for the Owner's approval. The Design Development Documents shall illustrate and describe the development of the approved Schematic Design Documents and shall consist of drawings and other documents including plans, sections, elevations, typical construction details, and diagrammatic layouts of building systems to fix and describe the size and character of the Project as to architectural, structural, mechanical and electrical systems, and such other elements as may be appropriate. The Design Development Documents shall also include outline specifications that identify major materials and systems and establish in general their quality levels.

§ 3.3.2 The Architect shall update the estimate of the Cost of the Work.

§ 3.3.3 The Architect shall submit the Design Development documents to the Owner, advise the Owner of any adjustments to the estimate of the Cost of the Work, and request the Owner's approval.

§ 3.4 CONSTRUCTION DOCUMENTS PHASE SERVICES

§ 3.4.1 Based on the Owner's approval of the Design Development Documents, and on the Owner's authorization of any adjustments in the Project requirements and the budget for the Cost of the Work, the Architect shall prepare Construction Documents for the Owner's approval. The Construction Documents shall illustrate and describe the further development of the approved Design Development Documents and shall consist of Drawings and

Specifications setting forth in detail the quality levels of materials and systems and other requirements for the construction of the Work. The Owner and Architect acknowledge that in order to construct the Work the Contractor will provide additional information, including Shop Drawings, Product Data, Samples and other similar submittals, which the Architect shall review in accordance with Section 3.6.4.

§ 3.4.2 The Architect shall incorporate into the Construction Documents the design requirements of governmental authorities having jurisdiction over the Project.

§ 3.4.3 During the development of the Construction Documents, the Architect shall assist the Owner in the development and preparation of (1) bidding and procurement information that describes the time, place and conditions of bidding, including bidding or proposal forms; (2) the form of agreement between the Owner and Contractor; and (3) the Conditions of the Contract for Construction (General, Supplementary and other Conditions). The Architect shall also compile a project manual that includes the Conditions of the Contract for Construction and Specifications and may include bidding requirements and sample forms.

§ 3.4.4 The Architect shall update the estimate for the Cost of the Work.

§ 3.4.5 The Architect shall submit the Construction Documents to the Owner, advise the Owner of any adjustments to the estimate of the Cost of the Work, take any action required under Section 6.5, and request the Owner's approval.

§ 3.5 BIDDING OR NEGOTIATION PHASE SERVICES
§ 3.5.1 GENERAL
The Architect shall assist the Owner in establishing a list of prospective contractors. Following the Owner's approval of the Construction Documents, the Architect shall assist the Owner in (1) obtaining either competitive bids or negotiated proposals; (2) confirming responsiveness of bids or proposals; (3) determining the successful bid or proposal, if any; and, (4) awarding and preparing contracts for construction.

§ 3.5.2 COMPETITIVE BIDDING
§ 3.5.2.1 Bidding Documents shall consist of bidding requirements and proposed Contract Documents.

§ 3.5.2.2 The Architect shall assist the Owner in bidding the Project by
    .1   procuring the reproduction of Bidding Documents for distribution to prospective bidders;
    .2   distributing the Bidding Documents to prospective bidders, requesting their return upon completion of the bidding process, and maintaining a log of distribution and retrieval and of the amounts of deposits, if any, received from and returned to prospective bidders;
    .3   organizing and conducting a pre-bid conference for prospective bidders;
    .4   preparing responses to questions from prospective bidders and providing clarifications and interpretations of the Bidding Documents to all prospective bidders in the form of addenda; and
    .5   organizing and conducting the opening of the bids, and subsequently documenting and distributing the bidding results, as directed by the Owner.

§ 3.5.2.3 The Architect shall consider requests for substitutions, if the Bidding Documents permit substitutions, and shall prepare and distribute addenda identifying approved substitutions to all prospective bidders.

§ 3.5.3 NEGOTIATED PROPOSALS
§ 3.5.3.1 Proposal Documents shall consist of proposal requirements and proposed Contract Documents.

§ 3.5.3.2 The Architect shall assist the Owner in obtaining proposals by
    .1   procuring the reproduction of Proposal Documents for distribution to prospective contractors, and requesting their return upon completion of the negotiation process;
    .2   organizing and participating in selection interviews with prospective contractors; and
    .3   participating in negotiations with prospective contractors, and subsequently preparing a summary report of the negotiation results, as directed by the Owner.

§ 3.5.3.3 The Architect shall consider requests for substitutions, if the Proposal Documents permit substitutions, and shall prepare and distribute addenda identifying approved substitutions to all prospective contractors.

---

§ 3.6 CONSTRUCTION PHASE SERVICES
§ 3.6.1 GENERAL
§ 3.6.1.1 The Architect shall provide administration of the Contract between the Owner and the Contractor as set forth below and in AIA Document A201™–2007, General Conditions of the Contract for Construction. If the Owner and Contractor modify AIA Document A201–2007, those modifications shall not affect the Architect's services under this Agreement unless the Owner and the Architect amend this Agreement.

§ 3.6.1.2 The Architect shall advise and consult with the Owner during the Construction Phase Services. The Architect shall have authority to act on behalf of the Owner only to the extent provided in this Agreement. The Architect shall not have control over, charge of, or responsibility for the construction means, methods, techniques, sequences or procedures, or for safety precautions and programs in connection with the Work, nor shall the Architect be responsible for the Contractor's failure to perform the Work in accordance with the requirements of the Contract Documents. The Architect shall be responsible for the Architect's negligent acts or omissions, but shall not have control over or charge of, and shall not be responsible for, acts or omissions of the Contractor or of any other persons or entities performing portions of the Work.

§ 3.6.1.3 Subject to Section 4.3, the Architect's responsibility to provide Construction Phase Services commences with the award of the Contract for Construction and terminates on the date the Architect issues the final Certificate for Payment.

§ 3.6.2 EVALUATIONS OF THE WORK
§ 3.6.2.1 The Architect shall visit the site at intervals appropriate to the stage of construction, or as otherwise required in Section 4.3.3, to become generally familiar with the progress and quality of the portion of the Work completed, and to determine, in general, if the Work observed is being performed in a manner indicating that the Work, when fully completed, will be in accordance with the Contract Documents. However, the Architect shall not be required to make exhaustive or continuous on-site inspections to check the quality or quantity of the Work. On the basis of the site visits, the Architect shall keep the Owner reasonably informed about the progress and quality of the portion of the Work completed, and report to the Owner (1) known deviations from the Contract Documents and from the most recent construction schedule submitted by the Contractor, and (2) defects and deficiencies observed in the Work.

§ 3.6.2.2 The Architect has the authority to reject Work that does not conform to the Contract Documents. Whenever the Architect considers it necessary or advisable, the Architect shall have the authority to require inspection or testing of the Work in accordance with the provisions of the Contract Documents, whether or not such Work is fabricated, installed or completed. However, neither this authority of the Architect nor a decision made in good faith either to exercise or not to exercise such authority shall give rise to a duty or responsibility of the Architect to the Contractor, Subcontractors, material and equipment suppliers, their agents or employees or other persons or entities performing portions of the Work.

§ 3.6.2.3 The Architect shall interpret and decide matters concerning performance under, and requirements of, the Contract Documents on written request of either the Owner or Contractor. The Architect's response to such requests shall be made in writing within any time limits agreed upon or otherwise with reasonable promptness.

§ 3.6.2.4 Interpretations and decisions of the Architect shall be consistent with the intent of and reasonably inferable from the Contract Documents and shall be in writing or in the form of drawings. When making such interpretations and decisions, the Architect shall endeavor to secure faithful performance by both Owner and Contractor, shall not show partiality to either, and shall not be liable for results of interpretations or decisions rendered in good faith. The Architect's decisions on matters relating to aesthetic effect shall be final if consistent with the intent expressed in the Contract Documents.

§ 3.6.2.5 Unless the Owner and Contractor designate another person to serve as an Initial Decision Maker, as that term is defined in AIA Document A201–2007, the Architect shall render initial decisions on Claims between the Owner and Contractor as provided in the Contract Documents.

§ 3.6.3 CERTIFICATES FOR PAYMENT TO CONTRACTOR
§ 3.6.3.1 The Architect shall review and certify the amounts due the Contractor and shall issue certificates in such amounts. The Architect's certification for payment shall constitute a representation to the Owner, based on the Architect's evaluation of the Work as provided in Section 3.6.2 and on the data comprising the Contractor's Application for Payment, that, to the best of the Architect's knowledge, information and belief, the Work has

progressed to the point indicated and that the quality of the Work is in accordance with the Contract Documents. The foregoing representations are subject (1) to an evaluation of the Work for conformance with the Contract Documents upon Substantial Completion, (2) to results of subsequent tests and inspections, (3) to correction of minor deviations from the Contract Documents prior to completion, and (4) to specific qualifications expressed by the Architect.

§ 3.6.3.2 The issuance of a Certificate for Payment shall not be a representation that the Architect has (1) made exhaustive or continuous on-site inspections to check the quality or quantity of the Work, (2) reviewed construction means, methods, techniques, sequences or procedures, (3) reviewed copies of requisitions received from Subcontractors and material suppliers and other data requested by the Owner to substantiate the Contractor's right to payment, or (4) ascertained how or for what purpose the Contractor has used money previously paid on account of the Contract Sum.

§ 3.6.3.3 The Architect shall maintain a record of the Applications and Certificates for Payment.

§ 3.6.4 SUBMITTALS

§ 3.6.4.1 The Architect shall review the Contractor's submittal schedule and shall not unreasonably delay or withhold approval. The Architect's action in reviewing submittals shall be taken in accordance with the approved submittal schedule or, in the absence of an approved submittal schedule, with reasonable promptness while allowing sufficient time in the Architect's professional judgment to permit adequate review.

§ 3.6.4.2 In accordance with the Architect-approved submittal schedule, the Architect shall review and approve or take other appropriate action upon the Contractor's submittals such as Shop Drawings, Product Data and Samples, but only for the limited purpose of checking for conformance with information given and the design concept expressed in the Contract Documents. Review of such submittals is not for the purpose of determining the accuracy and completeness of other information such as dimensions, quantities, and installation or performance of equipment or systems, which are the Contractor's responsibility. The Architect's review shall not constitute approval of safety precautions or, unless otherwise specifically stated by the Architect, of any construction means, methods, techniques, sequences or procedures. The Architect's approval of a specific item shall not indicate approval of an assembly of which the item is a component.

§ 3.6.4.3 If the Contract Documents specifically require the Contractor to provide professional design services or certifications by a design professional related to systems, materials or equipment, the Architect shall specify the appropriate performance and design criteria that such services must satisfy. The Architect shall review shop drawings and other submittals related to the Work designed or certified by the design professional retained by the Contractor that bear such professional's seal and signature when submitted to the Architect. The Architect shall be entitled to rely upon the adequacy, accuracy and completeness of the services, certifications and approvals performed or provided by such design professionals.

§ 3.6.4.4 Subject to the provisions of Section 4.3, the Architect shall review and respond to requests for information about the Contract Documents. The Architect shall set forth in the Contract Documents the requirements for requests for information. Requests for information shall include, at a minimum, a detailed written statement that indicates the specific Drawings or Specifications in need of clarification and the nature of the clarification requested. The Architect's response to such requests shall be made in writing within any time limits agreed upon, or otherwise with reasonable promptness. If appropriate, the Architect shall prepare and issue supplemental Drawings and Specifications in response to requests for information.

§ 3.6.4.5 The Architect shall maintain a record of submittals and copies of submittals supplied by the Contractor in accordance with the requirements of the Contract Documents.

§ 3.6.5 CHANGES IN THE WORK

§ 3.6.5.1 The Architect may authorize minor changes in the Work that are consistent with the intent of the Contract Documents and do not involve an adjustment in the Contract Sum or an extension of the Contract Time. Subject to the provisions of Section 4.3, the Architect shall prepare Change Orders and Construction Change Directives for the Owner's approval and execution in accordance with the Contract Documents.

§ 3.6.5.2 The Architect shall maintain records relative to changes in the Work.

§ 3.6.6 PROJECT COMPLETION

§ 3.6.6.1 The Architect shall conduct inspections to determine the date or dates of Substantial Completion and the date of final completion; issue Certificates of Substantial Completion; receive from the Contractor and forward to the Owner, for the Owner's review and records, written warranties and related documents required by the Contract Documents and assembled by the Contractor; and issue a final Certificate for Payment based upon a final inspection indicating the Work complies with the requirements of the Contract Documents.

§ 3.6.6.2 The Architect's inspections shall be conducted with the Owner to check conformance of the Work with the requirements of the Contract Documents and to verify the accuracy and completeness of the list submitted by the Contractor of Work to be completed or corrected.

§ 3.6.6.3 When the Work is found to be substantially complete, the Architect shall inform the Owner about the balance of the Contract Sum remaining to be paid the Contractor, including the amount to be retained from the Contract Sum, if any, for final completion or correction of the Work.

§ 3.6.6.4 The Architect shall forward to the Owner the following information received from the Contractor: (1) consent of surety or sureties, if any, to reduction in or partial release of retainage or the making of final payment; (2) affidavits, receipts, releases and waivers of liens or bonds indemnifying the Owner against liens; and (3) any other documentation required of the Contractor under the Contract Documents.

§ 3.6.6.5 Upon request of the Owner, and prior to the expiration of one year from the date of Substantial Completion, the Architect shall, without additional compensation, conduct a meeting with the Owner to review the facility operations and performance.

ARTICLE 4   ADDITIONAL SERVICES

§ 4.1 Additional Services listed below are not included in Basic Services but may be required for the Project. The Architect shall provide the listed Additional Services only if specifically designated in the table below as the Architect's responsibility, and the Owner shall compensate the Architect as provided in Section 11.2.
(Designate the Additional Services the Architect shall provide in the second column of the table below. In the third column indicate whether the service description is located in Section 4.2 or in an attached exhibit. If in an exhibit, identify the exhibit.)

Additional Services		Responsibility *(Architect, Owner or Not Provided)*	Location of Service Description *(Section 4.2 below or in an exhibit attached to this document and identified below)*
§ 4.1.1	Programming		
§ 4.1.2	Multiple preliminary designs		
§ 4.1.3	Measured drawings		
§ 4.1.4	Existing facilities surveys		
§ 4.1.5	Site Evaluation and Planning (B203™–2007)		
§ 4.1.6	Building information modeling		
§ 4.1.7	Civil engineering		
§ 4.1.8	Landscape design		
§ 4.1.9	Architectural Interior Design (B252™–2007)		
§ 4.1.10	Value Analysis (B204™–2007)		
§ 4.1.11	Detailed cost estimating		
§ 4.1.12	On-site project representation		
§ 4.1.13	Conformed construction documents		
§ 4.1.14	As-designed record drawings		
§ 4.1.15	As-constructed record drawings		
§ 4.1.16	Post occupancy evaluation		
§ 4.1.17	Facility Support Services (B210™–2007)		
§ 4.1.18	Tenant-related services		
§ 4.1.19	Coordination of Owner's consultants		
§ 4.1.20	Telecommunications/data design		

Additional Services		Responsibility *(Architect, Owner or Not Provided)*	Location of Service Description *(Section 4.2 below or in an exhibit attached to this document and identified below)*
§ 4.1.21	Security Evaluation and Planning (B206™–2007)		
§ 4.1.22	Commissioning (B211™–2007)		
§ 4.1.23	Extensive environmentally responsible design		
§ 4.1.24	LEED® Certification (B214™–2007)		
§ 4.1.25	Fast-track design services		
§ 4.1.26	Historic Preservation (B205™–2007)		
§ 4.1.27	Furniture, Finishings, and Equipment Design (B253™–2007)		
§ 4.1.28	Other		

§ 4.2 Insert a description of each Additional Service designated in Section 4.1 as the Architect's responsibility, if not further described in an exhibit attached to this document.

§ 4.3 Additional Services may be provided after execution of this Agreement, without invalidating the Agreement. Except for services required due to the fault of the Architect, any Additional Services provided in accordance with this Section 4.3 shall entitle the Architect to compensation pursuant to Section 11.3 and an appropriate adjustment in the Architect's schedule.

§ 4.3.1 Upon recognizing the need to perform the following Additional Services, the Architect shall notify the Owner with reasonable promptness and explain the facts and circumstances giving rise to the need. The Architect shall not proceed to provide the following services until the Architect receives the Owner's written authorization:

    .1   Services necessitated by a change in the Initial Information, previous instructions or approvals given by the Owner, or a material change in the Project including, but not limited to, size, quality, complexity, the Owner's schedule or budget for Cost of the Work, or procurement or delivery method;

    .2   Services necessitated by the Owner's request for extensive environmentally responsible design alternatives, such as unique system designs, in-depth material research, energy modeling, or LEED® certification;

    .3   Changing or editing previously prepared Instruments of Service necessitated by the enactment or revision of codes, laws or regulations or official interpretations;

    .4   Services necessitated by decisions of the Owner not rendered in a timely manner or any other failure of performance on the part of the Owner or the Owner's consultants or contractors;

    .5   Preparing digital data for transmission to the Owner's consultants and contractors, or to other Owner authorized recipients;

    .6   Preparation of design and documentation for alternate bid or proposal requests proposed by the Owner;

    .7   Preparation for, and attendance at, a public presentation, meeting or hearing;

    .8   Preparation for, and attendance at a dispute resolution proceeding or legal proceeding, except where the Architect is party thereto;

    .9   Evaluation of the qualifications of bidders or persons providing proposals;

    .10   Consultation concerning replacement of Work resulting from fire or other cause during construction; or

    .11   Assistance to the Initial Decision Maker, if other than the Architect.

§ 4.3.2 To avoid delay in the Construction Phase, the Architect shall provide the following Additional Services, notify the Owner with reasonable promptness, and explain the facts and circumstances giving rise to the need. If the Owner subsequently determines that all or parts of those services are not required, the Owner shall give prompt written notice to the Architect, and the Owner shall have no further obligation to compensate the Architect for those services:

.1   Reviewing a Contractor's submittal out of sequence from the submittal schedule agreed to by the Architect;

.2   Responding to the Contractor's requests for information that are not prepared in accordance with the Contract Documents or where such information is available to the Contractor from a careful study and comparison of the Contract Documents, field conditions, other Owner-provided information, Contractor-prepared coordination drawings, or prior Project correspondence or documentation;

.3   Preparing Change Orders and Construction Change Directives that require evaluation of Contractor's proposals and supporting data, or the preparation or revision of Instruments of Service;

.4   Evaluating an extensive number of Claims as the Initial Decision Maker;

.5   Evaluating substitutions proposed by the Owner or Contractor and making subsequent revisions to Instruments of Service resulting therefrom; or

.6   To the extent the Architect's Basic Services are affected, providing Construction Phase Services 60 days after (1) the date of Substantial Completion of the Work or (2) the anticipated date of Substantial Completion identified in Initial Information, whichever is earlier.

§ 4.3.3 The Architect shall provide Construction Phase Services exceeding the limits set forth below as Additional Services. When the limits below are reached, the Architect shall notify the Owner:

.1   (   ) reviews of each Shop Drawing, Product Data item, sample and similar submittal of the Contractor

.2   (   ) visits to the site by the Architect over the duration of the Project during construction

.3   (   ) inspections for any portion of the Work to determine whether such portion of the Work is substantially complete in accordance with the requirements of the Contract Documents

.4   (   ) inspections for any portion of the Work to determine final completion

§ 4.3.4 If the services covered by this Agreement have not been completed within   (   ) months of the date of this Agreement, through no fault of the Architect, extension of the Architect's services beyond that time shall be compensated as Additional Services.

ARTICLE 5 OWNER'S RESPONSIBILITIES

§ 5.1 Unless otherwise provided for under this Agreement, the Owner shall provide information in a timely manner regarding requirements for and limitations on the Project, including a written program which shall set forth the Owner's objectives, schedule, constraints and criteria, including space requirements and relationships, flexibility, expandability, special equipment, systems and site requirements. Within 15 days after receipt of a written request from the Architect, the Owner shall furnish the requested information as necessary and relevant for the Architect to evaluate, give notice of or enforce lien rights.

§ 5.2 The Owner shall establish and periodically update the Owner's budget for the Project, including (1) the budget for the Cost of the Work as defined in Section 6.1; (2) the Owner's other costs; and, (3) reasonable contingencies related to all of these costs. If the Owner significantly increases or decreases the Owner's budget for the Cost of the Work, the Owner shall notify the Architect. The Owner and the Architect shall thereafter agree to a corresponding change in the Project's scope and quality.

§ 5.3 The Owner shall identify a representative authorized to act on the Owner's behalf with respect to the Project. The Owner shall render decisions and approve the Architect's submittals in a timely manner in order to avoid unreasonable delay in the orderly and sequential progress of the Architect's services.

§ 5.4 The Owner shall furnish surveys to describe physical characteristics, legal limitations and utility locations for the site of the Project, and a written legal description of the site. The surveys and legal information shall include, as applicable, grades and lines of streets, alleys, pavements and adjoining property and structures; designated wetlands; adjacent drainage; rights-of-way, restrictions, easements, encroachments, zoning, deed restrictions, boundaries and contours of the site; locations, dimensions and necessary data with respect to existing buildings, other improvements and trees; and information concerning available utility services and lines, both public and private, above and below grade, including inverts and depths. All the information on the survey shall be referenced to a Project benchmark.

§ 5.5 The Owner shall furnish services of geotechnical engineers, which may include but are not limited to test borings, test pits, determinations of soil bearing values, percolation tests, evaluations of hazardous materials, seismic evaluation, ground corrosion tests and resistivity tests, including necessary operations for anticipating subsoil conditions, with written reports and appropriate recommendations.

§ 5.6 The Owner shall coordinate the services of its own consultants with those services provided by the Architect. Upon the Architect's request, the Owner shall furnish copies of the scope of services in the contracts between the Owner and the Owner's consultants. The Owner shall furnish the services of consultants other than those designated in this Agreement, or authorize the Architect to furnish them as an Additional Service, when the Architect requests such services and demonstrates that they are reasonably required by the scope of the Project. The Owner shall require that its consultants maintain professional liability insurance as appropriate to the services provided.

§ 5.7 The Owner shall furnish tests, inspections and reports required by law or the Contract Documents, such as structural, mechanical, and chemical tests, tests for air and water pollution, and tests for hazardous materials.

§ 5.8 The Owner shall furnish all legal, insurance and accounting services, including auditing services, that may be reasonably necessary at any time for the Project to meet the Owner's needs and interests.

§ 5.9 The Owner shall provide prompt written notice to the Architect if the Owner becomes aware of any fault or defect in the Project, including errors, omissions or inconsistencies in the Architect's Instruments of Service.

§ 5.10 Except as otherwise provided in this Agreement, or when direct communications have been specially authorized, the Owner shall endeavor to communicate with the Contractor and the Architect's consultants through the Architect about matters arising out of or relating to the Contract Documents. The Owner shall promptly notify the Architect of any direct communications that may affect the Architect's services.

§ 5.11 Before executing the Contract for Construction, the Owner shall coordinate the Architect's duties and responsibilities set forth in the Contract for Construction with the Architect's services set forth in this Agreement. The Owner shall provide the Architect a copy of the executed agreement between the Owner and Contractor, including the General Conditions of the Contract for Construction.

§ 5.12 The Owner shall provide the Architect access to the Project site prior to commencement of the Work and shall obligate the Contractor to provide the Architect access to the Work wherever it is in preparation or progress.

ARTICLE 6 COST OF THE WORK

§ 6.1 For purposes of this Agreement, the Cost of the Work shall be the total cost to the Owner to construct all elements of the Project designed or specified by the Architect and shall include contractors' general conditions costs, overhead and profit. The Cost of the Work does not include the compensation of the Architect, the costs of the land, rights-of-way, financing, contingencies for changes in the Work or other costs that are the responsibility of the Owner.

§ 6.2 The Owner's budget for the Cost of the Work is provided in Initial Information, and may be adjusted throughout the Project as required under Sections 5.2, 6.4 and 6.5. Evaluations of the Owner's budget for the Cost of the Work, the preliminary estimate of the Cost of the Work and updated estimates of the Cost of the Work prepared by the Architect, represent the Architect's judgment as a design professional. It is recognized, however, that neither the Architect nor the Owner has control over the cost of labor, materials or equipment; the Contractor's methods of determining bid prices; or competitive bidding, market or negotiating conditions. Accordingly, the Architect cannot and does not warrant or represent that bids or negotiated prices will not vary from the Owner's budget for the Cost of the Work or from any estimate of the Cost of the Work or evaluation prepared or agreed to by the Architect.

§ 6.3 In preparing estimates of the Cost of Work, the Architect shall be permitted to include contingencies for design, bidding and price escalation; to determine what materials, equipment, component systems and types of construction are to be included in the Contract Documents; to make reasonable adjustments in the program and scope of the Project; and to include in the Contract Documents alternate bids as may be necessary to adjust the estimated Cost of the Work to meet the Owner's budget for the Cost of the Work. The Architect's estimate of the Cost of the Work shall be based on current area, volume or similar conceptual estimating techniques. If the Owner requests detailed cost estimating services, the Architect shall provide such services as an Additional Service under Article 4.

§ 6.4 If the Bidding or Negotiation Phase has not commenced within 90 days after the Architect submits the Construction Documents to the Owner, through no fault of the Architect, the Owner's budget for the Cost of the Work shall be adjusted to reflect changes in the general level of prices in the applicable construction market.

§ 6.5 If at any time the Architect's estimate of the Cost of the Work exceeds the Owner's budget for the Cost of the Work, the Architect shall make appropriate recommendations to the Owner to adjust the Project's size, quality or budget for the Cost of the Work, and the Owner shall cooperate with the Architect in making such adjustments.

§ 6.6 If the Owner's budget for the Cost of the Work at the conclusion of the Construction Documents Phase Services is exceeded by the lowest bona fide bid or negotiated proposal, the Owner shall

.1   give written approval of an increase in the budget for the Cost of the Work;

.2   authorize rebidding or renegotiating of the Project within a reasonable time;

.3   terminate in accordance with Section 9.5;

.4   in consultation with the Architect, revise the Project program, scope, or quality as required to reduce the Cost of the Work; or

.5   implement any other mutually acceptable alternative.

§ 6.7 If the Owner chooses to proceed under Section 6.6.4, the Architect, without additional compensation, shall modify the Construction Documents as necessary to comply with the Owner's budget for the Cost of the Work at the conclusion of the Construction Documents Phase Services, or the budget as adjusted under Section 6.6.1. The Architect's modification of the Construction Documents shall be the limit of the Architect's responsibility under this Article 6.

ARTICLE 7   COPYRIGHTS AND LICENSES

§ 7.1 The Architect and the Owner warrant that in transmitting Instruments of Service, or any other information, the transmitting party is the copyright owner of such information or has permission from the copyright owner to transmit such information for its use on the Project. If the Owner and Architect intend to transmit Instruments of Service or any other information or documentation in digital form, they shall endeavor to establish necessary protocols governing such transmissions.

§ 7.2 The Architect and the Architect's consultants shall be deemed the authors and owners of their respective Instruments of Service, including the Drawings and Specifications, and shall retain all common law, statutory and other reserved rights, including copyrights. Submission or distribution of Instruments of Service to meet official regulatory requirements or for similar purposes in connection with the Project is not to be construed as publication in derogation of the reserved rights of the Architect and the Architect's consultants.

§ 7.3 Upon execution of this Agreement, the Architect grants to the Owner a nonexclusive license to use the Architect's Instruments of Service solely and exclusively for purposes of constructing, using, maintaining, altering and adding to the Project, provided that the Owner substantially performs its obligations, including prompt payment of all sums when due, under this Agreement. The Architect shall obtain similar nonexclusive licenses from the Architect's consultants consistent with this Agreement. The license granted under this section permits the Owner to authorize the Contractor, Subcontractors, Sub-subcontractors, and material or equipment suppliers, as well as the Owner's consultants and separate contractors, to reproduce applicable portions of the Instruments of Service solely and exclusively for use in performing services or construction for the Project. If the Architect rightfully terminates this Agreement for cause as provided in Section 9.4, the license granted in this Section 7.3 shall terminate.

§ 7.3.1 In the event the Owner uses the Instruments of Service without retaining the author of the Instruments of Service, the Owner releases the Architect and Architect's consultant(s) from all claims and causes of action arising from such uses. The Owner, to the extent permitted by law, further agrees to indemnify and hold harmless the Architect and its consultants from all costs and expenses, including the cost of defense, related to claims and causes of action asserted by any third person or entity to the extent such costs and expenses arise from the Owner's use of the Instruments of Service under this Section 7.3.1. The terms of this Section 7.3.1 shall not apply if the Owner rightfully terminates this Agreement for cause under Section 9.4.

§ 7.4 Except for the licenses granted in this Article 7, no other license or right shall be deemed granted or implied under this Agreement. The Owner shall not assign, delegate, sublicense, pledge or otherwise transfer any license granted herein to another party without the prior written agreement of the Architect. Any unauthorized use of the

Instruments of Service shall be at the Owner's sole risk and without liability to the Architect and the Architect's consultants.

ARTICLE 8   CLAIMS AND DISPUTES
§ 8.1 GENERAL

§ 8.1.1 The Owner and Architect shall commence all claims and causes of action, whether in contract, tort, or otherwise, against the other arising out of or related to this Agreement in accordance with the requirements of the method of binding dispute resolution selected in this Agreement within the period specified by applicable law, but in any case not more than 10 years after the date of Substantial Completion of the Work. The Owner and Architect waive all claims and causes of action not commenced in accordance with this Section 8.1.1.

§ 8.1.2 To the extent damages are covered by property insurance, the Owner and Architect waive all rights against each other and against the contractors, consultants, agents and employees of the other for damages, except such rights as they may have to the proceeds of such insurance as set forth in AIA Document A201–2007, General Conditions of the Contract for Construction. The Owner or the Architect, as appropriate, shall require of the contractors, consultants, agents and employees of any of them similar waivers in favor of the other parties enumerated herein.

§ 8.1.3 The Architect and Owner waive consequential damages for claims, disputes or other matters in question arising out of or relating to this Agreement. This mutual waiver is applicable, without limitation, to all consequential damages due to either party's termination of this Agreement, except as specifically provided in Section 9.7.

§ 8.2 MEDIATION

§ 8.2.1 Any claim, dispute or other matter in question arising out of or related to this Agreement shall be subject to mediation as a condition precedent to binding dispute resolution. If such matter relates to or is the subject of a lien arising out of the Architect's services, the Architect may proceed in accordance with applicable law to comply with the lien notice or filing deadlines prior to resolution of the matter by mediation or by binding dispute resolution.

§ 8.2.2 The Owner and Architect shall endeavor to resolve claims, disputes and other matters in question between them by mediation which, unless the parties mutually agree otherwise, shall be administered by the American Arbitration Association in accordance with its Construction Industry Mediation Procedures in effect on the date of the Agreement. A request for mediation shall be made in writing, delivered to the other party to the Agreement, and filed with the person or entity administering the mediation. The request may be made concurrently with the filing of a complaint or other appropriate demand for binding dispute resolution but, in such event, mediation shall proceed in advance of binding dispute resolution proceedings, which shall be stayed pending mediation for a period of 60 days from the date of filing, unless stayed for a longer period by agreement of the parties or court order. If an arbitration proceeding is stayed pursuant to this section, the parties may nonetheless proceed to the selection of the arbitrator(s) and agree upon a schedule for later proceedings.

§ 8.2.3 The parties shall share the mediator's fee and any filing fees equally. The mediation shall be held in the place where the Project is located, unless another location is mutually agreed upon. Agreements reached in mediation shall be enforceable as settlement agreements in any court having jurisdiction thereof.

§ 8.2.4 If the parties do not resolve a dispute through mediation pursuant to this Section 8.2, the method of binding dispute resolution shall be the following:
(Check the appropriate box. If the Owner and Architect do not select a method of binding dispute resolution below, or do not subsequently agree in writing to a binding dispute resolution method other than litigation, the dispute will be resolved in a court of competent jurisdiction.)

☐   Arbitration pursuant to Section 8.3 of this Agreement

☐   Litigation in a court of competent jurisdiction

☐   Other *(Specify)*

§ 8.3 ARBITRATION

§ 8.3.1 If the parties have selected arbitration as the method for binding dispute resolution in this Agreement, any claim, dispute or other matter in question arising out of or related to this Agreement subject to, but not resolved by, mediation shall be subject to arbitration which, unless the parties mutually agree otherwise, shall be administered by the American Arbitration Association in accordance with its Construction Industry Arbitration Rules in effect on the date of this Agreement. A demand for arbitration shall be made in writing, delivered to the other party to this Agreement, and filed with the person or entity administering the arbitration.

§ 8.3.1.1 A demand for arbitration shall be made no earlier than concurrently with the filing of a request for mediation, but in no event shall it be made after the date when the institution of legal or equitable proceedings based on the claim, dispute or other matter in question would be barred by the applicable statute of limitations. For statute of limitations purposes, receipt of a written demand for arbitration by the person or entity administering the arbitration shall constitute the institution of legal or equitable proceedings based on the claim, dispute or other matter in question.

§ 8.3.2 The foregoing agreement to arbitrate and other agreements to arbitrate with an additional person or entity duly consented to by parties to this Agreement shall be specifically enforceable in accordance with applicable law in any court having jurisdiction thereof.

§ 8.3.3 The award rendered by the arbitrator(s) shall be final, and judgment may be entered upon it in accordance with applicable law in any court having jurisdiction thereof.

§ 8.3.4 CONSOLIDATION OR JOINDER

§ 8.3.4.1 Either party, at its sole discretion, may consolidate an arbitration conducted under this Agreement with any other arbitration to which it is a party provided that (1) the arbitration agreement governing the other arbitration permits consolidation; (2) the arbitrations to be consolidated substantially involve common questions of law or fact; and (3) the arbitrations employ materially similar procedural rules and methods for selecting arbitrator(s).

§ 8.3.4.2 Either party, at its sole discretion, may include by joinder persons or entities substantially involved in a common question of law or fact whose presence is required if complete relief is to be accorded in arbitration, provided that the party sought to be joined consents in writing to such joinder. Consent to arbitration involving an additional person or entity shall not constitute consent to arbitration of any claim, dispute or other matter in question not described in the written consent.

§ 8.3.4.3 The Owner and Architect grant to any person or entity made a party to an arbitration conducted under this Section 8.3, whether by joinder or consolidation, the same rights of joinder and consolidation as the Owner and Architect under this Agreement.

ARTICLE 9   TERMINATION OR SUSPENSION

§ 9.1 If the Owner fails to make payments to the Architect in accordance with this Agreement, such failure shall be considered substantial nonperformance and cause for termination or, at the Architect's option, cause for suspension of performance of services under this Agreement. If the Architect elects to suspend services, the Architect shall give seven days' written notice to the Owner before suspending services. In the event of a suspension of services, the Architect shall have no liability to the Owner for delay or damage caused the Owner because of such suspension of services. Before resuming services, the Architect shall be paid all sums due prior to suspension and any expenses incurred in the interruption and resumption of the Architect's services. The Architect's fees for the remaining services and the time schedules shall be equitably adjusted.

§ 9.2 If the Owner suspends the Project, the Architect shall be compensated for services performed prior to notice of such suspension. When the Project is resumed, the Architect shall be compensated for expenses incurred in the interruption and resumption of the Architect's services. The Architect's fees for the remaining services and the time schedules shall be equitably adjusted.

§ 9.3 If the Owner suspends the Project for more than 90 cumulative days for reasons other than the fault of the Architect, the Architect may terminate this Agreement by giving not less than seven days' written notice.

§ 9.4 Either party may terminate this Agreement upon not less than seven days' written notice should the other party fail substantially to perform in accordance with the terms of this Agreement through no fault of the party initiating the termination.

§ 9.5 The Owner may terminate this Agreement upon not less than seven days' written notice to the Architect for the Owner's convenience and without cause.

§ 9.6 In the event of termination not the fault of the Architect, the Architect shall be compensated for services performed prior to termination, together with Reimbursable Expenses then due and all Termination Expenses as defined in Section 9.7.

§ 9.7 Termination Expenses are in addition to compensation for the Architect's services and include expenses directly attributable to termination for which the Architect is not otherwise compensated, plus an amount for the Architect's anticipated profit on the value of the services not performed by the Architect.

§ 9.8 The Owner's rights to use the Architect's Instruments of Service in the event of a termination of this Agreement are set forth in Article 7 and Section 11.9.

ARTICLE 10 MISCELLANEOUS PROVISIONS
§ 10.1 This Agreement shall be governed by the law of the place where the Project is located, except that if the parties have selected arbitration as the method of binding dispute resolution, the Federal Arbitration Act shall govern Section 8.3.

§ 10.2 Terms in this Agreement shall have the same meaning as those in AIA Document A201–2007, General Conditions of the Contract for Construction.

§ 10.3 The Owner and Architect, respectively, bind themselves, their agents, successors, assigns and legal representatives to this Agreement. Neither the Owner nor the Architect shall assign this Agreement without the written consent of the other, except that the Owner may assign this Agreement to a lender providing financing for the Project if the lender agrees to assume the Owner's rights and obligations under this Agreement.

§ 10.4 If the Owner requests the Architect to execute certificates, the proposed language of such certificates shall be submitted to the Architect for review at least 14 days prior to the requested dates of execution. If the Owner requests the Architect to execute consents reasonably required to facilitate assignment to a lender, the Architect shall execute all such consents that are consistent with this Agreement, provided the proposed consent is submitted to the Architect for review at least 14 days prior to execution. The Architect shall not be required to execute certificates or consents that would require knowledge, services or responsibilities beyond the scope of this Agreement.

§ 10.5 Nothing contained in this Agreement shall create a contractual relationship with or a cause of action in favor of a third party against either the Owner or Architect.

§ 10.6 Unless otherwise required in this Agreement, the Architect shall have no responsibility for the discovery, presence, handling, removal or disposal of, or exposure of persons to, hazardous materials or toxic substances in any form at the Project site.

§ 10.7 The Architect shall have the right to include photographic or artistic representations of the design of the Project among the Architect's promotional and professional materials. The Architect shall be given reasonable access to the completed Project to make such representations. However, the Architect's materials shall not include the Owner's confidential or proprietary information if the Owner has previously advised the Architect in writing of the specific information considered by the Owner to be confidential or proprietary. The Owner shall provide professional credit for the Architect in the Owner's promotional materials for the Project.

§ 10.8 If the Architect or Owner receives information specifically designated by the other party as "confidential" or "business proprietary," the receiving party shall keep such information strictly confidential and shall not disclose it to any other person except to (1) its employees, (2) those who need to know the content of such information in order to perform services or construction solely and exclusively for the Project, or (3) its consultants and contractors whose contracts include similar restrictions on the use of confidential information.

Init. /

15

ARTICLE 11 COMPENSATION

§ 11.1 For the Architect's Basic Services described under Article 3, the Owner shall compensate the Architect as follows:
(Insert amount of, or basis for, compensation.)

§ 11.2 For Additional Services designated in Section 4.1, the Owner shall compensate the Architect as follows:
(Insert amount of, or basis for, compensation. If necessary, list specific services to which particular methods of compensation apply.)

§ 11.3 For Additional Services that may arise during the course of the Project, including those under Section 4.3, the Owner shall compensate the Architect as follows:
(Insert amount of, or basis for, compensation.)

§ 11.4 Compensation for Additional Services of the Architect's consultants when not included in Section 11.2 or 11.3, shall be the amount invoiced to the Architect plus   percent (%), or as otherwise stated below:

§ 11.5 Where compensation for Basic Services is based on a stipulated sum or percentage of the Cost of the Work, the compensation for each phase of services shall be as follows:

Schematic Design Phase:	percent (	%)
Design Development Phase:	percent (	%)
Construction Documents Phase:	percent (	%)
Bidding or Negotiation Phase:	percent (	%)
Construction Phase:	percent (	%)
Total Basic Compensation	one hundred percent (	100.00%)

§ 11.6 When compensation is based on a percentage of the Cost of the Work and any portions of the Project are deleted or otherwise not constructed, compensation for those portions of the Project shall be payable to the extent services are performed on those portions, in accordance with the schedule set forth in Section 11.5 based on (1) the lowest bona fide bid or negotiated proposal, or (2) if no such bid or proposal is received, the most recent estimate of the Cost of the Work for such portions of the Project. The Architect shall be entitled to compensation in accordance with this Agreement for all services performed whether or not the Construction Phase is commenced.

§ 11.7 The hourly billing rates for services of the Architect and the Architect's consultants, if any, are set forth below. The rates shall be adjusted in accordance with the Architect's and Architect's consultants' normal review practices.
(If applicable, attach an exhibit of hourly billing rates or insert them below.)

§ 11.8 COMPENSATION FOR REIMBURSABLE EXPENSES

§ 11.8.1 Reimbursable Expenses are in addition to compensation for Basic and Additional Services and include expenses incurred by the Architect and the Architect's consultants directly related to the Project, as follows:

- .1 Transportation and authorized out-of-town travel and subsistence;
- .2 Long distance services, dedicated data and communication services, teleconferences, Project Web sites, and extranets;
- .3 Fees paid for securing approval of authorities having jurisdiction over the Project;
- .4 Printing, reproductions, plots, standard form documents;
- .5 Postage, handling and delivery;
- .6 Expense of overtime work requiring higher than regular rates, if authorized in advance by the Owner;
- .7 Renderings, models, mock-ups, professional photography, and presentation materials requested by the Owner;
- .8 Architect's Consultant's expense of professional liability insurance dedicated exclusively to this Project, or the expense of additional insurance coverage or limits if the Owner requests such insurance in excess of that normally carried by the Architect's consultants;
- .9 All taxes levied on professional services and on reimbursable expenses;
- .10 Site office expenses; and
- .11 Other similar Project-related expenditures.

§ 11.8.2 For Reimbursable Expenses the compensation shall be the expenses incurred by the Architect and the Architect's consultants plus   percent(%) of the expenses incurred.

§ 11.9 COMPENSATION FOR USE OF ARCHITECT'S INSTRUMENTS OF SERVICE

If the Owner terminates the Architect for its convenience under Section 9.5, or the Architect terminates this Agreement under Section 9.3, the Owner shall pay a licensing fee as compensation for the Owner's continued use of the Architect's Instruments of Service solely for purposes of completing, using and maintaining the Project as follows:

§ 11.10 PAYMENTS TO THE ARCHITECT

§ 11.10.1 An initial payment of   Dollars
($) shall be made upon execution of this Agreement and is the minimum payment under this Agreement. It shall be credited to the Owner's account in the final invoice.

§ 11.10.2 Unless otherwise agreed, payments for services shall be made monthly in proportion to services performed. Payments are due and payable upon presentation of the Architect's invoice. Amounts unpaid
(   ) days after the invoice date shall bear interest at the rate entered below, or in the absence thereof at the legal rate prevailing from time to time at the principal place of business of the Architect.
(Insert rate of monthly or annual interest agreed upon.)

§ 11.10.3 The Owner shall not withhold amounts from the Architect's compensation to impose a penalty or liquidated damages on the Architect, or to offset sums requested by or paid to contractors for the cost of changes in the Work unless the Architect agrees or has been found liable for the amounts in a binding dispute resolution proceeding.

§ 11.10.4 Records of Reimbursable Expenses, expenses pertaining to Additional Services, and services performed on the basis of hourly rates shall be available to the Owner at mutually convenient times.

ARTICLE 12   SPECIAL TERMS AND CONDITIONS

Special terms and conditions that modify this Agreement are as follows:

ARTICLE 13   SCOPE OF THE AGREEMENT

§ 13.1 This Agreement represents the entire and integrated agreement between the Owner and the Architect and supersedes all prior negotiations, representations or agreements, either written or oral. This Agreement may be amended only by written instrument signed by both Owner and Architect.

§ 13.2 This Agreement is comprised of the following documents listed below:

.1   AIA Document B101™–2007, Standard Form Agreement Between Owner and Architect

.2   AIA Document E201™–2007, Digital Data Protocol Exhibit, if completed, or the following:

.3   Other documents:
(List other documents, if any, including Exhibit A, Initial Information, and additional scopes of service, if any, forming part of the Agreement.)

This Agreement entered into as of the day and year first written above.

_____   _____
OWNER *(Signature)*   **ARCHITECT** *(Signature)*

_____   _____
(Printed name and title)   *(Printed name and title)*

AIA® Document B101™ – 2007 Exhibit A

Initial Information

for the following PROJECT:
(Name and location or address)

THE OWNER:
(Name and address)

This document has important legal consequences. Consultation with an attorney is encouraged with respect to its completion or modification.

THE ARCHITECT:
(Name and address)

This Agreement is based on the following information.
(Note the disposition for the following items by inserting the requested information or a statement such as "not applicable," "unknown at time of execution" or "to be determined later by mutual agreement.")

ARTICLE A.1   PROJECT INFORMATION
§ A.1.1 The Owner's program for the Project:
(Identify documentation or state the manner in which the program will be developed.)

§ A.1.2 The Project's physical characteristics:
(Identify or describe, if appropriate, size, location, dimensions, or other pertinent information, such as geotechnical reports; site, boundary and topographic surveys; traffic and utility studies; availability of public and private utilities and services; legal description of the site; etc.)

§ A.1.3 The Owner's budget for the Cost of the Work, as defined in Section 6.1:
(Provide total, and if known, a line item break down.)

§ A.1.4 The Owner's other anticipated scheduling information, if any, not provided in Section 1.2:

§ A.1.5 The Owner intends the following procurement or delivery method for the Project:
(Identify method such as competitive bid, negotiated contract, or construction management.)

§ A.1.6 Other Project information:
(Identify special characteristics or needs of the Project not provided elsewhere, such as environmentally responsible design or historic preservation requirements.)

ARTICLE A.2 PROJECT TEAM

§ A.2.1 The Owner identifies the following representative in accordance with Section 5.3:
(List name, address and other information.)

§ A.2.2 The persons or entities, in addition to the Owner's representative, who are required to review the Architect's submittals to the Owner are as follows:
(List name, address and other information.)

§ A.2.3 The Owner will retain the following consultants and contractors:
(List discipline and, if known, identify them by name and address.)

§ A.2.4 The Architect identifies the following representative in accordance with Section 2.3:
(List name, address and other information.)

§ A.2.5 The Architect will retain the consultants identified in Sections A.2.5.1 and A.2.5.2.
(List discipline and, if known, identify them by name and address.)

§ A.2.5.1 Consultants retained under Basic Services:
    .1   Structural Engineer

    .2   Mechanical Engineer

    .3   Electrical Engineer

§ A.2.5.2 Consultants retained under Additional Services:

§ A.2.6 Other Initial Information on which the Agreement is based:
(Provide other Initial Information.)

B Standard Form of Agreement Between Owner and Contractor

AIA DOCUMENT A101–2007

AIA® Document A101™ – 2007

Standard Form of Agreement Between Owner and Contractor *where the basis of payment is a Stipulated Sum*

AGREEMENT made as of the   day of
in the year
(In words, indicate day, month and year)

BETWEEN the Owner:
(Name, address and other information)

This document has important legal consequences. Consultation with an attorney is encouraged with respect to its completion or modification.

AIA Document A201™–2007, General Conditions of the Contract for Construction, is adopted in this document by reference. Do not use with other general conditions unless this document is modified.

and the Contractor:
(Name, address and other information)

for the following Project:
(Name, location, and detailed description)

The Architect:
(Name, address and other information)

The Owner and Contractor agree as follows.

Init.

/

1

TABLE OF ARTICLES

ARTICLE 1   THE CONTRACT DOCUMENTS

The Contract Documents consist of this Agreement, Conditions of the Contract (General, Supplementary and other Conditions), Drawings, Specifications, Addenda issued prior to execution of this Agreement, other documents listed in this Agreement and Modifications issued after execution of this Agreement, all of which form the Contract, and are as fully a part of the Contract as if attached to this Agreement or repeated herein. The Contract represents the entire and integrated agreement between the parties hereto and supersedes prior negotiations, representations or agreements, either written or oral. An enumeration of the Contract Documents, other than a Modification, appears in Article 9.

ARTICLE 2   THE WORK OF THIS CONTRACT

The Contractor shall fully execute the Work described in the Contract Documents, except as specifically indicated in the Contract Documents to be the responsibility of others.

ARTICLE 3   DATE OF COMMENCEMENT AND SUBSTANTIAL COMPLETION

§ 3.1 The date of commencement of the Work shall be the date of this Agreement unless a different date is stated below or provision is made for the date to be fixed in a notice to proceed issued by the Owner.
(Insert the date of commencement if it differs from the date of this Agreement or, if applicable, state that the date will be fixed in a notice to proceed.)

If, prior to the commencement of the Work, the Owner requires time to file mortgages and other security interests, the Owner's time requirement shall be as follows:

§ 3.2 The Contract Time shall be measured from the date of commencement.

§ 3.3 The Contractor shall achieve Substantial Completion of the entire Work not later than
(   ) days from the date of commencement, or as follows:
(Insert number of calendar days. Alternatively, a calendar date may be used when coordinated with the date of commencement. If appropriate, insert requirements for earlier Substantial Completion of certain portions of the Work.)

, subject to adjustments of this Contract Time as provided in the Contract Documents.
(Insert provisions, if any, for liquidated damages relating to failure to achieve Substantial Completion on time or for bonus payments for early completion of the Work.)

ARTICLE 4   CONTRACT SUM
§ 4.1 The Owner shall pay the Contractor the Contract Sum in current funds for the Contractor's performance of the Contract. The Contract Sum shall be
Dollars ($), subject to additions and deductions as provided in the Contract Documents.

§ 4.2 The Contract Sum is based upon the following alternates, if any, which are described in the Contract Documents and are hereby accepted by the Owner:
(State the numbers or other identification of accepted alternates. If the bidding or proposal documents permit the Owner to accept other alternates subsequent to the execution of this Agreement, attach a schedule of such other alternates showing the amount for each and the date when that amount expires.)

§ 4.3 Unit prices, if any:
(Identify and state the unit price; state quantity limitations, if any, to which the unit price will be applicable.)

Item	Units and Limitations	Price Per Unit

§ 4.4 Allowances included in the Contract Sum, if any:
(Identify allowance and state exclusions, if any, from the allowance price.)

Item	Price

ARTICLE 5   PAYMENTS
§ 5.1 PROGRESS PAYMENTS
§ 5.1.1 Based upon Applications for Payment submitted to the Architect by the Contractor and Certificates for Payment issued by the Architect, the Owner shall make progress payments on account of the Contract Sum to the Contractor as provided below and elsewhere in the Contract Documents.

§ 5.1.2 The period covered by each Application for Payment shall be one calendar month ending on the last day of the month, or as follows:

§ 5.1.3 Provided that an Application for Payment is received by the Architect not later than the
(   ) day of a month, the Owner shall make payment of the certified amount to the Contractor not later than the
(   ) day of the (   ) month. If an Application for Payment is received by the Architect after the application date fixed above, payment shall be made by the Owner not later than
(   ) days after the Architect receives the Application for Payment.
(Federal, state or local laws may require payment within a certain period of time.)

§ 5.1.4 Each Application for Payment shall be based on the most recent schedule of values submitted by the Contractor in accordance with the Contract Documents. The schedule of values shall allocate the entire Contract Sum among the various portions of the Work. The schedule of values shall be prepared in such form and supported by such data to substantiate its accuracy as the Architect may require. This schedule, unless objected to by the Architect, shall be used as a basis for reviewing the Contractor's Applications for Payment.

§ 5.1.5 Applications for Payment shall show the percentage of completion of each portion of the Work as of the end of the period covered by the Application for Payment.

§ 5.1.6 Subject to other provisions of the Contract Documents, the amount of each progress payment shall be computed as follows:
- **.1** Take that portion of the Contract Sum properly allocable to completed Work as determined by multiplying the percentage completion of each portion of the Work by the share of the Contract Sum allocated to that portion of the Work in the schedule of values, less retainage of percent (%). Pending final determination of cost to the Owner of changes in the Work, amounts not in dispute shall be included as provided in Section 7.3.9 of AIA Document A201™–2007, General Conditions of the Contract for Construction;
- **.2** Add that portion of the Contract Sum properly allocable to materials and equipment delivered and suitably stored at the site for subsequent incorporation in the completed construction (or, if approved in advance by the Owner, suitably stored off the site at a location agreed upon in writing), less retainage of percent (%);
- **.3** Subtract the aggregate of previous payments made by the Owner; and
- **.4** Subtract amounts, if any, for which the Architect has withheld or nullified a Certificate for Payment as provided in Section 9.5 of AIA Document A201–2007.

§ 5.1.7 The progress payment amount determined in accordance with Section 5.1.6 shall be further modified under the following circumstances:
- **.1** Add, upon Substantial Completion of the Work, a sum sufficient to increase the total payments to the full amount of the Contract Sum, less such amounts as the Architect shall determine for incomplete Work, retainage applicable to such work and unsettled claims; and
(Section 9.8.5 of AIA Document A201–2007 requires release of applicable retainage upon Substantial Completion of Work with consent of surety, if any.)
- **.2** Add, if final completion of the Work is thereafter materially delayed through no fault of the Contractor, any additional amounts payable in accordance with Section 9.10.3 of AIA Document A201–2007.

§ 5.1.8 Reduction or limitation of retainage, if any, shall be as follows:
(If it is intended, prior to Substantial Completion of the entire Work, to reduce or limit the retainage resulting from the percentages inserted in Sections 5.1.6.1 and 5.1.6.2 above, and this is not explained elsewhere in the Contract Documents, insert here provisions for such reduction or limitation.)

§ 5.1.9 Except with the Owner's prior approval, the Contractor shall not make advance payments to suppliers for materials or equipment which have not been delivered and stored at the site.

§ 5.2 FINAL PAYMENT

§ 5.2.1 Final payment, constituting the entire unpaid balance of the Contract Sum, shall be made by the Owner to the Contractor when

 .1 the Contractor has fully performed the Contract except for the Contractor's responsibility to correct Work as provided in Section 12.2.2 of AIA Document A201–2007, and to satisfy other requirements, if any, which extend beyond final payment; and

 .2 a final Certificate for Payment has been issued by the Architect.

§ 5.2.2 The Owner's final payment to the Contractor shall be made no later than 30 days after the issuance of the Architect's final Certificate for Payment, or as follows:

ARTICLE 6 DISPUTE RESOLUTION
§ 6.1 INITIAL DECISION MAKER

The Architect will serve as Initial Decision Maker pursuant to Section 15.2 of AIA Document A201–2007, unless the parties appoint below another individual, not a party to this Agreement, to serve as Initial Decision Maker.
(If the parties mutually agree, insert the name, address and other contact information of the Initial Decision Maker, if other than the Architect.)

§ 6.2 BINDING DISPUTE RESOLUTION

For any Claim subject to, but not resolved by, mediation pursuant to Section 15.3 of AIA Document A201–2007, the method of binding dispute resolution shall be as follows:
(Check the appropriate box. If the Owner and Contractor do not select a method of binding dispute resolution below, or do not subsequently agree in writing to a binding dispute resolution method other than litigation, Claims will be resolved by litigation in a court of competent jurisdiction.)

 ☐ Arbitration pursuant to Section 15.4 of AIA Document A201–2007

 ☐ Litigation in a court of competent jurisdiction

 ☐ Other *(Specify)*

ARTICLE 7 TERMINATION OR SUSPENSION

§ 7.1 The Contract may be terminated by the Owner or the Contractor as provided in Article 14 of AIA Document A201–2007.

§ 7.2 The Work may be suspended by the Owner as provided in Article 14 of AIA Document A201–2007.

ARTICLE 8 MISCELLANEOUS PROVISIONS

§ 8.1 Where reference is made in this Agreement to a provision of AIA Document A201–2007 or another Contract Document, the reference refers to that provision as amended or supplemented by other provisions of the Contract Documents.

§ 8.2 Payments due and unpaid under the Contract shall bear interest from the date payment is due at the rate stated below, or in the absence thereof, at the legal rate prevailing from time to time at the place where the Project is located. *(Insert rate of interest agreed upon, if any.)*

§ 8.3 The Owner's representative:
(Name, address and other information)

§ 8.4 The Contractor's representative:
(Name, address and other information)

§ 8.5 Neither the Owner's nor the Contractor's representative shall be changed without ten days written notice to the other party.

§ 8.6 Other provisions:

ARTICLE 9 ENUMERATION OF CONTRACT DOCUMENTS

§ 9.1 The Contract Documents, except for Modifications issued after execution of this Agreement, are enumerated in the sections below.

§ 9.1.1 The Agreement is this executed AIA Document A101–2007, Standard Form of Agreement Between Owner and Contractor.

§ 9.1.2 The General Conditions are AIA Document A201–2007, General Conditions of the Contract for Construction.

§ 9.1.3 The Supplementary and other Conditions of the Contract:

Document	Title	Date	Pages

§ 9.1.4 The Specifications:
(Either list the Specifications here or refer to an exhibit attached to this Agreement.)

Section	Title	Date	Pages

§ 9.1.5 The Drawings:
(Either list the Drawings here or refer to an exhibit attached to this Agreement.)

Number	Title	Date

§ 9.1.6 The Addenda, if any:

Number	Date	Pages

Portions of Addenda relating to bidding requirements are not part of the Contract Documents unless the bidding requirements are also enumerated in this Article 9.

§ 9.1.7 Additional documents, if any, forming part of the Contract Documents:
.1   AIA Document E201™–2007, Digital Data Protocol Exhibit, if completed by the parties, or the following:

.2   Other documents, if any, listed below:
(List here any additional documents that are intended to form part of the Contract Documents. AIA Document A201–2007 provides that bidding requirements such as advertisement or invitation to bid, Instructions to Bidders, sample forms and the Contractor's bid are not part of the Contract Documents unless enumerated in this Agreement. They should be listed here only if intended to be part of the Contract Documents.)

ARTICLE 10   INSURANCE AND BONDS

The Contractor shall purchase and maintain insurance and provide bonds as set forth in Article 11 of AIA Document A201–2007.
(State bonding requirements, if any, and limits of liability for insurance required in Article 11 of AIA Document A201–2007.)

This Agreement entered into as of the day and year first written above.

_____   _____
OWNER *(Signature)*   **CONTRACTOR** *(Signature)*

_____   _____
(Printed name and title)   *(Printed name and title)*

General Conditions of the Contract for Construction

AIA DOCUMENT A201–2007

AIA® Document A201™ – 2007

General Conditions of the Contract for Construction

for the following PROJECT:
(Name and location or address)

This document has important legal consequences. Consultation with an attorney is encouraged with respect to its completion or modification.

THE OWNER:
(Name and address)

THE ARCHITECT:
(Name and address)

TABLE OF ARTICLES

Init.

/

5

ARTICLE 1   GENERAL PROVISIONS
§ 1.1 BASIC DEFINITIONS
§ 1.1.1 THE CONTRACT DOCUMENTS
The Contract Documents are enumerated in the Agreement between the Owner and Contractor (hereinafter the Agreement) and consist of the Agreement, Conditions of the Contract (General, Supplementary and other Conditions), Drawings, Specifications, Addenda issued prior to execution of the Contract, other documents listed in the Agreement and Modifications issued after execution of the Contract. A Modification is (1) a written amendment to the Contract signed by both parties, (2) a Change Order, (3) a Construction Change Directive or (4) a written order for a minor change in the Work issued by the Architect. Unless specifically enumerated in the Agreement, the Contract Documents do not include the advertisement or invitation to bid, Instructions to Bidders, sample forms, other information furnished by the Owner in anticipation of receiving bids or proposals, the Contractor's bid or proposal, or portions of Addenda relating to bidding requirements.

§ 1.1.2 THE CONTRACT
The Contract Documents form the Contract for Construction. The Contract represents the entire and integrated agreement between the parties hereto and supersedes prior negotiations, representations or agreements, either written or oral. The Contract may be amended or modified only by a Modification. The Contract Documents shall not be construed to create a contractual relationship of any kind (1) between the Contractor and the Architect or the Architect's consultants, (2) between the Owner and a Subcontractor or a Sub-subcontractor, (3) between the Owner and the Architect or the Architect's consultants or (4) between any persons or entities other than the Owner and the Contractor. The Architect shall, however, be entitled to performance and enforcement of obligations under the Contract intended to facilitate performance of the Architect's duties.

§ 1.1.3 THE WORK
The term "Work" means the construction and services required by the Contract Documents, whether completed or partially completed, and includes all other labor, materials, equipment and services provided or to be provided by the Contractor to fulfill the Contractor's obligations. The Work may constitute the whole or a part of the Project.

§ 1.1.4 THE PROJECT
The Project is the total construction of which the Work performed under the Contract Documents may be the whole or a part and which may include construction by the Owner and by separate contractors.

§ 1.1.5 THE DRAWINGS
The Drawings are the graphic and pictorial portions of the Contract Documents showing the design, location and dimensions of the Work, generally including plans, elevations, sections, details, schedules and diagrams.

§ 1.1.6 THE SPECIFICATIONS
The Specifications are that portion of the Contract Documents consisting of the written requirements for materials, equipment, systems, standards and workmanship for the Work, and performance of related services.

§ 1.1.7 INSTRUMENTS OF SERVICE
Instruments of Service are representations, in any medium of expression now known or later developed, of the tangible and intangible creative work performed by the Architect and the Architect's consultants under their respective professional services agreements. Instruments of Service may include, without limitation, studies, surveys, models, sketches, drawings, specifications, and other similar materials.

§ 1.1.8 INITIAL DECISION MAKER
The Initial Decision Maker is the person identified in the Agreement to render initial decisions on Claims in accordance with Section 15.2 and certify termination of the Agreement under Section 14.2.2.

§ 1.2 CORRELATION AND INTENT OF THE CONTRACT DOCUMENTS
§ 1.2.1 The intent of the Contract Documents is to include all items necessary for the proper execution and completion of the Work by the Contractor. The Contract Documents are complementary, and what is required by one shall be as binding as if required by all; performance by the Contractor shall be required only to the extent consistent with the Contract Documents and reasonably inferable from them as being necessary to produce the indicated results.

§ 1.2.2 Organization of the Specifications into divisions, sections and articles, and arrangement of Drawings shall not control the Contractor in dividing the Work among Subcontractors or in establishing the extent of Work to be performed by any trade.

§ 1.2.3 Unless otherwise stated in the Contract Documents, words that have well-known technical or construction industry meanings are used in the Contract Documents in accordance with such recognized meanings.

§ 1.3 CAPITALIZATION
Terms capitalized in these General Conditions include those that are (1) specifically defined, (2) the titles of numbered articles or (3) the titles of other documents published by the American Institute of Architects.

§ 1.4 INTERPRETATION
In the interest of brevity the Contract Documents frequently omit modifying words such as "all" and "any" and articles such as "the" and "an," but the fact that a modifier or an article is absent from one statement and appears in another is not intended to affect the interpretation of either statement.

§ 1.5 OWNERSHIP AND USE OF DRAWINGS, SPECIFICATIONS AND OTHER INSTRUMENTS OF SERVICE
§ 1.5.1 The Architect and the Architect's consultants shall be deemed the authors and owners of their respective Instruments of Service, including the Drawings and Specifications, and will retain all common law, statutory and other reserved rights, including copyrights. The Contractor, Subcontractors, Sub-subcontractors, and material or equipment suppliers shall not own or claim a copyright in the Instruments of Service. Submittal or distribution to meet official regulatory requirements or for other purposes in connection with this Project is not to be construed as publication in derogation of the Architect's or Architect's consultants' reserved rights.

§ 1.5.2 The Contractor, Subcontractors, Sub-subcontractors and material or equipment suppliers are authorized to use and reproduce the Instruments of Service provided to them solely and exclusively for execution of the Work. All copies made under this authorization shall bear the copyright notice, if any, shown on the Instruments of Service. The Contractor, Subcontractors, Sub-subcontractors, and material or equipment suppliers may not use the Instruments of Service on other projects or for additions to this Project outside the scope of the Work without the specific written consent of the Owner, Architect and the Architect's consultants.

§ 1.6 TRANSMISSION OF DATA IN DIGITAL FORM
If the parties intend to transmit Instruments of Service or any other information or documentation in digital form, they shall endeavor to establish necessary protocols governing such transmissions, unless otherwise already provided in the Agreement or the Contract Documents.

ARTICLE 2 OWNER
§ 2.1 GENERAL
§ 2.1.1 The Owner is the person or entity identified as such in the Agreement and is referred to throughout the Contract Documents as if singular in number. The Owner shall designate in writing a representative who shall have express authority to bind the Owner with respect to all matters requiring the Owner's approval or authorization. Except as otherwise provided in Section 4.2.1, the Architect does not have such authority. The term "Owner" means the Owner or the Owner's authorized representative.

§ 2.1.2 The Owner shall furnish to the Contractor within fifteen days after receipt of a written request, information necessary and relevant for the Contractor to evaluate, give notice of or enforce mechanic's lien rights. Such information shall include a correct statement of the record legal title to the property on which the Project is located, usually referred to as the site, and the Owner's interest therein.

§ 2.2 INFORMATION AND SERVICES REQUIRED OF THE OWNER
§ 2.2.1 Prior to commencement of the Work, the Contractor may request in writing that the Owner provide reasonable evidence that the Owner has made financial arrangements to fulfill the Owner's obligations under the Contract. Thereafter, the Contractor may only request such evidence if (1) the Owner fails to make payments to the Contractor as the Contract Documents require; (2) a change in the Work materially changes the Contract Sum; or (3) the Contractor identifies in writing a reasonable concern regarding the Owner's ability to make payment when due. The Owner shall furnish such evidence as a condition precedent to commencement or continuation of the Work or the portion of the Work affected by a material change. After the Owner furnishes the evidence, the Owner shall not materially vary such financial arrangements without prior notice to the Contractor.

§ 2.2.2 Except for permits and fees that are the responsibility of the Contractor under the Contract Documents, including those required under Section 3.7.1, the Owner shall secure and pay for necessary approvals, easements, assessments and charges required for construction, use or occupancy of permanent structures or for permanent changes in existing facilities.

§ 2.2.3 The Owner shall furnish surveys describing physical characteristics, legal limitations and utility locations for the site of the Project, and a legal description of the site. The Contractor shall be entitled to rely on the accuracy of information furnished by the Owner but shall exercise proper precautions relating to the safe performance of the Work.

§ 2.2.4 The Owner shall furnish information or services required of the Owner by the Contract Documents with reasonable promptness. The Owner shall also furnish any other information or services under the Owner's control and relevant to the Contractor's performance of the Work with reasonable promptness after receiving the Contractor's written request for such information or services.

§ 2.2.5 Unless otherwise provided in the Contract Documents, the Owner shall furnish to the Contractor one copy of the Contract Documents for purposes of making reproductions pursuant to Section 1.5.2.

§ 2.3 OWNER'S RIGHT TO STOP THE WORK
If the Contractor fails to correct Work that is not in accordance with the requirements of the Contract Documents as required by Section 12.2 or repeatedly fails to carry out Work in accordance with the Contract Documents, the Owner may issue a written order to the Contractor to stop the Work, or any portion thereof, until the cause for such order has been eliminated; however, the right of the Owner to stop the Work shall not give rise to a duty on the part of the Owner to exercise this right for the benefit of the Contractor or any other person or entity, except to the extent required by Section 6.1.3.

§ 2.4 OWNER'S RIGHT TO CARRY OUT THE WORK
If the Contractor defaults or neglects to carry out the Work in accordance with the Contract Documents and fails within a ten-day period after receipt of written notice from the Owner to commence and continue correction of such default or neglect with diligence and promptness, the Owner may, without prejudice to other remedies the Owner may have, correct such deficiencies. In such case an appropriate Change Order shall be issued deducting from payments then or thereafter due the Contractor the reasonable cost of correcting such deficiencies, including Owner's expenses and compensation for the Architect's additional services made necessary by such default, neglect or failure. Such action by the Owner and amounts charged to the Contractor are both subject to prior approval of the Architect. If payments then or thereafter due the Contractor are not sufficient to cover such amounts, the Contractor shall pay the difference to the Owner.

ARTICLE 3 CONTRACTOR
§ 3.1 GENERAL
§ 3.1.1 The Contractor is the person or entity identified as such in the Agreement and is referred to throughout the Contract Documents as if singular in number. The Contractor shall be lawfully licensed, if required in the jurisdiction where the Project is located. The Contractor shall designate in writing a representative who shall have express authority to bind the Contractor with respect to all matters under this Contract. The term "Contractor" means the Contractor or the Contractor's authorized representative.

§ 3.1.2 The Contractor shall perform the Work in accordance with the Contract Documents.

§ 3.1.3 The Contractor shall not be relieved of obligations to perform the Work in accordance with the Contract Documents either by activities or duties of the Architect in the Architect's administration of the Contract, or by tests, inspections or approvals required or performed by persons or entities other than the Contractor.

§ 3.2 REVIEW OF CONTRACT DOCUMENTS AND FIELD CONDITIONS BY CONTRACTOR
§ 3.2.1 Execution of the Contract by the Contractor is a representation that the Contractor has visited the site, become generally familiar with local conditions under which the Work is to be performed and correlated personal observations with requirements of the Contract Documents.

Init.

/

11

§ **3.2.2** Because the Contract Documents are complementary, the Contractor shall, before starting each portion of the Work, carefully study and compare the various Contract Documents relative to that portion of the Work, as well as the information furnished by the Owner pursuant to Section 2.2.3, shall take field measurements of any existing conditions related to that portion of the Work, and shall observe any conditions at the site affecting it. These obligations are for the purpose of facilitating coordination and construction by the Contractor and are not for the purpose of discovering errors, omissions, or inconsistencies in the Contract Documents; however, the Contractor shall promptly report to the Architect any errors, inconsistencies or omissions discovered by or made known to the Contractor as a request for information in such form as the Architect may require. It is recognized that the Contractor's review is made in the Contractor's capacity as a contractor and not as a licensed design professional, unless otherwise specifically provided in the Contract Documents.

§ **3.2.3** The Contractor is not required to ascertain that the Contract Documents are in accordance with applicable laws, statutes, ordinances, codes, rules and regulations, or lawful orders of public authorities, but the Contractor shall promptly report to the Architect any nonconformity discovered by or made known to the Contractor as a request for information in such form as the Architect may require.

§ **3.2.4** If the Contractor believes that additional cost or time is involved because of clarifications or instructions the Architect issues in response to the Contractor's notices or requests for information pursuant to Sections 3.2.2 or 3.2.3, the Contractor shall make Claims as provided in Article 15. If the Contractor fails to perform the obligations of Sections 3.2.2 or 3.2.3, the Contractor shall pay such costs and damages to the Owner as would have been avoided if the Contractor had performed such obligations. If the Contractor performs those obligations, the Contractor shall not be liable to the Owner or Architect for damages resulting from errors, inconsistencies or omissions in the Contract Documents, for differences between field measurements or conditions and the Contract Documents, or for nonconformities of the Contract Documents to applicable laws, statutes, ordinances, codes, rules and regulations, and lawful orders of public authorities.

§ **3.3 SUPERVISION AND CONSTRUCTION PROCEDURES**
§ **3.3.1** The Contractor shall supervise and direct the Work, using the Contractor's best skill and attention. The Contractor shall be solely responsible for, and have control over, construction means, methods, techniques, sequences and procedures and for coordinating all portions of the Work under the Contract, unless the Contract Documents give other specific instructions concerning these matters. If the Contract Documents give specific instructions concerning construction means, methods, techniques, sequences or procedures, the Contractor shall evaluate the jobsite safety thereof and, except as stated below, shall be fully and solely responsible for the jobsite safety of such means, methods, techniques, sequences or procedures. If the Contractor determines that such means, methods, techniques, sequences or procedures may not be safe, the Contractor shall give timely written notice to the Owner and Architect and shall not proceed with that portion of the Work without further written instructions from the Architect. If the Contractor is then instructed to proceed with the required means, methods, techniques, sequences or procedures without acceptance of changes proposed by the Contractor, the Owner shall be solely responsible for any loss or damage arising solely from those Owner-required means, methods, techniques, sequences or procedures.

§ **3.3.2** The Contractor shall be responsible to the Owner for acts and omissions of the Contractor's employees, Subcontractors and their agents and employees, and other persons or entities performing portions of the Work for, or on behalf of, the Contractor or any of its Subcontractors.

§ **3.3.3** The Contractor shall be responsible for inspection of portions of Work already performed to determine that such portions are in proper condition to receive subsequent Work.

§ **3.4 LABOR AND MATERIALS**
§ **3.4.1** Unless otherwise provided in the Contract Documents, the Contractor shall provide and pay for labor, materials, equipment, tools, construction equipment and machinery, water, heat, utilities, transportation, and other facilities and services necessary for proper execution and completion of the Work, whether temporary or permanent and whether or not incorporated or to be incorporated in the Work.

§ **3.4.2** Except in the case of minor changes in the Work authorized by the Architect in accordance with Sections 3.12.8 or 7.4, the Contractor may make substitutions only with the consent of the Owner, after evaluation by the Architect and in accordance with a Change Order or Construction Change Directive.

§ 3.4.3 The Contractor shall enforce strict discipline and good order among the Contractor's employees and other persons carrying out the Work. The Contractor shall not permit employment of unfit persons or persons not properly skilled in tasks assigned to them.

§ 3.5 WARRANTY

The Contractor warrants to the Owner and Architect that materials and equipment furnished under the Contract will be of good quality and new unless the Contract Documents require or permit otherwise. The Contractor further warrants that the Work will conform to the requirements of the Contract Documents and will be free from defects, except for those inherent in the quality of the Work the Contract Documents require or permit. Work, materials, or equipment not conforming to these requirements may be considered defective. The Contractor's warranty excludes remedy for damage or defect caused by abuse, alterations to the Work not executed by the Contractor, improper or insufficient maintenance, improper operation, or normal wear and tear and normal usage. If required by the Architect, the Contractor shall furnish satisfactory evidence as to the kind and quality of materials and equipment.

§ 3.6 TAXES

The Contractor shall pay sales, consumer, use and similar taxes for the Work provided by the Contractor that are legally enacted when bids are received or negotiations concluded, whether or not yet effective or merely scheduled to go into effect.

§ 3.7 PERMITS, FEES, NOTICES, AND COMPLIANCE WITH LAWS

§ 3.7.1 Unless otherwise provided in the Contract Documents, the Contractor shall secure and pay for the building permit as well as for other permits, fees, licenses, and inspections by government agencies necessary for proper execution and completion of the Work that are customarily secured after execution of the Contract and legally required at the time bids are received or negotiations concluded.

§ 3.7.2 The Contractor shall comply with and give notices required by applicable laws, statutes, ordinances, codes, rules and regulations, and lawful orders of public authorities applicable to performance of the Work.

§ 3.7.3 If the Contractor performs Work knowing it to be contrary to applicable laws, statutes, ordinances, codes, rules and regulations, or lawful orders of public authorities, the Contractor shall assume appropriate responsibility for such Work and shall bear the costs attributable to correction.

§ 3.7.4 Concealed or Unknown Conditions. If the Contractor encounters conditions at the site that are (1) subsurface or otherwise concealed physical conditions that differ materially from those indicated in the Contract Documents or (2) unknown physical conditions of an unusual nature that differ materially from those ordinarily found to exist and generally recognized as inherent in construction activities of the character provided for in the Contract Documents, the Contractor shall promptly provide notice to the Owner and the Architect before conditions are disturbed and in no event later than 21 days after first observance of the conditions. The Architect will promptly investigate such conditions and, if the Architect determines that they differ materially and cause an increase or decrease in the Contractor's cost of, or time required for, performance of any part of the Work, will recommend an equitable adjustment in the Contract Sum or Contract Time, or both. If the Architect determines that the conditions at the site are not materially different from those indicated in the Contract Documents and that no change in the terms of the Contract is justified, the Architect shall promptly notify the Owner and Contractor in writing, stating the reasons. If either party disputes the Architect's determination or recommendation, that party may proceed as provided in Article 15.

§ 3.7.5 If, in the course of the Work, the Contractor encounters human remains or recognizes the existence of burial markers, archaeological sites or wetlands not indicated in the Contract Documents, the Contractor shall immediately suspend any operations that would affect them and shall notify the Owner and Architect. Upon receipt of such notice, the Owner shall promptly take any action necessary to obtain governmental authorization required to resume the operations. The Contractor shall continue to suspend such operations until otherwise instructed by the Owner but shall continue with all other operations that do not affect those remains or features. Requests for adjustments in the Contract Sum and Contract Time arising from the existence of such remains or features may be made as provided in Article 15.

§ 3.8 ALLOWANCES

§ 3.8.1 The Contractor shall include in the Contract Sum all allowances stated in the Contract Documents. Items covered by allowances shall be supplied for such amounts and by such persons or entities as the Owner may direct, but the Contractor shall not be required to employ persons or entities to whom the Contractor has reasonable objection.

§ 3.8.2 Unless otherwise provided in the Contract Documents,

   .1   allowances shall cover the cost to the Contractor of materials and equipment delivered at the site and all required taxes, less applicable trade discounts;

   .2   Contractor's costs for unloading and handling at the site, labor, installation costs, overhead, profit and other expenses contemplated for stated allowance amounts shall be included in the Contract Sum but not in the allowances; and

   .3   whenever costs are more than or less than allowances, the Contract Sum shall be adjusted accordingly by Change Order. The amount of the Change Order shall reflect (1) the difference between actual costs and the allowances under Section 3.8.2.1 and (2) changes in Contractor's costs under Section 3.8.2.2.

§ 3.8.3 Materials and equipment under an allowance shall be selected by the Owner with reasonable promptness.

§ 3.9 SUPERINTENDENT

§ 3.9.1 The Contractor shall employ a competent superintendent and necessary assistants who shall be in attendance at the Project site during performance of the Work. The superintendent shall represent the Contractor, and communications given to the superintendent shall be as binding as if given to the Contractor.

§ 3.9.2 The Contractor, as soon as practicable after award of the Contract, shall furnish in writing to the Owner through the Architect the name and qualifications of a proposed superintendent. The Architect may reply within 14 days to the Contractor in writing stating (1) whether the Owner or the Architect has reasonable objection to the proposed superintendent or (2) that the Architect requires additional time to review. Failure of the Architect to reply within the 14 day period shall constitute notice of no reasonable objection.

§ 3.9.3 The Contractor shall not employ a proposed superintendent to whom the Owner or Architect has made reasonable and timely objection. The Contractor shall not change the superintendent without the Owner's consent, which shall not unreasonably be withheld or delayed.

§ 3.10 CONTRACTOR'S CONSTRUCTION SCHEDULES

§ 3.10.1 The Contractor, promptly after being awarded the Contract, shall prepare and submit for the Owner's and Architect's information a Contractor's construction schedule for the Work. The schedule shall not exceed time limits current under the Contract Documents, shall be revised at appropriate intervals as required by the conditions of the Work and Project, shall be related to the entire Project to the extent required by the Contract Documents, and shall provide for expeditious and practicable execution of the Work.

§ 3.10.2 The Contractor shall prepare a submittal schedule, promptly after being awarded the Contract and thereafter as necessary to maintain a current submittal schedule, and shall submit the schedule(s) for the Architect's approval. The Architect's approval shall not unreasonably be delayed or withheld. The submittal schedule shall (1) be coordinated with the Contractor's construction schedule, and (2) allow the Architect reasonable time to review submittals. If the Contractor fails to submit a submittal schedule, the Contractor shall not be entitled to any increase in Contract Sum or extension of Contract Time based on the time required for review of submittals.

§ 3.10.3 The Contractor shall perform the Work in general accordance with the most recent schedules submitted to the Owner and Architect.

§ 3.11 DOCUMENTS AND SAMPLES AT THE SITE

The Contractor shall maintain at the site for the Owner one copy of the Drawings, Specifications, Addenda, Change Orders and other Modifications, in good order and marked currently to indicate field changes and selections made during construction, and one copy of approved Shop Drawings, Product Data, Samples and similar required submittals. These shall be available to the Architect and shall be delivered to the Architect for submittal to the Owner upon completion of the Work as a record of the Work as constructed.

§ 3.12 SHOP DRAWINGS, PRODUCT DATA AND SAMPLES

§ 3.12.1 Shop Drawings are drawings, diagrams, schedules and other data specially prepared for the Work by the Contractor or a Subcontractor, Sub-subcontractor, manufacturer, supplier or distributor to illustrate some portion of the Work.

§ 3.12.2 Product Data are illustrations, standard schedules, performance charts, instructions, brochures, diagrams and other information furnished by the Contractor to illustrate materials or equipment for some portion of the Work.

§ 3.12.3 Samples are physical examples that illustrate materials, equipment or workmanship and establish standards by which the Work will be judged.

§ 3.12.4 Shop Drawings, Product Data, Samples and similar submittals are not Contract Documents. Their purpose is to demonstrate the way by which the Contractor proposes to conform to the information given and the design concept expressed in the Contract Documents for those portions of the Work for which the Contract Documents require submittals. Review by the Architect is subject to the limitations of Section 4.2.7. Informational submittals upon which the Architect is not expected to take responsive action may be so identified in the Contract Documents. Submittals that are not required by the Contract Documents may be returned by the Architect without action.

§ 3.12.5 The Contractor shall review for compliance with the Contract Documents, approve and submit to the Architect Shop Drawings, Product Data, Samples and similar submittals required by the Contract Documents in accordance with the submittal schedule approved by the Architect or, in the absence of an approved submittal schedule, with reasonable promptness and in such sequence as to cause no delay in the Work or in the activities of the Owner or of separate contractors.

§ 3.12.6 By submitting Shop Drawings, Product Data, Samples and similar submittals, the Contractor represents to the Owner and Architect that the Contractor has (1) reviewed and approved them, (2) determined and verified materials, field measurements and field construction criteria related thereto, or will do so and (3) checked and coordinated the information contained within such submittals with the requirements of the Work and of the Contract Documents.

§ 3.12.7 The Contractor shall perform no portion of the Work for which the Contract Documents require submittal and review of Shop Drawings, Product Data, Samples or similar submittals until the respective submittal has been approved by the Architect.

§ 3.12.8 The Work shall be in accordance with approved submittals except that the Contractor shall not be relieved of responsibility for deviations from requirements of the Contract Documents by the Architect's approval of Shop Drawings, Product Data, Samples or similar submittals unless the Contractor has specifically informed the Architect in writing of such deviation at the time of submittal and (1) the Architect has given written approval to the specific deviation as a minor change in the Work, or (2) a Change Order or Construction Change Directive has been issued authorizing the deviation. The Contractor shall not be relieved of responsibility for errors or omissions in Shop Drawings, Product Data, Samples or similar submittals by the Architect's approval thereof.

§ 3.12.9 The Contractor shall direct specific attention, in writing or on resubmitted Shop Drawings, Product Data, Samples or similar submittals, to revisions other than those requested by the Architect on previous submittals. In the absence of such written notice, the Architect's approval of a resubmission shall not apply to such revisions.

§ 3.12.10 The Contractor shall not be required to provide professional services that constitute the practice of architecture or engineering unless such services are specifically required by the Contract Documents for a portion of the Work or unless the Contractor needs to provide such services in order to carry out the Contractor's responsibilities for construction means, methods, techniques, sequences and procedures. The Contractor shall not be required to provide professional services in violation of applicable law. If professional design services or certifications by a design professional related to systems, materials or equipment are specifically required of the Contractor by the Contract Documents, the Owner and the Architect will specify all performance and design criteria that such services must satisfy. The Contractor shall cause such services or certifications to be provided by a properly licensed design professional, whose signature and seal shall appear on all drawings, calculations, specifications, certifications, Shop Drawings and other submittals prepared by such professional. Shop Drawings and other submittals related to the Work designed or certified by such professional, if prepared by others, shall bear such professional's written approval when submitted to the Architect. The Owner and the Architect shall be entitled to rely upon the adequacy, accuracy and completeness of the services, certifications and approvals performed or provided by such design professionals, provided the Owner and Architect have specified to the Contractor all performance and design criteria that such services must satisfy. Pursuant to this Section 3.12.10, the Architect will review, approve or take other appropriate action on submittals only for the limited purpose of checking for conformance with information given and the design concept expressed in the Contract Documents. The Contractor shall not be responsible for the adequacy of the performance and design criteria specified in the Contract Documents.

§ 3.13 USE OF SITE

The Contractor shall confine operations at the site to areas permitted by applicable laws, statutes, ordinances, codes, rules and regulations, and lawful orders of public authorities and the Contract Documents and shall not unreasonably encumber the site with materials or equipment.

§ 3.14 CUTTING AND PATCHING

§ 3.14.1 The Contractor shall be responsible for cutting, fitting or patching required to complete the Work or to make its parts fit together properly. All areas requiring cutting, fitting and patching shall be restored to the condition existing prior to the cutting, fitting and patching, unless otherwise required by the Contract Documents.

§ 3.14.2 The Contractor shall not damage or endanger a portion of the Work or fully or partially completed construction of the Owner or separate contractors by cutting, patching or otherwise altering such construction, or by excavation. The Contractor shall not cut or otherwise alter such construction by the Owner or a separate contractor except with written consent of the Owner and of such separate contractor; such consent shall not be unreasonably withheld. The Contractor shall not unreasonably withhold from the Owner or a separate contractor the Contractor's consent to cutting or otherwise altering the Work.

§ 3.15 CLEANING UP

§ 3.15.1 The Contractor shall keep the premises and surrounding area free from accumulation of waste materials or rubbish caused by operations under the Contract. At completion of the Work, the Contractor shall remove waste materials, rubbish, the Contractor's tools, construction equipment, machinery and surplus materials from and about the Project.

§ 3.15.2 If the Contractor fails to clean up as provided in the Contract Documents, the Owner may do so and Owner shall be entitled to reimbursement from the Contractor.

§ 3.16 ACCESS TO WORK

The Contractor shall provide the Owner and Architect access to the Work in preparation and progress wherever located.

§ 3.17 ROYALTIES, PATENTS AND COPYRIGHTS

The Contractor shall pay all royalties and license fees. The Contractor shall defend suits or claims for infringement of copyrights and patent rights and shall hold the Owner and Architect harmless from loss on account thereof, but shall not be responsible for such defense or loss when a particular design, process or product of a particular manufacturer or manufacturers is required by the Contract Documents, or where the copyright violations are contained in Drawings, Specifications or other documents prepared by the Owner or Architect. However, if the Contractor has reason to believe that the required design, process or product is an infringement of a copyright or a patent, the Contractor shall be responsible for such loss unless such information is promptly furnished to the Architect.

§ 3.18 INDEMNIFICATION

§ 3.18.1 To the fullest extent permitted by law the Contractor shall indemnify and hold harmless the Owner, Architect, Architect's consultants, and agents and employees of any of them from and against claims, damages, losses and expenses, including but not limited to attorneys' fees, arising out of or resulting from performance of the Work, provided that such claim, damage, loss or expense is attributable to bodily injury, sickness, disease or death, or to injury to or destruction of tangible property (other than the Work itself), but only to the extent caused by the negligent acts or omissions of the Contractor, a Subcontractor, anyone directly or indirectly employed by them or anyone for whose acts they may be liable, regardless of whether or not such claim, damage, loss or expense is caused in part by a party indemnified hereunder. Such obligation shall not be construed to negate, abridge, or reduce other rights or obligations of indemnity that would otherwise exist as to a party or person described in this Section 3.18.

§ 3.18.2 In claims against any person or entity indemnified under this Section 3.18 by an employee of the Contractor, a Subcontractor, anyone directly or indirectly employed by them or anyone for whose acts they may be liable, the indemnification obligation under Section 3.18.1 shall not be limited by a limitation on amount or type of damages, compensation or benefits payable by or for the Contractor or a Subcontractor under workers' compensation acts, disability benefit acts or other employee benefit acts.

ARTICLE 4 ARCHITECT
§ 4.1 GENERAL
§ 4.1.1 The Owner shall retain an architect lawfully licensed to practice architecture or an entity lawfully practicing architecture in the jurisdiction where the Project is located. That person or entity is identified as the Architect in the Agreement and is referred to throughout the Contract Documents as if singular in number.

§ 4.1.2 Duties, responsibilities and limitations of authority of the Architect as set forth in the Contract Documents shall not be restricted, modified or extended without written consent of the Owner, Contractor and Architect. Consent shall not be unreasonably withheld.

§ 4.1.3 If the employment of the Architect is terminated, the Owner shall employ a successor architect as to whom the Contractor has no reasonable objection and whose status under the Contract Documents shall be that of the Architect.

§ 4.2 ADMINISTRATION OF THE CONTRACT
§ 4.2.1 The Architect will provide administration of the Contract as described in the Contract Documents and will be an Owner's representative during construction until the date the Architect issues the final Certificate For Payment. The Architect will have authority to act on behalf of the Owner only to the extent provided in the Contract Documents.

§ 4.2.2 The Architect will visit the site at intervals appropriate to the stage of construction, or as otherwise agreed with the Owner, to become generally familiar with the progress and quality of the portion of the Work completed, and to determine in general if the Work observed is being performed in a manner indicating that the Work, when fully completed, will be in accordance with the Contract Documents. However, the Architect will not be required to make exhaustive or continuous on-site inspections to check the quality or quantity of the Work. The Architect will not have control over, charge of, or responsibility for, the construction means, methods, techniques, sequences or procedures, or for the safety precautions and programs in connection with the Work, since these are solely the Contractor's rights and responsibilities under the Contract Documents, except as provided in Section 3.3.1.

§ 4.2.3 On the basis of the site visits, the Architect will keep the Owner reasonably informed about the progress and quality of the portion of the Work completed, and report to the Owner (1) known deviations from the Contract Documents and from the most recent construction schedule submitted by the Contractor, and (2) defects and deficiencies observed in the Work. The Architect will not be responsible for the Contractor's failure to perform the Work in accordance with the requirements of the Contract Documents. The Architect will not have control over or charge of and will not be responsible for acts or omissions of the Contractor, Subcontractors, or their agents or employees, or any other persons or entities performing portions of the Work.

§ 4.2.4 COMMUNICATIONS FACILITATING CONTRACT ADMINISTRATION
Except as otherwise provided in the Contract Documents or when direct communications have been specially authorized, the Owner and Contractor shall endeavor to communicate with each other through the Architect about matters arising out of or relating to the Contract. Communications by and with the Architect's consultants shall be through the Architect. Communications by and with Subcontractors and material suppliers shall be through the Contractor. Communications by and with separate contractors shall be through the Owner.

§ 4.2.5 Based on the Architect's evaluations of the Contractor's Applications for Payment, the Architect will review and certify the amounts due the Contractor and will issue Certificates for Payment in such amounts.

§ 4.2.6 The Architect has authority to reject Work that does not conform to the Contract Documents. Whenever the Architect considers it necessary or advisable, the Architect will have authority to require inspection or testing of the

Work in accordance with Sections 13.5.2 and 13.5.3, whether or not such Work is fabricated, installed or completed. However, neither this authority of the Architect nor a decision made in good faith either to exercise or not to exercise such authority shall give rise to a duty or responsibility of the Architect to the Contractor, Subcontractors, material and equipment suppliers, their agents or employees, or other persons or entities performing portions of the Work.

§ 4.2.7 The Architect will review and approve, or take other appropriate action upon, the Contractor's submittals such as Shop Drawings, Product Data and Samples, but only for the limited purpose of checking for conformance with information given and the design concept expressed in the Contract Documents. The Architect's action will be taken in accordance with the submittal schedule approved by the Architect or, in the absence of an approved submittal schedule, with reasonable promptness while allowing sufficient time in the Architect's professional judgment to permit adequate review. Review of such submittals is not conducted for the purpose of determining the accuracy and completeness of other details such as dimensions and quantities, or for substantiating instructions for installation or performance of equipment or systems, all of which remain the responsibility of the Contractor as required by the Contract Documents. The Architect's review of the Contractor's submittals shall not relieve the Contractor of the obligations under Sections 3.3, 3.5 and 3.12. The Architect's review shall not constitute approval of safety precautions or, unless otherwise specifically stated by the Architect, of any construction means, methods, techniques, sequences or procedures. The Architect's approval of a specific item shall not indicate approval of an assembly of which the item is a component.

§ 4.2.8 The Architect will prepare Change Orders and Construction Change Directives, and may authorize minor changes in the Work as provided in Section 7.4. The Architect will investigate and make determinations and recommendations regarding concealed and unknown conditions as provided in Section 3.7.4.

§ 4.2.9 The Architect will conduct inspections to determine the date or dates of Substantial Completion and the date of final completion; issue Certificates of Substantial Completion pursuant to Section 9.8; receive and forward to the Owner, for the Owner's review and records, written warranties and related documents required by the Contract and assembled by the Contractor pursuant to Section 9.10; and issue a final Certificate for Payment pursuant to Section 9.10.

§ 4.2.10 If the Owner and Architect agree, the Architect will provide one or more project representatives to assist in carrying out the Architect's responsibilities at the site. The duties, responsibilities and limitations of authority of such project representatives shall be as set forth in an exhibit to be incorporated in the Contract Documents.

§ 4.2.11 The Architect will interpret and decide matters concerning performance under, and requirements of, the Contract Documents on written request of either the Owner or Contractor. The Architect's response to such requests will be made in writing within any time limits agreed upon or otherwise with reasonable promptness.

§ 4.2.12 Interpretations and decisions of the Architect will be consistent with the intent of, and reasonably inferable from, the Contract Documents and will be in writing or in the form of drawings. When making such interpretations and decisions, the Architect will endeavor to secure faithful performance by both Owner and Contractor, will not show partiality to either and will not be liable for results of interpretations or decisions rendered in good faith.

§ 4.2.13 The Architect's decisions on matters relating to aesthetic effect will be final if consistent with the intent expressed in the Contract Documents.

§ 4.2.14 The Architect will review and respond to requests for information about the Contract Documents. The Architect's response to such requests will be made in writing within any time limits agreed upon or otherwise with reasonable promptness. If appropriate, the Architect will prepare and issue supplemental Drawings and Specifications in response to the requests for information.

ARTICLE 5 SUBCONTRACTORS
§ 5.1 DEFINITIONS
§ 5.1.1 A Subcontractor is a person or entity who has a direct contract with the Contractor to perform a portion of the Work at the site. The term "Subcontractor" is referred to throughout the Contract Documents as if singular in number and means a Subcontractor or an authorized representative of the Subcontractor. The term "Subcontractor" does not include a separate contractor or subcontractors of a separate contractor.

§ 5.1.2 A Sub-subcontractor is a person or entity who has a direct or indirect contract with a Subcontractor to perform a portion of the Work at the site. The term "Sub-subcontractor" is referred to throughout the Contract Documents as if singular in number and means a Sub-subcontractor or an authorized representative of the Sub-subcontractor.

§ 5.2 AWARD OF SUBCONTRACTS AND OTHER CONTRACTS FOR PORTIONS OF THE WORK

§ 5.2.1 Unless otherwise stated in the Contract Documents or the bidding requirements, the Contractor, as soon as practicable after award of the Contract, shall furnish in writing to the Owner through the Architect the names of persons or entities (including those who are to furnish materials or equipment fabricated to a special design) proposed for each principal portion of the Work. The Architect may reply within 14 days to the Contractor in writing stating (1) whether the Owner or the Architect has reasonable objection to any such proposed person or entity or (2) that the Architect requires additional time for review. Failure of the Owner or Architect to reply within the 14-day period shall constitute notice of no reasonable objection.

§ 5.2.2 The Contractor shall not contract with a proposed person or entity to whom the Owner or Architect has made reasonable and timely objection. The Contractor shall not be required to contract with anyone to whom the Contractor has made reasonable objection.

§ 5.2.3 If the Owner or Architect has reasonable objection to a person or entity proposed by the Contractor, the Contractor shall propose another to whom the Owner or Architect has no reasonable objection. If the proposed but rejected Subcontractor was reasonably capable of performing the Work, the Contract Sum and Contract Time shall be increased or decreased by the difference, if any, occasioned by such change, and an appropriate Change Order shall be issued before commencement of the substitute Subcontractor's Work. However, no increase in the Contract Sum or Contract Time shall be allowed for such change unless the Contractor has acted promptly and responsively in submitting names as required.

§ 5.2.4 The Contractor shall not substitute a Subcontractor, person or entity previously selected if the Owner or Architect makes reasonable objection to such substitution.

§ 5.3 SUBCONTRACTUAL RELATIONS

By appropriate agreement, written where legally required for validity, the Contractor shall require each Subcontractor, to the extent of the Work to be performed by the Subcontractor, to be bound to the Contractor by terms of the Contract Documents, and to assume toward the Contractor all the obligations and responsibilities, including the responsibility for safety of the Subcontractor's Work, which the Contractor, by these Documents, assumes toward the Owner and Architect. Each subcontract agreement shall preserve and protect the rights of the Owner and Architect under the Contract Documents with respect to the Work to be performed by the Subcontractor so that subcontracting thereof will not prejudice such rights, and shall allow to the Subcontractor, unless specifically provided otherwise in the subcontract agreement, the benefit of all rights, remedies and redress against the Contractor that the Contractor, by the Contract Documents, has against the Owner. Where appropriate, the Contractor shall require each Subcontractor to enter into similar agreements with Sub-subcontractors. The Contractor shall make available to each proposed Subcontractor, prior to the execution of the subcontract agreement, copies of the Contract Documents to which the Subcontractor will be bound, and, upon written request of the Subcontractor, identify to the Subcontractor terms and conditions of the proposed subcontract agreement that may be at variance with the Contract Documents. Subcontractors will similarly make copies of applicable portions of such documents available to their respective proposed Sub-subcontractors.

§ 5.4 CONTINGENT ASSIGNMENT OF SUBCONTRACTS

§ 5.4.1 Each subcontract agreement for a portion of the Work is assigned by the Contractor to the Owner, provided that

.1   assignment is effective only after termination of the Contract by the Owner for cause pursuant to Section 14.2 and only for those subcontract agreements that the Owner accepts by notifying the Subcontractor and Contractor in writing; and

.2   assignment is subject to the prior rights of the surety, if any, obligated under bond relating to the Contract.

When the Owner accepts the assignment of a subcontract agreement, the Owner assumes the Contractor's rights and obligations under the subcontract.

§ 5.4.2 Upon such assignment, if the Work has been suspended for more than 30 days, the Subcontractor's compensation shall be equitably adjusted for increases in cost resulting from the suspension.

§ 5.4.3 Upon such assignment to the Owner under this Section 5.4, the Owner may further assign the subcontract to a successor contractor or other entity. If the Owner assigns the subcontract to a successor contractor or other entity, the Owner shall nevertheless remain legally responsible for all of the successor contractor's obligations under the subcontract.

ARTICLE 6   CONSTRUCTION BY OWNER OR BY SEPARATE CONTRACTORS
§ 6.1 OWNER'S RIGHT TO PERFORM CONSTRUCTION AND TO AWARD SEPARATE CONTRACTS
§ 6.1.1 The Owner reserves the right to perform construction or operations related to the Project with the Owner's own forces, and to award separate contracts in connection with other portions of the Project or other construction or operations on the site under Conditions of the Contract identical or substantially similar to these including those portions related to insurance and waiver of subrogation. If the Contractor claims that delay or additional cost is involved because of such action by the Owner, the Contractor shall make such Claim as provided in Article 15.

§ 6.1.2 When separate contracts are awarded for different portions of the Project or other construction or operations on the site, the term "Contractor" in the Contract Documents in each case shall mean the Contractor who executes each separate Owner-Contractor Agreement.

§ 6.1.3 The Owner shall provide for coordination of the activities of the Owner's own forces and of each separate contractor with the Work of the Contractor, who shall cooperate with them. The Contractor shall participate with other separate contractors and the Owner in reviewing their construction schedules. The Contractor shall make any revisions to the construction schedule deemed necessary after a joint review and mutual agreement. The construction schedules shall then constitute the schedules to be used by the Contractor, separate contractors and the Owner until subsequently revised.

§ 6.1.4 Unless otherwise provided in the Contract Documents, when the Owner performs construction or operations related to the Project with the Owner's own forces, the Owner shall be deemed to be subject to the same obligations and to have the same rights that apply to the Contractor under the Conditions of the Contract, including, without excluding others, those stated in Article 3, this Article 6 and Articles 10, 11 and 12.

§ 6.2 MUTUAL RESPONSIBILITY
§ 6.2.1 The Contractor shall afford the Owner and separate contractors reasonable opportunity for introduction and storage of their materials and equipment and performance of their activities, and shall connect and coordinate the Contractor's construction and operations with theirs as required by the Contract Documents.

§ 6.2.2 If part of the Contractor's Work depends for proper execution or results upon construction or operations by the Owner or a separate contractor, the Contractor shall, prior to proceeding with that portion of the Work, promptly report to the Architect apparent discrepancies or defects in such other construction that would render it unsuitable for such proper execution and results. Failure of the Contractor so to report shall constitute an acknowledgment that the Owner's or separate contractor's completed or partially completed construction is fit and proper to receive the Contractor's Work, except as to defects not then reasonably discoverable.

§ 6.2.3 The Contractor shall reimburse the Owner for costs the Owner incurs that are payable to a separate contractor because of the Contractor's delays, improperly timed activities or defective construction. The Owner shall be responsible to the Contractor for costs the Contractor incurs because of a separate contractor's delays, improperly timed activities, damage to the Work or defective construction.

§ 6.2.4 The Contractor shall promptly remedy damage the Contractor wrongfully causes to completed or partially completed construction or to property of the Owner, separate contractors as provided in Section 10.2.5.

§ 6.2.5 The Owner and each separate contractor shall have the same responsibilities for cutting and patching as are described for the Contractor in Section 3.14.

§ 6.3 OWNER'S RIGHT TO CLEAN UP

If a dispute arises among the Contractor, separate contractors and the Owner as to the responsibility under their respective contracts for maintaining the premises and surrounding area free from waste materials and rubbish, the Owner may clean up and the Architect will allocate the cost among those responsible.

ARTICLE 7   CHANGES IN THE WORK

§ 7.1 GENERAL

§ 7.1.1 Changes in the Work may be accomplished after execution of the Contract, and without invalidating the Contract, by Change Order, Construction Change Directive or order for a minor change in the Work, subject to the limitations stated in this Article 7 and elsewhere in the Contract Documents.

§ 7.1.2 A Change Order shall be based upon agreement among the Owner, Contractor and Architect; a Construction Change Directive requires agreement by the Owner and Architect and may or may not be agreed to by the Contractor; an order for a minor change in the Work may be issued by the Architect alone.

§ 7.1.3 Changes in the Work shall be performed under applicable provisions of the Contract Documents, and the Contractor shall proceed promptly, unless otherwise provided in the Change Order, Construction Change Directive or order for a minor change in the Work.

§ 7.2 CHANGE ORDERS

§ 7.2.1 A Change Order is a written instrument prepared by the Architect and signed by the Owner, Contractor and Architect stating their agreement upon all of the following:

.1   The change in the Work;
.2   The amount of the adjustment, if any, in the Contract Sum; and
.3   The extent of the adjustment, if any, in the Contract Time.

§ 7.3 CONSTRUCTION CHANGE DIRECTIVES

§ 7.3.1 A Construction Change Directive is a written order prepared by the Architect and signed by the Owner and Architect, directing a change in the Work prior to agreement on adjustment, if any, in the Contract Sum or Contract Time, or both. The Owner may by Construction Change Directive, without invalidating the Contract, order changes in the Work within the general scope of the Contract consisting of additions, deletions or other revisions, the Contract Sum and Contract Time being adjusted accordingly.

§ 7.3.2 A Construction Change Directive shall be used in the absence of total agreement on the terms of a Change Order.

§ 7.3.3 If the Construction Change Directive provides for an adjustment to the Contract Sum, the adjustment shall be based on one of the following methods:

.1   Mutual acceptance of a lump sum properly itemized and supported by sufficient substantiating data to permit evaluation;
.2   Unit prices stated in the Contract Documents or subsequently agreed upon;
.3   Cost to be determined in a manner agreed upon by the parties and a mutually acceptable fixed or percentage fee; or
.4   As provided in Section 7.3.7.

§ 7.3.4 If unit prices are stated in the Contract Documents or subsequently agreed upon, and if quantities originally contemplated are materially changed in a proposed Change Order or Construction Change Directive so that application of such unit prices to quantities of Work proposed will cause substantial inequity to the Owner or Contractor, the applicable unit prices shall be equitably adjusted.

§ 7.3.5 Upon receipt of a Construction Change Directive, the Contractor shall promptly proceed with the change in the Work involved and advise the Architect of the Contractor's agreement or disagreement with the method, if any, provided in the Construction Change Directive for determining the proposed adjustment in the Contract Sum or Contract Time.

§ 7.3.6 A Construction Change Directive signed by the Contractor indicates the Contractor's agreement therewith, including adjustment in Contract Sum and Contract Time or the method for determining them. Such agreement shall be effective immediately and shall be recorded as a Change Order.

§ 7.3.7 If the Contractor does not respond promptly or disagrees with the method for adjustment in the Contract Sum, the Architect shall determine the method and the adjustment on the basis of reasonable expenditures and savings of those performing the Work attributable to the change, including, in case of an increase in the Contract Sum, an amount for overhead and profit as set forth in the Agreement, or if no such amount is set forth in the Agreement, a reasonable amount. In such case, and also under Section 7.3.3.3, the Contractor shall keep and present, in such form as the Architect may prescribe, an itemized accounting together with appropriate supporting data. Unless otherwise provided in the Contract Documents, costs for the purposes of this Section 7.3.7 shall be limited to the following:

.1 Costs of labor, including social security, old age and unemployment insurance, fringe benefits required by agreement or custom, and workers' compensation insurance;

.2 Costs of materials, supplies and equipment, including cost of transportation, whether incorporated or consumed;

.3 Rental costs of machinery and equipment, exclusive of hand tools, whether rented from the Contractor or others;

.4 Costs of premiums for all bonds and insurance, permit fees, and sales, use or similar taxes related to the Work; and

.5 Additional costs of supervision and field office personnel directly attributable to the change.

§ 7.3.8 The amount of credit to be allowed by the Contractor to the Owner for a deletion or change that results in a net decrease in the Contract Sum shall be actual net cost as confirmed by the Architect. When both additions and credits covering related Work or substitutions are involved in a change, the allowance for overhead and profit shall be figured on the basis of net increase, if any, with respect to that change.

§ 7.3.9 Pending final determination of the total cost of a Construction Change Directive to the Owner, the Contractor may request payment for Work completed under the Construction Change Directive in Applications for Payment. The Architect will make an interim determination for purposes of monthly certification for payment for those costs and certify for payment the amount that the Architect determines, in the Architect's professional judgment, to be reasonably justified. The Architect's interim determination of cost shall adjust the Contract Sum on the same basis as a Change Order, subject to the right of either party to disagree and assert a Claim in accordance with Article 15.

§ 7.3.10 When the Owner and Contractor agree with a determination made by the Architect concerning the adjustments in the Contract Sum and Contract Time, or otherwise reach agreement upon the adjustments, such agreement shall be effective immediately and the Architect will prepare a Change Order. Change Orders may be issued for all or any part of a Construction Change Directive.

§ 7.4 MINOR CHANGES IN THE WORK

The Architect has authority to order minor changes in the Work not involving adjustment in the Contract Sum or extension of the Contract Time and not inconsistent with the intent of the Contract Documents. Such changes will be effected by written order signed by the Architect and shall be binding on the Owner and Contractor.

ARTICLE 8 TIME
§ 8.1 DEFINITIONS

§ 8.1.1 Unless otherwise provided, Contract Time is the period of time, including authorized adjustments, allotted in the Contract Documents for Substantial Completion of the Work.

§ 8.1.2 The date of commencement of the Work is the date established in the Agreement.

§ 8.1.3 The date of Substantial Completion is the date certified by the Architect in accordance with Section 9.8.

§ 8.1.4 The term "day" as used in the Contract Documents shall mean calendar day unless otherwise specifically defined.

§ 8.2 PROGRESS AND COMPLETION

§ 8.2.1 Time limits stated in the Contract Documents are of the essence of the Contract. By executing the Agreement the Contractor confirms that the Contract Time is a reasonable period for performing the Work.

§ 8.2.2 The Contractor shall not knowingly, except by agreement or instruction of the Owner in writing, prematurely commence operations on the site or elsewhere prior to the effective date of insurance required by Article 11 to be furnished by the Contractor and Owner. The date of commencement of the Work shall not be changed by the effective date of such insurance.

§ 8.2.3 The Contractor shall proceed expeditiously with adequate forces and shall achieve Substantial Completion within the Contract Time.

§ 8.3 DELAYS AND EXTENSIONS OF TIME

§ 8.3.1 If the Contractor is delayed at any time in the commencement or progress of the Work by an act or neglect of the Owner or Architect, or of an employee of either, or of a separate contractor employed by the Owner; or by changes ordered in the Work; or by labor disputes, fire, unusual delay in deliveries, unavoidable casualties or other causes beyond the Contractor's control; or by delay authorized by the Owner pending mediation and arbitration; or by other causes that the Architect determines may justify delay, then the Contract Time shall be extended by Change Order for such reasonable time as the Architect may determine.

§ 8.3.2 Claims relating to time shall be made in accordance with applicable provisions of Article 15.

§ 8.3.3 This Section 8.3 does not preclude recovery of damages for delay by either party under other provisions of the Contract Documents.

ARTICLE 9 PAYMENTS AND COMPLETION
§ 9.1 CONTRACT SUM

The Contract Sum is stated in the Agreement and, including authorized adjustments, is the total amount payable by the Owner to the Contractor for performance of the Work under the Contract Documents.

§ 9.2 SCHEDULE OF VALUES

Where the Contract is based on a stipulated sum or Guaranteed Maximum Price, the Contractor shall submit to the Architect, before the first Application for Payment, a schedule of values allocating the entire Contract Sum to the various portions of the Work and prepared in such form and supported by such data to substantiate its accuracy as the Architect may require. This schedule, unless objected to by the Architect, shall be used as a basis for reviewing the Contractor's Applications for Payment.

§ 9.3 APPLICATIONS FOR PAYMENT

§ 9.3.1 At least ten days before the date established for each progress payment, the Contractor shall submit to the Architect an itemized Application for Payment prepared in accordance with the schedule of values, if required under Section 9.2., for completed portions of the Work. Such application shall be notarized, if required, and supported by such data substantiating the Contractor's right to payment as the Owner or Architect may require, such as copies of requisitions from Subcontractors and material suppliers, and shall reflect retainage if provided for in the Contract Documents.

§ 9.3.1.1 As provided in Section 7.3.9, such applications may include requests for payment on account of changes in the Work that have been properly authorized by Construction Change Directives, or by interim determinations of the Architect, but not yet included in Change Orders.

§ 9.3.1.2 Applications for Payment shall not include requests for payment for portions of the Work for which the Contractor does not intend to pay a Subcontractor or material supplier, unless such Work has been performed by others whom the Contractor intends to pay.

§ 9.3.2 Unless otherwise provided in the Contract Documents, payments shall be made on account of materials and equipment delivered and suitably stored at the site for subsequent incorporation in the Work. If approved in advance by the Owner, payment may similarly be made for materials and equipment suitably stored off the site at a location agreed upon in writing. Payment for materials and equipment stored on or off the site shall be conditioned upon compliance by the Contractor with procedures satisfactory to the Owner to establish the Owner's title to such materials and equipment or otherwise protect the Owner's interest, and shall include the costs of applicable insurance, storage and transportation to the site for such materials and equipment stored off the site.

§ 9.3.3 The Contractor warrants that title to all Work covered by an Application for Payment will pass to the Owner no later than the time of payment. The Contractor further warrants that upon submittal of an Application for Payment all Work for which Certificates for Payment have been previously issued and payments received from the

Owner shall, to the best of the Contractor's knowledge, information and belief, be free and clear of liens, claims, security interests or encumbrances in favor of the Contractor, Subcontractors, material suppliers, or other persons or entities making a claim by reason of having provided labor, materials and equipment relating to the Work.

§ 9.4 CERTIFICATES FOR PAYMENT

§ 9.4.1 The Architect will, within seven days after receipt of the Contractor's Application for Payment, either issue to the Owner a Certificate for Payment, with a copy to the Contractor, for such amount as the Architect determines is properly due, or notify the Contractor and Owner in writing of the Architect's reasons for withholding certification in whole or in part as provided in Section 9.5.1.

§ 9.4.2 The issuance of a Certificate for Payment will constitute a representation by the Architect to the Owner, based on the Architect's evaluation of the Work and the data comprising the Application for Payment, that, to the best of the Architect's knowledge, information and belief, the Work has progressed to the point indicated and that the quality of the Work is in accordance with the Contract Documents. The foregoing representations are subject to an evaluation of the Work for conformance with the Contract Documents upon Substantial Completion, to results of subsequent tests and inspections, to correction of minor deviations from the Contract Documents prior to completion and to specific qualifications expressed by the Architect. The issuance of a Certificate for Payment will further constitute a representation that the Contractor is entitled to payment in the amount certified. However, the issuance of a Certificate for Payment will not be a representation that the Architect has (1) made exhaustive or continuous on-site inspections to check the quality or quantity of the Work, (2) reviewed construction means, methods, techniques, sequences or procedures, (3) reviewed copies of requisitions received from Subcontractors and material suppliers and other data requested by the Owner to substantiate the Contractor's right to payment, or (4) made examination to ascertain how or for what purpose the Contractor has used money previously paid on account of the Contract Sum.

§ 9.5 DECISIONS TO WITHHOLD CERTIFICATION

§ 9.5.1 The Architect may withhold a Certificate for Payment in whole or in part, to the extent reasonably necessary to protect the Owner, if in the Architect's opinion the representations to the Owner required by Section 9.4.2 cannot be made. If the Architect is unable to certify payment in the amount of the Application, the Architect will notify the Contractor and Owner as provided in Section 9.4.1. If the Contractor and Architect cannot agree on a revised amount, the Architect will promptly issue a Certificate for Payment for the amount for which the Architect is able to make such representations to the Owner. The Architect may also withhold a Certificate for Payment or, because of subsequently discovered evidence, may nullify the whole or a part of a Certificate for Payment previously issued, to such extent as may be necessary in the Architect's opinion to protect the Owner from loss for which the Contractor is responsible, including loss resulting from acts and omissions described in Section 3.3.2, because of

.1   defective Work not remedied;

.2   third party claims filed or reasonable evidence indicating probable filing of such claims unless security acceptable to the Owner is provided by the Contractor;

.3   failure of the Contractor to make payments properly to Subcontractors or for labor, materials or equipment;

.4   reasonable evidence that the Work cannot be completed for the unpaid balance of the Contract Sum;

.5   damage to the Owner or a separate contractor;

.6   reasonable evidence that the Work will not be completed within the Contract Time, and that the unpaid balance would not be adequate to cover actual or liquidated damages for the anticipated delay; or

.7   repeated failure to carry out the Work in accordance with the Contract Documents.

§ 9.5.2 When the above reasons for withholding certification are removed, certification will be made for amounts previously withheld.

§ 9.5.3 If the Architect withholds certification for payment under Section 9.5.1.3, the Owner may, at its sole option, issue joint checks to the Contractor and to any Subcontractor or material or equipment suppliers to whom the Contractor failed to make payment for Work properly performed or material or equipment suitably delivered. If the Owner makes payments by joint check, the Owner shall notify the Architect and the Architect will reflect such payment on the next Certificate for Payment.

§ 9.6 PROGRESS PAYMENTS

§ 9.6.1 After the Architect has issued a Certificate for Payment, the Owner shall make payment in the manner and within the time provided in the Contract Documents, and shall so notify the Architect.

§ **9.6.2** The Contractor shall pay each Subcontractor no later than seven days after receipt of payment from the Owner the amount to which the Subcontractor is entitled, reflecting percentages actually retained from payments to the Contractor on account of the Subcontractor's portion of the Work. The Contractor shall, by appropriate agreement with each Subcontractor, require each Subcontractor to make payments to Sub-subcontractors in a similar manner.

§ **9.6.3** The Architect will, on request, furnish to a Subcontractor, if practicable, information regarding percentages of completion or amounts applied for by the Contractor and action taken thereon by the Architect and Owner on account of portions of the Work done by such Subcontractor.

§ **9.6.4** The Owner has the right to request written evidence from the Contractor that the Contractor has properly paid Subcontractors and material and equipment suppliers amounts paid by the Owner to the Contractor for subcontracted Work. If the Contractor fails to furnish such evidence within seven days, the Owner shall have the right to contact Subcontractors to ascertain whether they have been properly paid. Neither the Owner nor Architect shall have an obligation to pay or to see to the payment of money to a Subcontractor, except as may otherwise be required by law.

§ **9.6.5** Contractor payments to material and equipment suppliers shall be treated in a manner similar to that provided in Sections 9.6.2, 9.6.3 and 9.6.4.

§ **9.6.6** A Certificate for Payment, a progress payment, or partial or entire use or occupancy of the Project by the Owner shall not constitute acceptance of Work not in accordance with the Contract Documents.

§ **9.6.7** Unless the Contractor provides the Owner with a payment bond in the full penal sum of the Contract Sum, payments received by the Contractor for Work properly performed by Subcontractors and suppliers shall be held by the Contractor for those Subcontractors or suppliers who performed Work or furnished materials, or both, under contract with the Contractor for which payment was made by the Owner. Nothing contained herein shall require money to be placed in a separate account and not commingled with money of the Contractor, shall create any fiduciary liability or tort liability on the part of the Contractor for breach of trust or shall entitle any person or entity to an award of punitive damages against the Contractor for breach of the requirements of this provision.

§ 9.7 FAILURE OF PAYMENT

If the Architect does not issue a Certificate for Payment, through no fault of the Contractor, within seven days after receipt of the Contractor's Application for Payment, or if the Owner does not pay the Contractor within seven days after the date established in the Contract Documents the amount certified by the Architect or awarded by binding dispute resolution, then the Contractor may, upon seven additional days' written notice to the Owner and Architect, stop the Work until payment of the amount owing has been received. The Contract Time shall be extended appropriately and the Contract Sum shall be increased by the amount of the Contractor's reasonable costs of shut-down, delay and start-up, plus interest as provided for in the Contract Documents.

§ 9.8 SUBSTANTIAL COMPLETION

§ **9.8.1** Substantial Completion is the stage in the progress of the Work when the Work or designated portion thereof is sufficiently complete in accordance with the Contract Documents so that the Owner can occupy or utilize the Work for its intended use.

§ **9.8.2** When the Contractor considers that the Work, or a portion thereof which the Owner agrees to accept separately, is substantially complete, the Contractor shall prepare and submit to the Architect a comprehensive list of items to be completed or corrected prior to final payment. Failure to include an item on such list does not alter the responsibility of the Contractor to complete all Work in accordance with the Contract Documents.

§ **9.8.3** Upon receipt of the Contractor's list, the Architect will make an inspection to determine whether the Work or designated portion thereof is substantially complete. If the Architect's inspection discloses any item, whether or not included on the Contractor's list, which is not sufficiently complete in accordance with the Contract Documents so that the Owner can occupy or utilize the Work or designated portion thereof for its intended use, the Contractor shall, before issuance of the Certificate of Substantial Completion, complete or correct such item upon notification by the Architect. In such case, the Contractor shall then submit a request for another inspection by the Architect to determine Substantial Completion.

§ 9.8.4 When the Work or designated portion thereof is substantially complete, the Architect will prepare a Certificate of Substantial Completion that shall establish the date of Substantial Completion, shall establish responsibilities of the Owner and Contractor for security, maintenance, heat, utilities, damage to the Work and insurance, and shall fix the time within which the Contractor shall finish all items on the list accompanying the Certificate. Warranties required by the Contract Documents shall commence on the date of Substantial Completion of the Work or designated portion thereof unless otherwise provided in the Certificate of Substantial Completion.

§ 9.8.5 The Certificate of Substantial Completion shall be submitted to the Owner and Contractor for their written acceptance of responsibilities assigned to them in such Certificate. Upon such acceptance and consent of surety, if any, the Owner shall make payment of retainage applying to such Work or designated portion thereof. Such payment shall be adjusted for Work that is incomplete or not in accordance with the requirements of the Contract Documents.

§ 9.9 PARTIAL OCCUPANCY OR USE

§ 9.9.1 The Owner may occupy or use any completed or partially completed portion of the Work at any stage when such portion is designated by separate agreement with the Contractor, provided such occupancy or use is consented to by the insurer as required under Section 11.3.1.5 and authorized by public authorities having jurisdiction over the Project. Such partial occupancy or use may commence whether or not the portion is substantially complete, provided the Owner and Contractor have accepted in writing the responsibilities assigned to each of them for payments, retainage, if any, security, maintenance, heat, utilities, damage to the Work and insurance, and have agreed in writing concerning the period for correction of the Work and commencement of warranties required by the Contract Documents. When the Contractor considers a portion substantially complete, the Contractor shall prepare and submit a list to the Architect as provided under Section 9.8.2. Consent of the Contractor to partial occupancy or use shall not be unreasonably withheld. The stage of the progress of the Work shall be determined by written agreement between the Owner and Contractor or, if no agreement is reached, by decision of the Architect.

§ 9.9.2 Immediately prior to such partial occupancy or use, the Owner, Contractor and Architect shall jointly inspect the area to be occupied or portion of the Work to be used in order to determine and record the condition of the Work.

§ 9.9.3 Unless otherwise agreed upon, partial occupancy or use of a portion or portions of the Work shall not constitute acceptance of Work not complying with the requirements of the Contract Documents.

§ 9.10 FINAL COMPLETION AND FINAL PAYMENT

§ 9.10.1 Upon receipt of the Contractor's written notice that the Work is ready for final inspection and acceptance and upon receipt of a final Application for Payment, the Architect will promptly make such inspection and, when the Architect finds the Work acceptable under the Contract Documents and the Contract fully performed, the Architect will promptly issue a final Certificate for Payment stating that to the best of the Architect's knowledge, information and belief, and on the basis of the Architect's on-site visits and inspections, the Work has been completed in accordance with terms and conditions of the Contract Documents and that the entire balance found to be due the Contractor and noted in the final Certificate is due and payable. The Architect's final Certificate for Payment will constitute a further representation that conditions listed in Section 9.10.2 as precedent to the Contractor's being entitled to final payment have been fulfilled.

§ 9.10.2 Neither final payment nor any remaining retained percentage shall become due until the Contractor submits to the Architect (1) an affidavit that payrolls, bills for materials and equipment, and other indebtedness connected with the Work for which the Owner or the Owner's property might be responsible or encumbered (less amounts withheld by Owner) have been paid or otherwise satisfied, (2) a certificate evidencing that insurance required by the Contract Documents to remain in force after final payment is currently in effect and will not be canceled or allowed to expire until at least 30 days' prior written notice has been given to the Owner, (3) a written statement that the Contractor knows of no substantial reason that the insurance will not be renewable to cover the period required by the Contract Documents, (4) consent of surety, if any, to final payment and (5), if required by the Owner, other data establishing payment or satisfaction of obligations, such as receipts, releases and waivers of liens, claims, security interests or encumbrances arising out of the Contract, to the extent and in such form as may be designated by the Owner. If a Subcontractor refuses to furnish a release or waiver required by the Owner, the Contractor may furnish a bond satisfactory to the Owner to indemnify the Owner against such lien. If such lien remains unsatisfied after payments are made, the Contractor shall refund to the Owner all money that the Owner may be compelled to pay in discharging such lien, including all costs and reasonable attorneys' fees.

§ **9.10.3** If, after Substantial Completion of the Work, final completion thereof is materially delayed through no fault of the Contractor or by issuance of Change Orders affecting final completion, and the Architect so confirms, the Owner shall, upon application by the Contractor and certification by the Architect, and without terminating the Contract, make payment of the balance due for that portion of the Work fully completed and accepted. If the remaining balance for Work not fully completed or corrected is less than retainage stipulated in the Contract Documents, and if bonds have been furnished, the written consent of surety to payment of the balance due for that portion of the Work fully completed and accepted shall be submitted by the Contractor to the Architect prior to certification of such payment. Such payment shall be made under terms and conditions governing final payment, except that it shall not constitute a waiver of claims.

§ **9.10.4** The making of final payment shall constitute a waiver of Claims by the Owner except those arising from

.1   liens, Claims, security interests or encumbrances arising out of the Contract and unsettled;

.2   failure of the Work to comply with the requirements of the Contract Documents; or

.3   terms of special warranties required by the Contract Documents.

§ **9.10.5** Acceptance of final payment by the Contractor, a Subcontractor or material supplier shall constitute a waiver of claims by that payee except those previously made in writing and identified by that payee as unsettled at the time of final Application for Payment.

ARTICLE 10   PROTECTION OF PERSONS AND PROPERTY
§ 10.1 SAFETY PRECAUTIONS AND PROGRAMS
The Contractor shall be responsible for initiating, maintaining and supervising all safety precautions and programs in connection with the performance of the Contract.

§ 10.2 SAFETY OF PERSONS AND PROPERTY
§ **10.2.1** The Contractor shall take reasonable precautions for safety of, and shall provide reasonable protection to prevent damage, injury or loss to

.1   employees on the Work and other persons who may be affected thereby;

.2   the Work and materials and equipment to be incorporated therein, whether in storage on or off the site, under care, custody or control of the Contractor or the Contractor's Subcontractors or Sub-subcontractors; and

.3   other property at the site or adjacent thereto, such as trees, shrubs, lawns, walks, pavements, roadways, structures and utilities not designated for removal, relocation or replacement in the course of construction.

§ **10.2.2** The Contractor shall comply with and give notices required by applicable laws, statutes, ordinances, codes, rules and regulations, and lawful orders of public authorities bearing on safety of persons or property or their protection from damage, injury or loss.

§ **10.2.3** The Contractor shall erect and maintain, as required by existing conditions and performance of the Contract, reasonable safeguards for safety and protection, including posting danger signs and other warnings against hazards, promulgating safety regulations and notifying owners and users of adjacent sites and utilities.

§ **10.2.4** When use or storage of explosives or other hazardous materials or equipment or unusual methods are necessary for execution of the Work, the Contractor shall exercise utmost care and carry on such activities under supervision of properly qualified personnel.

§ **10.2.5** The Contractor shall promptly remedy damage and loss (other than damage or loss insured under property insurance required by the Contract Documents) to property referred to in Sections 10.2.1.2 and 10.2.1.3 caused in whole or in part by the Contractor, a Subcontractor, a Sub-subcontractor, or anyone directly or indirectly employed by any of them, or by anyone for whose acts they may be liable and for which the Contractor is responsible under Sections 10.2.1.2 and 10.2.1.3, except damage or loss attributable to acts or omissions of the Owner or Architect or anyone directly or indirectly employed by either of them, or by anyone for whose acts either of them may be liable, and not attributable to the fault or negligence of the Contractor. The foregoing obligations of the Contractor are in addition to the Contractor's obligations under Section 3.18.

§ 10.2.6 The Contractor shall designate a responsible member of the Contractor's organization at the site whose duty shall be the prevention of accidents. This person shall be the Contractor's superintendent unless otherwise designated by the Contractor in writing to the Owner and Architect.

§ 10.2.7 The Contractor shall not permit any part of the construction or site to be loaded so as to cause damage or create an unsafe condition.

§ 10.2.8 INJURY OR DAMAGE TO PERSON OR PROPERTY

If either party suffers injury or damage to person or property because of an act or omission of the other party, or of others for whose acts such party is legally responsible, written notice of such injury or damage, whether or not insured, shall be given to the other party within a reasonable time not exceeding 21 days after discovery. The notice shall provide sufficient detail to enable the other party to investigate the matter.

§ 10.3 HAZARDOUS MATERIALS

§ 10.3.1 The Contractor is responsible for compliance with any requirements included in the Contract Documents regarding hazardous materials. If the Contractor encounters a hazardous material or substance not addressed in the Contract Documents and if reasonable precautions will be inadequate to prevent foreseeable bodily injury or death to persons resulting from a material or substance, including but not limited to asbestos or polychlorinated biphenyl (PCB), encountered on the site by the Contractor, the Contractor shall, upon recognizing the condition, immediately stop Work in the affected area and report the condition to the Owner and Architect in writing.

§ 10.3.2 Upon receipt of the Contractor's written notice, the Owner shall obtain the services of a licensed laboratory to verify the presence or absence of the material or substance reported by the Contractor and, in the event such material or substance is found to be present, to cause it to be rendered harmless. Unless otherwise required by the Contract Documents, the Owner shall furnish in writing to the Contractor and Architect the names and qualifications of persons or entities who are to perform tests verifying the presence or absence of such material or substance or who are to perform the task of removal or safe containment of such material or substance. The Contractor and the Architect will promptly reply to the Owner in writing stating whether or not either has reasonable objection to the persons or entities proposed by the Owner. If either the Contractor or Architect has an objection to a person or entity proposed by the Owner, the Owner shall propose another to whom the Contractor and the Architect have no reasonable objection. When the material or substance has been rendered harmless, Work in the affected area shall resume upon written agreement of the Owner and Contractor. By Change Order, the Contract Time shall be extended appropriately and the Contract Sum shall be increased in the amount of the Contractor's reasonable additional costs of shut-down, delay and start-up.

§ 10.3.3 To the fullest extent permitted by law, the Owner shall indemnify and hold harmless the Contractor, Subcontractors, Architect, Architect's consultants and agents and employees of any of them from and against claims, damages, losses and expenses, including but not limited to attorneys' fees, arising out of or resulting from performance of the Work in the affected area if in fact the material or substance presents the risk of bodily injury or death as described in Section 10.3.1 and has not been rendered harmless, provided that such claim, damage, loss or expense is attributable to bodily injury, sickness, disease or death, or to injury to or destruction of tangible property (other than the Work itself), except to the extent that such damage, loss or expense is due to the fault or negligence of the party seeking indemnity.

§ 10.3.4 The Owner shall not be responsible under this Section 10.3 for materials or substances the Contractor brings to the site unless such materials or substances are required by the Contract Documents. The Owner shall be responsible for materials or substances required by the Contract Documents, except to the extent of the Contractor's fault or negligence in the use and handling of such materials or substances.

§ 10.3.5 The Contractor shall indemnify the Owner for the cost and expense the Owner incurs (1) for remediation of a material or substance the Contractor brings to the site and negligently handles, or (2) where the Contractor fails to perform its obligations under Section 10.3.1, except to the extent that the cost and expense are due to the Owner's fault or negligence.

§ 10.3.6 If, without negligence on the part of the Contractor, the Contractor is held liable by a government agency for the cost of remediation of a hazardous material or substance solely by reason of performing Work as required by the Contract Documents, the Owner shall indemnify the Contractor for all cost and expense thereby incurred.

§ 10.4 EMERGENCIES

In an emergency affecting safety of persons or property, the Contractor shall act, at the Contractor's discretion, to prevent threatened damage, injury or loss. Additional compensation or extension of time claimed by the Contractor on account of an emergency shall be determined as provided in Article 15 and Article 7.

ARTICLE 11 INSURANCE AND BONDS
§ 11.1 CONTRACTOR'S LIABILITY INSURANCE

§ 11.1.1 The Contractor shall purchase from and maintain in a company or companies lawfully authorized to do business in the jurisdiction in which the Project is located such insurance as will protect the Contractor from claims set forth below which may arise out of or result from the Contractor's operations and completed operations under the Contract and for which the Contractor may be legally liable, whether such operations be by the Contractor or by a Subcontractor or by anyone directly or indirectly employed by any of them, or by anyone for whose acts any of them may be liable:

.1   Claims under workers' compensation, disability benefit and other similar employee benefit acts that are applicable to the Work to be performed;

.2   Claims for damages because of bodily injury, occupational sickness or disease, or death of the Contractor's employees;

.3   Claims for damages because of bodily injury, sickness or disease, or death of any person other than the Contractor's employees;

.4   Claims for damages insured by usual personal injury liability coverage;

.5   Claims for damages, other than to the Work itself, because of injury to or destruction of tangible property, including loss of use resulting therefrom;

.6   Claims for damages because of bodily injury, death of a person or property damage arising out of ownership, maintenance or use of a motor vehicle;

.7   Claims for bodily injury or property damage arising out of completed operations; and

.8   Claims involving contractual liability insurance applicable to the Contractor's obligations under Section 3.18.

§ 11.1.2 The insurance required by Section 11.1 shall be written for not less than limits of liability specified in the Contract Documents or required by law, whichever coverage is greater. Coverages, whether written on an occurrence or claims-made basis, shall be maintained without interruption from the date of commencement of the Work until the date of final payment and termination of any coverage required to be maintained after final payment, and, with respect to the Contractor's completed operations coverage, until the expiration of the period for correction of Work or for such other period for maintenance of completed operations coverage as specified in the Contract Documents.

§ 11.1.3 Certificates of insurance acceptable to the Owner shall be filed with the Owner prior to commencement of the Work and thereafter upon renewal or replacement of each required policy of insurance. These certificates and the insurance policies required by this Section 11.1 shall contain a provision that coverages afforded under the policies will not be canceled or allowed to expire until at least 30 days' prior written notice has been given to the Owner. An additional certificate evidencing continuation of liability coverage, including coverage for completed operations, shall be submitted with the final Application for Payment as required by Section 9.10.2 and thereafter upon renewal or replacement of such coverage until the expiration of the time required by Section 11.1.2. Information concerning reduction of coverage on account of revised limits or claims paid under the General Aggregate, or both, shall be furnished by the Contractor with reasonable promptness.

§ 11.1.4 The Contractor shall cause the commercial liability coverage required by the Contract Documents to include (1) the Owner, the Architect and the Architect's Consultants as additional insureds for claims caused in whole or in part by the Contractor's negligent acts or omissions during the Contractor's operations; and (2) the Owner as an additional insured for claims caused in whole or in part by the Contractor's negligent acts or omissions during the Contractor's completed operations.

§ 11.2 OWNER'S LIABILITY INSURANCE

The Owner shall be responsible for purchasing and maintaining the Owner's usual liability insurance.

§ 11.3 PROPERTY INSURANCE

§ 11.3.1 Unless otherwise provided, the Owner shall purchase and maintain, in a company or companies lawfully authorized to do business in the jurisdiction in which the Project is located, property insurance written on a builder's risk "all-risk" or equivalent policy form in the amount of the initial Contract Sum, plus value of subsequent Contract Modifications and cost of materials supplied or installed by others, comprising total value for the entire Project at the site on a replacement cost basis without optional deductibles. Such property insurance shall be maintained, unless otherwise provided in the Contract Documents or otherwise agreed in writing by all persons and entities who are beneficiaries of such insurance, until final payment has been made as provided in Section 9.10 or until no person or entity other than the Owner has an insurable interest in the property required by this Section 11.3 to be covered, whichever is later. This insurance shall include interests of the Owner, the Contractor, Subcontractors and Sub-subcontractors in the Project.

§ 11.3.1.1 Property insurance shall be on an "all-risk" or equivalent policy form and shall include, without limitation, insurance against the perils of fire (with extended coverage) and physical loss or damage including, without duplication of coverage, theft, vandalism, malicious mischief, collapse, earthquake, flood, windstorm, falsework, testing and startup, temporary buildings and debris removal including demolition occasioned by enforcement of any applicable legal requirements, and shall cover reasonable compensation for Architect's and Contractor's services and expenses required as a result of such insured loss.

§ 11.3.1.2 If the Owner does not intend to purchase such property insurance required by the Contract and with all of the coverages in the amount described above, the Owner shall so inform the Contractor in writing prior to commencement of the Work. The Contractor may then effect insurance that will protect the interests of the Contractor, Subcontractors and Sub-subcontractors in the Work, and by appropriate Change Order the cost thereof shall be charged to the Owner. If the Contractor is damaged by the failure or neglect of the Owner to purchase or maintain insurance as described above, without so notifying the Contractor in writing, then the Owner shall bear all reasonable costs properly attributable thereto.

§ 11.3.1.3 If the property insurance requires deductibles, the Owner shall pay costs not covered because of such deductibles.

§ 11.3.1.4 This property insurance shall cover portions of the Work stored off the site, and also portions of the Work in transit.

§ 11.3.1.5 Partial occupancy or use in accordance with Section 9.9 shall not commence until the insurance company or companies providing property insurance have consented to such partial occupancy or use by endorsement or otherwise. The Owner and the Contractor shall take reasonable steps to obtain consent of the insurance company or companies and shall, without mutual written consent, take no action with respect to partial occupancy or use that would cause cancellation, lapse or reduction of insurance.

§ 11.3.2 BOILER AND MACHINERY INSURANCE
The Owner shall purchase and maintain boiler and machinery insurance required by the Contract Documents or by law, which shall specifically cover such insured objects during installation and until final acceptance by the Owner; this insurance shall include interests of the Owner, Contractor, Subcontractors and Sub-subcontractors in the Work, and the Owner and Contractor shall be named insureds.

§ 11.3.3 LOSS OF USE INSURANCE
The Owner, at the Owner's option, may purchase and maintain such insurance as will insure the Owner against loss of use of the Owner's property due to fire or other hazards, however caused. The Owner waives all rights of action against the Contractor for loss of use of the Owner's property, including consequential losses due to fire or other hazards however caused.

§ 11.3.4 If the Contractor requests in writing that insurance for risks other than those described herein or other special causes of loss be included in the property insurance policy, the Owner shall, if possible, include such insurance, and the cost thereof shall be charged to the Contractor by appropriate Change Order.

§ 11.3.5 If during the Project construction period the Owner insures properties, real or personal or both, at or adjacent to the site by property insurance under policies separate from those insuring the Project, or if after final payment property insurance is to be provided on the completed Project through a policy or policies other than those insuring the Project during the construction period, the Owner shall waive all rights in accordance with the terms of Section 11.3.7 for damages caused by fire or other causes of loss covered by this separate property insurance. All separate policies shall provide this waiver of subrogation by endorsement or otherwise.

§ **11.3.6** Before an exposure to loss may occur, the Owner shall file with the Contractor a copy of each policy that includes insurance coverages required by this Section 11.3. Each policy shall contain all generally applicable conditions, definitions, exclusions and endorsements related to this Project. Each policy shall contain a provision that the policy will not be canceled or allowed to expire, and that its limits will not be reduced, until at least 30 days' prior written notice has been given to the Contractor.

§ 11.3.7 WAIVERS OF SUBROGATION

The Owner and Contractor waive all rights against (1) each other and any of their subcontractors, sub-subcontractors, agents and employees, each of the other, and (2) the Architect, Architect's consultants, separate contractors described in Article 6, if any, and any of their subcontractors, sub-subcontractors, agents and employees, for damages caused by fire or other causes of loss to the extent covered by property insurance obtained pursuant to this Section 11.3 or other property insurance applicable to the Work, except such rights as they have to proceeds of such insurance held by the Owner as fiduciary. The Owner or Contractor, as appropriate, shall require of the Architect, Architect's consultants, separate contractors described in Article 6, if any, and the subcontractors, sub-subcontractors, agents and employees of any of them, by appropriate agreements, written where legally required for validity, similar waivers each in favor of other parties enumerated herein. The policies shall provide such waivers of subrogation by endorsement or otherwise. A waiver of subrogation shall be effective as to a person or entity even though that person or entity would otherwise have a duty of indemnification, contractual or otherwise, did not pay the insurance premium directly or indirectly, and whether or not the person or entity had an insurable interest in the property damaged.

§ **11.3.8** A loss insured under the Owner's property insurance shall be adjusted by the Owner as fiduciary and made payable to the Owner as fiduciary for the insureds, as their interests may appear, subject to requirements of any applicable mortgagee clause and of Section 11.3.10. The Contractor shall pay Subcontractors their just shares of insurance proceeds received by the Contractor, and by appropriate agreements, written where legally required for validity, shall require Subcontractors to make payments to their Sub-subcontractors in similar manner.

§ **11.3.9** If required in writing by a party in interest, the Owner as fiduciary shall, upon occurrence of an insured loss, give bond for proper performance of the Owner's duties. The cost of required bonds shall be charged against proceeds received as fiduciary. The Owner shall deposit in a separate account proceeds so received, which the Owner shall distribute in accordance with such agreement as the parties in interest may reach, or as determined in accordance with the method of binding dispute resolution selected in the Agreement between the Owner and Contractor. If after such loss no other special agreement is made and unless the Owner terminates the Contract for convenience, replacement of damaged property shall be performed by the Contractor after notification of a Change in the Work in accordance with Article 7.

§ **11.3.10** The Owner as fiduciary shall have power to adjust and settle a loss with insurers unless one of the parties in interest shall object in writing within five days after occurrence of loss to the Owner's exercise of this power; if such objection is made, the dispute shall be resolved in the manner selected by the Owner and Contractor as the method of binding dispute resolution in the Agreement. If the Owner and Contractor have selected arbitration as the method of binding dispute resolution, the Owner as fiduciary shall make settlement with insurers or, in the case of a dispute over distribution of insurance proceeds, in accordance with the directions of the arbitrators.

§ 11.4 PERFORMANCE BOND AND PAYMENT BOND

§ **11.4.1** The Owner shall have the right to require the Contractor to furnish bonds covering faithful performance of the Contract and payment of obligations arising thereunder as stipulated in bidding requirements or specifically required in the Contract Documents on the date of execution of the Contract.

§ **11.4.2** Upon the request of any person or entity appearing to be a potential beneficiary of bonds covering payment of obligations arising under the Contract, the Contractor shall promptly furnish a copy of the bonds or shall authorize a copy to be furnished.

ARTICLE 12   UNCOVERING AND CORRECTION OF WORK
§ 12.1 UNCOVERING OF WORK

§ **12.1.1** If a portion of the Work is covered contrary to the Architect's request or to requirements specifically expressed in the Contract Documents, it must, if requested in writing by the Architect, be uncovered for the Architect's examination and be replaced at the Contractor's expense without change in the Contract Time.

§ **12.1.2** If a portion of the Work has been covered that the Architect has not specifically requested to examine prior to its being covered, the Architect may request to see such Work and it shall be uncovered by the Contractor. If such Work is in accordance with the Contract Documents, costs of uncovering and replacement shall, by appropriate Change Order, be at the Owner's expense. If such Work is not in accordance with the Contract Documents, such costs and the cost of correction shall be at the Contractor's expense unless the condition was caused by the Owner or a separate contractor in which event the Owner shall be responsible for payment of such costs.

§ 12.2 CORRECTION OF WORK
§ 12.2.1 BEFORE OR AFTER SUBSTANTIAL COMPLETION
The Contractor shall promptly correct Work rejected by the Architect or failing to conform to the requirements of the Contract Documents, whether discovered before or after Substantial Completion and whether or not fabricated, installed or completed. Costs of correcting such rejected Work, including additional testing and inspections, the cost of uncovering and replacement, and compensation for the Architect's services and expenses made necessary thereby, shall be at the Contractor's expense.

§ 12.2.2 AFTER SUBSTANTIAL COMPLETION
§ **12.2.2.1** In addition to the Contractor's obligations under Section 3.5, if, within one year after the date of Substantial Completion of the Work or designated portion thereof or after the date for commencement of warranties established under Section 9.9.1, or by terms of an applicable special warranty required by the Contract Documents, any of the Work is found to be not in accordance with the requirements of the Contract Documents, the Contractor shall correct it promptly after receipt of written notice from the Owner to do so unless the Owner has previously given the Contractor a written acceptance of such condition. The Owner shall give such notice promptly after discovery of the condition. During the one-year period for correction of Work, if the Owner fails to notify the Contractor and give the Contractor an opportunity to make the correction, the Owner waives the rights to require correction by the Contractor and to make a claim for breach of warranty. If the Contractor fails to correct nonconforming Work within a reasonable time during that period after receipt of notice from the Owner or Architect, the Owner may correct it in accordance with Section 2.4.

§ **12.2.2.2** The one-year period for correction of Work shall be extended with respect to portions of Work first performed after Substantial Completion by the period of time between Substantial Completion and the actual completion of that portion of the Work.

§ **12.2.2.3** The one-year period for correction of Work shall not be extended by corrective Work performed by the Contractor pursuant to this Section 12.2.

§ **12.2.3** The Contractor shall remove from the site portions of the Work that are not in accordance with the requirements of the Contract Documents and are neither corrected by the Contractor nor accepted by the Owner.

§ **12.2.4** The Contractor shall bear the cost of correcting destroyed or damaged construction, whether completed or partially completed, of the Owner or separate contractors caused by the Contractor's correction or removal of Work that is not in accordance with the requirements of the Contract Documents.

§ **12.2.5** Nothing contained in this Section 12.2 shall be construed to establish a period of limitation with respect to other obligations the Contractor has under the Contract Documents. Establishment of the one-year period for correction of Work as described in Section 12.2.2 relates only to the specific obligation of the Contractor to correct the Work, and has no relationship to the time within which the obligation to comply with the Contract Documents may be sought to be enforced, nor to the time within which proceedings may be commenced to establish the Contractor's liability with respect to the Contractor's obligations other than specifically to correct the Work.

§ 12.3 ACCEPTANCE OF NONCONFORMING WORK
If the Owner prefers to accept Work that is not in accordance with the requirements of the Contract Documents, the Owner may do so instead of requiring its removal and correction, in which case the Contract Sum will be reduced as appropriate and equitable. Such adjustment shall be effected whether or not final payment has been made.

ARTICLE 13 MISCELLANEOUS PROVISIONS

§ 13.1 GOVERNING LAW

The Contract shall be governed by the law of the place where the Project is located except that, if the parties have selected arbitration as the method of binding dispute resolution, the Federal Arbitration Act shall govern Section 15.4.

§ 13.2 SUCCESSORS AND ASSIGNS

§ 13.2.1 The Owner and Contractor respectively bind themselves, their partners, successors, assigns and legal representatives to covenants, agreements and obligations contained in the Contract Documents. Except as provided in Section 13.2.2, neither party to the Contract shall assign the Contract as a whole without written consent of the other. If either party attempts to make such an assignment without such consent, that party shall nevertheless remain legally responsible for all obligations under the Contract.

§ 13.2.2 The Owner may, without consent of the Contractor, assign the Contract to a lender providing construction financing for the Project, if the lender assumes the Owner's rights and obligations under the Contract Documents. The Contractor shall execute all consents reasonably required to facilitate such assignment.

§ 13.3 WRITTEN NOTICE

Written notice shall be deemed to have been duly served if delivered in person to the individual, to a member of the firm or entity, or to an officer of the corporation for which it was intended; or if delivered at, or sent by registered or certified mail or by courier service providing proof of delivery to, the last business address known to the party giving notice.

§ 13.4 RIGHTS AND REMEDIES

§ 13.4.1 Duties and obligations imposed by the Contract Documents and rights and remedies available thereunder shall be in addition to and not a limitation of duties, obligations, rights and remedies otherwise imposed or available by law.

§ 13.4.2 No action or failure to act by the Owner, Architect or Contractor shall constitute a waiver of a right or duty afforded them under the Contract, nor shall such action or failure to act constitute approval of or acquiescence in a breach there under, except as may be specifically agreed in writing.

§ 13.5 TESTS AND INSPECTIONS

§ 13.5.1 Tests, inspections and approvals of portions of the Work shall be made as required by the Contract Documents and by applicable laws, statutes, ordinances, codes, rules and regulations or lawful orders of public authorities. Unless otherwise provided, the Contractor shall make arrangements for such tests, inspections and approvals with an independent testing laboratory or entity acceptable to the Owner, or with the appropriate public authority, and shall bear all related costs of tests, inspections and approvals. The Contractor shall give the Architect timely notice of when and where tests and inspections are to be made so that the Architect may be present for such procedures. The Owner shall bear costs of (1) tests, inspections or approvals that do not become requirements until after bids are received or negotiations concluded, and (2) tests, inspections or approvals where building codes or applicable laws or regulations prohibit the Owner from delegating their cost to the Contractor.

§ 13.5.2 If the Architect, Owner or public authorities having jurisdiction determine that portions of the Work require additional testing, inspection or approval not included under Section 13.5.1, the Architect will, upon written authorization from the Owner, instruct the Contractor to make arrangements for such additional testing, inspection or approval by an entity acceptable to the Owner, and the Contractor shall give timely notice to the Architect of when and where tests and inspections are to be made so that the Architect may be present for such procedures. Such costs, except as provided in Section 13.5.3, shall be at the Owner's expense.

§ 13.5.3 If such procedures for testing, inspection or approval under Sections 13.5.1 and 13.5.2 reveal failure of the portions of the Work to comply with requirements established by the Contract Documents, all costs made necessary by such failure including those of repeated procedures and compensation for the Architect's services and expenses shall be at the Contractor's expense.

§ 13.5.4 Required certificates of testing, inspection or approval shall, unless otherwise required by the Contract Documents, be secured by the Contractor and promptly delivered to the Architect.

§ 13.5.5 If the Architect is to observe tests, inspections or approvals required by the Contract Documents, the Architect will do so promptly and, where practicable, at the normal place of testing.

§ 13.5.6 Tests or inspections conducted pursuant to the Contract Documents shall be made promptly to avoid unreasonable delay in the Work.

§ 13.6 INTEREST

Payments due and unpaid under the Contract Documents shall bear interest from the date payment is due at such rate as the parties may agree upon in writing or, in the absence thereof, at the legal rate prevailing from time to time at the place where the Project is located.

§ 13.7 TIME LIMITS ON CLAIMS

The Owner and Contractor shall commence all claims and causes of action, whether in contract, tort, breach of warranty or otherwise, against the other arising out of or related to the Contract in accordance with the requirements of the final dispute resolution method selected in the Agreement within the time period specified by applicable law, but in any case not more than 10 years after the date of Substantial Completion of the Work. The Owner and Contractor waive all claims and causes of action not commenced in accordance with this Section 13.7.

ARTICLE 14   TERMINATION OR SUSPENSION OF THE CONTRACT
§ 14.1 TERMINATION BY THE CONTRACTOR

§ 14.1.1 The Contractor may terminate the Contract if the Work is stopped for a period of 30 consecutive days through no act or fault of the Contractor or a Subcontractor, Sub-subcontractor or their agents or employees or any other persons or entities performing portions of the Work under direct or indirect contract with the Contractor, for any of the following reasons:

　　.1　Issuance of an order of a court or other public authority having jurisdiction that requires all Work to be stopped;

　　.2　An act of government, such as a declaration of national emergency that requires all Work to be stopped;

　　.3　Because the Architect has not issued a Certificate for Payment and has not notified the Contractor of the reason for withholding certification as provided in Section 9.4.1, or because the Owner has not made payment on a Certificate for Payment within the time stated in the Contract Documents; or

　　.4　The Owner has failed to furnish to the Contractor promptly, upon the Contractor's request, reasonable evidence as required by Section 2.2.1.

§ 14.1.2 The Contractor may terminate the Contract if, through no act or fault of the Contractor or a Subcontractor, Sub-subcontractor or their agents or employees or any other persons or entities performing portions of the Work under direct or indirect contract with the Contractor, repeated suspensions, delays or interruptions of the entire Work by the Owner as described in Section 14.3 constitute in the aggregate more than 100 percent of the total number of days scheduled for completion, or 120 days in any 365-day period, whichever is less.

§ 14.1.3 If one of the reasons described in Section 14.1.1 or 14.1.2 exists, the Contractor may, upon seven days' written notice to the Owner and Architect, terminate the Contract and recover from the Owner payment for Work executed, including reasonable overhead and profit, costs incurred by reason of such termination, and damages.

§ 14.1.4 If the Work is stopped for a period of 60 consecutive days through no act or fault of the Contractor or a Subcontractor or their agents or employees or any other persons performing portions of the Work under contract with the Contractor because the Owner has repeatedly failed to fulfill the Owner's obligations under the Contract Documents with respect to matters important to the progress of the Work, the Contractor may, upon seven additional days' written notice to the Owner and the Architect, terminate the Contract and recover from the Owner as provided in Section 14.1.3.

§ 14.2 TERMINATION BY THE OWNER FOR CAUSE

§ 14.2.1 The Owner may terminate the Contract if the Contractor

　　.1　repeatedly refuses or fails to supply enough properly skilled workers or proper materials;

　　.2　fails to make payment to Subcontractors for materials or labor in accordance with the respective agreements between the Contractor and the Subcontractors;

　　.3　repeatedly disregards applicable laws, statutes, ordinances, codes, rules and regulations, or lawful orders of a public authority; or

　　.4　otherwise is guilty of substantial breach of a provision of the Contract Documents.

§ 14.2.2 When any of the above reasons exist, the Owner, upon certification by the Initial Decision Maker that sufficient cause exists to justify such action, may without prejudice to any other rights or remedies of the Owner and after giving the Contractor and the Contractor's surety, if any, seven days' written notice, terminate employment of the Contractor and may, subject to any prior rights of the surety:

    .1   Exclude the Contractor from the site and take possession of all materials, equipment, tools, and construction equipment and machinery thereon owned by the Contractor;

    .2   Accept assignment of subcontracts pursuant to Section 5.4; and

    .3   Finish the Work by whatever reasonable method the Owner may deem expedient. Upon written request of the Contractor, the Owner shall furnish to the Contractor a detailed accounting of the costs incurred by the Owner in finishing the Work.

§ 14.2.3 When the Owner terminates the Contract for one of the reasons stated in Section 14.2.1, the Contractor shall not be entitled to receive further payment until the Work is finished.

§ 14.2.4 If the unpaid balance of the Contract Sum exceeds costs of finishing the Work, including compensation for the Architect's services and expenses made necessary thereby, and other damages incurred by the Owner and not expressly waived, such excess shall be paid to the Contractor. If such costs and damages exceed the unpaid balance, the Contractor shall pay the difference to the Owner. The amount to be paid to the Contractor or Owner, as the case may be, shall be certified by the Initial Decision Maker, upon application, and this obligation for payment shall survive termination of the Contract.

§ 14.3 SUSPENSION BY THE OWNER FOR CONVENIENCE
§ 14.3.1 The Owner may, without cause, order the Contractor in writing to suspend, delay or interrupt the Work in whole or in part for such period of time as the Owner may determine.

§ 14.3.2 The Contract Sum and Contract Time shall be adjusted for increases in the cost and time caused by suspension, delay or interruption as described in Section 14.3.1. Adjustment of the Contract Sum shall include profit. No adjustment shall be made to the extent

    .1   that performance is, was or would have been so suspended, delayed or interrupted by another cause for which the Contractor is responsible; or

    .2   that an equitable adjustment is made or denied under another provision of the Contract.

§ 14.4 TERMINATION BY THE OWNER FOR CONVENIENCE
§ 14.4.1 The Owner may, at any time, terminate the Contract for the Owner's convenience and without cause.

§ 14.4.2 Upon receipt of written notice from the Owner of such termination for the Owner's convenience, the Contractor shall

    .1   cease operations as directed by the Owner in the notice;

    .2   take actions necessary, or that the Owner may direct, for the protection and preservation of the Work; and

    .3   except for Work directed to be performed prior to the effective date of termination stated in the notice, terminate all existing subcontracts and purchase orders and enter into no further subcontracts and purchase orders.

§ 14.4.3 In case of such termination for the Owner's convenience, the Contractor shall be entitled to receive payment for Work executed, and costs incurred by reason of such termination, along with reasonable overhead and profit on the Work not executed.

ARTICLE 15   CLAIMS AND DISPUTES
§ 15.1 CLAIMS
§ 15.1.1 DEFINITION
A Claim is a demand or assertion by one of the parties seeking, as a matter of right, payment of money, or other relief with respect to the terms of the Contract. The term "Claim" also includes other disputes and matters in question between the Owner and Contractor arising out of or relating to the Contract. The responsibility to substantiate Claims shall rest with the party making the Claim.

§ 15.1.2 NOTICE OF CLAIMS
Claims by either the Owner or Contractor must be initiated by written notice to the other party and to the Initial Decision Maker with a copy sent to the Architect, if the Architect is not serving as the Initial Decision Maker.

Claims by either party must be initiated within 21 days after occurrence of the event giving rise to such Claim or within 21 days after the claimant first recognizes the condition giving rise to the Claim, whichever is later.

§ 15.1.3 CONTINUING CONTRACT PERFORMANCE

Pending final resolution of a Claim, except as otherwise agreed in writing or as provided in Section 9.7 and Article 14, the Contractor shall proceed diligently with performance of the Contract and the Owner shall continue to make payments in accordance with the Contract Documents. The Architect will prepare Change Orders and issue Certificates for Payment in accordance with the decisions of the Initial Decision Maker.

§ 15.1.4 CLAIMS FOR ADDITIONAL COST

If the Contractor wishes to make a Claim for an increase in the Contract Sum, written notice as provided herein shall be given before proceeding to execute the Work. Prior notice is not required for Claims relating to an emergency endangering life or property arising under Section 10.4.

§ 15.1.5 CLAIMS FOR ADDITIONAL TIME

§ 15.1.5.1 If the Contractor wishes to make a Claim for an increase in the Contract Time, written notice as provided herein shall be given. The Contractor's Claim shall include an estimate of cost and of probable effect of delay on progress of the Work. In the case of a continuing delay, only one Claim is necessary.

§ 15.1.5.2 If adverse weather conditions are the basis for a Claim for additional time, such Claim shall be documented by data substantiating that weather conditions were abnormal for the period of time, could not have been reasonably anticipated and had an adverse effect on the scheduled construction.

§ 15.1.6 CLAIMS FOR CONSEQUENTIAL DAMAGES

The Contractor and Owner waive Claims against each other for consequential damages arising out of or relating to this Contract. This mutual waiver includes

.1 damages incurred by the Owner for rental expenses, for losses of use, income, profit, financing, business and reputation, and for loss of management or employee productivity or of the services of such persons; and

.2 damages incurred by the Contractor for principal office expenses including the compensation of personnel stationed there, for losses of financing, business and reputation, and for loss of profit except anticipated profit arising directly from the Work.

This mutual waiver is applicable, without limitation, to all consequential damages due to either party's termination in accordance with Article 14. Nothing contained in this Section 15.1.6 shall be deemed to preclude an award of liquidated damages, when applicable, in accordance with the requirements of the Contract Documents.

§ 15.2 INITIAL DECISION

§ 15.2.1 Claims, excluding those arising under Sections 10.3, 10.4, 11.3.9, and 11.3.10, shall be referred to the Initial Decision Maker for initial decision. The Architect will serve as the Initial Decision Maker, unless otherwise indicated in the Agreement. Except for those Claims excluded by this Section 15.2.1, an initial decision shall be required as a condition precedent to mediation of any Claim arising prior to the date final payment is due, unless 30 days have passed after the Claim has been referred to the Initial Decision Maker with no decision having been rendered. Unless the Initial Decision Maker and all affected parties agree, the Initial Decision Maker will not decide disputes between the Contractor and persons or entities other than the Owner.

§ 15.2.2 The Initial Decision Maker will review Claims and within ten days of the receipt of a Claim take one or more of the following actions: (1) request additional supporting data from the claimant or a response with supporting data from the other party, (2) reject the Claim in whole or in part, (3) approve the Claim, (4) suggest a compromise, or (5) advise the parties that the Initial Decision Maker is unable to resolve the Claim if the Initial Decision Maker lacks sufficient information to evaluate the merits of the Claim or if the Initial Decision Maker concludes that, in the Initial Decision Maker's sole discretion, it would be inappropriate for the Initial Decision Maker to resolve the Claim.

§ 15.2.3 In evaluating Claims, the Initial Decision Maker may, but shall not be obligated to, consult with or seek information from either party or from persons with special knowledge or expertise who may assist the Initial Decision Maker in rendering a decision. The Initial Decision Maker may request the Owner to authorize retention of such persons at the Owner's expense.

§ 15.2.4 If the Initial Decision Maker requests a party to provide a response to a Claim or to furnish additional supporting data, such party shall respond, within ten days after receipt of such request, and shall either (1) provide a response on the requested supporting data, (2) advise the Initial Decision Maker when the response or supporting data will be furnished or (3) advise the Initial Decision Maker that no supporting data will be furnished. Upon receipt of the response or supporting data, if any, the Initial Decision Maker will either reject or approve the Claim in whole or in part.

§ 15.2.5 The Initial Decision Maker will render an initial decision approving or rejecting the Claim, or indicating that the Initial Decision Maker is unable to resolve the Claim. This initial decision shall (1) be in writing; (2) state the reasons therefor; and (3) notify the parties and the Architect, if the Architect is not serving as the Initial Decision Maker, of any change in the Contract Sum or Contract Time or both. The initial decision shall be final and binding on the parties but subject to mediation and, if the parties fail to resolve their dispute through mediation, to binding dispute resolution.

§ 15.2.6 Either party may file for mediation of an initial decision at any time, subject to the terms of Section 15.2.6.1.

§ 15.2.6.1 Either party may, within 30 days from the date of an initial decision, demand in writing that the other party file for mediation within 60 days of the initial decision. If such a demand is made and the party receiving the demand fails to file for mediation within the time required, then both parties waive their rights to mediate or pursue binding dispute resolution proceedings with respect to the initial decision.

§ 15.2.7 In the event of a Claim against the Contractor, the Owner may, but is not obligated to, notify the surety, if any, of the nature and amount of the Claim. If the Claim relates to a possibility of a Contractor's default, the Owner may, but is not obligated to, notify the surety and request the surety's assistance in resolving the controversy.

§ 15.2.8 If a Claim relates to or is the subject of a mechanic's lien, the party asserting such Claim may proceed in accordance with applicable law to comply with the lien notice or filing deadlines.

§ 15.3 MEDIATION
§ 15.3.1 Claims, disputes, or other matters in controversy arising out of or related to the Contract except those waived as provided for in Sections 9.10.4, 9.10.5, and 15.1.6 shall be subject to mediation as a condition precedent to binding dispute resolution.

§ 15.3.2 The parties shall endeavor to resolve their Claims by mediation which, unless the parties mutually agree otherwise, shall be administered by the American Arbitration Association in accordance with its Construction Industry Mediation Procedures in effect on the date of the Agreement. A request for mediation shall be made in writing, delivered to the other party to the Contract, and filed with the person or entity administering the mediation. The request may be made concurrently with the filing of binding dispute resolution proceedings but, in such event, mediation shall proceed in advance of binding dispute resolution proceedings, which shall be stayed pending mediation for a period of 60 days from the date of filing, unless stayed for a longer period by agreement of the parties or court order. If an arbitration is stayed pursuant to this Section 15.3.2, the parties may nonetheless proceed to the selection of the arbitrator(s) and agree upon a schedule for later proceedings.

§ 15.3.3 The parties shall share the mediator's fee and any filing fees equally. The mediation shall be held in the place where the Project is located, unless another location is mutually agreed upon. Agreements reached in mediation shall be enforceable as settlement agreements in any court having jurisdiction thereof.

§ 15.4 ARBITRATION
§ 15.4.1 If the parties have selected arbitration as the method for binding dispute resolution in the Agreement, any Claim subject to, but not resolved by, mediation shall be subject to arbitration which, unless the parties mutually agree otherwise, shall be administered by the American Arbitration Association in accordance with its Construction Industry Arbitration Rules in effect on the date of the Agreement. A demand for arbitration shall be made in writing, delivered to the other party to the Contract, and filed with the person or entity administering the arbitration. The party filing a notice of demand for arbitration must assert in the demand all Claims then known to that party on which arbitration is permitted to be demanded.

§ **15.4.1.1** A demand for arbitration shall be made no earlier than concurrently with the filing of a request for mediation, but in no event shall it be made after the date when the institution of legal or equitable proceedings based on the Claim would be barred by the applicable statute of limitations. For statute of limitations purposes, receipt of a written demand for arbitration by the person or entity administering the arbitration shall constitute the institution of legal or equitable proceedings based on the Claim.

§ **15.4.2** The award rendered by the arbitrator or arbitrators shall be final, and judgment may be entered upon it in accordance with applicable law in any court having jurisdiction thereof.

§ **15.4.3** The foregoing agreement to arbitrate and other agreements to arbitrate with an additional person or entity duly consented to by parties to the Agreement shall be specifically enforceable under applicable law in any court having jurisdiction thereof.

§ 15.4.4 CONSOLIDATION OR JOINDER

§ **15.4.4.1** Either party, at its sole discretion, may consolidate an arbitration conducted under this Agreement with any other arbitration to which it is a party provided that (1) the arbitration agreement governing the other arbitration permits consolidation, (2) the arbitrations to be consolidated substantially involve common questions of law or fact, and (3) the arbitrations employ materially similar procedural rules and methods for selecting arbitrator(s).

§ **15.4.4.2** Either party, at its sole discretion, may include by joinder persons or entities substantially involved in a common question of law or fact whose presence is required if complete relief is to be accorded in arbitration, provided that the party sought to be joined consents in writing to such joinder. Consent to arbitration involving an additional person or entity shall not constitute consent to arbitration of any claim, dispute or other matter in question not described in the written consent.

§ **15.4.4.3** The Owner and Contractor grant to any person or entity made a party to an arbitration conducted under this Section 15.4, whether by joinder or consolidation, the same rights of joinder and consolidation as the Owner and Contractor under this Agreement.

APPENDIX

D Performance and Payment Bonds

AIA Document A312–1984

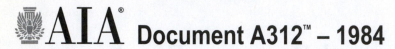

Document A312™ – 1984

Performance Bond

CONTRACTOR *(Name and Address)*: **SURETY** *(Name and Principal Place of Business)*:

OWNER *(Name and Address)*:

Any singular reference to Contract, Surety, Owner or other party shall be considered plural where applicable.

CONSTRUCTION CONTRACT
Date:

Amount:

Description *(Name and Location)*:

BOND
Date *(Not earlier than Construction Contract Date)*:

Amount:

Modifications to this Bond: ☐None   ☐ See page 4

CONTRACTOR AS PRINCIPAL	**SURETY**
Company: *(Corporate Seal)*	Company: *(Corporate Seal)*
Signature: _____	Signature: _____
Name and Title:	Name and Title:

(Any additional signatures appear on page 4)

(FOR INFORMATION ONLY - Name, Address and Telephone)
AGENT or **BROKER**:   **OWNER'S REPRESENTATIVE** *(Architect, Engineer or other party)*:

§ 1 The Contractor and the Surety, jointly and severally, bind themselves, their heirs, executors, administrators, successors and assigns to the Owner for the performance of the Construction Contract, which is incorporated herein by reference.

§ 2 If the Contractor performs the Construction Contract, the Surety and the Contractor shall have no obligation under this Bond, except to participate in conferences as provided in Section 3.1.

§ 3 If there is no Owner Default, the Surety's obligation under this Bond shall arise after:
§ 3.1 The Owner has notified the Contractor and the Surety at its address described in Section 10 below that the Owner is considering declaring a Contractor Default and has requested and attempted to arrange a conference with the Contractor and the Surety to be held not later than fifteen days after receipt of such notice to discuss methods of performing the Construction Contract. If the Owner, the Contractor and the Surety agree, the Contractor shall be allowed a reasonable time to perform the Construction Contract, but such an agreement shall not waive the Owner's right, if any, subsequently to declare a Contractor Default; and

§ 3.2 The Owner has declared a Contractor Default and formally terminated the Contractor's right to complete the contract. Such Contractor Default shall not be declared earlier than twenty days after the Contractor and the Surety have received notice as provided in Section 3.1; and

§ 3.3 The Owner has agreed to pay the Balance of the Contract Price to the Surety in accordance with the terms of the Construction Contract or to a contractor selected to perform the Construction Contract in accordance with the terms of the contract with the Owner.

§ 4 When the Owner has satisfied the conditions of Section 3, the Surety shall promptly and at the Surety's expense take one of the following actions:
§ 4.1 Arrange for the Contractor, with consent of the Owner, to perform and complete the Construction Contract; or

§ 4.2 Undertake to perform and complete the Construction Contract itself, through its agents or through independent contractors; or

§ 4.3 Obtain bids or negotiated proposals from qualified contractors acceptable to the Owner for a contract for performance and completion of the Construction Contract, arrange for a contract to be prepared for execution by the Owner and the contractor selected with the Owner's concurrence, to be secured with performance and payment bonds executed by a qualified surety equivalent to the bonds issued on the Construction Contract, and pay to the Owner the amount of damages as described in Section 6 in excess of the Balance of the Contract Price incurred by the Owner resulting from the Contractor's default; or

§ 4.4 Waive its right to perform and complete, arrange for completion, or obtain a new contractor and with reasonable promptness under the circumstances:
  .1   After investigation, determine the amount for which it may be liable to the Owner and, as soon as practicable after the amount is determined, tender payment therefor to the Owner; or
  .2   Deny liability in whole or in part and notify the Owner citing reasons therefor.

§ 5 If the Surety does not proceed as provided in Section 4 with reasonable promptness, the Surety shall be deemed to be in default on this Bond fifteen days after receipt of an additional written notice from the Owner to the Surety demanding that the Surety perform its obligations under this Bond, and the Owner shall be entitled to enforce any remedy available to the Owner. If the Surety proceeds as provided in Section 4.4, and the Owner refuses the payment tendered or the Surety has denied liability, in whole or in part, without further notice the Owner shall be entitled to enforce any remedy available to the Owner.

§ 6 After the Owner has terminated the Contractor's right to complete the Construction Contract, and if the Surety elects to act under Section 4.1, 4.2, or 4.3 above, then the responsibilities of the Surety to the Owner shall not be greater than those of the Contractor under the Construction Contract, and the responsibilities of the Owner to the Surety shall not be greater than those of the Owner under the Construction Contract. To the limit of the amount of this Bond, but subject to commitment by the Owner of the Balance of the Contract Price to mitigation of costs and damages on the Construction Contract, the Surety is obligated without duplication for:
§ 6.1 The responsibilities of the Contractor for correction of defective work and completion of the Construction Contract;

§ 6.2 Additional legal, design professional and delay costs resulting from the Contractor's Default, and resulting from the actions or failure to act of the Surety under Section 4; and

---

§ 6.3 Liquidated damages, or if no liquidated damages are specified in the Construction Contract, actual damages caused by delayed performance or non-performance of the Contractor.

§ 7 The Surety shall not be liable to the Owner or others for obligations of the Contractor that are unrelated to the Construction Contract, and the Balance of the Contract Price shall not be reduced or set off on account of any such unrelated obligations. No right of action shall accrue on this Bond to any person or entity other than the Owner or its heirs, executors, administrators or successors.

§ 8 The Surety hereby waives notice of any change, including changes of time, to the Construction Contract or to related subcontracts, purchase orders and other obligations.

§ 9 Any proceeding, legal or equitable, under this Bond may be instituted in any court of competent jurisdiction in the location in which the work or part of the work is located and shall be instituted within two years after Contractor Default or within two years after the Contractor ceased working or within two years after the Surety refuses or fails to perform its obligations under this Bond, whichever occurs first. If the provisions of this Paragraph are void or prohibited by law, the minimum period of limitation available to sureties as a defense in the jurisdiction of the suit shall be applicable.

§ 10 Notice to the Surety, the Owner or the Contractor shall be mailed or delivered to the address shown on the signature page.

§ 11 When this Bond has been furnished to comply with a statutory or other legal requirement in the location where the construction was to be performed, any provision in this Bond conflicting with said statutory or legal requirement shall be deemed deleted here from and provisions conforming to such statutory or other legal requirement shall be deemed incorporated herein. The intent is that this Bond shall be construed as a statutory bond and not as a common law bond.

§ 12 DEFINITIONS
§ 12.1 Balance of the Contract Price: The total amount payable by the Owner to the Contractor under the Construction Contract after all proper adjustments have been made, including allowance to the Contractor of any amounts received or to be received by the Owner in settlement of insurance or other claims for damages to which the Contractor is entitled, reduced by all valid and proper payments made to or on behalf of the Contractor under the Construction Contract.

§ 12.2 Construction Contract: The agreement between the Owner and the Contractor identified on the signature page, including all Contract Documents and changes thereto.

§ 12.3 Contractor Default: Failure of the Contractor, which has neither been remedied nor waived, to perform or otherwise to comply with the terms of the Construction Contract.

§ 12.4 Owner Default: Failure of the Owner, which has neither been remedied nor waived, to pay the Contractor as required by the Construction Contract or to perform and complete or comply with the other terms thereof.

§ 13 MODIFICATIONS TO THIS BOND ARE AS FOLLOWS:

(Space is provided below for additional signatures of added parties, other than those appearing on the cover page.)

CONTRACTOR AS PRINCIPAL		**SURETY**	
Company:	*(Corporate Seal)*	Company:	*(Corporate Seal)*
Signature: _____		Signature: _____	
Name and Title:		Name and Title:	
Address:		Address:	

AIA Document A312™ – 1984. The American Institute of Architects.

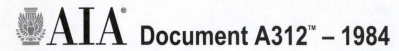

Document A312™ – 1984

Payment Bond

CONTRACTOR *(Name and Address)*:   **SURETY** *(Name and Principal Place of Business)*:

OWNER *(Name and Address)*:

Any singular reference to Contract, Surety, Owner or other party shall be considered plural where applicable.

CONSTRUCTION CONTRACT
Date:

Amount:

Description *(Name and Location)*:

BOND
Date *(Not earlier than Construction Contract Date)*:

Amount:

Modifications to this Bond: ☐None   ☐ See page 4

CONTRACTOR AS PRINCIPAL   **SURETY**
Company: *(Corporate Seal)*   Company: *(Corporate Seal)*

Signature: _____   Signature: _____
Name and Title: Name and Title:

(Any additional signatures appear on page 4)

(FOR INFORMATION ONLY - Name, Address and Telephone)
AGENT or **BROKER**:   **OWNER'S REPRESENTATIVE** *(Architect, Engineer or other party)*:

§ 1 The Contractor and the Surety, jointly and severally bind themselves, their heirs, executors, administrators, successors and assigns to the Owner to pay for labor, materials and equipment furnished for use in the performance of the Construction Contract, which is incorporated herein by reference.

§ 2 With respect to the Owner, this obligation shall be null and void if the Contractor:
§ 2.1 Promptly makes payment, directly or indirectly, for all sums due Claimants, and

§ 2.2 Defends, indemnifies and holds harmless the Owner from claims, demands, liens or suits by any person or entity whose claim, demand, lien or suit is for the payment for labor, materials or equipment furnished for use in the performance of the Construction Contract, provided the Owner has promptly notified the Contractor and the Surety (at the address described in Section 12) of any claims, demands, liens or suits and tendered defense of such claims, demands, liens or suits to the Contractor and the Surety, and provided there is no Owner Default.

§ 3 With respect to Claimants, this obligation shall be null and void if the Contractor promptly makes payment, directly or indirectly, for all sums due.

§ 4 The Surety shall have no obligation to Claimants under this Bond until:
§ 4.1 Claimants who are employed by or have a direct contract with the Contractor have given notice to the Surety (at the address described in Section 12) and sent a copy, or notice thereof, to the Owner, stating that a claim is being made under this Bond and, with substantial accuracy, the amount of the claim.

§ 4.2 Claimants who do not have a direct contract with the Contractor:
  .1 Have furnished written notice to the Contractor and sent a copy, or notice thereof, to the Owner, within 90 days after having last performed labor or last furnished materials or equipment included in the claim stating, with substantial accuracy, the amount of the claim and the name of the party to whom the materials were furnished or supplied or for whom the labor was done or performed; and
  .2 Have either received a rejection in whole or in part from the Contractor, or not received within 30 days of furnishing the above notice any communication from the Contractor by which the Contractor has indicated the claim will be paid directly or indirectly; and
  .3 Not having been paid within the above 30 days, have sent a written notice to the Surety (at the address described in Section 12) and sent a copy, or notice thereof, to the Owner, stating that a claim is being made under this Bond and enclosing a copy of the previous written notice furnished to the Contractor.

§ 5 If a notice required by Section 4 is given by the Owner to the Contractor or to the Surety, that is sufficient compliance.

§ 6 When the Claimant has satisfied the conditions of Section 4, the Surety shall promptly and at the Surety's expense take the following actions:
§ 6.1 Send an answer to the Claimant, with a copy to the Owner, within 45 days after receipt of the claim, stating the amounts that are undisputed and the basis for challenging any amounts that are disputed.

§ 6.2 Pay or arrange for payment of any undisputed amounts.

§ 7 The Surety's total obligation shall not exceed the amount of this Bond, and the amount of this Bond shall be credited for any payments made in good faith by the Surety.

§ 8 Amounts owed by the Owner to the Contractor under the Construction Contract shall be used for the performance of the Construction Contract and to satisfy claims, if any, under any Construction Performance Bond. By the Contractor furnishing and the Owner accepting this Bond, they agree that all funds earned by the Contractor in the performance of the Construction Contract are dedicated to satisfy obligations of the Contractor and the Surety under this Bond, subject to the Owner's priority to use the funds for the completion of the work.

§ 9 The Surety shall not be liable to the Owner, Claimants or others for obligations of the Contractor that are unrelated to the Construction Contract. The Owner shall not be liable for payment of any costs or expenses of any Claimant under this Bond, and shall have under this Bond no obligations to make payments to, give notices on behalf of, or otherwise have obligations to Claimants under this Bond.

§ 10 The Surety hereby waives notice of any change, including changes of time, to the Construction Contract or to related subcontracts, purchase orders and other obligations.

---

§ 11 No suit or action shall be commenced by a Claimant under this Bond other than in a court of competent jurisdiction in the location in which the work or part of the work is located or after the expiration of one year from the date (1) on which the Claimant gave the notice required by Section 4.1 or Section 4.2.3, or (2) on which the last labor or service was performed by anyone or the last materials or equipment were furnished by anyone under the Construction Contract, whichever of (1) or (2) first occurs. If the provisions of this Paragraph are void or prohibited by law, the minimum period of limitation available to sureties as a defense in the jurisdiction of the suit shall be applicable.

§ 12 Notice to the Surety, the Owner or the Contractor shall be mailed or delivered to the address shown on the signature page. Actual receipt of notice by Surety, the Owner or the Contractor, however accomplished, shall be sufficient compliance as of the date received at the address shown on the signature page.

§ 13 When this Bond has been furnished to comply with a statutory or other legal requirement in the location where the construction was to be performed, any provision in this Bond conflicting with said statutory or legal requirement shall be deemed deleted herefrom and provisions conforming to such statutory or other legal requirement shall be deemed incorporated herein. The intent is that this Bond shall be construed as a statutory bond and not as a common law bond.

§ 14 Upon request by any person or entity appearing to be a potential beneficiary of this Bond, the Contractor shall promptly furnish a copy of this Bond or shall permit a copy to be made.

§ 15 DEFINITIONS
§ 15.1 Claimant: An individual or entity having a direct contract with the Contractor or with a subcontractor of the Contractor to furnish labor, materials or equipment for use in the performance of the Contract. The intent of this Bond shall be to include without limitation in the terms "labor, materials or equipment" that part of water, gas, power, light, heat, oil, gasoline, telephone service or rental equipment used in the Construction Contract, architectural and engineering services required for performance of the work of the Contractor and the Contractor's subcontractors, and all other items for which a mechanic's lien may be asserted in the jurisdiction where the labor, materials or equipment were furnished.

§ 15.2 Construction Contract: The agreement between the Owner and the Contractor identified on the signature page, including all Contract Documents and changes thereto.

§ 15.3 Owner Default: Failure of the Owner, which has neither been remedied nor waived, to pay the Contractor as required by the Construction Contract or to perform and complete or comply with the other terms thereof.

§ 16 MODIFICATIONS TO THIS BOND ARE AS FOLLOWS:

(Space is provided below for additional signatures of added parties, other than those appearing on the cover page.)

CONTRACTOR AS PRINCIPAL		**SURETY**	
Company:	*(Corporate Seal)*	Company:	*(Corporate Seal)*
Signature: _____		Signature: _____	
Name and Title:		Name and Title:	
Address:		Address:	

Standard Form of Agreement Between Contractor and Subcontractor

AIA DOCUMENT A401–2007

AIA® Document A401™ – 2007

Standard Form of Agreement Between Contractor and Subcontractor

AGREEMENT made as of the   day of
in the year
(In words, indicate day, month and year)

BETWEEN the Contractor:
(Name, address and other information)

and the Subcontractor:
(Name, address and other information)

This document has important legal consequences. Consultation with an attorney is encouraged with respect to its completion or modification. AIA Document A201™–2007, General Conditions of the Contract for Construction, is adopted in this document by reference.

The Contractor has made a contract for construction (hereinafter, the Prime Contract) dated:

with the Owner:
(Name, address and other information)

for the following Project:
(Name, location and detailed description)

The Prime Contract provides for the furnishing of labor, materials, equipment and services in connection with the construction of the Project. A copy of the Prime Contract, consisting of the Agreement Between Owner and Contractor (from which compensation amounts may be deleted) and the other Contract Documents enumerated therein, has been made available to the Subcontractor.

The Architect for the Project:
(Name, address and other information)

The Contractor and the Subcontractor agree as follows.

Init.

/

1

TABLE OF ARTICLES

ARTICLE 1   THE SUBCONTRACT DOCUMENTS

§ 1.1 The Subcontract Documents consist of (1) this Agreement; (2) the Prime Contract, consisting of the Agreement between the Owner and Contractor and the other Contract Documents enumerated therein; (3) Modifications issued subsequent to the execution of the Agreement between the Owner and Contractor, whether before or after the execution of this Agreement; (4) other documents listed in Article 16 of this Agreement; and (5) Modifications to this Subcontract issued after execution of this Agreement. These form the Subcontract, and are as fully a part of the Subcontract as if attached to this Agreement or repeated herein. The Subcontract represents the entire and integrated agreement between the parties hereto and supersedes prior negotiations, representations or agreements, either written or oral. An enumeration of the Subcontract Documents, other than Modifications issued subsequent to the execution of this Agreement, appears in Article 16.

§ 1.2 Except to the extent of a conflict with a specific term or condition contained in the Subcontract Documents, the General Conditions governing this Subcontract shall be the AIA Document A201™–2007, General Conditions of the Contract for Construction.

§ 1.3 The Subcontract may be amended or modified only by a Modification. The Subcontract Documents shall not be construed to create a contractual relationship of any kind (1) between the Architect and the Subcontractor, (2) between the Owner and the Subcontractor, or (3) between any persons or entities other than the Contractor and Subcontractor.

§ 1.4 The Contractor shall make available the Subcontract Documents to the Subcontractor prior to execution of this Agreement, and thereafter, upon request, but the Contractor may charge the Subcontractor for the reasonable cost of reproduction.

ARTICLE 2 MUTUAL RIGHTS AND RESPONSIBILITIES

The Contractor and Subcontractor shall be mutually bound by the terms of this Agreement and, to the extent that the provisions of AIA Document A201–2007 apply to this Agreement pursuant to Section 1.2 and provisions of the Prime Contract apply to the Work of the Subcontractor, the Contractor shall assume toward the Subcontractor all obligations and responsibilities that the Owner, under such documents, assumes toward the Contractor, and the Subcontractor shall assume toward the Contractor all obligations and responsibilities which the Contractor, under such documents, assumes toward the Owner and the Architect. The Contractor shall have the benefit of all rights, remedies and redress against the Subcontractor that the Owner, under such documents, has against the Contractor, and the Subcontractor shall have the benefit of all rights, remedies and redress against the Contractor that the Contractor, under such documents, has against the Owner, insofar as applicable to this Subcontract. Where a provision of such documents is inconsistent with a provision of this Agreement, this Agreement shall govern.

ARTICLE 3 CONTRACTOR
§ 3.1 SERVICES PROVIDED BY THE CONTRACTOR

§ 3.1.1 The Contractor shall cooperate with the Subcontractor in scheduling and performing the Contractor's Work to avoid conflicts or interference in the Subcontractor's Work and shall expedite written responses to submittals made by the Subcontractor in accordance with Section 4.1 and Article 5. Promptly after execution of this Agreement, the Contractor shall provide the Subcontractor copies of the Contractor's construction schedule and schedule of submittals, together with such additional scheduling details as will enable the Subcontractor to plan and perform the Subcontractor's Work properly. The Contractor shall promptly notify the Subcontractor of subsequent changes in the construction and submittal schedules and additional scheduling details.

§ 3.1.2 The Contractor shall provide suitable areas for storage of the Subcontractor's materials and equipment during the course of the Work. Additional costs to the Subcontractor resulting from relocation of such storage areas at the direction of the Contractor, except as previously agreed upon, shall be reimbursed by the Contractor.

§ 3.1.3 Except as provided in Article 14, the Contractor's equipment will be available to the Subcontractor only at the Contractor's discretion and on mutually satisfactory terms.

§ 3.2 COMMUNICATIONS

§ 3.2.1 The Contractor shall promptly make available to the Subcontractor information, including information received from the Owner, that affects this Subcontract and that becomes available to the Contractor subsequent to execution of this Subcontract.

§ 3.2.2 The Contractor shall not give instructions or orders directly to the Subcontractor's employees or to the Subcontractor's Sub-subcontractors or material suppliers unless such persons are designated as authorized representatives of the Subcontractor.

§ 3.2.3 The Contractor shall permit the Subcontractor to request directly from the Architect information regarding the percentages of completion and the amount certified on account of Work done by the Subcontractor.

§ 3.2.4 If hazardous substances of a type of which an employer is required by law to notify its employees are being used on the site by the Contractor, a subcontractor or anyone directly or indirectly employed by them (other than the Subcontractor), the Contractor shall, prior to harmful exposure of the Subcontractor's employees to such substance, give written notice of the chemical composition thereof to the Subcontractor in sufficient detail and time to permit the Subcontractor's compliance with such laws.

§ 3.2.5 The Contractor shall furnish to the Subcontractor within 30 days after receipt of a written request, or earlier if so required by law, information necessary and relevant for the Subcontractor to evaluate, give notice of or enforce mechanic's lien rights. Such information shall include a correct statement of the record legal title to the property, usually referred to as the site, on which the Project is located and the Owner's interest therein.

§ 3.2.6 If the Contractor asserts or defends a claim against the Owner that relates to the Work of the Subcontractor, the Contractor shall promptly make available to the Subcontractor all information relating to the portion of the claim that relates to the Work of the Subcontractor.

§ 3.3 CLAIMS BY THE CONTRACTOR

§ 3.3.1 Liquidated damages for delay, if provided for in Section 9.3 of this Agreement, shall be assessed against the Subcontractor only to the extent caused by the Subcontractor or any person or entity for whose acts the Subcontractor may be liable, and in no case for delays or causes arising outside the scope of this Subcontract.

§ 3.3.2 The Contractor's claims for the costs of services or materials provided due to the Subcontractor's failure to execute the Work shall require

.1 seven days' written notice prior to the Contractor's providing services or materials, except in an emergency; and

.2 written compilations to the Subcontractor of services and materials provided by the Contractor and charges for such services and materials no later than the fifteenth day of the month following the Contractor's providing such services or materials.

§ 3.4 CONTRACTOR'S REMEDIES

If the Subcontractor defaults or neglects to carry out the Work in accordance with this Agreement and fails within five working days after receipt of written notice from the Contractor to commence and continue correction of such default or neglect with diligence and promptness, the Contractor may, by appropriate Modification, and without prejudice to any other remedy the Contractor may have, make good such deficiencies and may deduct the reasonable cost thereof from the payments then or thereafter due the Subcontractor.

ARTICLE 4 SUBCONTRACTOR
§ 4.1 EXECUTION AND PROGRESS OF THE WORK

§ 4.1.1 For all Work the Subcontractor intends to subcontract, the Subcontractor shall enter into written agreements with Sub-subcontractors performing portions of the Work of this Subcontract by which the Subcontractor and the Sub-subcontractor are mutually bound, to the extent of the Work to be performed by the Sub-subcontractor, assuming toward each other all obligations and responsibilities that the Contractor and Subcontractor assume toward each other and having the benefit of all rights, remedies and redress each against the other that the Contractor and Subcontractor have by virtue of the provisions of this Agreement.

§ 4.1.2 The Subcontractor shall supervise and direct the Subcontractor's Work, and shall cooperate with the Contractor in scheduling and performing the Subcontractor's Work to avoid conflict, delay in or interference with the Work of the Contractor, other subcontractors, the Owner, or separate contractors.

§ 4.1.3 The Subcontractor shall promptly submit Shop Drawings, Product Data, Samples and similar submittals required by the Subcontract Documents with reasonable promptness and in such sequence as to cause no delay in the Work or in the activities of the Contractor or other subcontractors.

§ 4.1.4 The Subcontractor shall furnish to the Contractor periodic progress reports on the Work of this Subcontract as mutually agreed, including information on the status of materials and equipment that may be in the course of preparation, manufacture, or transit.

§ 4.1.5 The Subcontractor agrees that the Contractor and the Architect each have the authority to reject Work of the Subcontractor that does not conform to the Prime Contract. The Architect's decisions on matters relating to aesthetic effect shall be final and binding on the Subcontractor if consistent with the intent expressed in the Prime Contract.

§ 4.1.6 The Subcontractor shall pay for all materials, equipment and labor used in connection with the performance of this Subcontract through the period covered by previous payments received from the Contractor, and shall furnish satisfactory evidence, when requested by the Contractor, to verify compliance with the above requirements.

§ 4.1.7 The Subcontractor shall take necessary precautions to protect properly the work of other subcontractors from damage caused by operations under this Subcontract.

§ 4.1.8 The Subcontractor shall cooperate with the Contractor, other subcontractors, the Owner, and separate contractors whose work might interfere with the Subcontractor's Work. The Subcontractor shall participate in the preparation of coordinated drawings in areas of congestion, if required by the Prime Contract, specifically noting and advising the Contractor of potential conflicts between the Work of the Subcontractor and that of the Contractor, other subcontractors, the Owner, or separate contractors.

§ 4.2 PERMITS, FEES, NOTICES, AND COMPLIANCE WITH LAWS

§ 4.2.1 The Subcontractor shall give notices and comply with applicable laws, statutes, ordinances, codes, rules and regulations, and lawful orders of public authorities bearing on performance of the Work of this Subcontract. The Subcontractor shall secure and pay for permits, fees, licenses and inspections by government agencies necessary for proper execution and completion of the Subcontractor's Work, the furnishing of which is required of the Contractor by the Prime Contract.

§ 4.2.2 The Subcontractor shall comply with Federal, state and local tax laws, social security acts, unemployment compensation acts and workers' compensation acts insofar as applicable to the performance of this Subcontract.

§ 4.3 SAFETY PRECAUTIONS AND PROCEDURES

§ 4.3.1 The Subcontractor shall take reasonable safety precautions with respect to performance of this Subcontract, shall comply with safety measures initiated by the Contractor and with applicable laws, statutes, ordinances, codes, rules and regulations, and lawful orders of public authorities for the safety of persons and property in accordance with the requirements of the Prime Contract. The Subcontractor shall report to the Contractor within three days an injury to an employee or agent of the Subcontractor which occurred at the site.

§ 4.3.2 If hazardous substances of a type of which an employer is required by law to notify its employees are being used on the site by the Subcontractor, the Subcontractor's Sub-subcontractors or anyone directly or indirectly employed by them, the Subcontractor shall, prior to harmful exposure of any employees on the site to such substance, give written notice of the chemical composition thereof to the Contractor in sufficient detail and time to permit compliance with such laws by the Contractor, other subcontractors and other employers on the site.

§ 4.3.3 If reasonable precautions will be inadequate to prevent foreseeable bodily injury or death to persons resulting from a hazardous material or substance, including but not limited to asbestos or polychlorinated biphenyl (PCB), encountered on the site by the Subcontractor, the Subcontractor shall, upon recognizing the condition, immediately stop Work in the affected area and promptly report the condition to the Contractor in writing. When the material or substance has been rendered harmless, the Subcontractor's Work in the affected area shall resume upon written agreement of the Contractor and Subcontractor. The Subcontract Time shall be extended appropriately and the Subcontract Sum shall be increased in the amount of the Subcontractor's reasonable additional costs of demobilization, delay and remobilization, which adjustments shall be accomplished as provided in Article 5 of this Agreement.

§ 4.3.4 To the fullest extent permitted by law, the Contractor shall indemnify and hold harmless the Subcontractor, the Subcontractor's Sub-subcontractors, and agents and employees of any of them from and against claims, damages, losses and expenses, including but not limited to attorneys' fees, arising out of or resulting from performance of the Work in the affected area if in fact the material or substance presents the risk of bodily injury or death as described in Section 4.3.3 and has not been rendered harmless, provided that such claim, damage, loss or expense is attributable to bodily injury, sickness, disease or death, or to injury to or destruction of tangible property (other than the Work itself) except to the extent that such damage, loss or expense is due to the fault or negligence of the party seeking indemnity.

§ 4.3.5 The Subcontractor shall indemnify the Contractor for the cost and expense the Contractor incurs 1) for remediation of a material or substance brought to the site and negligently handled by the Subcontractor or 2) where the Subcontractor fails to perform its obligations under Section 4.3.3, except to the extent that the cost and expense are due to the Contractor's fault or negligence.

§ 4.4 CLEANING UP

§ 4.4.1 The Subcontractor shall keep the premises and surrounding area free from accumulation of waste materials or rubbish caused by operations performed under this Subcontract. The Subcontractor shall not be held responsible for conditions caused by other contractors or subcontractors.

§ 4.4.2 As provided under Section 3.3.2, if the Subcontractor fails to clean up as provided in the Subcontract Documents, the Contractor may charge the Subcontractor for the Subcontractor's appropriate share of cleanup costs.

§ 4.5 WARRANTY

The Subcontractor warrants to the Owner, Architect, and Contractor that materials and equipment furnished under this Subcontract will be of good quality and new unless the Subcontract Documents require or permit otherwise. The Subcontractor further warrants that the Work will conform to the requirements of the Subcontract Documents and will be free from defects, except for those inherent in the quality of the Work the Subcontract Documents require or permit. Work, materials, or equipment not conforming to these requirements may be considered defective. The Subcontractor's warranty excludes remedy for damage or defect caused by abuse, alterations to the Work not executed by the Subcontractor, improper or insufficient maintenance, improper operation, or normal wear and tear under normal usage. If required by the Architect and Contractor, the Subcontractor shall furnish satisfactory evidence as to the kind and quality of materials and equipment.

§ 4.6 INDEMNIFICATION

§ 4.6.1 To the fullest extent permitted by law, the Subcontractor shall indemnify and hold harmless the Owner, Contractor, Architect, Architect's consultants, and agents and employees of any of them from and against claims, damages, losses and expenses, including but not limited to attorney's fees, arising out of or resulting from performance of the Subcontractor's Work under this Subcontract, provided that any such claim, damage, loss or expense is attributable to bodily injury, sickness, disease or death, or to injury to or destruction of tangible property (other than the Work itself), but only to the extent caused by the negligent acts or omissions of the Subcontractor, the Subcontractor's Sub-subcontractors, anyone directly or indirectly employed by them or anyone for whose acts they may be liable, regardless of whether or not such claim, damage, loss or expense is caused in part by a party indemnified hereunder. Such obligation shall not be construed to negate, abridge, or otherwise reduce other rights or obligations of indemnity which would otherwise exist as to a party or person described in this Section 4.6.

§ 4.6.2 In claims against any person or entity indemnified under this Section 4.6 by an employee of the Subcontractor, the Subcontractor's Sub-subcontractors, anyone directly or indirectly employed by them or anyone for whose acts they may be liable, the indemnification obligation under Section 4.6.1 shall not be limited by a limitation on the amount or type of damages, compensation or benefits payable by or for the Subcontractor or the Subcontractor's Sub-subcontractors under workers' compensation acts, disability benefit acts or other employee benefit acts.

§ 4.7 REMEDIES FOR NONPAYMENT

If the Contractor does not pay the Subcontractor through no fault of the Subcontractor, within seven days from the time payment should be made as provided in this Agreement, the Subcontractor may, without prejudice to any other available remedies, upon seven additional days' written notice to the Contractor, stop the Work of this Subcontract until payment of the amount owing has been received. The Subcontract Sum shall, by appropriate Modification, be increased by the amount of the Subcontractor's reasonable costs of demobilization, delay and remobilization.

ARTICLE 5 CHANGES IN THE WORK

§ 5.1 The Owner may make changes in the Work by issuing Modifications to the Prime Contract. Upon receipt of such a Modification issued subsequent to the execution of the Subcontract Agreement, the Contractor shall promptly notify the Subcontractor of the Modification. Unless otherwise directed by the Contractor, the Subcontractor shall not thereafter order materials or perform Work that would be inconsistent with the changes made by the Modification to the Prime Contract.

§ 5.2 The Subcontractor may be ordered in writing by the Contractor, without invalidating this Subcontract, to make changes in the Work within the general scope of this Subcontract consisting of additions, deletions or other revisions, including those required by Modifications to the Prime Contract issued subsequent to the execution of this Agreement, the Subcontract Sum and the Subcontract Time being adjusted accordingly. The Subcontractor, prior to the commencement of such changed or revised Work, shall submit promptly to the Contractor written copies of a claim for adjustment to the Subcontract Sum and Subcontract Time for such revised Work in a manner consistent with requirements of the Subcontract Documents.

§ 5.3 The Subcontractor shall make all claims promptly to the Contractor for additional cost, extensions of time and damages for delays or other causes in accordance with the Subcontract Documents. A claim which will affect or become part of a claim which the Contractor is required to make under the Prime Contract within a specified time period or in a specified manner shall be made in sufficient time to permit the Contractor to satisfy the requirements of the Prime Contract. Such claims shall be received by the Contractor not less than two working days preceding the time by which the Contractor's claim must be made. Failure of the Subcontractor to make such a timely claim shall bind the Subcontractor to the same consequences as those to which the Contractor is bound.

ARTICLE 6 MEDIATION AND BINDING DISPUTE RESOLUTION
§ 6.1 MEDIATION

§ 6.1.1 Any claim arising out of or related to this Subcontract, except those waived in this Subcontract, shall be subject to mediation as a condition precedent to binding dispute resolution.

§ 6.1.2 The parties shall endeavor to resolve their claims by mediation which, unless the parties mutually agree otherwise, shall be administered by the American Arbitration Association in accordance with its Construction Industry Mediation Procedures in effect on the date of the Agreement. A request for mediation shall be made in writing, delivered to the other party to this Subcontract and filed with the person or entity administering the mediation. The request may be made concurrently with the filing of binding dispute resolution proceedings but, in such event, mediation shall proceed in advance of binding dispute resolution proceedings, which shall be stayed pending mediation for a period of 60 days from the date of filing, unless stayed for a longer period by agreement of the parties or court order. If an arbitration is stayed pursuant to this Section, the parties may nonetheless proceed to the selection of the arbitrators(s) and agree upon a schedule for later proceedings.

§ 6.1.3 The parties shall share the mediator's fee and any filing fees equally. The mediation shall be held in the place where the Project is located, unless another location is mutually agreed upon. Agreements reached in mediation shall be enforceable as settlement agreements in any court having jurisdiction thereof.

§ 6.2 BINDING DISPUTE RESOLUTION
For any claim subject to, but not resolved by mediation pursuant to Section 6.1, the method of binding dispute resolution shall be as follows:
(Check the appropriate box. If the Contractor and Subcontractor do not select a method of binding dispute resolution below, or do not subsequently agree in writing to a binding dispute resolution method other than litigation, claims will be resolved by litigation in a court of competent jurisdiction.)

☐   Arbitration pursuant to Section 6.3 of this Agreement

☐   Litigation in a court of competent jurisdiction

☐   Other *(Specify)*

§ 6.3 ARBITRATION
§ 6.3.1 If the Contractor and Subcontractor have selected arbitration as the method of binding dispute resolution in Section 6.2, any claim subject to, but not resolved by, mediation shall be subject to arbitration which, unless the parties mutually agree otherwise, shall be administered by the American Arbitration Association in accordance with its Construction Industry Arbitration Rules in effect on the date of the Agreement. A demand for arbitration shall be made in writing, delivered to the other party to the Subcontract, and filed with the person or entity administering the arbitration. The party filing a notice of demand for arbitration must assert in the demand all claims then known to that party on which arbitration is permitted to be demanded.

§ 6.3.2 A demand for arbitration shall be made no earlier than concurrently with the filing of a request for meditation but in no event shall it be made after the date when the institution of legal or equitable proceedings based on the claim would be barred by the applicable statute of limitations. For statute of limitations purposes, receipt of a written demand for arbitration by the person or entity administering the arbitration shall constitute the institution of legal or equitable proceedings based on the claim.

§ 6.3.3 Either party, at its sole discretion, may consolidate an arbitration conducted under this Agreement with any other arbitration to which it is a party provided that (1) the arbitration agreement governing the other arbitration permits consolidation; (2) the arbitrations to be consolidated substantially involve common questions of law or fact; and (3) the arbitrations employ materially similar procedural rules and methods for selecting arbitrator(s).

§ 6.3.4 Either party, at its sole discretion, may include by joinder persons or entities substantially involved in a common question of law or fact whose presence is required if complete relief is to be accorded in arbitration, provided that the party sought to be joined consents in writing to such joinder. Consent to arbitration involving an additional person or entity shall not constitute consent to arbitration of a claim not described in the written consent.

§ 6.3.5 The Contractor and Subcontractor grant to any person or entity made a party to an arbitration conducted under this Section 6.3, whether by joinder or consolidation, the same rights of joinder and consolidation as the Contractor and Subcontractor under this Agreement.

§ 6.3.6 This agreement to arbitrate and any other written agreement to arbitrate with an additional person or persons referred to herein shall be specifically enforceable under applicable law in any court having jurisdiction thereof. The award rendered by the arbitrator or arbitrators shall be final, and judgment may be entered upon it in accordance with applicable law in any court having jurisdiction thereof.

ARTICLE 7 TERMINATION, SUSPENSION OR ASSIGNMENT OF THE SUBCONTRACT
§ 7.1 TERMINATION BY THE SUBCONTRACTOR
The Subcontractor may terminate the Subcontract for the same reasons and under the same circumstances and procedures with respect to the Contractor as the Contractor may terminate with respect to the Owner under the Prime Contract, or for nonpayment of amounts due under this Subcontract for 60 days or longer. In the event of such termination by the Subcontractor for any reason which is not the fault of the Subcontractor, Sub-subcontractors or their agents or employees or other persons performing portions of the Work under contract with the Subcontractor, the Subcontractor shall be entitled to recover from the Contractor payment for Work executed and for proven loss with respect to materials, equipment, tools, and construction equipment and machinery, including reasonable overhead, profit and damages.

§ 7.2 TERMINATION BY THE CONTRACTOR
§ 7.2.1 If the Subcontractor repeatedly fails or neglects to carry out the Work in accordance with the Subcontract Documents or otherwise to perform in accordance with this Subcontract and fails within a ten-day period after receipt of written notice to commence and continue correction of such default or neglect with diligence and promptness, the Contractor may, by written notice to the Subcontractor and without prejudice to any other remedy the Contractor may have, terminate the Subcontract and finish the Subcontractor's Work by whatever method the Contractor may deem expedient. If the unpaid balance of the Subcontract Sum exceeds the expense of finishing the Subcontractor's Work and other damages incurred by the Contractor and not expressly waived, such excess shall be paid to the Subcontractor. If such expense and damages exceed such unpaid balance, the Subcontractor shall pay the difference to the Contractor.

§ 7.2.2 If the Owner terminates the Contract for the Owner's convenience, the Contractor shall promptly deliver written notice to the Subcontractor.

§ 7.2.3 Upon receipt of written notice of termination, the Subcontractor shall
    .1   cease operations as directed by the Contractor in the notice;
    .2   take actions necessary, or that the Contractor may direct, for the protection and preservation of the Work; and
    .3   except for Work directed to be performed prior to the effective date of termination stated in the notice, terminate all existing Sub-subcontracts and purchase orders and enter into no further Sub-subcontracts and purchase orders.

§ 7.2.4 In case of such termination for the Owner's convenience, the Subcontractor shall be entitled to receive payment for Work executed, and costs incurred by reason of such termination, along with reasonable overhead and profit on the Work not executed.

§ 7.3 SUSPENSION BY THE CONTRACTOR FOR CONVENIENCE
§ 7.3.1 The Contractor may, without cause, order the Subcontractor in writing to suspend, delay or interrupt the Work of this Subcontract in whole or in part for such period of time as the Contractor may determine. In the event of suspension ordered by the Contractor, the Subcontractor shall be entitled to an equitable adjustment of the Subcontract Time and Subcontract Sum.

§ 7.3.2 An adjustment shall be made for increases in the Subcontract Time and Subcontract Sum, including profit on the increased cost of performance, caused by suspension, delay or interruption. No adjustment shall be made to the extent that
    .1   performance is, was or would have been so suspended, delayed or interrupted by another cause for which the Subcontractor is responsible; or
    .2   an equitable adjustment is made or denied under another provision of this Subcontract.

§ 7.4 ASSIGNMENT OF THE SUBCONTRACT
§ 7.4.1 In the event the Owner terminates the Prime Contract for cause, this Subcontract is assigned to the Owner pursuant to Section 5.4 of A201–2007 provided the Owner accepts the assignment.

§ 7.4.2 Without the Contractor's written consent, the Subcontractor shall not assign the Work of this Subcontract, subcontract the whole of this Subcontract, or subcontract portions of this Subcontract.

ARTICLE 8   THE WORK OF THIS SUBCONTRACT
The Subcontractor shall execute the following portion of the Work described in the Subcontract Documents, including all labor, materials, equipment, services and other items required to complete such portion of the Work, except to the extent specifically indicated in the Subcontract Documents to be the responsibility of others.
(Insert a precise description of the Work of this Subcontract, referring where appropriate to numbers of Drawings, sections of Specifications and pages of Addenda, Modifications and accepted alternates.)

ARTICLE 9   DATE OF COMMENCEMENT AND SUBSTANTIAL COMPLETION
§ 9.1 Subcontract Time is the period of time, including authorized adjustments, allotted in the Subcontract Documents for Substantial Completion of the Work described in the Subcontract Documents. The Subcontractor's date of commencement is the date from which the Subcontract Time of Section 9.3 is measured; it shall be the date of this Agreement, as first written above, unless a different date is stated below or provision is made for the date to be fixed in a notice to proceed issued by the Contractor.
(Insert the date of commencement, if it differs from the date of this Agreement or, if applicable, state that the date will be fixed in a notice to proceed.)

§ 9.2 Unless the date of commencement is established by a notice to proceed issued by the Contractor, or the Contractor has commenced visible Work at the site under the Prime Contract, the Subcontractor shall notify the Contractor in writing not less than five days before commencing the Subcontractor's Work to permit the timely filing of mortgages, mechanic's liens and other security interests.

§ 9.3 The Work of this Subcontract shall be substantially completed not later than
(Insert the calendar date or number of calendar days after the Subcontractor's date of commencement. Also insert any requirements for earlier substantial completion of certain portions of the Subcontractor's Work, if not stated elsewhere in the Subcontract Documents.)

, subject to adjustments of this Subcontract Time as provided in the Subcontract Documents.
(Insert provisions, if any, for liquidated damages relating to failure to complete on time.)

§ 9.4 With respect to the obligations of both the Contractor and the Subcontractor, time is of the essence of this Subcontract.

§ 9.5 No extension of time will be valid without the Contractor's written consent after claim made by the Subcontractor in accordance with Section 5.3.

ARTICLE 10   SUBCONTRACT SUM
§ 10.1 The Contractor shall pay the Subcontractor in current funds for performance of the Subcontract the Subcontract Sum of   Dollars
($), subject to additions and deductions as provided in the Subcontract Documents.

§ 10.2 The Subcontract Sum is based upon the following alternates, if any, which are described in the Subcontract Documents and have been accepted by the Owner and the Contractor:
(Insert the numbers or other identification of accepted alternates.)

Init.

/

§ 10.3 Unit prices, if any:
(Identify and state the unit price, and state the quantity limitations, if any, to which the unit price will be applicable.)

Item	Units and Limitations	Price per Unit

§ 10.4 Allowances included in the Subcontract Sum, if any:
(Identify allowance and state exclusions, if any, from the allowance price.)

Item	Price

ARTICLE 11 PROGRESS PAYMENTS

§ 11.1 Based upon applications for payment submitted to the Contractor by the Subcontractor, corresponding to applications for payment submitted by the Contractor to the Architect, and certificates for payment issued by the Architect, the Contractor shall make progress payments on account of the Subcontract Sum to the Subcontractor as provided below and elsewhere in the Subcontract Documents. Unless the Contractor provides the Owner with a payment bond in the full penal sum of the Contract Sum, payments received by the Contractor and Subcontractor for Work properly performed by their contractors and suppliers shall be held by the Contractor and Subcontractor for those contractors or suppliers who performed Work or furnished materials, or both, under contract with the Contractor or Subcontractor for which payment was made to the Contractor by the Owner or to the Subcontractor by the Contractor, as applicable. Nothing contained herein shall require money to be placed in a separate account and not commingled with money of the Contractor or Subcontractor, shall create any fiduciary liability or tort liability on the part of the Contractor or Subcontractor for breach of trust or shall entitle any person or entity to an award of punitive damages against the Contractor or Subcontractor for breach of the requirements of this provision.

§ 11.2 The period covered by each application for payment shall be one calendar month ending on the last day of the month, or as follows:

§ 11.3 Provided an application for payment is received by the Contractor not later than the   day of a month, the Contractor shall include the Subcontractor's Work covered by that application in the next application for payment which the Contractor is entitled to submit to the Architect. The Contractor shall pay the Subcontractor each progress payment no later than seven working days after the Contractor receives payment from the Owner. If the Architect does not issue a certificate for payment or the Contractor does not receive payment for any cause which is not the fault of the Subcontractor, the Contractor shall pay the Subcontractor, on demand, a progress payment computed as provided in Sections 11.7, 11.8 and 11.9.

§ 11.4 If the Subcontractor's application for payment is received by the Contractor after the application date fixed above, the Subcontractor's Work covered by it shall be included by the Contractor in the next application for payment submitted to the Architect.

§ 11.5 The Subcontractor shall submit to the Contractor a schedule of values prior to submitting the Subcontractor's first Application for Payment. Each subsequent application for payment shall be based upon the most recent schedule of values submitted by the Subcontractor in accordance with the Subcontract Documents. The schedule of values shall allocate the entire Subcontract Sum among the various portions of the Subcontractor's Work and be prepared in such form and supported by such data to substantiate its accuracy as the Contractor may require. This schedule, unless objected to by the Contractor, shall be used as a basis for reviewing the Subcontractor's applications for payment.

§ 11.6 Applications for payment submitted by the Subcontractor shall indicate the percentage of completion of each portion of the Subcontractor's Work as of the end of the period covered by the application for payment.

§ 11.7 Subject to the provisions of the Subcontract Documents, the amount of each progress payment shall be computed as set forth in the sections below.

§ 11.7.1 Take that portion of the Subcontract Sum properly allocable to completed Work as determined by multiplying the percentage completion of each portion of the Subcontractor's Work by the share of the total Subcontract Sum allocated to that portion of the Subcontractor's Work in the schedule of values, less that percentage actually retained, if any, from payments to the Contractor on account of the Work of the Subcontractor. Pending final determination of cost to the Contractor of changes in the Work that have been properly authorized by the Contractor, amounts not in dispute shall be included to the same extent provided in the Prime Contract, even though the Subcontract Sum has not yet been adjusted;

§ 11.7.2 Add that portion of the Subcontract Sum properly allocable to materials and equipment delivered and suitably stored at the site by the Subcontractor for subsequent incorporation in the Subcontractor's Work or, if approved by the Contractor, suitably stored off the site at a location agreed upon in writing, less the same percentage retainage required by the Prime Contract to be applied to such materials and equipment in the Contractor's application for payment;

§ 11.7.3 Subtract the aggregate of previous payments made by the Contractor; and

§ 11.7.4 Subtract amounts, if any, calculated under Section 11.7.1 or 11.7.2 that are related to Work of the Subcontractor for which the Architect has withheld or nullified, in whole or in part, a certificate of payment for a cause that is the fault of the Subcontractor.

§ 11.8 Upon the partial or entire disapproval by the Contractor of the Subcontractor's application for payment, the Contractor shall provide written notice to the Subcontractor. When the basis for the disapproval has been remedied, the Subcontractor shall be paid the amounts withheld.

§ 11.9 SUBSTANTIAL COMPLETION

When the Subcontractor's Work or a designated portion thereof is substantially complete and in accordance with the requirements of the Prime Contract, the Contractor shall, upon application by the Subcontractor, make prompt application for payment for such Work. Within 30 days following issuance by the Architect of the certificate for payment covering such substantially completed Work, the Contractor shall, to the full extent allowed in the Prime Contract, make payment to the Subcontractor, deducting any portion of the funds for the Subcontractor's Work withheld in accordance with the certificate to cover costs of items to be completed or corrected by the Subcontractor. Such payment to the Subcontractor shall be the entire unpaid balance of the Subcontract Sum if a full release of retainage is allowed under the Prime Contract for the Subcontractor's Work prior to the completion of the entire Project. If the Prime Contract does not allow for a full release of retainage, then such payment shall be an amount which, when added to previous payments to the Subcontractor, will reduce the retainage on the Subcontractor's substantially completed Work to the same percentage of retainage as that on the Contractor's Work covered by the certificate.

ARTICLE 12 FINAL PAYMENT

§ 12.1 Final payment, constituting the entire unpaid balance of the Subcontract Sum, shall be made by the Contractor to the Subcontractor when the Subcontractor's Work is fully performed in accordance with the requirements of the Subcontract Documents, the Architect has issued a certificate for payment covering the Subcontractor's completed Work and the Contractor has received payment from the Owner. If, for any cause which is not the fault of the Subcontractor, a certificate for payment is not issued or the Contractor does not receive timely payment or does not pay the Subcontractor within seven days after receipt of payment from the Owner, final payment to the Subcontractor shall be made upon demand.

(Insert provisions for earlier final payment to the Subcontractor, if applicable.)

§ 12.2 Before issuance of the final payment, the Subcontractor, if required, shall submit evidence satisfactory to the Contractor that all payrolls, bills for materials and equipment, and all known indebtedness connected with the Subcontractor's Work have been satisfied. Acceptance of final payment by the Subcontractor shall constitute a waiver of claims by the Subcontractor, except those previously made in writing and identified by the Subcontractor as unsettled at the time of final application for payment.

ARTICLE 13 INSURANCE AND BONDS

§ 13.1 The Subcontractor shall purchase and maintain insurance of the following types of coverage and limits of liability as will protect the Subcontractor from claims that may arise out of, or result from, the Subcontractor's operations and completed operations under the Subcontract:

§ 13.2 Coverages, whether written on an occurrence or claims-made basis, shall be maintained without interruption from the date of commencement of the Subcontractor's Work until the date of final payment and termination of any coverage required to be maintained after final payment to the Subcontractor, and, with respect to the Subcontractor's completed operations coverage, until the expiration of the period for correction of Work or for such other period for maintenance of completed operations coverage as specified in the Prime Contract.

§ 13.3 Certificates of insurance acceptable to the Contractor shall be filed with the Contractor prior to commencement of the Subcontractor's Work. These certificates and the insurance policies required by this Article 13 shall contain a provision that coverages afforded under the policies will not be canceled or allowed to expire until at least 30 days' prior written notice has been given to the Contractor. If any of the foregoing insurance coverages are required to remain in force after final payment and are reasonably available, an additional certificate evidencing continuation of such coverage shall be submitted with the final application for payment as required in Article 12. If any information concerning reduction of coverage is not furnished by the insurer, it shall be furnished by the Subcontractor with reasonable promptness according to the Subcontractor's information and belief.

§ 13.4 The Subcontractor shall cause the commercial liability coverage required by the Subcontract Documents to include: (1) the Contractor, the Owner, the Architect and the Architect's consultants as additional insureds for claims caused in whole or in part by the Subcontractor's negligent acts or omissions during the Subcontractor's operations; and (2) the Contractor as an additional insured for claims caused in whole or in part by the Subcontractor's negligent acts or omissions during the Subcontractor's completed operations.

§ 13.5 The Contractor shall furnish to the Subcontractor satisfactory evidence of insurance required of the Contractor under the Prime Contract.

§ 13.6 The Contractor shall promptly, upon request of the Subcontractor, furnish a copy or permit a copy to be made of any bond covering payment of obligations arising under the Subcontract.

§ 13.7 Performance Bond and Payment Bond:
(If the Subcontractor is to furnish bonds, insert the specific requirements here.)

§ 13.8 PROPERTY INSURANCE

§ 13.8.1 When requested in writing, the Contractor shall provide the Subcontractor with copies of the property and equipment policies in effect for the Project. The Contractor shall notify the Subcontractor if the required property insurance policies are not in effect.

§ 13.8.2 If the required property insurance is not in effect for the full value of the Subcontractor's Work, then the Subcontractor shall purchase insurance for the value of the Subcontractor's Work, and the Subcontractor shall be reimbursed for the cost of the insurance by an adjustment in the Subcontract Sum.

§ 13.8.3 Property insurance for the Subcontractor's materials and equipment required for the Subcontractor's Work, stored off site or in transit and not covered by the Project property insurance, shall be paid for through the application for payment process.

§ 13.9 WAIVERS OF SUBROGATION
The Contractor and Subcontractor waive all rights against (1) each other and any of their subcontractors, sub-subcontractors, agents and employees, each of the other, and (2) the Owner, the Architect, the Architect's consultants, separate contractors, and any of their subcontractors, sub-subcontractors, agents and employees for damages caused by fire or other causes of loss to the extent covered by property insurance provided under the Prime Contract or other property insurance applicable to the Work, except such rights as they may have to proceeds of such insurance held by the Owner as a fiduciary. The Subcontractor shall require of the Subcontractor's Sub-subcontractors, agents and employees, by appropriate agreements, written where legally required for validity, similar waivers in favor of the parties enumerated herein. The policies shall provide such waivers of subrogation by endorsement or otherwise. A waiver of subrogation shall be effective as to a person or entity even though that person or entity would otherwise have a duty of indemnification, contractual or otherwise, did not pay the insurance premium directly or indirectly, and whether or not the person or entity had an insurable interest in the property damaged.

ARTICLE 14   TEMPORARY FACILITIES AND WORKING CONDITIONS
§ 14.1 The Contractor shall furnish and make available at no cost to the Subcontractor the Contractor's temporary facilities, equipment and services, except as noted below.

§ 14.2 Specific working conditions:
(Insert any applicable arrangements concerning working conditions and labor matters for the Project.)

ARTICLE 15   MISCELLANEOUS PROVISIONS
§ 15.1 Where reference is made in this Subcontract to a provision of another Subcontract Document, the reference refers to that provision as amended or supplemented by other provisions of the Subcontract Documents.

§ 15.2 Payments due and unpaid under this Subcontract shall bear interest from the date payment is due at such rate as the parties may agree upon in writing or, in the absence thereof, at the legal rate prevailing from time to time at the place where the Project is located.
(Insert rate of interest agreed upon, if any.)

§ 15.3 Retainage and any reduction thereto is as follows:

§ 15.4 The Contractor and Subcontractor waive claims against each other for consequential damages arising out of or relating to this Subcontract, including without limitation, any consequential damages due to either party's termination in accordance with Article 7.

ARTICLE 16 ENUMERATION OF SUBCONTRACT DOCUMENTS

§ 16.1 The Subcontract Documents, except for Modifications issued after execution of this Subcontract, are enumerated in the sections below.

§ 16.1.1 This executed AIA Document A401–2007, Standard Form of Agreement Between Contractor and Subcontractor.

§ 16.1.2 The Prime Contract, consisting of the Agreement between the Owner and Contractor dated as first entered above and the other Contract Documents enumerated in the Owner-Contractor Agreement.

§ 16.1.3 The following Modifications to the Prime Contract, if any, issued subsequent to the execution of the Owner-Contractor Agreement but prior to the execution of this Agreement:

Modification	Date

§ 16.1.4 Additional Documents, if any, forming part of the Subcontract Documents:

.1   AIA Document E201™–2007, Digital Data Protocol Exhibit, if completed by the parties, or the following:

.2   Other documents:
(List here any additional documents that are intended to form part of the Subcontract Documents. Requests for proposal and the Subcontractor's bid or proposal should be listed here only if intended to be made part of the Subcontract Documents.)

This Agreement entered into as of the day and year first written above.

_____
CONTRACTOR *(Signature)*

_____
(Printed name and title)

_____
SUBCONTRACTOR *(Signature)*

_____
(Printed name and title)

CAUTION: You should sign an original AIA Contract Document, on which this text appears in RED. An original assures that changes will not be obscured.

Init.

/

14

Construction Industry Dispute Resolution Procedures (Including Mediation and Arbitration Rules)

American Arbitration Association

Revised and in Effect on September 1, 2007

American Arbitration Association
Dispute Resolution Services Worldwide

Construction Industry Arbitration Rules and Mediation PROCEDURES
(Including Procedures for Large, Complex Construction Disputes)
Amended and Effective September 1, 2007

Summary of Changes

TABLE OF CONTENTS

F-8. Proceedings on Documents
F-9. Date, Time, and Place of Hearing
F-10. The Hearing
F-11. Time of Award
F-12. Time Standards
F-13. Arbitrator's Compensation

PROCEDURES FOR LARGE, COMPLEX CONSTRUCTION DISPUTES
L-1. Administrative Conference
L-2. Arbitrators
L-3. Preliminary Hearing
L-4. Management of Proceedings

ADMINISTRATIVE FEES
Fees
Refund Schedule
Hearing Room Rental

NATIONAL CONSTRUCTION DISPUTE RESOLUTION COMMITTEE

Representatives of the more than 30 organizations listed below constitute the National
Construction Dispute Resolution Committee (NCDRC). This Committee serves as an advisory
body to the American Arbitration Association on mediation and arbitration procedures.

American Association of Airport Executives
American Bar Association - Construction Forum
American Bar Association - Construction Litigation Committee
American Bar Association - Public Contract Law Section
American College of Construction Lawyers
American Consulting Engineers Council
American Institute of Architects
American Public Works Association
American Road and Transportation Builders Association
American Society of Civil Engineers
American Subcontractors Association
Associated Builders & Contractors, Inc.
Associated General Contractors of America
American Specialty Contractors, Inc.
Buildings Future Council
Construction Specifications Institute
Construction Management Association of America
Design Build Institute of America
Engineers Joint Contract Documents Committee
National Association of Home Builders
National Association of Minority Contractors
National Society of Professional Engineers
National Utility Contractors Association
Victor O. Schinnerer
Women Construction Owners & Executives, USA

IMPORTANT NOTICE

These rules and any amendment of them shall apply in the form in effect at the time the
administrative filing requirements are met for a demand for arbitration or submission agreement

received by the AAA. To insure that you have the most current information, see our Web site at www.adr.org.

INTRODUCTION

Each year, many thousands of construction transactions take place. Occasionally, disagreements develop over these transactions. Many of these disputes are resolved by arbitration, the voluntary submission of a dispute to a disinterested person or persons for final and binding determination. Arbitration has proven to be an effective way to resolve disputes privately, promptly, and economically.

The American Arbitration Association (AAA) is a public-service, not-for-profit organization offering a broad range of dispute resolution services to business executives, attorneys, individuals, trade associations, unions, management, consumers, families, communities, and all levels of government. Services are available through AAA headquarters in New York City and through offices located in major cities throughout the United States and Europe. Hearings may be held at locations convenient for the parties and are not limited to cities with AAA offices. In addition, the AAA serves as a center for education and training, issues specialized publications, and conducts research on all forms of out-of-court dispute settlement.

Mediation

Because of the increasing popularity of mediation, especially as a prelude to arbitration, the Association has combined its mediation procedures and arbitration rules into a single brochure.

By agreement, the parties may submit their dispute to mediation before arbitration under the mediation procedures in this brochure. Mediation involves the services of one or more individuals, to assist parties in settling a controversy or claim by direct negotiations between or among themselves. The mediator participates impartially in the negotiations, guiding and consulting the various parties involved. The result of the mediation should be an agreement that the parties find acceptable. The mediator cannot impose a settlement, but can only guide the parties toward achieving their own settlement.

The AAA will administer the mediation process to achieve orderly, economical, and expeditious mediation, utilizing to the greatest possible extent the competence and acceptability of the mediators on the AAA's Construction Mediation Panel. Depending on the expertise needed for a given dispute, the parties can obtain the services of one or more individuals who are willing to serve as mediators and who are trained by the AAA in the necessary mediation skills. In identifying those persons most qualified to mediate, the AAA is assisted by the NCDRC.

The AAA itself does not act as mediator. Its function is to administer the mediation process in accordance with the agreement of the parties, to teach mediation skills to members of the construction industry, and to maintain the National Roster from which topflight mediators can be chosen.

There is no additional administrative fee where parties to a pending arbitration attempt to mediate their dispute under the AAA's auspices. Procedures for mediation cases are described in Sections M-1 through M-17.

Arbitration

Regular Track Procedures:

The rules contain Regular Track Procedures which are applied to the administration of all arbitration cases, unless they conflict with any portion of the Fast Track Procedures or the Procedures for Large, Complex Construction Disputes whenever these apply. In the event of a conflict, either the Fast Track procedures or the Large, Complex Construction Disputes procedures apply.

The highlights of the Regular Track Procedures are:

- party input into the AAA's preparation of lists of proposed arbitrators;
- express arbitrator authority to control the discovery process;
- broad arbitrator authority to control the hearing;
- a concise written breakdown of the award and, if requested in writing by all parties prior to the appointment of the arbitrator or at the discretion of the arbitrator, a written explanation of the award;
- arbitrator compensation, with the AAA to provide the arbitrator's compensation policy with the biographical information sent to the parties;
- a demand form and an answer form, both of which seek more information from the parties to assist the AAA in better serving the parties.

Fast Track Procedures:

The Fast Track Procedures were designed for cases involving claims of no more than $75,000. The highlights of this system are:

- a 60-day "time standard" for case completion;
- establishment of a special pool of arbitrators who are pre-qualified to serve on an expedited basis;
- an expedited arbitrator appointment process, with party input;
- presumption that cases involving $10,000 or less will be decided on a documents only basis;
- requirement of a hearing within 30 calendar days of the arbitrator's appointment;
- a single day of hearing in most cases;
- an award in no more than 14 calendar days after completion of the hearing.

Procedures for Large, Complex Construction Disputes:

Unless the parties agree otherwise, the Procedures for Large, Complex Construction Disputes, which appear in this pamphlet, will be applied to all cases administered by the AAA under the Construction Arbitration Rules in which the disclosed claim or counterclaim of any party is at least $500,000 exclusive of claimed interest, arbitration fees and costs.

The key features of these procedures include:

- mandatory use of the procedures in cases involving claims of $500,000 or more;
- a highly qualified, trained Panel of Neutrals, compensated at their customary rates;
- a mandatory preliminary hearing with the arbitrators, which may be conducted by telephone;
- broad arbitrator authority to order and control discovery, including depositions;
- presumption that hearings will proceed on a consecutive or block basis.

The National Roster

The AAA has established and maintains as members of its National Roster individuals competent to hear and decide disputes administered under the Construction Industry Arbitration Rules. The AAA considers for appointment to the construction industry roster persons recommended by Regional Panel Advisory Committees as qualified to serve by virtue of their experience in the construction field. The majority of neutrals are actively engaged in the construction industry. Attorney neutrals generally devote at least half of their practice to construction matters. Neutrals serving under these rules must also attend periodic training.

The services of the AAA are generally concluded with the transmittal of the award. Although there is voluntary compliance with the majority of awards, judgment on the award can be entered in a court having appropriate jurisdiction if necessary.

Administrative Fees

The AAA charges a filing fee based on the amount of claim or counterclaim. This fee information, which is contained with these rules, allows the parties to exercise control over their administrative fees.

The fees cover AAA administrative services; they do not cover arbitrator compensation or expenses, if any, reporting services, hearing room rental or any post-award charges incurred by the parties in enforcing the award.

ADR Clauses

Mediation

If the parties elect to adopt mediation as a part of their contractual dispute settlement procedure, they can insert the following mediation clause into their contract in conjunction with a standard arbitration provision.

> *If a dispute arises out of or relates to this contract, or the breach thereof, and if the dispute cannot be settled through negotiation, the parties agree first to try in good faith to settle the dispute by mediation administered by the American Arbitration Association under its Construction Industry Mediation Procedures before resorting to arbitration, litigation, or some other dispute resolution procedure.*

If the parties choose to use a mediator to resolve an existing dispute, they can enter into the following submission.

> *The parties hereby submit the following dispute to mediation administered by the American Arbitration Association under its Construction Industry Mediation Procedures. (The clause may also provide for the qualifications of the mediator(s), method of payment, locale of meetings, and any other item of concern to the parties.)*

Arbitration

When an agreement to arbitrate is included in a construction contract, it might expedite peaceful settlement without the necessity of going to arbitration at all. Thus, an arbitration clause is a form of insurance against loss of good will. The parties can provide for arbitration of future disputes by inserting the following clause into their contracts.

> *Any controversy or claim arising out of or relating to this contract, or the breach thereof, shall be settled by arbitration administered by the American Arbitration Association under its Construction Industry Arbitration Rules, and judgment on the award rendered by the arbitrator(s) may be entered in any court having jurisdiction thereof.*

Arbitration of existing disputes may be accomplished by use of the following.

> *We, the undersigned parties, hereby agree to submit to arbitration administered by the American Arbitration Association under its Construction Industry Arbitration Rules the following controversy: (cite briefly). We further agree that the above controversy be submitted to (one)(three) arbitrator(s). We further agree that we will faithfully observe this agreement and the rules, that we will abide by and perform any award rendered by the arbitrator(s), and that a judgment of the court having jurisdiction may be entered on the*

award.

For further information about the AAA's Construction Dispute Avoidance and Resolution Services, as well as the full range of other AAA services, contact the nearest AAA office or visit our Web site at www.adr.org.

CONSTRUCTION INDUSTRY MEDIATION PROCEDURES

M-1. Agreement of Parties

Whenever, by stipulation or in their contract, the parties have provided for mediation or conciliation of existing or future disputes under the auspices of the American Arbitration Association (AAA) or under these procedures, the parties and their representatives, unless agreed otherwise in writing, shall be deemed to have made these procedural guidelines, as amended and in effect as of the date of filing of a request for mediation, a part of their agreement and designate the AAA as the administrator of their mediation.

The parties by mutual agreement may vary any part of these procedures including, but not limited to, agreeing to conduct the mediation via telephone or other electronic or technical means.

M-2. Initiation of Mediation

Any party or parties to a dispute may initiate mediation under the AAA's auspices by making a request for mediation to any of the AAA's regional offices or case management centers via telephone, email, regular mail or fax. Requests for mediation may also be filed online via WebFile at www.adr.org.

The party initiating the mediation shall simultaneously notify the other party or parties of the request. The initiating party shall provide the following information to the AAA and the other party or parties as applicable:

  i.   A copy of the mediation provision of the parties' contract or the parties' stipulation to mediate.
 ii.   The names, regular mail addresses, email addresses, and telephone numbers of all parties to the dispute and representatives, if any, in the mediation.
iii.   A brief statement of the nature of the dispute and the relief requested.
 iv.   Any specific qualifications the mediator should possess.

Where there is no preexisting stipulation or contract by which the parties have provided for mediation of existing or future disputes under the auspices of the AAA, a party may request the AAA to invite another party to participate in "mediation by voluntary submission". Upon receipt of such a request, the AAA will contact the other party or parties involved in the dispute and attempt to obtain a submission to mediation.

M-3. Representation

Subject to any applicable law, any party may be represented by persons of the party's choice. The names and addresses of such persons shall be communicated in writing to all parties and to the AAA.

M-4. Appointment of the Mediator

Parties may search the online profiles of the AAA's Panel of Mediators at www.aaamediation.com in an effort to agree on a mediator. If the parties have not agreed to the appointment of a mediator and have not provided any other method of appointment, the mediator shall be appointed in the following manner:

  i.   Upon receipt of a request for mediation, the AAA will send to each party a list of mediators from the AAA's Panel of Mediators.

The parties are encouraged to agree to a mediator from the submitted list and to advise the AAA of their agreement.

ii. If the parties are unable to agree upon a mediator, each party shall strike unacceptable names from the list, number the remaining names in order of preference, and return the list to the AAA. If a party does not return the list within the time specified, all mediators on the list shall be deemed acceptable. From among the mediators who have been mutually approved by the parties, and in accordance with the designated order of mutual preference, the AAA shall invite a mediator to serve.

iii. If the parties fail to agree on any of the mediators listed, or if acceptable mediators are unable to serve, or if for any other reason the appointment cannot be made from the submitted list, the AAA shall have the authority to make the appointment from among other members of the Panel of Mediators without the submission of additional lists.

M-5. Mediator's Impartiality and Duty to Disclose

AAA mediators are required to abide by the Model Standards of Conduct for Mediators in effect at the time a mediator is appointed to a case. Where there is a conflict between the Model Standards and any provision of these Mediation Procedures, these Mediation Procedures shall govern. The Standards require mediators to (i) decline a mediation if the mediator cannot conduct it in an impartial manner, and (ii) disclose, as soon as practicable, all actual and potential conflicts of interest that are reasonably known to the mediator and could reasonably be seen as raising a question about the mediator's impartiality.

Prior to accepting an appointment, AAA mediators are required to make a reasonable inquiry to determine whether there are any facts that a reasonable individual would consider likely to create a potential or actual conflict of interest for the mediator. AAA mediators are required to disclose any circumstance likely to create a presumption of bias or prevent a resolution of the parties' dispute within the time-frame desired by the parties. Upon receipt of such disclosures, the AAA shall immediately communicate the disclosures to the parties for their comments.

The parties may, upon receiving disclosure of actual or potential conflicts of interest of the mediator, waive such conflicts and proceed with the mediation. In the event that a party disagrees as to whether the mediator shall serve, or in the event that the mediator's conflict of interest might reasonably be viewed as undermining the integrity of the mediation, the mediator shall be replaced.

M-6. Vacancies

If any mediator shall become unwilling or unable to serve, the AAA will appoint another mediator, unless the parties agree otherwise, in accordance with section M-4.

M-7. Duties and Responsibilities of the Mediator

i. The mediator shall conduct the mediation based on the principle of party self-determination. Self-determination is the act of coming to a voluntary, uncoerced decision in which each party makes free and informed choices as to process and outcome.

ii. The mediator is authorized to conduct separate or *ex parte* meetings and other communications with the parties and/or their representatives, before, during, and after any scheduled mediation conference. Such communications may be conducted via telephone, in writing, via email, online, in person or otherwise.

iii. The parties are encouraged to exchange all documents pertinent to the relief requested. The mediator may request the exchange of memoranda on issues, including the underlying interests and the history of the parties' negotiations. Information that a party wishes to keep confidential may be sent to the mediator, as necessary, in a separate communication with the mediator.

iv. The mediator does not have the authority to impose a settlement on the parties but will attempt to help them reach a satisfactory resolution of their dispute. Subject to the discretion of the mediator, the mediator may make oral or written recommendations for settlement to a party privately or, if the parties agree, to all parties jointly.

v. In the event a complete settlement of all or some issues in dispute is not achieved within the scheduled mediation session(s), the mediator may continue to communicate with the parties, for a period of time, in an ongoing effort to facilitate a complete settlement.

vi. The mediator is not a legal representative of any party and has no fiduciary duty to any party.

M-8. Responsibilities of the Parties

The parties shall ensure that appropriate representatives of each party, having authority to consummate a settlement, attend the mediation conference.

Prior to and during the scheduled mediation conference session(s) the parties and their representatives shall, as appropriate to each party's circumstances, exercise their best efforts to prepare for and engage in a meaningful and productive mediation.

M-9. Privacy

Mediation sessions and related mediation communications are private proceedings. The parties and their representatives may attend mediation sessions. Other persons may attend only with the permission of the parties and with the consent of the mediator.

M-10. Confidentiality

Subject to applicable law or the parties' agreement, confidential information disclosed to a mediator by the parties or by other participants (witnesses) in the course of the mediation shall not be divulged by the mediator. The mediator shall maintain the confidentiality of all information obtained in the mediation, and all records, reports, or other documents received by a mediator while serving in that capacity shall be confidential.

The mediator shall not be compelled to divulge such records or to testify in regard to the mediation in any adversary proceeding or judicial forum.

The parties shall maintain the confidentiality of the mediation and shall not rely on, or introduce as evidence in any arbitral, judicial, or other proceeding the following, unless agreed to by the parties or required by applicable law:

   i.   Views expressed or suggestions made by a party or other participant with respect to a possible settlement of the dispute;
   ii. Admissions made by a party or other participant in the course of the mediation proceedings;
   iii. Proposals made or views expressed by the mediator; or
   iv. The fact that a party had or had not indicated willingness to accept a proposal for settlement made by the mediator.

M-11. No Stenographic Record

There shall be no stenographic record of the mediation process.

M-12. Termination of Mediation

The mediation shall be terminated:

   i.   By the execution of a settlement agreement by the parties; or
   ii. By a written or verbal declaration of the mediator to the effect that further efforts at mediation would not contribute to a resolution of the parties' dispute; or
   iii. By a written or verbal declaration of all parties to the effect that the mediation proceedings are terminated; or
   iv. When there has been no communication between the mediator and any party or party's representative for 21 days following the conclusion of the mediation conference.

M-13. Exclusion of Liability

Neither the AAA nor any mediator is a necessary party in judicial proceedings relating to the mediation. Neither the AAA nor any mediator shall be liable to any party for any error, act or omission in connection with any mediation conducted under these procedures.

M-14. Interpretation and Application of Procedures

The mediator shall interpret and apply these procedures insofar as they relate to the mediator's duties and responsibilities. All other procedures shall be interpreted and applied by the AAA.

M-15. Deposits

Unless otherwise directed by the mediator, the AAA will require the parties to deposit in advance of the mediation conference such sums of money as it, in consultation with the mediator, deems necessary to cover the costs and expenses of the mediation and shall render an accounting to the parties and return any unexpended balance at the conclusion of the mediation.

M-16. Expenses

All expenses of the mediation, including required traveling and other expenses or charges of the mediator, shall be borne equally by the parties unless they agree otherwise. The expenses of participants for either side shall be paid by the party requesting the attendance of such participants.

M-17. Cost of the Mediation

There is no filing fee to initiate a mediation or a fee to request the AAA to invite parties to mediate.

The cost of mediation is based on the hourly mediation rate published on the mediator's AAA profile. This rate covers both mediator compensation and an allocated portion for the AAA's services. There is a four-hour minimum charge for a mediation conference. Expenses referenced in Section M-16 may also apply.

If a matter submitted for mediation is withdrawn or cancelled or results in a settlement after the agreement to mediate is filed but prior to the mediation conference the cost is $250 plus any mediator time and charges incurred.

The parties will be billed equally for all costs unless they agree otherwise.

If you have questions about mediation costs or services visit our website at www.adr.org or contact your local AAA office.

Conference Room Rental

The costs described above do not include the use of AAA conference rooms. Conference rooms are available on a rental basis. Please contact your local AAA office for availability and rates.

CONSTRUCTION INDUSTRY ARBITRATION RULES

REGULAR TRACK PROCEDURES

R-1. Agreement of Parties

(a) The parties shall be deemed to have made these rules a part of their arbitration agreement whenever they have provided for arbitration by the American Arbitration Association (hereinafter AAA) under its Construction Industry Arbitration Rules. These rules and any amendment of them shall apply in the form in effect at the time the administrative requirements are met for a demand for arbitration or submission agreement received by the AAA. The parties, by written agreement, may vary the procedures set forth in these rules. After appointment of the arbitrator, such modifications may be made only with the consent of the arbitrator.

(b) Unless the parties or the AAA determines otherwise, the Fast Track Procedures shall apply in any case in which no disclosed claim or counterclaim exceeds $75,000, exclusive of interest and arbitration fees and costs. Parties may also agree to use these procedures in larger cases. Unless the parties agree otherwise, these procedures will not apply in cases involving more than two parties. The Fast Track Procedures shall be applied as described in Sections F-1 through F-13 of these rules, in addition to any other portion of these rules that is not in conflict with the Fast Track Procedures.

(c) Unless the parties agree otherwise, the Procedures for Large, Complex Construction Disputes shall apply to all cases in which the disclosed claim or counterclaim of any party is at least $500,000, exclusive of claimed interest, arbitration fees and costs. Parties may also agree to use these procedures in cases involving claims or counterclaims under $500,000, or in nonmonetary cases. The Procedures for Large, Complex Construction Disputes shall be applied as described in Sections L-1 through L-4 of these rules, in addition to any other portion of these rules that is not in conflict with the Procedures for Large, Complex Construction Disputes.

(d) All other cases shall be administered in accordance with Sections R-1 through R-55 of these rules.

R-2. AAA and Delegation of Duties

When parties agree to arbitrate under these rules, or when they provide for arbitration by the AAA and an arbitration is initiated under these rules, they thereby authorize the AAA to administer the arbitration. The authority and duties of the AAA are prescribed in the agreement of the parties and in these rules, and may be carried out through such of the AAA's representatives as it may direct. The AAA may, in its discretion, assign the administration of an arbitration to any of its offices.

R-3. National Roster of Neutrals

In cooperation with the National Construction Dispute Resolution Committee the AAA shall establish and maintain a National Roster of Construction Arbitrators ("National Roster") and shall appoint arbitrators as provided in these rules. The term "arbitrator" in these rules refers to the arbitration panel, constituted for a particular case, whether composed of one or more arbitrators, or to an individual arbitrator, as the context requires.

R-4. Initiation under an Arbitration Provision in a Contract

(a) Arbitration under an arbitration provision in a contract shall be initiated in the following manner.

   (i) The initiating party (the "claimant") shall, within the time period, if any, specified in the contract(s), give to the other party (the "respondent") written notice of its intention to arbitrate (the "demand"), which demand shall contain a statement setting forth the nature of the dispute, the names and addresses of all other parties, the amount involved, if any, the remedy sought, and the hearing locale requested.

   (ii) The claimant shall file at any office of the AAA two copies of the demand and two copies of the arbitration provisions of the contract, together with the appropriate filing fee as provided in the schedule included with these rules.

   (iii) The AAA shall confirm notice of such filing to the parties.

(b) A respondent may file an answering statement in duplicate with the AAA within 15 calendar days after confirmation of notice of filing of the demand is sent by the AAA. The respondent shall, at the time of any such filing, send a copy of the answering statement to the claimant. If a counterclaim is asserted, it shall contain a statement setting forth the nature of the counterclaim, the amount involved, if any, and the remedy sought. If a counterclaim is made, the party making the counterclaim shall forward to the AAA with the answering statement the appropriate fee provided in the schedule included with these rules.

(c) If no answering statement is filed within the stated time, respondent will be deemed to deny the claim. Failure to file an answering statement shall not operate to delay the arbitration.

(d) When filing any statement pursuant to this section, the parties are encouraged to provide descriptions of their claims in sufficient detail to make the circumstances of the dispute clear to the arbitrator.

R-5. Initiation under a Submission

Parties to any existing dispute may commence an arbitration under these rules by filing at any office of the AAA two copies of a written submission to arbitrate under these rules, signed by the parties. It shall contain a statement of the matter in dispute, the names and addresses of the parties, any claims and counterclaims, the amount involved, if any, the remedy sought, and the hearing locale requested, together with the appropriate filing fee as provided in the schedule included with these rules. Unless the parties state otherwise in the submission, all claims and counterclaims will be deemed to be denied by the other party.

R-6. Changes of Claim

A party may at any time prior to the close of the hearing increase or decrease the amount of its claim or counterclaim. Any new or different claim or counterclaim, as opposed to an increase or decrease in the amount of a pending claim or counterclaim, shall be made in writing and filed with the AAA, and a copy shall be mailed to the other party, who shall have a period of ten calendar days from the date of such mailing within which to file an answer with the AAA.

After the arbitrator is appointed no new or different claim or counterclaim may be submitted to the arbitrator except with the arbitrator's consent.

R-7. Consolidation or Joinder

If the parties' agreement or the law provides for consolidation or joinder of related arbitrations, all involved parties will endeavor to agree on a process to effectuate the consolidation or joinder.

If they are unable to agree, the Association shall directly appoint a single arbitrator for the limited purpose of deciding whether related arbitrations should be consolidated or joined and, if so, establishing a fair and appropriate process for consolidation or joinder. The AAA may take reasonable administrative action to accomplish the consolidation or joinder as directed by the arbitrator.

R-8. Jurisdiction

(a) The arbitrator shall have the power to rule on his or her own jurisdiction, including any objections with respect to the existence, scope or validity of the arbitration agreement.

(b) The arbitrator shall have the power to determine the existence or validity of a contract of which an arbitration clause forms a part. Such an arbitration clause shall be treated as an agreement independent of the other terms of the contract. A decision by the arbitrator that the contract is null and void shall not for that reason alone render invalid the arbitration clause.

(c) A party must object to the jurisdiction of the arbitrator or to the arbitrability of a claim or counterclaim no later than the filing of the answering statement to the claim or counterclaim that gives rise to the objection. The arbitrator may rule on such objections as a preliminary matter or as part of the final award.

R-9. Mediation

At any stage of the proceedings, the parties may agree to conduct a mediation conference under the

Construction Industry Mediation Procedures in order to facilitate settlement. The mediator shall not be an arbitrator appointed to the case. Where the parties to a pending arbitration agree to mediate under the AAA's rules, no additional administrative fee is required to initiate the mediation.

R-10. Administrative Conference

At the request of any party or upon the AAA's own initiative, the AAA may conduct an administrative conference, in person or by telephone, with the parties and/or their representatives. The conference may address such issues as arbitrator selection, potential mediation of the dispute, potential exchange of information, a timetable for hearings and any other administrative matters.

R-11. Fixing of Locale

The parties may mutually agree on the locale where the arbitration is to be held. If any party requests that the hearing be held in a specific locale and the other party files no objection thereto within fifteen calendar days after notice of the request has been sent to it by the AAA, the locale shall be the one requested. If a party objects to the locale requested by the other party, the AAA shall have the power to determine the locale, and its decision shall be final and binding.

R-12. Appointment from National Roster

If the parties have not appointed an arbitrator and have not provided any other method of appointment, the arbitrator shall be appointed in the following manner:

(a) Immediately after the filing of the submission or the answering statement or the expiration of the time within which the answering statement is to be filed, the AAA shall send simultaneously to each party to the dispute an identical list of 10 (unless the AAA decides that a different number is appropriate) names of persons chosen from the National Roster, unless the AAA decides that a different number is appropriate. The parties are encouraged to agree to an arbitrator from the submitted list and to advise the AAA of their agreement. Absent agreement of the parties, the arbitrator shall not have served as the mediator in the mediation phase of the instant proceeding.

(b) If the parties are unable to agree upon an arbitrator, each party to the dispute shall have 15 calendar days from the transmittal date in which to strike names objected to, number the remaining names in order of preference, and return the list to the AAA. If a party does not return the list within the time specified, all persons named therein shall be deemed acceptable. From among the persons who have been approved on both lists, and in accordance with the designated order of mutual preference, the AAA shall invite the acceptance of an arbitrator to serve. If the parties fail to agree on any of the persons named, or if acceptable arbitrators are unable to act, or if for any other reason the appointment cannot be made from the submitted lists, the AAA shall have the power to make the appointment from among other members of the National Roster without the submission of additional lists.

(c) Unless the parties agree otherwise when there are two or more claimants or two or more respondents, the AAA may appoint all the arbitrators.

R-13. Direct Appointment by a Party

(a) If the agreement of the parties names an arbitrator or specifies a method of appointing an arbitrator, that designation or method shall be followed. The notice of appointment, with the name and address of the arbitrator, shall be filed with the AAA by the appointing party. Upon the request of any appointing party, the AAA shall submit a list of members of the National Roster from which the party may, if it so desires, make the appointment.

(b) Where the parties have agreed that each party is to name one arbitrator, the arbitrators so named must meet the standards of Section R-18 with respect to impartiality and independence unless the parties have specifically agreed pursuant to Section R-18(a) that the party-appointed arbitrators are to be non-neutral and need not meet those standards.

(c) If the agreement specifies a period of time within which an arbitrator shall be appointed and any party fails to make the appointment within that period, the AAA shall make the appointment.

(d) If no period of time is specified in the agreement, the AAA shall notify the party to make the appointment. If within 15 calendar days after such notice has been sent, an arbitrator has not been appointed by a party, the AAA shall make the appointment.

R-14. Appointment of Chairperson by Party-Appointed Arbitrators or Parties

(a) If, pursuant to Section R-13, either the parties have directly appointed arbitrators, or the arbitrators have been appointed by AAA, and the parties have authorized them to appoint a chairperson within a specified time and no appointment is made within that time or any agreed extension, the AAA may appoint the chairperson.

(b) If no period of time is specified for appointment of the chairperson and the party-appointed arbitrators or the parties do not make the appointment within 15 calendar days from the date of the appointment of the last party-appointed arbitrator, the AAA may appoint the chairperson.

(c) If the parties have agreed that their party-appointed arbitrators shall appoint the chairperson from the National Roster, the AAA shall furnish to the party-appointed arbitrators, in the manner provided in Section R-12, a list selected from the National Roster, and the appointment of the chairperson shall be made as provided in that Section.

R-15. Nationality of Arbitrator in International Arbitration

Where the parties are nationals of different countries, the AAA, at the request of any party or on its own initiative, may appoint as arbitrator a national of a country other than that of any of the parties. The request must be made before the time set for the appointment of the arbitrator as agreed by the parties or set by these rules.

R-16. Number of Arbitrators

If the arbitration agreement does not specify the number of arbitrators, the dispute shall be heard and determined by one arbitrator, unless the AAA, in its discretion, directs that three arbitrators be appointed. A party may request three arbitrators in the demand or answer, which request the AAA will consider in exercising its discretion regarding the number of arbitrators appointed to the dispute.

R-17. Disclosure

(a) Any person appointed or to be appointed as an arbitrator shall disclose to the AAA any circumstance likely to give rise to justifiable doubt as to the arbitrator's impartiality or independence, including any bias or any financial or personal interest in the result of the arbitration or any past or present relationship with the parties or their representatives. Such obligation shall remain in effect throughout the arbitration.

(b) Upon receipt of such information from the arbitrator or another source, the AAA shall communicate the information to the parties and, if it deems it appropriate to do so, to the arbitrator and others.

(c) In order to encourage disclosure by arbitrators, disclosure of information pursuant to this Section R-17 is not to be construed as an indication that the arbitrator considers that the disclosed circumstances is likely to affect impartiality or independence.

R-18. Disqualification of Arbitrator

(a) Any arbitrator shall be impartial and independent and shall perform his or her duties with diligence and in good faith, and shall be subject to disqualification for

   (i) partiality or lack of independence,

   (ii) inability or refusal to perform his or her duties with diligence and in good faith, and

   (iii) any grounds for disqualification provided by applicable law. The parties may agree in writing, however, that arbitrators directly appointed by a party pursuant to Section R-13 shall be non-neutral, in which case such arbitrators need not be impartial or independent and shall not be subject to disqualification for partiality or lack of independence.

(b) Upon objection of a party to the continued service of an arbitrator, or on its own initiative, the AAA shall determine whether the arbitrator should be disqualified under the grounds set out above, and shall inform the parties of its decision, which decision shall be conclusive.

R-19. Communication with Arbitrator

(a) No party and no one acting on behalf of any party shall communicate ex parte with an arbitrator or a candidate for arbitrator concerning the arbitration, except that a party, or someone acting on behalf of a party, may communicate ex parte with a candidate for direct appointment pursuant to Section R-13 in order to advise the candidate of the general nature of the controversy and of the anticipated proceedings and to discuss the candidate's qualifications, availability or independence in relation to the parties or to discuss the suitability of the candidates for selection as a third arbitrator where the parties or party-designated arbitrators are to participate in that selection.

(b) Section R-19(a) does not apply to arbitrators directly appointed by the parties who, pursuant to Section R-18(a), the parties have agreed in writing are non-neutral. Where the parties have so agreed under Section R-18(a), the AAA shall as an administrative practice suggest to the parties that they agree further that Section R-19(a) should nonetheless apply prospectively.

R-20. Vacancies

(a) If for any reason an arbitrator is unable to perform the duties of the office, the AAA may, on proof satisfactory to it, declare the office vacant. Vacancies shall be filled in accordance with the applicable provisions of these rules.

(b) In the event of a vacancy in a panel of neutral arbitrators after the hearings have commenced, the remaining arbitrator or arbitrators may continue with the hearing and determination of the controversy, unless the parties agree otherwise.

(c) In the event of the appointment of a substitute arbitrator, the panel of arbitrators shall determine in its sole discretion whether it is necessary to repeat all or part of any prior hearings.

R-21. Preliminary Hearing

(a) At the request of any party or at the discretion of the arbitrator or the AAA, the arbitrator may schedule as soon as practicable a preliminary hearing with the parties and/or their representatives. The preliminary hearing may be conducted by telephone at the arbitrator's discretion.

(b) During the preliminary hearing, the parties and the arbitrator should discuss the future conduct of the case, including clarification of the issues and claims, a schedule for the hearings and any other preliminary matters.

R-22. Exchange of Information

(a) At the request of any party or at the discretion of the arbitrator, consistent with the expedited nature of arbitration, the arbitrator may direct

    (i) the production of documents and other information, and

    (ii) the identification of any witnesses to be called.

(b) At least five business days prior to the hearing, the parties shall exchange copies of all exhibits they intend to submit at the hearing.

(c) The arbitrator is authorized to resolve any disputes concerning the exchange of information.

(d) There shall be no other discovery, except as indicated herein or as ordered by the arbitrator in extraordinary cases when the demands of justice require it.

R-23. Date, Time, and Place of Hearing

The arbitrator shall set the date, time, and place for each hearing and/or conference. The parties shall respond to requests for hearing dates in a timely manner, be cooperative in scheduling the earliest practicable date, and adhere to the established hearing schedule. The AAA shall send a notice of hearing to the parties at least ten calendar days in advance of the hearing date, unless otherwise agreed by the parties.

R-24. Attendance at Hearings

The arbitrator and the AAA shall maintain the privacy of the hearings unless the law provides to the contrary. Any person having a direct interest in the arbitration is entitled to attend hearings. The arbitrator shall otherwise have the power to require the exclusion of any witness, other than a party or other essential person, during the testimony of any other witness. It shall be discretionary with the arbitrator to determine the propriety of the attendance of any person other than a party and its representative.

R-25. Representation

Any party may be represented by counsel or other authorized representative. A party intending to be so represented shall notify the other party and the AAA of the name and address of the representative at least three calendar days prior to the date set for the hearing at which that person is first to appear. When such a representative initiates an arbitration or responds for a party, notice is deemed to have been given.

R-26. Oaths

Before proceeding with the first hearing, each arbitrator may take an oath of office and, if required by law, shall do so. The arbitrator may require witnesses to testify under oath administered by any duly qualified person and, if it is required by law or requested by any party, shall do so.

R-27. Stenographic Record

Any party desiring a stenographic record shall make arrangements directly with a stenographer and shall notify the other parties of these arrangements at least three days in advance of the hearing. The requesting party or parties shall pay the cost of the record. If the transcript is agreed by the parties, or determined by the arbitrator to be the official record of the proceeding, it must be provided to the arbitrator and made available to the other parties for inspection, at a date, time, and place determined by the arbitrator.

R-28. Interpreters

Any party wishing an interpreter shall make all arrangements directly with the interpreter and shall assume the costs of the service.

R-29. Postponements

The arbitrator for good cause shown may postpone any hearing upon agreement of the parties, upon request of a party, or upon the arbitrator's own initiative.

R-30. Arbitration in the Absence of a Party or Representative

Unless the law provides to the contrary, the arbitration may proceed in the absence of any party or representative who, after due notice, fails to be present or fails to obtain a postponement. An award shall not be made solely on the default of a party. The arbitrator shall require the party who is present to submit such evidence as the arbitrator may require for the making of an award.

R-31. Conduct of Proceedings

(a) The claimant shall present evidence to support its claim. The respondent shall then present evidence supporting its defense. Witnesses for each party shall also submit to questions from the arbitrator and the adverse party. The arbitrator has the discretion to vary this procedure, provided that the parties are treated with equality and that each party has the right to be heard and is given a fair opportunity to present its case.

(b) The arbitrator, exercising his or her discretion, shall conduct the proceedings with a view to expediting the resolution of the dispute and may direct the order of proof, bifurcate proceedings, and direct the parties to focus their presentations on issues the decision of which could dispose of all or part of the case. The arbitrator shall entertain motions, including motions that dispose of all or part of a claim, or that may expedite the proceedings, and may also make preliminary rulings and enter interlocutory orders.

(c) The parties may agree to waive oral hearings in any case.

R-32. Evidence

(a) The parties may offer such evidence as is relevant and material to the dispute and shall produce such evidence as the arbitrator may deem necessary to an understanding and determination of the dispute. Conformity to legal rules of evidence shall not be necessary.

(b) The arbitrator shall determine the admissibility, relevance, and materiality of the evidence offered. The arbitrator may request offers of proof and may reject evidence deemed by the arbitrator to be cumulative, unreliable, unnecessary, or of slight value compared to the time and expense involved. All evidence shall be taken in the presence of all of the arbitrators and all of the parties, except where: 1) any of the parties is absent, in default, or has waived the right to be present, or 2) the parties and the arbitrators agree otherwise.

(c) The arbitrator shall take into account applicable principles of legal privilege, such as those involving the confidentiality of communications between a lawyer and client.

(d) An arbitrator or other person authorized by law to subpoena witnesses or documents may do so upon the request of any party or independently.

R-33. Evidence by Affidavit and Post-hearing Filing of Documents or Other Evidence

(a) The arbitrator may receive and consider the evidence of witnesses by declaration or affidavit, but shall give it only such weight as the arbitrator deems it entitled to after consideration of any objection made to its admission.

(b) If the parties agree or the arbitrator directs that documents or other evidence be submitted to the arbitrator after the hearing, the documents or other evidence, unless otherwise agreed by the parties and the arbitrator, shall be filed with the AAA for transmission to the arbitrator. All parties shall be afforded an opportunity to examine and respond to such documents or other evidence.

R-34. Inspection or Investigation

An arbitrator finding it necessary to make an inspection or investigation in connection with the arbitration shall direct the AAA to so advise the parties. The arbitrator shall set the date and time and the AAA shall notify the parties. Any party who so desires may be present at such an inspection or investigation. In the event that one or all parties are not present at the inspection or investigation, the arbitrator shall make an oral or written report to the parties and afford them an opportunity to comment.

R-35. Interim Measures

(a) The arbitrator may take whatever interim measures he or she deems necessary, including injunctive relief and measures for the protection or conservation of property and disposition of perishable goods.

(b) Such interim measures may be taken in the form of an interim award, and the arbitrator may require security for the costs of such measures.

(c) A request for interim measures addressed by a party to a judicial authority shall not be deemed incompatible with the agreement to arbitrate or a waiver of the right to arbitrate.

R-36. Closing of Hearing

When satisfied that the presentation of the parties is complete, the arbitrator shall declare the hearing closed.

If documents or responses are to be filed as provided in Section R-33, or if briefs are to be filed, the hearing shall be declared closed as of the final date set by the arbitrator for the receipt of documents, responses, or briefs. The time limit within which the arbitrator is required to make the award shall commence to run, in the absence of other agreements by the parties and the arbitrator, upon the closing of the hearing.

R-37. Reopening of Hearing

The hearing may be reopened on the arbitrator's initiative, or by direction of the arbitrator upon application of a party, at any time before the award is made. If reopening the hearing would prevent the making of the award within the specific time agreed to by the parties in the arbitration agreement, the matter may not be reopened unless the parties agree to an extension of time. When no specific date is fixed by agreement of the parties, the arbitrator shall have 30 calendar days from the closing of the reopened hearing within which to make an award.

R-38. Waiver of Rules

Any party who proceeds with the arbitration after knowledge that any provision or requirement of these rules has not been complied with and who fails to state an objection in writing shall be deemed to have waived the right to object.

R-39. Extensions of Time

The parties may modify any period of time by mutual agreement. The AAA or the arbitrator may for good cause extend any period of time established by these rules, except the time for making the award. The AAA shall notify the parties of any extension.

R-40. Serving of Notice

(a) Any papers, notices, or process necessary or proper for the initiation or continuation of an arbitration under these rules; for any court action in connection therewith, or for the entry of judgment on any award made under these rules, may be served on a party by mail addressed to the party or its representative at the last known address or by personal service, in or outside the state where the arbitration is to be held, provided that reasonable opportunity to be heard with regard thereto has been granted to the party.

(b) The AAA, the arbitrator and the parties may also use overnight delivery or electronic facsimile transmission (fax) to give the notices required by these rules. Where all parties and the arbitrator agree, notices may be transmitted by electronic mail (email), or other methods of communication.

(c) Unless otherwise instructed by the AAA or by the arbitrator, any documents submitted by any party to the AAA or to the arbitrator shall simultaneously be provided to the other party or parties to the arbitration.

R-41. Majority Decision

When the panel consists of more than one arbitrator, unless required by law or by the arbitration agreement, a majority of the arbitrators must make all decisions.

R-42. Time of Award

The award shall be made promptly by the arbitrator and, unless otherwise agreed by the parties or specified by law, no later than 30 calendar days from the date of closing the hearing, or, if oral hearings have been waived, from the date of the AAA's transmittal of the final statements and proofs to the arbitrator.

R-43. Form of Award

(a) Any award shall be in writing and signed by a majority of the arbitrators. It shall be executed in the manner required by law.

(b) The arbitrator shall provide a concise, written breakdown of the award. If requested in writing by all parties prior to the appointment of the arbitrator, or if the arbitrator believes it is appropriate to do so, the arbitrator shall provide a written explanation of the award.

R-44. Scope of Award

(a) The arbitrator may grant any remedy or relief that the arbitrator deems just and equitable and within the scope of the agreement of the parties, including, but not limited to, equitable relief and specific performance of a contract.

(b) In addition to the final award, the arbitrator may make other decisions, including interim, interlocutory, or partial rulings, orders, and awards. In any interim, interlocutory, or partial award, the arbitrator may assess and apportion the fees, expenses, and compensation related to such award as the arbitrator determines is appropriate.

(c) In the final award, the arbitrator shall assess fees, expenses, and compensation as provided in Sections R-50, R-51, and R-52. The arbitrator may apportion such fees, expenses, and compensation among the parties in such amounts as the arbitrator determines is appropriate.

(d) The award of the arbitrator may include interest at such rate and from such date as the arbitrator may deem appropriate; and an award of attorneys' fees if all parties have requested such an award or it is authorized by law or their arbitration agreement.

R-45. Award Upon Settlement

If the parties settle their dispute during the course of the arbitration and if the parties so request, the arbitrator may set forth the terms of the settlement in a "consent award." A consent award must include an allocation of arbitration costs, including administrative fees and expenses as well as arbitrator fees and expenses.

R-46. Delivery of Award to Parties

Parties shall accept as notice and delivery of the award the placing of the award or a true copy thereof in the mail addressed to the parties or their representatives at the last known address, personal or electronic service of the award, or the filing of the award in any other manner that is permitted by law.

R-47. Modification of Award

Within twenty calendar days after the transmittal of an award, the arbitrator on his or her initiative, or any party, upon notice to the other parties, may request that the arbitrator correct any clerical, typographical, technical or computational errors in the award. The arbitrator is not empowered to redetermine the merits of any claim already decided.

If the modification request is made by a party, the other parties shall be given ten calendar days to respond to the request. The arbitrator shall dispose of the request within twenty calendar days after transmittal by the AAA to the arbitrator of the request and any response thereto.

If applicable law provides a different procedural time frame, that procedure shall be followed.

R-48. Release of Documents for Judicial Proceedings

The AAA shall, upon the written request of a party, furnish to the party, at its expense, certified copies of any papers in the AAA's possession that may be required in judicial proceedings relating to the arbitration.

R-49. Applications to Court and Exclusion of Liability

(a) No judicial proceeding by a party relating to the subject matter of the arbitration shall be deemed a waiver of the party's right to arbitrate.

(b) Neither the AAA nor any arbitrator in a proceeding under these rules is a necessary or proper party in judicial proceedings relating to the arbitration.

(c) Parties to these rules shall be deemed to have consented that judgment upon the arbitration award may be entered in any federal or state court having jurisdiction thereof.

(d) Parties to an arbitration under these rules shall be deemed to have consented that neither the AAA nor any arbitrator shall be liable to any party in any action for damages or injunctive relief for any act or omission in connection with any arbitration under these rules.

R-50. Administrative Fees

As a not-for-profit organization, the AAA shall prescribe filing and other administrative fees and service charges to compensate it for the cost of providing administrative services. The fees in effect when the fee or charge is incurred shall be applicable.

The filing fee shall be advanced by the party or parties, subject to final apportionment by the arbitrator in the award.

The AAA may, in the event of extreme hardship on the part of any party, defer or reduce the administrative fees.

R-51. Expenses

The expenses of witnesses for either side shall be paid by the party producing such witnesses. All other expenses of the arbitration, including required travel and other expenses of the arbitrator, AAA representatives, and any witness and the cost of any proof produced at the direct request of the arbitrator, shall be borne equally by the parties, unless they agree otherwise or unless the arbitrator in the award assesses such expenses or any part thereof against any specified party or parties.

R-52. Neutral Arbitrator's Compensation

Arbitrators shall be compensated a rate consistent with the arbitrator's stated rate of compensation.

If there is disagreement concerning the terms of compensation, an appropriate rate shall be established with the arbitrator by the Association and confirmed to the parties.

Any arrangement for the compensation of a neutral arbitrator shall be made through the AAA and not directly between the parties and the arbitrator.

R-53. Deposits

The AAA may require the parties to deposit in advance of any hearings such sums of money as it deems necessary to cover the expense of the arbitration, including the arbitrator's fee, if any, and shall render an accounting to the parties and return any unexpended balance at the conclusion of the case.

R-54. Interpretation and Application of Rules

The arbitrator shall interpret and apply these rules insofar as they relate to the arbitrator's powers and duties. When there is more than one arbitrator and a difference arises among them concerning the meaning or application of these rules, it shall be decided by a majority vote. If that is not possible, either an arbitrator or a party may refer the question to the AAA for final decision. All other rules shall be interpreted and applied by the AAA.

R-55. Suspension for Nonpayment

If arbitrator compensation or administrative charges have not been paid in full, the AAA may so inform the parties in order that one of them may advance the required payment. If such payments are not made, the arbitrator may order the suspension or termination of the proceedings. If no arbitrator has yet been appointed, the AAA may suspend the proceedings.

FAST TRACK PROCEDURES

F-1. Limitation on Extensions

In the absence of extraordinary circumstances, the AAA or the arbitrator may grant a party no more than one seven-day extension of the time in which to respond to the demand for arbitration or counterclaim as provided in Section R-4.

F-2. Changes of Claim or Counterclaim

A party may at any time prior to the close of the hearing increase or decrease the amount of its claim or counterclaim. Any new or different claim or counterclaim, as opposed to an increase or decrease in the amount of a pending claim or counterclaim, shall be made in writing and filed with the AAA, and a copy shall be mailed to the other party, who shall have a period of five calendar days from the date of such mailing within which to file an answer with the AAA. After the arbitrator is appointed no new or different claim or counterclaim may be submitted to that arbitrator except with the arbitrator's consent.

If an increased claim or counterclaim exceeds $75,000, the case will be administered under the Regular procedures unless: 1) the party with the claim or counterclaim exceeding $75,000 agrees to waive any award exceeding that amount; or 2) all parties and the arbitrator agree that the case may continue to be processed under the Fast Track Procedures.

F-3. Serving of Notice

In addition to notice provided by Section R-40, the parties shall also accept notice by telephone. Telephonic notices by the AAA shall subsequently be confirmed in writing to the parties. Should there be a failure to confirm in writing any such oral notice, the proceeding shall nevertheless be valid if notice has, in fact, been given by telephone.

F-4. Appointment and Qualification of Arbitrator

Immediately after the filing of (a) the submission or (b) the answering statement or the expiration of the time within which the answering statement is to be filed, the AAA will simultaneously submit to each party a listing and biographical information from its panel of arbitrators knowledgeable in construction who are available for service in Fast Track cases. The parties are encouraged to agree to an arbitrator from this list, and to advise the Association of their agreement, or any factual objections to any of the listed arbitrators, within seven calendar days of the AAA's transmission of the list. The AAA will appoint the agreed-upon arbitrator, or in the event the parties cannot agree on an arbitrator, will designate the arbitrator from among those names not stricken for factual objections. Absent agreement of the parties, the arbitrator shall not have served as the mediator in the mediation phase of the instant proceeding.

The parties will be given notice by the AAA of the appointment of the arbitrator, who shall be subject to disqualification for the reasons specified in Section R-18. Within the time period established by the AAA, the parties shall notify the AAA of any objection to the arbitrator appointed. Any objection by a party to the arbitrator shall be for cause and shall be confirmed in writing to the AAA with a copy to the other party or parties.

F-5. Preliminary Telephone Conference

Unless otherwise agreed by the parties and the arbitrator, as promptly as practicable after the appointment of the arbitrator, a preliminary telephone conference shall be held among the parties or their attorneys or representatives, and the arbitrator.

F-6. Exchange of Exhibits

At least two business days prior to the hearing, the parties shall exchange copies of all exhibits they intend to submit at the hearing. The arbitrator is authorized to resolve any disputes concerning the exchange of exhibits.

F-7. Discovery

There shall be no discovery, except as provided in Section F-6 or as ordered by the arbitrator in extraordinary cases when the demands of justice require it.

F-8. Proceedings on Documents

Where no party's claim exceeds $10,000, exclusive of interest and arbitration costs, and other cases in which the parties agree, the dispute shall be resolved by submission of documents, unless any party requests an oral hearing or conference call, or the arbitrator determines that an oral hearing or conference call is necessary. The arbitrator shall establish a fair and equitable procedure for the submission of documents.

F-9. Date, Time, and Place of Hearing

In cases in which a hearing is to be held, the arbitrator shall set the date, time, and place of the hearing, to be scheduled to take place within 30 calendar days of confirmation of the arbitrator's appointment. The AAA will notify the parties in advance of the hearing date.

F-10. The Hearing

(a) Generally, the hearing shall not exceed one day. Each party shall have equal opportunity to submit its proofs and complete its case. The arbitrator shall determine the order of the hearing, and may require further submission of documents within two business days after the hearing. For good cause shown, the arbitrator may schedule one additional hearing day within seven business days after the initial day of hearing.

(b) Generally, there will be no stenographic record. Any party desiring a stenographic record may arrange for one pursuant to the provisions of Section R-27.

F-11. Time of Award

Unless otherwise agreed by the parties, the award shall be rendered not later than fourteen calendar days from the date of the closing of the hearing or, if oral hearings have been waived, from the date of the AAA's transmittal of the final statements and proofs to the arbitrator.

F-12. Time Standards

The arbitration shall be completed by settlement or award within 60 calendar days of confirmation of the arbitrator's appointment, unless all parties and the arbitrator agree otherwise or the arbitrator extends this time in extraordinary cases when the demands of justice require it. The Association will relax these time standards in the event the arbitration is stayed pending mediation.

F-13. Arbitrator's Compensation

Arbitrators will receive compensation at a rate to be suggested by the AAA regional office.

PROCEDURES FOR LARGE, COMPLEX CONSTRUCTION DISPUTES

L-1. Administrative Conference

Prior to the dissemination of a list of potential arbitrators, the AAA shall, unless the parties agree otherwise, conduct an administrative conference with the parties and/or their attorneys or other representatives by conference call. The conference call will take place within 14 days after the commencement of the arbitration. In the event the parties are unable to agree on a mutually acceptable time for the conference, the AAA may contact the parties individually to discuss the issues contemplated herein. Such administrative conference shall be conducted for the following purposes and for such additional purposed as the parties or the AAA may deem appropriate:

  (a) to obtain additional information about the nature and magnitude of the dispute and the anticipated length of hearing and scheduling;

  (b) to discuss the views of the parties about the technical and other qualifications of the arbitrators;

  (c) to obtain conflicts statements from the parties; and

  (d) to consider, with the parties, whether mediation or other non-adjudicative methods of dispute resolution might be appropriate.

L-2. Arbitrators

(a) Large, Complex Construction Cases shall be heard and determined by either one or three arbitrators, as may be agreed upon by the parties. If the parties are unable to agree upon the number of arbitrators and a claim or counterclaim involved at least $1,000,000, then three arbitrator(s) shall hear and determine the case. If the parties are unable to agree on the number of arbitrators and each claim and counterclaim is less than $1,000,000, then one arbitrator shall hear and determine the case.

(b) The AAA shall appoint arbitrator(s) as agreed by the parties. If they are unable to agree on a method of appointment, the AAA shall appoint arbitrator from the Large, Complex Construction Case Panel, in the manner provided in the Regular Construction Industry Arbitration Rules. Absent agreement of the parties, the arbitrator (s) shall not have served as the mediator in the mediation phase of the instant proceeding.

L-3. Preliminary Hearing

As promptly as practicable after the selection of the arbitrator(s), a preliminary hearing shall be held among the parties and/or their attorneys or other representatives and the arbitrator(s). Unless the parties agree otherwise, the preliminary hearing will be conducted by telephone conference call rather than in person.

At the preliminary hearing the matters to be considered shall include, without limitation:

(a) service of a detailed statement of claims, damages and defenses, a statement of the issues asserted by each party and positions with respect thereto, and any legal authorities the parties may wish to bring to the attention of the arbitrator(s);

(b) stipulations to uncontested facts;

(c) the extent to which discovery shall be conducted;

(d) exchange and premarking of those documents which each party believes may be offered at the hearing;

(e) the identification and availability of witnesses, including experts, and such matters with respect to witnesses including their biographies and expected testimony as may be appropriate;

(f) whether, and the extent to which, any sworn statements and/or depositions may be introduced;

(g) the extent to which hearings will proceed on consecutive days;

(h) whether a stenographic or other official record of the proceedings shall be maintained;

(i) the possibility of utilizing mediation or other non-adjudicative methods of dispute resolution; and

(j) the procedure for the issuance of subpoenas.

By agreement of the parties and/or order of the arbitrator(s), the pre-hearing activities and the hearing procedures that will govern the arbitration will be memorialized in a Scheduling and Procedure Order.

L-4. Management of Proceedings

(a) Arbitrator(s) shall take such steps as they may deem necessary or desirable to avoid delay and to achieve a just, speedy and cost-effective resolution of Large, Complex Construction Cases.

(b) Parties shall cooperate in the exchange of documents, exhibits and information within such party's control if the arbitrator(s) consider such production to be consistent with the goal of achieving a just, speedy and cost effective resolution of a Large, Complex Construction Case.

(c) The parties may conduct such discovery as may be agreed to by all the parties provided, however, that the arbitrator(s) may place such limitations on the conduct of such discovery as the arbitrator(s) shall deem appropriate. If the parties cannot agree on production of document and other information, the arbitrator(s), consistent with the expedited nature of arbitration, may establish the extent of the discovery.

(d) At the discretion of the arbitrator(s), upon good cause shown and consistent with the expedited nature of arbitration, the arbitrator(s) may order depositions of, or the propounding of interrogatories to such persons who may possess information determined by the arbitrator(s) to be necessary to a determination of the matter.

(e) The parties shall exchange copies of all exhibits they intend to submit at the hearing 10 business days prior to the hearing unless the arbitrator(s) determine otherwise.

(f) The exchange of information pursuant to this rule, as agreed by the parties and/or directed by the arbitrator(s), shall be included within the Scheduling and Procedure Order.

(g) The arbitrator is authorized to resolve any disputes concerning the exchange of information.

(h) Generally hearings will be scheduled on consecutive days or in blocks of consecutive days in order to maximize efficiency and minimize costs.

ADMINISTRATIVE FEES

The administrative fees of the AAA are based on the amount of the claim or counterclaim. Arbitrator compensation is not included in this schedule. Unless the parties agree otherwise, arbitrator compensation and administrative fees are subject to allocation by the arbitrator in the award.

Fees

An initial filing fee is payable in full by a filing party when a claim, counterclaim or additional claim is filed.

A case service fee will be incurred for all cases that proceed to their first hearing. This fee will be payable in advance at the time that the first hearing is scheduled. This fee will be refunded at the conclusion of the case if no hearings have occurred.

However, if the Association is not notified at least 24 hours before the time of the scheduled hearing, the case service fee will remain due and will not be refunded.

These fees will be billed in accordance with the following schedule:

Amount of Claim	Initial Filing Fee	Case Service Fee
Above $0 to $10,000	$750	$200
Above $10,000 to $75,000	$950	$300
Above $75,000 to $150,000	$1,800	$750
Above $150,000 to $300,000	$2,750	$1,250
Above $300,000 to $500,000	$4,250	$1,750
Above $500,000 to $1,000,000	$6,000	$2,500
Above $1,000,000 to $5,000,000	$8,000	$3,250
Above $5,000,000 to $10,000,000	$10,000	$4,000
Above $10,000,000	*	*
Nonmonetary Claims**	$3,250	$1,250

Fee Schedule for Claims in Excess of $10 Million .

The following is the fee schedule for use in disputes involving claims in excess of $10 million. If you have any questions, please consult your local AAA office or case management center.

Claim Size	Fee	Case Service Fee
$10 million and above	Base fee of $ 12,500 plus .01% of the amount of claim above $ 10 million.	$6,000
	Filing fees capped at $65,000	

**This fee is applicable when a claim or counterclaim is not for a monetary amount. Where a monetary claim is not known, parties will be required to state a range of claims or be subject to the highest possible filing fee.

Fees are subject to increase if the amount of a claim or counterclaim is modified after the initial filing date. Fees are subject to decrease if the amount of a claim or counterclaim is modified before the first hearing.

The minimum fees for any case having three or more arbitrators are $2,750 for the filing fee, plus a $1,250 case service fee.

Fast Track Procedures are applied in any case where no disclosed claim or counterclaim exceeds $75,000, exclusive of interest and arbitration costs.

Parties on cases held in abeyance for one year by agreement, will be assessed an annual abeyance fee of $300. If a party refuses to pay the assessed fee, the other party or parties may pay the entire fee on behalf of all parties, otherwise the matter will be closed.

Refund Schedule

The AAA offers a refund schedule on filing fees. For cases with claims up to $75,000 a minimum filing fee of $300 will not be refunded. For all other cases, a minimum fee of $500.00 will not be refunded. Subject to the minimum fee requirements, refunds will be calculated as follows:

' 100% of the filing fee, above the minimum fee, will be refunded if the case is settled or withdrawn with five calendar days of filing.

' 50% of the filing fee will be refunded if the case is settled or withdrawn between six and 30 calendar days of filing.

' 25% of the filing fee will be refunded if the case is settled or withdrawn between 31 and 60 calendar days of filing.

No refund will be made once an arbitrator has been appointed (this includes one arbitrator on a three arbitrator panel). No refunds will be granted on awarded cases.

Note: The date of receipt of the demand for arbitration with the AAA will be used to calculate refunds of filing fees for both claims and counterclaims.

Hearing Room Rental

The fees described above do not cover the rental of hearing rooms, which are available on a rental basis. Check with the AAA for availability and rates.

AAA047-10M 7/03

- AAA MISSION & PRINCIPLES
- PRIVACY POLICY
- TERMS OF USE
- TECHNICAL RECOMMENDATIONS
- ©2007 AMERICAN ARBITRATION ASSOCIATION. ALL RIGHTS RESERVED

Standard Form of Agreement Between Owner and Engineer for Professional Services

Engineers Joint Contract Documents Committee (EJCDC)

Reprinted by permission of the National Society of Professional Engineers. For information on ordering these and other EJCDC documents, call 1-800-417-0348 or visit www.nspe.org.

Exhibits C (Compensation Formulas for Engineer and Resident Project Representative), D (Providing Resident Project Engineer Services), E (Notice of Acceptability of Work), G (Insurance), J (Special Provisions), and K (Amendments) Omitted

This document has important legal consequences; consultation with an attorney is encouraged with respect to its use or modification. This document should be adapted to the particular circumstances of the contemplated Project and the Controlling Law.

STANDARD FORM OF AGREEMENT BETWEEN OWNER AND ENGINEER FOR PROFESSIONAL SERVICES

Prepared by

ENGINEERS JOINT CONTRACT DOCUMENTS COMMITTEE

and

Issued and Published Jointly By

PROFESSIONAL ENGINEERS IN PRIVATE PRACTICE
a practice division of the
NATIONAL SOCIETY OF PROFESSIONAL ENGINEERS

AMERICAN COUNCIL OF ENGINEERING COMPANIES

AMERICAN SOCIETY OF CIVIL ENGINEERS

This Agreement has been prepared for use with the Standard General Conditions of the Construction Contract (No. C-700, 2002 Edition) of the Engineers Joint Contract Documents Committee. Their provisions are interrelated, and a change in one may necessitate a change in the other. For guidance on the completion and use of this Agreement, see EJCDC User's Guide to the Owner-Engineer Agreement, No. E-001, 2002 Edition.

National Society of Professional Engineers
1420 King Street, Alexandria, VA 22314

American Council of Engineering Companies
1015 15th Street N.W., Washington, DC 20005

American Society of Civil Engineers
1801 Alexander Bell Drive, Reston, VA 20191

TABLE OF CONTENTS

STANDARD FORM OF AGREEMENT
BETWEEN OWNER AND ENGINEER
FOR
PROFESSIONAL SERVICES

THIS IS AN AGREEMENT effective as of _____ , _____ ("Effective Date") between

_____ ("Owner") and

_____ ("Engineer").

Owner intends to _____

_____

_____

_____ ("Project").

Owner and Engineer agree as follows:

ARTICLE 1 - SERVICES OF ENGINEER

1.01   Scope

A.   Engineer shall provide, or cause to be provided, the services set forth herein and in Exhibit A.

ARTICLE 2 - OWNER'S RESPONSIBILITIES

2.01   General

A.   Owner shall have the responsibilities set forth herein and in Exhibit B.

B.   Owner shall pay Engineer as set forth in Exhibit C.

C.   Owner shall be responsible for, and Engineer may rely upon, the accuracy and completeness of all requirements, programs, instructions, reports, data, and other information furnished by Owner to Engineer pursuant to this Agreement. Engineer may use such requirements, programs, instructions, reports, data, and information in performing or furnishing services under this Agreement.

ARTICLE 3 - SCHEDULE FOR RENDERING SERVICES

3.01   Commencement

A.   Engineer shall begin rendering services as of the Effective Date of the Agreement.

3.02   Time for Completion

A.   Engineer shall complete its obligations within a reasonable time. Specific periods of time for rendering services are set forth or specific dates by which services are to be completed are provided in Exhibit A, and are hereby agreed to be reasonable

B.   If, through no fault of Engineer, such periods of time or dates are changed, or the orderly and continuous progress of Engineer's services is impaired, or Engineer's services are delayed or suspended, then the time for completion of Engineer's services, and the rates and amounts of Engineer's compensation, shall be adjusted equitably.

C.   If Owner authorizes changes in the scope, extent, or character of the Project, then the time for completion of

Engineer's services, and the rates and amounts of Engineer's compensation, shall be adjusted equitably.

D.   Owner shall make decisions and carry out its other responsibilities in a timely manner so as not to delay the Engineer's performance of its services.

E.   If Engineer fails, through its own fault, to complete the performance required in this Agreement within the time set forth, as duly adjusted, then Owner shall be entitled to the recovery of direct damages resulting from such failure.

ARTICLE 4 - INVOICES AND PAYMENTS

4.01   Invoices

A.   *Preparation and Submittal of Invoices.* Engineer shall prepare invoices in accordance with its standard invoicing practices and the terms of Exhibit C. Engineer shall submit its invoices to Owner on a monthly basis. Invoices are due and payable within 30 days of receipt.

4.02   Payments

A.   *Application to Interest and Principal.* Payment will be credited first to any interest owed to Engineer and then to principal.

B.   *Failure to Pay.* If Owner fails to make any payment due Engineer for services and expenses within 30 days after receipt of Engineer's invoice, then:

1.   amounts due Engineer will be increased at the rate of 1.0% per month (or the maximum rate of interest permitted by law, if less) from said thirtieth day; and

2.   Engineer may, after giving seven days written notice to Owner, suspend services under this Agreement until Owner has paid in full all amounts due for services, expenses, and other related charges. Owner waives any and all claims against Engineer for any such suspension.

C.   *Disputed Invoices.* If Owner contests an invoice, Owner may withhold only that portion so contested, and must pay the undisputed portion.

D.   *Legislative Actions.* If after the Effective Date of the Agreement any governmental entity takes a legislative action that imposes taxes, fees, or charges on Engineer's services or compensation under this Agreement, then the Engineer may invoice such new taxes, fees, or charges as a Reimbursable Expense to which a factor of 1.0 shall be applied. Owner shall pay such invoiced new taxes, fees, and charges; such payment shall be in addition to the compensation to which Engineer is entitled under the terms of Exhibit C.

ARTICLE 5 - OPINIONS OF COST

5.01   Opinions of Probable Construction Cost

A.   Engineer's opinions of probable Construction Cost are to be made on the basis of Engineer's experience and qualifications and represent Engineer's best judgment as an experienced and qualified professional generally familiar with the construction industry. However, since Engineer has no control over the cost of labor, materials, equipment, or services furnished by others, or over contractors' methods of determining prices, or over competitive bidding or market conditions, Engineer cannot and does not guarantee that proposals, bids, or actual Construction Cost will not vary from opinions of probable Construction Cost prepared by Engineer. If Owner wishes greater assurance as to probable Construction Cost, Owner shall employ an independent cost estimator as provided in Exhibit B.

5.02   Designing to Construction Cost Limit

A.   If a Construction Cost limit is established between Owner and Engineer, such Construction Cost limit and a statement of Engineer's rights and responsibilities with respect thereto will be specifically set forth in Exhibit F, "Construction Cost Limit," to this Agreement.

5.03   Opinions of Total Project Costs

A.   The services, if any, of Engineer with respect to Total Project Costs shall be limited to assisting the Owner in collating the various cost categories which comprise Total Project Costs. Engineer assumes no responsibility for the accuracy of any opinions of Total Project Costs.

EJCDC E-500 Standard Form of Agreement Between Owner and Engineer for Professional Services

ARTICLE 6 - GENERAL CONSIDERATIONS

6.01 Standards of Performance

A.   The standard of care for all professional engineering and related services performed or furnished by Engineer under this Agreement will be the care and skill ordinarily used by members of the subject profession practicing under similar circumstances at the same time and in the same locality. Engineer makes no warranties, express or implied, under this Agreement or otherwise, in connection with Engineer's services.

B.   Owner shall not be responsible for discovering deficiencies in the technical accuracy of Engineer's services. Engineer shall correct any such deficiencies in technical accuracy without additional compensation except to the extent such corrective action is directly attributable to deficiencies in Owner-furnished information.

C.   Engineer may employ such Consultants as Engineer deems necessary to assist in the performance or furnishing of the services, subject to reasonable, timely, and substantive objections by Owner.

D.   Subject to the standard of care set forth in paragraph 6.01.A, Engineer and its Consultants may use or rely upon design elements and information ordinarily or customarily furnished by others, including, but not limited to, specialty contractors, manufacturers, suppliers, and the publishers of technical standards.

E.   Engineer and Owner shall comply with applicable Laws and Regulations and Owner-mandated standards that Owner has provided to Engineer in writing. This Agreement is based on these requirements as of its Effective Date. Changes to these requirements after the Effective Date of this Agreement may be the basis for modifications to Owner's responsibilities or to Engineer's scope of services, times of performance, and compensation.

G.   Engineer shall not be required to sign any documents, no matter by whom requested, that would result in the Engineer having to certify, guarantee, or warrant the existence of conditions whose existence the Engineer cannot ascertain. Owner agrees not to make resolution of any dispute with the Engineer or payment of any amount due to the Engineer in any way contingent upon the Engineer signing any such documents.

H.   The General Conditions for any construction contract documents prepared hereunder are to be the "Standard General Conditions of the Construction Contract" as prepared by the Engineers Joint Contract Documents Committee (No. C-700, 2002 Edition) unless both parties mutually agree to use other General Conditions by specific reference in Exhibit J.

I.   Engineer shall not at any time supervise, direct, or have control over Contractor's work, nor shall Engineer have authority over or responsibility for the means, methods, techniques, sequences, or procedures of construction selected or used by Contractor, for security or safety at the Site, for safety precautions and programs incident to the Contractor's work in progress, nor for any failure of Contractor to comply with Laws and Regulations applicable to Contractor's furnishing and performing the Work.

J.   Engineer neither guarantees the performance of any contractor nor assumes responsibility for any contractor's failure to furnish and perform the Work in accordance with the Contract Documents.

K.   Engineer shall not be responsible for the acts or omissions of any contractor, subcontractor, or supplier, or of any of their agents or employees or of any other persons (except Engineer's own employees and its Consultants) at the Site or otherwise furnishing or performing any Work; or for any decision made on interpretations or clarifications of the Contract Documents given by Owner without consultation and advice of Engineer.

6.02 Design without Construction Phase Services

A.   If Engineer's Basic Services under this Agreement do not include Project observation, or review of the Contractor's performance, or any other Construction Phase services, then (1) Engineer's services under this Agreement shall be deemed complete no later than the end of the Bidding or Negotiating Phase; (2) Engineer shall have no design or shop drawing review obligations during construction; (3) Owner assumes all responsibility for the application and interpretation of the Contract Documents, contract administration, construction observation and review, and all other necessary Construction Phase engineering and professional services; and (4) Owner waives any claims against the Engineer that may be connected in any way thereto.

EJCDC E-500 Standard Form of Agreement Between Owner and Engineer for Professional Services

6.03 Use of Documents

A. All Documents are instruments of service in respect to this Project, and Engineer shall retain an ownership and property interest therein (including the copyright and the right of reuse at the discretion of the Engineer) whether or not the Project is completed. Owner shall not rely in any way on any Document unless it is in printed form, signed or sealed by the Engineer or one of its Consultants.

B. A party may rely that data or information set forth on paper (also known as hard copies) that the party receives from the other party by mail, hand delivery, or facsimile, are the items that the other party intended to send. Files in electronic media format of text, data, graphics, or other types that are furnished by one party to the other are furnished only for convenience, not reliance by the receiving party. Any conclusion or information obtained or derived from such electronic files will be at the user's sole risk. If there is a discrepancy between the electronic files and the hard copies, the hard copies govern.

C. Because data stored in electronic media format can deteriorate or be modified inadvertently or otherwise without authorization of the data's creator, the party receiving electronic files agrees that it will perform acceptance tests or procedures within 60 days, after which the receiving party shall be deemed to have accepted the data thus transferred. Any transmittal errors detected within the 60-day acceptance period will be corrected by the party delivering the electronic files.

D. When transferring documents in electronic media format, the transferring party makes no representations as to long term compatibility, usability, or readability of such documents resulting from the use of software application packages, operating systems, or computer hardware differing from those used by the documents' creator.

E. Owner may make and retain copies of Documents for information and reference in connection with use on the Project by Owner. Engineer grants Owner a license to use the Documents on the Project, extensions of the Project, and other projects of Owner, subject to the following limitations: (1) Owner acknowledges that such Documents are not intended or represented to be suitable for use on the Project unless completed by Engineer, or for use or reuse by Owner or others on extensions of the Project or on any other project without written verification or adaptation by Engineer; (2) any such use or reuse, or any modification of the Documents, without written verification, completion, or adaptation by Engineer, as appropriate for the specific purpose intended, will be at Owner's sole risk and without liability or legal exposure to Engineer or to Engineer's Consultants; (3) Owner shall indemnify and hold harmless Engineer and Engineer's Consultants from all claims, damages, losses, and expenses, including attorneys' fees, arising out of or resulting from any use, reuse, or modification without written verification, completion, or adaptation by Engineer; (4) such limited license to Owner shall not create any rights in third parties.

F. If Engineer at Owner's request verifies or adapts the Documents for extensions of the Project or for any other project, then Owner shall compensate Engineer at rates or in an amount to be agreed upon by Owner and Engineer.

6.04 Insurance

A. Engineer shall procure and maintain insurance as set forth in Exhibit G, "Insurance." Engineer shall cause Owner to be listed as an additional insured on any applicable general liability insurance policy carried by Engineer.

B. Owner shall procure and maintain insurance as set forth in Exhibit G, "Insurance." Owner shall cause Engineer and Engineer's Consultants to be listed as additional insureds on any general liability or property insurance policies carried by Owner which are applicable to the Project.

C. Owner shall require Contractor to purchase and maintain general liability and other insurance in accordance with the requirements of paragraph 5.04 of the "Standard General Conditions of the Construction Contract," (No. C-700, 2002 Edition) as prepared by the Engineers Joint Contract Documents Committee and to cause Engineer and Engineer's Consultants to be listed as additional insureds with respect to such liability and other insurance purchased and maintained by Contractor for the Project.

D. Owner and Engineer shall each deliver to the other certificates of insurance evidencing the coverages indicated in Exhibit G. Such certificates shall be furnished prior to commencement of Engineer's services and at renewals thereafter during the life of the Agreement.

E. All policies of property insurance relating to the Project shall contain provisions to the effect that Engineer's and Engineer's Consultants' interests are covered and that in the event of payment of any loss or damage the insurers will have no rights of recovery against Engineer or its Consultants, or any insureds or additional insureds thereunder.

EJCDC E-500 Standard Form of Agreement Between Owner and Engineer for Professional Services

F.   At any time, Owner may request that Engineer or its Consultants, at Owner's sole expense, provide additional insurance coverage, increased limits, or revised deductibles that are more protective than those specified in Exhibit G. If so requested by Owner, and if commercially available, Engineer shall obtain and shall require its Consultants to obtain such additional insurance coverage, different limits, or revised deductibles for such periods of time as requested by Owner, and Exhibit G will be supplemented to incorporate these requirements.

6.05   Suspension and Termination

A.   *Suspension.*

By Owner: Owner may suspend the Project upon seven days written notice to Engineer.

By Engineer: If Engineer's services are substantially delayed through no fault of Engineer, Engineer may, after giving seven days written notice to Owner, suspend services under this Agreement.

B.   *Termination.* The obligation to provide further services under this Agreement may be terminated:

1.   For cause,

a.   By either party upon 30 days written notice in the event of substantial failure by the other party to perform in accordance with the terms hereof through no fault of the terminating party.

b.   By Engineer:

1)   upon seven days written notice if Owner demands that Engineer furnish or perform services contrary to Engineer's responsibilities as a licensed professional; or

2)   upon seven days written notice if the Engineer's services for the Project are delayed or suspended for more than 90 days for reasons beyond Engineer's control.

3)   Engineer shall have no liability to Owner on account of such termination.

c.   Notwithstanding the foregoing, this Agreement will not terminate under paragraph

6.05.B.1.a if the party receiving such notice begins, within seven days of receipt of such notice, to correct its substantial failure to perform and proceeds diligently to cure such failure within no more than 30 days of receipt thereof; provided, however, that if and to the extent such substantial failure cannot be reasonably cured within such 30 day period, and if such party has diligently attempted to cure the same and thereafter continues diligently to cure the same, then the cure period provided for herein shall extend up to, but in no case more than, 60 days after the date of receipt of the notice.

2.   For convenience,

a.   By Owner effective upon Engineer's receipt of notice from Owner.

C.   *Effective Date of Termination.* The terminating party under paragraph 6.05.B may set the effective date of termination at a time up to 30 days later than otherwise provided to allow Engineer to demobilize personnel and equipment from the Site, to complete tasks whose value would otherwise be lost, to prepare notes as to the status of completed and uncompleted tasks, and to assemble Project materials in orderly files.

D.   Payments Upon Termination.

1.   In the event of any termination under paragraph 6.05, Engineer will be entitled to invoice Owner and to receive full payment for all services performed or furnished and all Reimbursable Expenses incurred through the effective date of termination. Upon making such payment, Owner shall have the limited right to the use of Documents, at Owner's sole risk, subject to the provisions of paragraph 6.03.E.

2.   In the event of termination by Owner for convenience or by Engineer for cause, Engineer shall be entitled, in addition to invoicing for those items identified in paragraph 6.05.D.1, to invoice Owner and to payment of a reasonable amount for services and expenses directly attributable to termination, both before and after the effective date of termination, such as reassignment of personnel, costs of terminating contracts with Engineer's Consultants, and other related close-out costs, using methods and rates for Additional Services as set forth in Exhibit C.

EJCDC E-500 Standard Form of Agreement Between Owner and Engineer for Professional Services

6.06   Controlling Law

A.   This Agreement is to be governed by the law of the state in which the Project is located.

6.07   Successors, Assigns, and Beneficiaries

A.   Owner and Engineer each is hereby bound and the partners, successors, executors, administrators and legal representatives of Owner and Engineer (and to the extent permitted by paragraph 6.07.B the assigns of Owner and Engineer) are hereby bound to the other party to this Agreement and to the partners, successors, executors, administrators and legal representatives (and said assigns) of such other party, in respect of all covenants, agreements, and obligations of this Agreement.

B.   Neither Owner nor Engineer may assign, sublet, or transfer any rights under or interest (including, but without limitation, moneys that are due or may become due) in this Agreement without the written consent of the other, except to the extent that any assignment, subletting, or transfer is mandated or restricted by law. Unless specifically stated to the contrary in any written consent to an assignment, no assignment will release or discharge the assignor from any duty or responsibility under this Agreement.

C.   Unless expressly provided otherwise in this Agreement:

1.   Nothing in this Agreement shall be construed to create, impose, or give rise to any duty owed by Owner or Engineer to any Contractor, Contractor's subcontractor, supplier, other individual or entity, or to any surety for or employee of any of them.

2.   All duties and responsibilities undertaken pursuant to this Agreement will be for the sole and exclusive benefit of Owner and Engineer and not for the benefit of any other party.

3.   Owner agrees that the substance of the provisions of this paragraph 6.07.C shall appear in the Contract Documents.

6.08   Dispute Resolution

A.   Owner and Engineer agree to negotiate all disputes between them in good faith for a period of 30 days from the date of notice prior to invoking the procedures of Exhibit H or other provisions of this Agreement, or exercising their rights under law.

B.   If the parties fail to resolve a dispute through negotiation under paragraph 6.08.A, then either or both may invoke the procedures of Exhibit H. If Exhibit H is not included, or if no dispute resolution method is specified in Exhibit H, then the parties may exercise their rights under law.

6.09   Environmental Condition of Site

A.   Owner has disclosed to Engineer in writing the existence of all known and suspected Asbestos, PCBs, Petroleum, Hazardous Waste, Radioactive Material, hazardous substances, and other Constituents of Concern located at or near the Site, including type, quantity, and location.

B.   Owner represents to Engineer that to the best of its knowledge no Constituents of Concern, other than those disclosed in writing to Engineer, exist at the Site.

C.   If Engineer encounters an undisclosed Constituent of Concern, then Engineer shall notify (1) Owner and (2) appropriate governmental officials if Engineer reasonably concludes that doing so is required by applicable Laws or Regulations.

D.   It is acknowledged by both parties that Engineer's scope of services does not include any services related to Constituents of Concern. If Engineer or any other party encounters an undisclosed Constituent of Concern, or if investigative or remedial action, or other professional services, are necessary with respect to disclosed or undisclosed Constituents of Concern, then Engineer may, at its option and without liability for consequential or any other damages, suspend performance of services on the portion of the Project affected thereby until Owner: (1) retains appropriate specialist consultant(s) or contractor(s) to identify and, as appropriate, abate, remediate, or remove the Constituents of Concern; and (2) warrants that the Site is in full compliance with applicable Laws and Regulations.

E.   If the presence at the Site of undisclosed Constituents of Concern adversely affects the performance of Engineer's services under this Agreement, then the Engineer shall have the option of (1) accepting an equitable adjustment in its compensation or in the time of completion, or both; or (2) terminating this Agreement for cause on 30 days notice.

F.   Owner acknowledges that Engineer is performing professional services for Owner and that Engineer is not and shall not be required to become an "arranger," "operator," "generator," or "transporter" of hazardous substances, as defined in the Comprehensive Environmental Response, Compensation, and Liability Act (CERCLA), as amended, which are or may be encountered at or near the Site in connection with Engineer's activities under this Agreement.

6.10 Indemnification and Mutual Waiver

A.   *Indemnification by Engineer.* To the fullest extent permitted by law, Engineer shall indemnify and hold harmless Owner, and Owner's officers, directors, partners, agents, consultants, and employees from and against any and all claims, costs, losses, and damages (including but not limited to all fees and charges of engineers, architects, attorneys, and other professionals, and all court, arbitration, or other dispute resolution costs) arising out of or relating to the Project, provided that any such claim, cost, loss, or damage is attributable to bodily injury, sickness, disease, or death, or to injury to or destruction of tangible property (other than the Work itself), including the loss of use resulting therefrom, but only to the extent caused by any negligent act or omission of Engineer or Engineer's officers, directors, partners, employees, or Consultants. The indemnification provision of the preceding sentence is subject to and limited by the provisions agreed to by Owner and Engineer in Exhibit I, "Allocation of Risks," if any.

B.   *Indemnification by Owner.* To the fullest extent permitted by law, Owner shall indemnify and hold harmless Engineer, Engineer's officers, directors, partners, agents, employees, and Consultants from and against any and all claims, costs, losses, and damages (including but not limited to all fees and charges of engineers, architects, attorneys, and other professionals, and all court, arbitration, or other dispute resolution costs) arising out of or relating to the Project, provided that any such claim, cost, loss, or damage is attributable to bodily injury, sickness, disease, or death or to injury to or destruction of tangible property (other than the Work itself), including the loss of use resulting therefrom, but only to the extent caused by any negligent act or omission of Owner or Owner's officers, directors, partners, agents, consultants, or employees, or others retained by or under contract to the Owner with respect to this Agreement or to the Project.

C.   *Environmental Indemnification.* In addition to the indemnity provided under paragraph 6.10.B of this Agreement, and to the fullest extent permitted by law, Owner shall indemnify and hold harmless Engineer and its officers, directors, partners, agents, employees, and Consultants from and against any

and all claims, costs, losses, and damages (including but not limited to all fees and charges of engineers, architects, attorneys and other professionals, and all court, arbitration, or other dispute resolution costs) caused by, arising out of, relating to, or resulting from a Constituent of Concern at, on, or under the Site, provided that (i) any such claim, cost, loss, or damage is attributable to bodily injury, sickness, disease, or death, or to injury to or destruction of tangible property (other than the Work itself), including the loss of use resulting therefrom, and (ii) nothing in this paragraph shall obligate Owner to indemnify any individual or entity from and against the consequences of that individual's or entity's own negligence or willful misconduct.

D.   *Percentage Share of Negligence.* To the fullest extent permitted by law, a party's total liability to the other party and anyone claiming by, through, or under the other party for any cost, loss, or damages caused in part by the negligence of the party and in part by the negligence of the other party or any other negligent entity or individual, shall not exceed the percentage share that the party's negligence bears to the total negligence of Owner, Engineer, and all other negligent entities and individuals.

E.   *Mutual Waiver.* To the fullest extent permitted by law, Owner and Engineer waive against each other, and the other's employees, officers, directors, agents, insurers, partners, and consultants, any and all claims for or entitlement to special, incidental, indirect, or consequential damages arising out of, resulting from, or in any way related to the Project.

6.11 Miscellaneous Provisions

A.   *Notices.* Any notice required under this Agreement will be in writing, addressed to the appropriate party at its address on the signature page and given personally, by facsimile, by registered or certified mail postage prepaid, or by a commercial courier service. All notices shall be effective upon the date of receipt.

B.   *Survival.* All express representations, waivers, indemnifications, and limitations of liability included in this Agreement will survive its completion or termination for any reason.

C.   *Severability.* Any provision or part of the Agreement held to be void or unenforceable under any Laws or Regulations shall be deemed stricken, and all remaining provisions shall continue to be valid and binding upon Owner and Engineer, who agree that the Agreement shall be reformed to replace such

EJCDC E-500 Standard Form of Agreement Between Owner and Engineer for Professional Services
Copyright ©2002 National Society of Professional Engineers for EJCDC. All rights reserved.

stricken provision or part thereof with a valid and enforceable provision that comes as close as possible to expressing the intention of the stricken provision.

D. *Waiver.* A party's non-enforcement of any provision shall not constitute a waiver of that provision, nor shall it affect the enforceability of that provision or of the remainder of this Agreement.

E. *Accrual of Claims.* To the fullest extent permitted by law, all causes of action arising under this Agreement shall be deemed to have accrued, and all statutory periods of limitation shall commence, no later than the date of Substantial Completion.

ARTICLE 7 - DEFINITIONS

7.01 Defined Terms

A. Wherever used in this Agreement (including the Exhibits hereto) terms (including the singular and plural forms) printed with initial capital letters have the meanings indicated in the text above or in the exhibits; in the following provisions; or in the "Standard General Conditions of the Construction Contract," prepared by the Engineers Joint Contract Documents Committee (No. C-700, 2002 Edition):

1. *Additional Services*—The services to be performed for or furnished to Owner by Engineer in accordance with Exhibit A, Part 2, of this Agreement.

2. *Basic Services*—The services to be performed for or furnished to Owner by Engineer in accordance with Exhibit A, Part 1, of this Agreement.

3. *Construction Cost*—The cost to Owner of those portions of the entire Project designed or specified by Engineer. Construction Cost does not include costs of services of Engineer or other design professionals and consultants, cost of land, rights-of-way, or compensation for damages to properties, or Owner's costs for legal, accounting, insurance counseling or auditing services, or interest and financing charges incurred in connection with the Project, or the cost of other services to be provided by others to Owner pursuant to Exhibit B of this Agreement. Construction Cost is one of the items comprising Total Project Costs.

4. *Constituent of Concern*—Any substance, product, waste, or other material of any nature whatsoever (including, but not limited to, Asbestos, Petroleum, Radioactive Material, and PCBs) which is or becomes listed, regulated, or addressed pursuant to [a] the Comprehensive Environmental Response, Compensation and Liability Act, 42 U.S.C. §§9601 et seq. ("CERCLA"); [b] the Hazardous Materials Transportation Act, 49 U.S.C. §§1801 et seq.; [c] the Resource Conservation and Recovery Act, 42 U.S.C. §§6901 et seq. ("RCRA"); [d] the Toxic Substances Control Act, 15 U.S.C. §§2601 et seq.; [e] the Clean Water Act, 33 U.S.C. §§1251 et seq.; [f] the Clean Air Act, 42 U.S.C. §§7401 et seq.; and [g] any other federal, state, or local statute, law, rule, regulation, ordinance, resolution, code, order, or decree regulating, relating to, or imposing liability or standards of conduct concerning, any hazardous, toxic, or dangerous waste, substance, or material.

5. *Consultants*—Individuals or entities having a contract with Engineer to furnish services with respect to this Project as Engineer's independent professional associates, consultants, subcontractors, or vendors.

6. *Documents*—Data, reports, Drawings, Specifications, Record Drawings, and other deliverables, whether in printed or electronic media format, provided or furnished in appropriate phases by Engineer to Owner pursuant to this Agreement.

7. *Drawings*—That part of the Contract Documents prepared or approved by Engineer which graphically shows the scope, extent, and character of the Work to be performed by Contractor. Shop Drawings are not Drawings as so defined.

8. *Laws and Regulations; Laws or Regulations*—Any and all applicable laws, rules, regulations, ordinances, codes, and orders of any and all governmental bodies, agencies, authorities, and courts having jurisdiction.

9. *Reimbursable Expenses*—The expenses incurred directly by Engineer in connection with the performing or furnishing of Basic and Additional Services for the Project.

10. *Resident Project Representative*—The authorized representative of Engineer, if any, assigned to assist Engineer at the Site during the Construction Phase. The Resident Project Representative will be Engineer's agent or

employee and under Engineer's supervision. As used herein, the term Resident Project Representative includes any assistants of Resident Project Representative agreed to by Owner. The duties and responsibilities of the Resident Project Representative, if any, are as set forth in Exhibit D.

11. *Specifications*—That part of the Contract Documents consisting of written technical descriptions of materials, equipment, systems, standards, and workmanship as applied to the Work and certain administrative details applicable thereto.

12. *Total Project Costs*—The sum of the Construction Cost, allowances for contingencies, and the total costs of services of Engineer or other design professionals and consultants, together with such other Project-related costs that Owner furnishes for inclusion, including but not limited to cost of land, rights-of-way, compensation for damages to properties, Owner's costs for legal, accounting, insurance counseling and auditing services, interest and financing charges incurred in connection with the Project, and the cost of other services to be provided by others to Owner pursuant to Exhibit B of this Agreement.

ARTICLE 8 - EXHIBITS AND SPECIAL PROVISIONS

8.01   Exhibits Included

A.   Exhibit A, "Engineer's Services," consisting of _____ pages.

B.   Exhibit B, "Owner's Responsibilities," consisting of _____ pages.

C.   Exhibit C, "Payments to Engineer for Services and Reimbursable Expenses," consisting of _____ pages.

D.   Exhibit D, "Duties, Responsibilities and Limitations of Authority of Resident Project Representative," consisting of _____ pages.

E.   Exhibit E, "Notice of Acceptability of Work," consisting of _____ pages.

F.   Exhibit F, "Construction Cost Limit," consisting of _____ pages.

G.   Exhibit G, "Insurance," consisting of _____ pages.

H.   Exhibit H, "Dispute Resolution," consisting of _____ pages.

I.   Exhibit I, "Allocation of Risks," consisting of _____ pages.

J.   Exhibit J, "Special Provisions," consisting of _____ pages.

K.   Exhibit K, "Amendment to Standard Form of Agreement," consisting of _____ pages.

8.02   Total Agreement

A.   This Agreement (consisting of pages 1 to ___ inclusive, together with the exhibits identified above) constitutes the entire agreement between Owner and Engineer and supersedes all prior written or oral understandings. This Agreement may only be amended, supplemented, modified, or canceled by a duly executed written instrument based on the format of Exhibit K to this Agreement.

8.03   Designated Representatives

A.   With the execution of this Agreement, Engineer and Owner shall designate specific individuals to act as Engineer's and Owner's representatives with respect to the services to be performed or furnished by Engineer and responsibilities of Owner under this Agreement. Such individuals shall have authority to transmit instructions, receive information, and render decisions relative to the Project on behalf of each respective party.

IN WITNESS WHEREOF, the parties hereto have executed this Agreement, the Effective Date of which is indicated on page 1.

Owner: Engineer:

_____   _____

By: _____   By: _____

Title: _____   Title: _____

Date Signed: _____   Date Signed: _____

                                            Engineer License or Certificate No.
                                            State of:

Address for giving notices: Address for giving notices:

_____   _____

_____   _____

_____   _____

Designated Representative (see paragraph 8.03.A):   Designated Representative (see paragraph 8.03.A):

_____   _____

Title: _____   Title: _____

Phone Number: _____   Phone Number: _____

Facsimile Number: _____   Facsimile Number: _____

E-Mail Address: _____   E-Mail Address: _____

EJCDC E-500 Standard Form of Agreement Between Owner and Engineer for Professional Services

SUGGESTED FORMAT
(for use with E-500, 2002 Edition)

This is **EXHIBIT A**, consisting of ___ pages, referred to in and part of the **Agreement between Owner and Engineer for Professional Services** dated ___, ___.

Engineer's Services

Article 1 of the Agreement is amended and supplemented to include the following agreement of the parties. Engineer shall provide Basic and Additional Services as set forth below.

PART 1 – BASIC SERVICES

A1.01   Study and Report Phase

A. Engineer shall:

1.   Consult with Owner to define and clarify Owner's requirements for the Project and available data.

2.   Advise Owner of any need for Owner to provide data or services of the types described in Exhibit B which are not part of Engineer's Basic Services.

3.   Identify, consult with, and analyze requirements of governmental authorities having jurisdiction to approve the portions of the Project designed or specified by Engineer, including but not limited to mitigating measures identified in the environmental assessment.

4.   Identify and evaluate [insert specific number or list here] alternate solutions available to Owner and, after consultation with Owner, recommend to Owner those solutions which in Engineer's judgment meet Owner's requirements for the Project.

5.   Prepare a report (the "Report") which will, as appropriate, contain schematic layouts, sketches, and conceptual design criteria with appropriate exhibits to indicate the agreed-to requirements, considerations involved, and those alternate solutions available to Owner which Engineer recommends. For each recommended solution Engineer will provide the following, which will be separately itemized: opinion of probable Construction Cost; proposed allowances for contingencies; the estimated total costs of design, professional, and related services to be provided by Engineer and its Consultants; and, on the basis of information furnished by Owner, a summary of allowances for other items and services included within the definition of Total Project Costs.

6.   Perform or provide the following additional Study and Report Phase tasks or deliverables: [here list any such tasks or deliverables]

7.   Furnish ___ review copies of the Report and any other deliverables to Owner within ___ calendar days of authorization to begin services and review it with Owner. Within ___ calendar days of receipt, Owner shall submit to Engineer any comments regarding the Report and any other deliverables.

8.   Revise the Report and any other deliverables in response to Owner's comments, as appropriate, and furnish ___ copies of the revised Report and any other deliverables to the Owner within ___ calendar days of receipt of Owner's comments.

B. Engineer's services under the Study and Report Phase will be considered complete on the date when the revised Report and any other deliverables have been delivered to Owner.

A1.02   Preliminary Design Phase

A. After acceptance by Owner of the Report and any other deliverables, selection by Owner of a recommended solution and indication of any specific modifications or changes in the scope, extent, character, or design requirements of the Project desired by Owner, and upon written authorization from Owner, Engineer shall:

1.   Prepare Preliminary Design Phase documents consisting of final design criteria, preliminary drawings, outline specifications, and written descriptions of the Project.

2.   Provide necessary field surveys and topographic and utility mapping for design purposes. Utility mapping will be based upon information obtained from utility owners.

3.   Advise Owner if additional reports, data, information, or services of the types described in Exhibit B are necessary and assist Owner in obtaining such reports, data, information, or services.

4.   Based on the information contained in the Preliminary Design Phase documents, prepare a revised opinion of probable Construction Cost, and assist Owner in collating the various cost categories which comprise Total Project Costs.

5.   Perform or provide the following additional Preliminary Design Phase tasks or deliverables: *[here list any such tasks or deliverables]*

6.   Furnish ____ review copies of the Preliminary Design Phase documents and any other deliverables to Owner within ____ calendar days of authorization to proceed with this phase, and review them with Owner. Within ____ calendar days of receipt, Owner shall submit to Engineer any comments regarding the Preliminary Design Phase documents and any other deliverables.

7.   Revise the Preliminary Design Phase documents and any other deliverables in response to Owner's comments, as appropriate, and furnish to Owner ____ copies of the revised Preliminary Design Phase documents, revised opinion of probable Construction Cost, and any other deliverables within ____ calendar days after receipt of Owner's comments.

B. Engineer's services under the Preliminary Design Phase will be considered complete on the date when the revised Preliminary Design Phase documents, revised opinion of probable Construction Cost, and any other deliverables have been delivered to Owner.

A1.03   Final Design Phase

A. After acceptance by Owner of the Preliminary Design Phase documents, revised opinion of probable Construction Cost as determined in the Preliminary Design Phase, and any other deliverables subject to any Owner-directed modifications or changes in the scope, extent, character, or design requirements of or for the Project, and upon written authorization from Owner, Engineer shall:

1.   Prepare final Drawings and Specifications indicating the scope, extent, and character of the Work to be performed and furnished by Contractor. If appropriate, Specifications shall conform to the 16-division format of the Construction Specifications Institute.

2.   Provide technical criteria, written descriptions, and design data for Owner's use in filing applications for permits from or approvals of governmental authorities having jurisdiction to review or approve the final design of the Project; assist Owner in consultations with such authorities; and revise the Drawings and Specifications in response to directives from such authorities.

3.   Advise Owner of any adjustments to the opinion of probable Construction Cost known to Engineer.

4.   Perform or provide the following additional Final Design Phase tasks or deliverables: *[here list any such tasks or deliverables]*

5.   Prepare and furnish Bidding Documents for review by Owner, its legal counsel, and other advisors, and assist Owner in the preparation of other related documents. Within ___ days of receipt, Owner shall submit to Engineer any comments and, subject to the provisions of paragraph 6.01.G, instructions for revisions.

6.   Revise the Bidding Documents in accordance with comments and instructions from the Owner, as appropriate, and submit ___ final copies of the Bidding Documents, a revised opinion of probable Construction Cost, and any other deliverables to Owner within ___ calendar days after receipt of Owner's comments and instructions.

B. Engineer's services under the Final Design Phase will be considered complete on the date when the submittals required by paragraph A1.03.A.6 have been delivered to Owner.

C. In the event that the Work designed or specified by Engineer is to be performed or furnished under more than one prime contract, or if Engineer's services are to be separately sequenced with the work of one or more prime Contractors (such as in the case of fast-tracking), Owner and Engineer shall, prior to commencement of the Final Design Phase, develop a schedule for performance of Engineer's services during the Final Design, Bidding or Negotiating, Construction, and Post-Construction Phases in order to sequence and coordinate properly such services as are applicable to the work under such separate prime contracts. This schedule is to be prepared and included in or become an amendment to Exhibit A whether or not the work under such contracts is to proceed concurrently.

D. The number of prime contracts for Work designed or specified by Engineer upon which the Engineer's compensation has been established under this Agreement is ___. If more prime contracts are awarded, Engineer shall be entitled to an equitable increase in its compensation under this Agreement.

A1.04   Bidding or Negotiating Phase

A. After acceptance by Owner of the Bidding Documents and the most recent opinion of probable Construction Cost as determined in the Final Design Phase, and upon written authorization by Owner to proceed, Engineer shall:

1.   Assist Owner in advertising for and obtaining bids or proposals for the Work and, where applicable, maintain a record of prospective bidders to whom Bidding Documents have been issued, attend pre-Bid conferences, if any, and receive and process contractor deposits or charges for the Bidding Documents.

2.   Issue Addenda as appropriate to clarify, correct, or change the Bidding Documents.

3.   Provide information or assistance needed by Owner in the course of any negotiations with prospective contractors.

4.   Consult with Owner as to the acceptability of subcontractors, suppliers, and other individuals and entities proposed by prospective contractors for those portions of the Work as to which such acceptability is required by the Bidding Documents.

5.   Perform or provide the following additional Bidding or Negotiating Phase tasks or deliverables: *[here list any such tasks or deliverables]*

---

6.   Attend the Bid opening, prepare Bid tabulation sheets, and assist Owner in evaluating Bids or proposals and in assembling and awarding contracts for the Work.

B. The Bidding or Negotiating Phase will be considered complete upon commencement of the Construction Phase or upon cessation of negotiations with prospective contractors (except as may be required if Exhibit F is a part of this Agreement).

A1.05   Construction Phase

A. Upon successful completion of the Bidding and Negotiating Phase, and upon written authorization from Owner, Engineer shall:

1.   *General Administration of Construction Contract.* Consult with Owner and act as Owner's representative as provided in the General Conditions. The extent and limitations of the duties, responsibilities, and authority of Engineer as assigned in the General Conditions shall not be modified, except as Engineer may otherwise agree in writing. All of Owner's instructions to Contractor will be issued through Engineer, which shall have authority to act on behalf of Owner in dealings with Contractor to the extent provided in this Agreement and the General Conditions except as otherwise provided in writing.

2.   *Resident Project Representative (RPR).* Provide the services of an RPR at the Site to assist the Engineer and to provide more extensive observation of Contractor's work. Duties, responsibilities, and authority of the RPR are as set forth in Exhibit D. The furnishing of such RPR's services will not limit, extend, or modify Engineer's responsibilities or authority except as expressly set forth in Exhibit D. [If *Engineer will not be providing the services of an RPR, then delete this paragraph 2 by inserting the word "DELETED" after the paragraph title, and do not include Exhibit D.]*

3.   *Selecting Independent Testing Laboratory.* Assist Owner in the selection of an independent testing laboratory to perform the services identified in Exhibit B, paragraph B2.01.0.

4.   *Pre-Construction Conference.* Participate in a Pre-Construction Conference prior to commencement of Work at the Site.

5.   *Schedules.* Receive, review, and determine the acceptability of any and all schedules that Contractor is required to submit to Engineer, including the Progress Schedule, Schedule of Submittals, and Schedule of Values.

6.   *Baselines and Benchmarks.* As appropriate, establish baselines and benchmarks for locating the Work which in Engineer's judgment are necessary to enable Contractor to proceed.

7.   *Visits to Site and Observation of Construction.* In connection with observations of Contractor's Work while it is in progress:

a.   Make visits to the Site at intervals appropriate to the various stages of construction, as Engineer deems necessary, to observe as an experienced and qualified design professional the progress and quality of Contractor's executed Work. Such visits and observations by Engineer, and the Resident Project Representative, if any, are not intended to be exhaustive or to extend to every aspect of Contractor's Work in progress or to involve detailed inspections of Contractor's Work in progress beyond the responsibilities specifically assigned to Engineer in this Agreement and the Contract Documents, but rather are to be limited to spot checking, selective sampling, and similar methods of general observation of the Work based on Engineer's exercise of professional judgment as assisted by the Resident Project Representative, if any. Based on information obtained during such visits and observations, Engineer will determine in general if the Work is proceeding in accordance with the Contract Documents, and Engineer shall keep Owner informed of the progress of the Work.

b.   The purpose of Engineer's visits to, and representation by the Resident Project Representative, if any, at the Site, will be to enable Engineer to better carry out the duties and responsibilities assigned to and undertaken by Engineer during the Construction Phase, and, in addition, by the exercise of Engineer's efforts as an experienced and qualified design professional, to provide for Owner a greater degree of confidence that the completed Work will conform in general to the Contract Documents and that Contractor has implemented and maintained the integrity of the design concept of the completed Project as a functioning whole as indicated in the Contract Documents. Engineer shall not, during such visits or as a result of such observations of Contractor's Work in progress, supervise, direct, or have control over Contractor's Work, nor shall Engineer have authority over or responsibility for the means, methods, techniques, sequences, or procedures of construction selected or used by Contractor, for security or safety on the Site, for safety precautions and programs incident to Contractor's Work, nor for any failure of Contractor to comply with Laws and Regulations applicable to Contractor's furnishing and performing the Work. Accordingly, Engineer neither guarantees the performance of any Contractor nor assumes responsibility for any Contractor's failure to furnish and perform the Work in accordance with the Contract Documents.

8.   *Defective Work.* Recommend to Owner that Contractor's Work be rejected while it is in progress if, on the basis of Engineer's observations, Engineer believes that such Work will not produce a completed Project that conforms generally to the Contract Documents or that it will threaten the integrity of the design concept of the completed Project as a functioning whole as indicated in the Contract Documents.

9.   *Clarifications and Interpretations; Field Orders.* Issue necessary clarifications and interpretations of the Contract Documents as appropriate to the orderly completion of Contractor's work. Such clarifications and interpretations will be consistent with the intent of and reasonably inferable from the Contract Documents. Engineer may issue Field Orders authorizing minor variations in the Work from the requirements of the Contract Documents.

10.   *Change Orders and Work Change Directives.* Recommend Change Orders and Work Change Directives to Owner, as appropriate, and prepare Change Orders and Work Change Directives as required.

11.   *Shop Drawings and Samples.* Review and approve or take other appropriate action in respect to Shop Drawings and Samples and other data which Contractor is required to submit, but only for conformance with the information given in the Contract Documents and compatibility with the design concept of the completed Project as a functioning whole as indicated by the Contract Documents. Such reviews and approvals or other action will not extend to means, methods, techniques, sequences, or procedures of construction or to safety precautions and programs incident thereto. Engineer shall meet any Contractor's submittal schedule that Engineer has accepted.

12.   *Substitutes and "or-equal."* Evaluate and determine the acceptability of substitute or "or-equal" materials and equipment proposed by Contractor, but subject to the provisions of paragraph A2.02.A.2 of this Exhibit A.

13.   *Inspections and Tests.* Require such special inspections or tests of Contractor's work as deemed reasonably necessary, and receive and review all certificates of inspections, tests, and approvals required by Laws and Regulations or the Contract Documents. Engineer's review of such certificates will be for the purpose of determining that the results certified indicate compliance with the Contract Documents and will not constitute an independent evaluation that the content or procedures of such inspections, tests, or approvals comply with the requirements of the Contract Documents. Engineer shall be entitled to rely on the results of such tests.

14.   *Disagreements between Owner and Contractor.* Render formal written decisions on all duly submitted issues relating to the acceptability of Contractor's work or the interpretation of the requirements of the Contract Documents pertaining to the execution, performance, or progress of Contractor's Work; review each duly submitted Claim by Owner or Contractor, and in writing either deny such Claim in whole or in part, approve such Claim, or decline to resolve such Claim if Engineer in its

discretion concludes that to do so would be inappropriate. In rendering such decisions, Engineer shall be fair and not show partiality to Owner or Contractor and shall not be liable in connection with any decision rendered in good faith in such capacity.

15. *Applications for Payment.* Based on Engineer's observations as an experienced and qualified design professional and on review of Applications for Payment and accompanying supporting documentation:

a. Determine the amounts that Engineer recommends Contractor be paid. Such recommendations of payment will be in writing and will constitute Engineer's representation to Owner, based on such observations and review, that, to the best of Engineer's knowledge, information and belief, Contractor's Work has progressed to the point indicated, the quality of such Work is generally in accordance with the Contract Documents (subject to an evaluation of the Work as a functioning whole prior to or upon Substantial Completion, to the results of any subsequent tests called for in the Contract Documents, and to any other qualifications stated in the recommendation), and the conditions precedent to Contractor's being entitled to such payment appear to have been fulfilled in so far as it is Engineer's responsibility to observe Contractor's Work. In the case of unit price work, Engineer's recommendations of payment will include final determinations of quantities and classifications of Contractor's Work (subject to any subsequent adjustments allowed by the Contract Documents).

b. By recommending any payment, Engineer shall not thereby be deemed to have represented that observations made by Engineer to check the quality or quantity of Contractor's Work as it is performed and furnished have been exhaustive, extended to every aspect of Contractor's Work in progress, or involved detailed inspections of the Work beyond the responsibilities specifically assigned to Engineer in this Agreement and the Contract Documents. Neither Engineer's review of Contractor's Work for the purposes of recommending payments nor Engineer's recommendation of any payment including final payment will impose on Engineer responsibility to supervise, direct, or control Contractor's Work in progress or for the means, methods, techniques, sequences, or procedures of construction or safety precautions or programs incident thereto, or Contractor's compliance with Laws and Regulations applicable to Contractor's furnishing and performing the Work. It will also not impose responsibility on Engineer to make any examination to ascertain how or for what purposes Contractor has used the moneys paid on account of the Contract Price, or to determine that title to any portion of the Work in progress, materials, or equipment has passed to Owner free and clear of any liens, claims, security interests, or encumbrances, or that there may not be other matters at issue between Owner and Contractor that might affect the amount that should be paid.

16. *Contractor's Completion Documents.* Receive, review, and transmit to Owner maintenance and operating instructions, schedules, guarantees, bonds, certificates or other evidence of insurance required by the Contract Documents, certificates of inspection, tests and approvals, Shop Drawings, Samples and other data approved as provided under paragraph A1.05.A.11, and the annotated record documents which are to be assembled by Contractor in accordance with the Contract Documents to obtain final payment. The extent of such review by Engineer will be limited as provided in paragraph A1.05.A.11.

17. *Substantial Completion.* Promptly after notice from Contractor that Contractor considers the entire Work ready for its intended use, in company with Owner and Contractor, conduct an inspection to determine if the Work is substantially complete. If after considering any objections of Owner, Engineer considers the Work substantially complete, Engineer shall deliver a certificate of Substantial Completion to Owner and Contractor.

18. *Additional Tasks.* Perform or provide the following additional Construction Phase tasks or deliverables: [*here list any such tasks or deliverables*].

19. *Final Notice of Acceptability of the Work.* Conduct a final inspection to determine if the completed Work of Contractor is acceptable so that Engineer may recommend, in writing, final payment to Contractor. Accompanying the recommendation for final payment, Engineer shall also provide a notice in the form attached hereto as Exhibit E (the "Notice of Acceptability of

Work") that the Work is acceptable (subject to the provisions of paragraph A1.05.A.15.b) to the best of Engineer's knowledge, information, and belief and based on the extent of the services provided by Engineer under this Agreement.

B. *Duration of Construction Phase.* The Construction Phase will commence with the execution of the first construction Contract for the Project or any part thereof and will terminate upon written recommendation by Engineer for final payment to Contractors. If the Project involves more than one prime contract as indicated in paragraph A1.03.C, Construction Phase services may be rendered at different times in respect to the separate contracts. Subject to the provisions of Article 3, Engineer shall be entitled to an equitable increase in compensation if Construction-Phase services are required after the original date for final completion of the Work as set forth in the construction Contract.

C. *Limitation of Responsibilities.* Engineer shall not be responsible for the acts or omissions of any Contractor, or of any subcontractors, suppliers, or other individuals or entities performing or furnishing any of the Work. Engineer shall not be responsible for the failure of any Contractor to perform or furnish the Work in accordance with the Contract Documents.

A1.06   Post-Construction Phase

A. Upon written authorization from Owner, Engineer, during the Post-Construction Phase, shall:

1.   Provide assistance in connection with the adjusting of Project equipment and systems.

2.   Assist Owner in training Owner's staff to operate and maintain Project equipment and systems.

3.   Assist Owner in developing procedures for control of the operation and maintenance of, and record keeping for Project equipment and systems.

4.   Together with Owner, visit the Project to observe any apparent defects in the Work, assist Owner in consultations and discussions with Contractor concerning correction of any such defects, and make recommendations as to replacement or correction of Defective Work, if present.

5.   Perform or provide the following additional Post-Construction Phase tasks or deliverables: _____.

6.   In company with Owner or Owner's representative, provide an inspection of the Project within one month before the end of the Correction Period to ascertain whether any portion of the Work is subject to correction.

B. The Post-Construction Phase services may commence during the Construction Phase and, if not otherwise modified in this Exhibit A, will terminate at the end of the Construction Contract's correction period.

PART 2 — ADDITIONAL SERVICES

A2.01   Additional Services Requiring Owner's Written Authorization

A. If authorized in writing by Owner, Engineer shall furnish or obtain from others Additional Services of the types listed below.

1.   Preparation of applications and supporting documents (in addition to those furnished under Basic Services) for private or governmental grants, loans, or advances in connection with the Project; preparation or review of environmental assessments and impact statements; review and evaluation of the effects on the design requirements for the Project of any such

statements and documents prepared by others; and assistance in obtaining approvals of authorities having jurisdiction over the anticipated environmental impact of the Project.

2.   Services to make measured drawings of or to investigate existing conditions or facilities, or to verify the accuracy of drawings or other information furnished by Owner or others.

3.   Services resulting from significant changes in the scope, extent, or character of the portions of the Project designed or specified by Engineer or its design requirements including, but not limited to, changes in size, complexity, Owner's schedule, character of construction, or method of financing; and revising previously accepted studies, reports, Drawings, Specifications, or Contract Documents when such revisions are required by changes in Laws and Regulations enacted subsequent to the Effective Date of this Agreement or are due to any other causes beyond Engineer's control.

4.   Services resulting from Owner's request to evaluate additional Study and Report Phase alternative solutions beyond those identified in paragraph A1.01.A.4.

5.   Services required as a result of Owner's providing incomplete or incorrect Project information to Engineer.

6.   Providing renderings or models for Owner's use.

7.   Undertaking investigations and studies including, but not limited to, detailed consideration of operations, maintenance, and overhead expenses; the preparation of feasibility studies, cash flow and economic evaluations, rate schedules, and appraisals; assistance in obtaining financing for the Project; evaluating processes available for licensing, and assisting Owner in obtaining process licensing; detailed quantity surveys of materials, equipment, and labor; and audits or inventories required in connection with construction performed by Owner.

8.   Furnishing services of Engineer's Consultants for other than Basic Services.

9.   Services attributable to more prime construction contracts than specified in paragraph A1.03.C.

10.   Services during out-of-town travel required of Engineer other than for visits to the Site or Owner's office.

11.   Preparing for, coordinating with, participating in and responding to structured independent review processes, including, but not limited to, construction management, cost estimating, project peer review, value engineering, and constructibility review requested by Owner; and performing or furnishing services required to revise studies, reports, Drawings, Specifications, or other Bidding Documents as a result of such review processes.

12.   Preparing additional Bidding Documents or Contract Documents for alternate bids or prices requested by Owner for the Work or a portion thereof.

13.   Determining the acceptability of substitute materials and equipment proposed during the Bidding or Negotiating Phase when substitution prior to the award of contracts is allowed by the Bidding Documents.

14.   Assistance in connection with Bid protests, rebidding, or renegotiating contracts for construction, materials, equipment, or services, except when such assistance is required by Exhibit F.

15.   Providing construction surveys and staking to enable Contractor to perform its work other than as required under paragraph A1.05.A.5, and any type of property surveys or related engineering services needed for the transfer of interests in real property; and providing other special field surveys.

16.   Providing Construction Phase services beyond the original date for final completion of the Work.

17.   Providing assistance in responding to the presence of any Constituent of Concern at the Site, in compliance with current Laws and Regulations.

18.   Preparing and furnishing to Owner Record Drawings showing appropriate record information based on Project annotated record documents received from Contractor.

19.   Preparation of operation and maintenance manuals.

20.   Preparing to serve or serving as a consultant or witness for Owner in any litigation, arbitration, or other dispute resolution process related to the Project.

21.   Providing more extensive services required to enable Engineer to issue notices or certifications requested by Owner.

22.   Other services performed or furnished by Engineer not otherwise provided for in this Agreement.

A2.02   Additional Services Not Requiring Owner's Written Authorization

A. Engineer shall advise Owner that Engineer is commencing to perform or furnish the Additional Services of the types listed below. For such Additional Services, Engineer need not request or obtain specific advance written authorization from Owner. Engineer shall cease performing or furnishing such Additional Services upon receipt of written notice from Owner.

1.   Services in connection with Work Change Directives and Change Orders to reflect changes requested by Owner.

2.   Services in making revisions to Drawings and Specifications occasioned by the acceptance of substitute materials or equipment other than "or-equal" items; and services after the award of the Construction Contract in evaluating and determining the acceptability of a substitution which is found to be inappropriate for the Project or an excessive number of substitutions.

3.   Services resulting from significant delays, changes, or price increases occurring as a direct or indirect result of materials, equipment, or energy shortages.

4.   Additional or extended services during construction made necessary by (1) emergencies or acts of God endangering the Work, (2) the presence at the Site of any Constituent of Concern, (3) Work damaged by fire or other cause during construction, (4) a significant amount of defective, neglected, or delayed work by Contractor, (5) acceleration of the progress schedule involving services beyond normal working hours, or (6) default by Contractor.

5.   Services (other than Basic Services during the Post-Construction Phase) in connection with any partial utilization of any part of the Work by Owner prior to Substantial Completion.

6.   Evaluating an unreasonable claim or an excessive number of claims submitted by Contractor or others in connection with the Work.

7.   Services during the Construction Phase rendered after the date stated in A1.05.B.

SUGGESTED FORMAT
(for use with E-500, 2002 Edition)

This is **EXHIBIT B**, consisting of _____ pages, referred to in and part of the **Agreement between Owner and Engineer for Professional Services** dated _____, _____.

Owner's Responsibilities

Article 2 of the Agreement is amended and supplemented to include the following agreement of the parties.

B2.01   In addition to other responsibilities of Owner as set forth in this Agreement, Owner shall at its expense:

A. Provide Engineer with all criteria and full information as to Owner's requirements for the Project, including design objectives and constraints, space, capacity and performance requirements, flexibility, and expandability, and any budgetary limitations; and furnish copies of all design and construction standards which Owner will require to be included in the Drawings and Specifications; and furnish copies of Owner's standard forms, conditions, and related documents for Engineer to include in the Bidding Documents, when applicable.

B. Furnish to Engineer any other available information pertinent to the Project including reports and data relative to previous designs, or investigation at or adjacent to the Site.

C. Following Engineer's assessment of initially available Project information and data and upon Engineer's request, furnish or otherwise make available such additional Project related information and data as is reasonably required to enable Engineer to complete its Basic and Additional Services. Such additional information or data would generally include the following:

1.   Property descriptions.

2.   Zoning, deed, and other land use restrictions.

3.   Property, boundary, easement, right-of-way, and other special surveys or data, including establishing relevant reference points.

4.   Explorations and tests of subsurface conditions at or contiguous to the Site, drawings of physical conditions in or relating to existing surface or subsurface structures at or contiguous to the Site, or hydrographic surveys, with appropriate professional interpretation thereof.

5.   Environmental assessments, audits, investigations, and impact statements, and other relevant environmental or cultural studies as to the Project, the Site, and adjacent areas.

6.   Data or consultations as required for the Project but not otherwise identified in the Agreement or the Exhibits thereto.

D. Give prompt written notice to Engineer whenever Owner observes or otherwise becomes aware of the presence at the Site of any Constituent of Concern, or of any other development that affects the scope or time of performance of Engineer's services, or any defect or nonconformance in Engineer's services, the Work, or in the performance of any Contractor.

E. Authorize Engineer to provide Additional Services as set forth in Part 2 of Exhibit A of the Agreement as required.

F. Arrange for safe access to and make all provisions for Engineer to enter upon public and private property as required for Engineer to perform services under the Agreement.

G. Examine all alternate solutions, studies, reports, sketches, Drawings, Specifications, proposals, and other documents presented by Engineer (including obtaining advice of an attorney, insurance counselor, and other advisors or consultants as Owner deems appropriate with respect to such examination) and render in writing timely decisions pertaining thereto.

H. Provide reviews, approvals, and permits from all governmental authorities having jurisdiction to approve all phases of the Project designed or specified by Engineer and such reviews, approvals, and consents from others as may be necessary for completion of each phase of the Project.

I. Provide, as required for the Project:

1. Accounting, bond and financial advisory, independent cost estimating, and insurance counseling services.

2. Legal services with regard to issues pertaining to the Project as Owner requires, Contractor raises, or Engineer reasonably requests.

3. Such auditing services as Owner requires to ascertain how or for what purpose Contractor has used the moneys paid.

4. Placement and payment for advertisement for Bids in appropriate publications.

J. Advise Engineer of the identity and scope of services of any independent consultants employed by Owner to perform or furnish services in regard to the Project, including, but not limited to, cost estimating, project peer review, value engineering, and constructibility review.

K. Furnish to Engineer data as to Owner's anticipated costs for services to be provided by others (including, but not limited to, accounting, bond and financial, independent cost estimating, insurance counseling, and legal advice) for Owner so that Engineer may assist Owner in collating the various cost categories which comprise Total Project Costs.

L. If Owner designates a construction manager or an individual or entity other than, or in addition to, Engineer to represent Owner at the Site, define and set forth as an attachment to this Exhibit B the duties, responsibilities, and limitations of authority of such other party and the relation thereof to the duties, responsibilities, and authority of Engineer.

M. If more than one prime contract is to be awarded for the Work designed or specified by Engineer, designate a person or entity to have authority and responsibility for coordinating the activities among the various prime Contractors, and define and set forth the duties, responsibilities, and limitations of authority of such individual or entity and the relation thereof to the duties, responsibilities, and authority of Engineer as an attachment to this Exhibit B that is to be mutually agreed upon and made a part of this Agreement before such services begin.

N. Attend the pre-bid conference, bid opening, pre-construction conferences, construction progress and other job related meetings, and Substantial Completion and final payment inspections.

O. Provide the services of an independent testing laboratory to perform all inspections, tests, and approvals of Samples, materials, and equipment required by the Contract Documents, or to evaluate the performance of materials, equipment, and facilities of Owner, prior to their incorporation into the Work with appropriate professional interpretation thereof.

P. Provide Engineer with the findings and reports generated by the entities providing services to Owner pursuant to this paragraph.

Q. Perform or provide the following additional services: _____.

EJCDC E-500 Standard Form of Agreement Between Owner and Engineer for Professional Services

SUGGESTED FORMAT
(for use with E-500, 2002 Edition)

This is **EXHIBIT F**, consisting of ___ pages, referred to in and part of the
Agreement between Owner and Engineer for Professional Services
dated ___, ___.

Construction Cost Limit

Paragraph 5.02 of the Agreement is amended and supplemented to include the following agreement of the parties:

F5.02   Designing to Construction Cost Limit

A. Owner and Engineer hereby agree to a Construction Cost limit in the amount of Dollars ($___).

B. A bidding or negotiating contingency of ___ percent will be added to any Construction Cost limit established.

C. The acceptance by Owner at any time during Basic Services of a revised opinion of probable Construction Cost in excess of the then established Construction Cost limit will constitute a corresponding increase in the Construction Cost limit.

D. Engineer will be permitted to determine what types and quality of materials, equipment and component systems are to be included in the Drawings and Specifications. Engineer may make reasonable adjustments in the scope, extent, and character of the Project to the extent consistent with the Project requirements and sound engineering practices, to bring the Project within the Construction Cost limit.

E. If the Bidding or Negotiating Phase has not commenced within three months after completion of the Final Design Phase, or if industry-wide prices are changed because of unusual or unanticipated events affecting the general level of prices or times of delivery in the construction industry, the established Construction Cost limit will not be binding on Engineer. In such cases, Owner shall consent to an adjustment in the Construction Cost limit commensurate with any applicable change in the general level of prices in the construction industry between the date of completion of the Final Design Phase and the date on which proposals or Bids are sought.

F. If the lowest bona fide proposal or Bid exceeds the established Construction Cost limit, Owner shall (1) give written approval to increase such Construction Cost limit, or (2) authorize negotiating or rebidding the Project within a reasonable time, or (3) cooperate in revising the Project's scope, extent, or character to the extent consistent with the Project's requirements and with sound engineering practices. In the case of (3), Engineer shall modify the Contract Documents as necessary to bring the Construction Cost within the Construction Cost Limit. Owner shall pay Engineer's cost to provide such modification services, including the costs of the services of Engineer's Consultants, all overhead expenses reasonably related thereto, and Reimbursable Expenses, but without profit to Engineer on account of such services. The providing of such services will be the limit of Engineer's responsibility in this regard and, having done so, Engineer shall be entitled to payment for services and expenses in accordance with this Agreement and will not otherwise be liable for damages attributable to the lowest bona fide proposal or Bid exceeding the established Construction Cost limit.

SUGGESTED FORMAT
(for use with E-500, 2002 Edition)

This is **EXHIBIT H**, consisting of _____ pages, referred to in and part of the **Agreement between Owner and Engineer for Professional Services** dated _____, _____.

Dispute Resolution

Paragraph 6.09 of the Agreement is amended and supplemented to include the following agreement of the parties:

[NOTE: Select one of the two alternatives provided]

H6.09   Dispute Resolution

A. Mediation. Owner and Engineer agree that they shall first submit any and all unsettled claims, counterclaims, disputes, and other matters in question between them arising out of or relating to this Agreement or the breach thereof ("Disputes") to mediation by *[insert name of mediator, or mediation service]*. If such mediation is unsuccessful in resolving a Dispute, then (a) the parties may mutually agree to a dispute resolution of their choice, or (b) either party may seek to have the Dispute resolved by a court of competent jurisdiction.

[or]

A. Arbitration. All Disputes between Owner and Engineer shall be settled by arbitration in accordance with the *[here insert the name of a specified arbitration service or organization]* rules effective at the Effective Date of the Agreement, subject to the conditions stated below. This agreement to arbitrate and any other agreement or consent to arbitrate entered into in accordance with this paragraph H6.09.A will be specifically enforceable under prevailing law of any court having jurisdiction.

1.   Notice of the demand for arbitration must be filed in writing with the other party to the Agreement and with the *[specified arbitration service or organization]*. The demand must be made within a reasonable time after the Dispute has arisen. In no event may the demand for arbitration be made after the date when institution of legal or equitable proceedings based on such Dispute would be barred by the applicable statute of limitations.

2.   All demands for arbitration and all answering statements thereto which include any monetary claims must contain a statement that the total sum or value in controversy as alleged by the party making such demand or answering statement is not more than $___ (exclusive of interest and costs). The arbitrators will not have jurisdiction, power, or authority to consider, or make findings (except in denial of their own jurisdiction) concerning any Dispute if the amount in controversy in such Dispute is more than $___ (exclusive of interest and costs), or to render a monetary award in response thereto against any party which totals more than $___ (exclusive of interest and costs). Disputes that are not subject to arbitration under this paragraph may be resolved in any court of competent jurisdiction.

3.   The award rendered by the arbitrators shall be in writing, and shall include: (i) a precise breakdown of the award; and (ii) a written explanation of the award specifically citing the Agreement provisions deemed applicable and relied on in making the award.

4.   The award rendered by the arbitrators will be consistent with the Agreement of the parties and final, and judgment may be entered upon it in any court having jurisdiction thereof, and will not be subject to appeal or modification.

5.   If a Dispute in question between Owner and Engineer involves the work of a Contractor, subcontractor, or consultants to the Owner or Engineer (each a "Joinable Party"), either Owner or Engineer may join each Joinable Party as a party to the arbitration between Owner and Engineer hereunder, and Engineer or Owner, as appropriate, shall include in each contract with each such Joinable Party a specific provision whereby such Joinable Party consents to being joined in an arbitration between Owner and Engineer involving the work of such Joinable Party. Nothing in this paragraph H6.09.A.5 nor in the provision of such contract consenting to joinder shall create any claim, right, or cause of action in favor of the Joinable Party and against Owner or Engineer that does not otherwise exist.

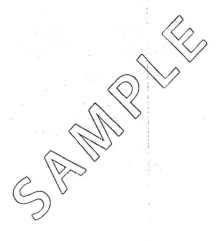

SUGGESTED FORMAT
(for use with E-500, 2002 Edition)

This is **EXHIBIT I**, consisting of _____ pages, referred to in and part of the
Agreement between Owner and Engineer for Professional Services
dated _____, _____.

Allocation of Risks

Paragraph 6.11 of the Agreement is amended and supplemented to include the following agreement of the parties:

16.11.B   Limitation of Engineer's Liability

[NOTE: *Select one of the three alternatives listed below for 16.11.B.*]

1.   *Engineer's Liability Limited to Amount of Engineer's Compensation.* To the fullest extent permitted by law, and notwithstanding any other provision of this Agreement, the total liability, in the aggregate, of Engineer and Engineer's officers, directors, partners, employees, agents, and Engineer's Consultants, and any of them, to Owner and anyone claiming by, through, or under Owner for any and all claims, losses, costs, or damages whatsoever arising out of, resulting from or in any way related to the Project or the Agreement from any cause or causes, including but not limited to the negligence, professional errors or omissions, strict liability or breach of contract, or warranty express or implied of Engineer or Engineer's officers, directors, partners, employees, agents, or Engineer's Consultants, or any of them, shall not exceed the total compensation received by Engineer under this Agreement.

[*or*]

1.   *Engineer's Liability Limited to Amount of Insurance Proceeds.* Engineer shall procure and maintain insurance as required by and set forth in Exhibit G to this Agreement. Notwithstanding any other provision of this Agreement, and to the fullest extent permitted by law, the total liability, in the aggregate, of Engineer and Engineer's officers, directors, partners, employees, agents, and Engineer's Consultants, and any of them, to Owner and anyone claiming by, through, or under Owner for any and all claims, losses, costs, or damages whatsoever arising out of, resulting from or in any way related to the Project or the Agreement from any cause or causes, including but not limited to the negligence, professional errors or omissions, strict liability or breach of contract, or warranty express or implied, of Engineer or Engineer's officers, directors, partners, employees, agents, or Engineer's Consultants, or any of them (hereafter "Owner's Claims"), shall not exceed the total insurance proceeds paid on behalf of or to Engineer by Engineer's insurers in settlement or satisfaction of Owner's Claims under the terms and conditions of Engineer's insurance policies applicable thereto (excluding fees, costs and expenses of investigation, claims adjustment, defense, and appeal). If no such insurance coverage is provided with respect to Owner's Claims, then the total liability, in the aggregate, of Engineer and Engineer's officers, directors, partners, employees, agents, and Engineer's Consultants, and any of them to Owner and anyone claiming by, through, or under Owner for any and all such uninsured Owner's claims shall not exceed $___.

[*or*]

1.   *Engineer's Liability Limited to the Amount of $___.* Notwithstanding any other provision of this Agreement, and to the fullest extent permitted by law, the total liability, in the aggregate, of Engineer and Engineer's officers, directors, partners, employees, agents, and Engineer's Consultants, and any of them, to Owner and anyone claiming by, through, or under Owner for

---

any and all claims, losses, costs, or damages whatsoever arising out of, resulting from, or in any way related to the Project or the Agreement from any cause or causes, including but not limited to the negligence, professional errors or omissions, strict liability or breach of contract, or warranty express or implied of Engineer or Engineer's officers, directors, partners, employees, agents, or Engineer's Consultants, or any of them, shall not exceed the total amount of $\_\_\_$.

[NOTE: If appropriate and desired, include 16.11.B.2 below]

2. *Exclusion of Special, Incidental, Indirect, and Consequential Damages.* To the fullest extent permitted by law, and notwithstanding any other provision in the Agreement, consistent with the terms of paragraph 6.11.E the Engineer and Engineer's officers, directors, partners, employees, agents, and Engineer's Consultants, or any of them, shall not be liable to Owner or anyone claiming by, through, or under Owner for any special, incidental, indirect, or consequential damages whatsoever arising out of, resulting from, or in any way related to the Project or the Agreement from any cause or causes, including but not limited to any such damages caused by the negligence, professional errors or omissions, strict liability, breach of contract, or warranties, express or implied, of Engineer or Engineer's officers, directors, partners, employees, agents, or Engineer's Consultants, or any of them, and including but not limited to:

> [NOTE: list here particular types of damages that may be of special concern because of the nature of the project or specific circumstances, e.g., cost of replacement power, loss of use of equipment or of the facility, loss of profits or revenue, loss of financing, regulatory fines, etc. If the parties prefer to leave the language general, then end the sentence after the phrase "or any of them."]

> [NOTE: the above exclusion of consequential and other damages can be converted to a limitation on the amount of such damages, following the format of paragraph 16.11.B.1 above, by providing that "Engineer's total liability for such damages shall not exceed $\_\_\_$."]

> [NOTE: If appropriate and desired, include 16.11.B.3 below] •

> [NOTE: The foregoing provisions may be included as a supplement to paragraph 6.11.E, which contains a mutual waiver of damages applicable to the benefit of both Owner and Engineer.]

3. *Agreement Not to Claim for Cost of Certain Change Orders.* Owner recognizes and expects that certain Change Orders may be required to be issued as the result in whole or part of imprecision, incompleteness, errors, omissions, ambiguities, or inconsistencies in the Drawings, Specifications, and other design documentation furnished by Engineer or in the other professional services performed or furnished by Engineer under this Agreement ("Covered Change Orders"). Accordingly, Owner agrees not to sue and otherwise to make no claim directly or indirectly against Engineer on the basis of professional negligence, breach of contract, or otherwise with respect to the costs of approved Covered Change Orders unless the costs of such approved Covered Change Orders exceed ___% of Construction Cost, and then only for an amount in excess of such percentage. Any responsibility of Engineer for the costs of Covered Change Orders in excess of such percentage will be determined on the basis of applicable contractual obligations and professional liability standards. For purposes of this paragraph, the cost of Covered Change Orders will not include any costs that Owner would have incurred if the Covered Change Order work had been included originally without any imprecision, incompleteness, error, omission, ambiguity, or inconsistency in the Contract Documents and without any other error or omission of Engineer related thereto. Nothing in this provision creates a presumption that, or changes the professional liability standard for determining if, Engineer is liable for the cost of Covered Change Orders in excess of the percentage of Construction Cost stated above or for any other Change Order. Wherever used in this paragraph, the term Engineer includes Engineer's officers, directors, partners, employees, agents, and Engineer's Consultants.

[NOTE: The parties may wish to consider the additional limitation contained in the following sentence.]

[Owner further agrees not to sue and otherwise to make no claim directly or indirectly against Engineer with respect to any Covered Change Order not in excess of such percentage stated above, and Owner agrees to hold Engineer harmless from and against any suit or claim made by the Contractor relating to any such Covered Change Order.

Suggested Form of Agreement Between Owner and Contractor for Construction Contract (Stipulated Price)

Engineers Joint Contract Documents Committee (EJCDC)

Reprinted by permission of the National Society of Professional Engineers.
For information on ordering these and other EJCDC documents, call
1-800-417-0348 or visit www.nspe.org.

This document has important legal consequences; consultation with an attorney is encouraged with respect to its use or modification. This document should be adapted to the particular circumstances of the contemplated Project and the controlling Laws and Regulations.

SUGGESTED FORM OF AGREEMENT
BETWEEN OWNER AND CONTRACTOR FOR
CONSTRUCTION CONTRACT (STIPULATED PRICE)

Prepared by

ENGINEERS JOINT CONTRACT DOCUMENTS COMMITTEE

and

Issued and Published Jointly by

AMERICAN COUNCIL OF ENGINEERING COMPANIES

———————————

ASSOCIATED GENERAL CONTRACTORS OF AMERICA

———————————

AMERICAN SOCIETY OF CIVIL ENGINEERS

———————————

PROFESSIONAL ENGINEERS IN PRIVATE PRACTICE
A Practice Division of the
NATIONAL SOCIETY OF PROFESSIONAL ENGINEERS

Endorsed by

CONSTRUCTION SPECIFICATIONS INSTITUTE

This Suggested Form of Agreement has been prepared for use with the Standard General Conditions of the Construction Contract (EJCDC C-700, 2007 Edition). Their provisions are interrelated, and a change in one may necessitate a change in the other. The language contained in the Suggested Instructions to Bidders (EJCDC C-200, 2007 Edition) is also carefully interrelated with the language of this Agreement. Their usage is discussed in the Narrative Guide to the 2007 EJCDC Construction Documents (EJCDC C-001, 2007 Edition).

Copyright © 2007 National Society of Professional Engineers
1420 King Street, Alexandria, VA 22314-2794
(703) 684-2882
www.nspe.org

American Council of Engineering Companies
1015 15th Street N.W., Washington, DC 20005
(202) 347-7474
www.acec.org

American Society of Civil Engineers
1801 Alexander Bell Drive, Reston, VA 20191-4400
(800) 548-2723
www.asce.org

Associated General Contractors of America
2300 Wilson Boulevard, Suite 400, Arlington, VA 22201-3308
(703) 548-3118
www.agc.org

The copyright for this EJCDC document is owned jointly by the four

EJCDC sponsoring organizations and held in trust for their benefit by NSPE.

INTRODUCTION

This Suggested Form of Agreement between Owner and Contractor for Construction Contract (Stipulated Price) ("Agreement") has been prepared for use with the Suggested Instructions to Bidders for Construction Contracts ("Instructions to Bidders") (EJCDC C-200, 2007 Edition); the Suggested Bid Form for Construction Contracts ("Bid Form") (EJCDC C-410, 2007 Edition); and the Standard General Conditions of the Construction Contract ("General Conditions") (EJCDC C-700, 2007 Edition). Their provisions are interrelated, and a change in one may necessitate a change in the others. See also the Guide to the Preparation of Supplementary Conditions (EJCDC C-800, 2007 Edition), and the Narrative Guide to the 2007 EJCDC Construction Documents (EJCDC C-001, 2007 Edition).

This Agreement form assumes use of a Project Manual that contains the following documentary information for a construction project:

- Bidding Requirements, which include the advertisement or invitation to bid, the Instructions to Bidders, and the Bid Form that is suggested or prescribed, all of which provide information and guidance for all Bidders; and

- Contract Documents, which include the Agreement, performance and payment bonds, the General Conditions, the Supplementary Conditions, the Drawings, and the Specifications.

The Bidding Requirements are not Contract Documents because much of their substance pertains to the relationships prior to the award of the Contract and has little effect or impact thereafter, and because many contracts are awarded without going through the bidding process. In some cases, however, the actual Bid may be attached as an exhibit to the Agreement to avoid extensive rekeying. (The definitions of terms used in this Agreement, including "Bidding Documents," "Bidding Requirements," and "Contract Documents," are set forth Article 1 of the General Conditions.)

Suggested provisions are accompanied by "Notes to User" to assist in preparing the Agreement. The provisions have been coordinated with the other forms produced by EJCDC. Much of the language should be usable on most projects, but modifications and additional provisions will often be necessary. When modifying the suggested language or writing additional provisions, the user must check the other documents thoroughly for conflicts and coordination of terms and make appropriate revisions in all affected documents.

All parties involved in construction projects benefit significantly from a standardized approach in the location of subject matter throughout the documents. Experience confirms the danger of addressing the same subject matter in more than one location: doing so frequently leads to confusion and unanticipated legal consequences. When preparing documents for a construction project, careful attention should be given to the guidance provided in the Uniform Location of Subject Matter (EJCDC N-122).

EJCDC has designated Section 00520 for this Agreement. If this convention is used, the first page of the Agreement would be numbered 00520-1. If CSI's MasterFormat 04™ is being used for the Project Manual, consult MasterFormat 04 for the appropriate section number and number the pages accordingly.

For brevity, paragraphs of the Instructions to Bidders are referenced with the prefix "I," those of the Bid Form are referenced with the prefix "BF," and those of this Agreement are referenced with the prefix "A."

NOTE: EJCDC publications may be purchased from any of the organizations listed on the page immediately following the cover page of this document.

SUGGESTED FORM OF AGREEMENT
BETWEEN OWNER AND CONTRACTOR
FOR CONSTRUCTION CONTRACT (STIPULATED PRICE)

THIS AGREEMENT is by and between _____ ("Owner") and

_____ ("Contractor").

Owner and Contractor hereby agree as follows:

ARTICLE 1 – WORK

1.1   Contractor shall complete all Work as specified or indicated in the Contract Documents. The Work is generally described as follows:

ARTICLE 2 – THE PROJECT

2.1   The Project for which the Work under the Contract Documents may be the whole or only a part is generally described as follows:

ARTICLE 3 – ENGINEER

3.1   The Project has been designed by _____ (Engineer), which is to act as Owner's representative, assume all duties and responsibilities, and have the rights and authority assigned to Engineer in the Contract Documents in connection with the completion of the Work in accordance with the Contract Documents.

ARTICLE 4 – CONTRACT TIMES

4.1   *Time of the Essence*

   A. All time limits for Milestones, if any, Substantial Completion, and completion and readiness for final payment as stated in the Contract Documents are of the essence of the Contract.

4.2   *Dates for Substantial Completion and Final Payment*

    A. The Work will be substantially completed on or before _____, and completed and ready for final payment in accordance with Paragraph 14.07 of the General Conditions on or before _____.

[or]

4.2   *Days to Achieve Substantial Completion and Final Payment*

    A. The Work will be substantially completed within _____ days after the date when the Contract Times commence to run as provided in Paragraph 2.03 of the General Conditions, and completed and ready for final payment in accordance with Paragraph 14.07 of the General Conditions within _____ days after the date when the Contract Times commence to run.

4.3   *Liquidated Damages*

    A. Contractor and Owner recognize that time is of the essence as stated in Paragraph 4.01 above and that Owner will suffer financial loss if the Work is not completed within the times specified in Paragraph 4.02 above, plus any extensions thereof allowed in accordance with Article 12 of the General Conditions. The parties also recognize the delays, expense, and difficulties involved in proving in a legal or arbitration proceeding the actual loss suffered by Owner if the Work is not completed on time. Accordingly, instead of requiring any such proof, Owner and Contractor agree that as liquidated damages for delay (but not as a penalty), Contractor shall pay Owner $_____ for each day that expires after the time specified in Paragraph 4.02 above for Substantial Completion until the Work is substantially complete. After Substantial Completion, if Contractor shall neglect, refuse, or fail to complete the remaining Work within the Contract Time or any proper extension thereof granted by Owner, Contractor shall pay Owner $_____ for each day that expires after the time specified in Paragraph 4.02 above for completion and readiness for final payment until the Work is completed and ready for final payment.

NOTE TO USER

If failure to reach a Milestone on time is of such consequence that the assessment of liquidated damages for failure to reach one or more Milestones on time is to be provided, appropriate amending or supplementing language should be inserted here.

ARTICLE 5 – CONTRACT PRICE

5.1   Owner shall pay Contractor for completion of the Work in accordance with the Contract Documents an amount in current funds equal to the sum of the amounts determined pursuant to Paragraphs 5.01.A, 5.01.B, and 5.01.C below:

    A. For all Work other than Unit Price Work, a lump sum of: $_____

All specific cash allowances are included in the above price in accordance with Paragraph 11.02 of the General Conditions.

B. For all Unit Price Work, an amount equal to the sum of the established unit price for each separately identified item of Unit Price Work times the actual quantity of that item:

<u>UNIT PRICE WORK</u>

Item No.	Description	Unit	Estimated Quantity	Bid Unit Price	Bid Price

Total of all Bid Prices (Unit Price Work) $_____

The Bid prices for Unit Price Work set forth as of the Effective Date of the Agreement are based on estimated quantities. As provided in Paragraph 11.03 of the General Conditions, estimated quantities are not guaranteed, and determinations of actual quantities and classifications are to be made by Engineer as provided in Paragraph 9.07 of the General Conditions.

C. For all Work, at the prices stated in Contractor's Bid, attached hereto as an exhibit.

NOTES TO USER

1. If adjustment prices for variations from stipulated Base Bid quantities have been agreed to, insert appropriate provisions.

2. Depending upon the particular project bid form used, use 5.01.A alone, 5.01.A and 5.01.B together, 5.01.B alone, or 5.01.C alone, deleting those not used and renumbering accordingly. If 5.01.C is used, Contractor's Bid is attached as an exhibit and listed as a Contract Document in A-9.

ARTICLE 6 – PAYMENT PROCEDURES

6.1 *Submittal and Processing of Payments*

A. Contractor shall submit Applications for Payment in accordance with Article 14 of the General Conditions. Applications for Payment will be processed by Engineer as provided in the General Conditions.

6.2 *Progress Payments; Retainage*

A. Owner shall make progress payments on account of the Contract Price on the basis of Contractor's Applications for Payment on or about the _____ day of each month during performance of the Work as provided in Paragraph 6.02.A.1 below. All such payments will be measured by the schedule of values established as provided in Paragraph 2.07.A of the General Conditions (and in the case of Unit Price Work based on the number of units completed) or, in the event there is no schedule of values, as provided in the General Requirements.

1. Prior to Substantial Completion, progress payments will be made in an amount equal to the percentage indicated below but, in each case, less the aggregate of payments previously made and less such amounts as Engineer may determine or Owner may withhold, including but not limited to liquidated damages, in accordance with Paragraph 14.02 of the General Conditions.

    a. _____ percent of Work completed (with the balance being retainage). If the Work has been 50 percent completed as determined by Engineer, and if the character and progress of the Work have been satisfactory to Owner and Engineer, then as long as the character and progress of the Work remain satisfactory to Owner and Engineer, there will be no additional retainage; and

    b. _____ percent of cost of materials and equipment not incorporated in the Work (with the balance being retainage).

B. Upon Substantial Completion, Owner shall pay an amount sufficient to increase total payments to Contractor to _____ percent of the Work completed, less such amounts as Engineer shall determine in accordance with Paragraph 14.02.B.5 of the General Conditions and less _____ percent of Engineer's estimate of the value of Work to be completed or corrected as shown on the tentative list of items to be completed or corrected attached to the certificate of Substantial Completion.

NOTE TO USER
Typical values used in Paragraph 6.02.B are 100 percent and 200 percent respectively.

6.3   *Final Payment*

A. Upon final completion and acceptance of the Work in accordance with Paragraph 14.07 of the General Conditions, Owner shall pay the remainder of the Contract Price as recommended by Engineer as provided in said Paragraph 14.07.

ARTICLE 7 – INTEREST

7.1   All moneys not paid when due as provided in Article 14 of the General Conditions shall bear interest at the rate of _____ percent per annum.

ARTICLE 8 – CONTRACTOR'S REPRESENTATIONS

8.1   In order to induce Owner to enter into this Agreement, Contractor makes the following representations:

A. Contractor has examined and carefully studied the Contract Documents and the other related data identified in the Bidding Documents.

B. Contractor has visited the Site and become familiar with and is satisfied as to the general, local, and Site conditions that may affect cost, progress, and performance of the Work.

C. Contractor is familiar with and is satisfied as to all federal, state, and local Laws and Regulations that may affect cost, progress, and performance of the Work.

D. Contractor has carefully studied all: (1) reports of explorations and tests of subsurface conditions at or contiguous to the Site and all drawings of physical conditions relating to existing surface or subsurface structures at the Site (except Underground Facilities), if any, that have been identified in Paragraph SC-4.02 of the Supplementary Conditions as containing reliable "technical data," and (2) reports and drawings of Hazardous Environmental Conditions, if any, at the Site that have been identified in Paragraph SC-4.06 of the Supplementary Conditions as containing reliable "technical data."

NOTE TO USER
Modify the above paragraph if there are no such reports or drawings.

E. Contractor has considered the information known to Contractor; information commonly known to contractors doing business in the locality of the Site; information and observations obtained from visits to the Site; the Contract Documents; and the Site-related reports and drawings identified in the Contract Documents, with respect to the effect of such information, observations, and documents on (1) the cost, progress, and performance of the Work; (2) the means, methods, techniques, sequences, and procedures of construction to be employed by Contractor, including any specific means, methods, techniques, sequences, and procedures of construction expressly required by the Contract Documents; and (3) Contractor's safety precautions and programs.

NOTE TO USER
If the Contract Documents do not identify any Site-related reports and drawings, modify this paragraph accordingly.

F. Based on the information and observations referred to in Paragraph 8.01.E above, Contractor does not consider that further examinations, investigations, explorations, tests, studies, or data are necessary for the performance of the Work at the Contract Price, within the Contract Times, and in accordance with the other terms and conditions of the Contract Documents.

G. Contractor is aware of the general nature of work to be performed by Owner and others at the Site that relates to the Work as indicated in the Contract Documents.

H. Contractor has given Engineer written notice of all conflicts, errors, ambiguities, or discrepancies that Contractor has discovered in the Contract Documents, and the written resolution thereof by Engineer is acceptable to Contractor.

I. The Contract Documents are generally sufficient to indicate and convey understanding of all terms and conditions for performance and furnishing of the Work.

ARTICLE 9 – CONTRACT DOCUMENTS

9.1   *Contents*

A. The Contract Documents consist of the following:

1. This Agreement (pages 1 to __, inclusive).

2. Performance bond (pages _____ to _____, inclusive).

3. Payment bond (pages _____ to _____, inclusive).

4. Other bonds (pages _____ to _____, inclusive).

    a. _____ (pages _____ to _____, inclusive).

    b. _____ (pages _____ to _____, inclusive).

    c. _____ (pages _____ to _____, inclusive).

5. General Conditions (pages _____ to _____, inclusive).

6. Supplementary Conditions (pages _____ to _____, inclusive).

7. Specifications as listed in the table of contents of the Project Manual.

8. Drawings consisting of _____ sheets with each sheet bearing the following general title: _____ [or] the Drawings listed on attached sheet index.

9. Addenda (numbers _____ to _____, inclusive).

10. Exhibits to this Agreement (enumerated as follows):

    a. Contractor's Bid (pages _____ to _____, inclusive).

    b. Documentation submitted by Contractor prior to Notice of Award (pages _____ to _____, inclusive).

    c. *[List other required attachments (if any), such as documents required by funding or lending agencies].*

11. The following which may be delivered or issued on or after the Effective Date of the Agreement and are not attached hereto:

    a. Notice to Proceed (pages _____ to _____, inclusive).

    b.   Work Change Directives.

    c.   Change Orders.

NOTE TO USER

If any of the items listed are not to be included as Contract Documents, remove such item from the list and renumber the remaining items.

B. The documents listed in Paragraph 9.01.A are attached to this Agreement (except as expressly noted otherwise above).

C. There are no Contract Documents other than those listed above in this Article 9.

D. The Contract Documents may only be amended, modified, or supplemented as provided in Paragraph 3.04 of the General Conditions.

ARTICLE 10 – MISCELLANEOUS

10.1 *Terms*

A. Terms used in this Agreement will have the meanings stated in the General Conditions and the Supplementary Conditions.

10.2 *Assignment of Contract*

A. No assignment by a party hereto of any rights under or interests in the Contract will be binding on another party hereto without the written consent of the party sought to be bound; and, specifically but without limitation, moneys that may become due and moneys that are due may not be assigned without such consent (except to the extent that the effect of this restriction may be limited by law), and unless specifically stated to the contrary in any written consent to an assignment, no assignment will release or discharge the assignor from any duty or responsibility under the Contract Documents.

10.3 *Successors and Assigns*

A. Owner and Contractor each binds itself, its partners, successors, assigns, and legal representatives to the other party hereto, its partners, successors, assigns, and legal representatives in respect to all covenants, agreements, and obligations contained in the Contract Documents.

10.4 *Severability*

A. Any provision or part of the Contract Documents held to be void or unenforceable under any Law or Regulation shall be deemed stricken, and all remaining provisions shall continue to be valid and binding upon Owner and Contractor, who agree that the Contract Documents shall be reformed to replace such stricken provision or part thereof with a valid and enforceable provision that comes as close as possible to expressing the intention of the stricken provision.

10.5   *Contractor's Certifications*

A. Contractor certifies that it has not engaged in corrupt, fraudulent, collusive, or coercive practices in competing for or in executing the Contract. For the purposes of this Paragraph 10.05:

1. "corrupt practice" means the offering, giving, receiving, or soliciting of any thing of value likely to influence the action of a public official in the bidding process or in the Contract execution;

2. "fraudulent practice" means an intentional misrepresentation of facts made (a) to influence the bidding process or the execution of the Contract to the detriment of Owner, (b) to establish Bid or Contract prices at artificial non-competitive levels, or (c) to deprive Owner of the benefits of free and open competition;

3. "collusive practice" means a scheme or arrangement between two or more Bidders, with or without the knowledge of Owner, a purpose of which is to establish Bid prices at artificial, non-competitive levels; and

4. "coercive practice" means harming or threatening to harm, directly or indirectly, persons or their property to influence their participation in the bidding process or affect the execution of the Contract.

10.6   *Other Provisions*

NOTES TO USER

1. If Owner intends to assign a procurement contract (for goods and services) to the Contractor, see Notes to User at Article 23 of Suggested Instructions to Bidders for Procurement Contracts (EJCDC P-200, 2000 Edition) for provisions to be inserted in this Article.

2. Insert other provisions here if applicable.

IN WITNESS WHEREOF, Owner and Contractor have signed this Agreement. Counterparts have been delivered to Owner and Contractor. All portions of the Contract Documents have been signed or have been identified by Owner and Contractor or on their behalf.

NOTE TO USER
See I-21 and correlate procedures for format and signing of the documents.

This Agreement will be effective on _____ (which is the Effective Date of the Agreement).

NOTE TO USER
The Effective Date of the Agreement and the dates of any Construction Performance Bond (EJCDC C-610) and Construction Payment Bond (EJCDC C-615) should be the same, if possible. In no case may the date of any bonds be earlier then the Effective Date of the Agreement.

OWNER: CONTRACTOR

_____   _____

By: _____   By: _____

Title: _____   Title: _____

                                                             (If Contractor is a corporation, a partnership,
                                                             or a joint venture, attach evidence of authority
                                                             to sign.)

Attest: _____   Attest: _____

Title: _____   Title: _____

Address for giving notices: Address for giving notices:

_____   _____

_____   _____

_____   _____

                                                             License No.: _____
                                                                           (Where applicable)

(If Owner is a corporation, attach evidence   *NOTE TO USER: Use in those states or other*
of authority to sign. If Owner is a public body, *jurisdictions where applicable or required.*
attach evidence of authority to sign and resolution
or other documents authorizing execution   Agent for service of process:
of this Agreement.)

                                                             _____

Standard General Conditions
of the Construction Contract

Engineers Joint Contract Documents Committee (EJCDC)

Reprinted by permission of the National Society of Professional Engineers.
For information on ordering these and other EJCDC documents, call
1-800-417-0348 or visit www.nspe.org.

This document has important legal consequences; consultation with an attorney is encouraged with respect to its use or modification. This document should be adapted to the particular circumstances of the contemplated Project and the controlling Laws and Regulations.

STANDARD GENERAL CONDITIONS OF THE CONSTRUCTION CONTRACT

Prepared by

ENGINEERS JOINT CONTRACT DOCUMENTS COMMITTEE

and

Issued and Published Jointly by

AMERICAN COUNCIL OF ENGINEERING COMPANIES

ASSOCIATED GENERAL CONTRACTORS OF AMERICA

AMERICAN SOCIETY OF CIVIL ENGINEERS

PROFESSIONAL ENGINEERS IN PRIVATE PRACTICE
A Practice Division of the
NATIONAL SOCIETY OF PROFESSIONAL ENGINEERS

Endorsed by

CONSTRUCTION SPECIFICATIONS INSTITUTE

These General Conditions have been prepared for use with the Suggested Forms of Agreement Between Owner and Contractor (EJCDC C-520 or C-525, 2007 Editions). Their provisions are interrelated and a change in one may necessitate a change in the other. Comments concerning their usage are contained in the Narrative Guide to the EJCDC Construction Documents (EJCDC C-001, 2007 Edition). For guidance in the preparation of Supplementary Conditions, see Guide to the Preparation of Supplementary Conditions (EJCDC C-800, 2007 Edition).

Copyright © 2007 National Society of Professional Engineers
1420 King Street, Alexandria, VA 22314-2794
(703) 684-2882
www.nspe.org

American Council of Engineering Companies
1015 15th Street N.W., Washington, DC 20005
(202) 347-7474
www.acec.org

American Society of Civil Engineers
1801 Alexander Bell Drive, Reston, VA 20191-4400
(800) 548-2723
www.asce.org

Associated General Contractors of America
2300 Wilson Boulevard, Suite 400, Arlington, VA 22201-3308
(703) 548-3118
www.agc.org

The copyright for this EJCDC document is owned jointly by the four
EJCDC sponsoring organizations and held in trust for their benefit by NSPE.

STANDARD GENERAL CONDITIONS OF THE CONSTRUCTION CONTRACT

TABLE OF CONTENTS

Page

ARTICLE 1 – DEFINITIONS AND TERMINOLOGY

1.01   *Defined Terms*

A. Wherever used in the Bidding Requirements or Contract Documents and printed with initial capital letters, the terms listed below will have the meanings indicated which are applicable to both the singular and plural thereof. In addition to terms specifically defined, terms with initial capital letters in the Contract Documents include references to identified articles and paragraphs, and the titles of other documents or forms.

1. *Addenda*—Written or graphic instruments issued prior to the opening of Bids which clarify, correct, or change the Bidding Requirements or the proposed Contract Documents.

2. *Agreement*—The written instrument which is evidence of the agreement between Owner and Contractor covering the Work.

3. *Application for Payment*—The form acceptable to Engineer which is to be used by Contractor during the course of the Work in requesting progress or final payments and which is to be accompanied by such supporting documentation as is required by the Contract Documents.

4. *Asbestos*—Any material that contains more than one percent asbestos and is friable or is releasing asbestos fibers into the air above current action levels established by the United States Occupational Safety and Health Administration.

5. *Bid*—The offer or proposal of a Bidder submitted on the prescribed form setting forth the prices for the Work to be performed.

6. *Bidder*—The individual or entity who submits a Bid directly to Owner.

7. *Bidding Documents*—The Bidding Requirements and the proposed Contract Documents (including all Addenda).

8. *Bidding Requirements*—The advertisement or invitation to bid, Instructions to Bidders, Bid security of acceptable form, if any, and the Bid Form with any supplements.

9. *Change Order*—A document recommended by Engineer which is signed by Contractor and Owner and authorizes an addition, deletion, or revision in the Work or an adjustment in the Contract Price or the Contract Times, issued on or after the Effective Date of the Agreement.

10. *Claim*—A demand or assertion by Owner or Contractor seeking an adjustment of Contract Price or Contract Times, or both, or other relief with respect to the terms of the Contract. A demand for money or services by a third party is not a Claim.

11. *Contract*—The entire and integrated written agreement between the Owner and Contractor concerning the Work. The Contract supersedes prior negotiations, representations, or agreements, whether written or oral.

12. *Contract Documents*—Those items so designated in the Agreement. Only printed or hard copies of the items listed in the Agreement are Contract Documents. Approved Shop Drawings, other Contractor submittals, and the reports and drawings of subsurface and physical conditions are not Contract Documents.

13. *Contract Price*—The moneys payable by Owner to Contractor for completion of the Work in accordance with the Contract Documents as stated in the Agreement (subject to the provisions of Paragraph 11.03 in the case of Unit Price Work).

14. *Contract Times*—The number of days or the dates stated in the Agreement to: (i) achieve Milestones, if any; (ii) achieve Substantial Completion; and (iii) complete the Work so that it is ready for final payment as evidenced by Engineer's written recommendation of final payment.

15. *Contractor*—The individual or entity with whom Owner has entered into the Agreement.

16. *Cost of the Work*—See Paragraph 11.01 for definition.

17. *Drawings*—That part of the Contract Documents prepared or approved by Engineer which graphically shows the scope, extent, and character of the Work to be performed by Contractor. Shop Drawings and other Contractor submittals are not Drawings as so defined.

18. *Effective Date of the Agreement*—The date indicated in the Agreement on which it becomes effective, but if no such date is indicated, it means the date on which the Agreement is signed and delivered by the last of the two parties to sign and deliver.

19. *Engineer*—The individual or entity named as such in the Agreement.

20. *Field Order*—A written order issued by Engineer which requires minor changes in the Work but which does not involve a change in the Contract Price or the Contract Times.

21. *General Requirements*—Sections of Division 1 of the Specifications.

22. *Hazardous Environmental Condition*—The presence at the Site of Asbestos, PCBs, Petroleum, Hazardous Waste, or Radioactive Material in such quantities or circumstances that may present a substantial danger to persons or property exposed thereto.

23. *Hazardous Waste*—The term Hazardous Waste shall have the meaning provided in Section 1004 of the Solid Waste Disposal Act (42 USC Section 6903) as amended from time to time.

24. *Laws and Regulations; Laws or Regulations*—Any and all applicable laws, rules, regulations, ordinances, codes, and orders of any and all governmental bodies, agencies, authorities, and courts having jurisdiction.

25. *Liens*—Charges, security interests, or encumbrances upon Project funds, real property, or personal property.

26. *Milestone*—A principal event specified in the Contract Documents relating to an intermediate completion date or time prior to Substantial Completion of all the Work.

27. *Notice of Award*—The written notice by Owner to the Successful Bidder stating that upon timely compliance by the Successful Bidder with the conditions precedent listed therein, Owner will sign and deliver the Agreement.

28. *Notice to Proceed*—A written notice given by Owner to Contractor fixing the date on which the Contract Times will commence to run and on which Contractor shall start to perform the Work under the Contract Documents.

29. *Owner*—The individual or entity with whom Contractor has entered into the Agreement and for whom the Work is to be performed.

30. *PCBs*—Polychlorinated biphenyls.

31. *Petroleum*—Petroleum, including crude oil or any fraction thereof which is liquid at standard conditions of temperature and pressure (60 degrees Fahrenheit and 14.7 pounds per square inch absolute), such as oil, petroleum, fuel oil, oil sludge, oil refuse, gasoline, kerosene, and oil mixed with other non-Hazardous Waste and crude oils.

32. *Progress Schedule*—A schedule, prepared and maintained by Contractor, describing the sequence and duration of the activities comprising the Contractor's plan to accomplish the Work within the Contract Times.

33. *Project*—The total construction of which the Work to be performed under the Contract Documents may be the whole, or a part.

34. *Project Manual*—The bound documentary information prepared for bidding and constructing the Work. A listing of the contents of the Project Manual, which may be bound in one or more volumes, is contained in the table(s) of contents.

35. *Radioactive Material*—Source, special nuclear, or byproduct material as defined by the Atomic Energy Act of 1954 (42 USC Section 2011 et seq.) as amended from time to time.

36. *Resident Project Representative*—The authorized representative of Engineer who may be assigned to the Site or any part thereof.

37. *Samples*—Physical examples of materials, equipment, or workmanship that are representative of some portion of the Work and which establish the standards by which such portion of the Work will be judged.

38. *Schedule of Submittals*—A schedule, prepared and maintained by Contractor, of required submittals and the time requirements to support scheduled performance of related construction activities.

39. *Schedule of Values*—A schedule, prepared and maintained by Contractor, allocating portions of the Contract Price to various portions of the Work and used as the basis for reviewing Contractor's Applications for Payment.

40. *Shop Drawings*—All drawings, diagrams, illustrations, schedules, and other data or information which are specifically prepared or assembled by or for Contractor and submitted by Contractor to illustrate some portion of the Work.

41. *Site*—Lands or areas indicated in the Contract Documents as being furnished by Owner upon which the Work is to be performed, including rights-of-way and easements for access thereto, and such other lands furnished by Owner which are designated for the use of Contractor.

42. *Specifications*—That part of the Contract Documents consisting of written requirements for materials, equipment, systems, standards and workmanship as applied to the Work, and certain administrative requirements and procedural matters applicable thereto.

43. *Subcontractor*—An individual or entity having a direct contract with Contractor or with any other Subcontractor for the performance of a part of the Work at the Site.

44. *Substantial Completion*—The time at which the Work (or a specified part thereof) has progressed to the point where, in the opinion of Engineer, the Work (or a specified part thereof) is sufficiently complete, in accordance with the Contract Documents, so that the Work (or a specified part thereof) can be utilized for the purposes for which it is intended. The terms "substantially complete" and "substantially completed" as applied to all or part of the Work refer to Substantial Completion thereof.

45. *Successful Bidder*—The Bidder submitting a responsive Bid to whom Owner makes an award.

46. *Supplementary Conditions*—That part of the Contract Documents which amends or supplements these General Conditions.

47. *Supplier*—A manufacturer, fabricator, supplier, distributor, materialman, or vendor having a direct contract with Contractor or with any Subcontractor to furnish materials or equipment to be incorporated in the Work by Contractor or Subcontractor.

48. *Underground Facilities*—All underground pipelines, conduits, ducts, cables, wires, manholes, vaults, tanks, tunnels, or other such facilities or attachments, and any encasements containing such facilities, including those that convey electricity, gases, steam, liquid petroleum products, telephone or other communications, cable television, water, wastewater, storm water, other liquids or chemicals, or traffic or other control systems.

49. *Unit Price Work*—Work to be paid for on the basis of unit prices.

50. *Work*—The entire construction or the various separately identifiable parts thereof required to be provided under the Contract Documents. Work includes and is the result of performing or providing all labor, services, and documentation necessary to produce such construction, and furnishing, installing, and incorporating all materials and equipment into such construction, all as required by the Contract Documents.

51. *Work Change Directive*—A written statement to Contractor issued on or after the Effective Date of the Agreement and signed by Owner and recommended by Engineer ordering an

addition, deletion, or revision in the Work, or responding to differing or unforeseen subsurface or physical conditions under which the Work is to be performed or to emergencies. A Work Change Directive will not change the Contract Price or the Contract Times but is evidence that the parties expect that the change ordered or documented by a Work Change Directive will be incorporated in a subsequently issued Change Order following negotiations by the parties as to its effect, if any, on the Contract Price or Contract Times.

1.02   *Terminology*

A. The words and terms discussed in Paragraph 1.02.B through F are not defined but, when used in the Bidding Requirements or Contract Documents, have the indicated meaning.

B. *Intent of Certain Terms or Adjectives:*

    1. The Contract Documents include the terms "as allowed," "as approved," "as ordered," "as directed" or terms of like effect or import to authorize an exercise of professional judgment by Engineer. In addition, the adjectives "reasonable," "suitable," "acceptable," "proper," "satisfactory," or adjectives of like effect or import are used to describe an action or determination of Engineer as to the Work. It is intended that such exercise of professional judgment, action, or determination will be solely to evaluate, in general, the Work for compliance with the information in the Contract Documents and with the design concept of the Project as a functioning whole as shown or indicated in the Contract Documents (unless there is a specific statement indicating otherwise). The use of any such term or adjective is not intended to and shall not be effective to assign to Engineer any duty or authority to supervise or direct the performance of the Work, or any duty or authority to undertake responsibility contrary to the provisions of Paragraph 9.09 or any other provision of the Contract Documents.

C. *Day:*

    1. The word "day" means a calendar day of 24 hours measured from midnight to the next midnight.

D. *Defective:*

    1. The word "defective," when modifying the word "Work," refers to Work that is unsatisfactory, faulty, or deficient in that it:

        a. does not conform to the Contract Documents; or

        b. does not meet the requirements of any applicable inspection, reference standard, test, or approval referred to in the Contract Documents; or

        c. has been damaged prior to Engineer's recommendation of final payment (unless responsibility for the protection thereof has been assumed by Owner at Substantial Completion in accordance with Paragraph 14.04 or 14.05).

E. *Furnish, Install, Perform, Provide:*

1. The word "furnish," when used in connection with services, materials, or equipment, shall mean to supply and deliver said services, materials, or equipment to the Site (or some other specified location) ready for use or installation and in usable or operable condition.

2. The word "install," when used in connection with services, materials, or equipment, shall mean to put into use or place in final position said services, materials, or equipment complete and ready for intended use.

3. The words "perform" or "provide," when used in connection with services, materials, or equipment, shall mean to furnish and install said services, materials, or equipment complete and ready for intended use.

4. When "furnish," "install," "perform," or "provide" is not used in connection with services, materials, or equipment in a context clearly requiring an obligation of Contractor, "provide" is implied.

F. Unless stated otherwise in the Contract Documents, words or phrases that have a well-known technical or construction industry or trade meaning are used in the Contract Documents in accordance with such recognized meaning.

ARTICLE 2 – PRELIMINARY MATTERS

2.01 *Delivery of Bonds and Evidence of Insurance*

A. When Contractor delivers the executed counterparts of the Agreement to Owner, Contractor shall also deliver to Owner such bonds as Contractor may be required to furnish.

B. *Evidence of Insurance.* Before any Work at the Site is started, Contractor and Owner shall each deliver to the other, with copies to each additional insured identified in the Supplementary Conditions, certificates of insurance (and other evidence of insurance which either of them or any additional insured may reasonably request) which Contractor and Owner respectively are required to purchase and maintain in accordance with Article 5.

2.02 *Copies of Documents*

A. Owner shall furnish to Contractor up to ten printed or hard copies of the Drawings and Project Manual. Additional copies will be furnished upon request at the cost of reproduction.

2.03 *Commencement of Contract Times; Notice to Proceed*

A. The Contract Times will commence to run on the thirtieth day after the Effective Date of the Agreement or, if a Notice to Proceed is given, on the day indicated in the Notice to Proceed. A Notice to Proceed may be given at any time within 30 days after the Effective Date of the Agreement. In no event will the Contract Times commence to run later than the sixtieth day after the day of Bid opening or the thirtieth day after the Effective Date of the Agreement, whichever date is earlier.

2.04   *Starting the Work*

    A. Contractor shall start to perform the Work on the date when the Contract Times commence to run. No Work shall be done at the Site prior to the date on which the Contract Times commence to run.

2.05   *Before Starting Construction*

    A. *Preliminary Schedules:* Within 10 days after the Effective Date of the Agreement (unless otherwise specified in the General Requirements), Contractor shall submit to Engineer for timely review:

        1. a preliminary Progress Schedule indicating the times (numbers of days or dates) for starting and completing the various stages of the Work, including any Milestones specified in the Contract Documents;

        2. a preliminary Schedule of Submittals; and

        3. a preliminary Schedule of Values for all of the Work which includes quantities and prices of items which when added together equal the Contract Price and subdivides the Work into component parts in sufficient detail to serve as the basis for progress payments during performance of the Work. Such prices will include an appropriate amount of overhead and profit applicable to each item of Work.

2.06   *Preconstruction Conference; Designation of Authorized Representatives*

    A. Before any Work at the Site is started, a conference attended by Owner, Contractor, Engineer, and others as appropriate will be held to establish a working understanding among the parties as to the Work and to discuss the schedules referred to in Paragraph 2.05.A, procedures for handling Shop Drawings and other submittals, processing Applications for Payment, and maintaining required records.

    B. At this conference Owner and Contractor each shall designate, in writing, a specific individual to act as its authorized representative with respect to the services and responsibilities under the Contract. Such individuals shall have the authority to transmit instructions, receive information, render decisions relative to the Contract, and otherwise act on behalf of each respective party.

2.07   *Initial Acceptance of Schedules*

    A. At least 10 days before submission of the first Application for Payment a conference attended by Contractor, Engineer, and others as appropriate will be held to review for acceptability to Engineer as provided below the schedules submitted in accordance with Paragraph 2.05.A. Contractor shall have an additional 10 days to make corrections and adjustments and to complete and resubmit the schedules. No progress payment shall be made to Contractor until acceptable schedules are submitted to Engineer.

        1. The Progress Schedule will be acceptable to Engineer if it provides an orderly progression of the Work to completion within the Contract Times. Such acceptance will not impose on Engineer responsibility for the Progress Schedule, for sequencing, scheduling, or progress of

the Work, nor interfere with or relieve Contractor from Contractor's full responsibility therefor.

2. Contractor's Schedule of Submittals will be acceptable to Engineer if it provides a workable arrangement for reviewing and processing the required submittals.

3. Contractor's Schedule of Values will be acceptable to Engineer as to form and substance if it provides a reasonable allocation of the Contract Price to component parts of the Work.

ARTICLE 3 – CONTRACT DOCUMENTS: INTENT, AMENDING, REUSE

3.01 *Intent*

A. The Contract Documents are complementary; what is required by one is as binding as if required by all.

B. It is the intent of the Contract Documents to describe a functionally complete project (or part thereof) to be constructed in accordance with the Contract Documents. Any labor, documentation, services, materials, or equipment that reasonably may be inferred from the Contract Documents or from prevailing custom or trade usage as being required to produce the indicated result will be provided whether or not specifically called for, at no additional cost to Owner.

C. Clarifications and interpretations of the Contract Documents shall be issued by Engineer as provided in Article 9.

3.02 *Reference Standards*

A. Standards, Specifications, Codes, Laws, and Regulations

1. Reference to standards, specifications, manuals, or codes of any technical society, organization, or association, or to Laws or Regulations, whether such reference be specific or by implication, shall mean the standard, specification, manual, code, or Laws or Regulations in effect at the time of opening of Bids (or on the Effective Date of the Agreement if there were no Bids), except as may be otherwise specifically stated in the Contract Documents.

2. No provision of any such standard, specification, manual, or code, or any instruction of a Supplier, shall be effective to change the duties or responsibilities of Owner, Contractor, or Engineer, or any of their subcontractors, consultants, agents, or employees, from those set forth in the Contract Documents. No such provision or instruction shall be effective to assign to Owner, Engineer, or any of their officers, directors, members, partners, employees, agents, consultants, or subcontractors, any duty or authority to supervise or direct the performance of the Work or any duty or authority to undertake responsibility inconsistent with the provisions of the Contract Documents.

3.03 *Reporting and Resolving Discrepancies*

A. *Reporting Discrepancies:*

1. *Contractor's Review of Contract Documents Before Starting Work*: Before undertaking each part of the Work, Contractor shall carefully study and compare the Contract Documents and check and verify pertinent figures therein and all applicable field measurements. Contractor shall promptly report in writing to Engineer any conflict, error, ambiguity, or discrepancy which Contractor discovers, or has actual knowledge of, and shall obtain a written interpretation or clarification from Engineer before proceeding with any Work affected thereby.

2. *Contractor's Review of Contract Documents During Performance of Work*: If, during the performance of the Work, Contractor discovers any conflict, error, ambiguity, or discrepancy within the Contract Documents, or between the Contract Documents and (a) any applicable Law or Regulation , (b) any standard, specification, manual, or code, or (c) any instruction of any Supplier, then Contractor shall promptly report it to Engineer in writing. Contractor shall not proceed with the Work affected thereby (except in an emergency as required by Paragraph 6.16.A) until an amendment or supplement to the Contract Documents has been issued by one of the methods indicated in Paragraph 3.04.

3. Contractor shall not be liable to Owner or Engineer for failure to report any conflict, error, ambiguity, or discrepancy in the Contract Documents unless Contractor had actual knowledge thereof.

B. *Resolving Discrepancies*:

1. Except as may be otherwise specifically stated in the Contract Documents, the provisions of the Contract Documents shall take precedence in resolving any conflict, error, ambiguity, or discrepancy between the provisions of the Contract Documents and:

   a. the provisions of any standard, specification, manual, or code, or the instruction of any Supplier (whether or not specifically incorporated by reference in the Contract Documents); or

   b. the provisions of any Laws or Regulations applicable to the performance of the Work (unless such an interpretation of the provisions of the Contract Documents would result in violation of such Law or Regulation).

3.04   *Amending and Supplementing Contract Documents*

A. The Contract Documents may be amended to provide for additions, deletions, and revisions in the Work or to modify the terms and conditions thereof by either a Change Order or a Work Change Directive.

B. The requirements of the Contract Documents may be supplemented, and minor variations and deviations in the Work may be authorized, by one or more of the following ways:

1. A Field Order;

2. Engineer's approval of a Shop Drawing or Sample (subject to the provisions of Paragraph 6.17.D.3); or

3. Engineer's written interpretation or clarification.

3.05 *Reuse of Documents*

A. Contractor and any Subcontractor or Supplier shall not:

1. have or acquire any title to or ownership rights in any of the Drawings, Specifications, or other documents (or copies of any thereof) prepared by or bearing the seal of Engineer or its consultants, including electronic media editions; or

2. reuse any such Drawings, Specifications, other documents, or copies thereof on extensions of the Project or any other project without written consent of Owner and Engineer and specific written verification or adaptation by Engineer.

B. The prohibitions of this Paragraph 3.05 will survive final payment, or termination of the Contract. Nothing herein shall preclude Contractor from retaining copies of the Contract Documents for record purposes.

3.06 *Electronic Data*

A. Unless otherwise stated in the Supplementary Conditions, the data furnished by Owner or Engineer to Contractor, or by Contractor to Owner or Engineer, that may be relied upon are limited to the printed copies (also known as hard copies). Files in electronic media format of text, data, graphics, or other types are furnished only for the convenience of the receiving party. Any conclusion or information obtained or derived from such electronic files will be at the user's sole risk. If there is a discrepancy between the electronic files and the hard copies, the hard copies govern.

B. Because data stored in electronic media format can deteriorate or be modified inadvertently or otherwise without authorization of the data's creator, the party receiving electronic files agrees that it will perform acceptance tests or procedures within 60 days, after which the receiving party shall be deemed to have accepted the data thus transferred. Any errors detected within the 60-day acceptance period will be corrected by the transferring party.

C. When transferring documents in electronic media format, the transferring party makes no representations as to long term compatibility, usability, or readability of documents resulting from the use of software application packages, operating systems, or computer hardware differing from those used by the data's creator.

ARTICLE 4 – AVAILABILITY OF LANDS; SUBSURFACE AND PHYSICAL CONDITIONS; HAZARDOUS ENVIRONMENTAL CONDITIONS; REFERENCE POINTS

4.01 *Availability of Lands*

A. Owner shall furnish the Site. Owner shall notify Contractor of any encumbrances or restrictions not of general application but specifically related to use of the Site with which Contractor must comply in performing the Work. Owner will obtain in a timely manner and pay for easements for permanent structures or permanent changes in existing facilities. If Contractor and Owner are unable to agree on entitlement to or on the amount or extent, if any, of any adjustment in the

Contract Price or Contract Times, or both, as a result of any delay in Owner's furnishing the Site or a part thereof, Contractor may make a Claim therefor as provided in Paragraph 10.05.

B. Upon reasonable written request, Owner shall furnish Contractor with a current statement of record legal title and legal description of the lands upon which the Work is to be performed and Owner's interest therein as necessary for giving notice of or filing a mechanic's or construction lien against such lands in accordance with applicable Laws and Regulations.

C. Contractor shall provide for all additional lands and access thereto that may be required for temporary construction facilities or storage of materials and equipment.

4.02   *Subsurface and Physical Conditions*

A. *Reports and Drawings:* The Supplementary Conditions identify:

  1. those reports known to Owner of explorations and tests of subsurface conditions at or contiguous to the Site; and

  2. those drawings known to Owner of physical conditions relating to existing surface or subsurface structures at the Site (except Underground Facilities).

B. *Limited Reliance by Contractor on Technical Data Authorized:* Contractor may rely upon the accuracy of the "technical data" contained in such reports and drawings, but such reports and drawings are not Contract Documents. Such "technical data" is identified in the Supplementary Conditions. Except for such reliance on such "technical data," Contractor may not rely upon or make any claim against Owner or Engineer, or any of their officers, directors, members, partners, employees, agents, consultants, or subcontractors with respect to:

  1. the completeness of such reports and drawings for Contractor's purposes, including, but not limited to, any aspects of the means, methods, techniques, sequences, and procedures of construction to be employed by Contractor, and safety precautions and programs incident thereto; or

  2. other data, interpretations, opinions, and information contained in such reports or shown or indicated in such drawings; or

  3. any Contractor interpretation of or conclusion drawn from any "technical data" or any such other data, interpretations, opinions, or information.

4.03   *Differing Subsurface or Physical Conditions*

A. *Notice:* If Contractor believes that any subsurface or physical condition that is uncovered or revealed either:

  1. is of such a nature as to establish that any "technical data" on which Contractor is entitled to rely as provided in Paragraph 4.02 is materially inaccurate; or

  2. is of such a nature as to require a change in the Contract Documents; or

3. differs materially from that shown or indicated in the Contract Documents; or

4. is of an unusual nature, and differs materially from conditions ordinarily encountered and generally recognized as inherent in work of the character provided for in the Contract Documents;

then Contractor shall, promptly after becoming aware thereof and before further disturbing the subsurface or physical conditions or performing any Work in connection therewith (except in an emergency as required by Paragraph 6.16.A), notify Owner and Engineer in writing about such condition. Contractor shall not further disturb such condition or perform any Work in connection therewith (except as aforesaid) until receipt of written order to do so.

B. *Engineer's Review:* After receipt of written notice as required by Paragraph 4.03.A, Engineer will promptly review the pertinent condition, determine the necessity of Owner's obtaining additional exploration or tests with respect thereto, and advise Owner in writing (with a copy to Contractor) of Engineer's findings and conclusions.

C. *Possible Price and Times Adjustments:*

1. The Contract Price or the Contract Times, or both, will be equitably adjusted to the extent that the existence of such differing subsurface or physical condition causes an increase or decrease in Contractor's cost of, or time required for, performance of the Work; subject, however, to the following:

   a. such condition must meet any one or more of the categories described in Paragraph 4.03.A; and

   b. with respect to Work that is paid for on a unit price basis, any adjustment in Contract Price will be subject to the provisions of Paragraphs 9.07 and 11.03.

2. Contractor shall not be entitled to any adjustment in the Contract Price or Contract Times if:

   a. Contractor knew of the existence of such conditions at the time Contractor made a final commitment to Owner with respect to Contract Price and Contract Times by the submission of a Bid or becoming bound under a negotiated contract; or

   b. the existence of such condition could reasonably have been discovered or revealed as a result of any examination, investigation, exploration, test, or study of the Site and contiguous areas required by the Bidding Requirements or Contract Documents to be conducted by or for Contractor prior to Contractor's making such final commitment; or

   c. Contractor failed to give the written notice as required by Paragraph 4.03.A.

3. If Owner and Contractor are unable to agree on entitlement to or on the amount or extent, if any, of any adjustment in the Contract Price or Contract Times, or both, a Claim may be made therefor as provided in Paragraph 10.05. However, neither Owner or Engineer, or any of their officers, directors, members, partners, employees, agents, consultants, or subcontractors shall be liable to Contractor for any claims, costs, losses, or damages (including but not limited to all fees and charges of engineers, architects, attorneys, and other

professionals and all court or arbitration or other dispute resolution costs) sustained by Contractor on or in connection with any other project or anticipated project.

4.04   *Underground Facilities*

A.   *Shown or Indicated:*   The information and data shown or indicated in the Contract Documents with respect to existing Underground Facilities at or contiguous to the Site is based on information and data furnished to Owner or Engineer by the owners of such Underground Facilities, including Owner, or by others. Unless it is otherwise expressly provided in the Supplementary Conditions:

1.   Owner and Engineer shall not be responsible for the accuracy or completeness of any such information or data provided by others; and

2.   the cost of all of the following will be included in the Contract Price, and Contractor shall have full responsibility for:

   a.   reviewing and checking all such information and data;

   b.   locating all Underground Facilities shown or indicated in the Contract Documents;

   c.   coordination of the Work with the owners of such Underground Facilities, including Owner, during construction; and

   d.   the safety and protection of all such Underground Facilities and repairing any damage thereto resulting from the Work.

B.   *Not Shown or Indicated:*

1.   If an Underground Facility is uncovered or revealed at or contiguous to the Site which was not shown or indicated, or not shown or indicated with reasonable accuracy in the Contract Documents, Contractor shall, promptly after becoming aware thereof and before further disturbing conditions affected thereby or performing any Work in connection therewith (except in an emergency as required by Paragraph 6.16.A), identify the owner of such Underground Facility and give written notice to that owner and to Owner and Engineer. Engineer will promptly review the Underground Facility and determine the extent, if any, to which a change is required in the Contract Documents to reflect and document the consequences of the existence or location of the Underground Facility. During such time, Contractor shall be responsible for the safety and protection of such Underground Facility.

2.   If Engineer concludes that a change in the Contract Documents is required, a Work Change Directive or a Change Order will be issued to reflect and document such consequences. An equitable adjustment shall be made in the Contract Price or Contract Times, or both, to the extent that they are attributable to the existence or location of any Underground Facility that was not shown or indicated or not shown or indicated with reasonable accuracy in the Contract Documents and that Contractor did not know of and could not reasonably have been expected to be aware of or to have anticipated. If Owner and Contractor are unable to agree on entitlement to or on the amount or extent, if any, of any such adjustment in Contract Price

or Contract Times, Owner or Contractor may make a Claim therefor as provided in Paragraph 10.05.

4.05 *Reference Points*

A. Owner shall provide engineering surveys to establish reference points for construction which in Engineer's judgment are necessary to enable Contractor to proceed with the Work. Contractor shall be responsible for laying out the Work, shall protect and preserve the established reference points and property monuments, and shall make no changes or relocations without the prior written approval of Owner. Contractor shall report to Engineer whenever any reference point or property monument is lost or destroyed or requires relocation because of necessary changes in grades or locations, and shall be responsible for the accurate replacement or relocation of such reference points or property monuments by professionally qualified personnel.

4.06 *Hazardous Environmental Condition at Site*

A. *Reports and Drawings:* The Supplementary Conditions identify those reports and drawings known to Owner relating to Hazardous Environmental Conditions that have been identified at the Site.

B. *Limited Reliance by Contractor on Technical Data Authorized:* Contractor may rely upon the accuracy of the "technical data" contained in such reports and drawings, but such reports and drawings are not Contract Documents. Such "technical data" is identified in the Supplementary Conditions. Except for such reliance on such "technical data," Contractor may not rely upon or make any claim against Owner or Engineer, or any of their officers, directors, members, partners, employees, agents, consultants, or subcontractors with respect to:

1. the completeness of such reports and drawings for Contractor's purposes, including, but not limited to, any aspects of the means, methods, techniques, sequences and procedures of construction to be employed by Contractor and safety precautions and programs incident thereto; or

2. other data, interpretations, opinions and information contained in such reports or shown or indicated in such drawings; or

3. any Contractor interpretation of or conclusion drawn from any "technical data" or any such other data, interpretations, opinions or information.

C. Contractor shall not be responsible for any Hazardous Environmental Condition uncovered or revealed at the Site which was not shown or indicated in Drawings or Specifications or identified in the Contract Documents to be within the scope of the Work. Contractor shall be responsible for a Hazardous Environmental Condition created with any materials brought to the Site by Contractor, Subcontractors, Suppliers, or anyone else for whom Contractor is responsible.

D. If Contractor encounters a Hazardous Environmental Condition or if Contractor or anyone for whom Contractor is responsible creates a Hazardous Environmental Condition, Contractor shall immediately: (i) secure or otherwise isolate such condition; (ii) stop all Work in connection with such condition and in any area affected thereby (except in an emergency as required by

Paragraph 6.16.A); and (iii) notify Owner and Engineer (and promptly thereafter confirm such notice in writing). Owner shall promptly consult with Engineer concerning the necessity for Owner to retain a qualified expert to evaluate such condition or take corrective action, if any. Promptly after consulting with Engineer, Owner shall take such actions as are necessary to permit Owner to timely obtain required permits and provide Contractor the written notice required by Paragraph 4.06.E.

E. Contractor shall not be required to resume Work in connection with such condition or in any affected area until after Owner has obtained any required permits related thereto and delivered written notice to Contractor: (i) specifying that such condition and any affected area is or has been rendered safe for the resumption of Work; or (ii) specifying any special conditions under which such Work may be resumed safely. If Owner and Contractor cannot agree as to entitlement to or on the amount or extent, if any, of any adjustment in Contract Price or Contract Times, or both, as a result of such Work stoppage or such special conditions under which Work is agreed to be resumed by Contractor, either party may make a Claim therefor as provided in Paragraph 10.05.

F. If after receipt of such written notice Contractor does not agree to resume such Work based on a reasonable belief it is unsafe, or does not agree to resume such Work under such special conditions, then Owner may order the portion of the Work that is in the area affected by such condition to be deleted from the Work. If Owner and Contractor cannot agree as to entitlement to or on the amount or extent, if any, of an adjustment in Contract Price or Contract Times as a result of deleting such portion of the Work, then either party may make a Claim therefor as provided in Paragraph 10.05. Owner may have such deleted portion of the Work performed by Owner's own forces or others in accordance with Article 7.

G. To the fullest extent permitted by Laws and Regulations, Owner shall indemnify and hold harmless Contractor, Subcontractors, and Engineer, and the officers, directors, members, partners, employees, agents, consultants, and subcontractors of each and any of them from and against all claims, costs, losses, and damages (including but not limited to all fees and charges of engineers, architects, attorneys, and other professionals and all court or arbitration or other dispute resolution costs) arising out of or relating to a Hazardous Environmental Condition, provided that such Hazardous Environmental Condition: (i) was not shown or indicated in the Drawings or Specifications or identified in the Contract Documents to be included within the scope of the Work, and (ii) was not created by Contractor or by anyone for whom Contractor is responsible. Nothing in this Paragraph 4.06.G shall obligate Owner to indemnify any individual or entity from and against the consequences of that individual's or entity's own negligence.

H. To the fullest extent permitted by Laws and Regulations, Contractor shall indemnify and hold harmless Owner and Engineer, and the officers, directors, members, partners, employees, agents, consultants, and subcontractors of each and any of them from and against all claims, costs, losses, and damages (including but not limited to all fees and charges of engineers, architects, attorneys, and other professionals and all court or arbitration or other dispute resolution costs) arising out of or relating to a Hazardous Environmental Condition created by Contractor or by anyone for whom Contractor is responsible. Nothing in this Paragraph 4.06.H shall obligate Contractor to indemnify any individual or entity from and against the consequences of that individual's or entity's own negligence.

I. The provisions of Paragraphs 4.02, 4.03, and 4.04 do not apply to a Hazardous Environmental Condition uncovered or revealed at the Site.

ARTICLE 5 – BONDS AND INSURANCE

5.01   *Performance, Payment, and Other Bonds*

A. Contractor shall furnish performance and payment bonds, each in an amount at least equal to the Contract Price as security for the faithful performance and payment of all of Contractor's obligations under the Contract Documents. These bonds shall remain in effect until one year after the date when final payment becomes due or until completion of the correction period specified in Paragraph 13.07, whichever is later, except as provided otherwise by Laws or Regulations or by the Contract Documents. Contractor shall also furnish such other bonds as are required by the Contract Documents.

B. All bonds shall be in the form prescribed by the Contract Documents except as provided otherwise by Laws or Regulations, and shall be executed by such sureties as are named in the list of "Companies Holding Certificates of Authority as Acceptable Sureties on Federal Bonds and as Acceptable Reinsuring Companies" as published in Circular 570 (amended) by the Financial Management Service, Surety Bond Branch, U.S. Department of the Treasury. All bonds signed by an agent or attorney-in-fact must be accompanied by a certified copy of that individual's authority to bind the surety. The evidence of authority shall show that it is effective on the date the agent or attorney-in-fact signed each bond.

C. If the surety on any bond furnished by Contractor is declared bankrupt or becomes insolvent or its right to do business is terminated in any state where any part of the Project is located or it ceases to meet the requirements of Paragraph 5.01.B, Contractor shall promptly notify Owner and Engineer and shall, within 20 days after the event giving rise to such notification, provide another bond and surety, both of which shall comply with the requirements of Paragraphs 5.01.B and 5.02.

5.02   *Licensed Sureties and Insurers*

A. All bonds and insurance required by the Contract Documents to be purchased and maintained by Owner or Contractor shall be obtained from surety or insurance companies that are duly licensed or authorized in the jurisdiction in which the Project is located to issue bonds or insurance policies for the limits and coverages so required. Such surety and insurance companies shall also meet such additional requirements and qualifications as may be provided in the Supplementary Conditions.

5.03   *Certificates of Insurance*

A. Contractor shall deliver to Owner, with copies to each additional insured and loss payee identified in the Supplementary Conditions, certificates of insurance (and other evidence of insurance requested by Owner or any other additional insured) which Contractor is required to purchase and maintain.

B. Owner shall deliver to Contractor, with copies to each additional insured and loss payee identified in the Supplementary Conditions, certificates of insurance (and other evidence of insurance requested by Contractor or any other additional insured) which Owner is required to purchase and maintain.

C. Failure of Owner to demand such certificates or other evidence of Contractor's full compliance with these insurance requirements or failure of Owner to identify a deficiency in compliance from the evidence provided shall not be construed as a waiver of Contractor's obligation to maintain such insurance.

D. Owner does not represent that insurance coverage and limits established in this Contract necessarily will be adequate to protect Contractor.

E. The insurance and insurance limits required herein shall not be deemed as a limitation on Contractor's liability under the indemnities granted to Owner in the Contract Documents.

5.04 *Contractor's Insurance*

A. Contractor shall purchase and maintain such insurance as is appropriate for the Work being performed and as will provide protection from claims set forth below which may arise out of or result from Contractor's performance of the Work and Contractor's other obligations under the Contract Documents, whether it is to be performed by Contractor, any Subcontractor or Supplier, or by anyone directly or indirectly employed by any of them to perform any of the Work, or by anyone for whose acts any of them may be liable:

 1. claims under workers' compensation, disability benefits, and other similar employee benefit acts;

 2. claims for damages because of bodily injury, occupational sickness or disease, or death of Contractor's employees;

 3. claims for damages because of bodily injury, sickness or disease, or death of any person other than Contractor's employees;

 4. claims for damages insured by reasonably available personal injury liability coverage which are sustained:

    a. by any person as a result of an offense directly or indirectly related to the employment of such person by Contractor, or

    b. by any other person for any other reason;

 5. claims for damages, other than to the Work itself, because of injury to or destruction of tangible property wherever located, including loss of use resulting therefrom; and

 6. claims for damages because of bodily injury or death of any person or property damage arising out of the ownership, maintenance or use of any motor vehicle.

B. The policies of insurance required by this Paragraph 5.04 shall:

1. with respect to insurance required by Paragraphs 5.04.A.3 through 5.04.A.6 inclusive, be written on an occurrence basis, include as additional insureds (subject to any customary exclusion regarding professional liability) Owner and Engineer, and any other individuals or entities identified in the Supplementary Conditions, all of whom shall be listed as additional insureds, and include coverage for the respective officers, directors, members, partners, employees, agents, consultants, and subcontractors of each and any of all such additional insureds, and the insurance afforded to these additional insureds shall provide primary coverage for all claims covered thereby;

2. include at least the specific coverages and be written for not less than the limits of liability provided in the Supplementary Conditions or required by Laws or Regulations, whichever is greater;

3. include contractual liability insurance covering Contractor's indemnity obligations under Paragraphs 6.11 and 6.20;

4. contain a provision or endorsement that the coverage afforded will not be canceled, materially changed or renewal refused until at least 30 days prior written notice has been given to Owner and Contractor and to each other additional insured identified in the Supplementary Conditions to whom a certificate of insurance has been issued (and the certificates of insurance furnished by the Contractor pursuant to Paragraph 5.03 will so provide);

5. remain in effect at least until final payment and at all times thereafter when Contractor may be correcting, removing, or replacing defective Work in accordance with Paragraph 13.07; and

6. include completed operations coverage:

   a. Such insurance shall remain in effect for two years after final payment.

   b. Contractor shall furnish Owner and each other additional insured identified in the Supplementary Conditions, to whom a certificate of insurance has been issued, evidence satisfactory to Owner and any such additional insured of continuation of such insurance at final payment and one year thereafter.

5.05 *Owner's Liability Insurance*

A. In addition to the insurance required to be provided by Contractor under Paragraph 5.04, Owner, at Owner's option, may purchase and maintain at Owner's expense Owner's own liability insurance as will protect Owner against claims which may arise from operations under the Contract Documents.

5.06 *Property Insurance*

A. Unless otherwise provided in the Supplementary Conditions, Owner shall purchase and maintain property insurance upon the Work at the Site in the amount of the full replacement cost thereof (subject to such deductible amounts as may be provided in the Supplementary Conditions or required by Laws and Regulations). This insurance shall:

1. include the interests of Owner, Contractor, Subcontractors, and Engineer, and any other individuals or entities identified in the Supplementary Conditions, and the officers, directors, members, partners, employees, agents, consultants, and subcontractors of each and any of them, each of whom is deemed to have an insurable interest and shall be listed as a loss payee;

2. be written on a Builder's Risk "all-risk" policy form that shall at least include insurance for physical loss or damage to the Work, temporary buildings, falsework, and materials and equipment in transit, and shall insure against at least the following perils or causes of loss: fire, lightning, extended coverage, theft, vandalism and malicious mischief, earthquake, collapse, debris removal, demolition occasioned by enforcement of Laws and Regulations, water damage (other than that caused by flood), and such other perils or causes of loss as may be specifically required by the Supplementary Conditions.

3. include expenses incurred in the repair or replacement of any insured property (including but not limited to fees and charges of engineers and architects);

4. cover materials and equipment stored at the Site or at another location that was agreed to in writing by Owner prior to being incorporated in the Work, provided that such materials and equipment have been included in an Application for Payment recommended by Engineer;

5. allow for partial utilization of the Work by Owner;

6. include testing and startup; and

7. be maintained in effect until final payment is made unless otherwise agreed to in writing by Owner, Contractor, and Engineer with 30 days written notice to each other loss payee to whom a certificate of insurance has been issued.

B. Owner shall purchase and maintain such equipment breakdown insurance or additional property insurance as may be required by the Supplementary Conditions or Laws and Regulations which will include the interests of Owner, Contractor, Subcontractors, and Engineer, and any other individuals or entities identified in the Supplementary Conditions, and the officers, directors, members, partners, employees, agents, consultants and subcontractors of each and any of them, each of whom is deemed to have an insurable interest and shall be listed as a loss payee.

C. All the policies of insurance (and the certificates or other evidence thereof) required to be purchased and maintained in accordance with this Paragraph 5.06 will contain a provision or endorsement that the coverage afforded will not be canceled or materially changed or renewal refused until at least 30 days prior written notice has been given to Owner and Contractor and to each other loss payee to whom a certificate of insurance has been issued and will contain waiver provisions in accordance with Paragraph 5.07.

D. Owner shall not be responsible for purchasing and maintaining any property insurance specified in this Paragraph 5.06 to protect the interests of Contractor, Subcontractors, or others in the Work to the extent of any deductible amounts that are identified in the Supplementary Conditions. The risk of loss within such identified deductible amount will be borne by Contractor, Subcontractors, or others suffering any such loss, and if any of them wishes property insurance coverage within the limits of such amounts, each may purchase and maintain it at the purchaser's own expense.

E. If Contractor requests in writing that other special insurance be included in the property insurance policies provided under this Paragraph 5.06, Owner shall, if possible, include such insurance, and the cost thereof will be charged to Contractor by appropriate Change Order. Prior to commencement of the Work at the Site, Owner shall in writing advise Contractor whether or not such other insurance has been procured by Owner.

5.07   *Waiver of Rights*

A. Owner and Contractor intend that all policies purchased in accordance with Paragraph 5.06 will protect Owner, Contractor, Subcontractors, and Engineer, and all other individuals or entities identified in the Supplementary Conditions as loss payees (and the officers, directors, members, partners, employees, agents, consultants, and subcontractors of each and any of them) in such policies and will provide primary coverage for all losses and damages caused by the perils or causes of loss covered thereby. All such policies shall contain provisions to the effect that in the event of payment of any loss or damage the insurers will have no rights of recovery against any of the insureds or loss payees thereunder. Owner and Contractor waive all rights against each other and their respective officers, directors, members, partners, employees, agents, consultants and subcontractors of each and any of them for all losses and damages caused by, arising out of or resulting from any of the perils or causes of loss covered by such policies and any other property insurance applicable to the Work; and, in addition, waive all such rights against Subcontractors and Engineer, and all other individuals or entities identified in the Supplementary Conditions as loss payees (and the officers, directors, members, partners, employees, agents, consultants, and subcontractors of each and any of them) under such policies for losses and damages so caused. None of the above waivers shall extend to the rights that any party making such waiver may have to the proceeds of insurance held by Owner as trustee or otherwise payable under any policy so issued.

B. Owner waives all rights against Contractor, Subcontractors, and Engineer, and the officers, directors, members, partners, employees, agents, consultants and subcontractors of each and any of them for:

1. loss due to business interruption, loss of use, or other consequential loss extending beyond direct physical loss or damage to Owner's property or the Work caused by, arising out of, or resulting from fire or other perils whether or not insured by Owner; and

2. loss or damage to the completed Project or part thereof caused by, arising out of, or resulting from fire or other insured peril or cause of loss covered by any property insurance maintained on the completed Project or part thereof by Owner during partial utilization pursuant to Paragraph 14.05, after Substantial Completion pursuant to Paragraph 14.04, or after final payment pursuant to Paragraph 14.07.

C. Any insurance policy maintained by Owner covering any loss, damage or consequential loss referred to in Paragraph 5.07.B shall contain provisions to the effect that in the event of payment of any such loss, damage, or consequential loss, the insurers will have no rights of recovery against Contractor, Subcontractors, or Engineer, and the officers, directors, members, partners, employees, agents, consultants and subcontractors of each and any of them.

5.08 *Receipt and Application of Insurance Proceeds*

A. Any insured loss under the policies of insurance required by Paragraph 5.06 will be adjusted with Owner and made payable to Owner as fiduciary for the loss payees, as their interests may appear, subject to the requirements of any applicable mortgage clause and of Paragraph 5.08.B. Owner shall deposit in a separate account any money so received and shall distribute it in accordance with such agreement as the parties in interest may reach. If no other special agreement is reached, the damaged Work shall be repaired or replaced, the moneys so received applied on account thereof, and the Work and the cost thereof covered by an appropriate Change Order.

B. Owner as fiduciary shall have power to adjust and settle any loss with the insurers unless one of the parties in interest shall object in writing within 15 days after the occurrence of loss to Owner's exercise of this power. If such objection be made, Owner as fiduciary shall make settlement with the insurers in accordance with such agreement as the parties in interest may reach. If no such agreement among the parties in interest is reached, Owner as fiduciary shall adjust and settle the loss with the insurers and, if required in writing by any party in interest, Owner as fiduciary shall give bond for the proper performance of such duties.

5.09 *Acceptance of Bonds and Insurance; Option to Replace*

A. If either Owner or Contractor has any objection to the coverage afforded by or other provisions of the bonds or insurance required to be purchased and maintained by the other party in accordance with Article 5 on the basis of non-conformance with the Contract Documents, the objecting party shall so notify the other party in writing within 10 days after receipt of the certificates (or other evidence requested) required by Paragraph 2.01.B. Owner and Contractor shall each provide to the other such additional information in respect of insurance provided as the other may reasonably request. If either party does not purchase or maintain all of the bonds and insurance required of such party by the Contract Documents, such party shall notify the other party in writing of such failure to purchase prior to the start of the Work, or of such failure to maintain prior to any change in the required coverage. Without prejudice to any other right or remedy, the other party may elect to obtain equivalent bonds or insurance to protect such other party's interests at the expense of the party who was required to provide such coverage, and a Change Order shall be issued to adjust the Contract Price accordingly.

5.10 *Partial Utilization, Acknowledgment of Property Insurer*

A. If Owner finds it necessary to occupy or use a portion or portions of the Work prior to Substantial Completion of all the Work as provided in Paragraph 14.05, no such use or occupancy shall commence before the insurers providing the property insurance pursuant to Paragraph 5.06 have acknowledged notice thereof and in writing effected any changes in coverage necessitated thereby. The insurers providing the property insurance shall consent by endorsement on the policy or policies, but the property insurance shall not be canceled or permitted to lapse on account of any such partial use or occupancy.

ARTICLE 6 – CONTRACTOR'S RESPONSIBILITIES

6.01 *Supervision and Superintendence*

A. Contractor shall supervise, inspect, and direct the Work competently and efficiently, devoting such attention thereto and applying such skills and expertise as may be necessary to perform the Work in accordance with the Contract Documents. Contractor shall be solely responsible for the means, methods, techniques, sequences, and procedures of construction. Contractor shall not be responsible for the negligence of Owner or Engineer in the design or specification of a specific means, method, technique, sequence, or procedure of construction which is shown or indicated in and expressly required by the Contract Documents.

B. At all times during the progress of the Work, Contractor shall assign a competent resident superintendent who shall not be replaced without written notice to Owner and Engineer except under extraordinary circumstances.

6.02 *Labor; Working Hours*

A. Contractor shall provide competent, suitably qualified personnel to survey and lay out the Work and perform construction as required by the Contract Documents. Contractor shall at all times maintain good discipline and order at the Site.

B. Except as otherwise required for the safety or protection of persons or the Work or property at the Site or adjacent thereto, and except as otherwise stated in the Contract Documents, all Work at the Site shall be performed during regular working hours. Contractor will not permit the performance of Work on a Saturday, Sunday, or any legal holiday without Owner's written consent (which will not be unreasonably withheld) given after prior written notice to Engineer.

6.03 *Services, Materials, and Equipment*

A. Unless otherwise specified in the Contract Documents, Contractor shall provide and assume full responsibility for all services, materials, equipment, labor, transportation, construction equipment and machinery, tools, appliances, fuel, power, light, heat, telephone, water, sanitary facilities, temporary facilities, and all other facilities and incidentals necessary for the performance, testing, start-up, and completion of the Work.

B. All materials and equipment incorporated into the Work shall be as specified or, if not specified, shall be of good quality and new, except as otherwise provided in the Contract Documents. All special warranties and guarantees required by the Specifications shall expressly run to the benefit of Owner. If required by Engineer, Contractor shall furnish satisfactory evidence (including reports of required tests) as to the source, kind, and quality of materials and equipment.

C. All materials and equipment shall be stored, applied, installed, connected, erected, protected, used, cleaned, and conditioned in accordance with instructions of the applicable Supplier, except as otherwise may be provided in the Contract Documents.

6.04   *Progress Schedule*

A.   Contractor shall adhere to the Progress Schedule established in accordance with Paragraph 2.07 as it may be adjusted from time to time as provided below.

1.   Contractor shall submit to Engineer for acceptance (to the extent indicated in Paragraph 2.07) proposed adjustments in the Progress Schedule that will not result in changing the Contract Times. Such adjustments will comply with any provisions of the General Requirements applicable thereto.

2.   Proposed adjustments in the Progress Schedule that will change the Contract Times shall be submitted in accordance with the requirements of Article 12. Adjustments in Contract Times may only be made by a Change Order.

6.05   *Substitutes and "Or-Equals"*

A.   Whenever an item of material or equipment is specified or described in the Contract Documents by using the name of a proprietary item or the name of a particular Supplier, the specification or description is intended to establish the type, function, appearance, and quality required. Unless the specification or description contains or is followed by words reading that no like, equivalent, or "or-equal" item or no substitution is permitted, other items of material or equipment or material or equipment of other Suppliers may be submitted to Engineer for review under the circumstances described below.

1.   *"Or-Equal" Items:* If in Engineer's sole discretion an item of material or equipment proposed by Contractor is functionally equal to that named and sufficiently similar so that no change in related Work will be required, it may be considered by Engineer as an "or-equal" item, in which case review and approval of the proposed item may, in Engineer's sole discretion, be accomplished without compliance with some or all of the requirements for approval of proposed substitute items. For the purposes of this Paragraph 6.05.A.1, a proposed item of material or equipment will be considered functionally equal to an item so named if:

a.   in the exercise of reasonable judgment Engineer determines that:

1)   it is at least equal in materials of construction, quality, durability, appearance, strength, and design characteristics;

2)   it will reliably perform at least equally well the function and achieve the results imposed by the design concept of the completed Project as a functioning whole; and

3)   it has a proven record of performance and availability of responsive service.

b.   Contractor certifies that, if approved and incorporated into the Work:

1)   there will be no increase in cost to the Owner or increase in Contract Times; and

2)   it will conform substantially to the detailed requirements of the item named in the Contract Documents.

2. *Substitute Items:*

   a. If in Engineer's sole discretion an item of material or equipment proposed by Contractor does not qualify as an "or-equal" item under Paragraph 6.05.A.1, it will be considered a proposed substitute item.

   b. Contractor shall submit sufficient information as provided below to allow Engineer to determine if the item of material or equipment proposed is essentially equivalent to that named and an acceptable substitute therefor. Requests for review of proposed substitute items of material or equipment will not be accepted by Engineer from anyone other than Contractor.

   c. The requirements for review by Engineer will be as set forth in Paragraph 6.05.A.2.d, as supplemented by the General Requirements, and as Engineer may decide is appropriate under the circumstances.

   d. Contractor shall make written application to Engineer for review of a proposed substitute item of material or equipment that Contractor seeks to furnish or use. The application:

      1) shall certify that the proposed substitute item will:

         a) perform adequately the functions and achieve the results called for by the general design,

         b) be similar in substance to that specified, and

         c) be suited to the same use as that specified;

      2) will state:

         a) the extent, if any, to which the use of the proposed substitute item will prejudice Contractor's achievement of Substantial Completion on time,

         b) whether use of the proposed substitute item in the Work will require a change in any of the Contract Documents (or in the provisions of any other direct contract with Owner for other work on the Project) to adapt the design to the proposed substitute item, and

         c) whether incorporation or use of the proposed substitute item in connection with the Work is subject to payment of any license fee or royalty;

      3) will identify:

         a) all variations of the proposed substitute item from that specified, and

         b) available engineering, sales, maintenance, repair, and replacement services; and

4) shall contain an itemized estimate of all costs or credits that will result directly or indirectly from use of such substitute item, including costs of redesign and claims of other contractors affected by any resulting change.

B. *Substitute Construction Methods or Procedures:* If a specific means, method, technique, sequence, or procedure of construction is expressly required by the Contract Documents, Contractor may furnish or utilize a substitute means, method, technique, sequence, or procedure of construction approved by Engineer. Contractor shall submit sufficient information to allow Engineer, in Engineer's sole discretion, to determine that the substitute proposed is equivalent to that expressly called for by the Contract Documents. The requirements for review by Engineer will be similar to those provided in Paragraph 6.05.A.2.

C. *Engineer's Evaluation:* Engineer will be allowed a reasonable time within which to evaluate each proposal or submittal made pursuant to Paragraphs 6.05.A and 6.05.B. Engineer may require Contractor to furnish additional data about the proposed substitute item. Engineer will be the sole judge of acceptability. No "or equal" or substitute will be ordered, installed or utilized until Engineer's review is complete, which will be evidenced by a Change Order in the case of a substitute and an approved Shop Drawing for an "or equal." Engineer will advise Contractor in writing of any negative determination.

D. *Special Guarantee:* Owner may require Contractor to furnish at Contractor's expense a special performance guarantee or other surety with respect to any substitute.

E. *Engineer's Cost Reimbursement*: Engineer will record Engineer's costs in evaluating a substitute proposed or submitted by Contractor pursuant to Paragraphs 6.05.A.2 and 6.05.B. Whether or not Engineer approves a substitute so proposed or submitted by Contractor, Contractor shall reimburse Owner for the reasonable charges of Engineer for evaluating each such proposed substitute. Contractor shall also reimburse Owner for the reasonable charges of Engineer for making changes in the Contract Documents (or in the provisions of any other direct contract with Owner) resulting from the acceptance of each proposed substitute.

F. *Contractor's Expense*: Contractor shall provide all data in support of any proposed substitute or "or-equal" at Contractor's expense.

6.06   *Concerning Subcontractors, Suppliers, and Others*

A. Contractor shall not employ any Subcontractor, Supplier, or other individual or entity (including those acceptable to Owner as indicated in Paragraph 6.06.B), whether initially or as a replacement, against whom Owner may have reasonable objection. Contractor shall not be required to employ any Subcontractor, Supplier, or other individual or entity to furnish or perform any of the Work against whom Contractor has reasonable objection.

B. If the Supplementary Conditions require the identity of certain Subcontractors, Suppliers, or other individuals or entities to be submitted to Owner in advance for acceptance by Owner by a specified date prior to the Effective Date of the Agreement, and if Contractor has submitted a list thereof in accordance with the Supplementary Conditions, Owner's acceptance (either in writing or by failing to make written objection thereto by the date indicated for acceptance or objection in the Bidding Documents or the Contract Documents) of any such Subcontractor, Supplier, or other individual or entity so identified may be revoked on the basis of reasonable objection after due investigation. Contractor shall submit an acceptable replacement for the rejected Subcontractor, Supplier, or other individual or entity, and the Contract Price will be adjusted by the difference in the cost occasioned by such replacement, and an appropriate Change Order will be issued. No acceptance by Owner of any such Subcontractor, Supplier, or other individual or entity, whether initially or as a replacement, shall constitute a waiver of any right of Owner or Engineer to reject defective Work.

C. Contractor shall be fully responsible to Owner and Engineer for all acts and omissions of the Subcontractors, Suppliers, and other individuals or entities performing or furnishing any of the Work just as Contractor is responsible for Contractor's own acts and omissions. Nothing in the Contract Documents:

1. shall create for the benefit of any such Subcontractor, Supplier, or other individual or entity any contractual relationship between Owner or Engineer and any such Subcontractor, Supplier or other individual or entity; nor

2. shall create any obligation on the part of Owner or Engineer to pay or to see to the payment of any moneys due any such Subcontractor, Supplier, or other individual or entity except as may otherwise be required by Laws and Regulations.

D. Contractor shall be solely responsible for scheduling and coordinating the Work of Subcontractors, Suppliers, and other individuals or entities performing or furnishing any of the Work under a direct or indirect contract with Contractor.

E. Contractor shall require all Subcontractors, Suppliers, and such other individuals or entities performing or furnishing any of the Work to communicate with Engineer through Contractor.

F. The divisions and sections of the Specifications and the identifications of any Drawings shall not control Contractor in dividing the Work among Subcontractors or Suppliers or delineating the Work to be performed by any specific trade.

G. All Work performed for Contractor by a Subcontractor or Supplier will be pursuant to an appropriate agreement between Contractor and the Subcontractor or Supplier which specifically binds the Subcontractor or Supplier to the applicable terms and conditions of the Contract Documents for the benefit of Owner and Engineer. Whenever any such agreement is with a Subcontractor or Supplier who is listed as a loss payee on the property insurance provided in Paragraph 5.06, the agreement between the Contractor and the Subcontractor or Supplier will contain provisions whereby the Subcontractor or Supplier waives all rights against Owner, Contractor, Engineer, and all other individuals or entities identified in the Supplementary Conditions to be listed as insureds or loss payees (and the officers, directors, members, partners, employees, agents, consultants, and subcontractors of each and any of them) for all losses and damages caused by, arising out of, relating to, or resulting from any of the perils or causes of loss covered by such policies and any other property insurance applicable to the Work. If the insurers on any such policies require separate waiver forms to be signed by any Subcontractor or Supplier, Contractor will obtain the same.

6.07   *Patent Fees and Royalties*

   A. Contractor shall pay all license fees and royalties and assume all costs incident to the use in the performance of the Work or the incorporation in the Work of any invention, design, process, product, or device which is the subject of patent rights or copyrights held by others. If a particular invention, design, process, product, or device is specified in the Contract Documents for use in the performance of the Work and if, to the actual knowledge of Owner or Engineer, its use is subject to patent rights or copyrights calling for the payment of any license fee or royalty to others, the existence of such rights shall be disclosed by Owner in the Contract Documents.

   B. To the fullest extent permitted by Laws and Regulations, Owner shall indemnify and hold harmless Contractor, and its officers, directors, members, partners, employees, agents, consultants, and subcontractors from and against all claims, costs, losses, and damages (including but not limited to all fees and charges of engineers, architects, attorneys, and other professionals, and all court or arbitration or other dispute resolution costs) arising out of or relating to any infringement of patent rights or copyrights incident to the use in the performance of the Work or resulting from the incorporation in the Work of any invention, design, process, product, or device specified in the Contract Documents, but not identified as being subject to payment of any license fee or royalty to others required by patent rights or copyrights.

   C. To the fullest extent permitted by Laws and Regulations, Contractor shall indemnify and hold harmless Owner and Engineer, and the officers, directors, members, partners, employees, agents, consultants and subcontractors of each and any of them from and against all claims, costs, losses, and damages (including but not limited to all fees and charges of engineers, architects, attorneys, and other professionals and all court or arbitration or other dispute resolution costs) arising out of or relating to any infringement of patent rights or copyrights incident to the use in the performance of the Work or resulting from the incorporation in the Work of any invention, design, process, product, or device not specified in the Contract Documents.

6.08   *Permits*

   A. Unless otherwise provided in the Supplementary Conditions, Contractor shall obtain and pay for all construction permits and licenses. Owner shall assist Contractor, when necessary, in obtaining such permits and licenses. Contractor shall pay all governmental charges and inspection fees necessary for the prosecution of the Work which are applicable at the time of opening of Bids, or, if there are no Bids, on the Effective Date of the Agreement. Owner shall pay all charges of utility owners for connections for providing permanent service to the Work.

6.09   *Laws and Regulations*

   A. Contractor shall give all notices required by and shall comply with all Laws and Regulations applicable to the performance of the Work. Except where otherwise expressly required by applicable Laws and Regulations, neither Owner nor Engineer shall be responsible for monitoring Contractor's compliance with any Laws or Regulations.

   B. If Contractor performs any Work knowing or having reason to know that it is contrary to Laws or Regulations, Contractor shall bear all claims, costs, losses, and damages (including but not limited to all fees and charges of engineers, architects, attorneys, and other professionals and all court or arbitration or other dispute resolution costs) arising out of or relating to such Work. However, it shall not be Contractor's responsibility to make certain that the Specifications and Drawings are in accordance with Laws and Regulations, but this shall not relieve Contractor of Contractor's obligations under Paragraph 3.03.

C. Changes in Laws or Regulations not known at the time of opening of Bids (or, on the Effective Date of the Agreement if there were no Bids) having an effect on the cost or time of performance of the Work shall be the subject of an adjustment in Contract Price or Contract Times. If Owner and Contractor are unable to agree on entitlement to or on the amount or extent, if any, of any such adjustment, a Claim may be made therefor as provided in Paragraph 10.05.

6.10   *Taxes*

A. Contractor shall pay all sales, consumer, use, and other similar taxes required to be paid by Contractor in accordance with the Laws and Regulations of the place of the Project which are applicable during the performance of the Work.

6.11   *Use of Site and Other Areas*

A. *Limitation on Use of Site and Other Areas:*

1. Contractor shall confine construction equipment, the storage of materials and equipment, and the operations of workers to the Site and other areas permitted by Laws and Regulations, and shall not unreasonably encumber the Site and other areas with construction equipment or other materials or equipment. Contractor shall assume full responsibility for any damage to any such land or area, or to the owner or occupant thereof, or of any adjacent land or areas resulting from the performance of the Work.

2. Should any claim be made by any such owner or occupant because of the performance of the Work, Contractor shall promptly settle with such other party by negotiation or otherwise resolve the claim by arbitration or other dispute resolution proceeding or at law.

3. To the fullest extent permitted by Laws and Regulations, Contractor shall indemnify and hold harmless Owner and Engineer, and the officers, directors, members, partners, employees, agents, consultants and subcontractors of each and any of them from and against all claims, costs, losses, and damages (including but not limited to all fees and charges of engineers, architects, attorneys, and other professionals and all court or arbitration or other dispute resolution costs) arising out of or relating to any claim or action, legal or equitable, brought by any such owner or occupant against Owner, Engineer, or any other party indemnified hereunder to the extent caused by or based upon Contractor's performance of the Work.

B. *Removal of Debris During Performance of the Work:* During the progress of the Work Contractor shall keep the Site and other areas free from accumulations of waste materials, rubbish, and other debris. Removal and disposal of such waste materials, rubbish, and other debris shall conform to applicable Laws and Regulations.

C. *Cleaning:* Prior to Substantial Completion of the Work Contractor shall clean the Site and the Work and make it ready for utilization by Owner. At the completion of the Work Contractor shall remove from the Site all tools, appliances, construction equipment and machinery, and surplus materials and shall restore to original condition all property not designated for alteration by the Contract Documents.

D. *Loading Structures:* Contractor shall not load nor permit any part of any structure to be loaded in any manner that will endanger the structure, nor shall Contractor subject any part of the Work or adjacent property to stresses or pressures that will endanger it.

6.12 *Record Documents*

A. Contractor shall maintain in a safe place at the Site one record copy of all Drawings, Specifications, Addenda, Change Orders, Work Change Directives, Field Orders, and written interpretations and clarifications in good order and annotated to show changes made during construction. These record documents together with all approved Samples and a counterpart of all approved Shop Drawings will be available to Engineer for reference. Upon completion of the Work, these record documents, Samples, and Shop Drawings will be delivered to Engineer for Owner.

6.13 *Safety and Protection*

A. Contractor shall be solely responsible for initiating, maintaining and supervising all safety precautions and programs in connection with the Work. Such responsibility does not relieve Subcontractors of their responsibility for the safety of persons or property in the performance of their work, nor for compliance with applicable safety Laws and Regulations. Contractor shall take all necessary precautions for the safety of, and shall provide the necessary protection to prevent damage, injury or loss to:

1. all persons on the Site or who may be affected by the Work;

2. all the Work and materials and equipment to be incorporated therein, whether in storage on or off the Site; and

3. other property at the Site or adjacent thereto, including trees, shrubs, lawns, walks, pavements, roadways, structures, utilities, and Underground Facilities not designated for removal, relocation, or replacement in the course of construction.

B. Contractor shall comply with all applicable Laws and Regulations relating to the safety of persons or property, or to the protection of persons or property from damage, injury, or loss; and shall erect and maintain all necessary safeguards for such safety and protection. Contractor shall notify owners of adjacent property and of Underground Facilities and other utility owners when prosecution of the Work may affect them, and shall cooperate with them in the protection, removal, relocation, and replacement of their property.

C. Contractor shall comply with the applicable requirements of Owner's safety programs, if any. The Supplementary Conditions identify any Owner's safety programs that are applicable to the Work.

D. Contractor shall inform Owner and Engineer of the specific requirements of Contractor's safety program with which Owner's and Engineer's employees and representatives must comply while at the Site.

E. All damage, injury, or loss to any property referred to in Paragraph 6.13.A.2 or 6.13.A.3 caused, directly or indirectly, in whole or in part, by Contractor, any Subcontractor, Supplier, or any other individual or entity directly or indirectly employed by any of them to perform any of the Work, or anyone for whose acts any of them may be liable, shall be remedied by Contractor (except damage or loss attributable to the fault of Drawings or Specifications or to the acts or omissions of Owner or Engineer or anyone employed by any of them, or anyone for whose acts any of them may be liable, and not attributable, directly or indirectly, in whole or in part, to the fault or negligence of Contractor or any Subcontractor, Supplier, or other individual or entity directly or indirectly employed by any of them).

F. Contractor's duties and responsibilities for safety and for protection of the Work shall continue until such time as all the Work is completed and Engineer has issued a notice to Owner and Contractor in accordance with Paragraph 14.07.B that the Work is acceptable (except as otherwise expressly provided in connection with Substantial Completion).

6.14 *Safety Representative*

A. Contractor shall designate a qualified and experienced safety representative at the Site whose duties and responsibilities shall be the prevention of accidents and the maintaining and supervising of safety precautions and programs.

6.15 *Hazard Communication Programs*

A. Contractor shall be responsible for coordinating any exchange of material safety data sheets or other hazard communication information required to be made available to or exchanged between or among employers at the Site in accordance with Laws or Regulations.

6.16 *Emergencies*

A. In emergencies affecting the safety or protection of persons or the Work or property at the Site or adjacent thereto, Contractor is obligated to act to prevent threatened damage, injury, or loss. Contractor shall give Engineer prompt written notice if Contractor believes that any significant changes in the Work or variations from the Contract Documents have been caused thereby or are required as a result thereof. If Engineer determines that a change in the Contract Documents is required because of the action taken by Contractor in response to such an emergency, a Work Change Directive or Change Order will be issued.

6.17 *Shop Drawings and Samples*

A. Contractor shall submit Shop Drawings and Samples to Engineer for review and approval in accordance with the accepted Schedule of Submittals (as required by Paragraph 2.07). Each submittal will be identified as Engineer may require.

1. *Shop Drawings:*

    a. Submit number of copies specified in the General Requirements.

    b. Data shown on the Shop Drawings will be complete with respect to quantities, dimensions, specified performance and design criteria, materials, and similar data to show Engineer the services, materials, and equipment Contractor proposes to provide and to enable Engineer to review the information for the limited purposes required by Paragraph 6.17.D.

2. *Samples:*

    a. Submit number of Samples specified in the Specifications.

    b. Clearly identify each Sample as to material, Supplier, pertinent data such as catalog numbers, the use for which intended and other data as Engineer may require to enable Engineer to review the submittal for the limited purposes required by Paragraph 6.17.D.

B. Where a Shop Drawing or Sample is required by the Contract Documents or the Schedule of Submittals, any related Work performed prior to Engineer's review and approval of the pertinent submittal will be at the sole expense and responsibility of Contractor.

C. *Submittal Procedures:*

    1. Before submitting each Shop Drawing or Sample, Contractor shall have:

        a. reviewed and coordinated each Shop Drawing or Sample with other Shop Drawings and Samples and with the requirements of the Work and the Contract Documents;

        b. determined and verified all field measurements, quantities, dimensions, specified performance and design criteria, installation requirements, materials, catalog numbers, and similar information with respect thereto;

        c. determined and verified the suitability of all materials offered with respect to the indicated application, fabrication, shipping, handling, storage, assembly, and installation pertaining to the performance of the Work; and

        d. determined and verified all information relative to Contractor's responsibilities for means, methods, techniques, sequences, and procedures of construction, and safety precautions and programs incident thereto.

    2. Each submittal shall bear a stamp or specific written certification that Contractor has satisfied Contractor's obligations under the Contract Documents with respect to Contractor's review and approval of that submittal.

    3. With each submittal, Contractor shall give Engineer specific written notice of any variations that the Shop Drawing or Sample may have from the requirements of the Contract Documents. This notice shall be both a written communication separate from the Shop

Drawings or Sample submittal; and, in addition, by a specific notation made on each Shop Drawing or Sample submitted to Engineer for review and approval of each such variation.

D. *Engineer's Review:*

1. Engineer will provide timely review of Shop Drawings and Samples in accordance with the Schedule of Submittals acceptable to Engineer. Engineer's review and approval will be only to determine if the items covered by the submittals will, after installation or incorporation in the Work, conform to the information given in the Contract Documents and be compatible with the design concept of the completed Project as a functioning whole as indicated by the Contract Documents.

2. Engineer's review and approval will not extend to means, methods, techniques, sequences, or procedures of construction (except where a particular means, method, technique, sequence, or procedure of construction is specifically and expressly called for by the Contract Documents) or to safety precautions or programs incident thereto. The review and approval of a separate item as such will not indicate approval of the assembly in which the item functions.

3. Engineer's review and approval shall not relieve Contractor from responsibility for any variation from the requirements of the Contract Documents unless Contractor has complied with the requirements of Paragraph 6.17.C.3 and Engineer has given written approval of each such variation by specific written notation thereof incorporated in or accompanying the Shop Drawing or Sample. Engineer's review and approval shall not relieve Contractor from responsibility for complying with the requirements of Paragraph 6.17.C.1.

E. *Resubmittal Procedures:*

1. Contractor shall make corrections required by Engineer and shall return the required number of corrected copies of Shop Drawings and submit, as required, new Samples for review and approval. Contractor shall direct specific attention in writing to revisions other than the corrections called for by Engineer on previous submittals.

6.18   *Continuing the Work*

A. Contractor shall carry on the Work and adhere to the Progress Schedule during all disputes or disagreements with Owner. No Work shall be delayed or postponed pending resolution of any disputes or disagreements, except as permitted by Paragraph 15.04 or as Owner and Contractor may otherwise agree in writing.

6.19   *Contractor's General Warranty and Guarantee*

A. Contractor warrants and guarantees to Owner that all Work will be in accordance with the Contract Documents and will not be defective. Engineer and its officers, directors, members, partners, employees, agents, consultants, and subcontractors shall be entitled to rely on representation of Contractor's warranty and guarantee.

B. Contractor's warranty and guarantee hereunder excludes defects or damage caused by:

1. abuse, modification, or improper maintenance or operation by persons other than Contractor, Subcontractors, Suppliers, or any other individual or entity for whom Contractor is responsible; or

2. normal wear and tear under normal usage.

C. Contractor's obligation to perform and complete the Work in accordance with the Contract Documents shall be absolute. None of the following will constitute an acceptance of Work that is not in accordance with the Contract Documents or a release of Contractor's obligation to perform the Work in accordance with the Contract Documents:

1. observations by Engineer;

2. recommendation by Engineer or payment by Owner of any progress or final payment;

3. the issuance of a certificate of Substantial Completion by Engineer or any payment related thereto by Owner;

4. use or occupancy of the Work or any part thereof by Owner;

5. any review and approval of a Shop Drawing or Sample submittal or the issuance of a notice of acceptability by Engineer;

6. any inspection, test, or approval by others; or

7. any correction of defective Work by Owner.

6.20 *Indemnification*

A. To the fullest extent permitted by Laws and Regulations, Contractor shall indemnify and hold harmless Owner and Engineer, and the officers, directors, members, partners, employees, agents, consultants and subcontractors of each and any of them from and against all claims, costs, losses, and damages (including but not limited to all fees and charges of engineers, architects, attorneys, and other professionals and all court or arbitration or other dispute resolution costs) arising out of or relating to the performance of the Work, provided that any such claim, cost, loss, or damage is attributable to bodily injury, sickness, disease, or death, or to injury to or destruction of tangible property (other than the Work itself), including the loss of use resulting therefrom but only to the extent caused by any negligent act or omission of Contractor, any Subcontractor, any Supplier, or any individual or entity directly or indirectly employed by any of them to perform any of the Work or anyone for whose acts any of them may be liable .

B. In any and all claims against Owner or Engineer or any of their officers, directors, members, partners, employees, agents, consultants, or subcontractors by any employee (or the survivor or personal representative of such employee) of Contractor, any Subcontractor, any Supplier, or any individual or entity directly or indirectly employed by any of them to perform any of the Work, or anyone for whose acts any of them may be liable, the indemnification obligation under Paragraph 6.20.A shall not be limited in any way by any limitation on the amount or type of damages, compensation, or benefits payable by or for Contractor or any such Subcontractor,

Supplier, or other individual or entity under workers' compensation acts, disability benefit acts, or other employee benefit acts.

C. The indemnification obligations of Contractor under Paragraph 6.20.A shall not extend to the liability of Engineer and Engineer's officers, directors, members, partners, employees, agents, consultants and subcontractors arising out of:

1. the preparation or approval of, or the failure to prepare or approve maps, Drawings, opinions, reports, surveys, Change Orders, designs, or Specifications; or

2. giving directions or instructions, or failing to give them, if that is the primary cause of the injury or damage.

6.21 *Delegation of Professional Design Services*

A. Contractor will not be required to provide professional design services unless such services are specifically required by the Contract Documents for a portion of the Work or unless such services are required to carry out Contractor's responsibilities for construction means, methods, techniques, sequences and procedures. Contractor shall not be required to provide professional services in violation of applicable law.

B. If professional design services or certifications by a design professional related to systems, materials or equipment are specifically required of Contractor by the Contract Documents, Owner and Engineer will specify all performance and design criteria that such services must satisfy. Contractor shall cause such services or certifications to be provided by a properly licensed professional, whose signature and seal shall appear on all drawings, calculations, specifications, certifications, Shop Drawings and other submittals prepared by such professional. Shop Drawings and other submittals related to the Work designed or certified by such professional, if prepared by others, shall bear such professional's written approval when submitted to Engineer.

C. Owner and Engineer shall be entitled to rely upon the adequacy, accuracy and completeness of the services, certifications or approvals performed by such design professionals, provided Owner and Engineer have specified to Contractor all performance and design criteria that such services must satisfy.

D. Pursuant to this Paragraph 6.21, Engineer's review and approval of design calculations and design drawings will be only for the limited purpose of checking for conformance with performance and design criteria given and the design concept expressed in the Contract Documents. Engineer's review and approval of Shop Drawings and other submittals (except design calculations and design drawings) will be only for the purpose stated in Paragraph 6.17.D.1.

E. Contractor shall not be responsible for the adequacy of the performance or design criteria required by the Contract Documents.

ARTICLE 7 – OTHER WORK AT THE SITE

7.01 *Related Work at Site*

    A. Owner may perform other work related to the Project at the Site with Owner's employees, or through other direct contracts therefor, or have other work performed by utility owners. If such other work is not noted in the Contract Documents, then:

        1. written notice thereof will be given to Contractor prior to starting any such other work; and

        2. if Owner and Contractor are unable to agree on entitlement to or on the amount or extent, if any, of any adjustment in the Contract Price or Contract Times that should be allowed as a result of such other work, a Claim may be made therefor as provided in Paragraph 10.05.

    B. Contractor shall afford each other contractor who is a party to such a direct contract, each utility owner, and Owner, if Owner is performing other work with Owner's employees, proper and safe access to the Site, provide a reasonable opportunity for the introduction and storage of materials and equipment and the execution of such other work, and properly coordinate the Work with theirs. Contractor shall do all cutting, fitting, and patching of the Work that may be required to properly connect or otherwise make its several parts come together and properly integrate with such other work. Contractor shall not endanger any work of others by cutting, excavating, or otherwise altering such work; provided, however, that Contractor may cut or alter others' work with the written consent of Engineer and the others whose work will be affected. The duties and responsibilities of Contractor under this Paragraph are for the benefit of such utility owners and other contractors to the extent that there are comparable provisions for the benefit of Contractor in said direct contracts between Owner and such utility owners and other contractors.

    C. If the proper execution or results of any part of Contractor's Work depends upon work performed by others under this Article 7, Contractor shall inspect such other work and promptly report to Engineer in writing any delays, defects, or deficiencies in such other work that render it unavailable or unsuitable for the proper execution and results of Contractor's Work. Contractor's failure to so report will constitute an acceptance of such other work as fit and proper for integration with Contractor's Work except for latent defects and deficiencies in such other work.

7.02 *Coordination*

    A. If Owner intends to contract with others for the performance of other work on the Project at the Site, the following will be set forth in Supplementary Conditions:

        1. the individual or entity who will have authority and responsibility for coordination of the activities among the various contractors will be identified;

        2. the specific matters to be covered by such authority and responsibility will be itemized; and

        3. the extent of such authority and responsibilities will be provided.

    B. Unless otherwise provided in the Supplementary Conditions, Owner shall have sole authority and responsibility for such coordination.

7.03   *Legal Relationships*

    A. Paragraphs 7.01.A and 7.02 are not applicable for utilities not under the control of Owner.

    B. Each other direct contract of Owner under Paragraph 7.01.A shall provide that the other contractor is liable to Owner and Contractor for the reasonable direct delay and disruption costs incurred by Contractor as a result of the other contractor's wrongful actions or inactions.

    C. Contractor shall be liable to Owner and any other contractor under direct contract to Owner for the reasonable direct delay and disruption costs incurred by such other contractor as a result of Contractor's wrongful action or inactions.

ARTICLE 8 – OWNER'S RESPONSIBILITIES

8.01   *Communications to Contractor*

    A. Except as otherwise provided in these General Conditions, Owner shall issue all communications to Contractor through Engineer.

8.02   *Replacement of Engineer*

    A. In case of termination of the employment of Engineer, Owner shall appoint an engineer to whom Contractor makes no reasonable objection, whose status under the Contract Documents shall be that of the former Engineer.

8.03   *Furnish Data*

    A. Owner shall promptly furnish the data required of Owner under the Contract Documents.

8.04   *Pay When Due*

    A. Owner shall make payments to Contractor when they are due as provided in Paragraphs 14.02.C and 14.07.C.

8.05   *Lands and Easements; Reports and Tests*

    A. Owner's duties with respect to providing lands and easements and providing engineering surveys to establish reference points are set forth in Paragraphs 4.01 and 4.05. Paragraph 4.02 refers to Owner's identifying and making available to Contractor copies of reports of explorations and tests of subsurface conditions and drawings of physical conditions relating to existing surface or subsurface structures at the Site.

8.06   *Insurance*

    A. Owner's responsibilities, if any, with respect to purchasing and maintaining liability and property insurance are set forth in Article 5.

8.07   *Change Orders*

    A. Owner is obligated to execute Change Orders as indicated in Paragraph 10.03.

8.08   *Inspections, Tests, and Approvals*

   A. Owner's responsibility with respect to certain inspections, tests, and approvals is set forth in Paragraph 13.03.B.

8.09   *Limitations on Owner's Responsibilities*

   A. The Owner shall not supervise, direct, or have control or authority over, nor be responsible for, Contractor's means, methods, techniques, sequences, or procedures of construction, or the safety precautions and programs incident thereto, or for any failure of Contractor to comply with Laws and Regulations applicable to the performance of the Work. Owner will not be responsible for Contractor's failure to perform the Work in accordance with the Contract Documents.

8.10   *Undisclosed Hazardous Environmental Condition*

   A. Owner's responsibility in respect to an undisclosed Hazardous Environmental Condition is set forth in Paragraph 4.06.

8.11   *Evidence of Financial Arrangements*

   A. Upon request of Contractor, Owner shall furnish Contractor reasonable evidence that financial arrangements have been made to satisfy Owner's obligations under the Contract Documents.

8.12   *Compliance with Safety Program*

   A. While at the Site, Owner's employees and representatives shall comply with the specific applicable requirements of Contractor's safety programs of which Owner has been informed pursuant to Paragraph 6.13.D.

ARTICLE 9 – ENGINEER'S STATUS DURING CONSTRUCTION

9.01   *Owner's Representative*

   A. Engineer will be Owner's representative during the construction period. The duties and responsibilities and the limitations of authority of Engineer as Owner's representative during construction are set forth in the Contract Documents.

9.02   *Visits to Site*

   A. Engineer will make visits to the Site at intervals appropriate to the various stages of construction as Engineer deems necessary in order to observe as an experienced and qualified design professional the progress that has been made and the quality of the various aspects of Contractor's executed Work. Based on information obtained during such visits and observations, Engineer, for the benefit of Owner, will determine, in general, if the Work is proceeding in accordance with the Contract Documents. Engineer will not be required to make exhaustive or continuous inspections on the Site to check the quality or quantity of the Work. Engineer's efforts will be directed toward providing for Owner a greater degree of confidence that the completed Work will conform generally to the Contract Documents. On the basis of such visits and observations, Engineer will keep Owner informed of the progress of the Work and will endeavor to guard Owner against defective Work.

B. Engineer's visits and observations are subject to all the limitations on Engineer's authority and responsibility set forth in Paragraph 9.09. Particularly, but without limitation, during or as a result of Engineer's visits or observations of Contractor's Work, Engineer will not supervise, direct, control, or have authority over or be responsible for Contractor's means, methods, techniques, sequences, or procedures of construction, or the safety precautions and programs incident thereto, or for any failure of Contractor to comply with Laws and Regulations applicable to the performance of the Work.

9.03   *Project Representative*

A. If Owner and Engineer agree, Engineer will furnish a Resident Project Representative to assist Engineer in providing more extensive observation of the Work. The authority and responsibilities of any such Resident Project Representative and assistants will be as provided in the Supplementary Conditions, and limitations on the responsibilities thereof will be as provided in Paragraph 9.09. If Owner designates another representative or agent to represent Owner at the Site who is not Engineer's consultant, agent or employee, the responsibilities and authority and limitations thereon of such other individual or entity will be as provided in the Supplementary Conditions.

9.04   *Authorized Variations in Work*

A. Engineer may authorize minor variations in the Work from the requirements of the Contract Documents which do not involve an adjustment in the Contract Price or the Contract Times and are compatible with the design concept of the completed Project as a functioning whole as indicated by the Contract Documents. These may be accomplished by a Field Order and will be binding on Owner and also on Contractor, who shall perform the Work involved promptly. If Owner or Contractor believes that a Field Order justifies an adjustment in the Contract Price or Contract Times, or both, and the parties are unable to agree on entitlement to or on the amount or extent, if any, of any such adjustment, a Claim may be made therefor as provided in Paragraph 10.05.

9.05   *Rejecting Defective Work*

A. Engineer will have authority to reject Work which Engineer believes to be defective, or that Engineer believes will not produce a completed Project that conforms to the Contract Documents or that will prejudice the integrity of the design concept of the completed Project as a functioning whole as indicated by the Contract Documents. Engineer will also have authority to require special inspection or testing of the Work as provided in Paragraph 13.04, whether or not the Work is fabricated, installed, or completed.

9.06   *Shop Drawings, Change Orders and Payments*

A. In connection with Engineer's authority, and limitations thereof, as to Shop Drawings and Samples, see Paragraph 6.17.

B. In connection with Engineer's authority, and limitations thereof, as to design calculations and design drawings submitted in response to a delegation of professional design services, if any, see Paragraph 6.21.

C. In connection with Engineer's authority as to Change Orders, see Articles 10, 11, and 12.

D. In connection with Engineer's authority as to Applications for Payment, see Article 14.

9.07   *Determinations for Unit Price Work*

A. Engineer will determine the actual quantities and classifications of Unit Price Work performed by Contractor. Engineer will review with Contractor the Engineer's preliminary determinations on such matters before rendering a written decision thereon (by recommendation of an Application for Payment or otherwise). Engineer's written decision thereon will be final and binding (except as modified by Engineer to reflect changed factual conditions or more accurate data) upon Owner and Contractor, subject to the provisions of Paragraph 10.05.

9.08   *Decisions on Requirements of Contract Documents and Acceptability of Work*

A. Engineer will be the initial interpreter of the requirements of the Contract Documents and judge of the acceptability of the Work thereunder. All matters in question and other matters between Owner and Contractor arising prior to the date final payment is due relating to the acceptability of the Work, and the interpretation of the requirements of the Contract Documents pertaining to the performance of the Work, will be referred initially to Engineer in writing within 30 days of the event giving rise to the question.

B. Engineer will, with reasonable promptness, render a written decision on the issue referred. If Owner or Contractor believes that any such decision entitles them to an adjustment in the Contract Price or Contract Times or both, a Claim may be made under Paragraph 10.05. The date of Engineer's decision shall be the date of the event giving rise to the issues referenced for the purposes of Paragraph 10.05.B.

C. Engineer's written decision on the issue referred will be final and binding on Owner and Contractor, subject to the provisions of Paragraph 10.05.

D. When functioning as interpreter and judge under this Paragraph 9.08, Engineer will not show partiality to Owner or Contractor and will not be liable in connection with any interpretation or decision rendered in good faith in such capacity.

9.09   *Limitations on Engineer's Authority and Responsibilities*

A. Neither Engineer's authority or responsibility under this Article 9 or under any other provision of the Contract Documents nor any decision made by Engineer in good faith either to exercise or not exercise such authority or responsibility or the undertaking, exercise, or performance of any authority or responsibility by Engineer shall create, impose, or give rise to any duty in contract, tort, or otherwise owed by Engineer to Contractor, any Subcontractor, any Supplier, any other individual or entity, or to any surety for or employee or agent of any of them.

B. Engineer will not supervise, direct, control, or have authority over or be responsible for Contractor's means, methods, techniques, sequences, or procedures of construction, or the safety precautions and programs incident thereto, or for any failure of Contractor to comply with Laws and Regulations applicable to the performance of the Work. Engineer will not be responsible for Contractor's failure to perform the Work in accordance with the Contract Documents.

C. Engineer will not be responsible for the acts or omissions of Contractor or of any Subcontractor, any Supplier, or of any other individual or entity performing any of the Work.

D. Engineer's review of the final Application for Payment and accompanying documentation and all maintenance and operating instructions, schedules, guarantees, bonds, certificates of inspection, tests and approvals, and other documentation required to be delivered by Paragraph 14.07.A will only be to determine generally that their content complies with the requirements of, and in the case of certificates of inspections, tests, and approvals that the results certified indicate compliance with, the Contract Documents.

E. The limitations upon authority and responsibility set forth in this Paragraph 9.09 shall also apply to the Resident Project Representative, if any, and assistants, if any.

9.10 *Compliance with Safety Program*

A. While at the Site, Engineer's employees and representatives shall comply with the specific applicable requirements of Contractor's safety programs of which Engineer has been informed pursuant to Paragraph 6.13.D.

ARTICLE 10 – CHANGES IN THE WORK; CLAIMS

10.01 *Authorized Changes in the Work*

A. Without invalidating the Contract and without notice to any surety, Owner may, at any time or from time to time, order additions, deletions, or revisions in the Work by a Change Order, or a Work Change Directive. Upon receipt of any such document, Contractor shall promptly proceed with the Work involved which will be performed under the applicable conditions of the Contract Documents (except as otherwise specifically provided).

B. If Owner and Contractor are unable to agree on entitlement to, or on the amount or extent, if any, of an adjustment in the Contract Price or Contract Times, or both, that should be allowed as a result of a Work Change Directive, a Claim may be made therefor as provided in Paragraph 10.05.

10.02 *Unauthorized Changes in the Work*

A. Contractor shall not be entitled to an increase in the Contract Price or an extension of the Contract Times with respect to any work performed that is not required by the Contract Documents as amended, modified, or supplemented as provided in Paragraph 3.04, except in the case of an emergency as provided in Paragraph 6.16 or in the case of uncovering Work as provided in Paragraph 13.04.D.

10.03   *Execution of Change Orders*

A. Owner and Contractor shall execute appropriate Change Orders recommended by Engineer covering:

1. changes in the Work which are: (i) ordered by Owner pursuant to Paragraph 10.01.A, (ii) required because of acceptance of defective Work under Paragraph 13.08.A or Owner's correction of defective Work under Paragraph 13.09, or (iii) agreed to by the parties;

2. changes in the Contract Price or Contract Times which are agreed to by the parties, including any undisputed sum or amount of time for Work actually performed in accordance with a Work Change Directive; and

3. changes in the Contract Price or Contract Times which embody the substance of any written decision rendered by Engineer pursuant to Paragraph 10.05; provided that, in lieu of executing any such Change Order, an appeal may be taken from any such decision in accordance with the provisions of the Contract Documents and applicable Laws and Regulations, but during any such appeal, Contractor shall carry on the Work and adhere to the Progress Schedule as provided in Paragraph 6.18.A.

10.04   *Notification to Surety*

A. If the provisions of any bond require notice to be given to a surety of any change affecting the general scope of the Work or the provisions of the Contract Documents (including, but not limited to, Contract Price or Contract Times), the giving of any such notice will be Contractor's responsibility. The amount of each applicable bond will be adjusted to reflect the effect of any such change.

10.05   *Claims*

A. *Engineer's Decision Required*: All Claims, except those waived pursuant to Paragraph 14.09, shall be referred to the Engineer for decision. A decision by Engineer shall be required as a condition precedent to any exercise by Owner or Contractor of any rights or remedies either may otherwise have under the Contract Documents or by Laws and Regulations in respect of such Claims.

B. *Notice:* Written notice stating the general nature of each Claim shall be delivered by the claimant to Engineer and the other party to the Contract promptly (but in no event later than 30 days) after the start of the event giving rise thereto. The responsibility to substantiate a Claim shall rest with the party making the Claim. Notice of the amount or extent of the Claim, with supporting data shall be delivered to the Engineer and the other party to the Contract within 60 days after the start of such event (unless Engineer allows additional time for claimant to submit additional or more accurate data in support of such Claim). A Claim for an adjustment in Contract Price shall be prepared in accordance with the provisions of Paragraph 12.01.B. A Claim for an adjustment in Contract Times shall be prepared in accordance with the provisions of Paragraph 12.02.B. Each Claim shall be accompanied by claimant's written statement that the adjustment claimed is the entire adjustment to which the claimant believes it is entitled as a result of said event. The opposing party shall submit any response to Engineer and the claimant within 30 days after receipt of the claimant's last submittal (unless Engineer allows additional time).

C. *Engineer's Action*: Engineer will review each Claim and, within 30 days after receipt of the last submittal of the claimant or the last submittal of the opposing party, if any, take one of the following actions in writing:

1. deny the Claim in whole or in part;

2. approve the Claim; or

3. notify the parties that the Engineer is unable to resolve the Claim if, in the Engineer's sole discretion, it would be inappropriate for the Engineer to do so. For purposes of further resolution of the Claim, such notice shall be deemed a denial.

D. In the event that Engineer does not take action on a Claim within said 30 days, the Claim shall be deemed denied.

E. Engineer's written action under Paragraph 10.05.C or denial pursuant to Paragraphs 10.05.C.3 or 10.05.D will be final and binding upon Owner and Contractor, unless Owner or Contractor invoke the dispute resolution procedure set forth in Article 16 within 30 days of such action or denial.

F. No Claim for an adjustment in Contract Price or Contract Times will be valid if not submitted in accordance with this Paragraph 10.05.

ARTICLE 11 – COST OF THE WORK; ALLOWANCES; UNIT PRICE WORK

11.01 *Cost of the Work*

A. *Costs Included:* The term Cost of the Work means the sum of all costs, except those excluded in Paragraph 11.01.B, necessarily incurred and paid by Contractor in the proper performance of the Work. When the value of any Work covered by a Change Order or when a Claim for an adjustment in Contract Price is determined on the basis of Cost of the Work, the costs to be reimbursed to Contractor will be only those additional or incremental costs required because of the change in the Work or because of the event giving rise to the Claim. Except as otherwise may be agreed to in writing by Owner, such costs shall be in amounts no higher than those prevailing in the locality of the Project, shall not include any of the costs itemized in Paragraph 11.01.B, and shall include only the following items:

1. Payroll costs for employees in the direct employ of Contractor in the performance of the Work under schedules of job classifications agreed upon by Owner and Contractor. Such employees shall include, without limitation, superintendents, foremen, and other personnel employed full time on the Work. Payroll costs for employees not employed full time on the Work shall be apportioned on the basis of their time spent on the Work. Payroll costs shall include, but not be limited to, salaries and wages plus the cost of fringe benefits, which shall include social security contributions, unemployment, excise, and payroll taxes, workers' compensation, health and retirement benefits, bonuses, sick leave, vacation and holiday pay applicable thereto. The expenses of performing Work outside of regular working hours, on

Saturday, Sunday, or legal holidays, shall be included in the above to the extent authorized by Owner.

2. Cost of all materials and equipment furnished and incorporated in the Work, including costs of transportation and storage thereof, and Suppliers' field services required in connection therewith. All cash discounts shall accrue to Contractor unless Owner deposits funds with Contractor with which to make payments, in which case the cash discounts shall accrue to Owner. All trade discounts, rebates and refunds and returns from sale of surplus materials and equipment shall accrue to Owner, and Contractor shall make provisions so that they may be obtained.

3. Payments made by Contractor to Subcontractors for Work performed by Subcontractors. If required by Owner, Contractor shall obtain competitive bids from subcontractors acceptable to Owner and Contractor and shall deliver such bids to Owner, who will then determine, with the advice of Engineer, which bids, if any, will be acceptable. If any subcontract provides that the Subcontractor is to be paid on the basis of Cost of the Work plus a fee, the Subcontractor's Cost of the Work and fee shall be determined in the same manner as Contractor's Cost of the Work and fee as provided in this Paragraph 11.01.

4. Costs of special consultants (including but not limited to engineers, architects, testing laboratories, surveyors, attorneys, and accountants) employed for services specifically related to the Work.

5. Supplemental costs including the following:

    a. The proportion of necessary transportation, travel, and subsistence expenses of Contractor's employees incurred in discharge of duties connected with the Work.

    b. Cost, including transportation and maintenance, of all materials, supplies, equipment, machinery, appliances, office, and temporary facilities at the Site, and hand tools not owned by the workers, which are consumed in the performance of the Work, and cost, less market value, of such items used but not consumed which remain the property of Contractor.

    c. Rentals of all construction equipment and machinery, and the parts thereof whether rented from Contractor or others in accordance with rental agreements approved by Owner with the advice of Engineer, and the costs of transportation, loading, unloading, assembly, dismantling, and removal thereof. All such costs shall be in accordance with the terms of said rental agreements. The rental of any such equipment, machinery, or parts shall cease when the use thereof is no longer necessary for the Work.

    d. Sales, consumer, use, and other similar taxes related to the Work, and for which Contractor is liable, as imposed by Laws and Regulations.

    e. Deposits lost for causes other than negligence of Contractor, any Subcontractor, or anyone directly or indirectly employed by any of them or for whose acts any of them may be liable, and royalty payments and fees for permits and licenses.

f. Losses and damages (and related expenses) caused by damage to the Work, not compensated by insurance or otherwise, sustained by Contractor in connection with the performance of the Work (except losses and damages within the deductible amounts of property insurance established in accordance with Paragraph 5.06.D), provided such losses and damages have resulted from causes other than the negligence of Contractor, any Subcontractor, or anyone directly or indirectly employed by any of them or for whose acts any of them may be liable. Such losses shall include settlements made with the written consent and approval of Owner. No such losses, damages, and expenses shall be included in the Cost of the Work for the purpose of determining Contractor's fee.

g. The cost of utilities, fuel, and sanitary facilities at the Site.

h. Minor expenses such as telegrams, long distance telephone calls, telephone service at the Site, express and courier services, and similar petty cash items in connection with the Work.

i. The costs of premiums for all bonds and insurance Contractor is required by the Contract Documents to purchase and maintain.

B. *Costs Excluded:* The term Cost of the Work shall not include any of the following items:

1. Payroll costs and other compensation of Contractor's officers, executives, principals (of partnerships and sole proprietorships), general managers, safety managers, engineers, architects, estimators, attorneys, auditors, accountants, purchasing and contracting agents, expediters, timekeepers, clerks, and other personnel employed by Contractor, whether at the Site or in Contractor's principal or branch office for general administration of the Work and not specifically included in the agreed upon schedule of job classifications referred to in Paragraph 11.01.A.1 or specifically covered by Paragraph 11.01.A.4, all of which are to be considered administrative costs covered by the Contractor's fee.

2. Expenses of Contractor's principal and branch offices other than Contractor's office at the Site.

3. Any part of Contractor's capital expenses, including interest on Contractor's capital employed for the Work and charges against Contractor for delinquent payments.

4. Costs due to the negligence of Contractor, any Subcontractor, or anyone directly or indirectly employed by any of them or for whose acts any of them may be liable, including but not limited to, the correction of defective Work, disposal of materials or equipment wrongly supplied, and making good any damage to property.

5. Other overhead or general expense costs of any kind and the costs of any item not specifically and expressly included in Paragraphs 11.01.A.

C. *Contractor's Fee:* When all the Work is performed on the basis of cost-plus, Contractor's fee shall be determined as set forth in the Agreement. When the value of any Work covered by a Change Order or when a Claim for an adjustment in Contract Price is determined on the basis of Cost of the Work, Contractor's fee shall be determined as set forth in Paragraph 12.01.C.

D. *Documentation:* Whenever the Cost of the Work for any purpose is to be determined pursuant to Paragraphs 11.01.A and 11.01.B, Contractor will establish and maintain records thereof in accordance with generally accepted accounting practices and submit in a form acceptable to Engineer an itemized cost breakdown together with supporting data.

11.02 *Allowances*

A. It is understood that Contractor has included in the Contract Price all allowances so named in the Contract Documents and shall cause the Work so covered to be performed for such sums and by such persons or entities as may be acceptable to Owner and Engineer.

B. *Cash Allowances:*

1. Contractor agrees that:

   a. the cash allowances include the cost to Contractor (less any applicable trade discounts) of materials and equipment required by the allowances to be delivered at the Site, and all applicable taxes; and

   b. Contractor's costs for unloading and handling on the Site, labor, installation, overhead, profit, and other expenses contemplated for the cash allowances have been included in the Contract Price and not in the allowances, and no demand for additional payment on account of any of the foregoing will be valid.

C. *Contingency Allowance:*

1. Contractor agrees that a contingency allowance, if any, is for the sole use of Owner to cover unanticipated costs.

D. Prior to final payment, an appropriate Change Order will be issued as recommended by Engineer to reflect actual amounts due Contractor on account of Work covered by allowances, and the Contract Price shall be correspondingly adjusted.

11.03 *Unit Price Work*

A. Where the Contract Documents provide that all or part of the Work is to be Unit Price Work, initially the Contract Price will be deemed to include for all Unit Price Work an amount equal to the sum of the unit price for each separately identified item of Unit Price Work times the estimated quantity of each item as indicated in the Agreement.

B. The estimated quantities of items of Unit Price Work are not guaranteed and are solely for the purpose of comparison of Bids and determining an initial Contract Price. Determinations of the actual quantities and classifications of Unit Price Work performed by Contractor will be made by Engineer subject to the provisions of Paragraph 9.07.

C. Each unit price will be deemed to include an amount considered by Contractor to be adequate to cover Contractor's overhead and profit for each separately identified item.

D. Owner or Contractor may make a Claim for an adjustment in the Contract Price in accordance with Paragraph 10.05 if:

1. the quantity of any item of Unit Price Work performed by Contractor differs materially and significantly from the estimated quantity of such item indicated in the Agreement; and

2. there is no corresponding adjustment with respect to any other item of Work; and

3. Contractor believes that Contractor is entitled to an increase in Contract Price as a result of having incurred additional expense or Owner believes that Owner is entitled to a decrease in Contract Price and the parties are unable to agree as to the amount of any such increase or decrease.

ARTICLE 12 – CHANGE OF CONTRACT PRICE; CHANGE OF CONTRACT TIMES

12.01 *Change of Contract Price*

A. The Contract Price may only be changed by a Change Order. Any Claim for an adjustment in the Contract Price shall be based on written notice submitted by the party making the Claim to the Engineer and the other party to the Contract in accordance with the provisions of Paragraph 10.05.

B. The value of any Work covered by a Change Order or of any Claim for an adjustment in the Contract Price will be determined as follows:

1. where the Work involved is covered by unit prices contained in the Contract Documents, by application of such unit prices to the quantities of the items involved (subject to the provisions of Paragraph 11.03); or

2. where the Work involved is not covered by unit prices contained in the Contract Documents, by a mutually agreed lump sum (which may include an allowance for overhead and profit not necessarily in accordance with Paragraph 12.01.C.2); or

3. where the Work involved is not covered by unit prices contained in the Contract Documents and agreement to a lump sum is not reached under Paragraph 12.01.B.2, on the basis of the Cost of the Work (determined as provided in Paragraph 11.01) plus a Contractor's fee for overhead and profit (determined as provided in Paragraph 12.01.C).

C. *Contractor's Fee:* The Contractor's fee for overhead and profit shall be determined as follows:

1. a mutually acceptable fixed fee; or

2. if a fixed fee is not agreed upon, then a fee based on the following percentages of the various portions of the Cost of the Work:

   a. for costs incurred under Paragraphs 11.01.A.1 and 11.01.A.2, the Contractor's fee shall be 15 percent;

   b. for costs incurred under Paragraph 11.01.A.3, the Contractor's fee shall be five percent;

c. where one or more tiers of subcontracts are on the basis of Cost of the Work plus a fee and no fixed fee is agreed upon, the intent of Paragraphs 12.01.C.2.a and 12.01.C.2.b is that the Subcontractor who actually performs the Work, at whatever tier, will be paid a fee of 15 percent of the costs incurred by such Subcontractor under Paragraphs 11.01.A.1 and 11.01.A.2 and that any higher tier Subcontractor and Contractor will each be paid a fee of five percent of the amount paid to the next lower tier Subcontractor;

d. no fee shall be payable on the basis of costs itemized under Paragraphs 11.01.A.4, 11.01.A.5, and 11.01.B;

e. the amount of credit to be allowed by Contractor to Owner for any change which results in a net decrease in cost will be the amount of the actual net decrease in cost plus a deduction in Contractor's fee by an amount equal to five percent of such net decrease; and

f. when both additions and credits are involved in any one change, the adjustment in Contractor's fee shall be computed on the basis of the net change in accordance with Paragraphs 12.01.C.2.a through 12.01.C.2.e, inclusive.

12.02 *Change of Contract Times*

A. The Contract Times may only be changed by a Change Order. Any Claim for an adjustment in the Contract Times shall be based on written notice submitted by the party making the Claim to the Engineer and the other party to the Contract in accordance with the provisions of Paragraph 10.05.

B. Any adjustment of the Contract Times covered by a Change Order or any Claim for an adjustment in the Contract Times will be determined in accordance with the provisions of this Article 12.

12.03 *Delays*

A. Where Contractor is prevented from completing any part of the Work within the Contract Times due to delay beyond the control of Contractor, the Contract Times will be extended in an amount equal to the time lost due to such delay if a Claim is made therefor as provided in Paragraph 12.02.A. Delays beyond the control of Contractor shall include, but not be limited to, acts or neglect by Owner, acts or neglect of utility owners or other contractors performing other work as contemplated by Article 7, fires, floods, epidemics, abnormal weather conditions, or acts of God.

B. If Owner, Engineer, or other contractors or utility owners performing other work for Owner as contemplated by Article 7, or anyone for whom Owner is responsible, delays, disrupts, or interferes with the performance or progress of the Work, then Contractor shall be entitled to an equitable adjustment in the Contract Price or the Contract Times, or both. Contractor's entitlement to an adjustment of the Contract Times is conditioned on such adjustment being essential to Contractor's ability to complete the Work within the Contract Times.

C. If Contractor is delayed in the performance or progress of the Work by fire, flood, epidemic, abnormal weather conditions, acts of God, acts or failures to act of utility owners not under the control of Owner, or other causes not the fault of and beyond control of Owner and Contractor, then Contractor shall be entitled to an equitable adjustment in Contract Times, if such adjustment is essential to Contractor's ability to complete the Work within the Contract Times. Such an adjustment shall be Contractor's sole and exclusive remedy for the delays described in this Paragraph 12.03.C.

D. Owner, Engineer, and their officers, directors, members, partners, employees, agents, consultants, or subcontractors shall not be liable to Contractor for any claims, costs, losses, or damages (including but not limited to all fees and charges of engineers, architects, attorneys, and other professionals and all court or arbitration or other dispute resolution costs) sustained by Contractor on or in connection with any other project or anticipated project.

E. Contractor shall not be entitled to an adjustment in Contract Price or Contract Times for delays within the control of Contractor. Delays attributable to and within the control of a Subcontractor or Supplier shall be deemed to be delays within the control of Contractor.

ARTICLE 13 – TESTS AND INSPECTIONS; CORRECTION, REMOVAL OR ACCEPTANCE OF DEFECTIVE WORK

13.01 *Notice of Defects*

A. Prompt notice of all defective Work of which Owner or Engineer has actual knowledge will be given to Contractor. Defective Work may be rejected, corrected, or accepted as provided in this Article 13.

13.02 *Access to Work*

A. Owner, Engineer, their consultants and other representatives and personnel of Owner, independent testing laboratories, and governmental agencies with jurisdictional interests will have access to the Site and the Work at reasonable times for their observation, inspection, and testing. Contractor shall provide them proper and safe conditions for such access and advise them of Contractor's safety procedures and programs so that they may comply therewith as applicable.

13.03 *Tests and Inspections*

A. Contractor shall give Engineer timely notice of readiness of the Work for all required inspections, tests, or approvals and shall cooperate with inspection and testing personnel to facilitate required inspections or tests.

B. Owner shall employ and pay for the services of an independent testing laboratory to perform all inspections, tests, or approvals required by the Contract Documents except:

1. for inspections, tests, or approvals covered by Paragraphs 13.03.C and 13.03.D below;

2. that costs incurred in connection with tests or inspections conducted pursuant to Paragraph 13.04.B shall be paid as provided in Paragraph 13.04.C; and

3. as otherwise specifically provided in the Contract Documents.

C. If Laws or Regulations of any public body having jurisdiction require any Work (or part thereof) specifically to be inspected, tested, or approved by an employee or other representative of such public body, Contractor shall assume full responsibility for arranging and obtaining such inspections, tests, or approvals, pay all costs in connection therewith, and furnish Engineer the required certificates of inspection or approval.

D. Contractor shall be responsible for arranging and obtaining and shall pay all costs in connection with any inspections, tests, or approvals required for Owner's and Engineer's acceptance of materials or equipment to be incorporated in the Work; or acceptance of materials, mix designs, or equipment submitted for approval prior to Contractor's purchase thereof for incorporation in the Work. Such inspections, tests, or approvals shall be performed by organizations acceptable to Owner and Engineer.

E. If any Work (or the work of others) that is to be inspected, tested, or approved is covered by Contractor without written concurrence of Engineer, Contractor shall, if requested by Engineer, uncover such Work for observation.

F. Uncovering Work as provided in Paragraph 13.03.E shall be at Contractor's expense unless Contractor has given Engineer timely notice of Contractor's intention to cover the same and Engineer has not acted with reasonable promptness in response to such notice.

13.04 *Uncovering Work*

A. If any Work is covered contrary to the written request of Engineer, it must, if requested by Engineer, be uncovered for Engineer's observation and replaced at Contractor's expense.

B. If Engineer considers it necessary or advisable that covered Work be observed by Engineer or inspected or tested by others, Contractor, at Engineer's request, shall uncover, expose, or otherwise make available for observation, inspection, or testing as Engineer may require, that portion of the Work in question, furnishing all necessary labor, material, and equipment.

C. If it is found that the uncovered Work is defective, Contractor shall pay all claims, costs, losses, and damages (including but not limited to all fees and charges of engineers, architects, attorneys, and other professionals and all court or arbitration or other dispute resolution costs) arising out of or relating to such uncovering, exposure, observation, inspection, and testing, and of satisfactory replacement or reconstruction (including but not limited to all costs of repair or replacement of work of others); and Owner shall be entitled to an appropriate decrease in the Contract Price. If the parties are unable to agree as to the amount thereof, Owner may make a Claim therefor as provided in Paragraph 10.05.

D. If the uncovered Work is not found to be defective, Contractor shall be allowed an increase in the Contract Price or an extension of the Contract Times, or both, directly attributable to such uncovering, exposure, observation, inspection, testing, replacement, and reconstruction. If the parties are unable to agree as to the amount or extent thereof, Contractor may make a Claim therefor as provided in Paragraph 10.05.

13.05 *Owner May Stop the Work*

A. If the Work is defective, or Contractor fails to supply sufficient skilled workers or suitable materials or equipment, or fails to perform the Work in such a way that the completed Work will conform to the Contract Documents, Owner may order Contractor to stop the Work, or any portion thereof, until the cause for such order has been eliminated; however, this right of Owner to stop the Work shall not give rise to any duty on the part of Owner to exercise this right for the benefit of Contractor, any Subcontractor, any Supplier, any other individual or entity, or any surety for, or employee or agent of any of them.

13.06 *Correction or Removal of Defective Work*

A. Promptly after receipt of written notice, Contractor shall correct all defective Work, whether or not fabricated, installed, or completed, or, if the Work has been rejected by Engineer, remove it from the Project and replace it with Work that is not defective. Contractor shall pay all claims, costs, losses, and damages (including but not limited to all fees and charges of engineers, architects, attorneys, and other professionals and all court or arbitration or other dispute resolution costs) arising out of or relating to such correction or removal (including but not limited to all costs of repair or replacement of work of others).

B. When correcting defective Work under the terms of this Paragraph 13.06 or Paragraph 13.07, Contractor shall take no action that would void or otherwise impair Owner's special warranty and guarantee, if any, on said Work.

13.07 *Correction Period*

A. If within one year after the date of Substantial Completion (or such longer period of time as may be prescribed by the terms of any applicable special guarantee required by the Contract Documents) or by any specific provision of the Contract Documents, any Work is found to be defective, or if the repair of any damages to the land or areas made available for Contractor's use by Owner or permitted by Laws and Regulations as contemplated in Paragraph 6.11.A is found to be defective, Contractor shall promptly, without cost to Owner and in accordance with Owner's written instructions:

1. repair such defective land or areas; or

2. correct such defective Work; or

3. if the defective Work has been rejected by Owner, remove it from the Project and replace it with Work that is not defective, and

4. satisfactorily correct or repair or remove and replace any damage to other Work, to the work of others or other land or areas resulting therefrom.

B. If Contractor does not promptly comply with the terms of Owner's written instructions, or in an emergency where delay would cause serious risk of loss or damage, Owner may have the defective Work corrected or repaired or may have the rejected Work removed and replaced. All claims, costs, losses, and damages (including but not limited to all fees and charges of engineers, architects, attorneys, and other professionals and all court or arbitration or other dispute resolution costs) arising out of or relating to such correction or repair or such removal and replacement (including but not limited to all costs of repair or replacement of work of others) will be paid by Contractor.

C. In special circumstances where a particular item of equipment is placed in continuous service before Substantial Completion of all the Work, the correction period for that item may start to run from an earlier date if so provided in the Specifications.

D. Where defective Work (and damage to other Work resulting therefrom) has been corrected or removed and replaced under this Paragraph 13.07, the correction period hereunder with respect to such Work will be extended for an additional period of one year after such correction or removal and replacement has been satisfactorily completed.

E. Contractor's obligations under this Paragraph 13.07 are in addition to any other obligation or warranty. The provisions of this Paragraph 13.07 shall not be construed as a substitute for, or a waiver of, the provisions of any applicable statute of limitation or repose.

13.08 *Acceptance of Defective Work*

A. If, instead of requiring correction or removal and replacement of defective Work, Owner (and, prior to Engineer's recommendation of final payment, Engineer) prefers to accept it, Owner may do so. Contractor shall pay all claims, costs, losses, and damages (including but not limited to all fees and charges of engineers, architects, attorneys, and other professionals and all court or arbitration or other dispute resolution costs) attributable to Owner's evaluation of and determination to accept such defective Work (such costs to be approved by Engineer as to reasonableness) and for the diminished value of the Work to the extent not otherwise paid by Contractor pursuant to this sentence. If any such acceptance occurs prior to Engineer's recommendation of final payment, a Change Order will be issued incorporating the necessary revisions in the Contract Documents with respect to the Work, and Owner shall be entitled to an appropriate decrease in the Contract Price, reflecting the diminished value of Work so accepted. If the parties are unable to agree as to the amount thereof, Owner may make a Claim therefor as provided in Paragraph 10.05. If the acceptance occurs after such recommendation, an appropriate amount will be paid by Contractor to Owner.

13.09 *Owner May Correct Defective Work*

A. If Contractor fails within a reasonable time after written notice from Engineer to correct defective Work, or to remove and replace rejected Work as required by Engineer in accordance with Paragraph 13.06.A, or if Contractor fails to perform the Work in accordance with the Contract Documents, or if Contractor fails to comply with any other provision of the Contract Documents, Owner may, after seven days written notice to Contractor, correct, or remedy any such deficiency.

B. In exercising the rights and remedies under this Paragraph 13.09, Owner shall proceed expeditiously. In connection with such corrective or remedial action, Owner may exclude Contractor from all or part of the Site, take possession of all or part of the Work and suspend Contractor's services related thereto, take possession of Contractor's tools, appliances, construction equipment and machinery at the Site, and incorporate in the Work all materials and equipment stored at the Site or for which Owner has paid Contractor but which are stored elsewhere. Contractor shall allow Owner, Owner's representatives, agents and employees, Owner's other contractors, and Engineer and Engineer's consultants access to the Site to enable Owner to exercise the rights and remedies under this Paragraph.

C. All claims, costs, losses, and damages (including but not limited to all fees and charges of engineers, architects, attorneys, and other professionals and all court or arbitration or other dispute resolution costs) incurred or sustained by Owner in exercising the rights and remedies under this Paragraph 13.09 will be charged against Contractor, and a Change Order will be issued incorporating the necessary revisions in the Contract Documents with respect to the Work; and Owner shall be entitled to an appropriate decrease in the Contract Price. If the parties are unable to agree as to the amount of the adjustment, Owner may make a Claim therefor as provided in Paragraph 10.05. Such claims, costs, losses and damages will include but not be limited to all costs of repair, or replacement of work of others destroyed or damaged by correction, removal, or replacement of Contractor's defective Work.

D. Contractor shall not be allowed an extension of the Contract Times because of any delay in the performance of the Work attributable to the exercise by Owner of Owner's rights and remedies under this Paragraph 13.09.

ARTICLE 14 – PAYMENTS TO CONTRACTOR AND COMPLETION

14.01 Schedule of Values

A. The Schedule of Values established as provided in Paragraph 2.07.A will serve as the basis for progress payments and will be incorporated into a form of Application for Payment acceptable to Engineer. Progress payments on account of Unit Price Work will be based on the number of units completed.

14.02 Progress Payments

A. *Applications for Payments:*

1. At least 20 days before the date established in the Agreement for each progress payment (but not more often than once a month), Contractor shall submit to Engineer for review an Application for Payment filled out and signed by Contractor covering the Work completed as of the date of the Application and accompanied by such supporting documentation as is required by the Contract Documents. If payment is requested on the basis of materials and equipment not incorporated in the Work but delivered and suitably stored at the Site or at another location agreed to in writing, the Application for Payment shall also be accompanied by a bill of sale, invoice, or other documentation warranting that Owner has received the materials and equipment free and clear of all Liens and evidence that the materials and equipment are covered by appropriate property insurance or other arrangements to protect Owner's interest therein, all of which must be satisfactory to Owner.

2. Beginning with the second Application for Payment, each Application shall include an affidavit of Contractor stating that all previous progress payments received on account of the

Work have been applied on account to discharge Contractor's legitimate obligations associated with prior Applications for Payment.

3. The amount of retainage with respect to progress payments will be as stipulated in the Agreement.

B. *Review of Applications:*

1. Engineer will, within 10 days after receipt of each Application for Payment, either indicate in writing a recommendation of payment and present the Application to Owner or return the Application to Contractor indicating in writing Engineer's reasons for refusing to recommend payment. In the latter case, Contractor may make the necessary corrections and resubmit the Application.

2. Engineer's recommendation of any payment requested in an Application for Payment will constitute a representation by Engineer to Owner, based on Engineer's observations of the executed Work as an experienced and qualified design professional, and on Engineer's review of the Application for Payment and the accompanying data and schedules, that to the best of Engineer's knowledge, information and belief:

   a. the Work has progressed to the point indicated;

   b. the quality of the Work is generally in accordance with the Contract Documents (subject to an evaluation of the Work as a functioning whole prior to or upon Substantial Completion, the results of any subsequent tests called for in the Contract Documents, a final determination of quantities and classifications for Unit Price Work under Paragraph 9.07, and any other qualifications stated in the recommendation); and

   c. the conditions precedent to Contractor's being entitled to such payment appear to have been fulfilled in so far as it is Engineer's responsibility to observe the Work.

3. By recommending any such payment Engineer will not thereby be deemed to have represented that:

   a. inspections made to check the quality or the quantity of the Work as it has been performed have been exhaustive, extended to every aspect of the Work in progress, or involved detailed inspections of the Work beyond the responsibilities specifically assigned to Engineer in the Contract Documents; or

   b. there may not be other matters or issues between the parties that might entitle Contractor to be paid additionally by Owner or entitle Owner to withhold payment to Contractor.

4. Neither Engineer's review of Contractor's Work for the purposes of recommending payments nor Engineer's recommendation of any payment, including final payment, will impose responsibility on Engineer:

   a. to supervise, direct, or control the Work, or

    b. for the means, methods, techniques, sequences, or procedures of construction, or the safety precautions and programs incident thereto, or

    c. for Contractor's failure to comply with Laws and Regulations applicable to Contractor's performance of the Work, or

    d. to make any examination to ascertain how or for what purposes Contractor has used the moneys paid on account of the Contract Price, or

    e. to determine that title to any of the Work, materials, or equipment has passed to Owner free and clear of any Liens.

5. Engineer may refuse to recommend the whole or any part of any payment if, in Engineer's opinion, it would be incorrect to make the representations to Owner stated in Paragraph 14.02.B.2. Engineer may also refuse to recommend any such payment or, because of subsequently discovered evidence or the results of subsequent inspections or tests, revise or revoke any such payment recommendation previously made, to such extent as may be necessary in Engineer's opinion to protect Owner from loss because:

    a. the Work is defective, or completed Work has been damaged, requiring correction or replacement;

    b. the Contract Price has been reduced by Change Orders;

    c. Owner has been required to correct defective Work or complete Work in accordance with Paragraph 13.09; or

    d. Engineer has actual knowledge of the occurrence of any of the events enumerated in Paragraph 15.02.A.

C. *Payment Becomes Due:*

1. Ten days after presentation of the Application for Payment to Owner with Engineer's recommendation, the amount recommended will (subject to the provisions of Paragraph 14.02.D) become due, and when due will be paid by Owner to Contractor.

D. *Reduction in Payment:*

1. Owner may refuse to make payment of the full amount recommended by Engineer because:

    a. claims have been made against Owner on account of Contractor's performance or furnishing of the Work;

    b. Liens have been filed in connection with the Work, except where Contractor has delivered a specific bond satisfactory to Owner to secure the satisfaction and discharge of such Liens;

    c. there are other items entitling Owner to a set-off against the amount recommended; or

    d. Owner has actual knowledge of the occurrence of any of the events enumerated in Paragraphs 14.02.B.5.a through 14.02.B.5.c or Paragraph 15.02.A.

  2. If Owner refuses to make payment of the full amount recommended by Engineer, Owner will give Contractor immediate written notice (with a copy to Engineer) stating the reasons for such action and promptly pay Contractor any amount remaining after deduction of the amount so withheld. Owner shall promptly pay Contractor the amount so withheld, or any adjustment thereto agreed to by Owner and Contractor, when Contractor remedies the reasons for such action.

  3. Upon a subsequent determination that Owner's refusal of payment was not justified, the amount wrongfully withheld shall be treated as an amount due as determined by Paragraph 14.02.C.1 and subject to interest as provided in the Agreement.

14.03 *Contractor's Warranty of Title*

  A. Contractor warrants and guarantees that title to all Work, materials, and equipment covered by any Application for Payment, whether incorporated in the Project or not, will pass to Owner no later than the time of payment free and clear of all Liens.

14.04 *Substantial Completion*

  A. When Contractor considers the entire Work ready for its intended use Contractor shall notify Owner and Engineer in writing that the entire Work is substantially complete (except for items specifically listed by Contractor as incomplete) and request that Engineer issue a certificate of Substantial Completion.

  B. Promptly after Contractor's notification, Owner, Contractor, and Engineer shall make an inspection of the Work to determine the status of completion. If Engineer does not consider the Work substantially complete, Engineer will notify Contractor in writing giving the reasons therefor.

  C. If Engineer considers the Work substantially complete, Engineer will deliver to Owner a tentative certificate of Substantial Completion which shall fix the date of Substantial Completion. There shall be attached to the certificate a tentative list of items to be completed or corrected before final payment. Owner shall have seven days after receipt of the tentative certificate during which to make written objection to Engineer as to any provisions of the certificate or attached list. If, after considering such objections, Engineer concludes that the Work is not substantially complete, Engineer will, within 14 days after submission of the tentative certificate to Owner, notify Contractor in writing, stating the reasons therefor. If, after consideration of Owner's objections, Engineer considers the Work substantially complete, Engineer will, within said 14 days, execute and deliver to Owner and Contractor a definitive certificate of Substantial Completion (with a revised tentative list of items to be completed or corrected) reflecting such changes from the tentative certificate as Engineer believes justified after consideration of any objections from Owner.

  D. At the time of delivery of the tentative certificate of Substantial Completion, Engineer will deliver to Owner and Contractor a written recommendation as to division of responsibilities pending final payment between Owner and Contractor with respect to security, operation, safety, and protection of the Work, maintenance, heat, utilities, insurance, and warranties and guarantees. Unless Owner and Contractor agree otherwise in writing and so inform Engineer in writing prior to Engineer's issuing the definitive certificate of Substantial Completion, Engineer's aforesaid recommendation will be binding on Owner and Contractor until final payment.

E. Owner shall have the right to exclude Contractor from the Site after the date of Substantial Completion subject to allowing Contractor reasonable access to remove its property and complete or correct items on the tentative list.

14.05 *Partial Utilization*

A. Prior to Substantial Completion of all the Work, Owner may use or occupy any substantially completed part of the Work which has specifically been identified in the Contract Documents, or which Owner, Engineer, and Contractor agree constitutes a separately functioning and usable part of the Work that can be used by Owner for its intended purpose without significant interference with Contractor's performance of the remainder of the Work, subject to the following conditions:

1. Owner at any time may request Contractor in writing to permit Owner to use or occupy any such part of the Work which Owner believes to be ready for its intended use and substantially complete. If and when Contractor agrees that such part of the Work is substantially complete, Contractor, Owner, and Engineer will follow the procedures of Paragraph 14.04.A through D for that part of the Work.

2. Contractor at any time may notify Owner and Engineer in writing that Contractor considers any such part of the Work ready for its intended use and substantially complete and request Engineer to issue a certificate of Substantial Completion for that part of the Work.

3. Within a reasonable time after either such request, Owner, Contractor, and Engineer shall make an inspection of that part of the Work to determine its status of completion. If Engineer does not consider that part of the Work to be substantially complete, Engineer will notify Owner and Contractor in writing giving the reasons therefor. If Engineer considers that part of the Work to be substantially complete, the provisions of Paragraph 14.04 will apply with respect to certification of Substantial Completion of that part of the Work and the division of responsibility in respect thereof and access thereto.

4. No use or occupancy or separate operation of part of the Work may occur prior to compliance with the requirements of Paragraph 5.10 regarding property insurance.

14.06 *Final Inspection*

A. Upon written notice from Contractor that the entire Work or an agreed portion thereof is complete, Engineer will promptly make a final inspection with Owner and Contractor and will notify Contractor in writing of all particulars in which this inspection reveals that the Work is incomplete or defective. Contractor shall immediately take such measures as are necessary to complete such Work or remedy such deficiencies.

14.07   *Final Payment*

A.   *Application for Payment:*

1.   After Contractor has, in the opinion of Engineer, satisfactorily completed all corrections identified during the final inspection and has delivered, in accordance with the Contract Documents, all maintenance and operating instructions, schedules, guarantees, bonds, certificates or other evidence of insurance, certificates of inspection, marked-up record documents (as provided in Paragraph 6.12), and other documents, Contractor may make application for final payment following the procedure for progress payments.

2.   The final Application for Payment shall be accompanied (except as previously delivered) by:

   a.   all documentation called for in the Contract Documents, including but not limited to the evidence of insurance required by Paragraph 5.04.B.6;

   b.   consent of the surety, if any, to final payment;

   c.   a list of all Claims against Owner that Contractor believes are unsettled; and

   d.   complete and legally effective releases or waivers (satisfactory to Owner) of all Lien rights arising out of or Liens filed in connection with the Work.

3.   In lieu of the releases or waivers of Liens specified in Paragraph 14.07.A.2 and as approved by Owner, Contractor may furnish receipts or releases in full and an affidavit of Contractor that: (i) the releases and receipts include all labor, services, material, and equipment for which a Lien could be filed; and (ii) all payrolls, material and equipment bills, and other indebtedness connected with the Work for which Owner might in any way be responsible, or which might in any way result in liens or other burdens on Owner's property, have been paid or otherwise satisfied. If any Subcontractor or Supplier fails to furnish such a release or receipt in full, Contractor may furnish a bond or other collateral satisfactory to Owner to indemnify Owner against any Lien.

B.   *Engineer's Review of Application and Acceptance:*

1.   If, on the basis of Engineer's observation of the Work during construction and final inspection, and Engineer's review of the final Application for Payment and accompanying documentation as required by the Contract Documents, Engineer is satisfied that the Work has been completed and Contractor's other obligations under the Contract Documents have been fulfilled, Engineer will, within ten days after receipt of the final Application for Payment, indicate in writing Engineer's recommendation of payment and present the Application for Payment to Owner for payment. At the same time Engineer will also give written notice to Owner and Contractor that the Work is acceptable subject to the provisions of Paragraph 14.09. Otherwise, Engineer will return the Application for Payment to Contractor, indicating in writing the reasons for refusing to recommend final payment, in which case Contractor shall make the necessary corrections and resubmit the Application for Payment.

C.   *Payment Becomes Due:*

1. Thirty days after the presentation to Owner of the Application for Payment and accompanying documentation, the amount recommended by Engineer, less any sum Owner is entitled to set off against Engineer's recommendation, including but not limited to liquidated damages, will become due and will be paid by Owner to Contractor.

14.08 *Final Completion Delayed*

A. If, through no fault of Contractor, final completion of the Work is significantly delayed, and if Engineer so confirms, Owner shall, upon receipt of Contractor's final Application for Payment (for Work fully completed and accepted) and recommendation of Engineer, and without terminating the Contract, make payment of the balance due for that portion of the Work fully completed and accepted. If the remaining balance to be held by Owner for Work not fully completed or corrected is less than the retainage stipulated in the Agreement, and if bonds have been furnished as required in Paragraph 5.01, the written consent of the surety to the payment of the balance due for that portion of the Work fully completed and accepted shall be submitted by Contractor to Engineer with the Application for such payment. Such payment shall be made under the terms and conditions governing final payment, except that it shall not constitute a waiver of Claims.

14.09 *Waiver of Claims*

A. The making and acceptance of final payment will constitute:

1. a waiver of all Claims by Owner against Contractor, except Claims arising from unsettled Liens, from defective Work appearing after final inspection pursuant to Paragraph 14.06, from failure to comply with the Contract Documents or the terms of any special guarantees specified therein, or from Contractor's continuing obligations under the Contract Documents; and

2. a waiver of all Claims by Contractor against Owner other than those previously made in accordance with the requirements herein and expressly acknowledged by Owner in writing as still unsettled.

ARTICLE 15 – SUSPENSION OF WORK AND TERMINATION

15.01 *Owner May Suspend Work*

A. At any time and without cause, Owner may suspend the Work or any portion thereof for a period of not more than 90 consecutive days by notice in writing to Contractor and Engineer which will fix the date on which Work will be resumed. Contractor shall resume the Work on the date so fixed. Contractor shall be granted an adjustment in the Contract Price or an extension of the Contract Times, or both, directly attributable to any such suspension if Contractor makes a Claim therefor as provided in Paragraph 10.05.

15.02 *Owner May Terminate for Cause*

A. The occurrence of any one or more of the following events will justify termination for cause:

1. Contractor's persistent failure to perform the Work in accordance with the Contract Documents (including, but not limited to, failure to supply sufficient skilled workers or suitable materials or equipment or failure to adhere to the Progress Schedule established under Paragraph 2.07 as adjusted from time to time pursuant to Paragraph 6.04);

2. Contractor's disregard of Laws or Regulations of any public body having jurisdiction;

3. Contractor's repeated disregard of the authority of Engineer; or

4. Contractor's violation in any substantial way of any provisions of the Contract Documents.

B. If one or more of the events identified in Paragraph 15.02.A occur, Owner may, after giving Contractor (and surety) seven days written notice of its intent to terminate the services of Contractor:

1. exclude Contractor from the Site, and take possession of the Work and of all Contractor's tools, appliances, construction equipment, and machinery at the Site, and use the same to the full extent they could be used by Contractor (without liability to Contractor for trespass or conversion);

2. incorporate in the Work all materials and equipment stored at the Site or for which Owner has paid Contractor but which are stored elsewhere; and

3. complete the Work as Owner may deem expedient.

C. If Owner proceeds as provided in Paragraph 15.02.B, Contractor shall not be entitled to receive any further payment until the Work is completed. If the unpaid balance of the Contract Price exceeds all claims, costs, losses, and damages (including but not limited to all fees and charges of engineers, architects, attorneys, and other professionals and all court or arbitration or other dispute resolution costs) sustained by Owner arising out of or relating to completing the Work, such excess will be paid to Contractor. If such claims, costs, losses, and damages exceed such unpaid balance, Contractor shall pay the difference to Owner. Such claims, costs, losses, and damages incurred by Owner will be reviewed by Engineer as to their reasonableness and, when so approved by Engineer, incorporated in a Change Order. When exercising any rights or remedies under this Paragraph, Owner shall not be required to obtain the lowest price for the Work performed.

D. Notwithstanding Paragraphs 15.02.B and 15.02.C, Contractor's services will not be terminated if Contractor begins within seven days of receipt of notice of intent to terminate to correct its failure to perform and proceeds diligently to cure such failure within no more than 30 days of receipt of said notice.

E. Where Contractor's services have been so terminated by Owner, the termination will not affect any rights or remedies of Owner against Contractor then existing or which may thereafter accrue. Any retention or payment of moneys due Contractor by Owner will not release Contractor from liability.

F. If and to the extent that Contractor has provided a performance bond under the provisions of Paragraph 5.01.A, the termination procedures of that bond shall supersede the provisions of Paragraphs 15.02.B and 15.02.C.

15.03 *Owner May Terminate For Convenience*

A. Upon seven days written notice to Contractor and Engineer, Owner may, without cause and without prejudice to any other right or remedy of Owner, terminate the Contract. In such case, Contractor shall be paid for (without duplication of any items):

1. completed and acceptable Work executed in accordance with the Contract Documents prior to the effective date of termination, including fair and reasonable sums for overhead and profit on such Work;

2. expenses sustained prior to the effective date of termination in performing services and furnishing labor, materials, or equipment as required by the Contract Documents in connection with uncompleted Work, plus fair and reasonable sums for overhead and profit on such expenses;

3. all claims, costs, losses, and damages (including but not limited to all fees and charges of engineers, architects, attorneys, and other professionals and all court or arbitration or other dispute resolution costs) incurred in settlement of terminated contracts with Subcontractors, Suppliers, and others; and

4. reasonable expenses directly attributable to termination.

B. Contractor shall not be paid on account of loss of anticipated profits or revenue or other economic loss arising out of or resulting from such termination.

15.04 *Contractor May Stop Work or Terminate*

A. If, through no act or fault of Contractor, (i) the Work is suspended for more than 90 consecutive days by Owner or under an order of court or other public authority, or (ii) Engineer fails to act on any Application for Payment within 30 days after it is submitted, or (iii) Owner fails for 30 days to pay Contractor any sum finally determined to be due, then Contractor may, upon seven days written notice to Owner and Engineer, and provided Owner or Engineer do not remedy such suspension or failure within that time, terminate the Contract and recover from Owner payment on the same terms as provided in Paragraph 15.03.

B. In lieu of terminating the Contract and without prejudice to any other right or remedy, if Engineer has failed to act on an Application for Payment within 30 days after it is submitted, or Owner has failed for 30 days to pay Contractor any sum finally determined to be due, Contractor may, seven days after written notice to Owner and Engineer, stop the Work until payment is made of all such amounts due Contractor, including interest thereon. The provisions of this Paragraph 15.04 are not intended to preclude Contractor from making a Claim under Paragraph 10.05 for an adjustment in Contract Price or Contract Times or otherwise for expenses or damage directly attributable to Contractor's stopping the Work as permitted by this Paragraph.

ARTICLE 16 – DISPUTE RESOLUTION

16.01   *Methods and Procedures*

A. Either Owner or Contractor may request mediation of any Claim submitted to Engineer for a decision under Paragraph 10.05 before such decision becomes final and binding. The mediation will be governed by the Construction Industry Mediation Rules of the American Arbitration Association in effect as of the Effective Date of the Agreement. The request for mediation shall be submitted in writing to the American Arbitration Association and the other party to the Contract. Timely submission of the request shall stay the effect of Paragraph 10.05.E.

B. Owner and Contractor shall participate in the mediation process in good faith. The process shall be concluded within 60 days of filing of the request. The date of termination of the mediation shall be determined by application of the mediation rules referenced above.

C. If the Claim is not resolved by mediation, Engineer's action under Paragraph 10.05.C or a denial pursuant to Paragraphs 10.05.C.3 or 10.05.D shall become final and binding 30 days after termination of the mediation unless, within that time period, Owner or Contractor:

   1. elects in writing to invoke any dispute resolution process provided for in the Supplementary Conditions; or

   2. agrees with the other party to submit the Claim to another dispute resolution process; or

   3. gives written notice to the other party of the intent to submit the Claim to a court of competent jurisdiction.

ARTICLE 17 – MISCELLANEOUS

17.01   *Giving Notice*

A. Whenever any provision of the Contract Documents requires the giving of written notice, it will be deemed to have been validly given if:

   1. delivered in person to the individual or to a member of the firm or to an officer of the corporation for whom it is intended; or

   2. delivered at or sent by registered or certified mail, postage prepaid, to the last business address known to the giver of the notice.

17.02   *Computation of Times*

A. When any period of time is referred to in the Contract Documents by days, it will be computed to exclude the first and include the last day of such period. If the last day of any such period falls on a Saturday or Sunday or on a day made a legal holiday by the law of the applicable jurisdiction, such day will be omitted from the computation.

17.03 *Cumulative Remedies*

A. The duties and obligations imposed by these General Conditions and the rights and remedies available hereunder to the parties hereto are in addition to, and are not to be construed in any way as a limitation of, any rights and remedies available to any or all of them which are otherwise imposed or available by Laws or Regulations, by special warranty or guarantee, or by other provisions of the Contract Documents. The provisions of this Paragraph will be as effective as if repeated specifically in the Contract Documents in connection with each particular duty, obligation, right, and remedy to which they apply.

17.04 *Survival of Obligations*

A. All representations, indemnifications, warranties, and guarantees made in, required by, or given in accordance with the Contract Documents, as well as all continuing obligations indicated in the Contract Documents, will survive final payment, completion, and acceptance of the Work or termination or completion of the Contract or termination of the services of Contractor.

17.05 *Controlling Law*

A. This Contract is to be governed by the law of the state in which the Project is located.

17.06 *Headings*

A. Article and paragraph headings are inserted for convenience only and do not constitute parts of these General Conditions.

Subject Index

Case Index